HANDBOOK OF
RADIOTHERAPY PHYSICS

Theory and Practice

HANDBOOK OF
RADIOTHERAPY PHYSICS
Theory and Practice

Edited by

P MAYLES
Clatterbridge Centre for Oncology, UK

A NAHUM
Clatterbridge Centre for Oncology, UK

J C ROSENWALD
Institute Curie, Paris, France

Taylor & Francis
Taylor & Francis Group
New York   London

Taylor & Francis is an imprint of the
Taylor & Francis Group, an informa business

CRC Press
Taylor & Francis Group
6000 Broken Sound Parkway NW, Suite 300
Boca Raton, FL 33487-2742

© 2007 by Taylor & Francis Group, LLC
CRC Press is an imprint of Taylor & Francis Group, an Informa business

No claim to original U.S. Government works
Printed in the United States of America on acid-free paper
10 9 8 7

International Standard Book Number-10: 0-7503-0860-5 (Hardcover)
International Standard Book Number-13: 978-0-7503-0860-1 (Hardcover)

Library of Congress Cataloging-in-Publication Data

Handbook of radiotherapy physics : theory and practice / edited by P. Mayles, A. Nahum, and J.C. Rosenwald.
  p. ; cm.
  Includes bibliographical references and index.
  ISBN-13: 978-0-7503-0860-1 (hardcover : alk. paper)
  ISBN-10: 0-7503-0860-5 (hardcover : alk. paper)
  1. Medical physics--Handbooks, manuals, etc. 2. Cancer--Radio-therapy--Handbooks, manuals, etc.
I. Mayles, P. (Philip) II. Nahum, Alan E. III. Rosenwald, J. C. (Jean-Claude)
  [DNLM: 1. Radiotherapy--methods. 2. Radiation Dosage. 3. Radiation Effects. WN 250 H236 2007]

  R896.7.H36 2007
  610.1'53--dc22   2006029753

Visit the Taylor & Francis Web site at
http://www.taylorandfrancis.com

and the CRC Press Web site at
http://www.crcpress.com

PREFACE

Radiotherapy is a comprehensive and fast-moving discipline which plays a major role in cancer care. Safe and effective radiotherapy requires close collaboration between radiation oncologists, radiation technologists and medical physicists, and all must have an understanding of each others' disciplines. Our aim has been to provide a comprehensive text providing most of the theoretical and practical knowledge that medical physicists need to know, including the essential underlying radiation biology. Although principally aimed at practising medical physicists, this book will also be useful to all other professionals involved in radiation therapy, whether they be students (master's and PhD level), university teachers, researchers, radiation oncologists, or radiation technologists.

The book is organised into 13 Parts, each dealing with a major self-contained subject area. Each part begins with an introduction by the editors and is subdivided into chapters mostly written by a single author. References are collected together at the end of each part. In order to cover in detail all aspects of radiotherapy physics and biology, a high level of expertise was required. Contributions have been brought together from eminent specialists in the field, mostly from Europe, but also some from North America. The editors have, where necessary, combined contributions from different authors in order to provide a logical flow—as far as possible details of who wrote what are shown on the title page of each chapter.

Parts A through C provide the fundamentals of the underlying physics, radiobiology and technology respectively. Parts D through H provide the practical information needed to support external-beam radiotherapy: dose measurements, properties of clinical beams, patient dose computation, treatment planning and quality assurance. An appendix to Part D gives complementary details to enable a thorough understanding of the methods and data used for absolute dose measurement. Part I seeks to capture the exciting new developments in the subject including those in particle therapy, thus providing a basis for the reader to understand the ever expanding literature in this area. Parts J and K deal with brachytherapy using sealed and unsealed sources respectively. The framework of radiation protection is covered in Part L including an appendix describing the detailed application of UK legislation. Part M contains useful tables of physical constants, and electron and photon interaction data.

In a multi-author book of this length there will inevitably be a certain unevenness of style and level of detail and also some repetition; we see this as a strength rather than a weakness although we as editors have sought to ensure consistency. We wish to thank our many authors for their high-class contributions and not least for their patience during the time it has taken to bring together this work. It is unavoidable that some chapters are even more up-to-the-minute than others which were written more promptly. We would also like to thank Don Chapman and Helen Mayles who have read and commented on some of the chapters.

Finally, we have tried to make the content as international as possible by taking into account different practices and terminology in different parts of the world. For example we have used the words *Radiation Oncologist* and *Radiation Technologist* to describe the medical staff who prescribe radiotherapy and who operate the treatment machines respectively. Colleagues who carry out computer planning are referred to as *Dosimetrists*.

It is our hope that our readers will learn as much from reading this book as we have from editing it.

Philip Mayles
Alan Nahum
Jean-Claude Rosenwald

THE EDITORS

Philip Mayles Philip Mayles was born in 1947. He graduated from Gonville and Caius College, Cambridge with a BA in natural sciences in 1968. He completed a master of science degree in radiation physics at St Bartholomew's Hospital Medical School, following which he joined the staff of Guy's Hospital, where he worked until 1986. During this time he obtained a PhD from the Faculty of Medicine in London University. In 1986 he was appointed head of clinical radiotherapy physics at the Royal Marsden Hospital Surrey Branch under Professor Bill Swindell.

In 1994 he moved to Clatterbridge Centre for Oncology near Liverpool, as head of the Physics Department. During his time there he has overseen the expansion of the Radiotherapy Department from having five linear accelerators with conventional collimators to nine machines all equipped with multileaf collimators and portal imaging devices. The department uses modern imaging technology to enable 3D treatment planning for a high proportion of its patients and has been one of the pioneers of intensity modulated radiotherapy and image guided radiotherapy in the UK.

As chairman of the Radiotherapy Topic Group of the Institute of Physics and Engineering in Medicine he was instrumental in producing Report 75 on the design of radiotherapy treatment facilities and Report 81 on quality assurance in radiotherapy. He has an active interest in research especially in improving the physical basis of radiotherapy treatment and treatment planning.

In 1992 he and Alan Nahum established the Royal Marsden Course in Radiotherapy Physics which has been running successfully ever since, now under the direction of Margaret Bidmead and Jim Warrington. This course provided the inspiration for this book.

He and his wife, Helen, also a medical physicist, met at Guy's Hospital, London and have two daughters.

Alan Nahum was born in Manchester in 1949. He read physics at Worcester College, Oxford University. He gained his PhD in 1975 on theoretical radiation dosimetry using Monte Carlo methods at Edinburgh University under the supervision of Professor John Greening.

Alan then trained and worked as a school science teacher and worked as a gas cutter in a Volvo factory in Arvika, Sweden, before re-entering the medical physics world in 1979 as forskarassistent at the Radiofysik department, University of Umeå, where he taught and organised courses in radiation dosimetry, worked on ion-chamber response and on dosimetry codes of practice with Professor Hans Svensson, and where his two daughters Rebecka and Marie were born. He took a sabbatical (in Spring 1983) in Ottawa at the National Research Council of Canada working with Dave Rogers and Alex Bielajew on Monte Carlo simulation of ion-chamber response. He became a docent of Umeå University

Medical Faculty in 1983. He published *The Computation of Dose Distributions in Electron Beam Radiotherapy* just before leaving Umeå.

In 1985 Alan joined the Joint Department of Physics at the Institute of Cancer Research and Royal Marsden Hospital, Sutton, UK where his interests were redirected towards conformal (radio)therapy and subsequently to radioiological modelling. He co-directed (with Ralph Nelson and Dave Rogers) a course on Monte Carlo Transport of Electrons and Photons in Erice, Sicily in 1987 also co-editing a book with this title. In 1992 he and Philip Mayles started an annual 2×1-week course in radiotherapy physics which is still running. His PhD students at ICR were Charlie Ma, Mike Lee, Richard Knight, Paul Mobit, John Fenwick, Francesca Buffa, Mark Atthey and Margarida Fragoso. Postdoctoral scientists on his team included Beatriz Sanchez-Nieto, Frank Verhaegen, Cephas Mubata, Stefano Gianolini and João Seco, working on Monte Carlo simulation applied to dosimeter response and treatment planning, on biological modelling, and on building a database for analysis of clinical trials in conformal therapy. In 1997 he became Reader in Medical Physics of London University. He served as an associate editor for the journal *Medical Physics* between 1997 and 2004.

Alan left the ICR/Marsden in 2001 and after brief spells as a visiting scientist in Philadelphia, Reggio Emilia and Copenhagen, joined Clatterbridge Centre for Oncology, in June 2004, as head of physics research, becoming a Visiting Professor of Liverpool University in 2005. He was a member of the teaching faculty on the ESTRO course on IMRT and other conformal techniques from 1998 to 2005. His main current research is on using TCP and NTCP models in treatment plan optimisation.

Recreational interests include foreign languages, cooking, cricket and classical music.

Jean-Claude Rosenwald was born in 1945 in Neuilly, close to Paris. After earning an engineering degree in electronics, nuclear physics and computing sciences obtained in Nancy in 1967, he began his career as a computer scientist developing dose calculation programs in brachytherapy at the Institut Gustave Roussy in Villejuif, under the supervision of Andrée Dutreix. He obtained his PhD on this subject in 1976 from the University of Nancy.

He was appointed medical physicist at the Institut Gustave Roussy from 1971–1975 and then moved in 1976 to the Institut Curie in Paris, as head of the Physics Department. In 1996 he obtained an Habilitation à Diriger les Recherches from the Université Paul Sabattier in Toulouse in recognition of his capacity to coordinate research programmes. Altogether, more than 14 PhD and 60 master's students have undertaken research programmes under his supervision.

Dr. Rosenwald has a particular interest in the use of computers in radiation therapy, has participated in several international conferences, has been a co-author of several reports on this subject and was involved in the development of commercial solutions for treatment planning both for external-beam radiotherapy and brachytherapy. He has also promoted the use of proton beams in radiotherapy and played a major role in the development of the Centre de Protonthérapie d'Orsay. He has contributed to the expansion of the Radiotherapy Department at the Institut Curie, which is one of the leading centres in France, possessing modern equipment and practising modern radiotherapy techniques, based on advanced imaging devices and including intensity modulated radiotherapy, proton therapy and tomotherapy.

He served as president of the French Society for Medical Physics (today SFPM) from 1979–1982, as chairman of the Scientific Committee of the European Federation of Medical Physics (EFOMP) from 1990–1993 and chaired the Scientific Committee for Medical Physics at the International Conference of Bioengineering and Medical Physics held in Nice in 1997. He is a member of the editorial board of *Radiotherapy and Oncology*.

CONTRIBUTORS

Edwin Aird
Medical Physics Department
Mount Vernon Hospital
Northwood, Middlesex, United Kingdom

Gudrun Alm Carlsson
Department of Radiation Physics, IMV
Linköping University
Linköping, Sweden

Pedro Andreo
International Atomic Energy Agency
Vienna, Austria
and
Medical Radiation Physics
University of Stockholm–Karolinska Institute
Stockholm, Sweden

Mark Atthey
Medical Physics Department
CancerCare Manitoba
Winnipeg, Manitoba, Canada

Margaret Bidmead
Joint Department of Physics
Institute of Cancer Research
and
Royal Marsden NHS Foundation Trust
London, United Kingdom

Alex Bielajew
Nuclear Engineering and Radiological Sciences
University of Michigan
Ann Arbor, Michigan

Peter Blake
Institute of Cancer Research
and
Royal Marsden NHS Foundation Trust
London, United Kingdom

John N. H. Brunt
Physics Department
Clatterbridge Centre for Oncology
    NHS Foundation Trust
Wirral
Merseyside, United Kingdom

Jean Chavaudra
Service de Physique Médicale
Institut Gustave Roussy
Villejuif, France

Peter Childs
Joint Department of Physics
Institute of Cancer Research
and
Royal Marsden NHS Foundation Trust
London, United Kingdom

Roger Dale
Department of Radiation Physics and
    Radiobiology
Hammersmith Hospitals NHS Trust
and
Imperial College Faculty of Medicine
Charing Cross Hospital
London, United Kingdom

David Dance
Joint Department of Physics
Institute of Cancer Research
and
Royal Marsden NHS Foundation Trust
London, United Kingdom

Philip Evans
Joint Department of Physics
Institute of Cancer Research
and
Royal Marsden NHS Foundation Trust
London, United Kingdom

Maggie Flower
Joint Department of Physics
Institute of Cancer Research
and
Royal Marsden NHS Foundation Trust
London, United Kingdom

Tony Greener
Medical Physics Department
Guy's and St Thomas's NHS Foundation Trust
London, United Kingdom

Vibeke Nordmark Hansen
Joint Department of Physics
Institute of Cancer Research
and
Royal Marsden NHS Foundation Trust
London, United Kingdom

Dorothy Ingham
Medical Physics Department
Royal Devon and Exeter NHS Foundation Trust
Exeter, United Kingdom

Oliver Jäkel
Department of Medical Physics in Radiation Oncology
German Cancer Research Center
Heidelberg, Federal Republic of Germany

Colin Jones
Joint Department of Physics
Institute of Cancer Research
and
Royal Marsden NHS Foundation Trust
London, United Kingdom

Vincent Khoo
Academic Unit of Radiotherapy
Royal Marsden NHS Foundation Trust
and
Institute of Cancer Research
London, United Kingdom

Gerald Kutcher
Department of Radiation Oncology
Babies Hospital North
New York, New York

Christine Lord
Department of Medical Physics
Royal Surrey County Hospital NHS Trust
Guildford, United Kingdom

Les Loverock
Formerly with
Medical Physics Department
Royal Surrey County Hospital NHS Trust
Guildford, United Kingdom

C.-M. Charlie Ma
Department of Radiation Oncology
Fox Chase Cancer Center
Philadelphia, Pennsylvania

Ginette Marinello
Unité de Radiophysique et Radioprotection
Hôpital Henri Mondor
Créteil, France

Helen Mayles
Physics Department
Clatterbridge Centre for Oncology
    NHS Foundation Trust
Wirral
Merseyside, United Kingdom

Philip Mayles
Physics Department
Clatterbridge Centre for Oncology
    NHS Foundation Trust
Wirral
Merseyside, United Kingdom

Alejandro Mazal
Service de Physique Médicale
Institut Curie
Paris, France

David McKay
Radiotherapy Department
Royal Preston Hospital
Preston, United Kingdom

Alan McKenzie
Radiotherapy Physics Unit, Bristol Oncology Centre
United Bristol Healthcare NHS Trust
Bristol, United Kingdom

Cephas Mubata
Radiation Oncology
Columbia St-Mary's Hospital
Milwaukee, Wisconsin

Alan Nahum
Physics Department
Clatterbridge Centre for Oncology
    NHS Foundation Trust
Wirral
Merseyside, United Kingdom

Anthony Neal
Royal Surrey County Hospital NHS Trust
Guildford, Surrey, United Kingdom

Mark Oldham
Radiation Oncology and Biomedical Engineering
Duke University Medical Center
Durham, North Carolina

Ivan Rosenberg
Department of Radiotherapy Physics
University College London Hospitals
London, United Kingdom

Mike Rosenbloom
Joint Department of Physics
Institute of Cancer Research
and
Royal Marsden NHS Foundation Trust
London, United Kingdom

Jean-Claude Rosenwald
Service de Physique Médicale
Institut Curie
Paris, France

Roland Sabattier
Service d'Oncologie-Radiothérapie
Centre Hospitalier Régional d'Orléans
Orléans, France

John Sage
Department of Medical Physics and Bio-Engineering
University Hospitals of Coventry NHS Trust
Walsgrave Hospital
Coventry, United Kingdom

John Saunders
Formerly with
Medical Physics Department
Guy's and St Thomas's NHS Foundation Trust
London, United Kingdom

Glyn Shentall
Radiotherapy Department
Royal Preston Hospital
Preston, United Kingdom

Gordon Steel
Academic Department of Radiotherapy
Institute of Cancer Research
and
Royal Marsden NHS Foundation Trust
London, United Kingdom

David Thwaites
Department of Medical Physics and Engineering
Yorkshire Cancer Centre

Leeds University Teaching Hospitals
Leeds, United Kingdom

Jim Warrington
Joint Department of Physics
Institute of Cancer Research
and
Royal Marsden NHS Foundation Trust
London, United Kingdom

Steve Webb
Joint Department of Physics
Institute of Cancer Research
and
Royal Marsden NHS Foundation Trust
London, United Kingdom

Peter Williams
North Western Medical Physics Department
Christie Hospital NHS Trust
Manchester, United Kingdom

Jamal Zweit
Radiochemical Targeting and Imaging Department
Paterson Institute for Cancer Research
Manchester, United Kingdom

TABLE OF CONTENTS

PART A

FUNDAMENTALS

Editors: Jean-Claude Rosenwald and Alan Nahum

INTRODUCTION

This Part introduces the fundamental concepts underlying radiotherapy physics. It moves from the structure of matter and radioactivity (Chapter 1 and Chapter 2) to an explanation of the various interactions between radiation and matter (Chapter 3 and Chapter 4). A description of the Monte Carlo method is given in Chapter 5, as it illustrates the way in which knowledge of the detailed interactions can lead to an understanding of the transport of radiation in matter. Finally, in Chapter 6, the main dosimetric quantities are defined and their relationship is discussed in order to provide a more thorough understanding of dose-measurement methods presented in Part D and dose-calculation methods presented in Part E and Part F. Numerical values of many useful quantities are given in Part M. For complementary data on fundamental radiation physics used for medical applications the reader can refer to other more detailed textbooks (Goodwin and Rao 1977; Halliday and Resnic 1988; Bushberg et al. 1994; Cherry, Sorenson and Phelps 2003).

CHAPTER 1

Structure of Matter

Jean Chavaudra

CONTENTS

1.1   THE CONCEPT OF THE ATOM

The concept of matter composed of empty space filled with small indivisible particles had been proposed as early as several centuries BC, in particular by the Greek Demokritos. However, the Aristotelean concept of continuous matter prevailed for centuries until the development, circa 1800, of modern quantitative chemistry by scientists including A.L. de Lavoisier, J. Dalton, J.L. Gay-Lussac, A. Avogadro, and L.J. Proust. Their work developed the principle of the composition of compound materials from well-defined proportions of chemical elements.

Nevertheless, the most convincing evidence of the structure of matter was associated with the discovery of radioactivity, which revealed valuable information about the atomic structure.

1.2   THE ATOMIC STRUCTURE

1.2.1   BUILDING UP THE MODELS

The discovery of the electron and the ratio of its electric charge to its mass by J.J. Thomson in 1897 not only explained the nature of electric current, but also provided a basis for J. Perrin and J.J. Thomson's first hypothesis in 1901–1902 that atoms were made of negative and positive charges, globally neutral and structured according to some kind of planetary system.

The confirmation of this hypothesis was obtained through experiments done by E. Rutherford, and by H. Geiger and E. Marsden using α particle scattering in gold leaf. The results suggested that atoms were made of a heavy nucleus exhibiting a positive electric charge, surrounded by orbiting electrons bound to the nucleus by the electrostatic force of attraction, and exhibiting an equivalent negative charge.

This model was not satisfactory, as it could not explain how the electrons could avoid losing radiant energy and consequently, following the classical physics rules, being at last captured by the nucleus.

The currently accepted model for the description of atoms relies on the 1913 Bohr model. This model adds two postulates that contradict principles of classical physics, but it is further reinforced by the contributions of the pioneers of quantum-wave-mechanics, especially those of M. Planck, A. Sommerfeld, W. Pauli, P.A.M. Dirac, M. Born, W. Heisenberg, E. Schrödinger, and L. de Broglie.

The classical rules of mechanics and electricity could not explain atomic stability or the discontinuities observed in the results of early experiments on light emission and absorption by atoms. Explaining these phenomena required the adoption of Planck's quantum mechanical principles:

- Electrons revolve about the nucleus only in orbits with radii such that the relationship $M = nh/2\pi$ is satisfied, in which M is the angular momentum of the electron, h is Planck's constant and n is an integer. This means that the angular momentum (electron-mass×speed-along-orbit×radius) must be a multiple of $nh/2\pi$, and that only well defined orbits are possible
- Electrons do not gain or lose energy when they remain in a given orbit. They only exchange energy when they move from one orbit to another.

1.2.2   SCHEMATIC DESCRIPTION OF THE ATOMIC STRUCTURE

1.2.2.1   *The Nucleus*

Most of the mass of the atom is concentrated in the nucleus, which has a very high density (diameter on the order of a few fermis, or about $10^{-4}\times$ the atom's diameter).

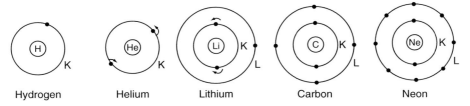

FIGURE 1.1
Schematic diagram showing the electron structure for hydrogen, helium, lithium, carbon and neon. (From Johns, H. E. and Cunningham, J. R., *The Physics of Radiology*, Charles C. Thomas., 4th Ed., Springfield, IL, 1983.)

The nucleus may be considered as made up of Z protons and N neutrons (nucleons).

The *protons* are responsible for the positive electric charge of the nucleus. Their electric charge has the same absolute value as the charge of the electron, $e = 1.602\ 176\ 53(14) \times 10^{-19}\ C^*$ and a rest mass $m_p = 1.672\ 621\ 71(29) \times 10^{-27}\ kg$ (Mohr and Taylor 2005).

To compensate for the Z *electric charges* of the nucleus, Z electrons should be present in the atom; Z is called the *atomic number* of the atom. The atoms classically considered as *natural* have atomic numbers ranging from 1 (hydrogen) to 92 (uranium).

The *neutrons* have no electric charge. Their rest mass is very close to the proton mass. The number of neutrons in a nucleus is close to the number of protons. Outside the nucleus, neutrons are unstable, dividing into protons, electrons and antineutrinos.

The mass of the nucleus is slightly smaller than the sum of the masses of the Z protons and N neutrons, due to their nuclear bonding energy corresponding, according to Einstein, to the use of a tiny proportion of the nuclear mass, called the mass defect (see Section 1.3.2.1).

1.2.2.2 The Peripheral Electrons/Electronic Shells

The *electrons* have a negative electric charge $e = 1.602\ 176\ 53(14) \times 10^{-19}\ C$ and a rest mass $m_e = 9.109\ 3826(16) \times 10^{-31}\ kg$.

The Z peripheral electrons revolve about the nucleus in well defined orbits, called electronic K, L, M, N,…shells according to their rank, from the nucleus to the periphery of the atom.

According to the principles of the atomic model, the number of electrons present in a given shell is limited. The innermost shell, called the K shell, has the rank of 1. Any shell with a rank n has a maximum allowed number of electrons of $2n^2$.

Hydrogen is the simplest atom, consisting of one electron revolving about one proton on the K shell (Figure 1.1). The next simplest atom, helium, has two electrons, saturating the K shell, and spinning in opposite directions. The next atom, lithium, has 3 electrons and the additional electron is alone on the L shell. The same evolution can be observed from carbon to neon atoms, the latter presenting a saturated L shell.

The most peripheral electrons, those belonging to the outermost shell (valence electrons), are directly linked to the chemical properties of atoms and molecules, and so determine the chemical elements or nuclides reported in the Mendeleyev table. Helium and neon atoms, presenting a saturated outer shell, are remarkably chemically stable.

---

* These values are given in the so-called *concise form* in which the number in brackets is the *standard uncertainty*.

1.2.2.3 The Global Atom

Mass Number: The mass number A is defined as the total number of nucleons in the atom. It ranges from 1 (hydrogen) to more than 200 for the heaviest nuclei.

Atomic Mass: As seen previously, the mass of a given atom is close to the sum of the masses of the nucleons and the electrons. Remember that the mass of an electron is about 1840 times smaller than the mass of a nuclear particle. The actual mass of an atom X, $m(X)$, expressed in terms of kilograms, the international unit of mass, is very small and not of much practical use. Consequently, another approach is often used, by defining a special unit of *atomic mass* (a.m.u. or u).

The international unit of *atomic mass* is approximately the mass of a nuclear particle, permitting the atomic mass of an atom to be expressed with numbers close to the mass number. The official definition of the atomic mass unit (unified atomic mass m_u) is related to 1/12 of the mass of the carbon atom with 6 protons and 6 neutrons, i.e. $1 u = 1.660\,538\,86(28) \times 10^{-27}$ kg. Because the mass number of a carbon atom is exactly 12, its atomic mass is also 12 u. However, for oxygen 16, for example, 1 atom corresponds to 15.9991 u. Atomic masses range from 1 u (hydrogen) to more than 200 u for the heaviest nuclei.

Another definition of atomic mass is used in chemistry, with the same symbol A, but in this case, it refers to the so-called atomic weights. The difference from the definition above is that, for a given element, this definition takes the naturally occurring mixture of nuclides into account (with the isotopes) so the value of A may be different from the mass number. We may also refer to molar masses, so that if, for instance, a mole (mol) of carbon atoms is made of N_A (Avogadro's number $= 6.022\,1415(10) \times 10^{23}$ mol^{-1}) atoms, a mole of carbon atoms has a mass of exactly 12 g. Thus, the molar mass $M(X)$ of any other element X can be derived from the carbon molar mass, in the same way as the atomic masses[*].

1.2.3 Atomic Structure Interpretation according to the Wave-Mechanical Model

In this model, the classical fundamental law, $\vec{F} = m\vec{\gamma}$ in which the acceleration γ of a particle with mass m is obtained by applying a force $\vec{F}$ to the particle, is replaced by the Schrödinger equation in which the particle is associated with a probability of presence at a given place, at a given time, thus extending the meaning of electron orbits or shells. Furthermore, in addition to the classical parameters qualifying the behaviour of a particle (mass, 3D coordinates and speed components) an intrinsic angular momentum, the spin, is added. Particles with spins that are odd multiples of $\frac{1}{2}h/2\pi$ are called fermions and include in particular the *classical* components of matter (electrons and nucleons), also called leptons. Particles with spins that are multiples of $h/2\pi$ are called bosons and include photons.

---

[*] "The mole is the amount of substance of a system that contains as many elementary entities as there are atoms in 0.012 kg of carbon 12; its symbol is "mol". When the mole is used, the elementary entities must be specified, and may be atoms, molecules, ions, electrons, other particles or specified groups of such particles." (Bureau International des Poids et Mesures 1998).

1.2.3.1 Peripheral Electrons

According to the model, the peripheral electrons of the atoms are each characterised by four quantum numbers:

The *principal quantum number*, n, defines the shell within which the electron resides. As stated above, n can have the values 1, 2, 3... corresponding to K, L, M ... shells.

The *azimuthal quantum number*, l, describes the electron's angular momentum and provides information on the elliptical characteristics of the orbit. l can have integer values ranging from 0 to $n-1$.

The *magnetic quantum number*, m_e, represents the orientation of the electron's magnetic moment in a magnetic field and provides information on the orbit orientation with respect to a given reference direction. The values of this quantum number, for a given orbit l, are the integer values in the range $-l \leq m_e \leq l$.

The *spin quantum number*, m_s, defines the direction of the electron's spin upon its own axis with respect to a given reference direction, and can have two values, $+1/2$ and $-1/2$.

1.2.3.2 Electronic Status

The Pauli exclusion principle states that a given set of quantum numbers can characterise only one electron.

This allows us to derive the maximum number of electrons in a given shell. For example:

In the K shell, $n=1$,

therefore $l=0$ and $m_e=0$.

m_s can have values of $+1/2$ or $-1/2$.

Only 2 electrons can reside in the K shell.

The same rule leads to 8 possibilities for the L shell, 18 for the M shell, etc.

1.2.3.3 The Nucleus

Two main models have been proposed for the nucleus. The *liquid drop model* assumes that the nucleus is made up of closely packed nucleons in constant motion. This model is compatible with the explanation of interactions of heavy particles with the nucleus, but is not compatible with the explanation for discrete nuclear energy states revealed in interactions with light particles. Because of this, a shell model of the nucleus has been proposed, which is similar to the shell model for the peripheral electrons. In this model, each nucleon is characterised by four quantum numbers, with the same meanings, replacing *electron* with *nucleon*:

The *principal quantum number*, n, related to each nucleon shell (protons or neutrons);

The *orbital quantum number*, l, characterising the orbital motion of the nucleons inside the nucleus (from 0 to $n-1$);

The *spin quantum number*, m_s, equal to $\pm 1/2$;

The *magnetic quantum number*, m_e (from $-l$ to l).

More elaborate models have been proposed in order to fit better with the latest experimental data, but they will not be considered here.

1.2.4 NOMENCLATURE

To characterise a given nuclide, the following symbolism is used: $^A_Z X$, in which X is the element nuclide symbol, Z the atomic number and A the mass number.

Isotones are atoms that have the same number of neutrons, such as $^{51}_{23}$V, $^{52}_{24}$Cr, $^{54}_{26}$Fe.

Isotopes are atoms that have the same number of protons (Z) with different number of neutrons. They are therefore different versions of the same element. Many elements are made of a mixture of several isotopes, with a fairly stable composition. The atomic mass A for such elements may then be calculated and is not necessarily an integer.

Isobars are atoms that have the same number of nucleons (value of A). They belong to different elements as they have different values of Z (e.g. $^{40}_{18}$A, $^{40}_{19}$K, $^{40}_{20}$Ca). For the so-called natural nuclides, A ranges from 1 to 238 $\left(^{238}_{92}\text{U}\right)$.

1.3 BINDING ENERGIES IN ATOMS AND MOLECULES

1.3.1 ENERGY AND MATTER

There are various forms of energy, mechanical, kinetic, and electrical. The fundamental laws of physics include the conservation of the total energy of a given system, whatever the transformations in it. The energy unit, in the International System of Units is the joule (J), but because this quantity is too large when applied to particle energies, the electron volt (eV) is often used instead.

This unit, representing the energy acquired by an electron accelerated through a potential difference of 1 V, is such that:

$$1 \text{ eV} \approx 1.6 \times 10^{-19} \text{ J}$$

$$1 \text{ keV} \approx 1.6 \times 10^{-16} \text{ J}$$

$$1 \text{ MeV} \approx 1.6 \times 10^{-13} \text{ J}$$

1.3.1.1 Energy of Photons

Photons have no rest mass. Considering the frequency v of the electromagnetic wave associated with a given photon (behaving as a particle in the energy range of radiotherapy- i.e. above tens of keV), the amount of energy E carried by the photon can be obtained, according to Einstein, from the equation:

$$E = hv \quad \text{or} \quad E = \frac{hc}{\lambda}$$

where E is given in eV, Planck's constant h is given as $4.135\,667\,43 \times 10^{-15}$ eV s, v is in Hz, c is the velocity of light in vacuo* (299 792 458 ms^{-1}) and λ is the wavelength in m.

After Duane and Hunt, a numerical value of E can be obtained from:

$$E(\text{eV}) = \frac{1240}{\lambda(\text{nm})}$$

It will be seen that photon energies increase when their wavelength decreases. Also, although photons have no rest mass, they have a momentum equal to hv/c and a virtual dynamic mass $m = h/\lambda c$

---

* In fact the symbol should in principle be written c_0 (to refer to velocity *in vacuo*).

TABLE 1.1

Relationship between Kinetic Energy, Relative Mass and Relative Velocity for Electrons

T Kinetic Energy (MeV)	m_v/m_0	Relative Velocity
0	1	0
0.051	1.1	0.416
0.511	2	0.866
1.022	3	0.942
5.11	11	0.996
51.1	101	0.99995

Source: From Dutreix, J. et al., *Biophysique des Radiations et Imagerie Médicale*, 3rd Ed., Masson, Paris, France, 1993.

1.3.1.2 Energy of Particles with Mass

According to the relativity principle, the mass defined above from classical physics principles is not constant when the particle velocity v varies. It is given by:

$$m = \frac{m_0}{\sqrt{1 - \frac{v^2}{c^2}}}$$

where m_0 is the rest mass ($v=0$), and c is the velocity of light in vacuo.

The particle momentum is given by $\overrightarrow{m \cdot v} = \overrightarrow{m_0 \cdot v_0}/\sqrt{1-(v^2/c^2)}$

Also, the rest mass m_0 is considered as a particular form of energy, according to the Einstein equation:

$$E = m_0 c^2$$

For example, the energy corresponding to one u is 931 MeV, and the energy corresponding to the rest mass of an electron is 511 keV.

When a particle is moving, its total energy is the sum of the energy corresponding to its rest mass, and of the translation kinetic energy. So, the total energy of the particle becomes $E=mc^2$, according to Einstein, in such a way that $mc^2=m_0c^2+T$, where T is the translation kinetic energy, i.e. the additional energy due to the particle movement. T can also be expressed as equal to $m_0c^2((1/\sqrt{1-v^2/c^2})-1)$. When the particle velocity becomes small, the above equation becomes close to the classical value of $T=\frac{1}{2}mv^2$.

This theory shows that, for instance, in the energy range of the electrons considered in medical radiological physics, their very small rest mass can nevertheless allow large kinetic energies obtained with large velocities approaching the light velocity in vacuo (Table 1.1).

1.3.2 BINDING ENERGIES IN ATOMS

Matter appears to be made of large numbers of particles associated through bonds resulting from forces with a wide range of intensity. The nature of these forces is not yet fully understood. Four kinds of interactions have been considered to explain such remote interactions, involving quantum energy exchanges between elementary particles: gravity, electromagnetic interactions, weak interactions and strong interactions. For example, in the nucleus, the strong interactions are believed to be between the elementary particles called quarks; nucleons are believed to be made up of three quarks. The quantum exchanged in such interactions is called the gluon, which has no mass.

The strength of the bonds existing between subgroups of particles depends upon whether the forces are within the nucleus, inside the atom, between atoms or between molecules. The rest of this Chapter will deal with atomic and molecular structures. The structure of the nucleus will be considered in Chapter 2.

The bonds associated with different matter structures can be quantified with the definition of *binding energy*: the energy required to dissociate a given structure or substructure. It is usually denoted by W. Table 1.2 provides examples of binding energies corresponding to various structures.

1.3.2.1 Mass Defect

The energy required to create a bond in a system is lost by the system and induces a mass decrease. So, the total mass of the system is smaller than the sum of the masses of its individual components. This is called the *mass defect* and is about 7 MeV per nucleon in the helium atom.

1.3.2.2 Electron Binding Energy and Energy Levels of the Atomic Shells

According to the models above, an electron of an inner shell of an atom is attracted by the nucleus with an electrostatic force greater than the force applied by the nucleus to an electron of an outer shell. The binding energy required to extract a given electron from an atom depends on the shell considered and the electric charge Z of the nucleus, according to the approximate Moseley rule:

$$W = 13.6 \frac{(Z-b)^2}{n^2}$$

in which W is the binding energy of the electron (in eV), Z is the atomic number of the atom, n is the electron shell number and b is a constant used to correct for the electrostatic *screening effect* due to electrons situated between the nucleus and the electron considered.

Obviously, the outer shells are less dependent on the Z number of the atoms, due to increasing values of b. Consequently the outer shells correspond to a binding energy ranging from 1 to 16 eV, whatever the value of Z. This corresponds to the first ionisation energy of the atoms.

Note that the value of n is related to the principal energy level of a given shell (usually $n=1$ means the K shell, $n=2$ the L shell etc). The binding energies of sublevels are close to the binding energy of the principal energy level, but cannot be calculated with the empirical Moseley formula.

TABLE 1.2

Orders of Magnitude of Various Binding Energies

Action	Energy Required (eV)
To extract one alcohol molecule from a water solution through distillation	13
To break the covalent link between H and O in H_2O using electrolysis	5
To extract one electron from	
A hydrogen atom	13
Molecules of biological materials	15
The M shell of a tungsten atom	2500
The K shell of a tungsten atom	70 000
To dissociate the helium nucleus into 4 nucleons, for each nucleon	7 000 000

Source: From Dutreix, J. et al., *Biophysique des Radiations et Imagerie Médicale*, 3rd Ed., Masson, Paris, France, 1993.

Also, when all electrons belong to shells in such a way that all binding energies are at the maximum value, the atom is said to be in the fundamental state. This state also corresponds to the minimum internal energy.

1.3.3 BINDING ENERGIES IN MOLECULES

In molecules, the links between atoms rely on sharing outer shell electrons. Consequently, the binding energies of such electrons are changed. The magnitude of the changes depends on the molecules, and on the chemical species considered. The binding energies of electrons belonging to inner shells are less affected, or may remain unchanged.

In crystals, the situation is different, as the number of electrons shared is usually large, depending on the crystal structure and the atom. Consequently, electrons can be found at many different energy levels, in an energy range comparable to an energy band, called the *valence band*.

If, in the fundamental state, the crystal absorbs external energy, electrons can reach a higher energy level (an excited state), and behave as they do in conductors of electricity. The corresponding energy band is then the *conduction band*.

When the *conduction band* overlaps the *valence band*, the crystal is a conductor of electricity. If the bands do not overlap, a *forbidden band* of a few eV between the two bands ensures that the crystal behaves as an insulating material.

1.4 PERTURBATION OF BINDING ENERGIES

1.4.1 EXCITATION

If a given atom absorbs external energy at a level smaller than any electron binding energy, an electron may be moved from one shell to another, farther from the nucleus. This corresponds to a higher level of internal energy. The atom is then said to be excited.

One can assume that, if the electron moves, for example, from the L shell to the M shell, it is because the absorbed energy ΔW is such that:

$$\Delta W = W_M - W_L$$

W_M and W_L being binding energies of M and L shells, respectively.

If the energy given to the atom, increasing its internal energy, is defined as positive, then the binding energies should be considered as negative. For instance, for Tungsten:

$$W_L = -11\,280 \text{ eV} \quad \text{and} \quad W_M = -2\,810 \text{ eV}$$

$$\text{So:} \quad \Delta W = -2\,810 + 11\,280 = 8\,470 \text{ eV}.$$

As in the fundamental state all inner shell electron positions are occupied, the electron transitions are most often observed between outer shells, involving the peripheral electrons, which are also responsible for the chemical characteristics of the atom. As such, peripheral electrons have a weak binding energy, and excitations can be produced with low energy photons (UV or visible). Conversely, excitations linked to internal shells require higher energy photons.

1.4.2 IONISATION

If a given atom absorbs external energy at a level equal to or higher than an electron binding energy, an electron becomes free because its link with the atom has been broken. As a result, the electrical equilibrium in the atom is no longer maintained, and the atom becomes a *positive ion*. For a given electron to be removed from the atom, the energy transfer must be higher than the binding energy of this electron. The excess of energy is, in principle, shared between the ionised atom and the electron as kinetic energy. Since particle momentum is conserved, most of the kinetic energy is given to the electron, because of the very large difference in masses.

1.4.3   EQUILIBRIUM RECOVERY: FLUORESCENCE

After receiving a given amount of energy, leading to excitation or ionisation, an atom has an excess of internal energy, becomes unstable, and tends to return to its fundamental state. This recovery of the fundamental state is associated with re-emission of energy. In the *fluorescence* process, the energy re-emission is made through prompt emission of one or several photons (after a delay of the order of 10^{-6} s).

The mechanism of this process is described as follows: after excitation or ionisation, vacancies or *holes* appear in electron shells, and are promptly filled by electrons cascading from energy levels corresponding to shells farther from the nucleus. As the vacancies are filled, energy is released (for instance through photon emission) and the internal energy of the atom is reduced.

A single photon is emitted if the original event is a single ionisation and if the position of the electron removed from the atom is re-occupied by an external free electron. The photon energy is then equal to the binding energy of the electron removed by ionisation (W).

Several photons are emitted if the return to the fundamental state is made through successive transitions of different electrons, from the inner part of the atom to the peripheral shell, where a free external electron can be captured. The global energy emitted through the photons is still equal to W, the binding energy of the electron initially removed by the ionisation.

If the original event is an excitation, the energy available is the difference in the binding energies corresponding to the shells involved in the electron transition (Figure 1.2a and Figure 1.2b).

The energy of the emitted fluorescence photons is closely related to the mechanism of their production. This energy is therefore characteristic of the energy structure of the atoms and molecules involved in the production process and the emitted photons are called *characteristic x-rays*. The fluorescence spectrum is made up of lines allowing the characteristic radiation to be identified and associated with a specific atom. Fluorescence emission is usually described according to the destination of the cascading electron, as, for instance, K characteristic radiation, or K fluorescence.

K shell is a general name for a family of sub-shells with close energy levels, in such a way that the K characteristic radiation is made of a number of lines with close energies, with subfamilies described as Kα, Kβ... according to the original energy level of the electron (an example is given for tungsten in Figure 1.3). Depending on the energy levels in the atoms, the fluorescence photons can belong to the infrared, visible, UV or x-ray part of the electromagnetic spectrum.

1.4.4   EQUILIBRIUM RECOVERY: AUGER EFFECT

Occasionally, the recovery energy may be used to eject a second electron instead of a photon. This ejected electron is called an Auger electron. Like fluorescence photons, Auger electrons have well-defined energies, depending on whether the energy is effectively transferred to more external electrons or whether the emission of the Auger electron is due to a free electron filling the initial vacancy (Figure 1.2c and Figure 1.2d).

The probability of the Auger effect is higher for low-Z biological media than the probability of fluorescence, close to 1 for $Z < 10$ and about 0.1 for $Z > 80$.

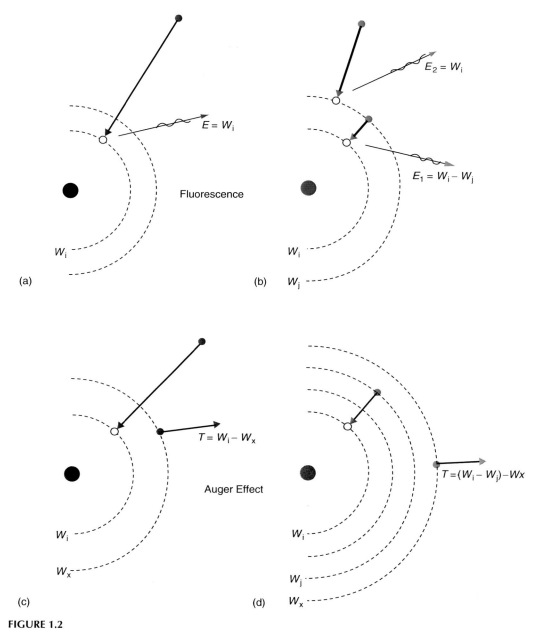

FIGURE 1.2
Description of the mechanisms of fluorescence and the Auger Effect: (a) an i shell vacancy is filled with a free external electron, (b) an i shell vacancy is filled through two successive transitions involving the j shell, (c) after an i shell vacancy is filled with an external electron, an electron of the x-shell is ejected as an Auger electron, (d) similar to c, but the vacancy filling is due to a transition between the j and i shells. (From Dutreix, J. et al., *Biophysique des Radiations et Imagerie Médicale*, 3rd Ed., Masson, Paris, France, 1993.)

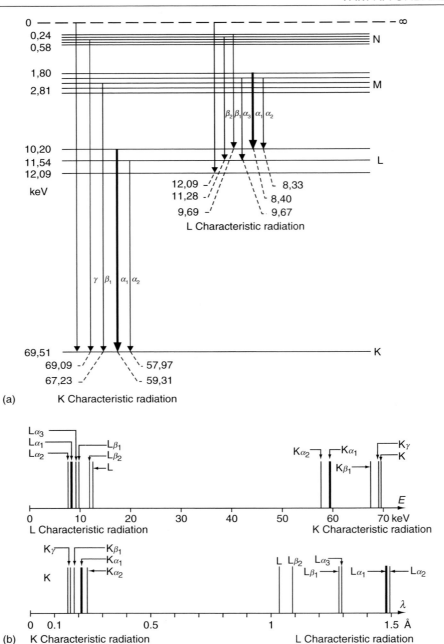

FIGURE 1.3
Fluorescence in Tungsten (a) Schematic description of the electron energy levels in tungsten atoms: The thick lines correspond to the highest probabilities of emission. The M fluorescence has an energy of a few keV and is usually not visible in the spectra from x-ray tubes, (b) Spectral distributions of fluorescence lines for tungsten atoms: Families of lines can be shown as a function of energy or of wavelength. (From Dutreix, J. et al., *Biophysique des Radiations et Imagerie Médicale*, 3rd Ed., Masson, Paris, France, 1993.)

After an Auger effect transition, the atom is still ionised, and another process can occur, leading to fluorescence or to a new Auger electron.

1.5 EXAMPLES OF ATOMS AND MOLECULES OF INTEREST FOR RADIATION PHYSICS

TABLE 1.3

Examples of Electron Binding Energies for the Main Components of Biological Tissues and for Materials of Interest in Radiology

Materials	Atoms	Electron Binding Energies (keV)	
		K Shell	Outer Shell
Biological tissues	$_1$H	0.0136	0.0136
	$_6$C	0.283	0.0113
	$_8$O	0.532	0.0136
	$_{15}$P	2.142	0.011
	$_{20}$Ca	4.038	0.0061
X-Ray tube anodes	$_{74}$W	69.51	0.008
Radiographic films	$_{35}$Br	13.48	
	$_{47}$Ag	25.53	

Source: From Dutreix, J. et al., *Biophysique des Radiations et Imagerie Médicale*, 3rd Ed., Masson, Paris, France, 1993.

Table 1.3 gives binding energy data for some materials of interest in radiology and radiotherapy.

CHAPTER 2

RADIOACTIVITY

Jean Chavaudra

CONTENTS

2.1 STABLE NUCLEUS: NUCLEAR ENERGY STRUCTURE

2.1.1 NUCLEAR ENERGY LEVELS

According to the shell model of the nucleus, the forces existing between nucleons determine a structure supported by discrete strong binding energies as illustrated in Figure 2.1.

This could appear paradoxical because the nucleus is made up of a mixture of protons and neutrons despite the repulsive electrostatic forces between protons. In fact, the nuclear energy structure can be explained by considering the following principal forces between the nucleons:

- The postulated attractive force ensuring the cohesion of the nucleus is called the *nuclear force* or *strong force*. It is supposed to be effective when the distance between nucleons is smaller than the diameter of the nucleus, and it is associated with the exchange of gluons between the quarks constituting the nucleons.
- The *electrostatic repulsive force* expected to take place between the protons is effective when the distance between the protons is greater than the diameter of the nucleus. Its strength is much smaller than the strength of the nuclear force (10^{-2} to 10^{-6}), and it is believed to be associated with the exchange of photons.
- The *weak force* is a very weak force, about 10^{-10} times smaller than the nuclear force, believed to explain some β disintegrations, and it also involves electrons and neutrinos. It is associated with the exchange of bosons (so called W and Z particles which have large mass but cannot be observed directly).
- The *gravitational force* is a general component of the universe, but it is very tiny when small masses are considered, of the order of 10^{-40} times smaller than the nuclear force, in relation to both protons and neutrons.

The binding energy of nucleons can be considered similarly to the principles presented for the peripheral electrons. The major difference is the magnitude of this binding energy that is of the order of 1 MeV per nucleon, i.e. about 10^6 times the binding energy of the peripheral electrons. This binding energy is obviously linked to the content of the nucleus, and it is the result of the combination of the different forces described above. Figure 2.2 shows the variation of the mean nuclear binding energy per nucleon as a function of the total number of nucleons *A*.

It appears that the maximum binding energy per nucleon (about 9 MeV per nucleon) occurs for values of *A* around 60 to 70 (Fe, Ni, Co, Ca, etc.), corresponding to very stable nuclei. When the number of nucleons, *A*, becomes higher, the binding energy decreases

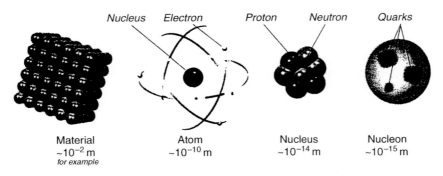

FIGURE 2.1
The different components of the structure of matter, progressively magnified. (From Gambini, D. J. and Granier, R., *Manuel Pratique de Radioprotection*, Technique & Documentation collection, Lavoisier Editions, Cachan, France, 1997.)

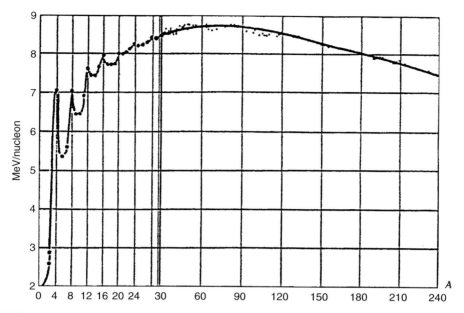

FIGURE 2.2
Mean nuclear binding energy per nucleon as a function of the mass number (also the total number of nucleons A).
(From Gambini, D. J. and Granier, R., *Manuel Pratique de Radioprotection*, Technique & Documentation collection,
Lavoisier Editions, Cachan, France, 1997.)

because of the larger distances between nucleons and the increasing influence of Coulomb forces. Discontinuities can be observed for a number of atoms where the saturation of nuclear shells (according to the shell model of the nucleus) is associated with a remarkable atomic stability (e.g. $_2^4$He, $_6^{12}$C, and $_8^{16}$O).

It can be predicted that the fusion of two light-nuclei, resulting in a heavier one, or the fission of a very heavy nucleus into lighter ones will allow the mean binding energy per nucleon to increase, leading to more stable atoms. This is associated with the release of energy and a greater global mass defect. This energy release can be very high. For example, $_{92}^{235}$U fission produces near 10^{11} J/g, whereas, the fusion of $_1^3$H and $_1^2$H produces 3.1011 J/g.

2.1.2 ABUNDANCE OF STABLE NUCLEI AS A FUNCTION OF THE NUMBER OF PROTONS AND NEUTRONS

The possible number of nuclear configurations leading to stable atoms is only a small fraction of all possible configurations for a given set of nucleons. Approximately 300 different stable atoms naturally exist on earth, but more than 2000 other atoms have already been observed, at least for very short times during the many experiments that have taken place over recent years.

Figure 2.3 is interesting to consider as it shows the number of neutrons as a function of the number of protons for a wide range of existing atomic mass numbers. It appears that, for light atoms, the number of neutrons is about equal to the number of protons. For mass numbers above approximately 40, the number of neutrons becomes higher than the number of protons to allow these existing elements to remain stable for a significant time. If a given nucleus has a neutron or a proton content differing from the optimal structure, it would be expected that unstable configurations will arise during the constant changes linked to the orbits of the nucleons. Most of the nuclei with atomic numbers greater than approximately 80 are unstable, but unstable nuclei also exist with smaller atomic numbers.

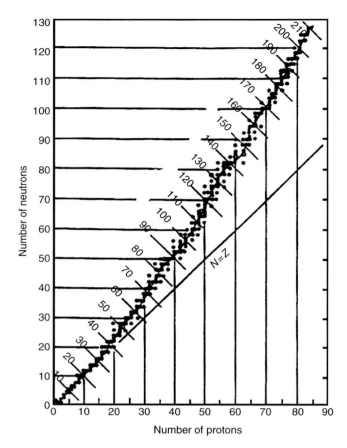

FIGURE 2.3
Number of neutrons as a function of the number of protons for existing atoms for mass numbers *A* in the range 0–210. (From Dutreix, J. et al., *Biophysique des Radiations et Imagerie Médicale*, 3rd Ed., Masson, Paris, France, 1993.)

2.1.3 INFLUENCE OF N/Z ON STABILITY

Excepting catastrophic events such as fusion or fission, more stable structures can be obtained through changes in the number or in the nature of nucleons for a given nucleus. For instance, a nucleus with an excess of neutrons can eject a negative charge, changing a neutron into a proton (β^- emission), whereas, a nucleus with an excess of protons can eject a positive charge, changing a proton into a neutron (β^+ emission). Very heavy nuclei that contain too many protons and neutrons can eject a group of two protons and two neutrons (i.e. a helium nucleus) which forms one of the most stable structures (α emission).

2.2 NUCLEAR INSTABILITY: RADIOACTIVITY

2.2.1 DEFINITION OF RADIOACTIVITY

Discovered by Henri Becquerel in 1896 in the form of emissions from uranium, radio-activity is commonly described as the possibility that a given atomic nucleus will spontaneously emit particles through disintegration, leading to possible change in its physical and chemical properties. In 1898 Pierre and Marie Curie announced that they had identified two hitherto unknown elements, polonium and radium. These elements were much more radio-active than uranium and formed a very small fraction of the uranium ore, pitchblende.

The complex phenomena involved in the radioactivity process required at least ten years to be properly explained with the initial work made by Henri Becquerel, Pierre and Marie Curie, Ernest Rutherford, and Frederick Soddy at the beginning of the twentieth century. Another main step was the discovery of *artificial* radioactivity by Irène and Frédéric Joliot in 1934, introducing the production of a great number of *radionuclides* and the world of nuclear energy. The process involved in the radioactivity phenomenon is the same for natural and artificial radionuclides.

Based on the current view of the energy structure of the atomic nucleus (Blanc and Portal 1999) the principal different, spontaneous radioactive transformations can be categorised according to whether they are associated with strong interactions, electrostatic interactions or weak interactions. They are described in Section 2.2.2 through Section 2.2.4 according to these classifications.

2.2.2 RADIOACTIVE TRANSFORMATIONS ASSOCIATED WITH STRONG INTERACTIONS

2.2.2.1 α Radioactivity

As previously stated, one of the most strongly bound structures is the helium nucleus ${}^{4}_{2}\text{He}$. It is emitted as alpha particles by a number of nuclei with high atomic numbers (>80), and it was first observed by H. Becquerel and described by E. Rutherford around 1900.

An alpha transition can be described as follows:

$$ {}^{A}_{Z}\text{X} \rightarrow {}^{A-4}_{Z-2}\text{Y} + {}^{4}_{2}\text{He} $$

where the nuclide, X, changes into another nuclide, Y. The daughter nuclide frequently is in an excited energy state, leading to a gamma photon emission associated with the return to the basic energy state.

It can be observed that the sums of the mass numbers and the atomic numbers of the daughter nuclide and the alpha particle equal the mass number and the atomic number, respectively, of the parent nuclide. The discrete transition energy is usually shared between the alpha particle kinetic energy and the gamma photon. For this reason, alpha particles are emitted with discrete energies.

A classical alpha decay is the decay of ${}^{226}_{88}\text{Ra}$:

$$ {}^{224}_{88}\text{Ra} \rightarrow {}^{222}_{86}\text{Rn} + {}^{4}_{2}\text{He} $$

The corresponding radioactive decay scheme is represented in Figure 2.4.

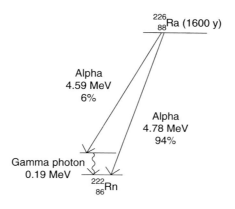

FIGURE 2.4
Radioactive-decay scheme for alpha decay of ^{226}Ra. (From Hendee, W. R., *Medical Radiation Physics*, 2nd Ed., Yearbook Medical Publishers, Inc., Chicago, IL, 1979.)

2.2.2.2 *Spontaneous Fission*

In spontaneous fission, the nucleus splits into two or more pieces with the simultaneous emission of fast neutrons. This phenomenon happens with the heaviest nuclei, and it seems to be the reason for an upper limit of the atomic numbers of approximately 110. An example of such radioactive sources used in radiation therapy is $^{252}_{98}Cf$ that produces α and γ particles, and neutrons with a mean energy of 2.35 MeV.

2.2.3 RADIOACTIVE TRANSFORMATIONS ASSOCIATED WITH THE ELECTROSTATIC FORCE

2.2.3.1 *Nuclear Isomerism (or γ Radioactivity)*

So-called *pure gamma radioactivity**** is really nuclear isomerism. A nucleus is artificially elevated to an excited energy state, and it then returns to the fundamental state through well-defined gamma emissions. This may be represented by:

$$^{Am}_{Z}X \rightarrow {}^{A}_{Z}X + \gamma$$

A classical example in the medical field of this type of decay is

$$^{99m}_{43}Tc \rightarrow {}^{99}_{43}Tc + \gamma$$

that results in gamma photons of 140 keV.

2.2.3.2 *γ Emission and Internal Conversion*

γ photon emission is often observed during radioactive transformations as the daughter nucleus is formed in an excited, metastable state. The γ photons, first observed by Villard in 1900, directly result from the return of the daughter nucleus to ground energy state, and they are produced according to the discrete energies available, corresponding to characteristic lines in the energy spectrum. As a result, a large number of radionuclides with different decay schemes include γ emissions in the resulting emission spectrum. When the nuclear transition energy available is sufficiently high, the return to ground energy level may require several steps with cascades of γ-rays and β particles.

Sometimes, the energy release is produced by internal conversion, corresponding to an electron being ejected from the shells close to the nucleus. This electron receives kinetic energy equal to the energy released minus the binding energy of the electron. This process could appear as an interaction of a γ photon with the electronic shells. The result of such a vacancy in an electronic shell is a cascade of electron transitions to fill the vacancy with emission of characteristic x-rays and Auger electrons.

The probability of internal conversion increases rapidly with atomic number and with the lifetime of the metastable state of the nucleus.

---

* Photons emitted from the nucleus are called γ photons, whereas, the photons produced outside the nucleus are called x-rays.

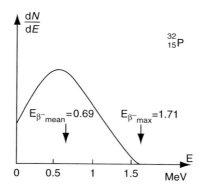

FIGURE 2.5

β^- spectrum of $^{32}_{15}P$. (From Gambini, D. J. and Granier, R., *Manuel Pratique de Radioprotection*, Technique & Documentation collection, Lavoisier Editions, Cachan, France, 1997.)

2.2.4 RADIOACTIVE TRANSFORMATIONS ASSOCIATED WITH THE WEAK INTERACTION

2.2.4.1 β^- Radioactivity

β^- *radioactivity** can be observed when neutrons are in excess in the nucleus. The following process occurs through weak interactions where a neutron becomes a proton with the release of a negative electron (*negatron*) and of an electronic antineutrino (no electric charge and rest mass <15 eV/c^2):

$$^{1}_{0}n \rightarrow {}^{1}_{1}p + {}^{0}_{-1}e + \overline{\nu}_e$$

Generally, the transition energy is equally shared between the electron (β^- particle) and the neutrino, according to probabilities leading to a continuous energy spectrum as shown on the Figure 2.5 for $^{32}_{15}P$, even though the transition energies are discrete. The energy transmitted to the nucleus is negligible.

The result for the atom is as follows:

$$^{A}_{Z}X \rightarrow {}^{A}_{Z+1}Y + \beta^- + \overline{\nu}_e$$

As the atomic number A does not change, this transition is called *isobaric*.

β^- decay is possible whatever the atomic mass as is also the case for β^+ decay below.

2.2.4.2 β^+ Radioactivity

β^+ decay can be observed when protons are in excess in the nucleus. The same type of process as shown above is produced with

$$^{1}_{1}p \rightarrow {}^{1}_{0}n + {}^{0}_{+1}e + \nu_e$$

---

* Electrons emitted from the nucleus are called β particles, whereas, electrons arising from the electron shells are not β particles (they can, for instance, be Auger electrons).

where $_{+1}^{0}e$ is an antielectron, called a *positron*. ν_e is an electron-neutrino. Again, the transition energy is shared between the β^+ particles and the neutrinos in such a way that a continuous energy spectrum is observed for the β^+ particles.

The β^+ particles do not permanently exist, and they initially behave like the β^- particles with respect to interactions with matter. Eventually, when they have almost reached rest energy, a recombination of the positron and a negatron of the matter takes place.

$$_{-1}^{0}e + _{+1}^{0}e \rightarrow 2 \text{ annihilation photons} \quad (0.511 \text{ MeV})$$

This means that nuclei that cannot produce at least 1.02 MeV for a given transition do not decay through positron emission.

With this kind of radioactivity, the result for the atom is

$$_{Z}^{A}X \rightarrow _{Z-1}^{A}Y + \beta^+ + \nu_e$$

This transition is also isobaric.

2.2.4.3 General Aspects of β Decay

There are a large number of radionuclides that are β emitters. The energy spectrum of β radioactivity is such that the mean energy of the β particles is of the order of one third of the maximum energy, with differences from one nucleus to another, and between β^- and β^+. It is possible for a number of radionuclides to be simultaneously β^- and β^+ emitters with a well-defined ratio between both emission probabilities.

2.2.4.4 Electron Capture

This process is another way for a given atom to increase its n/p ratio, especially when the transition energy is not high enough to allow a β^+ decay. It can be described as a *capture* by the nucleus of an electron belonging to the shells, leading to:

$$_{1}^{1}p + _{-1}^{0}e \rightarrow _{0}^{1}n + \nu_e$$

The result for the atom is;

$$_{Z}^{A}X + _{-1}^{0}e \rightarrow _{Z-1}^{A}Y + \nu_e$$

In this case, the neutrinos produced are monoenergetic. After such a capture, most often of a K shell electron, a vacancy is produced. This is then filled by an electron cascading from an energy level farther from the nucleus with emission of characteristic x-rays. The energy spectrum of such x-rays is then made up of well-defined lines.

A typical radionuclide used in Brachytherapy and that decays through electron capture is ^{125}I with photon emissions corresponding to the characteristic x-rays of the Iodine atom.

2.2.5 ARTIFICIAL RADIOACTIVITY

Unstable nuclei, leading to radioactive transformations, can be obtained by bombarding stable nuclei with a number of particles such as neutrons, high energy protons, deuterons,

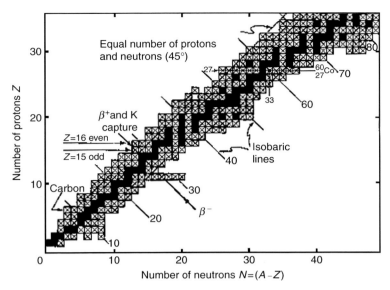

FIGURE 2.6
Chart showing the proportions of protons and neutrons in nuclei. Stable nuclei are represented by solid squares; radioactive nuclei by crosses. Nuclei with equal numbers of neutrons and protons lie along the line $N = Z$. Isotopes appear along horizontal lines and isobars along lines at 45°. (From Johns, H. E. and Cunningham, J. R., *The Physics of Radiology*, 4th Ed., Charles C. Thomas, Springfield, IL, 1983.)

α particles or γ-rays. During a collision of the particles with the nucleus, the particles can either be absorbed or they can eject a nucleon from the nucleus. The result is the production of a new nuclide with a nuclear content that may allow radioactive transformations. A very large number of artificial radioactive isotopes* have been produced with the wide range of high-energy particle-accelerators now available, and most of the radionuclides used in medicine are man-made (Figure 2.6).

The classical example of the production of artificial radioactivity is the experiment that allowed Frédéric and Irène Joliot-Curie to discover this process. They bombarded aluminium foil with α particles produced by a ^{210}Po source (5.3 MeV). Aluminium produced positrons with an exponential decay, suggesting a radioactive process. In fact, the following processes were involved:

$$^{27}_{13}\text{Al} + {}^{4}_{2}\text{He} \rightarrow {}^{30}_{15}\text{P} + {}^{1}_{0}\text{n}$$

$$^{30}_{15}\text{P} \rightarrow {}^{30}_{14}\text{Si} + \beta^{+} + \nu_e$$

Through chemical analysis, they were able to confirm that phosphorus had actually been produced. A number of radionuclides are also produced through the fission process, the fission fragments corresponding to nuclei situated outside the stability conditions. The mathematical description of radioactive decay is considered in Section 2.3 and the artificial production of radionuclides in Section 2.4.

---

* The term *isotopes* is frequently used to designate *radionuclides* (i.e. radioactive nuclides)

2.3 QUANTIFICATION OF RADIOACTIVITY

2.3.1 ACTIVITY: QUANTITY AND UNIT

According to ICRU (ICRU 1998): "the activity A of an amount of radioactive nuclide in a particular energy state at a given time is the quotient of dN/dt, where dN is the expectation value of the number of spontaneous nuclear transitions from that energy state in the time interval dt".

$$A = \frac{dN}{dt}$$

The unit of activity is s^{-1}. The special name for the unit of activity is becquerel (Bq)

$$1\ Bq = 1\ s^{-1}$$

This unit replaces the former curie (Ci) such that

$$1\ Ci = 3.7 \times 10^{10}\ Bq\ (exactly)$$

2.3.2 RADIOACTIVE DISINTEGRATION AND DECAY

The principles of radioactive decay are discussed here. Examples of decay characteristics for radionuclides of interest in radiotherapy will be found in Table M.4.

2.3.2.1 Law of Radioactive Decay

The radioactive decay law was experimentally established in 1902 by Rutherford and Soddy in Great Britain. According to this law, the number dN of spontaneous disintegrations happening in a given amount of radioactive material during an infinitesimal time interval dt is proportional to:

- The number of radioactive atoms included in this amount of material at the time considered
- The time interval dt
- A constant λ, called the decay constant. It represents, for a given nucleus of a given radionuclide, the quotient, dP/dt, where dP is the probability that the nucleus will undergo a spontaneous nuclear transition from the original energy state during the time interval dt. This constant is a specific characteristic of the nuclide and is not influenced by its physico-chemical status.

So,

$$dN = -\lambda N\ dt \tag{2.1}$$

Starting with an initial number N_0 of radioactive atoms, the number, N_t, of atoms present at time t will be

$$N_t = N_0 \exp(-\lambda t) = N_0\, e^{-\lambda t} \tag{2.2}$$

From the above definition, it follows that the activity A is represented by:

$$A = \left| \frac{dN}{dt} \right| = \lambda N \tag{2.3}$$

and by following a similar argument to the derivation of N_t:

$$A_t = A_0 \exp(-\lambda t) = A_0 e^{-\lambda t} \tag{2.4}$$

2.3.2.2 Half-Life of a Radioactive Nuclide

The principles of exponential decay lead to the fact that, in a given time interval, a given proportion of radioactive atoms disintegrate.

To illustrate the decay rate of a given radionuclide, the time required for half of the existing radioactive atoms to disintegrate has been chosen as a reference and is called the half-life, T^*. The *half-life* can be introduced into the above equations as follows:

$$0.5 = \frac{N}{N_0} = \exp(-\lambda T) = e^{-\lambda T}$$

hence:

$$T = \frac{\ln 2}{\lambda} \approx \frac{0.693}{\lambda} \tag{2.5}$$

$$A_t = A_0 \exp\left(-\frac{\ln 2 \times t}{T}\right) \tag{2.6}$$

Sometimes, the *mean life*, τ, is used instead of the half-life T.

$$\tau = \frac{1}{\lambda} = \frac{T}{\ln 2} \approx 1.44T$$

2.3.2.3 Specific Activity

When, in a given amount of material, stable and radioactive atoms of the same element are present, the *specific activity* of this radionuclide is given by:

$$A_{\text{spec}} = \frac{N_{\text{ra}}}{N_{\text{st}} + N_{\text{ra}}} \tag{2.7}$$

where N_{ra} is the number of radioactive atoms, and N_{st} the number of stable atoms. It represents the activity per unit mass of that radionuclide. It is different from the *mass specific activity* that, in a material containing a radionuclide, is the activity per unit mass of that material.

2.3.2.4 Equilibrium with Radioactive Daughter Products

When daughter nuclides produced by a radioactive process are also radioactive, the decay of the daughter nuclides in a mixture of original/daughter nuclides may appear different from the decay expected for the daughter nuclides alone.

A radionuclide with a decay constant λ_1, and the daughter with a decay constant λ_2, larger than λ_1 (shorter half-life) is now considered. According to the above equations,

$$dN_1 = -\lambda_1 N_1 dt$$

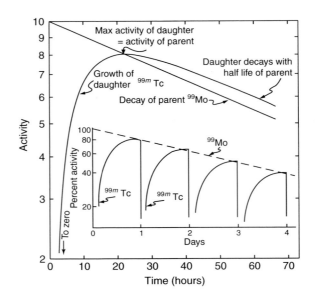

FIGURE 2.7
Illustration of activity decay of parent ^{99}Mo and activity growth of daughter ^{99m}Tc as a function of time, assuming that it is started from a pure source of parent. The figure insert shows the same data when the daughter is *milked* from the parent once every day. The actual situation is somewhat more complicated because only about 86% of the parent ^{99}Mo nuclei decay to the daughter ^{99m}Tc. This means that the activity of the ^{99m}Tc is about 14% less than indicated on the graph. (From Johns, H. E. and Cunningham, J. R., *The Physics of Radiology*, 4th Ed., Charles C. Thomas, Springfield, IL, 1983. With permission.)

$$dN_2 = \lambda_1 N_1 dt - \lambda_2 N_2 dt \tag{2.8}$$

The corresponding activities are then

$$A_{1t} = A_{1t_0} e^{-\lambda_1 t}$$

$$A_{2t} = \frac{\lambda_2}{\lambda_2 - \lambda_1} A_{1t_0} \left[e^{-\lambda_1 t} - e^{-\lambda_2 t} \right] = \frac{\lambda_2}{\lambda_2 - \lambda_1} A_{1t_t} \left(1 - e^{-(\lambda_2 - \lambda_1)t} \right) \tag{2.9}$$

A_1 decays exponentially, whereas, A_2 increases then decays after a maximum value t_m such that

$$t_{\mathrm{m}} = \frac{\ln(\lambda_2/\lambda_1)}{\lambda_2 - \lambda_1} \tag{2.10}$$

For this example where $T_1 > T_2$, after a long time, a *transient equilibrium**** is obtained as the daughter nuclide decreases with an apparent half-life equal to the half-life of the parent. The production of daughter atoms is slower than the disintegration rate of the daughter itself. At

---

* *Editors note*: The use of the terms *transient* and *secular* (which means "very long lasting") equilibrium is controversial (Hendee and Bednarek 2004). The *equilibrium* usually described in textbooks is equilibrium between the activity of the daughter and the parent. This ratio becomes constant as in Equation 2.11 so this *equilibrium* is *secular* whether $T_1 \gg T_2$ or not. We could also consider the equilibrium in the balance between rate of increase in activity of the daughter and its rate of decay which occurs when $dN_2/dt = 0$, i.e. when $\lambda_1 N_1$ is equal to $\lambda_2 N_2$ as can be seen from Equation 2.8. This is transient in both cases, but if T_1 is very large the situation may *appear* secular. More obviously, the *equality* of activity between daughter and parent is *transient* unless $T_1 \gg T_2$.

this point, the exponential term in Equation 2.9 tends to zero and:

$$\frac{A_1}{A_2} = \frac{\lambda_2 - \lambda_1}{\lambda_2} \qquad (2.11)$$

The ratio between the activities of parent and daughter remains constant. This principle is used in generators for radioactive nuclides used in nuclear medicine. For instance, for the production of ^{99m}Tc ($T=6.03$ h) from ^{99}Mo ($T=66.7$ h), a quasi-transient equilibrium can be obtained after less than one day (Figure 2.7).

If $T_1 \gg T_2$, a *secular equilibrium* is obtained, and the activities of the parent and daughter radionuclides become equal after several half-lives of the daughter.

On the other hand, if $T_1 \ll T_2$, the production of daughter atoms is faster than the daughter decay, and the daughter activity increases and, after a long delay, the parent has decayed to a negligible activity so that the daughter decay can be observed alone.

In practice, since the situation depends upon the individual half-life of the parent and daughter radionuclide, it becomes much more complex if the decay scheme involves several successive daughter radionuclides.

2.4 PRODUCTION OF RADIOACTIVE SOURCES THROUGH THE ACTIVATION PROCESS

As previously stated, artificial radioactivity can be induced by bombarding stable nuclei with various particles, including neutrons, protons, α particles or γ-rays, with appropriate energy.

2.4.1 STANDARD PRODUCTION OF ARTIFICIAL RADIONUCLIDES

Production of artificial radionuclides takes place in nuclear reactors, either as by-products of nuclear fission (e.g. ^{137}Cs, ^{90}Sr, ^{131}I) or by neutron bombardment, through (n, γ) reactions (e.g. ^{60}Co, ^{198}Au, ^{192}Ir). Cyclotrons are also used for the production of radionuclides used in nuclear medicine.

The principles and the quantitative aspects of the production of radioactive atoms can be summarised as follows:

A thin target (thickness dx) with an area being equal to 1 cm^2 is considered. This target contains n atoms involved in the activation process.

The production of radioactive atoms is a function of the following parameters:

- The duration dt of the bombardment
- The number of atoms under bombardment. (n dx for this target)
- The neutron fluence rate $\dot{\Phi}$
- The cross section σ of the nuclear reactions, leading to neutron capture in the target material. This cross section depends on the material and on the neutron energy. It is expressed in cm^2 per atom.

So, if the activation process starts at t_0, with n initial number N_0 of radioactive atoms in the target equal to zero, during dt, the following can be observed:

$$\frac{dN}{dt} = \sigma \dot{\Phi} \, \text{n} \, dx - \lambda N \qquad (2.12)$$

λ being the decay constant of the radionuclide produced, decaying during dt.

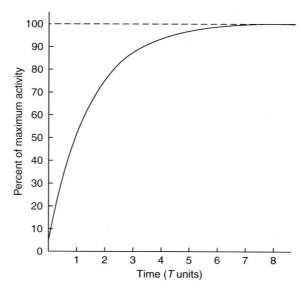

FIGURE 2.8
Production of artificial radionuclides: Variation of activity in the target as a function of time expressed in half-life units.

Hence, the total number of radioactive atoms at time t is:

$$N_t = \sigma \dot{\Phi} \, \mathrm{n} \, \mathrm{d}x \frac{(1 - e^{-\lambda t})}{\lambda} \tag{2.13}$$

The radionuclide activity after a bombarding time t is:

$$A_t = \lambda N_t = \sigma \dot{\Phi} \, \mathrm{n} \, \mathrm{d}x (1 - e^{-\lambda t}) \tag{2.14}$$

This increase of activity in the target as a function of time is represented on Figure 2.8.

It appears that, for a given neutron fluence rate, the activity reaches a limit value even for an infinite bombarding time. It is called saturation activity and given by:

$$A_s = \sigma \dot{\Phi} \, \mathrm{n} \, \mathrm{d}x \tag{2.15}$$

In this situation, the atom disintegration rate is equal to the atom production rate.

As higher activities are required (i.e. for radiotherapy), the target mass or the neutron fluence rate must increase. Because the source size is usually constrained by the application, an increase in activity can only be achieved by increasing the neutron fluence rate.

For the production of cobalt-60, the (n, γ) reaction transforms an atom of atomic mass A into an isotope of atomic mass $A + 1$:

$$^{59}\mathrm{Co} + {}^{1}_{0}\mathrm{n} \rightarrow {}^{60}\mathrm{Co} \qquad (^{59}\mathrm{Co} \text{ is the natural cobalt element})$$

The neutron spectrum should, of course, be optimised in order to get the highest value possible for σ.

To reach 75% of the saturation activity in a given reactor neutron fluence rate, it is necessary to wait for two times the half-life. In the case of the production of $^{60}\mathrm{Co}$ ($T = 5.271$ years), this would represent more than ten years, which is not pratically possible.

TABLE 2.1

Examples of Radionuclides Possibly Produced in Linear Accelerators by a (γ, n) Reaction

Element	Reaction	Resulting Radionuclide and Half-Life	Threshold (MeV)
^{12}C	(γ, n)	^{11}C—20 min	18.6
^{14}N	(γ, n)	^{13}N—10 min	10.5
^{16}O	(γ, n)	^{15}O—2 min	15.6
^{63}Cu	(γ, n)	^{62}Cu—10 min	10.9
^{65}Cu	(γ, n)	^{64}Cu—13 h	9.8
^{54}Fe	(γ, n)	^{53}Fe—8.5 min	13.6
^{204}Pb	(γ, n)	^{203}Pb—6.1 s/52 h	8.2

Source: From *Radiation Protection Design Guide Lines for 0.1–100 MeV particle Accelerators Facilities,* NCRP Report no. 51, Washington, 1977

2.4.2 Unintentional Activation

Unintentional activation can occur in high energy accelerators, and this activation should be taken into consideration for radiation protection. For electron accelerators, the process first involves the production of photons in the target or in various parts of the unit (contributing to photon leakage around the machine). It then involves the production of neutrons through a photonuclear interaction. This photonuclear interaction occurs when the photon energy has the same order of magnitude as the binding energy of the nucleons. When the photon is absorbed by the nucleus, the additional internal energy is generally lost through a neutron or a proton emission (usually presented as (γ, n) or (γ, p) reactions). Except for light atoms, the most common reaction is (γ, n).

Such a reaction appears only above a given energy threshold that depends on the material. The induced activity depends on the material (cross-section value) and on the irradiation conditions (energy, dose rate, and irradiation time).

A few examples of activation reactions are given in Table 2.1. This table shows that there are parts of the linac made of copper that can be activated when the electron/photon energies are higher than approximately 10 MeV. For higher energies, some activation occurs in air (and in the patient). Therefore, the global radiation protection problem for the higher energy range used in radiotherapy is linked to both the neutron production during irradiation (suitable absorbing materials must be present in the walls and in the door of treatment rooms) and the photons emitted during, and even some time after, irradiation. The latter are due to activation and must be taken into consideration for maintenance and repair activities. More severe radiation protection problems may occur in the use of protons, neutrons or heavy ions in radiation therapy (see Chapter 46 and Chapter 49).

CHAPTER 3

INTERACTIONS OF CHARGED PARTICLES WITH MATTER

Alan Nahum

CONTENTS

3.1 INTRODUCTION

Charged particles are fundamental to the medical use of radiation. Even if the primary radiation is a photon beam, it is the charged particles, known as secondary radiation in such cases, that cause the biological effect, whether it be cell killing or other changes that may eventually induce cancer. In fact, charged particles are often termed *ionising* radiation, and photons (and neutrons) termed *non-ionising* or *indirectly ionising*. Furthermore, a precise knowledge of the *spatial* distribution of the absorbed dose is crucial to radiotherapy treatment planning and delivery (and in certain cases, to radiation protection considerations), and this can only be obtained if the transport of the energy by the charged particles (overwhelmingly electrons) can be modelled. In many cases, the ranges of, for example, the Compton electrons (see Section 4.3.2) generated by megavoltage x-ray beams are appreciable (up to several cm) and must, therefore, be accurately modelled. The generation of x-rays, i.e. bremsstrahlung, is a charged-particle interaction. Alternatively, radiotherapy is sometimes delivered by primary charged particle beams, usually megavoltage electrons, where electron interactions with matter are obviously crucial. However, increasingly, proton beams are coming into therapeutic use (see Chapter 46), and even so-called heavy ions such as carbon are used (see Chapter 49). Mention can also be made of unsealed source therapy (Part K) with, for example, β-emitting radionuclides; these electrons (or positrons) can also have ranges up to a centimetre. The subject of radiation dosimetry (see Chapter 6 and Part D) depends on an intimate knowledge of the interactions of both non-ionising and directly ionising (i.e. charged) particles (e.g. the Bragg–Gray cavity principle) as will be made clear in Chapter 6. At the microdosimetric level, a fundamental understanding of the action of radiation on cells can only come through studying the track structure of particle tracks in relation to the relevant targets (i.e. the DNA in the cell nucleus). Again, this requires knowledge of the charged-particle interactions. Perhaps the only use of radiation in radiotherapy that does not rely heavily on charged particle interactions is imaging by diagnostic x-rays.

In this chapter, the emphasis is on electrons. However, much of the material applies with little modification to protons and other ions, and this will be indicated where appropriate. There are primarily three interaction mechanisms of importance for electrons in the energy range from a few hundred eV up to 50 MeV[*]: first, *collisions* with bound atomic electrons (Møller scattering); second, *bremsstrahlung* or radiative losses; and third, *elastic scattering* largely because of the heavy, positively charged nucleus. For heavy particles only (inelastic) collision losses and elastic scattering are important (see Chapter 47 and Chapter 49). In all cases, the consequences of the interactions are twofold: modifications of the incident (direct or secondary) charged particles in terms of energy loss and direction, and transfer of energy to matter, resulting in energy absorption and dose deposition. This second issue will be considered in more detail in Chapter 6.

3.2 COLLISION LOSSES

3.2.1 THEORY

Coulomb interactions with the bound atomic electrons are the principal way that charged particles (electrons, protons, etc.) lose energy in the materials and energies of interest in radiotherapy. The particle creates a trail of ionisations and excitations along its path. Occasionally, the energy transfer to the atomic electron is sufficient to create a so-called *delta ray* (or δ-ray)

---

[*] See Section 5.4 for some historical references related to the fundamental electrons and positrons interaction processes.

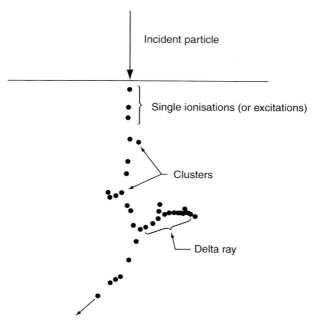

FIGURE 3.1
Diagrammatic representation of the track of a charged particle in matter. (From ICRU, *Linear Energy Transfer, Report 16*, ICRU, Bethesda, MD, 1970.)

which is a (secondary) electron with an appreciable range of its own. This is schematically illustrated in Figure 3.1. A fuller and very readable account of the theory of inelastic collisions between fast charged particles and atomic electrons can be found in Evans (1955).

The physical model of the Coulomb interaction between the fast charged particle and a bound electron in the medium is shown in Figure 3.2. The electron is assumed to be free, and its binding energy assumed to be negligible compared to the energy it receives. The primary particle imparts a net impulse to the bound electron in a direction perpendicular to its path.

Using classical, non-relativistic collision theory, as Bohr once did, from Newton's second law, (i.e. the change in momentum is equal to the impulse [the time integral of the force]) and from the Coulomb law for the force between charged particles, it can be shown that the energy transfer Q is given by:

$$Q = \frac{2k^2z^2e^4}{mb^2v^2} \tag{3.1}$$

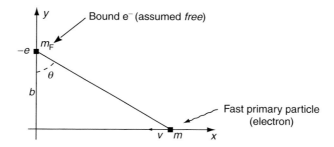

FIGURE 3.2
Interaction between a fast primary charged particle and a bound electron. The incoming electron is moving at a speed v in a direction opposite to axis x. (From Nahum, A. E., *The Computation of Dose Distributions in Electron Beam Radiotherapy*, Medical Physics Publishing, Madison, WI, 1985.)

where b, the distance of closest approach, is known as the *impact parameter*; m is the mass of the electron; z is the charge on the primary particle (in units of the electronic charge e); v is the velocity of the primary particle; and the constant, k, is defined below. It should be noted that the mass of the primary particle does not enter into Equation 3.1, which equally applies to electrons, protons (that both have $z=1$), and other heavier charged particles (see Chapter 46 and Chapter 49). Equation 3.1 leads to the following classical expression for the cross-section per electron, differential in the energy transfer Q:

$$\frac{d\sigma}{dQ} = \frac{2\pi z^2 e^4 k^2}{m v^2} \frac{1}{Q^2} \tag{3.2}$$

The full relativistic, quantum-mechanical cross-section for Coulomb interactions between free electrons, due to Møller (1932), is:

$$\frac{d\sigma}{d\varepsilon} = \frac{2\pi e^4 k^2}{T m_e v^2} \left[\frac{1}{\varepsilon^2} + \frac{1}{(1-\varepsilon)^2} + \left(\frac{\tau}{\tau+1}\right)^2 - \frac{2\tau+1}{(\tau+1)^2} \frac{1}{\varepsilon(1-\varepsilon)} \right] \tag{3.3}$$

where

    T is the electron kinetic energy (k.e.)
    $\varepsilon = Q/T$ is the energy transfer in units of the electron k.e.
    $\tau = T/m_e c^2$ is the k.e. in units of the electron rest mass
    v is the electron velocity
    $k = 8.9875 \times 10^9 \, \text{Nm}^2 \, \text{C}^{-2}$ (constant appearing in the Coulomb-force expression)

The first term in Equation 3.3 dominates. Ignoring the remaining terms and changing the variable from ε to the energy transfer Q, the classical result given above (Equation 3.2) is obtained. The $1/Q^2$ dependence clearly demonstrates that small losses predominate; the average energy loss in low atomic-number materials is of the order of 60 eV (ICRU 1970). The Møller expression (Equation 3.3) is valid provided that the electron energy is much greater than the binding energies of the atoms in the medium. The binding energies also set a lower limit to the energy transfer possible (see Section 3.2.2).

The terms *soft* and *hard* are often used to describe the different types of collisions. So-called *soft* collisions are said to occur when the fast particle passes an atom at a relatively large distance, and the Coulomb force field affects the atom as a whole, distorting it, with possibly an excitation or ionisation of a valence-shell (i.e. outer) electron. Only a small amount of energy can be transferred of the order of eV. These soft collisions are by far the most numerous type of collision. Cerenkov radiation can result from soft collisions (see Attix 1986); it is entirely negligible as a fraction of the total energy lost in soft collisions. The density or polarization effect is also concerned with soft or distant collisions (see Section 3.2.3).

When the fast particle passes relatively close to the atom (i.e. of the order of the atomic dimensions), then one can properly speak of an interaction with a single bound electron (Figure 3.2). Such hard collisions result in electrons being ejected with appreciable kinetic energy, and these are known as *knock-on* electrons or δ-rays. Hard collisions are naturally much rarer than soft ones, but the contributions of hard and soft collisions to the total energy loss are comparable in magnitude.

If an inner-shell electron is ejected as a result of a hard collision, then the atom will return to its ground state either by the emission of characteristic x-rays or by Auger electrons in the same manner as for the photoelectric effect (see Section 4.3.1).

3.2.2 COLLISION STOPPING POWER

The occurrence of very frequent, small energy losses along the path of any charged particle in matter (in marked contrast to the way that photons transfer energy, see Chapter 4) leads naturally to the concept of *stopping power*, defined as the average energy loss, dE, per unit distance, ds, along the track of the particle. This is usually expressed as the *mass* collision stopping power, written $(1/\rho)(dE/ds)_{col}$ or S_{col}/ρ, which is calculated from

$$\frac{1}{\rho}\left(\frac{dE}{ds}\right)_{col} = N_A\left\langle\frac{Z}{A}\right\rangle\int_{Q_{min}}^{Q_{max}} Q\frac{d\sigma}{dQ}\,dQ \tag{3.4}$$

where N_A is Avogadro's number, and Z and A have their usual meaning (see Chapter 1). Q_{max} for an electron with kinetic energy E_0 is equal to $E_0/2^*$. Note that strictly (dE/ds) is a negative quantity as it expresses energy *loss* not energy *gain* as the distance s increases. However, this negative sign has generally been omitted in what follows.

The evaluation of the minimum energy transfer Q_{min} represents a major difficulty. The integral for stopping power (Equation 3.4) can alternatively be cast in terms of the impact parameter b and then integrating out to $b = \infty$ yields an infinite stopping power because of the large number of soft collisions at large distances. The first approximate solution to this far from trivial problem is due to Niels Bohr (1913, 1948). Bohr's first attempt at this was in 1913 before he had conceived of quantized energy levels for the atomic electrons. Bohr treated the problem in terms of the *time of collision*, τ, and the natural frequency of the atomic electron, ν, such that the electron, if displaced, oscillated about its natural equilibrium position with a period of $1/\nu$. If the collisions are such that τ ($\approx b/v$ i.e. impact parameter/velocity) is short (i.e. $\tau \ll 1/\nu$), then the electron would behave as though it were free and accept the impulse (corresponding to close collisions). If, on the other hand, the collision time was relatively long, such that $\tau \gg 1/\nu$, then the electron acts as though bound; its orbit is distorted or deformed by the passage of the charged particle, but no net energy transfer takes place. This is sometimes referred to as adiabatic behaviour. Bohr showed that the maximum impact parameter $b_{max} = 1.123v/2\pi\nu$ and, consequently, a finite energy loss per unit pathlength was obtained that included the geometric mean value $\bar{\nu}$ of the Z individual atomic frequencies that are characteristic of different atoms

$$\left(\frac{dE}{ds}\right)_{classical}^{Bohr} = \frac{4\pi z^2 e^4}{m_e v^2} NZ \ln\frac{1.123 m_e v^3 M}{2\pi\bar{\nu}ze^2(M + m_e)} \quad \text{ergs/cm}$$

where M is the mass of the fast charged particle and N is the number of atoms per unit volume.

The full quantum-mechanical expression for the electron mass collision stopping power (Berger and Seltzer 1964; ICRU 1984a, 1984b) is given by

$$\frac{1}{\rho}\left(\frac{dE}{ds}\right)_{col} = \frac{2\pi r_e^2 m_e c^2 N_A}{\beta^2}\left\langle\frac{Z}{A}\right\rangle\left\{\ln\left[\frac{\tau^2(\tau + 2)}{2(I/m_e c^2)^2}\right] + F(\tau) - \delta\right\} \tag{3.5}$$

where

$$F(\tau) = 1 - \beta^2 + [\tau^2/8 - (2\tau + 1)\ln 2]/(\tau + 1)^2 \tag{3.6}$$

and the extra quantities not defined so far are

$m_e c^2$, rest mass energy of the electron

$\beta =$ v/c

---

* Two electrons are indistinguishable after the collision; therefore, it cannot be determined which was the incident electron. Arbitrarily, *the faster electron after the collision is taken to be the incident one*; this results in $Q_{max} = T/2$. For heavy particles, it can be shown from kinematics that $Q_{max} = T\{1 + (2Mc^2/T)\}/\{1 + (M + m_0)^2 c^2/2m_0 T\}$ (e.g. Evans 1955).

r_e, electron radius ($=e^2/m_e c^2 = 2.818 \times 10^{-15}$ m)

I, mean excitation energy

δ, density-effect correction (see Section 3.2.3)

The above expression can be somewhat simplified if it is expressed in the virtually universally employed units of MeV cm^2 g^{-1} (ICRU 1984a):

$$\frac{1}{\rho}\left(\frac{dE}{ds}\right)_{col} = 0.1535 \frac{1}{\beta^2} \left\langle \frac{Z}{A} \right\rangle \left\{ \ln\left[\frac{\tau^2(\tau+2)}{2(I/m_e c^2)^2}\right] + F(\tau) - \delta \right\} \text{ MeV cm}^2 \text{ g}^{-1} \quad (3.7)$$

The mean excitation energy or potential, I, is an average of the transition energies E_i weighted by their oscillator strengths f_i according to the following:

$$Z \ln I = \sum_i f_i \ln E_i \quad (3.8)$$

It is effectively the geometric mean of all the ionisation and excitation potentials of the atoms in the absorbing medium; it is, of course, the more exact counterpart of Bohr's mean characteristic frequency that was discussed above. In general, I cannot be derived theoretically except in the simplest cases such as monoatomic gases. Instead, it must be derived from measurements of stopping power or range. The most recent values of I, based largely on experimental data, are given in ICRU (1984b). For example the best current estimate of the I-value for water is 75.0 eV. Generally, the I-value increases as Z increases (see Table 3.1).

The correspondence between classical and quantum-mechanical treatments of the energy loss spectrum is illustrated in Figure 3.3, taken from Evans (1955). The quantity $\sigma(Q)/\sigma_0(Q)$ on the ordinate is the ratio of the effective cross-section to the classical one. I is the effective minimum excitation potential of the sth electron. Q is the adiabatic limit of minimum classical energy transfer to the sth electron when the impact parameter has its maximum effective value. In the quantum-mechanical treatment, the energy losses in soft collisions correspond to a type of resonance phenomenon. The fast incident particle has a finite probability of transferring energy to that particular atom. However, either the energy transferred is zero or it is equal to the

TABLE 3.1

Mean Excitation Energies, I, and Other Quantities Relevant to the Evaluation of the Collision Stopping Power of Selected Human Tissues and Other Materials of Dosimetric Interest

Material	I (eV)	$\langle Z/A \rangle$	Density (g cm^{-3})
Adipose tissue (ICRP)	63.2	0.558468	0.920
Air (dry)	85.7	0.499190	1.205×10^{-3}
Bone, compact (ICRU)	91.9	0.530103	1.850
Bone, cortical (ICRP)	106.4	0.521299	1.850
Ferrous-sulphate dosimeter solution	76.3	0.553282	1.024
Lithium fluoride	94.0	0.462617	2.635
Muscle, skeletal (ICRP)	75.3	0.549378	1.040
Muscle, striated (ICRU)	74.7	0.550051	1.040
Photographic emulsion	331.0	0.454532	3.815
PMMA (lucite, perspex)	74.0	0.539369	1.190
Polystyrene	68.7	0.537680	1.060
Water (liquid)	75.0	0.555087	1.000

Source: Adapted from Nahum, A. E., *The Computation of Dose Distributions in Electron Beam Radiotherapy,* Medical Physics Publishing, Madison, WI, pp. 27–55,1985. With permission; Data taken from ICRU (International Commission on Radiation Units and Measurements), Report 37, *Stopping Powers for Electrons and Positrons,* ICRU, Bethesda, MD, 1984. With permission.

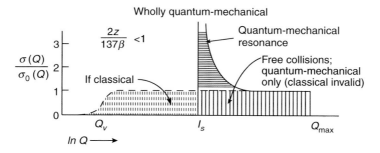

FIGURE 3.3
The ratio of the effective cross-section $\sigma(Q)$ to the classical one $\sigma_0(Q)$ for high particle velocities. The classical theory predicts too large an energy loss as the adiabatic limit, Q_ν, becomes much smaller than the allowed minimum energy loss, I_s, as β increases. (From Evans, R. D., *The Atomic Nucleus*, McGraw Hill, New York, 1955. With permission.)

excitation or ionisation energy of the atom. Therefore, quantum mechanics replaces the classical multitude of small losses to each of the atoms by larger losses to only a few of the atoms.

The essential features of the mass collision stopping power are retained in the following simplified expression:

$$\frac{S_{col}}{\rho} \propto \left\langle \frac{Z}{A} \right\rangle \frac{1}{v^2} [f(\tau) - 2 \ln I - \delta] \tag{3.9}$$

Comparing this expression with Equation 3.5 or Equation 3.7, the increase at decreasing sub-relativistic energies due to the $(1/v^2)$ factor can be identified. This is simply explained by the fact that *slow* electrons spend more time going past an atom than do *fast* ones, and consequently the impulse (see above) is greater and thus more energy is lost. At relativistic energies, there is a more gradual increase in the stopping power which is known as the *relativistic rise*. Figure 3.4 illustrates these two main features.

The explanation for the relativistic rise in stopping power is as follows. The electric field of the moving primary particle undergoes a *Lorentz contraction* at relativistic energies. In the forward and backward directions the field is *weakened* by the factor γ whereas in the transverse directions it is *strengthened* by the factor γ (see footnote); the field-line pattern is changed from spherical to *pancake*-shaped. This *contraction* means that the field of the moving particle acts on the bound atomic electrons for a shorter time (i.e. the collision time, τ, is reduced). Following Bohr's reasoning, this leads to an increase in b_{max}*. The net result of a decrease in τ, a lateral broadening of the field, and an increase in the maximum impact parameter is an increase in the stopping power, approximately logarithmic with energy, as v approaches c, as illustrated in Figure 3.4.

3.2.3 DENSITY EFFECT

The density or polarization effect (Fermi 1940; Sternheimer 1961; ICRU 1984a) reduces the value of S_{col} at relativistic energies in condensed media via the term δ in Equation 3.5 and Equation 3.9. It is connected to the relativistic rise in the stopping power. The increase in the maximum value of the impact parameter, b_{max}, discussed in the previous section, implies an increase in the volume of the cylinder around the path of the fast charged particle where energy transfers are possible. However, if the stopping medium has a high density, (i.e. condensed media as opposed to gases) then the electric field seen by the atoms distant from the fast particle track is reduced due to the polarization of the intervening atoms (as illustrated in Figure 3.5). Consequently, the contribution of these distant collisions to the stopping power will be reduced.

---

* b_{max} increases by the factor $\gamma = 1/\sqrt{1 - v^2/c^2}$.

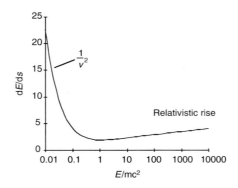

FIGURE 3.4
Collision energy loss as a function of (log) electron kinetic energy (in units of electron rest mass). (From Nahum, A. E., *The Computation of Dose Distributions in Electron Beam Radiotherapy*, Medical Physics Publishing, Madison, WI, 1985.)

This reduction in collision stopping power is known as the *polarization* or *density effect*. Fermi (1940) gave the first theoretical treatment. Sternheimer (1961) subsequently worked out the theory in more detail and gave numerical recipes to calculate the correction factor δ for a number of different materials.

The density effect is of particular importance in the dosimetry of megavoltage electron and photon beams as will become apparent when the water-to-air stopping-power ratio is discussed in Section D.1 of Appendix D. Figure 3.6 shows the variation of S_{col}/ρ with energy for air and water, two substances with similar atomic compositions and similar I-values; note that the energy scale is linear, not logarithmic as in Figure 3.4. The relativistic rise in the collision stopping power in the condensed medium is much less pronounced compared to that in the gas because of the density effect. Consequently, the ratio of mass stopping powers, water to air, is strongly energy dependent above around 0.5 MeV. It will be seen in Section 6.7 and Section 18.2.4 that it is precisely this ratio that determines the energy variation of the response (in terms of dose to water) of an ionisation chamber placed in water, and the energy region above 0.5 MeV is precisely the one that is relevant for the megavoltage photon and electron beams used in radiotherapy.

3.2.4 ELECTRON STOPPING-POWER DATA FOR SUBSTANCES OF MEDICAL INTEREST

The material-dependent terms in Equation 3.5 are $\langle Z/A \rangle$, the mean excitation energy I, and the density-effect correction δ. Table 3.1 lists the relevant parameters for various human tissues

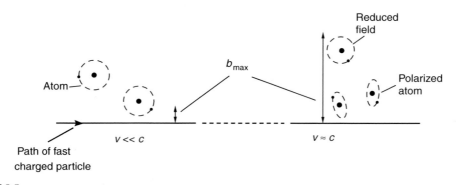

FIGURE 3.5
Schematic explanation of the mechanism of the density (or polarization) effect. (From Nahum, A. E., *The Computation of Dose Distributions in Electron Beam Radiotherapy*, Medical Physics Publishing, Madison, WI, 1985.)

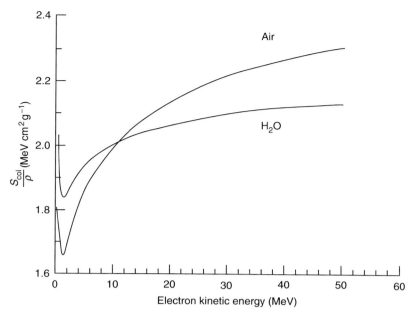

FIGURE 3.6
The variation of mass collision stopping power with electron kinetic energy for air and for water; note the linear energy scale. (From Nahum, A. E., Annual Meeting of the Canadian Association of Physicists, Quebec, 1983.)

and some other substances of dosimetric interest taken from ICRU (1984a) as well as for water as a comparison. It can be seen that the I-values all fall between 73 eV and 75 eV with the exception of adipose tissue (high hydrogen content) and bone with its high calcium content. In fact, the I-value is approximately proportional to the mean atomic number (e.g. the values for aluminium and lead are 166 eV and 823 eV, respectively). Given the similarity of the values of I, $\langle Z/A \rangle$, and the (mass) density (the latter being involved in the density-effect correction δ) in the table, the values of (S_{col}/ρ) must also be very similar. This is very convenient, as it means that the electron energy loss over a given distance in the body can be derived from that in water by simply multiplying by the density, assuming that radiation losses are also very similar, which will generally be the case below around 20 MeV (see Section 3.3.2).

The values of the (mass) stopping-power ratio, $s_{med,air}$, for various substances of interest in medical dosimetry, as a function of electron kinetic energy in the megavoltage region are shown in Figure 3.7. The quantity $\langle Z/A \rangle_{med}/\langle Z/A \rangle_{water}$ has been added along the right hand axis in order to show that for all these substances except bone, where the atomic number is appreciably higher, and air, where the density effect plays the dominant role (see Section 3.2.3), $(S_{col}/\rho)_{med} \propto \langle Z/A \rangle_{med}$ (Equation 3.9). In fact, the curve labelled *Air* also follows this pattern at energies below about 1 MeV where the density effect is negligible. It can also be noted that the ratio is virtually independent of energy except for that of air; this is very convenient for dosimeter response evaluation and treatment planning purposes. These medium-to-water stopping-power ratios are likely to find direct application in the conversion of Monte-Carlo-derived dose distributions in patients (see Chapter 28) to *water-equivalent* doses (Siebers et al. 2000).

3.2.5 RESTRICTED STOPPING POWER

Use is frequently made of the *restricted stopping power*. This means that only energy transfers below a certain value Δ are included; it is calculated by setting Q_{max} equal to Δ in Equation 3.4. The full expression is again given by Equation 3.5 and Equation 3.6, but with the $F(\tau)$ term now modified to:

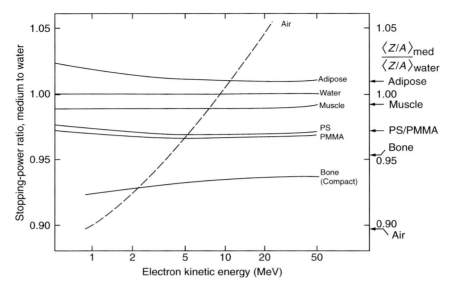

FIGURE 3.7
Ratios of mass collision stopping powers, medium to water, for various substances of medical and dosimetric interest. PS stands for polystyrene and PMMA for polymethyl methacrylate (perspex). The quantity $\langle Z/A \rangle_{med}/\langle Z/A \rangle_{water}$ for the various media is given on the right-hand axis. (From Nahum, A. E., *The Computation of Dose Distributions in Electron Beam Radiotherapy*, Medical Physics Publishing, Madison, WI, 1985.)

$$F(\tau,\varDelta) = -1 - \beta^2 + \ln\{4\varDelta(\tau-\varDelta)\tau^{-2}\} + \tau/(\tau-\varDelta) + \{\varDelta^2/2 + (2\tau+1)$$
$$\ln(1-\varDelta/\tau)\}(\tau+1)^{-2} \tag{3.10}$$

It should be noted that the expression for the restricted stopping power still includes the density-effect correction factor δ (i.e. it is implicitly assumed that all the losses that do not appear because of the density effect are below $\varDelta$ in energy). This is reasonable as it is precisely the so-called distant (i.e. soft) collisions that are affected by the density effect.

The restricted stopping power is often written $L_\varDelta$; it is required in the evaluation of the Spencer–Attix form of the (mass) stopping-power ratio (see Section 6.7.4). The ratio $L_\varDelta/S_{col}$ as a function of $\varDelta$ is plotted in Figure 3.8 for 1 MeV and 10 MeV electrons in water; very similar plots would have been obtained for other low-Z substances. The slow decrease from unity with decreasing $\varDelta$ demonstrates that small energy transfers dominate in the overall energy transfer process as has been previously emphasised.

3.2.6 COLLISION STOPPING POWER FOR HEAVY CHARGED PARTICLES

The stopping-power formula for a singly charged particle heavier than an electron is (Attix 1986; ICRU 1984a)

$$\frac{1}{\rho}\left(\frac{dE}{ds}\right)_{col} = \frac{2\pi N_A r_e^2 m_e c^2 z^2}{\beta^2} \left\langle \frac{Z}{A} \right\rangle \left\{ \left\{ \ln\left(\frac{2m_e c^2 \beta^2 T'_{max}}{I^2(1-\beta^2)}\right) \right\} - 2\beta^2 - 2\frac{C}{Z} - \delta \right\} \tag{3.11}$$

where T'_{max} is the maximum energy that can be transferred in a head-on collision with an (unbound) atomic electron (see footnote). For kinetic energy $T < M_0 c^2$, $T'_{max} \approx 2m_0 c^2(\beta^2/1-\beta^2)^*$. Following Attix (1986), the above expression can then be

---

* For a 10 MeV proton, $T'_{max} = 20\text{keV}$; this can be contrasted with a maximum delta ray energy of 5 MeV for a 10 MeV electron.

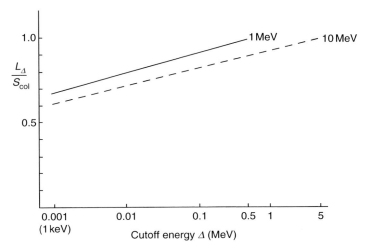

FIGURE 3.8
The ratio of restricted collision stopping power, L_Δ, to unrestricted collision stopping power, S_{col}, for 1 MeV and 10 MeV electrons in water. (From Nahum, A. E., *The Computation of Dose Distributions in Electron Beam Radiotherapy*, Medical Physics Publishing, Madison, WI, 1985.)

simplified to

$$\frac{1}{\rho}\left(\frac{dE}{ds}\right)_{col} = 0.1535\frac{z^2}{\beta^2}\left\langle\frac{Z}{A}\right\rangle\left\{27.675 + 2\ln\left(\frac{\beta^2}{1-\beta^2}\right) - 2\beta^2 - 2\ln I - 2\frac{C}{Z} - \delta\right\} \quad (3.12)$$

Compared to the corresponding formula for electron collision stopping power (Equation 3.6 and Equation 3.7), the above expression contains the so-called shell correction term C/Z. This is necessary as the Born approximation assumption (i.e. $2Zz/137 \ll \beta$) is no longer satisfied when particle velocities approach those of the bound electrons in the stopping medium. K-shell electrons will be affected first as they have the highest (orbital) velocities, then the L-shell and so on. The shell correction approximately accounts for the resulting error in the stopping-power formula. As particle velocities approach those of the various shells, (dE/ds) decreases as those electrons can no longer take part in the energy transfer process. This term may be as large as 10% at low energies, but it is negligible for electrons > 0.1 MeV in energy. The term C/Z is the same for all charged particles of the same velocity. Note that concerning the use of heavy particles in radiotherapy (see Chapter 46 and Chapter 49), the density-effect term δ is almost always negligible as the beam energies necessary for adequate penetration generally fall far short of the relativistic region.

Figure 3.9 compares the collision stopping power in water for protons with that for electrons. It can be seen that the basic shapes are the same, but the proton curve is shifted in energy such that the particle velocities are roughly comparable. Therefore, the minimum for the proton case occurs at around 2000 MeV which is a factor 2000 greater than that for electrons and is basically the ratio of the rest masses.

Heavy charged particles with relatively high energies (100 MeV or greater) can also lose energy through interacting with the nucleus (see Chapter 46 and Chapter 49). This can give rise to neutrons and γ-rays. The magnitude is generally small. As an example, around 2.5% of the energy in a 100 MeV proton beam in a carbon absorber is lost because of nuclear interactions and therefore does not appear as excitations or ionisations (Attix 1986). The effect of nuclear interactions is conventionally not included in the stopping power (or the range) of heavy charged particles. However, in accurate computations of the dose deposited by proton beams, for example using Monte-Carlo simulation, nuclear interactions

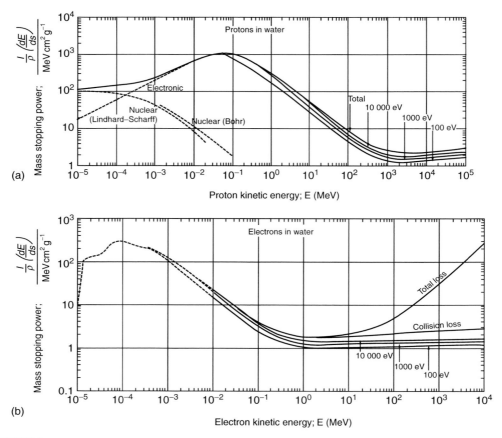

FIGURE 3.9

Mass stopping power in water for protons (a) and electrons (b). Note the similarities in the shapes of the collision stopping power curves, with the minima close to the respected particles' rest mass energies. The curves labelled with energy numbers refer to mass restricted stopping power for different value of Δ—see Section 3.2.5. The curve for the total stopping power (see Section 3.4.1) is also shown. (From ICRU, *Linear Energy Transfer, Report 16*, ICRU, Bethesda, MD, 1970.)

must be taken account of by decreasing the number of particles in the beam as the depth increases; this corresponds to a decrease in the *planar* fluence of the beam with depth (see Section 6.3.5).

3.3 RADIATIVE LOSSES (BREMSSTRAHLUNG)

3.3.1 THEORY

The acceleration of the (very light) electrons in the strong electric field of a nucleus leads to the production of *bremsstrahlung*. The acceleration is proportional to the nuclear charge, Z, divided by the mass, m, of the moving particle (i.e. $\propto Z/m$). The intensity of radiation produced is then proportional to $(Z/m)^2$. One sees immediately that this is a relatively unimportant energy loss mechanism below about 10 MeV in low-Z materials, and it is completely negligible for heavy charged particles. The cross-section, σ_{rad}, for this totally *non-classical* process is extremely complicated. One significant feature is that, very approximately:

$$\frac{d\sigma_{rad}}{dh\nu} \propto \frac{1}{h\nu} \tag{3.13}$$

Therefore, on average, the losses will be appreciably larger than for collisions (cf. Equation 3.2). This means that one can expect considerable energy-loss straggling due to radiation losses (see Section 3.4.2). Figure 3.10 is a plot of the variation of the product of the differential cross-section (as in the expression above) and the photon energy (i.e. the intensity) with photon energy (expressed as a fraction of the electron kinetic energy) for the case of lead. These curves correspond to the x-ray spectrum that would be produced by monoenergetic electrons beam hitting a thin target, integrated over all directions of the emitted photons.

3.3.2 RADIATION STOPPING POWER

In an exactly analogous fashion to that for collision losses in the previous section, one can define a radiative stopping power, $(dE/ds)_{rad}$ or S_{rad}, and also a mass radiation stopping power (S_{rad}/ρ). The general form of the mass radiative stopping power for high energies (complete screening: $\tau \gg 1/\alpha\, Z^{1/3}$) is given by:

$$(S/\rho)_{rad} = \frac{4r_e^2\alpha}{\beta^2} N_A \frac{Z(Z+1)}{M_A}(\tau + 1)m_e c^2 \ln(183 Z^{-1/3} + 1/18) \tag{3.14}$$

where α is the fine structure constant ($\alpha \approx 1/137$). From an inspection of Equation 3.14, it can be seen that the radiative stopping power increases almost linearly with kinetic energy

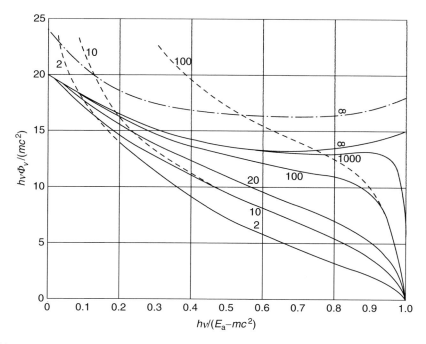

FIGURE 3.10

Energy distribution of the radiation emitted by monoenergetic electrons hitting a thin target. Ordinate—intensity of radiation per unit frequency interval; abcissa—energy of emitted quantum as a fraction of the kinetic energy of the emitting electron. The numbers on the curves indicate the energy of the electron in units of mc^2. Solid curves are for lead, including the effect of screening. Dotted curves are without screening, valid for all Z. (From Evans, R. D., *The Atomic Nucleus*, McGraw Hill, New York, 1955.)

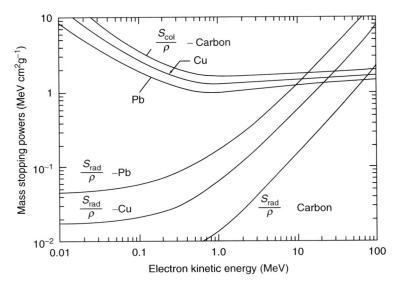

FIGURE 3.11
A comparison of the mass radiative and mass collision stopping powers, S_{rad}/ρ and S_{col}/ρ respectively, for carbon, copper, and lead as a function of the electron kinetic energy. (From Attix, F. H., *Introduction to Radiological Physics and Radiation Dosimetry*, Wiley, New York, 1986.)

in the MeV region, in contrast to the weak logarithmic energy dependence of the collision stopping power (see Section 3.2). Approximately, one can write:

$$\frac{S_{rad}}{\rho} \propto \frac{Z^2}{A} EB \qquad (3.15)$$

where B is a very slowly varying function of E and Z. The factor Z^2/A causes an increase in S_{rad}/ρ for higher Z (cf. S_{col}/ρ that decreases slowly with increasing Z). The combined effect of these influences is also evident in Figure 3.11.

From a comparison of Equation 3.9 and Equation 3.14, and also by examining the plots of S_{col}/ρ in Figures 3.4 and 3.9 which show that S_{col}/ρ is approximately constant at the energies where S_{rad}/ρ is non-negligible, it is apparent that $S_{rad}/S_{col} \propto ZE$. A useful quantitative approximation is:

$$\frac{(S_{rad}/\rho)}{(S_{col}/\rho)} = \frac{ZE}{1600 m_e c^2} \qquad (3.16)$$

Figure 3.12 illustrates the variation of S_{rad}/S_{col} with electron kinetic energy E for various media of interest in medical dosimetry as well as for aluminium. As the mean atomic number increases, from adipose tissue to bone, S_{rad}/S_{col} increases. It is less than 0.1 below 5 MeV, even for aluminium.

The bremsstrahlung process should be clearly distinguished from the much more frequent *elastic* scattering of charged particles in the Coulomb field of the nucleus that is responsible for almost all the changes in direction of electrons and is dealt with in Section 3.5.

3.3.3 RADIATION YIELD

A useful quantity is the fraction of the initial electron energy, E_0, that is lost to bremsstrahlung in slowing down to rest. This fraction is known as the *Radiation Yield*, $Y(E_0)$, and is given by:

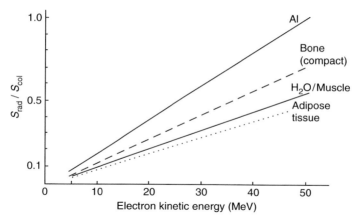

FIGURE 3.12
The energy and material dependence of the ratio of the radiative to the collision stopping power for electrons above 5 MeV. (From Nahum, A. E., *The Computation of Dose Distributions in Electron Beam Radiotherapy*, Medical Physics Publishing, Madison, WI, 1985.)

$$\Upsilon(E_0) = \frac{1}{E_0} \int_0^{E_0} \frac{S_{\text{rad}}(E)}{S_{\text{col}}(E) + S_{\text{rad}}(E)} \, \mathrm{d}E \tag{3.17}$$

The dependence of $\Upsilon(E_0)$ on E_0 and on Z is approximately linear, which closely corresponds to the relation between S_{rad}/ρ, E, and Z discussed above.

The radiation yield is involved in calculating the dosimetric quantity, g, which is the fraction of energy transferred (by photons) to a medium in the form of electron kinetic energy that is subsequently re-radiated as bremsstrahlung (see Section 6.3.2).

3.3.4 ANGULAR DISTRIBUTION OF BREMSSTRAHLUNG PHOTONS

The angular distribution of the emitted photons is very strongly forward-peaked at relativistic electron energies with a mean value $\theta \approx mc^2/E$ where E is the total energy of the electrons. This forward-peaking is the reason for the flattening filter in a linear accelerator treatment head. It can be noted that, at the much lower electron energies involved in largely diagnostic but also some therapeutic kilovoltage x-ray machines, the angular distribution is much closer to isotropic; hence, the steep angle of x-ray targets (see Section 10.3.1.3).

Chapter 9 in Attix (1986) should be consulted for more details on all aspects of the bremsstrahlung process.

3.4 TOTAL ENERGY LOSSES

3.4.1 TOTAL STOPPING POWER

The potentially confusing convention in this field is to term the inelastic electron–electron Coulomb interactions *collisions*; whereas bremsstrahlung losses are always termed *radiative losses*. The collision and radiative stopping powers are frequently summed to give the total stopping power, written $(\mathrm{d}E/\mathrm{d}s)_{\text{tot}}$ or S_{tot}:

$$\left(\frac{\mathrm{d}E}{\mathrm{d}s}\right)_{\text{tot}} = \left(\frac{\mathrm{d}E}{\mathrm{d}s}\right)_{\text{col}} + \left(\frac{\mathrm{d}E}{\mathrm{d}s}\right)_{\text{rad}} \tag{3.18}$$

or

$$S_{tot} = S_{col} + S_{rad}$$

Figure 3.9b shows the total mass stopping power (labelled "Total Loss"), mass collision stopping power, and several restricted mass collision stopping powers ($\Delta = 10$ keV, 1 keV and 100 eV) for water against electron kinetic energy E for values between 10^{-5} MeV and 10^4 MeV. It can be seen that $(S/\rho)_{tot}$ varies slowly with E over the energy range of primary interest in radiotherapy (from 1.937 MeV cm^2 g^{-1} at 4 MeV to only 2.459 MeV cm^2 g^{-1} at 25 MeV).

Several features should be noted:

- Radiation losses only become important above around 10 MeV in water
- The relativistic rise in the collision losses is small because of the density effect
- Collision losses restricted to $\Delta < 10$ keV only result in a modest reduction in stopping power compared to the unrestricted S_{col} which emphasizes the predominance of very small losses
- The approximate value for the electronic stopping power in water in the MeV region is around 2 MeV cm^{-1}; the value in tissue is very similar.

3.4.2 ENERGY-LOSS STRAGGLING

It is important to realize that stopping power is an average value for the energy loss per unit distance. Fluctuations will occur about this mean value in any real situation. This gives rise to what is known as *energy-loss straggling*. Figure 3.13 illustrates the idea; E_0 is the electron energy incident on a thin absorber and the distribution of electron energies emerging from the foil are shown where ΔE_{tot} would be predicted by the product of the thickness of the absorber and the (total) stopping power, S_{tot}. The width Γ is related to the distribution of the individual energy losses. If all energy loss is much smaller than the electron kinetic energy, then Γ will be narrow. This implies that most of the straggling results from the rare, large-energy losses (see Berger and Wang 1988 for a more in-depth treatment).

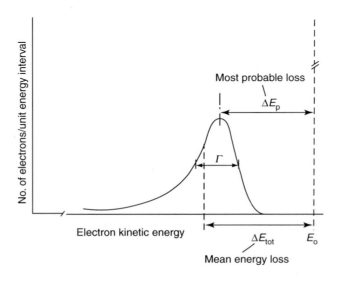

FIGURE 3.13
Energy broadening because of energy-loss straggling after the passage of a monoenergetic electron beam (energy E_0) through a thin absorber. (From Nahum, A. E., *The Computation of Dose Distributions in Electron Beam Radiotherapy*, Medical Physics Publishing, Madison, WI, 1985.)

3.4.3 Continuous-Slowing-Down-Approximation (CSDA) Range

Charged particles lose energy in a quasi-continuous fashion along their tracks in matter, eventually coming to rest. This means that, unlike photons with their exponential attenuation (see Chapter 4), charged particles do have a finite, reasonably well-defined range. Mathematically it has been found convenient to define the so-called continuous-slowing-down-approximation (csda) range, r_0 in the following fashion:

$$r_0 = \int_0^{E_0} \frac{1}{S_{\text{tot}}(E)} \, dE \qquad (3.19)$$

This represents the average pathlength travelled in coming to rest by a charged particle having kinetic energy, E_0. Note that for electrons, as opposed to heavy charged particles, this will always be considerably greater than the average penetration depth because of the marked angular deflections that electrons suffer in slowing down (see Section 3.5). The csda range r_0 is approximately proportional to E_0 in the therapeutic energy range because of the relatively slow variation of S_{tot} in this energy range (see Figure 3.9b). The quantity $r_0\rho$ is analogous to the mass stopping power, (S/ρ), and has the units of mass per unit area (e.g. g cm^{-2}). Other definitions of electron range of particular use in radiotherapy with electron beams, such as the practical range, R_{p}, and R_{50}, are discussed elsewhere in the book (see Section D.3.2 and Section 24.1).

3.4.4 Tabulated Stopping-Power Data

ICRU (1984a) provides total mass stopping powers, mass collision stopping powers, mass radiative stopping powers, csda ranges, radiation yields, and density-effect corrections for electrons for a very wide range of materials, including many of interest in radiotherapy dosimetry. A selection of stopping-power and range data can be found in Tables M.2, at the end of this book.

3.5 ELASTIC NUCLEAR SCATTERING

When a charged particle passes close to the atomic nucleus, at a distance much smaller than the atomic radius, the Coulomb interaction will now be between the fast particle and the nuclear charge rather than with one of the bound electrons (see Section 3.2). In the case of electrons, this causes appreciable changes of direction, but almost never (with the exception of the bremsstrahlung process—see Section 3.3) any change in energy. The scattering is basically *elastic*, the energy lost being the negligible amount required to satisfy momentum conservation between the very light electron and the positively charged nucleus (at least two thousand times heavier). Figure 3.14 gives an indication of the importance of this process which accounts for almost all the changes in electron direction.

This interaction process is essentially Rutherford scattering with differential cross-section:

$$\frac{d\sigma}{d\theta} = \frac{\pi e^4 Z^2 (1 - \beta^2)}{2m^2 v^4} \frac{\sin\theta}{\sin^4(\theta/2)} \qquad (3.20)$$

A correction for screening is essential to prevent the cross-section from becoming infinite when integrated over all angles. A more accurate cross-section that accounted for electron spin was later derived by Mott (see Berger and Wang 1988).

In order to be able to treat the very large number of (single) scatterings occurring in even thin absorbers, analytical theories for multiple scattering (i.e. the combined effect of many

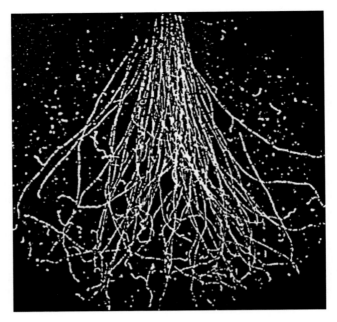

FIGURE 3.14
Bubble chamber picture of a narrow 9.3 MeV electron beam in propane. The increasing multiple scatter toward the end of the electron range is clearly seen. (From Harder, D., Harigel, G., and Schultz, K., *Strahlentherapie*, 115,1, 1961.)

individual single-scattering events) were developed (Rossi 1952; Andreo 1985; Berger and Wang 1988; Fernández-Varea et al. 1993). The simplest multiple scattering theory, though of limited accuracy, is the Gaussian small-angle one (Rossi 1952) that can be expressed as

$$P(\theta)\mathrm{d}\theta = \frac{2}{\overline{\theta^2}}\,\theta\,\exp\left(-\frac{\theta^2}{\overline{\theta^2}}\right)\mathrm{d}\theta \qquad (3.21)$$

which is the probability of finding an electron with a direction between θ and $\theta + \mathrm{d}\theta$ after traversal through an absorber. The mean square scattering angle, $\overline{\theta^2}$, is given by the product of the absorber thickness, s, and the scattering power, defined as $(\mathrm{d}\overline{\theta^2}/\mathrm{d}s)$ and usually (and somewhat confusingly cf. kinetic energy) denoted by T:

$$\overline{\theta^2} = s\left(\frac{\mathrm{d}\overline{\theta^2}}{\mathrm{d}s}\right) = sT \qquad (3.22)$$

The quantity scattering power is analogous to stopping power. The mass scattering power is similarly written T/ρ. It increases with decreasing electron energy and increasing atomic number. Scattering power is dealt with in detail in Kase and Nelson (1978) and in ICRU (1984a) which gives tables of T/ρ for materials and energies of relevance in dosimetry. Li and Rogers (1994) should be consulted for an excellent discussion of the validity and application of the concept of electron mass scattering power.

Following Rossi (1952) and Kase and Nelson (1978), through the superposition of small and independent events, the mass scattering power (expressed for instance in radians^2 cm^{-1}) is given by

$$\frac{1}{\rho}\frac{d\overline{\theta^2}}{ds} \equiv T = \frac{N_0}{A}\int_{\theta_1}^{\theta_2}\theta^2\frac{d\sigma}{d\omega}d\omega \qquad (3.23)$$

which becomes

$$\frac{T}{\rho} = 16\pi N_0 \frac{Z^2}{A}r_0^2\left(\frac{mc^2}{p\beta c}\right)^2 \ln[196(Z/A)^{1/6}Z^{-1/3}] \qquad (3.24)$$

where the Rutherford single-scattering cross-section has been used (Equation 3.20), p is the electron momentum, the screening angle θ_1 is given by $\alpha Z^{1/3}(m/p)$, the large-angle limit θ_2 by $280\,A^{-1/3}\,(m/p)$, and the small-angle assumption $\sin\theta \approx \theta$ has been made. Equation 3.24 illustrates the dependence of T/ρ on the electron kinetic energy and on Z.

An idea of the magnitude of electron scattering in low-atomic number materials is given in Figure 3.15, where the Gaussian multiple-scattering distribution (Equation 3.21) has been evaluated in the framework of small-angle Fermi–Eyges transport theory (Chapter 27) for 11.8 MeV electrons in water.

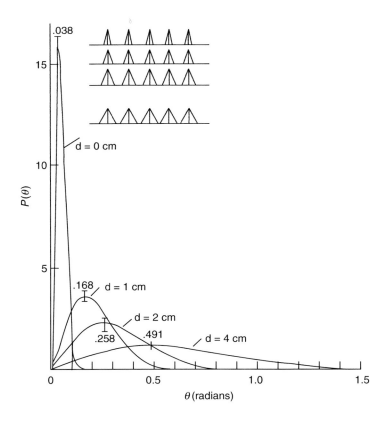

FIGURE 3.15
The angular distribution of electrons ($E_0 = 11.8$ MeV) vs. depth, d, in water based on Fermi–Eyges transport theory which uses simple Gaussian multiple scattering; note that a small angular spread was assumed at the surface. (From Hogstrom, K. R. and Almond, P. R., *The Effect of Electron Multiple Scattering on Dose Measured in Non-Water Phantoms*, AAPM Annual Meeting, 1982.)

3.6 APPLICATION TO AN ELECTRON DEPTH–DOSE CURVE

Figure 3.16 illustrates the physics of electron interactions as they apply to electron beams used in radiotherapy (see also Nahum and Brahme 1985); it shows three different depth–dose curves obtained through Monte-Carlo simulation (see Chapter 5), corresponding to different approximations about electron transport physics for a 30 MeV broad, monoenergetic, and parallel electron beam in water (see also Section 24.2 for a detailed description of electron depth–dose curves).

The curve labelled "CSDA straight ahead" corresponds to straight tracks and shows the Bragg Peak, normally associated with heavy charged particles (see Chapter 46); this extremely simple approximation illustrates very clearly the behaviour of the total stopping power S_{tot} as the electron energy gradually decreases with depth (see Section 6.6 for the relation between absorbed dose, electron fluence, and mass stopping power). The gradual decrease in dose with depth mirrors the decrease in total stopping power with falling electron energy. At an energy close to that of the electron rest mass (0.511 MeV), however, the collision stopping power goes through a minimum and then rises rapidly (principally because of the $1/\beta^2$ term in Equation 3.5.

The "CSDA multiple scattering" curve involves directional changes through multiple scattering (Equation 3.20 through Equation 3.23), but does not involve any secondary particle transport or any simulation of energy-loss straggling (see Section 3.4.2). The increase in dose away from the surface is entirely due to the increasing average obliquity of the electron tracks with depth and the fact that the beam is *broad* (i.e. there is lateral scattering equilibrium); this is

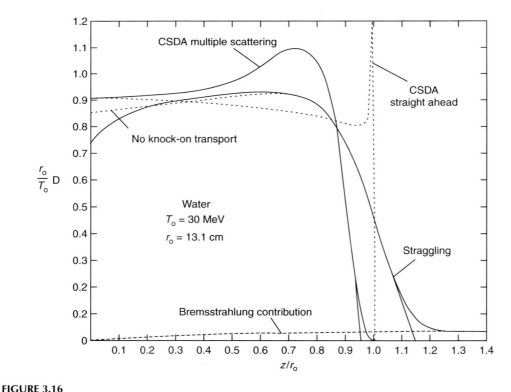

FIGURE 3.16

The effect of various approximations on the electron depth-dose curve for a broad, 30 MeV electron beam in water (csda range $r_o = 13.1$ cm), illustrating the physics of electron interactions covered in this chapter. (From Seltzer, S. M., Hubbel, J. H., and Berger, M. J., *National and International Standardization of Radiation Dosimetry*, Vol. 2, IAEA, Vienna, 1978.)

sometimes known as *scatter buildup*). At around $z/r_0 = 0.7$, the planar fluence (see Section 6.3.5) starts to decrease as electron tracks begin to reach their end. The maximum occurs as a result of a competition between the scatter build-up and the decrease in the planar fluence because of electrons reaching the end of their range.

The "No knock-on transport" curve does not include the generation and transport of knock-on electrons or delta rays, but it does include radiative losses (i.e. bremsstrahlung [see Section 3.3]) and one sees the so-called *bremsstrahlung tail* beyond the practical range. Also, now the slope of the dose falloff is much reduced; this is primarily due to the incorporation of energy-loss straggling (see Section 3.4.2). Finally, the unlabelled full curve corresponds to a simulation including the full electron transport physics (of most relevance in this energy region). The effect of simulating δ-ray transport is clearly seen in the build-up close to the surface; this is analogous to the much more pronounced one in megavoltage photon beams where the ranges of the mainly Compton electrons are significantly greater than those of the predominantly low-energy δ-rays.

In the case of heavy charged particles, the depth-dose curve (for a broad, approximately parallel beam) has a totally different appearance. The well-known Bragg Peak is the most prominent feature; in fact, this is the reason why proton beams are so well-suited to radio-therapy (see Chapter 46). In marked contrast to electron beams, the angular deflections in this case are much less pronounced due to these particles being at least 2000 times more massive than electrons. Consequently, all heavy-particle tracks are approximately straight except at the very end of their range; therefore, the appearance of depth-dose (and the fluence) curves most resembles that labelled "CSDA straight ahead" in Figure 3.16. Furthermore, there is negligible bremsstrahlung, and consequently much less energy-loss straggling, and also a gradual *increase* in the stopping power and hence the dose with decreasing (sub-relativistic) particle kinetic energy (i.e. increasing depth).

CHAPTER 4

INTERACTIONS OF PHOTONS WITH MATTER

David Dance and Gudrun Alm Carlsson

CONTENTS

4.1 INTRODUCTION

Electromagnetic radiation consists of electromagnetic waves that propagate with their electric and magnetic field vectors perpendicular to each other and to the direction of propagation. An electromagnetic wave is characterized by its wavelength, λ, and moves in a vacuum at the speed of light, c. Therefore, the number of oscillations per unit time, the frequency ν, is given by $\nu = c/\lambda$. Electromagnetic radiation is classified as ionising when it is capable of ionising atoms and molecules. In this case, the interactions with matter are most suitably described by considering the electromagnetic wave as a stream of particles, the photons. Each photon carries energy $h\nu$ where h is Planck's constant.

Photon interactions are stochastic (i.e. random) by nature. Unlike electrons, they may undergo a few, one, or no interactions as they pass through matter. In this chapter, the basic interaction processes and the probability of each interaction occuring when a beam of photons enters a medium are described. In each interaction, secondary ionising particles are created. These may be charged particles (usually electrons) or uncharged particles (usually photons). The charged particles deposit their energies close to the interaction site and contribute to the local energy deposition, whereas, secondary photons may be transported some distance before interacting. The transfer of energy to charged particles has particular significance in dosimetry, and interaction coefficients quantifying this energy transfer will also be introduced.

Secondary photons are important because they contribute to the photon fluence inside and around an irradiated body and to dose when they interact and produce secondary electrons. The relative importance of secondary photons depends on the energies of the primary photons. In external beam therapy using megavoltage beams, the dominant contribution to the absorbed doses within the patient is due to primary photons. This is also the case with low energy photons (<40 keV) such as those emitted from radioactive nuclides, e.g. ^{125}I, used in brachytherapy where photoelectric absorption is the dominant interaction process. In the intermediate range of photon energies, 50 keV to 200 keV, a considerable fraction of the absorbed dose may, depending on the depth in the body, be due to scattered photons. This is particularly so at photon energies in the interval 60 keV to 100 keV where multiple scattering becomes important within tissue depths relevant to radiotherapy.

This chapter gives an outline treatment of photon interactions. For more detailed information, the reader is referred to Evans (1955), Attix (1986) Hubbell (1999), and other references cited[*].

4.2 PHOTON INTERACTION CROSS-SECTIONS

4.2.1 INTERACTION CROSS-SECTIONS

Photons interact with various target entities such as atomic electrons, nuclei, atoms or molecules. The probability of interaction with a target entity is usually expressed in terms of the cross-section σ. This may be interpreted as the cross-sectional area that the target presents in the plane normal to the incident photon direction and can be used to estimate the probability of interaction in the following way. Consider a beam with one photon randomly incident on an absorber containing one target entity per unit p area measured in a plane normal to the photon direction. Further, assume that the photon will interact provided it passes through the cross-sectional area of the target. The probability, p, of interaction is then given by the fraction of the unit area that is occupied by the target area. This is numerically equal to σ.

---

[*] See also Section 5.3 for some historical references related to the fundamental photon interaction process.

Photon interactions can be characterized as absorption or scattering processes. In a full absorption process, the incoming photon loses all its energy and the energy is transferred to the target entity. Secondary particles are emitted during or subsequently to the interaction. In a full scattering process, an incoming photon interacts with a target entity and its direction of motion, energy and momentum may be changed because of this interaction. The photon, however, is not absorbed, and changes of energy and momentum are governed by the laws of relativistic kinematics. The main absorption processes are photoelectric (pe) absorption, pair (pair), and triplet (trip) production. The main scattering processes are coherent (coh) and incoherent (incoh) scattering. Nuclear photo-effect (phn) is an absorption process that is mostly neglected but needs to be considered in some cases. The total interaction cross-section, independent of which process occurs, is the sum of the cross-sections for the individual processes:

$$\sigma = \sigma_{pe} + \sigma_{coh} + \sigma_{incoh} + \sigma_{pair} + \sigma_{trip} + \sigma_{phn} \qquad (4.1)$$

As noted above, the target entity may be an atomic electron, a nucleus, an atom or a molecule. Care must be taken to distinguish between these targets. In the following, the type of target for the interaction is marked, when necessary, by adding an index to σ. Therefore, $_e\sigma$ and $_a\sigma$ designate the cross-section per electron and per atom, respectively. The relation between them is given by $_a\sigma = Z \times {}_e\sigma$ where Z is the atomic number of the atom.

The unit of cross-section is m^2. Although it does not belong to the International System of Units, the barn is still frequently used (including in this chapter):

$$1 \text{ barn} = 10^{-24} \text{ cm}^2 = 10^{-28} \text{ m}^2$$

4.2.2   DIFFERENTIAL SCATTERING CROSS-SECTIONS

In a scattering process, the distribution of scattered photons may not be isotropic, but may instead be anisotropic in some fashion related to the direction of the incoming photon and its polarisation. In order to quantify such effects, the cross-section is regarded as a function of the solid angle Ω in the direction of the scattered photon and the concept of the differential cross-section $d\sigma/d\Omega$ is introduced. The differential cross-section is defined in a way analogous to the total cross-section with $(d\sigma/d\Omega)d\Omega$ related to the probability that the photon scatters into solid angle $d\Omega$ (Figure 4.1). It follows that:

$$\sigma = \int_{4\pi} \frac{d\sigma(\theta,\phi)}{d\Omega} d\Omega \qquad (4.2)$$

where θ is the scattering (polar) angle, and ϕ is an azimuthal angle. In many situations, the scattering will, on average, have no azimuthal dependence, and the equation can then be

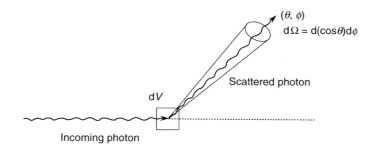

FIGURE 4.1
Scattering angles and solid angle. A photon incident on a small volume element dV with one target entity per unit area normal to its direction of motion has probability $(d\sigma/d\Omega)d\Omega$ to scatter into the solid angle element $d\Omega$ (spherical polar angles θ and ϕ.)

written

$$\sigma = \int_0^\pi \frac{d\sigma(\theta)}{d\Omega} 2\pi \sin\theta d\theta \qquad (4.3)$$

For isotropic scattering $d\sigma/d\Omega = \sigma/4\pi$.

Equation 4.3 may also be written

$$\sigma = \int_0^\pi \frac{d\sigma(\theta)}{d\theta} d\theta \qquad (4.4)$$

where

$$\frac{d\sigma(\theta)}{d\theta} = 2\pi \sin\theta \frac{d\sigma(\theta)}{d\Omega} \qquad (4.5)$$

The quantity $d\sigma/d\theta$ is also referred to as a differential cross-section. Indeed, this concept can be further generalised. For example, cross-sections differential in the energy of the scattered photon, $d\sigma/d(h\nu')$, can also be of utility.

4.3 PHOTON INTERACTION PROCESSES

4.3.1 PHOTOELECTRIC ABSORPTION

Photoelectric absorption is illustrated in Figure 4.2. In this process, an incoming photon interacts with an atom and is absorbed. An atomic electron is ejected with kinetic energy T from one of the atomic shells. Its kinetic energy is given by:

$$T = h\nu - E_B \qquad (4.6)$$

Here, $h\nu$ is the energy of the interacting photon, and E_B is the binding energy of the atomic electron. The process cannot occur with a free electron. The atom is needed in order to conserve momentum. Because of the heavy mass of the nucleus, the energy transferred to the atom is negligible.

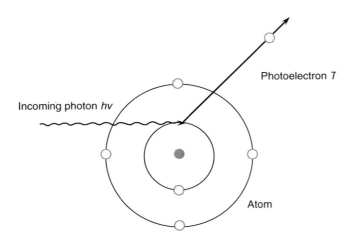

FIGURE 4.2
Photoelectric absorption. An incoming photon with energy $h\nu$ interacts with the atom and ejects a photoelectron with kinetic energy T.

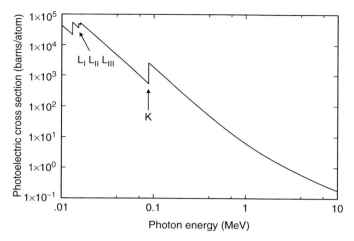

FIGURE 4.3
The total photoelectric absorption cross-section for lead as a function of photon energy. The positions of the L_I, L_{II}, L_{III}, and K absorption edges are indicated. (Data taken from Berger, M. J. and Hubbell, J. H., *XCOM: Photon Cross Sections on a Personal Computer*, 87–3597, NBS, Washington, DC, 1987.)

In general, the cross-section σ_{pe} for photoelectric absorption increases strongly with decreasing photon energy. Figure 4.3 shows this cross-section for lead over the photon energy range 10 keV to 10 MeV. The cross-section displays a series of discontinuities at energies corresponding to the binding energies of the electrons in the atomic shells. These discontinuities are known as absorption edges. Below the absorption edge, the photon does not have sufficient energy to liberate an electron from the shell. At energies just above the edge, the photon has sufficient energy to liberate the electron. Therefore, the cross-section abruptly increases because the number of electrons that can take part in the absorption process increases. The larger the binding energy, the larger the increase of the cross-section at the absorption edge. The absorption edge is most pronounced at the K shell in a high atomic number material. The L-shell has three sub-shells and, correspondingly, three absorption edges are seen in Figure 4.3 at the energies 13.04 keV, 15.20 keV, and 15.86 keV of the L sub-shells in lead. At energies above the K absorption edge, about 80% of the interactions take place in the K shell (see also Section 1.3.2.2).

The cross-section for photoelectric absorption depends strongly on atomic number. Above the K absorption edge, the cross-section per atom as a function of photon energy and atomic number is approximately given by

$$_a\sigma_{ph} \cong kZ^4/(h\nu)^3 \tag{4.7}$$

The cross-section increases as the fourth power of the atomic number and is inversely proportional to the third power of photon energy. This points to the strong impact of this process at low photon energies, particularly, at high atomic numbers. It also explains why bones offer high beam attenuation in soft tissue radiography.

The angular distribution of the photoelectrons is peaked at angles of $\pi/2$ to the forward direction at low photon energies, but it becomes increasingly forward directed as the photon energy increases.

After photoelectric absorption, a vacancy is left in the atomic shell. This vacancy is subsequently filled with an electron from an outer shell. The energy released is equal to the difference in the binding energies of an electron in the two shells (e.g. $E_K - E_L$ in a transition from the L to the K shell). The energy released is carried away either by the emission of a

TABLE 4.1

K-Fluorescence Yield and K X-Ray Energies for Selected Elements

Element	ω_K	E_K (keV)
C	0.003	0.3
Al	0.04	1.5
Cu	0.45	8.0
Ag	0.83	22.1
Pb	0.96	74.2

Source: From Storm, E. and Israel, H. I., *Nuclear Data Tables*, 7, 6, 1970; Hubbell, J. H., et al., *J. Phys. Chem. Ref. Data*, 23, 2, 339–364, 1994.

photon or an electron. The photon is known as a characteristic x-ray because of its fixed energy determined by the atomic number of the atom and the shells involved. Characteristic x-rays are isotropically emitted. At energies immediately above an absorption edge, they may carry a substantial fraction of the incident photon energy. They will have a lower photoelectric cross-section than the primary photon as their energy will be below that of the absorption edge. Electrons emitted after electronic rearrangement are known as Auger electrons (see Section 1.4.4). They are also isotropically emitted. The kinetic energy of an Auger electron is equal to the energy released in the transition minus its binding energy (e.g. an Auger electron emitted from the L shell after filling a K shell vacancy with an electron from the L shell receives kinetic energy $E_K - 2E_L$).

The probability of emission of a characteristic x-ray is called the fluorescence yield, ω, leaving probability $1 - \omega$ for emission of an Auger electron. Values of the K-fluorescence yield ω_K and energies of characteristic x-rays are given for some elements in Table 4.1.

The fluorescence yield is large when filling a vacancy in the K-shell of a high atomic number material and strongly decreases as the binding energy of the electrons in the shell containing the vacancy decreases. Therefore, the K-fluorescence yield strongly decreases with decreasing atomic number, and the fluorescence yields of the L- and higher shells are considerably lower than those of the K shell. As a vacancy has been filled, one or two new vacancies are created in outer shells. A cascade of events follow that may finally leave the atom in a highly ionized condition, particularly after the ejection of a K shell electron in a high atomic number atom (see also Section 1.4).

4.3.2 COMPTON INTERACTION AND SCATTERING PROCESSES

In a scattering process, the photon changes its direction of motion. If its energy is reduced, the scattering is called incoherent. The scattering may also occur without energy loss and is then referred to as coherent scattering. The terms *elastic* and *Rayleigh scattering* have also been used for this process.

For photon energies that considerably exceed the binding energies of the atomic electrons, the kinematics of the scattering process is usually described by considering the target electron to be free and at rest at the moment of collision. In this case, the scattering is incoherent because the photon will lose energy upon being scattered. At lower photon energies, the binding energies of the atomic electrons cannot be neglected. The photon can then scatter from individual bound electrons (incoherent scattering) or from all the bound electrons together, scattering in phase (coherent scattering). In the latter case, the whole atom takes part in the scattering process to conserve momentum.

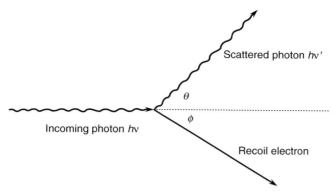

FIGURE 4.4
Scattering angles and energies for Compton scatter. An incoming photon of energy $h\nu$ scatters to produce a scattered photon of energy $h\nu'$ and a recoil electron.

4.3.2.1 Incoherent Scattering

Compton Scattering by Free Electrons: In incoherent scattering, the photon transfers part of its energy to an atomic electron that is ejected from the atomic shell. The process was first described by Compton who assumed the electron to be free and at rest at the moment of collision. In this approximation, the process is also known as Compton scattering. The kinematics of Compton scattering is illustrated in Figure 4.4.

The incoming photon with energy $h\nu$ is scattered through the angle θ (the scattering angle). The energy $h\nu'$ of the scattered photon is given by the Compton equation

$$h\nu' = \frac{h\nu}{1 + \alpha(1-\cos\theta)} \tag{4.8}$$

where $\alpha = h\nu/(m_0c^2)$ and m_0 is the rest mass of the electron.

The equation shows that the photon loses no energy when scattered in the forward direction ($\theta = 0$). For 180° scattering (backward scattering), the photon loses the largest fraction of its energy. This fraction increases with increasing photon energy, and the energies of backscattered photons approach the limiting value of $m_0c^2/2$ (256 keV) at high photon energies. For 90° scattering, the corresponding limiting value is m_0c^2 (511 keV).

Kinetic energy $T_e = h\nu - h\nu'$ is transferred to the electron. This energy is zero when the photon scatters in the forward direction and takes its largest value, $T_e = (h\nu 2\alpha)/(1+2\alpha)$, when the photon is backscattered. Figure 4.5 shows the maximum kinetic energy of the electron expressed as a fraction of the incident photon energy. The figure shows that at low photon energies, the energy transferred is small, but at high energies, the energy transfer is almost complete.

The angle ϕ between the directions of the incoming photon and the electron (Figure 4.4) is given by

$$\cot\phi = (1 + \alpha)\tan\frac{\theta}{2} \tag{4.9}$$

This equation shows that $\phi \leq \pi/2$, and the recoil electron is never emitted in the backward direction.

The cross-section for Compton scattering is named after Klein and Nishina who first derived an expression for its value. The differential Klein–Nishina cross-section per electron

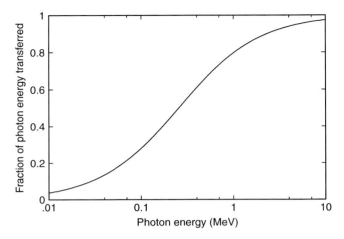

FIGURE 4.5
Fraction of the incident photon energy that is transferred to the recoil electron for Compton collisions where the photon is backscattered at 180°.

is given by

$$\frac{d_e\sigma_{KN}(\theta)}{d\Omega} = \frac{r_e^2}{2}\left(\frac{h\nu'}{h\nu}\right)^2\left[\frac{h\nu'}{h\nu} + \frac{h\nu}{h\nu'} - \sin^2\theta\right] \tag{4.10}$$

At low energies ($h\nu \rightarrow 0$), this reduces to

$$\frac{d_e\sigma_{Th}(\theta)}{d\Omega} = \frac{r_e^2}{2}(1 + \cos^2\theta) \tag{4.11}$$

This cross-section is known as the classical Thomson differential cross-section. In Equation 4.11, r_e is the classical electron radius given by $r_e = e^2/m_0c^2 = 2.8179 \times 10^{-15}$ m.

The total Klein–Nishina cross-section per electron may be obtained by integrating Equation 4.10, substituting for $h\nu'$ using Equation 4.8. The result is:

$$_e\sigma_{KN} = 2\pi r_e^2\left(\frac{1+\alpha}{\alpha^2}\left[\frac{2(1+\alpha)}{1+2\alpha} - \frac{\ln(1+2\alpha)}{\alpha}\right] + \frac{\ln(1+2\alpha)}{2\alpha} - \frac{1+3\alpha}{(1+2\alpha)^2}\right) \tag{4.12}$$

The total cross-section per atom is $_a\sigma_{KN} = Z_e\sigma_{KN}$.
Similarly, the total Thomson scattering cross-section is

$$_e\sigma_{Th} = \int_0^\pi \frac{r_e^2}{2}(1 + \cos^2\theta)2\pi\sin\theta d\theta = \frac{8}{3}\pi r_e^2 = 0.665 \text{ barns} \tag{4.13}$$

The energy transferred to the Compton electron will be subsequently deposited close to the point of interaction (within the electron range), whereas, the secondary photon may travel much farther. It is interesting to partition the total cross-section into absorption, σ_a, and scatter, σ_s, components

$$\sigma = \sigma_a + \sigma_s \tag{4.14}$$

where

$$\sigma_a = \frac{\langle T\rangle}{h\nu}\sigma$$

and

$$\sigma_s = \frac{\langle h v' \rangle}{h v} \sigma$$

Here, $\langle T \rangle$ and $\langle h v' \rangle$ are the expectation values of the energies of the Compton electron and the scattered photon, respectively. Equation 4.8 and Equation 4.10 may be combined to yield the following expression for $_e\sigma_s$:

$$_e\sigma_s = \pi r_e^2 \left[\frac{\ln(1 + 2\alpha)}{\alpha^3} + \frac{2(2\alpha^3 - 3\alpha - 1)}{\alpha^2(1 + 2\alpha)^2} + \frac{8\alpha^2}{3(1 + 2\alpha)^3} \right] \tag{4.15}$$

The differential Klein–Nishina cross-section is shown in Figure 4.6a; the total cross-section and the partial cross-sections $_e\sigma_a$ and $_e\sigma_s$ in Figure 4.6b.

At low photon energies, the Compton scattering angular distribution is almost symmetrical about $\cos \theta = 0$ (Figure 4.6a). With increasing energy, the photons are increasingly scattered into the forward direction, and an increasing fraction of the energy is transferred to the Compton electron (Figure 4.6b). At 0.01 MeV, only 1.87% of the photon energy is converted to electron kinetic energy. This fraction increases to 50% at 1.7 MeV and reaches 79.4% at 100 MeV. These fractions are, of course, limited by those shown in Figure 4.5.

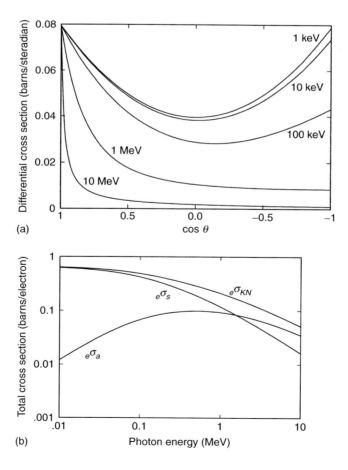

(a)

(b)

FIGURE 4.6
Cross-sections for Compton scattering from free electrons: (a) Variation of differential cross-section with $\cos \theta$ and the energy of the incident photon, (b) Variation of the total cross-section, KN (Klein Nishina) and the partial cross-sections for scattered energy, s, and energy absorption, a, with the energy of the incident photon.

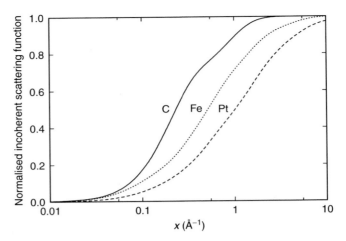

FIGURE 4.7
Variation of the incoherent scattering function with the momentum transfer parameter x^*. Results are shown for carbon (C), iron (Fe), and platinum (Pt). In each case, the incoherent scattering function has been normalised to the atomic number. (Data taken from Hubbell, J. H., et al., *J. Phys. Chem. Ref. Data*, 23(2), 339–364, 1994.)

Scattering from Bound Electrons: The influence of electron binding on the incoherent scattering cross-section is usually quantified by the incoherent scattering function $S(x,Z)$. The differential scattering cross-section for incoherent scattering per atom is then given by

$$\frac{d_a\sigma_{incoh}(\theta)}{d\Omega} = \frac{d_e\sigma_{KN}(\theta)}{d\Omega} S(x,Z) \tag{4.16}$$

The incoherent scattering function is generally assumed to be a function of the momentum transfer and the atomic number, Z. It is tabulated in terms of the momentum transfer related quantity x given by

$$x = \frac{\sin(\theta/2)}{\lambda} \tag{4.17}$$

where λ is the wavelength of the primary photon.

The incoherent scattering function is related to the probability that the electron will be ejected from the atomic shell as a result of the momentum transfer x to the atom. It takes its lowest value at $\theta = 0$ where $S(x,Z) = 0$. It monotonically increases with increasing x (increasing photon energy and/or scattering angle) and approaches Z, the number of electrons in the atom, as x approaches infinity. The cross-section then reduces to the free electron cross-section. At high photon energies, the cross-section for Compton scattering is approached at all scattering angles. The incoherent scattering function is shown in Figure 4.7 for the elements carbon, iron, and platinum. To facilitate comparison, the function has been normalized to the atomic number. It can be seen that the momentum transfer required for the incoherent scattering function to approach its maximum value increases with increasing atomic number.

---

* The momentum transfer is expressed in units of inverse ångström (Å^{-1}) to match the units used in the standard reference cited (1 ångström $= 10^{-10}$m).

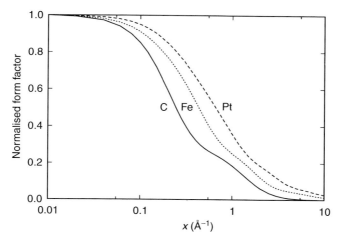

FIGURE 4.8
Variation of the form factor for coherent scattering with the momentum transfer parameter x*. Results are shown for carbon (C), iron (Fe), and platinum (Pt). In each case, the form factor has been normalized to the atomic number. (Data taken from Hubbell, J. H. and Øverbø, I., *J. Phys. Chem. Ref. Data*, 9, 69, 1979.)

4.3.2.2 Coherent Scattering

In coherent scattering, the photon is collectively scattered by the atomic electrons. Essentially, no energy is lost by the photon as it transfers momentum x to the atom while being scattered through the angle θ. The scattering from the different electrons is in phase, and the resultant angular deflection is determined by an interference pattern characteristic of the atomic number of the atom. The differential cross-section for coherent scattering is obtained as the product of the differential Thomson scattering cross-section and the atomic form factor F squared

$$\frac{\mathrm{d}_e \sigma_{coh}(\theta)}{\mathrm{d}\Omega} = \frac{\mathrm{d}_e \sigma_{Th}(\theta)}{\mathrm{d}\Omega} F^2(x,Z) \tag{4.18}$$

The atomic form factor is, like the incoherent scattering function, a universal function of x. Figure 4.8 shows this form factor for the elements carbon, iron, and platinum. To facilitate comparison, the form factor is normalised to the atomic number.

The form factor takes its maximum value in the forward direction ($\theta = 0$) where $F(0,Z) = Z$. It decreases to zero as x increases; with increasing momentum transfer x, it gets increasingly difficult for all electrons to scatter in phase without absorbing energy. However, for a given momentum transfer, the normalised form factor F/Z increases with increasing atomic number.

As noted above, the form factor F increases with decreasing x (Figure 4.8). Corresponding to this increase, the incoherent scattering function S decreases (Figure 4.7) and, to some extent, compensates for the increase in F. In hydrogen with only one atomic electron, the compensation is exact, and $S(x, 1) + F^2(x, 1) = 1$ at all values of x. In higher atomic number media, for small values of x, coherent scattering more than compensates for the reduced incoherent scattering. This is illustrated in Figure 4.9 where differential cross-sections for coherent scattering, incoherent scattering, and the sum of coherent and incoherent scattering are compared to the free electron differential scattering cross-section for carbon at a photon energy of 50 keV. The data in the figure is plotted just for the range $\cos \theta = 1.0$–0.6 in order to show the peaking of the coherent and total differential scattering cross-section in the forward direction.

---

* The momentum transfer is expressed in units of inverse ångström (Å^{-1}) to match the units used in the standard reference cited (1 ångström $= 10^{-10}$m).

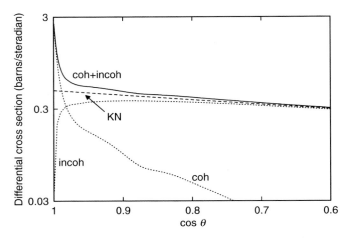

FIGURE 4.9
The differential cross-sections for coherent scattering (coh), incoherent scattering (incoh), total scattering (coh+incoh), and free electron scattering (KN) from carbon for the $\cos \theta$ range 1.0–0.6. (Data taken from Hubbell, J. H. and Øverbø, I., *J. Phys. Chem. Ref. Data*, 9, 69, 1979; Hubbell, J. H., et al., *J. Phys. Chem. Ref. Data*, 23(2), 339–364, 1994.)

At low photon energies, x is close to zero at all scattering angles, and the total cross-section for coherent scattering is then

$$_a\sigma_{coh} \cong e\sigma_{Th}Z^2 \tag{4.19}$$

At higher energies, x becomes larger, and F is small for much of the energy range. The cross-section then decreases with increasing energy, approximately as

$$\sigma_{coh} \propto (h\nu)^{-2} \tag{4.20}$$

4.3.3 PAIR AND TRIPLET PRODUCTION

Pair production is illustrated in Figure 4.10. In pair production, the photon is absorbed in the electric field of the nucleus. An electron (negatron)-positron pair is created and emitted with the sum of their kinetic energies, $T^- + T^+$, being determined by the requirement for

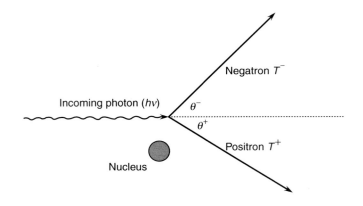

FIGURE 4.10
Pair production. An incoming photon with energy $h\nu$ is absorbed in the electric field of a nucleus. An electron (negatron)—positron pair is created with kinetic energies T^- and T^+, respectively.

conservation of energy

$$T^- + T^+ = h\nu - 2m_0c^2 \tag{4.21}$$

Here, m_0c^2 is the energy equivalent of the electron rest mass m_0.

As in photoelectric absorption, the nucleus is needed to conserve momentum. Also in this case, the energy associated with the recoil of the atom is negligible. From the above equation, it is clear that the process has a threshold value of $2m_0c^2$(1.02 MeV), the minimum energy required to create two electrons. On average, the electron–positron pair about equally shares the kinetic energy available. The probability for either particle to absorb most of the energy is small.

The two electrons tend to be ejected in the forward direction. At energies well above the threshold energy, the average angle of *departure* from the photon direction, $\bar{\theta}$, is

$$\bar{\theta} \approx \frac{m_0c^2}{\bar{T}}; \quad \bar{T} = \frac{1}{2}(h\nu - 1.02 \text{ MeV}) \tag{4.22}$$

The process of pair production may also occur in the electric field of an atomic electron. The atomic electron will recoil with sufficient energy to be ejected from the atomic shell. Three electrons appear as a result of the interaction and, accordingly, the process is called *triplet production*. Triplet production has an energy threshold at $4m_0c^2$ (2.04 MeV).

The cross-section for pair production in the nuclear field is zero below threshold. It then rapidly increases with increasing energy and, well above threshold, varies approximately as the square of the nuclear charge Z, i.e.

$$_a\sigma_{Pair} \propto Z^2 \tag{4.23}$$

The cross-section for triplet production, at energies above threshold, approximately varies as Z

$$_a\sigma_{Pair} \propto Z \tag{4.24}$$

Triplet production is as important as pair production in hydrogen ($Z=1$), but it gets increasingly less important, compared to pair production, with increasing atomic number. In high atomic number media, Z dependence of these processes becomes weaker because of screening of the electric fields of the target entities by the surrounding atomic electrons.

4.3.4 NUCLEAR PHOTOEFFECT

When the photon energy exceeds that of the binding energy of a nucleon, it can be absorbed in a nuclear reaction (see Section 2.1.1). As a result of the reaction, one or more nucleons (neutrons and/or protons) are ejected. The cross-section for the nuclear photoeffect depends, in a complicated manner, on both the atomic number, Z, and the atomic mass, A, and thus on the isotopic abundance in a sample of a given element. Because of these irregularities, it is not readily given in tabulated form. The cross-section has an energy threshold, and it is shaped as a giant resonance peak. The peak occurs between 5 and 40 MeV, depending on the element, and it can contribute between 2% (high-Z element) and 6% (low-Z element) to the total cross-section.

4.3.5 THE TOTAL ATOMIC CROSS-SECTION

The total atomic cross-section and its partial cross-sections are given in Figure 4.11 for the elements carbon ($Z=6$) and lead ($Z=82$). In both elements, photoelectric absorption is the

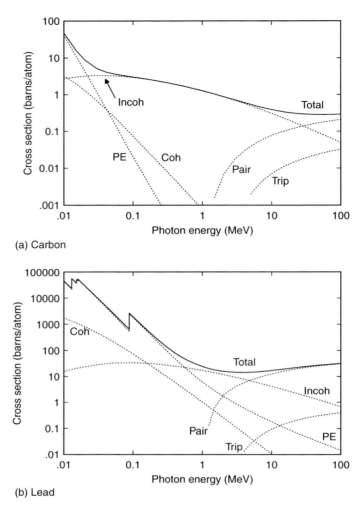

FIGURE 4.11
The total and partial cross-sections for carbon (a) and lead (b) for photon energies from 10 keV to 100 MeV. (Data taken from Berger, M. J. and Hubbell, J. H., *XCOM: Photon Cross Sections on a Personal Computer*, 87-3597, NBS, Washington, DC, 1987.)

dominating interaction at low photon energies (<50 keV in carbon, <800 keV in lead). With increasing photon energy the next process to become dominant is incoherent scattering (although the absolute amount of incoherent scattering actually decreases as the energy increases - see Figure 4.11). In a low atomic number medium such as carbon, this process remains the most important over a wide range of energies, 100 keV to 20 MeV. In a high atomic number element, the predominance of incoherent scattering is limited to a much smaller energy interval, 800 keV to 5 MeV in lead. This is due to the onset of pair production at energies above 1.02 MeV and the strong dependence of this process on atomic number. The cross-section for pair production approaches a constant value at 100 MeV in both elements. The larger relative contribution of triplet production in the low atomic number element is clearly seen. Nuclear photoeffect significantly contributes to the total cross-section in a narrow energy interval around the giant resonance peak. The threshold for nuclear photoeffect is above 10 MeV in carbon, but it occurs at a lower photon energy in lead. Coherent scattering is never the dominating interaction. At low energies, it is more important than incoherent scattering; however, at these energies, photoelectric absorption is, by far, the most important process.

4.4 MACROSCOPIC BEHAVIOUR

4.4.1 BEAM ATTENUATION AND ATTENUATION COEFFICIENTS

Photons incident on an absorber will either interact in it (producing secondary electrons and/or scattered photons) or else pass through it without interacting. The number of photons transmitted undisturbed through an absorber of thickness t of a given element and density can, for mono-energetic photons, be derived in the following way (see Figure 4.12). The number of primary photons, $d\Phi$, interacting in a thin layer dx at depth x is proportional to the thickness of the layer and the number of photons incident on the layer so that

$$d\Phi = -\mu dx \Phi(x) \tag{4.25}$$

The linear attenuation coefficient μ is a property of the material and depends on photon energy. The minus sign indicates that photons are removed from the beam. Integrating the equation from $x=0$ to $x=t$ gives the number, $\Phi(t)$, of primary photons that are transmitted through the absorber. This number decreases exponentially with increasing thickness t according to

$$\Phi(t) = \Phi_0 e^{-\mu t} \tag{4.26}$$

with $\Phi_0 = \Phi(0)$ the number of incident photons.

The linear attenuation coefficient is the probability per unit length for interaction and is related to the total atomic cross-section, σ_{tot}, through the relation

$$\mu = N\sigma_{tot} \tag{4.27}$$

where N is the number of target entities per unit volume. It is given by

$$N = \frac{N_A}{A}\rho \tag{4.28}$$

where N_A is the Avogadro's number ($N_A = 6.022 \times 10^{23}$ atoms/mol), A is the relative atomic mass of the target element (normalized to the mass of an atom of the nuclide ^{12}C), and ρ its density.

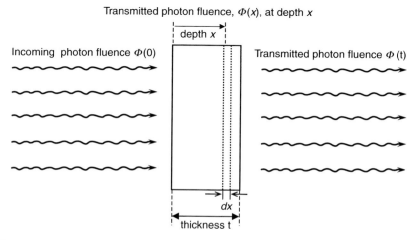

FIGURE 4.12
Calculation of photon transmission through a slab of matter. The incident fluence of primary photons $\Phi(0)$ is reduced to a fluence $\Phi(x)$ of primary photons at depth x in the slab.

The mass attenuation coefficient, μ/ρ, obtained by dividing μ with ρ, is independent of the actual density of the absorber and makes this quantity attractive for use in compilations.

The penetration power of a photon beam is commonly expressed by means of the mean free path. This is defined as the average distance, $\bar{x}$, travelled by the photon before it interacts. For mono-energetic photons it is given by

$$\bar{x} = \int_0^\infty x e^{-\mu x} \mathrm{d}x = \frac{1}{\mu} \tag{4.29}$$

The mean free path equals the thickness of an absorber that reduces the primary photons to a fraction $1/e = 37\%$ of their initial number.

The mass attenuation coefficient for mixtures and compounds such as water can be calculated as the weighted sum of the mass attenuation coefficients of their components.

$$\frac{\mu}{\rho} = \sum w_i \left(\frac{\mu}{\rho}\right)_i \tag{4.30}$$

Here, w_i is the fraction by weight of the element i in the compound, e.g. for water

$$\left(\frac{\mu}{\rho}\right)_{H_2O} = \frac{2}{18}\left(\frac{\mu}{\rho}\right)_H + \frac{16}{18}\left(\frac{\mu}{\rho}\right)_O \tag{4.31}$$

4.4.2 ENERGY TRANSFER AND ENERGY ABSORPTION COEFFICIENTS

The energy transferred to kinetic energy of secondary electrons in a photon interaction is of particular interest in dosimetry because the electrons will impart their energies close to the point where they were released (within their maximum range). The mass energy transfer coefficient μ_{tr}/ρ has been defined to allow calculations of this energy and is defined as

$$\frac{\mu_{tr}}{\rho} = \frac{\mu}{\rho}\frac{\langle T \rangle}{h\nu} \tag{4.32}$$

where $\langle T \rangle$ is the expectation value of the energy converted to kinetic energy of secondary electrons in an interaction. Considering each type of interaction i separately, the right side is obtained as the weighted sum

$$\frac{\mu_{tr}}{\rho} = \sum_i f_i \frac{\mu_i}{\rho} \tag{4.33}$$

The weighting factors, f, for the different processes (photo electric, incoherent, coherent, pair production) are given respectively by

$$f_{pe} = 1 - \frac{\delta}{h\nu} \tag{4.34a}$$

$$f_{incoh} = 1 - \frac{\langle h\nu' \rangle + \delta}{h\nu} \tag{4.34b}$$

$$f_{coh} = 0 \tag{4.34c}$$

$$f_{pp} = 1 - \frac{2m_0 c^2}{h\nu} \tag{4.34d}$$

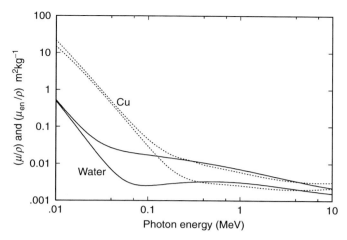

FIGURE 4.13

Mass attenuation coefficient μ/ρ, and mass energy absorption coefficients μ_{en}/ρ, for water (solid curves) and copper (dotted curves). For each material the mass attenuation coefficient lies above the mass absorption coefficient. (Data taken from Hubbell and Seltzer 1995).

Here, δ^* is the average energy of characteristic x-rays emitted subsequent to an ionization event (photoelectric absorption or incoherent scattering) because of rearrangements of vacancies in the atomic shells and $\langle h\nu' \rangle$ is the average energy of an incoherently scattered photon.

In slowing down, the electrons may lose energy to secondary photons (mainly bremsstrahlung, see Section 3.3.3). The mass energy absorption coefficient μ_{en}/ρ allows for these losses

$$\frac{\mu_{en}}{\rho} = \sum_i (1 - g_i) f_i \frac{\mu_i}{\rho} = (1 - g) \frac{\mu_{tr}}{\rho} \qquad (4.35)$$

Here, g is the fraction of the kinetic energy lost to photons during the complete slowing down of secondary electrons. It is usual to allow for bremsstrahlung when estimating g (e.g. ICRU 1980); however, account should also be taken of positron annihilation in flight and characteristic x-rays (Seltzer 1993).

In Figure 4.13, values of μ/ρ and μ_{en}/ρ are compared for water and copper. In a low atomic number medium such as water, the mass energy absorption coefficient comes close to the mass attenuation coefficient at low photon energies where photoelectric absorption prevails. Both fluorescence yields and energies of the characteristic x-rays are low so that δ (the average energy emitted as fluorescence radiation) becomes a negligibly small fraction of the photon energy. In copper, however, for photon energies immediately above the K absorption edge, the high fluorescence yield and energies of characteristic x-rays produce a significant deviation between the mass energy absorption and the mass attenuation coefficients. At higher photon energies, δ will be an increasingly smaller fraction of the photon energy and the deviation between the mass energy absorption and mass attenuation coefficients is mainly due to the fraction of energy carried away by incoherently scattered photons. At still higher energies, bremsstrahlung and annihilation in flight additionally contribute to the deviation between the two coefficients.

---

[*] The symbol delta used here should not be confused with the symbol used for *delta-rays* (electrons) discussed in Section 3.2.1., nor with the *density effect correction factor* discussed in Section 3.2.3.

As the photon energy increases, the energies and ranges of their secondary electrons also increase. In particular, the quotient, μR, between the maximum range of the secondary electrons, R, and the mean free path of the photons, $1/\mu$, increases, and the applicability of the mass energy absorption coefficient needs to be carefully considered.

4.5 SOURCES OF INTERACTION DATA

The overall fluence of photons inside and around a body consists of photons that have not interacted (primary photons) and secondary photons of various generations. With knowledge of the actual irradiation geometry including the energy spectrum of photons emitted from the source, the attenuation coefficients for relevant materials, and the differential cross-sections for scattering processes, the photon field may be calculated using analytical or Monte Carlo methods. For high atomic number media, the energies of absorption edges, characteristic x-rays and fluorescence yields also need to be known.

The cross-section data needed for such calculations are available in published tabulations and are known with a high degree of accuracy. In the photon energy range of most interest in medical applications, 5 keV–50 MeV, attenuation coefficients are known within 1–2% (one standard deviation; see review by Hubbell 1999). Mass attenuation coefficients can be obtained from the XCOM program first published by Berger and Hubbell (1987) and available on line at http://physics.nist.gov/PhysRefData/Xcom/Text/XCOM.html. They are tabulated in Part M for energies 0.001 MeV up to 50 MeV and for several materials of interest in radiotherapy physics (e.g. biological tissues, detectors components, radiation protection barriers). Information on mass attenuation and mass energy absorption coefficients is available from the NISTIR compilation (Hubbell and Seltzer 1995). The compilation by Hubbell et al. (1975) provides information on the incoherent scattering function, $S(x,Z)$ that may be used to derive differential scattering cross-sections for incoherent scattering. Differential cross-sections for coherent scattering may be derived from the compilation of relativistic form factors, $F(x,Z)$, by Hubbell and Øverbø (1979), replacing the atomic form factors in Hubbell et al. (1975).

Data on absorption edges in different materials and energies of characteristic x-rays are available from Storm and Israel (1970). Fluorescence yields are tabulated in Hubbell et al. (1994).

CHAPTER 5

THE MONTE CARLO SIMULATION OF RADIATION TRANSPORT

Alex Bielajew

CONTENTS

5.1 INTRODUCTION

As applied to radiation transport applications in radiotherapy and dosimetry, the Monte Carlo method provides a numerical solution to the Boltzmann transport equation (e.g. Kase and Nelson 1978; Duderstadt and Martin 1979) that directly employs the fundamental microscopic physical laws of electron-atom and photon-atom interactions. Monte Carlo simulation faithfully reproduces the individual particle tracks, in a statistical sense, within current knowledge of the physical laws: the scattering and absorption cross-sections. The radiation fields' macroscopic features (e.g. the average track-length per incident photon in a given volume of space) are computed as an average over many individual particle simulations or histories. If the true average $\bar{x}$ exists and the distribution in x has a true finite variance, σ_x^2, the *Central Limit Theorem* (Lindeberg 1922; Feller 1967) for energies 0.001 MeV up to 50 MeV guarantees that the Monte Carlo estimator for $\bar{x}$, that is referred to here as $\langle x \rangle$, can be made arbitrarily close to $\bar{x}$ by increasing the number, N, of particle histories simulated. Moreover, the *Central Limit Theorem* predicts that the distribution of $\langle x \rangle$ is Gaussian, characterised by a variance $\sigma_{\langle x \rangle}^2$ that may be simply estimated in the simulation. The *Central Limit Theorem* also predicts that in the limit $N \to \infty$, $\sigma_{\langle x \rangle}^2 \to 0$. This limiting result is also proven by the *Strong Law of Large Numbers* (Feller 1967).

It is these facts that have been partly responsible for the rapid increase in the use of the Monte Carlo method in radiotherapy and dosimetry applications (Nahum 1988), a trend that has continued since Nahum's review article[*]. If one knows the governing physical laws to sufficient accuracy and has access to sufficient computing resources, then the answer to any well-posed physical question may be computed. Fortunately, the physical laws required for most applications in radiotherapy and dosimetry are well known. They are the results of Quantum Electrodynamics (QED) (Bjorken and Drell 1965; Sakurai 1967), one of the most successful theories of theoretical physics. Additionally, the computer resources required for most of our applications are modest, and they may be executed to sufficient accuracy on affordable desktop computers and workstations. Today's powerful PCs, generally running under Linux, effectively function as Monte Carlo calculation engines. This confluence of theory and computational ability puts the Monte Carlo method into the standard toolbox of the medical physicists, especially if they are involved in research.

Of course, there are significant technical obstacles to be overcome, but these are surmountable. The point of this chapter is to demonstrate the power and utility of the Monte Carlo method without the detail that would be considered in a fuller discussion. Fortunately, there are some valuable detailed reviews on this topic (Raeside 1976; Mackie 1990; Rogers and Bielajew 1990; Andreo 1991; Zaidi and Sgouros 2003). The reader should realise that an application with tens of thousands of volume elements and many different materials to describe a simulation geometry is only slightly more complicated than a simple two-medium interface problem. More organisation and computer coding is required, but the technical ability to solve the simpler problem empowers one with the ability to solve the geometrically more complex problem.

5.2 A BRIEF HISTORY OF MONTE CARLO

Usually, the first reference to the Monte Carlo method is that of the Comte de Buffon (Buffon 1777) who proposed a Monte Carlo-like method to evaluate the probability of tossing a needle onto a ruled sheet. This reference goes back to 1777, well before automatic calculating machines. Buffon calculated that a needle of length L randomly tossed on a plane ruled with parallel lines of distance d apart, where $d > L$, would have a probability:

$$p = \frac{2L}{\pi d} \tag{5.1}$$

---

[*] A more recent paper by Carrier et al. (2004) clearly illustrates how widespread Monte Corlo simulation has become in Medical (radiation) Physics, and it also demonstrates the impressive choice of sophisticated public domain Monte Corlo codes today's hospital physicist may access.

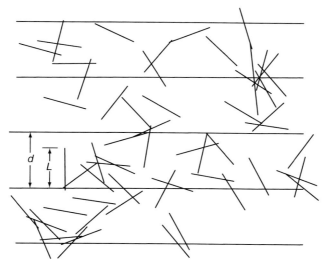

FIGURE 5.1
A computer simulation of the buffon needle problem. The distance d between lines is larger than the length L of each needle ($L/d=3/4$). 50 needles are randomly tossed. The probability that a needle will cross one line is $(2/\pi)\times(3/4) \approx 1/2$ (i.e. 25 needles crossing a line).

A computer simulation of 50 needles (where $L/d=\frac{3}{4}$) on a finite grid of 5 lines is shown in Figure 5.1.

Much later, Laplace (1886) suggested that this procedure could be employed to determine the value of π, albeit slowly. Several other historical uses of Monte Carlo predating computers are cited by Kalos and Whitlock (1986). The modern Monte Carlo age was ushered in by von Neumann and Ulam during the initial development of thermonuclear weapons[*]. Ulam and von Neumann coined the phrase *Monte Carlo*, and they were pioneers in the development of the Monte Carlo technique and its realisations on digital computers[†].

5.3 PHOTON INTERACTION PROCESSES

The photon interaction processes that should be modelled by a Monte Carlo code designed for applications in radiotherapy and dosimetry are:

- Pair production in the nuclear and atomic fields (Davies et al. 1954; Motz et al. 1969; Tsai 1974)
- Compton scattering from atomic electrons (incoherent scattering) (Klein and Nishina 1929; Compton and Allison 1935)
- Photoelectric absorption and photoelectron production (Sauter 1931)
- Rayleigh scattering from atomic and molecular fields (coherent scattering) (Rayleigh 1871)

These processes are reviewed in systematic detail in Chapter 4.

---

[*] The two books by Richard Rhodes, *The Making of the Atomic Bomb* and *Dark Sun* are excellent historical sources for this period.
[†] Despite their history, thermonuclear weapons have never been deployed in conflict. Therefore, Monte Carlo calculations have not been employed for destructive purposes.

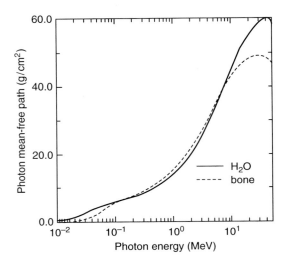

FIGURE 5.2
Mean-free path of photons in water and bone. (Data extracted from EGS4 code system: Nelson W. R. et al., *The EGS4 Code System Report SLAC-265*, Stanford Linear Accelerator Center, Stanford, California, 1985.)

The relative importance of the various processes involved for materials common to radiotherapy applications are now considered. For water and bone, the mean-free path, i.e. the average distance a photon travels before interacting via one of the interaction processes, is shown in Figure 5.2 as a function of energy. When presented in units of grams per square centimetre, the mean-free path in bone is almost the same as that in water between 100 keV and 10 MeV, but it is higher outside of this range. These differences may be quite easily modelled with Monte Carlo methods.

Figure 5.2 shows that photon's interaction distances in the energy range $10 \text{ keV} \leq E_\gamma \leq 40 \text{ MeV}$ are of the order of 20 cm for common low-Z materials (i.e. the photons will interact only a few times in macroscopic objects such as a radiation measuring device, a tank of water, or a human being undergoing treatment with radiotherapy)[*]. This means that if one considers the transport of photons alone, it is feasible, with modest computational resources, to simulate hundreds of millions of particle histories.

Figure 5.3 is a schematic illustration of an actual photon history that includes the tracks of the secondary particles. The history begins at position 1 in a vacuum; photons are indicated by sinusoidal tracks and the secondary electrons by straight lines.

The simulation consists of the following steps (here, steps 1 to 7 are explicitly labelled in the figure):

1. Choose photon energy, direction, and starting position based on sampling from distribution of incident photons, and transport photon to first boundary
2. Choose distance to first interaction (see Section 5.6.2) and transport photon to this interaction point
3. Choose the type of interaction (Compton scatter, photo-electric, pair-production, Rayleigh scatter)
4. Choose direction, energy, etc. of new particles (such as Compton electrons by sampling from the Klein–Nishina differential cross-section; characteristic photons; Auger electrons). Put them on the stack (i.e. a list of secondary particles to be followed later)

---

[*] Monte Carlo based radiotherapy treatment planning requires of the order of 10^8 to 10^9 photon histories (see Chapter 28).

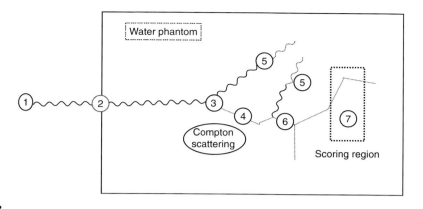

FIGURE 5.3
A schematic illustration of a Monte Carlo photon history (Reproduced courtesy of F. Verhaegen, private communication). Electron transport is dealt with in much greater detail in Section 5.4 and Section 5.5.)

5. Transport scattered photon until it either leaves the geometry, or it reaches some predetermined energy cut-off (e.g. PCUT in the EGS code system)

6. Transport secondary electron. Keep track of any δ electrons and bremsstrahlung photons produced

7. Score deposited energy, fluence spectra, etc. in region of interest

8. Repeat steps 1–7 for many more particles until scored quantities reach a sufficiently low statistical uncertainty

Figure 5.4 shows a flow diagram for a photon Monte Carlo simulation.

The techniques involved in sampling from distributions, including random number generation as well as the trigonometric relationships required in ray-tracing, are dealt with briefly in Section 5.6.

5.4 ELECTRON INTERACTION PROCESSES

The electron and positron interaction processes that should be modelled by a Monte Carlo code designed for applications in radiotherapy and dosimetry are

- Møller scattering of electrons from atomic electrons (Møller 1932)
- Bhabha scattering of positrons from atomic electrons (Bhabha 1936)
- Bremsstrahlung photon creation in the nuclear and atomic fields (Koch and Motz 1959; Tsai 1974)
- Positron annihilation with atomic electrons (Heitler 1954)
- Elastic scattering of electrons and positrons from nuclei (Mayol and Salvat 1997)
- Excitation of atoms and molecules by electrons and positrons

The most important of the above processes are dealt with in some detail in Chapter 3.

Figure 5.5 presents the electron mean-free paths for the elastic, ionisation, excitation, and bremsstrahlung interactions in oxygen. The distance to an interaction in the relativistic region (greater than, say, 1 MeV) is noted as between 10^{-5} and 10^{-4} g/cm^2. The range of a 10 MeV electron in oxygen is 5.6 g/cm^2. This means, for example, that a relativistic electron must undergo 10^5 to 10^6 interactions before slowing down. A typical electron is completely

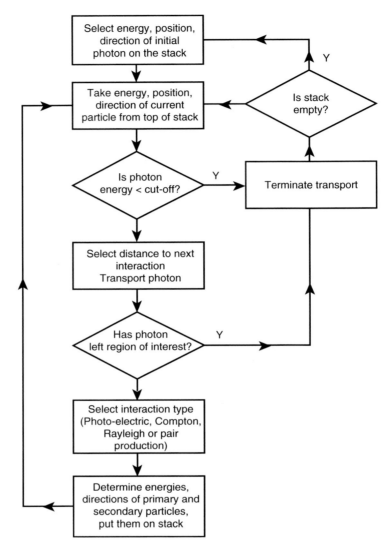

FIGURE 5.4
Flow chart showing the essential steps in the Monte-Carlo transport of photons; the transport of electrons is not dealt with here. (Reproduced courtesy of F. Verhaegen, Private communication).

slowed down in a typical simulation geometry. This means that each one of the 10^5 to 10^6 interactions would have to be simulated for each electron history. This form of calculation is called *analogue* simulation, and it is always done for photons (see Section 5.3). However, it requires Teraflop computational resources for most practical problems involving electron transport.

Fortunately, there is a practical solution to this problem that was pioneered by Berger (1963). Berger called his technique *Condensed History Electron Transport*. It is based on the realisation that, whereas electrons undergo many interactions, relatively few of these interactions cause a great deal of energy loss or directional change. Therefore, the effect of most of these interactions is small, involving little energy loss or small angular deflections. Therefore, one can combine the effect of these small-effect interactions into single virtual large-effect interactions. These large-effect interactions can be theoretically predicted through cumulative-event theories. For energy losses, there is the continuous slowing down approximation

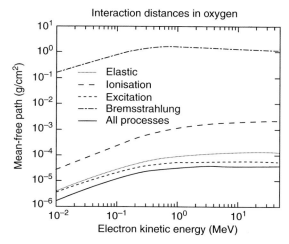

FIGURE 5.5
Electron elastic, ionisation, excitation, and bremsstrahlung mean-free path in oxygen (Data derived from the Lawrence Livermore Evaluated Electron Data Library, Perkins, S. T., et al., *Lawrence Livermore National Laboratory Report UCRL-50400*, 31, 1991.)

(CSDA) method with the energy loss being characterised by the stopping power[*] (Bethe 1930, 1932; Bloch 1933; ICRU 1984a and 1984b) or by distributions that are a function of the length of the electron path (Landau 1944; Vavilov 1957). The effect of cumulative elastic scattering events is predicted by several *small-angle* theories (Molière 1947; Eyges 1948; Molière 1948; Bethe 1953); although, these are being superseded by the *any-angle* theory of Goudsmit and Saunderson (1940a, 1940b). All of these theories require single-event elastic scattering models. The Goudsmit–Saunderson approach is favoured because it can be employed to use any cross-section. In particular, elastic cross-section calculations using partial-wave analysis are becoming quite sophisticated (Mayol and Salvat 1997) and are being adopted in advanced Monte Carlo algorithms. Fortunately, the elastic scattering that produces the angular deflections produces no energy loss. Additionally, the energy-loss process produces very little angular deflection. Therefore, these two processes nearly decouple, making the theoretical development somewhat simpler and the construction of algorithms somewhat easier[†].

The *Condensed History Method* (CHM) has been put on a stronger theoretical footing by Larsen (1992) who showed that the CHM converges at the exact solution of the Boltzmann transport equation in the limit of small electron path-lengths. This is actually a warning that the CHM is an approximation, and its use can lead to calculation artefacts when applied outside of its range of validity (Bielajew et al. 1985; Bielajew and Rogers 1989; Seltzer 1991). This has led to the development of high-accuracy CHM (Bielajew and Rogers 1987; Seltzer 1991; Fernández-Varea et al. 1993; Kawrakow and Bielajew 1998a). Research in this area remains quite active. Nahum (1999) has provided a review of the breadth of applications of the CHM in medical physics.

The benefits of adopting the CHM are evident in the partial mean-free-paths depicted in Figure 5.6. In this figure, the mean-free-path to a bremsstrahlung interaction and a Møller interaction (δ-creation event) are shown for secondary particle thresholds of 1, 10, and 100 keV. The analogue Monte Carlo method is employed for interactions that produce

---

[*] See Chapter 3 for a detailed treatment of electron collision stopping power.
[†] High accuracy approaches have to consider this coupling and treat it carefully (Kawrakov and Bielajew 1998a, 1998b).

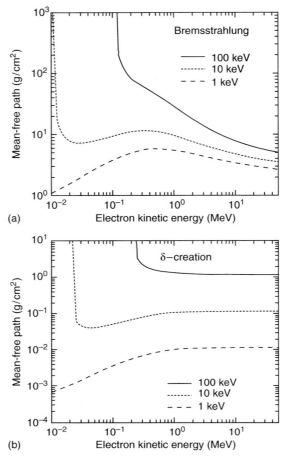

FIGURE 5.6
Mean-free-path to a bremsstrahlung interaction (a) or a Møller interaction (b) for different thresholds. (Data extrated from the EGS4 code system: Nelson W. R. et al., *The EGS4 Code System Report SLAC-265*, Stanford Linear Accelerator Center, Stanford, California, 1985.)

secondary particles above the thresholds. The *Continuous Slowing Down* (CSD) or other energy-loss methods are employed for the sub-threshold events. It should be noted that higher thresholds produce longer mean-free paths and more efficient calculation. This is traded for a loss of information on particle fluences below the thresholds. Selection of these thresholds (and other parameters like electron path-length) becomes one of the skills of the Monte Carlo user. Further information on these parameters is best obtained from the extensive literature describing the various general-purpose Monte Carlo codes available such as EGS4 (Nelson et al. 1985; Bielajew et al. 1994), ETRAN (Seltzer 1989, 1991), ITS (Halbleib and Mehlhorn 1984; Halbleib 1989; Halbleib et al. 1992), MCNP (Briesmeister 1986, 1993, 1997), and PENELOPE (Baró et al. 1995; Salvat et al. 1996; Sempau et al. 1997).

5.5 COUPLED ELECTRON–PHOTON TRANSPORT

The process by which coupled electron–photon transport is carried out is described in Figure 5.7.

In Figure 5.7a, the simulation geometry is depicted by the outer rectangular box. Photons are represented by straight dashed lines; electrons and positrons by solid curves. A photon

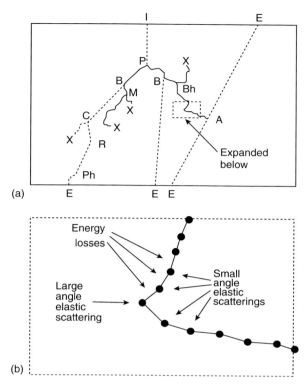

FIGURE 5.7
(a) Global picture of electron transport. Photons are represented by straight dashed lines; electrons and positrons by solid curves (I, insertion; E, escape; P, pair production; B, bremsstrahlung; M, Møller; X, end-of-range; C, Compton; R, Rayleigh; Ph, photoelectric; Bh, Bhabha; A, annihilation). (b) Expanded version of the dashed box in (a).

starts the simulation by insertion, I, into the simulation geometry. It undergoes a pair interaction, P, producing an e^-e^+ pair. The electron track (on the left) undergoes a bremsstrahlung interaction at B. Following the photon, it can be seen that there is a Compton interaction at C. The resulting electron is then transported until it runs out of energy at X. The photon undergoes a Rayleigh interaction at R, and it is then absorbed in a photoelectric event at Ph. The photoelectron escapes the simulation geometry at E. Returning to the initial electron, it undergoes a Møller interaction at M. Both the electron and its knock-on are then transported until their end-of-range at the X's. The positron that was born in the pair interaction at P undergoes a bremsstrahlung interaction at B, followed by a Bhabha interaction at Bh. The electron from B reaches its end-of-range at X. The positron eventually annihilates, producing two back-to-back annihilation quanta that escape the simulation geometry at their respective Es. The details of the CHM depicted in Figure 5.7b are now the focus. At each one of the vertices, represented by a solid circle, the positron changes direction. (Electron and positron transport in the CHM are essentially the same.) The deposition of energy according to the CSD or other models can be considered to happen anywhere along the track segments.

5.6 MATHEMATICAL METHODS OF MONTE CARLO

Before concluding this chapter, some of the basic mathematics of the Monte Carlo method will be briefly covered. A complete treatment would involve a detailed discussion of random number generation, sampling theory, displacements and rotations, estimating means and variances, and geometry. References are provided to allow a more in-depth study.

5.6.1   RANDOM NUMBER GENERATION

It will be assumed that a good source of random numbers is being used. Much mathematical study has been devoted to random number generation (Ehrman 1981; Knuth 1981; James 1988). These latter three references are excellent reviews of random number generation theory and methods up to about 1987. The following references contain more modern material: Marsaglia et al. (1990); Marsaglia and Zaman (1991); James (1994); Lüscher (1994); and Knuth (1981). It should also be noted that random number generation continues to be an area of active research; therefore, information that was given last year may be proven to be misleading this year. The best way to stay in tune is to track the discussions concerning this topic on the web sites of organisations where random number generation is critical. A particularly good example is that of the European Laboratory for Particle Physics (CERN: www.cern.ch).

5.6.2   ELEMENTARY SAMPLING THEORY

There are a number of sampling techniques well-known to all constructors of Monte Carlo codes. Only the *Cumulative Probability Distribution Function Method* is dealt with here.

Assume that a probability function $p(x)$ that is normalised over some range between a and b is used; that is:

$$\int_a^b dx' p(x') = 1 \tag{5.2}$$

Its cumulative probability distribution function is now constructed:

$$c(x) = \int_a^x dx' p(x') \tag{5.3}$$

that is monotonically increasing and bounded between 0 and 1. It can be shown that a variable x is distributed randomly according to $p(x)$ if x values are selected by inverting $c(x)$ according to:

$$x = c^{-1}(R) \tag{5.4}$$

where R is a uniformly distributed random number between 0 and 1.

For example, it can be shown that the distance to an interaction is given by:

$$x = -\frac{1}{\mu_{tot}} \log_e(1 - R) \tag{5.5}$$

where μ_{tot} is the attenuation coefficient (or macroscopic cross-section) for photons of that energy in the medium concerned[*]. This follows directly from the probability distribution for interaction distances:

$$p(x) = \mu \exp(-\mu x) \quad 0 \leq x \leq \infty \tag{5.6}$$

It is important to note that the technique of inverting a function implicit in Equation 5.4 and Equation 5.5 may not always be analytically possible, but it is always numerically possible. A treatment of other sampling techniques can be found elsewhere (e.g. Raeside 1976; Nelson et al. 1985; Turner 1985; Bielajew 1993).

---

[*] Written here as μ_{tot} to emphasise that it is the sum of the interaction coefficients for the separate processes (i.e. pair + Compton + photoelectric + Rayleigh); $1/\mu_{tot}$ is equal to the mean-free path between interactions. (μ_{tot} is simply denoted μ in Section 4.4.1).

5.6.3 Displacements and Rotations

Ray tracing (displacement and rotation) is carried out using very simple geometrical constructs. Given that a particle has a position $\vec{x}_0$ and direction $\vec{u}_0$ and distance to travel s, the new position, $\vec{x}$, is given by:

$$\vec{x} = \vec{x}_0 + \vec{u}_0 s \tag{5.7}$$

where $\vec{u}_0 = (u_0, v_0, w_0) = (\sin\theta_0\cos\phi_0, \sin\theta_0\sin\phi_0, \cos\theta_0)$, θ_0 and ϕ_0 are the polar and azimuthal angles, respectively.

Rotation after a scattering by polar angle Θ and azimuthal angle Φ is carried out via the rotation

$$u = \sin\theta\cos\phi = u_0\cos\Theta + \sin\Theta(w_0\cos\Phi\cos\phi_0 - \sin\Phi\sin\phi_0)$$
$$v = \sin\theta\sin\phi = v_0\cos\Theta + \sin\Theta(w_0\cos\Phi\sin\phi_0 + \sin\Phi\cos\phi_0) \tag{5.8}$$
$$w = \cos\theta = w_0\cos\Theta - \sin\Theta\sin\theta_0\cos\Phi$$

that provides the new direction for the particle after a scattering.

5.6.4 Estimating Means and Variances

The conventional approach to calculating the estimated error is as follows:

- Assume that the calculation calls for the simulation of N particle histories.
- Assign and accumulate the value x_i for the score associated with the ith history where $1 \leq i \leq N$. As well, assign the square of the score x_i^2 for the ith history.
- Estimate the mean value of x

$$\langle x \rangle = \frac{1}{N}\sum_{i=1}^{N}x_i \tag{5.9}$$

- Estimate the variance associated with the distribution of the x_i

$$s_x^2 = \frac{1}{N-1}\sum_{i=1}^{n}(x_i - \langle x\rangle)^2 = \frac{1}{N-1}\sum_{i=1}^{n}(x_i^2 - \langle x\rangle^2) \tag{5.10}$$

- The estimated variance of $\langle x \rangle$ is the standard variance of the mean

$$s_{\langle x\rangle}^2 = \frac{s_x^2}{N} \tag{5.11}$$

Note that it is the error in $\langle x \rangle$ that is sought, not the spread of the distribution of the x_i
- Report the final result as $\langle x \rangle = \pm s_{\langle x\rangle}$.

The *Central Limit Theorem* interpretation is that $\langle x \rangle$ satisfies $|\bar{x} - \langle x\rangle| < s_{\langle x\rangle}$ 68% of the time.

5.6.5 Geometry

Geometrical elements are made up of regions of space bounded by planes and quadric surfaces (spheres, cylinders, etc.). The essential problem is to determine the smallest intercept distance along the forward flight direction to one of the bounding surfaces. A well-defined prescription for doing this (Nelson and Jenkins 1987; Bielajew 1995) is summarised below.

Borrowing from the notation of Olmsted (1947), an arbitrary quadric surface in $3(x,y,z)$-space* can be represented by

$$f(\vec{x}) = \sum_{i,j=0}^{3} a_{ij}x_i x_j = 0 \qquad (5.12)$$

The a_{ij} symbols are arbitrary constants, and the 4-vector x_i has components $(1, x, y, z)$. The 0th component is unity by definition, allowing a very compact representation and a_{ij} is symmetric with respect to the interchange of i and j, that is $a_{ij}=a_{ji}$. Equation 5.12 is very general and encompasses a wide variety of possibilities, including solitary planes (e.g. only a_{0i} non-zero), intersecting planes (e.g. only a_{11} and a_{22} non-zero), cylinders (circular, elliptical, parabolic, and hyperbolic), spheres, spheroids and ellipsoids, cones (circular and elliptical), hyperboloids of one and two sheets, and elliptic and hyperbolic paraboloids. These surfaces can be combined to make geometrical objects of great complexity and are extremely useful in Monte Carlo modelling of physical objects.

Despite having apparently ten independent constants, Equation 5.12 represents only ten independent real surfaces (including the simple plane), unique after a translation and rotation to standard position. The three cross terms (a_{ij} for $i \neq j$ and $i, j \geq 1$) can be eliminated by rotation. The resultant equation then only involves terms like x_i^2 and x_i. In addition, providing that a given variable's quadratic constant is non-zero, the linear terms can be eliminated by a translation. The result is that there are only two generic forms:

$$f(\vec{x}) = \sum_{i=1}^{3} a_i x_i^2 + c = 0 \qquad (5.13)$$

and

$$f(\vec{x}) = \sum_{i=1}^{2} a_i x_i^2 + bx_3 = 0 \qquad (5.14)$$

Equation 5.13 and Equation 5.14 describe only ten distinct possibilities with real solutions

1. Ellipsoids: $a_1^2 x_1^2 + a_2^2 x_2^2 + a_3^2 x_3^2 - c^2 = 0$
2. Cones: $a_1^2 x_1^2 + a_2^2 x_2^2 - a_3^2 x_3^2 = 0$
3. Cylinders: $a_1^2 x_1^2 + a_2^2 x_2^2 - c^2 = 0$
4. Hyperboloids of one sheet: $a_1^2 x_1^2 + a_2^2 x_2^2 - a_3^2 x_3^2 - c^2 = 0$
5. Hyperboloids of two sheets: $a_1^2 x_1^2 + a_2^2 x_2^2 - a_3^2 x_3^2 + c^2 = 0$
6. Elliptic paraboloids: $a_1^2 x_1^2 + a_2^2 x_2^2 + a_3 x_3 = 0$
7. Hyperbolic paraboloids: $a_1^2 x_1^2 - a_2^2 x_2^2 + a_3 x_3 = 0$
8. Hyperbolic cylinders: $a_1^2 x_1^2 - a_2^2 x_2^2 + c^2 = 0$
9. Parabolic cylinders: $a_1^2 x_1^2 + a_3 x_3 = 0$
10. Simple planes: $a_3 x_3 + c = 0$

There are other possible imaginary surfaces such as imaginary ellipsoids ($a_1^2 x_1^2 + a_2^2 x_2^2 + a_3^2 x_3^2 + c^2 = 0$) or quadrics that can be made up of two independent planes in various

---

* The only variance with Olmsted's notation is that the 4th component is labelled as the 0th component in this text.

orientations (e.g. intersection planes $a_1^2 x_1^2 - a_2^2 x_2^2 = 0$, parallel planes $a_1^2 x_1^2 - c^2 = 0$, and coincident planes $a_1^2 x_1^2 = 0$).

For more information on the reduction to canonical form, the reader is encouraged to read Olmsted's book (1947). Olmsted also gives the classification of the surfaces and lists the entire set of 17 canonical quadric forms.

To calculate the intercept distance to an arbitrary quadric surface, consider that the particle's trajectory is given by:

$$\vec{x} = \vec{p} + \vec{\mu}s \qquad (5.15)$$

where the starting position of the particle is $\vec{p} = (p_x, p_y, p_z)$. A positive value of s expresses a distance along the direction that the particle is going (forward trajectory), and a negative value is associated with a distance along the direction that the particle came from (backward trajectory). Therefore, negative solutions that are found for s below will be rejected.

In Monte Carlo particle transport calculations as well as ray-tracing algorithms, a common problem is to find the distance a particle has to travel in order to intersect a surface. This is done by substituting for $\vec{x}$ from Equation 5.15 in Equation 5.12 to give:

$$s^2 \left(\sum_{i,j=0}^{3} a_{ij} \mu_i \mu_j \right) + 2s \left(\sum_{i,j=0}^{3} a_{ij} p_i \mu_j \right) + \left(\sum_{i,j=0}^{3} a_{ij} p_i p_j \right) = 0 \qquad (5.16)$$

where the convention that $\mu_0 = 0$ and $p_0 = 1$ has been adopted. This is a quadratic equation in s of the form $A(\vec{\mu})s^2 + 2B(\vec{\mu}, \vec{p})s + C(\vec{p}) = 0$ where $A(\vec{\mu}) = \sum_{i,j=0}^{3} a_{ij} \mu_i \mu_j$, $B(\vec{\mu}, \vec{p}) = \sum_{i,j=0}^{3} a_{ij} p_i \mu_j$ and $C(\vec{p}) = \sum_{i,j=0}^{3} a_{ij} p_i p_j$.

The intercept distance is given by the smallest positive solution of the above quadratic equation.

5.7 CONCLUSION

This chapter merely skims the surface of the Monte Carlo method as applied to radiotherapy and dosimetry. However, the simplicity of the approach or the power of the method cannot be overstated. With sufficient computational resources, any well-posed question can be answered as accurately as the knowledge of the basic cross-sections allows. Other approaches suffer from systematic uncertainties whose magnitude may be impossible to estimate[*]. However, these methods may be useful if the computational time for a Monte Carlo approach is prohibitive. If a computation is feasible by Monte Carlo methods, one is not required to estimate systematic errors, and one may proceed with some confidence to tackle other problems. If one factors in the human effort, ultimately, the Monte Carlo approach may prove to be the most productive.

---

[*] A relevant example is the variety of analytical (i.e. non-Monte Carlo approaches) to computing the dose distributions from megavoltage photon or electron beams required in radiotherapy treatment planning (see Part G) in extremely heterogeneous regions of patient anatomy such as the thorax and the head and neck; this is dealt with in Part F where the limitations of the various methods (pencil-beam, 3D superposition-convolution, etc.) are discussed, and it is then shown (in Chapter 28) how Monte Carlo simulation can overcome these more or less serious approximations involved though generally at the cost of a substantial increase in computation time.

CHAPTER 6

PRINCIPLES AND BASIC CONCEPTS IN RADIATION DOSIMETRY

Alan Nahum

CONTENTS

6.1 INTRODUCTION

The accurate determination of absorbed dose is crucial to the success of radiotherapy because of the relatively steep sigmoidal dose–response curves for both tumour control and normal-tissue damage. A difference of only a few percent in the dose (to the tumour) may separate uncomplicated tumour eradication from either failure to control the tumour through underdosage or serious normal tissue damage through overdosage (see Section 37.4.2). There are many different steps involved in the determination of the absorbed dose distribution in the patient. One of the most important of these involves measurements with a detector (often termed a *dosimeter*) in a phantom (often water, sometimes water-like plastic) placed in the radiation field. Such measurements include determining the absolute dose at a reference depth in a reference-size field (Chapter 18), relative doses at many positions in the phantom in order to map out a complete dose distribution (Chapter 19), and so-called in-vivo doses on the patient's skin during treatment (Section 40.2). In all cases the detector will, via a calibration factor, yield the dose D_{det} to its own sensitive material from a quantity of charge, light, film blackening, etc. Generally, the dose is required at a position r in the medium in the absence of the detector, $D_{med}(r)$. The conversion of D_{det} to D_{med} is, therefore, a fundamental step and requires a knowledge of the theoretical aspects of radiation dosimetry. The same is true of the calculation of the dose distribution inside the complicated inhomogeneous geometry of the patient (see Part F).

This chapter covers the fundamental ideas and principles involved in radiation dosimetry, independent of the particular detector being used. The characteristics of the different detectors or dosimeters used in radiotherapy, as well as the numerical values of the relevant quantities, are covered principally in Part D. Comprehensive treatments of the theoretical aspects of radiation dosimetry are given in Greening (1981) and Attix (1986). Kase and Nelson (1978), Johns and Cunningham (1983), Rajan (1992) and Metcalfe et al. (1997) are also very useful references.

It may be thought from the formal definitions of the quantities such as *absorbed dose* and *kerma*, given below, that radiation dosimetry is an old, well-established discipline. In fact, it has a rather tortuous history. Following Röntgen's discovery of x-rays in 1895, the first radiation unit proposed was based on its ionising power. Christen (1914) was ahead of his time in advocating the quantity *dose* defined as "radiant energy per unit volume". In 1928, the International x-ray Unit Committee defined the *Röntgen* as, "the quantity of x-radiation which, when secondary electrons are fully utilised and the wall effect of the chamber is avoided, produces in one cubic centimetre of atmospheric air at 0°C and 76 cm mercury pressure such a degree of conductivity that one electrostatic unit of charge is measured at saturation current". This quantity later became known as *exposure* (see Section 6.3.2).

It was not until 1950 that the International Commission on Radiation Units and Measurements formalised the definition of *dose*, "in terms of the quantity of energy absorbed per unit mass (ergs per gram) of irradiated material at the point of interest" (ICRU 1951). In 1954 the ICRU finally approved the term *absorbed dose*, with the unit being the *rad* which is defined as "100 ergs per gram". The modern unit, the *gray*, is equal to 100 rad. The meaning and formal definition of absorbed dose was further refined and its modern form (see Section 6.3.1) was given in ICRU (1980) and carried through to ICRU (1998).

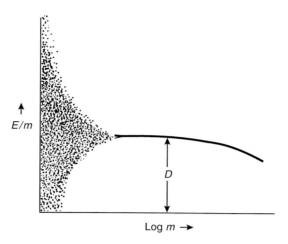

FIGURE 6.1
Energy deposited per unit mass, plotted against the mass, m, of the scoring volume, as this volume is gradually changed in size for a given incident radiation fluence. The shaded portion represents the range where statistical fluctuations become important as the volume (i.e. m) is getting smaller. (From Rossi, H. H., *Radiation Dosimetry*, Vol. 1, Academic Press, New York, 1968.)

6.2 THE STOCHASTIC NATURE OF ENERGY DEPOSITION

The absorbed dose arises from the deposition of energy in matter by radiation tracks. This is an inherently random process, as Chapter 3 and Chapter 4 make clear. Consider a situation where the energy, E, deposited in ever decreasing volumes centred on the same point in a uniform medium is determined for a series of irradiations of identical duration. A plot is then made of E, divided by the mass m of these volumes for each separate irradiation. Figure 6.1 reflects the anticipated result.

For large volumes (i.e. large log m) E/m has the same value each time[*]. Below a certain value of m, fluctuations start to appear which increase as m decreases. These are due to the fact that energy is deposited through interactions along particle tracks; some volumes, if small enough, will contain few or no tracks while others may contain many. Therefore, for a given particle fluence (see Section 6.3.3), at some size of the elementary volume in which we determine the dose in this experiment, the randomness of energy deposition will become apparent.

The absorbed dose discussed in what follows is *non-stochastic*, i.e. it is assumed to be determined in a mass element sufficiently large for the fluctuations to be negligible. The subject of fluctuations in the dose is generally dealt with under the heading of *microdosimetry* (ICRU 1983; Goodhead 1987) and falls outside the scope of this chapter.

6.3 DEFINITIONS OF DOSIMETRIC QUANTITIES

6.3.1 ABSORBED DOSE

ICRU (1980, 1998) defines *absorbed dose* as the quotient of $d\bar{\varepsilon}$ by dm, where $d\bar{\varepsilon}$ is the mean energy imparted by ionising radiation to matter of mass dm:

$$D = \frac{d\bar{\varepsilon}}{dm} \tag{6.1}$$

---

[*] The fall off at large values of m or log m corresponds to attenuation of the radiation when the path length across the volume becomes comparable with the mean free path for radiation interactions and is not relevant in this discussion.

The unit of absorbed dose is the *gray* which is 1 Joule per kilogram (J kg^{-1}); the old unit is the *rad* which is 10^{-2} gray (sometimes referred as a centigray).

The *energy imparted*, ε, by ionising radiation to the matter in a volume is defined by ICRU (1980, 1998) as:

$$\varepsilon = R_{in} - R_{out} + \sum Q \tag{6.2}$$

where

> R_{in} is the sum of the *energies* (excluding rest mass energies) of all those charged and uncharged ionising particles that enter the volume (known as the *radiant* energy)
>
> R_{out} is the sum of the *energies* (*excluding* rest mass energies) of all those charged and uncharged ionising particles that leave the volume, and
>
> $\sum Q$ is the sum of all *changes* (decreases: positive sign, increases: negative sign) of the rest mass energy of nuclei and elementary particles in any nuclear transformations that occur in the volume.

Figure 6.2 illustrates the concept of energy imparted. In the left part of the figure which represents a Compton interaction within the volume V, the energy imparted is given by

$$\varepsilon = h\nu_1 - (h\nu_2 + h\nu_3 + T') \tag{6.3}$$

where T' is the kinetic energy of the charged particle—of initial kinetic energy T—upon leaving the volume V. Note that the photon $h\nu_4$ does not appear as this is not emitted within the volume V. The $\sum Q$ term is not involved here.

The volume on the right in the figure involves γ-ray emission ($h\nu_1$) from a radioactive atom, pair production (kinetic energies T_1 and T_2), and annihilation radiation as the positron comes to rest. The energy imparted in this case is given by

$$\varepsilon = 0 - 1.022 \text{ MeV} + \sum Q \tag{6.4}$$

where

$$\sum Q = h\nu_1 - 2m_0c^2 + 2m_0c^2 = h\nu_1 \tag{6.5}$$

In Equation 6.4, the zero is the R_{in} term; '1.022 MeV' is the R_{out} term comprising the two annihilation γ-rays. The γ-ray energy $h\nu_1$ arises from a *decrease* in the rest mass of the nucleus; the $-2m_0c^2$ term is due to the creation of an electron–positron pair, i.e. an *increase* in rest mass

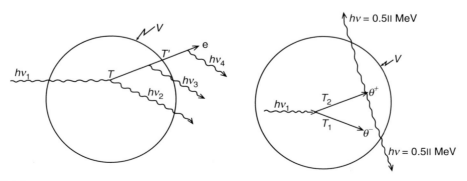

FIGURE 6.2
An illustration of the concept of the energy imparted to an elementary volume by radiation. (From Attix, F. H., *Introduction to Radiological Physics and Radiation Dosimetry*, Wiley, New York, 1986. With permission.)

energy. Finally the $+2m_0c^2$ term is due to the annihilation of an electron and a positron. Careful thought is required in this example.

Note that the quantity, ε, is stochastic, although we require its mean value, which is nonstochastic, in the definition of absorbed dose.

6.3.2 KERMA (AND EXPOSURE)

The quantity *kerma*, can be thought of as a step towards absorbed dose. It is conceptually very close to *exposure*, the first radiation quantity to be formally defined (see Section 6.1 and Greening 1981). All practising hospital physicists will come across kerma (in air in a cobalt-60 γ-ray beam) in the context of calibrating ionisation chambers at a National Standards Laboratory (see Section 18.2). The formal definition (ICRU 1980, 1998) follows:

The *kerma*, K, is the quotient dE_{tr} by dm, where dE_{tr} is the sum of the initial kinetic energies of all the charged ionising particles liberated by uncharged ionising particles in a material of mass dm:

$$K = \frac{dE_{tr}}{dm} \tag{6.6}$$

The units of kerma are the same as for absorbed dose, i.e. J kg^{-1} or gray (Gy). Kerma applies only to indirectly ionising particles which, for our purposes, almost always mean photons, although neutrons also fall into this category.

Exposure is conceptually closely related to *air kerma*. Exposure, usually denoted by X, is the quotient of dQ by dm where dQ is the absolute value of the total charge of the ions of one sign produced in air when all the electrons (negatrons and positrons) liberated by photons in air of mass dm are completely stopped in air. Until the late 1970s, all ionisation chambers were calibrated in terms of exposure; subsequently this was replaced by air kerma.

Figure 6.3 illustrates the concept of kerma (and exposure). It is the initial kinetic energies that are involved; the eventual fate of the charged particles (i.e. if they do or do not leave the elementary volume), does not affect kerma. In the volume in the figure, the initial kinetic energies of the two electrons labelled e_1 contribute to the kerma, as both were generated in the volume. The fact that one of these electrons leaves the volume with a residual kinetic energy T is irrelevant. None of the kinetic energy of the electron *entering* the volume with kinetic energy T contributes to kerma as this electron was generated *outside* the volume.

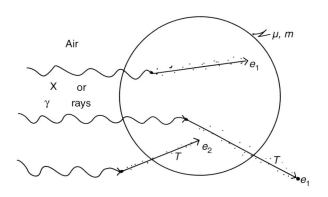

FIGURE 6.3
Illustration of the concepts of kerma and exposure. (Taken from Attix, F. H., *Introduction to Radiological Physics and Radiation Dosimetry*, Wiley, New York, 1986. With permission.)

It is important to realise that kerma includes the energy that the charged particles will eventually re-radiate in the form of bremsstrahlung photons. Kerma can be partitioned as follows (Attix 1979):

$$K = K_c + K_r \qquad (6.7)$$

where c refers to collision losses and r to radiation losses. The collision kerma K_c is related to the (total) kerma by

$$K_c = K(1 - g) \qquad (6.8)$$

The quantity g is the fraction of the initial kinetic energy of the electrons that is re-radiated as bremsstrahlung (in the particular medium of interest)[*].

Exposure and air kerma can be related to each other as follows. Multiplying the charge dQ (refer to the definition of exposure given above) by the mean energy required to produce one ion pair, divided by the electron charge, i.e. W/e (see Section 15.1), yields the collision part of the energy transferred, i.e. $dE_{tr}(1 - g)$ and therefore

$$X(W/e) = K_{air}(1 - g) \qquad (6.9)$$

or

$$X(W/e) = K_{air,c} \qquad (6.10)$$

6.3.3 Particle Fluence

To calculate absorbed dose we require quantities that describe the radiation field; these are known as 'field quantities' (Greening 1981). Particle fluence is a very important basic quantity, involving the number of particles per unit area. The concept is illustrated in Figure 6.4.

Let N be the expectation value of the number of particles striking a finite sphere surrounding point P (during a finite time interval). If the sphere is reduced to an infinitesimal one at P with a cross sectional area of dA, then the fluence Φ is given by

$$\Phi = \frac{dN}{dA} \qquad (6.11)$$

which is usually expressed in units of m^{-2} or cm^{-2} (ICRU 1980, 1998). Fluence is a *scalar* quantity—the direction of the radiation is not taken into account.

We will be meeting fluence, differential in energy, often written as Φ_E:

$$\Phi_E = \frac{d\Phi}{dE} \qquad (6.12)$$

in which case the (total) fluence is given by:

$$\Phi = \int_0^{E_{max}} \Phi_E \, dE \qquad (6.13)$$

It should be noted that fluence can also be expressed as the quotient of the sum of the track

---

[*] $g \approx 0.003$ in the case of cobalt-60 γ-rays in air (IAEA 1997).

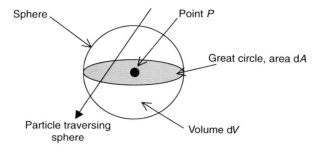

FIGURE 6.4
Characterization of the radiation field at a point P in terms of the radiation traversing a sphere centred at P. (From Attix, F. H., *Introduction to Radiological Physics and Radiation Dosimetry*, Wiley, New York, 1986.)

lengths Δs of the particles crossing the elementary sphere and the volume of the sphere[*] (Chilton 1978):

$$\Phi = \frac{\sum \Delta s}{dV} \qquad (6.14)$$

This form is extremely useful when considering so-called *cavity integrals* (see Section 6.7) which involve evaluating the fluence averaged over a volume.

6.3.4 ENERGY FLUENCE

Energy fluence is simply the product of fluence and particle energy. Let R be the expectation value of the total energy (excluding rest mass energy) carried by all the N particles in Figure 6.4. Then the *energy fluence* Ψ is given by (ICRU 1980, 1998)

$$\Psi = \frac{dR}{dA} \qquad (6.15)$$

If only a single energy E of particles is present, then $R = EN$ and $\Psi = E\Phi$.

In a similar fashion, one can define the above quantities per unit of time, i.e. fluence rate and energy fluence rate.

6.3.5 PLANAR FLUENCE

Planar fluence is the number of particles crossing a fixed plane in either direction (i.e., summed by scalar addition) per unit area of the plane. One can also define a vector quantity corresponding to *net flow* but this is of little use in dosimetry because it requires scalar, not vector, addition of the effects of individual particles.

Planar fluence is a particularly useful concept when dealing with beams of charged particles. In certain situations, e.g. at small depths in a parallel electron beam, one can say that the planar fluence remains constant as the depth increases; whereas, fluence generally increases due to the change in direction of the electron tracks (see Section 3.6). The difference between the two quantities is illustrated in Figure 6.5.

---

[*] This can be proved by noting that, for a sphere of volume V and external surface S, the mean chord length is $4V/S$.

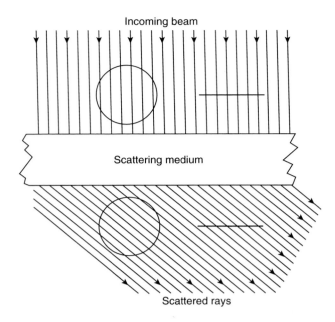

FIGURE 6.5
Schematic 2D illustration of the concept of planar fluence. The number of particles crossing the two horizontal lines (of equal length) is the same. This shows that the planar fluence in the original beam direction remains constant, whereas the fluence, illustrated by the total track length in the two circles, is much greater on the downstream side of the scattering medium.

6.4 RELATIONS BETWEEN FLUENCE AND DOSIMETRIC QUANTITIES FOR PHOTONS

This section will relate quantities concerning energy *transferred to* (i.e. kerma) or *deposited in* (i.e. absorbed dose) a medium to the *fluence*, which is a basic description of the radiation field. This will first be done for indirectly ionising particles (i.e. photons) starting with the relation between fluence and kerma (Section 6.4.1), then moving on to fluence and absorbed dose (Section 6.4.2) via the important concept of *charged particle equilibrium* (Section 6.5). Subsequently, the corresponding fluence dose relation for electrons will be derived (Section 6.6). From these fluence absorbed dose relationships we will then be in a position to derive important expressions for detector or dosimeter response, often known as *cavity theory*.

6.4.1 RELATION BETWEEN FLUENCE AND KERMA

Consider the schematic Figure 6.6 showing N photons, each of energy E, crossing perpendicularly a thin layer (of material *med*) of thickness dl and area dA. To extract energy from particle tracks and transfer it to the medium, we require an interaction coefficient. We can use the ICRU (1980, 1998) definition of the mass energy-transfer coefficient μ_{tr}/ρ (see Section 4.4.2):

$$\frac{\mu_{tr}}{\rho} = \frac{1}{\rho dl} \frac{dR_{tr}}{R} \qquad (6.16)$$

Identifying the fraction of incident radiant energy dR_{tr}/R as $dE_{tr}/(N \times E)$, and making a simple rearrangement, we have

$$dE_{tr} = \mu_{tr} \, dl \, NE \qquad (6.17)$$

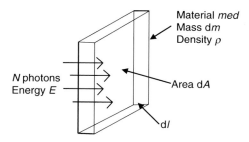

FIGURE 6.6
Illustration of N radiation tracks (photons) of energy E crossing a thin layer of material *med*, thickness dl, area dA, mass dm, density ρ.

Then, dividing both sides by the mass of the layer dm and rearranging again:

$$\frac{dE_{tr}}{dm} = \mu_{tr}E\left[\frac{N\,dl}{dm}\right] \tag{6.18}$$

Replacing dm by $\rho \times dV$ we can write:

$$\frac{dE_{tr}}{dm} = \frac{\mu_{tr}}{\rho}E\left[\frac{N\,dl}{dV}\right] \tag{6.19}$$

where we recognise the left hand side as kerma (in medium *med*) and the expression in the square brackets as the sum of the track lengths divided by the volume [i.e. fluence (see Section 6.3.3)], resulting in

$$K_{med} = \left(\frac{\mu_{tr}}{\rho}\right)_{med} E\Phi \tag{6.20}$$

or in terms of energy fluence Ψ:

$$K_{med} = \left(\frac{\mu_{tr}}{\rho}\right)_{med} \Psi \tag{6.21}$$

Note that the perpendicular incidence in Figure 6.6 was assumed for simplicity; Equation 6.20 and Equation 6.21 are valid for arbitrary angles of incidence*.

Up to this point, we have confined ourselves to particles crossing the thin layer with a *single* energy. In the more practical case of a spectrum of energies, described by the fluence differential in energy, Φ_E, we evaluate the K_{med} from:

$$K_{med} = \int_0^{E_{max}} E\Phi_E\left(\frac{\mu_{tr}(E)}{\rho}\right)_{med} dE \tag{6.22}$$

where the energy dependence of $(\mu_{tr}/\rho)_{med}$ has been shown explicitly. Therefore, we have now arrived at a relationship connecting kerma and fluence for photons.

It will also be necessary to calculate the collision kerma, K_c from photon fluence (see Section 6.3.2). To do this, one replaces the mass energy *transfer* coefficient μ_{tr}/ρ by the mass energy *absorption* coefficient μ_{en}/ρ, where *energy absorbed* is defined to exclude that part of the initial kinetic energy of charged particles converted to bremsstrahlung photons (see also

---

* This can most easily be seen by considering that fluence is the sum of the track lengths per unit volume, irrespective of their directions (see Section 6.3.3).

Section 4.4.2). The two coefficients are related by:

$$\mu_{en} = \mu_{tr}(1 - g) \tag{6.23}$$

which is naturally the same factor that relates K_c and K (see Section 6.3.2). It therefore follows that:

$$(K_c)_{med} = \left(\frac{\mu_{en}}{\rho}\right)_{med} E\Phi \tag{6.24}$$

and, similarly, for the integral over Φ_E in the case of a spectrum of incident photons:

$$(K_c)_{med} = \int_0^{E_{max}} E\Phi_E \left(\frac{\mu_{en}(E)}{\rho}\right)_{med} dE \tag{6.25}$$

6.4.2 Relation between Kerma and Absorbed Dose

Having established a relationship between kerma and fluence in the previous section, if absorbed dose can be related to kerma then a relationship will finally be established between absorbed dose and fluence for photons. However, the absorbed dose D_{med} in medium *med* concerns the (mean) value of energy *imparted* to an elementary volume, whereas kerma concerns energy *transferred* as the charged particles can leave the elementary volume (or thin layer), taking a fraction of the initial kinetic energy with them. This is illustrated in Figure 6.7. Note that the quantity denoted in the figure by E_{tr}^n is the *net* energy transferred and excludes that part of the initial kinetic energy converted into bremsstrahlung photons. It is equal to $E_{tr}(1 - g)$ as we have seen above.

In Figure 6.7, let the energy *imparted* to the layer be denoted by ε, the (net) kinetic energy leaving the layer be denoted by E_{out}^n and the (net) kinetic energy entering the layer on charged particles be denoted by E_{in}^n. Then, from Equation 6.2, we have:

$$\varepsilon = E_{tr}^n - E_{out}^n + E_{in}^n \tag{6.26}$$

If now the electron track that leaves the layer is replaced by an identical track that enters the layer we can write:

$$E_{in}^n = E_{out}^n$$

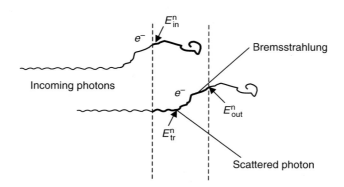

FIGURE 6.7
Schematic illustration of how secondary electrons created by photons can transfer (kinetic) energy E_{tr}^n to a thin layer, leave that layer, and also enter the layer from outside. If the energy leaving the layer (E_{out}^n) is exactly replaced by energy entering (E_{in}^n), then *charged particle equilibrium* is said to exist and absorbed dose can be equated to collision kerma. This is schematically represented on the figure, where the full length of the track of a secondary electron is exactly equal to the partial tracks of electrons (in bold) within the layer limits.

It then follows that:

$$\varepsilon = E_{tr}^{n} \qquad (6.27)$$

The equality between energy leaving and energy entering on charged particles is known as *charged particle equilibrium* (CPE). This can be realised under certain circumstances as is explained in Section 6.5.

Dividing both sides of Equation 6.27 by the mass of the layer or volume element, and changing from stochastic to average quantities, we can write:

$$D_{med} \overset{CPE}{=} (K_c)_{med} \qquad (6.28)$$

which is a very important result: under the special condition of charged particle equilibrium, the absorbed dose is equal to the *collision kerma*.

Consequently, replacing kerma by absorbed dose in Equation 6.24 and Equation 6.25, for monoenergetic photons it follows that

$$D_{med} \overset{CPE}{=} \left(\frac{\mu_{en}}{\rho}\right)_{med} E\Phi \qquad (6.29)$$

and, similarly, for the integral over Φ_E in the case of a spectrum of incident photons:

$$D_{med} \overset{CPE}{=} \int_{0}^{E_{max}} E\Phi_E \left(\frac{\mu_{en}(E)}{\rho}\right)_{med} dE \qquad (6.30)$$

Equation 6.29 and Equation 6.30 are very important relationships in radiation dosimetry; they will be exploited in Section 6.7.2.

6.5 CHARGED PARTICLE EQUILIBRIUM

Charged particle equilibrium (CPE), also known as *electronic equilibrium**, is said to exist in a volume V in an irradiated medium *if each charged particle of a given type and energy leaving V is replaced by an identical particle of the same energy entering V*.

There are, however, many situations where there is a lack of CPE. Strictly speaking, it is impossible to divide the lack of CPE into separate components (i.e. *longitudinal* or *lateral* disequilibrium). However, this distinction may be useful for a better understanding, especially for high energy photons where the secondary electrons are primarily peaked forward.

Figure 6.8 illustrates schematically how charged particle equilibrium can actually be achieved in a photon beam. Naturally, the figure is a huge oversimplification as in practice there will be a whole spectrum of secondary electron energies and directions. However, the arguments are not essentially altered by showing only one electron starting in each voxel, (labelled A to G in the figure) and travelling in a straight line.

In each voxel, one electron is generated and therefore, the kerma will be constant as it is assumed here that there is no or negligible photon attenuation. Only a fraction of the electron track deposits energy in voxel A; therefore, the dose is low and there is clearly no replacement for the part of the track that leaves the volume. In voxel B a *new* electron starts but here there is also part of the electron track which started *upstream* in voxel A. Hence the dose is higher than in voxel A; in voxel C the dose is higher still. However, in voxel D, where the electron which

---

* Other charged particles are involved but, in radiotherapy photon beams, the role of secondary electrons predominates.

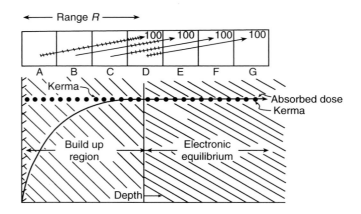

FIGURE 6.8
Diagram showing the build up to charged particle equilibrium for the idealised case of no attenuation of the photon beam and one straight electron track generated in each slab labelled A to G. (Taken from Johns, H. E. and Cunningham, J. R., *The Physics of Radiology*, 4th Ed., Charles, C., Ed., Thomas Publisher, Springfield, IL, 1983. With permission.)

started in voxel A comes to rest, all the sections of an electron track are present. This means that the sum of the kinetic energies leaving this volume must be exactly balanced by the sum of the kinetic energies entering and remaining in the volume, i.e. CPE is first attained in D. Subsequent voxels (E, F, G, etc.) will contain patterns of electron tracks identical to those in D and, therefore, CPE must also apply in these volumes. At the depth of voxel D, the absorbed dose is now equal to kerma (strictly this ought to be *collision* kerma) and this will also be the case in the subsequent voxels E, F, G, etc., in the absence of photon attenuation.

In the situation of a photon beam true CPE is strictly impossible to achieve in practice. Attenuation means that the photon fluence does not remain constant and, therefore, the number of secondary particles (electrons) starting at different depths also cannot be constant. Table 6.1 illustrates the degree of photon attenuation in water thicknesses ensuring *transient* electronic equilibrium for photon beams of different energies. The degree of CPE failure increases as the photon energy increases. Consequently, experimental determination of

TABLE 6.1

Approximate Thickness of Water Required to Establish Transient Charged Particle Equilibrium.

Maximum Energy of Photons (MeV)	Approximate Thickness of Water for Equilibrium (mm)	Approximate Photon Attenuation (%)
0.3	0.1	0.03
0.6	0.4	0.1
1	0.8	0.3
2	2.5	0.8
3	8	2
6	15	4
8	25	6
10	30	7
15	50	9
20	60	11
30	80	13

For bremsstrahlung beams of different maximum energies; the final column gives approximately the attenuation of photons in that thickness of water.
Source: From Greening, J. R., *Fundamentals of Radiation Dosimetry*, Adam Hilger, Bristol, 1981. With permission.

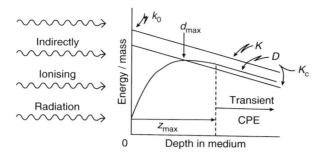

FIGURE 6.9
Variation of kerma, K, collision kerma, K_c, and dose, D, with depth in a beam of indirectly ionising radiation such as a photon beam. (Taken from Attix, F. H., *Introduction to Radiological Physics and Radiation Dosimetry*, Wiley, New York, 1986. With permission.)

exposure (now air kerma) is not attempted at maximum photon energies above approximately 3 MeV (see Section 15.3). Even below this energy, small corrections have to be made for the effect of photon attenuation.

Even though strict CPE may not exist, in many situations it is very well approximated, such as at depths beyond the dose maximum in media irradiated by photons below around 1 MeV in energy. At higher energies, the equals sign in Equation 6.28 can be replaced by a proportionality sign. This is then termed *transient charged particle equilibrium* (TCPE):

$$D \overset{\text{TCPE}}{\propto} K_c$$

The quantities K, K_c and dose D are plotted as a function of depth in Figure 6.9. The build-up region and the region of TCPE, where the D and K_c curves become parallel to each other are illustrated.

If radiative interactions and scattered photons are ignored it can be shown (Greening 1981) that:

$$D \overset{\text{TCPE}}{\approx} K_c(1 + \mu\bar{x}) \tag{6.31}$$

where μ is the common slope of the D, K and K_c curves and $\bar{x}$ is the mean distance the secondary charged particles carry their energy in the direction of the primary rays while depositing it as dose (Attix 1986). This constant of proportionality between dose and collision kerma is usually denoted by β, i.e.

$$D \overset{\text{TCPE}}{=} \beta K_c \tag{6.32}$$

Another situation where the dose is different from the collision kerma because CPE is not achieved is for beams with very small cross sections. Figure 6.10 and Figure 6.11 show the results from a Monte-Carlo simulation (using the EGSnrc code—see Section 28.2) of depth–dose variation for a cylindrical cobalt-60 γ-ray beam (energies 1.17 and 1.33 MeV, equally weighted) incident perpendicularly on the end of a cylinder of water with radius 20 cm and length 50 cm. The dose has been scored along the central axis in cylindrical columns of increasing radii, r, with 1 mm depth intervals for the first centimetre then 0.5 cm down to 10 cm depth (the results at greater depths are not shown)[*].

---

[*] In Figure 6.10 and Figure 6.11, the doses, in units of gray per unit incident photon fluence (Gy cm²), have been renormalised so that they can be compared for each of the scoring radii, r, 0.05 cm, 0.5 cm, and 10.0 cm, by multiplying the doses for the latter two radii by $\{r(\text{cm})/(0.05)\}^2$. This converts all three dose distributions, via the so-called reciprocity theorem (e.g. see Nahum 1976; Bielajew 1993), into three mathematically equivalent geometries of beams equal to the actual respective scoring radii, with the dose scored in regions with a radius equal to the actual beam radius. Note however, that the numerical values of dose shown in Figures 6.10 and 6.11 have only been rescaled in a relative and not an absolute sense to correspond to these equivalent geometries.

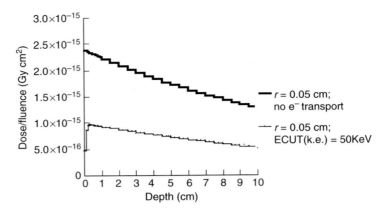

FIGURE 6.10
Cobalt-60 γ-ray beam in water incident on a water cylinder (radius 20 cm, length 50 cm): depth–dose curves obtained from Monte-Carlo simulation (10^8 photon histories for each case) when the *equivalent* beam radius, r, is 0.05 cm. A complete description of the geometry of the simulation geometry is given in the text and the normalisation is explained in the footnote. The upper curve, with no transport of secondary electrons, corresponds to the *kerma*. The lower curve includes full electron transport and shows clearly the lack of CPE. (Lasse Rye Aarup, private communication.)

In Figure 6.10 the radius of the scoring volume is very small, 0.05 cm. The upper curve corresponds to *no electron transport* (which is achieved by setting the electron transport cutoff to 0.5 MeV—see Section 5.4); thus the quantity scored corresponds exactly to (water) *kerma*. Upon switching electron transport back on (by simply reducing the electron transport cutoff to a suitably low value: 50 keV was chosen) the buildup of dose in the first 3 mm, due to electron transport, can now be observed. At the same time, however, the absolute value of the dose has fallen by more than a factor 2. Why should this be? The radius of only half a millimetre is in fact equal to only a fraction of the ranges of the highest energy secondary electrons created (virtually 100% of these will be due to Compton interaction—see Section 4.3.2) which is insufficient for the establishment of CPE or, put another way, the electrons will predominantly *leave* the scoring regions without being balanced by an equal number *entering* them, as the beam is so narrow. The absorbed dose is consequently much lower than the kerma in this very special geometry.

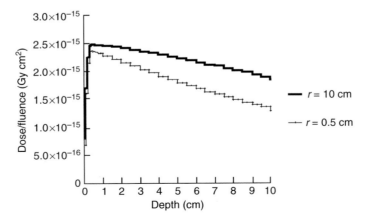

FIGURE 6.11
Depth–dose curves obtained for cobalt-60 γ-ray beams from Monte-Carlo simulation (10^8 photon histories) with equivalent radii, r, of 0.5 and 10.0 cm incident on a water cylinder (radius 20 cm, length 50 cm) with full electron transport (ECUT − KE = 50 keV). The increased dose and reduced effective attenuation for the larger radius is due to photon (Compton) scatter. (Lasse Rye Aarup, private communication.)

In Figure 6.11, the equivalent beam radii, r, are 0.5 cm and 10.0 cm, respectively. The depth–dose curve for the larger radius lies considerably above that for the smaller one. The latter was chosen to be large enough for the establishment of (partial) CPE. The reduced effective attenuation at the larger beam radius is explained by an increasing build-up of Compton-scattered photons with depth. The dependence of depth–dose curves on field size is a well-known and important phenomenon in external-beam radiotherapy (see Chapter 22).

It must be stressed that for large collimated beams where CPE is achieved on the beam axis, it is still partially lost at the edge, which contributes to enlarging the penumbra region (see Section 22.3.3.2).

6.6 RELATION BETWEEN FLUENCE AND DOSE FOR ELECTRONS

6.6.1 STOPPING POWER AND CEMA

The previous two sections have concerned indirectly ionising particles, i.e. photons. In this section, equivalent relationships will be derived between charged particle fluence (i.e. electrons) and absorbed dose. Referring back to Figure 6.6, consider that we now have N electron tracks incident perpendicularly on the thin layer of medium *med* and thickness dl, etc. Instead of the mass energy-transfer coefficient for photons, for charged particles the quantity of relevance is *stopping power*, the energy lost per unit track length (see Chapter 3) and we denote this energy by dE_l to distinguish it from dE_{tr}, which is used for indirectly ionising radiation. We are interested in energy locally deposited in the thin layer so it is clearly appropriate to employ the collision stopping power, S_{col}, rather than the total stopping power as the latter would include the energy lost in the form of bremsstrahlung that would escape the thin layer (this is analogous to the difference between kerma and collision kerma). Thus we can write

$$dE_l = S_{col}\, dl\, N \tag{6.33}$$

Note that, unlike the analogous Equation 6.17 for photons, we do not need the energy of the particles. Dividing both sides by the mass of the layer dm and expressing this as $\rho \times dV$ on the right hand side, we obtain

$$\frac{dE_l}{dm} = \frac{S_{col}\, N\, dl}{\rho\, dV} \tag{6.34}$$

which can be rearranged to give:

$$\frac{dE_l}{dm} = \frac{S_{col}}{\rho} \left[\frac{N\, dl}{dV} \right] \tag{6.35}$$

where, as in the case of indirectly ionising radiation, the quantity in the square brackets is the fluence Φ and therefore:

$$\frac{dE_l}{dm} = \frac{S_{col}}{\rho} \Phi \tag{6.36}$$

Until recently, there was no equivalent of *kerma* for the case of charged particles. However, the quantity *cema*, converted energy per unit mass, was proposed by Kellerer et al. (1992). ICRU (1998) defines *cema* as the energy lost by charged particles, excluding secondary electrons, in electronic collisions in a mass dm of a material. 'Secondary electrons' refer to the delta rays generated by the incident primary electrons (see Section 3.2), and their kinetic energies have already been included in dE_l. Therefore, cema is equal to dE_l/dm and, consequently, to the product of electron fluence and mass collision stopping power.

Cema is not necessarily equal to *absorbed dose*, as some of the delta rays can leave the thin layer, just as secondary electrons can do in the case of the primary radiation being photons (Figure 6.7). To involve absorbed dose, it must follow that any charged particle kinetic energy *leaving* the thin layer or elementary volume is replaced by an exactly equal amount *entering* the layer and being deposited in it or imparted to it. Consequently, it must be assumed that there is *delta-ray equilibrium* in order to be able to equate cema with absorbed dose and therefore we can write, for a medium m:

$$D_m \overset{\delta-\text{eqm}}{=} \Phi \left(\frac{S_{\text{col}}}{\rho} \right)_m \tag{6.37}$$

or, in the case of polyenergetic electron radiation:

$$D_m \overset{\delta-\text{eqm}}{=} \int_0^{E_{\text{max}}} \Phi_E \left(\frac{S_{\text{col}}(E)}{\rho} \right)_m dE \tag{6.38}$$

where Φ_E is the fluence, differential in energy (see Section 6.3.3).

6.6.2 DELTA-RAY EQUILIBRIUM

Naturally, delta-ray equilibrium must always exist if charged particle equilibrium exists. However, where the primary radiation consists of charged particles CPE can never be achieved except in the rather special case of uniformly distributed β sources in a large medium. For the beams of high-energy electrons used in radiotherapy, the energy of the primary electrons decreases continuously with depth and hence there cannot be equilibrium. However, the ranges of the delta rays are predominantly extremely short (see Section 3.2) and almost all the energy transferred through collision losses, i.e. the cema, is deposited locally. One need only look at how close the ratio L_Δ/S_{col} is to unity for very small values of Δ to be convinced of the above (Figure 3.8), i.e. most of the collisions result in very small energy losses and electrons with these low energies have extremely short ranges. Thus, delta-ray equilibrium is generally fulfilled to a high degree in media irradiated by electron beams.

One situation where delta-ray equilibrium is definitely not a good approximation, however, is very close to the phantom surface in an electron beam. The appreciable range in the forward direction of the most energetic delta rays results in a small but discernible *delta-ray build up* (see Figure 3.16)[*].

---

[*] As illustrated in Figure 3.16 and Figure 6.5, the increase in primary electron fluence due to increasing obliquity of the tracks is the main reason for the dose increase with depth for an electron beam.

6.7 CAVITY THEORY

6.7.1 GENERAL

When a measurement is made with a detector, the detector material will, in general, differ from that of the medium into which it is introduced. As mentioned in Section 6.1, the signal from a radiation detector will generally be proportional to the energy absorbed in its sensitive material and thus to the absorbed dose in this material, D_{det}. The step from the *raw* detector signal to D_{det} at some reference radiation quality, generally known as *calibration*, for the various detectors of interest, is dealt with in Part D and will not be covered further here.

The detector can be thought of as a *cavity* introduced into the uniform medium of interest; this name stems from the fact that gas-filled ionisation chambers dominated the development of the subject (Greening 1981) and the associated theory, which relates D_{det} to D_{med}, is known as *cavity theory*. In its most general form, the aim of cavity theory is to determine the factor f_Q given by

$$f_Q = \left(\frac{D_{med}}{D_{det}}\right)_Q \tag{6.39}$$

for an arbitrary detector 'det', in an arbitrary medium 'med', and in an arbitrary radiation quality Q (here, photons or electrons). Figure 6.12 illustrates the situation schematically.

This section contains the key ideas and expressions as far as applying cavity theory to practical radiotherapy dosimetry is concerned. The treatment relies heavily on the results obtained in the previous sections, and is very much based on the concept of particle fluence. For a much more detailed treatment following more traditional lines, Burlin (1968) makes very interesting reading.

The two most important situations where it is possible to derive an exact expression for f_Q will now be considered. The first of these, presented below, is that of a so-called large detector in a photon-irradiated medium. The term large refers to its size in relation to the ranges of the secondary particles, i.e. electrons. The well-known *Bragg–Gray* cavity situation then follows in Section 6.7.3.

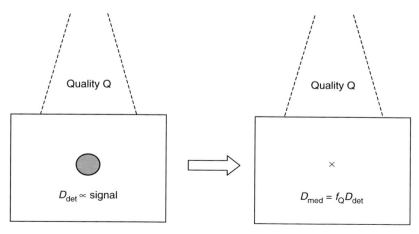

FIGURE 6.12
The general situation of a detector introduced into a medium (left), yielding D_{det} for a given exposure to radiation of quality Q and then being converted into the dose D_{med} at $\times$ in the absence of the detector by multiplying by the cavity theory factor f_Q.

6.7.2 CAVITY THEORY FOR LARGE PHOTON DETECTORS

Let us imagine that the radiation beam in Figure 6.12 consisting for simplicity of mono-energetic photons is incident on a phantom of material, med, and has energy fluence ψ at the depth of interest z. Provided that this depth is sufficient for CPE to be established, then using Equation 6.29 the dose in the medium at depth z can be written as

$$D_{med,z} \stackrel{CPE}{=} \Psi_{med,z}\left(\frac{\mu_{en}}{\rho}\right)_{med} \tag{6.40}$$

Suppose now that a detector is placed with its centre at depth z and that the sensitive material of the detector, denoted by 'det', is large enough for there to be charged particle equilibrium in this material, i.e. that its extent is greater than the maximum range of the secondary electrons generated in this material. Figure 6.13 illustrates the situation schematically.

The photons interact with the detector material and produce secondary electrons tracks (changes in the photon direction/energy have been ignored for simplicity in the figure). Within a narrow rind region on either side of the detector wall (labelled 'boundary') there is no CPE, as electrons there are generated partly in the surrounding medium, partly in the material of the detector wall (which may or may not be present) and partly in the detector itself. In the detector material away from the *boundary* CPE will be re-established. The absorbed dose in the detector, with the bar indicating that this is an average over the detector volume, will be given by:

$$\bar{D}_{det} \stackrel{CPE}{=} \Psi_{det,z}\left(\frac{\mu_{en}}{\rho}\right)_{det} \tag{6.41}$$

where the energy fluence is that in the detector material (strictly an average over the detector volume) and the mass–energy absorption coefficient applies to the detector material 'det'. It is clear that the large detector condition is fulfilled when the volume of the no-CPE rind is only a small fraction of the total volume of the sensitive material of the detector.

Combining Equation 6.40 and Equation 6.41 we have

$$\frac{D_{med}}{\bar{D}_{det}} = \frac{\Psi_{med,z}(\mu_{en}/\rho)_{med}}{\Psi_{det,z}(\mu_{en}/\rho)_{det}} \tag{6.42}$$

Assuming further that the detector does not disturb the photon (energy) fluence existing in the medium, i.e. $\Psi_{det,z}=\Psi_{med,z}$ then the ratio of the absorbed doses, or the cavity theory factor f_Q,

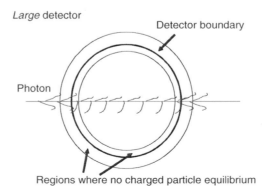

FIGURE 6.13
The deposition of energy by photons (simplified to a single photon unusually represented by a straight line) in a *large* detector. The short tracks represent secondary electrons with ranges much less than any dimension of the detector.

is given by

$$f_Q = \frac{D_{\text{med},z}}{\bar{D}_{\text{det}}} = \frac{(\mu_{\text{en}}/\rho)_{\text{med}}}{(\mu_{\text{en}}/\rho)_{\text{det}}} \tag{6.43}$$

In all practical situations, there will be a spectrum of photon energies. This will be true even for a monoenergetic photon source as scattered photons of lower energy will be present at a depth in the medium. The dose ratio is then given by:

$$\frac{D_{\text{med},z}}{\bar{D}_{\text{det}}} = \frac{\displaystyle\int_0^{E_{\text{max}}} E\frac{\mathrm{d}\Phi_{\text{med},z}}{\mathrm{d}E}\left(\frac{\mu_{\text{en}}(E)}{\rho}\right)_{\text{med}} \mathrm{d}E}{\displaystyle\int_0^{E_{\text{max}}} E\frac{\mathrm{d}\Phi_{\text{med},z}}{\mathrm{d}E}\left(\frac{\mu_{\text{en}}(E)}{\rho}\right)_{\text{det}} \mathrm{d}E} \tag{6.44}$$

This ratio, often written in shorthand form as

$$\frac{D_{\text{med}}}{D_{\text{det}}} = (\bar{\mu}_{\text{en}}/\rho)_{\text{med},\text{det}} \tag{6.45}$$

is the well-known mass–energy absorption coefficient ratio.

In general, the photon fluence spectrum at a depth in the medium can only be obtained by using Monte-Carlo simulation (see Chapter 5 and Chapter 28).

It should be noted that the detector should not be large with respect to the attenuation of photons within it; i.e. to the photon mean free path $s = 1/\mu$, otherwise, it would be impossible to fulfil the requirement that the energy fluence in the detector is negligibly different from that in the medium. In such cases, it will be necessary to add a correction or perturbation factor for the differences in photon attenuation due to the detector thickness (i.e. its dimension in the beam direction) and an equal thickness of the medium. Mobit et al. (2000) have made a detailed Monte-Carlo based study of this problem.

An example of a large detector situation is the use of small thermoluminescent dosimeters (TLDs) (see Section 16.2) in kilovoltage photon fields; the maximum secondary electron range for kV x-rays in the common TLD material lithium fluoride is of the order of 0.5 mm (i.e. less than the dosimeter thickness). Figure 6.14 shows how the mass–energy absorption coefficient

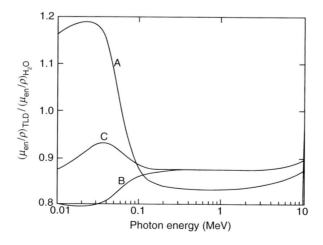

FIGURE 6.14
Variation of the mass–energy absorption coefficient ratio relative to water for different TLD materials. (A) Lithium fluoride, (B) Lithium tetraborate $Li_2B_4O_7$, (C) $Li_2B_4O_7 + 0.3\%$ Mn. (From Greening, J. R., *Fundamentals of Radiation Dosimetry*, Adam Hilger, Bristol, 1981. With permission.)

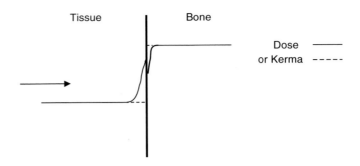

FIGURE 6.15
Variation of absorbed dose at an interface between soft tissue and bone, irradiated by kilovoltage photons but assuming negligible photon attenuation. There is charged particle equilibrium in both media except very close to the interface, where electron scattering effects will occur. The dashed lines indicate kerma where it differs from absorbed dose.

ratio (Equation 6.43) varies with photon energy for three different thermoluminescent materials. The rapid variation below around 100 keV is due to the marked dependence of the photoelectric effect on atomic number (see Section 4.3.1). The virtually flat portion of the curves corresponds to the energy region where the Compton effect dominates, i.e. the interaction coefficients only depend upon electron density (see Section 4.3.2.2).

The large photon cavity result can be exploited to calculate the change in the absorbed dose close to an interface, e.g. going from tissue to bone in a patient irradiated by a low-energy x-ray beam. Figure 6.15 is only schematic; the effect of photon attenuation, which is appreciable at low energies, has not been shown. In Figure 6.15, the *equilibrium* dose levels in the tissue and the bone will be proportional to $(\bar{\mu}_{en}/\rho)_{tissue}$ and $(\bar{\mu}_{en}/\rho)_{bone}$, respectively (evaluated over the photon spectrum). Extremely close to (i.e. within an electron range of) the interface there will not be CPE; the increased electron backscattering from the bone, due to its higher effective atomic number (see Section 3.5), produces the complicated behaviour indicated. Kerma, which is unaffected by electron transport, is also shown. The reader will understand after studying Section 6.7.3 that the ratio of the doses *exactly* on either side of the interface will be given by the (mass) stopping-power ratio as here it is the *electron* fluences which must be equal. Johns and Cunningham (1983) is recommended for more details on interface effects in photon beams.

In the megavoltage energy range, the appreciable ranges of the secondary electrons make it impossible for radiation detectors to fulfil the *large* photon detector condition without becoming impractically large. Nevertheless, the (μ_{en}/ρ)-ratio will be encountered again in, Chapter 15 and Chapter 18 as a component in the correction factor for the wall materials of ionisation chambers.

6.7.3   BRAGG–GRAY CAVITY THEORY

We will now focus on a very different class of detector, one that is *small* compared to the ranges of the secondary electrons. Consider the situation shown in Figure 6.16. Photons irradiate the uniform medium on the left and the tracks of several secondary electrons are shown. On the right a detector has been introduced which is small enough so as not to disturb the electron tracks.

In this case, the extent of the detector in any direction represents only a tiny fraction of the buildup depth required for the establishment of charged particle equilibrium (cf. Figure 6.8). Therefore, we cannot derive the absorbed dose in this small detector from the product of photon energy fluence and $(\mu_{en}/\rho)_{det}$. Instead, it is much more logical to use the relationship between electron fluence and absorbed dose. In this small-detector case, it is precisely the *electron* fluence that connects the detector and undisturbed-medium situations.

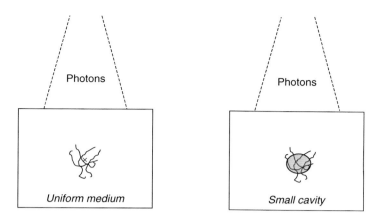

FIGURE 6.16
An illustration of the behaviour of a small detector in a photon-irradiated medium. On the left, some of the secondary electron tracks close to the point of interest x are shown. The introduction of the small detector (on the right) does not disturb these tracks. Such detectors are known as *Bragg–Gray cavities*.

It therefore follows from Equation 6.37 that at the depth z in the medium:

$$\frac{D_{\text{med},z}}{\bar{D}_{\text{det}}} = \frac{\Phi_{\text{med},z}\,(S_{\text{col}}/\rho)_{\text{med}}}{\Phi_{\text{det}}\,(S_{\text{col}}/\rho)_{\text{det}}} \tag{6.46}$$

If we assume that the introduction of the detector does not disturb the electron fluence existing in the undisturbed medium, i.e.

$$\Phi_{\text{det}} = \Phi_{\text{med},z} \tag{6.47}$$

then the fluences in Equation 6.46 cancel out, yielding

$$\frac{D_{\text{med}}}{\bar{D}_{\text{det}}} = \frac{(S_{\text{col}}/\rho)_{\text{med}}}{(S_{\text{col}}/\rho)_{\text{det}}} \tag{6.48}$$

The above expression is known as the (mass) *stopping-power ratio*, often written simply as $s_{\text{med,det}}$.

In all practical cases there will be a spectrum of electron energies and the stopping-power ratio must be evaluated from

$$\frac{D_{\text{med}}}{\bar{D}_{\text{det}}} = s_{\text{med,det}}^{\text{BG}} = \frac{\displaystyle\int_0^{E_{\text{max}}} (\Phi_E)_{\text{med},z}\,(S_{\text{col}}(E)/\rho)_{\text{med}}\,dE}{\displaystyle\int_0^{E_{\text{max}}} (\Phi_E)_{\text{med},z}\,(S_{\text{col}}(E)/\rho)_{\text{det}}\,dE} \tag{6.49}$$

where the energy dependence of the collision stopping power has been made explicit and it is understood that $(\Phi_E)_{\text{med},z}$ refers to the medium *in the absence of the detector* in both the numerator and the denominator. It must be stressed that this is the fluence of *primary* electrons only; no delta-rays are involved (see Section 6.7.4). For reasons that will become apparent in the next paragraph, it is convenient to denote the stopping-power ratio evaluated according to Equation 6.49 by $s_{\text{med,det}}^{\text{BG}}$ (Nahum 1978; ICRU 1984b).

Detectors that behave in the above fashion in photon-irradiated media, i.e. that respond to, but do not disturb, the electron fluence that exists in the absence of any detector, are known as *Bragg–Gray cavities* after the pioneering work of Bragg (1912) and Gray (1929, 1936). As we have seen in the introduction to this chapter, gas-filled detectors dominated all the early radiation measurement techniques. Bragg presented qualitative arguments and Gray reasoned in a more quantitative way that a small gas-filled cavity ought to negligibly modify the number, energy, and direction of electrons present in a photon-irradiated medium (there were no electron beams at the time). Gray showed that the ratio of energy *lost* per unit mass in the medium to that in the gas was equal to the ratio of mass stopping powers $s_{\text{med,gas}}$ and then assumed further that ratio of energy *absorbed* in the respective media was also equal to $s_{\text{med,gas}}$. This is equivalent to assuming delta-ray equilibrium and is discussed further in Section 6.7.4.

In radiotherapy we also deal with electron beams. In this case, the *primary* particles are electrons and there are no photon-generated secondary electrons (ignoring the small bremsstrahlung contribution). Thus the Bragg–Gray idea can easily be extended to detectors in electron beams—see below.

In order for a detector to be treated as a Bragg–Gray (B–G) cavity there is really only one condition which must be fulfilled:

> *The cavity must not disturb the charged particle fluence (including its distribution in energy) existing in the absence of the cavity.*

In practice this means that the cavity must be small compared to the electron ranges, and in the case of photon beams, only *gas*-filled cavities, i.e. ionisation chambers fulfil this.

Generally a second condition is added:

> *The absorbed dose in the cavity is deposited entirely by the charged particles crossing it.*

Now this means that any contribution to the dose due to photon interactions in the cavity must be negligible—clearly this only applies to photon beams. In many ways it is a corollary to the first condition. If the cavity is *small* enough to fulfil the first condition then the buildup of dose due to interactions in the cavity material itself has to be negligible; if this is not the case then the charged particle fluence will differ from that in the undisturbed medium for this very reason.

A third condition is sometimes added:

> *Charged particle equilibrium must exist in the absence of the cavity.*

This condition is only of historical interest; Greening (1981) stated that Gray's original theory required this. In fact, provided that the stopping-power ratio is evaluated over the charged particle (i.e. electron) spectrum *at the position of the detector*, then CPE is *not* required. Nor in the electron beams used in radiotherapy is there ever CPE as has already mentioned. Gray and other early workers required this condition as they did not have the theoretical tools to evaluate Φ_E in Equation 6.49, except in the CPE situation, as Attix (1986) points out.

The air-filled ionisation chamber of the dimensions used in radiotherapy (see Chapter 15) irradiated by a *megavoltage* photon beam is a clear case of a Bragg–Gray cavity. The same is true for electron beams. But can we say the same for kilovoltage x-ray beams? (see Section 18.6) Table 6.2 gives values of $F_{\text{air}}^{\text{spec}}$, the ratio of the dose to the air due to photon interactions (in the air cavity) to the equilibrium dose* in the cavity for a plane-parallel ionisation chamber of typical dimensions, calculated by Monte-Carlo simulation (Ma and Nahum 1991). In a Bragg–Gray cavity, $F_{\text{air}}^{\text{spec}}$ ought to be close to zero. It can be seen in the table that this is clearly true for the ^{60}Co γ-ray and 4 MV x-ray qualities but is definitely not the case for any of the kilovoltage x-ray

---

* The dose corresponding to a *thick* wall, i.e. for which full buildup is achieved by the wall alone cf. ion chamber buildup caps (see Section 15.3.2.2).

TABLE 6.2

Calculated Dose Ratio F_{air}^{spec}, Direct Photon Interactions/Total Cavity Dose, for an Air Cavity Exposed to Different Photon Beam Qualities

Incident Beam			F_{air}^{spec} (5 cm depth in water)
Tube Potential	HVL (mm Al)	F_{air}^{spec} (in vacuum)	
50 kV	1.62	0.26	0.18
150 kV	5.01	0.27	0.29
240 kV	7.30	0.27	0.27
240 kV	17.4	0.18	0.23
^{60}Co		0.0037	0.0060
4 MV		0.0014	0.0098

The air cavity is a 6mm diameter and 6mm thick cylinder, located either in vacuum or at a depth of 5cm in water. F_{air}^{spec} is $\ll$ in a Bragg-Gray Cavity.
Source: From Ma, C.-M. and Nahum, A. E., *Phys. Med. Biol.*, 36, 413–428, 1991. With permission.

qualities. Thus it is not possible to say that air-filled ion chambers behave as true *Bragg–Gray* cavities in any kilovoltage x-ray beam.

6.7.4 THE SPENCER–ATTIX MODIFICATION OF BRAGG–GRAY THEORY

In Section 6.7.3, we neglected the question of delta-ray equilibrium, which is a pre-requisite for the strict validity of the stopping-power ratio as evaluated in Equation 6.49. The original Bragg–Gray theory effectively assumed that all collision losses resulted in energy deposition within the cavity; another way of expressing this is that the charged particles should lose energy by *continuous slowing down*, i.e. in a large number of extremely small energy-loss events. In fact, Greening (1981) made this one of the conditions for Gray's cavity theory.

Experiments by Attix et al. (1958) amongst others with flat air-filled ion chambers with walls made of C, Al, Cu, Sn and Pb, ranging from low and air-like to high atomic numbers, showed clearly that the ionisation density (i.e. per unit mass of gas) depended on the wall separation (that was varied from tenths of a millimetre to 10 mm) in the 412 keV γ-rays from ^{198}Au. Such a dependence should not exist in Bragg–Gray cavities. At the time it was suggested that the reason for this was delta ray production.

Spencer and Attix (1955) proposed an extension of the Bragg–Gray idea that took account of, in an approximate manner, the effect of the finite delta ray ranges. In their theory all the electrons above a cutoff energy Δ, whether primary or delta rays, were considered to be part of the fluence spectrum incident on the cavity. All energy losses in the cavity below Δ in energy were assumed to be *local* to the cavity and all losses above Δ were assumed to escape entirely. The size of Δ is related to the size of the cavity; Spencer and Attix suggested that Δ be set equal to the energy of electrons with a range (in the cavity material, i.e. air) just sufficient to cross the cavity[*]. The local energy loss was calculated by using the collision stopping power restricted to losses less than Δ, $(dE/dS)_{col,\Delta}$, usually denoted by L_Δ (see Section 3.2.5).

---

[*] This is impossible to define uniquely even if the cavity is spherical, given the various directions that electrons will enter the gas volume. Generally, however, $s_{water,air}^{SA}$ only varies slowly with Δ, which is conventionally set to 10 keV for the ionisation chambers used in radiotherapy. However, Buckley et al. (2003) found that $\Delta = 16$ keV corresponded to the mean chord length for a cylindrical ion chamber of Farmer dimensions (air cavity 6.3 mm in diameter and 2 cm in length) and that this value provided significantly better agreement between Spencer–Attix theory and Monte-Carlo simulation than assuming $\Delta = 10$ keV.

This approximate 2-component model leads to a stopping-power ratio given by (Nahum 1978, ICRU 1984b):

$$\frac{\bar{D}_{\text{med}}}{\bar{D}_{\text{det}}} = \frac{\displaystyle\int_{\Delta}^{E_{\text{max}}} \Phi_E^{\text{tot}}(L_\Delta(E)/\rho)_{\text{med}}\, dE + \left[\Phi_E^{\text{tot}}(\Delta)(S_{\text{col}}(\Delta)/\rho)_{\text{med}}\Delta\right]}{\displaystyle\int_{\Delta}^{E_{\text{max}}} \Phi_E^{\text{tot}}(L_\Delta(E)/\rho)_{\text{det}}\, dE + \left[\Phi_E^{\text{tot}}(\Delta)(S_{\text{col}}(\Delta)/\rho)_{\text{det}}\Delta\right]} \quad (6.50)$$

The electron fluence has been denoted by Φ^{tot} to emphasise that (all generations of) delta rays must be included. The term in square brackets accounts for the energy deposition by those electrons falling below Δ in energy, the so-called *track-end term* (Nahum 1976, 1978; ICRU 1984b); $\Phi_E^{\text{tot}}(\Delta)$ is the fluence, differential in energy, evaluated at energy Δ^*. The shorthand form established in ICRU (1984b) for the Spencer–Attix stopping-power ratio is $s_{\text{med,det}}^{\text{SA}}$.

The use of $s_{\text{med,det}}^{\text{SA}}$ instead of the conventional Bragg–Gray ratio based on the unrestricted stopping power, $s_{\text{med,det}}^{\text{BG}}$, was shown to result in improved agreement with experiment (Greening 1957; Attix et al. 1958; Burlin 1962). However, for cases where the atomic composition, and hence the mean atomic number, Z, of the cavity and the medium are quite similar (e.g. air and water), there is only a small difference between $s_{\text{med,det}}^{\text{SA}}$ and the simpler $s_{\text{med,det}}^{\text{BG}}$. For example, for $\Delta = 10$ keV, Nahum (1978) showed that $s_{\text{water,air}}^{\text{SA}}$ was generally about 1% higher than $s_{\text{water,air}}^{\text{BG}}$ for megavoltage photon and electron beams.

The use of the Spencer–Attix expression implies that the electron fluence in the cavity and in the medium are identical only down to energy Δ, but not necessarily below this. This is in principle an easier condition to fulfil than the original Bragg–Gray one where the fluences must be identical down to the lowest electron energies present in the fluence spectrum. In fact, at energies below Δ the electron fluence in the cavity material will be largely determined by delta rays generated in the cavity material itself as their ranges will be smaller than the cavity dimensions.

Should we expect that $s_{\text{med,det}}^{\text{SA}}$ tends towards $s_{\text{med,det}}^{\text{BG}}$ for large cavities (i.e. large Δ) or for small cavities (i.e. small Δ)? In fact, it is large cavities that most closely correspond to the simpler unrestricted BG ratio (Nahum 1976; Attix 1986). This is reasonable as it is largely *primary* electrons which deposit energy in *large* cavities, most delta rays having ranges which are small compared to the cavity size which corresponds better to the implicit BG assumption of negligible delta ray range. If the cavity is very small, however, then delta rays generated in the surrounding medium will be responsible for a much greater fraction of the cavity dose. However, in the case of photon radiation, the cavity size cannot be allowed to be so large that it violates the Bragg–Gray principle that a negligible fraction of the dose arises from photon interactions in the cavity. Nevertheless, Borg et al. (2000) concluded from their Monte-Carlo study that Spencer–Attix cavity theory with $\Delta = 10$ keV was applicable within 0.5% for photon energies at 300 keV or above. The behaviour of cavities which are too large for the Bragg–Gray assumptions to apply is discussed separately under *general cavity theory* in Section 6.7.6.

All modern dosimetry protocols for megavoltage photon and electron beams (see Chapter 18 and Appendix D) use values of $s_{\text{water,air}}^{\text{SA}}$ evaluated according to Equation 6.50, as part of the conversion of the reading of an ion chamber to absorbed dose to water. Considerable detail can

---

* It is instructive to read Spencer and Attix (1955) as the formalism they used is very different from that in Equation 6.50; they did not explicitly include a track-end term, but instead evaluated the stopping power in a different way in the energy interval between 2Δ and Δ.

be found on the evaluation of $s_{\text{water,air}}^{\text{SA}}$ in Appendix D, Section D.1.2 (electron beams) and Appendix D, Section D.1.3 (photon beams).

6.7.5 DEPARTURES FROM PERFECT BRAGG–GRAY BEHAVIOUR IN SMALL CAVITIES

The expression relating the dose in the (undisturbed) medium to that in a detector behaving as a Bragg–Gray cavity can be written most simply as

$$D_{\text{med}} = D_{\text{det}} s_{\text{med,det}}$$

where it assumed that $s_{\text{med,det}}$ has been evaluated using Spencer–Attix theory from the total electron fluence spectrum at the appropriate depth in the medium. In practice not even small gas-filled ionisation chambers in high-energy electron or photon beams truly fulfil the Bragg–Gray conditions. One clear reason for this is the effect of the chamber wall material in the photon-beam case. If the wall is not medium-equivalent in terms of radiation interactions, then some of the secondary electrons in the cavity will have originated from photon interactions in the wall material rather than in the surrounding medium, thereby violating the identical electron fluence condition. This will be the case, despite the fact that the air cavity itself is small enough to function as an *electron* detector.

In the case of electron beams, there are no photon-generated electrons to consider. The response of TLD chips and semiconductor diodes in megavoltage electron beams, for example, should be predicted by the stopping-power ratio. However, the assumption that the fluence must not be appreciably disturbed by the *cavity* limits the applicability of the above theory to detectors that are either small compared to the range of the primary electrons or to materials which scatter electrons in a similar way to the medium.

Departures from Bragg–Gray assumptions are generally treated as *perturbations* (see Nahum 1996) and a perturbation factor, p, (or the product of several factors) is generally added to the above expression, which becomes simply

$$D_{\text{med}} = D_{\text{det}} s_{\text{med,det}} p \qquad (6.51)$$

The subject of perturbation factors is an extensive and important one in the use of ionisation chambers in radiotherapy due to the high demands placed on accuracy; Section 18.5.3 and Appendix D, Section D.2 should be consulted for more details.

6.7.6 GENERAL CAVITY THEORY

We have so far looked at two extreme cases:

1. Detectors that are small compared to the electron ranges and that do not disturb the electron fluence (Bragg–Gray cavities)
2. Detectors that are large compared to the electron ranges, and in which, therefore, CPE is established (photon radiation only).

Many situations involve measuring the dose from photon (or neutron) radiation using detectors which fall into neither of the above categories. In such cases there is no exact theory. However, so-called *General Cavity Theory* has been developed as an approximation (Burlin 1966). Burlin was the first to formally tackle this problem and he proposed a factor, which is a weighted mean of the stopping-power ratio and the mass–energy absorption coefficient ratio:

$$\frac{\overline{D_{\text{det}}}}{D_{\text{med}}} = d\, s_{\text{det,med}} + (1-d)\left(\frac{\mu_{\text{en}}}{\rho}\right)_{\text{det,med}} \tag{6.52}$$

where d is a weighting factor that varies between unity for small (or Bragg–Gray) cavities and zero for large cavities (or photon detectors). Burlin provided a very approximate theory to estimate d based on the exponential attenuation of the electron fluence entering the cavity through the wall, balanced by the exponential buildup of the cavity-generated electron fluence. Attix (1986) discusses in detail the Burlin theory and also subsequent attempts to develop a more sophisticated theory by Horowitz and Dubi (1982), Janssens (1983), Kearsley (1984), and others. Attix concluded that analytical theories were unlikely to be able to compete in accuracy with Monte-Carlo simulation in the future (Mobit et al. 1997).

6.7.7 THE FANO THEOREM

Fano (see Greening 1981; Attix 1986) was able to prove in a formal mathematical way that:

"In a medium of given composition exposed to a uniform fluence of primary radiation (such as x-rays or neutrons) the fluence of secondary radiation is also uniform and independent of the density of the medium as well as of the density variations from point to point"

Figure 6.17 illustrates schematically the situation to which the Fano theorem applies. A medium large enough for CPE to be established is irradiated by indirectly ionising radiation which gives rise to electron tracks (shown as straight lines for simplicity). The central region has a higher density but exactly the same atomic composition as the region surrounding it. More electron tracks are started per unit volume in this central region but each track is correspondingly shorter due to the higher stopping power. Fano's theorem then predicts that *the electron fluence (total track length per unit volume) in the central region will be exactly the same as that in the outer region*, irrespective of the size of the central region.

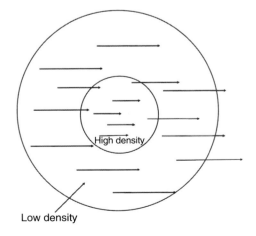

Electron tracks in medium with density variation

High density

Low density

FIGURE 6.17
Illustration of the Fano theorem: the total track length of secondary electrons per unit volume, i.e. the electron fluence, remains the same although the density of the photon-irradiated medium may vary. The electron tracks are shown as straight lines for simplicity.

A practical example of the use of the Fano theorem is in the dosimetry of low-energy x-ray beams where it is not possible to make a gas cavity small enough to fulfil the Bragg–Gray condition (see Section 6.7.3). If the low-density cavity has the same elemental condition as the surrounding medium, then there is no limitation on the size of the cavity (e.g. air surrounded by air-equivalent walls).

The theorem assumes that the mass stopping power of a medium is independent of its density. At relativistic energies (i.e. above around 0.5 MeV for electrons) this is no longer the case as a result of the *density effect* (see Section 3.2.3). The consequences of this violation of the Fano theorem conditions may, however, have no practical importance.

PART A: FUNDAMENTALS
REFERENCES

Alm Carlsson, G., Theoretical basis for dosimetry, in *The Dosimetry of Ionising Radiation*, Kase, K. R., Bjarngard, B. E., and Attix, F. H., Eds., Vol. 1, Academic Press, London, pp. 1–75, 1985.

Andreo, P., The interaction of electrons with matter: II. Scattering, in *The Computation of Dose Distributions in Electron Beam Radiotherapy*, Nahum, A. E., Ed., Medical Physics Publishing, Madison, WI, pp. 56–71, 1985.

Andreo, P., Stopping-power ratios for dosimetry, in *Monte-Carlo Transport of Electrons and Photons*, Jenkins, T. M., Nelson, W. R., Rindi, A., Nahum, A. E., and Rogers, D. W. O., Eds., Plenum, New York, pp. 485–501, 1988.

Andreo, P., Monte Carlo techniques in medical radiation physics, *Phys. Med. Biol.*, 36, 861–920, 1991.

Attix, F. H., The partition of kerma to account for bremsstrahlung, *Health Phys.*, 36, 347–354, 1979.

Attix, F. H., *Introduction to Radiological Physics and Radiation Dosimetry*, Wiley, New York, 1986.

Attix, F. H., De La Vergne, L., and Ritz, V. H., Cavity ionisation as a function of wall material, *J. Res. NBS*, 60, 235, 1958.

Barø, J., Sempau, J., Fernández-Varea, J. M., and Salvat, F., PENELOPE: An algorithm for Monte Carlo simulation of the penetration and energy loss of electrons and positrons in matter, *Nucl. Instrum. Methods*, B100, 31–46, 1995.

Berger, M. J., Monte Carlo calculation of the penetration and diffusion of fast charged particles, *Methods Comput. Phys.*, 1, 135–215, 1963.

Berger, M. J. and Hubbell, J. H., *XCOM: Photon cross sections on a personal computer NBSIR 87-3597*, National Bureau of Standards, Washington, DC, 1987.

Berger, M. J. and Seltzer, S. M., *Tables of energy losses and ranges of electrons and positrons NASA SP-3012*, National Aeronautics and Space Administration, Washington, DC, 1964.

Berger, M. J. and Wang, R., Multiple-scattering angular deflections and energy-loss straggling, in *Monte Carlo Transport of Electrons and Photons*, Jenkins, T. M., Nelson, W. R., Rindi, A., Nahum, A. E., and Rogers, D. W. O., Eds., Plenum, New York, pp. 21–56, 1988.

Bethe, H. A., Theory of passage of swift corpuscular rays through matter, *Ann. Physik*, 5, 325, 1930.

Bethe, H. A., Scattering of electrons, *Z. für Physik*, 76, 293, 1932.

Bethe, H. A., Molière's theory of multiple scattering, *Phys. Rev.*, 89, 1256–1266, 1953.

Bhabha, H. J., The scattering of positrons by electrons with exchange on Dirac's theory of the positron, *Proc. R. Soc.*, A154, 195, 1936.

Bielajew, A. F., Efficiency, statistics and sampling, *National Research Council of Canada Report PIRS-0395*, National Research Council of Canada, Ottawa, 1993.

Bielajew, A. F., HOWFAR and HOWNEAR: Geometry modeling for Monte Carlo particle transport, *National Research Council of Canada Report PIRS-0341*, National Research Council of Canada, Ottawa, 1995.

Bielajew, A. F. and Rogers, D. W. O., PRESTA: The parameter reduced electron-step transport algorithm for electron Monte Carlo transport, *Nucl. Instrum. Methods.*, B18, 165–181, 1987.

Bielajew, A. F. and Rogers, D. W. O., Electron step-size artefacts and PRESTA, in *Monte Carlo Transport of Electrons and Photons*, Jenkins, T., Nelson, W., Rindi, A., Nahum, A., and Rogers, D., Eds., Plenum Press, New York, pp. 115–137, 1989.

Bielajew, A. F., Hirayama, H., Nelson, W. R., and Rogers, D. W. O., History, overview and recent improvements of EGS4, *National Research Council of Canada Report PIRS-0436*, National Research Council of Canada, Ottawa, 1994.

Bielajew, A. F., Rogers, D. W. O., and Nahum, A. E., Monte Carlo simulation of ion chamber response to ^{60}Co-Resolution of anomalies associated with interfaces, *Phys. Med. Biol.*, 30, 419–428, 1985.

Bjorken, J. D. and Drell, S. D., *International Series in Pure and Applied Physic: Relativistic Quantum Fields*, McGraw-Hill, New York, 1965.

Blanc, D. and Portal, G., *Précis de Physique Nucléaire (in French)*, 2nd ed., Dunod, Paris, 1999.

Bloch, F., Stopping power of atoms with several electrons, *Z. für Physik*, 81, 363, 1933.

Bohr, N., On the theory of the decrease of velocity of moving electrified particles on passing through matter, *Philos. Mag. and J. Sci.*, 25, 10–31, 1913.

Bohr, N., The penetration of atomic particles through matter, *PK. Danske Vidensk. Selsk., Mat.-Fys. Medd.*, 18(8), 1–144, 1948.

Borg, J., Kawrakow, I., Rogers, D. W. O., and Seuntjens, J. P., Monte Carlo study of correction factors for Spencer–Attix cavity theory at photon energies at or above 100 keV, *Med. Phys.*, 27, 1804–1813, 2000.

Bragg, W. H., *Studies in Radioactivity*, Macmillan, New York, 1912.

Brahme, A. and Svensson, H., Specification of electron beam quality from the central-axis depth absorbed-dose distribution, *Med. Phys.*, 3, 95, 1976.

Briesmeister, J. F., MCNP-A general purpose Monte Carlo code for neutron and photon transport, *Version 3A Los Alamos National Laboratory Report LA-7396-M*, Los Alamos National Laboratory, Los Alamos, New Mexico, 1986.

Briesmeister, J. F., MCNP-A general Monte Carlo N-particle transport code, *Los Alamos National Laboratory Report LA-12625-M*, Los Alamos National Laboratory, Los Alamos, New Mexico, 1993.

Briesmeister, J. F., MCNP-A general Monte Carlo N-particle transport code, *Los Alamos National Laboratory Report LA-12625-M, Version 4B*, Los Alamos National Laboratory, Los Alamos, New Mexico, 1997.

Buckley, A. L., Kawrakow, I., and Rogers, D. W. O., An EGSnrc investigation of cavity theory for ion chambers measuring air kerma, *Med. Phys.*, 30, 1211–1218, 2003.

Buffon, G.-L. C., *Essai d'arithmétique morale*, Vol. 4, Supplément à l'Histoire Naturelle, Imprimerie Royale, 1777.

Bureau International des Poids et Mesures (BIPM), *The International System of Units (with suppl. 2000)*, 7th ed., Organisation Intergouvernementale de la Convention du Mètre, Sèvres, France, 1998.

Burlin, T. E., An experimental examination of theories relating the absorption of gamma-ray energy in a medium to the ionisation produced in a cavity, *Phys. Med. Biol.*, 6, 33–53, 1961.

Burlin, T. E., A general theory of cavity ionisation, *Br. J. Radiol.*, 39, 727–734, 1966.

Burlin, T. E., Cavity-chamber theory, in *Radiation Dosimetry*, Attix, F. H. and Roesch, W. C., Eds., Vol. 1, Academic Press, New York, pp. 331–392, 1968.

Bushberg, J. T., Seibert, J. A., Leidholt, E.M., Jr., and Boone, J. M., *The Essential Physics of Medical Imaging*, William and Wilkins, Baltimore, MD, 1994.

Carrier, J.-F., Archambault, L., Beaulieu, L., and Roy, R., Validation of GEANT4, an object-oriented Monte Carlo toolkit for simulations in medical physics, *Med. Phys.*, 31, 484–492, 2004.

Cherry, S. R., Sorenson, J. A., and Phelps, M. E., *Physics in Nuclear Medicine*, 3rd ed., W. B. Saunders, New York, 2003.

Chilton, A. B., A note on the fluence concept, *Health Phys.*, 34, 715–716, 1978.

Christen, T., Radiometry, *Arch. Roentgen Ray*, 19, 210–219, 1914.

Compton, A. H. and Allison, S. K., *X-rays in Theory and Experiment*, D. Van Nostrand Co. Inc., New York, 1935.

Davies, H., Bethe, H. A., and Maximon, L. C., Theory of bremsstrahlung and pair production. II. Integral cross sections for pair production, *Phys. Rev.*, 93, 788, 1954.

Duderstadt, J. J. and Martin, W. R., *Transport Theory*, Wiley, New York, 1979.

Dutreix, J., Desgrez, A., Bok, B., and Vinot, J. H., *Biophysique des Radiations et Imagerie Médicale (in French)*, 3rd ed., Masson, Paris, France, 1993.

Ehrman, J. R., The care and feeding of random numbers, *SLAC VM Notebook, Module 18*, SLAC Computing Services, 1981.

Evans, R. D., *The Atomic Nucleus*, McGraw Hill, New York, 1955.

Eyges, L., Multiple scattering with energy loss, *Phys. Rev.*, 74, 1534, 1948.

Feller, W., *An Introduction to Probability Theory and its Applications*, Vol. 1, 3rd ed., Wiley, New York, 1967.

Fermi, E., The ionisation loss of energy in gases and in condensed materials, *Phys. Rev.*, 57, 485–493, 1940.

Fernández-Varea, J. M., Mayol, R., Baró, J., and Salvat, F., On the theory and simulation of multiple elastic scattering of electrons, *Nucl. Instrum. Methods*, B73, 447–473, 1993.

Goodhead, D. T., Relationship of microdosimetric techniques to applications in biological systems, in *The Dosimetry of Ionising Radiation*, Kase, K. R., Bjärngard, B. E., and Attix, F. H., Eds., Vol. 2, Academic Press, Orlando, FL, pp. 1–89, 1987.

Goodwin, P. N. and Rao, D. V., *An Introduction to the Physics of Nuclear Medicine*, Charles C. Thomas, Springfield, IL, 1977.

Goudsmit, S. A. and Saunderson, J. L., Multiple scattering of electrons, *Phys. Rev.*, 57, 24–29, 1940a.

Goudsmit, S. A. and Saunderson, J. L., Multiple scattering of electrons II, *Phys. Rev.*, 58, 36–42, 1940b.

Gray, L. H., The absorption of penetrating radiation, *Proc. R. Soc.*, A122, 647–668, 1929.

Gray, L. H., An ionisation method for the absolute measurement of gamma-ray energy, *Proc. R. Soc.*, A156, 578–596, 1936.

Greening, J. R., An experimental examination of theories of cavity ionisation, *Br. J. Radiol.*, 30, 254–262, 1957.

Greening, J. R., *Fundamentals of Radiation Dosimetry*, Adam Hilger, Bristol, 1981.

Halbleib, J., Structure and operation of the ITS code system, in *Monte Carlo Transport of Electrons and Photons*, Jenkins, T., Nelson, W., Rindi, A., Nahum, A., and Rogers, D., Eds., Plenum Press, New York, pp. 249–262, 1989.

Halbleib, J. A. and Mehlhorn, T. A., ITS: The integrated TIGER series of coupled electron/photon Monte Carlo transport codes, *Sandia Report SAND84-SA0573*, 1984.

Halbleib, J. A., Kensek, R. P., Mehlhorn, T. A., Valdez, G. D., Seltzer, S. M., and Berger, M. J., ITS Version 3.0: The integrated TIGER series of coupled electron/photon Monte Carlo transport codes, *Sandia Report SAND91-1634*, 1992.

Halliday, D. and Resnic, R., *Fundamental of Physics*, 3rd ed., Wiley, New York, 1988.

Harder, D., Harigel, G., and Schultz, K., Bahnspuren schneller Elektronen, *Strahlentherapie*, 115, 1, 1961.

Heitler, W., *The Quantum Theory of Radiation*, Clarendon Press, Oxford, 1954.

Hendee, W. R., *Medical Radiation Physics*, 2nd ed., Year book Medical Publishers Inc., Chicago, 1979.

Hogstrom, K. R. and Almond, P. R., The effect of electron multiple scattering on dose measured in non-water phantoms, AAPM annual meeting, available from authors, 1982.

Horowitz, Y. P., Photon general cavity theory, *Radiat. Prot. Dosim.*, 9, 5–16, 1984.

Horowitz, Y. S. and Dubi, A., A proposed modification of Burlin's general theory for photons, *Phys. Med. Biol.*, 27, 867–870, 1982.

Hubbell, J. H., Review of photon interaction cross section data in the medical and biological context, *Phys. Med. Biol.*, 44, R1–R22, 1999.

Hubbell, J. H. and Øverbø, I., Relativistic atomic form factors and photon coherent scattering cross sections, *J. Phys. Chem. Ref. Data.*, 9, 69, 1979.

Hubbell, J. H. and Seltzer, S. M., *Tables of X-ray Mass Attenuation Coefficients and Mass Energy-absorption Coefficients 1 keV to 20 MeV for Elements Z=1 to 92 and 48 Additional Substances of Dosimetric Interest*. NISTIR 5632, US Department of Commerce, Gaithersburg, MD, 1995.

Hubbell, J. H., Trehan, P. N., Singh, N., Chand, B., Mehta, D., Garg, M. L., Garg, R. R., Singh, S., and Puri, S., A review, bibliography and tabulation of K, L and higher atomic shell X-ray fluorescence yields, *J. Phys. Chem. Ref. Data.*, 23(2), 339–364, 1994.

Hubbell, J. H., Veigele, W. J., Briggs, E. A., Brown, R. T., Cromer, D. T., and Howerton, R. J., Atomic form factors, incoherent scattering functions, and photon scattering cross sections, *J. Phys. Chem. Ref. Data.*, 4, 471–538, 1975. Erratum. 1977. 6:615–616

IAEA (International Atomic Energy Agency), *Absorbed dose determination in Photon and Electron Beams—an International Code of Practice*, 2nd ed., TRS. No. 277, IAEA, Vienna, 1997.

ICRU (International Commission on Radiation Units and Measurements), Report 6, NBS Handbook 47, *Br. J. Radiol.*, 24, 54, 1951.

ICRU (International Commission on Radiation Units and Measurements), *Report 16: Linear Energy Transfer*, ICRU, Bethesda, MD, 1970.

ICRU (International Commission on Radiation Units and Measurements), *Report 33: Radiation Quantities and Units*, ICRU, Bethesda, MD, 1980.

ICRU (International Commission on Radiation Units and Measurements), *Report 36: Microdosimetry*, ICRU, Bethesda, MD, 1983.

ICRU (International Commission on Radiation Units and Measurements), *Report 37: Stopping Powers for Electrons and Positrons*, ICRU, Bethesda, MD, 1984a.

ICRU (International Commission on Radiation Units and Measurements), *Report 35: Radiation Dosimetry; Electron Beams with Energies Between 1 and 50 MeV*, ICRU, Bethesda, MD, 1984b.

ICRU (International Commission on Radiation Units and Measurements), *Report 49: Stopping Powers for Electrons and Positrons*, ICRU, Bethesda, MD, 1993.

ICRU (International Commission on Radiation Units and Measurements), *Report 60: Fundamental Quantities and Units for Ionising Radiation*, ICRU, Bethesda, MD, 1998.

James, F., A review of pseudorandom number generators, *CERN-Data Handling Division, Report*, DD/88/22, 1988.

James, F., RANLUX: A FORTRAN implementation of the high-quality pseudorandom number generator of Lüscher, *Comput. Phys. Commun.*, 79, 111–114, 1994.

Johns, H. E. and Cunningham, J. R., *The Physics of Radiology*, 4th ed., Charles C. Thomas, Springfield, IL, 1983.

Kalos, M. H. and Whitlock, P. A., *Monte Carlo Methods*, Vol. I: Basics, Wiley, New York, 1986.

Kase, K. R. and Nelson, W. R., *Concepts of Radiation Dosimetry*, Pergamon Press, Oxford, 1978.

Kawrakow, I. and Bielajew, A. F., On the condensed history technique for electron transport, *Nucl. Instrum. Methods*, B142, 253–280, 1998a.

Kawrakow, I. and Bielajew, A. F., On the representation of electron multiple elastic-scattering distributions for Monte Carlo calculations, *Nucl. Instrum. Methods.*, B134, 325–336, 1998b.

Kearsley, E. E., A new general cavity theory, *Phys. Med. Biol.*, 29, 1179–1187, 1984.

Kellerer, A. M., Hahn, K., and Rossi, H. H., Intermediate dosimetric quantities, *Radiat. Res.*, 130, 15–25, 1992.

Klein, O. and Nishina, Y., Über die Streuung von Strahlung durch freie Elektronen nach der neuen relativistischen Quantendynamik von Dirac, *Z. Physik*, 52, 853–868, 1929.

Knuth, D. E., *Seminumerical Algorithms, Vol. II: The Art of Computer Programming*, Addison Wesley, Reading, Mass, 1981.

Koch, H. W. and Motz, J. W., Bremsstrahlung cross-section formulas and related data, *Rev. Mod. Phys.*, 31, 920–955, 1959.

Landau, L., On the energy loss of fast particles by ionisation, *J. Phys. U.S.S.R*, 8, 201, 1944.

Laplace, P. S., Théorie analytique des probabilités, Livre 2, in *Oeuvres complètes de Laplace,* Vol. 7, Part 2, L'Académie des Sciences, Paris, pp. 365–366, 1886.

Larsen, E. W., A theoretical derivation of the condensed history algorithm, *Ann. Nucl. Energy*, 19, 701–714, 1992.

Lindeberg, J. W., Eine neue Herleitung des Exponentialgesetzes in der Wahrscheinlichkeitrechnung, *Mathematische Zeitschrift*, 15, 211–225, 1922.

Li, X. A. and Rogers, D. W. O., Electron mass scattering powers: Monte Carlo and analytical calculations, *Med. Phys.*, 22, 531–541, 1995.

Lüscher, M., A portable high-quality random number generator for lattice field theory simulations, *Comput. Phys. Commun.*, 79, 100–110, 1994.

Ma, C.-M. and Nahum, A. E., Bragg–Gray theory and ion chamber dosimetry for photon beams, *Phys. Med. Biol.*, 36, 413–428, 1991.

Mackie, T. R., Applications of the Monte Carlo method in radiotherapy, in *Dosimetry of Ionising Radiation*, Vol. III, Kase, K., Björngard, B., and Attix, F. H., Eds., Academic Press, New York, pp. 541–620, 1990.

Marsaglia, G. and Zaman, A., A new class of random number generators, *Ann. Appl. Probab.*, 1, 462–480, 1991.

Marsaglia, G., Zaman, A., and Tsang, W. W., Toward a universal random number generator, *Stat. Probab. Lett.*, 8, 35–39, 1990.

Mayol, R. and Salvat, F., Total and transport cross sections for elastic scattering of electrons by atoms, *Atom. Data Nucl. Data Tables*, 65, 55–154, 1997.

Metcalfe, P., Kron, T., and Hoban, P., *The Physics of Radiotherapy X-rays from Linear Accelerators*, Medical Physics Publishing, Madison, WI, 1997.

Mobit, P. N., *Monte Carlo and Experimental Studies of Dose Measurements in Radiotherapy Beams with LiF TLDs and other Solid State Dosimeters*, PhD thesis, (Available from University of London as ICR-PHYS-2/96), 1996.

Mobit, P. N., Nahum, A. E., and Mayles, P., An EGS4 Monte Carlo examination of general cavity theory, *Phys. Med. Biol.*, 42, 1319–1334, 1997.

Mobit, P. N., Sandison, G. A., and Nahum, A. E., Photon fluence perturbation correction factors for solid state detectors irradiated in kilovoltage photon beams, *Phys. Med. Biol.*, 45, 267–277, 2000.

Mohr, P. J. and Taylor, B. N., CODATA recommended values of the fundamental physical constants, *2002 Rev. Mod. Phys.*, 77, 1, 2005.

Molière, G. Z., Theorie der Streuung schneller geladener Teilchen. I. Einzelstreuung am abgeschirmten Coulomb-Field, *Z. Naturforsch.*, 2a, 133–145, 1947.

Molière, G. Z., Theorie der Streuung schneller geladener Teilchen.II. Mehrfach- und Vielfachstreuung, *Z. Naturforsch.*, 3a, 78–97, 1948.

Møller, C., Zur Theorie des Durchgangs schneller Elektron durch Materie, *Ann. Phys.*, 14, 531–585, 1932.

Motz, J. W., Olsen, H. A., and Koch, H. W., Pair production by photons, *Rev. Mod. Phys.*, 41, 581–639, 1969.

Nahum, A. E., *Calculations of Electron Flux Spectra in Water Irradiated with Megavoltage Electron and Photon Beams with Applications to Dosimetry*, PhD Thesis, University of Edinburgh, July 1975, (Available from University Microfilms International: Order No 7770,006), 1976.

Nahum, A. E., Water/air mass stopping-power ratios for megavoltage photon and electron beams, *Phys. Med. Biol.*, 23, 24–38, 1978.

Nahum, A. E., *Stopping Powers and Dosimetry*, Annual Meeting of the Canadian Association of Physicists, Quebec, 18–23 June, 1983, (Available as NRCC report: PXNR2653), 1983.

Nahum, A. E., The interactions of electrons with matter, Vol. I. Energy losses, stopping power and range, in *The Computation of Dose Distributions in Electron Beam Radiotherapy*, Nahum, A. E., Ed., Medical Physics Publishing, Madison, WI, pp. 27–55, 1985.

Nahum, A. E., Overview of photon and electron Monte Carlo, in *Monte Carlo transport of Electrons and Photons*, Jenkins, T., Nelson, W., Bindi, A., Nahum, A., and Rogers, D., Eds., Plenum Press, New York, pp. 3–20, 1989.

Nahum, A. E., Perturbation effects in dosimetry: Part I: Kilovoltage x-rays and electrons, *Phys. Med. Biol.*, 41, 1531–1580, 1996.

Nahum, A. E., Condensed-history Monte-Carlo simulation for charged particles: What can it do for us? *Radiat. Environ. Biophys.*, 38, 163–173, 1999.

Nahum, A. E. and Brahme, A., Electron depth-dose distributions in uniform and nonuniform media, in *The Computation of Dose Distributions in Electron Beam Radiotherapy*, Nahum, A. E., Ed., Medical Physics Publishing, Madison, WI, pp. 98–127, 1985.

Nelson, W. R. and Jenkins, T. M., Writing subroutine HOWFAR for EGS4, *Report SLAC-TN-87-4*, Stanford Linear Accelerator Center, Stanford, CA, 1987.

Nelson, W. R., Hirayama, H., and Rogers, D. W. O., The EGS4 code system, *Report SLAC-265*, Stanford Linear Accelerator Center, Stanford, CA, 1985.

Olmsted, J. M. H., *Solid Analytic Geometry*, Appleton-Century-Crofts, Inc., New York, 1947.

Perkins, S. T., Cullen, D. E., and Seltzer, S.M., Tables and graphs of electron-interaction gross sections from 10 eV to 100 GeV derived from the LLNL Evaluated Electron Data Library (EEDL), $Z=1$–100, *Lawrence Livermore National Laboratory Report UCRL-50400*, Vol. 31, Lawrence Livermore National Laboratory, Livermore, CA, 1991.

Raeside, D. E., Monte Carlo principles and applications, *Phys. Med. Biol.*, 21, 181–197, 1976.

Rajan, G., *Advanced Medical Radiation Dosimetry*, Prentice-Hall of India Private Ltd, New Delhi, 1992.

Rayleigh, L., On the light from the sky, its polarization and colour, *Philos. Mag.*, 41, 107–120, see also pp 274–279, 1871.

Rogers, D. W. O. and Bielajew, A. F., Monte Carlo techniques of electron and photon transport for radiation dosimetry, in *The Dosimetry of Ionising Radiation,* Vol. III, Kase, K., Bjängard, B., and Attix, F., Eds., Academic Press, New York, pp. 427–539, 1990.

Rossi, B., in *High Energy Particles*, Prentice Hall, Englewood Cliffs, pp. 63–77, 1952.

Rossi, H. H., Microscopic energy distribution in irradiated matter, in *Radiation Dosimetry*, Attix, F. H. and Roesch, W. C., Eds., Vol. 1, Academic Press, New York, pp. 43–92, 1968.

Sakurai, J. J., *Advanced Quantum Mechanics*, Addison-Wesley, Reading, MA, 1967.

Salvat, F., Fernández-Varea, J. M., Barø, J., and Sempau, J., PENELOPE, an algorithm and computer code for Monte Carlo simulation of electron-photon showers, *University of Barcelona Ciemat (Centro de Investigaciones Energéticas, Medioambientales y Tecnológicas)*, Report 799, 1996.

Sauter, F., Über den atomaren Photoeffekt in der K-Schale nach der relativistischen Wellenmechanik Diracs, *Ann. Physik*, 11, 454–488, 1931.

Seltzer, S. M., An overview of ETRAN Monte Carlo methods, in *Monte Carlo Transport of Electrons and Photons*, Jenkins, T., Nelson, W., Rindi, A., Nahum, A., and Rogers, D., Eds., Plenum Press, New York, pp. 153–182, 1989.

Seltzer, S. M., Electron–photon Monte Carlo calculations: The ETRAN code, *Int. J. Appl. Radiat. Isot.*, 42, 917–941, 1991.

Seltzer, S. M., Calculation of photon mass energy-transfer and mass energy-absorption coefficients, *Radiat. Res.*, 136, 147–170, 1993.

Seltzer, S. M., Hubbel, J. H., and Berger, M. J., Some theoretical aspects of electron and photon dosimetry, IAEA-SN-222/05, *National and International Standardization of Radiation Dosimetry*, Vol. 2, IAEA, Vienna, pp. 3–43, 1978.

Sempau, J., Acosta, E., Barø, J., Fernández-Varea, J. M., and Salvat, F., An algorithm for Monte Carlo simulation of coupled electron–photon showers, *Nucl. Instrum. Methods*, B132, 377–390, 1997.

Shultis, J. K. and Faw, R. E., *Radiation Shielding*, Prentice Hall, Upper Saddle River, 1996.

Siebers, J. V., Keall, P. J., Nahum, A. E., and Mohan, R., Converting absorbed dose to medium to absorbed dose to water for Monte Carlo based photon beam dose calculations, *Phys. Med. Biol.*, 45, 983–995, 2000.

Spencer, L. V. and Attix, F. H., A theory of cavity ionisation, *Radiat. Res.*, 3, 239–254, 1955.

Sternheimer, R. M., in *Methods of Experimental Physics*, Yuan, L. C. L. and Wu, C. S., Eds., Vol. 5A, Academic Press, New York, pp. 1–89, 1961.

Storm, E. and Israel, H. I., Photon cross sections from 1 keV to 100 MeV for elements $Z=1$ to $Z=100$, *Nucl. Data Tables*, 7(6), 565–681, 1970.

Tsai, Y. S., Pair production and bremsstrahlung of charged leptons, *Rev. Mod. Phys.*, 46, 815, 1974.

Turner, J. E., Wright, H. A., and Hamm, R. N., A Monte Carlo primer for health physicists, *Health Phys.*, 48, 717–733, 1985.

Vavilov, P. V., Ionisation losses of high-energy heavy particles, *Sov. Phy. JETP.*, 5, 749, 1957.

Zaidi, H. and Sgouros, G., Eds., *Therapeutic Applications of Monte Carlo Calculations in Nuclear Medicine*, Institute of Physics, Bristol, 2003.

PART B

RADIOBIOLOGY

Editors: Alan Nahum and Philip Mayles[*]

[*] The editors are grateful to Don Chapman who read and commented on the text.

INTRODUCTION

Physical approaches to the optimisation of radiation therapy increasingly require attention to biological factors. The name of the game, as with any form of curative cancer treatment, is *selectivity*: how to maximise effects on tumours without incurring serious damage to normal tissues. This is largely achieved in the case of radiation therapy by careful attention to dose distributions, but biological effects depend not only on the *spatial* but also on the *temporal* distribution of dose, and selectivity depends on biological response processes. All these aspects will be covered in Part B.

It is fair to say that radiobiology is currently undergoing something of a renaissance with today's unprecedented ability to shape dose distributions through 3-D conformal and intensity-modulated therapy (Chapter 43 and Chapter 44), stereotactic techniques (Chapter 45), and heavy charged-particle therapy (Chapter 46 and Chapter 49) acting as the catalyst. With such dose-shaping tools available, one has many alternative treatment plans to choose from, and tools such as *dose-volume histograms* have become routine. The field is moving toward the use of so-called biological models of Tumour Control Probability (TCP) and Normal-Tissue Complication Probability (NTCP) (see Chapter 36). These models introduce the biological response on the top of the physical dose distribution; hence, radiobiological concepts are entering into the vocabulary of the radiotherapy physicist with renewed vigour (Glatstein 2002).

Many current issues in modern radiotherapy need radiobiological input in order to be used optimally. In recent years intriguing phenomena such as low-dose hypersensitivity (see section 7.13.3) and the possibility of low α/β in prostate and breast (see Section 9.3) have been reported and their impact needs to be studied. Radiobiology is also needed to guide extracranial stereotaxy with large fraction sizes in lung and liver (Chapter 45; Blomgren et al. 1995; Fowler et al. 2004; Grau et al. 2006; Lax et al. 2006; SBRT 2006). PET-guided dose painting to combat hypoxia (Chao et al. 2001; Chapman et al. 2003) and also to target regions of particularly high clonogen density, e.g. in the lung (van der Wiele et al. 2003; Paulino and Johnstone 2004) has become a possibility. In the prostate, Nutting et al. (2002) have proposed that images from magnetic resonance spectroscopy (MRS) could be used to target dose more precisely. All the above require radiobiological knowledge in addition to radiation physics. Physicists' involvement in radiotherapy even extends to radiobiologically based analyses of the clinical effects of waiting times for treatment (Wyatt et al. 2003) and on the effect of the time interval between surgery and postoperative radiotherapy (Al-Dweri et al. 2004).

The field of radiation biology is now vast, comprising not only of applications to therapy but also to radiation protection (dealt with in Part L; see also Sachs and Brenner 2005). Many areas covered in this Part are amplified in the book *Basic Clinical Radiobiology* (Steel 2002a) which also deals with topics not covered here, including the recent emphasis on molecular approaches. The review articles by Fowler (1984, 2006) and Peters et al. (1990) are also very useful.

CHAPTER 7

RADIOBIOLOGY OF TUMOURS

*Gordon Steel**

CONTENTS

---

* With contributions by Don Chapman and Alan Nahum.

This chapter begins with the basic concepts of cell killing by radiation. Although this is discussed in the context of killing tumour cells, most of these ideas equally apply to the normal tissues at risk in radiotherapy.

7.1 CONCEPT OF CLONOGENIC CELLS

The maintenance of tissue size, and therefore of tissue function, in the normal renewal tissues of the body depends on the existence of a small number of primitive *stem cells*. These are cells with unlimited proliferative capacity which are at the base of the hierarchy of cells that make up the normal epithelial and haemopoietic tissues of the body. Carcinomas are derived from such hierarchical tissues, and the ability to recognise this in histological sections derives from the fact that the tumours often maintain many of the features of differentiation of the tissue within which they arose. Well-differentiated tumours do this to a greater extent than do anaplastic tumours. It follows that not all the cells in a tumour are *neoplastic stem cells*; some have embarked on an irreversible process of differentiation. In addition, carcinomas contain many normal host cells that make up the stroma (fibroblasts, endothelial cells, macrophages, etc.). To some readers it may come as a surprise that most of the cells that make up the volume of many tumour types are not malignant. If these non-malignant cells were the only cells left intact at the end of therapy, then the tumour would not regrow. Tumour growth is driven by the neoplastic stem cells, and it is these that must be eradicated.

When a tumour regrows after non-curative treatment, it does so because some stem cells were not killed. Radiobiologists have recognised that the key to understanding tumour response is to determine how many tumour stem cells are left. It is almost impossible to recognise tumour stem cells in situ, and assays have been developed which allow these cells to be detected after removal from the tumour. These assays generally detect stem cells by their ability to form a colony within some growth environment. These cells are called *clonogenic* or *colony-forming* cells.

After exposure to a therapeutic dose of radiation, damaged cells do not immediately die, and they may produce a modest family of descendants. This is illustrated in Figure 7.1. The growth of single mouse L-cells was observed under the microscope, and one selected colony was irradiated with **200** *roentgens* (about 2 Gy) of x-rays at the 4-cell stage. Subsequent

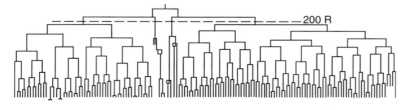

FIGURE 7.1
Pedigree of a clone of mouse L-cells irradiated with an exposure of 200 roentgens at the 4-cell stage, illustrating the concept of surviving and non-surviving clonogenic cells. (From Trott, K. R., *Curr. Top. Radiat. Res.*, 7, 336–337, 1972. With permission.)

growth was carefully recorded, and in the figure each vertical line indicates the time from birth at mitosis to subsequent division of a daughter cell. The two irradiated cells on the left and the right of this figure produced continuously expanding colonies, although, some daughter cells had a long intermitotic time. The other two irradiated cells fared badly; they underwent a number of irregular divisions including a tripolar mitosis. But note that at the end of the experiment cells are present from each of the original four cells: the difference is that two produced expanding colonies while the other two did not. The two irradiated cells on the left and right of this diagram were *surviving clonogenic cells*, and the other two are usually described as having been *killed* by radiation. More precisely, they should be described as having *lost their colony-forming ability*; this is what irradiation of tumour cells seeks to achieve.

These examples of the kinetics of cell death after irradiation indicate the process of *proliferative* cell death. Another mechanism of cell death that is more prompt is *apoptosis*, or *programmed cell death*, which will be described later in this chapter.

7.2 CLONOGENIC ASSAYS

Clonogenic assays have formed the basis of cellular response studies in tumours as well as in some normal tissues. The basic idea is to remove cells from the tumour, place them in a defined growth environment, and test for their ability to produce a defined colony of descendants. Many types of assay have been described; here, the principle is illustrated by a simple assay in tissue-culture that is analogous to a microbiological assay.

A single-cell suspension of tumour cells is prepared and divided into two parts. One is irradiated, the other kept as an unirradiated control. The two suspensions are then plated out in separate tissue culture dishes under identical conditions, but since it is expected that irradiation has killed some cells, a larger number of the irradiated cells will have to be plated. Plating 100 control cells and 400 irradiated cells is typical. After allowing growth over a suitable period of incubation, the colonies are *scored*. To do this, it must first be decided what number of cells should be classified as a viable colony. The usual criterion is 50 cells, equivalent to 5–6 generations of cell division. If in the control dishes an average of approximately 20 colonies is found, the plating efficiency (PE) is $20/100 = 0.2$. If the irradiated cells yield an average of approximately eight colonies, then their plating efficiency is lower: $8/400 = 0.02$. The *surviving fraction* is calculated as the ratio of these plating efficiencies:

$$\text{Surviving fraction} = \frac{\text{PE}_{\text{treated}}}{\text{PE}_{\text{control}}} = \frac{0.02}{0.2} = 0.1$$

The assumption here is that $\text{PE}_{\text{control}}$ indicates the detection efficiency of clonogenic cells, and is unaffected by irradiation. Using this equation, the colony counts from the irradiated cells can be corrected for this detection efficiency. Surviving fraction is often given as a percentage (10% in this case).

The above description started with a suspension of tumour cells. In order to measure in vivo cell survival, two groups of experimental tumours (i.e. in mice or rats) are taken. One is irradiated and the other is kept as a control. At some time after irradiation cell suspensions are made from both groups, treated identically, and plated as before. The difference here is that the cells are irradiated under in vivo conditions.

Irradiation not only reduces the colony numbers; it also produces non-lethal effects that shift the colony-size distribution to smaller values. Some of these small colonies may represent clones that eventually die out; others may arise from cells that, as a result of non-lethal injury, have a reduced colony growth rate. Unless they reach the usual cut-off of 50 cells, they will not be counted. However, the implication of these partially viable colonies for the evaluation of radiation effects on tumours may be worthy of greater attention.

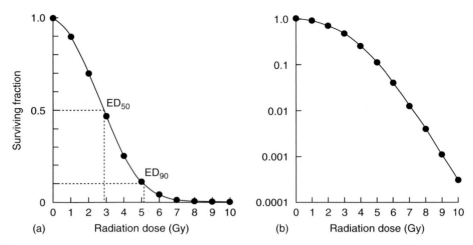

FIGURE 7.2
A typical cell survival curve for cells irradiated in tissue-culture, plotted (a) on a linear survival scale; (b) shows the same data plotted on a logarithmic scale.

7.3 CELL-SURVIVAL CURVES

A cell survival curve is a plot of surviving fraction against dose (of radiation, cytotoxic drug, or other cell-killing agent). Figure 7.2a shows that when plotted on *linear* scales, the survival curve for cells irradiated in tissue culture is often sigmoid: there is a shoulder followed by a curve which asymptotically approaches zero survival. To indicate the sensitivity of the cells to radiation, the ED_{50} or ED_{90} values could be read off from the graph. (ED_{90} is the dose necessary to kill 90% of the cells.) In doing this no assumptions are made about the shape of the curve.

There are two reasons why survival curves are usually plotted on a *logarithmic* scale of survival as in Figure 7.2b. If cell killing is the result of random *single-hit* radiation events then survival will be an *exponential* function of dose. This will be a straight line on a semi-log plot[*]. Secondly, a logarithmic scale more easily allows effects at very low survival levels to be shown and compared. This is important because, as indicated in the next section, the cure of a tumour requires many orders of magnitude of cell kill.

7.4 THE RELATIONSHIP BETWEEN CELL SURVIVAL AND GROSS TUMOUR RESPONSE

The objective of clonogenic cell survival studies is to be able to understand, or to make predictions about, the main features of tumour response to therapy: these are tumour growth delay (or in clinical terms *duration of remission*) and local tumour control.

7.4.1 TUMOUR GROWTH DELAY

Incomplete treatment of a tumour leads to a temporary phase of tumour regression which is subsequently followed by tumour recurrence. This pattern is illustrated in Figure 7.3.

---

[*] The linear–quadratic model of cell killing involves both *single-hit* and *double-hit* mechanisms and will be introduced later in Section 7.13.

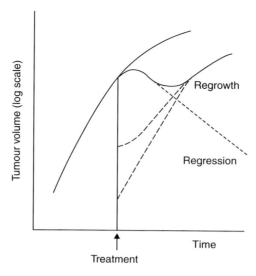

FIGURE 7.3
The volume response of an uncontrolled tumour is the resultant of two processes: regression and regrowth. Repopulation during the period of regression may take place at a rate that may differ from the growth rate of the untreated tumour.

Regression is due to the death and disappearance of cells killed by radiation and also of differentiated (i.e. maturing) cells of limited life-span which would have been produced by the killed stem cells.

The rate of tumour regression differs widely from one tumour to another. Some tumours shrink during a course of radiation therapy, whereas others regress very slowly. This process strongly influences the patient's and the clinician's perception of the effectiveness of the treatment. But it is important to stress (as illustrated by Figure 7.3) that the effectiveness of treatment (as judged by the time-scale of tumour recurrence) depends on the *regrowth component* and not on the regression component.

The regrowth component in Figure 7.3 is due to repopulation by surviving clonogenic cells. The speed of regrowth varies considerably from one tumour to another, and the upper of the two broken lines in the figure illustrates the possibility of a lag period before repopulation fully gets under way. There is also evidence that once it begins, the speed of repopulation may resemble the higher growth rate of a very small untreated tumour. This is sometimes known as *accelerated repopulation*.

7.4.2 LOCAL TUMOUR CONTROL

The eradication of every clonogenic tumour cell will lead to tumour cure; however, this is a very difficult objective. Every gram of tumour may contain 10^9 cells of which as few as 1%, may be clonogenic. A human tumour at presentation will have a mass of tens or hundreds of grams, and thus the total number of clonogenic cells could exceed 10^9. Figure 7.4 illustrates the response of a tumour that initially contains 10^{10} clonogenic cells to radiotherapy. Each radiation dose is assumed to result in a surviving fraction of roughly 0.5, thus approximately halving the number of clonogenic cells. Consequently, cell kill is exponential with cumulative dose. This assumption implies *constant effect per fraction*, for which there is some experimental evidence. In this example around 30 fractions of 2 Gy are required to reduce the initial number of clonogenic cells to ten surviving cells ($0.5^{30} \approx 10^{-9}$). When treatment has reduced survival to 1% of the original number of clonogenic cells, this might, in a rapidly shrinking tumour,

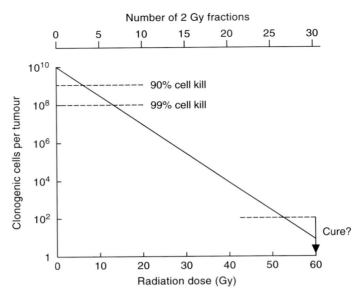

FIGURE 7.4
Illustrating the treatment of a tumour that initially contains 10^{10} clonogenic cells with a sequence of 2 Gy radiation doses; each roughly reduces survival by half.

lead to a complete disappearance of all *visible* tumour. The diagram indicates that four or five times this dose is expected to be needed in order to eradicate the last surviving clonogenic tumour cell thereby reaching a curative level.

Whether it is always necessary to eradicate the last clonogenic tumour cell in order to achieve local tumour control has been a matter of intense debate. During the 1970s there was a widespread belief that the immune responses of patients against their tumours might be strong, and this gave rise to many attempts at immunotherapy. Unfortunately, it has subsequently been realised that the animal tumours used to support the optimistic claims for immunotherapy were not truly syngeneic (i.e. genetically identical to their hosts). They had developed an artificially strong immune response against the hosts they were growing in, and consequently the results obtained with them were misleading. Although it is possible that weak tumour-directed immune responses may exist within cancer patients, perhaps sufficient to eradicate a few of the surviving tumour cells, this cannot be relied on.

The key point that is illustrated in Figure 7.4 is that a course of radiotherapy which kills 99% of clonogenic cells and which might lead to complete tumour regression is very far from achieving local tumour control. Hence the adage, "99% tumour cell kill is complete failure" (see Section 36.1 on the mathematical modelling of tumour control probability).

7.4.3   SELECTIVITY IS THE NAME OF THE GAME

The very considerable *log cell kill* that is required to control a tumour has to be achieved without exceeding the tolerance of critical normal tissues which are within or adjacent to the high-dose volume. This is a daunting task for it is known that organs such as the small intestine will fail (for instance, by ulceration of the mucosa) if the proportion of stem cells is reduced below around 1% (see Figure 8.5). This implies that ~10 logs of cell kill in the tumour must be achieved without exceeding ~2 logs in the normal tissues. This requires a factor of 5 in dose selectivity, i.e. the dose to the tumour must be *five times* the dose to the sensitive normal tissue. Radiotherapy goes a long way toward achieving this by physically *focusing* the radiation and reducing the amount of normal structures which are within the high-dose volume. If this is not achieved, radiotherapy will almost always fail. Incidentally, this is the predicament of cancer chemotherapy. When it is given as a systemic treatment, critical normal tissues such as the bone marrow and intestines receive a full drug dose and success is only possible when (as is rarely the case) the drugs or drug combinations have a very high level of selectivity for tumour cell kill (see also Wilson et al. 2006).

7.5 WHY CELLS DIE WHEN THEY ARE IRRADIATED

Exposure of biological tissues to ionising radiation immediately leads to ionisation and excitation of their constituent atoms. The molecules where these atoms reside then tend to fall apart, resulting in so-called *free radicals*. Water is the most prevalent molecule within the cell so most of the free radicals are produced by the radiolysis of water. Free radicals are highly unstable. They react with other nearby molecules, thereby transferring chemical damage to them; these free-radical processes are usually complete within a millisecond under physiological conditions. All components of the cell will be damaged in this way: proteins, enzymes, membrane components, etc. However, such molecules are present in vast numbers in every cell and damage to a few of them probably has little impact on the cell's viability; they will be quickly regenerated. But there is one cellular component which is almost unique: DNA. DNA is a very long double-helix molecule consisting of a repeated sequence of bases, and every chromosome has approximately 200 million bases. Groups of bases form the genes that contain instructions for proteins and thus for all aspects of cellular function. There is usually some duplication of genes, but even so there is a serious risk that radiation damage may lead to the loss (or modification) of some genes and thus to a loss of specific functions (some of which may be essential for survival). This is the reason why DNA is the most vulnerable part of a cell to radiation damage. Cells have evolved to withstand such damage, however; they have a formidable armoury of sophisticated repair enzymes which are continually monitoring the integrity of the DNA, recognising damage, and repairing it. A radiation dose of 1 Gy produces, in every cell nucleus, roughly 2×10^5 ionisations, leading to around 1000 single-strand breaks in DNA and possibly 40 double-strand breaks. But repair processes are so efficient that, in spite of all this damage, most cells survive.

Direct evidence that damage to DNA is in most cases the critical event for cell survival has come from experiments showing that when short-range Auger-emitting isotopes are incorporated into DNA they are far more toxic than when the same type and amount of radioactivity is bound to other parts of the cell.

When they are fatally damaged by radiation most cells do not immediately die; after a dose-dependent delay in progression through the cell cycle they usually proceed to mitosis. Damaged cells often fail to complete mitosis, or they may go through one or more further cell cycles before getting stuck at a subsequent cell division (Figure 7.1). However, lymphocytes as well as some other cell types die before reaching mitosis. This is called *intermitotic cell death*. It is closely related to what has come to be called *programmed cell death* or *apoptosis* which has been the focus of much cancer research, the aim being to manipulate this process for therapeutic benefit (Haimovitz-Friedman et al. 1996).

7.6 CELLULAR RECOVERY FROM RADIATION DAMAGE

Experimental methods have been devised to detect the recovery of cells and tissues from radiation exposure, and these recovery processes are of direct relevance to clinical radiotherapy.

1. *Split-dose experiments.* The effect of a given dose of radiation is less if it is split into two fractions that are delivered a few hours or more apart. This effect has been termed recovery from *sublethal damage* (SLD), or *Elkind recovery*. Almost all cells show this effect. The basis for

this is that if after giving one pulse of damage to a cell, a time delay is allowed before a second pulse, repair will enable the cell to withstand the second pulse more readily. Giving both together may, in contrast, produce an unrepairable level of damage.

2. *Delayed-plating experiments.* If cells are irradiated in a non-growing state and left for increasing periods of time before assaying for survival, an increase in survival is often observed. This has been termed recovery from *potentially lethal damage* (PLD). The rationale is that assay involves cell proliferation and cells which are made to replicate using a non-repaired genome are more likely to die. The kinetics of PLD recovery and SLD recovery are similar.

3. *Dose-rate effect.* The sparing of radiation damage as the dose rate is reduced to around 1 Gy/h is primarily due to cellular recovery. What happens is that at low dose rates each cell that suffers a particle traversal will, on average, experience a considerable delay before a second particle comes along. As in the split-dose situation, this will lead to less overall damage.

4. *Fractionation.* The sparing effect of *fractionating* radiation treatment within a relatively short overall time is primarily due to recovery (see Section 9.1). With daily dose fractions, there is a gap of ~24 h after each fraction during which repair can occur. This is the reason why, in clinical radiotherapy, the total radiation dose required for tumour control must be increased as the number of fractions increases (see Section 9.2).

The speed of cellular recovery has been measured in many cell- and tissue types. There is evidence, mainly from DNA repair studies, that recovery is a multi-exponential process with a first component half-time of a few minutes, a main component whose half-time may be approximately 1 hour, and perhaps also an even slower component. The clinical implication of this is that when treatment is given with multiple fractions per day, it is important to allow full recovery between fractions. This may require an inter-fraction gap of at least 6 hours (see Section 9.4).

7.7 VARIATION OF CELL KILLING THROUGH THE CELL CYCLE

The radiosensitivity of cells varies considerably as they pass through the cell cycle. A cell's life is conventionally divided into four phases: its birth at mitosis, a period of DNA synthesis known as the S-phase, preceded by G1, and followed by G2 (see Figure 7.5). Although this has not been studied in a large number of cell lines, there seems to be a general tendency for cells in

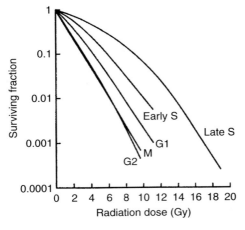 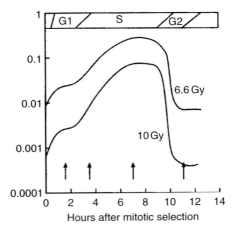

FIGURE 7.5
The variation of radiosensitivity with the phase of the cells in the cell cycle; the surviving fraction varies by a factor of almost 100. (From Sinclair, W. K. and Morton, R. A., *Biophys. J.*, 5, 1–25, 1965.)

the S-phase, in particular the latter part of the S-phase, to be the most resistant and for cells in G2 and mitosis to be the most sensitive. The reason for the resistance in S may be related to the *conformation* of DNA at that time. The sensitivity in G2 probably results from the fact that those cells have little time to repair radiation damage before being called upon to divide. The analysis of the radiation response of synchronized cells by the linear-quadratic model could shed additional light on the underlying mechanisms (Chapman 2003).

The effect of the variation in sensitivity during the cell cycle is that a degree of *synchrony* is created in the cells that survive irradiation. Immediately after a dose of x-rays, all the cells will still be at precisely the same point in the cell cycle as they were before irradiation; some will have lost their reproductive integrity, and the greatest number of viable cells will tend to be in the S-phase. The implication for clinical radiotherapy is that when radiation doses are given close together in time, the extent of cell-cycle progression will influence the overall amount of cell kill. If surviving cells have time to move out of their resistant phase into a more sensitive one, then the effect of a second radiation dose will be greater. This process is called *reassortment* or *redistribution*.

7.8 THE 5 Rs OF RADIOTHERAPY

The biological factors that influence the response of normal and neoplastic tissues to fractionated radiotherapy can be summarised as:

Repair   As evidenced by cellular recovery during the few hours after exposure

Reassortment   Cell-cycle progression effects. Cells that survive a first dose of radiation will tend to be in a resistant phase of the cell cycle, and within a few hours, they may progress into a more radiosensitive phase

Repopulation   During a 4- to 6-week course of radiotherapy, tumour cells that survive irradiation may proliferate and thus increase the number of cells which must be killed

Reoxygenation   In a tumour the radioresistant hypoxic cells will selectively survive after a dose fraction, but thereafter, when their oxygen supply improves, their radiosensitivity will increase

Radiosensitivity   For a given fractionation course (or for single-dose irradiation), the haemopoietic system shows a greater response than the kidney, even allowing for the different timing of response. Similarly, some tumours are more radioresponsive than others to a particular fractionation schedule, and this is largely due to differences in radiosensitivity

Note that two of these processes (*repair* and *repopulation*) will tend to make the tissue more *resistant* to a second dose of radiation; two others (reassortment and reoxygenation) tend to make it more *sensitive*. The first four factors given above were christened the *4 Rs of Radiotherapy* by Withers (1975).

7.9 THE IMPORTANCE OF OXYGEN

The response of cells to ionising radiation is strongly dependent upon oxygen. This is illustrated in Figure 7.6 for mammalian cells irradiated in culture. Cell surviving-fraction is shown as a function of radiation dose administered either under normal aerated conditions or under hypoxia which is generally achieved by passing nitrogen gas over the cell suspension for

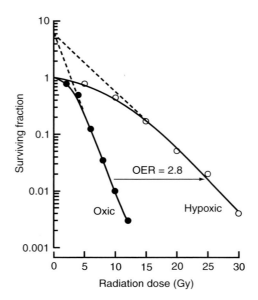

FIGURE 7.6
Survival curves for cultured mammalian cells exposed to x-rays under oxic or hypoxic conditions. Note that the dose level is ≈ 10 Gy, SF ≈ 0.015 where the OER value equals 2.8, i.e. far in excess of the clinically relevant 2 Gy.

a period of 15–30 min. The enhancement of radiation damage by oxygen is *dose-modifying*, i.e. the isoeffective dose is reduced by the same factor at all levels of survival (note that in Figure 7.6 the broken lines extrapolate to the same point on the Y-axis). This allows calculation of an *oxygen enhancement ratio* (OER) that is simply the ratio of the radiation dose in hypoxia to the dose in air needed to achieve the same biological effect. For most cells, the OER for x-rays is around 3.0. However, certain studies have shown a *lower* OER for the single-hit mechanism relative to the double-hit mechanism of the linear-quadratic model so that at radiation doses of 3 Gy or less, the OER is actually *reduced* (Palcic and Skarsgard 1984; Chapman 2003; Nahum et al. 2003). This is an important finding because this is the dose range for the overwhelming majority of fractionated treatments.

It has been demonstrated from rapid-mix studies that the oxygen effect only occurs if oxygen is present either during irradiation or within a few milliseconds thereafter. This occurs because the effect of oxygen is to modify free-radical reactions, leading to greater fixation of DNA damage. The dependence of OER on oxygen tension is shown in Figure 7.7. By definition, the OER under anoxic conditions is 1.0. As the oxygen level increases, there is a corresponding increase in radiosensitivity and therefore in the OER. The greatest change occurs from 0 to about 20 mm Hg; further increase in oxygen concentration, even up to that seen in air (155 mm Hg) or to 100% oxygen (760 mm Hg), has little influence on radiosensitivity. Also shown in Figure 7.6 are the oxygen partial pressures typically found in arterial and venous blood. Therefore, from a radiobiological standpoint, most normal tissues can be considered to be well-oxygenated, although it is recognised that the response of some normal tissues may increase if their oxygen supply is improved.

7.10 HYPOXIA IN TUMOURS

Oxygen also plays an important role in the radiation response of tumours. Solid tumour growth requires the induction of new blood vessels, a process that is referred to as *angiogenesis*. This new blood supply is primitive in nature, and it may be inadequate for meeting all the needs of the growing tumour. Nutrient-deprived and oxygen-deprived regions often develop,

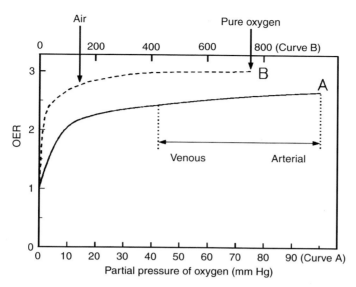

FIGURE 7.7
Variation of oxygen enhancement ratio (OER) with oxygen tension. The dashed curve shows the same curve plotted on the upper scale. The range of blood oxygen tensions is indicated. (From Denekamp, J., in *The Biological Basis of Radiotherapy*, 2nd ed., Steel, G. G., Adams, G. E., and Horwich, A., Eds., Elsevier Science, Amsterdam, 1989. With permission.)

yet the hypoxic cells existing in these areas may still be viable. The first real indication that hypoxia could exist in tumours was made in 1955 by Thomlinson and Gray from their observations on histological sections of fresh specimens from carcinoma of the bronchus (Thomlinson and Gray 1955). They observed viable tumour regions surrounded by vascular stroma where the tumour cells obtained their nutrient and oxygen requirements. As these regions expanded, areas of necrosis appeared at the centre. The thickness of the resulting 'shell' of viable tissue was found to be similar to the calculated diffusion distance of oxygen in respiring tissues; it was suggested that as oxygen diffused from the stroma, it was consumed by the cells. Although those cells beyond the diffusion distance were unable to survive, cells immediately bordering the necrotic regions might be viable, but hypoxic.

Tannock (1968) made similar observations in mouse mammary tumours. The extent of necrosis in these tumours was much greater, and each patent blood vessel was surrounded by a 'cord' of viable tumour cells outside which there was necrosis. Cells at the edge of the cords are thought to be hypoxic and are often called *chronically hypoxic cells*. Tannock showed that because the cell population of the cord is in a dynamic state of cell turnover, these cells will have a short lifespan. More recently, it has been suggested that some tumour blood vessels may open and close, leading to *transient* or *acute* hypoxia. The mechanisms responsible for intermittent blood flow in tumours are not entirely clear. They might include vessel-plugging by white blood cells, red blood-cell *rouleaux* or circulating tumour cells, vessels collapsing in regions of high tumour interstitial pressure, or spontaneous vasomotion in incorporated host arterioles affecting blood flow in downstream capillaries.

Since hypoxic cells are resistant to radiation, their presence in tumours is critical in determining the response of tumours to single treatments with large doses of radiation. The presence of such cells in experimental tumours can be easily demonstrated as seen in Figure 7.8. This shows (from left to right) the radiation survival response of EMT6 mouse mammary tumours irradiated (a) as single cells in monolayer culture under oxic conditions; (b) in situ in air-breathing mice (having high or low haemoglobin (Hb) levels); or (c) hypoxic (in nitrogen-asphyxiated mice). Cell survival was estimated immediately after irradiation using an in vitro assay. The figure shows that the survival curves in air-breathing mice were biphasic;

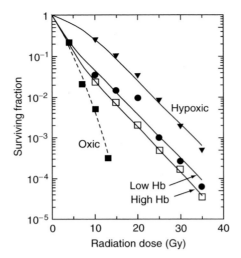

FIGURE 7.8
The survival of an EMT6 mouse sarcoma irradiated under aerobic or hypoxic conditions. The dashed curve shows the in vitro survival of oxic cells. (From Hill, R. P., et al., *Br. J. Radiol.*, 44, 299–304, 1971. With permission.)

for doses under 5 Gy, the curves followed that for oxic cells, but above this dose level, the curves are parallel to the hypoxic curve. High single doses of radiation wipe out the oxic cells, and the surviving cells are almost all hypoxic.

7.10.1 HYPOXIC FRACTION

The term *hypoxic fraction* refers to the fraction of clonogenic tumour cells that have a radiosensitivity which is characteristic of hypoxic cells. Figure 7.8 shows the standard way in which this has been measured. In air-breathing animals, the survival curve will have the biphasic form mentioned above, reflecting the radiosensitivity of a mixture of oxic and hypoxic cells. In the killed animals, only the hypoxic component will be seen. The fact that the two survival curves are parallel confirms that they are both characteristic of hypoxic cells. If a vertical line is drawn at any point where the lines are parallel, then the ratio (survival in air-breathing mice)/(survival in dead mice) gives the hypoxic fraction of the tumours in air-breathing mice. The example shown in Figure 7.8 also illustrates the therapeutically important conclusion that the hypoxic fraction of a tumour may be greater in patients whose blood has a reduced haemoglobin content. Smoking reduces blood oxygen tension and is particularly inadvisable for patients undergoing radiotherapy.

7.10.2 REOXYGENATION

The time-course of changes in the hypoxic fraction of a tumour before and after irradiation is illustrated in Figure 7.9. Tumour nodules whose diameters are less than about 1 mm have been found to be fully oxygenated. Above this size, the hypoxic fraction increases until it reaches between 10% and 50% of clonogenic cells. Irradiation of a tumour will inevitably kill more oxic than hypoxic cells, and after a single large radiation dose, the hypoxic fraction may approach 100%. The *number* of surviving hypoxic cells will be low, but their *fraction* of all surviving clonogenic cells will be high. The term *reoxygenation* refers to the process by which these hypoxic cells become better supplied with oxygen. Reoxygenation has been studied by measuring the hypoxic fraction (as described above) at various times after a

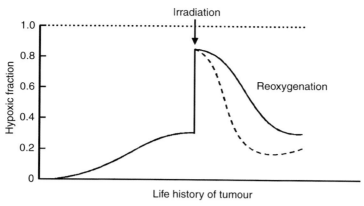

FIGURE 7.9
The time-course of changes in the hypoxic fraction during the life history and response to irradiation of a tumour.

priming dose of radiation. The experiments are difficult because they involve assaying the increase in cell kill from adding one radiation dose to another. The results of these studies have indicated that some tumours reoxygenate quickly (the hypoxic fraction falling from 100% to 10% within approximately one day), whereas, other tumours reoxygenate slowly (taking a week or more).

During the 1960s and 1970s, tumour hypoxia was perceived to be one of the principal causes of failure in radiation therapy, and much research effort was directed at developing ways of selectively killing hypoxic cells: high-LET radiations (see Chapter 49), hyperbaric oxygen, chemical radiosensitizers, etc. The disappointing results from these approaches have led to a widespread view that reoxygenation may be quite efficient in many clinical situations. Tumour cells that are hypoxic at the time of one radiation-dose fraction may be oxic by the time a subsequent dose fraction comes along. Therefore, provided fractionated radiotherapy is given over a conventional period of time, the presence of hypoxia may be less serious than had been previously believed. When the overall treatment time is shortened, as with brachytherapy or accelerated external-beam treatment, hypoxia may still be a significant problem.

Several clinical studies performed throughout the 1990s with the Eppendorf pO_2 micro-electrodes have confirmed that hypoxia identified prior to the start of radiotherapy does significantly predict for the treatment failure of cervical, head-and-neck, and prostate carcinomas and of sarcomas (Hockel et al. 1993; Nordsmark et al. 1996; Movsas et al. 2002). Indeed, the use of the hypoxic-cell radiosensitizer Nimorazole has been shown to increase local control in supraglottic larynx and pharynx carcinoma in a phase-III clinical trial (Overgaard et al. 1998).

7.11 TUMOUR CONTROL PROBABILITY

The main factor determining success in clinical radiotherapy is radiation dose. Low doses are ineffective, but if it is possible to give a very large total radiation dose then, in principle, any tumour can be (locally) controlled. Between these extremes is a probability of tumour control which varies with dose according to a sigmoid relationship (see Figure 8.2 and Figure 8.3; see also Section 36.1 on the mathematical modelling of tumour control probability). For any particular type of cancer, the characteristics of this curve are crucial to the success of therapy, and this is stressed in Section 8.5.

The position of such a dose-response curve on the dose scale depends on the various biological factors that have been discussed above. Its steepness is also clinically important and depends on:

1. An underlying Poisson relationship (Section 8.6).
2. Variation in the possibility of cure among tumours of the same type, arising from differences in bulk, cellular radiosensitivity, repopulation, hypoxia, etc. All forms of variability will tend to make dose-response curve less steep (cf. the parameter σ_α in Section 36.1 on TCP modelling).
3. Inter-patient variation in the actual radiation dose delivered.

Quantification of the steepness of dose-response curves is dealt with in Section 8.6.1 and Section 36.2.1.3.

7.12 EXPERIMENTAL TUMOUR SYSTEMS

Research into the radiation response of tumours has been performed on a wide variety of tumour cell systems, including the following:

In-vitro tumour cell lines. These are cells derived from rodent or human tumours that have been induced to grow in tissue culture. They usually grow rapidly and have to be *passaged* (i.e. a small sample of the culture is transferred to new culture medium) every few weeks. They may be stored for long periods in a tissue bank, and they are useful for model experiments where simplicity and good reproducibility are important.

Multicellular spheroids. Small fragments of a tumour tissue culture can be grown under culture conditions where the cells stay together and grow as spherical balls of cells. As the spheroids grow, they develop a hypoxic centre. They have been used as in-vitro models of solid tumours.

Transplanted tumours in experimental animals. Within a highly inbred strain of mice or rats (i.e. syngeneic animals), tissues can be readily transplanted from one animal to another. A tumour that has arisen within such animals can be indefinitely *passaged*. For any one experiment, a large number of identical tumours can be produced and this greatly aids the experimental design. A disadvantage is that such tumours tend to drift away from their original characteristics, often growing more rapidly. Both the tumour and the mouse strain may drift genetically which leads to the development of immune rejection mechanisms and to potential artefacts.

Primary animal tumours. These may spontaneously arise or be induced by chemicals. They are real tumours with less artificiality than the systems described above, but they are expensive, often available only in small numbers, and are less easy to use than transplanted tumours.

Human tumour xenografts. Xenograft refers to transplantation across a species barrier. This is facilitated by the production of mice that almost completely lack an immune response. There are congenitally immune-deficient mouse strains such as the nude or SCID mice, but it is also possible using whole-body irradiation and immuno suppressive drugs to make conventional mice receptive to the subcutaneous transplantation of human tumour fragments. Xenografts maintain many of the biological characteristics of the source tumour and are useful for some aspects of tumour radiobiology. But the artificiality of the immunological situation of xenografts may lead to artefacts.

Tumours in patients. All experimental models of human cancer have drawbacks, and in many respects, the ideal is to work directly on cancer patients. However, practical and ethical considerations seriously limit the scope for such research. Clinical trials as the endpoint of research development are, of course, established. The quantification of treatment response, both of tumours and normal tissues, is an important research area. There are also some opportunities to perform experimental therapy on small groups of patients in order to test the feasibility of using radiosensitizers, hyperthermia, etc.

These experimental tumour systems have been listed in order of increasing closeness to cancer in patients. As we move down the list we come closer to the reality that we seek to understand but the scope for precise and sophisticated experimentation decreases.

7.13 THE RADIOSENSITIVITY OF HUMAN TUMOUR CELLS

7.13.1 THE INITIAL SLOPE OF THE CELL-SURVIVAL CURVE

Two key questions in radiation biology applied to radiotherapy are: how radiosensitive are human tumour cells, and how does this relate to clinical radiocurability? Prior to 1980, there was little information on these issues, but interest was aroused by Fertil and Malaise (1981) who surveyed the published literature on in-vitro human tumour cell survival curves. They found evidence for a correlation with clinical response. This survey was repeated by Deacon et al. (1984) who summarised data on 51 non-HeLa cell lines. These covered 17 different histopathological tumour types that were placed into five categories of local tumour radiocurability: A: lymphoma, myeloma, neuroblastoma; B: medulloblastoma, small-cell lung cancer; C: breast, bladder, cervix carcinoma; D: pancreas, colo-rectal, squamous lung cancer; and E: melanoma, osteosarcoma, glioblastoma, renal carcinoma. The placing of tumour types in this list is somewhat controversial, and the underlying clinical data do not allow this to be unequivocally done. However, the ranking A–E broadly reflects clinical experience.

The published in-vitro data for each cell line were analysed to determine the surviving fraction at 2 Gy (which Deacon et al. termed SF_2): chosen as a measure of the initial slope of the cell survival curves. The result is shown in Figure 7.10. Within each category of clinical

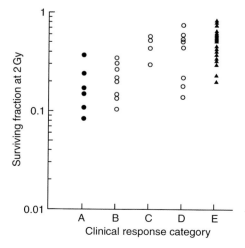

FIGURE 7.10
Surviving fraction at 2 Gy for 51 human tumour cell lines arranged in five categories of clinical radioresponsiveness.

radioresponsiveness there is a considerable scatter (not surprising in view of the many different sources, cell lines, and techniques used). In confirmation of Fertil and Malaise's conclusions, there was a significant trend in the data towards Group A having lower and Group E having higher SF_2 values. This important conclusion underlies the belief that the steepness of the initial slope is a significant factor in the clinical response of tumours to radiotherapy. One should note that, while there is some evidence of a trend for average tumour radiosensitivity, the intra-group heterogeneity is so large that the most radioresistant tumour cells in Group A (the sensitive group) are more resistant than the most sensitive tumour cells in Group E (the resistant group).

7.13.2 Cell-Survival Curves for Human Tumour Cells

The survival curves shown in Figure 7.11 illustrate the range of radiosensitivity commonly found among human tumour cell lines. The range of doses that correspond to a survival of 1% of cells varies by approximately a factor of three. The range of steepness of the initial slopes of these curves is even wider. The full lines in Figure 7.11 are *linear-quadratic* curves fitted to the data

$$\text{Surviving fraction} = e^{-\alpha d - \beta d^2} \tag{7.1}$$

where d is the radiation dose. With these and other data sets, the fit is good; there is a clear initial slope, and the data are consistent with a *continuously*-bending curve. This is usually the case when cells of *homogeneous* radiosensitivity are irradiated at *low* temperatures where the repair of sublethal damage is minimized (Chapman 2003).

Because there may be a molecular basis for regarding the separate terms of the linear-quadratic equation as mechanistically distinct, it is interesting to examine their relative

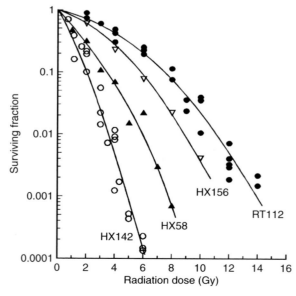

FIGURE 7.11
Cell survival curves for four representative human tumour cell lines irradiated at high dose rate. HX142 neuroblastoma; HX58 pancreatic; HX156 cervix; RT112 bladder carcinoma. (From Steel, G. G., *Radiother. Oncol.*, 20, 71–83, 1991. With permission.)

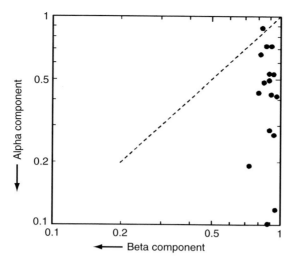

FIGURE 7.12

Contributions to the surviving fraction at 2 Gy from the linear and quadratic components of radiation cell killing for 17 human tumour cell lines. The dotted line indicates equality of the two components. (From Steel, G. G. and Peacock, J. H., *Radiother. Oncol.*, 15, 63–72, 1989. With permission.)

contributions to cell killing. Therefore, they are separated as:

$$\text{Linear or alpha component} = \exp(-\alpha d)$$

$$\text{Quadratic or beta component} = \exp(-\beta d^2)$$

Having fitted the data with the linear–quadratic (LQ) expression (Equation 7.1), as shown in Figure 7.11, the values for α and β are known, and for any chosen dose, these two components can be calculated. This has been done in Figure 7.12 for 17 human tumour cell lines for a radiation dose of 2 Gy, a typical dose per fraction in clinical radiotherapy. It can be seen that the calculated points lie down the right-hand border of this diagram; the dispersion in radiosensitivity among these cell lines at 2 Gy is entirely due to differences in the steepness of the *linear* component of cell killing. The small amounts of cell kill because of the *beta* component (indicated by the distance of the point from the right-hand boundary) do not seem to correlate with the *alpha* sensitivity. The conclusion is one that is derived from most of the models of radiation cell killing: *cell killing at clinically realistic doses per fraction is dominated by the steepness of the linear component of the cell-survival curve.* An excellent review of cell-killing mechanisms in radiation therapy has been written by Chapman (2003).

7.13.3 DEPARTURES FROM THE LQ MODEL FOR DOSES BELOW 1 GY

There is now a large body of experimental evidence (e.g. Marples and Joiner 1993; Short et al. 1999; Joiner et al. 2001) that for many mammalian cell lines, the surviving fraction at doses below around 1.0 Gy is not well predicted by the LQ expression (Equation 7.1). This low-dose behaviour is illustrated in Figure 7.13, which shows the survival of *asynchronous* T98G human glioma cells between 0 Gy and 6.0 Gy measured by Short et al. (1999).

As the figure clearly shows, the cells exhibit increased radiosensitivity at doses up to around 1.0 Gy, and this phenomenon has become known as *low-dose hyper-radiosensitivity* (HRS). This important behaviour has only come to light relatively recently because of the difficulty of making measurements at doses less than 1 Gy; effective techniques involve either a so-called fluorescence-activated cell sorter (FACS) to plate a predetermined number of cells, or microscopic scanning to identify an exact number of cells after plating (Joiner 2003; see also Section 7.1.3).

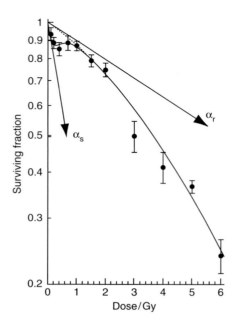

FIGURE 7.13
Experimental data on the survival of human glioma cells showing hypersensitive behaviour, clearly departing from the linear–quadratic expression (dashed curve), at doses below around 1.0 Gy. The radiation quality used was 240 kV x-rays. The solid line is a fit of the Induced Repair Model (IndRep) to the data (see Equation 7.2). The initial slope parameter of the IndRep model is indicated as α_s that is clearly much greater than that of the LQ model, α_r. The surviving fraction between ≈ 0.2 Gy and ≈ 1.1 Gy is approximately constant at ≈ 0.87; whereas, the LQ model would predict that it should vary between ≈ 0.98 and ≈ 0.87. (From Short, S. C., et al., *Int. J. Radiat. Biol.*, 75, 847–855, 1999. With permission.)

A modified form of the LQ expression has been proposed (Joiner 2002a) which is known as the induced repair (IndRep) model

$$\text{Surviving Fraction} = \exp\left\{-\alpha_r D\left[1 + (\alpha_s/\alpha_r - 1) \times \exp(-D/D_c)\right] - \beta D^2\right\} \qquad (7.2)$$

where D_c (about 0.2 Gy) is the dose at which the transition from hypersensitive behaviour to the increased radioresistance (IRR) associated with following the LQ model begins (see Figure 7.13). Inspection of Equation 7.2 shows that for $D \gg D_c$, $\exp(-D/D_c) \to 0$, and the right-hand side reduces to the LQ expression (Equation 7.1) with parameters α_r and β. At very low doses, $(D \ll D_c)$, $\exp(-D/D_c) \to 1$, and the expression also reduces to the LQ model with parameters α_s and β.

The mechanism behind this intriguing low-dose behaviour of the surviving fraction has not yet been fully elucidated. Joiner et al. (2001) suggest that an increase in DNA repair capacity occurs in the IRR region. Other radiobiologists have hypothesized that it is a manifestation of the widely differing radiosensitivities of cells in different phases of the cell cycle in an asynchronous cell population (see Section 7.7); the much steeper initial slope corresponds to the response of these highly radiosensitive cells undergoing mitosis (Chapman, private communication). However, simple modeling of this variable radiosensitivity in the cell cycle yields too small a deviation from LQ behaviour.

Is there any evidence that the hypersensitivity observed from in-vitro experiments also exists in vivo? Joiner et al. (2001) concluded that such evidence does exist; the total dose required to produce a given level of damage in mouse skin, kidney and lung *decreases* when the fraction size is reduced below 1 Gy. They termed this a *reverse fractionation* effect. Short et al. (2001) found that a 3-hour interfraction interval was sufficient for HRS recovery between fractions. Joiner et al. (2001) have discussed how HRS might be clinically exploited, e.g. by giving dose fractions around 0.5 Gy (which they termed *ultrafractionation*). Clearly a greater excess of radiosensitivity is required in the tumour than in the surrounding organ(s)-at-risk for there to be a therapeutic gain (see Section 8.5).

Fractional doses of the order of 0.2 Gy to 1.0 Gy correspond to the 10–50% isodoses in a treatment plan, (for a 2 Gy per fraction prescription dose) and consequently the presence of HRS in the cells in certain organs-at-risk could have consequences for complication rates. In particular, the rapidly emerging field of predicting complication probability by analysing dose distributions and DVHs in organs at risk (normal-tissue complication probability modeling— see Section 36.3) may need to take account of HRS. To give a simple example, if a certain conformal or intensity-modulated technique *reduces* the volume of normal tissue irradiated at around 1 Gy per fraction, but at the same time *increases* the normal-tissue volume irradiated at around 0.2 Gy per fraction, and if the cells involved exhibit HRS, then this may not lead to any net reduction in cell killing at all and thus not reduce the complication probability. Honoré (2002) has explored scenarios of this nature.

7.14 THE EFFECT OF DIFFERENT RADIATION QUALITIES OR LET

There has so for been very little mention of the effect of the *quality* of the radiation employed, i.e. whether this be photons (x- or γ-rays), electrons (linac-produced or β-rays), neutrons, protons, α particles, etc. or of their energy or energy spectrum. This is a complex and important issue, but it is possible to simplify it inasmuch as the radiobiological effect over the radiation-quality range employed in modern radiotherapy, i.e. overwhelmingly megavoltage beams from linear accelerators or cobalt-60 γ-radiation, exhibits entirely negligible variations (for the same total dose delivered in a given fractionation pattern at a given dose rate). This is also true of the high-energy end of the kilovoltage x-ray range (approx. 50 kV and above) and is very nearly the case even for high-energy proton beams. All of the above radiation qualities are classed as *low-LET*. This refers to the fact that the ionisation density along the tracks of the secondary charged particles (overwhelmingly electrons) is relatively low. LET stands for *Linear Energy Transfer* which is another term for (collision) Stopping Power (see Chapter 3). However, this constancy of biological effect over a wide quality range does not extend to neutrons or charged particles heavier than protons which are often known as *light ions* (see Chapter 46 and Chapter 49). Such qualities are classed as high LET. In fact, one could justifiably add a 6th R to the 5 Rs previously discussed: *Radiation Quality*.

The considerable difference between low- and high-LET charged-particle tracks is illustrated in Figure 7.14. An absorbed dose of 1 Gy corresponds to around 1000 electron tracks but only about 4 from α-particles (in a volume equal to that of the cell nucleus). This shows how much more *energy* is transferred *per unit distance* along the high-LET α-particle tracks compared to that transferred by the low-LET electrons.

A huge number of cell irradiation experiments have been carried out over the years with different qualities of radiation (e.g. Barendsen 1968; Raju 1980; Fowler 1981; Joiner 2002b). From these experiments, a clear picture has emerged. As the LET increases, more cell killing is generally produced per Gy. Figure 7.15 shows experimentally derived surviving fraction vs. dose curves for human kidney cells irradiated by radiation ranging from 250 kV x-rays, representing low-LET radiation (LET = 2 keV/μm), through to 2.5 MeV α-particles at the high-LET end (LET = 165 keV/μm)[*]. Several things can be observed in Figure 7.15. First, the curves

---

[*] To characterize a radiation quality by a single value for LET is a gross simplification (ICRU 1970). All radiations give rise to a broad spectrum of charged-particle energies in irradiated media; this is especially true of megavoltage photons or electrons, which produce electron energies from several MeV down to tens of eV, with S_{col} values in water (see Chapter 3), i.e. with LET in water, ranging between 2 MeV/cm (0.2 keV/μm) and 250 MeV/cm (25 keV/μm). An example of the complete electron-fluence spectra produced in water by megavoltage radiation can be found in Nahum (1976, 1999).

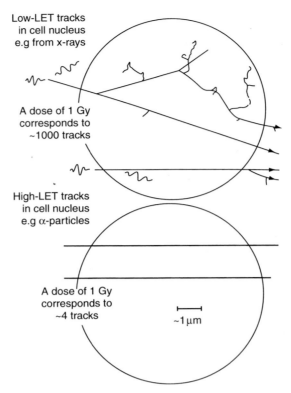

FIGURE 7.14

Particle track structure for low-LET radiation (upper picture) and for α-particles (lower picture). The circles represent the size of the nucleus of a typical mammalian cell. The tortuous nature of the (low-LET) secondary electron tracks are in complete contrast to the high-LET particles, of which only around *four* are required to deposit a dose of 1 Gy in that small volume. (From Goodhead, D. T., *Health Phys.*, 55, 231–240, 1988. With permission.)

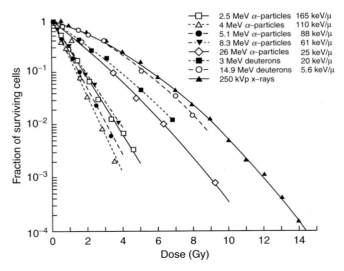

FIGURE 7.15

Surviving fraction measured in vitro for human kidney cells exposed to radiations with a wide range of (mean) LET. (From Barendsen, G. W., *Curr. Top. Radiat. Res. Q.*, 4, 293–356, 1968. With permission.)

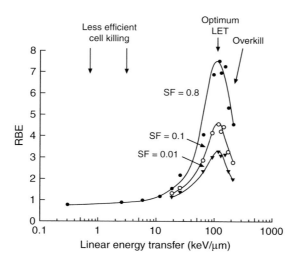

FIGURE 7.16
The relationship between RBE, for cell killing, and (mean) LET. The three curves correspond to different levels of cell survival, showing how RBE depends on exactly how it was determined. The low-LET qualities overwhelmingly employed in radiotherapy correspond to the mean values of keV/μm <1.0 where the RBE is essentially constant. (From Barendsen, G. W., *Curr. Top. Radiat. Res. Q.*, 4, 293–356, 1968. With permission.)

become progressively steeper with increasing LET, corresponding to increased radiosensitivity. Second, these curves also become progressively straighter (or there is less *shoulder*). This corresponds to higher and higher α/β-ratios. Or, in other words, more of the cell killing is by the α- or single-hit (non-repairable) mechanism (see Section 7.13.2 and Chapman 2003) and correspondingly less by the β- or repairable mode. This can be understood from the very different track structures illustrated in Figure 7.14; a hit on the DNA in the cell nucleus from a densely ionising high-LET track will almost always cause major, irreparable damage, whereas, the vast majority of the low-LET hits are much less damaging because much less energy is transferred. Such considerations fall under the scope of *Microdosimetry* (e.g. ICRU 1983; Goodhead 1988).

The radiotherapeutic implications of the higher α/β-ratio for high-LET radiation will be understood by referring to Chapter 9; briefly, for a given total dose, one would expect much *less* dependence of either tumour control probability or normal-tissue complication rate on *fraction size*. This has certainly been observed, at least in mice and rats, for the case of neutron radiation (Withers et al. 1982) which is a high-LET quality because the secondary charged particles are low-energy protons and these are high-LET particles (unlike the very high-energy proton beams required for proton radiotherapy).

The differing biological effectiveness of various radiation qualities is generally expressed through the quantity RBE (*relative biological effectiveness*). RBE is defined as

$$\text{RBE} = \frac{\text{dose of reference radiation}}{\text{dose of radiation under investigation}} \qquad (7.3)$$

to give the same biological effect (i.e. in this case, surviving fraction)[*]. The standard reference radiation is usually a low-LET quality such as 250 kV x-rays.

Figure 7.16 shows the relationship between RBE and (mean) LET. Over a range of low LET values, between 0.3 and about 5.0, there is little change. The maximum is a characteristic

---

[*] Another frequently encountered endpoint is cancer induction (see Part L; Sachs and Brenner 2005); RBE values can be very different for the two endpoints of cell killing and carcinogenesis.

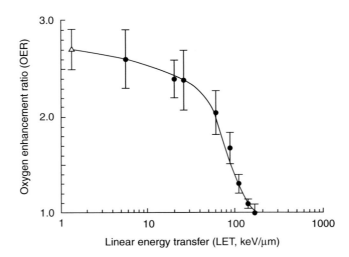

FIGURE 7.17

The relationship between OER and (mean) LET; the closed circles correspond to monoenergetic α-particles and deuterons and the open triangle to 250 kV x-rays; the dose levels where these OER were determined was not stated. Note that OER ≈ 2 at the mean LET of therapeutically useful carbon ions. (From Barendsen, G. W., *Curr. Top. Rad. Res. Q.*, 4, 293–356, 1968. With permission.)

feature of such curves and corresponds to a type of saturation, or overkill, effect. It should be noted that different values result from determining the RBE at different cell survival levels; SFs of 0.01, 0.1, and 0.8 were used by Barendsen (1968) in constructing Figure 7.16.

Another important radiobiological difference between radiations of differing LET concerns the effect of oxygen on cellular radiosensitivity. In Section 7.9, the concept of Oxygen Enhancement Ratio (OER) was introduced in the context of the low-LET qualities commonly employed in radiotherapy. As pointed out in Section 7.9 and Section 7.10, OER values are sufficiently high in low-LET radiation (between ≈ 2 and ≈ 3) that one would not expect to achieve uncomplicated tumour control in situations where the tumour contains an appreciable fraction of hypoxic clonogens. There is clinical evidence that this is, in fact, the case (e.g. Brizel et al. 1999; Movsas et al. 2002). With high-LET radiation, there is evidence from cell-survival curves obtained in vitro that the OER is reduced below its low-LET value with the degree of reduction being dependent on mean LET as Figure 7.17 illustrates.

However, as Dasu and Denekamp (1998), Nahum et al. (2003), and others have pointed out, the effective OER for the 2-Gy fraction size commonly used in megavoltage photon or electron (i.e. low-LET radiotherapy) is closer to 2 than to 3. Therefore, the theoretical advantage of high-LET over low-LET radiotherapy in hypoxic tumours (e.g. through the use of high-energy carbon-ion beams, which have a mean LET of around 80 keV/μm in the spread-out Bragg peak region) (Suzuki et al. 2000; see also Chapter 49), may turn out to be negligible in practice.

CHAPTER 8

RADIOBIOLOGY OF NORMAL TISSUES

Gordon Steel

CONTENTS

8.1 NORMAL-TISSUE REACTIONS TO RADIOTHERAPY

The damage caused by radiation treatment in the various normal tissues of the body varies widely, both in type and in severity. Many patients testify to their experiences that the morbidity associated with cancer therapy is very distressing, and it substantially detracts from the therapeutic benefits. Some tissues, such as the lungs, intestine, and bone marrow, are vital to life; severe radiation damage to these tissues may lead to death. Damage to other tissues, such as the skin, limbs, or gonads, is not usually life-threatening; nevertheless, it can lead to significant impairment of life.

8.2 WHAT DETERMINES THE SEVERITY OF NORMAL-TISSUE DAMAGE?

Many factors are involved in determining the severity of normal-tissue damage; these may be divided into two categories: controllable factors and uncontrollable factors.

8.2.1 CONTROLLABLE FACTORS

The term *controllable factors* principally refers to total radiation dose and treatment volume. Radiation effects are always more severe or more likely to occur as the radiation dose is increased. They also become more frequent as the *size* of the radiation field is increased; this is called the *volume effect* (Section 8.9). Related to this is the question of which normal structures are contained within the radiation field, as enlargement of the field may result in the inclusion of a structure that otherwise would not have been irradiated. A number of other controllable factors influence radiation damage to normal tissues. The choice of fractionation parameters is important: damage tends to increase with increasing dose per fraction (for a given total dose) and is often greater for short overall treatment times (see Section 9.3 and Section 9.4). If the interval between fractions is reduced below 24 h (and especially to below 6 h), then repair of radiation damage may be incomplete and the damage is increased. Concomitant therapy, especially with cytotoxic drugs, also often leads to increased normal-tissue complications.

8.2.2 UNCONTROLLABLE FACTORS

Uncontrollable factors include the age and clinical status of the patient, concurrent disease, genetic make-up, and to some extent, lifestyle. Patients who are older, or in poor health, often have reduced tolerance of radiation treatment. Smoking and excessive alcohol consumption are examples of lifestyle factors that increase the effect of radiation on normal tissue. The issue of genetic make-up is highly topical at the present time. Some radiotherapy patients *burn* more easily than others, just as some individuals react more to sunlight than others. There are a number of recognised inherited syndromes that are associated with an increased risk of damage by radiotherapy, especially ataxia-telangiectasia, Fanconi's anaemia, and Cockayne's syndrome. Genes associated with these and other syndromes are now being identified, and the range of responses seen within the *normal* population may be partly because some people carry these genes. The term *heterozygote* means a person who is carrying a copy of a particular gene in a *recessive* mode, i.e. without expressing the disease itself. Heterozygotes for a radiosensitivity gene do not show the same level of sensitivity as the *homozygote* (who carries two identical copies of a particular gene), but they may show a tendency towards increased radiation damage. Efforts are being made to develop screening procedures for genes that could influence radiation sensitivity, and this might lead to tests that enable unduly sensitive individuals to be identified prior to radiation treatment.

8.3 THE PROLIFERATIVE STRUCTURE OF TISSUES

Most tissues consist of a complex variety of cell types. Often there is a principal functional cell type that is referred to as the *parenchymal* component of the tissue. Invariably, there is also a connective-tissue component made up of fibroblasts, blood vessels, nerve cells, and mobile cells such as macrophages. The epithelial tissues that cover the outside of the body (i.e. epidermis) and the lining of the gut are structurally the easiest to understand. Here the parenchymal component is a sheet of cells, one or more cells thick, lying on supportive connective-tissue structures. When this type of tissue is exposed to radiation, all cell types are damaged and this damage may be expressed in different ways. Some epithelial tissues are in a state of rapid cell turnover in which the structure shown diagramatically in Figure 8.1 can clearly be seen. There are *stem cells* that have the dual ability to renew themselves and also to *differentiate* into mature cells. *Differentiation* means changing into a special type of cell with the ability to perform a specialised function. In the small intestine, this function is to absorb nutrients from the gut; in the skin, the function is to act as a waterproof and physical barrier.

When stem cells begin to differentiate, they usually proliferate faster and tend to be more sensitive to proliferation-dependent cytotoxic agents and may be more radiosensitive. Irradiation tends to stop cell proliferation, and if the dose is high enough, the production of new cells may fail to keep up with cell loss. The parenchymal cell population will, therefore, decline, and the tissue may eventually break down. In the epidermis, this leads to the loss of the superficial layers of the skin, *moist desquamation*, and even ulceration. Skin that has been damaged in this way may heal (if the dose is not too high) but problems may develop at later times as a result of the radiation damage to connective tissues underlying the epidermis. Damaged blood vessels may become permanently swollen and visible to the eye as the disfiguring appearance of *telangiectasia*. The skin may become hard and less flexible. In an extreme case, the tissue may break down into a deep ulcer (i.e. necrosis), a serious clinical problem. In tissues other than the skin, more than one manifestation of radiation injury may also develop, again attributable to damage to the various cell types that are present.

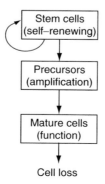

FIGURE 8.1
Schematic outline of the structure of a hierarchical tissue such as the skin.

8.4 EARLY- AND LATE-RESPONDING TISSUES

The effects of ionising radiation on the various tissues of the body vary considerably, both in terms of the radiation dose required to produce damage and in the timing of the expression of damage. Broadly speaking, tissues are divided into two categories: early-responding tissues and late-responding tissues.

8.4.1 EARLY-RESPONDING TISSUES

Early-responding tissues show the effects of radiation damage within a few weeks of being irradiated. Examples of such tissues are skin, oral mucosa, intestines, bone marrow, and testis. Each of these tissues contains functional cells (*epithelial* cells in the case of the surface tissues and *haemopoietic* cells in the case of the marrow) and also connective-tissue cells. The early reactions are due to damage to these parenchymal components which usually have a short functional life-span.

8.4.2 LATE-RESPONDING TISSUES

Late-responding tissues show their response to radiation damage over months to years after exposure. Examples of late-responding tissues are lung, kidney, and spinal cord. Damage to these tissues may often be the result of damage to connective tissues, especially blood vessels. Furthermore, some of the early-responding tissues mentioned above may subsequently show a late response that is either due to direct damage to connective-tissue cells or to connective-tissue damage that arises as a result of a very severe loss of parenchymal cells (the so-called *consequential late effects*).

Whereas damage to early-responding tissues tends to heal, damage to the late-responding tissues tends to be more permanent. Although patients can die from severe early reactions, these are usually transient and can be managed by appropriate treatment and nursing. It is the possibility of irreversible late reactions that is usually perceived by the radiation oncologist to be the main limiting factor on treatment. Therefore, it is these reactions that determine the maximum radiation dose that can be given. For instance, damage to the spinal cord appears as paralysis. This is nonrecoverable and a serious impediment to health, and the radiation oncologist will seek to avoid it at all costs.

The distinction between early- and late-responding tissues has become more important as a result of the recognition that time–dose relationships are systematically different between them (see Section 9.3).

8.5 CONCEPTS OF NORMAL-TISSUE TOLERANCE AND THERAPEUTIC GAIN

Basic to curative radiotherapy is the concept of *treating to tolerance*. As the radiation dose is increased, the effects on the tumour and on normal tissues both increase. The greater the dose of radiation, the greater the likelihood of controlling the tumour. But some types of radiation-induced morbidity are so serious that they must be avoided, and the risk of these also increases with radiation dose. How should the radiation oncologist decide where to stop? In principle, the first step is to decide what level of normal-tissue damage (early or late) the patient can tolerate and the radiation dose is then increased until this level is reached. The level of tumour response will be determined by this maximum dose.

The probability of local tumour control by radiotherapy increases with dose according to a sigmoid relationship (see Figure 8.2a, right hand curve). The probability of normal-tissue damage also increases with dose (see Figure 8.2b, right hand curve). This curve may also be

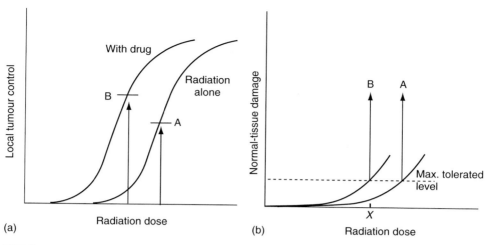

FIGURE 8.2
Illustrating the procedure by which an improvement in therapeutic index might reliably be identified as a result of adding chemotherapy to radiotherapy.

sigmoid in shape, but for serious complications only the bottom part of the curve is of interest. The radiation oncologist must decide the maximum level of complications that can be permitted (the *tolerance* level). This, in turn, fixes the level of tumour control that can be achieved (i.e. point A in Figure 8.2a).

Figure 8.2 also illustrates a very important principle: the concept of *therapeutic index* or *therapeutic gain*. For example, if chemotherapy is added to treatment that is initially by radiotherapy, how can it be known if a significant gain has been made? This is a complicated question that in clinical practice is seldom satisfactorily answered. The complication comes from the fact that chemotherapy, when given with radiotherapy, commonly increases not only the tumour response but also many types of radiation-induced normal-tissue damage. Therefore, an observed increase in tumour control is often at the expense of increased radiation morbidity. How do we know whether the same gain could not also have been obtained just by increasing the radiation dose?

The strategy to deal with this problem is illustrated in Figure 8.2, and it goes as follows: In the patients who are being given the combined (drug+radiation) treatment, a range of radiation doses is explored in order to identify the radiation dose (X) which, when combined with chemotherapy, gives the *same* incidence of treatment complications as chosen for radiotherapy alone (the assumed tolerance level). The expectation shown in Figure 8.2 is that this will lead to a level of tumour control at point B. If it can be statistically shown that point B is higher than point A, then there has been a proven *therapeutic gain* by adding the chemotherapy. Tumour response has been increased without increasing the treatment-induced morbidity (for radiation dose X), and such a gain could *not* have been obtained by merely increasing the radiation dose.

This concept is clear in theoretical terms. Unfortunately it is very difficult to put into practice. In the clinic it is impractical and unethical to explore a range of radiation doses in order to plot out the curves shown in Figure 8.2. This can be, and has been, done in experimental animals where the ethical constraints are not (yet) as stringent. Furthermore, the numbers of patients available for such a study is invariably too small to allow more than a three-arm clinical trial, and the precision that can be expected on the data (i.e. incidence of tumour control and complications) at each dose level is usually too low to allow it to be clearly shown that Point B is different from point A. But if this cannot be done, then there is no way of knowing whether a claimed benefit of adding chemotherapy could not have also been gained just by increasing the radiation dose. In addition, the large variations in tumour parameters

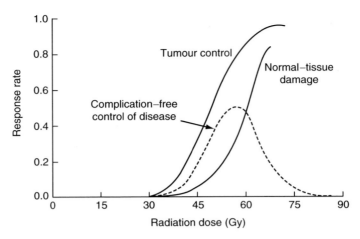

FIGURE 8.3
Illustrating the idea of complication-free tumour control. See text for a warning about this concept.

(volume, site, histology, gene expressions, etc.) of patients recruited to clinical trials will further confound the outcomes. The broader aspects of the combination of radiotherapy and chemotherapy are dealt with in Steel (2002a, b), Stewart and Bartelink (2002) and Wilson et al. (2006).

Figure 8.3 shows a similar type of representation of therapeutic index. From the dose–response curves for tumour control and normal-tissue damage respectively, the *probability of complication-free tumour control* has been calculated as shown by the dashed line. This is a bell-shaped curve against radiation dose. Although this is an interesting notion, it does not provide an appropriate approach for clinical practice: it would be wrong to suppose that the correct treatment policy is to prescribe a radiation dose that is at the peak of the dashed curve because, in some situations, this might be a dose that greatly exceeds an acceptable level of normal-tissue damage. Furthermore, calculation in this way assumes that the therapeutic *value* of tumour control can be equated to the *penalty* of a complication. There is no practical basis for such an assumption, and it is better to recognise that a key part of the judgement of a radiation oncologist in any individual case is to balance the risks and severity of radiation responses, both in the tumour and in the surrounding normal tissues.

8.6 STEEPNESS OF DOSE–RESPONSE CURVES

As previously demonstrated, if the total dose given to a patient is increased, it is expected that both local tumour control and normal-tissue damage will increase. Whether there will be a therapeutic advantage in doing this depends on the steepness of the dose–response curves for these two endpoints. Dose–response curves are, therefore, fundamental to clinical radiobiology.

Dose–response curves are usually sigmoid in shape (see Figure 8.3). At a low dose, such a curve is shallow and there is only a slow increase in the effect with dose. At a high dose, the same is also true. The steepness of the curve reaches a maximum at the point of 50% response. The theory and significance of dose–response curves has been further described by Brahme et al. (1988) and Bentzen (2002).

8.6.1 BASIC PROPERTIES OF DOSE–RESPONSE CURVES

In an ideal model situation where there is no variability between tumours or patients and no host response against residual cancer cells, tumour control would be achieved when the last

TABLE 8.1

Steepness of Dose–Response Curves from Studies in Clinical Radiotherapy

Normal-Tissue Damage		Tumour Control	
Type of Damage	γ_{50}	Tumour Type	γ_{37}
Frozen shoulder	4.8	Head and neck	2.8
Larynx oedema	4.3	Larynx	1.7–2.8
Rectal complications	0.4	Nasopharynx	1.6–1.8
Subcutaneous fibrosis	3.7	Neck nodes	1.4
Telangiectasia	2.2–3.5	Base of tongue	0.5, 0.9
		Supraglottic larynx	0.6, 0.8

Steepness is indicated by γ-values (see text). Ranges show the range of mean values from a number of clinical studies. From Bentzen (1997), where error estimates and original references will be found.

clonogenic cell is sterilized. The probability of cure will then be based on Poisson statistics (see Equation 36.5). If m is the mean number of clonogenic cells that survive the treatment, then the tumour control probability (TCP) is given by the zero-order term of a Poisson series:

$$TCP = e^{-m} \tag{8.1}$$

If there is a cell-survival curve that is accurate down to such small survival levels, then how m depends on dose is known and a TCP vs. dose curve can be calculated (see Section 36.2.1.2 for a detailed explanation of how this can be done). Such a curve is steep, but not infinitely steep. A similar argument applies to dose–response curves for normal-tissue damage where the endpoint is a reduction in the number of stem cells (or, more accurately, tissue-rescuing units) below a critical threshold.

The conventional way of expressing the steepness of a dose–response curve is to indicate, at the steepest point of the curve, the *percentage increase in effect for 1% increase in dose*. This dimensionless quantity is called the γ-value (Brahme et al. 1988). To be more precise, γ_{50} is the steepness at the 50% effect level, γ_{37} at the 37% level, and the appropriate choice between these depends on the form of the dose–response curve. Under the ideal circumstances just described, the maximum value for γ is around 7. Some values calculated from actual clinical data are shown in Table 8.1. For the head and neck tumours, γ_{37} ranges from 0.5 to 2.8. For various types of normal-tissue complication, the γ_{50} values range from below 1 to 4.8. Of course, these values are subject to various types of experimental error, but the following conclusions can be drawn from the body of data so far accumulated:

- Dose–response curves for complications of radiotherapy are often very steep with γ-values exceeding 3. There is a tendency for γ-values to be higher, and for the curves to be steeper, for *late* responses than for damage to *early*-responding tissues.
- The steepness of tumour control curves is often less than that for late complications.

8.6.2 What Determines the Steepness of Dose–Response Curves?

In practice, a tumour control curve has to be obtained by recording success or failure in a large number of patients treated with different radiation doses. Such a clinical study cannot, for ethical reasons, be planned to obtain these data in a well controlled way; the data usually have to be accumulated from trials with different doses performed over a clinical *learning period*. Therefore, the results are subject to the sources of error that always accompany such noncontrolled trials.

More importantly, inter-patient and inter-treatment variation will always tend to make the dose–response curve shallower[*]. If, for example, there is a mixture of sensitive and more resistant tumours, then the sensitive subgroup will have a dose–response curve that lies to the left of that for the more resistant tumours. The overall dose–response curve will, therefore, be a composite of steep curves at different points on the dose scale, and it will be flatter than any single one of them. Factors which may vary from one patient to another and which will contribute to such flattening include the quality of dose delivery and variation in various biological factors: cellular radiosensitivity, tumour hypoxia, speed of repopulation, etc. There is a growing realisation that these biological factors may, to a greater or lesser extent, be under genetic control and that, by genetic testing, it might be possible to identify sensitive and resistant subgroups of patients. If this could be done, then the dose–response curve for each subgroup would be steeper than for the patient group as a whole and there might be the possibility of improved therapeutic management.

The value of γ, either for tumour or treatment complications, also depends on whether, it is the *fraction number* or the *dose per fraction* that is kept constant as the dose is increased (Bentzen 2002). If it is the number of fractions then as the total dose increases so does the dose per fraction, and because *biological effectiveness* increases with dose per fraction (see Section 9.3), this will make the dose–response curve steeper. This is a manifestation of what has come to be called *double trouble*.

8.7 RADIATION PATHOLOGY

Each major tissue of the body shows a characteristic series of reactions to radiation exposure. They may be classified as stochastic or nonstochastic (also called deterministic, see Section 59.1.1).

8.7.1 STOCHASTIC EFFECTS

Stochastic effects are those for which the *incidence*, but not the *severity*, of damage increases with radiation dose. A prime example is radiation carcinogenesis, i.e. the induction of second tumours by radiation. The key biological event is neoplastic transformation, the probability of which is radiation-dose dependent, but the eventual development of a tumour is an *all-or-nothing* event. The induction of tumours by radiation was long thought to follow a bell-shaped curve against dose because the probabilities of transformation and the killing of transformed cells both increase in a nonlinear way with radiation dose. High radiation doses produce more transformants, but even more of these will be killed, and the actual frequency of viable transformants was thought to eventually decrease with increasing dose. However, Sachs and Brenner (2005) have put forward an alternative hypothesis which predicts a slow continuous increase with dose at high doses.

8.7.2 NONSTOCHASTIC EFFECTS

Nonstochastic effects are the more common and more clinically important effects, in which the *severity* of damage increases with radiation dose. Whether the *incidence* also increases with dose depends on how the data are presented. A common practice is to count the number of cases where the severity of a reaction exceeds a defined level. When this is done, it will usually be found that the incidence (e.g. of grade-3 effects) increases with dose for such nonstochastic effects. The following are examples of nonstochastic effects in three selected tissues.

---

[*] The σ_{α} parameter was introduced by Webb and Nahum (1993) into their TCP model precisely in order to account for such *interpatient* differences in radiosensitivity (Section 36.2.1.2).

8.7.2.1 Skin and Mucosae

Effects in the skin are easily visible to the eye, both to the doctor and to the patient. Therefore these attract attention. As radiation dose is increased, the first evidence of damage is reddening (*erythema*), shortly followed by scaling (*dry desquamation*), and then by weeping (*moist desquamation*). After high radiation doses, the skin may finally break down in *necrosis*. The mucosal lining of the mouth cavity is a stratified epithelium, like the skin, and damage to this mucosal lining is a significant limiting factor in head-and-neck radiotherapy. Skin damage begins to appear at the end of the second week of radiotherapy, and reaches a peak between four and eight weeks after the start of treatment (see Figure 8.4a). This is an *early reaction* as described in Section 8.4. Skin damage will eventually heal unless it is very severe, but it may be followed by more persistent types of late damage. The skin may become hard and inflexible (i.e. fibrosis), and there may be long-term damage to blood vessels as evidenced by the visible and disfiguring appearance of *telangiectasia*.

8.7.2.2 Lung

The lungs are among the most radiosensitive of anatomical structures and are essential to life. The first evidence of damage is *pneumonitis*, which appears three to six months after irradiation. Lung function is impaired in a dose-dependent manner to an extent that may

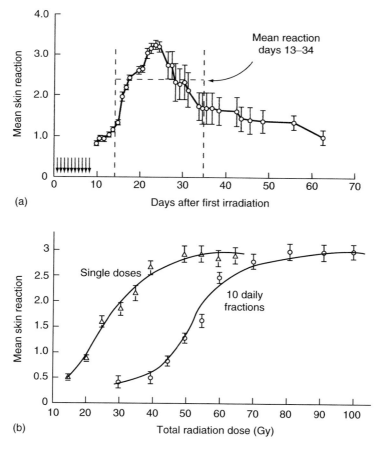

(a)

(b)

FIGURE 8.4

(a) The time-course of radiation-induced skin damage in mice irradiated with ten fractions of 6 Gy. (From Brown, J. M., Goffinet, D. R., Cleaver, J. E., and Kallman, R. F., *J. Natl. Cancer Inst.*, 47, 75–89, 1971.) (b) Dose–response curves calculated from such data as in panel A for irradiation with a single dose or with ten fractions.

lead to death. Patients who survive this first phase may suffer a second phase of damage a year or more after irradiation. This is due to a progressive and irreversible fibrosis of the lung that also manifests itself in breathing problems.

8.7.2.3 Brain and Spinal Cord

These are tissues which have a very low rate of cell turnover, and damage is only manifested many months to years after irradiation. In less severe cases there may be neurological effects that eventually resolve, but above a critical dose threshold the damage progresses to permanent radiation myelopathy or necrosis. In the cord this presents as paralysis, and the severity of this effect on the patient is such that radiation oncologists take great care to reduce the risk of it happening to below 1%.

8.8 QUANTIFICATION OF NORMAL-TISSUE DAMAGE

Radiobiologists have developed a wide range of techniques for quantifying the effects of radiation on normal tissues. These include visual scoring methods, assays of tissue function, and stem-cell cloning techniques; most have been developed in the context of laboratory studies on mice or rats, but some are applicable to humans.

8.8.1 VISUAL SCORING METHODS

Each manifestation of damage to a tissue such as the skin can be scored, by eye, using an arbitrary scale. For instance, for the skin one may set erythema as 1.0 and necrosis as 4.0 with intermediate scores for other types of damage. Figure 8.4a illustrates effects on mouse skin measured in this way. The ordinate is the average reaction score obtained in a group of mice given a dose of 60 Gy in 10 daily fractions. The reaction starts around days 7–10 (probably days 14–17 in humans); it rises to a peak after 4 weeks and then declines slowly over the next couple of months. During this *wave* of response, the character of the skin reaction changes: erythema is an early factor, and necrosis, if reached, comes in later and takes a long time to heal. Radiobiologists have quantified the effects on the skin by calculating the area under such a curve within a defined time interval (13–34 days in this example). By performing experiments at various radiation doses it is then possible to build up a *dose–response curve*. Figure 8.4b shows two such curves: for single-fraction and multiple-fraction irradiation. These curves are sigmoid in shape, and they clearly show the sparing effect of fractionation. Visual scoring may seem 'low-tech' and of dubious precision, but it has proved to be a valuable method which has made an important contribution to normal-tissue radiobiology.

8.8.2 ASSAYS OF TISSUE FUNCTION

In tissues where a relevant assay of tissue function is available, this may be a valuable endpoint for radiobiological studies. The lung is a good example. During the expression of radiation pneumonitis there is a decrease in lung function that can be detected by standard clinical lung-function tests. In mice, there is a rise in breathing rate above the normally very fast rate of around 350 breaths per minute. This can be detected in an enclosed chamber by a microphone, displayed as breathing rate on a pen recorder, and quantified electronically. A wide range of radiobiological studies has been made in this way.

The kidney is another example of a tissue where physiological function can easily be measured. A common technique is to study the excretion rate of a radioactively labelled tracer, such as ^{51}Cr-labelled EDTA (ethylenediaminetetraacetic acid).

Damage to the bladder has been documented by measuring urination frequency. Mice are placed in a cage that has a grid base; paper towels on a roller are automatically passed at a constant rate under the cage, and urine spots on the towel are counted or sized. Damage to the large intestine and rectum has been documented by measurements of faecal size, shape, and composition. The ingenuity of radiobiologists to invent clever and sensitive (but unsavoury) endpoints of normal-tissue damage knows no bounds.

8.8.3   STEM-CELL CLONING TECHNIQUES

In experimental animals, it is possible to look deeper into the biology of damaged tissues. For the skin, intestine and a few other tissues methods have been developed which allow colony formation by stem cells (Section 8.3) to be detected. Many early developments in normal-tissue radiobiology were made using these methods. For some tissues, it is possible to transplant irradiated cells into genetically identical recipient animals as a way of studying the radiosensitivity of stem cells. Best-known are the applications to blood-forming tissues. In classic studies on mouse bone marrow, Till, McCulloch, and their colleagues in the Ontario Cancer Institute intravenously transplanted marrow cells from test-irradiated animals into mice whose endogenous marrow had been destroyed by whole-body irradiation (Till and McCulloch 1961). The transplanted marrow stem cells grew to form colonies in the spleen that could be counted by eye. The number of colonies exponentially decreased with the radiation dose that the test cells had received, and it was possible to conclude that the mean inactivation dose for marrow stem cells was around 1 Gy.

An example of cloning studies on the mouse small intestine is shown in Figure 8.5. Hornsey (see Alper 1973) irradiated mice under a variety of conditions and measured both the effects on the whole animal (i.e. lethality) as well as the survival of intestinal stem cells using a colony assay.

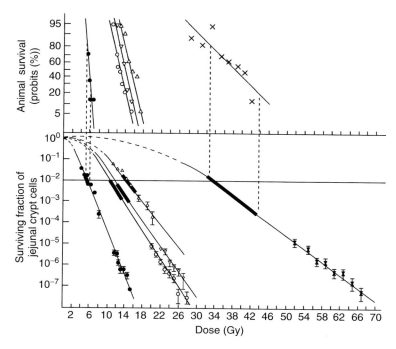

FIGURE 8.5
Survival of mice five days after irradiation (upper panel) matched with survival curves for intestinal stem-cells. From left to right, the irradiations were with neutrons, electrons (high and low dose rate), x-rays, and electrons under hypoxic conditions. (Reproduced from Alper, T., *Br. Med. Bull.*, 29, 3–6, 1973. With permission.)

The figure shows a remarkable observation: for each condition of irradiation, the radiation dose that caused around 50% lethality (upper panel) reduced the survival of intestinal stem cells to around 10^{-2} (i.e. 99% cell kill). This not only shows that the stem-cell assay produces realistic estimates of radiation damage, but it also indicates that, as a general rule, the intestine can tolerate the loss of no more than 99% of its stem cells. Less damage than this can be made up by rapid repopulation after irradiation; greater damage will lead to intestinal failure. As indicated in Section 7.4.3, this observation illustrates the predicament of the radiation oncologist: to achieve tumour control requires perhaps nine or ten decades of killing of tumour cells, but this must be achieved without exceeding around two decades in the critical normal tissues.

8.9 THE VOLUME EFFECT

8.9.1 INTRODUCTION

The severity of radiation-induced normal-tissue damage depends, to a very large extent, on the volume of tissue irradiated. *Volume effects* differ among the various organs of the body, depending on their structural organisation (see left-hand side of Figure 8.6) and on the migration characteristics of the surviving stem cells. Some tissues, in spite of having cells that are inherently radiosensitive, may nevertheless have a large volume effect and, therefore, are resistant to partial irradiation. Examples of this are the kidney or lung, both of which are among the most sensitive organs when the whole of the organ is irradiated but where partial volumes can be treated to far higher radiation doses. This occurs because there is considerable reserve capacity in these organs, and only one quarter to one third of the functional organ volume is required to sustain life under normal physiological conditions. In contrast, other more tube-like structures such as the spinal cord have a different volume effect: inactivation of a short segment causes loss of function of the whole organ (i.e. paralysis). Such tissues are sometimes said to have a *serial* organisation (lung and kidney would be described as having a *parallel* organisation).

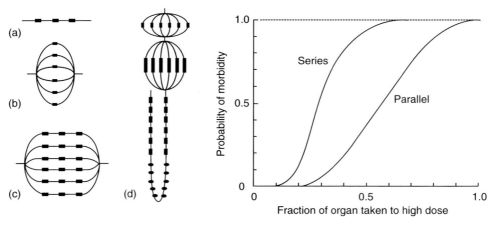

FIGURE 8.6
Clinical benefit from a reduction in field size may depend upon how morbidity increases with field size. The *left-hand* picture illustrates the concepts of (a) series (or serial), e.g. the spinal cord; (b) parallel, e.g. the lungs; (c) serial-parallel, e.g. the heart; and (d) a combination of parallel and serial structures, e.g. a nephron (From ICRU, *Prescribing, Recording and Reporting Photon Beam Therapy*, Report 62, ICRU, Bethesda, MD, 1999). The form of the complication probability vs. partial high-dose volume relationship can be expected to differ between series- and parallel-type tissues as indicated in the right-hand picture.

Intermediate between the *parallel* or *serial* organisation types are tissues composed of specialised elements, each carrying out very specific functions such as parts of the brain. Radiation damage even to a small area of this tissue leads to a permanent deficit for that particular function since the uninjured components may be unable to take over the functions of the affected elements. As a consequence, the *tolerance* dose of such tissues is only influenced to a small extent by the volume of the tissue irradiated. However, irradiating a larger volume may result in a more severe functional failure because of the elimination of a larger number of functional elements.

The magnitude of the volume effect is of critical importance to the development and success of *conformal radiotherapy* (see Chapter 43 and Part I generally). If there is a large volume effect, then reducing the volume of normal tissue taken to a high radiation dose should allow a considerable escalation in dose with the expectation of an improved tumour response. If the volume effect is low, then there is little to be gained in this way. Models for normal-tissue complication (see Section 36.3) must accurately reflect the volume effects of the various organs if they are to make reliable predictions of complication probability for arbitrary dose distributions.

8.9.2 BASIS OF THE VOLUME EFFECT

In spite of considerable attention by researchers, the biological basis for the volume effect is still unclear in most tissues. A major limitation has been the fact that most laboratory research has been on small animals (rats and mice) in which clinical-sized radiation fields cannot be used, and there is a serious scaling problem. A mouse is roughly 1/20 the *height* of a man: can it be assumed that changing a field diameter from 5 mm to 10 mm in the mouse is equivalent to changing from 10 cm to 20 cm in the clinic? Such scaling is unlikely to be valid because one component of the volume effect may be cell migration to replace damaged cells, and although this could occur faster in mice, it is unlikely to be twenty times faster. There may be other (e.g. vascular) factors that also do not scale with body size. The *fractional* volume of an organ that is irradiated may be important, but this also may not scale well over a wide range of body sizes.

Figure 8.7 shows data on the volume effect in pig skin (Hopewell 1986). The pig was chosen for these studies because the structure of its skin is similar to that of man. Strontium-90

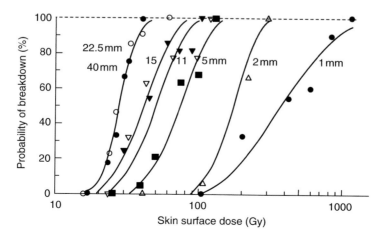

FIGURE 8.7

The volume effect in pig skin. The skin was irradiated with circular ^{90}Sr sources ranging in diameter from 1.0 to 40 mm. The percentage of skin fields showing moist desquamation is shown as a function of the skin surface dose. (From Hopewell, J. W., *Br. J. Radiol.* (Suppl. 19), 39–51, 1986. With permission.)

plaques of differing diameter were placed on the skin, and dose–response curves were generated as shown. As the plaque diameter was decreased from 40 mm to 2 mm, the skin dose that was required to produce moist desquamation increased from around 26 Gy to over 150 Gy. This indicates that there is a very large volume effect in skin over this small range of field sizes. Such an effect may well be due to the ability of undamaged epithelial cells to migrate into the irradiated area from its periphery and repopulate it, an effect that will clearly be more effective over millimetre than centimetre distances. Changing the field size in the clinical situation from 20 cm to 15 cm may show a volume effect, but *ingrowth* from the edges of the field could be a minor factor and is probably ineffective at these large field sizes. More important could be the discomfort, pain, and potential for infection of a large radiation burn compared with a smaller one. In other tissues such as the spinal cord, marked volume effects have also been seen in small animals over millimetre dimensions (Hopewell et al. 1987).

The basis of the volume effect may differ from one normal tissue to another, and in a body region such as the pelvis, a number of different organs will be involved. A basic characteristic may be how the probability of morbidity depends upon the *fraction* of the organ that is irradiated (Figure 8.6). *Series*-type tissues (e.g. spinal cord) will show a steeper curve in this type of plot than in *parallel*-type tissues (e.g. lung), and for the latter, there could be greater benefit from volume reduction. However the concept of *morbidity*, as used here, will depend not only on measurable biological factors but also on patient perception. The reader should refer to Section 36.3 on modelling normal-tissue complications where attempts to formulate a mathematical relationship between dose-volume and complication probability for different organs are described.

CHAPTER 9

DOSE FRACTIONATION IN RADIOTHERAPY

*Gordon Steel**

CONTENTS

---

* With contributions by Alan Nahum.

9.1 WHY WE FRACTIONATE IN RADIOTHERAPY

During the last twenty years, important new insights have been gained into the principles of dose fractionation in radiotherapy. The linear–quadratic (LQ) approach (see Section 7.13) provides a new understanding of why fractionation works and how it may be optimised; this approach has now become the standard method of calculation in modern radiotherapy departments, and it will be described in detail below. The *practical* basis of fractionation was established many years ago, well before the introduction of the LQ approach, through the skill, observation, and wisdom of radiotherapists in France, Austria, and other countries. These historical aspects have been well described by Thames and Hendry (1987), Bentzen (2002) and Fowler (2006).

The most direct reason why fractionation is routinely used in clinical radiotherapy is that it was *empirically* established. The early radiation oncologists began using single-dose treatments, but they soon realised that giving daily doses over a period of weeks also resulted in good tumour control with less severe side effects. In retrospect, this must have occurred because the normal-tissue reactions that limit radiation therapy have a greater recovery or repair capacity than most tumours.

Since those early days there have been some decades of intensive laboratory and clinical research into the scientific basis of radiation fractionation. This is a key area in clinical radiotherapy, and there is evidence that the optimisation of time–dose relationships is crucial to the achievement of a good clinical outcome.

9.2 HISTORICAL APPROACHES TO FRACTIONATION

9.2.1 THE APPROACHES OF STRANDQVIST AND ELLIS

The attempt to provide a quantitative basis for changes in dose fractionation began with the work of Strandqvist and Cohen (among others) in the 1940s. They documented their success or failure in controlling skin tumours by radiotherapy and recorded the incidence of skin necrosis. They found that the total dose required for tumour control increased with the number of fractions. At that time, it was the custom to treat with five daily doses per week, and it was not possible to separate the effects of time from the number of fractions. For individual patients with skin cancer, Strandqvist plotted, using double-logarithmic coordinates, the total radiation dose against the overall treatment time. Using different graph symbols he recorded the outcome in terms of local control or skin necrosis. He was able to draw straight lines that broadly divided the two types of symbol: these lines sloped upward (total dose increasing with overall time), and they are often referred to as *Strandqvist curves* (Strandqvist 1944). This was the first quantitative demonstration of recovery in tumours and normal tissues. Strandqvist curves are early examples of *isoeffect* curves which have an important place in thinking about radiotherapy optimisation. There was some uncertainty about the slope of the Strandqvist curves; Strandqvist believed it was 0.22, but with additional data, Cohen (1952) found a value of 0.33, giving rise to what was known as the *cube root law*. If a radiation oncologist wished to change the overall treatment time, he could read off from the Strandqvist curve a new total dose that should give the same treatment outcome.

Working in Oxford, Ellis (1969) developed the hypothesis that data such as those used by Strandqvist and Cohen could be summarised in a simple formula. Their straight lines on logarithmic coordinates implied a power-law relationship between total dose and overall time for a constant level of effect. Although accurate data on the relative roles of fraction number (N) and overall treatment time (T) were not available, Ellis used his intuition to separate the exponents for N and T as follows:

$$D = NSD \times N^{0.24} \times T^{0.11} \tag{9.1}$$

where D is the (isoeffective) total dose, and NSD (the *nominal standard dose*) determines the *intensity* of treatment and is assumed constant for any particular clinical situation. Note that Ellis had divided the 0.33 slope of Cohen into two exponents that roughly give the same result for the case of daily treatments. He included an adjustment to allow for daily (i.e. five) treatments per week.

The NSD formula was used for many years as the standard method for adjusting total dose in response to a change in fractionation, but it is an astonishing thought that its scientific basis was very weak. The original data that the slopes of Strandqvist curves were determined from showed considerable scatter, and there was uncertainty in how to represent single-dose data on a logarithmic scale of time. The original data related only to the treatment of skin tumours, but it later began to be used more widely in the absence of adequate alternative data. It gradually became clear that the formula only worked for certain well-defined conditions and within a limited range of fraction numbers. Examples of poor therapeutic results began to accumulate, especially when a fraction number greater than approximately 30 or less than approximately 10 was chosen. Detailed animal studies were performed which showed that a power-law relationship was usually not satisfactory in representing the data, and the slopes sometimes differed sustantially from those in the original NSD formula.

As a result, efforts were made to adjust the NSD formula. Variants of it (known as CRE and TDF) were introduced and widely used, and these may have been easier to use and safer. Around 1980, it began to be appreciated that the basic power-law relationship was inappropriate to represent time–dose relationships in radiotherapy. Furthermore, there was a pressing need to tailor the formulae to individual clinical situations. A critique of the Ellis formula and the background to these developments have been described by Bentzen (2002).

9.2.2 THE CONTRIBUTION OF EXPERIMENTAL RADIOBIOLOGY

Ethical and logistical considerations inevitably preclude clinical experiments exploring a wide range of fractionation regimes. This has led to extensive research using laboratory animals which has borne fruit in improved clinical dose schedules. Of course, laboratory animals have limitations mainly because the time-scale of tumour and normal-tissue effects in rodents is usually much shorter than in man.

An important observation in animals was that Strandqvist curves are usually not straight. Good examples of this are shown in Figure 9.1. In mouse skin, it is possible to make

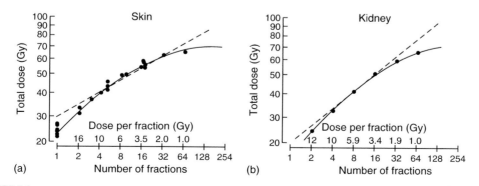

FIGURE 9.1
Isoeffect curves in the skin and kidney of the mouse. (a) Acute skin reactions (Douglas, B. G. and Fowler, J. F., *Radiat. Res.*, 66, 401–426, 1976.); (b) late injury in the kidney (Stewart et al., *Radiat. Res.*, 98, 407–420, 1984). Note that the relationship for kidney is steeper than for skin. The broken lines are NSD formulae fitted to the central part of each data set. The solid lines show the LQ model, from which the guide to the dose per fraction has been calculated.

fractionation experiments covering fraction numbers from 1 to 64, all delivered within eight days. Dose-response experiments were performed for each schedule, and this allowed the total radiation dose that gave an arbitrarily chosen isoeffect to be identified (clearly an experiment that is impracticable in the clinic). These are the doses that are plotted in Figure 9.1a. The data define an isoeffect curve that is convex upwards. As indicated in the figure, the data falls below the NSD line both for small and large fraction numbers. Why the NSD formula was unsafe in these conditions is immediately seen: the formula predicts a higher isoeffective dose than is actually tolerable, and treatment with such doses would lead to increased normal-tissue damage.

Figure 9.1b shows the results of similar experiments in the kidney, a late-responding tissue. Here again, for each fraction number, dose-response experiments were performed that allowed the total radiation dose for a selected isoeffect to be read off. These were large experiments that have produced very precise data. Once again, the isoeffect curve is clearly convex upwards.

9.3 THE FRACTIONATION RESPONSE OF EARLY- AND LATE-RESPONDING TISSUES

A major development in thinking about fractionation was the realisation by Thames et al. (1982) that in fractionation studies in mice, there is a systematic difference between *early*-responding and *late*-responding normal tissues (see Section 8.4). They summarised published isoeffect curves for fractionation effects on normal tissues in mice, considering only those studies where the overall time was kept short in order to minimise the effects of repopulation on the data. Their key result is shown in Figure 9.2. This shows a collection of isoeffect curves for various tissues as a function of the dose-per-fraction plotted from *right* to *left* (note that fraction number increases from left to right along the abscissa as in the case of the Strandqvist

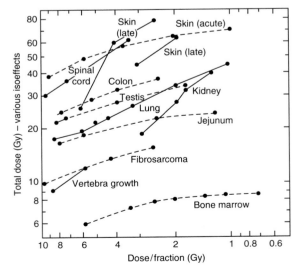

FIGURE 9.2
A collection of isoeffect curves for damage to normal tissues of the mouse following fractionated radiation treatment. Number of fractions increases to the right along the abscissa; dose per fraction, therefore, decreases. Full lines are for late-responding tissues; dashed lines for early-responding tissues. (From Thames, H. D., et al., *Int. J. Radiat. Oncol. Biol. Phys.*, 8, 219–226, 1982.)

curves). Many of these lines show evidence for the upward convexity shown by the data in Figure 9.1. But the most significant observation by the authors was that there is a systematic tendency for late-responding tissues to have *steeper* isoeffect lines than the early-responding tissues (i.e. effects on late-responding tissues are more sensitive to changes in fraction size). The position of the lines on the vertical scale covers a wide range of dose and this reflects two main factors. First, some tissues are more radiosensitive than others for any chosen dose per fraction; second, for each tissue studied, an isoeffect has been chosen, and the level of cell kill necessary to produce this effect differs among the tissues.

The collection of data shown in Figure 9.2 was discussed by Thames et al. (1990) in terms of the linear-quadratic model of cell killing (see Section 7.13). They went on to show that the form of isoeffect relationships for normal tissues is usually consistent with this model and that the difference in steepness of isoeffect curves between early- and late-responding tissues can be described by a single parameter: the α/β ratio. Their approach is described in what follows.

9.3.1   THE LINEAR–QUADRATIC APPROACH TO FRACTIONATION

It is believed that the effect (E) of a single radiation dose (d) is given by

$$E = \alpha d + \beta d^2 \qquad (9.2)$$

This is a linear-quadratic equation that can be considered to derive from a cell-survival relationship of the form: surviving fraction, $S = \exp(-\alpha d - \beta d^2)$, i.e. identical to Equation 7.2, where *log cell kill* is regarded as the *effect* of irradiation, i.e. $E = -\log_e[S]$.

For n fractions

$$E = n(\alpha d + \beta d^2) = \alpha D + \beta dD$$

where the total dose $D = nd$. This can be rearranged in the form

$$E/\alpha = D\left(1 + \frac{d}{\alpha/\beta}\right) \qquad (9.3)$$

E/α is a quantity that has the units of dose (E is dimensionless, and α has units Gy^{-1}). Note that in the limit $d \to 0$, $E/\alpha \to D$. Therefore, E/α can be described as the *Extrapolated Response Dose* (i.e. the dose that would produce the effect E when given in many small-dose fractions).

Fowler (1989a) has suggested that E/α be called the *Biologically Effective Dose* (BED). It is a measure of effect (E), in dose units, and for a given biological tissue (where α and β are constant). BED values indicate how much *damage* a particular fractionation regime will do; it can be written as

$$BED = D \times RE \qquad (9.4)$$

where BED is the total dose which, if given in *infinitely small fractions*, is equivalent to the actual fractionated regimen with fraction size d and total dose D, and

$$RE \text{ is the } Relative\ Effectiveness = \left(1 + \frac{d}{\alpha/\beta}\right)$$

Consider now an isoeffect calculation where it is asked: for a given tissue, if the dose per fraction is changed, how must the total dose be adjusted? It is presumed that a reference schedule is used where the total dose is D_{ref} and the dose per fraction is d_{ref}. For *isoeffect*,

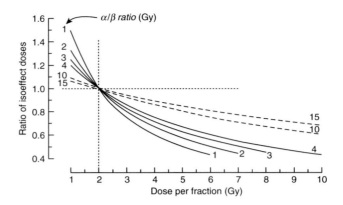

FIGURE 9.3
Compared with a reference treatment using 2 Gy per fraction, the diagram shows how the total dose must be changed in order to maintain a constant level of effect when dose per fraction is modified. Full lines are for low α/β ratios (as in late-responding normal tissues), and broken lines are for early-responding normal tissues or for most tumours.

E (Equation 9.3) is a constant, as are α and β for a given tissue type and given endpoint. Therefore

$$D\left(1 + \frac{d}{\alpha/\beta}\right) = D_{\mathrm{ref}}\left(1 + \frac{d_{\mathrm{ref}}}{\alpha/\beta}\right)$$

where D is the new total dose. The resulting simple *isoeffect* relationship:

$$\frac{D}{D_{\mathrm{ref}}} = \frac{d_{\mathrm{ref}} + (\alpha/\beta)}{d + (\alpha/\beta)} \tag{9.5}$$

was first proposed by Withers et al. (1983); the form of this relationship is shown in Figure 9.3*.

9.3.2 RATIONALE OF THE LQ APPROACH

The rationale underlying the linear-quadratic approach to fractionation can be stated as follows. Normal tissues and tumours respond differently to a change in dose fractionation. In part, this may be due to differences in repopulation, but if this is ruled out by suitable choice of overall treatment times, such differences probably arise largely from different intrinsic radio-sensitivities related to different α-components. These differences can be seen in data of the type shown in Figure 9.2. As can be seen from Equation 9.4, they can be described in terms of a single parameter: the α/β ratio. Late-responding normal tissues show greater changes in sensitivity in response to a change in dose per fraction than do early-responding tissues (i.e. they have a higher *fractionation sensitivity*), and this is consistent with a lower α/β ratio (or lower α-coefficients). The survival curves for target cells in early- and late-responding normal tissues can be envisaged to have systematic differences in shape as illustrated in Figure 9.4. The α-value determines the initial slope of this curve, whereas, β determines the degree of curvature; a lower α/β ratio means a more *bendy* curve.

---

* It should be noted that the use of the α and β parameters in these analyses is far removed from their original intent to define specific mechanisms of cell killing for homogeneous cell populations (Section 7.13) though it may be possible to identify the effective value of α/β with certain 'target' cells in the normal tissue in question.

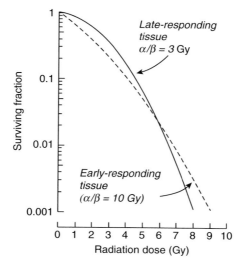

FIGURE 9.4
Cell survival curves for the putative target cells in early- and late-responding normal tissues. A lower α/β ratio produces a more *bendy* curve.

In the years since the original publication by Thames et al. (1982), a considerable literature has appeared, making it is clear that the linear-quadratic approach to dose fractionation is more reliable than the NSD approach (see Section 9.2.1). It must be emphasised that NSD should no longer be used and should be regarded as being of historic interest only. Values for the α/β ratio have been obtained for many tissues in experimental animals, some also in man, and values are given in Table 9.1 (Bentzen and Baumann 2002; Joiner and Bentzen 2002).

There is a general tendency for the α/β ratio to be high in early-responding tissues and to be low in late-responding tissues. Fowler (1989a) has expressed the view that because the

TABLE 9.1

Some Examples of α/β Ratios for Human Normal Tissues and Tumours

Tissue reaction	α/β ratio (Gy)
Early reactions	
Skin desquamation	~11
Mucosa	8–15
Late reactions	
Spinal cord	<3.3
Brachial plexopathy	~2–3
Skin, telangiectasia	2.5–3
Skin, subcutaneous fibrosis	~1.7
Lung, pneumonitis	3.3
Tumours:	
Head and neck	7–16
Skin	~8.5
Melanoma, liposarcoma	~0.5

Source: From Bentzen, S. M. and Baumann, M., in *Basic Clinical Radiobiology,* Steel, G. G. Ed., 3rd ed., Arnold, London, pp. 134–146, 2002. With permission.

uncertainties in α/β ratios are considerable and because the values for some tissues have not been shown to be different from those of others, for the time being, it is logical to assume standard values for α/β of 3 and 10 Gy for most late-responding and early-responding tissues, respectively. A prudent approach to these uncertainties is to make calculations for a range of α/β ratios, seeking to get a feel for what effect the uncertainty in α/β has on dose prescription. And if the variation in β-kill is much less than that for α-kill (as shown for tumour cells in Figure 7.12), then research should be directed towards more clearly defining and then manipulating the underlying mechanisms of this α-kill process (Chapman 1980, 2003).

Data on human tumours is sparse, but it is generally believed that most tumours resemble early-responding normal tissues in their fractionation sensitivity, thus having high α/β ratios. However, there are exceptions to this general rule: melanomas, some sarcomas, and possibly prostate tumours (Brenner and Hall 1999; Fowler et al. 2001; Brenner 2003) seem to have very low α/β ratios; although, this low prostate α/β has been challenged by Nahum et al. (2003) and others. The basis of the Nahum et al. case is that prostate tumours are not infrequently hypoxic (Movsas et al. 2002; Parker et al. 2004), and it is this, rather than low α/β, that explains the approximately equal biochemical control rates for external-beam radiotherapy and low dose-rate brachytherapy (for which β can be set approximately equal to zero; see Section 9.7 and also Chapter 55). In general it is prudent to be cautious in areas where there is insufficient clinical data or where these data are open to more than one interpretation.

9.3.3 HYPOFRACTIONATION

Hypofractionation means the use of a *reduced* number of fractions, or a *larger* dose per fraction. Equation 9.5, illustrated in Figure 9.3, indicates that if the dose per fraction is *increased* above the reference value of 2 Gy, then the total dose (for an isoeffect) must be *reduced* as indicated by the appropriate line in this chart. Many clinical studies have confirmed that such a dose reduction is required. The curves for late reactions (low α/β ratio) are steeper than for early reactions or for tumours (high α/β ratio). As an example, if the dose per fraction is increased to 5 Gy and the α/β ratio for late reactions is 3 Gy, then the total dose must be reduced to roughly 0.625 of its reference value (i.e. by 37.5%). If the tumour has an α/β ratio of 10 Gy, then at 5 Gy per fraction, the isoeffective total dose, say 70 Gy, should only be reduced by 20% to 56 Gy. A reduction by 37.5% is underdosing the tumour by 17.5% and this will almost certainly result in a significant loss in local control. The general rule is that the use of a large dose per fraction is radiobiologically sub-optimal (accepting that α/β is appreciably lower for late complications than for tumour–cell kill). Such large fraction sizes are widely used in palliative radiotherapy where factors of cost and convenience may predominate over radiobiological principles.

However, the use of large fraction sizes in situations other than palliative is on the increase, despite the above. So-called *extra-cranial* or *body* stereotaxy, with lung tumours often as the target, is one such area (Blomgren et al. 1995; Shiu et al. 2003 and Section 45.13); a summary of this work is given by Fowler et al. (2004b), Grau et al. (2006) and SBRT (2006). Shiu et al. (2003) treated paraspinal metastases in 5 fractions to 30 Gy, stating that their superior dose distribution (7–9 intensity-modulated beams) under CT image guidance was responsible for limiting the spinal cord dose to less than 10 Gy. The use of fraction sizes larger than 2 Gy is under intensive exploration also for treating the breast (Ash et al. 1995; Shelley et al. 2000; Whelan et al. 2002). The FAST trial has been launched in the United Kingdom (Yarnold 2004) and will test five 5.7 Gy or 6.0 Gy fractions to total doses of 28.5 and 30 Gy, respectively, in a total duration of either 35 or 15 days. The control arm in this prospective randomised clinical trial is 50.0 Gy in 25 fractions of 2.0 Gy in 35 days. The FAST trial is based on a recently completed trial that showed no difference in tumour control or normal tissue complications between 39 Gy in either 3 or 3.3 Gy fractions and 50 Gy in 25×2 Gy fractions, which corresponds to $\alpha/\beta = 4.0$ Gy for tumour control (Owen et al. 2006).

The contention that the α/β ratio is of the order of 1.5 for prostate clonogens (Brenner 2003) has led several investigators to propose hypofractionation schedules for prostate radiotherapy given that the α/β ratio for the principal organ at risk, the rectum, is believed to be considerably higher than 1.5. Therefore, the standard situation of low α/β for normal tissues, high α/β for tumours is effectively reversed, and giving a large number of small fractions is, at least theoretically, a disadvantage (i.e. is clearly suboptimal). Proposals for prostate hypofractionation schedules have been worked out by Fowler et al. (2003): assuming that $\alpha/\beta = 1.5$ Gy, they calculate that 10 fractions of 4.69 Gy should yield bNED* = 89.6% compared to only 69.2% for 72 Gy conventionally delivered in 36×2 Gy fractions for isoeffective normal-tissue response using $\alpha/\beta = 3.0$ Gy. Amer et al. (2003) have carried out a similar exercise, but they use, as a reference, the schedule of 50 Gy in 16 fractions of 3.13 Gy that has long been standard at the Christie Radiotherapy Department in Manchester. However, if $\alpha/\beta \approx 8.3$ and a significant proportion of the more advanced (pre-teatment PSA > 20 ng ml^{-1}) prostate tumours are hypoxic as Nahum et al. (2003) claim, then these hypofraction schedules may actually lead to poorer control rates than those currently obtained with 2 Gy fractions rather than the considerably improved outcome predicted by Fowler et al. and others.

9.4 EFFECT OF OVERALL TREATMENT TIME

A conventional overall treatment time of 5–7 weeks is long enough to allow considerable cell proliferation to take place, both in the tumour and in the critical normal tissues.

Early reactions. The early-responding normal tissues (see Section 8.4.1) usually have a relatively high rate of cell proliferation, and part of their ability to withstand irradiation is due to the extensive repopulation that they undergo over a period of weeks. If the overall treatment time is shortened, then there will be less time for repopulation to occur and the early reactions will be more severe.

Late reactions. The target cells for late reactions usually proliferate slowly. Changes in overall time have less effect on late than on early reactions. The usual advice is that for modest changes in overall time, it is unnecessary to adjust the total radiation dose.

Tumours. Repopulation rates in tumours vary widely. In view of the fact that most human tumours grow very slowly (volume doubling times for carcinomas average around three months), it might be supposed that there would be little growth within a 5–7 week period. This is not so. It seems that when a tumour is damaged and begins to shrink, there is usually a considerable increase in repopulation rate, and doubling times for repopulation have been estimated often to be less than seven days. Consequently overall treatment time is very important for tumour response. A one-week extension (for instance, because of a gap in treatment) may significantly reduce the chance of tumour control. Conversely, even a modest shortening of overall time, if it can be achieved without lowering the total dose, could improve the chance of local control.

The strongest evidence for tumour cell repopulation has come from analysis of clinical data where overall time has been changed. A classic example is the review by Withers et al. (1988) of data on the local control of head and neck cancers (Figure 9.5). As overall time is increased, there is a corresponding increase in the *isoeffective* total dose. The slope of the full lines in this figure is 0.6 Gy per day. This is a remarkably high number for it implies that every 2 Gy daily fractional dose is counteracted by a loss of 0.6 Gy because of repopulation, so that the effective dose is only 1.4 Gy.

---

* bNED means no biochemical evidence of disease and is based on monitoring the patient's PSA value; it is often used as a surrogate for local control.

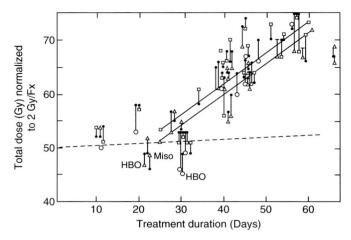

FIGURE 9.5

Tumour control dose (TCD$_{50}$) in head and neck tumours as a function of treatment duration. (From Withers et al., *Acta Oncol.*, 27, 131–146, 1988. With permission.)

The lag period. Both in early-responding tissues and in tumours, it takes time for repopulation to start. Lag periods vary and are seldom accurately known, but there is evidence that they may (in normal tissues or tumours) be in the region of 2–3 weeks. This is only of practical importance for very short treatment times, but it is one of the rationales for the CHART protocol* (Dische et al. 1997; Saunders et al. 1997, 1999) where treatment is over a 12-day period.

At the present time, the speed and time-course of repopulation in normal tissues and tumours are poorly understood (but see Fenwick 2006). They also vary between different tissues and from one patient to another. There is no simple equation to correct for the effect of a change in overall time. The lack of such an equation is sometimes used as a criticism of the LQ approach as compared with the NSD equation (see Section 9.2.1). This criticism is unfounded for although the Ellis formula allowed the effects of overall time to be calculated, it is now thought that these predictions were often inaccurate and misleading. In the commonest situation where late reactions are regarded as the dose-limiting factor, the usual recommendation is *not* to correct for overall time. Any correction will always be small and to ignore it is, in many cases, the safest option.

9.5 HYPERFRACTIONATION AND ACCELERATED FRACTIONATION

Two new approaches to radiotherapy fractionation were developed during the 1980s. Both involve the delivery of multiple fractions per day, but they are based on very different radiobiological principles.

9.5.1 HYPERFRACTIONATION

Hyperfractionation means using an increased number of fractions so that the dose per fraction is reduced below the conventional level of 1.8–2.0 Gy. In order to achieve the required

---

* Continuous Hyperfractionated Accelerated Radiotherapy.

total dose without extending the overall time, it is necessary to give more than one radiation dose per day (usually 2 or 3 fractions per day); the prostate study by Valdagni et al. (2005) is an example of this.

The aim of hyperfractionation is to exploit the greater recovery capacity of late-responding normal tissues. This is illustrated, on the basis of the LQ model, by Figure 9.3. As indicated in Section 9.3, the isoeffective total dose for late-responding normal tissues rapidly changes with change in dose per fraction. Increasing the dose per fraction above 2 Gy (i.e. hypofractionation) leads to a decrease in therapeutic index; however, by the same argument, there should be a therapeutic gain if dose per fraction is reduced below 2 Gy. This occurs because as dose per fraction is reduced, the isoeffective total dose for an α/β ratio of 3 Gy representing normal tissue increases faster than it does along the curve for α/β ratio $= 10$ Gy representing tumour. The dose difference between these curves indicates the magnitude of the potential therapeutic gain for any chosen dose per fraction.

9.5.2 ACCELERATED FRACTIONATION

Accelerated fractionation means reducing the overall duration of radiation therapy, using a conventional dose per fraction. The aim is to reduce the impact of tumour-cell proliferation during treatment. As with hyperfractionation, in order to deliver the prescribed total dose within a short overall time (without increasing dose per fraction), it is necessary to give more than one dose-fraction per day.

Both of these novel approaches to radiotherapy fractionation are showing some evidence of clinical benefit in current trials, and hybrid schedules between hyperfractionation and accelerated fractionation are also being used. As an example, Dische et al. (1997) obtained similar local control in head and neck cancer with the CHART schedule of 54 Gy given in 36 fractions over 12 days and with the 66 Gy given in 33 fractions over 6.5 weeks. Overgaard et al. (2003) reported on a comparison between five and six fractions per week of conventional (2 Gy fraction size) radiotherapy for squamous-cell carcinoma of the head and neck. 1476 patients were randomly assigned to one schedule or the other at the same total dose and fraction number. Their findings are shown in Figure 9.6.

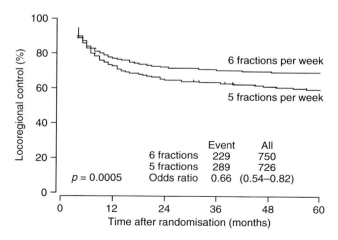

FIGURE 9.6
Primary locoregional tumour control as a function of the number of fractions per week in squamous-cell carcinoma of the head and neck, 66 Gy in 33 fractions or 68 Gy in 34 fractions, with all patients receiving the hypoxic radio-sensitiser nimorazole. (From Overgaard, J., et al., *Lancet*, 362, 933–940, 2003. With permission.)

This constitutes convincing evidence for tumour-cell proliferation and for the advantage of an accelerated schedule.

9.5.3 WHAT INTERFRACTION INTERVAL FOR MULTIPLE FRACTIONS PER DAY?

Accelerated- and hyper-fractionation both require the delivery of multiple fractions per day. It is essential that sufficient time is allowed between fractions to allow full repair of sublethal damage. The reason for this is that the rate of repair in late-responding normal tissues tends to be slower than in tumour cells. If fractions are given close together, repair may be less complete in the normal tissue than in the tumour, and this is a therapeutic disadvantage. It is usually recommended that dose fractions not be given closer than six hours apart, but this may be too close together when tissues such as the spinal cord are at risk.

9.6 HOW TO RESPOND TO GAPS IN TREATMENT

A substantial proportion of radiotherapy patients suffer an interruption in their prescribed course of treatment. This may arise for many reasons: some that derive from the patient (e.g. severe acute reactions, concurrent disease), and others that are unconnected with the patient (e.g. machine breakdown, transport difficulties, public holidays, etc.). How should the rest of the course be adjusted in response to such a gap? There are a number of options with some that are more advisable than others:

a. Ignore the missed fractions. This would be the worst response because it would reduce the total dose and seriously compromise local tumour control.

b. Add the missed fractions at the end of the course. This is frequently done, but its disadvantage is that the overall treatment time is extended, allowing greater tumour-cell repopulation. The usual practice is to make a correction for the loss of effectiveness that results from the gap. This can be done by adding one or more fractions (of the same or a similar size) or by increasing the dose per fraction, but neither is ideal.

c. Increase the dose per fraction for the rest of the course in order to complete treatment in the prescribed overall time. This avoids the repopulation problem but suffers from the radio-biological disadvantage of an increased dose per fraction (see Section 9.3).

d. Add the missed fractions to the remainder of the course either by treating at weekends or by giving multiple fractions per day. This option seeks to maintain the dose per fraction and overall treatment time as prescribed. It is thought to be the best option, but if the gap occurs toward the end of the treatment course, it may be logistically difficult to implement. If 2 or 3 fractions have to be given on a single day, they should be separated by at least six hours as indicated above.

The radiobiological and other considerations in the response to treatment gaps have been considered (in the United Kingdom) by a committee of the Royal College of Radiologists (Hendry et al. 1996).

9.7 LOW DOSE RATE: RATIONALE FOR BRACHYTHERAPY

A low dose rate can be viewed as the limiting case of hyperfractionation: many small fractions are given very frequently. Brachytherapy (i.e. radiation treatment using implanted radioactive sources) has distinct radiobiological advantages as well as potential disadvantages (see Chapter 55).

Advantages:

1. At a sufficiently low dose rate, this ultimate form of hyperfractionation maximally spares late-responding normal tissues.

2. It is usually achieved within a short overall treatment time, giving minimal time for tumour-cell repopulation.

3. The geometric form of the radiation field exploits the volume effect in normal-tissue damage; compared with external-beam treatment, the volume of normal tissue taken to high dose is usually smaller.

4. Cell killing around an implanted source is very intense and provides a form of tumour debulking.

Disadvantages:

1. The very non-uniform dose distribution risks the *geometric miss* of some tumour cells.

2. The short overall treatment time may be inadequate to allow full re-oxygenation of hypoxic tumour cells.

3. Particularly with pulsed high dose-rate brachytherapy, there is a temptation to treat too fast, losing the advantage listed as (1) above.

The effect of a single, low dose-rate exposure will differ from that of the same dose given at high dose rate because the extended treatment time allows a number of biological processes to take place. These are the 4 Rs of radiotherapy listed in Section 7.8. Over what range of dose rates will these biological processes act? This depends on their speed. A fast recovery process can compete with rapid infliction of damage and will produce effects at high dose rate. As a rough guide, the time the process takes (for example, after a brief radiation exposure) can be compared with the exposure time at low dose rate. Repair is a fast process with a half-time in the region of 1 h in mammalian cells. At a dose rate of 100 cGy/min or more (typical of high dose-rate exposures in clinical and experimental radiotherapy), a dose of, for example, 2 Gy is delivered in 2 min, and little recovery can occur during exposure. Therefore, dose rate will have little effect. Lowering the dose rate to 10 cGy/min increases the exposure time for 2 Gy to 20 min, and some recovery will occur. Reduction by a further factor of 10, to 1 cGy/min, gives an exposure time of over 3 h for 2 Gy. At this dose rate, recovery will be almost complete.

In contrast, proliferation is a much slower process. Cell cycle times for human cells are in the region of 2 to 4 days, and only when the exposure time becomes a day or more will this produce a significant effect. Figure 9.7 illustrates this in the form of calculated isoeffect curves for a cell population that has an α/β ratio of 3.7 Gy. The full line shows the sparing effect of repair (half-time 0.85 h) which has its main effect over the dose-rate range from around 20 cGy/min down to 0.2 cGy/min. The dashed lines show the effect of proliferation at various rates; these indicate the substantial sparing that may occur at very low dose rates in proliferating tissues.

The effect of changing the dose rate on cell survival is shown in Figure 9.8. For any given dose (imagine a vertical line in this figure), cell survival increases as the dose rate is reduced[*]. This reflects the effect of cellular recovery during irradiation. The shouldered survival curves at high dose rate become straight at a dose rate of around 1 cGy/min. This occurs because the time-dependent processes that produce the shoulder can then proceed during exposure and even go to completion.

---

[*] Wang et al. (2003) and Fowler et al. (2004a) have discussed the dose-rate issue in the context of the relatively prolonged delivery of certain intensity-modulated external-beam techniques; Fowler et al. concluded that any technique requiring more than 30 min for delivery could lead to a decrease in biological effect and tumour control probability.

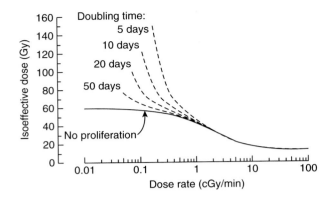

FIGURE 9.7
Theoretical isoeffect curves, showing the effect of changing dose rate for a single exposure either in the absence of proliferation or assuming proliferation with a doubling time in the range 5 to 50 days.

Wide variation is seen among different tumour cell lines in their sensitivity to low-dose-rate irradiation. This occurs because the low-dose-rate sensitivity is largely determined by the magnitude of the α component in the LQ equation, and this widely varies from one cell line to another (see Figure 7.11 and Figure 7.12). Studies on experimental animals have also found that the magnitude of the dose-rate effect differs from one normal tissue to another. Many tissues recover well from radiation damage and show a strong dose-rate effect. The bone-marrow is perhaps the extreme example of a tissue whose stem cells show little dependence on dose-rate since α-killing dominates (see Section 7.13; Chapman 2003).

In evaluating the therapeutic implications of these observations, it must be realized that the clinical use of low-dose-rate irradiation (in interstitial or intracavitary therapy) is not primarily for radiobiological reasons, but because these approaches lead to a better dose distribution within the tumour. Furthermore, the clinical choice is not between a single exposure at high dose rate or at low dose rate, but between fractionated high-dose-rate treatment and a single

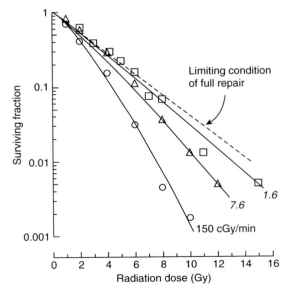

FIGURE 9.8
Cell survival curves for a human melanoma cell line irradiated at dose rates of 150, 7.5, or 1.6 cGy/min, illustrating the effect of dose rate on radiosensitivity.

(or a few) low-dose-rate exposures. As indicated above, both of these ways of protracting radiation treatment exploit the same biological processes (principally recovery and repopulation) and in broad terms tissues that are well spared by one will also be spared by the other[*]. Fowler (1989b) has calculated, on the basis of the linear-quadratic model, equivalence relationships between fractionated and low-dose-rate irradiation. For a tumour with an α/β ratio of 10 Gy and a late-responding normal tissue with $\alpha/\beta = 3$ Gy, the relative effects of continuous exposure at approximately 50 cGy/h will be equivalent to 30 high-dose-rate fractions of 2 Gy. The advantage of the continuous exposure is that it would be complete in only five days. Low-dose-rate radiotherapy is the most efficient way of maximizing recovery while keeping the treatment duration as short as possible. It is theoretically attractive where tumour-cell repopulation is thought to be rapid. For a more detailed description of dose-rate effects in brachytherapy, see Steel (2002a) and also Chapter 55.

---

[*] This equivalence between continuous low dose rate and small-fraction (at high dose rate) treatment is expanded upon in Chapter 55.

PART B: RADIOBIOLOGY
References

Al-Dweri, F. M. O, Guirado, D., Lallena, A. M, and Pedraza, V., Effect on tumour control of time interval between surgery and postoperative radiotherapy: an empirical approach using Monte Carlo simulation, *Phys. Med. Biol.*, 49, 2827–2839, 2004.

Alper, T., Relevance of experimental radiobiology to radiotherapy, *Br. Med. Bull.*, 29, 3–6, 1973.

Amer, A. M., Mott, J., Mackay, R., Williams, P. C., Livsey, J., Logue, J. P., and Hendry, J. H., Prediction of the benefits from dose-escalated hypofractionated intensity-modulated radiotherapy for prostate cancer, *Int. J. Radiat. Oncol. Biol. Phys.*, 56, 199–207, 2003.

Ash, D. V., Benson, E. A., Sainsbury, J. R., Round, C., and Head, C., Seven-year follow-up on 334 patients treated by breast conserving surgery and short course radical postoperative radiotherapy: A report of the Yorkshire Breast Cancer Group, *Clin. Oncol. (R. Coll. Radiol.)*, 7, 93–96, 1995.

Barendsen, G. W., Responses of cultured cells, tumours and normal tissues to radiations of different linear energy transfer, *Curr. Top. Radiat. Res. Q.*, 4, 293–356, 1968.

Bentzen, S. M., Dose-response relationships in radiotherapy, in *Basic Clinical Radiobiology*, Steel, G. G., Ed., 3rd ed., Arnold, London, 2002.

Bentzen, S. M. and Baumann, M., The L-Q model in clinical practice, in *Basic Clinical Radiobiology*, Steel, G. G., Ed., 3rd ed., Arnold, London, pp. 134–146, 2002.

Blomgren, H., Lax, I., Naslund, I., and Svanstrom, R., Stereotactic high dose fraction radiation therapy of extracranial tumors using an accelerator. Clinical experience of the first thirty-one patients, *Acta Oncol.*, 34, 861–870, 1995.

Brahme, A., Chavaudra, J., Landberg, T., McCullough, E. C., Nüsslin, F., Rawlinson, J. A., Svensson, G., and Svensson H., Accuracy requirements and quality assurance of external beam therapy with photons and electrons, *Acta Oncol.*, 15(Suppl. 1), 1–76, 1988.

Brenner, D. J., Hypofractionation for prostate cancer radiotherapy—what are the issues? *Int. J. Radiat. Oncol. Biol. Phys.*, 57, 912–914, 2003.

Brenner, D. J. and Hall, E. J., Fractionation and protraction for radiotherapy of prostate carcinoma, *Int. J. Radiat. Oncol. Biol. Phys.*, 43, 1095–1101, 1999.

Brizel, D. M., Dodge, R. K., Clough, R. W., and Dewhirst, M. W., Oxygenation of head and neck cancer: changes during radiotherapy and impact on treatment outcome, *Radiother. Oncol.*, 53, 113–117, 1999.

Brown, J. M., Goffinet, D. R., Cleaver, J. E., and Kallman, R. F., Preferential radiosensitization of mouse sarcoma relative to normal skin by chronic intra-arterial infusion of halogenated pyrimidine analogs, *J. Natl. Cancer Inst.*, 47, 75–89, 1971.

Chapman, J. D., Biophysical models of mammalian cell inactivation by radiation, in *Radiation Biology in Cancer Research*, Meyn, R. E. and Withers, H. R., Eds., Raven Press, New York, 1980.

Chapman, J. D., The single-hit mechanism of tumor cell killing by radiation, *Int. J. Radiat. Biol.*, 79, 71–81, 2003.

Chapman, J. D., Bradley, J. D., Eary, J. F., Haubner, R., Larson, S. M., Michalski, J. M., Okunieff, P. G., Strauss, H. W., Ung, Y. C., and Welch, M. J., Molecular (functional) imaging for radiotherapy applications: an RTOG symposium, *Int. J. Radiat. Oncol. Biol. Phys.*, 55, 294–301, 2003.

Chao, K. S., Bosch, W. R., and Mutic, S., A novel approach to overcome hypoxic tumor resistance: Cu-ATSM-guided intensity-modulated radiation therapy, *Int. J. Radiat. Oncol. Biol. Phys.*, 49, 1171–1182, 2001.

Cohen, L., Radiotherapy in breast Cancer 1. The dose N time relationship: Theoretical considerations, *Br. J. Radiol.*, 25, 636–642, 1952.

Dasu, A. and Denekamp, J., New insights into factors influencing relevant oxygen enhancement ratio, *Radiother. Oncol.*, 46, 269–277, 1998.

Deacon, J., Peckham, M. J., and Steel, G. G., The radioresponsiveness of human tumours and the initial slope of the cell survival curve, *Radiother. Oncol.*, 2, 317–323, 1984.

Denekamp, J., Physiological hypoxia and its influence on radiotherapy, in *The Biological Basis of Radiotherapy*, Steel, G. G., Adams, G. E., and Horwich, A., Eds. 2nd ed., Elsevier Science, Amsterdam, The Netherlands, pp. 115–134, 1989.

Dische, S., Saunders, M., Barrett, A., Harvey, A., Gibson, D., and Parmar, M., A randomised multicentre trial of CHART versus conventional radiotherapy in head and neck cancer, *Radiother. Oncol.*, 44, 123–136, 1997.

Douglas, B. G. and Fowler, J. F., The effect of multiple small doses of x rays on skin reactions in the mouse and a basic interpretation, *Radiat. Res.*, 66, 401–426, 1976.

Ellis, F., Dose, time and fractionation. A clinical hypothesis, *Clin. Radiol.*, 20, 1–7, 1969.

Fenwick, J. D., Delay differential equations and the dose-time dependence of early radiotherapy reactions, *Med. Phys.*, 33, 3526–3540, 2006.

Fertil, B. and Malaise, E. P., Inherent sensitivity as a basic concept for human tumour radiotherapy, *Int. J. Radiat. Oncol. Biol. Phys.*, 7, 621–629, 1981.

Fowler, J. F., *Nuclear Particles in Cancer Treatment*, Adam Hilger, Bristol, 1981.

Fowler, J. F., 40 years of radiobiology: its impact on radiotherapy, *Phys. Med. Biol.*, 29, 97–113, 1984.

Fowler, J. F., The linear-quadratic formula and progress in fractionated radiotherapy, *Br. J. Radiol.*, 62, 679–694, 1989a.

Fowler, J. F., Dose-rate effects in normal tissues. Brachytherapy 2, in *Proceedings of the 5th International SELECTRON Users' Meeting*, Mould, R. F., Ed., Nucletron International, Eersum, The Netherlands, pp. 26–40, 1989b.

Fowler, F., Development of radiobiology for oncology—a personal view, *Phys. Med. Biol.*, 51. R263–R286, 2006.

Fowler, J. F., Chappell, R. J., and Ritter, M. A., Is α/β for prostate tumours really low? *Int. J. Radiat. Oncol. Biol. Phys.*, 50, 1021–1031, 2001.

Fowler, J. F., Ritter, M. A., Chappell, R. J., and Brenner, D. J., What hypofractionated protocols should be tested for prostate cancer?, *Int. J. Radiat. Oncol. Biol. Phys.*, 56, 1093–1094, 2003.

Fowler, J. F., Welsh, J. S., and Howard, S. P., Loss of biological effect in prolonged fraction delivery, *Int. J. Radiat. Oncol. Biol. Phys.*, 59, 242–249, 2004a.

Fowler, J. F., Tome, W. A., Fenwick, J. D., and Mehta, M. P., A challenge to traditional radiation oncology, *Int. J. Radiat. Oncol. Biol. Phys.*, 60, 1241–1256, 2004b.

Glatstein, E., Intensity-modulated radiation therapy: the inverse, the converse, and the perverse, *Semin. Radiat. Oncol.*, 12, 272–281, 2002.

Goodhead, D. T., Estimation of a single property of low LET radiations which correlates with biological effectiveness, *Phys. Med. Biol.*, 28, 485–492, 1983.

Goodhead, D. T., Spatial and temporal distribution of energy, *Health Phys.*, 55, 231–240, 1988.

Grau, C., Høyer, M., Lindegaard, J., and Overgaard, J., The emerging evidence for Stereotactic Body Radiotherapy, *Acta Oncol.*, 45, 771–774, 2006.

Haimovitz-Friedman, A., Kolesnick, R. N., and Fuks, Z., Modulation of the apoptotic response: potential for improving the outcome in clinical radiotherapy, *Semin. Radiat. Oncol.*, 6, 273–283, 1996.

Hendry, J. H., Bentzen, S. M., Dale, R. G., Fowler, J. F., Wheldon, T. E., Jones, B., Munro, A. J., Slevin, N. J., and Robertson, A. G., A modelled comparison of the effects of using different ways to compensate for missed treatment days in radiotherapy, *Clin. Oncol.*, 8, 297–307, 1996.

Hill, R. P., Bush, R. S., and Yeung, P., The effect of anaemia on the fraction of hypoxic cells in an experimental tumour, *Br. J. Radiol.*, 44, 299–304, 1971.

Hockel, M., Knoop, C., Schlenger, K., Vorndran, B., Baussman, E., Mitze, M., Knapstein, P. G., and Vaupel, P., Intratumoral pO_2 predicts survival in advanced cancer of the uterine cervix, *Radiother. Oncol.*, 26, 45–50, 1993.

Honoré, H. B., Models for Optimization of Modern Radiotherapy, PhD thesis, Århus University, Århus, Denmark, 2002.

Hopewell, J. W., Mechanisms of the action of radiation on skin and underlying tissues, *Br J. Radiol.*, Suppl. 19, 39–51, 1986.

Hopewell, J. W., Morris, A. D., and Dixon-Brown, A., The influence of field size on the late tolerance of the rat spinal cord to single doses of x-rays, *Br. J. Radiol.*, 60, 1099–1108, 1987.

ICRU (International Commission on Radiation Units and Measurements), *Report 16. Linear Energy Transfer*, ICRU, Bethesda, MD, 1970.

ICRU (International Commission on Radiation Units and Measurements), *Report 36. Microdosimetry*, ICRU, Bethesda, MD, 1983.

ICRU (International Commission on Radiation Units and Measurements), *Report 62. Prescribing, Recording and Reporting Photon Beam Therapy*. Supplement to ICRU Report 50, ICRU, Bethesda, MD, 1999.

Joiner, M. C., Models of radiation cell killing, in *Basic Clinical Radiobiology*, Steel, G. G., Ed., 3rd ed., Arnold, London, pp. 64–70, 2002a.

Joiner, M. C., Particle beams in radiotherapy, in *Basic Clinical Radiobiology*, Steel, G. G., Ed., 3rd ed., Arnold, London, pp. 205–216, 2002b.

Joiner, M. C. and Bentzen, S. M., Time-dose relationships: the linear-quadratic approach, in *Basic Clinical Radiobiology*, Steel, G. G., Ed., 3rd ed., Arnold, London, pp. 120–133, 2002.

Joiner, M. C., Marples, B., Lambin, P., Short, S. C., and Turesson, I., Low-dose hypersensitivity: current status and possible mechanisms, *Int. J. Radiat. Oncol. Biol. Phys.*, 49, 379–389, 2001.

Lax, I., Panettieri, V., Wennberg, B., Amor Duch, M., Näslund, I., Baumann, P., and Gagliardi, G., Dose distributions in SBRT of lung tumors: comparisons between two different treatment planning algorithms and Monte-Carlo simulation including breathing motions, *Acta Oncol*, 45, 978–988, 2006.

Marples, B. and Joiner, M. C., The response of Chinese hamster V79 cells to low radiation doses: evidence of enhanced sensitivity of the whole cell population, *Radiat. Res.*, 133, 41–51, 1993.

Movsas, B., Chapman, J. D., Hanlon, A. L., Horwitz, E. M., Greenberg, R. E., Stobbe, C., Hanks, G. E., and Pollack, A., Hypoxic prostate/muscle pO$_2$ ratio predicts for biochemical failure in patients with prostate cancer: preliminary findings, *Urology*, 60, 634–639, 2002.

Nahum, A. E., *Calculations of Electron Flux Spectra in Water Irradiated with Megavoltage Electron and Photon Beams with Applications to Dosimetry*, PhD thesis (University of Edinburgh, July 1975). (available from Univ. Microfilms Int., 300 N ZEEB RD. Ann Arbor, MI 48106 U.S.A; Order No 77-70,006), 1976.

Nahum, A. E., Condensed-history Monte-Carlo simulation for charged particles: what can it do for us?, *Radiat. Environ. Biophys.*, 38, 163–173, 1999.

Nahum, A. E., Dearnaley, D. P., and Steel, G. G., Prospects for proton-beam radiotherapy, *Eur. J. Cancer*, 30A, 1577–1583, 1994.

Nahum, A. E., Movsas, B., Horwitz, E. M., Stobbe, C. C., and Chapman, J. D., Incorporating clinical measurements of hypoxia into tumor local control modeling of prostate cancer: implications for the α/β ratio, *Int. J. Radiat. Oncol. Biol. Phys.*, 57, 391–401, 2003.

Nordsmark, M., Overgaard, M., and Overgaard, J., Pretreatment oxygenation predicts radiation response in advanced squamous cell carcinoma of head and neck, *Radiother. Oncol.*, 41, 31–39, 1996.

Nutting, C. M., Corbishley, C. M., Sanchez-Nieto, B., Cosgrove, V. P., Webb, S., and Dearnaley, D. P., Potential improvements in the therapeutic ratio of prostate cancer irradiation: dose escalation of pathologically identified tumour nodules using intensity modulated radiotherapy, *Br. J. Radiol.*, 75, 151–161, 2002.

Overgaard, J., Hansen, H. S., Overgaard, M., Bastholt, L., Berthelsen, A., Specht, L., Lindelov, B., and Jorgensen, K., A randomized double-blind phase III study of nimorazole as a hypoxic radiosensitizer of primary radiotherapy in supraglottic larynx and pharynx carcinoma. Results of the Danish head and neck cancer study (DAHANCA). Protocol 5–85, *Radiother. Oncol.*, 46, 135–146, 1998.

Overgaard, J., Hansen, H. S., Specht, L., Overgaard, M., Grau, C., Andersen, E., Bentzen, J., Bastholt, L., Hansen, O., Johansen, J., Andersen, L., and Evensen, J. F., Five compared with six fractions per week of conventional radiotherapy of squamous-cell carcinoma of head and neck: DAHANCA 6 and 7 randomised controlled trial, *Lancet*, 362, 933–940, 2003.

Owen, J. R., Ashton, A., Bliss, J. M., Homewood, J., Harper, C., Hanson, J., Haviland, J., Bentzen, S. M., and Yarnold, J. R., Effect of radiotherapy fraction size on tumour control in patients with early-stage breast cancer after local tumour excision: Long-term results of a randomised trial, *Lancet Oncol.*, 7, 467–471, 2006.

Palcic, B. and Skarsgard, L. D., Reduced oxygen enhancement ratio at low doses of ionizing radiation, *Radiat. Res.*, 100, 328–339, 1984.

Parker, C., Milosevic, M., Toi, A., Sweet, J., Panzarella, T., Bristow, R., Catton, C., Catton, P., Crook, J., Gospodarowicz, M., McLean, M., Warde, P., and Hill, R. P., Polarographic electrode study of tumor oxygenation in clinically localized prostate cancer, *Int. J. Radiat. Oncol. Biol. Phys.*, 58, 750–757, 2004.

Paulino, A. C. and Johnstone, P. A. S., FDG-PET in radiotherapy treatment planning: pandora's box? *Int. J. Radiat. Oncol. Biol. Phys.*, 59, 4–5, 2004.

Peters, L. J., Brock, W. A., and Travis, E. L., Radiation biology at clinically relevant fractions, in *Important Advances in Oncology*, DeVita, V. T., Hellman, S., and Rosenberg, S., Eds., J.B. Lippincott, Philadelphia, PA, pp. 65–83, 1990.

Raju, M. R., *Heavy Particle Radiotherapy*, Academic Press, New York, 1980.

Sachs, R. K and Brenner, D. J., Solid tumor risks after high doses of ionizing radiation, *Proc. Natl. Acad. Sci.*, U.S.A., 102, 13040–13045, 2005.

Saunders, M. I., Dische, S., Barrett, A., Harvey, A., Gibson, D., and Parmar, M., Continuous hyperfractionated accelerated radiotherapy (CHART) vs. Conventional radiotherapy in non-small-cell lung cancer: a randomized multicentre trial, *Lancet*, 350, 161–166, 1997.

Saunders, M. I., Dische, S., Barrett, A., Harvey, A., Griffiths, D., and Palmar, M., Continuous hyperfractionated accelerated radiotherapy (CHART) vs. Conventional radiotherapy in non-small-cell lung cancer: mature data from the randomized multicentre trial, *Radiother. Oncol.*, 52, 137–148, 1999.

SBRT, 3rd Acta Oncologica Symposium on Stereotactic Body Radiotherapy, June 15–17, *Acta Oncol.*, Special Issue, 45, 2006.

Shelley, W., Brundage, M., Hayter, C., Paszat, L., Zhou, S., and Mackillop, W., A shorter fractionation schedule for postlumpectomy breast cancer patients, *Int. J. Radiat. Oncol. Biol. Phys.*, 47, 1219–1228, 2000.

Short, S. C., Mayes, C., Woodcock, M., Johns, H., and Joiner, M. C., Low dose hypersensitivity in the T98G human glioblastoma cell line, *Int. J. Radiat. Biol.*, 75, 847–855, 1999.

Short, S. C., Kelly, J., Mayes, C. R., Woodcock, M., and Joiner, M. C., Low-dose hypersensitivity after fractionated low-dose irradiation in vitro, *Int. J. Radiat. Biol.*, 77, 655–664, 2001.

Shiu, A., Chang, E. L., Lii, M. J., and Ye, J., Near simultaneous CT image-guided stereotactic spine radiotherapy: a novel paradigm for achieving true extracranial stereotaxy, *Int. J. Radiat. Oncol. Biol. Phys.*, 57, S427–S428, 2003.

Sinclair, W. K. and Morton, R. A., X-Ray and ultraviolet sensitivity of synchronized Chinese hamster cells at various stages of the cell cycle, *Biophys. J.*, 5, 1–25, 1965.

Steel, G. G., Cellular sensitivity to low dose-rate irradiation focuses the problem of tumour radioresistance, *Radiother. Oncol.*, 20, 71–83, 1991.

Steel, G. G., Ed., *Basic Clinical Radiobiology*, 3rd ed., Arnold, London, 2002a.

Steel, G. G., Combination of radiotherapy and chemotherapy: principles, in *Basic Clinical Radiobiology*, Steel, G. G., Ed., 3rd ed., Arnold, London, 2002b.

Steel, G. G. and Peacock, J. H., Why are some human tumours more radiosensitive than others? *Radiother. Oncol.*, 15, 63–72, 1989.

Stewart, F. A. and Bartelink, H., The combination of radiotherapy and chemotherapy, in *Basic Clinical Radiobiology*, Steel, G. G., Ed., 3rd ed., Arnold, London, 2002.

Stewart, F. A., Soranson, J. A., Alpen, E. L., Williams, M. V., and Denekamp, J., Radiation-induced renal damage: the effects of hyperfractionation, *Radiat. Res.*, 98, 407–420, 1984.

Strandqvist, M., Studien über die kumulative Wirkung der Roentgenstrahlen bei Fraktionierung, *Acta Radiol.*, Suppl. 55, 1–300, 1944.

Tannock, I. F., The relation between cell proliferation and the vascular system in a transplanted mouse mammary tumour, *Br. J. Cancer*, 22, 258–273, 1968.

Thames, H. D. and Hendry, J. H., *Fractionation in Radiotherapy*, Taylor & Francis, London, 1987.

Thames, H. D., Withers, H. R., Peters, L. J., and Fletcher, G. H., Changes in early and late radiation responses with altered dose fractionation: implications for dose survival relationships, *Int. J. Radiat. Oncol. Biol. Phys.*, 8, 219–226, 1982.

Thames, H. D., Mentzen, S. M., Turesson, I., Overgaard, J., and van den Bogaert, W., Time-dose factors in radiotherapy: a review of the human data, *Radiother. Oncol.*, 19, 219–235, 1990.

Thomlinson, R. H. and Gray, L. H., The histological structure of some human lung cancers and the possible implications for radiotherapy, *Br. J. Cancer*, 9, 539–549, 1955.

Till, T. E. and McCulloch, E. A., A direct measurement of the radiation sensitivity of normal mouse bone marrow, *Radiat. Res.*, 14, 213–222, 1961.

Trott, K. R., Relation between division delay and damage expressed in later generations, *Curr. Top. Radiat. Res.*, 7, 336–337, 1972.

Valdagni, R., Italia, C., Montanaro, P., Lanceni, A., Lattuada, P., Magnani, T., Fiorino, C., and Nahum, A., Is the alpha-beta ratio of prostate cancer really low? A prospective non-randomized trial comparing standard and hyperfractionated conformal radiation therapy, *Radiother. Oncol.*, 75, 74–82, 2005.

Van de Wiele, C., Lahorte, C., Oyen, W., Boerman, O., Goethals, I., Slegers, G., and Dierckx, R. A., Nuclear medicine imaging to predict response to radiotherapy: a review, *Int. J. Radiat. Oncol. Biol. Phys.*, 55, 5–15, 2003.

Wang, J. Z., Li, X. A., D'Souza, W. D., and Stewart, R. D., Impact of prolonged delivery times on tumour control: a note of caution for intensity-modulated radiation therapy (IMRT), *Int. J. Radiat. Oncol. Biol. Phys.*, 57, 543–552, 2003.

Webb, S. and Nahum, A. E., A model for calculating tumour control probability in radiotherapy including the effects of inhomogeneous distributions of dose and clonogenic cell density, *Phys. Med. Biol.*, 38, 653–666, 1993.

Whelan, T., MacKenzie, R., Julian, J., Levine, M., Shelley, W., Grimard, L., Lada, B., Lukka, H., Perera, F., Fyles, A., Laukkanen, E., Gulavita, S., Benk, V., and Szechtman, B., Randomized trial of breast irradiation schedules after lumpectomy for women with lymph node-negative breast cancer, *J. Natl. Cancer Inst.*, 94, 1143–1150, 2002.

Wilson, G. D., Bentzen, S. M., and Harari, P. M., Biologic basis for combining drugs with radiation, *Semin. Radiat. Oncol.*, 16, 2–9, 2006.

Withers, H. R., The four R's of radiotherapy, in *Adv. Radiat. Biol.*, Lett, J. T. and Adler, H., Eds., Academic Press, New York, 5, 241–271, 1975.

Withers, H. R., Biological aspects of conformal therapy, *Acta Oncol.*, 39, 569–577, 2000.

Withers, H. R., Thames, H. D., and Peters, L. J., Biological bases for high RBE values for late effects of neutron irradiation, *Int. J. Radiat. Oncol. Biol. Phys.*, 8, 2071–2076, 1982.

Withers, H. R., Thames, H. D., and Peters, L. J., A new isoeffect curve for change in dose per fraction, *Radiother. Oncol.*, 1, 187–191, 1983.

Withers, H. R., Taylor, J. M. G., and Maciejewski, B., The hazard of accelerated tumor clonogen repopulation during radiotherapy, *Acta Oncol.*, 27, 131–146, 1988.

Wyatt, R. M., Beddoe, A. H., and Dale, R. G., The effect of delays in radiotherapy treatment on tumour control, *Phys. Med. Biol.*, 48, 139–155, 2003.

Yarnold, J. R., FAST prospective randomised clinical trial testing 5.7 Gy and 6.0 Gy fractions of whole breast radiotherapy in terms of late normal tissue responses and tumour control, *Protocol*, Academic Radiotherapy Department, Royal Marsden NHS Foundation Trust, Sutton, U.K., 2004.

PART C

EQUIPMENT

Editors: Philip Mayles and Jean-Claude Rosenwald

INTRODUCTION

This part deals with the equipment used in external beam radiotherapy. The treatment devices used for radiation production are either *kilovoltage x-ray units*, or *megavoltage linear accelerators*, or *cobalt-60 machines*. *Kilovoltage x-ray units*, described in Chapter 10 were historically the first to be used and they played an important role in the early development of radiotherapy, mainly between 1910 and 1950. Although still useful for the treatment of superficial lesions, they have been superseded for most treatments by *linear electron accelerators*, emitting megavoltage x-ray beams and possibly electron beams. These accelerators, which started to be used clinically in the late 1960s, are described in Chapter 11 with a brief reference to *betatrons* and *microtrons*, which were other types of electron accelerator developed for radiotherapy. *Cobalt-60 machines*, described in Chapter 12, provide penetrating γ-ray beams emitted from a radioactive source. They were introduced as a replacement for kilovoltage x-ray units in the 1950s. It is only in the last decade of the past century that the superiority of linear accelerators became evident and that the number of cobalt-60 machines started to decrease all over the world, except in developing countries where they are still considered as the best option, mainly because of their reliability. Equipment for radiotherapy using proton or high LET particle beams are dealt with in Chapter 46 and Chapter 49 respectively.

For the treatment devices considered in this Part, a description is given of the physical principles involved and of practical considerations relating to machine construction. For a radiation beam to be clinically useful, the beam must be collimated and filtered in order to provide the radiation characteristics appropriate for use in radiotherapy treatment. These aspects are only touched on briefly in this Part, but they will be dealt with more fully in Part E.

Beside the equipment used to treat patients, specialist imaging devices are required to perform treatment preparation and verification. These are the so called *simulators* and *portal imaging devices* dealt with in Chapter 13 and Chapter 14 respectively. Standard diagnostic imaging devices, which also make an essential contribution to treatment preparation (see Chapter 30 and Chapter 31) are not described in this book and the reader interested in this subject should refer to other books dedicated to imaging equipment.

CHAPTER 10

KILOVOLTAGE X-RAY UNITS

Tony Greener

CONTENTS

10.1 INTRODUCTION

Kilovoltage x-rays have been used in radiotherapy from the earliest attempts at external beam treatments. Their use has gradually diminished over the years because of the widespread availability of high energy photons, and more recently, of electrons generated from high energy linear accelerators. Nevertheless, kilovoltage x-rays still provide an important option in the range of treatments available within many departments.

In principle, kilovoltage x-rays are generated by stopping electrons accelerated across the potential difference between a cathode and an anode. The energy or *quality* of the beam can generally be described in terms of the magnitude of this potential difference. However, the spectrum of radiation can be significantly altered by placing metal filters in the beam, and beams of different penetrative quality can be produced from the same accelerating potential. For this reason, it is customary to define the beam quality in terms of the thickness of a given material (usually aluminium or copper) that will reduce the beam intensity to half of its initial value. This is called the *half value layer* or HVL. This is discussed further in Section 21.2.

The kilovoltage range covers x-ray beams generated between 10 kV and 400 kV and is usually subdivided into categories according to increasing beam penetration. These reflect the type of treatments that each range is suited for as well as other physical considerations such as equipment design, beam characteristics, and the method employed in determining the dose rate (Klevenhagen et al. 2000).

1. *Grenz rays* (10 kV to 20 kV, 0.02 mm to 0.15 mm Al HVL) lie between the hardest ultraviolet rays and x-rays and are named from the German word *Grenze* (meaning *boundary*). They are sometimes also called *border* x-rays. At such energies, absorption in air can be appreciable. This range is rarely used in modern radiotherapy.

2. *Short distance* or *contact therapy.* (10 kV to 60 kV, 0.02 mm to 3.3 mm Al HVL) was originally developed to reproduce the conditions of surface Radium treatment. The very short source skin distance (SSD) of typically 1.5 cm to 5.0 cm ensures a rapid fall-off in depth dose, even over a wide range of qualities, providing a useful treatment depth up to several millimetres (BIR 1996).

3. *Superficial therapy* (50 kV to 150 kV, 1 mm to 8 mm Al HVL). This energy range has probably been least affected by the introduction of high energy linear accelerators providing the treatment of choice for many superficial lesions, and in many other cases, an adequate alternative to electrons. The beam characteristics enable lesions to a depth of around 5 mm to be encompassed by 90% of the surface dose.

4. *Orthovoltage therapy* or *deep therapy* (150 kV to 400 kV). Most clinical units in this range are operated with generating potentials from 160 kVp to 300 kVp corresponding to 0.5 mm to 4 mm Cu HVL for typical filtrations. The 90% dose in this range lies 1 cm to 2 cm beneath the incident skin surface at the usual treatment distance of 50 cm SSD. The percentage depth dose gradually varies with applied tube potential in this range. It changes by 2% or less of the local dose down to 50% of the peak dose per 50 kV change in tube voltage (BIR 1996). Multi-energy units covering both the *superficial* and *orthovoltage* ranges are now used in many departments.

10.2 PRINCIPLES OF X-RAY GENERATION

10.2.1 X-RAY PRODUCTION

As discussed in Chapter 3, when a high speed electron penetrates a target material, as explained in Chapter 3, three main interactions can occur.

1. The electron is subjected to small deflections by the electron cloud of the target material. These energy losses cause excitation and heat production, and they account for the majority of interactions.

2. The incoming electron interacts with an inner shell orbital electron and has sufficient energy to eject it. The resultant vacancy is filled by an electron from an outer shell that emits a photon of energy equal to the difference in energy between the two shells. This is termed *characteristic radiation* as the energy difference between orbital shells depends on atomic number and is characteristic of the target material. An electron from a shell still farther from the nucleus will then fill this new vacancy. This process is repeated, leading to the emission of a series of discrete lines of characteristic x-rays. Characteristic radiation accounts for only a small percentage of the x-rays produced. An electron must have an energy of 69.5 keV to eject a K shell electron from tungsten. The principal K-line emission for tungsten is 59.3 keV (range 58 keV to 69.1 keV).

3. *Bremsstrahlung (braking* radiation) accounts for the majority of x-ray photons emitted, and it occurs when the incoming electron passes close enough to the positive charge of the target nucleus to be attracted and, consequently, decelerated. The energy lost because of this braking effect is emitted as an x-ray photon. The electron can lose all of its incident energy in a single collision with a nucleus.

10.2.2 THE X-RAY SPECTRUM

The bremsstrahlung energy spectrum emerging from the target shows a continuous distribution of energies with characteristic x-rays at discrete energies superimposed upon this. In the absence of any filtration, the calculated energy spectrum emerging from a thick target will be a straight line given by the following equation (Johns and Cunningham 1983)

$$I(E) = CZ(E_{max} - E)$$

where $I(E)$ is the bremsstrahlung intensity at energy E, C is a constant that will include the number of electrons bombarding the target (proportional to tube current), and Z is the atomic number of the target material. The maximum energy of emitted photons (E_{max}) equals the maximum energy lost by the bombarding electrons that is determined by the peak kV (kVp) applied across the tube. To increase beam intensity a high-atomic-number target material is required that should also be capable of withstanding the very high temperatures generated. Tungsten satisfies this criterion ($Z=74$, melting point 3380°C). As shown in Figure 10.1, the maximum intensity of CZE_{max} at $E=0$ decreases linearly to zero at $E=E_{max}$. The total emitted x-ray energy given by the area under the line increases

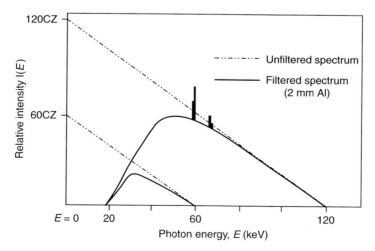

FIGURE 10.1
Unfiltered and filtered bremsstrahlung spectra from a tungsten target ($Z=74$) for peak applied tube potentials of 60 kVp and 120 kVp. The K-line characteristic emissions are superimposed on the filtered spectrum.

according to E_{max}^2 and with *applied kVp²*. For filtered beams, this proportionality may be raised to an even higher power of kVp, highlighting the importance of stable and reproducible kV control for accurate dose delivery. In practice, the emergent spectrum is attenuated by *inherent* and added filtration that preferentially absorb the lower energy components producing the filtered spectrum as shown in Figure 10.1. The thickness and type of filtration determines the minimum energy. As before, the maximum photon energy (keV) equals the applied kVp.

10.3 PRACTICAL X-RAY GENERATORS

10.3.1 THE X-RAY TUBE

In recent years, there have been numerous developments in the equipment associated with kV x-ray units. These include the widespread replacement of the glass enveloped x-ray tube by the more efficient and robust metal–ceramic tube, the introduction of high frequency HT generators, and the use of microprocessors for monitoring and controlling generator parameters.

Conventional kilovoltage x-ray units are designed around a stationary anode tube of which there are several possible designs depending on the age and type of equipment. The standard glass enveloped stationary anode tube has largely been superseded by the metal–ceramic tube (Figure 10.2) that offers numerous advantages over its predecessor. The basic radiotherapy x-ray tube consists of a cathode and anode assembly inside an evacuated tube. The cathode is a tungsten filament that, when heated by passing a current through it, emits electrons by the process of thermionic emission. The filament is set into a focusing cup that concentrates the electrons onto a small area of the target several centimetres away called the focal spot. The anode consists of a tungsten target embedded within a block of copper. A high voltage difference applied between the cathode and anode accelerates the electrons emitted from the cathode toward the anode where they deposit their energy in the target as outlined above. Most of this energy is deposited as heat, but a small percentage (ranging from 0.5% at 60 kVp to 1% at 200 kVp [Meredith and Massey 1977]) is converted to x-rays.

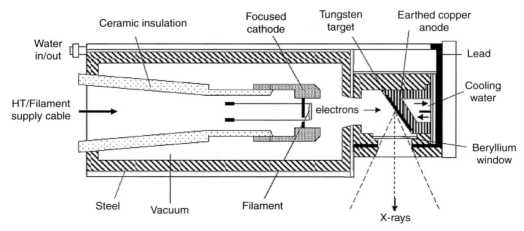

FIGURE 10.2
Schematic cross-section of a metal–ceramic x-ray tube (Comet MXR 161) for superficial therapy.

10.3.1.1 Metal–Ceramic Tube Design

The earthed anode is welded to the metal tube envelope to form a robust unit, facilitating simple and efficient direct cooling. The target is recessed within the copper anode, providing an integral anode hood that absorbs unwanted stray x-rays. The negative HT supply to the cathode is made by a connector from the HT generator that locates into a ceramic receptacle, reducing the likelihood of electrical discharge across this extended insulating surface. Metal–ceramic tubes are manufactured with an integral beryllium window (2 mm to 5 mm thick) that has a low atomic number ($Z=4$) and physical density. For the same tube operating conditions, this lower inherent filtration results in increased output compared with glass envelope tubes. Consequently, metal–ceramic tubes can be run at lower tube currents and allow lower beam qualities to be produced for a given tube kVp.

10.3.1.2 Contact Therapy Tubes

The design of contact therapy tubes is based on bringing the target as close as possible to the treatment surface. This leads to quite different configurations that have involved the use of both transmission targets (Chaoul unit) and reflection targets (Philips unit) (Birch and Blowes 1990). This equipment is not commonly used today, and the reader is referred to earlier texts for more information.

10.3.1.3 Target Angle and Radiation Distribution in Air

The inverse-square law and the amount of oblique filtration through attenuating materials in the beam path will govern the directional distribution of the beam. The beam profile perpendicular to the anode–cathode direction should be symmetric. This is unlikely to be the case in the anode–cathode direction where x-rays will be differentially absorbed within the target depending on the angle where they emerge. This *heel effect* can result in increased or reduced beam intensity on the anode side of the beam axis compared with that on the cathode side (Klevenhagen et al. 2000). The magnitude of this effect depends on both the target angle and applied kV, and it is additionally constrained by the need for a large useful beam at relatively short SSDs. The target angle, defined as the angle between the incident electrons and the normal to the target, is typically 40° for superficial x-rays and 30° for orthovoltage equipment. For multi-energy equipment, the larger target angle is chosen so that the useful field size at the superficial energy is not significantly reduced by the heel effect.

10.3.2 X-Ray Tube Housing

The x-ray tube housing consists of a metal shield that encloses and protects the tube providing shielding from unwanted x-rays. Permissible leakage levels through the shield at 1 m from the focal spot are 1 mGy/h and 10 mGy/h for equipment running up to 150 kVp and 300 kVp, respectively (IEC 1987). A filter holder and applicator mount are attached to the tube housing at the beam exit aperture. The filter holder usually contains one or more micro-switches to enable the system control unit to recognise coded external metal filters. For orthovoltage equipment with applicators of SSD above 40 cm, a radiation monitor should be provided to indicate the tube output rate (IEC 1987). This typically takes the form of a transmission ionisation chamber placed downstream of the filter holder. Such equipment can be calibrated to deliver the intended dose in terms of monitor units rather like a linear accelerator (Gerig et al. 1994).

10.3.3 Tube Cooling System

The requirement for tube cooling results from the inefficiency of x-ray production in the kilovoltage energy range with about 99% of the energy of the electrons striking the target being dissipated as heat. Heat is conducted through the target into the copper anode block and, in the case of the metal–ceramic tube, efficiently removed by direct water (or oil) cooling of the copper block. Some units may also incorporate secondary cooling of the pumped coolant using a remote heat exchanger. Interlocks prevent the beam from running if the coolant temperature gets too high or if its flow rate drops below an acceptable level.

10.3.4 HT GENERATOR AND RECTIFICATION

The HT step-up transformer, filament transformer, and any rectification circuitry are located in a separate oil filled earthed metal tank. The HT transformer has a fixed turns ratio whose voltage output is directly proportional to the input. Modern systems supply a high frequency (25 kHz) sine wave input voltage to the step-up transformer. The high frequency HT output from this transformer is then fed into a voltage multiplier circuit that consists of a network of diodes and capacitors. The HT voltage across the output capacitors of this multiplier network is essentially constant (less than 1% ripple) and can be dynamically monitored and regulated with high precision (0.1%) by external control circuits. The HT supplied to the x-ray tube from the voltage multiplier network requires no further rectification. A low voltage transformer is used to supply the filament current because the filament is at the cathode potential relative to earth and must be isolated from the primary input voltage.

10.4 ANCILLARY EQUIPMENT

10.4.1 TUBE SUPPORT STAND

A good tube support stand is vital for accurate radiotherapy, and it may be floor or ceiling mounted. Treatments should not be compromised because the tube stand is unable to support the tube in the desired position. Independent lockable movements should enable most beam orientations to be attained. Fine adjustment of the tube position should be possible. Because it is normal for the applicators to be touching the patient skin, it is essential that fail-safe mechanisms exist to interrupt motor driven tube movements in the event of a faulty drive switch. In addition all movements must be locked in case of a power failure.

10.4.2 SYSTEM CONTROL UNIT AND USER INTERFACE

The system control unit must allow complete and safe control of important generator parameters and system timers, and it will normally include a start, stop, and emergency shut-off switch. Modern microprocessor based control systems are interfaced to the HT generator and dynamically monitor parameters such as tube kV and filament current. When an exposure is started, the voltage supplied across the tube ramps to the desired value over a period of up to several seconds. Some control units are able to detect unusual ramp-up cycles, and they terminate the beam before the full tube potential is applied. During this ramp-up period, the dose rate from the x-ray tube will increase from zero to its steady state value. For timer-based systems, the treatment timer will be started as soon as the start button is depressed or at some time during the ramp-up stage. As it is usual for the steady-state dose rate to be used to calculate treatment times, the ramp-up region could lead to a dose deficit or excess, depending on the precise point where the timer was started. This correction is called the *timer error* and should be dosimetrically determined for all combinations of kV and mA (Heales et al. 1998 and see Section 38.8.4). For machines delivering dose based on monitor ionisation chambers, this is not a problem as the treatment is defined and terminated by the control unit according to transmitted dose (Gerig et al. 1994; Aukett et al. 1996). The control unit will also be linked into treatment unit and room safety interlocks only permitting treatment when these are all successfully completed. Display requirements for the user interface, which may be integral to or independently communicate with the control unit, are outlined in IEC 601-2-8 (1987). Modern microprocessor-based user interfaces provide more functionality, including automated tube run-up routines and recording of patient-specific treatment parameters.

10.4.3 MACHINE INTERLOCKS

As with all radiotherapy equipment, necessary interlocks must be provided to ensure correct and safe operation of the equipment and to reduce exposure of the operator and patient to any undue hazard (IEC 1987). The main interlocks and safety features are as follows: emergency stop, start, and stop switch; last person out/treatment door closed; cooler over-temperature; cooler flow-rate; filter in place; filter selection confirmation; HT generator interlocks (over-current, over-voltage, power limit); x-ray warning; ready lights; and audible alarm.

10.5 BEAM FILTRATION

The spectrum of radiation emitted from an x-ray tube includes a considerable low-energy component. These lower-energy photons will be quickly attenuated in tissue and contribute little dose at greater depths. Depending on the intended depth of treatment, this low-energy component needs to be substantially reduced as it increases dose in superficial tissue without clinical benefit. In order to achieve preferential absorption of the lower-energy component, metal filters are placed in the path of the beam. In this energy range, the probability of a photon undergoing a photoelectric interaction decreases with increasing energy according to an approximate inverse cube relationship (see Equation 4.7). The lower-energy photons are attenuated more than those at a higher energy, so the mean energy of the spectrum will increase or *harden* (Figure 10.3). The filter will attenuate the spectrum at all energies, but it should not lead to an unacceptable overall reduction in transmitted intensity. The metal should have no photoelectric absorption edge occurring within the required useful energy range as this will

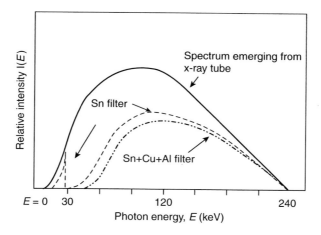

FIGURE 10.3
Beam hardening effect of a single Sn (0.4 mm) and composite Sn (0.4 mm) + Cu (0.25 mm) + Al (0.5 mm) filter on the intensity spectrum emerging from an x-ray tube at 240 kVp applied generating potential. Note that the curves are to illustrate the effect only—more accurate curves are shown in Figure 21.1.

lead to preferential absorption of higher energy photons in this region. The resultant filter must be mechanically stable, easy to handle, and of uniform thickness. It is important to verify the thickness and integrity of all filters at commissioning as irregularities in thickness across the filter could have an undesirable effect on the uniformity of the transmitted beam.

10.5.1 SUPERFICIAL ENERGIES

Although metals of high atomic number tend to harden the beam more efficiently, the resultant filter is often too thin to be mechanically stable, and an absorption edge is likely to be present within the required energy range. Aluminium is normally used (0 mm to 2.5 mm thick, depending on the amount of inherent filtration). Copper and aluminium composite filters may be used at the higher end of this range (120 kVp to 150 kVp).

10.5.2 ORTHOVOLTAGE ENERGIES

Higher atomic number filters of adequate thickness can be used at orthovoltage energies. Because of the production of characteristic radiation and incomplete removal of some parts of the lower energy spectrum by these filters, composite filters are used. These consist of additional metal filters of lower atomic number placed immediately downstream of the first filter. These filters preferentially remove the residual lower energy photons with a minimal effect on overall beam intensity. Examples include $Cu + Al$ or $Sn + Cu + Al$ (Thoraeus) filters (Figure 10.3). The order of composite filters is vital, and it must not be possible to administer treatment with the filter upside down.

10.5.3 FILTER BOX

Most units have a filter storage box that contains interchangeable filters that can be inserted beneath the beam exit aperture to produce beams of different qualities. Mechanical shaping of the filter and their holders within the box can limit individual filters to a unique place in the box. The box can be interlocked so that the system control unit can detect the removal of any filter and indicate an error condition if two or more filters are removed. A thick lead filter is normally provided for times when the tube is being run up or is not in use.

10.6 BEAM COLLIMATION AND APPLICATORS

The primary collimator usually consists of a conical hole within a block of lead, or other heavy metal, set into the tube housing and/or the x-ray tube. Although distant collimating systems with a projected light field are also used, secondary collimation is normally accomplished using a set of interchangeable applicators covering the range of available treatment sizes (Figure 10.4). The main collimation from an applicator is provided by a diaphragm of the correct size and shape that slides beneath the filter holder and is thick enough to reduce the unwanted beam to less than 1% of its incident intensity. The size of the divergent beam passing through the diaphragm must closely match the treatment field size as defined by the end of the applicator. The finite focal spot size means there will be a penumbra region at the beam edge that may be *trimmed* by making the edges of the divergent beam graze the applicator walls just before the end of the applicator. The *cone* part of the applicator consists of a rigid structure normally made from stainless steel. No primary radiation should strike the applicator walls, and any scattered radiation will usually be incident on the wall at an oblique angle as shown in Figure 10.4. Applicator walls can be made thinner than the diaphragm, keeping their overall weight down. The measurement technique and permissible levels of leakage radiation through

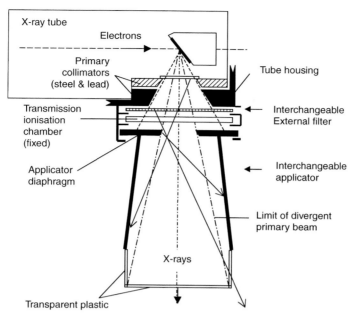

FIGURE 10.4
Schematic cross-section of the beam collimation system on an orthovoltage unit. Superficial units will be similar, but they will have no transmission ionisation chamber or plastic applicator end plate. The scattered radiation is schematically represented by solid-line arrows.

applicators are given in IEC 601-2-8 (1987). For a 10 cm circular applicator, this corresponds to 2% of the air-kerma rate on the radiation beam axis at a lateral distance between 2.5 cm and 4.5 cm from the distal·end of the applicator falling to 0.5% beyond this. It is common for the last few centimetres of an applicator to be made of clear plastic, allowing accurate positioning of the applicator against visible guide marks or surface lesions.

10.6.1 SUPERFICIAL APPLICATORS

Typical SSDs for superficial applicators range from 10 cm to 35 cm with field sizes from 1 cm to 20 cm diameter. Applicators are open-ended, ensuring the maximum surface dose rate is available to treat superficial lesions.

10.6.2 ORTHOVOLTAGE APPLICATORS

The typical SSD for orthovoltage applicators is 50 cm with rectangular or circular field sizes from 4 cm to 20 cm. Applicators in this range normally have closed ends to enable compression of the patient's skin. This brings the treated tissue closer to the applicator surface and so increases the percentage depth dose at the tumour with respect to the skin dose.

Applicators also accurately define the treatment distance and the dose rate. Actual applicator SSDs should be determined at commissioning as small differences can lead to unexplained step changes in plots of in-air dose rate as a function of field size.

CHAPTER 11

LINEAR ACCELERATORS

Les Loverock*

CONTENTS

* With contributions from Peter Williams (Section 11.4.2 and Section 11.4.3), David Thwaites and Alan McKenzie
(Section 11.4.5) and Philip Mayles (Section 11.4.1 and Section 11.4.4).

11.1 INTRODUCTION

Kilovoltage x-ray beams are useful for the treatment of skin lesions and shallow tumours, but for deep-seated tumours, the dose that can be delivered is limited by the high skin dose. Megavoltage beams are not only more penetrating, but they have the major benefit that the maximum dose is delivered below the skin surface (see Section 22.3.1). In addition, because the principal interaction with tissue is through the Compton effect, the locally absorbed dose is not dependent on the atomic number of the tissue, and the dose to bone is not enhanced.

It is not practicable to use transformer-based high voltage x-ray generators for energies above 300 kV, and another method of accelerating the electrons must be used. Higher energy beams can be produced in a number of ways. Methods of accelerating electrons to high energies include *linear accelerators*, *betatrons*, and *microtrons*. Betatrons were widely used in Europe and can produce very high energy electron beams (up to 50 MeV). However, they are bulky; the beam current is limited, and these machines are now obsolete. The *microtron* (from Scanditronix) is able to produce high energy electron beams at higher currents, but the high cost of the equipment has limited sales outside of Scandinavia. The linear accelerator is now the principal means of generating megavoltage beams in the developed world. Space constraints limit the electron energy that can be achieved to about 22 MeV, but in practice, this energy is sufficient for satisfactory treatments. The electron beams from linear accelerators can be converted to x-rays or can alternatively be directly used for patient treatment. A useful review of the history of clinical linear accelerators has been written by Thwaites and Tuohy (2006).

11.2 PRINCIPLES OF LINEAR ACCELERATORS

In a linear accelerator (*linac*) designed for radiotherapy, electrons gain energy by interacting with a synchronised radio-frequency electromagnetic field rather than by acceleration by direct potential. In free air, electromagnetic waves travel at the speed of light, but in a suitably designed waveguide, the speed of propagation of the waves can be substantially reduced. The *accelerating waveguide* (or accelerator structure) consists of a long cylindrical tube (see Figure 11.1), containing a series of circular baffles. These are designed so that speed of propagation of the microwaves increases in the first part of the accelerating tube until it eventually reaches velocities close to the speed of light. Bunches of electrons generated in the gun are injected into the guide in synchronism with pulsed microwave radiation and are carried down the guide in a manner analogous to riding the crest of a wave (i.e. surfing).

The high energy electron beams, typically 6 MeV or above, can be directly used for therapy and have a number of advantages compared to kilovoltage x-ray beams. Their depth dose curves are characterised by initial skin sparing followed by several centimetres of uniform dose (depending on energy) and then a rapid fall-off in dose (see Chapter 24). Electron beams from linear accelerators are useful for treatment of tumours up to about 70 mm deep, but for more deep-seated tumours, it is better to use photon beams (see Chapter 22). If the electron beam is to be used for therapy, the originally narrow beam of electrons must be broadened by scattering the electrons.

In photon mode, the electrons are focused onto a high-atomic-number thick target, and their energy loss is converted into bremsstrahlung radiation. At megavoltage energies, the principal direction of bremsstrahlung emission is in the forward direction. In the simplest accelerators, the target is fixed and the accelerating structure is coaxial with the emerging x-ray beam (i.e. it is parallel to the direction of travel of the electrons so that no bending of the electrons takes place) and is perpendicular to the cranio-caudal axis of the patient. However, for energies above 6 MeV, the length of the accelerator tube is such that it makes this impracticable. In order for the radiation beam to be brought in to irradiate the patient from any angle, it becomes necessary to bend the beam. Electrons are easily deflected in a magnetic field, and it is convenient to bend them through about 90°.

The challenge for accelerator design is to produce a stable monoenergetic high current electron beam concentrated onto a small focal spot that will ensure that a sharply focussed x-ray beam can be produced. The x-ray beam must be modified to allow uniform irradiation of the intended treatment area of the patient, which can vary from a very small area up to a

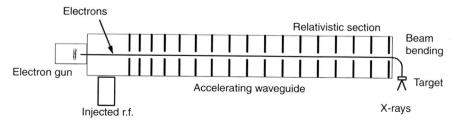

FIGURE 11.1
Schematic diagram of a linear accelerator designed for radiotherapy and used in x-ray mode. The electrons are injected with an electron gun in a section where an electromagnetic field is obtained from a radiofrequency (r.f.) microwave power source. At the exit of the relativistic section the electron beam may be bent before reaching the x-ray target.

maximum of about 40 cm × 40 cm. It is also of considerable benefit to be able to choose the beam energy to suit the tumour being treated.

Modern accelerators are the result of a complex design process with different emphases being placed by different manufacturers. In the following pages, the issues relating to the different components will be considered.

11.3 THE LINAC COMPONENTS

11.3.1 THE MICROWAVE POWER SOURCE

In the 2 MW to 10 MW power range that is necessary for pulsed operation of medical accelerators, there are two radiofrequency (r.f.) power sources available: the magnetron and the klystron. In lower energy machines (4 MeV to 6 MeV), magnetrons providing peak outputs up to 3 MW are used, but in dual energy linacs, requiring peak r.f. powers in excess of 5 MW, some manufacturers favour klystrons.

Medical linacs normally operate in the S-band either at 2856 MHz or 2998 MHz frequencies. The corresponding wavelengths in free space are 10.5 cm and 10 cm, respectively. This represents a compromise between the high power capability of large L-band (390 MHz to 1550 MHz) wave guides and the compactness needed in therapy applications.

11.3.1.1 The Magnetron

The *magnetron* (see Figure 11.2 [Greene and Williams 1997]) is an r.f. oscillator that extracts microwave energy from electrons by means of a slow wave resonant structure in a strong magnetic field. Negative 50 kV voltage pulses, lasting up to 5 μs, are applied to a heated cylindrical cathode. Thermionic electrons are then attracted toward a radially symmetric anode, comprising an array of cylindrical cavities. A magnetic field parallel to the cathode axis causes electrons to follow curved trajectories. At a critical field strength, the electrons sweep close to the anode before spiralling back to the cathode. In doing so, two things happen; first, induction currents are set up in the anode cavities, creating oscillating electric fields that will subsequently modulate later spiralling electrons; and second, the returning electrons will liberate more electrons by secondary emission at the cathode. This latter feature requires the cathode heater current to be reduced during the high-voltage pulse to avoid an avalanche situation developing. The cavity oscillations cause electrons to form bunches or spokes at each alternate cavity in a slowly rotating space charge cloud. A quasi-stable equilibrium state is reached where the rotating cloud of electrons transfer energy to the oscillating cavities. R.f. power is extracted at the output transition via wires from two adjacent cavities that are fed out beneath an evacuated glass dome to a loop antenna that directly feeds into a transverse-magnetic mode rectangular waveguide.

The dimensions of the resonant cavities determine the operating frequency of the magnetron, although an adjustable tuning probe can vary this over a small range (see discussion of automatic frequency control later—Section 11.7.1). The maximum power available is determined by the electron emission from the cathode, the magnitude of the applied high-voltage pulse, the transverse magnetic field strength, and the ability to shed heat from the water-cooled anode assembly.

As the r.f. power source essentially determines the maximum energy of an accelerator, some manufacturers of linacs offering dual x-ray operation use long anode, so-called Boot magnetrons (named after an early designer H. Boot). These have adjustable electromagnets, and when they are supplied with variable pulse voltages, they permit operation in low and high energy modes and can deliver peak outputs of 5 MW.

Magnetrons are compact devices and are usually mounted in the gantry arm. This avoids the need for expensive rotational r.f. couplers and lengthy transmission waveguide elements.

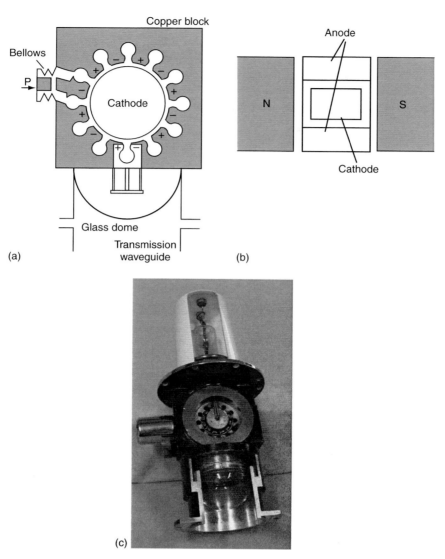

FIGURE 11.2
(a) Diagram of a tunable magnetron. This is placed in a magnetic field with lines of force normal to the plane of the diagram. (b) A section through the magnetron normal to that in (a), showing the poles of the magnet. (c) A photograph of a magnetron that has been opened to show the anode–cathode structure.

However, despite the high power levels present in magnetrons, they are very sensitive to magnetic field effects, and gantry mounting does require that corrections are made for different orientation within the earth's magnetic field as a function of gantry angle. For this reason, Siemens employed a fixed mounting position in their smaller Mevatron accelerators, avoiding geomagnetic variations and other causes of instability.

11.3.1.2 The Klystron

One of the earliest examples of a *klystron* was built by the Varian brothers in 1937, but since then, the design and application range has extended enormously. Unlike the magnetron, the klystron is not an r.f. oscillator, but it is an r.f. amplifier. It requires an r.f. input signal to excite it that it then amplifies from a few dB to nearly 100 dB, depending on the design. Functionally, klystrons can be regarded as inverse accelerators in that they transfer energy from an electron beam by velocity modulation to power an intense r.f. microwave output.

Figure 11.3 schematically shows the main components of a *Pulsed Klystron Amplifier*. Mounted vertically, an *electron gun* injects thermionically emitted electrons through an aperture in the anode into the r.f. section above. At the gun end, these electrons are confined to a narrow axial beam by electrostatic or magnetic focusing and are kept on axis throughout the r.f. section by large solenoidal coils. Finally, at the top of the klystron, the electrons are captured in the walls of the *collector*.

R.f. operation begins at the input or buncher cavity. Usually, an input loop from a coaxial transmission line, fed by a low power microwave oscillator, excites this first cavity. Induced field effects result in bunching the electrons as they pass through, some being slightly accelerated and some retarded. This is known as velocity modulation. After passing through a field-free region or drift tube, the now bunched electrons will induce r.f. oscillations in a similarly designed resonator cavity. This can be repeated at successive cavities, building up r.f. power at the expense of that present in the original electron beam. Finally, at the output cavity or *catcher*, the r.f. power can be propagated through a coupling iris and fed to a transmission waveguide for subsequent use in the accelerator waveguide.

At this stage, it is worth noting some of the parameters involved in a modern high power klystron. Typical high voltage pulses lie in the range 100 kV to 150 kV, electron currents

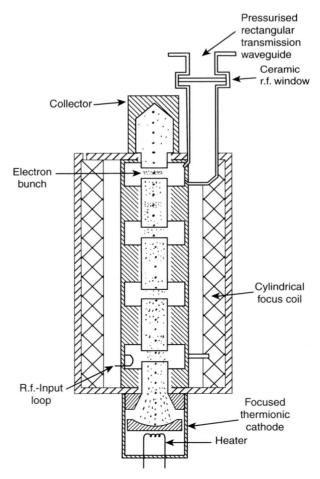

FIGURE 11.3
Schematic representation of a Klystron.

during these pulses are of the order of 100 A, and the r.f. power conversion gain is typically 50%. Three design requirements should be highlighted:

1. Electrons reaching the collector will generate bremsstrahlung x-rays (up to 100 keV) and need some significant shielding.
2. Even with a low duty cycle (1:1000), tens of kilowatts of heating will occur at the collector and need substantial cooling.
3. To reduce the risk of electrical arcing and to keep the overall dimensions small, at least part of the klystron should be oil-immersed. In practice, as the high voltage is confined to the gun end of the tube (negatively pulsed cathode) with the remainder at or close to earth potential, it is only the lower portion that is required to be mounted in an oil tank.

Linac klystrons and their associated shielding are heavy and bulky items. For this reason, they are always mounted in a fixed position in the drive stand and not in the gantry arm. As discussed in reference to the magnetron, this incurs a higher cost because of the need for more extensive r.f. plumbing.

11.3.2 HIGH VOLTAGE PULSED POWER SUPPLY OR MODULATOR

The *modulator* supplies high-voltage pulses lasting a few microseconds to the *microwave source* and *electron gun*. Details of the many circuit configurations employed are outside the scope of this work. The interested reader is referred to Karzmark et al. (1993) and service sheets supplied by accelerator manufacturers. Figure 11.4 shows the important components of a linac modulator in a much simplified form, following the treatment of Greene and Williams (1997). A high-voltage, three phase power supply (with two output levels in dual energy applications) typically delivers 10 kV to the smoothing capacitor C_1. This can be monitored at the voltage divider R_1 and R_2 where $R_1 \gg R_2$. The *pulse forming network*

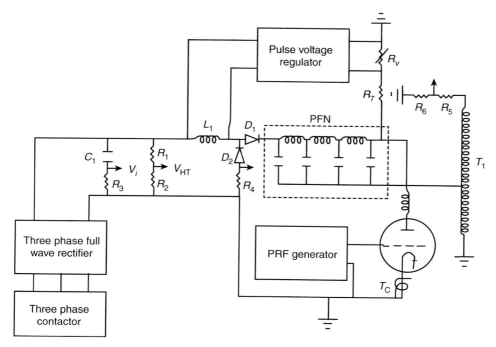

FIGURE 11.4
Simplified modulator circuit.

(PFN) is charged through inductance L_1 and diode D_1 while the *hydrogen thyratron* T_c is in its non-conducting state. The inductance L_1 and the combined capacity of the PFN act as a series resonant circuit (the inductance of $L_1 \gg$ PFN), and the voltage across the PFN will swing up to twice that of the power supply. Diode D_1 holds the voltage at its maximum level until the high power *thyratron* is fired to discharge the PFN. The resulting current pulse is fed to the primary windings of the pulse transformer T_1. This has its low end close to earth potential, and the high voltage ends go to the microwave generator and electron gun of the accelerator. Pulse lengths of 3 µs to 6 µs are common, and peak currents in excess of 500 A flow through the thyratron. The PRF (pulse repetition frequency) is set by the *PRF generator* connected to the thyratron grid and is usually adjustable in the range 50 Hz to 1000 Hz.

11.3.3   THE ACCELERATING WAVEGUIDE

At a basic level, this core component of a linac looks like a copper pipe. Internally, *irises* or *washer-like disks* divide the structure into *cavities* throughout its length. Powerful r.f. waves enter the guide and establish intense electric and magnetic fields in each cavity. These fields that accelerate the electrons along the guide axis are set up by induction currents created in the surface of the cavity walls. To keep resistive losses low, high purity copper is the most common material used. The microwave power dissipated P in the walls of a guide for an electron energy gain V and guide length L is given by

$$P = V^2/ZL \qquad (11.1)$$

where Z is the shunt impedance per unit length of the accelerator and provides a simple measure of waveguide efficiency.

There are two distinct types of guide: the *travelling waveguide* favoured by Elekta (formerly also ABB/Dynaray and Philips) and the *standing waveguide* used by Mitsubishi, Siemens, and Varian (formerly also CGR/GE). Both systems launch electrons from an electron gun into an initial accelerating buncher section in phase with an intense r.f. waveform. In the former, the wave is propagated along the guide axis, and in the latter, it is stationary.

Detailed coverage of waveguide design is beyond the scope of this text; interested readers are referred to the references Lapostolle and Septier (1970), Karzmark et al. (1993), and Greene and Williams (1997).

11.3.3.1   TW or Travelling Waveguides

These guides have relatively low shunt impedances compared to standing waveguide systems; therefore, they need to be physically longer to achieve the same output energy. However, they were the first in clinical use and their construction is much simpler. Figure 11.5b shows a section through part of a modern Elekta Travelling Waveguide. Electrons from the *gun* end enter from the bottom of the figure with a velocity of 0.8 c (at 80 kV) where c is the velocity of light. After the first 30 cm, they are travelling at velocities close to c. This first part of the guide is called the *buncher section* (notice the reduced spacing of the irises here and their tapering diameters). Thereafter, further energy gain results in a relativistic mass increase, and the iris separation remains constant. In the *buncher*, as the name implies, the electrons are brought together in space, phase and velocity as they acquire energy from the electromagnetic fields. These are set up by a synchronised r.f. pulse entering the guide through a mode transformer near the gun.

An overview of a typical microwave system used in the acceleration process is schematically shown in Figure 11.6. Rectangular *transmission guides*, pressurised with high dielectric gas (e.g. sulphur hexafluoride (SF_6) or freon), are used to transport r.f. power from the magnetron to the accelerating guide. These gases not only help to reduce r.f. arcing, but they also cool the output glass dome of the magnetron. At both ends of the accelerator, *door-knob mode*

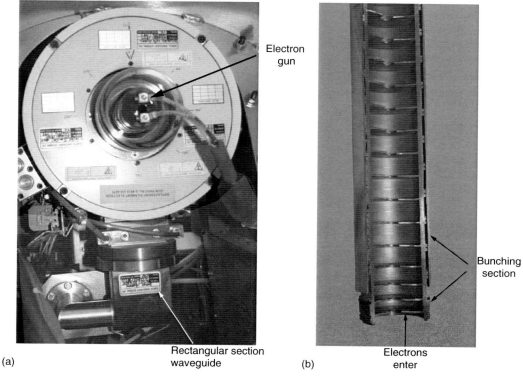

FIGURE 11.5

(a) View of the gun of an Elekta accelerator looking towards the accelerating waveguide from the rear. The rectangular waveguide at the bottom of the image brings the microwaves from the magnetron. They pass along this guide and upwards (in the picture) to where a junction is made with the circular accelerating waveguide involving a further turn through 90° to pass down the accelerating waveguide (located behind the glass gun assembly at the centre of the circular area). (b) Cut away section of the circular accelerating waveguide—electrons are accelerated in the direction from the bottom towards the top of the page as indicated by the arrow. Note the baffles in the waveguide are closer together at the bottom. This is the buncher section in which electrons are accelerated to relativistic velocities.

transformers (input and output) are used to couple the rectangular transmission sections to the accelerating guide. Sintered ceramic *r.f. windows* separate the pressurised guide regions from the highly evacuated (using *ion pumps*) *accelerating structure*. At the output end of the guide, unused r.f. power is either dumped into an absorbing water load or returned through a *re-circulator circuit* to augment the available power at the input.

    Also shown in the figure are external *solenoidal focus and centring (steering) coils* that are used to confine the electron beam and keep it directionally on axis. These help maximise the guide's efficiency and improve beam definition at the exit window.

11.3.3.2 SW or Standing Waveguides

    If each end of an iris loaded waveguide is terminated with a conducting disc, microwave power will be reflected with a $\pi/2$ phase change at the ends, allowing *standing waves* to build up. A simple account is given by Greene and Williams (1997) and will be followed here. Figure 11.7a shows a schematic *standing waveguide* with the arrows in cavities 1, 3, and 5 indicating the axial electric field at an instant. The fundamental mode (at least in this model) is shown by the solid curve in (b) with nodes at 2, 4, and 6. Clearly, electrons present at this moment in 2, 4, and 6 will receive no acceleration. One half cycle later, the situation has reversed and is shown by the dotted curve in (b). If the time required for an electron to travel from cavity 1 to cavity 3 equals half the r.f. wave period, then it will gain energy in both cavities.

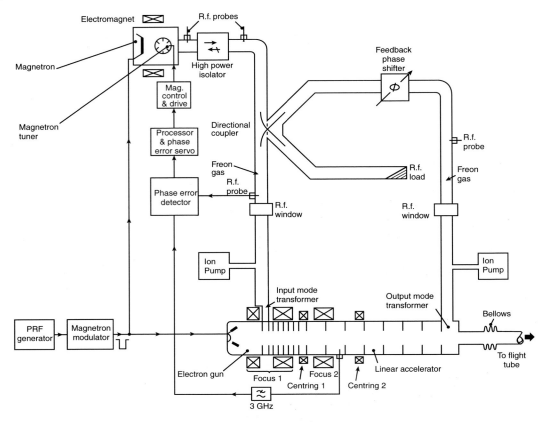

FIGURE 11.6
Full microwave system in a travelling wave accelerator. (Reproduced by permission of Elekta Oncology Systems.)

This is the basis of a standing waveguide. In practice, this can be implemented a number of ways, depending on the energy and particular application. For single energy x-ray applications (4 MV to 6 MV), both Siemens and Varian have adopted the *side coupled cavity* design shown in Figure 11.7c. Here, the non-accelerating cavities are taken out to the side, leaving only accelerating cavities on axis. Figure 11.8 shows a detailed sectional view of a Varian side cavity system. Such a longitudinally compact design allows these guides to be mounted *straight through* down the collimator axis and so avoid the need for a bending magnet as shown in Figure 11.9. In dual energy high powered guides, where their length would always preclude such an arrangement, the guide's transverse dimensions can become a problem.

With injection voltages of typically 50 kV, standing waveguides also require an initial *buncher section* where the electrons are gaining velocity. Here, it is necessary to shorten the first few cavities to allow transit times to be kept to one half cycle. Because the guide is soon filled with standing waves, it no longer matters where the r.f. power is fed in. This allows designers to avoid perturbing field effects at the gun end and to choose a more convenient central location.

11.3.3.3 Comparison of Travelling and Standing Waveguides

Shunt impedances in *standing* waveguide systems may be twice as high as in corresponding *travelling* waveguides. Inter-cavity coupling is usually magnetic rather than electric and more intense fields are set up. This does place higher demands on manufacturing tolerances, but in turn, results in a much tighter beam both in energy and spatial position. Where guide length is critical, standing wave systems, especially of the side cavity type, can offer a fourfold length

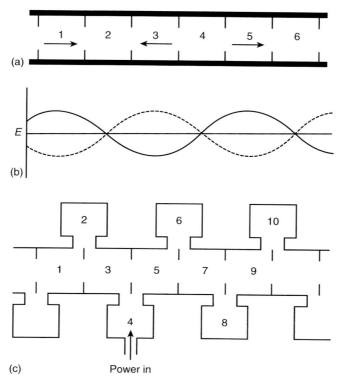

FIGURE 11.7
Standing wave accelerator: (a) Shows schematically the arrangement of the waveguide. (b) Illustrates the standing waves. (c) Shows *a side-coupled cavity.* (From Greene, D. and Williams, P. C., *Linear Accelerators for Radiation Therapy,* Institute of Physics, Bristol, 1997. With permission.)

reduction. However, maximum beam currents can be much higher in travelling waveguides, up to 2 A compared with 0.5 A in standing waveguides. In electron treatment modes where the required beam currents are very low compared with those for x-ray production, small changes in *frequency* in travelling waveguides can be used to change the electron energy. Standing waveguides, on the other hand, are much higher Q^* devices and must be driven very close to their design frequency. The electron energy here is determined by the *r.f. power* available and its *phase* relationship to the cavity oscillations.

11.3.4 THE ELECTRON GUN

The *electron gun* is the source of the electrons that are to be accelerated. As in a conventional x-ray tube, the electrons are generated from a heated cathode by the process of thermionic emission. A pulsed electron beam is produced by the application of a voltage pulse of about 10 kV. This produces an instantaneous beam current of the order of milliamps. Details of electron gun construction differ between manufacturers.

In *travelling* waveguides with their larger iris apertures (necessary for adequate electric field coupling) and the lower field strengths present, the system is more spatially tolerant. This allows simple heated tungsten wire coils to be used. Not only does this reduce the

---

* The Quality Factor Q in this context refers to the narrow performance window for electron acceleration in a waveguide. In a resonant circuit, it is usual to measure Q as the ratio of the resonant frequency to that of the difference between the upper and lower half power frequencies.

FIGURE 11.8
(See colour insert following page 590.) A Varian 4 MV side cavity standing waveguide.

complexity and the cost of the gun assembly, but it also has the additional benefit of providing a simple method of dose-rate control. As later demonstrated, this is a crucial requirement of any clinical accelerator and can be performed in these guides by controlling the gun current provided that it responds rapidly enough. The thermal capacities of these guns are very small, and they meet this requirement. A picture of the cathode for such a gun can be seen in Figure 11.10.

Standing waveguides are much more critical structures and cannot accept the possible spatial fluctuations of a gun filament as it ages or distorts with gantry rotation. Instead, indirectly heated guns are used that comprise a heating element embedded in a solid matrix supporting a high thermionic emissive cathode front surface. The whole forms part of an electrostatic focusing assembly precisely aligned with the anode aperture. For even

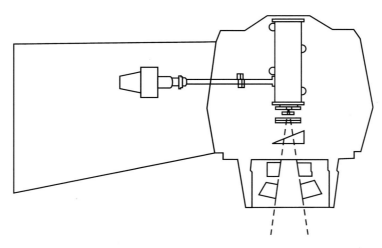

FIGURE 11.9
Magnetron-driven straight-through accelerator head. (From Greene, D. and Williams, P. C., *Linear Accelerators for Radiation Therapy*, Institute of Physics, Bristol, 1997. With permission.)

FIGURE 11.10
Cathode assembly from an Elekta SL series linac. Note the spiral cathode at the centre.

better control, a gridded element can be added in front of the cathode surface, creating a triode gun assembly and is schematically shown in Figure 11.11. The precise and constant geometry of these indirectly heated guns is favoured in standing wave machines, but because of their relatively high thermal capacities, they necessitate alternative methods of dose-rate control.

For normal operation, the peak beam current drawn from the gun is fixed at a precise level. This, together with the design of waveguide, the phase relationship, and r.f. power level

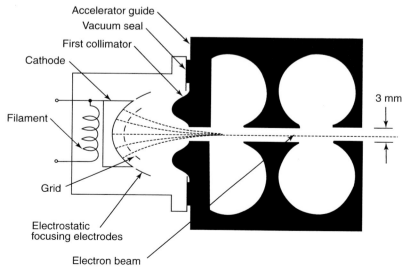

FIGURE 11.11
Triode electron gun assembly. (Reproduced by permission of Varian Medical Systems.)

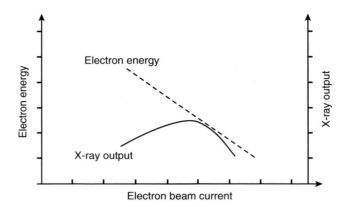

FIGURE 11.12
The relationship between electron beam current, electron energy, and x-ray output in a linear accelerator. (From Greene, D. and Williams, P. C., *Linear Accelerators for Radiation Therapy*, Institute of Physics, Bristol, 1997. With permission.)

available, will determine the output energy. In both travelling and standing wave systems, it is important to realise that small variations in gun current alone will affect the energy of the accelerated beam. Figure 11.12 shows how the electron energy falls as beam current rises in an exactly analogous way to a transformer load line. Clearly, any departure in beam current from its intended level will give rise to a consequential energy change and this indicates how important a parameter it is to control.

11.3.5 BEAM STEERING AND FOCUSING

In both travelling and standing waveguides, it is necessary to constrain (focus) the accelerating electron bunches close to the central axis of the guide. Throughout the guide, but especially in the buncher section, mutual repulsion would otherwise cause the electron beam to diverge. These losses would not only result in a reduced beam current, but they could also seriously damage the guide structure. Again, the standing-wave tolerances are tighter, and as the waveguides tend to be wider (especially those using side cavities), more powerful and correspondingly larger solenoidal focus coils are necessary.

Because a narrow beam of electrons travelling parallel to the axis of the waveguide is required, it should not be surprising that all systems employ some beam steering elements in their design. So-called steering coils, usually a pair of Helmholtz coils[*], are mounted at critical points along the guide to steer the beam *radially* R (i.e. along the treatment couch) or *transversely* T (i.e. across the couch). Adjustment of currents through coil pairs mounted near the gun end of the accelerating guide (commonly referred to as 1R or 1T for their radial or transverse effects, respectively)[†] are made during installation or following a major repair. Coils mounted at or near the bending chamber (2R and 2T) are used to adjust beam centring either manually during servicing or automatically via feedback circuitry from any unflatness signals monitored by the transmission ion chamber (see Section 11.7.3.2). This latter dynamic aspect of accelerator control is critical for safe operation. Any beam

---

[*] Helmholtz coil pairs are commonly used to obtain a region of uniform magnetic field between them. The arrangement comprises two equal coaxial circular coils set parallel to one another with their centres at a distance apart equal to the radius of either coil. With the same current passing through each coil and wound in the same direction, a uniform magnetic field is established between them. For steering or deflection purposes, it is sufficient to vary the current in either coil.

[†] The 'R' and 'T' designations are manufacturer dependent and other conventions may be used for the same purpose.

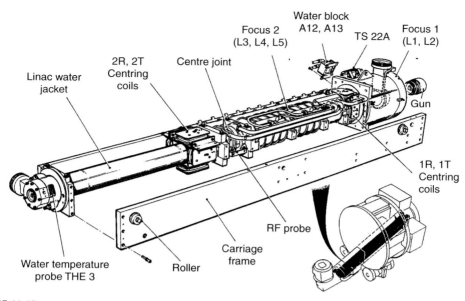

FIGURE 11.13
Focus and centring coil positions for a travelling waveguide. The insert at the bottom of the figure shows how this waveguide is mounted on the accelerator gantry. (Reproduced by permission of Elekta Oncology Systems.)

misalignment would result in the beam striking the target at the wrong angle, leading to a non-uniform x-ray beam with possible effects on beam energy. Figure 11.13 shows the focus and centring coil positions on a travelling waveguide (notice that it has not been necessary to have solenoidal focus coils around the second half of the waveguide).

11.3.6 THE BENDING SYSTEM

Ideally, the accelerating tube would be mounted with its axis parallel to the central axis of the radiation beam. However, except for low energy linacs with standing waveguides, the resulting height of the linac would be prohibitive. It is necessary to mount the accelerating waveguide approximately at right angles to the direction of the radiation beam. This requires that the electron beam is bent through an angle of approximately 90°.

Electrons moving in a magnetic field will be bent in a trajectory dependent on the energy of the individual electron. The requirement for the generation of satisfactory clinical beams is that all the electrons shall exit from the accelerating tube within a small focal spot and at the correct angle. Ideally, the electron beam at the input to the bending chamber will have a very narrow energy spread so that all the electrons will automatically follow the same trajectory. Such precise control of beam energy is difficult to achieve and a more stable solution is to design a so-called *achromatic* bending system where electrons of different energies exit at the same point and in the same direction in spite of their having followed different trajectories. *Non-achromatic* bending systems do, however, have the advantage that the bending magnet acts as a very precise system for ensuring that the energy is correct. In achromatic systems, some other means must be found to monitor beam energy (see Section 11.7.2.3). In all systems, it is of the utmost importance to set the magnet currents at a very precise and stable value, depending on the energy, because it will have a direct influence on the electron trajectories.

Many systems have found commercial success over the years, each having their advantages and disadvantages. Five of them will be briefly discussed, but for a fuller coverage, the reader is referred to Karzmark et al. (1993).

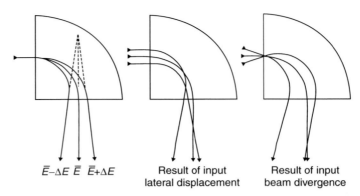

$\bar{E}-\Delta E$ $\bar{E}$ $\bar{E}+\Delta E$ Result of input   Result of input
                                      lateral displacement   beam divergence

FIGURE 11.14
Simple 90° non-achromatic bending system.

11.3.6.1 *90° Bending Magnet*

The simplest bending system provides a 90° bend using a simple dipole magnet. This was the bending system employed on many early linacs. Such a system is not achromatic. In Figure 11.14, the effect of energy spread lateral displacement, and beam divergence on electron trajectories through such a magnet can be seen. Precise control of beam energy is difficult to achieve and as a result, target diameters in excess of 2 cm were common, and even with large flattening filters, fluctuations in x-ray uniformity were considerable. Exit window *slits* were tried in order to trim the electron beam, but these became secondary sources of x-rays, further broadening the effective focal spot.

11.3.6.2 *A 270° Magnet Using Hyperbolic Pole Faces*

Figure 11.15 shows how, by shaping the pole faces, more energetic electrons (larger radii) enter closer spaced regions with higher bending fields, and less energetic electrons encounter lower bending fields between the wider pole spaces. In principle, all electrons that entered on

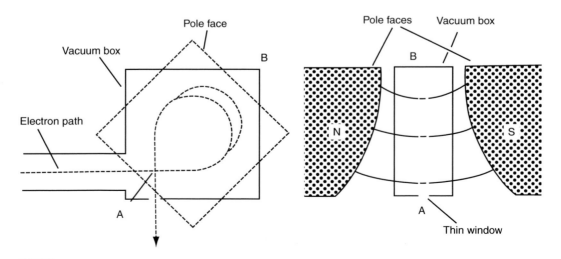

FIGURE 11.15
Bending chambers with hyperbolic pole faces. (From Greene, D. and Williams, P. C., *Linear Accelerators for Radiation Therapy*, Institute of Physics, Bristol, 1997. With permission.)

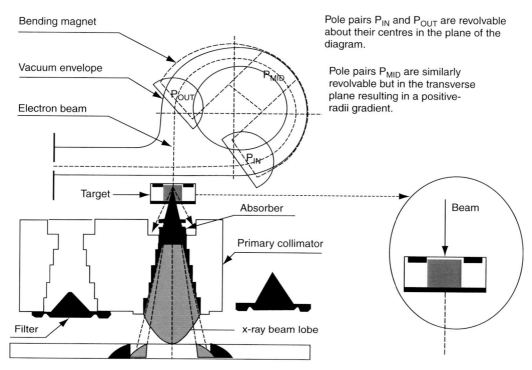

Bending magnet

Vacuum envelope

Electron beam

Pole pairs P_{IN} and P_{OUT} are revolvable about their centres in the plane of the diagram.

Pole pairs P_{MID} are similarly revolvable but in the transverse plane resulting in a positive-radii gradient.

P_{MID}

P_{OUT}

P_{IN}

Target

Absorber

Beam

Primary collimator

Filter

x-ray beam lobe

FIGURE 11.16
Locally tilted pole pieces bending chamber and target assembly. (Reproduced by permission of Siemens Medical Engineering.)

axis at 0° should converge again at the same point at 270°. The deflection is without dispersion with energy (i.e. achromatic). This system was used in the ABB/Dynaray units, but it requires stringent manufacturing tolerances in the pole face shaping and is a costly component.

11.3.6.3 A 270° Magnet with Locally Tilted Pole Pieces

Figure 11.16 shows the configuration of locally tilted pole pieces that are used in Siemens Mevatron and Primus accelerators. In this system, both uniform and non-uniform field regions are combined in the same magnet sector. This focuses a range of entrant momenta and a range of lateral displacements and divergences to be geometrically equivalent at its output and rotated by 270°. It is achieved by having two adjustable pole sections of the magnet providing an increased radial field gradient. At higher radii, more energetic electrons encounter a narrower spaced higher bending field and are focused to a single spot at the x-ray target. Entrant and exit pole faces are adjustable for optimal radial and transverse focusing.

11.3.6.4 A 270° Three Sector System

The 270° three sector design, favoured by Varian, is shown in Figure 11.17. It comprises three uniform dipole magnets separated by short drift spaces. A narrow pair of *slits* limits the transmitted energy to within 3% of that intended and where any bremsstrahlung is produced, it is directed away from the useful beam into the shielded head. The whole assembly is enclosed in an evacuated bending chamber which also includes the target sections and radial and transverse steering coils. Notice the target position lies above the accelerator guide axis, which allows a lower isocentre position for what would otherwise be a large magnet assembly.

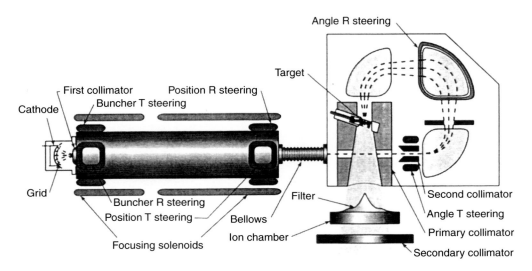

FIGURE 11.17
Three sector bending chamber. Also shown are the guide and chamber steering coil positions. (Reproduced by courtesy of Varian Medical Systems.)

11.3.6.5 A 112.5° Double Focusing (Slalom Bend) System

The slalom bending system (Figure 11.18) pioneered by Philips (now Elekta) is intended to minimise the vertical height associated with other achromatic bending systems. The purpose is to reduce the isocentre height of the overall machine. Maximum energy dispersion occurs between magnets M2 and M3, but the energy slit S is located at the entrance to the second dipole magnet to minimise the x-ray shielding that would otherwise be required closer to the treatment head. In the lower figure, the transverse or Y-field is shown being weakly focused at the exit plane. This is achieved by having the entry and exit pole faces of each dipole magnet slightly rotated. With better beam control systems in the early part of the waveguide, it has proved possible to dispense with the energy slit.

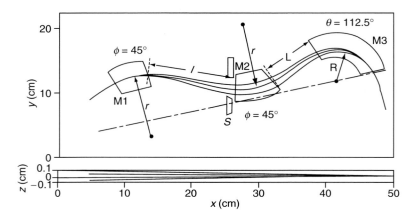

FIGURE 11.18
Slalom bending magnet system. (From Botman, J. I. M., Bates, T., and Hagedoorn, H. L., *Nuclear Instruments and Methods in Physics Research*, B10/11, North Holland, Amsterdam, 1985, pp. 796–798. With permission.)

11.4 THE HEAD ASSEMBLY

The accelerating and bending structure described above will deliver a narrow focussed beam of electrons that would be of little use for patient treatment. In order to generate a beam that can be used for patient treatment, this beam must either be converted to photons or the narrow electron beam must be appropriately scattered. An example of the features contained in the head is shown in Figure 11.19.

11.4.1 PHOTON BEAM GENERATION

11.4.1.1 The X-Ray Target

Just as with kilovoltage x-rays, megavoltage x-rays are generated by bremsstrahlung from a high energy electron beam striking a high atomic number metal target, usually tungsten or a copper–tungsten laminate. However, as the energy is increased, the bremsstrahlung emission becomes increasingly directed forward, so that at megavoltage energies, the target acts as a *transmission target*. The target is water cooled, but because of the higher efficiency of x-ray production, the target assembly is sufficiently compact so that it can be moved out of the beam to enable an electron beam to be delivered.

11.4.1.2 The Flattening Filter

The bremsstrahlung generated when a high energy electron beam strikes a target is primarily directed forward as shown in Figure 11.20a. For this reason, it is necessary to

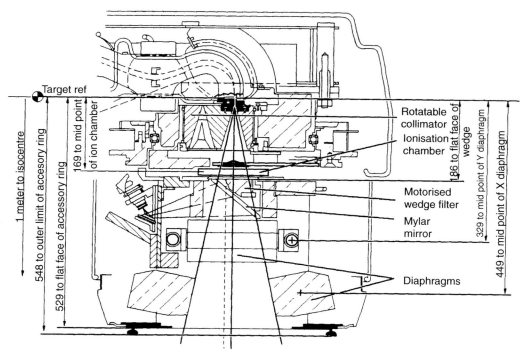

FIGURE 11.19
Diagram of the dual energy treatment head of an Elekta SL series accelerator showing the position of the primary collimator, flattening filter, wedge filter, and secondary collimators. Note that for the high energy beam, the high energy flattening filter (shown to the left of the beam axis in the diagram) is used in conjunction with the low energy flattening filter, which is the solid object just above the ionisation chamber. The numbers are distances expressed in mm. (From Elekta Oncology Systems.)

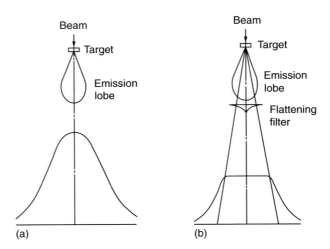

FIGURE 11.20
Linear accelerator beam without (a) and with (b) flattening filter.

insert a flattening filter in the beam as indicated in Figure 11.19. The effect of this is shown in Figure 11.20b. The use of a flattening filter makes it possible to compensate for the lack of scatter at the edge of the field by deliberately designing a profile that increases toward the edges. The flattening filter is generally circularly symmetric; although, additional material may be needed especially for machines with non-achromatic bending systems. The design of the flattening filter can have a considerable impact on the radiation beam, and this will be discussed further in Section 22.3.3.2.

11.4.1.3 Visual Beam Position Indication

In order to enable the treatment staff to see where the radiation beam will be located, a light beam is generated that coincides with the useful limits of the beam. This is obtained by a high intensity light bulb containing a small filament that must be optically coincident with the source position. It can be mounted on a *moving mechanism* or, as shown in Figure 11.19, can be placed outside the radiation field using a *mirror* to ensure that the image of the light source appears to be at the same point as the radiation source. A graticule is also included in the optical path so that the centre of the radiation beam can be indicated as a cross.

11.4.2 Collimation of Rectangular Fields

The radiation beam must be constrained in some way to ensure that only the required part of the patient is irradiated. In order to meet the requirements for very low dose rates at a large distance beyond the edge of the radiation field, accelerators have a *circular primary collimator* close to the source. This limits the field to a circular shape. On most accelerators, this circle is slightly smaller than the diagonal of the largest square field so that the corners of the field will be cut off (see, for example, Prasad and Bassano 1991). In the Elekta SL25, the maximum field size of 400 mm square at 100 cm distance from the target has a diagonal of 565 mm, but the primary collimator limits the field to a maximum circle of 500 mm. The Varian 2100C has the same maximum field size and a maximum circle of 495 mm. Other manufacturers' current equipment is similar. It is important to be aware of any particular machine's limitations in this respect.

The *secondary collimators* (sometimes also called *diaphragms*) is usually made of flat-faced blocks of lead that move in an arc in order that the face of the block shall be aligned with the field's divergent edge. Inevitably, one of the collimator jaws will be closer to the source than the other, and as a result, the sharpness of the edges of the beam will be different. Various approaches have been used to try to make the two collimators more nearly equivalent. In the

CGR head, an extra penumbra trimmer was attached around the outside of the upper jaw. This has the effect of reducing the penumbra by around 1 mm. Such methods may increase the size of the head or restrict jaw movement across the centre. However, it is more common to use a single block for each collimator jaw as this facilitates the generation of asymmetric fields and large fields while minimising the size of the collimator assembly. The thickness of the jaws and the relationship between the source size and the distance of the front face of the collimators must be such as to achieve the desired beam sharpness. This is discussed further in Section 22.3.3.

The geometric basis of the penumbra (see Section 22.3.3.1 and Section 22.5.1) dictates that the 50% field size is defined by the geometric edge of the field. However, optical systems may indicate a slightly larger field size than the useful radiation beam, which can be rectified using optical penumbra trimmers. These usually are thin strips of metal attached to the collimator so that they obscure the optical field while not blocking the radiation beam. It used to be common practice to use optical penumbra trimmers to make the optical field demonstrate the 80% field margins (i.e. the limits of that part of the field where the dose is greater than 80% of the dose on the central axis). Although this had the advantage of discouraging radiation oncologists from applying too narrow field margins, it does lead to difficulties when matching fields, especially using asymmetric collimation. Optical penumbra trimmers can also lead to confusion in setting up the field because it is very easy for them to become asymmetrically set, thereby masking other problems. For all these reasons, it is better to avoid their use and accept a slight mismatch in the field size. The standard specification is that the optical field must agree with the radiation field to within 2 mm, although, most centres will aim for agreement within 1 mm.

The collimation system and head shielding is required by IEC 601 (IEC 1998) to ensure that:

> "In a plane circular surface of radius 2 m centred on and perpendicular to the axis of the beam at the normal treatment distance [isocentre] and outside the useful beam, the absorbed dose due to leakage radiation (excluding neutron radiation) shall not exceed a maximum of 0.2% and an average of 0.1% of the absorbed dose measured at the point of intersection of the axis of the beam and the plane surface. These values refer to measurements which have already been averaged over 100 cm^2. Except in the area defined above, the absorbed dose due to leakage radiation (excluding neutron radiation) at 1 m from the path of the electrons between the electron gun and the Target or electron window [i.e. the target] shall not exceed 0.5% of the maximum absorbed dose on the axis of the beam at the normal treatment distance."

Provided that the collimation system has been correctly designed, any problems in meeting this specification are likely to be caused by poor beam steering resulting in the electron beam striking the edges of the waveguide and generating x-rays at a point other than the target.

Conventionally, jaws are labelled X1, X2 and Y1, Y2. The current IEC specification known as IEC1217 (1996) defines the orientation of these jaws as follows: With the gantry vertical and the head at 0°, Y2 is at the gantry end, and X2 is on the right side (i.e. supine patient's left) as viewed facing the gantry. Clockwise rotations as viewed from the isocentre are positive. This is illustrated in Figure 11.21.

As an illustration of the conventions and limitations which can be found on current machines, on Elekta accelerators that conform to IEC1217:

1. The built-in wedge toe points towards Y1
2. The X jaws are closest to the source and have a poorer penumbra
3. The X jaws will cross the centre-line by 125 mm (asymmetric head only). (100 mm on the Varian 2100C)
4. The Y jaws will only move as far as the centre-line (asymmetric head only).

On older Philips accelerators, the X and Y jaw labelling is reversed. Older Varian machines used A and B rather than X and Y. In spite of the IEC1217 recommendations, there are still

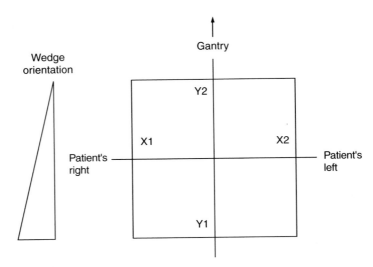

FIGURE 11.21
Labelling of collimator jaws according to IEC1217 (supine patient, head towards the gantry). Wedge orientation shown is also consistent with IEC1217 and is used for the Elekta built-in automatic wedge.

differences according to the manufacturer or even to the accelerator model in the method used to name the jaws and also in the orientation of the wedges. This is further discussed in Section 11.5.4 and illustrated in Figure 11.31. In all cases the collimator dimensions are displayed to be representative of the field size at the distance of the isocentre (i.e. usually at 100 cm see Section 11.5.3.1).

11.4.3 MULTILEAF COLLIMATORS

Conventional collimators are only able to constrain the radiation to a rectangular shape. In order to provide more flexibility, multileaf collimators have been introduced. Instead of a single block of metal, these have up to 80 pairs of leaves that can move independently, allowing any beam shape to be produced subject to the width of the leaves.

There are many variants of the multileaf collimator; some are commercially available, and others are designed and built for specific applications by research groups. The principle is the same in each case. A detailed description of systems available up to 1993 can be found in Chapter 5 of "The Physics of Three Dimensional Radiation Therapy" Webb (1993).

Modern multileaf collimators for use in radiotherapy started to appear in the 1980s. Early developments of multileaf collimators took place in Japan. The first European development took place in Scandinavia where specialised MLCs were designed to fit onto the Scanditronix race-track microtron. Philips (now operating as Elekta) and Varian introduced commercially available MLCs in 1990 in Europe and in the USA. Siemens and GE were, at that time, a few years behind.

There are three main types of MLC:

- *Type A* (e.g. Scanditronix and Siemens): The MLC provides all the collimation (except for the primary collimator) and completely replaces the standard collimation system
- *Type B* (e.g. Elekta): The MLC provides field shaping but additional shielding is provided by backup collimators. The MLC (including the backup collimators) replaces the standard collimation system
- *Type C* (e.g. Varian): The MLC provides field shaping in addition to the standard collimation system. Although part of the head assembly, it is externally mounted and complements the standard collimation system.

In addition to these distinctions, there are several ways of measuring and controlling the position of each leaf. This will be discussed later, but it should be noted that some type C high resolution research systems have been constructed without any form of motor control. In these, the leaves were positioned manually. One way of doing this was to produce a template of the proposed field, scaled to the position of the leaf banks. Each leaf was then pushed to be in contact with the template before its position was locked. This has been used mainly for the mini-MLC (see Section 11.4.3.2).

Table 11.1 summarises the specification of the Varian, Elekta, and Siemens MLCs. As specifications are subject to change, current manufacturers' literature should be consulted for up-to-date definitive specifications.

11.4.3.1 *Design Features and Relationship to Performance*

The following description centres mainly on the Elekta type of system (Jordan and Williams 1994), but most of the considered points are applicable to the other configurations. More information can be found in Galvin et al. (1992) for Varian and in Boyer et al. (2001).

Figure 11.22 shows the arrangement of the jaws in the Elekta head. In this example, the two banks of tungsten leaves are mounted inside the treatment head immediately below the flattening filter and the housing of the motorised wedge. In this position, the leaves are at approximately 1/3 of the distance from the x-ray source to the isocentre and are approximately 3 mm wide in order to project to a leaf width of 10 mm in the isocentric plane at 100 cm from the target. Manufacturing tolerances are very tight in this design. MLCs with the leaves nearer the patient will not require such tight tolerances. The leaves are tapered so that each of their edges can be focused toward the target. The leaf thickness of about 7 cm is sufficient to reduce the intensity of the primary beam to 1%. Additional attenuation is provided by the backup collimators, the requirements of which will become apparent after further consideration of the leaves. In this design, the leaves are mounted on rollers that allow them to be moved across the beam by lead-screw mechanisms driven by small DC motors. There has to be a mechanical clearance between the leaves to permit easy movement, and this results in leakage between adjacent leaves that is minimised by the use of stepped and overlapped leaf sections as shown in Figure 11.23 and by keeping the gap as small as reasonably possible. The Varian design has two steps, forming a tongue and groove joint that has the same effect.

As long as the overlap is maintained, the maximum leakage between leaves will be approximately 10%, but in practice, lower leakage is achieved because the gap, being very small, does not subtend the full focal spot from points in the isocentric plane. The average leakage through banks of leaves designed in this way is about 2% with local maxima of up to 5%. Figure 11.24 shows the leakage and transmission measured through leaves alone (i.e. without backup collimators) for the prototype of this collimator. The peaks are on a 10 mm pitch as expected.

Such levels of leakage over the full area exposed by the primary collimator are not acceptable, and the backup collimator moving in the same direction as the leaves provides attenuation by a further factor of about 10 so that areas shielded by both leaves and backup collimators receive a dose of less than 0.5% of the dose in the beam. A more difficult leakage problem arises between the opposing banks of leaves. When a pair of opposing leaves is required to be closed, the gap for mechanical clearance is exacerbated by the curved profile, shown in Figure 11.25, the need for which will be explained below. The projection of this gap onto the isocentric plane can be slightly reduced by setting the closed position away from the central axis, but the leakage will approach 100%. Therefore, the backup collimator moving normal to the leaves must provide attenuation to about 1% of the open beam intensity. The intensity in the shielded areas outside a shaped beam is clearly dependent on which shielding elements are called into play. Table 11.2 gives the transmission through each of the elements of the Elekta MLC. Transmission at each point can be calculated from the products corresponding to various combinations. Note that where MLCs are used in addition to normal collimators, both the X and Y collimator will have similar transmission at less than 1%.

TABLE 11.1

Parameters of Multileaf Collimators Available in 2006 (Manufacturers' Specifications)

	Elekta	Elekta Beam Modulator	Varian MLC-80	Varian MLC-120	Siemens 82 leaf	Siemens 160 leaf
Type	B	A [a]	C	C	A	A
No. of leaf pairs	40	40	40	60	41	80
Leaf pitch at isocentre [b]	10 mm	4 mm	10 mm	40×5 mm+20×10 mm	41×10 mm [c]	80×5 mm
Field size at isocentre [d]	40×40 cm	16×21 cm	40×40 cm	40×40 cm	40×40 cm	40×40 cm
Interdigitation possible	No	Yes	Yes	Yes	No	Yes
Maximum over-travel	12.5 cm	11 cm	20 cm [e]	20 cm [e]	10 cm	20 cm
Leaf transmission	<2%	<1%	<2.5%	<2.5%	<1%	
Inter-leaf leakage	<5%	<1.7%	<4%	<3%	<2% (typically 1.3%)	
Positional accuracy	1 mm	0.5 mm	1 mm	1 mm	1 mm	0.5 mm
Leaf speed	20 mm/s		25 mm/s	30 mm/s		40 mm/s
Head to isocentre clearance	45 cm		41.5 cm	41.5 cm	42.8 cm (including fixed block tray)	42.8 cm (including fixed block tray)

[a] The Elekta Beam Modulator does not have jaws orthogonal to the MLC (Patel et al. 2005).

[b] Note that the length of a field need not be an integer multiple of the leaf pitch. It can be trimmed with the backup or standard collimator except in the case of the Elekta Beam Modulator.

[c] Siemens have developed a system called HD270 for moving the couch during treatment to produce a virtual leaf width from 10 to 2 mm. Unlike the other manufacturers, the centre of the centre leaf is on the central axis rather than the junction of two leaves.

[d] Elekta's MLC aperture is constrained to have rounded corners at primary field diameter of 50 cm.

[e] Subject to max separation of 15 cm for leaves belonging to the same bank.

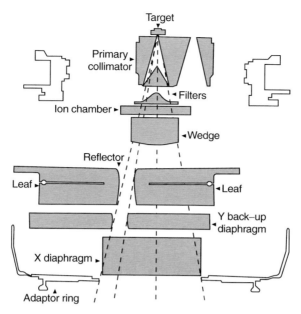

FIGURE 11.22
Diagram of Elekta MLC head. The tungsten leaves are mounted between the motorised wedge and the back-up diaphragm.

Of the three manufactures, only Siemens drives the leaves along arcs centred on the focal spot and, therefore, achieves a double focussed design with each leaf having a flat leading edge. Elekta and Varian use a curved leaf profile, shown in Figure 11.25, that provides a method of minimising the penumbra without the complex mechanical arrangements necessary to keep the flat edge of a conventional collimator aligned with the x-ray source. The curvature allows some penetration of the beam through the edge of each leaf, but this is the same for all field sizes. The penumbra (i.e. the distance between the points receiving 80% and 20% of the dose on the beam axis) is increased by penetration through the collimator. The primary beam intensity is attenuated to 80% along a ray passing through only approximately 7 mm of tungsten. This small depth is achieved within a fraction of a millimetre of the leading edge of the leaf. This is almost independent of the form of the leading edge. The position of the ray attenuated to 20%, having passed through 50 mm of tungsten, is more dependent on the form and can be several millimetres from the leading edge. Use of a circular form ensures that the chords along which 20% transmission occurs are at equal depths: the depth of penetration corresponding to the additional width of the geometric penumbra compared to a perfectly focused system. Because this is true for all chords, irrespective of the angle

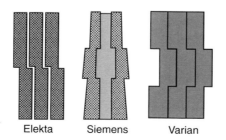

Elekta   Siemens   Varian

FIGURE 11.23
Stepped leaves to reduce leakage showing the method employed by different manufacturers. (From Huq, M. S., Das, J. J., Steinberg, T., and Galvin, J. M., *Phys. Med. Biol.*, 47, N159–N170, 2002. With permission.)

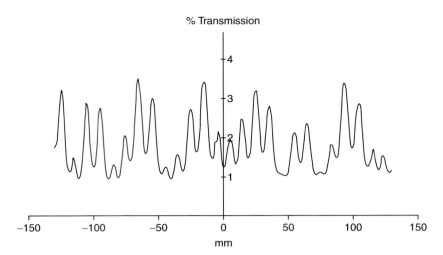

FIGURE 11.24
Example of leakage between leaves of the Elekta multileaf collimator (without back-up jaws).

of incidence, the degradation of geometric penumbra is independent of the position of the leaf. The penumbra width obtained by this simple design feature is 7 mm, which is only 1 mm greater than that of the standard collimator jaw. Because the position of the beam edge is defined by a different part of the leaf, depending on where it is relative to the central axis, the field size does not change linearly with leaf movement. Therefore, it is necessary to have a look-up table to correct for this.

Clearly, the positional control of up to 84 moving elements of a multileaf collimator (i.e. up to 80 leaves and 4 backup collimators)[*] is somewhat more complicated than for simple symmetric or independent systems. The simplest way to measure the position and to provide control signals for positional control is to use potentiometers and/or shaft encoders. Varian and Siemens use both on their MLCs. Scanditronix and Elekta use another system, the principle of which is shown in Figure 11.26.

In the Elekta system, each of the leaves is fitted with a reflector that is illuminated by the main optical beam. In addition, there are four fixed reflectors mounted on the collimator frame. These are then viewed by a CCD video camera via an optical beam splitter. The video image is captured 12.5 times per second (a sub-multiple of the video frame rate) and processed within the computer to identify the four reference reflectors and each of the 80 leaves. Failure to find a reflector within predefined limits of position and brightness causes the system to interrupt and the *MLC ready* interlock to the accelerator to be opened. If any spurious reflections are detected, this also interrupts the process.

The calculation of the position of each leaf from the measured positions of each of the reflectors in the video image is based on a stored calibration table. The measurement of the position of the reference reflectors provides a continuous check that the stored calibration data and the algorithms for the calculation are valid. The measured position of each leaf is displayed on a graphic display so that the radiographer can visualise the shape of the field defined by the 80 leaves. The accuracy of leaf positioning has been shown to be well within specification with a root-mean-square positional error of between 0.1 mm and 0.2 mm.

Movement of the leaf positions is by miniature DC motors, each driving a small lead screw through a nut fixed in each leaf. Control is achieved by servo control where the actual position of each leaf is compared with the prescribed position.

---

[*] Up to 120 or 160 leaves for the more recent designs of Varian and Siemens MLC respectively.

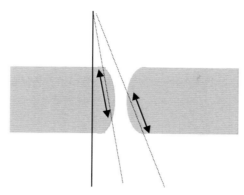

FIGURE 11.25
The penetration through curved leaves is independent of leaf position.

11.4.3.2 Mini and Micro Multileaf Collimators

The reduction in treatment volume afforded by the use of standard MLCs is comparable with that of customised divergent blocks, but the precise delineation of the edge of the treatment volume is, in many cases, compromised by the projected width of the MLC leaves. The stepping of the field edge, caused by an MLC with a projected leaf width of 10 mm, has been identified as a potential limitation in the restriction of morbidity in up to 19% of cases (Galvin et al. 1992). The limitations imposed by the leaf width have been addressed by smoothing techniques, and the development of MLCs with narrower leaves that are described as *mini* or *micro* MLCs. The distinction between mini and micro is arbitrary, but it has been suggested that MLCs with leaves projecting to between 2 mm and 5 mm should be considered as mini MLCs and those below 2 mm as micro MLCs (Schlegel et al. 1992; Bortfeld et al. 2000). It should be noted that by this definition, the higher resolution integrated MLCs available from each of the linac manufacturers shown in Table 11.1 are mini multileaf collimators (at least for part of the field).

Field edge smoothing can be achieved by the superposition of multiple fields, each of whose centres are shifted slightly relative to the next (Galvin et al. 1996), and this is implemented commercially in the Siemens HD270 facility. An alternative method of smoothing is to superimpose multiple fields rotated about the collimator axis. Superposition techniques provide smoothing, but not increased, resolution; they are severely limited in cases where it is necessary to shape a field with features smaller than 1 cm in the direction normal to that of leaf movement. A further disadvantage is that movement of the patient between fields increases the complexity of verification compared to that required for treatment with a single field.

Improved resolution can be achieved by the use of MLCs with narrower leaves. These are generally tertiary collimators mounted as accessories below standard collimators

TABLE 11.2

Elekta MLC Transmission at 20 MV (X and Y as defined in Figure 11.21)

Collimator Section	Transmission
Leaves (X direction)	2% (mean)
Gaps between leaves	4.3% (max)
Y jaw	0.5%
X jaw (back-up)	12%

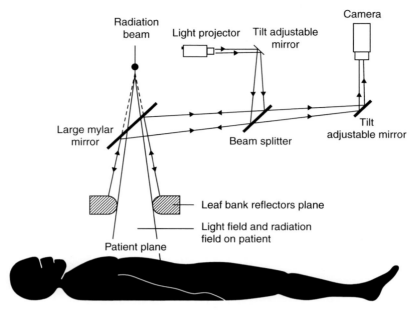

FIGURE 11.26
The optical system for measurement of leaf position for the Elekta MLC.

(or below conventional MLCs that replace the standard collimator). The geometric features of four such MLCs are listed in Table 11.3.

The number of leaves and leaf width determine the maximum field size, which in all cases is much smaller than for a conventional MLC. This has been justified if the main application is in stereotactic radiosurgery where it has been estimated that 95% of cases can be treated within the range of field size from 1 to 6 cm (Schlegel et al. 1997). In other applications, if precise shielding of a critical structure adjacent to the periphery of a large field or to a field that is significantly offset relative to the central axis is desired, these devices will have limitations. If they are to be used to provide precisely defined boosts to fields larger than the maximum covered by the mMLC, it would be necessary to attach the mMLC to the treatment machine between delivery of the main and boost fields to the patient. This would not be practicable in day-to-day use.

TABLE 11.3

Parameters of Mini and Micro MLCs not Integrated into the Treatment Head

	MRC (now Siemens)	MRC (now Siemens)	BrainLAB m3	Radionics	3D Line (IBA-Wellhöfer)
Leaf width (mm) (at isocenter)	1.6	2.5	3.0–5.5	4.0	4.5
Number of leaf pairs	40	40	26	30	20
Maximum field size (mm×mm)	73×64	120×100	100×100	100×120	90×90
Central axis overtravel (mm)	24	60	50	14	25
Reference	Schlegel et al. (1997)		Xia et al. (1999)	Shiu et al. (1997)	Meeks et al. (1999)

FIGURE 11.27
Automatic wedge filter as used in Elekta linacs. When the wedge is to be inserted into the beam the wedge filter is driven by the motor and worm drive to a position beneath the mirror. (The mirror is used to produce the field light. In this figure the beam direction is upwards.)

11.4.4 Beam Modifiers

11.4.4.1 Mechanical Fixed or Motorized Wedge Filters

As discussed in Section 33.3.3, there are frequent occasions when, instead of a uniform intensity across the beam profile, a wedge-shaped intensity profile is required. *Traditional wedge filters* are simply wedge-shaped pieces of brass, aluminium, or lead that are placed in front of the collimators. This simple design is still an option offered with Varian and Siemens accelerators. Filters of different maximum thickness are used to provide different wedge angles; commonly, angles of 15°, 30°, 45°, and 60° are provided. These wedges are mounted below the collimator assembly. Philips (now Elekta) and CGR/GE have introduced the concept of the *automatic* (or *motorized*) wedge. This is mounted inside the head (Figure 11.19) and is automatically moved into the beam for part of the treatment. Figure 11.27 shows a picture of an automatic wedge. By adjusting the proportion of the treatment for which the wedge is inserted into the beam, the effective wedge angle can be adjusted. Some authors have published formulae for calculating the wedge angle (Petti and Siddon 1985; Zwicker et al. 1985), but it is probably not necessary to precisely know the angle. Design and dosimetry considerations for mechanical wedge filters are discussed in Section 22.6.1.

11.4.4.2 Dynamic Wedges

The *automatic* wedge has the potential disadvantage that it increases the distance between the source and the front face of the accelerator head. This is a particular disadvantage for machines that use either an in-line accelerator tube or a 270° bending system. Because of the inconvenience of having to enter the treatment room to insert a manual wedge, there was a need to provide some form of automation without decreasing the clearance between the machine and the patient. Varian was the first to solve this problem by developing the *dynamic* or *flying* wedge

(Leavitt et al. 1990) that is now available from other manufacturers as well. To achieve a wedged beam profile, one of the jaws is kept stationary, and the other is moved during the treatment. This technique is potentially more flexible than fixed wedges because, in principle, it is possible to design any shape of beam profile that is required, but it is most often limited to one jaw pair because of the requirement for the jaw (or leaf) to cross the central axis of the beam.

11.4.4.3 Accessory Holders

Also mounted beneath the treatment head is an accessory holder. This can be used to add a *shielding tray* where lead or low melting point alloy blocks or a tissue compensating filter can be placed. The accessories should be linked to the treatment machine verification system so that confirmation is provided that the correct accessory is being used (see Chapter 41).

11.4.5 ELECTRON BEAM GENERATION

11.4.5.1 The Initial Beam

For clinical electron beam production, the x-ray target and flattening filter are withdrawn, under interlock control, and the accelerated electron beam emerges from the vacuum system through a thin metal vacuum window. This is typically chosen to be a lower atomic number material and is designed to produce minimal scatter and bremsstrahlung. The exiting beam is narrow, typically a few mm in diameter, and is essentially Gaussian in profile.

The narrow pencil electron beam is not suitable for clinical use as reasonably uniform distributions are required over fields of up to 25 cm × 25 cm. Two methods are used to produce this, either *scattering foils* or *scanned beams*.

11.4.5.2 Scattering Foil Systems

Most linear accelerator designs use *scattering foils* to widen the beam. In this case, thin foils of higher atomic number metals are placed in the beam, usually positioned on the same interlocked slide or carousel system that also carries the x-ray beam flattening filter. The foil thicknesses are chosen to be a compromise between being thick enough to produce acceptable levels of electron scattering while being thin enough to ensure that energy loss and additional bremsstrahlung contamination of the beam are not too high. As this balance of electron interactions is energy-dependent, each electron energy requires its own foil, although, in practice, this is not always the case. The electron distribution produced across the field from a single foil is generally not flat; rather, it is a significantly-widened Gaussian distribution. The clinical beam flatness is then improved towards the beam edges by making use of in-scattered electrons from the sides of applicators to increase the electron fluence. This can be adjusted by changing the x-ray collimator positions, and the optimum settings can be selected and pre-programmed in for each electron-energy/applicator size combination. Most scattering foil designs are based on a *dual-foil system* with two sets of foils a few cm apart. The first is a uniform thickness higher atomic number foil as in the single foil design. It is followed by the secondary foil that is a roughly cone-shaped stack of lower atomic number foils. This produces a better balance of effects and can give a flatter beam across the required width. The in-scatter required from the applicator is then less. Scattering foil systems have the advantage that they are simpler and safer in use than the alternative. However, they produce more energy loss for a given spectrum beam and larger bremsstrahlung contamination of the beam.

11.4.5.3 Scanned Beam Systems

The alternative to produce the clinically required beam is to scan the pencil electron beam across the field in a *raster* pattern using a magnet system controllable in two orthogonal directions. Various patterns are used to produce uniform coverage. The spatial resolution requirement does not need to be too demanding as the pencil beam widens in travelling through intervening material such as the dose-monitoring chamber and the air. To produce a uniform distribution along the raster path and, therefore, across the field, the raster line frequencies should be low (typically around a few Hz) compared to the pulse repetition rate of the accelerator (typically tens to hundreds of Hz). Scanned beam systems produce cleaner beams with less bremsstrahlung contamination and narrower spectra. This results in rather deeper therapeutic ranges and steeper gradient depth-dose curves. However, they do not provide as much flexibility to get optimal field flatness, especially for lower energies, because at low energies the scanning system under-doses the outer part of the beam and this is more difficult to compensate than with a shaped scattering foil. Furthermore, the system is significantly more complex than scattering foils, and it has the potential for malfunction, posing a safety concern. In addition, the high local beam intensities over short periods while the pencil scans near any particular point mean that ionisation chamber dosimetry requires significantly higher recombination corrections. Scanning systems are not favoured on recent linacs.

11.4.5.4 *Applicators*

As electron scatter is so significant, applicators are necessary to delineate the field sharply close to the patient. In addition, the applicator is designed to help to achieve a flat beam over the required widths as discussed above. Single scattering foil designs required applicators with an upper section of solid wall to provide in-scatter. However, this requirement is less stringent in modern designs, typically resulting in lighter more open applicators and an example of which is shown in Figure 11.28.

The lower sections are made up of successive collimation layers that absorb electrons that would scatter out of the field. The final layer that defines the field size typically lies at between 0 and 5 cm from the patient surface. *Applicators* are constructed of low Z materials (e.g. aluminium, plastic, etc.) to minimise x-ray contamination. They are generally supplied as a set of squares, circles, and rectangles that attach and are interlocked to the accelerator via the accessory plate. As noted above, each applicator for each specific energy will have pre-set x-ray collimator positions assigned to it that should be automatically safety checked as part of the modality verification. A frame is generally provided near to or at the final section, into which irregular or non-standard size *cut-outs* can be positioned for tertiary field shaping.

Some manufacturers offer *continuously variable electron trimmers* that attach to the head and provide more flexibility in varying field shape with a reduced need for cut-outs.

11.5 THE GANTRY

To allow patients to be treated with beams from different directions, there is a need to be able to rotate the treatment head through 360° about the patient. The accelerator waveguide and the head assembly must be mounted on a *gantry* that enables the beam to be rotated around the patient. There are two distinct mechanisms for supporting linear accelerators now being used: the *drive stand* and the *drum gantry* supports.

11.5.1 DRIVE STAND GANTRY SUPPORT

The *drive stand* (Figure 11.29) is popular with all manufacturers for low energy machines and at high energy for those using standing-wave accelerator guides. Here, an upright stand is bolted to a base frame embedded in the floor to which is attached an "L" shaped support via a

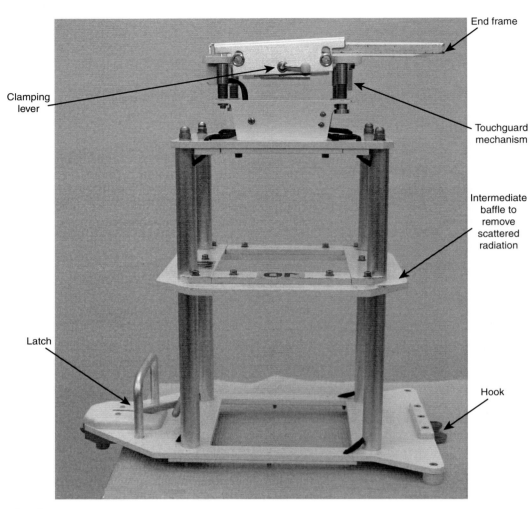

FIGURE 11.28
Open applicator (Elekta) suitable for a dual scattering foil system.

massive thrust bearing. It is usual to add weights to this support to counterbalance the mass of the waveguide and treatment head about the rotation axis. In some cases, this can be extended forward to act as a primary beam stop for use in rooms where there is a radiation shielding problem. The main attraction of the drive stand support is its compactness, and in recent years, much design effort has gone toward space reduction, enabling some manufacturers to avoid the need to have a separate modulator enclosure.

11.5.2 Drum Gantry Support

Manufacturers using high-energy travelling wave guides favour the *drum gantry mounting* (Figure 11.30). The guide length would otherwise require an excessively long gantry arm and place unacceptable out-of-plane loads on the central thrust bearings. With a drum mounting, the gantry arm can extend through on both sides of the support structure, allowing only enough overhang on the patient's side to cover at least half the length of the longest patient. Further, the counterbalancing load can be placed towards the rear of the drum and reduce the tipping out-of-plane moment as well. It is common with drum systems to have only the gantry

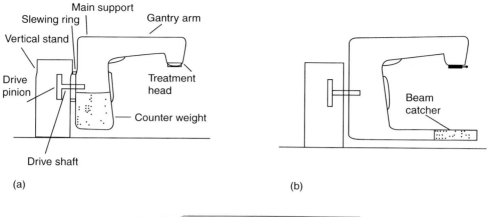

(a)   (b)

FIGURE 11.29
Drive stand gantry. (From Greene, D. and Williams, P. C., *Linear Accelerators for Radiation Therapy*, Institute of Physics, Bristol, 1997. With permission.) The counterweight can either be built into the base of the gantry as in (a), or a beam catcher can be used. The picture shows a Varian 2100C (without beam catcher) without its covers. The klystron can be seen at the extreme left of the picture.

arm on view protruding out of the surface of a wall panel. This allows the working components to be left exposed (behind safety interlocked doors in the panel) that greatly facilitates access for servicing.

11.5.3 THE ISOCENTRE

The intercept between the beam central axis and the gantry rotation axis (also the collimator and couch rotation axes) defines the isocentre of a linac, which should ideally be a fixed point in space. A typical dual energy accelerator weighs nearly ten tons with more than one ton in the gantry arm. In addition, to allow treatments to any part of an adult's anatomy, longitudinal access space between the isocentre and the vertical gantry structure needs to be in excess of 110 cm. This results in a large out-of-plane torque that is exaggerated further by the mass of the bending chamber, head shielding, and collimator. The engineering problems are therefore considerable. In general, the centre of rotation will not be a single point, but the locus of the centre of rotation will be contained within a sphere whose diameter defines the size of the isocentre. For geometrically accurate treatments, there is a need to obtain tight isocentre

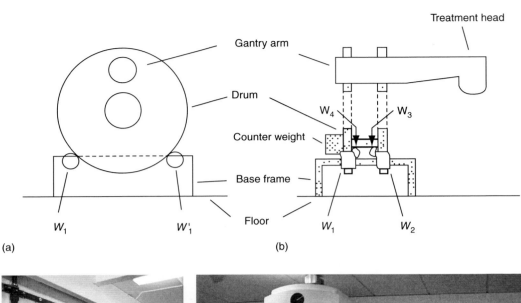

 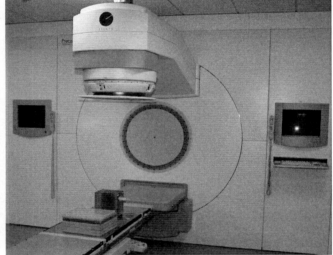

FIGURE 11.30
Drum gantry. Diagram (a) shows a section normal to the rotation axis, and (b) shows a section through the axis and the beam arm. The picture on the left shows the drum mounting from the rear. The picture on the right shows the accelerator as seen from the treatment room. (From Greene, D. and Williams, P. C., *Linear Accelerators for Radiation Therapy*, Institute of Physics, Bristol, 1997. With permission.)

tolerances. Some manufacturers have recently relaxed their specifications in machines for general use, but most can be set up with isocentric error surfaces having radii less than 1 mm.

11.5.3.1 *Isocentre Distance, Isocentre Clearance and Isocentre Height*

All current linac manufacturers employ 100 cm for their *isocentre distance* (i.e. the distance from the isocentre to the x-ray target). The *isocentre clearance* is the actual space available for the patient (i.e. the distance between the isocentre and the treatment head covers or accessory tray position). The *isocentre height* is simply the height of the isocentre above the floor. It is this last item that features rather large in accelerator sales information. Here it is argued (albeit correctly) that a high isocentre will force the treatment staff to use ladders or stools in order to carefully set the patient up (unless the staff are tall), costing time and fatigue.

To see what options are available to the designer, three combinations of accelerator and bending system will be considered:

(i) *A Low Energy Straight-Through Guide.* The benefits of the compact side cavity standing wave-guide, requiring no bending or beam steering elements, have already been described. Below the target, as with all machines, there must be a flattening filter, a dose-monitoring ion chamber, and collimation jaws. In practice, the only significant compaction available here is in the choice of wedge system: choosing dynamic jaw closure over mechanical wedges will save 5 cm height. However, at a gantry angle of 180° (undercouch), the balance of the overall guide length must be accounted for. Varian offers a retractable floor pit option that automatically operates when the gantry is within ±30° of 180°. This enables the isocentre to be kept moderately low for such a system: 133 cm for a 6 MV guide.

(ii) *A High Energy Standing Waveguide with a 270° Bend Magnet.* Above 6 MeV, it is no longer practical to use a straight-through guide mounting. Here, it is the vertical bulk of the various *Pretzel** bending magnets and the larger transverse dimensions of the standing waveguide and solenoid focus coils that the designers must contend with. In an earlier design, AECL chose to employ in-line cavities in their larger standing waveguides, and these, though longer, were correspondingly narrower than side cavity guides and had their isocentre heights reduced down to 131 cm. Other improvements can be made in the balance between isocentre height and clearance by careful positioning of the target in relation to the bending chamber. Varian mounts its x-ray targets within their large evacuated bending chamber at a position above the waveguide axis and achieves a 129 cm isocentre height as a result.

(iii) *A Travelling Waveguide Using 112.5° Slalom Magnets.* Characteristic of the travelling wave-guide is the reduced demand on beam focussing and narrower guide dimensions. If coupled with a vertically compact bending system, this combination should result in the lowest isocentre machines. At present, only Elekta produces high-energy machines of this type and can offer 125 cm isocentre heights. In principle, this could be further reduced if dynamic jaw closure was used instead of mechanical wedging.

11.5.4 STANDARD COORDINATE SYSTEMS

The IEC has defined the way in which the machine coordinates should be specified (IEC 1996). The details of this coordinate system are shown in Figure 11.31.

Not all machines (or even planning equipment) conform to this standard which can be problematic for networked departments (see Chapter 41 and Chapter 42). A common variant is the previous standard (IEC 601) where the couch rotation is reversed, and the X and Y jaws are differently labelled.

11.6 ANCILLARY EQUIPMENT

11.6.1 VACUUM SYSTEM

For clinical use, modern accelerators are expected to have a high up-time (i.e. they must rarely fail). Manufacturers now quote up-time figures as high as 98% on their new machines. One of the most important reasons for this is the systems' ability to provide a good vacuum. Typical vacuum levels of 10^{-7} torr are used (1 torr = 1 mm Hg = 133.3 Pa, 1 atm = 760 torr). This is to avoid arc discharges in any residual gas when exposed to high transient electric fields. This is particularly important in the accelerating waveguide where the phenomena of

---

* The electron trajectory, through typical 270° bend chambers, resembles that of the savoury Pretzel; hence, the name is often taken in describing such systems.

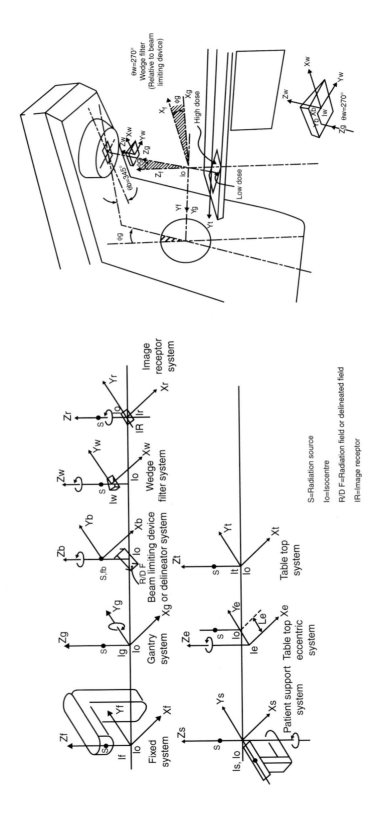

FIGURE 11.31
Coordinate system specified by IEC 1217. (From International Electrotechnical Commission, *IEC Publication 1217*, IEC, Geneva, 1996.)

multipactering may occur. This is where collisions of electrons with the guide wall are multiplied in intense field regions. When a scattered electron impacts the guide wall or iris, secondary emission electrons are liberated. These may then be swept up in localised field vortices and continue to impact the guide wall. An avalanche situation soon occurs and can lead to serious damage if ignored.

The minimum requirement of a vacuum system for a linac is that the accelerated electrons should not be deflected by collisions with gas molecules. Alternatively, the mean-free path between collisions needs to be long compared to the overall flight path. This distance is typically between 1 m and 3 m. The corresponding vacuum to satisfy this condition is roughly 10^{-5} torr. Early accelerators operated at these levels with the vacuum being provided by *oil diffusion pumps*.

11.6.1.1 Ion Pumps

The development of improved *ion pumps* was necessary to allow accelerators to be operated at higher r.f. field strengths and to increase gun (cathode) life. Figure 11.32 shows both *diode* and *triode* type ion pumps. Strong external permanent magnets (0.1 tesla to 0.3 tesla) and a cathode voltage of 5 kV are applied. Electrons caught in perpendicular electric and magnetic fields spiral toward the anode, increasing their path and chance of ionising gas atoms. This makes it possible to run a self-sustaining discharge at low pressures. Ionised atoms reaching the cathode liberate further electrons and keep the discharge going. Titanium is used for the cathode which sputters readily when bombarded by ions. Sputtered titanium atoms are deposited on the pump walls. Gas atoms striking the sputtered layer are adsorbed onto its surface and are subsequently buried in the next layer. The anode in the triode pump is held at the same potential as the wall, allowing the wall to collect low energy ions unable to sputter at the cathode. Clearly, an ion pump's life is determined by the consumption of the cathode material, typically 5000 h at 10^{-5} torr and 50 000 h at 10^{-6} torr.

Modern dual energy accelerators often employ three separate ion pumps: a large capacity (20 L/s) triode pump for the body of the waveguide and a smaller (2 L/s to 5 L/s) diode pump for the gun region. A third, usually a diode pump, is used to provide the vacuum for the klystron.

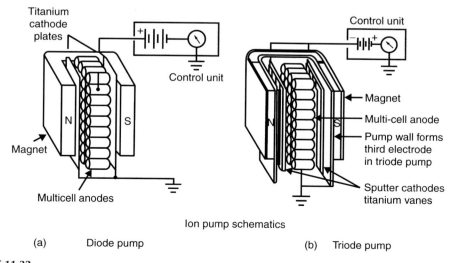

FIGURE 11.32

Ion pump schematics (a) shows a diode pump and (b) a triode pump. (Reproduced by permission of Varian Medical Systems.)

11.6.2 CIRCULATING WATER SUPPLIES

As already discussed, many accelerator components need cooling to shed heat, including the solenoidal focus coils, the waveguide, and the x-ray target. In addition, many of these components need the temperature to be maintained at a steady level. Linac *chiller* supplies circulate cooling water in a closed circuit through all the main components and return it to the secondary coils of a large *heat exchanger*. Flow rates at different points need to be carefully regulated to ensure a stable working system and can usually maintain temperatures to within 5°C.

11.7 THE LINAC CONTROL SYSTEMS

In order to deliver a stable radiation beam, *control circuits* are required. Beam characteristics are affected, among other things, by gantry angle, cooling water temperature, and ambient temperature as well as by long term changes such as component aging. Consequently, systems must exist both to correct for changes and to monitor errors. For multienergy machines, there are the safety issues of ensuring that the desired energy and modality is being delivered.

11.7.1 AUTOMATIC FREQUENCY CONTROL

Automatic frequency control (AFC) is employed in all linacs to keep them operating at their intended frequencies. In magnetron-powered systems, frequency changes may be necessary to correct for expansion effects at the anode where large amounts of heat are produced. Also, in gantry-mounted versions, different orientations in the earth's magnetic field will give rise to small frequency perturbing effects. Stability to a few MHz is needed, and this is provided via the AFC. In its simplest form, a *phase comparator* monitors the forward and reflected r.f. power and drives the *magnetron tuning plunger* to compensate. Traditionally, the tuning plunger has been driven by a rotary drive. A recent innovation is the so-called fast tuning magnetron where the tuning plunger is driven by a solenoid, allowing very rapid changes of frequency with consequent improvements in beam startup characteristics.

Frequency variations in klystron machines are more likely to result from guide temperature changes, and here, it is the frequency of the r.f. driver output that is adjusted to compensate.

11.7.2 ENERGY SELECTION AND CONTROL

The energy of the accelerated electrons is determined by the *r.f. power and phase* and by the *electron beam current*. How the energy is changed will depend on the accelerator system in use.

11.7.2.1 Magnetron Driven Travelling Wave Systems

For dual-x-ray use where the beam current needs to be high, energy selection is achieved by varying the *microwave power* supplied to the accelerating guide. Long cathode magnetrons with electromagnets allow both the magnetic field and the size of the HT pulse to be changed. As a result, outputs at 2.5 MW and 5.0 MW peak power are achievable. Low- or high-power operation is selected at the console, and different tappings are made on a high voltage transformer that feeds the *pulse forming network* discussed earlier.

High energy machines often provide the option to employ electron beams directly for treatment. Here, the beam currents are at least two orders of magnitude lower than for x-ray modes, and small *frequency adjustments* of the magnetron are sufficient to change the electron energy. A phase shifter, together with the AFC (see Section 11.7.1), maintains the frequency precision necessary.

11.7.2.2 *Klystron Driven Standing Wave Systems*

In klystron-driven systems, energy selection is also initiated by selecting high- or low-power modes in the modulator. This applies either 100 kV or 130 kV pulses to the klystron (compared to 40 kV or 50 kV pulses to a magnetron). These smaller pulse differences reflect the standing waveguide's inability to cope with the possible doubling of peak power levels that travelling waveguides can tolerate. Instead, extra measures must be taken to allow dual-x-ray operation. Figure 11.33 shows the main microwave power components of a standing wave machine. The output power pulse from the klystron is fed via a *directional coupler* (this prevents serious reflections getting back that could damage the klystron) to a *four port circulator*. This, in turn, directs r.f. power to the *shunt Tee assembly* that dumps a preset portion into a *water load* (port 2) where it is absorbed. The remaining r.f. then passes through a *rotary joint* that links the fixed klystron to the moveable gantry that houses the accelerating guide. The circulator also directs any reflected r.f. power from the accelerator to another water load (port 4).

Additionally (not shown), an *energy switch*, mounted in the accelerating guide, is inserted when the low energy mode is activated. This introduces a dielectric strip that changes the phase of the r.f. beam in the downstream portion of the guide, rendering it non-accelerating. In essence, for lower energy applications, these measures collectively allow the system to run at less than the optimum power levels for which the machine had been designed.

11.7.2.3 *Energy Monitoring*

As mentioned in Section 11.3.6, the path travelled by electrons within the bending system is precisely dependent on their energy. In 90° bending systems, the bending magnet current is an accurate measure of the energy, but in achromatic systems, it is necessary to insert *energy detecting slits* into the bending chamber. These offer the possibility to monitor the energy and to improve the electron spectrum because the electrons that do not have the proper energy are removed from the useful beam. However, there is the risk that large intensities would strike the slits and cause damage. An alternative method is to use the fact that the ratio between the dose at the centre and the periphery of a photon beam is energy dependent. This ratio can be measured by the transmission dose monitoring ionisation chamber. This so-called *hump detector* is then used to control the energy.

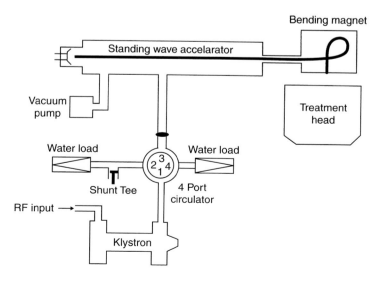

FIGURE 11.33
Microwave circuit for a high-energy klystron powered machine. (From Greene, D. and Williams, P. C., *Linear Accelerators for Radiation Therapy*, Institute of Physics, Bristol, 1997. With permission.)

11.7.3 CONTROL OF TOTAL DOSE, DOSE RATE, BEAM POSITION AND UNIFORMITY

As explained briefly in Section 11.4 and shown on Figure 11.19 and Figure 11.22, a *transmission ionisation chamber* (also called *dose-monitoring chamber* or simply *monitor*) is always present in the head of a linear accelerator. It is placed as close to the patient as possible but for practical reasons it is fixed and therefore located just above the rotating collimator assembly. It is designed to perform a number of different functions, aiming at delivering a precise, uniform radiation dose and at the correct energy. In use, the following are required:

- Monitoring and control of dose delivery whether in x-ray or electron beam mode
- Dose rate monitoring with rapid servo compensation
- Beam uniformity or symmetry and flatness monitoring also with dynamic correction
- Automatic treatment termination control if any of the above should lie outside acceptable tolerances.

Figure 11.34 shows the internal structure of a typical parallel plate linac monitor chamber. It consists of two *collector plates*, each divided into zones sandwiched between three uniform *field plates*. Notice that this creates two separate ion chambers mounted back to back, and it ensures that there is dual dosimetry operation for safety reasons. The lateral view that shows the plate stack has been expanded for clarity. To avoid sensitivity changes resulting from fluctuations in temperature and pressure, these chambers may be sealed. If they are not sealed, it is necessary to monitor the temperature and pressure of the chamber and to make appropriate corrections. Because of the pulsed nature of linac beams, high field strengths must be used to ensure operation under saturation conditions. Plate gaps of the order of 1 mm with polarising voltages of 500 V are sufficient and nearly a factor of two below dielectric breakdown levels. The chambers need to be wide enough to straddle the entire beam and are usually positioned immediately below the x-ray flattening filter. This is to reduce the diameter necessary and to be beyond the last beam perturbing element. As these devices are in-line monitors, they need to attenuate the beam as little as possible; therefore, a further requirement is that they should be thin and use low atomic number materials for their entry and exit windows.

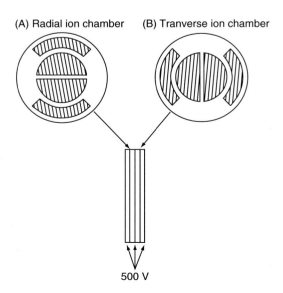

(A) Radial ion chamber   (B) Tranverse ion chamber

500 V

FIGURE 11.34
Typical multiplate ion chamber.

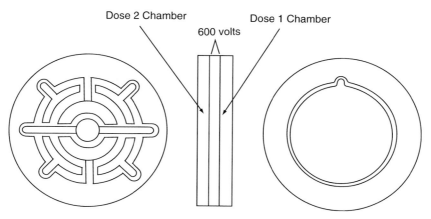

FIGURE 11.35
Alternative ion chamber plate configuration. (Reproduced by permission of Siemens Medical Engineering.)

Different manufacturers employ different collector plate configurations as shown in Figure 11.35 and Figure 11.36 which are examples of designs employed by Siemens and Elekta respectively. Clearly, the more structured or zoned the design, the more possibilities there are for monitoring beam flatness, etc. Two functionally distinct areas are important and are considered in the next two sections.

11.7.3.1 Dose Monitoring

With the beam on, the collector current in both chambers should be related to the actual dose rate being delivered to the patient. However, it is only possible to calibrate the chamber for one particular set of circumstances at a given position in the treatment beam (typically that at 100 cm for an unwedged 10 cm × 10 cm field at the depth of maximum depth-dose). The integrated current will, in principle, give a *measure of actual dose* delivered under these standard conditions. The integrated current associated with 1 cGy in the standard conditions is defined as one Monitor Unit (MU), (see Section 20.1.2) At the operating console, the number of monitor units required to deliver a treatment dose is set and the integrated current signal from the primary dosimetry channel is used to terminate

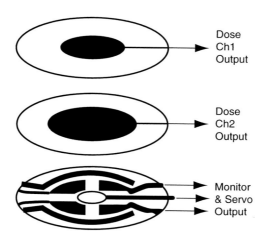

FIGURE 11.36
Design of ion chamber used by Elekta. Note the different size of the plates for the two dose channels. (Reproduced by permission of Elekta Oncolgy Systems.)

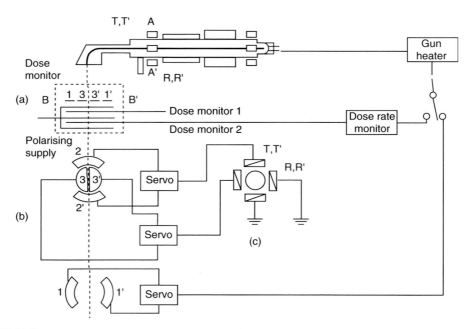

FIGURE 11.37
Dose monitoring and control system showing the dose-rate, beam-uniformity and beam-symmetry feedback controls achieved from the different zones of the monitor chambers. (From Greene, D. and Williams, P. C., *Linear Accelerators for Radiation Therapy*, Institute of Physics, Bristol, 1997. With permission.)

the beam delivery. The second dosimetry channel is typically allowed to over-run by a small percentage before it too would terminate the beam in the event that the primary channel should fail. With the design of Figure 11.36, the two dosimetry plates are sampling different parts of the beam so that a dose channel difference may be related to a change in beam energy rather than to a change in the calibration of the monitor ion chamber. Figure 11.37, shows how the signal from the dose monitor can be used to ensure a *constant dose rate* through feedback control of the gun heater.

11.7.3.2 Beam Uniformity and Beam Position Monitoring

Figure 11.37 also shows the links from the zoned collector plate structure to beam position and angle control circuit elements. Figure 11.38 illustrates how the plate positions can monitor changes in electron beam centring and angulation from variations in x-ray field uniformity. A simple two dimensional arrangement of four plate zones for monitoring the radial beam direction is shown. The two inner zones are labelled G_i and T_i for the Gun and Target directions, and the two outer zones are labelled G_o and T_o.

(i)   Ideal conditions with the electron beam both centrally and vertically incident over the target. Here signals from $G_i = T_i$ and $G_o = T_o$

(ii) The situation where, because of a steering error, the beam is no longer central. Here signals $G_i = T_i$ but $G_o > T_o$

(iii) The case of an inclined beam most commonly found in the radial plane. Here $G_i > T_i$ and $G_o > T_o$.

Imbalance in signals from any paired zones is a strong indicator of beam unflatness or asymmetry. Modern accelerator designs only allow very small excursions before interrupting a beam to ensure tight tolerances. Feedback signals can be sent to beam control elements, adjusting the currents in the radial (R) or transverse (T) steering coils to compensate

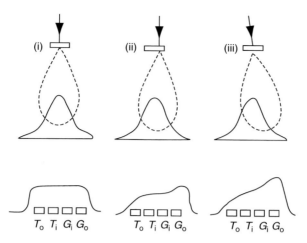

FIGURE 11.38
Segmented ion chamber response to 3 different beam conditions in the gun-target (i.e. *radial*) plane.

(Figure 11.37). Depending on the machine and operating conditions (i.e. energy, x-ray or electron mode) the feedback loop may be close or open. In the latter situation, any measurement out of the tolerance would immediately stop irradiation.

11.7.3.3 Dosimetry Interlocks

It is of fundamental importance that the correct dose is delivered to the patient, and a number of *interlocks* are used to ensure that this is the case. As previously mentioned, ionisation chambers are required to have two channels so that if one fails, the second will terminate treatment. The *accelerator control system* also monitors the difference between the two channels during treatment. It will interrupt treatment if the values exceed preset limits. Additionally, a *backup timer* is required that will terminate the beam after an elapsed time if the ionisation chambers have both failed. (The setting of this timer is usually transparent to the operator.) Ionisation chambers require a bias voltage to be applied, and an interlock circuit checks to ensure that this voltage is present with independent systems for both chambers.

CHAPTER 12

COBALT MACHINES

John Saunders

CONTENTS

12.1 INTRODUCTION

Teletherapy cobalt-60 units were first used for patient treatment in 1951 in Canada (Green and Errington 1952; Johns et al. 1952). Although megavoltage x-ray units had been available for some years prior to this date, they were not in widespread use, whereas, cobalt units became the mainstay of external beam therapy for the next 30 years or so world-wide.

Cobalt-60 is manufactured by irradiating cobalt-59 in a high neutron flux nuclear reactor. The main reasons for its suitability for teletherapy are the availability of relatively small, high specific activity, sources that minimise the beam penumbra (see Section 22.3.3.1); its relatively long half-life (5.27 years); and the almost monochromatic high-energy photon emission (photons of 1.173 MeV and 1.333 MeV in equal quantity).

Compared to modern linear accelerators, cobalt units offer poorer geometrical precision in treatment because of a larger penumbra and greater mechanical inaccuracy. Coupled with the fact that cobalt beams are less penetrating than those from linacs and that cobalt units lack the flexibility in the control of the radiation output offered by linacs, cobalt can no longer provide the basis for contemporary sophisticated radiation therapy. However, commercially available cobalt units may be quite adequate for many non-radical treatments, and there are even some applications where cobalt sources have been used in specially designed equipment to give a performance that could be argued to be superior to that obtainable from a conventional linac, for example stereotactic radiosurgery with the Elekta Gamma Knife (Leksell 1971; Walton et al. 1987; Lindquist 1995) and total body irradiation with custom-designed extended source skin distance (SSD) units (Leung et al. 1981 and see Section 47.4.1).

In countries where facilities for the maintenance of linear accelerators are lacking, cobalt-60 therapy may be the most appropriate choice for radiotherapy (IAEA 1998b). Many of the major techniques and advances in the physics of external beam therapy were developed on cobalt units including arc therapy (Jones et al. 1956); conformal therapy (Davy et al. 1975); transmission dosimetry (Fedoruk and Johns 1957); the development and measurement of tissue air ratios and the subsequent derivation of scatter air ratios (Gupta and Cunningham 1966); and differential SARs and the associated algorithms for treatment planning based on the separation of primary and secondary radiation. More recently, Poffenbarger and Podgorsak (1998) have investigated the possibility of using an isocentric unit for stereotactic radiosurgery, and Warrington and Adams (2001) have shown that conformal therapy, and even IMRT, could be adequately delivered with a cobalt-60 unit except for the most deep-seated tumours.

12.2 BASIC CONSTRUCTION AND FEATURES

A major advantage of cobalt units is the basic simplicity of their design and construction that generally makes them inherently reliable. Nevertheless, care must be taken to ensure that they are robustly built and carefully maintained to minimise the potential hazards from the high activity source and the associated heavy shielding. A sectional drawing of a modern cobalt unit treatment head is shown in Figure 12.1.

As with any item of radiotherapy equipment, it is essential that the physicists responsible for its performance understand the way that the unit is constructed and operates so that they will be aware of its capabilities and limitations. This knowledge will also help them to appreciate what routine safety measures should be put in place and also the possible causes when there are malfunctions or the performance deteriorates.

12.2.1 THE SOURCE

The cobalt source usually consists of a series of either millimetre-size cylindrical pellets (Jaffray et al. 1991) or thin, metallic discs sealed within a double capsule of stainless steel and manufactured to meet certain standards regarding protection from impact, corrosion, and heat (BSI 1976; ISO 1999). Active source diameters are generally in the region of 15 mm to 20 mm, being a compromise between achieving a high enough activity to obtain a reasonable output and keeping the source small enough to minimise the beam penumbra. The source length may be longer than the diameter, but this is not such a critical dimension because the photons from the rear of the source contribute relatively less to the clinical beam than those from the front. In a typical source, about 25% of the primary photons are lost due to self-attenuation. The space

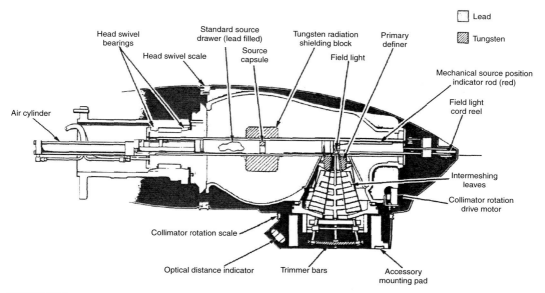

FIGURE 12.1
Diagram of the head of a Theratron 780 treatment unit showing the source drawer mechanism, collimator and shielding.

behind the active cobalt material in the capsule must be tightly packed with blank discs to eliminate movement. The low-energy β-ray emission from the source is filtered out by the capsule walls.

12.2.2 THE HEAD

The head of a cobalt unit has three basic functions: to shield the source, to expose the source as required, and to collimate the beam to the correct size. Shielding is achieved by surrounding the source and exposure mechanism with lead and, in many designs, with alloys of a higher density metal such as tungsten in order to reduce the volume. In some earlier designs, depleted uranium alloy (with a high percentage of U-238, density approximately 19.0×10^3 kg m^{-3}) was used, but this has been discontinued because of the problems with stability of the alloy as some of it became powdery and also because of the difficulties associated with eventual disposal.

The *source exposure mechanism* is usually one of two types where either the source is moved between a safe and exposed position (as shown in Figure 12.1) or where the source remains stationary, and a moving shutter opens or closes the beam. The latter solution, implemented on earlier machines is now obsolete. In the former solution, the source movement is either a translation (as in Figure 12.1) or a rotation. Various beam collimator designs exist to give variable rectangular fields with sides ranging in length, typically, from 4 cm to 30 cm or even up to 40 cm on isocentric units with a source axis distance (SAD) of 100 cm. Each of the four collimator leaves is usually focused on the edge of the source proximal to it so as to avoid cut-off of the primary beam and minimise penumbra. Distances from the source to the far edge of the collimators are typically between 40 cm and 50 cm for machines designed for 80 cm SSD, but this distance may be increased by *penumbra trimmers* that are particularly desirable when the machine is to be used for 100 cm SSD treatment.

With careful design of the collimation system and a 15 mm diameter source, a penumbra of no more than 10 mm (distance between the 20% and 80% decrement lines) may be achieved at 5 cm depth for field sizes with an area of less than 400 cm^2.

12.2.3 GANTRY

Modern cobalt units are manufactured in a standard *isocentric* configuration, with a source axis distance of 80 cm being the most common, to obtain a reasonable compromise between output, depth dose, and clearance around the patient. Units with an SAD of 100 cm are also practical if high activity sources can be afforded and offer the advantages of greater depth dose, larger field sizes, greater clearance around the patient, and geometrical compatibility with linacs. To increase their versatility, isocentric units have often been made with the ability to *swivel* the head about a horizontal axis through the source. This swivel motion keeps the beam axis in a vertical plane, and when used with an appropriate gantry angle, may be useful for extended SSD treatments or for treating immobile patients in a bed or chair. Some non-isocentric cobalt units have been manufactured with the head held by a *yoke on a vertical stand*, and these are particularly useful for giving single field palliative treatments. Where an isocentric unit has the swivel facility, great care must be taken to ensure that its position is accurately reset before the equipment is used for normal isocentric use since the slightest angulation of the head swivel will create a large deviation of the beam axis from the mechanical isocentre (see Section 38.2.1). Some isocentric units use slip rings for the supply of all power and control signals to the gantry and this allows continuous gantry rotation. Coupled with the guaranteed constant output from the source as the unit rotates, such a versatile and simple rotation mechanism provides an ideal unit for arc or full rotation therapy when this technique is required.

12.2.4 BEAM MODIFIERS

Although beam flattening filters can, in principle, be successfully used on cobalt units, they are rarely employed in practice because of the consequent reduction in output. Without a flattening filter, the beam homogeneity is relatively poor across a large area beam, especially at greater depths, because of scattering in the tissue and the greater distance from the source to the field edges compared to the centre for a constant depth. Users should be aware of this fact when using such fields, even for palliative applications, as the 90% decrement level will be much farther inside the beam than for a similar field on a linac.

Fixed wedge filters may be inserted into the beam below the end of the collimator to give a range of wedge angles, typically between 15° and 60°. These wedges will usually incorporate a certain amount of beam flattening, giving a somewhat straighter isodose curve within the central portion of the beam than for an open field. Wedge filters inevitably reduce the dose rate at the centre of the beam and this is more of a problem with a cobalt machine than with a linear accelerator. The reduction in output may be minimised by making the wedge shape always begin at the field edge by linking the toe end of the wedge to the collimator so that wedging always begins at the edge of the beam. With this system, the wedge factor will be different for each field size, so modern units usually have a fixed relationship between the centre of the wedge and the centre of the field with perhaps different wedges for different ranges of field size. In principle, there is no reason why a virtual wedge could not be created by dynamically moving one collimator during treatment in a similar fashion to that employed on linacs. Such a system has not yet been commercially produced and would probably require a different design for the collimators to allow them to travel across the beam axis, but it would have the advantage of minimising the increase in treatment time associated with the use of a wedge.

Other beam modifiers and accessories such as the *beam-defining light*, *range finder*, *accessory tray*, *laser positioning system*, *front* and *back pointers*, etc., will be very similar to those employed on linacs. However, on a cobalt-60 unit the beam-defining light may physically exchange places with the source as in Figure 12.1.

12.3 SAFETY

Many of the safety features and requirements for cobalt units will be identical to those for other megavoltage machines, so only those items where cobalt requires special attention will be dealt with here.

The IEC have published specific requirements for the safety of γ beam units (IEC 1997). The control of the source and its exposure is paramount. Since the output of the beam is accurately predictable, a dose monitoring system is not required, and dose to the patient can be controlled simply by the exposure time alone. *Dual independent timers* are required to cover the possibility of one timer failing, and it is also required that the pre-selected time for irradiation remains on display for the duration of the exposure. Units in which the timer is set are nomally min (and 1/100 min) rather than seconds. This must be clearly recognised since mixing-up minutes and seconds may be the origin of severe accidents (IAEA 1998b). It is also necessary to have a system that enables the operator to obtain the true exposure time, following a complete system failure, for example, after a mains power supply interruption, so that the dose received by the patient may be calculated. The source exposure system must be designed so that under all appropriate fault conditions, including mains power failure or any other major control system failure, the source exposure mechanism returns it to a safe state by an independent means (e.g. spring, compressed air, battery, etc.). In addition, a *manual system* must be provided to allow the unit to be restored to a safe state should the automatic system fail.

It is usual for one of the two timers to be arranged as the primary timer and for it to be activated by switches that sense when the source is in the fully exposed position and for the other, secondary timer to be activated by switches that sense when the source moves from or returns to the fully safe position. The difference between the two timers gives an accurate measure of the source or shutter transit time. This transit time should be regularly monitored to verify that the exposure mechanism is operating satisfactorily. Because of the finite transit time (typically 0.5 s to 2.0 s depending on the unit's design), it is necessary to measure a timer correction that must be applied to each treatment irradiation time to allow for the extra or, depending on the equipment type, occasionally reduced dose that would be given to the patient compared to what would be expected from the known dose rate and set time. The timer correction is most easily measured with a dose meter in the main beam using the dual exposure method (see Section 38.8.4). It is also a requirement that the total transition time from beam off, to on, and off again must not exceed a total of 5 s.

A *door interlock* mechanism must be provided and connected to the machine in such a way that (i) it is impossible to turn the beam on as long as the door is open, and (ii) if the beam is on, the source returns to the safe position as soon as the door opens.

The IEC standard requires that a *manual beam control system* is provided that directly operates on the source exposure mechanism that may be used in an emergency to return it to the beam-off condition. The status of the beam must be clearly indicated by *lights* on the head of the unit and by a mechanical system directly coupled to the exposure mechanism. An *audible warning* must also indicate when the source is not in the fully safe state.

It is essential that a robust interlock arrangement for wedge filters and, if used, flattening filters is provided. Once a filter has been chosen in the treatment room and inserted into the head, the type of filter must be re-confirmed at the external treatment control panel without the selected item's identity having been previously displayed. The orientation of any filter must also be clearly visible.

12.3.1 LEAKAGE

The IEC standard places limits on leakage radiation from the head both for *beam-on* and *beam-off* conditions. *In the beam-off condition*, the absorbed dose rate at one metre from the source must not exceed 0.02 mGy h^{-1}, and at a distance of 5 cm from the protective surface of the head, must not exceed 0.2 mGy h^{-1}. These measurements may be averaged over areas of up to 100 cm^2 and 10 cm^2, respectively and apply to a source of the maximum-rated activity. In some countries the 0.02 mGy h^{-1} figure is reduced to 0.01 mGy h^{-1}.

Leakage radiation *during irradiation* is measured with the collimator closed and any residual open area blocked by an attenuator thick enough to reduce penetration to below 1%. Leakage dose rates are measured at the standard treating distance and expressed as percentages of the dose rate at 5 mm depth for a 10 cm×10 cm field at this distance. Within the area enclosed by the limits of the maximum field size, whose area is A cm^2, the leakage dose rate for square fields (averaged over 8 points shielded by only one collimator and 4 points shielded by both) must be less than 1000/A % (e.g. 1% for a 32 cm×32 cm maximum field) and individual collimators must not allow more than 2% transmission. Leakage outside the limits defined by the maximum field size in the plane of the patient (within a circle of radius 2 m) is limited to a maximum of 0.2% (average 0.1%) of the dose on the beam axis. Elsewhere, the leakage dose rate, measured at one metre from the source must not exceed 0.5% of the maximum dose rate measured at one metre. The latter figure also applies for all positions outside of the maximum field size during source or shutter transit.

12.3.2 AREA RADIATION MONITORS

The International Basic Safety Standards (IAEA 1996) recommend that an *independently operated radiation monitor* should be installed in the treatment room to give a clear indication that the source is in a safe position. A suitable monitor would normally have an adjustable threshold alarm level so that an audible and visual warning is given when the beam is on. Some centres will choose to install the monitor in such a way that when the barrier interlock is closed, the audible warning will be suppressed. Such a monitor should be powered from a separate mains supply from that used for the treatment unit plus a battery back-up. It is also advisable to keep in the vicinity of the treatment room, but not in it, a *portable radiation monitor* with ranges extending from approximately 1 mSv h^{-1} to at least 1 Sv h^{-1}*.

Personal electronic radiation monitors that emit an audible indication of dose rate may be worn as an alternative (or preferably, in addition) to fixed monitors (IPEM 2001). It is desirable that such monitors should also record the cumulative dose received by the wearer so that a good estimate of any dose received in an emergency procedure can be quickly evaluated.

12.3.3 WIPE TESTS

It is also necessary to perform *wipe tests* on the equipment at least every two years to verify that there is no risk of contamination from a leaking source capsule. This test is most easily accomplished during a regular service of the unit when the covers are removed, and access to components such as the source mechanism and primary collimator may be achievable in order to wipe surfaces as close as possible to the source. If an activity of greater than 200 Bq is detected on the swab (IPEM 2001), the source should be considered as leaking, and steps should be taken to confirm the result followed by removal of the source and subsequent decontamination of the equipment.

12.3.4 LOCAL RULES

Local radiation safety rules must be written, and these must take into account the circumstances relating to each specific installation (see also Part L). For cobalt teletherapy units,

---

* Sv stands for sievert. It represents the unit of *Equivalent Dose* used for radiation protection purposes (see Section 59.1.2). In practice, for γ-radiation, 1 Sv=1 Gy.

a clearly documented *procedure* must be produced for implementation in the event of an emergency arising from a failure of the source exposure mechanism to return the source to a safe state. If a patient is in the beam at the time, this will probably make it necessary for staff to enter the room to remove the patient without being exposed to unnecessary radiation. Prior to entering the room, any remotely controlled emergency beam-off mechanisms should be tried, and if possible, the collimator should be closed down to a minimum using controls that should be available outside the room for such an emergency. These procedures must be routinely rehearsed so that the staff know exactly what to do when required. It is usually not advisable to attempt to make the source safe manually in these circumstances as this can result in further delay with the resultant additional unplanned dose to the patient. Such action should be performed only by authorised persons after the patient has been removed from the room.

12.3.5 MECHANICAL SAFETY

Because of the emphasis placed on the hazard from radiation, it could be easy to over look the importance of insuring adequate mechanical robustness. The considerable weight of the source shield and the collimation system requires that all of the supporting components in the head and gantry are carefully designed, manufactured, and maintained to avoid mechanical weakness that could result in inaccurate treatment delivery and possible injury from mechanical failure.

A mechanical lock should also be provided on the head to make sure that the source cannot be exposed when the unit is not under proper supervision. There is also a requirement that the room shall be physically secure when unattended to prevent potential terrorist activity (see Appendix L.1.5).

12.4 INSTALLATION OF COBALT SOURCES

Because of the huge activities involved, the installation of a cobalt-60 source is a potentially dangerous procedure that requires considerable care. It is usually necessary to bring a container with the new source into close contact with the head assembly, sometimes after having dismounted the collimator. The container is normally fitted with two *drawers* which can be activated from a distance, an empty one to collect the old source and the other one containing the new source. Proper mechanisms and interlocks ensure that after source exchange, the container can be safely decoupled from the treatment head and transported according to the current regulations. On the Alcyon cobalt machine, the whole head assembly is exchanged and used as the transportation container. In any case the operation must be performed by licensed source handlers under the control of the supplier. However, some of the particular issues that arise and that must be considered by the local staff are covered in the following paragraphs.

12.4.1 SOURCE SPECIFICATION

The two main characteristics that the user must decide upon before ordering a source are the *diameter* and the *output* at the treatment distance. The smaller the diameter, the better will be the penumbra, but this will also limit the maximum activity that can be obtained and will limit the output. Typically, sources with a diameter of 15 mm will be available with an output at depth of maximum dose for a 100 cm^2 field area of up to 220 cGy min^{-1} at 80 cm SSD or 140 cGy min^{-1} at 100 cm SSD and will have an activity in the order of 300 TBq. For sources with a diameter of 20 mm, these maximum output figures will increase to 360 cGy min^{-1} and 230 cGy min^{-1}, respectively, with an activity in the order of 500 TBq. Before ordering the source, checks must be made with the supplier that the head of the unit is designed to accommodate the type of source being considered and that the room shielding will also be adequate. The user should also ensure that he has the necessary authorisation from the

Regulatory Authority*. This must permit the user to have, at the hospital site, the total activity of both the new and old sources combined even though this may be only for a short period. Commercial suppliers of sources can usually give details of the approximate activity for a given source diameter to achieve a specified output. Although it is now recommended to express the output of a cobalt machine in air-kerma at a reference distance (in Gy h^{-1}). Suppliers may still express the output of a source in terms of the old *Röntgen* (R) unit of exposure per minute at one metre from its front face, commonly abbreviated to *RMM* (or Röntgen per hour at one metre abbreviated to *RHM*) (ICRU 1970). Purchasers should also be careful to note if the output measurement conditions stated by the source supplier refer to a scatter-free measurement or to a measurement with the source in the head of the machine. The added scattering from the collimator and phantom for a large field size could appear to enhance the figure by 10% to 20%. The R unit will, in any case, give a numerically larger value than the equivalent absorbed dose rate in water in cGy.

12.4.2 PREPARATION

It is essential that a detailed *quality plan* is prepared well in advance of the installation or source change by the radiotherapy physicist and that this is agreed to by all of the relevant parties. This plan will need to cover the transport arrangements, including local unloading and loading of the source container, local hospital security precautions, liaison with the radiotherapy department staff, and a clear definition of the responsibilities of the supplier and the local staff. It will also be necessary to clarify the arrangements for handing over the responsibility for the treatment room to the supplier's representative during the installation or source change. These should include any restrictions the hospital may wish to impose during this time. Similarly, it is better to agree well in advance to the arrangements for handing back the equipment to the hospital's control and to specify the tests the user requires to be successfully performed before accepting it back. If not involved directly, the person locally responsible for radiation protection must be consulted at all stages of the planning procedure. The hospital should also request the supplier to provide a *risk assessment* of the operations involving the sources and a corresponding contingency plan to deal with any reasonably foreseeable accident. Likewise, the hospital should also develop a *contingency plan* to cope with possible mishaps such as a source becoming stuck and irradiating an area of the room that is not fully protected. It should be noted that the IEC standard (IEC 1997) requires the equipment to be designed in such a way that the staff performing a source change receive no more than a dose of 1 mSv.

The quality plan should include taking the opportunity for local staff to become acquainted with the source exposure mechanism while there is no radioactivity present. If a source change is being made, a thorough inspection and service of the mechanism should be scheduled along with a wipe test as any maintenance of this part of the unit will, subsequently, be very expensive if it requires the source to be temporarily transferred to a separate container.

12.4.3 INSTALLATION

An experienced radiotherapy physicist must be nominated to co-ordinate and oversee all operations apart from those that have been clearly designated as the supplier's responsibility. Even then, the physicist must be available in the case of an unexpected event that may need intervention by hospital staff. Clear warning signs and barriers should be placed around the treatment room during source manipulation procedures to restrict access and indicate who is in control. The physicist must be informed of when the actual source transfer is to take place so that he can monitor dose rates in the vicinity. He should also be informed of when the first exposures are made with the new source for the same reason.

---

* The Environment Agency in the U.K.

Before accepting a new source, the supplier should demonstrate the following items as a recommended minimum set to satisfy the user that the specified performance has been met. It is also desirable that these checks are made before the old source leaves the site in case the wrong source has been delivered in error.

- Dose rate at d_{max} for specified field size and treatment distance
- Radiation leakage levels around the head
- Source/shutter transit time
- Radiation field size versus indicators across full range
- Radiation beam penumbra for specified field size and depth
- Correspondence between radiation field and light field edges across full range
- Symmetry of radiation field edges about collimator axis
- Isocentric accuracy of radiation beam axis
- No contamination from source
- Optical distance indicator

It is much preferable if the method of measurement and the acceptable tolerance levels for each item are agreed in writing beforehand by including them in the quality plan if not specified elsewhere. In the case of new equipment being installed, other acceptance tests will be necessary, but these will be similar to those for a linac except specific attention may need to be given to

- Timers–method of operation, accuracy, and linearity
- Operation of exposure control system
- Arc/rotational therapy control system
- Head swivel (if fitted) and precision of zero degree position

Quality control procedures for the new cobalt unit should be established prior to clinical operation (IPEM 1999).

12.4.4 Definitive Calibration

A definitive calibration must be performed whenever a new source is installed (IPEM 1999, Appendix A). This should be performed according to the *codes of practice* described in Chapter 18. Once the output has been measured an attempt should be made to reconcile the measured output with that stated on the source certificate provided by the supplier after making allowances as far as possible for the inevitable differing measurement conditions. *The source certificate must never be directly used to derive the reference dose rate for clinical use.*

From the definitive calibration, it will be possible to derive the output figures required for daily treatments. Some centres may work with a single output (e.g. that for a 10 cm$\times$10 cm field, 0.5 cm depth in water at the isocentre) from which the output for all other field sizes and depths is derived by tables of factors. This has the advantage of having to change only one figure to allow for source decay. Alternatively, other users may calculate the output for a range of field sizes at a particular reference depth each time. The usual interval for re-calculating a new output to allow for decay is normally one month, and this requires a correction of almost exactly one percent. The output is usually calculated to be correct for the middle of the month.

As part of the definitive calibration procedure, the timer correction to account for the time for the source to become fully exposed must be determined. In fact, if a calibration method is used where the absorbed dose is measured for a set time on the unit's timer, it will be necessary to allow for this timer correction in order to obtain the true absorbed dose rate (see Section 38.8.4).

CHAPTER 13

SIMULATORS

Jean-Claude Rosenwald

CONTENTS

13.1 INTRODUCTION

The simulator was introduced in the late 1960s as an additional device to assist with the preparation of external radiation therapy (Green et al. 1964; Karzmark and Rust 1972). The original concept came from the necessity to check prior to treatment both that the plan can be delivered in practice and that the relationship of the beam set-up to the patient's anatomical features is correct. This could be done with the therapy machine using film or electronic imaging devices (see Chapter 14 and Section 40.1), but there are then two major drawbacks:

1. The image quality of films performed with high energy x-rays is poor compared to the quality in the radiodiagnostic range

2. Using the treatment machine to simulate the actual treatment is not very efficient in terms of patient throughput.

From these considerations came the idea of designing a specific piece of equipment, mimicking the geometry of the therapy machine, located in a different room and fitted with a radio-diagnostic-type x-ray tube. It has been called a *simulator*.

Ideally, a simulator should have exactly the same mechanical and geometrical characteristics as the simulated treatment machine. This includes the patient couch and the external dimensions of the housing including any accessory equipment. It also means that the same movements, scales, and geometric accuracy should be available. There would then be one simulator for each treatment machine, and that would not be economically efficient. For this reason, the simulators presently available are designed to be universal and to match, as closely as possible, the characteristics of most existing therapy machines. For example, to cover the standard cobalt machines as well as the standard linear accelerators, the focus-isocentre distance can be changed from 80 cm to 100 cm; although as accelerators become more universal, with 100 cm as the standard distance, this facility is no longer universally provided (see Figure 13.1). This range can be even larger to cover special needs. However, it must be fully realised that significant differences exist between simulator and treatment machine. The most important issues are inconsistency of scales, inability to predict collisions or beam interception by the table structure, inconsistency of accessories or of their attachment, and differences in rigidity. In general, it must be ensured that the accuracy of the simulator is at least as good as the best treatment machine available on the site.

Simulators have the ideal characteristics to perform x-ray images in geometrically controlled conditions. Therefore, they are frequently used not only as a simulation device but also as a patient data acquisition device in conjunction with a mechanical pantograph for external contour measurements. These two practices are often combined, resulting in different schemes that will be discussed in Part F. Since the requirements for patient data acquisition and treatment simulation or verification are similar, a single treatment of the issues will suffice.

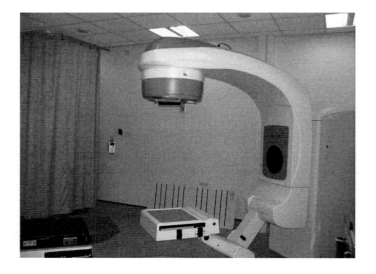

FIGURE 13.1
(See colour insert following page 590.) Photograph of a fixed SAD simulator showing a digital imaging system. A slot located behind the detector allows insertion of conventional film cassettes. The imager arm has considerable freedom of movement (focus-film distance in the Z direction and X-Y translation).

13.2 DESCRIPTION OF THE STANDARD SIMULATOR

The various components of a modern simulator can be seen in Figure 13.1.

13.2.1 X-RAY TUBE AND GENERATOR

The x-ray tube must be designed to perform high quality radiographs and to be used in fluoroscopic mode. Therefore, the requirements are not different from those for a basic diagnostic apparatus with a modest workload, although the tube and generator rating must be large enough to accommodate large distances between focus and detector (up to 170 cm) and long exposure times in fluoroscopic mode (McCullough and Earl 1979). The *focus* must be small enough to give sharp images of the anatomical structures and of radio-opaque markers that are used, for instance, to define the field limits. A typical recommended diameter is in the range 0.4 mm to 0.6 mm for the fine focus.

Unlike a diagnostic machine, the knowledge and stability of the geometrical position of the focus are of crucial importance. If two foci are available (large and small), it must be ensured that they are at the same position (even switching from radiographic to fluoroscopic mode) or preferably, provided that it doesn't limit the practical use of the equipment for thicker patients, access should be restricted to the small focus with an appropriate interlock. This focus must be aligned on the mechanical axis of diaphragm assembly. As for treatment machines, a light source must be placed at the same position. For practical reasons, this is always done through mirror reflection with adjustments for angle and distance.

13.2.2 DIAPHRAGM ASSEMBLY

It is not necessary to mount a heavy collimator on the tube housing; instead, *metal wires* are used to represent the collimator edges on the radiological images as well as on the light projected onto the skin. One advantage of using wires is the possibility to see the radiological structures inside and outside of the field limits. There should be one X pair and one Y pair covering the whole possible range of field sizes, preferably with provision for asymmetry.

A *variable x-ray diaphragm* is not strictly essential, but it is recommended in order to improve radiation protection of the patient. Minimising the x-ray aperture will also improve image quality by reducing the amount of scattered radiation. The diaphragms may be used to allow exposures with different openings to be added together to enhance the contrast for thicker parts of the body (i.e. lateral pelvis). The diaphragm opening must always be larger than the field delimited by the wires. It is, therefore, advisable to have access to rectangular shapes with independent control of each blade if asymmetric fields are simulated. On sophisticated equipment, there could be a servo mechanism to keep a constant margin between the field wire and the corresponding diaphragm limit, but it is important that this mechanism can be deactivated.

The central axis of the beam must be made visible on the skin and on radiological images. This is usually done by engraving a *cross-hair* on a polymethylmethacrylate (PMMA, also known as Perspex or Lucite) sheet that slides in at the base of the diaphragm assembly. By filling the grooves with some lead material, they become radio-opaque. In addition, radio-opaque scales are often engraved along the cross-hairs in order to be able to measure distances directly on the films (as shown in Figure 13.2). Distances measured using this scale will indicate lengths in a single plane orthogonal to the beam axis at a fixed distance from the focus (usually 100 cm). If measurements are to be valid at another distance instead (e.g. 80 cm) exchangeable perspex graticules will be needed together with a system to ensure that it is clear which graticule was used.

In order to provide a full simulation of a treatment machine collimator, the whole assembly must be able to rotate around the beam axis. A visual display of the current angular rotation of the collimator must be provided.

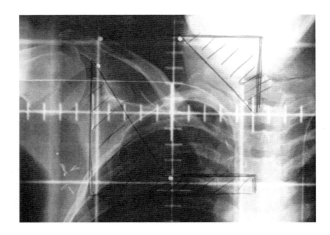

FIGURE 13.2
Example of a simulation film of the supra-clavicular region. The graticule is clearly seen with its centimetric scale (at SAD) on both axes. The small white circles correspond to lead beads used as skin markers. Additional shielding blocks are manually drawn on the same field through the skin markers.

13.2.3 GANTRY

As for a treatment machine, the source must rotate around the patient, keeping the beam axis in a vertical plane but passing through a fixed point called the *isocentre*. As discussed above, the source-isocentre or source-axis distance (SAD) is often variable and is implemented by allowing the tube mounting to slide along the rotating arm parallel to the beam axis. The SAD value must be recorded and displayed, and it is recommended that the field size indicators are automatically adjusted according to the SAD in order to display the field dimensions as recalculated at the SAD. For mechanical and practical reasons, additional movements such as head tilt or head swivel that were sometimes found on the previous generations of cobalt machines are not available on standard simulators.

The relatively light weight of the tube and diaphragm assembly should allow an overall rigidity, giving a high accuracy of the isocentric movement. However, the existence of a variable SAD, combined with the wish to reduce the cost as much as possible, could lead to a degradation of this accuracy as compared to a treatment machine. This should not be accepted.

13.2.4 IMAGE DETECTION

On the gantry, opposite to the x-ray tube, an image detection system must be installed. Although not strictly needed for true simulations, an *image intensifier* is useful when the field definition is radiologically performed, on the spot, without a predefined plan (see Section 32.2). The intensifier should be large enough to analyse a large portion of the field. On the other hand, if it is too bulky, the risk of collision is increased, and movements may be severely restricted. A reasonable compromise is found with image intensifiers of diameters in the range 30 cm to 35 cm. It is important to stress that if a field limit is too far from the centre to be projected onto the intensifier entrance screen, the practice of moving the patient after delineating the field with a radio-opaque marker placed on the skin must be strictly forbidden because it gives a wrong projection. Instead, the intensifier must be mounted on an arm with provision for X and Y translations perpendicular to the beam axis. Some of the limitations of intensifier size and image distortion will be removed by the advent of *digital imaging panels* (see Figure 13.1). (In the following text *imager* will be used to include both types of imaging device.)

An additional Z translation, parallel to the beam axis, is also important to bring the image plane as close as possible to the patient in order to cover a larger portion of the field and improve the image quality (with a sharper penumbra).

The imager housing is generally equipped with a special holder for film cassettes, incorporating an anti-scatter grid. The holder should accommodate large film sizes (more than 35 cm × 35 cm), depending on the treatment techniques in clinical use. The grid can be focused, but it should not be too sharp because it has to accommodate various distances to the focus. It is very useful to record and display the film-isocentre distance that can be calculated from the Z translation of the imager. This distance can be used to calculate the film magnification.

In order to limit the consequences of collisions that are likely to happen with the imager housing, mechanical or electronic sensors are generally installed. If the sensors detect a collision, all the movements of the simulator and table are stopped except those that are necessary to restore the minimum distance between the imager and the rest of the equipment. Some simulators now have anti-collision software programs, but do not guarantee that collisions are impossible.

13.2.5 TABLE

A good quality *couch* is an essential component of a modern simulator. Its general characteristics are required to be very similar to those of a treatment table.

The amplitude and scales of the linear movements (longitudinal, lateral, vertical) should be the same as for treatment tables. The rotations are less useful because the presence of the imager and the problems of image interpretation make it very difficult to simulate fields with table rotations (i.e. non-coplanar fields). However, table rotation is required to facilitate patient set-up or to easily move the patient from the simulator to a mechanical contouring device.

The table top must be radio-transparent, and its flatness and rigidity must be as good as that of the treatment couch. However, the material used could be thicker because there is no concern about preserving the skin sparing.

The table must be equipped to accept the same patient positioning accessories as for treatment. However, lateral metallic rails can be an obstacle when performing some imaging procedures such as Computerized Tomography (CT) (see Section 13.3.3).

13.2.6 CONTROL CONSOLE

As for a treatment machine, it should be possible to control all the movements of the simulator and of the table, while in the simulator room, using, for instance, a hand pendant. However, it is also important to be able to control these movements in fluoroscopic mode in order to perform radiological adjustment of the field limits. For this, a control console must be available in a protected area behind a radiological screen or in an adjacent room. The position of the console in relationship to the simulator must allow a clear vision of the patient (to enable collisions to be avoided). Obviously, at the same location, the operator must have full access to the x-ray control and to the monitoring of the fluoroscopic images. The possibility to freeze the fluoroscopic images is beneficial to allow detailed analysis without excessive dose to the patient.

13.3 SPECIAL FEATURES

As well as the features common to standard simulators, additional functions can be found on modern machines, especially in connection with the general trend toward better integration in the computerised environment of radiation therapy (see Chapter 32 and Chapter 41).

13.3.1 COMPUTERIZED CONTROL, NETWORKING AND AUTO SETUP

On modern simulators, the control of the machine is handled by a computer that allows movement selection and display. Therefore, even if the encoding of the various movements is done through potentiometers, the calibration of the corresponding scales can be achieved through software correction tables that make the procedure easier. Another interesting feature is the possibility to achieve relative x, y, z translations of the treatment couch coordinates in relationship to a given reference position where a new origin could be reset.

As soon as the machine coordinates are in digital form, they can be *imported* from or *exported* to other external equipment. This feature is of special interest when the simulator is part of a network, including check and confirm functions (see Chapter 41 and Chapter 42). When the simulator is used to confirm the geometry of a plan previously designed on a computerised treatment planning system (TPS), it is useful to import data into the simulator from the TPS to limit the risk of transcription errors. On the other hand, if the initial setup is decided during the simulation process, the machine parameters could advantageously be exported to the TPS or directly to the treatment machine.

On computerised simulators, an *auto setup* function is frequently provided where, as soon as the geometrical parameters are defined (by importation from external equipment, direct input from the simulator console, or by recall of a predefined position), a simple confirmation through a push button initiates and sequences the corresponding movements. For safety reasons, it is required to keep pressing a switch (known as a *dead man's switch*) while the machine is moving. The auto setup feature is also useful for loading or unloading the patient or to prepare the machine for specific imaging modalities such as CT scanning (see Section 13.3.3).

13.3.2 DIGITAL IMAGING

Access to simulator x-ray images in digital form is useful for a number of reasons such as storage and retrieval facilities, network data exchange and image processing. One major advantage is that it is possible to transfer a reference image to an electronic portal imaging system, and this can then be compared, either manually or automatically, to digital portal images of the same field acquired during the treatment course (see Section 40.1).

One simple approach is to use a *film scanner* to digitise conventional films. However, this process is time consuming and expensive, and there is image degradation because of a loss in resolution. In addition the dynamic range is reduced because of a combination of the limitations of film and of the film scanning device. The rapid development of *Digital Luminescent Radiography* (also known as *Computed Radiography* or CR) provides an alternative to conventional film with direct access to images in digital form.

Alternatively, it is also possible to use the video signal from the image intensifier and to convert it into a digital image. The problems with this approach are essentially the risk of image distortion (Morton et al. 1991a) and the size limitation directly linked to the entrance screen of the image intensifier. In order to reach an acceptable geometrical accuracy, it is necessary to have access to hardware and/or software solutions for distortion correction and to calibrate the whole system periodically. In addition, special software has been devised to reconstruct a single image from multiple adjacent images acquired after translation of the image intensifier over the field area.

The development of *amorphous silicon* and *amorphous selenium* allows direct digital imaging. The drawback of this technology is its high cost, but it is likely that the cost will decrease.

13.3.3 CT OPTION

Another additional feature, often available as an option on modern simulators, is the possibility of reconstruction of *computed tomograms* (CT). The original idea (Harrison and

Farmer 1976) comes from the fact that several major components of a CT scanner are already present on any standard simulator (i.e. x-ray tube, rotational movement, image detector). It is tempting to use the detector output signal acquired while rotating the gantry to derive an image of the patient central slice by back-projection.

The first attempts used analogue processing of the video signal taken from the output of the image intensifier (Webb et al. 1977; Harrison and Farmer 1978). However, the image quality was very poor with many artefacts and significant background noise. A slight improvement was obtained with digital processing of the same signal, but it was only after using other detection systems that the image quality became satisfactory.

Currently, there are essentially two approaches that are implemented on commercial equipment: one of them takes advantage of the image intensifier with replacement of the standard fluoroscopic camera by a special CCD camera (line of detectors) measuring the transmission profiles as seen on the output screen of the image intensifier; the other makes use of the same detection system as a conventional CT scanner attached at the intensifier's entrance surface. Each of these methods has several advantages and drawbacks. With the first approach, there is a risk of distortion of the transmission profiles as the gantry rotates, mostly because of the influence of the earth's magnetic field on the electronic deflection within the image intensifier. This problem can be solved by the addition of a compensating coil where the current is a function of the gantry angle. Another problem is the restricted field of view that is limited by the diameter of the intensifier entrance screen. Finally, it is necessary to add a collimating slit on the image intensifier surface to block as much scatter as possible from the patient. The second approach requires the integration of a bulky device, including the anti-scatter collimator, that is not easily fixed or retracted and is also more expensive and difficult to maintain.

In both cases, a significant improvement in the image quality is obtained by adding a collimating slit at the exit of the tube assembly and by monitoring the tube output and using this as a reference for transmission measurements. It is also usual to enlarge the field of view by moving the detector laterally to the edge of the central axis in such a way that half of the image is acquired with a 180° rotation and the other half with the remaining 180° (Figure 13.3). For this situation, a shaped compensating filter is frequently inserted into the x-ray beam, being thicker on the external side of the body where the transmission is expected to be higher as shown in Figure 13.3. This is sometimes referred to as a *bow-tie filter*.

More recently, a new generation of devices based on digital image acquisition using *flat panels* and *cone beam reconstruction* (Cho et al. 1995; Jaffray 2002) has become commercially available. With a single rotation of the gantry, the data for a patient cross section about 15 cm long can be acquired, and from this, several adjacent CT slices of thickness 5 mm to 10 mm can be extracted. However, unless special image processing algorithms are used, the image quality is significantly deteriorated by the presence of a large amount of scattered radiation. With the advent of amorphous silicon panels (see Section 14.3.2.2), this technology is now mature, and good quality images with equal resolution in all three directions can be obtained. Such devices are also available on linacs.

Whatever technique is used, the resulting image quality is not quite as good as for a diagnostic CT scanner. Also, the acquisition time per slice is limited by the rotational speed of the gantry that is typically around 1 min for 360°. However, provided that the intention is to acquire only a limited number of slices (typically less than 10)—or for cone beam CT a limited length of the patient—and making sure that the heat capacity of the tube assembly is high enough, the CT option is very useful for 2D planning. First of all, the digital images can be directly transferred to the treatment planning systems without any need for mechanical devices for external contours. Second, the patient is truly in the treatment position as can be easily verified by a conventional simulation performed on the same machine in combination with the CT image acquisition. Also, the diameter of the virtual tunnel of this equipment can be as much as 90 cm, instead of 60 cm to 73 cm usually found on diagnostic CT machines and this allows for accessories such as arm rests or wedged boards. Finally, until it becomes more common to have CT scanners belonging to radiotherapy departments, it is very convenient to have access

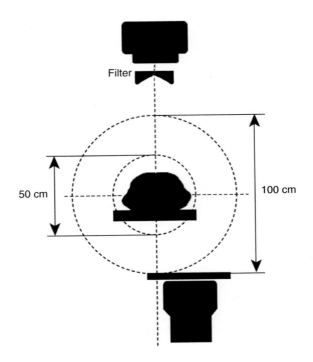

FIGURE 13.3
To obtain a full CT reconstruction with an image detector that is not large enough, the detector can be laterally shifted in order to cover half of the body for a fixed gantry position but the totality for a 360° rotation. A filter is usually added to compensate for variable thickness of the body. In this example, the field of reconstruction is 50 cm diameter, and the aperture is 100 cm diameter (redrawn with permission from Mecaserto documentation.)

to CT imaging on the spot directly managed by the technologists who specialise in radiation therapy.

In spite of these advantages, it must be clearly understood that the CT option of simulators cannot be used for conformal radiotherapy where a series of many thin adjacent slices is required for volume reconstruction (Dahl 1996; Rosenwald 1999).

13.3.4   Virtual Simulation and Image Guided Radiotherapy

The term *virtual simulation* is used to characterise an approach where the simulation process takes place without it being necessary for the patient to be present (Sherouse et al. 1990). Instead, starting from a volume acquisition on a CT scanner, perhaps complemented by other imaging modalities, the simulation is performed on a computer console offering advanced imaging capabilities (Kessler et al. 1995). The beams are interactively generated and handled on the console until considered satisfactory on the basis of geometrical, anatomical, and/or dosimetric criteria. More details on this approach can be found in Section 32.4.

Virtual simulators are becoming more widely used (Aird and Conway 2002; Mutic et al. 2003). They do not allow direct interaction with the patient, which means, for instance, that skin marking or accounting for patient internal movements is difficult. This is ideally done with a CT scanner dedicated to radiotherapy that is equipped with movable lasers to mark the isocentre position on the patient at the time of CT scanning based on a rapid definition of the target volume. It is still questionable if virtual simulation will completely replace, or just complement, conventional simulators (Webb 1990). A CT scanner suitable for virtual simulation should ideally have a wide bore and a radiotherapy-type couch as shown in Figure 13.4. *Wide-bore scanners* are now available from all the major CT manufacturers,

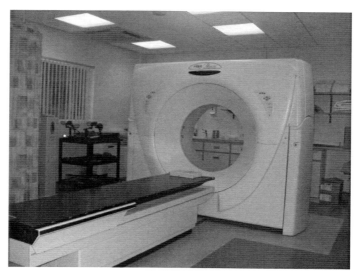

FIGURE 13.4
(See colour insert following page 590.) Photograph of a wide-bore CT scanner designed especially for virtual simulation. Note the standard flat-top couch and the lateral lasers.

and there is a steady trend toward CT simulation rather than the use of conventional simulators.

A more recent development is the so-called *image guided radiotherapy*. This approach consists of having access to imaging capabilities directly on the treatment machine and is briefly referred to in Section 40.1.5 and in the Introduction to Part I. Alternatively, the imaging system can be part of a treatment machine specifically designed to combine imaging and treatment as in tomotherapy (Mackie et al. 1993) (see also Section 43.3.4 and Section 44.3.6).

CHAPTER 14

PORTAL IMAGING DEVICES

Cephas Mubata

CONTENTS

14.1 INTRODUCTION

When a patient is treated, it is just as important that the radiation should be delivered to the correct anatomical site as it is that the correct dose should be delivered. With the advent of conformal radiotherapy, the requirements for positional accuracy are increased. Traditionally, quality assurance of patient positioning has been achieved using industrial x-ray film, but film needs to be processed, and the information is only available some time after the treatment fraction has been delivered. In order to provide an instantaneous verification of the patient setup, *electronic portal imaging devices* (EPIDs) have been developed that are attached to the gantry of the linear accelerator.

A major difficulty of portal imaging in megavoltage photon beams is the lack of tissue contrast inherent to the energy range because the photoelectric interactions responsible for high quality images at diagnostic energies are almost non-existent.

14.2 FILM-BASED METHODS

Until quite recently, film has been the traditional way of verifying patient position. In this section, the physical aspects of film used for patient setup verification are reviewed (see Section 17.2 for a description of the detailed physical properties of film for dose measurement). The difficulties encountered when using film are also addressed.

Radiographic films normally have light-sensitive emulsion coated on both sides. In treatment positioning verification, the film is normally sandwiched between two metal or fluorescent screens. Typical conventional metal screen combinations include two sheets of lead of thickness 0.15 mm and 0.3 mm for the front and back screen, respectively, or a 1.0 mm thick copper front screen in conjunction with a 0.25 mm lead back screen. The front screen serves a dual purpose. First, the screen absorbs the electrons from the patient and the intervening air column to reduce blurring in the image while generating recoil electrons through photon interactions that directly expose the film (see Figure 14.1a). A lead back screen is usually used to intensify the fluence of the back-scattered electrons. The back screen has been reported to reduce the dose or exposure per given density by up to 50%.

The detection efficiency of the film–cassette geometry can be increased by placing the film in contact with a fluorescent phosphor screen (Figure 14.1b). With phosphor screens, the recoil electrons from the metal front and back screens interact with the heavy metal ions to produce optical photons that expose the film. The process of light photon production is described in more detail in Barrett and Swindell (1981). The thickness of the phosphor has to be optimised for best results. Although thick screens will result in an increase in light output, this tends to degrade the system resolution by increasing the spread in the light photons when they reach the film. The phosphor thickness has to be minimised so that the generated photons reach the film with minimal lateral spread, but it must be thick enough to increase the light output. Calcium tungstate has long been used as a phosphor because of its excellent mechanical properties. More recently, rare-earth oxysulphides such as gadolinium, lanthanum, yttrium and other halides such as lanthanum oxybromide and barium fluorochloride have been explored.

Haus et al. (1997) have demonstrated a 3.5 times increase in film contrast with a combination of 1.0 mm thick front copper screen plus front and back gadolinium oxysulphide intensifying screen compared to conventional metal screen-only combinations This cassette–film

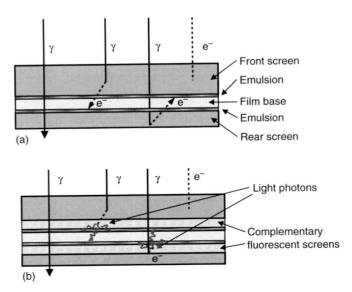

(a)

(b)

FIGURE 14.1
(See colour insert following page 590.) Image formation in film with (a) metal only and (b) metal/fluorescent screens.

combination is marketed by Kodak as the Kodak EC-L (enhanced contrast for localisation) fast cassette and Kodak EC-V slow cassette for verification. The improvement in contrast in the metal–phosphor–screen–film combination was attributed to both the screen combination and film granularity.

Film granularity gives an objective and quantitative measure of the film emulsion's graininess. A developed film appears *grainy* or *mottled* as a result of discrete grain structure of the randomly distributed silver grains within the clear gelatine of the emulsion. A comparison of electron photomicrographs for different film morphologies was provided by Haus et al. (1997). In their article, they showed that the enhanced contrast film with small grain sizes and a small size variation, has granularity that is about three times smaller than conventional films. Because the film sensitivity is proportional to the grain size, a narrow grain size distribution is exposed with a narrow range of exposure. This results in higher contrast than for films with large grain size.

Film is effectively a non-real-time imaging method. Some clinics will wait for the few minutes it takes to develop and read the film; this is especially true in difficult cases. Most of the time, it is used retrospectively. It is not possible to continuously monitor the patient position throughout the treatment (i.e. study patient movement). Similarly, gross errors such as missing blocks, inverted wedges, or wrongly set collimators will not be detected until it is too late. Film cannot be used to check the dynamically-varying beam parameters such as found in rotation therapy, dynamic wedges, or moving multileaf-collimator blades. For all these reasons, films tend to be progressively replaced by electronic portal imaging.

14.3 ELECTRONIC PORTAL IMAGING DEVICES

EPIDs have been developed to acquire and display portal images in as short a time as possible. The image can be analysed instantly for on-line verification, or more images can be acquired per treatment field without the need to enter the treatment room. One advantage of

TABLE 14.1

Properties of First Generation Electronic Portal Imaging Devices

EPID Model Supplier	Portal Vision Varian	SRI 100 Philips	Theraview/Target View Infimed	Beamview Siemens
Type	SLIC	VEPID	VEPID	VEPID
Detector	Ion chambers	Gd_2O_2S	Gd_2O_2S	Gd_2O_2S
SSD (cm)	Variable	160	Variable	140
Detector size (cm)	32.5×32.5	40×30	40×40	41×33
Detector holder	Robotic arm	Rigid, demountable	Motorised vertical and horizontal	Collapsible
Camera	—	CCD	Plumbicon	Newvicon
Detector matrix	256×256	512×256	512×512	512×480
Pixel size (mm)	1.27×1.27	0.78×1.17	0.78×0.78	0.65×0.52
Platform	PC	PC	SPARC station	SPARC station
Software tools	Comprehensive	Limited	Comprehensive	Limited
Check field size & shape	Yes	No	Yes	No
Image registration	Yes	No	Yes	No
Compression available	Yes	No	No	No
f_{50} at isocentre (lp/mm)	0.30	0.21–0.24	0.27	0.21–0.29

Note: The EPID types are either Scanning Liquid-filled Ionisation Chambers systems (SLIC) or Video-based Electronic Portal Imaging Device (VEPID).

the electronic devices is that the image is directly available in digital format for digital processing, archival and communication with other image handling modalities. A number of EPIDs have been developed based on different philosophies. These systems fall into three main categories: *video-* or *fluorescent-screen-based*; *matrix*; and *scanning diode* systems. A detailed review of such systems is provided by Antonuk (2002). The matrix devices can be subdivided into two groups: van Herk and Meertens' (1988) liquid filled ionisation chamber and the solid state systems of Antonuk et al. (1991) and Munro and Bouis (1998). The video and the matrix systems are available commercially, whereas, the scanning systems are still confined to the individual development groups. Table 14.1 gives an overview of the characteristics of some commercially available systems. The f_{50} values that provide a measure of the spatial resolution of the system will be discussed in more detail in Section 38.11.1.2.

14.3.1 FLUORESCENT-SCREEN-BASED SYSTEMS

Several authors (Benner et al. 1962; Bailey et al. 1980; Leong 1986, Shalev et al. 1989; Munro et al. 1990; Visser et al. 1990) have described systems based on imaging a fluorescent screen. The systems described in the latter three articles have been commercially developed by Siemens, Infimed, Eliav and Elekta. The principle is shown in Figure 14.2. The detector is a large metal screen that is bonded to a thick fluorescent screen. The patient image is formed by high-energy x-ray photons that undergo Compton scattering within the metal screen. The recoil electrons pass into the phosphor screen where each one produces a large number of optical photons, typically ten thousand. The optical image thus produced is relayed to a video camera via a 45° folding mirror; this bend in the optical path keeps the camera out of the main beam and allows the camera to be located in a more convenient place at or near the gantry. The signal from the video camera is then available for processing, analysis, storage and display; there is a wide range of special-purpose and general-purpose software available.

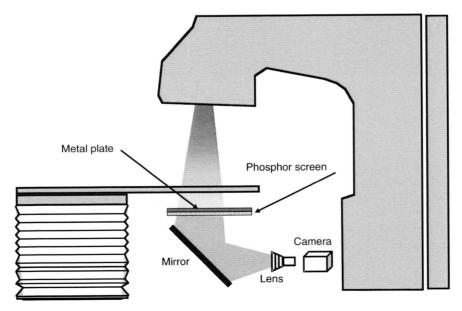

FIGURE 14.2
Configuration of a video-based EPID mounted on a linac. Light photons from the phosphor screen are reflected by the 45° mirror to the camera via the lens.

The *metal detection screen* must be thick enough to remove the vast majority of all the electrons that are scattered within the patient and patient-support system. This eliminates the patient-scattered electrons as a source of image-contrast degradation. A thickness of 1 g cm^{-2} is typically used. There is no point in having a thicker screen to absorb more x-ray photons because, by the same token, only secondary electrons produced within a distance equal to the stopping range of an electron in the metal plate will escape and help generate the optical image. At best, these metal screens have a quantum efficiency of about 1%.

The *phosphor screen* should be thick enough to remove most of the energy of the secondary electrons produced in the metal screen, and a thickness of 400 mg cm^{-2} is typical. Phosphor screens thicker than this have also been used with the aim of increasing the absorption of the x-ray photons, using the phosphor material itself. A problem arises over the efficiency with which the optical photons are transmitted to the video camera. Although many thousands of optical photons P_2 are produced on average for each detected x-ray photon, less than one in a thousand actually arrive at the photo detector because of the very small solid angle subtended by the camera lens from a position within the viewing screen. The optical coupling efficiency P_3 is given by Swindell (1991):

$$P_3 = \frac{T}{[4nF(1 + 1/m)]^2} \tag{14.1}$$

where m the optical magnification is typically 0.05; F is the f-number of the lens and is usually no less than about 0.6 (the smaller the better); and T is a factor usually very close to unity that describes transmission and vignetting losses within the camera lens. The refractive index of the scintillator is denoted by n. The importance of these factors can be seen from the expression that describes the overall transmission of signal-to-noise ratio through the system (after Swindell 1991):

$$\text{SNR}_{\text{out}}^2 = \text{SNR}_{\text{in}}^2 \left(1 + 1/(P_2 P_3 P_4)\right)^{-1} \tag{14.2}$$

where P_4, the quantum efficiency of the video camera, is typically 30% for a vidicon or orthicon type of device. CCD cameras have intrinsically high quantum efficiency, greater than 90%, but this is reduced to an effective value of sometimes only half of this figure because the chip has a large fraction of its surface used for non-detecting functions. The term $P_2 P_3 P_4$ is the average number of detected optical photons per detected x-ray photon, and provided this exceeds some small number, such as five, no significant additional noise is introduced by inefficiencies in the optical system. Most systems manage to achieve this desirable state of affairs. Of course, this product should be as large as possible for the purpose of maximising the signal level, thereby improving the overall SNR with respect to noise sources that occur elsewhere in the system such as in the camera. The above expression that is often quoted is derived on the assumption that the optical-photon density distribution is Poisson distributed. This turns out to be acceptable in this context.

These systems are efficient from the viewpoint that the detector is two-dimensional in nature, allowing the entire portal aperture to be sampled at once; but with a quantum efficiency of 1%, there are insufficient x-ray photons contributing to a video image's single frame. Some form of integration is required in order to reduce photon-noise degradation to an acceptable level. This can be done in software after the image has been digitised by the frame grabber, but Munro et al. (1990) have shown the advantage of integrating on the photo cathode of the camera in order to reduce read-out noise. Most commercial systems have photon-limited images and typically need to integrate over about five seconds of beam time (at 4 Gy/min incident on the patient) before acceptable quality images are produced.

These systems are typically large and bulky, and special handling techniques have been developed to rapidly remove the imager from the treatment zone around the patient.

14.3.2 MATRIX DEVICES

14.3.2.1 *The Matrix Ionisation Chamber*

The *matrix ionisation chamber* system was developed at the Netherlands Cancer Institute in Amsterdam (Meertens et al. 1985; van Herk and Meertens 1987, 1988) and is shown in Figure 14.3. In the clinical prototype, the chamber was made from two fibreglass circuit boards, each etched with 128 linear electrodes on a pitch of 2.5 mm to form an array of 128 by 128 separate ionisation chambers. Therefore, the field of view at the detector was 320 mm by 320 mm. The 1 mm gap between the boards was filled with iso-octane to act as the ionising medium. The detector's spatial resolution has since been improved by increasing the number of electrodes to 256×256 while maintaining the same detector area. As a result, the electrodes are now 0.6 mm wide with 0.6 mm gaps, giving an effective pitch of $1.27 \text{ mm} \times 1.27 \text{ mm}$ pixels. The image is acquired by sequentially applying a high voltage to each of the high voltage electrodes and then measuring the signal generated in each of the 256 signal electrodes. On the first commercial detectors, a 250 V potential was applied to the switches for approximately 20 ms. This acquired an image in 5.5 s in the standard mode or 1.37 s in the reduced spatial resolution mode. The switching speed was limited by switching transients that have now been eliminated on the improved commercial version of the detector (van Herk et al. 1996). On the newer detector, the voltage was increased from 250 V to 500 V in order to increase acquisition speed. Increasing the voltage reduces the dwell or the on-time for each high voltage line, thereby reducing the acquisition time. With the more recent detector, image acquisition is possible in between 0.6 s and 2.0 s under clinical conditions. This reduces the required monitor units by up to 10 times compared to the original system.

Ideally, the detector should use all available radiation to improve image quality. As with most scanning EPIDs, the system is intrinsically inefficient because only one high voltage line is active at any one time. The physics of signal generation in the octane (2-2-4-trimethylpentane) considerably improves the quantum utilisation of this chamber. The concentration of ion pairs is stationary in the octane with a slow ion recombination rate. Calculations have shown that a latent image is formed over the entire irradiated region of the chamber after just 0.5 s of irradiation (van Herk et al. 1996). The signal measured by an electrode depends, therefore, on the electrode's previous irradiation history and the formation of new ions during the

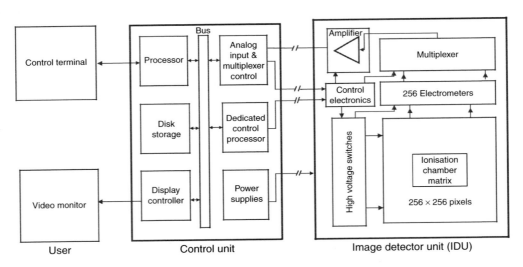

FIGURE 14.3
Diagrammatic layout of the matrix ionisation chamber system, showing the detector unit and layout of the control circuitry. The high-voltage switches and multiplex electronics are close to the field of view edges, so care has to be taken to reduce excess exposure.

measurement interval is negligible. The charge integration in the octane increases the measured signal. There is still, however, a large fraction of ignored radiation quanta not utilised in image acquisition. The faster timing electronics with the new detector compensate for this inefficiency so as to reduce the higher doses that would be required to generate the images. One area that still needs to be addressed is the detector's susceptibility to dose-rate-variation induced artefacts. There is no correction for the dose-rate variations during the signal collection, so a stable beam is assumed.

The system is relatively lightweight making it radiographer friendly. The detector unit is rigid and geometrically reliable, so there are no geometric distortions in the images, and mechanical problems are minimal.

14.3.2.2 Amorphous Silicon (a-Si) Devices

The most recent imaging detector to be developed is the *amorphous silicon array* detector unit (Antonuk et al. 1991, 1993, 1994, 1997; Munro and Bouius 1998). The a-Si electronic portal imaging detector consists of a scattering metal plate to produce Compton electrons and a phosphor layer to absorb the high energy electrons and emit light photons that are detected by the large-area photodiode array (as shown in Figure 14.4).

Principles of operation:

1. Compton electrons are produced in the 1.5 mm copper plate.
2. The electrons produce light photons in the phosphor material. The phosphor materials that can be used include gadolinium oxysulphide (Gd_2O_2S) and thallium-doped caesium iodide (CsI:Tl). The screen is either glued using transparent glue or mechanically pressed onto the light sensor system.
3. The light sensor detector pixels are comprised of photodiodes and TFT transistors that are connected to the readout and scanning electronics. The diodes initially are at 5 V bias before irradiation. During irradiation, the TFTs are non-conducting; the light photons discharge the diodes. On readout, the TFTs are made conducting to recharge the diodes. The charging is carried out row-by-row, and the charge required to re-bias the diodes is proportional to the light reaching the photodiode.

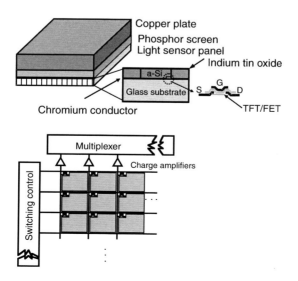

FIGURE 14.4

Schematic diagram of the amorphous Si detector. Photodiodes that make the individual pixels accumulate the signal generated by the light photons. When each row is switched, the TFT/FET act as switches for the individual diodes to allow current to flow along the data line to the charge amplifiers.

TABLE 14.2

Parameters of the Varian and Elekta Amorphous Silicon Imager Panels

Model/Supplier	Portal Vision aS500/Varian[a]	iViewGT/Elekta[b]
Detector area	40.14 cm×30.11 cm	40.96 cm×40.96 cm
Pixel format	512×384	512×512
Pixel pitch	0.784 mm	0.80 mm
Maximum image acquisition rate	15 fps	3 fps
Image display and storage rate	2 s per image	~0.3 s per image
Pixel depth	14 bits	16 bits
Metal plate thickness	1mm Cu	1mm Cu
Scintillator	Gd_2O_2S:Tb	Gd_2O_2S:Tb
f_{50}	0.45 lp/mm	0.46 lp/mm
Software tools	Comprehensive	Comprehensive

[a] The Varian panel potentially has twice the resolution of the Perkin Elmer panel, but in the aS500 version, the signals are combined to reduce the resolution.
[b] The Elekta panel is similar to the Siemens panel; both are based on Perkin Elmer detectors.
Source: From Menon, G. V. and Sloboda, R. S., *Med. Dosim.*, 29, 11–17, 2004.

This is a self-scanning system that is very compact and flat, providing very high spatial resolution images that have been comparable to film (Antonuk 1992; Munro 1995). The spatial resolution of these detectors is governed by the size of the sensors and the focal spot size of the linear accelerator (Antonuk 2002). The size of the sensors varies from 0.4 mm to 0.78 mm for the currently available amorphous silicon panels. The read-out speeds range from 3 frames per second (fps) for RID 1640 Perkin Elmer detector that is currently supplied by Siemens and Elekta to 15 fps for the aS500 (Varian Medical systems). These panels are tiled together to provide a wide area detector. Typically, the panels are 41 cm×41 cm in dimension for the Perkin Elmer detector and 40 cm×30 cm for the aS500. These details are summarized in Table 14.2.

These devices are well-suited to *transit dosimetry* because of the linear response to dose and their compactness that make them easy to position near the patient or between heterogeneous media. For this usage their properties and acquisition parameters must be fully understood and controlled (Greer 2005; Berger et al. 2006, see also Section 40.1.4). They have gained market acceptance in spite of their significantly higher cost of ownership.

14.3.3 OTHER SYSTEMS

Several groups (Taborsky et al. 1982; Lam et al. 1985; Morton and Swindell 1988; Morton et al. 1991b) have developed systems that produce images by mechanically scanning a row of detectors along a path at right angles to the long dimension of the row. The Johns Hopkins group (Taborsky et al. 1982; Lam et al. 1985) used high-voltage rectifier diodes underneath a 1.1 mm plate of lead. The image is acquired by moving the scanning array across the field in 2 mm incremental steps. The Royal Marsden group (Morton and Swindell 1988) has used a double row of photo-diodes in conjunction with zinc tungstate scintillating crystals. More details of these scanning systems are provided in the review by Boyer et al. (1992).

A method based on the photo-stimulable phosphor plate, the Fuji system (Sonoda et al. 1983), has been further described by Wilenzick et al. (1987). The image receptor is a flexible plate supporting a 1 mm thick layer that contains europium-doped barium fluoro-halide. The phosphor contains electron energy traps that are filled when the plate is exposed to ionising radiation. The plate is read out in a separate reader where a red, scanning laser beam releases the stored energy in the form of visible photons that are then transmitted to a photomultiplier via a light guide. The system gives very high quality images that can be digitally stored and processed in all the usual ways. It has gained wide acceptance in radiology as a replacement for films with the double advantage of eliminating the need for chemical processing and producing images directly in digital form. It is also more and more frequently used for radiotherapy portal imaging when EPID systems cannot be used (e.g. because of the limited size of detecting area or for special beam arrangements).

Although the lens/mirror system in video-based imagers is the most common method of coupling phosphor screens to photo-detectors such as CCD cameras, the use of fibre optic reducers has also been reported (Wong et al. 1991a and 1991b). The reducers are usually in the form of optical fibres of constant diameter that have been fused together to form a light guide. The fibre optic bundle is tapered to accomplish the required demagnification. The bundle is also bent through 90° in order to locate the photo-detector out of the radiation beam.

An excellent review of the principles of operation and some constructional details of the different types of existing EPID has been provided by Boyer et al. (1992).

PART C: EQUIPMENT
REFERENCES

Aird, E. G. and Conway, J., CT simulation for radiotherapy treatment planning, *Br. J. Radiol.*, 75, 937–949, 2002.

Antonuk, L. E., Electronic portal imaging devices: A review and historical perspective of contemporary technologies and research, *Phys. Med. Biol.*, 47, R31–R65, 2002.

Antonuk, L. E., Yorkston, J., Kim, C. W., Huang, W., Morton, E. J., Longo, M. J., and Street, R. A., Light-response characteristics of amorphous silicon arrays for megavoltage and diagnostic imaging, *Mater. Res. Soc. Symp. Proc.*, 219, 531–536, 1991.

Antonuk, L. E., Yorkston, J., Huang, W., Boudry, J., Morton, E. J., and Street, R. A., Radiation response characteristics of amorphous silicon arrays for megavoltage radiotherapy imaging, *IEEE Trans. Nucl. Sci.*, 39, 1069–1074, 1992.

Antonuk, L. E., Yorkston, J., Kim, C. W., Huang, W., Morton, E. J., Longon, M. J., and Street, R. A., Light-response characteristics of amorphous silicon arrays for megavoltage and diagnostic imaging, *Mater. Res. Soc. Symp. Proc.*, 219, 531–536, 1993.

Antonuk, L. E., El-Mohri, Y., Huang, W., Siewerdsen, J. H., Yorkston, J., and Street, R. A., A large area, high resolution a-Si:H array for x-ray imaging, *Mater. Res. Soc. Symp. Proc.*, 336, 855–860, 1994.

Antonuk, L. E., El-Mohri, Y., Siewerdsen, J. H., Yorkston, J., Huang, W., Scarpine, V. E., and Street, R. A., Empirical investigation of the signal performance of a high-resolution, indirect detection, active matrix flat-panel imager (AMFPI) for fluoroscopic and radiographic operation, *Med. Phys.*, 24, 51–70, 1997.

Aukett, R. J., Thomas, D. W., Seaby, A. W., and Gittins, J. T., Performance characteristics of the Pantak DXT-300 kilovoltage x-ray treatment machine, *Br. J. Radiol.*, 69, 726–734, 1996.

Bailey, N. A., Horn, R. A., and Kampp, T. D., Fluoroscopic visualisation of megavoltage therapeutic x-ray beams, *Int. J. Radiat. Oncol. Biol. Phys.*, 6, 935–939, 1980.

Barrett, H. H. and Swindell, W., *Radiological Imaging: The Theory of Image Formation, Detection, and Processing*, Vol. 1, Academic Press, New York, 1981.

Benner, S., Rosengren, B., Wallman, H., and Netteland, O., Television monitoring of a 30 MV x-ray beam, *Phys. Med. Biol.*, 7, 29–34, 1962.

Berger, L., Francois, P., Gaboriaud, G., and Rosenwald, J.-C., Performance optimization of the Varian aS500 EPID system, *J. Appl. Clin. Med. Phys.*, 7(1), 105–114, 2006.

BIR (British Institute of Radiology), Central axis depth dose data for use in radiotherapy, *Br. J. Radiol.*, (Suppl. 25), 1996.

Birch, M. J. and Blowes, R. W., A contact x-ray therapy unit for intracavitary irradiation, *Phys. Med. Biol.*, 35, 275–280, 1990.

Bortfeld, T., Oelfke, U., and Nill, S., What is the optimum leaf width of a multileaf collimator? *Med. Phys.*, 27, 2494–2502, 2000.

Botman, J. I. M., Bates, T., and Hagedoorn, H. L., A double focusing magnet system for a medical linear accelerator, *Nucl. Instrum. Methods Phys. Res., Sect. B*, 10–11 Part 2, 796–798, 1985.

Boyer, A. L., Antonuk, L., Fernster, A., van Herk, M., Meertens, H., Munro, P., Reistein, L., and Wong, J., A review of electronic portal imaging devices (EPIDS), *Med. Phys.*, 19(1), 1–16, 1992.

Boyer, A., Biggs, P., Galvin, J., Klein, E., LoSasso, T., Law, D., Mah, K., and Yu, C., Basic applications of multileaf collimators, *AAPM Report 72*, American Institute of Physics by Medical Physics Publishing, Madison, WI, 2001, http://www.aapm.org/pubs/reports/rpt_72.pdf.

BSI (British Standards Institution), BS 5288, *Specification. Sealed radioactive sources*, BSI, London, 1976.

Cho, P. S., Johnson, R. H., and Griffin, T. W., Cone-beam CT for radiotherapy applications, *Phys. Med. Biol.*, 40, 1863–1883, 1995.

Dahl, O., Kardamakis, D., Lind, B., and Rosenwald, J.-C., *Current status of conformal radiotherapy*, Acta. Oncol., 35(Suppl. 8), 41–57, 1996.

Davy, T. J., Johnson, P. H., Redford, R., and Williams, J. R., Conformation therapy using the tracking cobalt unit, *Br. J. Radiol.*, 48, 122–130, 1975.

Fedoruk, S. O. and Johns, H. E., Transmission dose measurement for cobalt 60 radiation with special reference to rotation therapy, *Br. J. Radiol.*, 30, 190–195, 1957.

Galvin, J. M., Smith, A. R., Moeller, R. D., Goodman, R. L., Powlis, W. D., Rubenstein, J., Solin, L. J., Michael, B., Needham, M., Huntzinger, C. J. et al., Evaluation of multileaf collimator design for a photon beam, *Int. J. Radiat. Oncol. Biol. Phys.*, 23, 789–801, 1992.

Galvin, J. M., Leavitt, D. D., and Smith, A. A., Field edge smoothing for multileaf collimators, *Int. J. Radiat. Oncol. Biol. Phys.*, 35, 89–94, 1996.

Gerig, L., Soubra, M., and Salhani, D., A translating-bed technique for total-body irradiation, *Phys. Med. Biol.*, 39, 1377–1392, 1994.

Green, D. T. and Errington, R. F., Design of a cobalt 60 beam therapy unit, *Br. J. Radiol.*, 25, 309–313, 1952.

Greene, D., Nelson, K. A., and Gibb, R., The use of a linear accelerator *simulator* in radiotherapy, *Br. J. Radiol.*, 37, 394–397, 1964.

Greene, D. and Williams, P. C., *Linear Accelerators for Radiation Therapy*, Institute of Physics, Bristol, 1997.

Greer, P. B., Correction of pixel sensitivity variation and off-axis response for amorphous silicon EPID dosimetry, *Med. Phys.*, 32, 3558–3568, 2005.

Gupta, S. K. and Cunningham, J. R., Measurement of tissue-air ratios and scatter functions for large field sizes for cobalt 60 gamma radiation, *Br. J. Radiol.*, 39, 7–11, 1966.

Harrison, R. M. and Farmer, F. T., Possible application of a radiotherapy simulator for imaging of body cross sections, *Br. J. Radiol.*, 49, 813, 1976.

Harrison, R. M. and Farmer, F. T., The determination of anatomical cross-sections using a radiotherapy simulator, *Br. J. Radiol.*, 51, 448–453, 1978.

Haus, A. G., Dickerson, R. E., Huff, K. E., Monte, S., Schlager, B. A., Atanas, M., and Matloubieh, A., Evaluation of a cassette-screen-film combination for radiation therapy portal localization imaging with improved contrast, *Med. Phys.*, 24, 1605–1608, 1997.

Heales, J. C., Harrett, A., and Blake, S., Timer error and beam quality variation during "ramp-up" of a superficial x-ray therapy unit, *Br. J. Radiol.*, 71, 1306–1309, 1998.

Huq, M. S., Das, I. J., Steinberg, T., and Galvin, J. M., A dosimetric comparison of various multileaf collimators, *Phys. Med. Biol.*, 47, N159–N170, 2002.

IAEA (International Atomic Energy Agency), International Basic Safety Standards for protection against ionizing radiation and for the safety of radiation sources, *IAEA Safety Series*, Number 115, IAEA, Vienna, 1996.

IAEA (International Atomic Energy Agency), Accidental overexposure of radiotherapy patients in San José, Costa Rica, *Special Report Series*, IAEA, Vienna, 1998 (http://www-pub.iaea.org/MTCD/publications/PDF/P027_scr.pdf)

IAEA (International Atomic Energy Agency), Design and implementation of a radiotherapy programme: Clinical, medical physics, radiation protection and safety aspects, TECDOC-1040, IAEA, Vienna, 1998. (Revision entitled *Setting-up a radiotherapy programme: Clinical, medical physics, radiation protection and safety aspects*, 2007.)

ICRU (International Commission on Radiation Units), Report 18, *Specification of High Activity Gamma-ray Sources*, ICRU, Bethesda, MD, 1970.

IEC (International Electrotechnical Commission), *Medical Electrical Equipment Specification for Therapeutic X-Ray Generators*, IEC Publication 601-2-8, IEC, Geneva, 1987.

IEC (International Electrotechnical Commission), *Radiotherapy Equipment—Coordinates, Movements and Scales*, IEC Publication 1217, IEC, Geneva, 1996.

IEC (International Electrotechnical Commission), *Medical Electrical Equipment. Particular Requirements for Safety. Specification for Gamma Beam Therapy Equipment*, IEC Publication 60601-2-11, IEC, Geneva, 1997.

IPEM (Institute of Physics and Engineering in Medicine), Report 81, *Physics Aspects of Quality Control in Radiotherapy*, IPEM, York, 1999.

IPEM (Institute of Physics and Engineering in Medicine), *Medical and Dental Guidance Notes. A Good Practice Guide to Implement Ionizing Radiation Protection Legislation in the Clinical Environment*, IPEM, York, 2001.

ISO (International Standards Organisation), *Radiation Protection—Sealed Radioactive Sources—General Requirements and Classification*, ISO, Geneva, 1999.

Jaffray, D. A., Munro, P., and Battista, J. J., Activity distribution of a cobalt-60 teletherapy source, *Med. Phys.*, 18, 288–291, 1991.

Jaffray, D. A., Siewerdsen, J. H., Wong, J. W., and Martinez, A. A., Flat-panel cone-beam computed tomography for image-guided radiation therapy, *Int. J. Radiat. Oncol. Biol. Phys.*, 53, 1337–1349, 2002.

Johns, H. E. and Cunningham, J. R., *The Physics of Radiology*, 4th ed., Charles C. Thomas, Springfield, IL, pp. 63–65, 1983.

Johns, H. E., Bates, L. M., and Watson, T. A., 1000 Curie cobalt units for radiation therapy, *Br. J. Radiol.*, 25, 296–302, 1952.

Jones, D. E. A., Gregory, C., and Birchall, I., Dosage distribution in rotational cobalt 60 therapy, *Br. J. Radiol.*, 29, 196–201, 1956.

Jordan, T. J. and Williams, P. C., The design and performance characteristics of a multileaf collimator, *Phys. Med. Biol.*, 39, 231–251, 1994.

Karzmark, C. J. and Rust, D. C., Radiotherapy simulators and automation, *Radiology*, 105, 157–161, 1972.

Karzmark, C. J., Nunan, C. S., and Tanabe, E., *Medical Electron Accelerators*, McGraw-Hill, Springfield, IL, 1993.

Kessler, M. L., McShan, D. L., and Fraass, B. A., A computer-controlled conformal radiotherapy system. III: Graphical simulation and monitoring of treatment delivery, *Int. J. Radiat. Oncol. Biol. Phys.*, 33, 1173–1180, 1995.

Klevenhagen, S. C., D'Souza, D., and Bonnefoux, I., Complications in low energy x-ray dosimetry caused by electron contamination, *Phys. Med. Biol.*, 36, 1111–1116, 1991.

Klevenhagen, S. C., Thwaites, D. I., and Aukett, R. J., Kilovoltage x-rays, in *Radiotherapy Physics in Practice*, Williams, J. R. and Thwaites, D. I., Eds. 2nd ed., University Press, Oxford, 2000. Chapter 6

Lam, K. S., Partowmah, M., and Lam, W. C., An on-line electronic portal imaging system for external beam radiotherapy, *Br. J. Radiol.*, 59, 1007–1013, 1985.

Lapostolle, P. M. and Septier, A. L., *Linear Accelerators*, North Holland Publishing, Amsterdam, 1970.

Leavitt, D. D., Martin, M., Moeller, J. H., and Lee, W. L., Dynamic wedge field techniques through computer-controlled collimator motion and dose delivery, *Med. Phys.*, 17, 87–91, 1990.

Leksell, L., *Stereotaxis and Radiosurgery*, Charles C. Thomas, Springfield, IL, 1971.

Leong, J., Use of digital fluoroscopy as an on-line verification device in radiation therapy, *Phys. Med. Biol.*, 31, 985–992, 1986.

Leung, P. M. K., Rider, W. D., Webb, H. P., Aget, H., and Johns, H. E., Cobalt-60 therapy unit for large field irradiation, *Int. J. Radiat. Oncol. Biol. Phys.*, 7, 705–771, 1981.

Lindquist, C., Gamma knife radiosurgery, *Semin. Radiat. Oncol.*, l5, 197–202, 1995.

Mackie, T. R., Holmes, T., Swerdloff, S., Reckwerdt, P., Deasy, J. O., Yang, J., Paliwal, B., and Kinsella, T., Tomotherapy: A new concept for the delivery of dynamic conformal radiotherapy, *Med. Phys.*, 20, 1709–1719, 1993.

McCullough, E. C. and Earle, J. D., The selection, acceptance testing, and quality control of radiotherapy treatment simulators, *Radiology*, 131, 221–230, 1979.

Meeks, S. L., Bova, F. J., Kim, S., Tome, W. A., Buatti, J. M., and Friedman, W. A., Dosimetric characteristics of a double-focused miniature multileaf collimator, *Med. Phys.*, 26, 729–733, 1999.

Meertens, H., van Herk, M., and Weeda, J., A liquid ionisation detector for digital radiography of therapeutic megavoltage photon beams, *Phys. Med. Biol.*, 30, 313–321, 1985.

Menon, G. V. and Sloboda, R. S., Quality assurance measurements of a-Si EPID performance, *Med. Dosim.*, 29, 11–17, 2004.

Meredith, W. J. and Massey, J. B., *Fundamental Physics of Radiology*, 3rd ed., John Wright and Sons Ltd, Bristol, p. 54, 1977.

Morton, E. J. and Swindell, W., A digital system for the production of radiotherapy verification images, in *The Use of Computers in Radiation Therapy*, Bruinvis, I. A. D. et al., Eds., North Holland, Amsterdam, pp. 375–377, 1988.

Morton, E. J., Evans, P. M., Ferraro, M., Young, E. F., and Swindell, W., Development of video frame store and distortion correction facilities for an external-beam radiotherapy treatment simulator, *Br. J. Radiol.*, 64, 747–750, 1991a.

Morton, E. J., Swindell, W., Lewis, D. G., and Evans, P. M., A linear array: Scintillation crystal photodiode detector for megavoltage imaging, *Med. Phys.*, 18, 681–691, 1991b.

Munro, P., Portal imaging technology: Past, present and future, *Semin. Radiat. Oncol.*, 5, 115–133, 1995.

Munro, P. and Bouius, D. C., X-ray quantum limited portal imaging using amorphous silicon flat panel arrays, *Med. Phys.*, 25, 689–702, 1998.

Munro, P., Rawlinson, J. A., and Fenster, A., A digital fluoroscopic imaging device for radiotherapy localisation, *Int. J. Radiat. Oncol. Biol. Phys.*, 18, 641–649, 1990.

Mutic, S., Palta, J. R., Butker, E. K., Das, I. J., Huq, M. S., Loo, L. D., Salter, B. J., Mccollough, C., and Van Dyk, J., Quality assurance for computed-tomography simulators and the computed-tomography-simulation process: Report of the AAPM Radiation Therapy Committee Task Group No. 66, *Med. Phys.*, 30, 2762–2792, 2003.

Patel, I., Glendinning, A. G., and Kirby, M. C., Dosimetric characteristics of the Elekta beam modulator™, *Phys. Med. Biol.*, 50, 5479–5492, 2005.

Petti, P. L. and Siddon, R. L., Effective wedge angles with a universal wedge, *Phys. Med. Biol.*, 30, 985–991, 1985.

Poffenbarger, B. A. and Podgorsak, E. B., Viability of an isocentric cobalt-60 teletherapy unit for stereotactic radiosurgery, *Med. Phys.*, 25, 1935–1943, 1998.

Prasad, S. C. and Bassano, D. A., Corner transmission in several linear accelerator photon beams, *Med. Phys.*, 18, 763–764, 1991.

Rosenwald, J.-C., Gaboriaud, G., and Pontvert, D., Conformal radiotherapy: principles and classification (in French), *Cancer Radiother.*, 3, 367–377, 1999.

Schlegel, W., Pastyr, O., Bortfeld, T., Becker, G., Schad, L., Gademann, G., and Lorenz, W. J., Computer systems and mechanical tools for stereotactically guided conformation therapy with linear accelerators, *Int. J. Radiat. Oncol. Biol. Phys.*, 24, 781–787, 1992.

Schlegel, W., Pastyr, O., Kubesch, R., Stein, J., Diemar, T., Hover, K. H., and Rhein, B., A computer controlled micro-multileaf collimator for stereotactic conformal radiotherapy, in *Proceedings of the XIIth International Conference on Computers in Radiotherapy*, Leavitt, D. D. and Starkschall, G., Eds., Medical Physics Publishing, Madison, WI, pp. 79–82, 1997.

Shalev, S., Lee, T., Leszczynski, K., Cosby, S., Chu, T., and Stein, L., Video techniques for on-line portal imaging, *Comput. Med. Imaging Graph.*, 13, 217–226, 1989.

Sherouse, G. W., Bourland, J. D., Reynolds, K., McMurry, H. L., Mitchell, T. P., and Chaney, E. L., Virtual simulation in the clinical setting: Some practical considerations, *Int. J. Radiat. Oncol. Biol. Phys.*, 19, 1059–1065, 1990.

Shiu, A. S., Kooy, H. M., Ewton, J. R., Tung, S. S., Wong, J., Antes, K., and Maor, M. H., Comparison of miniature multileaf collimation (MMLC) with circular collimation for stereotactic treatment, *Int. J. Radiat. Oncol. Biol. Phys.*, 37, 679–688, 1997.

Sonoda, M., Akano, M., Miyahara, J., and Kato, H., Computed radiography utilising scanning laser stimulated luminescence, *Radiology*, 148, 833–838, 1983.

Swindell, W., The lens coupling efficiency in megavoltage imaging, *Med. Phys.*, 18, 1152–1153, 1991.

Taborsky, S. C., Lam, W. C., Sterner, R. E., and Skarda, G. M., Digital imaging for radiation therapy verification, *Opt. Eng.*, 21, 888–893, 1982.

van Herk, M. and Meertens, H., A digital imaging system for portal verification, in *The Use of Computers in Radiation Therapy*, Bruinvis, I. A. D. et al., Eds., North Holland, Amsterdam, pp. 371–373, 1987.

van Herk, M. and Meertens, H., A matrix ionisation chamber imaging device for on-line patient set up verification during radiotherapy, *Radiother. Oncol.*, 11, 369–378, 1988.

van Herk, M., Boellard, R., Brugmans, M., and van Dalen, A., Feasibility of the portalvision system for verification of dynamic therapy, *Proceedings of the 4th International Workshop on Electronic Portal Imaging*, Amsterdam, 1996.

Visser, A. G., Huizenga, H., Althof, V. G. M., and Swanenburg, B. N., Performance of a prototype fluoroscopic radiotherapy imaging system, *Int. J. Radiat. Oncol. Biol. Phys.*, 18, 43–50, 1990.

Walton, L., Bomford, C. K., and Ramsden, D., The Sheffield stereotactic radiosurgery unit: Physical characteristics and principles of operation, *Br. J. Radiol.*, 60, 897–906, 1987.

Warrington, A. P. and Adams, E. J., Conformal and intensity modulated radiotherapy using cobalt-60 and 6 MV x-ray beams: A treatment planning comparison of different sites, *Radiother. Oncol.*, 61(Suppl. 1), S73–S74, 2001.

Webb, S., *The Physics of Three Dimensional Radiation Therapy: Conformal Radiotherapy, Radiosurgery and Treatment Planning*, IOPP, Bristol, 1993.

Webb, S., Non-standard CT scanners: Their role in radiotherapy, *Int. J. Radiat. Oncol. Biol. Phys.*, 19, 1589–1607, 1990.

Webb, S., Lillicrap, S. C., Steere, H., and Speller, R. D., Application of a radiotherapy simulator for imaging of body cross sections, *Br. J. Radiol.*, 50, 152–153, 1977.

Wilenzick, R. M., Merritt, C. R. B., and Balter, S., Megavoltage portal films using computed radiographic imaging with photostimulable phosphors, *Med. Phys.*, 14, 389–392, 1987.

Wong, J. W., Binns, W. R., Cheng, A. Y., Epstein, J. W., and Klarmann, J., Development of a high resolution fiber-optic radiotherapy imaging system, *Med. Phys.*, 18, 609, 1991a.

Wong, J. W., Graham, M. L., Halverson, K. S., and Leung, T. C., Study of treatment variation using a fiber-optic on-line radiotherapy imaging system, *Med. Biol. Eng. Comput.*, 29(Suppl. 2), 858, 1991b.

Xia, P., Geis, P., Xing, L., Ma, C., Findley, D., Forster, K., and Boyer, A., Physical characteristics of a miniature multileaf collimator, *Med. Phys.*, 26, 65–70, 1999.

Zwicker, R. D., Shahabi, S., Wu, A., and Sternick, E. S., Effective wedge angles for 6 MV wedges, *Med. Phys.*, 12, 347–349, 1985.

PART D

DOSE MEASUREMENT

Editors: Jean-Claude Rosenwald and Philip Mayles

INTRODUCTION

Dose measurement is essential in radiation therapy. The dose delivered to the patient is required to be known with an accuracy, at worst, of the order of 5 % to 7 %, taken to be the 95 % confidence level (ICRU 1976; Dutreix 1984; Mijnheer et al. 1987) as discussed further in Section 37.4. The dose finally delivered to the patient is the end result of a complex sequence, involving several steps and many processes and parameters, and each contributing component must have an appropriately higher accuracy to achieve this overall value. In particular, the beams generated from the treatment machines should be calibrated with an accuracy approaching 1 %. In this Part, Chapter 15 to Chapter 17 describe the dosimeters that are commonly used in radiation therapy departments and in Chapter 18 and Chapter 19 the methodology of beam calibration and the acquisition of the experimental data needed for patient dose calculations is discussed. A minimum knowledge of clinical beam properties is useful for a better understanding of the methodology for dose measurement, for which it will occasionally be useful to refer to Part E (Clinical Beams).

In principle, any radiation effect may be used as the basis of dose measurement provided that it can be suitably linked to a useful dosimetric quantity. Systems that respond to radiation but in a way that does not give quantitative information on dose or a related quantity are termed *detectors*. They only indicate the presence or absence of significant radiation and are not calibrated in a dosimetric quantity. To be used for dose measurements the response of a dosimeter requires various corrections, depending on the measurement system and measurement conditions and a transfer step to convert dose to the medium of its sensitive volume to dose to the medium of interest (generally water or soft tissue for clinical applications). This requires relative radiation interaction information as input data that introduces additional uncertainties.

Absorbed dose can never be *measured* in a direct sense. In practice, a quantity such as a charge, a temperature change, or some chemical change is measured from which the dose is *derived*. However, in deference to accepted usage, the expression *dose measurement* has been retained for the title and throughout the rest of this Part.

Some general requirements of dosimeters may be listed that have consequences for the choice of the system and the construction and operation of the instrument.

- Repeatability: stability for repeated measurements over a short period of time
- Reproducibility: stability of material, construction, etc. over a long period of time
- Accuracy, precision*: the derivation of the dose from the dosimeter response must be possible with minimum uncertainty, but the requirements may differ for different applications
- Sensitivity: must be high enough for low dose rate measurements. If the sensitivity is too high, it may cause rapid saturation at high dose rate
- Adequate dose range and (preferably) linearity of the response as a function of accumulated dose
- Energy response: preferably independence of response as a function of energy
- Insensitivity of response to influence quantities (dose rate, temperature, pressure, directional effect, accumulated dose, etc.): response should be independent, or variation should be known or measurable in order to perform adequate correction.

Other requirements may also need to be taken into account such as compatibility with phantom materials and detector size, both being linked to considerations of perturbation of the

---

* *Precision* encompasses the statistical reproducibility of measurements and the resolution of the measuring instrument. The *accuracy* of a measuring system reflects how closely the measured value lies to the true value of the given quantity of interest (e.g. absolute dose).

radiation field and the latter also to spatial resolution. In any situation where a dosimetry system has been chosen, the user must be familiar with its limitations to ensure that the system is appropriate to the application.

The type of direct or indirect radiation effect used for dose measurement can be ionisation in gas (ionisation chambers in Chapter 15) or in solid material (diodes in Chapter 16). It can be the emission of light (thermoluminescent materials in Chapter 16), an increasing opacity or blackening in the case of photographic film (Chapter 17), or a chemical change such as in the case of the Fricke or Ferrous Sulphate dosimeter. Provided strictly controlled experimental conditions can be achieved, generally only possible in specialised calibration laboratories, the heat increase resulting from radiation energy absorption in materials like graphite or water can also be used to derive the dose (Ross and Klassen 1996; Huq and Andreo 2004). This method of measurement is called calorimetry. It will not be further addressed here, but it is discussed briefly in connection with proton dosimetry in Section 46.4.3.

Only three types of dosimeters can be regarded as independently absolute, i.e. may be used without reference to others. These are calorimeters, ionisation chambers, and Fricke chemical dosimeters. However, they can only be used for absolute dose determination in a certain configuration and under certain limiting conditions, and conversion and correction factors must still be applied to the readings. All other systems must be calibrated against one of these absolute instruments. Ionisation chambers and calorimeters form the basis of primary standards (at international and national levels), and generally, ionisation chambers are used as secondary standards (at regional or national levels). Hospital field-class ionisation chamber instruments are then calibrated against a secondary standard. Any other dosimetry system used in the hospital must be calibrated against the calibrated field instrument in conditions that are as close as possible to those to be used in the clinical application.

The introduction of special radiotherapy techniques such as those presented in Part I (Intensity-Modulated Radiotherapy, stereotactic techniques, proton therapy, etc.) pose new challenges for dose measurements, and the specific requirements for each technique must be considered (i.e. McKerracher and Thwaites 1999; Medin and Grusell 2000; Moran 2003). Some of these issues will be addressed in Part I; others will be discussed in the following chapters. However, several relatively new types of dosimeters, some of which are beginning to be used in conjunction with these special techniques, will not be dealt with in this book, and readers are referred to specific publications to get additional information. Of particular importance among these new dosimeters are

- Diamond detectors (Rustgi 1995; Laub et al. 1999; Bucciolini et al. 2003).
- Metal Oxide Silicon Field Effect Transistors (MOSFETs) (Chuang et al. 2002; Kron et al. 2002; Halvorsen 2005).
- Electron Paramagnetic Resonance (EPR) or Electron Spin Resonance (ESR) alanine dosimeters (Wielopolski et al. 1987; Kuntz et al. 1996; Sharpe et al. 1996; De Angelis et al. 2005).

In addition, a promising new type of two-dimensional dosimeter is the electronic portal imaging device (see Chapter 14) when used for transit dosimetry. This will be dealt with in Section 40.1.4.4.

CHAPTER 15

IONISATION CHAMBERS

Pedro Andreo, Alan Nahum, and David Thwaites

CONTENTS

15.1 INTRODUCTION

Ionisation is the creation of ion pairs. This results from interactions that transfer energy from the radiation to the atomic system. For energy transfers greater than electron binding energies, electrons are ejected (see Section 1.4.2), leaving behind a positively charged *ion*. The ability of ionising radiation to *ionise* a gas was one of its first properties to be observed; and indeed, it is the defining property. The measurement of ionisation in a gas was also the basis for the first radiation detectors.

Ionisation (or ion) chambers have remained central not only to the detection of radiation but also to the precise measurement of absorbed dose. They are the reference instruments used in the radiotherapy clinic, both for absolute and relative dose measurements (see Chapter 18 and Chapter 19 respectively), and it is fair to say that more is known about the response of ionisation chambers than about any other radiation-detecting instrument. Because of the key role that ionisation chambers play in radiotherapy dosimetry, it is vital that practising physicists and others involved in radiation measurements in radiotherapy beams have a thorough knowledge of their characteristics. As will be made clear, though widely used and well understood, ionisation chambers are far from simple in their response or even in their mode of operation and do not possess the ideal properties of a radiation detector for *dose measurement*.

First, the sensitive medium is a gas, usually air, at atmospheric pressure that has a density around 700 times less than that of water. Hence, the signal strength from an ionisation chamber will always be relatively low compared to so-called solid-state detectors (Chapter 16 and Chapter 17). Furthermore, air being a gas is not particularly close to being equivalent to water[*] or to any other commonly used phantom material such as polymethyl methacrylate (PMMA, known as Perspex or Lucite) or polystyrene, and the quantity measured, charge, has first to be related to dose to the gas D_g (or D_{air}) before cavity theory (Section 6.7) is used to derive the dose to the medium D_{med}.

In order to be able to accurately calculate D_{med} from D_g, the ionisation chamber must behave as a Bragg–Gray cavity in a photon or electron beam that means that the range of the electrons in the gas (secondary electrons in the case of a primary photon beam) must be considerably greater than the dimensions of the cavity; in this case, it can be said that the gas cavity does not disturb the fluence of electrons present in the medium as Figure 15.1 illustrates. This will generally be the case for all high-energy or megavoltage radiotherapy beams[†]. Consequently, the ratio D_{med}/D_g can be equated to the (mass collision) stopping-power ratio $s_{med,g}$ (see Equation 6.48 and Section 18.5.1). It should be noted, however, that the Bragg–Gray principle does not apply to the so-called free air chamber that is also dealt with in this chapter (see Section 15.3.1).

If an ionisation chamber is to be used to make an absolute determination of the dose, then the charge per unit mass of gas, Q/m, also known as the mass ionisation J_g, must be determined. The gas density ρ will generally be accurately known from its pressure and temperature (see Section 15.4.4); a precise knowledge of the volume V containing the gas is also required as $m = \rho \times V$. Then, the dose to gas is derived by multiplying J_g by the so-called W-value that is the mean energy required to create an ion pair. This leads to the following equation:

---

[*] It should be stressed that in spite of this large difference in density, the mass attenuation and mass absorption coefficients of air are not very different from those of water (see tables in Part M) which makes the relative response air/water approximately constant across a wide range of photon energies. Furthermore, (W_{air}/e) is constant across a wide range of photon and electron beam qualities (see Section 18.5.4).

[†] In practical chamber designs, there are, however, important departures from Bragg–Gray behaviour principally because of the non-medium equivalence of the chamber wall material and the finite size of the air cavity; these deviations, termed *perturbations*, are dealt with in detail in Section 18.5.3 and in Section D.2 of Appendix D.

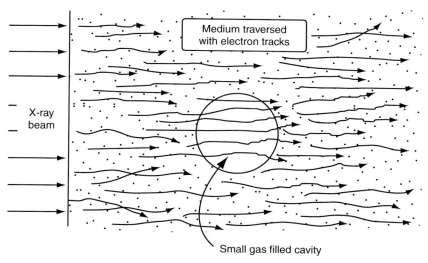

FIGURE 15.1

A Bragg–Gray cavity in a medium irradiated by an x-ray beam; the secondary electron tracks are undisturbed by the small gas-filled cavity. (Reproduced from Johns, H. E. and Cunningham, J. R., *The Physics of Radiology*, 4th Ed., Charles C. Thomas, Eds., Springfield, Illinois, 1983. With permission.)

$$D_{\mathrm{g}} = \frac{Q}{\rho V} \left(W_{\mathrm{g}}/e \right) \tag{15.1}$$

and for the common practical case of water as the medium and air as the gas, we have:

$$D_{\mathrm{water}} = \frac{Q}{\rho V} \left(W_{\mathrm{air}}/e \right) s_{\mathrm{water,\ air}} \tag{15.2}$$

To make an absolute determination of the dose to water with an ionisation chamber, therefore, each of the quantities in Equation 15.2 must be known accurately: the charge Q collected during the radiation exposure; the chamber volume, V; the ratio of the stopping power in water to that in air, $s_{\mathrm{water,air}}$, which is in principle dependent on radiation quality[*]; and the mean energy required to produce an ion pair, divided by the electron charge, (W_{air}/e). (W_{air}/e) which has the precisely determined value of 33.97 eV/ion pair (or joules/coulomb) (see Section 18.5.4) and is independent of radiation quality except at photon or electron energies well below the range of interest in radiotherapy (ICRU 1979). The charge Q can usually be determined with high accuracy, including correction for any recombination losses (see Section 18.5.1), and a great deal of attention has been paid to the accurate computation of stopping-power ratios (see Section 18.5.1). In practice, the difficulty lies in determining the chamber volume with sufficient precision. This is the principal reason why ionisation chambers used for absolute dose determination in radiotherapy, where accuracy of the order of 1% is required, are provided with a calibration factor which accounts for the volume of the chamber as well as for other factors depending on the *Code of Practice* (see Chapter 18). Strictly, ionisation chambers should not be classed as absolute dose-measuring instruments in the radiotherapy context.

---

[*] The quantity $s_{\mathrm{water,air}}$ has a weak dependence on the chamber properties through the size of the cutoff $\varDelta$ in the Spencer-Attix formulation of the stopping-power ratio (see Section 6.7.4) which is related to the dimensions of the air cavity.

15.2 PHYSICAL AND OPERATIONAL PRINCIPLES

15.2.1 INFLUENCE OF THE APPLIED VOLTAGE—THE IONISATION REGION

Ionisation chambers are, in principle, the simplest of all gas-filled detectors. They consist of a gas-filled receptacle with two electrodes between which a voltage is applied. As the gas is theoretically a perfect insulator, no electric current will flow between the electrodes. The radiation interactions ionise the gas, and the electric field causes the movement of the electric charges toward the electrodes of opposite sign, creating a current that can be measured with an appropriate device called an *electrometer*.

At low voltages, the charge collected from the gas is relatively small because considerable *ion recombination* occurs. Then, as the voltage is increased, the signal increases because a greater proportion of the charge is collected before it can recombine. Finally, the signal reaches a saturation value where all the ion pairs created from radiation are collected*; the detector is then said to operate in the *ionisation chamber region*. Increasing the voltage to values beyond this region causes the electrons produced in the gas to be accelerated to such an extent that they produce additional ionisation of the gas, increasing the charge collected beyond that corresponding to I_{sat} (see Figure 15.2). Successive increases in the applied voltage cause the detector to operate successively in what are called the *proportional region*, *limited proportionality region*, *Geiger region* and *continuous discharge region* (Figure 15.2).

This typical behaviour, in addition to the possible electrical breakdown of insulators or distortion of the detector volume, gives rise to a practical limit to the voltage applied to an ionisation chamber, corresponding to the position of the first vertical dashed line in Figure 15.2. A slightly lower limit, at which the chamber can be safely used, is generally specified in the manufacturer's recommendations. Typical voltage values for ionisation chambers used in radiotherapy dosimetry are in the range 200 V to 400 V, with some parallel-plate chambers where the distance between electrodes is very small, having recommended values lower than this (see, for instance, Thwaites et al. 2003); this also applies to certain very small diameter cylindrical chambers.

Multiple designs of ionisation chambers have been developed for use in different situations. These are discussed in Section 15.3 where it is shown that the ionisation chamber is the most versatile radiation detector, capable of operating over a wide range of radiation intensities. For the determination of absorbed dose to water in radiotherapy beams, the most common types are cylindrical (thimble) and plane-parallel (parallel plate or coin) chambers; and air is the usual gas (see Section 15.3.2). The international standard IEC 60731 (1997) on dosimetry with ionisation chambers for radiotherapy provides recommended limits for the performance characteristics of these devices. The International Electrotechnical Commission (IEC) defines an ionisation chamber as "a detector consisting of a chamber filled with air, in which an electric field insufficient to produce gas multiplication is provided for the collection at the electrodes of charges associated with the ions and the electrons produced in the measuring volume of the detectors by ionising radiation".

In general, the chamber itself consists of the measuring volume, the central collecting electrode and the outer electrode (which consists of the chamber wall and possibly a conductive coating). A build-up cap or water-proof housing may be added if needed. The collecting volume of typical cavity chambers used for radiotherapy dosimetry are in the range 0.1 cm^3 to 1 cm^3 (see Section 15.3.2). For dose rates that can vary from 0.1 Gy/min to 10 Gy/min, the typical currents created from air ionisation in the chamber cavity are in the range 6 pA to 6 nA. Such currents are very small and difficult to measure and require high quality connectors and

---

* The limitations in charge collection of radiotherapy ion chambers will be discussed in Section 15.4.

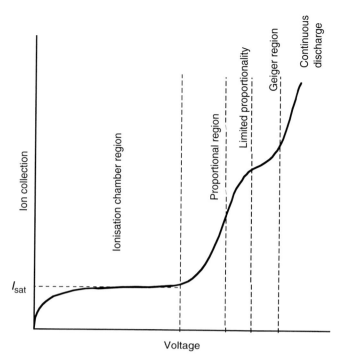

FIGURE 15.2
Variation of the charge collected, as a function of the applied voltage, in a gas detector. The different operating regions of the detector are indicated; the plateau corresponds to the so-called *ionisation chamber region* where a saturation current I_{sat} is achieved.

amplifiers. In radiation protection where the dose rates are much lower, the typical volume of the ionisation chambers used is several hundred cubic centimetres; even with such large volumes, only dose rates above about 10 µGy/h can be measured.

15.2.2 INSULATORS AND GUARD RINGS

The charge (or current) created in the air chamber is conducted to the measuring device (the electrometer) through leads connected to the electrodes. These leads pass through the chamber stem that is attached to the chamber cavity and designed to hold the whole assembly. Because the charge circulating between the electrodes should be restricted to that created within the air cavity, there must be no other conducting path between the electrodes except that of the ionised air, and an insulator must be provided between them. Although, in the past, various naturally occurring substances were used as insulators (amber, quartz and ceresin wax), today synthetic plastics can be used. These are polymethyl methacrylate (PMMA), polystyrene, polyethylene, Nylon, polytetrafluoroethylene (PTFE, Teflon), Mylar film and polytrifluoro-monochloroethylene (Kel-F) (Boag 1966). Surface leakage because of humidity is resisted well by Teflon; however, Teflon is more prone to radiation damage than polystyrene, for example.

As a result of the electric field between the electrodes, a leakage current can be created in the insulator that adds to the measured signal due to the radiation. This leakage is increased by the presence of moisture or small dirt particles; this is the reason why insulators should not be breathed upon or touched with the fingers. To minimize the voltage across the insulator, and therefore, the leakage current, the insulator is usually segmented into two parts with a conducting guard ring; one part separates the ring from the negative electrode, and another separates it from the positive electrode (see Figure 15.3a). The guard ring extends through the insulator assembly in the chamber stem and connects to the inner braid in the cable. The guard

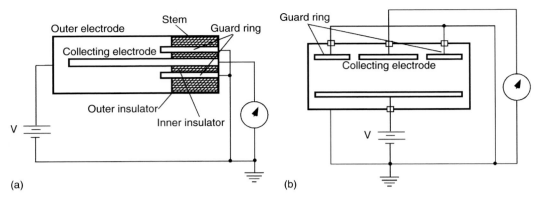

FIGURE 15.3
Set-up of the guard ring in a cylindrical ionisation chamber (a) to avoid leakage current in the insulator. In a plane-parallel chamber, (b) the guard ring defines the effective collection volume between the collecting electrode where the electric field is uniform.

and the collector are held at the same potential, so any charge produced in the cable is collected between outer and inner braids and does not affect the measurement. Different guard ring designs are implemented in plane-parallel chambers where they are used to define the effective collection volume (See Figure 15.3b).

15.2.3 ELECTROMETERS

The electronics required to measure the charge (or current) is provided by the electrometer which, in most cases, includes a voltage supply and a display unit. For further details on electrometers, readers are referred to the compilation given by Attix (1986). In the early days of radiation dosimetry physicists used a gold-leaf electroscope equipped with a quantitative scale as an electrometer. Today, most electrometers are based on *negative-feedback operational amplifiers*.

An operational amplifier, represented by the triangle in Figure 15.4, has a high open-loop gain ($>10^4$) and high input impedance ($>10^{12}$ ohm) and consequently a low potential (<100 mV) is maintained between its positive and negative inputs. Negative feedback means that some or all of the amplifier's output is fed back to its negative input. Total feedback results in a gain of unity. When a capacitor or a resistor is connected in the feedback loop of an operational amplifier whose negative input is connected to the chamber collection electrode and the positive input to the polarizing voltage or to the case ground, the voltage measured at the output of the amplifier is proportional to the charge collected and is equal to the voltage across the capacitor or resistor. If a capacitor C is placed in the feedback loop and the voltage measured across the capacitor is V, then the charge collected, Q, is equal to CV. This is known as capacitive feedback circuit mode, and unity gain ensures that exactly the amount of charge collected by the chamber collection electrode is deposited in the feedback capacitor. Similarly, if instead of the capacitor a resistor R is placed in the loop, the voltage V measured across the resistor allows the current I (related to the charge rate) to be determined as V/R. The reproducibility of a measurement depends almost entirely on the quality and stability of the capacitor or the resistor. Desired electrometer characteristics are the following:

- Fast warm-up (5–10 min)
- Short equilibration time after changing the high voltage across the electrodes
- Negligible ($<10^{-13}$ A) pre- and post-irradiation leakage currents
- Sensitivity independent of ambient conditions, time, or scale position
- Ability to change the bias voltage in magnitude and polarity.

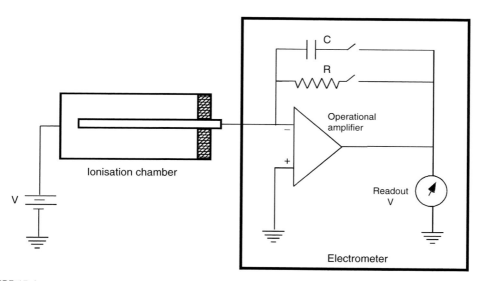

FIGURE 15.4
The principle of a *negative-feedback operational-amplifier electrometer*. The capacitor, *C*, or the resistor, *R*, are the feedback elements. The use of the capacitor or the resistor corresponds, respectively, to an electrometer operating in integrating or rate mode.

Most modern readout devices are digital voltmeters capable of achieving reproducibility of the reading over the entire range. They usually have maximum readings of 1999 or 19 999.

15.2.4 CABLES AND CONNECTORS

The cables connecting the chamber and the electrometer are an important component of the dosimetry system (Humphries and Slowey 1986). High-quality low-noise triaxial cables having three separated conductive coatings are the obvious choice; the microphonic noise and dielectric-response time in cables provided by different manufacturers can vary significantly. These usually have a minor influence on the response of the system, but a more important aspect is the radiation-induced signal, which varies with the beam energy and, as shown by Humphries and Slowey, is often rather different between high-energy photons and electrons and from one manufacturer to another. The desired cable characteristics are the following: short equilibration time (less than one minute) when the high voltage is changed; low radiation-induced signal; low microphonic noise; low leakage (less than 10^{-14} A); low capacitance; pliability; and easy and robust connector installation.

The connectors between the chamber lead and the electrometer have not been given much attention in the dosimetry literature. However they can cause numerous problems for users, e.g. the effects of dirt, moisture or small metal fragments, bad contacts, misaligned pins, breaks in connection of the cable leads to the connector contacts, mismatches of connectors, slightly different sizes of the same nominal connector, strain of the cable where it meets the connector, etc. If the two devices are from the same manufacturer they usually match each other without further problems but this is not necessarily the case when ionisation chambers and electrometers have different origins. In addition *male* and *female* connectors must be considered. There is a range of connector types used for commercially available chambers and electrometers (Nette and Czap 1994), some of which are shown in Figure 15.5.

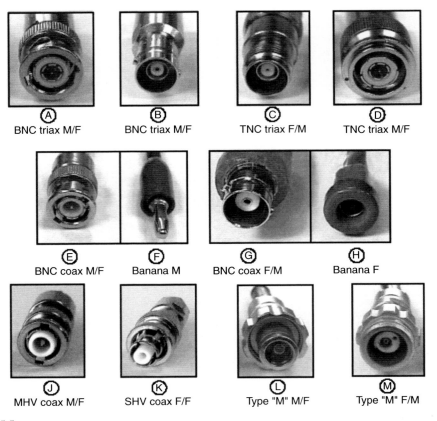

FIGURE 15.5
Different types of connectors between ionisation chambers and electrometers (courtesy of Standard Imaging Inc, WI, USA). BNC coaxial (PTW, Exradin); BNC triaxial (Keithley); Miniquick coaxial (Nuclear Enterprise); TNC triaxial (Nuclear Enterprise, Capintec); PTW biaxial (PTW, type "M").

The term *dosimeter* describes the entire set-up, including chamber, electrometer and cables; although some manufacturers use this term solely to refer to the electrometer. IEC 60731 (1997) includes in the term *radiotherapy dosimeter* not only the chamber and measuring assemblies, but also stability check sources and phantoms.

15.3 TYPES OF CHAMBERS

In what follows, those types of ionisation chambers that are used in radiotherapy for determining the dose in the treatment field will be further discussed. The discussion will not include chambers designed for radiation protection purposes, which generally have a relatively large gas volume in order to increase their sensitivity because the measured doses are generally orders of magnitude lower than those relevant to treating tumours. In the radiotherapy clinic, essentially only one type of chamber is employed. This is known as a *cavity chamber* as its air volume is generally small (see Section 15.3.2). However, there is another type of chamber of great relevance to accurate dosimetry in radiotherapy, namely the free-air ionisation chamber that is used at Standards Dosimetry Laboratories (SDLs) in order to realise air kerma standards that are the basis for the majority of calibrations of cavity chambers.

TABLE 15.1

Approximate CSDA Ranges in Air of Mean Energy Secondary Electrons for Different X-Ray Qualities Ranging from 30 to 300 kV and for ^{60}Co γ-Rays

Radiation Quality (kVp)	Average Initial Secondary Electron Kinetic Energy (keV)[a]	CSDA Range in Air (cm)[b]
30	10	0.24
50	13	0.36
80	12	0.33
100	11	0.25
200	11	0.25
300	18	0.64
^{60}Co γ-rays	590	210

[a] The average initial secondary electron kinetic energy was estimated from Table A-3a in Johns and Cunningham (1983), on the basis of the mean x-ray energy in the spectrum approximated as 37% of the maximum energy (i.e. kVp)

[b] The CSDA ranges were taken from ICRU Report 37 (1984).

Sources: From Johns, H. E. and Cunningham, J. R., *The Physics of Radiology,* 4th Ed., Charles C., Thomas, Eds., Springfield, Illinois, 1983; ICRU, Report 37, ICRU, Bethesda, Maryland, 1984.

15.3.1 FREE-AIR CHAMBERS

The so-called *free-air ionisation chamber* is a specialist instrument usually only to be found at Primary Standard Dosimetry Laboratories (PSDLs). It is called a free-air chamber because, in principle, the walls of the chamber do not play any role in its response. The basic idea underlying the design is to produce perfect *charged particle equilibrium* (Section 6.15) in the defined collecting volume for a particular photon beam and hence to measure directly exposure (or air kerma)[*]. This necessitates a consideration of the maximum range in air of the secondary electrons for various photon-beam qualities, as the dimensions of the chamber in all directions from the interaction volume must be at least as large as this for the energy of the particular beam under consideration. Table 15.1 contains estimates of this quantity. It can be seen that for megavoltage qualities (i.e. for x-ray qualities above around 300 kV) it becomes totally impractical to build such instruments.

Figure 15.6 is a schematic illustration of a free-air chamber. The ionisation to be measured is that produced by secondary electrons created in the interaction volume V. Note that the collector, length l, is guarded by plates G. The two illustrative electron tracks e_3 and e_2 are shown balancing each other to indicate that there is *Charged Particle Equilibrium* over the complete inter-electrode collecting volume.

The mass of air irradiated is derived from the product of the cross-sectional area A of the beam aperture, and the length l of the collecting electrodes in the beam direction ($V=Al$) multiplied by the density ρ of the air. The exposure rate $\Delta X/\Delta t$ is then given by

$$\frac{\Delta X}{\Delta t} = \frac{Q}{Al\rho} \tag{15.3}$$

where Q is the charge collected in a specified time interval Δt. Small corrections must be made for the influence of, for example, the attenuation by the air and photons scattered in the chamber. Details of these can be found in Attix (1986).

---

[*] *Exposure* is mostly used in this section since its original definition was closely related to the measurements that could be performed with a free-air ionisation chamber; *Air Kerma* is now preferred and can be considered as the modern replacement quantity for exposure (see Section 6.3.2).

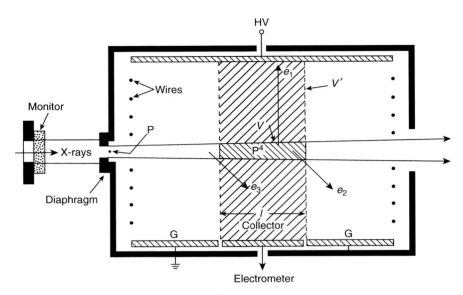

FIGURE 15.6
Schematic cross-sectional view of a typical standard free-air ionisation chamber. The ionisation desired for an exposure measurement is that produced by the electrons such as e_1 originating in interaction volume V (volume defined by the intersection of the beam and the collecting volume V', in turn defined by the collector electrode). The measured ionisation is that collected from volume V'. Charged Particle Equilibrium ensures that if a particle e_2 escapes from the interaction volume V, it is compensated by a particle such as e_3 collected in V'. (Taken from Attix, F. H., *Introduction to Radiological Physics and Radiation Dosimetry*, Wiley, New York, 1986. With permission.)

The modern quantity required is no longer exposure but air kerma, K_{air}, which is obtained from

$$K_{air} = \frac{X}{(1-g)}(W_{air}/e) \qquad (15.4)$$

where g is the fraction of the initial secondary electron kinetic energy re-radiated as bremsstrahlung (see Section 6.3.2) in air, and (W_{air}/e) is the mean energy required to produce an ion pair in air, divided by the electron charge (and has units of J C^{-1}). The unit of time has been omitted from both sides of the expression for simplicity. Therefore, it is clearly seen that all determinations of air kerma rely directly on the value of (W/e) for air; this key quantity is discussed in Section 18.5.4 and in ICRU (1979).

Free-air chambers of different dimensions are used at SDLs to determine air kerma in beams of x-rays between 10 kV and 300 kV. At higher energies, the secondary electron range in air becomes too large to make it practical to build such a chamber. At photon qualities of ^{60}Co γ-rays and above, air kerma is determined through the use of so-called cavity ionisation chambers.

15.3.2 CAVITY CHAMBERS

Cavity chambers are small chambers with walls enclosing the air volume. They are used both in SDLs and in radiotherapy clinics. The key distinction is that the design of the cavity chambers employed at SDLs must enable highly precise determinations to be made of the sensitive air volume, and that the materials used in the walls and central electrode must be as close to homogeneous as possible in order for the correction factors to be known as precisely as possible.

There are basically two types of cavity chambers: cylindrical (sometimes known as *thimble*) chambers and plane-parallel (also known as *parallel-plate*, *flat*, or *coin*) chambers. Both of these types are designed to behave as Bragg–Gray cavities in megavoltage photon and electron beams (see Section 6.7.3) and their design features can best be understood in the light of this. Somewhat different considerations apply to chambers used in kilovoltage x-ray beams (see Section 18.6). In all cases, the gas in the cavity is air, and the cavity should not be sealed in order for the air pressure to equilibrate with the ambient pressure (see Section 15.4.4).

15.3.2.1   Dosimetric Requirements for Cavity Chambers

Requirements for fulfilling Bragg–Gray conditions: For a chamber to behave as a good approximation to a perfect Bragg–Gray cavity, it should have the following features:

1. The air cavity should be small in order for the sensitive volume to be as well defined in space as possible, ideally being small in relation to dose gradients in any direction. In practice, the requirements that a potential difference of some 300 V be maintained between the electrodes and that the sensitivity (which is proportional to the air volume) not be too low preclude the lengths of any dimension being less than around 2 mm and the cavity volume being much less than 0.1 cm^3.

2. The wall (and central electrode in the case of thimble chambers) should be made out of materials that are as homogeneous as possible and also as water- (or medium-) equivalent as possible; in practice, this is seldom possible.

3. The walls should be as thin as possible, especially in the beam direction on the upstream side of the cavity, consistent with the demands of robustness (though in principle, this would not matter if the walls were perfectly medium-equivalent and no measurements were required close to the surface of the phantom).

Departures from Bragg–Gray behaviour: Departure from the Bragg–Gray cavity theory (see Section 6.7.3 for the theoretical derivation of this important condition) is an important issue that implies that correction factors must be used for accurate measurement of the dose to the medium. These factors will be discussed in detail in Section 18.5.3 and Section 2 of Appendix D. A short description of the various perturbations is, however, useful here in order to appreciate their importance in connection with the design of ionisation chambers.

Referring to conditions 2 and 3 above, the chambers wall(s) should either have negligible thickness or be continuous with or equivalent to the surrounding medium as it is only then correct to assume that the secondary electrons in the cavity all originate in the medium or in medium-equivalent material. In practice, one cannot make chamber walls from perfectly water-equivalent materials. Also, for the minimum wall thickness compatible with robustness (i.e. around 0.5 mm), some 50% of the ionisation measured with a typical cylindrical chamber in a cobalt-60 beam is due to electrons generated in the wall, falling to no lower than around 20% at the highest photon energies used in radiotherapy (Andreo et al. 1986; IAEA 1997a). Consequently, a correction for the effect of the wall will generally be necessary, and this is expressed through the wall perturbation factor p_{wall}; generally, this is within 2% of unity for the wall materials used in practice (see Table 15.2). Note that this effect is considered to be significant only in photon beams, although, an approximate theory for wall effects in electron beams (Nahum 1988) predicts effects of the order of 0.5%.

The second important departure from Bragg–Gray behaviour is due to the finite size of the air cavity (Condition 1). This is generally* dealt with by assigning to the chamber an effective point of measurement, denoted by P_{eff}, that, in general, lies *upstream* from the centre C of the

---

* Another approach, presented in Chapter 18 and Appendix D, consists in using a multiplicative correction factor p_{dis} that accounts for the displacement of the medium by the air cavity.

TABLE 15.2

Examples of Common Cylindrical Ionisation Chambers

Ionisation Chamber Type[a]	Cavity Volume (cm³)	Cavity Length (mm)	Cavity Radius (mm)	Wall Material	Wall Thickness (g/cm²)	Buildup Cap Material[b]	Buildup Cap Thickness[b] (g/cm²)	Central Electrode Material[b]	Waterproof
Capintec PR-05 mini	0.14	11.5	2.0	C-552	0.220	Polystyrene	0.568	C-552	N
Capintec PR-06C/G Farmer	0.65	22.0	3.2	C-552	0.050	PMMA[c]	0.547	C-552	N
Exradin T2 Spokas (4 mm cap)	0.53	11.4	4.8	A-150	0.113	A-150	0.451	A-150	Y
Exradin T1 mini Shonka (4 mm cap)	0.05	5.7	2.0	A-150	0.113	A-150	0.451	A-150	Y
Exradin A12 Farmer	0.65	24.2	3.1	C-552	0.088	C-552	0.493	C-552	Y
Far West Tech IC-18	0.1	9.5	2.3	A-150	0.183	A-150	0.386	A-150	N
Nuclear Assoc. 30-749	0.08	4.0	3.0	C-552	0.068			C-552	Y
Nuclear Assoc. 30-716	0.25	10.0	3.0	C-552	0.068			C-552	Y
Nuclear Assoc. 30-752 Farmer	0.69	23.0	3.1	Graphite	0.072	Delrin	0.560	Al	Y
NE 2515/3 Farmer	0.2	7.0	3.2	Graphite	0.066	PMMA[c]	0.543	Al	N
NE 2505/3, 3A Farmer	0.6	24.0	3.2	Graphite	0.065	PMMA[c]	0.551	Al	N
NE 2571 Farmer	0.6	24.0	3.2	Graphite	0.065	Delrin	0.551	Al	N
NE 2581 Farmer (PMMA cap)	0.6	24.0	3.2	A-150	0.041	PMMA	0.584	A-150	N
NE 2561/2611 Sec. Std	0.33	9.2	3.7	Graphite	0.090	Delrin	0.600	Al (hollow)	Y
PTW 23323 micro	0.1	12.0	1.6	PMMA[d]	0.197	PMMA	0.357	Al	N
PTW 23332 rigid	0.3	18.0	2.5	PMMA[d]	0.054	PMMA	0.357	Al	N
PTW 30001 Farmer	0.6	23.0	3.1	PMMA[d]	0.045	PMMA	0.541	Al	N
SNC 100740 Farmer	0.6	24.4	3.5	Graphite	0.085	PMMA	0.536	Al	N
Victoreen 30-351	0.6	23.0	3.1	PMMA[d]	0.060	PMMA	0.360		N
Scanditronix-Wellhöfer IC 25	0.25	10.0	3.0	C-552	0.068			C-552	Y
Scanditronix-Wellhöfer IC 70 Farmer	0.67	23.0	3.1	Graphite	0.068	POM[e]	0.560	Al	Y

[a] As obtained from the manufacturers.

[b] Blanks correspond to no information available.

[c] Polymethyl methacrylate ($C_5H_8O_2$), also known as acrylic. Trade names are Lucite, Plexiglass, or Perspex.

[d] Like most chamber types with non-conductive plastic walls, the chamber wall has an inner conductive layer made of graphite. For this chamber type, the thickness and density of the graphite layer is supplied in the chamber specifications.

[e] Poly Oxy Methylene (CH_2O). A trade name is Delrin.

Source: Taken from International Atomic Energy Agency, *Technical Report Series no. 398*, Table 3, IAEA, Vienna, 2000. With permission.

cavity; the magnitude of the $(P_{eff}-C)$ shift is non-negligible in most practical chamber designs. More detail is given on P_{eff} in Appendix D. The effect is similar in both photon and electron beams.

There is a further effect due to the non-negligible cavity size that only affects measurements in electron beams and is due to the large differences in density between the gas and the wall or medium. This results in more electrons being scattered *into* the cavity than being scattered *out*. For certain cavity shapes in relatively low-energy electron beams, this results in an appreciable increase in the electron fluence in the cavity compared to that in the (undisturbed) medium at the equivalent position (Harder 1968; ICRU 1984a; Nahum 1996; IAEA 1997b). This fluence and signal increase is corrected for, by introducing a perturbation factor less than unity usually denoted by p_{cav}. There is essentially no corresponding effect in photon beams because of the existence of (partial) charged particle equilibrium, i.e. $p_{cav}=1$ in Equation 15.5 (see Appendix D, Section D.2.4 for an explanation).

The expression to obtain the dose to the medium, D_{med}, from the dose to the air measured by the ionisation chamber is modified for departures from Bragg–Gray behaviour (Equation 15.2) to read

$$D_{med}(P_{eff}) = D_{air}\, s_{med,air}\, p_{wall}\, p_{cav}\, p_{cel} \qquad (15.5)$$

where it is to be understood that the measurement relates to position P_{eff} in the medium; a correction, p_{cel} for the effect of the central electrode material (see Appendix D, Section D.2.3) has also been added for completeness.

Chambers designed to be air-equivalent: Some designs of ionisation chamber for use in radiotherapy are intended to fulfill other criteria than Bragg–Gray ones. In kilovoltage x-ray beams, practical chambers do not behave as Bragg–Gray cavities because of the lower-energy secondary electrons generated with their much reduced ranges (see Section 6.7.2). Here, the primary concern is that the air kerma calibration factor, N_K, (see Section 18.2) should show as little variation as possible over a wide range of radiation quality, from say 50 kV (Half Value Layer ≈ 1 mm Al) up to 300 kV (Half Value Layer ≈ 20 mm Al) and sometimes even extending to cobalt-60 (with a build-up cap) as it was often desirable that the same chamber, e.g. a graphite-walled *Farmer* (see Section 15.3.2.2), be provided with N_K factors for the complete range of qualities existing in the radiotherapy clinic. Empirically, it was found, for example, that a graphite wall in combination with an aluminium central electrode resulted in a variation in N_K of no more than around 5% (Humphries and Slowey 1986).

The extent to which these various requirements are fulfilled is now discussed in relation to the two main chamber types, cylindrical and plane-parallel. For completeness, the so-called monitor ionisation chambers found in accelerator treatment heads are also briefly described.

15.3.2.2   Cylindrical Chambers

The classic *Farmer* chamber (Farmer 1955; Aird and Farmer 1972), a very common design of cylindrical chamber for radiotherapy, is shown in Figure 15.7. The cylindrical outer wall acts as one of the electrodes and is made of graphite. The other electrode is a thin aluminium central rod. The nominal cavity volume is $0.6\ \text{cm}^3$. The cavity length is 24 mm with an inner diameter of 6.25 mm. The graphite outer wall is 0.5 mm in thickness.

There are several other types of *Farmer* chambers commercially available with walls made variously of A-150 (muscle-tissue equivalent conducting plastic), C-552 (air equivalent conducting plastic), Delrin, Nylon 66, PMMA[*]. The closest of these to water equivalence

---

[*] When PMMA is used, it is coated on the inside with a conducting layer, usually *dag* that is basically a suspension of graphite in alcohol.

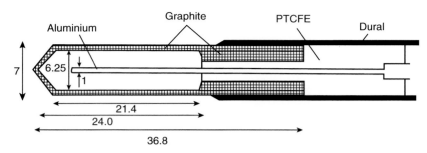

FIGURE 15.7
A Farmer-type ionisation chamber (redrawn from Aird and Farmer 1972), dimensions in millimetres.

in terms of atomic composition is PMMA. There are also models where the central electrode is made of A-150 instead of aluminium.

Other chamber designs may be smaller in diameter and length in order to achieve a higher spatial resolution (particularly important in situations where the dose is changing rapidly, e.g. in the penumbra region) but at the cost of a smaller signal per unit dose. Table 15.2 shows a representative selection of cylindrical chamber designs.

Most cylindrical chambers are supplied with a build-up cap. This is placed around the chamber wall in order to achieve Charged Particle Equilibrium when the chamber is irradiated free-in-air in a ^{60}Co γ-ray beam as is done when providing the chamber with an air-kerma calibration factor, N_K, at a Standards Dosimetry Laboratory (see Section 18.2). The build-up cap is *not* used when the chamber is irradiated in a phantom at the Standards Dosimetry Laboratory for a direct calibration in dose to water $N_{D,w}$ (see Section 18.3) or in a phantom in the clinic. However, if the chamber is not waterproof, it is necessary to place a thin water-proof sheath around it before using it in a water phantom.

Cylindrical chambers such as those given in Table 15.2 can be used for either absolute or relative dose measurements. For absolute dose measurements, they must be supplied with a calibration factor (N_K or $N_{D,w}$) in order to determine the dose in absolute units (Gy) at the reference point in water by following a Code of Practice. For relative dose measurements, other types of cylindrical chambers may be used in conjunction with computer controlled beam-data acquisition systems. They are then preferably considerably smaller in order to improve the spatial resolution (Section 19.1.4). They should be waterproof or easily made waterproof. Most manufacturers make such chambers with cavities with volumes of the order of 0.1 cm^3, lengths of 10 mm or less, and inner diameters of around 4 mm (see Table 15.2).

15.3.2.3 Plane-Parallel Chambers

The *plane-parallel* or *parallel-plate* chamber design appears to be very different from that of the cylindrical chamber, although, both have cylindrical air volumes. The cylindrical or thimble chamber is irradiated with its central axis *perpendicular* to the beam direction, whereas, the plane-parallel chamber is irradiated with its axis *parallel* to the beam direction.

The chamber design may be different if the chamber is to be used in kilovoltage x-ray beams; the designs discussed up to this point are intended to be used in photon or electron beams from linear accelerators. For kilovoltage x-rays, it is preferable to have a very thin entrance wall, e.g. 0.004 mm of conducting mylar. It is also necessary to choose the wall and body materials with care in order to avoid significant energy dependence of the response for low-energy x-rays. For electron or megavoltage-photon beams, the entrance wall may be thicker as shown in Figure 15.8.

Plane-parallel chambers are preferred to cylindrical chambers in situations where the dose gradient is steep in the direction of the beam, for example, in the build-up region in a

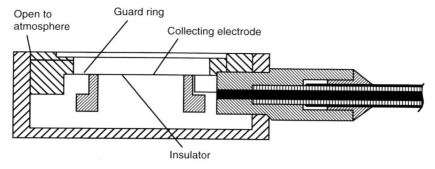

FIGURE 15.8
The NACP plane-parallel design of ionisation chamber. (From Nordic Association of Clinical Physics, *Acta Radiol. Oncol.*, 20, 401–415, 1981. With permission.)

megavoltage photon beam. Their use is recommended for absolute dose determination in electron beams of energy below 15 MeV. This is partly because of their improved spatial resolution (generally 2 mm or less cavity depth), but is mainly because the perturbation effect of *in-scattering* (see Appendix D, Section D.2.4) can be eliminated with a sufficiently wide guard ring (IAEA 1997b; Nahum 1996). Furthermore, the position of the effective point of measurement is well defined at the centre of the inside surface of the front window (see Appendix D, Section D.2.1). Table 15.3 lists desirable properties for a plane-parallel chamber in electron radiation.

The NACP chamber (Figure 15.8) is an example of a successful design of a flat chamber. It has a thin front window of 0.6 mm thickness, made of mylar foil and graphite, to enable measurements to be made at small depths. The back wall is made of graphite, and the side walls of rexolite. The collecting electrode is a very thin layer (>0.1 mm) of graphited rexolite mounted on a thin insulating layer in order to give a negligible polarity effect (see Section 15.4.2). There is a 3 mm guard ring in order to exclude from the signal electrons scattered in from the side walls. The distance between the electrodes is 2 mm.

Table 15.4 lists the characteristics of plane-parallel chamber types in common use. It should be noted that relatively large values of the polarity effect have been observed with certain plane-parallel chamber types, and particular care should be taken to investigate this property when choosing a chamber. The width of the guard rings in Table 15.4 are at least 2.5 mm except for that of the Markus chamber; the latter, 0.2 mm, is insufficient to ensure a negligible perturbation from electrons scattered into the cavity through the side walls.

TABLE 15.3

Desirable Properties for a Plane-Parallel Chamber in Electron Beams

Chamber dimensions	
Front window thickness	≤ 1 mm
Collecting electrode diameter	≤ 20 mm
Ratio of guard ring width to cavity height	≥ 1.5
Cavity height	≤ 2 mm
In-scattering perturbation effect, p_{cav}	$<1\%$
Wall and backscattering perturbation effects, p_{wall}	$<1\%$
Polarity effect	$<1\%$
Leakage current	$<10^{-14}$ A
Long term stability	$\pm 0.5\%$[a]

[a] Response variation per year for any radiation quality in the rated range.
Source: From International Atomic Energy Agency, *Technical Report Series no. 381*, IAEA, Vienna, 1997. With permission.

TABLE 15.4

Characteristics of Plane-Parallel Chambers in Common Use

Ionisation Chamber Type[a]	Materials	Window Thickness	Electrode Spacing	Collecting Electrode Diameter	Guard Ring Width	Recommended Phantom Material
NACP01 (Scanditronix) Calcam-1 (Dosetek)	Graphite window Graphited rexolite electrode Graphite body (back wall) Rexolite housing	90 mg/cm^2 0.5 mm	2 mm	10 mm	3 mm	Polystyrene, Graphite, Water (with water-proof housing)
NACP02 (Scanditronix)	Mylar foil and graphite window Graphited rexolite electrode Graphite body (back wall) Rexolite housing	104 mg/cm^2	2 mm	10 mm	3 mm	Water, PMMA
Calcam-2 (Dosetek)		0.6 mm				
Markus chamber PTW 23343 NA 30-329 NE 2534	Graphited polyethylene foil window Graphited polystyrene collector PMMA body PMMA cap	102 mg/cm^2 0.9 mm (incl. Cap)	2 mm	5.3 mm	0.2 mm	Water, PMMA
Scanditronix-Wellhöfer PPC 05	Window and body C-552 Graphited (PEEK[b]) electrode	176 mg/cm^2 1 mm	0.5 mm	10 mm	3.5 mm	Water
Holt chamber (Memorial) NA 30-404	Graphited polystyrene wall and electrode Polystyrene body	416 mg/cm^2 4 mm	2 mm	25 mm	5 mm	Polystyrene (Integral phantom)
Capintec PS-033	Aluminised mylar foil window Carbon impregnated air equivalent, plastic electrode Polystyrene body	0.5 mg/cm^2 0.004 mm	2.4 mm	16.2 mm	2.5 mm	Polystyrene

Chamber	Material					
Exradin 11	Conducting plastic wall and electrodes, Model P11 : polystyrene equivalent Model A11 : C-552, air equivalent Model T11 : A-150, tissue equivalent	P11: 104 mg/cm² 1 mm	2 mm	20 mm	5.1 mm	P11: Polystyrene, Water
Roos chamber PTB FK6 PTW 34001	PMMA, graphited electrodes	118 mg/cm² 1 mm	2 mm	16 mm	4 mm	Water, PMMA
Scanditronix -Wellhöfer PPC 35 PPC 40						
Attix chamber RMI 449	Kapton conductive film window, graphited polyethylene collector, solid water body	4.8 mg/cm² 0.025 mm	1 mm (0.7 mm reported)	12.7 mm	13.5 mm	Solid water

a Some of the chambers listed in this table fail to meet the requirements described in Table 15.3. They have been included here because they are used in certain clinics.

b Polyetheretherketone ($C_{19}H_{18}O_3$) 1.265 g/cm³.

Source: Taken from International Atomic Energy Agency, *Technical Report Series no. 398*, Table 4, IAEA, Vienna, 2000. With permission.

The use of plane-parallel chambers for absolute dose determination in electron beams based on an air-kerma calibration (see Section 18.2) requires a knowledge of the factor p_{wall} in ^{60}Co γ-radiation. In general, it has proved difficult to determine $p_{wall,Co60}$ accurately because of the difficulty of making such chambers with homogeneous wall materials. Furthermore, it has been found that for certain chamber designs, $p_{wall,Co60}$ varies appreciably between chambers of exactly the same type (IAEA 1997b). Therefore, the use of plane-parallel chambers in photon beams for absolute dose measurements is discouraged.

Plane-parallel chambers are also used for dose determination in low-energy x-ray beams (see Section 18.6). At these qualities, the entrance window must be extremely thin in order to minimise photon attenuation. The chamber is mounted with the window flush with the surface of a phantom; in a so-called *Grenz-ray* chamber, the phantom forms part of the chamber body (Grenz rays are x-rays at or below 20 kVp). Commercial examples are the PTW M23342 and M23344 with cavity volumes of 0.02 cm^3 and 0.2 cm^3, respectively, and the NE 2532/3A and 2536/3A of volume 0.03 cm^3 and 0.3 cm^3, all with 2.5 mg cm^{-2} polyethylene entrance windows.

15.3.2.4 *Monitoring Ionisation Chambers*

A further example of a type of ionisation chamber, usually referred to as a *monitor chamber*, can be found in the treatment heads of linear accelerators. Monitor chambers are essentially used to control the total dose, the dose rate and the beam position; they are described in Section 11.7.3. Their principle and design have some similarities with plane-parallel chambers, but, since they are used as *transmission* chambers, their diameter is much larger, and their electrodes (i.e. entrance and exit windows) are very thin to prevent significant perturbation of the beam characteristics. Monitor chambers are generally sealed in order to keep the enclosed mass of air constant, independently of the pressure and (machine head) temperature. However, keeping the mass of air accurately constant is difficult to achieve in practice, and the monitor response must be checked frequently to prevent any drift. Alternatively, some monitor chambers (e.g. in *Elekta* machines) are unsealed, and their response is automatically corrected for changes in pressure and temperature (see Section 15.4.4) directly measured in the machine head. This does not, however, remove the need for frequent checks (see Section 38.8.1).

15.4 DETERMINATION OF THE CHARGE PRODUCED

Ionisation chambers must be used in the voltage range where the collected charge reaches, or asymptotically approaches, a saturation value (Figure 15.2). However, because of the physics of ion transport and chamber electrical design, the charge collected can be different from the charge produced in the chamber gas by the passage of radiation. The most important phenomena affecting the charge collected are ion recombination, polarity effects, and leakage and stem effects. An additional important issue arises at the time of determining the charge per unit mass of air in the chamber cavity when a correction for the local pressure and temperature is usually needed.

15.4.1 RECOMBINATION

15.4.1.1 *Initial and General Recombination*

The recombination of positive and negative ions within the air cavity reduces the amount of charge collected. Two main mechanisms for the disappearance of ions have been described (Boag 1966, 1987) that are related to the structure of radiation tracks and their distribution; these are termed *initial* and *general* recombination.

In the process of *initial* recombination, positive and negative ions in the track of a single particle recombine; this process is also called *columnar* recombination. Because only one single particle track is involved for a given field strength, this intra-track process is independent of the number of tracks formed per time interval (i.e. the dose rate). In practice, this effect is only significant for tracks with a high ionisation density, i.e. high-LET particles such as heavy ions (see Chapter 49).

General recombination proceeds once diffusion and ion drift have modified the initial track structure, and positive and negative ions of different tracks recombine as they drift toward the chamber electrodes. The process is also known as *volume* recombination, and it increases as the density of ions increases, being dependent on dose rate. This is the most significant process in the majority of radiotherapy beams.

Approximate empirical relationships have been proposed to describe initial and general recombination where so-called *Jaffé plots* ($1/q$ as a function of $1/V$ or $1/V^2$ where q is the measured charge and V the polarizing voltage) are made to represent the complete saturation curve. *Initial recombination* is described by

$$1/q = 1/q_0 + \text{constant}/V \qquad (15.6)$$

whereas *general recombination* follows the relationship

$$1/q = 1/q_0 + \text{constant}/V^2 \qquad (15.7)$$

Practical limitations of these expressions have been discussed in the context of plane-parallel chambers for electron dosimetry in TRS-381 (IAEA 1997b).

15.4.1.2 Collection Efficiency: Boag's Theory

For *general recombination*, Boag (1966, 1987) defined the collection efficiency f of an ionisation chamber as the ratio of the measured charge to the ideal saturation charge. It should be noted that the recombination correction factor, p_s, commonly used in dosimetry protocols, is the inverse of the collection efficiency. In his comprehensive review, Boag developed step-by-step the theory for the collection efficiency in continuous radiation, pulsed beams and pulsed scanning beams for different types of ionisation chambers. Here, a summary is given for the case of the pulsed beams commonly found in clinical accelerators; for fuller details, readers can consult Boag (1987). A more condensed, practical approach is given in ICRU (1982).

In pulsed beams, the collection efficiency is given by

$$f(u) = \frac{1}{u}\ln(1 + u) \qquad (15.8)$$

where the variable u is defined as

$$u = \mu \frac{q_i}{V} A_I \qquad (15.9)$$

and the coefficient μ is a constant that depends only on properties of the ions and their mobility (3.02×10^{10} V m C^{-1}); q_i is the initial charge density, V the polarizing voltage and A_I is the solution of a double integral that, in the case of a plane-parallel chamber, is equal to the square of the distance between the electrodes. In the case of a cylindrical chamber, A_I is given by

$$A_I = (a - b)^2 \frac{(a + b)}{(a - b)} \frac{\ln(a/b)}{2} \qquad (15.10)$$

where a and b are, respectively, the radii of the external and internal electrodes.

A problem with Equation 15.8 is that it depends on q_i, the initial charge density that is not known until the amount of recombination has been determined, whereas, the measured charge density q_m is known. However, $q_m = f q_i$, and Boag introduced an auxiliary variable $v = fu = \ln(1 + u)$ with the result that

$$f = \frac{v}{e^v - 1} \tag{15.11}$$

This equation allows f to be determined from the measured charge as $v = \left[\dfrac{\mu A_I}{V}\right] q_m$ from Equation 15.9.

A simpler approximate equation, essentially a fit to calculated values from the complete expressions, has been given in IAEA (1997b); for plane-parallel chambers the recombination correction factor p_s is given by

$$p_s = 1 + 0.54 d \frac{s^2}{V} \tag{15.12}$$

where d is the average dose per pulse in mGy, s the electrode spacing in millimetres and V the polarizing voltage in volts. The average dose per pulse refers to the absorbed dose to air. A similar equation can be written for cylindrical chambers, replacing s^2 by the corresponding value A_I from Equation 15.10.

Some analyses of practical cylindrical chambers such as the NPL design standard chamber (Kemp 1972; Burns and Rosser 1990) or the Farmer chamber (Havercroft and Klevenhagen 1993) have used Boag's approach and taken into account both the cylindrical body and the end of the chamber. Expressions have been given for both these chamber designs that can be expressed as

$$p_s = 1 + n \frac{d}{V} \tag{15.13}$$

where d is again the dose per pulse in mGy, and V is the polarizing voltage in volts. The values of n are 4.6 for the NPL designed chamber and 5.75 for the Farmer chamber. Both the simpler equations given above in terms of dose per pulse (Equation 15.12 for plane-parallel chambers and Equation 15.13 for the Farmer design) are in close agreement with the illustrative values given in UK Codes of Practice for electron beams (IPEMB 1996a; Thwaites et al. 2003) and also with experimental measurements using a range of chambers recommended in those codes (Nisbet and Thwaites 1998a).

15.4.1.3 Collection Efficiency: The Two-Voltage Method

Situations can be encountered when an ionisation chamber does not reach saturation at the highest polarization voltage that can safely be used. However, a method based on making a second measurement at a much lower voltage has been described by Boag and Currant (1980). If the collected charges Q_1 and Q_2 have been obtained at the voltages V_1 and V_2, and Q_i is the desired saturation value, then

$$Q_1 = Q_i f_1 = Q_i \frac{1}{u_1} \ln(1 + u_1)$$

$$Q_2 = Q_i f_2 = Q_i \frac{1}{u_2} \ln(1 + u_2) \tag{15.14}$$

and their ratio becomes

$$\frac{Q_1}{Q_2} = \frac{u_2 \ln(1 + u_1)}{u_1 \ln(1 + u_2)} = \frac{V_1 \ln(1 + u_1)}{V_2 \ln(1 + u_2)} \tag{15.15}$$

because $u_1/u_2 = V_2/V_1$ from Equation 15.9. Equation 15.15 can be solved for u_1 numerically or graphically because both the ratios Q_1/Q_2 and V_1/V_2 are known. Weinhous and Meli (1984) carried out this calculation and fitted the results for the recombination correction factor p_s to the function

$$p_s = a_0 + a_1 \left(\frac{Q_1}{Q_2} \right) + a_2 \left(\frac{Q_1}{Q_2} \right)^2 \tag{15.16}$$

where the coefficients a_0, a_1 and a_2 were given as a function of V_1/V_2 for pulsed and for pulsed-scanned beams. These solutions have been implemented in the majority of the dosimetry protocols.

Provided that the correction required is small, the quadratic term in Equation 15.16 can be ignored, and the percentage correction reduces to the *percentage change in reading* divided by (*voltage ratio* − 1) (IPEMB 1996a). In the simplest case, if a voltage ratio of 2 is selected, the percentage recombination correction is equal to the percentage change in reading. This method does, however, rely on the linearity of the correction, and it is wise to verify this by plotting the full curve for each chamber at least once. Typical voltage values for cylindrical chambers may be $V_1 = 400$ V and $V_2 = 200$ V. These voltages are generally in the linear portion of the correction curve. For plane-parallel chambers, the deviation from linearity may occur at a smaller voltage value (e.g. 100 V). An example of the response of one particular plane-parallel ionisation chamber is shown in Figure 15.9.

15.4.2 POLARITY EFFECT

Reversing the polarity on an ionisation chamber (e.g. making the central electrode +V with respect to the wall and then changing it to −V) may yield different readings; this phenomenon is termed the polarity effect. For most chamber types, the effect is practically negligible in high-energy photon beams, but in electron beams, notably at low energy, the effect may be significant. Several explanations have been given for this process (e.g. see Attix 1986); the effect

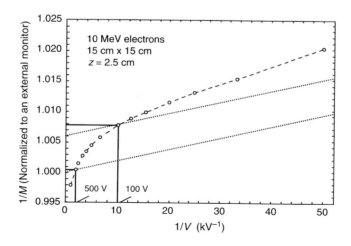

FIGURE 15.9

Response of one particular plane-parallel ionisation chamber when the bias voltage is changed. *M* is the chamber reading, and *V* the polarizing voltage in the so-called *Jaffé plot*. The deviation from linearity at the highest voltages indicates that this is not a true saturation curve but, instead, describes the overall response of the chamber. To obtain a measure of the true saturation correction using the two-voltage technique, the readings at the two voltages, corrected by Boag's theory (intercepts of the dotted lines on the ordinate) should be used instead of the uncorrected readings (solid-line intercepts). (Reproduced from International Atomic Energy Agency, *Technical Report Series no. 381*, IAEA, Vienna, 1997. With permission.)

is better understood in the case of plane-parallel chambers in electron beams (IAEA 1997b) where it is considered to be essentially a charge-balance effect on electrodes with asymmetric design. The magnitude of this effect depends on the energy and angular distribution of the incident radiation, the measurement depth in a phantom and also on the field size. By examining the pattern of charge deposition by electrons in a phantom, it can also be seen that the sign of the polarity effect may change with depth.

In situations where there is a measurable polarity effect, the true chamber reading is taken to be the mean of the absolute values of readings taken at each polarity. In the routine use of a given ionisation chamber, a single polarizing potential and polarity are normally employed. The effect on the chamber reading of using polarizing potentials of opposite polarity for each user beam quality can be accounted for by using a correction factor

$$k_{\mathrm{pol}} = \frac{|Q_+| + |Q_-|}{2Q} \tag{15.17}$$

where Q_+ and Q_- are the electrometer readings obtained at positive and negative polarity, respectively, and Q is the electrometer reading obtained with the polarity used routinely (positive or negative). The readings Q_+ and Q_- should be made after ensuring that the chamber reading is stable following any change in polarity (some chambers can take up to 20 min to stabilize).

When the chamber is sent for calibration, a decision is normally made, either by the user or by the Standards Dosimetry Laboratory, on the polarizing potential and polarity to be adopted for the routine use of the chamber. The calibration should be carried out at this same polarizing potential (and polarity if only one polarity is used for the calibration). The Standards Dosimetry Laboratory may or may not correct for the polarity effect at the calibration quality. This should be stated in the calibration certificate. When the calibration laboratory has already corrected for the polarity effect, then the user must apply the correction factor k_{pol} derived using Equation 15.17 to all measurements made using the routine polarity. When the calibration laboratory has *not* corrected for the polarity effect, the subsequent treatment of the polarity effect depends on the facilities available to the user and on what beam qualities must be measured. Detailed procedures on how to perform such corrections can be found, for instance, in IAEA (2000). The polarity effect will vary in different electron beams and at different depths, but it should be less than 1% for well designed chambers (Nisbet and Thwaites 1998a).

15.4.3 Leakage and Stem Effect

Electric leakage in the ionisation chamber and electrometer may occur as a consequence of the irradiation of insulators and chamber parts, cables, and electronics of the measuring equipment. This is termed *post-irradiation* leakage, an effect that continues after the irradiation has ceased and usually decreases exponentially with time. IEC 60731 (1997) recommends that within 5 s after the end of a 10 min irradiation, the leakage current shall have decreased to $\pm 1.0\%$ of the ionisation current produced in the measuring volume during the irradiation[*]. Another kind of leakage is the so-called *stray radiation* leakage that is produced by scattered radiation in those parts of the chamber assembly and pre-amplifier that are usually outside the radiation beam (but within the radiation room). Mechanical stress on cable insulators can also cause a leakage current, and for this reason, bending or twisting cables should be avoided.

---

[*] Similarly, the so-called *pre-irradiation* leakage, obtained with the maximum rated polarizing voltage applied to the chamber, shall not exceed $\pm 0.5\%$ of the ionisation current produced by the minimum effective dose rate.

In the case of insulators, leakage may occur either through the volume or across the surface of insulators. The mechanisms of radiation effects on insulators are complicated and numerous (Attix 1986). The electrons produced during the ionisation of atoms in the insulators may be displaced a considerable distance from the ionised atom, requiring some time to recombine. As a result of the applied voltage, the electron-hole dipoles become aligned in a common direction, and their relaxation induces charges in the electrodes connected to the insulator forming a leakage current. Another effect in insulators that received considerable attention in the mid-eighties (Galbraith et al. 1984; Mattsson and Svensson 1984; Thwaites 1984) was charge accumulation in non-conductive plastic phantoms; this can give rise to a very large electric field around the chamber which then directs the flow of electrons toward the chamber cavity, yielding an increased signal. This is discussed further in Section 18.8.3 for phantom materials, but similar effects can occur in the insulators of ionisation chambers.

The effect of irradiating the chamber stem is also a complicated phenomenon which, fortunately, has been minimized in modern ionisation chambers. Two mechanisms have been described by IEC 60731 (1997), namely *stem scatter* and *stem leakage*. The stem scatter effect arises from scattered radiation arising from interactions in the stem that may be different from those in the phantom material and that may contribute differently to the ionisation in the cavity volume. This effect can be determined using a *dummy stem*. The chamber is irradiated successively with and without the presence of the dummy stem, and the ratio of the readings allows a correction factor to be determined. The stem leakage arises as a consequence of the irradiation of the stem and the insulators and cables in it. The effect is usually determined by irradiating a chamber twice with the same narrow rectangular field, parallel and perpendicular to the long side, and a correction factor is derived as above[*].

15.4.4 CORRECTION FOR THE CHANGES IN THE MASS OF AIR WITHIN THE CAVITY

The mean absorbed dose to the air in the cavity of an ionisation chamber, $\overline{D}_{air}$, is derived from the basic relation

$$\overline{D}_{air} = \frac{Q}{m_{air}}(W_{air}/e) \qquad (15.18)$$

where Q is the charge produced in the air cavity having a mass m_{air} and (W_{air}/e) is the mean energy required to produce an ion pair in air, divided by the electron charge.

The mass of air is equal to $V\rho_{air}$ where V is the cavity volume and ρ_{air} the air density inside the chamber. The number of interactions and consequently the charge produced is proportional to the mass of air, which in turn is proportional to its density. As most ionisation chambers are vented to the ambient atmosphere, ρ_{air} is a function of the atmospheric pressure, temperature and humidity.

It is common practice to fix the value of ρ_{air} for certain conditions and convert Q to those conditions, and not the reverse. The SDLs take the reference air density value as 1.2930 kg m^{-3}, which is the value for dry air at 0°C and 101.325 kPa. They then measure charge in the reference beams at 20°C, 101.325 kPa, and 50% relative humidity (common reference conditions). They correct to dry air and to 0°C and 101.325 kPa, using the usual density corrections for dry air. The conversion from charge to absorbed dose to air also involves the

---

[*] When performing such measurements, it is recommended to turn the detector through 90° rather than rotating the collimator or exchanging the X and Y dimensions of the collimator; doing this prevents the influence of the collimator settings on the monitor chamber being interpreted as a stem leakage effect (see also Section 19.2.5 and Section 23.2.1.13).

value for (W_{air}/e), and for consistency in Equation 15.18, the value to be used is that for dry air, i.e. 33.97 J C^{-1}.

When calibrating an ionisation chamber at the Standards Dosimetry Laboratory, the charge measured by a vented chamber depends on air temperature and pressure (and also humidity); therefore, the calibration factor must be given for stated reference values of these parameters. Most SDLs supply calibration factors to the user that have been corrected to $\theta_0 = 20°C^*$ and $P_0 = 101.325$ kPa without humidity correction. Instead, the relative humidity during calibration is controlled within the range 45–55%, so the calibration factor applies for relative humidity around 50% (in fact, in the range from 20 to 70%, no correction is required[†]). A detailed treatment of the effect of humidity is given in Appendix D, Section D.4.

In the user beam, the correction factor

$$k_{\theta P} = \frac{(273.2 + \theta)}{(273.2 + \theta_0)} \frac{P_0}{P} \tag{15.19}$$

should be applied to convert the measured charge to the reference conditions used for the chamber calibration at the Standards Dosimetry Laboratory. P and θ(°C) are the cavity air pressure and temperature at the time of the measurements, and P_0 and θ_0(°C) are the reference values (usually 101.325 kPa and 20°C). The air temperature in a chamber cavity should be taken to be that of the phantom. This is not necessarily the same as the temperature of the surrounding air, and if a significant difference is observed, the phantom temperature must be checked regularly, or preferably, waiting a sufficient time should allow the phantom to reach the room temperature. In addition, for measurements in a water phantom, the chamber waterproof sleeve should be vented to the atmosphere in order to achieve rapid equilibrium between the ambient air and the air in the chamber cavity.

---

[*] In some countries, the reference temperature is 22°C.
[†] In practice, a correction is required only in tropical countries.

CHAPTER 16

RADIOTHERMOLUMINESCENT DOSIMETERS AND DIODES

Ginette Marinello

CONTENTS

16.1 INTRODUCTION

Among the wide choice of solid state detectors that can be used for applications in radio-therapy, brachytherapy, diagnostic radiology and radiation protection of the patient, radiothermoluminescent dosimeters (TLD) and diodes are currently the most used (see Figure 16.1). The purpose of this chapter is to give a brief description of the underlying theoretical principles, the materials used, and their main dosimetric characteristics, leaving the reader to refer to the specialized literature on the subject to get the in-depth information that is impossible to provide in a generalist handbook. The practical use of these detectors for in-beam measurements and for quality control of beam characteristics will be presented in Section 19.1.4 and Section 38.4 respectively. Their use for in-vivo dose measurements in external radiotherapy and brachytherapy and for dosimetric comparisons betwen centres will be presented in Section 40.2.

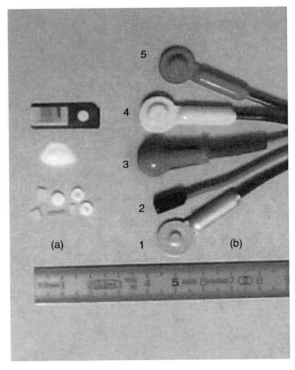

FIGURE 16.1
(See colour insert following page 590.) Examples of solid state detectors of current use: (a) radiothermoluminescent dosimeters made of LiF or $Li_2B_4O_7$ doped with Mn or Cu and (b) diodes (1: Sun Nuclear-QED 111200 for electrons; 2 and 3: Scanditronix EDD-5 and EDP-5 for organs at risks and cobalt 60, respectively; 4 and 5: Sun Nuclear QED 111500 and 111600 for 6–12 and 15–25 MV, respectively).

16.2 THERMOLUMINESCENT DOSIMETRY

Thermoluminescent (TL) dosimetry has now become a well-established technique. TL materials are readily available commercially and do not require to be linked with a cable to the reading equipment. They also have the advantage of providing a very sensitive dosimeter with a small volume and which when correctly chosen, is equivalent to the different human tissues. The time required for readout is a drawback, but the time can be considerably decreased by choosing an automatic reader and a good methodology.

16.2.1 PRINCIPLES OF TL DOSIMETRY

Thermoluminescent dosimetry (TLD) is based on the ability of certain imperfect crystals to absorb and store the energy of ionising radiation that, upon heating, is re-emitted in the form of light. The light is detected, and the light output is correlated to the absorbed dose previously received by them. One of the possible mechanisms to explain it (McKinlay 1981; McKeever 1985) refers to the band theory of the multiatomic crystalline structures.

When *irradiated*, free electrons and holes are produced in the TL material. The electrons are free to travel through the conduction band for a short time. Three outcomes are then possible: they may be trapped at defects (i.e. in a *metastable energy state*); they may fall back into the valence band and recombine with holes either radiatively (fluorescence) or non-radiatively; or they may be captured at luminescence centres already activated by holes as a result of the irradiation and deactivate the centre with the emission of light.

When *heated*, the electrons trapped in the metastable energy states are given sufficient thermal energy to escape from the trap into the conduction band again, where they are free to travel and, as before, have three possible outcomes: either be retrapped at defects, fall into the valence band and recombine radiatively or non-radiatively with holes, or radiatively recombine at a hole-activated luminescence centre. The light emission phenomenon from the last process is called *thermoluminescence* (TL). The thermoluminescent light emission (or glow curve) varies with the mode of heating (linear or isothermal) and the heating temperature (Marinello et al. 1992). Corresponding to the different energy traps in the crystal, the glow curve exhibits several peaks (see Figure 16.2). The unstable low temperature peaks are

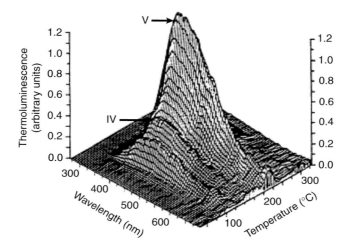

FIGURE 16.2

Readout thermoluminescence spectra of Bicron–Harshaw LiF 600 dosimeters, previously irradiated by x-rays. Peaks IV and V are stable peaks of light emitted with a wavelength around 450 nm when the dosimeter is heated at temperature around 100 to 200°C used for TL dosimetry. (From Townsend, P. D. et al., *Radiati. Effects*, 72, 245–257, 1983.)

eliminated by preheating. The more stable high temperature peaks, IV and V, are used for dosimetry with readers which allow temperatures up to 400°C and peaks VI and VII can be extracted at higher temperatures. An indication of the emission temperatures and wavelengths of stable peaks of TL materials of current use are shown in Table 16.1.

After readout, the TL material either returns to its original state or requires a special heating treatment called *annealing* in order to restore it. For instance, $Li_2B_4O_7$:Mn or $Li_2B_4O_7$:Cu can be used and re-used many times without thermal treatments between the successive irradiations and readouts. On the contrary, $CaSO_4$, CaF_2 and LiF require annealing procedures depending on the manufacture of the TL material and on the dose previously received by it (Toivonen 1993; McKeever et al. 1995). The recommended procedure for Harshaw-LiF:Mg,Ti is to carry out 4 cycles of 1 h at 400°C + controlled cooling + 20 h at 80°C prior to the first use (*initialisation*). This cycle is also used as an annealing process to allow reuse after irradiation if a dose greater than 0.5 Gy has been given (Driscoll et al. 1986). Some readers include an annealing cycle after each readout, but this lengthens the readout process by 40 s to 60 s per reading, depending on the system.

16.2.2 DETECTORS

TL detectors are either generated naturally or by doping phosphors with a very small percentage of *activators* (LiF:Mg–Ti is lithium fluoride doped with magnesium and titanium, $Li_2B_4O_7$:Cu is lithium borate doped with copper, etc.). It should be noted that the characteristics of the pure phosphor dosimeters may differ considerably from those of the composite and depend widely on the percentage of activators (Wall et al. 1982). They are available in the form of powder, solid dosimeters made of polycrystalline extrusions (rods, sintered pellets, or extremity chips) or powder bonded to a polypropylene and plastic substrate (Martin et al. 2000). Some examples are shown in Figure 16.1a.

The TL material to be used is chosen by comparing the parameters of tissue and TLD material that are relevant to the main interactions taking place for the radiation being measured:

For photon beams, these are the effective atomic number for the photoelectric effect or pair production or the electron density for the Compton effect depending on the photon energy range (see Chapter 4). Table 16.2 could be used to choose the TL material with the closest characteristics to those of the studied tissue.

For electron beams, the mass collision stopping ratio (ICRU 1984) is more relevant (see Chapter 3). On this basis, LiF and $Li_2B_4O_7$ can be used for measurements in soft tissues and $CaSO_4$ and CaF_2 in bones because the variation of their mass collision stopping power ratio to that of these tissues is less than a few percent in the energy range from 200 keV to 50 MeV.

Although very useful for choosing the appropriate TL material, these theoretical data must not be used for accurate energy correction because the shape and size of the dosimeter and surrounding material may have a significant influence on its energy response (see Section 16.2.4.9).

16.2.3 READERS

Only commercially available readers used for routine applications in radiation physics and patient dosimetry will be considered in this chapter. Most of them are listed in Table 16.3.

A reader has the following components (see also Figure 16.3):

Heating system housed in a readout chamber: Depending on the reader, the heating system may consist of a metallic support (planchet) heated by an electric current, isothermal fingers, hot nitrogen gas, an infrared sensor using an intense light pulse, or of a laser beam

TABLE 16.1

Emission Temperatures and Wavelengths of Dosimetric Peaks of Different TL Materials

Emission Temperature (°C)	Modal Wavelength (nm)								
	300	368	380	400	450	478 and 571	480 and 577	500	600
80–90									CaSO$_4$:Sm
200								CaSO$_4$:Mn	
200–240	LiF:Mg,Cu,P [GR 200]								
210–220				LiF:Mg,Ti [Harshaw]			CaF$_2$:Dy		Li$_2$B$_4$O$_7$:Mn
220–250					CaSO$_4$:Tm	CaSO$_4$:Dy			
240–270		Li$_2$B$_4$O$_7$:Cu							
260			CaF$_2$:Nat						
300								CaF$_2$:Mn	

The peak sensitivity of usual bialkali photocathodes is around 400 nm. As shown in this table, the nature of the doping material (indicated after the colon) has a great influence on the wavelength of the TL emitted light, and therefore, on the type of optical filters to be placed in front of the tube of the photomultiplier (PM) tube.

TABLE 16.2

Comparison between TL, Diode Detectors and Human Tissues Irradiated by Photon Beams of Different Energies

Material	Photoelectric Effect Z_{eff}	Compton Effect e^-/g	Pair Production Z_{eff}	Density g/cm^3
Silicon (diodes)[a]	14	3×10^{23}	—	2.33
LiF (Mg,Ti)[b]	8.14	2.79×10^{23}	7.50	2.64
LiF (Mg,Ti,Na)[b]	8.14	2.79×10^{23}	7.50	2.64
$Li_2B_4O_7$:Mn[b]	7.4	2.92×10^{23}	6.90	2.30
$Li_2B_4O_7$:Cu[c]	7.4	2.92×10^{23}	6.90	2.30
$CaSO_4$:Mn[b]	15.3	3.02×10^{23}	—	2.61
$CaSO_4$:Dy[b]	15.3	3.03×10^{23}	—	2.61
CaF_2:Mn[b]	16.3	2.95×10^{23}	—	3.18
CaF_2:Dy[b]	16.3	2.95×10^{23}	—	3.18
Air[d]	7.64	3.03×10^{23}	7.36	1.293×10^{-3}
Water[d]	7.42	3.34×10^{23}	6.60	1.00
Fat[d]	5.92	3.48×10^{23}	5.2	0.91
Muscle[d]	7.42	3.36×10^{23}	6.60	1.04
Bone[d]	14	3×10^{23}	10	1.01–1.60

[a] Hall (1994).
[b] McKeever (1985).
[c] Visocekas et al. (1985).
[d] Tubiana et al. (1963).

(see Table 16.3). It must be possible to heat the TL dosimeter to two different temperatures: the preheating temperature used to clear unstable peaks and the readout temperature used to collect the information from dosimetric peaks. Stability and reproducibility of these temperatures must be carefully controlled, requiring close contact between the TLD and its support when a planchet or isothermal fingers are used.

The heating kinetics of TLD depend on the system and have an influence on the dosimetric properties. Heating can be linear (TL material progressively heated to preheating and readout temperatures) or isothermal (TL material quasi-instantaneously heated to both these temperatures). Readers that are designed for the readout of a large number of dosimeters in a short time generally have isothermal heating kinetics (Marinello et al. 1992) or heat the TL dosimeter in hot gas.

The readout chamber must be continuously flushed with nitrogen gas in order to reduce spurious phenomena (McKeever 1985).

Light detection system: The luminous flux emitted by the TL dosimeter is collected and guided by a light guide into a photomultiplicator (PM) tube with a bialkali photocathode (peak sensitivy around 400 nm). This is well adapted to the blue emission of LiF:Mg,Ti or $Li_2B_4O_7$:Cu, but not necessarily optimal for other TL materials (see Table 16.1). Optical filters placed in front of the PM window considerably improve the response, provided that they are adapted to both the spectral response of the PM and the wavelength of the light emitted by the TL material. A system enabling interchange of the associated filters is a benefit when practical applications requiring the use of different TL materials have to be performed.

Signal integrator: The signal proportional to the light emission is either amplified and fed to an integrator (*d.c. operation continuous integration regimen*) or converted into pulses and fed to a scaler (*pulse counting regimen*). In most readers, glow curves are also displayed during dose measurements. Irrespective of the regimen, the voltage of the PM tube has to be correctly stabilised in order to achieve good reproducibility of measurements.

TABLE 16.3

Main Characteristics of TLD Readers Commercially Available in 2006

TLD Reader Companies	Model	Manual or Automatic	Heating Procedure	Maximum Temperature (°C)	Light Detection	TLD Nature	TLD Form
Thermo-Electron (USA)	Harshaw 3500	Manual	Planchet	400	Hamamatsu bialkali PMT fixed optical filter	LiF, CaSO$_4$	Powder and solid detectors
Thermo-Electron (USA)	Harshaw 4500	Manual	Planchet and hot nitrogen gas	400 (Planchet) or 600 (Optional) 400 (Gas)	Hamamatsu bialkali PMT fixed optical filter	LiF, CaSO$_4$	Powder, solid detectors, capton cards
Thermo-Electron (USA)	Harshaw 5500	Automatic 50 in/25 min	Hot nitrogen gas	400	Hamamatsu bialkali PMT fixed optical filter	LiF, CaSO$_4$	Solid dosimeters
FIMEL (France)	LTM	Manual	Planchet	500	Hamamatsu bialkali PMT+interchange-able optical filters	All TL material	Powder and solid detectors
FIMEL (France)	PCL 3	Automatic 45 in/15 min	2 Isothermal fingers	600	Hamamatsu bialkali PMT+interchange-able optical filters	All TL material	Powder and solid detectors

All readers allow the preheating of TLD, and some of them annealing after readout.

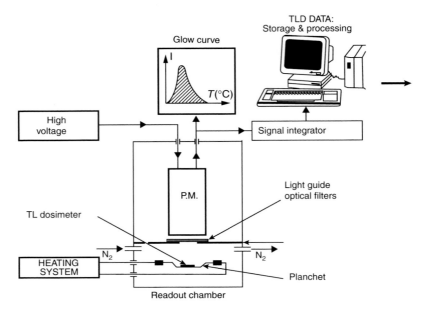

FIGURE 16.3
Different components of a TLD reader based on a planchet heating system.

PC and Associated Software: Results, which have to be converted to absorbed dose, are either read out and stored by the operator, or most often, automatically saved on a PC. Generally the PC both controls the reading process and allows data management. Computer programs are very different from one manufacturer to another. They may offer direct dose calculations provided individual calibration factors are entered. They may also perform statistical analysis (mean, standard deviation, etc.) and allow processing, display and printout of experimental results. In order to facilitate dosimeter management, some readers are mounted with a bar-code scanner that allows rapid identification of bar-coded TL dosimeters (see example on the top left corner of Figure 16.1).

16.2.4 DOSIMETRIC PROPERTIES AND INFLUENCE FACTORS

Dosimetric properties should be established for the readout conditions used in practice because they depend upon the heating kinetics. Moreover, characteristics described in this section are only valid for TL materials in their original state or restored by correct annealing (see Section 16.2.1).

16.2.4.1 Reproducibility

Achievable reproducibility is very dependent on the quality of the TL material, the type of reader used, and the purity of the nitrogen gas circulating in the readout chamber (use of N_2 containing less than 5 μL/L of O_2 is recommended to avoid non-dose-related luminescence). In addition, the operator must take care that the adjustments of the reader are correctly selected and check that the shape of the glow curve is normal (McKeever 1985; McKeever et al. 1995; Mayles et al. 2000).

The reproducibility can be evaluated by randomly taking ten samples of TL powder or dosimeters of the same batch, inserting them into a phantom made of tissue-equivalent material, and delivering them the same dose in the same irradiation conditions.

After readout (and annealing procedure when necessary), the operation is repeated several times. When readout parameters have been optimised, a standard deviation of $\pm 2\%$ or less can routinely be obtained with either manual or automatic readers of good quality associated with reliable TL materials (Kirby et al. 1992; Marinello et al. 1992; Toivonen 1993).

16.2.4.2   Background Signal

The background signal is the dose recorded with unirradiated dosimeters. It comes from several sources:

- The dark current of the photomultiplier, which can be very low if the PM power supply is permanently on and if the photomultiplier is not exposed to ambient light (if it happens accidentally, the instrument should be left for several hours to settle down)
- The background luminescence because of chemical interactions at the surface of the dosimeters or, with powder, to movements in the crystal lattice (tribothermoluminescence). It is reduced by reading out the dosimeters in an atmosphere of oxygen-free nitrogen (see Section 16.2.3)
- Residual signals from previous irradiations that can be removed by the annealing procedure (see Section 16.2.1).

The background signal should be evaluated with unirradiated TL samples read out with the same reader. Although for most measurements in radiotherapy, it is not a significant problem; it is preferable to subtract background signal from measurements particularly when low doses are expected. Some readers allow this to be done automatically.

16.2.4.3   Detection Threshold

The detection threshold is the minimum dose that can be detected by the TL material. It depends on the sensitivity of the material and on the background signal. As an indication, the detection threshold is approximately $10\ \mu Gy$ for LiF:Mg,Ti and $1\ \mu Gy$ for LiF:Mg,Cu,P (Wang et al. 1993; Zha et al. 1993) provided that they are read out with readers equipped with a PM tube and associated filters adapted to the wavelength of the emitted light. For lithium borate TLDs doped with Mn and Cu, the thresholds are approximately $40\ \mu Gy$ and $10\ \mu Gy$, respectively (Wall et al. 1982).

16.2.4.4   Fading

Fading is the decrease of TL response because of the loss of some of the charges initially trapped, between the end of the irradiation and the readout. It can be caused by the influence of heat, even at room temperature (*thermal fading*), or by an unwanted exposure to light (*optical fading*).

Thermal fading can be considerably reduced by appropriate preheating, but it can never be totally eliminated (see Table 16.4). In practice it is important:

- To avoid the use of TL materials that are subject to significant thermal fading (for instance CaF_2:Mn) and to evaluate thermal fading with the reader together with the TL material intended to be used. If it is greater than the expected value indicated in the table (i.e. 1% per month or less for the different preparations of LiF or 1% per week or less for $Li_2B_4O_7$, depending upon the doping), it means that optimal readout and annealing conditions have not been reached;
- To introduce a fading correction when a long delay separates irradiation from readout.

Optical fading can be avoided by manipulating the dosimeters in a room illuminated with incandescent light and wrapping them in opaque containers or envelopes when used in rooms illuminated with fluorescent light.

TABLE 16.4

Dosimetric Properties of Some TL Material

TL Material	Linear Zone (Gy)	Saturation (Gy)	Thermal Fading
LiF:Mg,Ti[a]	5×10^{-5} to 1	10^3	5% to 10% per year
LiF:Mg,P,Cu[c] [GR 200]	Up to 12	> 40	No evidence over 100 days
$Li_2B_4O_7$:Mn[a]	10^{-4} to 3	3×10^4	2.5% per month
$Li_2B_4O_7$:Cu[b]	5×10^{-4} to 120	10^3	4% per month
$CaSo_4$:Dy[a]	10^{-6} to 30	10^3	1% to 5% per month
$CaSo_4$:Mn[a]	10^{-7} to 30	10^2	10% per month
CaF_2:Dy[a]	10^{-5} to 10	10^4	25% per month
CaF_2:Mn[a]	10^{-5} to 10	10^3	7% per day

Except $Li_2B_4O_7$:Cu all TLD are distributed by Thermo-Electron (USA).
[a]   McKeever (1985).
[b]   Wang et al. (1993).
[c]   Visocekas et al. (1985).

16.2.4.5 Mass

TL Powder: Because the TL signal is generally proportional to the mass, samples of equal weight must be used or a correction factor taking into account the mass of each sample introduced. It should be also noted that some TL material such as $Li_2B_4O_7$:Cu or some LiF powders have a response that can be considered as independent of mass, within a certain range of masses, when they are read out with readers using isothermal kinetics (Marinello 1994). In this case, it is not necessary to weigh the samples; a simple volumetric measurement is sufficient.

Solid TL Dosimeters: For manufacturing reasons, variations in sensitivity within a batch of dosimeters are unavoidable. Two methods can be used to limit the effect of these variations (Van Dam and Marinello 1994).

One consists of individually identifying each dosimeter, irradiating them all together in the same geometrical conditions, reading them out and attributing to each of them a sensitivity factor S_i equal to R_i/R where R_i is the TL readout from dosimeter number i and R the mean of all values of R_i. Sensitivity factors should be periodically checked in order to take into account a possible loss of material occurring when TL dosimeters are not handled carefully.

The other consists of separating the TL dosimeters into sensitivity groups (i.e. groups of dosimeters with a response less than ± 1 or $\pm 2\%$ from the group mean response) and using a larger number of dosimeters used for each point of measurement. The distribution of sensitivities within a group should be periodically checked for the same reasons as explained above.

Both methods can achieve similar accuracy, the second one being faster when a large number of measurements have to be performed particularly if an automatic reader is available.

16.2.4.6 Dose

Generally, the response variation curve of TL materials with dose comprises a linear, a supralinear, and a sublinear region followed by saturation (Figure 16.4a). The importance of the different regions depends on the TL material, the nature and the energy of the particles used to irradiate it and the readout parameters. For LiF, for instance, when the heating kinetics and maximum temperature of the readout system allow the extraction of nearly all the TL peaks, the dose curve is linear over a much greater range than when the reader allows the extraction of peaks IV and V only (see Section 16.2.1 and Figure 16.4b). With such optimal readout conditions, it has been shown that most lithium fluorides and borates irradiated with a cobalt-60 beam have a dose response curve that can be considered as linear up to 12 Gy

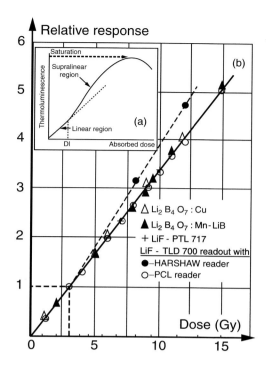

FIGURE 16.4

(a) General shape of a typical TL curve as a function of absorbed dose. (b) Results of measurements performed with different TL materials and different readers. The dose response curve (normalised to 1 for 3 Gy) for LiF-TLD 100 material obtained with the BICRON–HARSHAW reader, which uses isothermal heating kinetics, is linear (open circles) whereas it is supralinear in the same dose range with the FIMEL–PCL reader (closed circles and dashed line), which uses adiabatic heating kinetics. (From Marinello, G. et al., *Radiother. Oncol.*, 25, 63–66, 1992.)

(Marinello et al. 1992). Because of this dependance on the exact readout conditions, the dose response curve must be established with the reader and the TL material that are intended to be used. Data shown in Table 16.4 is given as an indication only and may be not valid for modern readers that allow for greater readout temperatures.

In practice, it is recommended that TL dosimeters are used in the linear region. When this is not the case, a correction should be applied to the signal from a curve established locally and this should be checked periodically. TL dosimeters must not be used in the sublinear region approaching saturation.

16.2.4.7 Dose-Rate

TL dosimeters are, to a large extent, dose-rate independent. As shown by Tochilin and Goldstein (1966) and Goldstein (1972), LiF and $Li_2B_4O_7$:Mn are independent of the dose-rate up to 45 and 1000 Gy per pulse of 0.1 µs, respectively. That means that no correction is needed in practice to take into account the dose-rate variations produced by beam modifiers, SSD variations, patient thicknesses, etc. This consideration also makes TLD very useful for dose measurements in intensity modulated beams.

16.2.4.8 Temperature

As the temperature required to obtain the light signal out of the TL crystal is high compared to room or patient temperature, the response of TL dosimeters is independent of temperature in usual conditions of use. However, care should be taken not to store the dosimeters close to a heating device (such as a radiator).

16.2.4.9 Energy

As mentioned in Section 16.2.2, it is very important to choose a TL material in agreement with the intended use.

Photon energies below 300 keV: For experiments performed in water or tissue-equivalent material irradiated with photon energies below 300 keV, TL dosimeters should be very thin and used without a build-up cap. Lithium borates are preferred to LiF and a fortiori to other TL materials because their effective atomic number is comparable to that of tissue in the photoelectric range (see Table 16.2). Theoretical data such as the mass energy absorption coefficient ratio to tissue can be used whenever the TL dosimeter is sufficiently small (Jayachandran 1970). For very low photon energies (below 50 keV), a theoretical calculation of the energy response is no longer possible because the shape and dimensions of the detector can induce considerable response variations within the dosimeter volume (Bassi et al. 1976; McKeever 1985; Davis et al. 2003). Moreover, the nature of the activator may also cause large differences in the response of TL materials in this energy range (Wall et al. 1982). The only solution consists of directly comparing the response of the TL dosimeters to a calibrated ionisation chamber specially designed and calibrated for low energy x-rays (see Section 18.6).

High energy photon beams: For dose measurements in media irradiated with high energy photon beams, all TL materials can be used except with very high energy photon beams that are sometimes contaminated by neutrons. LiF enriched with ^{6}Li and $Li_2B_4O_7$ responds to slow neutrons via reactions with ^{6}Li and ^{10}B, and for such high energy photons it is preferable to use LiF enriched with ^{7}Li that is insensitive to neutrons. Except for superficial measurements (see Section 40.2.3.1), TL dosimeters should be surrounded by a suitable build-up cap corresponding to the energy and geometrical irradiation conditions, in order to ensure electronic equilibrium (Van Dam and Marinello 1994 and Section 40.2.3.2). When the build-up cap is made of tissue-equivalent material, it is theoretically possible to evaluate the absorbed dose in TL dosimeters knowing the relative variation of mass energy absorption coefficient between the TL material and water as a function of photon energy. In practice, because of the influence of the surrounding material (build-up cap and patient), the sizes and shapes of the TL dosimeters may modify the expected results by a few per cent (Mobit et al. 1996a). The heating conditions may also modify the results slightly (McKeever 1985). The most reliable method consists of directly comparing the TL dosimeter and its associated build-up cap to a calibrated ionisation chamber.

High energy electron beams: When the irradiation is performed with high energy electron beams, it is theoretically possible to evaluate the absorbed dose in TL dosimeters knowing the variation of the ratio of the mass collision stopping power of the TL material to that of tissue or water as a function of energy. This variation is less than 2% for LiF and less than 5% for $Li_2B_4O_7$ in the energy range from 200 keV to 50 MeV (ICRU 1984). In practice, and for the same reasons as for photon beams, it is preferable to make a direct comparison between the response of the TL dosimeters to be used the dose measured with a calibrated ionisation chamber. The validity of the method has been verified by different authors (Holt et al. 1975; Bagne 1977; Mobit et al. 1996b; Marre et al. 2000).

16.2.4.10 Directional Effect

Irrespective of the conditions of use, no correction for directional effect is necessary except if the dosimeter container and associated build-up cap have an asymmetric shape (e.g. flat dosimeters should be oriented normal to the incident beam).

16.3 DIODE DOSIMETRY

Diodes are useful for dosimetric purposes (Huyskens et al. 2001; AAPM 2005) mainly because they have a high sensitivity within a small volume (see Figure 16.1), a good spatial resolution, and a real-time response. They can yield accurate measurements provided that their response is regularly monitored and corrected by factors taking into account both the aging of the diode and the irradiation conditions (field size, dose-rate, etc.).

16.3.1 PRINCIPLES OF DIODE DOSIMETRY

Most diodes are made of silicon, which is a crystalline material in which atomic electrons are arranged in bands of energy. During irradiation, electron–hole pairs are created. Conduction occurs by movement of electrons in the conduction band and by motion of hole states in the valence band. The gap between the valence and the conduction band is large enough to prevent rapid thermal repopulation of electrons (or holes). The number of charge carriers is controlled by doping the material with impurities like phosphorus, arsenic, or boron. Depending on the element used, the crystal is given either an excess or a deficit of free electrons that can carry electrical current. An excess of electrons carries a negatively charged current, giving an *n-type* semiconductor; whereas, an electron deficit (often described as an excess of electron *holes*) carries a positively charged current giving a *p-type* semiconductor.

A diode is typically formed by implanting a high density of p-type atoms into a lightly doped n-type substrate. Using silicon as a substrate, there are two major classes of detectors: the Charge Coupled Device (CCD) that is based on arrays of Metal-Oxide-Silicon (MOS) capacitors, and the p–n junction diodes that will be the only one considered in this section. Details for both types of detectors can be found in Hall (1994) and Barthe (2001).

16.3.2 DETECTORS

Commercially available diodes are generally made of a small quantity of silicon doped with phosphorus (n-type diodes) or boron (p-type diodes). They are encapsulated in caps of various natures, shapes and dimensions. Their design depends on the manufacturer and is different for photon beams or electron beams (Figure 16.1b). The diode cap ensures mechanical protection and acts as build-up material for photon beams. For diodes dedicated to in vivo measurements, the cap is generally made of tissue-equivalent material (polystyrene or epoxy), but stainless steel or tantalum are sometimes used in the construction in order to reduce its bulkiness and to provide filtration, making diode response less dependent of photon energy. In practice, it is important to know the diode cap's characteristics before undertaking measurements and to check that its shape and thickness are well adapted to the irradiation conditions and do not cause too much perturbation of the dose distribution (Sen et al. 1996; Marre and Marinello 2004). The nature and thickness of the caps of most of the diodes that are commercially available in 2006 are shown in Table 16.5. Intracavitary detectors that consist of diodes mounted within flexible probes and specifically designed for intracavitary measurements are not listed in the table.

Diodes from different manufacturers are generally easily identified by their shape and colour. As relatively large differences may occur between responses of diodes of the same type, it is recommended that each diode should be individually labelled with a number attached to its cable.

16.3.3 ELECTROMETERS

In order to minimize leakage, diodes for dosimetry are operated without external bias voltage, their signals are measured either in open-circuit mode (voltage) or in short-circuit mode (current). This last mode, most often encountered in commercial electrometers, offers

TABLE 16.5

Main Characteristics of Diode Systems Dedicated to Patient Monitoring (In Vivo Dosimetry) and Commercially Available in 2006

Company	Electrometers	Detector		Radiation Quality	Build-Up Cap		Pre-irradiation Dose
		Name	Type		Material	Thickness (g/cm^2)	
Cardinal Health Radiation Management Services	Veridose PDMQC^a (5 channels)	30-471	n or p	1–4 MV	Brass	0.732	20 kGy with high energy electrons
		30-472	n or p	5–11 MV	Brass	1.369	
		30-473	n or p	12–17 MV	Tungsten	2.606	
		30-474	n or p	18–25 MV	Tungsten	3.574	
		30-475	n or p	5–25 MeV elec.	None	0.284	
PTW	VIVODOS	T60010L	P	1–5 MV	Titanium	1	Not pre-irradiated
		T60010M	P	5–15 MV	Lead	2	
		T60010H	P	15–25 MV	Tungsten	3	
		T60010E	P	4–30 MeV elec.	PMMA	0.35	
IBA (Scanditronix-Wellhöfer)	DPD 3 (3 channels) DPD 510 (10 channels) DPD-12pc (12 channels)	EDD-5	P	Organs at risks	PVC+epoxy	0.5	8 kGy delivered with 10 MeV electrons
		EDP-5	P	Cobalt-60	Polyst+epoxy	0.5	
		EDP-10	P	4–8 MV	Stainless steel	1	
		EDP-15	P	6–12 MV	Stainless steel	1.5	
		EDP-20	P	10–20 MV	Stainless steel	2	
		EDP-HL	P	≥16 MV	Tantal	1.7	
		EDD-2	P	Electrons	Epoxy	0.2	
Sun Nuclear	IVD or Rf-IVD (3 channels)	QED 111300	n or p^b	70 kV and up	None	0.11	10 kGy delivered with 10 MeV electrons
		QED 111400	n or p	1–4 MV	Aluminium	1.03	
		QED 111500	n or p	6–12 MV	Brass	1.85	
		QED 111600	n or p	15–25 MV	Brass	3.04	
		QED 111200	n or p	Electrons	Acrylic	0.30	

Intracavitary detectors are excluded from this table.
a   Can also be connected to a beam uniformity quality control device.
b   QED diodes were p-type from 1997 to 2002 and n-type from 2003.

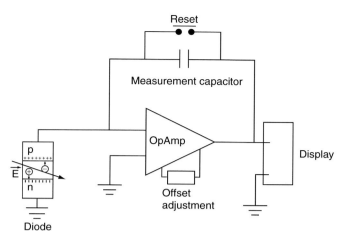

FIGURE 16.5
Principle of diode detection in short-circuit mode.

the advantage of producing a linear relationship between the charge generated in the diode and the dose (Figure 16.5).

Two types of equipment are generally available.

The devices dedicated to *patient dose monitoring* consist of multichannel electrometers (number of channels varying from 2 to 12), allowing the use of several individual diodes at the same time (Table 16.5). Most electrometers provide zero level adjustments for each diode, and also allow individual calibration factors to be applied to the diode signals so that readings can be directly displayed as absorbed dose. Some systems also provide a correction factor that accounts for differences between calibration and clinical conditions.

The devices dedicated to *quality control of linac beams* consist of electrometers linked to special phantoms in which diodes are embedded at fixed positions. At least five diodes are used to measure beam output as well as symmetry and uniformity. More recently, specific devices including arrays of several hundreds of diodes have been developed for quality control of intensity modulated beams (see Section 38.9.3).

Some electrometers are compatible with both of these applications.

Irrespective of their type, electrometers can often be interfaced with a standard printer that allows output of report including the date and time at which measurements were made, descriptions of the measurement conditions, and the results in terms of the dose delivered to the patient or linac quality control data. Most of the recent devices are computer controlled or can be interfaced with a PC providing long term storage and further processing of data.

16.3.4 DOSIMETRIC PROPERTIES AND INFLUENCE FACTORS

Measurements with diodes should be performed under conditions of electronic equilibrium. If the built-in build-up is not sufficient for the particular energy, an additional build-up cap can be placed on the diodes or they can be inserted into a special phantom drilled with holes of the appropriate dimensions with enough material in front of them. It should also be noted that dosimetric properties are not only dependent on the diode design, but they are also influenced by the doses accumulated by the diode so that they must be determined for each diode and regularly checked.

16.3.4.1 *Repeatability and Reproducibility*

The attainable repeatability and reproducibility depend on the quality of both the diodes and the associated electrometer. Repeatability can be evaluated by measuring the signal for at least ten consecutive irradiations with the same dose delivered in the same conditions of irradiation. Assessment of reproducibility requires that measurements are repeated over a longer period of time (e.g. daily during one week). The standard deviation for both of them should be less than 1%.

16.3.4.2 Fading

Fading is the drift of signal that can be observed after irradiation. It can be evaluated by measuring the signal at increasing delays after irradiation with the same dose delivered in the same geometrical conditions. The diodes and the associated electrometer should have a drift less than 1% in 1 h.

When diodes are intended to be used for continuous measurements during a long period of time such as when measuring the patient dose for low-dose-rate total body irradiations, the checks should be extended over a period of several hours.

16.3.4.3 Detection Threshold

The detection threshold depends both on the electrometer and on the active volume and doping level of the semiconductor detector. The pre-irradiation dose and the doses subsequently accumulated also have an influence on it. For commercially available systems and new pre-irradiated diodes, the detection threshold varies from 0.1 cGy to some tenths of cGy.

16.3.4.4 Background Signal

The background signal is due to thermally generated charge carriers. It causes problems only if the electrometer's input offset voltage is not zero. The effect is strongly temperature-dependent and appears to be greater for n-type diodes than for p-type diodes. Although the current can be zeroed out before measurements are taken, several minutes are necessary to obtain a satisfactory background measurement, and any subsequent change in temperature will upset the balance of the circuit and cause a change in diode response. This problem should be taken into account when diodes are used for in vivo measurements, because the temperature of the patient's body is different from the room temperature (see Section 16.3.4.7).

16.3.4.5 Dose

All silicon diodes show a loss of sensitivity as a function of accumulated dose because of radiation damage that displaces silicon atoms from their lattice sites. This introduces recombination centres that capture charge carriers, leading to a reduction in sensitivity and increased dose-rate dependence (Rikner and Grusell 1983). The effect depends on the doping level of the diode and is more pronounced for *n-type* than for *p-type* diodes because holes are more easily trapped than electrons (Grusell and Rikner 1984, 1993; Van Dam et al. 1990). It is also more important when the irradiation is performed with electron beams than with photon beams: after Rikner and Grusell (1987) the amount of damage for a given dose is 20 times more important with 20 MeV electrons than 8 MV photons and increases as a function of energy for both the particles. Moreover Rikner and Grusell (1983) have shown that the steep decrease in sensitivity observed for new diodes was considerably reduced after an irradiation of 20 kGy with 18 MV x-ray beams or 5 kGy with 20 MeV electron beams. This is the reason why most companies sell pre-irradiated diodes (Table 16.5). It is recommended that diodes that have not been pre-irradiated by the manufacturer are given a high dose before use. Once this has been done, if the electrometer is used in integration mode, the variation of response of diodes as a function of dose is generally linear in the range 1 Gy to 11 Gy (Van Dam and Marinello 1994).

16.3.4.6 Dose-Rate

For the reasons previously given in Section 16.3.4.5, diode sensitivity increases as dose rate or dose per pulse increases (Grusell and Rikner 1984; Van dam et al. 1990). The effect is essentially dependent on the doping level and not always more pronounced for n-type than for p-type diodes (Shi et al. 2003). Although reduced by an adequate pre-irradiation of the diode, it remains significant in the dose range encountered in practice and results either in a non-linearity of dose response with dose-rate for n-type detectors or in a limited linear dependence for p-type ones (Grusell and Rikner 1993). As pointed out by Van Dam and Marinello (1994), the dose-rate dependence of a diode should be taken into account for linacs that offer the choice of different monitor unit rates by variation of pulse intensity (not of pulse repetition frequency) or those where the field flatness is achieved by magnetically scanning an electron pencil beam of high intensity over the irradiation field. The dose-rate dependence of a diode also deserves special attention when depth-dose or dose-profile measurements are carried out. When wedge filters, compensators, or shielding blocks are used, they induce simultaneous variations in dose rate and beam quality. Similar problems can be encountered when diodes are used with intensity modulated radiation therapy. Diode response changes of several percent can be observed in such cases. Consequently, it is recommended that diodes are always calibrated in the dose-rate range for which they will be used and it is not advisable to use them when the dose-rate is likely to vary considerably as a function of time.

16.3.4.7 Temperature

Many authors (Grusell and Rikner 1986, 1993; Van Dam et al. 1990; Heukelom et al. 1991; Welsh and Reinstein 2001; Saini and Zhu 2002; Marre and Marinello 2004) have reported an increase of diode response with temperature probably because of an increase of the energy of the charge carriers that escape from recombination. The increase in sensitivity with temperature is reduced, or even eliminated, when diodes are correctly pre-irradiated. In practice, it is recommended that the increase in sensitivity with temperature is determined for each new diode and checked periodically. For in vivo measurements, as long as the variation remains less than 0.1% per °C, it is not necessary to correct for the difference between the room (where the calibration takes place) and the patient's skin temperature. It must, however, be ensured that the electrometer is zeroed out some minutes after the diodes have been placed on the patient. If it exceeds 0.4% per °C, the diode must be eliminated or correction factors must be introduced.

16.3.4.8 Energy

Because of their high effective atomic number compared to that of water or human tissues (Table 16.5), diodes are not tissue equivalent for photon energies below ^{60}Co. This was experimentally confirmed by Heukelom et al. (1991) and Edwards et al. (1997) and is of particular concern for conventional x-rays or scattered radiation.

Compensated diodes as described by Rikner and Grusell (1985) can be used without problem for dosimetry in water or tissue-equivalent material irradiated by high energy photon beams because they are shielded from low energy scattered radiation. Nevertheless, care must be taken when making measurements at depths or in fields where low energy scattered radiation contributes a substantial fraction of the dose, as for instance when measuring doses under shielding blocks or beyond the edge of the irradiation field.

In the case of measurements performed in high energy electron beams, the stopping power ratios of silicon to water vary less than 3% between 3 MeV and 50 MeV (ICRU 1984). That means that diodes can give good results in practice provided their build-up cap is very thin and made of tissue-equivalent material (Eveling et al. 1999; Marre and Marinello 2004).

In any case, it is advisable to measure the energy response of a diode in the beam or energy range in which it is intended to be used.

16.3.4.9 Directional Effect

This effect is related to the shape and construction of the diodes which do not always have a symmetrical shape in all directions (Figure 16.1b). Because of the lack of symmetry, the response of the detector can change according to the beam angle when the diode is on the beam axis, or when off-axis measurements are performed. The directional effect also depends on the beam parameters such as energy and field size, so its importance should be evaluated in the expected experimental or clinical conditions and with diodes that will be used for measurement. Generally, the correction can be neglected for incidence angles less than $\pm 30°$, but it can reach more than 5% for some detectors for incidence angles larger than 50° (Van Dam and Marinello 1994; Marre and Marinello 2004). Moreover Mayles et al. (1993) have pointed out the asymmetrical response of domed diodes that can be as much as 15% when irradiated by mistake on the flat side instead of the domed one. They also mentioned that the symmetrical response of cylindrical detectors can be reduced by about 15% if irradiated from the tip.

16.4 CONCLUSIONS

Both TLD and diodes have the advantage, compared to ionisation chambers, that they do not need to be connected to a high voltage supply; but they both require correction for energy response. Diodes have an advantage over TLD because of their real time response but are inferior with respect to their need for a cable connection and their response variation as a function of accumulated dose, dose rate, and temperature and directional effects. Both detectors should be handled by physicists or operators who know the correction factors to be applied; otherwise, the results obtained will not be accurate. The equipment should also be checked regularly in order to maintain its performance. A typical list of checks to be performed can be found in the book chapter of Mayles et al. (1993) or in other specialized publications.

CHAPTER 17

RADIATION SENSITIVE FILMS AND GELS

Mark Oldham *

CONTENTS

---

* Thanks to David Jaffray, Mike Sharpe, Greg Edmundson, Jeff Siewerdsen, and Daniel Letourneau for proofreading this chapter.

17.1   INTRODUCTION

Radiographic film is the dosimeter of choice for a wide range of therapeutic and diagnostic applications in radiation-therapy departments. It is particularly useful for qualitative measurements of ionising radiation beams (e.g. high and low energy x-rays and beams of electrons and protons) and the verification of clinical radiation treatments both prior to and during therapy. For many years, film has been used to verify light-field and radiation-field coincidence and beam homogeneity of a linear accelerator. Films are also often used to verify the correct set-up of a megavoltage therapeutic beam with respect to the patient by means of a portal image of the beam as it exits the patient (see Section 40.1). These applications all benefit from the primary strengths of film that include high contrast (image quality), high spatial resolution and high sensitivity. The limitations of radiographic film include a variation of response as a function of photon energy, a high sensitivity to processing conditions*, and the fact that film is only useable in the form of thin two-dimensional (2D) sheets that do not easily lend themselves to the examination of complex dose-distributions in 3D. The advent of multi-detector arrays (Jursinic and Nelms 2002; Letourneau et al. 2004) which can be convenient, are near-real time, do not require processing, and can directly interface to computer read-out, has led to a decline in the use of film in many clinics. Recently, this trend has somewhat reversed as many clinics implement Intensity Modulated Radiation Therapy where it is common practice to compare 2D computed dose distributions with film measurements in the same conditions (Dogan et al. 2002; Bucciolini et al. 2004). Ideally, IMRT would be verified through a comprehensive measurement of the distribution in 3D, and this has led to the development of radiation sensitive gels and plastics that respond and can be imaged in 3D, giving more complete information for complex distributions. Gel-dosimetry is a relatively new development to the field of dosimetry and is discussed further in the second part of this chapter (see Section 17.3).

17.2   FILM DOSIMETRY

17.2.1   THE RADIOGRAPHIC PROCESS

The basic physical mechanisms of image formation and development are the same for both radiographic and photographic film. The active component of radiographic film consists of a gelatin emulsion that suspends small radiation sensitive silver-halide crystals (e.g. silver-bromide AgBr). The emulsion is evenly coated onto both sides of a polyester base that provides the stiffness and mechanical stability of the film. Stiffness is required for handling purposes (e.g. viewing on a light-box), and stability is required to avoid film expansion during exposure to heat and chemicals during development. Adhesive is used to ensure that the emulsion remains attached to the base throughout development, and the whole film is coated with a strong protective gelatin super coat that protects the soft emulsion from scratches and skin oils during handling. Details of film manufacture and the active chemical ingredients can be found, for instance, in Chapter 19 of Carlton and Adler (2005).

17.2.2   THE LATENT IMAGE

The key mechanism utilised by radiographic film is the formation of a latent image in the emulsion of the film by incident ionising radiation. The latent image is then amplified and

---

* With the increased use of digital radiography leading to less use of film for other purposes, maintaining the stability of processors is becoming more of a challenge. Radiochromic film (see Section 17.2.7) provides a useful alternative to traditional silver based film. Computed radiography (CR) systems such as the Kodak 2000RT may also provide a partial solution.

rendered visible by chemical development. Gurney and Mott (1938) initially proposed the theory of latent image formation, and although some areas are still not fully understood, this theory remains largely unchallenged today. Each radiation sensitive crystal, or grain, in the emulsion can be thought of as a cubic lattice containing $\sim 10^{10}$ atoms. The silver-bromide molecules orientate such that the halide ions (bromine and iodine) congregate on the surface of the grain, presenting a negatively charged exterior. The centre of the grain has a net positive charge because of the preponderance of silver ions. Both electrons and silver atoms (or ions) are able to drift through the lattice.

Incident photons generate free electrons inside the grain through the photoelectric, Compton and pair production effects. These electrons get trapped on impurity molecules known as sensitivity specks that are incorporated in the grain surface and may protrude out from the grain surface. The sensitivity speck becomes negatively charged and attracts mobile silver ions that then recombine to form silver atoms on the surface of the sensitivity speck. A single incident photon may free thousands of electrons for deposition at the sensitivity speck. Not all free silver atoms are deposited at the speck, but it is estimated that at least three must be deposited for the entire grain to be rendered developable. The latent image corresponds to the conglomeration of silver ions near the speck and to the deposition of, at least, three silver atoms on the sensitivity speck that stabilise the speck and allow it to act as an electron gate into the grain during the development of the film.

17.2.3   FILM PROCESSING

Film processing involves rendering the latent image visible by chemical amplification. It involves the following steps:

(1) *Developing*—The alkaline, electron-rich reducing agents in the developer fluid supply electrons to the interior of the grain through the sensitivity speck that acts as an electron gate. Grains that were exposed to high doses of radiation have larger sensitivity specks that correspond to more efficient electron gates. When developing is stopped, the amount of black metallic silver deposited in the grain is in proportion to the size of the sensitivity speck gate. It is this proportionality that leads to the exquisite grey-level contrast in film images. If a film is left in the developer, electrons will eventually get into even those grains without a sensitivity speck gate. The presence of the gate differentiates between irradiated and un-irradiated grains, but only within the short term. Further details of latent image formation and amplification can be found in Chapter 3 of Herz (1969).

(2) *Fixing*—the acidic environment of the fixing fluid (or stop bath stage) quickly stops the developing process of the latent image. Clearing agents (usually sodium thiosulphate) bond with undeveloped silver halide molecules, removing them from the emulsion. The darkened appearance of irradiated film is caused by black metallic silver deposited during latent image formation and amplified during developing that remains unaffected during fixing of the film.

(3) *Washing*—the film is then washed in pure water to remove most of the fixer and developer from the film, and it is then air-dried. Poorly washed films will suffer silver and/or yellow emulsion staining under long-term residual chemical action.

17.2.4   FILM BLACKENING AND OPTICAL DENSITY

The degree of film blackening after irradiation and processing can be quantified by determining how much a beam of light is attenuated as it passes through the film. The degree of blackening thus measured is known as the light opacity and is defined as I_0/I where I_0 is the light intensity measured in the absence of the film, and I is the intensity transmitted through the film in a perpendicular manner. A more useful quantity is the optical density (OD) of the film, defined as the $\log_{10}$ of the light opacity (see below), because it is proportional to dose within limiting conditions. Theoretically, this linearity can be understood with the following simplified model. If the average area obscured by a single grain of silver (i.e. the cross-section of a single developed silver bromide grain) is a, and N is the number of developed grains per cm^3, then the intensity $I(z)$ of light after passing through a depth z of film will follow Beer's law

$$I(z) = I_0 e^{-aNz} \tag{17.1}$$

where I_0 is the light intensity without film.

If it is assumed that a single photon hit renders a grain of silver bromide developable and that all grains present the same area, a, to the fluence, then the number of grains that are made developable is $N = N_{AgBr} a\phi$, providing $N \ll N_{AgBr}$, where N_{AgBr} is the initial number of undeveloped grains in the unirradiated film, and ϕ is the photon fluence. Substituting this into Equation 17.1 and taking the logarithm, the optical density is defined as:

$$OD = \log_{10}\left(\frac{I_0}{I(z)}\right) = a^2 N_{AgBr} z\phi \times 0.4343 \tag{17.2}$$

For low fluence where $N \ll N_{AgBr}$, the optical density is proportional to the fluence, ϕ, and hence, the dose. In practice, the optical density is linear with dose only in the limited range before saturation effects occur. Equation 17.2 demonstrates that the most effective way to alter the speed of the film is to increase a, the cross-section of the grain and then to increase the film's thickness. Extensions of this simplified theory are given in Chapter 4 of Shani (2000).

The instrument used to measure the optical density of a film is called a *densitometer*. A simple densitometer consists of a calibrated light source, a stage on which to place the film to be measured, a light aperture to control the amount of light from the source and an optical sensor on a moveable arm. The densitometer is calibrated to read zero on an unirradiated part of the film prior to reading optical density values in the irradiated film. The amount of light transmitted through the film is measured and displayed by the densitometer, which normally gives a read-out calibrated to optical density values. More complex scanning densitometers automatically move the light source and sensor over the film in a Cartesian grid to build up the 2D image of the optical density variations across the film. In all cases, it is necessary to check the densitometer response using a calibrated optical wedge consisting of a film exposed to multiple known light intensities resulting in known optical density steps. Calibrated optical wedges are generally supplied with new densitometers. They can also be produced locally by exposing a piece of film to a special illuminating device called a *sensitometer*. Films exposed with sensitometers are used to check the constancy of film batches and film processing.

17.2.5 CHARACTERISTIC FILM CURVES

The responses of different films vary with many factors including the manufactures, the radiation energy, and daily fluctuations in the relative chemical components in the developer unit. Manufacturers customarily use a characteristic curve to describe their film performance. Characteristic curves are commonly referred to as H&D curves after Hurter and Driffield who first described one in 1890. The H&D curve plots *optical density* versus *$log_{10}(dose)$* (Figure 17.1) and describes the film performance in terms of resolution, speed, contrast and latitude. It should not be used as a calibration curve because it does not account for the local processing conditions.

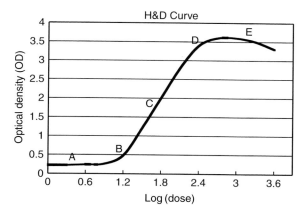

FIGURE 17.1

Typical H&D curve of x-ray film indicating (A) base+fog, (B) toe, (C) linear portion, (D) shoulder and (E) reversal or solarization region. The exact shape of this curve depends on the particular film type. The gradient in the straight line region is a measure of the inherent contrast of the film and is historically referred to as gamma.

17.2.6 Measuring a Dose Distribution with Film

Film has been used to measure relative dose distributions in radiotherapy since the 1950s (Gramke et al. 1952; Fleeman and Frantz 1954). It became popular in the 1960s, especially for dose measurements in electron beams (Dutreix and Dutreix 1969). For photon beams, the published results were somewhat contradictory, and the accuracy of film dosimetry was considered to be not as good as that of other detectors (Danciu et al. 2001). However, it has regained popularity in the last ten years, as a relative dosimetry Quality Assurance tool, in conjunction with the development of Intensity Modulated Radiotherapy.

In order to derive dose distributions from films, the first step is to determine a calibration (or sensitometric) curve of optical density versus dose (Figure 17.2) that describes the response

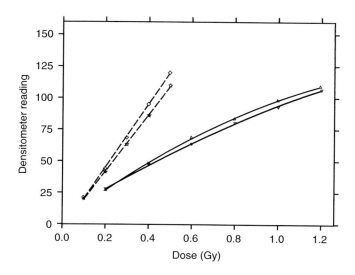

FIGURE 17.2

Calibration curves (6 MV radiation) for two different makes of therapy film; solid lines are Kodak X-OMAT V film, and the dashed lines are CEA films (Cheng and Das 1996). The steeper line for each film corresponds to radiation incident perpendicular to the plane of the film. The shallower line for each film corresponds to radiation incident parallel to the plane of the film. The ordinate axis is in relative units of optical density (OD). Note that the drop in film response when radiation is incident non-normally to the plane of the film is more pronounced for lower energy beams (Herz 1969). (Figure taken from Oldham, M. and Webb, S., *Med. Phys.*, 24, 827–836, 1997. With permission.)

of the film as a function of dose (Williamson et al. 1981). It is important to note that the calibration curve is linear before flattening out as it approaches saturation. This linearity makes relative dosimetry straightforward as the scanned optical density values represent the relative dose distribution without the need for the calibration constant to be known. Calibration films should be irradiated in the same orientation with respect to incident photons as the experimental film, to maintain consistent dose response.

The main problems in film dosimetry relate to the high atomic number of silver that results in an increase in film sensitivity compared to water-equivalent detectors for energies lower than about 400 keV. The calibration curves can be expected to change when the photon spectrum changes as a function of depth or field size or under blocks and at the edge of or outside the beam. Changes in the calibration curve are larger in the energy range 4 MV to 10 MV (and for cobalt-60 beams) than in the range 10 MV to 25 MV. Some variations have also been observed, depending on the film orientation with respect to the beam incidence (see Figure 17.2). These effects have been experimentally investigated and theoretical attempts are made to provide correction factors (Zhu et al. 2003; Palm et al. 2004).

To obtain accurate results, the dosimetric film is usually pressed firmly into a solid phantom* consisting of slabs of water equivalent material, in order to eliminate dose distorting air pockets near the film surface. Film can be cut into shapes in a darkroom to fit into small phantoms, and opaque sealing tape should be used to protect from light exposure at the edges when necessary. Various solutions have been suggested to minimize the influence of spectral changes such as introduction of high atomic number filtering slabs within the phantom (Burch et al. 1997; Yeo et al. 1997) or the use of multiple sensitometric curves (Stern et al. 2004). If radiation incidence is parallel to the film plane, then sensitivity to beam hardening effects can be minimised by tilting the film a few degrees away from the incident direction.

The sensitivity of the film to be used must be such that the optical density remains within an acceptable range (i.e. below 3) for doses of clinical relevance. This is especially important for the verification of Intensity Modulated Radiotherapy where it is safer to check the dose distribution in situations resembling the treatment situation as much as possible. Therefore, specific types of films have been investigated in replacement of the very popular Kodak XV film. Examples are the CEA TVS film (Cheng and Das 1996) and the Kodak EDR2 film (Olch et al. 2002; Zhu et al. 2002) that yield an optical density of approximately 1.0 for a dose around 2 Gy.

After exposure, the dosimetric film is developed and may be digitised with a scanning densitometer. The variation of the overall film response is slightly dependent on the type of densitometer used and on the film batch (typical variation being 5–15%). However, the main variation comes from the processing conditions that can yield differences greater than 30%. Such differences can be overcome for relative measurements by using a normalised calibration curve (Bos et al. 2002; Georg et al. 2003), and for absolute measurements, by ensuring accurate quality control of the performance and stability of the processor and of the consistency between different films in the same batch. Under these conditions, the uncertainty may be reduced to about 2–3% for relative dose determination and to 3–5% for absolute dose determination.

17.2.7 RADIOCHROMIC FILM

A promising and relatively new approach in film technology is radiochromic or Gafchromic film (Niroomand Rad et al. 1998; Rink et al. 2005), a transparent film that responds to

---

* Most films can also be placed directly in water if required.

ultraviolet light and ionising radiation by turning blue and exhibiting absorption bands at 670 nm and 610 nm. The radiation induced colour change (a dye forming or polymerisation process) is formed directly, without the need for any chemical processing, and is stable at temperatures up to 60°C. The image can be read out by a He–Ne laser scanning densitometer (Reinstein and Gluckman 1997) or a regular transmission flat-bed-scanner (Devic et al. 2005). An attractive feature is that the photon mass energy absorption coefficients and electron mass collision stopping powers are very similar to those of water and skeletal muscle, making the film tissue equivalent and much less sensitive to spectral hardening than radiographic film. Furthermore, the film is easy to handle as it is insensitive to visible light. Historic challenges to radiochromic film were sensitivity (about a factor of 30 less than standard radiographic film) and non-uniformity of response, corresponding to the manufacturing-inhomogeneity of the dye layer (Cheng and Das 1996). However recent advances and the improved sensitivity of new formulations, associated with price reduction, are expected to lead to substantially increased use of radiochromic film for clinical applications (Devic et al. 2004; Rink et al. 2005).

17.2.8 PRACTICAL CONSIDERATIONS WHEN USING FILM AS A DOSIMETER

(1) *Energy response*: Typically, film is 10 times more sensitive to 0.1 MeV photons than to 1 MeV photons because of the strong photoelectric interaction in AgBr below 150 keV (see Figure 17.3).

(2) *Dose-rate*: an important rule concerning photographic exposure to x-rays and γ-rays is the reciprocity rule. This states that the photographic response is independent of dose rate. Without the reciprocity rule, film dosimetry would hardly be possible at all as one would always need to know the dose rate. Thankfully, provided the film is processed within a few days of irradiation, the reciprocity rule holds in essentially all film applications in radiation-therapy and dose-rate effects can be ignored. The latent image may fade in irradiated films left for long periods prior to processing (Chapter 4 of Herz 1969).

(3) *Dose range*: Kodak EDR2 film: range 25 cGy to 400 cGy; saturates at 700 cGy. Kodak XV-2: range 0 to 100 cGy; saturates at 200 cGy. XTL film: range 1 cGy to 15 cGy; saturates at 30 cGy. Kodak PPL film: range 0.25 cGy to 5 cGy; saturates at 10 cGy. Older gafchromic film (e.g. MD55): range 0 to 30 Gy. More recent gafchromic EBT film: 0 to 6 Gy.

(4) The main advantages of film as a dosimeter are its robustness and flexibility of application, high contrast and high spatial resolution (≤0.1 mm in principle, but in practice, usually limited by the aperture of the scanning densitometer to ≥0.3 mm).

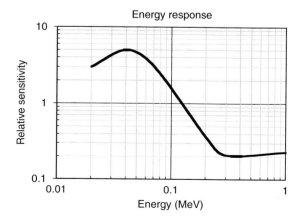

FIGURE 17.3
Sensitivity of a typical radiographic x-ray film (irradiated without any additional screen) as a function of photon energy. (From Shani, G., *Radiation Dosimetry Instrumentation and Methods*, CRC Press Inc., Boca Raton, Florida, 1991.)

(5) The problems of using film to measure absolute dose and beam data are rooted in (a) film production and processing fluctuations (for example, variations of 1°C in the developer bath can affect the optical density by as much as 10%), and (b) lack of tissue equivalence that causes variation in film sensitivity with the energy of incident radiation.

(6) Film can be used to measure dose distributions from electron and neutron beams and is much less energy dependent with these modalities[*].

17.3 RADIATION SENSITIVE GELS

Traditional dosimeters such as ionisation chambers, radiographic films, thermolumines-cent dosimeters and solid state detectors have limitations for the measurement of complex distributions produced by modern techniques such as intensity-modulated-radiation-therapy (see Section 43.3.8). One of the main limitations is that none of these dosimeters are truly 3D in that they measure dose at a single point or in a single plane. Ideally, a new technology is required that can produce an accurate high resolution 3D map of the delivered dose distri-bution. To this end, much recent effort has been invested into developing radiation sensitive gel-dosimeters that can capture the 3D distribution via chemical processes that are radiation dependent. Many chemical dosimeters have been found in nature, including ferrous sulphate (4 Gy to 400 Gy), ceric sulphate (400 Gy to 4×10^6 Gy), chlorinated solutions (1 kGy), oxalic acid based solutions and polyacrylamide gels (PAG) (2 Gy to 30 Gy). Because of dose-range and ease of manufacturing, only the Fricke (ferrous sulphate) and PAG gel dosimeters have found applications in radiation therapy departments, although, new reports indicate a new material, PRESAGE, to be a promising alternative (Guo et al. 2006a, 2006b).

17.3.1 FRICKE GELS

Fricke and Hart originally proposed the ferrous-sulphate dosimeter in 1966 (Fricke and Hart 1966). The initial formulation was not a gel but an aerated solution containing 1 mmol/L ferrous sulphate ($FeSO_4$), 1 mmol/L NaCl and 0.4 mmol/L sulphuric acid. When irradiated, Fe^{2+} ions are oxidised to Fe^{3+} along the following pathways:

$$Fe^{2+} + OH \rightarrow Fe^{3+} + OH^-$$

$$H + O_2 \rightarrow HO_2$$

$$Fe^{2+} + HO_2 \rightarrow Fe^{3+} + HO_2^-$$

$$Fe^{2+} + H_2O_2 \rightarrow Fe^{3+} + OH + OH^-$$

The sensitivity of the Fricke solution to impurities is reduced by the presence of NaCl, but it is still necessary to use triple distilled water and pure high-grade chemicals. Since ferrous sulphate gradually oxidises with time, the solution must be freshly prepared before use.

17.3.2 CHEMICAL YIELD AND THE G VALUE

The radiation chemical yield of Fe^{3+} ions produced in solution is known as the G value and is measured either by chemical titration, or often more conveniently, by absorption

---

[*] However, for neutron, proton, or ion beams, the film response as a function of dose may be misleading because, for high LET radiation, changes in LET (e.g. as a function of depth) may cause changes in optical density including saturation effects (see Section 46.4.3).

spectroscopy at 304 nm (ICRU 1969). The latter requires only 1 cm^3 of solution in a thermally equilibrated optical cell. The G value is the number of Fe^{3+} ions produced by 100 eV of absorbed dose and has been determined to be 15.5 ± 0.4 for megavoltage radiation (ICRU 1969). A G value should be measured at the local institution as a calibration of local equipment. The dose absorbed in a dosimetric Fricke solution is then determined from

$$D(Gy) = \frac{\Delta M}{\rho G} \times 0.964 \times 10^6 \qquad (17.3)$$

where ΔM (moles/L) is the change in concentration of Fe^{3+} ions, and ρ is the density of the solution.

Fricke solutions have limited use as the spatial distribution of dose is not preserved. However, Gore et al. (1984) showed that Fe^{3+} ions produced in aqueous Fe^{2+} solution sufficiently affect the magnetic resonance (MR) T_1 relaxation time of water protons such that dose differences of 0–40 Gy can be resolved. They suggested that fixing the ferrous ions in a gel matrix could preserve spatial information and this was later demonstrated by Appelby et al. (1987). Subsequently, investigators have attempted to model the Fricke gel response (Podgorsak and Schreiner 1992; Audet and Schreiner 1997) and study performance issues associated with composition (Olsson et al. 1989; Schulz et al. 1990; Olsson et al. 1990; Hazle et al. 1991) and MR imaging (Johansson et al. 1998; Shortt 2001). Fricke gels have been applied to study many dosimetric situations encountered in the clinic including 3D conformal treatment planning (Chan and Ayyanger 1995), radiosurgery (Olsson et al. 1992; Schulz et al. 1993), brachytherapy (Olsen and Hellesness 1994; Schreiner et al. 1994), and to protons (Gambarini et al. 1997). The T_1 image of an irradiated gel is converted to a dose map by the application of a calibration curve of T_1 values for known dose in the same batch of Fricke gel. Two problems have hindered routine clinical application of Fricke gels. First, the Fe^{3+} ions are not well fixed in the gel matrix and start to gradually diffuse through the gel immediately after irradiation; and second, access to expensive MR imaging facilities is required to obtain the T_1 image and the dose distribution. The diffusion of ions through the gel leads to the loss of penumbra effects in a matter of hours (Balcom et al. 1995), so that a Fricke dosimeter gel must be imaged soon after irradiation. Some progress has been made to reduce the diffusion by changing the gelling environment (Chu et al. 2000).

17.3.3 POLYACRYLAMIDE GELS

Maryanski et al. (1993, 1994, 1996a) proposed a new type of radiation sensitive poly acrylamide gel (PAG). PAGs utilize the mechanism of radiation induced polymerisation of monomers where small monomer molecules join together under the influence of ionising radiation. The resultant long polymer chains are too big to move through the gel lattice eliminating image degradation due to diffusion. The initial PAG composition was called the BANG™*, gel as it constituents by weight were Bis (3%) Acrylamide (3%) and Nitrogen and Gelatin (5%) mixed with water. Subsequent development led to the BANG-2 gel where acrylamide was replaced with acrylic acid and 1% of sodium. On irradiation, polymerisation and cross-linking occurs between the acrylic and bis monomers. Regions of the gel exposed to dose undergo polymerisation in proportion to the dose, and polymerised regions affect both the MR relaxivity of water protons and also the optical scattering power of the gel. The 3D-polymerisation distribution in the gel can be imaged either by taking a relaxivity image ($R_2 = 1/T_2$) in an MR scanner (Figure 17.4) (Maryanski et al. 1996b) or by performing an optical CT scan (Gore et al. 1996; Kelly et al. 1998; Oldham et al.

---

* MGS Research Inc., PO Box 581, Guilford, Connecticut, 06437-0581 USA.

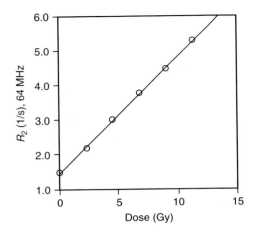

FIGURE 17.4
Dose response curve for BANG-2 gel irradiated with 6 MV x-rays. The relaxivity R_2 data are fitted with a linear function $R_2 = 1.45 + 0.34D$. (From Maryanski M. J. et al., *Med. Phys.*, 23, 699–705, 1996b. With permission.)

2001). A slice through the dose distribution delivered by a 3 isocentre radiosurgery treatment as imaged by an optical-CT scanner is illustrated in (Figure 17.5). PAG dosimetry is a relatively recent technique, and there is still much to understand concerning the theoretical relations between dose and polymerisation and the corresponding changes in R_2 or optical scattering (Lepage et al. 2001).

A relaxivity R_2 image of an irradiated dosimetric gel is converted to dose by application of a calibration curve (Oldham et al. 1998). Because of the difficulty of manufacturing gels with the same radiation sensitivity, the calibration gel must be from the same batch as the dosimetric gel, and the thermal and temporal histories should be as identical as possible. To date, most work on PAG gels has involved characterisation of performance with respect to composition and MR imaging (Maryanski et al. 1997; Baldock et al. 1998; De Deene et al. 1998; Mc Jury et al. 2000),

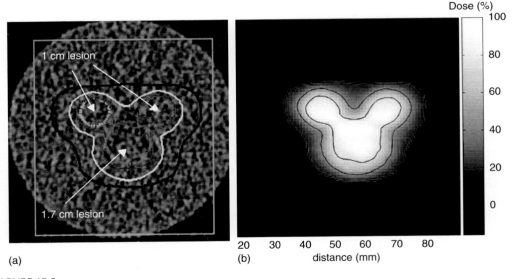

FIGURE 17.5
(See colour insert following page 590.) Illustration of the use of a gel-dosimeter in conjunction with optical-CT scanning to verify a 3 isocentre radiosurgery treatment. The left hand image is the dose-distribution calculated by the treatment planning system (isodoses 80%, 50% and 20%) superimposed onto the CT scan of the head-phantom. The right hand image is a dose map obtained from optical CT scanning through the central plane of the distribution (lines are also 80, 50 and 20%). (From Oldham, M. et al., *Med. Phys.*, 28,1436–1445, 2001. With permission.)

but the optical CT approach is attractive as it is more amenable for high resolution dosimetry and obviates the need for expensive MR access and imaging time (Oldham et al. 2001). PAGs have been applied to conformal therapy (De Deene et al. 1998), tomotherapy (Gore et al. 1996), brachytherapy (Mc Jury et al. 1999; De Deene et al. 2001) and radiosurgery (Oldham et al. 2001).

17.3.4 COMPARISON OF FRICKE WITH PAG GELS

PAG gels have both advantages and disadvantages when compared with Fricke gels. PAGs can be difficult to manufacture, sensitive to light, oxygen and thermal history, and are toxic*. Although a small long-term background post-irradiation polymerisation does occur with PAG gels (De Deene et al. 2000), they are much more stable than Fricke gels, allowing greater freedom to schedule imaging time. PAGs can measure doses in the range from 0 Gy to 10 Gy, whereas, Fricke gels are less sensitive with a useful range from 10 Gy to 40 Gy. The signal to noise ratio of Fricke gels is reduced by the intrinsically high electrical conductivity of the gel that attenuates the radiofrequency field. In contrast, PAGs show insignificant radiofrequency attenuation and a higher signal to noise ratio. The accuracy of relative dose measurements for a spatial resolution between 1 mm and 2 mm is reported as between 3% and 5% for PAG gels (Baldock et al. 1999; Mc Jury et al. 2000) and between 3% and 10% for Fricke gels (Hazle et al. 1991; Johansson et al. 1998). PAGs become increasingly opaque when irradiated, potentially allowing for optical CT scanning that would have major time and cost advantages (Oldham 2004).

17.3.5 PRACTICAL CONSIDERATIONS WHEN USING RADIATION-SENSITIVE GELS AS A DOSIMETER

(1) Energy response: both Fricke and PAG gels show no significant dependence of sensitivity (i.e. <1%) on radiation energy within the therapy range (Maryanski et al. 1996b; Schulz et al. 1990; Shani 1991).

(2) Dose rate: both Fricke and PAG gels show no dependence on dose-rate within the range normally produced by linear accelerators (Olsson et al. 1989; Schulz et al. 1990; Ibbott et al. 1997; Johanson et al. 1998; Shortt 2001). New formulations of BANG-3™ gels have been found to exhibit some non-linearity at high dose rates.

(3) Dose range: PAGs are currently available that measure maximum doses of anywhere between 2 Gy and 20 Gy. As Fricke gels are less sensitive, they are generally used to measure doses of between 25 Gy and 30 Gy.

(4) PAGs, prior to irradiation, may be sensitive to both the presence of light and oxygen. For this reason, if the gel container is made of oxygen porous material (e.g. polymethylmethacrylate [PMMA, Known as Perspex]) then it must be irradiated within a few hours after it has set or the dose-sensitivity will be reduced because of oxygen contamination. Containers can also be made from *Barex* (acrylonitrile) plastic.

(5) For absolute dosimetry measurements, each gel batch must be independently calibrated.

(6) MR imaging of the T_1 distribution (Fricke) or T_2 ($=1/R_2$) distribution (PAGs) is a complex process. Image quality and accuracy of measured dose values depend on MR sequencing parameters and techniques and local scanner characteristics (e.g. distortion). Careful selection of sequences and parameters and good understanding of the characteristics of the local MR machine are required (De Deene et al. 2001).

---

* A PAG formulation was recently proposed that is less sensitive to oxygen exposure during the manufacturing process, rendering the gels considerably easier to make (Fong et al. 2001).

CHAPTER 18

ABSOLUTE DOSE DETERMINATION UNDER REFERENCE CONDITIONS

*Pedro Andreo and Alan Nahum**

CONTENTS

---

* With contributions from David Thwaites, who wrote Section 18.8.

18.1 INTRODUCTION

This chapter is concerned with the accurate determination of the absorbed dose in gray at the reference depth in a water phantom. This is sometimes known as *calibrating the beam* and is expressed in terms of a number of monitor units per gray for a given standard field size, Source to Surface Distance (SSD) and beam quality. For a better understanding of some of the concepts discussed in this chapter, the reader should consult Part E on the general characteristics of photon and electrons beams.

Whereas it is possible to determine the absolute value of the absorbed dose for the various clinical beam qualities from first principles, this would be an elaborate procedure involving expensive instrumentation; this is both impractical and inappropriate in a radiotherapy clinic. In practice, a *code of practice* or *dosimetry protocol* is followed, based on the use of an ionisation chamber having a calibration factor, traceable either directly to a Primary Standard Dosimetry Laboratory (PSDL) or to a Secondary Standard Dosimetry Laboratory (SSDL), itself, in turn, linked to a PSDL. Codes of Practice are generally issued by national or regional organisations such as AAPM (North America), SEFM (Spain), or IPEM (U.K.), but they are also established by international bodies such as the IAEA. This procedure ensures a high level of consistency in dose determination between different radiotherapy clinics in a given country and also between one country and another. This chapter deals with the theory and practice involved in the various codes, but it is emphasised that in any given clinical situation one particular code should be followed. Therefore, comprehensive sets of data such as stopping-power ratios and chamber correction factors have deliberately not been given. The reader should refer to the relevant codes of practice (either national or international) for the methodology to be followed and the data to be used.

The emphasis here is on the principles of the codes of practice for the dosimetry of megavoltage photon (including ^{60}Co γ-ray) and electron beams discussed in Section 18.2 through Section 18.5. There are separate sections on kilovoltage x-rays (Section 18.6) and heavy charged particle beams (Section 18.7). Although the recommended reference medium is water, the use of other phantoms is discussed in Section 18.8. More details on each of the components involved in the dose determination can be found in Appendix D.

All codes of practice are based on the air-filled ionisation chamber as the dose measuring instrument (see Chapter 15). An ionisation chamber, often of a standard *reference* design, is provided with a calibration factor N_{S,Q_0} in terms of the radiation quantity S in a beam of a standard quality Q_0, and it is subsequently placed at the reference depth in water in the user's beam, of quality Q. A general form of the equation for calibration is then:

$$D_{w,Q} = M_Q \, N_{S,Q_0} \, f_{Q,Q_0}^{D,S} \qquad (18.1)$$

where M_Q is the instrument (i.e. electrometer) reading, suitably corrected (see Section 15.4) in the user's beam, and $f_{Q,Q_0}^{D,S}$ is any overall factor necessary to convert both from the calibration quantity S to dose D and from the calibration quality Q_0 to the user's quality Q.

Until relatively recently, all megavoltage photon and electron beam codes were based on calibrating an ionisation chamber in a ^{60}Co γ-ray beam[*] in terms of air kerma (e.g. IAEA 1987). This approach is still in use in many parts of the world, e.g. SEFM (1984, 1987), NCS (1986), IAEA (1987, 1997a, 1997b), IPEMB (1996a). However, in recent years there has been a concerted effort to move toward calibrations in terms of the quantity actually required, absorbed dose to water (e.g. IPSM 1990, AAPM 1999, IAEA 2000). The field is currently in a transition period, and consequently both approaches are described here: air kerma (see Section 18.2) and absorbed dose to water (see Section 18.3), as well as the relationship between the two (see Section 18.4) and quantities common to both approaches (see Section 18.5).

18.2 CODES OF PRACTICE BASED ON AIR-KERMA CALIBRATION

18.2.1 BACKGROUND

The ionisation chamber has been the instrument of choice for many years for the accurate determination of absorbed dose on account of its widespread availability, excellent reproducibility and stability. The long history of its use means that its properties are well understood (see Chapter 15). Originally, ionisation chambers were used for the dosimetry of kilovoltage x-ray beams, and Standards Dosimetry Laboratories (SDLs) provided calibrations in terms of exposure (charge per unit mass of air) over a range of radiation qualities, characterised by their half-value layer (HVL) in aluminium or copper (see Section 21.2.2). This method was subsequently extended to megavoltage photon beams (Greene and Massey 1966; ICRU 1969; HPA 1969) through the concept of the C_λ factor which corresponds exactly to $f_{Q,Q_0}^{D,S}$ in Equation 18.1, Q_0 here being ^{60}Co γ-rays.

The quantity *air kerma*, K_{air}, subsequently replaced *exposure* as the calibration quality realised at the Standards Dosimetry Laboratories; kerma has the same units as absorbed dose, i.e. gray (see Section 6.3.2).

The next two sections explain the principles of codes of practice based on air kerma calibration. The formalism, established by the Nordic Code (NACP 1980) and subsequently adopted by the IAEA (IAEA 1987, 1997a) and others (e.g. IPEMB 1996a), will follow.

There are three steps involved in an air-kerma based code: first, the determination of the N_K factor for the reference chamber at the SDL; second, the calculation of the $N_{D,air}$ factor; and finally, the absorbed dose to water determination in the user's beam. These steps are outlined in the following sections.

---

[*] In the U.K., a 2 MV Van der Graaf generator was used.

18.2.2   From Air Kerma to the N_K Factor

The starting point in the complete dosimetry chain is the realisation of air kerma at a specified point in a ^{60}Co γ-ray beam at the Standards Dosimetry Laboratory (e.g. Boutillon and Niatel 1973); this is illustrated on Figure 18.1. The so-called *primary standard chamber (a)* has an air volume known very precisely. Various correction factors are applied, principally for the attenuation and scattering by the chamber wall and also for its lack of air equivalence (Boutillon and Niatel 1973; Bielajew 1990; Bielajew and Rogers 1992). The Figure 18.1b illustrates the calibration of the *user's reference chamber*. For a chamber (including build-up cap) positioned with its centre at a point where the air kerma K_{air} is known and giving a reading (corrected to standard temperature and pressure) M_{Q_0}, the calibration factor, denoted by N_K, is given by

$$N_K = \frac{K_{air}}{M_{Q_0}} \qquad (18.2)$$

It is this factor that is supplied by the Standards Dosimetry Laboratory and that constitutes the first step in the procedure. It may be noted that for many years the National Physical Laboratory in the U.K. would only provide a N_K factor for the designated secondary standard chamber, the NE2611, including buildup cap[*]; the IPEMB (1996a) Code of Practice only considered this chamber. Generally, however, Standards Dosimetry Laboratories will supply any cylindrical chamber together with its appropriate build-up cap, such as a Farmer chamber (see Table 15.2), with an N_K factor.

18.2.3   From the N_K to the $N_{D,\text{AIR}}$ Factor

The second step in the procedure is the determination of the $N_{D,air}$ factor; this is defined by:

$$N_{D,air} = \frac{\bar{D}_{air}}{M_{Q_0}} \qquad (18.3)$$

where $\bar{D}_{air}$ is the dose to the air averaged over the chamber cavity. This is identical in concept to N_{gas} in AAPM (1983)[†].

$N_{D,air}$ is a calibration factor for the chamber to convert its corrected reading to mean absorbed dose in the sensitive air volume. Some insight into the nature of $N_{D,air}$ can be gained as follows. D_{air}, is equal to the product of the charge per unit mass, J_g, (known as the *mass ionisation*) and the mean energy required to produce an ion pair in air, divided by the electron charge, (W_{air}/e):

$$\bar{D}_{air} = J_g(W_{air}/e) \qquad (18.4)$$

which can be expressed in terms of the volume of the air cavity, V, the air density, ρ, and the charge measured, Q:

$$\bar{D}_{air} = \frac{Q}{\rho V}(W_{air}/e) \qquad (18.5)$$

Combining Equation 18.3 and Equation 18.5 leads to

$$N_{D,air} = \frac{Q}{M_{corr}} \frac{1}{\rho V}(W_{air}/e) \qquad (18.6)$$

---

[*] Previously, the NE2561, taken to be equivalent to the NE2611 in this chapter.
[†] $N_{D,air}$ was originally written as N_D, e.g. in NACP(1980) and IAEA (1987, 1997a), but *air* has recently been explicitly added (e.g. IPEMB 1996a; IAEA 1997b) in order to clearly distinguish it from the absorbed-dose-to-water factor $N_{D,w}$.

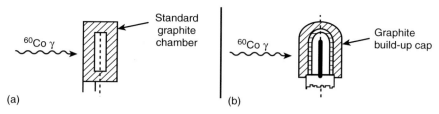

FIGURE 18.1
Air kerma determination (a) and dissemination (b) at a Primary Standard Dosimetry Laboratory (PSDL) in a ^{60}Co γ beam.

where M_{corr} is the reading (M_{Q_0}) corrected for recombination, polarity effect, temperature and pressure (see Section 15.4). Now, clearly the measured signal M_{corr} must be proportional to the charge Q, and the density ρ is a constant. Furthermore, there is very good evidence that (W_{air}/e) does not vary with radiation quality (except at electron energies below a few kiloelectron volts) (ICRU 1979). Consequently, $N_{D,air}$ is simply inversely proportional to the volume of the air in the cavity, V, i.e. it is simply related to a fixed geometrical property of the chamber and *will therefore be constant irrespective of the radiation quality.*

The air kerma standard depicted in Figure 18.1 is realised by using a thick-walled cavity chamber, i.e. full build-up is achieved in the chamber wall, and there is charged particle equilibrium (CPE) in the cavity gas (in photon beams of lower energies, CPE is achieved by the use of a free-air chamber, e.g. Attix 1986) (see Section 15.3). Ideally, the build-up material should be air-equivalent. Furthermore, it should not attenuate or scatter the incident beam. In practice, therefore, correction factors are required. Referring to the Figure 18.1b, the mean absorbed dose $\bar{D}_{air}$ in the air cavity of the user's ionisation chamber in a cobalt-60 beam is related to the air kerma (free in air), K_{air}, by

$$\bar{D}_{air} = K_{air}(1-g)k_{att}\, k_m\, k_{cel} \tag{18.7}$$

where g is the fraction of the initial kinetic energy of the secondary electrons generated by ^{60}Co photons that is converted to bremsstrahlung in air (see Section 6.3.2).

The correction factors k refer to the effect of the chamber on the radiation field (e.g. IAEA 1997a, 1997b and Appendix D, Section D.2):

- k_m takes into account the lack of air equivalence of the ionisation chamber wall materials and build-up cap at the calibration in the ^{60}Co γ-ray beam;
- k_{att} takes into account the attenuation and scatter of the photons in the ionisation chamber materials and build-up cap;
- k_{cel} takes into account the non-air equivalence of the central electrode of a cylindrical ionisation chamber.

Dividing both D_{air} and K_{air} by the corrected meter reading for the irradiation in a cobalt-60 beam (Figure 18.1), it follows from Equation 18.2 and Equation 18.3 that[*]

$$N_{D,air} = N_K(1-g)k_{att}\, k_m\, k_{cel} \tag{18.8}$$

---

[*] Note that the corresponding expression in IAEA (1987) did not include k_{cel} and, therefore, the N_D in that code of practice did not relate solely to the geometrical characteristics of the chamber (Equation 18.6). The factor k_{cel} was included in a global factor p_{cel} to account for the combined effect of the central electrode, both during the calibration of the chamber in air in a cobalt-60 beam and during subsequent measurements in photon and electron beams in a phantom. This was remedied in IAEA (1997b).

An expression for k_m for a single material, e.g. chamber wall and build-up cap both made of graphite, can be derived from cavity theory; k_m is then given by Equation 18.9 where $\alpha = 1$ (Bielajew 1986). However, in many cases, the cap and the wall materials differ, and the following approximate 2-component formula, based on similar ideas as for p_{wall} (see Appendix D, Section D.2.2), must be used:

$$k_m = \alpha\, s_{air,wall}(\mu_{en}/\rho)_{wall,air} + (1-\alpha)\, s_{air,cap}(\mu_{en}/\rho)_{cap,air} \tag{18.9}$$

where α is the fraction of ionisation because of electrons generated in the chamber wall.

This expression was evaluated for many materials of interest by Andreo et al. (1986) using a consistent set of stopping-power data. Values range from 0.965 to 1.006 for homogeneous cap and wall made of A-150 and of C-552 (approx. air equivalent) material, respectively.

The factor k_{att} has been obtained by Monte-Carlo simulation (Nath and Schulz 1981; Rogers et al. 1985) for a large number of chamber and cap combinations. Andreo et al. (1986) compared the product $k_m k_{att}$ for a number of chamber and cap combinations in common use with experimental values from Mattsson and Johansson (1984); the latter had been determined by using the all-graphite wall and cap *SSI* chamber (i.e. constructed by the Swedish National Radiation Protection Institute) as reference (using the theoretical value $k_m k_{att} = 0.989$) and then determining the chamber-volume ratio in a high-energy electron beam where the effect of the different walls was assumed to be negligible. The agreement between theory and experiment was generally within 0.7%. A comprehensive table of $k_m k_{att}$ is given in IAEA (1987) based on the data in Andreo et al. (1986).

The factor k_{cel} for cylindrical chambers with graphite and aluminium central electrodes was not explicitly included in IAEA (1987). Monte-Carlo simulations by Ma and Nahum (1993) showed that $k_{cel} = 1.006$ for both the Farmer NE2571 (solid Al central electrode) and the U.K. secondary standard NE2611 (hollow Al central electrode).

For a given ionisation chamber, Equation 18.6 shows that the $N_{D,air}$ factor is a constant provided that W_{air}/e does not vary with energy. Therefore, the $N_{D,air}$ factor derived at the cobalt-60 quality Q_0 is also valid at the user quality Q. The factor $N_{D,air}$ then allows the determination of the mean absorbed dose within the air cavity at the user beam quality Q

$$\bar{D}_{air,Q} = M_Q\, N_{D,air} \tag{18.10}$$

where M_Q is the corrected reading in that beam.

18.2.4 USING THE $N_{D,AIR}$ FACTOR TO DERIVE THE ABSORBED DOSE TO WATER

The third, and final, step in the determination of the absorbed dose to water at the reference depth in the user's beam is illustrated in Figure 18.2. The chamber is used without its build-up cap, but a thin waterproof sheath of polymethylmethacrylate (PMMA, known as perspex or lucite) or polystyrene is advisable for non-waterproof chambers.

The absorbed dose to water, $D_{w,Q}$, at a point in a phantom where the effective point of measurement of the chamber is positioned, is obtained from the dose to air ($M_Q N_{D,air}$) using the Bragg–Gray principle

$$D_{w,Q}(P_{eff}) = M_Q N_{D,air}(s_{w,air})_Q\, p_Q \tag{18.11}$$

where M_Q is the dosimeter reading at the beam quality Q corrected for influence quantities; $s_{w,air}$ is the stopping power ratio, water to air; p_Q is the overall perturbation factor of the ionisation chamber for in-phantom measurements at a beam quality Q; and P_{eff} is the effective point of

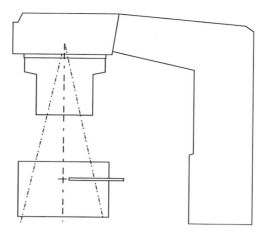

FIGURE 18.2
The basic Bragg–Gray situation of a thimble chamber in a phantom irradiated by a megavoltage photon or electron beam. The geometrical center C of the chamber cavity is aligned on the beam axis. The effective point of measurement P_{eff} is on the same axis, slightly displaced from C in the source direction.

measurement of the chamber which is generally not situated at the chamber centre* (see Appendix D, Section D.2.1). The effective point of measurement P_{eff} and the perturbation factor p_Q are required because the chamber does not behave as a perfect Bragg–Gray cavity, i.e. the electron fluence in the air cavity is *not* identical to that at the reference depth in the undisturbed medium. These two quantities are chamber specific. The remainder of any air-kerma based code of practice is essentially occupied with the correct choice of these quantities and that of the stopping-power ratio. These are dealt with briefly in Section 18.5 and in more detail in Appendix D.

In principle, there is no difference between the procedures for photon and electron beams as the Bragg–Gray principle can be applied equally to either modality (see Section 6.7.3). In practice, however, plane-parallel chambers tend to be preferred to cylindrical ones for electron beams (see Section 15.3.2.3 and Appendix D).

18.3 CODES OF PRACTICE BASED ON ABSORBED DOSE TO WATER CALIBRATIONS

18.3.1 INTRODUCTION

All codes of practice concern the determination of the quantity *absorbed dose to water*. It is, therefore, logical to provide ionisation chambers with a calibration factor in terms of this quantity, rather than in terms of *air kerma* if at all possible.

In 1988, the development of standards of absorbed dose to water at PSDLs became a major recommendation of the *Comité Consultatif pour les Etalons de Mesure des Rayonnements Ionisants (Section I—CCEMRI(I) 1988)*. Although only the water calorimeter allows the direct determination of the absorbed dose to water in a water phantom, the required conversion and perturbation factors for other procedures (the ionisation method, chemical dosimetry and

---

* It should be noted that an alternative procedure used in some codes of practice is to position the centre of the chamber at the point of interest and use a modified p_Q that also includes a displacement correction. In U.K. codes of practice the effective point of measurement correction is not explicitly included because the National Physical Laboratory includes this correction in $N_{D,w}$, effectively defining the effective point of measurement as the centre of the chamber.

graphite calorimetry) are now well known at many laboratories (ICRU 2001). Many PSDLs already provide $N_{D,w}$ calibrations for ^{60}Co γ-ray beams, and some laboratories have extended these calibration procedures to high-energy photon and electron beams (Burns 1994; McEwen et al. 1998) while others are developing the necessary techniques.

The procedure for the determination of absorbed dose to water based on standards of absorbed dose to water has been implemented in the national dosimetry recommendations by the IPSM (1990) based on calibrations over a range of megavoltage qualities, by DIN 6800-2 (1997) based on cobalt-60 calibrations and more recently, by the AAPM (1999) and the IAEA (2000).

The most significant advantages of calibrations in terms of absorbed dose to water and dosimetry procedures using these calibration factors can be summarised as follows:

- *Reduced uncertainty.* Measurements based on calibration in air in terms of air kerma require chamber-dependent conversion factors to determine absorbed dose to water (see Section 18.2). Conversion factors such as perturbation corrections and stopping-power ratios are derived from complex measurements or calculations based on theoretical models with uncertainties that are often difficult to estimate. In addition, they do not account for differences between individual chambers of a particular type. In contrast, $N_{D,w}$ calibrations can be performed under similar conditions to those in the user's beam, thus accounting for the response of each individual chamber.

- *A more robust system of primary standards.* Although international comparisons of air-kerma standards have shown very good agreement, they are all based on the use of ionisation chambers as illustrated in Figure 18.1, and are therefore subject to the same uncertainties. In contrast, primary standards of absorbed dose to water are based on a number of different physical principles with no assumptions or correction factors common to all of them. Therefore, the good agreement among these standards (see Figure 18.3) gives much greater confidence.

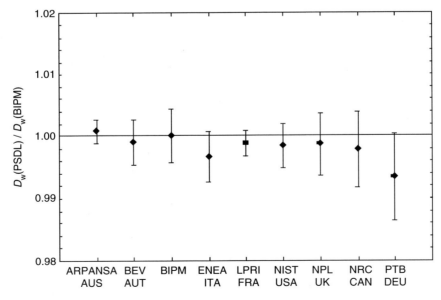

FIGURE 18.3

Comparisons of standards of absorbed dose to water in the BIPM cobalt-60 beam. The results are relative to the BIPM determination and are those for the most recent comparison for each PSDL, the oldest dating from 1989. The uncertainty bars represent the relative standard uncertainty of the D_w determination at each institute. Laboratories contributing were as follows: ARPANSA (Australian Radiation Protection and Nuclear Safety Agency); BEV (Bundesamt für Eich- und Vermessungswesen, Austria); BIPM (Bureau International des Poids et Mesures); ENEA (Italian National Agency for New Technologies, Energy and the Environment); LPRI (Laboratoire Primaire des Rayonnements Ionisants, France); NIST (National Institute of Standards and Technology, U.S.A.); NPL (National Physical Laboratory, U.K.); NRC (National Research Council, Canada); PTB (Physikalisch-Technische Bundesanstalt, Germany). (From Allisy-Roberts, P. J. and Burns, D. T., Report CCRI(I)/99-1, BIPM, Sèvres, France, 1999.)

- *Use of a simple formalism.* The formalism given in most air-kerma based dosimetry protocols is based on the application of several coefficients and correction factors to convert from air-kerma in cobalt-60 beam to in-phantom absorbed dose to water in the user's beam, i.e. Equation 18.8 and Equation 18.11. Information about physical characteristics of the chamber used, e.g. wall and build-up cap material, is also required. The simpler procedure starting from a calibration factor in terms of absorbed dose to water, and applying correction factors for all influence quantities, reduces the possibility of mistakes in the determination of absorbed dose to water (see Section 18.3.2).

18.3.2 $N_{D,\text{w}}$-BASED FORMALISM

The formalism used here follows that in IAEA (2000), originally developed by Hohlfeld (1988) and extended by Andreo (1992) and Rogers (1992a).

The absorbed dose to water at the reference depth z_{ref} in water for a reference beam of quality Q_0 and in the absence of the chamber is given by

$$D_{\text{w},Q_0} = M_{Q_0} N_{D,\text{w},Q_0} \qquad (18.12)$$

where M_{Q_0} is the reading of the dosimeter under the reference conditions used at the SDL, and N_{D,w,Q_0} is the calibration factor in terms of absorbed dose to water of the dosimeter obtained from the SDL. The chamber centre is positioned at the reference depth. When a dosimeter is used in a beam of the same quality as that used at its calibration, Q_0, the absorbed dose to water is simply given by Equation 18.12 where the dosimeter reading M_{Q_0} is corrected to the reference values of pressure and temperature (and other relevant influence quantities) at which the calibration factor is valid. In general, the calibration or reference quality is ^{60}Co γ-rays, i.e. the same as that for the air-kerma standard. Ideally, an SDL would possess a linear accelerator and be able to calibrate the user's chamber at all high-energy photon and electron qualities, Q, used in a radiotherapy clinic:

$$D_{\text{w},Q} = M_Q N_{D,\text{w},Q} \qquad (18.13)$$

where $N_{D,\text{w},Q}$ pertains to the user's quality.

18.3.2.1 *Correction for the Radiation Quality of the Beam, k_{Q,Q_0}*

The commonest situation, that of $N_{D,\text{w}}$ only being provided for a single reference quality, will now be dealt with. When a dosimeter is used in a beam of quality Q different from that used at its calibration, Q_0, the absorbed dose to water is given by

$$D_{\text{w},Q} = M_Q N_{D,\text{w},Q_0} k_{Q,Q_0} \qquad (18.14)$$

where the factor k_{Q,Q_0} corrects for the differences in chamber response between the reference beam quality Q_0 and the user quality Q. The beam-quality correction factor k_{Q,Q_0} is defined as the ratio of the calibration factors of the ionisation chamber, in terms of absorbed dose to water, at the qualities Q and Q_0,

$$k_{Q,Q_0} = \frac{N_{D,\text{w},Q}}{N_{D,\text{w},Q_0}} = \frac{D_{\text{w},Q}/M_Q}{D_{\text{w},Q_0}/M_{Q_0}} \qquad (18.15)$$

The most common reference quality Q_0 used for the calibration of ionisation chambers is ^{60}Co γ-radiation where the simplified symbol k_Q is commonly used for the beam-quality correction factor.

At some PSDLs, high-energy photon and electron beams are directly used for calibration purposes, and the symbol k_{Q,Q_0} is used in these cases with Q_0 to be specified. Ideally, the beam-quality correction factor should be measured directly for each chamber at the same quality as the user's beam. However, this is not possible in most standards laboratories. Such measurements can be performed only in laboratories with access to the appropriate beam qualities; for this reason, the technique is, at present, restricted to a few PSDLs in the world as the procedure also requires the availability of an energy-independent dosimetry system, such as a calorimeter, operating at these qualities.

When no experimental data are available or when it is difficult to directly measure k_{Q,Q_0} in clinical beams, the correction factors can, in many cases, be calculated theoretically. By comparing Equation 18.13 with the $N_{D,air}$ formalism given in Equation 18.6 and Equation 18.11, k_{Q,Q_0} can be written (Andreo 1992)

$$k_{Q,Q_0} = \frac{(s_{w,air})_Q}{(s_{w,air})_{Q_0}} \frac{(W_{air})_Q}{(W_{air})_{Q_0}} \frac{p_Q}{p_{Q_0}} \tag{18.16}$$

which is valid for all types of high-energy beams. It involves ratios, at the qualities Q and Q_0 respectively, of water/air stopping-power ratios, $s_{w,air}$ (see Section 18.5.1 and Appendix D, Section D.1) of the mean energy expended in air per ion pair formed, W_{air}, and of the perturbation factors p_Q and p_{Q_0} for departures from the ideal Bragg–Gray detector conditions (see Section 18.5.3 and Appendix D, Section D.2).

In therapeutic electron and photon beams, the assumption of $(W_{air})_Q = (W_{air})_{Q_0}$[*] yields the simpler equation for k_{Q,Q_0} (Hohlfeld 1988)[†]

$$k_{Q,Q_0} = \frac{(s_{w,air})_Q}{(s_{w,air})_{Q_0}} \frac{p_Q}{p_{Q_0}} \tag{18.17}$$

The calculations of k_{Q,Q_0} are based on exactly the same data that are used in the calculations in the air-kerma based approach (see Section 18.2), but the parameters ($s_{w,air}$ and p_Q in Equation 18.11) are used as ratios that have reduced uncertainties as compared to the individual values.

In the case of kilovoltage x-ray beams, Bragg–Gray conditions do not apply; therefore, Equation 18.17 cannot be used. In addition, the chamber-to-chamber variation in response is usually rather large. Kilovoltage absolute dose determination is covered in Section 18.6.

18.3.2.2 A Modified k_{Q,Q_0} for Electron Beams

For electron beams, when the calibration quality of the dosimeter Q_0 is cobalt-60 the situation is the same as discussed above, i.e. for the user's electron beam quality Q, the beam quality correction factor k_Q is given by Equation 18.16 (or Equation 18.17). The alternative is the direct calibration of chambers in electron beams (Thwaites et al. 2003) which is not presently offered by many PSDLs (McEwen et al. 1998). However, the ongoing development of electron-beam primary standards will enable calibration at a series of electron beam qualities Q. In such cases the equation is simply

$$D_{w,Q} = M_Q N_{D,w,Q} \tag{18.18}$$

A third possibility (which in the absence of direct calibration in electron beams is the

---

[*] Note that this is the same assumption as for the beam-quality independence of $N_{D,air}$ (see Equation 18.6).
[†] It is emphasised that when experimental and theoretical values of k_{Q,Q_0} are compared, it is strictly the full Equation 18.16 that is relevant rather than Equation 18.17. A possible weak energy variation of W_{air}, has been suggested (cf. Svensson and Brahme 1986).

preferred choice) in some $N_{D,w}$ based dosimetry protocols (AAPM 1999; IAEA 2000) is the cross calibration of a plane-parallel chamber against a calibrated cylindrical chamber in a high-energy electron beam of quality Q_{cross}. Well designed plane-parallel chambers are preferred in electron beams because of the negligible perturbation they exhibit as discussed in Section 15.3.2.3 and Section 18.5.3. The $N_{D,w}$ factor for the plane-parallel chamber is then given by

$$N^{pp}_{D,w,Q_{cross}} = \frac{M^{ref}_{Q_{cross}}}{M^{pp}_{Q_{cross}}} N^{ref}_{D,w,Q_0} k^{ref}_{Q_{cross},Q_0} \tag{18.19}$$

where $M^{ref}_{Q_{cross}}$ and $M^{pp}_{Q_{cross}}$ are dosimeter readings for the reference (cylindrical) chamber and the plane-parallel chamber, respectively and are corrected for influence quantities (temperature, pressure, etc.); *ref* and *pp* refer to the cylindrical and plane-parallel chambers, respectively.

Unfortunately, unless the quality Q_{cross} is specified beforehand, then for each chamber type a *two dimensional* table of $k_{Q,Q_{cross}}$ factors is required for the subsequent use of this chamber in an electron beam of quality Q. A practical solution, adopted both in AAPM (1999) and in IAEA (2000), is to present the required data in a single table by introducing an arbitrary *intermediate* electron beam quality Q_{int}*. The required $k_{Q,Q_{cross}}$ factor is evaluated from

$$k_{Q,Q_{cross}} = \frac{k_{Q,Q_{int}}}{k_{Q_{cross},Q_{int}}} \tag{18.20}$$

The factor $(k_{Q_{cross},Q_{int}})^{-1}$ converts the actual chamber calibration factor $N_{D,w,Q_{cross}}$ into a calibration factor that applies at the intermediate quality Q_{int}. The factor $k_{Q,Q_{int}}$ corrects this latter calibration factor into one that applies at Q so that the general Equation 18.13 for $D_{w,Q}$ can be applied. The expressions for $k_{Q,Q_{int}}$ and $k_{Q_{cross},Q_{int}}$ follow from Equation 18.17, and the stopping-power ratios and perturbation factors at Q_{int} can be eliminated†.

18.3.3 $N_{D,w}$ AND k_{Q,Q_0} OPTIONS AVAILABLE IN PRACTICE

As already indicated, many Standards Dosimetry Laboratories provide calibrations of ionisation chambers in terms of absorbed dose to water in a cobalt-60 γ-ray beam, and some laboratories have developed calibration procedures for high-energy photon and electron beams. The options available can be summarised as follows:

1. Users are provided with a calibration factor $N_{D,w}$ at a reference beam quality Q_0, usually cobalt-60. For additional qualities Q, the SDL provides directly measured beam-quality correction factors k_{Q,Q_0} *for that particular chamber*.

2. Users are provided with a series of $N_{D,w,Q}$ calibrations of the user's ionisation chamber at beam qualities Q.

Options 1 and 2 are very similar. The main advantage of either approach is that the individual chamber response in a water phantom irradiated by various beam types and qualities is intrinsically taken into account. For option 1, there is an advantage in presenting the data in the form of a single calibration factor N_{D,w,Q_0} together with directly measured values of k_{Q,Q_0}. Once directly measured values of k_{Q,Q_0} for a particular chamber have been

---

* No measurements are made at Q_{int}. It is only a device to simplify the presentation of the data.

† This approach has another advantage. For a given Q and Q_{cross}, the value for $k_{Q,Q_{cross}}$ is the same for all well guarded plane-parallel chamber types (as it only depends on a ratio of stopping-power ratios); for cylindrical chamber types, it depends also on the chamber radius r_{cyl}. The value chosen for Q_{int} (usually $R_{50} = 7.5$ g cm^{-2}) minimises the differences for cylindrical chambers of different r_{cyl} over the range of beam qualities for which cylindrical chambers are used.

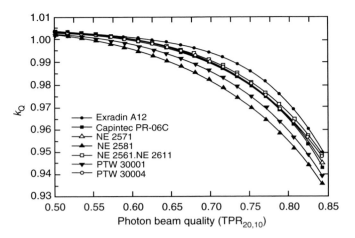

FIGURE 18.4
Calculated values of k_Q for various cylindrical ionisation chambers commonly used for reference dosimetry as a function of photon beam quality (TPR$_{20,10}$). Open symbols correspond to graphite-walled ionisation chambers, solid symbols to plastic-walled chambers. (From International Atomic Energy Agency, Technical Report Series no. 398, IAEA, Vienna, 2000. With permission.)

obtained, it may not be necessary for the user to re-calibrate the chamber at all qualities Q; it may be sufficient to do this only at the single reference quality Q_0; the full calibration at all qualities to verify the quality dependence of that chamber can be performed less often. Furthermore, the calibration at the single reference quality need to be performed at the laboratory where the k_{Q,Q_0} values were measured (usually a PSDL). A possible limitation, common to options 1 and 2, lies in the difference between the beam qualities used at the SDL and at the user's facility, which is of special relevance for high-energy beams (c.f. Andreo 1993) and which is still the subject of studies at some SDLs. These two approaches are the preferred alternatives as they are based on the individual calibration of the user's chamber. However, it is acknowledged that for beam qualities other than cobalt-60, such possibilities are currently restricted to a small number of PSDLs. Two further approaches are possible:

3. Users are provided with a N_{D,w,Q_0} calibration factor for the ionisation chamber at the reference quality cobalt-60, and the *code of practice* gives theoretically derived beam-quality correction factors k_{Q,Q_0} for other beam qualities Q *for that chamber type* using Equation 18.17. This method, however, ignores chamber-to-chamber variations in response with energy for a given chamber type, and the calculations rely on chamber specifications provided by manufacturers. Nevertheless, this approach is currently the most common one.

4. Users are provided with a single measured N_{D,w,Q_0} for a given chamber, obtained at a selected reference quality, together with *generic** experimental values of k_{Q,Q_0} for that ionisation chamber type. In common with 3, this option does not take into account possible chamber-to-chamber variations for a given chamber type. IAEA (2000) recommends that such generic experimental k_Q or k_{Q,Q_0} values should be determined at an SDL.

Figure 18.4 shows calculated k_Q for a number of cylindrical chambers in common use. Figure 18.5 shows a comparison between measured and calculated k_Q.

---

* In the present context, *generic* means factors common to a specific ion-chamber type, in a specific form, as constructed by a given manufacturer.

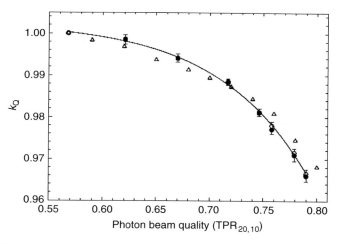

FIGURE 18.5

Comparison of calculated values of k_Q (triangles) and mean values of k_Q measured at the NPL (Sharpe 1999) for Secondary Standard ionisation chambers of the type NE 2561 (open circles) and NE 2611 (filled circles). The solid line is a sigmoidal fit to the experimental data. The uncertainty bars represent chamber-to-chamber variations, determined as the standard deviations of samples of 13 NE 2561 (upper half of the error bar) and 11 NE 2611 (lower half of the error bar) chambers. The values of k_Q are normalized to a TPR$_{20,10}$ of 0.568 (cobalt-60 beam at the NPL). Note that the calculated values do not distinguish between the two types of chamber. (From International Atomic Energy Agency, Technical Report Series no. 398, IAEA, Vienna, 2000. With permission.)

18.4 INTER-RELATIONSHIP BETWEEN N_K-$N_{D,\mathrm{AIR}}$ AND $N_{D,\mathrm{W}}$ CODES OF PRACTICE

In this section, the inter-relationship between the air-kerma and the absorbed-dose-to-water formalisms is discussed. In comparing Equation 18.11 and Equation 18.12 for the same reference beam quality Q_0, it follows that

$$N_{D,\mathrm{w},Q_0} = N_{D,\mathrm{air}}(s_{\mathrm{w,air}})_{Q_0}\, p_{Q_0} \qquad (18.21)$$

or in expanded form,

$$N_{D,\mathrm{w},Q_0} = [N_K(1-g)k_{\mathrm{att}}k_{\mathrm{m}}k_{\mathrm{cel}}]_{^{60}\mathrm{Co}}(s_{\mathrm{w,air}})_{Q_0}[p_{\mathrm{cav}}\,p_{\mathrm{dis}}\,p_{\mathrm{wall}}\,p_{\mathrm{cel}}]_{Q_0} \qquad (18.22)$$

where Q_0 usually refers to cobalt-60 γ-rays. The separate components of the overall perturbation factor p_{Q_0} shown in Equation 18.22 are listed in Section 18.5.3 and discussed in Appendix D, Section D.2.

The adoption of absorbed-dose-to-water-based codes of practice inevitably introduces small differences in the value of the absorbed dose to water determined in clinical beams compared with air-kerma based codes. However, any conclusions drawn from comparisons between these two types of protocols must take account of differences between primary standards. For a given primary standard, the results of a comparison will depend on the type and quality of the beam and on the type of ionisation chamber.

For a cobalt-60 beam, which is generally better characterised than other modalities, determination of D_{w} based on the two different types of protocol differs by typically 1%. Figure 18.6 shows the ratio of absorbed dose to water in cobalt-60 determined using $N_{D,\mathrm{w}}$ to that using N_K following the IAEA TRS-277 *Code of Practice* (IAEA 1987) for a number of individual ionisation chambers of various types commonly used in clinical dosimetry. Discrepancies of this order can be expected when $N_{D,\mathrm{w}}$ and N_K calibrations, traceable to the BIPM and to most PSDLs, are used in hospitals and SSDLs. The change may be greater or smaller when calibrations are traceable to laboratories lying at the extremes of the distributions shown

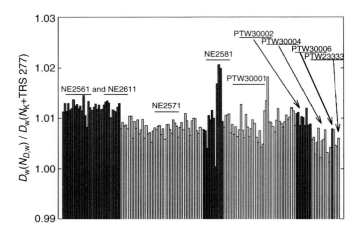

FIGURE 18.6
The ratio of the absorbed dose to water in cobalt-60 determined with calibration factors in terms of absorbed dose to water, $N_{D,w}$ (IAEA 2000), to the absorbed dose determined with calibration factors in terms of air kerma, N_K (IAEA 1987), for a large number of cylindrical ionisation chambers commonly used in radiotherapy dosimetry. Both calibration factors are traceable to the BIPM. Results for a given chamber type show chamber-to-chamber variations. The large variation for NE 2581 chambers is considered to be caused by the hygroscopic properties of the A150 plastic walls. (Reproduced from International Atomic Energy Agency, Technical Report Series no. 398, IAEA, Vienna, 2000. With permission.)

in Figure 18.6. Any systematic discrepancy between the two methods is probably due to inaccuracies in the numerical factors and expressions (for example, k_m, p_{wall} etc.) used in the N_K-based formalism; in addition, there is the possibility of a systematic error in the air-kerma primary standards (Bielajew and Rogers 1992).

In the case of high-energy photon and electron beam calibrations, only the situations involving theoretically derived values of k_Q (category 3 in Section 18.3.3) are discussed here. The change in D_w at cobalt-60 is simply propagated through to high-energy beams as most of the coefficients and factors involved in the calculation of k_Q factors are approximately the same as in previous protocols, and no other differences are expected. For electron beams, in addition to the propagated change in D_w at cobalt-60 quality, the second largest contribution will be due to the implementation of stopping-power ratios $s_{w, air}^{SA}$ for *clinical* beam qualities, as the basic data (i.e. $s_{w, air}^{SA}$ for mono-energetic beams) has remained practically the same (see Section 18.5.1 and Appendix D, Section D.1); this will result in changes of the order of 0.5%. As with cobalt-60, the differences lie within the combined standard uncertainty of the two types of codes of practice (see Appendix D, Section D.5).

18.5 QUANTITIES COMMON TO BOTH APPROACHES

The air-kerma based formalism for the determination of absorbed dose to water in reference conditions presented in Section 18.2 includes stopping-power ratios and correction factors for so-called perturbation effects, the latter being detector-dependent. Some of the analytical models employed for the calculation of perturbation correction factors also include mass energy-absorption coefficient ratios. Concerning the absorbed-dose-to-water formalism presented in Section 18.3, the approach most commonly used today relies on theoretically determined beam quality factors k_Q that are also based on stopping-power ratios and perturbation correction factors. These latter two quantities, common to the two formalisms, are briefly discussed here. The stopping-power ratios are discussed in Section 18.5.1, the mass energy-absorption coefficient ratios in section 18.5.2, and the perturbation factors in Section 18.5.3. Another important quantity common to both approaches is the mean energy required to produce an ion pair in air, divided by the electron charge, (W_{air}/e), which is dealt with in Section 18.5.4. Finally, in order to be able to select the appropriate values of stopping-power ratio and the perturbation factors, the user's beam quality must be known. Therefore, beam quality specification, a crucial component of any code of practice, is covered in Section 18.5.5. More detailed treatments of each of these components are given in Appendix D.

18.5.1 Stopping-Power Ratios

18.5.1.1 General

The determination of the absorbed dose in a medium using an ionisation chamber is based on the Bragg–Gray principle (see Section 6.7.3 and Section 18.2.4) which relates the absorbed dose at a point in the medium (water), D_w, to the mean absorbed dose in the detector (air), $\bar{D}_{air}$, through the mean (collision) stopping power ratio, water/air

$$D_w = \bar{D}_{air} s_{w,air} \tag{18.23}$$

The key Bragg–Gray assumption is that the electron fluence present in the detector is identical to that in the (undisturbed) medium at the position of interest (see Section 6.7.3). The gas-filled ionisation chamber behaves to a good approximation as a Bragg–Gray detector in megavoltage photon or electron beams. Any deviations from perfect Bragg–Gray behaviour are corrected for by perturbation factors which are the subject of Section 18.5.3 and Appendix D, Section D.2.

The stopping-power ratio applies to the electron spectrum at the point of interest in the *undisturbed* medium and is independent of the detector*. The determination of this stopping-power ratio is the task of cavity theory and is described in Appendix D, Section D.1.

18.5.1.2 Stopping-Power Ratios for Electron Beams

The most important characteristic of electron stopping-power ratios is their strong dependence on energy and depth (see Figure 18.7), which is due to a combination of the considerable differences in energy spectra at the various depths and the strong energy dependence of the quantity $[S_{col}(E)/\rho]_{water}/[S_{col}(E)/\rho]_{air}$ in the megavoltage energy range†.

The selection of the appropriate stopping-power ratio for the user's beam at a particular depth (e.g. $(s_{w,air})_Q$ in Equation 18.11) is dealt with in Appendix D, Section D.1.2.

18.5.1.3 Stopping-Power Ratios for Photon Beams

Stopping-power ratios in high-energy beams are calculated in the same way as for high-energy electron beams, i.e. by averaging $[S_{col}(E)/\rho]_{water}/[S_{col}(E)/\rho]_{air}$ over the spectrum of (secondary) electrons present at the depth of interest. In contrast to the electron-beam case, the depth dependence of $s_{w,air}$ for photon beams is negligible except possibly in the build-up region close to the surface; this lack of dependence on depth is essentially due to the establishment of quasi charged-particle equilibrium beyond the depth of maximum dose. Therefore, only one value of $s_{w,air}$ per beam quality is required. A much more detailed treatment is given in Appendix D, Section D.1.3.

Figure 18.8 shows the variation of $s_{w,air}$ with photon beam quality, characterised here by the widely established parameter $TPR_{20,10}$ (see Section 18.5.5 and Appendix D, Section D.3.1);

---

* There is a minor dependence on air-cavity size expressed through the cutoff Δ in the Spencer-Attix stopping-power ratio.
† This strong dependence on electron kinetic energy E is, in turn, due to the influence on $[S_{col}(E)/\rho]_{water}$ of the density or polarisation effect at relativistic energies (see Section 3.2.3).

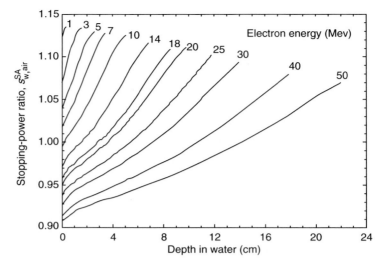

FIGURE 18.7
Depth variation of the Spencer-Attix water/air stopping-power ratio, $s_{w,air}^{SA}$, for $\Delta = 10$ keV, derived from Monte-Carlo generated electron spectra for broad, monoenergetic, plane-parallel electron beams. (From Andreo, P., *Nucl. Instr. Meth. B*, 51, 107–121, 1990; and International Atomic Energy Agency, Technical Report Series no. 381, IAEA, Vienna, 1997. With permission.)

low values of $TPR_{20,10}$ correspond to low photon (megavoltage) energies and vice versa. One can see from Figure 18.8 that the total variation in $s_{w,air}$ over the quality range from ^{60}Co γ-rays to 50 MV x-ray beams is only around 6%, i.e. much less than that in electron beams (Figure 18.7).

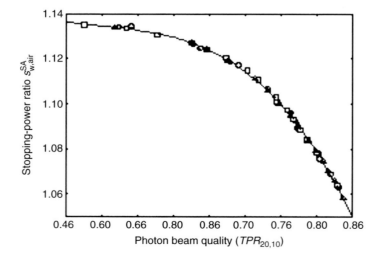

FIGURE 18.8
Spencer-Attix ($\Delta = 10$ keV) water/air stopping-power ratios at the reference depth for clinical photon beams versus the photon beam quality specifier $TPR_{20,10}$. Circles correspond to spectra published by different authors (Table 2 in Andreo and Brahme 1986), and squares represent the calculated spectra in the same reference. The solid line is a fit of the stopping-power ratios to the beam quality specifier $TPR_{20,10}$. (From Andreo, P., *Med. Phys.*, 27, 434–440, 2000.)

18.5.2 MASS ENERGY ABSORPTION COEFFICIENT RATIOS

The role of mass energy-absorption coefficient ratios in modern dosimetry protocols is mainly restricted to their use in calculating perturbation (principally p_{wall}) and other correction factors for ionisation chambers in cobalt-60 γ-ray and high-energy photon beams*. In general they are associated with the fraction of energy deposited within a detector due to electrons generated by photon interactions in the detector material itself (i.e. the large cavity component—see Section 6.7.2).

Depending on the medium, the photon fluence spectrum may change appreciably with depth or material thickness, and it also depends on the field size of the incident beam. Cunningham et al. (1986) investigated the effects of spectral changes within a phantom on the mean mass energy-absorption coefficients using Monte-Carlo techniques. They showed that only for large field sizes or for low-energy photon beams is there more than a 0.5% variation in $(\bar{\mu}_{en}/\rho)_{w,m}$, the mass energy absorption coefficient ratio of water, w, to medium, m, for tissue-like materials.

Spectrum-averaged mass energy-absorption coefficient ratios are calculated according to

$$(\bar{\mu}_{en}/\rho)_{w,m} = \frac{\int_0^{E_{max}} \Phi_{E,w}(\mu_{en}(E)/\rho)_w E\, dE}{\int_0^{E_{max}} \Phi_{E,w}(\mu_{en}(E)/\rho)_m E\, dE} \tag{18.24}$$

where $\Phi_{E,w}$ is the photon fluence, differential in energy at the point of measurement, assumed to be the same both in water, w, and in material, m; and $\mu_{en}(E)/\rho$ is the mass energy-absorption coefficient at energy E for each of the two materials respectively (see Section 6.7.2). For materials other than water, a similar relationship is applied.

The same set of bremsstrahlung spectra used by Andreo and Brahme (1986) to calculate stopping-power ratios has been used as input for calculations of μ_{en}/ρ-ratios by Cunningham (c.f. IAEA 1987) to produce a consistent set of data for photon dosimetry that is used in most dosimetry protocols.

18.5.3 PERTURBATION CORRECTION FACTORS

For a detector to behave as a Bragg–Gray cavity, the electron fluence in the sensitive medium of the detector must be identical to that at a specified point in the uniform medium. The only possible true Bragg–Gray detector would be a small air bubble; all the codes of practice for absolute dose determination are, in fact, based on air-filled ionisation chambers.

For megavoltage photon radiation, the Bragg–Gray conditions (see Section 6.7.3 and Section 15.3.2.1) are adequately fulfilled for air cavities of the size encountered in practical ionisation chambers, i.e. the ranges in air of the secondary electrons generated in megavoltage photon beams are much greater than the cavity dimensions (Ma and Nahum 1991). However, as Figure 18.9 illustrates, an ionisation chamber does not only consist of an air cavity. There will always be a wall that, in general, will not be perfectly medium-equivalent (and is often made of graphite), whereas, the medium is usually water. For cylindrical chambers there is a central electrode which is frequently aluminium. There may be other materials around

---

* The mass energy-absorption coefficients play a key role in photon dosimetry with so-called thick-walled chambers where the thickness of the chamber wall (or wall plus build-up cap) is greater than the range of the photon-generated secondary electrons (see Appendix D, Section D.2.2). Under the assumption that secondary charged-particle equilibrium exists (e.g. in the chamber wall), the absorbed dose to a medium, D_m, for a given photon fluence differential in energy Φ_E is given by (Section 6.4.2) $D_m = \int \Phi_E(\mu_{en}/\rho)_m E dE$.

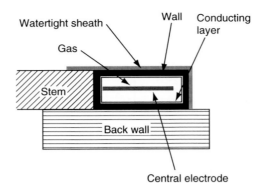

FIGURE 18.9
An imaginary ionisation chamber. (From Nahum, A. E., *Phys. Med. Biol.*, 41, 1531–1580, 1996. With permission.)

the chamber such as a stem for cylindrical chambers and a back wall in the case of plane-parallel designs. All of these features can introduce deviations from perfect Bragg–Gray behaviour.

These deviations are generally dealt with by introducing one or more correction factors, often known as *perturbation* factors, into the expression for the absorbed dose, i.e. the factor p_Q in Equation 18.11. This overall factor is often written as the product of a number of perturbation factors, each one accounting for a different effect, assumed to be independent of the others (Equation 18.22). The word *perturbation* is used in the sense of a perturbation by the detector of the electron fluence present at the position of interest P in uniform medium, $\Phi_{\mathrm{med}}(P)$. Naturally, it is the *fluence in the detector*, inevitably a mean value over a finite volume, $\bar{\Phi}_{\mathrm{det}}$, that gives rise to the signal, i.e. the fluence in the air in the case of an ionisation chamber. Therefore, one can define the overall perturbation factor p with complete generality through (Nahum 1996)

$$\Phi_{\mathrm{med}}(P) = \bar{\Phi}_{\mathrm{det}}\, p \qquad (18.25)$$

The different components of the overall perturbation factor are the following:

- p_{dis} accounts for the displacement of a certain volume of medium by the air of the cavity, and the corresponding modification of the beam attenuation at the measuring point assumed to be located at the centre of the cavity. The displacement effect is most often treated by referring the measured dose to an effective point of measurement, P_{eff}, which is shifted toward the source of radiation by a distance expressed as a fraction of the detector radius, r, for cylindrical chambers (e.g. 0.6 r) or located on the inside of the entrance wall for plane-parallel chambers

- p_{wall} accounts for the fact that the nature and composition of the chamber wall, which is responsible for a large proportion of secondary electrons contributing to the chamber signal, is different from water. This effect is considered to be negligible in electron beams

- p_{cel} accounts for the lack of *air* equivalence of the central electrode of cylindrical chambers

- p_{cav} accounts for the perturbation of the electron fluence in the air of the cavity; this effect is important for electron beams but is negligible for photon beams

The values of the perturbation factors which should be used in practical dose determination depend on the nature and quality of the radiation and on the characteristics of the detector. They are given in each code of practice. More details on the methods that have been used to obtain these factors and their order of magnitude can be found in Appendix D, Section D.2. The general conclusions are presented here.

In both photon and electron beams, account must be taken of the displacement effect. This is most conveniently done by assigning the chamber reading to an effective point of measurement P_{eff}. For photon beams, the effect of the chamber wall is important, and an approximate formula for p_{wall} is generally used in air-kerma based codes of practice (e.g. AAPM 1983; IAEA 1987, 1997a); this is acceptably accurate for cylindrical, but not for plane-parallel chambers. For electron beams, the *in-scattering* effect is important for cylindrical chambers at energies E_z below around 10 MeV and experimentally derived values for p_{cav} are given in the codes. Plane-parallel chambers can be designed which minimize this perturbation. Wall effects are generally unimportant in electron beams, and *in-scattering* is unimportant in photon beams. The effect of the central electrode in cylindrical chambers can be separately taken into account through p_{cel}, which is significantly different from unity for aluminium central electrodes in low-energy photon beams.

18.5.4   Mean Energy Expended in Air per Ion Formed, (W_{air}/e)

It is generally assumed that a constant value of W_{air} can be used for the complete photon and electron energy range used in radiotherapy dosimetry, and therefore, $(W_{air})_Q = (W_{air})_{Q_0}$. There is a lot of indirect support for such an assumption. However, the data available has been obtained from measurements at a narrow range of photon qualities, ^{60}Co and ^{137}Cs γ-ray beams and 2 MV x-rays. The value $(W_{air}/e) = 33.85 \pm 0.15$ J/C, recommended by ICRU report 31 (1979), was derived from a weighted mean value of the available experimental data, mainly obtained from absorbed dose measurements that use a graphite calorimeter and a graphite ionisation chamber in a graphite phantom where

$$(W_{air}/e) = \frac{D_{calorimetry}}{J_{air}\, s_{graphite,air}} \qquad (18.26)$$

This method of evaluation requires a change in (W_{air}/e) if the stopping-power ratio $s_{graphite,air}$ is changed (Andreo et al. 1987). Following the introduction of new electron stopping power data by ICRU (1984b), the value of (W_{air}/e) has been modified by the CCEMRI(I) (1985) to $(W_{air}/e) = 33.97 \pm 0.06$ J/C, based on the work by Boutillon and Perroche-Roux (1987).

Svensson and Brahme (1986) analysed the then available experimental data at higher energies, mainly for electron beams, and concluded that a small but non-negligible energy dependence in (W_{air}/e) could not be excluded, but experimental uncertainties and the use of different stopping-power ratios over the years do not allow a definitive conclusion to be reached.

18.5.5   Beam Quality Specification

A key issue in practical dosimetry is how to characterise the beam qualities available in radiotherapy clinics so that appropriate values of the energy-dependent stopping-power ratios and perturbation factors can be chosen. Ideally, these quantities should be available for the user beam and detector configuration in the measurement conditions at the particular hospital. In the near future, however, Monte-Carlo calculations may become available for every possible user configuration; the only option widely available today is to correlate the relevant quantities to beam-quality specifiers or indices related to the penetration properties of clinical beams. The most common methods of quality specification used in the various codes of practice are the following:

- For megavoltage photon beams, instead of specifying the quality in terms of *nominal energy* (in MV), it is clearly recognised that it is much better to use the beam *attenuation* in water[*] as a means to select the best values for $s_{w,air}$ and for the perturbation effects. This attenuation is generally expressed by the quantity $TPR_{20,10}$ defined as the ratio of absorbed dose to water on the beam axis at the depths of 20 cm and 10 cm in a water phantom, obtained with a constant source-detector distance and a $10 \text{ cm} \times 10 \text{ cm}$ field size at the position of the detector.

- For electron beams, the indices most commonly used are the mean energy at the surface, $\bar{E}_0$, and the mean energy at depth z, $\bar{E}_z$, calculated from the ranges obtained from depth dose curves (see Section 24.2) from the following relationships:

$$\bar{E}_0[\text{MeV}] = C R_{50} \tag{18.27}$$

where $C = 2.33 \text{ MeV cm}^{-1}$ and R_{50} is the depth where the absorbed dose has decreased to 50% of its maximum value and is expressed in centimetres.

$$\bar{E}_z = \bar{E}_0(1 - z/R_p) \tag{18.28}$$

where R_p is the practical range defined as the depth where the tangent at the steepest point (the inflection point) on the almost straight descending portion of the depth–dose curve meets the extrapolated bremsstrahlung background.

More details on the specification of radiation quality can be found in Appendix D, Section D.3.

18.6 KILOVOLTAGE X-RAY BEAMS

18.6.1 INTRODUCTION

During the past 30 years or so, there has been a great deal of development in the dosimetry of high-energy, i.e. megavoltage, photon and electron beams as the above sections clearly demonstrate. However, the dosimetry of low-energy or *orthovoltage* (below 400 kV) x-ray beams (referred to in what follows as *kilovoltage* and described in Chapter 21) remained more or less static until the late 1980s. As an example, in the U.K. Code of Practice for photon dosimetry (HPA 1983) so-called *medium-energy x-radiation* was confined to a short appendix. Despite this, kilovoltage beams are still in widespread use for the treatment of superficial lesions (Ma et al. 1996; Poen 1999). Furthermore, radiobiologists have continued to use such qualities in cell-irradiation experiments, and the validity of RBE values, for example, depend on the ability to determine the dose accurately in kilovoltage x-ray beams, often in much more complicated geometries than are encountered in clinical radiotherapy. The IAEA (1987) Code of Practice devoted a separate, detailed section to kilovoltage x-rays, setting out a new air-kerma based formalism, and this was followed by national codes on kilovoltage x-ray dosimetry (DIN 1988, 1996; IPEMB 1996b; NCS 1997; Ma et al. 2001 for the AAPM).

The emphasis here is on *air-kerma* based dosimetry. However, with kilovoltage x-rays, as with all other external-beam radiotherapy modalities, there is a move toward absorbed-dose-to-water standards (see Section 18.6.7).

---

[*] This concept is somewhat similar to the Half Value Layer approach for kilovoltage x-rays, but water is preferred to layers of metal for megavoltage beams because the variation of the attenuation coefficient as a function of the energy is more rapid, and therefore, yields a more accurate index.

18.6.2 Some Basic Physics

The dosimetry of kilovoltage x-rays was originally based on the concept of treating the dosimeter as an *exposure meter* (e.g. Attix 1986), i.e. the chamber was calibrated in terms of exposure at the quality of interest, and its reading in the phantom was interpreted as measuring exposure at the chamber centre. Air kerma has now replaced exposure, but the concept remains identical. The original kV dosimetry formalism involved so-called F-factors that converted from roentgen, the unit of exposure, to rad, the old unit of dose (1 Gy = 100 rads); F is essentially the product of $(\mu/\rho)_{w,air}$ and (W_{air}/e) (see, for example, Nahum 1999b). This *exposure meter* approach was carried over into megavoltage photon dosimetry through the C_λ concept (e.g. ICRU 1969) though it was subsequently replaced by the $N_{D,air}$ formalism (see Section 18.2) which is conceptually much clearer because the ionisation chamber can be explicitly treated as a Bragg–Gray cavity.

In the case of kilovoltage x-rays, however, the secondary electrons have extremely short ranges because of their much lower initial energy coupled with the rapid increase of the collision stopping power at sub-relativistic energies (see Section 3.2). The Bragg–Gray principle (Section 6.7.3) can no longer be applied to such qualities, i.e. the electron fluence in the air cavity of the ionisation chamber is not exclusively determined by the electron interactions in the surrounding medium. This had been understood for many years; Ma and Nahum (1991) used Monte-Carlo simulation to estimate the fraction, F_{air}, of the ionisation in an air cavity that originated from photon interactions in the air itself. Their results are given in Table 18.1.

The numbers in this table show that at 240 kV, almost 30% of the signal in the chamber comes from photon interactions in the air, compared to less than 1% for cobalt-60 γ-rays and above. Clearly, one cannot base a formalism for kilovoltage x-ray dosimetry on Bragg–Gray assumptions and therefore on stopping-power ratios. However, if it is a disadvantage that ionisation chambers cannot be treated as Bragg–Gray cavities in kilovoltage x-ray beams, then one has the compensating advantage that dose can be equated to kerma to a very good approximation, and also g, the fraction of kerma re-radiated as bremsstrahlung, can be assumed to be zero (see Section 6.3.2).

Another important difference between kilovoltage and megavoltage x-ray dosimetry lies in the nature of the calibration factor. Air-kerma based megavoltage dosimetry (see Section 18.2) employs the factor N_K which is *realised* only at one quality, ^{60}Co γ-rays. The reason for this is that it is increasingly difficult to determine air kerma (originally exposure) for any higher photon energies because of the increase in electron ranges, making the establishment of charged particle equilibrium (CPE) virtually impossible; very thick build-up caps would be required with correspondingly large correction factors (k_m, k_{att}, etc.). Then, as has been

---

TABLE 18.1

Origin of the Ionisation in a Cavity for a Range of Photon Energies

Incident Beam		F_{air}	
Tube Potential	HVL (mm Al)	In Vacuum	5 cm Depth in Water
50 kV	1.62	0.26	0.18
150 kV	5.01	0.27	0.29
240 kV	7.30	0.27	0.27
240 kV	17.4	0.18	0.23
Cobalt-60		0.0037	0.0060
4 MV		0.0014	0.0098

F_{air}, is the ratio of the ionisation due to photon interactions in the air itself to the total ionisation measured in the air of the cavity for a range of photon beam qualities. The cavity size is 6 mm in thickness and diameter. The kilovoltage photon beams are characterized by both tube potential and first half-value layer (HVL).
Source: From Ma, C.-M. and Nahum, A. E., *Phys. Med. Biol.*, 36, 413–428, 1991. With permission.

mentioned in earlier sections, stopping-power ratios are required to convert from cobalt-60 to other qualities, based on Bragg–Gray principles. However, at kilovoltage qualities, air kerma can *always* be realised; CPE can be established as the electron ranges are so much shorter (see Table 6.1 in Section 6.5). In fact, in kilovoltage beams, air kerma is determined at PSDLs using free-air chambers (see Section 15.3.1) where CPE is actually established in the air itself, rather than in the thick wall of a cavity chamber as it is in a cobalt-60 beam (as depicted in Figure 18.1). Consequently, SDLs provide N_K for the user's chamber over a range of kilovoltage photon qualities.

A third difference lies in the way that kilovoltage beam quality is specified. This is done in terms of the *Half-Value Layer* (HVL, see Section 21.2.2), generally expressed in millimetres of aluminium or, at the top end of the energy range copper. The HVL is that thickness of material that reduces the air kerma to half its value. It must be measured in narrow-beam geometry such that the influence of photon scatter generated in the attenuating material is negligible (see Section 21.2.2). It should be noted that beams with widely differing generating potentials can have similar HVL because of the marked effect of different filtrations (e.g. Knight and Nahum 1994 and Table 21.1). Therefore, the user must determine the HVL of the beam of interest and then choose N_K values for the calibrated chamber for that beam using the calibration curve supplied by the SDL.

18.6.3 BACKGROUND TO AIR-KERMA BASED CODES OF PRACTICE

In the case of kilovoltage x-rays, codes of practice generally recommend two entirely separate methods: one to be used at *medium* energies that involves placing the calibrated chamber in a water phantom at a specified depth (as for megavoltage beams), and the other to be used at *low* energies that is based on a measurement *free-in-air*, i.e. with no phantom involved at all. In the German and U.K. protocols, there is even a third method at *very low* energies.

The IAEA (1987) Code of Practice established an air-kerma based formalism for kilovoltage x-rays with clear expressions containing all the factors. Furthermore, IAEA (1987) also contained new values of the backscatter coefficient B and the correction factors for the in-phantom ionisation chamber at medium energies; previously, the latter had been implicitly assumed to be equal to unity in the exposure-based codes of practice (ICRU 1973; HPA 1983).

If one compares the dose derived following the IAEA (1987) Code of Practice to that derived following the ICRU (1973) Code or equally well the HPA (1983) Code, one finds a difference of nearly 10% in the dose at 5 cm depth in water at ≈ 100 kV. For a beam with an HVL of 0.44 mm of Al and a 10 cm diameter field the difference is nearly 6% in the dose at the surface. These differences, which were due to an error in IAEA (1987) (see Section 18.6.5), were largely responsible for a *renaissance* in the dosimetry of low-energy x-rays. New national protocols specifically on kilovoltage x-ray dosimetry have appeared (DIN 1988, 1996; IPEMB 1996b; NCS1997; Ma et al. 2001 for the AAPM). They are all quite similar in their formalisms and have been summarised and compared by Peixoto and Andreo (2000). The basic ideas common to these protocols are given in the following sections.

18.6.4 THE BACKSCATTER METHOD (*LOW* ENERGIES)

Clinically, the dose is most often prescribed on the skin surface (strictly just below the skin where CPE is established). This has led to the most widely used method of determining the absorbed dose. The principle is straightforward. A chamber is positioned free-in-air, i.e. with no phantom involved, at a position corresponding to the centre of the field on the

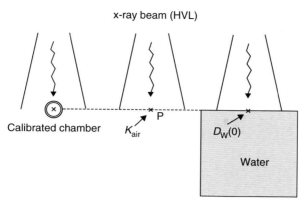

FIGURE 18.10
A schematic illustration of the *backscatter method* for determining the dose at the surface of a water phantom.

patient's skin surface. The reading of the (calibrated) chamber yields air kerma. This is then converted into dose to water at the field size of interest by the application of a backscatter factor B that is a function of field size and the HVL of the beam (Figure 18.10). The energy, or quality range, for this method differs slightly from code to code, but all of them denote it by the term *low energy*; in IPEMB (1996b), for example, this refers to HVLs in the range 1 mm to 8 mm in aluminium, covering approximately 50 kV to 160 kV beams; in IAEA (1987), this range is 10 kV to 160 kV.

In IPEMB (1996b), NCS (1997) and AAPM (2001), the following theoretical route has been chosen: K_{air} is converted into water kerma, K_w, (through the mass-energy absorption coefficient ratio, water to air) but still under *free-in-air* conditions, i.e. for the primary beam spectrum. This has the advantage that $(\bar{\mu}/\rho)_{w,air}$ is independent of field size. Then the backscatter factor B_w converts $K_{w,primary}$ into the water kerma at the surface of a water phantom* (Nahum and Knight 1994).

Following the nomenclature in IPEMB (1996b), the absorbed dose to water at the surface is given by

$$D_{w,z=0} = M N_K B_w \left[\left(\frac{\bar{\mu}_{en}}{\rho} \right)_{w,air} \right]_{air} \qquad (18.29)$$

where M is the instrument reading corrected to standard conditions, N_K is the air-kerma chamber calibration factor for the HVL of the user's beam, B_w is the backscatter factor (see above) for the field size, HVL, and Source to Surface Distance of the user's beam and $[(\bar{\mu}_{en}/\rho)_{w,air}]_{air}$ is the mass energy-absorption coefficient ratio, water to air, for the *free-in-air* (i.e. *primary*) spectrum. The factor B_w is written with the subscript w to emphasise that it is a ratio of *water* kermas (explicitly for the spectrum at the water-phantom surface to that of the primary beam).

IAEA (1987) gave values of the backscatter factor that differed considerably (by several percent for certain combinations of HVL and field size) from the values given in BIR (1983) that had been recommended in earlier codes such as HPA (1983). These new IAEA factors

---

* The backscatter factor B_w is here a ratio of *water* kermas. This differs from IAEA (1987) in which the backscatter factor is a ratio of *air* kermas (though this was never made explicit), and the mass-energy absorption coefficient ratio is field-size dependent; the two *routes* are formally equivalent. Numerically, there is very little difference between B_w and B_{air} but conceptually the distinction is important.

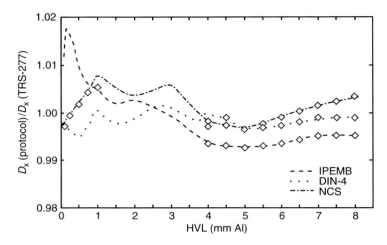

FIGURE 18.11
Ratios of absorbed dose to water in reference conditions at the surface of the phantom for x-rays in the range 10 kV to 160 kV determined using the protocols IPEMB (1996b) DIN-4 (1988) and NCS (1997) normalized to that determined using IAEA (1997a). Symbols correspond to extrapolated data (the first symbol in each extrapolated region corresponds to the last value tabulated in the protocol). (Reproduced from Peixoto J. G. P. and Andreo, P., *Phys. Med. Biol.*, 45, 563–575, 2000. With permission.)

were computed by Grosswendt (1984) using Monte-Carlo simulation, and subsequently confirmed by Knight (1993, 1996). Measurements of B by Klevenhagen (1989), using a specially constructed thin-window parallel-plate chamber, and by Coudin and Marinello 1998, using thin lithium borate TLD chips, generally agreed much better with the Monte-Carlo values than with those given in BIR (1983).

In principle, the type of ionisation chamber has no significance when one utilizes the backscatter method to determine the surface dose; one is merely using the chamber as a means of transferring the air kerma from the Standards Dosimetry Laboratory to the user's beam. In practice, however, the chamber should show a small, preferably negligible, variation in N_K with *HVL* over the full quality range. In the IPEMB Code of Practice, the U.K. secondary standard instrument, formerly the NE2561 now replaced by the NE2611, is to be used over the complete *medium* and *low* energy range. The IAEA (1987) code recommended a thin-window plane-parallel chamber for the *low* energy range.

A comparison of the absorbed dose to water at the surface determined using four different kV codes, with the IAEA 1997a code as a reference, is given in Figure 18.11. The differences are mostly within 1%.

18.6.5 THE IN-PHANTOM METHOD (*MEDIUM* ENERGIES)

For so-called *medium* energy x-ray beams, defined in IPEMB (1996b) as being in the range of approximately 160 kV to 300 kV*, 0.5 mm to 4 mm HVL in copper or >8 mm in aluminium, it is recommended that the dose be determined at a reference depth in a water phantom. The various codes differ in their specification of this reference depth. IAEA (1987) specifies a depth of 5 cm. In IPEMB (1996b), NCS (1997) and Ma et al. (2001) this is 2 cm, which is more representative of clinical practice.

---

* IAEA (1987) specified this as 100 kV to 300 kV.

Using again the IPEMB (1996b) nomenclature, the absorbed dose to water at 2 cm depth, $D_{w,z=2}$, is given by

$$D_{w,z=2} = M \, N_K k_{ch} \left[\left(\frac{\bar{\mu}_{en}}{\rho} \right)_{w,air} \right]_{z=2,\phi} \qquad (18.30)$$

where M is the instrument reading corrected to standard conditions, N_K is the air-kerma chamber calibration factor for the HVL of the user's beam, $[(\bar{\mu}_{en}/\rho)_{w,air}]_{z=2,\phi}$ is the mass energy-absorption coefficient ratio, water-to-air, for the photon spectrum at 2 cm depth in water and for field size ϕ of the user's beam, and k_{ch} is an overall correction factor[*].

The factor k_{ch} corrects for several different effects. It converts from *air kerma* at the centre of a *hole* in the phantom made by the outer surface of the chamber wall, corresponding to the free-in-air calibration situation, to the air kerma at the centre of the water volume that replaces the chamber. Then $(\bar{\mu}_{en}/\rho)_{w,air}$ converts from *air kerma* to water kerma, and finally, dose to water is equated to water kerma as there is CPE (Nahum and Knight 1994). The different effects in k_{ch} include the displacement of water by the chamber, the change in photon energy and angular distributions between the *in-air* and *in-phantom* situations and chamber stem effects (e.g. Hohlfeld 1996).

IAEA (1987) effectively gave k_{ch} values (denoted by p_u) that steadily increased with decreasing kV or HVL to 1.10 at 100 kV (4 mm Al); these were for Farmer-like cylindrical chambers such as the NE2571 (see Table 15.2 in Chapter 15).

Following the publication of IAEA (1987), the factor k_{ch} was extensively investigated, mostly for the NE2611 and NE2571 cylindrical chambers (e.g. Ma and Nahum 1994, 1995a, 1995b; Hohlfeld 1996; Seuntjens and Verhaegen 1996; Seuntjens et al. 1999). There was agreement that the IAEA (1987) values were incorrect. The IAEA officially withdrew their 1987 numbers, and IAEA (1997a) includes a revised table of p_u values that vary from 1.0 at around HVL = 3 mm Cu to 1.03 at 0.2 mm Cu with the uncertainty estimated to be 2% (1 SD). There is now a reasonable consensus on the values of the overall correction factor as given in IAEA (1997a) and IPEMB (1996b); these are for the NE2571 and NE2611 chambers as a function of the HVL in Al and Cu. Additionally, the AAPM code (Ma et al. 2001) gives values for a number of other chambers, including the Capintec PR-06C, Exradin A12 and NE2611. All of these go through a maximum around 0.5 mm to 0.6 mm Cu. This consensus is within 1.5%; none of the values given in any of the codes is below 0.99 or above 1.04.

The factor $(\bar{\mu}_{en}/\rho)_{w,air}$ has been evaluated by Monte-Carlo simulation by, for example, Knight (Knight and Nahum 1994; Knight 1996). The values by various investigators agree within 1% or so; the factor $(\bar{\mu}_{en}/\rho)_{w,air}$ only depends weakly on field size or depth.

Figure 18.12 shows how the various codes (but not including AAPM—Ma et al. 2001) differ in dose determinations at 2 cm depth in water; the range is within 1.5% at all qualities of interest.

18.6.6 *Very Low* Energies

IPEMB (1996b) has defined a third energy range: *very low* (0.035 mm to 1.0 mm Al, approximately 8 kV to 50 kV); this category was necessary for several reasons. Firstly, in the U.K. (and also in France and Sweden, for example), such low energies are used for certain treatments (e.g. so-called Buckytherapy or Grenz-ray therapy). Secondly, the secondary standard chamber cannot be used at these energies as its response, i.e. the N_K vs HVL calibration curve, is no longer approximately flat. Instead, a thin-window parallel-plate

---

[*] In IAEA (1987) this factor is denoted by the product of k_u and p_u, and by $P_{Q,cham}$ in AAPM (2001).

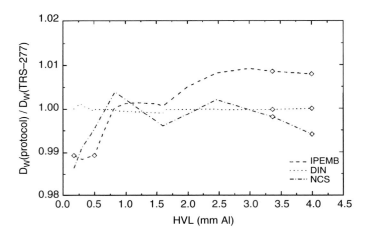

FIGURE 18.12
Ratios of absorbed dose to water in reference conditions at 2 cm depth in water for x-rays in the range 100 kV to 300 kV determined using the protocols IPEMB (1996b) DIN-5 (1996) and NCS (1997), normalized to that determined using IAEA (1997a). Symbols correspond to extrapolated data (the first symbol in each extrapolated region corresponds to the last value tabulated in the protocol). (Reproduced from Peixoto J. G. P. and Andreo, P., *Phys. Med. Biol.*, 45, 563–575, 2000. With permission.)

chamber is recommended. Furthermore, the backscatter method may be invalid for the very small field sizes sometimes clinically employed in such low-energy beams, i.e. the field size can be insufficient to completely cover the chamber, and consequently the value of the product $M \times N_K$ will no longer yield the correct value for air kerma free-in-air in the user's beam.

For this *very low* energy region, IPEMB (1996b) recommends that a parallel-plate chamber be placed at the surface of a water phantom and the dose at the surface be determined; DIN (1988) also specifies this method. The relevant expression is identical to that for the in-phantom *medium-energy* method (see Section 18.6.5) except that, now, the factor k_{ch} refers to the specific parallel-plate chamber employed and pertains to the surface dose rather than that at 2 cm depth and is consequently designated $k_{pp,z=0}$. At the time of the publication of the IPEMB code, no values of $k_{pp,z=0}$ had been determined either experimentally or theoretically, and therefore this factor was simply assumed to be equal to unity. However, this cannot be correct as it assumes that the scatter from the body of the chamber is negligible (cf. the hole involved in determining k_{ch} for a cylindrical chamber in Section 18.6.5); effectively $k_{pp,z=0}$ has to *put this scatter back*. Ipe et al. (1999) have since experimentally determined $k_{pp,z=0}$ and found that it varies from about 1.01 to 1.08 for a field diameter of 5.4 cm (at a focal distance of 50 cm) depending on the chamber, beam quality and phantom.

Perrin et al. (2001) determined k_{ch} for three different chamber types: PTW 23344 large-volume and PTW 23342 small-volume soft x-ray chambers and a Markus parallel-plate electron chamber, as a function of FSD, applicator size and beam quality; their experimental method involved effectively extending the energy range of the backscatter factor method (see Section 18.6.4). For each of the chamber types, an additional 0.1 mm polythene sheet was placed over the entrance window to increase the (total) window thickness to greater than 8.5 mg cm^{-2} as recommended in IPEMB (1996b). No variation of k_{ch} was found with FSD. For the PTW 23344 chamber, k_{ch} increased approximately linearly with HVL from 1.01 ± 0.005 at 0.04 mm Al to 1.075 ± 0.005 at 0.98 mm Al for a 50 mm diameter field; Ipe et al. (1999) had obtained values approximately 1% higher. For the PTW23342 chamber, Perrin et al. found very similar k_{ch} values in the 0.04 mm to 0.6 mm Al range and $k_{ch} \approx 1.06$ at 0.98 mm Al. The variation with applicator size/field diameter was found to be much less than with HVL, k_{ch} decreasing by around 1% as the field diameter was increased from 2 mm to 15 mm for both PTW chamber designs, but a clear trend was difficult to discern. For the

Markus chamber, the trend was toward increasing k_{ch} as the field size increased; however, this chamber type should not be used for absolute dose determinations at very low energies as its N_K varies by more than the 5% over this quality range. They concluded that the use of $k_{ch} = 1$ given provisionally in the IPEMB kV code (IPEMB 1996b) could result in a significant *underestimation* of the dose in this *very low* energy region.

18.6.7 ABSORBED DOSE TO WATER CALIBRATION

All of the above applies to air-kerma calibrations. Standards of absorbed dose to water in the kilovoltage x-ray range are not generally available. However, it is possible to derive calibration factors in terms of absorbed dose to water from air-kerma calibration factors using one of the accepted codes of practice (see IAEA 2000). In this way any calibration laboratory with standards of air kerma can provide (derived) calibration factors in terms of absorbed dose to water. Even though this is formally equivalent to the user obtaining an air-kerma calibration and individually applying the same air-kerma code of practice as used by the calibration laboratory, it has the advantage of extending the use of the unified methodology of absorbed dose to water standards to kilovoltage x-rays for which such standards have yet to be developed.

18.7 PROTONS AND HEAVY CHARGED PARTICLES

The use of proton and heavy-ion beams in radiotherapy has been increasing in recent years (see Chapters 46 and 49). Practical dosimetry at these qualities is also based on the use of ionisation chambers that can be provided with calibrations both in terms of air kerma and in terms of absorbed dose to water. This section presents an overview of the different procedures for the calibration of clinical proton and heavy-ion beams, extending the review of proton beam dosimetry provided by Medin (1997) to all types of heavy charged particles.

Although protons have been used for radiotherapy since the mid-1950s, no dosimetry recommendations were issued until 1986. Proton and heavy-charged particle beam dosimetry was largely based on Faraday cups and calorimeters used together with ad hoc procedures for ionometry. This situation contrasts with the much longer tradition of codes of practice for high-energy photon and electron beam dosimetry. The pioneer dosimetry protocol issued by AAPM (1986) was addressed to heavy charged particles in general. As few data were available at that time, this protocol concentrated on a discussion of the different dosimetry methods available and retained Faraday cups and calorimeters as the preferred choice. The subsequent dosimetry protocols for protons issued by the European Clinical Heavy Particle Dosimetry Group, ECHED (Vynckier et al. 1991, 1994), represented an improvement with regard to the data and practical procedures provided. Although calorimetry and Faraday cups were still the primary dosimetry recommendation in the 1991 protocol, the use of ionisation chamber dosimetry received more attention than in AAPM (1986). This was in parallel with neutron dosimetry, and A-150 was the preferred choice of material both for calorimetry and for the construction of ionisation chamber walls. The 1994 protocol updated proton stopping powers, adopting those from ICRU (1993), and allowed the use of other types of ionisation chambers even if no detailed chamber data were supplied*.

The ICRU published a report (no.59) on proton dosimetry (ICRU 1999) where, for the first time, the recommended choice of reference dosimeter was a thimble ionisation chamber. The use of water-based calorimeters is, however, encouraged. ICRU Report 59 is the most comprehensive document on proton dosimetry published to date, and includes both the *air-kerma* and *absorbed-dose-to-water* based formalisms for the determination of absorbed

---

* The user was, instead, referred to IAEA (1987a) and AAPM (1983) to derive chamber-specific factors.

dose to water using ionisation chambers. Unfortunately, instead of providing a unified form-alism, this ICRU publication mixes the notation of AAPM (1983), ICRU (1984a) and IAEA (1987), resulting in unnecessary complications for the user.

For all types of heavy charged particles, the *air-kerma based formalism* for absorbed dose determination using an ionisation chamber is the same as has been described above (see Section 18.2) except for the lack of constancy of the mean energy required to produce an ion pair in air, W_{air}. When the reference quality Q_0 is cobalt-60 γ-rays, the $N_{D,air}$ chamber factor, at this quality, is identical to that given in Equation 18.8, but the expression for the absorbed dose to water is modified for the user's heavy charged particle beam quality Q by the ratio of W_{air}-values for this quality Q and for cobalt-60 according to

$$D_{w,Q}(P_{eff}) = M_Q \, N_{D,air,Q_0} \left[(W_{air})_Q/(W_{air})_{Q_0}\right] (s_{w,air})_Q \, p_Q \qquad (18.31)$$

where, as for electron and photon beams, the factor p_Q includes the product of different perturbation correction factors of the ionisation chamber at the user's heavy charged particle beam quality Q (see Section 18.5.3 and Appendix D, Section D.2). These factors are assumed to be equal to unity for most ionisation chambers used in charged-particle beam dosimetry although the shift of the effective point of measurement for a cylindrical ionisation chamber is still required for heavy ions. As in the dosimetry of all other types of radiation, the values for W_{air} and for the stopping-power ratios must be consistent as most determinations of W_{air} are based on the knowledge of stopping-power data. Figure 18.13 illustrates the level of consist-ency achieved in absorbed dose to water determinations based on the IAEA and ICRU N_K formalisms using different cylindrical ionisation chambers (Medin et al. 2000).

The IAEA Code of Practice, based on standards of *absorbed dose to water* (IAEA 2000), has provided recommended procedures and updated data for the dosimetry of proton and heavy-ion beams based on the $N_{D,w}$-formalism (see Section 18.3.2). Because of the lack of standards for this type of radiation, the only current possibility is to use ionisation chambers calibrated in a cobalt-60 beam. The absorbed dose to water in reference conditions is then determined according to Equation 18.14 with k_Q values determined according to Equation 18.16 (Medin et al. 1995). The beam quality for protons is defined in terms of the *residual range*, R_{res}. This is the difference in a water medium between practical range R_p (defined as for electrons—see Appendix D, Section D.3.2) minus the reference depth of measurement (the middle of the spread out Bragg peak, see Section 46.3.11). Note that for proton and

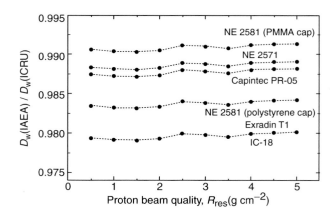

FIGURE 18.13
Ratios of absorbed dose to water as a function of proton beam quality for various cylindrical ionisation chambers determined according to the N_K-based formalisms of IAEA (1987, 1997a) and ICRU (1999). Note that the data for the Exradin T1 and the IC-18 chambers overlap. The main contributions to the discrepancies shown are the different values adopted for W_{air} and $s_{w,air}$ and the formalism in ICRU that ignores the influence of the chamber build-up cap. (Reproduced from Medin, J., et al., *Phys. Med. Biol.*, 45, 3195–3212, 2000. With permission.)

heavy-ion beams, the use of the approximate Equation 18.17 is not possible because of the lack of constancy of W_{air}.

Whereas the dosimetry of proton beams currently has an acceptable degree of accuracy and only the perturbation factors are still the subject of investigations (c.f. Medin and Andreo 1997; Palmans et al. 1999), considerable effort is required to improve heavy-ion beam dosimetry. The complexity of the physical processes involved, mainly because of projectile and target fragmentation for which detailed cross-sections are not yet available, means that the determination of stopping-powers and W_{air}-values relies on rather crude approximations (c.f. IAEA 2000). It is also important to emphasise that in the case of heavy-ion beams, for which the *Relative Biological Effectiveness* (RBE) varies with depth and with the dose delivered to tissue, there is a considerable difference between the so-called *biological dose* and the *physical dose* (c.f. IAEA 2000). Dosimetry protocols are restricted to the determination of the physical dose, ensuring that consistency is achieved in the first step of clinical dosimetry.

18.8 PHANTOMS FOR DOSE DETERMINATION IN REFERENCE CONDITIONS

18.8.1 INTRODUCTION

The quantity to be determined as the basis of a radiotherapy beam calibration is dose to water in reference conditions. All of the above material is based on this assumption, and the formalism presented in the previous sections of this chapter has this objective. Any conversions to dose to other materials are generally undertaken as part of the treatment-planning process, if at all. Therefore, the basic recommendation in dosimetry protocols for all these situations is to use water as the primary reference phantom material. Most ionisation chambers of interest either have built-in waterproofing or can be water-proofed using thin plastic sleeves, ideally no more than 1 mm thick (see Section 15.3.2.2), so this does not generally present any impossible practical demands. Phantoms should be large enough to provide sufficient side-scatter and back-scatter. As a general recommendation, this requires the phantom to be at least 5 cm wider than each field edge at the depth of measurement and to have at least 5 cm of phantom material beyond the maximum depth of measurement (10 cm in the case of medium energy kilovoltage x-ray beams), although, the exact recommendations may vary with beam type and quality and the particular dosimetry protocol. Where possible, the beam should be brought into the phantom directly through the water surface, but where this is not possible, then the phantom wall lying over the chamber position should be made of plastic and should be less than 5 mm thick. The water/equivalent thickness (thickness times density in $g\ cm^{-2}$) of this wall should be taken into account when positioning the chamber at the required depth. For very thin windows in the wall, as present on some commercial phantoms, *bowing* can occur because of water pressure. This should be carefully monitored as it can change with time and can therefore alter the depth of water between the surface and the chamber.

Although water is the preferred phantom material, there are situations where using a solid plastic phantom may be more convenient or may provide better positional accuracy, particularly for lower-energy electron beams and for lower-energy kilovoltage x-ray beams. In addition, solid phantoms are more convenient for many relative measurement and routine checks because of the ease of set-up and the reproducibility of chamber depth. Ideally, a solid phantom should be *water-equivalent*, i.e. it should exactly mimic water in all the relevant physical properties over the range of radiation qualities to be met in the particular situation. In practice, no material meets this requirement. Practical phantoms are, at best, an approximation to water. Their use will be limited to the qualities and energies for which they approximate more-or-less to water. In general, each will require the use of some correction factors. However, these correction factors may be close to unity for some materials and for certain irradiation situations. Various materials are discussed in ICRU (1989). The materials that have been widely used have been acrylic plastic (PMMA, known as perspex or lucite) and poly-styrene. More recently, a number of commercial phantom materials, based on epoxy-resin formulations, have become available, specifically designed for dosimetry purposes.

It should be noted that the use of any material other than water will increase the measure-ment uncertainties (although it may decrease positional uncertainties), and this is why the primary recommendation is to use water wherever possible for beam calibration. Other materials should be tested on purchase to ensure that their density and slab or sheet thicknesses are as expected. They should be checked for flatness as warped sheets can introduce significant air gaps into the radiation path. In addition, the user should demonstrate that the phantom really is *water-equivalent* by comparing the results of measurements made in the phantom material to those made in real water. If necessary, corrections for differences in density may be applied.

18.8.2 MEGAVOLTAGE X-RAY AND ^{60}Co γ-RAY BEAMS

For these beams, no significant problems arise in the use of water phantoms. Dosimetry protocols all recommend water, and they discuss the waterproofing and positioning require-ments needed. The AAPM (1983) protocol allowed the use of polystyrene and acrylic plastics and discussed the depth scaling and transfer correction factors required although this protocol has now been superseded.

Where other materials might be useful for relative or routine measurements, it may be noted that:

a. Depth-scaling factors can be taken as proportional to electron density

b. Measurements in the epoxy-resin-based materials are typically within around $\pm 0.5\%$ of water over the energy range from ^{60}Co γ-rays to 16 MV x-rays (Tello et al. 1995; Allahverdi et al. 1999). Monotonic trends are observed with beam quality, rising for some materials, falling for others. Therefore, dosimetry differences between different phantom materials can be up to 1% in this energy range if no corrections are made

c. Measurements in polystyrene can show differences from water of up to 3% (Christ 1995), depending on photon beam quality and depth

d. All the epoxy-resin phantom materials yield reasonably good relative measurements (depth doses, Tissue Phantom Ratio, Tissue Maximum Ratio, etc., generally to within $\pm 1\%$ of water and Quality Index to within around $\pm 0.5\%$ of water).

18.8.3 ELECTRON BEAMS

Non-water phantom materials are allowed in most dosimetry protocols for lower energy electron beams (generally, less than around 10 MeV) where dose gradients are steeper, depth-dose distributions extend to smaller overall distances and parallel-plate chambers are preferred to cylindrical chambers to minimise the perturbation in the direction of the beam (see Section 15.3.2.3 and Appendix D, Section D.2). For these chambers, positioning problems can become more significant, and some designs cannot be easily water-proofed. Generally, PMMA and polystyrene (white and/or clear) have been included as allowed phantom materials with epoxy-resin based plastics (White et al. 1977; Constantinou et al. 1982; Ho and Paliwal 1986; Tello et al. 1995) being introduced in more recent protocols (IPEMB 1996a; IAEA 1997b, 2000). The following problems are recognised in the use of non-water phantoms:

a. Depths must be scaled to the equivalent depth in water such that the spectra are similar, and therefore, the same $s_{w,air}$ is applicable. Various approaches have been used, e.g. scaling by electron density, by stopping power or range, and by 50% depth. A discussion of these methods can be found in AAPM (1991) and IPEMB (1996b). Recommended scaling factors can be found in IPEMB (1996b) and IAEA (1997, 2000). Where scaling factors are given in terms of linear dimensions, they require modification if the density of the plastic sample is non-standard (IPEMB 1996b). Similar scaling is required where non-water phantoms are used to determine electron beam quality or to measure electron depth doses.

b.   Even when appropriately depth-scaled, the electron fluence will typically be different to that in water as the scattering powers will not be the same in the two materials. The carbon content in typical plastics (PMMA, clear polystyrene) produces a lower effective atomic number, and therefore, a lower scattering power when the stopping power is matched. This results in a lower electron fluence at an equivalent depth. Plastics with the addition of some higher-Z material partially redress this (e.g. white, or high-impact, polystyrene, having a few percent of TiO_2 added or the epoxy-resin formulations). A fluence-ratio correction, h_m, generally >1, is required to correct for this (Thwaites 1985; Bruinvis et al. 1985; Thomadsen 1995; Tello et al. 1995; Ding et al. 1996; Nisbet and Thwaites 1998b). The specific values depend on beam energy (falling as energy increases), the angular and energy distributions incident on the phantom, and the exact composition of the phantom material. Therefore, there can be significant uncertainties associated with the data. Representative values are given in IPEMB (1996b) and IAEA (1997b, 2000) for depths close to the reference depths. For clear polystyrene, these correction factors can be up to around 1.03 at an effective mean energy below about 3 MeV, falling linearly to unity with increasing energy at a mean energy of around 20 MeV. For white polystyrene, the maximum correction is around 1.02; for a number of commercial epoxy-resin-based plastics (e.g. RMI[*] *solid water* 451, 457; WT1[†]), it is around 1.01; and for WTe[‡] and plastic water[¶], it is around 1.00, i.e. these two formulations have negligible fluence-ratio corrections to within experimental uncertainties. It is assumed that similar values to those tabulated (e.g. IPEMB 1996) can be applied at greater depths, selecting for the appropriate mean energy at that depth[§] and that the values at shallower depths than the depth of maximum dose can be linearly interpolated to unity at the phantom surface. However, this assumes that differences in back-scatter are negligible. There are significant differences in the literature for PMMA. Figure 18.14 shows the spread in h_m found by different investigators.

c.   For phantoms which are electrical insulators, problems can arise from *charge storage* where the stopped electrons remain in the phantom producing high-strength electric fields that can modify the electron fluence incident on the chamber and the measured ionisation during that irradiation and during subsequent irradiations (Galbraith et al. 1984; Thwaites 1984). These effects can be large, particularly in thick slabs of phantom material (Galbraith et al. 1984; Pitchford and Thwaites 1985). They are unpredictable, depending on the accumulated dose, the time, dose and radiation modality history of the phantom and the specific insulating properties of the material both under normal conditions and under the action of radiation beams. Effects can be large for cylindrical chambers, but they appear to be negligible for parallel-plate chambers[**]. Effects can be minimised by using thin sheets, no more than 2 cm thick, and sleeves of the same material around the chamber (Thwaites 1984; IPEMB 1996b). The magnitude of these effects can be checked by comparing measurements against those in a conducting phantom, e.g. water. The epoxy-resin-based plastics do not appear to show charge storage effects, implying that their conductivity under the action of radiation beams is sufficiently high to allow the charge to leak away.

d.   Variations in all of the above effects have been observed between different samples of the same nominal material from different manufacturers, mixes, or batches. For example, there are well-

---

[*] Radiation Measurements Incorporated, Wisconsin; solid water mixes 451 and 457.

[†] WT1 is the original White photon formulation, available from Radiation Physics, St Bartholomew's Hospital, London.

[‡] WTe is a later electron/photon formulation, also available from Radiation Physics, St Bartholomew's Hospital, London.

[¶] CIRS, Norfolk, VA.

[§] Where only single values are given, e.g. IAEA 2000, the reference depth value is assumed to apply to greater depths.

[**] The use of cylindrical chambers in electron beams to cross-calibrate parallel-plate chambers could introduce problems into parallel-plate chamber calibration factors.

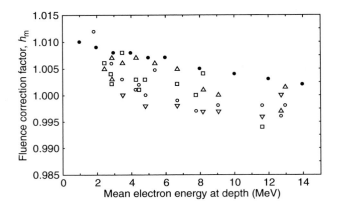

FIGURE 18.14

Factor h_m to correct for the difference in the electron fluence in a PMMA phantom compared to that in water at equivalent depths. The filled circles correspond to the data given in IPEMB (1996) and IAEA (1997). The rest of the data has been measured by participants in an IAEA research project that uses different ionisation chambers: NACP (triangles), Roos (circles), Markus plane-parallel chambers (squares) and a Farmer NE 2571 cylindrical chamber (inverted triangles). (Reproduced from DuSautoy, A., et al., in Review of data and methods recommended, in *IAEA TRS-381, The Use of Plane Parallel Ionisation Chambers in High Energy Electron and Photon Beams*. *TECDOC-1173*, 1–10, IAEA, Vienna, 2000. With permission.)

established differences between clear and white polystyrene or between different water-substitute plastics from different manufacturers or between different mixes (sometimes with the same commercial name) from the same manufacturer. There can be other, subtler batch differences because of different contaminants, material density, etc. This may explain the differences in h_m, etc., reported in the literature. In general, it is recommended that, as a minimum, a check of the density should be carried out and a correction * made if it is different from the generally accepted value. Other checks are also required to ensure that the behaviour of a specific sample of material is as expected (see Section 18.8.1).

18.8.4   KILOVOLTAGE X-RAY BEAMS

At kilovoltage qualities, small differences between different phantom materials are likely to be magnified as compared to megavoltage beams. For use with parallel-plate chambers in phantom for low or very low energy beams, PMMA and epoxy-resin plastic phantoms have been recommended[†]. However, the measurement in this situation is on the surface, so no correction factors are required for depth. The phantom is intended to provide back-scatter. There is also at least one commercial polyethylene-based material formulated for kilovoltage x-ray beams. For *medium* energy beams, cylindrical chambers are used, and water phantoms are recommended for beam calibration. For routine and relative measurements where improved positional accuracy or reproducibility is required, epoxy-resin plastic phantoms have been investigated (Nisbet et al. 1998). The comparisons reported indicate that larger differences are observed than for megavoltage x-rays, but that agreement is obtained within a few percent for the photon epoxy-resin plastic formulations. The electron formulations have a greater content of higher-Z material and are expected to yield larger differences.

---

* The more straightforward correction consists in scaling depth according to material density provided that the distance from the source to the detector remains unchanged or that its variation is accounted for.

† As various protocols recommend in-air measurements for these beams, this depends on dosimetry protocol and approach (see Section 18.6).

18.8.5 PROTONS AND HEAVY CHARGED PARTICLES

For proton and heavy-ion beams, there is little or no information available on the equivalent effects such as h_m values, and water is recommended for measurements. Any use of non-water phantoms would require careful assessment in the user's own beam relative to water.

18.9 SUMMARY AND EXPECTED FUTURE DEVELOPMENTS

A fundamental issue pertaining to the development of methods for the determination of the dose in reference conditions is the accuracy that can be currently achieved. The uncertainties attached to the various codes of practice in given conditions are discussed in Appendix D. In general, these are now below 2%. This is the result of considerable efforts made in the last twenty years or so during which the determination of many of the quantities and correction factors used in external-beam dosimetry with ionisation chambers has evolved considerably. The most important changes, of the order of several percent for some of the quantities involved in dose determination as well as a dramatic improvement in current knowledge and interpretation of physical phenomena, occurred during the 1980s as a result of the extended use of Monte-Carlo calculations. The 1990s have been mainly characterised by a refinement of the data at the level of one percent or so.

An important effort to promote the TRS 398 Code of Practice (IAEA 2000), based on absorbed dose to water, has been undertaken by the IAEA. Several concerted projects have been launched to test its implementation and provide guidelines to SSDLs and clinical physicists (IAEA 2005). In the near future, it can be expected that the formalisms involving stopping-power ratios and perturbation correction factors will not be needed for reference dose determination because calibrations in terms of absorbed dose to water and experimental k_Q factors will be available for most configurations of clinical interest. The uncertainty in dose determination will then all be similar to that available today for a cobalt-60 beam using the $N_{D,w}$ calibration, which is below the one-percent level.

Some issues remain unsolved, however, such as the difference between the dose determined using the $N_K - N_{D,air}$ formalism and with an $N_{D,w}$ calibration even for a cobalt-60 beam; this is a question to be solved at the level of PSDLs. This has also a direct relation to the differences between primary standards in different countries which cause *1 Gy in one country to be different from 1 Gy in another*. This problem can be of importance in geographical regions where a common dosimetry protocol is used and yet users obtain different calibration factors from different laboratories; the inconsistency may be even greater in countries without a PSDL where it is common practice to purchase instruments calibrated in other countries. This can result in a situation where, within a given country or even within one institution, two instruments may yield slightly different values for the same quantity, and in certain cases, these may even be national secondary standards. Until a kind of averaged mean value is adopted, as proposed by Boutillon and Andreo (1997), international harmonization for the determination of absorbed dose will not have been achieved, and dosimetry will remain one step behind other branches of physics where one metre or one second are the same everywhere in the world.

CHAPTER 19

RELATIVE DOSE MEASUREMENTS AND COMMISSIONING

Ivan Rosenberg

CONTENTS

19.1 INTRODUCTION

19.1.1 RATIONALE FOR ABSOLUTE AND RELATIVE DOSE MEASUREMENTS

The use of therapy beams for patient treatment requires that the dose can be determined at any point within the patient. Because it is necessary to know beforehand what the dose distribution will be, and it is impossible to perform measurements within the patient's body, one has to predict the dose from calculations. Methods for calculating the patient dose at any point require a good understanding of the properties of clinical beams, and these issues are presented in Part E. The methods for calculating the dose with a computer system and the corresponding dose modelling are presented in Part F. To provide basic data for the necessary dose models, a number of preliminary measurements must be performed on the treatment machine. In practice, it is convenient to make a distinction between *absolute* and *relative* dose measurements.

The *absolute dose* determination in reference conditions is based on published codes of practice discussed in Chapter 18. It is important to stick closely to the recommendations of these codes of practice because they guarantee accuracy and consistency of the dose determination throughout the world. The reference conditions are chosen to provide a point at a depth of clinical interest where electronic equilibrium can be relied on (for photon beams).

However, in practice, the actual treatment situation is different, and it is essential to measure the dose for a wide range of conditions representative of the clinical use. To be able to relate these measurements to absolute dose, they should be normalized to the reference conditions and are called *relative dose* measurements. The detailed procedures for performing these measurements are described in this chapter. Performing these measurements requires an understanding of the properties of clinical beams and of the definition of the corresponding dosimetric parameters presented in Part E. Where appropriate, cross references are provided to the relevant sections.

19.1.2 PRE-MEASUREMENT PREPARATION

The period between delivery of a new radiation treatment machine and its release to clinical use is the best, and sometimes only, time that physicists have to characterise the behaviour of the radiation beams and of the various modifiers (such as wedges, blocks, and multi-leaf collimator (MLC)) that will be used with it. It is very important to ensure that measurements taken in this period are accurate and comprehensive. A measure of aforethought and preparation will make the task smoother.

This initial period is divided into the acceptance phase and commissioning. *Acceptance* consists of verifying that the performance of the equipment meets the minimum requirements outlined in the purchase contract and in the manufacturer's specifications. *Commissioning* is the process through which all the possible machine characteristics relevant to clinical use are investigated, measured, and recorded. Commissioning of a treatment unit should only start after a thorough acceptance process has established its mechanical and radiation consistency (see Section 38.2 and Section 38.3). In preparation for both the acceptance and commissioning measurements, all in-house dosimetry equipment should be checked for calibration, accuracy, and availability. For example, if a computerised water phantom is to be used, its mechanical motions and the data collection software need to be checked for accuracy; electrometers need to be checked for noise, leakage, and proper polarising voltage; ionisation chambers and diodes need to be checked for leakage and waterproofing if used; a survey should be made of available solid phantoms to establish their slab thicknesses and also of detectors that can be accommodated in them.

The minimum set of measurements needed for commissioning would correspond to the requirements for monitor unit (MU) calculations and the beam data input specifications of the treatment planning system (TPS) to be used at the institution (see Table 19.1).

TABLE 19.1

Example of Beam Data Input Required for Different Types of Treatment Planning System (TPS)

Treatment Planning System	XiO	Masterplan	Plato	Pinnacle	Eclipse
Treatment machine name	Y	Y	Y	Y	Y
Modality	Y	Y	Y	Y	Y
Energy	Y	Y	Y	Y	Y
Couch, gantry and collimator limits and direction of rotation	Y	Y	Y	Y	Y
Jaw labelling	Y	Y	Y	Y	Y
Available setups, SSD limits	Y	Y	Y	Y	Y
Source to collimator distance	Y	Y	Y	Y	Y
Source to wedge distance	Y	Y	Y	Y	N
Wedge dimensions	Y	N	N	N	N
Source to monitor chamber	N	Y	N	N	N
Source to flattening filter	Y	Y	Y	Y	N
Source to block tray	Y	Y	N	N	N
Primary collimator opening	N	Y	N	N	N
Percent depth dose for a full range of field sizes open and wedge fields	Y	Y	Y	Verification only	Y
Primary axis profiles for a full range of field sizes open and wedge fields	Y	Y	Y	Verification only	Y
In air outputs	Y (Square/open)	Y (Square/length Open/wedged)	Y (Open/wedged)	Y	Y
In water output	Y	Y	Y	Y	Y
SSD and reference depth	Optional	Y (90 cm SSD 10 cm deep, 10 cm×10 cm field size)	Optional	Optional	Optional
Absolute calibration dose	Y	Y	Y	Y	Y
Wedge factor	Y	Y	Y	Y	Y
Wedge type	Y	Y	Y	Y	Y
Wedge angle(s)	Y	Y	Y	Y	Y
Block/wedge/collimator density and composition	Y	Y	Y	Y	N
Block thickness	Y	Y	Y	Y	N
Collimator transmission	Y	N	Y	N	N

This table is intended to illustrate the type of data needed. Although it has been updated by the editors, it should not be used to determine TPS requirements which may be different for a particular version of software. Data required for the MLC are not listed.

Source: Adapted from Sally Morgan-Fletcher et al., at Linac commissioning—Changing the Culture, IPEM meeting, BIR London, 15 November 2001.

However, a much larger range of measurements than these should be performed to check the accuracy of the TPS calculations in widely different situations, to understand the influence of beam modifiers and of asymmetric geometries, and to be able to answer likely clinical questions such as bolusing and skin sparing, which may not be modelled by the TPS (AAPM 1994a; Mayles et al. 1998). A plan of all the measurements to be taken should be created in advance, specifying which equipment would be most suitable for the task and with a thought to the efficient use of the various resources. For example, once a water phantom is properly aligned, the maximum set of measurements should be performed before disturbing its geometry.

19.1.3   CHOICE OF PHANTOMS

The choice of phantoms for absolute dose measurements in reference conditions has been discussed in Section 18.8. For relative dose measurements, the main difference is that measurements must be repeated at many points and for many different situations. In addition, the measurements are always relative to some reference situation, which gives more flexibility without loss of accuracy. The alternative is essentially between a remotely controlled water phantom and a solid plastic phantom, and the choice is usually obvious.

A modern computer-controlled water phantom (Figure 19.1) can position a detector within a large tank of water to a precision of 1 mm or better, can acquire data at each point with good signal-to-noise ratio in under a second*, and can adjust its dwell time or step size, or both, depending on the local radiation field properties. Most models can be programmed to acquire specific sets of scans in an appropriate sequence, display the results on a monitor in real-time, and, crucially for transfer to a TPS, store the results in digital form. Some models even offer the option of varying and measuring the water level in the tank during the measurements; however, the water ripples limit the precision of this method.

Water phantoms are ideal for measuring dose distributions varying in space (but not in time) and for fixed source-to-phantom geometry. On the other hand, they are cumbersome and require a large setup time for accurate work. Checks are needed prior to final scanning to ensure that:

- The central axis of the beam (CAX) lies close to the centre of the tank to enable the maximum range of travel in all directions
- The detector's axes of motion are parallel, in depth, to the CAX and, transversally, to the water surface, to a precision of 1 mm over the range of travel
- The origin of the scanning coordinate system (where the CAX intersects the water surface) is correctly identified and recorded to better than 1 mm accuracy, taking into account the effective point of measurement of the detector
- The orientation of the detector is appropriate relative to the direction of scanning
- The required measurement spatial range does not produce collisions between the detector and the tank sides
- The reference detector does not perturb the measured dose distribution
- The gain of the electrometers is appropriate to avoid saturation at all measurement points
- The source-to-surface distance is correct and constant over the time of data collection (mechanical sag of the tank support system and evaporation are the main culprits).

In light of the above, there are many instances when it is preferable to use a solid phantom to a water phantom. If motion of the detector is not required as in relative output measurements at the same depth or even to measure tissue phantom ratios where the detector is at a constant

---

* A fast response is especially important if the signal acquisition is performed in continuous mode with a fixed detector used as a reference to correct for dose-rate fluctuations. It is then the ratio of the signal between the moving detector and the reference detector which is used to represent the spatial distribution of the dose.

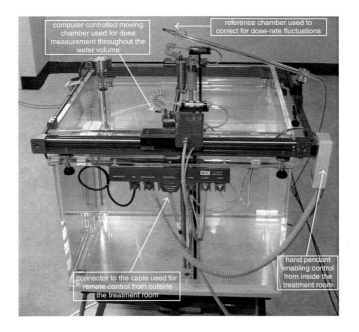

FIGURE 19.1
A remotely controlled water phantom system (Scanditronix-Wellhöfer blue phantom). Detectors can be positioned within the water tank and moved under computer control. The system is most often used with the beam pointing vertically downwards, striking the open water surface; other orientations are possible such as horizontal beams striking the PMMA lateral walls of the tank, which are sometimes equipped with thin entrance windows.

distance from the source and varying amounts of phantom material are added above it, solid phantom slabs are faster, easier, and more reproducible to work with. Relative output measurements are also, usually, better acquired with a high precision, energy independent integrating system such as a Farmer dosimeter (Farmer 1955), and these ionisation chambers are not easy to waterproof. Another example of the preferred use of solid phantoms is in the measurement of doses near the surface that should be performed with a parallel plate (or extrapolation) chamber with a very thin front wall (IAEA 1997). The waterproofing of such a chamber in a water phantom is often difficult, and both water ripples and surface tension so close to the water surface would prevent accurate measurements.

The use of solid phantoms raises the question of their composition and the relationship of the measured dose distributions to those in the universal reference material, water. In the past, materials with effective atomic number and electron density reasonably close to water such as Mix-D™, polymethylmethacrylate (PMMA, also known as Perspex™ or Lucite™), and polystyrene have been used, and correction factors for their effective attenuation relative to water are given in various protocols (IAEA 1987; IPEMB 1996). On the other hand, special plastic compositions specifically designed to have the same linear attenuation as water for various energy ranges have been commercially available* (White 1977; Constantinou 1982; ICRU 1992) and are recommended for commissioning work. Care must be taken to use the correct material for the radiation under investigation, as different compounds exist for photon and electron water equivalence, and to apply any residual correction factors to the results. When used with electron beams, care should also be taken to avoid accumulating charges in large nonconductive plastic blocks as these will distort the readings (Galbraith et al. 1984; Thwaites 1984) (see also Section 18.8.3).

---

* For example, St. Bartholomew's Hospital, London, U.K., or MRI, Middleton, Wisconsin, U.S.A.

19.1.4 Choice of Detectors

The radiation detectors used most commonly in commissioning work are air-filled ionisation chambers and semiconductor diodes. Films are also used, especially for radiation leakage measurements of MLC and with electron beams, and thermoluminescent dosimeters (TLD) also play a part in special situations such as whole body dosimetry.

Ion chambers have long-term stability, high sensitivity, and very small energy dependence in response to photon beams. To maximise the advantages of good spatial resolution (Sibata et al. 1991), the logical detectors to use for photons in water phantoms are small volume water-proof ionisation chambers, 0.1 cm^3 thimble for profiles (Figure 19.2 and Figure 19.3), and plane-parallel for depth doses (Figure 19.4). For relative output measurements in solid phantoms, the high precision 0.6 cm^3 Farmer-type chambers (Farmer 1955) are the most popular choice (see Section 15.3.2.2), but plane-parallel chambers with small active diameters are more accurate for small field sizes and close to the surface.

Ionisation chamber response to electron beams, on the other hand, is highly energy-dependent, albeit in a known way, and stopping-power-ratio corrections need to be applied to the readings where the mean electron energy is changing, either with depth or in the penumbra. In addition, the cavity perturbation effects on electron beams can also be significant, and chambers with small volumes and simple geometry such as plane-parallel chambers (Markus 1976; Mattsson et al. 1981; AAPM 1994b; IAEA 1997a, 1997b) are recommended for relative and absolute dosimetry (see also Section 15.3.2.3).

When using an ionisation chamber for dose measurements the cavity displacement effects of ionisation chambers can be corrected for in two ways (see Appendix D, Section D.2.1). The preferred method for air-kerma-based protocols is to define an effective point of measurement P_{eff} for the detector that is not coincident with the geometrical centre of the chamber (IAEA 1987; IPEMB 1996). For a cylindrical chamber, this point is taken to be shifted by a certain

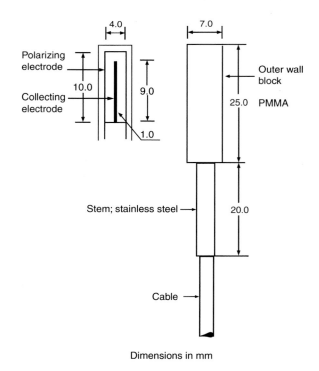

Dimensions in mm

FIGURE 19.2
Schematics and dimensions (in mm) of a typical small volume scanning ionisation chamber (0.12 cm^3 RK model from Scanditronix).

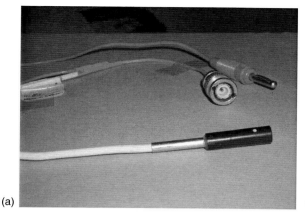
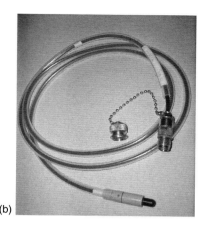

FIGURE 19.3

Picture of two waterproof scanning chambers with cables and connectors. (a) the Scanditronix RK chamber depicted in Figure 19.2 equipped with BNC plug for signal and banana plug for high-voltage. (b) the Wellhöfer IC10 chamber shown in Figure 19.1, equipped with a triaxial plug for both signal and high-voltage. These chambers are vented and the communication with ambient air is through the additional waterproof sleeve around the cable.

fraction (0.6 for photons and 0.5 for electrons, IAEA 2000) of its radius towards the radiation source, and for a plane parallel chamber it is defined at the inside surface of its front wall . Alternatively, some photon dosimetry protocols (particularly absorbed-dose-to-water-based codes of practice) take the effective point of measurement as the centre of the chamber. A perturbation correction p_{dis} is then applied that depends on the beam energy and the size of the detector. When making *relative* depth dose measurements, whatever code of practice is used for absolute dose measurements, care should always be taken to correct for the effective point of measurement (see Figure D.6). In the build up-region a better accuracy can be achieved if a plane parallel chamber is being used (IAEA 2000). When measuring relative off-axis dose profiles, it is usual to relate the measurement to the geometrical center of the ionisation chamber but one should realize that the size of ionisation the chamber could lead to an overestimation of the penumbra width. It is therefore recommended to use small-volume cylindrical chambers with their axis perpendicular to both the beam axis and the movement direction.

Diodes (see Section 16.3) have much smaller measuring volumes than even the smallest practical ionisation chambers and are ideal candidates for scanning in a water phantom (Sibata et al. 1991). Their sensitivity is not constant, and it depends on their radiation history. Therefore, they are invariably pre-irradiated to ensure some medium term stability.

FIGURE 19.4

Two typical waterproof plane-parallel chambers. The 0.16 cm³ Scanditronix/Wellhöfer NACP-02 chamber (left figure) has a graphite front window 0.5 mm thick covered with a 0.1 mm thick mylar foil. The 0.35 cm³ PTW Roos chamber (right figure) has an acrylic entrance window 1 mm thick. (see also Figure 15.8 and Table 15.4).

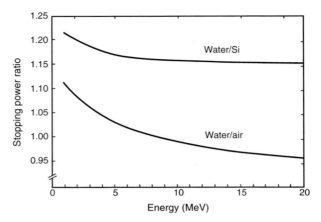

FIGURE 19.5
Plots of the water to silicon and water to air stopping power ratios as a function of electron energy. (From Berger, M. J. and Seltzer, S. M., *Stopping Powers and Ranges of Electrons and Positrons,* 2nd ed., US Department of Commerce, National Bureau of Standards, Washington, DC, 1983.)

When used for relative dosimetry such as water phantom scanning, this variation in sensitivity is not a problem.

The response of diodes to electron beams above 4 MeV is practically energy independent because of the small variation with energy of the water to silicon stopping power ratio (Berger and Seltzer 1983; Rikner 1985a) (Figure 19.5). The same is not true, however, for photon beams where diodes exhibit an energy dependent response, both at large depths and at the edges of the beam. To overcome these problems, energy-compensated diodes for photon measurements have been designed with a high-density material applied to the back (Rikner 1985b; Rikner and Grüssel 1985) (Figure 19.6). This leads to a directional dependence in response,

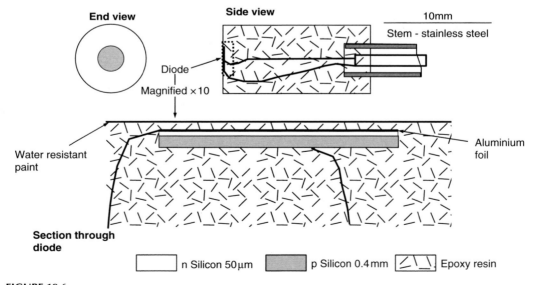

FIGURE 19.6
Diagram of an energy-compensated diode suitable for photon measurements. A thin n-Si silicon layer is overlaying a thicker p-Si silicon layer. The aluminium foil provides an energy compensation. The robustness is ensured by the epoxy resin covered with water-resistant paint. (From Rikner, G. and Grusell, E., *Acta Radiol. Oncol.,* 24, 65–69, 1985. With permission.)

making them unsuitable for general dosimetry. When used in a water phantom, however, these diodes are always mounted with the front facing the incident beam, thereby minimising the effect of their directionality.

19.1.5 BEAM QUALITY

The quality or energy spectrum of the radiation beams produced by linear accelerators, based on manufacturers' specifications, is checked as part of the initial acceptance procedure. This test is usually performed with a water phantom and looks at certain characteristics of the depth-dose curves for both photon and electron beams. It is highly recommended to check the constancy of the beam quality throughout the long commissioning process and even more so during clinical use. To this end, an easy method of measurement is desirable.

For photon beams, a widely accepted measure of beam quality is the quality index, QI (also known as *ionisation ratio* in the U.S.A.). The QI is presented in Appendix D.3.1. It is defined as the ratio $TPR_{20,10}$ of doses at two depths in water (20 cm and 10 cm) at the same distance from the source (usually 100 cm) for a standard field (10 cm $\times$ 10 cm). Various dosimetry protocols base their choice of energy-dependent factors on the value of the QI. Simple water-filled (or water substitute) phantoms can be built to facilitate quick and accurate measurement of this quantity. However, QI is not very sensitive to beam quality changes; 0.3% variation in its value reflects significant changes in energy spectrum.

The beam quality of teletherapy cobalt-60 beams is naturally more stable and predictable than that of accelerator-produced beams. However, it is a good idea to measure the QI even for cobalt units at acceptance and occasionally afterwards to ensure the integrity of the source.

The energy of electron beams is defined by reference to the 50% depth dose R_{50} and the practical range R_p (see Appendix D.3.2). These quantities are derived from depth-dose measurements in a water phantom. For a quick and easy constancy check on the energy, though, an approach similar to the QI can be used for electron beams by recording and checking the ratio of readings at two depths in a small solid phantom. In this case, these depths need to be carefully chosen for each energy; the first should be near the location of dose maximum and the other close to R_{50}. The values of these ratios then become a very sensitive measure of beam-energy constancy; for example, a shift of 1 mm in the depth of the 50% depth dose can produce a 15% change in this ratio for a 6 MeV electron beam and a 4% change at 20 MeV.

19.2 PHOTON MEASUREMENTS

19.2.1 VIRTUAL SOURCE POSITION

The concept of a virtual source is described in Section 23.2.1.16. The measurement of its position needs to be done in air as only the primary beam is expected to follow the inverse-square law. As with all in-air measurements with high-energy photons, an appropriate build-up cap needs to be added to the detector to achieve electronic equilibrium and to eliminate contaminating electrons. These caps can be made of plastic or high density material, or a mini-phantom can be used (see Figure 19.8). To establish the position of the virtual source, the dose rate must be measured at several nominal distances from the target. The range of distances should vary from about 80 cm to about 120 cm, and this can be achieved either by moving the couch on which the detector stands (making sure that scatter from the couch is minimal) or by remotely adjusting the depth in an empty water phantom. This test should be repeated for several field sizes, spanning the available range. The resulting variation of dose rate with distance can then be analysed as discussed in Section 23.2.1.16.

19.2.2 RELATIVE-DEPTH DOSES

The variation of dose with depth along the central axis (CAX), known as relative or percent depth dose (PDD), is one of the fundamental parameters that characterise a radiation beam (see Section 22.3.2). PDD should be measured for open fields and for all available physical wedges because of the beam hardening produced by the wedge attenuation. It is best obtained using a remotely controlled water phantom and either a small volume ionisation chamber or a photon-compensated diode. The PDD should be collected for several square field sizes, from the smallest to the largest available, as well as for some rectangular fields and with the water surface at the standard Source to Surface Distance (SSD) (usually 100 cm). Scanning should start with the detector at the deepest possible position and moving toward the surface, possibly overshooting the origin by a few millimetres. This procedure minimises the effect of water ripples and gives an independent check on the position of the surface where the PDD will abruptly change gradient. If the scanning steps can be varied, it is desirable to sample more densely from around 5 cm deep through the depth of maximum dose to the surface and less densely at deeper depths.

19.2.3 BUILD-UP REGION

The relative depth dose of a high-energy photon beam increases from the surface to a maximum, at a depth dependent on its energy, before decreasing in a quasi-exponential fashion (see Section 22.3.1). The layer between the surface and the depth of dose maximum is known as the build-up region, and knowledge of its characteristics is very important in order to estimate the amount of clinical skin sparing or, conversely, the risk of under-dosing a superficial lesion. Although the shape of the dose distribution in this region can be obtained from the relative-depth-dose measurements in water described above (Section 19.2.2), those results can be distorted by the thickness of the ionisation chamber waterproofing and the ripples and surface tension effects caused by the motion of the detector close to the water surface. To obtain a more accurate characterisation of the build-up region, a thin-walled plane-parallel chamber should be used in a solid phantom providing full backscatter. Measurements are performed by adding slabs of well-known thickness and density above the chamber, ranging from zero to just beyond the depth of dose maximum. The set-up geometry for these measurements should correspond to the relative-depth-dose data; that is, a constant SSD should be used. This can be achieved by ensuring that the support on which the phantom is resting is moved every time a build-up slab is added or removed. Alternatively, and more accurately, at the start of the measurement all slabs to be used in the build-up measurements can be placed below the detector as part of the backscattering material. Slabs are then transposed from below the chamber to above it or vice versa, and the source-to-surface distance will stay constant. However, the alternative approach of leaving the chamber at a fixed distance from the source is compatible with tissue maximum ratio (TMR) data, and conversion from one geometry to the other is simple and accurate within the distances involved (see Section 23.2.2.3). Build-up curves should be obtained for several field sizes, spanning the available range because scatter from the collimators will affect the surface dose. For the same reason, additional measurements should be done with solid wedges in position as these filters will modify the scatter contribution to the surface layers.

The presence of a tray, especially if close to the skin, could also affect the build-up curve and the surface dose. Build-up measurements should be performed with all the available trays to evaluate their effect on skin sparing.

19.2.4 OFF-AXIS PROFILES

The variation of dose with distance from the CAX of beam but at the same depth, known as *off-axis profiles* or *off-axis ratios* (OAR), is also a fundamental parameter required to characterise a radiation beam (see Section 22.3.3 and Section 23.2.1.15). OARs should be measured for open fields and for all available physical wedges. The shape of the OAR is a function of depth, and profiles at several depths are required. They are best obtained using a remotely controlled water phantom with its surface at the same SSD as for the PDD measurements and either a small volume thimble ionisation chamber or a photon compensated diode. Care should be taken to adjust the electrometer gain to avoid saturation at the thin side of wedged profiles. The depths at which the profiles should be scanned will be determined by the requirements of the treatment planning system, but they should include the depth of maximum dose and the deepest possible position of the detector in the tank. The width of the scans perpendicular to the beam will be a function of both the field size and the depth (because of divergence) and should be large enough to include not only the beam edges but also at least 5 cm beyond the geometrical edge of the beam. For large field sizes, the dimensions of the water phantom may prevent a full scan. As this usually only occurs with unwedged symmetrical fields, the water phantom can be offset to one side and only one half of the beam scanned to the appropriate distance from the CAX.

19.2.5 OUTPUT FACTORS

Measurements of the dose rate (or dose per MU) in phantom as a function of field size is a necessary step in the commissioning process. A large number of measurements are required because the dose per MU to a fixed point in a phantom depends on the size of the beam at that point, generally increasing monotonically with field size. The measured in-phantom *field output factors* (FOF) are assumed to be the product of two independent effects: *phantom scatter factor* (S_p) and *collimator (or head) scatter factor* (S_c). That is

$$\mathrm{FOF} = S_\mathrm{p} \times S_\mathrm{c} \tag{19.1}$$

where all three quantities are normalised to unity for the reference field (Khan 1984; van Gasteren et al. 1991). As discussed in Section 23.2.1.10, phantom scatter depends only on the scatter geometry within the phantom or patient; this can be modified by beam shaping, SSD, and patient shape. Collimator (or head) scatter, on the other hand, is independent of the phantom position, but depends on the collimator settings and the presence of additional filters (see Section 23.2.1.12). S_c also includes the effect of the monitor chamber response with collimator settings.

Measurements of FOF should be performed with an ionisation chamber, either a Farmer-type or a small area parallel plate, in a water (or water substitute) phantom (see Figure 19.7). The measurements should be at the depth and distance from the source, corresponding to the reference conditions used for calibration (see Chapter 18). Measurements should never be made at the depth of dose maximum (only a few centimetres below the surface), but instead should be made at a

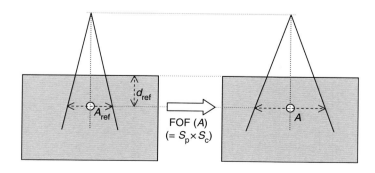

FIGURE 19.7
Diagram illustrating the measurement of the *field output factor* (FOF) in a full scatter phantom. A_ref is the reference field size at the measurement depth d_ref and A is the current field size.

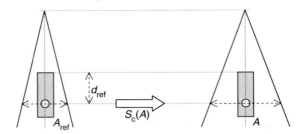

FIGURE 19.8
Diagram illustrating the measurement of the *collimator* (or *head*) *scatter factor* S_c (A) in a mini-phantom. A_{ref} and A are the reference and the current set field sizes respectively (see also Table 23.1). The dimensions of the mini-phantom (and especially its cross-section) must be just large enough to provide electron equilibrium. The depth of measurement d_{ref}, should be the same as for the measurement of the FOF (see Figure 19.7).

larger depth such as 5 cm or 10 cm where the influence of electrons scattered from the collimator is negligible. The measured dose rates could then be converted to values at the depth of maximum dose by applying a PDD or TMR correction (see Section 23.2.2).

Several square and rectangular field sizes should be measured spanning the available range. Rectangles should be measured with both the upper and lower jaws defining the long side to determine the *collimator exchange effect* (see Section 23.2.1.13). The measured dose rates are normalised to the reference field, usually a $10 \text{ cm} \times 10 \text{ cm}$ square.

It is desirable to be able to separate the two contributions to FOF, and several methods have been proposed to achieve this. These approaches can be divided into two categories: those that attempt to measure S_c separately and those that try to estimate S_p. In both cases, the total output variation in phantom, FOF, also needs to be measured. The third quantity can then be derived by application of Equation 19.1.

To measure S_c, an ionisation chamber should be placed at the depth corresponding to the FOF measurements (that could be at 5 cm or 10 cm) but with minimum side scatter contribution. This can be achieved with a mini-phantom* (van Gasteren et al. 1991; Dutreix et al. 1997) whose cross-section is just large enough for lateral electron equilibrium (Figure 19.8). For energy ranges from 4 MV to 25 MV, a $4 \text{ cm} \times 4 \text{ cm}$ cross-section is considered appropriate (Ahnesjö et al. 1987; Dutreix et al. 1997). In this mini-phantom, the scatter contribution to the ionisation chamber is constant for field sizes larger than its cross-section, and values of S_c can be derived from the measurements. It is preferable to make the measurements with the ionisation chamber at the same distance from the source as in the FOF setup. However, because collimator scatter is assumed to be independent of distance, a larger SSD could be used for the mini-phantom to extend the range of field sizes that totally cover it. Squares and rectangles (in both orientations) should be measured because the collimator exchange effect originates from collimator and head geometry. This method fails for field sizes comparable to or smaller than the mini-phantom cross-section.

On the other hand, S_p can be calculated from knowledge of the tissue air ratio (as a function of equivalent square) at the reference depth (see Section 23.2.2.6).

19.2.6   WEDGE DOSIMETRY

The availability of wedge filters that attenuate the beam differentially across their width requires additional measurements. The need to collect wedges profiles, relative depth doses, and

---

* The symbols S_c and FOF given here are the most frequenty encountered in the literature. In the ESTRO booklet, Dutreix et al. (1997) suggested the use of O_0 instead of S_c and O_R instead of FOF. However they kept the symbol S_p for the *phantom scatter factor*.

build-up curves has already been mentioned. It is also essential to characterise the beam outputs when wedges are present, using the same geometry employed for the FOF measurements and to relate them to the reference condition. The ratio of the dose rate with the wedge present to that without the wedge, for the same beam parameters, is known as the wedge transmission factor (WF). Because of the steep gradients in the beam profiles, particular care needs to be taken to ensure that the detector is aligned with its narrowest dimension along the wedged width and that it is truly on the axis of collimator rotation (CAX). The latter condition can be easily checked by measuring the dose rate through the steepest wedge at two opposite wedge orientations by rotating the collimator by 180°; the readings should vary by less than 1%.

Wedge transmission factors should be obtained for several field sizes spanning the available range because asymmetrical scatter in the phantom will affect the CAX dose. Rectangles as well as squares should be measured to investigate if the variation in wedge outputs can be approximated by a function of field width or of a combination of width and length (e.g. equivalent square see Section 23.2.1.8). If the formalism separating FOFs in S_c and S_p is to be employed, measurements of these quantities should also be performed for all wedges as described in Section 19.2.5.

19.2.7 SECONDARY BLOCKING AND ATTENUATORS

Most teletherapy equipment has accessories that enable the positioning of secondary shielding blocks between the machine head and the patient to modify the shape of the radiation field. These blocks are made of lead or high density alloy, and they are held in position by attaching them to a *tray* (usually clear plastic but sometimes metal with a hole pattern) that covers the entire beam. These blocks can be located either on the target side or the patient side of the tray. Several dosimetry issues are raised by the presence of the blocks and tray, and they need to be addressed by appropriate measurements.

First, trays attenuate the radiation, and their transmission needs to be quantified. The transmission factor is the ratio of the doses (at the depth and distance from the source corresponding to the reference condition) with and without the tray in position. This measurement needs to be performed for all available trays and also for any other absorber that could sometimes be in the path of the beam such as head supports, breast boards, or the couch top. For attenuators with uniform thickness across the beam, this factor is not usually a function of field size so that only the reference field need be used.

As previously mentioned, the presence of a tray, especially if close to the skin, could also affect the build-up curve and the surface dose. Build-up measurements should be performed with all the available trays to evaluate their effect on skin sparing.

The introduction of shielding blocks will alter the effective equivalent square of the beam entering the patient and the value of S_p, and they could also change the value of head scatter factor, S_c. Additional measurements should be made to evaluate the influence of beam shaping on the dose rate and to check the output factor calculation formalisms.

The beam penumbra, defined by the blocks, could be different from the one measured for the adjustable collimators, especially if the blocks are much farther from the source or if their internal faces are not aligned with the beam divergence. Additional profile measurements to define the beam edges should be taken using typical blocks.

19.2.8 ASYMMETRIC INDEPENDENT JAWS

Many modern accelerators offer the option of moving each jaw in a set of collimators independently from the other so as to produce fields asymmetric relative to the beam CAX. Although this facility offers added flexibility in clinical practice, it also increases the possible permutations of field shapes and the amount of data needed to characterise the beams. The variation of output factors and of S_c and S_p, separately, needs to be investigated for asymmetric as well as symmetric fields. The shape of the profiles across asymmetric collimators may or may not be predictable by simply splicing two symmetrical half-profiles together. Additional profiles should be measured to verify this assumption. On the other hand, the penumbra produced by each jaw should not be influenced by whether the opposite jaw is at the same distance from the CAX or not.

In addition to their radiation properties, the mechanical behaviour of asymmetric fields needs to be investigated more carefully than for the simple symmetric version. Their symmetry about the CAX when in symmetrical mode needs to be checked on a routine basis as well as the accuracy and reproducibility of independently repositioning each jaw. Of particular concern in this respect is the accurate positioning at the centre line because asymmetric half-field techniques are sometimes used to abut orthogonal fields at the isocentre.

19.2.9 MULTI-LEAF COLLIMATORS

Modern accelerators also offer the option of shaping the beam through the use of adjustable multi-leaf collimators (MLC). These devices consist of several narrow blocks with substantial attenuation along the beam rays that can be positioned independently of each other under computer control to form any desired beam shape (see Section 11.4.3). MLCs have obvious advantages over custom-made blocks in terms of preparation, handling, and storage, but they require additional mechanical and dosimetry checks, both at commissioning and in a routine quality assurance program (see Section 38.3).

The conversion of a required manually drawn beam shape into MLC positions or the transfer of MLC shapes from a treatment planning computer needs to be checked for accuracy. The mechanical accuracy and reproducibility of leaf positions need to be verified by extensive operation of the MLC system at regular intervals. Preventive maintenance should be scheduled for the mechanical components of the MLC system.

If the MLC are used as replacements for secondary blocks in conjunction with two orthogonal sets of collimators, the effect of beam shaping on the output needs to be investigated as for blocks. In some accelerator models, the MLC actually replaces one of the sets of jaws. In this case, the position of each leaf will have a significant effect on the overall *head scatter factor*, S_c, and more extensive measurements will be needed to fully characterise this quantity.

The penumbra, both along and across the MLC leaves, needs to be determined carefully as it will depend on the leaf design and position from the CAX. This information will be needed to determine the required margin between the MLC leaf positions and the planned target volume projection in *beam's-eye-view* (BEV)-based conformal treatment plans.

Radiation leakage within an MLC system is an additional quantity that needs to be evaluated. Leakage can occur though the small gap between two touching opposing leaves. It is worse when they abut on the central plane of the beam because of a lack of divergence, and it decreases as the junction is moved to one side. In clinical use, the position of the junction between *closed* leaves should be forced to be outside the field. Leakage also occurs between adjacent leaves and could also be a function of their distance from the centre line. In practice, this leakage can be minimised by enclosing the MLC shape as tightly as possible with the collimators. These variations in leakage need to be quantified as a function of distance from the centre. They are best studied either with a diode scanned in a water phantom or by exposing a film perpendicular to the beam direction while sandwiched in a solid phantom at the depth of maximum dose.

19.2.10 DYNAMIC TREATMENTS

The term *dynamic treatment* refers to several different irradiation modes where some machine parameters vary while the beam is on, from standard arc therapy to full-blown dynamic intensity-modulated radiation therapy (IMRT). Because the dose distribution is changing in time as well as in space, some of the standard measurement techniques previously mentioned need to be modified. Various commonly available modalities will be discussed in detail with some general guidelines on the more esoteric developments.

19.2.10.1 Arc Therapy

Most teletherapy units equipped with an isocentric gantry can deliver radiation while the gantry rotates around the isocentre. The equipment's ability to deliver the required MUs within the specified arc range needs to be validated. To check the TPS predictions, output measurements should be performed within cylindrical or anthropomorphic phantoms of known density that should be introduced into the TPS through CT scanning. These checks are best carried out with Farmer-type chambers in integration mode, located at pre-drilled holes in the phantom at the isocentre and other points of interest.

19.2.10.2 Dynamic Wedges

Some accelerators can produce wedge-shaped profiles by moving one independent jaw across the field while the beam is on. In this case, there is no beam hardening to worry about, so there is no pressing need to measure relative depth doses under those conditions. Many treatment planning systems can model the wedge shape directly using a knowledge of the jaw movements. In this case it is only necessary to make a few verification measurements. However, if this facility is not available, profiles need to be collected at several field sizes for each wedge angle - more than for fixed wedges. This was also necessary with early Varian implementations of the dynamic wedge where jaw motion was field size dependent.

These requirements are made even more time-consuming because of the difficulties posed by the time-varying nature of the exposure. The use of a remotely controlled water phantom moving a single detector to measure profiles becomes problematical. One solution consists of using an array of large numbers of detectors in integration mode suspended in the water phantom (Sidhu 1999; Martens et al. 2001). To extend the range and resolution of this array, it can be shifted using the water phantom controls with the full programmed exposure repeated several times. Alternatively, a large number of films need to be exposed and carefully scanned, taking into account the nonlinearity of the film's response over the large range of doses produced by a steep wedge.

Another consequence of the custom programming of the motion pattern for each field size mentioned above is that the output on the CAX (i.e. the effective WF) could be a strong and unpredictable function of the field dimension along the jaw motion (width). In that case, measurements of effective WF need to be made at every programmable field width so that there is no need to rely on interpolation.

Considering the large number of measurements required for dynamic wedges, alternative solutions have been investigated where the dynamic data are obtained from calculated super-position of static beam data (Papatheodorou et al. 1999) However, even in this case, an experimental verification for a limited number of conditions covering the whole range of clinical situations is required.

19.2.10.3 Other Dynamic Treatment Modalities

With the advent of fully three-dimensional (3D) planning systems incorporating BEV facilities and inverse planning, a large variety of complex dynamic treatments can be planned and delivered on modern accelerators (see Section 43.3). Examples include dynamic arc therapy where the MLC shape is adjusted to the BEV projection of the target every few degrees of rotation without pause in the irradiation; sequential IMRT that is obtained by a programmed succession of differently shaped fixed fields at each beam orientation; and dynamic IMRT where the shape of the field is varied while the beam is on, moving each MLC leaf at a programmed differential speed.

The basic commissioning of such treatment modalities as far as the dosimetry of each of its separate static components is concerned, is not different in principle to that for normal MLC treatments. The difficulty arises in attempting to verify that the complex sequence of configurations that constitute the complete treatment is being correctly performed as planned. Measuring the integrated dose at the centre of the treated volume with an ionisation chamber is an obvious minimum requirement. However, as the aim of these techniques is the delivery of a very specific dose distribution in three dimensions, checking the dose at only one point is not a sufficient check. Films (see Section 17.2) or multidetector-arrays (Letourneau et al. 2004) are of great value to assess the 2D fluence pattern for each beam individually. In addition, the task of validating dynamic treatments will be facilitated by the development of truly 3D and tissue-equivalent dosimeters such as Fricke or BANG gels that can be scanned in an MR imager to yield a 3D picture of absorbed dose (see Section 17.3).

19.3 ELECTRON MEASUREMENTS

19.3.1 VIRTUAL SOURCE POSITION AND ANGULAR SPREAD

The *effective* source position for electron beams depends somewhat on how this information is intended to be used. To estimate the dose rate at extended distances, a value for the virtual source position derived from inverse square analysis of output at different distances as described for photons is appropriate (see Section 19.2.1 and Section 23.2.1.16). In contrast to the latter, measurements should be performed at the depth of maximum dose in a water phantom[*]. In addition, the virtual source position for electrons is a strong function of the beam energy, jaw settings, field size of the applicator and endframe area, and extensive data need to be collected.

Some implementations of the *pencil beam* approach to calculating electron dose distributions (Hogstrom et al. 1981) require the input of an effective source position as well as an initial angular spread $\sigma_{\theta x}$ (see also Section 27.2.4.4). Both these quantities as well as the effective collimator position can be extracted from measurements of several beam profiles in air at different distances from the source. These measurements can be performed using a small ionisation chamber or a diode in a remotely controlled water phantom by scanning profiles at different distances but without filling the tank with water. Alternatively, films exposed in air perpendicular to the beam direction, supported by expanded polystyrene blocks at various distances, can be scanned to obtain the profile and penumbra information. A plot of the beam width (that is, the distance between the two points that receive 50% of the CAX dose) against the nominal distance from the source will extrapolate back to zero width at the *virtual* source position. A similar plot of the distance between the 90% and 10% penumbra levels against the nominal distance from the source will extrapolate back to zero penumbra at the *effective* collimator position. The slope of this regression line will be proportional to the initial angular spread (see Section 27.2.4.4).

19.3.2 RELATIVE DEPTH DOSES AND OFF-AXIS PROFILES

The same procedures for photon measurements described above should be followed for electron depth doses and profiles. If an ionisation chamber is employed to measure relative depth doses, the readings will have to be converted from ionisation to dose by applying energy-dependent corrections at each depth in the scan. In contrast to photons, electron depth doses have definite ranges beyond which only a bremsstrahlung contamination dose will be present (see Section 24.2.2). Depth scans need not reach much deeper than about 5 to 10 cm beyond the range of each electron energy. This depth can be estimated by the well-known rule of thumb that the range in water, in centimetres, is half the nominal energy in MeV.

---

[*] Such measurements are not trivial since the depth of maximum dose varies with field size and slightly with distance (see Section 24.2.4 and Section 24.2.6). However, in practice, the measuring depth could be kept constant for a large range of situations and be adjusted only for smaller field sizes or extreme SSDs.

Another consequence of the finite range of electron beams is that the depths for profiles need to be carefully chosen at each energy, restricting them to values smaller than the practical range. A possible profile measurement scheme would be to scan at depths beyond the maximum corresponding to the 100%, 90%, 70%, 50%, 30%, and 10% relative depth dose values as well as at one depth relatively close to the surface, for instance 0.5 cm. The shape of the profiles and the width of the penumbra in electron beams are strongly dependent on the distance from the end of the applicators to the surface. Depth doses and profiles should be measured at several source-to-surface distances covering the expected clinical range.

19.3.3 OUTPUT FACTORS

The measurement and behaviour of output factors for electron beams differ greatly from those of photon beams.

Because of the sharp fall-off of the electron depth-dose curves, measurements at depths larger than the depth of maximum dose (d_{max}) would introduce unacceptable uncertainties with even small variations in detector position. The point of output measurement is invariably at the depth of d_{max}. For the same reason, care must be taken to determine this depth for all measured field sizes, as it can change for smaller fields, and reposition the detector accordingly. Alternatively, if the same depth of measurement corresponding to the depth of maximum for the reference field is used for all field sizes, the value at each d_{max} needs to be calculated using the corresponding depth dose curve. The optimal detector for electron output factors is a small volume plane-parallel chamber, both because of the unambiguity of its effective position (at the inside surface of its front wall) and the virtual absence of perturbation corrections. However, as previously mentioned, the ionisation-to-dose conversion factors depend on the effective electron energy and on the depth of measurement. These corrections need to be applied before normalisation to the reference field. Measurement of output factors for very small fields where the field size is comparable to the chamber size need special consideration.

The variation of output dose rate with field size for electron beams depends on accelerator design. For units with a single variable electron applicator consisting of four *trimmers* attached to the photon jaws, the output is a smooth function of applicator size. More often, field shaping is achieved by separate *applicators* for different field size ranges with the provision to add *endframes* to achieve intermediate or irregular fields. These applicators often have additional scattering rings within them to improve the beam flatness near the edge of the beam. These scattered electrons also contribute to the dose rate at the centre of the beam. As a consequence, the variation of output with applicator size and with endframe area is not obviously predictable just on the basis of the beam shape at the surface. Therefore, dose rates need to be measured for each applicator and for a range of endframe shapes within each applicator to be able to determine the dosimetric characteristics of such electron beams. For some accelerators, the position of the photon jaws is adjusted to achieve a preliminary collimation upstream from the applicator. Therefore, the field size delimited by the photon jaws is both dependent on the size of the applicator and on the beam energy. Because the opening of these jaws has a significant influence on the beam output, care should be taken to ensure that commissioning measurements and clinical use are performed in a consistent way.

Because a large number of measurements at different depths for different energies and field sizes are required, it can take a long time. The careful use of a remotely controlled water phantom can speed up the acquisition of output factors at different depths of maximum dose, as an alternative to measurements at a fixed depth in solid phantoms corrected to the individual d_{max} through use of previously acquired depth dose curves.

SUPPLEMENTARY DETAILS ON CODES OF PRACTICE FOR ABSOLUTE DOSE DETERMINATION

Pedro Andreo and Alan Nahum

CONTENTS

This appendix is intended to provide supplementary information on the codes of practice for absolute dose determination which are described in Chapter 18. The origin of the various factors involved in dose determination and the questions which are still to be resolved are discussed. Only megavoltage photon and electron beams are dealt with. The proper choice for stopping-power ratios and perturbation factors are dealt with in Section D.1 and Section D.2, respectively. Section D.3 is dedicated to the specification of the beam quality. Section D.4 considers the issue of humidity correction for ionisation chambers. Finally, Section D.5 contains an estimation of the overall uncertainties in absolute dose determination in reference conditions. For a better understanding of the concepts discussed in this appendix, it is recommended that the reader familiarise her/himself with the codes of practice presented in Chapter 18 and with the general characteristics of photon and electron beams presented in Chapter 20, Chapter 22 and Chapter 24.

D.1 STOPPING-POWER RATIOS

D.1.1 THEORY

The determination of the absorbed dose in a medium using an ionisation chamber is based on the Bragg–Gray principle, which relates the absorbed dose at a point in the medium (water), D_w, to the mean absorbed dose in the detector (air), $\bar{D}_{air}$, through a proportionality factor that has been classically identified as the ratio of the mass (collision) stopping powers, water/air:

$$D_w = \bar{D}_{air}\, s_{w,air} \qquad (D.1)$$

The key Bragg–Gray assumption is that the electron fluence present in the detector is identical to that in the (undisturbed) medium at the position of interest. The gas-filled ionisation chamber in a high-energy photon or electron beam behaves to a good approximation as a Bragg–Gray detector. Any deviations from perfect Bragg–Gray behaviour are taken care of by perturbation factors which are dealt with in Section D.2.

Thus the stopping-power ratio applies to the electron spectrum at the point of interest in the undisturbed medium and is thus independent of the detector (except for the minor influence of the Spencer–Attix cutoff—see below). The determination of this stopping-power ratio is the task of cavity theory (see Section 6.7.3 for a derivation of the Bragg-Gray result) and will now be described.

To understand the reasons for the existence of some of the theories and approximations described below, ones needs to be aware of the calculation limitations some decades ago, when the determination of the so-called *slowing-down spectrum* (which today is called the *electron fluence spectrum*) by numerical methods was a considerable undertaking (see, for instance, the review given in NCRP [1961]). Approximations had to be developed to make the analytical calculations feasible, and some of these approximations have persisted for many years. Today, it is actually more straightforward to carry out accurate calculations using Monte Carlo simulation, as will be seen. In what follows, the terminology and formulation given by ICRU (1984a) will be used.

For an ideal detector that does not perturb the fluence of electrons in the medium, and assuming that electrons lose their energy *continuously* and *locally*, the *Bragg–Gray stopping-power ratio* is given by

$$s_{w,air}^{BG} = \frac{\int_0^{E_{max}} (\Phi_{E,w})_p (S/\rho)_{col,w}\, dE}{\int_0^{E_{max}} (\Phi_{E,w})_p (S/\rho)_{col,air}\, dE} \qquad (D.2)$$

where $(\Phi_{E,w})_p$ is the fluence of primary electrons in water, differential in energy, at the point of measurement, and $(S/\rho)_{col}$ is the unrestricted mass collision stopping-power of an electron with kinetic energy E in water or air*. In this context, *primary* electrons are either *incident electrons* or *photon-generated electrons*, e.g. Compton electrons; subsequent generations of electrons, i.e. δ-rays, are termed *secondaries*. The assumption of continuous and local energy loss ignores the finite ranges of δ-rays whose energy is assumed to be deposited at their site of production. Thus these δ-rays do not contribute to the primary electron fluence, $(\Phi_{E,w})_p$. Note that the numerator and denominator of Equation D.2 can be identified with the quantity *cema*, defined by ICRU (1998). The use of the *Bragg–Gray* (i.e. unrestricted) stopping-power ratio can only strictly be justified when δ-ray equilibrium conditions are realised due to a complete build-up of the δ-ray electron spectrum, which in practice never exists. However, approximate δ-ray equilibrium can be achieved through the use of a detector wall (or a thin lining on the inner side of the wall, e.g. graphite dag) which is approximately equivalent (in the sense of radiation interaction coefficients) to the cavity material itself (air). Bragg–Gray stopping-power ratios were used in ICRU Report 14 for photon dosimetry (ICRU 1969); they were evaluated using a primary electron fluence estimated from the reciprocal of the total stopping-power. Such methods are discussed in Johns and Cunningham (1983).

The *Spencer–Attix* stopping-power ratio (Spencer and Attix 1955) takes approximate account of the finite ranges of δ-rays with regard to energy deposition *in* the detector by δ-rays generated *outside* the cavity (however, the detector walls are assumed equivalent to the surrounding medium). Quoting NCRP (1961), "the Spencer–Attix theory is the only higher approximation that has been developed to the point where comparison with experiment is generally possible. The theory is not a rigorous one. At the risk of oversimplifying the physical picture, rigor was reduced to the point where numerical calculations became feasible." Nahum (1976, 1978) recast the original Spencer–Attix formulation into a more practical one involving (depth-dependent) electron fluence spectra.

The Spencer–Attix stopping-power ratio is given by (Nahum 1978; ICRU 1984a)

$$s_{w,air}^{SA} = \frac{\int_{\Delta}^{E_{max}} \Phi_{E,w}(L/\rho)_{\Delta,w}\, dE + \Phi_E(\Delta)_w(S(\Delta)/\rho)_w \Delta}{\int_{\Delta}^{E_{max}} \Phi_{E,w}(L/\rho)_{\Delta,air}\, dE + \Phi_E(\Delta)_w(S(\Delta)/\rho)_{air}\Delta} \tag{D.3}$$

where $\Phi_{E,w}$ is the total fluence of electrons in water, differential in energy (at the point of interest), and $(L/\rho)_{\Delta}$ is the mass collision stopping-power in water or air *restricted to energy losses less than* Δ. In analogy with Equation D.2, the integrals in the numerator and denominator can be identified with the quantity *restricted cema*, defined by ICRU (1998). The quantities $\Phi_E(\Delta)_w$ and $(S(\Delta)/\rho)_w$ are, respectively, the total electron spectrum and *unrestricted* collision stopping-power evaluated at energy Δ; the product of these two gives approximately the number of electrons falling below Δ in energy which, when multiplied by energy Δ, is equal to the total energy dissipated by the so-called *track-ends*. The energy cut-off, Δ, is related to the dimensions of the cavity, being equal to the energy of an electron whose range is equal to the mean chord

---

* A simple approximation for the *Bragg–Gray* stopping-power ratio has been given by Harder (1965a):

$$s_{w,air}^{H} = \frac{(S(\bar{E})/\rho)_{col,w}}{(S(\bar{E})/\rho)_{col,air}}$$

where the unrestricted mass collision stopping-powers are evaluated at the mean energy $\bar{E}$ of the primary electron spectrum at the point of measurement. This approximation for $s_{w,air}^{BG}$ is only acceptable when $(S(E)/\rho)_{col,w}/(S(E)/\rho)_{col,air}$ varies *linearly* with energy over the range of energies in the primary electron spectrum, as is the case when the detector and the medium are similar materials, and presupposes that the mean energy $\bar{E}$ is correctly estimated. For electron beams Nahum (1976, 1978) found discrepancies of up to 3.5% between the Harder approximation using a very crude expression for $\bar{E}$ and a rigorous evaluation of $s_{w,air}^{BG}$ using Monte-Carlo generated $\Phi(z)_{E,w}$.

length across the cavity. Electrons with energy greater than Δ are assumed to originate *outside* the cavity and to deposit their energy locally *inside* the cavity. It can be mentioned that some early determinations of Spencer–Attix stopping-power ratios omitted the track-end term (see, for example, Nahum [1976, 1978]).

The expression for $s_{\mathrm{w,air}}^{\mathrm{SA}}$ given in Equation D.3, with a cut-off, Δ, of 10 keV (approximately valid for most ionisation chambers in clinical use), has been used for the determination of practically all stopping-power ratio data available today for electron and photon beam dosimetry. The calculation of electron energy spectra in different media produced by photon and electron beams has relied extensively on the use of the Monte-Carlo method*. An alternative procedure to that of first calculating electron energy spectra and then using this to evaluate the cavity integrals is to compute *directly* the stopping-power-ratios at all depths during a Monte Carlo simulation. During the transport of charged particles in a medium, the product of the electron step-length (see Section 5.4) and the (restricted) stopping-power for any medium, evaluated at the mean energy of the step-length, yields the energy deposition in the material where the transport is simulated. There are two main advantages in this procedure, namely, it is independent of the number and size of the energy bins (as opposed to the calculation of electron spectra differential in energy) and it allows a straightforward determination of the statistical uncertainty of the calculated stopping-power ratios. This *on-the-fly* procedure for the computation of stopping-power ratios has been used for photon, electron and proton beams (see, for example, Andreo [1991, 1994], Malamut et al. [1991], and Medin and Andreo [1997]).

D.1.2 STOPPING-POWER RATIOS FOR ELECTRON BEAMS

Berger and Seltzer (1969) were the pioneers in using Monte Carlo simulation to derive electron fluence spectra as a function of depth in monoenergetic electron beams; subsequently they applied this to computing $s_{\mathrm{w,air}}^{\mathrm{SA}}$ (Berger et al. 1975; see also ICRU 1984a). This was improved upon by Nahum (1976, 1978) who included the track-ends in the evaluation of $s_{\mathrm{w,air}}^{\mathrm{SA}}$, as mentioned above. Andreo (1990a) calculated a comprehensive set of ICRU-37 (ICRU 1984b) based stopping-power ratios for monoenergetic electrons with different Monte Carlo codes, finding a level of agreement that provided confidence in the data used for clinical dosimetry. The most important characteristic of electron stopping-power ratios is their strong dependence on energy and depth (see Figure D.1), mainly resulting from the considerable differences in energy spectra at the various depths.

Until recently, the selection of stopping-power ratios, $s_{\mathrm{w,air}}$, for the user's beam in electron dosimetry protocols was based on monoenergetic-beam data, using a procedure based on the characterization of the electron beam via the mean electron energy *at the phantom surface*, $\bar{E}_{\mathrm{o}}$, together with the depth of measurement, z (IAEA 1987; IPEMB 1996a; IAEA 1997b;)[†]. Clinical beams are, however, far from monoenergetic and monodirectional at the phantom surface. The validity of the $s_{\mathrm{w,air}}(\bar{E}_{\mathrm{o}}, z)$ selection procedure has been reviewed in detail in the IAEA Code of Practice for plane–parallel ionisation chambers (IAEA 1997b); it was concluded that even for beams with large energy and angular spread the maximum error produced by such a procedure was always within 1%. For most beams used in clinical practice, even for those with a certain degree of photon contamination, the agreement was within the estimated uncertainty of the calculated stopping-power ratios, which is of the order of 0.6%. As illustrated in Figure D.2, differences can be greater at small depths for beams with considerable energy and angular spread. It is worth emphasizing that curves of $s_{\mathrm{w,air}}$ vs. depth for clinical and for monoenergetic beams have different slopes, crossing at a depth (greater than the depth of maximum absorbed dose, d_{max}) which varies with the contamination of the incident electron

---

* See Andreo (1988) for a review of this technique applied to the determination of stopping-power ratios.
[†] See Section D.3.2 for a discussion on electron beam quality specification.

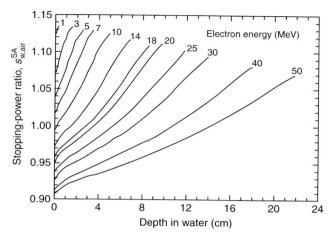

FIGURE D.1
Depth variation of the Spencer–Attix water/air stopping-power ratio, $s_{w,air}^{SA}$, for $\varDelta = 10$ keV, derived from Monte-Carlo generated electron spectra for monoenergetic, plane–parallel, broad electron beams. (From Andreo (1990a); IAEA (International Atomic Energy Agency), *The use of plane–parallel ionisation chambers in high-energy electron and photon beams. An International Code of Practice for Dosimetry, Technical Report Series no. 381*, IAEA, Vienna, 1997.)

beam. This supports the recommendation of the use of a depth greater than d_{max} as the reference depth for electron beams (ICRU 1984a; IAEA 1987, 1997b).

The work by Andreo and Fransson (1989), including stopping-power ratios for the *realistic* electron beams obtained by Udale (1988), who simulated in detail the treatment head of some clinical accelerators, has been extended by Ding et al. (1995) to other accelerators using the Monte-Carlo code BEAM (c.f. Rogers et al. 1995). Their results have confirmed the conclusions given above; thus no dramatic changes are expected in electron-beam dosimetry solely due to this improvement in the calculation of stopping-power ratios. The new stopping-power ratio data have been fitted empirically by Burns et al. (1996) to a ratio of polynomials,

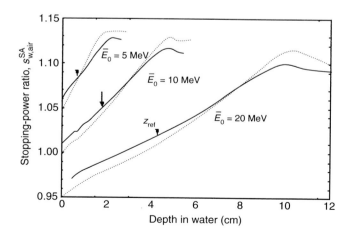

FIGURE D.2
Water/air stopping-power ratios of highly *contaminated* electron beams calculated directly with the Monte Carlo method (solid lines), compared with the $s_{w,air}^{SA}(\bar{E}_o = 2.33\ R_{50},\ z)$ method used in most N_K-based dosimetry protocols (dashed lines). The *contaminated* beams, with initial energies of 5, 10, and 20 MeV, include the energy and angular spread of electrons and contaminant photons produced in single scattering lead foils, the beam defining system, air, and water; their contamination is stronger than that existing in most clinical accelerators. The arrows indicate the reference depths recommended for beam calibration, usually different from the depth of maximum dose. (From IAEA (International Atomic Energy Agency), *The use of plane–parallel ionisation chambers in high-energy electron and photon beams. An International Code of Practice for Dosimetry, Technical Report Series no. 381*, IAEA, Vienna, 1997.)

constituting the set of data in common use today in the AAPM (1999) and IAEA (2000) dosimetry protocols based on standards of absorbed dose to water.

D.1.3 Stopping-Power Ratios for Photon Beams

For high-energy photon beams, Nahum (1976, 1978) first performed calculations of stopping-power ratios based on Monte Carlo derived electron slowing-down or fluence spectra at different depths. The technique has been used by Andreo and Nahum (1985) to compute $s_{w,air}$ values for a set of monoenergetic photon beams and then weight these to obtain stopping-power ratios for bremsstrahlung, i.e. polyenergetic spectra. This technique has been used also by Kosunen and Rogers (1993), who found exceptionally good agreement in the calculated stopping-power ratio data (at the 0.1% level), even when the Monte Carlo codes were different. As in the case of electron beams, agreement between different Monte Carlo calculations provides confidence in the data used for photon-beam dosimetry.

The most important characteristic of the depth variation of the stopping-power ratios of monoenergetic photons is that they are almost constant beyond the depth of transient electronic equilibrium, as Figure D.3 clearly shows. The range of variation of the stopping-power ratio data with energy is also much smaller than in the case of electrons with similar energies. In the case of bremsstrahlung spectra, interesting phenomena can be observed (see Figure D.4 and Figure D.5). The constancy of the stopping-power ratio is reached at shallower depths due to the presence of low-energy photons in the spectrum; furthermore, due to the higher values of the stopping-power ratio at lower energies, there is a *build-down* of the final stopping-power ratio, instead of the build-up shown for monoenergetic photon beams (Nahum 1982). This typical build-down pattern at small depths in a pure bremsstrahlung photon beam almost disappears in practice (except at very high energies) due to the presence of electron contamination (Figure D.5); the stopping-power of these electrons is usually lower than that of the photons. Thus, the two effects tend to cancel each other out and the constancy of the stopping-power ratio in a real photon beam is thus reached at smaller depths than in the case of a *clean* photon beam (Figure D.5).

Stopping-power ratios correlated in a consistent manner with the quality of clinical photon beams were calculated by Andreo and Brahme (1986) for a large number of bremsstrahlung spectra, including also the effect on $s_{w,air}$ of the electron contamination in photon beams and their dependence with depth and field size. These results have been used to calculate stopping-power

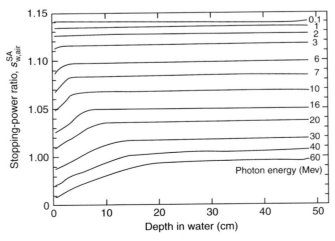

FIGURE D.3
Depth variation of the Spencer–Attix water/air stopping-power ratio, $s_{w,air}^{SA}$, for $\Delta = 10$ keV, derived from Monte Carlo generated electron spectra for monoenergetic, plane–parallel photon beams. (Adapted from Andreo, P. and Nahum, A. E., *Phys. Med. Biol.*, 30, 1055–1065, 1985.)

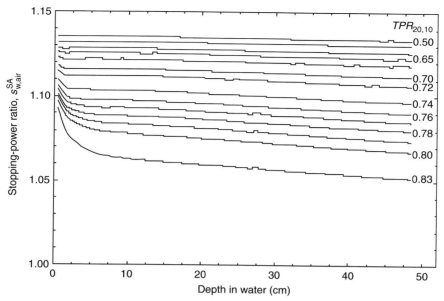

FIGURE D.4

Depth dependence of the water/air stopping-power ratios for uncontaminated plane–parallel (i.e. non-divergent) bremsstrahlung x-ray spectra as a function of the quality of the beams, $TPR_{20,10}$ (defined in Section D.3.1). The range of values for $s_{w,air}^{SA}$ should be compared with that for monoenergetic photons in Figure D.3. (Adapted from Andreo, P. and Brahme, A., *Phys. Med. Biol.*, 31, 839–858, 1986.)

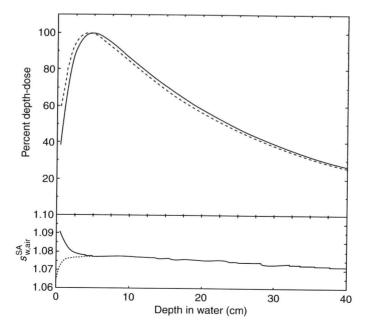

FIGURE D.5

The influence of electron contamination in depth-dose and water/air stopping-power ratio for a 25 MV linac spectrum. The solid lines show distributions for the *pure* photon bremsstrahlung beams, and the dashed lines show the final distributions when electron contamination is taken into account. (Adapted from Andreo, P. and Brahme, A., *Phys. Med. Biol.*, 31, 839–858, 1986.)

ratios for different materials and derive the ionisation-chamber correction factors used today in the majority of dosimetry protocols (Andreo et al. 1986). Improved calculations of stopping-power ratios using different Monte Carlo codes, and their correlation to the quality of photon beams based on the convolution of Monte Carlo calculated kernels, have also been made (Andreo 1991, 1994) but in most cases differences found were negligible, thus providing a firm consensus and lending reliability to the existing set of data. These have also been included in the IAEA (2000) dosimetry protocol based on standards of absorbed dose to water. The set of data calculated by Kosunen and Rogers (1993) have been used by the AAPM (1999) protocol but, as already mentioned, these stopping-power ratios are practically identical to those used by other protocols, the only difference residing in the use of a different photon-beam quality specifier to select stopping-power ratios. This aspect will be discussed in Section D.3.1.

D.2 PERTURBATION FACTORS

The perturbation factors are related to all the corrections which are required to take into account the fact that the Bragg–Gray conditions are not fulfilled when a real ionisation chamber is introduced into the medium. In what follows there is no further subdivision into photon and electron beams, though some of the effects are much more important for one modality than the other. In the case of kilovoltage x-ray beams, the formalism in the codes of practice is not based on Bragg–Gray behaviour and therefore the correction factors involved, such as k_{ch}, are not strictly perturbations and are not discussed here. In fact, one cannot logically talk about *perturbation* unless there is a quantity to be *perturbed*. Detailed treatments on perturbation have been given by Andreo (1991) and Nahum (1994, 1996). The emphasis here is on the *physics* of these correction factors; again, it is stressed that numerical values for the particular chamber and radiation quality of interest should be taken from the particular *code of practice* being followed.

D.2.1 THE DISPLACEMENT EFFECT (P_{eff} OR P_{dis})

When an ionisation chamber measurement is performed, a certain volume of the medium is *displaced* by the detector. Even if it is assumed that the wall is medium-equivalent, one still has the volume occupied by the air cavity. In general, for cylindrical chambers, the dimensions of this volume are not negligible compared to any gradients in the radiation field and hence in the dose distribution, e.g. the dose may change by a few percent over a distance equal to the diameter of the chamber. Clearly the chamber reading will then be affected by this *missing* medium. In simple terms one can expect that the reduced attenuation, in the case of photon beams, will result in a higher chamber reading, compared to that in a vanishingly small *air bubble* situated at the centre of the detector. However, there is another effect: the missing material means that there is less scatter. This will counterbalance the first effect. The net result is still generally an *increase* in the signal and a factor known as the displacement factor is required, usually denoted by p_{dis}, which will therefore be less than unity.

The value of p_{dis} will in general depend on both the radiation quality and the physical dimensions of the air cavity in the direction of the beam, as well as on the depth of measurement. In photon beams, p_{dis} will be practically constant beyond the depth of dose maximum, due to the exponential fall-off in dose. However, in the buildup region, it will vary in a complicated fashion with depth. For a Farmer chamber, which has an internal radius of 2.95 mm, the value is close to 0.988 in a ^{60}Co beam beyond d_{max} (Johansson et al. 1978).

The correction for displacement can be viewed in an alternative way. Instead of applying a factor to correct the chamber reading, assumed to be positioned so that its centre is at the depth of interest, a *shift* in the position of the chamber can be made. For a cylindrical chamber the electrons enter the wall at various depths, generally upstream of its centre, and hence the electron fluence in the air cavity is representative of that existing at some point in the uniform

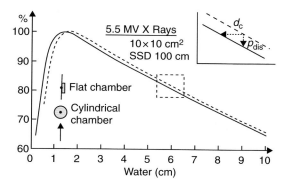

FIGURE D.6

Depth-ionisation curves measured at 100 cm Source to Surface Distance (SSD) with a cylindrical chamber (dashed line) or with a parallel-plate chamber (solid line). The depths are referred to the geometrical centre of the cylindrical chamber and to the inner surface of the front window of the parallel-plate chamber. Coincidence between the two curves can be achieved by translating the dashed line upstream by a constant shift, d. (From Dutreix, A. and Bridier, A., in *The Dosimetry of Ionizing Radiation*, Vol. 1, Kase, K. R., Bjärngard, B. E., Attix F. H., Eds., Academic Press, Orlando, FL, 163–228, 1985.)

medium shifted upstream of the chamber centre. In fact, it was found (Dutreix 1976; Johansson et al. 1978) that the (relative) readings of different chambers could be brought into coincidence with one another by performing shifts depending on the chamber dimensions. Thus the concept of the *effective point of measurement*, P_{eff}, was developed. This is illustrated in Figure D.6.

Assuming that the secondary electrons travel in the *forward*, i.e. beam direction, and that the number of ion pairs created is proportional to the chord length of the electrons in the cavity, it can be shown that the shift is $8r/3\pi(=0.85r)$ for a cylindrical cavity and $3r/4$ for a spherical cavity, both of radius r (Dutreix and Bridier 1985). More detailed treatments have been attempted, which take into account backscatter and the angular distribution of the secondary electrons (Svensson and Brahme 1986) but still fall well short of an exact theoretical analysis. Various experimental determinations of the *upstream* shift of P_{eff} from the cavity centre have been carried out (e.g. Hettinger et al. 1967; Dutreix 1976; Johansson et al. 1978); the readings of cylindrical chambers with different radii were generally compared with those of a plane–parallel chamber (cf. Figure D.6).

The newer absorbed-dose-to-water-based codes of practice favour the p_{dis} approach for the reference calibration point[*]. However, air-kerma-based codes of practice utilise the P_{eff} concept in preference to p_{dis}[†]. IAEA (1987) recommended a shift of $0.5r$ for ^{60}Co γ-rays, increasing to $0.75r$ for all higher energy photon beams; this is the quantity d_c in Figure D.7. More recent reviews of the experimental evidence on the magnitude of the shift (Andreo 1993; IAEA 1996) led the IAEA to recommend a single value of $0.6r$ for all high-energy photon beams (IAEA 1997a, 1997b).

In electron beams the use of p_{dis} is impractical as the depth-dose curve is very irregular in shape in total contrast to the quasi-exponential decrease in photon beams at depths beyond the buildup region, i.e. p_{dis} would vary rapidly and in an irregular way with depth in an electron beam. Thus in this case the P_{eff} concept is universally employed. For cylindrical chambers the recommended shift is $0.5r$ (IAEA 1987, 1997a, 1997b). For plane–parallel chambers P_{eff} is assumed to be situated at the centre of the inside face of the front wall, as illustrated in Figure D.7 (IAEA 1987,1997b;

---

[*] It must be stressed that, although both approaches are possible and yield equivalent results for the absolute beam calibration at the reference point, the relative dose measurements (e.g. depth dose variation) must always be based on the P_{eff} concept, i.e. assigning the measured dose to the effective point of measurement.

[†] In the case where a p_{dis} coefficient is used at the PSDL to obtain a calibration factor at the *centre* of the ionisation chamber, the same factor is strictly valid only if the user's beam falls off at approximately the same rate as the calibration beam.

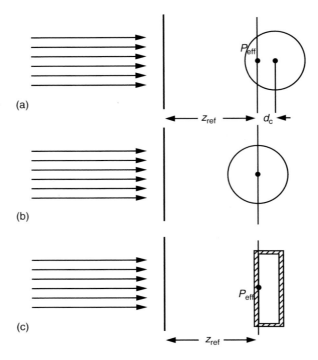

FIGURE D.7

In most N_K-based dosimetry protocols the effective point of measurement of a cylindrical ionisation chamber is positioned at the reference depth, z_{ref}, where the absorbed dose is required; the chamber centre is deeper than z_{ref} a distance d_c equal to the shift of P_{eff} (for example 0.6 r_{cyl} for photon beams in TRS-277 IAEA 1987/1997a) (a). Except in electron and heavy-ion beams, in $N_{D,w}$-based protocols, the centre of a cylindrical chamber is positioned at the reference depth, z_{ref}, and the absorbed dose is determined at this position (b). For plane–parallel chambers all protocols position the effective point of measurement (front of the air cavity) at the reference depth, z_{ref} (c).

IPEMB 1996a); this is logical as, in a well-guarded chamber, it can be assumed that all the electrons entering the sensitive air volume do so through the front window.

D.2.2   THE EFFECT OF THE CHAMBER WALL (p_{wall})

Bragg–Gray behaviour implies that the electron fluence in the sensitive volume of the detector is identical (strictly both in magnitude and in energy and angular distribution) to that present in the undisturbed medium at the position of interest. However, an ionisation chamber has a wall which in general is not made of medium-equivalent material. In the case of *photon beams*, the electron fluence in the air cavity in an ionisation chamber, assumed cylindrical, with a wall of a typical thickness, will consist partly of electrons generated in the (uniform) medium surrounding the wall and which have travelled through the wall, and partly of electrons generated by photon interactions in the wall material. Quite clearly, the number and energy distribution of these wall-generated secondary electrons will be characteristic of photon interactions with the material of the wall and not of the medium as demanded by Bragg–Gray conditions.

It can be shown (e.g. Attix 1986; Bielajew 1986) for the case of an (air-filled) ionisation chamber with a so-called *thick* wall, i.e. of a thickness equal to the buildup depth in the wall material, and where this material is a single substance, e.g. graphite or aluminium, that the relationship between the (mean) dose to air and the (mean) dose to medium is given, to a good approximation, by

$$D_{med} = \bar{D}_{air} \left(\frac{\bar{\mu}_{en}}{\rho} \right)_{wall}^{med} s_{wall,air} \qquad (D.4)$$

This expression is based on assuming that there is charged particle equilibrium in the wall at the position of the air cavity, i.e. using the *large detector* result $[D_{wall} = \psi\,(\mu_{en}/\rho)_{wall}]$, and then the Bragg–Gray result to relate $\bar{D}_{air}$ to D_{wall}. By comparing it to

$$D_{med} = \bar{D}_{air}\, s_{med,\,air}\, p_{wall} \tag{D.5}$$

the perturbation factor $p_{thick\text{-}wall}$ can be written as

$$p_{thick-wall} = \left(\frac{\bar{\mu}_{en}}{\rho}\right)_{wall}^{med} s_{wall,med} \tag{D.6}$$

For a graphite wall in a water medium, $p_{thick\text{-}wall}$ is practically constant at approximately 0.982 over the megavoltage range, and only begins to increase at the high-energy end, $TPR_{20,10} > 0.70^{*}$, where pair-production becomes important; for a nylon wall, the corresponding value of $p_{thick\text{-}wall} \approx 1.022$. Nahum (1994) has shown that over the broad (megavoltage) energy range where the Compton effect dominates, the material dependence of $p_{thick\text{-}wall}$ is essentially proportional to the difference between the mean excitation energies $(I_{med} - I_{wall})$.

For ionisation chambers in practical use in radiotherapy the wall thicknesses are much less than the buildup depths, and thus the above expression is not valid. Almond and Svensson (1977) wrote down an approximate two-component expression for the case of a wall of intermediate thickness, which is equivalent to:

$$p_{wall} = \frac{\alpha\, s_{wall,air}(\mu_{en}/\rho)_{med,wall} + (1-\alpha)s_{water,air}}{s_{water,air}} \tag{D.7}$$

where α is the fraction of the dose to the air in the cavity due to electrons generated in the chamber wall; thus if this is zero then p_{wall} reduces to unity, as expected. In Equation D.7 the medium is water; this is the expression given in some air-kerma-based codes of practice such as IAEA (1987).

The very simple Equation D.7 cannot in any sense be *derived* for the practically important intermediate-wall-thickness case; it is a purely empirical approach to an extremely complex problem. Figure D.8 illustrates schematically the origins of the different contributions to the dose in the air cavity:

- The spectrum of the water- or more generally, medium-generated electrons will be modified by their passage through the wall material, thus modifying the effective stopping-power ratio, $s_{med,air}$, for these electrons
- The δ-rays generated in the wall by the electrons originating in the medium will also affect $s_{med,air}$
- The μ_{en}/ρ ratio in Equation D.7 applies only to wall thicknesses sufficient for charged particle equilibrium which does not exist in the intermediate wall case
- The stopping-power ratio for the wall-generated electrons, $s_{wall,air}$, should reflect the fact that this spectrum is only partially *built-up*, i.e. similar to that existing in the buildup region in a high-energy photon beam (Andreo and Nahum 1985)

In a more complete analytical treatment all of the above factors would need to be taken into account. Only Monte-Carlo simulation can yield reliable theoretical values for p_{wall}.

To evaluate p_{wall} from Equation D.7, the wall-generated fraction of the ionisation, α, must be estimated. Lempert et al. (1983) measured α by adding successively thicker sheaths to a thin-walled chamber until full build-up was achieved at a number of different qualities. Codes of practice such as IAEA (1987, 1997a) give graphs of α as a function of beam quality, $TPR_{20,10}^{*}$, for different wall thicknesses, expressed in units of g cm^{-2} and it is

---

* $TPR_{20,10}$ is an index representative of beam quality of a megavoltage photon beam; it is defined in Section D.3.1 and increases with beam energy.

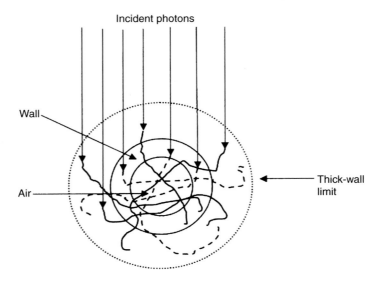

FIGURE D.8
The influence of the chamber wall on the secondary electron fluence in the air cavity; the dashed tracks are wall-generated electrons, the full lines are medium-generated.

then assumed that these α values are valid for any low-Z wall material. Values of $s_{wall,air}$ and $(\mu_{en}/\rho)_{water,wall}$ as a function of $TPR_{20,10}$ are also given in the IAEA code. Figure D.9 shows the values of p_{wall} using Equation D.7 for a range of commonly used chamber-wall materials, of thickness 0.5 mm. An additional small correction has been made for a 0.5 mm thick water-proofing sleeve of PMMA, using the extension of Equation D.7 to a three-component model, with a third term $\tau s_{sheath,air}(\mu_{en}/\rho)_{water,sheath}$ where $\tau s_{sheath,air}$ is the fraction of the ionisation due to electrons generated in the sheath (Gillin et al. 1985; Hanson and Dominguez-Tinoco 1985).

Experimental determinations of p_{wall} have been made by assuming that, for two chambers of different wall materials, the ratio of their air volumes is given by the ratio of the signals in an

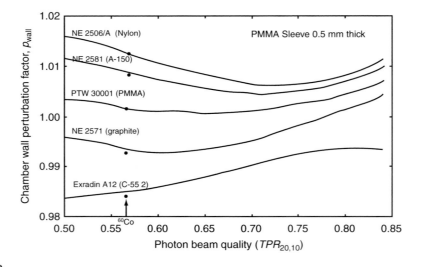

FIGURE D.9
The chamber wall perturbation factor, p_{wall}, as a function of the quality of photon beams for different ionisation chambers (wall materials). It is assumed that all the chambers are used in a water phantom with a waterproofing sleeve of PMMA 0.5 mm thick. (Adapted from Andreo, P., Nahum, A. E., and Brahme, A., *Phys. Med. Biol.*, 31, 1189–1199, 1986.)

electron beam, i.e. that the wall effects are negligible in an electron beam (see below). Using this method, Johansson et al. (1978) and Mattsson (1984) obtained values for Farmer-like chambers which were reasonably consistent with the Almond–Svensson two-component theory (Equation D.7); Nahum (1994) should be consulted for a more detailed discussion. A more direct test of Equation D.7 was carried out by Nahum et al. (1985) using a Farmer chamber, irradiated in a ^{60}Co γ-ray beam, with a wall of aluminium coated on the inside with successively thicker layers of dag, which is essentially graphite. Clear deviations from the simple two-component model were found, though aluminium represents an extreme test, having an atomic number more different from water than the wall materials used in practical chambers.

The formalism for evaluating p_{wall} described above was developed for cylindrical chambers, e.g. the Farmer chamber. However, plane–parallel chambers, which are specifically recommended for use in electron beams, especially at energies below about 10 MeV, are sometimes also used in photon beams. One particular instance of this is to determine the $N_{D,air}$ factor for a plane–parallel chamber by cross-calibrating it against a cylindrical chamber in a ^{60}Co beam (NACP 1981; IAEA 1997b). In this case, p_{wall} for the plane–parallel chamber must be known. However, it has been found that Equation D.7 gives values that are inconsistent with experiment for a number of plane–parallel designs (Mattsson 1984; Wittkämper et al. 1992; Nyström and Karlsson 1993). This is almost certainly due to the fact that several different materials are generally used in their construction. A Monte-Carlo simulation of plane–parallel chamber response (Rogers 1992b) showed that the energy deposition in the air cavity depends strongly on the material *behind* the cavity, probably due to electron backscattering. Such effects cannot be accounted for by the simple two- or three-component model. Consequently, it is no longer recommended that plane–parallel chambers be used in photon beams except for *relative* dose measurements (IAEA 1997b); however even here there can be perturbation effects in the buildup region unless there is a wide guard ring (Nilsson and Montelius 1986).

In the case of *electron beams*, it is generally assumed that the effect of the chamber wall is negligible. Certainly the physics here is different from the photon-beam case because there is no counterpart to the generation of secondary electrons characteristic of the wall material which can provide a large fraction of the chamber signal (through the factor α in Equation D.7). However, δ-rays are generated in the chamber wall which could, in principle, affect the chamber signal. Nahum (1988) developed an approximate theory to take this into account; this predicts that for walls of practical thicknesses and materials (graphite, A-150, C-552 plastic) the effect is always less than 0.5% and virtually independent of electron energy. Nahum (1996) should be consulted for a more detailed treatment.

D.2.3   The Effect of the Central Electrode (p_{cel})

Cylindrical chambers have a central electrode which is usually made of aluminium but can be made of graphite. A correction for the lack of air-equivalence of this is in principle necessary and is denoted by p_{cel}, though in some cases it has been included in p_{wall}. Mattsson (1984) measured the ratio of signals of cylindrical chambers with aluminum and graphite central electrodes at different photon beam qualities; he found a 0.9% effect at cobalt-60 on the assumption that there was no effect in the high-energy electron beam used to derive the ratio of the chamber volumes; Kristensen (1983) obtained 0.8% using a more direct method, without involving any assumptions in electron beams. Ma and Nahum (1993) carried out Monte-Carlo simulations of the effect of the electrode compared to a chamber with no electrode and their values are now the generally accepted ones: in photon beams, p_{cel} for a 1 mm diameter graphite electrode is negligibly different from unity at all qualities investigated (cobalt-60 to $TPR_{20,10} = 0.80$) but decreases from 1.008 to 1.004 for a 1 mm diameter aluminium electrode. Ma and Nahum

(1993) showed that the effect was negligible for graphite in electron beams, and never greater than 0.2% at any energy (5 MeV to 20 MeV) or depth for a 1 mm diameter aluminium electrode.

D.2.4 The Electron Fluence (In-)Scattering Effect (p_{cav})

An ionisation chamber introduces a low-density heterogeneity into a medium. In principle this is no different from the general problem of the influence of heterogeneities on the dose distribution in treatment planning. In an *electron beam*, density changes can cause *hot* or *cold* spots as a result of electron scattering. Harder (1968) analysed the problem of the effect of small air cavities on the electron fluence and produced an approximate theory using small-angle scattering which predicted an increase in the chamber signal. The reason for this is clear from the elegant figure due to Harder (see Figure D.10). As a result of (elastic nuclear) scattering, in an electron beam the angular distribution of electrons broadens with depth in the wall (or phantom material) whereas this broadening will be negligible in a gas cavity due to the very low density. Consequently, more electrons are scattered *into* a low-density cavity than are scattered *out*, resulting in an *increase* in the electron fluence towards the *downstream* end of the cavity, compared to the fluence in a uniform medium at that depth.

The magnitude of this *in-scattering* effect must be referred to a particular depth in the undisturbed medium. Thus a perturbation factor, p_{cav}, which corrects for the effect of the air cavity is defined as

$$\Phi_{med}(P_{eff}) = \bar{\Phi}_{cav}\, p_{cav} \tag{D.8}$$

where Φ_{med} (P_{eff}) and $\bar{\Phi}_{cav}$ refer to the primary electron fluence in the undisturbed medium (at the depth of the effective point of measurement P_{eff}) and in the cavity (averaged over the volume), respectively. It can be noted that this definition of the perturbation correction factor is only consistent with the expression for the absorbed dose to water (Equation D.4) if the energy distribution of the electron fluence in the cavity is very similar to that at P_{eff} in the uniform medium. This makes it possible to use the standard computations

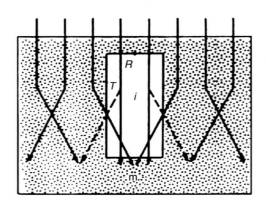

FIGURE D.10
Perturbation of the electron fluence caused by a gas-filled cavity in a solid or liquid phantom irradiated by an electron beam. Electron tracks are idealized to emphasize the effects being shown. The dashed lines represent tracks which would exist if the cavity was not present. The dominance of in-scattering over out-scattering gives rise to an increase in the fluence towards the back of the cavity and hence an increase in the chamber signal. (From Harder, D., Fano's therorem and the multiple scatter correction, in *Proceedings of the 4th Symposium on Microdosimetry*, Booz, H. G. E. J., Eickel, R., Waker, A., Eds., Euratom, Brussels, 677–693, 1974.)

of the stopping-power ratio, which are based on the Bragg–Gray requirements of identical fluences, differential in energy, in the medium and in the detector material.

The in-scattering effect has been experimentally demonstrated for cylindrical chambers by several workers (Harder 1968; Johansson et al. 1978). Johansson et al. (1978) used PMMA-walled chambers of diameters 3, 5, and 7 mm. The ratios of the sensitive air volumes in the chambers were assumed to be given by the signal ratios at d_{max} in photon beams. The measurements in the electron beams were all made in a PMMA phantom at the peak of the depth-ionisation curves to eliminate any displacement effects (see Section D.2.1). The energy at this depth z, E_z, varied between 2.5 and 22 MeV. For Farmer-type chambers (by interpolation) $p_{\text{cav}} = 0.995$ at $E_z = 20$ MeV, but decreases to 0.970 at 6 MeV. Johansson et al. (1978) found a linear relation between p_{cav} and cavity radius in contrast to the non-linear dependence on radius predicted by the Harder (1968) theory. All modern air-kerma-based dosimetry protocols (IAEA 1987; NCS 1989; IPEMB 1996a) include values of the perturbation factor determined experimentally by Johansson et al. (1978). Palm and Mattsson (2000) made another experimental investigation of p_{cav} using the same PMMA phantom and PMMA-walled cylindrical chambers as had been originally used by Johansson et al. (1978). Only minor deviations from the 1978 values were found. Additionally, Palm and Mattsson measured p_{cav} for graphite-walled Farmer-like chambers and found values at the lowest energies (E_z between 5 and 10 MeV) about 1% higher than those given in IAEA (1997b). In all cases, the linear relationship between p_{cav} and cavity radius was confirmed. Nahum (1994, 1996) and IAEA (1997b) should be consulted for a more detailed treatment of the *in-scattering* effect.

The magnitude of the in-scattering perturbation exceeds 3% for Farmer-type chambers for E_z below 8 MeV. This is one of the principal reasons why plane–parallel (or parallel-plate) chambers are recommended in low-energy electron beams (NACP 1981; AAPM 1994b; IPEMB 1996a; IAEA 1997b). A plane–parallel chamber has a completely different geometry to a cylindrical or thimble chamber. The diameter of the air cavity (typically between 13 and 20 mm) is deliberately made much greater than its thickness (the electrode spacing), which is 2 mm in almost all commercial designs. Thus most of the electrons enter the air cavity through the front face of the chamber and only a small fraction through the side walls. Furthermore, well-designed plane–parallel chambers have a relatively wide guard ring—3 mm or more—which ensures that almost no electrons entering through the short side walls can contribute to the chamber signal. Consequently, *in-scattering* is virtually

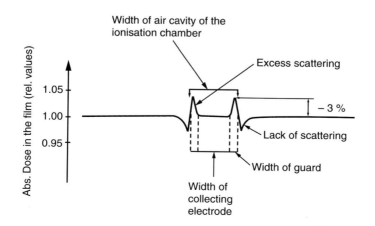

FIGURE D.11

Film measurements across the front surface of an air cavity in a PMMA phantom at the depth of maximum dose in a 6 MeV electron beam; the effect of the guard ring in eliminating the perturbation is clearly demonstrated (From Mattsson, L. O., Johansson, K.-A., Svensson, H., *Acta Radiol. Oncol.*, 20, 385–399, 1981.)

eliminated. Figure D.11 shows how any perturbation of the dose distribution is confined to the guard-ring region. The electron fluence in the sensitive volume of such a chamber is therefore that existing in the uniform medium at the depth of the inside face of the front window, which is the position of the effective point of measurement P_{eff} as illustrated in Figure D.7c.

There is a substantial body of experimental evidence that in-scattering perturbation is negligible for well-designed parallel-plate chambers such as the NACP and Roos chambers (e.g. Mattsson et al. 1981; Wittkämper et al. 1991; IAEA 1997b). A good example of this evidence is the study by Roos (1993) who compared the response of chambers with different guard-ring widths, each with a 2-mm cavity depth, in a 6-MeV electron beam. He showed that for guard-ring widths greater than 3 mm there was no change in the chamber responses compared to a chamber with an ultrabroad (8 mm) guard ring (IAEA 1997b). However, for one commercial parallel-plate design in particularly widespread use, the *Markus chamber*, several series of measurements have demonstrated a perturbation effect of the order of 1% at $E_z = 5$ MeV, increasing to 2% at 3 MeV (this is summarized in IAEA 1997b). This was to be expected given that the Markus chamber has a collecting electrode diameter of only 5.3 mm, a cavity depth of 2 mm, and a guard ring of only 0.2 mm.

When investigating the response of an ionisation chamber, the perturbation factor obtained experimentally is naturally an *overall* factor and may well include contributions from effects other than the primary one that is believed to dominate, e.g. in-scattering in electron beams. An example of a possible secondary effect in plane–parallel chambers is differences in electron backscattering from the material behind the air cavity compared to that of the (uniform) medium (Hunt et al. 1988; Klevenhagen 1991; Nilsson et al. 1996). The back wall of the Roos chamber is PMMA and is only 1.3 mm thick, whereas the NACP chamber has about 7 mm of graphite. And yet the NACP/Roos response ratio has been shown to be independent of electron energy (Roos 1993) and given that the designs of these two chambers almost certainly ensure that there is a negligible in-scattering effect, this strongly suggests that backscattering effects are also negligible (IAEA 1997b)

In *photon beams*, should a perturbation of the electron fluence in the air cavity due to the in-scattering effect also be expected? The effect in electron beams is due to the change in the *angular* distribution of the electron fluence with *depth*. However, in photon beams there is generally charged particle equilibrium (or a very good approximation to it), and there will therefore be no (or negligible) change in either the energy or angular distribution of the secondary electrons with position in the irradiated medium. Harder (1974) also reasoned that any such effect would be inconsistent with Fano's theorem (Attix 1986). Johansson et al. (1978) tried to detect such an effect by measuring the response ratios of cylindrical chambers with radii of 3, 5, and 7 mm at photon qualities from ^{60}Co γ-radiation to 42 MV x-rays with the chamber centres at d_{max} to eliminate the displacement effect (see Appendix D.2.1). A constant ratio of the readings was obtained despite the very different secondary electron energies over the photon quality range; it was thus concluded that the effect was negligible in photon beams. However, in the buildup region in photon beams, where there is incomplete charged particle equilibrium, perturbation effects have been demonstrated (Velkley et al. 1975; Nilsson and Montelius 1986).

D.2.5   CONCLUDING REMARKS

Experimental determinations of perturbation factors often involve questionable assumptions, e.g. that a particular effect is negligible in the beam quality used to normalise the responses of the different chambers, and none of the analytical treatments for any of these effects is wholly satisfactory. Only Monte-Carlo simulation of chamber response can be expected to yield highly accurate values of the different perturbation factors in the various chamber types (Rogers 1992b, 1993; Nahum 1999a; Kawrakow 2000; Verhaegen et al. 2006).

D.3 BEAM-QUALITY SPECIFICATION

D.3.1 PHOTON BEAM QUALITY SPECIFIERS

Prior to the use of modern practices for specifying photon beam quality, the nominal accelerator potential was the parameter most commonly used in photon-beam dosimetry. Measured absorbed-dose or ionisation ratios were first used as a beam-quality index in the dosimetry recommendations of the Nordic Association of Clinical Physicists (NACP, 1972; NACP, 1980). The measured dose ratio was, however, associated with the nominal accelerating potential, which was then used for the selection of conversion factors. A first attempt at improving the NACP procedure was made in the AAPM TG-21 protocol (AAPM 1983) where stopping-power and mass energy-absorption coefficient ratios, were associated in graphical form with measured ionisation ratios. However, these data were still given numerically as a function of the nominal MV, which involved a unique correspondence between MV and ionisation ratios, similar to the drawback pointed out for the NACP recommendations. Andreo and Brahme (1986) showed that the use of only the nominal accelerator potential, ignoring the actual penetration properties of a clinical beam, could yield variations of up to 1.5% in the stopping-power ratio. It is mainly for this reason that the use of a quantity representative of the beam attenuation in water and derived from the Tissue Phantom Ratio (TPR defined in Section 23.2.1.6), $TPR_{20,10}$ was considered to be a more appropriate choice for radiotherapy beams than the nominal accelerating potential. Most dosimetry protocols, based on standards of air kerma and on standards of absorbed dose to water, have recommended $TPR_{20,10}$, as the specifier of the quality of high-energy photon beams (AAPM 1983; SEFM 1984, 1987; NCS 1986; IAEA 1987; AIFB 1988; IPSM 1990; DIN 1997). $TPR_{20,10}$ is defined as the ratio of water absorbed doses on the beam axis at the depths of 20 cm and 10 cm in a water phantom, obtained with a constant source-detector distance (e.g. 100 cm) and a 10 cm × 10 cm field size at the position of the detector. The parameter $TPR_{20,10}$ is a measure of the effective attenuation coefficient, and describes the approximately exponential decrease of a photon depth-dose curve beyond the depth of maximum dose[*]; more importantly, it is independent of the electron contamination in the incident beam.

Other beam-quality specifiers have been proposed for photon-beam dosimetry which are, in most cases, related to the depth of maximum absorbed dose and can, therefore, be affected by

---

[*] A photon depth dose distribution can be described by the functional form (c.f. Brahme and Svensson 1979; Greening 1981; Brahme and Andreo 1986):

$$D(z) \approx D_0(e^{-\mu_p z} - \nu\, e^{-\mu_e z})$$

where μ_p is the effective attenuation coefficient of the photon spectrum and μ_e is the effective attenuation coefficient of photon-produced electrons plus contaminant electrons (ν is an adjusting parameter). At large depths, where the electron component can be ignored, the ratio of dose at two depths, z_1 and z_2, is given by,

$$\frac{D(z_2)}{D(z_1)} \approx e^{-\mu_p(z_2 - z_1)}$$

from where

$$\mu_p \approx \frac{1}{z_1 - z_2} \ln \frac{D(z_2)}{D(z_1)}$$

Thus, the ratio of absorbed dose at two depths is related to the effective attenuation coefficient of the photon spectrum. This relationship of the dose ratio with the effective attenuation coefficient is valid both for Tissue-Phantom Ratio and for Percent Depth Dose distributions although the μ_p will be of a different kind (for a parallel or for a divergent beam). The relationship is not valid when the dose is specified at one depth only, referred to d_{max}, where the distribution has a shoulder and it is not exponential.

the electron contamination at this depth[*]. Based on percentage depth-dose distributions, a widely disseminated recommendation for specifying the quality of high-energy photon beams was made in Supplement 17 of the British Journal of Radiology (British Institute of Radiology 1983); the parameter d_{80} was defined as the depth of the 80% depth dose (i.e. 80% of the dose maximum) for a 10 cm × 10 cm field size at a Source to Surface Distance (SSD) of 100 cm. In this supplement it was pointed out that electron contamination should be considered as a practical shortcoming of the method. The use of d_{80} as a photon beam quality index has also been endorsed in Supplement 25 of the *British Journal of Radiology* (British Institute of Radiology 1996), although other beam quality specifiers, like the Percent Depth Dose (PDD defined in Section 23.2.1.1) at a given reference depth, are also considered. In its conclusions, Supplement 25 also referred to contaminating electrons as the greatest problem for normalization at d_{max}, as by changing the dose at this depth electron contamination can alter the apparent beam quality[†].

The parameter PDD(10), the percentage depth-dose at 10 cm depth, determined under the same conditions of field size and source to surface distance (SSD) as d_{80}, has the same limitation with regard to the effect of electron contamination as d_{80} (see Section 23.2). This parameter has been commonly used by accelerator manufacturers, associating it with an effective accelerator potential. The work by LaRiviere (1989), proposing a relation between the beam quality specified in terms of MV and PDD(10), has been used by manufacturers to justify the use of this parameter. This has, however, produced the paradoxical situation where an accelerator could have an effective MV larger than the accelerator electron energy (see Section 22.2.2). LaRiviere also proposed a relation between PDD(10) and the dose-weighted mean energy of the photon spectrum, which was suggested as an alternative beam quality index. That proposal led Kosunen and Rogers (1993) to investigate the relation of $PDD(10)_x$ in a *pure photon beam* (i.e. without electron contamination) to stopping-power ratios. Based on the linearity of the relation obtained, they proposed extending the use of $PDD(10)_x$ to specify the quality of photon beams and to select conversion and correction factors. Kosunen and Rogers referred to the problem of the difference between absorbed-dose and ionisation measurements with cylindrical ionisation chambers due to the use of a displacement (or replacement) factor, and also emphasized that electron contamination should be removed from the photon beam for measuring $PDD(10)_x$. According to these authors the latter can be achieved using a thin lead foil as filter, which has become the method recommended by the AAPM TG-51 dosimetry protocol (AAPM 1999). Limitations of this attempt to remove contaminant electrons have been stressed by Andreo (2000).

It is important to emphasize that, in principle, stopping-power ratios and k_Q values can be related to any parameter indicating the penetration characteristics of photon beams, as both stopping-power ratios and dose distributions are usually determined in a correlated manner using a Monte Carlo calculation. Figure D.12 illustrates the variation of water/air stopping-power ratios with different photon beam quality specifiers, $TPR_{20,10}$, PDD(10) and d_{80}, using the beam quality parameters given in Table 5.iii of Supplement 25 of the *British Journal of Radiology* (British Institute of Radiology 1996) for the latter two specifiers. However,

---

[*] Stopping-power ratios at the depth of d_{max} are usually affected by electron contamination, but this occurs in a way such that the depth-variation of $s_{w,air}$ is practically constant after shallow depths (c.f. Andreo and Brahme 1986); pure photon beams have a build-up effect that makes $s_{w,air}$ different at shallow and at large depths. In addition, the use of ionisation distributions measured with thimble-type ion chambers is problematic, since the displacement perturbation factor has to be taken into account in converting ionisation into dose at depths other than the maximum. The alternative concept of the effective point of measurement is to be preferred. Plane–parallel ion chambers solve this problem, although they are not commonly used in photon beam dosimetry.

[†] It is worth noting that ever since the publication of BJR-11 (BJR 1972), the problem of electron contamination and the need to normalize dose-distributions at depths greater than d_{max} has been addressed by different BJR Supplements, but alternatives for a specifier independent of electron contamination have not been recommended in this series of publications.

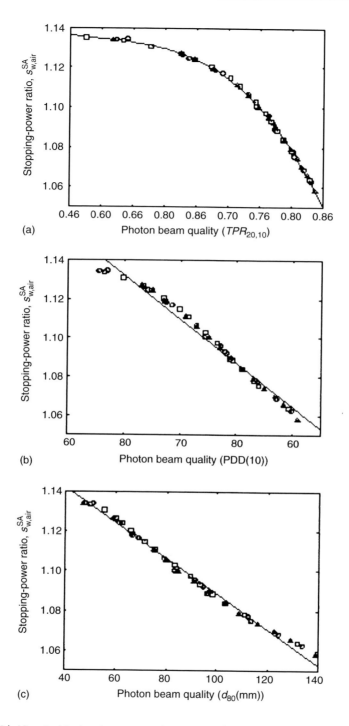

FIGURE D.12

Spencer–Attix ($\Delta = 10$ keV) water/air stopping-power ratios, $s_{w,air}$, at the reference depth for clinical photon beams versus different photon beam quality specifiers: (a) $TPR_{20,10}$; (b) PDD(10); and (c) d_{80}. Circles correspond to spectra published by different authors (c.f., Table 2 in Andreo and Brahme 1986) and squares represent the calculated spectra in the same reference. The triangles correspond to the qualities given in the *British Journal of Radiology*, Suppl. 25, 1996. The solid lines represent fits of the stopping-power ratios to each beam quality specifier; note that the left-most data point of (a) has not been included in the linear fits of (b) and (c). (Adapted from Andreo, P., *Med. Phys.*, 27, 434–440, 2000.)

calculating stopping-power ratios directly as a function of PDD(10) or d_{80} for realistic beams is not possible due to the lack of information on the spectra of contaminant electrons, which would be required as input to the calculations. Instead, the parameter $PDD(10)_x$, i.e. PDD(10) for *pure photon beams*, has been recommended to select stopping-power ratios (Kosunen and Rogers 1993; Rogers 1996). The problem is that in reality *pure photon beams* do not exist. Therefore, beam quality specification must rely either on a simple practical parameter which is truly related to the intrinsic physics of photon interactions (such as the practical attenuation coefficient discussed in relation with $TPR_{20,10}$) or the problems associated with electron contamination and attempts to remove it will outweigh any possible theoretical advantage in the use of an apparently more attractive simple linear fit.

Advantages and limitations of the different photon-beam quality specifiers have been discussed by Andreo (2000). The general conclusion is that there is no unique beam-quality specifier that works satisfactorily in all possible conditions for the entire range of photon energies used in radiotherapy and all possible accelerators used in hospitals and in standards laboratories. Most recent dosimetry protocols or codes of practice, based on the calibration of ionisation chambers in terms of absorbed dose to water, use a photon beam quality specifier in terms of $TPR_{20,10}$ (IPSM 1990; DIN 1997; IAEA 2000); the AAPM TG-51 protocol (AAPM 1999) uses PDD(10). Although both sets of data, PDD and TPR, are generally available for routine clinical use, there are more practical problems in measuring $PDD(10)_x$ than $TPR_{20,10}$, and errors in determining the beam quality index may have in general more serious consequences with $PDD(10)_x$ than with $TPR_{20,10}$. The overall impact on clinical photon-beam dosimetry resulting from the use of different photon-beam quality specifiers to select k_Q values is that they are not expected to yield a significant change (i.e. 0.5% at most and in most cases within 0.2%, c.f. Huq et al. [1999]) in the value of the absorbed dose to water in reference conditions for most clinical beams. This difference is considerably smaller than the combined uncertainty of the different factors and coefficients used in photon dosimetry and it is also less than the uncertainties quoted by standard dosimetry laboratories for calibrations in terms of absorbed dose to water (typically around 0.6%).

D.3.2 ELECTRON-BEAM QUALITY SPECIFIERS

For more than two decades, following the early recommendations of NACP (1980), the quality of electron beams has been specified in practically all dosimetry protocols by the mean electron energy, $\bar{E}_0$, of the incident spectrum at the phantom surface. This specifier was derived from the measurement of the half-value depth, R_{50}, defined as the depth at which the absorbed dose has decreased to 50% of its maximum value (ICRU 1984a). An empirical relationship between $\bar{E}_0$ and R_{50}, derived by Brahme and Svensson (1976), is

$$\bar{E}_0[\text{MeV}] = CR_{50} \tag{D.9}$$

where $C = 2.33$ MeV cm^{-1} and R_{50} is in cm. This approximate relationship was derived from measurements on clinical accelerators in common use during the seventies. Strictly speaking, it is valid only for large field sizes (*broad beams*), energies between 5 and 30 MeV, and for R_{50} determined from depth-dose distributions measured in water with a constant source-chamber distance (i.e. equivalent to a plane-parallel beam) (Brahme and Svensson 1976; ICRU 1984a). The criterion for a *broad beam* is when the depth-dose distribution can be considered to be relatively independent of the field size, which is approximately achieved with at least 12 cm $\times$ 12 cm for energies up to about 15 MeV, and at least 20 cm $\times$ 20 cm for higher energies.

When the dose distribution has been obtained with a constant source-surface distance this relation is not valid. As an alternative, IAEA (1997b) has provided the relations

$$\bar{E}_0[\text{MeV}] = 0.818 + 1.935R_{50}^I + 0.040(R_{50}^I)^2 \tag{D.10}$$

for R_{50}^J determined from a depth-ionisation curve, and

$$\bar{E}_o[\text{MeV}] = 0.656 + 2.059R_{50}^D + 0.022(R_{50}^D)^2 \qquad (\text{D}.11)$$

for the case of an R_{50}^D determined from a depth-dose curve.

There has been much controversy around the use of Equation D.9 and consequently new energy-range relationships between $\bar{E}_o$ and R_{50}, based on Monte-Carlo calculations for mono-energetic electron beams, have been developed (see for instance Rogers and Bielajew 1986; Andreo 1993; Ding et al. 1996); all yield $\bar{E}_o$ values higher than that those obtained using Equation D.9. This would result in *lower* stopping-power ratios at the reference depth compared to those obtained with the $s_{w,air}(\bar{E}_o, z)$ method and $\bar{E}_o$ from Equation D.9. It has been shown (see IAEA 1997a) that accurate determinations of $s_{w,air}$ based on beam quality derived from detailed Monte Carlo simulations of accelerator treatment heads yield values of $s_{w,air}$ at the reference depth which are closer to those derived from Equation D.9 than from the use of new energy-range relationships. Further details on the energy-range relationship in connection with the selection of $s_{w,air}$ have been given in an appendix of IAEA (1997). It should be stressed that accurate experimental determinations of the mean energy at the surface are extremely difficult. Energy determinations based on spectrometric measurements have been reported (André et al. 1985; Sorcini and Rosander 1993) in which detailed straggling distributions around the most probable energy have been obtained. However, precise measurements of the low-energy tail of a straggling distribution are inherently limited by the experimental procedure, and therefore experimental determinations of the mean energy of the spectrum will in general overestimate the true value.

The mean energy at depth, $\bar{E}_z$ is a related quantity in widespread use in electron-beam dosimetry. An empirical relation, originally developed by Harder (1965b), for the most probable energy of an electron spectrum has been recommended by many dosimetry protocols for estimating $\bar{E}_z$:

$$\bar{E}_z = \bar{E}_o(1 - z/R_p) \qquad (\text{D}.12)$$

where R_p is the practical range defined as the depth where the tangent at the steepest point (the inflection point) on the almost straight, descending portion of the depth-dose curve meets the extrapolated bremsstrahlung background (ICRU 1984a) (see Figure 24.1). Other expressions for the calculation of $\bar{E}_z$, based on the continuous-slowing-down approximation, have been given in ICRU (1984a). Using the Monte Carlo method it has been shown by various authors that Equation D.12 is an acceptable approximation of the mean energy at depth only for electron beams with incident energies less than 10 MeV or for small depths at higher energies (Andreo and Brahme 1981; Ding et al. 1996; Fernandez-Varea et al. 1996). Much of the available data for electron dosimetry, which were originally given in terms of $\bar{E}_z$ determined according to Equation D.12, have been recast to Monte-Carlo determined $\bar{E}_z$ in recent dosimetry protocols (IPSM 1996; IAEA 1997b, 2000).

It must be emphasized that the use of $\bar{E}_o$ and $\bar{E}_z$ in electron-beam dosimetry is only an approximation for the practical purpose of selecting dosimetric quantities and correction factors, and is not intended to be an accurate statement of the energy parameters of an electron beam. This misunderstanding has resulted in numerous publications questioning the validity of the available relations and data, but the reality is that electron-beam dosimetry has not been improved by more than 1%–2% due to the more accurate determination of these beam parameters. Whereas the scientific interest of the new calculations is unquestionable, on the practical side the real improvements have been due to the availability of more accurate stopping-power data and perturbation correction factors included in recent dosimetry protocols, as well as the improved understanding of the physical processes involved.

A recent suggestion for electron beam quality specification effectively puts an end to the controversy of energy-based relationships. The proposal by Burns et al. (1996) for directly using R_{50} as a beam-quality specifier for selecting stopping-power ratios (and reference depths) parallels the longstanding practice in photon dosimetry where beam qualities are expressed in terms of the penetration of the beam. The choice of R_{50} as the beam quality index is a change from the current practice of specifying beam quality in terms of the mean energy at the phantom surface; because $\bar{E}_o$ is normally derived from R_{50}, this change in beam quality index is merely a simplification which avoids the need for a conversion to energy. The recent dosimetry protocols based on standards of absorbed dose to water, AAPM (1999) and IAEA (2000), have endorsed this approach and all data are expressed in terms of R_{50}, measured with a constant SSD of 100 cm and a field size at the phantom surface of at least $10 \text{ cm} \times 10 \text{ cm}$ for $R_{50} \leq 7$ cm ($\bar{E}_o < 16$ MeV) and at least $20 \text{ cm} \times 20 \text{ cm}$ for $R_{50} > 7$ cm ($\bar{E}_o > 16$ MeV). Stopping-power ratios for electron beams as a function of R_{50} are given in Figure D.13, which shows a comparison between the values in IAEA (2000) which include the effect of the clinical accelerator head on the beam quality and those in IAEA (1997b) for monoenergetic electrons, recast in terms of R_{50}.

When the half-value depth is determined from a depth-ionisation distribution in water, $R_{50,\text{ion}}$, the half-value of the depth-dose distribution in water R_{50} is obtained using

$$R_{50} = 1.029 R_{50,\text{ion}} - 0.06 \text{ cm} \ (R_{50,\text{ion}} \leq 10 \text{ cm})$$
$$R_{50} = 1.059 R_{50,\text{ion}} - 0.37 \text{ cm} \ (R_{50,\text{ion}} > 10 \text{ cm})$$

(D.13)

The reference depth for beam calibration is also expressed in terms of R_{50} as

$$z_{\text{ref}} = 0.6 R_{50} - 0.1 \text{ cm} \ (R_{50} \text{ in cm})$$

(D.14)

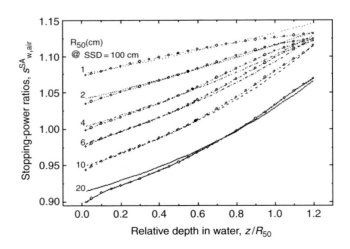

FIGURE D.13

A comparison between the water/air stopping-power ratios, $s_{w,air}^{SA}$, for electron beams given in IAEA (1997) (open circles) and those recommended in IAEA (2000) (lines without symbols). The latter set of data originate from Monte Carlo simulations which included details of the accelerator heads of clinical accelerators for a variety of accelerator types (Ding et al. 1995; Burns et al. 1996). The values in IAEA (1997), for monoenergetic electrons, have been recast to electron beam quality specified in terms of R_{50} and depths scaled according to R_{50}. The filled circles correspond to $s_{w,air}$ values at the reference depth. The agreement is very good except at large depths, which is consistent with the discussion on beam contamination. Note that an unexplained large difference occurs for an energy close to 50 MeV ($R_{50} = 20$ cm approximately), even if in both cases the beams are practically monoenergetic. (Reproduced from Dusautoy, et al., in *The Use of Plane Parallel Ionisation Chambers in High Energy Electron and Photon Beams.* TECDOC-1173 IAEA, Vienna, 1–10, 2000.)

This depth is close to the depth of the absorbed-dose maximum d_{max} at beam qualities $R_{50} < 4$ cm ($E_0 < 10$ MeV), but at higher beam qualities is greater than d_{max}. It has been recognized that this choice of reference depth may be less convenient than that recommended in previous dosimetry protocols because, for a given accelerator, no two reference beams will have the same reference depth. However, the new depth has been shown to significantly reduce machine-to-machine variations in chamber calibration factors (c.f. Burns et al. 1996) and the accuracy gained justifies its use. For completeness, and to allow the re-casting of existing dosimetry data, an expression for the mean energy at the reference depth in terms of R_{50} has been given in IAEA (2000):

$$\bar{E}_{z_{ref}} = 0.07 + 1.027 R_{50} - 0.0048(R_{50})^2 \tag{D.15}$$

which is a fit to the Monte-Carlo data used in previous IAEA Codes of Practice (IAEA 1987, 1997) at the reference depths given by Equation D.14.

D.4 HUMIDITY CORRECTIONS FOR IONISATION CHAMBERS

It is known that the W_{air}/e value for air at 20°C, 101.325 kPa and 50% relative humidity is 0.6% lower than that for dry air at the same temperature and pressure. That is, it has a value of 33.77 J C^{-1} instead of 33.97 J C^{-1} (c.f. Figure 5.15 in ICRU 1979). Thus, for the same amount of energy available for creating charge, 0.6% more charge will be created in air at 50% relative humidity than in dry air (at 20°C and 101.325 kPa). However, the response of a cavity ionisation chamber will also depend on the degree to which the incident charged particles deposit energy, i.e. on the stopping power, which is different for ambient air than for dry air. For the secondary electrons generated in a ^{60}Co beam, for example, the combined effect is that 0.3% more charge is created in air at 20°C, 101.325 kPa and 50% relative humidity than in dry air at this temperature and pressure. Thus for ^{60}Co, the humidity correction factor, k_h, which corrects the ambient-air charge, Q_h, to the dry-air charge, Q_{dry}, is 0.997 (c.f. Figure 5.14 in ICRU (1979).

By definition, the air kerma is independent of air temperature, pressure, and humidity. However, all primary standards for air kerma are based on ionisation in air and the measured charge depends on these parameters. The air-kerma calibration factor, N_K, is defined as

$$N_{K,h} = \frac{K_{air}}{Q_h} \tag{D.16}$$

where K_{air} is the air kerma, free in air (and independent of humidity), and Q_h is the chamber charge measured at 50% relative humidity, corrected to 20°C and to 101.325 kPa. The subscript h has been added to N_K to indicate the charge is measured at 50% relative humidity. When the $N_K - N_{D,air}$ formalism is used, one derives the chamber factor $N_{D,air}$, which strictly should be written as

$$N_{D,air} = N_{K,h}(1 - g)k_{att} k_m k_{cel} \frac{1}{k_h} \tag{D.17}$$

This equation does two things. First, it converts K_{air} into the mean absorbed dose to the air in the cavity, $\bar{D}_{air}$. If, as is usual, the electron stopping power ratios and mass energy-absorption coefficient ratios entering into k_m are the dry-air values, then $\bar{D}_{air}$ is also the dry-air value. Secondly, by including k_h in the denominator, Equation D.17 converts the air charge at 50% relative humidity Q_h to the dry-air value Q_{dry}. Clearly then the chamber factor can be written unambiguously as $N_{\text{D,dry air}}$*. The absorbed dose to water in the user's beam u is then derived using

$$D_{\text{w,u}} = Q_{\text{dry,u}} N_{\text{D,dry air}} (s_{\text{w,dry air}})_u p_u \qquad (D.18)$$

Note here that the dry-air value $(s_{\text{w,dry air}})_u$ must be used for consistency with the dry air electron stopping powers used in k_m in Equation D.17. Note also that the charge, $Q_{\text{dry,u}}$, must be the dry-air value. Thus the measured value for ambient air, $Q_{\text{h,u}}$ must be corrected for the effect of humidity. Combining Equation D.17 and Equation D.18 yields the final equation for the determination of absorbed dose to water using N_K referred to air at 50% relative humidity,

$$D_{\text{w,u}} = Q_{\text{h,u}} k_{\text{h,u}} N_{K,\text{h}} (1-g) k_{\text{att}} k_m k_{\text{cel}} \frac{1}{k_\text{h}} (s_{\text{w,dry air}})_u p_u \qquad (D.19)$$

where all the physical constants are referred to dry air, and the humidity correction $k_{\text{h,u}}$ in the user's beam cancels, at least to first order, with the ^{60}Co- value k_h. The importance of consistency in the use (or omission) of k_h becomes obvious from this expression. The approximate cancellation of $k_{\text{h,u}}$ and k_h in Equation D.19 is the reason why dosimetry protocols do not include a humidity correction in the expressions for $N_{\text{D,air}}$ and $D_{\text{w,u}}$ when N_K refers to air at 50% relative humidity. When N_K is referred to dry air, as already stated, k_h in the denominator is omitted and the remaining humidity correction to the charge measured must be applied.

D.5   UNCERTAINTIES IN REFERENCE DOSIMETRY

Uncertainties are generally evaluated following the recommendations given by ISO (1995). They are expressed as relative standard uncertainties classified into type A and type B. The method of evaluation of a type-A standard uncertainty is by statistical analysis of a series of observations, whereas the method of evaluation of a type-B standard uncertainty is based on means other than statistical analysis of a series of observations. It is emphasized that the classification rests exclusively on the procedure used to evaluate the uncertainty, and does not relate to the origin of the uncertainty. It is therefore incorrect to identify type-A and type-B uncertainties as systematic and random uncertainties, respectively; the use of the latter terms is discouraged. Type-A and type-B uncertainties are combined in quadrature to yield the combined standard uncertainty, u_c, and an expanded uncertainty (previously called *overall uncertainty*) can be built by multiplying u_c by a factor, which must always be stated; this *coverage factor*, k, is typically in the range 2 to 3.

Whereas uncertainties of type A are evaluated in a straightforward manner as the standard deviation of the mean value, uncertainties of type B are in many cases difficult to estimate and the ISO guide has provided examples to help in the determination of this type of uncertainty. A good example of an application to dosimetry is the case when there is no specific knowledge about the possible values of a variable X_i within an interval; then one can only assume that the variable X_i lies within a uniform rectangular distribution with an expected value x_i at the midpoint of the interval and an associated type-B uncertainty, $u_B = a/\sqrt{3}$, where a is the half-width of the interval. This should be compared to the common *overestimation* made when, for example, two different values are available for a given quantity or correction factor and the uncertainty is taken to be equal to the difference of the values ($u_B = 2a$). A common mistake made for the expanded uncertainty is to ignore the coverage factor used and

---

* It should be noted that if the reference conditions for the air-kerma calibration factor include dry air (that is, $N_{K,\text{dry}}$ is supplied rather than $N_{K,\text{h}}$), then the factor k_h has already been applied at the standards laboratory and so must be removed from Equation D.17.

compare, for instance, the standard uncertainty ($k = 1$) for one value of a quantity, with the 95% confidence limit ($k = 2$) for another. It is for this reason that the comparison of uncertainty estimates becomes particularly problematic and requires extreme care; in any case, subjective type-B evaluations are always the largest source of discrepancy between different estimates.

Uncertainties in external-beam radiotherapy dosimetry can be classified into four major groups according to:

- Absolute dose determination at the reference point in a water phantom (beam calibration), performed with a calibrated ionisation chamber according to a code of practice or dosimetry protocol (see Chapter 18)
- Procedures involving relative beam dosimetry (field size output dependence, influence of beam modifiers like wedges, secondary collimation and blocks, etc.), performed with any type of detector suitable for measurements in water or in a plastic phantom (see Chapter 19)
- The calculation of the dose delivered to the patient (monitor units or irradiation time to deliver the prescribed dose) and its distribution, usually performed with a computerized treatment planning system (see Chapter 23 and Part F)
- The process of treatment delivery throughout a complete treatment, which accounts for daily variations both in patient and machine setup, patient movements and machine instability during several weeks of treatment (see Part H)

The first of these uncertainties is of concern here. Following the work by Loevinger and Loftus (1977), estimates of the combined standard uncertainty of the dose distribution in the different steps of a treatment with external photon and electron beams have been given, for example, by Brahme et al. (1988). These provided the basis for the uncertainty estimates included in IAEA (1987) for the dose determination in reference conditions. Updates have been given by Andreo (1990b, 1992) and Thwaites (1994) showing that the uncertainty component due to basic physical quantities is considerably smaller than previously estimated. A comprehensive new estimate of the uncertainties in dose determination with all types of beams has been made in IAEA (2000) to which the reader is referred for details; IAEA(2000) takes into account the issues raised above on the evaluation of type-B uncertainties, as well as the cancellation of common terms in some contributing factors. These estimates are considered to be an update of the more conservative estimates given in Andreo (1992) and IAEA (1997b) for photon and electron beams, respectively.

A summary of the estimated uncertainties for the calibration of photon and electron beams using the $N_K - N_{D,air}$ and $N_{D,w}$ formalisms is presented in what follows; it focuses only on aspects related to measurements in the user's beam or factors selected by the user. The values provided are intended only to be a guide and refer only to some of the cases found in clinical use. Dose determinations in low-energy electron beams made in plastic phantoms, for example, are not considered even if these determinations are considered to be one of the biggest problems in dosimetry, but any uncertainty estimation would be too approximate. Because different alternatives are available for the calibration of the user's reference detector, it will be assumed here, except when stated otherwise, that calibrations are made at a secondary standards laboratory with a typical average uncertainty for the relevant calibrations*.

---

[*] It is of interest to mention that the uncertainty of an N_K or $N_{D,w}$ calibration factor for the user's chamber does not change significantly (i.e. more than a linearly added 0.2% or so) if the calibration is carried out at a primary or at a secondary standards laboratory. This is because the largest part of this uncertainty is directly associated with the primary standards used to realize the relevant quantity. The dissemination of this quantity to a secondary standard and then to the user's chamber does not involve any appreciable increase of the final uncertainty after the uncertainties of the different steps at a secondary standards laboratory (mainly related to reproducing the reference conditions) have been combined in quadrature.

D.5.1 AIR-KERMA BASED FORMALISM

For the $N_K - N_{D,air}$ formalism, the first step made by the user is the determination of the chamber factor $N_{D,air}$ based on an N_K calibration. Table D.1 shows the estimated relative uncertainties for the various contributing factors, as well as the relative combined standard uncertainty for $N_{D,air}$ which is close to 1% and thus approximately double that of the uncertainty in the N_K calibration factor. The largest contribution is due to the factor k_m, which is mainly caused by the uncertainty of the stopping-power ratios and of the approximate theoretical model used in the calculation of k_m. No attempt is made to include the contribution due to assuming generic values of k_m and k_{att} for all chambers of a specific type by a given manufacturer, as information on manufacturing tolerances (dimensions and thicknesses) is in general rather scarce.

At this point it is of interest to refer to the uncertainty of stopping-power ratios, which also enters in later steps of the dose-determination process. This estimate has been the subject of numerous studies and controversy. Uncertainties in the calculated stopping-power ratios are almost entirely type-B uncertainties because type-A uncertainties can be easily reduced to 0.1% or less, usually by a Monte-Carlo calculation. The uncertainty quoted by ICRU (1984b) for the basic electron stopping-power data is 1% to 2% at the level of two standard deviations, but it is difficult to estimate the uncertainty in the *ratios* of stopping powers, for which some components cancel out due to correlation. At the level of one standard deviation, Berger et al. (1975) estimated an uncertainty in the *ratios* of 0.6%, and more recently Berger (1988) reduced this to 0.25% to 0.5%. Andreo and Fransson (1989, 1992) used statistical procedures to argue that the uncertainty of stopping-power ratios is 0.4% to 0.6%. Any comparison among the estimations given in different references becomes even more problematic because uncertainties are usually given using different probability coverage factors, and these are not always clearly stated. Changes in the electron stopping-power data in the past typically led to changes in the stopping-power ratios of about the same amount, and the entire variation has sometimes been used as an estimate of the uncertainty (see the comment given above on the overestimation of type B uncertainties). Table D.2 reproduces a survey of estimations made by different authors which provides support to the figures used in this section.

The second step to be considered is that of the measurements carried out by the user in the clinical beam, and these estimates are the same for either type of formalism, i.e. $N_{D,air}$ or $N_{D,w}$. The estimations can be considered to be typical when modern and reliable equipment is used. The components in this step, in addition to the long-term stability of the user's dosimeter, are

TABLE D.1

Estimated Relative Standard Uncertainty of the Chamber Factor $N_{D,air}$ and of the Various Factors Entering in Its Definition

Factors in $N_{D,air}$	Standard Uncertainty		
N_k	0.4%		
g	0.02%		
k_m	0.7%	Factors in k_m	
		α	0.1%
		$s_{med,air}$[a]	0.5%
		$(\mu_{en}/\rho)_{med,air}$	0.2%
		k_m empirical model	0.4%
k_{att}	0.2%		
Combined uncertainty in $N_{D,air}$	0.8%		

[a] med refers to the wall and build-up cap of the ionisation chamber.

TABLE D.2

Survey of Estimates of the Uncertainties of Stopping-Power Ratios for Photon and Electron Beam Dosimetry

Reference	Uncertainty Quoted as	Standard Combined Uncertainty, u_c
ICRU-14 (1969)	2% (95% confidence limit)	1%
Berger et al. (1975)	0.55% (1std dev)	0.55%
Loevinger and Loftus (1977)	1.5% (95% confidence limit)	0.55%[a]
ICRU-35 (1984a)	1.5% (3 std dev)	0.5%
Boutillon and Perroche (1987, 1991)		0.2%–0.3%
Berger (1988)	0.5%–1% (2 std dev)	0.25%–0.5%
Brahme et al. (1988)		1.5%
Andreo (1990b)		0.4%–0.6%
Rogers (1991)		1% or larger
ICRU-37 (1984b)	1%–2% (2 std dev)	0.5%–1%

The middle column shows the statement of uncertainties quoted or inferred from the corresponding reference. The uncertainties given in the last row are those quoted by ICRU report 37 (1984b) for the collision stopping-powers of electrons above 10 keV, rather than for stopping-power ratios. The values in the last column should be considered as standard combined uncertainties, u_c, according to ISO (1993).

[a] The original statement of 1.5% (95% confidence limit) in this reference, or 0.75% (1 std dev), included also a component resulting from the uncertainty in the specification of the quality of the beam, approximately equal to 0.5% (1 std dev), which has been subtracted in quadrature here.

Source: Adapted from Andreo, P., The status of high-energy photon and electron beam dosometry five years after the publication of the IAEA code of practice in the Nordic countries, *Acta Oncologica*, 32, 483–500, 1993.

the establishment of the experimental set-up in the clinical conditions, the uncertainty of the readings of the dosimeter relative to a timer or beam monitor, and the necessary corrections for the *influence quantities*. The latter include polarity effect, temperature and pressure, humidity, stem effect, and recombination. The combined uncertainty varies with the type of beam and ionisation chamber, being typically between 0.6% for the use of a cylindrical chamber in a ^{60}Co beam and 1% for a plane–parallel chamber in a low-energy electron beam. For completeness, the details on these uncertainty values have been included in step 2 of Tables D.3 and D.4, which summarize the uncertainty estimates for the entire beam calibration process performed with the two formalisms.

The third step in the uncertainty estimate of the dose determination corresponds to the values of quantities and perturbation correction factors selected by the user for the dosimeter-beam combination. This is without any doubt the most controversial step and here type-B uncertainties play the major role. Especially for the perturbation correction factors, in some cases there exist experimental values determined by multiple authors without a clear uncertainty estimate. In others these are calculated using very approximate empirical equations; yet others are simply assumed based on some theory, etc. Table D.3, step 3, gives the most up-to-date estimates available for the factors entering into the N_K–$N_{D,air}$ formalism. In the case of a plane–parallel chamber cross-calibrated in a high-energy electron beam against a cylindrical chamber (calibrated in a ^{60}Co beam), due to the uncertainties of the different perturbation factors for cylindrical and plane–parallel chambers, the uncertainty in the $N_{D,air}$ factor (step 1) increases considerably with respect to the $N_{D,air}$ of the reference cylindrical chamber. Quoting every single reference used for the uncertainty estimates of all the factors in step 3 would result in a considerable list and the interested reader is referred to IAEA (2000) and references therein for details. The combined uncertainty of the product of all quantities and perturbation correction factors ranges from 0.8% for a cylindrical chamber in a ^{60}Co beam to 1.0% for a cylindrical chamber in an electron beam.

When the three steps are combined, the relative standard uncertainty in the determination of the absorbed dose to water at the reference depth in ^{60}Co gamma and high-energy photon

TABLE D.3

Estimated Relative Standard Uncertainty (%) in the Determination of Absorbed Dose to Water at the Reference Depth in ^{60}Co and High-Energy Photon and Electron Beams Using the $N_K - N_{D,air}$ Formalism

Type of Ionisation Chamber	^{60}Co Beams	Mega-voltage Photon Beams	Electrons with $R_{50} > 4$ g cm^{-2} ($\bar{E}_o > 10$ MeV)	Electron Beams of Any Energy
		Cylindrical		Plane–parallel[a]
Step 1: User chamber calibration factor				
Combined uncertainty in $N_{D,air}$	0.8	0.8	0.8	1.3
Step 2: User beam measurements				
Long-term stability of user dosimeter	0.3	0.3	0.3	0.4
Establishment of reference conditions	0.5	0.4	0.4	0.6
Dosimeter reading relative to timer or beam monitor	0.1	0.6	0.6	0.6
Correction for influence quantities	0.3	0.4	0.4	0.5
Combined uncertainty in dosimeter reading	0.6	0.9	0.9	1.0
Step 3: Quantities and perturbation factors for the user beam				
$s_{w,air}$	0.5	0.5	0.5	0.5
Assignment of $s_{w,air}$ to beam quality	0.1	0.2	0.2	0.2
p_{dis} or p_{eff}	0.3	0.2	0.3	0.1
p_{wall}	0.5	0.5	0.5	0.3
p_{cav} (or global p_Q in plane parallel chambers)	<0.1	<0.1	0.5	<0.1
p_{cel}	0.2	0.2	0.1	-
W/e (relative to ^{60}Co for high-energy beams)	0.2	0.3	0.3	0.3
Combined uncertainty in quantities and p_Q factors	0.8	0.9	1.0	0.7
Combined uncertainty in D_w	**1.3**	**1.5**	**1.6**	**1.9**

[a]   The plane–parallel chamber is cross-calibrated in a high-energy electron beam against a cylindrical chamber calibrated in a ^{60}Co beam.

and electron beams using the $N_K - N_{D,air}$ formalism ranges between 1.3% and 1.9% for the cases discussed here. Note that these values are considerably smaller than the figures given for example in IAEA (1987), which vary between 2.5% and 3.7%. As already mentioned, the changes are justified not only in terms of the more accurate values for some physical quantities and perturbation correction factors available today, and more up-to-date uncertainty estimates, but also in the way in which many of the uncertainty components have been estimated and combined. It is emphasized that many of the new values used in this analysis are not included in any of the $N_K - N_{D,air}$ protocols currently in use.

D.5.2   Absorbed-Dose-to-Water Formalism

For the $N_{D,w}$ formalism there are multiple options available for the calibration of the user's reference dosimeter, and for simplicity only the most common cases of cylindrical chambers calibrated in ^{60}Co beams and plane–parallel chambers cross-calibrated in high-energy electrons will be included in the discussion.

For the first step, the $N_{D,w}$ calibration factor, typical uncertainties are around 0.6% for cylindrical chambers and 1.4% for cross-calibrated plane–parallel chambers. Uncertainties associated with the measurements in the user's beam, the second step, are identical to those quoted above for the $N_K - N_{D,air}$ formalism. The only component in step 3 is that of the beam-quality correction factor, k_Q, which is not relevant in the case of ^{60}Co beams. It has already been mentioned that the most common approach available today is based on the use of

TABLE D.4

Estimated Relative Standard Uncertainty (%) in the Determination of Absorbed Dose to Water at the Reference Depth in ^{60}Co and High-Energy Photon and Electron Beams Using the $N_{D,w}$ Formalism, Based on a Cylindrical Chamber Calibration in a ^{60}Co Beam

Type of Ionisation Chamber	^{60}Co Beams	Megavoltage Photon Beams	Electrons with $R_{50} > 4$ g cm^{-2} ($\bar{E}_o > 10$ MeV)	Electron Beams of Any Energy
		Cylindrical		Plane–Parallel[a]
Step 1: User chamber calibration factor				
Combined uncertainty in $N_{D,w}$	0.6	0.6	0.6	1.4
Step 2: User beam measurements				
Long-term stability of user dosimeter	0.3	0.3	0.3	0.4
Establishment of reference conditions	0.5	0.4	0.4	0.6
Dosimeter reading relative to timer or beam monitor	0.1	0.6	0.6	0.6
Correction for influence quantities	0.3	0.4	0.4	0.5
Combined uncertainty in dosimeter reading	0.6	0.9	0.9	1.1
Step 3: Beam quality correction				
3a k_{Q,Q_o} (calculated)		0.9	1.1	0.6[b]
3b k_{Q,Q_o} (measured at a standards laboratory)		0.7	0.7	0.8
Combined uncertainty in D_w (calculated k_{Q,Q_o})	**0.9**	**1.4**	**1.6**	**1.9**
Combined uncertainty in D_w (measured k_{Q,Q_o})	**0.9**	**1.3**	**1.3**	**1.9**

[a] The plane–parallel chamber is cross-calibrated in a high-energy electron beam against a cylindrical chamber calibrated iin a ^{60}Co beam. If the plane–parallel chamber is calibrated directly in an electron beam at a standards laboratory, them $u_c(N_{D,w}) = 0.7\%$, yielding $u_c(D_w) = 1.4\%$.

[b] The k_{Q,Q_o} factor is in this case calculated for a reference electron beam, whereas for the other two cases the reference beam is ^{60}Co. Using ^{60}Co as reference quality for the calculation of k_{Q,Q_o} for plane–parallel chambers would increase its uncertainty up to 1.7%.

theoretically determined k_Q factors, which include ratios, at the user's quality Q and at ^{60}Co of water/air stopping-power ratios, $s_{w,air}$, the mean energy expended in air per ion pair formed, W_{air}, and the perturbation factors p_Q. The uncertainty of calculated k_Q factors is, therefore, obtained by a combination of the uncertainties of these quantities and perturbation factors given in Table D.3, but correlations between the ratios of various parameters must be taken into account. The estimates made in IAEA (2000), which should be consulted for the details, are reproduced in Table D.4, step 3. The uncertainties for a cylindrical chamber are 0.9% for megavoltage photons and 1.1% for high-energy electron beams; for a plane–parallel chamber the uncertainty with reference to an electron beam is 0.6% (this value increases to 1.7% if ^{60}Co is used as reference). These uncertainties are larger than the corresponding values in step 3 for the $N_K - N_{D,air}$ case, but they are lower than that which would result from a simple combination in quadrature of the different ratios.

The relative standard uncertainty in the determination of the absorbed dose to water at the reference depth in ^{60}Co and high-energy photon and electron beams using the $N_{D,w}$ formalism, based on a cylindrical chamber calibrated in ^{60}Co and a plane–parallel chamber cross-calibrated in high-energy electrons, is, upon the combination of the uncertainties in the different steps, smaller for photon beams and about the same for electrons compared to the corresponding values for the $N_K - N_{D,air}$ formalism, ranging between 0.9% for ^{60}Co and 1.9% for low-energy electron beams. Note that, for ^{60}Co beams in particular, it is the first time that the uncertainty in dose determination has been brought below 1%. It is also worth pointing out that, even if the uncertainty for low-energy electrons is about the same as for the $N_K - N_{D,air}$ formalism, this value would reduce to 1.4% if the plane–parallel chamber were calibrated directly in an electron beam (see footnote a in the table). The case when k_Q factors are directly measured for the user's chamber in high-energy photon and electron beams at standards laboratories has also been included in Table D.4 for completeness (see step 3b). In the case of electron beams this is a situation of reduced application at present because of the limited availability of such experimental determinations (McEwen et al. 1998).

PART D: DOSE MEASUREMENT
REFERENCES

AAPM (American Association of Physicists in Medicine), Task Group 21: A protocol for the determination of absorbed dose from high-energy photon and electron beams, *Med. Phys.*, 10, 741–771, 1983.

AAPM (American Association of Physicists in Medicine), Report 16, Task Group 20: Protocol for Heavy Charged-Particle Therapy Beam Dosimetry, AAPM, New York, 1986.

AAPM (American Association of Physicists in Medicine), Task Group 25: Clinical electron-beam dosimetry, *Med. Phys.*, 18, 73–109, 1991.

AAPM (American Association of Physicists in Medicine), Task Group 40: Comprehensive QA for radiation oncology, *Med. Phys.*, 21, 581–618, 1994a.

AAPM (American Association of Physicists in Medicine), Task Group No. 39: The calibration and use of plane-parallel ionisation chambers for dosimetry of electron beams: An extension of the 1983 AAPM protocol, *Med. Phys.*, 21, 1251–1260, 1994b.

AAPM (American Association of Physicists in Medicine), Task Group 51: Protocol for clinical reference dosimetry of high-energy photon and electron beams, *Med. Phys.*, 26, 1847–1870, 1999.

AAPM (American Association of Physicists in Medicine), Task Group 62: Diode in vivo dosimetry for patients receiving external beam irradiation therapy. Medical physics publishing, Madison WI, 2005.

Ahnesjö, A., Andreo, P., and Brahme, A., Calculation and application of point spread functions for treatment planning with high energy photon beams, *Acta Oncol.*, 26, 49, 1987.

AIFB (Associazione Italiana di Fisica Biomedica), Protocollo per la dosimetria di base nella radioterapia con fasci di fotoni ed elettroni con E_{max} fra 1 e 40 MeV, *Notiziario della A.I.F.B.*, VI(2), 1988.

Aird, E. G. and Farmer, F. T., The design of a thimble chamber for the Farmer dosemeter, *Phys. Med. Biol.*, 17, 169–174, 1972.

Allahverdi, M., Nisbet, A., and Thwaites, D. I., An evaluation of epoxy resin phantom materials for megavoltage photon dosimetry, *Phys. Med. Biol.*, 44, 1125–1132, 1999.

Allisy-Roberts, P. J. and Burns, D. T., Comparisons and calibrations at the BIPM, Report CCRI(I)/99-1, BIPM, Sevres, France, 1999.

Alm-Carlsson, G., Theoretical basis for dosimetry, in *The Dosimetry of Ionizing Radiation*, Kase, K. R., Bjarngard, B. E., and Attix, F. H., Eds., Academic Press, New York, 1985.

Almeida, C. E. D., Perroche-Roux, A. M., and Boutillon, M., Perturbation correction of a cylindrical thimble-type chamber in a graphite phantom for ^{60}Co gamma rays, *Phys. Med. Biol.*, 34, 1443–1449, 1989.

Almond, P. R. and Svensson, H., Ionisation chamber dosimetry for photon and electron beams: Theoretical considerations, *Acta Radiol. Ther. Phys. Biol.*, 16, 177–186, 1977.

Almond, P. R., Andreo, P., Mattsson, O., Nahum, A. E., and Roos, M., The use of plane-parallel ionisation chambers in high-energy electron and photon beams. *An International Code of Practice for Dosimetry*, IAEA Technical Report Series no. 381, IAEA, Vienna, 1997.

André, L., Born, J., Mini, R., and Poretti, G., Energy distribution of electron beams from clinical accelerators measured with a magnetic spectrometer, *Med. Biol. Eng. Comp.*, (Suppl. 1), 1169, 1985.

Andreo, P., Stopping-power ratios for dosimetry, in *Monte Carlo Transport of Electrons and Photons*, Jenkins, T. M., Nelson, W. R., Rindi, A., Nahum, A. E., and Rogers, D. W. O., Eds., Plenum Press, New York, pp. 485–501, 1988.

Andreo, P., Depth-dose and stopping-power data for monoenergetic electron beams, *Nucl. Instr. Methods*, B51, 107–121, 1990a.

Andreo, P., Uncertainties in dosimetric data and beam calibration, *Int. J. Radiat. Oncol. Biol. Phys.*, 19, 1233–1247, 1990b.

Andreo, P., Dosimetry of high-energy photon beams. Present status and future prospects for the data used in ionisation chamber dosimetry, Report RI 1991-03 Radiation Department, Stockholm, University of Stockholm, 1991a.

Andreo, P., Absorbed dose beam quality factors for the dosimetry of high-energy photon beams, *Phys. Med. Biol.*, 37, 2189–2211, 1991b.

Andreo, P., The status of high-energy photon and electron beam dosimetry five years after the publication of the IAEA code of practice in the Nordic countries, *Acta Oncol.*, 32, 483–500, 1993.

Andreo, P., Improved calculations of stopping-power ratios and their correlation with the quality of therapeutic photon beams, in *Measurement Assurance in Dosimetry*, IAEA Proceedings Series, Vienna, pp. 335–359, 1994.

Andreo, P., On the beam quality specification of high-energy photons for radiotherapy dosimetry, *Med. Phys.*, 27, 434–440, 2000.

Andreo, P. and Brahme, A., Mean energy in electron beams, *Med. Phys.*, 8, 682–687, 1981.

Andreo, P. and Brahme, A., Stopping power data for high energy photon beams, *Phys. Med. Biol.*, 31, 839–858, 1986.

Andreo, P. and Fransson, A., Stopping-power ratios and their uncertainties for clinical electron beam dosimetry, *Phys. Med. Biol.*, 34, 1847–1861, 1989.

Andreo, P. and Fransson, A., Estimation of uncertainties in stopping-power ratios using Monte Carlo methods, *Appl. Radiat. Isotop.*, 43, 1425–1426, 1992.

Andreo, P. and Nahum, A. E., Stopping power ratio for a photon spectrum as a weighted sum of the values for monoenergetic photon beams, *Phys. Med. Biol.*, 30, 1055–1065, 1985.

Andreo, P., Nahum, A. E., and Brahme, A., Chamber-dependent wall correction factors in dosimetry, *Phys. Med. Biol.*, 31, 1189–1199, 1986.

Andreo, P., Nahum, A. E., and Svensson, H., Recent developments in basic dosimetry, *Radiother. Oncol.*, 10, 117–126, 1987.

Andreo, P., Lindborg, L., and Medin, J., On the calibration of plane-parallel ionisation chambers using ^{60}Co beams, *Med. Phys.*, 18, 326–327, 1991.

Andreo, P., Burns, D. T., Hohlfeld, K., Huq, M. S., Kanai, T., Laitano, F., Smyth, V., and Vynckier, S., *Absorbed Dose Determination in External Beam Radiotherapy: An International Code of Practice for Dosimetry Based on Standards of Absorbed Dose to Water*, IAEA Technical Reports Series no. 398, IAEA, Vienna, 2000.

Appleby, A., Christman, E. A., and Leghrouz, A., Imaging of spatial radiation dose distribution in agarose gels using magnetic resonance, *Med. Phys.*, 14, 382–384, 1987.

Attix, F. H., *Introduction to radiological physics and radiation dosimetry*, Wiley, New York, 1986.

Audet, C. and Schreiner, L. J., Multiple-site fast exchange model for spin-lattice relaxation in the Fricke-gelatin dosimeter, *Med. Phys.*, 24, 201–209, 1997.

Awschalom, M., Rosenberg, I., and Ten Haken, R. K., A new look at displacement factor and point of measurement corrections in ionisation chamber dosimetry, *Med. Phys.*, 10, 307–313, 1983.

Bagne, F., A comprehensive study of LiF TL response to high energy photon and electrons, *Radiology*, 123, 753–760, 1977.

Balcom, B. J., Lees, T. J., Sharp, A. R., Kulkarni, N. S., and Wagner, G. S., Diffusion in Fe(II/III) radiation dosimetry gels measured by MRI, *Phys. Med. Biol.*, 40, 1665–1676, 1995.

Baldock, C., Burford, R. P., Billingham, N., Wagner, G. S., Patval, S., Badawi, R. D., and Keevil, S. F., Experimental procedure for the manufacture and calibration of polyacrylamide gel (PAG) for magnetic resonance imaging (MRI) radiation dosimetry, *Phys. Med. Biol.*, 43, 695–702, 1998.

Baldock, C., Murry, P., and Kron, T., Uncertainty analysis in polymer gel dosimetry, *Phys. Med. Biol.*, 44, N243–N246, 1999.

Barthe, J., Electronic dosimeters based on solid state detectors, *Nucl. Instrum. Methods Phys. Res., Sect. B*, 184, 158–189, 2001.

Bassi, P., Busuoli, G., and Raimondi, O., Calculated energy dependence of some RTL and RPL detectors, *Int. J. Appl. Radiat. Isot.*, 27, 291–305, 1976.

Baustert, I. C., Oldham, M., Smith, T. A. D., Hayes, C., Webb, S., and Leach, M. O., Optimised MR imaging for polyacrylamide gel dosimetry, *Phys. Med. Biol.*, 45, 847–858, 2000.

Berger, M. J., Electron stopping-powers for transport calculations, in *Monte Carlo Transport of Electrons and Photons*, Jenkins, T. M., Nelson, W. R., Rindi, A., Nahum, A. E., and Rogers, D. W. O., Eds., Plenum Press, New York, pp. 57–80, 1988.

Berger, M. J. and Seltzer, S. M., *Tables of Energy Losses and Ranges of Electrons and Positrons*, 1133, National Academy of Sciences—National Research Council, Washington, DC, pp. 205–268, 1964.

Berger, M. J. and Seltzer, S. M., Calculation of energy and charge deposition and of the electron flux in a water medium bombarded with 20 MeV electrons, *Ann. New York Acad. Sci.*, 161, 8–23, 1969.

Berger, M. J. and Seltzer, S. M., *Stopping Powers and Ranges of Electrons and Positrons*, 2nd ed., US Department of Commerce, National Bureau of Standards, Washington, DC, 1983.

Berger, M. J., Seltzer, S. M., Domen, S. R., and Lamperti, P. J., Stopping-power ratios for electron dosimetry with ionisation chambers, in *Biomedical Dosimetry*, IAEA Proceedings Series, Vienna, pp. 589–609, 1975.

Bielajew, A. F., Ionisation cavity theory—a formal derivation of perturbation factors for thick-walled ionisation chambers in photon beams, *Phys. Med. Biol.*, 30, 1261–1270, 1986.

Bielajew, A. F., Correction factors for thick-walled ionisation chambers in point-source photon beams, *Phys. Med. Biol.*, 35, 517–538, 1990.

Bielajew, A. F. and Rogers, D. W. O., Implications of new correction factors on primary air kerma standards in Co-60 beams, *Phys. Med. Biol.*, 37, 1283–1291, 1992.

Bjärngard, B. E. and Kase, K. R., Replacement correction factors for photon and electron dose measurements, *Med. Phys.*, 12, 785–787, 1985.

Boag, J. W., Ionisation chambers, in *Radiation Dosimetry*, Attix, F. H. and Roesch, W. C., Eds., Academic Press, New York, pp. 1–72, 1966.

Boag, J. W., Ionisation chambers, in *The Dosimetry of Ionizing Radiation*, Kase, K. R., Bjärngard, B. E., and Attix, F. H., Eds., Academic Press, New York, pp. 169–243, 1987.

Boag, J. W. and Currant, J., Current collection and ionic recombination in small cylindrical ionisation chambers exposed to pulsed radiation, *Br. J. Radiol.*, 53, 471–478, 1980.

Bos, L. J., Danciu, C., Cheng, C. W., Brugmans, M. J., van der Horst, A., Minken, A., and Mijnheer, B. J., Inter-institutional variations of sensitometric curves of radiographic dosimetric films, *Med. Phys.*, 29, 1772–1780, 2002.

Boutillon, M., Perturbation correction for the ionometric determination of absorbed dose in a graphite phantom for ^{60}Co gamma rays, *Phys. Med. Biol.*, 28, 375–388, 1983.

Boutillon, M. and Andreo, P., Some thoughts on the dissemination of ND, w calibration factors in Co-60 gamma radiation. Report CCEMRI(I)/97-1, BIPM, Sevres, France, 1997.

Boutillon, M. and Niatel, M. T., A study of a graphite cavity chamber for absolute exposure measurements of Co-60 gamma rays, *Metrologia*, 9, 139–146, 1973.

Boutillon, M. and Perroche-Roux, A. M., Ionometric determination of absorbed dose in water for cobalt-60 gamma rays. Report CCEMRI(I)/91-1, BIPM, Sevres, France, 1991.

Boutillon, M. and Perroche-Roux, A. M., Re-evaluation of the W for electrons in dry air, *Phys. Med. Biol.*, 32, 213–219, 1987.

Brahme, A. and Andreo, P., Dosimetry and quality specification of high energy photon beams, *Acta Radiol. Oncol.*, 25, 213–223, 1986.

Brahme, A. and Svensson, H., Specification of electron beam quality from the central axis depth absorbed dose distribution, *Med. Phys.*, 3, 95–102, 1976.

Brahme, A. and Svensson, H., Radiation beam characteristics of a 22 MeV microtron, *Acta Radiol. Oncol.*, 18, 244–272, 1979.

Brahme, A., Chavaudra, J., Landberg, T., McCullough, E. C., Nüsslin, F., Rawlinson, J. A., Svensson, G., and Svensson, H., Accuracy requirements and quality assurance of external beam therapy with photons and electrons, *Acta Oncol.*, (Suppl. 1), 1988.

British Institute of Radiology, Central axis depth dose data for use in radiotherapy, *Br. J. Radiol.*, (Suppl. 17), The British Institute of Radiology, London, 1983.

British Institute of Radiology, Central axis depth dose data for use in radiotherapy, *Br. J. Radiol.*, (Suppl. 25), The British Institute of Radiology, London, 1996.

Bruinvis, I. A. D., Heukelom, S., and Mijnheer, B. J., Comparison of ionisation measurements in water and polystyrene for electron beam dosimetry, *Phys. Med. Biol.*, 30, 1043–1053, 1985.

Bucciolini, M., Banci Buonamici, F., Mazzocchi, S., De Angelis, C., Onori, S., and Cirrone, G. A. P., Diamond detector versus silicon diode and ion chamber in photon beams of different energy and field size, *Med. Phys.*, 30, 2149–2154, 2003.

Bucciolini, M., Banci Buonamici, F., and Casati, M., Verification of IMRT fields by film dosimetry, *Med. Phys.*, 31, 161–168, 2004.

Burch, S. E., Kearfott, K. J., Trueblood, J. H., Sheils, W. C., Yeo, J. I., and Wang, C. K., A new approach to film dosimetry for high energy photon beams: Lateral scatter filtering, *Med. Phys.*, 24, 775–783, 1997.

Burns, J. E., Absorbed-dose calibrations in high-energy photon beams at the National Physical Laboratory: conversion procedure, *Phys. Med. Biol.*, 39, 1555–1575, 1994.

Burns, J. E. and Rosser, K. E., Saturation correction for the NE 2560/1 dosemeter in photon dosimetry, *Phys. Med. Biol.*, 35, 687–693, 1990.

Burns, D. T., Ding, G. X., and Rogers, D. W. O., R_{50} as a beam quality specifier for selecting stopping-power ratios and reference depths for electron dosimetry, *Med. Phys.*, 23, 383–388, 1996.

Carlton, R. R. and Adler, A. M., Radiographic film, Chapter 19 in *The Principles of Radiographic Imaging*, 4th ed., Carlton, R. R. and Adler, A. M., Eds., Thomson Delmar Learning, New York, 2005.

CCEMRI(I) Comité Consultatif pour les Etalons de Mesure des Rayonnements Ionisants (Section I), Report to the Comite International des Poids et Mesures (S C Ellis, Rapporteur), 8th Meeting, Bureau International des Poids et Mesures, Sevres, France, 1985.

CCEMRI(I) Comité Consultatif pour les Etalons de Mesure des Rayonnements Ionisants (Section I), Report to the Comité International des Poids et Mesures (N J Hargrave, Rapporteur), 9th Meeting, Bureau International des Poids et Mesures, Sevres, France, 1988.

Chan, M. F. and Ayyanger, K. M., Confirmation of target localisation and dosimetry for 3D conformal radiotherapy treatment planning by MR imaging of a ferous sulphate head phantom, *Med. Phys.*, 22, 1171–1175, 1995.

Cheng, C. W. and Das, I. J., Dosimetry of high energy photon and electron beams with CEA films, *Med. Phys.*, 23, 1225–1232, 1996.

Christ, G., White polystyrene as a substitute for water in high energy photon dosimetry, *Med. Phys.*, 22, 2097–2100, 1995.

Chu, K. C., Jordan, K. J., Battista, J. J., Van Dyk, J., and Rutt, B. K., Polyvinyl alcohol-Fricke hydrogel and cryogel: two new gel dosimetry systems with low Fe^{3+} diffusion, *Phys. Med. Biol.*, 45, 955–969, 2000.

Chuang, C. F., Verhey, L. J., and Xia, P., Investigation of the use of MOSFET for clinical IMRT dosimetric verification, *Med. Phys.*, 29, 1109–1115, 2002.

Constantinou, C., Attix, F. H., and Paliwal, B. R., A solid water phantom material for radiotherapy x-ray and γ-ray beam calibrations, *Med. Phys.*, 9, 436–441, 1982.

Coudin, D. and Marinello, G., Lithium borate TLD for determining the backscatter factors for low energy x-rays: Comparison with chamber-based and Monte-Carlo derived values, *Med. Phys.*, 25, 347–353, 1998.

Cunningham, J. R., Woo, M., Rogers, D. W. O., and Bielajew, A. F., The dependence of mass energy absorption coefficient ratios on beam size and depth in a phantom, *Med. Phys.*, 13, 496–502, 1986.

Danciu, C., Proimos, B. S., Rosenwald, J.-C., and Mijnheer, B. J., Variation of sensitometric curves of radiographic films in high energy photon beams, *Med. Phys.*, 28, 966–974, 2001.

Davis, S. D., Ross, C. K., Mobit, P. N., van der Zwan, L., Chase, W. J., and Short, K. R., The response of LiF thermoluminescence dosemeters to photon beams in the energy range from 30 kV x-rays to ^{60}Co gamma rays, *Radiat. Prot. Dosimetry*, 106, 33–42, 2003.

De Angelis, C., De Coste, V., Fattibene, P., Onori, S., and Petetti, E., Use of alanine for dosimetry intercomparisons among Italian radiotherapy centers, *Appl. Radiat. Isot.*, 62, 261–265, 2005.

De Deene, Y., de Wagter, C., van Duyse, B., Derycke, S., de Neve, W., and Achten, E., Three-dimensional dosimetry using polymer gel and magnetic resonance imaging applied to the verification of conformal radiation therapy in head-and-neck cancer, *Radiother. Oncol.*, 48, 283–291, 1998.

De Deene, Y., Hanselaer, P., De Wagter, C., Achten, E., and De Neve, W., An investigation of the chemical stability of a monomer/polymer gel dosimeter, *Phys. Med. Biol.*, 45, 859–878, 2000.

De Deene, Y., Reynaert, N., and De Wagter, C., On the accuracy of monomer/polymer gel dosimetry in the proximity of a high-dose-rate ^{192}Ir source, *Phys. Med. Biol.*, 46, 2801–2825, 2001.

Devic, S., Seuntjens, J., Hegyi, G., Podgorsak, E. B., Soares, C. G., Kirov, A. S., Ali, I., Williamson, J. F., and Elizondo, A., Dosimetric properties of improved GafChromic films for seven different digitizers, *Med. Phys.*, 31, 2392–2401, 2004.

Devic, S., Seuntjens, J., Sham, E., Podgorsak, E. B., Schmidtlein, C. R., Kirov, A. S., and Soares, C. G., Precise radiochromic film dosimetry using a flat-bed document scanner, *Med. Phys.*, 32, 2245–2253, 2005.

DIN (Deutsches Institut für Normung), Klinische Dosimetrie: Anwendung von Röntgenstrahlen mit Röhrenspannungen von 10 bis 100 kV in der Strahlentherapie und in der Weichteildiagnostik (DIN6809-4). DIN, Berlin, 1988.

DIN (Deutsches Institut für Normung), Klinische Dosimetrie: Anwendung von Röntgenstrahlen mit Röhrenspannungen von 10 bis 100 kV in der Strahlentherapie (DIN6809-5), DIN, Berlin, 1996.

DIN (Deutsches Institut für Normung), Dosismessverfahren nach der Sondenmethode für Photonen- und Elektronenstrahlung, Teil 2: Ionisationsdosimetrie (DIN 6800-2), DIN, Berlin, 1997.

Ding, G. X., Rogers, D. W. O., and Mackie, T. R., Calculation of stopping-power ratios using realistic clinical electron beams, *Med. Phys.*, 22, 489–501, 1995.

Ding, G. X., Rogers, D. W. O., and Mackie, T. R., Mean energy, energy–range relationships and depth-scaling factors for clinical electron beams, *Med. Phys.*, 23, 361–376, 1996.

Ding, G., Rogers, D. W. O., Cygler, J. E., and Mackie, T. R., Electron fluence correction factors for conversion of dose in plastic to dose in water, *Med. Phys.*, 24, 161–176, 1997.

Dogan, N., Leybovich, L. B., and Sethi, A., Comparative evaluation of Kodak EDR2 and XV2 films for verification of intensity modulated radiation therapy, *Phys. Med. Biol.*, 47, 4121–4130, 2002.

Driscoll, C. M. H., Barthe, J. R., Obertofer, M., Busuoli, G., and Hickman, C., Annealing procedures for commonly used radiothermoluminescent materials, *Radiat. Prot. Dosimetry*, 14, 17–32, 1986.

Dusautoy, A., Roos, M., Svensson, H., and Andreo, P., Review of data and methods recommended in the international code of practice for dosimetry, in *The use of plane parallel ionisation cambers in high energy electron and photon beams*, IAEA Technical Report Series No. 381, IAEA, Vienna, pp. 1–10, 2000.

Dutreix, A., Problems of high-energy x-ray dosimetry, in *High-Energy Photons and Electrons: Clinical Applications in Cancer Management*, Kramer, N. S. and Kramer, G. F. Z. S., Eds., Wiley, New York, pp. 202–214, 1976.

Dutreix, A., When and how can we improve precision in radiotherapy? *Radiother. Oncol.*, 2, 275–292, 1984.

Dutreix, A. and Bridier, A., Dosimetry for external beams of photon and electron radiation, in *The Dosimetry of Ionizing Radiation*, Kase, K. R., Bjärngard, B. E., and Attix, F. H., Eds., Vol. I, Academic Press, Orlando, pp. 163–228, 1985.

Dutreix, J. and Dutreix, A., Film dosimetry of high-energy electrons, *Ann. New York Acad. Sci.*, 161, 33–43, 1969.

Dutreix, A., Bjarngard, B., Bridier, A., Minjnheer, B., Shaw, J., and Svensson, H., Monitor unit calculation for high energy photon beams, ESTRO Booklet 3, Garant, Leuven, Belgium, 1997.

Edwards, C. R., Green, S., Palethorpe, J. E., and Mountford, P. J., The response of a MOSFET, p-type semiconductor and LiF TLD to quasi-monoenergetic x-rays, *Phys. Med. Biol.*, 42, 2383–2391, 1997.

Eveling, J. N., Morgan, A. M., and Pitchford, W. G., Commissioning a p-type silicon diode for use in clinical electron beams, *Med. Phys.*, 26, 100–107, 1999.

Farmer, F. T., A sub-standard x-ray dosimeter, *Br. J. Radiol.*, 28, 304–310, 1955.

Fernandez-Varea, J. M., Andreo, P., and Tabata, T., Detour factors in water and plastic phantoms and their use for range and depth-dose scaling in electron beam dosimetry, *Phys. Med. Biol.*, 41, 1119–1139, 1996.

Fleeman, J. and Frantz, F. S., Film dosimetry of electrons in the energy range 0.5 to 1.4 MeV, *Am. J. Roentgenol. Radium Ther. Nucl. Med.*, 71, 1049–1055, 1954.

Fricke, H. and Hart, E. J., Chemical dosimetry, in *Radiation Dosimetry*, Vol. 2, Attix, F. H. and Roesch, W. C., Eds., Academic Press, New York, pp. 167–239, 1966.

Galbraith, D. M., Rawlinson, J. A., and Munro, P., Dose errors due to charge storage in electron irradiated plastic phantoms, *Med. Phys.*, 11, 197–203, 1984.

Gambarini, G., Monti, D., Fumagalli, M. L., Birattari, C., and Salvadori, P., Phantom dosimeters examined by NMR analysis: A promising technique for 3-D determinations of absorbed dose, *Appl. Radiat. Isot.*, 48, 1477–1484, 1997.

Georg, D., Kroupa, B., Winkler, P., and Potter, R., Normalized sensitometric curves for the verification of hybrid IMRT treatment plans with multiple energies, *Med. Phys.*, 30, 1142–1150, 2003.

Gillin, M. T., Kline, R. W., Niroomand-Rad, A., and Grimm, D. F., The effect of thickness of the waterproofing sheath on the calibration of photon and electron beams, *Med. Phys.*, 12, 234–236, 1985.

Goldstein, N., Dose-rate dependence of lithium fluoride for exposures above 15,000 R per pulse, *Health Phys.*, 22, 90–92, 1972.

Gore, J. C., Kang, Y. S., and Schulz, R. J., Measurement of radiation dose distributions by nuclear magnetic resonance imaging, *Phys. Med. Biol.*, 29, 1189–1197, 1984.

Gore, J. C., Ranade, M., Maryanski, M. J., and Schulz, R. J., Radiation dose distributions in three dimensions from tomographic optical density scanning of polymer gels: I. Development of an optical scanner, *Phys. Med. Biol.*, 41, 2695–2704, 1996.

Gramke, R. C., Wright, K. A., Evans, W. W., Nelson, J. E., and Trump, J. G., The film method of tissue dose studies with 2.0 MeV roentgen rays, *Am. J. Roentgenol. Radium Ther. Nucl. Med.*, 28, 207–216, 1952.

Greene, D. and Massey, J. B., The use of Farmer-Baldwin and Victrometer ionisation chambers for dosimetry of high-energy X-radiation, *Phys. Med. Biol.*, 11, 569–575, 1966.

Greening, J. R., Fundamentals of radiation dosimetry, *Medical Physics Handbooks*, Adam Hilger Ltd, Bristol, 1981.

Grimbergen, T. W. M., Aalbers, A. H. L., Mijnheer, B. M., Seuntjens, J. et al., The NCS code of practice for dosimetry of low and medium-energy x-rays, in *Kilovoltage X-Ray Beam Dosimetry for Radiotherapy and Radiobiology*, Ma, C. -M. and Seuntjens, J. P., Eds., Medical Physics Publishing, Wisconsin, pp. 55–68, 1999.

Grosswendt, B., Backscatter factors for x-rays generated at voltages between 10 and 100 kV, *Phys. Med. Biol.*, 29, 579–591, 1984.

Grusell, E. and Rikner, G., Radiation damage induced dose-rate non linearity in an n-type silicon detector, *Acta Radiol. Oncol.*, 23, 465–469, 1984.

Grusell, E. and Rikner, G., Evaluation of temperature effects in p-type silicon detectors, *Phys. Med. Biol.*, 31, 527–534, 1986.

Grusell, E. and Rikner, G., Linearity with dose-rate of low resistivity p-type silicon semiconductor detectors, *Phys. Med. Biol.*, 38, 785–792, 1993.

Guo, P., Adamovics, J. A., and Oldham, M., Characterization of a new radiochromic 3D dosimeter, *Med. Phys.*, 33, 1338–1345, 2006a.

Guo, P., Adamovics, J. A., and Oldham, A., A practical 3D dosimetry system for radiation therapy, *Med. Phys.*, 33, 3962–3972, 2006b.

Gurney, R. W. and Mott, N. F., The theory of the photolysis of silver bromide and the photographic latent image, *Proc. Roy. Soc. London, Ser. A*, 164, 151–167, 1938.

Hall, G., Semiconductor detectors, in *Medical Radiation Detectors*, Kember, N. F., Ed., Institute of Physics Publishing, Bristol/Philadelphia, pp. 47–56, 1994.

Halvorsen, P. H., Dosimetric evaluation of a new design MOSFET in vivo dosimeter, *Med. Phys.*, 32, 110–117, 2005.

Hanson, W. F. and Dominguez-Tinoco, J. A., Effects of plastic protective caps on the calibration of therapy beams in water, *Med. Phys.*, 12, 243–248, 1985.

Harder, D., Berechnung der Energiedosis aus Ionisationsmessungen bei Sekundär-elektronen-Gleichgewicht, in *Symposium on high energy electrons—Montreux,* 1964, Zuppinger, A. and Poretti, G., Eds., Springer, Berlin, pp. 40–48, 1965a.

Harder, D., Energiespektren schneller Elektronen in verschiedenen Tiefen, in *Symposium on High Energy Electrons—Montreux,* 1964, Zuppinger, A. and Poretti, G., Eds., Springer, Berlin, pp. 26–33, 1965b.

Harder, D., Einfluss der Vielfachstreuung von Elektronen auf die Ionisation in gasgefüllten Hohlräumen, *Biophysik*, 5, 157–164, 1968.

Harder, D., Fano's theorem and the multiple scattering correction, in *Proc. 4th Symp. Microdosimetry (1973)*, Booz, H. G. E. J., Eickel, R., and Waker, A., Eds., Verbania Pallanza, Italy, pp. 677–693, 1974.

Havercroft, J. M. and Klevenhagen, S. C., Ion recombination corrections for plane-parallel and thimble chambers in electron and photon radiation, *Phys. Med. Biol.*, 38, 25–38, 1993.

Hazle, J. D., Hefner, L., Nyerick, C. E., Wilson, L., and Boyer, A. L., Dose–response characteristics of a ferrous-sulphate-doped gelatin system for determining radiation absorbed dose distributions by magnetic resonance imaging (Fe MRI), *Phys. Med. Biol.*, 36, 1117–1125, 1991.

Herz, R. H., *The Photographic Action of Ionising Radiation* (Chapters 3 and 4), Wiley Interscience, New York, 1969.

Hettinger, G., Pettersson, C., and Svensson, H., Displacement effect of thimble chambers exposed to a photon or electron beam from a betatron, *Acta Radiol. Ther. Phys. Biol.*, 6, 61–64, 1967.

Heukelom, S., Lanson, J. H., and Mijnheer, B. J., Comparison of entrance and exit dose measurements using ionisation chambers and silicon diodes, *Phys. Med. Biol.*, 36, 47–59, 1991.

Ho, A. and Paliwal, B., Stopping-power and mass energy-absorption coefficient ratios for solid water, *Med. Phys.*, 13, 403–404, 1986.

Hogstrom, K., Mills, M., and Almond, P., Electron beam dose calculations, *Phys. Med. Biol.*, 26, 445–459, 1981.

Hohlfeld, K., The standard DIN 6800: Procedures for absorbed dose determination in radiology by the ionisation method (IAEA-SM-298/31), in *Dosimetry in Radiotherapy*, Vol. 1, IAEA Proceedings Series, Vienna, 1988.

Hohlfeld, K., In-phantom measurement of absorbed dose to water in medium energy x-ray beams, in *Review of data and methods recommended in the international code of practice IAEA Technical Reports Series No. 277, Absorbed Dose Determinations in Photon and Electron Beams*, IAEA-TECDOC Series no. 897, IAEA, Vienna, pp. 47–67, 1996.

Holt, J. G., Edelstein, G. R., and Clark, T. E., Energy dependence of the response of lithium fluoride TLD rods in high energy electron fields, *Phys. Med. Biol.*, 20, 559–570, 1975.

HPA (Hospital Physicists Association), A code of practice for the dosimetry of 2 to 35 MV x-ray and caesium-137 and cobalt-60 gamma ray beams, *Phys. Med. Biol.*, 13, 1–8, 1969.

HPA (Hospital Physicists Association), Revised Code of Practice for the dosimetry of 2 to 25 MV x-ray, and of caesium-137 and cobalt-60 gamma-ray beams, *Phys. Med. Biol.*, 28, 1097–1104, 1983.

Humphries, L. J. and Slowey, T. W., Dosimetry instrumentation, in *Radiation Oncology Physics (AAPM Summer School)*, Kereiakes, J. G., Elson, H. R., and Born, C. G., Eds., AAPM, New York, pp. 110–138, 1986.

Hunt, M. A., Kutcher, G. J., and Buffa, A., Electron backscatter corrections for parallel-plate chambers, *Med. Phys.*, 15, 96–103, 1988.

Huq, M. S. and Andreo, P., Advances in the determination of absorbed dose to water in clinical high-energy photon and electron beams using ionisation chambers, *Phys. Med. Biol.*, 49, R49–R104, 2004.

Huq, M. S., Hossain, M., and Andreo, P., A comparison of the AAPM TG51 protocol and the IAEA absorbed-dose-to-water based Code of Practice for dosimetry calibration of high energy photon beams, *Med. Phys.*, 26, 1153, 1999.

Huyskens, D., Bogaerts, R., Verstraete, J., Lööf, M., Nyström, H., Fiorino, C., Broggi, S., Jornet, N., Ribas, M., and Thwaites, D., Partical guidelines for the implementation of in vivo dosimetry with diodes in external radiotherapy with photon beams (entrance dose), ESTRO Estro Booklet on Physics for clinical radiotherapy N° 5, 2001, [http://estro.be]

IAEA (International Atomic Energy Agency), *Absorbed dose determination in photon and electron beams: An international code of practice*, Technical Report Series no. 277, IAEA, Vienna, 1987.

IAEA (International Atomic Energy Agency), *Review of data and methods recommended in the International Code of Practice, Absorbed dose determination in photon and electron beams IAEA Technical Reports Series No. 277, Proceedings of a consultants meeting*, Vienna, 8–11, December, 1992. IAEA TECDOC Series no. 897, Vienna, 1996.

IAEA (International Atomic Energy Agency), *Absorbed dose determination in photon and electron beams: An International Code of Practice*, Technical Report Series no. 277, 2nd ed., IAEA, Vienna, 1997a.

IAEA (International Atomic Energy Agency), *The use of plane-parallel ionisation chambers in high-energy electron and photon beams: An International Code of Practice for dosimetry*, Technical Report Series no. 381, IAEA, Vienna, 1997b.

IAEA (International Atomic Energy Agency), *Absorbed dose determination in external beam radiotherapy: An international Code of Practice for dosimetry based on standards of absorbed dose to water*, Technical Report Series no. 398, IAEA, Vienna, 2000.

IAEA (International Atomic Energy Agency), *Implementation of the international Code of Practice on dosimetry in radiotherapy (TRS398): Review of testing results*, IAEA TECDOC Series no. 1455, Vienna, 2005.

Ibbott, G. S., Maryanski, M. J., Eastman, P., Holcomb, S. D., Zhang, Y. S., Avison, R. G., Sanders, M., and Gore, J. C., 3D visualisation and measurement of conformal dose-distributions using MRI of BANG-gel dosimeters, *Int. J. Rad. Oncol. Biol. Phys.*, 38, 1097–1103, 1997.

ICRU (International Commission on Radiation Units and Measurements), *Radiation dosimetry: X-rays and gamma rays with maximum photon energies between 0.6 and 50 MeV*, ICRU Report 14. ICRU, Bethesda, MD, 1969.

ICRU (International Commission on Radiation Units and Measurements), *Measurement of absorbed dose in a phantom irradiated by a single beam of x or gamma rays*, ICRU Report 23, ICRU, Bethesda, MD, 1973.

ICRU (International Commission on Radiation Units and Measurements), *Determination of absorbed dose in a patient irradiated by beams of x or gamma rays in radiotherapy procedures*, ICRU Report 24, ICRU, Bethesda, MD, 1976.

ICRU (International Commission on Radiation Units and Measurements), *Average energy required to produce an ion pair*, ICRU Report 31, ICRU, Bethesda, MD, 1979.

ICRU (International Commission on Radiation Units and Measurements), *The dosimetry of pulsed radiation*, ICRU Report 34, ICRU, Bethesda, MD, 1982.

ICRU (International Commission on Radiation Units and Measurements), *Radiation dosimetry: Electron beams with energies between 1 and 50 MeV*, ICRU Report 35, ICRU, Bethesda, MD, 1984a.

ICRU (International Commission on Radiation Units and Measurements), *Stopping powers for electrons and positrons*, ICRU Report 37, ICRU, Bethesda, MD, 1984b.

ICRU (International Commission on Radiation Units and Measurements), *Tissue substitutes in radiation dosimetry and measurement*, ICRU Report 44, ICRU, Bethesda, MD, 1989.

ICRU (International Commission on Radiation Units and Measurements), *Photon, electron, proton, and neutron interaction data for body tissues*, ICRU Report 46, ICRU, Bethesda, MD, 1992a.

ICRU (International Commission on Radiation Units and Measurements), *Phantoms and computational models in therapy, diagnosis and protection*, ICRU Report 48, ICRU, Bethesda, MD, 1992b.

ICRU (International Commission on Radiation Units and Measurements), *Stopping powers and ranges for protons and alpha particles*, ICRU Report 49, ICRU, Bethesda, MD, 1993.

ICRU (International Commission on Radiation Units and Measurements), *Fundamental quantities and units for ionizing radiation*, ICRU Report 60, ICRU, Bethesda, MD, 1998.

ICRU (International Commission on Radiation Units and Measurements), *Clinical proton dosimetry, Part I: Beam production, beam delivery and measurement of absorbed dose*, ICRU Report 59, ICRU, Bethesda, MD, 1999.

ICRU (International Commission on Radiation Units and Measurements), *Dosimetry of high-energy photon beams based on standards of absorbed dose to water*, ICRU Report 64, ICRU, Bethesda, MD, 2001.

IEC, *Medical electrical equipment: Dosimeters with ionisation chambers as used in radiotherapy Standard IEC-60731*, International Electrotechnical Commission, Geneva, 1997.

Ipe, N. E., Rosser, K. E., Moretti, C. J., Manning, J. W., Palmer, M. J., Air kerma calibration factors and k_{ch} values for the PTW soft x-ray, NACP and Roos ionisation chambers at very low x-ray energies (0.035mm–1.0 mm al HVL). SLAC-PUB-8099, Radiation Physics Department, Stanford Linear Accelerator Center, Stanford, CA, 1999, [www.slac.Stanford.edu/pubs/slacpubs/8000/slac-pub-8099.html].

IPEMB (Institute of Physics and Engineering in Medicine and Biology), The IPEMB code of practice for electron dosimetry for radiotherapy beams of initial energy from 2 to 50 MeV based on air kerma calibration, *Phys. Med. Biol.*, 41, 2557–2603, 1996a.

IPEMB (Institute of Physics and Engineering in Medicine and Biology), The IPEMB code of practice for the determination of absorbed dose for x-rays below 300 kV generating potential (0.035 mm Al–4 mm Cu HVL; 10–300 kV generating potential), *Phys. Med. Biol.*, 41, 2605–2625, 1996b.

IPSM (Institute of Physical Sciences in Medicine), IPSM Report 54, *Commissioning and quality assurance of linear accelerators*, IPEM, York, UK, 1988.

IPSM (Institute of Physical Sciences in Medicine), Code of Practice for high-energy photon therapy dosimetry based on the NPL absorbed dose calibration service, *Phys. Med. Biol.*, 35, 1355–1360, 1990.

IPSM (Institute of Physical Sciences in Medicine), Report of the Institute of Physical Sciences in Medicine Working Party on low and medium energy x-ray dosimetry, *Phys. Med. Biol.*, 36, 1027–1038, 1991.

ISO (International Organization for Standardization), *Guide to the expression of uncertainty in measurement*, 2nd ed., Published by ISO in the name of BIPM, IEC, IFCC, IUPAC, IUPAP and OIML, ISO, Geneva, 1995.

Jayachandran, C. A., The response of thermoluminescent dosimetric borates equivalent to air, water and soft tissue and of Lif TLD-100 to low energy x-rays, *Phys. Med. Biol.*, 15, 325–334, 1970.

Johansson, K. A., Mattsson, L. O., Lindborg, L., and Svensson, H., Absorbed-dose determination with ionisation chambers in electron and photon beams having energies between 1 and 50 MeV, in *National and International Standardization of Radiation Dosimetry*, Vol. 2, IAEA Proceedings Series, Vienna, pp. 243–270, 1978.

Johansson Bäck, S. A., Magnusson, P., Fransson, A., Olsson, L. E., Montelius, A., Holmberg, O., Andreo, P., and Mattsson, S., Improvements in absorbed dose measurements for external radiation therapy using ferrous dosimeter gel and MR imaging (FeMRI), *Phys. Med. Biol.*, 43, 261–276, 1998.

Johns, H. E. and Cunningham, J. R., *The Physics of Radiology*, 4th Ed., Charles C. Thomas, Springfield, IL, 1983.

Jursinic, P. A. and Nelms, B. E., A 2-D diode array and analysis software for verification of intensity modulated radiation therapy delivery, *Med. Phys.*, 29, 1687–1692, 2002.

Kawrakow, I., Accurate condensed history Monte Carlo simulation of electron transport. II. Application to ionisation chamber response simulations, *Med. Phys.*, 27, 499–513, 2000.

Kelly, B. G., Jordan, K. J., and Battista, J. J., Optical CT reconstruction of 3D dose distributions using the ferrous-benzoic-xylenol (FBX) gel dosimeter, *Med. Phys.*, 25, 1741–1750, 1998.

Kemp, L. A., The National Physical Laboratory secondary standard therapy-level x-ray exposure meter, *Br. J. Radiol.*, 45, 775–778, 1972.

Khan, F., *The Physics of Radiation Therapy*, Williams and Wilkins, Baltimore, 1984.

Kirby, T. H., Hanson, W. F., and Johnston, D. A., Uncertainty analysis of absorbed dose calculations from thermo-luminescence dosimeters, *Med. Phys.*, 19, 1427–1433, 1992.

Klevenhagen, S. C., Experimentally determined backscatter factors for x-rays generated at voltages between 16 and 140 kV, *Phys. Med. Biol.*, 34, 1871–1882, 1989.

Klevenhagen, S. C., Implications of electron backscattering for electron dosimetry, *Phys. Med. Biol.*, 36, 1013–1018, 1991.

Klevenhagen, S. C., *Physics and Dosimetry of Therapy Electron Beams*, Medical Physics Publishing, Madison, WI, 1993.

Knight, R. T., Backscatter factors for low and medium energy x-rays calculated by the Monte Carlo method, ICR-PHYS-1/93, Available from: Physics Dept., Royal Marsden NHS Trust, Sutton SM2 5PT, UK, 1993.

Knight, R. T., Absorbed dose conversion factors for therapeutic kilovoltage and megavoltage x-ray beams calculated by the Monte Carlo method, Ph.D. Thesis, London University, ICR-PHYS-1/96, Available from: Physics Dept., Royal Marsden NHS Trust, Sutton SM2 5PT, UK, 1996.

Knight, R. T. and Nahum, A. E., Depth and field-size dependence of ratios of mass-energy absorption coefficient, water-to-air, for kV x-ray dosimetry, IAEA-SM-330/17, in *Measurement Assurance in Dosimetry*, IAEA Proceedings Series, Vienna, pp. 361–370, 1994.

Kosunen, A. and Rogers, D. W. O., Beam quality specification for photon beam dosimetry, *Med. Phys.*, 20, 1181–1188, 1993.

Kristensen, M., Measured influence of the central electrode diameter and material on the response of a graphite ionisation chamber to cobalt-60 gamma rays, *Phys. Med. Biol.*, 28, 1269–1278, 1983.

Kron, T., Rosenfeld, A., Lerch, M., and Bazley, S., Measurements in radiotherapy beams using on-line MOSFET detectors, *Radiat. Prot. Dosimetry*, 101, 445–448, 2002.

Kuntz, F., Pabst, J. Y., Delpech, J. P., Wagner, J. P., and Marchioni, E., Alanine-ESR in vivo dosimetry: A feasibility study and possible applications, *Appl. Radiat. Isot.*, 47, 1183–1188, 1996.

LaRiviere, P. D., The quality of high-energy x-ray beams, *Br. J. Radiol.*, 62, 473–481, 1989.

Laub, W. U., Kaulich, T. W., and Nüsslin, F., A diamond detector in the dosimetry of high-energy electron and photon beams, *Phys. Med. Biol.*, 44, 2183–2192, 1999.

Lempert, G. D., Nath, R., and Schulz, R. J., Fraction of ionisation from electrons arising in the wall of an ionisation chamber, *Med. Phys.*, 10, 1–3, 1983.

Lepage, M., Whittaker, A. K., Rintoul, L., Back, S. A., and Baldock, C., Modelling of post-irradiation events in polymer gel dosimeters, *Phys. Med. Biol.*, 46, 2827–2839, 2001.

Letourneau, D., Gulam, M., Yan, D., Oldham, M., and Wong, J. W., Evaluation of a 2D diode array for IMRT quality assurance, *Radiother. Oncol.*, 70, 199–206, 2004.

Loevinger, R. and Loftus, T. P., Uncertainty in the delivery of absorbed dose, *Proceedings of the International Course on Ionizing Radiation Metrology*, Casnati, E., Ed., Editrice Compositori, Bologna, pp. 459–273, 1977.

Ma, C. -M. and Nahum, A. E., Bragg–Gray theory and ionisation chamber dosimetry for photon beams, *Phys. Med. Biol.*, 36, 413–428, 1991.

Ma, C. -M. and Nahum, A. E., Effect of the size and composition of the central electrode on the response of cylindrical ionisation chambers in high-energy photon and electron beams, *Phys. Med. Biol.*, 38, 267–290, 1993.

Ma, C. -M. and Nahum, A. E., Monte-Carlo calculated correction factors for a NE2571 chamber in medium-energy photon beams (IAEA-SM-330/5), in *Measurement Assurance in Dosimetry* (Proc. Symp. Vienna, 1993), IAEA Proceedings Series, Vienna, pp. 371–381, 1994.

Ma, C. -M. and Nahum, A. E., Calculations of ionisation chamber displacement effect corrections for medium-energy x-ray dosimetry, *Phys. Med. Biol.*, 40, 45–62, 1995a.

Ma, C. -M. and Nahum, A. E., Monte Carlo calculated stem effect corrections for NE2561 and NE2571 chambers in medium-energy x-ray beams, *Phys. Med. Biol.*, 40, 63–72, 1995b.

Ma, C. -M., Coffey, C. W., DeWerd, L. A., Liu, C., Nath, R., Seltzer, S. M., and Seuntjens, J., Status of kilovoltage x-ray beam dosimetry in radiotherapy, NRCC Report PIRS-0533, Available from Ionizing Radiation Standards, INMS, National Research Council Canada, Ottawa, Canada K1A 0R6, 1996.

Ma, C. -M., Li, X. A., and Seuntjens, J., Study of dosimetry consistency for kilovoltage x-ray beams, in *Kilovoltage x-ray Beam Dosimetry for Radiotherapy and Radiobiology*, Ma, C. -M. and Seuntjens, J. P., Eds., Medical Physics Publishing, Wisconsin, pp. 69–87, 1999.

Ma, C. -M., Coffey, C. W., DeWerd, L. A., Liu, C., Nath, R., Seltzer, S. M., and Seuntjens, J. P., AAPM protocol for 40–300 kV x-ray beam dosimetry in radiotherapy and radiobiology, *Med. Phys.*, 28, 868–893, 2001.

Malamut, C., Rogers, D. W. O., and Bielajew, A. F., Calculation of water/air stopping-power ratios using EGS4 with explicit treatment of electron-positron differences, *Med. Phys.*, 18, 1222–1228, 1991.

Marinello, G., Thermoluminescence dosimetry applied to quality assurance in radiotherapy, brachytherapy and radiodiagnostic, *IAEA TECDOC Series no.* 896, 267–280, 1994.

Marinello, G., Barthe, J., Pollack, J., and Portal, G., PCL, a new automatic fast reader suitable for *in vivo* dosimetry, *Radiother. Oncol.*, 25, 63–66, 1992.

Markus, B., Eine Parallelplatten-Kleinkammer zur Dosimetrie schneller Elektronen, *Strahlentherapie*, 150, 307–320, 1976.

Marre, D. and Marinello, G., Comparison of p-type commercial electron diodes for in vivo dosimetry, *Med. Phys.*, 31, 50–56, 2004.

Marre, D., Ferreira, I. H., Bridier, A., Björeland, A., Svensson, H., Dutreix, A., and Chavaudra, J., Energy correction factors of LiF powder TLDs irradiated in high-energy electron beams and applied to mailed dosimetry for quality assurance networks, *Phys. Med. Biol.*, 45, 3657–3674, 2000.

Martens, C., de Wagter, C., and de Neve, W., The value of the LA48 linear ion chamber array for characterization of intensity-modulated beams, *Phys. Med. Biol.*, 46, 1131–1148, 2001.

Martin, S., Lisbona, A., Richard, J., Morteau, S., Denizot, B., and Bardies, M., Production of new thermoluminescent mini-dosimeters, *Phys. Med. Biol.*, 45, 479–494, 2000.

Maryanski, M. J., Gore, J. C., Kennan, R. P., and Schulz, R. J., NMR relaxation enhancement in gels polymerised and cross-linked by ionising radiation: A new approach to 3D dosimetry by MRI, *Magn. Res. Imaging*, 11, 253–258, 1993.

Maryanski, M. J., Schulz, R. J., Ibbott, G. S., Gatenby, J. C., Xie, J., Horton, D., and Gore, J. C., Magnetic resonance imaging of radiation dose distributions using a polymer-gel dosimeter, *Phys. Med. Biol.*, 39, 1437–1455, 1994.

Maryanski, M. J., Zastavker, Y. Z., and Gore, J. C., Radiation dose distributions in three dimensions from tomographic optical density scanning of polymer gels: II. Optical properties of the BANG polymer gel, *Phys. Med. Biol.*, 41, 2705–2717, 1996a.

Maryanski, M. J., Ibbott, G. S., Eastman, P., Schulz, R. J., and Gore, J. C., Radiation therapy using magnetic resonance imaging of polymer gels, *Med. Phys.*, 23, 699–705, 1996b.

Maryanski, M. J., Audet, C., and Gore, J. C., Effects of crosslinking and temperature on the dose response of a BANG polymer gel dosimeter, *Phys. Med. Biol.*, 42, 303–311, 1997.

Mattsson, L. O., Application of the water calorimeter, Fricke dosimeter and ionisation chamber in clinical dosimetry, Ph.D. Thesis, University of Gothenburg, 1984.

Mattsson, L. O. and Johansson, K. A., Experimentally determined wall correction factors, k_m and k_{att}, for cylindrical ionisation chambers used in high energy photon and electron beam dosimetry. RADFYS 84:04 (Department of Radiation Physics, University of Göteborg), 1984.

Mattsson, L. O. and Svensson, H., Charge build-up effects in insulating phantom materials, *Acta Radiol. Oncol.*, 23, 393–399, 1984.

Mattsson, L. O., Johansson, K. A., and Svensson, H., Calibration and use of plane-parallel ionisation chambers for the determination of absorbed dose in electron beams, *Acta Radiol. Oncol.*, 20, 385–399, 1981.

Mayles, W. P. M., Heisig, S., and Mayles, H. M. O., Treatment verification and in vivo dosimetry, in *Radiotherapy Physics*, 2nd ed.,Williams, J. R. and Thwaites, D. I., Eds., Oxford University Press, Oxford, pp. 220–246, 2000.

Mayles, W. P. M., Lake, R., McKenzie, A. L., Macaulay, E. M., Morgan, H. M., Jordan, T. J., and Powley, S. K., *Physics aspects of quality control in radiotherapy*. Report 81, IPEM, York, UK, 1998.

McEwen, M. R., DuSautoy, A. R., and Williams, A. J., The calibration of therapy level electron beam ionisation chambers in terms of absorbed dose to water, *Phys. Med. Biol.*, 43, 2503–2519, 1998.

McJury, M., Oldham, M., Leach, M. O., and Webb, S., Dynamics of polymerisation in polyacrylamide gel (PAG) dosimeters: (I) Aging and long-term stability, *Phys. Med. Biol.*, 44, 1863–1873, 1999.

McJury, M., Oldham, M., Cosgrove, V. P., Murphy, P. S., Doran, S., Leach, M. O., and Webb, S., Radiation dosimetry using polymer gels: Methods and applications, *Br. J. Radiol.*, 73, 919–929, 2000.

McKeever, S. W. S., *Thermoluminescence of Solids*, University Press, New York, 1985.

McKeever, S. W. S., Moscovitch, M., and Townsend, P. D., *Thermoluminescence Dosimetry Materials: Properties and Uses*, Nuclear Technology Publishing, Ashford, UK, 1995.

McKerracher, C. and Thwaites, D., Assessment of new small-field detectors against standard field detectors for practical stereotactic beam data acquisition, *Phys. Med. Biol.*, 44, 2143–2160, 1999.

McKinlay, A. F., *Thermoluminescence dosimetry*, Medical Physics Handbooks, Adam Hilger Ltd, Bristol, 1981.

Medin, J., Studies of clinical proton dosimetry using Monte Carlo simulation and experimental techniques. Ph.D. Thesis, University of Stockholm, 1997.

Medin, J. and Andreo, P., Monte Carlo calculated stopping-power ratios water/air for clinical proton dosimetry (50–250 MeV), *Phys. Med. Biol.*, 42, 89–105, 1997.

Medin, J. and Grusell, E., General characteristics of the use of silicon diode detectors for clinical dosimetry in proton beams, *Phys. Med. Biol.*, 45, 2382–2573, 2000.

Medin, J., Andreo, P., Grusell, E., Mattsson, O., Montelius, A., and Roos, M., Ionisation chamber dosimetry of proton beams using cylindrical and plane-parallel chambers N_w versus N_K ionisation chamber calibrations, *Phys. Med. Biol.*, 40, 1161–1176, 1995.

Medin, J., Andreo, P., and Vynckier, S., Comparison of dosimetry recommendation for clinical proton beams, *Phys. Med. Biol.*, 45, 3195–3212, 2000.

Mijnheer, B. J., Battermann, J. J., and Wambersie, A., What degree of accuracy is required and can be achieved in photon and neutron therapy? *Radiother. Oncol.*, 8, 237–252, 1987.

Mobit, P. N., Mayles, P., and Nahum, A. E., The quality dependence of LiF TLD in megavoltage photon beams: Monte Carlo simulation and experiments, *Phys. Med. Biol.*, 41, 387–398, 1996a.

Mobit, P. N., Mayles, P., and Nahum, A. E., The energy correction factor of LiF thermoluminescent dosimeters in megavoltage electron beams: Monte Carlo simulation and experiments, *Phys. Med. Biol.*, 41, 979–993, 1996b.

Moran, J., *Dosimetry metrology for IMRT* Proceedings AAPM Summer School on Intensity-Modulated Radiation Therapy: The state of art. Medical Physics Monograph No. 29, Colorado, June 22–26, 2003.

NACP (Nordic Association of Clinical Physics), Procedures in radiation therapy dosimetry with 5 to 50 MeV electrons and roentgen and gamma rays with maximum photon energies between 1 and 50 MeV, *Acta Radiol. Ther. Phys. Biol.*, 11, 603–624, 1972.

NACP (Nordic Association of Clinical Physics), Procedures in external radiation therapy dosimetry with electron and photon beams with maximum energies between 1 and 50 MeV, *Acta Radiol. Oncol.*, 19, 55–79, 1980.

NACP (Nordic Association of Clinical Physics), Supplement to the recommendations of NACP (1980): Electron beams with mean energies at the phantom surface below 15 MeV, *Acta Radiol. Oncol.*, 20, 401–415, 1981.

Nahum, A. E., Calculations of electron flux spectra in water irradiated with megavoltage electron and photon beams with applications to dosimetry. Ph.D. Thesis, University of Edinburgh. University Microfilms, Ann Arbor, International Order no. 77-70006, 1976.

Nahum, A. E., Water/air mass stopping power ratios for megavoltage photon and electron beams, *Phys. Med. Biol.*, 23, 24–38, 1978.

Nahum, A. E., Monte Carlo calculations of dosimetrical parameters for photon beams. *Proceedings of 6th International Congress on Medical Physics*, Hamburg, 5–11, September 1982, MPBE, Hamburg, pp. 27–92, 1982.

Nahum, A. E., An extension of Spencer-Attix cavity theory to the 3-media situation for electron beams, in *Dosimetry in Radiotherapy*, IAEA Proceedings Series, Vienna, pp. 87–115, 1988.

Nahum, A. E., Perturbation effects in dosimetry, ICR-PHYS-1/94, Available from: Physics Dept., Royal Marsden NHS Trust, Sutton SM2 5PT, UK, 1994.

Nahum, A. E., Perturbation effects in dosimetry, Part I: Kilovoltage x-rays and electrons, *Phys. Med. Biol.*, 41, 1531–1580, 1996.

Nahum, A. E., Condensed-history Monte-Carlo simulation for charged particles: What can it do for us? *Radiat. Environ. Biophys.*, 38, 163–173, 1999a.

Nahum, A. E., kV x-ray dosimetry: Current status and future challenges, in *Kilovoltage X-ray Beam Dosimetry for Radiotherapy and Radiobiology*, Ma, C. -M. and Seuntjens, J. P., Eds., Medical Physics Publishing, Wisconsin, pp. 7–26, 1999b.

Nahum, A. E. and Knight, R. T., Consistent formalism for kV x-ray dosimetry, in *Measurement Assurance in Dosimetry*, IAEA Proceedings Series, Vienna, pp. 451–459, 1994.

Nahum, A. E., Henry, W. H., and Ross, C., Response of carbon- and aluminium-walled thimble chambers in Co-60 and 20 MeV electron beams, *XIV ICMBE and VII ICMP Vol Suppl. Part I*, Med. Biol. Eng. Comp. 23. Espoo, Finland, pp. 612–613, 1985.

Nath, R. and Schulz, R. J., Calculated response and wall correction factors for ionisation chambers exposed to ^{60}Co gamma-rays, *Med. Phys.*, 8, 85–93, 1981.

NCRP (National Committee on Radiation Protection and Measurements), NCRP Report 27, *Stopping powers for use with cavity chambers*, NCRP, Washington, DC, 1961.

NCS (Nederlandse Commissie voor Stralingsdosimetrie), NCS Report 2, *Code of practice for the dosimetry of high-energy photon beams*, NCS, Delft, 1986.

NCS (Nederlandse Commissie voor Stralingsdosimetrie), NCS Report 10, *Dosimetry for low and medium energy x-rays: A code of practice in radiotherapy and radiobiology*, NCS, Delft, 1997.

Nette, P. and Czap, L., *Cable adapters* (Working material for the IAEA/WHO SSDL network), Internal Report DMRP-9400, International Atomic Energy Agency, Vienna, 1994.

Nilsson, B. and Montelius, A., Fluence perturbation in photon beams under nonequilibrium conditions, *Med. Phys.*, 13, 191–195, 1986.

Nilsson, B., Montelius, A., and Andreo, P., Wall effects in plane-parallel ionisation chambers, *Phys. Med. Biol.*, 41, 609–623, 1996.

Niroomand-Rad, A., Blackwell, C. R., Coursey, B. M., Gall, K. P., Galvin, J. M., McLaughlin, W. L., Meigooni, A. S., Nath, R., Rodgers, J. E., and Soares, C. G., Radiochromic film dosimetry: Recommendations of AAPM radiation therapy committee task group 55, *Med. Phys.*, 25, 2093–2115, 1998.

Nisbet, A. and Thwaites, D. I., Polarity and ion recombination correction factors for ionization chambers employed in electron beam dosimetry, *Phys. Med. Biol.*, 43, 435–443, 1998a.

Nisbet, A. and Thwaites, D. I., An evaluation of epoxy resin phantom materials for electron dosimetry, *Phys. Med. Biol.*, 43, 1523–1528, 1998b.

Nisbet, A., Aukett, R., and Thwaites, D. I., The use of epoxy resin phantom materials in kV x-ray dosimetry, *Radiother. Oncol.*, 48(Suppl. 1), S51, 1998.

Nyström, H. and Karlsson, M., Correction factors applied to plane-parallel ionisation chambers, *Phys. Med. Biol.*, 38, 311–322, 1993.

Olch, A. J., Dosimetric performance of an enhanced dose range radiographic film for intensity-modulated radiation therapy quality assurance, *Med. Phys.*, 29, 2159–2168, 2002.

Oldham, M., Optical-CT scanning of polymer gels, *J. Phys. Conf. Ser.*, 3, 293–329, 2004.

Oldham, M. and Webb, S., Intensity modulated radiotherapy by means of static tomotherapy: A planning and verification study, *Med. Phys.*, 24, 827–836, 1997.

Oldham, M., McJury, M., Baustert, I. B., Webb, S., and Leach, M. O., Improving calibration accuracy in gel-dosimetry, *Phys. Med. Biol.*, 43, 2709–2720, 1998.

Oldham, M., Siewerdsen, J. H., Shetty, A., and Jaffray, D. A., High resolution gel-dosimetry by optical-CT and MR scanning, *Med. Phys.*, 28, 1436–1445, 2001.

Olsen, D. R. and Hellesnes, J., Absorbed dose distribution measurements in brachytherapy using ferrous sulphate gel and magnetic resonance imaging, *Br. J. Radiol.*, 67, 1121–1126, 1994.

Olsson, L. E., Petersson, S., Ahlgren, L., and Mattsson, S., Ferrous sulphate gels for determination of absorbed dose distributions using MRI technique: Basic studies, *Phys. Med. Biol.*, 34, 43–52, 1989.

Olsson, L. E., Fransson, A., Ericsson, A., and Mattsson, S., MR imaging of absorbed dose distributions for radio-therapy using ferrous sulphate gels, *Phys. Med. Biol.*, 35, 1623–1631, 1990.

Olsson, L. E., Arndt, J., Fransson, A., and Nordell, B., Three-dimensional dose mapping from gamma knife treatment using a dosimeter gel and MR-imaging, *Radiother. Oncol.*, 24, 82–86, 1992.

Palm, Å., and Mattsson, O., Experimental determination of p_{cav} factors for cylindrical ionisation chambers in electron beams. IAEA TECDOC Series no. 1173, Vienna, pp. 81–89, 2000.

Palm, Å., Kirov, A. S., and LoSasso, T., Predicting energy response of radiographic film in a 6 MV x-ray beam using Monte Carlo calculated fluence spectra and absorbed dose, *Med. Phys.*, 31, 3168–3178, 2004.

Palmans, H., Verhaegen, F., Denis, J. M., Vynckier, S., and Thierens, H., Experimental study of perturbation correction factors for ionisation chambers in a 75 MeV clinical proton beam, *Radiother. Oncol.*, 51(Suppl. 1), S39, 1999.

Papatheodorou, S., Zefkili, S., and Rosenwald, J.-C., The equivalent wedge implementation of the Varian Enhanced Dynamic Wedge (EDW) into a treatment planning system, *Phys. Med. Biol.*, 44, 509–524, 1999.

Peixoto, J. G. P. and Andreo, P., Determination of absorbed dose to water in reference conditions for radiotherapy kilovoltage x-rays between 10 and 300 kV: A comparison of the data in the IAEA, IPEMB, DIN and NCS dosimetry protocols, *Phys. Med. Biol.*, 45, 563–575, 2000.

Perrin, B. A., Whitehurst, P., Cooper, P., and Hounsell, A. R., The measurement of k_{ch} factors for application with the IPEMB very low energy dosimetry protocol, *Phys. Med. Biol.*, 46, 1985–1995, 2001.

Pitchford, W. G. and Thwaites, D. I., Effect on dose of charge storage in electron irradiated phantoms. (*Proceedings of VII Int. Conf. on Medical Physics, Espoo, 1985*), *Med. Biol. Eng. Comput.*, 23(Suppl. 1), 610–611, 1985.

Podgorsak, M. B. and Schreiner, L. J., Nuclear magnetic relaxation characterization of irradiated Fricke solution, *Med. Phys.*, 19, 87–95, 1992.

Poen, J. C., Clinical applications of orthovoltage radiotherapy: Tumors of the skin, endorectal therapy, and intraoperative radiation therapy, in *Kilovoltage X-Ray Beam Dosimetry for Radiotherapy and Radiobiology*, Ma, C. -M. and Seuntjens, J. P., Eds., Medical Physics Publishing, Wisconsin, pp. 1–5, 1999.

Reinstein, L. E. and Gluckman, G. R., Comparison of dose response of radiochromic film measured with He–Ne laser, broadband, and filtered light densitometers, *Med. Phys.*, 24, 1531–1533, 1997.

Rikner, G., Characteristics of a p-Si detector in high energy electron fields, *Acta Radiol. Oncol.*, 3, 279–283, 1985a.

Rikner, G., Characteristics of a selectively shielded p-Si detector in Co-60 and 8 and 16 MV Roentgen radiation, *Acta. Radiol. Oncol.*, 24, 205–208, 1985b.

Rikner, G. and Grusell, E., Effect of radiation damage on p-type silicon detectors, *Phys. Med. Biol.*, 28, 1261–1267, 1983.

Rikner, G. and Grusell, E., Selective shielding of a p-Si detector for quality independence, *Acta Radiol. Oncol.*, 24, 65–69, 1985.

Rikner, G. and Grusell, E., General specifications for silicon semiconductors for use in radiation dosimetry, *Phys. Med. Biol.*, 32, 1109–1117, 1987.

Rink, A., Vitkin, J. A., and Jaffray, D. A., Characterization and real-time optical measurements of the ionizing radiation dose response for a new radiochromic medium, *Med. Phys.*, 32, 2510–2516, 2005.

Rogers, D. W. O., The role of Monte Carlo simulation of electron transport in radiation dosimetry, *Appl. Radiat. Isot.*, 42, 965–974, 1991.

Rogers, D. W. O., The advantages of absorbed-dose calibration factors, *Med. Phys.*, 19, 1227–1239, 1992a.

Rogers, D. W. O., Calibration of parallel-plate chambers: Resolution of several problems by using Monte Carlo calculations, *Med. Phys.*, 19, 889–899, 1992b.

Rogers, D. W. O., How accurately can EGS4/PRESTA calculate ion-chamber response? *Med. Phys.*, 20, 319–323, 1993.

Rogers, D. W. O., Fundamentals of dosimetry based on absorbed-dose standards, in *Teletherapy Physics, Present and Future*, Palta, J. R. and Mackie, T. R., Eds., AAPM, Washington, DC, pp. 319–356, 1996.

Rogers, D. W. O. and Bielajew, A. F., Differences in electron depth-dose curves calculated with EGS and ETRAN and improved energy–range relationships, *Med. Phys.*, 13, 687–694, 1986.

Rogers, D. W. O., Bielajew, A. F., and Nahum, A. E., Ionisation chamber response and A_{wall} correction factors in a ^{60}Co beam by Monte Carlo simulation, *Phys. Med. Biol.*, 30, 429–443, 1985.

Rogers, D. W. O., Faddegon, B. A., Ding, G. X., Ma, C. -M., We, J., and Mackie, T. R., BEAM: A Monte Carlo code to simulate radiotherapy treatment units, *Med. Phys.*, 22, 503–524, 1995.

Roos, M., *The state of the art in plane-parallel chamber hardware with emphasis on the new Roos and Attix chambers*, Work commissioned by the IAEA plane-parallel working group, December 1993. Unpublished (available from the author at PTB, Lab. 6.43, Postfach 33 45, D-38023 Braunschweig, Germany), 1993.

Ross, C. K. and Klassen, N. V., Water calorimetry for radiation dosimetry, *Phys. Med. Biol.*, 41, 1–29, 1996.

Rustgi, S. N., Evaluation of the dosimetric characteristics of a diamond detector for photon beam measurements, *Med. Phys.*, 22, 567–570, 1995.

Saini, A. S. and Zhu, C. T., Temperature dependence of commercially available diode detectors, *Med. Phys.*, 29, 622–630, 2002.

Schreiner, L. J., Crooks, I., Evans, M. D., Keller, B. M., and Parker, W. A., Imaging of HDR brachytherapy dose distributions using Fricke-gelatin dosimetry, *NMR Magn. Reson. Imaging*, 12, 901–907, 1994.

Schulz, R. J., deGuzman, A. F., Nguyen, D. B., and Gore, J. C., Dose–response curves for Fricke-infused agarose gels as obtained by nuclear magnetic resonance, *Phys. Med. Biol.*, 35, 1611–1622, 1990.

Schulz, R. J., Maryanski, G. S., Ibbott, G. S., and Bond, J. E., Assessment of the accuracy of stereotactic radiosurgery using Fricke-infused gels and MRI, *Med. Phys.*, 20, 1731–1734, 1993.

SEFM (Sociedad Española de Física Médica), Report SEFM 84-1, Procedimientos recomendados para la dosimetría de fotones y electrones de energías comprendidas entre 1 MeV y 50 MeV en radioterapia de haces externos, SEFM, Madrid, 1984.

SEFM (Sociedad Española de Física Médica), Report SEFM 87-1, Suplemento al Documento 84-1: Procedimientos recomendados para la dosimetría de fotones y electrones de energías comprendidas entre 1 MeV y 50 MeV en radioterapia de haces externos, SEFM, Madrid, 1987.

Sen, A., Parsei, E. I., McNeeley, S. W., and Ayyangar, K. M., Quantitative assessment of beam perturbations caused by silicon diodes used for in vivo dosimetry, *Int. J. Radiat. Oncol. Biol. Phys.*, 36(1), 205–211, 1996.

Seuntjens, J. and Verhaegen, F., Dependence of overall correction factor of a cylindrical ionisation chamber on field size and depth in medium-energy x-ray beams, *Med. Phys.*, 23, 1789–1796, 1996.

Seuntjens, J. P., van der Zwan, L., and Ma, C. -M., Type dependent correction factors for cylindrical chambers for in-phantom dosimetry in medium-energy x-ray beams, in *Kilovoltage X-Ray Beam Dosimetry for Radiotherapy and Radiobiology*, Ma, C. -M. and Seuntjens, J. P., Eds., Medical Physics Publishing, Wisconsin, pp. 159–174, 1999.

Shani, G., *Radiation Dosimetry Instrumentation and Methods*, CRC Press, Inc., Boca Raton, MA, 1991.

Sharpe, P. H., Rajendran, K., and Sephton, J. P., Progress towards an alanine/ESR therapy level reference dosimetry service at NPL, *Appl. Radiat. Isot.*, 47, 1171–1175, 1996.

Shi, S., Simon, W. E., and Zhu, T. C., Modelling the instantaneous dose rate dependence of radiation dose detectors, *Med. Phys.*, 30, 2509–2519, 2003.

Shortt, K., Ross, C., Seuntjens, J., Delaunay, F., Ostrowsky, A., Gross, P., and Leroy, E., Comparison of dosimetric standards of Canada and France for photons at ^{60}Co and higher energies, *Phys. Med. Biol.*, 46, 2119–2142, 2001.

Sibata, C., Mota, H., Beddar, A., Higgins, P., and Shin, K., Influence of detector size in photon beam profile measurements, *Phys. Med. Biol.*, 36, 621–631, 1991.

Sidhu, N. P., Interfacing a linear diode array to a conventional water scanner for the measurement of dynamic dose distributions and comparison with a linear ion chamber array, *Med. Dosimetry*, 24, 57–60, 1999.

Sorcini, B. B. and Rosander, S., Energy distributions from a racetrack microtron measured with a magnetic spectrometer, *Med. Phys.*, 20, 695, 1993.

Spencer, L. V. and Attix, F. H., A theory of cavity ionisation, *Radiat. Res.*, 3, 239–254, 1955.

Stern, R. L., Kurylo, J., Siantar, C. H., Lehmann, J., and Goldberg, Z., Film dosimetry in the peripheral region using multiple sensitometric curves, *Med. Phys.*, 31, 327–332, 2004.

Svensson, H. and Brahme, A., Recent advances in electron and photon dosimetry, in *Radiation Dosimetry, Physical and Biological Aspects*, Orton, C. G., Ed., Plenum Press, New York, pp. 87–170, 1986.

Tarte, B. J., Jardine, P. A., and van Doorn, T., Laser scanned agarose gel sections for radiation field mapping, *Int. J. Radiather. Oncol. Biol. Phys.*, 36, 175–179, 1996.

Tello, V. M., Tailor, R. C., and Hanson, W. F., How water equivalent are water-equivalent solid materials for output calibration of photon and electron beams? *Med. Phys.*, 22, 1177–1189, 1995.

Thomadsen, B., Constantinou, C., and Ho, A., Evaluation of water-equivalent plastics as phantom material for electron-beam dosimetry, *Med. Phys.*, 22, 291–296, 1995.

Thwaites, D. I., Charge storage effect on dose in insulating phantoms irradiated with electrons, *Phys. Med. Biol.*, 29, 1153–1156, 1984.

Thwaites, D. I., Measurements of ionisation in water, polystyrene and a 'solid water' phantom material for electron beams, *Phys. Med. Biol.*, 30, 41–53, 1985.

Thwaites, D. I., Uncertainties at the end point of the basic dosimetry chain (IAEA-SM-330/18), in *Measurement Assurance in Dosimetry*, IAEA Proceedings Series, IAEA, Vienna, pp. 239–255, 1994.

Thwaites, D. I., DuSautoy, A. R., Jordan, T., McEwen, M. R., Nisbet, A., Nahum, A. E., and Pitchford, W. G., The IPEM code of practice for electron dosimetry for radiotherapy beams of initial energy from 4 to 25 MeV based on an absorbed dose to water calibration, *Phys. Med. Biol.*, 48, 2929–2970, 2003.

Tochilin, E. and Goldstein, N., Dose-rate and spectral measurements from a pulsed x-ray generator, *Health Phys.*, 12, 1705, 1966.

Toivonen, M. J., Improving the accuracy of TLD systems in clinical applications, *Radiat. Prot. Dosimetry*, 47, 497–503, 1993.

Townsend, P. D., Ahmed, K., Chandler, P. J., McKeever, S. W. S., and Whitlow, H., Measurements of the emission spectra of LiF during thermoluminescence, *Radiat. Eff.*, 72, 245–257, 1983.

Tubiana, M., Dutreix, J., Dutreix, A., and Jockey, P., *Bases physiques de la radiothérapie et de la radiobiologie*, Masson Editeurs, Paris, pp. 117–118, 1963.

Udale, M., A Monte Carlo investigation of surface doses for broad electron beams, *Phys. Med. Biol.*, 33, 939–954, 1988.

Van Dam, J. and Marinello, G., Methods for in vivo dosimetry in external radiotherapy, *ESTRO Estro Booklet on Physics in Radiotherapy*, Garant, Leuven-Apeldoorn, 1994, (http://www.estro.be).

Van Dam, J., Leunens, G., and Dutreix, A., Correlation between temperature and dose rate dependence of semiconductor response: Influence of accumulated dose, *Radiother. Oncol.*, 13, 345–351, 1990.

van Gasteren, J., Heukelom, S., van Kleffens, M., van der Laarse, R., Venselaar, J., and Westermann, C., The determination of phantom and collimator scatter components of the output of megavoltage photon beams: Measurements of the collimator scatter part with a coaxial narrow cylindrical phantom, *Radiother. Oncol.*, 20, 250–257, 1991.

Verhaegen, F., Zakikhani, R., Dusautoy, A., Palmans, H., Bostock, G., Shipley, D., and Seuntjens, J., Perturbation correction factors for the NACP-02 plane-parallel ionization chamber in water in high-energy electron beams, *Phys. Med. Biol.*, 51, 1221–1235, 2006.

Visocekas, R., Lorrain, S., and Marinello, G., Evaluation of a preparation of $Li_2B_4O_7$:Cu for thermoluminescence dosimetry, *Nucl. Sci.*, 22, 61–66, 1985.

Vynckier, S., Bonnett, D. E., and Jones, D. T. L., Code of practice for clinical proton dosimetry, *Radiother. Oncol.*, 20, 53–63, 1991.

Vynckier, S., Bonnett, D. E., and Jones, D. T. L., Supplement to the code of practice for clinical proton dosimetry, *Radiother. Oncol.*, 32, 174–179, 1994.

Wall, B. F., Driscoll, C. M. H., Strong, J. C., and Fischer, E. S., The suitability of different preparations of thermoluminescent lithium borate for medical dosimetry, *Phys. Med. Biol.*, 2, 1023–1034, 1982.

Wang, S., Wang, Y., Cai, G., Wang, S., and Zha, Z., A new Tl detector developed for multiple applications, *Radiat. Prot. Dosimetry*, 47, 223–225, 1993.

Weinhous, M. S. and Meli, J. A., Determining P_{ion}, the correction factor for recombination losses in an ionisation chamber, *Med. Phys.*, 11, 846–849, 1984.

Welsch, K. T. and Reinstein, L. E., The thermal characteristics of different diodes on *in vivo* patient dosimetry, *Med. Phys.*, 28, 844–849, 2001.

White, D. R., The formulation of tissue substitute materials using basic interaction data, *Phys. Med. Biol.*, 22, 889–899, 1977.

White, D., Martin, R., and Darlison, R., Epoxy resin-based tissue substitutes, *Br. J. Radiol.*, 50, 814–821, 1977.

Wielopolski, L., Maryanski, M., Ciesielski, B., Forman, A., Reinstein, L. E., and Meek, A. G., Continuous three-dimensional radiation dosimetry in tissue-equivalent phantoms using electron paramagnetic resonance in L-alpha-alanine, *Med. Phys.*, 14, 646–652, 1987.

Williamson, J. F., Khan, F. M., and Sharma, S. C., Film dosimetry of megavoltage photon beams: A practical method of isodensity-to-isodose curve conversion, *Med. Phys.*, 8, 94–98, 1981.

Wittkämper, F. W., Thierens, H., van der Plaetsen, A., de Wagter, C., and Mijnheer, B. J., Perturbation correction factors for some ionisation chambers commonly applied in electron beams, *Phys. Med. Biol.*, 36, 1639–1652, 1991.

Wittkämper, F. W., Aalbers, A. H. L., and Mijnheer, B., Experimental determination of wall correction factors part II: NACP and Markus plane-parallel ionisation chambers, *Phys. Med. Biol.*, 37, 995–1004, 1992.

Yeo, I. J., Wang, C. K., and Burch, S. E., A filtration method for improving film dosimetry in photon radiation therapy, *Med. Phys.*, 24, 1943–1953, 1997.

Zha, Z., Wang, W. S., Zhu, J., and Cai, G., Preparation and characteristics of LiF:Mg,Cu,P thermoluminescent material, *Radiat. Prot. Dosimetry*, 47, 111–118, 1993.

Zhu, X. R., Jursinic, P. A., Grimm, D. F., Lopez, F., Rownd, J. J., and Gillin, M. T., Evaluation of Kodak EDR2 film for dose verification of intensity modulated radiation therapy delivered by a static multileaf collimator, *Med. Phys.*, 29, 1687–1692, 2002.

Zhu, X. R., Yoo, S., Jursinic, P. A., Grimm, D. F., Lopez, F., Rownd, J. J., and Gillin, M. T., Characteristics of sensitometric curves of radiographic films, *Med. Phys.*, 30, 912–919, 2003.

Zoetelief, J., Engels, A. C., and Broerse, J. J., Effective measuring point of ionisation chambers for photon dosimetry in phantoms, *Br. J. Radiol.*, 53, 580–583, 1980.

Zoetelief, J., Engels, A. C., and Broerse, J.J., Displacement corrections for spherical ionisation chambers in phantoms irradiated with neutron and photon beams (IAEA-SM-249/38), in *Biomedical Dosimetry: Physical Aspects, Instrumentation, Calibration*, IAEA Proceedings Series, Vienna, pp. 125–138, 1981.

Note: Many IAEA documents can be freely downloaded from the IAEA web site: http://www.iaea.org/publications/

PART E

CLINICAL BEAMS

Editors: Jean-Claude Rosenwald and Philip Mayles

INTRODUCTION

The equipment used to deliver the correct amount of dose to the patient has been introduced in Part C. The methods by which absolute dose and relative dose distributions are measured have been dealt with in Part D. It is then necessary to devise a methodology where these measurements are used to make accurate predictions of the dose distribution delivered to the patients, taking into account the beam characteristics as well as the patient characteristics.

The main issues related to this methodology are addressed in Chapter 20, where it is demonstrated that a detailed knowledge of the beam properties is required. The characteristics of photon beams adapted to clinical needs and generated either from kilovoltage or from megavoltage equipment (consisting both of cobalt units and linear accelerators) are presented in Chapter 21 and Chapter 22 respectively. The quantities and the methods used to calculate the dose using a spreadsheet or pocket calculator (often referred to as *manual calculations*) at a specified depth for a patient treated with photon beams are presented in Chapter 23. Finally Chapter 24 covers the case of clinical electron beams.

CHAPTER 20

FROM MEASUREMENTS TO CALCULATIONS

Philip Mayles

CONTENTS

20.1 CONTROL OF DOSE DELIVERY

Whatever the method employed for a particular type of equipment, there is a need to convert an elapsed time or a reading on a dose meter into the absolute dose delivered. For modern x-ray treatment machines, it is usual to have an ionisation chamber situated in the beam close to the x-ray source (see Section 11.7.3.1). This produces a reading proportional to the dose delivered by the beam. By its nature, the reading will generally have a relationship to the dose delivered that is dependent on various beam parameters. For this reason, the term *Monitor Unit* (MU) has been introduced. In principle, it is not necessary that there is a one-to-one correspondence between monitor units and dose, but it is usual to calibrate the machine so that a MU corresponds to 1 cGy at a specified point in a phantom and with a specified field size.

20.1.1 KILOVOLTAGE X-RAY UNITS

The dose rate delivered by a kilovoltage x-ray unit is likely to be fairly constant for a given combination of kilovoltage and filtration, but will change significantly if either the kilovoltage or the filter is changed. Older units were controlled by a timer, and the absolute dose could then be calculated using a dose rate chart. Modern units mostly have a transmission ionisation chamber. The relationship between the reading of this chamber and the dose delivered to the patient will depend on the applicator in use (especially the distance from the source to the end of the applicator) and the quality of the beam. It is usual to choose a standard applicator such as a 10 cm circle and to relate the doses for other applicators to this. Because of the rapid fall off in dose with depth, it is usually the dose at the surface of the phantom that is used for reference; although, the method of calibration may involve dose measurements in air or at a depth in the phantom (see Section 18.6).

20.1.2 LINEAR ACCELERATORS

For linear accelerators, the standard practice is to choose a 10 cm square field, but the position of the specified point and the set-up of the phantom can be different. The most common definitions of the MU are

(a) The monitor chamber reads 1 MU when 1 cGy is delivered to a point at the depth of dose maximum in a water equivalent phantom whose surface is at the isocentre of the machine (i.e. usually at 100 cm from the source) with a field size at the surface of 10 cm×10 cm.

(b) The monitor chamber reads 1 MU when 1 cGy is delivered to a point at a given depth in the phantom with the surface of the phantom positioned so that the specified point is at the isocentre of the machine and the field size is 10 cm×10 cm at the isocentre.

Whether definition (a) or definition (b) is used in a particular department is largely historical, but definition (a) is most appropriate for percentage depth dose based calculations and method (b) for tissue phantom based calculations (see Section 23.3). Many linear accelerators can be used at different energies, and the treatment beam can be either photons or electrons. It is usually possible to establish a different monitor chamber calibration factor for each energy and modality so that the relationship between the monitor unit and the absolute dose in the standard conditions is not dependent on modality or energy.

The calibration procedure must be carefully performed following one of the protocols described in Chapter 18. It is essential to maintain the stability of the calibration throughout the clinical use of the accelerator. Therefore, the stability of the monitor response must be ensured by setting up a Quality Assurance programme that guarantees that there is no drift beyond acceptable limits (see Section 38.8).

20.1.3 COBALT UNITS

Cobalt teletherapy units are usually controlled by a timer. Calibration of the unit involves measuring the dose rate with a standard setup as for linear accelerators. The mechanical nature of the source transfer mechanism means that the dose delivered in one minute will generally be greater than the dose rate measured with the source fully exposed. Timers are often set to measure the time for which the source is fully exposed, and a timer end correction will be necessary to compensate for the dose delivered while the source is in transit at the beginning and end of the exposure. Unlike for kilovoltage equipment or linear accelerators, there is no risk that the dose rate fluctuates during irradiation. Provided that the timer is accurate and that an appropriate correction is performed for the decay of the source activity, the calibration is accurately maintained over a long period of time with a minimal quality assurance program. Apart from this singularity, the treatment time is not conceptually different from the MU, and in what follows, no distinction will be drawn. The term *dose rate* (or sometimes, simply, *dose*) will be used to refer to the absorbed dose at a specific point in specific conditions for one monitor unit. *Absolute dose rate* will be used when it is expressed, for instance, in cGy/MU, or *relative dose rate* will be used when it is expressed as a percentage and normalized to an arbitrary reference point in some reference conditions.

20.2   RELATIVE DOSE DISTRIBUTIONS

The methods for measuring relative dose distributions, as part of the commissioning of a new treatment machine or as a supplementary procedure for a machine already in use, have been described in Chapter 19. In the present chapter the emphasis is on the link between absolute and relative dose measurements and some factors that are necessary to an understanding of the application of clinical beams are considered. Detailed definitions and relationships between the quantities briefly mentioned here will be found in Chapter 23.

Having established the relationship between the monitor unit and the absolute dose delivered to a reference point under specific conditions, doses at all other points and in all other circumstances can be calculated by making measurements relative to this standard situation. However, the range of different situations for patient treatment is very wide, and it is necessary to establish a system of measurements that will allow the relationship between absolute dose and monitor units to be calculated for any treatment situation. For a given beam energy, the dose delivered to the patient will be affected by:

- The distance of the skin surface from the source of radiation
- The depth inside the patient
- The field size and shape (and the method of shaping)
- The distance of the point of interest from the centre of the beam
- The composition of the patient tissue and, particularly, its density
- Oblique incidence of the beam on the skin surface
- Any accessories used such as wedge filters and other beam attenuators.

The system of measurements must allow all these factors to be considered. The details of the methods used to apply the necessary corrections depend on the quality and type of radiation being used, but some of the broad principles are common to all radiation beams. Allowance for the effect of all these factors can be carried out with a pocket calculator or a spreadsheet based on these broad principles, although some of the calculations are more appropriately carried out with a treatment planning computer. These basic principles will be considered in the following sections.

20.2.1   VARIATION OF DOSE WITH DISTANCE FROM THE SOURCE

Unattenuated dose rates at a distance from a point source of radiation vary as the inverse square of the distance from the source. This is a direct consequence of the fact that if radiation is emitted isotropically from a point source, the total number of particles crossing the surface of a sphere of radius r centred on the source, is independent of r. Therefore the particle fluence at a distance r, obtained by dividing this number of particles by the area of the surface of the sphere, varies proportionally to r^{-2}. Since the dose is proportional to the particle fluence (see Section 6.4 and Section 6.6), it varies also as a function of the inverse square of r. This variation is generally referred to as the *inverse square law*. It is useful to bear this in mind when determining if the inverse square law is likely to be obeyed in particular circumstances. For example, in situations where a beam is scattered, the inverse square law will not be followed exactly in proportion to the amount of scatter. It may also be deduced that on the beam axis of a radiation treatment unit, the dose rate fall-off in air will follow closely the inverse square law if the energy is such that the attenuation and scattering in air may be neglected (in practice, photon energies above some tens of kV and electron energies above some MeV) and if the distance from the scattering materials (treatment head, room walls, and floor) is large enough.

20.2.2 VARIATION OF DOSE WITH DEPTH WITHIN THE PATIENT

In a solid medium, the effect of the inverse square law will be added to that of the absorption and scattering of radiation in the material. For a polyenergetic photon beam the primary component would be exponentially attenuated (see Section 4.4.1) if it were not for the fact that the low energy components of the beam will be attenuated more than the high energy components, making the primary beam progressively more penetrating. However, the scattered radiation is not lost to the beam, and so with increasing depth, there is an increasing scatter contribution. Examples of depth-dose curves for photon beams are shown in Figure 20.1. Depth-dose curves are easily measured in a water tank with a remotely controlled detector (see Section 19.1.3). Using the depth dose curve, it is easy to calculate the dose at a depth if the dose at the normalisation point is known. However, with megavoltage equipment where the source-skin distance (SSD) is not fixed, it is necessary to make a correction to the depth dose curve measured at another SSD. An alternative approach is to measure the variation of the Tissue Maximum Ratio (TMR) or Tissue Phantom Ratio (TPR) (see Section 23.2.1.6) with depth. The TMR or TPR are measured by keeping the source to detector distance constant and varying the depth of the detector in the phantom (by adding water).

Because the TMR and TPR are almost independent of SSD, they effectively allow a separation of the effect of distance from the source from the effect of attenuation and scatter. Figure 20.2 shows the variation of TMR with depth. The difference between the

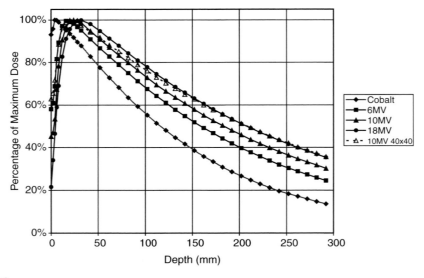

FIGURE 20.1
Percentage depth dose curves in water for a cobalt-60 beam, 6 MV, 10 MV, and 25 MV x-ray beams for a 10 cm square field. Also shown is the depth dose curve for a 10 MV beam with a 40 cm square field.

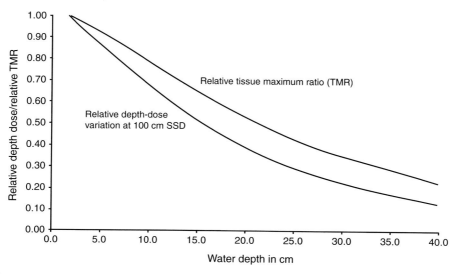

FIGURE 20.2
Comparison of the variation of percentage depth dose (x 0.01), at 100 cm SSD and tissue maximum ratio with depth for a 6 MV x-ray beam with a 10 cm square field. Graph drawn from data given in British Journal of Radiology, Supplement 25. (From BIR, *Br. J. Radiol.,* Suppl. 25, 1996.)

depth-dose curve and the depth TMR curve shows the contribution of the inverse square law to the depth dose curve. The details of calculations using TMRs will be dealt with in Chapter 23. In practice, making direct measurements of TMR or TPR in water is not easy because the water level must be accurately controlled (i.e. using a pump and a depth sensor) while the detector is kept in fixed position. It is much easier to change the position of the detector by moving it with a computer controlled system. Alternatively, TMR or TPR can be measured in solid phantoms. On the other hand, in spite of their outstanding advantages for dose calculation, TMR or TPR do not exactly reflect the depth dose variation that occurs within the patient. For these reasons, it is common practice to measure depth-dose curves at a fixed SSD and to use the formalism in Chapter 23 to calculate the TMR or TPR.

20.2.3 EFFECT OF FIELD SIZE AND SHAPE

As the field size increases, the amount of scatter generated within the patient and reaching a point on the central axis of the beam increases, and consequently, the dose increases, especially at larger depths. In addition, the amount of radiation reaching the patient from the machine head components also increases because of scatter from the collimators. With photon beams from linear accelerators, an additional (and more significant) factor comes into play; more of the radiation scattered out of the primary beam in the flattening filter is able to reach the patient (Chaney et al. 1994). Where customized shielding is used to shape the radiation beam, this will have a different effect depending on where the shielding is placed in relation to the flattening filter. For this reason, it is usual for photon beams to separate the effect of field size into phantom scatter and collimator scatter factors. The collimator scatter factor may be measured using relative measurements made in air. The phantom scatter is indirectly measured by measuring the variation in dose per monitor unit for different field sizes to obtain an output factor and using the relationship

$$\text{Output factor} = \text{Collimator Scatter Factor} \times \text{Phantom Scatter Factor}$$

When the TMR/TPR approach is used, the output factor is often measured at a depth beyond the depth of dose maximum. This depth is taken as the *reference depth*. Even when depth dose-based calculation methods are used, it is better to make the relative output measurements at a depth and make appropriate corrections to calculate the output factor at the depth of dose maximum in air or in a Mini-phantom (see also Section 19.2.5 and Chapter 23).

20.2.4   EFFECT OF DISTANCE FROM THE CENTRAL AXIS

As the distance of the point of interest from the central axis of the beam increases, the dose rate compared to that on the central axis will change. The dose rate is usually (but not always) highest on the central axis. In order to account for this in dose calculations, an *off-axis factor*, defined as the ratio of the dose at the appropriate distance from the axis to the dose on the central axis at the same depth, can be measured. The off-axis dose distributions for megavoltage photon beams are discussed in Section 22.3.3. Further details of the use of off-axis factors in dose calculation are presented in Section 23.2.1.15, and examples of off-axis curves (also called *dose profiles*) are shown in Figure 23.9. For a given field size, dose profiles are essentially determined by the depth and the distance from the source. Because the beam diverges from a small source, the value of the off-axis factor is found to vary least if a fan-beam geometry is assumed. This means that the off-axis factors along a ray line directed back to the source are approximately constant. This forms the basis of the Bentley–Milan model of a photon radiation beam that is described in more detail in Section 26.2.1.1.

20.2.5   COMPOSITION OF PATIENT TISSUE

The effect of the composition of patient tissue depends on the nature and energy of the beam. At orthovoltage energies, the principal method of interaction with tissue is through the photoelectric effect that has a dependence on the atomic number of the tissue according to Z^4, and this gives rise to differential absorption in bone (see Section 4.3.1). In megavoltage beams, the effect is less because it is primarily the density of the tissue that affects the amount of attenuation. However, lung tissue has a major effect on the attenuation of the beam, and methods of calculating the effect of this are needed. The effect of inhomogeneities is even larger for electron beams due to their influence on the electron scattering processes. Only a rough calculation is possible with pocket calculator-based methods, and more sophisticated computer calculations are required (see Part F).

20.3   REPRESENTATION OF DOSE DISTRIBUTIONS: ISODOSE CURVES AND SURFACES

Although depth-dose curves and beam profiles can be used to calculate the dose at any point in the beam, it is useful to have a representation of the effect of the radiation beam as a whole. For this purpose, it is helpful to represent dose distributions in the form of isodose curves. In a given plane, the isodose curves (or, simply, *isodoses*) represent lines of equal dose rather like the isobars on a weather map. The isodose lines can be relative isodoses that are normalised to some point (e.g. the point of maximum dose on the beam axis) within the plane of interest, or they can be absolute isodoses that are lines representing a particular dose value for a given machine setting. Examples of isodose curves are shown in Figure 21.8 for a kilovoltage x-ray machine, in Figure 22.11 for a cobalt machine and a linear accelerator and in Figure 24.7 for electron beams. In three dimensions, isodose lines become isodose surfaces.

CHAPTER 21

KILOVOLTAGE X-RAYS

Philip Mayles

CONTENTS

21.1 INTRODUCTION

Kilovoltage x-rays were the first form of external beam therapy to be used. Because of their inadequate penetrative qualities it was necessary to use multiple overlapping beams to deliver a high dose to the tumour while keeping within tissue tolerance. At kilovoltage energies the photoelectric effect makes a significant contribution to the absorbed dose especially in higher atomic number materials such as bone. The soft tissue within trabecular bone receives an enhanced dose (see Section 6.7.3), which is normally undesirable. Kilovoltage x-rays are used mostly for superficial lesions and for palliative treatments. In modern radiotherapy many superficial lesions could also be treated with electrons. However there are some disadvantages in using electrons for superficial lesions. For instance, with low energy electrons it is difficult to obtain an accurate and uniform dose for small fields or for an irregular patient surface (see Chapter 24 and Chapter 34) and they may not be ideal for eye treatments (see Section 21.5).

21.2 BEAM QUALITY

21.2.1 ENERGY SPECTRUM AND ROLE OF FILTRATION

The penetrative quality of an x-ray beam will depend on the peak kilovoltage used to generate it and the filtration applied. As discussed in Section 10.2.2, the spectrum of an x-ray beam is the bremsstrahlung spectrum arising from electrons used to produce it. The resulting x-ray energies typically range from the energy of the electrons reaching the anode down to a minimum energy determined by the filtration of the x-ray tube. X-ray tubes can be either glass walled with an equivalent inherent filtration of between 0.6 mm and 1 mm of aluminium or have a beryllium window between 1 mm and 2 mm thick. The mean energy of the beam is typically one third of the maximum x-ray energy. A typical medium energy spectrum is shown in Figure 21.1.

The energy fluence shown in Figure 21.1 must be multiplied by the mass energy-transfer coefficient in air to obtain the resulting air kerma in order to approximate the variation with energy of the dose that will be delivered. When this is done the spectrum shown in Figure 21.2 is obtained. Without filtration a very high dose would be delivered to the skin with very little penetration and the lower energies must be filtered out in order to produce a more penetrating beam*. Filtration aims to reduce the low energy radiation while allowing high energy radiation to pass through. This result can be achieved because the photoelectric effect has a higher cross section for low energies than for high energies. As can be seen from the two figures very different penetration characteristics can be expected depending on the combination of generating potential and filtration. The effect on the depth-dose curve is shown in Figure 21.3.

For energies up to 100 kV a simple aluminium filter is used. At higher energies, to maximise the photoelectric cross section, a high atomic number material may need to be used. However, there is then the problem that the attenuation of the material will be sharply reduced below the K-absorption edge, and additionally the photoelectric effect will produce radiation at this characteristic energy. For this reason, at higher energies it is common to use a combination of metals so that the lower atomic number material will remove the radiation below the K-absorption edge. A much-used combination is the Thoraeus filter in which tin, copper and aluminium are combined (the K-absorption edge for tin occuring at 29 keV, while that of copper is at 9 keV). The original Thoraeus filter was composed of 0.4 mm of tin, 0.25 mm of copper and 1.0 mm of aluminium, but different combinations of these materials have been used since to produce different spectra. Care must be taken to ensure that the correct filter is in place. Ideally machines should be used for one beam quality only, but where multiple energies are used, a system of filter interlocks must be provided to ensure that the appropriate filter is selected.

21.2.2 HALF VALUE LAYER

By a judicious choice of filter it is possible for beams of different penetrating properties to be produced by a single generating voltage. It is not practicable to measure the full spectrum on

---

* With very low energy x-rays an undesirably high surface dose can be delivered and this can be rectified by the use of a thin sheet of Mylar film which absorbs the lowest energy x-rays.

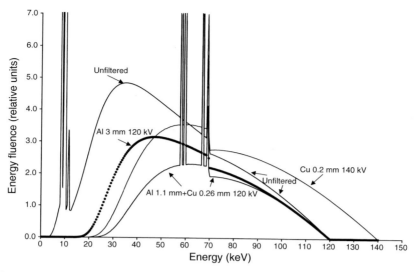

FIGURE 21.1
Spectra for a superficial x-ray unit with different combinations of filtration and generating potential. The *unfiltered* curve shows the output with only the inherent filtration of the x-ray tube for which the mean energy is 32 keV. With 3 mm of aluminium filtration the mean energy of the beam is increased to 54 keV. A more penetrating beam is obtained with 1.1 mm of aluminium and 0.26 mm of copper, for which the mean energy is 62 keV. Increasing the tube potential to 140 kV with 0.2 mm of copper reduces the penetrative quality in spite of the increase in mean energy to 65 keV. Spectra produced and mean energies calculated using the spectrum generator contained in IPEM Report 78. (IPEM, Scientific Report Series No. 78, *Catalogue of diagnostic x-ray and other spectra*, IPEM, York, 1997.)

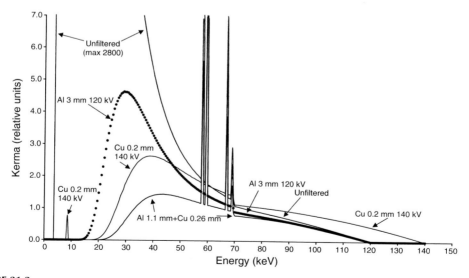

FIGURE 21.2
Air Kerma spectra for a superficial x-ray unit with different combinations of filtration and generating potential. These results are derived from those in Figure 20.1 by applying the air energy-transfer correction factor which decreases rapidly as the photon energy increases. The unfiltered curve goes off scale illustrating the dangers of treating without a filter. Note the contribution from the K-absorption edge of copper at around 9 keV that would be part of the beam if no aluminium were used in the filter. Spectra produced and mean energies calculated using the spectrum generator contained in IPEM Report 78. (IPEM, Scientific Report Series No. 78, *Catalogue of diagnostic x-ray and other spectra*, IPEM, York, 1997.)

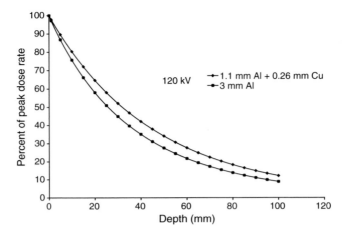

FIGURE 21.3
Depth dose curves for kV x-ray beams generated with 120 kV, but with different filtration.

a routine basis, and therefore some method must be found to define the penetrative quality of the beam in an unambiguous way. This is done by measuring the attenuation of the beam by a metallic absorber in scatter-free conditions (see Figure 21.4). In such narrow-beam conditions, if the radiation is monoenergetic the attenuation will be exponential so that, for an absorber thickness t the dosemeter reading I will be related to the reading without the attenuator I_o by the equation:

$$I = I_o e^{-\mu t} \tag{21.1}$$

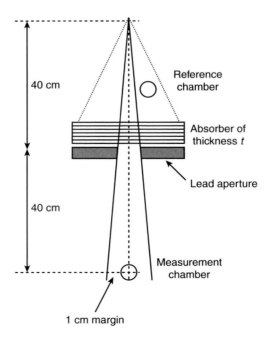

FIGURE 21.4
Measurement arrangement in narrow-beam conditions for HVL measurement. The aperture is just large enough to expose uniformly the measurement ionisation chamber (using about 1 cm margin) and placed conveniently at about 40 cm from the source. The measurement chamber is located at twice this distance. A reference chamber can be placed above the aperture to allow correction for instability of the machine output.

The linear attenuation coefficient, μ, could be used to characterise the penetrative beam quality, but it is more usual to express this quality as the thickness of material required to reduce the dose rate measured by the dosemeter to half of its original value. This is called the *half value layer* (HVL). Since clinical beams are not monoenergetic, the increasing thickness of the absorber will remove the lower energies preferentially, and consequently the beam will become more penetrating (or *harder*). As a result, the attenuation will not follow the exponential relationship of Equation 21.1, and the thickness required to reduce the dose rate to a quarter of its original value will be greater than twice the amount required to reduce it to 50% (see Figure 21.5). When specifying the penetrative quality of the beam using the HVL, if not otherwise specified, it is the first HVL that is quoted.

It is customary to use aluminium for low energies (below 140 kV) and copper for higher energies. When making these measurements the beam should be collimated to an aperture that is just large enough to ensure that the whole of the ionisation chamber is covered by the beam with a margin of about 10 mm (see Figure 21.4). Ideally, two ionisation chambers should be used. The first chamber, used to monitor the constancy of the radiation, is placed above the aperture at such a position that it is not in the part of the beam which passes through the aperture. The second chamber is placed a long distance beyond the aperture. It is

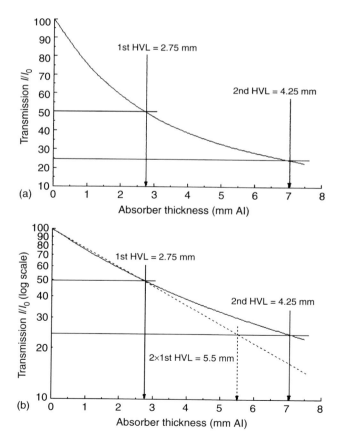

FIGURE 21.5
Variation of relative dose rate (i.e. percent transmission) with thickness of absorber, measured as shown in Figure 21.4. Transmission is plotted with a linear scale in (a) and with a logarithmic scale in (b). Since the absorber removes preferentially the low energy components, the second HVL is larger than the first one. If the beam were monoenergetic, the attenuation would be exponential (straight dotted line on graph (b)) and the second HVL would be equal to the first HVL.

TABLE 21.1

Mean Energy and Penetrative Quality for the Beams of Figure 21.1

Generating Potential (kV)	Filtration	Mean Energy (keV)	HVL of Aluminium (mm)
120	Inherent	32	0.08
120	3 mm Al	54	4.3
120	1.1 mm Al + 0.26 mm Cu	62	7.7
140	0.2 mm Cu	65	7.3

The data are calculated with the spectrum generator in IPEM Report 78, but although this is designed for diagnostic x-rays the results closely match measured results with actual therapeutic x-ray machines.

appropriate for the aperture to be about 400 mm from the source and the measurement chamber at about 800 mm. Thin sheets of the appropriate metal are placed to cover the aperture and the effect on the attenuation measured. The chemical purity of the metal sheets used is particularly important. It is also important to obtain measurements for absorber thicknesses close to the HVL and not to try to fit an exponential curve through experimental points.

Beams generated with the same generating potential may have different penetrative qualities. Table 21.1 shows the mean energy and HVL for the four different beams shown in Figure 21.1. It has been found that the HVL together with the generating potential provides a satisfactory measure of the quality of the beam with regard to penetrative ability and the amount of scattered radiation. The first HVL measurement, together with a dose measurement in some specified conditions, should therefore be sufficient to enable dose calculations for all other conditions of that beam using standard data (i.e. depth dose curves for the same HVL). Although this is an approximation it nevertheless works well in practice.

21.3 DEPTH DOSE CHARACTERISTICS

As discussed in Section 20.2.1, the radiation intensity due to primary photons originating from a point source falls off as the inverse square of the distance from the source. However, the inverse-square law will break down in the event of scattering or absorption of the radiation. An x-ray beam measured in air will conform closely to the inverse-square law with respect to the distance from the measurement point to the source (i.e. the target). In a more dense medium there will be more significant interactions; some of the beam energy will be absorbed either through the photoelectric effect or as secondary Compton electrons which are themselves absorbed close to the site of the original interaction. The Compton interactions will also give rise to scattered photons which may have sufficient energy to travel a significant distance from the site of the interaction. In order to predict the variation of dose rate with depth in the medium we consider a detector placed on the beam central axis at different depths. For a sufficiently narrow beam the dose rate reduction as a function of depth will be caused by the loss of both the photons absorbed in the medium and the photons scattered outside of the narrow-beam which will not reach the central axis detector. This reduction is approximately exponential. The geometrical dispersion of the primary photons with the increasing distance from the source also contributes to the reduction in dose rate according to the inverse-square law. Thus in tissue the dose rate will fall off according to the relationship:

$$I = I_{\mathrm{o}} \frac{(\mathrm{SSD} + d)^2}{\mathrm{SSD}^2} f(d,A) \qquad (21.2)$$

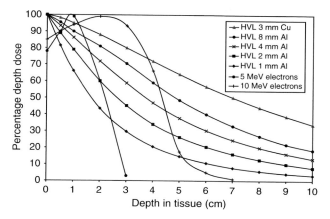

FIGURE 21.6
Central axis depth dose curves for kilovoltage x-ray beams of different energies. Field size is a 10 cm diameter circle. The SSD for all except the 3 mm Cu HVL curve is 25 cm: for the 3 mm Cu HVL curve it is 50 cm. As the energy increases or the SSD decreases the depth dose curve becomes more dominated by the inverse square law. For comparison depth dose curves for 5 and 10 MeV electrons (see Chapter 24) are also shown. Graph drawn from data given in British Journal of Radiology Supplement 25. (From BIR, *Br. J. Radiol.*, Suppl. 25, 1996.)

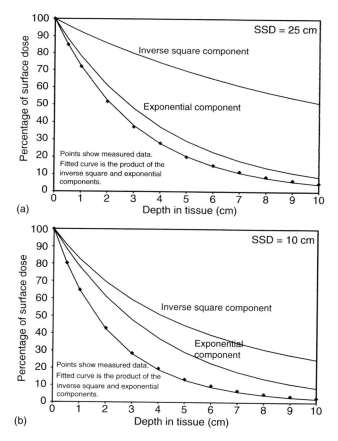

FIGURE 21.7
Comparison of the contribution to the depth dose curve for an x-ray beam of HVL 2 mm of Al and a 2 cm diameter field size. (a) shows the result for 25 cm SSD and (b) for 10 cm SSD. Graph drawn from data given in British Journal of Radiology supplement 25 with the effect of SSD calculated by the author. (From BIR, *Br. J. Radiol.*, Suppl. 25, 1996.)

where *SSD* is the distance of the surface of the phantom from the source, *d* is the depth within the phantom, *A* is the field size, and I_o is the dose rate at the surface of the phantom. For small field sizes and heavily filtered x-ray beams an exponential relationship (i.e. $f(d,A) = e^{-\mu d}$) is a good mathematical approximation to the change in dose with depth. However, for clinical beams, the polyenergetic nature of x-rays and the increase of scatter as a function of depth ensure that an exponential relationship is not correct (i.e. the value of μ is both depth and field size dependent). Some example depth dose curves are shown in Figure 21.6. For shorter source skin distances (SSDs) and higher energies the inverse-square law becomes more important. This is illustrated in Figure 21.7, that shows the relative contributions of the inverse-square law to the reduction in dose for 10 cm SSD and 25 cm SSD. The serious consequences of not using data that are appropriate to the SSD are apparent.

21.4 DOSE DISTRIBUTION

A sharp beam edge at all depths characterizes kilovoltage x-ray dose distributions, but scattered radiation causes radiation to be present outside the geometrically divergent beam as shown in Figure 21.8. This figure also illustrates the standard way of displaying radiation dose distributions as isodose curves. The off-axis dose reduction, illustrated by the rounded isodoses, is caused partly by the increased distance from the source, from the beam axis to the edge of the beam (3% for a 25 cm square field at 50 cm SSD) and partly by the reduction in scattered radiation. In an inhomogeneous medium the beam will be significantly more attenuated in high atomic number materials—this effect being greater the lower the energy.

The profile of dose rate across a kilovoltage x-ray beam will be also affected by the heel effect (see Section 10.3.1.3). This will result in a variation in the dose rate of about 5% in the cathode to anode direction with the dose rate being lower at the anode end. Although in principle it would be possible to compensate for the heel effect with a filter, this is not normally done. Isodose curves from kilovoltage units are usually shown plotted in the orthogonal direction so that this effect is masked.

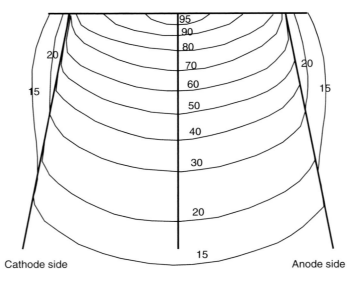

FIGURE 21.8
Isodose curve for 120 kV x-rays with HVL of 8 mm of Al. Applicator is a 10 cm circle at 25 cm SSD. Isodoses are shown relative to the surface dose in the centre of the beam. Note the slight asymmetry associated with the heel effect.

21.5 BEAM SHAPING

X-ray applicators are used to limit the radiation to the desired area as described in Section 10.6. Further shaping can be achieved with a lead cut-out placed on the patient's skin. This should be 0.5 mm thick up to 90 kV and 1 mm thick up to 140 kV. For higher energies the weight of lead required makes detailed shaping in this way an option of last resort. Lead may also be placed behind the target volume (e.g. the ear lobe or the lip) to prevent further penetration, but in this case it is necessary to put a layer of wax, about 1 mm thick, on the lead to prevent an increased dose due to electrons back-scattered from the lead.

A uniform beam within 3% can be achieved up to field sizes of about 24 cm at 50 cm SSD in the direction orthogonal to the anode–cathode direction. However, in the anode–cathode direction the heel effect limits the field size over which the uniformity is 3% to 12 cm. The maximum beam divergence achievable is around 15°. For this large field size (26 cm diameter at 50 cm SSD) a variation across the field to between 10% and 20%, depending on the orientation can be expected.

Protection of the eye when treating tumours close to the eye can be achieved with a silver-coated lead internal eye shield that is placed beneath the lid. Such shields are very effective in reducing the eye dose and the possibility of shielding the eye may make ortho-voltage x-rays preferable to electron beams when irradiating skin tumours close to the eye (Amdur et al. 1992). This is because with electrons the dose below the shield is increased by bremsstrahlung generated in the lead.

Scattered radiation will be emitted from an applicator as a result of interactions with the tube, tube housing, filter and applicator, and will be equipment dependent. Its effect on beam characteristics such as beam output, depth dose, and fall off in dose-rate with stand off from the applicator face must be determined therefore by measurement. Care must be taken when performing dose measurements with thin parallel-plate ionisation chambers; secondary photoelectrons generated in the collimation system may have sufficient energy to penetrate the thin chamber window leading to erroneous interpretation of dosimetric data if measurements are done in air without any overlying material (Klevenhagen et al. 1991; Lee and Chan 2001). Lee and Chan suggest the use of a piece of paper to eliminate these secondary photoelectrons.

The additional thickness of the plastic end-plate reduces the percentage dose at all depths compared with an open ended or diaphragm limited field. Depth dose data are provided in British Journal of Radiology Supplement 25 (BIR 1996) for both types of applicator and this publication includes a method of correcting the data according to end-plate thickness. For other field sizes, including those of irregular shape, the beam must be further collimated by the use of lead cut-outs taped to the skin.

21.6 DOSE CALCULATIONS FOR KILOVOLTAGE X-RAYS

The dose delivered to the patient with a given x-ray energy is affected by two factors: the applicator used and the size of the area being treated. To facilitate dose calculation a table of applicator factors and back-scatter factors will be created at the time of commissioning. These two factors are described below.

21.6.1 BACK-SCATTER FACTORS

If dose rates are measured in scatter-free conditions at the end of an applicator, the dose rate measured will be lower than that measured at the surface of a phantom. This is because x-rays interacting at other points within the phantom will produce scattered secondary x-rays by means of Compton scattering. These secondary x-rays will be scattered in all directions so

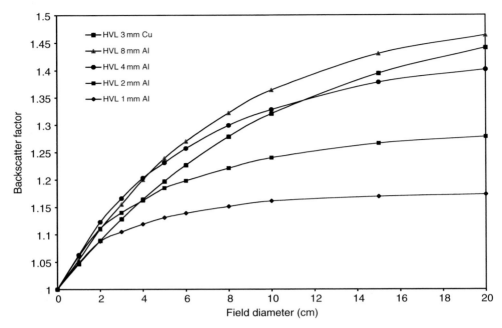

FIGURE 21.9
Variation of back-scatter factor with field size. Data from IPSM (1991). Data are for 30 cm SSD for all except the 3 mm Cu for which the SSD is 50 cm.

that the radiation detector will be irradiated not only by the radiation coming directly from the radiation source, but also by secondary radiation from within the phantom. The amount of back-scattered radiation will depend on the area of the field. To quantify this the *back-scatter factor* (BSF) is defined as follows:

$$BSF = \frac{\text{Dose rate at the surface of the phantom for the given field size}}{\text{Dose rate in air for the same field size}}$$

Back-scatter factors are usually tabulated for equivalent circular fields or equivalent square fields. Accurate measurement of back-scatter is difficult to achieve because of the finite size of ionisation chambers (Klevenhagen et al. 1991). An alternative approach is to use back-scatter data calculated by Monte Carlo methods (e.g. Groswendt 1990; IPSM 1991; Knight 1993). This is the approach recommended by the IAEA (1997a) and the Institute of Physics and Engineering in Medicine (IPEM) (IPEMB 1996a). For higher energies the back-scatter factors vary slightly with SSD. Figure 21.9 shows how backscatter varies with field size and HVL.

21.6.2 APPLICATOR FACTORS

The dose rate normally will be measured for one standard applicator following the appropriate protocol (see Section 18.6). For other applicators the dose rate may be specified relative to the standard applicator by means of an *applicator factor*. Applicator factors are measured at the time of commissioning the unit, following the protocol for each applicator in turn. This may involve a measurement in water or a measurement in air. In the latter case it is then necessary to multiply by the appropriate BSF in order to obtain the dose rate in tissue. *Applicator factors* can be defined either as "the factor by which the dose per monitor unit for the standard applicator must be multiplied to give the dose per monitor unit for the applicator in question", or as "the factor by which the number of monitor units calculated with the standard output must be multiplied to give the required dose". These two approaches give inverted results, and therefore it is important that tables of output factors define clearly which of the two approaches is being taken. It is also important to specify systematically whether *in-air* or *at-surface* applicator factors are being used. Note that for two applicators being used at the same SSD the *at-surface* applicator factor will approximate to the ratio of the backscatter-factors for the two field sizes. However, because of the different amount of scatter from the side walls of the applicators, this calculation will not be exact. Then it will be necessary to measure the applicator factors directly. Applicator factors should be checked from time to time to verify that they have remained constant.

21.6.3 Cut-out Factors

It is often the case that the shape of a surface lesion to be treated is not the same as that of any of the applicators provided with the machine. In this event it is necessary to make a lead cut-out shaped to fit the patient's skin. The dose rate in this cut-out area will be less than that for the open applicator by the ratio of the BSF for the cut-out to the BSF for the applicator. For irregular shaped cut-outs the method of Clarkson integration may be applied (Clarkson 1941). The irradiated area is divided into sectors, and the equivalent diameter of each sector is measured. Using a table of backscatter factors against equivalent diameters the average backscatter factor can be calculated. The Clarkson method is illustrated diagrammatically in Figure 23.5 (p. 490).

21.6.4 Example of Monitor Units (or Treatment Time) Calculation

A basal cell carcinoma is 5 mm thick and has an elliptical shape with major axes of 1.8 cm and 2.3 cm and a surface dose of 45 Gy in 15 fractions has been prescribed*. It is decided to treat it with a 140 kV beam with an HVL of 4 mm of Aluminium. (This has a depth dose of 87% at 5 mm and 75% at 10 mm.) The reference situation for this machine is a 10 cm diameter circular applicator for which the standard SSD is 25 cm. The applicator used must be larger than the cut-out so a 3 cm circle is chosen for which the SSD is also 25 cm. With this applicator it is found that because of the shape of the area being treated there is a standoff of 10 mm between the surface of the applicator and the lesion (i.e. the lesion is 10 mm further from the source than the applicator end). For the reference situation 100 MU is measured as giving a surface dose of 1.21 Gy. The *at-surface* applicator factor[†] for the 3 cm circle is 0.885. The backscatter factor for a 3 cm circle is 1.16. The equivalent diameter of the cut-out is 2.05 cm and the backscatter factor is 1.12. Therefore the surface dose rate may be calculated as:

Dose per MU at surface = Reference Surface Dose Rate × Applicator Factor
× Backscatter Correction Factor × Standoff Factor

$$= \frac{1.21}{100} \times 0.885 \times \frac{1.12}{1.16} \times \frac{25^2}{(25+1)^2} = 0.00956 \text{ Gy/MU}$$

The number of monitor units for a prescribed surface dose of 3 Gy/fraction is then
$$N = 3/0.00956 = 314 \text{ MU}$$

Note that if the unit is sufficiently stable it may be appropriate to amalgamate the output factor with the applicator factor and have a table of applicator factors.

---

* The common practice for kilovoltage x-rays is to prescribe the dose at the patient surface. This is different in megavoltage radiation where the dose is usually prescribed at a specified depth.
† Applicator factors in this example are defined as multiplying factors for the dose per monitor unit.

CHAPTER 22

MEGAVOLTAGE PHOTON BEAMS

Philip Mayles and Peter Williams

CONTENTS

22.1   INTRODUCTION

Megavoltage* photon beams are the most widely used radiation in radiotherapy because of their greater penetration compared to kilovoltage beams. Early megavoltage radiotherapy was carried out with cobalt-60 teletherapy units, and these are still the most suitable where the technical infrastructure is not adequate for the support of linear accelerators (IAEA 1998) Linear accelerators have the advantage of more penetrating beams, the versatility of the choice of beam energy, a smaller penumbra at the edge of the beam, and they deliver a higher dose rate. They also have the advantage that the source does not need to be replaced and the old source disposed of. A major advantage of megavoltage beams is the effect of skin sparing brought about by dose build-up (see Section 22.3.1). This effect is greater the higher the energy.

22.2   BEAM QUALITY

22.2.1   COBALT UNITS

Cobalt unit γ-rays are emitted at two well-defined energies (1.173 MeV and 1.332 MeV). Additional Compton emission generated in the source results in the beam incident on the patient having a continuum of energies below this, with a mean energy less than 1 MeV. At cobalt-60 energies the Compton effect is the principal interaction mechanism in the patient, which generates further low-energy scattered radiation. This results in the spectrum shown in Figure 22.1.

22.2.2   LINEAR ACCELERATORS

The spectrum of the beam from a linear accelerator is similar in shape to that of a kilovoltage x-ray beam (see Figure 21.1). A nominal 6 MV beam will have a mean x-ray energy of around 2 MeV, and the range of the secondary electrons produced by interaction in tissue will be about 16 mm.

As explained in Section 11.4.1.2, a flattening filter is used to convert the forward-peaked bremsstrahlung distribution into a clinically useful beam. The choice of the material with which to make the flattening filter depends on the energy of the beam (Luxton and Astrahan 1988). Figure 22.2 shows the variation of transmission with beam energy for steel and tungsten. At 6 MV either steel or tungsten would provide a satisfactory filter. However, at higher energies there is benefit to using steel rather than tungsten (in spite of the fact that the filter needs to be physically larger). This is illustrated in Figure 22.3, which shows the spectrum of energies

---

* The term *megavoltage* refers to the range of energy of photon beams generated from linear accelerators (typically between 4 and 25 MV). In principle it is not appropriate for cobalt machines. However, since the photon beams emitted from cobalt-60 have similar characteristics, the same term is used here for both types of equipment.

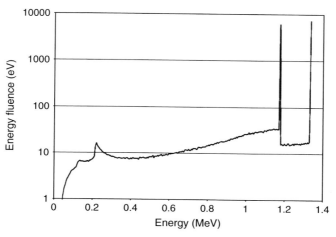

FIGURE 22.1
Spectrum of radiation in a cobalt-60 beam at 5 cm deep in water. Data were derived by Monte Carlo calculation. (Data provided by Price, R. A., private communication.)

transmitted by two filters of similar shape. Since more high energy radiation will be transmitted by a steel filter, the resulting beam will be more penetrating (see Section 22.3.2.1). There is also a significant problem with neutron activation of the flattening filter which is reduced if stainless steel is used (Rawlinson et al. 2002).

For practical radiotherapy the penetration of the radiation beam is the factor of interest, rather than the energy of the accelerated electrons hitting the target. The Half Value Layer concept as used for kilovoltage beams (see Section 21.2.2) would require the use of high atomic number materials and proves not to be very sensitive to quality. For this reason it is usual to specify quality in terms of the variation of dose with depth. This could be by specifying the dose at 5 cm or 10 cm as a percentage of the dose at the depth of dose maximum (d_{max}). However, measurements at d_{max} are subject to unpredictable variations because of the effect of electrons and scattered x-rays produced in the collimating system. For this reason it is usual to use the ratio of the doses at two greater depths such as 5 cm and 15 cm or 10 cm and 20 cm. The latter leads to the *Quality Index* (see Appendix D.3.1) which is defined as the ratio of the

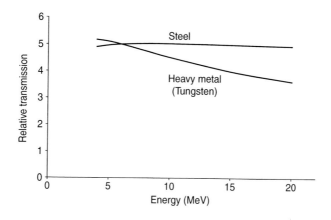

FIGURE 22.2
Variation of attenuation of steel and tungsten with energy. Graphs are based on linear attenuation coefficient data from the National Institute of Standards and Technology (NIST) web site (http://physics.nist.gov/PhysRefData/XrayMass-Coef/tab3.html). Transmission is normalised to an arbitrary value of 5 at 6 MeV at which point the absolute transmission through steel is 1.8 times that of the same thickness of tungsten.

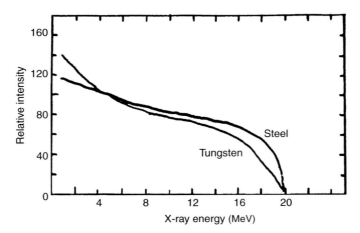

FIGURE 22.3
Effect of the flattening filter material on the spectrum of transmitted radiation for a 20 MeV maximum electron energy.
(From information provided by Elekta Oncology Systems.)

tissue phantom ratio at 10 cm and 20 cm , or $TPR_{20,10}$. Andreo (2000) gives a detailed discussion of the reasons for choosing this measure of quality. The AAPM (Almond 1999) uses a different measure as described in Section 22.3.1. It is, however, traditional to refer to, say, a 25 MV beam—this effective energy being based on the penetrative qualities of accelerator beams in the 1960s. This does not necessarily relate to the energy of the electrons in a modern accelerator, since the use of steel as a flattening filter material allows a more penetrating beam to be produced from the same initial electron energy. Care must be taken in applying the concept of the Quality Index since it is still possible to have two beams with the same Quality Index, but with different spectra—especially when comparing research accelerators to clinical ones. For such non-standard accelerators, knowledge of the beam spectrum or some other element of the beam energy is necessary when transferring the calibration of an ionisation chamber from a standards laboratory to a hospital.

22.3 CHARACTERISTICS OF PHOTON BEAMS

22.3.1 DOSE BUILD-UP

The depth dose curve for cobalt-60 and all megavoltage photon beams exhibits the phenomenon called *build-up*. The dose is higher a short distance below the skin compared to the dose at the surface. This is the result of the lack of *electronic equilibrium* at shallow depth, explained in Section 6.5 and illustrated in Figure 22.4. Absorbed dose is delivered not directly by the photons but by the secondary electrons generated by their interactions with tissues. At these higher energies the secondary electrons are principally Compton electrons directed in the forward direction and give up their energy further away from the point of interaction than is the case with low energy x-rays. The number of electrons passing through each layer of the phantom will progressively increase until the point at which equilibrium is reached. Since electrons lose energy almost uniformly along their path, the dose deposited will progressively increase until electron equilibrium is attained at a depth approximating to the range of the electrons. This depth is about 5 mm for a cobalt-60 beam. For a megavoltage beam the depth expressed in cm is approximately 1/4 of the maximum energy expressed in MeV (see also Table 6.1). On the other hand the kerma, is at a maximum at the skin surface and reduces continuously with increasing depth (see Figure 6.9).

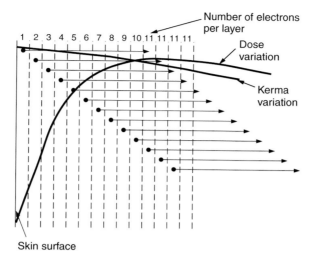

FIGURE 22.4
Idealised representation of build-up of dose in a high energy beam. Electrons are generated in equal numbers in each layer of the patient (assuming no attenuation takes place). Build-up of the number of electrons traversing a layer continues until slice 10 at which point equilibrium is reached (10 slices approximately representing the range of the electrons). In practice, the number of electrons generated begins to decrease as photon attenuation takes place and the electrons do not travel in straight lines, but this is not shown in this idealised diagram.

The depth dose curve at shallow depths will also be affected by the secondary electrons produced in the collimator system (Nilsson and Brahme 1986), and this will result in differences in the build-up region of the depth-dose curves for different accelerators. The region of the depth dose curve affected by secondary electrons from the collimator extends beyond the depth of dose maximum. It is principally for this reason that reference dose measurements for high energy beams are made at depths below the depth of dose maximum (5 cm, 7 cm, or 10 cm deep depending on the dosimetry code being employed) rather than directly at the depth of dose maximum (see Section 18.1).

The value of the dose actually at the surface of the patient depends both on beam energy and field size. One might expect that the surface dose would be smaller the higher the energy. However, this is only the case for small field sizes. For field sizes above 10 cm square, the surface dose may be greater at very high energies. The reason for this is that the surface dose is actually determined by the electron contamination of the beam which derives partly from the collimators, but mainly from the flattening filter (Nilson and Brahme 1986). The contamination is higher for large fields and more energetic beams. For a 5 cm×5 cm field the surface dose is typically 5% of the maximum dose, whereas for a 40 cm×40 cm it is typically 40%. The use of a perspex shadow tray in the beam to hold lead blocks or other beam modifiers increases the surface dose. A multileaf collimator defined field does not suffer from this problem. A metal (e.g. lead) sheet can be used to remove electron contamination (Leung and Johns 1977; Attix et al. 1983; Parthasaradhi et al. 1989; Rogers 1999) and when a fixed wedge is used, the surface dose can be reduced by a third because the metal wedge stops some of the electron contamination. A careful study of surface doses from Varian accelerators has been carried out by Kim et al. (1998a) and from Elekta accelerators by Klein et al. (2003). Lamb and Blake (1998a) have developed a model to predict the surface dose in various conditions.

In the AAPM TG51 code of practice for dosimetry (Almond 1999) the definition of photon beam energy is based on the percentage depth dose rather than $TPR_{20,10}$. Because depth dose measurements would be affected by the electron contamination of the beam for energies of 10 MV and above, they recommend the use of a 1 mm lead sheet to remove the electron contamination. The measure of beam quality used is then the depth dose at 10 cm deep (100 cm Source Skin Distance) measured with such a lead sheet placed at 50 cm from the surface. A small amount of electron contamination remains due to the lead sheet itself, and a formula is provided to calculate the magnitude of this.

22.3.2 VARIATION OF DOSE WITH DEPTH

Beyond the point of dose maximum, and in addition to the dose rate decrease as a function of the distance to the source according to the inverse square law (see Section 20.2.1), the dose rate in the absorbing medium will eventually fall off in an approximately exponential manner. This fall-off is due to the attenuation of the primary photons which are either absorbed or scattered out of the primary beam. The overall depth-dose variation will depend on the energy of the beam, the field size, and the source-to-surface distance as well as on the composition of the medium.

22.3.2.1 *Effect of Energy*

The attenuation of the primary photon beam will be determined by the mass energy attenuation coefficient and by the density of the attenuating material at the relevant energy. In biological tissue, for energies up to about 15 MV, attenuation will be primarily through the Compton effect and will therefore depend on the electron density (approximately analogous to the actual density). At higher energies pair production begins to occur and there will therefore be a dependence on the atomic number of the medium (Bradley et al. 1998). As explained in Section 22.2.2, the depth dose for a given electron energy incident on the target will depend on the design and the material of the flattening filter. This is illustrated in Figure 22.5.

22.3.2.2 *Effect of Field Size*

The Compton effect has two components: transfer of the photon energy to an electron and scattering of the photon. Both components contribute to attenuation of the beam. As the field size increases, the dose rate on the beam axis will increase because it will include a contribution from the Compton photons scattered from the rest of the irradiated volume. For small field sizes, the scatter contribution is small and does not change much as a function of depth. For larger field

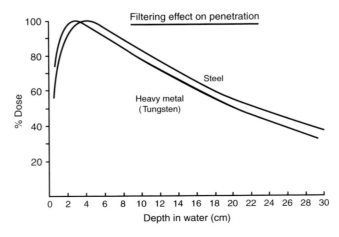

FIGURE 22.5
Percentage depth dose for a beam flattened with a steel filter compared to a tungsten filter. (Data provided courtesy of Elekta Oncology Systems.)

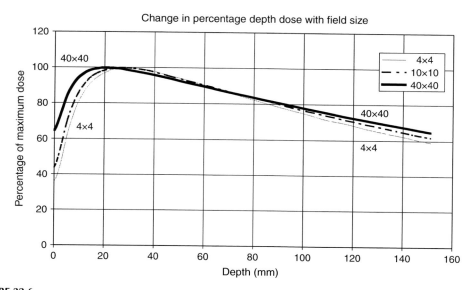

FIGURE 22.6
Graph showing percentage depth dose variation for 3 different field sizes (4 cm×4 cm, 10 cm×10 cm, and 40 cm× 40 cm) for a 15 MV beam (from a Varian 2100C). Note that generally the percentage depth dose is greater for larger fields, but that close to the surface this may not be the case.

sizes this contribution is greater and, since the field divergence is also greater, it will increase more rapidly as a function of depth and consequently the effective attenuation will be less.

As pointed out in Section 22.3.1, the field size will also affect the build-up region, and the depth of the maximum will be smaller as the field size increases. As a consequence, in spite of the smaller effective attenuation for larger field sizes for the higher energies, the whole depth-dose curve will be shifted towards the patient surface resulting in an apparent decrease of the depth dose for some depths (see Figure 22.6).

22.3.2.3 Effect of Source Skin Distance

The Source Skin Distance (SSD) is a significant determinant of the change of dose rate with depth. The variation of dose with depth can be represented approximately by Equation 22.1:

$$D_d = D_{max} \frac{(SSD + d_{max})^2}{(SSD + d)^2} f(d, A) \tag{22.1}$$

Here D_d is the dose rate at depth d and D_{max} is the dose rate at the depth of dose maximum, d_{max}. $f(d, A)$ is an approximately exponential function of the depth d, for the field A, and represents the beam attenuation beyond the maximum. A detailed derivation of the effect of the change of SSD is given in Section 23.2.2.4. In Equation 22.1, the beam attenuation is enhanced by a factor including the ratio of the square of the distances from the source to d_{max} and to the point of interest, respectively. In contrast with the dose rate which decreases as the distance to the source increases, the fall-off in dose with depth is smaller at large distances than at short ones. For example, with a cobalt-60 beam the percentage depth dose at 10 cm deep (i.e. relative to dose at d_{max}) is 56.4% at 80 cm SSD and 61.4% at 140 cm SSD (BIR 1996). Thus a cobalt-60 beam at 140 cm SSD would have the same effective penetration as a 5 MV x-ray beam at 80 cm SSD.

22.3.3 OFF-AXIS DOSE DISTRIBUTION

The off-axis dose distribution at the edge of the beam in linear accelerators or cobalt-60 beams is largely dependent upon the design of the collimator which is used to delineate the beam (see Section 22.5.1). In the central part of the beam, the off-axis distribution for an accelerator depends on the design of the flattening filter.

22.3.3.1 Cobalt Beams

The shape of the cobalt-60 beam profile is affected by the large source size and by the effect of scatter. Unlike in an accelerator beam for which a flattening filter is used, the dose rate decreases towards the edge of the beam because of the reduced contribution of scatter dose as the beam edge approaches. This results in a rounded beam profile. At the longer distances associated with megavoltage units, the difference in SSD between the centre and the edge of the beam makes only a minor contribution to the fall-off in dose at the beam edge (3.5% for a 30 cm square field at 80 SSD).

The shape of the profile at the edge of the beam is determined largely by the size of the source. For a cobalt-60 beam the half value thickness of lead is 10 mm, so for beam shaping, lead collimators around 70 mm thick are needed to reduce the dose to 1%. Because of the large size of the source (usually 15 mm or 20 mm in diameter), the beam edge is less sharp than for a kilovoltage unit. This is caused by the fact that at the edge of the beam there is a penumbra region where part of the source will be shielded by the collimators and part will not (see Figure 22.7). The *geometric penumbra*, p_{geom}, at a given distance from the source is defined as the width of the region where the source is partly shielded by the collimator. It can be calculated from the following expression:

$$p_{geom} = d_{source} \frac{(f - SCD)}{SCD} \tag{22.2}$$

where d_{source} is the source diameter, f is the distance from the source to the plane of interest, and SCD is the distance from the source to the collimator end. The penumbra width is smaller

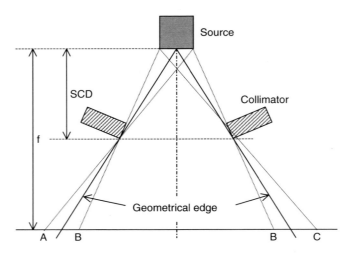

FIGURE 22.7
Diagram showing the geometric and transmission penumbra for a cobalt-60 beam. A and B represent the external and internal limits of the geometric penumbra respectively. C is in the region where the transmission through the collimator is incomplete.

if the beam is collimated further from the source and increases as the distance to the collimator is greater.

If the collimator were made of only two separate orthogonal blocks, the penumbra from the collimator jaw closest to the source would be unacceptably large. So it is usual for this jaw to be split into at least two parts, one above and one below the orthogonal collimator pair. To produce a sharper beam an additional strip of lead, or tungsten, may be attached beneath the main collimator. This is called a penumbra trimmer. However, penumbra trimmers reduce the clearance between the treatment head and the patient, and many centres accept the increased penumbra rather than use them. If they are used, it is essential that they are precisely aligned, as otherwise the penumbra width could actually be increased. Since the collimators are usually slightly further from the source than half way between the source and the patient, the geometric penumbra width will be slightly smaller than the diameter of the source.

The overall penumbra is usually defined as the distance, p_{20-80}, separating the 20% and 80% of the dose on the beam axis at the depth of interest. This overall penumbra accounts also for the fact that some photons emitted by the distal parts of the source are only partially attenuated by the collimator, resulting in a *transmission* penumbra illustrated in Figure 22.7. It should be noted that when additional shielding blocks are used (see 22.5.3.1), the *geometric* penumbra at the edge of these blocks, which are closer to the patient skin, gets smaller; however, if they are not focused towards the source, the *transmission* penumbra gets larger. For cobalt-60 beams, at depth d_{max}, p_{20-80} is usually slightly smaller than the geometric penumbra at the same distance and typically of the order of 10 mm to 15 mm. As the depth increases, p_{20-80} increases. This is due to the increase of both the geometric penumbra and the scatter contribution in tissue and a larger *safety margin* must be used to ensure a uniform dose distribution within the target volume.

22.3.3.2 Linear Accelerators

Beam edge: The source size of a linear accelerator is the size of the electron spot incident on the target, which is typically of the order of a few millimetres. This results in a sharper beam edge than with a cobalt-60 unit, and p_{20-80} is typically about 6 mm at d_{max}. This is larger than the geometric penumbra predicted from Equation 22.2 using source sizes quoted by manufacturers. Actually the flattening filter acts as a source of secondary photons, effectively broadening the focal spot. Secondly, the effect of scattered photons, which is greater at larger depth, leads to a broadening of the penumbra as the depth increases. Thirdly, the role of secondary electrons transported outside of the beam edge without being compensated for by electrons coming in becomes significant at energies above 10 MV (Day et al. 1990). For the larger energies (i.e. 20 MV to 25 MV), the penumbra width at the reference depth may even be the same order of magnitude as for a cobalt-60 beam.

Electron transport is also a problem for very small fields such as those used in stereotactic arc therapy where lateral electron equilibrium may not be achieved anywhere in the beam, thus reducing drastically the dose on the beam axis. It becomes important to use tissue equivalent dosemeters in this situation since the assumptions made in routine dosimetry are no longer valid. Computer algorithms designed for larger field sizes may be significantly in error for such fields. If real data for very small fields are input to the planning computer, this may cause a problem for larger fields because the algorithm is forced to extrapolate from a field size where there is equilibrium to a field size where there is not. In such circumstances it may be appropriate to omit the very small field sizes. Specially designed algorithms, or beam data files, should be used for the very small fields.

Beam centre: The shape of the central part of the beam profile will depend on the design of the flattening filter. The use of a flattening filter makes it possible to compensate for the lack of

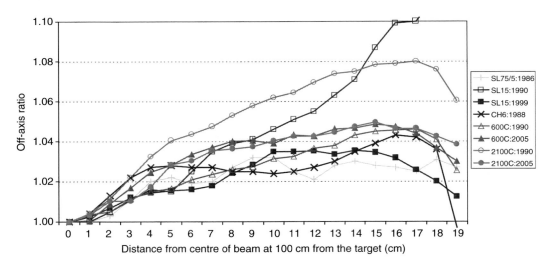

FIGURE 22.8
(See colour insert following page 590.) Profiles at peak depth for different low energy accelerators including a Philips SL75/5, two Elekta SL15s (one installed in 1990 and the other in 1999), an ABB CH6, two Varian 600Cs and two Varian 2100Cs installed at different times. Note the more pronounced increase in dose towards the edge for the SL15 and 2100C installed around 1990.

scatter at the edge of the field by deliberately designing a profile which increases towards the edges at shallow depth. The beam profile will therefore be very dependent on the detailed design of the flattening filter. Because of the flattening filter, the energy spectrum and consequently the Quality Index of the beam varies between the centre and periphery of the beam (Hanson et al. 1980, 1980a; Zefkili et al. 1994).

Differences between the design aims of different flattening filter designers can result in very different beam profiles for accelerators from different manufacturers—and indeed for different machines from the same manufacturer (Figure 22.8). For example, older Philips (now Elekta) accelerators such as the SL75/5 were designed to produce a flat beam at 5 cm deep. This resulted in a maximum off-axis ratio at the depth of peak build-up of 1.03. However, the flattening filter for the SL series accelerators was designed to produce a flat beam (within 3%) at 10 cm deep. For a typical SL series machine at 6 MV, the maximum off-axis ratio at the depth of peak build-up was 1.10. The International Electrotechnical Commission (IEC 1989a, 1989b) specifies that the ratio of the maximum to the minimum dose at the depth of dose maximum should be less than 1.07 over the central

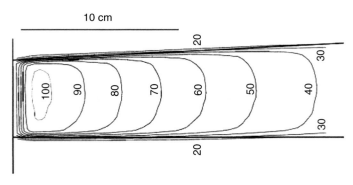

FIGURE 22.9
Isodose curve for a half blocked field from an SL series (Elekta) accelerator.

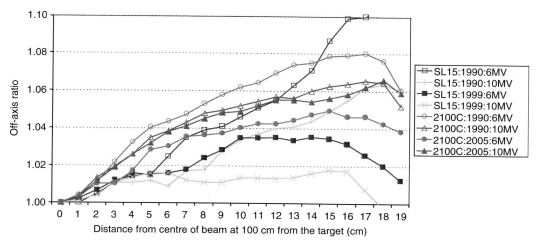

FIGURE 22.10
(See colour insert following page 590.) Flatness profiles at peak depth for Varian (2100C) and Elekta (SL15) 6 MV beams compared to 10 MV beams. Two machines from each manufacturer installed at different times are compared. Note the higher doses at the beam edge for the earlier 6 MV beams which were designed for flatness at 10 cm deep.

80% of the beam width for field sizes between 5 cm and 30 cm and less than 1.09 up to the maximum square field[*]. Note that IEC 976/977 is a disclosure standard—i.e. it defines how things should be specified—and, although maximum recommended values are given, they are not mandatory. The European Community (EC 1977) has specified acceptance criteria for accelerators which are mandatory in Europe (see Section 38.1.3), but these do not specify flatness at the peak.

When asymmetric fields are used, the scatter is no longer at a maximum on the central axis of the accelerator. This results in a profile for a half-blocked beam that is not symmetrical about the centre line of the open field (see Figure 22.9). It could be argued that, in striving to achieve flatter symmetric fields, the comparability of the beam profiles of asymmetric and symmetric fields is compromised.

For energies above 6 MV, there is less scatter and consequently less difference between the beam profile at the peak and that at 10 cm depth. For this reason the beam profile at the depth of peak build-up can be less peaked than that at 6 MV (see Figure 22.10) Because of the much thicker filter required for high energy beams the low energy filter may be used in conjunction with the high energy filter, as can be seen in Figure 11.19.

22.3.4 ISODOSE DISTRIBUTIONS

Isodose distributions from typical cobalt-60 and linear accelerator beams are shown in Figure 22.11. Note the wider penumbra and more rounded isodoses of the cobalt-60 beam. The isodoses shown are those obtained in a uniform water phantom. In non-uniform media the isodoses may be significantly changed (see below, Section 22.4.2).

22.4 INFLUENCE OF PATIENT SHAPE AND COMPOSITION

The isodose distributions shown in Figure 22.11 apply to a uniform water phantom. Real patients have irregular surfaces and inhomogeneous internal composition. For most regions of

---

[*] Most manufacturers consider 35 cm to be the maximum size for this purpose because of the cut-off corners.

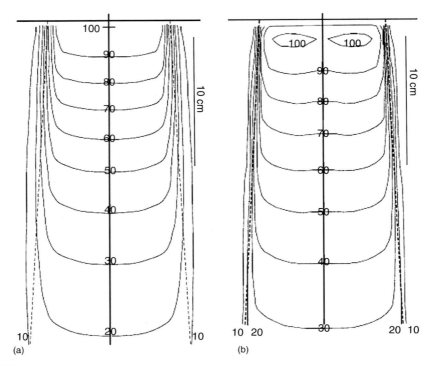

FIGURE 22.11
Isodose lines in the central plane (a) from an 80 SSD cobalt-60 beam and (b) from a 6 MV linear accelerator. The isodose lines close to the entrance surface are not represented. The dashed lines represent the edge of the geometrical field. Note the influence of the flattening filter and the tighter penumbra for the 6 MV beam.

the body it is appropriate to treat the patient as water equivalent. A more accurate patient representation can be obtained using a CT scan of the patient to show the patient shape and the real variation in density. Because CT scans are obtained with 120 kV beams (the attenuation of which is strongly affected by photoelectric interaction cross sections), it is necessary to make a conversion between CT numbers and relative electron density to obtain a density map of the patient (see Section 39.2.1). Methods of correction for oblique incidence and patient inhomogeneities are briefly presented here. They are dealt with in more detail in Section 26.2.3.

22.4.1 EFFECT OF OBLIQUE INCIDENCE

When a beam is incident obliquely on the surface of the patient, the isodoses will be changed compared to a beam incident on a flat surface (Garrett and Jones 1962). In this case the isodoses effectively become wedged. It might be expected that the isodose lines would remain parallel to the patient surface. This is not the case because part of the fall-off of the dose in depth is associated with the inverse-square law, independently of the tissue attenuation. To predict the effect it is possible to use the *isodose shift* method described by Barry (1970) and van der Giessen (1973). For a cobalt-60 unit at 80 cm SSD, the isodose curve is shifted by two-thirds of the distance between the line perpendicular to the skin entry point on the central axis and the skin surface. For linacs, particularly as the energy is increased, the shift is reduced (see Table 22.1 and ICRU 1976).

The wedging effect due to the skin obliquity is usually not what is wanted and it is then necessary to compensate in some way. In some situations (such as breast treatments) this can be achieved with a wedge filter (see Section 22.6.1) whose angle should be about two-thirds of the

TABLE 22.1

Isodose Shift from Lines Perpendicular to the Beam Axis, Expressed as Fraction of the Air Gap between This Plane and the Skin, for Various Energies and SSDs

	Cobalt-60	5 MV	10 MV
100 cm SSD	0.69	0.67	0.59
80 cm SSD	0.67	0.64	0.55

These values are calculated from the ratio between depth-dose fall-off due to tissue attenuation alone and total depth-dose fall-off (combining both inverse square of the distance and attenuation).

angle of the skin surface for a cobalt-60 beam and correspondingly shallower for a linear accelerator. This is illustrated in Figure 22.12.

22.4.2 INHOMOGENEITIES

Where the tissue is not water, equivalent distortions to the beam profiles will occur. If the density is substantially less than that of water, as in lung or in air cavities, the attenuation will be reduced. The effect of this may be very approximately accounted for by considering the water equivalent path length, d_{eff}, calculated by correcting the geometrical depth d with the ratio of the density of the inhomogeneity to that of water. In this case, d_{eff} is then used instead of d in the $f(d,A)$ term in Equation 22.1 (Section 22.3.2.3). While this approximation works well for points that are well beyond the inhomogeneity, it does not represent the dose variation within the inhomogeneity or just beyond it. This is for two different reasons: firstly, the equivalent path length approximation does not consider the modifications of the scatter component which have a greater influence within or close to the inhomogeneity; secondly, at the interface, there is a loss of electronic equilibrium and the inverse of the build-up effect occurs just inside the inhomogeneity. Just beyond the inhomogeneity the dose builds up again for the same reasons as is illustrated in Figure 22.13.

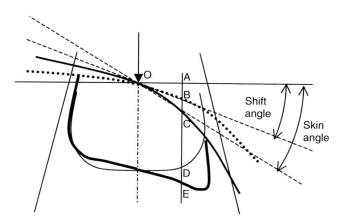

FIGURE 22.12

Effect of oblique incidence on isodose distributions. The patient surface is represented by a thick solid line. The isodose lines are parallel to the dotted line which is intermediate between the skin and a line perpendicular to the beam axis. On each line parallel to the beam axis, the isodose shift (DE=AB) is a constant fraction of the gap AC between the line perpendicular to the beam axis and the patient surface. For an accelerator at high energy (around 15 MV–20 MV) and 100 cm SSD, this fractional shift, AB/AC, is about 0.5. For a cobalt-60 beam, at 80 cm SSD, the shift is about 2/3. To compensate for this effect on the isodose lines and to restore them perpendicular to the beam axis, a wedge filter can be used with a wedge angle defined by the tangent to the dotted line representing the isodose shift— referred to as the *shift angle* in the figure.

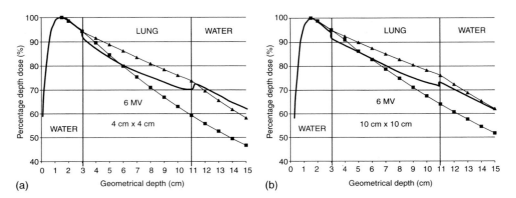

FIGURE 22.13
Comparison of the actual depth dose curve (thick solid line without symbol), the curve that would be obtained in a homogeneous medium (solid squares) and the result of an equivalent path length calculation (solid triangles) for 6 MV beam energy. The left side shows a 4 cm × 4 cm field and the right side shows a 10 cm × 10 cm field. The actual depth dose curve is based on the Monte Carlo calculations from Arnfield, M.R. et al., *Med. Phys.*, 27, 1266–1274, 2000 and the homogeneous dose values are taken from BIR, *Br. J. Radiol.*, Suppl. 25, 1996.

Figure 22.13 shows the effect on the depth-dose curve of an 8 cm thick block of lung for a 6 MV beam. The effect is slightly greater for the 4 cm × 4 cm field than for the 10 cm × 10 cm field and the rebuild-up effect is more marked. However, at 18 MV (Figure 22.14) there is a marked difference between the 4 cm × 4 cm field and the 10 cm × 10 cm field. For a 6 MV beam the depth of dose maximum is at 1.5 cm, so on the central axis electronic equilibrium is established for both field sizes. However for the 18 MV beam the depth of dose maximum is 3.2 cm, so there is a complete loss of electronic equilibrium for the small field.

In addition to the effect on the depth-dose curve, a low density region will also have an effect on the beam penumbra. The effect of electron transport will be particularly acute for high energy beams in low density media such as lung. The lack of electronic equilibrium in this situation means that there is a zone at the edge of the beam between the 50% and the 90% isodose, sometimes referred to as the beam fringe (Ekstrand and Barnes 1990; Miller et al. 1998; Engelmans et al. 2001; Tsiakalos et al. 2004), where the dose-rate is significantly reduced. The penumbra width and the beam fringe may increase by a factor greater than 2.5 in lung compared to a water-like medium when the energy is about 20 MV (see Figure 22.15).

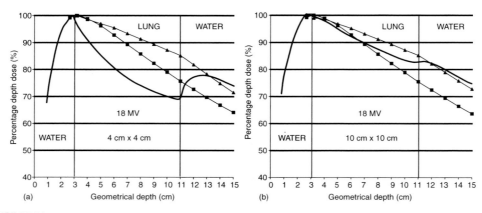

FIGURE 22.14
The same comparison as in Figure 22.13 but with a beam energy of 18 MV. At this energy the depth of dose maximum is at 3.2 cm in water, and for a 4 cm × 4 cm field (a) there is a loss of electronic equilibrium in the lung and after the lung-water interface. (Data from Arnfield, M.R. et al., *Med. Phys.*, 27, 1266–1274, 2000 and BIR, *Br. J. Radiol.*, Suppl. 25, 1996. With permission.)

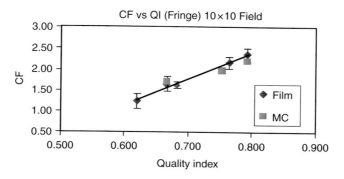

FIGURE 22.15

Increase of the beam fringe (distance from 50% to 90% of the central axis dose) as a function of the beam Quality Index (QI) in lung material of density 0.3 g cm^{-3} for a 10 cm × 10 cm field size. The correction factor CF represents the ratio of the fringe within lung material to the fringe in water-equivalent tissue obtained both from film measurements and Monte Carlo computation (MC). It increases from 1.2 to 2.6 as the energy increases from 4 MV (QI = 0.620) to 20 MV (QI = 0.794). (From Tsiakalos, M. F., Theodorou, K., Kappas, C., Zefkili, S., and Rosenwald, J. C., *Med. Phys.*, 31, 943–949, 2004.)

The equivalent-path-length method used to calculate the dose in the presence of inhomogeneities will be discussed again in Section 23.3.4.5 with the corresponding equations. Other methods frequently used in computer based dose calculations will be presented in Section 26.2.3.2.

22.5 BEAM DELINEATION

22.5.1 STANDARD COLLIMATORS

Most accelerators have standard collimators designed to produce a rectangular field and to reduce the unwanted dose to about 1% of the central axis dose. The geometric edge of the radiation beam so defined is obtained by considering the geometric projection of the distal end of the collimator from the centre of the source. For cobalt-60 beams, the point source to be considered is the centre of the proximal surface of the source relative to the collimator. For cobalt-60 and linear accelerator beams it is found that, whatever the distance and the depth, this geometrical edge coincides with the 50% dose relative to the central axis at the relevant depth. For cobalt-60 beams, the maximum aperture of the standard collimators is typically 30 cm × 30 cm at the treatment distance of 80 cm. For linear accelerators it is typically 40 cm × 40 cm at the treatment distance of 100 cm. For some accelerators the circular primary collimator limits the diameter of the maximum field so that a markedly lower dose is delivered to the corners of a 40 cm × 40 cm field (see also Section 11.4.2). This needs to be considered when using large fields as it may not be correctly modelled by the treatment planning system. One possibility is to place additional shielding in the corners of such fields so that some tissue is not partially irradiated (as is done automatically by the Elekta Multileaf Collimators).

22.5.2 REQUIREMENTS FOR ENHANCED BEAM SHAPING

The need for further field shaping arises as, in general, the projection of the radiotherapy target volume in the *Beam's-Eye View* (BEV) is not rectangular. The observation that rectangular beams were a limitation was pointed out as early as 1959 by Green et al. (1960) who stated that target volumes "are not composed of cubes and cylinders", which are the volumes generated by the intersection of beams with rectangular cross sections. Clearly it did not take 40 years for the radiotherapy community to take notice of this observation, and the shaping

TABLE 22.2

Volumes of Regular Shapes

		Volume (cm³)
Cube	8 cm	512
Cylinder	8 cm×8 cm	402
Sphere	8 cm	268

of beams has been carried out at differing levels of sophistication for as long as radiotherapy has been available.

The use of beam shaping has two main purposes:

- It reduces the volume of normal tissue that is irradiated and therefore allows the radiation oncologist to exploit the inverse relationship between normal-tissue-tolerance dose and target volume irradiation. Empirical understanding of this relationship on the basis of clinical experience predates the development of radiobiological understanding and the concepts of Tumour Control Probability (TCP) and Normal Tissue Complication Probability (NTCP) (see Chapter 36).

- It reduces the dose to critical structures which are close to the boundary of the planned target volume to levels where unacceptable morbidity is avoided.

For a single beam the reduction in volume of tissue irradiated by a shaped beam is obvious, but is particularly marked when considering the use of multiple shaped fields which intersect to form the planned target volume. Beam shaping is very effective in reducing the treated volume below that achieved by the use of rectangular fields.

This is illustrated in Table 22.2 which shows the relative volume of three simple shapes all with major dimensions of 8 cm. A sphere can be formed by the intersection of multiple non-coplanar circular beams, a cylindrical volume by isocentric rotation of a square (or rectangular) beam and a cube by the intersection of two orthogonal pairs of square beams.

Assuming a partial volume effect relationship of the irradiated healthy tissue, the radiobiological consequences of these volume reductions according to the Lyman–Burman model of NTCP (see Section 36.3.2) can be evaluated from Equation 22.3:

$$TD_2 = TD_1 \left(\frac{V_2}{V_1}\right)^{-n} \tag{22.3}$$

where TD_1 is the 50% tolerance dose (i.e. the dose delivered uniformly to the volume V_1 which would lead to complications in 50% of the population), and TD_2 the 50% tolerance dose corresponding to a uniform irradiation of V_2. The exponent n is characteristic of the partial irradiation tolerance of each organ and is near unity if the volume effect is large as for parallel type tissue (see Section 8.9).

This relationship is shown graphically in Figure 22.16 for $n=0.16$. If this relationship could be shown to be valid in clinical practice then a reduction of the target volume by 30% would allow for a 5% escalation in dose without incurring extra morbidity, and a 50% reduction would allow for a 10% dose escalation. If these were achieved then significant improvements in tumour control could be expected.

22.5.3 BLOCKS

22.5.3.1 *Simple Beam Blocking*

As part of the planning process the shape of each beam is determined. The shape can then be described by marks applied to the patient's skin or by a template that can be placed in the optical beam projected through the collimation system. In each case beam shaping is achieved

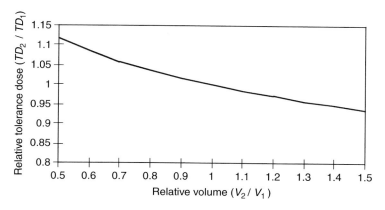

FIGURE 22.16
Variation of tolerance dose with volume assuming a partial volume effect according to Equation 22.3 and taking a *n* value equal to 0.16.

by placing lead blocks on the shadow tray. These blocks are then aligned so that the optical beam from the accelerator projects a shadow to coincide with the area to be shielded. This method is limited for many reasons:

- It is time consuming.
- In its simplest form it uses blocks with parallel sides (compromising the beam penumbra).
- It requires ingenuity in supporting blocks on the tray when it is not horizontal.
- The blocks are heavy, which is a serious health and safety problem for both patients and radiographers.

22.5.3.2 Customised Blocking

Some of the limitations in simple blocking can be overcome by manufacturing customised blocks for each patient. A customised block can be pre-mounted on a slide fitted to the shadow tray and can therefore be repositioned very accurately and used at any gantry orientation.

These customised blocks are cast from a low melting point alloy of lead and other metals. One such alloy is Cerrobend, or Wood's Metal, which can be cast into an expanded polystyrene mould formed with a hot wire cutter. Several simple commercial devices are available which allow the shape of the target to be traced from an outline drawn on a simulator radiograph and demagnified so that the resulting block is correctly scaled to the position of the shadow tray. Alternative *high tech* solutions to block manufacture include computer controlled milling machines linked to treatment planning systems.

Blocks cast with these devices have edges focused towards the target in order to achieve the best penumbra. In fact, the geometric penumbra (arising from the finite size of the focal spot) is somewhat smaller for a customised block than it is for the conventional rectangular collimator as the block is nearer to the patient than the collimator.

In many respects the customised block is the gold standard for blocking, but the method is not perfect for the following reasons:

- The blocks are still heavy and need to be handled manually (a few elaborate handling devices have been designed to overcome this problem but they have not been widely used).
- The blocks are cast from alloys which need careful handling as they contain toxic materials[*].

---

[*] In the UK handling of such materials are covered by the Control of Substances Hazardous to Health (COSHH) regulations.

- The density of these alloys is usually lower than lead and therefore they need to be thicker.
- As with the simple blocks they need to be mounted near to the patient and secondary electrons generated in the shadow tray significantly increase the surface dose (Hounsell and Adams (1995) found that the surface dose increased from 25% to up to 50% of the peak dose).
- The manufacturing process is time consuming and as a result changes in blocking, which may be required during a course of treatment, are difficult to implement.

22.5.4 MULTILEAF COLLIMATION

22.5.4.1 Standard Multileaf Collimators

The multileaf collimator (MLC) was designed as a way of overcoming most of these difficulties. Details of the construction of multileaf collimators have been discussed in Section 11.4.3.

As compared with a conventional collimator, which has 4 collimators arranged as 2 orthogonal pairs, the MLC consists of 2 banks of leaves which can be individually positioned. The radiation field can therefore be shaped to match the projection of the target volume from each beam direction within the resolution of the leaves. Figure 22.17 illustrates the principle and shows the stepped beam edge which would be avoided with a customised block.

In considering the performance characteristics of a multileaf collimator the following considerations are important:

1. Radiation transmission and leakage:
   a. Radiation transmitted through the leaf material
   b. Radiation leakage between adjacent leaves
2. Resolution:
   a. The size of the steps defining the beam edge
   b. The maximum size of the field
   c. The maximum projection of the leaves across the central axis
3. Penumbra:
   a. How well the leaves are focused

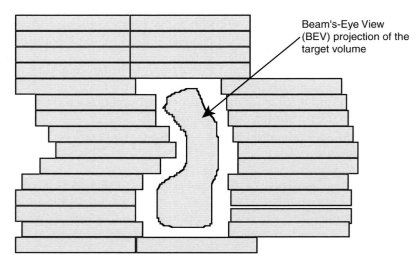

FIGURE 22.17
Leaves shaping the beam to match the target with a margin.

4. Precision:
   a. The accuracy of the leaf positioning
5. Beam modification:
   a. How the MLC affects the clinical dosimetry of the beam (percentage depth dose, field size factors, etc.)

The position of the leaves is of some interest, but only as a surrogate for the position of the isodose curves. Clearly the precise position of each isodose depends on the geometry of the multileaf collimator and the radiation scattering processes that occur in the treatment head and in the patient or phantom. Figure 22.18 shows the isodoses on a plane normal to the central axis of a beam with a relatively simple shape.

It can be seen that even where the stepping of the beam edge is most evident the isodoses are relatively smooth. These isodoses were measured for a single field at d_{max}. The effects of scatter in the patient will smooth the curves even further and for multiple beams the undulations from one field will not necessarily register with those from the other fields. However, this does not mean that lower level isodoses are unimportant. If the planned target volume is defined by the 90% isodose, then an undulating contribution at the 10% level from one field can add to a relatively smooth distribution which has reached the 85% level from other fields. It can also have a significant effect on the final position of the defining isodose.

There is no evidence that an MLC significantly alters the percentage depth dose as compared to customised blocks, but measurable changes to head scatter do occur (see Section 23.2.1.12). These result in changes to field area factors and have to be taken into account in accurate planning. The degree to which the collimator scatter factor is affected by the MLC will depend on the position of the MLC relative to the collimator jaws and monitor chamber. Thus the effect of the leaf positions on the head scatter factor will be greater with the Elekta design than with the Varian design.

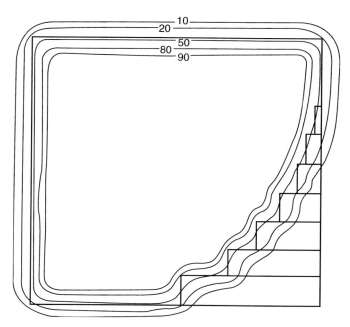

FIGURE 22.18
Typical isodose lines at d_{max} for an MLC shaped field.

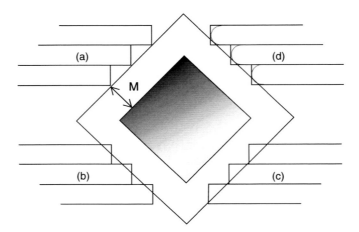

FIGURE 22.19
Leaf fitting strategies. A diamond-shaped planning target volume (shaded) is expanded uniformly along each of its edges by an MLC margin M to create an expanded BEV contour. Leaves are shown fitted to the expanded contour using touch (a) inner edge, (b) centre, (c) outer edge, (d) circle strategies. (From Fenwick, J. D. et al., *Phys. Med. Biol.*, 49, 1505–1519, 2004. With permission.)

As the edge of an MLC field is stepped, it is necessary to have some rules about where leaves should be positioned relative to the boundary of the expanded target volume. Figure 22.19 illustrates the problem. The leaves can be set so that:

(a) Their tips touch the boundary.
(b) The leaf centre is a best fit to the boundary.
(c) The troughs touch the boundary.
(d) Rounded tips touch the boundary.

The rule applied will depend on the main intention of the field shaping since the best compromise must be reached between irradiating the target correctly (a) and ensuring a good protection of healthy tissue (c). In clinical practice, the intermediate solution (b) is often preferred.

This is not a problem unique to the MLC. The geometric projection of any collimator is insufficient to describe the dose distribution. Any rules relating geometric edges of beams to the position of a particular isodose are only rules of thumb. If it is necessary to know where the beam edge is to a precision better than a few millimetres, the full 3D dose distribution must be considered. In the case of an MLC, it is important to establish that the planning system chosen can cope adequately with the stepped beam edges. It may be necessary to combine a knowledge of the geometric shape of the isodoses relative to the leaf positions obtained by measurement, with the knowledge of the relative contribution of the other beams.

The issue relative to fitting of MLC leaves for clinical situations is discussed further in Section 32.4.4.

22.5.4.2 *Mini and Micro Multileaf Collimators*

The reduction in treatment volume afforded by the use of standard MLCs is comparable with that of customised divergent blocks, but the precise delineation of the edge of the treatment volume is in many cases compromised by the projected width of the MLC leaves. The stepping of the field edge, caused by an MLC with a projected leaf width of 10 mm, has been identified as a potential limitation in the restriction of morbidity in up to 19% of cases (Galvin et al. 1992). The limitations imposed by the leaf width have been addressed by smoothing techniques and the development of MLCs with narrower leaves, described as mini or micro

MLCs. The distinction between *mini* and *micro* is arbitrary, but it has been suggested that MLCs with leaves projecting to between 2 mm and 5 mm should be considered as mini-MLCs and those below 2 mm as micro-MLCs (Bortfeld et al. 2000). It should be noted that by this definition the Varian 120 leaf MLC is a mini-MLC for fields up to 20 cm long.

Field edge smoothing can be achieved by the superposition of multiple fields, each of whose centres are shifted slightly relative to the next (Galvin et al. 1996). An alternative method of smoothing is to superimpose multiple fields rotated about the collimator axis, but the need to use a wedge filter may limit this approach. Superposition techniques provide smoothing but not increased resolution, and they are severely limited in cases where it is necessary to shape a field with features less than 1 cm in the direction normal to that of leaf movement. A further disadvantage is that movement of the patient between fields increases the complexity of verification compared to that required for treatment with a single field.

The improved resolution that is achievable by the use of MLCs with narrower leaves was discussed in Section 11.4.3.2. The number of leaves and the leaf width determine the maximum field size, which in all cases is much smaller than for a conventional MLC. This has been justified if the main application is in stereotactic radiosurgery, where it has been estimated that 95% of cases can be treated within the range of field sizes from 1 cm to 6 cm (Schlegel et al. 1997). In other applications, if precise shielding of a critical structure adjacent to the periphery of a large field, or a field which is significantly offset relative to the central axis is required, these devices will have limitations. Specifically, their use to provide precisely-defined boosts to fields larger than the maximum covered by the micro-MLC presents the logistical difficulties of having to attach the micro-MLC to the treatment machine between the delivery of the main and boost fields to each patient.

22.6 BEAM MODULATION TECHNIQUES

Although beam blocking is very effective in conforming the projection of each beam to the target volume, it does nothing to ensure uniformity of dose within the volume to be treated. This is important because, for a given mean dose, the maximum Tumour Control Probability is achieved with a uniform dose distribution within the target volume. However, a deliberately non-uniform dose may sometimes be desirable close to a sensitive normal tissue. More often, non uniform dose distributions are achieved on each individual beam of a composite treatment plan, either to compensate for surface irregularities or to obtain a resulting uniform dose from the addition of several modulated beams.

22.6.1 MECHANICAL FIXED OR MOTORISED WEDGE FILTERS

In order to produce a wedge shaped isodose distribution as shown in Figure 22.20 wedge filters may be introduced into the beam. The different types of wedge filters currently used have been described in Part C (see Section 11.4.4.1 for accelerators and Section 12.2.4 for cobalt-60 machines).

A wedge is characterised by its *wedge angle*. This is defined as the angle between the wedge isodose lines and a line perpendicular to the central axis of the beam (see Figure 22.20a). According to the ICRU Report 24 (1976) this should be defined at a depth of 10 cm. The older definition of the angle of the 50% isodose is not recommended because it is more appropriate to cobalt-60 beams than to the more penetrating megavoltage beams. At greater depths the isodose angle will be smaller because the increased scattered radiation will reduce the wedge effect.

Another useful quantity is the *wedge factor* (*WF*) defined as the ratio of the dose with and without wedge for a specified field size and at a specified depth on the beam axis. This factor is mostly dependent on the on-axis transmission of the wedge for the primary beam. Therefore,

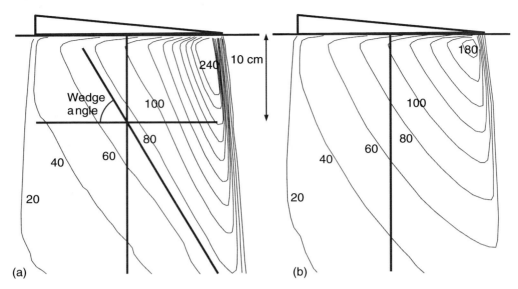

FIGURE 22.20
Dose distribution resulting from the introduction of a wedge filter into the beam. The results of two wedge designs is shown—(a) designed to produce isodoses that are at the wedge angle all the way across the field (Elekta SL75/5) and (b) a simple triangular shape (Elekta SL15). The definition of the wedge angle is illustrated in (a). Note that the symbol of the wedge does not represent the actual shape of the wedge (see Figure 22.21).

for a fixed mechanical wedge, *WF* is not very sensitive to changes of depth or field size. *WF* is used principally to calculate the number of monitor units (see 22.6.1.2 and Chapter 23).

If a limited number of mechanical filters are available, intermediate angles may be obtained from a weighted addition of beams with and without wedge. The proportion of the treatment for which the wedge is inserted into the beam in order to obtain the desired effective wedge angle can be calculated from the following expression (Petti and Siddon 1985):

$$\frac{\text{on-axis dose with wedge}}{\text{on-axis total Dose}} = \frac{\tan(\text{desired wedge angle})}{\tan(\text{maximum wedge angle})} \tag{22.4}$$

The same principle is applied in the case of automatically-moving (or motorised) wedges. Table 22.3 illustrates how intermediate angles may be obtained from the 60° motorised wedge filter of a GE Saturne accelerator.

TABLE 22.3

Example of Intermediate Wedge Angles Obtained for the 6MV Beam of a GE Saturne Accelerator Equipped with a 60°Motorised Wedge

Desired wedge angle	$\frac{\tan(\text{wedge angle})}{\tan(60°)}$	Weight of the 60° wedged beam (dose)	Weight, K_{MU}, of the 60° wedged beam (monitor units)	Equivalent wedge factor (WF_{eq})
0°	0	0%	0%	1.00
20°	0.21	21%	40%	0.76
30°	0.33	33%	55%	0.67
40°	0.48	48%	70%	0.58
50°	0.69	69%	85%	0.49
60°	**1.00**	**100%**	**100%**	**0.40**

As can be observed from Table 22.3 it is important to avoid confusion between the weight expressed in terms of *dose* (third column) and in terms of *monitor units* (fourth column). The former will be used to combine beams with and without wedge in a treatment planning system. The latter will be used at the treatment console or in a record and verify system to set the number of monitor units to be actually delivered by the accelerator. The fifth column represents the *equivalent wedge factor*, WF_{eq}, defined as the ratio of the total dose resulting from the combination of wedged and open field to the dose which would be obtained for the same number of MU in an open field. WF_{eq} is a useful practical quantity, as a replacement of WF, to calculate the number of MU for a preset combination of open and wedged filter (see Section 23.3.4.2).

The relationships between the quantities presented in Table 22.3 may be established as follows: Let $\dot{D}_o$ be the dose per monitor unit in the open field at an on-axis reference point. For a given combination, the total dose at the reference point is:

$$D_t = D_o + D_w = (MU_o \times \dot{D}_o) + (MU_w \times \dot{D}_o \times WF) \tag{22.5}$$

where MU_o and MU_w are the number of monitor units used to deliver the dose D_o and D_w without and with wedge filter respectively. This can also be written in terms of the equivalent wedge factor:

$$D_t = \dot{D}_o(MU_o + MU_w) \times WF_{eq} \tag{22.6}$$

The *wedged MU factor*, defined as the proportion of MU with wedge (4th column of table 22.3) may be written as:

$$K_{MU} = \frac{MU_w}{MU_o + MU_w} \tag{22.7}$$

Combining Equation 22.6 and Equation 22.7, we can calculate the *equivalent wedge factor* as:

$$WF_{eq} = \frac{D_t}{\dot{D}_o (MU_o + MU_w)} = 1 - K_{MU}(1 - WF) \tag{22.8}$$

The *wedged dose weighting factor* of Equation 22.4 (3rd column of table 22.3) is then:

$$\frac{D_w}{D_t} = K_{MU} \times \frac{WF}{WF_{eq}} \tag{22.9}$$

Since WF is slightly field size and depth dependent the *wedged dose weighting factor* and the *equivalent wedge factor* may vary to some extent for a specified value of the *wedged MU factor* However, fixed average values as shown in Table 22.3 are generally acceptable within 2% or 3%. Instead of being applied to preset combinations of wedged and open beams, Equation 22.4 to Equation 22.9 may be used in a continuous way to generate any angle intermediate between 0° and the maximum angle defined by the mechanical wedge.

It should be noted that the names given to the quantities discussed above have not gained general acceptance. Therefore, various manufacturers or medical physicists may use different terminologies or concepts. A thorough understanding of the local definitions is essential to avoid potential serious errors which could result from an improper use of all these wedge related factors.

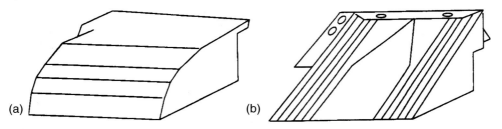

FIGURE 22.21
Wedge shapes used to produce the wedge profiles shown in Figure 22.20. The appearance of a rounded shape (a) is illusory as in fact there are two distinct values of the slope of the wedge face which changes on the central axis. In the design (b) — also pictured in Figure 11.27 — the slope is constant across the face of the wedge but the thickness varies in the unwedged direction to compensate both for the oblique path through the wedge and the less penetrating off-axis radiation. (Courtesy of Elekta Oncology Systems.)

22.6.1.1 Design of Mechanical Wedge Filters

The design of a manual mechanical wedge filter is based on an isodose curve of the largest open field size that is required to be wedged (Aron and Scapicchio 1966). The aim is to produce a series of isodose lines crossing a line perpendicular to the axis of the beam at 10 cm depth at the appropriate wedge angle. An idealised wedged isodose curve can be drawn on a template based on the central axis depth doses as indicated in Figure 22.22. If we assume that the wedge will not affect the depth dose characteristics of the beam, we can draw the central axis of the beam and, using the open field depth-dose data, mark off the points at which the isodose lines cross the central axis at, say, 10% intervals. Angled lines are drawn through these points. Figure 22.22 shows such an idealised curve for a wedge angle of 60°. The ray line M-M' represents the ray line at the tip of the wedge where there is no attenuation required. The dotted lines represent the unwedged isodose line and the wedged isodose line passing through the point A. Point A is the point at which the line through the dose maximum (100%) on the central axis crosses the line at 10 cm depth[*]. The transmission at A and beyond must be calculated by consideration of the open field dose at the point M', where the wedged isodose through A intersects the zero attenuation ray line. In an idealised world the open field depth dose at M' where the line through A meets the projection of the ray line through M is about 28% compared to the dose rate at 10 cm deep (70%)[†]. We may deduce that the transmission required on the ray line through A is $28/70 = p$. A similar process can be carried out for other points to the right of A, for each point searching for the intersection with the ray line through M. It is easier to calculate the relative transmission required at points to the left of A. Considering point B, it will be seen that it is the 40% isodose that must pass through B so that if the transmission at A is p%, the transmission at B must be $p \times 40/100$. The wedge factor can be calculated by considering the transmission on the central axis at C. The transmission here is $p \times 70/100$, i.e. 28%, which is comparable with the wedge factor measured in practice. To determine the shape of the wedge it is necessary to consider its position in the head of the treatment machine. Let us suppose that the wedge is to be placed 30 cm from the source. Point A is 4.9 cm from central axis and this corresponds to $(4.9 \times 30)/110 = 1.34$ cm from the central axis at the wedge distance. Point B is equivalent to a point 1.93 cm from the central axis. The thickness of the wedge at the points can be calculated for the required attenuation based on the linear attenuation factor of the material of the wedge for the current energy spectrum.

---

[*] For a shallower wedge or a narrower field this point may lie outside the beam.
[†] As indicated by the dotted curves the real isodoses will be rounded towards the edge of the field, but using the points M and M' provides sufficient accuracy at this stage.

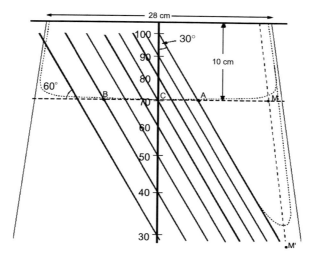

FIGURE 22.22
Wedge filter design for a 60° wedge for a 28 cm wide field and a 6 MV beam. See text for explanation.

A number of other factors must be considered when designing a real wedge. The idealised approach above assumes that the unwedged profile is perfectly flat. If at point A the off-axis factor was 1.02 the attenuation required along that ray line could be increased by 2%. The linear attenuation factor will decrease as the thickness of the wedge increases because of beam hardening as is shown in Figure 22.23. The attenuation factor is also dependent on the radial distance from the centre of the beam because of the beam-hardening effect of the flattening filter (see Section 22.2.2). The simplistic approach taken above is based on primary beam attenuation and takes no account of scatter (Mageras et al. 1991; Heukelom et al. 1994a). The effect of scatter will be to reduce the wedge effect and will require a thicker wedge. Calculating the effect of scatter is not easy and in practice it is simpler to make a prototype wedge and to measure the resulting isodose curves. Corrections can then be

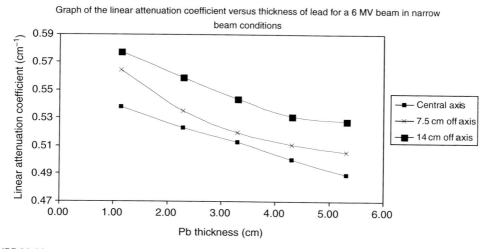

FIGURE 22.23
Variation of lead effective attenuation coefficient with lead thickness and position within the field. (Data measured at Clatterbridge Centre for Oncology.)

calculated to improve the design. Finally account must be taken of the effect of the different energies if the wedge is to be used in a multi-energy accelerator. In practice this is not as great a problem as it would appear because both the value of the attenuation coefficient and the equivalent depth of water increase roughly in proportion for energies where Compton scattering dominates. The wedge isodose curves shown in Figure 22.20a would be produced if the above method of wedge design is followed.

Although this wedge design meets the specification, there are some problems with this approach. With the advent of asymmetric collimators, there has been a demand for larger and larger wedged fields and for these to be used in asymmetric mode. A large field size is particularly important in the unwedged direction. If half-blocked beams are being used to facilitate field matching, an unwedged field length up to 36 cm is required for a half field length of 18 cm, which would be typical of the size required for a breast treatment. The assumption made in most planning systems is that the beam profile in the unwedged direction is unaffected by the presence of the wedge. In order for this to be the case, the wedge needs to be thinner at the edges than at the centre because of the change in half-value thicknesses at different radial distances from the centre. When asymmetric fields are used in the wedged direction, the effect of scatter will change depending on whether the thin end or the thick end of the wedge is being used. There will therefore be a marked difference in the isodoses when the heel of the wedge (the thick part) or the toe (the thin part) is used. For this reason a wedge shape which has a uniform slope is preferred. The large wedge currently installed in Elekta linacs (Figure 11.27) conforms to the design shown in Figure 22.20b and Figure 22.21b.

22.6.1.2 Dosimetry Considerations for Wedge Filters

Wedged fields will have a number of differences from open fields. The electron contamination of the beam is likely to be less, resulting in lower skin dose (Klein and Esthappen 2003). The hardening of the beam will also result in the beam being more penetrating than the open field. Thus it is important to measure the depth dose curve for the wedge directly and not to assume that it is the same as for the unwedged field (Knöös and Wittgren 1991; Heukelom et al. 1994b).

The introduction of a wedge into the beam will clearly reduce the dose rate on the central axis of the beam. The ratio of the dose rate at the depth of dose maximum with the wedge in place to the dose rate with an unwedged field is called the *wedge factor*. The wedge factor may also be defined as the ratio of the doses with and without the wedge at some different specified depth. In principle the magnitude of the wedge factor relates to the thickness of the physical wedge on the central axis of the beam, but because of the effects of scatter (see above) and of beam hardening the wedge factor will also vary with depth and field size. The wedge factor may be measured directly at the calibration depth making an exposure with and without the wedge, but, depending on the exact definition adopted for the wedge factor, it will be necessary to make a correction for the different percentage depth doses of the wedge and open fields (Knöös and Wittgren 1991). The wedge factor for a given wedge will depend on the maximum permitted field width in the wedged direction. This is because a greater width of the wedge also implies a greater thickness of material at the centre of the beam. The wedge factor will be smaller (i.e. the reduction in dose rate will be greater) as the wedge angle increases. For a typical $60°$ wedge allowing a field width up to 30 cm, the wedge factor will be about 0.3. For a given wedge it will also be a function of the actual field size used and wedge factors must therefore be measured over the full range of field sizes. On a cobalt-60 unit where the dose rate is already relatively low, it is common practice to have different wedges for small and large fields. Another technique that has been used is to attach the wedge by a pin to the collimator jaw on the thin side of the wedge so that the wedge thickness in the beam is the minimum necessary for the jaw opening. This has the drawback that the wedge factor varies rapidly with field size. It increases the risk of error and has proved difficult to model in treatment planning computers. In measuring the wedge factor it is important to take the mean of the two values obtained when the wedge is rotated through 180° and to ensure that, if a thimble ion chamber is used, the thimble is parallel to the unwedged direction (see Section 19.2.6).

Another factor to be considered in the design of automatic wedges is the direction of wedge movement and the arrangements for ensuring that the wedge is correctly positioned. Ideally the wedge should slide along runners that are parallel to the unwedged direction. In this arrangement a small positioning error has no effect on the wedge factor. This arrangement is also to be preferred for manually inserted wedges, although it might be in contradiction with the IEC recommendation on wedge orientation (IEC 1996) or with practical ease of use. Wedges can be used to compensate for missing tissue or to allow beams to be brought in from two angled directions. In correcting for skin obliquity the wedge angle should be reduced in the proportion indicated in Table 22.1 compared to the angle between the skin surface and a line perpendicular to the beam (see Figure 22.12). In combining two beams the wedge angle should be half the complement of the angle between them (see Section 33.3.5). Often the reason for use of the wedge is a combination of the two, and in a simplistic way the two angles can be added together. It is not essential to use the wedge angle defined by this method, which may in any case not be optimal, but it does provide a good starting point. For example, rather than use two 45° wedges a 60° and a 30° wedge may be used. However, in going away from the natural choice care must be taken that the balance of the distribution is not affected.

22.6.2 Dynamic Wedges

The idea of moving the collimator jaws dynamically (i.e. with the beam on) is not new. The disadvantage of not having an automatic wedge system and the increase in the distance from the source to the front of the treatment head that would be involved in designing one has led Varian to pursue the development of the *dynamic* or *flying* wedge (Leavitt et al. 1990). To achieve a wedged beam profile one of the jaws is kept stationary and the other is moved during the treatment. When the dynamic wedge was first introduced each wedge angle used a different series of jaw positions with the jaw starting to move at the start of the beam and moving more or less slowly depending on the wedge angle. Verification and calculation of the wedge factor in such circumstances was difficult. To simplify things the enhanced dynamic wedge (EDW) was developed. This is similar to the automatic wedge described in Section 22.6.1 in that an open field portion is first delivered followed by a wedged portion equivalent to a 60° wedge. There is thus in principle only one wedge that is required to be measured. In a treatment planning system the wedge can either be simulated by a combination of an equivalent solid 60° wedge and an open field, or the jaw positions used (contained in the so-called *Golden Segmented Treatment Table* [STT]) can be simulated directly. The wedge factor for the dynamic wedge depends almost entirely on the field size in the wedged direction (and varies by a much greater amount than fixed or automatic wedges). It is therefore mandatory that a planning system that uses an equivalent solid wedge can be set up to consider the effect of each pair of jaws on the output independently rather than use an equivalent square field. In the Siemens implementation the wedge angle is sensitive to the calibration of the secondary dose rate channel (see Section 11.7.3.1), but the wedge factor is within 2% of 1.0. Special consideration must be also given to asymmetric fields. It is possible to make direct use of the golden STT in dose calculation (Papatheodorou 1999). Another method for calculating dynamic wedge factors for use in hand calculations which is also able to cope with asymmetric fields is described by Yu (2002). This is based on the number of monitor units for which the point of interest is exposed during the dynamic delivery, i.e.:

$$\text{Wedge Factor} = \frac{\text{Dose for wedged field}}{\text{Dose for open field}}$$

$$= \frac{\text{Monitor units open field}}{\text{Monitor units by wedged field to the point}} \quad (22.10)$$

However, this formula* does not take into account the change in scatter conditions when the jaw is moving. In practice the scatter dose to the point when the point is nominally shielded counterbalances the loss of scatter caused by the reduction in field size when the point is unshielded. To compensate for this, Yu proposes the revised formula:

$$\text{Wedge Factor} = \frac{GS(Y_0 + \alpha)}{GS(Y_F)} \quad (22.11)$$

where $GS(Y)$ is the golden STT table value for the jaw position, Y_0 represents the jaw position as it passes the measurement point, Y_F represents the fixed jaw position, and α is an empirical factor which is found to have a value 0.6 cm.

The dynamic wedge has the advantage that there is no beam-hardening effect. The depth dose can therefore be considered as independent of the presence of the wedge.

The dynamic wedge is potentially more flexible than fixed wedges because in principle it is possible to design any shape of beam profile that is required. In practice, however, the use of such concepts is limited by the difficulties of quality assurance and treatment planning. Convery and Rosenbloom (1992) have shown how any beam profile can be produced if both the jaws can be moved simultaneously with the minimum beam on time. Another suggestion was that a field could be made up in segments (Cardarelli 1991). The problem with such methods of beam formation is that they increase the total beam-on time and this in turn increases the dose delivered to areas of the patient that are not intended to be treated. This approach can be considered as a 1D precursor of 2D intensity modulation (see Section 22.6.4 and Chapter 43).

22.6.3 TISSUE COMPENSATORS

Although wedges provide adequate compensation for missing tissue in many situations, there are areas, such as the head and neck, for which it is very difficult to achieve a uniform dose distribution because of the irregular patient contours. In such situations a custom tissue compensator can be used. Many different approaches to the design of compensators exist. The simplest is to measure the distance to the skin surface from a line perpendicular to the beam. A fixed ratio of thickness of lead (Cardarelli et al. 1991; ICRU 1976) or aluminium (Ellis et al. 1959) to tissue is then used as a compensator (see Figure 22.24). Wax has also been used since, in principle, it is appropriate simply to scale down the compensator geometrically in the direction perpendicular to the beam axis to allow for beam divergence. However, higher atomic number materials cause less electron contamination (ICRU 1976). In order to obtain the data needed for the design of such missing tissue compensators, it is convenient to use a device to take multiple contours of the patient. This can be conveniently done using a laser outlining system, and the necessary measurements for the compensator can then be carried out automatically in the computer. Thus the inaccuracies associated with making a large number of physical measurements on the patient or the patient's cast are reduced. The problem with using a fixed relationship between tissue thickness and compensator thickness is that the relationship varies with depth and field size. Rather than use a table of equivalent

---

* It must be stressed that Equation 22.10 is not valid for motorised wedges where the number of monitor units for a wedged beam must be increased, compared to the open beam monitor units, in order to compensate for the beam attenuation (see Equation 22.5 and Table 22.3).

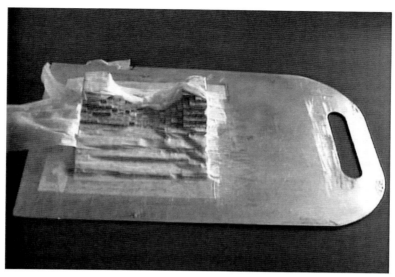

FIGURE 22.24
Missing tissue compensator composed of aluminium blocks taped on a tray at distance from the patient skin.

thicknesses, a more satisfactory approach is to calculate the attenuation required and then to apply the appropriate thickness of compensating material. Even this method is not entirely accurate, as it does not account for the effect of scatter or the effect of increasing thickness on the effective attenuation (as illustrated in Figure 22.23). However, it does allow inhomogeneities inside the patient to be taken into consideration. In using this approach it is convenient to select a plane within the patient at which to create a uniform distribution (Valdagni et al. 1992). For example several centres have been trying to improve dose homogeneity in treatment of carcinoma of the breast (Solin et al. 1991; Valdagni et al. 1992). Here it is particularly important to allow for the reduced attenuation through the lung. Evans et al. (1995) used the transmission dose measured with a portal imaging device to design an appropriate compensator which automatically included a correction for the lung attenuation.

Such techniques are appropriate for simple single field treatments or for parallel opposed fields. For combinations of beams from several directions, dose homogeneity within the target volume may be improved by considering the combined effect of all the beams and specifically setting out to achieve a homogeneous dose distribution as a result of the combination of beams rather than compensating each beam independently. In the TARGET treatment planning system (developed by GE) an algorithm was developed in the early 1980s to calculate the necessary thickness of an appropriate attenuating material such as lead to produce a homogeneous distribution within the defined target volume. Maps were then printed out to show the compensator shape. In this case it is convenient to use thin lead sheets for the compensator, as these can be cut to the appropriate shape. A more recent development of this idea is to identify the target and the sensitive normal tissues and to assign the desired dose to each. An inverse planning method (see Section 43.2) is then applied to design the most appropriate fluence profile for each beam in order to optimise the dose distribution according to the specified criteria. Such an approach was adopted in the Nucletron Plato planning system which derives the thicknesses using an inverse planning algorithm. Although this approach was originally designed for use with multileaf collimators, it can also be applied to physical compensators where the quality control difficulties are less. Compensators designed in this way are likely to have a much greater dynamic range, sometimes requiring 4 or 5 half-value thicknesses of the compensator material. In calculating the appropriate thickness of the compensator material it becomes particularly important to consider the effect of the departure from exponential attenuation (see Figure 22.23) and to include the effect of the scatter from the compensating material (Castellanos and Rosenwald 1998).

Fabrication of compensators can be a time-consuming process. Automatic compensator production devices have been developed which use a milling machine to fabricate the compensator. These require the input of outline data for a number of slices, which can be obtained from multiple CT slices or from an automatic outlining device. These devices often cut negative formers out of polystyrene which must then be filled with an appropriate absorbing material. Cerrobend could be used but, because of its high density, both the former and its filler have to be more accurate than is easy to achieve. Consequently lower density fillers such as granulate of stainless steel (van Santvoort et al. 1995), loaded plaster of Paris, and tin loaded wax (Jursinic et al. 1994) have been developed.

22.6.4 INTENSITY MODULATION WITH THE MLC

The flying or dynamic wedge concept, originally developed for conventional 4 block collimators can easily be extended for use with the MLC. By judicious prescription of the leaf trajectories the wedge angle need not be the same for each leaf pair and the direction of the wedge gradient can be chosen to be at any angle relative to the major axes of the collimator.

The ultimate delivery system for conformal therapy is the dynamically controlled MLC which is able to produce fields equivalent to compensated fields. Several methods of modulation are available and are further discussed in Section 43.3.3. Figure 22.25 illustrates the method described by Convery and Rosenbloom (1992). Each pair of leaves generates a *sliding window* which is scanned across the field. As the window moves its width is varied so that points in the treatment plane are exposed for variable lengths of time and thus receive varying doses.

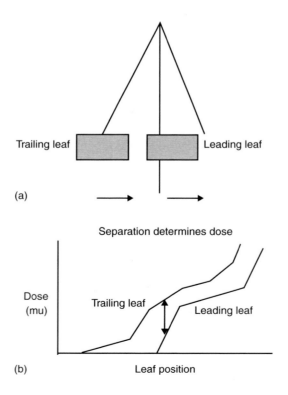

FIGURE 22.25
Beam modulation using the sliding window dynamic multileaf collimation technique (a) leaf separation (b) beam modulation. (From Convery, D. J. and Rosenbloom, M. E., *Phys. Med. Biol.*, 37, 1359–1374, 1992.)

As discussed in Section 43.3, delivery of a modulated beam can be either by full dynamic control of the MLC where the leaves move during irradiation or by the sequential irradiation of multiple fixed beams, often called *step and shoot* delivery. Various algorithms, *interpreters,* and *sequencers* used to calculate leaf trajectories and sequences have to be integrated with the treatment planning process as, to achieve a particular dose distribution, account must be taken of the head and phantom scatter in addition to the primary beam fluence delivered by the MLC. Determination of trajectories must also take into account machine limitations including the maximum leaf speed, the startup characteristics of the linear accelerator, and any mechanical restrictions necessary to avoid collisions between opposing leaves.

The main difference between intensity modulation for compensation and intensity modulation for enhanced conformation comes from the method used to calculate the so-called *fluence map*. For conformal radiotherapy it is necessarily obtained from inverse planning involving all beams simultaneously, whereas for compensation more simple algorithms can be used on a beam per beam basis (i.e. surface compensation or effective transmission required to get a uniform dose distribution in a given plane). Once the fluence map is determined, the conversion into MLC segments will be carried out using exactly the same interpreter.

CHAPTER 23

MANUAL DOSE CALCULATIONS IN PHOTON BEAMS

Ivan Rosenberg

CONTENTS

23.1 INTRODUCTION

Quantitative characterisation of clinical beams is needed in order to undertake the calculation of dose distributions within patients. This calculation is a two-step process where the starting point is the absolute dose calibration in standard conditions, followed by the derivation of the dose at other points in other conditions. Such derivation requires a deep understanding of the beam characteristics presented in Chapter 22 and makes use of a number of quantities and parameters which must be part of a self-consistent system. Many of the parameters defined and discussed in this chapter were originally intended to describe the behaviour of orthovoltage photon beams (see Section 21.6) and have been adapted or expanded to fit the properties of high-energy (megavoltage) radiation. Only the latter will be considered in what follows. The treatment given here is based on the methodology of Khan (1984).

23.2 PARAMETERS FOR DOSE CALCULATIONS

23.2.1 DEFINITIONS

Given the wide variety in the terminology and symbols used in the literature, a careful definition of the quantities used for photon beam calculations is necessary for the sake of clarity. Notations and conventions used in this chapter are given in Table 23.1 and related ones in Table 23.2. All quantities discussed below are intrinsically a function of beam quality (energy spectrum).

23.2.1.1 Percent Depth Dose

This is one of the basic measured quantities from which many of the other parameters are derived (Burns 1983a). Referring to Figure 23.1 it is defined as

$$\mathrm{PDD}(d, A_s, \mathrm{SSD}) = 100 D(d, A_s, \mathrm{SSD})/D(d_{max}, A_s, \mathrm{SSD}) \tag{23.1}$$

TABLE 23.1

Notations and Conventions Used for the Definition of the Dosimetric Quantities Described in This Chapter

d	Depth in tissue from surface, measured parallel to beam axis
d_{max}	Depth of dose maximum (*peak*)
d_{ref}	Depth of dose reference point
x	Off-axis distance from collimator rotation axis (CAX)
SAD	Source to isocentre distance (axis of gantry rotation)
SSD	Source to surface distance
A	Field size at isocentre (set field)
A_s	Field size at surface, if not the same as above
A_m	Field size at d_{max}, if not the same as above
A_d	Field size at point of calculation, if not the same as above
A_{ref}	Reference field size for calibration (usually 10 cm × 10 cm)
ESQ	Equivalent square field at point of calculation
W	Wedge filter present
D_0	Calibration Dose/MU at the reference point for the reference field size (usually 1 Gy/100 MU except for isotope teletherapy machines)

The concept of percentage depth dose (PDD) was the obvious choice for early orthovoltage clinical beams, where the distance from the source to the skin was fixed by the use of an applicator. It is most suited for measurements using a computer controlled water phantom, as the phantom geometry is stationary and the detector moves (see Chapter 19). The field size

TABLE 23.2

Summary of the Related Dosimetric Quantities Defined and Used throughout the Present Chapter

PDD(d, A_s, SSD)	Percent Depth Dose at Depth d in tissue for field size A_s at surface at SSD
TAR(d, A_d)	Tissue Air Ratio at depth d for field size A_d at calculation point
TMR(d, A_d)	Tissue Maximum Ratio at depth d for field size A_d at calculation point
TPR(d, A_d)	Tissue Phantom Ratio at depth d for field size A_d at calculation point
PSF(A)	Peak Scatter Factor for field size A, defined in tissue at d_{max}
NPSF(A)	Normalised Peak Scatter Factor, defined in tissue at d_{max}
$S_p(d_{ref}, ESQ)$	Phantom Scatter Factor, for given equivalent square (ESQ) at d_{ref}, defined in tissue at d_{ref}
$S_c(A)$	Collimator Scatter Factor for set field A, also called Air Output Factor
FOF(d_{ref}, A, ESQ)	Field Output Factor, for set field A and equivalent square (ESQ) at d_{ref}, defined in tissue at d_{ref}
OAR(x, d, A_d)	Off-Axis Ratio at Off-axis distance x and depth d in tissue for field size A_d at calculation point
WF(d_{ref}, A_d)	Wedge Factor for field size A_d: ratio of output with and without wedge, measured in tissue at d_{ref}
D(air, A, SAD)	Absorbed Dose to water (per MU[a]) on the CAX in air at a distance SAD from the source, for field size A at isocentre
$D(d_{ref}, A_d, SSD)$	Absorbed Dose (per MU[a]) on the CAX at depth d_{ref} in tissue with surface at SSD for field size A_d at calculation point
$D(d_{max}, A_d, SSD)$	Absorbed Dose (per MU[a]) on the CAX at depth d_{max} in tissue with surface at SSD for field size A_d at calculation point ($=D_{max}$)
$D(d, A_s, SSD, W)$	Absorbed Dose (per MU[a]) on the CAX at depth d in tissue with surface at SSD for field size A_s at surface with (possibly) a wedge filter present
$D(d, A_d, SSD, W)$	Absorbed Dose (per MU[a]) on the CAX at depth d in tissue with surface at SSD for field size A_d at calculation point with (possibly) a wedge filter present
$D(x, d, A_d, SSD, W)$	Absorbed Dose (per MU[a]) at a point x cm away from the CAX at depth d in tissue with surface at SSD for field size A_d at calculation point with (possibly) a wedge filter present

[a] For definition of monitor units (MU) see Section 20.1.2.

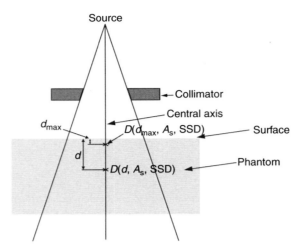

FIGURE 23.1
Percent depth dose is 100 $D(d, A_s, SSD)/D(d_{max}, A_s, SSD)$, where d is any depth and d_{max} is the reference depth of maximum dose.

parameter A_s refers to the value at the surface of the phantom (or, alternatively, at the depth of dose maximum), not at the point of measurement. Also, as the PDD is a ratio of doses at two different points in space, its value is dependent on the source surface distance (SSD) at measurement. We will discuss below how to convert PDD from one SSD set-up to another.

23.2.1.2 Tissue Air Ratio

With reference to Figure 23.2, tissue air ratio (TAR) is defined as the ratio of dose in water measured at depth on the beam central axis in a large phantom, to the dose to water measured in air at the same point (i.e. at same distance from the source):

$$TAR(d, A_d) = D(d, A_d, SSD)/D(air, A_d, SAD) \qquad (23.2)$$

where in this case, $SSD = SAD - d$, to emphasise that the doses refer to the same point in space.

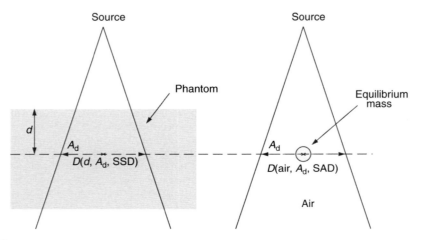

FIGURE 23.2
Illustration of the Tissue Air Ratio as defined in Equation 23.2.

In contrast to PDD, the field size parameter A_d always refers to the value at the point of measurement or calculation.

The original definition of TAR was given by ICRU 23 (1973) as, "the ratio of the absorbed dose at a given point in a phantom to the absorbed dose which would be measured at the same point in free air within a volume of the phantom material just large enough to provide *the maximum electronic build-up* at the point of reference". In ICRU 24 (1976) the words in italic were replaced with *electronic equilibrium*.

Burns (1983b) and others pointed out several difficulties with this definition; mainly, the fact that, for high energy beams, the build-up volume becomes so large that the denominator $D(air, A_d, SAD)$ is not free of scattered contribution. Additional problems arise when we consider the TAR for field sizes smaller than the minimum diameter of the build-up material, and with the definition of peak scatter factor (PSF) (see below). An alternative definition replaces the denominator in Equation 23.2 with *absorbed dose at d_{max} due to primary radiation only*. This definition makes the task of measuring TAR directly even more difficult than the original version. Measurements in air using ionisation chambers with a build-up cap are fraught with difficulties, while separating primary from scattered radiation is a task best left to Monte Carlo simulations. In spite of this, the TAR concept is still useful to relate dose in air, which is assumed to follow the *inverse square law* (see Section 23.2.1.16), to dose in tissue. For this reason, Burns (1983b) proposed an empirical, albeit somewhat circular, definition for TAR in terms of more easily measurable quantities:

$$TAR(d, A_d) = TMR(d, A_d) \ PSF(A_d) \qquad (23.3)$$

where the other quantities are defined below. In practice, TAR values are frequently derived from measured PDD curves, as discussed in the next section.

23.2.1.3 Peak Scatter Factor

The definition of peak scatter factor (PSF) was adapted from the original definition of back scatter factor (BSF) that was defined at the surface for orthovoltage beams (see Section 21.6.1). The most recent definition refers to the depth of dose maximum (*peak*):

$$PSF = \frac{\text{absorbed dose in tissue at depth of dose maximum}}{\text{absorbed dose at same point due to primary radiation only}} \qquad (23.4)$$

Thus, PSF is a special case of TAR, i.e. $PSF(A_d) = TAR(d_{max}, A_d)$, and as such suffers from the same difficulties in measurement. Values for high-energy beams have been published, based on Monte Carlo calculations (Li 1999). In practice, for high-energy beams, it is the ratio of two PSF's that is utilised. This is called the *normalised peak scatter factor* (NPSF) as discussed in Section 23.2.1.11. Note that, by definition, $PSF(0) = 1$.

23.2.1.4 Zero Field Tissue Air Ratio (TAR0)

An ancillary definition is the zero-field size TAR: $TAR0(d) = TAR(d, 0)$, a quantity that represents the attenuation of the primary beam in tissue with no scatter contribution. Its behaviour beyond the depth of maximum dose can be approximated by a simple exponential attenuation curve: $\exp[-\mu(d - d_{max})]$ where μ is the effective linear attenuation coefficient for the beam in the given phantom. The value of μ is energy dependent and requires some averaging for a polyenergetic beam. It will also change for fan lines at an angle from the central axis, because of the differential filtration of the flattening filter. This quantity can be measured using a column of water or solid phantom in a narrow beam geometry similar to HVL measurements for orthovoltage beams (Karlsson et al. 1993; Zefkili et al. 1994). More often, it is derived from extrapolation of the TAR values for finite field sizes. TAR0 is used mainly to separate the primary and scatter components of the radiation beam. Note that, by definition, $TAR0(d_{max}) = 1$.

23.2.1.5 Scatter-Air Ratio

Cunningham (1972) introduced the scatter-air ratio (SAR), defined as "that part of the TAR that is due to radiation scattered solely within the phantom" (ICRU 1974) and calculated from the following expression:

$$SAR(d, A_d) = TAR(d, A_d) - TAR0(d) \qquad (23.5)$$

The SAR can be considered as independent of the distance to the source. We will encounter this quantity again when discussing scatter summation for irregular fields (see Section 23.2.1.8).

23.2.1.6 Tissue Phantom Ratio TPR (Tissue Maximum Ratio TMR)

Because of the above mentioned difficulties with in air measurements for high energy beams, a new quantity was introduced (Karzmark et al. 1965; Khan 1984) that relates two doses, both defined in phantom. The general form of this quantity, TPR, is defined as the ratio of the dose at a given point on the beam central axis in phantom to the dose at the same point at a fixed reference depth d_{ref}, all other machine parameters being constant, i.e.:

$$TPR(d, A_d) = \frac{D(d, A_d, SSD)}{D(d_{ref}, A_d, SSD + d - d_{ref})} \qquad (23.6)$$

When d_{ref} is chosen as d_{max}, the TPR becomes the TMR (Figure 23.3).

These quantities are measured in water or water-substitute solid phantoms by keeping the detector at a constant distance from the source and varying the overlying depth of material. As small a detector as possible (such as a flat chamber with small gap) should be used when measuring close to d_{max}, to avoid errors due to the displacement of water by the chamber cavity. As TPR is a ratio of two doses at the same point for the same field size, it is practically independent of the distance of the point in question from the source (as is true for TAR). Note that, by definition, both $TPR(d_{ref}, A_d)$ and $TMR(d_{max}, A_d)$ are equal to 1.

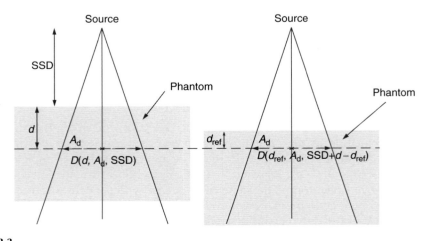

FIGURE 23.3
Diagram illustrating the definitions of tissue phantom ratio (TPR) and tissue maximum ratio (TMR). TPR is defined as in Equation 23.6. If d_{ref} is the depth of maximum dose then the TPR becomes the tissue maximum ratio (TMR).

23.2.1.7 *Zero Field Tissue Maximum Ratio (TMR0)*

An ancillary definition is the zero-field size TMR: $\mathrm{TMR0}(d) = \mathrm{TMR}(d, 0)$, a quantity that is analogous (actually, numerically identical) to the zero-field size TAR discussed in Section 23.2.1.4. The definition and derivation of scatter maximum ratio (SMR) (see Section 23.2.2.5) will have to wait until some necessary quantities are defined below.

23.2.1.8 *Equivalent Square*

The equivalent square (ESQ) concept is invaluable in clinical dose calculations for photon beams (Day and Aird 1983). As it is not practical to measure every single possible field size and shape, interpolations need to be made between measured (tabulated) values of quantities that are a function of scattered dose as well as of the primary radiation. Most of these quantities (TAR, TPR, PDD, NPSF, etc.) are measured for a series of square fields on accelerators with orthogonal collimators. Thus other field shapes need to be expressed in terms of an equivalent square for interpolation purposes. The definition of ESQ is the side of the square field that has the same SAR at the calculation point as the field in question. For example, circular fields of radius r have the same SAR as squares of the same area, i.e. $\mathrm{ESQ} = r\sqrt{\pi}$. There are a number of ways to calculate ESQ for more irregular fields:

- *Table look up* (for rectangular fields). Tables of ESQ for rectangular fields can be found in BJR Supplement 25 (Day and Aird 1983). These tables were originally obtained by application of Clarkson's method (Clarkson 1941) for a particular energy, but are accurate for all clinically used beam qualities.

- *Sterling's formula* (Sterling et al. 1964). A reasonably accurate and simple method relates the ESQ to the ratio of the area to the perimeter of the field:

$$\mathrm{ESQ} = \frac{4\mathrm{Area}}{\mathrm{Perimeter}} \tag{23.7}$$

This expression is mostly used for rectangular fields where it then becomes

$$\mathrm{ESQ} = \frac{2WL}{(W + L)} \tag{23.8}$$

for a rectangle of width W and length L. This gives reasonably good agreement with BJR Supplement 25. A useful discussion of this formula and others is given by McDermott (1998).

- *Summation of sub rectangles*. For reasonably regular field shapes, such as in Figure 23.4, the ESQ can be obtained as the average of rectangles centred on the calculation point. This approach was first proposed by Day (1950), and it is also useful in calculating the scatter contribution to points not in the centre of the field (see also Figure 23.15).

- *Approximate equivalent rectangle*. For many odd-shaped fields, a useful rule of thumb consists in determining the area of the unshielded field, choosing one of the original sides of the set field (depending on the position of the blocks) and then calculating the other side of an equivalent rectangle to give the same area. The ESQ can then be derived from this equivalent rectangle.

- *Clarkson's Scatter Integration*. Clarkson's scatter integration (Clarkson 1941) is the most accurate, albeit the most labour-intensive method (unless a computer programme is used). The open part of any irregularly shaped field can be subdivided into N equal narrow sectors of arc, as shown in Figure 23.5. Each sector can be thought of as a sector of a circle, and the SAR for each radius can be summed together:

$$\mathrm{SAR(ESQ)} = \frac{\sum \mathrm{SAR}(r_i)}{N} \tag{23.9}$$

ESQ is then determined by interpolation of tabulated SAR for squares.

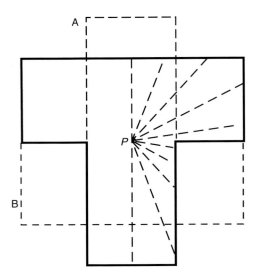

FIGURE 23.4
Calculation of the equivalent square for point *P* both by Clarkson's integration method illustrated also in Figure 23.5, and by summation of sub-rectangles. The scatter contribution from the T shape field at point P is half the contribution of rectangle A summed with half the contribution of rectangle B. (From Khan, F. M., *The Physics of Radiation Therapy*, Williams and Wilkins, Baltimore, 1984.)

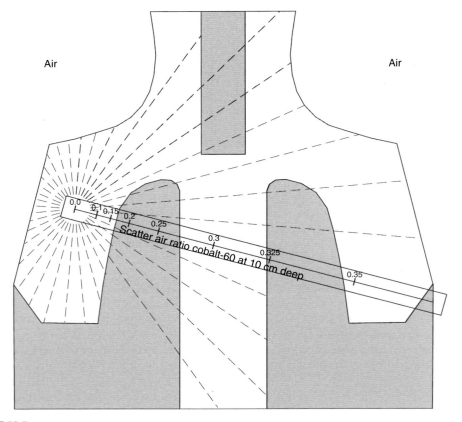

FIGURE 23.5
Clarkson's scatter integration over a highly irregular field. The cursor is scaled to the SAR at 10 cm deep for a cobalt-60 beam—different depths can be used as appropriate. The mean value of the SAR is then taken and used to look up the equivalent square field that would give this value.

23.2.1.9 Field Output Factor

The field output factor (FOF) is defined as the ratio of the output at the reference depth d_{ref} in a full scatter phantom for a given field size to that for the reference field geometry. FOF can be measured directly by placing a detector at d_{ref} in a phantom at standard geometry and varying the field size (see Section 19.2.5). Because of electron contamination (as discussed in Section 22.3.1), in order to obtain the FOF at d_{max}, it is preferable to make the measurements at a depth d_{ref} larger than d_{max} such as 5 cm or 10 cm (depending on the energy of the beam) and then to relate the measured outputs to d_{max} through the use of PDD or TMR (see Section 23.2.1.1 and Section 23.2.1.6).

For open square and rectangular fields at the standard distance, FOF is the only quantity that would be needed for MU calculations. The value of FOF for those fields not originally measured can be derived by interpolation using the equivalent square concept (Day and Aird 1983). For rectangular fields, care must be taken to distinguish between the upper and lower jaws when interpolating, because of the collimator exchange effect (see Section 23.2.1.13). However, when beams are modified by blocks or MLC, or for non-standard geometries, the direct use of FOF becomes inadequate and can lead to large errors. This happens because the effects of the machine head geometry are then decoupled from the scatter conditions within the phantom. To address these situations, it is commonly assumed that the FOF can be described as the product of two independent factors, the collimator scatter factor and the phantom scatter factor (Figure 23.6) (Khan 1984) (defined in Section 23.2.1.12 and Section 23.2.1.10):

$$\text{FOF}(d_{ref}, A) = S_c(A)\, S_p(d_{ref}, A_d) = S_c(A)\, S_p(d_{ref}, \text{ESQ}) \qquad (23.10)$$

Parameter ESQ is used in the above equation as distinct to A, to emphasise that if the beam is modified in any way, such as extended distance or blocks, two different values of equivalent square need to be considered, one for the head geometry and one for the phantom scatter (Kim et al. 1998b).

23.2.1.10 Phantom Scatter Factor S_p

The phantom scatter factor S_p essentially quantifies the variation of scatter contribution with field size to the dose in tissue at the reference depth, normalised to the reference field. Thus, it can be measured directly by varying the effective field size at the phantom, while keeping all other machine settings constant (Figure 23.7) (Khan 1984). Alternatively, it can be derived from the ratio of measurements of output factors in a full scatter phantom and *in air* (see Section 23.2.1.12). S_p is mostly a function of the beam quality (energy spectrum) and thus

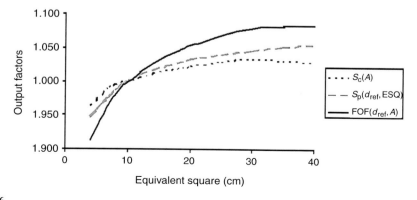

FIGURE 23.6
Typical representation of field output factor (FOF) as a function of the equivalent square field ESQ. FOF can be considered as the product of the collimator scatter factor S_c and the phantom scatter factor S_p.

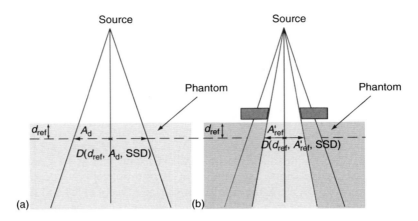

FIGURE 23.7
Diagrams to illustrate definition of S_p. (a) Dose in phantom at reference depth for a given field. (b) Dose at the same collimator opening. Note that A' has been used on the right hand side (b) to indicate that the field size is reduced by blocking to the reference field size rather than using the collimator.

has little or no dependence on machine head geometry. It is also a function of depth and of the effective field size at the point of calculation (ESQ), including the effect of SSD and blocks, and not only of the position of the collimators. Values of phantom scatter factors at 10 cm deep for high energy beams as a function of beam quality have been published (Storchi and Van Gasteren 1996). Note that $S_p(d_{ref}, A_{ref}) = 1$ by definition. The relationship of S_p to TAR will be discussed in Section 23.2.2.6.

23.2.1.11   Normalised Peak Scatter Factor

The normalised peak scatter factor (NPSF) is defined as the ratio of the PSF for the given field size to that for the reference field size (usually a 10 cm × 10 cm square field).

$$\text{NPSF}(A) = \text{PSF}(A)/\text{PSF}(A_{ref}) \qquad (23.11)$$

This quantity is coincident with S_p when the depth of reference is chosen at d_{max}, as is the case for PDD or TMR based calculations. NPSF is also mostly a function of the beam quality (energy spectrum) and independent of particular accelerator design.

Note that, while by definition $\text{PSF}(0) = 1$, it is $\text{NPSF}(A_{ref})$ that equals unity by definition, and that $\text{NPSF}(0)$ is always less than unity. Values for NPSF, averaged from the literature, are given in BJR Supplement 25 for several beam qualities. Values for PSF can be derived from these published values by extrapolating the NPSF to zero field size and applying the above definitions.

23.2.1.12   Collimator or Head Scatter Factor S_c

The collimator scatter factor (also known as head scatter factor or in air output factor), is defined as the ratio of the output in air for a given field size to that for the reference field, i.e.:

$$S_c = \frac{D(\text{air}, A)}{D(\text{air}, A_{ref})} \qquad (23.12)$$

Thus, it could in principle be measured directly using an ion chamber in air with an adequate build-up cap. However, this method suffers from the difficulties mentioned above for TAR measurements, in as much as it fails for field sizes less than the build-up cap diameter. This minimum diameter, on the other hand, is not always sufficient to prevent electron

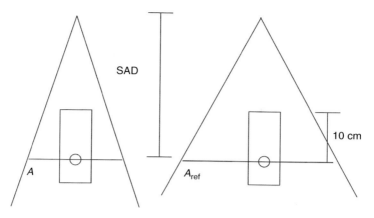

FIGURE 23.8

Direct measurement of the collimator scatter factor S_c using a mini-phantom. (Based on Dutreix, A., Bjärngard, B. E., Bridier, A., Minjnheer, B., Shaw, J. E., and Svensson, H., *ESTRO Booklet 3*, Garant, Leuven, Belgium, 1997. With permission.)

contamination produced in the machine head from reaching the ion chamber cavity. To overcome these difficulties, a new phantom geometry has been proposed for measurements of collimator scatter factors (Figure 23.8) (Van Gasteren et al. 1991; Dutreix et al. 1997). It is based on a *mini-phantom* which consists of a solid column of water-substitute material of cross-section sufficient to achieve lateral scatter equilibrium, and of thickness sufficient to position the detector at a depth where electron contamination becomes negligible. Typical dimensions for such a mini-phantom are 4 cm×4 cm cross-section, a depth to the detector of 10 cm with 1 cm backscatter (Vadash and Bjärngard 1993). In this mini-phantom, the scatter contribution to the ion chamber is constant for field sizes larger than its cross-section. The dose at depth d in the mini-phantom is then:

$$D(A, d, \text{mini}) = D_0 \, S_c(A) \, S_p(d, \text{ESQ}_\text{m}) \qquad (23.13)$$

where A is the set field and ESQ_m is the equivalent square corresponding to the mini-phantom cross-section. Then, values of S_c can be derived from measurements of dose in the mini-phantom as a function of field size, as long as the beam fully covers it. For beam cross-sections smaller than that of the mini-phantom, $S_p(d, \text{ESQ}_\text{m})$ becomes close to $S_p(d, A)$, except for the lack of full backscatter. This method, therefore, fails for field sizes comparable or smaller than the mini-phantom cross-section.

In these cases, it is preferable to derive S_c from the ratio of FOFs in a full scatter phantom and the corresponding S_p (see Section 23.2.1.10), as long as S_p can be determined independently (i.e. by extrapolation to very small field sizes or by Monte Carlo computation).

The value of S_c is assumed to be independent of distance from the source. This is not strictly true, especially for the contribution to S_c due to radiation scattered from the head components. This assumption is also equivalent to assuming that the effective source position for the inverse square law behaviour of output in air (see below) is independent of field size. This can be seen from Equation 23.12. If the virtual source position varied with field size, this ratio could not be the same at different distances.

The origin of the variation of S_c with field size is a subject of debate (Huang et al. 1987; Dunzenli et al. 1993). The major contributions are believed to come from (a) back scatter from the field defining jaws into the monitor chambers; (b) small angle scatter from the internal faces of the jaws; and (c) the shielding of the scattered dose from the flattening filter. Ideally, each of the three components causing the variation of S_c with collimator opening should be treated separately, since they are not affected similarly by the presence of filters, such as wedges or blocks, between the machine head and the phantom. However, in most situations the assumption that the value of S_c depends only on the setting of the jaws can be considered acceptable.

23.2.1.13   Collimator Exchange Effect

The values of S_c for rectangular fields depend on which collimator (upper or lower) is used to define the long side. This is known as the collimator exchange effect, and is dependent on the design of each particular accelerator. It is assumed to originate mostly from the different amount of backscatter into the monitor chambers from upper and lower jaws defining the same width at the isocentre, since this contribution is more sensitive to the position of the jaws than the two others. Its importance is markedly accelerator-dependent. Based on this effect, a modified Sterling's formula for equivalent square (Equation 23.8) has been proposed for rectangular fields to correlate the Collimator Exchange Effect uniquely with square fields (Yu et al. 1995; Meyer et al. 1999):

$$ESQ' = \frac{(1+f)UL}{(U+fL)} \tag{23.14}$$

where the factor f is related to the distances of the upper (U) and lower (L) jaws from the monitor chambers and from the isocentre. In practice, the value of f is derived from a least square fit to the measured set of collimator scatter factors.

23.2.1.14   Wedge Transmission Factor

The wedge transmission factor (WF) (sometimes simply called wedge factor) is defined as the ratio of the outputs for a given field size, at the reference depth d_{ref} in a full scatter phantom at standard geometry, with and without the presence of a wedge filter (see also Section 22.6.1). Because of the differential attenuation through the wedge, the beam quality, and therefore the PDD or TMR for the wedged beam will be different to that of the corresponding unfiltered beam. It follows that the value of WF depends on the choice of d_{ref}. Generally, the value of WF defined as above also varies monotonically with field width. For fixed mechanical wedges the attenuation of the primary component of the radiation at the central axis is obviously constant, and independent of the position of the jaws. The fact that WF varies with field size, then, could be a consequence of the different scatter conditions within the phantom due to both beam quality changes (that is, a change in S_p) and asymmetric primary fluence attenuation, as well as to modifications to collimator scatter (S_c) due to the presence of the wedge. In other words, the definition of WF is given by:

$$D(d_{ref}, A_d, SSD, W) = WF(d_{ref}, A_d) \times D(d_{ref}, A_d, SSD, op) \tag{23.15}$$

where W indicates the wedged beam and op the open one.

By analogy with Equation 23.13:

$$D(d_{ref}, A_d, SSD, op) = D_0 \, S_c(A) \, S_p(d_{ref}, A_d) \tag{23.16}$$

where D_0 is the dose at the reference depth for the reference condition.

It is possible, in principle, to account separately for the variation in beam quality and in collimator scatter by determining values $S_{p,w}$ and $S_{c,w}$ appropriate for the wedged conditions. This could be done by the methods described above, e.g. by measuring $S_{c,w}$ in a mini-phantom and by measuring $S_{p,w}$ in a full scatter phantom using secondary blocking (or by using the ratio of TAR derived for the wedged beam quality). The remaining variation of wedged output with field size, WF', would then relate only to the asymmetric attenuation of the primary fluence and consequent scatter contribution to the central axis:

$$D(d_{\text{ref}}, A_d, \text{SSD}, W) = D_0\,\text{WF}'(d_{\text{ref}}, A_d)\,S_{c,w}(A)\,S_{p,w}(d_{\text{ref}}, A_d) \qquad (23.17)$$

The dependence of WF' on field size would be different from that of the measured WF. This analysis may seem more complex than necessary for regular wedged beams, but may be necessary for improved accuracy when dealing with asymmetric beams, where the quantities S_c and S_p could change as a function of the distance from the central axis as well as with the presence of the wedge. WF' is generally a slowly varying function of equivalent square at the point of calculation.

The situation is quite different with dynamic wedges (see Section 22.6.2). Some accelerators can produce wedge-shaped profiles by moving one independent jaw across the field while the beam is on. In this case, there is no beam hardening to worry about, and no significant expected influence on collimator scatter. The corresponding wedge transmission factor DWF is thus not expected to depend significantly on the choice of d_{ref}, and use of Equation 23.15 would be justified. However, as mentioned in Section 22.6.2, the DWF could be a strong function of the field width (the moving jaw direction), not of ESQ.

23.2.1.15   Off-Axis Ratio

The off-axis ratio (OAR) has been previously discussed in Section 22.3.3. It is measured in planes perpendicular to the beam axis, and is defined as the ratio of dose at a point away from the central axis to the dose at the point on the central axis at the same depth:

$$\text{OAR}(x, d, A_d) = D(x, d, A_d, \text{SSD}, W)/D(0, d, A_d, \text{SSD}, W) \qquad (23.18)$$

where W accounts for the possible presence of a wedge filter.

A plot of the variation of OAR with distance x from the central axis is known as a beam profile. These profiles are also easily measured using a computer-controlled scanning tank, by moving a small detector in the water. An example of open profiles at several depths is shown in Figure 23.9a, plotted against actual distance and scaled according to the PDD, as they would be measured. Because of the definition of OAR, the same profiles are shown in Figure 23.9b normalised to 100% at the central axis. It can be seen more clearly in Figure 23.9b that the shape of the profiles changes with depth. First of all, because of the beam divergence, the profiles become progressively wider with depth. In fact, the definition of field width is intimately linked with the shape of these profiles: in most institutions, the width of the 50% decrement line (the distance between the 50% OAR values at each depth) is taken as the definition of field width. The geometrical field edge (defined by the centre of the source and the collimators) and the light field edge are made to coincide with these 50% points. This definition has many advantages, not least when matching abutting fields: by matching the geometric edge, the two fields are matched at the 50% OAR, thus providing the smoothest dose distribution across the match. The fact that the 50% OAR closely follows the geometric edge is illustrated in Figure 23.9c, where the same profiles are plotted as a function of normalised off-axis distance. In this plot, points with the same ordinate lie on the same fan line (a diverging line from the centre of the source) at all depths. The profiles all overlap close to the 50% OAR, at the geometric edge, where $2x/A_d = 1$. Profiles at shallow depths show the horns produced by the flattening filter. These horns gradually decrease with depth, even along fan lines. This effect is due partly to an in-scatter/out-scatter imbalance at the edge of the beam and partly to the fact that the quality of the radiation becomes progressively softer (the attenuation coefficient increases) away from the central axis, because of the effect of the flattening filter shape.

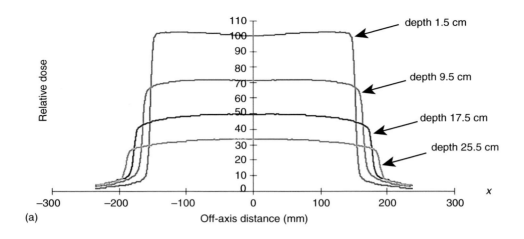

(a)

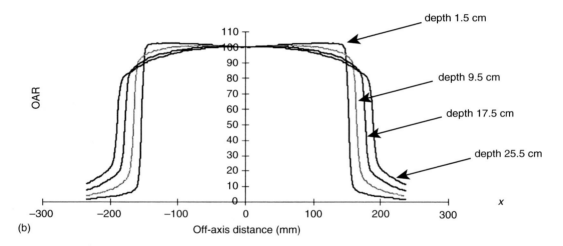

(b)

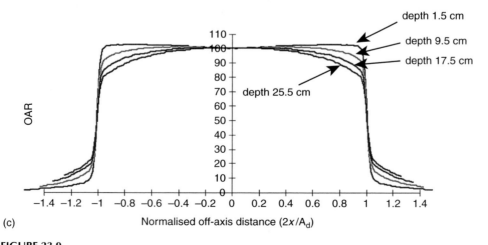

(c)

FIGURE 23.9

Plot of off-axis curves for a 30 cm × 30 cm field size at 100 cm SSD with 3 different scalings: (a) without normalization. (b) OAR with dose normalisation on the beam axis. (c) same as (b) but plotted as a function of the normalised off-axis distance $2x/A_d$.

23.2.1.16 *Inverse Square Law*

As discussed in Section 20.2.1 isotropic radiation from a point source will fall off in a vacuum according to the inverse square law:

$$\frac{D(L_1)}{D(L_2)} = \left(\frac{L_2}{L_1}\right)^2 \tag{23.19}$$

where L_1 and L_2 are distances measured from the point source. In the case of radiation from an accelerator, the source (target) is not a point, and radiation is scattered in the beam direction by the flattening filter and the primary collimators, as well as by the field defining jaws, monitor chambers and other objects. Nevertheless, the output in air follows the inverse square law reasonably well.

For more accurate calculations, an effective source position can be determined by measuring the output in air at several distances from the isocentre used as a reference to establish the distances to the *physical* target (Figure 23.10a). Once a set of dose rates $D(L)$ as a function of nominal distance to this target are obtained, the results are plotted in the

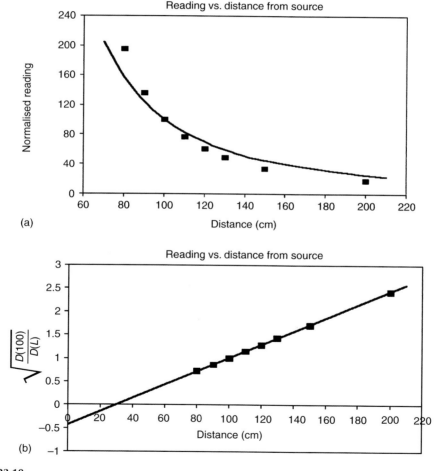

(a)

(b)

FIGURE 23.10

Determination of the effective source position by measuring and plotting the dose rate at different distances from an arbitrary point (in principle taken coincident with the *physical* source). The solid line is a fit to the inverse square law and the filled rectangles are the measured points. (a) The quantity plotted is the normalised dosimeter reading; there is a small deviation of the experimental values compared to the inverse square law. (b) The quantity plotted is the normalized inverse square root of the reading; the experimental points are aligned and the extrapolation to zero (i.e. infinite dose) yields the effective source position.

form $y = \sqrt{D(L_0)}/\sqrt{D(L)}$ vs. distance to obtain a linear regression (Figure 23.10b). The intercept of the linear fit with the abscissa axis ($y=0$) will yield the shift of the virtual source from whatever value is taken for the reference distance L_0.

The reason for this derivation can be seen by following the algebra. Call S_v the virtual source shift upstream from the *physical* source. Then:

$$D(L) = D(L_0)\frac{(L_0+S_v)^2}{(L + S_v)^2}$$

and therefore, by rearranging to get a linear relationship ($y=mx+c$):

$$\frac{\sqrt{D(L_0)}}{\sqrt{D(L)}} = \frac{1}{(L_0 + S_v)}L + \frac{S_v}{(L_0 + S_v)} = 0 \quad \text{at} \quad L = -S_v$$

A negative value of S_v (interception for $L>0$) implies that the effective source of the radiation is downstream from the assumed position of the *physical* target.

23.2.2 RELATIONSHIPS AND DERIVATIONS

Armed with the preceding definitions and assumptions, we can now proceed to discuss the relationships among the above quantities and some more derivations. These relationships are useful, for instance, to convert measured quantities into other quantities more appropriate for dose calculations. Many of the derivations are based on the inverse square assumption.

23.2.2.1 PDD from TAR

Referring to Figure 23.11, the dose in air at depth d is related to the dose in air at depth d_{max} by the inverse square law:

$$D(\text{air}, A, \text{SSD} + d) = D(\text{air}, A, \text{SSD} + d_{max}) \left(\frac{\text{SSD} + d_{max}}{\text{SSD} + d}\right)^2 \qquad (23.20)$$

The dose in tissue at depth d is then given by the application of TAR:

$$D(d, A_d, \text{SSD}) = D(\text{air}, A, \text{SSD} + d)\,\text{TAR}(d, A_d) \qquad (23.21)$$

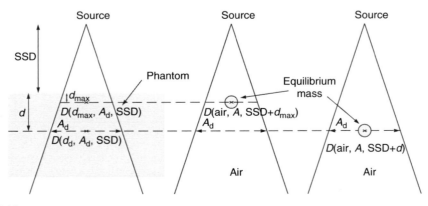

FIGURE 23.11
Relationship between PDD and TAR.

where:

$$A_{\mathrm{d}} = A\left(\frac{\mathrm{SSD}+d}{\mathrm{SAD}}\right) \tag{23.22}$$

Similarly, at $d_{\max}$, the dose in tissue is derived from the PSF:

$$D(d_{\max},\ A_{\mathrm{m}},\ \mathrm{SSD}) = D(\mathrm{air},\ A,\ \mathrm{SSD}+d_{\max})\,\mathrm{PSF}(A_{\mathrm{m}}) \tag{23.23}$$

where:

$$A_{\mathrm{m}} = A\left(\frac{\mathrm{SSD}+d_{\max}}{\mathrm{SAD}}\right) \tag{23.24}$$

Thus, the PDD is given by:

$$\mathrm{PDD}(d,\ A_{\mathrm{s}},\ \mathrm{SSD}) = 100\,\frac{D(d,\ A_{\mathrm{d}},\ \mathrm{SSD})}{D(d_{\max},\ A_{\mathrm{m}},\ \mathrm{SSD})}$$

$$= 100\,\frac{\mathrm{TAR}(d,\ A_{\mathrm{d}})}{\mathrm{PSF}(A_{\mathrm{m}})}\left(\frac{\mathrm{SSD}+d_{\max}}{\mathrm{SSD}+d}\right)^{2} \tag{23.25}$$

Usually, PDD is the measured quantity, and the inverse relationship (Burns 1983a) is used to derive TAR, given the values of PSF. These could be extracted from published PSF tables, as they vary slowly with beam quality.

23.2.2.2 TPR and TMR from TAR

Referring to Figure 23.2 and Figure 23.3, the TPR or TMR can be derived from:

$$\mathrm{TPR}(d, A_{\mathrm{d}}) = \mathrm{TAR}(d, A_{\mathrm{d}})/\mathrm{TAR}(d_{\mathrm{ref}}, A_{\mathrm{d}}) \tag{23.26}$$

and, in particular

$$\mathrm{TMR}(d, A_{\mathrm{d}}) = \mathrm{TAR}(d, A_{\mathrm{d}})/\mathrm{PSF}(A_{\mathrm{d}}) \tag{23.26a}$$

as the ratio is of doses at the same point. This last relation is the reverse of Burn's definition of TAR (Burns 1983a) (See Equation 23.3 above).

23.2.2.3 PDD from TMR

The derivation of percentage depth dose from the tissue maximum ratio follows from the above two relations (Equation 23.25 and Equation 23.26a):

$$\mathrm{PDD}(d, A_{\mathrm{s}}, \mathrm{SSD}) = 100\,\mathrm{TMR}\,(d, A_{\mathrm{d}})\frac{\mathrm{PSF}(A_{\mathrm{d}})}{\mathrm{PSF}(A_{\mathrm{m}})}\left(\frac{\mathrm{SSD}+d_{\max}}{\mathrm{SSD}+d}\right)^{2} \tag{23.27}$$

As with TAR, it is usually the inverse relation that is used to derive TMR from measured PDD. Often, the ratio of PSFs is ignored in the derivation, as it is close to unity for high-energy beams.

23.2.2.4 PDD from One SSD to Another (TAR Method)

When changing the SSD from SSD_1 to SSD_2, the PDD_1 measured at SSD_1 will be modified, because of both a difference in scatter conditions and a difference in distance relations (Figure 23.12). From the above Equation 23.25, for the same field size A_{s} at the surface (which implies different set fields at isocentre):

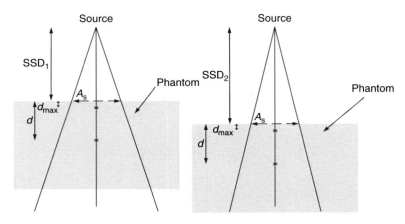

FIGURE 23.12
Relationship between PDD at one SSD and another. In both cases the field size at the surface A_s and depth d are the same.

$$PDD_1(d, A_s, SSD_1) = 100 \frac{TAR(d, A_{d1})}{PSF(A_{m1})} \left(\frac{SSD_1 + d_{max}}{SSD_1 + d} \right)^2$$

where A_{d1} and A_{m1} are defined as above (Equation 23.22 and Equation 23.24). Similarly:

$$PDD_2(d, A_s, SSD_2) = 100 \frac{TAR(d, A_{d2})}{PSF(A_{m2})} \left(\frac{SSD_2 + d_{max}}{SSD_2 + d} \right)^2$$

Thus, the relation between the two PDDs is:

$$PDD_2(d, A_s, SSD_2) = PDD_1(d, A_s, SSD_1) \frac{TAR(d, A_{d2})}{PSF(A_{m2})} \frac{PSF(A_{m1})}{TAR(d, A_{d1})}$$

$$\left(\frac{SSD_2 + d_{max}}{SSD_2 + d} \right)^2 \left(\frac{SSD_1 + d}{SSD_1 + d_{max}} \right)^2 \qquad (23.28)$$

Rearranging the above expression, and neglecting the ratio of $PSF(A_{m1})/PSF(A_{m2})$ as being very close to unity:

$$PDD_2(d, A_s, SSD_2) = PDD_1(d, A_s, SSD_1) \frac{TAR(d, A_{d2})}{TAR(d, A_{d1})}$$

$$\left\{ \left(\frac{SSD_2 + d_{max}}{SSD_1 + d_{max}} \right) \left(\frac{SSD_1 + d}{SSD_2 + d} \right) \right\}^2 \qquad (23.29)$$

The expression in the { } brackets is known as Mayneord's F factor (Mayneord and Lamerton 1944). This factor is larger than unity if SSD_2 is greater than SSD_1. This purely geometrical inverse square factor is thus modified by the ratio of the TARs for the two projected field sizes at depth. A similar relationship, derived by Burns (1983a) for the case when the field size is the same at the depth of calculation (and therefore not the same at the surface or the isocentre) can also be derived from the above full relationship (Equation 23.28). An alternative formulation using NPSFs is given by Knight and Mayles (1991).

23.2.2.5 Scatter Maximum Ratio

The scatter maximum ratio (SMR) is the analogous scatter component for TMR as SAR was for TAR (see Section 23.2.1.5). It can be used in determining ESQ instead of SAR, thus avoiding having to bring in air doses into the discussion (Johns and Cunningham 1983; Khan 1984).

Thus, SMR should be defined similarly to Equation 23.5 above:

$$SMR(d, A_d) = TMR(d, A_d) - TMR0(d)$$

However, this simple analogy is incorrect (even though it has been used in several text books). Given that $TMR(d_{max}, A_d) = TMR0(d_{max}) = 1$ (see Section 23.2.1.6), it would force $SMR(d_{max}, A_d)$ to be identically zero for all field sizes, implying that there is no scatter contribution at d_{max}. Following Kahn (1984), we can define SMR as:

$$SMR(d, A_d) = SAR(d, A_d)/TMR0(d_{max}) \tag{23.30}$$

and, rewriting Equation 23.5 in terms of TMR:

$$SAR(d, A_d) = TMR(d, A_d)\, PSF(A_d) - TMR0(d) \tag{23.31}$$

remembering that $TMR0(d) = TAR0(d)$. Further, from the definition of NPSF ($= S_p$ at d_{max}), (Equation 23.11), it can be shown that:

$$PSF(A_d) = NPSF(A_d)/NPSF(0) = S_p(d_{max}, A_d)/S_p(d_{max}, 0) \tag{23.32}$$

Combining Equation 23.30 through Equation 23.32, we finally obtain the correct relationship between TMR and SMR (Khan 1984):

$$SMR(d, A_d) = TMR(d, A_d)[S_p(d_{max}, A_d)/S_p(d_{max}, 0)] - TMR0(d) \tag{23.33}$$

23.2.2.6 Phantom Scatter Factor S_p from TAR

Consider the step-by-step derivation of dose at d_{ref} for an arbitrary beam, starting from the standard reference dose in tissue D_0 produced by reference field A_{ref} at reference depth d_{ref} at isocentre. Each step relies on the definition of the quantity used.

(i)   The dose *in air* at isocentre for the reference field A_{ref} is given by:

$$D(\text{air}, A_{ref}, SAD) = D_0/TAR(d_{ref}, A_{ref}) \tag{23.34}$$

(ii)   The dose *in air* at isocentre for any other field A is given by:

$$D(\text{air}, A, SAD) = D(\text{air}, A_{ref}, SAD)\, S_c(A) \tag{23.35}$$

(iii)   The dose in tissue at depth d_{ref} at isocentre ($SSD = SAD - d_{ref}$) for field A is given by:

$$D(d_{ref}, A_d, SSD) = D(\text{air}, A, SAD)\, TAR(d_{ref}, A_d) \tag{23.36}$$

(iv)   Substituting from Equation 23.34 and Equation 23.35 into Equation 23.36 gives:

$$D(d_{ref}, A_d, SSD) = D_0\, S_c(A)\, \frac{TAR(d_{ref}, A_d)}{TAR(d_{ref}, A_{ref})} \tag{23.37}$$

(v)   On the other hand, from the definition of FOF and its components Equation 23.10:

$$D(d_{ref}, A_d, SSD) = D_0\, FOF(d_{ref}, A_d) = D_0\, S_c(A)\, S_p(d_{ref}, A_d) \qquad (23.38)$$

Therefore, comparing Equation 23.37 and Equation 23.38, the relationship between S_p and TAR is given by:

$$S_p(d_{ref}, A_d) = TAR(d_{ref}, A_d)/TAR(d_{ref}, A_{ref}) \qquad (23.39)$$

It follows that, when d_{ref} is chosen to be at d_{max}:

$$S_p(d_{max}, A_d) = PSF(A_d)/PSF(A_{ref}) = NPSF(A_d) \qquad (23.39a)$$

23.3 MONITOR UNIT CALCULATIONS

23.3.1 GENERAL METHODOLOGY

In spite of the almost universal use of computers for carrying out calculations in radiotherapy, there is still a need for physicists to be able to do quick point dose calculations by hand, both to verify the computer plans and to perform double checks of the number of monitor units used clinically for patients. This practice is also invaluable as a teaching tool to understand the physics behind computerised systems and to acquire a sound judgement about the results of computation.

Various methods for calculating monitor units for standard and complex situations have been proposed (Khan et al. 1986; Rosenberg et al. 1995). Some of these procedures involve a large number of correction factors, often predicated on a large amount of additional measurements (e.g. with a mini-phantom (Van Gasteren et al. 1991; Dutreix et al. 1997; Mijnheer et al. 2001)), to improve accuracy in non-standard situations. While these efforts are useful, especially if incorporated in computerised planning, there is a danger that the link between the calculations and the basic beam data can be lost among the numerous correction factors. This would defeat the main purpose of performing hand calculations. These calculations should be easily traceable to measured beam data acquired during commissioning (as described in Chapter 19), or to straightforward derivations from these measured data (as described in Section 23.2.2), and should be able to deal simply with the variety of modifications to the measured beam geometry necessitated by clinical needs. This approach is followed throughout this chapter.

Let us apply the quantities discussed above to calculate the dose at a point at depth d in a phantom with a given SSD using available tabulated data. What we want to calculate in each case is the dose per monitor unit (MU), or conversely the MU for a required dose. To be able to do this, though, we must first discuss the various ways a treatment machine can be calibrated (see also Section 20.1.2).

23.3.2 CALIBRATION CONVENTIONS

The process of calibration consists in setting and maintaining (or, in the case of an isotope teletherapy machine, determining and verifying) the dose output (or dose rate) D_0 per MU at the reference point for the reference field size and geometry. Several options for the reference geometry are or have been in general use (Figure 23.13). Although it used to be common practice to measure D_0 according to any of the three setups shown in Figure 23.13, direct measurements of D_0 *in air* or at d_{max} are no longer recommended. Nevertheless, it is still useful to deal with these calibration methods since the corresponding quantities are often used in manual or computer based calculations.

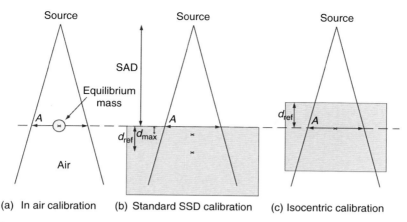

FIGURE 23.13
Possible options for calibration quantities: (a) In Air (b) SSD (c) Isocentric. Actual measurements should be perfomed with SSD or isocentric set up at a reference depth larger than d_{max} (typically 5 cm or 10 cm).

23.3.2.1   In Air Calibration

In air calibration was used for isotope teletherapy units (^{60}Co and ^{137}Cs). The reference field size was usually a 10 cm$\times$10 cm square field, and the reference point was located in air at the isocentre (usually the centre of rotation of the gantry if so mounted), at the source to axis distance SAD. The dose output D_0 was measured in air with an appropriate build-up cap. As the radiation is produced by radioactive decay, the output will decrease with time following a predictable exponential function. Output factors, namely the in air collimator scatter factor $S_c(A)$, would also be defined at the reference point, i.e. at the isocentre (Figure 23.13a).

23.3.2.2   Standard SSD Calibration at d_{max}

The definition of D_0 as the dose rate at d_{max} at a standard SSD was the ideal choice where most treatments were delivered at a fixed source to surface distance (SSD), but is still often used for historical reasons. It is well adapted to dose calculation based on the percentage depth dose. The reference field size is usually a 10 cm$\times$10 cm square field, and the reference point is located at the depth of maximum dose (d_{max}) in a phantom with a standard SSD equal to the SAD. The distance from the source to the reference point is thus equal to SAD$+d_{max}$. Output factors, namely FOF$(A)=S_c(A)S_p(d_{max}, A_m)$, would also be defined at the reference point, where S_c is the collimator scatter factor and S_p is the phantom scatter factor, which at d_{max} is equal to NPSF(A_m). The PSF is strictly defined for the field size A_m but it is commonly tabulated according to A (Figure 23.13b).

23.3.2.3   Isocentric Calibration

This method is the most convenient for institutions where most treatments are delivered isocentrically, i.e. with the centre of the target volume at the machine isocentre. It is well adapted to dose calculation based on tissue maximum ratios (TMR) or tissue phantom ratios (TPR). The reference field size is usually a 10 cm$\times$10 cm square field, and the reference point is located at the reference depth (d_{ref}) in phantom at a distance from the source equal to the SAD. A number of institutions still use $d_{ref}=d_{max}$, although larger depths such as 5 cm or 10 cm are recommended. Output factors, namely FOF$(A)=S_c(A_{ref})S_p(d_{ref}, A_{ref})$, would also be defined at the reference point, i.e. at the isocentre (Figure 23.13c).

TABLE 23.3

Relationship between the Recommended Measured Dose Rate D_{ref} and the Calibration Quantity D_0

Calibration ref. quantity D_0	Relationship with the Measured Quantity D_{ref}
In air (cobalt)	$D_0 = D_{ref}/TAR(d_{ref}, A_{ref})$
SSD at d_{max}	$D_0 = D_{ref}/PDD(d_{ref}, A_{ref}, SSD)$
Isocentric at d_{max}	$D_0 = D_{ref}/TMR(d_{ref}, A_{ref})$
Isocentric or SSD at d_{ref}	$D_0 = D_{ref}$

Note: These relationships can be used to derive the number of MU required to deliver a given dose at a specific point in defined conditions.

23.3.2.4 *Relationships with the Recommended Protocols for Absolute Dose Determination*

Although still corresponding to the local practice in a number of institutions, some of the calibration methods described above differ significantly from the recommended procedure described in Chapter 18 for absolute dose determination in reference conditions. However, this is not in contradiction provided that the measurement is actually performed within the phantom at depth d_{ref} and that the D_0 calibration reference quantity is recalculated from the relationships given previously. For clarification, the corresponding relationships are recalled in Table 23.3. This methodology is essential to ensure consistency and to guarantee that recalculation in reference conditions (using *exactly* the same quantities) will return accurately the measured dose rate. Care must be taken to ensure that distances to the measuring point and field size specifications are unambiguous and self consistent.

23.3.3 CALCULATIONS FOR RECTANGULAR FIELDS AT THE STANDARD DISTANCE

To illustrate the process of calculation based on the above approaches, we describe here systematically, for both SSD and isocentric beam set up, the possible solutions of using a given calibration method, in conjunction with each of the principal dosimetric parameters (TAR, PDD, or TMR/TPR). Although all combinations are possible, some are easier to implement and Table 23.4 illustrates the recommended choices. A step by step derivation is used, based on the previous definitions for each parameter. This is intended to illustrate the relationships between calibration conditions and dosimetric quantities. It should be noted that, in all of the following, the dose in air is the only quantity assumed to follow the inverse square law. Therefore, it needs to be included in the derivation for most cases.

23.3.3.1 *SSD Beams*

SSD beams are set up with the surface of the patient at the isocentre, i.e. SSD = SAD. For a point at depth d on the central axis we want to calculate $D(d, A_s, SSD) = D(d, A_s, SAD)$.

In Air Calibration: D_0 is here defined as the output in air at SAD.

TABLE 23.4

Dosimetric Parameters for Dose Calculation

Beam Set Up	In Air (Cobalt)	Calibration SSD (at d_{max})	Isocentric
SSD	**TAR** or PDD or TMR	TAR or **PDD** or TMR	TAR or PDD or **TMR**
Isocentric	**TAR** or PDD or TMR	TAR or PDD or **TMR**	TAR or PDD or **TMR**

Recommended choice for the use of appropriate dosimetric parameters when combining various possibilities of beam set up and calibration method. All possibilities are listed, but the recommended solutions are shown in bold type. In all cases the tissue phantom ratio (TPR) could be used advantageously instead of the tissue maximum ratio (TMR).

(a) TAR Method

From Equation 23.18:

$$D(\text{air}, A_{\text{ref}}, \text{SAD}+d) = D_0\left(\frac{\text{SAD}}{\text{SAD}+d}\right)^2$$

From the definition of S_c in Equation 23.12:

$$D(\text{air}, A, \text{SAD}+d) = D(\text{air}, A_{\text{ref}}, \text{SAD}+d)\, S_c(A)$$

From the definition of TAR in Equation 23.2:

$$D(d, A_s, \text{SAD}) = D(\text{air}, A, \text{SAD}+d)\, \text{TAR}(d, A_d)$$
$$= D_0 S_c(A)\, \text{TAR}(d, A_d)\left(\frac{\text{SAD}}{\text{SAD}+d}\right)^2 \quad (23.40)$$

where $A_d = A(\text{SSD}+d)/\text{SAD}$.

(b) PDD Method

From Equation 23.18:

$$D(\text{air}, A, \text{SAD}+d_{max}) = D_0\left(\frac{\text{SAD}}{\text{SAD}+d_{max}}\right)^2$$

From the definition of S_c in Equation 23.12:

$$D(\text{air}, A, \text{SAD}+d_{max}) = D(\text{air}, A_{\text{ref}}, \text{SAD}+d_{max})\, S_c(A)$$

From the definition of PSF in Equation 23.4:

$$D(d_{max}, A_s, \text{SSD}) = D(\text{air}, A, \text{SSD}+d_{max})\, \text{PSF}(A_m)$$

From the definition of PDD in Equation 23.1:

$$D(d, A_s, \text{SAD}) = D(d_{max}, A_s, \text{SAD})\, \text{PDD}(d, A_s, \text{SSD})$$
$$= D_0\, S_c(A)\, \text{PSF}(A_m)\, \text{PDD}(d, A_s, \text{SSD})\left(\frac{\text{SAD}}{\text{SAD}+d_{max}}\right)^2 \quad (23.41)$$

where $A_m = A(\text{SSD}+d_{max})/\text{SAD}$.

(c) TMR Method

From Equation 23.18:

$$D(\text{air}, A_{\text{ref}}, \text{SAD} + d) = D_0 \left(\frac{\text{SAD}}{\text{SAD} + d} \right)^2$$

From the definition of S_c in Equation 23.12:

$$D(\text{air}, A, \text{SAD} + d) = D(\text{air}, A_{\text{ref}}, \text{SAD} + d) \; S_c(A)$$

From the definition of PSF in Equation 23.4:

$$D(d_{\text{max}}, A_d, \text{SAD} + d - d_{\text{max}}) = D(\text{air}, A, \text{SAD} + d) \; \text{PSF}(A_d)$$

The definition of TMR (Section 23.2.1.6) leads to:

$$D(d, A_s, \text{SSD}) = D(d_{\text{max}}, A_d, \text{SAD} + d - d_{\text{max}}) \; \text{TMR}(d, A_d)$$

$$= D_0 \, S_c(A) \, \text{PSF}(A_d) \, \text{TMR}(d, A_d) \; \left(\frac{\text{SAD}}{\text{SAD} + d} \right)^2 \qquad (23.42)$$

where $A_d = A \, (\text{SSD} + d) \, / \, \text{SAD}$.

SSD Calibration at d_{max}: D_0 is here defined as the output in tissue at d_{max} at distance $\text{SAD} + d_{\text{max}}$.

(a) TAR Method

From the definition of PSF in Equation 23.4:

$$D(\text{air}, A_{\text{ref}}, \text{SAD} + d_{\text{max}}) = \frac{D_0}{\text{PSF}(A_{\text{m, ref}})}$$

From Equation 23.18:

$$D(\text{air}, A_{\text{ref}}, \text{SAD} + d) = D(\text{air}, A_{\text{ref}}, \text{SAD} + d_{\text{max}}) \left(\frac{\text{SAD} + d_{\text{max}}}{\text{SAD} + d} \right)^2$$

From the definition of S_c in Equation 23.12:

$$D(\text{air}, A, \text{SAD} + d) = D(\text{air}, A_{\text{ref}}, \text{SAD} + d) \; S_c(A)$$

From the definition of TAR in Equation 23.2:

$$D(d, A_s, \text{SSD}) = D(\text{air}, A, \text{SAD} + d) \; \text{TAR}(d, A_d)$$

$$= \frac{D_0}{\text{PSF}(A_{\text{m, ref}})} S_c(A) \; \text{TAR}(d, A_d) \left(\frac{\text{SAD} + d_{\text{max}}}{\text{SAD} + d} \right)^2 \qquad (23.43)$$

where $A_{\text{m,ref}} = A_{\text{ref}} \, (\text{SSD} + d_{\text{max}}) \, / \, \text{SAD}$

(b) PDD Method

Using the definition of field output factor (Section 23.2.1.9):

$$D(d_{\text{max}}, A_s, \text{SSD}) = D_0 \text{FOF}(A) = D_0 S_c(A) \; S_p(d_{\text{max}}, A_m)$$

as can be seen from Equation 23.10. The definition of percentage depth dose in Equation 23.1 leads to:

$$D(d, A_s, \text{SSD}) = D(d_{\max}, A_s, \text{SSD}) \ \text{PDD}(d, A_s, \text{SSD})$$

$$= D_0 \ S_c(A) \ S_p(d_{\max}, A_m) \text{PDD}(d, A_s, \text{SSD}) \tag{23.44}$$

(c) TMR Method

From the definition of PSF in Equation 23.4:

$$D(\text{air}, A_{\text{ref}}, \text{SAD} + d_{\max}) = \frac{D_0}{\text{PSF}(A_{m, \text{ref}})}$$

From Equation 23.18:

$$D(\text{air}, A_{\text{ref}}, \text{SAD} + d) = D(\text{air}, A_{\text{ref}}, \text{SAD} + d_{\max}) \left(\frac{\text{SAD} + d_{\max}}{\text{SAD} + d} \right)^2$$

From the definition of S_c in Equation 23.12:

$$D(\text{air}, A, \text{SAD} + d) = D(\text{air}, A_{\text{ref}}, \text{SAD} + d) \ S_c(A)$$

From the definition of PSF in Equation 23.4:

$$D(d_{\max}, A_d, \text{SAD} + d - d_{\max}) = D(\text{air}, A, \text{SAD} + d) \ \text{PSF}(A_d)$$

The definition of TMR (Section 23.2.1.6) leads to:

$$D(d, A_s, \text{SSD}) = D(d_{\max}, A_d, \text{SAD} + d - d_{\max}) \ \text{TMR}(d, A_d)$$

$$= \frac{D_0}{\text{PSF}(A_{m, \text{ref}})} \left(\frac{\text{SAD} + d_{\max}}{\text{SAD} + d} \right)^2 S_c(A) \ \text{PSF}(A_d) \ \text{TMR}(d, A_d)$$

$$= D_0 \ S_c(A) \ S_p(d_{\max}, A_d) \ \text{TMR}(d, A_d) \left(\frac{\text{SAD} + d_{\max}}{\text{SAD} + d} \right)^2 \tag{23.45}$$

by combining the previous equations and the definition of S_p in Section 23.2.1.10.

Isocentric Calibration at $d_{\max}$: D_0 is here defined as the output in tissue at $d_{\max}$ with the detector at a distance SAD from the source.

(a) TAR Method

From the definition of PSF in Equation 23.4:

$$D(\text{air}, A_{\text{ref}}, \text{SAD}) = \frac{D_0}{\text{PSF}(A_{\text{ref}})}$$

From Equation 23.18:

$$D(\text{air}, A_{\text{ref}}, \text{SAD} + d) = D(\text{air}, A_{\text{ref}}, \text{SAD}) \left(\frac{\text{SAD}}{\text{SAD} + d} \right)^2$$

From the definition of S_c in Equation 23.12:

$$D(\text{air}, A, \text{SAD} + d) = D(\text{air}, A_{\text{ref}}, \text{SAD} + d) \ S_c(A)$$

From the definition of TAR in Equation 23.2:

$$D(d, A_s, \text{SSD}) = D(\text{air}, A, \text{SAD} + d) \ \text{TAR}(d, A_d)$$

$$= \frac{D_0}{\text{PSF}(A_{\text{ref}})} S_{\text{c}}(A) \, \text{TAR}(d, A_{\text{d}}) \left(\frac{\text{SAD}}{\text{SAD} + d} \right)^2 \qquad (23.46)$$

by combining the previous equations.

(b) PDD Method

From the definition of PSF in Equation 23.4:

$$D(\text{air}, A_{\text{ref}}, \text{SAD}) = \frac{D_0}{\text{PSF}(A_{\text{ref}})}$$

From Equation 23.18:

$$D(\text{air}, A_{\text{ref}}, \text{SAD} + d_{\text{max}}) = D(\text{air}, A_{\text{ref}}, \text{SAD}) \left(\frac{\text{SAD}}{\text{SAD} + d_{\text{max}}} \right)^2$$

From the definition of S_{c} in Equation 23.12:

$$D(\text{air}, A_{\text{m}}, \text{SAD} + d_{\text{max}}) = D(\text{air}, A_{\text{ref}}, \text{SAD} + d_{\text{max}}) \, S_{\text{c}}(A)$$

From the definition of PSF in Equation 23.4:

$$D(d_{\text{max}}, A_{\text{m}}, \text{SSD}) = D(\text{air}, A_{\text{m}}, \text{SAD} + d_{\text{max}}) \, \text{PSF}(A_{\text{m}})$$

The definition of percentage depth dose in Equation 23.1 leads to:

$$D(d, A_s, \text{SSD}) = D(d_{\text{max}}, A_{\text{m}}, \text{SSD}) \, \text{PDD}(d, A_s, \text{SSD})$$

$$= D_0 \, S_{\text{c}}(A) \, S_{\text{p}}(d_{\text{max}}, A_{\text{m}}) \, \text{PDD}(d, A_s, \text{SSD}) \left(\frac{\text{SAD}}{\text{SAD} + d_{\text{max}}} \right)^2 \qquad (23.47)$$

by combining the previous equations and the definition of S_{p} in Section 23.2.1.10.

(c) TMR Method

From the definition of PSF in Equation 23.4:

$$D(\text{air}, A_{\text{ref}}, \text{SAD}) = \frac{D_0}{\text{PSF}(A_{\text{ref}})}$$

From Equation 23.18:

$$D(\text{air}, A_{\text{ref}}, \text{SAD} + d) = D(\text{air}, A_{\text{ref}}, \text{SAD}) \left(\frac{\text{SAD}}{\text{SAD} + d} \right)^2$$

From the definition of S_{c} in Equation 23.12:

$$D(\text{air}, A, \text{SAD} + d) = D(\text{air}, A_{\text{ref}}, \text{SAD} + d) \, S_{\text{c}}(A)$$

From the definition of PSF in Equation 23.4:

$$D(d_{\text{max}}, A_{\text{d}}, \text{SAD} + d - d_{\text{max}}) = D(\text{air}, A, \text{SAD} + d) \, \text{PSF}(A_{\text{d}})$$

The definition of TMR (Section 23.2.1.6) leads to:

$$D(d, A_s, \text{SSD}) = D(d_{\text{max}}, A_{\text{d}}, \text{SAD} + d - d_{\text{max}}) \, \text{TMR}(d, A_{\text{d}})$$

$$= \frac{D_0}{\text{PSF}(A_{\text{ref}})} \left(\frac{\text{SAD}}{\text{SAD} + d} \right)^2 S_{\text{c}}(A) \, \text{PSF}(A_{\text{d}}) \, \text{TMR}(d, A_{\text{d}})$$

$$= D_0 S_c(A)\, S_p(d_{max},\, A_d)\, \mathrm{TMR}(d,\, A_d)\left(\frac{\mathrm{SAD}}{\mathrm{SAD} + d}\right)^2 \tag{23.48}$$

by combining the previous equations and the definition of S_p in Section 23.2.1.10.

23.3.3.2 Isocentric Beams

Isocentric beams are set up with the target point of calculation at the isocentre, i.e. $\mathrm{SSD}=\mathrm{SAD}\text{-}d$ for a point at depth d on the central axis. We want to calculate $D(d,\, A_d,\, \mathrm{SSD})=D(d,\, A_d,\, \mathrm{SAD}\text{-}d)$.

In Air Calibration: D_0 is here defined as the output in air at SAD.

(a) TAR Method

From the definition of S_c in Equation 23.12:

$$D(\mathrm{air},\, A,\, \mathrm{SAD}) = D_0 S_c(A)$$

From the definition of TAR in Equation 23.2:

$$D(d,\, A_d,\, \mathrm{SSD}) = D(\mathrm{air},\, A,\, \mathrm{SAD})\, \mathrm{TAR}(d,\, A_d)$$

$$= D_0\, S_c(A)\, \mathrm{TAR}(d,\, A_d) \tag{23.49}$$

(b) PDD Method

From Equation 23.18:

$$D(\mathrm{air},\, A_{ref},\, \mathrm{SAD}-d+d_{max}) = D_0\left(\frac{\mathrm{SAD}}{\mathrm{SAD}-d+d_{max}}\right)^2$$

From the definition of S_c in Equation 23.12:

$$D(\mathrm{air},\, A,\, \mathrm{SAD}-d+d_{max}) = D(\mathrm{air},\, A_{ref},\, \mathrm{SAD}-d+d_{max})\, S_c(A)$$

From the definition of PSF in Equation 23.4:

$$D(d_{max},\, A_d,\, \mathrm{SSD}) = D(\mathrm{air},\, A,\, \mathrm{SAD}-d+d_{max})\, \mathrm{PSF}(A_m)$$

From the definition of PDD in Equation 23.1:

$$D(d,\, A_d,\, \mathrm{SSD}) = D(d_{max},\, A_d,\, \mathrm{SSD})\, \mathrm{PDD}_m(d,\, A_s,\, \mathrm{SSD})$$

$$= D_0 S_c(A)\, \mathrm{PSF}(A_m)\, \mathrm{PDD}_m(d,\, A_s,\, \mathrm{SSD})\left(\frac{\mathrm{SAD}}{\mathrm{SSD} + d_{max}}\right)^2 \tag{23.50}$$

where $\mathrm{PDD}_m(d, A_s, \mathrm{SSD})$ represents the PDD modified for the new $\mathrm{SSD}=\mathrm{SAD}\text{-}d$ according to Equation 23.29.

(c) TMR Method

From the definition of S_c in Equation 23.12:

$$D(\mathrm{air},\, A,\, \mathrm{SAD}) = D_0\, S_c(A)$$

From the definition of PSF in Equation 23.4:

$$D(d_{max}, A_d, SSD) = D(air, A, SAD)\, PSF(A)$$

The definition of TMR (Section 23.2.1.6) leads to:

$$D(d, A_d, SSD) = D(d_{max}, A_d, SSD)\, TMR(d, A_d)$$

$$= D_0\, S_c(A)\, PSF(A)\, TMR(d, A_d) \qquad (23.51)$$

SSD Calibration at d_{max}: D_0 is here defined as the output in tissue at d_{max} at distance $SAD + d_{max}$.

(a) TAR Method

From the definition of PSF in Equation 23.4:

$$D(air, A_{ref}, SAD + d_{max}) = \frac{D_0}{PSF(A_{m,\, ref})}$$

From Equation 23.18:

$$D(air, A_{ref}, SAD) = D(air, A_{ref}, SAD + d_{max}) \left(\frac{SAD + d_{max}}{SAD}\right)^2$$

From the definition of S_c in Equation 23.12:

$$D(air, A, SAD) = D(air, A_{ref}, SAD)\, S_c(A)$$

From the definition of TAR in Equation 23.2:

$$D(d, A_s, SSD) = D(air, A, SAD)\, TAR(d, A_d)$$

$$= \frac{D_0}{PSF(A_{m,\, ref})} S_c(A)\, TAR(d, A_d) \left(\frac{SAD + d_{max}}{SAD}\right)^2 \qquad (23.52)$$

(b) PDD Method

From the definition of PSF in Equation 23.4:

$$D(air, A_{ref}, SAD + d_{max}) = \frac{D_0}{PSF(A_{m,\, ref})}$$

From Equation 23.18:

$$D(air, A_{ref}, SSD + d_{max}) = D(air, A_{ref}, SAD + d_{max}) \left(\frac{SAD + d_{max}}{SSD + d_{max}}\right)^2$$

From the definition of S_c in Equation 23.12:

$$D(air, A_{ref}, SSD + d_{max}) = D(air, A_{ref}, SSD + d_{max})\, S_c(A)$$

From the definition of PSF in Equation 23.4:

$$D(d_{max}, A_d, SSD) = D(air, A_d, SSD + d_{max})\, PSF(A_m)$$

The definition of percentage depth dose in Equation 23.1 leads to:

$$D(d, A_d, SSD) = D(d_{max}, A_d, SSD)\, PDD_m(d, A_s, SSD)$$

$$= \frac{D_0}{PSF(A_{m,\, ref})} S_c(A)\, PSF(A_m)\, PDD_m(d, A_s, SSD) \left(\frac{SAD + d_{max}}{SSD + d_{max}}\right)^2$$

$$= D_0 S_c(A) \, S_p(d_{max}, A_m) \, \text{PDD}_m(d, A_s, \text{SSD}) \left(\frac{\text{SAD} + d_{max}}{\text{SSD} + d_{max}} \right)^2 \quad (23.53)$$

Where $\text{PDD}_m(d, A_s, \text{SSD})$ represents the PDD modified for the new $\text{SSD} = \text{SAD-}d$ according to Equation 23.29.

(c) TMR Method

From the definition of PSF in Equation 23.4:

$$D(\text{air}, A_{ref}, \text{SAD} + d_{max}) = \frac{D_0}{\text{PSF}(A_{m, ref})}$$

From Equation 23.18:

$$D(\text{air}, A_{ref}, \text{SAD}) = D(\text{air}, A_{ref}, \text{SAD} + d_{max}) \left(\frac{\text{SAD} + d_{max}}{\text{SAD}} \right)^2$$

From the definition of S_c in Equation 23.12:

$$D(\text{air}, A, \text{SAD}) = D(\text{air}, A_{ref}, \text{SAD}) \, S_c(A)$$

From the definition of PSF in Equation 23.4:

$$D(d_{max}, A_d, \text{SAD} - d_{max}) = D(\text{air}, A, \text{SAD}) \, \text{PSF}(A_d)$$

The definition of TMR (Section 23.2.1.6) leads to:

$$D(d, A_d, \text{SSD}) = D(d_{max}, A_d, \text{SAD} - d_{max}) \, \text{TMR}(d, A_d)$$

$$= \frac{D_0}{\text{PSF}(A_m)} \left(\frac{\text{SAD} + d_{max}}{\text{SAD}} \right)^2 S_c(A) \, \text{PSF}(A_d) \, \text{TMR}(d, A_d)$$

$$= D_0 S_c(A) \, S_p(d_{max}, A_d) \, \text{TMR}(d, A_d) \left(\frac{\text{SAD} + d_{max}}{\text{SAD}} \right)^2 \quad (23.54)$$

by combining the previous equations and the definition of S_p in Section 23.2.1.10.

Isocentric Calibration at d_{ref}: D_0 is here defined as the output in tissue at d_{ref} with the detector at a distance SAD from the source.

(a) TAR Method

From the definition of TAR in Equation 23.2:

$$D(\text{air}, A_{ref}, \text{SAD}) = \frac{D_0}{\text{TAR}(d_{ref}, A_{ref})}$$

From the definition of S_c in Equation 23.12:

$$D(\text{air}, A, \text{SAD}) = D(\text{air}, A_{ref}, \text{SAD}) \, S_c(A)$$

From the definition of TAR in Equation 23.2:

$$D(d, A_d, \text{SSD}) = D(\text{air}, A, \text{SAD}) \, \text{TAR}(d, A_d)$$

$$= \frac{D_0}{\text{TAR}(d_{ref}, A_{ref})} S_c(A) \, \text{TAR}(d, A_d)$$

$$= D_0 \, S_c(A) \, S_p(d_{ref}, A_d) \frac{TAR(d, A_d)}{TAR(d_{ref}, A_d)} \tag{23.55}$$

by combining the previous equations and Equation 23.39. This can also be written in terms of the TPR as in Section 23.2.12.6:

$$= D_0 \, S_c(A) \, S_p(d_{ref}, A_d) \, TPR(d, A_d) \tag{23.56}$$

(b) PDD Method

From the definition of TAR in Equation 23.2:

$$D(air, A_{ref}, SAD) = \frac{D_0}{TAR(d_{ref}, A_{ref})}$$

From Equation 23.18:

$$D(air, A_{ref}, SSD + d_{max}) = D(air, A_{ref}, SAD)\left(\frac{SAD}{SSD + d_{max}}\right)^2$$

From the definition of S_c in Equation 23.12:

$$D(air, A, SSD + d_{max}) = D(air, A_{ref}, SSD + d_{max}) \, S_c(A)$$

From the definition of PSF in Equation 23.4:

$$D(d_{max}, A_d, SSD) = D(air, A, SSD + d_{max}) \, PSF(A_m)$$

The definition of percentage depth dose in Equation 23.1 leads to:

$$D(d, A_d, SSD) = D(d_{max}, A_d, SSD) \, PDD_m(d, A_s, SSD)$$

$$= \frac{D_0}{TAR(d_{ref}, A_{ref})} S_c(A) \, PSF(A_m) \, PDD_m(d, A_s, SSD) \left(\frac{SAD}{SAD + d_{max}}\right)^2 \tag{23.57}$$

by combining the previous equations and the definition of S_p in Section 23.2.1.10. Here $PDD_m(d, A_s, SSD)$ represents the PDD modified for the new SSD=SAD-d as according to Equation 23.29.

(c) TPR Method

From the definition of S_c in Equation 23.12:

$$D(d_{ref}, A_d, SSD) = D_0 \, FOF(A)$$

$$= D_0 \, S_p(d_{ref}, A_d) \, S_c(A)$$

from Equation 23.9. The definition of TPR (see Section 23.2.1.6) leads to:

$$D(d, A_d, SSD) = D(d_{ref}, A_d, SSD) \, TPR(d, A_d)$$

$$= D_0 \, S_c(A) \, S_p(d_{ref}, A_d) \, TPR(d, A_d) \tag{23.58}$$

by combining the previous equations and the definition of S_p in Section 23.2.1.10. Here SSD=SAD-d.

As already indicated in Table 23.4 it is clear from the above exercise that, although any system of dosimetric calculations can be used with any calibration convention and any patient set-up, some parameters are naturally linked to particular calibration conventions and beam geometry. Thus, an institution that prefers to treat all its patients at a constant SSD would be better off to calibrate its accelerators at the standard SSD also, and to use a system based on

PDD. Conversely, where most patient treatments are given isocentrically, an isocentric calibration and a dosimetry system based on TPR/TMR would be more appropriate.

23.3.4 CALCULATIONS FOR MODIFIED BEAMS

The basic principles included in the above formulae can be transposed to cover a number of beam modifications. Some possibilities are discussed in this section.

23.3.4.1 Blocked Fields

Take, for example, an isocentric beam treated on a machine with isocentric calibration. Using the TPR method (see Section 23.3.3.2.3), the dose at the target point is given by:

$$D(d, \mathrm{ESQ}, \mathrm{SSD}) = D_0\, S_\mathrm{c}(A)\, S_\mathrm{p}(d_\mathrm{ref}, \mathrm{ESQ})\, \mathrm{TPR}(d, \mathrm{ESQ})\, \mathrm{TF} \qquad (23.59)$$

where ESQ is the equivalent square field determined at the calculation point, after taking account of the blocks. Thus, both S_p and TPR are modified by the presence of the block, while S_c is assumed to depend only on the set field. In some accelerators models, the MLC actually replace one of the sets of jaws. In this case, S_c will depend on the MLC shape and will also vary with ESQ (S_c (ESQ)). Also there is a possibility, depending on each particular design, that the presence of blocks or MLC between the machine head and the patient will modify S_c by filtering (Meyer et al. 1999) or, conversely, amplifying the radiation scattered from the jaws. This variation could be studied by measuring S_c in mini-phantom conditions and parametrising it as a function of the block shape and geometry. However, in keeping with the overall simplicity of the present approach, it is generally acceptable to ignore such variations, which often amount to less than 1% (in particular for additional blocks). If the blocks are supported in the beam by a tray, a tray attenuation factor (TF) needs to be added to the above expression. If the blocking is achieved with an MLC assembly, the TF will be unity.

23.3.4.2 Wedge Filters

The presence of a wedge filter in the beam modifies the beam profiles. It will also introduce an attenuation characterised by the wedge transmission factor (WF) at the central axis, which may be a function of field size. In addition, the wedge may harden the beam by filtering out some of the lower energies in the spectrum (Knöös and Wittgren 1991) (see Section 22.6.1). For an SSD beam treated on a machine with SSD calibration, the dose at depth d on the beam axis will be given by the expressions given in Section 23.2.1.14 and Section 23.3.3.1.2:

$$D(d, A_\mathrm{s}, \mathrm{SSD}) = D_0\, \mathrm{FOF}(A)\, \mathrm{PDD_w}(d, A_\mathrm{s}, \mathrm{SSD})\, \mathrm{WF}(A) \qquad (23.60)$$

Again, the output factors and wedge factors could be dissected, as in the *mini-phantom* methodology (Van Gasteren et al. 1991; Dutreix et al. 1997) into wedge dependent $S_\mathrm{p,w}$ and $S_\mathrm{c,w}$, *primary wedge attenuation factor*, and other factors, in order to account for the modifications in beam energy and fluence produced by the presence of the wedge. In practice, though, the variation of PDD (TPR) for wedged fields (Knöös and Wittgren 1991) is often ignored in hand calculations, and an average WF is used for all field sizes, unless the variation is more than 3%. Some accelerators can produce wedge-shaped profiles by moving one independent jaw across the field while the beam is on (*dynamic wedges*, see Section 22.6.2). In this case there is no beam hardening to worry about, and the open field PDD (TPR) can properly be used. The effective collimator scatter factor $S_\mathrm{c,\,DW}$, averaged over the many positions of the moving jaw, could well be different from the value of S_c for the starting field. This variation is usually ignored and its effects are incorporated into the overall wedge factor. On the other hand, the effective WF can be a strong and unpredictable function of the variable field dimension (in contrast to the equivalent field size), and care must be taken in interpolating for the wedge factors. (see Section 22.6.2)

23.3.4.3 Non-Standard SSD

There are a couple of cases where neither PDD nor TPR as such are sufficient by themselves, when patients are treated at an SSD different from the standard. On the one hand, larger distances from the source can sometimes be used to obtain a larger field size at the skin. On other occasions, even though the SSD is shorter than the SAD, the prescription point is not at the isocentre. In both cases, the inverse square law needs to be invoked. The dose rate at a point at an arbitrary depth d and non-standard source to skin distance SSD_2 will be calculated in two ways.

SSD Calibration, PDD System: This is analogous to Example given in Section 23.3.3.1.2(b) (SSD set up, SSD calibration, PDD method), if SSD is set to SSD_2 instead of SAD-d. Thus:

$$D\,(d, A, SSD_2) = D_0\, S_c(A)\, S_p(d_{max}, A_2)\, PDD_2(d, A_2, SSD_2)\left(\frac{SAD + d_{max}}{SSD_2 + d_{max}}\right)^2$$

where A_2 is the field size at the surface of the extended SSD_2: $A_2 = A\, SSD_2/SAD$ and $PDD_2(d, A_2, SSD_2)$ represents PDD modified for the new SSD (see Equation 23.29).

$$PDD_2(d, A_2, SSD_2) = PDD_1(d, A_2, SSD_1)\left[\frac{TAR(d, A_{d2})}{TAR(d, A_{d1})}\right]\left\{\left(\frac{SSD_2 + d_{max}}{SSD_1 + d_{max}}\right)\left(\frac{SSD_1 + d}{SSD_2 + d}\right)\right\}^2$$

Rearranging and simplifying, and neglecting the ratio of TAR in the square brackets:

$$D(d, A, SSD_2) = D_0\, S_c(A)\, S_p(d_{max}, A_2)\, PDD_1(d, A_2, SSD_1)\left(\frac{SSD_1 + d}{SSD_2 + d}\right)^2 \qquad (23.61)$$

Isocentric Calibration, TPR System: This is analogous to the example given in Section 23.3.3.1.3(c) (SSD set up, isocentric calibration, TMR method), if SSD is set to SSD_2 instead of SAD:

$$D(d, A, SSD_2) = D_0\, S_c(A)\, S_p(d_{ref}, A_2)\, TPR(d, A_2)\left(\frac{SAD}{SSD_2 + d}\right)^2 \qquad (23.62)$$

with the same definition of A_2 as above.

23.3.4.4 Asymmetric Fields and Off-Axis Calculation Points

Modern accelerators are equipped with independent jaws capable of producing fields which are asymmetric about the central axis of the beam (CAX, defined as the axis of rotation of the collimator assembly). While this option makes adjusting fields on patients much easier for the radiation oncologist it complicates the calculation for the physicist. The prescription point can now be in the centre of the open field (COF), rather than on the CAX, or it can still be on the CAX, but with a different scatter geometry than with symmetric fields. This innovation has stimulated a number of works discussing the dosimetry of such fields (e.g. Khan et al. 1986; Rosenberg et al. 1995). A relatively simple approach, suitable for hand calculations, makes use of the concept of effective or equivalent field (Rosenberg et al. 1995). Referring to Figure 23.14, the dose output (Gy/MU) at any point $P(x, d)$ in the open portion of an asymmetric field AS (potentially modified by a wedge W) is given by a

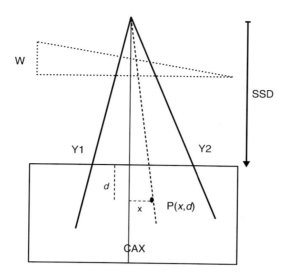

FIGURE 23.14
Calculation of the dose rate for an off-axis point *P* with an asymmetric field in presence of wedge filter. (From Rosenberg, I. et al., *Med. Phys.*, 22, 55–61, 1995. With permission.)

modification of Equation 23.38:

$$D_p(x, d, \text{SSD}, \text{AS}, \text{W}) = D_0 \, S_c(\text{eff}) \, S_p(d_{\text{ref}}, \text{eff}) \, \text{WF} \, \text{TPR}(d, \text{eff}) \, \text{OAR}(x, d, \text{W}) \, \text{G} \quad (23.63)$$

where OAR (x, d, W) denotes an OAR potentially modified by a wedge and G is the geometrical inverse square law factor consistent with the machine calibration convention. In Equation 23.63, eff, the effective field size at the calculation point for which the field size dependent factors, S_c, S_p, and TPR, are to be calculated, can be obtained (following Day 1950) from an average of four rectangular fields centered on the point (see Figure 23.15). Note that this definition of effective field size makes the values for these three parameters,

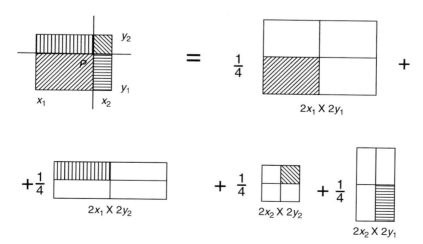

FIGURE 23.15
The Day method allows calculation of the scatter phantom factor S_p at any point *P* for a symmetric or asymmetric rectangular field by considering separately the contribution of each quadrant as being the quarter of the contribution of the corresponding centred symmetric field. (Adapted from Day, M. J., *Br. J. Radiol.*, 23, 368, 1950. With permission.)

dependent not only on the position of the individual jaws or blocks, but also on the location of the calculation point relative to the field edges.

In this simple approach to asymmetric dosimetry several factors are not specifically taken into account. The beam quality, on which S_c, S_p, and TPR all depend, will change with off-axis distance, due to differential hardening by the flattening filter. The *primary* OAR at a given off-axis distance may also be a function of the asymmetric jaw positions. The *mini-phantom* methodology could, in principle, measure the variation of those quantities. However, if parameterization were required for points at both COF and CAX for different degrees of asymmetry, the number of measurements would be unmanageable. The simple expression given above, by using OAR(x, d, W), that is, off-axis ratios measured at depth, partially accounts for off-axis beam quality variations and differential attenuation in phantom.

23.3.4.5 Corrections for Inhomogeneities

In contrast to the water phantoms in which the basic beam information is acquired, patients contain structures with different densities. The task of calculating the full effect of these inhomogeneities on the dose distribution is covered in Chapter 26 and is mostly beyond the scope of hand calculations. A simple method for estimating the approximate effect of an intervening inhomogeneity is the equivalent path length correction, where an effective depth is used to calculate the depth dependent parameters.

Referring to Figure 23.16, the effective depth d_{eff} to point P in, say, Equation 23.45, would be changed from $d = (d_1 + d_2 + d_3)$ to $d_{eff} = (d_1 + d_3) + d_2\rho_e$, where ρ_e is the electron density of the inhomogeneity relative to that of water. Thus:

$$D(d, A_s, \text{SSD}) = D_0\, S_c(A)\, S_p(d_{max}, A_d)\, \text{TMR}(d_{eff}, A_d)\left(\frac{\text{SAD} + d_{max}}{\text{SAD} + d}\right)^2 \qquad (23.64)$$

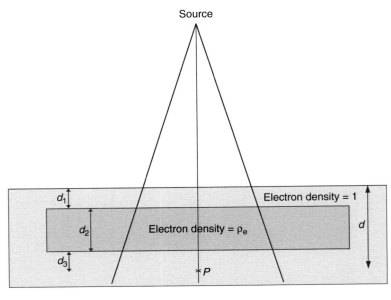

FIGURE 23.16
Schematic diagram showing a water equivalent phantom containing an inhomogeneity of electron density ρ_e relative to that of water. P is the point of dose calculation.

Geometrical factors, such as the inverse square law and field size divergence A_d, would still be based on the actual distance $SSD + d$. Thus, care needs to be taken not to use PDD calculation methods directly with this effective depth correction.

This method is reasonably accurate when the calculation point is well beyond the boundary of the inhomogeneity. At short distances from the boundary, and within the inhomogeneous volume, however, the equivalent path correction is no longer valid because of the importance of the modification of scattered radiation for the lower energies, and because of the perturbations of the secondary electron transport for the higher energies. More sophisticated computer based methods (see Section 26.2.3.2) are then required to obtain accurate results.

CHAPTER 24

ELECTRON BEAMS

David Thwaites and Alan McKenzie

CONTENTS

24.1 INTRODUCTION

The electron beams most commonly available in radiotherapy departments have energies of between 4 MeV and 25 MeV as produced by standard clinical linear accelerators, although some microtrons provide higher energies. Electron beams have advantages for a variety of clinical situations due to the characteristics of their depth-dose curves (Figure 24.1). They deliver acceptably uniform doses to a relatively well-defined region extending from the surface to the therapeutic range (taken as the distal 85% or 90%), which can be altered to fit the clinical situation by varying the beam energy, followed by a steep fall in relative dose that spares underlying structures. The general approach to electron-beam measurement and dosimetry is similar to that for megavoltage x-ray beams. However, there are a number of significant differences:

- Electrons, being charged particles, lose energy continuously as they pass through matter. Therefore, the mean energy decreases continuously with depth in the phantom.
- Electrons are light particles and undergo significant scatter. Therefore, the angular distribution of an electron beam also changes with its passage through matter. Scatter effects cause large perturbations around patient surface irregularities and inhomogeneities. Beam characteristics change when scatter equilibrium is lost. Scatter from collimators and other structures in the accelerator head has a marked effect on the detail of the dose distribution in the region around the surface and extending at least to d_{max} and also on the link between dose and monitor units(MU). As a consequence, some of the detail of beam properties can depend much more strongly on the design of the machine head than is the case for megavoltage x-ray beams.

The basic physics relating to electron interactions has been discussed previously (Chapter 3) and additional material can be found in ICRU (1984) and Klevenhagen (1985, 1993). In this chapter, the characteristics of clinical electron beams are described.

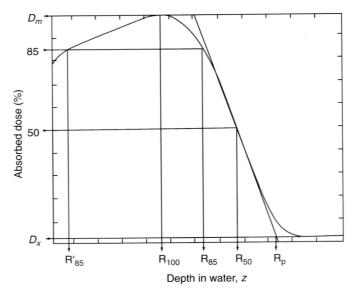

FIGURE 24.1
An electron beam depth-dose curve in water illustrating the significant characteristic parameters. Depending on local practice either the distal 90% (R_{90}) or the distal 85% (R_{85}) is taken to be the therapeutic range. R_{100} is the depth of maximum, R_{50} is the 50% depth, D_x is the bremsstrahlung tail percentage and R_p is the extrapolated range. (From ICRU, Report 35, International Commission on Radiation Units and Measurements, Bethesda, MD, 1984. With permission.)

Section 11.4.5 describes how electron beams are generated. Chapter 18 and Section 19.3 present the methods for dose measurement in electron beams considering the absolute dose calibration and the experimental determination of the relative dose distribution respectively. Chapter 27 and Chapter 28 describe electron treatment planning algorithms and Monte Carlo approaches to electron-beam modelling, whereas Chapter 34 deals with the techniques that make use of these beams in clinical situations.

24.2 DEPTH–DOSE CHARACTERISTICS OF ELECTRON BEAMS

24.2.1 SPECTRUM AT THE PATIENT SURFACE

The electron-beam spectrum at the patient or phantom surface will be reduced in energy and broadened as compared to that emerging from the vacuum window, with a more significant skewing towards lower energies (Figure 24.2), due to the cumulative effect of the interactions experienced in passing through scattering foils, monitor chamber, air, etc. It is characterised by a *most probable energy at the surface*, $E_{p,0}$, and a lower *mean energy at the surface*, $\bar{E}_0$, where the former is used to characterise the dose distribution and should approximately coincide with the energy stated by the manufacturer and the latter has generally been used in selecting appropriate correction coefficients for dose measurements. Both are determined practically by measuring depth–dose characteristics.

As the beam travels through the patient or phantom, all of the electrons undergo scattering and energy-loss interactions. The spectrum gradually shifts to lower energies and continues to broaden (Figure 24.2), until the electrons have lost all their energy. It is

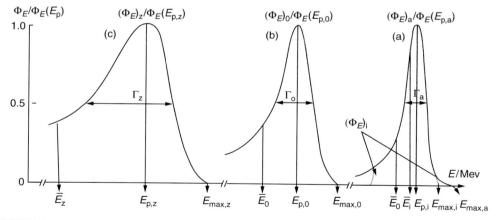

FIGURE 24.2

Illustrative representation of electron-beam spectra for 3 different locations on the beam path: (a) the *initial beam exiting the accelerator* (right-hand side: the subscript *a* represents the intrinsic accelerator beam, whereas *i* represents the initial electron beam; any difference is due to energy-defining slits); (b) the *beam incident at the surface of the phantom* (middle: represented by the subscript *0*); and (c) *at depth z* (left-hand side, represented by z). The abscissa represents nominal energies and illustrates the maximum (subscript *max*), most probable (subscript *p*) and mean energy ($\bar{E}$) for each case; the ordinate is the electron fluence differential in energy, normalised to the peak value. (From ICRU, Report 35, International Commission on Radiation Units and Measurements, Bethesda, MD, 1984. With permission.)

characterised by the mean energy at depth z, $\bar{E}_z$. As a rough estimate of this, the mean energy loss is approximately 2 MeV/cm in tissue or water.

24.2.2 GENERAL SHAPE OF DEPTH–DOSE CURVE

Figure 24.1 shows the general shape of an electron beam depth-dose curve in a reasonably large field. It illustrates the relatively uniform dose region from surface to *therapeutic range*, typically taken to be the depth of the distal 90% or 85% dose depending on local practice, and the rapidly falling part of the curve beyond that. The therapeutic range can be roughly estimated in cm of water or soft tissue as one-third of the beam energy in MeV.

The *surface dose* (D_s), conventionally stated at 0.5 mm depth, generally lies between around 75% or 80% for lower-energy beams (4 MeV to 6 MeV) and between 90% and 100% for higher-energy beams (20 MeV to 25 MeV). The rise in relative dose from the surface is due to the increasing obliquity of the electron paths as a result of scatter as the electrons penetrate the material. The electron paths are deflected through increasing mean angles from the original incident direction. This, in turn, increases the mean pathlength travelled for fixed increments of depth into the material and increases the electron fluence with depth (see Section 6.3.5). Thus, the energy deposited in successive layers of the material increases (see Figure 6.5). This continues until the beam is essentially fully diffused, i.e. the mean scattering angle does not increase further, in which case the depth-dose curve becomes flat. At increasing depths, it continues until electrons begin to be lost from the beam, in which case the depth-dose curve begins to fall. The depth of dose maximum (R_{100}) and the shape of the peak of the depth–dose curve is governed by the balance of these scatter effects with eventual loss of electrons from the beam. It therefore depends not only on beam energy, but also on the details of the accelerator head design because this determines the angular distribution of the incident electrons. The difference in relative value between D_s and 100% at R_{100} is governed by the scatter properties of the beam and is therefore larger for lower energies where scatter is more significant, i.e. D_s is lower at lower energies.

The fall-off of the curve at depth is described by the dose *gradient*, which becomes less steep as energy increases, significantly so for beams having initial energies above approximately 20 MeV, but also depending on machine design. Scanning beam machines generally produce deeper therapeutic range and steeper fall-off than scattering foil machines. A *bremsstrahlung (x-ray) tail* is observed beyond the region where primary electrons are directly depositing energy, typically being around 1% of maximum dose for lower-energy beams to around 5% for 20 MeV to 25 MeV beams. Most of this originates in the machine head, with around half that contribution from the scattering foils where they are present. The *practical range*, R_p*, representing those electrons that have travelled through the material with the minimum of scatter deviations from the original direction producing a straight or almost straight path, is obtained from the intersection of the extrapolated x-ray tail with the tangent to the steepest part of the falling curve. It can be estimated in cm of water or soft tissue as roughly half the beam energy in MeV. The *50% depth*, R_{50}, is the depth at which the relative dose falls to half the maximum value and lies approximately midway between therapeutic range and practical range.

Figure 24.1 also illustrates an alternative general rule for estimating the depth of significant doses on the central axis of electron beams. This is called the *2, 3, 4, 5 rule* and it is used to find the central-axis depth (in mm) of the 100%, 90% and 50% doses and the practical range by multiplying the incident beam energy (in MeV) by 2, 3, 4 and 5, respectively. Although it is only an approximate rule, particularly with regard to the depth of 100%

---

* Although the subscript p is always used in the symbol, the same quantity is also frequently called *extrapolated range* because it is obtained by extrapolation from the depth–dose curve.

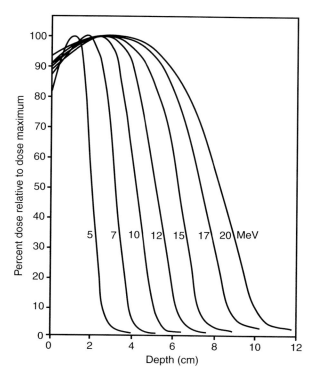

FIGURE 24.3
Typical central axis percentage depth-dose curves for beam energies from 5 MeV to 20 MeV.

dose, which breaks down above about 15 MeV, it is nevertheless a convenient, practical aid in the clinic.

24.2.3 VARIATION WITH BEAM ENERGY

Figure 24.3 shows some representative electron-beam depth–dose curves for a range of energies from 5 MeV to 20 MeV. General features of these curves are that, as energy increases:

- The relative surface dose increases.
- The depth of dose maximum[*] increases over the lower energies, but less strongly, if at all, at higher energies. This characteristic can vary with machine and does not necessarily vary in a monotonic way with energy.
- The penetration increases, as reflected in the values of therapeutic range, R_{50}, and practical range, R_p.
- The steepness of the falling part of the curve is roughly constant at the lower energies, but decreases at the higher energies.
- The x-ray tail increases.

Three energy values are used to specify an electron beam spectrum, two of them being used mostly for the purposes of dose measurements (see Section 18.5.5). They are obtained from

---

[*] The depth of maximum dose, R_{100}, is not always well defined in electron beams since, except for the lower energies, the depth–dose curve is rather flat in this region. Therefore it could be more relevant to define R_{100} as the mid-depth between distal and proximal 99% or 98%.

empirical relationships between energy (expressed in MeV) and the measured values of particular characteristics of the depth-attenuation curves in water (expressed in cm).

1. The mean energy at the surface, $\bar{E}_0$, is obtained from the 50% depth. A number of different approaches have been employed. The most common relation is:

$$\bar{E}_0 = 2.33\,R_{50} \tag{24.1}$$

   which is taken to be valid from 5 MeV to 35 MeV. Strictly, this applies to the R_{50} from depth–dose curves for monoenergetic broad parallel beams. However, it is often more convenient to use values from depth–dose curves obtained at fixed Source Skin Distance. Various dosimetry protocols give data linking $\bar{E}_0$ with R_{50} obtained in different ways (e.g. IPEMB 1996b; IAEA 1997b).

2. The most probable energy at the surface, $E_{p,0}$, is obtained from the practical range. There are a number of empirical expressions in the literature linking $E_{p,0}$ and R_p. The most widely used relation is

$$E_{p,0} = 0.22 + 1.98R_p + 0.0025R_p^2 \tag{24.2}$$

3. The mean energy at depth z, $\bar{E}_z$. A widely used relation for beams having a mean surface energy of $\bar{E}_0$ is

$$\bar{E}_z = \bar{E}_0(1 - z/R_p) \tag{24.3}$$

where z and R_p are measured for the same material. This approximation is close to more accurate values only for lower energy beams or for higher energies at depths close to the surface and close to R_p.

For any of the above, the determination of R_{50} and R_p must be carried out in a sufficiently large field, such that the values obtained are independent of field size, i.e. scatter equilibrium is obtained. The energies obtained are then the appropriate value to be used for any irradiation conditions in that beam, including smaller field sizes. Up to 15 MeV, at least $12\,\text{cm} \times 12\,\text{cm}$ fields are preferable. Larger field sizes, e.g. $20\,\text{cm} \times 20\,\text{cm}$, are appropriate at higher energies. All the energy relationships given above are only approximations to the true values due to the simplifications involved that are not applicable to the whole range of possible clinical beam spectra.

24.2.4 VARIATION WITH FIELD SIZE

At medium and large field sizes, central-axis depth doses do not vary significantly in their basic characteristics, although the exact value of surface dose and the detail of the depth–dose curve between the surface and dose maximum may vary. This is the case as long as the field is large enough to produce scatter equilibrium at the central axis. Below these dimensions, more electrons are scattered away from the central axis, reducing the fluence and the dose deposited, than are scattered inward. This imbalance results in the high-value depth doses moving progressively towards the surface as the field size reduces, with an increase in the surface dose relative to maximum dose. The maximum penetration of electrons stays the same, as it must because it essentially represents the electron range. Thus the practical range is hardly changed. Therefore, the steepness of the depth–dose curve is progressively reduced. This is illustrated in Figure 24.4 for two electron energies. This loss of side-scatter equilibrium occurs for field dimensions that are less than some critical value that is dependent on electron energy, but is also affected by machine design. In practice, it is often found to be approximately the same as R_p.

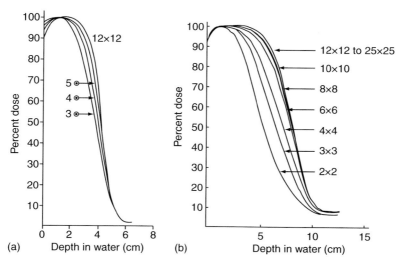

FIGURE 24.4
Typical variation of the central axis percentage depth-dose curves with field size at 100 cm Source Skin Distance for (a) a 10 MeV (3 cm, 4 cm, and 5 cm circles and 12 cm square) and (b) a 20 MeV electron beam. Square field sizes expressed in cm × cm.

24.2.5 VARIATION WITH FIELD SHAPE

Where the field shape has all dimensions greater than the scatter equilibrium value, the depth–dose characteristics will be the same as for medium to large square fields. Where any dimension is smaller than the critical value, the depth dose will be modified in a similar way to that discussed above. The exact depth dose will depend on the shape and the sizes involved. Various methods have been proposed for finding equivalent fields. For example, for rectangular fields of sides $a \times b$ it has been suggested (Hogstrom et al. 1981) that the percentage depth dose (PDD) at a given depth can be estimated from square field information, using:

$$\text{PDD}(a,b) = [\text{PDD}(a,a) \cdot \text{PDD}(b,b)]^{1/2} \qquad (24.4)$$

provided that the variation in collimator scatter is negligible. However, the concept of a single equivalent field has limited application and will vary with the specific machine and irradiation situation; therefore, any such simple models should be checked experimentally for the particular beams in use. More sophisticated models, such as pencil-beam approaches (see Section 27.2), are required to describe the real situation, particularly when the shape is irregular.

24.2.6 VARIATION WITH SOURCE SKIN DISTANCE (SSD)

In general, electron treatments are carried out at SSDs close to the nominal value. Except in special cases, shorter SSDs are not possible because of the applicator. Longer SSDs are dictated where surface shape prevents the applicator being positioned close to the surface. To deal with this situation, a virtual source may be defined as the point that is closest to the intersection of the back projections of the mean directions of electrons passing through the plane perpendicular to the central axis at the isocentre. This point will be different for different beam energies and can also vary with field size. It can be determined by taking measurements at different distances and plotting the reciprocal of the square root of the readings against distance (see Figure 23.10). The virtual source position is determined either by the distance-intercept of the resulting line (see Section 19.3.1 and Section 23.2.1.16) or by the inverse of the gradient of the line (Khan 1978; AAPM

1991). An alternative method is to plot field widths measured at various distances against distance and again examine the intercept of the resulting line. ICRU (1984) and AAPM (1991) give further discussion of these methods. The virtual source position defines the effective SSD that can be used in conjunction with an inverse square law correction for that beam energy and applicator size. In general, the effective SSD is shorter for smaller field sizes for a given energy, as the contribution of scattered electrons to the central axis from x-ray collimators and applicator is more significant. For very small field sizes, this can be quite marked. It also tends to be shorter at lower energies where scatter is more important.

After the virtual source position is determined, corrections can be made to the depth–dose curve. In general, because the penetration of electron beams is not very large, an Inverse Square Law correction to depth–dose curves is not very significant, particularly for changes in SSD of up to 10 cm, which is often the limit met in practice. Changes might be expected close to the surface and possibly extending to around the depth of therapeutic range. This is particularly marked if collimator scatter is used to significantly flatten the beam. Relative doses close to the surface decrease, but the depth of therapeutic range increases (Figure 24.5). The main problem with increased SSD is that of flatness and penumbra (see Section 24.3.5).

24.2.7 VARIATION WITH OBLIQUE INCIDENCE

Oblique incidence has an effect on the depth–dose distribution due to changes in the effective penetration of the electrons and changes in scatter. A simplistic approach to correct for this is to apply the inverse square law to calculate the change of dose due to changing distance from the source at oblique angles. This approach assumes that at a point where the SSD is increased compared to the SSD on the central axis, the dose rate is reduced in proportion to the inverse square of the ratio of the SSDs and vice versa for a reduced SSD. (ICRU 1984). However, in 1979, McKenzie divided the field into elemental beams to show that scatter effects can work in the opposite direction, i.e. near to an oblique surface that is increasing in distance from the normal FSD; the spreading of the elemental beams due to scatter can result in an increase in dose at larger distances. This increase is at shallower depths, with a lower dose beneath this at greater depths. Later work confirmed these findings

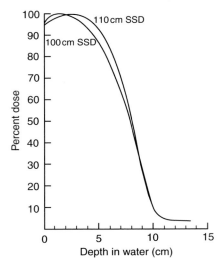

FIGURE 24.5
The effect of extended SSD on the percentage depth-dose curve for a 6 cm × 6 cm 20 MeV electron beam. Both curves are normalised to the maximum dose.

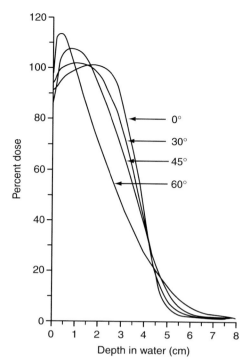

FIGURE 24.6
Variation of central axis depth doses with angle of incidence, for a 10 MeV electron beam. All the curves are normalised to the maximum dose for a normal incidence.

(Ekstrand and Dixon 1982; Khan et al. 1985). In clinical practice, this increase in dose close to the surface commonly compensates or even overcompensates for the reduction due to inverse-square effects.

Thus, as the angle between the electron-beam central axis and a perpendicular to the surface increases, the depth–dose curve pulls towards the surface in the region of the therapeutic range. The dose at the depth of maximum dose increases and becomes greater than the zero angle maximum when the angle becomes steep, due to scatter from upstream excess of tissue. Conversely, scatter from downstream missing tissue penetrates to depths larger than the practical range, increasing the residual dose. Figure 24.6 illustrates this with depth doses measured along the central axis with increasing angle of incidence.

24.3 ISODOSES

24.3.1 GENERAL CHARACTERISTICS

Typical isodose distributions are presented in Figure 24.7 for the same field size for both a low energy and a high energy electron beam, showing the isodose lines from 90% to 10% of the maximum central axis dose. These distributions illustrate the typical features. The relatively uniform dose region between the surface and the therapeutic range and the rapid fall-off with depth beyond this are evident, with only minor changes across the central part of the field. However there are significant changes towards the field edges due to scatter. At the surface, the penumbra is narrow, as the field is sharply defined by the applicator close to the surface. As the beam penetrates to increasing depths, the 50% isodose line initially approximately follows

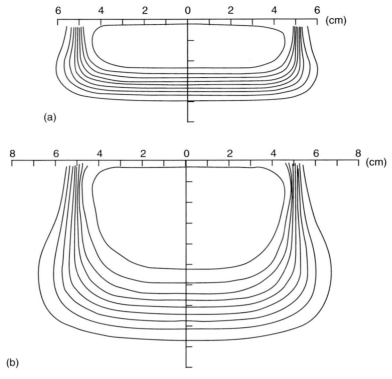

FIGURE 24.7
Typical central plane electron isodose curves in water for (a) a 7.5 MeV and (b) a 17 MeV beam for a 10 cm×10 cm applicator placed in contact with the surface. The isodose lines are ranging from 90% to 10% of the maximum central axis dose value with a 10% increment.

the geometric beam edge. This defines the field size and corresponds to the dimensions of the applicator projected downwards. Lower value isodose lines bulge significantly outwards due to scatter, producing significant dose outside the field edge. Correspondingly, the higher-value isodose lines are pulled in at depth, reflecting the loss of dose from the out-scattered electrons. This has the consequence that the high-dose volume, e.g. that encompassed by the 90% isodose, is reduced in width as depth increases and can be considerably narrower at the depth corresponding to the therapeutic range than at the surface. The width of the 90% isodose will depend on beam energy, field size and machine design, and will increase towards the diagonal positions at the field corners. The clinical consequences of these characteristics are discussed in Section 34.3.

24.3.2 ELECTRON BEAM FLATNESS AND SYMMETRY

Field flatness and symmetry are specified over some stated central region either in terms of particular beam profiles at particular depths or in planes at stated reference depths. The specified central region may be defined in terms of a proportion of the field width, e.g. within the central 80% of the beam, or in terms of fixed distances from the geometric beam edge or the 50% isodose line on the principal and diagonal axes. For electron beams, the International Electrotechnical Commission (IEC 1989) defines the *standard measurement depth* as half the depth of the distal 80% dose on the central axis; and the *base depth* as the depth of the 90% dose on the central axis. Figure 24.8 illustrates the general shapes and the position of these planes. For flatness and symmetry criteria, measurements are required at these

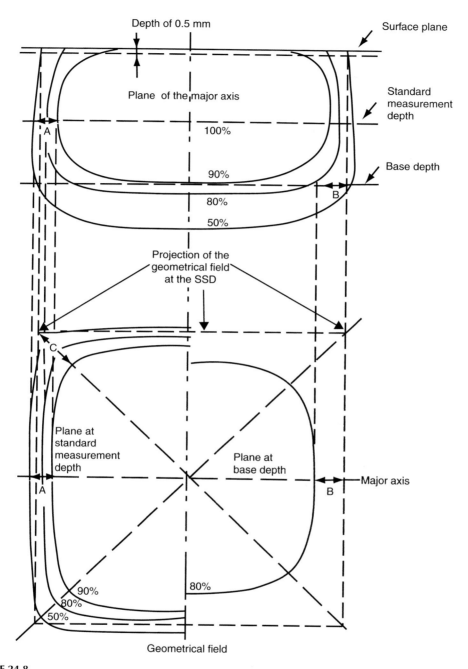

FIGURE 24.8

Schematic electron isodose distributions to illustrate the general form, the specified measurement planes and the International Electrotechnical Commission (IEC)-specified flatness-defining distances. (From IEC, *Medical Electrical Equipment—Medical Electron Accelerators in the Range 1 to 50 MeV—Functional Performance Characteristics*, IEC No 976, International Electrotechnical Commission, Geneva, 1989.)

depths and also at 0.5 mm depth and at the depth of dose maximum. The IEC specification includes:

1. The maximum distance between the 90% isodose contour and the edge of the projection of the geometric field (from the surface and parallel to the beam radiation axis) on both major axes at the standard measurement depth should not exceed 10 mm (Figure 24.8, distance *A*).

2. The maximum distance between the 80% contour and the edge of the projection of the geometric field on both major axes at the base depth should not exceed 15 mm (Figure 24.8, distance *B*).

3. The maximum distance between the 90% contour and the corner of the geometric field (diagonal axes) at the standard measurement depth should not exceed 20 mm (Figure 24.8, distance *C*).

4. The ratio of the highest absorbed dose anywhere in the beam at the standard measurement depth to the dose on the central axis at depth of dose maximum should not exceed 1.03 [*].

5. In an area 1 cm inside the 90% isodose contour at the standard measurement depth, the ratio of doses at any two points equidistant from the central axis should not exceed 1.05.

24.3.3 ELECTRON-BEAM PENUMBRA

The penumbra for electron beams is defined either in terms of the distance between two isodose values on a beam profile at the depth of maximum dose (or at the standard measurement depth), or indirectly in terms of distances between specified isodoses and the geometric field edge under stated conditions as above. If the former, then generally the 20%–80% width is expected to be 10 mm to12 mm for electron beams below 10 MeV, and 8 mm to10 mm for electron beams between 10 MeV and 20 MeV. These values apply for applicators with the final collimation stage at 5 cm or less from the skin, but for greater seperation between the applicator and the skin the penumbra will increase.

24.3.4 VARIATION OF ISODOSES WITH FIELD SIZE

As the field size reduces, the central uniform portion of the isodose distribution reduces in width, until for small field sizes the isodoses become essentially two penumbral regions and the high-dose volume is very constricted in width at depth (Figure 24.9). The central axis distribution reflects the small-field depth dose behaviour, as discussed above.

Similar constrictions occur in any regions where the dimensions are small enough, e.g. small applicators, small shaped fields and larger fields where one dimension, or part of the irradiated area, is sufficiently small (Figure 24.10).

24.3.5 VARIATION OF ISODOSES WITH SSD

The variation of depth dose with SSD was discussed above. In addition, the distribution across the beam can be altered significantly. In particular, larger gaps result in greater penumbra widths, as head-scattered electrons are allowed to diverge more before being incident on the patient or phantom surface. Hence, increasing SSD does not generally produce wider high-dose regions; instead, these may remain approximately constant or even be reduced in width. Beam flatness may also be altered by increased SSD, particularly in those linac designs in which electron scatter off the collimator system is used to improve

---

[*] It may be noted that this specification does not preclude *cold spots* within the treatment field at, or close to, the depth of dose maximum being as low as 90% of the maximum dose. IPEM have proposed an additional specification to address this issue (IPEM 1999).

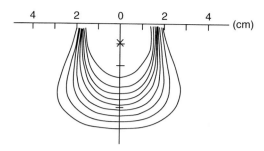

FIGURE 24.9
Effect of small beam area on electron isodose curves: a 3.5 cm circular applicator on a 10 MeV beam, at close distance from the distal end of the cone.

flatness. A sharp penumbra can be regained by using field shaping at the patient surface, but the maximum field size for acceptably uniform dose is still determined by the scatter behaviour. Thus, unlike photons, extending SSD does not usually allow larger areas to be treated uniformly.

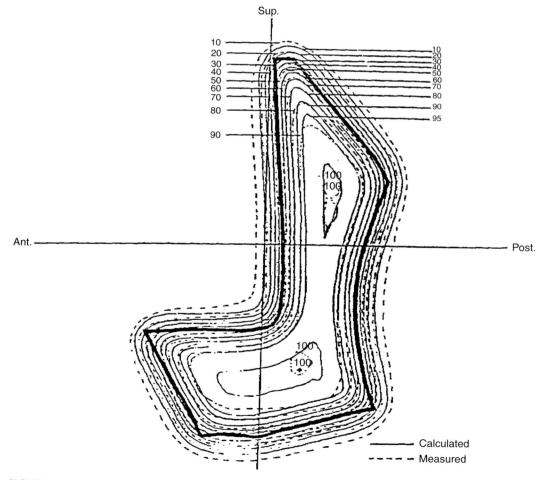

FIGURE 24.10
Typical dose distribution perpendicular to the beam axis, at depth of dose maximum, for an irregular-field 9-MeV electron beam.

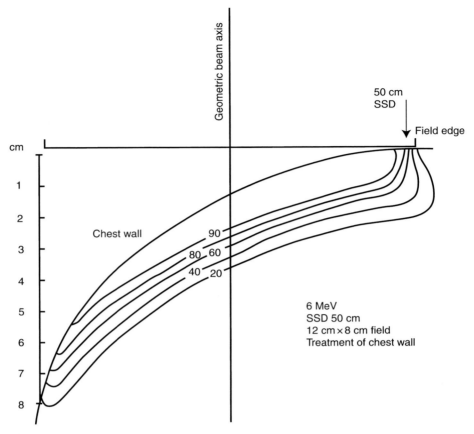

FIGURE 24.11
A 6 MeV electron beam incident on a chest wall illustrating the changing effects on isodose as the obliquity increases. This irradiation would not be ideal as the beam is not in best apposition. However, the beam direction may be dictated by adjacent treatment.

24.3.6 VARIATION OF ISODOSES WITH OBLIQUE INCIDENCE

Isodose lines generally tend to roughly follow surface shape up to incident angles of around 30°. As obliquity increases above this, the electron paths become more tangential, air gaps increase and scatter changes become more significant. The isodoses move increasingly closer to the surface and the penumbral region becomes increasingly wide (Figure 24.11). At larger angles, scatter effects, as discussed for oblique incidence depth doses, can produce increased dose hot-spots (see Section 24.6).

24.4 FIELD SHAPING/SHIELDING

24.4.1 FIELD SHAPING

Most accelerators have a set range of electron applicators (or cones) that allow limited flexibility in modifying field shape. For other sizes, or for irregular fields, shaping can be achieved relatively simply using cut-outs of lead or low-melting-point alloy attached to, or close to, the end of the applicator (cone) or on the patient surface. As a rule of thumb, the minimum thickness of lead, in mm, is about half the nominal beam energy, in MeV, e.g. 10 mm for 20 MeV. These thicknesses should be increased by a factor of 1.2 for standard low-melting-point alloy.

Generally, these give adequate shielding with less than 5% bremsstrahlung residual dose at low energies which increases to about 10% at 20 MeV. The highest doses under the shielding are on the skin. If excessively thin shielding is used, this can result in skin dose being increased as compared to the open-field values due to the production of forward-scattered electrons and bremsstrahlung. Sharp field definition can be maintained by using shielding in close proximity to the patient surface. However, some compromise may be necessary, depending on the weight, the applicator design and the standard air gap. Normal electron field edge effects occur at a shielding edge; therefore, scatter outside the shielding should be taken into consideration.

24.4.2 INTERNAL SHIELDING/BACKSCATTERED ELECTRONS

One particular specialised application is to position shielding material inside anatomical structures to spare tissues lying below the target volume, for example in treatments involving eyelids, lip, or cheek. Given space restrictions, the minimum thickness to provide the required shielding should be used for the field size and depth involved. Any shielding material will produce backscatter that can significantly increase the dose at the tissue/shield interface. This excess dose increases as the atomic number of the scattering material increases and as the mean energy at the position of the shield ($\bar{E}_s$) decreases (Figure 24.12). For the specific example of lead, Klevenhagen et al. (1982) give the empirical relationship

$$\text{EBF} = 1 + 0.735 \exp(-0.052\bar{E}_s) \qquad (24.5)$$

where the electron backscatter factor (EBF) gives the relative increase in dose at the surface of the backscattering shield compared to the dose without shield. Dose enhancement can be up to approximately 60% for $\bar{E}_s$ of 3 MeV or 4 MeV.

Lambert and Klevenhagen (1982) have reported the transmission of backscattered electrons through tissue or water substitute materials, with exponential fits to the data (Figure 24.13). Acceptable dose distributions can be achieved by using appropriate layers of such low-Z material between the shield and the overlying tissue to attenuate the backscatter. More details and a clinical example are given in Section 34.4.2.

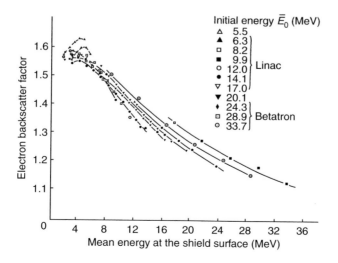

FIGURE 24.12

Variation of electron backscatter factor (EBF) with beam energy $\bar{E}_s$ at the shield surface, for lead of sufficient thickness to provide shielding of underlying tissue. Each curve is for one beam of initial energy E_0. As the depth of the shield changes, the beam energy at the shield surface changes. (From Klevenhagen, S. C. et al., *Phys. Med. Biol.*, 27, 363–373, 1982. With permission.)

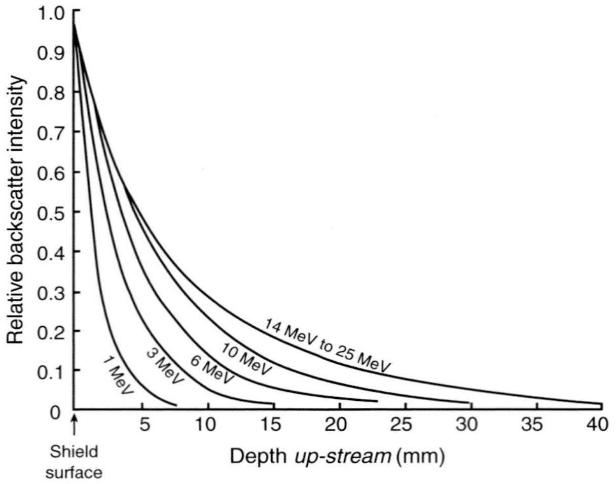

FIGURE 24.13
Penetration of electrons (backscattered from underlying lead) in polystyrene in the *up-stream* direction of the primary beam. The energies indicated on the curves are the mean energies of the primary electrons incident on the lead. As the energy increases, a thicker layer of polystyrene is required to stop most of the backscattered electrons. (From Lambert, G. D. and Klevenhagen, S. C., *Phys. Med. Biol.*, 27, 721–725, 1982.)

24.5 BOLUS/ENERGY DEGRADERS

In various clinical situations (see Section 34.4), it is necessary to add tissue-similar material at the base of the applicator and/or in contact with the patient's skin. One particular situation is the modification of the depth dose to reduce the therapeutic range to match a specific clinically required depth that is not achievable with a standard beam energy. Alternatively, for superficial target volumes, low-energy beams may give excessively low surface doses and a higher energy beam may be selected in conjunction with an energy degrader whose thickness is chosen to bring the therapeutic range back to the desired depth, but with the low dose part of the build-up occurring in the degrader such that by the time the beam is incident on the skin the patient surface dose is higher. Suitable materials include wax or commercial flexible tissue substitutes. Alternatively, sheets of plastic, e.g. polymethyl methacrylate (PMMA, known as Perspex or Lucite), can be used, provided that any air gaps are not too large. The thicknesses should be chosen taking into account the electron density differences between the material used and water or soft tissue. Lambert et al. (1999), and others, have explored the use of higher-Z materials to increase surface doses.

24.6 ELECTRON SCATTER EFFECTS

Electrons scatter very easily and strong scatter effects are observed in all electron-beam situations. The fundamental physics of the electron scatter processes has been described in Section 3.5. Scatter perturbations cause significant effects in dose measurements and scatter determines a number of the unique characteristics of electron-beam distributions, as discussed above. Scatter perturbations also produce significant effects in clinical dose distributions, some of which will be discussed in Chapter 34.

A scatter perturbation is produced when the junction of two materials of different composition is struck tangentially, or at a large oblique angle, by an electron beam. More electrons are scattered away from the higher density or higher-atomic-number material towards a lower density or lower-atomic-number material than are scattered back. This imbalance gives rise to increased electron fluence and therefore increased dose (hot-spot) scatter lobes into or under the lower-density or lower-Z medium. There are correspondingly lower fluences and lower dose areas in or under the higher-density or higher-Z region, reflecting the relative loss of the out-scattered electrons. As an extreme example, if an electron field overlaps the edge of a phantom that is parallel to the beam axis, then the distribution at the edge of the phantom will be qualitatively similar to that at the edge of a normal electron isodose distribution, with high-dose isodose lines curving inwards due to loss of electrons scattered out into the adjacent air. Such situations arise clinically where fields overlap the edges of normal or abnormal body structures (which may then require added side-scatter to achieve the required distribution), at surface irregularities, at edges of energy degraders (bolus) in the field if they are not tapered, at edges of shielding or collimation, or at internal interfaces between tissues (e.g. tissue/lung, tissue/bone, tissue/air cavity, etc.) that are lying roughly in the same direction as the beam. These can give large local changes in dose distribution. Where small inhomogeneities or irregularities are present, the effects may be reinforced by scatter perturbations from more than one interface overlapping. Similar effects occur around dosimeters, e.g. the fluence perturbation around ion chambers (see Section D.2.4 and Figure D.10).

Sophisticated methods are required to deal with these effects in detail, i.e. at least the use of pencil-beam methods (Section 27.2) and ideally Monte Carlo modeling (Section 28.4). However, simple methods can be used as a means of providing a quick estimate of the magnitude of effects. For example, Pohlit and Manegold (1976) showed experimentally that the effects are maximum at an angle α (Figure 24.14), which is approximately 60° at 5 MeV, 30° at 10 MeV and 15° at 20 MeV, where the energies are mean values at the depth of the interface.

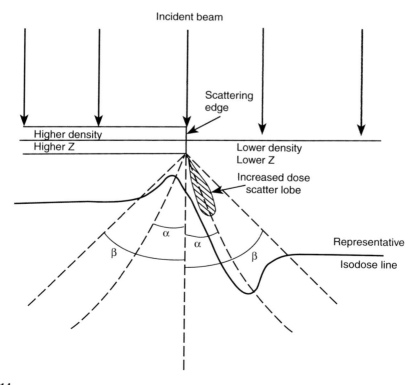

FIGURE 24.14
The general form of the scatter perturbation effect on dose distributions beneath an interface. (After Pohlit, W. and Manegold, K. H., *High-Energy Photons and Electrons*, Wiley, London, 1976. With permission.)

The effects decrease with lateral distance from the interface to become negligible at a second angle β, beyond which interface or edge effects are not significant and the distribution is only influenced by differences in absorption of the electrons passing through the different materials. β is approximately 2.2 to 2.5 times α in the energy range from 10 MeV to 30 MeV. Maximum changes in dose, whether reductions or increases as compared to the dose at the equivalent point in a homogeneous water phantom, may be estimated along either line at angle α. They are found to increase with increasing energy at the depth of the interface, for example varying from approximately 4% at 5 MeV to 8% at 10 MeV and 14% at 20 MeV for a bone/water interface. For air/water or lead/water interfaces, the figures are approximately 1.8 times these values. Thus, air cavities give rise to greater effects than similar-sized bone cavities. Figure 24.15 provides some examples of the general form and magnitude of the scatter perturbation effects in a few typical situations (Thwaites 2000).

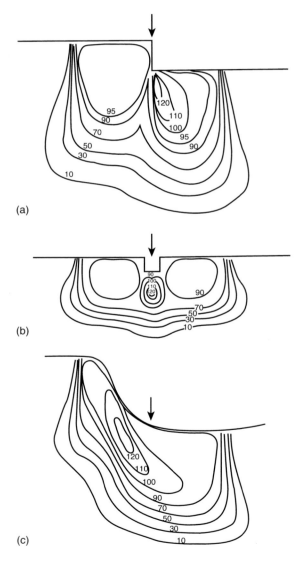

FIGURE 24.15
Illustrative examples of scatter perturbation effects on dose distributions. (a) A 20 MeV beam incident on a sharp irregularity; (b) a 10 MeV beam incident on a narrow air cavity, (c) a 17 MeV beam incident on a neck/chin contour. (From Thwaites, D. I., *Radiotherapy Physics in Practice*, 2nd Ed., Williams J. R. and Thwaites, D. I., Eds., Oxford University Press, Oxford, p. 332, 2000. With permission.)

In addition to side-scatter from interfaces parallel to, or at shallow angles to, the beam, backscatter will also occur where an interface between materials of different density and atomic number lies across an electron beam. Some information on the dependence on energy and atomic number of the material is presented above as it relates to backscatter from internal shielding (see Section 24.4.2). However, similar effects will increase doses upstream from any higher atomic number material, e.g. bone, metallic implants, etc. and the information presented in Figure 24.12 can be used to estimate the magnitude of effects, although absolute values will be lower for lower-atomic-number back-scattering materials. Similarly, backscatter is reduced upstream from an interface with lung or air.

24.7 OUTPUT FACTORS AND MONITOR UNIT CALCULATIONS

The output factors, in dose/MU, are a function of applicator size and SSD. For the nominal SSD, applicator (cone) factors relative to a standard applicator size (usually 10 cm × 10 cm) should be measured at depth of dose maximum for each field size. The depth of dose maximum will vary for smaller field sizes and the output factor is not simply a ratio of ion chamber readings in this case, but is a ratio of reading times stopping power ratio, as the stopping power ratios vary with depth (see also Section D.1.2). If the machine is an older design that uses a fixed side-scattering cone mount or a fixed x-ray collimator setting, with a range of add-on field-defining applicators, the factors usually vary relatively smoothly with field size at a given energy. For machines that have differing pre-set x-ray collimator positions for each combination of applicator and energy, the variation of dose/MU with field size may be less predictable. For variable trimmer systems, the output varies relatively smoothly with field size so that a limited set of measurements allows interpolation to intermediate settings.

The dose/MU at dose maximum on the central axis varies with the electrons scattered off the x-ray collimators and those scattered off the applicator, as well as phantom scatter. If further field shaping is used, there may be additional scatter from the edges of the shield. The relative contributions of these components may vary with: the beam energy, the depth of dose maximum (itself energy and field size dependent), the particular standard applicator involved, the x-ray collimator position and the overall size and shape of any modified field, including the distance from the central axis of shaping material. This represents a complex combination and R_{100} and output factors must be determined for a sufficiently wide range of fields and cut-outs. Output factors for cut-outs must relate to the dose maximum of the modified field; they may be expressed relative to the output of the open applicator at its dose maximum or relative to a reference output for a standard applicator. Measurements should be carried out under the conditions to be used clinically. Thus, in the event that the applicator design and the position of field shaping material require changes in the nominal treatment distance, this is included in the factor. If any field dimension is less than the range of the electrons, side scatter equilibrium at the central axis will be lost and the dose per monitor unit will be less. This effect is more pronounced for lower energies. Similar effects will also be present for fields where shielding is used to reduce the size of one dimension to within this range. However, provided that there is no change in SSD and the field shaping does not reduce any dimensions below that necessary for scatter equilibrium on the central axis, the output factor will not differ significantly from the unshaped field output factor. For shaped fields, various methods have been applied to predicting output factors, including equivalent area, scatter integration, pencil beam and Monte Carlo. Because of the various components contributing to output when cut-outs are used, the relative output factors may not be smooth or monotonic in behaviour and prediction for unmeasured fields must be carried out with care and verified by measurement.

Electron-beam treatments are most commonly carried out at the standard treatment distance defined by the applicator end, usually with a short stand-off distance. As discussed above, extending the SSD generally worsens the dose distribution; however, some clinical situations dictate longer SSD due to problems in positioning, surface shapes, etc. In these cases, the dose/MU will be reduced with increased distance. The calculation of the dose/MU for a SSD change could then be based on the inverse square law relative to the *virtual*, or *effective*, point source location, which will generally be variable with both energy and field size (see Section 24.2.6).

PART E: CLINICAL BEAMS
REFERENCES

Almond, P. R., Biggs, P. J., Coursey, B. M., Hanson, W. F., Huq, M. S., Nath, R., and Rogers, D. W., AAPM TG-51 protocol for clinical reference dosimetry of high-energy photon and electron beams, *Med. Phys.*, 26, 1847–1870, 1999.

Amdur, R. J., Kalbaugh, K. J., Ewald, L. M., Parsons, J. T., Mendenhall, W. M., Bova, F. J., and Million, R. R., Radiation therapy for skin cancer near the eye: Kilovoltage x-rays versus electrons, *Int. J. Radiat. Oncol. Biol. Phys.*, 23, 769–779, 1992.

AAPM (American Association of Physicists in Medicine), Task Group 25: Clinical electron beam dosimetry, *Med. Phys.*, 18, 73–109, 1991.

Andreo, P., On the beam quality specification of high-energy photons for radiotherapy dosimetry, *Med. Phys.*, 27, 434–440, 2000.

Arnfield, M. R., Siantar, C. H., Siebers, J., Garmon, P., Cox, L., and Mohan, R., The impact of electron transport on the accuracy of computed dose, *Med. Phys.*, 27, 1266–1274, 2000.

Aron, B. S. and Scapicchio, M., Design of a universal wedge filter system for a cobalt-60 unit, *Am. J. Roentgenol. Radium Ther. Nucl. Med.*, 96, 70–74, 1966.

Attix, F. H., Lopez, F., Owolabi, S., and Paliwal, B. R., Electron contamination in ^{60}Co gamma-ray beams, *Med. Phys.*, 10, 301–306, 1983.

Barry, L. M. and Rapley, L. F., A simple method of making the "isodose shift" obliquity correction, *Br. J. Radiol.*, 43, 910–911, 1970.

BIR (British Institute of Radiology), Central axis depth dose data for use in radiotherapy, *Br. J. Radiol.*, (Suppl. 25), 1996.

Bortfeld, T., Oelfke, U., and Nill, S., What is the optimum leaf width of a multileaf collimator? *Med. Phys.*, 27, 2494–2502, 2000.

Bradley, J., Reft, C., Goldman, S., Rubin, C., Nachman, J., Larson, R., and Hallahan, D. E., High-energy total body irradiation as preparation for bone marrow transplantation in leukemia patients: Treatment technique and related complications, *Int. J. Radiat. Oncol. Biol. Phys.*, 40, 391–396, 1998.

Burns, J. E., Conversion of percentage depth dose for photon beams from one SSD to another and calculation of TAR, TMR and TPR, *Br. J. Radiol.*, (Suppl. 17) 115–119, 1983a.

Burns, J. E., Definition of tissue-air ratio, *Br. J. Radiol.*, (Suppl. 17) 137–147, 1983b.

Cardarelli, G. A., Rao, S. N., and Cail, D., Investigation of the relative surface dose from Lipowitz-metal tissue compensators for 24- and 6-MV photon beams, *Med. Phys.*, 18, 282–287, 1991.

Castellanos, M. E. and Rosenwald, J.-C., Evaluation of the scatter field for high-energy photon beam attenuators, *Phys. Med. Biol.*, 43, 277–290, 1998.

Chaney, E. L., Cullip, T. J., and Gabriel, T. A., A Monte Carlo study of accelerator head scatter, *Med. Phys.*, 21, 1383–1390, 1994.

Chui, C. S., Mohan, R., and Fontenla, D., Dose computations for asymmetric fields defined by independent jaws, *Med. Phys.*, 15, 92–95, 1988.

Clarkson, J. R., A note on depth doses in fields of irregular shapes, *Br. J. Radiol.*, 14, 255, 1941.

Convery, D. J. and Rosenbloom, M. E., The generation of intensity-modulated fields for conformal radiotherapy by dynamic collimation, *Phys. Med. Biol.*, 37, 1359–1374, 1992.

Convery, D. J. and Rosenbloom, M. E., Treatment delivery accuracy in intensity-modulated conformal radiotherapy, *Phys. Med. Biol.*, 40, 979–999, 1995.

Cunningham, J. R., Scatter-air ratios, *Phys. Med. Biol.*, 17, 42–51, 1972.

Day, M. J., A note on the calculation of dose in X-ray fields, *Br. J. Radiol.*, 23, 368, 1950.

Day, M. J. and Aird, E. G. A., The equivalent-field method for dose determination in rectangular fields, *Br. J. Radiol.*, (Suppl. 17), 17, 105–114, 1983.

Day, M. J., Lambert, G. D., and Locks, S. M., The effect of secondary electron spread on the penumbra in high energy photon beam therapy, *Br. J. Radiol.*, 63, 278–285, 1990.

Dunzenli, C., McLean, B., and Field, C., Backscatter into the beam monitor chamber: Implications for dosimetry of asymmetric collimators, *Med. Phys.*, 20, 363–367, 1993.

Dutreix, A., Bjärngard, B. E., Bridier, A., Minjnheer, B., Shaw, J. E., and Svensson, H., Monitor unit calculation for high energy photon beams, ESTRO Booklet 3, Garant, Leuven, Belgium, 1997.

EC (European Commission), *Radiation protection 91.Criteria for acceptability of radiological (including radiotherapy) and nuclear medicine installations*. Chapter 7, ISBN 92-828-1140-9, Office for Official Publications of the European Communities, Luxembourg, 1997.

Ekstrand, K. E. and Barnes, W. H., Pitfalls in the use of high energy X rays to treat tumors in the lung, *Int. J. Radiat. Oncol. Biol. Phys.*, 18, 249–252, 1990.

Ekstrand, K. E. and Dixon, R. L., The problem of obliquely incident beams in electron-beam treatment planning, *Med. Phys.*, 9, 276–278, 1982.

Ellis, F., Hall, E. J., and Oliver, R., A compensator for variations in tissue thickness for high energy beams, *Br. J. Radiol.*, 32, 421–422, 1959.

Engelsman, M., Damen, E. M., Koken, P. W., van't Veld, A. A., van Ingen, K. M., and Mijnheer, B. J., Impact of simple tissue inhomogeneity correction algorithms on conformal radiotherapy of lung tumours, *Radiother. Oncol.*, 60, 299–309, 2001.

Evans, P. M., Hansen, V. N., Mayles, W. P., Swindell, W., Torr, M., and Yarnold, J. R., Design of compensators for breast radiotherapy using electronic portal imaging, *Radiother.Oncol.*, 37, 43–54, 1995.

Fenwick, J. D., Temple, S. W. P., Clements, R. W., Lawrence, G. P., Mayles, H. M. O., and Mayles, W. P. M., Geometric leaf placement strategies, *Phys. Med. Biol.*, 49, 1505–1519, 2004.

Gaballa, H. E., Mitev, G., Zwicker, R. D., Ting, J. Y., and Tercilla, O. F., Tissue compensation using dynamic collimation on a linear accelerator, *Int. J. Radiat. Oncol. Biol. Phys.*, 32, 723–731, 1995.

Galvin, J. M., Smith, A. R., Moeller, R. D., Goodman, R. L., Powlis, W. D., Rubenstein, J., Solin, L. J., Michael, B., Needham, M., and Huntzinger, C. J., Evaluation of multileaf collimator design for a photon beam, *Int. J. Radiat. Oncol. Biol. Phys.*, 23, 789–801, 1992.

Galvin, J. M., Leavitt, D. D., and Smith, A. A., Field edge smoothing for multileaf collimators, *Int. J. Radiat. Oncol. Biol. Phys.*, 35, 89–94, 1996.

Garrett, J. H. and Jones, D. E. A., Obliquity problems in megavoltage radiotherapy, *Br. J. Radiol.*, 35, 739–742, 1962.

Green, A., Jennings, W. A., and Christie, H. M., Radiotherapy by tracking the spread of disease, *Trans. IXth Int. Cong. of Radiology* (Munich 1959), Urban and Schwarzenberg, Berlin, pp. 766–772, 1960.

Hanson, W. F., Berkley, L. W., and Peterson, M., Off-axis beam quality change in linear accelerator x-ray beams, *Med. Phys.*, 7, 145–146, 1980a.

Hanson, W. F., Berkley, L. W., and Peterson, M., Calculative technique to correct for the change in linear accelerator beam energy at off-axis points, *Med. Phys.*, 7, 147–150, 1980b.

Hogstrom, K. R., Mills, M. D., and Almond, P. R., Electron beam dose calculation, *Phys. Med. Biol.*, 26, 445–459, 1981.

Heukelom, S., Lanson, J. H., and Mijnheer, B. J., Wedge factor constituents of high-energy photon beams: Head and phantom scatter dose components, *Radiother. Oncol.*, 32, 73–83, 1994a.

Heukelom, S., Lanson, J. H., and Mijnheer, B. J., Wedge factor constituents of high energy photon beams: Field size and depth dependence, *Radiother. Oncol.*, 30, 66–73, 1994b.

Hounsell, A. R. and Adams, E. J., Changes to dose in the build-up region when using multi-leaf collimators in place of lead blocks supported on an accessory tray, *Radiother. Oncol.*, 37, 225–229, 1995.

Huang, P.-H., Chu, J., and Bjärngard, B. E., The effect of collimator backscatter radiation on photon output of linear accelerators, *Med. Phys.*, 14, 268–269, 1987.

IAEA (International Atomic Energy Agency), *Absorbed dose determination in photon and electron beams: An international code of practice*, Technical Report Series 277, IAEA, Vienna, 1997a.

IAEA (International Atomic Energy Agency), *The use of plane parallel ionization chambers in high-energy electron and photon beams: An international code of practice*, Technical Report Series 381, IAEA, Vienna, 1997b.

IAEA (International Atomic Energy Agency), *Design and implementation of a radiotherapy programme: Clinical, medical physics, radiation protection and safety aspects*, TECDOC Series No. 1040, IAEA, Vienna, 1998 (being revised).

ICRU (International Commission on Radiation Units and Measurements), *Measurement of absorbed doses in a phantom irradiated by a single beam of X or gamma rays*, ICRU Report 23, ICRU, Bethesda, MD, 1973.

ICRU (International Commission on Radiation Units and Measurements), *Determination of absorbed dose in a patient irradiated by beams of X or gamma rays in radiotherapy procedures*, ICRU Report 24, ICRU, Bethesda, MD, 1976.

ICRU (International Commission on Radiation Units and Measurements), *Radiation dosimetry: Electron beams with energies between 1 and 50 MeV*, ICRU Report 35, ICRU, Bethesda, MD, 1984.

IEC (International Electrotechnical Commission), *Medical electrical equipment—medical electron accelerators in the range 1 MeV to 50 MeV—functional performance characteristics*, IEC No 976, IEC, Geneva, 1989a.

IEC (International Electrotechnical Commission), *Medical electrical equipment—medical electron accelerators in the range 1 MeV to 50 MeV—guidelines for functional performance characteristics*, IEC No. 977, IEC, Geneva, 1989b.

IEC (International Electrotechnical Commission), *Radiotherapy equipment—coordinates, movements and scales*, IEC Publication 1217, IEC, Geneva, 1996.

IPEM (Institute of Physics and Engineering in Medicine), Scientific Report Series No. 78, *Catalogue of diagnostic x-ray and other spectra*, IPEM, York, 1997.

IPEM (Institute of Physics and Engineering in Medicine), Report 81: *Physics Aspects of Quality Control in Radiotherapy*, IPEM, York, 1999.

IPEMB (Institution of Physics and Engineering in Medicine and Biology), The IPEMB code of practice for the determination of absorbed dose for x-rays below 300 kV generating potential (0.035 mm Al–4 mm Cu HVL; 10–300 kV generating potential), *Phys. Med. Biol.*, 41, 2605–2626, 1996a.

IPEMB (Institution of Physics and Engineering in Medicine and Biology), The IPEMB code of practice for electron dosimetry for radiotherapy beams of initial energy from 2 to 50 MeV, *Phys. Med. Biol.*, 41, 2557–2603, 1996b.

IPSM (Institute of Physical Sciences in Medicine), Report of the IPSM working party on low- and medium-energy x-ray dosimetry. Institute of Physical Sciences in Medicine, *Phys. Med. Biol.*, 36, 1027–1038, 1991.

Johns, H. E. and Cunningham, J. R., *The Physics of Radiology*, 4th ed., C.C. Thomas, Springfield, IL, 1983.

Jordan, T. J. and Williams, P. C., The design and performance characteristics of a multileaf collimator, *Phys. Med. Biol.*, 39, 231–251, 1994.

Jursinic, P. A., Podgorsak, M. B., and Paliwal, B. R., Implementation of a three-dimensional compensation system based on computed tomography generated surface contours and tissue inhomogeneities, *Med. Phys.*, 21, 357–365, 1994.

Karlsson, M., Nystrom, H., and Svensson, H., Photon beam characteristics on the MM50 racetrack microtron and a new approach for beam quality determination, *Med. Phys.*, 20, 143–149, 1993.

Karzmark, C. J., Deubert, A., and Loevinger, R., Tissue-phantom-ratios—an aid to treatment planning, *Br. J. Radiol.*, 38, 158–159, 1965.

Kase, K. R. and Svensson, G. K., Head scatter data for several linear accelerators (4–18 MV), *Med. Phys.*, 13, 530–532, 1986.

Khan, F. M., *The Physics of Radiation Therapy*, Williams and Wilkins, Baltimore, 1984.

Khan, F. M., Dosimetry of wedged fields with asymmetric collimation, *Med. Phys.*, 20, 1447–1451, 1993.

Khan, F. M., Sewchand, W., and Levitt, S. H., Effect of air space on depth dose in electron beam therapy, *Radiology*, 126, 249–253, 1978.

Khan, F. M., Deibel, F. C., and Soleimani-Meigooni, A., Obliquity correction for electron beams, *Med. Phys.*, 12, 749–753, 1985.

Khan, F. M., Gerbi, B. J., and Deibel, F. C., Dosimetry for asymmetric x-ray collimators, *Med. Phys.*, 13, 936–941, 1986.

Kim, S., Liu, C. R., Zhu, T. C., and Palta, J. R., Photon beam skin dose analyses for different clinical setups, *Med. Phys.*, 25, 860–866, 1998a.

Kim, S., Palta, J. R., and Zhu, T. C., The equivalent square concept for the head scatter factor based on scatter from flattening filter, *Phys. Med. Biol.*, 43, 1593–1604, 1998b.

Klein, E. E., Esthappan, J., and Li, Z., Surface and buildup dose characteristics for 6, 10, and 18 MV photons from an Elekta Precise linear accelerator, *J. Appl. Clin. Med. Phys.*, 4(1), 1–7, 2003.

Klevenhagen, S. C., *Physics of Electron Beam Therapy*, Adam Hilger, Bristol, 1985.

Klevenhagen, S. C., *Physics and Dosimetry of Therapy Electron Beams*, Medical Physics Publishing, Inc., Madison, WI, 1993.

Klevenhagen, S. C., Lambert, G. D., and Arbabi, A., Backscattering in electron beam therapy for energies between 3 and 35 MeV, *Phys. Med. Biol.*, 27, 363–373, 1982.

Klevenhagen, S. C., Aukett, R. J., Burns, J. E., Harrison, R. M., Knight, R. T., Nahum, A. E., and Rosser, K. E., Memorandum from the Institute of Physical Sciences in Medicine. Back-scatter and F-factors for low- and medium-energy X-ray beams in radiotherapy. Working Party of the Institute of Physical Sciences in Medicine, *Br. J. Radiol.*, 64, 836–841, 1991.

Knight, R. T., Backscatter factors for low and medium energy x-rays calculated by the Monte Carlo method, ICRPHYS-1/93, Available from: Physics Dept., Royal Marsden NHS Trust, Sutton SM2 5PT, UK, 1993.

Knight, R. T. and Mayles, W. P., An application of a computer spreadsheet to checking dose plans in radiotherapy planning, *Phys. Med. Biol.*, 36, 655–658, 1991.

Knöös, T. and Wittgren, L., Which depth dose data should be used for dose planning when wedge filters are used to modify the photon beam? *Phys. Med. Biol.*, 36, 255–267, 1991.

Lamb, A. and Blake, S., Investigation and modelling of the surface dose from linear accelerator produced 6 and 10 MV photon beams, *Phys. Med. Biol.*, 43, 1133–1146, 1998.

Lambert, G. D. and Klevenhagen, S. C., Penetration of backscattered electrons in polystyrene for energies between 1 and 25 MeV, *Phys. Med. Biol.*, 27, 721–725, 1982.

Lambert, G. D., Richmond, N. D., Kermode, R. H., and Porter, D. J., The use of high density metal foils to increase surface dose in low-energy clinical electron beams, *Radiother. Oncol.*, 53, 161–166, 1999.

Leavitt, D. D., Martin, M., Moeller, J. H., and Lee, W. L., Dynamic wedge field techniques through computer-controlled collimator motion and dose delivery, *Med. Phys.*, 17, 87–91, 1990.

Lee, C. H. and Chan, K. K., Electron contamination from the lead cutout used in kilovoltage radiotherapy, *Phys. Med. Biol.*, 45, 1–8, 2000.

Leung, P. M. and Johns, H. E., Use of electron filters to improve the buildup characteristics of large fields from cobalt-60 beams, *Med. Phys.*, 4, 441–444, 1977.

Li, X. A., Peak scatter factors for high energy photon beams, *Med. Phys.*, 26, 962–966, 1999.

Loshek, D. D. and Keller, K. A., Beam profile generator for asymmetric fields, *Med. Phys.*, 15, 604–610, 1988.

Luxton, G. and Astrahan, M. A., Characteristics of the high-energy photon beam of a 25-MeV accelerator, *Med. Phys.*, 15, 82–87, 1988.

Mageras, G. S., Mohan, R., Burman, C., Barest, G. D., and Kutcher, G. J., Compensators for three-dimensional treatment planning, *Med. Phys.*, 18, 133–140, 1991.

Marinello, G. and Dutreix, A., A general method to perform dose calculations along the axis of symmetrical and asymmetrical photon beams, *Med. Phys.*, 19, 275–281, 1992.

Mayneord, W. V. and Lamerton, L. F., A survey of depth dose data, *Br. J. Radiol.*, 14, 255, 1944.

McDermott, P. N., The physical basis for empirical rules used to determine equivalent fields for phantom scatter, *Med. Phys.*, 25, 2215–2219, 1998.

McKenzie, A. L., Air-gap correction in electron treatment planning, *Phys. Med. Biol.*, 24, 628–635, 1979.

Meeks, S. L., Bova, F. J., Kim, S., Tome, W. A., Buatti, J. M., and Friedman, W. A., Dosimetric characteristics of a double-focused miniature multileaf collimator, *Med. Phys.*, 26, 729–733, 1999.

Meyer, K., Skubic, S. E., Baird, L. C., Virudachalam, R., and Asche, D., Clinically shaped fields: Relative outputs for Varian multileaf collimators and tray-mounted cerrobend blocks, *Med. Phys.*, 26, 1257–1259, 1999.

Mijnheer, B., Bridier, A., Garibaldi, C., Torzsok, K., and Venselaar, J., *Monitor unit calculation for high energy photon beams—practical examples*, ESTRO Booklet no. 6, ESTRO, Brussels, 2001, Available at http://www.estroweb.org/ESTRO/upload/pdfs/booklet6.pdf

Miller, R. C., Bonner, J. A., and Kline, R. W., Impact of beam energy and field margin on penumbra at lung tumor-lung parenchyma interfaces, *Int. J. Radiat. Oncol. Biol. Phys.*, 41, 707–713, 1998.

Nilsson, B. and Brahme, A., Electron contamination from photon beam collimators, *Radiother. Oncol.*, 5, 235–244, 1986.

Papatheodorou, S., Zefkili, S., and Rosenwald, J.-C., The equivalent wedge implementation of the Varian Enhanced Dynamic Wedge (EDW) into a treatment planning system, *Phys. Med. Biol.*, 44, 509–524, 1999.

Parthasaradhi, K., Prasad, S. G., Rao, B. M., Lee, Y., Ruparel, R., and Garces, R., Investigation on the reduction of electron contamination with a 6-MV x-ray beam, *Med. Phys.*, 16, 123–125, 1989.

Petti, P. L. and Siddon, R. L., Effective wedge angles with a universal wedge, *Phys. Med. Biol.*, 30, 985–991, 1985.

Pohlit, W. and Manegold, K. H., *High-Energy Photons and Electrons*, Wiley, London, 1976.

Rawlinson, J. A., Islam, M. K., and Galbraith, D. M., Dose to radiation therapists from activation at high-energy accelerators used for conventional and intensity-modulated radiation therapy, *Med. Phys.*, 29, 598–608, 2002.

Rogers, D. W., Correcting for electron contamination at dose maximum in photon beams, *Med. Phys.*, 26, 533–537, 1999.

Rosenberg, I., Chu, J. C. H., and Saxena, V., Calculation of monitor units for a linear accelerator with asymmetric jaws, *Med. Phys.*, 22, 55–61, 1995.

Schlegel, W., Pastyr, O., Kubesch, R., Stein, J., Diemar, T., Hover, K. H., and Rhein, B., A computer controlled mico-multileaf collimator for stereotactic conformal radiotherapy, in *Proceedings of the XXIIth ICCR*, Leavitt, D. D. and Starkschall, G., Eds., Medical Physics Publishing, Madison, WI, pp. 79–82, 1997.

Shiu, A. S., Wong, J., Ewton, J. E., Rittichier, H. E., Tung, S. S., and Dong, L., Computer-controlled miniature multileaf collimator, in *Proceedings of the XXIIth ICCR*, Leavitt, D. D. and Starkschall, G., Eds., Medical Physics Publishing, Madison, WI, pp. 83–85, 1997.

Solin, L. J., Chu, J. C., Sontag, M. R., Brewster, L., Cheng, E., Doppke, K., Drzymala, R. E., Hunt, M., Kuske, R., and Manolis, J. M., Three-dimensional photon treatment planning of the intact breast, *Int. J. Radiat. Oncol. Biol. Phys.*, 21, 193–203, 1991.

Sterling, T. D., Perry, H., and Katz, L., Derivation of a mathematical expression for the percent depth dose surface of cobalt-60 beams and visualization of multiple field dose distributions, *Br. J. Radiol.*, 37, 544–550, 1964.

Storchi, P. and van Gasteren, J. J. M., A table of phantom scatter factors of photon beams as a function of the quality index and field size, *Phys. Med. Biol.*, 41, 563–571, 1996.

Thomas, S. J. and Thomas, R. L., A beam generation algorithm for linear accelerators with independent collimators, *Phys. Med. Biol.*, 35, 325–332, 1990.

Thwaites, D. I., Electron beam treatment-planning techniques, in *Radiotherapy Physics in Practice*, Williams, J. R. and Thwaites, D. I., Eds. 2nd ed., Oxford University Press, Oxford, pp. 220–246, 2000.

Tsiakalos, M. F., Theodorou, K., Kappas, C., Zefkili, S., and Rosenwald, J.-C., Analysis of the penumbra enlargement in lung versus the Quality Index of photon beams: A methodology to check the dose calculation algorithm, *Med. Phys.*, 31, 943–949, 2004.

Vadash, P. and Bjärngard, B., An equivalent square formula for head scatter factor, *Med. Phys.*, 20, 733–734, 1993.

Valdagni, R., Ciocca, M., Busana, L., Modugno, A., and Italia, C., Beam modifying devices in the treatment of early breast cancer: 3-D stepped compensating technique, *Radiother. Oncol.*, 23, 192–195, 1992.

Van der Giessen, P. H., A method of calculating the isodose shift in correcting for oblique incidence in radiotherapy, *Br. J. Radiol.*, 46, 978–982, 1973.

Van Gasteren, J. J. M., Heukelom, S., van Kleffens, M. J., van der Laarse, R., Vensellar, J. L. H., and Westermann, C. F., The determination of phantom and collimator scatter components of the output of megavoltage photon beams: Measurements of the collimator scatter part with a coaxial narrow cylindrical phantom, *Radiother. Oncol.*, 20, 250–257, 1991.

Van Santvoort, J. P., Binnekamp, D., Heijmen, B. J., and Levendag, P. C., Granulate of stainless steel as compensator material, *Radiother. Oncol.*, 34, 78–80, 1995.

Webb, S., The Physics of Three Dimensional Radiation Therapy: Conformal Radiotherapy, Radiosurgery and Treatment Planning, Institute of Physics Publishing, Bristol, 1993.

Wu, Q., Manning, M., Schmidt-Ullrich, R., and Mohan, R., The potential for sparing of parotids and escalation of biologically effective dose with intensity-modulated radiation treatments of head and neck cancers: A treatment design study, *Int. J. Radiat. Oncol. Biol. Phys.*, 46, 195–205, 2000.

Xia, P., Geis, P., Xing, L., Ma, C., Findley, D., Forster, K., and Boyer, A., Physical characteristics of a miniature multileaf collimator, *Med. Phys.*, 26, 65–70, 1999.

Yu, M. K., Analytical representation of enhanced dynamic wedge factors for symmetric and asymmetric photon fields, *Med. Phys.*, 29, 2606–2610, 2002.

Yu, M. K., Murray, B., and Sloboda, R., Parameterisation of head scatter factors for rectangular photon fields using an equivalent square formalism, *Med. Phys.*, 22, 1329–1332, 1995.

Zefkili, S., Kappas, C., and Rosenwald, J.-C., On-axis and off-axis primary dose component in high energy photon beams, *Med. Phys.*, 21, 799–808, 1994.

PART F

PATIENT DOSE COMPUTATION METHODS

Editors: Jean-Claude Rosenwald and Philip Mayles

INTRODUCTION

Calculation of the dose at a single point on the central axis of a beam under well-defined conditions can be performed using beam data tables and a calculator (see Chapter 23 for photons and Chapter 24 for electrons). Calculating the distribution of dose in a patient is not so straightforward. Before computerized treatment planning systems became widely available, dose distributions were calculated manually by the addition of percentages estimated from the superposition of isodose charts (such as those shown in Figure 22.11). The methods used involved empirically derived corrections to account for patient shape and inhomogeneities (see Section 22.4). So-called *hand planning* was time consuming and relied heavily on the experience of the planner; even so, complex planning methods were attempted at many centres. Modern computer technology has allowed increasingly sophisticated techniques to be routinely applied to treatment planning and has widened the aims and scope of treatment planning itself.

In the 1970s and 1980s, treatment planning algorithms were developed by individuals or groups working in, or attached to radiotherapy institutions. Commercial systems based on published algorithms or collaboration with research institutes are now widely used, and the number of one-off hospital-based systems is relatively small. As the power of computers has grown, so has the complexity of the planning algorithms. Some insight into the impressive development that took place during the last 40 years can be obtained from browsing through the series of proceedings of the International Conference on the Use of Computers in Radiation Therapy (ICCR); these meetings typically have been held every three years since 1961 (see references).

Although photon and electron algorithms have much in common (Chapter 25), there are significant differences which will be addressed separately in Chapter 26 and Chapter 27. Also, the general approach of dose computation principles covers a wide range from models that are mostly a representation of the experimental dose distribution to physically based models that model the transport of elementary particles from their production to their energy deposition in the patient. The Monte Carlo algorithms that can be used for photons as well as for electrons belong to the latter category, and they will be presented separately in Chapter 28.

CHAPTER 25

PRINCIPLES OF PATIENT DOSE COMPUTATION

Jean-Claude Rosenwald

CONTENTS

25.1 GENERAL REQUIREMENTS

The dose calculation algorithms, as discussed here, primarily are intended to be integrated into clinically used treatment planning software. Their ultimate goal is to predict accurately the dose delivered at any single point in the patient in order to decide if a given treatment plan is acceptable or to choose the best plan from several alternative ones. The corresponding requirements are therefore to obtain rapidly an accurate and comprehensive result in all situations with a maximum of flexibility. The issues related to comprehensiveness and flexibility are dependent mostly on the practical implementation of the algorithm within the treatment planning software. They are discussed in Chapter 39. Other issues are essentially algorithm dependent and are discussed in what follows.

25.1.1 ACCURACY

There has been extensive discussion in the literature on the accuracy required when treating a patient. This is covered in detail in Section 37.4 and Section 37.5. Since the dose computation is only one element of a complex chain (see Figure 43.1), the general principle is that it should not add a significant uncertainty to the final outcome. An interesting analysis based on this principle has been made by Ahnesjö and Aspradakis (1999) and is illustrated in Table 25.1. This table lists the uncertainty attached to each link of the chain starting from the absolute dose determination (see Chapter 18) and ending with the patient dose calculation. Two situations are considered the *present* technique (i.e. in 1999) and the future development. Several hypothetical levels of dose calculation uncertainty (from 0.5% to 5%) are considered. For each level the overall uncertainty is calculated from quadratic combination of individual uncertainties and compared to the overall uncertainty excluding dose calculation. It turns out that a dose calculation uncertainty of about 2% or 3% does not lead to a significant increase in the overall uncertainty at *present*, whereas 1% calculation accuracy might be required in the future.

Some indications on how to assess the accuracy of a dose planning algorithm are given later (see Section 39.3.4.2), but it must be recognised that it is very difficult to state the accuracy range for a given dose computation algorithm independently of the context in which it is used. The reasons are twofold: (i) the parameters used by the algorithm (basic data) have a huge influence on the result itself and hence on the corresponding accuracy, and (ii) the accuracy can be very good for some situations (e.g. a water medium) and very poor for others (e.g. inhomogeneous media). For these reasons, simple models can turn out to be more accurate than *sophisticated* models, at least in some instances. The important point is to be aware of the cases where the error is outside acceptable limits.

25.1.2 SPEED

The increase in the speed of computers has made life much easier than before when planning a patient's treatment. However, with more powerful computers has come an increase in the demands placed upon them; a full 3D dose distribution (i.e. in the whole treated area) for a large number of complex beams is often demanded, together with

TABLE 25.1

Determination of the Accuracy Required for Dose Calculation

	Present Technique $100 \times \Delta D(1\sigma)/D$	Future Development $100 \times \Delta D(1\sigma)/D$
Absorbed dose determination at the calibration point	2.0	1.0
Additional uncertainty for other points	1.1	0.5
Monitor stability	1.0	0.5
Beam flatness	1.5	0.8
Patient data uncertainties	1.5	1.0
Beam and patient set-up	2.5	1.6
Overall excluding dose calculation	**4.1**	**2.4**
Dose calculation	1.0 **2.0 3.0** 4.0 5.0	0.5 **1.0** 2.0 3.0 4.0
Resulting overall uncertainty	4.2 **4.6 5.1** 5.7 6.5	2.4 **2.6** 3.1 3.8 4.7

Uncertainties are expressed as percentages. The *present* technique refers to 1999.
Source: From Ahnesjö, A. and Aspradakis, M. M., *Phys. Med. Biol.*, 44, R99–R155, 1999.

searching for the best plan through an interactive trial-and-error procedure*. Typically, for such cases where the planner uses the graphical display to look at the dose changes resulting from beam modifications, the overall response time should not be more than 10 s to 20 s. With the advent of intensity modulated radiotherapy (IMRT) the speed problem is even more crucial since the inverse planning approach requires that the dose computation be repeated for many beams, many times at many points, as part of an automated iterative process (see Chapter 43).

It is easily understood that speeding up the calculation generally means a decrease in accuracy (for given computer hardware). Therefore, the choice of the dose calculation algorithm on a given occasion is largely a compromise which should take into account the clinical impact of a greater inaccuracy in dose. Although somewhat difficult to organise in practice, a good solution is to have access to several algorithms and make the appropriate choice depending on the circumstances (e.g. have a faster algorithm for inverse planning and then confirm the dose distribution with a slower but more accurate one before the start of treatment). However, if used for inverse planning, the fast algorithm must have sufficient accuracy to ensure that its limitations do not have a significant effect on the outcome of optimisation.

25.1.3 GENERAL ASPECTS

The calculation of the dose distribution in a patient implies that relevant beam data are combined with individualised patient data. In this process, it is therefore essential that the dose calculation algorithm takes into account the characteristics both of the patient anatomy and of the clinical beam. As seen in Part E, factors affecting the delivered dose are quite complex, and they include the influence of all accessories (i.e. wedge filters, compensators, blocks, etc.). They are even more complex when special techniques are being used (see Part I).

As far as the patient anatomy is concerned, the most effective representation is a 3D matrix of voxels derived from a CT scan of the patient in the treatment position (Chapter 30). Ideally, the exact tissue density and composition of each of these voxels should be taken into account.

In practice, however, many simplifications have to be made for both the patient representation and the beam description. Again this is mostly a trade-off between the current possibilities of the algorithm, ease of access to the relevant beam and patient data, speed of calculation and accuracy.

25.2 COORDINATE SYSTEMS AND GEOMETRICAL ISSUES

In order to compute the dose at any point within the patient it is necessary to know the corresponding geometry. As a first approximation, it could be sufficient to calculate the overlying thickness of tissue along the line joining the source to the calculation point. This would typically allow an accurate calculation of the primary component. This thickness z, generally referred to as the *depth* of the calculation point, can be expressed as a geometrical distance (as if the patient were entirely made of water) or be corrected for tissue inhomogeneity, using the concept of *equivalent path length* (sometimes called *radiological* or *water equivalent* thickness). This concept is rather intuitive (Figure 25.1). It is based on the fact that the product $(t_m \times \rho_m)$ of a given thickness t_m of material m and the *density* ρ_m of this material gives approximately the same attenuation for photons (and the same penetration for electrons) as the product $(t_w \times \rho_w)$, where w stands for water ($\rho_w = 1$). It can be demonstrated that this equivalence is perfect for the primary component of a polyenergetic photon beam, provided that density is expressed as

---

* Such a trial-and-error procedure has also become known as *forward planning* in contrast to *inverse planning* (see Chapter 43).

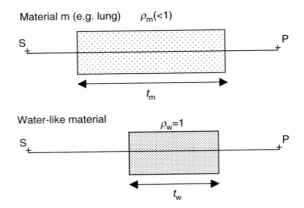

FIGURE 25.1
Illustration of the equivalent path length method: the attenuation at point P for a thickness t_m of a material of electronic density relative to water ρ_m is assumed to be the same as for a *scaled* thickness t_w of water where $t_w = t_m \cdot \rho_m$. This is rigorous for primary photons when only Compton interactions with peripheral electrons are considered (same total number of electrons/cm^2 interposed between the source and point P).

electronic density relative to water and that the Compton effect is the only interaction process to be considered (see also Section 22.4 and Section 23.3.4.5). For a given photon spectrum, whatever interaction process of the primary component is considered, this equivalence is also rigorous using the mass density under the assumption that the *mass* attenuation coefficient is independent of the material (see Section 4.4.1).

In order to achieve a higher accuracy, it is necessary to consider also the influence of the scattered photons and secondary electrons. As far as photons are concerned, the equivalent path length method can be extended to the lateral dimensions by application of the O'Connor theorem (O'Connor 1957). With this theorem it can be shown that, when considering two media of different densities but the same atomic composition exposed to the same beam, the dose at corresponding points in the two media will be the same provided that all geometric distances in the two media, including field sizes, are scaled inversely with density (Figure 25.2) (Bjärngard 1987; Ahnesjö and Aspradakis 1999).

The practical application of the O'Connor theorem is not straightforward. In general, however, understanding some of the fundamentals and some of their related principles is very useful in gaining an insight into the possibilities and limitations of the various dose

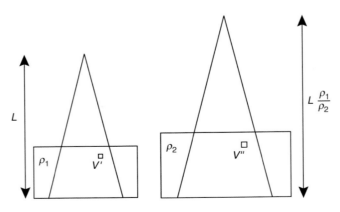

FIGURE 25.2
Illustration of the conditions required for O'Connor theorem to be valid. The figure illustrates a situation where ρ_1 is larger than ρ_2. (From Ahnesjö, A. and Aspradakis, M. M., *Phys. Med. Biol.*, 44, R99–R155, 1999. With permission.)

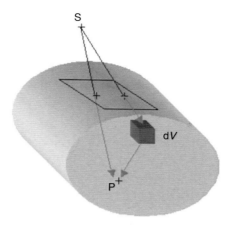

FIGURE 25.3
Schematic drawing of a beam incident on a patient. The dose at point P is influenced principally by the tissues lying between the source S and P but also by the scattered radiation from the rest of the patient volume.

calculation algorithms. Although some simplifications are often acceptable, it should always be remembered that the problem is a 3D problem and that one should take into account not only what happens along the line joining the source to the calculation point but also the contributions to the dose at that point from the scattered radiation (both photons and electrons) arising in the rest of the medium (see Figure 25.3).

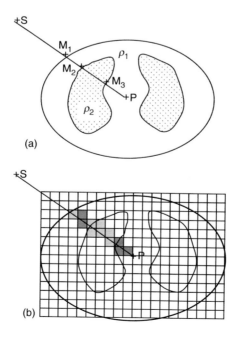

FIGURE 25.4
Illustration of two methods for the calculation of equivalent path length t_{eq}: (a) For *contour-based* patient representation, t_{eq} is calculated from the (3D) intersections of a line joining the source S to the calculation point P with the external body surface (at M_1) and with internal structures such as lung (at M_2 and M_3): $t_{eq} = M_1 M_2 . \rho_1 + M_2 M_3 . \rho_2 + M_3 P . \rho_1$. (b) For *voxel-based* patient representation, t_{eq} is calculated from the sum of the elementary path lengths through each individual image voxel: $t_{eq} = \sum t_i \rho_i$ where t_i is the geometrical distance through voxel i and ρ_i the density of voxel i.

For electron beams (and for the secondary electrons generated in photon beams), similar scaling considerations can be useful and provide an often acceptable approximation (see also the description of Fano's theorem in Section 6.7.3).

In practice, the implementation of clinical dose calculation methods requires the definition of unambiguous coordinate systems. It also generally requires dealing simultaneously or sequentially with *vector-based* (i.e. contours) and *image-based* representations (i.e. pixels and voxels) of the patient, with algorithms for transforming one representation into the other. The depth calculations could be handled in either of these two modes (Figure 25.4) and thus may require, as a first step, a coordinate transformation in which a point linked to the patient would be expressed in relation to a Cartesian coordinate system linked to the beam. Thanks to this transformation which takes into account the beam geometrical characteristics such as 3D position, orientation and divergence, the geometrical problem is very much simplified (Siddon 1985), and the source-to-point distances and point depths become directly available. In subsequent chapters, a coordinate system with the z axis coincident with the beam axis and the x and y directions, parallel to the width and length of the field, respectively, will be considered. The depth of the point will be referred to as d or z and the source-to-point distance as f. We will use the quantities and symbols defined in Section 23.2.1.

25.3 ABSOLUTE AND RELATIVE DOSE

The dose calculation algorithm ought to be able to yield the absolute dose distribution (i.e. in Gy) within the patient, for a given number of monitor units (MU). However, the current practice is to calculate relative dose distributions, which is relative to the dose at a given arbitrary point in the open part of the beam. This point, called the *weight point* or *beam normalisation point*, is frequently coincident with some reference point used for patient set-up or for dose prescription (i.e. the isocentre, the ICRU reference point or the depth of maximum dose d_{max}). Though often overlooked in the past, the link between relative and absolute dose is now recognised as fundamental (Ahnesjö et al. 1992a; Mackie et al. 1995, 1996; Fraas et al. 1998; Ahnesjö and Aspradakis 1999; Liu et al. 2000; Jiang et al. 2001; Mijnheer et al. 2004).

The point is that one should take full advantage of the sophistication of modern algorithms to assess precisely the relationship between the dose at a given reference point and the number of monitor units. The need for such an approach is even more evident when special techniques are used, e.g. intensity modulated radiation therapy (see Chapter 43), where traditional methods of calculating monitor units, such as those described in Chapter 23, are not applicable. The solution to the above problem is far from evident. It depends very much on the type of algorithm and on its clinical implementation. However, the general principle can be described.

Whatever kind of dose modelling is being used, the dose at any point is directly or indirectly linked to an internal quantity Q_0 used as a reference (i.e. used for beam normalisation purposes). If, for instance, the dose is simply obtained from storage and retrieval of dose measurements performed on a Cartesian grid for different field sizes at a given SSD, then Q_0 could be the dose at the normalisation point of each grid (e.g. at constant depth) and it could therefore be field-size and depth dependent. As seen later, Q_0 could also be the dose in air $D_{air} = D(air, A, SAD)$ (or in a mini-phantom) as defined in Chapter 23. D_{air} is independent of the phantom and is easily calculated at different distances. At a given distance f it depends almost exclusively upon the collimator opening A. The collimator opening A is also the only significant *uncontrolled** parameter of influence if Q_0 represents the primary photon energy

---

* We use the term *uncontrolled* to characterize the parameters for which the influence on Q_0 is not predictable by the calculation algorithm. The variation of Q_0 as a function of such parameters should be measured instead.

fluence at a reference distance in the so-called kernel-based convolution/superposition method (see Section 26.3). For methods such as Monte-Carlo simulation, where the particles emitted from the source (photons or electrons) are transported through the treatment head and through the patient, Q_0 could be the number of primary particles used for simulation (see Section 28.6.1).

In all cases, once the *uncontrolled* parameter(s) of influence of Q_0 is (are) clearly identified, a reference set-up is decided arbitrarily with a simple geometrical phantom (i.e. used for beam calibration). For this set-up, the parameter(s) *param* is (are) varied systematically and both dose *measurements* (for a given number of MU) and dose *calculations* (for given value of Q_0) are performed. In principle one could then derive the quantity of interest $(Q_0/MU)(param)$ using the following equation:

$$\frac{Q_0}{MU}(param) = \frac{\frac{D_{\text{meas}}(param)}{MU}}{\frac{D_{\text{calc}}(param)}{Q_0}} \tag{25.1}$$

where $D_{\text{meas}}(param)$ and $D_{\text{calc}}(param)$ are the measured and calculated doses for a particular value of the parameter(s) *param* respectively.

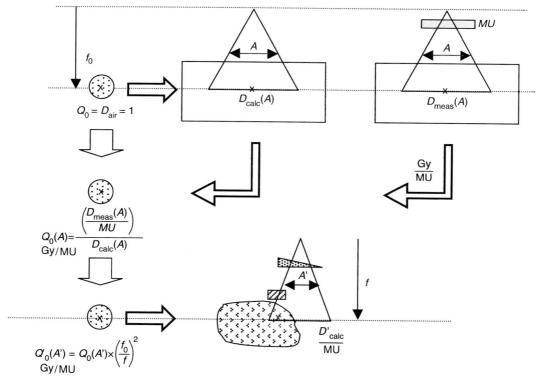

FIGURE 25.5
The principle of the *internal calibration* of a dose calculation algorithm allowing the calculation of the absolute dose (D_{calc}) per MU at any point and for any situation when D_{air} is the reference quantity Q_0 (left column of the figure), the field size A being the *uncontrolled* parameter. The upper part of the figure describes the reference set-up (both measurements and calculation). $D_{\text{calc}}(A)$ is first calculated keeping Q_0 equal to unity. $D_{\text{calc}}(A)$ is then used as a normalisation factor of the measured dose $D_{\text{meas}}(A)$ to derive the quantity $Q_0(A)$ that must be used in all subsequent calculations. The lower part of the figure corresponds to calculation for the clinical situation (field size A', distance f, including beam modifiers and patient characteristics). In this example, the calculated quantity $Q_0(A)$ is proportional to the collimator-scatter factor S_c (see Section 23.2.1.12) but it is related to the absolute dose and directly derived from in-phantom measurements performed in reference conditions.

It is then easy to calculate the *absolute dose per* MU at any point and under any conditions, provided that the algorithm makes use of the quantity $(Q_0/MU)(param)$ instead of Q_0.

This overall procedure can be considered as an *internal calibration* of the dose calculation algorithm. It is illustrated in Figure 25.5 when the dose is calculated from the reference quantity $Q_0 = D_{air}$, using the Clarkson–Cunningham integration of the scatter component (see Section 26.3.2). The detail of the method for other formalisms can be found in (Ahnesjö and Aspradakis 1999) and in Section 28.6.1.

Another point of discussion is the definition of the dose in a non-water like medium. Actually, for a given photon energy fluence the local dose is dependent on the surrounding medium (see Chapter 6). For instance for low energy x-rays the absorbed dose in bone is much higher than in soft tissue. In the current algorithms used for clinical planning, this is not considered and even if the photon (and/or electron) fluence is corrected for the influence of inhomogeneities, the local dose is always referred to a water-like medium. This may change as Monte-Carlo methods come into clinical use (du Plessis et al. 1998; Siebers et al. 2000b; and Section 28.6.2). It is therefore important to state clearly to which medium the calculated dose is related.

25.4 ALGORITHM CLASSIFICATION

There is no current consensus on a classification methodology for dose calculation algorithms. ICRU (1987) distinguished between tabular (or matrix) formats, beam generating functions, separation of primary and scattered radiation and representations using basic principles. Mackie et al. (1995, 1996) suggested distinguishing between correction-based and model-based methods. This was related mostly to the handling of beam modifiers and inhomogeneity corrections; in the former case the open-beam dose distribution in water is computed and then corrected for the beam modifiers and patient characteristics, whereas in the latter case the interaction in the patient is computed directly. The former approach could be considered as based principally on correction factors extracted from measurement and the latter on more physical modelling of beam interactions. However, this conceptual distinction is

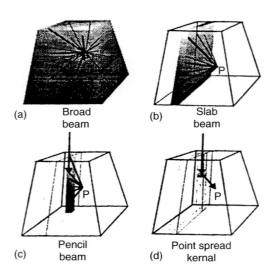

FIGURE 25.6
Integration of scatter from various types of kernels. (From Battista, J. J. et al., in *XIIth International Conference on Computers in Radiotheraphy*, Leavitt, D. D. and Starkshall, G., Eds., Medical Physics Publishing, Salt Lake City, UT, pp. 29–42, 1997. With permission.)

difficult to apply strictly since there is a wide range of possibilities between measurement-oriented and fundamental methods.

Another suggestion was made by Battista et al. (1997), who referred to the fact that most modern methods used clinically make use of some integration or superposition of elementary *kernels* throughout the patient volume (Figure 25.6) (see Section 26.3). They made a distinction according to the number of dimensions over which the integration is carried out: the absence of integration was referred to as the *broad-beam* approach, because the beam is treated as a whole rather than being broken up into its component segments. The Monte-Carlo approach (see Chapter 5 and Chapter 28), where each particle is tracked in 3D, does not fit exactly with this scheme. It could be considered as the ultimate form of the 3D integration and treated separately.

This provides a systematic approach to the frequent classification into 1D, 2D or 3D (or 2.5D, etc.) models. However, as explained by Battista et al., one must also consider how the anatomical dimensions are taken into account in the implementation of the model. Typically for a 1D algorithm the dose at a point P depends only on the accessories and patient characteristics found on the line joining the source to P; a 2D algorithm takes into account the modifications in the transverse plane, excluding the changes in the cranio-caudal direction, whereas a 3D algorithm accounts for any change in the whole volume (see also Figure 11 and Table 7 in AAPM 2004). This classification of dose algorithms should be differentiated from the classification of Treatment Planning Systems where the possibilities of 3D beam orientation and 3D display must also be considered (see Chapter 39).

Although discussed in the literature mostly with reference to photons, all the previous considerations are very similar for electrons. In fact the methods for dose calculation, which originally were quite different, have certain unifying features (Bloch 1988). In what follows, nevertheless they have been kept separate. Photon beam are dealt with in Chapter 26 and electron beams in Chapter 27. The application of the Monte-Carlo method, which is common to both photons and electrons, and which is just starting to be used for clinical planning, is covered in Chapter 28.

It should be noted that, whatever method is used, some empirical tuning of the model with respect to the actual beam characteristics is needed to a greater or lesser extent. Such tuning is of the utmost importance for the validity of the results (see Section 39.3.6).

CHAPTER 26

PATIENT DOSE COMPUTATION FOR PHOTON BEAMS

Jean-Claude Rosenwald, Ivan Rosenberg, and Glyn Shentall[*]

CONTENTS

---

[*] With contributions from David McKay.

26.1 PHYSICAL BACKGROUND

As seen in Part A and Part E, the mechanisms of dose deposition in the patient by a clinical photon beam are quite complex. They are summarised in Figure 26.1. Note that we refer in this chapter implicitly to megavoltage photon beams which were dealt with in Chapter 22; detailed dose computation in patients is not generally carried out for kilovoltage x-ray beams which were dealt with in Chapter 21.

In this figure, four main dose components have been identified:

- The primary dose dominates and represents generally more that 70% of the total dose
- The phantom scatter dose is the second-largest contribution, and can represent up to 30% of the total dose

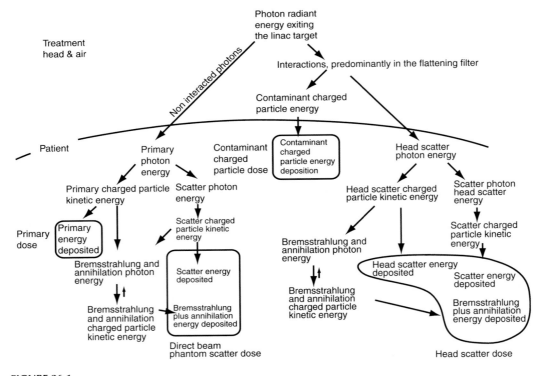

FIGURE 26.1

Schematic representation of the main interactions and dose components for clinical photon beams. (From Ahnesjö, A. and Aspradakis, M. M., *Phys. Med. Biol.*, 44, R99–R155, 1999. With permission.)

- The head scatter dose is of less importance but it can amount to 5% to 10% of the total dose
- The contaminant charged-particle energy deposition has a large influence, especially for high-energy photons, but only at small depths (typically less than 4 cm, i.e. only in the buildup region; externally incident electrons cannot reach any further).

Depending on the sophistication of the model, these components, although always implicitly taken into account in contributing to the total dose, may or may not be considered explicitly. In the following distinction between models, we will consider separately the *broad beam* methods (also referred to as *empirical* methods), which do not make any distinction between these components and the *superposition* methods, which have many variants according to the type of components taken into account and to the integration order of these components. Excellent general descriptions of photon beam dose calculations methods have been given by Mackie et al. (1996) and Ahnesjö and Aspradakis (1999). These have been drawn from extensively in what follows.

26.2 BROAD-BEAM (EMPIRICAL) METHODS

26.2.1 SIMPLE BEAMS IN WATER

The dose representation of *simple beams* (i.e. rectangular fields limited by the main collimator without any accessories) in water can be further subdivided into two types:

- Representations that use *tabulated beam data* dose distributions for a number of beams measured under reference conditions are stored in tabular form and are interpolated by the treatment planning system (TPS) during calculation.
- Representations that use an *analytical approach* (beam generating functions): depth-dose and profile characteristics are modelled using mathematical functions. Some of these approaches are purely mathematical, while others more closely reflect the physics of broad beam interactions.

26.2.1.1 Tabulated Beam Data Representations

Many early dose calculation algorithms were based on this principle (see ICRU 1987). For the sake of simplicity only the most frequently implemented one is described here.

In the 1970s Bentley and Milan developed a small digital computer system for radiotherapy treatment planning (Bentley and Milan 1971). This was a forerunner of the modern TPSs. The need for a fast and accurate system had a strong influence on the model they developed and the Bentley–Milan model, as it is now known, was still commonly used until recently. The beam data are stored in a tabular form as central axis depth-dose data and off-axis ratios (Milan and Bentley 1974).

Depth-dose data depend on the width and length of the field under consideration. It would have been impractical to store depth-dose curves for all possible open and wedge fields so *equivalent squares* (see Section 23.2.1.8) were used to reduce the required data to square fields only. Each square field had depth-dose data stored at 17 depths, equally spaced, starting at the depth of dose maximum and extending as far as the user decides is necessary. This is usually to a depth of between 30 cm and 35 cm. The dose in the build-up region was modelled very crudely considering that the dose at a depth, d, smaller than the depth of maximum dose, d_{max}, could be calculated from a linear interpolation between the surface dose and the maximum dose assigned to a modified depth, d_B, calculated as

$$d_B = d_{max} - \frac{(d_{max} - d)^3}{d_{max}^2} \qquad (26.1)$$

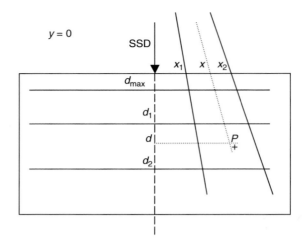

FIGURE 26.2
Tabular method for dose calculation at a point P located in the central patient cross-section ($y=0$) at depth d (parallel to the beam axis) and at a distance x from the central axis (defined at the patient surface). The dose at P can be obtained from linear interpolation between the four closest points of the grid (lines x_1, x_2, d_1, d_2).

Off-axis ratios depend mainly on the width of the field. The effect of the length of the field on the profile can be ignored, so only square field profiles need to be stored. The off-axis ratios will also depend on depth but if the effect of divergence is removed the shape of the profile only changes slowly. This is due to the different amounts of scatter present at different depths. Therefore, by adopting a system of fan lines that diverge from the radiation source, the off-axis ratios need only be stored at 5 depths. The depths were chosen to correspond to measurement points 1, 5, 9, 13 and 17 on the depth-dose curve. Each profile was covered by 47 fan lines, divided up so that number 24 lies on the central axis of the beam with 23 others on either side. The spacing of the fan lines was originally constant but could also depend on the field size and be non-uniform so that more fan lines could be placed in the penumbra region where the dose is changing rapidly. The off-axis ratio on the central fan line was unity by definition.

The dose at any point in the central cross-section of a patient (i.e. in a transverse plane passing through the beam axis) is then calculated after co-ordinate transformation by linear interpolation as a function of depth d and off-axis distance x (see Figure 26.2). In upper and lower patient cross-sections, the dose can also be interpolated as a function of y, assuming that the behaviour is the same as that in the x direction.

26.2.1.2 Analytical Beam Representations

A wide variety of mathematical functions has been developed over the years to model beams (see Appendix A in ICRU 1987). Generally the mathematical representation of the beam is split into two parts: central axis depth-dose values, and off-axis ratios. Each part is approximated by an analytical function. The dose is found by multiplying the two parts together. The range of energies and field sizes for which the expressions hold must be determined and their use restricted to this range.

One analytical approach that was used in many centres and also in at least one commercial TPS was that of van de Geijn (1965). Over a number of years he published analytical models and refinements that provided means of fitting depth dose and profile data over a wide range of energies (van de Geijn 1970, 1972). One advantage of this method was that depth-dose tables for a particular beam energy could be characterised by taking only seven measurements. Each of the parameters used in the fitting process was linked to the physical properties of broad beams.

26.2.2 IMPROVEMENT OF OFF-AXIS CALCULATIONS AND CORRECTIONS FOR WEDGE FILTERS

Tabular or analytical beam representations, may be used for points located not only off-centre (in the x direction), but also off-plane (in the y direction). The more straightforward approach consists in assuming that, for a flat homogeneous medium, the dose at any (x,y,z) point can be obtained from the dose at $(0,0,z)$ by multiplying it by two off-axis functions, $g(x)$ $g(y)$. However, some corrections are required, especially for megavoltage radiation where a flattening filter is present, to account for the radial changes of the off-axis curves as a function of depth in case of both symmetric and asymmetric fields (Chui and Mohan 1986; Storchi and Woudstra 1995).

With the broad-beam representation, the wedge filters could be treated as part of the initial beam description or as an additional accessory requiring specific correction. The first approach simply involves including a number of dose distributions with wedge filters in the original experimental data set. The second approach requires a dose profile for a large field to be acquired in the wedge direction with and without the wedge filter. Then a *transmission* profile is derived from these two curves, tabulated for computing purposes and used as a correction factor. Alternatively this transmission profile can be calculated from the physical dimensions and composition of the filter using the appropriate attenuation coefficient. It must then be verified experimentally that this coefficient, which is slightly dependent upon the wedge thickness, gives acceptable results.

The concept of a wedge transmission profile is not restricted to the broad beam approach and has been implemented in more recent algorithms. As discussed in Section 22.6.1.2, although correction factors may be added (Yao and Ranganathan 1994; Myler and Szabo 2002), considering only the effect of the wedge on the fluence incident on the surface contains several simplifications since it implicitly assumes that the beam quality does not change as a function of off-axis position and that the scatter component behaves in the same way as the primary one.

26.2.3 CORRECTION FOR PATIENT SHAPE AND INHOMOGENEITIES

In the discussion so far it has been implicitly assumed that the measurements and beam models have dealt with the *reference situation*, i.e. a beam incident normally on a flat surface of homogeneous, water-equivalent material. Needless to say, the human body is not like this; therefore dose calculations in a patient must include correction factors to account for the shape of the patient and the shape and position of inhomogeneities. The accuracy and complexity of the correction method employed depend on circumstances but it should be remembered that the more complex (and presumably therefore more accurate) the correction the longer it takes to compute. Correction factors are normally expressed as a ratio of the corrected dose to the uncorrected dose.

26.2.3.1 Patient Shape

One method of correcting for the patient shape in regions not too far from the central axis (i.e. ignoring off-axis correction) is the effective source-skin distance (SSD) method (Figure 26.3a). This applies an inverse square law correction to the measured depth-dose data. If the SSD on the central axis is represented by SSD, the increase in SSD for a particular off-axis point is h, and the corresponding point is at a depth d, then the dose at the same depth on the central axis must be multiplied by the correction factor C, defined by

$$C = \left(\frac{SSD + d}{SSD + h + d} \right)^2 \qquad (26.2)$$

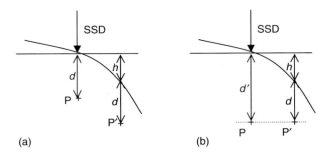

FIGURE 26.3
Correction for patient surface obliquity using either the effective SSD method (a) or the TAR/TPR ratio method (b). In (a), the dose at P' can be obtained from the dose at P by correcting for the increase in distance from (SSD + d) to (SSD + h + d) keeping the same attenuation (for depth d). In (b), P and P' are at the same distance (SSD + d') from the source but the depth is decreased from d', or d + h, to d. In both cases, the beam divergence and off-axis corrections have been omitted for simplicity (see also ICRU 1976).

Bentley and Milan used this method to correct for patient shape. However in practice, in order to make it more general and usable at any SSD, they modified the stored beam data so that it was effectively *infinite* SSD data by multiplying it by

$$\left(\frac{SSD + d}{SSD}\right)^2$$

The correction applied by the algorithm to the dose on the beam axis at a depth d and at *infinite* SSD is then simplified to

$$\left(\frac{SSD}{SSD + h + d}\right)^2$$

The tissue–air ratio or tissue–phantom ratio (TAR or TPR) method (see Section 23.2.1 for definitions) is an alternative way of correcting for shape (Figure 26.3b). As the name suggests the correction factor can either be a ratio of two TARs or two TPRs. The ratios are a function both of depth and of field size at that depth. They are not dependent on SSD. If the increase in SSD is h and the depth of the point under consideration is d, as before, then the dose at the point of interest is the dose on the central axis at a depth of $d + h$ (i.e. at the same distance from the source) multiplied by the correction factor

$$C = \frac{T(d, A_d)}{T(d + h, A_d)} \tag{26.3}$$

where T is the TAR or TPR at depth d with the field size at that depth being A_d.

26.2.3.2 *Patient Inhomogeneities*

Up to this point, it has been assumed that the patient is composed of water-equivalent tissue and thus that depths in the patient correspond to the same depths in water. This is never exactly the case and in practice, especially for lung treatments, corrections are needed to take into account changes in attenuation and scatter due to the presence of an inhomogeneity (see AAPM 2004 for a complete review of the inhomogeneity corrections methods).

Effective-Depth or Equivalent Path Length Method: The simplest method to correct for an inhomogeneity is the equivalent path length method (see Section 25.2). The *effective depth* to a calculation point P is the thickness of water-equivalent tissue that would attenuate the radiation by the same amount as the actual tissue along a fan line between the surface and point P. It is a one dimensional correction that ignores changes in scatter. If the radiation passes through n different tissues, each of thickness t_i and density ρ_i to reach the calculation point then the *effective* (or *radiological*) depth d_{eff} is

$$d_{eff} = \sum_{i=1}^{n} t_i \rho_i \qquad (26.4)$$

The dose at a point below the inhomogeneity can be obtained from the uncorrected dose by applying a correction factor, C, given by

$$C = \frac{PDD(d_{eff})}{PDD(d)} \times \left(\frac{SSD + d_{eff}}{SSD + d} \right)^2 \qquad (26.5)$$

where $PDD(d_{eff})$ is the percentage depth dose for the corrected depth, $PDD(d)$ is the percentage depth dose for the actual depth and the SSD is SSD.

Note that the inverse square law correction is also needed. This is needed because $PDD(d_{eff})$ assumes the distance from the radiation source to the point under consideration is $SSD + d_{eff}$ when it is actually $SSD + d$.

TAR Ratio (or TPR Ratio) Method: The TAR ratio method is a slightly more precise implementation of the equivalent path length approach. The correction factor is calculated as a ratio of TARs or TPRs. The correction factor to be applied to the uncorrected dose is

$$C = \frac{T(d_{eff}, A_d)}{T(d, A_d)} \qquad (26.6)$$

where T is the TAR or TPR and A_d is the field size at depth d.

Again, this correction is strictly true only for the primary radiation and does not apply to scatter. In some situations, especially at some distance from inhomogeneities, both corrections (i.e. Equation 26.5 and Equation 26.6) are equivalent and will give an acceptable estimate of the dose. In others, more sophisticated corrections are needed that account for size and position of the inhomogeneity.

Power-Law Correction (Batho Correction): The power-law correction is again a ratio of two TARs (or TPRs) but in this instance they are raised to a power that depends on the density of the surrounding material (Batho 1964) as shown in Equation 26.7. For a point located in a slab of density ρ_1 overlaid by a slab of density ρ_2, the correction factor is

$$C = \frac{TAR(l_2, A_d)^{(\rho_2 - \rho_1)}}{TAR(l_1, A_d)^{(1 - \rho_1)}} \qquad (26.7)$$

where l_1 and l_2 are the distances from the point to the upper (i.e. closest to the source) boundaries of inhomogeneities 1 and 2, respectively (Figure 26.4). The distances, l, are distances from the boundaries of the inhomogeneity, so this correction accounts for the position of the inhomogeneity but not for its overall size and shape. It produces better agreement with measurements than the effective depth or TAR ratio methods. However, it predicts dose discontinuities at interfaces between the normal tissue and the inhomogeneity.

Batho's original suggestion to use TAR has been shown to exaggerate the correction for a point lying in a low density inhomogeneity since it ignores the backscatter from underlying material; the accuracy was improved by using Tissue Maximum Ratio (TMR) or TPR instead (El-Khatib and Battista 1984; Kappas and Rosenwald 1985).

For a series of superimposed inhomogeneous slabs a more general form for this correction (Sontag and Cunningham 1977; Parker et al. 1980) can be written as

$$C = \prod_{i=1}^{n} TPR(l_i, A_d)^{(\rho_i - \rho_{i-1})} \qquad (26.8)$$

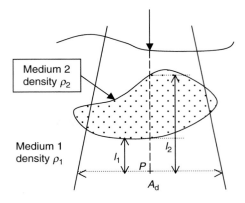

FIGURE 26.4
Illustration of the power law correction method applied at point P located in a medium of density ρ_1 below an inhomogeneity of density ρ_2: assuming a slab geometry, the correction factor at point P can be calculated from Equation 26.7.

where l_i is the distance from the dose calculation point to the anterior part of the ith inhomogeneity (starting from inhomogeneity 1, where the point lies), ρ_i is the relative electron density of the ith inhomogeneity and $\rho_0 = 1$.

Beam Subtraction Method: The power-law correction, which indirectly accounts for (first) scatter modification (Wong and Henkelman 1982), fails to give accurate results when the lateral dimensions of the inhomogeneity are smaller than the field size. The beam subtraction method makes use of a smaller *virtual* beam which exactly covers the inhomogeneity (Lulu and Bjängard 1982; Kappas and Rosenwald 1982, 1985) (see Figure 26.5). For this virtual beam, the power law correction is satisfactory. The additional scatter from the outer part of the original beam is generated in water-equivalent material. It can therefore be calculated by subtracting the dose due to the virtual beam from that due to the original beam to yield a correction factor

$$C = 1 + \frac{T(d, A_d^*)}{T(d, A_d)} \times (C^* - 1) \tag{26.9}$$

where T is the TAR or TPR, A_d is the field size at depth d, A_d^* is the virtual field size at depth d and C^* is the power law correction factor for the virtual field size.

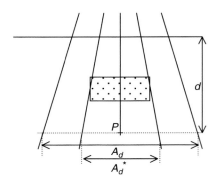

FIGURE 26.5
Illustration of the beam-subtraction method for an inhomogeneity smaller than the beam size: The Batho correction factor is applied to the beam A_d^* that has the same dimensions as the inhomogeneity whereas the scatter from the outer part is generated in water and calculated by subtracting the dose in water for the virtual beam A_d^* from the dose in water for the whole beam A_d.

After experimental verification, the 3D version of this method has been implemented in some clinical planning systems (Kappas and Rosenwald 1986a).

Equivalent Tissue Air Ratio (ETAR) Correction: The ETAR method is another way to account for the 3D shape of the inhomogeneity. It was developed by Sontag and Cunningham (1978), who suggested that the TAR ratio method described above (Equation 26.6) could be improved by scaling not only the depth, but also the field size. The general expression for the correction factor is then

$$C = \frac{T(d_{\text{eff}}, A'_d)}{T(d, A_d)} \tag{26.10}$$

where A_d in the numerator has been replaced by A'_d obtained by scaling A_d (actually the radius of the equivalent circular field) according to the density of the surrounding inhomogeneities. In practice, Sontag and Cunningham scaled the field size using an effective density, $\tilde{\rho}$ different from the one used for scaling the depth d_{eff} (see Equation 26.4). The quantity $\tilde{\rho}$ is calculated as

$$\tilde{\rho} = \frac{\sum_i \sum_j \sum_k \rho_{ijk} W_{ijk}}{\sum_i \sum_j \sum_k W_{ijk}} \tag{26.11}$$

using the densities ρ_{ijk} for each voxel ijk of a series of CT images, where the W_{ijk} are weighting factors describing the relative contribution of each voxel to the effective density at the calculation point. These weighting factors are different for each calculation point.

Although the theory behind this method appears to be reasonable, in practice it takes too long to perform a full summation over the entire irradiated volume. To overcome this limitation imposed by the calculation time the three dimensional density array (from CT) is reduced to two dimensions by coalescing pixels with the same x,y coordinates into a single slice at a distance Z_{eff} from the central slice. Z_{eff} is calculated from the weighted sum of the Z values of individual slices, where each weighting factor, which represents the scatter contribution of each slice to the calculation point, is calculated from the circular beam scatter-air ratio difference for two different radii depending upon the field size and slice position. After this simplification the W_{ij} (replacing W_{ijk}) are calculated from the probability of scattered photons from the density element ρ_{ij} reaching the point of calculation. Although this method was implemented on clinically used TPSs, the associated simplifications have been shown to cause some inconsistencies (Wong and Henkelman 1983; Carrasco et al. 2004). The main inherent limitation of this method is that the scatter is considered as a whole, without explicit ray tracing for calculation of the scatter dose (Wong and Purdy 1990). Further refinements in the calculation of the correction factor, are out of the scope of the *broad beam* methods. They are discussed briefly in the last paragraph of Section 26.3.2.2.

Of the above methods for inhomogeneity correction, the effective depth, the TAR (or TPR) ratio and the standard power law methods can be carried out by hand or using a computer. The more advanced application of power law, beam subtraction and ETAR methods can only be implemented with a computer. For all these methods, the density information can be obtained either from contours or from voxels (see Figure 25.4). However, using methods designed for relatively large inhomogeneities (such as the power-law correction) in situations involving voxel-sized inhomogeneities is likely to be significantly inaccurate and should be avoided.

26.2.4 Limitations and Drawbacks of Broad-Beam Methods

Beginning with the dose calculation under reference conditions, we have seen how it is necessary to apply many corrections to the original measured data in order for it to be accurate in real situations involving patients. Although the corrections applied are only approximations, they give reasonable results in the vast majority of clinical situations. However this broad-beam approach, which nevertheless requires a lot of experimental effort or somewhat arbitrary curve-fitting procedures, is inadequate as soon as the situation is more complex or when greater accuracy is required. This is, for instance, the case for non-rectangular or intensity-modulated fields (see Chapter 43 and Chapter 44) or when dose-escalation protocols are implemented. Because the speed of computers has increased dramatically since the start of computerized treatment planning, the broad-beam approach has now been superseded in clinical treatment planning by more sophisticated and computer-resource intensive algorithms based on the super-position principle.

26.3 SUPERPOSITION METHODS

26.3.1 THE SUPERPOSITION PRINCIPLE

The superposition principle is summarised in Figure 26.6.

If we consider separately the primary photons and the secondary particles (photons and/or electrons), the dose at a point P (x, y, z) can be considered as the sum of the contributions of the energy *launched* at a distance from P, in volume elements dV (x', y', z'). This elementary energy originates from the energy fluence $p(x', y', z')$ of the *primary* photons impinging on dV. Let $s(x, x', y, y', z, z')$ be the scatter energy per unit primary photon fluence launched at dV and reaching P. Then the dose at P is

$$D_P(x,y,z) = \iiint_V p(x',y',z')\, s(x,x',y,y',z,z')\, dV \qquad (26.12)$$

where the limits of integration V correspond to the limits of the actual scattering volume.

As discussed previously (see Figure 25.6) there are several alternative ways of performing this 3D integration, including reducing it to 1D or 2D integration in order to speed up the calculations. These possibilities, which imply different levels of underlying assumptions, have been extensively exploited in clinical treatment planning algorithms. Some of them will be discussed in what follows.

Referring to Figure 26.1 above, the dose at P is the sum of several components which involve photon and electron interactions and these could be treated separately. These interactions take place at a distance from point P. The electrons travel only a short path and

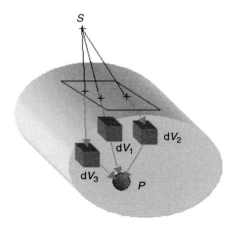

FIGURE 26.6
The superposition principle: the dose at P can be considered as the sum of the contributions of energy deposited in a small volume surrounding P by particles (electrons, photons) originating from primary photon interactions in distant volume elements dV_i.

therefore contribute to the dose at points which are typically at a maximum distance of a few mm (for ^{60}Co γ-rays), to between 3 cm and 4 cm (for 25 MV photons). By contrast, photons travel large distances without interacting and must therefore be considered, whether or not they are close to point P. In a homogeneous medium (i.e. water equivalent), even for high primary photon energy, electronic (or charged-particle) equilibrium is achieved for most of the points and the dose is, to a good approximation, only dependent on the local photon fluence (or kerma) (see Section 6.5). This is no longer true at tissue interfaces (e.g. patient skin, edges of inhomogeneities) where, especially at high energy, electron disequilibrium occurs and should be taken into account. It is also not true when the lateral dimensions of the beam are smaller than the range of the secondary electrons in the medium, e.g. for 25 MV photons in the lung ($\rho \approx 0.3$ g/cm^3) for field sizes smaller than ≈ 10 cm $\times$ 10 cm. If electron transport is faithfully modelled, however, then the effects of the lack of charged-particle equilibrium are automatically taken into account.

The three methods presented below are currently in fairly widespread use. They can be seen as different ways of combining the dose components on the one hand and the order of integration on the other hand. They all require some means to calculate the fluence (or the dose) for the primary photons. Although they are all largely based on experimental data, the nature and use of these data varies significantly according to the method and its practical implementation.

26.3.2 PRIMARY-SCATTER SEPARATION

26.3.2.1 Principle of the Method

The primary-scatter separation method as applied to computerised treatment planning was originally developed by Cunningham to solve the problem of dose calculation in irregular fields such as mantle fields (Cunningham et al. 1972). This idea originated from Clarkson's scatter integration method presented in Section 23.2.1.8. As seen in that section, the method makes use of the Scatter-Air Ratio, defined as the ratio of the *dose at a point due to the scattered radiation only* to the *dose in free space at the same point* (Cunningham 1972).

$$SAR(z,A_z) = TAR(z,A_z) - TAR0(z) \qquad (26.13)$$

where $TAR(z, A_z)$ is the tissue-air ratio at a depth z in the field of size A_z and $TAR0(z)$ is the tissue-air ratio at the same depth but in a field of zero area. $TAR0(z)$ is taken to represent the primary radiation. The reason for this is that, as the irradiated area is reduced, the scattering volume approaches zero and if there is no scattering volume there can be no scattered radiation, only primary radiation. The zero-area TAR is a mathematical abstraction*; it cannot be measured directly but it can be obtained by extrapolating TARs for small field sizes back to zero field size. Beyond the build-up region, $TAR0(z)$ can also be obtained by performing measurements in scatter-free conditions, typically using large distances between source and detector, placing in-between variable thicknesses of water equivalent material and collimating the beam both on the source side and on the detector side to reduce scatter as much as possible. In megavoltage beams, qualitative changes occur for off-axis points (see Section 22.3.3.2) and therefore, $TAR0(z)$ may have to be corrected, either selecting different tables according to the off-axis radius or applying a radial correction factor (Zefkili et al. 1994; Lee 1997).

---

* Strictly, the dose in free space or the dose in zero field size would be zero as there would be no charged-particle equilibrium (CPE); one has to imagine that in both these situations there is enough phantom-like material surrounding the point to establish CPE.

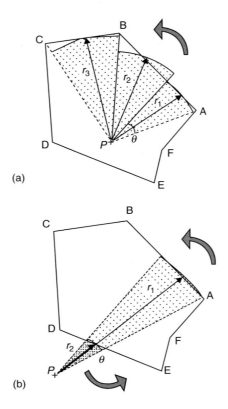

FIGURE 26.7
Clarkson's scatter-integration method: For a field such as ABCDEF, and a calculation point P, the assumption is made that the scatter from the whole area can be calculated by addition of the scatter contributions of angular sectors of angle θ and radii such as r_1, r_2, etc. The contribution from each of these sectors, i, to the total scatter dose at P is a fraction $\theta/2\pi$ of the scatter $S(z,r_i)$ at the centre of a circular field of radius r_i. If P is outside the field (as in b) a similar summation is possible, but in this case the contribution that would have arisen from the parts of the sectors that are outside the field must be subtracted from the total. This is achieved by proceeding around the edge of the contour as before, but assigning a positive sign to sectors for which the direction of rotation is anticlockwise (i.e. from A to B) and a negative sign to those for which it is clockwise (i.e. from D to E) so that the contribution of the small sector (r_2) is subtracted from the large sector (r_1) leaving the contribution from inside the field area.

The scatter component for any field can be calculated using Clarkson's method (Figure 26.7). This method uses sector integration; so the field must be divided up into n sectors centred on the point (x,y,z) at which the dose is required. If the ith sector has an angular width $\Delta\theta_i$ and radius r_i and the SARs are tabulated as functions of depth and field radius at that depth then its SAR can be found by interpolating from the tables and weighting it according to its angular width. The total scatter dose, $D_S(x,y,z)$ is found by summing over all sectors and is given by

$$D_S(x,y,z) = D_A(z) \times \sum_i S(z,r_i) \frac{\Delta\theta_i}{2\pi} \qquad (26.14)$$

where $D_A(z)$ is dose in air* (i.e. without attenuation) at position $(0,0,z)$, $S(z,r_i)$ is the scatter-air ratio for a depth z in a circular beam of radius, r_i.

---

* $D_A(z)$ at the distance f from the source and for the actual collimator opening A is easily calculated from the dose in air $D(\text{air}, A_{\text{ref}}, \text{SAD})$ at the isocentre distance SAD and for the reference field A_{ref} by multiplying this quantity by $(\text{SAD}/f)^2$ and by the collimator scatter factor (or air output factor) $S_c(A)$ (see Table 23.1 and Table 23.2).

The dose, $D_P(x, y, z)$ due to the primary radiation is

$$D_P(x, y, z) = D_A(z) \times T(z,0) \times f(x,y) \tag{26.15}$$

where $T(z,0)$ is the zero area TAR at depth z and $f(x,y)$ is an off axis factor in air that corrects for the beam profile.

The total dose at the point is then

$$D(x, y, z) = D_P(x, y, z) + D_S(x, y, z) \tag{26.16}$$

26.3.2.2 *Application to Cobalt-60 and Medium-Energy X-Ray Beams*

In-Phantom Basic Measurements: The application of primary-scatter separation for ^{60}Co γ-ray beams or medium energy x-ray beams (typically <6 MV) is relatively straightforward. TAR tables for square fields, including TAR0, are easily obtained from published tables or preferably from measurements. The *in air* reference dose D_A is obtained with an ionisation chamber with a 5 mm thick build-up cap for cobalt-60 (or thicker at higher energy). $S(z,r)$ tables are then derived using Equation 26.13 and the equivalence between a circular and a square field (see Section 23.2.1.8):

$$r = \frac{1}{\sqrt{\pi}} \times a \tag{26.17}$$

where r is the radius of the circular field equivalent to the square field $a \times a$.

In-Air Beam Profiles: The *in air* primary component must account for the beam profile as described by the correction factor $f(x,y)$ (see Equation 26.15). This factor is close to unity in the central region, decreases slowly for off-axis points as the oblique distance to the source increases, and decreases sharply at the edge of the collimator in the penumbra region. For rectangular beams ($2X \times 2Y$), Cunningham (Johns and Cunningham 1983) has suggested computing $f(x,y)$ as $g(x) \times g(y)$ where:

$$
\begin{aligned}
g(x) &= 0.5e^{(-\alpha_{ext}(x-X))} \quad \text{if } x > X \\
g(x) &= 1 - 0.5e^{(-\alpha_{int}(X-x))} \quad \text{if } x < X
\end{aligned}
\tag{26.18}
$$

a similar expression being used for $g(y)$, replacing x with y and X with Y.

α_{int} and α_{ext} are two *collimation* coefficients, characteristic of the penumbra width for the interior and the exterior of the beam respectively. The higher these coefficients, the sharper is the profile in the penumbra region. They can be adjusted according to the experimental in-air off-axis curves and shown to be practically proportional to the inverse of the width of the geometrical penumbra. A further correction to the above equation is required to account for the residual collimator transmission.

For x-ray beams, it may be necessary to account for the influence of the flattening filter which usually produces a profile in free air or at a shallow depth that is higher towards the edges than at the centre. This is necessary to produce a flat profile at larger depths (see Section 22.3.3.2). This is easily performed by introducing an additional correction with a multiplicative radial function representing the in-air measurement, normalised to the value on beam axis and measured along the diagonal of the largest field.

For irregular fields, Equation 26.18 cannot be applied as such. A modification (Cunningham et al. 1972) consists in assuming that the source is represented by an *intensity function* which can be imagined as a large radioactive disk seen through the collimator opening, with a local *density* which is infinite on the beam axis and decreases exponentially from the centre to the periphery. The detailed expression is given as a footnote[*]. In this expression, β is a collimation coefficient very similar to the α coefficient defined for rectangular fields. Within the Clarkson integration process, the β value may be different for the various field-delimiting segments. This offers the possibility of accounting for the differences in the penumbra, depending on whether it is generated by the main jaws or by blocks at a given point. Because β (or α) is directly correlated with the width of the computed penumbra, it is easy to adjust these to match experimental in-air profiles.

Intensity Modulation: Up to now we have implicitly assumed that the primary fluence (i.e. the dose in air) was uniform over the field area and that the patient surface was flat and normal to the beam axis, so that the scatter from an angular sector could be considered as a fraction of the scatter from a circular field at the depth of the calculation point (Figure 26.8a).

However, if the surface is oblique or irregular or if the primary beam is non-uniform (i.e. modulated by insertion of a wedge filter or compensator), it is possible, using the same $S(z,r)$ tables, to replace the 1D angular Clarkson integration by a 2D integration of elementary *pencil-like* scatter elements.

$$\Delta S_{ij}(z_{ij}, \theta_i) = \frac{\Delta \theta_i}{2\pi} \times (S(z_{ij}, r_{ij} + \Delta r) - S(z_{ij}, r_{ij})) \qquad (26.19)$$

where the scattering pencil beam is referenced to a polar co-ordinate system centred on the calculation point P, located at a distance r_{ij} along the radius defined by angle θ_i, and z_{ij} is the thickness of tissue above P as calculated along the pencil beam (see Figure 26.8b). In addition, when performing the integration over the field area, each scattering pencil beam must be weighted according to the primary fluence at point ij. When extended to higher energy photons (see below), this method gives satisfactory results for intensity modulated radiotherapy performed with multileaf collimators (Papatheodorou et al. 2000).

Figure 26.8c illustrates the so-called *differential scatter-air ratio* (dSAR) method (in a polar coordinate system) that is an extension of the Clarkson method to a 3D integration of scatter (Beaudoin 1968; Cunningham 1972).

Inhomogeneities: Separation of the medium into elementary volumes makes it possible to refine the corrections for inhomogeneities described previously in Section 26.2.3.2, provided that the primary-beam intensity reaching a voxel accounts for the transmission through the inhomogeneous medium and that the scatter from a voxel is weighted by its electronic density and also corrected by the presence of adjacent inhomogeneities. This was the basis of the dSAR method. However, due to the complex behaviour of multiple scatter, this method turned out to be inaccurate and had to be improved. This was attempted with the delta volume method (DV) proposed by Wong and Henkelman (1983) and with the dTAR method proposed by Kappas and Rosenwald (1986b). The ETAR method described by Sontag and Cunningham (1978) which also considers primary and scatter components separately, but which has been shown to be inaccurate in some situations (see Section 26.2.3.2) has been significantly improved by Redpath and Thwaites (1991). They allowed for the effective depth above the inhomogeneities and made a slight change to the method by which the equivalent density, $\tilde{\rho}$, was calculated, taking into account the scatter changes with respect to depth. They also optimised the calculation time and obtained quite good results for several benchmark examples.

---

[*] For a circular field and a calculation point on the beam axis, the normalised apparent *density* seen through the projected opening of radius R is:

$$Dens\,(R) = \frac{\beta^2}{2\pi} \int_0^{2\pi} \int_0^{R} e^{-\beta r} r\, dr\, d\theta = 1 - (1 + \beta R)e^{-\beta r}$$

The Clarkson integration method is then applied to this expression for any point x,y:

$$f(x,y) = \frac{\Delta \theta}{2\pi} \sum_{i=0}^{n} Dens(r_i)$$

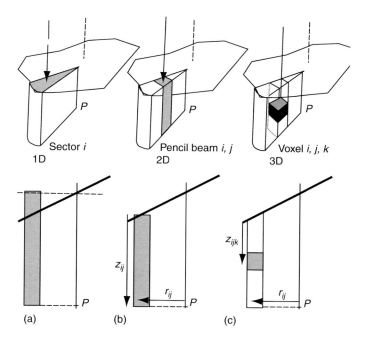

FIGURE 26.8
Some of the possibilities for medium decomposition to obtain the total scatter at point P for an irregular field. The upper views are 3D representations of the elementary volumes. The lower views are 2D cuts through the angular sectors in the case of surface obliquity. (a) Conventional Clarkson's 1D angular integration; the scatter at P is calculated assuming that the depth of all scattering elements is the same as P (shaded scattering column or *pencil*). (b) Pencil-like, 2D angular + radial integration; the scatter from each scattering pencil (θ_i, r_{ij}) is calculated taking into account the surface obliquity, with the possibility of weighting the scatter according to the primary-beam intensity at (θ_i, r_{ij}). (c) Voxel-based, 3D integration (dSAR method) which requires the calculation of the attenuation of the primary before reaching the voxel element i,j,k.

26.3.2.3 *Extension to Higher Energies*

As the photon energy increases, the quantities related to the *in-air* reference become irrelevant because of the need for large build-up caps to ensure electronic equilibrium and because of the importance of the attenuation and scatter contribution in such caps. The TAR must therefore be replaced by other quantities (see Section 23.2.1). However, the same general approach is still possible with the introduction of the mini-phantom concept (see Section 23.2.1.12; Drouard et al. 1986; van Gasteren et al. 1991; Dutreix et al. 1997). The attenuation of the *primary* component is then represented by the tissue-phantom ratio in the mini-phantom, $\mathrm{TPR}(z, \mathrm{ESQ_m})$, where $\mathrm{ESQ_m}$ is the equivalent square corresponding to the mini-phantom geometry[†]. By extension of Equation 26.13, the scatter air ratio used in the Clarkson integration is replaced by SAR^* calculated from the following equation at depth z for the field A_z:

$$\mathrm{SAR}^*(z, A_z) = \mathrm{TAR}^*(z, A_z) - \mathrm{TPR}(z, \mathrm{ESQ_m}) \tag{26.20}$$

where TAR^*, the replacement quantity for TAR, can be written as:

$$\mathrm{TAR}^*(z, A_z) = \mathrm{TPR}(z, A_z) \times \frac{S_p(z_{\mathrm{ref}}, A_z)}{S_p(z_{\mathrm{ref}}, \mathrm{ESQ_m})} \tag{26.21}$$

[†] To obtain a quantity that accurately represents the primary component, one should correct the TPR for the scatter from the mini-phantom (Zefkili et al. 1994). This could be necessary for accurate calculation of the dose under small blocks.

In this expression, S_p is the phantom scatter factor defined in Section 23.2.1.10. In addition, the *in-air* reference dose D_A of Equation 26.14 and Equation 26.15 has to be replaced by the dose in the mini-phantom at depth z_{ref} for the current collimator opening A.

This method provides accurate results for rectangular or irregular fields of dimensions larger than ESQ_m. However, it implicitly assumes that the path length of secondary electrons is negligible and that the *primary* penumbra can be accurately modelled by adjustment of the collimation coefficients described above, independently of depth and field size. Nevertheless, at higher photon energies, secondary electron path lengths cannot be ignored and specific adaptations must be performed to account properly for electron transport in water-equivalent media (Rosenwald et al. 1987; Woo et al. 1990).

The advantage of the primary-scatter separation is that it allows one to account accurately for the influence of beam modifiers (modulation of the primary and even of the scatter component) and for the volume of tissues included within the beam limits. Its main limitation is related to the accuracy of inhomogeneity corrections which are normally performed according to one of the methods described above (see Section 26.2.3.2 and Section 26.3.2.2). In particular, these methods cannot accurately handle the situations where there is a lack of electronic equilibrium at the beam edge or on the axis of small fields, e.g. high energy photon beams through lungs (Mohan and Chui 1985; Mackie et al. 1985; AAPM 2004).

26.3.3 PENCIL-BEAM APPROACH

26.3.3.1 *General*

The first mention of using the pencil-beam approach for photon beams can be ascribed to Schoknecht (1971). The method was in fact developed first for electron beams (see Chapter 27). A detailed model for photon beams was first described by Mohan and Chui (1987). Since that time, the literature has grown and one can mention papers by Ahnesjö et al. (1992b), Bourland and Chaney (1992), and Bortfeld et al. (1993), that deal with implementation aspects, and other papers that discuss the limitations of the pencil-beam method (e.g. Knöös et al. 1995) in certain inhomogeneous situations.

In essence, convolving a pencil beam over the field area defined by the final collimator is similar to performing the 2D integration described in Figure 26.8b. For example, one can easily model any variation in incident beam intensity, due, say, to a wedge or compensator. This approach is also suited to modelling complex treatments such as dynamic wedges or intensity-modulated radiation therapy, where the incident fluence is modulated over time. Furthermore, the beam spectrum variations, either off-axis or due to filtration by a wedge or any arbitrary compensator can, in principle, be taken into account by changing the pencil-beam quality according to its entrance position. The main difference between the two approaches is the following: In the primary-scatter separation, the scatter component is the only one which is scanned over the field area whereas in the pencil-beam approach described below, the *pencil* is made of the total energy (from secondary electrons and scattered photons) generated by the primary photons and deposited at a distance. This energy is characterised by a *pencil-beam convolution kernel*, K_{PB}, which represents the energy distribution released from the pencil beam and normalised to unit fluence at the penetration point on the patient surface (see Figure 26.9).

The general expression for the dose computed from pencil-beam *kernels* is therefore:

$$D(x,y,z) = \iint \frac{\mu}{\rho} \Psi_E(x',y') K_{PB}(x-x', y-y', z) \mathrm{d}x' \mathrm{d}y' \qquad (26.22)$$

where the 2D integration is restricted to the field area; $\Psi_E(x',y')$ is the *entrance* energy fluence at point P' (J m^{-2}); $\frac{\mu}{\rho} \Psi_E(x',y')$ is the total energy (secondary electrons and scattered photons) per unit mass released from P', known as *terma* (J kg^{-1} or Gy), μ/ρ being the mass attenuation

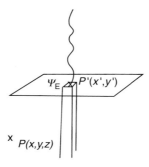

FIGURE 26.9
Principle of the pencil-beam approach. The dose at point P is computed from the integration over the field area of pencil kernels weighted by the primary fluence impinging on the patient surface.

coefficient ($m^2 \ kg^{-1}$) for medium at P' (see Chapter 4); $K_{PB} \ (x-x', y-y', z)$ is the value of the kernel at point $P(x, y, z)$ for a pencil beam impinging on the patient at point $P'(x', y')$; it represents the fractional energy deposition per unit mass at P due to primary energy fluence entering the patient at P'.

In the above expression, the pencil-beam kernels are in principle unique for each position of both P and P'. However, in a homogeneous water-like medium, one can consider that, at a given depth z and distance between P and the pencil-beam axis, they are as a first approximation independent of the pencil-beam position (x', y'). In that case Equation 26.22 is actually a convolution between the incident energy fluence and the pencil-beam convolution kernels. In this case, as discussed in Section 26.3.3.4, the computation can be speeded up by means of fast Fourier transform (FFT) convolution algorithms. If this assumption is considered as unacceptable (e.g. because of off-axis spectral variations due to the flattening filter) or if inhomogeneities are considered, then the kernel is no longer spatially invariant and the mathematical solution is a *superposition* where the FFT approach is no longer applicable.

In both cases, the dose calculation problem involves the determination of the entrance energy fluence on the one hand and of pencil-beam kernels on the other hand.

26.3.3.2 Determination of Ψ_E, the Entrance Primary Energy Fluence

In the absence of beam modifiers, the primary fluence is basically uniform within the field aperture and drops to a small residual value outside. The method of calculation is generally based on geometrical considerations complemented by calculation of transmission through the main collimator or convolution with a source function, to account for the finite size of the source (Mohan and Chui 1987). As seen in Section 26.3.2.2, the influence of the flattening filter can be further accounted for by application of a radial function derived from dose profile measurements along the diagonal of the largest field, performed at shallow depth (just large enough to avoid electron contamination) (Storchi and Woustra 1995). The extrafocal radiation (i.e. radiation not originating in the x-ray target) generated as scatter mainly in the primary collimator and in the flattening filter should be included (Ahnesjö and Trepp 1991; Jaffray et al. 1993; Papanikolaou et al. 1993). A frequent approach, which can be used also for irregular fields, consists in calculating the scattered dose from the head by integration of an extended source model over the parts of the head visible from the point of calculation (Ahnesjö 1994; Sharpe et al. 1995; Liu et al. 1997a).

The dose at the edge of collimating blocks can be calculated with the same method as for the main collimator (Mohan and Chui 1987; Storchi and Woustra 1996). When additional beam modifiers are used (wedges, compensators), the primary beam is attenuated but also some scatter is generated which can be accounted for by performing scatter integration over the modifying device (Ahnesjö et al. 1995; Islam and Van Dyk 1995; Castellanos and Rosenwald 1998).

Altogether, the methods for calculation of the primary fluence distribution are very similar to the methods described in the Section 26.3.2.2 (*in air* profiles). They can also be thought of as the product of correction matrices, each of them representing the beam attenuation or modulation due to each specific beam limiting or modifying device. Ideally, such calculations should also account for the spectral modifications while passing through the modifiers, e.g. wedges, compensators (Liu et al. 1997b). It should be realised that with the development of intensity-modulated radiotherapy, it is becoming very important to include an accurate value for the transmission of the individual leaves of multileaf collimators.

Another issue is the fact that, in principle, for each position of P' the local pencil-beam kernel should be modified to compensate for the changes in the energy spectrum owing to off-axis *softening* as the flattening filter gets thiner and to beam filtration from additional beam modifiers (Tailor et al. 1998). This is generally ignored.

26.3.3.3 *Determination of* K_{PB}, *the Pencil-Beam Kernels*

There are several methods of obtaining the relevant pencil-beam kernels. Mohan and Chui (1987) performed direct Monte Carlo (MC) calculations, for a number of monoenergetic pencil beams as well as for polyenergetic linear-accelerator pencil beams. In practice, polyenergetic kernels can be computed by simple addition of monoenergetic kernels, each of them being weighted according to the local x-ray spectrum. However, such an approach requires the spectrum of the photons emitted by the accelerator to be known (ideally at any point on the patient surface). This spectrum cannot be easily measured in a clinical environment. It can, however, be computed by the Monte Carlo method from the characteristics of electrons hitting the target and details of the head characteristics (Mohan et al. 1985; Lovelock et al. 1995; Liu et al. 1997a) (see also Section 28.3). It can also be derived from attenuation measurements (Huang et al. 1983; François et al. 1997; Nisbet et al. 1998) or adjusted empirically to minimise the deviations between computed and measured depth-dose distributions (Ahnesjö and Andreo 1989; Starkschal et al. 2000).

Other approaches for pencil-beam kernel determination include integration of pre-calculated point kernels (see Section 26.3.4), deconvolution from broad beam measurements (Chui and Mohan 1988), or experimental estimation using scatter-factor differentiation (Ceberg et al. 1996; Storchi and Woustra 1996).

In what follows, we will describe some examples of application of the pencil-beam method, including short descriptions of different *tricks* which have been used for clinical implementations.

26.3.3.4 *Some Practical Implementations*

As indicated previously, under certain circumstances, the fast Fourier transform can speed up the computation enormously (e.g. Boyer 1984). This was successfully exploited by Mohan and Chui (1987). They demonstrated the power of the pencil-beam method in predicting dose distributions for irregular fields in a uniform medium with a flat incident surface (Figure 26.10). They obtained their elementary pencil beams for cobalt-60, 6 MV and 18 MV qualities by running the EGS4 Monte Carlo code. They used the FFT method assuming that the pencil-beam *kernel* was invariant with respect to its position (i.e. they neglected the consequences of their off-axis spectral variations). Nevertheless, Mohan and Chui (1987) made the point that their method relied only on basic principles of physics, without resorting to empirical assumptions or employing arbitrary analytical functions to describe the source distribution or the shape of the boundary. They stated that the reason for this was that the Monte Carlo generated pencil beams automatically took into account the transport (in water) of scattered photons and secondary electrons.

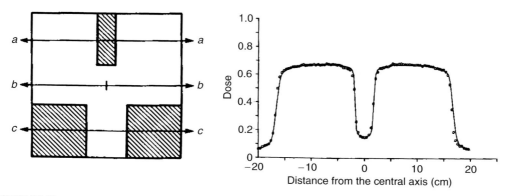

FIGURE 26.10

Pencil-beam FFT convolutions vs. measurements for a blocked 6 MV beam. The right-hand side figure shows a profile at level (a) of the irregular field displayed on the left-hand side taken at 10 cm deep. On the graph the points represent measurements and the line represents the convolution calculation. (From Mohan, R. and Chui, C.-S., *Medical Phys.*, 14, 70–77, 1987. With permission.)

Bortfeld et al. (1993) maximised the efficiency of the pencil-beam method by decomposing the pencil-beam kernel into three separate terms and thereby managed to reduce the required number of 2D convolutions; they also employed the fast Hartley transform. They reported a calculation time for the three convolutions of less than 10 s, out of a total time of only 40 s, for a 3D dose calculation of a 10 cm ×10 cm field (using a DEC VAXstation 3100). For the 15 MV beam, excellent agreement with measurement was demonstrated for the case of various irregular fields in a uniform water medium.

Ahnesjö et al. (1992b) presented a complete pencil-beam model for clinical use. They employed polyenergetic pencil-beam kernels, obtained from in-depth convolution of Monte Carlo-derived energy deposition point kernels (see Section 26.3.4.3) with the fluence variation of the primary photons (Ahnesjö and Andreo 1989). They found that these pencil beams could be represented analytically with high accuracy by a sum of two exponentials over the radius:

$$K_{PB}(r,z) = \frac{A_z e^{-a_z r}}{r} + \frac{B_z e^{-b_z r}}{r} \qquad (26.23)$$

where r is the cylindrical radius from the pencil-beam axis and the four coefficients are fitting parameters which depend on the depth, z.

In the Ahnesjö et al. (1992b) pencil-beam model, for all dose components except (photon) scatter, the effects of tissue variations and patient contour are modelled by depth scaling, i.e. by calculating the dose using pencil-beam parameters taken at the radiological depth (i.e. the equivalent depth in water). For the scatter dose, the pencil-beam parameters at the geometrical depth are used and the kernel integration is followed by the application of a special scatter correction factor described in their paper. Knöös et al. (1994) reported on a dosimetric verification of the Ahnesjö et al. (1992b) pencil-beam model as implemented at that time in the *TMS/HELAX* TPS but this study was confined to homogeneous water phantoms, where its performance was found to be excellent.

Another pencil-beam algorithm implementation was performed by Storchi and Woustra (1996) who based their model parametrisation on a limited number of experimental data and introduced both *scatter* and *boundary* kernels. The comparison between measurements and calculations was performed for rectangular and irregular fields, with and without wedge filters, yielding deviations less than 2% except for 45° wedges or under blocks where it could be between 4% and 5%. This algorithm was incorporated into the Cadplan (Dosetek–Varian) TPS and migrated more recently into the Eclipse (Varian) system.

26.3.3.5 Inhomogeneities and Other Limitations

Bortfeld et al. (1993) stressed the inherent limitations of the pencil-beam approach: that high accuracy is in principle only achievable for homogeneous phantoms with flat surfaces. The very same point was made by Mohan and Chui (1987). Nor do Ahnesjö et al. (1992b) make any exaggerated claims for the performance of their particular pencil-beam model in inhomogeneous situations.

Knöös et al. (1995) made a detailed study of the limitations of the pencil-beam approach in the case of lung tissue. They investigated the effects of scaling operations for heterogeneity corrections, such as the method described in Ahnesjö et al. (1992b), by comparisons of the pencil-beam model in the *TMS/HELAX* system with Monte Carlo calculations in the admittedly difficult geometry of the mediastinum. Deviations in low-density volumes were found to increase with increasing beam energy from around 3% at 4 MV to 14% at 18 MV; this was ascribed to electron disequilibrium. They concluded that the magnitude of this limitation of the pencil-beam model in heterogeneous situations was similar to that found in *conventional* inhomogeneity correction models (see Section 26.2.3.2 and Section 26.3.2.2).

The pencil-beam approach has been quite extensively implemented in commercial planning systems. One reason for this is that it represents an acceptable compromise between speed and accuracy for most of the clinical situations and that it is particularly well suited for intensity modulation techniques (including inverse planning: see Chapter 43). However, the principle of the method is very similar to primary-scatter double decomposition; its quality depends strongly of the detailed implementation and it suffers inherently from serious limitations in handling inhomogeneity corrections properly since the pencil beam is generally treated as a whole, i.e. not separated into primary and scatter components. For more accurate results in the presence of inhomogeneities, so-called *differential* pencil-beam approaches, usually referred to as *point-kernel convolution/superposition methods*, are preferable. These are covered in the following section.

26.3.4 POINT KERNEL CONVOLUTION/SUPERPOSITION

26.3.4.1 General

The idea of the point kernel convolution/superposition[*]approach to dose calculation for treatment planning was put forward by several independent groups in 1984 (Ahnesjö 1984; Boyer and Mok 1984; Chui and Mohan 1984; Mackie and Scrimger 1984). This was prior to the first attempts to use pencil-beam kernels for photons (see Section 26.3.3). Actually, in contrast to what happened with electron beams (see Chapter 27), the photon pencil-beam approach was developed essentially as a subset of the more general photon point-kernel approach. Although much of the material pertaining to pencil-beam kernels holds also for point kernels, a full description is justifiable for a method which has been implemented in several commercial planning systems and represents a good compromise between the limitations of the other superposition methods and the potentially highly accurate but computer-resource intensive Monte Carlo computations (see Chapter 28).

The principle of the method is illustrated in Figure 26.11.

For an homogeneous medium, this can be expressed by the following equation:

$$D(x, y, z) = \iiint \frac{\mu}{\rho} \, \Psi(x', y', z') \, K(x - x', y - y', z - z') \, dV' \qquad (26.24)$$

---

[*] The fundamental mathematical operation is *superposition*, but if the kernel is spatially invariant, *convolution* can be used as a means of rapid calculation of the superposition (see Section 26.3.4.6).

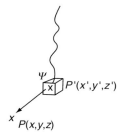

FIGURE 26.11
Principle of point kernel convolution: the dose at P is calculated from 3D integration of the energy released from voxels such as dV' centred on P' (see also Figure 26.6).

The 3D integration is performed over the patient volume; $\Psi(x',y',z')$ is the energy fluence at point P' ($\mathrm{J\,m^{-2}}$); $dV' = dx'dy'dz'$ is an elementary volume (approximated by a voxel) around P'; $\frac{\mu}{\rho}\Psi(x',y',z')$ is the total energy (secondary electrons and scattered photons) per mass released from dV, known as *terma* ($\mathrm{J\,kg^{-1}}$ or Gy), μ/ρ being the mass attenuation coefficient ($\mathrm{m^2\,kg^{-1}}$) for the medium at P'; $K(x-x',y-y',z-z')$ is the *energy deposition point kernel* (sometimes called *dose-spread array* or *point spread function* or *differential pencil-beam kernel*) which represents the fraction of energy released from dV' that is deposited at P, when primary photons interact at P'.

A more general form of the above equation for a clinical polyenergetic beam consists in performing an overall integration of the expression according to the energy bins of the local beam spectrum, where Ψ should be replaced by the energy fluence differential in energy, μ/ρ and K being both energy dependent. Simplifying the expression by the use of an equivalent mean energy is not accurate enough but a limited number of bins can be used as demonstrated by Boyer et al. (1989) and Zhu and Van Dyk (1995). Some other practical solutions will be discussed below.

In common with the pencil-beam approach, the dose computation requires that the energy fluence (or the terma) and the point kernels are first determined.

26.3.4.2 Determination of the Energy Fluence and Terma

The determination of the 2D energy fluence at the patient surface has been discussed above as part of the pencil-beam approach. This fluence is then transported through the patient by ray tracing along paths diverging from the source position, applying the inverse square-law correction and a tissue attenuation correction according to the patient composition integrated between the patient surface and P'. For extra-focal radiation this is an approximation;. however it is acceptable as long as this contribution is small. In principle the attenuation calculation should account for the depth-hardening effect on the primary beam as well as the off-axis softening due to the influence of the flattening filter (Zefkili et al. 1994; Lee 1997).

The terma at any point is then obtained by multiplication of the energy fluence and mass attenuation coefficient, the latter depending on the local energy spectrum and tissue composition (Metcalfe et al. 1990; Papanikolaou et al. 1993).

26.3.4.3 Determination of the Energy Deposition Point Kernels, K

The standard method for determination of K for monoenergetic beams in water is Monte Carlo computation in a phantom made of small, adjacent voxels, where the primary photons are forced to interact at the voxel at the origin (Figure 26.12).

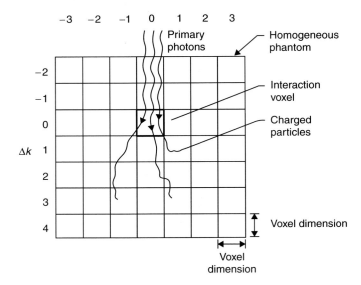

FIGURE 26.12
Generation of an energy deposition kernel in a homogeneous phantom by Monte Carlo computation. (From Mackie, T. R. et. al., *Med. Phys.*, 12, 188–195, 1985. With permission.)

Spherical geometry is usually chosen, using polar coordinates with set radial and angular intervals in order to obtain an accurate representation of energy deposition at any distance from the interaction point (Mackie et al. 1988; Mohan et al. 1986; Ahnesjö et al. 1987). The graphical representation of the kernels for various energies either as isolevels or profiles is illustrative of the energy-deposition processes and helps in understanding the principle of the method (Figure 26.13).

Figure 26.13 also illustrates the possibility of computing separately *primary* kernels (related to the energy deposited by the charged particles set in motion by primary photons) and *scatter* kernels (related to the energy deposited by all charged particles set in motion by scattered photons). Such a decomposition could be further extended to different orders of scatter, bremsstrahlung and annihilation radiation and may prove to be useful for simplification and/or acceleration of computation. In addition, the decrease in the contribution of scattered photons and the increase in the importance of electron transport as the energy increases are clearly seen in the figure, as well as the directional behaviour.

Although other possibilities for normalisation have been suggested (Ahnesjö and Aspradakis 1999), the kernel K in Equation 26.24 is the fraction of incident energy per unit volume. It can be therefore expressed as cm^{-3} and should be consistent with the equation

$$\iiint_{\infty} K(x,y,z)\, dV \equiv 1$$

As pointed out by Mackie et al. (1988) and Boyer (1988):

$$\iiint_{\infty} K_p(x,y,z)\, dV = \frac{\mu_{en}}{\mu} \quad \text{and} \quad \iiint_{\infty} K_s(x,y,z)\, dV = \frac{\mu - \mu_{en}}{\mu} \qquad (26.25)$$

where K_p is the *primary* kernel; K_s the *scatter* kernel; μ is the linear attenuation coefficient; and μ_{en} is the linear energy-absorption coefficient (see Chapter 4).

This turns out to be useful for verification of the generated kernels and for the application of relevant corrections for beam-quality variations.

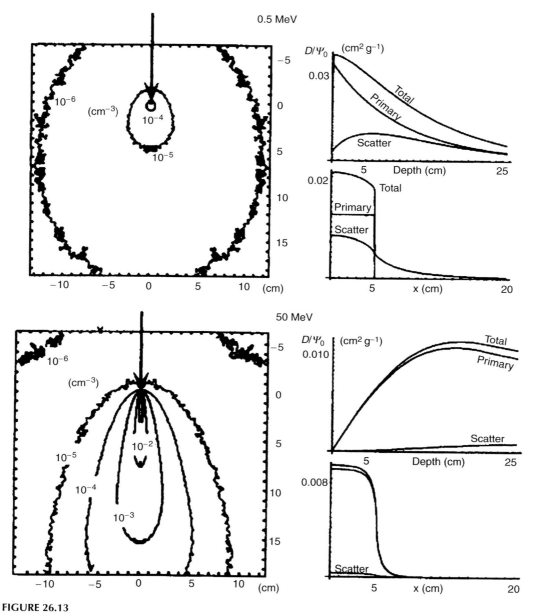

FIGURE 26.13
Graphical representation of energy deposition kernels for 0.5 MeV (top) and 50 MeV photons (bottom), both in isolevel format (left) and profiles format (right). (From Ahnesjö, A. and Aspradakis, M. M., *Phys. Med. Biol.*, 44, R99–R155, 1999. With permission.)

When modelling clinical beams, it is important to account as faithfully as possible for their spectral composition and the spatial variations thereof. The most straightforward method is to generate polyenergetic kernels by weighted addition of precalculated monoenergetic kernels according to the normalised incident fluence spectrum. This spectrum must first be obtained by one of the methods explained above for the pencil-beam approach. However, the fluence spectrum changes with depth and lateral position. Because it is impractical to use different kernels for different positions of P and because the overall dose per unit fluence does not change much, it has been considered as acceptable to make corrections according to Equation 26.25. Thus the primary kernel is multiplied by a factor equal to the ratio between μ_{en} at point P' and

μ_{en} at the surface and the scatter kernel is multiplied by a similar ratio of energy-scatter coefficients $(\mu-\mu_{en})$ (Metcalfe et al. 1990; Papanikolaou et al. 1993; Hoban 1995).

Instead of using tables of precalculated kernels, Ahnesjo and Mackie (1987) have suggested an analytical fit of the form:

$$K(r) = \frac{A_\theta\, e^{-a_\theta} + B_\theta\, e^{-b_\theta}}{r^2} \qquad (26.26)$$

where r is the distance between P and P'; A_θ, a_θ, B_θ and b_θ are fitting parameters depending on the scattering angle, θ.

In this formulation, the first term mainly describes the primary and the second one mainly the scatter-energy deposition fraction.

26.3.4.4 Beam Divergence and Tilting of the Kernels

In addition to a modification of the photon fluence, the beam divergence introduces a modification of the kernel directions which are *tilted* accordingly. Therefore the same kernel should strictly not be used for different off-axis distances. In practice, it has been found acceptable in most clinical cases to neglect this effect (Sharpe and Battista 1993), especially if it is compensated for by applying the inverse square-law correction at dose calculation sites instead of at each primary interaction site (Papanikolaou et al. 1993). This issue has also been addressed by Liu et al. (1997c).

26.3.4.5 Kernel Scaling according to Tissue Density and Tissue Inhomogeneities

If the density of the medium is different from unity, then according to the O'Connor theorem the same dose distribution would be obtained by scaling appropriately all phantom dimensions (see Section 25.2). This is equivalent to using a scaled distance for fluence computation at P' and scaling the kernels in all directions according to the inverse of the density. This is, in turn equivalent to replacing all distances in Figure 26.11 and Equation 26.24 by the equivalent radiological distances. In Figure 26.11 we have to consider separately the density ρ_1 of the medium between surface and P' which modifies $\Psi(x',y',z')$, the density ρ of the medium surrounding P' which modifies the *terma* term according to μ/ρ and the density of ρ_2 of the medium between P' and P which changes the kernel value to $K[\rho_2(x\text{-}x'),\ \rho_2(y\text{-}y'),\ \rho_2(z\text{-}z')]$. This leads to a generalisation of Equation 26.24 for an inhomogeneous medium where all the distances can be scaled according to the relevant radiological equivalent path lengths with respect to the voxel-density distribution and the positions of P and P'.

It is clear however that, as soon as the kernel K can no longer be considered as spatially invariant, Equation 26.24 is no longer a convolution and the integration must be conducted by the superposition method instead. Therefore, one expects much longer computation time if inhomogeneities are taken into account. Zhu and Boyer (1990) have presented an alternative way to compute the dose in 3D for inhomogeneous materials. The kernels remain spatially invariant but the fluence is computed by ray tracing through the 3D CT matrix of inhomogeneous densities; In this way the possibility of using Fourier space with its attendant speed advantages is retained. Zhu and Boyer (1990) show that the method compares well with measurements for fairly simple inhomogeneous phantoms. Obviously it will fail in the region of large mismatches at interfaces where there is no electronic equilibrium, i.e. in exactly those situations where one might expect significant advantages from the point-kernel approach compared to the pencil-beam approach. Even when the kernels are scaled, Equation 26.24 in an inhomogeneous medium is only approximate. In particular, the densities to be considered should include not only the voxels between P' and P but also the other surrounding voxels. Another approximation comes from the fact that rectilinear scaling is not exact with respect to multiple scattering of secondary electrons in an inhomogeneous medium. Therefore

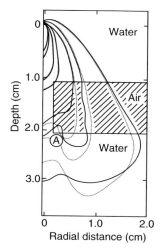

FIGURE 26.14

Graphical comparison of a density scaled kernel (dotted line) and a Monte Carlo kernel (full line) in the presence of an air ring; it can be assumed that the Monte Carlo-derived kernel correctly shows the effect of the inhomogeneities. (From Woo, M. K. and Cunningham, J. R., *Med. Phys.*, 17, 187–194, 1990. With permission.)

a number of refinements have been suggested, which inevitably lead to an increase in computation time (Keall and Hoban 1995; Sauer 1995; Yu et al. 1995). In spite of the inherent limitations and approximations of the scaling method, it seems that in most situations there is an *averaging* effect of the surrounding kernels which gives acceptable results (Woo and Cunningham 1990) (Figure 26.14). Significant discrepancies occur only in extreme situations according to Arnfield et al. (2000). It should also be remembered that the results can be very different depending on the practical implementation of the method.

26.3.4.6   Practical Implementation

Direct summation (superposition) of the dose using an equation such as Equation 26.24, implies repeating the elementary computation at n^3 deposition voxels from n^3 interaction voxels. The resulting order of calculation is n^6. If we apply a scaling with respect to the positions of P and P', this is likely to further increase the order to n^7. This is very demanding and efforts have been made to reduce the computation time.

When convolution is possible (i.e. for homogeneous medium), the use of FFT drastically reduces the number of operations. Instead of n^6 operations for full superposition, a 3D FFT algorithm would generate a number of operations proportional to $n^3 \log_2 n$ (Ahnesjö and Aspradakis 1999).

Another method for reducing the computation time consists of avoiding having to account systematically for all n^3 voxels which theoretically contribute to the dose at any specific single point. This has been done in the so called *collapsed cone convolution* (CCC) method which takes advantage of the fact that at large distances the kernel values fall off rapidly (Mackie et al. 1987; Ahnesjö 1989). Consequently, if we draw in 3D a set of m lines (say for instance 3 orthogonal and $6+4$ diagonal) from the deposition site, they will intersect the neighbouring voxels preferentially and avoid more distant voxels (see Figure 26.15). If we calculate only the contribution from the voxels along these m lines, there will be n.m voxels sampled for each of the n^3 deposition voxels and the number of operations is reduced to the order $n^4 m$. (instead of n^6). It is however necessary to make sure that the total transported energy remains the same as if all voxels were sampled. Therefore some normalising procedure is required which *collapses* the energy emitted from all neighbouring voxels onto the average direction defined by the sampling lines. This is where the name "collapsed cone" comes from.

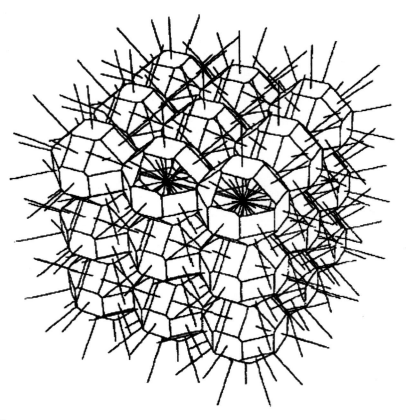

FIGURE 26.15
Schematic representation of 26 possible discrete directions of particle transport (13 cone axes) from each interaction voxel of a lattice made of $3\times3\times3$ voxels. (Reproduced from Ahnesjö, A. and Aspradakis, M. M., *Phys. Med. Biol.*, 44, R99–R155, 1999. With permission.)

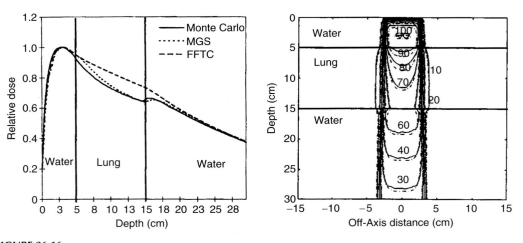

FIGURE 26.16
Comparison between a full Monte Carlo calculation (solid line) and a derivation of the CCC method called MGS (dashed line) for a 5 cm×5 cm 15 MV beam with a lung slab (density 0.31) 10 cm thick. (From Miften, M. et al., *Phys. Med. Biol.*, 45, 817–833, 2000. With permission.)

The CCC method or other methods derived from it have been successfully implemented in several clinical treatment planning systems: Philips-Pinnacle (Mackie et al. 1985; Bedford et al. 2003), Helax-TMS which has become Oncentra Masterplan (Ahnesjö et al. 2005), CMS Focus which is now called XiO (Miften et al. 2000), and the DKFZ system (Scholz et al. 2003). These implementations account, for example, with reasonable accuracy for both the observed penumbra broadening and dose reduction on the axis of a high energy photon beam passing through lung (i.e. a low-density inhomogeneity) (Figure 26.16). Both of these effects are due to secondary electron transport and cannot be accounted for by any of the simpler super-position methods discussed in Section 26.3.2 and Section 26.3.3.

26.4 CONCLUDING REMARKS

A very wide variety of methods for photon dose computation in external-beam therapy has been implemented in treatment planning systems. These range from very basic dose representation methods (Section 26.2) to sophisticated methods involving convolution/superposition of precalculated kernels (Section 26.3). Although most of them perform well in simple clinical situations, there are still many cases where approximations have to be made for some or all of the different algorithms. These include the dose representation in the build-up region, under blocks, at beam edges and in the presence of inhomogeneities. The superposition methods with all their variants are the most promising. Nevertheless, it is necessary to find a compromise between speed and accuracy and to recognise the limitations of the various algorithms, especially for situations where a lack of electronic equilibrium can be of clinical importance such as in the lung particularly at high photon energies. It is emphasised that there is no completely correct way of scaling the water-based point kernels/spread functions to account for the effect of inhomogeneities (Figure 26.14) short of using Monte Carlo simulation (Chapter 28).

CHAPTER 27

PATIENT DOSE COMPUTATION FOR ELECTRON BEAMS

Alan Nahum

CONTENTS

27.1 INTRODUCTION

Methods of calculating the dose distribution in patients undergoing external-beam radio-therapy have gradually evolved over the years. The emphasis has justifiably been on photon beams (see Chapter 25 and Chapter 26). The division of a *broad* beam into a large number of narrow beams, or *pencils*, would seem to be a natural line of attack, given that this corresponds to the way the particles, either photons or electrons, are incident on the phantom or patient. However, it was not until the early 1980s that pencil-beam methods were first exploited, and the first models were developed for electrons (Brahme et al. 1981; Hogstrom et al. 1981; Lax et al. 1983) rather than for photons. This was possibly due to the fact that the inadequacies of so-called *broad-beam* approaches (see Section 26.2) in modelling irregular fields and inhomo-geneities are more dramatic for electrons than for photons (e.g. Figure 27.1).

The overall concept is the same regardless of the incident particle type, i.e. integration or *convolution* with the incident fluence at the collimator, but the physical forms of the elementary photon and electron pencil-beam dose distributions are very different. This is due to the difference between the physics of photon and electron transport in matter, at least as far as the primaries are concerned (see Chapter 3 and Chapter 4). Naturally, the photons produce electrons which then must, in principle, be transported in the same way as if they were the primary particles.

For photons, the *excitement* since the mid 80s has not been about pencil-beam methods, but about the *convolution* of *point spread dose functions (kernels)* with the spatial distribution of primary-photon fluence (see Section 26.3.4). For photons, pencil beams represent one level below this in complexity but with the advantage of shorter computation times (see Section 26.3.3). On the other hand, for electrons there is *no* level beyond that of the pencil beam. It makes no sense to discuss electron point spread dose functions because there is no *primary* component; all the electrons interact immediately upon entering the medium or the patient; in other words, *the scatter component is 100% at all depths* (see Chapter 3).

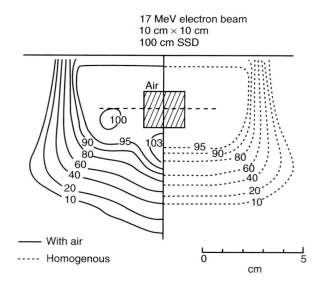

FIGURE 27.1
Perturbation of electron-beam dose distribution due to a low-density heterogeneity. The distribution in a homogeneous water phantom is shown on the right side. The introduction of an air cavity (2 cm×2 cm), shown on the left side, causes an alteration of the shapes of the isodoses including a *hot* region of 103% relative to the maximum dose on the left side. (Reproduced from ICRU, Report 71, International Commission on Radiation Units and Measurement, Bethesda, MD, 2004. With permission.)

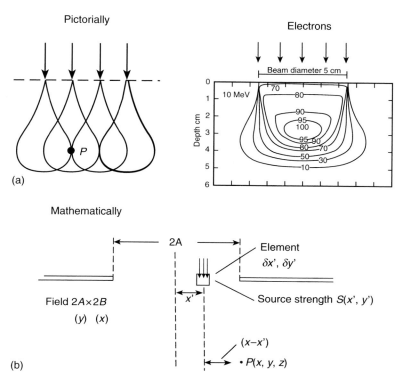

FIGURE 27.2
Decomposing a broad beam into narrow pencils and summing them up to give the dose at a point. (Reproduced from Nahum, A. E., in *The Computation of Dose Distributions in Electron Beam Radiotherapy*, Nahum, A. E., Ed., Medical Physics Publishing, Madison, WI, 1985, pp. 151–184.)

Up until the early 1980s, electron-beam patient dose-calculation methods had mirrored the simplest of those for photon beams (see Section 26.2), following a 1-D approach where the effect of inhomogeneities was accounted for by simply shifting the isodoses by an amount depending only on the extent and density of the inhomogeneity along the ray line passing through the dose point.

Figure 27.1 shows how the presence of an air cavity produces a so-called *hot spot* downstream from the cavity; there is no way that the 1-D isodose-shift method can predict such a distribution (e.g. especially regions with dose greater than 100%).

A natural alternative to the 1-D rayline approach is to decompose the beam into narrow pencils. The dose at a point then becomes a sum over the pencils comprising the beam. Figure 27.2 illustrates the idea. The dose distribution due to a narrow electron beam is shown in Figure 27.3.

27.2 THE FERMI–EYGES PENCIL-BEAM ALGORITHM IMPLEMENTATION (HOGSTROM MODEL)

27.2.1 GENERAL

Hogstrom et al. (1981) showed that the Fermi–Eyges analytical form of the pencil beam for a geometry consisting of inhomogeneous layers is capable of predicting the existence and approximate position of *perturbations* in dose distributions caused by inhomogeneities

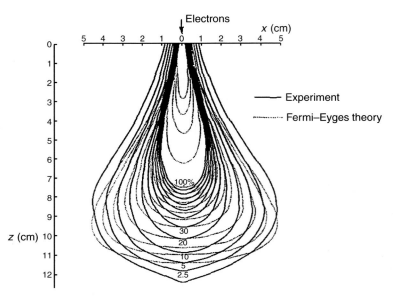

FIGURE 27.3
Narrow-beam dose distribution for 3 mm diameter 22.5 MeV electron beam in water. (Adapted from Brahme A., Lax, I., and Andreo, P., *Acta Radiol. Oncol.*, 20, 147–158, 1981.)

(though not necessarily their magnitude; see later sections). Furthermore, the effects of a lack of lateral scattering equilibrium on oblique incidence are correctly accounted for. For irregular fields the pencil-beam approach is the natural choice, as it clearly lends itself to integration over any shape of aperture or collimator.

The first implementations of the pencil-beam approach were limited to two-dimensional descriptions of inhomogeneities. This was solely due to the practical requirement for reasonable calculation times. Calculations were generally carried out for points defined by the plane of a particular CT slice (Figure 27.5). The two-dimensional approach implies that the inhomogeneities present in the slice are assumed to extend infinitely in the direction perpendicular to the CT-slice plane. This enables the dose summation to be carried out over a series of *strip beams* rather than pencils. It has been shown that the 2D model can introduce serious errors for cases where the inhomogeneity actually extends only a short distance perpendicular to the CT plane (Cygler et al. 1987; Mah et al. 1989). Ideally, the pencil-beam model should be fully 3D; this is generally the case with the pencil-beam models in treatment planning systems today. However, despite its apparent sophistication, there are still major limitations to a purely analytical pencil-beam approach (see in particular Section 27.3).

The basic ideas behind the electron pencil-beam approach will be covered in the remainder of this section, with an emphasis on the Hogstrom model developed at the M.D. Anderson Hospital (Hogstrom et al. 1981; Nahum 1985; Jette 1995; Hogstrom and Steadham 1996).

27.2.2 THEORY OF THE FERMI–EYGES PENCIL-BEAM MODEL

An analytical solution to the transport equation for charged particles in the small-angle approximation, first developed by Fermi, was extended by Eyges (1948) to inhomogeneous media and to take account of energy loss. The Fermi–Eyges solution involves both angular and spatial variables. In what follows, the angular dependence has been removed by integration over all possible directions of particle motion.

FIGURE 11.8
A Varian 4 MV side cavity standing waveguide.

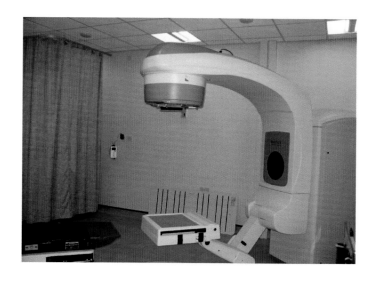

FIGURE 13.1
Photograph of a fixed SAD simulator showing a digital imaging system. A slot located behind the detector allows insertion of conventional film cassettes. The imager arm has considerable freedom of movement (focus-film distance in the Z direction and X-Y translation).

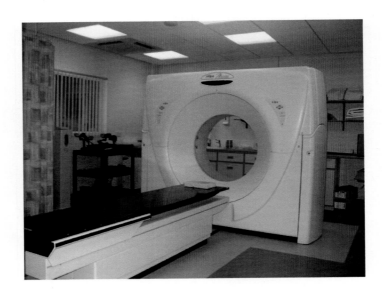

FIGURE 13.4
Photograph of a wide-bore CT scanner designed especially for virtual simulation. Note the standard flat-top couch and the lateral lasers.

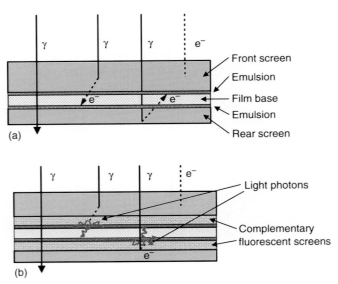

FIGURE 14.1
Image formation in film with (a) metal only and (b) metal/fluorescent screens.

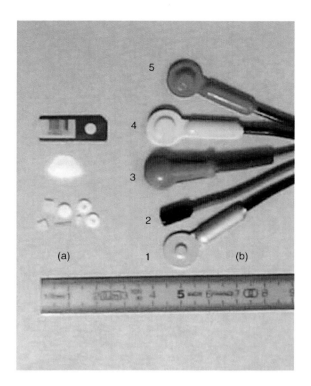

FIGURE 16.1
Examples of solid state detectors of current use: (a) radiothermoluminescent dosimeters made of LiF or $Li_2B_4O_7$ doped with Mn or Cu and (b) diodes (grey: Sun Nuclear-QED 111200 for electrons; black and blue: Scanditronix EDD-5 and EDP-5 for organs at risks and cobalt-60, respectively; yellow and red: Sun Nuclear QED 111500 and 111600 for 6–12 and 15–25 MV, respectively).

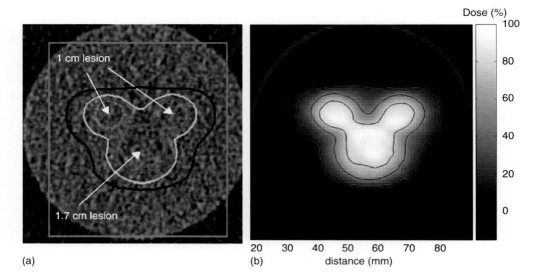

(a)   (b)   distance (mm)

FIGURE 17.5
Illustration of the use of a gel-dosimeter in conjunction with optical-CT scanning to verify a 3 isocentre radiosurgery treatment. The left hand image is the dose-distribution calculated by the treatment planning system (isodoses 80%, 50% and 20%) superimposed onto the CT scan of the head-phantom. The right hand image is a dose map obtained from optical CT scanning through the central plane of the distribution (lines are also 80, 50 and 20%). (From Oldham, M. et al., *Med. Phys.*, 28,1436–1445, 2001. With permission.)

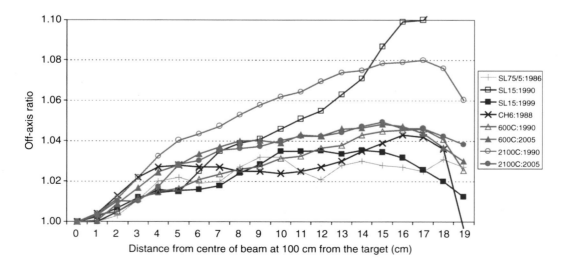

FIGURE 22.8
Profiles at peak depth for different low energy accelerators including a Philips SL75/5, two Elekta SL15s (one installed in 1990 and the other in 1999), an ABB CH6, two Varian 600Cs and two Varian 2100Cs installed at different times. Note the more pronounced increase in dose towards the edge for the SL15 and 2100C installed around 1990.

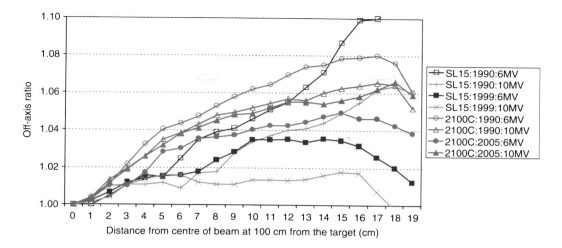

FIGURE 22.10
Flatness profiles at peak depth for Varian (2100C) and Elekta (SL15) 6 MV beams compared to 10 MV beams. Two machines from each manufacturer installed at different times are compared. Note the higher doses at the beam edge for the earlier 6 MV beams which were designed for flatness at 10 cm deep.

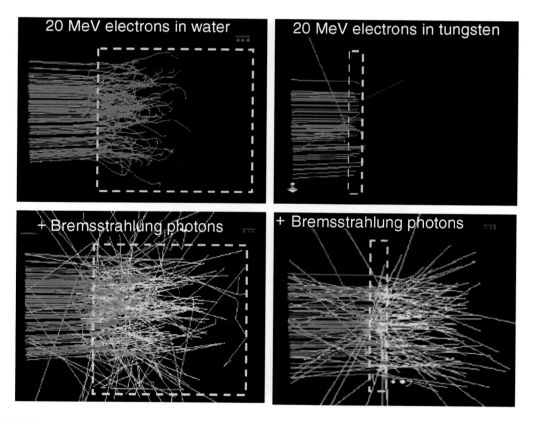

FIGURE 28.3
20 MeV electron tracks generated by the EGS4 code system incident from a vacuum on phantoms of water (left, top and bottom) and tungsten (right, top and bottom). In the top figures, only the charged particle tracks are shown. In the bottom figures, the bremsstrahlung photon tracks have been added. (Courtesy of the EGS development group at the National Research Council of Canada.)

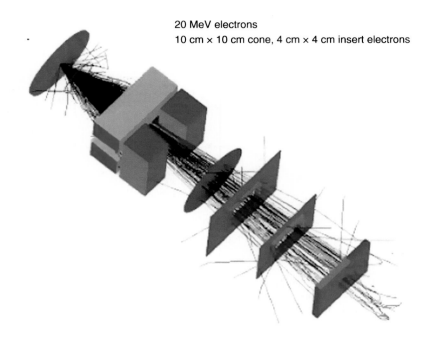

20 MeV electrons
10 cm × 10 cm cone, 4 cm × 4 cm insert electrons

FIGURE 28.6
Graphical representation of the simulation geometry used in the BEAM Monte-Carlo system to model a Varian Clinac 2100C in electron mode; a number of electron tracks are shown. (Courtesy of the National Research Council of Canada, http://www.irs.inms.nrc.ca/BEAM/egs_windows/egs_windows.html).

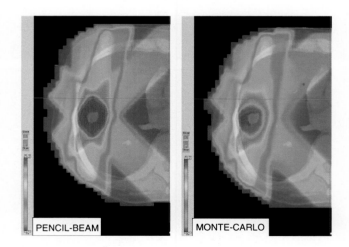

PENCIL-BEAM   MONTE-CARLO

FIGURE 28.10
A comparison of patient CT-based 6 MV photon-beam dose distributions computed at Copenhagen University Hospital. The *pencil-beam* calculation from the Eclipse TPS (Varian-Dosetek) is on the left, and that from the Monte-Carlo code system BEAMnrc is on the right; the linear accelerator is a Varian 2300EX. MC parameters: ECUT = 0.70 MeV, PCUT = 0.10 MeV, voxel size = 2.5 mm, 9×10^7 histories per beam ($\approx 3 \times 10^6$ particles in the *phase-space* file for each beam); runtime = 0. 6 h on 12 parallel CPUs. (Lasse Rye Aarup, personal communication).

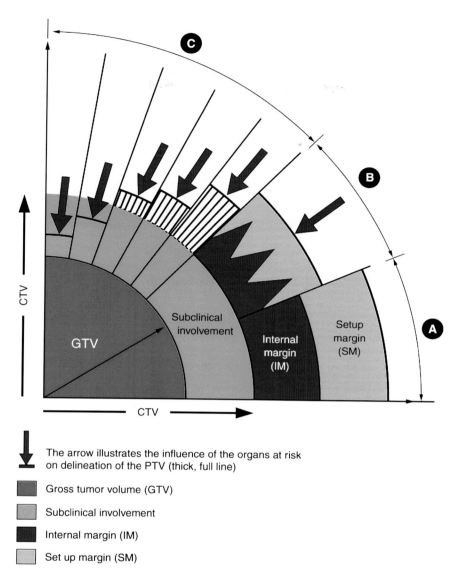

FIGURE 29.3
Schematic representation of different possibilities to combine uncertainties to define the PTV from the GTV: (A) linear addition of margins; (B) probabilistic addition of IM and SM; (C) definition of a *global* safety margin based on an empirical compromise between adequate coverage of GTV and unacceptable irradiation of OARs. (From ICRU, Report No. 62, International Commission on Radiation Units and Measurements, Bethesda, MD, 1999. With permission.)

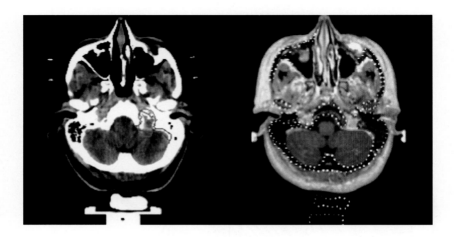

FIGURE 30.3

A side-by-side display of a CT image (left) and an MR image (right) of the skull. The bony structures cannot be seen on the MR image, where the dotted line corresponds to the registered bony outline of the CT image. The tumour delineation (blue) is much easier to tell on the MR image. The red outline on the left hand image is the volume that would have been delineated using CT alone. (From Khoo, V. S. et al., *International Journal of Radiation Oncology Biology Physics*, 46, 1309–1317, 2000. With permission.)

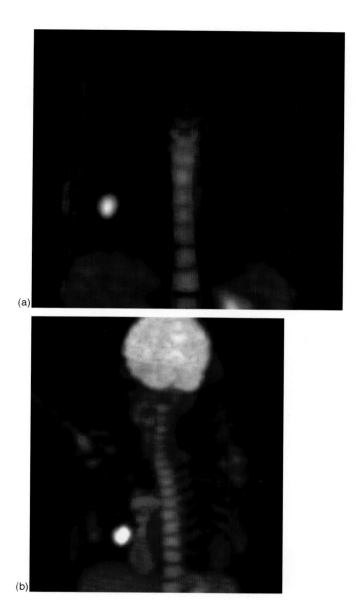

(a)

(b)

FIGURE 30.4

An isolated right lung metastasis demonstrated by fluorodesoxyglucose (FDG) PET imaging. The coloured PET image is blended with a registered grey scale CT image, both acquired on a PET–CT machine. The PET image is usually displayed in colour. The upper image (a) is a coronal slice through the metastasis and the lower image (b) is a 3D reconstruction in maximum-intensity projection (MIP) mode. (Courtesy of the Institut Curie, Paris.)

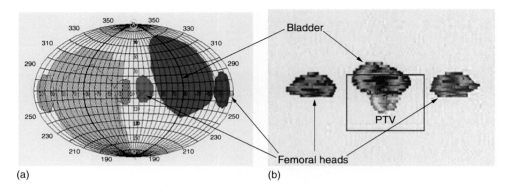

(a)   (b)

FIGURE 32.1
(a) A spherical view produced from Voxelplan. The gantry angle is depicted as meridians and the couch angle as latitudes. (b) The 3D Beam's eye view (BEV) (or Observer's Eye View) of the same patient showing planning target volume (PTV) (yellow), bladder (green), right (blue), and left (red) femoral head. This view is from 0° gantry and 0° couch. The yellow cross at the top of the spherical view (a) indicates the couch and gantry angle for this BEV.

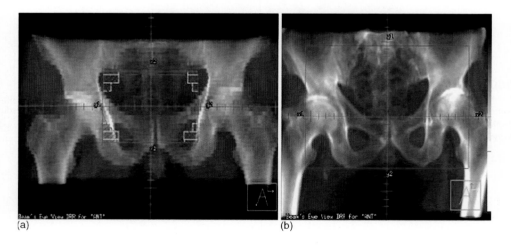

(a)   (b)

FIGURE 32.3
Examples of digitally reconstructed radiographs (DRRs) generated by the Pinnacle planning system; in (a) the CT data set had 5 mm slice separation and a coarse reconstruction was used, whereas in (b) the CT data set had 2.5 mm slice separation and the DRR algorithm was smooth; therefore, the DRR looks very much like a simulator radiograph.

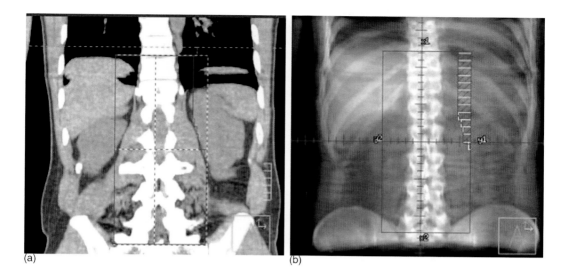

FIGURE 32.4
(a) A coronal view, showing the position of the kidneys relative to the field borders, and (b) a DRR on which the field size and the individual MLC leaves can be positioned.

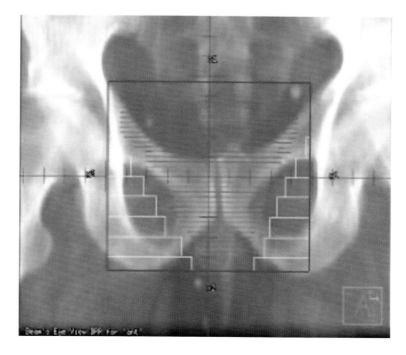

FIGURE 32.5
The DRR shown in Figure 32.3b with the prostate PTV contours (purple) and the MLC projection overlaid.

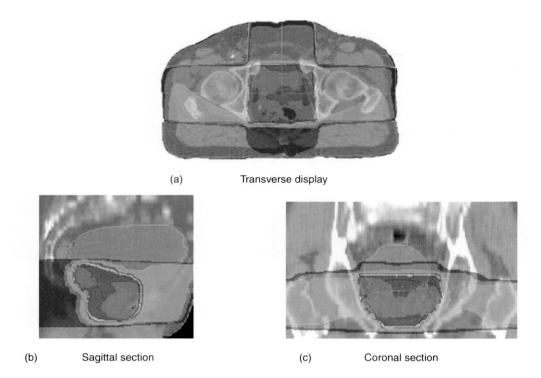

(a) Transverse display

(b) Sagittal section

(c) Coronal section

FIGURE 35.1
Planar *colour-wash* representation of the dose distribution in three different calculation planes through the patient: each colour corresponds to a range of dose according to a predefined scale. The corresponding isodose lines can be seen at the frontier of adjacent colours. This display may be easier to understand than the standard isodose line representation.

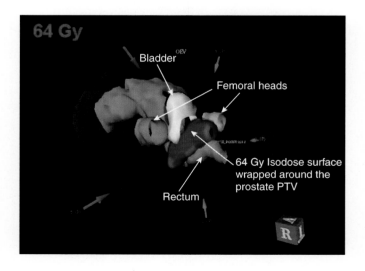

FIGURE 35.2
3D surface rendering showing anatomical structures (rectum, purple; bladder, white; femoral heads, blue and yellow) and 64 Gy isodose surface (red) for a five-beam prostate treatment plan.

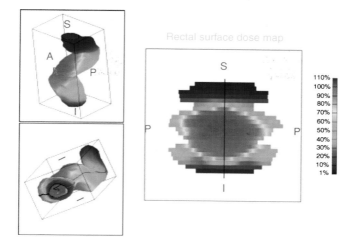

FIGURE 36.27
A rectal dose-surface map calculated from a 3D treatment plan for prostate cancer (Sanchez-Nieto et al. 2001b); spatial information of the high-dose region is preserved unlike in dose-volume (or dose-surface) histograms. (Reproduced from GUINESS, a treatment plan analysis software tool developed by Stefano Gianolini at the Institute of Cancer Research, Sutton.)

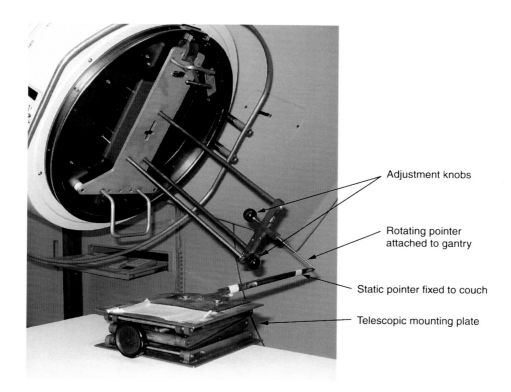

FIGURE 38.2
Precision front pointer suitable for mechanical isocentre measurement. The pointer is attached to the front of the collimator and the screw adjustments allow it to be centred on the collimator axis of rotation (see text). The telescopic device that is placed on the treatment couch allows the position of the fixed pointer to be moved precisely.

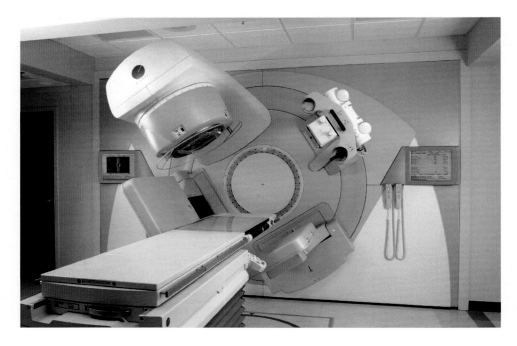

FIGURE 40.11
The Elekta synergy image guided radiotherapy solution. A similar concept is available from Varian. (Photo courtesy of Elekta Oncology Systems.)

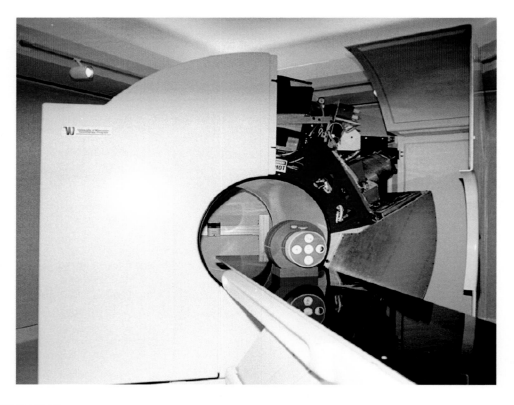

FIGURE 43.22
The UW tomotherapy research unit. (Courtesy of Prof. Rock Mackie.)

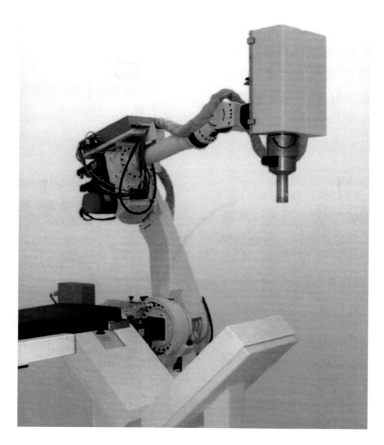

FIGURE 43.24
The Accuray Cyberknife.

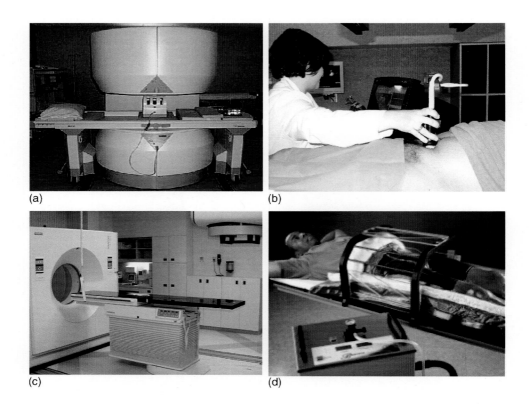

(a) (b)

(c) (d)

FIGURE 44.1
Imaging, immobilization, patient positioning and target localization devices used for IMRT at Fox Chase Cancer Center, Philadelphia, Pennsylvania; (a) a low-field open MRI scanner for structure segmentation; (b) a BAT ultrasound system for soft-tissue target localization; (c) a CT scanner in the treatment room for target localization; (d) a body localizer with vacuum fixation for patient immobilization and repositioning.

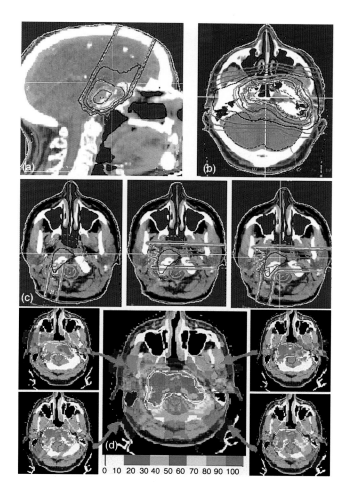

0 10 20 30 40 50 60 70 80 90 100

FIGURE 46.13

Treatment planning systems and dosimetry of intracranial targets with high energy proton beams: (a) noncoplanar beams, ray-tracing algorithm; (b) combined treatment with photons and protons, pencil beam algorithm; (c) "patching" techniques to boost targets surrounding a critical structure; (d) intensity modulated proton therapy with four coplanar scanned proton beams. (Figures (a–c) courtesy of Centre de Protontherapie d'Orsay with Isis system, Paris; Figure (d) is courtesy of Trofimov, A. and Bortfeld, T., Boston with KONRAD system, German Cancer Research Center (DKFZ), Heidelberg.)

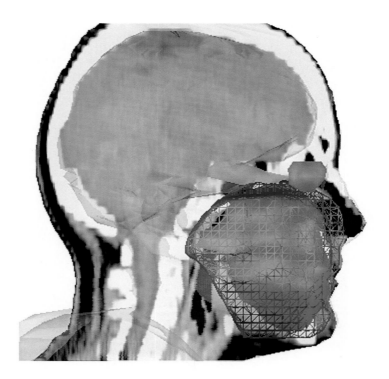

FIGURE 49.3
Example of 3D dose distribution for a large tumour of the maxillary sinus with irregular fields and partially wedged neutron beams at the Orléans neutron therapy facility. The wireframe represents the prescribed dose distribution around the clinical target volume.

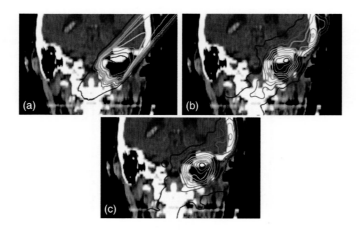

FIGURE 49.6
Example of the online PET imaging of the induced activity in a patient treated for a tumour of the skull base. The figure shows the absorbed dose distribution of one of the treatment fields (a), the expected activity as calculated from the treatment plan (b) and the measured PET activity (c). Due to *wash out* effects in soft tissue, the measured activity is in general somewhat lower than the calculated activity, while in the bony structures the measured and calculated activities agree very well. (Courtesy of W. Enghardt.)

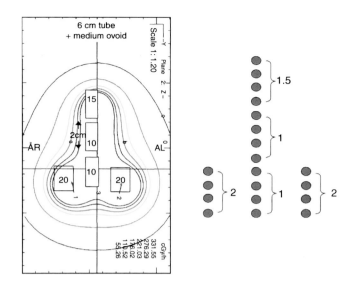

FIGURE 54.14
HDR source dwell position pattern with relative dwell weights to reproduce typical Manchester loading.

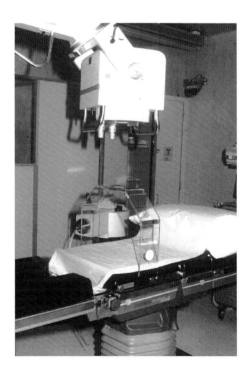

FIGURE 54.21
X-ray unit and localisation box used to reconstruct the 3D coordinates of sources. With such a box, it is not necessary to have an accurate x-ray set-up since the radioopaque markers of the box can be used to restore an orthogonal coordinate system.

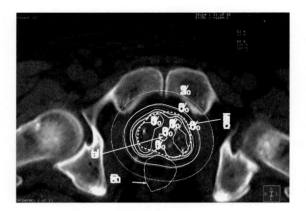

FIGURE 54.29

Seventeen-catheter prostate implant. The intended target volume is shown dotted and the thick line represents the prescription 100% isodose line. Note the sparing of the urethra in the centre of the prostate which is only given 80% of the dose.

Referring now to a Cartesian coordinate system (x,y,z) and an electron, incident perpendicularly on a medium at $(0,0,0)$ in the z direction, the Fermi–Eyges expression for the probability of finding the electron at depth z with displacement between x and $x+dx$, y and $y+dy$ is

$$p(x,y,z)dx\,dy = \frac{1}{2\pi\sigma_{MCS}^2}\exp\left[-\frac{x^2+y^2}{2\sigma_{MCS}^2}\right]dx\,dy \qquad (27.1)$$

where

$$\sigma_{MCS}^2 = \frac{1}{2}\int_0^z (z-u)^2 T(u)du \qquad (27.2)$$

The quantity $T(u)$ is the linear scattering power of the medium at depth u (see Section 3.5); this is evaluated at the mean energy of the electrons at the depth in question (MCS stands for multiple coulomb scattering).

The expression $p(x,y,z)$ can be separated into the product $p(x,z)\times p(y,z)$, where it can be seen that

$$p(x,z)dx = \frac{1}{\sqrt{2\pi}\sigma}\exp\frac{-x^2}{2\sigma^2}dx \qquad (27.3)$$

is a Gaussian, or normal distribution, and σ can be identified as the *standard deviation*, which is a measure of the width of the distribution. As the depth increases, σ_{MCS} increases and the pencil spreads out. The evaluation of σ_{MCS} through Equation 27.2 allows the medium to change in a layered manner, as indicated in Figure 27.4, with $T(u)$ appropriate to the layer in question. The pencil that would have resulted from water only is shown by the dashed outline.

The integration of a Gaussian can be expressed as

$$\text{erf}(x) \equiv \frac{2}{\sqrt{\pi}}\int_0^x e^{-t^2}dt$$

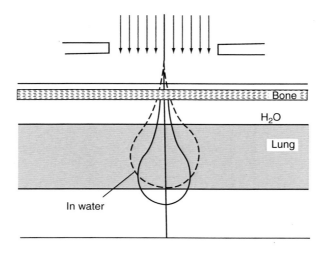

FIGURE 27.4
The layered geometry to which the Fermi–Eyges expression applies. (Adapted from Hogstrom, K. R., Mills, M. D., and Almond, P. R., *Phys. Med. Biol.*, 26, 445–459, 1981.)

which defines the error function erf(x). It should be noted that erf($-x$)=erf(x), erf(∞)=1 and erf(x)>0.995 for x>2; extensive use is made of tabulations of the error function in practical implementations. The above integral can be recast in a more directly useful form as

$$\int_0^A \exp\frac{-x^2}{2\sigma^2}\,dx = \sqrt{\frac{\pi}{2}}\sigma\,\mathrm{erf}\left(\frac{A}{\sqrt{2}\sigma}\right)$$ (27.4)

For a field of size $2A \times 2B$ and a uniform source strength, ignoring beam divergence, the total probability of finding an electron, or the *electron number*, at position (x,y,z) per unit area will be given by

$$N(x,y,z) = \frac{1}{2\pi\sigma_{MCS}^2}\int_{-B}^{B}\exp\left[\frac{-(y-y')^2}{2\sigma_{MCS}^2}\right]dy' \times \int_{-A}^{A}\exp\left[\frac{-(x-x')^2}{2\sigma_{MCS}^2}\right]dx'$$ (27.5)

which, with the help of Equation 27.4, reduces to

$$N(x,y,z) = \frac{1}{4}\left[\mathrm{erf}\left(\frac{A-x}{\sqrt{2}\sigma_{MCS}}\right) + \mathrm{erf}\left(\frac{A+x}{\sqrt{2}\sigma_{MCS}}\right)\right]\left[\mathrm{erf}\left(\frac{B-y}{\sqrt{2}\sigma_{MCS}}\right) + \mathrm{erf}\left(\frac{B+y}{\sqrt{2}\sigma_{MCS}}\right)\right]$$ (27.6)

Recalling that erf(x) is very close to unity for x>2, it can be seen that $N(x,y,z)=1$ if both x and y are more than about $3\sigma_{MCS}$ from the field edges, i.e. the dose profile is *flat* away from the edges of a broad beam, exactly as one would expect.

The above result also implies that $N(x,y,z)$ is constant with depth, i.e. the build-up due to scattering (see Section 24.2.2) is not predicted. This is because $N(x,y,z)$ is proportional to the *planar* fluence (Section 6.3.5; Nahum and Brahme 1985). Furthermore, Fermi–Eyges theory cannot account for electron loss. An empirical correction factor is necessary in order to reproduce measured depth–dose curves.

27.2.3 ADAPTATION OF FERMI–EYGES THEORY TO THE TREATMENT SITUATION

The application of the Fermi–Eyges theory to the treatment situation described here follows the approach developed by Hogstrom et al. (1981). The electron pencils are started at the level of the beam-defining device, or secondary collimator. Figure 27.5 explains the situation.

The electrons begin to spread out as they leave the vacuum window due to air scattering. The scattering-foil system broadens the beam in order to produce a uniform profile (see Section 11.4.5.2). The appreciable angular spread present in the beam at the secondary collimator level must be incorporated into the model. This is done by giving the pencils an initial angular distribution σ_{θ_x} (this is the *sigma* of the angular distribution projected onto the $x-z$ plane). As a result of this initial divergence, the elementary pencils started at the secondary collimator would continue to spread out even in the absence of any material further downstream. The resulting lateral spread σ_{air} at depth z follows from Equation 27.2 by putting $T(u)\Delta z = 2\sigma_{\theta_x}$:

$$\sigma_{air} = (z + L_0)\sigma_{\theta_x}$$ (27.7)

where L_0 is the distance between the secondary collimator and the $z=0$ plane (Figure 27.5); the small amount of extra scattering caused by the air over the distance L_0 has been neglected.

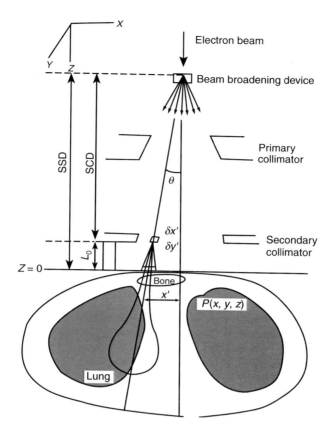

FIGURE 27.5
Schematic representation of an elementary pencil in the $x-z$ plane in the treatment situation. (Reproduced from Hogstrom, K. R., Mills, M. D., and Almond, P. R., *Phys. Med. Biol.*, 26, 445–459, 1981. With permission.)

The lateral spread of the beam at a depth in the patient is obtained by convolving one Gaussian with spread σ_{air} with another Gaussian with spread σ_{MCS}, which results in a Gaussian with spread σ_{med}:

$$\sigma^2_{med} = \sigma^2_{air} + \sigma^2_{MCS} \tag{27.8}$$

27.2.3.1 *Incorporation of Measured Depth–Dose Curves*

The Fermi–Eyges function $p(x,y,z)$ gives the spatial distribution of the *planar fluence* of electrons due to an electron pencil. To transform this to a (pencil) *dose* distribution $d(x,y,z)$, a weighting factor $g(z)$ is defined such that

$$d(x,y,z) = p(x,y,z)\,g(z) \tag{27.9}$$

Note that the weighting factor is a function of depth, z, only.

The full expression for the dose $\delta D(x,y,z)$ from a pencil originating from the small element $\delta x' \delta y'$ with an intensity weighting $W(x',y')$, for a given source-skin distance (SSD) (see Figure 27.5), now becomes:

$$\delta D(x,y,z) = W(x',y')\frac{1}{2\pi\sigma^2_{med}}\exp\left[-\frac{(x-x')^2 + (y-y')^2}{2\sigma^2_{med}}\right]g(z)\left[\frac{SSD}{SSD+z}\right]^2 \delta x' \delta y' \tag{27.10}$$

which should be compared with Equation 27.1.

An absolute requirement of the model is that it be able to reproduce measured central-axis depth–dose distributions, $CAXD(z)$, at different regular field sizes. This is effected by choosing the weighting factor $g(z)$ such that the dose as a function of depth on the central axis derived from integrating $\delta D(x,y,z)$ over a given field size $2A \times 2B$ exactly equals the measured $CAXD(z)$ for that field size. This measured $CAXD(z)$ is first corrected to infinite SSD and then the bremsstrahlung dose, assumed constant at all depths less than R_p, is subtracted (see Section 27.2.3.4) to yield $D_{\text{meas,e}^-}(0,0,z)$.

Integrating the expression in Equation 27.10 out to the projected field edges at depth z, but with no SSD term as $g(z)$ is defined for an infinite SSD beam, one arrives at:

$$g(z) = \frac{D_{\text{meas,e}^-}(0,0,z)}{\text{erf}\left[\frac{A(1+z/\text{SSD})}{\sqrt{2}\sigma_{\text{med}}(z)}\right]\text{erf}\left[\frac{B(1+z/\text{SSD})}{\sqrt{2}\sigma_{\text{med}}(z)}\right]} \qquad (27.11)$$

where the two pairs of error functions involving x and y, respectively, in Equation 27.6 are identical because $x=y=0$ and thus have been reduced to two single erfs; and consequently the factor $1/4$ in Equation 27.6 is multiplied by 4; and SSD refers to the value at which the reference $CAXD(z)$ was measured. The $\sigma_{\text{med}}(z)$ is evaluated from Equation 27.7 and Equation 27.8, where $\sigma_{\text{MCS}}(z)$ is for a uniform water medium in this case. Physically, $g(z)$ is the effective *broad beam*, *infinite SSD*, central-axis depth–dose distribution in water corresponding to a particular field size. In practice, $g(z)$ must be determined for a range of field sizes for a beam of a particular energy as the model can only predict the change in $CAXD(z)$ over a small range of field sizes (see Section 27.3).

27.2.3.2 The F_{air} Weighting Factor

The Fermi–Eyges theory is strictly valid only for a layered geometry (Figure 27.4); the $\sigma_{\text{MCS}}(z)$ for each pencil is evaluated along a ray line through the patient, with only the inhomogeneities along this ray line being accounted for. This approximation works best at small depths where the pencil has not spread out significantly. However, in order to model the increase in the penumbra with depth, the pencils are started at the level of the secondary collimator. As Figure 27.5 makes clear, the pencils already have a certain spread, or *width*, when they meet the patient surface. This means that the effect of inhomogeneities close to the surface will be underestimated as the contribution to σ_{med} due to changes in σ_{MCS} will be swamped by the value of σ_{air}. In order to remedy this, Hogstrom et al. (1981) split σ_{med} back into its *MCS* and *air* components (technically this is a deconvolution).

The (planar) fluence at position (x,y,z) in the absence of all matter below the secondary collimator is calculated by integrating the unmodified Fermi–Eyges pencils, with $\sigma_{\text{air}}(z)$ evaluated from Equation 27.7, over the projected collimator area at depth z, i.e. as for Equation 27.5, but replacing σ_{MCS} by σ_{air} and the field limits A, B by the collimator, or endframe shape for the field in question. This special fluence, which is denoted here by F_{air}, is then used to weight pencils that start at the level of the final collimator and which have a *width* given now by σ_{MCS} alone; these MCS-pencils do not begin to broaden out until the patient surface or any bolus is reached. In other words, the pencils have been *redefined* at the final collimator level (see Section 27.5 and Boyd et al. 2001c).

27.2.3.3 The Final Expression for the Elementary Pencil

The final expression for the dose $\delta D(x,y,z)$ due to an elementary pencil located at (x',y') then becomes

$$\delta D(x,y,z) = \frac{1}{2\pi\sigma_{\text{MCS}}^2(z)}\exp\left[-\frac{(x-x')^2+(y-y')^2}{2\sigma_{\text{MCS}}^2(z)}\right]W(x',y')\,F_{\text{air}}(x',y',z')\,g(z)\left[\frac{\text{SSD}}{\text{SSD}+z}\right]^2\delta x'\delta y'$$

$$(27.12)$$

where F_{air} is the air fluence weighting factor and W is the intensity weighting factor that accounts for beam non-uniformities. This is the expression used to calculate the dose distribution for an inhomogeneous patient and an arbitrary field aperture.

27.2.3.4 The Photon Dose

The dose due to bremsstrahlung was originally subtracted from the measured depth–dose curve. This must now be added back to the electron dose, after putting back the inverse square law dependence. It is assumed that the dose beyond the depth of the practical range (see Section 24.4.2) is entirely due to photons. The photon dose at a given depth is further assumed to be constant within the collimator and zero outside it. Thus the field size dependence and the shape of the penumbra are entirely due to the electron dose as given by Equation 27.12.

27.2.3.5 Practical Implementation for an Inhomogeneous Patient

The two most basic parameters in the model that depend on the depth are the weighting factor $g(z)$ and the mean electron energy at depth z, $\bar{E}_z$. The latter must be known in order to evaluate σ_{MCS}. The quantity $g(z)$ is determined for water (see above). It is scaled for the inhomogeneous patient by means of an effective water depth d_{eff}, i.e. the value $g(d_{eff})$ is used. This quantity d_{eff} is calculated according to

$$d_{eff}(z) = \int_0^z \frac{(S_{tot})_{med}}{(S_{tot})_{water}}\, dz \qquad (27.13)$$

where S_{tot} is the total (linear) stopping power (see Section 3.4.1). d_{eff} is that depth in water having approximately the same $\bar{E}_z$ as the geometrical depth z in the patient (see also Section 25.2). Values of $(S_{tot})_{water}$ as a function of energy are given in Table M.2. It is assumed that $(S_{tot})_{med}/(S_{tot})_{water}$ is independent of $\bar{E}_z$. The value of d_{eff} can be calculated from the CT numbers (Hogstrom et al. 1981).

The mean energy at depth in water, $\bar{E}_z$, is required. Hogstrom et al. (1981) used the approximate formula (see also Section 18.5.5):

$$\bar{E}_z = \bar{E}_0(1 - z/R_p) \qquad (27.14)$$

where $\bar{E}_0$ is the mean energy at the surface and R_p is the practical range. More accurate values, based on Monte Carlo simulation, have been given by Andreo and Brahme (1981).

The patient will exhibit density variations from point to point. Consequently the inhomogeneities along any ray line will in principle differ from those along any other ray line, i.e. $\sigma_{MCS}(z)$ will be a function of the x and y position. In 2D implementations the rectangular beam aperture is divided into strips of a constant length in the y direction for the evaluation of $\sigma_{MCS}(z)$ (Hogstrom et al. 1981; Nahum 1985). In a 3D model account is also taken of the variations in patient density (and surface contour) in the direction perpendicular to the dose calculation plane. Small rectangular sub-beams, with centre (x_k, y_k) at depth z, described by $\sigma_{MCS}(x_k, y_k, z)$ are therefore required instead of the 2D strip-beams.

The calculation of $\sigma_{MCS}(z)$ involves the medium- and energy-dependent scattering power $T_{med}(\bar{E}_z)$ at all depths up to z (Equation 27.2). It is generally assumed that T_{med}/T_{water} is energy independent. Values of $T_{water}(E)$ and T_{med}/T_{water} can be found in ICRU (1984); a conversion relation between CT number and T_{med}/T_{water} is given in Hogstrom et al. (1981).

27.2.4 Input Data

This section is not intended to be a detailed description of exactly how these quantities are to be determined for the user's beams, which will depend on the particular implementation and the manual of the user's particular treatment planning system should be consulted.

27.2.4.1 Measured Depth–Dose Distributions

Depth–dose distributions for electron beams of a given $\bar{E}_0$ are extremely dependent on the particular beam transport system (scattering foils or scanning system, applicator design. etc.) of the user's accelerator (see Section 24.2). The Fermi–Eyges model requires central-axis depth–dose distributions, $CAXD(z)$, for a number of different rectangular field sizes, as input data in order to determine the infinite-SSD, broad-beam weighting factor $g(z)$ which converts planar fluence into dose in water (see Section 27.2.3.1).

27.2.4.2 Dose Profiles

A dose profile is required in order to derive the weighting factor $W(x,y)$. This should be measured at or near d_{max} in a principal plane across the larger field dimension for each field for which the $CAXD(z)$ is required.

27.2.4.3 Mean Energy at the Surface, $\bar{E}_0$

The quantity $\bar{E}_0$ required in Equation 27.14 can be determined from the approximate formula:

$$\bar{E}_0(\text{MeV}) = 2.33 R_{50}(\text{cm}) \tag{27.15}$$

where R_{50} is the depth at which the dose has fallen to 50% of its maximum value. The dose distribution should be for a *broad beam*, corrected to infinite SSD. Alternative methods may be given in National and International dosimetry protocols (e.g. IPEMB 1996, IAEA 1987, see also Section D.3.2).

27.2.4.4 The Initial Angular Spread, σ_{θ_x}

This parameter can be determined from in-air measurements of the penumbra at different distances from the collimator using film (Hogstrom et al. 1981). Plots of the width of the penumbra measured as the distance between the 90% and the 10% isodoses against distance below the collimator for both the y-axis and the x-axis penumbras should be straight lines cutting the ordinate-axis at the effective collimator position. The angular sigma is then given by 0.391 multiplied by the slope of the line fitting the points, with all distances in the same units. An average of the two values obtained should be used for σ_{θ_x}.

27.2.4.5 The Penumbra Adjustment Factor, FMCS

The exact form of the penumbra at depth will vary from accelerator to accelerator. Further-more, Fermi-Eyges theory can only predict the penumbra approximately (see below) even for perfectly *clean* beams, i.e. not contaminated with electrons scattered from the side walls of the applicator. The predictions of the model should be compared to the measurements in the penumbra region (Figure 27.6) and improved somewhat by the use of a penumbra adjustment factor, *FMCS*, which multiplies the theoretical value of $(\sigma_{MCS})^2$. Suitable values are between 1.0 and 1.4. However, perfect agreement cannot be expected. A value of *FMCS* as close to unity as possible is preferable in order not to distort the predictions of the model in regions of inhomogeneities.

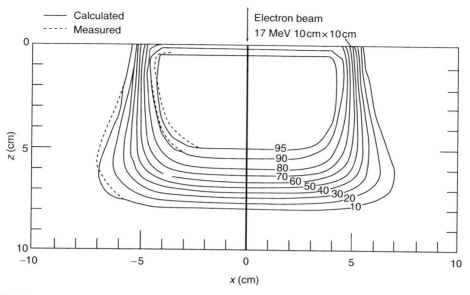

FIGURE 27.6

Comparison of calculated (full line) with measured (broken line) isodose curves for 10 cm × 10 cm field size at 100 cm SSD for a 17 MeV beam from a Therac 20 linear accelerator. (Reproduced from Hogstrom, K. R., Mills, M. D., and Almond, P. R., *Phys. Med. Biol.*, 26, 445–459, 1981. With permission.)

27.3 LIMITATIONS OF THE FERMI–EYGES PENCIL-BEAM MODEL

Electron-beam penetration in matter is extremely complex (see Chapter 3). The Fermi–Eyges pencil-beam model described above may appear to be sophisticated but in fact contains some severe approximations. Firstly the Fermi–Eyges model predicts that the pencil-beam *planar* fluence, and hence the dose (as the weighting factor $g(z)$ is only a function of depth not lateral position) has a Gaussian off-axis distribution at all depths. Figure 27.3 makes this point clearly by comparing a measured narrow-beam dose distribution with one calculated with Fermi–Eyges theory; there are considerable differences. It has been clearly demonstrated that the Gaussian prediction is only approximately true at small and large depths, with very considerable deviations in-between (Brahme et al. 1981; Lax et al. 1983); the Gaussian fall-off with off-axis distance is generally much too rapid. One consequence is that the Hogstrom/Fermi–Eyges model only predicts the change in shape of the central-axis depth–dose distribution over a narrow range of field sizes; this is why measured depth–dose distributions are required at several different field sizes. Lax et al. (1983) proposed therefore an alternative pencil-beam model where the elementary pencil was composed of three Gaussians fitted to a Monte-Carlo generated distribution (see Section 27.5).

A further factor that is not generally taken into account in pencil-beam models is scatter from the edges of customised low-melting-point alloy cut-outs (Bruinvis 1987; Ebert and Hoban 1995). This will affect the irregular field calculations, which should be checked against measurement.

The pencil-beam width as a function of depth, $\sigma_{MCS}(z)$, calculated using Equation 27.2, agrees well with experiment at small and moderate depths but then continues to increase, whereas in reality it goes through a maximum and finally decreases (Figure 27.7). This is due to the reduction of the number of electrons in the beam at large depths due to *range straggling* (see Section 3.4.2). Attempts have been made to modify the Fermi–Eyges model so

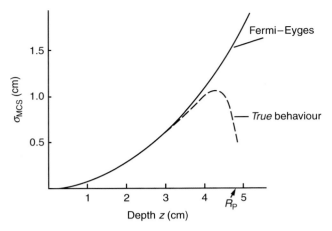

FIGURE 27.7
The true behaviour of $\sigma_{MCS}(z)$ compared to the predictions of unmodified Fermi–Eyges theory, for a 10 MeV beam in water. (Reproduced from Nahum, A. E., in *The Computation of Dose Distributions in Electron Beam Radiotherapy*, Nahum, A. E., Ed., Medical Physics Publishing, Madison, WI, 1985, pp. 151–184. With permission.)

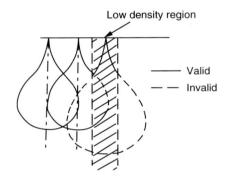

FIGURE 27.8
Experimental tests of accuracy of electron-beam treatment planning systems. Schematic representation of the errors introduced by the *central-ray approximation* in presence of inhomogeneities. (Reproduced from Nahum, A. E., in *The Computation of Dose Distributions in Electron Beam Radiotherapy*, Nahum, A. E., Ed., Medical Physics Publishing, Madison, WI, 1985, pp. 151–184. With permission.)

that it can predict electron loss (Bruinvis et al. 1989; Jette and Walker 1992; Jette 1995). In the Hogstrom model the electron loss is effectively incorporated through the factor $g(z)$.

The most significant approximation in the pencil-beam models concerns the effect of the so-called *central-ray approximation* on the treatment of inhomogeneities. Each pencil is corrected for inhomogeneities by computing $\sigma_{MCS}(z)$ according to the material along the central ray of the pencil. This is equivalent to assuming that the pencil is incident on a layered phantom (Figure 27.4). Some of the consequences of this approximation are illustrated schematically in Figure 27.8. The pencil farthest to the left is *correct* at *all* depths, but the other two are only correct at *small* depths. As a result, one cannot expect that the effect of inhomogeneities which are narrow in the beam direction will be accurately predicted at depths where the pencil has spread out appreciably*. In fact, this seemingly insoluble limitation (but see Section 27.5) is one of the main factors propelling the development of Monte Carlo approaches to electron dose distributions in inhomogeneous media (see Section 27.4 and Chapter 28).

---

* Lax (1986) should be consulted for a more complete discussion.

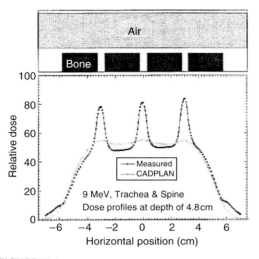
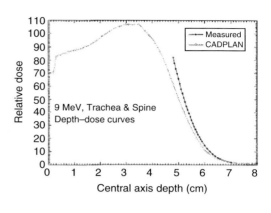

FIGURE 27.9
Calculated (CADPLAN pencil beam) and measured dose distributions for a 9 MeV beam from a Siemens KD2 linear accelerator incident on a trachea and spine phantom. The base of the phantom was a slab of solid water 15 cm $\times$ 15 cm $\times$ 4.7 cm. There was a long cylindrical hole, diameter 26 mm, with its long axis perpendicular to the central axis of the beam 2 mm below the top surface. Four 14 mm thick, 25 mm diameter hard-bone disks, separated by 5 mm, were situated 5 mm below the trachea with their axes parallel to the beam axis. The measured depth–dose curve starts at the bottom of the solid water phantom which was placed in water. (Reproduced from Ding, G. X. et al., *Med. Phys.*, 26, 2571–2580, 1999. With permission.)

27.4 EXPERIMENTAL TESTS OF ACCURACY OF ELECTRON-BEAM TREATMENT PLANNING SYSTEMS

Significant inaccuracies of the 2D version of the Hogstrom pencil-beam algorithm (see Section 27.2.1) have been demonstrated by Lax (1987) and by Cygler et al. (1987) amongst others. However, of greater interest are experimental tests of the more recent three-dimensional implementations of pencil-beam algorithms. Blomquist et al. (1996) evaluated the Lax et al. (1983) generalized Gaussian pencil beam algorithm as implemented in the Helax-TMS treatment planning system (version 2.10-J) at electron energies between 10 MeV and 50 MeV in a number of specially built phantoms including ones simulating a *neck* and a *nose*. The Helax pencil-beam results were mostly within $\pm 2\%$ of the measured doses, when averaged over volumes larger than a sphere of 15 mm diameter, or ± 2 mm in isodose position. More significant deviations were found for oblique incidences and below small air cavities.

A rigorous and comprehensive evaluation of the CADPLAN (version 2.7.9, Varian/Dosetek) electron-beam algorithm was made by Ding et al. (1999). They focussed on inhomogeneous phantoms and used both measurements and Monte Carlo calculations. A pencil beam size of 2.5 mm $\times$ 2.5 mm was used. Their trachea and spine phantom is a relatively complex 3D inhomogeneous geometry and represents a severe test for any analytical dose-calculation method. Figure 27.9 shows the significant underestimation of the dose peaks just behind the gaps between the bone disks.

Measured and calculated electron cut-out factors (ratio of doses at respective d_{max} for cut-out field to that for an open applicator) were also compared at 6 MeV, 12 MeV, and 18 MeV for an ELEKTA SL20 linear accelerator. The agreement was within between 1% and 2% except for very small fields. The authors point out that the CADPLAN algorithm does not take into account the dose contribution from electrons scattered from low-melting-point-alloy shaped cut-outs and thus all CADPLAN predicted cut-out factors are less than or equal to unity. Ding et al. (1999) concluded that their results were similar to those found by Samuelsson et al. (1998) for homogeneous phantoms, except that large discrepancies were found very close to the surface. Quite severe limitations of the CADPLAN pencil-beam algorithm were exposed when phantoms with high-density material such as aluminium and relatively complex 3D inhomogeneities such as the trachea and spine phantom were investigated.

27.5 OTHER ELECTRON-BEAM ALGORITHMS

An alternative to the Hogstrom pencil-beam model was developed by Lax et al. (1983). In their model, the dose profile at each depth is represented by the sum of three Gaussian distributions, with their parameters derived from Monte-Carlo calculated pencils. This model produces much better agreement with measurement than the Hogstrom-Fermi–Eyges one in the penumbra region (addressing, for example, the problems illustrated in Figure 27.3 and Figure 27.6). The Gaussians are scaled for layered media as described by Lax and Brahme (1985); however, the assumption that the central-ray of each pencil passes through material of infinite lateral extent is retained. This generalized Gaussian pencil-beam model was implemented in the CADPLAN treatment planning system, for example (Hyödynmaa 1991). Shiu and Hogstrom (1987) have described a *pencil-beam redefinition algorithm* (PBRA) which attempts to remove the *central-ray* limitation but at the expense of a substantial increase in calculation time. However, the *small-angle approximation*, which is invalid at large depths, is still retained and the input of a measured depth–dose distribution of a broad beam is still required. Further development of the PBRA is reported in Boyd et al. (2001b).

Another approach has been suggested by Dutreix and Briot (1985). It consists in extending to electron beams the concept of primary-scatter separation, which was originally developed for photon beams (see Section 26.3.2). For electron beams this separation has no strict physical meaning but, as a first approximation, the electrons undergoing very small deviations in the patient could be considered to behave as *primary* particles. Their fluence decreases very rapidly from the surface to the practical range according to a function which can be determined experimentally by extrapolation to a zero field area for a full range of energies, or modelled analytically. The difference between depth–dose curves for various circular (or square) fields and this primary depth–dose curve, all of them being normalised at the phantom surface where the electron fluence is not much influenced by the collimator opening, provides *scatter* functions. These functions could then be used to perform a 2D-pencil like (or even 3D) Clarkson integration (see Figure 26.8). A similar approach has been suggested by Van de Geijn et al. (1987) and a method to predict and accurately represent the penumbra for this model has been devised by Rosenwald et al. (1994). Since the calculation is based on a separation of components from experimental data, followed by a recombination of these components after corrections for patient shape and inhomogeneities, the result is quite accurate and this algorithm has been implemented in the ISIS and DOSIGRAY (now combined in the ISOGRAY) treatment-planning systems.

The 6D *phase-space-time-evolution* (PSTE) model (Huizenga and Storchi 1989; Morawska-Kaczynska and Huizenga 1992; Janssen et al. 1994) has been developed as a more accurate alternative to the pencil-beam model. The PSTE model is a numerical solution to the transport equation in phase space (see Section 5.1). The model is based on the various physical processes of electron interactions with matter, i.e. energy loss, scattering, secondary electron and bremsstrahlung production (see Chapter 3). Huizenga and Storchi (1989) showed that it could predict the depth–dose distribution of a broad, clinical, high-energy electron beam in a homogeneous medium, in contrast to the Fermi–Eyges model. Morawska-Kczynska and Huizenga (1992) extended the PSTE model to a multi-layered geometry and found excellent agreement with Monte-Carlo simulation, but in shorter calculation times. Korevaar et al. (2000) state that it takes 30 min (on a Hewlett-Packard 9000/B1000 workstation) for the PSTE model to calculate a 3D dose distribution with a 5 mm $\times$ 5 mm $\times$ 5 mm calculation grid. This can be compared with 15 min for the CADPLAN (Varian-Dosetek) pencil-beam model for the complete 3D dose distribution consisting of 30 (CT) slices, i.e. only a factor of 2 faster. However, the pencil-beam model can calculate the distribution in a *single slice* in around 30 sec, which is not an option with the PSTE model. Monte-Carlo calculations (EGS4/PRESTA) took 6 times longer to reach 2% statistical uncertainty and 25 times longer for 1% uncertainty (see Chapter 28).

27.6 CONCLUDING REMARKS

Pencil-beam models, including the Hogstrom model (described in some detail here) represent a significant advance over earlier *broad-beam* methods. The Hogstrom et al. (1981) and Lax et al. (1983) pencil-beam algorithms correctly predict the existence of *hot* and *cold* spots due to inhomogeneities which are narrow compared to the field width but are less successful in getting the *magnitude* of these dose perturbations correct. Some serious limitations remain, including that of the *central-ray approximation* and the inability to predict central-axis depth–dose distributions. The considerably more sophisticated PSTE method has not yet been implemented in a commercial treatment-planning system. Monte-Carlo simulation (Chapter 5 and Chapter 28) has demonstrated its high accuracy against measurements in heterogeneous geometries (e.g. Shortt et al. 1986) and is currently routinely used as a *gold standard* against which to compare analytical methods. Increases in calculation speed means that it is now feasible for Monte Carlo methods to be used to calculate electron-beam patient dose distributions for clinical applications (see Chapter 28).

CHAPTER 28

MONTE-CARLO BASED PATIENT DOSE COMPUTATION

Alan Nahum

CONTENTS

28.1 INTRODUCTION

The various *analytical* methods used to compute the dose distribution in patients irradiated by photon and electron beams have been described in Chapter 26 and Chapter 27. These methods attempt to account for the effect of irregular patient surface and inhomogeneities in the body with varying degrees of success. For photon beams, three-dimensional convolution of the point-spread function or *kernel* with the primary photon fluence is the most sophisticated analytical method to date (see Section 26.3.4). The kernels are derived from Monte-Carlo

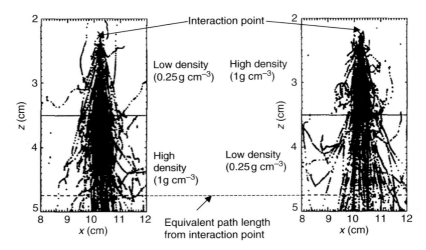

FIGURE 28.1
The limitation of *point kernel scaling* illustrated with an example of the electron tracks resulting from 100 5 MeV photons generated by Monte-Carlo simulation. The approximation involved is explained in the text. (Reproduced from Keall, P. J. and Hoban, P. W., *Med. Phys.*, 22, 1413–1418, 1995. With permission.)

(MC) simulation in water (Mackie et al. 1988). However, there is no exact analytical way to modify these water kernels to account for the non-water patient and thus approximate *scaling* methods have been devised. Figure 28.1 is an elegant illustration of the approximations involved in such kernel scaling. The pattern of electron tracks at an equivalent path length (indicated by the horizontal dashed line) *downstream* from the interaction point is clearly different depending on the order of the varying densities. In the situation on the right, at the level of the dashed line, the tracks are clearly more *spread out* than on the left. Therefore, using a *scaled kernel* (i.e. stretched according to the tissue density) to calculate the dose at the level of the dashed line is clearly unphysical.

Other approximations are also often employed in practical implementations of the 3D convolution method in treatment planning systems in order to reduce the otherwise considerable computation times, such as a spatially-invariant *polyenergetic* kernel and the so-called *collapsed cone* method (e.g. Ahnesjö and Aspradakis 1999; Figure 26.15). Despite the considerable improvements obtained from 3D point-kernel convolution as compared to primary-scatter separation or the pencil-beam approach, there are still some approximations, especially related to electron transport close to interfaces of different density or composition (Figure 28.1 being one example) which can yield inaccurate results (Mohan 1997; Arnfield et al. 2000; Krieger and Sauer 2005).

Radiotherapy in heterogeneous regions of the body such as the thorax can cause particular difficulties for analytical photon-beam algorithms due, for example, to the small fields used in stereotactic techniques (Solberg et al. 1995; Ayyangar and Jiang 1998; Solberg et al. 1998; Verhaegen et al. 1998) and the narrow beam *elements* of differing fluence in intensity-modulated fields. There may be a lack of charged-particle equilibrium on the central axes of these very small fields, especially at the higher end of the clinically used megavoltage range. Figure 28.2 is a graphic illustration of the effect of a low-density inhomogeneity, in this case lung, on the dose distribution in a 24 MV x-ray beam. The figure clearly demonstrates the large drop in dose at small field sizes close to and inside the heterogeneity due to lack of *lateral* electron equilibrium. Any analytical dose-calculation method that does not faithfully account for (secondary) electron transport will not correctly predict the dose in such situations (Carrasco et al. 2004).

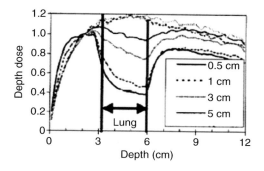

FIGURE 28.2
Depth-dose curves for a 24 MV photon beam showing the change in dose caused by a lung slab (density 0.26 g cm^{-3}) sandwiched in water, for different field sizes, calculated using the EGSnrc Monte Carlo code system. (Reproduced from Jones, A. O., Das, I. J., and Jones, F. L., *Med. Phys.*, 30, 296–300, 2003. With permission.)

For electron beams, the pencil-beam method (see Section 27.2), is well established in commercial planning systems. However, due principally to the unavoidable assumption that, for each individual *pencil*, only the inhomogeneities on its central axis can be accounted for (the *central-ray approximation*; see Figure 27.4 and Figure 27.8) the agreement with experiment is not satisfactory when large density differences (such as those encountered in lung and bone) are involved. This has been documented in several studies (Samulesson et al. 1998; Ding et al. 1999). It is fair to say that the 3D planning of electron treatments has been severely impeded by a general lack of confidence in current dose-calculation methods.

There is a calculation technique that can, in principle, avoid all the approximations referred to above: Monte-Carlo simulation of radiation transport. The basic principles of MC methods are explained in Chapter 5. Due to its accuracy in both homogeneous and heterogeneous media, MC simulation has found extensive application in medical radiation physics (Andreo 1991). This is especially the case in radiotherapy dosimetry where it has been used to compute quantities essential to accurate absolute-dose determination such as (water/air) stopping-power ratios and backscatter factors (see Chapter 18) and also the point-spread functions employed in the 3D convolution method for photon-beam dose calculation referred to above and discussed in detail in Section 26.3.4 (Nahum 1988; Mackie 1990; Rogers and Bielajew 1990; Andreo 1991; Rogers 1991; Nahum 1999; Bielajew 2001; Carrier et al. 2004).

The following figures illustrate some of the power and the pedagogical value of MC simulation in situations relevant to radiotherapy. Figure 28.3 is a beautiful example of how individual *histories* in an MC simulation can be plotted as *tracks* by recording the coordinates of the many interaction positions making up each history. In the upper part of the figure only the electron tracks are shown. The tortuous nature of their tracks in the medium and their finite ranges are apparent. In the lower part the bremsstrahlung photon tracks have been added and it can be seen that these are straight. The rare case of an electron backscattered from tungsten can be seen in the top right figure. One sees at a glance how much more efficient tungsten is, compared to water, at generating such photons.

The next example (Figure 28.4) involves a further processing of the raw output from a simulation, this time into isodose lines so familiar in planar views of radiotherapy treatment plans. Close to the beam axis and at depths less than around 5 cm, the isodoses are extremely well defined and show the pear-shaped electron distribution characteristic of electron penumbra, which is due to the strong scattering of megavoltage electrons in condensed media. At greater depths and off-axis distances, the increase in statistical noise becomes evident, as is inevitable at some low level of dose in any MC simulation.

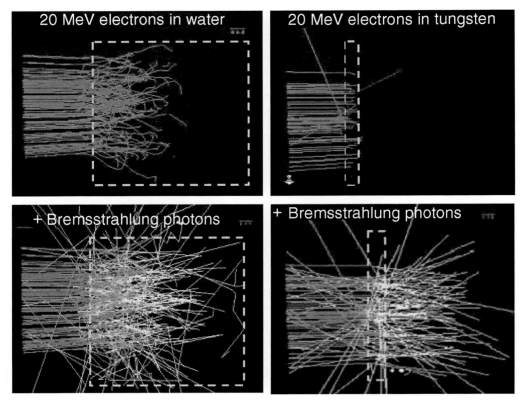

FIGURE 28.3
(See colour insert following page 590.) 20 MeV electron tracks generated by the EGS4 code system incident from a vacuum on phantoms of water (left, top and bottom) and tungsten (right, top and bottom). In the top figures, only the charged particle tracks are shown. In the bottom figures, the bremsstrahlung photon tracks have been added. (Courtesy of the EGS development group at the National Research Council of Canada.)

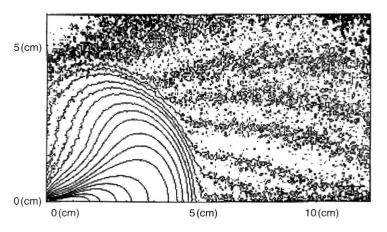

FIGURE 28.4
A simulation of a so-called electron pencil beam incident on a semi-infinite water medium. The electron energy is 10 MeV. The cylindrically symmetrical dose distribution has been transformed into isodose lines. (Courtesy of P. Andreo.)

A further example of the pedagogical value of Monte-Carlo simulation can be seen in Figure 3.17 at the end of Chapter 3 on charged-particle interactions. In that example, various electron-interaction phenomena have been successively switched off with dramatic effects on the shape of the depth-dose curve (Seltzer et al. 1978). In Figure 3.17, it can be seen that electrons would show a *Bragg peak*, familiar from heavy charged-particle physics (see Section 46.2.2), if the decrease of the electron fluence resulting from the large scattering effect did not take place. Further, the almost imperceptible build-up at small depths can be seen to be due to the forward transport of *delta rays*.

The main reason that MC methods are not already in widespread use for computing dose distributions in patients undergoing radiotherapy is the vast amount of computer-processing power necessary (Bielajew 1994; Mohan 1997). A little over a decade ago this was estimated to be of the order of hundreds of hours for a photon-beam radiotherapy treatment plan using the computer hardware then available to medical physicists (Nahum 1988), but nevertheless some groups made pioneering efforts (e.g. Manfredotti et al. 1987; Ito 1988). Today, however, due to the continuing reduction in the cost of ultra-fast processing power, MC-based treatment planning is perfectly feasible (e.g. Ma et al. 1999; Sempau et al. 2000; Fraass et al. 2003; Leal et al. 2003; Cygler et al. 2004; van der Zee et al. 2005).

28.2 CHOICE OF AVAILABLE MONTE-CARLO CODES

It is generally no longer necessary for radiation physicists to write their own MC radiation transport codes. Many decades of research and development have resulted in a number of very powerful software packages or code systems being made available, the majority of which are freely distributed, for example, at relevant websites. The following does not claim to be an exhaustive list but includes, in no particular order, the principal codes that are currently applied to problems in medical radiation physics, and in radiotherapy in particular. Where known, the names of the most recent versions and also the website from which the code may be downloaded are listed. In other cases, the potential user may have to enrol on a course in order to obtain the code in question, e.g. BEAMnrc at the National Research Council of Canada or MCDOSE/MCSIM at the Fox-Chase Cancer Center.

1. EGSnrc (the new version of the Electron Gamma Shower (EGS) code system, following on from EGS4 (Nelson et al. 1985) the most widely used code in medical physics, cited thousands of times in the medical physics literature; EGSnrc is the only code demonstrated to be able to simulate (gas-filled) ion chamber response in an entirely valid manner)
    a. Kawrakow 2000a; Kawrakow and Rogers 2003
    b. http://www.irs.inms.nrc.ca/inms/irs/EGSnrc/EGSnrc.html
2. BEAMnrc (technically not an independent code but a version of the EGS system designed for modelling radiotherapy treatment machines, primarily a *research* tool, as it is not optimised for speed)
    a. Rogers et al. 1995
    b. http://www.irs.inms.nrc.ca/inms/irs/BEAM/beamhome.html
3. MCNPX (includes neutron transport; very extensive use in nuclear power industry and increasingly in medical physics)
    a. Briesmeister 2000; MCNPX 2003
    b. http://mcnpx.lanl.gov/
4. GEANT4 (huge, many-particle MC toolkit; beginning to be used in medical physics)
    a. Agostinelli et al. 2003;
    b. http://geant4.web.cern.ch/geant4/

5. MMC (Macro MC: semi-numerical i.e. hybrid electron transport code for fast electron-beam radiotherapy treatment planning)
   a. Neuenschwander and Born 1992; Neuenschwander et al. 1995, 1997

6. PENELOPE (sophisticated electron transport; a *research* code)
   a. Salvat et al. 1996
   b. http://www.nea.fr/html/dbprog/penelope-2003.pdf

7. PEREGRINE (developed specifically for radiotherapy planning; available commercially through the NOMOS corporation)
   a. Hartmann Siantar et al. 2001; Heath et al. 2004; Boudreau et al. 2005
   b. http://www.llnl.gov/peregrine/

8. MCDOSE/MCSIM (developed specifically for fast treatment planning, based on EGS4/BEAM but optimized for speed due to, e.g. *electron-track repeating*; has user-friendly, measurement-based clinical beam commissioning system)
   a. Ma et al. 2002; Ma et al. 2004b
   b. http://www.fccc.edu/clinical/radiation_oncology/monte_carlo_course.html

9. Voxel-MC (VMC)++ (developed specifically for fast treatment planning; exploits *electron-track repeating*; is the MC dose engine in the MC option for electron-beam treatments in the Nucletron treatment planning system (TPS) *Oncentra/Masterplan*)
   a. Fippel 1999; Kawrakow and Fippel 2000; Cygler et al. 2004, 2005

10. Particle DMLC (further development of the EGS/BEAM-based MCV system involving fast particle transport through dynamic multileaf-collimator geometry specifically for intensity-modulated radiotherapy (IMRT)
    a. Siebers et al. 2000a; Keall et al. 2001; Siebers et al. 2002

11. DPM (developed specifically for fast radiotherapy treatment planning; electron transport scheme involves large condensed-history *steps* crossing medium boundaries)
    a. Sempau et al. 2000; Chetty et al. 2003
    b. http://www.upc.es/inte/downloads/dpm.htm

12. ORANGE (a fast MCNP-based dose engine for radiotherapy treatment planning)
    a. van der Zee et al. 2005

Although the MC simulation of radiation transport can take account of all the *physics* involved, nevertheless, some differences between different MC codes cannot be ruled out, especially where charged-particle transport is concerned (e.g. Nahum 1988, 1999; Rogers and Bielajew 1988; Jeraj et al. 1999; van der Zee et al. 2005). As is explained in Section 5.3, photon transport is simulated in an *analogue* manner (i.e. interaction by interaction), which means that this component of any code will be virtually identical (except for possible differences in the interaction cross sections). By contrast, it would be hopelessly inefficient to construct the tracks of charged particles in this analogue manner; consequently, all the above codes use the condensed-history approach (see Section 5.4) but in ways that can differ considerably from one code to another. It is thus imperative that careful tests (sometimes known as *bench-marking*) are carried out when a particular code is applied to a new situation where charged-particle transport is critical. One such example is illustrated in Figure 28.5. Here, the general-purpose-many-particle GEANT4 code has been compared to the specialist medical physics version of the EGS4 code using the PRESTA electron-transport scheme (Bielajew and Rogers 1987), in a standard test geometry designed by Rogers and Mohan (2000) specifically for treatment-planning purposes. Subsequently, van der Zee et al. (2005) used this same standard geometry and 18 MV beam quality to demonstrate equally good agreement between their MCNP-based code ORANGE and EGS4/PRESTA.

However, none of the above code systems are free from limitations; it will always be possible to find a class of problems that any given code fails to tackle in a valid manner. A good example of this is the failure of GEANT4 (and most other well-known MC codes) to yield

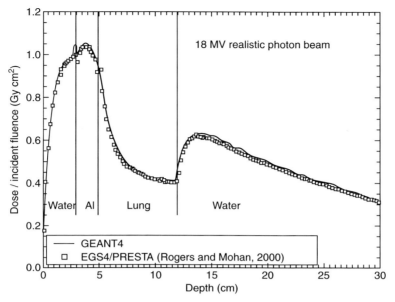

FIGURE 28.5
Depth-dose curve for a multislab phantom irradiated by a 18 MV *realistic* clinical beam, showing the excellent agreement between EGS4/PRESTA and GEANT4 Monte Carlo codes. (Reproduced from Carrier, J.-F., Archambault, L., and Beaulieu, L., *Med. Phys.*, 31, 484–492, 2004. With permission.)

correct results when simulating the response of an ionisation chamber (the radiation quality in this particular case (Poon et al. 2005) was a 1.25 MeV photon beam). This is in contrast to EGSnrc (Kawrakow and Rogers 2003), which has a highly sophisticated electron-transport scheme, with a user-controlled *switch* from *condensed-history* to *analogue* electron transport and back again, developed specifically to tackle the ion-chamber problem (Kawrakow 2000a, 2000b).

28.3 TREATMENT-MACHINE SIMULATION

As input data for each beam quality, most treatment planning systems require extensive measurements of dose distributions in water at different field sizes for open and wedged fields (see Table 19.1). MC simulation can only make use of such data in an indirect way. Instead, what is required is that the initial state of the particles in the beam, incident on the patient surface, be accurately known, i.e. the energies, directions and positions of the photons, electrons (and positrons). This data set is referred to as *phase-space**. There is no practical way to obtain such information by measurement; a portable spectrometer would be required with a degree of sophistication not yet available to radiotherapy clinics. Instead, the MC method itself has to be used, where one starts the simulation at the position where the electron beam leaves the vacuum window of the linear accelerator.

It was first demonstrated by Mohan (1988) for photon beams and by Udale (1988) for electron beams (which is a much more critical case) that only when such detail is incorporated into the simulation can good agreement with measurement in water phantoms be obtained. Consequently, there have been extensive efforts to simulate in full geometrical detail the passage of radiation through the complex treatment heads (target, flattening filter, monitor chamber,

---

* Phase-space is a standard mathematical term for a multi-dimensional array that describes a system.

collimating jaws, multileaf collimators, etc.) of modern clinical linear accelerators (Rogers et al. 1995; De Vlamynck et al. 1999; Lewis et al. 1999; Ma and Jiang 1999; Verhaegen and Das 1999; Jiang et al. 2000; Verhaegen and Seuntjens 2003). Rogers et al. (1995) developed an exquisitely detailed EGS4 usercode called BEAM for the purpose of detailed treatment-head modelling. BEAM produces a (very large) phase-space file of co-ordinates (particle type, energy, direction, position) above the level of the devices that define the beam shape for an individual treatment. This detailed modelling is only possible, however, if the accelerator manufacturers provide full information on the dimensions and composition of the key components in the treatment head of their machines. Figure 28.6 is an example of the kind of spectacular graphics that can be generated from detailed treatment-head simulations, in this case, using the EGS/BEAM code system (Rogers et al. 1995).

An alternative to the generation of a phase-space file from a simulation of the accelerator treatment-head geometry is the so-called *source model* approach (Ma and Rogers 1995; Ma et al. 1997). The first step is to devise an appropriate representation of the beam for a given design of treatment head. The beam representation is a concise mathematical description of the phase space, generally in terms of multiple (virtual) *sources*, such as the target, the primary collimator and the flattening filter (Ma et al. 1997). In the second step, the phase space is reconstructed from these separate *sources* and then fed into the dose-calculation code one particle at a time, thus eliminating the need to store large amounts of phase-space data. Figure 28.7 illustrates the general idea of source models. Deng et al. (2000) have further developed this multiple-source modelling approach and obtained excellent agreement (within two percent) in heterogeneous lung and bone phantoms when using the source model as input to the MC calculations instead of the complete phase-space. Fippel et al. (2003) have developed what they term a *virtual photon energy fluence* model based largely on measured dose distributions in water and air, with additional technical information on the linear accelerator in question. Fix et al. (2004) present a similarly detailed three-component source model for 6 MV and 18 MV photon beams that match the full phase-space data. Fippel et al. (2003) go so far as to state that "by comparison

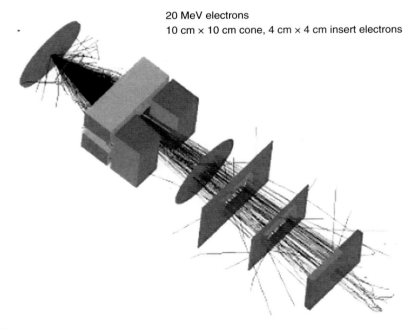

20 MeV electrons
10 cm × 10 cm cone, 4 cm × 4 cm insert electrons

FIGURE 28.6
(See colour insert following page 590.) Graphical representation of the simulation geometry used in the BEAM Monte-Carlo system to model a Varian Clinac 2100C in electron mode; a number of electron tracks are shown. (Courtesy of the National Research Council of Canada, http://www.irs.inms.nrc.ca/BEAM/egs_windows/egs_windows.html).

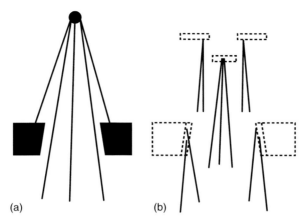

FIGURE 28.7
Schematic examples of source models: (a) the linac is replaced by a simple point source representing photons emerging from the target, with the jaws acting as perfect absorbers; (b) the treatment head is modeled as a number of sub-sources. (Reproduced from Verhaegen, F. and Seuntjens, J., *Phys. Med. Biol.*, 48, R107–R164, 2003. With permission.)

with results derived by the BEAM-NRC software system, it is demonstrated that time-consuming MC simulations of the whole accelerator head are unnecessary for radiation therapy planning purposes". Undoubtedly, such a method based on measured dose distributions is the most practical for a clinical MC system.

A further parameter must be known in order to simulate a treatment beam: the *energy* and possibly the *spectrum* of the electrons emerging from the vacuum window of the linear accelerator waveguide system. It cannot be assumed that, for example, the emerging electrons have an energy of exactly 6.0 MeV when the linac is set to produce a *nominal* 6 MV photon beam (see Section 22.2.2). The strategy generally adopted is to *deduce* the value of the electron energy that gives the best agreement between measured and simulated depth-doses in water by a process of trial and error. In the case of photon beams, it has been found (e.g. Verhaegen and Seuntjens 2003) that the central-axis depth-dose curve is relatively insensitive to small changes in electron energy, whereas the shape of the cross-beam *profiles* are much more sensitive, with the *horns** becoming more pronounced as the energy is lowered. It is generally also necessary to set up radial and angular intensity distributions of the incident electrons and then adjust these to produce the best agreement between measured and simulated depth-dose characteristics. For example, to achieve good agreement for an Elekta SL linac at (nominal) 6 MV, Seco et al. (2005) found that the best fit was obtained for electron energy of 6.0 MeV, an energy spread of 7% and a pencil-beam radius of 0.5 mm. Huang et al. (2005) found by measurement that, for a Varian Clinac 21EX linac, the electron focal spot was elliptical with a full width at half maximum (FWHM) between 1.69 mm and 2.24 mm and, additionally, a divergent (elementary) beam model was necessary to obtain good agreement with measurement for large field electron beams. There have also been instances when it has not been possible to obtain good agreement between measurement and MC simulation no matter how the initial electron distributions in energy, angle and space have been *adjusted*. The reason has generally been traced to incorrect or incomplete information provided by the linac manufacturers. An interesting example of this is how Bieda et al. (2002) deduced by MC-guided trial and error process how the scattering foil system had to be configured in a linac in electron mode, a form of MC *detective work*, and subsequently discovered that they had been correct.

---

* Horns are high-dose regions, relative to the beam's central axis, at the edge of the beam, when measured in air or close to the surface of a phantom (see Figure 22.8).

The MC modelling of electron beams from medical accelerators was reviewed in detail by Ma and Jiang (1999). A similarly comprehensive review of MC photon-beam modelling has been written by Verhaegen and Seuntjens (2003). It can be noted that in addition to mega-voltage photons from linear accelerators, cobalt-60 units (Han et al. 1987) and kilovoltage x-ray therapy machines have also been simulated using the MC method with a high degree of success (Verhaegen et al. 1999).

28.4 MONTE-CARLO SIMULATION OF THE DOSE DISTRIBUTION IN THE PATIENT

The simulation of radiation transport in the patient is generally started from the plane of the *phase-space* file, the particles being tracked through the particular beam-defining devices as illustrated schematically in Figure 28.8 (e.g. Hartmann Siantar et al. 1997; DeMarco et al. 1998). Note that the phase-space scoring plane must be above the level of the moving jaws. The patient geometry is represented by a 3D cartesian array of voxels, as for other 3D dose-calculation methods (see Section 25.2). Mora et al. (2001) recommend (cubic) voxel sizes of no larger than 3 mm for the head and 5 mm elsewhere in the body. Schemes have been developed for deriving appropriate parameters such as density and material type (tissue, lung, bone, air, etc.) from the CT numbers and associating these with the appropriate cross-section data required for the MC simulation (e.g. Schneider et al. 2000). This conversion from CT number to atomic composition is far from trivial, as the detailed analysis by Verhaegen and Devic (2005) makes clear.

Naturally, the key factor in all this is the central processor (CPU) time required to achieve an acceptably low uncertainty. For *electron* beams, runtimes of the order of a few minutes for 1% to 2% uncertainty (1σ) with a small voxel size (1 mm to 2 mm) have been demonstrated by Neuenschwander et al. (1997) using the macro Monte Carlo (MMC) approach on hardware with a performance comparable to a Pentium 200 MHz personal computer (PC). For primary *photon* beams (used far more often than electrons in radio-therapy) of the order of 2×10^8 histories are required (compared to about 10^7 for electron beams) to achieve 1 mm to 2 mm resolution; the PEREGRINE group first demonstrated this using a 16-CPU Pentium Pro-based system (Hartmann Siantar et al. 1997). Their runtimes

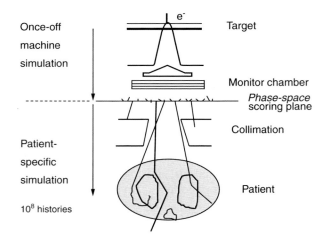

FIGURE 28.8
Schematic illustration of the two-step approach to Monte Carlo simulation in radiotherapy treatment planning. (Reproduced from Nahum, A. E., *Radiat. Environ. Biophys.*, 38, 163–73, 1999. With permission.)

were under 30 minutes. It can be noted that in a multi-beam plan the total number of histories required for a given level of uncertainty in dose in voxels of a given volume is essentially *independent of the number of beams* (Ma et al. 2005). This is because this uncertainty is determined solely by the number of energy-deposition events in the scoring volume, whether these arise from one or several beams. This is in marked contrast to the situation for analytical algorithms, where the total computation time generally increases approximately in proportion to the number of beams.

An approximate formula for the number of photon histories incident on the phantom/patient surface, N_γ, required to obtain a statistical uncertainty, δ, on the dose (in the central part of the beam) has been given by Spezi et al. (2002):

$$N_\gamma = \frac{1}{\delta^2 \mu_{en}^{eff}} \frac{A_{beam}}{V_{voxel}} \qquad (28.1)$$

where (μ_{en}^{eff}) is the *effective* linear energy-absorption coefficient for photons in water (*effective* because x-ray beams are not monoenergetic) for a beam of area A_{beam} and volume of dose-scoring voxel V_{voxel}. Though this expression is for a single beam, it can also be used to estimate the total number of histories required in the case of a multibeam treatment plan provided that each beam has a similar area (and quality). This total number is then divided up among the beams in proportion to the number of monitor units for each beam.

When employing a phase-space file, for reasons of finite storage size (≈ 2 Gb is around the current practical limit) the number of particles in the phase-space file may be smaller than the number of histories required to achieve the desired uncertainty in the phantom or patient; this will frequently be the case for photon beams. As an example, in simulations of a Varian Clinac 23EX performed in 2004 in the radiotherapy department of Copenhagen University Hospital, the 2-Gbyte phase-space files contained around 67 million histories. For a 6 MV photon beam with a 40 cm $\times$ 40 cm field size, approximately 3×10^9 histories were run (Jutemark 2005), i.e. the phase-space file *generated* this number of particles, which represents oversampling (i.e. reuse of a given history) by a factor 45.

This *recycling* of the particles in the phase-space in order to improve the statistics of the final dose calculations can have undesirable effects. Sheikh-Bagheri and Rogers (2002) state that the phase-space can be recycled up to 80 or 90 times, provided that the statistics in the phase-space file are good enough to represent all categories of particles (e.g. photons, electrons, primary, scattered, etc.). However, should this condition not be met, any *noise* in the phase space will be propagated through the system and still be present in phantom or patient simulations, irrespective of how many histories are run in the latter simulation. Fix et al. (2004) consider that the limited size of the phase-space file gives rise to what they term a *systematic error*, also known as *latent* or *remnant variance* (Sempau et al. 2001; Fippel et al. 2003). They point out that this limitation can be overcome through the use of source models and, as an alternative to analytical source models (see Section 28.3), they recommend a *histogram-based* source model based on the phase-space data (Schach von Wittenau et al. 1999). Through sampling the initial parameters of a particle from such histogram distributions no particle is reused, which reduces the latent variance.

28.5 MONTE-CARLO TREATMENT PLANNING IMPLEMENTATIONS

Several implementations of MC simulation in treatment planning systems have appeared in the literature: Kawrakow et al. (1996), Hartmann Siantar et al. (1997), Neuenschwander et al. (1997), DeMarco et al. (1998), Wang et al. (1998, 1999), Ma et al. (1999, 2002), Li et al.

(2000), Siebers et al. (2000a), and He (2002). A variety of different codes have been used, as indicated in Section 28.2.

The important subject of *variance reduction* in MC simulation has been introduced in Chapter 5. Techniques such as photon interaction forcing, particle splitting, Russian roulette, electron range rejection, correlated sampling and indeed the condensed-history method itself are commonly employed in applications of MC in medical radiation physics (e.g. Bielajew and Rogers 1988; Rogers and Bielajew 1990; Andreo 1991; Ma and Nahum 1993). However, researchers have gone even further in looking for ways of improving efficiency specifically for the MC simulation of radiotherapy treatment planning. The first of these was MMC (Neuenschwander and Born 1992; Neuenschwander et al. 1995, 1997), which used pre-calculated MC electron dose distributions in small spheres and led to electron-beam treatment plan computation times of the order of minutes. The VMC method (Kawrakow et al. 1996; Fippel 1999; Kawrakow and Fippel 2000) re-uses the same electron history in different patient regions. This is also known as *electron-track repeating* and has been exploited by several other research groups (Keall and Hoban 1996a; Li et al. 2000; Ma et al. 2002). There is also Super MC (Keall and Hoban 1996b; Keall et al. 1998), which combines the convolution/superposition method for photon beams (see Section 26.3.4) with MC electron transport. Figure 28.9 shows that the VMC method can achieve virtually identical results to the extensively bench-marked EGS4 MC code (Nelson et al. 1985) in a lung-water phantom, with a reduction in runtime by a factor of around 35. The limitations of the MD Anderson Hospital Hogstrom pencil-beam algorithm, discussed in Section 27.3, in this particular geometry are clearly shown.

Sempau et al. (2000) have developed DPM, an MC code specifically optimised for treatment planning calculations, principally through a highly efficient electron-transport scheme that permits long transport steps (of the order of 5 mm) which can straddle boundaries

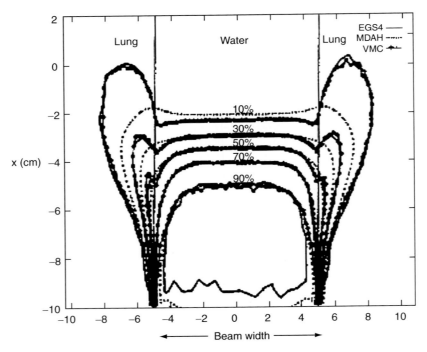

FIGURE 28.9
A comparison of the Voxel Monte-Carlo (VMC) method with EGS4 and with the MDAH/Hogstrom pencil-beam algorithm for 16 MeV electrons. The phantom consists of water and lung (0.26 g cm^{-3}), and the boundaries of the lung region coincide with the edge of the radiation field. (From Kawrakow, I., Fippel, M., and Friedrich, K., *Med. Phys.*, 23, 445–457, 1996. With permission.)

between different media. Sempau et al. (2000) stated that the simulation of 1 million 10 MeV electrons on a water phantom of 128^3 cubic voxels of 1 mm side length took approximately three minutes on a modern desktop workstation. Chetty et al. (2003) tested the accuracy of the DPM code for small field sizes and heterogeneous media (water/lung/water) for 6 MV and 15 MV photon beams; the phase-space files for a Varian 21EX linac were generated using the BEAMnrc code (Rogers et al. 1995). For the inhomogeneous phantom, DPM depth-dose calculations were within $\pm 1\%$ (based on the root-mean-square deviation) of measurements. Relative profile differences at depths within and beyond the lung were, on average, within $\pm$ 2% in the inner and outer beam regions. It was concluded that the DPM MC code was capable of accurate photon-beam dose calculations in situations of pronounced lateral electron disequilibrium.

It is imperative that developers of such special MC codes demonstrate that their modifications of the transport physics do not introduce errors that could be significant in the radiotherapy context; MC simulation would otherwise become 'just another dose-calculation method involving approximations'. An interesting discussion of these issues has been given by Ma et al. (2005).

Cygler et al. (2004, 2005) have reported on the first implementation of MC for electron-beams in a commercial treatment planning system (Nucletron Oncentra/Masterplan). The dose engine is VMC++ (Kawrakow and Fippel 2000). Runtimes (for 10^4 histories/cm^2) were typically around one minute on a 2.2 GHz Pentium 4 Xeon computer for voxel sizes of 3 mm$\times$3 mm$\times$3 mm ($\approx 500\,000$ voxels). Good agreement with measurement (relative doses) was obtained for most of the situations tested and for monitor unit calculations the agreement was described as excellent. The authors state that, as they are now more confident of the actual dose delivered to the patient, they calculate 3D dose distributions for a wider variety of electron-beam techniques and anatomical sites than has previously been their practice. Further, this has led to changes in treatment decisions.

Heath et al. (2004) have described the performance of the PEREGRINE code (Hartmann Siantar et al. 2001), which is the first MC code *commercially available* for photon-beam treatment plan dose calculations. Bar et al. (2003) found that MC dose calculation did not have a significant impact in head-and-neck planning. Boudreau et al. (2005) compared dose distributions calculated with the *finite-size-pencil-beam* algorithm (using an equivalent pathlength correction for heterogeneities) in the CORVUS TPS (Nomos corporation) with those computed with the PEREGRINE MC algorithm for 11 head-and-neck cases treated with a 6-MV IMRT technique. It was concluded that for head-and-neck cases the differences between the finite-size-pencil-beam plans and the MC plans "may not be significant enough to warrant the use of a MC based algorithm for treatment planning" but that sites located in the thoracic region "may be prime candidates for the use of MC based dose calculation". An example of the difference in dose distribution that can be obtained in the lung between (BEAMnrc) MC and a pencil-beam algorithm (here using the Varian ECLIPSE TPS) is shown in Figure 28.10. However, the differences between MC and the point-kernel collapsed-cone convolution algorithm are not as large (Carrasco 2004).

28.6 SPECIAL FEATURES OF MONTE-CARLO BASED TREATMENT PLANNING

28.6.1 ABSOLUTE DOSE PER MONITOR UNIT

A further feature of the MC approach is the accurate calculation of Monitor Units (MU) for each beam, taking into account the (heterogeneous) patient. Manual methods to calculate MU rely on a series of formulae which are largely based on the water-phantom

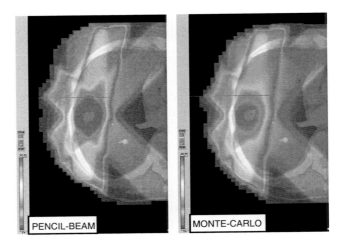

FIGURE 28.10
(See colour insert following page 590.) A comparison of patient CT-based 6 MV photon-beam dose distributions computed at Copenhagen University Hospital. The *pencil-beam* calculation from the Eclipse TPS (Varian-Dosetek) is on the left, and that from the Monte-Carlo code system BEAMnrc is on the right; the linear accelerator is a Varian 2300EX. MC parameters: ECUT = 0.70 MeV, PCUT = 0.10 MeV, voxel size = 2.5 mm, 9×10^7 histories per beam ($\approx 3 \times 10^6$ particles in the *phase-space* file for each beam); runtime = 0.6 h on 12 parallel CPUs. (Lasse Rye Aarup, personal communication).

situation (see Section 23.3). Adjustment for anatomical geometry and composition is subject to approximations which may introduce errors.

Monte-Carlo methods, used for the calculation of the absolute dose in terms of monitor units, rely on the correct calibration of monitor units for the reference field size by measurement (see also Section 25.3). Hence, the first step is the conventional *calibration of the beam* (see Chapter 18) by placing an ionisation chamber at the reference depth in a reference field size, A_{ref}, generally $10 \text{ cm} \times 10 \text{ cm}$, and determining the dose (to water) in Gy per monitor unit, $D(A_{\text{ref}})/MU$. A MC simulation of this exact geometry and beam quality is then carried out, starting from the phase-space plane, i.e. using particles from the phase-space file. This simulation yields $D_{\text{MC}}(A_{\text{ref}})$ *per simulated particle* (i.e. per number of histories run from the phase-space file) or $D_{\text{ref}}(A_{\text{ref}})/\text{history}$, which can be written as $D_{\text{ref,MC/hist}}(A_{\text{ref}})$. A *calibration factor* for this beam quality, F_{Q}, can then be derived:

$$F_{\text{Q}} = \frac{D_{\text{ref}}(A_{\text{ref}})/MU}{D_{\text{ref, MC/hist}}(A_{\text{ref}})} \qquad (28.2)$$

This MC calibration factor F_{Q} should only depend on the beam quality (e.g. 6 MV photons or 10 MeV electrons) and has dimensions of *particles per monitor unit*. The phantom is now replaced by the patient, and the field size is adjusted to what is specified in the treatment plan, say A_{plan}. A MC simulation of this new situation is done and yields, per history from the phase-space file, a dose $D_{\text{P}}(A_{\text{plan}})_{\text{MC/hist}}$ at some position of interest, P, in the patient. The dose in *Gy per monitor unit* at P will be given by

$$D_{\text{P}}(A_{\text{plan}})/MU = F_{\text{Q}} \times D_{\text{P}}(A_{\text{plan}})_{\text{MC/hist}}$$

or, the *absolute* dose at P, in Gy, is

$$D_{\text{P}}(A_{\text{plan}}) = MU \times F_{\text{Q}} \times D_{\text{P}}(A_{\text{plan}})_{\text{MC/hist}} \qquad (28.3)$$

for *MU* monitor units set on the treatment machine. Ma et al. (2004a) derived an equivalent expression, using a slightly different formalism, and extended this to cover intensity-modulated treatments.

The above expression assumes that changing the field size has no influence on the signal from the monitor ion chamber. However, there is a possibility that photons will be (*back*)-scattered from the collimator jaws into the monitor chamber. This can lead to a dependence of the monitor chamber signal on the settings of the collimator jaws, necessitating a correction to Equation 28.2. Verhaegen et al. (2000) studied this problem for the 6 MV and 10 MV photon beams from a Varian 2100C linac, finding a 2% to 3% increase in the monitor chamber signal due to backscatter for a field size of 0.5 cm$\times 0.5$ cm compared to 40 cm$\times 40$ cm. An empirical correction factor B_{mon} can be defined as

$$B_{mon}(A) = \frac{MU_{backscatter}(A_{ref})}{MU_{backscatter}(A)} \qquad (28.4)$$

where $MU_{backscatter}(A)$ is the ratio of the monitor ion chamber signal, including backscatter, to that excluding it, for a fixed beam current, for field size A. Thus, B varies between unity for large field sizes and ≈ 1.03 for very small field sizes. This modifies Equation 28.3, which becomes

$$D_P(A_{plan}) = MU \times F_Q \times B_{mon}(A_{plan}) \times D_P(A_{plan})_{MC/hist} \qquad (28.5)$$

It can be noted that if one started the simulations from a position *upstream* of the monitor chamber and thus automatically included the monitor chamber response in the simulation then this backscatter issue would disappear, but at the expense of significantly increased calculation time.

It has been demonstrated that *Output Factors* for irregular electron-applicator shapes (see Section 24.7) can be predicted within 2% or better (Kapur et al. 1998; Verhaegen et al. 2001). Thus MC holds out the prospect of high accuracy in *absolute* as well as *relative* dose determination. Ultimately, it may well be the MU computation that proves to be the most significant component in the improvement of dosimetric accuracy in radiotherapy due to Monte-Carlo simulation (Ma et al. 2005).

28.6.2 DOSE TO MEDIUM OR DOSE TO WATER?

One implicit feature of MC-based patient dose calculation is that the doses are computed in the actual media (i.e. one obtains dose to tissue, dose to lung, dose to bone, etc.). Treatment planning systems, however, give the dose distributions in (effective) dose to water whether corrections are made for heterogeneities or not. Siebers et al. (2000b) have explored this important issue. They strongly recommend that the MC doses be converted to dose to water by multiplying by the water-to-tissue, water-to-bone (etc.) mass absorption coefficient ratio or mass stopping-power ratio. Fortunately, however, this ratio is very close to unity except for the case of bone and air (but the dose to air cavities in the body has no clinical significance). For such cases, if this stopping-power ratio conversion is *not* made then there is a danger that comparisons between dose distributions based on MC and those based on analytical dose-calculation methods will be invalid, as one will be comparing *apples* with *pears*.

It should be borne in mind that all the experience of clinical radiation oncologists since the advent of radiotherapy at the turn of the previous century is based on dose to water, however imperfectly calculated. Thus, it can be argued that the dose distributions to be fed into *biological* models for Tumour Control Probability (TCP) and Normal Tissue Complication Probability (NTCP) (Chapter 36), which are currently in the form of dose-volume histograms (Section 35.3) ought to be in terms of dose to water. Any change in this philosophy would require a consensus in the worldwide radiation-therapy community. Similar issues arise when considering whether to change dose prescriptions as a result of changes in dose distributions brought about by the introduction into the clinic of a new (or updated) treatment planning system with a different (hopefully improved) non-MC-based dose calculation algorithm.

The re-analysis of the connection between lung doses and complications by De Jaeger et al. (2003) is a good example of this dilemma.

28.6.3 Computer Hardware Considerations

Shorter and shorter execution times to carry out simulations of the vast numbers of histories (of the order of 10^8) required for photon-beam treatment plans are being reported in the literature. One reason for this reduction in total simulation time is the use of variance-reduction techniques tailor-made for the treatment-planning situation, such as the two-stage phase-space file approach, *electron-track repeating* or MMC (see Figure 28.5). However, another important factor is the inherently *parallel* nature of MC simulation; each history is totally independent of any other history. This means that the total number of histories required can be divided up between a number of independent machines and the simulations run in parallel. Each machine, generally a *state-of-the-art* PC, runs the complete simulation, and therefore the complete software system must be loaded onto it. These days, the capacity of PCs is such that this is no problem at all, even for large, complex and comprehensive code systems such as MCNPX, GEANT and EGSnrc. It is now commonplace for a research group or radiotherapy clinic to purchase between 10 and 50 PC-CPUs and configure them in a cluster. One of these PCs acts as the server and copies the necessary input or job-initiation files to each PC in the cluster, together with *different* random-number seeds to ensure that there is no duplication of histories. Thereafter, no communication between the machines in the cluster is necessary. The results from each CPU in the cluster are then combined with weights proportional to the number of histories run on that machine. As an example, Seco et al. (2005) employed a *Linux Beowulf* cluster consisting of 30 CPUs with an average speed of 1.4 GHz. Increases in speed roughly in proportion to the number of machines in a cluster have been obtained (Jutemark 2005).

28.7 CONCLUDING REMARKS

The use of Monte-Carlo simulations to compute patient dose distributions for treatment planning appears to have many advantages, in terms of accuracy, in heterogeneous parts of the human anatomy. To date, however, while there are a growing number of commercial TPS with an electron MC option, there is currently no standard commercial TPS offering a photon MC dose calculation algorithm (apart from specialist Monte-Carlo planning systems such as Peregrine). This is despite some fifteen years of continuous development of MC applied to external-beam treatment planning. This slowness of clinical progress is possibly because there are still some important issues to be resolved before MC comes into widespread clinical use. One of these is whether the all-important characterisation of clinical beams is done by each user for the particular accelerators in the clinic or whether generic phase-space files or source models can be used for a particular type of accelerator. The paper by Fraass et al. (2003) contains a summary and recommendations from a workshop specifically convened to identify issues limiting the clinical use of MC in radiotherapy treatment planning. Among the key points in this paper are:

- The desirability of a procedure that would promote distribution of accurate MC machine modelling data from manufacturers
- The desirability of methods to validate phase-space models and/or simulations of accelerator-head geometry
- The need to study the sensitivity of MC dose calculations to uncertainties in machine design, machine tuning and changes in accelerator components known to occur over time
- How should radiation oncologists use this more accurate dose information? Retrospective MC-based dose assessments of already existing clinical complications and local control data should be carried out

- Additional study of variance-reduction techniques, improved computational techniques, including parallel processing and de-noising (Deasy et al. 2002; El Naqa et al. 2005) are all warranted

- Improved implementation of IMRT delivery techniques into the MC algorithms and integration of MC calculation algorithms into the inverse planning/optimisation process for IMRT in a time-effective way

- Education of physicists and physicians about MC principles and the clinical implications of MC algorithms and their implementation is crucial. Without such efforts, there is a risk that the potential superior accuracy of MC methods will be accepted as fact. In reality, they may be worse than existing clinical dose-calculation methods in clinics which do not properly implement and understand the method.

The use of MC simulations in radiotherapy treatment planning offers a number of new possibilities in addition to the potential increase in the accuracy of the calculated dose distributions. Among these are

- Computation of the photon or electron fluence spectra at any position thus allowing the computation of stopping-power or mass-energy-absorption-coefficient ratios for the actual beam quality for dosimetric purposes (see Section 18.5)

- *Benchmarking* of other algorithms (e.g. pencil beam or 3D convolution) to quantify the effect of their particular approximations with a minimum of measurements

- Investigation of the physics of dose delivery by allowing the effect of modifying the real physics to be observed (e.g. by changing the electron cutoff energies)

- Visualisation of the particle or photon tracks both as a teaching aid and to verify the code

Where MC dose calculation produces different results from traditional methods, this will potentially affect the expected results of treatment. This should be estimated via TCP and NTCP models (Chapter 36) so that the impact of MC dose calculation can be assessed.

Because of the time required for MC beam calculation it is likely for some time to come that a simpler algorithm will be needed to support *forward planning* of conventional beams. Recalculation of the dose distribution after small changes, e.g. beam angle or weight, are made needs to be in almost *real time* for this purpose. However, MC *recalculation* of the dose distribution once the optimum *forward* plan has been designed will be perfectly feasible. For intensity-modulated radiation therapy (IMRT), the *inverse-planning* approach attenuates the need for a *real time* response and MC algorithms could be used within the optimisation loop. Because inverse planning will modify the beam intensities based on the calculated dose distribution, including any imperfections in the calculation, it becomes more important that the calculation should be accurate at all points in the patient and not just at the ICRU reference point (see Section 37.6). However, *intensity modulation* means a large number of sub-beams of different intensity and small cross-section in addition to the problem of heterogeneities in the patient's body. This places even greater demands on beam calculation methods. MC is now being applied to IMRT and both static and dynamic delivery using multileaf collimators is being simulated (Laub et al. 2000; Ma et al. 2000; Keall et al. 2001; Pawlicki and Ma 2001; Leal et al. 2003; Seco et al. 2005). Furthermore, MC simulation has been applied to the radical new Tomotherapy machine (see Section 43.3.4) with its narrow-intensity-modulated beams delivered in a helical pattern (Jeraj et al. 2004) and has been shown to produce virtually perfect agreement with measurement. It can be expected that MC dose computation will have an important role to play in increasing our confidence in using these new treatment techniques.

Because of the difficulties in modelling the passage of the electrons from the accelerator gun to the target, MC dose calculation will probably always require some matching of the model to experimental measurement. However, provided that the physics of radiation transport has been correctly modelled and the interaction coefficient data used by the MC code are accurate, then uncertainties in MC-generated dose distributions are primarily statistical in nature and can be quantified by a standard deviation that will depend only on the size of the scoring volume and the number of particle histories (see Equation 28.1). This is in contrast to the uncertainties in analytical methods, which depend on the quality of experimental beam data and on approximations in the physics that may be impossible to quantify and cannot simply be reduced by changing, say, the grid size of the calculation. The day is finally in sight when the radiotherapy physicist can rely on an MC dose calculation to overcome the limitations of *model-based* algorithms.

PART F: PATIENT DOSE COMPUTATION METHODS
REFERENCES

AAPM (American Association of Physicists in Medicine). Report No. 85. Tissue inhomogeneity corrections for megavoltage photon beams. Report of AAPM Radiation Therapy Committee Task Group 65, Med. Phys. Publishing, Madison, WI, 2004.

Agostinelli, S., Allison, J., Amako, K., Apostolakis, J., Araujo, H. et al., GEANT4—a simulation toolkit, *Nucl. Instrum. Methods Phys. Res., Sect. A*, 506, 250–303, 2003.

Ahnesjö, A., Application of transform algorithms for calculation of absorbed dose in photon beams, *Computers in Radiation Therapy: Proc. VIIIth Int. Conf. on the use of Computers in Radiation Therapy*, IEEE Computer Society Press, Los Alamos, CA, pp. 17–20, 1984.

Ahnesjö, A., Invariance of convolution kernels applied to dose calculations for photon beams, in *The Use of Computers in Radiation Therapy*, Bruinvis, I. A. D., van der Giessen, P. H., van Kleffens, H. H., and Wittkämper, F. W., Eds., Elsevier Science Publishers BV, North Holland, pp. 99–102, 1987.

Ahnesjö, A., Collapsed cone convolution of radiant energy for photon dose calculation in heterogeneous media, *Med. Phys.*, 16, 577–592, 1989.

Ahnesjö, A., Photon beam characterisation for dose calculations from first principles, in *The Use of Computers in Radiation Therapy*, Hukku, S. and Iyer, P.S., Eds., Institute of Medical Sciences, Lucknow, pp. 64–67, 1990.

Ahnesjö, A., Analytic modeling of photon scatter from flattening filters in photon therapy beams, *Med. Phys.*, 21, 1227–1235, 1994.

Ahnesjö, A., Three-dimension dose calculation (I); application of convolution methods, in *Dreidimensionale Strahlentherapieplanung*, Schlegel, W., Bortfeld, T., and Stein, J., Eds., DKFZ, Heidelberg, pp. 55–72, 1995.

Ahnesjö, A. and Andreo, P., Determination of effective Bremsstrahlung spectra and electron contamination from therapeutic photon beam depth dose data, *Phys. Med. Biol.*, 34, 1451–1464, 1989.

Ahnesjö, A. and Aspradakis, M. M., Dose calculations for external photon beams in radiotherapy, *Phys. Med. Biol.*, 44, R99–R155, 1999.

Ahnesjö, A. and Mackie, R., Analytical description of Monte Carlo generated photon dose convolution kernels, in *The Use of Computers in Radiation Therapy*, Bruinvis, I. A. D., van der Giessen, P. H., van Kleffens, H. H., and Wittkämper, F. W., Eds., Elsevier Science Publishers BV, North Holland, pp. 197–200, 1987.

Ahnesjö, A. and Trepp, A., Acquisition of the effective lateral energy fluence distribution for photon beam dose calculations by convolution methods, *Phys. Med. Biol.*, 36, 973–985, 1991.

Ahnesjö, A., Andreo, P., and Brahme, A., Calculation and application of point spread functions for treatment planning with high energy photon beams, *Acta. Oncol.*, 26, 49–56, 1987.

Ahnesjö, A., Knöös, T., and Montelius, A., Application of the convolution method for calculation of the output factors for therapy photon beams, *Med. Phys.*, 19, 295–301, 1992a.

Ahnesjö, A., Saxner, M., and Trepp, A., A pencil beam model for photon dose calculation, *Med. Phys.*, 19, 263–273, 1992b.

Ahnesjö, A., Weber, L., and Nilsson, P., Modeling transmission and scatter for photon beam attenuators, *Med. Phys.*, 22, 1711–1720, 1995.

Ahnesjö, A., Weber, L., Murman, A., Saxner, M., Thorslund, I., and Traneus, E., Beam modeling and verification of a photon beam multisource model, *Med. Phys.*, 32, 1722–1737, 2005.

Al-Affan, I. A. M., A comparison of speeds of personal computers using an x-ray scattering Monte Carlo benchmark, *Phys. Med. Biol.*, 41, 309–313, 1996.

Andreo, P., Monte-Carlo techniques in medical radiation physics, *Phys. Med. Biol.*, 36, 861–920, 1991.

Andreo, P. and Brahme, A., Mean energy in electron beams, *Med. Phys.*, 8, 682–687, 1981.

Arnfield, M. R., Hartmann, S. C., Siebers, J., Garmon, L., Cox, L. et al., The impact of electron transport on the accuracy of computed dose, *Med. Phys.*, 27, 1266–1274, 2000.

Ayyangar, K. M. and Jiang, S. B., Do we need Monte Carlo treatment planning for Linac-based stereotactic radiosurgery?—A case study, *Med. Dosim.*, 23, 161–167, 1998.

Bar, W., Schwarz, M., Alber, M., Bos, L. J., Mijnheer, B. J., Rasch, C., Schneider, C., Nusslin, F., and Damen, E. M., A comparison of forward and inverse treatment planning for intensity-modulated radiotherapy of head and neck cancer, *Radiother. Oncol.*, 69, 251–258, 2003.

Batho, H., Lung corrections in cobalt-60 beam therapy, *J. Can. Assoc. Radiol.*, 15, 79–83, 1964.

Battista, J. J. and Sharpe, M., True three-dimensional dose computations for megavoltage x-ray therapy: A role for the superposition principle, *Aust. Phys. Eng. Sci. Med.*, 15, 159–178, 1992.

Battista, J. J., Mackie, T. R., El-Khatib, E., and Scrimger, J. W., Lung dose corrections for 6MV and 15MV XC-rays, *Anomalies Proceedings: Eighth International Conference on the use of Computers in Radiation Therapy, July 9–12, Toronto*, IEEE Computer Society Press, Silver Spring, pp. 31–35, 1984.

Battista, J. J., Sharpe, M., Wong, E., and van Dyk, J., A new classification scheme for photon beam dose algorithms, in *XIIth International Conference on Computers in Radiotherapy*, Leavitt, D. D. and Starkschall, G., Eds., Medical Physics Publishing, Salt Lake City, UT, pp. 39–42, 1997.

Beaudoin, L., Analytical approach to the solution of the dosimetry in heterogeneous media. MSc Thesis, University of Toronto, 1968.

Bentley, R. E. and Milan, J., An interactive digital computer system for radiotherapy planning, *Br. J. Radiol.*, 44, 826–833, 1971.

Bieda, M. R., Antolak, J. A., and Hogstrom, K. R., The effect of scattering foil parameters on electron beam Monte Carlo calculations, *Med. Phys.*, 28, 2527–2534, 2002.

Bielajew, A. F., Monte Carlo modelling in external electron beam radiotherapy—why leave it to chance? in *Proceedings of the XIth International Conference on the Use of Computers in Radiation Therapy*, Hounsell, A. R., Wilkinson, J. M., and Williams, P. C., Eds., Medical Physics Publishing, Madison, WI, pp. 2–5, 1994.

Bielajew, A. F., Some random thoughts on Monte Carlo electron and photon transport, in *Advanced Monte Carlo for Radiation Physics, Particle Transport Simulation and Applications*, Kling, A., Barao, F., Nakagawa, M., Távora, L., and Vaz, P., Eds., *Proceedings of the Monte Carlo 2000 Conference, Lisbon*, Springer, Berlin, pp. 1–6, 2001.

Bielajew, A. F. and Rogers, D. W. O., PRESTA: The parameter reduced electron-step algorithm for electron monte carlo transport, *Nucl. Instr. Meth. B*, 18, 165–181, 1987.

Bielajew, A. F. and Rogers, D. W. O., Variance-reduction techniques, in *Monte Carlo Transport of Electrons and Photons Below 50 MeV*, Jenkins, T. M., Nelson, W. R., Rindi, A., Nahum, A. E., and Rogers, D. W. O., Eds., Plenum Press, New York, pp. 407–419, 1988.

Bjängard, B. E., On Fano's and O'Connor's theorems, *Radiat. Res.*, 109, 84–89, 1987.

Bloch, P., A unified electron/photon dosimetry approach, *Phys. Med. Biol.*, 33, 373–377, 1988.

Blomquist, M., Karlsson, M., and Karlsson, M., Test procedures for verification of an electron pencil beam algorithm implemented for treatment planning, *Radiother. Oncol.*, 39, 271–286, 1996.

Bortfeld, T., Boesecke, R., Schlegel, W., and Bohsung, J., 3D dose calculation using 2D convolutions and ray tracing methods, in *The Use of Computers in Radiation Therapy*, Hukku, S. and Iyer, P. S., Eds., Institute of Medical Sciences, Lucknow, pp. 238–241, 1990.

Bortfeld, T., Schlegel, W., and Rhein, B., Decomposition of pencil beam kernels for fast dose calculations in three-dimensional treatment planning, *Med. Phys.*, 20, 311–318, 1993.

Boudreau, C., Heath, E., Seuntjens, J., Ballivy, O., and Parker, W., IMRT head and neck treatment planning with a commercially available Monte Carlo based planning system, *Phys. Med. Biol.*, 50, 879–890, 2005.

Bourland, J. D. and Chaney, E. L., A finite-size pencil beam model for photon dose calculations in three dimensions, *Med. Phys.*, 19, 1401–1412, 1992.

Boyd, R. A., Hogstrom, K. R., and Rosen, I. I., Effects of using an initial polyenergetic spectrum with the pencil beam redefinition algorithm for electron dose calculations in water, *Med. Phys.*, 25, 2176–2218, 1998.

Boyd, R. A., Hogstrom, K. R., Antolak, J. A., and Shiu, A. S., A measured data set for evaluating electron-beam dose algorithms, *Med. Phys.*, 28, 950–958, 2001a.

Boyd, R. A., Hogstrom, K. R., and Starkschall, G., Electron pencil-beam redefinition algorithm dose calculations in the presence of heterogeneities, *Med. Phys.*, 28, 2096–2104, 2001b.

Boyd, R. A., Hogstrom, K. R., White, R. A., and Starkschall, G., Modeling pencil beam divergence with the electron pencil-beam redefinition algorithm, *Phys. Med. Biol.*, 46, 2841–2856, 2001c.

Boyd, R. A., Hogstrom, K. R., White, R. A., and Antolak, J. A., The use of an extra-focal electron source to model collimator-scattered electrons using the pencil-beam redefinition algorithm, *Med. Phys.*, 29, 2571–2583, 2002.

Boyer, A. L., Shortening the calculation time of photon dose distributions in an inhomogeneous medium, *Med. Phys.*, 11, 552–554, 1984.

Boyer, A. L., Mok, E. C., Photon beam modelling using fourier transform techniques, *International Conference on the Use of Computers in Radiation Therapy*, VIII ICCR, Toronto, Canada, 14–16, IEEE Computer Society Press, Los Alamos, CA, 1984.

Boyer, A. L. and Mok, E. C., A photon dose distribution model employing convolution calculations, *Med. Phys.*, 12, 169–177, 1985.

Boyer, A. L. and Mok, E. C., Calculation of photon dose distributions in an inhomogeneous medium using convolutions, *Med. Phys.*, 13, 503–509, 1986.

Boyer, A. L., Wackwitz, R., and Mok, E. C., A comparison of the speeds of three convolution algorithms, *Med. Phys.*, 15, 224–227, 1988.

Boyer, A. L., Zhu, Y., Wang, L., and François, P., Fast fourier transform convolution calculations of x-ray isodose distributions in homogeneous media, *Med. Phys.*, 16, 248–253, 1989.

Brahme, A., Lax, I., and Andreo, P., Electron beam dose planning using discrete gaussian beams, *Acta Radiol. Oncol.*, 20, 147–158, 1981.

Briesmeister, J. F., *MCNP—a general Monte Carlo n-particle transport code*, Version 4C, *Technical Report LA-13709-M*, Los Alamos National Laboratory, Los Alamos, NM, 2000.

Bruinvis, I. A. D., *Electron Beams in Radiation Therapy*, University of Amsterdam, Amsterdam, 1987.

Bruinvis, I. A. D., Mathol, W. A. F., and Andreo, P., Inclusion of electron range straggling in the Fermi-Eyges multiple-scattering theory, *Phys. Med. Biol.*, 34, 491–550, 1989.

Buffa, F. M. and Nahum, A. E., Monte-Carlo dose calculations and radiobiological modelling: Analysis of the effect of the statistical noise of the dose distribution on the probability of tumour control, *Phys. Med. Biol.*, 45, 3009–3023, 2000.

Carrasco, P., Jornet, N., Duch, M. A., Weber, L., Ginjaume, M. et al., Comparison of dose calculation algorithms in phantoms with lung equivalent heterogeneities under conditions of lateral electronic disequilibrium, *Med. Phys.*, 31, 2899–2911, 2004.

Carrier, J.-F., Archambault, L., and Beaulieu, L., Validation of GEANT4, an object-oriented Monte Carlo toolkit for simulations in medical physics, *Med. Phys.*, 31, 484–492, 2004.

Castellanos, M. E. and Rosenwald, J.-C., Evaluation of the scatter field for high-energy photon beam attenuators, *Phys. Med. Biol.*, 43, 277–290, 1998.

Ceberg, C. P., Bjarngard, B. E., and Zhu, T. C., Experimental determination of the dose kernel in high-energy x-ray beams, *Med. Phys.*, 23, 505–511, 1996.

Chetty, I. J., Charland, P. M., Tyagi, N., McShan, D., Fraass, D. et al., Photon beam relative dose validation of the DPM Monte Carlo code in lung-equivalent media, *Med. Phys.*, 30, 563–573, 2003.

Chui, C. S. and Mohan, R., Differential pencil beam dose computation model (abstract), *Med. Phys.*, 11, 392, 1984.

Chui, C. S. and Mohan, R., Off-center ratios for three-dimensional dose calculations, *Med. Phys.*, 13, 409–412, 1986.

Chui, C. S. and Mohan, R., Extraction of pencil beam kernels by the deconvolution method, *Med. Phys.*, 15, 138–144, 1988.

Cunningham, J. R., Scatter-air ratios, *Phys. Med. Biol.*, 17, 42–51, 1972.

Cunningham, J. R., Tissue inhomogeneity corrections in photon-beam treatment planning, in *Progress in Medical Radiation Physics: 1*, Orton, C. G., Ed., Plenum Press, New York, pp. 103–131, 1982.

Cunningham, J. R., Shrivastava, P. N., and Wilkinson, J. M., Program IRREG—calculation of dose from irregularly shaped radiation beams, *Comput. Programs Biomed.*, 2, 192–199, 1972.

Cygler, J. E., Battista, J. J., Scrimger, J. W., Mah, E., and Antolak, J., Electron dose distributions in experimental phantoms: A comparison with 2-D pencil beam calculations, *Phys. Med. Biol.*, 32, 1073–1086, 1987.

Cygler, J. E., Daskalov, G. M., Chan, G. H., and Ding, G. X., Evaluation of the first commercial Monte Carlo dose calculation engine for electron beam treatment planning, *Med. Phys.*, 31, 142–153, 2004.

Cygler, J. E., Lochrin, C., Daskalov, G. M., Howard, M., Zohr, R. et al., Clinical use of a commercial Monte Carlo treatment planning system for electron beams, *Phys. Med. Biol.*, 50, 1029–1034, 2005.

De Jaeger, K., Hoogeman, M. S., Engelsman, M., Seppenwoolde, Y., Damen, E. M. et al., Incorporating an improved dose-calculation algorithm in conformal radiotherapy of lung cancer: Re-evaluation of dose in normal lung tissue, *Radiother. Oncol.*, 69, 1–10, 2003.

De Vlamynck, K., Palmans, H., Verhaegen, F., de Wagter, C., de Neve, W., and Thierens, H., Dose measurements compared with Monte Carlo simulations of narrow 6 MV multileaf collimator shaped photon beams, *Med. Phys.*, 26, 1874–1882, 1999.

Deasy, J. O., Wickerhauser, M. V., and Picard, M., Accelerating Monte Carlo simulations of radiation therapy dose distributions using wavelet threshold de-noising, *Med. Phys.*, 29, 2366–2377, 2002.

DeMarco, J. J., Solberg, T. D., and Smathers, J. B., A CT-based Monte-Carlo simulation tool for dosimetry planning and analysis, *Med. Phys.*, 25, 1–11, 1998.

Deng, J., Jiang, S. B., Kapur, A., Li, J., Pawlicki, T., and Ma, C. M., Photon beam charcterization and modelling for Monte Carlo treatment planning system, *Phys. Med. Biol.*, 45, 411–427, 2000.

Ding, G. X., Cygler, J. E., Zhang, G. G., and Yu, M. K., Evaluation of a commercial three-dimensional electron beam treatment planning system, *Med. Phys.*, 26, 2571–2580, 1999.

Doucet, R., Olivares, M., DeBlois, F., Podgorsak, E. B., Kawrakow, I., and Seuntjens, J., Comparison of measured and Monte Carlo calculated dose distributions in inhomogeneous phantoms in clinical electron beams, *Phys. Med. Biol.*, 48, 2239–2254, 2003.

Drouard, J., Rosenwald, J.-C., and Simonian. M., Generation of primary and scatter tables for dose computation in high energy photon beams. *Proceedings of 5th ESTRO Meeting on Radiotherapy and Oncology*, Baden–Baden, 1986.

du Plessis, F. C., Willemse, C. A., Lotter, M. G., and Goedhals, L., The indirect use of CT numbers to establish material properties needed for Monte Carlo calculation of dose distributions in patients, *Med. Phys.*, 25, 1195–1201, 1998.

Dutreix, A., Bjärngard, B.E., Bridier, A., Mijnheer, B., Shaw, J., and Svensson H., 1997 *Monitor unit calculation for high energy photon beams*, ESTRO Booklet 3, ESTRO, Garant Publishers, Leuven, 1997.

Dutreix, A. and Briot, E., The development of a pencil-beam algorithm for clinical use at the Institut Gustave Roussy, in *The Computation of Dose Distributions in Electron Beam Radiotherapy* Nahum A. E., Ed., pp. 242–270 Available from Medical Physics Publishing Madison, WI, or Medical Radiation Physics Dept, Umeå University, 90 187 Umeå, Sweden, 1985.

Ebert, M. A. and Hoban, P. W., A model for electron beam applicator scatter, *Med. Phys.*, 22, 1419–1429, 1995.

El Naqa, I., Kawrakow, I., Fippel, M., Siebers, J. V., Lindsay, P. E. et al., A comparison of Monte Carlo dose calculation denoising techniques, *Phys. Med. Biol.*, 50, 909–922, 2005.

El-Khatib, E. and Battista, J. J., Improved lung dose calculation using tissue-maximum ratios in the Batho correction, *Med. Phys.*, 11, 279–286, 1984.

Eyges, L., Multiple scattering with energy loss, *Phys. Rev.*, 74, 1534, 1948.

Field, C. and Battista, J. J., Photon dose calculations using convolution in real and Fourier space: Assumptions and time estimates, in *The Use of Computers in Radiation Therapy*, Bruinvis, I. A. D., van der Giessen, P. H., van Kleffens, H. H., and Wittkämper, F. W., Eds., Elsevier Science Publishers BV, North Holland, pp. 103–106, 1987.

Fippel, M., Fast Monte Carlo dose calculation for photon beams based on the VMC electron algorithm, *Med. Phys.*, 26, 1466–1475, 1999.

Fippel, M., Haryanto, F., Dohm, O., and Nüsslin, F., A virtual photon energy fluence model for Monte Carlo dose calculation, *Med. Phys.*, 30, 301–311, 2003.

Fix, M. K., Keall, P. J., Dawson, K., and Siebers, J. V., Monte Carlo source model for photon beam radiotherapy: Photon source characteristics, *Med. Phys.*, 31, 3106–3121, 2004.

Fraas, B., Doppke, K., Hunt, M., Kutcher, G., Starkschall, G., Stern, R., and Dyke, J. V., American Association of Physicists in Medicine Radiation Therapy Committee Task Group 53: Quality assurance for clinical radiotherapy treatment planning, *Med. Phys.*, 25, 1773–1829, 1998.

Fraass, D., Deye, J., and Smathers, J., Summary and recommendations of a National Cancer Institute workshop on issues limiting the clinical use of Monte Carlo dose calculation algorithms for megavoltage external beam radiation therapy, *Med. Phys.*, 30, 3206–3216, 2003.

François, P., Coste, F., Bonnet, J., and Caselles, O., Validation of reconstructed Bremsstrahlung spectra between 6 MV and 25MV from measured transmission data, *Med. Phys.*, 24, 769–773, 1997.

Han, K., Ballon, D., Chui, C., and Mohan, R., Monte Carlo simulation of a cobalt-60 beam, *Med. Phys.*, 14, 411–419, 1987.

Hartmann Siantar, C. L., Bergstrom, P. M., Chandler, W. P., Chase, L. et al., Lawrence livermore national laboratory's PEREGRINE project, in *Proceedings of the XIIth International Conference on the Use of Computers in Radiation Therapy*, Leavitt, D. D. and Starkschall, G. et al., Eds., Medical Physics Publishing, Madison, WI, pp. 19–22, 1997.

Hartmann Siantar, C. L., Walling, R. S., Daly, T. P., Faddegon, B., Albright, N. et al., Description and dosimetric verification of the PEREGRINE Monte Carlo dose calculation system for photon beams incident on a water phantom, *Med. Phys.*, 28, 1322–1337, 2001.

He, T. T., Implementation of a Monte Carlo based inverse planning model for clinical IMRT with MCNP code. PhD thesis, Wayne State University, Detroit, MI, available from the author—het@kci.wayne.edu, 2002.

Heath, E., Seuntjens, J., and Sheikh-Bagheri, D., Dosimetric evaluation of the clinical implementation of the first commercial IMRT Monte Carlo treatment planning system at 6 MV, *Med. Phys.*, 31, 2771–2779, 2004.

Hoban, P. W., Accounting for the variation in collision kerma-to-terma ratio in polyenergetic photon beam convolution, *Med. Phys.*, 22, 2035–2044, 1995.

Hogstrom, K. R., Dosimetry of electron heterogeneities, in *Medical Physics Monograph No. 9: Advances in Radiation Therapy Treatment Planning*, Wright, A. and Boyer, A., Eds., American Institute of Physics, New York, pp. 223–243, 1983.

Hogstrom, K. R. and Almond, P. R., Comparison of experimental and calculated dose distributions, Electron beam dose planning at the M.D. Anderson Hospital., *Acta Radiol. Suppl.*, 364, 89–99, 1983.

Hogstrom, K. R., Mills, M. D., and Almond, P. R., Electron beam dose calculations, *Phys. Med. Biol.*, 26, 445–459, 1981.

Hogstrom, K. R. and Steadham, M. S., Electron beam dose computation, in *Teletherapy: Present and Future, Proceedings of the 1996 AAPM Summer School*, Mackie, T. R. and Palta, J., Eds., Advanced Medical Publishing, Madison, WI, pp. 137–174, 1996.

Huang, P. H., Kase, K. R., and Bjarngard, B. E., Reconstruction of 4-MV Bremsstrahlung spectra from measured transmission data, *Med. Phys.*, 10, 778–785, 1983.

Huang, V. W., Seuntjens, J., Devic, S., and Verhaegen, F., Experimental determination of electron source parameters for accurate Monte Carlo calculation of large field electron therapy, *Phys. Med. Biol.*, 50, 779–786, 2005.

Huizenga, H. and Storchi, P. R., Numerical calculation of energy deposition by broad high-energy electron beams, *Phys. Med. Biol.*, 34, 1371–1396, 1989.

Hyödynmaa, S., Implementations of the generalised Gausssian pencil beam algorithm for three-dimensional electron beam dose planning, PhD Thesis, University of Kuopio, Espoo: Technical Research Centre of Finland, Publication 74, 1991.

IAEA (International Atomic Energy Agency), *Absorbed dose determination in photon and electron beams—an International Code of Practice*. Technical Reports Series No. 277, IAEA, Vienna, 1987.

ICRU (International Commission on Radiation Units and Measurements). *Determination of absorbed dose in a patient irradiated by beams of X or gamma rays in radiotherapy procedures*. Report No. 24, ICRU, Bethesda, MD, 1976.

ICRU (International Commission on Radiation Units and Measurements), *Radiation dosimetry: Electron beams with energies between 1 and 50 MeV*. Report 35, ICRU, Bethesda, MD, 1984.

ICRU (International Commission on Radiation Units and Measurements), *Use of computers in external beam radiotherapy procedures with high-energy photons and electrons*. Report No. 42, ICRU, Bethesda, MD, 1987.

ICRU (International Commission on Radiation Units and Measurements), *Prescribing, recording, and reporting electron beam therapy*. Report No. 71, ICRU, Bethesda, MD, 2004.

IPEMB (Institute of Physics and Engineering in Medicine and Biology), The IPEMB Code of Practice for the electron dosimetry for radiotherapy beams of initial energy from 2 to 50 MeV based on an air-kerma calibration, *Phys. Med. Biol.*, 41, 2557–2603, 1996.

Islam, M. K. and van Dyk, J., Effects of scatter generated by beam-modifying absorbers in megavoltage photon beams, *Med. Phys.*, 22, 2075–2081, 1995.

Ito, A., Three-dimensional dose calculations for total body irradiation, in *Monte Carlo Transport of Electrons and Photons below 50 MeV*, Jenkins, T. M., Nelson, W. R., Rindi, A., Nahum, A. E., and Rogers, D. W. O., Eds., Plenum Press, New York, pp. 573–598, 1988.

Jaffray, D. A., Battista, J. J., Fenster, A., and Munro, P., X-ray sources of medical linear accelerators: Focal and extrafocal radiation, *Med. Phys.*, 20, 1417–1427, 1993.

Janssen, J. J., Riedeman, D. E. J., Morawska-Kaczynska, M., Storchi, P. R. M., and Huizenga, H., Numerical calculation of energy deposition by high-energy electron beams: III. Three-dimensional heterogeneous media, *Phys. Med. Biol.*, 39, 1351–1366, 1994.

Jeraj, R., Keall, P. J., and Ostwald, P. M., Comparisons between MCNP, EGS4 and experiment for clinical electron beams, *Phys. Med. Biol.*, 44, 705–717, 1999.

Jeraj, R., Mackie, T. R., Balog, J., Olivera, G., Pearson, D., et al., Radiation characteristics of helical tomotherapy, *Med. Phys.*, 31, 396–404, 2004.

Jette, D., Electron beam dose calculations, in *Radiation Therapy Physics*, Smith, A. R., Ed., Springer-Verlag, Heidelberg, Germany, pp. 95–121, 1995.

Jette, D. and Walker, S., Electron dose calculation using multiple-scattering theory: Evaluation of a new model for inhomogeneities, *Med. Phys.*, 19, 11,241–11,253, 1992.

Jiang, S. B., Kapur, A., and Ma, C. M., Electron beam modelling and commissioning for Monte Carlo treatment planning, *Med. Phys.*, 27, 180–191, 2000.

Jiang, S. B., Boyer, A. L., and Ma, C.-M., Modeling the extrafocal radiation and monitor chamber backscatter for photon beam dose calculation, *Med. Phys.*, 28, 55–66, 2001.

Johns, H. E. and Cunningham, J. R., in *The Physics of Radiology*, 4th ed., Charles C. Thomas, Springfield, IL, pp. 369–376, 1983.

Jones, A. O., Das, I. J., and Jones, F. L., A Monte Carlo study of IMRT beamlets in inhomogeneous media, *Med. Phys.*, 30, 296–300, 2003.

Jutemark, B., Monte Carlo based investigation of the influence of accelerator-head geometry on megavolt photon beam quality in radiotherapy. *MSc Thesis*, Lund University, Sweden. LUJI-RADFY-EX-1/2005, 2005.

Kappas, K. and Rosenwald, J.-C., Inhomogeneity corrections when lateral dimensions are smaller than field dimensions, *Proceedings of World Congress on Medical Physics and Biomedical Engineering*, Hamburg, 1982.

Kappas, K. and Rosenwald, J.-C., Calcul des doses en radiothérapie en présence d'hétérogénéités de petites dimensions, *J. Eur. Radiother.*, 6, 35–45, 1985.

Kappas, K. and Rosenwald, J.-C., A 3-D beam subtraction method for inhomogeneity correction in high energy x-ray radiotherapy, *Radiother. Oncol.*, 5, 223–233, 1986a.

Kappas, K. and Rosenwald, J.-C., Theoretical and experimental analysis of scatter from inhomogeneous slabs in a ^{60}Co beam: The differential tissue-air ratio method (DTAR), *Phys. Med. Biol.*, 31, 1211–1228, 1986b.

Kapur, A., Ma, C. M., Mok, E. C., Findley, D. O., and Boyer, A. L., Monte Carlo calculations of electron beam output factors for a medical linear accelerator, *Phys. Med. Biol.*, 43, 3479–3494, 1998.

Kawrakow, I., Accurate condensed history Monte Carlo simulation of electron transport: I. EGSnrc, the new EGS4 version, *Med. Phys.*, 27, 485–498, 2000a.

Kawrakow, I., Accurate condensed history Monte Carlo simulation of electron transport: II. Application to ion chamber response simulations, *Med. Phys.*, 27, 499–513, 2000b.

Kawrakow, I. and Fippel, M., VMC++, a fast MC algorithm for radiation treatment planning, in *The Use of Computers in Radiation Therapy*, XIIIth Int'l, Schlegel, W. and Bortfeld, T., Eds., Springer, Heidelberg, pp. 126–128, 2000.

Kawrakow, I., Fippel, M., and Friedrich, K., 3D electron dose calculation using a Voxel based Monte Carlo algorithm (VMC), *Med. Phys.*, 23, 445–457, 1996.

Kawrakow, I. and Rogers., D. W. O., The EGSnrc code system: Monte Carlo simulation of electron and photon transport. NRC Report PIRS-701, 4th printing, National Research Council of Canada, Ottawa, 2003.

Keall, P. J. and Hoban, P. W., Accounting for primary electron scatter in x-ray beam convolution calculations, *Med. Phys.*, 22, 1413–1418, 1995.

Keall, P. J. and Hoban, P. W., Superposition dose calculation incorporating Monte Carlo generated electron track kernels, *Med. Phys.*, 23, 479–485, 1996a.

Keall, P. J. and Hoban, P. W., Super-Monte Carlo: A 3D electron beam dose calculation algorithm, *Med. Phys.*, 23, 2023–2034, 1996b.

Keall, P. J., Hoban, P. W., and West, M. P., Super-Monte Carlo: A photon/electron dose calculation algorithm for radiotherapy, *Radiat. Phys. Chem.*, 53, 275–281, 1998.

Keall, P. J., Siebers, J. V., Jeraj, R., and Mohan, R., The effect of dose calculation uncertainty on the evaluation of radiotherapy plans, *Med. Phys.*, 27, 478–484, 2000.

Keall, P. J., Siebers, J. V., Arnfield, M., Kim, J. O., and Mohan, R., Monte Carlo dose calculations for dynamic IMRT treatments, *Phys. Med. Biol.*, 46, 929–941, 2001.

Knöös, T., Ceberg, C., Weber, L., and Nilsson, P., Dosimetric verification of a pencil beam based treatment planning system, *Phys. Med. Biol.*, 39, 1609–1628, 1994.

Knöös, T., Ahnesjö, A., Nilsson, P., and Weber, L., Limitation of a pencil beam approach to photon dose calculations in lung tissue, *Phys. Med. Biol.*, 40, 1411–1420, 1995.

Korevaar, E. W., Akhiat, A., Heijmen, B. J. M., and Huizenga, H., Accuracy of the phase space evolution dose calculation model for clinical 25 MeV electron beams, *Phys. Med. Biol.*, 45, 2931–2945, 2000.

Krieger, T. and Sauer, O. A., Monte Carlo- versus pencil beam-/collapsed-cone-dose calculation in a heterogeneous multi-layer phantom, *Phys. Med. Biol.*, 50, 859–868, 2005.

Laub, W., Alber, M., Birkner, M., and Nusslin, F., Monte Carlo dose computation for IMRT optimization, *Phys. Med. Biol.*, 45, 1741–1754, 2000.

Lax, I., Inhomogeneity corrections in electron-beam dose planning. Limitations with the semi-infinite slab approximation, *Phys. Med. Biol.*, 31, 879–892, 1986.

Lax, I., Accuracy in clinical electron beam dose planning using pencil beam algorithms, *Radioth. Oncol.*, 10, 307–319, 1987.

Lax, I. and Brahme, A., Electron beam dose planning using Gaussian beams: Energy and spatial scaling with inhomogeneities, *Acta Radiol. Oncol.*, 24, 75–85, 1985.

Lax, I., Brahme, A., and Andreo, P., Electron beam dose planning using Gaussian beams: Improved radial dose profiles, *Acta Radiol. Suppl.*, 364, 49–59, 1983.

Leal, A., Sánchez-Doblado, F., Arráns, R., Roselló, J., Carrasco Pavón, E. et al., Routine IMRT verification by means of an automated Monte Carlo simulation system, *Int. J. Radiat. Oncol. Ther. Phys.*, 56, 58–68, 2003.

Lee, P. C., Monte Carlo simulations of the differential beam hardening effect on a flattening filter on a therapeutic x-ray beam, *Med. Phys.*, 24, 1485–1489, 1997.

Lewis, R. D., Ryde, S. J. S., Hancock, D. A., and Evans, C. J., An MCNP-based model of a linear accelerator x-ray beam, *Phys. Med. Biol.*, 44, 1219–1230, 1999.

Li, J. S., Pawlicki, T., Deng, J., Jiang, S. B., Mok, E., and Ma, C.-M., Validation of a Monte Carlo dose calculation tool for radiotherapy treatment planning, *Phys. Med. Biol.*, 45, 2969–2985, 2000.

Liu, H. H., Mackie, T. R., and McCullough, E. C., A dual source photon beam model used in convolution/superposition dose calculations for clinical megavoltage x-ray beams, *Med. Phys.*, 24, 1960–1974, 1997a.

Liu, H. H., Mackie, T. R., and McCullough, E. C., Calculating dose and output factors for wedged photon radiotherapy fields using a convolution/superposition method, *Med. Phys.*, 24, 1714–1728, 1997b.

Liu, H. H., Mackie, T. R., and McCullough, E. C., Correcting kernel tilting and hardening in convolution/superposition dose calculations for clinical divergent and polychromatic photon beams, *Med. Phys.*, 24, 1729–1741, 1997c.

Liu, H. H., Mackie, T. R., and McCullough, E. C., Modeling photon output caused by backscattered radiation into the monitor chamber from collimator jaws using a Monte Carlo technique, *Med. Phys.*, 27, 737–744, 2000.

Lovelock, D. M., Chui, C. S., and Mohan, R., A Monte Carlo model of photon beams used in radiation therapy, *Med. Phys.*, 22, 1387–1394, 1995.

Lulu, B. A. and Bjängard, B. E., Batho's correction factor combined with scatter summation, *Med. Phys.*, 9, 372–377, 1982.

Ma, C.-M. and Nahum, A. E., Calculation of absorbed dose ratios using correlated Monte Carlo sampling, *Med. Phys.*, 20, 1189–1199, 1993.

Ma, C.-M. and Rogers, D. W. O., Beam characterization: A multiple-source model, *National Research Council of Canada Report* PIRS-0509(D), NRCC, Ottawa,1995.

Ma, C.-M. and Jiang, S. B., Monte Carlo modelling of electron beams from medical accelerators, *Phys. Med. Biol.*, 44, R157–R189, 1999.

Ma, C.-M., Faddegon, B. A., Rogers, D. W. O., and Mackie, T. R., Accurate characterization of Monte Carlo calculated electron beams for radio-therapy, *Med. Phys.*, 24, 401–416, 1997.

Ma, C.-M., Mok, E., Kapur, A., Pawlicki, T., Findley, D. et al., Clinical implementation of a Monte Carlo treatment planning system, *Med. Phys.*, 26, 2133–2143, 1999.

Ma, C.-M., Pawlicki, P., Jiang, S. B., Li, J. S., and Deng, J., MCDOSE: A Monte Carlo dose calculation tool for radiation therapy planning, in *The Use of Computers in Radiation Therapy, XIIIth Int. Conf. (ICCR)*, Schlegel, W. and Bortfeld, T., Eds., Springer, Heidelberg, pp. 123–125, 2000a. .

Ma, C.-M., Pawlicki, P., Jiang, S. B., Li, J. S., and Deng, J., Monte Carlo verification of IMRT dose distributions from a commercial treatment planning optimisation system, *Phys. Med. Biol.*, 45, 2483–2495, 2000b.

Ma, C.-M., Li, J. S., Pawlicki, T., Jiang, S. B., Deng, J. et al., A Monte Carlo dose calculation tool for radiotherapy treatment planning, *Phys. Med. Biol.*, 47, 1671–1689, 2002.

Ma, C.-M., Price, R. A., Li, J. S., Chen, L., Wang, L. et al., Monitor unit calculation for Monte Carlo treatment planning, *Phys. Med. Biol.*, 49, 1671–1687, 2004.

Ma, C.-M., Li, J. S., Pawlicki, T., Jiang, S. B., Deng, J. et al., MCSIM: A Monte Carlo dose calculation tool for radiation therapy, in *Proc. 14th Int. Conf. On the Use of Computers in Radiation Therapy (ICCR)*, Yi, B. Y., Ahn, S. D., Choi, E. K., and Ha, S. W., Eds., Jeong Publishing, Seoul, pp. 515–519, 2004.

Ma, C.-M., Li, J. S., Jiang, S. B., Pawlicki, T., Xiong, W. et al., Effect of statistical uncertainties on Monte Carlo treatment planning, *Phys. Med. Biol.*, 50, 891–907, 2005.

Mackie, T. R., Applications of the Monte Carlo method in radiotherapy, in *The Dosimetry of Ionizing Radiation*, Kase, K. R., Bjärngard, B. E., and Attix, F. H., Eds., Vol. III, Academic Press, San Diego, CA, pp. 541–620, 1990.

Mackie, T. R. and Scrimger, J. W., Computing radiation dose for high energy x-rays using a convolution method. *Proceedings of Int. Conf on the use of computers in radiation therapy, VIII (ICCR)*, 36–40. Toronto: IEEE Computer Society Press. 1984.

Mackie, T. R., El-Khatib, E., Battista, J. J., and Scrimger, J. W., Lung dose corrections for 6MV and 15MV x-rays, *Med. Phys.*, 12, 327–332, 1985.

Mackie, T. R., Ahnesjö, A., Dickof, P., and Snider, A., Development of a convolution/superposition method for photon beams, in *The Use of Computers in Radiation Therapy*, Bruinvis, I. A. D., van der Giessen, P. H., van Kleffens, H. H., and Wittkämper, F. W., Eds., Elsevier Science Publishers BV, North Holland, pp. 107–110, 1987.

Mackie, T. R., Bielajew, A. F., Rogers, D. W. O., and Battista, J. J., Generation of photon energy deposition kernels using the EGS Monte Carlo Code, *Phys. Med. Biol.*, 33, 1–20, 1988.

Mackie, T. R., Reckwerdt, P. J., Holmes, T. W., and Kubsad, S. S., Review of convolution/superposition methods for photon beam dose computation. in *The Use of Computers in Radiation Therapy*, Hukku, S. and Iyer, P. S., Eds., Institute of Medical Sciences, Lucknow, PP. 20–23, 1990a.

Mackie, T. R., Reckwerdt, P. J., Gehring, M. A., Holmes, T. W., Kubsad, S. S., Thomadsen, B. R., Sanders, C. A., Paliwal, B. R. and Kinsella, T. J., Clinical implementation of the convolution/superposition method, in *The Use of Computers in Radiation Therapy*, Hukku, S. and Iyer, P. S., Eds., Institute of Medical Sciences, Lucknow, pp. 322–325. 1990b.

Mackie, T. R., Reckwerdt, P. J., Gehring, M. A., Holmes, T. W., Kubsad, S. S., Sanders, C. A., Bhudatt, R. P., and Kinsella, T. J., A fast convolution/superposition algorithm for radiation therapy treatment planning, *Proc. 1st Biennial ESTRO Meeting on Physics in Clinical Radiotherapy, Budapest 14–17 Oct.*, 1991, 93, 1991.

Mackie, T. R., Reckwerdt, P. J., Wells, C. M., Yang, J. N., Deasy, J. O., Podgorski, M., Holmes, M. A. et al., The Omega project: Comparison among EGS4 electron beam simulations, 3-D fermi-Eyges calculations, and dose measurements, In *Proceedings of the XIth International Conference on the Use of Computers in Radiation Therapy, 20–24 March 1994, Manchester*, Hounsell, A. R., Wilkinson, J. M., and Williams, P. C., Eds., NW Medical Physics Dept., Christie Hospital NHS Trust, Manchester, pp. 152–157, 1994.

Mackie, T. R., Reckwerdt, P. J., and Papanikolaou, N., 3-D photon beam dose algorithms. in *3-D radiation treatment planning and conformal therapy*, Purdy J. A. and Emami, B., Eds., Medical Physics Publishing, Madison, WI, 1995.

Mackie, T. R., Reckwerdt, P., McNutt, T., Gehring, M., and Sanders, C., Photon beam dose computations, in *Teletherapy: Present and future*, Palta, J. and Mackie, T. R., Eds., American Associations of Physicists in Medicine, College Park, MD, pp. 103–135, 1996.

Mah, E., Antolak, J., Scrimger, J. W., and Battista, J. J., Experimental verification of a 2D and 3D electron pencil beam algorithm, *Phys. Med. Biol.*, 34, 1179–1194, 1989.

Manfredotti, C., Nastasi, U., Ragona, R., and Anglesio, S., Comparison of threedimensional Monte Carlo simulation and the pencil beam algorithm for an electron beam from a linear accelerator, *Nucl. Instr. Meth. in Phys. Res.*, A255, 355–359, 1987.

MCNPX. *MCNPX User's Manual, Version 2.4.0*, LA-CP-02-408. Los Alamos, NM: Los Alamos National Laboratory, 2003.

Metcalfe, P. E., Hoban, P. W., Murray, D. C., and Round, W. H., Beam hardening of 10 MV radiotherapy x-rays, Analysis using a convolution/superposition method, *Phys. Med. Biol.*, 35, 1533–1549, 1990.

Miften, M., Wiesmeyer, M., Monthofer, S., and Krippner, K., Implementation of FFT convolution and multigrid superposition models in the FOCUS RTP system, *Phys. Med. Biol.*, 45, 817–833, 2000.

Mijnheer, B., Olszewska, A., Fiorino, C., Hartmann, G., Knöös, T., Rosenwald, J.-C., and Welleweerd, H., *Quality Assurance of Treatment Planning Systems—Practical Examples for External Photon Beams*. ESTRO Booklet No. 7, ESTRO, Garant Publishers, Leuven, 2004.

Milan, J. and Bentley, R., The storage and manipulation of radiation dose data in a small digital computer, *Br. J. Radiol.*, 47, 115–121, 1974.

Mohan, R., Monte Carlo simulation of radiation treatment machine heads, in *Monte Carlo Transport of Electrons and Photons below 50 MeV*, Jenkins, T. M., Nelson, W. R., Rindi, A., Nahum, A. E., and Rogers, D. W. O., Eds., Plenum Press, New York, pp. 453–468, 1988.

Mohan, R., Why Monte Carlo? in *Proceedings of the XIIth International Conference on the Use of Computers in Radiation Therapy*, Leavitt, D. D. and Starkschall, G., Eds., Medical Physics Publishing, Madison, WI, pp. 16–18, 1997.

Mohan, R. and Chui, C. S., Validity of the concept of separating primary and scatter dose, *Med. Phys.*, 12, 726–730, 1985.

Mohan, R. and Chui, C. S., Use of fast Fourier transforms in calculating dose distributions for irregularly shaped fields for three-dimensional treatment planning, *Med. Phys.*, 14, 70–77, 1987.

Mohan, R., Chui, C. S., and Lidofsky, L., Energy and angular distribuiton of photons from medical linear accelerators, *Med. Phys.*, 12, 592–597, 1985.

Mohan, R., Chui, C. S., and Lidofsky, L., Differential pencil beam dose computation model for photons, *Med. Phys.*, 13, 64–73, 1986.

Mora, G., Pawlicki, T., Maio, A., and Ma, C. M., Effect of voxel size on Monte Carlo dose calculations for radiotherapy treatment planning, in *Advanced Monte Carlo for Radiation Physics, Particle Transport Simulation and Applications (Proceedings of the Monte Carlo 2000 Conference, Lisbon)*, Kling, A., Barao, F., Nakagawa, M., Távora, L., and Vaz, P., Eds., Springer, Berlin, 2001.

Morawska-Kaczynska, M. and Huizenga, H., Numerical calculation of energy deposition by broad high-energy electron beams: II: Multi-layered geometry, *Phys. Med. Biol.*, 37, 2103–2116, 1992.

Myler, U. and Szabo, J. J., Dose calculation along the nonwedged direction for externally wedged beams: Improvement of dosimetric accuracy with comparatively moderate effort, *Med. Phys.*, 29, 748–754, 2002.

Nahum, A. E., The MDAH pencil-beam algorithm, in *The Computation of Dose Distributions in Electron Beam Radiotherapy*, Nahum, A. E., Ed., pp. 151–184, Available from Medical Physics Publishing, Madison, WI. Radiation Physics Dept, Umeå University, 901 87 Umeå, Sweden, 1985.

Nahum, A. E., Overview of photon and electron Monte Carlo, in *Monte Carlo Transport of Electrons and Photons below 50 MeV*, Jenkins, T. M., Nelson, W. R., Rindi, A., Nahum, A. E., and Rogers, D. W. O., Eds., Plenum Press, New York, pp. 3–20, 1988.

Nahum, A. E., 3D dose calculation (III): Application of Monte-Carlo methods, in *Dreidimensionale Strahlentherapieplanung, 27–29 April, Heidelberg*, Schlegel, W., Bortfeld, T., and Stein, J., Eds., DKFZ, Heidelberg, pp. 89–97, 1995.

Nahum, A. E., Condensed-history Monte-Carlo simulation for charged particles: What can it do for us? *Radiation and Environmental Biophysics*, 38, 163–173, 1999.

Nahum, A. E. and Brahme, A., Electron depthdose distributions in uniform and nonuniform media, in *The Computation of Dose Distributions in Electron Beam Radiotherapy*, Nahum, A. E., Ed., pp. 98–127, Available from Medical Physics Publishing, Madison, WI,. Medical Radiation Physics Dept., Umeå University, 901 87 Umeå, Sweden, 1985.

Nelson, W. R., Hirayama, A., and Rogers, D. W. O., *The EGS4 Code System Stanford Linear Accelerator Center Report SLAC265*, SLAC, Stanford, CA, 1985.

Neuenschwander, H. and Born, E. J., A macro Monte Carlo method for electron beam dose calculations, *Phys. Med. Biol.*, 37, 107–125, 1992.

Neuenschwander, H., Mackie, T. R., and Reckwerdt, P. J., MMC—a high-performance Monte Carlo code for electron beam treatment planning, *Phys. Med. Biol.*, 40, 543–574, 1995.

Neuenschwander, H., Volken, W., Cris, C., Mini, R., and Schwab, P., Fast Monte Carlo algorithms for electron beam treatment planning, in *Proceedings of the XIIth International Conference on the Use of Computers in Radiation Therapy*, Leavitt, D. D. and Starkschall, G., Eds., Medical Physics Publishing, Madison WI, pp. 23–26, 1997.

Nilsson, M. and Knöös, T., Application of the Fano theorem in inhomogeneous media using a convolution algorithm, *Proceedings of the 1st Biennial ESTRO Meeting on Physics on Clinical Radiotherapy, Budapest, 14–17 Oct., 1991*, 70, 1991.

Nisbet, A., Weatherburn, H., Fenwick, J. D., and McVey, G., Spectral reconstruction of clinical megavoltage photon beams and the implications of spectral determination on the dosimetry of such beams, *Phys. Med. Biol.*, 43, 1507–1521, 1998.

O'Connor, J. E., The variation of scattered x-rays with density in an irradiated body, *Phys. Med. Biol.*, 1, 352–369, 1957.

Papanikolaou, N., Mackie, T. R., Meger-Wells, C., Gehring, M., and Reckwerdt, P., Investigation of the convolution method for polyenergetic spectra, *Med. Phys.*, 20, 1327–1336, 1993.

Papatheodorou, S., Rosenwald, J.-C., Zefkili, S., Murillo, M. C., Drouard, J., and Gaboriaud, G., Dose calculation and verification of intensity modulation generated by dynamic multileaf collimators. *Med. Phys.*, 27, 960–971, 2000.

Parker, R. P., Contier de Freitas, L., Cassell, M., Webb, S., and Hobday, P., A method of implementing inhomogeneity corrections in radiotherapy treatment planning, *J. Eur. Radiother*, 1, 93–100, 1980.

Pawlicki, T. and Ma, C.-M., Monte Carlo dose modeling for MLC-based IMRT, *Med. Dosim.*, 26, 157–168, 2001.

Poon, E., Seuntjens, J., and Verhaegen, F., Consistency test of the electron transport algorithm in the GEANT4 Monte Carlo code, *Phys. Med. Biol.*, 50, 681–694, 2005.

Redpath, A. T. and Thwaites, D. I., A 3-dimensional scatter correction algorithm for photon beams, *Phys. Med. Biol.*, 36, 779–798, 1991.

Rogers, D. W. O., The role of Monte Carlo simulation of electron transport in radiation dosimetry, *Int'l J. of Appl. Radiat. Isotopes*, 42, 965–974, 1991.

Rogers, D. W. O. and Bielajew, A. F., A comparison of EGS and ETRAN, in *Monte Carlo Transport of Electrons and Photons below 50 MeV*, Jenkins, T. M., Nelson, W. R., Rindi, A., Nahum, A. E., and Rogers, D. W. O., Eds., Plenum Press, New York, pp. 323–342, 1988.

Rogers, D. W. O. and Bielajew, A. F., Monte Carlo techniques of electron and photon transport in radiation dosimetry. in *The Dosimetry of Ionizing Radiation*, Vol. III, Kase, K. R., Bjärngard, B. E., and Attix, F. H., Eds., Academic Press, NewYork, PP. 427–539, 1990.

Rogers, D. W. O. and Mohan, R., Questions for comparison of clinical Monte Carlo codes, in *The Use of Computers in Radiation Therapy, XIIIth Int'l Conf.*, Schlegel, W. and Bortfeld, T., Eds., Springer, Heidelberg, pp. 120–122, 2000.

Rogers, D. W. O., Faddegon, B. A., Ding, G. X., Ma, C.-M., and Wei, J., BEAM: A Monte Carlo code to simulate radiotherapy treatment units, *Med. Phys.*, 22, 503–524, 1995.

Rosenwald, J.-C., Drouard, J., and Simonian, M., Representation of the lack of electronic equilibrium in high energy photon beams using modified scatter tables. *Proceedings of the 9th International Conference on the Use of Computers in Radiation Therapy*, Amsterdam, Elsevier, pp. 327–330, 1987.

Rosenwald, J.-C., Oozeer, R., Belshi, R., Drouard, J., and Mazal, A., Penumbra representation for primary-scatter decomposition of electron and proton beams, in *Proceedings of the XIth International Conference on the Use of Computers in Radiation Therapy*, Hounsell, A. R., Wilkinson, J. M., and Williams, P. C., Eds., NW Medical Physics Dept., Christie Hospital NHS Trust, Manchester, pp. 128–129, 1994.

Salvat, F., Fernández-Varea, J. M., Baró, J., and Sempau, J., PENELOPE, an algorithm and Computer Code for Monte Carlo simulation of electron-photon showers. *Informes Tecnicos CIEMAT Report* No. 799, CIEMAT, Madrid, 1996.

Samuelsson, A., Hyödynmaa, S., and Johansson, K.-A., Dose accuracy check of the 3D electron beam algorithm in a treatment planning system, *Phys. Med. Biol.*, 43, 1529–1544, 1998.

Sauer, O. A., Calculation of dose distributions in the vicinity of high-Z interfaces for photon beams, *Med. Phys.*, 22, 1685–1690, 1995.

Schach von Wittenau, A. E., Cox, L. J., Bergstrom, P. M., Chandler, W. P., Hartmann-Siantar, C. L. et al., Correlated histogram representation of Monte Carlo derived medical accelerator photon-output phase space, *Med. Phys.*, 26, 1196–1211, 1999.

Schneider, W., Bortfeld, T., and Schlegel, W., Correlation between CT number and tissue parameters needed for Monte Carlo simulations of clinical dose distributions, *Phys. Med. Biol.*, 45, 459–478, 2000.

Schoknecht, G., Die Beschreubung von Strahlenfeldern durch Separierung von Primär- und Streustrahlung IV: Berechnung von Streuverteilungen für parallele Photonen-Strahlenfelder, *Strahlentherapie*, 141, 326–331, 1971.

Scholz, C., Schulze, C., Oelfke, U., and Bortfeld, T., Development and clinical application of a fast superposition algorithm in radiation therapy, *Radiother. Oncol.*, 69, 79–90, 2003.

Seco, J., Adams, E., Bidmead, M., Partridge, M., and Verhaegen, F., Head-and-neck IMRT treatments assessed with a Monte Carlo dose calculation engine, *Phys. Med. Biol.*, 50, 817–830, 2005.

Seltzer, S. M., Hubbel, J. H., and Berger, M. J., Some theoretical aspects of electron and photon dosimetry, IAEA-SN-222/05, *National and International Standardization of radiation Dosimetry,*, Vol. 2, IAEA, Vienna, PP. 3–43, 1978.

Sempau, J., Wilderman, S. J., and Bielajew, A. F., DPM—a fast, accurate Monte Carlo code optimized for photon and electron radiotherapy treatment planning dose calculations, *Phys. Med. Biol.*, 45, 2263–2291, 2000.

Sempau, J., Sanchez-Reyes, F., Salvat, F., ben Tahar, O., Jiang, S. B. et al., Monte Carlo simulation of electron beams from an accelerator head using PENELOPE, *Phys. Med. Biol.*, 46, 1163–1186, 2001.

Seuntjens, J., van der Plaetsen, A., and Thierens, H., Comparison of measured and calculated dose distributions in lung after electron beam treatment of the chest wall, *Med. Phys.*, 21, 1959–1968, 1994.

Sharpe, M. B. and Battista, J. J., Dose calculations using convolution and superposition principles: The orientation of dose spread kernels in divergent x-ray beams, *Med. Phys.*, 20, 1685–1694, 1993.

Sharpe, M. B., Jaffray, D. A., Battista, J. J., and Munro, P., Extrafocal radiation: A unified approach to the prediction of beam penumbra and output factors for megavoltage x-ray beams, *Med. Phys.*, 22, 2065–2074, 1995.

Sheikh-Bagheri, D. and Rogers, D. W. O., Sensitivity of megavoltage photon beam Monte Carlo simulations to electron beam parameters, *Med. Phys.*, 29, 379–390, 2002.

Shiu, A. S. and Hogstrom, K. R., A pencil-beam redefinition algorithm for electron dose distributions, in *The Use of Computers in Radiation Therapy*, Bruinvis, I. A. D., van der Giessen, p. h., van Kleffens, H. H., and Wittkämper, F. W., Eds., Elsevier Science Publishers BV, North Holland, pp. 69–72, 1987.

Shiu, A. S., Tung, S., Hogstrom, K. R., Wong, J. W., Gerber, R. L. et al., Verification data for electron beam dose algorithms, *Med. Phys.*, 19, 623–636, 1992.

Shortt, K. R., Ross, C. K., Bielajew, A. F., and Rogers, D. W. O., Electron beam dose distributions near standard inhomogeneities, *Phys. Med. Biol.*, 31, 235–249, 1986.

Shu, Y. and Boyer, A., X-ray dose computations in heterogeneous media using 3-dimensional FFT convolution, *Phys. Med. Biol.*, 35, 351–368, 1990.

Siddon, R. L., Fast calculation of exact radiological path for a three-dimensional CT array, *Med. Phys.*, 12, 252–255, 1985.

Siebers, J. V., Keall, P. J., and Mohan, R., The impact of Monte Carlo dose calculations on intensity-modulated radiation therapy, in *Advanced Monte Carlo for Radiation Physics, Particle Transport Simulation and Applications (Proceedings of the Monte Carlo 2000 Conference, Lisbon)*, Kling, A., Barao, F., Nakagawa, M., Távora, L., and Vaz, P., Eds., Springer, Berlin, pp. 205–210, 2001.

Siebers, J. V., Keall, P. J., Kim, J. O., and Mohan, R., Performance benchmarks of the MCV Monte Carlo system, in *Proceedings of the XVIII Int. Conf. on the Use of Computers in Radiation Therapy*, Schlegel, W. and Bortfeld, T., Eds., Springer, Berlin, pp. 129–131, 2000a.

Siebers, J. V., Keall, P. J., Nahum, A. E., and Mohan, R., Converting absorbed dose to medium to absorbed dose to water for Monte Carlo based photon beam dose calculations, *Phys. Med. Biol.*, 45, 983–995, 2000b.

Siebers, J. V., Keall, P. J., Kim, J. O., and Mohan, R., A method for photon beam Monte Carlo multileaf collimator particle transport, *Phys. Med. Biol.*, 47, 3225–3249, 2002.

Solberg, T. D., Holly, F. E., DeSalles, A. A. F., and Smathers, J. B., Implications of tissue heterogeneity for radiosurgery in head and neck tumors, *Int. J. Radiation Oncology Biol. Phys.*, 32, 235–239, 1995.

Solberg, T. D., DeMarco, J. J., Holly, F. E., Smathers, J., and DeSalles, A. A. F., Monte Carlo treatment planning for stereotactic radiosurgery, *Radiotherapy and Oncology*, 49, 73–84, 1998.

Sontag, M. R. and Cunningham, J. R., Corrections to absorbed dose calculations for tissue inhomogeneities, *Med. Phys.*, 4, 431–436, 1977.

Sontag, M. R. and Cunningham, J. R., The equivalent tissue-air ratio method for making absorbed dose calculations in heterogeneous medium, *Radiology*, 129, 787–794, 1978.

Spezi, E., Lewis, D. G., and Smith, C. W., A DICOM-RT-based toolbox for the evaluation and verification of radiotherapy plans, *Phys. Med. Biol.*, 47, 4223–4232, 2002.

Starkschall, G., Steadham, R. E., Popple, R. A., Ahmad, S., and Rosen, I. I., Beam-commissioning methodology for a three-dimensional convolution/superposition photon dose algorithm, *J. Appl. Clin. Med. Phys.*, 1, 8–27, 2000.

Storchi, P. and Woudstra, E., Calculation models for determining the absorbed dose in water in off-axis planes of rectangular fields for open and wedged photon beams, *Phys. Med. Biol.*, 40, 511–527, 1995.

Storchi, P. and Woudstra, E., Calculation of the absorbed dose distribution due to irregularly shaped photon beams using pencil beam kernels derived from basic beam data, *Phys. Med. Biol.*, 41, 637–656, 1996.

Tailor, R. C., Tello, V. M., Schroy, C. B., Vossler, M., and Hanson, W. F., A generic off-axis energy correction for linac photon beam dosimetry, *Med. Phys.*, 25, 662–667, 1998.

Thomas, S. J., A modified power-law formula for inhomogeneity corrections in beams of high-energy x-rays, *Med. Phys.*, 18, 719–723, 1991.

Udale, M., A Monte Carlo investigation of surface doses for broad electron beams, *Phys. Med. Biol.*, 33, 939–954, 1988.

van de Geijn, J., The computation of two- and three-dimensional dose-distributions in cobalt-60 teletherapy, *Br. J. Radiol.*, 38, 369–377, 1965.

van de Geijn, J., A computer program for 3-D planning in external beam radiation therapy: EXTDOS, *Comput. Programs Biomed.*, 1, 47–57, 1970.

van de Geijn, J., EXTDOS 71. Revised and expanded version of EXTDOS. A program for treatment planning in external beam therapy, *Comput. Programs Biomed.*, 2, 169–177, 1972.

van de Geijn, J., Chin, B., Pochobradsky, J., and Miller, R. W., A new model for computerized clinical electron beam dosimetry, in *The Use of Computers in Radiation Therapy*, Bruinvis, I. A. D., van der Giessen, P. H., van Kleffens, H. H., and Wittkämper, F. W., Eds., Elsevier Science Publishers BV, North Holland, pp. 141–144, 1987.

van der Zee, W., Hogenbirk, A., and van der Marck, S. C., ORANGE: A Monte Carlo dose engine for radiotherapy, *Phys. Med. Biol.*, 50, 625–641, 2005.

van Gasteren, J. J. M., Heukelom, S., van Kleffens, H. J., van der Laarse, R., Venselaar, J. L. M., and Westermann, C. F., The determination of phantom and collimator scatter components of the output of megavoltage photon beams: Measurement of the collimator scatter part with a beam-coaxial narrow cylindrical phantom, *Radiother. Oncol.*, 20, 250–257, 1991.

Verhaegen, F. and Das, I. J., Monte Carlo modelling of a virtual wedge, *Phys. Med. Biol.*, 44, 251–259, 1999.

Verhaegen, F. and Seuntjens, J., Monte Carlo modelling of external radiotherapy photon beams, *Phys. Med. Biol.*, 48, R107–R164, 2003.

Verhaegen, F. and Devic, S., Sensitivity study for CT image use in Monte Carlo treatment planning, *Phys. Med. Biol.*, 50, 937–946, 2005.

Verhaegen, F., Das, I. J., and Palmans, H., Monte Carlo dosimetry study of a 6MV stereotactic radiosurgery unit, *Phys. Med. Biol.*, 43, 2755–2768, 1998.

Verhaegen, F., Nahum, A. E., van de Putte, S., and Namito, Y., Monte Carlo modelling of radiotherapy kV x-ray units, *Phys. Med. Biol.*, 44, 1767–1789, 1999.

Verhaegen, F., Symonds-Tayler, R., Liu, H. H., and Nahum, A. E., Backscatter towards the monitor ion chamber in high-energy photon and electron beams: Charge integration versus Monte Carlo simulation, *Phys. Med. Biol.*, 45, 3159–3170, 2000.

Verhaegen, F., Mubata, C., Pettingell, J., Bidmead, A. M., Rosenberg, I. et al., Monte Carlo calculation of output factors for circular, rectangular and square fields of electron accelerators (6–20 MeV), *Med. Phys.*, 28, 938–949, 2001.

Wang, L., Chui, C. S., and Lovelock, M., A patient-specific Monte Carlo dose-calculation method for photon beams, *Med. Phys.*, 25, 867–878, 1998.

Wang, L., Lovelock, M., and Chui, C.-S., Experimental verification of a CT-based Monte Carlo dose-calculation method in heterogeneous phantoms, *Med. Phys.*, 26, 2626–2634, 1999.

Webb, S., *The Physics of Three-Dimensional Radiation Therapy*, IOP Publishing Ltd, Bristol, UK., 1993.

Wong, J. W. and Henkelman, R. M., Reconsideration of the power-law (Batho) equation for inhomogeneity corrections, *Med. Phys.*, 9, 521–530, 1982.

Wong, J. W. and Henkelman, R. M., CT pixel-based dose calculations in heterogeneous media, *Med. Phys.*, 10, 199–208, 1983.

Wong, J. W. and Purdy, J. A., On methods of inhomogeneity corrections for photon transport, *Med. Phys.*, 17, 807–814, 1990.

Woo, M. K. and Cunningham, J. R., The validity of the density scaling method in primary electron transport for photon and electron beams, *Med. Phys.*, 17, 187–194, 1990.

Woo, M. K., Cunningham, J. R., and Jezioranski, J. J., Extending the concept of primary and scatter separation to the condition of electronic desequilibrium, *Med. Phys.*, 17, 588, 1990.

Yao, J. Y. and Ranganathan, G., On three-dimensional dose calculation of photon beam with wedge filters, *Med. Phys.*, 21, 809–816, 1994.

Yu, C. X., Mackie, T. R., and Wong, J. W., Photon dose calculation incorporating explicit electron transport, *Med. Phys.*, 22, 1157–1165, 1995.

Zefkili, S., Kappas, K., and Rosenwald, J.-C., On axis and off-axis primary dose component in high energy photon beams, *Med. Phys.*, 21, 799–808, 1994.

Zhu, Y. and Boyer, A., X-ray dose computations in heterogeneous media using 3-dimensional FFT convolution, *Phys. Med. Biol.*, 35, 351, 1990.

Zhu, Y. and van Dyk, J., Accuracy requirements of the primary x-ray spectrum in dose calculations using FFT convolution techniques, *Med. Phys.*, 22, 421–426, 1995.

INTERNATIONAL CONFERENCES ON THE USE OF COMPUTERS IN RADIOTHERAPY

The following is a list of the 15 International Conferences on the Use of Computers in Radiotherapy (ICCR) together with details of the document containing their published papers (if available)

I. Cambridge, England, 1966. *The Use of Computers in Therapeutic Tadiology*. Special Report No. 1, British Institute of Radiology, London, 1967.

II. Chicago, U.S.A., 1968. *Computers in Radiotherapy*, Cohen, M., Ed., Special Report No. 4, Br. J. Radiol. 43, 658–663, 1970.

III. Glasgow, Scotland, 1970. *Computers in Radiotherapy*, Glicksman, A. S., Cohen, M. and Cunningham, J. R., Eds., Special Report No. 5, British Insitute of Radiology, London, 1971.

IV. Uppsala, Sweden, 1972. *Computers in Radiation Therapy*. Radiofysikavdelningen, Akademiska Sjukhuset. S-75014, Uppsala, Sweden, 1972.

V. Hanover, NH, U.S.A., 1975. *Computer Applications in Radiation Oncology*, Sternick, E. S., Ed., University Press of New England, 1976.

VI. Göttingen, Germany, 1977. *Computers in Radiotherapy*, Rosenow, U., Ed., Strahlenabteilung Universitäts Frauenklinic, D-3400, Göttingen, Germany, 1978.

VII. Kawasaki and Tokyo, Japan, 1980. *Computers in Radiation Therapy*, Umegaki, Japan Radiological Soc., Tokyo, Japan, 1981.

VIII. Toronto, Canada, 1984. *Use of Computers in Radiation Therapy*. IEEE Computer Society Press, IEEE Los Angeles, U.S.A., 1984.

IX. Scheveningen, The Netherlands, 1987. *The Use of Computers in Radiation Therapy*, Bruinvis, I.A.D., van der Giessen, P.H., van Kleffens, H.H., and Wittkämper, F.W., Elsevier Science Publishers, North Holland, BV, 1987.

X. Lucknow, India, 1990. *The Use of Computers in Radiation Therapy*, Hukku, S. and Iyer. P. S., Eds., Institute of Medical Sciences, Lucknow, 1990.

XI. Manchester, U.K., 1994. *The Use of Computers in Radiation Therapy*, Hounsell, A. R., Wilkinson, J. M., and Williams, P. C., Eds., North Western Medical Physics Department of Christie Hospital NHS Trust, Manchester, U.K. and Medical Physics Publishing, Madison, WI, 1994.

XII. Salt Lake City, U.S.A., 1997. *The Use of Computers in Radiation Therapy*, Leavitt, D.D. and Starkschall, G. Eds., Medical Physics Publishing, Madison, WI, 1997.

XIII. Heidelberg, Germany, 2000. *The Use of Computers in Radiation Therapy*, Schlegel, W. and Bortfeld, T., Eds., Springer, Heidelberg, 2000.

XIV. Seoul, Korea, 2004. *The Use of Computers in Radiation Therapy*.

XV. Toronto, Canada, 2007.

PART G

Treatment Planning

Editors: Jean-Claude Rosenwald and Philip Mayles

INTRODUCTION

Treatment planning is the process of determining the most appropriate way to irradiate the patient. It is a combination of the following five essential steps:

1. Choosing an appropriate patient positioning and immobilisation method so that treatments will be reproducible;
2. Identifying the shape and the location of the tumour (i.e. the *target*) and of the neigbouring organs at risk;
3. Selecting a suitable beam arrangement;
4. Evaluating the resulting dose distribution;
5. Calculating the treatment machine settings to deliver the required absolute dose.

This results in the flow of data illustrated in Figure 39.1.

Prior to treatment planning, it will be necessary to determine the nature and extent of the tumour using various diagnostic techniques. Occasionally, it will be possible to conduct the diagnostic investigation in such a way as to provide appropriate data for treatment planning, but normally, it will be necessary to carry out a separate imaging study. This occurs because the requirements of diagnostic imaging and treatment planning are largely incompatible. For diagnosis, the priority is to obtain artefact-free images with appropriate enhancement of the tumour. Treatment planning requires a geometrically accurate three-dimensional model of the patient in the precise treatment position together with a means of transferring the three-dimensional coordinate system to the treatment machine. For accurate tissue inhomogeneity corrections, the electron density distribution of the patient is also required.

For common treatment situations, a standard beam arrangement will be established as a class solution and planning will consist of establishing the correct beam geometry followed by a dose computation to determine the monitor unit settings required. However, it is important to realise that the arrangement of the beams relative to each other can have a significant effect on the dose gradient at the edge of the target volume. Therefore, where the edge of the target volume is determined by the penumbra of only one beam, the dose gradient will be shallower than when more than one beam edge is involved—as in the superior or inferior direction for coplanar transverse beams. For this reason, purely geometric planning is not sufficient. In determining the field margins to be geometrically applied in such class solutions and for non-standard treatments, the relationship between the edge of the target volume and the edge of the dose distribution will need to be considered in detail.

The size of the target volume and the margins around what is visible on an image should be determined as suggested in ICRU 50 (ICRU 1993) and ICRU 62 (ICRU 1999). This is covered in Chapter 29. Chapter 30 and Chapter 31 describe how the various imaging methods can be used and combined to obtain a representation of the patient where the target volume and the organs at risks are properly identified. Chapter 32 considers the solutions for beam setup in relationship to patient anatomy. Chapter 33 and Chapter 34 look at ways of combining beams to achieve an appropriate dose distribution for photon and electron beams, respectively. It is possible to calculate the effect of simple beam combinations using manual graphical methods to estimate the summed contributions. Although such methods can be instructive and can help the student to understand the effect of combining beams, modern treatment planning is almost exclusively carried out using computer algorithms (see Part F); therefore, manual methods will not be discussed. Finally, in Chapter 35 and Chapter 36, methods of evaluating the dose distributions to establish the physical relationship with the target volume and to obtain an estimate of the biological effect of the treatment are considered.

Treatment planning requires a combination of a large amount of data and computer systems to produce the final result. It is essential that there is a detailed quality assurance programme to minimise the risk of error. This is covered in Part H, Chapter 39.

CHAPTER 29

TARGET DEFINITION

Anthony Neal

CONTENTS

29.1 INTRODUCTION

The therapeutic use of radiation (radiotherapy) is an established method of treating malignant tumours. Radiotherapy is a local treatment and is complementary to systemic treatments such as chemotherapy and hormone therapies. It is important to appreciate that the desired anti-cancer activity of radiotherapy is only seen in the tissues directly irradiated by the primary beam(s).

Definition of the tumour, adjacent organs at risk (OAR), and other anatomical structures are therefore an essential part of the planning process to ensure that the beam size, number, trajectory and weighting are optimised in relation to these structures. This will in turn ensure that the prescribed radiation dose is delivered to the appropriate volume of tumour-bearing tissue to achieve the endpoint of tumour cure or palliation (symptom control) while incurring an acceptable rate and severity of both early and late radiation morbidity.

In recent years, radiotherapy has become increasingly complex. The current trend is toward the implementation of conformal radiotherapy and intensity modulated radiotherapy (see Chapter 43). These aim to reduce normal tissue toxicity by limiting irradiation of normal tissues while still irradiating the tumour in its entirety. This should in turn produce reduced morbidity for a given absorbed dose compared with standard treatment techniques, or allow dose escalation for an equivalent level of morbidity. These technologically demanding treatments can only fulfil their promise if the therapeutic absorbed dose is delivered to the tumour reliably and reproducibly from day to day during a protracted course of treatment.

The aim of this chapter is to outline the basic principles of target definition, exploring the associated problems and uncertainties, and how they can be addressed in everyday clinical practice. The detailed procedures for acquiring the anatomical data needed for the definition of target volumes and other anatomical structures of interest will be described in Chapter 30 with special reference to computerised tomography (CT) imaging. The specific role of magnetic resonance (MR) imaging for treatment planning will be described in Chapter 31.

29.2 THE INTERNATIONAL COMMISSION ON RADIATION UNITS AND MEASUREMENTS VOLUMES

The International Commission on Radiation Units and Measurements (ICRU) 50 report (ICRU 1993) stipulates standard protocols for recording and reporting radiotherapy treatments of all degrees of complexity. These include the adoption of standard terminology to describe the volumes relevant to radiotherapy treatment planning. The adoption of these terms serves several purposes:

- Improves clarity of thought and encourages a logical approach to planning
- Promotes consistency in physics planning and clinical practice
- Allows standardisation of clinical trial protocols, particularly for complex, multi-phase treatments
- Facilitates communication between different centres and within clinical trials

Since the publication of the ICRU report 50, there has been considerable discussion in the literature about volume definition and related uncertainties. A comprehensive review and analysis of these issues can be found in a booklet published by the British Institute of Radiology (2003). The recommendations of ICRU Report 50 have been further developed in ICRU Report 62 (ICRU 1999) (see Section 29.4).

The volumes originally defined in the ICRU 50 report are described below (Figure 29.1).

29.2.1 GROSS TUMOUR VOLUME

The gross tumour volume (GTV) is the gross palpable, visible and demonstrable extent and location of the malignant growth. This is determined by physical examination by the oncologist and the results of radiological investigations relevant to the site of the tumour. As the term suggests, tumours have a length, breadth and depth, and the GTV must therefore be identified using orthogonal 2D or 3D imaging. It may be clearly delineated for some tumours (e.g. cerebral metastases or lung cancers) and very ill-defined for others (e.g. high grade gliomas of the brain or tumours diffusely involving an organ). Even the imaging

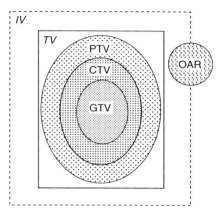

FIGURE 29.1
Schematic representation of the volumes defined in ICRU Report 50 demonstrating the relationship of different volumes to each other: GTV, gross tumour volume; CTV, clinical target volume; PTV, planning target volume; TV, treated volume; IV, irradiated volume; OAR, organ at risk. (Adapted from ICRU, Report No. 50, International Commission on Radiation Units and Measurements, Bethesda, MD, 1993.)

modalities with the best tissue contrast (CT and MR) may not resolve the tumour edge with precision, and indeed may not distinguish between tumour and inflammatory tissue reaction. The role of the various imaging techniques with respect to GTV delineation will be further discussed in Section 29.3.1, Chapter 30 and Chapter 31.

29.2.2   CLINICAL TARGET VOLUME

The clinical target volume (CTV) is a tissue volume encompassing the GTV but with an additional margin accounting for any microscopic extension of the primary tumour or regional lymph node spread. It is impossible to ascertain the degree of microscopic tumour invasion around a tumour unless it is removed completely. Therefore, this volume may not be defined separately but considered when defining the planning target volume (PTV) (see Section 29.2.3). When specified, the CTV is usually obtained from a somewhat empirical margin added concentrically to the GTV. In a limited number of cases, the results of clinico-pathological studies of primary tumours where microscopic extent could be measured, and correlated with tumour type and other pathological markers of invasiveness, are available (Holland et al. 1985; Giraud et al. 2000). The involved lymph nodes, which are often also to be irradiated, can be considered as a different type of CTV. In spite of efforts to standardise the delineation of such volumes (Martinez-Monge et al. 1999; Grégoire et al. 2000), individual clinical judgment often remains the basis for their delineation. The increasing use of positron emission tomography (PET) images (see Section 30.3.4) could provide additional data in the future. However, in all cases it is imperative that the CTV receives the prescribed radiation dose; otherwise, tumour cure is unlikely.

29.2.3   PLANNING TARGET VOLUME

The PTV is usually the final volume defined by the clinician during the planning process. It fully encompasses the GTV and CTV. It is in effect obtained from a safety margin added to the GTV/CTV to take account of the organ motion and of the uncertainties inherent in radiotherapy setup and treatment delivery (BIR 2003). Once again, a concentric margin around the GTV/CTV is often used. This margin should ideally be determined from uncertainty analysis. In practice, it is often the result of a compromise between two contradictory issues: making sure that the CTV will receive the prescribed dose while at the same time ensuring that OARs will not receive an excessive dose. Asymmetric margins are therefore used for certain tumours (e.g. prostate, in the posterior direction to protect the rectum). Since the problem of organ motion is quite different from setup uncertainties, these issues have been considered separately in a more recent ICRU report (see Section 29.4).

In contrast with the GTV and the CTV which are purely oncological concepts, independent of any therapeutic approach, the PTV is a purely geometric volume, which does not necessarily correspond to tissue or organ borders. It is used to select the appropriate beam sizes, shapes and arrangements to ensure that the prescribed dose is actually delivered to all parts of the CTV.

29.2.4 TREATED VOLUME

The treated volume (TV) is the volume of tissue enclosed by an isodose surface selected and specified by the clinician as being appropriate to achieve the aim of treatment, i.e. cure or palliation. For example, this may be the volume encompassed within the 95% isodose surface (with 100% in the centre of the PTV) for a curative treatment plan. The TV should not be significantly larger than the PTV. The use of 3D treatment planning and shaping the radiation fields to the shape of the PTV using conformal radiation delivery techniques ensures that the TV encloses the PTV with as narrow a margin as possible. This ensures minimal irradiation of surrounding OARs while coverage of the PTV is assured.

29.2.5 IRRADIATED VOLUME

The irradiated volume (IV) is the tissue volume receiving a radiation absorbed dose that is considered significant in relation to normal tissue tolerance. This concept is not often considered in practice but may be useful when comparing one or more competing treatment plans. Clearly, it would be preferable to accept the plan with the smallest IV, all else being equal.

29.2.6 ORGANS AT RISK

Organs at risk are organs adjacent to the PTV which are non-target, as they do not contain malignant cells. The aim should therefore be to minimise irradiation of OARs as they are often relatively sensitive to the effects of ionising radiation and, if damaged, may lead to substantial morbidity. The OARs to be considered will vary greatly according to the anatomical region being treated, the size of the PTV and the location of the PTV in these regions. The following are examples of the most common OARs that must be considered:

- Brain: lens of eye, optic chiasm, brain stem
- Head & neck: lens of eye, parotid glands
- Thorax: spinal cord, lungs
- Abdomen: spinal cord, large bowel, small bowel, kidneys
- Pelvis: bladder, rectum, femoral heads, large bowel, small bowel

Many OARs have a defined tolerance dose at which a given level of late radiation morbidity can be expected. These are summarised well by Emami et al. (1991a). Some OARs such as lung, brain and kidney demonstrate a volume effect with increasing loss of function as an increasing proportion of the organ is irradiated.

Controversy exists concerning the outlining of hollow OARs, e.g. the colon or rectum. Delineation of the outer surface of the wall of the structure is adequate for determining the anatomical position. However, when considering the expected side effects, it is the dose to the wall of the organ that is likely to be relevant. A dose volume histogram (DVH, see Section 35.3) of the whole organ may not be representative of the volume of the wall and the risk of side effects may be overestimated. Several approaches to this problem have been applied, with some centres outlining the outer and inner walls and others calculating *dose surface histograms* (see Section 35.3).

29.3 CLINICAL ISSUES IN TARGET DEFINITION

The acquisition of high quality images and the definition of the ICRU 50 volumes are integral parts of treatment planning. Outlining of these volumes (segmentation) is a time-consuming and tedious process, and therefore prone to random errors due to lapses in concentration and uncertainties in the definition of the tumour boundary. Other steps in the planning process may introduce other random and systematic errors. Failure to consider these issues will lead to errors being implemented during the patient's treatment. It is therefore vital to appreciate and if possible eliminate sources of error and uncertainty. Some recognised problems with volume definition are discussed below. Although many of the examples given are related to brain tumours, most issues are similar for other clinical sites. Specific issues relate to the type of OAR close to the tumour, to the imaging technique used and to the mobility of the organs.

29.3.1 POOR ORGAN/TUMOUR DEFINITION

For some malignancies, it is the usual practice to treat the whole of the organ from which the tumour originates (e.g. the prostate or the bladder). In these instances, the primary tumour may not be clearly seen, but this is irrelevant due to the large margin for error.

Some tumours are well circumscribed with an easily defined margin of demarcation from the surrounding normal tissue. For example, benign tumours (e.g. meningiomas of the brain and acoustic neuromas) have little propensity to invade the surrounding tissues and are often encapsulated, making the tumour stand out clearly from the tissue of origin. The same can be said for some malignant tumours such as brain metastases, low grade gliomas of the brain and some soft tissue sarcomas. In these examples, there will be little dispute regarding delineation of the GTV, and in turn the PTV.

Inevitably, there are some tumours that are far less well defined. This may be because of diffuse infiltration at the tumour periphery or because the tumour has a similar radiographic density compared to the surrounding normal tissue. Thus, it is possible that the tumour size and shape will not be fully appreciated by the clinician. This is a particular problem with high grade gliomas arising within the brain (Figure 29.2), where clinico-pathological studies have shown tumours extending for some centimetres beyond the edge of the main tumour mass. This is usually accounted for by allowing a large (e.g. 3 cm) margin between the GTV and PTV. Imaging with MR is deemed superior to CT due to the superior soft tissue contrast for neural tissue which is rich in lipids (and therefore protons), but it does not always allow clear tumour delineation.

Comparison of CT and MR volumes suggests that in some cases there is a substantial difference in the tumour volumes (Ten Haken et al. 1992). Similar issues are found in prostate cancer (Khoo et al. 1999; Rasch et al. 1999; Sannazzari et al. 2002).

The appreciation of the position and extent of the GTV and of OARs clearly benefits from the combination of several complementary imaging modalities. This requires that the various image data sets are mapped onto a common coordinate system. This process, called *image registration*, is mostly used to combine CT and MR images and has been developed initially for the brain (see Section 31.6.2). The CT images are used as the reference and the MR images are

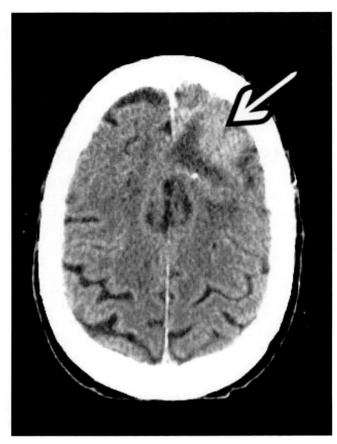

FIGURE 29.2
CT image of a high grade glioma of the brain (see arrow). Note the poor demarcation between tumour and the surrounding brain and in turn the difficulty in delineating the GTV. (Courtesy Institut Curie, Paris.)

recalculated and registered with respect to the CT coordinate system, using one of the methods described in Section 31.6.2.

The practical use of registered multimodality data sets requires specific tools which must be made available as part of a specific radiotherapy imaging console or be included in the treatment planning system used for calculation of dose distributions. One solution can be to display the reference and the registered images side by side and have a cursor linked pixel for pixel, so that the clinician can move from one imaging modality to the other, modifying the volume according to the image which demonstrates it best. The images can also be overlaid on the top of each other, with the possibility to switch instantaneously from one modality to the other. Finally, the two images can be blended into a single image with the possibility in some systems of interactively changing the relative weight of each of them. This process is commonly called image fusion.

29.3.2   INTER-OBSERVER VARIATION

Some tumours are well circumscribed, and in these instances, there can be little debate regarding the extent of the GTV. However, for more ill-defined tumours, there can be some considerable inter-observer variation. It is clear that if the same set of images were presented to a panel of specialists, one would expect some degree of concordance. In the study by

Leunens et al. (1993), CT images from several glioma patients were presented to a multi-disciplinary group of clinicians including radiation oncologists, diagnostic radiologists and neurosurgeons, each varying in seniority. Comparison of the GTVs and PTVs suggested considerable variation from person to person in the GTV contour, in some cases with a substantial difference in tumour volumes*. The only way to eliminate such subjectivity and improve consistency may be to use computer image analysis, although there are no data to suggest that this would be superior to the average oncologist.

29.3.3 INTERNAL ORGAN MOVEMENTS

For many years, it was assumed that internal organs did not move much from day to day, or hour to hour throughout the day. Accordingly, little attempt was made to account for these variations. This assumption is satisfactory for some sites where the tissue containing the tumour has very little scope for movement, e.g. the brain. However, with the emergence of precision radiation delivery techniques such as conformal radiotherapy, these issues have had to be considered. For example, conformal radiotherapy to the prostate ensures conformation of the shape of the TV to the PTV with as small a margin as possible. Therefore, there is little room for error, and slight movement of the prostate may result in under-treatment, resulting in failure to control the tumour, or over-treatment of an OAR, leading to excessive radiation morbidity. As conformal radiotherapy has been most widely used for treatment of the prostate, it is for this indication for which most data are available. The prostate is moderately fixed within the bony pelvis by a series of ligaments. The bladder lies anteriorly, a hollow, muscular sac that undergoes regular distension and contraction according to the volume of urine contained within it. The rectum lies posteriorly, a similarly hollow, muscular sac that also undergoes regular distension and contraction according to the amount of gas and faeces contained within it. Laterally lie the relatively immobile and inflexible bony wall of the pelvis and femoral heads. It is clear that unless care is taken to ensure consistency of bladder and rectal filling at the time of imaging, and subsequently at the time of treatment delivery, diurnal variations could lead not only to changes in the position of the prostate CTV, but also to intrusion of the bladder and rectal walls (each OARs) into the TV. This hypothesis has been elegantly demonstrated by Ten Haken et al. (1991) using beam's-eye-view images of the prostate, bladder and rectum and 3D dose distributions. Other investigators have since demonstrated the vectors and magnitudes of these movements using, for instance, radio-opaque marker seeds (Balter et al. 1995a, 1995b). Such information can be used to optimise the margin allowed between the CTV and the PTV.

Although it has been conventional practice to add a uniform margin around the GTV to yield a PTV for planning, some authors have advocated the use of non-uniform margins to take account of the particular treatment technique and anatomical relations between the PTV and OARs. For example, Pickett et al. (1995) have proposed the use of margins varying from less than 1 cm up to more than 2 cm for the prostate gland treated with a co-planar 6 field conformal radiotherapy technique. This approach may provide the best compromise between CTV coverage and avoidance of OARs. Similar data are available for other internal organs susceptible to motion. For example, in the abdomen the kidneys are often in close proximity to the radiation beam trajectory and occasionally are themselves susceptible to malignant disease. They are also very susceptible to the late effects of ionising radiation and essential for the day-to-day metabolic function of vertebrates. Being close to the thoracic cavity, the kidneys are susceptible to respiratory motion, moving inferiorly with inspiration and superiorly with expiration. The study by Schwartz et al. (1994) suggests that the inferior poles of the

---

* The variability of the results may have been greater than would have been obtained by a group of experienced oncologists alone because of the different criteria being used by the different groups.

kidneys may move by more than 4 cm in the superior–inferior direction between maximum inspiration and expiration (mean 1.7 cm), while other studies have indicated that the displacement may be even greater (Moerland et al. 1994).

It has been noted that respiration leads to measurable variations in the CT volume relationships of both the lungs and the liver, which can be appreciated by changes in the dose volume histograms for these organs (Balter et al. 1996). For those organs which move with respiration, it is of course impractical to plan and deliver treatment during a breath hold without some form of gating or respiration-monitoring system. In the past, the relatively slow image acquisition of CT images in transverse mode gave a time averaged picture which could be considered as representative of the situation during treatment. However, this is not any longer true with modern CTs for which each transverse slice is randomly acquired throughout the respiratory cycle. Empirical corrections in the superior–inferior direction can be used when planning intrapulmonary tumours to take account of respiratory excursion, but it is recognised that this may lead to increased normal tissue complication probabilities for the OARs (Ten Haken et al. 1997). Other options include the use of a stereotactic body frame to restrict lung movement or to employ a *respiratory-gated* treatment in which the CT scanner, and subsequently the therapeutic radiation beam, is switched on only at a predetermined phase of the respiratory cycle. Several techniques are used clinically, either based on active or passive breath-holding, or based on movement detection and synchronised irradiation. More details on these techniques are given in the introduction to Part I.

29.4 ICRU 62 ADDITIONS TO ICRU 50

To address some of these issues, ICRU Report 62 (1999) has been introduced as a supplement to ICRU Report 50. The definitions of GTV and CTV remain unchanged, as these are oncological concepts independent of any technical developments. The main relevant changes include:

- The margin between the CTV and PTV is segregated into an *internal margin* (IM) to take account of variations in the size, shape and position of the CTV relative to anatomical reference points, and a *setup margin* (SM) to take account of uncertainties related to setup errors, reproducibility of patient position and mechanical stability of the treatment unit. The addition of an IM to the CTV results in an *internal target volume* (ITV) that is intermediate between the CTV and the PTV. The ICRU report reflects the obvious uncertainty as to how these two margins should be added together to produce the composite margin and therefore allow definition of the PTV (see Figure 29.3). At least it may serve to clarify thought and lead to a custom-defined PTV that takes into account the relative vectors of these uncertainties, accepting that the whole process is a compromise that relies on the experience and judgement of the radiotherapy team.

- OARs are defined according to their pattern of response to radiation damage (see Section 8.9). *Serial* OARs have an architecture whereby damage to any one functional subunit in a linear chain leads to sudden organ failure and in turn manifestation of morbidity (e.g. spinal cord injury leading to radiation myelitis). *Parallel* OARs require damage to a critical number of functional subunits to manifest radiation damage, and increasing radiation damage leads to progressive loss of organ function (e.g. lung, kidney). *Serial-parallel* OARs demonstrate attributes of both types (e.g. heart).

- Introduction of the concept of *planning organ at risk volumes* (PRVs) which take account of changes in the size, shape, volume and position of OARs. The PRV's derivation from the OAR is therefore analogous to the PTV relative to the CTV. The PTV and PRV(s) may of course overlap, in which case judgement and experience must prevail, but it is usual to give more weight to giving a sufficient dose to the PTV than to sparing a margin around the OAR.

- Introduction of the concept of a *conformity index*, defined as the quotient of the TV and PTV, where the TV encompasses the PTV (see Section 35.4.2).

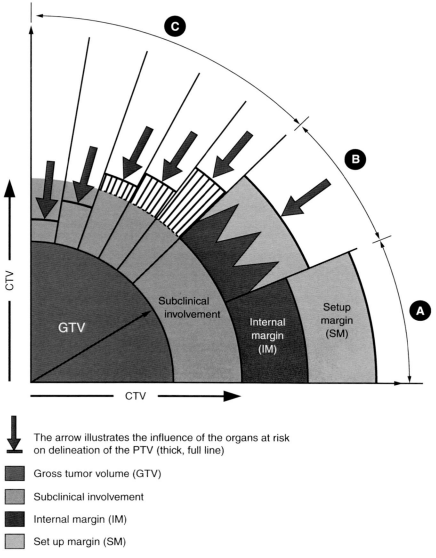

CTV

CTV

↓ The arrow illustrates the influence of the organs at risk
on delineation of the PTV (thick, full line)

▮ Gross tumor volume (GTV)

▮ Subclinical involvement

▮ Internal margin (IM)

▮ Set up margin (SM)

FIGURE 29.3
(**See colour insert following page 590.**) Schematic representation of different possibilities to combine uncertainties to define the PTV from the GTV: (A) linear addition of margins; (B) probabilistic addition of IM and SM; (C) definition of a *global* safety margin based on an empirical compromise between adequate coverage of GTV and unacceptable irradiation of OARs. (From ICRU, Report No. 62, International Commission on Radiation Units and Measurements, Bethesda, MD, 1999. With permission.)

29.5 CONCLUSIONS AND FUTURE TRENDS

The increasing sophistication of treatment planning and radiation delivery has outstripped our capability to delineate the ICRU volumes with precision and consistency. A comprehensive methodology to include systematic and random uncertainties into the volume definition and into the treatment planning process has been developed by a working party of the British Institute of Radiology (BIR 2003). They made a distinction between the various origins of the overall uncertainty: the systematic errors which can be either Gaussian (doctor's delineation; target position, shape and size; phantom transfer*; patient setup) or linear (patient breathing; treatment planning beam algorithm) and the treatment execution errors assumed to be Gaussian (daily setup; target position, shape and size). The methodology was illustrated for several clinical cases. This approach may be considered as too cumbersome. However, failure to consider uncertainties in daily practice will undermine the treatment and ultimately will lead to an undesirable shift in the balance between tumour control and normal tissue morbidity.

Future improvements in medical imaging such as functional imaging (PET or single photon emission computed tomography (SPECT)) and their incorporation into the planning process will contribute to delineation of the GTV for ill-defined tumours. More clinico-pathological data are necessary to relate tumour type, size and grade to the degree of microscopic tumour extension around the GTV. These data may be used to determine the optimum margins to be allowed around tumours to form the CTV; particularly for those tumours that have reduced in size due to treatment with chemotherapy or hormone therapy prior to radiotherapy. The acquisition of more data regarding internal organ motion may also permit a better estimation of the margin that is necessary between the CTV and ITV.

It must be recognised that using a common language in the definition of the volumes which are used throughout the treatment planning process constitutes the basis for consistency within one centre and between different centres for prescribing, recording and reporting the dose to the patient. The ICRU has played a significant role in promoting this approach and complementary reports addressing the problem of other treatment techniques have been produced, such as electron therapy (ICRU 2004), or are in preparation, such as proton or intensity modulated beams. The ICRU has also sought to reduce uncertainties relating to different approaches to the definition of dose and these are discussed in Section 37.6.

---

* In this context, phantom transfer encompasses all uncertainties associated with imaging, planning and treatment equipment or procedures which can be quantified using a phantom instead of a real patient.

CHAPTER 30

PATIENT DATA ACQUISITION

Anthony Neal

CONTENTS

30.1 INTRODUCTION

Patient data acquisition is the first step in the treatment planning process. It is required for three different purposes:

1. To assess the position and extent of the target volume in relation to the other anatomical structures, particularly the organs at risk (see Chapter 29)
2. To acquire the data required for accurate computation of the dose distribution (e.g. shape and composition of the body)
3. To acquire the information necessary for the accurate set up of the patient (e.g. landmarks or reference structures)

In all cases, it is essential that the patient position be defined at the beginning and maintained throughout the whole treatment planning and delivery process. This will be addressed in

Section 30.2. Patient data for treatment planning is often based on data acquired using a single imaging modality with the aim of achieving the goals listed above. The most common methods are described in Section 30.3. A full chapter is dedicated to the use of Magnetic Resonance (MR) imaging for treatment planning, which is becoming increasingly important (see Chapter 31). With some imaging modalities, such as positron emission tomography (PET), and to some extent MR, the geometric accuracy and the lack of electron density information makes it necessary to co-register the data with CT. The methods used to register image sets obtained from different modalities are described briefly in Section 31.6.2 in the context of the use of Magnetic Resonance Imaging in radiotherapy.

30.2 PATIENT POSITIONING AND IMMOBILISATION

An important issue in the radiotherapy process is the ability to reproduce accurately the same patient position throughout the treatment and to keep the patient still during each fraction. This fixed position must be decided before the acquisition of the patient data. It must enable the optimal beam directions to irradiate the planning target volume (PTV) without colliding with the treatment machine or its accessories.

Some of the patient movement can be prevented by positioning patients comfortably and sympathetically and explaining the importance of keeping still. The use of additional positioning accessories or immobilisation devices throughout planning and treatment is often recommended. Examples of such devices include the following:

- Simple sand/bolus bags for general support
- External arm supports and body *boards*, e.g. for breast and thoracic treatments
- Thermoplastic casts, e.g. for head and neck, and pelvic treatments
- Vacuum bags, e.g. for breast and pelvic treatments
- Thermochemical polystyrene devices (headrests and knee supports), e.g. head and neck, and pelvic treatments
- Stereotactic frames, e.g. for brain treatments

The standard devices that can be used for a series of patients can be either homemade or purchased. They are more and more frequently made of carbon fibre, which allows the beams to pass through the material without too much perturbation of the dose distribution[*]. Individual casts can also be used. These are kept only during the treatment course and must be disposed of afterwards. The casts can be made either of polyurethane foam (often referred to under the commercial name of *Alpha Cradle*) or consist of sheets of thermoplastic material that are moulded directly on the patient's body (e.g. head, thorax, limbs, etc.). An intermediate solution is to mould some parts of the body with vacuum bags containing small polystyrene beads. Such bags maintain their shape as long as there is no air input. They can be kept evacuated throughout the treatment and reused afterwards for other patients.

Unfortunately, these devices are often insufficient to ensure an accurate reproducibility of the treatment position. Even individual head casts or masks allow some movement of the patient's head inside. Therefore, it is fundamental that the patient set-up be performed by experienced therapy radiographers. Accurate set-up must still rely to some extent on skin

---

[*] When a photon treatment beam passes through part of an accessory used for patient set-up, in addition to the attenuation there is a risk that the build-up effect will be restored and result in significant skin reactions. Perturbations could also be important in electron beams. In all cases, the possible modifications of dose distribution must be carefully investigated before accepting that the beam is intercepted by a positioning device.

marks or bony landmarks, which must coincide with crosses projected by wall-mounted lasers and other optical aids. The radiographers should also be involved in the imaging procedures used for patient data acquisition.

The more accurate solution for head immobilisation is the use of a stereotactic frame screwed to the skull and used as a reference both for imaging and treatment (see Section 45.1). Intermediate, less-invasive solutions are also available and have been developed for the head and other parts of the body. They consist of devices that are firmly attached to some fixed landmarks (e.g. a moulded *bite block* fixed to the upper jaw, ear plugs, etc.) (see Section 44.3.6 and Section 45.5.2). For greater accuracy, it is also possible to define a number of fixed landmarks (anatomical or fiducial) that can be seen on images performed systematically before each fraction. These images are then compared with a reference image to enable the treatment position or beam parameters to be adjusted. Appropriate imaging procedures must be devised for such techniques.

30.3 IMAGING FOR TREATMENT PLANNING

The clinician must be able to define the size, shape, and location of the gross tumour volume and organs at risk in three-dimensional (3D) space. Small, superficial tumours (e.g. skin cancers and tumours of the oral cavity) are easily visualised and palpated and therefore may be fully appreciated by the clinician without requiring complex medical imaging methods. However, for less well-defined tumours, particularly those arising within the deeper parts of the body, accurate information can only be obtained by one or more methods of medical imaging. Quality assurance at each stage is essential to ensure integrity between the planning and implementation of the treatment.

The ideal imaging modality for radiotherapy treatment planning would have the following attributes:

- Well-tolerated and atraumatic
- Quick to acquire images (less movement artefact)
- Excellent soft tissue and bone contrast with high resolution
- No geographical image distortion
- Able to record a 3D imaging data-set
- Provides data that can be used for radiotherapy dose calculations
- Widely available and cost effective

Although for many years treatment plans have been prepared using patient external contours acquired with mechanical or optical devices (see Section 32.2.1), the direct use of images is the standard practice today. The most widely used imaging modalities will now be compared and contrasted with the context of radiation treatment planning.

30.3.1 COMPUTED TOMOGRAPHY

Computed tomography (CT) should be considered the *gold standard*, especially with regard to geometric accuracy. A description of the principles of CT imaging can be found in most basic textbooks on medical imaging. The clinical use of CT imaging in radiotherapy started in the late 1970s (Goitein 1979; Hobday et al. 1979; Battista et al. 1980). CT is now widely available at all cancer centres and many small *district* hospitals. An example of a CT image of the head is shown in Figure 30.1. The images can be acquired reasonably quickly and comfortably, so movement artefact should not be a problem, except in the thorax. The images of the tumour can be augmented by the injection of radio-opaque iodinated contrast

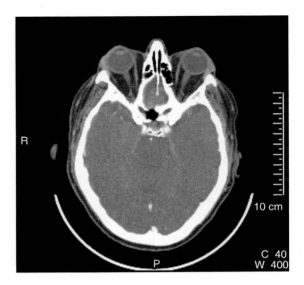

FIGURE 30.1
An example of a CT image of the head. The grey scale ranges from white (bones) to black (air). This scale is mapped onto a range of Hounsfield units (HU) within a window, W, corresponding to 400 HU centred around a value C equal to 40 HU (range = −160 to 240 HU). Other important parameters for 3D reconstruction are the slice position and thickness.

media; this is particularly useful for tumours of the central nervous system. Overall, soft tissue contrast is very good, with high resolution due to the small pixel size attainable. Bone contrast is excellent, a useful attribute when bone invasion must be visualised, e.g. for tumours arising within the base of the skull and those invading adjacent bone. The modest radiation exposure is not a problem for the cancer patient, but it does mean that children and anxious adults must remain unaccompanied during imaging.

CT produces a full 3D set of imaging data which are easily imported to and handled by all commercial treatment planning systems. The high resolution is important when small surface landmarks (e.g. a small radio-opaque point marker as a reference point) must be resolved. There is no inherent image distortion, and therefore the 3D positional data are reliable, provided that adequate quality assurance is performed (see Section 39.2.1). One particular advantage of CT is that the images are an indirect representation of the various tissue–electron densities within the images. Because megavoltage radiation interacts with tissue predominantly according to electron density (Compton scattering), the CT data can be used directly for radiation-absorbed dose calculations, provided a correction factor is applied to take into account the kilovoltage photon energy used for CT data acquisition. This is particularly relevant when the radiation beams traverse tissue inhomogeneities, such as lung and bone, in which the attenuation is very different from that of a homogeneous volume of soft tissue. The direct use of CT density data is far preferable to making manual corrections from densities approximately alloted to the different organs. The pixel information of a CT image is expressed in Hounsfield units, where an arbitrary value of −1000 is set to air and a value of zero is set to water. The CT scanner must be calibrated to measure the relationship between tissue density and Hounsfield units. A typical example of this relationship is shown in Figure 39.2.

CT is particularly useful for acquiring the 3D external contour of the patient, which would be very tedious to acquire manually (although there are a number of laser devices commercially available for this purpose). The external contour is essential for planning because it represents the interface between air and tissue and therefore allows the planning computer to take into account the variations in beam attenuation and obliquity of incidence to calculate dose distributions. Fortunately, automatic segmentation algorithms can be used on the 3D data to draw the external contour on each transverse image.

When acquiring treatment–planning images, CT is sufficiently versatile to permit imaging in the radiotherapy treatment position. This will entail substitution of the convex couch top that is normally used for diagnostic imaging with a flat one so that the set-up emulates a radiotherapy or treatment couch*. As with conventional simulator-based planning†, it is important that the patient be placed in treatment position by a therapy radiographer. This position should be both comfortable and reproducible, preferably following a standard departmental protocol employing the placement of radio-opaque fiducial reference points and the use of immobilisation devices as appropriate.

It has been customary to acquire data as a sequence of transverse images for which the cranio-caudal resolution is limited by the slice spacing. However, most modern scanners have the capability to use a faster helical data acquisition mode. The volume to be imaged will be determined by the anatomical region occupied by the tumour, the extent of the likely PTV, and the likely trajectories of the beams that will be used to cover it. Any organs at risk must be imaged in their entirety if dose volume histograms (DVHs) or biological endpoints such as normal tissue complication probability (NTCP) will be calculated (see Section 35.3 and Section 36.3). The imaging protocol will require that contiguous transverse images are captured to ensure that a full 3D appreciation of the anatomy is obtained. The image slice separation will not normally be greater than 10 mm; often it will be 5 mm or less, particularly when the tumour region itself is imaged and even more so when the gross tumour volume (GTV) is very small, e.g. in the brain. The use of narrower slice widths for image acquisition through the region of the tumour allows accurate delineation of the superior and inferior extent of the GTV, which in turn increases the accuracy of the final CTV and PTV. Acquisition of thin contiguous slices (e.g. with a thickness smaller than 5 mm) over a long superoinferior distance has the drawbacks of needing a large amount of computer storage space and the tedium of outlining the therapy-relevant outlines on a large number of individual images. On the other hand, because large series of thin images are useful for the generation of good quality digitally reconstructed radiographs (see Section 32.4.5), a reasonable compromise must be reached.

The restricted aperture and field of view of a conventional diagnostic CT scanner can be a problem for treatment planning, particularly for breast cancer patients who would normally be treated with the arm elevated. This problem can be overcome if the CT option on a simulator is used (Section 13.3.3). Older simulators were restricted by the time required for each slice, which required one rotation of the gantry, and the limitations of tube heating. The introduction of cone-beam CT algorithms has made it possible, with a single gantry rotation, to acquire a full 3D data set over a length of about 20 cm, albeit with poorer Hounsfield unit accuracy. Specialist CT scanners with a wider (85 cm) bore are now available (see Section 13.3.4), but there is a slight loss in image quality associated with the greater source detector distance. One other disadvantage of CT is seen when there is a metallic implant such as a joint prosthesis (see Figure 30.2) surgical clips, or large masses of dental amalgam. Such objects can lead to considerable streaking of the CT images, which may make it impossible to identify the tumour accurately. They also make it impossible to use the CT images directly to perform inhomogeneity corrections, since the image artefacts would be interpreted as density changes (Reft et al. 2003) (see also Section 39.2.2).

---

* This is sometimes done using an insert with a flat top. In this case it is important to ensure that it fits correctly so that the flat top of the insert is horizontal.
† Simulator-based planning consists of choosing the beam position with the assistance of a simulator (see Chapter 13) and without preliminary patient data acquisition. The corresponding procedure is described in Section 32.2.1.

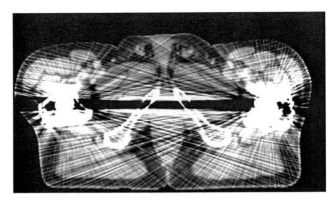

FIGURE 30.2
A CT image of the pelvis, with artefacts from a metallic hip prosthesis. (Courtesy of the Institut Curie, Paris.)

30.3.2 Magnetic Resonance Imaging

Magnetic resonance (MR) imaging provides an important added value to CT imaging, especially for target delineation. Although often presented in axial sections like CT images, MR images can also be obtained in coronal and sagittal sections. The appearance of an MR image varies widely depending on the acquisition sequence and the display settings. Because of the MR principle and acquisition method, the bony structures do not exhibit any signal (i.e. appear as black areas) while small differences in soft tissue composition can be easily seen. A comparison between registered CT and MR images of the skull is shown in Figure 30.3.

The increasing importance of MR imaging for treatment planning is such that a full chapter (Chapter 31) is devoted to this topic.

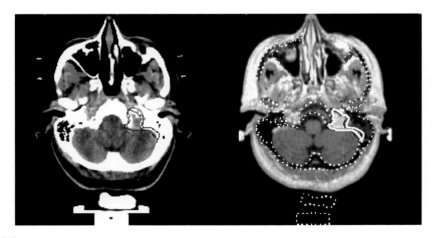

FIGURE 30.3
(See colour insert following page 590.) A side-by-side display of a CT image (left) and an MR image (right) of the skull. The bony structures cannot be seen on the MR image, where the dotted line corresponds to the registered bony outline of the CT image. The tumour delineation (shown as a continuous white line on the MR image) is much easier to tell on the MR image. On the left image the tumour is shown using the MR delineation (black) and the volume that would have been delineated using CT alone (grey). (From Khoo, V. S. et al., *International Journal of Radiation Oncology Biology Physics*, 46, 1309–1317, 2000. With permission.)

30.3.3 ULTRASOUND

Ultrasound is cheap, simple to perform, non-invasive, and well tolerated. It provides reasonable images of tumours, but only when they are in close proximity to the probe. Thus, ultrasound is unsuitable for tumours situated deep within the thorax or abdomen. Bone significantly limits the depth of ultrasound penetration and is itself not resolved with much detail. Generally, ultrasound images are both difficult for the non-expert to interpret and open to subjectivity. Image quality has been improved in recent years by the introduction of transducers that can be introduced into body cavities to visualise adjacent structures, e.g. rectal ultrasound probes to image the prostate (see Section 54.4). Because ultrasound relies on assumptions about the speed of sound in tissue, it may be subject to geometrical distortion.

Another limitation of ultrasound for radiotherapy is that it does not produce a full 3D data set for processing by a treatment planning computer. Instead, it is more likely to yield a succession of 2D images obtained at the discretion of the radiologist.

Apart from its use in prostate seed implants, the role of ultrasound is currently confined to providing staging information that complements the other imaging modalities and that can indirectly guide the clinician faced with the task of volume definition. Simple depth measurements of the chest wall are useful for planning electron boosts in breast cancer patients, as they help select the optimum electron energy. In the pelvic region, it is also sometimes used at the time of treatment delivery as a means for checking in real time the position of the target with respect to the CT image used for initial planning (Lattanzi et al. 2000; Artignan et al. 2004).

30.3.4 POSITRON EMISSION TOMOGRAPHY

Positron emission tomography (PET) uses positron-emitting and biologically active molecules. Thus, PET images reflect the biochemical function of tissues and organs rather than their structure, producing information complementary to that obtained from CT and MR. PET yields 3D data sets which can be imported into the treatment planning computer. It may have a useful role in helping to define the edge of ill-defined, metabolically highly active tumours such as gliomas of the brain (Gross et al. 1998). Both CT and MR imaging may not yield much information regarding the involvement of regional lymph nodes, and often involvement can only be inferred when the node is significantly enlarged. In contrast, PET may help to determine whether a node of borderline size is likely to be involved, and it can even detect a tumour in nodes that appeared normal using CT or MR. Therefore, PET imaging can provide a means for direct evaluation of the CTV, since it is likely to provide useful information on the spread of the microscopic disease. In addition to the investigation of brain tumours, PET is being used in radiotherapy (e.g. to assess the extent of the node involvement) in non-small-cell lung cancers and head and neck tumours (Bradley et al. 2004; Scarfone et al. 2004).

As with MR, the images must be registered to the CT data (Daisne et al. 2003). Because PET images are rather low in resolution compared to CT and MR, PET machines are frequently coupled with CT to provide a straightforward hardware registration of the two image sets (Ciernik et al. 2003; Costa et al. 2003). An additional benefit is that the CT densities obtained from the CT images can be used to apply a transmission correction that significantly improves the quality of the PET images. One of the main disadvantages, however, is the fact that PET imaging facilities are few and far between, and the technology is expensive both in terms of capital and revenue cost. Even fewer centres have a dedicated cyclotron for generating the precious short half-life positron-emitting isotopes. An example of a PET image is shown in Figure 30.4. An alternative to the PET scans available from dedicated PET scanners is the use of a double-headed gamma camera to obtain PET images. Unfortunately, these images have even poorer resolution. They can be adequate to identify active lung tumours, but they are being systematically replaced by dedicated PET imaging devices.

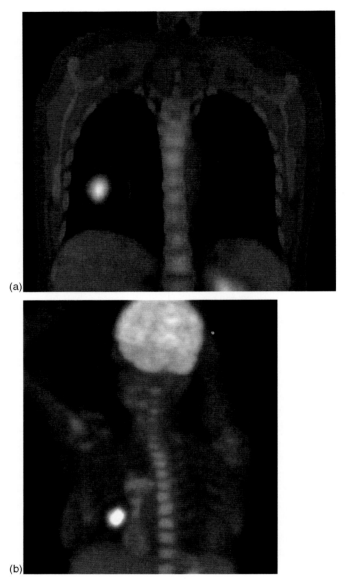

(a)

(b)

FIGURE 30.4
(**See colour insert following page 590.**) An isolated right lung metastasis demonstrated by fluorodesoxyglucose (FDG) PET imaging. The PET image is blended with a registered grey scale CT image, both acquired on a PET–CT machine. The PET image is usually displayed in colour. The upper image (a) is a coronal slice through the metastasis and the lower image (b) is a 3D reconstruction in maximum-intensity projecion (MIP) mode. (Courtesy of the Institut Curie, Paris.)

30.3.5 PLANAR RADIOGRAPHS

Planar radiographs have the advantage of being easy to obtain and interpret. However, since it is important that their geometry be the same as that of the treatment machine, the images should be obtained using a treatment simulator (see Chapter 13). Planar radiographs may be used alone to define primary tumours of the lung or secondary tumours of bone where there is sufficient contrast between the tumour and surrounding tissues. Reliance on such 2D images may be sufficient for the planning of non-conformal curative treatments (e.g. head and neck) or of non-curative, palliative treatments where a geographically inaccurate tumour

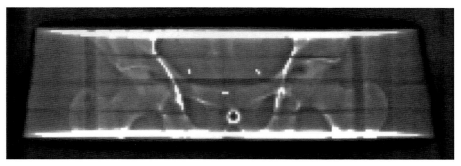

FIGURE 30.5
A CT topogram showing three gold seeds within the pelvis of a humanoid phantom.

delineation may not affect symptom relief and where normal tissue toxicity is not as much of an issue because lower treatment doses are used. A single film may be sufficient when an anterior–posterior parallel opposed pair of beams is to be used. Orthogonal films are used when the tumour has to be more precisely defined. An anterior film allows delineation of the tumour in the superior–inferior and left–right directions, while a lateral film allows delineation in the superior–inferior and anterior–posterior directions.

Contrast-enhanced radiographs are useful in non-3D planning when it is important to identify the position of a hollow organ. The following techniques may be used:

- Barium may be swallowed to outline the oesophagus, stomach, and small bowel
- Barium may be given as an enema to outline the rectum and lower colon
- Radio-opaque tampons may be inserted into the vagina to outline the structure
- Intravenous iodinated contrast may delineate the kidneys and allow them to be avoided when the abdomen is treated
- Iodinated contrast may be introduced into the bladder to outline it and suggest the position of the top of the prostate
- Iodinated contrast may be introduced into the urethra to suggest the position of the inferior extent (apex) of the prostate
- Metallic seeds may be introduced into the pelvis under guidance of trans-rectal ultrasound to define the prostatic apex and base of seminal vesicles (see Figure 30.5).

30.4 CONCLUSIONS

Acquisition of high-quality 3D imaging data is becoming increasingly necessary to exploit recent advances in the precision delivery of radiation. CT and MR are the cornerstones of any imaging protocol. The use of PET imaging, preferably combined with CT within the same piece of equipment, is likely to become standard practice in the future. Care must be taken to ensure consistency of patient set-up for each imaging modality and for each subsequent step of the planning process. This requires the implementation of a quality assurance program, as discussed in Chapter 39.

CHAPTER 31

MAGNETIC RESONANCE IMAGING IN TREATMENT PLANNING

Vincent S. Khoo

CONTENTS

31.1 INTRODUCTION

Computed tomography (CT) has made a dramatic impact on the accuracy of radiotherapy treatment planning in the past two decades by improving the localisation of target volumes. Improved target localisation will allow higher local control rates with less geographical misses and permit better awareness of the surrounding critical organs, thereby potentially minimising normal tissue complications (Goitein 1979; Goitein et al. 1979). The development of magnetic resonance (MR) has introduced several added imaging benefits that may confer an advantage over the use of CT in treatment planning.

31.2 PRINCIPLES OF MAGNETIC RESONANCE IMAGING

The basis of magnetic resonance techniques is the measurement of radiofrequency radiation, resulting from transitions induced between nuclear spin states of tissue hydrogen atoms (protons) in the presence of a strong external magnetic field. Whereas CT depends on x-ray attenuation by tissues that are a function of atomic number and electron density, relative pixel intensities in MR images are a function of proton densities and the different proton spin relaxation times in different tissues. Magnetic resonance imaging (MRI) does not use ionising radiation.

The major source of contrast in clinical MRI is the difference in relaxation times between different tissue types. The two parameters that are often used to characterise the behaviour of an MR signal and that can be used as a basis for generating MR contrast are the Spin–Lattice (T1) and Spin–Spin (T2) relaxation times. MR images are calculated from signals acquired by applying a particular imaging or pulse sequence comprised of one or more radiofrequency pulses and associated magnetic field gradients. The image itself represents the rate at which the transitions or relaxation times occur. The nature of the acquired MR signal and the image contrast depends on the relative timings of the applied pulse(s), gradients, and signal acquisition in relation to the proton density structure of the tissue(s) being imaged.

A feature of MRI is that there is no standard, universally applied imaging sequence; there are an infinite set of combinations of possible timings and arrangements of the various imaging sequence components. In practice, sequences are often designed to give images that are weighted according to a particular relaxation time (i.e. T1- or T2-weighted). Even when utilising a particular type of weighting, there is flexibility and variability in the imaging parameters used. Therefore, T1 weighted images acquired on different MR scanners are often obtained using slightly different values of the echo (TE) and repetition (TR) times, resulting in images with subtle differences in contrast. Pixel intensities also vary from measurement to measurement, depending on the instrument design and calibration. MRI has a greater flexibility in varying contrast compared to CT where essentially fixed imaging parameters are used.

The flexibility inherent in MRI can be used, to some extent, to tailor the image contrast to meet clinical requirements. Choice of contrast is then influenced by both clinical considerations and scanner constraints. Some of the clinical factors include the need for varying relaxation times, the administration of appropriate contrast agent, the choice of the optimal imaging plane (i.e. axial, sagittal, coronal, etc.), the scanning volume, and patient tolerance. Other factors that influence image quality include image signal-to-noise and contrast-to-noise ratios, the image resolution, and the scanning time. The acquired MR image is often a compromise between interdependent and mutually exclusive imaging requirements.

31.3 RATIONALE FOR THE USE OF MAGNETIC RESONANCE IMAGING IN TREATMENT PLANNING

The main advantage of MR compared to CT is its ability to better demonstrate and characterise tumours and soft tissues.

Contrast between healthy and malignant tissue can be obtained from differences in the T1 and T2 relaxation times exploited by using the appropriate imaging sequences. Imaging sequences can be conveniently characterised in terms of Spin-Echo (SE) or Gradient-Echo (GE) sequences. A common patient scanning protocol involves both T1-weighted spin-echo imaging (short TE and TR) for good anatomical delineation and T2-weighted spin-echo imaging (long TE and TR) to differentiate pathological from normal tissues. An example of this can be seen for the prostate gland (Figure 31.1).

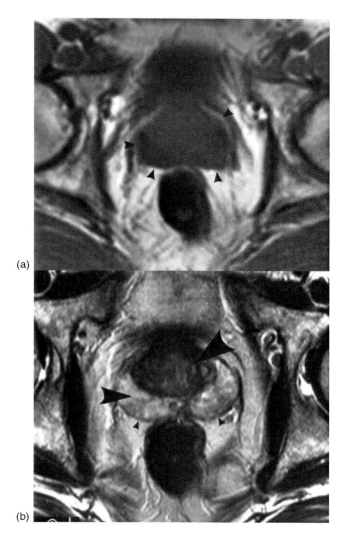

(a)

(b)

FIGURE 31.1

(a) T1 weighted MR image of the prostate. T1 weighted images are useful for highlighting the periprostatic tissues that lie between the prostatic capsule and the puborectalis muscle (indicated by small arrows). (b) T2 weighted MR image of the prostate. T2 weighted images show the internal prostatic structure better with the central zone appearing dark (indicated by large arrows on right side) and the peripheral zone appearing brighter (medium arrow on left side). The prostatic capsule is also easily visualised (small black arrow heads inferiorly).

Anatomical regions surrounded by thick bone will absorb x-rays from CT, reducing soft tissue image quality, producing artefacts, and decreasing the visualisation of nearby tumours. This effect is absent with MR because cortical bone does not provide an MR signal and appears as a dark area. Tumours within the posterior fossa or brainstem and tumours centred at bony prominences or enclosures such as the spinal cord are better defined.

The use of MR contrast agents such as Gadopentetate dimeglumine (Gd–DTPA) may further enhance visualisation of the tumour or disease process under investigation. MR has been shown to be superior to CT in the staging of soft tissue (Herrlin et al. 1990) and pelvic tumours (Hricak 1991), and it is sensitive in detecting and defining the extent of bone marrow disease (Kattapuram et al. 1990). It is the imaging modality of choice for brain, spinal cord, and some head and neck tumours (De Vries and Bydder 1988; Sze 1988).

Detailed MR images can be acquired in any orthogonal or non-orthogonal plane without loss of spatial resolution. This multiplanar capability provides greater functionality for 3D treatment planning compared to CT. Furthermore, 3D volumetric data sets can be obtained, avoiding the need for separate axial, sagittal, and coronal plane acquisitions. Recent developments have produced faster imaging and real time applications. MRI in cine mode display can provide beam's-eye-view capability (Ogino et al. 1993) and may be used to assess organ motion (Padhani et al. 1999). These developments may allow the development of *virtual* or *real-time* MR treatment simulation (Mizokawi et al. 1996).

By combining the high resolution anatomical information from MR with functional information available from positron emission tomography (PET) and single photon emission computed tomography (SPECT), a clinically useful structure-function map can be produced. This procedure has been used in glioma radiotherapy (Pardo et al. 1994) and applied with radiolabeled monoclonal antibodies or radioimmunotherapy to evaluate the 3D dosimetry of the target to surrounding normal tissue absorbed-dose ratios (Sgouros et al. 1993).

MR can also provide physiological and biochemical tumour information. MR angiography can assess cerebral vasculature, providing a non-invasive method to plan stereotactic radiotherapy for cerebral vascular malformations and angiomas. MR spectroscopy can non-invasively measure the behaviour of a range of clinically significant biochemical compounds and may offer new methods to study morphological alterations and physiological and metabolic changes non-invasively within tumours during and post-treatment (Leach 1994). This may allow individual modification of radiation dose or prompt alternative treatments for non-responders. MRI provides a much wider range of applications not only for treatment planning but also for cancer management. CT is limited in these aspects, showing only anatomical images with contrast based solely on differences in intrinsic or contrast agent generated x-ray attenuation. Differences between CT and MRI are listed in Table 31.1.

31.4 PROBLEMS WITH THE USE OF MAGNETIC RESONANCE IMAGING IN TREATMENT PLANNING

Although the imaging quality and versatility of MR is ideal for acquiring the necessary information for treatment planning, MR has not yet seriously challenged CT for treatment planning in most sites for the reasons described in the following sections.

31.4.1 ELECTRON DENSITY INFORMATION

Whereas CT depends on x-ray attenuation by tissues, which is a function of atomic number and electron density, relative pixel intensities in MR images are a function of proton densities and tissue relaxation times. For CT data, electron density can be automatically derived from Hounsfield units, but MR signal intensity has no such correlation. For dose calculations, MR images will either need to have absorption coefficients assigned to the relevant anatomical

TABLE 31.1

General Comparisons between Magnetic Resonance Imaging (MRI) and Conventional CT Imaging

Subject	Parameters	MRI	CT
Patient	Magnetic safety concerns	Present	Nil
	Radiofrequency heat deposition	Present	Nil
	Ionizing radiation dose	Nil	Present
	Claustrophobia in scan tube	More marked	Minimal
	Scanning noise	Moderate	Minimal
	Contrast materials allergy:		
	Iodinated contrast	Not applicable	Present
	Gd DTPA (gadolinium)	Minimal	Not applicable
Characteristics	Soft tissue contrast	Excellent	Moderate
	Cortical bone contrast	Poor	Excellent
	Detection of calcifications	Poor	Excellent
	Metallic artefacts:		
	Non-ferromagnetic material	Some	Marked
	Ferromagnetic material	Marked	Marked
Machine	Size of tunnel aperture	Smaller	Larger
	Image resolution	Good	Better
	Scanning time	Moderate[a]	Short[a]
	Electron density information	Nil	Present
	Functionality and technical sequences	Large	Limited
	Geometrical image accuracy	Distortion present[b]	Excellent
	Multiplanar imaging	Any plane	Limited
	Multiplanar reconstructions	Available	Available
	Cost	Higher	Lower
	Availability	Restricted	Widely available

[a] Scanning times are dependent on the imaging volume, number of slices, and for MRI on the pulse sequences used.
[b] Object, sequence and machine dependent.

structures and regions or to register the MR and CT images so that the superior tumour definition of MR can be transferred to the CT data for planning.

31.4.2 IMAGING OF BONE

Cortical bone in MRI is shown as regions of very low signal intensity. The presence, type (compact bone vs. spongy bone), shape, and extent of bone are important in assessing bone and tissue or bone and air inhomogeneities for dose calculations. Bony boundaries and landmarks are not clearly visible on Magnetic Resonance images, which can limit image registration (see Section 31.6.2).

31.4.3 MAGNETIC RESONANCE IMAGE DISTORTION

A highly uniform magnetic field with perfectly linear orthogonal gradients is desired for MRI. However, inhomogeneities of the main magnetic field and non-linearities in the gradients together with the presence of objects within the magnet field will cause both geometric and signal intensity distortion in clinical practice. The presence of any imaging distortion presents a major hindrance to the utilisation of MR images for treatment planning as it will provide inaccurate spatial information and preclude accurate image correlation. In contrast, CT data are spatially accurate and can be used directly in treatment planning systems without the need for correction. Sources of MR image distortion can be conveniently

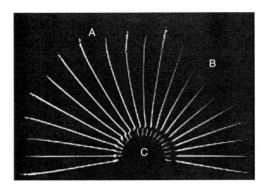

FIGURE 31.2
Various forms of MR-related distortion can be seen in this image of a coplanar array of water-filled tubes supported within a circular solid plastic block. System-related distortion effects are seen in the apparent curvature of the tubes at A and their disappearance at B that was due to warping distortion of the imaging plane.

considered as system-related distortions and object-induced (i.e. patient-dependent) effects, Figure 31.2. Magnetic susceptibility differences because of the presence of the plastic support block at C give rise to object-induced distortions in the form of discontinuities at the point where each tube enters the support block. The resultant displacements have occurred along the readout (horizontal) direction.

31.4.3.1 System-Related Distortions

System-related image distortions arise because of inhomogeneities in the main magnetic field, non-linearity of magnetic field gradients, and the presence of eddy-current effects. All magnetic fields possess inhomogeneities that result from imperfections in the magnet windings or stray external magnetic fields and that cause image-slice distortion and in-plane distortions in the read-out gradient direction. For a given magnet inhomogeneity, the magnitude of the resulting image distortion is inversely proportional to the gradient strength. The homogeneity of the static magnetic field (B_0) decreases with increasing distance from the centre of the magnet bore; therefore, the degree of distortion increases for larger fields-of-view.

MR images can also be distorted by non-linearities in the applied gradients and by the presence of eddy currents that cause both slice and in-plane distortion in both the read-out and phase-encoding directions. Eddy currents arise when the applied gradients are turned on or off, resulting in the presence of transient magnetic fields that contribute to image distortion. Newer MR scanners utilise shielded and compensated gradients to reduce eddy currents. In practice, these distortions remain constant if the imaging sequence and its parameters (TE, TR, etc.) are unchanged. Therefore, eddy current-induced distortions may be quantified and corrected provided the same imaging sequence is used for mapping distortions and imaging patients.

For brain imaging with head coils or similar fields-of-view (i.e. between 10 cm and 15 cm in each direction), investigators have reported distortion errors of 3 mm to 4 mm at the periphery and less than 1 mm at the centre of the image; these images were considered satisfactory for planning purposes (Schad et al. 1987a; Pötter et al. 1992; Hill et al. 1994). With larger fields-of-view between 20 cm and 40 cm, phantom studies have revealed point displacements of up to 10 mm or 16 mm, respectively at the edge of the image (Kokoves 1991; Finnigan et al. 1997). Uncorrected images with distortions of this magnitude cannot be used for the generation of accurate patient outlines in treatment planning. The system-related distortions are strongly dependent on the design of the MR system. They are also influenced by the imaging sequence, the size of the gradients employed, and the position within the field-of-view.

31.4.3.2   Object-Induced Distortions

Object-induced distortions principally arise from magnetic susceptibility and chemical shift effects. The magnetic susceptibility effect occurs because any object placed within a magnetic field changes the value of that field. This distortion is particularly marked at boundaries of structures having different intrinsic magnetic susceptibilities, and it tends to be most pronounced at air and tissue boundaries, for example, near the outline of the head and at internal cavities such as the sinuses. These effects can be complex as they depend on the orientation of the boundary to the main magnetic field and can influence the value of the field at substantial distances from the susceptibility interface.

Another major source of object-induced distortion is the chemical shift effect caused by the different resonance frequencies of protons in fat and water that results in an apparent positional shift of fatty tissues relative to water-based tissue along the read-out gradient direction. Chemical shift effects are more noticeable in the abdomen and pelvis because of the extensive presence of fatty tissues. They have also been documented in head and neck regions (Tien et al. 1991; Sakurai et al. 1992) but may not be as obvious for MR images of the central nervous system because of the relatively limited presence of fat.

The functional dependence of susceptibility and chemical shift effects is similar to that for main field inhomogeneities in that the resultant distortions are inversely proportional to gradient strength. Phantom studies of susceptibility effects at Royal Marsden Hospital have demonstrated shifts of the fiducial markers in the read-out gradient by between 2 mm and 3 mm with respect to the phantom (Finnigan et al. 1997). When the same markers were imaged on the pelvic skin surface of patients, the apparent shift had increased to 5 mm. This additional shift was attributed to the contribution of chemical shift from subcutaneous fatty tissues and correlated well with the expected chemical shift of 3.2 ppm (i.e. Hz/MHz) for fat.

31.4.3.3   Magnetic Resonance Correction of Image Distortion

The effect of MR distortions needs to be considered prior to utilising MR images for treatment planning. At the Royal Marsden Hospital, a protocol has been developed to correct for MR-related distortions (Finnigan et al. 1997; Tanner et al. 2000). Furthermore, this protocol incorporates measures to maintain the quality, accuracy, and safety of the imaging and correction procedures. System-related distortions may be corrected using dedicated MR phantoms of known geometry (Tanner et al. 2000). Studies using these phantoms have provided comprehensive maps of system distortions in three orthogonal planes, and by applying these measurements, observed distortions of up to 9 mm may be reduced to 1 mm or less (Finnigan et al. 1997). However, object- or patient-induced distortions cannot be corrected simply by phantom measurements. A procedure based on gradient reversals has been shown to be effective in correcting for object-induced distortions that include effects caused by the chemical shift (Finnigan et al. 1997). This correction procedure requires acquisition of one additional image for each slice acquired, and although this increases the overall patient scanning time, such correction is needed for accurate definition of target volumes, image correlation, and data registration for treatment planning. This distortion correction methodology has been further refined to improve the correction process for larger MR images that may allow imaging at extra-cranial sites and larger fields-of-view to be used (Doran et al. 2005; Reinsberg et al. 2005).

31.5 OTHER CONSIDERATIONS AFFECTING UTILISATION OF MAGNETIC RESONANCE IMAGING FOR TREATMENT PLANNING

31.5.1 GENERAL CONSIDERATIONS

MRI using closed MR systems that resemble narrow tunnels can be a formidable procedure for patients because of the noise and enclosed space of the magnet bore. Current generation closed MR systems have now been designed with flanged openings at either end of the magnet to create a better sense of space for the patient if he or she can be positioned toward the end of the magnet bore. Newer, open, C-shaped MRI systems may overcome the patient's claustrophobia to some extent, but are not as widely available as closed MRI systems. Alternatively, patient sedation may be provided, or it is possible to accompany and reassure the patient during MRI as ionising radiation is not used.

A small proportion of patients will be unsuitable for MRI by virtue of:

- Claustrophobia
- Presence of a cardiac pacemaker
- Previous history of metallic particle eye injury.

The latter two scenarios are excluded by careful screening with a questionnaire prior to patients entering the room housing the scanner. If doubt does exist regarding metallic foreign bodies in the eyes, plain radiographs of the orbits can help exclude the possibility of a very small ferro-magnetic object within the eye.

The reason for not allowing patients with ferro-magnetic objects close to critical structures to be scanned is that these objects may be moved by strong magnetic fields. External objects such as hair clips, scissors, and the like must also be excluded because of the potential danger from impact injuries as well as the introduction of distortions secondary to these objects within the system.

The MR scanning table must reproduce the flat bed design of the radiotherapy simulator and treatment couch. Any immobilisation device used in planning needs to be assessed for its magnetic safety as well as being able to fit into the patient tunnel and not interfere with the MR examination coils. Where this is not practical, e.g. for patients in stereotactic head frames, MR scans can be registered to planning CT scans.

31.5.2 MOTION

Any patient or organ motion will give rise to image blurring and artefacts, potentially masking pathology. Prolonged scan times will increase the potential for both patient and internal organ movement. Physiological movements such as cardiac and respiratory movements have a greater effect on the imaging of the thorax, abdomen, and pelvis compared to the limbs or the brain. Blood vessel motion can be significant and can result in severe artefacts that propagate through the images. Blood flow and motion of abdominal fat because of respiration can result in *phase mismapping* during data acquisition. This mismapping will give rise to image ghosting on image reconstruction (White et al. 1992). Peristaltic bowel movements and rectal and bladder filling are more subtle but can still alter the position of the target volumes within the pelvis during imaging.

Diagnostic methods exist to compensate for the problems of motion. However, it remains to be seen if these compensatory methods such as ECG gating, blood flow compensation techniques, presaturation methods, and respiratory gating methods will be applicable for treatment planning. Some methods such as respiratory gating have been less successful because the imaging times have substantially increased (Wood 1988). Techniques such as scan averaging and breath-holding imaging, using fast *turbo* sequences for example, considerably reduce artefacts because of respiration (Mirowitz et al. 1990; Flentje et al. 1993). Anti-muscarinic agents such as hyoscine butylbromide can be used to decrease gastrointestinal motility. Ultra-fast MR sequences such as Magnetisation Prepared Rapid Acquisition Gradient Echo (MP-RAGE), fast Spin-Echo, and Echo Planar Imaging (EPI) have much shorter scanning times and may produce acceptable image quality. Echo planar images have been obtained in under 100 ms, effectively freezing some physiological motion (Edelman et al. 1994). The possible distortion effects associated with these newer sequences need to be assessed before use.

31.6 METHODS TO ALLOW THE USE OF MAGNETIC RESONANCE IMAGING IN TREATMENT PLANNING

The lack of electron density information and the presence of MR distortion have meant that image processing methods must be employed if MRI is to be utilised for treatment planning. Once MR distortion and radiofrequency non-uniformity effects have been quantified and corrected, image segmentation and image registration or correlation techniques can be used to overcome the lack of electron density data from MR images. These procedures are not mutually exclusive as segmentation methods can be used to preprocess data for multimodality image registration. These concepts will only be briefly outlined.

31.6.1 IMAGE SEGMENTATION

One method of performing dose calculations on MR images is to segment the relevant tissues in order to assign the appropriate attenuation factors to these tissues. Manual segmentation methods are labour-intensive and time-consuming. Fully automatic segmentation methods can be more efficient, but the segmentation process needs to be accurate. Often these algorithms do not have adequate intelligence to distinguish reliably between subtle differences in soft tissue structures. Semi-supervised automatic methods are needed to ensure quality control for these procedures. A review of the many segmentation methods is not within the scope of this chapter, and further information may be obtained for instance from the review by Clarke et al. (1995).

In general, parameters such as MR pixel intensities can be used to threshold regions of interest and to define other features such as the edges and texture of anatomical organs to augment the segmentation process. However, there can be difficulty in determining the maximum contrast difference for the different soft tissues in order to select the optimum threshold for segmentation. Furthermore, non-uniformities of the radiofrequency field can vary the pixel intensities of a given tissue type within different regions of the same MR image, reducing the effectiveness of automatic pixel-based techniques. These issues may often require substantial operator input to provide reliable segmentation and may also result in both inter- and intra-observer variations. Attention needs to be paid to the increased noise associated with the use of faster modern MRI sequences that may result in reduced image quality and an increase in the variability of image grey scale shading that may result from radiofrequency non-uniformities and patient-induced distortion effects. MR segmentation methods appear to be successful for normal brain images (Clarke et al. 1993). The greater soft tissue contrast obtainable with MRI for regions such as the abdomen and pelvis may permit more reliable segmentation compared to CT (Khoo et al. 1999) as segmentation in these anatomical regions remains time-consuming (Neal et al. 1994). Segmentation algorithms for MRI need to minimise the level of experienced operator supervision and provide reliable accurate results before they can be adopted as standard procedure.

31.6.2 IMAGE REGISTRATION

The registration of MR images with CT images allows the use of CT-calculated isodose contours with the superior tumour definition of MR for improved treatment planning. Manual registration methods are time consuming and require a high level of skill and regular practice. Automated approaches may allow more efficient registration, but without the developed skill, errors in automatic registration may go unnoticed.

Simple 2D registration procedures include the use of a projection system with an appropriate magnification that allows MR images to be superimposed over simulation films or isodose plans. Bony landmarks such as vertebrae and pelvic bones are used as correlation points. Registration methods in 3D are more complex, and a multitude of parameters need to be considered in the matching procedure such as the reproducibility of the patient position, the angle of the imaging plane, the image contrast and resolution, and the volume of the image set, including the width and gap distance of each slice. Registration techniques may be applied either prospectively or retrospectively, and they may include point or line matching, iterative matching, principal axes, and surface matching. A review of current techniques has been published by van den Elsen et al. (1993).

There are a number of algorithms for 3D image registration, but generally, they fall into one of several categories:

- In *surface matching* (e.g. Levin et al. 1988), one or more 3D contours are outlined (e.g. the external contour and specific anatomical structures that are well visualised on the two image sets) and then brought together in 3D space until the best mathematical match is made. The acquisition of contiguous and narrow CT images ensures good spatial definition and optimises outcome. However, the boundaries of structures may not be identical in MR and CT and there is likely to be more distortion of the external contour in the MR image.

- In *point matching* (e.g. Hill et al. 1991), a number of (usually at least five) 3D anatomical reference points are nominated on each of the two image sets. These points are preferably easily identified, small, and immobile, e.g. bony landmarks. The more points nominated, the more accurate the final match of the images is likely to be. An alternative is to use external fiducial markers applied to the patient for each of the imaging sessions. Although it is easier to use a large number of points with this method compared with internal landmarks (which can be surprisingly difficult to identify), reproducibility of marker position between different imaging sessions on different days presents a challenge. Ideally the CT and MR should be carried out on the same day without removing the markers. If this is not possible, the marker positions must be documented with photographs, diagrams, and measurements of their position relative to reproducible anatomical landmarks. It is important that the points used for registration are well-separated in 3D, and it is advisable that when using external fiducials, some internal points are used as well. A well-designed registration algorithm will provide information on the mean and individual point errors so that the operator can concentrate on those points with the greatest uncertainty. The problems associated with external fiducial markers are eliminated for brain imaging by using a stereotactic frame (see Section 45.5) that is firmly and reproducibly located on the patient's head although this may restrict the MR coils that can be used. The frame is usually fixed by surgical pins that ensure that it does not move relative to the skull, but relocatable frames have been devised for treatment planning (Kooy et al. 1994).

- *Mutual-information-based registration* (e.g. Studholme et al. 1999) of several image series is becoming more frequently used. It is a completely automated process that makes use of the internal 3D distribution of the information (i.e. grey levels) within MR and CT images, respectively.

For a more detailed description of the various methodologies and issues, the reader is referred to Kessler et al. (1991). Whatever method is used, it is important that a careful visual check is carried out once the registration is complete to verify that it has been satisfactorily performed. For this purpose, it is useful to have access to specific tools designed for accurate comparison of several image sets. A moving window that allows one modality to be viewed on top of the other is a useful example.

Image registration does have limitations:

- It is time consuming.
- Patient positioning can be difficult to reproduce during acquisition of different types of image at different scanning sessions. For example, the degree of neck flexion can be very different for CT compared to MR and may present difficulties in registering images of the head and neck region when the body surfaces are used for matching the images to each other.
- There may be a problem with reproducibility of the position of different internal organs to each other between imaging sessions. This is a particular problem with the prostate, which may shift position, depending on differential filling of the bladder and rectum (see below). This may compromise the position of internal reference points.
- External fiducial markers (if used) must be large enough to be visualised but small enough to give a precise 3D reference point.
- An adequate number of common points must be identified to provide the registration algorithm with enough spatial information to perform an accurate geometric match between image sets. This can be difficult when internal anatomical reference points are used, as insufficient points may be identifiable as being common to the two sets of imaging data. Differences in slice orientation between CT and MR data sets can make it difficult to identify consistent internal reference points.

Assuming MR distortions have been corrected, brain registration methods have reported a mean correlative value of between 1 mm and 3 mm for CT–MR matching (Schad et al. 1987b; Pelizzari et al. 1989; Hill et al. 1994; Van Herk and Kooy 1994; Veninga et al. 2004). More difficulty can be expected for imaging of the thorax, abdomen, and pelvis where movement and distortion effects are greater.

31.7 CLINICAL SITES OF MAGNETIC RESONANCE IMAGING APPLICATIONS IN TREATMENT PLANNING

Magnetic resonance imaging has been extensively used for the brain, and it is the modality of choice for this site. Intracranial movement is minimal, the skull can be rigidly immobilised, and the head is easily imaged in the centre of the magnetic field; therefore, MR distortions are generally smaller compared to larger body regions. These factors coupled with effective head registration methods have allowed easier integration of MR and CT for treatment planning. MRI has been shown to improve target volume definition in between 62% and 82% of CT-based brain treatment plans (Pötter et al. 1992; Gademann et al. 1993; Heester et al. 1993) as well as providing complementary information not visualised by CT imaging (Khoo et al. 2000). MRI has been used alone for treatment planning by assuming a homogenous attenuation value inside the brain with an error of less than 2 percent compared to CT-based dosimetry (Schad et al. 1994). Co-registered MR and CT images have also been used for brain brachytherapy planning to reconstruct catheter or radioisotope seed arrays in 3D to enable calculation of the dose distribution (Hardy et al. 1992). High resolution 3D vascular and neuro-anatomical information for stereotactic radiotherapy of cerebral arterio-venous malformations can be obtained using flow compensated gradient echo and fast 3D MR sequences. This avoids the invasiveness of angiography and allows frequent follow-up scans to assess the occlusion of the treated feeding blood vessels (Ehricke and Schad 1992).

MR has provided improved delineation of target volumes for radiotherapy of nasopharyngeal and head and neck cancers (Curren et al. 1986; Kovacs et al. 1992). Fast breath-held MR sequences and coronal MR angiography techniques have provided similar improvements in supradiaphragmatic lymphoma treatment planning with better localisation and avoidance of partial volume imaging effects (Müller-Schimpfle et al. 1992). MRI has also been shown to provide improved definition of target volumes in prostate radiotherapy, especially in the localisation of the prostatic apex, compared to CT (Kagawa et al. 1997; Khoo et al. 1999). For further radiotherapy applications of MR, the reader is referred to reviews by Khoo et al. (1997) or Khoo and Joon (2006).

31.8 CONCLUSIONS

Magnetic resonance imaging has considerable potential for treatment planning. The superior soft tissue contrast provided by MR and the ability to vary contrast by manipulation of the imaging parameters facilitate optimal tumour evaluation. Together with its 3D multiplanar imaging capability, MR can provide advantages over reconstructed CT images. Used in conjunction with CT, MRI can provide both complementary and supplementary information for the localisation and characterisation of tumours and surrounding normal tissues.

MR distortions can be geometrically corrected to provide accurate spatial information. For distortion correction, it is important that both patient images and distortion data are acquired using the same set of imaging conditions since MR distortion is strongly influenced by the parameters of the imaging sequence.

The appropriate MRI parameters for optimum tumour visualisation and treatment planning need to be identified, and faster MR sequences will reduce the effect of motion artefacts. Non-operator dependent segmentation and image registration methods that are accurate and reliable are required so that MR can be either used alone or integrated with CT for dose calculations. Although the full potential of MR in treatment planning remains to be realised, these methods will greatly enhance the utilisation and integration of MR applications in treatment planning.

CHAPTER 32

BEAM DEFINITION—VIRTUAL SIMULATION

Vibeke Nordmark Hansen

CONTENTS

32.1 INTRODUCTION

The Planning Target Volume (PTV, see Section 29.2) must receive a dose large enough to reach the therapeutic goal. Therefore, it must be covered by the beams that are used for treating that volume. For skin tumours, the beam definition is straightforward; only one beam is used, and the beam shape directly reflects the tumour lesion plus a margin. Treating tumours not visible from the exterior of the patient requires the use of x-ray equipment or other medical imaging devices to define the location of the PTV (see Section 30.3). Treatment simulators (see Chapter 13) provide a diagnostic x-ray tube mounted in the same isocentric geometry as a treatment machine. Using a simulator, the bony anatomy is easy to identify and beam definition can be made relative to that. However, to obtain a precise match between the field shape and the shape of the PTV (together with an appropriate margin) onto a plane perpendicular to the beam direction is required. This may be achieved in one of three ways.

- The PTV and its margin may be implicitly estimated or explicitly drawn on radiographic images taken from the appropriate beam direction on the simulator. This is referred to as the *anatomical* approach.
- Planning may be based on a few transverse cross sections where the desired PTV is marked. The information about the position of the PTV may be derived from a number of sources (including planar radiographs). This is referred to as the *traditional* approach.
- Simulation may be performed on the computer using a full 3D data set usually from a CT scanner. The positioning of the beams is carried out on a computer simulation of the patient, and this is referred to as *virtual simulation*.

In the following discussion, these three methods of simulation will be covered.

32.2 BEAM DEFINITION USING ANATOMICAL SET-UP

32.2.1 DIRECT SIMULATION OF THE BEAM ARRANGEMENT

In cases where the precise shape of the PTV is not considered, the beam arrangement will be decided according to the diagnostic information that exists. This may be in the form of x-rays (possibly with contrast), CT scans, radionuclide imaging, ultrasound, MR scans, and possibly biological tests to determine the stage of the disease. The beam geometry is then defined during simulation on the basis of the anatomy that is identified using the fluoroscopic image.

A number of points need to be considered during simulation:

- Beam modality
- Beam energy
- Beam direction
- Beam divergence or the arrangements to avoid or compensate for beam divergence at one beam edge
- Beam shape or lead shielding

Some of these may have been decided prior to simulation, either on the basis of the specific patient or on the basis of protocols for the given tumour site.

In most radiotherapy departments, the available beam modalities are photons and electrons. Electron planning will be discussed in Chapter 34. In most departments, photons are available with different energies ranging from 4 MV up to about 20 MV. The choice of energy is dependent on patient separation (i.e. the distance between the beam entry and exit points) and the depth of tumour. If the skin or superficial tissue less than 0.5 cm deep is part of the target, bolus may be used to increase the dose to this region. The preferred energy range for targets close to and involving the skin is 4 MV to 6 MV. If, however, the target is deeply situated, higher energies offer better dose distributions. In some cases, a combination of, e.g. 20 MV and 6 MV will yield the best plan. Beam direction, shape, and blocking are site specific and will be briefly discussed for a limited number of sites in Section 32.2.2.

Finally, at simulation, all physical parameters required to calculate the dose need to be acquired. The field size and any associated shielding will be decided. It is also necessary to define the dose prescription point and any additional dose calculation points. For each of these, it is necessary to establish the depth and patient separation and the source to skin distances (SSD) for each beam. Some of these parameters may be indirectly acquired by taking a patient outline on which the beam axes are indicated. From the outline, the SSD and depth to the point of interest can be measured.

Outlines can be obtained in a number of ways, e.g. an outline may be manually acquired using the SSD indicator and polar graph paper, by using a pantograph, or with one of a number of commercial optical outlining devices. At least one outline should be taken in a transverse plane through the centre of the treated area. Further outlines are recommended if there is a significant change in the patient shape within the treatment area.

Optical devices generally make use of laser projections on the patient skin. The images are acquired by two (or more) fixed cameras (Wilks 1993) or by a single camera mounted on the simulator gantry and operated at different gantry angles. These images are converted into outlines that may be scaled and printed, or exported to a treatment planning computer. Because of patient curvature, several images from different angles need to be acquired and merged to obtain a surface outline. Further parallel outlines may be obtained by moving the couch. Satisfactory outlines of the patient's anterior surface, which is visible by the camera, can be obtained in this way. However, the delineation of the posterior surface of the patient is inaccurate because it is not visible by the camera, and it is often approximated by a straight line. Therefore, care must be taken if beams are to enter the patient from posterior angles. Measurements of the patient separation can be taken using callipers to check points on the contour. Systems for measuring surface shapes lack information on internal structures. One exception is the system described by Wilks (1993) that uses an algorithm to draw lungs on breast contours by using the depth of the lung seen on the radiograph of the simulated field. Alternatively systems can be based on stereo-video acquisition of light patterns projected on the patient skin (Bert et al. 2005).

If one has access to a simulator equipped with a CT option (see Chapter 13), the external outlines as well as the outlines of internal structures can be directly extracted from CT images transferred to the treatment planning computer.

32.2.2   SPECIFIC EXAMPLES OF ANATOMICAL SET-UP

Anatomical set-ups should be restricted to cases where it is considered acceptable to define the PTV on the basis of general diagnostic information without strict delineation of this volume on a series of axial slices. It must also be realised that such an approach is possible only if the beam arrangement is simple enough to be able to anticipate what the dose distribution will look like. Some typical examples are listed below. For a more comprehensive overview of radiotherapy treatment set-ups, see, for instance, Dobbs et al. (1999).

32.2.2.1   Parallel Opposed Simple Fields

A pair of parallel opposed fields with relatively simple shapes is frequently used, e.g. for palliative treatment of vertebral or brain metastases. In such cases, the field area will be defined at simulation according to bony landmarks. In some cases, contrast will be used to indicate the location of organs such as the bladder or rectum more clearly. For the location of the vagina, a radio-opaque tampon can be used.

32.2.2.2   Mantle Fields

It has been the practice to treat Hodgkins and non-Hodgkins lymphoma with large anterior and posterior fields covering the whole chest and neck. Because the aim is to treat the tumour and lymphatic drainage while avoiding critical structures such as the lung and heart, these fields need carefully designed shaping. Knowledge of the pre-chemotherapy extent of the solid tumour (generally in the form of a diagnostic CT scan of the patient) and information on the position of the lymph nodes (for which an anatomical atlas may be helpful) will enable the shielding to be determined. The field size required could be larger than the maximum size available on the linear accelerator at the standard SSD, so these treatments can be planned at an extended SSD that is typically between 120 cm and 140 cm. Extending the SSD has the added advantage of reducing the size and weight of the lung blocks. However, it requires that, for the posterior field, the table can be elevated to a distance of between 20 cm and 40 cm above the isocentre and this is not always possible. It also requires that the table top has an opening large enough to let the beam through or is made of a material, such as carbon fibre, that attenuates the beam minimally. If the required field length cannot be achieved with the gantry beneath the couch, it will be necessary to treat the patient prone, but this should be avoided if possible, because the relative positions of internal organs differ.

32.2.2.3   *Tangential Breast Beams*

To simulate tangential breast treatments, the gantry angle that covers the breast volume while minimising the amount of underlying lung tissue in the fields can be found by real-time fluoroscopic adjustment of the beam position. Lead markers can be placed on the skin at the margin of the palpable breast to facilitate the radiological set-up. The opposing beams should be angled such that their internal beam edges are non-divergent and include a minimal amount of lung. The relative gantry angles can be worked out by fluoroscopy or through trigonometry by using the field width.

For tangential breast beams, it is also advantageous to keep a non-divergent or vertical beam edge at the superior border in order to allow for the addition of an axillary field (even if that is not prescribed as primary treatment). This can be achieved either by using trigonometry to compensate by adjusting the couch angle and the collimator angle of the tangential fields, or one can use half beam blocks (asymmetric diaphragms) provided the field length is not restricted (see Section 33.9.3).

In most situations, an anatomical set-up does not dispense with the need to confirm the plan by computation of the dose distribution, using the data acquired during the simulation process. Small adjustments of the plan (i.e. beam weights, beam modifiers, shielded area) might be performed without resuming the simulation. However, large changes (i.e. beam directions) could necessitate a repeated simulation.

32.3   COMPUTER-BASED BEAM DEFINITION—THE TRADITIONAL APPROACH

Computerised planning based on one or more transverse sections of the patient is recommended when the choice of beam arrangement is too difficult to make an anatomical set-up practical. Beam characteristics will be defined during the dose planning process. In these cases, the simulation process is split into two parts:

- The acquisition of anatomical data (before dose planning)
- The verification simulation of the selected plan (after dose planning).

Information about the position of the tumour and organs at risk (OARs) can be obtained from any of the data acquisition methods that have been explained in Chapter 30 and Chapter 31. From a combination of these data sources, a set of external outlines, reference skin marks, and information on internal structures are obtained.

After transferring these data to the treatment planning system and indicating the limits of the PTV, the beams are graphically defined on transverse sections in order to cover the PTV and to avoid the OARs. Their characteristics (number, energy, incidence, position, field size, etc.) are interactively edited until a satisfactory dose distribution is obtained.

The beam parameters are finally transferred back to the simulator in order to check the plan's feasibility and consistency on the patient before progressing with treatment. At this stage, the beam entry points can be marked on the patient's skin (e.g. tattooed markers) and radiological images taken to serve as a reference.

This traditional approach is appropriate for a plan that makes use of a limited number of cross-sections (e.g. less than 10). When a large series of thin adjacent slices is acquired, a full 3D reconstruction can take place, and the beam definition can be made with the virtual simulation approach.

32.4 COMPUTER-BASED BEAM DEFINITION—VIRTUAL SIMULATION

32.4.1 PRINCIPLE OF VIRTUAL SIMULATION

When a full CT data set is available for a patient in the radiotherapy treatment position, the CT data serves as a 3D anatomical model of the patient. The CT data provides the basic electron density or physical density map that allows correction for tissue heterogeneities to be made in the dose calculation. In 3D planning systems (Goitein et al. 1983; Mohan et al. 1988; Galvin et al. 1995), the CT data can be viewed not only in transaxial (i.e. the original CT slices) but also in sagittal and coronal views. Generally, the sagittal and coronal views show a coarser resolution in the cranio-caudal direction; however, they allow a better appreciation of the basic anatomy and the 3D extent of the various anatomical structures.

The target and OARs should, as a minimum, be outlined on all slices. The projections of these outlines are shown on the transverse planes (where they were defined) and on the sagittal and coronal plans, either as contours or as colour wash. In many 3D planning systems, the user can toggle any structures on or off, which can be very helpful, as overlying structures can obscure each other and the underlying greyscale CT data. In addition to the three orthogonal 2D views, it is helpful to have a 3D surface rendered view of the structures outlined. In particular, it is useful to view the 3D volumes from different angles to decide on the beam directions.

The planning process for computer-based set-up is taken away from the simulator and performed at the planning computer. Presently, however, it is still common, as for the traditional approach, to send the patient to the simulator for a plan check, i.e. to get reference images of the planned fields and sometimes to mark the patient up for treatment, limiting the set-up time for the first treatment. In some departments, the simulator is interfaced to the record and verify system of the treatment machine. In this case the set-up parameters (except the monitor units) can be transferred automatically from the simulator to the accelerator. This saves time and reduces the risk of errors incurred by manual data entry (see Chapter 41). However, the definitive source of beam parameter data is the treatment planning system and it is better to transfer these data, including the monitor units, to the linear accelerator record and verify system. Where there is couch angulation, the simulation also checks for collision risks * as these are not usually considered in planning systems. Some groups have developed graphical simulation that enables collision situations between the gantry and couch to be

---

* In principle, ensuring that there will be no collision in the treatment room requires that the geometrical characteristics of the simulator, table, and accessories are the same as for the treatment equipment.

predicted (Kessler et al. 1995; Muthuswamy 1999; Nioutsikou et al. 2003). However, unless the patient anatomy is included in the graphical simulation, it does not allow for the risk of collision between patient and gantry. Assessing the collision risk is less difficult for coplanar beams.

32.4.2 BEAM AXIS DIRECTION

In some cases, standard beam angles can be used (e.g. the treatment of prostate tumours with between 3 and 6 coplanar fields), but in others, the incident directions for the beams are selected for individual patients by the treatment planner. The planner needs to find the beam direction that treats the PTV while best avoiding irradiation of the OAR. This may be done through the use of various computer visualisations.

A very useful representation, complementary to the visualisation of the transverse sections as used in the traditional approach, is the *beam's-eye-view* (BEV) (Goitein et al. 1983). It is generated by taking a plane perpendicular to the beam direction and showing the projected PTV and OARs relative to the beam edges (the rectangle on Figure 32.1b). As the beam direction and couch angle are interactively changed, the projections of the structures are shifted with respect to the beam edges until the best compromise between PTV coverage and OAR avoidance is achieved.

Another example of a helpful tool is the *spherical view* (Figure 32.1a). This represents a flattened globe with the geometric centre of the PTV, coincident with the treatment isocentre, at its centre and the OARs projected onto its surface. The colours of the OARs are as in the BEV shown in Figure 32.1b. The grey shaded areas represent forbidden angles for couch (latitude) and gantry (longitude) combinations. Beams with couch and gantry angles that are in the clear regions will not pass through an OAR in order to reach the PTV. The main restriction of the spherical view is that it only shows the projection of the central axis, and this shortcoming is illustrated in Figure 32.1b where part of the beam passes through the (green) bladder even though this is not apparent from the spherical view. The spherical view is not currently implemented in standard commercial planning systems.

Individual centres have developed software which semi-automatically optimises the beam direction and such software is now available in some commercial systems. However, as has been discussed in the literature, such optimisation will always be limited (Rowbottom et al. 1999a, 1999b; Bedford and Webb 2003).

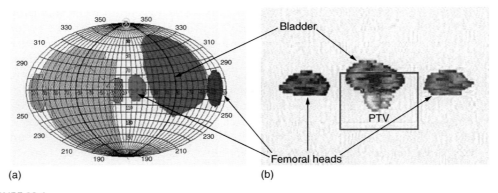

(a)   (b)

FIGURE 32.1
(**See colour insert following page 590.**) (a) A spherical view produced from Voxelplan. The gantry angle is depicted as meridians and the couch angle as latitudes. (b) The 3D Beam's eye view (BEV) (or Observer's Eye View) of the same patient showing planning target volume (PTV) (lighter region within the field rectangle), bladder (darker central region above PTV), right, and left femoral head. This view is from 0° gantry and 0° couch. The cross at the top of the spherical view (a) indicates the couch and gantry angle for this BEV.

32.4.3 Field Shaping

Having determined the incident beam axes, a BEV can be generated, showing the projected extent of the PTV and the OAR in a plane perpendicular to the beam axis. The field shape is then generated from the projection of the PTV by adding a margin to allow for the penumbra of the beam (Brewster et al. 1993). The ideal width of this margin depends on the field size and depth and whether the fields are coplanar or not. In a coplanar situation, the margin will need to be between 4 mm and 6 mm greater in the directions where all the field edges coincide. The shielding can either be a custom-made low-melting-point alloy block (see Section 22.5.3.2), or it can be achieved with a multi-leaf collimator (see Section 32.4.4). A custom-made block conforms to the field shape and is made either by sending the shape from the treatment planning system to a block cutter or having the blocks manually made using a printed template.

32.4.4 Fitting MLC Leaves in Beam's-Eye-Views

Multileaf collimators (MLC) do not produce smoothly rounded isodose distributions as discussed in Section 22.5.4. The geometric projection of the MLC leaves approximately defines the 50% isodose; hence, a margin to account for penumbra (~6mm) must first be applied to the PTV to establish the required field shape. It is this field shape that is fitted by the MLC, using one of several shape fitting algorithms. Generally, treatment planning systems support fitting leaves to the outside, the inside, or to the middle of the MLC leaf (see Figure 32.2a and Section 22.5.4). In practice, the corners of the MLC leaves may be considered to be cut off as shown in Figure 32.2b (Heisig et al. 1994) and this concept provides a good fit to the 50% isodose line.

In practice, the choice of model to use for fitting the MLCs will be restricted by what is available on the planning computer, and, for the sake of consistency, there should be a departmental policy on the margin size and fitting mode for different sites of treatment. How well the MLC conforms to the target shape will also depend on the collimator angle. Most planning systems provide the planner with a graph of either the total volume or the normal tissue volume irradiated plotted against collimator angle. This enables the planner to choose a collimator angle that results in the best fit of the MLCs to the PTV. If, however, a wedge is used, the collimator

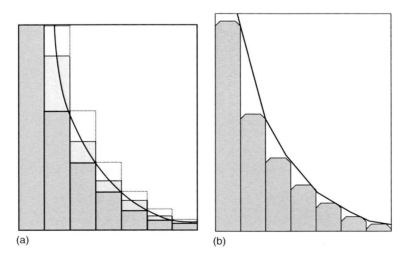

(a)   (b)

FIGURE 32.2
Multileaf collimators (MLC) fitting to the field shape. Here, a schematic corner of an elliptical PTV is shown. (a) Solid bold line-shaded dark grey: MLC fitting outside the field shape; light grey shaded: MLC fitting to middle of leaf; Dotted line: fitting to shield all outside of field shape. In (b) the fitting method of Heisig et al. (1994) is shown.

angle and the MLC orientation are determined by the wedge direction. In such cases, the MLC fit may be improved by shifting the isocentre for that beam (Fernandez et al. 1995).

The most usual MLC leaf width is 10 mm at the isocentre. However, standard or micro MLCs have also been designed with smaller leaf widths (see Section 11.4.3.2). The micro MLCs were originally designed for brain tumour treatment but are also shown to give a closer fit for prostate treatments (Kubo et al. 1999) sparing significantly more bladder and rectum. In all cases the conformity of the dose distribution to the PTV is the principal criterion in determining the optimal positions of the leaves.

32.4.5 DIGITALLY RECONSTRUCTED RADIOGRAPHS

Just as anatomical simulation uses fluoroscopic or conventional radiographs, so virtual simulation uses *digitally reconstructed radiographs* (DRRs) to enable treatment fields to be visualised in relation to the patient's anatomy. The advantage of a DRR produced from a therapy CT data set with the patient in the treatment position is that the data used for target localisation are derived from the same data set as that used for verification of the treatment beam direction. This means that the DRR can be used as the reference image so eliminating the simulation process. In addition, the PTV outlines can be directly projected onto the DRR (see Figure 32.3). However, in order to be useful tools in the clinic, DRRs must be of comparable quality to conventional radiographs. An understanding of how DRRs are produced will enable the physicist to ensure that best use is made of the software used to create and enhance DRRs.

DRRs are generated by summing the radiological thicknesses along each ray line from the virtual source of the beam through the 3D patient model to a point in the plane of the DRR. The geometry used to make the DRRs and the mathematical transformations required are described by Siddon (1981, 1985). The attenuation along the ray lines can be derived from the definition of the Hounsfield Unit (HU), as follows (Killoran et al. 2001):

$$HU = 1000\frac{\mu - \mu_{w}}{\mu_{w}} \tag{32.1}$$

Which can be rearranged in the form:

$$\mu = \mu_{w}\left(\frac{HU}{1000} + 1\right) \tag{32.2}$$

where μ and μ_{w} are the linear attenuation coefficients of the tissue and water, at the energy used to produce the CT, respectively.

The linear attenuation along a ray line will be the sum of the attenuation in the voxels in the path of the ray line. It is important to note that CT scans are acquired at about twice the energy used for conventional radiographs so that the ratio of the photoelectric effect to the Compton effect is much reduced. Consequently, bone contrast is inherently much lower in DRRs.

Using the relationship of linear attenuation coefficient with energy, it is possible to generate DRRs that resemble simulator images (photoelectric) and megavoltage images (Compton) by recalculating radiological thickness maps according to the beam quality. Modifications can also be made to the mapping between CT scanner Hounsfield Units and relative electron density to emphasise different anatomical features. For more details, see Sherouse et al. 1990; Cheng et al. 1987; Cullip et al. 1993; and Killoran et al. 2001.

For reference images, the photoelectric effect is most often used as that gives the highest contrast and best visualisation of bony landmarks. It is debatable if Compton images are of any use as treatment verification becomes a comparison of two low contrast images (Killoran et al. 2001). Radiological thickness maps contain the information required to make dose compensators (see Section 22.6.3) in situations where fields are opposed (e.g. tangential breast treatments).

DRR quality is influenced by a number of factors, related to both data acquisition and reconstruction algorithms.

- *The separation of the CT slices.* This determines the resolution in the longitudinal plane and provides an objective limit to DRR quality. Slice separations of 10 mm will result in poor DRRs. Images that are suitable for verification in the pelvic region can be produced with 5 mm slice separations (see Figure 32.3a) although a smaller separation gives a better image (see Figure 32.3b). In the chest, head, and neck region, a separation of 2 mm to 3 mm is always recommended (Dong et al. 1995; Gilhuijs et al. 1996). It is also important that the CT data set extends to the whole treatment area, which can be some distance beyond the tumour especially if non-coplanar beam arrangements are used.

- *The ray tracing mode.* This refers to the way the summation is performed. It either uses the nearest CT pixel only for a fast reconstruction, interpolates between the CT pixels of adjacent slices, or uses a full interpolation of all eight nearest pixels.

- *The step size.* This is the size of the increment used when summing up the DRR.

- *The number of pixels.* This is the number (rows × columns) that is in the output DRR.

- *The magnification of the DRR.* If print-outs are requested, DRRs will need to be rescaled to be the same size as the simulator verification image or the portal image. However, this increases the pixel size and reduces the spatial resolution.

- *The manipulation of the relative contrast of the structures in the DRR.* It is possible to enhance DRRs by selectively altering the visibility of different structures according to their relative densities. It is also possible to generate a DRR based exclusively on the voxels encountered within a given depth range and therefore to reinforce the contrast of the structures lying within this range. Several planning computers incorporate these features which, in principle, enable for instance airways and bones to be simultaneously enhanced, something which is impossible with conventional radiographs.

With the increased number of CT scanners available in radiotherapy departments, the use of CT scans for simple treatments such as parallel pairs is increasing as CT based planning offers

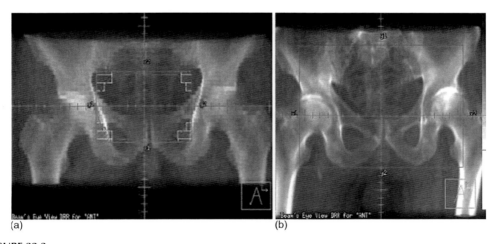

(a) (b)

FIGURE 32.3
(See colour insert following page 590.) Examples of digitally reconstructed radiographs (DRRs) generated by the Pinnacle planning system; in (a) the CT data set had 5 mm slice separation and a coarse reconstruction was used, whereas in (b) the CT data set had 2.5 mm slice separation and the DRR algorithm was smooth; therefore, the DRR looks very much like a simulator radiograph.

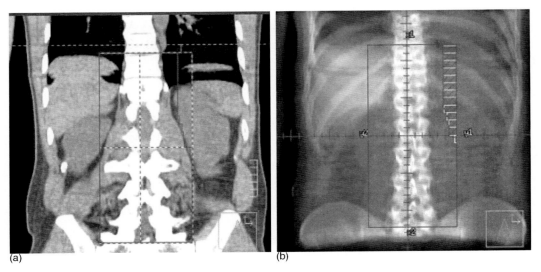

(a)   (b)

FIGURE 32.4
(See colour insert following page 590.) (a) A coronal view, showing the position of the kidneys relative to the field borders, and (b) a DRR on which the field size and the individual MLC leaves can be positioned.

increased accuracy to the treatment and often reduces the time the patient needs to spend in the radiotherapy planning department. For example, palliative lung cancer treatments where the lung tumour is medial, i.e. the tumour is partially in the shadow of the mediastinum, are often very difficult to simulate as the full extent cannot be seen using a conventional x-ray simulator. In these cases, a CT scan will reveal the extent of the tumour, and a combination of transverse and coronal views can be used to define the beams for the parallel pair.

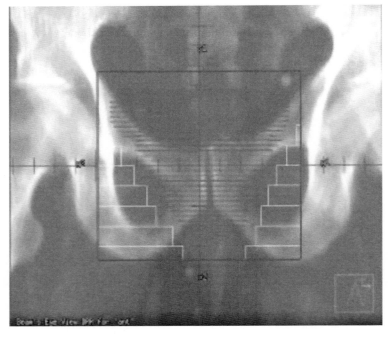

FIGURE 32.5
(See colour insert following page 590.) The DRR shown in Figure 32.3b with the prostate PTV contours (horizontal lines) and MLC projection overlaid.

CT scans can also be used to reduce the use of contrast agents administered to patients. One particular site is the treatment of the para-aortic nodes where the position of the kidneys relative to the treatment fields is crucial. Here, the patient benefits from the CT scan by avoiding the intravenous contrast that would be administered to show the extent of the kidneys. The planning is then performed by simultaneously looking at a coronal reconstruction (see Figure 32.4) and the DRR. This combination will give a full appreciation of the position of the kidneys (by scrolling through the coronal reconstructions) relative to the para-aortic nodes and allow for manual positioning of the MLC leaves to shield them.

If the PTV and other structures have been outlined on the CT data set, they can be overlaid on the DRR. The beam shape and conformal blocks or MLCs may also be added as overlays as shown in Figure 32.5. Using DRRs as the reference images for treatment verification elimin-ates the need to use the simulator to verify a computer plan and reduces one source of error. It is important that the patient has the same position at treatment as at CT scanning. This is facilitated in purpose-designed CT simulators that have rigid flat couch tops.

With the use of multislice CT scanners, slice seperations of 1 mm are readily achieved. With such small slice seperations, DRRs can almost match the spatial resolution of conven-tional radiographs with the additional possibility to enhance and manipulate the digital images and to benefit from the automatic overlay of the structures which were delineated in the transverse sections. It is important that DRRs are produced by skilled and experienced staff that have had the time to test the enhancing tools available and liaise with the clinicians who will use the images. Care must be taken to achieve an image quality that allows satisfactory delineation of the structures* that will be seen on the verification portal images which will be acquired during the course of the treatment.

---

* There are therefore two possibilities to obtain structures overlaid on a DRR: either drawn directly or projected from transverse contours. Then, the decision must be taken, for each anatomical site, about which mehod to use for which organ in order to achieve an optimal result.

CHAPTER 33

PHOTON-BEAM TREATMENT PLANNING TECHNIQUES

Peter Childs and Christine Lord

CONTENTS

33.1 INTRODUCTION

The goal of radiotherapy treatment planning is to design a beam configuration which will deliver a homogeneous dose to the specified planning target volume (PTV), ensuring that normal tissue receives a reasonably low dose and that critical organs receive less than their tolerance doses. ICRU 50 (ICRU 1993) suggests an acceptable dose variation of $+7\%/-5\%$ of the reference dose, the reference dose point being the centre of the PTV, the isocentre, mid-plane or some other suitable point (see Section 37.6). This is achieved by choice of treatment modality (photons or electrons), beam energy, beam arrangement, use of wedges, compensators or blocks, and methods such as conformal planning (see Section 43.1) as appropriate. The ICRU objectives are not always met and sometimes compromise may be necessary, depending on the aims of radiotherapy. For example, if the treatment is palliative, long-term radiation effects may not be a consideration and a simple beam arrangement will suffice. On the other hand, if the treatment is radical, i.e. aiming at local control of the tumour, long-term effects need to be taken into account and this may result in a more complex radiotherapy regime. Centres may also be limited in the choice of photon beam energy or the availability of electron therapy.

Single beams are easily visualised with the use of isodose curves, sets of which may be produced for a range of field sizes for each available treatment energy. This is particularly true for electron and kilovoltage treatments. In the past, when beams were combined, isodose curves were amalgamated manually by superposition of standard isodose curves with the help of tracing paper. This was a slow process and placed limits on the optimisation of treatment plans. Modern treatment planning computers provide a fast way of viewing dose distributions. The effect of adding and repositioning beams, changing beam weights and adding wedges can be seen almost instantaneously, thus facilitating the planning process and allowing the planner to investigate various options before deciding on the final plan. The ability to view dose distributions in three dimensions also adds to the armoury of the radiotherapy treatment planner, as does the use of dose volume histograms (see Section 35.3).

In this chapter, we will address the basic principles of photon-based treatment techniques that are currently used in radiation therapy to treat the most common types of cancer. Electron beam techniques will be dealt with in Chapter 34. Although the conformal approach is becoming the standard for a majority of curative treatments (see Chapter 43), it is still necessary, for safe and efficient treatment planning, to have a good knowledge of the more conventional techniques discussed below.

33.2 SINGLE BEAM TREATMENTS

33.2.1 KILOVOLTAGE BEAMS

Kilovoltage beams are mostly used to treat superficial disease, normally with single fields. With these energies, there is a large dose fall off with depth and the maximum dose is at or very near the surface. Typical doses at 1 cm depth are 80% for 120 kV, 90% for 150 kV and 95% for 300 kV. There is differential absorption of the dose in bone due to the photoelectric effect and lesions over bone or calcified cartilage should therefore be treated with electrons rather than with kilovoltage beams. (The exception is in the region of the eye because of the need for a corneal shield (see Section 21.5)).

33.2.2 MEGAVOLTAGE BEAMS

Single, high-energy photon fields provide skin sparing compared to kilovoltage beams due to the build-up effect (see Section 22.3.1) and increased dose with depth. Both the build-up

TABLE 33.1

Typical Beam Parameters for a 10 cm×10 cm Field at 100 cm SSD

Nominal Energy (MV)	d_{max} (cm)	% Depth Dose at 5 cm	% Depth Dose at 10 cm
Cobalt-60	0.5	80.4	58.7
4	1.0	83.9	63.0
6	1.5	86.9	67.5
8	2.0	89.6	71.0
10	2.3	91.4	73.0
15	2.9	94.5	77.0
25	3.8	98.5	83.0

Source: From British Institute of Radiology, *Central Axis Depth Dose Data for Use in Radiotherapy,* *Br. J. Radiol.,* Suppl. 25, 1996.

effect and the depth dose increase with energy. Energies available range from ^{60}Co sources to 50 MV linear accelerator beams, although more typical megavoltage energies would be between 4 MV and 25 MV (see Table 33.1 for typical beam parameters). Combinations of high- and low-energy photon beams can produce an equivalent intermediate energy if required.

The dose gradient across the PTV for a single photon beam is large, the fall off for a 6 MV beam being approximately 4% per cm. Therefore, dose homogeneity will probably fall short of the recommendations in ICRU 50. Single fields are therefore not suitable for treating deep-seated tumours.

The prescription may be made to a point at the required treatment depth, but this will result in a high dose at the point of dose maximum, d_{max}, and any normal tissue in the beam

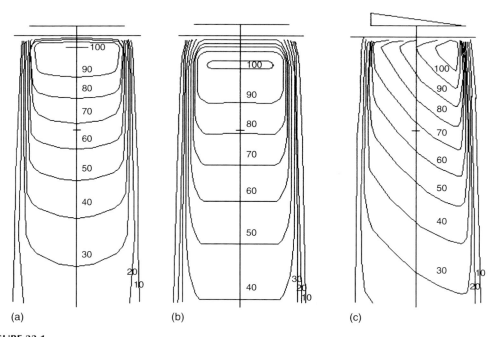

(a) (b) (c)

FIGURE 33.1

Dose distributions from single photon beams (a) 6 MV photon beam, (b) 25 MV beam, (c) 6 MV 45° wedged beam. 100% is at d_{max}, on the beam axis. 10 cm depth is indicated by a mark on the beam axis. The general shape of the isodose lines for a given energy may vary significantly depending on the design of the flattening filter (see Section 22.3.3.2).

overlying the PTV. The use of a higher energy will make the distribution more uniform (see Figure 33.1a and b). For volumes near the surface, bolus can be added to ensure that none of the PTV lies within the build up region.

The field size is normally defined by the 50% dose relative to the central axis, and a margin must therefore be added to the PTV when designing a beam to fit a target volume so that it does not lie within the beam's penumbral region. For x-rays (i.e. excluding cobalt) this margin will increase with energy, but a typical value is 6 mm at the isocentre.

Single fields provide a simple mode of treatment which may suit the clinical needs of the patient in particular cases. They are used to treat breast nodes (in combination with tangential fields to the breast, see below) and the lower neck and supraclavicular nodes (in combination with an opposed pair to the upper cervical nodes).

One example of the use of single beams is that of a long, narrow, posterior field in the treatment of the spine, with the anterior limit of the spinal cord being at approximately 5 cm deep. The variation in dose to the cord along the treatment field, due to the change in source surface distance (SSD) and depth, can be greater than 10%. It is possible to obtain a uniform dose by using purpose-built compensators: shaped blocks of aluminium or other suitable material, placed in the beam accessory tray, below the treatment head. Alternatively, this compensation of the single field may be achieved with smaller *top up* fields overlaid on the single field, using asymmetric collimator settings or with intensity modulation obtained with multileaf collimators.

33.3 TWO BEAM ARRANGEMENTS

33.3.1 PARALLEL OPPOSED PAIR

For deeper volumes, two opposed beams may be used, a *parallel opposed pair*. This will produce a more uniform dose in the PTV but will also give a high dose to overlying normal tissue in the path of the beam, the region of highest dose being near the surface. The shape of the isodoses in a parallel pair setup is that of an hourglass (see Figure 33.2a). For an antero-posterior pair, this is the case in both the transverse and sagittal planes. Parallel pair treatments are used widely in all areas of the body. The parallel pair may be used for the entire treatment course, especially when the intent is mainly palliative, or may form the first phase of a two-phase radical treatment. As the separation of the patient increases, the uniformity of the dose distribution decreases, with an increase in dose close to the surface compared to the mid-plane. This may lead to higher normal tissue doses or to an unacceptable dose inhomogeneity within the PTV. The non-uniformity can be reduced if a higher energy is chosen for treatment.

33.3.2 BEAM WEIGHTING

If the centre of the PTV is not at the centre of the patient's cross-section, i.e. in the mid-plane of the patient, then the opposed beams can be given different weightings (see Figure 33.2b). The definition of beam weights depends on the planning system or calculation technique used. It can be proportional to the *beam-on* time (i.e. set monitor units* for linear accelerators or time for cobalt-60 units), or proportional to the dose contribution of that beam to a specific point, usually the isocentre[†]. For a parallel pair with the isocentre at mid-plane, weighting to *set MU* is the same as weighting the dose to the isocentre (in the absence

---

* See Section 20.1.2 for a definition of monitor units (MU).
[†] In some planning systems the weight may be based on the calculated dose in a water medium or in conditions which may differ from the actual ones (e.g. without shielding blocks).

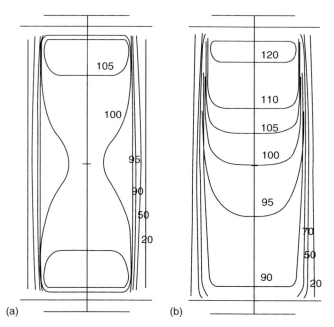

FIGURE 33.2
6 MV parallel pair distributions normalised to 100 at mid-plane, showing the effect of beam weighting; separation is 20 cm (a) equal weighting, (b) 2:1 weighting.

of inhomogeneities), so a 2:1 weighting will give the ratio of beam-on time and of contributions to the isocentre. However, weighting to a different point will result in different values of the beam weight, depending on which weighting system is in use. Increasing the weighting of the beam entering the side of the patient nearest the PTV will always increase the PTV dose relative to the normal tissue lying on the opposing side. Equal weights defined in terms of the dose administered to the centre of a PTV which is not at mid-plane would result in an overdosage of healthy tissues on the opposing side. With this definition of beam weight it is necessary to compensate by significantly increasing the weight of the beam entering the side nearest the PTV.

33.3.3 WEDGE FILTERS

Where opposed beams are applied to sloping surfaces, for example in larynx or breast treatments, it is necessary to use wedge filters to maintain a uniform dose distribution perpendicular to the beam direction. The effect of the wedge is to tilt the isodoses; in these situations it compensates for missing tissue. A detailed discussion of wedge filters can be found in Section 22.6.1 and Section 22.6.2.

For breast and larynx treatments the wedged direction is in the transverse plane, but in principle a wedge can be used for compensation in any orientation. An example of the use of the wedged direction in the sagittal plane is to compensate for the sloping surface of the chest in the upper thorax.

33.3.4 TANGENTIAL BEAMS

Radiotherapy of the breast and chest wall (post mastectomy) is usually carried out using two tangential fields. These may be set up so that the posterior borders of the two fields are aligned. This ensures that the beams do not diverge into the lung, which would result in an

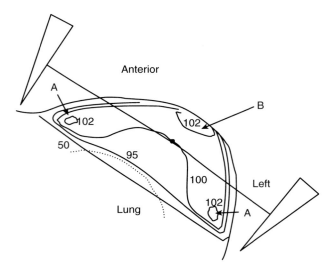

FIGURE 33.3
6 MV tangential breast fields, normalised to 100 at the isocentre. Note the indications of patient orientation (Anterior and Left) which should always be present on any treatment plan.

increase in lung dose. Alignment of the field edges may be achieved either by using half-blocked beams and placing the isocentre at the posterior edge of the field or by angling the gantry to make the posterior borders of the fields parallel. Wedges are added to achieve a more uniform dose distribution by compensating for the differential tissue thickness across the beam width (see Figure 33.3). The amount of lung in the beam also needs to be considered, both because of the sensitivity of lung tissue to radiation and because the lower density of lung tissue can lead to hot spots at points A in Figure 33.3. Not all treatment planning systems account for the lack of scatter in air, a further consideration when beams enter the patient obliquely. This may result in an overestimation of the dose at point B*. Normally equal weight is given to the two beams, but it may be necessary to vary beam weights to account for asymmetry of the breast or positioning of the isocentre to optimise the dose distribution.

Breast radiotherapy may involve node irradiation using an anterior field with a possible posterior boost field. Suitable collimator angles and couch twists can be set up on the tangential beams so that a vertical match plane is produced for the node fields. Alternatively, asymmetric fields can be matched by placing the isocentre at the superior border of the tangential fields to achieve a vertical match plane with no divergence into adjacent fields.

33.3.5 WEDGED PAIR BEAMS

Wedged pair beam arrangements are common in head and neck radiotherapy, e.g. parotid, oral cavity and maxillary antrum, with the angle between the beams, often referred to as the *hinge angle*, being less than 180°. In this case the wedge is used to compensate not only for missing tissue but also because the beam entry points are close together on one side of the patient. As a result the open field isodose lines would not be parallel to each other resulting in hot spots close to the surface between the beam entry points. The smaller the hinge angle, the

---

* Calculation of breast plans with older planning systems that did not allow for either the between decreased attenuation of lung or the decrease in dose at point B would lead to a requirement for a thicker wedge to achieve a homogeneous dose distribution. With a modern planning system it is usually found that a wedge angle between 15° and 20° is the maximum that is ever needed, whereas with older systems it was common to find 30° or even 45° wedges being used.

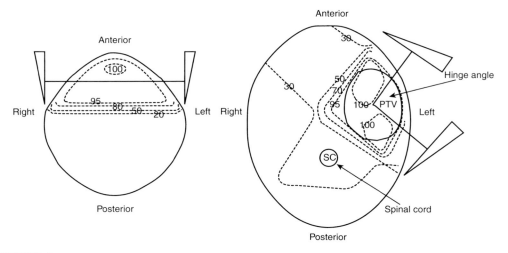

FIGURE 33.4
6 MV dose distributions normalised to 100 at the isocentre from two wedged fields. (a) A parallel opposed larynx treatment. (b) A typical beam arrangement for treatment of the parotid gland (labelled PTV).

larger the required wedge angles needed to obtain a uniform dose across the PTV. In an ideal case where no missing tissue compensation is required, the wedge angle should be (90°- 1/2 hinge angle). An example is shown in Figure 33.4. This is a good beam arrangement for volumes near the surface, as it minimises the dose to normal tissue.

33.3.6 PHOTON AND ELECTRON BEAMS

Photon and electron beams can be overlaid to treat the same site. The result of this arrangement is an increase in skin sparing compared to an electron-only treatment, and a lower dose at depth than that obtained from a photon-only treatment. Examples include treatment of the parotid gland (thus reducing the dose to the contralateral parotid and spinal cord) and the intra-mammary chain in the mediastinum.

33.4 MULTIPLE COPLANAR BEAMS

The use of three or more beams reduces the dose level to normal tissue within the irradiated volume compared with a parallel pair arrangement, as the entrance and exit doses do not coincide. This is important when the beam traverses a sensitive structure such as the bladder or rectum or a dose-limiting structure such as the spinal cord or lung. Beams converge on the PTV and, with the use of suitable beam angles, weights and wedges, produce a uniform dose within the PTV that is higher than in surrounding tissue. Beam weights and wedges can be chosen so that the dose to normal tissue in the path of the beams is spread equally, or to spare particular sensitive structures at the expense of giving a higher dose to other normal tissues. Selection of the best field arrangement for a specific site therefore depends upon many factors with varying degrees of clinical importance (Bedford et al. 1999).

33.4.1 THREE-BEAM ARRANGEMENTS

Three-beam treatments are used in the pelvis (prostate and bladder), abdomen and thorax as well as the brain.

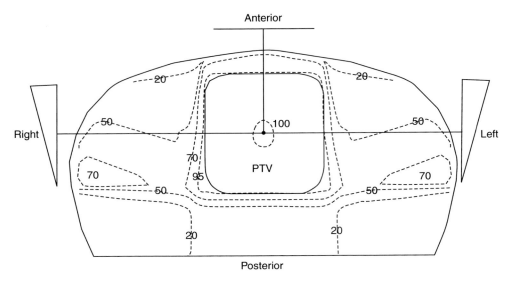

FIGURE 33.5
6 MV three-beam dose distribution normalised to 100 at the isocentre to treat the bladder (PTV).

The beam arrangement depends on the treatment site, but a typical configuration in the pelvic area is an anterior or posterior beam plus two lateral or lateral oblique fields (see Figure 33.5). In the pelvis (treatment of the prostate or bladder), the three-beam arrangement ensures that no beam enters through the rectum. Wedges on the lateral or lateral oblique beams ensure a uniform dose to the PTV, compensating for the uneven spacing of the beams around

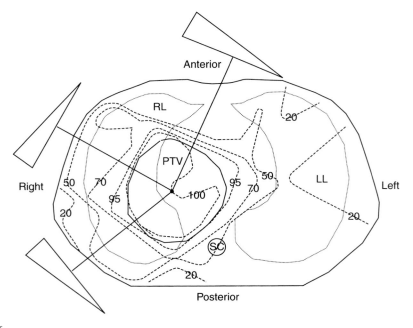

FIGURE 33.6
6 MV three-beam dose distribution normalised to 100 at the isocentre, for the treatment of a right bronchial tumour (PTV). The beams are arranged to avoid the contra-lateral lung and to keep dose to the spinal cord (SC) below tolerance. RL indicates the right lung and LL the left lung.

the patient. Figure 33.6 illustrates a three-beam bronchus treatment, with the beams arranged to avoid the spinal cord and the contralateral lung.

Oesophagus treatments may be delivered as two-phase treatments, with an anterior and posterior parallel pair as the first phase, followed by an anterior and two posterior oblique beams with wedges as the second phase. The first phase delivers minimum dose to the lungs, the second phase minimum dose to the spinal cord. The number of fractions for each phase is chosen so that the dose to the lungs and spinal cord is kept below tolerance.

It is possible to use different beam energies in multi-beam plans. A lower beam energy may be chosen for one beam to reduce the exit dose (e.g. the anterior beam which exits through the rectum in prostate treatments) at the expense of an increase in the surface and entrance doses.

33.4.2 FOUR-BEAM ARRANGEMENTS

Four-beam arrangements can be used in the pelvic area, consisting of an anterior, posterior and two lateral beams (see Figure 33.7). Wedges can be used to compensate for missing tissue. The beam weights may be different depending on the beam direction.

33.4.3 ARRANGEMENTS WITH MORE THAN FOUR BEAMS

Arrangements using more than four beams are less common. As the number of fields increases, the mean dose to normal tissue within the irradiated volume decreases, but the volume of normal tissue that is irradiated increases. Whilst there may be some advantages to using a large number of beams for specific cases, overall treatment time will be greater due to increased set up and delivery time.

Rotational or arc therapy is the ultimate endpoint for large numbers of beams, but this is seldom practised in conventional radiotherapy today. The main reason is that in conventional rotational therapy (i.e. with no dynamic conformal shielding), it is not possible to produce conformal dose distributions, so the technique is limited to cylindrical or spherical PTVs.

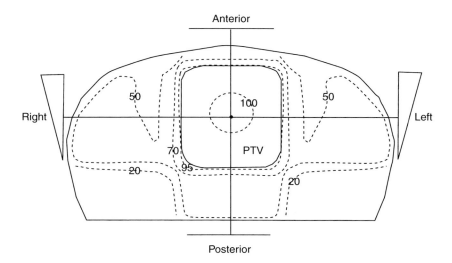

FIGURE 33.7
Dose distribution normalised to 100 at the isocentre for a 6 MV *four-beam brick* arrangement (also called *box* technique) in the pelvic area, treating the same patient and PTV as in Figure 33.5.

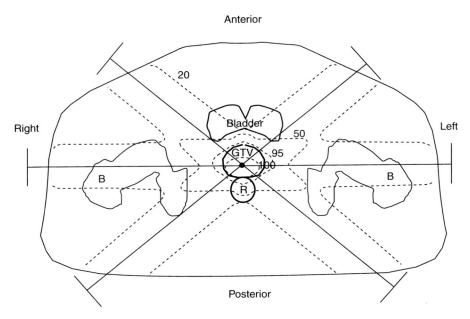

FIGURE 33.8
Dose distribution normalised to 100 at the isocentre for a Phase 2 boost treatment to the prostate (GTV) using six beams; R indicates the rectum and B the pelvic bones (and femoral heads). Note the lower dose to normal tissue relative to that in Figure 33.5 and Figure 33.7, but the consequent increase in the amount of normal tissue irradiated.

One example of a six-field treatment is that of prostate boost treatments, one of the arms of the RT01 trial (Sydes et al. 2004) (see Figure 33.8). This was the second phase of a two-phase prostate treatment. The first phase was a conventional three-field technique to the PTV with a prescribed dose of 64 Gy. The second phase consisted of two lateral beams with right and left anterior and posterior oblique fields to treat the GTV only to 10 Gy. Both phases included conformal blocking.

33.5 NON-COPLANAR BEAMS

In all of the above examples, the beam axes are contained in the transverse plane and are therefore coplanar. Even in relatively simple treatments such as treatment of carcinoma of the larynx by a wedged pair of anterior oblique beams, non-coplanar fields may be required. The couch may need to be rotated by 5° or 10° in order to avoid the beams passing through the shoulders. Such beams are non-coplanar. In traditional planning, where only a single slice was considered, these small angles were often ignored, as the error involved was minimal. However, modern 3D planning systems can easily model the angulation correctly, provided that they have multislice patient data. A 3D plan using the correct angles is essential if data are to be transferred electronically between the planning system and the treatment machine.

Although the use of multiple beams can significantly reduce the dose to normal tissue, it is often difficult to avoid the exit dose of one beam contributing to the entrance dose of another, opposing beam. More importantly, it may be impossible to direct several treatment beams at the PTV whilst avoiding critical structures if all the beam axes are in one plane (normally the transverse plane). To overcome these problems, non-coplanar beam arrangements can be employed. Below the neck, the possibilities of non-coplanar beams are limited by potential collisions between the gantry and either the patient or the patient support system. Any beams entering out of the transverse plane may have increased path lengths to reach the PTV, resulting in an increased irradiated volume. However, non-coplanar beams can be useful in treating small lung tumours, as they facilitate the avoidance of normal lung tissue.

The problem of access is reduced in the head, which can be seen as approximately spherical, although constraints such as the patient support system, be it a shell or stereotactic frame, can restrict the choice of beam arrangements. Because the critical organs can now potentially be irradiated from any direction, the ability to visualise these structures in three dimensions during the planning process rather than viewing single dose planes is essential. Analysis of the dose distribution must also be performed in three dimensions. As simply considering a transverse, sagittal and coronal dose plane through the isocentre may miss important information outside these planes, the use of tools such as dose volume histograms and 3D graphics are important in the planning process. The ideal four-beam arrangement, minimising field overlap, would be a tetrahedron. Constraints, for example not exiting into radiosensitive structures such as the thyroid and the eyes, can limit the freedom of beam orientations, while for conformally shaped beams, the beam configuration may be influenced as much by the optimum shielding of adjacent critical structures as by achieving a homogeneous dose distribution. It has been found that, when using cerrobend blocks to shape the field, increasing the number of static beams beyond four only marginally improves normal brain sparing (Perks et al. 1999) for lesions such as the pituitary. Larger numbers of conformal beams may be needed to further reduce the dose to specific adjacent normal tissues. Arcing non-coplanar fields have a role in treating spherical brain lesions (using circular fields), but where the PTV is of irregular shape, fixed conformal fields offer the best solution for normal tissue sparing (Laing et al. 1993). Stereotactic treatments of brain lesions are considered further in Chapter 45.

33.6 TREATMENTS IN INCLINED PLANES

Treatments of inclined cylindrical volumes, such as the oesophagus and some head and neck volumes, may be best planned with fields that are in the plane perpendicular to the longitudinal axis of the volume. By selecting the appropriate table, collimator and gantry angles, it is possible to arrange for beams from any angle to be coplanar in an inclined treatment plane. This method was originally described by Fleming and Orchard (1974) and further developed by Casebow (1976) and Siddon (1981).

In the general case, the long axis of the PTV is inclined to the horizontal and rotated with respect to the longitudinal axis of the patient. The projections of the two angles are measured

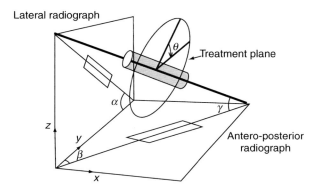

FIGURE 33.9
Illustration of a PTV inclined at an angle γ to the horizontal and showing the projection of γ onto lateral and AP radiographs, α and β. The treatment plane is normal to the axis of the PTV. The frame of reference of the linear accelerator is given by coordinates x, y, z, as defined by the IEC (see Figure 11.31). The angle θ represents the planned gantry angle in the treatment plane. (From Casebow, M. P., *Br. J. Radiol.*, 49, 278–280, 1976.)

on lateral and anterior radiographs. α is the projection of the PTV axis to the horizontal and β is the projection to the longitudinal axis. The true angle, γ, to the horizontal is given by: $\tan \gamma = \tan \alpha \tan \beta$. The sign of γ is important: viewed from the patient's left side (patient supine) clockwise rotations are positive. Let θ be the planned beam angle in the treatment plane (i.e. what would be the gantry angle if the plan were being performed in that plane). Figure 33.9, based on Casebow (1976), illustrates these angles. The gantry (GA), table (TA), and collimator (CA) angles of the treatment machine are given by:

$$\cos GA = \cos \gamma \cos \theta \tag{33.1}$$

$$\tan TA = (\sin \gamma / \tan \theta) - \beta \tag{33.2}$$

$$\tan CA = (\tan \gamma / \sin \theta) \tag{33.3}$$

If the PTV is in the sagittal plane, as is frequently the case, $\beta = 0$ and $\gamma = \alpha$.

33.7 FIXED SOURCE SURFACE DISTANCE BEAMS VS. ISOCENTRIC TREATMENTS

An essential component of any treatment plan is the set-up instructions that relate the dose distribution and field arrangement to the external reference marks of the patient. For multiple field arrangements, it is preferable to be able to set up the treatment at the outset, and to move only the gantry and collimator between fields, rather than adjusting the patient position and couch movements. This is illustrated in Figure 33.10 for a four-beam treatment in which the isocentric arrangement has been set up to the skin reference mark using the isocentre lasers and the optical distance indicator (ODI). The treatment unit lasers are centred on the isocentre and therefore intersect with the patient, enabling lateral marks to be used to ensure that the patient is not rotated prior to setting up the remaining fields. Although the ODI would not be necessary for setting up the remaining fields, it is useful as a secondary check, as the SSD for each beam can be obtained from the treatment plan. For the fixed-SSD set-up, the ODI and lasers must be set up to each beam entry point by moving the couch laterally and vertically, and separate measurements are required to ensure that the patient is not rotated. Isocentric plans are therefore easier to set up, as the isocentre only has to be set once, and this patient position then applies to all fields. In addition, the dose to the PTV is expected to be more accurate because (i) with isocentric set-up of opposing beams, an inaccurate patient positioning will have the opposite effect on the dose contributions of both beams; (ii) a change in patient thickness will have consequences on the tissue attenuation but not on the distance to the PTV, which remains unchanged.

It may be argued that maintaining a constant source-to-surface distance, SSD, in a multiple beam treatment simplifies the dose calculation, as the inverse square dependence of the beam output is constant and may be looked up directly from depth dose tables. However, using tissue phantom ratio or tissue maximum ratio tables for isocentric treatments is no more complex (see Section 23.3.3.2). With modern treatment planning computers, which can perform calculations accurately for any SSD, the need for non-isocentric setups for multiple convergent beam treatments is limited, and isocentric treatment is recommended for most situations. However, for 80 cm source axis distance (SAD) cobalt-60 machines, for thick patients, or when using lead shielding on a shadow tray, there may be difficulties with potential collisions or increase of surface dose due to electron contamination from the close perspex tray. This can be solved by the use of an SSD set-up.

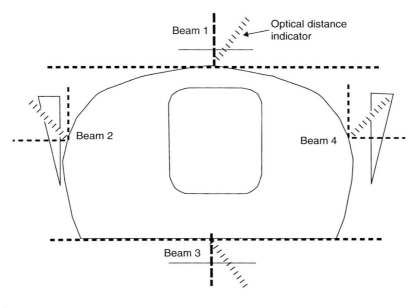

(a)

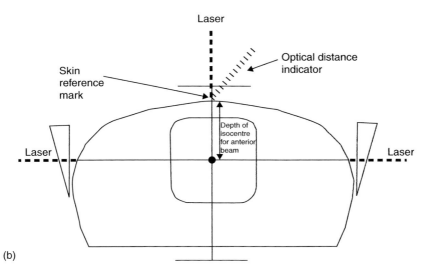

(b)

FIGURE 33.10
A comparison between a four-beam fixed SSD set-up and a four-beam isocentric set-up. For the fixed SSD set-up (a) the optical distance indicator (ODI) must be used for each of the four beams; for any of them and for the frequent situation where the SSD is taken equal to the SAD, the projection of the central axis on the skin coincides with the intersections or the projections of the wall mounted lasers (represented by thick dashed lines). For the isocentric set-up (b), the ODI could be used for the anterior beam to adjust or check the table vertical position; it could also be used to check the isocentre depth for any beam direction, but the set-up relies mostly on the laser projections and the table is not moved when rotating the gantry.

33.8 EXTENDED SSD

It may not be possible to cover larger PTVs with a field at the standard SSD, and an extended SSD is therefore required to increase the available field size on a particular treatment machine. Examples of long volumes include the spine in treatments of the whole central nervous system, and sarcomas, some of which extend from the pelvis to the knee and may be in excess of 50 cm long. Increasing the SSD increases the field size geometrically, with a dose output that falls by inverse square dependence. For example, a 40 cm long field will become 52 cm long at 130 cm SSD. Appropriate corrections should be made for the change in percentage depth dose, as described in Section 23.3.4.3. The option to treat at extended SSD depends upon the flexibility of the treatment unit itself, in that a large range of couch movements are required. Raising and lowering the couch will facilitate increased field sizes for vertical beams, but, for horizontal beams, the range of lateral couch movements available will usually be more restricted. As patients are generally treated while lying either supine or prone (to obtain a reproducible set-up), the possibilities for using extended SSD fields will be greater for antero-posterior beams.

Extended SSD fields can be set at oblique gantry angles, which by combining both couch lateral and vertical movements, provides the greatest SSD available for a given treatment machine. When oblique fields at extended SSD are employed, the machine isocentre is located considerably distant from the patient surface, and detailed set-up instructions are required to move the couch both vertically and laterally from a skin reference point. Figure 33.11 shows a beam configuration to treat a sarcoma in the thigh. The extent of the lesion in the transverse plane shown means that only a small volume of the normal tissue of the leg is spared from irradiation. The preservation of this small volume is essential in preventing amputation of the lower limb, so the beams are angled to achieve a non-diverging edge of the beam and therefore a sharp dose gradient from the PTV to the normal tissue. However, this means that the accuracy of the set-up is critical and demonstrates

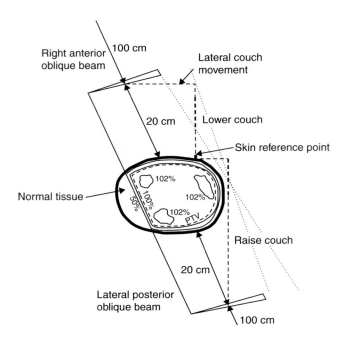

FIGURE 33.11
An extended SSD treatment of the thigh, showing the couch lateral and vertical movements required to set up each field individually from a skin reference point.

the importance of accurate and reproducible lateral and vertical couch movements and display scales over a range perhaps exceeding 80 cm.

33.9 MATCHING FIELDS

The need to match adjacent fields can arise when treating a volume greater in length than can be covered by a single field even at extended SSD. Another common need for beam matching is during treatment of an area abutting a previously treated volume. Perfect matching at all depths is usually difficult to achieve and it is important to establish whether overdosing or underdosing normal tissue is the deciding clinical factor when determining gaps between fields.

33.9.1 ABUTTING FIELDS WITH PARALLEL AXES

Where symmetric fields with parallel central axes (C/A) are to be abutted, it is necessary to calculate the separation required between the field edges at the skin surface to prevent an overdose or an underdose of the underlying tissue. Due to the divergence of the beam, exact matching can only occur at one depth (see Figure 33.12). Therefore when calculating gaps, a depth must be specified at which the fields will be matched. Frequently, the fields are matched at the midplane depth. Alternatively, matching may be to a critical structure (e.g. the spinal cord) and the gap calculated to ensure that no part of the structure receives more than 50% of dose from each field. The distance between the fields may be defined as a gap at the skin surface between the field edges, or by the distance between the central axes of the beams (Figure 33.12). The depth at which the field is defined must be known, i.e. how deep the isocentre is for each beam. Calculation of the gap is based on the use of similar triangles using the geometrical divergence of the field edge.

If a parallel opposed pair is to be matched to an adjacent opposed pair, all beam axes being parallel (e.g. at the junction of a supra- and infra-diaphragmatic treatment of lymph nodes) a compensation occurs between antero-posterior and postero-anterior divergences resulting in a fairly uniform depth dose, provided that the matching is at midplane depth and that the field sizes are not too different. When small fields are to be matched, it is possible to achieve almost perfect matching using asymmetric collimators to eliminate beam divergence at the junction and to enable the two fields to provide a uniform dose at all depths (Figure 33.12b)*. It may also be possible to match diverging field edges with asymmetric fields, as shown in Figure 33.12c, but this could be restricted by the limitations of the collimator travel and the maximum field size available on the treatment unit. The risk with matching fields is that a small error in the set-up can result in a large overlap volume in which dose is contributed from both fields. Another drawback in matching fields asymmetrically with no divergence at the junction is that the field edge not being abutted will diverge more and consequently irradiate more normal tissue. This effect is illustrated in Figure 33.11, in which the non-diverging field edge in this case is producing a sharp dose fall-off into the normal tissue region. However, in this treatment, the excess divergence arises in air, so there is no penalty. A non-diverging field edge is often used for breast treatments as shown in Figure 33.3. It is also used for matching neck irradiation with a single vertical beam to head irradiation with parallel opposed horizontal beams.

It should be understood that matching is usually to 50% of the central axis dose at the chosen depth. This is not the same as the 50% isodose line, as can be clearly seen in

---

* On Figure 33.12b, the lines indicated as central axes (C/A) are different from the collimator rotation axis. They represent a virtual axis defined from the centre of the asymmetric fields. Similar edge matching would be obtained with symmetric fields by rotating the gantry, but this would be restricted to the transverse plane.

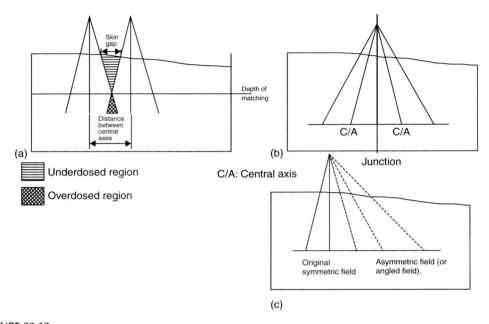

FIGURE 33.12
Matching fields, (a) at a specified depth, using skin gap or C/A separation, (b) using asymmetric collimators to eliminate divergence at the junction, or (c) using asymmetric collimators to match the divergence at the junction.

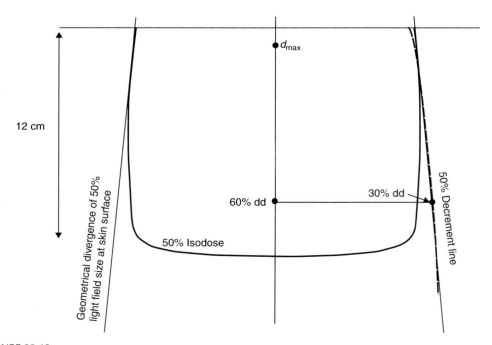

FIGURE 33.13
^{60}Co beam (80 cm SSD) showing 50% isodose and 50% decrement lines. The depth dose values (dd) are normalised at d_{max} on the beam axis.

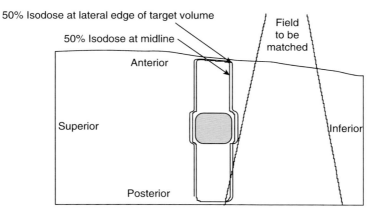

FIGURE 33.14
Matching a single field to a previously treated isocentric plan.

Figure 33.13, in which the 50% isodose of an 80 cm SSD Cobalt beam is shown. The line representing 50% of the central axis dose is called 50% decrement line[*]. The 50% decrement line follows the divergence of the beam almost exactly for all megavoltage beams.

33.9.2 MATCHING SINGLE FIELDS TO PARALLEL OPPOSED PAIRS

It is not possible to provide a satisfactory 50% match between a parallel opposed pair and a single abutting field due to the contradictory divergences of the opposed beams. The clinician must decide where the tissue may be overdosed or must be under-dosed before a gap can be determined. Figure 33.14 shows a field abutting the 50% level at the tumour but overdosing at the posterior field entry point.

33.9.3 MATCHING ADJACENT BEAMS IN INCLINED PLANES

When an inclined PTV is to be treated together with adjacent nodal regions, as in breast and head and neck treatments, it is important not to have a gap or an overlap between the field edges even though different field arrangements are used for the two volumes. If asymmetric jaws are available, it is often possible to match adjacent beams by using a single isocentre, but this reduces the field length available or may place the isocentre where it is difficult to set up. In these cases, it is possible to abut the field edges with the use of appropriate angulation (Siddon 1981; Casebow 1984). This methodology is widely used for matching tangential breast fields to supraclavicular and axilla nodal fields.

The fields to be matched are illustrated in Figure 33.15, in which the beam axes of the medial and lateral tangential fields are oblique in the cranio-caudal direction, rather than lying in a transverse section. The beam divergence, ε, is given by:

$$\varepsilon = \arctan\left(\frac{L}{2\mathrm{SAD}}\right)$$

---

[*] It used to be the practice to define the field size as the width of the 80% isodose line, and this is still sometimes done for cobalt units. In that case, geometric beam matching must take into consideration the distance between the optical field edge and the 50% point.

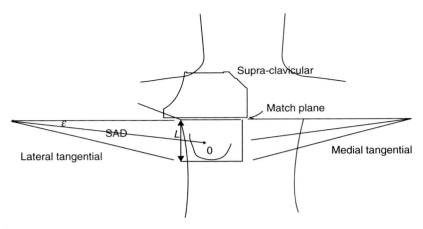

FIGURE 33.15
Diagram of a typical arrangement for a breast treatment where the tangential beams are angled superiorly and rotated so that their superior edges are parallel to each other and align with the inferior edge of the supraclavicular field. The distance from the target to the isocentre, SAD, is usually 100 cm.

where L is the beam length, and SAD is the source axis distance. ε is taken as positive for the edge of the beam nearest the gantry (i.e. it is positive for the superior edge of the beam and negative for the inferior edge). It is important that the tangential fields do not diverge into the supraclavicular field and that their posterior borders are aligned and do not diverge into the lung. To achieve this, the patient's back may be raised on a *breast board* so as to make the sternum horizontal and the plane of the tangential fields vertical. However, most CT scanners cannot accommodate a breast board and an alternative arrangement is that the patient lies flat and the tangential fields are in an inclined plane to which the supraclavicular field is matched. The following equations derived by Casebow (1984) (with the nomenclature adapted to be the same as in Section 33.6) ensure that the superior borders of the tangential fields are non-diverging and lie in a plane inclined at γ to the vertical, to which the inferior border of the supraclavicular field is matched.

$$GA = \arccos(\cos\theta\cos\gamma\cos\varepsilon - \sin\gamma\sin\varepsilon) \qquad (33.4)$$

$$CA = -\arcsin\left(\frac{\sin\varepsilon\cos GA + \sin\gamma}{\cos\varepsilon\sin GA}\right) \qquad (33.5)$$

$$TA = -\arcsin\left(\frac{\sin\gamma\cos GA + \sin\varepsilon}{\cos\gamma\sin GA}\right) \qquad (33.6)$$

Using the IEC1217 convention, 360° is added to negative gantry angles. The angle CA is added to the unrotated value of the collimator angle, which is defined by the wedge direction for the tangential fields. If the patient is on a breast board such that the match plane is vertical, γ becomes zero and the equations are simplified.

33.9.4 MATCHING ELECTRON TO PHOTON FIELDS

The shape of the electron beam isodoses makes exact matching impossible, although the 50% decrement line (i.e. the field size defined) will not be as bowed as the lower value decrement lines. A small gap of typically 2 mm offers the best compromise for most clinical electron energies, but depending on whether overdosing or under-dosing is the main clinical concern, no gap or up to a 5 mm gap at the skin surface may be employed. As with photon beam matching, concerns arising over the beam junctions may be alleviated by considering that modest doses may be involved, for example when a 20 Gy neck node electron field contributes to a total dose of 60 Gy. Matching of electron fields is discussed in Section 34.5.

CHAPTER 34

ELECTRON-BEAM TREATMENT PLANNING TECHNIQUES

Alan McKenzie and David Thwaites

CONTENTS

34.1 INTRODUCTION

The feature that makes electron beams very useful for a range of clinical applications is the shape of the depth dose curve. This allows for relatively uniform dose delivery to reasonably well-defined regions from the surface to the therapeutic range, with sparing of underlying tissue due to the rapid reduction of the dose towards the practical range. At about 20 MeV and above, the gradient of the depth dose curve begins to lose the characteristic steepness seen at lower energies and the separation of the high-dose and spared regions becomes increasingly blurred. Energies higher than this can be used for deep-seated tumours, but the techniques begin to approach those used for megavoltage photons. However, the majority of linear accelerators that provide clinical electron beams produce energies within the range 4 MeV to 20 MeV; this section concentrates primarily on their use. Electron beams of these energies deliver high doses (greater than about 90% depth dose) to depths of around 1cm to 6 cm. Common applications include skin and lip cancers, chest wall and peripheral lymphatic areas in breast cancer, additional boost doses to limited volumes such as scar areas and nodes, various head and neck cancers and other sites lying within these depths of the surface. The most commonly used beams are those of medium energy.

The basic physics of electron beams used in clinical practice has been described in Chapter 24. This section discusses how these principles may be applied to techniques and problem-solving in electron treatment planning. In the following discussion, the term *treatment planning* is taken to include treatment using single fields; indeed, such treatments constitute the bulk of electron therapy in clinical practice.

When presented with an electron treatment planning problem, it is helpful to use a checklist as an aid, and the following list contains items which should be considered.

34.2 CHECKLIST FOR ELECTRON TREATMENT PLANNING

Special Techniques. Does the problem warrant special techniques such as electron arcing?

Treatment Source Skin Distance (SSD). Does any part of the patient's anatomy, or do the available applicator sizes, prevent the treatment being given at the standard treatment distance (nominal SSD) on the central axis? This nominal SSD is generally 100 cm, which implies a gap of 50 mm between the applicator and the skin surface for many models of linear accelerator[*]. However, with Elekta linacs, it is a common practice to treat in skin apposition at 95 cm SSD.

Obliquity. If electrons are to enter any part of the patient with significant obliquity, then additional considerations apply, as described in Section 34.4.4.

Surface Irregularities. If surface irregularities are present in the area to be treated, side-scatter bolus may be required. Alternatively, bolus may be required to fill the irregularities and present a flat surface perpendicular to the beam.

Choice of Energy. The beam energy is determined, along with the amount of surface bolus, by considering both the required surface dose and the required depth of the treatment isodose beneath the surface.

Field Size. The decrease in dose near the field edges must be taken into account.

Calculation of Monitor Units (MU). The isodose to which the treatment is prescribed must be carefully specified (it should be 100% of the maximum dose) and the monitor unit calculation must take into account any field shaping or modification to the applicator.

Vulnerable Organs. Organs-at-risk may need to be shielded during treatment.

---

[*] This SSD is a nominal value, generally equal to the Source Axis Distance where the wall mounted lasers intersect, dictated by the geometry of the machine. The effective SSD will be variable with energy and field size as discussed in Section 24.2.6.

34.3 FIELD SIZE AND COVERAGE OF THE TARGET VOLUME

Electron applicator sizes, or insert (cut-out) sizes if the applicator carries an insert with an aperture cut in it, are generally quoted in terms of the geometrical size found by measuring the extent of the light beam projected through the applicator (or insert) onto the skin at the normal treatment distance. This will also be, to a close approximation, the size of the field defined by the 50% isodose at the skin surface.

The International Commission on Radiation Units and Measurements (ICRU 1993) recommends that planning target volumes should be covered by the 95% isodose or better, but to insist on this same criterion in electron therapy will often mean unacceptably large margins around the planning target volume. This difficulty is recognised and addressed by ICRU (2004). For electron therapy, the International Atomic Energy Agency (IAEA 1987) recommends an 85% coverage, and the American Association of Physicists in Medicine (AAPM 1991) recommends 90% coverage. Pragmatically, in electron planning, it is common practice to aim to cover the planning target volume with the 90% isodose. In some instances lower-value isodoses may be chosen, as for example, in chest wall irradiation, in which the 80% isodose may be chosen to lie at the lung surface in order to reduce penetration into the lung and hence overall lung dose. If any critical structure, such as cord, lies beneath the target volume, then the relative doses, both at the inner limit of the target and at this structure, must be carefully considered to make an appropriate clinical choice of the optimum beam energy.

As illustrated by Figure 34.1, the width of the *flat floor* of the 90% isodose surface is considerably less than the 50% field width at the tissue surface. The 90% isodose surface generally begins to flatten out at a distance of about $2E$ mm inside the 50% isodose (see Figure 34.1), where E is in MeV. Hence, to guarantee that a target volume is treated uniformly to the same 90% depth as the central axis, one would need to use a field size about $4E$ mm larger than the width of the target volume. In clinical practice, this is too large a margin, and, instead, the addition of 10 mm around the target volume is commonly used, which allows for some rounding of the bottom edges of the target volume. However, the significance of the $4E$ mm

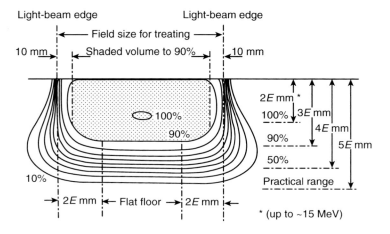

FIGURE 34.1
The flat floor of the 90% isodose curve is smaller than the field width by a margin of $4E$ mm, where E is the incident energy in MeV. In practice, a margin of 10 mm is commonly used, which will treat a target volume with rounded edges (shown shaded). Approximate depths of key percentage doses are also illustrated. Isodose values are shown in intervals of 10%.

dimension is that, for applicators smaller than $4E$ mm, there can be no flat floor to the 90% isodose surface. This means that the depth of the 90% isodose for applicators less than $4E$ mm wide is less than that achieved with a larger applicator, and great care should be exercised when planning with these relatively small applicators, or where any dimensions of a shaped field approach this critical value. Figure 34.1 also illustrates the 2, 3, 4, 5 rule as discussed in Section 24.2.2 to estimate the depths of various isodose parameters from the beam energy.

Rules should not be applied blindly, and it is always incumbent upon the treatment planner to take into account the clinical requirements for target coverage. Attention must be given to the dose outside the geometric beam edge, particularly when the beam is to be placed close to a critical structure (organ at risk). In this situation, some compromise on field size and positioning may be necessary.

34.4 EXAMPLES OF ELECTRON PLANNING

34.4.1 EXAMPLE A: SINGLE-ELECTRON FIELD WITH NO COMPLICATIONS

As an example of the application of the checklist to a relatively simple plan, consider the treatment of a parotid tumour (Figure 34.2) which is to be treated with 20 fractions of 2.75 Gy

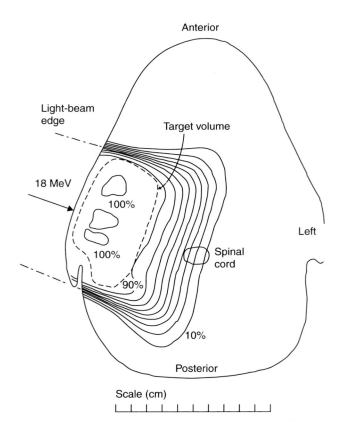

FIGURE 34.2
In this example of treating the parotid, a 90 mm wide field is sufficient to treat a target volume measuring 80 mm across. Isodose values are normalised to the maximum dose on the beam axis (at depth d_{max}) and shown in intervals of 10%.

in four weeks. The target volume measures 80 mm in the anterior–posterior direction, 80 mm in the superior–inferior direction and extends 49 mm beneath the skin surface.

Considering the checklist, no issues are raised by the need for special techniques, treatment SSD or obliquity. Other items on the checklist include:

Surface Irregularities. Because it is not the primary intention to treat the ear, it is unnecessary to provide side-scatter bolus, and the ear is merely taped as flat as possible. However, a wax ear-plug is required to prevent electrons tunnelling down the external auditory meatus.

Choice of Energy. The depth of the planning target volume is 49 mm, which will be reached by an 18 MeV beam, for which the 90% isodose is at 55 mm deep with a 200 mm × 200 mm applicator (see Table 34.1) and 53 mm for a 100 mm × 100 mm applicator. The surface dose of an 18 MeV beam is 88%, from Table 34.1. However, it is not the intention to treat to the skin surface, and the target volume does not come closer to the skin surface than 4 mm, which, from Table 34.1, is more than the 1 mm required to raise the dose to 90%.

Field Size. Using the criterion of a 10 mm margin around the planning target volume, the field size should be 100 mm × 100 mm. However, because the target volume has been drawn with rounded corners at depth, it may be treated with a 90 mm × 90 mm field in this case.

Calculation of Monitor Units. The dose prescription point is on the central axis of the treatment beam, at the depth of d_{max}, i.e. 22 mm for this linear accelerator. This means that a dose of 2.75 Gy will be given to 100% on the central axis, 22 mm deep.

If a 90 mm × 90 mm insert is not available, one may be produced by attaching lead strips onto the edges of a 100 mm × 100 mm insert to reduce the field size appropriately. The output (in Gy per monitor unit) of the newly fashioned insert should be measured, although experience shows that where the field edges are brought in only minimally, the measured output may not show significant change from that for the unmodified applicator.

Vulnerable Organs. The ear plug mentioned above is used to protect the external auditory meatus. The dose to the spinal cord is clinically acceptable.

34.4.2 EXAMPLE B: USE OF INTERNAL SHIELDING

A basal cell carcinoma on the cheek is to be treated with electrons using a circular applicator 40 mm in diameter. The full thickness of the cheek is to be treated; it measures 20 mm from the surface of the lesion to the inner surface of the cheek (Figure 34.3).

Surface Irregularities. The lesion itself is raised from the normal skin surrounding it. Wax bolus will be applied, which will eliminate the effects of the irregular surface (Figure 34.3).

Choice of Energy. From Table 34.1, it will be seen that, at normal incidence, the 90% depth dose of a 9 MeV beam is 27 mm. The surface of the lesion must be covered with 5 mm bolus (using Table 34.1) in order to raise the dose there to 90%. With this arrangement, 9 MeV is adequate for the purpose.

Field Size. Using the 4E rule, the minimum field size without affecting the central axis depth doses is 4 × 9 = 36 mm. Hence, the 4 cm diameter applicator should be appropriate, provided that the radiation oncologist accepts the rounded edges of the treatment volume at depth.

Vulnerable Organs. It is decided to protect the gums with a lead shield. However, because of backscatter from the lead, which would raise the dose on the inner mucosal surfaces of the

TABLE 34.1

Parameters for Electron Planning

Incident Beam Energy	6 MeV	9 MeV	12 MeV	15 MeV	18 MeV	22 MeV
Surface dose, $D_{0,s}$, at normal incidence relative to dose at d_{max} at normal incidence, 100 cm SSD	75%	86%	83%	87%	88%	88%
Surface dose, $D_{45,s}$, at 45° incidence, 105 cm SSD, relative to dose at d_{max} at normal incidence, 100 cm SSD	83%	85%	87%	89%	90%	90%
Surface build-up (i.e. bolus thickness) at normal incidence (in mm), 100 cm SSD, required to raise skin surface to 90% of dose at d_{max} at normal incidence	6 mm	5 mm	4 mm	1 mm	1 mm	1 mm
Surface build-up at 45° incidence, 105 cm SSD, required to raise skin surface to 90% of dose at d_{max} at normal incidence, 100 cm SSD	2 mm	2 mm	1 mm	1 mm	0 mm	0 mm
Depth, $d_{0,90\%}$, of dose at normal incidence, 100 cm SSD, which is 90% of dose at d_{max} at normal incidence	17 mm	27 mm	38 mm	45 mm	55 mm	63 mm
Depth, $d_{45,90\%}$, of dose at 45° incidence, 105 cm SSD, which is 90% of dose at d_{max} at normal incidence, 100 cm SSD	7 mm	11 mm	17 mm	23 mm	26 mm	30 mm
Depth of dose beneath skin surface, covered with build-up as in row 4, at 45° incidence, 105 cm SSD, which is 90% of dose at d_{max} at normal incidence, 100 cm SSD	5 mm	9 mm	16 mm	22 mm	26 mm	30 mm

For each energy shown, these data relate the skin surface dose and depth of the 90% isodose (at 45° incidence and 105 cm SSD), to the maximum dose (i.e. at d_{max}) measured at normal incidence and 100 cm SSD using a 200 mm × 200 mm applicator (see Figure 34.6). The bolus thicknesses required to bring the surface doses up to 90% have been included. While the parameters for individual machines and different angles of obliquity may vary from those listed above, the table serves as a first approach to designing a treatment plan.

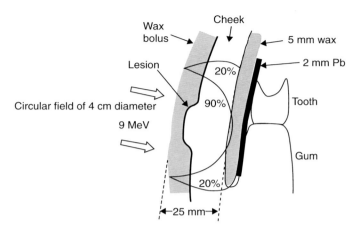

FIGURE 34.3
Illustration of the irradiation situation for the use of an internal shield behind the cheek as described in the example.

cheek, a layer of wax is needed between the lead and the cheek. Saunders and Peters (1974) have suggested that a lead shield covered with 5 mm wax to *dampen* the backscattering is sufficient for electrons with residual energy of 3.5 MeV or less at the lead interface. In practice, this arrangement is suitable for most applications, bearing in mind that internal lead shields will not generally be used for beams of incident energy much above 9 MeV.

A calculation for the combination of lead with 5 mm of wax is instructive. With the 5 mm wax protection, plus the 20 mm thickness of cheek and the 5 mm bolus on top, the lead is at a depth of 30 mm. For this rough calculation, it is sufficiently accurate to suppose a linear decrease of energy from the surface downwards. Thus, the energy of the electron beam when it hits the lead shield is $9(1-30/45)$ where the range of the electron beam is estimated to be 45 mm from the relation: practical range (mm) $= 5 \times$ incident energy (MeV) (see Figure 34.1). Hence, the electron energy at the lead shield is about 3 MeV.

Using the relation given by Klevenhagen et al. (1982), the electron backscatter factor (EBF) is

$$\text{EBF}_{(\text{at shield})} = 1 + 0.735 e^{-0.052E_s} = 1 + 0.735 e^{-0.052 \times 3} = 1.63 \qquad (34.1)$$

where E_s is the energy in MeV of the electrons at the shield. In other words, the backscatter fluence is 63% of the incident fluence at the lead surface. For energies at the shield of between 1 MeV and 9 MeV, Lambert and Klevenhagen (1982) give an exponential expression for the electron backscatter *upstream* from the shield

$$\text{EBF}_{(\text{at } t \text{ mm upstream})} = e^{(-kt)}$$

where k is given by the relation

$$k = 0.61 E_s^{-0.62}$$

Hence, in this case, $k = 0.31$ and the backscatter 5 mm upstream from the shield is $e^{-0.31 \times 5} = 0.21$, that is, 21% of the backscatter found at the shield itself. Hence, the backscatter 5 mm upstream is, finally, 21% of 63%, or 13% of the incident fluence at the shield. Since the incident fluence at the shield (at a depth of 30 mm) is 80% (from depth–dose curves), the mucosal surface of the cheek will experience $13\% \times 80\% = 10\%$ backscatter. However, the dose from the direct incident beam (at the depth of 25 mm) is 95% (from depth–dose tables), and so the total dose to the mucosa, with the lead shield plus 5 mm wax in place, is 105%, which is quite acceptable.

In general, incident beams will be of this energy or lower when shields need to be used, so that the 5 mm wax protection over the lead shield will usually be adequate. A 2 mm thickness of lead, by the same token, will also normally provide adequate shielding (in this case, it is perfectly able to absorb the residual 3 MeV energy).

34.4.3 EXAMPLE C: SINGLE-ELECTRON FIELD WITH SURFACE INHOMOGENEITIES

Figure 34.4 illustrates the treatment of the pinna of the ear. Referring to the check-list, there is no need for a special technique and there are no problems with SSD or obliquity. Other items on the list include:

Surface Irregularities. To avoid the problems of inhomogeneous dose distribution that would arise from irradiating the protruding pinna, a well-fitting wax block is fashioned to enclose the pinna on both the anterior and posterior surfaces. In practice, this may be arranged by producing the wax block in two halves so that the pinna is sandwiched between them. Vaseline helps to eliminate any remaining air gaps, but some centres prefer not to use it because it is messy.

Choice of Energy. The pinna extends 20 mm from the tip to the root. From Table 34.1, the minimum appropriate energy is 9 MeV, which has a 90% depth of 27 mm, and so the wax block must extend 7 mm beyond the tip of the pinna. This is also greater than the 5 mm required, from Table 34.1, to build up the dose to 90% on the tip of the pinna.

Field Size. The anterior–posterior dimension of the pinna to be treated is of the order of only 10 mm. However, from the $4E$ rule, a flat 90% isodose is only achieved in this case with surface field sizes of 4×9 mm $+ 10$ mm $= 46$ mm. In practice, a 60 mm $\times$ 60 mm applicator is used, and the pre- and post-auricular areas are shielded by lead. Since the lead is placed at the depth of the 90% isodose, the small aperture in the lead does not affect the 90% isodose itself, although it will constrain the lower-value isodoses of electrons transmitted through the aperture to take an almost semi-circular shape as shown in Figure 34.4.

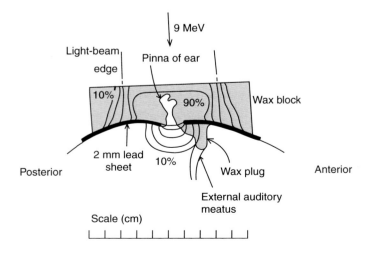

FIGURE 34.4
With the pinna in this position, the necessary side scatter is provided by a well-fitted wax plug. Lead sheet protects the surrounding skin, but a plug in the external auditory meatus is still necessary to prevent the possible electron tunnelling. Shown isodose values (both above and beneath the lead shielding) are 10%, 20%, 50%, 80%, and 90%.

Vulnerable Organs. Although lead has been used to shield tissue from the incident electron beam, it will be noticed in Figure 34.4 that electrons can leak through the aperture under the lead. It is therefore sensible to place a plug in the external auditory meatus to prevent possible tunnelling. The plug may be made from wax, or it could be produced by soaking cotton wool in olive oil. A thickness of 2 mm of lead will afford sufficient protection in this instance, because the overlying thickness of the wax block will have reduced the energy of the electrons to less than half that of the incident 9 MeV beam. In general, the thickness (in millimetres) of lead shielding required is given by half of the energy (in MeV) of the electrons incident upon the shielding.

34.4.4 EXAMPLE D: SINGLE-ELECTRON FIELD WITH OBLIQUITY

Consider now the treatment of the chest-wall after mastectomy. In Figure 34.5, the posterior surface of the planning target volume has not been extended as far as the surface of the lung in order to minimise the lung dose as discussed below. The lung extends relatively close to the skin surface, and the target volume extends considerably to the posterior of the patient, so that the technique of treating the chest wall with an opposed pair of tangential photon fields would mean that an unacceptably large portion of lung would be treated to a high dose, with the risk of radiation pneumonitis.

As before, we consider the points in the checklist in turn, and use them to illustrate general principles where appropriate. There are no surface irregularities.

Special Techniques. Ideally, this would be treated using electron arc therapy, although the varying radius of curvature of the breast does require complex planning (see Section 34.6). However, for reasons discussed later, this technique is not available in many centres, and so the simplest approach—a static beam—is considered in this case.

Treatment SSD. The applicator may be used at the standard distance of 50 mm on the central axis of the field. However, the curvature of the area means that the skin at the periphery of the field may lie considerably further from the applicator than the standard distance (see next point). In practice, it is the bolus applied over the skin surface that is placed at the standard distance of 50 mm.

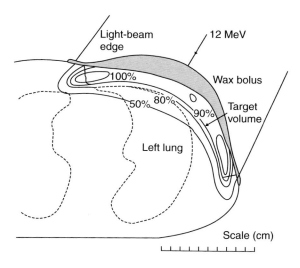

FIGURE 34.5

To treat a curved chest wall with a single-electron field, it is generally necessary to provide a bolus that is thick near the central axis of the field and tapers to the minimum required for surface build-up at the field edges. Shown isodose values, normalised to the maximum dose on the beam axis, are 50%, 80%, 90% and 100%. The maximum dose within the two *islands* defined by 100% isodoses at either edge of the field is 105%.

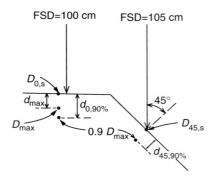

FIGURE 34.6
This illustrates the geometry used to construct Table 34.1. Measurements were made with a Varian Clinac 2500 linear accelerator using a 200 mm×200 mm applicator, and the data in Table 34.1 must be regarded only as a guide when applied to other linear accelerators.

Obliquity. The curvature of the area means that electrons at the edges of the field will enter with considerable obliquity, that is, at an angle that is significantly removed from the local perpendicular. Because the problems of obliquity become greater as the angle of obliquity increases (McKenzie 1979), it is nearly always best to arrange the angle of incidence of the central axis so that the angles of obliquity at the two extreme edges of the field, as shown in the transverse plane, are about equal, provided that this does not conflict with other clinical considerations. In this case, a gantry angle of 30° is appropriate, and the beam may be said to be *in apposition* to the skin.

Table 34.1 gives some representative data which are useful in a manual first pass approach to treatment planning with oblique fields. Figure 34.6 shows the geometry for those data.

Choice of Energy. In oblique incidence, it is generally the depth of the planning target volume beneath the skin at the edges of the field (16 mm in this case at the lateral edge of the field) that determines the required incident energy. The penetration of a 45° oblique beam at 12 MeV (bottom row of Table 34.1) is 16 mm from the skin to the 90% depth dose, and so this energy is used, together with 1 mm build up on the skin surface to raise the skin dose from 87% to 90% (again, using Table 34.1). At the central axis, the thickness of bolus needed is determined from Table 34.1 by subtracting the depth of the target volume on the central axis (16 mm) from the 90% penetration depth at normal incidence (38 mm from Table 34.1).

Using a treatment planning computer, and with these constraints on the bolus, it becomes a relatively easy matter to design the exact shape of the bolus required to shape the isodoses to conform more closely to the target volume, using perhaps one or two trial shapes. The shape of bolus in Figure 34.5 gives rise to the computed isodose distribution shown, which leaves an acceptably low incursion of high-value isodoses into the lung. Notice that the maximum dose at the field edges is 105%. If the increased distance from the source had not mitigated the effect of obliquity, this value would have been nearer 120%.

Field Size. At 107 cm SSD at the field edges, the geometric (50%) coverage afforded by a 200 mm×200 mm applicator is 214 mm. This is sufficient to provide the 200 mm coverage required between the 90% isodoses at the medial and lateral field edges.

Calculation of Monitor Units. It would normally be unnecessary to calculate the monitor units required to deliver the prescribed dose for this treatment because this should be computed automatically by the treatment planning system when producing the isodose distribution. Nevertheless, at 100 cm SSD to the surface of the bolus* on the central axis, the

---

* It may be easier to perform the SSD patient setup without bolus and to add the bolus afterwards. The monitor unit setting would then be slightly lower to compensate for the inverse square variation according to the effective source-bolus distance (typically -1% for a 5mm thick bolus at 100 cm SSD).

monitor unit setting will be given simply by that determined locally for the 200 mm × 200 mm applicator at the standard distance of 100 cm SSD. This setting may not agree exactly with that produced by the computer (because of the influence of the curved surface) but it will be acceptably close.

Vulnerable Organs. The lung inhomogeneities do not significantly influence the distribution of the isodoses that lie above them, but the low density of lung nevertheless plays a part in the design of the treatment plan. In this plan, a margin has been left between the target volume and the lung surface so that the 70% or 80% isodose is coincident with the lung surface rather than the 90%. If the 90% isodose were to lie only a few millimetres closer to the surface of the lung, this would push the 80% isodose considerably further into the lung volume, leaving little room for error and an increased risk of radiation pneumonitis.

34.5 FIELD MATCHING

Occasionally, situations arise in which adjacent fields must be matched in order to avoid an over- or under-dose. Examples of situations in which such beam matching may be required include:

- Providing varying penetration by using different energy electron beams for adjoining areas
- Treating a larger area than standard applicators allow
- Treating an area adjacent to a previously-treated area
- Reducing the extreme effects of obliquity when treating a curved surface

In beam matching, the join should be positioned away from critical areas. Staggered joins (i.e. shifting joins at different treatment fractions) may be considered if this is practical. One method of improving the matching is to utilise the broadened penumbra regions produced by extended SSD. This necessitates on-surface shaping of the other field edges to restore a sharp penumbra, and may well lead to a loss of beam flatness. Several solutions to the problem have emerged over the years, for example, Kalend et al. (1985), Kurup et al. (1992, 1993), and Feygelman (1994).

Ulnin and Palisca (1996) and Lachance et al. (1997) proposed techniques that broaden the penumbra on the edge to be matched. However, the applicator design modifications depend upon the applicator cones, and the overlap or gap that must be left between matching fields is again dependent upon the detail of the design. Furthermore, measurements are required for a range of applicator sizes and energies if computer treatment planning is to be used.

A simple alternative solution is to add a slab of material such as polymethylmethacrylate (PMMA) to cover completely the end of the electron applicator (McKenzie 1998). Such a slab is called a *spoiler* because it spoils the originally narrow penumbra by broadening it. With a spoiler:

- There is no dependence of the properties of the spoiler design upon the applicator cone, provided that the spoiler is large enough to cover the aperture completely.
- No design-dependent overlap or gap is required between adjacent fields, as matching is always to the light beam edge regardless of applicator size or SSD.
- Computer treatment planning programs can be used without modification or extra beam measurement to calculate the effect of the spoiler.

Figure 34.7 illustrates two important characteristics of the dose distribution in tissue irradiated with an electron beam through a spoiler. These characteristics are general in that they are independent of SSD, applicator size, electron energy incident upon the spoiler and spoiler thickness (provided that the thickness is not so great that the electron beam becomes totally diffuse, which is avoided in practice). In summary, these characteristics are:

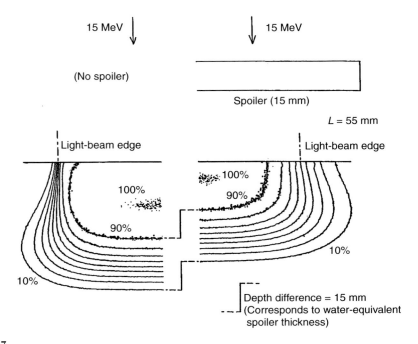

FIGURE 34.7
An illustration of characteristics of the isodose distribution under an electron beam spoiler. The light-beam edge marks the position of the 50% isodose at the surface, and depths of the flat portions of isodose surfaces are given by subtracting the effective thickness of the spoiler from the depths of corresponding isodoses in a beam with no spoiler. The dimension L is defined in Figure 34.8. Isodoses, normalised to the maximum dose on the beam axis, are shown at 10% intervals. (Note that the step indicated in the middle of the figure does not represent a continuous isodose line - the two halves of the figure are separate.) (From McKenzie, A. L., *Phys. Med. Biol.*, 43, 3456–3478, 1998. With permission.)

- The light-beam edge closely coincides with the 50% isodose at the tissue surface.
- The depths of the isodoses in the flat portion of the distribution in tissue irradiated through a spoiler may be found by subtracting the thickness of the spoiler from the depths of the corresponding isodoses in tissue irradiated without a spoiler.

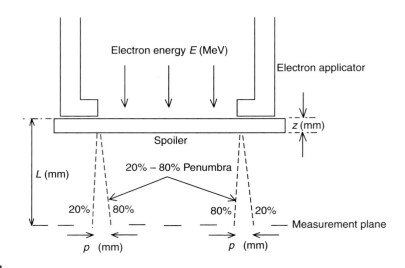

FIGURE 34.8
Illustration of the geometry used for beam-spoiler treatment in field matching and of the dimensions used in Equation 34.2. (From McKenzie, A. L., *Phys. Med. Biol.*, 43, 3456–3478, 1998.)

These simple features are particularly useful in planning electron therapy with spoilers. One direct consequence of the first characteristic is that electron beams should be matched by ensuring that the light-beam edges coincide at the tissue surface.

Another feature of using spoilers is that, in general, they increase the surface dose so that additional skin bolus may be unnecessary. A further advantage is that the number of monitor units required to deliver a given dose to the tissue is the same as that calculated for a treatment without a spoiler, except for a small inverse-square decrease to allow for the fact that the point of maximum dose within the tissue has moved closer to the surface by an amount equal to the thickness of the spoiler.

McKenzie (1998) has shown that the 20%–80% penumbra width, p, is given by

$$p = 1.2Lz^{1/2}E^{-1} \qquad (34.2)$$

where p is in mm, L is the distance in mm from the inner face of the spoiler (the one nearer the electron source) to the point where the penumbra is measured (see Figure 34.8), z is the thickness of the spoiler in mm and E is the energy (in MeV) of the electrons incident upon the front face of the spoiler. This dependence on L, z, and E is in accordance with theory predicting the broadening of pencil beams through thick absorbers. This equation can be used to indicate the tolerance of abutting fields to an inadvertent overlap or gap.

Figure 34.9 shows the isodose distribution at the junction of two 9 MeV beams transmitted through a 10 mm spoiler with a hinge angle of 0°. Although the two beams have been made to overlap by 5 mm, the measured hot spot is only 110%.

An advantage of using spoilers for matching is that the facilities required for treatment planning with spoiler beams, at least in two dimensions, are already available on planning computers that use the Fermi–Eyges theory (Hogstrom et al. 1981) or more recent variants (see Chapter 27). If the computer treatment planning program allows beam modifiers, then planning is straightforward with spoiler beams. If the beam modifier facility is not available,

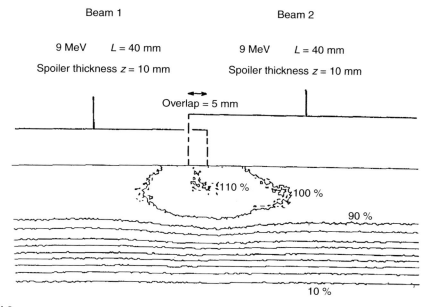

FIGURE 34.9
Even when two beams with a zero hinge angle are made to overlap (in this case by 5 mm), the overdosing is slight, because of the penumbra broadening introduced by the spoiler. Isodoses are normalised to the maximum dose on the beam axis and shown at 10% intervals. (From McKenzie, A. L., *Phys. Med. Biol.*, 43, 3456–3478, 1998. With permission.)

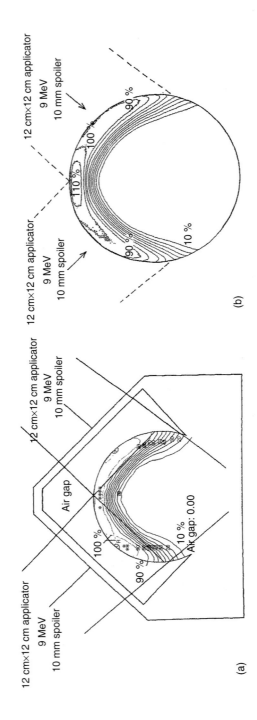

FIGURE 34.10

If the treatment planning computer does not have the facility to introduce beam modifiers, it is still possible to model the spoiler by the simple expedient of creating an air gap under a thickness of tissue equal to that of the beam spoiler. (a) This shows the distribution computed using the air-gap technique with a 160 mm diameter cylindrical phantom and the extremely large hinge angle of 90°. The maximum dose is 110%, just beneath the intersection of the two beam edges. (b) This shows the distribution obtained by measuring the arrangement in (a). The maximum dose (inside the 110% isodose curve) is 116%. Isodoses in both cases are normalised to the maximum dose on the beam axis and shown at 10% intervals. (From McKenzie, A. L., *Phys. Med. Biol.*, 43, 3456–3478, 1998. With permission.)

then it is nearly as simple to digitise an air gap under a thickness of tissue corresponding to that of the spoiler. This has been done in Figure 34.10a to predict the effect of two beams incident through 10 mm spoilers at the extreme hinge angle of 90°, treating a cylindrical phantom. The phantom diameter was 160 mm, taken to be typical of that of a scalp.

For comparison, Figure 34.10b shows the measured dose distribution for the arrangement in Figure 34.10a. Even at this extreme angle, it is noteworthy that the maximum dose measured in the phantom is only 116% (at a point inside the 110% isodose curve) and that the computer treatment plan is a relatively faithful representation of the measured dose.

If field matching is used in electron irradiation of the chest wall, the overdose at the matching point will generally be lower, partly because the hinge angles involved will be smaller, and partly because the larger radius of curvature of the chest wall means that matching points are further from the electron source, which reduces the overdose compared to the maximum dose on the central axes of the beams. However, in cases of field matching on curved surfaces, it is always important either to test the predicted doses in a phantom before treatment, or on the patient during treatment, because of the possibility that electrons from one field will be scattered well into the middle of the abutting field, contributing to the total dose.

When using spoilers, only the minimum thickness of spoiler necessary to produce an adequately wide penumbra should be used, and if that leads to too great a depth for the 90% dose, then skin bolus should be applied as well. A 10 mm spoiler works well with 9 MeV for chest wall irradiation. Since a spoiler broadens penumbrae at all edges of the field, consideration should be given to shielding tissue beyond field borders that are not involved in the matching.

34.6 ELECTRON ARC THERAPY

The problems of dose inhomogeneity associated with obliquity and beam matching in the electron treatment of curved surfaces are effectively eliminated with the use of narrow-beam electron arcing techniques. However, these techniques are not standard in every radiotherapy centre and, unlike other techniques that have been discussed earlier in this section, commissioning an electron arc therapy service requires considerably more preparation, starting from the technology, through data gathering to implementation of a system for dose computation. In view of these considerations, it is not practicable to provide enough information here to cover the topic completely, but only to highlight the key points to be considered in establishing electron arc therapy in a radiotherapy centre. Reviews of electron arc therapy can be found in Leavitt et al. (1985) and Leavitt (1996).

In electron arc therapy, it is important to try to minimise the variation in source-skin distance over the treatment arc, and this is achieved by placing the linear accelerator isocentre as close as possible to the centre of curvature of the arc of the patient's surface. The depth of the centre of curvature beneath the patient's surface should be at least R_p, the electron range, and preferably around twice this distance, in order to avoid possible overdosing from superposition of doses in the region of beam *crossfire*. In the treatment of surfaces with relatively large radii of curvature, such as the chest wall, it is clear that the treatment cannot be delivered through the electron applicator normally used to provide secondary collimation in stationary-beam therapy, because there would not be adequate clearance for the patient.

Secondary collimation[*] of the arc therapy beam therefore generally has to be positioned at a significant distance from the patient's skin, so that the beam profile in the plane of the arc extends well beyond the limits indicated by the light field (Leavitt et al. 1985; Lam et al. 1987).

---

[*] The term *primary beam collimation* is reserved here for that imposed upon the electron beam by the x-ray collimators.

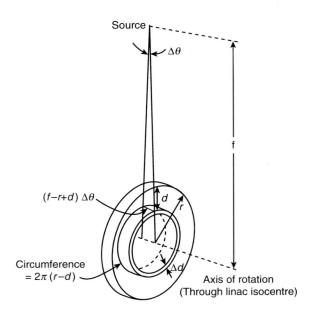

FIGURE 34.11
Schematic representation of a patient cross section (radius of curvature r) used to interpret its influence on the dose delivered during arc therapy. In a complete rotation, the dose delivered at depth d is proportional to $(r-d)^{-1}$ $(f-r+d)^{-1}$ (see text).

Because of this beam spread, the aim, at least for the middle sections of the arc, is for the complete beam profile to pass over any given point, so that the dose received by the point is integrated over the whole beam profile. This is more practicable with narrower beams, of around 5 cm width when projected to the isocentre, which will also minimise problems of obliquity at the edges of the beam profile.

To see the effect of the radius of surface curvature upon the dose delivered by arc therapy at a given depth, d, consider the irradiation of a cylindrical phantom of radius r. In particular, consider the disk-shaped volume (containing the central treatment plane) defined by a sector of the beam that extends an angle $\Delta\theta$ in the direction perpendicular to the plane of rotation (see Figure 34.11). At depth d, the electron energy delivered to an incremental depth Δd through the sector of angle $\Delta\theta$ during one complete rotation will be spread evenly within the volume $2\pi(r-d)(f-r+d)\Delta d\Delta\theta$. Because the dose will depend upon the reciprocal of the mass, and, hence, the volume throughout which the energy is distributed, the variation of dose $D(r,d)$ at depth d beneath a surface of radius r will be given by:

$$D(r,d) \propto (r-d)^{-1}(f-r+d)^{-1} \qquad (34.3)$$

It will be seen from this dependence of the dose upon the radius that, in the competing effects of the decrease in radius of curvature and the consequent increase in distance from the source, the former effect wins. In summary, a decrease in the radius of curvature leads to an increase in dose.

It may well be necessary to arrange for a treatment to be based on more than one centre of curvature, in order to accommodate the different skin curvatures in the central plane. In planes parallel to the central plane, lying superior or inferior to it, the skin curvature may again be different. The solution is, rather than use a rectangular treatment beam, to use a beam that varies in width at different distances superior or inferior to the central plane, depending upon how the skin curvature varies away from the central plane. The resultant beam shape may be fixed, and is generally trapezoidal in shape, or it may vary during treatment using multileaf collimators (Leavitt et al. 1989; Klein et al. 1996). Such field shaping with secondary collimation, whether static or dynamic, may also be used to neutralise the relatively poor beam profile in the superior–inferior direction (Leavitt et al. 1985).

The dose distribution at the ends of the arc is naturally compromised because of incomplete integration across the dose profile. The dose distribution may be sharpened by placing tertiary collimating lead or low-melting-point alloy shaped to the treatment field on the patient's skin and extending the treatment arc some 10° to 15° over the tertiary collimator. Alternatively, dynamic collimation may be used (Swalec et al. 1994), in which the jaws in the plane of the arc are progressively opened and closed at the beginning and end of the arc respectively, which mimics the effect of tertiary shielding.

Determination of the dose distribution in electron arc therapy treatment planning should be done using an appropriate pencil-beam or Monte-Carlo based computer algorithm. It is characteristic of electron arc therapy that the depth–dose curves tend to display a lower surface dose, a greater depth of maximum dose, d_{max}, and higher percentage depth–doses at given depths beyond d_{max} than they do for stationary electron beams of the same incident energy. This can largely be explained using the same reasoning underlying Equation 34.3: at greater treatment depths in arc therapy, the electrons are distributed over diminishing lengths of arc and so the fractional depth dose is greater than that for stationary beams.

The low surface doses in arc therapy can be an obstacle to achieving dose homogeneity, and Leavitt et al. (1990) have suggested combining electron energies, including using build-up on the lowest energy, to achieve a more uniform depth dose distribution.

In arc therapy, it is always important to remember that the bremsstrahlung contribution is additive around the isocentre (Kase and Bjärngard 1979). Clearly, as the angle of the treatment arc increases, so will the relative bremsstrahlung contribution, which may become a significant fraction of the treatment dose, particularly at higher incident electron energies.

34.7 INHOMOGENEITIES AND ELECTRON PLANNING ALGORITHMS

Inhomogeneities in composition and in shape modify dose distributions in two ways, both of which need to be considered in clinical situations. The first is the effect on absorption and the consequent shift in isodose lines, and is most pronounced through and beyond large inhomogeneities. The second is due to scatter differences between different materials (interface or edge effects) and is most pronounced for small inhomogeneities or near the edges of larger heterogeneities. The physics of the latter and some basic effects were discussed in Section 24.6. The resultant distribution depends on the size, shape and composition (atomic number and electron density) of the inhomogeneity, as well as on the energy and field size of the beam. In current practice they are best considered by using pencil beam algorithms, as available on most modern treatment planning systems. However, increasingly, Monte Carlo techniques are being applied to these problems. Simple manual methods can be used to provide initial rough estimates of the effects or approximate checks of more sophisticated approaches (e.g. see Klevenhagen 1985; Thwaites 2000).

Pencil beam algorithms are discussed and reviewed in Chapter 27 and Monte Carlo methods in Chapter 28. Other comprehensive reviews of the development of electron beam algorithms can be found in Jette (1996) and of Monte Carlo modelling of electron beams in Ma and Jiang (1999). A concise summary of electron beam algorithms is given in Thwaites (2000), along with a brief review of their accuracy (see also Chapter 28). The pencil-beam algorithms currently implemented have generally been shown to be accurate to within around $\pm 5\%$ or ± 5 mm over a range of applications, with the more recent algorithms showing improved performance. In simpler situations, accuracies of $\pm 2\%$ or ± 2 mm are achievable in accordance with the requirements laid down for photon algorithms (ICRU 1987). However, larger discrepancies on the order of 10% (and with the potential to be significantly larger) have been observed at larger depths for narrow inhomogeneities with long edges parallel to the beam. Pencil-beam algorithms are still currently the most widely implemented approach for electron treatment planning on commercial planning systems, although they are now beginning to be supplemented or replaced by Monte Carlo approaches (Kawrakow 2001; Cygler et al. 2004). There are a number of different pencil-beam algorithms and implementations in use and the user must ensure that the system is well tested locally and that any limitations are clearly understood. Commissioning recommendations and test methods are provided in Shiu et al. (1992), van Dyk et al. (1993), IPEMB (1994), AAPM (1998), IPEM (1999) and IAEA (2004).

34.8 TREATMENT PRESCRIPTION

Prescription depends on clinical requirements but is generally to the position of dose maximum, although other levels may be used. It should be noted that, in some circumstances, low-energy scatter components may give a dose maximum quite close to the surface, which may influence the position of the dose prescription. Electron prescription recording and reporting is dealt with in ICRU Report 29 (1978); this has been superseded by ICRU Report 71 (2004), in which newer recommendations for electron beams complement the recommendations included in ICRU 50 (1993) and ICRU 62 (1999). It is still recommended that the ICRU reference point used for reporting is the point of dose maximum on the beam axis. However, whereas for photon beams ICRU recommends that the dose variation within the Planning Target Volume should be constrained within +7% and −5% of the reference point dose, for electron beams −10% or −15% is acceptable as the 100% reference is by definition at the depth of dose maximum.

34.9 CURRENT AND FUTURE DEVELOPMENTS

Electron therapy is set to benefit from the same developments that are currently adding to the power of megavoltage photon therapy. Inverse treatment planning will increasingly be used to design solutions to difficult problems, and these solutions will be implemented in clinical practice using electron-beam shaping with multileaf collimators both in static and in dynamic mode, e.g. Klein (1998) and Korevaar et al. (1998). This same technology also offers the possibility of mixing the modalities of photon and electron beams, or different energy electron beams, to design individually tailored beam characteristics of depth dose, flatness, penumbra width and skin dose, e.g. Karlsson and Zackrisson (1997) and Korevaar et al. (1998).

Whilst pencil-beam based treatment planning algorithms will continue to develop, Monte Carlo programs that will handle complex dose distributions in acceptable times are now available commercially; these will be especially useful for dealing with the effects of inhomogeneities. Figure 34.4 was drawn freehand with an eye to the physics of backscattering from lead and diffusion through a narrow aperture. Monte Carlo programs will underpin the art of prediction in cases like this and will add significantly to the confidence placed in calculated distributions. This is turn will lead to an increase in the use of intensity-modulated electron beams and their combination with photon beams to tailor dose distributions in appropriate clinical situations.

CHAPTER 35

DOSE EVALUATION OF TREATMENT PLANS

Margaret Bidmead and Jean-Claude Rosenwald

CONTENTS

35.1 INTRODUCTION

The ideal radiotherapy treatment plan produces a uniform coverage of the target volume without giving a significant dose to surrounding normal tissue. The use of CT and MR in treatment planning, and fast computer hardware (for both planning and treatment delivery) has meant that 3D planning is a realistic option. More degrees of freedom in the use of noncoplanar beam arrangements and the introduction of intensity modulation both make plans more difficult to compare and evaluate. Good, fast plan evaluation tools are essential to simplify this process. Several such tools have been designed, and will be discussed in this chapter.

35.2 ISODOSE DISPLAY

Isodose information can be displayed on a 3D treatment plan as surfaces of equal dose, which correspond to the familiar isodose lines shown in a 2D display. Isodose surfaces can be labelled in terms of absolute dose (if the dose prescription has been incorporated into the treatment plan), or relative dose (where the dose is expressed as a percentage of a reference dose). One advantage of relative dose displays is that they are valid for any prescription dose, but this is less useful with computerised plans, as it is easy to change the prescription dose with rapid recalculation of the full dose distribution in terms of absolute dose. The type of display chosen depends on local protocols and may depend on whether more than one phase of radiotherapy treatment is to be delivered.

The calculation of a dose distribution requires a lattice of points, spread over the volume in which dosimetric information is required. These points are contained within the external surface of the patient and their number and spatial resolution are the result of a trade-off between the accuracy required and the speed of calculation, which also depends on the dose calculation algorithm used. The X and Y coordinates of these points are within the transverse planes, but the Z coordinate may require some interpretation since the information is limited by the spacing between slices*. The isodose contours in a given calculation plane or the isodose surfaces are usually produced by linear interpolation between this matrix of points. The separation of the points should therefore be small enough to allow accurate linear interpolation. If the dose calculation algorithm is too slow for quasi-real-time display, a low resolution may be used for rapid display during plan optimisation. A high resolution is then set for the final dose calculation. In this case, the final plan may be printed out and used to scrutinize the isodose curve chosen to cover the planning target volume (PTV), the isodoses through critical organs and the location of hot or cold spots in the dose distribution.

Although it is possible to display the dose distribution in terms of total absolute dose over the whole treatment course, some form of normalisation may be employed. The beam weights can be adjusted to give 100% to a predefined point, such as the ICRU reference point (ICRU 1993 and see Section 37.6). Independently of the method used to specify the weights or to combine the beams, the dose may be renormalized to the ICRU reference point at the end of the calculation. This facilitates the assessment of the dose uniformity within the PTV on the basis of a permitted percentage deviation from the prescribed dose (e.g. between 95% and 107%).

The dose distribution can be displayed in multiplanar views, as shown in Figure 35.1, particularly in transverse (a), sagittal (b), and coronal (c) sections. It can also be displayed as a 3D representation of anatomy and isodose surfaces (see Figure 35.2).

The possibility of choosing between different display methods for dose distribution and anatomical information should be a feature of any modern treatment planning system (Drzymala et al. 1994). Some useful features are listed as follows:

- Opaque *colour-wash* displays, which should be interchangeable with conventional isodose lines, as although they can give a good overall impression of the dose distribution, they can also obscure the target volume.

---

* The coordinate system used here is that used in the DICOM (Digital Imaging and Communications in Medicine) standard (Chapter 42) rather than that recommended by the International Electrotechnical Commission (see Section 11.5.4).

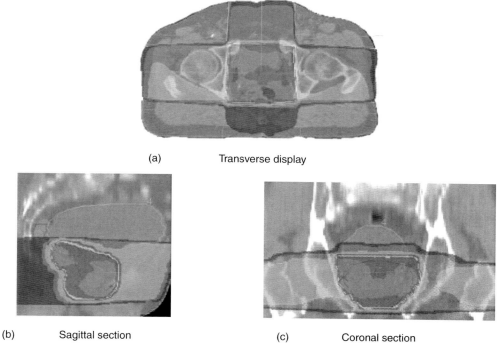

(a)   Transverse display

(b)   Sagittal section   (c)   Coronal section

FIGURE 35.1
(**See colour insert following page 590.**) Planar *colour-wash* representation of the dose distribution in three different calculation planes through the patient: each colour corresponds to a range of dose according to a predefined scale. The corresponding isodose lines can be seen at the frontier of adjacent colours. This display may be easier to understand than the standard isodose line representation.

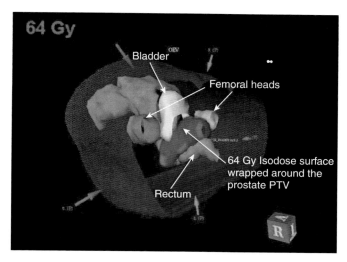

FIGURE 35.2
(**See colour insert following page 590.**) 3D surface rendering of anatomical structures and 64 Gy isodose surface for a five-beam prostate treatment plan.

- Translucent colour wash displays superimposed on CT grey-scale information which can be windowed.
- Zoom and pan facilities, with variable calculation resolution over the zoomed areas.
- 3D surface rendering of isodose surfaces, for instance as wire frames or translucent rendered surfaces, superimposed on organ volume surface displays, on a graphics device capable of real-time image manipulation (see Figure 35.2).
- *Volume of regret* representation, which is a dose-reduction technique applied to the dose distribution using surface displays: An acceptable window of dose level is chosen and portions of organs outside this window are highlighted. For critical organs this is a one-sided test, in which regions of dose above the acceptable level are displayed. For target volumes, regions of either excessive or insufficient dose can be shown. This is particularly useful when checking the coverage of PTV when using arcing beams, such as for stereotactic radiotherapy.

Ideally, plan evaluation should be done using real-time displays of the dose distribution. It is useful to display the transverse, sagittal, and coronal planes simultaneously on an interactive console to help overcome the limitations of a 3D display on a 2D medium. This requires special graphics hardware in which all of the images are loaded in graphics memory and rapidly accessed. The real-time capability gives the illusion of the user moving through the 3D display (Emami et al. 1991b). Other possibilities have been explored, such as real-time use of beam's-eye-views (BEVs) with the dose contour displayed (McShan et al. 1990) or colour wash surface dose display. It is also advantageous to be able to make visual comparisons between the dose distributions of rival plans using side-by-side display or other forms of graphical comparisons (i.e. dose difference displays).

35.3 DOSE VOLUME ANALYSIS

35.3.1 DEFINITION AND APPLICATION OF DOSE VOLUME HISTOGRAM

Volume calculations performed by 3D treatment planning systems provide a large quantity of dose information, which can be difficult to interpret and evaluate when it is displayed as isodose curves on several transverse, sagittal and coronal planes. It is much simpler to condense the 3D dose distribution data to a graph, which displays the radiation distribution within a specifically defined volume of interest, so that summarising and analysing the 3D data is possible (Chen 1988). Such a graphical representation is called a dose volume histogram (DVH).

The DVH can be expressed as

> The summed volume of elements receiving dose in a specified dose interval, against a set of equally spaced dose intervals.

This is a *differential* DVH and shows the absolute or relative volume in each dose interval (bin) directly.

More frequently used, however are *cumulative* dose volume frequency distributions, which are plots of the volume receiving a dose greater than, or equal to, a given dose, against dose. The volume accumulates starting at the highest dose bin continuing towards zero dose, eventually reaching 100% of the total volume (see Figure 35.3 and Figure 35.4). Normally volumes are expressed as a percentage of the total volume; however, in some situations the absolute volume may be more appropriate.

DVHs can be used during the planning process to check whether the dose is adequate and uniform throughout the target volume, and the extent and value of any hot spots in adjacent normal tissue. However, because they do not display positional information, they should not be the only method used. Their main use is as a plan evaluation tool. They can be used as a method of comparing different treatment plans on a single graph, for specifically identified structures and target volumes. For the PTV, DVHs should show a uniformly high dose

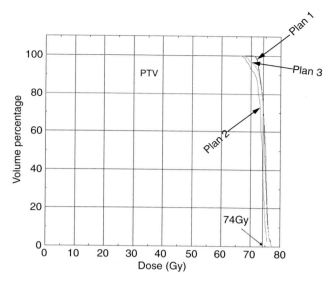

FIGURE 35.3
Dose volume histograms (DVHs) from 3 different plans for a prostate planning target volume (PTV where the prescribed dose is 74 Gy). Plan 1 provides a better PTV coverage than Plan 3 or Plan 2.

throughout the volume. The shape approximates to a step function and a steep slope shows that a large percentage of the volume has a similar dose. An ideal curve would show this step around the prescription dose, as in Figure 35.3, in which the DVHs for the PTV in three different treatment beam configurations are compared.

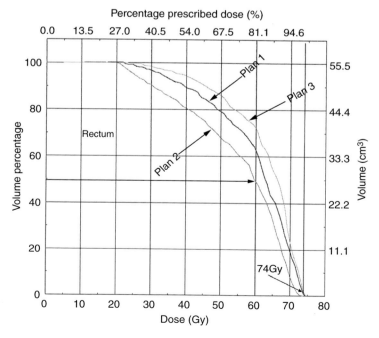

FIGURE 35.4
Dose volume histograms (DVHs) for the rectum for the same plans as shown in Figure 35.3. Plan 2 provides a better protection of the rectum than Plan 1 or Plan 3 but, as seen on Figure 35.3, this is at the detriment of the PTV coverage.

DVHs for organs-at-risk which are intended to be protected should preferably have a concave appearance: it may be acceptable either to deliver a relatively high dose to a small volume or a small dose to a large volume. Figure 35.4 shows comparative DVHs for a rectal volume for a prostate treatment where the prescribed dose is 74 Gy. It can be seen for example that 50% of the rectal volume is receiving more than 60 Gy, 63 Gy, or 66 Gy, depending on the treatment plan. It is also useful to display both the absolute dose and the absolute volume, together with normalised values, on the same graph. For the analysis of small differences in DVHs, it can be beneficial to use a non-linear volume axis.

DVHs were used originally for analysis of the effect of sophisticated treatment equipment (Chin et al. 1981) or for charged particle therapy (Goitein and Miller 1983; Chen et al. 1984; Austin-Seymour et al. 1986). They are now a method of choice in the evaluation and optimisation of all external-beam conformal plans (see also Chapter 43).

35.3.2 METHODS FOR DOSE VOLUME HISTOGRAM CALCULATION

35.3.2.1 *Sampling Methods*

There are essentially two approaches to calculate a DVH, depending on which method is used to sample the points within the structure of interest. These are the grid sampling method and the random sampling method (see Niemierko and Goitein 1990).

The grid method (Chen et al. 1984; Drzymala et al. 1991) consists of systematically calculating the dose at each point of a three dimensional grid of appropriate resolution that covers the structure of interest. This grid is normally derived from a sequence of CT scans in which cubic volume elements are equally spaced in the X and Y directions, but not necessarily in the Z direction. The volume of the voxel can then be defined by the product of the X and Y grid spacing and the slice thickness. To calculate the total volume of a particular structure, the volume elements (voxels) lying within the structure are summed. A reasonable resolution for sampling the volume of typical anatomical structures is a grid spacing of between 2 mm and 3 mm. The grid method is the natural extension of the dose calculation performed for isodose representation (see Section 35.2).

The random sampling method (Niemierko and Goitein 1990; Kooy et al. 1993) consists of setting a parallelepiped of dimensions around the structure of interest, large enough to cover the structure, and picking points randomly within this parallelepiped (i.e. X, Y, Z coordinates within the preset limits above). If N points have been picked with a uniform probability, each of them is representative of an elementary volume $\Delta V = V/N$, where V is the total volume of the parallelepiped (i.e. $V = (X_{max} - X_{min}) \times (Y_{max} - Y_{min}) \times (Z_{max} - Z_{min})$). If, among these N points, N' points have been found within the structure, the volume of the structure may be approximated by $N' \times \Delta V$. By calculating the dose at each of these N' points, a DVH may be obtained and plotted.

The advantages and disadvantages of these two methods have been the subject of controversy in the literature (Jackson et al. 1993; Lu and Chin 1993; Niemierko and Goitein 1993c). It is now clear that the random sampling technique is more efficient (i.e. a smaller number of calculation points[*] is required to reach the same level of representativity in the dose volume distribution). It is subject to less variability due to the shape or orientation of the structure. It is more flexible because it can be disconnected from the isodose calculation and launched *on the fly*[†]. For both methods, an indication of the validity of the calculation can be obtained from the comparison between the structure volume calculated from pure geometrical

---

[*] The number of dose calculation points required to reach a reasonable accuracy for a given structure is typically around 400. It should be increased to at least 1000 or 2000 if there are high gradient regions within the structure (for instance at the beam edge), especially for large complex volumes (e.g. lungs).

[†] Real-time dose volume histogram (DVH) calculation can be followed, and interrupted when it is stable.

considerations and the summation of the elementary volumes used in the DVH calculation, respectively.

35.3.2.2 Dose Binning and Graphical Representation

According to the definition given in Section 35.3.1, it is necessary to divide the range of expected dose values into equal intervals when constructing a histogram. For each interval, the volumes of the voxels receiving dose within that interval are accumulated in the appropriate element of an array, or bin. Cumulative DVHs are therefore obtained by adding the volume accumulated in each bin to the volumes in all bins corresponding to higher dose intervals. Differential DVHs treat the volume in each bin as a separate quantity, and a bin interval of between 2 Gy and 5 Gy is reasonable. For cumulative DVHs, the appropriate dose interval for the bins depends on the dose response curve for the structure of interest. An interval of 0.5 Gy has been shown to be reasonable, whereas 2 Gy is too wide an interval. The graphical representation of the resulting DVH is shown in Figure 35.5.

This approach means that the bin size has been carefully selected at the very beginning of the computation process or that the dose values to all points sampled within a structure are stored before distributing them into the proper bins. It has been suggested by Niemierko and Goitein (1994) that it is not the optimal solution, because it requires an unnecessary *dose* sampling process (i.e. dose binning) after having achieved the *volume* sampling. They suggested replacement of *dose volume histograms* (DVHs) by so-called *dose volume distributions* (DVDs), in which the calculation points are simply sorted according to dose values and are plotted by assigning the corresponding cumulated volume to each dose value (see Figure 35.6).

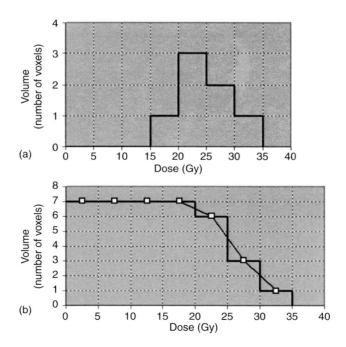

FIGURE 35.5
Dose volume histogram (DVH) plot, assuming that the dose has been calculated in seven points (i.e. seven voxels) evenly distributed throughout the structure volume, yielding dose values equal to 16, 31, 23, 24, 27, 28, 32 Gy. The abscissa axis represents the dose in Gy, using five Gy bins. The ordinate axis represents the volume in arbitrary units (i.e. number of voxels). The differential (a) and cumulative (b) forms are represented. The thinner solid line shows how it is possible to plot this histogram as a curve. (The number of points and the dose intervals have been reduced to illustrate the methodology.)

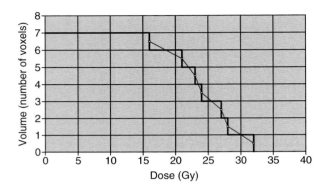

FIGURE 35.6
Dose volume distribution (DVD) plot for the same distribution as represented on Figure 35.5. Each point represents the exact dose at an individual dose calculation point which is assumed to represent the dose for the whole of the corresponding volume element. The graph can be plotted either as a continuous line through the points or with the familiar histogram steps (equally distributed along the *volume* axis rather than along the *dose* axis). In practice this distinction becomes irrelevant when a large number of points are plotted.

In practice, there is not much difference between these two approaches, but the DVD representation, which could be used in all situations, has particular advantages when there is a small number of sampling points.

35.3.2.3 *Boolean Operations on Structures for Dose Volume Histogram Calculations*

It is sometimes useful to have DVH information for structures that are completely or partially contained within another structure. Examples include the rectal wall, rather than the volume encompassed by a solid definition of rectum (i.e. all points lying inside of the external wall surface); a rectal tumour where normal rectum needs to be separated from tumour in the DVH calculation; or a combined DVH of both lungs. This cannot be performed by combining the individual DVHs assigned to each structure. To do this, a system of logical (boolean) operators is useful, such as *and*, *or*, *not*, *inside*, and *outside*. The DVH calculation then looks at each sampling point and determines within which logically specified region the point lies.

Another method, useful when one structure is contained within another, is to assign hierarchy numbers to each structure, the highest number representing the most important structure. Each element in the dose matrix is given a *tag* value, with the more important points given a higher value, so that any points occurring in both structures will have the highest priority for the structure of interest under calculation. For example, a structure within another can be subtracted out by assigning the inner structure a higher priority than the outer one.

35.3.3 CLINICAL USE AND LIMITATIONS OF DOSE VOLUME HISTOGRAMS

Care is required when comparing plans which have been computed with different dose computational parameters or using different algorithms because there is always some uncertainty related to DVH calculation and representation. Different dose calculation algorithms may not handle heterogeneities, or the penumbra, etc., in the same way; the dose sampling resolution or dose binning can also affect the results of the DVH, especially in regions of steep dose gradient. A good way to test the accuracy of dose calculations is to use test phantoms of known dimensions and dose. Useful phantom designs include a set of concentric cubes and circles of known volume (see Section 39.3.7).

DVHs provide a graphical summary of the dose distribution within a structure; they do not give spatial information. They show that hot and cold spots exist, but do not indicate where they occur, nor whether there are several small high- or low-dose areas or a just a large one. To overcome this difficulty, Cheng and Das (1999) have introduced the concept of spatial DVH (zDVH). Dose-area frequency distributions can be used for organ surfaces (Li et al. 1997; Tucker et al. 2004) and some use can be made of dose-length histograms for long thin structures (Drzymala et al. 1991). Interpretation of the DVH plot is fairly subjective and implications of small differences between DVHs are not well understood. This indicates the usefulness of an objective numeric score, such as tumour control probability (TCP) or normal tissue complication probability (NTCP) (see Chapter 36) to provide a ranking when comparing plans (Munzenrider et al. 1991). DVHs do not indicate the complexity of the field arrangements and will not show any use of couch or collimator angles, etc. that cannot be achieved in practice. DVHs must therefore be used in conjunction with other methods for treatment plan evaluation, especially those which will give spatial information.

The ordinate of the DVHs can be presented in terms of absolute volume or as the fraction of the total volume of the organ. If fractional volumes are being used, it is important that the whole of the organ used as reference has been delineated. This requires that sufficient CT slices are available to cover the whole organ. Whether absolute or relative DVHs are more appropriate will depend on the organ in question. For lung, for example, the percentage of the total volume of lung is usually regarded as most relevant. Yorke et al. (2002) found a good correlation between the incidence of radiation pneumonitis and V_{20} for the ipsilateral lung, where V_{20} is defined as the percentage of normal lung that receives at least 20 Gy (see also Graham et al. 1999; Allen et al. 2003; Gopal et al. 2003; Bradley et al. 2005). DVHs of the rectal wall have been used extensively to assess the risk of rectal bleeding in prostate irradiation (see for instance Boersma et al. 1998; Jackson et al. 2001; Cozzarini et al. 2003). For such an analysis, dose surface histograms may also be of benefit (Fenwick et al. 2001).

In summary, the DVH is a very useful tool for 3D plan comparisons and as an input parameter for TCP and NTCP calculations. TCP and NTCP are quantitative methods of predicting the likelihood of local control (see Chapter 36). DVHs and the predicted behaviour of the cell population in question are used as input data. There is, however, large uncertainty in the clinical data, due partly to the lack of consistency in the definition of anatomical volumes, and partly to the uncertainties in dose computation and data collection (Deasy et al. 2002; Fiorino et al. 2002; Cho et al. 2002). So absolute probabilities are not fully reliable, but relative probabilities can be used to assess the relative merits of different treatment plans.

35.4 DOSE STATISTICS AND PLAN SCORING

35.4.1 DOSES AT REFERENCE POINTS AND DATA DERIVED FROM DOSE VOLUME HISTOGRAMS

The simplest form of reporting the dose in a structure, either a target volume or an organ-at-risk, consists of choosing a point meant to be representative of the dose distribution within the structure. This was the approach chosen by ICRU (ICRU 1993) as a means to obtain a uniform method for dose reporting in a given centre or amongst several centres (see Section 37.6). It is still the method currently used on manual treatment chart or within the recording and verification systems to ensure that the dose accumulated at the reference point of

TABLE 35.1

Examples of Dose Statistics for a Given Structure

Statistic	Definition
Total volume	The sum of all dose volume histogram (DVH) voxels found within a set of boundary contours. It can be compared to the volume obtained directly from the contours as a mean to gain confidence in DVH calculation
Mean dose	The sum of the doses assigned to each voxel divided by the total number of voxels
Modal dose	The most probable dose, i.e. the dose which corresponds to the bin where the larger number of voxels is found. The modal dose is sometimes considered to be most representative of the organ irradiation, but it is dependent on the choice of the dose bin
Median dose	The dose level where the same number of voxels are found at higher and lower doses
Min dose, D_{min}	The minimum dose in the structure of interest. If this structure is the planning target volume (PTV), D_{min} should be appropriate to eradicate the tumour and ideally, the volume enclosed by the isodose surface D_{min} is the treated volume (TV)
Max dose, D_{max}	The maximum dose in the structure, sometimes excluding those hot spots which are associated to too-small volumes (e.g. <1 cm^3) (ICRU 1993)
Volume, V_{Dref}, corresponding to a reference dose, D_{ref}	The sum of all voxels receiving a dose greater than or equal to the user-supplied reference dose. For the PTV, the reference dose D_{ref} may be the prescription dose, or a dose expressed in Gy or as a percentage of the prescription dose. V_{Dref} is expressed either as an absolute volume (cm^3) or as a percentage of the total volume of the structure
Dose, D_{Vref}, corresponding to a volume equal to V_{ref}	The dose such that the volume receiving a dose higher than D_{Vref} is equal to V_{ref} (where V_{ref} is expressed either as an absolute volume or as a percentage of the total volume of the structure)

the PTV reaches the prescribed dose and that the dose to the organs at-risk remains below the predefined tolerance level (see Section 41.2.5). It is clear, however, that this method does not give any information about the dose homogeneity within the structures.

Dose statistics as outlined in Table 35.1 give a simplified view of the dose in a specified structure to compare 3D plans for evaluation purposes. They are obtained from DVHs, but because they are expressed directly as numerical information, they are easier to manipulate and can be used to define criteria for optimisation processes (see Section 44.3.4).

The parameters D_{Vref} or V_{Dref} allow the definition of one point on the cumulative DVH graph, where constraints can be set as a minimum for a PTV and as a maximum for an organ-at-risk.

35.4.2   CONFORMITY INDEX AND FIGURES OF MERIT

Considering the large number of parameters required to characterise a full 3D dose distribution with respect to the anatomical structures, it is attractive to try to define global quantities intended to summarize the overall dosimetic quality of a plan.

The conformity index (CI) is defined by ICRU (1999) as the quotient of the treated volume (TV, see Section 29.2.4) and the PTV. This definition implies that the TV totally encompasses the PTV. Then, if CI = 1, the conformality is optimum, with a TV covering exactly the PTV. If the TV is smaller than the PTV (underdosage), CI < 1. If the TV is too large for the PTV, causing potential damage to healthy tissue, CI > 1. Van't Riet et al. (1997) and Knöös et al. (1998) have used a similar concept to serve as an optimisation parameter. Other forms of CI have been used for stereotactic techniques (Leung et al. 1999; Theodorou et al. 2000).

In cases where a fraction of the PTV lies outside of the TV, a CI equal to 1 could still be obtained even though the conformality of the dose distribution was very poor. To overcome this problem the percentage of the PTV included in the TV should be reported (ICRU 2004). However, a more general definition of the CI for a given dose level D is the following (Oozeer et al. 2000):

$$CI(D) = IF_{PTV}(D) \times [1 - IF_{HT}(D)] \qquad (35.1)$$

where $IF_{PTV}(D)$ is the irradiation factor of the PTV, defined as the fraction of the PTV receiving a dose higher than D; $IF_{HT}(D)$ is the irradiation factor of healthy tissue, defined as the ratio of the volume of tissue outside the PTV receiving a dose greater than D to the volume of isodose D. Taking D as the dose used to define the TV:

- If the TV is included within the PTV, $IF_{PTV} < 1$ and $IF_{HT} = 0$, yielding $CI < 1$.
- If the TV encompasses the PTV, $IF_{PTV} = 1$ and $IF_{HT} < 1$, yielding $CI > 1$.
- If the TV is partially covering the PTV, $IF_{PTV} < 1$ and $IF_{HT} < 1$, yielding $CI < 1$.
- Only when TV is covering almost exactly the PTV, CI approaches 1.

In the expression of Equation 35.1, the importance of the PTV coverage and of the protection of healthy tissue are equally weighted. It is easy to adapt Equation 35.1 to reinforce the importance of either of these two constraints.

Conformity indices can be considered as simple forms of more general figures of merit or objective functions, which can then be used to score or optimise a treatment plan. For example, Jain and Khan (1992) and Jain et al. (1993) suggested a figure of merit calculated from inputs of TCP, NTCP, and oncologist preferences, and used to rank up to three competing treatment plans in order of merit by calculating a single numerical score for each treatment plan. A similar study was performed by Graham et al. (1996) for non-small-cell lung cancers.

Other more experimental methods of automating the decision process have been tried. Willoughby et al. (1996) described a system of evaluating and scoring radiotherapy treatment plans using an artificial neural network. Treatment plans were assigned a figure of merit by a radiation oncologist using a five-point rating scale. DVH data extracted from a large training set was correlated to the physician-generated figure of merit using an artificial neural network, and the net was tested on another set of plans. The accuracy of the neural net in scoring plans compared well with the reproducibility of the clinical scoring; the system is promising for the reliable generation of a clinically relevant figure of merit (see also Munley et al. 1999). Gopal and Starkschall (2002) have used a graphical representation, considering both the target coverage and the healthy tissue irradiation to rank the plans and help in the decision. The *receiver operating characteristics* (ROC) approach has also been used as a tool for plan scoring (Platoni et al. 1999; Dejean et al. 2001; Mavroidis et al. 2004).

The analysis of the three-dimensional dose volume distribution with respect to the target volumes and organs-at-risk forms the basis for the development of advanced radiotherapy techniques (see Chapter 43). Applied at the level of a patient population, it allows for an assessment of the margins to be used for optimal treatment (Van Herk et al. 2000). Provided that an adequate methodology is used, this allows for the analysis of clinical data and the improvement of the value of biological modeling (Deasy et al. 2002).

CHAPTER 36

BIOLOGICAL EVALUATION OF TREATMENT PLANS

Alan Nahum [*] *and Gerald Kutcher*

CONTENTS

[*] Alan Nahum wishes to acknowledge many useful and interesting discussions with Dr. Giovanna Gagliardi (Como and Stockholm) on the subject of NTCP modelling.

36.1 INTRODUCTION

Physicists working in radiotherapy spend a lot of their time measuring doses in phantoms and then calculating the dose distributions in patients due to a particular arrangement of beams. This is because, according to the present state-of-the-art practice, the radiation oncologist prescribes the treatment in terms of a (uniform) dose to the *target volume* accompanied by some sort of constraint on the dose to one or more *organs-at-risk*. However, the endpoints in radiotherapy that are truly of relevance are not dose distributions but the *probability of local control*, also known as the *Tumour Control Probability* (TCP) and the *Probability of Normal-Tissue Complications* (NTCP). This chapter deals with the modelling of TCP and NTCP with the emphasis on the *spatial* distribution of the absorbed dose within the target volume.

Some of the reasons why models for TCP and NTCP are desirable are listed below (the references cited are not intended to be exhaustive):

- Dose distributions in three dimension (3D) are inherently very complex and some way of assimilating this vast amount of information is needed (Mauro 1989; Goitein 1992).
- Biological models enable estimates to be made of the effect of uncertainties in dose and patient position on therapy outcome (Boyer and Schultheiss 1988; Mijnheer et al. 1989; Mackay et al. 1999; Zavgorodni 2004).
- The effect of non-uniformities in the tumour dose distribution can be approximately quantified (Brahme 1984; Sanchez-Nieto and Nahum 1999; Tomé and Fowler 2000).
- The values for α (especially important for tumours) and α/β from, for example, clonogenic assays can be both extracted from and fed into such models (Deacon et al. 1984; Peters et al. 1989; Mauro et al. 1989; West 1995; Bentzen 1997; Fenwick 1998; Sanchez-Nieto et al. 2001a; Buffa et al. 2001b; Levegrün et al. 2001, 2002; Wang et al. 2003a; Xiong et al. 2005; Carlone et al. 2006).
- Estimates can be made of the effects on local tumour control of hypoxia and other information derived from *functional* imaging (Poppel et al. 2002; Nahum et al. 2003; Ruggieri 2004; Nioutsikou et al. 2005; Ruggieri and Nahum 2006).
- The clinical effect of improvements in dose distributions through the use of new beam-delivery technology (e.g. MLCs, IMRT), 3D treatment planning systems, and new radiation modalities (brachytherapy, protons, light ions) can be approximately quantified (Webb 1993; Lee et al. 1994; Isacsson 1998; Gagliardi 1998, 2001; King et al. 2000; De Meerleer et al. 2000; Nahum and Glimelius 2001; Nutting et al. 2002).
- Optimisation/inverse planning is beginning to be done in terms of biological criteria such as highest TCP for a fixed low value of NTCP, equivalent uniform dose (EUD) etc. (Källman 1992; Nahum and Tait 1992; Mohan et al. 1992; Brahme 1999, 2001; De Gersem et al. 1999; Engelsman et al. 2001; Iori 2001; Sanchez-Nieto et al. 2001a; Schwarz et al. 2003; Peñagarícano et al. 2005; Kim and Tomé 2006; Hoffmann et al. 2006).

- Models for TCP and NTCP can serve as an aid to clarity of thought about radiotherapy (Dutreix et al. 1988).

36.2 TUMOUR CONTROL PROBABILITY

36.2.1 A RADIOBIOLOGICALLY-BASED TCP MODEL

36.2.1.1 General

So-called *dose–response curves* are believed to have a sigmoid shape. One approach is to fit a mathematical function to this curve (Walker and Suit 1981; Källman et al. 1992; Carlone et al. 2006). However, there is no natural way to accommodate changes in basic parameters such as tumour cell radiosensitivity, inhomogeneities in the dose distribution, variations in tumour volume, clonogenic cell density, etc. through empirical curve-fitting, such as the logistic model (e.g. Bentzen 2002). In the case of tumour control, in contrast to that for complications in normal tissues (see Section 36.3), it is possible to develop a model starting from the response of cells to radiation. Niemierko and Goitein (1993a) described such a model, which is similar to the one given here (Nahum and Tait 1992; Webb and Nahum 1993; Nahum and Sanchez-Nieto 2000).

36.2.1.2 A Model for TCP Based on Poisson Statistics

In this section we will derive a model for tumour control using the linear-quadratic model of cell survival. Numerous radiobiological experiments have demonstrated beyond doubt that, except at doses below about 0.6 Gy in certain cell lines (see Section 7.13.3) the killing of cells by radiation can be described by the linear-quadratic model (See Chapter 7 and also Fowler 1989; Steel 2002; Chapman 2003). This is generally written in the form

$$SF = \exp(-\alpha d - \beta d^2) \tag{36.1}$$

where SF is the surviving fraction after a (uniform) dose d^{*} of radiation to a population of cells. The parameters α and β characterise the initial slope and degree of curvature, respectively, of the survival curve.

When the irradiation is *fractionated* as in external-beam radiotherapy (Figure 36.1: full curve), for small fraction sizes such as 2 Gy, the effective slope of the resulting straight line is very nearly given by the value of α alone (assuming α/β is high). Replacing now SF by N_S/N_0, one can write:

$$N_s \approx N_0 \exp(-\alpha D) \tag{36.2}$$

where D is the total dose, N_0 is the *initial* number and N_s is the average number of clonogenic cells that survive, assumed here to be irradiated uniformly and to have uniform radiosensitivity α (Gy^{-1}); the latter is sometimes denoted by α_{eff}. The original formulation of this TCP model (Nahum and Tait 1992; Webb and Nahum 1993) was based on the approximate expression in Equation 36.2. However, in order to model fractionation effects correctly the β term must be included (Sanchez-Nieto and Nahum 1999); Mauro et al. (1989) and Niemierko and Goitein (1993a) include it in their model, for example. Thus for n fractions, each of dose d, and thus

---

* The convention adopted in this chapter is that dose is indicated by d except when it refers to total dose when D or sometimes D_{tot} is used.

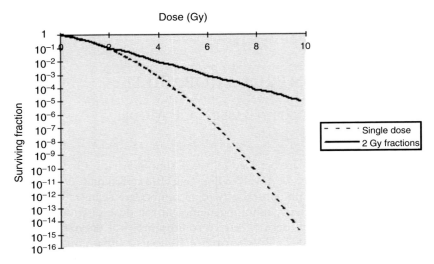

FIGURE 36.1
Semi-log plot showing the modification of the linear-quadratic cell-survival curve (dashed) when the radiation is given in 2Gy fractions (full curve). (From Nahum, A. E. and Sanchez-Nieto, B., *Physica. Medica.*, 17(2), 13–23, 2001. With permission.)

total dose D, the average number of surviving cells N_s is given by:

$$N_s = N_0 \exp\left[-\alpha D\left(1 + \frac{\beta}{\alpha}d\right)\right] \tag{36.3}$$

It can be observed that if $(\beta/\alpha)d \ll 1$ then Equation 36.3 reduces to Equation 36.2; this will be the case for small-fraction radiotherapy when the tumour clonogen α/β-ratio is very high and for continuous Low-Dose-Rate (LDR) brachytherapy irradiation for which $\beta \sim 0$ (see Section 9.4). Figure 36.2 shows how *SF* depends on total dose for $d=2$ Gy for four different α, β combinations.

The next step is to incorporate the endpoint, i.e. the eradication of the tumour, into the model. It will be assumed that a tumour is only controlled when every single clonogenic cell (i.e. cells with the potential for uncontrolled division) has been eliminated; this is known as the *target-cell hypothesis* of tumour control (Munro and Gilbert 1961). Thus, the quantity that we require is the *probability that no single clonogenic cell survives*. This can be obtained from the Poisson statistics result:

$$P(N,y) = \frac{e^{-N}N^y}{y!} \tag{36.4}$$

that gives the probability P of the occurrence of exactly y events when the mean number of events is N (e.g. Cooper 1969). Consequently, the desired probability corresponds to $y=0$, i.e.:

$$TCP = P(N,0) = e^{-N} \tag{36.5}$$

A beautifully pedagogical demonstration by Monte-Carlo simulation of the correctness of Equation 36.5 has been given by Bentzen (2002)[*].

---

[*] Porter (1980) discusses the mathematics of not just zero cells remaining, which we are assuming here, but also one-, two-, three cells etc; he concluded that they were "algebraic curiosities".

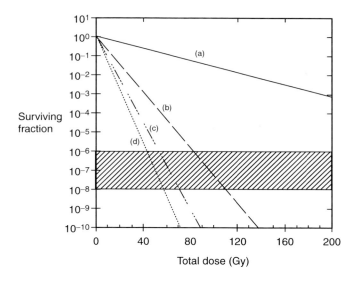

FIGURE 36.2
The surviving fraction as a function of the total dose delivered in 2 Gy fractions, according to Equation 36.3, for different values of α (Gy^{-1}) and β (Gy^{-2}): (a) $\alpha=0.036$ and $\beta=0$; (b) $\alpha=0.149$ and $\beta=0.00293$ corresponding to hypoxic prostate clonogens—see Section 36.1.5; (c) $\alpha=0.2603$ and $\beta=0$; (d) $\alpha=0.2603$ and $\beta=0.031$. The choices of α and β, taken from Nahum et al. (2003), are explained in Section 36.1.5. The hatched region, SF between 10^{-6} and 10^{-8}, indicates approximately the region where TCP can be expected to have a non-zero value. (Adapted from Nahum and Chapman 2004.)

Insight into the nature of the problem may be gained by considering that, if there is an average of one cell left after irradiation, then the probability of *no* cells remaining is $1/e$ or 37%. We can see that even if the irradiation succeeds in reducing the number of cells from 10^9 to only (a mean value of) 4, say, the resulting TCP is only e^{-4} or 1.8% despite the fact that the surviving fraction is only 4×10^{-9}; tumour *cure* is thus very difficult to achieve.

Combining Equation 36.1 and Equation 36.5, where $N=N_s=N_0 S$, we arrive at:

$$TCP = \exp[-N_0 \exp(-\alpha d - \beta d^2)] \tag{36.6}$$

for the case of a dose delivered in one fraction[*]. For the case of n fractions, each of identical dose d such that $D_{tot}=d\times n$, combining Equation 36.3 and Equation 36.5, the expression for *TCP* becomes:

$$TCP = \exp\left\{-N_0 \exp\left[-\alpha D_{tot}\left(1+\frac{\beta}{\alpha}d\right)\right]\right\} \tag{36.7}$$

A plot of the above expression for *TCP* as a function of dose D_{tot} produces the well-known sigmoidal curve. Using a realistic value for the number of initial clonogenic cells N_0 of the order of 10^9 (Porter 1980; Steel 2002), values of α from 0.1 Gy^{-1} to 1 Gy^{-1} (e.g. Deacon et al. 1984) and setting β to zero for simplicity, one obtains the family of curves shown in Figure 36.3.

Proliferation/repopulation can be approximately included in the expression for *TCP* by adding a term $\gamma(T-T_k)$ inside the square brackets in Equation 36.7 where $\gamma=\ln 2/T_d$, T_d is the average doubling time, T is the overall treatment time, and T_k is the time at which proliferation begins (e.g. Sanchez-Nieto and Nahum 2000). Tucker et al. (1990) and Zaider

---

[*] If β is negligible then the βD^2 term is simply removed as in Webb and Nahum (1993) and then Equation 36.6 is valid even for fractionated delivery where now $D=D_{tot}$.

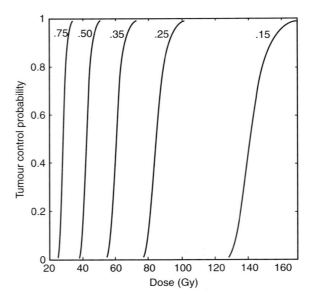

FIGURE 36.3
TCP curves from Equation 36.7 for clinically realistic values of $N_0 = 10^9$ and α ranging from 0.15 to 0.75 Gy^{-1} (as indicated on the curves) and with $\beta = 0$. (Reproduced from Nahum, A. E. and Sanchez-Nieto, B., *Physica Medica.*, 17(2), 13–23, 2001. With permission.)

and Minerbo (2000) have given a more rigorous treatment of the effect of including clonogen proliferation during irradiation.

36.2.1.3 *Consistency with Clinical Data*

Theoretical models must be compared to clinical data whenever possible. A number of Local Control vs Tumour Dose studies have been published (e.g. Morrison 1975; Batterman et al. 1981; Moore et al. 1983; Hanks et al. 1988, 1997; Stock et al. 1998; Martel et al. 1999; Jacob et al. 2005). Despite the limitations associated with such data, i.e. uncertainties in the dosimetry, inadequate patient numbers, imprecise clinical definition of local control, etc. there are almost no Tumour Control vs Dose curves with slopes anything like as steep as the ones in Figure 36.3. This has led some investigators to favour an empirical model to fit clinical dose–response curves, e.g. Mohan et al. (1992), Carlone et al. (2006).

Various hypotheses have been advanced over the years to explain the shallowness of the clinically observed Local Control vs Dose curves, e.g. Zagars et al. (1987) and Yaes et al. (1999). The most probable single explanation is *inter-patient heterogeneity* in the intrinsic radiosensitivity of the tumour cells, i.e. the α and β values (Deacon et al. 1984; Zagars et al. 1987; Boyer and Schultheiss 1988; Dutreix et al. 1988). This is in contrast to the possible heterogeneity of radiosensitivity of the clonogenic cells within any one tumour, i.e. *intra-patient heterogeneity* (e.g. Suit et al. 1992). Hypoxia (i.e. poor oxygenation of cells), has long been considered to be a major cause of failure to achieve local control in radiotherapy. It is now believed to play an important role only in certain tumours (Dasu and Denekamp 1998; Steel and Peacock 1989; Movsas et al. 2000, 2002; Buffa et al. 2001a; Poppel et al. 2002; Steel 2002; Chapman 2003; Nahum et al. 2003; see also Section 7.10 and Section 36.2.5). Explanations based on the effect of a very small hypoxic, and therefore radioresistant, fraction of cells (e.g. Källman et al. 1992) cannot, therefore, account for the shallow slopes in more than a small percentage of cases and only then for certain types of tumours (e.g. Yaes 1999; Steel 2002). In an analysis of clinical dose–response data (Brenner 1993) it was demonstrated that it was possible to explain the wide variations in the dose required to achieve local control for a number of different lesions solely in terms of variations in α from one lesion type to another *and* the variation in the number of clonogenic cells, assumed proportional to the volume of the lesion. The *bottom line* of this study was that it was not necessary to invoke any assumptions about, for example, the variation in the hypoxic cell fraction with tumour size. However, Brenner did not build into his analysis any inter-patient variation in radiosensitivity and as a consequence the number of clonogenic cells N_0 required to fit the clinical data came out as unrealistically small (Webb 1994; Webb and Nahum 1998).

The TCP model being developed in this section (see also Nahum and Sanchez-Nieto 2001) explicitly incorporates inter-patient variation by assuming that α is distributed normally amongst the patient population, with standard deviation σ_α. Thus the final *TCP* is calculated as the average of the *TCPs* of these patients with different radiosensitivities according to

$$\overline{TCP}(D,N_0,\bar{\alpha},\bar{\beta},\sigma_\alpha) = \sum_i g_i TCP(D,N_0,\alpha_i,\beta_i) \qquad (36.8)$$

where TCP $(D, N_0\,\alpha_i, \beta_i,)$ is given by Equation 36.7 and a fraction g_i of the patients have $\alpha = \alpha_i$ such that

$$g_i \propto \exp\left[-(\alpha_i - \bar{\alpha})^2/2\sigma_\alpha^2\right] \qquad (36.9)$$

where $\Sigma g_i = 1$. It is generally assumed that β also follows a normal distribution such that α/β remains constant (Sanchez-Nieto and Nahum 1999; Nahum and Sanchez-Nieto 2001; Xiong et al. 2005).

As the value of σ_α is increased, the slope of the TCP vs Dose curve decreases. One way of thinking about this is to regard this resulting TCP vs Dose curve as the sum of the curves for different α values in Figure 36.3. One can say that there is a group of patients with radio-resistant tumours (i.e. low α values who will never be cured (TCP=zero)), another group with radiosensitive tumours, i.e. high α values that will all be (locally) controlled, and a group with intermediate α values for which the term *stochastic fraction* applies, as the outcome for these patients is literally a matter of chance (Zagars et al. 1987).

This addition of a non-zero σ_α to the TCP model was first applied to the case of bladder tumours (Nahum and Tait 1992). A mean α-value of 0.35 Gy^{-1} was adopted, as determined by Deacon et al. (1984) for human bladder-tumour cells grown in vitro. The initial number of clonogenic cells, N_0, was estimated from the product $\rho_{cl} \times V_{tgt}$ with the clonogenic cell density ρ_{cl} taken to be 10^7 cm^{-3} (which corresponds to approximately 1 tumour cell in ten being clonogenic) and a mean value of the target volume, V_{tgt}, of 320 cm^3; this latter value was derived from an analysis of the actual target volumes, as outlined on CT, of patients entered into a clinical trial of conformal vs conventional pelvic radiotherapy (Tait et al. 1997). The steep sigmoid curve (dashed) in Figure 36.4 corresponds to $\sigma_\alpha=0$ and is equivalent to one of the curves in Figure 36.3. The full curve is a fit to clinical data from Batterman et al. (1981); the value $\sigma_\alpha=0.08$ Gy^{-1} was determined entirely empirically. The TCP at the actual clinical dose used, 64 Gy (32×2 Gy) came out at just below 0.5 for both curves. This 50% control rate was consistent with clinical findings and lent some confidence to the model.

The more usual way of describing the slope is by the parameter $\gamma=D(\Delta TCP/\Delta D)$ at either the 50% or 37% point on the TCP curve (Brahme 1984; Yaes 1999). Values between 1.7 and 2.0 are typical for human tumours (Fischer and Moulder 1975; Stewart and Jackson 1975), compared to around 7 for the conditions depicted in Figure 36.3. It was shown by Brahme (1984) that γ is related to the logarithm of the initial clonogen number N_0 through $\gamma=\ln N_0/e$. If σ_α is set to zero, the only way to obtain slopes similar to those consistent with clinical dose–response data is to set N_0 equal to an unrealistically small number, of the order of tens or hundreds, of clonogens (e.g. Brenner 1993; Webb 1994; Webb and Nahum 1998; Nahum et al. 2003; Wang et al. 2003).

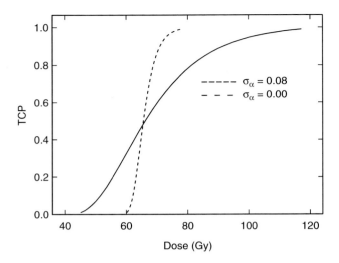

FIGURE 36.4
Tumour Control Probability (TCP) as a function of target dose, derived from Equation 36.7, Equation 36.8 and Equation 36.9, with $\alpha = 0.35$ Gy^{-1}, $\rho_{cl} = 10^7$ cm^{-3} for a volume = 320 cm^3, for $\sigma_\alpha = 0.0$ (steep curve) and for the clinically more realistic $\sigma_\alpha = 0.08$ Gy^{-1} (shallow curve); β was set equal to zero. (From Nahum, A. E. and Tait, D. M., *Advanced Radiation Therapy: Tumour Response Monitoring and Treatment Planning*, Springer, Heidelberg, 1992.)

36.2.1.4 *Inhomogeneous Dose Distributions*

Up to this point, it has been assumed that all cells receive exactly the same dose. Next, we incorporate a *dose distribution* into the TCP model. In order to do this, it is necessary to know the number of clonogenic cells $N_{0,i}$ that receive a (total) dose D_i. This is most conveniently obtained from a *dose–volume histogram*, or DVH, generated by the planning computer (see Section 35.3). Strictly, what is required is the *differential* dose–volume distribution or histogram, dV/dD from which the more familiar *cumulative* DVH is calculated. Thus, one generalises Equation 36.3 to

$$N_s = \sum_i^n N_{0,i} \exp\left[-\alpha D_i\left(1 + \frac{\beta}{\alpha}d_i\right)\right] \qquad (36.10)$$

where the summation is carried out over the *n bins* in the DVH.

The full expression for TCP can now be written as follows:

$$TCP = \frac{1}{\sigma_\alpha\sqrt{2\pi}} \int_0^\infty \left(\prod_i \exp\left[-\rho_{cl}V_i\exp\left\{-\alpha D_i\left(1 + \frac{\beta}{\alpha}d_i\right)\right\}\right]\right)\exp[-(\alpha - \bar{\alpha})^2/2\sigma_\alpha^2]d\alpha \qquad (36.11)$$

where the summation (of the indices of the exponential) over the *i dose bins* in the DVH (Equation 36.10) has been replaced by a *multiplication* of exponentials, i.e. of the control probabilities for the cells in each dose bin. The clonogenic cell density is denoted by ρ_{cl}, V_i is the volume of tissue in the *i*th dose bin, and $\bar{\alpha}$ is the mean value of the radiosensitivity in the inter-patient distribution, characterised by standard deviation σ_α*. It should be noted that this alternative way of computing the TCP is equivalent to writing the control probability for the *tumour* as the *product* of the control probabilities of all the elements, or *voxels*, comprising

---

* It is tacitly assumed here that the radiosensitivity coefficient β varies over the patient population in the same way as α; Xiong et al. (2005) show that σ_β becomes important for low α/β ratios.

the tumour, i.e.:

$$TCP = \prod_i VCP_i \qquad (36.12)$$

Writing the expression for TCP in this way can serve as a conceptual aid to thinking about the effect on TCP of, for example, boosting one part of the tumour, e.g. if the (pre-boost) control probability of the voxel to be boosted is 0.95 then the maximum possible increase in (overall) TCP would be 5%. Equally well, it shows immediately that if any *voxel* in the tumour receives a low dose, such that its control probability $(VCP) = 0.10$ say, then the (overall) TCP can never be higher than 0.10 no matter how high a dose be given to the clonogens in all the other voxels. Tomé and Fowler (2000) have modelled in detail the effect on the TCP of *boosting* subvolumes of a tumour.

Sanchez-Nieto and Nahum (1999) explored the idea of converting dose values at grid points in a treatment plan (corresponding to voxels of certain dimensions containing a given number of clonogens) into VCP values. However, for any clinically realistic values of the total dose, all the VCPs take values of the order of 0.999..., i.e. extremely close to unity. This type of *bioeffect* distribution (e.g. Wigg 2001) would not therefore be a useful way of representing a treatment plan.

Strictly speaking the relevant DVH is not that for the *target* (i.e. planning target volume (PTV)) but, instead, that for the *tumour* volume (i.e. gross tumour volume (GTV) in ICRU 50 terminology), i.e. one should not include the margin added to account for patient movement. However, it is currently generally not possible to be more precise about such issues (see below). A potentially equally serious current limitation is the implicit assumption that the clonogenic cell density ρ_{cl} is constant right out to the edges of the tumour or target volume. It can be argued that the clonogenic cell density must drop off significantly outside the GTV and that therefore any lower values of the dose located in the PTV–CTV–GTV edge region coincide with very low values of the clonogen density, i.e. relatively small numbers of clonogens. Consequently including possible *cold spots* located in this edge region in the DVH (e.g. that for the PTV) used to compute the TCP will artificially lower the overall TCP. Certain aspects of this difficult issue are discussed in more detail below.

36.2.2 THE EFFECT OF DOSE NONUNIFORMITY ON TCP

Brahme (1984) applied his TCP model to the question of the effect on TCP of both inhomogeneities in the target dose distribution and uncertainties in the absolute absorbed dose determination. Nahum and Sanchez-Nieto (2001) carried out a similar exercise (see Figure 36.5) using the model described above for the tumour parameters shown in Figure 36.4. The dose distribution in the target volume, consisting of 10^9 clonogenic cells of uniform radiosensitivity, was assumed to follow a *normal* distribution, i.e. $N_{0,i}$ was varied normally as a function of D_i (see Equation 36.10) with its width expressed in terms of σ_D/D (plotted along the *x*-axis in Figure 36.5). The four curves correspond to different values of σ_α. The mean dose was kept constant, at 60 Gy.

One immediate observation is that the TCP always decreases as σ_D/D increases, even though the mean dose does not change. Thus the maximum value of the TCP is for uniform dose, and *hot* and *cold* spots do not compensate each other, as has been pointed out elsewhere (e.g. Webb et al. 1994; Sanchez-Nieto and Nahum 1999). The figure shows that for a group of patients with tumours of exactly the same radiosensitivity, i.e. $\sigma_\alpha = 0.0$, relatively small dose inhomogeneities seriously lower the TCP; this corresponds to the very steep dose–response curves in Figure 36.3. A clinically more realistic value of σ_α, such as 0.10 Gy^{-1} results in a much less dramatic reduction in TCP as the dose inhomogeneity is increased. Thus the message here is that appreciable inter-patient variability in radiosensitivity reduces significantly the consequences of deviations from target-dose uniformity. The corollary of this is the conclusion

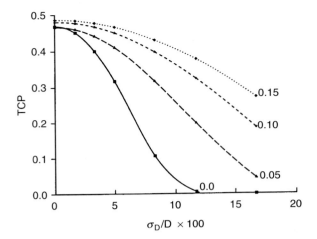

FIGURE 36.5
The effect on TCP of nonuniformity in the target dose distribution, expressed as σ_D/D, for values of the inter-patient radiosensitivity parameter σ_α from zero to 0.15 Gy^{-1}. The mean target dose is 60 Gy, the mean $\alpha = 0.35$ Gy^{-1}, $\beta = 0$ and $N_0 = 10^9$. (From Nahum, A. E. and Sanchez-Nieto, B., *Physica. Medica.*, 17(2), 13–23, 2001. With permission.)

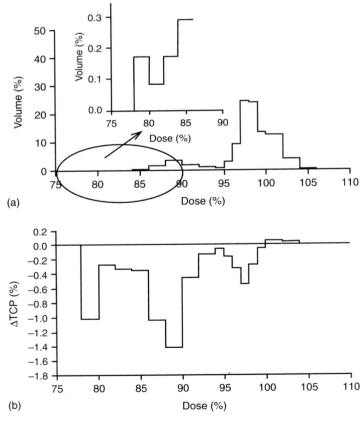

FIGURE 36.6
An example of the ΔTCP method applied to the differential DVH (a) of a simulated prostate GTV. The parameters in the ΔTCP calculation were $\alpha = 0.29$ Gy^{-1}, $\sigma_\alpha = 0.07$ Gy^{-1}, $\alpha/\beta = 10$ Gy; the TCP was 43.9%. The ΔTCP distribution (b) indicates that the TCP is reduced by 1.4% due to the volume in the 88–90% dose bin, and by 1.0% due to the 78–80% dose bin, etc. (From Sanchez-Nieto, B. and Nahum, A. E., *Int. J. Radiat. Oncol. Biol. Phys.*, 44, 369–380, 1999. With permission.)

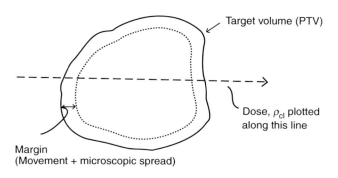

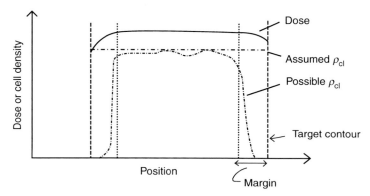

FIGURE 36.7
Schematic drawing illustrating the problem of variation of clonogenic cell density at the edge of the PTV; a hypothetical dose and cell density profile through the centre of the PTV is shown. (From Nahum, A. E. and Sanchez-Nieto, B., *Physica. Medica.*, 17(2), 13–23, 2001. With permission.)

reached by Brahme (1984)—that for certain classes of tumours with steep dose–response curves, notably in the larynx (normalised dose gradient $\gamma > 4$), only very small uncertainties in the absolute dose determination can be tolerated[*].

Sanchez-Nieto and Nahum (1999) have proposed a method whereby a TCP model could be used directly in a treatment planning system. This is done by assigning each dose bin in the (differential) DVH the quantity ΔTCP (or deltaTCP) which is the amount by which the TCP is decreased or increased because of the volume of tumour/target in this dose bin *not* being equal to the prescribed dose. In this way the effect of *cold* dose bins can be seen directly (though the volume in the bin will not necessarily be contiguous) (Figure 36.6).

36.2.3 VARIATION IN CLONOGENIC CELL DENSITY

If the model described here is applied to the DVH of the PTV (see Section 29.2.3) then it is implicitly being assumed that the clonogenic cell density is constant over the whole of the PTV, i.e. one calculates the number of clonogenic cells $N_{0,i}$ at dose D_i, in Equation 36.10 from the product of $V_{0,i}$ and ρ_{cl}. However, this assumption of constant ρ_{cl} is quite unrealistic as the PTV includes a margin for microscopic spread plus a second margin for geometrical inaccuracies. This point is illustrated in Figure 36.7.

---

[*] The calculations involved in producing Figure 36.5 have been repeated including a non-zero β value; this has the effect of slightly increasing the rate of decrease in TCP with increase in dose non-uniformity (W. Xiong, private communication).

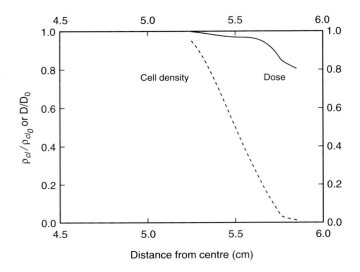

Distance from centre (cm)

FIGURE 36.8
The variation in the dose corresponding to the iso-VCP condition for a tumour with a hypothetical relative variation in clonogenic cell density (ρ_{cl}/ρ_{cl_0}) its centre to the periphery given by the dashed curve. The central target dose, D_0, is 64Gy and the mean value of α is 0.35 Gy. The solid line shows the relative dose variation, D/D_0, which is such that the voxel control probability (UCP) remains unchanged throughout the tomour. (From Webb, S. and Nahum, A. E., *Phys. Med. Biol.*, 38, 653–666, 1993. With permission.)

Although it is not known how the clonogen density varies throughout the PTV the model can be used to assess the effect that such variations might have on the local control probability. One way of looking at this is to calculate the change in dose that corresponds to a change in cell density ρ_{cl} when one requires that the *voxel control probability* (VCP) remains unchanged for a given volume element of cells. Figure 36.8 shows the iso-VCP dose corresponding to a decrease in clonogenic cell density given by the (hypothetical) dashed curve. One can see that for a considerable decrease in ρ_{cl} the *allowable* decrease in dose is very modest. Similar conclusions were drawn by Brahme and Ågren (1987).

36.2.4 Applying the TCP Model to Patient Data

The TCP model described above has been applied to the DVHs for the GTV for prostate patients in a clinical trial on conformal prostate therapy (Figure 36.9). Parameters appropriate to prostate tumours were obtained by fitting the TCP model to the Hanks et al. (1988) clinical data set (Sanchez-Nieto and Nahum 1999). Two curves were calculated, corresponding to (uniformly irradiated) tumour volumes of 10 cm^3 and 100 cm^3, which correspond approximately to stage B and C prostate cancer respectively; the parameters used to compute the curves shown in Figure 36.9 are given in the caption (Sanchez-Nieto and Nahum 1999). A much more detailed analysis of the relationship between prostate tumour radiotherapy outcome data and a TCP model has been made by Levegrün and colleagues (2000, 2001).

Figure 36.10 is a frequency distribution of the (population-averaged) TCP values for 30 patients treated in the non-dose-escalated arm of a clinical trial, at 64 Gy total dose (in 32 × 2-Gy fractions). The overall mean TCP came out at 0.64. The spread in values, from around 55% to 70% is due primarily to the different sizes of the GTVs (and hence different estimates of initial clonogen numbers); any non-uniformity in the dose distributions in these cases played only a minor role. It is important to realise that the TCP value for each patient represents a *mean* value for that particular treatment plan and patient anatomy, averaged over an α-distribution characterised by $\bar{\alpha} = 0.29$ Gy^{-1} and $\sigma_\alpha = 0.07$ Gy^{-1}; the individual α-value for the patient in question

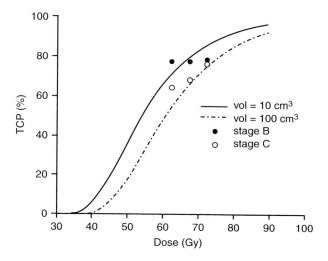

FIGURE 36.9

TCP curves resulting from the following parameters: $\alpha=0.29$ Gy^{-1}, $\sigma_\alpha=0.07$ Gy^{-1}, $\alpha/\beta=10$ Gy, $\rho_{cl}=10^7$ clonogens cm^{-3}, dose per fraction $=2$ Gy, volume $=10$ cm^3 or 100 cm^3. Solid and open circles represent 7-year recurrence-free clinical data for stage B and C respectively from Hanks et al. (1988). (From Sanchez-Nieto, B. and Nahum, A. E., *Int. J. Radiat. Oncol. Biol. Phys.*, 44, 369–380, 1999. With permission.)

is not known. Alternatively, the *individual patient* situation could be simulated by choosing a single α value for each patient by random sampling of the normal distribution for $\bar{\alpha}=0.29$ Gy^{-1} and $\sigma_\alpha=0.07$ Gy^{-1}. In this case, the range of TCP values obtained would be greater than that indicated in Figure 36.10 but would, be a better representation of the true situation.

One frequently encounters the statement that the probability of controlling a tumour depends on the *minimum* dose in the target volume. The model can be used to examine this hypothesis. Nahum and Sanchez-Nieto (2001) analysed 43 typical treatment plans for

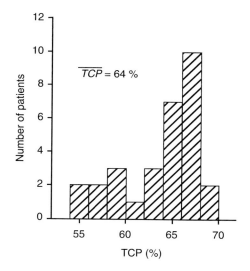

FIGURE 36.10

Frequency distribution of calculated TCP for the GTVs of 30 patients treated in the non-dose-escalated arm of the Royal Marsden prostate trial, using the following parameters: $\alpha=0.29$ Gy^{-1}, $\sigma_\alpha=0.07$ Gy^{-1}, $\alpha/\beta=10$ Gy, $\rho_{cl}=10^7$ clonogens cm^{-3}, dose per fraction $=2$ Gy, total dose 64 Gy. (From Nahum, A. E. and Sanchez-Nieto, B., *Physica. Medica.*, 17(2), 13–23, 2001.)

carcinoma of the bladder and calculated the TCP for the GTV of each plan using the same parameter values as for the *realistic* curve in Figure 36.4. For each plan the mean dose, the minimum dose and the effective dose D_{eff} were recorded. This latter dose is that value, which if uniform in the tumour, would give the same TCP as the actual dose distribution*; it follows from Figure 36.5 that it must lie between D_{min} and D_{mean}. Nahum and Sanchez-Nieto (2001) found that, on the average, D_{eff} was much closer to D_{mean} than to D_{min}. It can therefore be concluded that, at least for treatment plans where the tumour dose is reasonably uniform (as was the case), the *mean* dose is much better than the *minimum* dose as a single indicator of clinical effectiveness.

Terahara et al. (1999) analysed the effect of tumour-dose inhomogeneity on local control for 132 patients with skull-base chordoma treated with combined photons and protons, with the prescribed doses ranging from 66.6 to 79.2 Cobalt-Gray-Equivalent (see Chapter 46). Poor correlation was obtained between local control and measures such as mean dose and prescribed dose, but there was a significant correlation with the minimum dose (low p-value) and with the quantity *EUD*. The *EUD*, proposed by Niemierko (1997), is close in concept to TCP; it is the uniform dose in the target that gives the same level of cell kill and involves summing up the contributions of the dose bins in the DVH in the same way as in Equation 36.10; Ebert (2000) has compared the relative usefulness of *EUD* and TCP. The Terahara et al. study thus effectively provides clinical support for the effect of inhomogeneous dose distributions on TCP predicted by the model described here (and in Nahum and Sanchez-Nieto 2001).

36.2.5 THE EFFECT OF HYPOXIA ON LOCAL CONTROL: A CASE STUDY

Very low values of $\alpha/\beta \sim 1.5$ for prostate clonogens have been obtained from applying tumour local-control models to the clinical results obtained by external-beam therapy (XRT), on the one hand, and low-dose-rate (LDR) permanent brachytherapy seed implants (LDR brachy) on the other (Brenner and Hall 1999; King et al. 2000; Fowler et al. 2001; Lindsay et al. 2003; Wang et al. 2003b).

A study has also been carried out using the local control model described here but extended to incorporate hypoxic clonogens (Nahum et al. 2003a). A best set of prostate tumour-biology parameters was derived from averages of the literature values of the α- and β-inactivation coefficients for human prostate cancer cell lines, derived from in vitro irradiation; these are given in Table 36.1. The new aspect of this study was the explicit incorporation of hypoxia into the local control model. Movsas et al. (2000, 2002) had obtained, using an Eppendorf probe in vivo, median pO_2 values ≤ 1 mmHg in the prostates of brachytherapy patients in 21% of 115 cases. Oxygen Enhancement Ratios (OERs) of 1.75 and 3.25 for α- and $\sqrt{\beta}$-inactivation respectively were incorporated into the TCP model, following Chapman (2003), together with clonogen densities of $\sim 10^5$ cells/cm^3. Severe hypoxia, increasing with PSA level, was assumed in a proportion of tumours, in accordance with the Movsas et al. experimental findings.

The clinical-response data from both the Hanks' external-beam, dose–escalation study, stratified by PSA value, and the Stock et al. LDR brachytherapy study were well predicted by the Nahum et al. (2003a) model without assuming a low α/β; values of $\alpha/\beta = 8.5$ Gy and 50.3 Gy† were used for well-oxygenated and hypoxic clonogens respectively (Chapman 2003; Nahum and Chapman 2004). The fit to the Hanks external-beam data, stratified into three bands of PSA values, is shown in Figure 36.11. It should be noted that the local control vs dose curves

---

* D_{eff} is identical to the Equivalent Uniform Dose (*EUD*) proposed by Niemierko (1997).
$\dagger$ The value of $\alpha/\beta = 15.5$ Gy for hypoxic clonogens given in Nahum et al. (2003a) was found to be in error; it was subsequently corrected to 50.3 Gy in Nahum and Chapman (2004); this change did not affect the conclusions drawn from the study.

TABLE 36.1

The Tumour Biology Parameters Used in the Prostate Local-Control Study

$\alpha = 0.2603\ Gy^{-1}$
$\sigma_\alpha = 0.06\ Gy^{-1}$
$\beta = 0.0312\ Gy^{-2}$ (for XRT) $\beta = 0$ (for LDR brachy)
$OER_\alpha = 1.75$, i.e. $\alpha_{hypoxic} = \alpha_{oxic}/1.75$
$OER_{\sqrt\beta} = 3.25$, i.e. $\beta_{hypoxic} = \beta_{oxic}/(3.25)^2$
Clonogen number: $N_0 = 5 \times 10^6$ (LDR brachy); N_0 variable between 10^6 and 10^7 (XRT)
Fraction size: $d = 2\ Gy$ (for XRT)

Source: From Nahum A. E. et al., *Int. J. Radiat. Oncol. Biol. Phys.*, 57, 391–401, 2003a. With permission.

for the intermediate and high PSA-level patients are no longer sigmoid in shape due to the incorporation of hypoxia.

Nahum et al. (2003a) concluded that neither α/β ratios nor clonogen number need be extremely low to explain the response of prostate cancer to low dose rate (LDR) brachytherapy and external-beam therapy in contrast to other analyses (e.g. Brenner and Hall 1999; Fowler et al. 2001). It was strongly suggested that the overall cure-rate by radiation treatment is limited by severe hypoxia in the prostates of a non-negligible proportion of patients, this proportion increasing by PSA group.

36.2.6 CONCLUSIONS

The TCP model described here is mathematically simple and yet it is a reasonably complete description of the process of tumour eradication by radiotherapy, as it is able to deal with non-uniform dose distributions, variation in fraction size, proliferation (Tucker et al. 1990; Deasy 1996) and, in principle, variable clonogen density. The effects of hypoxia can be incorporated without any drastic modifications (Buffa et al. 2001a; Nahum et al. 2003a; see also Section 36.2.5), though modelling re-oxygenation is less straightforward (Poppel et al. 2002;

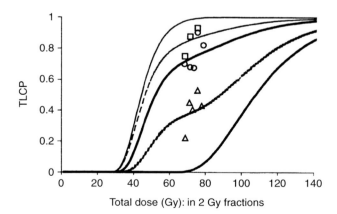

FIGURE 36.11
Tumour local control probability (TLCP) versus total external-beam radiotherapy dose delivered in 2 Gy fractions for prostate tumours whose clonogens have the same variable α (i.e. non-zero σ_α) and fixed β radiosensitivities as those in the table. The two sigmoidal curves forming the envelope correspond to 100% aerobic (left-most) and 100% hypoxic (right-most) tumours. Within the envelope, the three curves describe patient populations whose hypoxic tumour/-aerobic tumour ratios are 15/85 (dashed), 25/75 (full) and 60/40 (dashed-dotted) respectively. The data points are local-control rates (bNED) for prostate cancers of increasing stage with PSA levels of ≤ 10 ng/ml ($\square$), 10–20 ng/ml ($\triangle$) and ≥ 20 ng/ml ($\bigcirc$) from Hanks et al. (1997). (Reproduced from Nahum, A. E. and Chapman, J. D., *Int. J. Radiat. Onc. Biol. Phys.*, 58, 1637–1639, 2004. With permission.)

Ruggieri 2004; Ruggieri and Nahum 2006). Dose-rate effects can also be important in certain circumstances, principally when LDR or HDR brachytherapy is administered (see Section 9.4 and Chapter 55); there are well-established formulas available to estimate the decrease in the β coefficient as the dose rate is reduced (e.g. Dale 1985; Steel 2002; Wang et al. 2003a, 2003b; Chapter 55). There are also suggestions that the extended treatment times for external-beam therapy (as in some IMRT treatments) may reduce cell-killing and hence TCP (see Section 9.7).

The main limitations in the clinical use of the TCP model are the lack of clinical data on radiosensitivity and clonogenic cell density; a useful compilation of such data is found in Table 8.5 in Wigg (2001). Thus, the *absolute values* of predicted TCP must be treated with caution. There have been some recent attempts to use TCP models to fit clinical data on patient local control and/or survival and doses and hence to derive clinically reasonable parameter values, the most comprehensive of these being for prostate tumours (e.g. Sanchez-Nieto and Nahum 1999; Fowler et al. 2001; Levegrün et al. 2000, 2001; Lindsay et al. 2003; Nahum et al. 2003a) and there is a valuable study for the case of non-small cell lung tumours by Martel et al. (1999), further discussed by Fowler et al. 2004. There is enormous variation in the parameters obtained (especially in α and ρ_{cl}) depending on, for example, whether or not interpatient heterogeneity is incorporated (Brenner 1993; Webb 1994; Fenwick 1998; Webb and Nahum 1998; Brenner and Hall 1999; Buffa et al. 2000; Fowler et al. 2001; Nahum et al. 2003a; Wang et al. 2003a, 2003b; Carlone et al. 2006).

TCP modelling is a developing field. Refinements of the parameters and of the models themselves continue. Zavgorodni (2004) has looked at the radiobiological implications of inter-fraction dose fluctuations, due for example to patient set-up uncertainties, and derived an expression for the *equivalent constant dose* (ECD) producing the same cell survival fraction as a variable dose with a given standard deviation σ_d. The expression involves the mean dose, the number of fractions, σ_d and α/β. Keall et al. (2003) have modified a TCP model to enable predictions to be made for the combination of gene therapy and radiotherapy. The work of Antipas et al. (2004) represents a much more ambitious type of modelling; the growth of a tumour is simulated including cell division, changing oxygenation status, cell death, tumour shrinkage in response to irradiation, apoptosis, etc.

All the current tumour-response models make the implicit assumption that each cell responds to the absorbed dose at the location of the cell, i.e. to the *local* dose. However, the existence of so-called *bystander effects* has been established beyond any doubt (e.g. Mothersill et al. 2004), i.e. irradiated cells affect the cells in their immediate neighbourhood. The key question for radiotherapy is over what distance do these bystander effects act? If this is only fractions of a millimetre then probably the consequences are minimal. However, if the *action distance* is several millimetres or even centimetres then drastic revisions of many of our ideas will be necessary; the experiments by Suchowerska et al. (2005) suggest that it may well be centimetres. Certainly the predictions of current *local dose* TCP models regarding the effect of *hot* and *cold* spots will require revision. The rationale for the renaissance of so-called *grid therapy* in the megavoltage era is at least partly based on the belief that bystander effects are important in external-beam radiotherapy (Mohiuddin et al. 1999; Trapp et al. 2004; Zwicker et al. 2004).

Significant changes are underway in patient dose-calculation methods (Part F). Monte-Carlo simulation, because of its highly accurate treatment of the effect of heterogeneities, is gradually being introduced into treatment planning systems (see Chapter 28). However, a unique feature of Monte-Carlo generated dose distributions is the statistical noise, which can be thought of as a large number of artificial *hot* and *cold* regions. If these fluctuations are too large, due to too few *histories* being executed, then they could affect the prediction of TCP, yielding an artificially low value even though the mean value of the dose in the tumour will be correct. Analyses of this intriguing problem have appeared (Buffa and Nahum 2000; Keall et al. 2000); it has been shown, for example, that for a photon-beam plan, TCP varies strongly with voxel-size for 10^7 histories, but this variation (i.e. underestimation) is negligible for 10^8 histories.

There is software freely available that evaluates TCP (and NTCP) models including the one which has been described in detail above. One such package is BIOPLAN (Sanchez-Nieto and Nahum 2000); Warkentin et al. (2004) describe a similar system. El Naqa et al. (2005) offer a comprehensive software package which will not only compute TCP (and NTCP) but also fit these models to the user's own clinical data.

36.3 NORMAL TISSUE COMPLICATION PROBABILITY

36.3.1 GENERAL

The notion of using some type of biological index to characterize one or more features of a treatment plan was expounded upon early by Dritschilo who applied what he called a *complication probability factor* to rank rival treatment plans (Dritchillo et al. 1978). In spite of early attempts such as this, the increasing interest in NTCP modelling has, in large part, been a consequence of the development of 3D treatment planning systems and their use in conformal therapy. The potential of designing treatments with new and complex dose distributions to reduce normal tissue toxicity, and maybe escalate the dose to the tumour, lies behind attempts to characterize the merits of a treatment plan. This has become especially important because the limited clinical experience with innovative dose distributions makes it difficult to choose between two candidate treatment plans. NTCP is a particularly important index for ranking treatment plans. Like other quantitative indices (e.g. TCP), it reduces the large amount of data characterizing a three-dimensional dose distribution to a limited number of numerical indices, thus making the comparison of a number of rival plans more efficient. In addition, NTCP nominally represents an endpoint, which often determines the acceptability of a treatment plan, that is, whether the tolerance of normal tissues has been respected or exceeded.

The NTCP models currently in use stem from two main lines of development, both of which aim to predict the probability of a complication as a function of the dose (or biologically equivalent dose) and volume. These models can be distinguished by their descriptions of the *volume effect*, that is, how the probability of a complication changes with volume irradiated at a fixed dose. One of the lines of development has led to a phenomenological model which was first formulated by Lyman (1985), Lyman and Wolbarst (1987,1989a), and later augmented by Kutcher and Burman (1989) and Kutcher et al. (1991); this model is generally known as the L–K–B model (see Section 36.3.2). The L–K–B model seeks to describe complication probabilities using (clinical) tolerance doses for different irradiated volumes as input data. The tolerance doses for different volumes are related through a power law, and thus the clinical data are used to determine the magnitude of the volume effect, which can vary widely. In spite of this flexibility, the L–K–B model does not exhibit a threshold effect with volume and so would not strictly apply to endpoints like radiation pneumonitis. It is possible, however, to augment the model by introducing a fifth parameter, the so-called critical volume, below which there is no complication.

A second line of approach begins by making certain assumptions about the functional and anatomic organization of tissues and organs such that different volume effects are specific to different functional organizations. For example, *serial* (critical *element*) models assume that certain organs are arranged like links on a chain, and when one link (a functional subunit) is damaged the entire chain is broken and a complication ensues (Schultheiss et al. 1983; Wolbarst 1984; Withers et al. 1988; Niemierko and Goitein 1991). Organs with this architecture reveal a small volume effect. *Parallel* models (critical *volume* models) assume that a complication does not occur until a significant fraction of the independent functional sub-units (the *functional reserve*) of an organ has been incapacitated (Wolbarst et al. 1982; Källman et al. 1992; Jackson et al. 1993; Niemierko and Goitein 1993b; Yorke et al. 1993). Figure 36.12 is a schematic illustration of these ideas.

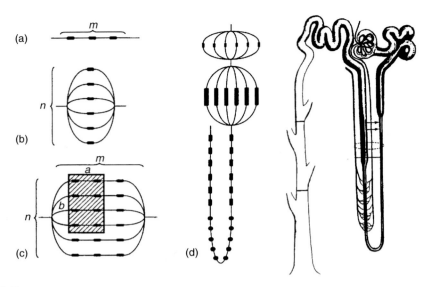

FIGURE 36.12
Schematic illustration of the concepts of (a) series, (b) parallel and (c) serial-parallel organisation of functional subunits in an organ; (d) is an example of the parallel-serial model applied to a functional subunit of the kidney, a nephron. (From Källman, P., et al., *Int. J. Radiat. Biol.*, 62, 249–262, 1992. With permission.)

The volume effect in these *parallel* tissues is necessarily large, since a complication does not occur if less than the functional reserve is irradiated. A model which makes explicit use of these ideas is the *relative seriality* model (Källman et al. 1992) described in Section 36.3.3.3.

36.3.2 Lyman–Kutcher–Burman (L–K–B) Empirical Model

36.3.2.1 *Homogeneous Irradiation*

Lyman (1985) argued that normal tissue complication probabilities depend upon volume as well as dose, and that they could be conveniently represented by an error function in dose and volume. For organs that are irradiated by a uniform dose to part of the volume and no dose to the rest, the probability of a complication can be represented by:

$$NTCP = \frac{1}{\sqrt{2\pi}} \int_{-\infty}^{t} e^{-t^2/2} dt \qquad (36.13)$$

$$t = \frac{D - TD_{50}(V/V_{\text{ref}})}{m\, TD_{50}(V/V_{\text{ref}})} \qquad (36.14)$$

$$TD_{50}(1) = TD_{50}(V/V_{\text{ref}})(V/V_{\text{ref}})^n \qquad (36.15)$$

The four parameters in this model are:

- $TD_{50}(1)$, the dose to the whole organ which would lead to complication in 50% of the population (note that $TD_{50}(V/V_{\text{ref}})$ is to be read as the TD_{50} at partial volume V/V_{ref});
- V_{ref}, a reference volume, which in many cases will be the (whole) organ volume;
- m, a parameter representing the steepness of the dose–response curve;

- n, the exponent of volume in the power law that relates the tolerance doses for uniform whole and uniform partial organ irradiation.

This latter parameter represents the *volume effect*: when n is near unity, the volume effect is large and when it is near zero, the volume effect is small. As we shall find, a large volume effect (n close to unity) implies that the NTCP correlates with the mean dose, while a small volume effect (n close to zero) implies that it correlates with the maximum dose in the organ. Moreover, referring to the architectural models which are described below, the analogue of a parallel architecture organ would be one with a large value of n, and for a serial organ, one with a small value of n. Figure 36.13 shows the NTCP function that is generated by Equation 36.13 (together with Equation 36.14 and Equation 36.15).

This figure was constructed by Lyman to represent then current estimates of partial-volume tolerance doses for the heart (Lyman 1985). It also serves to illustrate several features of the Lyman model. The sigmoid curve picked out in the figure corresponds to uniform irradiation of 50% of the heart volume, for which TD_5 is around 45 Gy and TD_{50} around 53 Gy. For a partial volume of unity the sigmoidal complication curve starts at around 30 Gy and reaches 100% at just over 50 Gy. As one moves towards *lower* values of the partial volume the sigmoid curve moves towards *higher* doses. In general, the more the curve moves (per unit decrease in partial volume) the greater is the volume effect, i.e. the larger the value of n (see also Figure 36.14).

The NTCP for partial organ irradiation in the Lyman model is based upon clinical esti-mates of organ tolerance for partial uniform irradiation at different doses and volumes. Such data are difficult to come by. Nevertheless, Emami et al. (1991a) estimated tolerance doses from the literature and unpublished experience. They chose to transform the data into one-third, two-thirds, and whole-organ tolerances, rather than using the fractions reported in the literature, and this forced them to interpolate the published data. Values of the parameters TD_{50}, m, and n, which provided the best fits of the Lyman model (Equation 36.13) to the Emami et al. partial organ tolerances are tabulated in Burman et al. (1991) for a large number of organs and endpoints. The values of n varied from near zero for some organs, e.g. the spinal cord, up to about 0.7 or greater for large organs like the liver and lung. Two examples of complication probabilities predicted by the Lyman model using the Burman et al. parameters are given in Figure 36.14.

It was clearly emphasised in Emami et al. (1991a) and Burman et al. (1991) that the partial organ tolerance doses themselves and hence the parameters derived from them were subject to considerable uncertainties. One important point to be made in this context is that these publi-cations date from the very beginning of the era of 3D treatment planning and hence there are

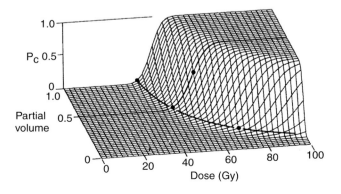

FIGURE 36.13
A 3D surface representation of the Lyman model (Equation 36.13) for complication probability P_C for uniform partial irradiation of an organ, the heart in this example, as a function of partial volume and dose. (From Lyman, J. T., *Radiat. Res.*, 104, S13–S19, 1985. With permission.)

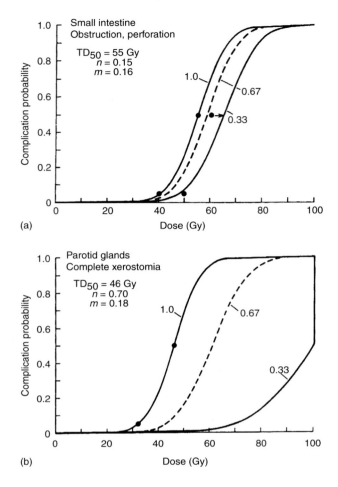

FIGURE 36.14
Complication probability, or NTCP, vs dose for two organs evaluated from the Lyman NTCP model (Equation 36.13) for uniform irradiation of the whole (1.0), two-thirds (0.67) and one-third (0.33) of the organ in question, for (a) obstruction or perforation in the small intestine, which has a *small* volume effect ($n = 0.15$) and (b) complete xerostomia in the parotid glands, which has a *large* volume effect ($n = 0.70$). All the parameters are taken from the fits to clinical data in Burman et al. (1991). (From Kutcher, G. J., et al., *Int. J. Radiat. Oncol. Biol. Phys.*, 21, 137–146, 1991. With permission.)

major limitations in the clinical data on which they were based. Since then, there have been a number of publications, which have suggested changes in the model parameters based upon new dose distributions, and toxicity data, which will be described in later sections (e.g. Ten Haken 2001).

36.3.2.2 Inhomogeneous Irradiation

Because of the uniform partial irradiation requirement the above approach will apply only to a limited number of treatments, namely those in which the organ is treated with parallel-opposed fields. For example, in head and neck cancer a large portion of the parotid glands are often irradiated to near uniform doses with opposed lateral fields, while the remainder receives a low (although nonzero) scatter dose. Nowadays, however, multi-field treatment techniques are the rule rather than the exception and these usually produce dose distributions that are inhomogeneous throughout the organ so that the above *uniform partial volume* modelling would not be applicable. Indeed, one of the difficulties with the data culled from the literature in the Emami compilation was that the condition of dose uniformity was not necessarily met.

It was necessary, therefore, to find a way of using the Lyman model in the more general case of inhomogeneous dose in the organ or tissue of interest. One way of doing this is to convert the organ's DVH into an *equivalent* uniform one using either interpolation (Lyman and Wolbarst 1989) or the so-called *effective volume* method (Kutcher and Burman 1989). If this DVH-conversion step is valid, then the dose and volume for the transformed uniform histogram can be used to read-off the complication probability from the Lyman representation of the uniform partial-organ data. In most instances both procedures lead to the same NTCP (Kutcher et al. 1991); here the latter method is described.

In the effective volume method, the DVH for the organ in question is transformed into one in which the volume V_{eff} (which is equal to or less than the whole organ volume) receives a dose equal to the *maximum* dose in the organ (D_{max}). One operates on each dose-bin of the *differential* DVH in turn, multiplying the fractional volume ΔV_i in that bin; of (mean) dose D_i, by $(D_i/D_{max})^{1/n}$ to yield the effective volume for that bin, ΔV_{eff}. Thus the effective volume is given by

$$V_{eff} = \Delta V_{max} + \Delta V_1(D_1/D_{max})^{1/n} + \Delta V_2(D_2/D_{max})^{1/n} + \ldots$$

i.e.

$$V_{eff} = \sum_{i=1}^{k} \Delta V_i(D_i/D_{max})^{1/n} \tag{36.16}$$

This is illustrated in Figure 36.15.

Moreover, there is a *family* of equivalent uniform DVHs with effective volume and dose related through the defining power-law relationship (Equation 36.15). Thus, there is an effective dose (D_{eff}) or *EUD* that, if applied to the whole organ (V_T), would yield the same NTCP:

$$D_{eff} = \sum_i \left[D_i^{1/n} \frac{V_i}{V_T} \right]^n \tag{36.17}$$

The effective volume transformation is self-consistent with the power-law model for uniform irradiation in that it can be derived from just two hypotheses: the organ is homogeneous in its response to radiation; and each element of the organ obeys the same power-law relationship as the whole organ (Kutcher and Burman 1989). The former condition is clearly problematic. Most normal organs are not homogeneous in response. For example, the

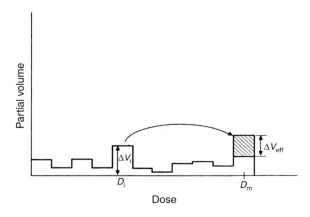

FIGURE 36.15
Illustration of *dose-volume histogram reduction* using the *effective volume* method; the DVH shown is in differential form, as opposed to the more familiar cumulative version. (From Kutcher, G. J., et al., *Int. J. Radiat. Oncol. Biol. Phys.*, 21, 137–146, 1991. With permission.)

nephrons of the kidney are not uniformly distributed, and the apex of the lung is less sensitive than the rest of that organ (e.g. Seppenwoolde et al. 2004).

It is instructive to consider the effect of different values of n on the calculation of V_{eff} and D_{eff}. For very small n, all terms $(D_i/D_{max})^{1/n}$ in Equation 36.16 will be negligible for all $D_i < D_{max}$ which will mean that all the volumes ΔV_i at lower doses will make a negligible contribution to the effective volume, which will therefore be essentially equal to the volume at maximum dose; in other words the complication probability will be solely determined by the fractional volume of the organ receiving doses very close to the tumour or prescription dose. For $n=1$, on the other hand, the effective dose (Equation 36.17) will be equal to the *mean* dose to the organ, and thus those parts of the organ receiving relatively low doses will play a significant role in contributing to the complication probability. Figure 36.16 is an example of how the NTCP of the parotid gland ($TD_{50}=46$ Gy, $n=0.70$) differs for two different treatment plans with quite different (cumulative) DVHs.

Niemierko has generalized the concept of EUD originally proposed for tumours (Niemierko 1997) to apply also to normal tissues and, in Niemierko (1999a), gave an expression for EUD identical to that for D_{eff} in Equation 36.17 above, except that the volume-effect exponent n was replaced by $1/a$. It was stated that a was a tissue-specific parameter, and that EUD was equal to the mean dose for $a=1$. Further, a was negative for tumours and positive for all normal structures. Building on this EUD concept for normal tissues, Niemierko (1999b) proposed a *generalized unified model*:

$$p = \Phi\left(\frac{\ln \text{EUD} - \ln D_{50}}{\sigma}\right) \qquad (36.18)$$

where $\Phi(x)$ is the standard normal distribution function, σ is the standard deviation of log-tolerance in a population and D_{50} is the median tolerance dose. Alternatively, this can be expressed as

$$p = \Phi\left[\gamma_{50} 2\pi (\ln \text{EUD} - \ln D_{50})\right] \qquad (36.19)$$

where γ_{50} is the slope of the normalized *NTCP* vs dose curve at $p=NTCP=50\%$ (see the definition of γ in Section 36.2.1.3). Niemierko stated that the fit of the model to several

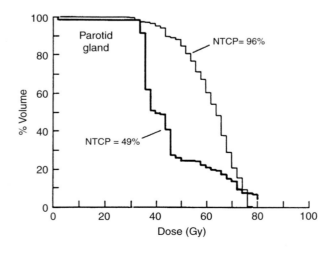

FIGURE 36.16
NTCP values for xerostomia according to the L–K–B model for two different parotid gland DVHs (from different treatment plans) using the Burman et al., 1991 parameter values given in Figure 36.14b. (From Kutcher, G. J. et al., *Int. J. Radiat. Oncol. Biol. Phys.*, 21, 137–146, 1991. With permission.)

tumour and normal-tissue data sets was at least as good as fits of "other more complicated models proposed in the literature".

36.3.3 ORGAN ARCHITECTURE MODELS

There have been a number of papers, quoted earlier, which have suggested that organ architecture is related to complication probability. Withers argued forcefully that the response of an organ to radiation might be related to the organisation of its more basic units (Withers et al. 1988). The simplest representation of an organ was that it was composed of independent *functional subunits* (FSUs) which could be defined either structurally (e.g. the nephrons of the kidney) or operationally (e.g. the FSUs of the skin). Moreover, it was suggested that the FSUs had either *serial* or *parallel* organizations. Since the FSUs are considered (by definition) to be independent of one another, at least in their response to radiation damage, the probability of a complication is amenable to statistical methods. In his paper, Withers applied these ideas to serial organs with uniform partial irradiation and derived the complication probability (see Section 36.3.3.1).

The NTCP for inhomogeneous dose distributions for both serial or parallel organs has been developed by Jackson et al. (1993) and Niemierko and Goitein (1991, 1993b) as well as by Källman et al. (1992)—see Section 36.3.3.3. Here, briefly, the reasoning of Jackson's solution is followed. Starting with a general inhomogenous dose distribution, Jackson derived a *risk histogram* (an extension of a DVH) which represented the family of fractional volumes of the organ in each of which the probability of eradicating its FSUs was equal to or greater than a certain probability. Using the risk histogram, he was able to calculate, by statistical methods, the probability $P(M)$ of eradicating M and only M FSUs. $P(M)$ is a complicated function of the organ architecture and the radiation conditions and the interested reader should refer to Jackson et al. for further details. An important point, however, is that once this function is known, it can be used to derive the complication probability for serial or parallel normal organs (as well as the control probability for tumours). For example, NTCP is calculated by summing $P(M)$ from a lower limit L (the minimum number of eradicated FSUs required to realize a complication) to N (the total number of FSUs in the organ):

$$NTCP = \sum_{M=L}^{N} P(M) \qquad (36.20)$$

If $L=1$, then we obtain a serial complication model where at least one FSU must be eradicated to realize a complication. If L is greater than unity, we have a parallel complication model in which at least L FSUs must be eradicated for a complication. Since the number of FSUs is usually quite large, typically 10^4–10^9 (Thames and Hendry 1987), the derived dose–response curves are very steep. If we interpret this result as the dose response of an organ in an individual, then it is possible to obtain more realistic dose–response curves by averaging over a *population* of patients (cf. the TCP model described in Section 36.2). For example, we can average the NTCPs over a distribution of functional reserves of an organ or we can average over intra- or inter-organ distributions of radiosensitivities. Such an approach leads to models with at least 4 parameters, two for the local response function, and two for each population distribution, although fewer are possible for a serial model (Niemierko and Goitein 1991).

36.3.3.1 Serial Organs

The description above is somewhat formal and general. We can obtain some insight by following Wither's argument for a serial chain model (e.g. Figure 36.12a). In the case of uniform irradiation of N of the FSUs and no radiation to the remainder, the probability of eradicating at least one FSU is given by one minus the probability of *not* eradicating any:

$$NTCP = 1 - (1-p)^N \qquad (36.21)$$

where p is the probability of eradicating a single FSU. This relationship yields a sigmoid curve of *NTCP* versus dose. As the number of irradiated FSUs increases (that is, as the volume of irradiated organ increases) the sigmoid response shifts to the left and the complication probability increases. In addition, the partial volume dose–response curves become steeper as the volume increases. Niemierko and Goitein (1991) have also shown that the serial model can be extended in a rather straightforward fashion for non-homogeneous irradiation, and that the serial and L–K–B models agree at low complication probability, that is in the clinically significant domain. Finally, for the serial case, the whole-organ dose–response curve can be shown to uniquely determine the volume effect of the organ.

36.3.3.2 Parallel Organs

In parallel-element tissues, which have also been modelled using binomial statistics for homogenous irradiation (Yorke et al. 1993) and for inhomogeneous irradiation (Jackson et al. 1993, Niemierko and Goitein 1993b), a complication occurs if the fraction of eradicated FSUs exceeds a threshold, the so-called *functional reserve* of the organ. The kidney, liver, and lung are believed to behave as parallel organs. Because the number of FSUs is always large in these organs (Thames and Hendry 1987), the functional reserve of an organ can be defined by the fraction rather than the number of eradicated FSUs. Moreover, as mentioned above, it is likely that this functional reserve varies over a population of patients. For example, liver function will differ depending, for example, on the age and alcohol consumption of the patients. It is also possible that radiosensitivities may vary over a population of patients (cf. Section 36.2.1.3) or that there are *intra*-organ variations, although Jackson et al. (1993) have shown that intra-organ distributions have negligible effects. For simplicity, if we assume that there is a population distribution of functional reserves only (i.e. not of radiosensitivities), then the observed NTCP would be proportional to the fraction of patients in which the local damage exceeds the functional reserve, that is

$$NTCP = H(f) \qquad (36.22)$$

where H is the cumulative distribution of functional reserves for the population and f is the fractional damage of FSUs, given by

$$f = \sum_i p(d_i)(V_i/V_T) \qquad (36.23)$$

where V_i is the volume of the organ receiving dose d_i with probability of damage $p(d_i)$.

Equation 36.22 and Equation 36.23 can be interpreted as follows. To obtain NTCP we first calculate the fractional damage from Equation 36.23 using the DVH for the organ and the local response function, $p(d)$. The NTCP is obtained by calculating the fraction of the population in which the fractional damage exceeds the functional reserve. This is nothing more than the area under the differential distribution of functional reserves from zero to the fractional damage, which is $H(f)$ (see Figure 36.17). To take one example, f might represent the fraction of *acini* of the liver that have been irradiated, while H would represent the fraction of the patients whose functional reserve for radiation hepatitis was fully depleted. In this representation we have a 4-parameter model where two of the parameters represent the distribution H (mean and width) and two parameters represent the $p(d_i)$.

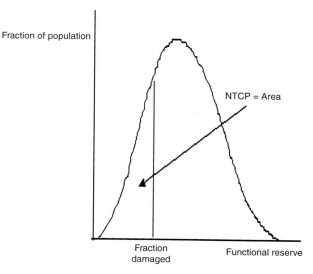

FIGURE 36.17
The calculation of NTCP in the parallel model in the limit where the distribution of functional reserves is used to obtain NTCP for a population.

36.3.3.3 *The Relative Seriality Model*

The so-called *relative seriality* model (Källman et al. 1992) represents an alternative attempt to construct a quasi-mechanistic model for NTCP which takes explicit account of the structure or *architecture* of an organ; the latter is assumed to consist of FSUs and/or *tissue rescuing units* (TRUs). These units are arranged in series, in parallel or in a combination of series and parallel as illustrated in Figure 36.12. The model includes a parameter s which reflects the degree to which the architecture of the FSUs is considered to be *serial* ($s = 1$) or *parallel* ($s \rightarrow 0$).

Källman et al. began by writing down the dose–response relationship for tumours (i.e. TCP) and normal tissues (NTCP), and included the probit, logit and Poisson models, the latter being expressed as

$$P(D_{\text{tot}}) = 2^{-\exp[e\gamma(1-D_{\text{tot}}/\text{TD}_{50})]} \tag{36.24}$$

where $P(D)$ is the probability of no cell surviving (and we retain here the upper-case Källman et al. notation for probability), D_{tot} is the total dose delivered (in small fractions), TD_{50} is the dose causing a 50% response rate (identical to TD_{50} in the L–K–B model), and γ is the normalised dose-response gradient (see Section 36.2.1.3). Källman et al. derived Equation 36.24 starting from the linear-quadratic model expression for the surviving fraction after fractionated dose delivery (Equation 36.3) and then use Poisson statistics to obtain the probability of eradicating all the cells (i.e. Equation 36.5)[*].

These authors then reasoned that the response, i.e. the NTCP, of an organ consisting purely of m serial subunits with local response P_i would be given by (cf. Equation 36.21)

$$NTCP = 1 - \prod_{i=1}^{m}(1 - P_i) \tag{36.25}$$

---

[*] Källman et al. show that $\gamma = (\ln N_0 + \beta\overline{D}^2/n)/e$ and $TD_{50} \approx \ln N_0/(\alpha + 2\beta\overline{D}/n)$ for n fractions and mean organ dose $\overline{D}$. Källman et al. use the notation D rather than D_{tot} and D_{50} rather than TD_{50}, but their notation has been changed here for consistency with the rest of the chapter.

whereas for an organ with n parallel subunits with local response P_j,

$$NTCP = \prod_{j=1}^{n} P_j \tag{36.26}$$

For an organ with an $n \times m$ matrix of parallel and serial subunits (e.g. Figure 36.12c), the overall complication probability would then be given by

$$NTCP = \prod_{j=1}^{n} \left[1 - \prod_{i=1}^{m}(1 - P_{ij}) \right] \tag{36.27}$$

Källman et al. now assumed that all the subunits had identical sensitivities and further that the dose distribution in the organ was uniform. Writing all $P_{ij} = P_\Delta$ one can then write Equation 36.27 as

$$NTCP = \left[1 - (1 - P_\Delta)^m \right]^n \tag{36.28}$$

from which follows, by inverting Equation 36.28, the key step of expressing the *local* response P_Δ in terms of the organ response, i.e. NTCP:

$$P_\Delta = 1 - (1 - NTCP^{1/n})^{1/m} \tag{36.29}$$

At this point in the derivation, it became necessary for the authors to introduce further approximations and the model became more empirical in nature. The identity of the elementary compartments changed from FSUs to simply *volume elements* or *voxels*. In particular, they pointed out that the response of a structured tissue or organ would depend critically on the geometrical alignment of structures relative to the beam direction(s), e.g. the parallel structures and the serial structures may be in different locations. Källman et al. then introduced the parameter s describing the *relative seriality* of the organ or tissue. It was shown that $s = 1/n$ and the response function of a subunit v of an organ, P_v, could be written

$$P_v = \left[1 - (1 - P(D_{tot})^s)^v \right]^{1/s} \tag{36.30}$$

where $P(D_{tot})$ is given by Equation 36.24 and v is the fractional volume of the subunit.

After some further algebraic manipulation, it was shown that for M identical volume elements of fractional volume Δv,

$$NTCP = [1 - (1 - (P_{\Delta v})^s)^M]^{1/s} \tag{36.31}$$

and finally, generalising Equation 36.30 to a *non-uniform* dose distribution, and expressing $P_{\Delta v}(D_i)$ in terms of $P(D_i)$ for the whole organ using Equation 36.29, the final expression for the *relative seriality* NTCP is obtained:

$$NTCP = \left[1 - \prod_{i=1}^{M} \left[1 - P(D_i)^s \right]^{\Delta v_i} \right]^{1/s} \tag{36.32}$$

where $P(D_i)$ is evaluated from Equation 36.24 and the fractional volumed Δv_i at doses D_i are most conveniently obtained from the M bins of the differential DVH for the organ in question. Therefore, no DVH reduction is required in this model.

Källman et al. (1992) give values of TD_{50}, γ and s for seven organs/endpoints based on the Emami et al. (1991a) compilation of clinical complication data. As examples, for the esophagus $s = 3.4$ which corresponds to an organ with a highly serial structure, and for the lung, $s = 0.0061$ which corresponds to a highly parallel structure.

The relative seriality model has been extensively employed in analysing treatment plans and in correlating clinical outcome with NTCP predictions (e.g. Gagliardi et al. 1996, 2000; Mackay et al. 1997; Gagliardi 1998). As an example, for the heart, for the endpoint of cardiac mortality, Gagliardi et al. (2001) give $TD_{50}=52.3$ Gy, $\gamma=1.28$ and $s=1.00$; these values were obtained from analysing patient data from two randomised breast radiotherapy trials.

36.3.3.4 Relationships between L–K–B, Serial and Parallel Models

Although the L–K–B and architecturally based models are derived in different ways they yield similar results for NTCP within certain limits. For example, as has already been mentioned, for small values of n the NTCP for the L–K–B model approaches that of the serial model. Thus we can think of organs with a small n as serial-like in their behaviour. Moreover, for small n the NTCP is determined primarily by the peak dose and is thus weakly correlated with the irradiated volume. As n increases so does the influence of volume on the probability of a complication. For $n=1$, the NTCP is a function of the mean dose delivered to the organ. For the two extremes, $n=0$ and $n=1$, to a first approximation, the NTCP is thus correlated with dosimetric variables. When evaluating treatment plans in this limit, we do not have to resort to making any further assumptions about the biological models and the values of the parameters. It therefore possible to use these two extremes as a starting point or rough approximation when evaluating a treatment plan. We might, for example, use the mean dose when the lung, liver, or parotid glands are irradiated, and the peak dose for the spinal cord or small bowel.

It is important to note that organs with serial or parallel architecture have very different volume effects, as shown in Figure 36.18. For serial organs, the NTCP increases linearly with the irradiated volume, while for parallel organs, the NTCP is zero up to a critical volume and then increases supralinearly.

Finally, Kwa et al. (1998a) have demonstrated a relationship between the L–K–B and parallel models. For the parallel model if we assume that for moderate doses the probability of killing an FSU can be represented by a power law in dose:

$$p(d) = \text{constant} \times d^{1/r} \qquad (36.33)$$

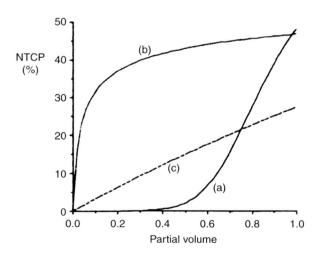

FIGURE 36.18
Comparison of the change in NTCP with volume at fixed dose (i.e. the volume effect), for parallel tissues (critical *volume*: curve (a)) and series (critical *element*) tissues with (curve (b): $\sigma=15\%$) and without (curve (c): $\sigma=0\%$) account being taken of interpatient variations in the model parameters. The NTCP for serial tissues show a linear response with volume irradiated, while for parallel tissues there is a threshold response. (From Niemierko, A. and Goitein, M., *Int. J. Radiat. Oncol. Biol. Phys.*, 25, 135–145, 1993b.)

then it is possible to show that D_{eff}, or the *EUD*, can be represented by:

$$D_{\text{eff}}^{\text{parallel}} = \sum_i [D_i^r V_i / V_T]^{1/r} \qquad (36.34)$$

This expression is the same as that for the L–K–B model (Equation 36.17), if $n = 1/r$. The assumption of a power law for $p(d)$ is quite reasonable in the limit where the dose delivered is smaller than that required for killing half the FSUs. We would expect that the above correspondence between parallel and L–K–B models would break down at dose levels where there is a very high level of cell killing.

36.3.4 THE EFFECT OF FRACTION SIZE

NTCP models ought to be able to predict the effect of different fraction sizes, consistent with our current radiobiological understanding (see Section 9.3). In the case of the TCP model described in Section 36.2 the influence of fraction size is automatically taken into account as the linear-quadratic expression for the surviving fraction of irradiated cells is used directly in the model, which requires as input the fraction size (strictly speaking all the fraction sizes in the differential DVH) and also α/β for the clonogens. The clinical data on which the Emami compilation of tolerance doses is based corresponds to 1.8–2.0 Gy fractions delivered to the tumour (Emami et al. 1991a). In the case of the L–K–B and relative seriality NTCP models, however, there is no obvious way to include fraction-size effects inside the theoretical expressions as neither fraction number nor the α/β-ratio explicitly appear anywhere.

Fowler (2001) stressed that for the several different NTCP models in the literature it was *essential* that the linear-quadratic formula be used to correct the total doses delivered at a range of different doses-per-fraction to the biological equivalent of 2-Gy fractions *in every voxel*. Fowler continued, "Fortunately these corrections are easy to make, involving basically a re-scaling of the dose scale of a DVH". We give here the expression as it appears in Wheldon et al. (1998):

$$LQED_2 = D_{tot,j}(\alpha/\beta + d_j)/(\alpha/\beta + 2) \qquad (36.35)$$

where $LQED_2$ denotes the *linear-quadratic equivalent dose for 2-Gy fractions* (Fowler 1989), $D_{tot,j}$ is the total dose and d_j the fraction size at that total dose for the jth dose bin in the DVH. The effect of applying Equation 36.35 to a DVH is shown in Figure 36.19. All volumes at doses below the total dose at which the fraction size is 2 Gy (≈ 60 Gy in this case) are shifted to lower $LQED_2$ doses and all volumes above this (prescription) dose are shifted to higher $LQED_2$ doses.

The effect on NTCP values of carrying out the above conversion of total doses to $LQED_2$ at the various fraction sizes in the normal-tissue DVH according to Equation 36.35 can be clearly seen in the example given in Table 36.2. It is supposed that the organ in question is the rectum and the Rancati et al. (2003) updated parameters for the endpoint *proctitis* have been used. The reference prescription is 74 Gy in 2 Gy fractions. The fraction size has been varied from 1.6 Gy to 5.9 Gy, keeping the total dose isoeffective for $\alpha/\beta = 3$ Gy. The NTCP values, which take into account the fraction-size variation, are approximately constant (7.3% to 8.0%) whilst the NTCP *without any biological correction of the doses* vary between 15.7% and 0.1%.

It can be noted that the fraction-size corrected NTCP values are not quite constant but decrease slowly with increasing fraction size. This is a consequence of the fact that the single dose d in the Withers formula for deriving the isoeffective total dose is conventionally set equal to the prescription fraction size (see Equation 9.5 and, e.g. Bentzen and Baumann 2002) whereas the *effective* dose in any normal tissue will always be lower than that. The Withers formula will therefore tend to *underestimate* the true (normal-tissue) isoeffective dose for

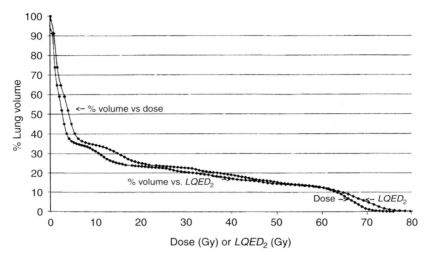

FIGURE 36.19
The conversion of a (cumulative) dose volume histogram into a 2 Gy per fraction equivalent DVH (using $\alpha/\beta=3.3$ Gy) for a typical lung in a 3D conformal plan. (From Yorke, E. D., *Semin. Radiat. Oncol.*, 11, 197–209, 2001. With permission.)

fraction sizes larger than 2 Gy and hence the NTCP values will be lower than might have been expected (Nahum et al. 2003a).

36.3.5 RELATIONSHIP OF MODELS TO CLINICAL COMPLICATIONS

36.3.5.1 *Radiation Proctitis*

Quantitative evidence for a volume effect in the rectum has changed rather significantly due to interest in conformal therapy and dose escalation in prostate cancer. Earlier, Emami estimated that the whole volume tolerance dose for 5% complications in the rectum was 60 Gy (Emami et al. 1991a). In addition, it was suggested that the same value of TD_5 be used no matter what volume was irradiated, because there was a minimal volume effect at best. In contrast, Benk et al. (1993) published a clinical study of 41 patients treated for prostate cancer with photons and protons, and found a statistically significant increase in the probability

TABLE 36.2

The Effect of the $LQED_2$ Correction to the DVH on NTCP for the Rectum/Proctitis

			$NTCP_{L-K-B}$ (%)	
Fraction Size (Gy)	Fraction Number	Total Prescription Dose (Gy)	No *Biological* Correction	With *LQED₂* Correction
1.606	50	80.32	15.7	8.0
2.0 (ref.)	37	74.0	9.0	7.9
3.055	20	61.10	2.1	7.6
4.252	12	51.02	0.5	7.5
5.923	7	41.46	0.1	7.3

Data evaluated using the L–K–B model, as the fraction size is varied at isoeffective ($\alpha/\beta=3$ Gy) prescription doses; the parameters used were $TD_{50}=81.9$ Gy, $m=0.19$; $n=0.23$. $LQED_2$ corrections to the (total) doses in the DVH assumed $\alpha/\beta=3$ Gy. The DVH used was the OAR.dvh file supplied with the BIOPLAN biological model evaluation software (Sanchez-Nieto and Nahum 2000; Rancati et al. 2003).

of Grade-2 rectal bleeding when more than 40% of the anterior portion of the rectal wall received at least 75 Cobalt-Gray-Equivalent (see Chapter 46). The prostate and pelvis were irradiated with photons, while the prostate was boosted with a single proton field (Shipley et al. 1979). Since for the latter phase the rectum was immobilised with a stent, it was possible to know the position of the rectum and therefore more accurately determine dose-volume histograms for the rectal wall.

The presence of a volume effect has also been confirmed in other studies. For example, the results of the dose escalation study by the group at Memorial Sloan-Kettering in New York makes a compelling case for a rectal volume effect. They were able to raise the dose to the prostate to 81 Gy and more while keeping rectal bleeding below levels commonly found at much lower doses (Leibel et al. 1994; Zelefsky et al. 1995). This was possible since the dose escalation was constrained by limiting rectal irradiation such that no more than 30% of the rectal wall exceeded 75 Gy (Kutcher et al. 1996). Moreover, it was also found that rectal bleeding was reduced when some of the rectum was blocked in the boost phase of the treatment (Schultheiss et al. 1995). Dearnaley et al. (1999) reported on the first randomised clinical trial of conformal vs conventional radiotherapy of the prostate in which a statistically significant reduction in the incidence of Grade-2 rectal bleeding was found in the conformal arm.

There remain a number of questions. How does rectal *motion* influence the correlation of dose and volume with rectal bleeding? What is the relative influence on rectal bleeding between increasing the length and increasing the fraction of the circumference of irradiated rectal wall? If we consider the rectum in two extremes, one in which the FSUs of the rectum are arranged as in Figure 36.20a, then full circumferential radiation would be required for a complication, while if the FSUs are arranged as in Figure 36.20b, then even partial circumferential irradiation would be sufficient. Dale et al. (1999) should be consulted for further discussion on this point.

Fenwick (1999) calculated rectal *dose-surface-histograms* for 79 prostate cancer patients, who received, using conformal or conventional radiotherapy techniques, a tumour (PTV) dose of 64 Gy (Tait et al. 1997; Dearnaley et al. 1999; Fenwick et al. 2001). Rectal bleeding complication data (dichotomised as Grade 0 versus Grade 1, 2, 3) was also available for these patients. The dependence of patients' complication status on rectal dose-distribution was statistically explored by fitting a parallel bleeding model to the dose-surface histogram and complication data (Fenwick et al. 2001). The *maximum-likelihood* fit of the model to the data suggests that local rectal-wall radiation damage increases sharply around a (physical) radiation dose of 58 Gy, although it should be noted that the dose-response is not defined very tightly by the data. The fit also indicates that patients' complication status depends significantly on the damage (calculated using the maximum-likelihood fit dose-response) totalled over the rectal wall: a reduction of 1% in the fraction of rectal wall receiving a dose

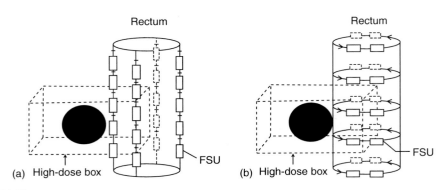

FIGURE 36.20
Two possible arrangements of FSUs for the rectum: (a) serially arranged along the length and (b) serially arranged around the circumference of the rectal wall; the prostate is represented by the black sphere. (From Dale, E., et al., *Int. J. Radiat. Oncol. Biol. Phys.*, 43, 385–391, 1999. With permission.)

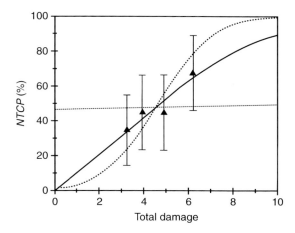

FIGURE 36.21
A plot of the maximum-likelihood fit curve (unbroken line) describing the dependence of the probability of Grade 1–3 bleeding on total damage (total number of radiation induced bleeding sites calculated as the fraction of the rectal surface area receiving a dose of more than 58 Gy), together with the steepest and shallowest sigmoid curves (dotted lines) which adequately fit the data. Also shown are points representing the observed complication rates (with 95% confidence intervals) and average calculated total damage for four cohorts of patients grouped by total damage. (From Fenwick, J. D., et al., *Int. J. Radiat. Oncol. Biol. Phys.*, 49, 473–480, 2001. With permission.)

of 58 Gy or more leads to a fall in the Grade 1,2,3 bleeding rate of about 1.1% (95% confidence interval [0.04%, 2.2%]) (Figure 36.21).

Studies such as those of Hartford et al. (1996), Boersma et al. (1998), Jackson et al. (2001), Fiorino et al. (2002a), Akimoto et al. (2004), Rancati et al. (2004), and Tucker et al. (2004) support the finding that the incidence of rectal bleeding decreases as the fractional area of rectal wall irradiated to intermediate or high dose-levels is reduced.

Rancati et al. (2003) have fitted late rectal-bleeding data using four different NTCP models; the data were the rectal dose-volume histograms of 547 prostate cancer patients pooled from five institutions previously analysed by Fiorino et al. (2003). All patients in the study were treated with one fraction per day, 5 days a week, with the fraction size in the range 1.8 Gy to 2.0 Gy. It should be noted that the volume contoured for which the DVH was computed always corresponded to that of the *solid* rectum, i.e. not the more logical dose-wall histogram or the dose-surface histogram (Lu et al. 1995) computed by Fenwick et al. (2001). The authors admit that this was a practical rather than an ideal solution[*]. The four NTCP models were the Lyman or probit model (Section 36.3.2) with the DVH *reduced* so that the whole organ volume was treated as receiving the EUD, using Equation 36.17, the relative seriality model (Section 36.3.3.3), the logistic model[†] and the Poisson model (Equation 36.24), both with the DVH reduced to the whole number *EUD*. The estimated best fit for the complete set of patients developing $\geq$ Grade 2 bleeding (38 out of 547 or 6.9%) in the case of the Lyman model, with the dose expressed in terms of the (whole-volume) *EUD*, is shown in Figure 36.22. In fact, the Lyman, relative seriality and logistic models yielded virtually indistinguishable NTCP values for whole, 2/3 and 1/3 organ irradiation over the range of EUDs encountered in clinical practice.

---

[*] See Fiorino et al. (2003) for a detailed study of this issue.
[†] The expression for NTCP for this model, also known as the logit formula, is $NTCP\,(D) = 1/[1 + (D_{50}/D)^k]$ where k is the slope of the NTCP vs dose curve at $D = TD_{50}$ and is related to the slope parameter m in the Lyman or probit formula through $k = 1.6/m$.

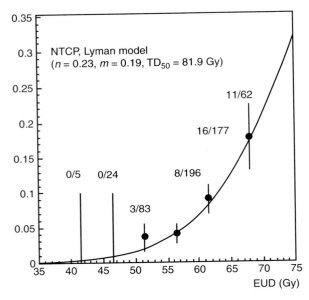

FIGURE 36.22
Incidence of late rectal bleeding as a function of EUD for the Lyman-EUD model: observed complication rates, shown as *number of complications/number of patients* in each EUD interval (solid circles) and predicted NTCP curve (full line, obtained using best estimates of parameters *n*, *m* and TD_{50} as shown—see Table 36.2. (From Rancati, T. et al., *Radiother. Oncol.*, 67, 275–283, 2003.)

The complete set of best-fit parameter values for the four different models, with all Grade ≥ 2 scored as *bleeders*, derived from the complete patient population (non-conformal and radical 3D-conformal as well as post-prostatectomy treatments, and prescription doses ranging between 64 Gy and 79.2 Gy), is given in Table 36.3. Note that the fitting exercise was performed both with and without applying corrections to the DVHs for fraction size, using the L-Q model with $\alpha/\beta = 3$ Gy (see Section 36.3.4).

When only severe (Grade 3) bleeding was considered, the dose-response curve had a steeper slope ($m = 0.06$ instead of 0.19) and the volume-effect parameter $n = 0.06$ instead of 0.23; this suggests that for this endpoint the rectum behaves in a more serial manner. However, the number of patients experiencing Grade 3 bleeding was only 10 out of the 547 and therefore this finding, however interesting, must be treated with caution.

The authors conclude by pointing out some *truths* concerning NTCP models. They emphasise that all the models used in their analysis are phenomenological in nature and thus the resulting dose-response curves cannot be considered to encompass the complexities of rectal radiobiology. Further, the good agreement as demonstrated in Figure 36.22 cannot be taken to imply that the Lyman-EUD model (in this case) used with the best-fit parameters will make reliable *predictions* when applied to arbitrary rectal DVHs. In fact, the predictions made by the models using the parameters obtained in the present study are to be tested in a separate ongoing *prospective* study on a large, independent patient population. Furthermore, this new study will include not only dose-volume histograms but also rectal surface dose maps (see Figure 36.27 in Section 36.3.6).

36.3.5.2 *Radiation Hepatitis (Radiation Induced Liver Disease)*

One of the most informative examples of how NTCP models and data can be successfully used is by Jackson and colleagues (1995) for radiation-induced liver disease (RILD also called radiation hepatitis). They used clinical outcomes in 93 patients treated for tumours of the liver, dose-volume histograms, the parallel model and the method of maximum likelihood to

TABLE 36.3

Estimated Parameter Values for the Four NTCP Models, with 1D-68% Confidence Intervals

Model	TD_{50} (Gy)	m or k or γ	n or s
Lyman-EUD			
No correction for fraction size	81.9 ± 1.8	0.19 ± 0.01	0.23 ± 0.05
LQED$_2$ correction ($\alpha/\beta=3$ Gy)	81.8 ± 2.3	0.22 ± 0.01	0.29 ± 0.06
Logit-EUD			
No correction for fraction size	82.2 ± 1.8	7.85 ± 0.5	0.24 ± 0.05
LQED$_2$ correction ($\alpha/\beta=3$ Gy)	83.5 ± 2.3	6.1 ± 0.4	0.30 ± 0.06
Poisson-EUD			
No correction for fraction size	87.7 ± 1.9	0.51 ± 0.07	0.23 ± 0.04
LQED$_2$ correction ($\alpha/\beta=3$ Gy)	88.2 ± 2.4	1.30 ± 0.06	0.29 ± 0.06
Relative seriality			
No correction for fraction size	83.1 ± 1.8	1.69 ± 0.13	0.49 ± 0.12
LQED$_2$ correction ($\alpha/\beta=3$ Gy)	83.6 ± 2.3	1.42 ± 0.11	0.50 ± 0.12

m applies to the Lyman model, k to the Logistic model and γ to the Poisson and RS models; n applies to the EUD DVH-reduction method, s to the RS model.
Source: From Rancati, T., et al., *Radiother. Oncol.*, 67, 275–283, 2003. With permission.

determine the model parameters (Lawrence et al. 1990; Lawrence et al. 1992; Jackson et al. 1995). They assumed that the organ had a functional reserve described by two parameters (whose values were obtained from the maximum likelihood fit) and that NTCP was given by the integral of the functional reserve up to the mean fraction of eradicated FSUs, that is, up to the fraction damaged. For each patient the fraction damaged was calculated by summing the product of the fractional volume of each voxel of the organ times the probability of damage. The latter was calculated using each patient's DVH and an assumed local-response function with two parameters (whose values were also obtained from the maximum likelihood fit).

The method of maximum likelihood was used by first assigning a best guess for the model parameters. The predicted probability of a complication for each patient was then compared against the observed grade of complication in that patient. The overall likelihood L of the observations was taken as:

$$L(\gamma_1,\gamma_2,...) = \prod_{\substack{m \\ \text{complication}}} O_m(t_m,\gamma_1,\gamma_2,...) \prod_{\substack{n \\ \text{no complication}}} (1 - O_n(t_n,\gamma_1,\gamma_2...)). \qquad (36.36)$$

where

$$O_m(t_m,\gamma_1,\gamma_2,...) = o(t_m)\text{NTCP}_m(\gamma_1,\gamma_2...), \qquad (36.37)$$

where L is the likelihood, $o(t_m)$ is the probability that a complication will manifest itself after the follow-up time t_m for the mth patient, calculated from complication and follow-up time data with the Kaplan–Meier method (Kaplan and Meier 1958), and $\gamma_1,\gamma_2,...$ indicates the model parameters. The likelihood is a product over all patients of the chance that there is a complication multiplied by the chance there is no complication, which is maximised with respect to the model parameters.

The results of applying such an analysis to the radiation hepatitis data are shown in Figure 36.23; the observed complication rate has been plotted as a function of calculated fraction of the liver, which is damaged. The observed complications are zero up to a threshold, followed by a steep response, exactly as predicted by the parallel model. The

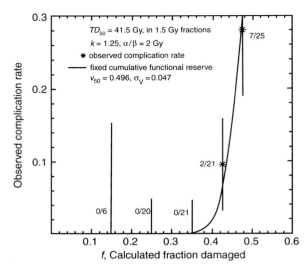

FIGURE 36.23
The observed complication rate for radiation hepatitis as a function of calculated fraction of damaged liver. (From Jackson et al., *Int. J. Radiat. Oncol. Biol. Phys.*, 31, 883–891, 1995. With permission.)

best fit for the functional reserve distribution (and confidence intervals) predicts that a threshold volume of about one-third must be exceeded before complications are observed. Moreover, complications were seen only in the cohort of patients that received whole-liver irradiation as part of the treatment course. The lack of complications among patients given only partial-volume irradiation suggests that an attempt to increase local control through dose escalation is feasible for these patients. Furthermore, the observation of complications in this subset of patients would be likely to reduce the correlated uncertainties found for the model parameters.

Dawson et al. (2002) carried out an analysis of radiation-induced liver disease using the L–K–B model. This was an updating of their earlier analysis (Lawrence et al. 1992), now extended from the original 79 patients to 203 patients, of which 19 developed Grade 3 or higher RILD. The patients had either unresectable intrahepatic cancer or metastic spread to the liver from colorectal carcinoma. All patients were treated on dose-escalation protocols, 61 with whole-liver irradiation, 20 of which had a boost to a partial liver volume, and 142 were treated with partial liver irradiation. Two fractions a day of 1.5–1.65 Gy/fraction (11 fractions/week) were delivered with a minimum inter-fraction interval of between 4 h and 6 h. Either fluorodeoxyuridine (FUDR) or bromodeoxyuridine (BUDR) was used as a radiation sensitizer for the first four weeks. Three-dimensional techniques were used throughout and the targets were expanded to account for both setup uncertainty and breathing motion. The GTV was subtracted from the organ to obtain *normal liver* DVHs for the NTCP analysis. The physical (total) dose values were converted to biologically isoeffective doses at 1.5 Gy/fraction according to Equation 36.35 using $\alpha/\beta = 2$ Gy, i.e. to $LQED_{1.5}$, before the DVHs were reduced to effective volumes V_{eff} at maximum dose (see Section 36.3.2.2). Table 36.4 gives the new, previous (Lawrence et al. 1992) and original (Burman et al. 1991) L–K–B model parameters for radiation-induced liver disease.

The differences between the three parameter sets in Table 36.4 are expressed in terms of ten-percent iso-NTCP curves in V_{eff}–*dose* space in Figure 36.24. The dashed curve corresponds to the original Burman 1991 parameter set, the intermediate curve to the Lawrence et al. (1992) set and the upper curve to the Dawson et al. (2002) set based on 203 patients with correspondingly narrow confidence limits. Quite clearly the Burman 1991 estimates were too conservative as the two upper curves indicate that considerably larger partial volumes than

TABLE 36.4

Radiation-Induced Liver Disease-Published Values of the Parameters for the Lyman–Kutcher–Burman NTCP Model

Source of Data	TD_{50} (Gy)	m	n
Burman et al. (1991)	40	0.15	0.32
Lawrence et al. (1992), 79 patients	45	0.15	0.67
Dawson et al. (2002)[a], 203 patients (95% C.I.)	43.3, (41.9–52.8)	0.18, (0.14–0.24)	1.1, (0.88–1.6)

[a]   The most recent parameter set can be considered as the most reliable—see text.

those indicated by dashed curve can be irradiated at all doses in the 40 Gy to 100 Gy range without exceeding 10% probability for radiation hepatitis.

The value $n = 1.1$ in the 209-patient fit suggested to Dawson et al. that there should be a strong correlation between RILD and *mean dose* in the normal liver. The observed complications were stratified into 4 Gy wide mean-dose bins and compared to the predictions of the L–K–B model using the authors' best-fit parameter set in Figure 36.25. No patient developed RILD at a mean liver dose below 30 Gy, which corresponds to a *threshold* value characteristic of a parallel organ. It should be noted that the L–K–B model cannot predict thresholds for RILD, unlike Jackson et al. (1995), and hence may overestimate the complication probability at small volumes of liver treated to high dose.

Dawson et al. also performed separate analyses on the patients with primary malignancies and those with liver metastases treated with FUDR. Respectively, they obtained $n = 0.90$ (0.66–2.0), $m = 0.09$ (0.05–0.23) and $TD_{50}(1)_{HB} = 39.6$ Gy (37.1–43.7) for the 84 patients with primary malignancies and $n = 1.27$ (0.70–>3), $m = 0.18$ (0.10–0.4) and $TD_{50}(1)_{LM} = 50.8$ Gy (45.4–99.0) for the 85 patients with liver metastases.

The Ann Arbor group (i.e. Ten Haken and colleagues) use the L–K–B NTCP model in an active way to plan the treatment of patients with unresectable intrahepatic malignancies; using

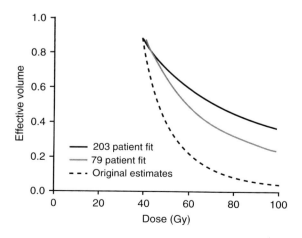

FIGURE 36.24

Ten percent iso-NTCP curves displayed in Effective Volume vs. Dose space (normalised to 1.5 Gy b.i.d.) for the three sets of parameters given in Table 36.3; the most recent 203-patient fit is the uppermost curve and suggests that the original Burman et al. estimate was far too pessimistic. (From Dawson L. et al., *Int. J. Radiat. Oncol. Biol. Phys.*, 53, 810–821, 2002. With permission.)

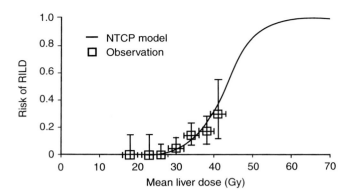

FIGURE 36.25
Observed NTCP for liver complications compared to predictions of the Lyman–Kutcher–Burman NTCP model with the patients grouped into mean-liver dose bins 4 Gy in width; the error bars correspond to 80% confidence intervals. The parameters used in calculating the NTCP curve were $n=1.1$, $m=0.18$ and $TD_{50}(1)=43.3$ Gy. (From Dawson L. A. et al., *Int. J. Radiat. Oncol. Biol. Phys.*, 53, 810–821, 2002. With permission.)

their revised set of L–K–B parameters (see Table 36.4) they determine the safe (defined as a 10% risk of RILD) prescription dose up to 90 Gy (McGinn et al. 1998).

36.3.5.3 *Radiation Pneumonitis*

The need to limit complications in the lung plays an important role in the treatment of several tumours in the thoracic region of the body, including lung tumours themselves, tumours in the breast and malignant lymphoma. Severe radiation pneumonitis in the first six months following radiotherapy can be fatal. It is therefore essential to know as much as possible about the relationship between the 3-D dose distribution and the incidence of radiation pneumonitis.

Convincing evidence for a large volume effect for pneumonitis in *mouse* lung was published by Liao, Travis, and Tucker (Liao et al. 1995). They demonstrated that mice with 70% of the lung irradiated to 20 Gy in one fraction were all dead after 28 weeks. In contrast, those with 40% of the lung irradiated to 20 Gy all survived. Additionally, lethality occurred at lower doses for mice given whole lung irradiation than for mice with 70% of the lung irradiated. These data strongly suggest that a parallel model may be appropriate for pneumonitis, since a tolerance dose for death after 28 weeks was not found for mice given 40% lung irradiation.

They have also argued that the base of mouse lung is more radiosensitive than the apex. If such a large difference were seen in (human) clinical data, then for equal numbers of tumours in the base and apex, when the overall complication rate is 20%, we would estimate that the complication rate for tumours in the base of the lung would be of the order of 30% higher that for tumours in the apex. Evidence to support this observation comes from Martel et al. (1994), who reported that all nine lung cancer patients with pneumonitis ($\geq$ Grade 1) had tumours in the middle or lower lobes, and Graham et al. (1995), who found that the incidence of $\geq$ Grade 3 pneumonitis was 6% in 49 patients with upper lobe tumours and 29% in 21 patients with middle or lower lobe tumours.

The clinical evidence for lung tolerance prior to conformal radiation therapy has been reviewed by Travis (1991) and by Marks (1994). They point out, for example, that whole lung volume tolerance doses that were established for patients treated with hemi-body irradiation (Van Dyk et al. 1981), after corrections for fractionation, are generally consistent with tolerance doses deduced from patients treated with various fractionation schedules to volumes in excess of 75% (Wara et al. 1973). Few quantitative data for partial volume tolerances were available before the advent of conformal treatment techniques. However, there were indications that the probability of pneumonitis was related to the volume of lung irradiated

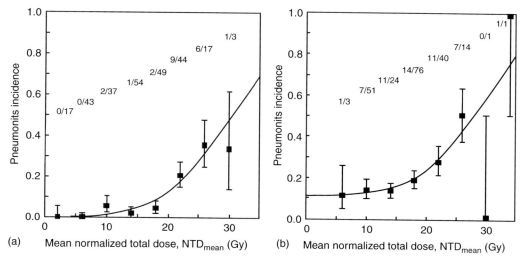

FIGURE 36.26

The incidence of radiation pneumonitis as a function of mean dose corrected for fractionation (NTD$_{mean}$) for a population of 504 patients from 5 institutions: (a) all lymphoma and breast patients and lung patients from one of the institutions ($N=264$); (b) all other lung patients ($N=276$). (From Kwa, S. L. S. et al., *Int. J. Radiat. Oncol. Biol. Phys.*, 42, 1–9, 1998b. With permission.)

(Brady et al. 1965a; Brady et al. 1965b; Perez et al. 1980; Rothwell et al. 1985; Seydel et al. 1985). Emami et al. (1991a) in their review of the literature concluded that the volume effect in the lung was very large, and the Burman et al. (1991) fit yielded a value for n of 0.87*.

Kwa et al. (1998b) looked at the incidence of pneumonitis in 540 patients from 12 institutions. Since the patients with lung cancer showed higher complication rates at lower doses, the patients were divided into those with lung cancer and others, mainly breast and lymphoma. Kwa et al. found that the rate of Grade 2 and higher complications for each group could be fitted to the Lyman model with $n=1$ (Section 36.3.2), suggesting that there was a large volume effect in lung, and that the complications were correlated with mean dose (see Figure 36.26).

Seppenwoolde and Lebesque (2001) reviewed the then current literature on how well the various models fitted or predicted the clinically observed incidence of radiation pneumonitis and concluded that there was no consensus on the best model. This led Seppenwoolde et al. (2003) to undertake a major new analysis of the data from 382 breast cancer, malignant lymphoma and inoperable non-small-cell lung cancer patients from two centres, the Netherlands Cancer Institute and the University of Michigan at Ann Arbor. The DVHs were reduced to a single parameter using various dose-effect relationships, e.g. mean lung dose and the volume of lung receiving more than some particular *threshold* dose such as V$_{20}$ (see Section 35.3.3). The two lungs (defined using CT by binary thresholding, thus excluding the GTV) were considered as one organ. 1 Gy dose bins were used in calculating the DVHs and the same treatment planning system, U-Mplan (University of Michigan) was used at both centres. Thus, maximum consistency in dose-volume information was ensured. Table 36.5 gives a comparison of the values of the parameters in various models in the literature values compared to those found by Seppenwoolde et al.

---

* It is essential to be aware that for the L–K–B model, and all the other lung complication models (see e.g. Table 36.5) except that of *relative seriality*, the lungs (minus tumour) are treated as a *single paired organ*.

TABLE 36.5

Values of the Parameters for Various Models of Radiation Pneumonitis in the Literature, Compared to Those Found in Seppenwoolde et al. (2003)

Model	Parameters				Log Likelihood
Parallel	TD_{50} (Gy)	m	$n = 1/k$	D_{50} (Gy)	
Seppenwoolde et al. (2003)	30.8[a]	0.37[a]	0.99[a]		−119.1
SPECT	31[a]	0.32[a]	0.67	63	−199.4
L–K–B	TD_{50} (Gy)	m	n		
Seppenwoolde et al. (2003)	30.8[a]	0.37[a]	0.99[a]		−119.1
Burman et al. (1991)[b]	24.5	0.18	0.87		n/a
Martel et al. (1994)	28	0.18	0.87		−141.7
MLD	TD_{50} (Gy)	m	n		
Seppenwoolde et al. (2003)	30.8[a]	0.37[a]	1		−119.1
Kwa et al. (1998b)	30.5	0.33	1		−119.9
V_D	rdV_{50} (%)	m		D_{th} (Gy)	
Seppenwoolde et al. (2003)	77[a]	0.44[a]		13[a]	−121.3
Seppenwoolde et al. (2003)	65[a]	0.46[a]		20	−123.4
Graham et al. (1999)	51.3	0.32		20	−130.2
Relative seriality	D_{50} (Gy)	γ	s		
Gagliardi et al. (2000)[c]	30.1	0.97	0.01		−119.3
Seppenwoolde et al. (2003)	34[a]	0.9[a]	0.06[a]		−118.1

D_{50}, local dose for 50% local damage; SPECT, single photon emission tomography; MLD, mean lung dose; V_{Dth}, lung volume receiving more than threshold dose D_{th}; rdV_{50}, relative damaged volume resulting in a 50% complication rate.

[a] Optimized in maximum likelihood analysis.
[b] No inhomogeneity corrections applied to the dose distributions.
[c] For radiation pneumonitis Grade ≥ 1, the lungs were considered as *separate* organs.

Source: From Seppenwoolde, Y. et al., *Int. J. Radiat. Oncol. Biol. Phys.*, 55, 724–35, 2003. With permission.

Seppenwoolde et al. (2003) concluded that the underlying dose-effect relationship for radiation pneumonitis was linear (i.e. mean lung dose fitted the data best) rather than a step function (i.e. a model based on the volume of lung receiving more than a certain dose D, V_D, as described in Section 35.3.3). However, there was a high correlation between these two parameters ($r^2 = 0.82$). Therefore, in clinical practice both the mean lung dose and V_{13Gy} (or V_{20Gy})could be used for predicting pneumonitis.

All analyses involving dose-volume data will be affected by the accuracy of the dose calculations themselves. In the case of treatment plans involving lung tissue, this can be particularly problematical because the average density of the lungs is around one-third of that of water; such situations represent a critical test for dose-calculation algorithms (see Part F). De Jaeger et al. 2003 re-evaluated published dose-effect relations for radiation pneumonitis using the more accurate 3-D convolution photon-beam algorithm (see Section 26.3.4) which takes the increased range of the secondary electrons in the low-density lung tissue into account. They found that the *equivalent-pathlength* (EPL) algorithm (see Section 26.2.3.2) consistently *overestimates* the dose at the lung-tumour boundary by about 10%. Re-evaluation using a *convolution-superposition* algorithm (see Section 26.3.4) showed that lung complications occur at a 12–14% *lower* dose. Thus, TD_{50} changed from 34.1 Gy to 29.2 Gy. This type of re-evaluation can be expected to become more common as more accurate dose-computation methods such as 3-D convolution and Monte-Carlo simulation (see Chapter 28) replace current less accurate ones.

36.3.6 Concluding Remarks

Two major methods of describing complication probabilities have been covered, namely, an empirical method and an *architectural* description. Underlying these NTCP models is the *volume effect*, which is central to how normal organs behave in response to inhomogeneous distributions of radiation and which therefore, has important implications for the design of treatment strategies. In organs with a small volume effect, it may be advantageous to use many fields in order to distribute the dose over a larger volume and thereby reduce the peak dose in the organ. Conversely, in parallel organs, it would be better to use a small number of fields to keep fractional damage below the organ's functional reserve. Intensity modulation (see Chapter 43) may prove useful in adjusting the volumetric distribution to normal organs while maintaining the intended dose to the large volume. In the future, we may find that organ architecture will play an important role in *designing*, in addition to evaluating treatment plans (Kutcher et al. 1994).

In spite of their importance, NTCP values should be used with caution when evaluating treatment plans, since the models suffer from a number of problems (Glatstein 2001). To begin with, they are generally quite crude and hardly even try to represent the multiple and inter-related toxicities observed in clinical practice (Deasy et al. 2002). In this respect, modelling of NTCP is much more complex than TCP because of the large and complex taxonomy of treatment toxicities (Niemierko and Goitein 1993b; Withers and Taylor 1993; Glatstein 2001; Schultheiss 2001; Travis 2001; Yorke 2001; Deasy et al. 2002). And even when clinical responses are reduced to the barest essentials, perhaps to one or a few critical endpoints, the paucity of clinical data and their large uncertainties makes it difficult to rely on the calculated complication probabilities (Deasy et al. 2001; Glatstein 2001). These uncertainties can lead not only to large variations in the *absolute* values of the calculated NTCPs (Lebesque et al. 1995) but can also affect the *relative ranking* of candidate treatment plans. However, this situation is changing as the foregoing sections have made clear; the increased use of 3D planning by many institutions is beginning to provide large databases of three-dimensional dose distributions potentially correlated with clinical endpoints. Although the endpoints characterising clinical complications, as is well known, are difficult to define and the data are painstaking to collect, nevertheless, there is a continuing accumulation of such data. There are also indications that the pooling of clinical data and the ability to share dose distributions between institutions electronically will provide databases that are more robust where models can be refined (Kwa et al. 1998b; Deasy et al. 2003; Rancati et al. 2003; El Naqa et al. 2006). Some consensus may then be reached on the models and their parameters, at least for some important organs and endpoints, e.g. Seppenwoolde et al. (2003) in the case of radiation pneumonitis in lung and Rancati et al. (2004) in the case of proctitis in the rectum. As a complement to the preceding sections, excellent summaries of the status of the NTCP models themselves and their application to a number of organs/endpoints can be found in Ten Haken (2001) and in Cattaneo et al. (2001). In the meantime, tools (such as BIOPLAN - Sanchez-Nieto and Nahum 2000 and that described in Warketin et al. 2004) are available which simplify the evaluation of the existing models (of both NTCP and TCP).

One potentially serious limitation of all the NTCP models described above is that information on the dose distribution is restricted to the dose-volume histogram for the organ or tissue concerned. Consequently no spatial information is retained (see Section 35.3.3); a contiguous region and separated *islands* of high dose with the same total volume cannot therefore be distinguished from each other. Nor can the *position* of the high-dose region in the organ, even if contiguous, be taken into account; this is known to be a limitation in the case of lung (e.g. Seppenwoolde et al. 2004). Due to the similarities between most clinically employed treatment techniques, this lack of spatial information is probably much less severe than it can seem, but ways of removing this limitation are nevertheless desirable. One way of doing this for the case of tubular organs such as the rectum is to effectively *unfold* the curved surface onto a flat surface; this can be compared to the way in which maps of the Earth's surface are reformatted in Mercator's projection. This idea, introduced by Sanchez-Nieto et al. (2001b) and further developed by Hoogeman et al. (2004), is shown in Figure 36.27. It would enable a clear distinction to be made, for example, between high-dose regions that wrap entirely around the circumference of the rectum and those that do not.

The development of further NTCP models is probably unnecessary. However, the generation of more dose-volume-complication data for the organs and endpoints of most interest is essential in order to derive more reliable values of the parameters for the currently established models such as the L–K–B (Section 36.3.2) and relative seriality (Section 36.3.3.3) formalisms. Furthermore, it is now necessary to start using the models and their associated parameters to make *predictions* of complications and to compare these predictions with the observed complication rates (Gagliardi et al. 2000). Only in this way will NTCP models come to be used with confidence in the clinic. Furthermore, for those organs/endpoints for which a reasonable amount of reliable dose-volume-complication data already exists (e.g. radiation pneumonitis in the lung; proctitis in the rectum; RILD in the liver), one can start *optimizing* treatment plans based on the calculated NTCP values. Two of the forms that this active use of NTCP modelling could take are

1. Start with a *relative-dose* treatment plan arrived at using, for example, dose-based criteria (e.g. PTV within 95–105% of D_{presc}, $V_{90\%}$ of OAR $<80\%$ of D_{presc}) and then adjust D_{presc} until $NTCP_{OAR}$ is equal to a value specified in the local clinical protocol, e.g. $NTCP_{proctitis} = 3\%$ (Nahum and Tait 1992; Ten Haken et al. 1993; McGinn et al. 1998; Sanchez-Nieto et al. 2001a).

2. Use NTCP and TCP as part of the *objective function* in the (computer) optimisation/inverse-planning process, thus allowing the mathematical and radiobiological properties of the models to drive the search for the optimum plan (e.g. Peñagarícano et al. 2005; Hoffmann et al. 2006).

Finally, it should not be forgotten that other variables such as fraction size, clonogen proliferation rate and the patient's performance status should ideally be incorporated into both the biological models and the optimisation process (Glatstein 2001; Bentzen 2004; Fowler et al. 2004; Nahum and Bentzen 2004). Low-dose hypersensitivity (LDHRS—see Section 7.13.3) may also play an important role in our understanding of how certain complications depend on dose and volume. One consequence of factoring LDHRS into NTCP

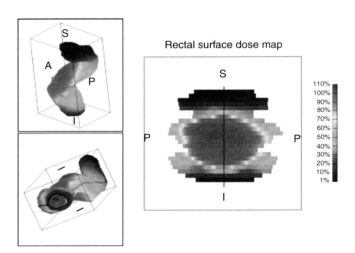

FIGURE 36.27
(See colour insert following page 590.) A rectal dose-surface map calculated from a 3D treatment plan for prostate cancer (Sanchez-Nieto et al. 2001b); spatial information of the high-dose region is preserved unlike in dose-volume (or dose-surface) histograms. (Reproduced from GUINESS, a treatment plan analysis software tool developed by Stefano Gianolini at the Institute of Cancer Research, Sutton.)

models, via a modification of the L-Q expression below around 0.6 Gy, will be an increased contribution to overall NTCP from those volumes of normal tissue receiving doses well below the tumour dose, i.e. at the 20% and lower isodoses. A significant increase in such volumes through the use of rotational techniques such as tomotherapy or many-field IMRT may be undesirable for certain organs and endpoints in which the killing of cells exhibiting LDHRS play a significant role in causing the complication. Finally we should mention the question of cancer *induction* by radiotherapy (see Section 61.5.2). Concerns have been raised (Glatstein 2002; Hall and Wuu 2003) that certain modern conformal techniques such as (many-field) IMRT, IMAT and tomotherapy (see Chapter 43) may increase cancer induction due to the increase in volumes irradiated at low doses compared to non-IMRT few-field techniques and especially compared to the use of proton beams (see Chapter 46). Hitherto cancer induction has not been classed as a complication due to the presumed low or extremely low frequency of occurrence. Models of (second) cancer induction probability (SCIP) which take into account details of the dose distribution in different organs and also the patient's age and general prognosis (Lindsay et al. 2001) would be desirable additions to the collection of tools for predicting the clinical effect of radiotherapy. Sachs and Brenner (2005) have produced a very useful analysis from which such SCIP models can now be developed.

REFERENCES

AAPM (American Association of Physicists in Medicine), Task Group 25: Clinical electron beam dosimetry, *Med. Phys.*, 18, 73–109, 1991.

AAPM (American Association of Physicists in Medicine), Task Group 53: Quality assurance for clinical radiotherapy treatment planning, *Med. Phys.*, 25, 1723–1829, 1998.

Akimoto, T., Muramatsu, H., Takahashi, M., Saito, J-I. , Kitamoto, Y. et al., Rectal bleeding after hypofractionated radiotherapy for prostate cancer: Correlation between clinical and dosimetric parameters and the incidence of grade 2 or worse rectal bleeding, *Int. J. Radiat. Oncol. Biol. Phys.*, 60, 1033–1039, 2004.

Allen, A. M., Henning, G. T., Ten Haken, R. K., Hayman, J. A., and Martel, M. K., Do dose-volume metrics predict pulmonary function changes in lung irradiation? *Int. J. Radiat. Oncol. Biol. Phys.*, 55, 921–929, 2003.

Antipas, V. P., Stamatakos, G. S., Uzunoglu, N. K., Dionysiou, D. D., and Dale, R. G., A spatio-temporal simulation model of the response of solid tumours to radiotherapy in vivo: Parametric validation concerning oxygen enhancement ratio and cell cycle duration, *Phys. Med. Biol.*, 49, 1485–1504, 2004.

Artignan, X., Smitsmans, M. H., Lebesque, J. V., Jaffray, D. A., van Herk, M. et al., Online ultrasound image guidance for radiotherapy of prostate cancer: Impact of image acquisition on prostate displacement, *Int. J. Radiat. Oncol. Biol. Phys.*, 59(2), 595–601, 2004.

Austin-Seymour, M. M., Chen, G. T. Y., Castro, J. R., Saunders, W. M., Pitluck, S., Woodruff, K. H., and Kessler, M., Dose volume histogram analysis of liver radiation tolerance, *Int. J. Radiat. Oncol. Biol. Phys.*, 12, 31–35, 1986.

Balter, J. M., Kwok, L. L., Sandler, H. M., Littles, J. F., Bree, R. L., and Ten Haken, R. K., Automated localisation of the prostate at the time of treatment using implanted radio-opaque markers: Technical feasibility, *Int. J. Radiat. Oncol. Biol. Phys.*, 33, 1281–1286, 1995a.

Balter, J. M., Sandler, H. M., Lam, K., Bree, R. L., Lichter, A. S., and Ten Haken, R. K., Measurement of prostate movement over the course of routine radiotherapy using implanted markers, *Int. J. Radiat. Oncol. Biol. Phys.*, 31, 113–118, 1995b.

Balter, J. M., Ten Haken, R. K., Lawrence, T. S., Lam, K. L., and Robertson, J. M., Uncertainties in CT-based radiation therapy treatment planning associated with patient breathing, *Int. J. Radiat. Oncol. Biol. Phys.*, 36, 167–174, 1996.

Batterman, J. J., Hart, A. A. M., and Breur, K., Dose-effect relations for tumour control and complication rate after fast neutron therapy for pelvic tumours, *Br. J. Radiol.*, 54, 899–904, 1981.

Battista, J., Computed tomography for radiotherapy planning, *Int. J. Radiat. Oncol. Biol. Phys.*, 6, 99–104, 1980.

Battista, J. J., Rider, W. D., and van Dyk, J., Computed tomography for radiotherapy planning, *Int. J. Radiat. Oncol. Biol. Phys.*, 6, 99–108, 1980.

Bedford, J. L. and Webb, S., Elimination of importance factors for clinically accurate selection of beam orientations, beam weights and wedge angles in conformal radiation therapy, *Med. Phys.*, 30, 1788–1804, 2003.

Bedford, J. L., Khoo, V. S., Oldham, M., Dearnaley, D. P., and Webb, S., A comparison of coplanar four-field techniques for conformal radiotherapy of the prostate, *Radiother. Oncol.*, 51, 225–235, 1999.

Benk, V. A., Adams, J. A., Shipley, W. U., Urie, M. M., McManus, P. L., Efird, J. T., Willett, J., and Goitein, M., Late rectal bleeding following combined x-rays and proton high dose irradiation for patients with stages T3–T4 prostate carcinoma, *Int. J. Radiat. Oncol. Biol. Phys.*, 26, 551–557, 1993.

Bentzen, S. M., Potential clinical impact of normal-tissue intrinsic radiosensitivity testing, *Radiother. Oncol.*, 43, 121–131, 1997.

Bentzen, S. M., Dose-response relationships in radiotherapy, in *Basic Clinical Radiobiology*, Steel, G. G., Ed., 3rd ed., Arnold, London, pp. 94–104, 2002.

Bentzen, S. M., High-tech in radiation oncology: Should there be a ceiling? *Int. J. Radiat. Oncol. Biol. Phys.*, 58, 320–330, 2004.

Bentzen, S. M. and Baumann, M., The L–Q model in clinical practice, in *Basic Clinical Radiobiology*, Steel, G. G., Ed., 3rd ed., Arnold, London, pp. 134–146, 2002.

Bentzen, S. M., Thames, H. D., and Overgaard, J., Does variation in the in vitro cellular radiosensitivity explain the shallow clinical dose-control curve for malignant melanoma? *Int. J. Radiat. Biol.*, 57, 117–126, 1990.

Bert, C., Metheany, K. G., Doppke, K., and Chen, G. T., A phantom evaluation of a stereo-vision surface imaging system for radiotherapy patient setup, *Med. Phys.*, 32(9), 2753–2762, 2005.

BIR (British Institute of Radiology), *Central Axis Depth Dose Data for Use in Radiotherapy*, Br. J. Radiol. Suppl. 25, British Institute of Radiology, London, 1996.

BIR (British Institute of Radiology), *Geometric uncertainties in radiotherapy*, British Institute of Radiology, London, 2003.

Boersma, L. J., van den Brink, M., Bruce, A. M., Shouman, T., Gras, L., te Velde, A., and Lebesque, J. V., Estimation of the incidence of late bladder and rectum complications after high-dose (70–78 Gy) conformal radiotherapy for prostate cancer, using dose-volume histograms, *Int. J. Radiat. Oncol. Biol. Phys.*, 41, 83–92, 1998.

Boyer, A. L. and Schultheiss, T., Effects of dosimetric and clinical uncertainty on complication-free local tumor control, *Radiother. Oncol.*, 11, 65–71, 1988.

Bradley, J., Thorstad, W. L., Mutic, S., Miller, T. R., Dehdashti, F., Siegel, B. A., Bosch, W., and Bertrand, R. J., Impact of FDG–PET on radiation therapy volume delineation in non-small-cell lung cancer, *Int. J. Radiat. Oncol. Biol. Phys.*, 59, 78–86, 2004.

Bradley, J., Graham, M. V., Winter, K., Purdy, J. A., Komaki, R., Roa, W. H., Ryu, J. K., Bosch, W., and Emami, B., Toxicity and outcome results of RTOG 9311: A phase I–II dose-escalation study using three-dimensional conformal radiotherapy in patients with inoperable non-small-cell lung carcinoma, *Int. J. Radiat. Oncol. Biol. Phys.*, 61, 318–328, 2005.

Brady, L. W., Cancer, L., Evans, G. C., and Forest, D. S., Carcinoma of the lung: Results of supervoltage radiation, *Arch. Surg.*, 90, 90–94, 1965.

Brady, L. W., German, P. A., and Cancer, L., The effects of radiation therapy on pulmonary function in carcinoma of the lung, *Radiology*, 85, 130–134, 1965.

Brahme, A., Dosimetric precision requirements in radiation therapy, *Acta Radiol. Oncol.*, 23, 379–391, 1984.

Brahme, A., Optimized radiation therapy based on radiobiological objectives, *Semin. Radiat. Oncol.*, 9, 35–47, 1999.

Brahme, A., Individualizing cancer treatment: Biological optimisation models in treatment and planning, *Int. J. Radiat. Oncol. Biol. Phys.*, 49, 327–337, 2001.

Brahme, A. and Ågren, A.-K., Optimal dose distribution for eradication of heterogeneous tumors, *Acta Oncol.*, 26, 377–385, 1987.

Brenner, D. J., Dose, volume and tumour-control predictions in radiotherapy, *Int. J. Radiat. Oncol. Biol. Phys.*, 26, 171–179, 1993.

Brenner, D. J. and Hall, E. J., Fractionation and protraction for radiotherapy of prostate carcinoma, *Int. J. Radiat. Oncol. Biol. Phys.*, 43, 1095–1101, 1999.

Brewster, L., Mageras, G. S., and Mohan, R., Automatic generation of beam apertures, *Med. Phys.*, 20, 1337–1342, 1993.

Buffa, F. M. and Nahum, A. E., Monte-Carlo dose calculations and radiobiological modelling: Analysis of the effect of the statistical noise of the dose distribution on the probability of tumour control, *Phys. Med. Biol.*, 45, 3009–3023, 2000.

Buffa, F. M., Fenwick, J. D., and Nahum, A. E., Letter to the editor: A realistic closed-form radiobiological model of clinical tumor-control data incorporating inter-tumor heterogeneity by Roberts and Hendry, *Int. J. Radiat. Onc. Biol. Phys.*, 41, 689–699, 1999.

Buffa, F. M., Fenwick, J. D., and Nahum, A. E., An analysis of the relationship between radiosensitivity and volume effects in tumour control probability modelling, *Med. Phys.*, 27, 1258–1265, 2000.

Buffa, F. M., West, C., Byrne, K., and Nahum, A. E., Radiation response and cure rate of human colon adenocarcinoma spheroids of different size: The significance of hypoxia on tumour control modelling, *Int. J. Radiat. Onc. Biol. Phys.*, 49, 1109–1118, 2001a.

Buffa, F. M., Davidson, S. E., Hunter, R. D., Nahum, A. E., and West, C. M. L., Incorporating biological measurements (SF2, CFE) in a tumor control probability model increases their prognostic significance: A study in cervical carcinoma treated with radiation therapy, *Int. J. Radiat. Onc. Biol. Phys.*, 50, 1113–1122, 2001b.

Burman, C., Kutcher, G. J., Emami, B., and Goitein, M., Fitting normal tissue tolerance data to an analytic function, *Int. J. Radiat. Oncol. Biol. Phys.*, 21, 123–135, 1991.

Carlone, M. C., Warkentin, B., and Stavrev, P., Fundamental form of a population TCP model in the limit of large heterogeneity, *Med. Phys.*, 33, 1634–1642, 2006.

Casebow, M. P., The angulation of radiotherapy machines in the treatment of inclined lesions, *Br. J. Radiol.*, 49, 278–280, 1976.

Casebow, M. P., Matching of adjacent radiation beams for isocentric radiotherapy, *Br. J. Radiol.*, 57, 735–740, 1984.

Cattaneo, G. M., Gagliardi, G., Nahum, A. E., Eds., The practical use of TCP and NTCP models for the evaluation of treatment plans in radiotherapy. Physica Medica 17 (Suppl. 2), 2001.

Chaney, E. L., Thorn, J. S., Tracton, G., Cullip, T. J., Rosenman, J. G., and Tepper, J. E., A portable software tool for computing digitally reconstructed radiographs, *Int. J. Radiat. Oncol. Biol. Phys.*, 32, 491–497, 1995.

Chang, H. and Fitzpatrick, J. M., A technique for accurate magnetic resonance imaging in the presence of field inhomogeneities, *IEEE Transactions on Medical Imaging*, 11, 319–329, 1992.

Chapman, J. D., Invited review: The single-hit mechanism of tumor cell killing by radiation, *Int. J. Radiat. Biol.*, 79, 71–81, 2003.

Chen, G. T. Y., Dose volume histograms in treatment planning, *Int. J. Radiat. Oncol. Biol. Phys.*, 14, 1319–1320, 1988.

Chen, G. T. Y., Austin-Seymour, M. M., Castro, J. R., Collier, J. M., Lyman, J. T., Pitluck, S., Saunders, W. M. and Zink, S. R., Dose volume histograms in treatment planning evaluation of carcinoma of the pancreas, *Proceedings of the Eighth International Conference on Uses of Computers in Radiation Therapy*, IEEE Computer Society Press, IEEE, Los Angeles, pp. 264–268, 1984.

Cheng, C. W. and Das, I. J., Treatment plan evaluation using dose-volume histogram (DVH) and spatial dose-volume histogram (zDVH), *Int. J. Radiat. Oncol. Biol. Phys.*, 43, 1143–1150, 1999.

Cheng, C. W., Chin, L. M., and Kijewski, P. K. A., Coordinate transfer of anatomical information from CT to treatment simulation, *Int. J. Radiat. Oncol. Biol. Phys.*, 13, 1559–1569, 1987.

Chin, L. M., Kijewski, P., Svensson, G. K., Chaffey, J. T., Levene, M. B., and Bjarngard, B., A computer controlled radiation therapy machine for pelvis and paraortic nodal areas, *Int. J. Radiat. Oncol. Biol. Phys.*, 7, 61–70, 1981.

Cho, B. C., van Herk, M., Mijnheer, B. J., and Bartelink, H., The effect of set-up uncertainties, contour changes, and tissue inhomogeneities on target dose-volume histograms, *Med. Phys.*, 29, 2305–2318, 2002.

Ciernik, I. F., Dizendorf, E., Baumert, B. G., Reiner, B., Burger, C., Davis, J. B., Lutolf, U. M., Steinert, H. C., and von Schulthess, G. K., Radiation treatment planning with an integrated positron emission and computer tomography (PET/CT): A feasibility study, *Int. J. Radiat. Oncol. Biol. Phys.*, 57, 853–863, 2003.

Clarke, L. P., Velthuizen, R. P., Phuphanich, S., Schellenberg, J. D., Arrington, J. A. et al., MRI: Stability of three supervised segmentation techniques, *Magn. Reson. Imaging*, 11, 95–106, 1993.

Clarke, L. P., Velthuizen, R. P., Camacho, M. A., Heine, J. J., Vaidyanathan, M. et al., MRI segmentation: Methods and application, *Magn. Reson. Imaging*, 13, 343–368, 1995.

Conway, J. and Robinson, M. H., CT virtual simulation, *Br. J. Radiol.*, Special Issue, S106–S118, Nov. 1997.

Cooper, B. E., *Statistics for Experimentalists*, Pergamon Press, Oxford, 1969.

Costa, D. C., Visvikis, D., Crosdale, I., Pigden, I., Townsend, C., Bomanji, J., Prvulovich, E., Lonn, A., and Ell, P. J., Positron emission and computed x-ray tomography: A coming together, *Nucl. Med. Commun.*, 24, 351–358, 2003.

Cozzarini, C., Fiorino, C., Ceresoli, G. L., Cattaneo, G. M., Bolognesi, A., Calandrino, R., and Villa, E., Significant correlation between rectal DVH and late bleeding in patients treated after radical prostatectomy with conformal or conventional radiotherapy (66.6–70.2 Gy), *Int. J. Radiat. Oncol. Biol. Phys.*, 55, 688–694, 2003.

Cullip, T. J., Symon, J. R., Rosenman, J. G., and Chaney, E. L., Digitally reconstructed fluoroscopy and other interactive volume visualisations in 3D treatment planning, *Int. J. Radiat. Oncol. Biol. Phys.*, 27, 145–151, 1995.

Curren, W. J., Hackney, D. B., Blitzer, P. H., and Bilaniuk, L., The value of magnetic resonance imaging in treatment planning of nasopharyngeal carcinoma, *Int. J. Radiat. Oncol. Biol. Phys.*, 12, 2189–2196, 1986.

Cygler, J., Daskalov, G., Chan, G., and Ding, G., Evaluation of the first commercial Monte Carlo dose calculation engine for electron beam treatment planning, *Med. Phys.*, 31, 142–153, 2004.

Dale, R. G., The application of the linear-quadratic dose-effect equation to fractionated and protracted radiotherapy, *Br. J. Radiol.*, 58, 515–528, 1985.

Dale, E., Olsen, D. R., and Fosså, S. D., Normal tissue complication probabilities correlated with late effects in the rectum after prostate conformal radiotherapy, *Int. J. Radiat. Oncol. Biol. Phys.*, 43, 385–391, 1999.

Daisne, J. F., Sibomana, M., Bol, A., Cosnard, G., Lonneux, M., and Gregoire, V., Evaluation of a multimodality image (CT, MRI and PET) coregistration procedure on phantom and head and neck cancer patients: accuracy, reproducibility and consistency, *Radiother. Oncol.*, 69, 237–245, 2003.

Dasu, A. and Denekamp, J., New insights into factors influencing the clinically relevant oxygen enhancement ratio, *Radiother. Oncol.*, 46, 269–277, 1998.

Dawson, L. A., Ten Haken, R. K., and Lawrence, T. S., Partial irradiation of the liver, *Semin. Radiat. Oncol.*, 11, 240–246, 2001.

Dawson, L. A., Normolle, D., Balter, J. M., McGinn, C. J., Lawrence, T. S. et al., Analysis of radiation-induced liver disease using the Lyman NTCP model, *Int. J. Radiat. Oncol. Biol. Phys.*, 53, 810–821, 2002.

De Gersem, W. R. T., Derycke, S., Colle, C. O., de Wagter, C., and de Neve, W. J., Inhomogeneous target-dose distributions: A dimension more for optimization? *Int. J. Radiat. Oncol. Biol. Phys.*, 44, 461–468, 1999.

De Jaeger, K., Hoogeman, M. S., Engelsman, M., Seppenwoolde, Y., Damen, E. M. F. et al., Incorporating an improved dose-calculation algorithm in conformal radiotherapy of lung cancer: Re-evaluation of dose in normal lung tissue, *Radiother. Oncol.*, 69, 1–10, 2003.

De Meerleer, G. O., Vakaet, L. A. M. L., de Gersem, W. R. T., de Wagter, C., De Naeyer, B., and De Neve, W., Radiotherapy of prostate cancer with or without intensity modulated beams: A planning comparison, *Int. J. Radiat. Oncol. Biol. Phys.*, 47, 639–648, 2000.

De Vries, L. S. and Bydder, G. M., Tumors of the central nervous system, in *Magnetic Resonance Imaging (MRI)*, Partain, C. C., Price, R. R., and Patton, J. A., Eds., W.B. Saunders, Philadelphia, pp. 144–168, 1988.

Deacon, J., Peckham, M. J., and Steel, G. G., The radioresponsiveness of human tumours and the initial slope of the cell survival curve, *Radiother. Oncol.*, 2, 317–323, 1984.

Dearnaley, D. D., Khoo, V. S., Norman, A., Meyer, L., Nahum, A. et al., Reduction of radiation proctitis by conformal radiotherapy techniques in prostate cancer: A randomized trial, *Lancet*, 353, 267–272, 1999.

Deasy, J., Poisson formulas for tumor control probability with clonogen proliferation, *Radiat. Res.*, 145, 382–384, 1996.

Deasy, J. O., Chao, K. S. C., and Markman, J., Uncertainties in model-based outcome predictions for treatment planning, *Int. J. Radiat. Oncol. Biol. Phys.*, 51, 1389–1399, 2001.

Deasy, J. O., Niemierko, A., Herbert, D., Yan, D., Jackson, A., Ten Haken, R. K., Langer, M., and Sapareto, S., Methodological issues in radiation dose-volume outcome analyses: Summary of a joint AAPM/NIH workshop, *Med. Phys.*, 29, 2109–2127, 2002.

Deasy, J. O., Blanco, A. I., and Clark, V. H., CERR: A computational environment for radiotherapy research, *Med. Phys.*, 30, 979–985, 2003.

Dejean, C., Lefkopoulos, D., Foulquier, J. N., Schlienger, M., and Touboul, E., Définition automatique de l'isodose de prescription pour les irradiations stéréotaxiques de malformations artérioveineuses, *Cancer Radiother.*, 5, 138–149, 2001.

Dobbs, J., Barrett, A., and Ash, D., *Practical Radiotherapy Planning*, 3rd ed., Edward Arnold, London, 1999.

Dong, L. and Boyer, A., An image correlation procedure for DRRs and electronic portal images, *Int. J. Radiat. Oncol. Biol. Phys.*, 33, 1052–1060, 1995.

Doran, S. J., Charles-Edwards, L., Reinsberg, S. A., and Leach, M. O., A complete distortion correction for MR images: I. Gradient warp correction, *Phys. Med. Biol.*, 50, 1343–1361, 2005.

Dritschilo, A., Chaffer, J. T., Bloumon, W. D., and Marck, A., The complication probability factor: A method of selection of treatment plans, *Br. J. Radiol.*, 51, 370–374, 1978.

Drzymala, R. E., Mohan, R., Brewster, M. S., Chu, J., Goitein, M., Harms, W., and Urie, M., Dose volume histograms, *Int. J. Radiat. Oncol. Biol. Phys.*, 21, 71–78, 1991.

Drzymala, R. E., Holman, M. D., Yan, D., Harms, W. B., Vain, N. L., Kahn, M. G., Emami, B., and Purdy, J., Integrated software tools for the evaluation of radiotherapy treatment plans, *Int. J. Radiat. Oncol. Biol. Phys.*, 30, 909–919, 1994.

Dutreix, J., Tubiana, M., and Dutreix, A., An approach to the interpretation of clinical data on the tumour control probability-dose relationship, *Radiother. Oncol.*, 11, 239–248, 1988.

Ebert, M. A., Viability of the EUD and TCP concepts as reliable dose indicators, *Phys. Med. Biol.*, 45, 441–457, 2000.

Edelman, R. R., Wielopolski, P., and Schmitt, F., Echo-planar MR imaging, *Radiology*, 192, 600–612, 1994.

Ehricke, H. H. and Schad, L. R., MRA-guided stereotactic radiation treatment planning for cerebral angiomas, *Comput. Med. Imaging Graph.*, 16, 65–71, 1992.

Eisbruch, A., Ship, J. A., Kim, H. M., and Ten Haken, R. K., Partial irradiation of the parotid gland, *Semin. Radiat. Oncol.*, 11, 234–239, 2001.

El Naqa, I., Suneja, G., Lindsay, P. E., Hope, A. J., Alaly, J. R., Vicic, M., Bradley, J. D., Apte, A., and Deasy, J. O., Dose response explorer: an integrated open-source tool for exploring and modelling radiotherapy dose-volume outcome relationships, *Phys. Med. Biol.*, 51, 5719–5735, 2006.

Emami, B., Lyman, J., Brown, A., Coia, L., Goitein, M. et al., Tolerance of normal tissue to therapeutic irradiation, *Int. J. Radiat. Oncol. Biol. Phys.*, 21, 109–122, 1991a.

Emami, B., Purdy, J. A., Manolis, J., Barest, G., Cheng, E. et al., Three-dimensional treatment planning for lung cancer, *Int. J. Radiat. Oncol. Biol. Phys.*, 21, 217–227, 1991b.

Engelsman, M., Remeijer, P., van Herk, M., Lebesque, J. V., Mijnheer, B., and Damen, E. M. F., Field size reduction enables Iso-NTCP escalation of tumor control probability for irradiation of lung tumours, *Int. J. Radiat. Oncol. Biol. Phys.*, 51, 1290–1298, 2001.

Fay, M., Tan, A., Fisher, R., Wirth, A., and Ball, D., Dose-volume histogram analysis as predictor of radiation pneumonitis in primary lung cancer patients treated with radiotherapy, *Int. J. Radiat. Oncol. Biol. Phys.*, 61, 1355–1363, 2005.

Fenwick, J. D., Predicting the radiation control probability of heterogeneous tumour ensembles: Data analysis and parameter estimation using a closed-form expression, *Phys. Med. Biol.*, 43, 2159–2178, 1998.

Fenwick, J. D., Biological modelling of pelvic radiotherapy: Potential gains from conformal techniques, PhD thesis, University of London, U.K., Available as internal report ICR/CCO/2000/1 from the author, Physics dept., Clatterbridge Centre for Oncology, Bebington CH63 4JY U.K., 1999.

Fenwick, J. D., Khoo, V. S., Nahum, A. E., Sanchez-Nieto, B., and Dearnaley, D. P., Correlations between dose-surface histograms and the incidence of long-term rectal bleeding following conformal or conventional radiotherapy treatment of prostate cancer, *Int. J. Radiat. Oncol. Biol. Phys.*, 49, 473–480, 2001.

Fernandez, E. M., Shentall, G. S., Mayles, W. P., and Dearnaley, D. P., The acceptability of a multileaf collimator as a replacement for conventional blocks, *Radiother. Oncol.*, 36, 65–74, 1995.

Feygelman, V., Mandelzweig, Y., and Baral, E., Matching electron beams without secondary collimation for treatment of extensive recurrent chest-wall carcinoma, *Med. Dosim.*, 19, 23–27, 1994.

Finnigan, D. I., Tanner, S. F., Dearnaley, D. P., Edser, E., Horwich, A. et al., Distortion-corrected magnetic resonance images for pelvic radiotherapy treatment planning, *Quant. Imaging Oncol.*, 3, 72–76, 1997.

Fiorino, C., Cozzarini, C., Vavassori, V., Sanguineti, G., Bianchi, C. et al., Relationships between DVHs and late rectal bleeding after radiotherapy for prostate cancer: Analysis of a large group of patients pooled from three institutions, *Radiother. Oncol.*, 64, 1–12, 2002a.

Fiorino, C., Vavassori, V., Sanguineti, G., Bianchi, C., Cattaneo, G. M., Piazzolla, A., and Cozzarini, C., Rectum contouring variability in patients treated for prostate cancer: Impact on rectum dose-volume histograms and normal tissue complication probability, *Radiother. Oncol.*, 63, 249–255, 2002b.

Fiorino, C., Gianolini, S., and Nahum, A. E., A cylindrical model of the rectum: Comparing dose-volume, dose-surface and dose-wall histograms in the radiotherapy of prostate cancer, *Phys. Med. Biol.*, 48, 2603–2616, 2003.

Fischer, J. J. and Moulder, J. E., The steepness of the dose-response curve in radiation injury, *Radiology*, 117, 179–184, 1975.

Fleming, J. S. and Orchard, P. G., Isocentric radiotherapy treatment planning where the treatment axis is not horizontal, *Br. J. Radiol.*, 47, 34–36, 1974.

Flentje, M., Zierhut, D., Schraube, P., and Wannenmacher, M., Integration of coronal magnetic resonance imaging (MRI) into radiation treatment planning of mediastinal tumors, *Strahlenther. Onkol.*, 169(6), 351–357, 1993.

Forman, J. D., Mesina, C. F., He, T., Devi, S. B., Ben-Josef, E. et al., Evaluation of changes in the location and shape of the prostate and rectum during a seven week course of conformal radiotherapy, *Int. J. Radiat. Oncol. Biol. Phys.*, 27, 222, 1993.

Fortier, G. A. and Pellitteri, M. R., MRI treatment planning for CNS neoplasm: Graphic display of target volume using boxed cursor, *Int. J. Radiat. Oncol. Biol. Phys.*, 19, 179–182, 1990.

Fowler, J. F., The linear quadratic formula and progress in fractionated radiotherapy, *Br. J. Radiol.*, 62, 679–694, 1989.

Fowler, J. F., Normal tissue complication probabilities: How well do the models work? *Physica Medica*, 17(Suppl. 2), 24–35, 2001.

Fowler, J. F., Chappell, R., and Ritter, M., Is α/β for prostate cancer really low? *Int. J. Radiat. Oncol. Biol. Phys.*, 50, 1021–1031, 2001.

Fowler, J. F., Tomé, W. A., Fenwick, J. D., and Mehta, M., A challenge to traditional radiation oncology, *Int. J. Radiat. Oncol. Biol. Phys.*, 60, 1241–1256, 2004.

Fraass, B. A., McShan, D. L., Diaz, R. F., Ten Haken, P. K., Aisen, A. et al., Integration of magnetic resonance imaging into radiotherapy treatment planning. I. Technical considerations, *Int. J. Radiat. Oncol. Biol. Phys.*, 13, 1897–1908, 1987.

Gademann, G., Schad, L.R., Schlegel, W., The definition of target volume in tumours of the brain, based on skull and facial area by means of MRI: Its impact on precision radiotherapy. Three Dimensional Treatment Planning, P. Minet. Liege, European Association of Radiology, pp. 47–55, 1993.

Gagliardi, G., Modeling heart and lung complication data in radiotherapy of the breast, PhD thesis, Stockholm University, 1998.

Gagliardi, G., Lax, I., Ottolenghi, A., and Rutqvist, L. E., Long-term cardiac mortality after radiotherapy of breast cancer—application of the relative seriality model, *Br. J. Radiol.*, 69, 839–846, 1996.

Gagliardi, G., Björhle, J., Lax, I., Ottolenghi, A., Eriksson, F. et al., Radiation pneumonitis after breast cancer irradiation: Analysis of the complication probability using the relative seriality model, *Int. J. Radiat. Oncol. Biol. Phys.*, 46, 373–381, 2000.

Gagliardi, G., Lax, I., and Rutqvist, L. E., Partial irradiation of the heart, *Semin. Radiat. Oncol.*, 11, 224–239, 2001.

Galvin, J. M., Sims, C., Dominiak, G., and Cooper, J. S., The use of digitally reconstructed radiographs for three-dimensional treatment planning and CT simulation, *Int. J. Oncol. Biol. Phys.*, 31, 935–942, 1995.

Gilhuijs, K. G., Drukker, K. A., Touw, A., van de Ven, P. J. H., and van Herk, M., Interactive three-dimensional inspection of patient setup in radiation therapy using digital portal images and computed tomography data, *Int. J. Oncol. Biol. Phys.*, 34, 873–885, 1996.

Giraud, P., Antoine, M., Larrouy, A., Milleron, B., Callard, P. et al., Evaluation of microscopic tumor extension in non-small-cell lung cancer for three-dimensional conformal radiotherapy planning, *Int. J. Radiat. Oncol. Biol. Phys.*, 48, 1015–1024, 2000.

Glatstein, E., Lichter, A. S., Fraass, B. A., and van de Geijn, J., The imaging revolution and radiation oncology: Use of CT, ultrasound and NMR for the localisation, treatment planning and treatment delivery, *Int. J. Radiat. Oncol. Biol. Phys.*, 11, 1299–1311, 1985.

Glatstein, E., Personal thoughts on normal tissue tolerance, or, what the textbooks don't tell you, *Int. J. Radiat. Oncol. Biol. Phys.*, 51, 1185–1189, 2001.

Glatstein, E., Intensity-modulated radiation therapy: The inverse, the converse, and the perverse, *Semin. Radiat. Oncol.*, 12, 272–281, 2002.

Goitein, M., The utility of computed tomography in radiation therapy: An estimate of outcome, *Int. J. Radiat. Oncol. Biol. Phys.*, 5, 1799–1807, 1979.

Goitein, M., Causes and consequences of inhomogeneous dose distributions in radiation therapy, *Int. J. Radiat. Oncol. Biol. Phys.*, 12, 701–704, 1986.

Goitein, M., The comparison of treatment plans, *Semin. Radiat. Oncol.*, 2, 246–256, 1992.

Goitein, M. and Miller, T., Planning treatments for eye with protons, *Med. Phys.*, 10, 275–283, 1983.

Goitein, M. and Niemierko, A., Intensity modulated therapy and inhomogeneous dose to the tumour: A note of caution, *Int. J. Radiat. Oncol. Biol. Phys.*, 36, 519–522, 1996.

Goitein, M., Abrams, M., Rowell, D., Pollari, H., and Wiles, J., Multidimensional treatment planning: II Beam's eye-view, back projection, and projection through CT sections, *Int. J. Radiat. Oncol. Biol. Phys.*, 9, 789–797, 1983.

Goitein, M., Wittenberg, J., Mendiondo, M., Doucette, J., Friedberg, C. et al., The value of CT scanning in radiation therapy treatment planning: A prospective study, *Int. J. Radiat. Oncol. Biol. Phys.*, 5, 1787–1798, 1979.

Goitein, M., Niemierko, A., and Okunieff, P., The probability of controlling an inhomogeneously irradiated tumour: A stratagem for improving tumour control through partial tumour boosting, in *Quantitative Imaging in Oncology—Proceedings of the 19th L.H. Gray Conference*, Kaulner, K., Carey, B., Crellin, A., Harrison, R. M., Eds., BIR Publishing, London, pp. 25–32, 1995.

Gopal, R. and Starkschall, G., Plan space: Representation of treatment plans in multidimensional space, *Int. J. Radiat. Oncol. Biol. Phys.*, 53, 1328–1336, 2002.

Gopal, R., Tucker, S. L., Komaki, R., Liao, Z., Forster, K. M., Stevens, C., Kelly, J. F., and Starkschall, G., The relationship between local dose and loss of function for irradiated lung, *Int. J. Radiat. Oncol. Biol. Phys.*, 56, 106–113, 2003.

Graham, M. and Purdy, J. A., Correspondence: In response to Dr. Lawrence et al., *Int. J. Radiat. Oncol. Biol. Phys.*, 41, 972–973, 1998.

Graham, M. V., Purdy, J. A., Emami, B., Matthews, J. W., and Harms, W. B., Preliminary results of a prospective trial using three dimensional radiotherapy for lung cancer, *Int. J. Radiat. Oncol. Biol. Phys.*, 33, 993–1000, 1995.

Graham, M. V., Jain, N. L., Kahn, M. G., Drzymala, R. E., and Purdy, J. A., Evaluation of an objective plan-evaluation model in the three dimensional treatment of nonsmall cell lung cancer, *Int. J. Radiat. Oncol. Biol. Phys.*, 34, 469–474, 1996.

Graham, M. V., Purdy, J. A., Emami, B., Harms, W., Bosch, W., Lockett, M. A., and Perez, C. A., Clinical dose-volume histogram analysis for pneumonitis after 3D treatment for non-small cell lung cancer (NSCLC), *Int. J. Radiat. Oncol. Biol. Phys.*, 45, 323–329, 1999.

Grégoire, V., Coche, E., Cosnard, G., Hamoir, M., and Reychler, H., Selection and delineation of lymph node target volumes in head and neck conformal radiotherapy. Proposal for standardizing terminology and procedure based on the surgical experience, *Radiother. Oncol.*, 56, 135–150, 2000.

Gross, M. W., Weber, W. A., Feldmann, H. J., Bartenstein, P., Schwaiger, M., and Molls, M., The value of F-18-fluorodeoxyglucose PET for the 3-D radiation treatment planning of malignant gliomas, *Int. J. Radiat. Oncol. Biol. Phys.*, 41, 989–995, 1998.

Hall, E. J. and Wuu, C. S., Radiation-induced second cancers: The impact of 3D-CRT and IMRT, *Int. J. Radiat. Oncol. Biol. Phys.*, 56, 83–88, 2003.

Hanks, G., Martz, K. L., and Diamond, J. J., The effect of dose on local control of prostate cancer, *Int. J. Radiat. Oncol. Biol. Phys.*, 15, 1299–1305, 1988.

Hanks, G. E., Schultheiss, T. E., Hanlon, A. L., Hunt, M., and Lee, W. R., Optimization of conformal radiation treatment of prostate cancer: Report of a dose escalation study, *Int. J. Radiat. Oncol. Biol. Phys.*, 37, 543–550, 1997.

Hardy, T. L., Brynildson, L. R., Gray, J. G., and Spurlock, D., Three-dimensional imaging for brachytherapy planning, *Stereotac. Funct. Neurosurg.*, 59(1–4), 179–181, 1992.

Hartford, A. C., Niemierko, A., Adams, J. A., Urie, M. M., and Shipley, W. U., Conformal irradiation of the prostate: Estimating long-term rectal bleeding risk using dose-volume histograms, *Int. J. Radiat. Oncol. Biol. Phys.*, 36, 721–730, 1996.

Hazle, J. D., Hefner, L., Nyerick, C. E., Wilson, L., and Boyer, A. L., Dose-response characteristics of a ferrous–sulphate–doped gelatin system for determining radiation absorbed dose distributions by magnetic resonance imaging (Fe MRI), *Phys. Med. Biol.*, 36(8), 1117–1125, 1991.

Heester, M. A., Wijrdeman, H. K., Strukmans, H., Witkamp, T., and Moerland, M. A., Brain tumor delineation based on CT and MR imaging, *Strahlenther. Onkol.*, 169(12), 729–733, 1993.

Heisig, S., Shentall, G. S., Mirza, K., Mayles, W. P. M., Application of the GE target planning computer to multi-leaf collimator treatments, Proceedings of XIth International Conference on Computers in Radiation Therapy, Hounsell, A. R., Wilkinson, J. M., and Williams, P. C., Eds., Medical Physics Publishing, Madison, WI, pp. 16–17, 1994.

Hendrick, R. E. and Raff, U., Image contrast and noise, in *Magnetic Resonance Imaging*, Stark, D. D. and Bradley, W. G., Eds., Mosby Year Book, St. Louis, pp. 109–145, 1992.

Herrlin, K., Bi-Ling, L., Pettersson, H., Willen, H., and Rydholm, A., Gadolinium-DPTA enhancement of soft tissue tumors in magnetic resonance imaging, *Acta Radiol.*, 31(3), 233–236, 1990.

Hill, D. L. G., Hawkes, D. J., Crossman, J. E., Gleeson, M. J., Cox, T. C. S., Bracey, E. E., Strong, A. J., and Graves, P., Registration of MR and CT images for skull base surgery using point-like anatomical features, *Br. J. Radiol.*, 64, 1030–1035, 1991.

Hill, D. L. G., Hawkes, D. J., Gleeson, M. J., Cox, T. C., Strong, A. J. et al., Accurate frameless registration of MR and CT images of the head: Applications in planning surgery and radiation therapy, *Radiology*, 191(2), 447–454, 1994.

Hobday, P., Hodson, J. J., Husband, J., and Parker, R. P., Computed tomography applied to radiotherapy treatment planning: Techniques and results, *Radiology*, 133, 477–482, 1979.

Hoffmann, A. L., Siem, A. Y., den Hertog, D., Kaanders, J. H., and Huizenga, H., Derivative-free generation and interpolation of convex Pareto optimal IMRT plans, *Phys. Med. Biol.*, 51, 6349–6369, 2006.

Hogstrom, K. R., Mills, M. D., and Almond, P. R., Electron beam dose calculations, *Phys. Med. Biol.*, 26, 445–449, 1981.

Holland, R., Veling, S. H., Mravunac, M., and Hendriks, J. H., Histologic multifocality of Tis, T1-2 breast carcinomas. Implications for clinical trials of breast-conserving surgery, *Cancer*, 56, 979–990, 1985.

Hoogeman, M. S., van Herk, M., de Bois, J., Muller-Timmermans, P., Koper, P. C. M. et al., Quantification of local rectal wall displacements by virtual rectum unfolding, *Radiother. Oncol.*, 70, 21–30, 2004.

Hricak, H., The role of imaging in the evaluation of pelvic cancer, in *Important Advances in Oncology*, DeVita, V. T., Hellmann, S., and Rosenberg, S. A., Eds., J.B. Lippincott, Philadelphia, pp. 103–133, 1991.

IAEA (International Atomic Energy Agency), Technical Report Series 277: *Absorbed dose determination in photon and electron beams*, IAEA, Vienna, 1987.

ICRU (International Commission on Radiation Units and Measurements), *Dose specification for reporting external beam therapy with photons and electrons*. Report No. 29, ICRU, Bethesda, MD, 1978.

ICRU (International Commission on Radiation Units and Measurements), *Radiation dosimetry: Electron beams with energies between 1 and 50 MeV*. Report No 35, ICRU, Bethesda, MD, 1984.

ICRU (International Commission on Radiation Units and Measurements). *Use of computers in external beam radiotherapy procedures with high-energy photons and electrons*. Report 42, ICRU, Bethesda, MD, 1987.

ICRU (International Commission on Radiation Units and Measurements), *Prescribing, recording and reporting photon beam therapy*. Report No. 50, ICRU, Bethesda, MD, 1993.

ICRU (International Commission on Radiation Units and Measurements), *Prescribing, recording and reporting photon beam therapy* (supplement to ICRU Report 50). Report No. 62, ICRU, Bethesda, MD, 1999.

ICRU (International Commission on Radiation Units and Measurements), *Prescribing, recording, and reporting electron beam therapy*. Report No. 71, Journal of the ICRU Vol. 4, Oxford University Press, Oxford, U.K., 2004.

Iori, M., Methods for physical and radiobiological optimization in radiotherapy with intensity modulation, *Physica Medica*, 17(Suppl. 2), 55–73, 2001.

IPEM (Institute of Physics and Engineering in Medicine), *Physics Aspects of Quality Control in Radiotherapy*, IPEM, York, 1999.

IPEMB (Institute of Physics and Engineering in Medicine and Biology), *A Guide to Commissioning and Quality Control of Treatment Planning Systems*, IPEM, York, 1994.

Isacsson, U., *Comparative treatment planning in external radiotherapy of malignant tumours*, PhD thesis, Department of Oncology, Uppsala University, 1998.

Jackson, A., Kutcher, G. J., and Yorke, E. D., Probability of radiation induced complications for normal tissues with parallel architecture subject to non-uniform irradiation, *Med. Phys.*, 20, 613–625, 1993.

Jackson, A., Ten Haken, R. K., Robertson, J. M., Kessler, M. L., Kutcher, G. J., and Lawrence, T. S., Analysis of clinical complication data for radiation hepatitis using a parallel architecture model, *Int. J. Radiat. Oncol. Biol. Phys.*, 31, 883–891, 1995.

Jackson, A., Skwarchuk, M. W., Zelefsky, M. J., Cowen, D. M., Venkatraman, E. S. et al., Late rectal bleeding after conformal radiotherapy of prostate cancer. II. Volume effects and dose-volume histograms, *Int. J. Radiat. Oncol. Biol. Phys.*, 49, 685–698, 2001.

Jacob, R., Hanlon, A. L., Horwitz, E. M., Movsas, B., Uzzo, R. G., and Pollack, A., Role of prostate dose escalation in patients with greater than 15% risk of pelvic lymph node involvement, *Int. J. Radiat. Oncol. Biol. Phys.*, 61, 695–701, 2005.

Jain, N. L. and Kahn, M. G., Ranking radiotherapy treatment plans using decision-analytic and heuristic techniques, *Comput. Biomed. Res.*, 25, 374–383, 1992.

Jain, N. L., Kahn, M. G., Drzymala, R. E., Emami, B., and Purdy, J. A., Objective evaluation of 3D radiation treatment plans: A decision-analytic tool incorporating treatment preferences of radiation oncologists, *Int. J. Radiat. Oncol. Biol. Phys.*, 26, 321–333, 1993.

Jette, D., Electron beam dose calculations, in *Radiation Therapy Physics*, Smith, A. R., Ed., Springer-Verlag, Berlin, pp. 95–121, 1996.

Judnick, J. W., Kessler, M. L., Fleming, T., Petti, P., and Castro, J. R., Radiotherapy technique integrates MRI into CT, *Radiol. Technol.*, 64(2), 82–89, 1992.

Just, M., Rosler, H. P., Higer, H. P., Kutzner, J., and Thelen, M., MRI-assisted radiation therapy planning of brain tumors—clinical experiences in 17 patients, *Magn. Reson. Imaging*, 9(2), 173–177, 1991.

Kagawa, K., Lee, W. R., Schultheiss, T. E., Hunt, M. A., Shaer, A. H. et al., Initial clinical assessment of CT-MRI image fusion software in localization of the prostate for 3D conformal radiation therapy, *Int. J. Radiat. Oncol. Biol. Phys.*, 38(2), 319–325, 1997.

Kalend, A. M., Zwicker, R. D., Wu, A., and Sternick, E. S., A beam edge modifier for abutting electron fields, *Med. Phys.*, 12, 793–798, 1985.

Källman, P., *Optimization of radiation therapy planning using physical and biological objective functions*, PhD thesis, Department of Radiation Physics, Stockholm University, 1992.

Källman, P., Ågren, A., and Brahme, A., Tumour and normal tissue responses to fractionated non-uniform dose delivery, *Int. J. Radiat. Biol.*, 62, 249–262, 1992.

Kaplan, E. L. and Meier, P., Nonparametric estimation from incomplete observations, *J. Am. Stat. Assoc.*, 53, 457–816, 1958.

Karlsson, M. and Zackrisson, B., Matching of electron and photon beams with a multi-leaf collimator, *Radiother. Oncol.*, 29(3), 317–326, 1993.

Karlsson, M. and Zackrisson, B., Exploration of new treatment modalities offered by high energy (up to 50 MeV) electrons and photons, *Radiother. Oncol.*, 43, 303–309, 1997.

Kase, K. R. and Bjärngard, B. E., Bremsstrahlung dose to patients in rotational electron therapy, *Radiology*, 133, 531–532, 1979.

Kattapuram, S. V., Khurana, J. C., Scott, J. A., and El Khoury, G. Y., Negative scintigraphy with positive magnetic resonance imaging in bone metastasis, *Skeletal Radiol.*, 19, 113–116, 1990.

Kawrakov, I., VMC + +, electron and photon Monte Carlo calculations optimised for radiation treatment planning, in *Advanced Monte Carlo for Radiation Physics, Particle Transport Simulation and Applications: Proceedings of the Monte Carlo 2000 Meeting, Lisbon*, Kling, A., Barao, F., Nakagawa, M., Távora, L., and Vaz, P., Eds., Springer, Berlin, pp. 229–236, 2001.

Keall, P. J., Siebers, J. V., Jeraj, R., and Mohan, R., The effect of dose calculation uncertainty on the evaluation of radiotherapy plans, *Med. Phys.*, 27, 478–484, 2000.

Keall, P. J., Lammering, G., Lin, P.-S., Winter, D., Chung, T. D. et al., Tumor control probability predictions for genetic radiotherapy, *Int. J. Radiat. Oncol. Biol. Phys.*, 57, 255–263, 2003.

Kessler, M. L., Pitluck, S., Petti, P., and Castro, J. R., Integration of multimodality imaging data for radiotherapy treatment planning, *Int. J. Radiat. Oncol. Biol. Phys.*, 21, 1653–1667, 1991.

Kessler, M. L., McShan, D. L., and Fraass, B. A., A computer-controlled conformal radiotherapy system, III: Graphical simulation and monitoring of treatment delivery, *Int. J. Radiat. Oncol. Biol. Phys.*, 33, 1173–1180, 1995.

Khan, F. M., *The Physics of Radiation Therapy*, Williams and Wilkins, Baltimore, 1992.

Khan, F. M., Fullerton, G. D., Lee, J. M. F., Moore, V. C., and Levitt, S. H., Physical aspects of electron-beam arc therapy, *Radiology*, 124, 497–500, 1977.

Khoo, V. S. and Joon, D. L., New developments in MRI for target volume delineation in radiotherapy, *Br. J. Radiol.*, 79, Special Issue 1, S2–S15, 2006.

Khoo, V. S., Adams, E. J., Saran, F., Bedford, J. L., Perks, J. R. et al., A comparison of planning volumes determined by CT versus MRI for meningiomas of the base of the skull, *Int. J. Radiat. Oncol. Biol. Phys.*, 46, 1309–1317, 2000.

Khoo, V. S., Dearnaley, D. P., Finnigan, D. J., Padhani, A., Tanner, S. F. et al., Magnetic resonance imaging (MRI): Considerations and applications in radiotherapy treatment planning, *Radiother. Oncol.*, 42(1), 1–15, 1997.

Khoo, V. S., Padhani, A. R., Tanner, S. F., Finnigan, D. J., Leach, M. O., and Dearnaley, D. P., Comparison of MRI with CT for the radiotherapy planning of prostate cancer: A feasibility study, *Br. J. Radiol.*, 72, 590–597, 1999.

Killoran, J. H., Baldini, E. H., Beard, C. J., and Chin, L., A technique for optimization of digitally reconstructed radiographs of the chest in virtual simulation, *Int. J. Radiat. Oncol. Biol. Phys.*, 49, 231–239, 2001.

Kim, Y. and Tomé, W., Risk-adaptive optimization: selective boosting of high-risk tumor subvolumes, *Int. J. Radiat. Oncol. Biol. Phys.*, 66, 1528–1542, 2006.

King, C. R., DiPetrillo, T. A., and Wazer, D. E., Optimal radiotherapy for prostate cancer: Predictions for conventional external beam, IMRT, and brachytherapy from radiobiologic models, *Int. J. Radiat. Oncol. Biol. Phys.*, 46, 165–172, 2000.

Klein, E. E., Modulated electron beams using multi-segmented multileaf collimation, *Radiother. Oncol.*, 48, 307–311, 1998.

Klein, E. E., Zuofeng, L., and Low, D. A., Feasibility study of multileaf collimated electrons with a scattering foil based accelerator, *Radiother. Oncol.*, 41, 189–196, 1996.

Klevenhagen, S. C., *Physics of Electron Beam Therapy*, Adam Hilger, Bristol, 1985.

Klevenhagen, S. C., Lambert, G. D., and Arbabi, A., Backscattering in electron beam therapy for energies between 3 and 35 MeV, *Phys. Med. Biol.*, 27, 363–373, 1982.

Knöös, T., Kristensen, I., and Nilsson, P., Volumetric and dosimetric evaluation of radiation treatment plans: Radiation conformity index, *Int. J. Radiat. Oncol. Biol. Phys.*, 42, 1169–1176, 1998.

Kokoves, L., *Assessment of geometrical distortion in MR images—investigation for the future applications in radiotherapy treatment planning*, MSc Thesis, University of Surrey, 1991.

Kooy, H. M., Nedzi, L. A., Alexander III, E., Loeffler, J. S., and Ledoux, R. J., Dose-volume histogram computations for small intracranial volumes, *Med. Phys.*, 20, 755–760, 1993.

Kooy, H. M., Dunbar, S. F., Tarbell, N. J., Mannarino, E., Ferarro, N., Shusterman, S., Bellerive, M., Finn, L., McDomough, C. V., and Loeffler, J. S., Adaptation and verification of the relocatable Gill–Thomas–Cosman frame in stereotactic radiotherapy, *Int. J. Radiat. Oncol. Biol. Phys.*, 30, 685–691, 1994.

Korevaar, E. W., van Vliet, R. J., Woudstra, E., Heijmen, B., and Huizenga, H., Sharpening the penumbra of high energy electron beams with low weight narrow photon beams, *Radiother. Oncol.*, 48, 213–220, 1998.

Kovacs, G., Pötter, P., Prott, F. J., Lenzen, B. and Knocke, T. H., The Münster experience with magnetic resonance imaging assisted treatment planning used for high dose rate after loading therapy of gynaecological and naso-pharyngeal cancer, *Advanced Radiation Therapy Tumour Response Monitoring and Treatment Planning*, Breit, A., Ed., Springer-Verlag, Berlin, pp. 661–665, 1992.

Kubo, H. D., Wilder, R. B., and Conrad, T. E., Impact of collimator leaf width on stereotactic radiosurgery and 3D conformal radiotherapy treatment plans, *Int. J. Radiat. Oncol. Biol. Phys.*, 44, 937–945, 1999.

Kurup, R. G., Wang, S., and Glasgow, G., Feasibility study of multileaf collimated electrons with a scattering foil based accelerator, *Phys. Med. Biol.*, 37, 145–153, 1992.

Kurup, R. G., Glasgow, G., and Leybovich, L. B., Design of electron beam wedges for increasing the penumbra of abutting fields, *Phys. Med. Biol.*, 38, 667–673, 1993.

Kutcher, G. J. and Burman, C., Calculation of complication probability factors for non uniform normal tissue irradiation: The effective volume method, *Int. J. Radiat. Oncol. Biol. Phys.*, 16, 1623–1630, 1989.

Kutcher, G. J., Burman, C., Brewster, L., Goitein, M., and Mohan, R., Histogram reduction method for calculating complication probabilities for three-dimensional treatment planning evaluations, *Int. J. Radiat. Oncol. Biol. Phys.*, 21, 137–146, 1991.

Kutcher, G. J., Niehaus, A., and Yorke, E. D., The effect of normal organ architecture on 3D conformal strategies, in *XIth International Conference on The Use of Computers in Radiotherapy, Manchester*, Hounsell, A. R., Wilkinson, J. M., and Williams, P. C., Eds., Medical Physics Publishing, Madison, WI, pp. 10–11, 1994.

Kutcher, G. J., Leibel, S. A., Ling, C. C., Zelefsky, M., and Fuks, Z., New wine in an old bottle? Dose escalation under dose-volume constraints: A model of conformal therpay of the prostate, *Int. J. Radiat. Oncol. Biol. Phys.*, 35, 415–416, 1996.

Kwa, S. L. S., Theuws, J. C. M., Wagenaar, A., Damen, E. M. F., Boersma, L. J., Baas, P., Muller, S. H., and Lebesque, J. V., Evaluation of two dose-volume histogram reduction models for the prediction of radiation pneumonitis, *Radiother. Oncol.*, 48, 33–44, 1998a.

Kwa, S. L. S., Lebesque, J. V., Theuws, J. C. M., Marks, L. B., Munley, M. T. et al., Radiation pneumonitis as a function of mean lung dose: An analysis of pooled data of 540 patients, *Int. J. Radiat. Oncol. Biol. Phys.*, 42, 1–9, 1998b.

Lachance, B., Tramblay, D., and Pouliot, J., A new penumbra generator for electron field matching, *Med. Phys.*, 24, 485–495, 1997.

Laing, R. W., Bentley, R. E., Nahum, A. E., Warrington, A. P., and Brada, M., Stereotactic radiotherapy of irregular targets: A comparison between static conformal beams and non-coplanar arcs, *Radiother. Oncol.*, 28, 241–246, 1993.

Lam, K. S., Lam, W. C., O'Neill, M. J., Lee, D. J., and Zinreich, E., Electron arc therapy: Beam data requirements and treatment planning, *Clin. Radiol.*, 38, 379–383, 1987.

Lambert, G. D. and Klevenhagen, S. C., Penetration of backscattered electrons in polystyrene for energies between 1 and 25 MeV, *Phys. Med. Biol.*, 27, 721–725, 1982.

Lattanzi, J., McNeeley, S., Hanlon, A., Schultheiss, T. E., and Hanks, G. E., Ultrasound-based stereotactic guidance of precision conformal external beam radiation therapy in clinically localized prostate cancer, *Urology*, 55(1), 73–78, 2000.

Lawrence, T. S., Tesser, R. J., and Ten Haken, R. K., An application of dose volume histograms to the treatment of intrahepatic malignancies with radiation therapy, *Int. J. Radiat. Oncol. Biol. Phys.*, 19, 1041–1047, 1990.

Lawrence, T. S., Ten Haken, R. K., Kessler, M. L., Robertson, J. M., Lyman, J. T. et al., The use of 3-D dose volume analysis to predict radiation hepatitis, *Int. J. Radiat. Oncol. Biol. Phys.*, 23, 781–788, 1992.

Leach, M. O., Magnetic resonance imaging and spectroscopy: An introduction to theory, hardware, current applications and safety, *J. Radiol. Prot.*, 12, 137–158, 1992.

Leach, M. O., Magnetic resonance spectroscopy applied to clinical oncology, *Technol. Health Care*, 2, 235–246, 1994.

Leavitt, D. D., Physics of electron arc therapy, in *Radiation Therapy Physics*, Smith, A. R., Ed., Springer-Verlag, Berlin, pp. 139–154, 1996.

Leavitt, D. D., Peacock, L. M., Gibbs, F. A., and Stewart, J. R., Electron arc therapy: Physical measurement and treatment planning techniques, *Int. J. Radiat. Oncol. Biol. Phys.*, 11, 987–999, 1985.

Leavitt, D. D., Stewart, J. R., and Earley, E., Improved dose homogeneity in electron arc therapy achieved by a multiple-energy technique, *Int. J. Radiat. Oncol. Biol. Phys.*, 19, 159–165, 1990.

Leavitt, D. D., Stewart, J. R., Moeller, J. H., and Earley, L., Optimization of electron arc therapy doses by multi-vane collimator control, *Int. J. Radiat. Oncol. Biol. Phys.*, 16, 489–496, 1989.

Lebesque, J. V., Bruce, A., Kroes, G., Shouman, T., and van Herk, M., Variation in volumes, dose-volume histograms and estimated normal tissue complication prbabilities of rectum and bladder during conorformal radiotherapy of T3 prostate cancer, *Int. J. Radiat. Oncol. Biol. Phys.*, 33, 251–257, 1995.

Lee, M., Wynne, C., Webb, S., Nahum, A. E., and Dearnaley, D., A comparison of proton and megavoltage x-ray treatment planning for prostate cancer, *Radiother. Oncol.*, 33, 239–253, 1994.

Leibel, S. A., Zelefsky, M. J., Kutcher, G. J., Burman, M., Kelson, S. M., and Fuks, Z., Three-dimsensional conformal radiation therapy in localised carcinoma of the prostate: Interim report of phase I dose-escalation study, *J. Urol.*, 152, 1792–1798, 1994.

Leunens, G., Menten, J., Weltens, C., Verstraete, J., and van der Schueren, E., Quality assessment of medical decision making in radiation oncology: Variability in target volume delineation for brain tumours, *Radiother. Oncol.*, 29, 169–175, 1993.

Leung, L. H., Chua, D. T., and Wu, P. M., A new tool for dose conformity evaluation of radiosurgery treatment plans, *Int. J. Radiat. Oncol. Biol. Phys.*, 45, 233–241, 1999.

Levegrün, S., Jackson, A., Zelefsky, M. J., Skwarchuk, M. W., Venkatraman, E. S., Schlegel, W. et al., Analysis of biopsy outcome after three-dimensional conformal radiation therapy of prostate cancer using dose-distribution variables and tumor control probability models, *Int. J. Radiat. Oncol. Biol. Phys.*, 47, 1245–1260, 2000.

Levegrün, S., Jackson, A., Zelefsky, M. J., Skwarchuk, M. W., Venkatraman, E. S., Schlegel, W. et al., Fitting tumor control probability models to biopsy outcome after three-dimensional conformal radiation therapy of prostate cancer: pitfalls in deducing radiobiologic parameters for tumors from clinical data, *Int. J. Radiat. Oncol. Biol. Phys.*, 51, 1064–1080, 2001.

Levin, D. N., Pelizzari, C. A., Chen, G. T. Y., Chen, C. T., and Cooper, M. D., Retrospective geometric correlation of MR, CT and PET images, *Radiology*, 169, 817–823, 1988.

Li, S., Boyer, A., Lu, Y., and Chen, G. T., Analysis of the dose-surface histogram and dose-wall histogram for the rectum and bladder, *Med. Phys.*, 24, 1107–1116, 1997.

Liao, Z. X., Travis, E. L., and Tucker, S. L., Damage and morbidity from pneumonitis after irradiation of partial volumes of mouse lung, *Int. J. Radiat. Oncol. Biol. Phys.*, 32, 1359–1370, 1995.

Lindsay, K. A., Wheldon, E. G., Deehan, C., and Wheldon, T. E., Radiation carcinogenesis modelling for risk of treatment-related second tumours following radiotherapy, *Br. J. Radiol.*, 74, 529–536, 2001.

Lindsay, P. E., Moiseenko, V. V., van Dyk, J., and Battista, J. J., The influence of brachytherapy dose heterogeneity on estimates of α/β for prostate cancer, *Phys. Med. Biol.*, 48, 507–522, 2003.

Ling, C. C., Rogers, C. C., and Morton, R. J., *Computed Tomography in Radiation Therapy*, Raven Press, New York, 1983.

Lohr, F., Schramm, O., Schraube, P., Sroka-Perez, G., Seeber, S., Schlepple, G., Schlegel, W., and Wannenmacher, M., Simulation of 3D-treatment plans in head and neck tumors aided by matching of digitally reconstructed radiographs (DRR) and on-line distortion corrected simulator images, *Radiother. Oncol.*, 45, 199–207, 1997.

Lu, X.-Q. and Chin, L. M., Sampling techniques for the evaluation of treatment plans, *Med. Phys.*, 20, 151–161, 1993.

Lu, Y., Li, S., Spelbring, D., Song, P., Vijayakumar, S. et al., Dose-surface histograms as treatment planning tool for prostate conformal therapy, *Med. Phys.*, 22, 279–284, 1995.

Lunsford, L. D., Martinez, A. J., and Latchaw, R. E., Stereotaxic surgery with a magnetic resonance and computerised tomography-compatible system, *J. Neurosurg.*, 64(6), 872–878, 1986.

Lyman, J. T., Complication probabilities as assessed from dose-volume histograms, *Radiat. Res.*, 104, S13–S19, 1985.

Lyman, J. T. and Wolbarst, A. B., Optimization of radiation therapy. III. A method for assessing complication probabilities from dose-volume histograms, *Int. J. Radiat. Oncol. Biol. Phys.*, 13, 103–109, 1987.

Lyman, J. T. and Wolbarst, A. B., Optimization of radiation therapy. IV. A dose volume reduction algorithm, *Int. J. Radiat. Oncol. Biol. Phys.*, 17, 433–436, 1989.

Ma, C.-M. and Jiang, S. B., Monte Carlo modelling of electron beams from medical accelerators, *Phys. Med. Biol.*, 44, R157–R189, 1999.

Mackay, R. I., Hendry, J. H., Moore, C. J., Williams, P. C., and Read, G., Predicting late rectal complications following prostate conformal radiotherapy using biologically effective doses and normalized dose-surface histograms, *Br. J. Radiol.*, 70, 517–526, 1997.

Mackay, R. I., Graham, P. A., Moore, C. J., Logue, J. P., and Sharrock, P. J., Animation and radiobiological analysis of 3D motion in conformal radiotherapy, *Radiother. Oncol.*, 52, 43–49, 1999.

Mackenzie, R. and Dixon, A. K., Measuring the effects of imaging: An evaluative framework, *Clin. Radiol.*, 50, 513–518, 1995.

Mageras, G. S., Podmaniczky, K. C., and Mohan, R., A model for computer-controlled delivery of 3-D conformal treatments, *Med. Phys.*, 19(4), 945–953, 1992.

Marchal, G., Bosmans, H., and van Fraeyenhoven, L., Intracranial vascular lesions: Optimisation and clinical evaluation of three-dimensional time-of-flight MR angiography, *Radiology*, 175, 443–448, 1990.

Marks, L. B., The pulmonary effects of thoracic irradiation, *Oncology*, 8, 89–100, 1994.

Martel, M. K., Ten Haken, R. K., Hazuka, M. B., Turrisi, A. T., Frass, B. A., and Lichter, A. S., Dose-volume histogram and 3-D treatment planning evaluation of patients with pneumonitis, *Int. J. Radiat. Oncol. Biol. Phys.*, 28, 575–581, 1994.

Martel, M. K., Ten Haken, R. K., Hazuka, M. B., Kessler, M. L., Strawderman, M. et al., Estimation of tumor control probability parameters from 3-D dose distributions of non-small cell lung cancer patients, *Lung Cancer*, 24, 31–37, 1999.

Martinez-Monge, R., Fernandes, P. S., Gupta, N., and Gahbauer, R., Cross-sectional nodal atlas: A tool for the definition of clinical target volumes in three-dimensional radiation therapy planning, *Radiology*, 211, 815–828, 1999.

Maryanski, M. J., Magnetic resonance imaging of radiation dose distributions using a polymer-gel dosimeter, *Phys. Med. Biol.*, 39, 1437–1455, 1994.

Mauro, F., Arcangeli, G., D'Angelo, L., Marino, C., and Benassi, M., Mathematical models of cell survival after ionising radiation: Application to radiotherapy planning, *Health Phys.*, 57, 355–361, 1989.

Mavroidis, P., Lind, B. K., Theodorou, K., Laurell, G., Fernberg, J. O., Lefkopoulos, D., Kappas, C., and Brahme, A., Statistical methods for clinical verification of dose-response parameters related to esophageal stricture and AVM obliteration from radiotherapy, *Phys. Med. Biol.*, 49, 3797–3816, 2004.

McGinn, C. J., Ten Haken, R. K., Ensminger, W. D., Walker, S., Wang, S. et al., Treatment of intrahepatic cancers with radiation doses based on a normal tissue complication probability model, *J. Clin. Oncol.*, 16, 2246–2252, 1998.

McKenzie, A. L., Air-gap correction in electron treatment planning, *Phys. Med. Biol.*, 24, 628–635, 1979.

McKenzie, A. L., A simple method for matching electron beams in radiotherapy, *Phys. Med. Biol.*, 43, 3456–3478, 1998.

McShan, D. L., Fraass, B. A., and Lichter, A. S., Full integration of the beam's eye view concept into computerized treatment planning, *Int. J. Radiat. Oncol. Biol. Phys.*, 18, 1485–1494, 1990.

Mijnheer, B. J., Batterman, J. J., and Wambersie, A., Reply to: Precision and accuracy in radiotherapy, *Radiother. Oncol.*, 14, 163–167, 1989.

Mirowitz, S. A., Lee, J. K. T., Brown, J. J., Eilenberg, S. S., Heiken, J. P. et al., Rapid acquisation spin-echo (RASE) MR imaging: A new technique for reduction of artefacts and acquisition time, *Radiology*, 175(1), 131–135, 1990.

Mizowaki, T., Nagata, Y., Okajima, K., Murata, R., Yamamoto, M., Kokubo, M., Hiraoka, M., and Abe, M., Development of an MR simulator: Experimental verification of geometric distortion and clinical application, *Radiology*, 199, 855–860, 1996.

Moerland, M. A., van den Bergh, A. C. M., Bhagwandien, R., Janssen, W. M., Bakker, C. J. G., Langendijk, J. J. W., and Battermann, J. J., The influence of respiration induced motion of the kidneys on the accuracy of radiotherapy treatment planning, a magnetic resonance imaging study, *Radiother. Oncol.*, 30, 150–154, 1994.

Mohan, R., Bareest, G., Brewster, L., Chui, C., Kutcher, G., Laughlin, J., and Fuks, Z., A comprehensive three-dimensional radiation treatment planning system, *Int. J. Radiat. Oncol. Biol. Phys.*, 15, 481–495, 1988.

Mohan, R., Mageras, G. S., Baldwin, B., Brewster, L. J., and Kutcher, G. J., Clinically relevant optimization of 3-D conformal treatments, *Med. Phys.*, 19, 933–944, 1992.

Mohiuddin, M., Fujita, M., Regine, W. F., Megooni, A. S., Ibbott, G. S., and Ahmed, M. M., High-dose spatially-fractionated radiation (GRID): a new paradigm in the management of advanced cancers, *Int. J. Radiat. Oncol. Biol. Phys.*, 45, 721–727, 1999.

Moore, J. V., Hendry, J. H., and Hunter, R. D., Dose incidence curves for tumor control and normal tissue injury in relation to the response of clonogenic cells, *Radiother. Oncol.*, 1, 143–157, 1983.

Morrison, R., The results of treatment of cancer of the bladder—a clinical contribution to radiobiology, *Clin. Radiol.*, 26, 67–75, 1975.

Mothersill, C. E., Moriarty, M. J., and Seymour, C. B., Radiotherapy and the exploitation of bystander effects, *Int. J. Radiat. Oncol. Biol. Phys.*, 58, 575–579, 2004.

Movsas, B., Chapman, J. D., Greenberg, R. E., Hanlon, A. L., Horwitz, E. M. et al., Increasing levels of hypoxia in human prostate carcinoma correlate significantly with increasing clinical stage and age: An Eppendorf pO2 study, *Cancer*, 89, 2018–2024, 2000.

Movsas, B., Chapman, J. D., Hanlon, A. L., Horwitz, E. M., Greenberg, R. E. et al., A hypoxic ratio of prostate pO2/muscle pO2 predicts for biochemical failure in prostate cancer patients, *J. Urol.*, 60, 634–639, 2002.

Müller-Schimpfle, M., Layer, G., Köster, A., Brix, G., Kimig, B. et al., MRI and MRA in treatment planning of subdiaphragmatic radiation therapy, *J. Comput. Assist. Tomogr.*, 16(1), 110–119, 1992.

Munley, M. T., Lo, J. Y., Sibley, G. S., Bentel, G. C., Anscher, M. S., and Marks, L. B., A neural network to predict symptomatic lung injury, *Phys. Med. Biol.*, 44, 2241–2249, 1999.

Munro, T. R. and Gilbert, C. W., The relationship between tumour lethal doses and the radiosensitivity of tumour cells, *Br. J. Radiol.*, 34, 246–251, 1961.

Munzenrider, J. E., Brown, A. P., Chu, J. C., Coia, L. R., Doppke, K. P., Emami, B., Kutcher, G. J., Mohan, R., Purdy, J. A., Shank, B. et al., Numerical scoring of treatment plans, *Int. J. Radiat. Oncol. Biol. Phys.*, 21, 147–163, 1991.

Muthuswamy, M. S., A method of beam-couch intersection detection, *Med. Phys.*, 26, 229–235, 1999.

Nahum, A.E., Converting dose distributions into tumour control probability, in *IAEA-TECDOC-896, Radiation Dose in Radiotherapy from Prescription to Delivery:* Proceedings of a Seminar Held in Rio de Janeiro, Brazil, 27–30 August 1994, IAEA, Vienna, pp. 27–40, 1996.

Nahum, A. E. and Bentzen, S. M., IMRT and hypofractionation: A double therapeutic gain if used with care, *Radiother. Oncol.*, 73(Suppl. 1), S174, 2004.

Nahum, A. E. and Chapman, J. D., Correspondence: In response to Dr. Colin Orton, *Int. J. Radiat. Onc. Biol. Phys.*, 58, 1637–1639, 2004.

Nahum, A. E. and Glimelius, B., Biological models applied to the comparison of proton and photon treatments, *Physica Medica*, 17(Suppl. 2), 126–130, 2001.

Nahum, A. E. and Sanchez-Nieto, B., Tumour control probability modelling: Basic principles and applications in treatment planning, *Physica Medica*, 17(Suppl. 2), 13–23, 2001.

Nahum, A. E. and Tait, D. M., in *Maximising Local Control by Customised Dose Prescription for Pelvic Tumours Advanced Radiation Therapy: Tumour Response Monitoring and Treatment Planning*, Breit, A., Ed., Springer, Heidelberg, pp. 425–431, 1992.

Nahum, A. E., Movsas, B., Horwitz, E. M., Stobbe, C. C., and Chapman, J. D., Incorporating clinical measurements of hypoxia into tumor local control modeling of prostate cancer: Implications for the α/β ratio, *Int. J. Radiat. Oncol. Biol. Phys.*, 57, 391–401, 2003a.

Nahum, A. E., Polico, R., Iori, M., Iotti, C., Paiusco, M. et al., Adapting the LQ expression for surviving fraction to organ-at-risk dose distributions: Why large fractions may make more sense for parallel OARs, *Radiother. Oncol.*, 68, S113, 2003b.

Neal, A. J., Sivewright, G., and Bentley, R., Evaluation of a region growing algorithm for segmenting pelvic computed tomography images during radiotherapy planning, *Br. J. Radiol.*, 67, 392–395, 1994.

Niemierko, A., Reporting and analysing dose distributions: A concept of equivalent uniform dose, *Med. Phys.*, 24, 103–110, 1997.

Niemierko, A., A generalized concept of equivalent uniform dose (EUD), *Med. Phys.*, 26, 1100, 1999a.

Niemierko, A., A unified model of tissue response to radiation, *Med. Phys.*, 26, 1100, 1999b.

Niemierko, A. and Goitein, M., Random sampling for evaluating treatment plans, *Med. Phys.*, 17, 753–762, 1990.

Niemierko, A. and Goitein, M., Calculation of normal tissue complication probability and dose-volume histogram reduction schemes for tissues with critical element architecture, *Int. J. Radiat. Oncol. Biol. Phys.*, 25, 135–145, 1991.

Niemierko, A. and Goitein, M., Implementation of a model for estimating tumor control probability for an inhomogeneously irradiated tumour, *Radiother. Oncol.*, 29, 140–147, 1993a.

Niemierko, A. and Goitein, M., Modeling of normal tissue response to radiation: The critical volume model, *Int. J. Radiat. Oncol. Biol. Phys.*, 25, 135–145, 1993b.

Niemierko, A. and Goitein, M., Comments on sampling techniques for the evaluation of treatment plans, *Med. Phys.*, 20, 1377–1380, 1993c.

Niemierko, A. and Goitein, M., Dose-volume distributions: A new approach to dose-volume histograms in three-dimensional treatment planning, *Med. Phys.*, 21, 3–11, 1994.

Nioutsikou, E., Bedford, J. L., and Webb, S., Patient-specific planning for prevention of mechanical collisions during radiotherapy, *Phys. Med. Biol.*, 48, N313–N321, 2003.

Nioutsikou, E., Partridge, M., Bedford, J. L., and Webb, S., Prediction of radiation-induced normal tissue complications in radiotherapy using functional image data, *Phys. Med. Biol.*, 50, 1035–1046, 2005.

Nutting, C. M., Corbishley, C. M., Sanchez-Nieto, B., Cosgrove, V. P., Webb, S. et al., Potential improvements in the therapeutic ratio of prostate cancer irradiation: Dose escalation of pathologically identified tumour nodules using intensity modulated radiotherapy, *Br. J. Radiol.*, 75, 151–161, 2002.

O'Donnell, M. and Edelstein, W. A., NMR imaging in the presence of magnetic field inhomogeneities and gradient field nonlinearities, *Med. Phys.*, 12, 20–26, 1985.

Ogino, T., Nawano, S., Shimizu, W., and Moriyama, N., Cine MRI in radiotherapy treatment planning of brain tumors, *Radiat. Med.*, 11(5), 201–205, 1993.

Ohara, K., Okumura, T., Akisada, M., Inada, T., Mori, T., Yokota, H., and Calaguas, M. J., Irradiation synchronised with respiration gate, *Int. J. Radiat. Oncol. Biol. Phys.*, 17, 853–857, 1989.

Olsen, D. R. and Hellesnes, J., Absorbed dose distribution measurements in brachytherapy using ferrous sulphate gel and magnetic resonance imaging, *Br. J. Radiol.*, 67, 1121–1126, 1994.

Oozeer, R., Chauvet, B., Garcia, R., Berger, C., Felix-Faure, C., and Reboul, F., Evaluation dosimétrique d'une radiothérapie conformationnelle: le facteur de conformation, *Cancer Radiother.*, 4, 207–216, 2000.

Padhani, A. R., Khoo, V.S, Suckling, J., Husband, J. E., Leach, M. O. et al., Evaluating the effect of rectal distension and movement on prostate gland position using cine MRI, *Int. J. Radiat. Oncol. Biol. Phys.*, 44, 525–533, 1999.

Pardo, F. S., Aronen, H. J., Kennedy, D., Moulton, G., Paiva, K. et al., Functional cerebral imaging in the evaluation and radiotherapeutic treatment planning of patients with malignant gliomas, *Int. J. Radiat. Oncol. Biol. Phys.*, 30(3), 663–669, 1994.

Pelizzari, C. A., Chen, G. T. Y., Spelbring, D. R., Weichselbaum, R. R., and Chen, C. T., Accurate three-dimensional registration of CT, PET and MR images of the brain, *J. Comput. Assist. Tomogr.*, 13, 20–27, 1989.

Peñagarícano, J. A., Papanikolaou, N., Wu, C., and Yan, Y., An assessment of biologically-based optimisation (BORT) in the IMRT era, *Med. Dosim.*, 30, 12–19, 2005.

Perez, C. A., Stanley, K., Rubin, P., Kramer, S., Brady, L. et al., A prospective randomized study of various irradiation doses and fractionation schedules in the treatment of inoperable non-oat-cell carcinoma of the lung, *Cancer*, 45, 2744–2753, 1980.

Perks, J. R., Jalali, R., Cosgrove, V. P., Adams, E. J., Shepherd, S. F., Warrington, A. P., and Brada, M., Optimization of stereotactically guided conformal treatment planning of sellar and parasellar tumours: The practical solution, *Int. J. Radiat. Oncol. Biol. Phys.*, 45, 507–513, 1999.

Peters, L. J., Brock, W. A., Chapman, J. D., Wilson, G., and Fowler, J. F., Response predictors in radiotherapy: A review of research into radiobiologically based assays, *Br. J. Radiol.*, Suppl. 22 69–108, 1989.

Petti, P. L. and Siddon, L. S., Effective wedge angles with a universal wedge, *Phys. Med. Biol.*, 30, 985–991, 1985.

Pickett, B., Roach III, M., Verhey, L., Horine, P., Malfatti, C., Akazawa, C. et al., The value of non-uniform margins for six field conformal irradiation of localised prostate cancer, *Int. J. Radiat. Oncol. Biol. Phys.*, 32, 211–218, 1995.

Platoni, K., Lefkopoulos, D., Grandjean, P., and Schlienger, M., Adaptation de l'analyse ROC pour l'évaluation quantitative des plans de traitement en radiothérapie stéréotaxique, *Cancer Radiother.*, 3, 494–502, 1999.

Poppel, R. A., Ove, R., and Shen, S., Tumor control probability for selective boosting of hypoxic subvolumes, including the effect of reoxygenation, *Int. J. Radiat. Oncol. Biol. Phys.*, 54, 921–927, 2002.

Porter, E. H., The statistics of dose/cure relationships for irradiated tumours Part II, *Br. J. Radiol.*, 53, 336–345, 1980.

Pötter, R., Heil, B., Schneider, L., Lenzen, H., Al-Dandashi, C. et al., Sagittal and coronal planes from MRI for treatment planning in tumors of brain, head and neck: MRI assisted simulation, *Radiother. Oncol.*, 23(2), 127–130, 1992.

Rancati, T., Ceresoli, G., Gagliardi, G., Schipani, S., and Cattaneo, G. M., Factors predicting radiation pneumonitis in lung cancer patients: A retrospective study, *Radiother. Oncol.*, 67, 275–283, 2003.

Rancati, T., Fiorino, C., Gagliardi, G., Cattaneo, G. M., Sanguineti, G. et al., Fitting late rectal bleeding data using different NTCP models: Results from an Italian multi-centric study (AIROPROS0101), *Radiother. Oncol.*, 73, 21–32, 2004.

Rasch, C., Barillot, I., Remeijer, P., Touw, A., van Herk, M., and Lebesque, J. V., Definition of the prostate in CT and MRI: A multi-observer study, *Int. J. Radiat. Oncol. Biol. Phys.*, 43, 57–66, 1999.

Reft, C., Alecu, R., Das, I. J., Gerbi, B. J., Keall, P. et al., Dosimetric considerations for patients with HIP prostheses undergoing pelvic irradiation. Report of the AAPM Radiation Therapy Committee Task Group 63, *Med. Phys.*, 30, 1162–1182, 2003.

Reinsberg, S. A., Doran, S. J., Charles-Edwards, E. M., and Leach, M. O., A complete distortion correction for MR images: II. Rectification of static-field inhomogeneities by similarity-based profile mapping, *Phys. Med. Biol.*, 50, 2651–2661, 2005.

Rothwell, R. I., Kelly, S. A., and Joslin, C. A. F., Radiation pneumonitis in patients treated for breast cancer, *Radiother. Oncol.*, 4, 9–14, 1985.

Rowbottom, C. G., Oldham, M., and Webb, S., Constrained customization of non-coplanar beam orientations in radiotherapy of brain tumours, *Phys. Med. Biol.*, 44, 383–399, 1999a.

Rowbottom, C. G., Oldham, M., and Webb, S., Is it possible to optimize a radiotherapy treatment plan? *Int. J. Oncol. Biol. Phys.*, 43, 698–699, 1999b.

Ruggieri, R., Hypofractionation in non-small cell lung cancer (NSCLC); suggestions from modelling both acute and chronic hypoxia, *Phys. Med. Biol.*, 49, 4811–4823, 2004.

Ruggieri, R. and Nahum, A. E., The impact of hypofractionation on simultaneous dose-boosting to hypoxic tumor subvolumes, *Med. Phys.*, 33, 4044–4055, 2006.

Sachs, R. K. and Brenner, D. J., Solid tumor risks after high doses of ionizing radiation, *Proc. Natl. Acad. Sci. U.S.A.*, 102, 13040–13045, 2005.

Sakurai, K., Fujita, N., Harada, K., Kim, S. W., Nakanishi, K. et al., Magnetic susceptibility artefact in spin-echo MR imaging of the pituitary gland, *Am. J. Neuroradiol.*, 13, 1301–1308, 1992.

Sanchez-Nieto, B. and Nahum, A. E., The Delta-TCP concept: A clinically useful measure of tumour control probability, *Int. J. Radiat. Oncol. Biol. Phys.*, 44, 369–380, 1999.

Sanchez-Nieto, B. and Nahum, A. E., *BIOPLAN:* Software for the biological evaluation of radiotherapy treatment plans, *Med. Dosim.*, 25, 71–76, 2000.

Sanchez-Nieto, B., Nahum, A. E., and Dearnaley, D. P., Individualisation of dose prescription based on normal-tissue dose-volume and radiosensitivity data, *Int. J. Radiat. Ther. Biol., Phys.*, 49, 487–499, 2001a.

Sanchez-Nieto, B., Fenwick, J. F., Nahum, A. E., and Dearnaley, D. P., Biological dose surface maps: Evaluation of 3D dose data for tubular organs, *Radiother. Oncol.*, 61, S52, 2001b.

Sannazzari, G. L., Ragona, R., Ruo Redda, M. G., Giglioli, F. R., Isolato, G., and Guarneri, A., CT-MRI image fusion for delineation of volumes in three-dimensional conformal radiation therapy in the treatment of localized prostate cancer, *Br. J. Radiol.*, 75, 603–607, 2002.

Saunders, J. E. and Peters, V. G., Back-scattering from metals in superficial therapy with high energy electrons, *Br. J. Radiol.*, 47, 467–470, 1974.

Scarfone, C., Lavely, W. C., Cmelak, A. J., Delbeke, D., Martin, W. H., Billheimer, D., and Hallahan, D. E., Prospective feasibility trial of radiotherapy target definition for head and neck cancer using 3-dimensional PET and CT imaging, *J. Nucl. Med.*, 45, 543–552, 2004.

Schad, L., Lott, S., Schmitt, F., Sturm, V., and Lorenz, W. J., Correction of spatial distortion in MR imaging: A prerequisite for accurate stereotaxy, *J. Comput. Assist. Tomogr.*, 11, 499–505, 1987a.

Schad, L., Boesecke, R., Schlegel, W., Hartmann, G. H., Sturm, V. et al., Three-dimensional image correlation of CT, MR and PET studies in radiotherapy treatment planning of the brain tumours, *J. Comput. Assist. Tomogr.*, 11, 948–1054, 1987b.

Schad, L. R., Bluml, S., Hawighorst, H., Wenz, F., and Lorenz, W. J., Radiosurgical treatment planning of brain metastases based on a fast three-dimensional MR imaging technique, *Magn. Reson. Imaging*, 12(5), 811–819, 1994.

Schultheiss, T. E., Orton, C. G., and Peck, R. A., Models in radiation therapy: Volume effects, *Med. Phys.*, 10, 410–415, 1983.

Schultheiss, T. E., Hanks, G. E., Hunt, M. A., and Lee, W. R., Incidence of and factors related to late complications in conformal and conventional radiation treatment of cancer of the prostate, *Int. J. Radiat. Oncol. Biol. Phys.*, 32, 643–649, 1995.

Schwartz, L. H., Richaud, J., Buffat, L., Touboul, E., and Schlienger, M., Kidney mobility during respiration, *Radiother. Oncol.*, 32, 84–86, 1994.

Schwarz, M., Lebesque, J. V., Mijnheer, B. J., and Damen, E. M. F., Sensitivity of treatment plan optimisation for prostate cancer using the equivalent uniform dose (EUD) with respect to the rectal wall volume parameter, *Radiother. Oncol.*, 73, 209–218, 2003.

Seppenwoolde, Y. and Lebesque, J. V., Partial irradiation of the lung, *Semin. Radiat. Oncol.*, 11, 247–258, 2001.

Seppenwoolde, Y., Lebesque, J. V., de Jaeger, K., Belderbos, J. S. A., Boersma, L. J. et al., Comparing different NTCP models that predict the incidence of radiation pneumonitis, *Int. J. Radiat. Oncol. Biol. Phys.*, 55, 724–735, 2003.

Seppenwoolde, Y., de Jaeger, K., Boersma, L. J., Belderbos, J. S. A., and Lebesque, J. V., Regional differences in lung radiosensitivity after radiotherapy for non-small-cell lung cancer, *Int. J. Radiat. Oncol. Biol. Phys.*, 60, 748–758, 2004.

Seydel, H. G., Diener-West, M., Urtasun, R., Podolsky, W. J., Cox, J. D. et al., Radiation therapy oncology group (RTOG): Hyperfractionation in the radiation therapy of unresectable non-oat cell carcinoma of the lung: Preliminary report of an RTOG pilot study, *Int. J. Radiat. Oncol. Biol. Phys.*, 11, 1841–1847, 1985.

Sgouros, G., Chiu, S., Pentlow, K. S., Brewster, L. J., Kalaigian, H. et al., Three-dimensional dosimetry for radioimmunotherapy treatment planning, *J. Nucl. Med.*, 34(9), 1595–1601, 1993.

Sherouse, G. S., Novins, K., and Chaney, E. L., Computation of digitally reconstructed radiographs for use in radiotherapy treatment design, *Int. J. Oncol. Biol. Phys.*, 18, 651–658, 1990.

Shipley, W. U., Tepper, J. E., Prout, G. R., Verhey, L. J., Mendiondo, O. A., Goitein, M., Koehler, A. M., and Suit, H. D., Proton radiation as boost therapy for localized prostatic carcinoma, *JAMA*, 241, 1912–1915, 1979.

Shiu, A. S., Tung, S., Hogstrom, K. R., Wong, J. W., Gerber, R. L. et al., Verification data for electron beam dose algorithms, *Med. Phys.*, 19, 623–636, 1992.

Siddon, R. L., Solution to treatment planning problems using coordinate transforms, *Med. Phys.*, 8, 766–774, 1981.

Siddon, R. L., Fast calculation of the exact radiological path for a three-dimensional CT array, *Med. Phys.*, 12, 252–255, 1985.

Steel, G. G., Ed., *Basic Clinical Radiobiology*, 3rd ed., Edward Arnold, London, 2002.

Steel, G. G. and Peacock, J. H., Why are some human tumours more radiosensitive than others?, *Radiother. Oncol.*, 15, 63–72, 1989.

Stewart, J. G. and Jackson, A. W., The steepness of the dose response curve both for tumor cure and normal tissue injury, *Laryngoscope*, 85, 1107–1111, 1975.

Stock, R. G., Stone, N. N., Tabert, A., Iannuzzi, C., and DeWyngaert, J. K., A dose-response study for I-125 prostate implants, *Int. J. Radiat. Oncol. Biol. Phys.*, 41, 101–108, 1998.

Suchowerska, N., Ebert, M. A., Zhang, M., and Jackson, M., *In vitro* response of tumour cells to non-uniform irradiation, *Phys. Med. Biol.*, 50, 3041–3051, 2005.

Suit, H., Skates, S., Taghian, A., Okunieff, P., and Efird, J. T., Clinical implications of heterogeneity of tumor response to radiation therapy, *Radiother. Oncol.*, 25, 251–260, 1992.

Swalec, J. J., Leavitt, D. D., and Moeller, J. H., Improved field edge definition in electron arc therapy with dynamic collimation techniques, *Int. J. Radiat. Oncol. Biol. Phys.*, 30, 205–210, 1994.

Sydes, M.R., Stephens, R.J., Moore, A.R., Edwin, G., Aird, E.G., Bidmead, A.M. et al., Implementing the U.K. Medical Research Council (MRC) RT01 trial (ISRCTN 47772397): Methods and practicalities of a randomised controlled trial of conformal radiotherapy in men with localised prostate cancer, *Radiother. Oncol.*, 72, 199–211, 2004.

Sze, G., Gadolinium-DPTA in spinal disease, *Radiol. Clin. North Am.*, 26, 1009–1024, 1988.

Tait, D. M., Nahum, A. E., Meyer, L., Law, M., Dearnaley, D. P., Horwich, A., Mayles, W. P., and Yarnold, J. R., Acute toxicity in pelvic radiotherapy: A randomized trial of conformal versus conventional treatment, *Radiother. Oncol.*, 42, 121–136, 1997.

Tanner, S. F., Finnigan, D. J., Khoo, V. S., Dearnaley, D. P., and Leach, M. O., Radiotherapy planning of the pelvis using distortion corrected MR images: The removal of system distortions, *Phys. Med. Biol.*, 45, 2117–2132, 2000.

Ten Haken, R.K., Ed., Partial organ irradiation, *Semin. Radiat. Oncol.*, 11, 181–267, 2001.

Ten Haken, R. K., Forman, J. D., Heimburger, D. K., Gerhardsson, A., McShan, D. L., and Perez-Tomayo, C., Treatment planning issues related to prostate movement in response to differential filling of the rectum and bladder, *Int. J. Radiat. Oncol. Biol. Phys.*, 20, 1317–1324, 1991.

Ten Haken, R. K., Thornton, A. F., Sandler, H. M., La Vigne, M. L., Quint, D. J., Fraass, B. A., Kessler, M. L., and McShan, D. L., A quantitative assessment of the addition of MRI to CT-based, 3-D treatment planning of brain tumors, *Radiother. Oncol.*, 25, 121–133, 1992.

Ten Haken, R. K., Martel, M. K., Kessler, M. L., Hazuka, M. B., Lawrence, T. S. et al., Use of v_{eff} and iso-NTCP in the implementation of dose escalation protocols, *Int. J. Radiat. Oncol. Biol. Phys.*, 27, 689–695, 1993.

Ten Haken, R. K., Balter, J. M., Marsh, L. H., Robertson, J. M., and Lawrence, T. S., Potential benefits of eliminating planning target volume expansions for patient breathing in the treatment of liver tumours, *Int. J. Radiat. Oncol. Biol. Phys.*, 38, 613–617, 1997.

Terahara, A., Niemierko, A., Goitein, M., Finkelstein, D., Hug, E. et al., Analysis of the relationship between tumor dose inhomogeneity and local control in patients with skull base chordoma, *Int. J. Radiat. Oncol. Biol. Phys.*, 45, 351–358, 1999.

Thames, H. D. and Hendry, J. H., *Fractionation in Radiotherapy*, Taylor and Francis, London (Philadelphia, PA), 1987.

Thames, H. D., Schultheiss, T. E., Hendry, J. H., Tucker, S. L., Dubray, B. M. et al., Can modest escalations of dose be detected as increased tumor control? *Int. J. Radiat. Oncol. Biol. Phys.*, 22, 241–246, 1992.

Theodorou, K., Platoni, K., Lefkopoulos, D., Kappas, C., Schlienger, M., and Dahl, O., Dose-volume analysis of different stereotactic radiotherapy mono-isocentric techniques, *Acta Oncol.*, 39, 157–163, 2000.

Thwaites, D. I., Electron beam treatment-planning techniques, in *Radiotherapy Physics in Practice*, Williams, J. R. and Thwaites, D. I., Eds., 2nd ed., Oxford University Press, Oxford, pp. 205–219, 2000.

Tien, R. D., Buxton, R. B., Schwaighofer, B. W., and Chu, P. K., Quantitative of structural distortion of the neural foramina in gradient-echo MR imaging, *J. Magn. Reson. Imaging*, 1, 683–687, 1991.

Tomé, W. A. and Fowler, J. F., Selective boosting of tumor subvolumes, *Int. J. Radiat. Oncol. Biol. Phys.*, 48, 593–599, 2000.

Trapp, J. V., Warrington, A. P., Partridge, M., Philps, A., Glees, J. et al., Measurement of the three-dimensional distribution of radiation dose in grid therapy, *Phys. Med. Biol.*, 49, N317–N323, 2004.

Travis, E. L., Lung morbidity of radiotherapy, in *Complications of Cancer Management*, Plowman, P. N., McElwin, T. J., and Meadows, A. T., Eds., Butterworth and Heinemann, Stonehan, MA, pp. 232–249, 1991.

Travis, E. L., Organizational response of normal tissues to irradiation, *Semin. Radiat. Oncol.*, 11, 184–196, 2001.

Tucker, S. L., Thames, H. D., and Taylor, J. M. G., How well is the probability of tumor cure after fractionated irradiation described by Poisson statistics? *Radiat. Res.*, 124, 273–282, 1990.

Tucker, S. L., Dong, L., Cheung, R., Johnson, J., Mohan, R., Huang, E. H., Liu, H. H., Thames, H. D., and Kuban, D., Comparison of rectal dose-wall histogram versus dose-volume histogram for modeling the incidence of late rectal bleeding after radiotherapy, *Int. J. Radiat. Oncol. Biol. Phys.*, 60, 1589–1601, 2004.

Ulnin, K. and Palisca, M., The scattering foil compensators in electron beam therapy, *Int. J. Radiat. Oncol. Biol. Phys.*, 35, 785–792, 1996.

van Dyk, J., Keane, T. J., Kan, S., Rider, W. D., and Fryer, C. J., Radiation pneumonitis following large single dose irradiation: A re-evaluation based on absolute dose to lung, *Int. J. Radiat. Oncol. Biol. Phys.*, 7, 461–467, 1981.

van Dyk, J., Barnett, R. B., Cygler, J. E., and Shragge, P. C., Commissioning and quality assurance of treatment planning computers, *Int. J. Radiat. Oncol. Biol. Phys*, 26, 261–273, 1993.

van Herk, M. and Kooy, H. M., Automatic three-dimensional correlation of CT–CT, CT–MRI and CT–SPECT using chamfer matching, *Med. Phys.*, 21, 1163–1178, 1994.

van Herk, M., Remeijer, O., Rasch, C., and Lebesque, J. V., The probability of correct target dosage: Dose population histograms for deriving treatment margins in radiotherapy, *Int. J. Radiat. Oncol. Biol. Phys.*, 47(4), 1121–1135, 2000.

van-den-Elsen, P. A., Pol, E. J., and Viergever, M. A., Medical image matching—A review with classification, *IEEE Trans. Biomed. Eng.*, 12, 26–39, 1993.

Van't Riet, A., Mak, A. C., Moerland, M. A., Elders, L. H., and van der Zee, W., A conformation number to quantify the degree of conformality in brachytherapy and external beam irradiation: Application to the prostate, *Int. J. Radiat. Oncol. Biol. Phys.*, 37, 731–736, 1997.

Veninga, T., Huisman, H., van der Maazen, R. W., and Huizenga, H., Clinical validation of the normalized mutual information method for registration of CT and MR images in radiotherapy of brain tumors, *J. Appl. Clin. Med. Phys.*, 5(3), 66–79, 2004.

Walker, A. M. and Suit, H. D., Choosing between two formulations of a dose/cure function, *Br. J. Radiol.*, 54, 1012–1013, 1981.

Wang, J. Z. and Allen Li, X., Evaluation of external beam radiotherapy and brachytherapy for localized prostate cancer using equivalent uniform dose, *Med. Phys.*, 30, 34–40, 2003a.

Wang, J. Z., Guerrero, M., and Allen Li, X., How low is the α/β ratio for prostate cancer? *Int. J. Radiat. Oncol. Biol. Phys.*, 55, 194–203, 2003b.

Wara, W. M., Phillips, T. L., Margolis, L. W., and Smith, V., Radiation pneumonitis: A new approach to the deterioration of time dose factors, *Cancer*, 32, 547–552, 1973.

Warkentin, B., Stavrev, P., Stavreva, N., Field, C., and Fallone, B. G., A TCP–NTCP estimation module using DVHs and known radiobiological models and parameter sets, *J. Appl. Clin. Med. Phys.*, 5(1), 50–63, 2004.

Webb, S., The effect on tumour control probability of varying the setting of a multileaf collimator with respect to the planning target volume, *Phys. Med. Biol.*, 38, 1923–1936, 1993.

Webb, S., Optimum parameters in a model for tumour control probability, including interpatient heterogeneity, *Phys. Med. Biol.*, 39, 1895–1914, 1994.

Webb, S. and Nahum, A. E., A model for calculating tumour control probability in radiotherapy, including the effects of inhomogeneous distributions of dose and clonogenic cell density, *Phys. Med. Biol.*, 38, 653–666, 1993.

Webb, S. and Nahum, A. E., Correspondence: Regarding, Wu, Chua, Sham, et al., *IJROBP* 37(4):913–920; 1997, *Int. J. Radiat. Oncol. Biol. Phys.*, 40, 1009–1010, 1998.

West, C. M., Intrinsic radiosensitivity as a predictor of patient response to radiotherapy, *Br. J. Radiol.*, 68, 827–837, 1995.

West, C. M. L., Hendry, J. H., Scott, D., Davidson, S. E., and Hunter, R. D., 25th Paterson Symposium—is there a future for radiosensitivity testing? *Br. J. Cancer*, 64, 197–199, 1991.

Wheldon, T. E., Deehan, C., Wheldon, E. G., and Barrett, A., The linear-quadratic transformation of dose-volume histograms in fractionated radiotherapy, *Radiother. Oncol.*, 46, 285–295, 1998.

White, R. D., Ehman, R. L., and Weinreb, J. C., Cardiovascular MR imaging: Current level of clinical activity, *J. Magn. Reson. Imaging*, 2, 365–370, 1992.

Wigg, D. R., *Applied Radiobiology and Bioeffect Planning*, Medical Physics Publishing, Madison, WI, 2001.

Wilks, R. J., An optical system for measuring surface shapes for radiotherapy planning, *Br. J. Radiol.*, 6, 351–359, 1993.

Willoughby, T. R., Starkschall, G., Janjan, N. A., and Rosen, I. I., Evaluation and scoring of radiotherapy treatment plans using an artificial neural network, *Int. J. Radiat. Oncol. Biol. Phys.*, 34, 923–930, 1996.

Withers, H. R. and Taylor, J. M., Critical volume model, *Int. J. Radiat. Oncol. Biol. Phys.*, 25, 151–152, 1993.

Withers, H. R., Taylor, J. M. G., and Maciejewski, B., Treatment volume and tissue tolerance, *Int. J. Radiat. Oncol. Biol. Phys.*, 14, 751–759, 1988.

Wolbarst, A. B., Optimization of radiation therapy. II. The critical voxel model, *Int. J. Radiat. Oncol. Biol. Phys.*, 10, 741–745, 1984.

Wolbarst, A. B., Chin, L. M., and Svensson, G. K., Optimization of radiation therapy: Integral-response of a model biological system, *Int. J. Radiat. Oncol. Biol. Phys.*, 8, 1761–1769, 1982.

Wood, M. L., Thoracic and abdominal motion artefacts, in *Magnetic Resonance Imaging*, Stark, D. D. and Bradley, W. G., Eds., Mosby, St. Louis, pp. 792–803, 1988.

Xiong, W., Li, J., and Ma, C.-M., Effect of patient variation on standard- and hypo-fractionated radiotherapy of prostate cancer, *Phys. Med. Biol.*, 50, 1483–1492, 2005.

Yaes, R. J., The slope of the sigmoid dose response curve for tumor control, *Int. J. Radiat. Oncol. Biol. Phys.*, 44, 470–471, 1999.

Yorke, E. D., Dose distributions in normal tissues, *Semin. Radiat. Oncol.*, 11, 197–209, 2001.

Yorke, E. D., Kutcher, G. J., Jackson, A., and Ling, C. C., Probability of radiation induced complications in normal tissues with parallel architecture under conditions of uniform whole or partial organ irradiation, *Radiother. Oncol.*, 26, 226–237, 1993.

Yorke, E. D., Jackson, A., Rosenzweig, K. E., Merrick, S. A., Gabrys, D. et al., Dose-volume factors contributing to the incidence of radiation pneumonitis in non-small-cell lung cancer patients treated with three-dimensional conformal radiation therapy, *Int. J. Radiat. Oncol. Biol. Phys.*, 54, 329–339, 2002.

Zagars, G. K., Schultheiss, T. E., and Peters, L. J., Inter-tumour heterogeneity and radiation dose-control curves, *Radiother. Oncol.*, 8, 353–362, 1987.

Zaider, M. and Minerbo, G. N., Tumour control probability: A formulation applicable to any temporal protocol of dose delivery, *Phys. Med. Biol.*, 45, 279–293, 2000.

Zavgorodni, S., The impact of inter-fraction dose variations on biological equivalent dose (BED): The concept of equivalent constant dose, *Phys. Med. Biol.*, 49, 5333–5345, 2004.

Zelefsky, M. J., Leibel, S. A., Kutcher, G. J., Kelson, S., Ling, C. C., and Fuks, Z., The feasibility of dose escalation with three dimensional conformal radiotherapy in patients with prostatic carcinoma, *Cancer J.*, 1, 142–150, 1995.

Zwicker, R. D., Meigooni, A., and Mohiuddin, M., Therapeutic advantage of grid irradiation for large single fractions, *Int. J. Radiat. Oncol. Biol. Phys.*, 58, 1309–1315, 2004.

PART H

QUALITY ASSURANCE

Editors: Philip Mayles and Jean-Claude Rosenwald

INTRODUCTION

Quality assurance is an essential part of the radiotherapy process. In recent years, it has become accepted that this is not just about ensuring that the treatment machines are correctly calibrated, but that it includes every part of the process. The ISO9000 standard has been used as the basis for such a system in a number of countries (Bleehen 1991; Leer et al. 1999). This book is not intended to be a quality assurance manual, so the treatment in this Part is restricted to the elements of the quality system that require the application of physics. It is the responsibility of the physicist to establish the physical and dosimetric framework underpinning the delivery of radiotherapy treatments. In Chapter 37, an introduction is given to some general principles underlying the quality assurance of this framework. In the remaining chapters, the physical aspects of quality control of each part of the process are considered. Chapter 38 deals with quality control of external beam treatment machines, including in Section 38.8 some practical guidance on dose calibration for photon treatments, the details of which have already been described in Part D. Chapter 39 deals with quality assurance in treatment planning including both the purchase and commissioning of the planning system and the ongoing quality control of the system and of individual patient plans. Quality control of treatment involves both portal imaging and in vivo dosimetry, and these are dealt with in Chapter 40. A modern radiotherapy department is heavily dependent on networked systems, and consideration needs to be given to the quality assurance of these. This is covered in Chapter 41. The Digital Imaging and Communications in Medicine (DICOM) standard is becoming universally used to transfer data in radiotherapy, and Chapter 42 is devoted to this particular aspect of network communication.

CHAPTER 37

RATIONALE AND MANAGEMENT OF THE QUALITY SYSTEM

Philip Mayles and David Thwaites[*]

CONTENTS

---

[*] With contribution from Jean-Claude Rosenwald.

Radiotherapy is a form of treatment that is directed towards the eradication of tumour cells and it has the inherent potential to damage normal tissue also. Calibration errors can lead to injury in a large number of patients. A systematic approach to the quality assurance of the whole radiotherapy process is therefore essential (WHO 1988; AAPM 1994a; Thwaites et al. 1995).

37.1 DEFINITIONS

Before discussing the requirements of a quality system, it is first necessary to set out some definitions. According to WHO (1988),

> "**Quality Assurance** is concerned with all those procedures that ensure consistency of the medical prescription and the safe fulfilment of that prescription as regards dose to the target volume, together with minimal dose to normal tissue, minimal exposure of personnel, and adequate patient monitoring aimed at determining the end result of treatment."

This is based on the general definition adopted by the International Organisation for Standardisation (ISO 1995) and also by the British Standards Institute (BSI) that:

> "**Quality Assurance (QA)** is defined as all those planned and systematic actions necessary to provide adequate confidence that a structure, system or component will perform satisfactorily in service", or will satisfy given requirements for quality.
>
> **Quality Control (QC)** is the regulatory process through which the actual quality performance is measured, compared with existing standards and finally the actions necessary to keep or regain conformance with the standard (ISO 1995).

So QC describes the operational techniques and activities that are used to fulfil the requirements for quality. It forms part of the wider quality assurance programme or system.

> **Quality Standards** are the set of accepted criteria against which the quality of the activity in question can be assessed.
>
> **Quality Audit** is an independent review of the quality assurance and quality control programmes, which is ideally external to the process or part of the process under review, i.e. performed using independent procedures and by independent persons who are not responsible for the performance of the product or process under review.

The aim of the introduction and development of these concepts in relation to radiotherapy is to produce and maintain consistent and continuing quality in treatment. The overall aim is to ensure that the clinical requirements on the quality of radiotherapy are met, to achieve optimum treatment in terms of maximising tumour control probability, whilst at the same time maintaining normal tissue injury to within clinically acceptable levels. As one part of this, the implementation of a quality assurance programme will minimise errors and accidents. However, the attainment of quality in treatment is a much more fundamental goal than that, being concerned with the general reduction of uncertainties throughout the whole radiotherapy process.

37.2 REQUIREMENTS OF A QUALITY SYSTEM

A systematic approach to quality assurance requires the following principal elements:

- Clear definition of responsibilities
- Documented procedures
- Accurate record keeping

- Control of system failures
- Internal and external audit of procedures
- Attention to training needs

ESTRO (Leer et al. 1999) have published a guide to the implementation of a quality system based around the requirements of the ISO9000:1994 quality system (ISO 1994). This quality standard has since been replaced by the ISO9000:2000 standard (ISO 2000) which reduces the requirement for extensive documentation and introduces the additional requirement to demonstrate continual improvement. The new standard requires that the whole process should be considered, but allows more freedom to define the level of documentation required.

It is good practice to begin by drawing a flow chart of the patient's journey and to use this to form the basis of an assessment of the QC procedures that are required to ensure that each stage of the process is completed accurately. This is summarised in Figure 37.1.

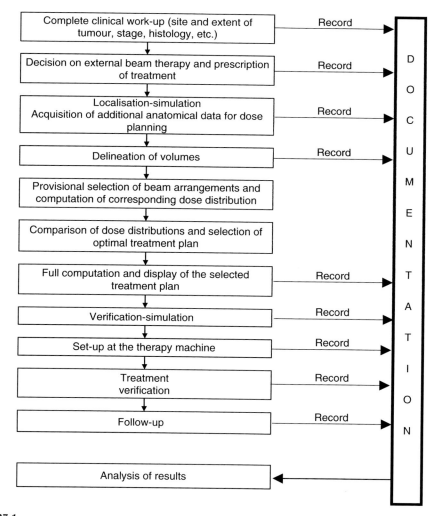

FIGURE 37.1
The radiotherapy process as defined in ICRU Report 50. (From International Commission on Radiological Units, Prescribing, recording and reporting photon beam therapy, ICRU Report 50, ICRU, Bethesda, MD, 1993.)

37.3 QUALITY-CONTROL PROCEDURES

A quality assurance system will require QC procedures to ensure that all equipment and processes are conforming to established specifications. A number of sources of such procedures are available. At the end of this Part, there is a full bibliography. Particularly useful are the AAPM reports from Task Groups 23, 40, and 45 (AAPM 1994a; 1994b; AAPM 1995, respectively), as well as IEC Publication 976 (IEC 1989) and IPEM Report 81 (IPEM 1999). It must be emphasised that these reports should not be treated as recipe books for carrying out QC. Rather, they provide a yardstick against which the quality system of the individual department can be judged. The quality procedures must be adapted to the particular circumstances, equipment and treatment techniques in use.

37.4 BASIS FOR TOLERANCE LIMITS

For each parameter on which QC is carried out, it is necessary to provide a tolerance limit. It is a fundamental tenet of a quality system that the basis for these tolerance limits should be the fitness for purpose of the process. In this case, the aim is to deliver radiotherapy with the required clinical accuracy, and to determine the tolerances necessary to evaluate these requirements. In some situations, the limiting factor may not be the clinical accuracy required, but the accuracy achievable at the present time. It is in these areas that the struggle for improvement should be concentrated.

37.4.1 STATEMENTS OF ACCURACY

The term *accuracy* in radiotherapy is often used loosely. As a starting point, absolute accuracy must be distinguished from *precision* or *reproducibility*. Overall uncertainty is a combination of random and systematic uncertainties. Random uncertainties can be estimated from repeated independent observations and can be expressed as a standard deviation (SD). Systematic uncertainties on the other hand can only be estimated by an analysis of the process under consideration, assigning reasonable variations to parameters where the uncertainties are not, by definition, exactly quantifiable. They can be expressed as effective SD, being taken as the estimate of the limits within which the correct value is expected to occur in around 70% of cases. Uncertainties of both types (SD or effective SD) can be combined in quadrature to provide an estimate of overall uncertainty[*] (BIPM 1980; Mijnheer et al. 1987; NIST 1994; BSI 1995; NIST 2000; ISO 2002). As is well recognised, systematic errors contribute to absolute deviations from the correct value in any situation whilst random errors are associated with precision. However, the distinction can become blurred in certain circumstances.

In radiotherapy, within one centre and one radiation modality, it is reproducibility of dose delivery that is critical. However in transferring experience from one centre to another or in intercomparing results between centres or between modalities, for example in clinical trials, some systematic uncertainties also become important. When the radiation oncologist prescribes a dose to a given patient, the prescription may be subject to different interpretations and, as a consequence, to large deviations in the dose actually delivered to the target volume for the same prescription. This cause of discrepancy will be further discussed in Section 37.6. In intercomparisons between a number of centres, some systematic uncertainties are effectively randomly distributed. On the other hand some systematic uncertainties are common to all participants and can be omitted from routine consideration, e.g. those associated with a given

---

[*] The NIST online reference (NIST 2000) provides a particularly useful resource on definition of uncertainties.

dosimetry protocol if all participants follow the same protocol, or those associated with basic physical data if the protocols involved are all based on the same consistent set. Such factors only require consideration when estimating absolute accuracy but play no part in estimations of dosimetric consistency (precision). Thus a clear understanding is required of which uncertainties require inclusion in any given situation. However in practice it is often difficult to separate the two types (Dutreix 1984). As a further point in aiding clarity of discussion of uncertainties, it should be noted that any stated values should also have their meaning defined. All values stated in this Chapter, whether for individual parameters or for cumulative values, where all involved uncertainties are combined in quadrature, are given as one SD or effective SD.

37.4.2 DOSE EFFECT RELATIONSHIPS AND ACCURACY

Two sets of dose relationships are of importance in any radiotherapy situation, the relationship between dose and tumour control (see Section 7.4.2) and that between dose and normal tissue complications (see Section 8.6). Each has the characteristic form shown in Figure 37.2, exhibiting threshold doses, relatively steep rises, and saturation at high doses. The most common practical radiotherapy situation is to find that the curves overlap along the dose axis such that the aim of radiotherapy is to attempt to maximise control of the tumour while maintaining normal tissue complications at an acceptable level.

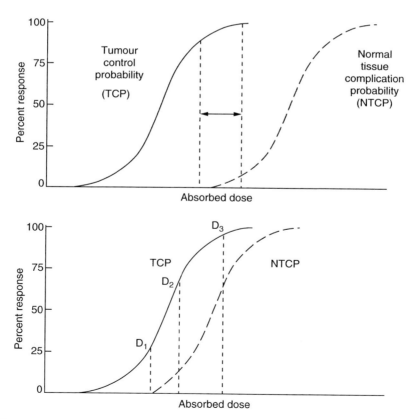

FIGURE 37.2

Dose effect curves for tumour control and normal tissue complications. The curves are sigmoid and are assumed here to have the same shape and steepness, for simplicity. The upper figure shows a favourable situation for radiotherapy where a dose selected within the range shown by a double arrow is satisfactory. The lower figure shows a less favourable one where none of the 3 dose levels indicated by D_1, D_2 or D_3 can achieve both a high TCP and low NTCP.

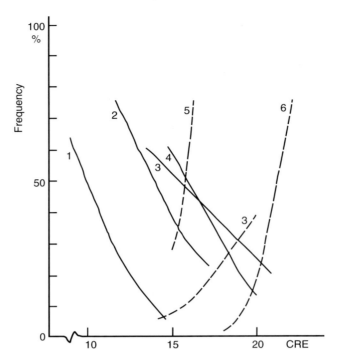

FIGURE 37.3
Dose response curves collected from the clinical literature for local recurrence frequency (solid line) and frequency of complications (broken line) versus the cumulative radiation effect (CRE). 1. Malignant lymphomas, Fuks and Kaplan (1973); 2. Nasopharynx, Moench and Phillips (1972); 3. Bladder, Morrison (1975); 4. Tonsil, Shukovsky and Fletcher (1974); 5. Skin reaction Turesson and Notter (1984); 6. Brachial plexus, Svensson and others (1975). (From Johansson, L. et al., *Int. J. Radiat. Biol. Relat. Stud. Phys. Chem. Med.*, 41, 411–420, 1982.)

The general requirements on accuracy are obtained by considering the steepness of the dose-effect relationships. It is obvious for relationships of this general type that if uncertainties on a delivered dose are large, then the probability of tumour control will be reduced or the incidence of normal tissue complications increased depending on the direction of the deviation from the stated dose. Thus clinical accuracy requirements must be such as to ensure that deviations are acceptable in these terms, with high accuracy on the steep part of the curves (Figure 37.2).

Clinical dose-effect data have been reviewed by a number of authors (e.g. Johansson 1982; Moore et al. 1983; Mijnheer et al. 1987; Brahme et al. 1988). Some examples are given in Figure 37.3 with the doses for different fractionation schedules corrected to an equivalent dose. These curves are subject to a number of uncertainties but give an indication of the steepness of the dose response curves. More detail is given in Part B.

37.4.3 REQUIRED ACCURACY IN ABSORBED DOSE DELIVERY

As a general requirement, it is necessary to base accuracy recommendations on the steeper dose-effect relationships encountered in routine clinical situations. These are generally for normal tissues. The International Commission on Radiation Units and Measurements (ICRU), in Report 24 (1976), reviewed the limited information available up to that time and considered that ±5% accuracy was required in the delivery of absorbed dose to the target volume, but that in critical situations ±2% may be required. However, it was recognised that the latter figure was not generally realisable in then-current practice. It was not made clear what uncertainty value the figures represented. Mijnheer and others (1987) obtained a figure for required accuracy by considering normal tissue effects. They considered the steepness of dose-effect curves in terms of the percentage increase in absorbed dose to produce a change in the probability of normal tissue complications from 25% to 50%. A representative value of 7% was taken for this relative gradient and it was concluded that any transfer of clinical information from one centre to another will involve unacceptable risks of complications if overall uncertainty in absorbed dose is larger than this value. This was then assigned to the 2 SD level, resulting in a value of 3.5%, one relative SD, as the general accuracy requirement on absorbed dose delivery. Brahme and others (1988) considered the effects of variations in dose on tumour control for typical values, showing that the largest loss in tumour control introduced by dosimetric inaccuracy is found at the highest level of tumour control probability. A general figure of 3% (relative SD) on the delivered absorbed dose to the patient was recommended as the tolerance level on accuracy in dose delivery, to keep variations in the probability of tumour control within acceptable limits.

Thus overall a figure of 3% SD can be taken as the currently recommended accuracy requirement, considered as one relative SD, on the value of the dose delivered to the patient at the dose specification point. This implies there is a 96% probability that changes will be clinically observable for dose changes at twice this level, in situations described by the steeper dose-effect relationships. This is also consistent with more anecdotal evidence on clinical observations following inadvertent dose changes due to dosimetric errors (Dutreix 1984).

37.4.4 REQUIRED ACCURACY IN ABSORBED DOSE DISTRIBUTIONS

Besides variations in the overall level of absorbed dose delivered, variations in the dose distribution across the target volume may be expected to affect the outcome of treatment and to contribute to the steepness of practical dose-effect relationships. Brahme (1984) has considered the effects of different distributions on uniform tumour volumes and argued that a SD of the dose distribution in the range of 3% to 5% was required. Combining this with the recommended uncertainty in delivered dose to the specification point gives a requirement that the accuracy of dose to all other points in the target volume should be around 4% to 6%. As a general recommendation, a figure of 5% can be taken, again being one relative SD.

37.4.5 REQUIRED GEOMETRIC ACCURACY IN RADIOTHERAPY

Geometric uncertainties arise for a variety of reasons, including treatment machine specifications and tolerances, simulation and treatment set-up, patient or organ movement during treatment and changes of patient shape between fractions. Clinical information is limited concerning the effects of such variations on treatment outcome. Geometric miss of tumour/target will obviously decrease tumour control probability, whilst overlap of fields with adjacent normal structures, particularly critical organs, will be detrimental in terms of complications. In general appropriate margins are allowed around the target volume to allow for these uncertainties so it is difficult to find definitive data on the effect of inaccuracies. For this reason estimates of the required accuracy must be based on modelling studies (e.g. Brahme et al. 1988).

Many studies were carried out when computerised tomography (CT) was introduced routinely to radiotherapy to look at the impact of using CT on tumour localisation as compared to conventional methods (e.g. Badcock 1983; Dobbs et al. 1983) and on the impact of CT or simulator CT on treatment planning (e.g. Redpath 1988). These unambiguously show the significant advantage of using CT methods for determining margins and for input to the treatment planning process and can provide information on the changes in dose variation across target volumes due to changes in determination of those volumes.

The estimation of required geometric accuracy is best carried out by considering the various sources of uncertainty, combining them to give an overall value, and using the best practically

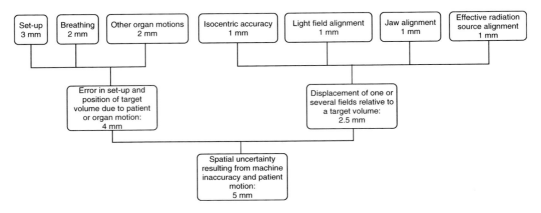

FIGURE 37.4

Cumulative geometric uncertainties in external beam therapy. (Redrawn from AAPM, Report 13, *Physical aspects of quality assurance in radiotherapy*, American Institute of Physics, New York, 1984. With permission.)

achievable figures as the level to be recommended. On this basis the AAPM (1984) arrived at a figure corresponding to 5 mm, one effective SD, as illustrated in Figure 37.4. Mijnheer and others (1987) considered a wider set of data and recommended an accuracy of positioning of field edges and shielding blocks of 4 mm, one SD. This is in general agreement with the philosophy underlying the IEC recommendations for geometric tolerances of radiotherapy equipment (Rassow et al. 1988).

37.4.6 OVERALL ACCURACY REQUIREMENTS

In summarising all the currently available evidence, general recommendations on accuracy in radiotherapy can be made:

- Three percent on the absorbed dose delivered at the specification point
- Five percent on the dose at all other points in the target volume
- Four millimetres on the position of field edges and shielding blocks in relation to the treatment isocentre

All these are given as one SD and all are general requirements for routine clinical practice. In certain cases (for example, palliative treatments) higher values may be acceptable. However it is generally impractical to work to different standards of accuracy for different treatment techniques and so these values must be widely applied. In some special cases, smaller uncertainties may be demanded if very steep complication curves are involved or tight geometric tolerances are required, as in stereotactic treatments (Chapter 45) or proton therapy (Chapter 46).

It must be appreciated that these figures are on final delivered dose to the patient at the end of all the contributing stages involved in the radiotherapy process and this implies that the accuracy of individual contributions to the overall delivered dose must be higher. The dosimetry chain begins with primary dosimetry standards and the required basic physical data. It leads through dosimeter calibration, treatment-beam calibration in reference conditions, relative dosimetry in all other conditions, and the treatment planning process, including acquisition of patient data and specification of treatment volume and dose. It ends with the delivery of the prescribed, planned, and accepted treatment to the patient under day-to-day conditions over the course of the treatment. The accuracy requirement on each part of the whole process must be significantly less than the overall recommendations to achieve the final values recommended.

37.5 ACCURACY CURRENTLY ACHIEVABLE

37.5.1 GENERAL

Uncertainties are introduced at all levels in the dosimetry chain. The overall cumulative uncertainty can be estimated in one of two ways, which complement each other: (1) by an a priori approach, considering potential sources of uncertainty at every stage, using all available information to assign reasonable practical values, or at worst making educated guesses in the absence of hard data; (2) by appropriate dosimetric intercomparisons or audits to assess cumulative uncertainties experimentally at various levels. The nearer to treatment conditions that the intercomparison is carried out, the more factors are included. Differences in uncertainties at different levels can provide information on those introduced in the intervening steps. Neither method is ideal. For example inter-centre intercomparisons cannot detect any systematic errors common to both the auditing system and the audited centre. In addition inevitably there are omissions of some clinical factors in inter-centre dosimetric intercomparisons, as these can only be used to investigate limited combinations of clinical treatment conditions (Thwaites et al. 1992; Thwaites et al. 2003b). However inter-centre comparisons can be complemented by internal audit using in vivo dose measurement and on-line portal imaging.

37.5.2 ESTIMATION OF ACHIEVABLE UNCERTAINTIES

Various attempts have been made to analyse the radiotherapy process to estimate cumulative uncertainties on delivered dose (e.g. Brahme 1984; Dutreix 1984; Mijnheer et al. 1987; Brahme et al. 1988). This has been done considering each major step, including

- The determination of the absorbed dose to a reference point in water
- The establishment of other doses relative to this
- Treatment planning
- Treatment delivery to the patient

These steps are then broken down into substeps and best estimated uncertainties at each level are assigned to each contributing factor. The overall cumulative uncertainties obtained in this way have ranged from 2.5% to 8.5%, as one effective SD. The different values in part reflect the judgements used by different authors, the number of separate sub-steps and parameters they included and whether they tended towards an optimal or a conservative approach. A figure of 5% (1 SD) could be taken to be representative of these types of estimates and they would be likely to produce smaller uncertainties for simpler treatments and larger for more complex treatments.

More recently, uncertainty estimates based on harder evidence from intercomparisons, audits and in vivo dosimetry have been made (e.g. Thwaites 1993; Thwaites 2003; Thwaites 2004). Table 37.1 provides an example for external beam megavoltage x-ray treatments, following U.K. procedures and codes of practice for basic dosimetry and utilising results from:

- The U.K. dosimetry intercomparisons and audits, all based on site visits and ion chamber dosimetry (Thwaites et al. 1992; Nisbet and Thwaites 1997)
- The Scottish+ audit group of the U.K. radiotherapy dosimetry audit network (Thwaites et al. 2003b), which includes intercomparisons in a number of phantoms simulating treatment sites, also based on site visits and ion chamber dosimetry, and some inter-centre in vivo intercomparisons
- Analysis of QC records and specific experimental evaluation of the uncertainties at many levels in the radiotherapy process in one department (Allahverdi and Thwaites 1998)
- Systematic in vivo (diode) dosimetry in one department over more than 10 years (Thwaites et al. 2003a)

TABLE 37.1

Uncertainties in External-Beam Megavoltage X-Ray Dosimetry Based on Experimental Determinations

Source/Level	Single Centre (different beams/times) (%)	Multi-Centre (%)
1. Dose at reference point in water phantom		
Uncertainties quoted on calibration factors by the U.K. standards lab (NPL) are 0.7% (1 effective SD)	0.7	0.7
Variation in reference dose determination	0.5	0.7–1.0
2. Dose to phantoms representing various treatment sites (at a range of points within target volumes; given relative to reference dose)	0.8–1.8	1.1–2.3
3. Patient dose at specification point (based on estimates from in vivo dosimetry; for a wide range of treatment sites and techniques; given relative to reference dose)	1.5–3.0	1.6–3.3
where lung is significantly involved	(5.0)	(5.2)
4. Estimated overall cumulative uncertainty on delivered patient dose at the specification point including standards lab uncertainty (i.e. relative to absolute Gy)	1.7–3.1	1.8–3.5
where lung is significantly involved	(5.1)	(5.3)

The beam/dosimetry calibration method follows the U.K. system and code of practice (based on a direct absorbed dose to water calibration covering that beam quality at the standards laboratory). It is assumed that uncertainties may be combined (and separated) in quadrature. All uncertainties are given as 1 effective SD.

The uncertainties in Table 37.1 are all given as effective SD values; 95% confidence levels can be taken to be approximately twice these values. It may be noted that the final cumulative patient dose uncertainties for a single centre are not very different from those for multi-centre intercomparisons. The values at level 2 indicate what is achievable in relative dosimetry comparisons between phantom measurements and planned doses, i.e. normalised to the local reference dose value. The values at level 3 similarly indicate what is achievable in comparing patient measurements and planned doses, i.e. again normalised to the local reference dose value. The values at level 4 indicate the overall uncertainties relative to absolute dose, i.e. including the basic dosimetry (standards laboratory) uncertainties.

The general overall cumulative uncertainties at the specification point, for many treatment sites and techniques other than when lung is significantly involved, are within—or very close to—the recommended required values of 3% (1 SD) (see Section 37.4.6). They are likely to be representative of fairly optimal situations in normal clinical practice. However, overall uncertainties will be larger if any steps or sub-steps have larger uncertainties. For example other more recent multi-centre dosimetry intercomparisons or audits carried out by ESTRO-EQUAL (Ferreira et al. 2003) and the U.S. RPC (Aguirre et al. 2003) using mailed thermoluminescent dosimeters (TLD) have shown larger variations than those quoted above where direct comparison is possible. For example, in determination of reference dose in megavoltage photon beams SD of 1.6% to 1.9% are observed. In part this reflects the wider international spread of such audits, involving different codes of practice and different standards laboratories in the dosimetry chain. In part, also it reflects the larger inherent uncertainties in the use of mailed TLDs compared to site visits using ion chambers.

The only significant exception to the general values is treatments where there is marked involvement of lung. Here in vivo measurements indicate that uncertainties may increase due both to movement and to the ability of planning systems to cope with such situations. This will depend on many factors, including the exact site, the beam directions, the planning algorithms involved (see Section 39.3.3), the approach used for lung correction, and so on.

The overall cumulative uncertainties are estimated for patient-delivered dose at the specification point. For other points within the target volume, additional uncertainties are present, which can be estimated by folding in uncertainties from multi-point measurements in phantom or from 2D (or 3D) verification of fields and treatments (here for example, experience is rapidly growing from the development and implementation of verification methods for intensity modulated radiotherapy (IMRT)). Optimal uncertainties in these points are likely to be similar to those at level 2 in the table, giving estimated cumulative uncertainties (1 SD) on the dose at other points in the target volume in the range of approximately 2% to 4.5%. These figures may also be interpreted in the context of the various levels recommended for assessing the accuracy of treatment planning systems, as discussed in Section 39.3.4.2 (and, for example, by Venselaar et al. 2001).

It may be noted that realistic phantom intercomparisons have shown uncertainties which are in general agreement with Table 37.1, for example, the START trial audit (Venables et al. 2003) reported SD of 1.2% and 1.3% in the results of phantom measurements at the specification point for 2D and 3D breast plans respectively, with the totality of individual measurements lying within the range -10% to $+12\%$ compared to locally planned values. The RTO1 trial audit (Moore et al. 2003) showed a 1.3% SD at the isocentre for multi-centre measurements at the specification point in a prostate phantom, comparing to locally planned values, with a 2% SD for measurements within the clinical target volume and a 4% SD for all measurements, including some outside the target volumes. There are also results beginning to appear from clinical trial QA groups carrying out systematic in vivo measurements on patients entered into the trials. For example, Venables and others (2004) have reported the distribution of measured/expected doses for more than 400 breast patients in the START trial, where the mean value observed was 0.99 and the SD 4%. Here the measurements used a mailed TLD system, with its own uncertainties reported as close to 3% (1 SD). When this is taken into account there is good agreement with the data used for Table 37.1.

Geometric uncertainties translate into dosimetric uncertainties, in part depending on considerations of choices of margins for target volumes and for organs at risk and also on correction strategies. A rapidly growing database of information is available from studies considering positioning uncertainties and organ motion (see Section 39.2 and Hurkmans et al. 2001). A major review (BIR 2003) provides detailed discussion of geometric uncertainties in general and also for particular treatment sites[*]. Overall geometric uncertainties have been reported from a few mm to between 1 cm and 2 cm (1 SD), depending on site, techniques, immobilisation, etc. and must be evaluated in each centre for the procedures in use.

37.6 ICRU DOSE SPECIFICATION AND REPORTING

As pointed out in Section 37.4.1, clarity of the prescription of the radiation oncologist is fundamental. The ICRU (1993, 1999, 2004) have published recommendations on the reporting of radiotherapy doses. The geometric part of these recommendations is discussed in Section 29.2, to which the reader is referred for the definition of the planning target volume (PTV). The various possibilities for dose specification are discussed briefly in Section 35.4. Here, the specification of the dose is considered in the context of the need to ensure consistency of dose specification and reporting.

The ICRU defines the ICRU reference point and recommends that the dose at this point be always reported. The recommendations relate to the recording and reporting of radiotherapy doses and do not make it a requirement that the dose is prescribed to the ICRU reference point. However, it improves consistency if all prescriptions are to this reference

---

[*] This is discussed further in Section 29.3 and Section 29.4.

point. Even if the prescription is to a different point it is sensible for dose distributions to be normalised to the ICRU reference point. It should then be apparent that a prescription is different because a prescription to the minimum dose within the PTV, say, will appear as a prescription to 95% of the reference point dose, in which case the *reported* dose would be about 5% larger than the *prescribed* dose.

The ICRU reference point is chosen according to the following criteria:

- The dose at the point should be clinically relevant and representative of the dose throughout the PTV.
- The point should not be in a region of steep dose gradient.
- It should be possible to calculate the dose at the point accurately.
- The definition of the point should be clear and unambiguous.

For simple plans with a single intersection point that is close to the centre of the PTV, the intersection point, usually the isocentre, is defined as the ICRU reference point. However, where this point is not representative of the dose to the PTV it is appropriate to define an alternative point. In selecting this point, the first of the above factors is fundamental. The aim is to select a point that is approximately at the middle of the tumour mass of the PTV. Examples of situations where a point other than the isocentre are appropriate are as follows:

1. *Glancing field breast treatments*. Here, the isocentre may be very close to the chest wall or to the skin surface. In either case this point is not representative of the dose to the majority of the PTV and in addition there are uncertainties in the absolute dose value because of loss of lateral scatter. A point half way between the skin surface and the ribs at a point midway between the beam entry points is appropriate.
2. *Asymmetric fields*. It is common practice to facilitate matching of fields to place the isocentre on the match plane. Here the reference point should be placed in the centre of the half field.
3. *Weighted parallel-opposed fields*. The centre of the PTV on the central axis of the beam is the obvious choice here.
4. *Electron beams*. For electron beams the reference point is always in the central part of the PTV and for single fields this should be on the beam axis at the level of the peak dose. Where adjacent fields are used a reference point should be defined for each field (ICRU 2004). For more complex techniques a corresponding point should be chosen.
5. *Simultaneous boost fields*. It is becoming common practice to treat the gross tumour volume (GTV) and nodal fields with a single dose distribution either by IMRT or by combining a number of matched fields. In this case there will be several PTVs and it is appropriate to define a reference point that is representative of the PTV dose for each.
6. *IMRT fields*. IMRT fields are inherently somewhat non-uniform even in the centre of the field and it may therefore be inappropriate to define a point dose. The median or modal dose in the PTV (see Section 35.4.1, Table 35.1) is a useful alternative, although this does not meet the requirement to be easily calculated.

In addition to the reference-point dose, the ICRU recommends that the maximum and minimum doses to the PTV also be reported. In considering the maximum dose, it is recognised that a very small area of high dose is unlikely to be clinically significant. It is therefore recommended that the maximum dose should be the average dose in a volume of 15 mm diameter. For the minimum dose there is no such volume criterion and it is the minimum point dose that should be reported.

Differences in prescription and dose reporting are a significant factor in the variability of doses in radiotherapy. A survey of the fractionation practice in the U.K. in 1990 (Hendry and Roberts 1991) showed that 20% of U.K. radiation oncologists prescribed doses (allowing for fractionation) more than 10% different from the mean. Subtle factors can also lead to different prescriptions. For example in Europe it is common practice to prescribe lung doses with corrections made for the lung inhomogeneities, whereas in the U.S.A. it has been common practice to prescribe the dose as calculated without lung corrections[*], although this may be changing with the advent of IMRT. Thus what may appear to be the same dose could be over 10% different for individual patients (Frank et al. 2003). For prostate cases the difference will be approximately 5% (Ginestet et al. 2000) because of attenuation by the femoral heads. This issue can lead to considerable confusion both in the analysis of historical studies and in present day prescription. Surprisingly ICRU Reports 50 and 62 (1993, 1999) do not make recommendations as to what should be done[†]. It is therefore essential that any reported dose includes a statement about whether inhomogeneity corrections have been made or not. In interpreting published data this issue must be given careful consideration.

37.7 ERRORS AND ACCIDENTS

Although it is clearly important to minimise the uncertainty in the patient dose associated with the combination of these individual uncertainties, there are occasionally errors made in patient treatment that are of a significantly greater order of magnitude. A quality assurance system must therefore spend at least as much effort in minimising the possibility of such errors. The IAEA publication on the lessons learnt from radiotherapy errors (IAEA 2000b) considered the causes of a number of such incidents around the world. It identified the problem of ensuring accuracy when a large number of tasks have to be performed many times every day differing only slightly from one patient to the next. When many people work together contributing a small part of the entire process and staff in several disciplines must interact in highly technical measurements and calculations the potential for mistakes is large. It must be acknowledged that all staff make mistakes and there is therefore a requirement for defence in depth and for checks to be made that require new thought. Particularly telling is the story of a physicist who checked the same calculation at weekly intervals and only on the third occasion spotted the error. Checks must therefore be devised that are as independent as possible of the previous work. It is also essential that working conditions are as conducive to careful work as possible, with attention paid to reducing interruptions and distractions, and that staff have the proper amount of time allocated to the work they have to do.

Critical steps were identified in the process. These were:

- Commissioning of the treatment machine and treatment planning computer
- Periodic calibration
- Identification of the patient
- Target volume localisation
- Preparation of the treatment plan
- Prescription of the dose and fractionation
- Simulation
- The first treatment session
- Daily positioning and parameter selection
- Accumulation of doses on the treatment sheet
- Maintenance of equipment

---

[*] The rationale for not correcting for inhomogeneities is that current doses are based on historical experience where no lung corrections were made. However, the difference between the real (inhomogeneity corrected) dose and the uncorrected dose will differ from one patient to another which will introduce increased variability. It is also argued that inhomogeneity correction algorithms do not all yield the same result and are in any case not 100% accurate.
[†] In both reports the example lung case is presented without lung corrections. In contrast, the sample prostate case in Report 50 is inhomogeneity corrected, whereas in Report 62 it is not.

Critical steps were also identified in brachytherapy (see Chapter 51), including

- Removing sources after brachytherapy
- Identification of sources
- Activity specification and measurement

To minimise the possibility of error, a number of prevention measures are identified. These include the quality system approaches of ISO9000, but identify also some less obvious factors. Particular attention must be taken to consider the unusual situations that may occur and to prepare staff for them. All procedures and training programmes should pay attention to the identification of those situations that could lead to an accident in an unusual or unplanned situation. One person, usually a medical physicist, should have overall responsibility for all aspects of the dosimetry system to ensure that the lines of communication concerning dosimetry of patients are working properly. This requirement requires that the physicist should be closely involved in all aspects of the treatment process so that misunderstandings do not occur.

The vital importance of communication is also identified. Problems include failure to transmit information, transmission of incorrect information and communication to the wrong person. Communication should be written wherever possible, and printed forms should be used to avoid the problems of unreadable handwriting. Safety-critical communications include:

- Prescription
- Input data for preparation of treatment plan
- Proposed treatment plan and calculation, prescribed dose
- Patient immobilisation and shielding aids
- Confirmed plan
- Equipment fault report
- Request for maintenance
- Instruction to discontinue treatment
- Maintenance complete

Responsibilities for these communications must be clearly assigned and formal procedures established, including the design of forms and checklists. Particular attention must be paid to the procedures for the transfer of treatment machines between staff groups. For example, after a repair the machine must be checked by a physicist and then handed over to the treatment staff. Care is also required after measurements or calibration changes have been made on the machine. These may have required a change in the operating conditions of the machine or the over-riding of an interlock.

Because of the impact of errors, as opposed to inaccuracies, as much attention must be paid to the establishment of proper procedures for reducing both their frequency and impact as to the accurate calibration and maintenance of the equipment. This attention to procedures is enshrined in European and national law (EC 1997a; HS 2000) which requires procedures to be written by the employing authority and followed by the staff.

37.8 EQUIPMENT QUALITY ASSURANCE

37.8.1 EQUIPMENT USED IN EXTERNAL BEAM RADIOTHERAPY

The equipment currently used in radiotherapy encompasses a large variety of devices, from very simple mechanical items to highly sophisticated computerised systems. A typical list of

TABLE 37.2

Typical Equipment Used in External Radiotherapy

Dosimetric equipment (ionisation chambers, water tanks, films, TLD, etc.)
Treatment machines (cobalt-60, accelerators)
Immobilisation and positioning devices (head holders, casts, masks, lasers, etc.)
Localising/imaging devices (simulator, CT scanner, etc.)
Computerised treatment planning systems
Beam limiting devices (blocks, multileaf collimator)
Beam modifying devices (wedges, compensators)
High energy imaging devices (films, electronic portal imaging)
Computerised record and verify systems (with or without network integration)

such equipment is given in Table 37.2. QC systems must be set up for each of these items of equipment, methods for some of which will be described in subsequent chapters.

37.8.2 STEPS IN EQUIPMENT QUALITY ASSURANCE

Between the acquisition of a new piece of equipment and its clinical use, several steps must be taken: the acquisition should always be based on a well-defined list of items, characteristics and features on which the user and the supplier must agree. This constitutes the *specification* phase that is required before any official order. After delivery and/or installation of the equipment, a series of *acceptance tests* must be conducted. Their aim is to make sure that the equipment meets the specifications previously agreed upon. During the acceptance procedure, the user is not expected to cover all aspects of clinical use. For sophisticated systems such as a treatment machine or a computerised treatment planning system, a significant time is still required to perform measurements, adjustments and tests before it can be clinically used. This is referred to as the *commissioning* phase, where the ability of the system to meet the clinical requirements in terms of possibilities and accuracy must be assessed.

After commissioning, the equipment must still be kept under control to make sure that the original characteristics are not changed inadvertently. This is done by setting up *periodic QC checks*, performed at regular intervals or after each major event such as repair or software updating.

CHAPTER 38

QUALITY CONTROL OF MEGAVOLTAGE EQUIPMENT

Edwin Aird, Philip Mayles, and Cephas Mubata

CONTENTS

38.1 INTRODUCTION

The tests described in this chapter are considered for each area of quality control (QC). It is not the purpose of this book to provide a detailed manual of QC. However, a brief description of tests that can be done to check a particular area is given together with some discussion about the testing and the issues that arise when parameters are found to be out of tolerance.

Details of these methods can also be found in other publications (IPEM 1999; AAPM 1994a) and some discussion about each area is also given.

38.1.1 QUALITY CONTROL OF LINEAR ACCELERATORS AND COBALT MACHINES

In the early days of megavoltage radiotherapy when cobalt-60 units were widespread, only limited QC checks were required. These included a monthly check of the dose rate and the mechanical alignment. The advent of linear accelerators required more frequent detailed QC.

Although at first the main concern was to maintain accurate output, it was soon recognised that, if overall accuracy of treatment was to be within the tolerances expected by radiation oncologists, other parameters needed regular checking. In summary, these tolerances (see Section 37.4) are

- Accuracy of delivered dose to the specification point ±3% (1 standard deviation (SD))
- Accuracy of delivered dose at all other points in the target volume ±5% (1 SD)
- Accuracy of positioning beam edges and shielding blocks in relation to the planning target volume (PTV) ±4 mm (1 SD)

A number of QC protocols have been written and these often differ as to the frequencies recommended (see, for example, Table 38.1). For each individual department, it is necessary for the local physicist to put together a coherent and comprehensive series of checks. In some countries there will be a legal requirement to carry out a minimum series of checks. In others the physicist professional body may have published recommendations that do not have the full force of law. It is important to read these recommendations carefully. In The Netherlands (NCS 1996), it was decided to define a minimum level that would be regarded as mandatory, whereas in the U.K. the frequencies quoted were largely based on the median frequencies resulting from a national survey. In deciding on the appropriate frequency for an individual department a number of factors will need to be considered:

- The treatment techniques and fractionation schedules being used in the department
- The equipment being used (e.g. age, stability based on experience, control system)
- The time and resources and equipment available for QC

TABLE 38.1

Frequency of Checks of Dose Monitoring System According to Different Protocols

Test	IPEM, U.K.	AAPM, USA	NACP, Nordic	NCS, Dutch	SFPH, France	DIN, Germany	IEC	WHO	ActaR, Nordic
Output calibration									
Photons	D	D	W	2W	D	W	W	D	D
Electrons	D	D/3D	W	2W	D	W	W	D/3D	D
Output stability									
During day			3M	—		A	M		M
Moving beam			3M	3M		M	M		
Reproducibility	M	A	M	—	6M		6M		6M
Proportionality	M	A	M		6M	A	6M	A	6M
Output variation with									
Gantry position	M		3M	A	M	A	6M		M
Arcing	M	A		3M	A	A	6M		6M
Factors									
Wedge factor	M	A		3M	M	6M	6M	A	
Tray factor		A			M			A	
Applicator factors	3M								

D, daily; 3D, every third day; M, monthly; 3M, every third month; 6M, every sixth month; A, annual.
Source: Adapted from Thwaites, D. I., *Radiotherapy from Prescription to Delivery* (Proceedings of an Interregional Seminar for Europe, the Middle East and Africa, Leuven, 16–20 September 1991), IAEA TECDOC Series No. 734, IAEA, Vienna, 1994.

The aim of the QC process is to ensure that patients are given accurate treatment. For example, if a machine is being used regularly for stereotactic single fraction high dose treatments the QC checks relating to the mechanical alignment and stability with arc rotation will need to be carried out more frequently, perhaps even before each treatment. On the other hand for treatments being given over a six-week period a dose inaccuracy of 3% for two or three fractions can be easily compensated and less frequent checks may be appropriate. In practice it may be difficult to adjust frequencies to this level of detail. An analysis of the issues associated with the frequency of checks has been carried out by McKenzie and others (2006). Some checks are very quick to perform and others may take several hours. Evidently it makes sense to consider carefully requirements for the latter. In some situations, such as for IMRT treatments (see Section 39.3.9.3) or electron treatments, it may be appropriate to carry out individual patient QC measurements.

As well as providing assurance of continuing accuracy of dose delivery, QC also provides the radiotherapy department with an invaluable record of the safety of treatment given by the centre so that in case of litigation, machine error can be eliminated.

The quality assurance (QA) documentation produced by professional bodies will be used by bodies involved with auditing, accreditation, or in legal cases as a guide to *best practice*. Whilst these are valuable documents, most are based on surveys of existing practice and are in danger of becoming standard practice set-in-stone, rather than allowing a process of continual review.

38.1.2 RESPONSIBILITIES

The EC directive (EC 1997a) requires that QC is the responsibility of a Medical Physics Expert (MPE). This requirement has been translated into the national legal framework in most European countries. The EC directive is based in turn on documents such as the IAEA Basic Safety Standards (IAEA 1996). This physicist (with the assistance of the maintenance engineer or technician) must co-ordinate the QC checks and interpret deviations in terms of the effect that they will have on patient treatment. There is also an important role for the radiographer/-technologist in daily QC and continued vigilance is required when operating the unit to make note of any error message or abnormal functioning.

38.1.3 TOLERANCES

Tolerances are given in most of the reports discussed above. A summary of European requirements (EC 1997b) is given in Table 38.2. Measurement of many of these parameters does not require special techniques and will not be discussed further. In what follows we will consider the assessment of those parameters where the interpretation of the results requires a greater degree of understanding

38.2 ISOCENTRIC TREATMENT MACHINES

Modern megavoltage machines use an isocentric mounting. The aim is that the rotation axes of the collimator, gantry and treatment table should pass through a single point called the isocentre. Because of mechanical limitations, this aim is not exactly achievable and so the *isocentre* is then defined as the centre of the sphere that contains the axes of rotation of these components for all gantry, collimator and couch angles. The diameter of this sphere (the *isosphere*) defines the size of the isocentre. The position of the isocentre is usually indicated optically by two or more orthogonal lasers, which should be set up to indicate the centre of the sphere.

Ultimately, the aim is that the centre of the target volume within the patient shall be placed at the isocentre. If the gantry, collimator and couch all rotate about this point, the radiation beam

TABLE 38.2

Criteria for Acceptability of Linear Accelerators and Cobalt Units as Required by the EC (1997b)

Test	Remedial Action Level
Gantry and collimator rotation indication	$\pm 1°$
Yoke rotation (cobalt only)	$\pm 0.2°$
Isocentre diameter	± 2 mm
Source distance and beam axis indicators	± 2 mm
Numerical field size indicators	± 2 mm
Light field compared to radiation field	± 2 mm
Treatment couch scale	2 mm
Couch deflection under load	5 mm
Immobilisation devices (e.g. casts etc.)	± 2 mm
Patient alignment devices	± 2 mm
Light field indication (field size)	± 2 mm
Light field indication (density measurements)	± 1 mm per edge
Dose calibration at reference point	$\pm 3\%$ for photons and $\pm 4\%$ for electrons
Output constancy (including accelerators, cobalt and orthovoltage)	$\pm 2\%$
Timer of cobalt unit	± 0.01 min
Electron/photon beam type	Correct type
Beam flatness and symmetry	$\pm 3\%$ (Photons and electrons and cobalt)
Orthovoltage beam symmetry	$\pm 6\%$
Transmission factor of wedge or compensator	$\pm 2\%$
Dose monitoring system	
Precision	$\pm 0.5\%$
Linearity	$\pm 1\%$
Dose rate effect	$\pm 2\%$
Stability	$\pm 2\%$
Variation with gantry angle	$\pm 3\%$

Source: From EC, *Radiation Protection 91. Criteria for Acceptability of Radiological (Including Radiotherapy) and Nuclear Medicine Installations,* Chapter 7, Office for Official Publications of the European Communities, Luxembourg, 1997.

can be made to come from a wide variety of directions by rotation of these components of the treatment machine and the beam will still be centred on the target. Therefore, in addition to the coincidence of these mechanical axes of rotation, it is necessary that the radiation source is on the axis of rotation of the collimator and that the direction at which the electron beam strikes the target is coincident with this axis. Furthermore, the collimators must be centred on this axis.

Because the radiation beam is invisible, optical systems have been designed which indicate its position and are used to align the patient. With modern accelerators with multileaf collimators (MLCs) it is no longer common practice to set the *borders* of the radiation field to skin marks and the aim of patient treatment set-up is to ensure that the *isocentre* of the machine is placed at the desired point within the tumour. Some years ago an isosphere diameter of 4 mm was the best that could be achieved, but the standard specification of modern accelerators is 2 mm. Accelerators optimised for stereotactic treatment can now achieve an isocentre diameter of 1 mm. The coincidence of the radiation and light field is less critical than it used to be for the purpose of patient set-up and in testing machine alignment more emphasis should be placed on ensuring the alignment of the radiation beam and of the collimators. Methods of assessing the radiation beam are discussed in Section 38.2.2.

Tests associated with the isocentre may have one of three aims:

- Setting up and optimising the isocentre at installation or in the event of a problem
- Investigating the isocentric accuracy of a machine on a routine basis
- Checking that the optical indication of the isocentre position is correct

At installation, or where the isocentric accuracy is found to be below the required specification, the actions that can be taken depend on the mechanical design of the accelerator. All machines will have some possibility to optimise the isocentre, but this should only be necessary at installation and will usually be carried out by the manufacturer's installation engineer. One or more of the following actions may be possible:

- The machine head may be slightly rotated or tilted with respect to the gantry by loosening or tightening the attachment bolts.
- On some machines, it is possible to adjust the axis of rotation of the gantry (e.g. by adjustment of the jockey wheels on a ring gantry).
- The gantry balance weights may be altered.
- The attachment of the accelerator to the floor may be adjusted with the help of shims.
- The couch position may be adjusted to bring its rotation axis into coincidence with the rotation axis of the collimator (with the gantry vertical).

Once the isocentre size has been minimised, the position of the isocentre, being the best point in space to give the least excursion with (couch) gantry and collimator rotation, is defined and the optical and mechanical indicators of the isocentre position are set up (see Section 38.2.1.2).

If it is considered impossible to bring the isocentre excursions within tolerance (because of wear on machine bearings), it may be necessary to restrict the use of the linear accelerator to certain types of treatment not requiring high accuracy or small fields.

Although the ultimate aim is to establish the position and size of the radiation isocentre, adjustments are most easily carried out without the use of the radiation beam and a systematic approach to mechanical and optical alignment should result in the correct alignment of the radiation beam. In the following section (Section 38.2.1) approaches to isocentre-related tests will be described that do not require the radiation beam.

38.2.1 MECHANICAL ALIGNMENT OF ISOCENTRIC MACHINES

38.2.1.1 Optical Alignment Checks

Routine checks of the alignment of the optical indicators are the simplest way to assess the mechanical accuracy of a treatment machine. However, if problems are identified by such methods, more elaborate methods are required (as described in Section 38.2.1.2). The projection of the graticule at the centre of the field should provide an accurate indication of the centre of the beam and hence of the centre of rotation of the collimator. This can be checked as follows:

- Place a white board, with a clear thin black cross marked on it on the treatment table at the level of the isocentre (i.e. 100 cm from the target).
- Set the gantry to be exactly vertical using a spirit level on the front face of the collimator (or by some other method). Align the cross with the projected light field centre, then rotate the collimator and note the excursion of the optical cross from the fixed cross every 90°. This should not exceed the stated tolerance (e.g. 1 mm)[*].
- Any movement of this optical beam centre with gantry angle can be checked by rotating the gantry to 90° and 270° having set a rotatable plate (see Figure 38.1) at the isocentre when the machine was vertical—this plate can also be used to check that the wall laser beams correctly indicate the isocentre.

---

[*] This test is often combined with a check on the optical field size and its rotation tolerances.

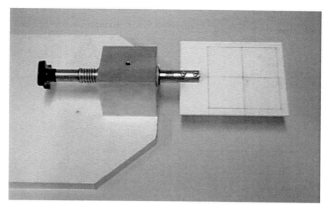

FIGURE 38.1
Rotatable plate for assessing optical isocentre accuracy. (Photograph courtesy of Clatterbridge Centre for Oncology.)

38.2.1.2 Definitive Mechanical Isocentre Assessment

If problems are identified by a quick optical check, a mechanical check of the isocentre is called for. The fundamental basis of an isocentric linear accelerator is the coincidence of the axis of rotation of the collimator and of the gantry. The aim of this test is to determine the mechanical relationship of these two axes.

A definitive assessment of the mechanical isocentre can be carried out using a mechanical front pointer attached to the head of the treatment machine (see Figure 38.2) together with a

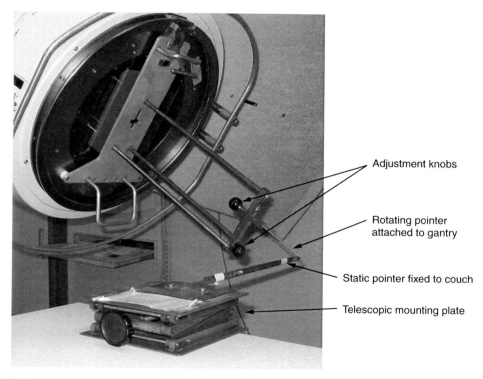

Adjustment knobs

Rotating pointer attached to gantry

Static pointer fixed to couch

Telescopic mounting plate

FIGURE 38.2
(**See colour insert following page 590.**) Precision front pointer suitable for mechanical isocentre measurement. The pointer is attached to the front of the collimator and the screw adjustments allow it to be centred on the collimator axis of rotation (see text). The telescopic device that is placed on the treatment couch allows the position of the fixed pointer to be moved precisely.

rigid pointer, such as a hypodermic needle, attached to the end of the treatment table. This static pointer is used to assess the movement of the moving pointer in space. The moving pointer must first be adjusted to be coincident with the axis of rotation of the collimator. This is done by adjusting the pointer until no movement of its tip can be detected as the collimator is rotated. This can usually be achieved without difficulty.

Once this has been done the rotation of the gantry can be assessed. As the gantry rotates, flexing of the gantry structure will occur which causes the tip of the pointer to move both parallel to the gantry axis and perpendicular to it. Initially the distance of the tip of the pointer from the head is adjusted so that its tip is at the same point with the gantry at 90° and 270°. The fixed pointer is then set to this point and its height adjusted to be half way between the position of the moving pointer tip with the gantry at 0° and 180°. The gantry is then rotated and the movement between the two pointers is observed. The position of the fixed pointer may be further adjusted so that it gives an optimum indication of the isocentre. Finally the lasers are adjusted to point to this position.

If the position of the fixed pointer determined as described in the previous paragraph is coincident with the position of the moving pointer with the gantry at 90° or 270° there is no problem. However, it is commonly found that the height of the moving pointer with the gantry at 90° or 270° is below this point* as illustrated in Figure 38.3. Since most treatments are given with the gantry at one of these four cardinal angles it may be appropriate to set the indication of the isocentre to the mid-point of the position of the moving pointer with the gantry at 90° or 270° is this will indicate the height of the beam centre for these beam directions. As a consequence of this choice, the source-isocentre distance will be slightly larger with the gantry at 0° (and smaller at 180°) than observed at 90° or 270°. Provided that the source axis distance change is smaller than 2 mm to 3 mm, this may be the best option (Figure 38.3).

This method can also be used to assess the accuracy of the indicated isocentre. In this case the fixed pointer is set to the indicated position and then for each gantry angle (every 30° or 45°) the distance between the pointers in both the transverse plane and the longitudinal plane must be measured and recorded. These can either be plotted in polar co-ordinates and compared with an acceptance baseline or a tolerance can be defined (e.g. 2 mm) which, if exceeded, triggers appropriate action.

38.2.1.3 Isocentric Couch Rotation Axis

If non-coplanar treatments or treatments in an inclined plane are carried out in the department it is important to include the couch rotation within the assessment of the isocentre. The couch rotation axis can be checked using the mechanical front pointer[†] and graph paper on the couch. The couch is rotated and the excursions of the front pointer noted. Care must be taken not to shift the couch laterally or longitudinally while rotating it. It is frequently the case that couch rotation is less accurately isocentric than collimator rotation.

38.2.1.4 Establishment of Vertical Axis

It is also important that the couch moves precisely vertically, and with the gantry vertical the optical indication of the beam should also be vertical. The vertical movement of the couch

---

* The reason for this is that the flexion of the supporting arm, in a plane perpendicular to the gantry axis, is usually more pronounced when the beam is horizontal than when it is vertical. However, in the direction parallel to the gantry axis, the flexion is opposite when the beam is rotated from 0° to 180° and it is commonly observed that the tip of the rotating pointer shifts slightly outwards from the gantry. The average position is found at 90° or 270° and constitutes here again the best compromise.

† Alternatively if the graticule has been shown to be at the centre of rotation of the collimator it may be more convenient to use this instead.

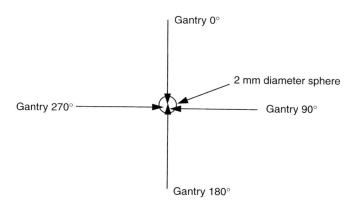

FIGURE 38.3
Figure indicating the possible position of the tip of the moving pointer at four gantry rotations. It may be appropriate to set the horizontal indication of the isocentre to be at the mean of the position with the gantry at 90° or 270° rather than at the centre of the 2 mm diameter sphere containing the isocentre.

can first be established using a plumb line held so that the pivot point from which it hangs does not move as the line is extended. The beam indication can then be checked as the couch is driven up and down.

38.2.1.5 *Considerations Relating to the Mechanical Isocentre*

If one of the optical isocentre measurements is out of tolerance it is important to recognise why and how drift has occurred. Adjustment of one optical indicator to align with another that is incorrect will lead to a progressively more unsatisfactory set-up. It is usually necessary to reassess the mechanical isocentre before proceeding with making any adjustments.

When deciding where to centre the isocentre, it is important that the radiographers/technologists understand how this relates to the wall laser settings. These lasers must be aligned with the centre of a horizontal beam throughout their length and not just at the isocentre. This cannot be achieved just by altering the angle of the laser beam and it is essential that the lasers are accurately centred at the correct position on the wall. A laser level that gives a horizontal line across the room from the couch top is useful for this (a water level using a clear plastic pipe can also be used).

An important point connected to the checks at gantry angles 90° and 270° is that the electronic indicators of these angles may have drifted. These should be checked against the mechanical angle indicator and, if there is still some uncertainty, against the true vertical and horizontal. A laser tool can also be used for this.

It is also important to check that the couch top is horizontal, whatever weight is added within representative values (see IEC specifications (1989)). Some deflection in the sagittal plane is inevitable, but it should be possible to set the couch top exactly level in the transverse plane.

38.2.2 RADIATION BEAM ALIGNMENT

38.2.2.1 *Radiation Isocentre*

Ultimately, it is the radiation beam alignment that determines the accuracy of treatment. This can be assessed directly using a star film or the Winston–Lutz test.

A star film is created by mounting an envelope wrapped film vertically in a poly-methylmethacrylate (PMMA, known as Perspex or Lucite) holder in the plane of rotation of the

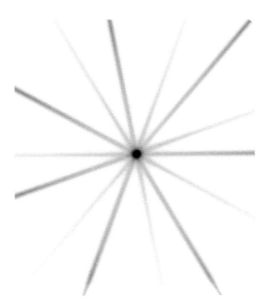

FIGURE 38.4
Star film exposed at 50° gantry increments. The film is placed in a PMMA holder in the plane, through the machine isocentre, that is perpendicular to the axis of rotation of the gantry.

gantry with its centre at the indicated position of the isocentre. As narrow a field as possible[*] is set and the film irradiated with the gantry at 0°, 120° and 240° (these directions are not prescriptive). The PMMA holder may have holes drilled at the centre and at 120° increments around the centre to coincide with the directions of irradiation. The holes are used to pierce the film to assist analysis. The image on the film will be a star shape and it is relatively easy to draw lines on the film indicating the centre of each beam. These should all meet at the marked isocentre. If the film is scanned into a computer, automatic analysis becomes possible. An example for which the increments were 50° is shown in Figure 38.4. It will be noted that this method gives no indication of errors in the direction orthogonal to the plane of rotation.

An alternative method developed originally for stereotactic radiotherapy is the Winston–Lutz test (Lutz et al. 1988). This was originally carried out using a circular collimator as shown in Figure 38.5. With a circular applicator attached to the head of the machine and a ball bearing suspended at the isocentre, the gantry may be rotated and films exposed at appropriate angles. The apparent movement of the image of the ball bearing relative to the circular field is easily observed. For stereotactic treatments, rotation of the couch and collimator can also be included among the different exposures. A circular field can be defined using an MLC and the similar set of exposures made. A small square field could also be defined. An electronic portal imaging device (EPID) can also be used.

38.2.2.2 Radiation and Light Field Coincidence

There are various methods of measuring coincidence between the radiation field and the light field, and at the same time checking the calibration of the field size indicators, by using film. The film must be placed on some back-scattering material, such as water equivalent plastic (which also provides a flat surface), and build-up of the appropriate thickness placed on top.

---

[*] The narrow field may require to be set using the MLC, because of restrictions in the minimum field size of standard collimators.

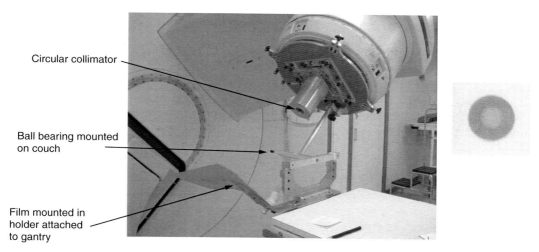

FIGURE 38.5
Set-up to carry out a Winston–Lutz test using a stereotactic collimator and head ring. The image on the right shows the image of the ball bearing coincident with the centre of the circular collimator.

Some test objects combine the two tests in one by having lead/tungsten markers at the field edges to which the optical field is aligned. Prisciandaro describe how an EPID can be used for radiation and light field coincidence measurement with such a test object (Prisciandaro et al. 2003).

It is usually more convenient and easier to interpret if the field is set to a particular value on the field indicator and the edges of the optical field are then marked onto the film either with a pressure mark or a pinprick*.

Following exposure to give an optical density of between 1.0 and 1.5, the radiation beam edge at the optical density representing 50% of the dose at the centre of the field can be found using a hand held densitometer or a film scanner. The distance between the optical edge and the radiation edge should not exceed tolerance (typically 2 mm). Ideally the film should be calibrated on each occasion but if the film has been checked to have a linear response over the range used and this is checked occasionally, it is not necessary to do a calibration every time. A slow film should be used which typically requires an exposure of about 0.5 Gy. This ensures that the resultant x-ray field represents the stable running condition instead of the initial switch-on condition—during the first few monitor units (MU)—when, particularly in bent beam accelerators, the servo systems are operating to centre the electron beam on the target. For monthly testing it is usual to measure at least the reference field size (10 cm × 10 cm), but it is nice to use different field sizes in rotation as well. Software is available to facilitate the analysis of scanned films.

38.2.2.3 Considerations Relating to Radiation Alignment of the Beam

The field size and the penumbra width can be measured when a scanning densitometer or film scanner is used. The most likely cause of any misalignment is due to problems with the

---

* With test objects containing lead markers the field has to be set visually to the appropriate size. For modern accelerators with assisted setup of the field shape this is therefore rarely the best option. However, as digital imaging takes over such methods may become essential.

optical beam; only rarely is it due to x-ray beam misalignment as a misalignment of the x-ray beam would probably be identified in assessing beam flatness (see Section 38.3.5). If an x-ray collimator has drifted out of calibration or shifted mechanically, both the x-ray beam and the optical beam will be affected. If the angle of the face of the x-ray collimator has tilted slightly with age or mechanical shock, the x-ray penumbra may be affected.

A second method for measuring all the field size parameters (i.e. x-ray field size versus optical field size and penumbra width) is to use an in-air scanner with the ion chamber set into a block with its centre of measurement at 100 cm from the source (Aird et al. 1995). The chamber is placed in a special white-faced plate in which the chamber's effective measuring point is about 2 mm below the surface of the plate and marked by a fine, black cross. The chamber can be moved so that the cross is aligned to particular points in the light field and its position noted. Build-up of the appropriate thickness is added to the block after each alignment has been made. For comparison of the x-ray and optical fields, the plate is first placed on the central axis of the beam and readings that represent 100% at the centre of beam taken. The cross on the plate is then aligned with the optical beam edge and readings taken that should represent 50%. The difference from 50% can be translated into a distance error in mm by comparison with the x-ray beam profile for the particular field size investigated.

The advantage of this approach where a jig is attached to the gantry is that beam parameters can be measured at any gantry angle. This is also possible with electronic systems containing arrays of ionisation chambers or diodes.

38.2.2.4 Interpretation of Alignment Checks

The results of some of the preceding tests may indicate a problem in alignment of the unit. Examples of possible alignment problems together with the symptoms that they will produce are described below. If one of the tests identifies a problem it is important to carry out all the associated tests before trying to make a diagnosis. The causes of misalignment can be quite complex and it is easy for misalignment of one component of the system to be compensated for by a misalignment of another component. The examples given below should not be regarded as an exhaustive list of possibilities. The first step in aligning the system is to ensure that the projection of the crosswires does not move when the collimator is rotated.

The Crosswire or the Light Source is Not on the Collimator Axis: The light source and the crosswire must both be located on the mechanical axis of rotation of the collimator (the collimator axis). The testing procedure and its interpretation would be somewhat different depending upon whether the light source is designed to rotate with the collimator (as in Elekta and Varian linacs) or not (as in Siemens linacs).

If the crosswire is not on the collimator axis, its projection will always move as the collimator rotates. If the light source is fixed, movement of the crosswire projection is always an indication that the crosswire is misplaced and it should be adjusted until its projection is static. If the light source rotates, it could also be responsible for the movement of the crosswire projection. Indeed, if both the crosswire and the light source are misplaced, it could happen exceptionally that at a given distance the projection does not move (see Figure 38.6) but this would not be true for another distance. One method to decide whether it is the light source or the crosswire that must be adjusted, is to measure the diameter of the projected circle at two different distances. If the diameter increases in proportion to the distance to the source, the crosswire is misplaced; if it increases in proportion to the distance to the crosswire, the source is not on the collimator axis. Another particularly effective method is to suspend a 50 cm focal length lens on the axis of the beam so that a sharp image of the light source is projected onto the floor (or a wall). The exact position is not critical. As the collimator is rotated this image of the light source will not move if the light source is on the axis of rotation.

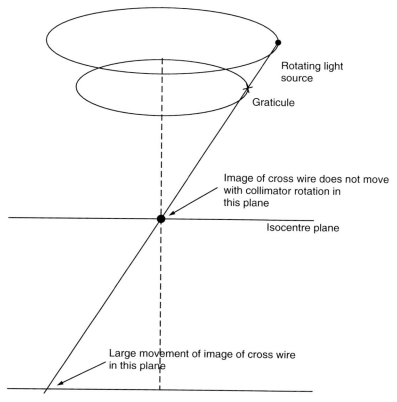

Rotating light source

Graticule

Image of cross wire does not move with collimator rotation in this plane

Isocentre plane

Large movement of image of cross wire in this plane

FIGURE 38.6
Misalignment of the crosswire or light source can be compensated for in one plane, but not in all.

When a light source that does not rotate is misplaced the projection will still not move if the crosswire is on the collimator axis. One solution to detect this situation is to set the collimator base to be horizontal with a spirit level (i.e. the gantry vertical) and to check with a plumb line if the optical axis is vertical.

The light source is not at the correct distance from the isocentre: In this case, the field size at extended source skin distance (SSD) will not be in correct proportion to the field size at the isocentre and coincidence of the radiation and the light field will not be correct at the extended SSD.

The light source mounting is loose: If the light source moves as the gantry is rotated the alignment of the light source will be different at different orientations of the gantry. This can be detected by checking that the coincidence of the radiation and light field remains the same as the gantry rotates*, for example at 0° and 180°.

The optical field trimmers protrude too much or too little into the beam or are asymmetric: Older machines used optical penumbra trimmers† to allow exact coincidence between the radiation and light field to be established. Optical trimmers can be very confusing as they can easily mask a more fundamental misalignment. Incorrectly set trimmers will be manifested by a difference in field size between the radiation and the light field. If they are set asymmetrically, the light

---

* If the light source rotates with the collimator, this may be more conveniently detected by checking the movement of the projection of the crosswires as the collimator is rotated with the gantry at 0° and 180°.
† This was particularly necessary if the optical field was to be set to show the 80% dose level. However, this field size convention is impractical with modern equipment because of the use of asymmetric fields and with MLCs optical trimmers are impractical.

field will be asymmetric and the asymmetric jaw will move round as the head is rotated through 180°. If there is doubt about this it is useful to adjust the light-field trimmers so that they do not protrude into the beam at all and then ensure that the correct radiation field size is set before adjusting the light field.

The radiation source is not on the central axis of the machine: The radiation source does not move with collimator rotation. If it is not on the axis of rotation of the collimator there will be an asymmetric radiation field compared to the light field. This asymmetry will be equal and opposite for opposing jaws that are themselves correctly set up and will be offset in the same direction as the collimator is rotated. This can be confirmed as described in Section 38.2.2.1. The misalignment of the radiation and light field will be emphasised by taking radiographs at different SSDs. A more sensitive test of a misaligned beam is the beam flatness (see Section 38.3.5) because if the radiation beam centre is not aligned with the beam flattening filter there will be a significant asymmetry in the middle of the beam.

The electron beam is striking the target at an angle other than 90°: If the electron beam strikes the target at the wrong angle this may result in an asymmetric beam. The flatness scans will also show a misalignment of the minimum in the centre of the beam with the geometric centre (provided that the scanner is set up correctly so that the indication of the central axis is correct) and poor beam flatness (see Section 38.3.5). There may also be an energy change associated with the need to compensate for the incorrect angle. These problems will result in an asymmetric beam profile at the edges. Because of this possibility it is important to check the beam profile before adjusting the optical field and to check the light and radiation field coincidence after any beam adjustment.

The collimators are asymmetric: The asymmetry of both the optical and the radiation field relative to the isocentre will change sides as the collimator is rotated through 180°.

The collimator sides are not aligned with respect to the source: If the collimators are not aligned with the edge of the diverging beam an increase in the width of the penumbra will be seen. It should also be possible to detect significant angulation by visually inspecting the collimator.

The gantry sags or there are distortions in the drum or gantry bearings: Gantry inaccuracy will be demonstrated by a longitudinal misalignment of the beam at 0° and 180°. This can be demonstrated with a mechanical front pointer or by marking up a film with the gantry at 0° and exposing it with the gantry at 180°. The former is the most effective way to demonstrate problems with the gantry bearings as sudden jumps of the pointer may be seen.

The collimator bearings are faulty so that it does not rotate correctly about its axis: Erratic movements of the central cross will demonstrate errors in the collimator bearings when the collimator is rotated.

The rotation axis of the gantry and collimator differ: This will result in a shift in the transverse direction (i.e. perpendicular to the gantry rotation axis) as the gantry is rotated through 180°.

38.3 ALIGNMENT AND QUALITY CONTROL OF MULTILEAF COLLIMATORS

MLCs require particular QC measures depending on their application. Some recommendations for these are given in IPEM Report 81 (1999), the report of AAPM Task Group 50 (Boyer et al. 2001) and in SFPM Report 40 (2003). For simple field shaping the requirements are less demanding than for intensity-modulated radiotherapy (IMRT, see also Chapter 43 and Chapter 44), where errors in leaf positioning can lead to errors in the delivered dose. With the growing importance of leaf calibration because of IMRT a number of authors have developed methods of testing involving EPIDs (Boyer and Li 1997; Samant et al. 2002; Vieira et al. 2002, 2003; Chang et al. 2004; Baker et al. 2005; Budgell et al. 2005b) and Wiezorek et al. (2005) describe the use of other electronic test tools for this purpose.

38.3.1 LEAF CALIBRATION

When establishing the leaf calibration there are two stages: setting the mean position of the whole set (or bank) of leaves and then setting the position of individual leaves (sometimes referred to as minor offsets). The relationship between the mean position of the leaves and the position of the backup jaws (if any) must also be established. Several factors must be considered when setting up the MLC leaf calibration:

- For MLCs with rounded leaf ends (as those of Varian and Elekta) the part of the leaf which defines the field edge varies as the leaf is tracked across the beam and it is therefore necessary to test leaf alignment at different positions relative to the centre of the beam (Graves et al. 2001).

- It is not possible to adjust the light field by using light field trimmers so that once the optics have been adjusted so that the (virtual) light source is on the central axis of the beam no further adjustment of the relationship of the optical field to the radiation field is possible.

- As discussed by Jordan and Williams (1994), when leaves are used to define rectangular fields in conjunction with backup jaws (as in the Elekta design) the position of the 50% isodose line will be different in relation to the leaf position compared to when the leaves define the edge of the field alone. For IMRT the most critical factor is the relationship between opposing leaves as this is fundamental to being able to create composite fields by the combination of multiple small segments by dynamic jaw movements.

A universally useful test is to expose a single film using a series of adjacent strip fields, usually 20 mm wide. Any error in leaf position will show up as a non-uniformity in the resulting film. The exact technique for setting up leaf calibration is dependent on the particular MLC. The procedure for Varian MLC's is described by Chui et al. (1996) and Mubata et al. (1997), for Elekta MLC's by Hounsell and Jordan (1997) and Sastre-Padro et al. (2004), and for Siemens MLC's by Samant et al. (2002) and Bayouth et al. (2003).

38.3.2 ALIGNMENT OF THE MLC TO THE TREATMENT MACHINE AXES

The MLC leaves must be aligned so that they are respectively parallel and perpendicular to the standard collimator jaws. In addition the centre of the leaf bank must be aligned with the central axis of the machine. This is principally an issue at commissioning, but with the Varian design where the leaves are mounted on a moving carriage it is possible that they may become misaligned during use. This can be checked on an occasional basis as described by Mubata and others (1997). When using a detachable mini MLC it is necessary to carry out this test each time the collimator is fitted. A rapid method of doing this is to set up a field in which the top left and bottom right quadrants of the field are open while the other two quadrants are shielded. A portal image is then taken, ideally using an EPID. After rotating the collimator through 180° a second image is taken. If using an EPID, this must then be added to the first image, but a film is simply left in place to record both images. The overlap of the two images indicates how well the MLC is aligned.

38.3.3 DYNAMIC LEAF MOVEMENTS

When dynamic leaf movements are to be used for IMRT it is necessary to verify the speed of movement of the leaves. If one leaf moves faster than is intended compared to the opposing leaf, the dose delivered will be either greater or less than intended depending on which leaf is moving too fast. Testing procedures involve trying to deliver a uniform dose to a film using different leaf speeds (Chui et al. 1996) or by measuring the dose delivered to an ionisation chamber by scanning a thin slit across the field (LoSasso et al. 1998, 2001; Chauvet et al. 2005).

Dynamic therapy carried out with a tomotherapy machine (see Section 43.3.4) has a number of additional special requirements which must be considered (Fenwick et al. 2004, 2005) including issues such as the ability to recover after a fault which occurs in mid treatment.

38.3.4 LEAKAGE BETWEEN LEAVES

Interleaf leakage is one of the important criteria in the design of an MLC (Jordan and Williams 1994). Where backup jaws are provided it could be argued that the interleaf leakage is less important, but at present only the Elekta machine is designed so that the backup jaws track the maximum leaf opening at all times. For the Siemens machine there are no backup jaws in the direction of leaf movement and for the Elekta *beam modulator* there are no backup jaws in either direction. For other MLCs it is usual for the main jaws to be left in a fixed position during IMRT treatment. MLC leakage is principally measured at commissioning, but it is also necessary to check from time to time that leakage has not increased due to wear (Hounsell and Jordan 1997). A procedure for measurement can be found in SFPM Report 40 (2003).

38.3.5 VARIATION WITH GANTRY ANGLE

The leaves in an MLC are heavy and it is potentially possible for a number of parameters to vary with gantry angle. These include leaf calibration, interleaf leakage and the speed of leaf movement. Occasional tests should therefore be carried out with the gantry at 90° or 270° and with different collimator rotations (Mubata et al. 1997; Hounsell and Jordan 1997). The dynamic slit test (Section 38.3.3) can also be carried out at different gantry angles (Chauvet et al. 2005).

38.4 FLATNESS AND SYMMETRY OF PHOTON BEAMS

The flatness and symmetry specification according to the IEC (1989) is as illustrated in Figure 38.7. *Flatness* is defined as $F = D_{max}/D_{min}$, where D_{max} and D_{min} are the maximum and minimum absorbed dose values in the central 80% of a dose profile in a plane transverse to the beam axis (the *flattened area*). Other definitions of flatness are used in some countries (e.g. NACP 1980).

For routine QC tests, it is normal to measure the flatness along the major axes of the largest field. These checks are usually combined with symmetry checks where *symmetry* is defined as: "The maximum value of the ratio of the higher to the lower absorbed dose at any two positions symmetrical to the radiation beam axis inside the flattened area" (IEC 1989).

When more extensive QC checks are made, for example post service and annual checks, it is advisable to extend the flatness and symmetry checks to include the diagonals of the largest field and all four major gantry angles (0°, 90°, 180° and 270°).

The most accurate measurement of flatness is made using an ion chamber in a water tank normally with the ion chamber at 10 cm depth (at 100 cm source-to-detector distance) and the surface of the water at 90 cm. However, this method is mainly used when rigorous testing is required and for annual QC.

A less rigorous method (which was used extensively in the Continuous Hyperfractionated Accelerated Radiotherapy (CHART) multi-centre trial QA program (Aird et al. 1995)) is to use a linear *in-air* scanner. The ion chamber is set into a block of water equivalent plastic. When this scanner is attached to the head of the accelerator it can be used at any gantry angle. The resulting profiles are very similar to those in water except for the edges of the beam that are not part of the definition of flatness.

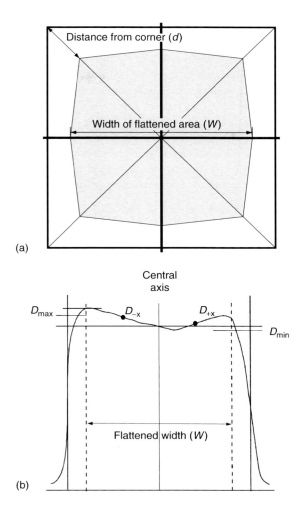

FIGURE 38.7
(a) IEC (IEC 1989) definition of flattened area (shaded). W is 80% of the field width for field sizes between 10 cm and 30 cm. For sizes below 10 cm it is 2 cm less than the field size and for sizes larger than 30 cm it is 6 cm less than the field size. Along the diagonal, the distance from the corner, d, is 20% of the field width for field sizes between 10 cm and 30 cm. For smaller sizes, it is allowed to be 2 cm, and for larger sizes it is 6 cm. (b) IEC definition of flatness and symmetry. Flatness is the ratio of D_{max} to D_{min} expressed as a percentage. Symmetry is the absolute maximum value of the ratio of D_{-x} to D_{+x} (where D_{-x} represents the value of the dose rate at a distance x on one side of the central axis of the beam and D_{+x} is the dose rate at the corresponding point on the other side of the beam axis). The maximum asymmetry may be at any point within the flattened area, i.e. it will not be at the point indicated in the diagram.

Some centres use this principle of *in-air* scanning but with less build-up material above and around the ion chamber. The resulting profiles are different from *in-water* scans, mainly in having larger *horns* at maximum dose due to the measurement reflecting mainly primary radiation flux. This type of measurement is satisfactory provided the results are interpreted against base-line measurements made at the same time as water-tank measurements.

Similarly, when film is used to measure flatness and symmetry, the interpretation must be made with care, particularly because film is less accurate than measurement with an ion chamber. However, film does have the advantage of producing a 2D map of dose across the entire transverse plane of the beam.

Excellent diode or ion chamber arrays are now available to measure flatness and symmetry. In order to measure the whole field a linear array is needed, but for day-to-day use it is sufficient to measure only the central part of the beam with the (usually) smaller 2D arrays.

The Sun nuclear profiler allows two 20 cm square arrays to be bolted together to enable the whole field to be measured if required. Software for such devices allows automatic interpretation of the results in terms of the IEC measures of flatness and symmetry. An electronic device also allows the variation of flatness with time, as the beam stabilises, to be observed. This aspect can be very important for situations requiring a fast beam rise time, such as IMRT. Some multi-detector arrays have a detector at an increased depth (achieved in one such device using a stainless steel absorber) to have a simultaneous measure of energy.

The portal imaging device can also in principle be used as a quick check of flatness. There are some problems with this as portal-imaging software is designed to correct imperfections in the beam and it is usual for the imager to be calibrated to show a uniform field. For the purposes of a quick constancy check or if the manufacturer specifically supports this use it can be very convenient.

38.4.1 CONSIDERATIONS RELATING TO BEAM FLATNESS AND SYMMETRY

Flatness and symmetry are important parameters to maintain within tolerance as they can have a major impact on the accuracy with which treatment can be given. They are also useful parameters diagnostically when determining faulty functioning of the accelerator, particularly with bent beam machines. Although the monitor ion chamber with its associated electronics is intended to maintain flatness by servo control in these machines (see Section 11.7.3.2), the ion chamber currents and control amplifiers can drift.

The depth at which flatness is measured is important to consider. If the check is to be made against the specification of the machine or a base-line measurement made at acceptance, then it is normal to make the routine check at 10 cm depth. However, some low energy accelerators may be used mainly for head and neck treatments, in which case a depth of 5 cm for beam flatness may be more appropriate.

Any adjustments of flatness or symmetry that may be needed should be made in collaboration between the maintenance engineer and the physicist. It is important that they are aware of the interaction of the various beam control systems (centring, steering and energy) that will affect the flatness and symmetry. Protocols should be written as part of the QA system to ensure that adjustments are made in the correct order to avoid mismanagement of these important parameters.

When assessing the cause of a beam asymmetry, it may be useful to carry out a scan with the flattening filter removed, but care must be taken to restore the filter to normal position, and to check that this is in place, before returning the equipment to clinical use. With a computer controlled accelerator, removing the flattening filter can often be achieved by typing a command at the keyboard. The resulting scan will show clearly whether the peaked dose distribution is correctly centred and whether it is symmetrical.

38.5 ENERGY CHECKS

While x-ray beam energy can be specified in terms of nominal MV, this is not a parameter that can be directly measured. As discussed in Section 18.5.5, beam quality is usually specified for dosimetry purposes in terms of the quality index or $TPR_{20,10}$. The percentage depth dose at a standard depth is useful as an energy specification for a particular accelerator type. The percentage depth dose is more easily related to the clinical data in a treatment-planning computer, but a measure related to the quality index may be easier to use as a quick check as it can be easily measured with high precision. Any convenient ratio of two ion chamber readings made with the ion chamber fixed but at two different depths can be used. The ideal system uses blocks of water equivalent plastic of 10 and 20 cm above the chamber (with at least 5 cm below the chamber) to measure the $TPR_{20,10}$ directly.

38.5.1   CONSIDERATIONS RELATING TO BEAM ENERGY

Measurement of quality index is a very sensitive and very precise test of energy (reproducibility should be around 0.2%). However, it is unusual to observe changes in energy of greater than 1% unless the testing follows a major service when beam control currents have been altered. Beam energy changes will also be noted when flatness is measured, because the flattening filter is optimised for one energy. If flatness changes at the beam edge are observed—particularly significant if scans are made at d_{max} instead of at 10 cm deep—it is very likely that the energy has changed.

38.6   QUALITY CONTROL OF ELECTRON BEAMS

38.6.1   ELECTRON FLATNESS AND SYMMETRY

The optimum way to study the flatness of electron beams is in a water tank. However, if flatness is to be measured regularly, possibly at various gantry angles, it is necessary to use a scanning jig (or array) rigidly attached to the treatment head, but normally with the measuring probe at a slightly extended distance, say 5 cm, from the end of the applicator. If an ion chamber is used in a water-equivalent plastic block at the depth of peak dose, the flatness profile is very similar to that in water. When changing beam energy, it is essential to remember to change the depth in the water-equivalent plastic to get the correct value for uniformity and symmetry. However, it is unusual for uniformity to change unless there is a large change in energy, so for routine measurements it is more important to concentrate on symmetry. This can be done using a compromise depth for several energies where the depth is always below or near the peak dose. The tolerance given by several groups is 3% for symmetry. The IEC definition of flatness refers to the ratio of the maximum dose in the field at the depth of dose maximum to the dose on the central axis rather than to the minimum dose (as in Figure 38.7). However, as pointed out by McKenzie in IPEM Report 81 (IPEM 1999), this does not satisfactorily deal with a situation where, for example, the dose maximum is on the central axis and the dose falls off rapidly at the beam edges. He therefore recommends that if the off-axis dose within the flattened area is less than that on the central axis the greatest difference from the central axis should be recorded.

38.6.2   ELECTRON ENERGY

Electron energy is of fundamental importance. It determines not only the treatment depth and depth of penetration of the electrons—which may be crucial when critical structures are close to the practical range of the beam—but also the ion chamber calibration factors used for absolute dose measurements.

38.6.2.1   Energy Selection

In the typical radiotherapy linear accelerator, the energy of the electron beam is determined by filament temperature, waveguide frequency and bending current. For a 90° beam bending system (e.g. Philips SL75-type machines) the 90° bend was used as a spectrometer, so that only electrons of a narrow spread of energy around the nominal energy selected by the bending current would escape from the bending system. For achromatic bending systems other methods of energy control are applied (see Section 11.3.6).

Prior to selection by the bending magnet, the electron current (governed by the filament temperature) and the waveguide frequency are adjusted so that electrons reaching the drift tube and entering the bending magnet have approximately the correct energy.

The energy reaching the surface of the patient is modified by the various objects in the beam and at the edges of the beam as it traverses the treatment head and exits through an applicator. The only true measure of energy relevant to treatment is the depth ionisation or depth dose curve in a water phantom. The energy can then be determined using the Equation D.10 or D.11 respectively given in Appendix D. Any check measurement used for QC must be performed on the same occasion so that a baseline can be established against the appropriate depth dose or depth ionisation curve.

38.6.2.2   *Quality-Control Energy Check*

There are various methods of checking energy constancy that use film or ion chambers. The simplest, that also gives an immediate answer not dependent on film development, is to use an ionisation chamber at 2 or 3 depths. It is not necessary to move the chamber between different depths since it is only the slope of the curve that is used to determine the energy, not the absolute depth dose. Using 3 depths can give more information and be more accurate. A further parameter that should be measured, at less frequent intervals, is the level of the contaminating bremsstrahlung at a depth beyond the electron range. Commercial devices are available with multiple detectors that give an instant reading of the energy.

Tolerance on energy can be critical. A 0.5-MeV energy variation will cause a depth dose shift of about 2 mm.

Each electron energy should be checked monthly at a gantry angle of 0°. In addition, energies should be checked annually at gantry angles of 90° and 270°.

38.6.3   APPLICATOR FACTORS

The measurement of applicator factors is more part of commissioning than QC, but more stress needs to be put on regular checking than is the case with photon-field-size factors, since there is more opportunity for electron-applicator factors to change. These factors are very sensitive to any change in material in the beam, such as the thickness of the plastic foil at the end of the treatment head, or to the position of the secondary diaphragms. A similar argument applies to electron end-plate factors. For most field sizes the ratio of the output at the standard calibration depth to the output at the same depth with the standard field (usually 10 cm × 10 cm) is sufficiently accurate. However, for small field sizes, especially for higher energies, it will be necessary to measure at the true position of dose maximum. These ratios should be checked annually, or following any maintenance or alterations to the head that might change the ratios. If there is uncertainty about this a quick check to decide whether to measure all sizes or not is to check the applicator or end-plate that gives the largest change. At the time of the annual check some checks of the depth of dose maximum should be made.

38.7 THE MEASUREMENT OF PERCENTAGE DEPTH DOSE

Measurements of percentage depth dose are also part of the commissioning process (see Section 19.2.2 and Section 19.3.2). However, they must be repeated from time to time, especially after accelerator maintenance. Some hints are given below.

Depth-dose measurements are made in a large water tank in which all movements of the measuring probe are remotely driven. If the operator is unused to a particular tank it is vital to become familiar with its operation and to know how to set it up accurately.

The following is a minimum list of checks that should be made before attempting to measure percentage depth dose:

- The field size calibration should be checked (using film).
- Is the scanning mechanism rigid in the tank and does the depth movement move truly vertically?
- Are the scanning movements, particularly the depth movements, smooth and accurate?
- Has the calibration of the movement been checked?
- Is the tank set horizontally after it was filled with water? It is best to allow time (15–30 min or longer if the water is not at room temperature) after filling the tank for the support system to settle.
- The vertical scale must be set so that the zero position is correct with respect to the water surface bearing in mind the position of the sensitive volume of the detector.
- An appropriate correction should be made in the software for the effective point of measurement (see Section 18.5.3).
- The water will gradually change its level during a day of measurements as it evaporates and the tank settles further. It is important to check the surface distance periodically during the day.
- Is the appropriate detector being used? (Ion chambers are best for the reference photon depth-dose measurements, but otherwise diodes are satisfactory and provide better resolution. For depth dose measurements with electrons the most appropriate detector is an unshielded diode.)
- Is the reference probe at an appropriate position (i.e. giving sufficient signal but not interfering with the measurement by attenuating the beam)?
- Is the gantry exactly at $0°$ so that the beam is coming out of the head travelling vertically down?
- Use simple profile scanning at depths to check the beam path.
- Even when everything looks perfect it is important to test that the measuring probe moves vertically down to 30 cm or 40 cm.
- For electron beams with ionisation chambers appropriate conversions must be made between ionisation and dose.

When the percentage depth dose has been measured, the dose at particular depths should be compared to those recorded at commissioning. For electron beams, automatic analysis software will usually be available to extract significant values such as R_{85} and R_{50}, etc. (see Section 24.2.2).

38.8 OUTPUT MEASUREMENT

The measurement of radiation output and output stability is fundamental to the QC of treatment units. Absolute dose determination will follow a calibration protocol (e.g. IPEMB 1996a, 1996b; IAEA 1997a, 1997b, 2000a; Almond et al. 1999; Ma et al. 2001; Thwaites et al. 2003c) as discussed in Chapter 18 and Appendix D. The protocol must be followed exactly when carrying out the definitive calibration of the unit, but simpler approaches may be defined for the purposes of day-to-day calibration. In this section we consider the practical aspects of maintaining correct calibration.

38.8.1 STRATEGIES FOR QUALITY CONTROL OF TREATMENT UNIT CALIBRATION

Three different levels of calibration can be defined:

- Definitive calibration: where there is no link to a previous calibration so that there is a possibility that a serious miscalibration may go unnoticed. Examples of such situations are when a new machine is commissioned, when a new monitor ionisation chamber is installed in the accelerator or a new source installed in a cobalt unit.

- Routine calibration: when the machine is calibrated using a procedure that provides traceability to a national standard.
- Constancy check: when a quick check device is used to establish that the machine output does not have a significant error.

Each of these situations will be considered in turn.

38.8.1.1 Definitive Calibration

In the U.K. the requirements for *definitive* calibration were laid down following an incident in which a cobalt source was incorrectly calibrated (Bleehen 1991; IPEM 2002). The calibration should be carried out by an experienced physicist, who should ensure that all other QC procedures that may have an impact on the calibration have been satisfactorily carried out. This is essential because parameters such as isocentre indication (Section 38.2.1) or beam flatness (Section 38.4 and Section 38.6.1 and also Section 11.7.3) can adversely affect the calibration. A national or international protocol (see Chapter 18) should be carefully followed and a dosimeter with a calibration traceable to a secondary standardisation laboratory should be used. The calibration should be repeated with an independent dosimeter and by an independent physicist, thus providing *defence in depth* (IAEA 2000b). The equipment used for the calibration should be removed completely between these two calibrations.

In the circumstances of a definitive calibration, the link with previous calibrations has been broken and it is therefore possible for a large error in the calibration to be made. Wherever possible appropriate cross checks should be made (IAEA 2000b). For example with a new cobalt source there will be a source certificate. The ratio between the quoted output for the old source and the standard calibration dose rate can be compared with the corresponding ratio for the new source. With a new accelerator it may be possible to compare the calibration settings in the accelerator with those of other accelerators of the same type. While these will probably not be identical they should be reasonably close. When no other cross check is available it is wise to make use of a mailed dosimetry service such as the thermoluminescent dosimetry (TLD) services run by the IAEA and ESTRO or the alanine service run by the National Physical Laboratory. Alanine has particular advantages being almost energy independent and having a calibration accuracy on the order of 1%, which is traceable to the national standard.

When deciding on a calibration protocol to follow it is advisable to follow the recommendations of the national professional body. These will have been made in conjunction with the national standardising laboratory and should be self-consistent. For example in the U.K. the National Physical Laboratory provide a calibration factor in absorbed dose to water as defined at the centre of the chamber at a depth in water. It would not be appropriate to include a correction for the effective point of measurement as specified in the IAEA air kerma based code (IAEA 1997a) when using a dosemeter calibrated in this way.

38.8.1.2 Routine Calibration

The second level of calibration is the routine calibration usually done at least weekly. This should be carried out in a way that allows traceability of the calibration, but may use a simplified set-up with appropriate corrections. Such corrections might include a correction for the use of a small PMMA phantom rather than a 30 cm cube of water. It is important that such corrections should not be arbitrary (i.e. it should be possible to estimate the correction factor from first principles) or significantly different from unity as otherwise there is the possibility for a systematic error. Staff carrying out such calibrations must have adequate training in the principles of dosimetry.

38.8.1.3 Constancy Checks

The third level of calibration is the *constancy check* that is usually carried out on a daily basis. This can be as simple as an instrument that shows a green light if the output is within the defined limits, but it is preferable to have a numeric output which will allow the monitoring of trends. The distinguishing factor of a *constancy check* is that a comparison must be made to the standard calibration method to establish the expected reading. This comparison should be carried out at the time of definitive calibration and repeated at appropriate intervals.

A number of constancy devices are available which may use diodes or ionisation chambers that can be rapidly placed in position. However, it is important that such devices should maintain their calibration with a degree of reliability as otherwise much time can be wasted in dealing with false indications that the treatment machine is out of calibration. Some would argue that it is very little extra work to perform a proper calibration with a thimble ionisation chamber rather than use a less precise method. However, many constancy check devices have more than one detector and can therefore provide information about the beam flatness as well. Following an incident in Poland where a power failure caused damage to a treatment machine the IAEA recommends that a constancy check should be carried out whenever an unexpected power failure occurs (IAEA 2004b) although it is acknowledged that the accelerator concerned did not meet current IEC requirements (IEC 1998).

38.8.1.4 Frequency of Checks

The frequency with which output checks are made varies from centre to centre. Most centres measure at least weekly. Each centre needs to review its findings regularly. If outputs are very stable then there is no need to change practice. If outputs seem to be drifting, more checks may be needed. In particular it may be important to investigate whether outputs are changing throughout the day. An action level of 3% is appropriate to use.

38.8.2 USE OF IONISATION CHAMBERS FOR OUTPUT CALIBRATION

Ionisation chambers are the most reliable method of measuring the dose delivered by a radiation beam and are the recommended instrument for this purpose. For absolute dose measurement in a photon beam, a chamber with a graphite cap is preferred. Each department should have a particular ion chamber and electrometer that is maintained as the departmental standard and calibrated against a secondary standard instrument with a calibration traceable to a standards laboratory. A strontium check source should be available for use with the chamber and this should be used to check its constancy on a regular basis whenever a definitive calibration is required (see Section 38.8.3).

Most ionisation chambers used in radiotherapy are unsealed, so measurements need to be corrected for temperature and pressure using the standard formula based on the *ideal gas* law (see Section 15.4.4):

$$\mathrm{CF}_{T\&P} = \frac{101.33}{P} \times \frac{\theta + 273}{293}$$

where P is the pressure in kPa and θ is the temperature in °C. Note that this assumes that the secondary standard dosimeter to which the chamber was compared was calibrated to read at 20°C. Care must be taken to ensure that the pressure used is the pressure at the height of the department and not corrected to mean sea level. A common practice is to ring the local airport to obtain a check of the pressure. However, this can lead to errors as the pressure may be different or be quoted at a different height and it is therefore better to use a properly calibrated aneroid barometer or a well-maintained Fortin barometer. It is also important to note that the temperature must be that of the chamber when it is placed in the phantom. If the phantom is kept in a cold cupboard and not in the treatment room, it is possible that there may be a considerable temperature difference, which is to be avoided. For accurate measurements, it is essential that the phantom and chamber be allowed to reach room temperature before measurements commence.

38.8.3 PRACTICAL ISSUES IN PHOTON CALIBRATIONS

For the calibration of photon beams the chamber is placed in reference conditions as defined in the protocol. This is at a depth beyond the depth at which electron contamination of the beam can occur—at least 5 cm for energies up to and including 10 MV and 7 cm for higher energies.

There is no universal agreement as to how the accelerator monitor unit should be defined (see Section 20.1.2). Some centres choose to define 1 MU as indicating 1 cGy at the isocentre at the reference depth (10 cm) for a 10 cm square field. Another common choice is to define 1 MU as delivering 1 cGy at the depth of maximum dose with the phantom surface at 100 cm from the radiation source. Even if this definition is chosen, the measurement must still be made in the reference conditions and the standard clinical depth dose tables used to convert the dose measured at the reference depth to the dose at the depth of maximum dose. Some may be tempted to use a measurement of the percentage depth dose for calibration purposes that is independent of the clinical data; but this is not a good idea.

A linear accelerator is required to have a dual dosimetry system in which there are two independent measures of the machine output. One of these channels will be responsible for terminating the treatment and the other will be a backup. The two channels should be independently calibrated rather than made equal to each other. Occasional checks should be carried out that the second channel will terminate the beam in the expected way. If changing the calibration factor of one of the channels can only do this it is best to change the secondary channel rather than the primary. Such a check should only be carried out under very careful supervision.

The monitor ion chamber may either be sealed or should have a built in temperature and pressure correction*. A sealed chamber may become unsealed and it is useful to compare both the fully temperature and pressure corrected reading with the uncorrected reading on a regular basis. If the calibration changes with the changes in atmospheric pressure it is a sign that the chamber has become unsealed. Recombination corrections (and polarity corrections for electrons) should also be applied to the readings of the measurement chamber (see Section 15.4.1, Section 15.4.2 and Section 38.9.1). If different dose rates are used, their influence on the monitor calibration must be assessed.

In addition to the weekly check, a check on the variation with gantry angle should be done at least annually with the variation between outputs at gantry $0°$ and gantry $\pm 90°$ not exceeding 2%. This should include checks of output factors with wedges. In addition a full calibration following the definitive calibration procedure should be carried out annually.

38.8.4 CALIBRATION OF UNITS CONTROLLED BY A TIMER

It is common practice for kilovoltage x-ray units and cobalt-60 units to be controlled by a timer rather than by a transmission dosemeter. Usually the dose rate will take a finite time to stabilise so that the dose delivered in 1 min will in general differ from the dose rate in mid exposure. This difference is called the *timer end error* (see Section 10.4.2) or, in the case of cobalt units, the *source-transit time* (see Section 12.3). Two approaches to the measurement of this correction are possible. In the dual exposure method the output dose is measured first for 1 min and then for 2 min. The instantaneous dose rate is then the difference between the two readings and the timer correction is the difference between the instantaneous dose rate

---

* Some older units had unsealed chambers with no such correction. In this case the variation may be reduced by correcting the readings to the average room conditions.

(in cGy/min) and the dose measured for a 1-min exposure. It is essential when using this method that the electrometer is able to record the 2-min exposure using the same range as the 1 min exposure and that periods of exactly 1 and 2 min are used. In the second method, the beam is first turned on and then the dose delivered in a fixed period is measured using the start and stop buttons on the electrometer. For some electrometers the dose-rate range is directly related to the dose range and can therefore be used to make a direct measurement of dose rate with the beam on.

38.8.5   ELECTRON OUTPUT

For the reasons explained in Part D, a plane-parallel chamber must be used for absolute electron measurements except at the higher energies when a cylindrical chamber can be used. Comparisons to an absorbed dose to water standard made at the NPL (du Sautoy et al. 2003) have resulted in a recommendation that only the Roos (Roos 1993) and NACP (Mattson et al. 1981; NACP 1981) chambers should be used for absolute dose determinations. It is necessary to ensure that the measurement is made at the correct depth and for this purpose it is useful to have a small water tank with a device to allow the chamber to be moved vertically by remote control. Making measurements in liquid water obviates the problems associated with charge storage in plastic phantoms. However, for routine calibrations a water equivalent plastic phantom selected to be water equivalent for electrons is convenient. This should be made of thin slabs rather than being a solid block.

Because of the rapid fall-off in dose with depth and because of the change in electron energy with depth which results in a change in the chamber calibration factor it is important that absolute calibrations are carried out at the correct depth as defined in the calibration protocol. However, for rapid checks of constancy it is convenient to use the same depth for a number of energies with a direct comparison being carried out to establish a *phantom factor*. For this purpose a cylindrical ionisation chamber may be used for all energies.

For small fields the depth of dose maximum may change significantly from the standard value for large fields and it is useful to be able to move the chamber remotely to find the depth of the maximum. The shape of the beam profile must also be considered as for small fields the dose often falls off towards the edge of the field. A chamber such as the Roos chamber may not be suitable in these circumstances because of its large area and it may be appropriate to use a Markus chamber that has a much smaller area.

At least each week a check should be made on the output for each electron energy used clinically. Some centres build this into their daily QC by measuring some of the range of available energies each day. The output check can be performed in a phantom that is also used for x-ray output checks so that no time is wasted changing phantoms and chambers. This can be achieved easily if the phantom is made of water equivalent plastic sections with holes at different depths and if a compromise depth, for example 1 cm, is used for all outputs. A standard 10 cm × 10 cm applicator is normally used. A combined calibration factor for the dosemeter and ion chamber should be determined for each energy on the same occasion that the definitive calibration was made with the appropriate ion chamber in water. It is important that a typical treatment level of dose is delivered, for example, 2 Gy or 3 Gy. The measured dose should be within 2% of the calibrated dose, although some centres use an action level of 3% (or even 5% for a daily constancy check).

38.9   CHECKS ON DOSIMETERS

38.9.1   IONISATION CHAMBERS

It is important to maintain regular checks and calibration data on all dosimeters used for output measurements or checks. Most ion chambers can be disconnected from their electrometer (although it is preferable for routine checks to leave the chamber connected at all times). It is vital to have appropriate labels made showing the calibration factors for the ion chamber serial number together with its electrometer serial number.

Strontium-90 checks are useful on a regular basis to check the maintenance of the sensitivity of the ion chamber/electrometer combination. Six monthly checks of this nature should show a consistency of better than 1%. However, it is important to recognise that the sensitivity of a chamber to ^{90}Sr beta particles may remain constant even when the sensitivity to photons, particularly at low energy, has changed markedly. Alternatively such checks could be made using a standard reproducible setup on a cobalt unit where available. In all cases, it is fundamental to correct the reading according to the decay of the reference source used (either ^{90}Sr or ^{60}Co).

Cross calibrations against the secondary standard should be made regularly at between 1 year and 3 year intervals (or if any doubt is raised about a particular ion chamber and/or electrometer). At the time of cross calibration, checks of polarity correction, recombination correction and the stem effect should also be made. Measurement of recombination correction can be carried out with the dual polarising voltage method (Boag and Currant 1980; Weinhous and Meli 1984), but it is advisable to verify that this works with the particular chamber by measuring at several different voltages—a linear plot should be obtained of the correction against the polarising voltage. The ion recombination correction is typically about 0.5%. This correction is only valid if the response of the ionisation chamber is close to linear with applied voltage. There is some evidence that this may not be the case with some parallel-plate chambers used at polarising voltages in excess of 100 V. Recombination corrections should, strictly speaking, be applied both at the time of dose measurement and at the time of comparison with the secondary standard. Alternatively if the dose rate (or dose per pulse) for the measurement will be the same the recombination correction appropriate to the secondary standard instrument can be applied to the field chamber at the time of measurement.

Visual checks and sensitive touch, of the thimble particularly, to ensure no breakage should be carried out before each use. Graphite thimbles have been known to crack when knocked and the crack only to become apparent some time later when the chamber falls apart.

38.9.2 THERMOLUMINESCENT DOSIMETERS

TLD can provide accurate dosimetry if an appropriate calibration method is employed. Details are contained in Section 16.2.4. The calibration scheme adopted will depend on the accuracy that is required of the measurements. Whatever system is adopted, it is vital to handle TLD materials with great care and attention to detail. The light output will change considerably if the disc or chip surface is contaminated with grease or any other substance. Chips can easily be damaged and are best handled with vacuum tweezers.

TLD work is notoriously fraught with problems. For a particular use it is vital to adopt good procedures for handling, read-out and annealing at the start and refine the calibration process. Otherwise the source of error is very difficult to find, since there are so many possibilities. Some of the points to keep in mind and check are:

- Oven temperature for readout and anneal
- Cleanliness of TLDs and trays
- Care with storage of TLDs

38.9.3 DIODES

Issues that must be considered in connection with the calibration of diodes are described in Section 16.3.4. Regular calibration is essential both for diodes used for patient dosimetry and for diodes used for QC measurements. Different calibration factors are required in different situations—for example for entrance and exit measurements—but the relative values can be expected to be constant over a long period of time. However, the standard dose calibration should be checked on a regular basis whose frequency is determined by the amount of dose delivered to the diode. The temperature correction factor for diodes used for in vivo dosimetry should also be checked regularly[*].

Arrays of diodes may be used for beam flatness measurements either for photons or electrons. In these situations it is only necessary to verify that each diode in the array maintains its sensitivity and that any radiation damage affects each diode by the same amount. However, the amount of radiation received by the central section of the array will inevitably be much greater over a period of time than the total dose received at the edges of the array. A regular check that should be made is to measure a beam with an intensity gradient (produced, for example, by a wedge), then rotate the array through 180° and check that the outputs of the two array positions are identical. An alternative approach is to use a method that shifts the whole array by one detector distance. Dose per pulse may also affect sensitivity and should be checked. A more stringent check that should be made at least annually is to check the sensitivity of each diode by using a narrow beam of radiation.

When using a diode to measure depth dose in photon beams it is vital to check the sensitivity with depth, field size and energy since the amount of scatter of different energies varies with these parameters. It is advisable to compare these depth doses with those from an ionisation chamber on each occasion when fundamental commissioning data are measured. It is inadvisable to use diode depth dose data as the primary measurement for photon beams except where it is essential, such as for very small field sizes. There are no such reservations for electron beams where the response of diodes is almost constant for 5 MeV to 25 MeV beams. However, it is still advisable to check the corrected ion chamber depth dose against the diode readings for a range of energies.

38.10 PHANTOMS AND PHANTOM MATERIALS

The particular dosimetry protocol used will state the preferred options for phantom materials. For the U.K. protocol water is preferred and the minimum phantom size 30 cm × 30 cm × 30 cm. The chamber should be fitted with a thin waterproofing sheath made from PMMA and not more than 1 mm thick.

For plastic materials (mainly PMMA and polystyrene) appropriate corrections must be made. Water-equivalent-plastics made from resin-based materials should be avoided for definitive calibrations. They are acceptable for relative measurements including routine output checks. However, even for these checks care must be taken to use the same block(s) of water equivalent plastic for QC, since there may be variation in composition and density between blocks.

Most modern protocols point out the problems of non-water phantoms, but some of the problems to be aware of are:

- Density and electron density of material will affect depth dose, stopping power etc.
- Different scattering properties affect electron fluence build-up.
- Many of these plastics are insulators and exhibit charge-storage effects which can modify the measured ionisation. The effects are minimised by using phantoms made from a number of sheets rather than a solid block.

---

[*] If diodes are used such as to require monthly recalibration, the temperature correction should be checked annually.

38.11 PORTAL IMAGING DEVICES

The QA of EPIDs used for verification of patient positioning and transit dosimetry should cover:

- Monitoring of the quality of image produced by such devices during the commissioning stage and their variations over time
- The monitoring of the mechanical aspects of the devices to ensure safety of patients and staff
- Checking the integrity of storage facilities for digital images

Procedures or protocols to monitor these three aspects of the electronic portal imaging system are necessary to ensure that the information obtained from the system is accurate.

With solid-state imaging systems, it is important to ensure that systems are in place to prevent irradiation of the electronics that can cause rapid deterioration of the images. Ideally, software to assist in this function should be provided by the manufacturer.

38.11.1 IMAGE QUALITY ASSESSMENT

A general definition of image quality addresses the effectiveness by which the image can be used for its intended task. The image acquired by the detector is a representation of the distribution of attenuating properties of the different structures of the human body or phantom under investigation. The performance of the detector is measured by the detectable contrast between these different tissues, the spatial resolution and detector noise. The performance of the EPID can be assessed by analysing images of known objects. The analysis can either involve the participation of human observers (subjective), or the use of mathematical models (objective) that are evaluated using computers.

38.11.1.1 Subjective Image Analysis

One of the earliest image quality testing tools was developed by Lutz and Bjarngard (1985). The test tool consists of a series of objects with different contrasts. The evaluation process involves identifying which objects are visible. The contrast detail phantom of Harrison and Lambert (1994) or the "Las Vegas" phantom (Leszcynski and Shalev 1989) consists of a block of aluminium with holes of different diameters and depth. A typical image obtained with the Las Vegas phantom is shown in Figure 38.8. However, unless specific software is being used, the analysis of images from these phantoms is not entirely objective. As one gets familiar with the hole pattern, it is easy to see holes that are not there. In case of direct visual analysis, a more comprehensive procedure requiring receiver operating characteristic (ROC) image analysis methodology (ICRU 1996) is required for the interpretation of the hole patterns in order to increase the objectivity of the results with the Las Vegas phantom.

38.11.1.2 Objective Image Analysis

Shalev and others (1997) have proposed and designed a phantom that provides a totally objective test. The images acquired are analysed by a computer program and the results compared to the manufacturer's specification or to a base line obtained at commissioning. The image quality index is therefore not prone to human perception.

An image obtained from a test phantom constructed according to this design is shown in Figure 38.9. The phantom consists vertically and horizontally layered slabs of PMMA and lead of different thickness chosen to provide spatial resolution in the range 0–1.5 line pairs/mm, and contrast ranging from 20% to 80%, respectively. The phantom is placed on the detector at an

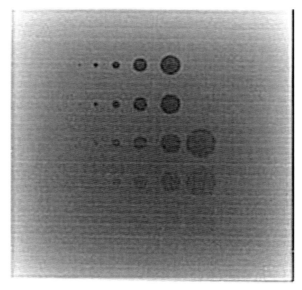

FIGURE 38.8
A typical image acquired with the "Las Vegas" phantom showing images of holes of different diameters and depths.

angle of 45° to avoid the artefacts associated with scanning electrodes in matrix chambers. Analysis of the images can be carried out using computer programs. Computer software such as PIPS of Shalev and others (1994), or in-house developed programs, can then be used to determine the frequency dependant square wave modulation transfer function (SWMTF) from which the frequency at 50% modulation, $f50$, is derived. The SWMTF is determined by the

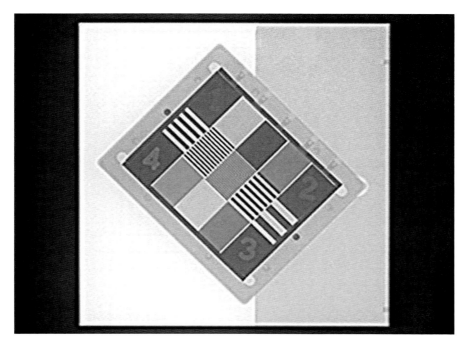

FIGURE 38.9
An image acquired with a test phantom constructed based on the Shalev and others (1997) design. The numbers 1–4 represent the regions with different lead and PMMA combinations to produce different contrasts. The spatial resolution is determined from the line pair patterns in the middle regions spanning lengthwise.

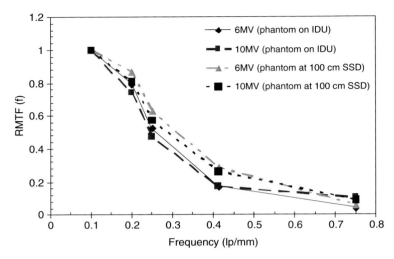

FIGURE 38.10
Typical RMTF curve for 6 and 10 MV photon beams, with the phantom at 100 cm SSD and on top of the image detector unit (IDU). The detector was positioned at 140 cm source detector distance.

MV method used by Droege (1983) and Rajapskshe (1996):

$$\text{SWMTF} = \frac{\Delta E(f)}{\Delta E_0}$$

where ΔE_0 and $\Delta E(f)$ are the modulation input and output from the system. The above equation can be modified to obtain a relative measure of the modulation transfer function (RMTF), just for monitoring purposes, given by:

$$\text{RMTF}(f) = \frac{\Delta E(f)}{\Delta E(f_1)}$$

In this case, $\Delta E(f_1)$ is taken as the highest output frequency modulation. The $f50$ is obtained from linear interpolation of the RMTF (f) versus frequency data.

This modulation transfer function provides a measure of the spatial detail transfer characteristics of the imaging system. The analysis is carried out on the original image with no signal manipulation. A typical RMTF curve is shown in Figure 38.10, for 6 MV and 10 MV photon beams, obtained using the Varian liquid-ionisation-matrix chamber. From Figure 38.10, the $f50$ values of 0.278 lp/mm and 0.252 lp/mm for this detector can be extracted. Typical values for other detector types are shown in Part C, Table 14.1. The $f50$ values can be measured over a period of time to monitor the stability of the system using this simple and reproducible test.

38.11.2 Mechanical Stability

QC procedures on the image detection unit assembly depend on holder mounting design and partly on image processing software. Part C, Table 14.1 lists the mounting designs for some of the commercially available detectors. For detectors that have to be attached to the gantry temporarily when in use, such as the now superseded Philips SRI-100, QA checks have to be carried out on the integrity of the interlocking system. This has to be carried out at various gantry angles to ensure safety of patients. Reproducibility of the positioning of the detectors is only critical if the geometric centre of the image detector unit is used during image analysis. This may be more critical when the absolute output at every pixel position is required as in exit and transit dosimetry. The calibration in this case would be affected by beam characteristics such as flatness.

38.11.3 QUALITY CONTROL OF IMAGE DATA

Portal imaging devices require a comprehensive image handling system that is used to manipulate, store, import and export images. The images are stored on the system in lossless compressed format. The integrity of the stored and retrieved data can be checked by monitoring the pixel values for an acquired and stored image over a period of time. This can be achieved using the image analysis tools supplied with the system. A region of interest can be chosen and the statistical results obtained using the system tools and recorded over a period of time. There are other methods of checking the accuracy of file storage and retrieval such as performing checksums on the image files.

CHAPTER 39

QUALITY ASSURANCE OF THE TREATMENT PLANNING PROCESS

Jean-Claude Rosenwald

CONTENTS

39.1 THE TREATMENT PLANNING PROCESS

39.1.1 DEFINITION OF TREATMENT PLANNING

The term *treatment planning* is often used in a restrictive manner to refer only to the production of isodose distributions associated with the choice of a treatment plan (*dose planning*). However, it is also used to describe the whole technical process, from patient data acquisition to treatment verification (Fraass et al. 1998). It is this second definition that is used here. The whole sequence of events is given, in a simplified form, in Table 39.1.

The sequence in Table 39.1 is subject to variations according to different practices. However, it is recognised as being the most consistent one, especially in conformal therapy (see Chapter 43.1) when one is aiming at a precise match between the shape of the target volume and the high-value isodose envelopes. It is often an iterative process since, in a number of instances, a correction to a previous step must be made as a result of a subsequent one. This is the case in the dose-planning phase when the beam parameters are subsequently modified to get the desired dose distribution. It is also the case when discrepancies are found at verification requiring review of some of the previous steps. The way in which this fits into the overall treatment process can be seen in Figure 37.1.

39.1.2 DATA EXCHANGE IN RADIOTHERAPY

The radiotherapy process involves the use of a number of different pieces of equipment, as listed in Table 37.2. For each piece of equipment, specific quality assurance (QA) procedures are designed as discussed in Chapter 38 through Chapter 41 but, in addition, special attention must be given to data exchanges between different pieces of equipment, such as shown in Figure 39.1.

TABLE 39.1

Main Steps of the Treatment Planning Process

Patient positioning and patient data acquisition
Definition of the treatment beams
Calculation and evaluation of the dose distribution
Verification before the actual plan implementation

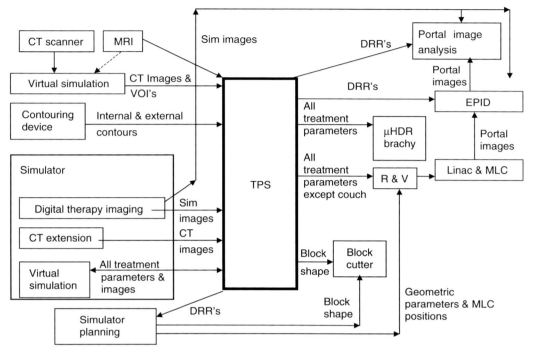

FIGURE 39.1
Example of communication between equipment in a radiotherapy department (slightly simplified from a real department).

It is essential to ensure a coherent quality-control system for the whole treatment planning and delivery process.

39.1.3 QUALITY ASSURANCE IN TREATMENT PLANNING

A comprehensive QA programme for the whole process implies taking into account many constituents, such as definition of clear aims, appropriate staffing level and qualifications, definition of responsibilities, existence of detailed written procedures, etc. (see Section 37.2). The main issues discussed here are related to QA of the treatment-planning equipment. It must also be ensured that, for individual patients, errors due to wrong data input, incorrect interpretation of results, or inadvertent/undetected failure of equipment will not appear. Therefore, QA procedures related to individual plans must also be devised.

For further detail, the reader is referred to the American Association of Physicists in Medicine Report of Task Group 53 (TG53 Fraass et al. 1998), the Institute of Physics and Engineering in Medicine Report 81 (IPEM 1999), the ESTRO booklet on QA of treatment planning systems (ESTRO 2004) and the IAEA report TRS 430 (IAEA 2004a).

39.2 PATIENT POSITIONING AND ACQUISITION OF ANATOMICAL DATA

The importance of QA in this phase is often overlooked. Some of the points that deserve special attention for the consistency and accuracy of the planning process are discussed below. However, one crucial point is the interpretation of the patient data by the treatment-planning system (TPS), which will be discussed in more detail in Section 39.3.5. Another important point is that the devices used for data acquisition have been properly calibrated.

39.2.1   QUALITY ASSURANCE OF EQUIPMENT USED FOR PATIENT DATA ACQUISITION

The equipment used for patient data acquisition has been described in Chapter 30. The data acquired can be classified as providing geometric or density information.

If the anatomical data consist of a series of contours obtained from mechanical or optical devices, eventually combined with some imaging data for internal structure, the important point is to make sure that these contours have accurate dimensions within a tolerance range on the order of 2–3 mm. The simplest solution to check it, is to use a standard solid block of simple geometry and known dimensions (e.g. a cube or a cylinder), ideally including internal inhomogeneities, and to acquire the corresponding contour(s) and internal structures as if it were a patient. The shape and dimensions should be the same as expected within the stated tolerance.

If CT density data are to be used for inhomogeneity corrections, then it is necessary to carry out a calibration to establish the relationship between the Hounsfield Units (HU) and the corresponding tissue densities (Constantinou et al. 1992; Schneider et al. 1996; Guan et al. 2002). This relationship is somewhat scanner-dependent. It could also, to some extent, depend on the type and quality of radiation used for treatment: for high-energy photon beams, where the Compton effect is prominent for most patient tissues, one should preferably use the electron density instead of the mass density (see Chapter 4). For particle beams (electrons and protons), the relevant quantity for *density corrections* should be the tissue stopping power or scattering power coefficient (see Chapter 3). However, in most cases, the direct use of the mass density results in sufficient accuracy to correct the dose for patient inhomogeneities. To perform the CT calibration, one can use commercial phantoms in which a series of materials of different known densities are embedded. The CT console is used to measure the average HU in a representative area of each of these sample materials. Typically, a curve similar to the one represented in Figure 39.2 is obtained. Actually, this curve is normally linear between the density of air (used as a reference for HU = −1000) and the density of water (used as a reference for HU = 0). Therefore, this part of the curve is

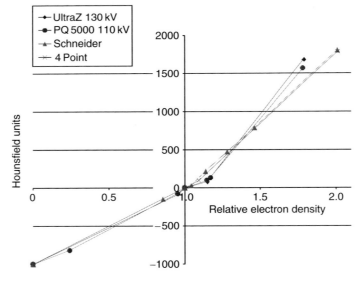

FIGURE 39.2
Example of the density calibration curve for two different CT scanners. The curve attributed to Schneider et al. (1996) corrects for the real composition of bone compared to the phantom materials used to measure the curve. The four-point curve is a simplified representation of the data. (After Guan, H. et al., *Phys. Med. Biol.*, 47, N223–N231, 2002.)

independent of the scanner type and CT acquisition parameters. This is not the case in the upper region of the curve, which is much more sensitive to these parameters and even to the phantom size or sample position within the phantom. The reason is that for high-atomic number material (typically bone), some photoelectric effect takes place during CT imaging which depends strongly on local x-ray quality*. Nevertheless, it is generally acceptable to consider that above roughly 100 HU, there is a second linear portion of curve with a steeper slope based on the average of measurements performed with bone equivalent materials.

Other CT parameters are important for ensuring an accurate patient reconstruction from a series of CT slices. These are essentially comprised of the following:

- Capability to reproduce actual dimensions of objects
- Absence of distortion
- Reproducibility and accuracy of slice position
- Accuracy of slice thickness
- Accuracy of gantry rotation and table swivel scales (preferably fixed at 0°)
- Alignment of laser positioning devices (if any)

The methods to check these points can be developed from the procedures used for mechanical checks of treatment machines and quality control of CT scanners for radiological purposes. Most of them are relatively simple to implement using simple solid phantoms with radio-opaque markers and performing appropriate measurements on the CT console display. The requirements are more stringent than for diagnostic applications, and the accuracy goal is typically between 1 mm and 2 mm or between 0.1° and 0.2° (Mutic et al. 2003).

39.2.2 PROCEDURES FOR PATIENT DATA ACQUISITION

The patient setup at the time of data acquisition must be such that the same position can be used in further steps of the process, i.e. simulation (if any) and treatment. A distinction could be made between *immobilisation* devices, which aim to ensure that the patient does not move during the corresponding procedure but do not guarantee a reproducible setup, and *repositioning* devices, which help to restore the patient's position at the time of anatomical data acquisition. The use of skin markers, combined with the skill of experienced radiographers, is a more primitive, although often rather efficient, system for patient repositioning. Systematic pictures (i.e. taken with a digital camera) can also be quite helpful. Ultimately, devices such as stereotactic frames ensure both efficient patient immobilisation and accurate repositioning. In all cases, the devices used for treatment should be used also for patient data acquisition, including the table top of the imaging system which should resemble, as much as possible, the table top of the treatment couch and positioning aids such as laser pointers.

Patients' internal movements are also a potential source of inaccuracy. Techniques are developing to take these movements into account, especially for treatment in the lung area. This may be done either by using the *deep inspiration breath hold* technique, where the beam is delivered only while the patient is holding a deep breath under passive or active control of a spirometer, or by using free-breathing *gated radiotherapy*. For this latter technique, the respiratory movements are analysed by an appropriate sensor which triggers the irradiation when the organ position or cycle is within a given range (Ford et al. 2002; Ozhasoglu and Murphy 2002). In such instances, the patient data acquisition should be consistent with the average organ position in the beam-on cycle (see the Introduction to Part I). In the standard situation, the patients are asked to breathe normally, as they would do during treatment, in order to get data

---

* For this reason, the higher-density inserts should be of equivalent chemical composition to the patient tissues that they represent.

representative of the whole patient anatomy in the treated region. In all cases, some evaluation of the magnitude of internal movements should be established in order to include them in the planning target volume (PTV) margins (see Chapter 29 and Chapter 30).

Another potential source of inaccuracy is the use of contrast media that change the local tissue density when the images are acquired, but are no longer there when the patient is treated. If contrast media are considered necessary and if the dose calculation algorithm makes direct use of the HU information, some corrective procedure must be implemented at the dose-planning stage.

39.3 TREATMENT PLANNING SYSTEM

Bringing a treatment planning system into clinical use can be implemented in five phases:

1. Specification and purchase of the system
2. Assessment of the system for completeness against the specification
3. Commissioning the beam data
4. Testing the calculation accuracy and algorithms
5. Clinical implementation

Each of these phases will be considered in turn, but it is first necessary to be clear about the components of the system.

39.3.1 COMPONENTS AND FUNCTIONS OF A TREATMENT PLANNING SYSTEM

Computerised treatment planning systems have evolved significantly as computer technology has become more and more sophisticated. However, since the 1970s—when the first computer with real time display and graphics possibilities became available—the main components have remained essentially the same. They are shown in Figure 39.3.

Besides the computer itself—which typically includes the processor, memory, and hard disk drive—a monitor is used to display contours, images, and dose distributions, allowing real-time interaction to perform the plan optimisation. The other pieces of equipment and their main characteristics are described below.

39.3.1.1 Curve Digitiser

Curve digitisers are used for input of patient contours as obtained from mechanical contouring devices (see Chapter 30). They are also useful in defining the shape of the field from radiographs for dose calculation purposes, or for designing the actual field shaping (for transfer to a block-cutting device or to the controller of a multileaf collimator (MLC)). Typically, it should be possible to acquire curves with accuracy better than 0.5 mm. With the increasing use of digital images and networking solutions, curve digitisers are progressively becoming obsolete.

39.3.1.2 Film Scanner

A high-quality film scanner used to be an essential part of a planning system, but images are increasingly available in digital format. The most important feature of such a scanner is geometric linearity. It is not recommended to use them to digitise CT images printed on films, since the images are often small and subject to geometrical distortion.

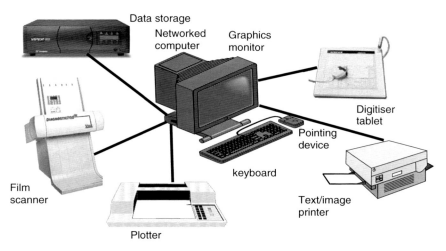

FIGURE 39.3
Main components of a typical treatment planning system. Most of the components shown in the figure can be networked and shared between several workstations. It is particularly useful to have a central patient database and a full connection to the rest of the equipment as indicated in Figure 39.1.

39.3.1.3   Plotter/Printer

Although most data and results can be archived on storage devices and displayed on monitors, it is still a common practice to use hard copies of the plans for easier transmission between the various people involved in the planning process and for filing purposes.

Various technologies are available, yielding differences in speed, image resolution, geometric accuracy, and stability. Some printers are limited to text, and others to text and lines (pen plotters). Many modern devices (e.g. inkjet or laser printers) support text, graphics, and images. They are controlled through appropriate drivers, and the compatibility with the planning software must be ensured.

As the paper printout is often used for direct geometrical measurements, the magnification factor must be known and should be controlled in both directions with accuracy better than 0.5 mm.

39.3.1.4   Storage Devices

Three-dimensional (3D) image manipulation places great demands on storage capacity. Storage devices are evolving rapidly, and the more recent technologies offer a wide choice of solutions. It is necessary to clearly define the needs, both for data backup and for archiving. In assessing these needs, it is necessary to:

- Differentiate clearly between the short term (i.e. while the patient is under treatment) and long-term requirements, and to evaluate the relevant needs for storage and access
- Decide whether it is necessary to keep all images used for contouring
- Decide at what stage and in which format (e.g. original CT, treatment planning system format, compressed or uncompressed)
- Define the procedure to restore archived data (e.g. where direct access to the images is required or whether it is sufficient to restore backup versions when the images are needed).

Alternatively, a connection with a picture archive and communication system (PACS) could be provided.

39.3.1.5 Network Interfaces

More and more frequently, the treatment planning system is part of a network which consists of several consoles accessing a common dedicated server, with special protocols for data exchange with other pieces of equipment (see Figure 39.1). To have good communication between these various devices, it is essential to optimise the physical and logical characteristics of the network. It is, therefore, necessary to establish contact with the relevant specialists who are in charge of the hospital network and to make sure that the network infrastructure is adequate for the treatment planning system requirements. If necessary, a local network could be partially isolated from the general hospital network through proper bridges.

39.3.1.6 Software Functions

Besides the hardware components, the software is obviously fundamental to obtaining the required functionality. Originally, the functions of a treatment planning system were mostly limited to dose calculation for a given beam arrangement, taking into account the beams and patient characteristics. The requirements have been considerably expanded and now include a wide variety of features necessary for the preparation of a clinical plan. A list of the main features needed is given in Table 39.2.

This table does not reflect the ease of use of the given features. A well-designed user interface is important. This should allow maximum flexibility to change any parameter at any time and immediately see the effect on the dose distribution.

Additional functions and tools are required: for instance, to prepare and check the basic beam data that are processed by the dose calculation algorithm in order to ensure consistency with the measured beam data.

TABLE 39.2

Typical Main Functions of a Modern Treatment Planning System

Patient anatomy
- Transfer of images from the CT scanner (and other imaging devices)
- Registration of images from different modalities
- Automatic delineation of the external surface of the body (for dose calculation)
- Automatic delineation of inhomogeneities and/or density mapping
- Manual or semi-automatic delineation of target volumes
- Manual or semi-automatic delineation of organs at risk
- Automatic addition of 3D margins around structures

Beams
- Manual or automatic beam positioning (coplanar, non coplanar)
- Manual or automatic delineation of beam aperture
- Inclusion of beam modifiers (blocks, wedges, bolus, compensators, etc.)
- Definition of beam contributions (beam weights)

Evaluation/optimisation of the dose distribution
- Dose computation for various points, planes and in 3D
- 2D isodose display with superimposition on anatomical images
- 3D surface rendering of anatomical structures and isodose surfaces
- Calculation/display of dose volume histograms for volumes of interest
- Calculation of biological indices
- Optimisation (i.e. inverse planning for intensity modulated radiotherapy)

Aids for patient setup
- Print-out and automatic transfer of beam parameters
- Print-out and automatic transfer of individual shielding block shapes
- Print-out and automatic transfer of multileaf collimator (MLC) settings
- Generation and export of digitally reconstructed radiographs (DRR)

Another important issue, which will be discussed later (see Section 39.3.4.2), is the accuracy with which the treatment planning system reproduces the actual dose distribution within the patient, considering both the geometric and dosimetric uncertainties.

39.3.2 SPECIFICATION AND PURCHASE PROCESS

The acquisition of a new treatment planning system should begin with the careful preparation of a specification. This specification should form the basis of the assessment of the performance of the equipment and the design of the QA system. Different arrangements may be appropriate, depending on the local organisation, but in any case, it is essential to appoint a group of people to prepare the specifications of the future system. This group should be experienced in treatment planning and directly involved in the future use of the system. It could consist, for instance, of a physicist, a dosimetrist, and a radiation oncologist. The basic requirements should be identified in conjunction with the existing or foreseen imaging and treatment devices and with the present or foreseen procedures applied in clinical practice (e.g. conventional photon beams, electron beams, isocentric or fixed Source Skin Distance (SSD) treatments, arc therapy, beam modifiers, stereotactic irradiation, brachytherapy, etc.). Table 39.2 is a useful reference to check the main characteristics that should be considered.

One must sometimes choose between a single multi-purpose treatment planning system and several independent systems dedicated to specialised techniques, such as brachytherapy or stereotaxy. While a multipurpose system may be a cheaper option, other important aspects must also be considered. The single system may be less well-adapted to the whole range of situations. There may be a requirement for a number of people to access different aspects of the system at the same time. This may be easier if there are dedicated systems or, conversely, a single system with multiple terminals may provide greater flexibility. There may also be an issue of accessibility. As a multi-terminal planning system expands, issues of maintaining and upgrading it without disruption can become more difficult.

Before preparing the detailed specification, it is useful to organise visits or demonstrations of some of the systems currently available on the market. It is then much easier to figure out what could be offered and to write specifications at the right level to cover local needs while also taking into account the existing commercial products.

As underlined in the AAPM Report of Task Group 53 (Fraass et al. 1998), creating the specification for a modern 3D treatment planning system is a huge task, and only some general guidelines can be given. The main items that should be considered are the following:

- Computer hardware description*:
  - Number of consoles
  - Input and output peripherals
  - Connections with other equipment
  - Storage capacity and access for backup data and for archiving

- Software requirements:
    These should be described with enough detail to avoid any ambiguous responses. Table 39.2 or the tables given in Chapter 4 of the TRS 430 report (IAEA 2004a) can be used as an aide memoire. The details must be adapted to local requirements. The questions should normally be phrased to require a yes-or-no answer, but in many situations, it is also useful to allow for some comments from the manufacturer.

---

* In most situations, it is not appropriate to specify a given type of computer or operating system, since the important point is to make sure that the system is capable of fulfilling its purpose, whatever technical solution is chosen by the manufacturer.

TABLE 39.3

Functions Requiring Particular Attention in Testing

Automatic delineation of structures
Automatic expansion of structures (margins)
Automatic isocentre positioning
Automatic field shaping
Dose calculation with shielding blocks
Dose calculation with inhomogeneities
Dose calculation with wedge filters
Dose calculation with asymmetric fields
Dose calculation with multileaf collimators

- Performance requirements:
    Requirements such as accuracy or speed should be expressed in such a way that they could be actually checked by running an appropriate benchmark test.

An example of a list of requirements for a relatively basic treatment planning system can also be found in Appendix C of IAEA TECDOC 1040 (IAEA 1998).

It is not practical to conduct an exhaustive test of a treatment planning system before purchase, and it is, therefore, important that a sufficiently detailed specification of the requirements is prepared so that the manufacturer has a clear commitment to provide what is expected. The manufacturer or the vendor may be asked to conduct preliminary tests (following, for instance, the published benchmark tests) and to demonstrate them to the user during the purchase or the acceptance phase (ESTRO 2004). Typical areas in which detailed specification and assessment are necessary are listed in Table 39.3.

39.3.3 ACCEPTANCE OF THE TREATMENT PLANNING SYSTEM

Acceptance testing involves testing the function of the hardware and assessment of the system for completeness against the specification. It is unreasonable to expect to carry out a full evaluation of the system at this stage, but sufficient tests should be conducted to identify major deficiencies. It is also wise to test connectivity to other equipment at this stage, but it is unlikely to be possible to test the calculation accuracy unless the manufacturer is able to supply standard beam data for the published test data given in AAPM Report 55 (1995) or Venselaar and Welleweerd (2001). It is similarly important to check the documentation provided with the system. This must include comprehensive details of the data required and the algorithms used, as well as the standard instructions for use. These instructions must be read carefully, as a treatment planning system can produce unexpected results if not used in the intended fashion (IAEA 2001). The manufacturer must also provide appropriate training particularly of the physicist responsible for the planning data.

39.3.4 OBJECTIVES OF TREATMENT PLANNING SYSTEM COMMISSIONING

39.3.4.1 Aim of Commissioning and Risk Analysis

Here, *commissioning* is understood as the steps required to prepare a new treatment planning system, a new software release, or a new set of beam data so that it may be used safely for patient planning. In contrast to the approach of IPEM Report 81 (IPEM 1999), the process of data collection and data entry into the planning system is treated as one process. Since there are many risks associated with the use of a treatment planning system, qualified and experienced staff must perform commissioning, and sufficient time must be allocated to carry out the whole process.

More and more frequently, the systems being sold and installed are subject to some kind of certification by accredited bodies. Examples are the Food and Drug Administration (FDA) system in the U.S.A. and the CE marking in Europe. Depending on local regulations, such certification is mandatory for clinical use. However, the testing required for certification is limited to ensuring that the system is well-documented and that it behaves as described. Such testing does not prevent large uncertainties or errors in local clinical practice (IAEA 2001). With the increasing complexity of treatment planning systems, not all situations can be covered, and computer glitches with severe implications can be found from time to time. Published recommendations often contain an exhaustive list of tests to detect possible problems. The author believes that, for commercial software, this extensive testing is primarily the responsibility of the manufacturer, with the help of clinical beta test sites (ESTRO 2004). Users should concentrate on the points where their own interaction can induce inaccurate or erroneous results. These are linked to parameter definition by the user on the one hand and to misinterpretation in using the software on the other hand. The full testing of in-house software or software which has not been released commercially remains the responsibility of the user. The possibility of undiscovered software programming errors (so-called bugs) will remain in spite of the most exhaustive testing, and a quality control system for individual plans is also necessary (see Section 39.3.9).

Any system requires some form of parameter definition. Some of the parameters are used only as *prefered options* and have no influence on the dose calculation. Some are fundamental, especially those related to machine and beam characteristics that must be consistent with the local data. The commissioning phase is the unique opportunity to adjust these parameters with great care and to check their value in terms of overall accuracy for a number of clinically relevant situations. This will be discussed in more detail in the following paragraphs.

Another aim of commissioning is to achieve a thorough understanding of the various functions of the treatment planning system and the underlying assumptions and limitations. It is also necessary to establish that the accuracy is within acceptable limits.

39.3.4.2 *Accuracy and Tolerances*

The accuracy of the planning system should be compared against an agreed-upon standard for both geometric and dosimetric accuracy. The goal is that the treatment planning system should not introduce significant additional errors to the errors already existing in the chain, starting from beam calibration and ending with treatment completion. This leads to figures which are typically between ± 2 mm and ± 3 mm for geometry and between ± 2 percent and ± 3 percent for doses (see Section 37.4 and Section 37.5). However, the accuracy required depends on the clinical objective. For instance, a plan for a stereotactic case would be much more demanding than one for a palliative treatment. In addition, for practical reasons, it may be acceptable that a larger tolerance is accepted in complex situations or specific regions (see below). Ultimately, it is the user's responsibility to decide in particular circumstances if the results are clinically acceptable.

The appropriate tolerances for different regions of the beam have been considered in several publications (Van Dyk et al. 1993; TG53: Fraass et al. 1998; Venselaar et al. 2001; ESTRO 2004). It is generally agreed that the regions of the beam should be divided up as illustrated in Figure 39.4 and described in Table 39.4.

In defining tolerances as percentages, it is necessary to be clear about the definition of the normalisation (100%) point. For the dose at the normalisation point, the only sensible way to define the tolerance is in relationship to the measured (or expected) absolute dose for a given number of monitor units (MU)[*]. At other points, a number of different

---

[*] For a discussion of the definition of MU, see Section 38.8.3 and Section 20.1.2.

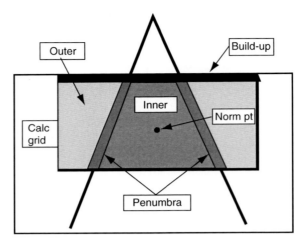

FIGURE 39.4
Regions of a single beam in for which tolerance levels can be set. More detail is contained in Table 39.5.
(From Fraass, B. A. et al., *Int. J. Radiat. Oncol. Biol. Phys.*, 42, 651–659, 1998.)

choices become possible. TG53 (Fraass et al. 1998) recommends that all criteria be a percentage of the central-axis normalisation dose. Venselaar et al. (2001) suggest that this can lead to an unrealistically lenient requirement in areas of low dose. They recommend that the normalisation is to the local dose (i.e. the measured dose at the point of interest) for the inner region and to the dose on the central axis at the same depth for low-dose regions (i.e. outside of the beam limits or under shielding blocks). It will be seen that these definitions will result in significantly different values of the assessment criteria, even if the recommendations actually agree in principle.

Table 39.5 gives an example of the possible criteria, based on the table by Venselaar et al. (2001). In an ideal world, the tolerance should not depend on the complexity of the treatment, being instead based on the clinical requirements, but the widening of the tolerances is necessary in order to be achievable. Note that the tolerance in the outer region is similar to that in the central region if expressed as a percentage of the

TABLE 39.4

Description of Regions of a Beam as Shown in Figure 39.4

Region Name	Definition	Dose Level	Dose Gradient
Build-up region	Depth < dose maximum	High	High
Inner region	Central high-dose part of beam	High	Low
Penumbra region	5 mm inside and outside the geometric limits of the beam	High	High
Beam fringe	50–90% region of penumbra	High	High
Outer region	Outside the penumbra or areas where the dose is <7% of the central axis dose	Low	Low
Central axis	Centre of open field	High	Low
Normalisation point	Point at which absolute dose is defined	High	Low

Source: Definitions following Fraass et al. (1998) and Venselaar et al. (2001).

TABLE 39.5

Example Criteria for Tolerances

Region	Homogenous, Simple Geometry	Complex Geometry (inhomogeneity, wedge, asymmetry, blocks or MLC)	More Complex Geometries (more than one complexity)
Build-up region	2 mm or 10%	3 mm or 15%	3 mm or 15%
Inner region	2%	3%	4%
Penumbra region	2 mm or 10%	3 mm or 15%	3 mm or 15%
Beam fringe	2 mm	3 mm	3 mm
Outer region	30% (3%)	40% (4%)	50% (5%)
Central axis	2%	3%	4%
Normalisation point	0.5%	1%	2%

For the outer region, tolerances are shown as the percentage of the local dose with the percentage of the central axis dose in brackets.
Source: Adapted from Venselaar et al. (2001) taking into account also Fraass et al. (1998).

central axis dose. When a point is in a high dose gradient, it is appropriate to use a distance criterion.

39.3.4.3 *Tolerance Specification in Non-Uniform Dose Distributions*

In areas where the dose distribution is varying in an unpredictable way, such as with Intensity Modulated Radiation Therapy (IMRT) fields, it may be misleading to use a strict dose percentage criterion. In these circumstances, a combination of a dose criterion and a percentage criterion may be appropriate. For this purpose, the concept of the *gamma index* has been proposed (Low et al. 1998; Van Esch et al. 2002; Depuydt et al. 2002; Bakai et al. 2003). Here, the tolerance is expressed as the maximum distance to a point of agreement or the maximum percentage difference, whichever is greater. To determine whether a tolerance in terms of gamma is exceeded at any point, the calculated dose is first compared with the measured dose, and the difference (Δ%) calculated. Next, the nearest measured dose point that matches the calculated dose is found, and the distance (δ mm) to this point is calculated. A gamma criterion of 3% or 3 mm is met (i.e. the gamma index is smaller than 1) if $\Delta < 3$% and $\delta < 3$ mm. The mathematical definition of the gamma index by Low et al. (1998) and a comprehensive presentation of the possible solutions to express the deviations between computed and measured dose and illustrative examples can be found in the ESTRO booklet (2004).

39.3.5 GEOMETRIC ASPECTS OF TREATMENT PLANNING SYSTEM COMMISSIONING

The nondosimetric aspects of treatment planning system commissioning have often been neglected in the past. Since the publication of TG53 (Fraass et al. 1998), they are considered to be an important part of the commissioning process, especially with the development of conformal radiotherapy, in which geometrical issues are fundamental. The tests listed here are purely an indication and are not exhaustive. The principal benefit of going through these different tests is to explore the various functions of the software. This helps the user to understand the possibilities, limitations, and pitfalls, rather than to discover the bugs in the software.

TABLE 39.6

Examples of Tests to Check the Validity of Anatomical Data Entry into the Treatment Planning System

	Item	Scope	Procedure
1.	Digitisation of contours	Check scale and linearity	Digitise simple shapes of known dimensions. Measure these with the TPS ruler functions. Apply beams to the shapes and verify that their sizes are as expected
2.	Identification	Check that image identification by the TPS is correct and unambiguous	Understand the identification structure of the imaging device (i.e. patient/examination/series/-image). Using all the available protocols, transfer images to the TPS and verify how these relate to the TPS image identification
3.	Series of images	Check consistency	Generate a series of scans with mixed characteristics (e.g. duplicated slices, different field of view, different orientation, with and without contrast, etc.) and make sure that the TPS rejects, generates warnings, or processes the inconsistencies
4.	Patient orientation	Check that orientation is correctly understood and processed by the TPS	Use a test object (e.g. a cube with engraved or protruding markers on different sides) and scan it in different patient orientations (e.g. supine/prone, head/feet first). Check how this is displayed in 2D or 3D reconstructions performed in the TPS
5.	Geometrical representation of images	Check that there is no significant modification of distances or distortion	Compare the distances measured on the imaging console to those measured on the TPS using appropriate tools (a solid phantom of known dimensions could also be used) as in item 1
6.	Tissue density for CT	Check that the density calculation by the TPS is consistent with original HU data	Compare original HU on the CT console to the HU measured on the TPS (preferably using a density phantom) and to the density derived manually from the HU to (electron) density curve (see Section 39.2.1)

A number of these tests could be performed on dummy clinical cases as part of the training programme.

39.3.5.1 Anatomical Data

The suggested tests for the accuracy of anatomical data are summarised in Table 39.6. The general precautions to be taken during patient data acquisition have already been discussed in Section 39.2. The object of these tests is to make sure that anatomical data are transferred and interpreted correctly by the treatment planning system.

During the planning process, these data are subject to further processing, which could be explored with tests such as those listed in Table 39.7.

TABLE 39.7

Examples of Tests to Check the Processing of Anatomical Data

	Item	Scope	Procedure
1.	Delineation of structures	Check that the various options of manual or automated delineation give accurate results	Using either patient or phantom images, practice 2D contouring using the different tools available. Look at the position of the resulting contour overlaid onto the original image. Check known dimensions on phantom images using measurement tools. Analyse the behaviour for bifurcated structures (i.e. the same structure separated into different contours on a single slice)
2.	Image reconstruction	Check if image reconstruction in planes, such as sagittal or coronal, is accurate	Using either patient or phantom images in which sufficient landmarks can be identified in each plane, check that distances are correctly reproduced and that the position of each plane is satisfactorily displayed. Check also the position of reconstructed contours in relationship to the reconstructed images
3.	Expansion of structures (Volume growing)	Check that adding margins around a given structure is a 3D process based on actual values of these margins	Starting from a given simple structure, create other structures with either symmetric or asymmetric margins for different slice thicknesses. Look at the result in different reconstruction planes (e.g. a point-like original structure with symmetric margins should expand to a sphere). The volume obtained by applying four successive 10 mm margins should be the same as that obtained by applying a 40 mm margin in one step. A beam's-eye view is also a useful way of checking 3D margining
4.	Bolus[a]	Check if the bolus coverage and thickness are correct and if they can be associated with individual beams	Use either patient or phantom data and create bolus of a given thickness. Check how it is displayed and how it affects the depth, SSD, and dose distribution of the associated beam(s)
5.	Image registration	Check if images from different modalities can be merged within a single consistent patient model	Use either patient or phantom images with sufficient landmarks identifiable in both modalities to compare their position in the two series after registration. The comparison is easier if the two series are viewed in different orientations (e.g. sagittal and coronal). It can also be useful to have two data sets with a known geometric relationship

[a] Bolus can be considered either as an addition to the patient *anatomy*, or as a beam accessory.

39.3.5.2 Machine Parameters

The treatment planning system should model both the geometric and dosimetric characteristics of the treatment machine. These characteristics include:

- Machine identification, available energies and modalities, scale definitions, accessories, etc.
- Dosimetric coefficients and tables for each beam configuration.

These data are stored in the machine beam data library. Software should be provided to facilitate the construction, verification, and maintenance of these data. This section considers the geometric parameters; the dosimetric parameters are considered in Section 39.3.6.

With the development of record-and-verify systems and networking, it is becoming more important to model the machine geometry exactly (see Section 41.1.3). If even one parameter is not correctly represented in the treatment planning system, it will not be possible to use digital data transfer. In spite of the efforts to reach an agreed-upon international standard (IEC 1996), there are still many departments where machines from different manufacturers, or even different generations of machine from the same manufacturer, have inconsistent scales. Depending on the treatment planning system, some degree of customisation will be possible. However, it can lead to confusion if a patient is treated on two differently-configured machines so that the scale conventions are different for different beams of the same plan*. Some compromises may be required in order to achieve a consistent representation within the planning system. The major areas of difficulty are collimator jaw settings (especially in asymmetric mode), wedge conventions, and table rotations.

Another important issue is the definition of the allowed range for these various data. Ideally, it should be impossible to prepare a plan that cannot be implemented. Time will be wasted if a plan is prepared which cannot be achieved on the machine. A potentially more-serious problem is that a machine facility may be selected which has not been validated, resulting in an incorrect treatment. However, it may not be possible to reproduce the limitations of couch movements exactly as the exact position of the patient on the treatment couch can vary. Each machine parameter should be examined to ensure that it has been correctly set up.

39.3.5.3 Beam Definition and Representation

For an individual plan, beam parameters have to be specified, including the following:

- Machine, mode, and energy
- Treatment technique (SSD, isocentric, arc)
- Beam orientation and position
- Collimator settings, including MLC leaves
- Additional beam-limiting devices or beam-modifying devices

Table 39.8 gives some examples of tests that should help to understand and check the treatment planning system behaviour for beam-related features. Some of these tests are relevant only if the corresponding technique is being used. More extensive testing may be appropriate where special techniques, such as conformal radiotherapy or intensity modulated radiation therapy, are implemented.

Although the primary aim is to investigate non-dosimetric characteristics, it is convenient to use the generic data supplied with the planning system to display the dose distribution

---

* With modern machines, it is usually possible to change existing machines to the new scale conventions retrospectively. This may involve some physical changes, but if a record-and-verify system is in use software changes are very simple. The added convenience and safety will more than repay the effort involved.

TABLE 39.8

Examples of Tests to Check the Processing of Beam Data

	Item	Scope	Procedure
1.	Beam positioning	Test the various means to define the beam position and orientation	Use a patient or phantom model to position the beam using various techniques, with different beam orientations and positions. Check the graphical or numerical position of the entry point (on the skin) and the isocentre depth. Check that movement of the beam in the longitudinal direction is correctly represented, including for different slice thicknesses. Simulate non-coplanar beams and check the graphical and numerical display of table rotation
2.	Collimator settings	Test the conventions used for collimator opening and rotation	For the different techniques and different SSDs, set a rectangular field of known dimensions and check the graphical and numerical display. Understand how the field size is defined (at what distance). Check maximum and minimum field sizes, symmetric and asymmetric modes, use of electron cones, etc. Check also how the display reproduces the changes induced by collimator rotation in different planes
3.	Wedge filters	Test the wedge selection and orientations	Add a wedge to an open beam. Change the wedge orientation or rotate the collimator. Select another machine/energy while the wedge is on. Check changes in display and wedge identification
4.	Conformal fields	Test the beam's-eye view design of field aperture	Prepare customised field shapes either manually or automatically, using the various available means (e.g. film digitising, manual or automatic contouring, editing of an existing shape), and check the shape and size of the resulting aperture. Check how it reacts to modifications of machine, technique, collimator rotation, distances, etc. Check how the information is transferred and used for construction of individualised blocks
5.	Multileaf collimator (MLC)	Test the internal and external consistency of generated MLC files	From given explicit or virtual field shapes (i.e. automatically obtained from BEV projection of structures), generate MLC leaf positions for different leaf fitting options and collimator rotations. Check graphically that leaf positions are as expected. Test the minimum and maximum aperture limits, including rules about interdigitation. Check the relationship to the primary jaws. Check consistency with main jaws or backup jaws position
6.	Beam weight[a]	Test the exact meaning of weight (generally associated with a *beam normalisation point*)	Investigate the meaning of beam weight for single and multiple beams. Establish how weights are related to total dose, fraction dose, and MUs. For percent and fractional weights, establish what is the associated reference value and whether beam characteristics such as wedge, shielding blocks and trays, inhomogeneities, etc. are included in the weight. Check the position of the beam weight (or normalisation) point and its relationship to the overall normalisation. Consider also the case of several weighting points in different slices

[a] This parameter is linked to the beam but it is also, strictly speaking, a dosimetric quantity. It has been however arbitrarily included in this list which deals with understanding how to handle the whole series of beam parameters.

for both photons and electrons, to check if it is qualitatively consistent with what is expected (see also Section 39.3.6.1). Furthermore, it is a good opportunity to understand the exact meaning of the planning system definition of beam weights and the relationship to the resulting dose distribution.

39.3.5.4 Transfer of Beam Data to Simulator or Treatment Machine

If simulator or treatment machines are under the supervision of a *record-and-verify* system (see Section 41.2.1), it is important that the plans can be transferred directly from the treatment planning system to the simulator and/or treatment machine(s). Transfer from the simulator to the treatment planning system may also be possible.

Data transfer usually involves translation between the data structure of the treatment planning system and that of the treatment machine. This may involve user-defined data translation tables. In addition to establishing the physical connection between the two computers, it is essential to verify that all the parameters discussed in Table 39.7 are correctly transferred. A major risk is that a systematic error will be made for all treatments on a given machine, rather than the occasional errors that might otherwise occur. Tests should include dose and MU parameters and MLC settings. It is important to vary these parameters within the clinically significant range to check the various interlock features. Checksum controls embedded in the transferred files have been found useful to confirm data integrity. This issue is discussed further in relation to the DICOM transfer protocol in Chapter 42, Section 2.7 and Section 4.2.

39.3.6 DOSIMETRIC ASPECTS OF TREATMENT PLANNING SYSTEM COMMISSIONING

39.3.6.1 Qualitative Check of Dose Computation Features

Some testing of the dosimetric possibilities and limitations of the treatment planning system software can be carried out using *generic* beam data that should normally be supplied with the system. As mentioned in Section 39.3.5.3, these data are useful to explore the treatment planning system functions related to beam definition and processing. They can also be used to assess the possibilities and limitations of the dose computation algorithm (Rosenwald 1995). For example, one can define a reference situation (e.g. using a flat semi-infinite water-equivalent phantom) and then change the phantom configuration or beam position to explore qualitatively if the expected change occurs in the dose distribution. This approach has been used successfully to assess the treatment planning system ability to correct for field blocking, missing tissues, or the presence of inhomogeneities (Simonian-Sauve and Smart 1998). Used in the framework of a European intercomparison, it has given unexpected results for systems that used dose algorithms believed to be sophisticated enough to handle relevant corrections (Panitsa et al. 1997).

Such investigations are useful to introduce the new user to the operation of the planning system, but can equally be completed by the manufacturer or by a users' group. If relying on such tests carried out by others, it is important to check that the same software version was used.

39.3.6.2 Basic Beam Data and the Construction of the Beam Data Library

The central part of the commissioning process is the establishment of a complete beam data set corresponding to the local treatment machines and the validation of the treatment planning system representation of these data. This is the function of the beam data library and associated utility software.

The dosimetric data which are included in the library are very much dependent on the dose calculation algorithm (see Part F), and have as their basis the minimum set of experimental data specified by the treatment planning system vendor. Even for fundamental models such as Monte Carlo algorithms, the basic experimental data will include depth dose curves (or tissue phantom ratios), off-axis profiles, and output factors. If not used as direct input to the algorithm, these data will be required to support the iterative process whereby dose calculation parameters (including energy spectra) are adjusted, until the computed doses match the experimental data (Starkschall 2000). Specific measurements may be required to enable the influence of the flattening filter, the shape of the collimating device, the transmission of blocks or trays, etc. to be modelled. The relative dose distribution must be linked to the absolute dose through the treatment time or MU calculation, and the appropriate data for this must be measured (see Chapter 23 and Chapter 26).

A computerised 3D water phantom is essential to perform these measurements, and the data can be transferred as electronic files to the treatment planning system (see Chapter 19). However, it is important that the whole set of measurements is self-consistent and representative of the *normal* behaviour of the treatment machine. Some re-normalisation and smoothing of the data may be necessary to achieve this. For example, small asymmetries in the measured data should normally be averaged out.

The process of measuring, entering, processing, and checking the beam data is time-consuming. To assist this process, a number of software tools are required to enable interactive adjustment of the coefficients to yield an acceptable match between the experimental and the computed dose distribution. The most straightforward approach to comparing these distributions consists of superimposing plots of computed and experimental depth-dose curves and profiles with appropriate scales. As discussed in Section 39.3.4.2, the associated tolerance levels are the responsibility of the user. Venselaar et al. (2001) have suggested that more stringent criteria (i.e. $\pm 1\%$ tolerance) be used when comparing the computed depth-dose data or profiles to those measurements that formed the basis of the beam dose-model. They also defined an *upper confidence limit* as

$$\Delta = \mathrm{AD} + 1.5\,\mathrm{SD}$$

where the average deviation, AD, is the average of the modulus of the difference between calculation and measurement for a number of data points and SD is the standard deviation of the differences. This methodology is sensible and the suggested tolerance values are realistic, but they may not always be achievable. The process of creating a beam model is dependent on the algorithm and on the degree to which the model can be controlled by adjustments of the various parameters. Because these parameters are generally interdependent, the final outcome tends always to be a compromise.

Great care must be exercised in carrying out this procedure, since the result will have a significant influence on the quality of all subsequent dose calculations. It must be conducted by well-trained staff who take full responsibility for the results. It must also be fully documented, since it will be used as a reference for future software releases or changes to treatment machine characteristics or accessories.

The number of basic measurements that are required to build the model is system-dependent. Measurements of dose distributions for rectangular fields and for different SSDs, collimation or beam modification devices, etc., may be required. If they are not part of the essential basic data, such measurements are still necessary to confirm the global validity of the dose calculations. These additional measurements are often called *reference data* instead of *basic data*.

39.3.6.3 *Quantitative Checks for Simple Situations Using Reference Data*

It is difficult to give precise recommendations for the extent of additional comparisons between measurement and computation. Some lists of cases have been suggested by various authors (Van Dyk et al. 1993; Fraass et al. 1998; IPEM 1999) without convincing justification for their choice. Their approach was to cover the range of clinical situations as fully as possible but, in order to limit the workload (and so accept a larger uncertainty), a *minimum* subset of situations measurements was specified. However, commissioning a treatment planning system is not a task that can be satisfactorily achieved by following a predefined list of actions. The rationale for the choice of such comparisons should be based on the following: "Is there a risk that my beam data library, as used by the dose calculation algorithm, will cause discrepancies in situations which have not been explored during the model-building process?" Rather than being prescriptive, Table 39.9 seeks as far as possible to answer this question. For this reason, the consideration of issues such as obliquity or inhomogeneity corrections has been excluded and will be addressed later (see Section 39.3.6.4). The items to test and the range of parameters should be chosen according to the existing or intended treatment techniques.

The format and the presentation of the measured data set (1D, 2D, 3D) is very much dependent on the tools and methodology used for comparison. Ideally, one would produce a series of 3D matrices of measured doses and have an automatic comparison with 3D computed dose distributions. Software tools may be limited to a series of 1D comparisons on the basis of depth dose curves and off-axis profiles, but the methodology is likely to evolve towards automated comparisons, using evaluation criteria such as the gamma index (Van Esch et al. 2002). Several automated 2D and 3D comparison tools using film or multiple diode dosemeters are becoming available (Jursinic and Nelms 2003; Renner et al. 2003; Létourneau et al. 2004). These are particularly necessary for IMRT dose distributions.

To assess the discrepancy between calculated and measured doses precisely, the comparison should ideally be in terms of absolute dose for a given number of MU. This approach—which includes, for example, the influence of the output factor or the transmission of accessories—is highly recommended. It can be implemented by calculating the MU required to deliver a given dose to the *weight point* and displaying the dose distribution as absolute dose per MU. However, the comparison of relative dose distributions (properly normalised) is easier to achieve and allows a better understanding of the reasons for discrepancies. The tolerance range could be the one suggested by Venselaar et al. (see Table 39.5). In that case, the additional discrepancy due to the MU calculation should be included in the assessment.

39.3.6.4 Quantitative Checks for Complex Situations

Here, the phrase *complex situations* means those cases of clinical interest which can present significant discrepancies between computation and reality, but are not in principle influenced by the adjustment of the beam data library. Such complex situations include tissue inhomogeneities, loss of lateral scatter, and oblique incidence. It is, of course, of interest to recognise the limitation and inaccuracy due to the algorithm, and several approaches have been suggested:

- One approach consists of using an extensive published reference data set, together with all the data needed to build a beam data library (AAPM 1995; SGSMP 1997; Venselaar and Welleweerd 2001; NCS 2006). The problem is then that the treatment planning system user would have to spend a huge amount of time adjusting the parameters of a beam data library that he will never use clinically. Such an approach is unrealistic for individual users, but could be carried out as a cooperative endeavour by groups of users of a given treatment planning system. Alternatively, the vendor could supply the corresponding beam data libraries, together with the series of test cases and ideally, with proper tools for analysis.

- Instead of using published data, the user could perform a series of measurements to create a similar data set for the same set of test situations as the published data. These data would then be consistent with the local treatment machines. However, such measurements are difficult, time-consuming, and perhaps unrealistic for most centres.

TABLE 39.9

An Example Reference-Beam Data Set

	Item	Scope	Range and Analysis
1.	Open fields (photons)	Check basic dose distribution and MU (or treatment time) calculation over clinical range	Minimum to maximum field sizes, including some rectangular fields. SAD and SSD technique, including extended SSD. Special attention should be paid to small fields Dose distribution in central planes or planes perpendicular to the beam axis. Variation with field size of the MU required to deliver a given dose to a point. Variation of MU with technique and distance
2.	Wedges	Check dose distribution and MU (or treatment time) calculation for wedges (including motorised and virtual wedges)	Dose distribution with wedges for at least one field size (preferably the same range as for open fields). Special attention should be paid to larger field sizes with different widths and lengths Dose distribution in wedged and unwedged direction. MU calculation for different field sizes with and without wedge (inverse of wedge factor). Special attention should be given to the relationship between wedge output and field width and length for virtual wedges
3.	Blocks	Check influence of blocks and tray	For each energy, dose under block for fields smaller and larger than block size, compared to dose without block Influence of tray (mostly MU calculation). Penumbra at the edge of block. Simplified mantle-type geometry could be used, making sure that blocks used for measurements are similar to blocks used clinically
4.	Asymmetric jaws	Check influence of asymmetry on dose distribution and MU (or treatment time) calculation	For a limited number of asymmetric fields, dose distribution in central or perpendicular plane(s) Changes in MU for a given dose on the field axis, as a given square field is scanned from the centre to the periphery (including diagonal)
5.	Multileaf collimator (MLC)	Check influence of MLC presence and shape on dose distribution and MU (or treatment time) calculation	Use some of the values for the open field, but delimited by MLC with 0° and 45° collimator rotation Dose distribution in planes perpendicular to beam axis. Penumbra region. Dose under the leaves, influence of main collimator, or backup jaws. MU calculation as for open field and asymmetric fields
6.	Electron beams	Check dose distribution and MU calculation	Same as open field for photons, with special attention, if cones and inserts are being used, to the position of x-ray jaws Special attention should be paid to small fields

In both cases, if any significant discrepancy is found (provided that the experimental data are fully validated), the cause is a priori related to a combination of beam-library data (which have already been validated in the so-called simple situations) and the dose calculation algorithm (which cannot be changed by the user).

An alternative approach has been proposed by Caneva et al. (2000), in which the ability of the algorithm to deal with inhomogeneities, etc. is tested independently of the beam data. The principle is to quantify the dose perturbation resulting after a modification (mostly of phantom shape or composition) from a reference situation. By measuring the perturbation in a number of beams of different energy, it can be expressed as a function of the beam quality index. If there is a smooth variation, independent of the other machine characteristics, the correction factor for any beam quality may be inferred. It is then straightforward for any user to simulate the same reference and modified situations on their treatment planning system, compute the corresponding perturbation factor, and compare to the perturbation factor expected for the same quality index. This approach has been implemented successfully for photon beams and used for treatment planning system intercomparisons for missing tissue perturbation (Caneva et al. 2000), electronic disequilibrium after an air slab (Caneva et al. 2006), and penumbra enlargement in a low-density medium (Tsiakalos et al. 2004). This approach is illustrated in Figure 39.5.

Another possibility is to use a computed reference data set instead of an experimental one. Such a possibility, which is more and more attractive as Monte Carlo algorithms are developing, is especially useful in cases where measurements turn out to be inaccurate or impractical. However, as a preliminary step, a comparison between such dose calculations and experimental values is always required for the *simple situations* referred to above, where measurements are still mandatory.

Due to the difficulty of exploring these complex situations, in many centres this will not be included in the commissioning process. Instead, the users will generally rely on the QA system

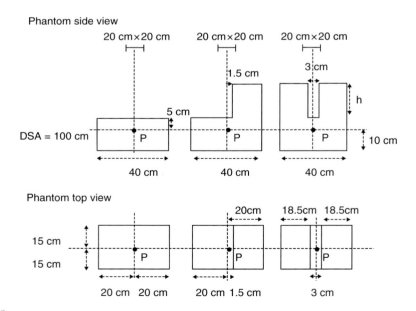

FIGURE 39.5
Experimental configurations designed to investigate the modification of the dose at the point of measurement, P, when there are changes in phantom lateral scatter. A series of measurements is made once for each of the phantoms and for a number of beam energies. These data allow the ratio of each test configuration to the reference configuration to be determined as a function of the beam quality index and these ratios have been published. For a beam quality, it is then possible to predict the value of this ratio and to compare it to the ratio computed with a given treatment planning system. (From Caneva, S. et al., *Med. Phys.*, 27, 1018–1024, 2000. With permission.)

implemented by the treatment planning system manufacturers with the support of their beta test sites. However, it remains the user's responsibility to ensure safe treatment of the patients in their centre. Users should not embark on complex treatments unless they have been able to establish that adequate verification of their version of the planning system algorithm has been properly tested for that treatment configuration. This is particularly true for IMRT treatments (see Section 39.3.9.3 and Section 44.3.5).

39.3.7 CLINICAL VERIFICATION

The last important step before allowing the treatment planning system to be used for clinical plans consists in conducting an overall verification of the whole chain.

If not done previously, this would start with phantom scanning or direct use of patient scans, including transfer to the treatment planning system. After that, a number of typical plans should be created which are as far as possible representative of the current clinical practice. This should be followed by data transfer to the external pieces of equipment (block cutters, milling machines, MLC, record and verify systems, etc.).

Overall verification of the absolute dose is particularly important. This consists, for instance, of preparing a plan for a simple phantom geometry, aiming to deliver a stated dose at a reference point (normally the *weight point*), and calculating the corresponding MU (or treatment time). The dose is then measured in the phantom and compared to the expected dose. Such measurements should be repeated for the various energies and for a significant range of beam parameters (especially with or without blocks, wedge filters, MLCs, etc.). Most planning systems allow a patient plan to be recalculated using a standard phantom, and it is convenient to use this facility for the first patients for whom plans are created on the new treatment planning system.

While performing the clinical verification, special attention should be given to the various displays and documents associated with the treatment plans. One should make sure, for instance, that the identification of all documents is unique and can be related unambiguously to the corresponding plan and the correct patient. The 2D and 3D dose distributions should be compared to ensure that they are consistent, together with the corresponding dose volume histograms (DVH). If there is any doubt, or if clinical decisions are to be based on the DVH, further tests can be conducted to assess the accuracy of DVH calculations, which are likely to be sensitive to such parameters as grid size, number of sampling points, or dose gradient in the investigated structure (Panitsa et al. 1998).

39.3.8 PERIODIC CHECKS OF THE TREATMENT PLANNING SYSTEM

39.3.8.1 Regular Quality Control Checks

After the treatment planning system has been fully commissioned and is in clinical use, the user must remain vigilant. There is not much risk that a change would appear spontaneously. One must, however, recognise the risk of drift of input or output devices such as the plotter or digitiser, which should be checked periodically and recalibrated if necessary.

As far as the computer itself is concerned, it is likely (but not certain) that any file corruption will result in a program crash or other evident error. It is unreasonable to set up a time-consuming systematic quality control programme for an incident which is very unlikely and which could happen at any time. However, an automatic checksum verification can be carried out with little effort if the necessary software is available.

It is, nevertheless, good practice to prepare a number of typical composite plans (*reference plans*), taking advantage as much as possible of the various options and situations currently encountered in clinical reality, and to use them as a reference for further tests. It is then quite easy to *replay* them systematically at given intervals (say, every 6 months) or in case of doubt.

These reference plans are also very useful to trace the treatment planning system history during its evolution.

39.3.8.2 After Changes to the Beam Data Library

From time to time, it is necessary to update the beam data library. The reason can be a major change, such as a new treatment machine installation, but it can be also the modification of an accessory (e.g. tray thickness) or a change in parameters to improve dosimetric accuracy. In any event, one must be very careful, since even minor changes have generated unexpected changes in the resulting dose or MU calculations. Ideally, a full recommissioning should be done, but this is unrealistic.

To avoid major errors, the reference plans referred to above can be repeated and any small differences between the original and the new version reconciled. Good documentation during commissioning is invaluable for this purpose. In any case, all changes to the beam data library and the results of the associated tests should be carefully documented and recorded.

39.3.8.3 After a New Software Release

The problem is somewhat similar after installation of a new software release. In the case of a major release, it is probably necessary to go through many of the steps of the acceptance and commissioning process. Some simplifications could be acceptable; for instance, if the dose calculation algorithm has been announced as unchanged, and if this is confirmed for some of the reference beams, it could be assumed that it will be the same for the other beams.

The number and types of tests to be done is very much a question of personal appreciation of the risks associated with the software evolution. The detailed list of the changes that have been made from the previous version should normally be supplied with each new release. It is very helpful to tentatively identify the critical points that deserve further attention. However, this does not prevent unexpected errors elsewhere, since treatment planning system software is highly complex and contains modules which are very much interdependent.

Again, it is essential to document and record systematically what is done during and after each new software release. This could also be extended to hardware problems, replacements, or upgrading. Finally, one must be careful when starting the clinical use of a new release where it is highly recommended that qualified physicists follow very closely the plans prepared with the new version. They should also be responsible for ensuring that the local procedures are updated to conform to any changes in the software.

39.3.9 Individual Plan Checks

39.3.9.1 Rationale

In spite of the QA programme set up by the manufacturer for the development of the treatment planning system, and in spite of the local commissioning efforts, there is a significant risk that some individual patient plans are erroneous. The two main causes of error are software error and data error.

A software error (bug) is likely to remain undetected until a specific sequence of operations leads to an untested part of the system. This will mostly happen when new people are using the software or when special clinical cases require the use of unexplored procedures or functions. In the best case, the system will crash and the error will be evident. In other cases, the plan will be completed but erroneous.

A data error will happen if an incorrect parameter is entered into the system. There are many causes for such errors, from a simple mistyping to an inadequate imaging procedure. The consequences of the errors can be very serious, and procedures should be implemented to try to detect them. While a totally error-free system is unachievable, the risk may be significantly decreased by implementing systematic review of individual plans and by applying an independent method for the most critical data.

39.3.9.2 Methodology

The plan review should be carried out by experienced specialists, who have a deep understanding of the influence of the various parameters and of the expected dose distributions for clinical cases. The review can be conducted on the printed documents or using the computer display, provided that an electronic signature mechanism is available. For complex plans, reviewing the plan on the computer may be essential.

The reviewer should make sure that the plan version being checked is the final one which has been approved or (preferably) is about to be submitted to the responsible physician. The patient and plan identification should also be checked. The review mainly consists of scrutinising all the available data, checking their internal consistency, and comparing all the associated documentation. Special attention should be given to the beam parameters (including the presence of accessories), to the prescription values (dose, weights and weight point positions, normalisation, etc.), and to the graphical displays of anatomy, beams, and isodoses.

The MU (treatment time) calculations deserve special consideration, since they will directly affect the final dose given to the patient. For this reason, it is recommended that an independent dose calculation is performed. This calculation may be less accurate than the planning system calculation, but with experience, the magnitude of such differences is predictable. If an independent calculation is not possible, reference to a table of expected values may suffice.

Altogether, the QA of the dose planning process is very much a question of state of mind. In addition to well defined procedures, one should always be questioning the validity of what one is doing. Computer output in particular should never be regarded as immune from error. Special attention should be paid to any situation that is not standard current practice.

39.3.9.3 Clinical Verification of IMRT Treatments

The QA of IMRT is discussed in Section 44.3.5, but it is important to contrast the QA of the complex situation described there with the considerations described in this chapter. The accuracy of IMRT treatments depends not only on the treatment planning system, but also on the treatment delivery (Kung and Chen 2000; Budgell et al. 2001; Dong et al. 2003; Budgell et al. 2005a). When calculating an IMRT dose distribution, a number of factors, such as small field dosimetry and MLC leaf modelling, become much more important than in standard therapy. In addition, there are often special parameters used in IMRT dose calculation that are not considered when standard plans are calculated. For this reason, the fact that the beam data library produces satisfactory results in the conditions described in Section 39.3.6.3 is no guarantee of satisfactory dose calculations for IMRT beams. It is, therefore, recommended, at least until consistent accurate results are found in the centre, to check individual plans by comparison between the planning system computation and measurements. This can be performed for each modulated beam, applying the beam setting used for the patient (but with vertical incidence) to both compute the dose distribution at a given depth of a flat phantom and measure it (Van Esch et al. 2002). Another solution is to compute and measure the dose for all beams together, keeping the beam directions but using a simple geometrical phantom. Such plans where the patient is replaced by a phantom are sometimes named *hybrid plans*. It must be recognised that an individual beam which only partially irradiates the detector may give significantly different doses from those calculated (Woo and Nico 2005). The combination of beams will, in general, smooth the differences. However, even for such a combination, the phantom should approximate the shape of the patient, because if the

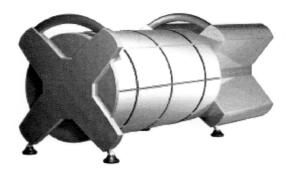

FIGURE 39.6
Delta4 phantom containing an array of diodes along the coronal and sagittal axes, used to measure 3D dose distributions in real time.

dimensions are very different, the relative contributions of the different beams may be so distorted that the dose is no longer uniform. It is a common experience that the measured dose differs for IMRT beams more than for standard beams (Low et al. 1998; Kung and Chen 2000; Mohan et al. 2000; Arnfield et al. 2001; Cadman et al. 2002) and a systematically low result by 1–2% has been reported (MacKenzie et al. 2002; Francescon et al. 2003; Budgell et al. 2005a).

There are two aspects to the verification of IMRT that are equally important: the absolute dose and the dose distribution. For the dose distribution, the position of the edge of the treated volume is particularly important, and film dosimetry is a convenient way of approaching this (Esthappen et al. 2002; Ju et al. 2002). Here, the gamma approach to simultaneous assessment of dosimetric and geometric accuracy is very useful (see Section 39.3.4.3). Software is available to compare the dose grid calculated by the planning computer with either film measurements or arrays of detectors (usually diodes). A new device (Delta4, manufactured by ScandiDos) has been developed under the EU Invorad project; it consists of an orthogonal array of 1069 diode detectors in a cylindrical phantom, enabling real-time measurements in three dimensions (see Figure 39.6). With this device, there is freedom to examine individual beams and the composite dose and to look at both absolute dose and the dose distribution all at the same time.

As experience grows with IMRT delivery, users will become more willing to use standard IMRT plans measured at regular intervals rather than individual patient dose measurements. Independent dose calculations are also now available to reduce the possibility of an unpredictable error in the treatment planning system calculation. However, in the initial phases of the introduction of IMRT, phantom measurements for a number of individual patient plans are essential. It is also essential to complement the quality control of the treatment planning system with detailed quality control checks of the linear accelerator. Leaf position accuracy (especially in relation to the other leaf of the pair) is particularly important, and for dynamic delivery, it should be tested in dynamic mode (Chauvet et al. 2005). For step and shoot delivery, the accuracy of dose for small numbers of MUs should also be checked. Niemierko (2004) and Budgell et al. (2005a) discuss the issues and criteria for transition from individual patient measurements to a system based on an independent calculation together with special machine quality control.

CHAPTER 40

QUALITY CONTROL OF TREATMENT DELIVERY

Philip Evans * *and Ginette Marinello* †

CONTENTS

---

* Section 40.1.
† Section 40.2.

40.1 ELECTRONIC PORTAL IMAGING DEVICES

40.1.1 INTRODUCTION

The technology of electronic portal imaging devices (EPIDs) was discussed in Chapter 14. The development of these devices has opened up new possibilities for quality assurance of treatment technique and for verification of treatment accuracy. The clinical usefulness of EPIDs depends on the availability of good quality images for analysis and the availability of methods of analysing the images obtained at the computer software level. The analysis should yield a quantitative verification of treatment quality. This verification may be in terms of patient positioning or dosimetry.

Here we consider the use of EPIDs for treatment verification. First, the use of image processing to enhance the information content in the image such that it may be used for treatment accuracy measurement is discussed. Then, the tools needed for investigations of positional accuracy are considered. Such tools allow the comparison of an electronic portal image taken on the treatment machine with a reference image. From this comparison, the positional accuracy of the treatment may be evaluated. Dosimetry with EPIDs is then discussed. Such applications often yield dose measurements or measures of patient thickness. Finally, other uses of EPIDs are considered such as in the verification of intensity-modulated treatments.

40.1.2 IMAGE ENHANCEMENT

Electronic portal images are generally of low contrast. This is mainly because the primary x-ray interaction process leading to image formation at radiotherapy treatment energies is Compton scattering. The cross-section for Compton scattering is proportional to the number of electrons available for scattering. For the low-Z atoms constituting biological material, the ratio of the atomic number to the mass number (Z/A) is constant. Therefore, the scattering probability is proportional to the density of the material. This situation should be contrasted with that for lower, diagnostic energies, where the photoelectric effect dominates. The photoelectric effect has a very high Z dependence, leading to large intrinsic contrast (for a given dose and spatial resolution), particularly between tissues such as soft tissue and bone.

Because of this low intrinsic contrast, it is often desirable to apply image enhancement techniques to electronic portal images to extract or enhance important features in the images, which may then be used for image matching, as discussed in Section 40.1.3. Image enhancement generally falls into three classes of operation, which will be described in the following sections.

40.1.2.1 Point Operations

Point operations are carried out on the histogram of image intensities. The most common example is windowing or contrast stretching. The relationship between intensity in the measured detector signal and displayed intensity is modified to accentuate an intensity range corresponding to features of interest. The simplest form of this is histogram equalisation, in which the goal is to obtain a uniform histogram for the output image, i.e. the number of pixels is the same for each displayed intensity range.

There are various methods that have been developed, based on this technique, which involve analysis of the histogram of the image on a region-by-region basis. These are known as local histogram equalisations. One of the most common of these is the adaptive histogram equalisation (AHE) (Pizer et al. 1984). In this technique the image is divided into a set of square regions of typically 16 pixels × 16 pixels. The histogram of each region is determined. The transformation to equalise the histogram in each region is then computed. For each point in the image, bilinear interpolation of the transformations for the four neighbouring regions is used to determine the output intensity. This is illustrated in Figure 40.1. There are variations on this technique, which are often used in electronic portal image processing. These include contrast limited adaptive histogram equalisation (CLAHE), in which intensity levels above a predefined clipping level are reduced. The purpose of this is to prevent the overenhancement of noise (Leszczynski et al. 1992).

40.1.2.2 Spatial Operations

Spatial operations are carried out on points or groups of points. Usually, a spatial filter is convolved with the image. The filter will usually be high-pass, low-pass, or band-pass.

Filters usually have kernel sizes of 3×3 or 5×5. High-pass filters are generally used to enhance edges, such as those in bony anatomy. Examples are the Laplacian and Sobel filters. Low-pass filters are generally used to reduce noise in the image. Examples are filters that

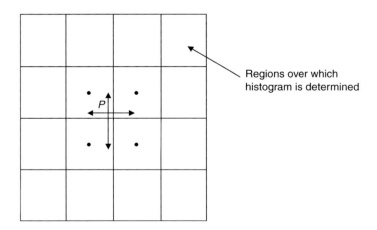

FIGURE 40.1
Illustration of image segmentation and interpolation in adaptive histogram equalisation. The image is segmented into 16 regions in this example. The histogram mapping for the point P is determined using bilinear interpolation of the mappings of the four closest regions, illustrated with the arrows. The four large dots indicate the centres of the closest regions.

average the values of adjacent pixels, median window filters (in which a pixel is replaced with the median of the surrounding region) and Gaussian filters.

40.1.2.3 Transform Operations

Transform operations involve taking a transform of an image and applying a filter to the transformed image. The inverse of the transform is then applied to yield a filtered image. The transform most commonly used is the Fourier transform. The frequency space filter used is generally a high-pass or low-pass filter.

Figure 40.2 shows an example of a mantle image with various filters applied to it.

40.1.3 IMAGE COMPARISON TECHNIQUES

Measurement of treatment setups using EPIDs generally consists of several distinct steps or components:

1. A reference image must be established to define the *ideal* setup.
2. The reference image and the treatment image are analysed to find the field edge and important anatomical detail.
3. The images are compared and the difference in positioning of the anatomy relative to the field edge is measured.

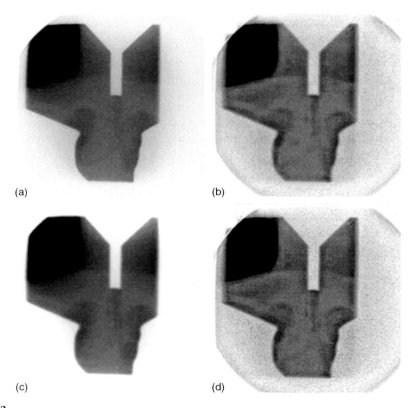

FIGURE 40.2
Illustration of filtering applied to electronic portal images: (a) original image; (b) application of adaptive histogram equalisation; (c) application of a 3×3 smoothing filter; (d) application of Sobel edge-enhancement filter in *x* and *y*. Note the similarity between (b) and (d): the Sobel filter enhances contrast by enhancing edges and the histogram equalisation boosts contrast between regions and hence edges.

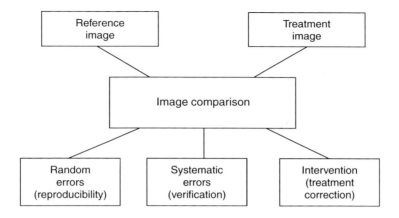

FIGURE 40.3
Illustration of the use of image comparison. A treatment image is compared with a reference image. The result is used to measure treatment reproducibility, for verification, or to correct treatment setup.

The results of such comparisons may be used in three ways:

1. *Treatment reproducibility can be measured* by comparing a series of images acquired at various times during the treatment course. The measured variations in patient positioning are often referred to as random errors.

2. *Treatment accuracy can be verified* by comparing a reference image with a treatment image. The measured differences between such images are often referred to as systematic errors.

3. *Treatment errors can be corrected* using intervention protocols. There are two ways intervention may be implemented. Each treatment may be corrected on-line. Alternatively, the measurement of the errors in one treatment may be used to modify future treatments.

The image comparison process is illustrated in Figure 40.3.

40.1.3.1 Reference Images

The reference image will come from one of three sources:

1. *A treatment image* could be used either from a digital detector or from a portal film. Film should ideally be digitised, to enable simple and quick image analysis with appropriate software.

2. *A simulator image* may be obtained from film (again with the possibility of digitisation) or from an x-ray image intensifier. Image intensifiers generally suffer from radial distortion, given by the expression:

$$r' = r + Dr^3 \qquad (40.1)$$

where r' is the radial distance of a point from the centre of the image intensifier field with distortion and r is the distance it would be if no distortion were present (i.e. its correct distance). D is a parameter describing the distortion. This distortion may be removed using a calibration process involving images of a test phantom containing a set of regularly spaced objects of known size and placement (Rudin et al. 1991). This distortion may well be dependent on the gantry angle used to obtain the simulator image[*]. Distortion is not a problem with solid-state imaging panels.

---

[*] The reason for the dependence of the distortion with respect to the gantry angle is the inevitable presence of the earth's magnetic field. However this may be compensated using corrective coils.

3. *A Digitally Reconstructed Radiograph (DRR)* can be computed from the CT data. Although simulation images are often taken as the reference for patient positioning studies, they clearly do not represent the true, planned position, since in effect a simulation setup is a special case of a treatment setup and hence prone to similar random errors as the treatment. The planned position is best represented by a DRR; a virtual simulation film generated from the treatment plan and the planning CT data (see Section 32.4.5 for more detail). This also has the added advantage that it is digital in nature. However, the spatial resolution of a DRR is limited by the slice separation in the CT data set. Hence there is the need for closely-spaced slices to produce a clinically useful DRR.

40.1.3.2 Field Edge Determination

The field edge is usually defined as the position of the 50% isodose contour. An estimate of this may be found simply by finding the 50% intensity level in the image (Evans et al. 1992). However, this will be inaccurate if the patient thickness varies across the field.

Other techniques developed include:

1. *A double exposure technique*. An image is taken of the treatment field. This is then divided by an *anatomical* image of a larger area of the patient. The net result is an image of the field edge (Meertens et al. 1990b).
2. *Histogram analysis*. The field edge may be defined as the region of maximum gradient. This corresponds to a minimum in the histogram. The histogram of the image contains several peaks and minima because of the various regions, i.e. the inside of the field, blocked regions and the area completely outside the field. A minimum corresponding to a first estimate of the field edge is found. The points in this first estimate are shifted to where the local gradient is maximal (Bijhold et al. 1991a).

40.1.3.3 Field Edge Matching

Various methods have been developed for automatic matching of field edges. Methods developed include the calculation of low-order moments in treatment and reference images. The images are moved and scaled until the moments match as closely as possible (Bijhold et al. 1992).

Another method involves the technique known as *chamfer matching* (Gilhuijs and van Herk 1993). In chamfer matching, the field edge in one image is turned into a wire-frame outline. In the other image, it is turned into a v-shaped valley, where the dips in the valley follow the contour of the field edge and the sides of the valley decrease in depth as distance from the field edge increases. The field edges are matched by fitting the wire-frame outline into the v-shaped valley. This is illustrated in Figure 40.4. The v-shaped valley is often created from the reference image, as this is available before treatment and thus may readily be processed prior to treatment. The field edge in the treatment image is generated using one of the automatic methods described above.

The techniques described above produce numerical results describing the goodness of fit and thus may be used to detect gross field definition errors, such as the misplacement of a block. For such an error, the moments will not match very well and the residual root-mean-square distance between points will be large.

40.1.3.4 Segmentation and Matching of Anatomy

Electronic portal images are generally of low contrast. Thus the process of segmenting them is far more difficult than that of finding the field edge. Both manual, operator-dependent and automatic methods have been developed. Some image matching techniques involve no

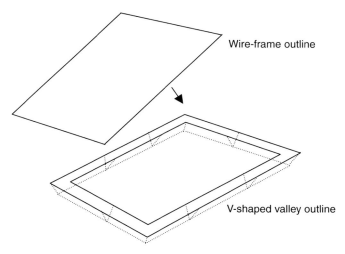

Wire-frame outline

V-shaped valley outline

FIGURE 40.4
Schematic illustration of chamfer matching. The best fit of the wire-frame into the v-shaped valley is determined.

formal segmentation step. Hence, in the following discussion, we shall consider segmentation and matching of anatomy together.

EPID images are, of course, two-dimensional, while the patient is three-dimensional. The differences in patient positioning between one imaging session and another (be they different treatments or a simulation and a treatment) are describable by six degrees of freedom, three translations and three rotations. Three of these, the two translations in the plane of the imager and rotation in the plane of the imager, are easily measured from portal images taken of each field. Translations in the plane perpendicular to the imager will show up as magnification changes. However, the sensitivity of the magnification to out-of-plane translations is very small. A large translation is required before the magnification changes significantly. The other two degrees of freedom, rotations out of the plane of the imager, result in spatial distortion of the image (Bijhold et al. 1991b). In general, it may be difficult to distinguish between such distortion and the effects of patient motion during treatment.

The first image matching method involves the use of fiducial reference markers, which are drawn on the prominent features on one image and the corresponding points in the other (Meertens et al. 1990a; Halverston et al. 1991; Evans et al. 1992; Bijhold 1993; McParland 1993). A method such as singular value decomposition (SVD) (Balter et al. 1992), is then used to find the transformation to achieve the best match between the points. This is usually expressed in terms of translation and rotation in the plane of the image and magnification out of the plane of the image.

Template overlays are sometimes used (Bijhold et al. 1991b; Evans et al. 1992). A wire-frame diagram is drawn around the anatomy in the reference image and either matched on the treatment image or the corresponding template from the treatment image.

The third set of methods falls in the category of visual registration. A two-image movie technique has been used. The treatment image is interactively translated and rotated relative to the reference image, until the best fit is obtained. A difference image technique has been used to compare two treatment images (Evans et al. 1992). Coloured overlays are used to indicate regions of significant intensity difference. One image is rotated and translated until the coloured region inside the field is minimised. Colour overlays have been used in which, for example, one image is displayed in green and the other is displayed in purple. If two maximum grey level regions in both images become superimposed, they appear white (Weinhous 1990; Bijhold et al. 1991b; Graham et al. 1991).

Various automatic image comparison methods have been developed. Gilhuijs and van Herk (1993) described a fully automatic method of segmentation and matching. Firstly, the

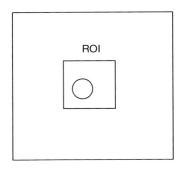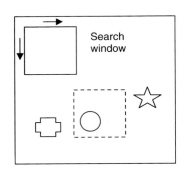

FIGURE 40.5
Illustration of correlation technique. On the left is a reference image with a region of interest (ROI) drawn around a feature (a circle in this schematic example). On the right is a treatment image. A search window is scanned across this image and the correlation with the ROI in the test image calculated. The best fit occurs when the correlation is maximal (dotted window).

positioning of anatomical edges is found in the reference image. The anatomical edges in the treatment image are then automatically extracted using a morphological transform that looks for features of a certain size and shape. Then chamfer matching is used to register the images as described above. They found their automatic method to yield a success rate in excess of 90% (Gilhuijs et al. 1995).

Jones and Boyer (1991) described a correlation technique allowing for automatic image registration (Figure 40.5). They found their technique to be most effective for most values of translation and small rotations in and out of plane. Other groups have developed similar techniques. In one such technique, the operator selects pairs of features in the reference and treatment image. A correlation technique is then used to register these pairs of features. The centres of the features are then taken as match points for use in the fiducial reference marker technique described above (Moseley and Munro 1994).

40.1.3.5 Measurement of Setup Errors

Once the field edge has been found and the anatomical detail delineated, one of two approaches may be used to calculate the setup error. The field edges may be aligned first and the residual displacement in anatomy equated to the setup inaccuracy; or the anatomy may be aligned first and the positioning of the field edges analysed.

If the first method is used, the quality of fit in the field edge matching procedure must be good; otherwise, the subsequent matching of anatomy will be prone to error. The transformation to match the field edges is also applied to the anatomy. Any residual difference in the positioning of anatomy between the two images is interpreted as a setup error.

If the anatomy is matched first, then the transformation parameters are used to register the reference image and the portal image. Any residual difference in field edge positioning is a setup error. It is important that the anatomy matching stage be accurate. Often the delineation of anatomical landmarks is difficult. Furthermore, image distortion and out-of-plane rotations will lead to inaccuracies.

40.1.3.6 Accuracy of Analysis Techniques

The accuracy of setup error measurement is limited by the accuracy of the techniques presented in Section 40.1.3.4. The determination and matching of the field edge should be highly accurate and reproducible, however, many of the anatomy matching methods require user interaction and thus are subject to inter-user variability.

Meertens et al. (1990a) studied the dependence of the technique using fiducial reference markers on the location of the markers. They found that the reproducibility of the results depended strongly on how far apart the markers were located. If the markers were too close together, then the small discrepancies inevitable in the identification of the markers by different users result in large uncertainties in the rotation required to match them.

Evans et al. (1992) tested several algorithms and found reproducibility of less than 1 mm and 1° for the fiducial reference marker, template overlay and visual registration techniques in a phantom study. For clinical images, only the visual registration technique remained reproducible to within 1 mm and 1°.

Michalski et al. (1993) tested the fiducial reference marker and template overlay technique. They found both methods to be reproducible to within 1 mm and 1° for phantom studies. For clinical images, they found the single user variation to be as small, but the multiple user variation was up to 2 mm and 3°.

40.1.3.7 *Visual Inspection of Images*

The methods described in Section 40.1.3.4 are all designed to measure the transformation to match treatment and reference images. However, much information may be obtained by visual inspection of images. For instance, a gross error due to the misplacement of a block will be shown. A set of images may be displayed in a movie loop. This may provide quantitative information on day-to-variations in setup and variations during the course of a single irradiation, due to breathing, etc. (Reinstein et al. 1992).

40.1.3.8 *Interpretation of Position Measurements*

Once a set of position measurements has been made for a set of patients, these may be analysed to extract information on the sizes of the random and systematic errors. Such measurements are the essential first stage in the use of EPIDs. Figure 40.6 illustrates how the accumulated measurements might look for a set of patients when measurements are made in the superior–inferior and left–right directions. Other planes will show similar behaviour.

This graph shows measured differences between a set of treatment images and the reference image for each patient. The data for each patient will fall within an ellipse, whose size indicates the random, day-to-day setup error. The eccentricity of the ellipse indicates the difference in accuracy of setup in superior–inferior and left–right directions. The distance of

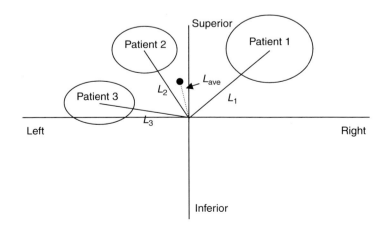

FIGURE 40.6
2D representation of a set of measurements of random and systematic errors for a hypothetical set of three patients. Here the systematic errors (L_1, L_2 and L_3) are larger than the random errors (shown by the ellipses).

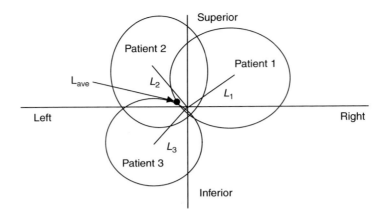

FIGURE 40.7
2D representation of a set of measurements of random and systematic errors for a hypothetical set of three patients. Here the random errors (ellipses) are larger than the systematic errors (individual L_1, L_2, L_3 and average, L_{ave}).

the ellipse centroid from the origin (L_1, etc.) is the difference between the mean treatment position and the position in the reference image. This is often referred to as the systematic error. This may be different for each patient (as in the example), indicating a random component in the error between simulation and treatment. The average of the L_1, L_2 measurements over a large group of patients (L_{ave}) is an estimate of the systematic difference between the simulated and treated position and should be close to zero.

In the example shown in Figure 40.6, the systematic errors are larger than the random errors, but the case may be the opposite (e.g. Figure 40.7). Whichever is the case, it is clearly most efficacious to try to minimise the larger source of error first and foremost.

The purpose of any method of improving treatment accuracy is to shrink the ellipses in Figure 40.6 and Figure 40.7 and to get them as close as possible to the origin.

40.1.3.9 Intervention in Treatment Delivery

Once the setup error is known, the treatment may be adjusted in order to correct the error. This may be done either on-line (during the irradiation) or retrospectively (at the next treatment fraction).

In the case of portal film, only retrospective correction is really feasible, owing to the time taken to develop and analyse the film. If the analysis of the film shows a significant setup error, then this is corrected at subsequent treatment fractions. To determine whether a correction can reasonably be made, it is necessary to have good information on the random and systematic errors associated with a given treatment technique. This is because retrospective correction should reduce systematic errors, but it can have no effect on day-to-day, random variations. Hence, if random errors are relatively large, this technique is likely to be of limited efficacy.

Various decision rules have been developed for retrospective correction. For a given fraction, the decision to adjust the treatment setup is based on the results of the random and systematic error values determined from the previous fractions. Often, these decision rules start with a relatively large threshold for correcting setup, which is gradually reduced as the treatment progresses, i.e. as better statistics on the random and systematic errors for the individual patient are acquired. Examples of decision rules may be found in the work of Bel et al. (1993) and van Herk (2004).

For on-line intervention, an image is taken at the start of the irradiation, using a few monitor units, and the beam is turned off. If the setup error is small, the irradiation is continued. If it is large, the setup is adjusted before the treatment recommences. Another image may be taken at the end of the treatment fraction to verify the correction. This technique reduces both types of error, but is time consuming and hence has significant logistic implications. Several groups have investigated the feasibility of on-line intervention (Ezz et al. 1991; De Neve et al. 1992; Gildersleve et al. 1994).

An alternative approach to intervention in treatment delivery is to monitor the random errors during the first few fractions and to modify the PTV based on the size of the random variations. This strategy has been called *adaptive radiation therapy* by its proponents (Yan et al. 1997; Martinez et al. 2001).

40.1.4 DOSIMETRIC APPLICATIONS

Portal films (Van Dam et al. 1992; Fiorino et al. 1993; Weltens et al. 1994) as well as EPIDs (Boellard et al. 1997a; Van Esch et al. 2004) have been used for in vivo dosimetry. The purpose of this application is to use an EPID to measure some dosimetric aspect of a treatment. To do this the intensity at each point in the image must be calibrated in a quantitative sense. In other words, the intensity must relate to dose in some way. Therefore,

1. The dose response of each pixel in the detector must be known.
2. The effects of scattered radiation in the detector must be understood.
3. The temporal stability of the detector must be known.

Knowledge of the temporal stability is needed in order that the detector may be calibrated frequently enough.

The information produced by dosimetric calibration may be used to verify the dose distribution or to design compensators to modify the dose distribution. Dose verification may be done by measuring the dose within a certain plane, or by reconstructing the dose distribution within the patient. Figure 40.8 illustrates the process.

40.1.4.1 *Dosimetric Calibration*

The relationship between intensity and dose may depend upon:

- Gantry angle
- Field size
- Air gap between patient exit surface and detector
- Other parameters

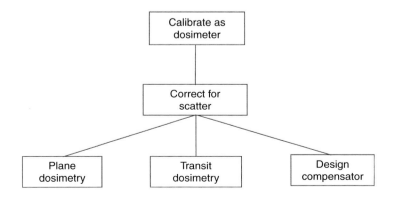

FIGURE 40.8
The steps required to extract dosimetric information from an EPID.

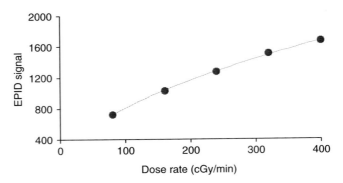

FIGURE 40.9
Response of the Liquid Ionisation Chamber Varian Portal Vision system as a function of the accelerator dose rate. The dots are measured points. The solid line is a fit using Equation 40.2.

Commercially-available systems fall into three groups detailed below.

Camera-Based Systems. These are generally linear in response up to a certain dose at which they start to saturate (Kirby and Williams 1993).

Liquid Ionisation Chambers. These usually have a dose response close to the square root of dose:

$$I = a\dot{D}^{1/2} + b\dot{D} \qquad (40.2)$$

where I is pixel intensity, $\dot{D}$ is the dose rate, and a and b are parameters. For dose rates commonly used in radiotherapy, a is large (over 90% of the contribution) and b is small. Earlier versions of these detectors are prone to variations with gantry angle, which can be corrected (Boellaard et al. 1996). Figure 40.9 illustrates a dosimetric calibration performed for the Varian Portal Vision system by varying the treatment machine dose rate. As can be seen, a square-root behaviour is a good representation of the detector's response.

Flat Panel Imagers. These are usually amorphous silicon arrays. They are generally linear with dose, but the acquisition parameters should be selected to avoid saturation (Berger et al. 2006). There is evidence that they show some *ghosting* effect, which may be a problem when they are applied to dosimetry. This ghost is a long-term image persistence that may last up to 30 min or longer (Siewerdsen and Jaffray 1999) although Van Esch et al. (2004) found that the effect of ghosting was not clinically relevant.

40.1.4.2 Scatter Correction

Scatter correction is needed to remove the effects of scatter from the patient (Swindell and Evans 1996) and, in the case of camera-based systems, to remove the effects of light scatter within the detector that lead to optical cross-talk between pixels in the image (Heijmen et al. 1995). Such corrections are needed to extract primary fluence for various dosimetric techniques and remove the influence of the air gap on the measured signal.

Scattered radiation from the patient adds a component to the EPID signal that will vary with patient thickness, field size and the air gap between the patient and the detector. Two approaches have been made to the determination of the effects of this scattered radiation. The first method, developed by Boellaard et al. (1997a) involves the measurement of the detector signal for various air gaps. For very large air gaps, the scatter signal is very small and hence, the detector intensity is mainly primary radiation. The relationship between primary radiation intensity and air gap is simply given by the inverse-square law. The scatter signal for smaller air gaps is determined by taking the difference between the measured signal and that expected by applying the inverse-square law correction to data measured with a large air gap.

These measurements are used to calculate a kernel, known as an exit dose spread function, EDSF. The second approach is to use Monte Carlo simulation to estimate the scatter signal (Swindell and Evans 1996). Studies of uniform, homogeneous objects have shown that the scatter signal is, to first order, proportional to the volume of scatterer in the field, with correction terms that are functions of the thickness of the object and the field area. Applying the same model to the calculation of kernels to describe scatter, Hansen et al. (1997) found that as the thickness of the scatterer is increased, the scatter contribution increases in magnitude up to a maximum thickness of typically 200 mm, beyond which the magnitude decreases. This is because as the thickness increases, the probability of scattering occurring increases. However, for large thickness, the probability of the scattered photons being absorbed becomes significant.

The light scattering inside camera-based systems has been extensively studied by Heijmen et al. (1995). They used images of light fields of various size to show there is optical cross-talk within the detector housing and none, or very little, in the CCD camera. From these measurements, they deduced a kernel to describe the optical scatter, which can be used for any field size and shape. Munro et al. (1998) modelled the optical cross-talk as a result of multiple reflection between mirror and phosphor screen. They showed that the distribution of the component of the image signal from light scatter might be very non-uniformly distributed, with a peak close to the junction between the mirror and the screen. This is particularly true for large field sizes.

40.1.4.3 Plane Dosimetry

As its name suggests, this involves using the EPID to deduce the dose in a certain plane. Most commonly this is the patient exit surface, the patient midplane, or the plane of the imaging system (what dose would be measured by an ion chamber with full build-up if it were positioned where the EPID is located) (Boellaard et al. 1997b).

This generally involves calibrating the detector for a large air gap and relating the signal measured in this plane to the dose in the desired plan. For instance, in the case of exit dosimetry, it is necessary to correct for the greater amount of patient scatter at the exit plane compared with that in the image plane (Boellaard et al. 1997a, 1998).

If dosimetry in the plane of the imager is performed, the EPID is usually calibrated in comparison with an ionisation chamber, which is the gold-standard for dosimetric measurement. The measured dose images are often referred to as portal dose images (PDIs). Kroonwijk et al. (1998) generated PDIs for a set of prostate cancer patients and compared their results with predicted PDIs from their treatment planning system. They found average agreement between the planning system and measurement of ~0.5%, with 2% to 3% standard deviation, but off-axis differences could be much larger (up to 15%) perhaps due to organ motion.

Huyskens et al. (1994) used semiconductor diodes to measure the entrance dose and film to measure the exit dose. They interpolated between entrance and exit dose to determine the midplane dose. In phantoms, they found agreement with ionisation chamber measurements to within a few percent.

40.1.4.4 Transit Dosimetry

This is a method of estimating the dose distribution within the whole patient using information from EPIDs. It involves a combination of measurement and dose planning. It requires CT information at time of treatment to enable the beam attenuation within the patient to be calculated. Generally this is not available so planning CT data are used. One transit dosimetry algorithm involves the following steps (Hansen et al. 1996):

1. Calibrate portal imager as dosimeter.
2. Subtract the scatter signal to yield a primary fluence map.

3. Back-project this information through the planning CT data to yield a primary fluence distribution.

4. Convolve this primary fluence distribution with dose deposition kernels to generate a dose distribution.

The dose distribution produced may be compared with the original planned distribution.

McNutt et al. (1996) generated PDIs from their planning system. They compared these with measured PDIs from their EPID. An iterative process was then used to adjust the primary fluence in their planning system model until it agreed with measurement. The suitability of amorphous silicon imagers for transit dosimetry is considered by Louwe et al. (2004).

40.1.4.5 Compensator Design

This requires the EPID to be calibrated for radiological thickness. Once a radiological thickness map is available, a planning protocol is used to determine what compensation is needed; e.g. compensation may be applied to equalise doses in the midplane.

EPIDs have been used to design compensators for lung (Yin et al. 1994) and for breast (Evans et al. 1995). One method of calibrating EPIDs for thickness is shown in Figure 40.10.

A set of uniform calibration blocks of known radiological thickness is imaged and, for each pixel in the detector, the relationship between pixel intensity and thickness is determined. This is done for a large field that covers the whole active region of the detector (A_{ref} in the left side of Figure 40.10). The relationship between intensity and thickness is given by

$$I = I_0 \exp\left(\alpha t + \beta t^2\right) \qquad (40.3)$$

where I is the measured intensity, I_0 is the intensity for no object in the field, t is the thickness and α and β are fitted parameters.

When a patient is imaged, (right side of Figure 40.10), the field size will, in general, be different. The intensity of the patient image (I_{meas}) is firstly corrected for the difference in field size, using a field size factor, F (equivalent to an output factor). This corrected intensity is then fed into Equation 40.3 to yield a first estimate of thickness, t. A correction for scatter within the patient is then made to yield an estimate of the intensity value (I_{ref}) if the patient were imaged with field size A_{ref}.

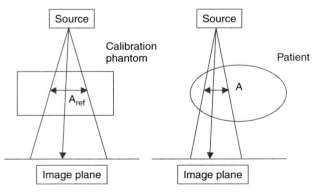

FIGURE 40.10
Method of calibrating an EPID for patient thickness.

$$I_{\text{ref}} = I_{\text{meas}} F\left(\frac{1 + \text{SPR}\left(A_{\text{ref}}, t\right)}{1 + \text{SPR}(A, t)}\right) \qquad (40.4)$$

The value for I_{ref} is then fed into Equation 40.3 to yield an improved estimate of the thickness, t.

40.1.5 OTHER APPLICATIONS

EPIDs have other applications than the measurement of treatment accuracy and dosimetry discussed above. These applications include regular QA of treatment machines and the verification of complex treatment techniques, such as intensity-modulated radiation therapy (IMRT).

Kirby and Williams (1995) showed how EPIDs may be used to verify multiple static field treatments using a multileaf collimator (MLC). Partridge et al. (1998) have developed a method of using an EPID, triggered by the linear accelerator pulses, to verify dynamic MLC treatments, in which the MLC leaves scan with the radiation on. A set of frames is obtained, from which the leaf positions as a function of dose delivered are extracted and compared with the prescribed parameters. The use of flat panel imagers to measure the positioning of gold grains implanted in the prostate has been shown to yield good results for localisation (Nederveen et al. 2001).

The use of portal imaging technology to obtain pretreatment megavoltage CT (MVCT) scans was first investigated in the 1980s (Swindell et al. 1983; Brahme et al. 1987). More recently cone-beam MVCT has been demonstrated (Mosleh-Shirazi et al. 1998).

Jaffray et al. (2002) attached a diagnostic x-ray set and flat panel imager to a radiotherapy gantry and have demonstrated CT images with good soft tissue delineation. This forms the basis of the approach known as Image Guided Radiotherapy. The commercial version of Jaffray's device is shown in Figure 40.11. Varian has a similar device with two

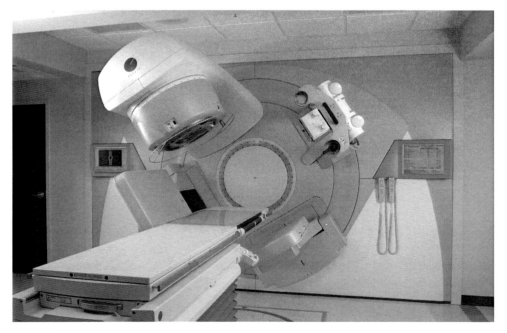

FIGURE 40.11
(**See colour insert following page 590.**) The Elekta synergy image guided radiotherapy solution. A similar concept is available from Varian. (Photo courtesy of Elekta Oncology Systems.)

motor driven arms. Siemens have taken the approach of mounting a CT scanner in the treatment room on a floor track which allows the scanner and the treatment machine to use the same couch, and have also developed megavoltage CT.

40.1.6 FURTHER READING

Several reviews are available. A review of the technology of imaging systems has been provided by Antonuk (2002). Herman et al. (2001) have presented a thorough set of recommendations for the clinical use of electronic portal imaging. Hurkmans et al. (2001) have provided a review of current clinical practice.

40.2 IN VIVO DOSIMETRY

40.2.1 INTRODUCTION

Historically, in vivo dosimetry began to be used at the same time as radiotherapy when skin erythema was the only form of dosimetry available. More than 30 years ago in vivo measurements were also recommended for brachytherapy of cancer of the uterine cervix because it was not possible to calculate the dose delivered to the bladder and rectum, which were limiting factors for the irradiation (IAEA 1967). Currently, in vivo dosimetry is most often considered a quality assurance tool useful to identify deviations in the delivery of standard (Van Dam and Marinello 1994; Huyskens et al. 2001; AAPM 2005) or complex treatments such as IMRT, to evaluate the dose to critical structures (lens, gonads, etc.) or when computer calculations are not possible or are questionable (limits of block shielding, junction of non-coplanar irradiation fields, dose at skin or within inhomogeneities, etc.). In vivo dosimetry can also be used to monitor the irradiation for special techniques such as total body irradiation (Chapter 47) or total skin electron irradiation (Chapter 48) etc.

Dosimetric investigations in special or anthropomorphic phantoms loaded with dosimeters and irradiated in the same conditions as patients can also be useful to check the validity of special techniques prior to routine practice, to point out problems related to suboptimal treatment planning systems, errors in irradiation technique or in dose calculations, or simply to validate the method used for in vivo dosimetry. They have also been used by cooperative groups for clinical and dosimetric intercomparison between centres (see Section 40.2.4.6). The problems to be solved being similar in the above cases, the object of this chapter is to make a review of the principal detectors that can be used and to present methodologies for some applications encountered in normal practice. Although the use of ionisation chambers is possible for in vivo dosimetry (IEC 1997), they will not be considered in this section as they are fragile and require a high voltage that may be dangerous for patients (except for condenser chambers, seldom used nowadays). Only solid state detectors will be considered.

40.2.2 CHOICE OF DETECTORS AND ACTION LEVELS

In the first place the material and method should be as simple as possible to reduce time, effort and cost (Kesteloot et al. 1993). Secondly, the tolerance and action levels should be previously established. Generally a maximum dose deviation level must be defined for each treatment situation, and all deviations exceeding this level must be investigated and explained. That means that considerable care needs to be taken in the choice of the detector and associated readout system, which must allow dose measurements with an overall uncertainty lower than the action level desired.

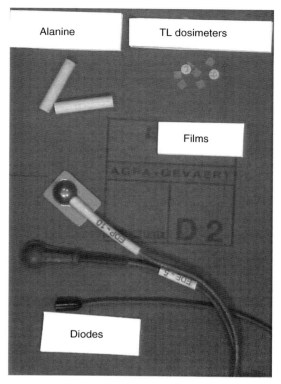

FIGURE 40.12
Comparison of some one or two-dimensional solid dosimeters which can be used for in vivo dosimetry.

Overall uncertainty (IAEA 1994) is evaluated by combining the type A uncertainties (i.e. the standard deviation of the mean value, which can, in principle, be reduced by increasing the number of individual readings) and the type B ones (i.e. uncertainties linked to the measuring instruments and methods that may affect the measured value in a systematic way). To minimise all the sources of type B uncertainties, it is recommended to calibrate the dosimeters in situations as close as possible to those used for the patient measurements.

The detectors that can be used to carry out in vivo dosimetry and anthropomorphic phantom measurements (see Figure 40.12) can be grouped into:

- *Point detectors*, which measure the dose at a point, such as alanine, diodes, MOSFETs, and thermoluminescent dosimeters
- *Two-dimensional detectors*, which measure the dose distribution in a plane, such as standard or radiochromic films, plastic scintillators and electronic portal imaging devices
- *Three-dimensional detectors*, such as gel dosimeters which are useful for IMRT verification because they can provide the dose distribution in complex volumes

The main properties of the different detectors yielding advantages or disadvantages for in vivo or anthropomorphic phantom dosimetry are summarized in Table 40.1. Details of detector characteristics are covered in Part D (Chapter 16 for TLD and diodes and Chapter 17 for film and gel dosimetry). EPID dosimetry has already been discussed in Section 40.1.4. For the choice of phantom material, the reader can refer to ICRU Report 44 (1989) and Report 48 (1992) and also to Section 18.8.

In practice, detectors that show an overall uncertainty higher than ±3% are either of poor quality or have been incorrectly handled. They can only be used to detect large errors such as

TABLE 40.1

Main Properties of Detectors That Can Be Used for In Vivo Dosimetry

Detector	Usability for Volume Dosimetry	Measurement Instrument	Cable Linked to Patient?	Readout	Destructive Readout?	Range of Response with Dose	Accumulated Dose Dependent?	Results Dose-Rate Dependent?	Temperature Dependent?	Tissue Equivalence	Directionally Dependent?
Diode	1D	Simple	Yes	Immediate	No	Up to 10 Gy	Yes	Yes	Yes	Limited range	Yes
TLD	1D/2D	Simple	No	Delayed	Yes	Up to 10^2 or 10^3 Gy[b]	No	No	No	Large range	No
MOSFET	1D	Simple	No	Immediate[a]	No	1–180 Gy	Yes	No	Yes	Limited range	Can be
Alanine	1D/2D	Specialised laboratory	No	Delayed	No	$10–10^5$ Gy	No	No	No	Large range	No
Plastic Scintillators	1D/2D	Specialised laboratory	No	Delayed[c]	Yes	1–10 Gy	No	No	No	Large range	No
Radiochromic films	2D	Simple	No	Delayed	No	0.5–40 Gy	No	No	Yes	Large range	No
Portal films	2D	Moderately simple	No	Delayed	No	A few cGy	No	No	No	Limited range	Yes
EPID	2D	EPID	No	Delayed	Yes	1–20 cGy	No	Yes	No	No	Yes
Gel dosimeters	3D	Use of MRI or optical CT scanner	No	Delayed	Yes	10–70 Gy	No	Not up to 4 Gy/min	Yes	Large range	No

a   However, not really *on line* because the gate voltage must be determined after the end of irradiation.
b   Depending on TL material and readout conditions.
c   Immediate data acquisition is possible.

the forgetting of a wedge filter, a bad position of a compensator, or a serious machine failure. The detection of small systematic errors or verification of the homogeneity of the dose distribution during treatment requires a tolerance and action level less than 5%. This implies the use of detectors of high quality and an experimental method well adapted to the intended application. Whenever possible, a preliminary check of the method in special or anthropomorphic phantoms is desirable.

40.2.3 METHODOLOGY

For obvious reasons, the direct measurement of the patient dose is only possible in a limited way and mostly restricted to skin or intracavitary dose measurements. However, it is possible to obtain information on the dose to the tumour or critical structures by combining data from entrance and exit dose measurements for techniques involving a limited number of homogeneous photon beams. The principle of the methodology to be applied is similar for all types of *point* detectors provided that they are correctly handled and their response corrected to take into account their particular characteristics. Use of two-dimensional or three-dimensional detectors requires a complex methodology not considered in this section (but that is discussed in Section 40.1.4). Nevertheless, the type of detector to be chosen depends on the application to be carried out and on the predefined action level. In vivo measurements must not interfere with the treatment and measurements close to metallic structures, such as dental fillings or a prosthesis, should be avoided because of the interface effect.

40.2.3.1 *Skin Dose Measurements*

Skin dose measurements are mostly used with electron beams either used alone, or joined at the patient skin to cover a large surface for breast or total skin electron irradiations (Chapter 48). They can also be used in some limited cases with photon beams in order to explain abnormal cutaneous reactions or to know the dose at a scar, etc.

The *skin dose* or *surface dose* is defined by convention at 0.5 mm below the surface irrespective of the nature and energy of the ionizing radiation used. That implies that it should be measured with very thin detectors such as TL chips or a thin layer of TL powder wrapped in plastic or kapton (Dupont polyimide film) pouches or envelopes made of thin opaque paper, monocoated photographic emulsions or radiochromic films. Diodes, even though they are very thin (such as diodes dedicated to electron beam measurements), must not be used in high-energy photon beams because their response is very much influenced by electron contamination existing in such beams (Sjögren and Karlson 1998). In all cases, the detector is stuck directly to the patient skin taking care to minimise the air space between the skin and the dosimeter wrapping envelope and no build-up cap is required.

If the water equivalent thickness of the detector (sensitive part plus cover) is less than 0.5 mm, a thin layer of tissue equivalent material must be added above it to reach an adequate thickness. On the other hand, when detectors of thickness greater than 0.5 mm are used, correction factors have to be applied to their response to take into account their effective point of measurement, which is not at 0.5 mm below their surface (Kron et al. 1993). The experimental determination of the correction factor is only possible for detectors of constant and known thickness.

40.2.3.2 *Entrance Dose Measurements*

Entrance dose measurements are made at an equivalent depth corresponding to the dose maximum on the beam axis of photon or electron beams either used alone or in combination with other beams (Huyskens et al. 2001). They are mainly used to check the reproducibility of patient setup and machine output during the different fractions of external beam therapy, the positioning of shielding in relation to the position of the patient or, when adjoining fields are

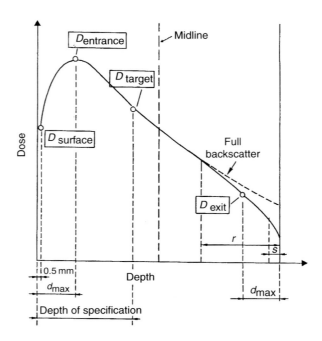

FIGURE 40.13

Dose distribution along the axis of a single photon beam passing through a phantom made of homogeneous tissue-equivalent material of limited thickness. The surface dose ($D_{surface}$) is defined at 0.5 mm below the entrance surface, the entrance dose ($D_{entrance}$) at d_{max}, the target dose (D_{target}) at the depth of dose specification, the exit dose (D_{exit}) at a distance of d_{max} from the exit surface on the beam axis. (After Van Dam, J. and Marinello, G., *Methods for in vivo dosimetry in external radiotherapy, ESTRO Booklet Number 1*, Garant, Leuven-Apeldoorn 1994.)

used, to check the field junction. In combination with exit dose measurements (see Section 40.2.3.3), they can allow the target or midline dose to be calculated using an appropriate protocol (Van Dam and Marinello 1994).

Photon Beams: The individual entrance dose of each beam contributing to the target dose and the case of two directly opposed beams will be considered in turn.

When a medium is irradiated by a *single photon beam*, the dose gradually increases from a low value at the surface up to a maximum value D_{max} at a depth d_{max} that depends on the energy, the collimator aperture, the source skin distance (SSD) and the presence or absence of modifying devices traversed by the beam or inhomogeneities close to the entrance surface (Figure 40.13). This depth can vary from a few millimetres to several centimetres. As the sensitive part of the detectors is generally very thin, entrance dose measurements cannot be carried out with bare detectors directly placed on the patient skin. A build-up cap made of a tissue-equivalent material must be placed in front of and around the bare detector to achieve electronic equilibrium. The dimensions of this build-up cap are very dependent on the irradiation conditions. For example a different thickness is desirable for open beams or under shielding, both because of the energy variation and the presence of the block tray. The appropriate thickness must be carefully determined for each type of treatment. The thickness of the build-up cap to be used to check external beam treatment will be between 2 mm and 3 mm with cobalt-60 and approximately 10 mm with 4 MV x-rays. For x-rays between 18 MV and 25 MV the required build-up cap thickness will increase from 20 mm to 35 mm as the field size is reduced from 30 cm × 30 cm to 5 cm × 5 cm (Van Dam and Marinello 1994). It should also be noted that the use of very thick build-up caps, although required to provide electronic equilibrium, can lead to an underdosage of the treatment volume together with loss of skin sparing in small areas (Nilsson et al. 1988). This effect can be reduced by using a slightly different detector position for the different treatment fractions, at the expense of some uncertainties in the dose determination. For this reason and to reduce the size of the detector generally, it is common practice to use a material such as stainless steel in the build-up cap instead of a tissue equivalent material.

When *opposed beams* are used both the contribution of the beams and their respective weights modify the depth d_{max}, and therefore the thickness of the build-up cap to be used. The presence of a large inhomogeneity traversed by the beams can also change the conditions of electronic equilibrium considerably (Marinello et al. 1982). Consequently the best solution involves the experimental determination of d_{max} in conditions as close as possible to those used for the clinical applications and to deduce from these measurements the optimal dimensions of the build-up cap. However, in practice commercial diodes provided a restricted choice of build-up cap (see Table 16.5 and Huyskens et al. 2001).

Electron Beams: Generally the entrance dose of electron beams is determined along the axis of a single beam used alone or adjacent to other beams to cover target volumes of limited thickness. In practice d_{max} is reached at depths varying from a few millimetres to several centimetres depending on the energy, the type of electron collimator (cone or tube collimator, or an open collimator, as shown in Figure 11.28), the source skin distance (SSD), the presence or absence of modifying devices, etc. To have a response representative of the dose at d_{max} the bare detector must be surrounded by a certain thickness of tissue-equivalent material but care must be taken because build-up caps more than a few millimetres in thickness can induce a relatively important scattering artefact and may alter the dose distribution substantially because of the high scattering properties of the electrons. Consequently it is preferable to derive the entrance dose from surface dose measurements (Section 40.2.3.1) using correction factors established in a phantom irradiated in the same irradiation conditions as for the patient. The correction factor to be used is the ratio of the response of the bare detector on the surface to that at the depth of the maximum dose on the beam axis.

40.2.3.3 Exit Dose Measurements

Exit dose measurements are primarily performed for photon beam irradiations, most often at the same time as the entrance dose measurement (except for portal transit dosimetry). In this region there is a *build-down* related to the lack of backscatter radiation from the air behind the patient (Figure 40.13). As the position at which the exit dose D_{exit} should be defined is less obvious than for $D_{entrance}$, it is generally taken at d_{max} from the exit surface, i.e. symmetrically with the entrance dose with respect to the midline. For reasons similar to those discussed for photons in Section 40.2.3.2, the detector must be covered with enough material behind and around it to ensure complete electron backscatter (otherwise it would be in too a high dose gradient, which would decrease the accuracy). The dimensions of the build-up cap required for exit dose measurements are in principle different from those used for entrance dose measurements, but it is usual in practice to use the same caps for both situations but to use different calibration factors which must be determined experimentally (Figure 40.14). This has the advantage that the same setup can be used for opposed beams with appropriate correction factors being used for each beam. When positioning the detectors for simultaneous entrance and exit dose measurements, it is necessary to avoid their shadowing effect by offsetting them slightly from each other.

40.2.3.4 Target and Midplane Doses

A very simplified approach which consists of considering the target dose equal to the mean of the entrance and exit doses can be acceptable in some practical conditions but may frequently induce errors of several percent. These errors can be attributed to one of the following reasons:

- Deviation from the linear approximation of dose variation with depth for lower energies and greater patient thickness

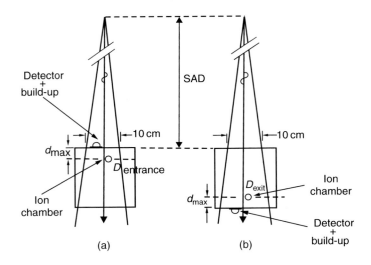

FIGURE 40.14
Example of method for determination of entrance and exit dose calibration factors of solid state detectors by comparison with an ion chamber. (After Van Dam, J. and Marinello, G., *Methods for In Vivo* Dosimetry in External Radiotherapy, ESTRO Booklet Number 1, Garant, Leuven-Apeldoorn, 1994.)

- Variation of response with change in energy spectrum
- Changes in the effect of the inverse square law

The validity of the method must be checked prior to use for in vivo dosimetry either experimentally or using a calculation method based on data established in irradiation conditions as close as possible to those used for patient treatments.

If the approximate method is not valid, it is necessary to derive the target dose from simultaneous entrance and exit dose measurements using one of the calculation methods published by different authors. Most of them are based on the symmetrical (with respect to the midline point) expansion or compression of the real patient to a thicker or thinner *water equivalent patient* and on the concept of midline and exit transmission curves (Rizzotti et al. 1985; Leunens et al. 1990). Accurate results can be obtained using these methods in regions of the human body which have a symmetrical disposition of the different types of tissue irradiated provided that midline and exit transmission curves are established from tissue-phantom ratios corresponding to the energy and field sizes used for irradiation and that the lack of exit backscatter is taken into account (Van Dam and Marinello 1994). Symmetry is generally adequate for lateral fields, but unfortunately, except for the brain and skull, the methods are not straightforwardly applicable for antero-posterior beam directions because of the lack of symmetry.

Some authors have proposed new methods to deduce the midplane dose directly in 2D using portal films or an electronic portal imaging device (see Section 40.1.4.3). They have shown that, under certain conditions, midplane doses could be estimated with a similar accuracy compared to existing conventional methods for in vivo dosimetry.

40.2.3.5 Intracavitary Dose Measurements

In vivo measurements within body cavities such as the mouth, nasopharynx, oesophagus, vagina, rectum, etc. can be performed either for external beam therapy, including IMRT, or brachytherapy. TL dosimeters sealed inside waterproof catheters are the most appropriate for such measurements (Engström et al. 2005). The position of the dosimeters must be checked radiographically using radio-opaque markers. It should be noted that the increase in scattered radiation resulting from the presence of high atomic number radio-opaque markers can cause uncertainties of a few per cent in the absorbed dose to the TL dosimeter, although this can be

FIGURE 40.15
Example of an anthropomorphic phantom including tissue inhomogeneities and allowing the insertion of films or solid dosimeters (left) and example of the results obtained with films (right) irradiated in conditions of total electron irradiation (8 MeV decelerated by lucite screens as shown in Part I, Figure 48.6).

allowed for. MOSFET detectors have also been successfully used for this purpose (Cygler et al. 2001; Marcié et al. 2005).

40.2.4 CLINICAL APPLICATION

Irrespective of the type of detector and the method intended to be practiced for in vivo measurements, the validity of the approach should first be checked by carrying out measurements in phantoms made of tissue-equivalent materials and irradiating them in conditions as close as possible to those used to treat patients. Details of suitable phantom materials can be found in ICRU Report 44 (1989) and Report 48 (1992). Body and limb cross sections can be simulated by elliptical, oval or cylindrical phantoms of the appropriate radii but it should be realised that homogeneous phantoms with regular shapes are not a good representation of real patients. It is preferable to confirm the method intended to be used for in vivo measurements with an anthropomorphic phantom (see Figure 40.15) drilled with holes and cut into sections allowing the insertion of dosimeters within it.

The accuracy of the measuring system and the different correction factors to be applied must be previously investigated by comparison with an ionisation chamber calibrated at an energy close to that of the measurements. This is the simplest way to compensate for factors such as dose-rate dependence, directional dependence, temperature effects, etc. which can influence the response of some detectors. If the detector is used for entrance and exit dose measurements, it has to be assigned separate entrance and exit dose calibration factors. One of the methods which can be used to determine them is shown in Figure 40.14.

Different methods for in vivo dosimetry which can be used for standard cases of external beam radiotherapy or in brachytherapy are briefly described below. For a more detailed description of them, the reader can refer to the work of Van Dam and Marinello (1994), the Report of Task Group 62 of the AAPM Radiation Therapy Committee (AAPM 2005) or to the extensive list of references relating to Part H.

40.2.4.1 Patient Dose Verification in External Beam Therapy

The large number of steps and persons involved leads to a large probability of errors in the preparation and execution of a radiation treatment (Leunens et al. 1992; Essers and Mijnheer 1999; Huyskens et al. 2001). Different ways can be used to assess the uncertainty in the dose delivery. The sequence of the different procedures involved in delivering a dose to a patient (dose calculation, treatment machine calibration, patient setup etc.) can be analysed. The total uncertainty in all the steps may then be considered as the maximum attainable accuracy in dose delivery. In practice the actual accuracy in dose delivery may be less than desirable for any of the following reasons:

- Errors in patient contours
- Patient mobility
- Inhomogeneities
- Organ motion
- Transfer of treatment data from simulator to the treatment unit
- Machine settings
- Positioning of modifiers, etc.

The ultimate check of the individual dose delivered to a patient can only be performed at the patient level, by means of in vivo dosimetry.

The method most commonly used for standard techniques consists of positioning *point* detectors on the patient's skin and measuring the entrance and/or exit doses in conditions adapted to the type and energy of the beam (see Section 40.2.3.2 and Section 40.2.3.3). By combination of the two (as discussed in Section 40.2.3.4), it is then possible to obtain the dose inside the patient (target or midline dose, for instance). The detectors usually employed are diodes and thermoluminescence dosimeters (Van Dam and Marinello 1994; Mayles et al. 2000). Some authors have also used portal films or electronic portal imaging devices as discussed in Section 40.1.4.

Although more demanding to carry out, in vivo dosimetry can be used to detect errors or to check the dose delivered using intensity modulated therapy (see Chapter 43). Depending upon the objectives to be achieved dosimeters can either be put at the entrance of the different beams at points situated in a low dose gradient area (Ling et al. 1996; Higgins et al. 2003) or put within the target volume using, for example, a nasogastric tube (Van Esch et al. 2002; Engström et al. 2005). When this is not possible, in vivo measurements done in high dose gradient regions are subject to question (LoSasso 2003).

Diodes: Diodes allow the detection of important systematic or human errors, with the patient still on the treatment couch which allows immediate corrective actions (Ciocca et al. 1991; Heukelom et al. 1991a, 1992; Mijnheer 1994; Noel et al. 1995; Loncol et al. 1996; Georg et al. 1999; Lanson et al. 1999; Verney and Morgan 2001). Because of their directional effect care must be taken when diodes are used for tangential fields (Heukelom et al. 1991b) and they should not be used in regions of high dose gradient. Some authors (Shackford and Bjarngard 1995) have also mentioned the possible disturbance of diode signals by radiofrequency radiation produced by linacs: although this should not be a problem with properly designed electronics. An overall uncertainty from 3 to 5% can be obtained for routine measurements provided that the constancy of diode response is regularly checked and a separate calibration carried out for each type of beam in which the diode is used. In practice a good solution consists of assigning an electrometer plus diodes to each treatment unit, and assigning them a correction factor for each irradiation condition (for instance a correction factor for standard fields and another one for large blocked fields used for Hodgkin disease etc.).

Thermoluminescent Dosimeters (TLD) [*]: TLD is also a good tool to detect important systematic or human errors in radiotherapy with an overall uncertainty of $\pm 3\%$ or less when correctly chosen and handled (Bascuas et al. 1977; Marinello et al. 1992; Loncol et al. 1996).

---

[*] A system similar to TLD, but with the possibility of rereading the detector is optically stimulated luminescence (OSL). This is also available as an on-line device connected via an optical fibre cable (Aznar et al. 2004).

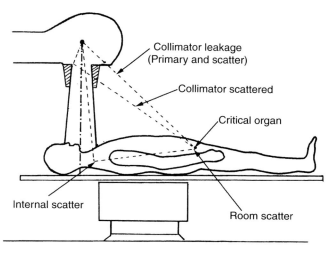

FIGURE 40.16
Measurements outside the irradiated volume: different components of the photon beam at the level of point of measurement.

Because of their limited size, their weak energy variation and their absence of directional effect, they are suitable for measurements in most clinical applications. They can be used for small or large fields with or without beam modifiers, irregularly shaped beams, tangential fields, zones of high dose gradient such as the penumbra region (although great care is needed with positioning accuracy), field junction zones, dose within inhomogeneities, IMRT, etc. Because of their insensitivity to dose-rate the dose variations produced by beam modifiers, SSD variations or patient anatomy do not change their response (even in scanned electron beams).

40.2.4.2   Dose to Organs at Risks Close to or Outside the Irradiation Field

A very common use of in vivo dosimetry is the determination of the dose to sensitive organs for which relatively small radiation doses may be undesirable. Among the most common such measurements are those of the dose delivered to the gonads and to the eyes.

Dose Delivered to Gonads: The treatment of seminoma, Hodgkin's disease and non-Hodgkin's lymphoma generally involves irradiation of sub-diaphragmatic nodal areas. Although outside the irradiation field, the gonads will be exposed to low doses of radiation. This is due both to scatter which is of lower energy than the primary beam and may be generated by the irradiated volume, by parts of the machine and the walls of the room, and to leakage through the machine shielding and collimator (Figure 40.16). As shown by different authors who have used them, thermoluminescent dosimeters are the most suitable dosimeters for in vivo measurements in this region because of their high sensitivity in a small volume, their absence of directional dependence and their good energy response in the relevant energy range (Sharma et al. 1981; Marcié et al. 1995). Diodes specially designed for measurements of doses to organs at risk situated outside the irradiation fields are also commercially available (see Table 16.5). The validity of their response must be evaluated using phantom measurements prior to in vivo use because of their directional effect and their enhanced response at low energy (Section 16.3.4.8 and Section 16.3.4.9).

Dose Delivered to Eyes: When the dose delivered to the eyes is to be assessed, it must be borne in mind that the sensitive tissue is the lens and not the eyelid, where doses are most easily measured. A phantom study carried out by Harnett et al. (1987) suggests that for lateral irradiation a dosimeter placed on the outer canthus may give a more accurate estimate of lens dose than one placed on the eyelid. If anterior beams are used (Galbraith et al. 1985), the surface dose just outside the beam edge actually overestimates the lens dose because of the electron contamination of the radiation reaching the dosimeter. Prior to in vivo measurements it is necessary to perform phantom measurements in irradiation conditions close to those used for patients to reduce the unavoidable inaccuracy of evaluation of dose to the lens. Because of their high sensitivity in a small volume and their dosimetric properties, TL dosimeters are again the most suitable dosimeters for such measurements. Diodes can also be used provided that care is taken in interpreting the results because of their directional effect (Section 16.3.4.9).

40.2.4.3 Intraoperative Irradiation

Intraoperative radiation therapy is a complex irradiation technique practiced during a surgical operation with either electron beams of energies ranging from 6 MeV to 17 MeV or x-rays of between 50 kV and 300 kV. The entire dose is delivered during a single irradiation in a complex configuration that makes dose control tricky but indispensable. A radiation dose mapping technique based on laser heating of thermoluminescent dosimetric plates has been developed allowing the measurement of a two-dimensional dose distribution of non-planar surfaces with a resolution of μm which is more satisfactory than film for this particular application (Beteille et al. 1996).

40.2.4.4 Brachytherapy

Currently used brachytherapy sources are either beta sources or low energy photon sources (see Section 51.2). They are characterised by a large variation in their energy spectrum and a high dose gradient existing around them. Consequently the detectors to be used for in vivo or phantom dosimetry must be of very small dimensions, have a response that varies little over the relevant energy ranges and be dose-rate independent. Among the possible solid detectors there are TLD, plastic scintillators, radiochromic film and polymer gels.

TLD. Thermoluminescent rods, chips, or powder included in small plastic tubes or catheters can be used. Care must be taken when using LiF because of its large variation of energy response at low photon energy which is not the case for lithium borate (see Section 16.2.4.9). TL measurements have been used since the 1970s for phantom or in vivo measurements intended to check the first computer programs (Joelsson et al. 1972). They have also been used by different teams for conservative breast treatments either to check the dose

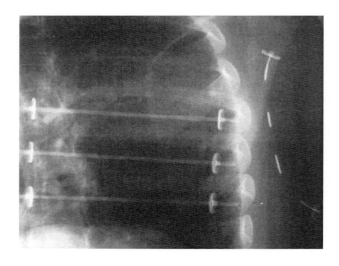

FIGURE 40.17
In vivo determination of the dose delivered to the axilla during endobrachytherapy of breast cancer using LiF powder enclosed in a thin catheter. As LiF is invisible on radiographs, lead markers are necessary to show the position of dosimeters. (After Marinello, G. et al. *J. Radiol. Electrol.*, 56, 791–796, 1975.)

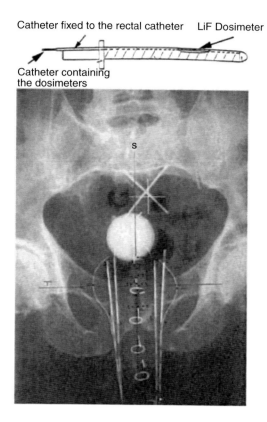

Catheter fixed to the rectal catheter   LiF Dosimeter

Catheter containing
the dosimeters

FIGURE 40.18
Rectal catheter used for in vivo measurements during HDR brachytherapy of prostate performed using the Paris System. (After Brune, D., Prostate, in *A Practical Manual of Brachytherapy*, Pierquin, B. and Marinello, G., Eds., Medical Physics Publishing, Madison, WI, 1997.)

delivered to the target volume (Hamers et al. 1993), to evaluate the skin dose in the vicinity of the implants (Mechakra et al. 1988) or the dose delivered to the axilla as shown in Figure 40.17 (Marinello et al. 1975). TL dosimeters or sheets have also been used for rectal dose determination during brachytherapy of the cervix (Iwata et al. 1992) or the prostate (Figure 40.18) and of rectal cancer.

Plastic Scintillators: Plastic scintillators have been shown to be very convenient to measure the dose distribution around ophthalmic applicators (Flühs et al. 1996) or to check the dose distribution of intravascular sources.

Radiochromic Film: Radiochromic films are useful for quality assurance purposes. For instance small samples of film included in vaginal cylinder applicators can be used to check the dose delivered during each fraction by HDR brachytherapy (Pai et al. 1998).

Polymer Gels: Polymer gels read out by MRI which have been found to be of the same accuracy as TLD measurements by Knusten et al. (1997) when used in a phantom study of dose distributions in brachytherapy of cervix.

Diodes: Diodes should be used cautiously (Piermattei et al. 1995; Alecu and Alecu 1999) but are not recommended because of their poor tissue equivalence in the involved energy range (Section 16.3.4.8).

Irrespective of the detectors chosen for in vivo dosimetry, it is necessary to calibrate the dosimeters in a water or a solid water phantom by comparison with an ionisation chamber of very small dimensions and calibrated for low energy photons or electrons, depending on the intended application. Both the chamber and TLD must be placed at a distance from the source representative of the practical application. The same phantom can be used for calibration at

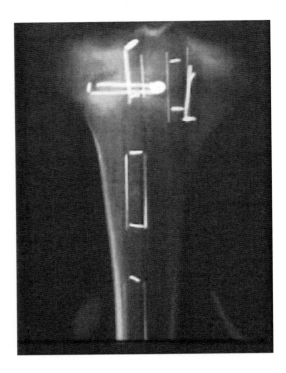

FIGURE 40.19
Use of alanine dosimeters to check the dose delivered to human bone allografts. Thin lead markers are necessary to see the position of the dosimeters within the bone structure because of the good tissue equivalence of alanine that make them invisible on a radiograph. (After Hernigou, P., Marce, D., Julieron, A., Marinello, G., and Dormont, D., *Revue Chirurgie Orthopédique*, 79, 445–451, 1993.)

different distances provided that the holes drilled in the phantom are not in front of each other to avoid the influence of an air cavity on the detector responses.

40.2.4.5 Prosthesis Measurements

Implantation of dried bone allografts after freezing and sterilisation by high doses of radiation (25 kGy to 35 kGy) is current practice in osseous surgery. Because of its good accuracy for the evaluation of doses from 10 Gy to 10^5 Gy (the saturation dose level being 5×10^5 Gy) and its dosimetric properties (Ciesielski et al. 1988, 1993; Olsen et al. 1990), alanine dosimetry is recommended to check the irradiation dose delivered to the prosthesis prior to the implant (Figure 40.19). It is simple to handle, with good long term stability, and its response is independent of the size and shape of the dosimeters. It can be considered as bone-equivalent in the photon and electron energy range used in practice. The relative variation of mass energy absorption coefficient between alanine and bone is less than 5% between 0.3 MeV and 8 MeV for photons. The relative variation of the mass collision stopping power of the material to that of compact bone is less than 2% between 0.2 MeV and 40 MeV for electrons (Mc Laughlin et al. 1991). As the readout equipment is expensive and rather complicated, most centres mail the dosimeters to specialized laboratories for readout.

40.2.4.6 Comparison between Centres: Mailed Dosimetry

Mailed dosimetry, most often practiced with thermoluminescence or alanine dosimeters, is extensively used by different international and national organisations (IAEA/WHO, Radiological Physics Centers in the U.S.A., ESTRO-EQUAL etc.) for quality assurance networks. Generally measurements are performed in a specially designed phantom and are intended to evaluate the uncertainties in the calibration of radiotherapy machines in the different radio-therapy centres. Since 1989 the European Organization for Research and Treatment of Cancer (EORTC) has also carried out pilot studies including either anatomical phantom measure-ments (Johansson et al. 1987; Hamers et al. 1991] or in vivo dosimetry in a limited number of patients. A good example is in vivo dosimetry with TLD in the conservative treatment of breast cancer patients treated with the EORTC protocol 22881 in which in vivo dosimetry has been used both for external therapy (Hamers et al. 1991) and for brachytherapy (Hamers et al. 1993).

CHAPTER 41

RECORDING AND VERIFICATION—NETWORKING

Margaret Bidmead

CONTENTS

41.1 RECORDING AND VERIFICATION IN EXTERNAL BEAM RADIOTHERAPY

To improve the accuracy of treatment delivery, by eliminating as many errors as possible, checking systems, whether human or computer-assisted are commonly used on radiotherapy treatment machines (Rosenbloom et al. 1977; Klein et al. 1998). Errors can either be caused by human fallibility or machine malfunction, so interlock and checking systems have been devised to help reduce these errors. In contrast with portal imaging and in vivo measurements, which are generally not used systematically for all patients and all fractions, record and verify (R and V) systems ensure that the machine settings will be repeated throughout the treatment course. They also ensure that the programmed total dose will not be exceeded. This places a considerable burden on the person entering the data for the first time, as any errors may be repeated throughout the treatment (see Section 41.2.2).

41.1.1 CONTROL OF THE TREATMENT MACHINE

All modern linear accelerators are computer controlled and have electronic monitoring circuits to limit machine malfunction. Parameters such as beam flatness, symmetry, output from dual monitor chambers, dose rate, and mechanical positional interlocks are all under computer control.

Features such as dynamic wedges and IMRT, which involve simultaneous control of motion of collimators and of dose-rate, require close and precise monitoring and verification systems. It is important for the operators and engineers to have a good appreciation of the function, trigger level, and importance of each interlock in the computer controlled systems—which ones are major and which ones may be overridden for later investigation. The issue of overriding interlocks will be discussed in Section 41.4.3 that addresses patient-related verification.

In addition to computer control of the accelerator, R and V systems have been designed to store the details of a patient's treatment and to help prevent errors. The original aim of an R and V system was to compare the daily treatment-setup parameters with the planned parameters, and to inhibit treatment or flag a warning signal if the two did not agree (Rosenbloom et al. 1997). Since then, systems have evolved through recording the parameters actually used, to storing the data required for treatment, controlling the linear accelerator user interface, allowing assisted setup and even automated treatment. At the end of the treatment session, in addition to recording the parameters actually used, the system will record any overrides along with the identification of the operator that performed them, the accumulated dose and the treated fraction.

41.1.2 NETWORKING

Record and verify systems can be linked together so that data can be transferred around the department electronically. The networking of the department is discussed further in Section 41.6 and in Chapter 42.

41.2 IMPLEMENTATION OF THE RECORD AND VERIFY SYSTEM

41.2.1 WORKFLOW IN A NETWORKED DEPARTMENT

As soon as a computerised system is installed, it becomes necessary to have a strictly regimented workflow. A typical sequence is as follows:

1. A patient plan is prepared on the treatment planning system.

2. Patient and treatment plan data are entered directly via a network link between the planning system and R and V system, and checked by an independent operator before the patient arrives for simulation of the plan (or the first treatment if verification simulation is omitted).

3. When the patient is simulated, the plan parameter data are recalled from the R and V storage system and used to set up each treatment field for simulation. Any changes at this stage are automatically acquired into the R and V system and passed back to the treatment planning system to check for any dosimetric changes.

4. If the patient requires blocks or multi-leaf collimation, data files containing the relevant parameters may be sent over the network to a block-cutter or the MLC workstation on the treatment machine.

5. When the patient attends for the first treatment the data are accessed by entering the patient's unique ID and the data are then displayed on the R and V console.

6. Each field is called up and the *planned* data are then displayed, together with the *actual* settings from the treatment machine.

7. The setting of different treatment fields is speeded up by use of the assisted setup facility that automates some of the machine's features, such as gantry, collimator, and jaw movements.

8. If, for some reason, some of the actual parameters are not within the tolerance assigned for the planned parameters, a warning is given that requires intervention before treatment is allowed to commence.

9. When treatment is finished, the system records to the database, all the set-up parameters and the doses actually used for the treatment.

The parameters that are verified in a typical system are listed in Table 41.1.

TABLE 41.1

Parameters for Verification

Patient identification	Patient number, Date of birth	
Prescription	Treatment plan phase	
	Dose	Total dose
		Cumulative dose
	Individual fraction data	Fraction number
		Contribution to target dose
		Reference point doses (optional)
Individual field data	Field identification	Beam energy and modality
		Static, dynamic, IMRT, arc
	Dose information	Monitor units—open field
		Monitor units—wedged field
		Dose rate
	Wedge data	Physical or dynamic
		Orientation of wedge
	Collimator settings	Jaw settings X1 and X2, Y1 and Y2
		Collimator rotation
		Individual MLC leaf positions
	Gantry setting	Gantry angle
	Couch settings	Lateral
		Longitudinal
		Height
		Isocentric rotation
		Pedestal rotation
	Accessories	Block tray (with optional coding)
		Electron applicator and insert

41.2.2 Data Entry and Checking

A separate terminal remote from the treatment machines, for data entry and/or checking, is a great advantage in a busy department. In some systems, the data can be acquired directly from the simulator or treatment planning system. It is important that data are entered carefully and then checked by an authorised second person, in a quiet atmosphere. Any mistakes at this stage could be carried through the whole treatment, and therefore have graver consequences than random mistakes made during the course of a treatment.

Machine breakdowns and emergency treatments will mean there are times when it is not possible to enter patient data before the first treatment. In this case, it is possible to treat without the verification system, or to use a direct data entry mode. A minimum of data necessary for defining a field needs to be entered. Some software has a facility whereby the only data to be typed in are patient and field identification, and the system automatically records the machine settings, if requested to, by automatic acquisition. This is not to be encouraged as a regular practice on a treatment machine, but can be useful in the event of a machine breakdown, provided the data are carefully checked before the next fraction of treatment.

After the patient and plan data have been entered into the system it may be useful to print the data so that a careful check can be carried out. There are several ways of doing this: some systems use specialised cards, bar codes, or simply A4 sheets, while some systems are customisable to individual department's requirements. However, hardcopy is often the weak part of the whole system, and the R and V printout is often not a suitable replacement for the existing treatment card system within a department. As R and V systems evolve and benefit from better integration with the hospital information system, it is likely that printed charts will progressively replace manual charts, which may ultimately be phased out altogether in favour of paperless departments.

41.2.3 Data Retrieval

The first step in treating a patient whose data has been entered and verified, is accessing their data using the patient's identity. The field is then selected by number and label, and the screen will display all the planned machine settings that are required. It is useful to have an additional monitor and keyboard in the treatment room, to allow data entry and display the settings. In some systems, it is obligatory to treat the fields in the order in which they have been entered and in all systems, the treatment of the same field twice over is inhibited.

41.2.4 Computer-Assisted Setup

Patient setup can be completed considerably more quickly by using the computer assisted setup facility. An auto button either on the handset or on the console can be activated taking the machine to all the required settings simultaneously. If this is not required, then the auto facility can still be used to set up each parameter individually. Various automated movements, such as couch rotation, can be *forbidden*, if required. The usefulness of assisted setup should not replace a visual check of the field positions on the patient before leaving the room. Users must also be aware of the possible danger of a collision and set parameters individually if there is any uncertainty. Some computer-assisted setups can be directed from outside the treatment room to facilitate conformal therapy or IMRT. Some systems can be programmed to deliver the whole treatment with one button press. The faster treatments achieved using computer-assisted setup, can not only improve throughput, but can also improve patient comfort and because there is less time for the patient to move during treatment, reproducibility.

41.2.5 RECORD KEEPING

The data entered at the pre-treatment stage is automatically recorded and stored. After the treatment of a field, the actual machine settings used, the monitor units delivered, any overrides, the accumulated dose from that field, and the number of fractions is immediately recorded. The system then sums the accumulated dose from all the fields and compares it with the prescribed value. Treatment is inhibited when either the accumulated dose has been reached or when the number of fractions equals the prescribed value. After a field has been recorded as treated in a particular session, a warning is displayed if it is accessed on the same day (unless explicitly declared as a hyperfractionated treatment). Provision to override the time interval is necessary to allow patients to be treated twice on one day to allow for gaps in treatment.

In an ideal world, the database of the R and V system should provide a complete record of treatment. However, it is common practice to maintain a manual system and to print the treatment details each day when all the patient fields have been treated. Paper systems have the advantage of being able to record signatures and check marks, facilities that are often not adequately catered for in the R and V software.

41.2.6 PATIENT VERIFICATION

The advent of electronic portal imaging has allowed routine evaluation and verification of the patient position on treatment and the accuracy of the treatment beam delivery. It also generates large amounts of image data, which need to be incorporated into a general image-viewing network. Images of treatment fields are easy to acquire as part of the individual field treatment delivery, but if the field is small, as in conformal therapy and IMRT, it is sometimes necessary to perform double exposure imaging, which gives extra radiation dose to the patient. It is useful to be able to incorporate this extra exposure into the record and verify system, both to program when images are to be taken and to add in the extra dose delivered to the patient. Control over the machine parameters and number of MU's set are also important as the larger field is not part of the prescribed treatment and needs to be monitored. Diode dosimetry for entrance and exit doses can also be incorporated into the R and V system, both in terms of a reminder as to when doses should be measured and to record the doses that have actually been delivered.

41.3 SYSTEM MANAGEMENT

A system manager should be nominated to manage the R and V system. The system manager should be very familiar with computers in general and the R and V software, in particular. The manager should also be able to customise the system to the users' requirements, and help to provide maximum use of all the features of a particular system.

41.3.1 PASSWORDS

Different levels of authorisation are password protected. Table 41.2 shows a typical example of how password access may be used. If security is to be effective, it is important that the user logs out when finished, unless the system does this automatically. The system should allow a user with a high level of access to enter their password temporarily to execute a particular task. Overriding parameters that are out of tolerance often requires a password, so is only allowed to be carried out by users with certain levels of access.

TABLE 41.2

Hierarchy of Duties Associated with R and V Systems

1. Treat patients without modifying the data
2. Enter patient data
3. Enter and verify patient data
   Delete patient data
   Override out of tolerance items
4. Backup and restore patient data using long term storage
   Edit and enter the tolerance table
5. Customise machine parameters
   Backup and restore machine parameters using long term storage
6. Troubleshooting involving direct access to the database

These duties may be ranked in order by passwords. It is assumed that each level includes authorisation to perform all the lower levels.

41.3.2 Tolerance Tables

The achievable accuracy of patient setup varies according to the treatment site. For example, head-and-neck tumours are usually treated with the patient in an immobilisation device, which is often fixed to the table in a reproducible manner, but there may be much more variation in patient position on the treatment couch for chest and abdomen patients. It is therefore necessary to assign different tolerances to the agreement between planned and actual machine parameters for different treatment sites. The accuracy required for particular treatments may also vary. Different *tolerance tables* are customisable by the user to specify the range over which parameters can differ from the prescribed value before treatment is inhibited. Tolerance tables are identified by number and need to be assigned to every field. Some systems use a default tolerance table when none is specified. In this case it is strongly recommended that the default table should be the tightest tolerance table, so that inadvertent choice of a wide tolerance will be avoided. Table 41.3 lists reasonable tolerance values for typical sites.

TABLE 41.3

Verification Parameters and Recommended Tolerance Tables for Routine Use

	Units	Tolerance Table Number			
		1	2	3	4
Gantry angle	Degrees	0.1	0.2	0.2	20.0
Collimator	Degrees	0.1	0.5	0.5	20.0
Field X and Y	mm	1	1	1	1
Couch Height	mm	1	2	10	100
Couch Lat & Long	mm	1	5	50	100
Couch Rotation	Degrees	0.1	0.5	2.0	10.0

Tolerance Table 1: Smallest tolerances: for accurate treatments with the patient immobilised on the couch (e.g. using a stereotactic frame).
Tolerance Table 2: Small tolerances on couch lateral and longitudinal movement: for treatments where patients are immobilised in a shell.
Tolerance Table 3: Tolerances suitable for pelvis/abdomen/chest where separations, etc, are checked daily.
Tolerance Table 4: Maximum flexibility: for static electron treatments.

41.3.3 Backup and Archiving

The large amount of data requires considerable storage capacity. Backup records of given treatments can be kept on optical disc or some form of data tape. It is useful if the backup system allows individual patient records to be retrieved as well as a full restore of the database. A daily backup of patients on treatment is usually performed (it should be possible to program this so that it occurs automatically during the night).

Old data are permanently removed from the system and transferred to long-term storage during the archival process. It is important that backups of archived data are also maintained. Ideally, archival should be controlled by the R and V system, so that a record of the patient's existence can be retained. Data storage should comply with national regulations relating to patient privacy.

41.3.4 System Resilience

Retrieval from tape is notoriously problematic, and any backup method should be fully checked (by retrieving patients) before it is actually needed. If data are stored centrally on a server, it is wise to have a backup server so that the system can be brought back online in the event of a serious failure. The backup server must be kept up to date. Transaction logs can be used as a method of restoring very recent data.

41.3.5 Fault Reporting

R and V systems are constantly evolving and it is important that software suppliers are made fully aware of any problems of use so that they can be rectified. For this purpose, it is helpful if the software stores an audit trail so that the source of any problems can be traced.

41.4 ERROR CONTROL

Although it is true that R and V systems have the potential for causing serious patient mistreatments when incorrect data are entered and not subsequently checked (Muller-Runkel and Watkins 1991; Barthelemy-Brichant et al. 1999), studies have shown that the rate of malfunctions is a fraction of the rate of mistakes prevented, and as systems are continuously improved, this fraction will decrease (Podmaniczky et al. 1985; Fraass et al. 1998). In a study of errors aggravated by the R and V system, Patton et al. (2003) identified 0.04% of treatment errors to be in this category. In order to keep these to a minimum, vigilance is required.

41.4.1 Types of Error and Measures to Avoid Them

Patton et al. (2003) identify four error types:

1. *Incorrect Patient*. Incorrect identification of the patient is surprisingly common. This may be because of two patients with the same name or because a patient is expecting their name to be called and responds to that of another patient. It is common practice for one member of staff to be interacting with the computer system while another prepares the patient for treatment. Asking the patient for their name and date of birth, each day, is a useful precaution, as is requiring the person who set up the patient to check the identity of the patient selected on the R and V system.

2. *Incorrect Data*. Incorrect data can arise in a number of different ways. A bug in the data-transfer software may result in data being incorrectly transferred. Problems may also be associated with human error. Incorrect data may also result from the use of the copy-settings facility referred to in Section 41.2.2.

3. *Incorrect Site*. It is common practice to use a reference point and to then carry out a move relative to this point. It is possible to forget to carry out the move, and if wide tolerances are set, the R and V system may not identify this as an error (Fontenla et al. 1996). The chances of this leading to an error are reduced if tolerances are kept as tight as possible (see Section 41.3.2).

4. *Incorrect Beam Modifier*. Beam modifiers such as bolus, lead blocks or compensators, are often not identified by the R and V system. This type of error is perhaps not directly related to the use of the R and V system, but where different coding is provided it should be used. It is also useful to put a reminder into the system that appears on the screen during treatment.

41.4.2 Training Requirements

Staff training is extremely important for the acceptance and efficient use of any new system, particularly as the R and V system is often part of a much larger package of new technology. It is important that staff who will be using the system on a daily basis have time and training to become familiar with all aspects of the system and have access to a trainer or the system manager for an immediate response to software and hardware problems. When data are to be entered by hand, it is important that the input information is in a standard format, to assist accuracy and speed. It is useful to have the same user interface for all linacs in a department, even when the linacs are from different manufacturers.

Commercial systems are often manufacturer-specific, and never exactly conform to the requirements of a particular department. Acceptability is dependent on the software design and the reliability of the computer hardware. The whole system should be both user-friendly and robust enough to cope with many operators. It is sensible to run the system initially with minimum functionality, and then to add to it once the system has become familiar. Initially, when an R and V system is introduced, patient throughput decreases until staff become familiar with the system. However, this is temporary, and patient throughput will soon rise, particularly if assisted setup is used.

41.4.3 Use of Override Facilities

There will be times when the settings for a field are modified at the time of treatment; for example, a change in daily dose or a reduced field size. In this case, an override facility allows the user to change parameters. Because the check system is disabled, the use of the override facility is a significant cause of errors. It is important that there is a human check of the overridden data. It is also important that the software makes clear which items are being overridden; overriding everything when half way through an assisted setup can lead to some items being overridden unintentionally. Provided that passwords are correctly used at all times, the system can provide a useful audit trail.

The R and V override facility may also be used when the machine control system is exhibiting a fault, but only with *extreme care*. It is not advisable to use it to speed up treatment when a change has been made.

41.5 QUALITY ASSURANCE

A quality assurance program should ensure that failures, such as corrupt data, power failures, etc. are kept to a minimum and do not significantly affect patient treatments. Individual QA programs will depend on the system in use. Table 41.4 gives a suggested list of checks that might be performed by trained personnel while the system is relatively new to the department. The frequency of some of the QA checks can be reduced as confidence in and familiarity with

TABLE 41.4

Quality Assurance Program

Daily checks:

1. **Before treatments commence**

   Treat a *test* patient, using the assisted setup (this has the advantage of familiarising the operator with the system), print out the daily record and check that the following are as prescribed:
   - The parameters obtained using the assisted setup
   - The Monitor Units delivered (look at screen and printout)
   - The number of fractions
   - The cumulative dose
   - The date on the printout

2. **After every treatment**
   - Look at the screen and daily printout (if done) and check it before going on to next field and signing it
   - Check next field before treating
   - Be especially careful with overridden parameters

3. **At the end of the day**
   - Check the record of that day's treatments either using printout or by examination of the data on screen
   - Backup and check patient database

Weekly checks:
   - Backup and verify system files
   - Check machine performance statistics from machine record printout

Monthly checks:
   - Archive or delete patient prescriptions from database one month after completion
   - Backup and verify customisation programs (such as tolerance table consistency)
   - Test and time the assisted setup of machine parameters
   - Test uploading of simulator parameters into the prescription

Checks at the end of a course of treatment:
   - Check that the number of fractions and the cumulative dose are as prescribed
   - Check that the R&V record is the same as the treatment sheet, if used

the system grows (Carson et al. 1992). The daily check (which can be carried out as part of the run-up process) and backup of the system are the most important. IPEM Report 93 (2006) provides further recommendations.

41.6 THE NETWORKED DEPARTMENT

To use the R and V system efficiently, networking is essential between simulator, treatment machines, and treatment planning systems. If the treatment machines are networked, it is then possible to switch patients from one machine to another without having to re-enter the data. If the planning system and the simulator are linked to the network, the need for typing in field parameters is removed, and patients are treated with the same data that has been simulated and planned.

There are different views on the best method of networking. Some systems have a master database on each unit, the data of which may be accessed by other units but not transferred to them. This ensures that a patient's data cannot exist on two units simultaneously. Others have a central master database of all patients stored, as well as separate databases on each of the units of the patients currently on treatment. A warning is displayed when treating a patient, if the two databases are not identical. The general trend is toward a single relational database accessible from anywhere. This puts a high demand on reliability and safety features. For a single database, record locking should ensure that it is not possible to access the data on two units simultaneously. A typical medium-sized radiotherapy network is shown in Figure 41.1.

For patient data transfer alone (i.e. mainly text files), a typical network in a small department would have a hub and star configuration, using 10-base-T twisted pair cabling. Image

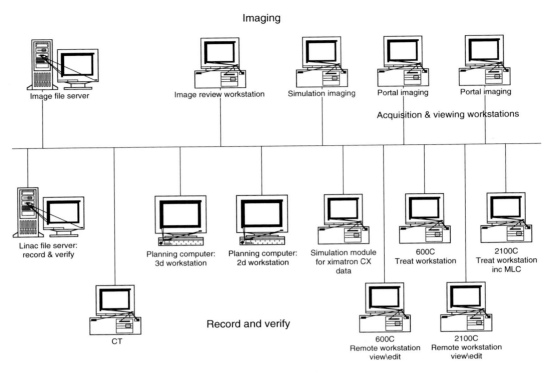

FIGURE 41.1
A typical radiotherapy network.

handling and transfer requires a much larger capacity cable and a storage system (fileserver) of much higher capacity. At present, 100-base-T cabling is essential for image transfer in reasonable time. Any network cabling over distances greater than 98 metres requires the use of fibre-optic cables.

A contentious issue is whether the network should be isolated from the main hospital network. Virus infections on a radiotherapy network can be disastrous and strict discipline needs to be exercised to prevent attacks as some system suppliers are reluctant to authorise the use of antivirus software on their equipment because of the problem of testing, and undesirable effects on real-time control systems. It is best practice to use a firewall or VLAN to isolate the R and V system from the hospital network. Some manufacturers now also firewall the treatment workstations from the radiotherapy network.

41.6.1 INTEGRATION WITH HOSPITAL INFORMATION SYSTEM

The next step is to have an integrated radiotherapy system where a central database contains all the patient data, including clinic appointments, test results etc, which is linked to all the treatment units, simulators and treatment planning computers and can handle images from these systems (Saunders 1990). This can be a costly setup, because not only does it need to have a powerful database and linked workstations situated at treatment units, in offices, and clinic rooms, but also a sophisticated backup system so that treatments can continue in the event of computer failure. It is important to have sufficiently powerful uninterruptible power supplies (UPSs) installed to maintain the system long enough to keep the full integrity of the data. For a completely integrated system, the computing power should be sufficient to handle image files, such as CT, MRI, portal images, simulator images, Beam's-Eye Views (BEVs) and Digitally Reconstructed Radiographs (DRRs), and connections to other hospital sites are extremely useful.

As these systems are introduced, staff instruction will need to include an emphasis on the interrelationship between all sections of staff. For example, secretaries may enter the general information about a new patient and simulator staff use the system to book simulation, mould room work, and treatment. Complete integration of the databases may not be possible, but it should be possible to ensure automatic transfer of data. Duplication of data entry should be avoided because it may lead to two different versions of the data. The simulator or treatment planning staff should be responsible for the entry of the treatment setup parameters. After treatment, follow-up data is added by different staff from those producing statistics, and it is important for the latter to make known to the former which are the vital data to ensure maximum usefulness of the system.

CHAPTER 42

DATA COMMUNICATION WITH
DICOM

John P. Sage, John N. H. Brunt, and W. Philip M. Mayles

CONTENTS

42.1 INTRODUCTION

Unless all the equipment being used in a department is from a single software supplier, it is a requirement that image and other data should be transferable between different computing systems in such a way that it can be interpreted by different software. The Digital Imaging and Communications in Medicine (DICOM) standard aims to provide a standard format and has now reached general acceptance as the standard for such data transfer. In this chapter, the radiotherapy requirements for data transfer are considered in the context of the DICOM standard.

The DICOM standard began in 1985 as the American College of Radiologists and National Electrical Manufacturers Association (ACR–NEMA) standard for the transfer of CT images. As digital imaging progressed the need for a broader standard was identified and the ACR-NEMA standard was expanded in scope to become the DICOM standard we have today. The standard itself can be found in digital form on the NEMA web site (http://medical.nema.org), but it is written in quite technical language, and less-technical web sites such as http://www.rsna.org/Technology/DICOM or http://www.dclunie.com may be more helpful. The lecture notes for a course run by the Swiss Society of Radiobiology and Medical Physics are available at http://www.sgsmp.ch/dicom. Published articles on DICOM (Bidgood et al. 1997; ECRI 2001; Mildenberger et al. 2002) are also a useful source of information.

This chapter is intended to give an introduction to some of the concepts that need to be understood by the radiotherapy physicist to support the use of the standard for radiotherapy communication. It begins with a description of the building blocks of the standard and then considers the application to radiotherapy.

42.2 ELEMENTS OF THE STANDARD

The DICOM standard has three main components: a data format, a definition of actions to be taken on the data and a protocol for the communication of the data*. The standard applies the concepts of object-oriented programming languages, defining a series of data objects and operations that may be done to them.

42.2.1 THE DATA FORMAT

As with any database, each real-world object can be described by an information object. Part 3 of the standard contains a series of information object definitions (IODs) which specify the structure of all DICOM objects. Part 10 of the standard describes a disk file format.

---

* This began primarily as a data transfer protocol in the same way as HTTP, but it is increasingly common for the data to be stored on disk in DICOM format.

A truly object-oriented data structure would contain within the object only those items of information that are related to that object. For example, the *information* about the patient—such as name and date of birth—would be contained in a *patient* object, and information about an image—such as the pixel size and image location—would be contained in a separate *image* object. Some of the simpler object types, such as print jobs or system messages, follow these principles, but items such as images are defined as composite objects. This means that the image format includes within its header, details of all objects that are relevant to the image, such as the patient details and the clinician's name, as well as details (or *attributes*) of the image, such as the pixel size and the location of the image. Because each object is sent in a separate transaction, or stored in a separate file, the use of composite objects is faster.

Each item or data element in a DICOM file consists of a tag indicating what sort of data it is, an indication of the length of the data and finally the data itself in the form:

Tag 1	Tag 2	VR	Value Length	Data
16 bit	16 bit	16 bit	16 bit	Variable

The *tag* consists of two 16-bit integers. For example, the tag (0010,0020) represents a patient identification number. The first of these numbers indicates the data group—group 0010 contains all tags relating to the patient*. A list of all the tags defined is contained in the data dictionary (Part 6 of the standard) and their application to a particular data object is contained in the IODs (Part 3). The next two integers provide a definition of the data representation that follows. In the original ACR–NEMA version of the standard, the type of data or *value representation* (VR) was looked up in the data dictionary and the *value length*, was a 32-bit integer. This method of data type definition is called *Implicit VR*. In the newer versions of the standard, a method called *Explicit VR* is preferred. In this, the first 16 bits define the data type, or VR and the second indicates the length. VRs use a two-letter code indicating the type of data. For example, the patient name is type PN and an integer is type IS. Provisions are also made for variable lengths. The DICOM standard aims to provide backwards compatibility, so both methods are permitted. The DICOM data dictionary defines three types of data. *Type-1 data* are required to be present and a blank value is not acceptable. *Type-2 data* elements are also required, but it is permitted for their value to be empty. *Type-3* data are optional. Both type-1 and type-2 data can be conditional (as in *type 1C*), indicating that their presence is required only if some condition is met.

Different computer systems use different ways of storing data that are larger than 8 bits (1 byte). Some computers, such as Windows™ based PCs, represent a two-byte number with the least significant byte being stored (or transmitted) first. This is referred to as the *little endian*[†] storage format. The other method, still used in many UNIX workstations, was to reverse the two bytes; this is called the *big endian* storage format. DICOM provides for both methods of representation, but little endian is the default.

To examine DICOM data, one must first get the data as a file in the part-10 format, which is not always easy; two systems may transfer data via DICOM but not use DICOM files internally. Many radiotherapy systems have an intermediate step where the data are stored as files until a user explicitly imports them. Otherwise, it is possible to set up a DICOM service on a PC

---

* The standard only specifies even-number groups. All odd-number groups are provided for companies to use for proprietary data not included in the DICOM object definitions. All such private groups are registered with the aim of ensuring that there is no misinterpretation of data between manufacturers. If the object is an image, the last tag (7fe0,0010) is the pixel data as a continuous stream. The information required to read the image data (image size, bit depth, etc.) is contained in group 0028 of the DICOM header.

[†] The term originates in *Gulliver's Travels*, where it referred to the issue of whether to open a soft-boiled egg at the big end or the little end.

which will receive DICOM data and store it as a file*. The text fields in the header of a DICOM file can, to some extent, be understood using a standard text editor, but a DICOM dump program (available from the ERL or OFFIS) will be needed for any thorough analysis.

42.2.2   DATA ORGANISATION

DICOM data structures are fundamentally based on a hierarchical data structure. At the top level is the patient. The patient's data are broken down into examinations, or "studies", that are further divided into "series". Thus, an MR study may consist of a T1 weighted series and a T2 weighted series. Because each slice of a multi-slice image is a separate composite object in DICOM, all of these related objects are grouped in the single series. As shall be seen later (Figure 42.2), DICOM RT objects, such as DICOM RT Plan, are stored at the series level.

42.2.3   DICOM ACTIONS (SERVICE CLASSES)

For two systems to communicate, a number of transactions must be defined. For composite information objects, such as images, the allowed transactions are STORE, GET, MOVE, FIND and ECHO. These elemental services are called *DICOM message service elements*, or DIMSEs. Any practical task may involve a sequence of these transactions, which is called a *service class*. A service class such as *Query/Retrieve* is made up of three elemental services: FIND, GET, and MOVE. Services are carried out on data objects, and the manner in which the service is performed is dependent on the data object on which it is performed. Thus, the service and the object form a *service-object pair class*, or SOP class. An example of an SOP class is CT image storage.

42.2.4   DICOM IDENTIFIERS (UNIQUE IDENTIFIERS)

The DICOM standard includes a system to enable objects to be identified in such a way that each system that generates data can create an identifier for the data that is guaranteed to be unique. This is called the unique identifier (UID). The UID can be used as a pointer that allows objects to be linked together. Each manufacturer registers a UID root unique to that manufacturer. It is then the responsibility of the manufacturer to ensure that the remainder of the numbers making up the full UID are also unique. Each SOP instance has its own UID stored in data element (0008,0018). If one object has been used as the basis to create a new object—for example, a structure set based on a CT image—then the structure set will reference the CT image UID within the DICOM header. There are also study and series UIDs through which a database can determine exactly which objects belong together. With the aid of UIDs, it is possible for related data objects to be sent by separate routes to an archive system, and the linkages can be restored on arrival.

42.2.5   DATA COMMUNICATION

The DICOM standard is built on the International Standards Organisation Open Systems Interconnection (ISO-OSI) seven-layer model in which data are passed through various layers from the application down to the physical transport layer. DICOM deals with the application layer, and some of its terminology, such as *application entity title (AE Title)*, is derived from this. For practical application, the standard is based on TCP/IP networking.

---

* Such software is available in the form of the Central Test Node (CTN) software which is available from the Electronic Radiology Laboratory of Mallinckrodt Institute of Radiology (http://wuerlim.wustl.edu) or from OFFIS (http://dicom.offis.de/dcmtk.php.en).

Each system has a fixed IP address, and service class providers (SCP) (see Section 42.2.5.1) will listen for DICOM associations on a specified TCP port. The standard port for DICOM communication is port 104, but provided that the port selected is not one dedicated to other network services, such as ping, ftp and telnet, any port can be chosen. A system may have different DICOM services operating simultaneously on different TCP ports.

42.2.5.1 Users and Providers

DICOM communication is always between two computers, one of which is the user (or client) and the other being the provider (or server) for the particular transaction. These are termed *Service Class User* (SCU) and *Service Class Provider* (SCP). Thus, a CT scanner wishing to transmit data is defined as the SCU, and the PACS* archive or treatment planning system is the SCP. A given computer may be both an SCP and an SCU. For example, the planning system may additionally be able to query the CT scanner database and retrieve images. For that transaction, the CT scanner is the Query/Retrieve SCP, and the planning system is the Query/Retrieve SCU.

A list is maintained in each system of all other systems with which it will associate. This list has three required fields: the IP address, the TCP port and the AE Title of the DICOM service. The AE Title is set within the software for each DICOM application. It can be anything, but it is wise to make sure each system has a different and representative AE Title. Some systems sort the list internally by AE Title, so multiple entries with identical AE Titles can cause serious problems.

42.2.5.2 DICOM Data Transfer

Transfer of data using DICOM consists of three stages. The first stage is a *handshaking* operation between the two computers, which is referred to as establishing an *association*. Each SCU and SCP has a defined AE title. The SCP listens on a defined port on the computer, so the SCU must know the SCP AE title, the port number and the IP address which identifies the computer. The SCU first requests an association with the SCP providing information on its own AE title and IP address. Depending on the degree of protection incorporated in the SCP, this association may be rejected if either the AE title or the SCU host IP address are not recognised, but the SCP may be set up to accept any association. The next step is to negotiate the action (SOP class) required and the transfer syntax (e.g. VR little endian) of the data. Communication may also be rejected at this stage of the negotiation.

Once it has been established that the data are potentially compatible, the data are transmitted. The SCP will now try to carry out the requested SOP class. It may be that this proves impossible because some data element is missing, in which case an error message may be returned. However, an error message is not necessarily an indication of failure of the data transfer, as it may refer to something that is not a requirement. Equally, an allegedly successful data transfer may not actually have succeeded because of some problem with the data. An example of this is that different systems use different methods to determine whether two images are different. If the recipient uses an optional DICOM element (see Section 42.2.1 and Section 42.2.6) for this purpose which is not transmitted by the sending system, succeeding images may overwrite each other.

42.2.6 THE DICOM CONFORMANCE STATEMENT

In view of the wide applicability of the DICOM standard, software that interacts with DICOM data is unlikely to support all the data types that are defined in the standard. For a

---

* PACS = Picture Archiving and Communication Systems.

given data object, certain mandatory parameters are defined which must be included if the data object is to be supported at all, but additional optional items can also be included. For this reason, each software supplier is required to produce a *DICOM conformance statement* in which the objects that are supported and the optional items that are used are defined. The conformance statement also lists actions that are taken and any specific requirements that have to be met. From the DICOM conformance statements of two pieces of software, it should be possible to determine whether the software systems will communicate with each other and the degree of interoperability that can be expected. In the example given in Section 42.2.5.2, images can be communicated between the systems, but interoperability is not achieved because the images are not stored properly. Another example of a failure of interoperability is where data about image magnification are not transmitted in the form used by the recipient system. By comparing two conformance statements, it is relatively easy to determine whether communication will be possible. However, because of the many optional items within DICOM and because of frequent misinterpretation of the standard, interoperability can only be confirmed reliably by a live data transfer test coupled with careful assessment of the transferred data.

42.2.7 DATA TRANSLATION AND QUALITY ASSURANCE

It is frequently thought that electronic data transfer is bound to be error-free, provided that an appropriate check (such as a checksum) is carried out by the system to ensure data integrity. This is not the case, because the process of data transfer is likely to involve translation of the data from one standard to another* (see, for example, Section 42.4.2). It is, therefore, essential that each DICOM interface be carefully tested to ensure that the transferred data are as expected.

42.3 DICOM APPLIED TO DIAGNOSTIC IMAGE DATA

Image data from a number of different sources are used in radiotherapy treatment planning. Data sources include CT, MR, PET, SPECT, ultrasound and planar x-ray images. The most mature part of the DICOM standard is that relating to CT data. As indicated above (Section 42.2.1), the DICOM image format consists of a header section, much of which is in ASCII format, which defines such things as the patient name, image identification, and details about the type and format of the image. Each slice of a CT data set is stored as a separate file, usually with the file extension *dcm* or *img*. In addition to transverse slice data, DICOM defines a secondary capture data type referred to as DICOM SC. This allows almost any 2D image type to be stored as a DICOM file. Some images will only be available as bitmap image format files, such as TIFF or BMP, and a means of translating such images into the DICOM SC format and attaching an appropriate header is useful. Viewers for DICOM image files are widely available, many of them at no cost (Escott and Rubinstein 2003).

42.3.1 DICOM PRINT FUNCTION

DICOM defines a standard for printing images which can be used to obtain hardcopy images on any printer that supports the standard, but which is particularly useful for a laser imager. Any software system that has a DICOM print output can then produce high quality images on the imager. This can be a useful feature for producing digitally reconstructed radiographs (DRRs) on x-ray film.

---

* There is an increasing trend towards using the DICOM standard for internal data storage. This will tend to reduce the requirement for data translation and thus improve security.

42.3.2   DICOM WORKLIST

The DICOM worklist provides a methodology to streamline the work of a diagnostic department and enables workloads for CT scanners or other image acquisition systems (*modalities*) to be more easily managed with close integration to a PACS or radiology information system (RIS). Where different modalities receive identical demographic information from the RIS via DICOM worklist, this eliminates problems of subsequently associating, in an archive, studies from the same patient which have accidental typing discrepancies in demographic data (see also Section 42.5). The *performed procedure step* allows the imaging system to record the start and completion of the requested imaging operation. A format also exists for the definition of an imaging report (Hussein et al., 2004). A later addition to the standard is the ability to request a storage commitment. This is useful when sending an image to an archive. The CT scanner sends a *storage commitment request* and the archive responds with confirmation that the data are satisfactorily committed to storage. The CT scanner can then safely delete the image.

42.3.3   MEDIA STORAGE

The standard includes definition of how images and other data are stored on various removable media, such as DVDs and CD-Rs. This enables data to be archived in a relatively future-proof manner in the absence of a PACS, to be easily transported to a different treatment centre and to peripheral clinics, or given to the patient.

42.4   RADIOTHERAPY DATA OBJECTS

Although the image itself is what is of principal interest in diagnostic radiology, in radiotherapy, the magnitude of the pixel data elements is also important. In addition, overlays indicating regions of interest need to be available as separate data objects so that they can be used in the design of treatment fields. For this reason, in 1997, the DICOM committee decided to add a set of radiotherapy objects to the standard (Neumann 2002). Figure 42.1 shows the flow of data in a radiotherapy department and the relevant DICOM objects that are transferred.

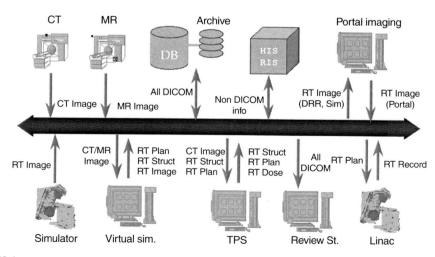

FIGURE 42.1

Diagram showing the transfer of DICOM data at different stages of the treatment planning and delivery process. A DICOM printer could also be added. Data transfer between devices implies that a connection has been specifically established between them. Alternatively, interfaces may be set up between each device and the archive, which can then provide a link between devices (although care must be taken to ensure that treatment plans are not made available to the linac until they are fully approved).

42.4.1 DICOM RT STRUCTURE SET

To carry out radiotherapy treatment planning, the target tissue and organs at risk are defined. This process of segmentation of the tomographic images leads to a set of structures, which are defined by the DICOM RT structure set object. The data structure allows for different contour types, including the body or *external* contour and the target as PTV, CTV or GTV. Structures can also be defined as *avoidance* structures. These structures have special meanings that must be maintained across the different computing systems. A Structure Set is linked to an image (see Figure 42.2 in Section 42.4.7). In addition to contours, the standard allows the definition of markers and isocentres, although interpretation of these may not be the same.

42.4.2 DICOM RT PLAN

The treatment planning process leads to a series of treatment beam definitions as defined by the DICOM RT plan object. This includes an indication of the position of the treatment isocentre, as well as the field sizes and beam orientations, and there is provision for definition of tolerance tables. The number of monitor units is defined, as well as methods of describing doses to the target and organs at risk and the planned fractionation. Couch movements can be relative or absolute. The DICOM standard requires that all beam definitions follow the IEC 1217 standard (IEC 1996, and see Section 11.5.4) apart from the definition of the coordinate system. In DICOM, the positive Z direction is along the couch towards the gantry, whereas in IEC1217, this is the Y direction. In IEC1217, the positive Z direction is anterior (for a supine patient), whereas in DICOM, the positive Y direction is posterior. The X direction is unchanged. This issue can lead to confusion.

Transmission of an RT plan may require a number of data translations. These may include the name of the treatment unit, the name of the wedge and of any electron applicators. It is, therefore, especially important to verify such data transfers with extreme care when the link is set up and to apply appropriate caution subsequently. For multi-segment treatments, such as IMRT beams and automatic wedge treatments, it may be necessary to change the number of segments. For example, the treatment planning system may consider a step-and-shoot IMRT treatment as consisting of four radiation segments, while the linear accelerator may treat this as a sequence with eight control points, one at the beginning and end of each segment, because the accelerator requires three extra *move-only* segments. DICOM itself uses the accelerator convention.

Provision is made for plan approval status, but it is not mandatory that this be implemented. Definitions also exist to define brachytherapy treatments, although these are not yet widely used.

42.4.3 DICOM RT DOSE

Treatment planning systems calculate the radiation dose distribution as a matrix of points with associated doses. These dose grid files are provided for in the DICOM RT dose object. Definitions also exist in the DICOM RT dose specification to store relationships between dose and structures through dose volume histograms and dose region of interest (ROI) statistics. Isodose curves may also be defined. This functionality is slowly being implemented.

42.4.4 DICOM RT IMAGE

In radiotherapy, planar x-ray images are closely associated with the geometric settings of the related treatment beam. The RT image is an extension of the planar x-ray format to give additional information about the exact geometry of the image, such as the distance to the source, and also to combine the image with the field definition modules of the DICOM RT plan object relating to that particular beam. In this way, the anatomical image may be overlaid by the shape of the treatment field with which that anatomy is to be treated, as well as any anatomical structures which are outlined. It is also possible to represent the field as a curve object, but this is less useful.

Such images may be generated in a number of ways. The treatment planning system will generate DRRs, which show the image that would be obtained from a radiotherapy treatment beam in a plane perpendicular to the beam beyond the patient. Additional sources of such images are the radiotherapy treatment simulator and the radiotherapy treatment set electronic portal imaging device. Associated with these images, but not currently covered by a DICOM object, are the associated field placement errors, which are calculated by comparison with the planning RT Images.

To provide backwards compatibility, many systems allow the overlay to be incorporated into the image pixel data and store the image in the DICOM SC format. The image may then be viewed on systems which do not interpret the RT Image elements. This is not a satisfactory solution, as image data are lost and the beam information is only retained in a visual form.

42.4.5   DICOM RT Treatment Record

Although not in widespread use, the DICOM RT treatment record object completes the list of available data objects under DICOM RT. Included are brachytherapy treatment records and treatment summaries. It is provided for completeness and for the purposes of future compatibility. It is more usual for this type of information to be interfaced directly with the electronic patient record.

42.4.6   Other Data Formats

Although each treatment planning manufacturer has their own internal proprietary data format and many can read and write data into the various DICOM objects, there is another data format that is widely used particularly in systems originating in the U.S.A. This is the RTOG format, which was designed for the purpose of transferring radiotherapy data to the Radiotherapy and Oncology Group data centre so that quality assurance of clinical trials could be performed. This data set provides access to all the data required for radiotherapy planning and is widely supported.

42.4.7   DICOM RT Dependencies

In the radiotherapy process, a number of DICOM objects are created. There is a system of dependencies in which each DICOM RT object contains references to the DICOM objects which preceded it. These are illustrated in Figure 42.2 and Figure 42.3.

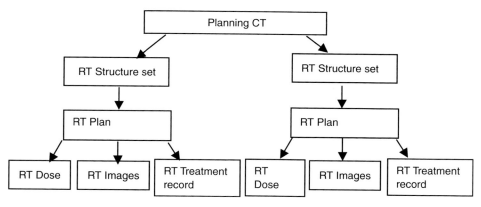

FIGURE 42.2
Example 1: CT Plan with two phases.

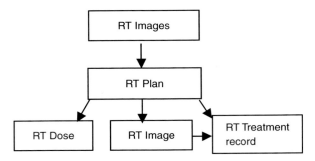

FIGURE 42.3
Example 2: Simulator plan.

42.5 DATA VISUALISATION IN THE RADIOTHERAPY PROCESS

While the requirements for viewing data in the diagnostic field are well-established, the viewing of radiotherapy data other than by using the proprietary software that generated the data is in its infancy. However, the standardisation provided by DICOM RT provides the opportunity for visualisation of radiotherapy data using generic visualisation tools that conform to the standard. In this section the way in which such facilities would be used is described.

Where visualisation of particular objects is not supported, it is essential that the storage system should accept all the objects without modification and be able to transmit them, unmodified, to other computer systems as required. Because some computer systems are unable to accept particular objects, the objects transmitted by the storage system must be selectable (i.e. it must be possible to transmit any combination of objects to the chosen recipient). It is also important that patient identifiers stored in different objects should be consistent. Recipient systems often require the ability to identify different DICOM objects as belonging to the same patient data set. Thus, it is essential that the specifications applied to issues such as the spaces between names and at the end of fields are identical for all the objects.

42.5.1 TARGET VOLUMES

Radiotherapy consultants often see patients in peripheral clinics away from the treatment planning computers. They want to be able to review and alter the volumes defined and to discuss them with colleagues in multidisciplinary teams. This requires the display of DICOM RT Structure Set overlays together with the possibility of editing them.

42.5.2 PLANNING RT IMAGE VIEWING

For simulator-planned patients, the radiation oncologist will need to view the RT Images acquired at simulation to establish that the plan is satisfactory. They will need to view the image with the associated beam overlay and be able to edit the collimator positions and any shielding structures. For CT-planned-patient viewing, the DRRs will form part of the review of the completed plan.

42.5.3 DOSE PLANNING

The availability of all the DICOM objects is essential for dose planning. To be able to use the defined contours, the structure sets are required to be available in the appropriate DICOM format.

The standardisation provided by DICOM RT allows interchange of data between planning systems and facilitates the independent verification of dose calculations (Spezi et al. 2002).

42.5.4   DOSE PLAN VIEWING

The dose plan will be generated by a specialist dose calculation computer. This dose plan will need to be viewed by the radiotherapy consultant and accepted (or otherwise) as suitable for treatment. It will also be desirable for these dose displays to be available for review whenever the patient is seen subsequently. A 3D view of the dose display and its relationship to the defined target and organ at risk volumes is required, as well as a 2D display superimposed on transverse images and on reformatted coronal and sagittal images.

42.5.5   PORTAL IMAGE VIEWING

Portal images are taken to be compared with simulator images or DRRs. Portal imaging software usually includes facilities for such image comparisons and for measuring the differences between the images or *field placement errors*. This is done by the use of overlays of the field shape and of patient anatomy. In order for medical staff to be able to review these registrations, it should be possible to display the images together with the overlays side-by-side. Ideally, it should be possible for the consultant to use the sort of facilities available on the imaging workstations to confirm or edit the registrations and to record approval or otherwise of the images. The values of the field placement errors need to be stored along with the images. One approach to this would be for the matched template images to be stored as DICOM-SC objects so that the consultant can see the end result of the matching process.

42.5.6   RECORDING APPROVAL

One of the main aims of RT object visualisation is for the clinical oncologist to review plans, dose and images, and give their approval to the plan that has been produced. Once treatment has started, the clinical oncologist may view the portal images in conjunction with planning reference images and give their approval for treatment to continue. The approval status of radiotherapy objects is very important and is included within the DICOM standard. It should be possible for clinicians to approve objects with an electronic signature and for this approval status to be clearly indicated.

42.6   SUMMARY

The DICOM standard provides a method to enhance interoperability between computer systems in radiotherapy. However, although image transfer is usually a seamless operation, the complexity of the radiotherapy extensions to the standard means that individual testing of each data link is essential. Advice on such testing can be found in IPEM Report 93 (IPEM 2006). Electronic transfer of data is more reliable than human data transfer by typing at a keyboard, but because of the many translations that occur during data transfer, accuracy cannot be guaranteed. Considered protocols must be in place to ensure that appropriate checks are carried out. Electronic methods of data verification that are independent of DICOM need to be developed. The DICOM standard is not yet fully mature and in the future we should expect that new standardised methods of viewing the data will facilitate wider availability of the electronic data. However, because of the potential for serious harm in radiotherapy treatment, caution should be applied before implementing a completely paperless system.

PART H: QUALITY ASSURANCE
REFERENCES

AAPM (American Association of Physicists in Medicine), Report 13: *Physical aspects of quality assurance in radiotherapy*, American Institute of Physics, New York, 1984.

AAPM (American Association of Physicists in Medicine), Comprehensive QA for radiation oncology: Report of AAPM Radiation Therapy Committee Task Group 40, *Med. Phys.*, 21, 581–618, 1994a.

AAPM (American Association of Physicists in Medicine), AAPM code of practice for radiotherapy accelerators: Report of AAPM Radiation Therapy Committee Task Group 45, *Med. Phys.*, 21, 1094–1121, 1994b.

AAPM (American Association of Physicists in Medicine), *Radiation treatment planning dosimetry verification*, AAPM Report 55 (Task Group 23), American Institute of Physics, New York, 1995.

AAPM (American Association of Physicists in Medicine), *Diode in vivo dosimetry for patients receiving external radiation therapy*, AAPM Report 87, American Institute of Physics, New York, 2005.

Aguirre, J. F., Tailor, R. C., Ibbott, G., Stovall, M., and Hanson, W. F., Thermoluminescence dosimetry as a tool for the remote verification of output for radiotherapy beams: 25 years of experience, in *Standards and Codes of Practice in Medical Dosimetry*, Vol. II, IAEA, Vienna, pp. 191–199, 2003.

Aird, E. G., Williams, C., Mott, G. T., Dische, S., and Saunders, M. I., Quality assurance in the CHART clinical trial, *Radiother. Oncol.*, 36, 235–244, 1995.

Alecu, R. and Alecu, M., In vivo rectal dose measurements with diodes to avoid misadministrations during intracavitary high dose rate brachytherapy for carcinoma of the cervix, *Med. Phys.*, 26, 768–770, 1999.

Allahverdi, M. and Thwaites, D. I., Achievable accuracy in radiotherapy dosimetry. Proc. 17th ESTRO meeting, Edinburgh, Sept. 1998, *Radiother. Oncol.*, 48(Suppl. 1), 187, 1998.

Almond, P. R., Biggs, P. J., Coursey, B. M., Hanson, W. F., Huq, M. S. et al., AAPM's TG-51 protocol for clinical reference dosimetry of high-energy photon and electron beams, *Med. Phys.*, 26, 1847–1870, 1999.

Antonuk, L. E., Electronic portal imaging devices: A review and historical perspective of contemporary technologies and research, *Phys. Med. Biol.*, 47, R31–R65, 2002.

Arnfield, M. R., Wu, Q., Tong, S., and Mohan, R., Dosimetric validation for multileaf collimator-based intensity-modulated radiotherapy: A review, *Med. Dosim.*, 26, 179–188, 2001.

Aznar, M. C., Andersen, C. E., Bøtter-Jensen, L., Bäck, S. A. J., Mattsson, S. et al., Real-time optical-fibre luminescence dosimetry for radiotherapy: Physical characteristics and applications in photon beams, *Phys. Med. Biol.*, 49, 1655–1669, 2004.

Badcock, P. C., The role of computed tomography in the planning of radiotherapy fields, *Radiology*, 147, 241–244, 1983.

Bakai, A., Alber, M., and Nüsslin, F., A revision of the γ-evaluation concept for the comparison of dose distributions, *Phys. Med. Biol.*, 48, 3543–3553, 2003.

Baker, S. J., Budgell, G. J., and Mackay, R. I., Use of an amorphous silicon electronic portal imaging device for multileaf collimator quality control and calibration, *Phys. Med. Biol.*, 50, 1377–1392, 2005.

Balter, J., Pelizzari, C., and Chen, T., Correlation of projection radiographs in radiation therapy using open curve segments and points, *Med. Phys.*, 19, 329–334, 1992.

Barthelemy-Brichant, N., Sabatier, J., Dewe, W., Albert, A., and Deneufbourg, J. M., Evaluation of frequency and type of errors detected by a computerized record and verify system during radiation treatment, *Radiother. Oncol.*, 53, 149–154, 1999.

Bascuas, J. L., Chavaudra, J., Vauthier, G., and Dutreix, J., Interest of systematic in vivo dosimetry in radiotherapy (Intérêt des mesures in vivo systématiques en radiothérapie), *J. Radiol. Electrol.*, 11, 701–708, 1977.

Bayouth, J. E., Wendt, D., and Morrill, S. M., MLC quality assurance techniques for IMRT applications, *Med. Phys.*, 30, 743–750, 2003.

Bel, A., van Herk, M., Bartelink, H., and Lebesque, J. V., A verification procedure to improve patient set-up accuracy using portal images, *Radiother. Oncol.*, 29, 253–260, 1993.

Berger, L., François, P., Gaboriaud, G., and Rosenwald, J.-C., Performance optimization of the Varian aS500 EPID system, *J. Appl. Clin. Med. Phys.*, 7(1), 105–114, 2006.

Beteille, D., Setzkorn, R., Prevot, H., Dusseau, L., Fesqet, J. et al., Laser heating of thermoluminescent plates: Application to intraoperative radiotherapy, *Med. Phys.*, 23, 1421–1424, 1996.

Bidgood, W. D., Horii, S. C., Prior, F. W., and van Syckle, D. E., Understanding and using DICOM, the data interchange standard for biomedical imaging, *J. Am. Med. Inform. Assoc.*, 4, 199–212, 1997.

Bijhold, J., Three-dimensional verification of patient placement during radiotherapy using digital portal images, *Med. Phys.*, 20, 347–356, 1993.

Bijhold, J., Gilhuijs, K. G. A., van Herk, M., and Meertens, H., Radiation field edge detection in portal images, *Phys. Med. Biol.*, 36, 1705–1710, 1991a.

Bijhold, J., van Herk, M., Vijlbrief, R., and Lebesque, J., Fast evaluation of patient set-up during radiotherapy by aligning features in portal and simulator images, *Phys. Med. Biol.*, 36, 1665–1679, 1991b.

Bijhold, J., Gilhuijs, K. G. A., and van Herk, M., Automatic verification of radiation field shape using digital portal images, *Med. Phys.*, 19, 1007–1014, 1992.

BIPM (Bureau International des Poids et Mesures), *Recommendation INC-1 by the Working Group on the Statement of Uncertainties*, BIPM, Sèvres, 1980.

BIR (British Institute of Radiology), *Geometric Uncertainties in Radiotherapy*, BIR, London, 2003.

Bleehen, N., Quality assurance in radiotherapy—Report of a Working Party of the Standing Subcommittee on Cancer of the Standing Medical Advisory Committee, May 1991, Department of Health, London, 1991.

Boag, J. W. and Currant, J., Current collection and ionic recombination in small cylindrical ionisation chambers exposed to pulsed radiation, *Br. J. Radiol.*, 53, 471–478, 1980.

Boellaard, R., van Herk, M., and Mijnheer, B. J., The dose response relationship of a liquid-filled electronic portal imaging device, *Med. Phys.*, 23, 1601–1611, 1996.

Boellaard, R., van Herk, M., and Mijnheer, B. J., A convolution model to convert transmission dose images to exit dose distributions, *Med. Phys.*, 24, 189–199, 1997a.

Boellaard, R., van Herk, M., Uiterwaal, H., and Mijnheer, B., Two-dimensional exit dosimetry using a liquid-filled electronic portal imaging device and a convolution model, *Radiother. Oncol.*, 44, 149–157, 1997b.

Boellaard, R., Essers, M., van Herk, M., and Mijnheer, B. J., New method to obtain the midplane dose using portal in vivo dosimetry, *Int. J. Radiat. Oncol. Biol. Phys.*, 41, 465–474, 1998.

Boyer, A. L. and Li, S., Geometric analysis of light-field position of a multileaf collimator with curved ends, *Med. Phys.*, 24, 757–762, 1997.

Boyer, A., Biggs, P., Galvin, J., Klein, E., LoSasso, T., et al., Basic applications of multileaf collimators, Report of the AAPM Radiation Therapy Committee Task Group No. 50, Medical Physics Publishing, Madison, WI, 2001.

Brahme, A., Dosimetric precision requirements in radiation therapy, *Acta Radiol. Oncol.*, 23, 379–391, 1984.

Brahme, A., Lind, B., and Nafstadius, P., Radiotherapeutic computed tomography with scanned photon beams, *Int. J. Radiat. Oncol. Biol. Phys.*, 13, 95–101, 1987.

Brahme, A., Chavaudra, J., Landberg, T., McCullough, E. C., Nüsslin, F. et al., Accuracy requirements and quality assurance of external beam therapy with photons and electrons, *Acta Oncol.*, 15(Suppl. 1), 1–76, 1988.

Brune, D., Prostate, in *A Practical Manual of Brachytherapy*, Pierquin, B. and Marinello, G., Eds., Medical Physics Publishing, Madison, WI, Chapter 12, 1997.

BSI (British Standards Institution), PD 6461–3: 1995 General metrology, Guide to the expression of uncertainty in measurement (GUM), BSI, London, 1995.

Budgell, G. J., Martens, C., and Claus, F., Improved delivery efficiency for step and shoot intensity modulated radiotherapy using a fast-tuning magnetron, *Phys. Med. Biol.*, 46, N253–N261, 2001.

Budgell, G. J., Perrin, B. A., Mott, J. H., Fairfoul, J., and Mackay, R. I., Quantitative analysis of patient-specific dosimetric IMRT verification, *Phys. Med. Biol.*, 50, 103–119, 2005a.

Budgell, G. J., Zhang, Q., Trouncer, R. J., and Mackay, R. I., Improving IMRT quality control efficiency using an amorphous silicon electronic portal imager, *Med. Phys.*, 32, 3267–3278, 2005b.

Cadman, P., Bassalow, R., Sidhu, N. P., Ibbott, G., and Nelson, A., Dosimetric considerations for validation of a sequential IMRT process with a commercial treatment planning system, *Phys. Med. Biol.*, 47, 3001–3010, 2002.

Caneva, S., Rosenwald, J.-C., and Zefkili, S., A method to check the accuracy of dose computation using quality index: Application to scatter contribution in high energy photon beams, *Med. Phys.*, 27, 1018–1024, 2000.

Caneva, S., Tsiakalos, M. F., Stathakis, S., Zefkili, S., Mazal, A., and Rosenwald, J.-C., Application of the quality index methodology for dosimetric verification of build-up effect beyond air-tissue interface in treatment planning system algorithms, *Radiother. Oncol.*, 79, 208–210, 2006.

Carson, H., Inglis, J., Maycock, P., and Porter, H., Quality assurance of computerised treatment verification and recording systems on radiotherapy accelerators, IPSM Conference on Quality in Radiotherapy York 1992. Abstract in *Scope*, December 1992, 1992.

Chang, J., Obcemea, C. H., Sillanpaa, J., Mechalakos, J., and Burman, C., Use of EPID for leaf position accuracy QA of dynamic multi-leaf collimator (DMLC) treatment, *Med. Phys.*, 31, 2091–2096, 2004.

Chauvet, I., Petitfils, A., Lehobey, C., Kristner, J. Y., Brunet, Y. et al., The sliding slit test for dynamic IMRT: A useful tool for adjustment of MLC related parameters, *Phys. Med. Biol.*, 50, 563–580, 2005.

Chui, C. S., Spirou, S., and LoSasso, T., Testing of dynamic multileaf collimation, *Med. Phys.*, 23, 635–641, 1996.

Ciesielski, B., Wielopolski, L., and Reinstein, L. E., The energy response of agar-alanine phantom dosimeter to gamma radiation, *Med. Phys.*, 15, 380–382, 1988.

Ciesielski, B., Reinstein, L. E., Meek, A. G., and Wielopolski, L., Energy response of agar-alanine free radical dosimetry to therapeutic electron beams, *Med. Phys.*, 20, 1453–1455, 1993.

Ciocca, M., Landoni, L., Italia, C., Montanaro, P., Canesi, P., and Valdagni, R., Quality control in the conservative treatment of breast cancer: Patient dosimetry using silicon detectors, *Radiother. Oncol.*, 22, 304–307, 1991.

Constantinou, C., Harrington, J. C., and DeWerd, L. A., An electron density calibration phantom for CT-based treatment planning computers, *Med. Phys.*, 19, 325–327, 1992.

Cygler, J., Saoudi, A., Wilkins, D., Morash, C., and Perry, C. E. G., Use of MOSFET detectors for in vivo dosimetry during permanent low-dose-rate prostate implants, *Med. Phys.*, 28, 1180, 2001.

De Neve, W., van den Heuvel, F., de Beukeleer, M., Coghe, M., Thon, L. et al., Routine clinical on-line portal imaging followed by immediate field adjustment using a tele-controlled patient couch, *Radiother. Oncol.*, 24, 45–54, 1992.

Depuydt, T., van Esch, A., and Huyskens, D. P., A quantitative evaluation of IMRT dose distributions: Refinement and clinical assessment of the gamma evaluation, *Radiother. Oncol.*, 62, 309–319, 2002.

Dobbs, H. J., Parker, R. P., Hodson, N. J., Hobday, P., and Husband, J. E., The use of CT in radiotherapy treatment planning, *Radiother. Oncol.*, 1, 133–141, 1983.

Dong, L., Antolak, J., Salehpour, M., Forster, K., O'Neill, L. et al., Patient-specific point dose measurement for IMRT monitor unit verification, *Int. J. Radiat. Oncol. Biol. Phys.*, 56, 867–877, 2003.

Droege, R. T., A practical method to routinely monitor resolution in digital images, *Med. Phys.*, 10, 337–343, 1983.

Du Sautoy, A., McEwen, M., Jordan, T., Nahum, A., Nisbet, A. et al., Current issues in electron dosimetry, *Radiother. Oncol.*, 68, S11, 2003.

Dutreix, A., When and how can we improve precision in radiotherapy?, *Radiother. Oncol.*, 2, 275–292, 1984.

EC, Council Directive 97/43/EURATOM, 30 June 1997, *Health protection of individuals against the dangers of ionizing radiation in relation to medical exposure*, Official Journal of the European Communities OJ No. L180/22, 1997a.

EC, Radiation Protection 91. *Criteria for Acceptability of Radiological (Including Radiotherapy) and Nuclear Medicine Installations*. Chapter 7, Office for Official Publications of the European Communities, Luxembourg, 1997b.

ECRI, DICOM reference guide, *Health Dev.*, 30, 5–30, 2001.

Engström, P. E., Haraldson, P., Landberg, T., Hansen, H. S., Elgelholm, S. A., and Nyström, H., In vivo dose verification of IMRT treated head and neck cancer patients, *Acta Radiologica*, 44, 572–578, 2005.

Escott, E. J. and Rubinstein, D., Free DICOM image viewing and processing software for your desktop computer: What's available and what it can do for you, *Radiographics*, 23, 1341–1357, 2003.

Essers, M. and Mijnheer, B. J., In vivo dosimetry during external photon beam therapy, *Int. J. Radiat. Oncol. Biol. Phys.*, 43, 245–259, 1999.

ESTRO (European Society for Radiotherapy and Oncology), *Quality assurance of treatment planning systems—practical examples for external photon beams*, Physics for clinical radiotherapy ESTRO booklet, ESTRO, Brussels, 2004.

Esthappan, J., Mutic, S., Harms, W. B., Dempsey, J. F., and Low, D. A., Dosimetry of therapeutic photon beams using an extended dose range film, *Med. Phys.*, 29, 2438–2445, 2002.

Evans, P. M., Gildersleve, J. Q., Morton, E. J., Swindell, W., Coles, R. et al., Image comparison techniques for use with megavoltage imaging systems, *Br. J. Radiol.*, 65, 701–709, 1992.

Evans, P. M., Hansen, V. N., Mayles, W. P. M., Swindell, W., Torr, M., and Yarnold, J. R., Design of compensators for breast radiotherapy using electronic portal imaging, *Radiother. Oncol.*, 37, 43–54, 1995.

Ezz, A., Munro, P., and Porter, A. T., Daily monitoring and correction of radiation field placement using a video-based portal imaging system: A pilot study, *Int. J. Radiat. Oncol. Biol. Phys.*, 22, 159–165, 1991.

Fenwick, J. D., Tome, W. A., Jaradat, H. A., Hui, S. K., James, J. A. et al., Quality assurance of a helical tomotherapy machine, *Phys. Med. Biol.*, 49, 2933–2953, 2004.

Fenwick, J. D., Tome, W. A., Kissick, M. W., and Mackie, T. R., Modelling simple helically delivered dose distributions, *Phys. Med. Biol.*, 50, 1505–1517, 2004.

Ferreira, I. H., Dutreix, A., Bridier, A., Svensson, H., ESTRO European assurance programme for radiation treatments (the EQUAL network), in *Standards and codes of practice in medical dosimetry*, Vol. II, IAEA, Vienna, pp. 157–165, 2003.

Fiorino, C., del Vecchio, A., Cattaneo, G. M., Fusca, M., Longobardi, B., Signorott, P., and Calandrino, R., Exit dose measurements by portal film dosimetry, *Radiother. Oncol.*, 29, 336–340, 1993.

Flühs, D., Heintz, M., Indenkämpen, F., and Wieczorek, C., Direct reading measurement of absorbed dose with plastic scintillators. The general concept and application to ophthalmic plaque dosimetry, *Med. Phys.*, 23, 427–434, 1996.

Fontenla, D. P., Curran, J., Yaparpalvi, R., and Vikram, B., Customization of a radiation management system to support in vivo patient dosimetry using diodes, *Med. Phys.*, 23, 1425–1429, 1996.

Ford, E. C., Mageras, G. S., Yorke, E., Rosenzweig, K. E., Wagman, R., and Ling, C. C., Evaluation of respiratory movement during gated radiotherapy using film and electronic portal imaging, *Int. J. Radiat. Oncol. Biol. Phys.*, 52, 522–531, 2002.

Fraass, B. A., Lash, K. L., Matrone, G. M., Volkman, S. K., McShan, D. L., Kessler, M. L., and Lichter, A. S., The impact of treatment complexity and computer-control delivery technology on treatment delivery errors, *Int. J. Radiat. Oncol. Biol. Phys.*, 42, 651–659, 1998.

Francescon, P., Cora, S., and Chiovati, P., Dose verification of an IMRT treatment planning system with the BEAM EGS4-based Monte Carlo code, *Med. Phys.*, 30, 144–157, 2003.

Frank, S. J., Forster, K. M., Stevens, C. W., Cox, J. D., Komaki, R., Liao, Z., Tucker, S. et al., Treatment planning for lung cancer: Traditional homogeneous point-dose prescription compared with heterogeneity-corrected dose-volume prescription, *Int. J. Radiat. Oncol. Biol. Phys.*, 56, 1308–1318, 2003.

Fuks, Z. and Kaplan, H. S., Recurrence rates following radiation therapy of nodular and diffuse malignant lymphomas, *Radiology*, 108, 675–684, 1973.

Galbraith, D. M., Aget, H., Leung, P. M. K., and Rider, W. D., Eye sparing in high energy x-ray beams, *Int. J. Radiat. Oncol. Biol. Phys.*, 11, 591–595, 1985.

Georg, D., de Ost, B., Hoornaert, M. T., Pillette, P., van Dam, J. et al., Build-up modification of commercial diodes for entrance dose measurements in higher energy photon beams, *Radiother. Oncol.*, 51, 249–256, 1999.

Gildersleve, J., Dearnaley, D. P., Evans, P. M., Law, M., Rawlings, C., and Swindell, W., A randomised trial of patient repositioning during radiotherapy using a megavoltage imaging system, *Radiother. Oncol.*, 31, 161–168, 1994.

Gilhuijs, K. G. A. and van Herk, M., Automatic on-line inspection of patient setup in radiation therapy using digital portal images, *Med. Phys.*, 20, 667–677, 1993.

Gilhuijs, K. G. A., Touw, A., van Herk, M., and Vijlbrief, R. E., Optimization of automatic portal image analysis, *Med. Phys.*, 22, 1089–1099, 1995.

Ginestet, C., Malet, C., Cohen, A., Lafay, F., and Carrie, C., Impact of tissues heterogeneities on monitor units calculation and ICRU dose point: Analysis of 30 cases of prostate cancer treated with 18-MV photons after three-dimensional planning, *Int. J. Radiat. Oncol. Biol. Phys.*, 48, 529–534, 2000.

Graham, M., Cheng, A., Geer, L., Binns, W., Vannier, M., and Wong, J., A method to analyze 2-dimensional daily radiotherapy portal images from an on-line fiber-optic imaging system, *Int. J. Radiat. Oncol. Biol. Phys.*, 20, 613–619, 1991.

Graves, M. N., Thompson, A. V., Martel, M. K., McShan, D. L., and Fraass, B. A., Calibration and quality assurance for rounded leaf-end MLC systems, *Med. Phys.*, 28, 2227–2233, 2001.

Guan, H., Yin, F. F., and Kim, J. H., Accuracy of inhomogeneity correction in photon radiotherapy from CT scans with different settings, *Phys. Med. Biol.*, 47, N223–N231, 2002.

Halverston, K., Leung, T., Pellet, J., Gerber, R., Weinhous, M., and Wong, J., Study of treatment variation in the radiotherapy of head and neck tumours using a fibre-optic on-line radiotherapy imaging system, *Int. J. Radiat. Oncol. Biol. Phys.*, 21, 1327–1336, 1991.

Hamers, H. P., Johansson, K. A., Venselaar, J. L., de Brouwer, P., Hansson, U., and Moudi, C., Entrance and exit TL-dosimetry in the conservative treatment of breast cancer: A pilot study for the EORTC-Radiotherapy Cooperative Group, *Radiother. Oncol.*, 22, 280–284, 1991.

Hamers, H. P., Johansson, K. A., Venselaar, J. L., de Brouwer, P., Hansson, U., and Moudi, C., In vivo dosimetry with TLD in conservative treatment of breast cancer patients treated with the EORTC protocol 22881, *Acta Oncologica*, 32, 435–443, 1993.

Hansen, V. N., Evans, P. M., and Swindell, W., The application of transit dosimetry to precision radiotherapy, *Med. Phys.*, 23, 713–721, 1996.

Hansen, V. N., Evans, P. M., and Swindell, W., Extraction of primary signal from EPIDs using only forward convolution, *Med. Phys.*, 24, 1477–1484, 1997.

Harnett, A. N., Hirst, A., and Plowman, P. N., The eye in acute leukemia 1. Dosimetric analysis in cranial radiation prophylaxis, *Radiother. Oncol.*, 10, 195–202, 1987.

Heijmen, B. J. M., Pasma, K. L., Kroonwijk, M., Althof, V. G. M., de Boer, J. C. J., Visser, A. G., and Huizenga, H., Portal dose measurement in radiotherapy using an electronic portal imaging device (EPID), *Phys. Med. Biol.*, 40, 1943–1955, 1995.

Hendry, J. H. and Roberts, S. A., The sensitivity of human tissues to changes in dose fractionation: Deductions from the RCR survey among U.K. radiotherapists, *Clin. Oncol. (R. Coll. Radiol.)*, 3, 22–27, 1991.

Herman, M. G., Balter, J. M., Jaffray, D. A., McGee, K. P., Munro, P. et al., Clinical use of electronic portal imaging: Report of AAPM Radiation Therapy Committee Task Group 58, *Med. Phys.*, 28, 712–737, 2001.

Hernigou, P., Marce, D., Julieron, A., Marinello, G., and Dormont, D., Bone sterilisation by irradiation and virus HIV (Stérilisation osseuse par irradiation et virus VIH), *Revue Chirurgie Orthopédique*, 79, 445–451, 1993.

Harrison, R. M. and Lambert, G. D., A method and contrast-detail phantom for the quantitative assessment of radiotherapy portal imaging systems, *Br. J. Radiol.*, 67, 384–388, 1994.

Heukelom, S., Lanson, J. H., and Mijnheer, B. J., Comparison of entrance and exit dose measurements using ionisation chambers and diodes, *Phys. Med. Biol.*, 36, 47–59, 1991.

Heukelom, S., Lanson, J. H., van Tienhoven, G., and Mijnheer, B. J., In vivo dosimetry during tangential breast treatment, *Radiother. Oncol.*, 22, 269–279, 1991.

Heukelom, S., Lanson, J. H., and Mijnheer, B. J., In vivo dosimetry during pelvic treatment, *Radiother. Oncol.*, 25, 111–120, 1992.

Higgins, P. D., Alaei, P., Gerbi, B. J., and Dusenbery, K. E., In vivo dosimetry for routine quality assurance in IMRT, *Med. Phys.*, 30, 3118–3123, 2003.

Hounsell, A. R. and Jordan, T. J., Quality control aspects of the Philips multileaf collimator, *Radiother.Oncol.*, 45, 225–233, 1997.

HS (Health and Safety), Statutory Instrument No. 1059: *The Ionising Radiation (Medical Exposures) Regulations*, HMSO, London, 2000.

Hurkmans, C. W., Remeijer, P., Lebesque, J. V., and Mijnheer, B. J., Set-up verification using portal imaging: Review of current clinical practice, *Radiother. Oncol.*, 58, 105–120, 2001.

Hussein, R., Engelmann, U., Schroeter, A., and Meinzer, H. P., DICOM Structured Reporting: Part 1. Overview and characteristics, *Radiographics*, 24, 891–896, 2004.

Huyskens, D., van Dam, J., and Dutreix, A., Midplane dose determination using in vivo dose measurements in combination with portal imaging, *Phys. Med. Biol.*, 39, 1089–1101, 1994.

Huyskens, D., Bogaerts, R., Verstraete, J., Lööf, M., Nyström, H., *Practical Guidelines for the Implementation of In vivo Dosimetry with Diodes in External Radiotherapy with Photon Beams*, ESTRO Booklet No. 5, ESTRO, Brussels, 2001, Available online at www.estroweb.org/ESTRO/upload/pdfs/booklet5.pdf

IAEA (International Atomic Energy Agency), Brachytherapy, in *Physical Aspects of Radioisotopes*, IAEA, Vienna, pp. 26–27, 1967.

IAEA (International Atomic Energy Agency), *Calibration of dosimeters used in radiotherapy*, IAEA Technical Report Series No. 374, IAEA, Vienna, 1994.

IAEA (International Atomic Energy Agency), *International Basic Safety Standards for protection against ionizing radiation and for the safety of radiation sources*, IAEA Safety Series No. 115, IAEA, Vienna, 1996.

IAEA (International Atomic Energy Agency), *Absorbed dose determination in photon and electron beams: An international code of practice*. 2nd ed., Technical Report Series 277, IAEA, Vienna, 1997a.

IAEA (International Atomic Energy Agency), *The use of plane-parallel ionisation chambers in high-energy electron and photon beams: An international code of practice*, IAEA Technical Report Series 381, IAEA, Vienna, 1997b.

IAEA (International Atomic Energy Agency), *Design and implementation of a radiotherapy programme: Clinical, medical physics, radiation protection and safety aspects*, TECDOC-1040, IAEA, Vienna, 1998. (Revision entitled *Setting-up a radiotherapy programme: Clinical, medical physics, radiation protection and safety aspects*. In press.)

IAEA (International Atomic Energy Agency), *Absorbed dose determination in External Beam Radiotherapy, An international code of practice for dosimetry based on standards of absorbed doses to water*, IAEA Technical Report Series 398, IAEA, Vienna, 2000a.

IAEA (International Atomic Energy Agency), Safety Report No. 17: *Lessons learned from accidental exposures in Radiotherapy*, IAEA, Vienna, 2000b.

IAEA (International Atomic Energy Agency), *Investigation of an Accidental Exposure of Radiotherapy Patients in Panama*, IAEA, Vienna, 2001.

IAEA (International Atomic Energy Agency), *Commissioning and quality assurance of computerized planning systems for radiation treatment of cancer*, IAEA Technical Report Series 430, IAEA, Vienna, 2004a.

IAEA (International Atomic Energy Agency), *Accidental Overexposure of Radiotherapy Patients in Białystok*, IAEA, Vienna, 2004b.

ICRU (International Commission on Radiation Units and Measurements), ICRU Report 24, *Determination of absorbed dose in a patient irradiated by beams of x or gamma rays in radiotherapy procedures*, ICRU, Bethesda, MD, 1976.

ICRU (International Commission on Radiation Units and Measurements), *Tissue substitutes in radiation dosimetry and measurement*, Report 44, ICRU, Bethesda, MD, 1989.

ICRU (International Commission on Radiation Units and Measurements), *Phantoms and computational models in therapy, diagnosis and protection*, ICRU Report 48, ICRU, Bethesda, MD, 1992.

ICRU (International Commission on Radiation Units and Measurements), *Prescribing, recording and reporting photon beam therapy*, ICRU Report 50, Landberg, T., (Chairman), Chavaudra, J., Dobbs, H. J., Hanks, G., Johansson, K. A., Möller, T., Purdy, J., ICRU, Bethesda, MD, 1993.

ICRU (International Commission on Radiation Units and Measurements), *Medical imaging—the assessment of image quality*, ICRU Report 54, ICRU, Bethesda, MD, 1996.

ICRU (International Commission on Radiation Units and Measurements), *Prescribing, recording and reporting photon beam therapy*, ICRU Report 62 (Supplement to ICRU Report 50), ICRU, Bethesda, MD, 1999.

ICRU (International Commission on Radiation Units and Measurements), *Prescribing, recording and reporting electron beam therapy*, ICRU Report 71, *Journal of the ICRU*, Vol. 4, No. 1, ICRU, Bethesda, MD, 2004.

IEC (International Electrotechnical Commission), *Medical electrical equipment—Medical electron accelerators. Functional performance characteristics*, IEC Publication 976, IEC, Geneva, 1989.

IEC (International Electrotechnical Commission), *Radiotherapy equipment—coordinates, movements and scales*, CEI/IEC 1217, First Edition, 1996–08, IEC, Geneva, 1996.

IEC (International Electrotechnical Commission), *Medical electrical equipment, Dosimeters with ionization chambers as used in radiotherapy*, Report 60731-ED.2, IEC, Geneva, 1997.

IEC (International Electrotechnical Commission), *Medical electrical equipment, Part 2: Particular requirements for the safety of electron accelerators in the range of 1–50 MeV*, IEC-60601-2-1, IEC, Geneva, 1998.

IPEM (Institute of Physics and Engineering in Medicine), Report 81: *Physics aspects of quality control in radiotherapy*, IPEM, York, 1999.

IPEM (Institute of Physics and Engineering in Medicine), *Medical and Dental Guidance Notes: A Good Practice Guide to Implementing Ionizing Radiation Protection Legislation in the Clinical Environment*, IPEM, York, 2002.

IPEM (Institute of Physics and Engineering in Medicine) Report 93. *Guidance for Commissioning and QA of a Networked Radiotherapy Department*, IPEM, York, 2006.

IPEMB (Institute of Physics and Engineering in Medicine and Biology), The IPEMB code of practice for electron dosimetry for radiotherapy beams of initial energy from 2 to 50 MeV based on an air kerma calibration, *Phys. Med. Biol.*, 41, 2557–2603, 1996a.

IPEMB (Institute of Physics and Engineering in Medicine and Biology), The IPEMB code of practice for the determination of absorbed dose for x-rays below 300 kV generating potential (0.035 mm Al-4 mm Cu HVL; 10–300 kV generating potential), *Phys. Med. Biol.*, 41, 2605–2625, 1996b.

ISO (International Standards Organisation), *Quality systems, Model for quality assurance in design, development, production, installation and servicing*, BS EN ISO 9001:1994 (being superseded by ISO 2000), British Standards Institution, London, 1994.

ISO (International Standards Organisation), *Quality management and quality assurance, Vocabulary*, BS EN ISO 8402:1995, BSI, London, 1995.

ISO (International Standards Organisation), *Quality management systems, Requirements*, BS EN ISO 9001:2000, BSI, London, 2000.

ISO (International Standards Organisation), *Guide for estimating uncertainties in dosimetry for radiation processing*, ISO/ASTM 51707:2002, ISO, Geneva, 2002.

Iwata, K., Yoshimura, H., Tsuji, Y., Shirai, A., Uto, F. et al., In vivo measurements of spatial dose distribution with thermoluminescent sheet around high dose-rate intracavitary source: Application to rectal cancer, *Int. J., Radiat. Oncol. Biol. Phys.*, 22, 1109–1115, 1992.

Jaffray, D. A., Siewerdsen, J. H., Wong, J. W., and Martinez, A. A., Flat panel cone-beam computed tomography for image-guided radiation therapy, *Int. J. Radiat. Oncol. Biol. Phys.*, 53, 1337–1349, 2002.

Joelsson, I., Rudén, B. I., Costa, A., Dutreix, A., and Rosenwald, J.-C., Determination of dose distribution in the pelvis by measurements and by computer in gynecologic radiation therapy, *Acta Radiologica*, 11, 289–304, 1972.

Johansson, L., Carlsson, J., and Nilsson, K., Radiosensitivity of human B-lymphocytic lymphomas in vitro, *Int. J. Radiat. Biol. Relat. Stud. Phys. Chem. Med.*, 41, 411–420, 1982.

Johansson, K. A., Horiot, J. C., and van der Schueren, E., Quality assurance control in the EORTC cooperative group of radiotherapy 3. Intercomparison in an anatomical phantom, *Radiother. Oncol.*, 9, 289–298, 1987.

Jones, S. and Boyer, A., Investigation of an FFT-based correlation technique for verification of treatment set-up, *Med. Phys.*, 18, 1116–1125, 1991.

Jordan, T. J. and Williams, P. C., The design and performance characteristics of a multileaf collimator, *Phys. Med. Biol.*, 39, 231–251, 1994.

Ju, S. G., Ahn, C. Y., Huh, S. J., and Yeo, I. J., Film dosimetry for intensity modulated radiation therapy: Dosimetric evaluation, *Med. Phys.*, 29, 351–355, 2002.

Jursinic, P. A. and Nelms, B. E., A 2-D diode array and analysis software for verification of intensity modulated radiation therapy delivery, *Med. Phys.*, 30, 870–879, 2003.

Kesteloot, K., Dutreix, A., and van der Schueren, E., A model for calculating the cost of in vivo dosimetry and portal imaging in radiotherapy departments, *Radiother. Oncol.*, 28, 108–117, 1993.

Kirby, M. C. and Williams, P. C., Measurement possibilities using an electronic portal imaging device, *Radiother. Oncol.*, 29, 237–243, 1993.

Kirby, M. C. and Williams, P. C., The use of an electronic portal imaging device for exit dosimetry and quality control measurements, *Int. J. Radiat. Oncol. Biol. Phys.*, 31, 593–603, 1995.

Klein, E. E., Morrison, A., Purdy, J. A., Graham, M. V., and Matthews, J., A volumetric study of measurements and calculations of lung density corrections for 6 and 18 MV photons, *Int. J. Radiat. Oncol. Biol. Phys.*, 37, 1163–1170, 1997.

Klein, E. E., Drzymala, R. E., Williams, R., Westfall, L. A., and Purdy, J. A., A change in treatment process with a modern record and verify system, *Int. J. Radiat. Oncol. Biol. Phys.*, 42, 1163–1168, 1998.

Knusten, B. H., Skretting, A., Hellebust, T. P., and Olsen, D. R., Determination of 3D dose distribution from intracavitary brachytherapy of cervical cancer by MRI of irradiated ferrous sulfate gel, *Radiother. Oncol.*, 43, 219–227, 1997.

Kron, T., Elliot, A., Wong, T., Showell, G., Clubb, B., and Metcalfe, P., X-ray surface dose measurements using TLD extrapolation, *Med. Phys.*, 20, 703–711, 1993.

Kroonwijk, M., Pasma, K. L., Quint, S., Koper, P. C. M., Visser, A. G., and Heijmen, B. J. M., In vivo dosimetry for prostate cancer patients using an electronic portal imaging device (EPID); demonstration of internal organ motion, *Radiother. Oncol.*, 49, 125–132, 1998.

Kung, J. H. and Chen, G. T., Intensity modulated radiotherapy dose delivery error from radiation field offset inaccuracy, *Med. Phys.*, 27, 1617–1622, 2000.

Lanson, J. H., Essers, M., Meijer, G. J., Minken, A. W. H., Uiterwaal, B. J., and Mijnheer, B. J., In vivo dosimetry during conformal radiotherapy. Requirements for and findings of a routine procedure, *Radiother. Oncol.*, 52, 51–59, 1999.

Leer, J. W. H., McKenzie, A. L., Scalliet, P., and Thwaites, D. I., Practical guidelines for the implementation of quality systems in radiotherapy, ESTRO Booklet No. 4, ESTRO, Brussels, 1999. Available online at www.estroweb.org/ESTRO/upload/publications/Qart.pdf

Leszcynski, K. and Shalev, S., Digital contrast enhancement for on-line portal imaging, *Med. Biol. Eng. Comput.*, 277, 507–512, 1989.

Leszczynski, K. W., Shalev, S., and Cosby, N. S., The enhancement of radiotherapy verification images by an automated edge detection technique, *Med. Phys.*, 19, 611–621, 1992.

Létourneau, D., Gulam, M., Yan, D., Oldham, M., and Wong, J. W., Evaluation of a 2D diode array for IMRT quality assurance, *Radiother. Oncol.*, 70, 199–206, 2004.

Leunens, G., van Dam, J., Dutreix, A., and van der Schueren, E., Quality assurance in radiotherapy. 2. Determination of the target absorbed dose, *Radiother. Oncol.*, 19, 73–87, 1990.

Leunens, G., Verstraete, J., van den Bogaert, W., van Dam, J., Dutreix, A., and van der Schueren, E., Human errors in data transfer during the preparation and delivery of radiation treatment affecting the final result: "Garbage in, garbage out", *Radiother. Oncol.*, 23, 217–222, 1992.

Lillicrap, S. C., Owen, B., Williams, J. R., and Williams, P. C., Code of Practice for high-energy photon therapy dosimetry based on the NPL absorbed dose calibration service, *Phys. Med. Biol.*, 35, 1355–1360, 1990.

Ling, C. C., Burman, C., Chui, C. S., Kutcher, G. J., Leibel, S. A. et al., Conformal radiation treatment of prostate cancer using inversely-planned intensity-modulated photon beams produced with dynamic multileaf collimation, *Int. J. Radiat. Oncol. Biol. Phys.*, 35, 721–730, 1996.

LoSasso, T., Chui, C. S., and Ling, C. C., Physical and dosimetric aspects of a multileaf collimation system used in the dynamic mode for implementing intensity modulated radiotherapy, *Med. Phys.*, 25, 1919–1927, 1998.

LoSasso, T., Chui, C. S., and Ling, C. C., Comprehensive quality assurance for the delivery of intensity modulated radiotherapy with a multileaf collimator used in the dynamic mode, *Med. Phys.*, 28, 2209–2219, 2001.

LoSasso, T. J., Quality assurance of IMRT, in *A practical guide to intensity modulated radiation therapy*, Medical Physics Publishing, Madison, WI, 2003.

Loncol, T. H., Greffe, J. L., Vinckier, S., and Scalliet, P., Entrance and exit dose with semi-conductors and thermo-luminescent dosimeters: A comparison of methods and in vivo results, *Radiother. Oncol.*, 41, 179–187, 1996.

Louwe, R. J., McDermott, L. N., Sonke, J. J., Tielenburg, R., Wendling, M. et al., The long-term stability of amorphous silicon flat panel imaging devices for dosimetry purposes, *Med. Phys.*, 31, 2989–2995, 2004.

Low, D. A., Harms, W. B., Mutic, S., and Purdy, J. A., A technique for the quantitative evaluation of dose distributions, *Med. Phys.*, 25, 656–661, 1998.

Low, D. A., Mutic, S., Dempsey, J. F., Gerber, R. L., Bosch, W. R., Perez, C. A., and Purdy, J. A., Quantitative dosimetric verification of an IMRT planning and delivery system, *Radiother. Oncol.*, 49, 305–316, 1998.

Lutz, W. R. and Bjarngard, B. E., A test object for evaluation of portal films, *Int. J. Radiat. Oncol. Biol. Phys.*, 11, 631–634, 1985.

Lutz, W., Winston, K. R., and Maleki, N., A system for stereotactic radiosurgery with a linear accelerator, *Int. J. Radiat. Oncol. Biol. Phys.*, 14, 373–381, 1988.

Ma, C. M., Coffey, C. W., DeWerd, L. A., Liu, C., Nath, R. et al., AAPM protocol for 40–300 kV x-ray beam dosimetry in radiotherapy and radiobiology, *Med. Phys.*, 28, 868–893, 2001.

MacKenzie, M. A., Lachaine, M., Murray, B., Fallone, B. G., Robinson, D., and Field, G. C., Dosimetric verification of inverse planned step and shoot multileaf collimator fields from a commercial treatment planning system, *J. Appl. Clin. Med. Phys.*, 3, 97–109, 2002.

Marcié, S., Costa, A., and Lagrange, J. L., Protection of testes during radiation treatment by irregular and focused fields of 25 MV x rays: In vivo evaluation of the absorbed dose, *Med. Dosim.*, 20, 269–273, 1995.

Marcié, S., Charpiot, E., Bensadoun, R-J., Ciais, G., Hérault, J., Costa, A., and Gérard, J-P., In vivo measurements with MOSFET detectors in oropharynx and nasopharynx intensity-modulated radiation therapy, *Int. J. Radiat. Oncol. Biol. Phys.*, 61, 1603–1606, 2005.

Marinello, G., Raynal, M., Brule, A. M., and Pierquin, B., Use of LiF in clinical dosimetry. Application to the measurement of dose deliverd to the region of the axilla by ^{192}Ir in brachytherpy of skin cancers (Utilisation du fluorure de lithium en dosimétrie clinique. Application à la mesure de la dose délivrées à la région axillaire par l'iridium 192 dans l'endocuriethérapie des cancers du sein), *J. Radiol. Electrol.*, 56, 791–796, 1975.

Marinello, G., Barrié, A. M., and Le Bourgeois, J. P., Measurement and calculation of lung dose in total body irradiation performed with cobalt-60, *J. Eur. Radiother.*, 4, 174–182, 1982.

Marinello, G., Barthe, J., Pollack, J., and Portal, G., "PCL", a new automatic fast reader suitable for in vivo dosimetry, *Radiother. Oncol.*, 25, 63–66, 1992.

Martinez, A. A., Yan, D., Lockman, D., Brabbins, D., Kota, K. et al., Improvement in dose escalation using the process of adaptive radiotherapy combined with three-dimensional conformal or intensity-modulated beams for prostate cancer, *Int. J. Radiat. Oncol. Biol. Phys.*, 50, 1226–1234, 2001.

Mattsson, L. O., Johansson, K.-A., and Svensson, H., Calibration and use of parallel-plate ionization chambers for the determination of absorbed dose in electron beams, *Acta. Radiol. Oncol.*, 20, 385–399, 1981.

Mayles, W. P. M., Heisig, S., and Mayles, H. M. O., Treatment verification and in vivo dosimetry, in *Radiotherapy Physics in Practice*, Williams, J. R. and Thwaites, D. I., Eds., Oxford University Press, Oxford, pp. 220–246, 2000.

McKenzie, A., Briggs, G., Buchanan, R., Harvey, L., Iles, A., et al., Balancing costs and benefits of checking in radiotherapy, IPEM Report 92, IPEM, York, 2006.

McLaughlin, W. L., Yung-Dong, C., and Soares, C. G., Sensitometry of the response of new radiochromic film dosimeter to gamma radiation and electron beams, *Nuclear Instruments and Methods in Physics Research*, A-302, 165–176, 1991.

McNutt, T. R., Mackie, T. R., Reckwerdt, P., and Paliwal, B. R., Modelling dose distributions from portal dose images using the convolution/superposition method, *Med. Phys.*, 23, 1381–1392, 1996.

McParland, B., Uncertainty analysis of field placement error measurements using digital portal and simulation image correlations, *Med. Phys.*, 20, 679–685, 1993.

Mechakra Tahiri, D. S., Gérard, J. P., Ginestet, C., and Berger, M., Dosimetry using TLD in breast cancer treated with ^{192}Ir wire (Dosimetrie par thermoluminescence dans les cancers mammaires traités par fils d'iridium), *J. Eur. Radiother.*, 9, 109–114, 1988.

Meertens, H., Bijhold, J., and Strackee, J., A method of the measurement of field placement errors in digital portal images, *Phys. Med. Biol.*, 35, 299–323, 1990a.

Meertens, H., van Herk, M., Bijhold, J., and Bartelink, H., First clinical experience with a newly developed electronic portal imaging device, *Int. J. Radiat. Oncol. Biol. Phys.*, 18, 1173–1181, 1990b.

Michalski, J. M., Wong, J. W., Bosch, W. R., Yan, D., Cheng, A. et al., An evaluation of two methods of anatomical alignment of radiotherapy portal images, *Int. J. Radiat. Oncol. Biol. Phys.*, 27, 1199–1206, 1993.

Mijnheer, B. J., Possibilities and limitations of in vivo dosimetry, in *Radiotherapy from prescription to delivery* (Proceedings of an Interregional Seminar for Europe, the Middle East and Africa, Leuven, 16–20 September 1991), IAEA TECDOC Series No. 734, IAEA, Vienna, 1994.

Mijnheer, B. J., Battermann, J. J., and Wambersie, A., What degree of accuracy is required and can be achieved in photon and neutron therapy?, *Radiother. Oncol.*, 8, 237–252, 1987.

Mildenberger, P., Eichelberg, M., and Martin, E., Introduction to the DICOM standard, *Eur. Radiol.*, 12, 920–927, 2002.

Moench, H. C. and Phillips, T. L., Carcinoma of the nasopharynx. Review of 146 patients with emphasis on radiation dose and time factors, *Am. J. Surg.*, 124, 515–518, 1972.

Mohan, R., Arnfield, M., Tong, S., Wu, Q., and Siebers, J., The impact of fluctuations in intensity patterns on the number of monitor units and the quality and accuracy of intensity modulated radiotherapy, *Med. Phys.*, 27, 1226–1237, 2000.

Moore, J. V., Hendry, J. H., and Hunter, R. D., Dose-incidence curves for tumour control and normal tissue injury, in relation to the response of clonogenic cells, *Radiother. Oncol.*, 1, 143–157, 1983.

Moore, A. R., Stanley, S., Bidmead, A. M., Dearnaley, D. P., Mayles, W. P. M., and Warrington, A. P., Quality assurance of dosimetry in U.K. MRC RT01 trial of dose escalation using conformal radiotherapy in prostate cancer (ISRCTC 47772397), *Clin. Oncol.*, 15, 58–59, 2003.

Morrison, R., The results of treatment of cancer of the bladder—a clinical contribution to radiobiology, *Clin. Radiol.*, 26, 67–75, 1975.

Moseley, J. and Munro, P., A semiautomatic method for registration of portal images, *Med. Phys.*, 21, 551–558, 1994.

Mosleh-Shirazi, M. A., Evans, P. M., Swindell, W., Webb, S., and Partridge, M., A cone-beam megavoltage CT scanner for treatment verification in conformal radiotherapy, *Radiother. Oncol.*, 48, 319–328, 1998.

Mubata, C. D., Childs, P., and Bidmead, A. M., A quality assurance procedure for the Varian multi-leaf collimator, *Phys. Med. Biol.*, 42, 423–431, 1997.

Muller-Runkel, R. and Watkins, S. S., Introducing a computerized record and verify system: Its impact on the reduction of treatment errors, *Med. Dosim.*, 16, 19–22, 1991.

Munro, P., Bouius, D. C., Moseley, J., Martin, L., Zhang, Y., Jaffray, D. A., et al., Glaring errors in transit dosimetry, Proceedings of EPI98, 5th International Workshop on Electronic Portal Imaging IX, Arizona, EPI98 Organising Committee, 128–129, 1998.

Mutic, S., Palta, J. R., Butker, E. K., Das, I. J., Huq, M. S. et al., Quality assurance for computed-tomography simulators and the computed-tomography-simulation process: Report of the AAPM Radiation Therapy Committee Task Group No. 66, *Med. Phys.*, 30, 2762–2792, 2003.

NACP (Nordic Association of Clinical Physics), Procedures in external radiation therapy dosimetry with electron and photon beams with maximum energies between 1 and 50 MeV, *Acta. Radiol. Oncol.*, 19, 55–79, 1980.

NACP (Nordic Association of Clinical Physics), Supplement to the recommendations of NACP 1980: Electron beams with mean energies at the phantom surface below 15 MeV, *Acta. Radiol. Oncol.*, 20, 401–415, 1981.

NCS (The Netherlands Commission on Radiation Dosimetry), Quality control of medical linear accelerators: Current practice and minimum requirements, NCS Report 9, NCS, Delft, 1996.

NCS (The Netherlands Commission on Radiation Dosimetry), Quality assurance of 3-D treatment planning systems for external photon and electron beams; practical guidelines for acceptance testing, commissioning, and periodic quality control of radiation therapy treatment planning systems, NCS Report 15, NCS, Delft, 2006.

Nederveen, A. J., Lagendijk, J. J. W., and Hofman, P., Feasibility of automatic marker detection with an a-Si flat panel imager, *Phys. Med. Biol.*, 46, 1219–1230, 2001.

Neumann, M., DICOM—current status and future developments for radiotherapy, *Z. Med. Phys.*, 12, 171–176, 2002.

Niemierko, A., Quality assurance and validation of individual IMRT treatment plans before delivery, *Med. Phys.*, 31, 421–422, 2004.

Nilsson, B., Rudén, B. I., and Sorcini, B., Characteristics of silicon diodes as patient dosimeters in external radiation therapy, *Radiother. Oncol.*, 11, 279–288, 1988.

Nisbet, A. and Thwaites, D. I., A dosimetric intercomparison of electron beams in U.K. radiotherapy centres, *Phys. Med. Biol.*, 42, 2393–2409, 1997.

NIST (National Institute of Standards and Technology), Technical Note 1297—Guidelines for Evaluating and Expressing the Uncertainty of NIST Measurement Results, NIST, Washington, D.C., Available online at www.physics.nist.gov/cuu/Uncertainty/international2.html, 1994.

NIST (National Institute of Standards and Technology), The NIST Reference on constants units and uncertainty, Uncertainty of measurement results, Online guide at http://physics.nist.gov/cuu/Uncertainty/index.html, 2000.

Noel, A., Aletti, P., Bey, P., and Malissard, L., Detection of errors in individual patients in radiotherapy by systematic in vivo dosimetry, *Radiother. Oncol.*, 34, 144–151, 1995.

Olsen, K. J., Hansen, J. W., and Wille, M., Response of the alanine radiation dosemeter to high-energy photon and electron beams, *Phys. Med. Biol.*, 35, 43–52, 1990.

Ozhasoglu, C. and Murphy, M. J., Issues in respiratory motion compensation during external-beam radiotherapy, *Int. J. Radiat. Oncol. Biol. Phys.*, 52, 1389–1399, 2002.

Pai, S., Reinstein, L., Gluckman, G., Xu, Z., and Weiss, T., The use of improved radiochromic film for in vivo quality assurance of high dose rate brachytherapy, *Med. Phys.*, 25, 1217–1221, 1998.

Panitsa, E., Rosenwald, J.-C., and Kappas, C., Intercomparison of treatment planning systems (TPS) used in radiotherapy: Progress report of the TPS group of the European Dynarad project, *Physica Medica*, 13(Suppl. 1), 139–143, 1997.

Panitsa, E., Rosenwald, J.-C., and Kappas, C., Quality control of dose volume histogram computation characteristics of 3D treatment planning systems, *Phys. Med. Biol.*, 43, 2807–2816, 1998.

Partridge, M., Evans, P. M., Mosleh-Shirazi, A., and Convery, D., Independent verification using portal imaging of intensity-modulated beam delivery by the dynamic MLC technique, *Med. Phys.*, 25, 1872–1879, 1998.

Patton, G. A., Gaffney, D. K., and Moeller, J. H., Facilitation of radiotherapeutic error by computerized record and verify systems, *Int. J. Radiat. Oncol. Biol. Phys.*, 56, 50–57, 2003.

Piermattei, A., Azario, L., Monaco, G., Soriani, A., and Arcovito, G., P-type silicon detector for brachytherapy dosimetry, *Med. Phys.*, 22, 835–839, 1995.

Pizer, S. M., Zimmerman, J. B., and Staab, E. V., Adaptive grey level assignment in CT scan display, *J. Comput. Assist. Tomogr.*, 8, 300–305, 1984.

Podmaniczky, K. C., Mohan, R., Kutcher, G. J., Kestler, C., and Vickram, B., Clinical experience with a computerised record and verify system, *Int. J. Radiat. Oncol. Biol. Phys.*, 11, 1529–1537, 1985.

Prisciandaro, J. I., Herman, M. G., and Kruse, J. J., Utilizing an electronic portal imaging device to monitor light and radiation field congruence, *J. Appl. Clin. Med. Phys.*, 4, 315–320, 2003.

Rajapakshe, R., Luchka, K., and Shalev, S., A quality control test for electronic portal imaging devices, *Med. Phys.*, 23, 1237–1244, 1996.

Rassow, J., Quality control of radiation therapy equipment, *Radiother. Oncol.*, 12, 45–55, 1988.

Redpath, A. T., An analysis of the changes to radiation dose distributions resulting from the use of simulator computed tomography, *Br. J. Radiol.*, 61, 1063–1065, 1988.

Reinstein, L. E., Pai, S., and Meek, A. G., Assessment of geometric treatment accuracy using time-lapse display of electronic portal images, *Int. J. Radiat. Oncol. Biol. Phys.*, 22, 1139–1146, 1992.

Renner, W. D., Sarfaraz, M., Earl, M. A., and Yu, C. X., A dose delivery verification method for conventional and intensity-modulated radiation therapy using measured field fluence distributions, *Med. Phys.*, 30, 2996–3005, 2003.

Rizzotti, A., Compri, C., and Garusi, G. F., Dose evaluation to patients irradiated by ^{60}Co, by means of direct measurement on the incident and on the exit surfaces, *Radiother. Oncol.*, 3, 279–283, 1985.

Roos, M., Derikum, K., and Lange, B., A new ionisation chamber construction for electron dosimetry (Eine neue Flachkammerkonstruktion für die Elektronendosimetrie). *Mediz. Physik 24 Wissenschaftliche Tagung der Deutschen Gesellschaft für Medicinische Physik (Erlangen, 1993)*, 364, Institut für Radiologie, Erlangen, 1993.

Rosenbloom, M. E., Killick, L. J., and Bentley, R. E., Verification and recording of radiotherapy treatments using a small computer, *Br. J. Radiol.*, 50, 637–644, 1977.

Rosenwald, J.-C., Quality control of 3D treatment planning systems used for conformal radiotherapy, in *Proceedings of Medizinische Physik 95*, Richter, J., Ed., DGMP, Würzburg, pp. 168–169, 1995.

Rudin, S., Bednarek, D. R., and Wong, R., Accurate characterization of image intensifier distortion, *Med. Phys.*, 18, 1145–1151, 1991.

Samant, S. S., Zheng, W., Parra, N. A., Chandler, J., Gopal, A. et al., Verification of multileaf collimator leaf positions using an electronic portal imaging device, *Med. Phys.*, 29, 2900–2912, 2002.

Sastre-Padro, M., van der Heide, U. A., and Welleweerd, H., An accurate calibration method of the multileaf collimator valid for conformal and intensity modulated radiation treatments, *Phys. Med. Biol.*, 49, 2631–2643, 2004.

Saunders, J. E., A radiotherapy management system. in *Proceedings of the 10th ICCR, Lucknow India*, Hukku, S. and Iyer, P. S., Eds., Alpana Arts, Lucknow, pp. 178–180, 1990.

Schneider, U., Pedroni, E., and Lomax, A., The calibration of CT Hounsfield units for radiotherapy treatment planning, *Phys. Med. Biol.*, 41, 111–124, 1996.

SFPM (Société Française de Physique Médicale), Quality control of multileaf collimators [Contrôle de qualité des collimateurs multilames], SFPM Report No 20, SFPM, Paris, Available online at www.bhpa.be/pages/publi/MLCfinal.pdf, 2003.

SGSMP (Swiss Society of Radiobiology and Medical Physics), Quality control of treatment planning systems for teletherapy, Recommendations No. 7, SGSMP, Lausanne, 1997.

Shackford, H. and Bjärngard, E., Disturbance of diode dosimetry by radiofrequency radiation, *Med. Phys.*, 22, 807, 1995.

Shalev, S., Gluhchev, G., and Chen, D., PIPS: A PC based processing system for portal images, *Med. Phys.*, 21, 964, 1994.

Shalev, S., Rajapakshe, R., Luchka, K., et al., Techniques for commissioning electronic portal imaging devices, XII International Conference on the use of Computers in Radiation Therapy, Salt Lake City, Utah, U.S.A, 1997.

Sharma, S. C., Williamson, J. F., Khan, F. M., and Lee, C. K., Measurement and calculation of ovary and fetus dose in extended field radiotherapy for 10 MV x-rays, *Int. J. Radiat. Oncol. Biol. Phys.*, 7, 843–846, 1981.

Shukovsky, L. J. and Fletcher, G. H., Time-dose and tumor volume relationships in the irradiation of squamous cell carcinoma of the tonsillar fossa, *Radiology*, 107, 621–626, 1973.

Siewerdsen, J. H. and Jaffray, D. A., A ghost story: Spatio-temporal response characteristics of an indirect-detection flat-panel imager, *Med. Phys.*, 26, 1624–1641, 1999.

Simonian-Sauve, M. and Smart, C., Evaluation tests of computer systems concerning 3-dimensional dose calculations. (Tests d'évaluation des systèmes informatiques concernant le calcul de la dose en trois dimensions), *Bulletin du Cancer/Radiothérapie*, 2, 53–62, 1998.

Sjögren, R. and Karlsson, M., Influence of electron contamination on in vivo surface dosimetry for high energy photon beams, *Med. Phys.*, 25, 916–921, 1998.

Soubra, M., Cygler, J., and Mackay, G., Evaluation of a dual bias dual MOSFET detector as a radiation dosimeter, *Med. Phys.*, 21, 567–572, 1994.

Spezi, E., Lewis, D. G., and Smith, C. W., A DICOM-RT-based toolbox for the evaluation and verification of radio-therapy plans, *Phys. Med. Biol.*, 47, 4223–4232, 2002.

Starkschall, G., Steadham, R. E., Popple, R. A., Ahmad, S., and Rosen, I. I., Beam-commissioning methodology for a three-dimensional convolution/superposition photon dose algorithm, *J. App. Clin. Med. Phy.*, 1(1), 8–27, 2000.

Svensson, H., Quality assurance in radiation therapy: Physical aspects, *Int. J. Radiat. Oncol. Biol. Phys.*, 10(Suppl. 1), 59–65, 1984.

Svensson, H., Westling, P., and Larsson, L. G., Radiation-induced lesions of the brachial plexus correlated to the dose-time-fraction schedule, *Acta Radiol. Ther. Phys. Biol.*, 14, 228–238, 1975.

Swindell, W. and Evans, P. M., Scattered radiation in portal images; a Monte-Carlo simulation and a physical model, *Med. Phys.*, 23, 63–73, 1996.

Swindell, W., Simpson, R. G., and Olsen, J. R., Computed tomography with a linear accelerator with radiotherapy applications, *Med. Phys.*, 10, 416–420, 1983.

Thwaites, D. I., Uncertainties at the endpoint of the basic dosimetry chain, in *Measurement assurance in dosimetry*, IAEA, Vienna, pp. 239–255, 1993.

Thwaites, D. I., Radiation dose, in *Radiotherapy from Prescription to Delivery* (Proceedings of an Interregional Seminar for Europe, the Middle East and Africa, Leuven, 16–20 September 1991), IAEA TECDOC Series No. 734, IAEA, Vienna, 1991.

Thwaites, D. I., Experience with the U.K. (IPEM) absorbed-dose-to-water radiotherapy dosimetry protocols for photons (1990) and electrons (2003), in *Standards and Codes of Practice in Medical Dosimetry*, Vol. I, IAEA, Vienna, pp. 243–256, 2003.

Thwaites, D. I., Williams, J. R., Aird, E. G., Klevenhagen, S. C., and Williams, P. C., A dosimetric intercomparison of megavoltage photon beams in U.K. radiotherapy centres, *Phys. Med. Biol.*, 37, 445–461, 1992.

Thwaites, D. I., Scalliet, P., Leer, J. W., and Overgaard, J., Quality assurance in radiotherapy (European Society for Therapeutic Radiology and Oncology advisory report to the Commission of the European Union for the Europe Against Cancer Programme), *Radiother. Oncol.*, 35, 61–73, 1995.

Thwaites, D. I., Blyth, C., Carruthers, L., Elliott, P. A., Kidane, G., et al., Experience with in vivo diode dosimetry for verifying radiotherapy dose delivery: The practical implementation of cost effective approaches (IAEA-CN-96/131P), in *Standards and Codes of Practice in Medical Dosimetry*, Vol. II, IAEA, Vienna, pp. 415–423, 2003a.

Thwaites, D. I., DuSautoy, A. R., Jordan, T., McEwen, M. R., Nisbet, A., Nahum, A. E., and Pitchford, W. G., The IPEM code of practice for electron dosimetry for radiotherapy beams of initial energy from 4 to 25 MeV based on an absorbed dose to water calibration, *Phys. Med. Biol.*, 48, 2929–2970, 2003b.

Thwaites, D. I., Powley, S., Nisbet, A., Allahverdi, M., et al., The United Kingdom's radiotherapy dosimetry audit network, In *Standards and codes of practice in medical dosimetry*, IAEA, Vienna, pp. 183–190, 2003c.

Thwaites, D. I., and Tuohy, J. B., Back to the future: the history and development of the clinical linear accelerator, *Phys. Med. Biol.*, 51, R343–R362, 2006.

Tsiakalos, M. F., Theodorou, K., Kappas, C., Zefkili, S., and Rosenwald, J.-C., Analysis of the penumbra enlargement in lung versus the Quality Index of photon beams: A methodology to check the dose calculation algorithm, *Med. Phys.*, 31, 943–949, 2004.

Turesson, I. and Notter, G., The influence of fraction size in radiotherapy on the late normal tissue reaction—II: Comparison of the effects of daily and twice-a-week fractionation on human skin, *Int. J. Radiat. Oncol. Biol. Phys.*, 10, 599–606, 1984.

van Dam, J., Marinello, G., Methods for *in vivo* dosimetry in external radiotherapy, ESTRO Booklet No. 1, Leuven, Apeldoorn, 1994.

van Dam, J., Vaerman, C., Blankaert, N., Leunens, G., Dutreix, A., and van der Schueren, E., Are port films reliable for in vivo exit dose measurements? *Radiother. Oncol.*, 25, 67–72, 1992.

van Dyk, J., Barnett, R. B., Cygler, J. E., and Shragge, P.C, Commissioning and quality assurance of treatment planning computers, *Int. J. Radiat. Oncol. Biol. Phys.*, 26, 261–273, 1993.

van Esch, A., Bohsung, J., Sorvari, P., Tenhunen, M., Paiusco, M. et al., Acceptance tests and quality control (QC) procedures for the clinical implementation of intensity modulated radiotherapy (IMRT) using inverse planning and the sliding window technique: Experience from five radiotherapy departments, *Radiother. Oncol.*, 65, 53–70, 2002.

van Esch, A., Depuydt, T., and Huyskens, D. P., The use of an aSi-based EPID for routine absolute dosimetric pre-treatment verification of dynamic IMRT fields, *Radiother. Oncol.*, 71, 223–234, 2004.

van Herk, M., Errors and margins in radiotherapy, *Semin. Radiat. Oncol.*, 14, 52–64, 2004.

Venables, K., Winfield, E. A., Aird, E. G., and Hoskin, P. J., Three-dimensional distribution of radiation within the breast: An intercomparison of departments participating in the START trial of breast radiotherapy fractionation, *Int. J. Radiat. Oncol. Biol. Phys.*, 55, 271–279, 2003.

Venables, K., Miles, E. A., Aird, E. G., and Hoskin, P. J., The use of in vivo thermoluminescent dosimeters in the quality assurance programme for the START breast fractionation trial, *Radiother. Oncol.*, 71, 303–310, 2004.

Venselaar, J. and Welleweerd, H., Application of a test package in an intercomparison of the photon dose calculation performance of treatment planning systems used in a clinical setting, *Radiother. Oncol.*, 60, 203–213, 2001.

Venselaar, J., Welleweerd, H., and Mijnheer, B., Tolerances for the accuracy of photon beam dose calculations of treatment planning systems, *Radiother. Oncol.*, 60, 191–201, 2001.

Verney, J. N. and Morgan, A. M., Evaluation of in vivo dose measurements for patients undergoing electron boost treatment, *Radiother. Oncol.*, 59, 293–296, 2001.

Vieira, S. C., Dirkx, M. L., Pasma, K. L., and Heijmen, B. J., Fast and accurate leaf verification for dynamic multileaf collimation using an electronic portal imaging device, *Med. Phys.*, 29, 2034–2040, 2002.

Vieira, S. C., Dirkx, M. L., Pasma, K. L., and Heijmen, B. J., Dosimetric verification of x-ray fields with steep dose gradients using an electronic portal imaging device, *Phys. Med. Biol.*, 48, 157–166, 2003.

Weinhous, M., Treatment verification using a computer workstation, *Int. J. Radiat. Oncol. Biol. Phys.*, 19, 1549–1554, 1990.

Weinhous, M. S. and Meli, J. A., Determining P_{ion}, the correction factor for recombination losses in an ionization chamber, *Med. Phys.*, 11, 846–849, 1984.

Weltens, C., van Dam, J., Leunens, G., Dutreix, A., and van der Schueren, E., Reliability of clinical port films for measuring dose inhomogeneities in radiotherapy for head and neck tumors, *Radiother. Oncol.*, 30, 167–170, 1994.

Wiezorek, T., Banz, N., Schwedas, M., Scheithauer, M., Salz, H. et al., Dosimetric quality assurance for intensity-modulated radiotherapy feasibility study for a filmless approach, *Strahlenther. Onkol.*, 181, 468–474, 2005.

WHO (World Health Organization), *Quality Assurance in Radiotherapy*, WHO, Geneva, 1988.

Woo, M. K. and Nico, A., Impact of multileaf collimator leaf positioning accuracy on intensity modulation radiation therapy quality assurance ion chamber measurements, *Med. Phys.*, 32, 1440–1445, 2005.

Yan, D., Wong, J., Vicini, F., Michalski, J., Pan, C. et al., Adaptive modification of treatment planning to minimize the deleterious effects of treatment setup errors, *Int. J. Radiat. Oncol. Biol. Phys.*, 38, 197–206, 1997.

Yin, F. F., Schell, M. C., and Rubin, P., A technique of automatic compensator design for lung inhomogeneity correction using an electronic portal imaging device, *Med. Phys.*, 21, 1729–1732, 1994.

FURTHER READING

European Community Documents

EC, Framework Directive 89/391/EEC, *Council directive on the introduction of measures to encourage improvements in the safety and health of workers*, 1989.

EC, Council Directive 93/42/EEC 12 July 1993 *Medical Devices Directive*, Official Journal of the European Communities OJ No. L169/1, 1993.

EC, Council Directive 96/29/Euratom of 13 May 1996 laying down basic safety standards for the protection of the health of workers and the general public against the dangers arising from ionizing radiation, Official journal NO. L 159, 29/06/1996 P, 0001–0114, 1996.

See also EC (1997a, 1997b) in main reference list.

Council Directives may be obtained from HMSO or downloaded via http://eur-lex.europa.eu/en/index.htm. Further information about the Medical Devices Directive and its application in the U.K. is available from: Medicines and Healthcare products Regulatory Agency (MHRA), 10-2 Market Towers, 1 Nine Elms Lane, London, SW8 5, telephone: 020 7084 2000. http://www.mhra.gov.uk.

British Government Documents

Current legal framework:

HS (Health and Safety), *The Ionising Radiations Regulations 1999*, HMSO (IRRRev), London, 1999.
HSE (Health and Safety Executive), *The Ionising Radiations Regulations 1999. Approved Code of Practice and Supporting Guidance*, HMSO (IRRRev ACOP), London, 1999.
HS (Health and Safety), *The Radiation (Emergency Preparedness and Public Information) Regulations*, HMSO (REPPIR), London, 1999.
HS (Health and Safety), *The Management of Health and Safety at Work Regulations 1992*, HMSO, London, 1992.
HS (Health and Safety), *The Radioactive Substances Act 1993*, HMSO, London, 1993.
HS (Health and Safety), *The Ionising Radiations (Outside Workers) Regulations*, HMSO, London, 1993.
HSE (Health and Safety Executive), Equipment used in connection with medical exposure, *Guidance Note PM77 (3rd edition)*, HMSO, London, 2006.

See also HS (2000) in main reference list.

HSE and HSC publications are available from HSE Books, PO Box 1999, Sudbury, Suffolk CO10 2WA Tel. +44 1787 881165 Fax Tel. +44 1787 313995. The HSE web page can be found at: www.hse.gov.uk.

Other publications:

DH (Department of Health), *Manual of cancer services standards*, HMSO, London, 2001, Downloadable by searching www.dh.gov.uk.
DHSS (Department of Health and Social Security), *Health Notice HN (Hazard) (78) 26*, Department of Health, London, 1978.
DTI, *Guide to Software Quality Management System Construction and Certification Using EN 29001*, HMSO, London, 1992.

See also Bleehen (1991) in main reference list.

International Standards

BSI (British Standards Institute), *Specification: Sealed radioactive sources*, BS 5288, BSI, London, 1976.
IEC (International Electrotechnical Commission), *Medical Electrical Equipment Dosimeters with Ionization Chambers as used in Radiotherapy*, IEC Publication 731, IEC, Geneva, 1986A.
IEC (International Electrotechnical Commission), *Medical Electrical Equipment Dosimeters with Ionization Chambers as used in Radiotherapy*, IEC Publication 977, IEC, Geneva, 1986B.
IEC (International Electrotechnical Commission), *Medical Electrical Equipment, Part 2: Particular requirements for the safety of gamma beam therapy equipment*, IEC Publication 601-2-11, IEC, Geneva, 1987.
IEC (International Electrotechnical Commission), *Medical Electrical Equipment, Part 2: Particular requirements for the safety of gamma beam therapy equipment. Amendment 1*, IEC Publication 601-2-11 Amdt 1, IEC, Geneva, 1988.
IEC (International Electrotechnical Commission), *Medical Electrical Equipment, Section 2.17 Specification for remote-controlled automatically driven gamma-ray afterloading equipment*, IEC Publication 601-2-17, IEC, Geneva, 1989.
IEC (International Electrotechnical Commission), *Medical Electrical Equipment, Part 2: Particular requirements for the safety of gamma beam therapy equipment. Amendment 2*, IEC Publication 601-2-11 Amdt 2, IEC, Geneva, 1993.
IEC (International Electrotechnical Commission), *Functional safety of electrical/electronic/programmable electronic safety-related systems, IEC Publication 61508*, IEC, Geneva, 2004.
IEC (International Electrotechnical Commission), *Software engineering. Guidelines for the application of ISO 9001:2000 to computer software*, ISO/IEC 90003:2004, IEC, Geneva, 2004. (Simplified version: www.praxiom.com/iso-90003.htm)

See also BSI (1995), IEC (1989), (1996), (1998), ISO (2000), NIST (2000) in main reference list.

These standards are available from the British Standards Institute, Linford Wood, Milton Keynes MK14 6LE. Lists of standards are available on the BSI website: www.bsonline.bsi-global.com. IEC standards can be obtained from the IEC website: www.iec.ch.

International Dosimetry Recommendations

ICRU (International Commission on Radiation Units and Measurements), *Report 14, Radiation Dosimetry: x-Rays and Gamma Rays with Maximum Photon Energies Between 0.6 and 50 MeV*, ICRU, Bethesda, MD, 1969.

ICRU (International Commission on Radiation Units and Measurements), *Determination of absorbed dose in a patient irradiated by beams of X and gamma rays in radiotherapy procedures*. ICRU Report 24, ICRU, Washington, D.C., 1976.

ICRU (International Commission on Radiation Units and Measurements), *Dose and volume specification for reporting intracavitary therapy in gynaecology*, ICRU Report 38, ICRU, Bethesda, MD, 1985.

ICRU (International Commission on Radiation Units and Measurements), *Use of computers in external beam radiotherapy procedures with high-energy photons and electrons*, ICRU Report 42, ICRU, Bethesda, MD, 1987.

IAEA (International Atomic Energy Agency), *Calibration of dosimeters used in radiotherapy: A manual*, Technical Report Series 374, IAEA, Vienna, 1994.

See also IAEA (1997a), (1997b), (2000), ICRU (1992), (1993), (1999) in main reference list.

ICRU publications are available from the International Commission on Radiation Units and Measurements, 7910 Woodmont Avenue, Bethesda, Maryland. 20814 U.S.A. IAEA publications are available from the International Atomic Energy Agency, Wagramerstrasse 5, P.O.Box 100, A-1400, Vienna, Austria. Their web site is: www.iaea.org/Publications.

AAPM Guidance

AAPM (American Association of Physicists in Medicine), A protocol for the determination of absorbed dose from high energy photon and electron beams, *Med. Phys.*, 10, 741, 1983.

AAPM (American Association of Physicists in Medicine), *Quality Assurance Methods and Phantoms for Magnetic Resonance Imaging*, AAPM Report No. 28, American Institute of Physics, New York, 1990.

AAPM (American Association of Physicists in Medicine)Purdy, J. A., Biggs, P. J., and Bowers, C., Medical accelerator safety considerations: Report of AAPM Task Group 35, *Med. Phys.*, 20, 1261–1275, 1993.

AAPM (American Association of Physicists in Medicine), Report of AAPM Task Group 41. *Remote afterloading technology*, American Institute of Physics, New York, 1993.

AAPM (American Association of Physicists in Medicine)Almond, P. R., Attix, F. H., Humphries, L. J., Kubo, H., Nath, R., Goetsch, S., and Rogers, D. W. O., The calibration and use of plane-parallel ionisation chambers for dosimetry of electron beams: An extension of the 1983 AAPM protocol report of AAPM Radiation Therapy Committee Task Group No. 39, *Med. Phys.*, 21, 1251–1260, 1994.

AAPM (American Association of Physicists in Medicine), Report 68. Permanent Prostate Seed Implant Brachytherapy, *Med. Phys.*, 26, 1999.

AAPM (American Association of Physicists in Medicine), TG-51 protocol for clinical reference dosimetry of high-energy photon and electron beams, *Med. Phys.*, 26, 1847–1870, 1999.

See also Almond (1999), AAPM (1984), (1994, 1995, 2004), Ma (2001) in main reference list.

AAPM publications are obtainable from AAPM, American Center for Physics, One Physics Ellipse, College Park, MD, 20740-3843. Their web site is located at: www.aapm.org/pubs.

BIR Guidance

BIR (British Institute of Radiology), Criteria and Methods for Quality Assurance in Medical X-Ray Diagnosis, British Journal of Radiology Supplement 18, British Institute of Radiology, London, 1985.

BIR (British Institute of Radiology), *Treatment Simulators*, British Journal of Radiology Supplement 23, British Institute of Radiology, London, 1989.

BIR (British Institute of Radiology), Aird, E. G. A., Jones, C. H., Joslin, C. A. F., Klevenhagen, S. C., Rossiter, M. J., Welsh, A. D., Wilkinson, J. M., Woods, M. J., and Wright, S.J., *Recommendations for Brachytherapy Dosimetry, Report of a Joint BIR/IPSM Working Party*, British Institute of Radiology, London, 1993.

BIR (British Institute of Radiology), Central axis depth dose data for use in radiotherapy, *Br. J. Radiol.* Suppl. 25, 1996.

BIR publications are obtainable from The British Institute of Radiology, 36 Portland Place, London W1N 4AT., Tel. +44 171 580 4085 Fax +44 171 255 3209. The BIR web site can be found at: www.bir.org.uk.

IPEM documents

HPA (Hospital Physicists Association), *The Physics of Radiodiagnosis*, HPA Scientific Report Series 6, IPEM, York, 1977.
HPA (Hospital Physicists Association), *Measurement of the performance characteristics of diagnostic x-ray systems used in medicine: Part 1, X-ray tubes and generators*, Topic Group Report 32 Part 1, IPEM, York, 1980.
IPEM (Institute of Physics and Engineering in Medicine), Report 91, *Recommended Standards for the Routine Performance Testing of Diagnostic X-ray Imaging Systems*, IPEM, York, 2005.
IPSM (Institute of Physical Sciences in Medicine), Supplement to Report 54, *Commissioning and Quality Assurance of Linear Accelerators*, IPEM, York, 1990.
IPSM (Institute of Physical Sciences in Medicine), Code of practice for high-energy photon therapy dosimetry based on the NPL absorbed dose calibration service, *Phys. Med. Biol.*, 35, 1355–1360, 1990.
IPSM (Institute of Physical Sciences in Medicine), Protocol for establishing and maintaining the calibration of medical radionuclide calibrators and their quality control. Report 65, In *Quality Standards in Nuclear Medicine*, Smith, A. H. and Hart, G. C., Eds..
IPEMB (Institution of Physics and Engineering in Medicine and Biology), Report 68, *A guide to commissioning and quality control of treatment planning systems*, Shaw, J. E., Ed., IPEMB, York, 1996.
IPSM (Institute of Physical Sciences in Medicine), The IPEMB code of practice for electron dosimetry for radiotherapy beams of initial energy from 2 to 50 MeV based on an air kerma calibration, *Phys. Med. Biol.*, 41, 2557–2603, 1996.
IPSM (Institute of Physical Sciences in Medicine), The IPEMB code of practice for the determination of absorbed dose for x-rays below 300 kV generating potential (0.035 mm Al-4 mm Cu HVL; 10–300 kV generating potential), *Phys. Med. Biol.*, 41, 2605–2626, 1996.
IPSM (Institute of Physical Sciences in Medicine), Report 74, *Application of the Medical Devices Directive, Guidance Notes*, IPEM, York, 1996.

See also IPEM (1999), IPEM (2002) in main reference list.

IPEM publications are obtainable from The Institute of Physics and Engineering in Medicine, PO Box 303, York YO1 2WR. Tel. +44 1904 610821 Fax +44 1904 612279. Web address: www.ipem.ac.uk.

Miscellaneous International Recommendations

ESTRO (European Society for Radiotherapy and Oncology), *Recommendations for a Quality Assurance Programme in External Radiotherapy*, Physics for clinical radiotherapy, ESTRO Booklet No. 2, Aletti, P. and Bey, P., Eds. ESTRO, Brussels, 1995.
IAEA (International Atomic Energy Agency), Safety Series No. 115, *International Basic Safety Standards for Protection against Ionizing Radiation and for the Safety of Radiation Sources*, IAEA, Vienna, 1996.
IAEA (International Atomic Energy Agency), Technical Document No. 1040, *Design and Implementation of a Radiotherapy Programme: Clinical, Medical Physics, Radiation Protection and Safety Aspects*, IAEA, Vienna, 1998 (Currently being revised).
ICRP (International Commission on Radiological Protection), Recommendation No. 86, *Prevention of accidents in radiotherapy*, Pergamon Press, Oxford, 2001.
ISO (International Organization for Standardization), *Quality management systems, Fundamentals and vocabulary*, ISO 9000:2000, ISO, Geneva, 2000.
ISO (International Organization for Standardization), ISO 9000:2000 *Quality management systems, Fundamentals and vocabulary*, ISO, Geneva, 2000.
NCRP (National Council on Radiation Protection and Measurements), Report 69, *Dosimetry of x-ray and gamma-ray beams for radiation therapy in the energy range 10 keV–50–MeV*. NCRP, Bethesda, MD, 1981, third reprinting, 1993.
NCRP (National Council on Radiation Protection and Measurements), Report 102, *Medical x-ray, electron beam and gamma ray protection for energies up to 50 MeV*, NCRP, Bethesda, MD, 1989.

NPL (National Physical Laboratory), *Guide to the measurement of pressure and vacuum*, Institute of Measurement and Control, London (Obtainable from the Institute of Measurement and Control, 87 Gower Street, London WC1E 6AF or from the National Physical Laboratory, www.npl.co.uk/pressure/publications), 1998.

See also Huyskens et al., (2001), IAEA (1994, 2000), NACP (1980), Leer et al. (1999), NCS (1996), SFPM (2003), SGSMP (1997), Thwaites, et al. (1995), van Dam and Marinello (1994) and WHO (1998) in main reference list.

NCS publications are obtainable via the website www.ncs-dos.org. SGSMP publications are obtainable via the website www.sgsmp.ch/sgsmp-e.htm.

PART I

SPECIAL TECHNIQUES

Editors: Alan Nahum and Philip Mayles

INTRODUCTION

In this Part we cover all those external-beam radiotherapy techniques that, for a variety of reasons, do not easily fit into conventional treatment planning (Part G). Today Conformal Therapy (CFRT)* is no longer considered to be a special technique, but it is used here as a foundation for introducing Intensity-Modulated Radiotherapy (IMRT) which is a particular form of CFRT (Chapter 43). The deceptively simple idea of introducing another degree of freedom in treatment planning, namely to create deliberate variations or modulations in the intensity across the beam, has spawned an unprecedented period of development in treatment machine technology (e.g. dynamically controlled multileaf collimators (MLC)), in treatment planning systems capable of so-called *inverse planning*, and also in the search for truly meaningful criteria by which to judge the merit of a treatment plan (see Part G, Chapter 36). A separate chapter is devoted to the *clinical* implementation of IMRT (Chapter 44).

Stereotactic techniques (Chapter 45) have come a long way since the initial idea of putting a small circular collimator on a linear accelerator as an alternative to the multisource cobalt-60 machine known as the Gamma Knife®. Superb dose distributions to treat small quasi-spherical lesions in the head can be achieved with these small circular collimators, but currently there are multileaf collimators with especially narrow leaves (miniMLCs) that can adjust to the projected shape of the tumour as the gantry rotates, enabling highly conformal dose distributions to be delivered also to irregularly shaped tumours. Additionally, stereotactic *body* radiotherapy, also known as *extracranial* stereotactic radiotherapy, is now a very active field of development (SBRT 2006).

Proton beams (Chapter 46) are the ultimate modality with which to do Conformal and Intensity Modulated Therapy, possessing far superior depth–dose characteristics than either megavoltage photon or electron beams; today, only the high initial capital cost of the accelerator (cyclotron or synchrotron), and more especially, the proton gantry, is hindering the widespread construction of proton therapy facilities.

In complete contrast to conformal techniques of increasing sophistication, there are situations where the aim is to treat the whole body with very large fields. The two principal techniques of this nature are described here: total body irradiation with photon beams (Chapter 47) and total skin irradiation with electrons (Chapter 48).

All the modalities mentioned thus far have one thing in common—they are *low-LET*[†] radiation, i.e. the ionisation density along the charged particle tracks is relatively low; this is the case even for protons except at the very end of their tracks. Chapter 49 deals with two, so-called high-LET modalities in clinical use, neutron beams, which have unfavourable depth–dose characteristics, and so-called *light-ion* beams such as carbon ions, which have depth–dose characteristics similar to those of proton beams. The important difference between high-LET and low-LET radiations is in their *radiobiological* properties (see Part B), and efforts are underway to exploit these differences to further improve the clinical efficacy of radiation treatment.

As indicated above, the pace of change in radiotherapy is considerable, and new techniques are constantly being developed. These techniques are evolving rapidly, and it is through the scientific (research) literature rather than textbooks that readers can keep up-to-date with such developments. The common thread in these new developments is to take advantage of the progress in tumour imaging to improve the accuracy of treatment delivery; the weak points in the current practice of radiotherapy are the uncertainties in patient set-up and in internal movements (Goitein 2004; Purdy 2004) as well as the uncertainty in delineating the tumour itself. For example PET imaging has become an essential part of tumour delineation

---

* The abbreviation CFRT is used here rather than 3DCRT which is commonly used in the U.S.A.
† For a further discussion of LET (linear energy transfer) see Section 7.14.

in non-small cell lung cancer (Nestle et al. 2006). These advances in tumour imaging allow margins to be reduced that, in turn, enable the prescription dose to be escalated where appropriate with the intention of increasing the probability of local tumour control (TCP) while maintaining an acceptably low probability of normal tissue complications (NTCP). It is even possible, in some cases, with the use of PET/CT and MR spectroscopy to identify particular parts of the tumour where a higher dose is required (Coleman 2002; De Santis et al. 2002; Perez et al. 2002; Bentzen 2005; Malinen et al. 2006); this leads to the concept of the *Biological Target Volume* (Ling et al. 2000; Grégoire et al. 2006). PET imaging is now a recommended part of planning for non-small cell lung cancer (Nestle et al. 2006).

One way to improve the accuracy of treatment delivery is to use advanced imaging methods throughout the process. To describe this, the term *Image Guided Radiotherapy* (IGRT) has been coined (Mackie et al. 2003; Bortfeld et al. 2006). However, IGRT is not a precise concept; indeed, most radical radiotherapy carried out in recent years could be described as *image guided*. IGRT as presently used implies that a 3D imaging system is provided on the treatment machine, allowing accurate assessment of the beam setup in relation to the tumour before each individual fraction. This follows the experience acquired in proton therapy where this approach has been carried out for many years, using fixed kilovoltage x-ray tubes mounted in the treatment room (see Section 46.2.2 and Figure 46.10). Elekta (see Figure 40.11) and Varian are both marketing accelerators with a conventional kilovoltage x-ray tube, orthogonally mounted to the megavoltage beam, which is capable of both standard x-ray and so-called *cone-beam* CT imaging. Siemens offers a CT scanner installed in the same room as the accelerator with a common couch that supports the patient both for treatment and for CT scanning. A group in Utrecht (Raaymakers et al. 2004; Raaijmakers et al. 2005) is investigating the use of MR imaging to provide image guidance in the treatment room, and this concept is being taken up by Elekta.

Tomotherapy (see Section 43.3.4) fully integrates imaging with treatment delivery (Fenwick et al. 2006). This solution, described originally by Mackie et al. (1993), consists of rotating a 5 MV linac around the patient, similarly to what is done with a kilovoltage x-ray tube on a CT scanner. Complemented with appropriate beam shutters and detectors and combined with longitudinal motion of the table, *helical* intensity modulated treatment as well as megavoltage CT imaging are carried out (using a 3.5 MV beam—Forrest et al. 2004).

The availability of on-line image data presents a challenge to the treatment staff who are required to make instant interpretations of these data in order to decide whether to correct the patient position before proceeding to treat the patient or not. As computer speed increases, it may become possible to carry out instant recalculations of treatment plans based on these images (Court et al. 2005; Mohan et al. 2005; Scherrer et al. 2005); techniques such as *target of the day* and *adaptive radiotherapy* (Yan et al. 1997; Hugo et al. 2007) are currently being evaluated. The use of implanted markers (King et al. 2003; Shirato et al. 2004) enables the position of the soft-tissue target volumes to be detected by planar kilovoltage x-ray imaging.

In contrast with the more conventional use of orthogonal portal images, 3D reconstruction (possibly aided by implanted markers) enables, in principle, the patient position to be corrected to ensure perfect consistency with the treatment plan provided that the six degrees of freedom can be used. Current systems for correcting the patient position are based on orthogonal movements of the couch and couch rotation, which do not allow accurate correction for couch tilt or *roll*. Therefore, efforts are being made to develop systems, often derived from industrial robotic techniques, that allow more flexibility in 3D patient positioning (Hornick et al. 1998). The use of an accelerator mounted on a robotic arm (Webb 1999) as achieved, for instance, with the Cyberknife (Schweikard et al. 2000; King et al. 2003; and Section 43.3.9), is another solution to correct for 3D misalignment of the beam with respect to the patient.

As an additional aid to patient positioning, video-based systems that record the external surface of the patient can be used (Johnson et al. 1999; Yan et al. 2002; Meeks et al. 2005; Djajaputra and Li 2005). A reference surface is either obtained from external contours derived from CT images or acquired during a reference session (i.e. combined with portal images), and this is used to ensure patient position reproducibility, either inter- or intra-fraction. Such a system may also be used to record the variation of the surface with the patient's respiratory movement.

This movement of the tumour because of respiration (or possibly the cardiac cycle) presents another opportunity for the reduction of treatment margins. This has become known as breathing-adapted radiotherapy (BART). The general principle of BART is that breathing is monitored or controlled, and treatment is suspended when the tumour moves out of the field. However, this process relies on reproducible motion of the tumour in synchrony with breathing, and this condition may not be satisfied in some patients. Different methods of monitoring respiration and controlling the beam are being developed (Kubo et al. 2000; Giraud et al. 2003; Dawson and Balter 2004; Mageras and Yorke 2004; Jiang 2006; Keall et al. 2006).

A first solution resides in asking patients to hold their breath at a fixed level of respiration while the beam is on (Rosenzweig et al. 2000). This is achieved with the help of a spirometric system connected to a patient mouthpiece that provides a real-time signal of the respiration displayed on a monitor or on video glasses and made visible to the patient. *Beam-on* is allowed only when the signal is within a preset range corresponding generally to medium inspiration with a tolerance around ± 0.1 L. In the Active Breathing Control (ABC) system, a valve blocks the respiratory flow when the preset level is reached (Wong et al. 1999). In the voluntary breath hold system, the patient keeps full control of the respiration, but the beam is stopped if the signal drifts out of the preset range (Garcia et al. 2002; Zhang et al. 2003). In both cases, patient training and coaching are necessary (Kini et al. 2003).

With synchronisation methods, the patient breathes normally and regularly. One of the most commonly used systems is the Real-Time Position Manager (RPM) system developed by Varian where a small cube equipped with reflecting circular spots is placed on the patient's chest and its movement followed by a fixed infra-red camera. The movement of the cube seen from the camera is converted into a signal that is used to switch the beam on and off (i.e. to gate the beam) in such a way that the radiation delivery occurs within a preset range of respiratory movement (Mageras et al. 2001; Ford et al. 2002; Hugo et al. 2002). This method requires some patient cooperation, training, and coaching to achieve a reproducible breathing cycle, but not as much as with spirometric techniques. An even more sophisticated approach is to move the MLC blades in synchrony with breathing (Neicu 2003; Murphy 2004) or to robotically track the tumour as in the Cyberknife (Schweikard et al. 2002 and Section 43.3.9).

Whichever BART approach is used, the CT images used for planning must be representative of the patient anatomy during delivery. Therefore, breathing monitoring, similar to that used for treatment, is required during CT acquisition. For the RPM, two solutions have been implemented: either *prospective* CT gating where images are acquired only in the gating window or *retrospectively* gated CT where a full series of images are acquired covering a full respiratory cycle with sufficient redundancy to allow reconstruction at any desired phase of the cycle. This latter solution, sometimes referred to as 4D CT, necessitates sorting of the slices according to each phase of the recorded respiratory cycle and subsequent selection of one or more series to serve as the reference for planning (Ford et al. 2003; Underberg et al. 2004; Giraud et al. 2005).

The benefit of BART is still to be fully assessed (Starkschall et al. 2004; Rosenzweig et al. 2005). When comparing breath-hold and gating techniques, there are two issues to be considered: margin reduction where both techniques may yield similar results, and improvement of the dose distribution with respect to planning target volume coverage and organ-at-risk protection where normal breathing is probably not as beneficial as forced breathing. However, the difficulties of reproducibility may limit the number of patients for whom BART is applicable. Fortunately, Bortfeld et al. (2004) have shown that the effect of movement on the accuracy of the dose distribution is not as great as it might at first appear, and therefore it may not be essential to eliminate movement. Whatever solution is adopted, the incorporation of internal organ movements and, ideally, of setup uncertainties into treatment planning is one of the most important developments taking place in precise conformal radiotherapy.

Whereas these solutions are perhaps more versatile than tomotherapy, they currently lack the systems integration that is offered by the tomotherapy machine. Dose distributions produced with the tomotherapy device present a considerable challenge to the linear accelerator manufacturers; it remains to be seen if techniques such as Intensity Modulated Arc Therapy (IMAT) (Yu et al. 2002), which is similar to tomotherapy but can be carried out on a conventional linac, can match these dose distributions. There is also some debate as to whether the *dose bath* (i.e. large volumes of normal tissue receiving relatively low dose) associated with tomotherapy and IMAT will result in more second cancers (see Section 61.5.2). This is also an issue in connection with cone-beam CT. The dose must always be limited to that which is strictly necessary for the purpose rather than seeking the highest quality images for their own sake.

All these new approaches require the physicist to have a good understanding of patient positioning techniques as well as of the imaging and image registration techniques used by the treatment staff. In addition, new algorithms are called for to allow computers to assist as much as possible in the process. A reappraisal is also required of the degree to which the equipment can be relied on and the degree to which manual checking is required. Without greater acceptance of the capabilities of the technology, it will be difficult to make further progress and thus to deliver the potential benefits of these more advanced techniques.

CHAPTER 43

CONFORMAL AND INTENSITY-MODULATED RADIOTHERAPY

Steve Webb

CONTENTS

43.1 GENERAL OVERVIEW

The rationale for conformal radiotherapy (CFRT) is straightforward to state. The goal is to achieve a tumorcidal high-dose volume which conforms to (i.e. wraps closely around) the planning target volume (PTV) whilst simultaneously the organs-at-risk (OARs) (which might lie tightly adjacent to the PTV) receive a dose sufficiently low as not to cause any complications. The concept behind this rationale is that if the difference between the relative dose received by the PTV and the OARs can be made as large as possible, then the dose to the PTV can be *escalated* with consequent expectation of a higher tumour control probability (TCP) without causing unwanted radiation damage to normal-tissues (i.e. a low normal-tissue complication probability (NTCP)). Today the words *conformal radiotherapy* have replaced the somewhat older term *conformation radiotherapy* coined by Takahashi (1965) over 35 years ago.

So what is new? Has this not always been the goal of radiotherapy? Of course it has (Mould 1993, 1995; Thomas et al. 1995). In that case, why is CFRT viewed as a modern development? The answer is simple. Radiation oncologists have always known that they wanted to *concentrate* the radiation at the target (Kieffer 1938) but because photons have to pass through normal tissues to arrive at the target and also scatter *en route* to deposit dose at unwanted locations, the goal has been thwarted. Additionally, before the advent of modern 3D imaging techniques (the first being x-ray computed tomography (CT) in 1972; see Webb 1990) radiation oncologists were unsure of the exact location of disease (Webb 1988). Also radiotherapy equipment for delivering photons generally formed a rectangular field, maybe with simple wedges or blocking, limiting the possible shape of the high-dose volume. Moreover, the patient geometry is known to change between fractions and factoring this in limits the precision of the delivery. In the very early days, the lack of high-energy beams with sufficient penetration was itself a limitation. Techniques to verify the patient position at the time of treatment were limited to the use of film and there were no methods of automatically correlating the at-treatment position with the at-simulation position. Finally, knowledge of radiobiology had to grow co-temporally with the development of radiotherapy techniques and therefore normal-tissue tolerances and dose-(tumour) response data had to be collected. Hence it may be observed that whilst radiation oncologists wanted to perform CFRT (they would not have used that term of course) from about 1896 onwards, they simply lacked the tools to do so.

It may be said that the reason CFRT is possible today is that there has been a simultaneous development in all the required elements of the *chain of radiotherapy*, i.e. one may observe that the CFRT process is like a chain having many links, all of which must be strong and reliable if CFRT is to succeed (Figure 43.1). An exemplary radiotherapy department is one in which there is emphasis given to the physical (and biological) basis of all links of the chain. CFRT has

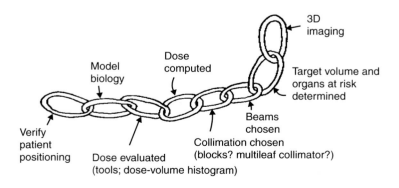

FIGURE 43.1
The chain of processes leading to conformal radiotherapy and illustrating that the physical basis of radiotherapy depends on good performance at each stage.

become a major specialty and has already been the subject of several substantial reviews (Webb 1993, 1997a, 1998a, 2000a, 2000b, 2003a, 2003b, 2004b; Smith 1995; Meyer and Purdy 1996; Sternick 1997; Van Dyk 1999; Schlegel and Mahr 2001; Palta and Mackie 2003; Bortfeld 2006; Ahnesjö et al. 2006). Techniques to shape beams, including the multileaf collimator, are covered in Section 22.5. Although beam-shaping is an essential aspect of conformal therapy, it can no longer be considered as a *special technique*. The emphasis in this chapter is on more advanced CFRT, principally intensity-modulated radiotherapy (IMRT).

The *chain of CFRT* comprises the following links:

- High-quality three-dimensional medical imaging to localise both the extent of disease and the anatomy of neighbouring normal tissue (Chapter 30 and Chapter 31). Ideally this imaging should comprise use of x-ray CT, magnetic resonance imaging (MRI), single-photon-emission-computed tomography (SPECT) and positron emission tomography (PET). All 3D imaging modalities should be networked to each other and to the planning workstation. Tools should exist to determine the PTV and the OARs automatically with the possibility of manual editing.

- The planning workstation should allow all tasks to be performed in 3D. It should include tools to optimise the beam directions and geometrical shapes, tools to optimise the beamweights, tools to optimise the fluence of radiation on a fine spatial scale (for IMRT) and tools to export the determined fields to the treatment-delivery device.

- The planning workstation should also be able to compute dose via accurate algorithms. These should ideally include both modern convolution algorithms and Monte Carlo dose calculation (see Part F).

- The planning workstation should include tools to evaluate the treatment plan and in particular competing treatment plans based on different human choices. These should include the computation and display of dose-volume histograms, of shaded colour dose planes, of isodose plots, of 3D surface dose distributions, of volumetric dose data and the prediction of biological outcome (TCP and NTCP) through suitable, albeit quasi-empirical, models (see Chapter 35 and Chapter 36).

- The patient position at the time of treatment should be verifiable via electronic portal imaging (EPI). Tools should exist to quantify discrepancies between treatment position and planned position and to evaluate consequences and make corrections (Section 40.1).

When is CFRT required? Ideally, the answer is *always*. There is clearly no point in irradiating normal tissues. However, the circumstances absolutely requiring CFRT are those tumour sites for which the cure rate would very probably be increased by increasing the prescription dose (i.e. the dose to the target volume), but where presently this dose increase is not possible due, for example, to OARs closely abutting the PTV and more specifically for those circumstances in which the PTV has a concave surface and an OAR lies in the concavity. Hence there has been a great deal of interest in CFRT of the prostate since this organ is concave and the adjacent rectum limits dose escalation (Dearnaley et al. 1999) (Figure 43.2). A second CFRT indication is for tumours where the cure rate is satisfactory but treatment-related morbidity, while not life-threatening, is highly unpleasant, e.g. xerostomia (or dryness of the mouth) from irradiation of the parotid gland. Head-and-neck tumours present a challenge to CFRT because many such tumours are in the vicinity of the parotid gland and the spinal cord, both of which require functional preservation. Some brain tumours have concave shapes with potential danger for irradiating the orbits, optic nerves, chiasm and normal brain. The breast requires CFRT to spare heart, lungs and brachial plexus. CFRT of the lung may spare other normal lung function and also the heart. There have been hundreds of papers on the physics of CFRT of these targets. Prostate in particular has been subject to numerous theoretical studies on improving treatment by CFRT including IMRT. However at the time of writing it should be emphasised that clinical trials to establish and quantify the clinical benefit of either *conventional* CFRT or advanced CFRT, i.e. IMRT, have been few. Amongst reasons for the slow introduction of CFRT and IMRT in many countries, the following may be identified:

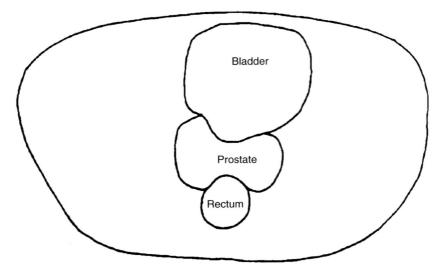

FIGURE 43.2
Why CFRT and IMRT are needed for the prostate. The concave prostate is shown abutting the rectum and the bladder.

- The argument that *proof of benefit* should precede change
- The requirement to quality-assure new techniques (particularly challenging for IMRT)
- The lack of the *complete chain* in every clinic

A U.K. study (Dobbs et al. 2002) has emphasised how essential it is that CFRT and IMRT become routinely established and the author's observation is that the year 2002 marked the change from IMRT being viewed as a research specialty to its acceptance as a practical clinical tool. In the U.S.A. IMRT is now well documented as a clinical practice (IMRTCWG 2001; Ezzel et al. 2003; Ten Haken and Lawrence 2006)

What are the physical means for achieving CFRT? The techniques divide into two broad classes:

1. Those involving only geometrical field shaping, sometimes known as *conventional* CFRT
2. IMRT techniques

Radiation delivered by the first method can only yield a convex 3D dose distribution whereas a concave 3D dose distribution can result from the latter. The main achievement of the techniques with only geometrical shaping is the shaping of the radiation field to the beam's-eye-view of the target. Although for decades this has been possible with cast metal blocks, the multileaf colimator (see Section 11.4.3 and Section 22.5.4) provides fast automatic verifiable beam shaping. IMRT may be delivered by many methods, some involving MLC-generated field-shaping combined with modulating the radiation fluence. This is appropriate for the estimated 30% of tumours which have concave surfaces. IMRT is a rapidly expanding subspecialty of CFRT. The various methods of IMRT delivery will be discussed in Section 43.3.

The necessary infrastructure for the development of CFRT is apparent from the above comments. Both conventional CFRT and IMRT are the product of synergistic development in computer technology, in electromedical equipment manufacture and in computer modelling. CFRT stands on the firm rock of clinical 3D medical imaging. Against this background, some heroic historical attempts at CFRT can be identified.

Takahashi in Japan developed rotation therapy with variable field shaping and blocking in the 1960s (Takahashi 1965). Also in the 1960s Proimos and colleagues in Boston developed rotational techniques with both variable field shaping and synchronous shielding (Proimos 1960). Jennings and colleagues in the 1960s at London's Royal Free Hospital developed the *tracking cobalt unit* which continuously varied the field shape to conform to the beam's-eye-view of the target (Jennings 1985). Note that all these developments took place before the invention of CT and before computerised control of equipment was possible. They also pre-dated any form of 3D treatment planning. Heroic indeed and holding a worthy place in history but never widely developed or copied and ultimately failing to be established clinically at other centres. The modern development of CFRT including IMRT is really the story of development in the late 1980s and the 1990s (Marinker and Peckham 1998; Webb 2003c; Webb and Evans 2006).

43.2 OPTIMISATION AND INVERSE PLANNING

43.2.1 WHY OPTIMISATION?

In the very early days of radiotherapy there were no methods of treatment planning. Radiation Oncologists flew by 'the seat of their pants'. They applied beams to the tumours until crude indices such as skin reddening showed that some particular dose had been delivered. There was no method of combining dose from different beam directions. By the 1940s the dose-deposition properties of single beams were beginning to be understood. Depth-dose distributions could be measured. They were transferred to acetate sheets. Then the combined dose distribution from several beams emanating from different directions could be *computed* by overlaying several sheets and adding up the doses to each point by hand. This became known as *hand planning* and even in the 1970s treatment planners were still using this method to learn their trade, and in many centres to plan real cases. Physicists designed devices to perform similar manual additions when the beams were not coplanar.

By the late 1960s, a number of pioneering groups were developing computer-based treatment planning. A computer was used to do the dose additions. The planning was two-dimensional. Perhaps the best known early system in the U.K. was the RAD8 designed by Roy Bentley and Jo Milan at the Royal Marsden Hospital and now something of a legend (Bentley and Milan 1971). By the mid 1980s, 3D treatment planning was well-established in research centres and commercial equipment was becoming available. Today, 3D treatment planning is regarded as a state-of-the art necessity for a strong link in the chain of radiotherapy. The planning stands on a firm bedrock of knowledge of the patient anatomy, of the dose-deposition characteristics of photon beams of different energy, size and shape, of the relationship between the position of the beams and that of the patient (Webb 1993, 1997a, 2000a).

Yet even in the early years of the new millennium, most treatment planning is what, in the jargon of the trade, has become known as *forward planning*, i.e. the treatment planner knows what dose distribution they would like to achieve in consultation with the radiation oncologist, making use of prior experience of similar cases, and then the task is to try out the positioning, shaping and weighting of a number of beams until the goal is reached. This is still the everyday technique of routine treatment planning. It relies on the expectation that patients are largely similar in geometry and so what works well for one patient will probably work well for another after some fine tuning. For many relatively simple situations it works satisfactorily.

43.2.2 PHILOSOPHY OF OPTIMISATION AND INVERSE PLANNING

The main objection to *forward planning* is that there is no way of knowing if the plan thus produced is the best which could be obtained. There is just not enough time for the human planner to explore all the options. The Holy Grail of planning is what has come to be called *optimisation* although, for the reasons to be discussed, the term *customisation* might be prefer-able. A second major objection to forward planning is that it simply cannot be done for complex IMRT. The number of variables in the planning problem is just too large. When several beam directions each with 2D intensity modulation are combined there is a requirement to determine hundreds of beam-weights. Hence from the earliest days of IMRT (and long before it was called by that name) *inverse planning* was developed (Lax and Brahme 1982; Brahme 1988). The 3D dose distribution was specified and the computer was programmed to arrive at the *optimum plan*, the set of beam element (bixel) beam-weights, which would deliver a 3D dose distribution as close to the prescription as the physics of photon-tissue interactions would allow. Inverse planning has become a cottage industry, a major specialty of radiotherapy physics, and today there are dozens of techniques, many variants on each other, and thousands of papers on the subject. Lengthy reviews have appeared (Webb 1993, 1997b, 2000a, 2003a; Ten Haken and Lawrence 2006). The principle is simple. The differences between the methods lie in the details.

Whilst inverse planning has grown up to serve IMRT, it need not be restricted to this. There is an equally large number of papers reporting the application of inverse planning to the determination of the beam-weights for a few open or wedged fields and for determining beam orientations. Inverse planning is rapidly being implemented into commercial treatment-plan-ning system, thus moving out of the research-tool arena and into everyday clinical practice.

Before reviewing some key techniques, a fundamental question should be addressed: does *inverse planning* as currently practised really produce the optimum treatment plan? The answer is *no* (Rowbottom et al. 1999). To arrive at an *optimum plan* the computer would have to examine an infinite range of beam orientations, an infinite number of beams, a wide range of beam energies, the combinations of beams with different energies, beams of different particles (photons, electrons, protons, heavy ions), an infinite flexibility on numbers of bixels and fluence quanta in IMRT. It does not do this. Instead all inverse planning techniques *fix* certain parameters and then seek to optimise the remaining free parameters, e.g. usually the beam energy is chosen a priori. A fixed number of beams is selected (whether for IMRT or non-IMRT), a set of predetermined orien-tations are chosen. No particle-modality combining is done. Then the computer is asked to *optimise* the beam-weights. The result is called the *optimum plan*. Yet if the initial assumptions were to be changed, a new and possibly better *optimum plan* might result. The same is true for the result of changing the cost-function being minimised by inverse planning. This would lead to statements of the kind that one of the two contenders was *more optimum* than the other, an English language violation. Yet the planners would know what they meant. It is proposed that the term for the plan resulting from inverse planning be called the *customised plan* and the term for the process be called *customisation* because this would overcome the difficulty. There can certainly be a *better customised plan*. The term *optimum plan* would refer to that in which the TCP were unity and the NTCP were zero, something probably unobtainable. However, it must reluctantly be recognised that the term *optimisation* has entered the language such that it is unlikely to be reversed. Hence, the terms *optimisation* and *optimised plan* will be used here with the above caveats.

43.2.3 PRINCIPLES OF MODERN INVERSE PLANNING

The 3D dose distribution, D, is related to the set of beam-weights, w_i, through a matrix, $\mathbf{M}$, via the operation

$$D = \mathbf{M}w_i \qquad (43.1)$$

where $\mathbf{M}$ is a matrix containing all the information linking beam-space to dose-space and there are i beams. When i is a small number, forward planning adjusts the beamweights, w_i, until the distribution, D, is *acceptable*. The process is one of informed trial and error. Inverse planning does the reverse and is particularly useful when i is large for IMRT and the weights w_i refer to small bixels (Webb 1995). The computer explores a range of options for w_i, deduces the delivered D and accepts or rejects options depending on whether some cost function decreases or increases with respect to that for other choices. The methods divide conveniently into those which are stochastically iterative and those which are deterministically iterative. The methods are characterised by the choice of cost function. Incidentally this also provides another reason to criticise the term *optimisation* because a change of cost function using the same technique would yield a different outcome.

The easiest way to understand cost functions is to consider the widely used quadratic cost function of dose:

$$C = \sum_i^N I_i (D_i - D_i^{\mathrm{P}})^2 \tag{43.2}$$

where D_i is the dose delivered to the ith dose voxel (at some stage in the optimisation) and D_i^{P} is the prescription dose to the ith voxel and there are N voxels. I_i is the importance factor for the ith voxel. Clearly this has the desirable property that minimising the cost function minimises the difference between the dose distribution obtained at some stage of the iteration and the prescribed dose distribution (in a least-squares sense). The presence of the *importance factor* (a simple dimensionless number) allows the user to assign more importance to certain regions of space rather than others. For example, given that photons deposit dose by the laws of physics, one cannot obtain arbitrarily high dose to a PTV and negligibly small dose to an adjoining OAR. The importance factor allows the planner to choose between the importance of obtaining the dose as close as possible to the prescription in the PTV whilst possibly compromising the OAR, or conversely, the clear protection of the OAR whilst possibly compromising the dose distribution in the PTV (conformal avoidance). Part of the art of inverse planning becomes that of selecting these importance factors; in turn, it is seen that this, yet again, gives an added flexibility to the solution which mitigates against calling the result the optimum plan. Critics of optimisation claim this human activity has simply replaced that of selecting beamweights by experience. To counter this, the generally adopted scheme is to determine *recommended importance factors* from test cases which are then applied to future inverse planning. Papers have appeared indicating ways to avoid presetting importance factors and these are reviewed in Chapter 5 of Webb (2000a).

Of course, although it is intuitive and possibly the first cost function used, the quadratic dose-based cost is only a *physical* cost function. It makes no distinction between under and overdosing. It takes no account of the location of dose discrepancies. For organs at risk it seems to require specification of a *required dose*, whereas critics would argue that this should be zero. Consequently, *dose-volume-based* cost functions have been investigated (Bortfeld et al. 1997). In these the volume of normal tissue to be raised to a certain dose is specified for a number of points on the dose-volume histogram—in principle, for all such points. Optimisation then attempts to force the OAR DVH within the constraints and makes no further improvements provided these can be met. It is claimed that such cost functions more closely model the way clinicians think.

A further option is to use biologically-based cost functions (see Chapter 36), e.g. if TCP (D_i) is the tumour control probability corresponding to the 3D dose distribution D_i at some stage of the inverse planning and NTCP (D_i) is the normal tissue complication probability at the same stage, then one could do any of the following (Webb 1997b):

- Maximise TCP subject to some maximally allowed NTCP
- Minimise NTCP subject to some minimally allowed TCP
- Maximise $P_+ = \mathrm{TCP}(1 - \mathrm{NTCP})$, the so-called probability of uncomplicated tumour control

These approaches each have their champions. They require, of course, a model for TCP and NTCP together with data to support the model. Herein lies the basis of much criticism of these approaches. It can be argued that the models and certainly the data on which they rely are somewhat arbitrary and insubstantial (see Chapter 36). It can also be argued that clinicians have a century of experience to guide them regarding requirements on dose and are less experienced in *biological modelling* and its predictions. There is also a halfway-house position in which the optimisation is conducted using dose-based cost functions but the results are reported also in terms of TCP and NTCP—the best of both worlds?

In its simplest form optimisation can be viewed as a computer recipe. The process states: make some change; observe the effect on the cost, either accept or reject the change and carry on until the cost converges to a minimum. Whatever the *state of the choices* is at this endpoint becomes *the solution*. Methods can be classified in terms of the instruction base which determines the directions of change. Stochastic instructions are slow but particularly easy to understand and very flexible. Deterministic instructions are generally faster but not always allowable. Hoffmann et al. (2006) have presented a particularly interesting analysis of this important area.

43.2.4 WHAT CAN BE OPTIMISED?

The following parameters have all been subject to optimisation schemes:

- Beamweights for open fields
- Beam directions for *few-field* irradiation
- Wedge angles for *few-field* irradiation
- Bixel weights in IMRT

For the first and last, an obvious constraint is that the weights must remain positive. This can be easily built into stochastic iterative optimisation. For some other schemes which generate negative beam-weights these have to be set to zero *a posteriori*, arguably damaging the *optimum* status of the solution. There have been several very detailed papers explaining how perfect dose distributions could be obtained with negative beam-weights but these papers are of academic interest only.

In summary, optimisation is entirely necessary for IMRT which can produce concave dose distributions such as those discussed in Section 43.1. It is also feasible for non-IMRT problems aiming to achieve convex dose distributions and particularly relevant to the optimisation of the number of beams, the beam directions and wedge angles.

There have been so many proposed variants on the optimisation scheme that they cannot be comprehensively reviewed in a short section such as this; readers requiring more details can consult Webb (1993, 1997b, 2000a, 2003a). In the next subsections a couple of specific methods are presented to give a flavour for what is done, how it is done and to provide some typical results.

43.2.5 OPTIMISATION BY SIMULATED ANNEALING

This technique was introduced into radiotherapy at the end of the 1980s (Webb 1989, 1991). At the time it was developed to minimise the quadratic-dose cost function. The methodology was based on that described by Barrett et al. (1984) for reconstructing nuclear medicine images. Since that time, it has become apparent that the convex cost function has only one global minimum (for beam-weight optimisation) and so the complexities of simulated annealing are *not* required. However the methodology is still sound (if slow) and it certainly *is* needed for optimising beam directions and for optimising problems using certain biologically-based cost functions where local minima are expected. It also has an intrinsic comprehensibility. It is easy to describe and to see exactly what is happening. Its variant, least-squares optimisation, is similarly transparent.

The method works as follows. It will be described as it applies to determining the 1D intensity-modulated beams (IMBs) for IMRT of a 2D dose slice. First, certain choices are fixed: the number of 1D IMBs and their orientation, the number of bixels per 1D IMB, the modality and beam energy, the deposition vector linking each bixel to each dose pixel, etc. For total transparency, imagine, for instance, 9 1D IMBs spaced at 40° azimuthal orientations with 20 bixels per 1D IMB, using photons from a spectrum of peak potential 6 MV as would be delivered by the NOMOS MIMiC operating at a fixed gantry orientation (see Section 43.3.4). The problem becomes that of determining the $20 \times 9 = 180$ bixel beam-weights (Figure 43.3).

Let $C(D_i)$ be the cost associated with the dose distribution D_i at the ith iteration. It is expected that the form of this cost function is such that it has several local minima. The goal is to arrive at the distribution of beam weights which yields the global minimum and avoids any trapping in local minima.

The process starts with, for example, empty bixels, i.e. zero applied fluence. A *grain* of fluence is offered to a randomly selected bixel (the selection is simply controlled by using the random number generator in the computer). The grain is very small in relation to the expected final *settled* distribution of bixel intensities, say 1% of the expected maximum. The 3D dose distribution is computed throughout the irradiated slice using Equation 43.1 and the corresponding cost is deduced from whatever function is represented by C. It could be Equation 43.2, although, as already stated, this strictly does not need simulated annealing.

Provided the cost function *decreases* as a result of the grain addition, the addition is accepted. One might imagine the converse to then be invoked, i.e. that an *increase* in the cost function would lead to the decision to reject the change. However, this is not so. Let ΔC_i be the change in the cost function at the ith iteration. When ΔC_i is negative the change is accepted. If ΔC_i is positive the change is accepted with a probability $\exp{-(\Delta C_i)/kT}$, where kT

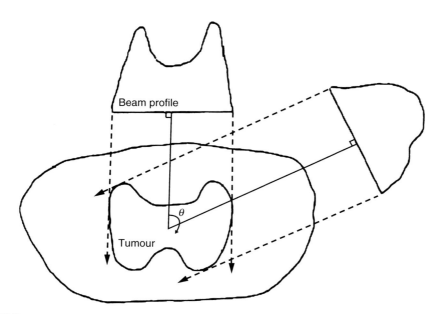

FIGURE 43.3
The concept of intensity-modulated beams. Two (of a large set of such beams with 1D intensity-modulation) are shown irradiating a 2D slice. The beams combine to create a high dose treatment volume spanning the PTV which has a concave outline in which OARs might lie. Such uniform high-dose treatment volumes cannot be achieved using beams without intensity modulation. Planning such treatments relies on tools such as simulated annealing (for cost functions with local minima) and projection–reprojection.

is a controlling parameter with the same dimensions as ΔC_i. T may be thought of as a tempera-ture and k is Boltzmann's constant. At the start of iteration the temperature is set high so many *wrong-way* (or *uphill*) changes are accepted. The temperature is gradually reduced during the iteration such that at the end of the process it is so low that no matter how small is ΔC_i the change is rejected. The process is subject to *cooling* and it can be shown that provided the cooling is slower than the reciprocal of the logarithm of the iteration number then the global minimum will be reached. This is known as the Metropolis condition after the discoverer of the process. The technique mimics the slow annealing of a metal or crystal, the crystalline state representing a global minimum potential whereas the higher local minimum potentials corre-spond to amorphous states. Thus the local minima are avoided and the iteration always reaches the global minimum.

Both positive and negative grains require to be added so that structure created early in the iteration can be removed if required. A random number randomly selects positive and negative grains. The positivity constraint on bixel beam-weights can easily be applied as the iteration progresses. The grain size can be reduced during the iteration so that the late changes are very small.

An analogy which is sometimes considered useful is to imagine that the progress of the iteration towards a minimum cost function is like a skier descending a hill. The starting line at the summit represents the cost associated with the initial conditions (e.g. empty beams). As the IMBs are constructed so the cost decreases, corresponding to the skier descending the *cost slope*. The goal is for the skier to arrive at the foot of the hill (the point of global minimum cost).

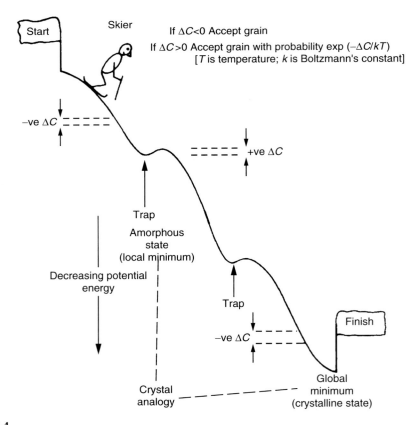

FIGURE 43.4
The analogy between minimising a cost function by simulated annealing and the technique of a skier descending a slope (see text for details).

However en-route there are small snow bumps creating local minima. If skiers only ever agree to go downhill (reducing cost), then they will be trapped behind such bumps, knowing that they are at a local, not global, minimum, but unable to *escape*. The acceptance of *wrong-way* moves is the mechanism by which the skier escapes, i.e. has to occasionally ski uphill which is counter-intuitive in a minimisation problem but absolutely essential to reaching the global minimum. This is illustrated in Figure 43.4.

The technique described is the so-called *classical simulated annealing mechanism*. An alternative is *fast simulated annealing*. In this, the temperature is reduced much faster but the grains are selected from a Cauchy distribution whose width is initially large and decreases as iteration progresses. This has the effect of allowing the system to jump out of a local minimum by *tunneling*.

Simulated annealing has been widely used together with its variant, least-squares mini-misation (when $T=0$ is set and there is only one minimum in the cost function). It is intuitive, easy to control, capable of much adaptation (e.g. the application of smoothing constraints), always gives positive IMBs but is computationally slow. This is one reason why faster semi-analytic optimisation techniques have been developed. The technique due to Bortfeld and colleagues will be described as an example. There are many variants of this too. The simulated annealing technique forms the basis of the inverse planning which is provided by PEACOCKPLAN or the CORVUS inverse-planning system marketed by the NOMOS Corporation (Sternick 1997).

43.2.6 FASTER OPTIMISATION BY PROJECTION-BACKPROJECTION

The projection-backprojection method was developed by Bortfeld and colleagues (1994a). The 3D solution creates a first estimate of the 2D fluence profile $\varphi_\theta^{(0)}(x_s, y_s)$ in the direction θ (x_s and y_s indicate coordinates in the plane normal to the central axis of the beam through the isocentre; x_s is in the transverse plane). This is done by *projecting* the 3D dose prescription $D^P(r)$ along fan lines but weighting it by a factor which applies a higher weight to rays that pass though regions of the target with a larger average distance from the source and vice-versa. The projections were filtered by a 1D function $h(x_s)$ and finally a positivity constraint C^+ was applied. Formally,

$$\varphi_\theta^{(0)}(x_s, y_s) = C^+[\{P_\theta[D^P(r)]/P_\theta^\Delta[D^P(r)]\}(P_\theta[D^P(r)]\otimes h(x_s))] \qquad (43.3)$$

Here, there are two projection operators. The first, P_θ, simply projects all the dose values at r along each ray line. The second, P_θ^Δ, projects the same values but each weighted by a function Δ that is the product of the tissue-maximum-ratio and the inverse-square-law factor. It may be seen that Equation 43.3 contains a ratio that applies a higher weight to rays that pass through regions of the target with a larger average distance from the source.

In turn, this leads to a first estimate of the 3D dose distribution by *backprojecting* the fluence weighted by photon attenuation and the inverse-square-law and summing over all the beams, i.e.

$$D^{(k)}(r) = \sum_\theta \varphi_\theta^{(k)}(x_s, y_s)\Delta \qquad (43.4)$$

The parameter k labels the iteration number; for the first time, $k=0$.

This first estimate of the 2D fluence profile was then iteratively refined using the projection of the difference between the 3D dose prescription $D^P(r)$ and the first estimate of the delivered 3D dose $D^{(k)}(r)$ at $k=0$, weighted by a function f_s that controlled the relative importance of constraints in the PTV and OAR. The resulting new fluence profile became the next estimate and the process cycled to convergence. Formally,

$$\varphi_\theta^{(k+1)}(x_s, y_s) = C^+[\varphi_\theta^{(k)}(x_s, y_s) - P_\theta^\Delta[D^{(k)}(r) - D^P(r)]f_s] \qquad (43.5)$$

The parameter f_s was unity in the PTV and a value between 1 and 10 in the OAR if, and only if, the dose was above a preset constraint limit and zero elsewhere.

As the delivered dose distribution finally approaches the prescription the cyclic operation of Equation 43.4 and Equation 43.5 will have diminishing effect. Whilst the authors have made ongoing modifications to this technique, the above represents the fundamental steps and is the basis of the inverse planning applied in the KONRAD treatment planning system (Bortfeld et al. 1997, see Section 43.2.8). There are many variants on this theme. The main observation is that this technique is much more rapid than simulated annealing. It is still based on minimising the difference between prescription dose and delivered dose at each iteration. Later variants have considered dose-volume constraints.

43.2.7 EXAMPLE OF IMRT-DELIVERED DOSE DISTRIBUTIONS COMPUTED BY INVERSE-PLANNING OPTIMISATION

Some results from the application of the above two methods of inverse-treatment planning are now presented. Bortfeld and colleagues (1994a) constructed a model problem in which the prostate, rectum and bladder were represented on a series of 2D contiguous slices. Inverse planning was performed for each slice using the method described in Section 43.2.6 and then the IMBs so determined were actually delivered to a phantom using the method known as multiple-static-field MLC delivery (see Section 43.3.2). Nine fixed fields were used spaced at 40° intervals. The spacing in the projection grid was 2.21 mm × 10 mm and the spacing in the calculation grid was 1 mm × 4.87 mm × 1 mm. The tolerance dose of the bladder was 85% of the maximum PTV dose and that of the rectum was 60% of the maximum PTV dose. The penalty value $f_s = 3$. Eleven iterations were performed. Figure 43.5 shows, for one such slice, the comparison between the computed isodose distribution and the measured one. The concave nature of the high-dose region is readily apparent and the 80% isodose contour was deemed to match well. The subject of the actual delivery will be discussed in Section 43.3. Here, the intention is just to show the planning outcome. This somewhat old reference is quoted because it was the first to make measurements in so-called MSF mode.

One slice of the same problem was planned using the simulated annealing technique (Webb 1994) described in Section 43.2.5. Figure 43.6 shows the results from this approach for two different arrangements of the number of beams (either 9 or 120) which represent different methods of delivering IMRT (see Section 43.3). Once again, the concave dose distribution so determined is readily apparent.

43.2.8 OPTIMISATION IN COMMERCIAL TREATMENT-PLANNING SYSTEMS

The serious growth phase of optimisation and inverse planning commenced in the late 1980s at a time when commercial treatment-planning systems were dedicated to *forward planning*. Twenty years later, the situation has changed radically; most commercial treatment-planning systems now have an inverse planning module. The first inverse-planning system was PEACOCKPLAN from the NOMOS Corporation, announced in 1992, which gradually evolved into the present system known as CORVUS. This made use of a simulated annealing algorithm based on the concepts expressed in Webb (1989, 1991). An example from this system is shown in Figure 43.7.

The second such system was KONRAD which emerged from DKFZ, Heidelberg, and was based on the projection-reprojection method. This was commercialised by MRC (now part of Siemens) and was subsequently incorporated into other commercial treatment-planning systems e.g. PLATO (Nucletron) to combine the advantages of forward and inverse planning in one treatment-planning system. Other manufacturers (e.g. Varian (Eclipse), Philips

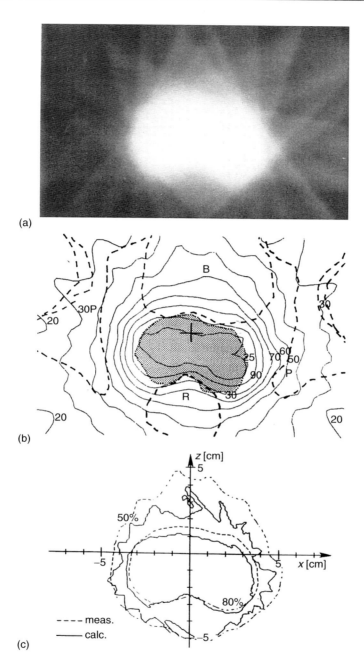

FIGURE 43.5

Representations of the dose distribution in the central slice of a model planning problem posed by Bortfeld. (a) Copy of the exposed and developed film (negative display). (b) Superposition of the measured isodose distribution on the anatomy; R, rectum; B, bladder; P, pelvis. Isodose values are in percent of the maximum global dose. (c) Comparison of measured and calculated isodose lines. (From Bortfeld, T. et al., *Int. J. Radiat. Oncol. Biol. Phys.*, 30, 899–908, 1994. With permission.)

(Pinnacle), CMS (Focus), Nucletron (Oncentra/Masterplan) have since incorporated inverse IMRT planning into their systems. Potential purchasers of treatment-planning systems are advised to ask manufacturers what optimisation tools are provided since commercial systems are continually evolving.

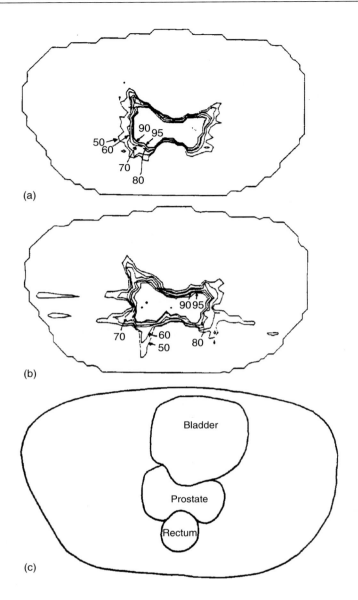

FIGURE 43.6
The isodose plot for the plan created according to two separate conditions: (a) 120 beams; (b) 9 beams. The cost function had importance parameters $I_{PTV}=200$ and $I_{OAR}=10$. The isodose distribution is tightly conformal in both cases. In case (a), corresponding to delivery by tomotherapy (see Section 43.3.4) the isodose lines are a little tighter than in case (b) that is nevertheless still an acceptable distribution. The isodose lines shown are 95%, 90%, 80%, 70%, 60%, and 50%. The ROIs are shown in (c). The bean-shaped ROI is the prostate; the anterior ROI is the bladder and the posterior ROI is the rectum.

In summary, optimisation has been the subject of enormous interest, particularly since the late 1980s, and the literature on it is vast. The papers selected here for reference are just a few examples of what one might cite. Optimisation could be said to be desirable for conventional radiotherapy but it is *essential* for IMRT planning. Inverse planning is the way forward; see Chapter 36 for a discussion on incorporating biological functions into the inverse-planning objective function. Virtually all manufacturers now have inverse-planning capabilities in their products.

Having established the methods by which optimised IMRT may be planned, the next section will describe how intensity-modulated radiotherapy can be delivered.

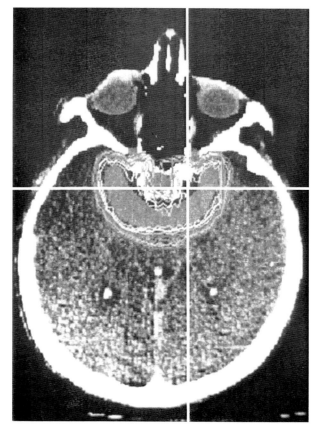

FIGURE 43.7
An example of the highly concave treatment volume which may be obtained from the NOMOS inverse-planning system for IMRT. The horseshoe-shaped tumour wraps around the sensitive structure in the brain which is correspondingly shielded.

43.3 INTENSITY-MODULATED BEAM DELIVERY

43.3.1 CLASSES OF IMRT DELIVERY TECHNIQUE AND HISTORY

IMRT offers the potential for a substantial leap in TCP at fixed normal-tissue complication probability in view of its ability to *sculpt* the high-dose treatment volume around the possibly concave target volume, and in other circumstances (e.g. breast radiotherapy) it can be used to create highly uniform coverage of target volumes. The planning of IMRT has been dealt with in the previous section. This section deals with the different ways to deliver and verify IMRT.

Before the early 1990s, there were virtually no practical methods to deliver IMRT. The methods available at present are (Webb 1993, 1997b, 1998a, 2000a, 2000b, 2003b; Williams 2003; Webb and Evans 2006):

- The metal compensator
- The MSF or *step & shoot* technique
- The dynamic multileaf collimator (DMLC) technique
- Tomotherapy
- Swept pencil beams

- The scanning attenuating bar method
- Intensity-modulated arc therapy (IMAT)
- The robotic cyberknife

Some of these techniques are still at the research and development phase. Others are being introduced clinically with varying levels of support, and still others have been found to be impractical. They each have advantages and disadvantages, champions and critics (Sternick 1997, Webb 1998a, 2000a, 2000b, 2003b; Williams 2003).

Historically the compensator is the oldest method of creating an IMB and has been available for much longer than any of the other techniques by a factor of at least 3 in time. Simply a piece of metal whose thickness in the direction of the beam varies with position in the beam, it creates an intensity-modulated beam (IMB) with the minimum of technological inventiveness. If a fluence $I(x,y)$ is required and the unattenuated fluence is I_0, then the thickness, x, is given by $x = (1/\mu) \ln [I_0/I(x, y)]$ where μ is the x-ray effective linear attenuation coefficient appropriate to the spectrum of the beam and the compensator thickness[*]. This equation clearly specifies only the primary fluence on the exit side of the compensator. The total fluence will be a complex mixture of primary and scatter and this could be determined by modelling the radiation transport using Monte Carlo simulation (see Chapter 5 and Chapter 28). There are many ways to make a compensator. These include:

- Pouring liquid metal into a mould of the appropriate shape
- Building a pixellated compensator using blocks of a finite size and different heights (like a LEGO^{TM} construction) (Ellis et al. 1959)
- Cutting thin (e.g. 0.5 mm) sheets of lead and sticking them together on a baseplate to make the required variable thickness (Figure 43.8)

The advantages of the compensator are:

- It is a physical device that is placed in the shadow tray on the linac. One can see it and easily check that the correct one is in place and that it is in the correct orientation.
- It has no moving parts, unlike many other methods.
- There are no constraints due to the linac collimation.
- There is no need for complex portal verification.
- There is no *tongue-and-groove* interleaf leakage (as there could be for MLC-based methods), etc.

The advocates of the technique are quick to point these out (Sherouse 1997). However there are disadvantages:

- The compensator has to be fabricated manually for every required field. Therefore, it is impractical for those methods requiring a large number of fields.
- It is cast at the beginning of the treatment and so the IMBs cannot be altered during the treatment (to account for changes in the target volume) without some difficulty.
- Its construction is a time-consuming process and somehow anachronistic today when one is able to control fieldshapes by computer.

It must be recognised that, in a field in which safety is paramount, there is a lot to be said for using tried-and-trusted methods, whatever their disadvantages. Consequently, given that the other methods also require more sophisticated technology, the compensator will not vanish

---

[*] The attenuation is not strictly exponential because there is a spectrum of photon energies; hence, the effective μ varies with thickness.

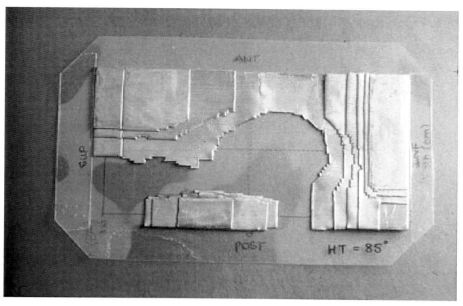

FIGURE 43.8
A photograph of a metal compensator created by cutting thin sheets of lead and gluing them together. (Courtesy of Dr P. Evans and the Breast Dosimetry Group, Royal Marsden NHS Foundation Trust.)

until the other methods have developed an equal following and the necessary quality assurance is achieved, and possibly not even then.

There have also been a few other heroic attempts at IMRT in the distant past (Webb 2003c). Of these the most-remembered is the synchronous field-shaping and organ shielding method developed by Basil Proimos and colleagues 40 years ago (Proimos 1960). The method is simple in principle. Radiotherapy is delivered by a rotation technique in which the beam sweeps a full circle around the patient. As it does so, the collimation is adjusted so the beam continuously adopts the beam's-eye-view of the target. This is achieved by fabricating special rotating collimators (Figure 43.9). At the same time the OARs are kept in the shadow of absorbers which rotate synchronously. There were many implementations of the method which was continually developed throughout the 1960s and Figure 43.9 shows one of them in which the upright patient, collimation and protection synchronously rotate in a stationary beam. There were other implementations in which the patient was horizontal and the collimation and protection was rotated by gravity with respect to the reoriented beam.

This method of delivering radiation is intensity-modulated in a very crude sense—the primary radiation is binary *on* or *off*. The method did not gain wide acceptability because:

- It was complicated (all the equipment had to be individually fabricated for each patient).
- The delivered dose was not uniform because the modulation was not constructed through an inverse-planning process.
- This was pre-CT and so there was opposition from those who did not believe the targets and normal structures were well enough localised.

However, the method has resurfaced (Danciu and Proimos 1999) from its original protagonist.

There now follows a brief review of modern techniques to deliver IMRT. The six methods outlined in the next sections were all developed co-temporally, largely in the 1990s (maybe

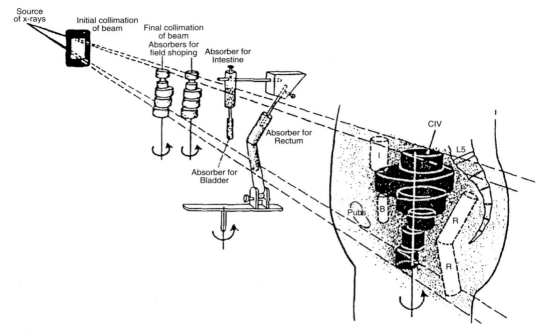

FIGURE 43.9
How synchronous absorbers for field shaping and synchronous protection can be arranged to provide a high-dose volume in the continuously irradiated volume (CIV) and protect the intestine (I), bladder (B), and rectum (R). (From Ilfield, D. N., Wright, K. A., and Salzman, F. A., *Am. J. Roentgenol.*, 112, 792–796, 1971. With permission.)

with a little activity at the end of the 1980s). There has, of course, been some rivalry and intercomparisons.

43.3.2 THE MULTIPLE-STATIC-FIELD (MSF) TECHNIQUE

The MSF technique relies on the observation that any 2D IMB can be created by a superposition of open fields with different fieldshapes. The fieldshapes are conveniently defined by a multileaf collimator (MLC). The principle is illustrated in Figure 43.10 for a 1-D IMB and even better by the many excellent computer simulations (as *.mpg* files) in Schlegel and Mahr (2001).

A 1-D IMB may be created by sequentially delivering open fields of different widths, each created by one leaf-pair of the MLC. Given that this may be done independently for any number of leaf pairs at the same time, the 2D IMB is built up from a series of 2D open field shapes. Figure 43.11 and Figure 43.12 show two possible configurations which yield the required primary fluence shown in Figure 43.10.

Thomas Bortfeld and Art Boyer (Bortfeld et al. 1994a, 1994b) were the originators of this method. An intuitive advantage is that the method is primarily a geometrical addition and so one can be easily convinced that the required 2D IMB is formed. The leaf-pairs can initially be considered to act independently and so the 3D dose-delivery problem requiring 2D IMBs breaks down into a set of 2D dose-delivery problems requiring 1D IMBs. Therefore, the delivery can be worked up from a series of 2D inverse-planning problems. This was how the method was initially conceived. However, very soon the connectivity between the 2D problems was recognised and genuine 3D inverse-planning was performed. The forward-dose calculations also took account of the 2D MLC shapes. Once conceived, the method was tested experimentally (Bortfeld et al. 1994a, 1994b) by laboriously delivering 2D IMBs

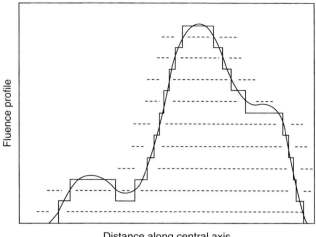

FIGURE 43.10
Shows how a 1D intensity modulation may be created for a radiotherapy beam profile. The horizontal axis is the distance along the direction of travel of the leaf, measured at the isocentre of the beam (called a central axis in the transaxial cross-section of the patient). The vertical axis is x-ray fluence. The solid line is the intensity modulation expressed as a continuous function of distance, interpolated from the discrete modulation resulting from some method of inverse planning. The horizontal dotted lines are the discrete intervals of fluence. Vertical lines are created where the dotted lines intersect the continuous profile thus giving a set of discrete distances at which discrete fluence increments or decrements take place. These are realised by setting the left and right leaves of a MLC leaf-pair at these distances in either the *close-in* or *leaf-sweep* technique. Note that all left leaf settings occur at positions where the fluence is increasing and all right leaf settings occur at positions where the fluence is decreasing.

at 9 gantry orientations each with between 21 and 30 subfields. At that time there was a significant time-penalty attached to resetting the leaf positions because the beam had to be switched off between each segment and then stabilise itself again at switch-on for new segments. Hence the experiment took many hours. However it is now regarded as a pioneering demonstration of the power of the MSF IMRT delivery technique.

One difficulty with the MSF technique is that the MLC imposes specific constraints on the leaf placements, these constraints varying between MLCs from different manufacturers. For example, the 10 mm Elekta MLC does not allow any leaf pair to close completely, nor does it allow a left leaf to come closer than a specified distance to its adjacent right leaf. Interdigitation, in which a left leaf travels to the right of an adjacent right leaf, is certainly disallowed (Figure 43.13). On the other hand, the Siemens MLC allows leaves to touch and the Varian MLC does allow interdigitation. The Elekta Beam Modulator MLC also allows interdigitation.

There are limitations on the allowed 2D configurations for MLCs not allowing interdigitation. It implies that one cannot simply solve the 1D leaf-placement problems independently and then put the solutions together. Another issue is that, when fields are shaped by an MLC, partial-depth leaf sides jut into the field components (Figure 43.13). Hence, unless the issue is specifically addressed, so-called *tongue-and-groove* underdoses can occur along the thin lines between leaves. Help is at hand because of the enormous number of ways in which 1D leaf segments can be put together to make any 2D IMB. This problem has been studied in detail and the number of configuration options established, whatever the number of local maxima and minima in the 1D IMB (Webb 1998b). One can then pose a formal optimisation problem of requesting the set of 2D field-shapes which, when added together, yield the required 2D IMB, without violating any of the MLC leaf-placement constraints, and minimising, preferably zeroing, the tongue-and-groove underdose. It has been shown that this formal optimisation problem cannot be solved due to the enormous computational effort required. However, it has also been shown that certain well-posed 2D

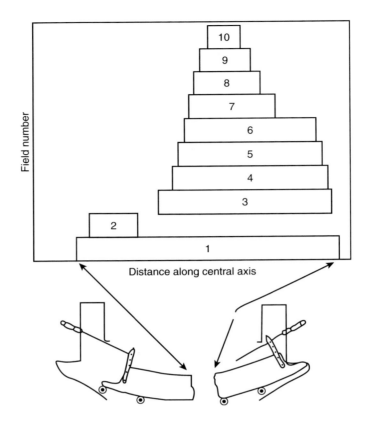

FIGURE 43.11
Shows the ten separate fields which when combined would give the distribution of fluence shown in Figure 43.10.
Each rectangle represents a field and the left vertical edge is the position of the left leaf and the right vertical edge is the
position of the right leaf. This method of setting the leaves is known as the *close-in* technique. A schematic of a pair of
MLC leaves is shown in the lower part of the figure with arrows indicating the correspondence with the field edges.

IMBs (Figure 43.14) *can* be delivered without violating MLC constraints and with zero
tongue-and-groove underdose (Webb 1998c).

It may be observed that the MSF method is less efficient than the delivery of the corresponding IMRT using a compensator when the 2D IMB has multiple extrema. There is also the
concern (above) over the *deadtime* between field resettings and field restabilisation. Manufacturers are continually working to reduce this intersegment deadtime; for example, the fast
tuning magnetron of Elekta has reduced the *deadtime* from about 5 s previously to 1 s.
However, apart from Varian, manufacturers have taken the view that step-and-shoot IMRT
is a simpler option and shown reluctance to implement dynamic leaf movement.

IMRT may also be delivered via the dynamic MLC (DMLC) method (see next section) in
which leaves sweep across the field and take up many positions. When this is executed with the
radiation *off* during leaf movement, the method reduces to the MSF IMRT technique albeit
with a huge number of segments.

Finally a word about nomenclature. The MSF method does not have to be inverse planned.
The IMRT required for breast radiotherapy may be delivered via metal compensators, or using
a few static IMRT fields (Evans et al. 1997). Several other centres, notably the University of
Michigan, the University of Ghent and The Netherlands Cancer Institute, are also delivering
what they call *segmented-field* therapy which is another form of the MSF method. In some cases
the 2D fields are very simply segmented into a couple of components, one with the beam's-eye-
view of the target only and the other with the beam's-eye-view of the target and organs at risk.

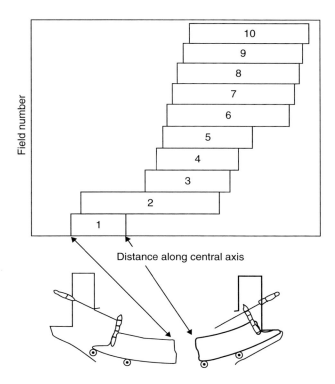

FIGURE 43.12
Shows the ten separate fields which when combined would give the distribution of fluence shown in Figure 43.10. Each rectangle represents a field and the left vertical edge is the position of the left leaf and the right vertical edge is the position of the right leaf. This method of setting the leaves is known as the *leaf-sweep* technique. A schematic of a pair of MLC leaves is shown in the lower part of the figure with arrows indicating the correspondence with the field edges.

The subfields are separately weighted (usually by an inverse-planning optimisation technique) and then separately delivered via MLC collimation. It may be appropriate to restrict the terminology *MSF method* to imply the (possibly not automated) delivery of a relatively small number of subfields at each gantry orientation and to use *step-and-shoot* IMRT to describe inverse-planned MSF IMRT.

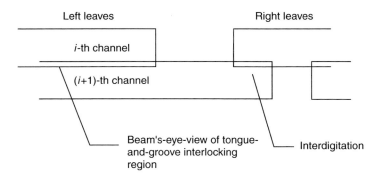

FIGURE 43.13
Two leaf-pairs of an MLC are shown, the ith and the $(i+1)$th channels. The leaves are shown interdigitating, an arrangement forbidden on the original 10 mm leaf-width Elekta MLC. The beam's-eye-view of the tongue-and-groove interlocking region is also illustrated and one might seek to minimise the underdose in this region by a judicious choice of 2D field components.

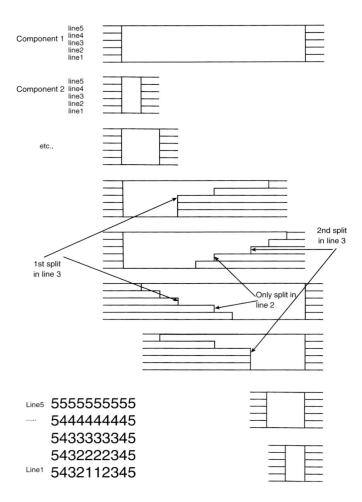

FIGURE 43.14
This shows the nine field components required to give the inverted half-pyramid 2D IMB fluence distribution shown in the lower left. Line 1 contains exactly 9 rising fluence increments and hence 9 subfields are required. However given that the largest fluence is 5 units the efficiency of delivery is only 5/9 for this problem. This optimum solution, found by an intelligent search is a symmetrical set of fields in which the first is entirely open and so may be verified by portal imaging. There is no leaf collision and no interdigitation and also zero tongue-and-groove underdose. This cannot be determined by simply looking at the components here but can easily be established by writing code to determine the fluence both in the line-of-sight of leaves and leaf intersections. (From Webb, S., *Phys. Med. Biol.*, 43, 1481–1495, 1998. With permission.)

43.3.3 The Dynamic Multileaf Collimator (DMLC) Technique

A related but technically quite different method to deliver IMRT is the DMLC technique. This, as its name suggests, also makes use of the multileaf collimator. Imagine the radiation switched on with all the leaves closed and at the far left of a field. The right leaves then start moving to the right exposing the field to radiation, followed at some later time by the commencement of movement of the left leaves. The right and left leaves move, changing their speeds from time to time until both sets come to rest in a closed position at the right of the field. To a first approximation the primary intensity, or fluence, delivered to some point in the field will be given by the difference between the time at which that point is uncovered by the leading leaf and the time at which it is once again covered by the trailing leaf, provided the fluence is delivered at a uniform rate. Therefore, any arbitrary 2D IMB can be generated by varying the velocity profile of each leaf pair. The principle is delightfully simple. Needless to say there are many practical complexities.

First, it is worth stating the laws which govern the leaf movement in an ideal situation. Imagine that the 2D IMB is created by a set of 1D IMBs each created by a leaf-pair. Then concentrate on that leaf-pair.

$$I(x) = t_1(x) - t_2(x) \qquad (43.6)$$

where $t_1(x)$ is the time of arrival at x of the trailing leaf and $t_2(x)$ is the time of arrival at x of the leading leaf. Intensity and time are measured in the same units. To generate an arbitrary 1D IMB, the leaf velocities are chosen such that

$$v_2(x) = v_{max}$$
$$v_1(x) = v_{max}/[1 + v_{max}(dI/dx)] \quad \text{if } dI/dx > 0$$

and

$$v_1(x) = v_{max}$$
$$v_2(x) = v_{max}/[1 - v_{max}(dI/dx)] \quad \text{if } dI/dx < 0 \qquad (43.7)$$

where v_{max} is the maximum leaf speed. This means that when the fluence profile is increasing the leading leaf moves at maximum speed and the trailing leaf provides the modulation. However, when the fluence profile is decreasing the trailing leaf moves at maximum speed and the leading leaf provides the modulation. This algorithm was simultaneously discovered and published by three groups in different countries (Stein et al. 1994; Svensson et al. 1994; Spirou and Chui 1994). It has been shown that this gives the most efficient delivery and it is possible to calculate the total delivery time as the sum of the time taken for the leaf-pairs to sweep the whole field plus the sum of the rising fluence parts (Stein et al. 1994). The trajectory diagrams then characterise the radiation delivery (Figure 43.15 and Figure 43.16). A particularly graphic demonstration of these equations (due to Art Boyer) is in Chapter 3 of Webb (2000a).

In practice there are a lot of instrumentational limitations to consider e.g. the equations imply that leaves will make sudden large accelerations. The experimentally possible accelerations may thus be exceeded but there are methods to cope with this. Versions of these equations exist which also take account of an upstream modulation in radiation primary fluence.

More importantly, the equations only express the movements required to obtain a required *primary* fluence. In practice there will also be a contribution to delivered fluence through leaf leakage and also through the head scatter and phantom scatter generated. Methods to cope with all of these exist and rely on an iterative process whereby the leaf movements are first generated for a required primary fluence. Then the actual delivered fluence is computed through some model. Then the primary fluence is modified so that after a second run of the equations the actual fluence more closely matches the required fluence and so on iteratively until convergence of actual to required fluence.

The above operations are sometimes called an *interpreter* because the mathematics interprets the IMB into a leaf pattern. However, as previously discussed in Section 43.3.2, it is not enough to interpret each 1-D IMB independently of its neighbours because when the separate interpreted leaf patterns are merged, this may lead to violations of the equipment limitations on interdigitation. It also can lead to a phenomenon known as the *tongue-and-groove effect* in which the beam's-eye-view of the interlocking leaf sides is underdosed with respect to the dose delivered to adjacent channels.

Several workers have shown that the tongue-and-groove problem is entirely soluble by a process known as leaf synchronisation (Van Santvoort and Heijmen 1996; Webb et al. 1997) which sets the underdose to exactly zero. Also, more intelligent interpreters have been developed which make use of both leaves and jaws and can thus create an island of zero

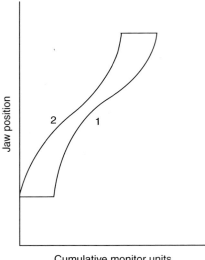

FIGURE 43.15
Showing the jaw or leaf positions as a function of cumulative monitor units for two jaws or leaves (the trailing jaw or leaf is number 1) in dynamic therapy. The horizontal axis is a measure of cumulative MUs representing time t. The vertical axis is a measure of position x. At any position x, the horizontal width between the two curves gives the intensity of the IMB in MU.

primary fluence (leakage ignored) in a field (Convery and Webb 1998). This can also arrange for zero tongue-and-groove underdose and can also avoid the interdigitation excluded by some (e.g. the standard Elekta) MLCs. The only price paid is a moderately complex algorithm and a slightly increased treatment delivery time. An example of one such component field is shown in Figure 43.17.

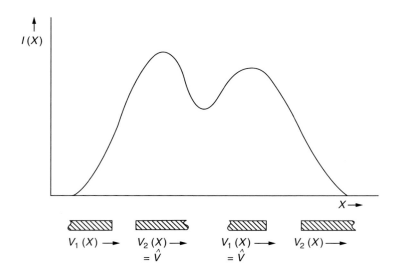

FIGURE 43.16
When the gradient dI/dx of the intensity profile, $I(x)$, is positive, the leading leaf (2) should move at the maximum velocity; conversely when the gradient dI/dx of the intensity profile $I(x)$ is negative the trailing leaf (1) should move at the maximum velocity. Equation 43.7 is illustrated by showing a schematic of the pair of leaves in two separate locations delivering the IMB profile shown in the upper part of the figure.

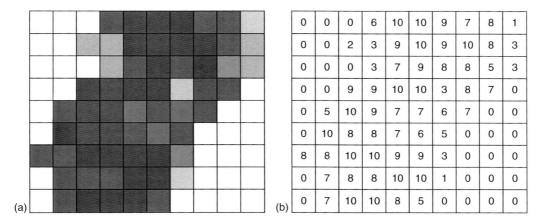

0	0	0	6	10	10	9	7	8	1
0	0	2	3	9	10	9	10	8	3
0	0	0	3	7	9	8	8	5	3
0	0	9	9	10	10	3	8	7	0
0	5	10	9	7	7	6	7	0	0
0	10	8	8	7	6	5	0	0	0
8	8	10	10	9	9	3	0	0	0
0	7	8	8	10	10	1	0	0	0
0	7	10	10	8	5	0	0	0	0

FIGURE 43.17
An example of a discrete IMB shown as (a) greyscale and (b) the matrix of relative intensities. The intensity distribution is for a posterior oblique field of an eight-field prostate plan with 10% intensity stratification and 10 mm by 10 mm beam elements at the isocentric plane planned with the CORVUS planning system. (From Convery, D. and Webb, S., *Phys. Med. Biol.*, 43, 2521–2538, 1998. With permission.)

Many different groups have generated different interpreters, all with slightly different aims. The reader is referred to the wider literature. IMRT by the DMLC technique has been promoted by the accelerator manufacturer Varian. The technique has been clinically implemented in many centres.

43.3.4 TOMOTHERAPY

The concept of tomotherapy represents a totally different technique for delivering IMRT. It does not make use of the conventional multileaf collimator. Instead the radiation is collimated to a narrow fan and the fan-beam rotates isocentrically around the patient. As it does so the 1-D IMB is created and continually modified by shooting, vanes, of attenuating tungsten into the aperture at right angles to the direction of the slit. There are two methods which, whilst related through this concept, have distinctive features. In the first the patient lies on a stationary table. Therefore, IMRT is performed for a small slice of the patient and to complete IMRT of the other slices the patient table must be incremented sequentially. In the other implementation, the patient table translates slowly through the rotating beam which then executes a helical motion relative to the patient considered stationary. The first method was proposed and developed by Mark Carol and the NOMOS Corporation* and the second by Rock Mackie and colleagues at the University of Wisconsin (Mackie et al. 1993). The first technique is illustrated in Figure 43.18 and the second in Figure 43.19.

Common to both realisations is the multivane slit collimator. The collimator was invented and patented by Swerdloff et al. (1994) in Wisconsin but was first constructed by the NOMOS Corporation who christened it the MIMiC (multivane intensity modulating collimator). The MIMiC comprises two banks of 20 vanes of tungsten which can be fired into and out of the (half) slit under electro-pneumatic control. Air pressure to the back of the vanes keeps them closed until a higher air pressure to their front opens them. They can be switched from open to closed state in 20 ms. The IMB is constructed from the varying dwell-times of the vanes in the aperture. The vanes are focused to the source axially, are 5 mm wide on the source side, 6 mm

---

* This was first presented at the American Society of Radiotherapy and Oncology meeting in 1992.

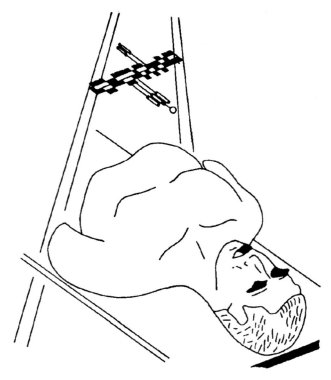

FIGURE 43.18
Shows the location of the MIMiC stubby-vane collimator in relation to the patient with some of the vanes closed and others open. The two back-to-back slit apertures each comprise 20 vanes.

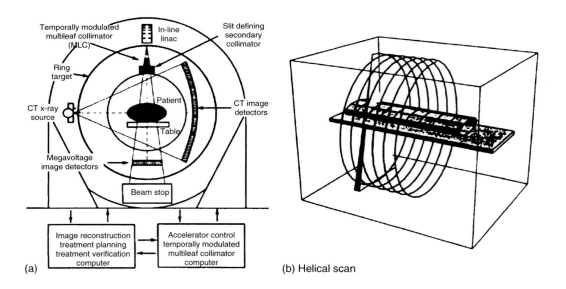

FIGURE 43.19
A schematic diagram of Mackie's original proposal for tomotherapy. (a) An in-line linac with a temporally modulated MLC is mounted on a gantry along with other components which will perform megavoltage CT and diagnostic CT. The equipment rotates continually as the patient is translated slowly through the beam, hence (b) the motion is helical with respect to the patient. In the commercial implementation the separate diagnostic CT scanning source has been omitted with CT images being created using the linear accelerator run at a lower energy. (Figure (b) from Mackie, T. R. et al., *Med. Phys.*, 20, 1709–1719, 1993.)

wide on the patient side, 8 cm deep and the *throw* can be adjusted to be 1 cm or 2 cm at the isocentre. The MIMiC creates two IMRT slices simultaneously (Figure 43.20). The MIMiC changes its 1-D IMB (per bank) every 5° of arc and generally rotates through 270° avoiding irradiation through the couch. The 1-D IMBs are computed via the inverse-planning process on PEACOCKPLAN (CORVUS) (see Section 43.2.8) and the plan is transferred on disk to an onboard computer. The apparatus looks a little cumbersome but has been designed to fit any accelerator and is used in many clinics, particularly in the U.S.A.

An issue of concern with the MIMiC technique is that of ensuring the matchline is accurate. There have been several studies which show that the couch increment must be made to 0.1 mm accuracy and NOMOS have provided the CRANE to achieve this, a device that permits accurate increments without backlash and with full stability. There is also a quite complex commissioning and quality-assurance procedure to follow before clinical use.

The NOMOS MIMiC IMRT technique has been a practical reality since 1992. The University of Wisconsin prototype tomotherapy apparatus whilst conceptualised in 1993, underwent a much longer period of development, and was first made commercially available by Tomotherapy, Inc. in 2002. This is partly because the scheme is more ambitious, intending to create a *radiotherapy system in a box*, doing away with almost all existing techniques. The group have also made an enormous number of careful studies of the concept including building a full laboratory mockup to study the process. Fuller details of these and indeed the other IMRT techniques are given in Webb (1997b, 2000a). Up-to-date reviews of commercial clinical spiral tomotherapy can be found in Palta and Mackie (2003) and in Fenwick et al. (2006). Figure 43.21 and Figure 43.22 show the modern Tomotherapy equipment. Quality control of tomotherapy is discussed briefly in Section 38.

43.3.5 SWEPT PENCIL BEAMS

The concept of *sweeping* a *pencil beam* of radiation is attractive, giving maybe the ultimate control over beam delivery. The idea was proposed some time ago using the Racetrack

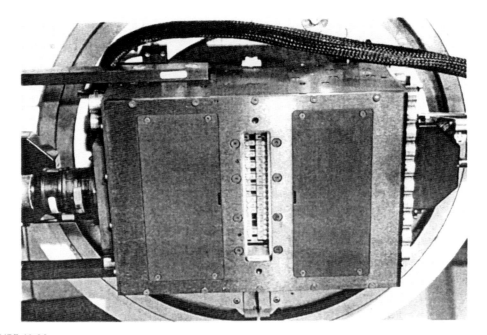

FIGURE 43.20
A view of the NOMOS MIMiC collimator attached to an Elekta accelerator at the Royal Marsden NHS Foundation Trust, showing a leaf pattern. (Loan—courtesy of the NOMOS Corporation.)

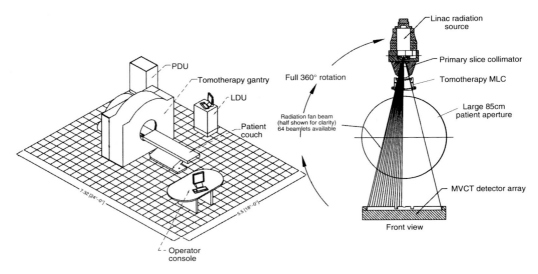

FIGURE 43.21
The tomotherapy system. (Courtesy of Prof. Rock Mackie.)

Microtron (Lind and Brahme 1995). However at that time the photon beam, far from being a *pencil*, was more like a *broom* with a FWHM of many centimetres and so the idea did not seriously rival the other IMRT methods. Recently it has been shown theoretically that a narrower beam can be generated using an optimised bremsstrahlung source and a short isocentric distance (Svensson et al. 1999).

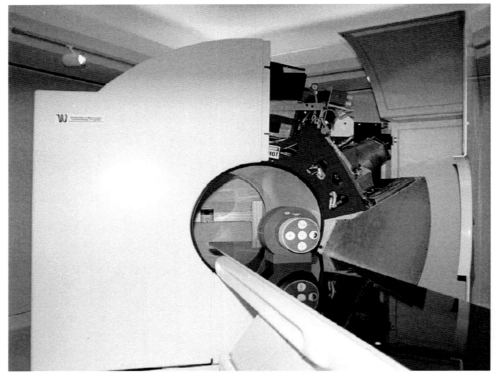

FIGURE 43.22
(See colour insert following page 590.) The UW tomotherapy research unit. (Courtesy of Prof. Rock Mackie.)

43.3.6 THE MOVING BAR TECHNIQUE

A 1-D IMB can be created by sweeping an attenuating bar through the field. By varying the dwell-time of the bar at different locations the 1D profile receives an intensity modulation. In its simplest form the algorithm requires the bar to dwell at location x_i for time t_i, where

$$t_i = T(1 - I(x_i)/I_u) \qquad (43.8)$$

I_u is the unattenuated fluence, and $I(x_i)$ is the fluence at location x_i and T is the total irradiation time. It may be appreciated that because of the condition:

$$\sum_i^N t_i \leq T \qquad (43.9)$$

not all modulations are achievable; the sum of the dwell times must clearly not come to more than the total irradiation time. It is clear that the method works best for shallow modulations, the exact opposite of the ideal conditions for the DMLC technique where shallow modulations would demand high leaf velocities. The moving bar technique involves many considerations. The spatial scale of the modulation is determined by the width of the bar in the direction of scanning. Therefore, there is a compromise between the accuracy of the IMB and the ability to actually provide the modulation at all. Detailed equations (see Webb 1997b) explain these concerns.

The technique has received attention because it has actually been engineered (Fiorino et al. 1992) and used, for example, to modulate the intensity to provide a missing-tissue compensator for head-and-neck treatments (Figure 43.23). In principle one could imagine creating a 2D collimator with this approach but then the complexity would rival that of the DMLC method. Hence this development is not expected.

43.3.7 INTENSITY-MODULATED ARC THERAPY (IMAT)

IMAT is a technique invented by Yu et al. (1995). It capitalises on the observation that any 1-D IMB can be created by a set of MLC-shaped fields in an enormous number of ways (Webb

FIGURE 43.23
The moving bar technique. The bar is translated across the radiation field spending a variable time shielding each part. In this way some (but not all) 1D IMBs can be constructed. The apparatus is motor controlled and resides in the blocking tray of the linear accelerator.

1998b, 1998c). Consider just one 2D slice irradiated by a set of 1-D IMBs arranged to be delivered by gantry rotation. The basis of the method is to decompose the 1-D IMBs into components and deliver the components by multiple gantry rotations. For each rotation the difference in field size between one orientation and another is minimised leading to an optimal delivery method. The method was implemented successfully in a prototype IMRT delivery and is now in use clinically.

43.3.8   VERIFICATION OF IMRT

The delivery of IMRT is a major change from the delivery of radiation with geometrical shaping alone. Many of the methods require collimating apparatus to be moving either whilst the radiation is on or between successive bursts of radiation. When clinical IMRT was first started (about 1994) there was consequently natural concern about its safety. Physicists have risen to the challenge of demonstrating that delivered fields match those planned within acceptable tolerances; this is known as IMRT verification. The years since 1994 have seen the steady growth of a variety of verification tools and now there is more confidence. It is a big subject, extensively reviewed by Webb (2000a) (see references therein), and here some of the available techniques for assuring quality are simply listed:

- Film may be strapped to the accelerator head and record the integrated fluence as leaves move in the DMLC technique. These films may then be compared with the corresponding maps from inverse-planning computers. By inspecting films taken on successive occasions the stability of leaf movement may be studied. Some groups have also used electronic pixellated detectors in the blocking tray with the same purpose.

- Electronic portal imaging devices can record the integrated fluence from the sequence of irradiations which make up the intensity-modulated beam in either the DMLC technique or the step-and-shoot technique. The detector developed at the Royal Marsden NHS Foundation Trust has the special property of recording the dose from each pulse of radiation. The positions of the leaf ends can also be extracted from EPID data and correlated with the expected positions. Movies of these motions are particularly instructive; e.g. sticky leaves can be identified. Other groups have developed *a posteriori* methods of correlating image frames from EPIDs to the planned geometrical patterns.

- Intensity-modulated beams may be replanned on to simple geometric or anthropomorphic phantoms to give a new plan. These beams are then delivered to such phantoms loaded with TLD chips and/or with film sandwiched between slices. The doses measured by TLD and film are then compared with the plans. It is usual to quote the dose error in regions of low-dose gradient and the distance error for selected isodoses in regions of high-dose gradient (see also Section 39.3.4.3).

- Polyacrylamide (BANG) gel dosimetry is growing in popularity as it is an integrating 3D dosimeter (see Chapter 17). There is a linear relationship between the $1/T_2$ image intensity of BANG gel and dose (Maryanski et al. 1996). Therefore, one can irradiate a phantom containing gel, read out the dose using MR and correlate with the plan. The gels may also be imaged optically.

- Megavoltage CT (MVCT) has been used to reconstruct the patient position during therapy to compare with the planned position. The University of Wisconsin tomotherapy device (Section 43.3.4) can actually make MVCT images with either *flashed on* open fields or even the leakage radiation from the modulated beams created by the collimator.

- NOMOS MIMIC IMRT (Section 43.3.4) has been verified through the use of a special box of films to which the modulated irradiations are applied followed by comparing measurements with computed predictions.

- Quality assurance programmes for DMLC have been designed based around the deliberate starting and stopping of leaves to create expected patterns which are then compared with measured patterns of dose.

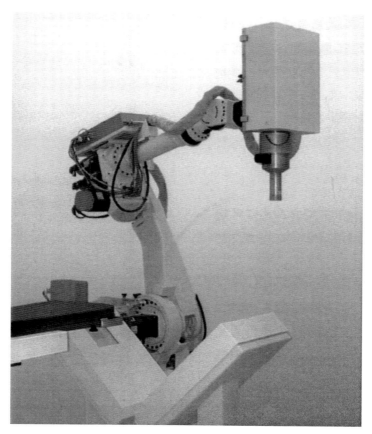

FIGURE 43.24
(See colour insert following page 590.) The Accuray Cyberknife.

43.3.9 THE FUTURE OF IMRT

IMRT is here to stay. It represents the future of improving the physical basis of radiotherapy. Its study has generated an enormous literature. Seven methods have been proposed and demonstrated at least in principle. From the literature the reader may draw their own conclusions about which are likely to succeed clinically; a technique which scores well for one clinical application may not be so good in others.

A radical new proposal has been put forward, namely that the ultimate IMRT may be delivered by a robotically controlled short-length linac (Webb 1999, 2000c), such as the Accuray Cyberknife (Figure 43.24). The inverse-planning technique has been completely worked out; it has been shown that complex 3D dose distributions can be delivered, albeit with very poor efficiency. Conversely, several radically new concepts for IMRT have been proposed but not yet experimentally realized (Webb 2001), e.g.:

- The shuttling MLC (Webb 2000d, 2001)
- *Jaws plus mask* IMRT (Webb 2002a, 2002b)
- The variable aperture collimator (Webb et al. 2003, 2004a)

The technical development of IMRT shows every sign of continuing (Webb 2004c; Bortfeld et al. 2006).

CHAPTER 44

INTENSITY-MODULATED THERAPY: PRACTICAL ASPECTS

C.-M. Charlie Ma[*]

CONTENTS

---

[*] With a contribution by Helen Mayles and Philip Mayles (Section 44.3.4).

44.1 INTRODUCTION

Clinical implementation of intensity-modulated radiotherapy (IMRT) needs all the work required to implement 3D conformal radiotherapy (CFRT) and more (see Chapter 39 and Chapter 40). Consideration must be given to important issues such as equipment and space requirements, time and personnel requirements, additional dosimetry verification and treatment quality assurance (QA), staff training and patient education and changes in treatment scheduling. It is essential for a clinic to designate an implementation team to think through the implications in advance, and to periodically update procedures as lessons are learned. Bringing the modality to the clinic is only the first step. Much effort is needed to keep it running smoothly and to keep pace with upgrades and future developments in IMRT technology. To achieve the full benefit from IMRT, various resources must be in place, and all persons involved in IMRT, not only physicists but also physicians and other personnel such as radiographers, service engineers and administrators, must be properly trained. Furthermore, IMRT is an integrated system. Every single technical/physical component and treatment step must be considered carefully. In this chapter, a brief description is provided, which details the requirements for clinical implementation and guidelines on IMRT treatments.

44.2 REQUIREMENTS FOR CLINICAL IMPLEMENTATION

44.2.1 EQUIPMENT, SPACE AND SHIELDING

It may be necessary to upgrade existing equipment to provide IMRT functionality. MLC-based IMRT may involve adding an MLC, upgrading an existing MLC to dynamic capability, or purchasing special add-on collimators. For compensator-based IMRT, special equipment, such as a computer-controlled milling machine, may be required. Similarly, special software for IMRT treatment planning must be acquired, which may include add-on modules for an existing treatment planning system, or a dedicated treatment optimisation system. Existing record and verification systems may need to be upgraded to accommodate IMRT treatments. Computer networks may need to be enlarged or improved in order to permit the necessary file transfers. Additional dosimetry equipment may be needed for the commissioning and ongoing QA of IMRT (e.g. it is important to have an efficient film-scanning system to accomplish these tasks). Additional phantoms may also be needed.

Extra space may be required for additional computer workstations, especially if IMRT planning is to be done on a dedicated system. Space may also be needed for additional equipment, such as add-on collimators or dosimetry devices. IMRT treatments require more beam-on time than conventional treatments, so room shielding should be re-evaluated. The increase in monitor units (MU) is 1 to 3 times for MLC-based IMRT treatments and up to 9 times for sequential tomotherapy delivery (see Section 44.3.4), depending on the number of gantry rotations involved. Compensator-based IMRT generally requires only a small fractional increase in MU. However, it is usually the secondary barriers, not the primary barriers, that may be affected (Mutic et al. 2001; Stathakis et al. 2005). The changes in neutron dose due to beam-intensity modulation basically scale with the increase of MU, but have limited clinical effect (Followill et al. 1997; Price et al. 2003a). However, these authors suggest that 18 MV should only be used for IMRT when treating older patients.

44.2.2 TIME AND PERSONNEL

Sufficient time and resources should be allocated to complete all the tasks involved in clinical implementation. Radiation oncologists and treatment planners (physicists or radiographers) will need to learn a very different approach to planning. Physicists must complete comprehensive measurements to assure that the treatment planning and treatment delivery systems are accurate. The implementation team should set up and test the processes used for individual patient treatments. QA procedures must be implemented. Special training should be arranged for personnel involved in the IMRT process. These tasks will probably require an initial investment of several person-months of work on the part of the physics staff and other members of the implementation team.

44.2.3 TREATMENT SCHEDULING

Due to the changes in patient immobilisation, target localisation, beam delivery and dosimetry verification, some IMRT treatments will take at least 50% longer than conventional treatments*. Only some of the treatment machines may be equipped for IMRT. New imaging studies and immobilisation devices may be required. Staff responsible for scheduling will need to be advised of these new requirements. They should be consulted early in the implementation process so that consequences of those changes can be anticipated, and adjustments made. Changes in charting procedures should also be considered, including documentation of daily treatment with many complex fields, documentation of QA procedures and records of dose summaries that adequately describe dose-volume goals and results.

44.2.4 STAFF TRAINING

IMRT is an integrated process which is much more complex than CFRT. Therefore, staff training and education should be an essential part of the clinical implementation of IMRT. Experience gained by the staff in 3D treatment planning and delivery (see Chapter 39) is helpful but not sufficient for IMRT. There are significant differences between the two that necessitate additional specialised training. Each member of the IMRT team should understand the whole process in order to be able to use this technology safely and effectively. IMRT differs from traditional radiation therapy enough that it should be considered as a special procedure, which necessitates the didactic training of key members before the new modality is implemented in a clinic. The training curriculum for each IMRT team member must include all of the critical steps in the IMRT process.

44.2.4.1 Radiation Oncologists

IMRT represents a significant departure from current radiotherapy practice. Treatment planning for IMRT is complex and non-intuitive. The optimisation process is generally completely computer-controlled (see Section 44.3.4), and its success in achieving the clinical goals is very much dependent on the set of parameters used as input to the computer algorithm. It is therefore difficult for the oncologist and planner to identify an optimal solution without having a complete understanding of the optimisation process and its limitations. Learning how to adjust the parameters to steer the results in the desired direction is essential to avoid treating a patient with a sub-optimal IMRT treatment plan.

A basic use of IMRT is to treat tumours that are in close proximity to critical, normal structures. This requires the oncologist to delineate between the target and the critical structures precisely and accurately, and to choose appropriate planning margins judiciously. This, in turn, requires training in image-guided treatment planning and a good understanding of treatment-planning and delivery uncertainties. IMRT often results in the gross tumour and regions of sub-clinical disease being given different doses per fraction concomitantly.

---

* The treatment time can also be reduced if replacing a complex non-IMRT treatment.

In addition, the dose distribution in the target volume is often much more inhomogeneous than is the case with (non-IMRT) conformal therapy. It is important that the radiation oncologists critically evaluate differential dose-fractionation schedules for IMRT in light of their clinical experience with conventional radiation therapy. This requires an understanding of the concept of biologically equivalent doses and tissue-tolerance doses (see Part B). Radiation oncologists whose residency training did not include IMRT should consider attending special training courses or workshops. Short-term clinical training at academic institutions that have active clinical IMRT programs can also be considered.

44.2.4.2 Physicists and Dosimetrists

Physicists play a much more significant and direct role in IMRT planning and delivery than in conventional radiation therapy. It is therefore essential that they (and dosimetrists) have a good understanding of the mathematical principles of dose optimisation (Section 43.2), computer-controlled delivery systems and dosimetry issues related to small and complex shaped radiation fields. They also need to have a better understanding of treatment setup, planning and delivery uncertainties and their impact on patients treated with IMRT. The QA procedures for IMRT are much more complex than for conventional radiation therapy (see Part H). Physicists are responsible for the design and testing of the QA procedures, and thus they should actively participate in the QA process. It is imperative that each physicist involved with IMRT has special training in the whole process of IMRT.

44.2.4.3 Treatment Radiographers

Implementing IMRT requires the active involvement of the radiographers. They should be involved in the design and testing of treatment procedures. If the IMRT delivery involves specialized equipment (e.g. an add-on collimating device or compensators), there will be the need to train the radiographers in its use and storage. They may also have responsibilities for basic maintenance and QA. Radiographers will need to be trained to use any new immobilisation or localisation systems, as implemented in the new procedures. Carrying out mock procedures with phantoms ought to be part of the process of testing the new procedures. Delivery details that escape the physicist's notice may be important to the radiographers. For example, the initial field shape for an IMRT treatment may obscure the light field or the crosshair, requiring that the patient be positioned before the MLC is programmed. Radiographers must be provided with the means of knowing that the treatment they are about to deliver is correct. Given the complexity of IMRT treatments, it is essential for the treatment delivery to be fully monitored by a record and verify system. Radiographers must nevertheless be trained to verify that the record and verify system is running correctly and this training must include how to respond to unplanned events. They need to know how to interrupt and restart a treatment, how to recover from a partial treatment that requires the console to be reprogrammed and how to recognise and act on new error messages and interlocks. Radiographers will need to be trained in any new procedures related to portal imaging as well as in daily QA tests. As with any QA procedure, clear instructions and action levels must be provided.

44.2.4.4 Service Engineers

Service engineers must have a good knowledge of all aspects of the IMRT beam delivery process and the equipment required. Compared to standard treatment techniques, it is much more difficult to recover cleanly from an interruption in dose delivery after an intensity-modulated treatment has started. Therefore, accelerators with a poor history of reliability are not suitable for this type of treatment, and expanded preventive maintenance programs are extremely important. Intensity modulated dose delivery can place demands on the MLC that far exceed the criteria used for the design of these systems. For example, some implementations can require several hundred MLC fields per patient, or many thousands of fields per treatment day, which may result in more frequent component failure. The accuracy of calibration of MLC leaves is more critical than for CFRT, and for 'step and shoot' IMRT (see Section 43.3.2), beam startup characteristics are more important because of the small doses for individual beam segments. Special QA procedures should be established to ensure proper calibration of leaf position and to avoid treatment interruptions. With the co-operation of the physicist, preventive maintenance programs should be implemented to meet the special needs of IMRT.

44.2.5 PATIENT EDUCATION

Patients treated with IMRT should be informed of several issues. These include the goal of the treatment, potential side effects, realistic estimates of the time required for each treatment, the immobilisation method to be used and the beam delivery process. IMRT can potentially improve local control and quality of life through dose escalation and normal-tissue sparing. This will be site- and protocol-specific. If IMRT is used to improve local control by escalating doses, then the potential for acute or chronic sequelae may increase. Parotid-sparing protocols may decrease the incidence of xerostomia, but increase acute mucositis, especially if target doses are more inhomogeneous than with conventional treatments. It is reasonable to expect that compared to CFRT, normal-tissue complications will be reduced by IMRT for the same target dose. For example, it is possible to achieve identical prostate dose coverage with a reduction in the dose delivered to the nearby critical structures such as rectum, bladder and femoral heads. It is necessary to discuss patients' expectations of IMRT. Some patients may desire to be treated with this new, highly advertised modality, whether or not it is advantageous or appropriate for their condition.

44.3 GUIDELINES ON IMRT TREATMENTS

44.3.1 GENERAL CONSIDERATIONS

Intensity modulation increases significantly the level of sophistication and complexity of the treatment planning and dose delivery process. Table 44.1 shows the key components of a general IMRT treatment process. Detailed discussions are given below on these key components.

44.3.2 IMAGE ACQUISITION

There is growing interest in image-guided radiation therapy (see Introduction to PART I), especially with the widespread use of IMRT. Image guidance is playing an increasingly important role in structure segmentation, target (treatment volume) determination, inter- and intra-fraction target localisation and/or target redefinition (Mah et al. 2002; Chapman et al. 2003; Grégoire et al. 2006). IMRT requires more precise information about the target and normal tissue structures for treatment planning. Hence, there will be increasing use of contrast agents for CT scans, as well as increasing registration of images from other modalities, such as MRI or PET scans, which both provide additional anatomical as well as functional information. This information can, in principle, make it possible to plan heterogeneous dose distributions within the target which account for variations in tumor cell density and oxygenation conditions

TABLE 44.1

Components of an IMRT Treatment Process

Patient immobilization
|
Image acquisition
|
Structure segmentation
|
Treatment planning
|
Dosimetry validation
|
Position verification/Target localization
|
Beam delivery

e.g. IMRT can be used for *dose painting* or *sculpting* (Ling et al. 2000; Webb 2003d; Bentzen 2005).

CT scanning for treatment planning will be performed with the patient in the treatment position in the immobilisation device. Chen et al. (2004) show that an MRI may also be used for treatment planning of abdominal and pelvic sites (Figure 44.1a). IMRT treatment planning needs 3D image data with more slices at a finer spacing than has previously been the norm,

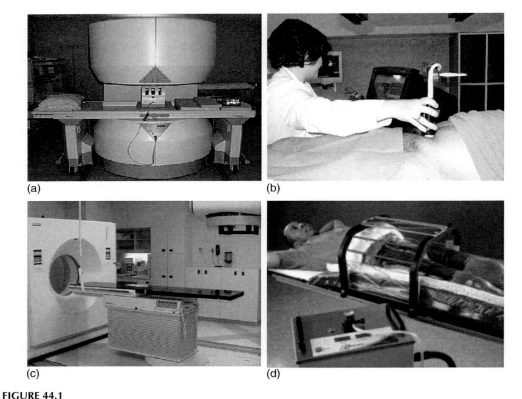

(a)   (b)

(c)   (d)

FIGURE 44.1
(See colour insert following page 590.) Imaging, immobilization, patient positioning and target localization devices used for IMRT at Fox Chase Cancer Center, Philadelphia, Pennsylvania; (a) a low-field open MRI scanner for structure segmentation; (b) a BAT ultrasound system for soft-tissue target localization; (c) a CT scanner in the treatment room for target localization; (d) a body localizer with vacuum fixation for patient immobilization and repositioning.

depending on the MLC leaf width. For example, a slice spacing of no more than 5 mm should be used for an MLC with 10 mm leaves, and finer spacing may be needed to generate digitally reconstructed radiographs of sufficient quality. It may be necessary to extend the range of slice acquisition for planning non-coplanar beam arrangements.

44.3.3 STRUCTURE SEGMENTATION AND TARGET DEFINITION

It is labour-intensive to delineate targets, critical organs and other reference structures for IMRT treatment planning. One important feature of inverse planning is that the high-dose region will be limited to the exact target volume defined, and only those normal-tissue structures specifically identified will be avoided. This requires the physicians and the planners to define structures in more detail. For example, implementing an IMRT protocol for breast patients would require the breast target volume together with the lung and the heart (and perhaps the contralateral breast) to be defined on each axial slice, with due consideration for margins. This can be more difficult than defining conventional tangential photon fields on simulator films. Other reference structures may be added to assist the optimisation process in improving target dose conformity (Price et al. 2003b). This also requires more effort from the physicians and the planners.

44.3.4 IMRT TREATMENT PLANNING

Treatment planning and plan optimisation for IMRT require a change of approach compared to conventional planning. In terms of clinical implementation, a key point is to allow time for the physicians and planners to develop their skills in using the system. Physicians have to prescribe dose-volume limits that define an acceptable plan, and planners need to learn how to control the dose distribution by modifying unfamiliar input parameters. There is a change from planning beams to give a uniform dose to a volume, to designing volumes for the target and organs at risk that will support the appropriate objectives to achieve the desired dose distribution. Tissue not specifically included in the target volume will probably get a low dose and sensitive normal tissues that are not specified as such may well receive a higher dose than is desirable. Clinics will need to develop tools to aid these tasks. New forms may be needed for radiation oncologists to communicate complex planning goals to the planners, for planners to keep track of the parameters that have been tried and for planners to report the final plan results. Since an IMRT plan is only *optimal* (or customised) in terms of the predefined beam directions and plan parameters, its dose distribution may not necessarily be superior to an alternative CFRT plan. IMRT is useful for those unusual cases where there are particular challenges, but in order to develop an understanding of the way in which the constraints available with a particular planning system affect the dose distribution, it is useful to begin with a common tumour site, such as prostate. In the initial implementation phase, plan comparisons between IMRT and conventional CFRT should be performed for each specific site and treatment protocol in order to maximise the benefit from the new technology. This information may be obtained from published results and the experiences of other institutions, but it is important to realise that the way in which IMRT objectives affect the outcome of optimisation is planning-system dependent. As a guide for such plan comparisons, calculation of relative TCP, NTCP and/or EUD (see Chapter 36) may be useful. Templates with favour- able beam arrangements and optimisation parameters, called *class solutions*, can be established for common treatment sites (such as prostate), to facilitate routine IMRT treatment planning.

The most obvious parameters available to the planner are the *maximum* and *minimum dose* to *target* tissues and the *maximum dose* to *organs at risk*. In addition there is the possibility of defining *dose volume objectives* so that a particular percentage of an organ-at-risk volume can be set to receive no more than a certain dose. The latter is usually called a dose-volume histogram (DVH) objective (see Section 35.3.1). Another useful option is a *uniformity* objective.

The perfect plan, with the prescribed dose delivered uniformly to the target volume and no dose outside the target, is unachievable and the optimisation algorithm has to balance the various objectives in order to achieve an approximation to the desired result. In order to guide the algorithm each objective can be assigned an *importance factor* or *weight*. If all the objectives were achievable there would be no need of weighting factors. Although experience shows that setting achievable objectives is more successful than setting unattainable goals, it is useful to be able to indicate which objectives are more important. Another concept is that of a particular goal that *has* to be met even if all other objectives are not achieved—this can be distinguished from other types of objective by describing it as a *constraint*, although this term is often used interchangeably with *objective*. IMRT optimisation essentially involves *moving* dose from one sensitive part of the patient to another part that is less sensitive. If no dose limit is specified for a particular volume of tissue it is possible that the dose will be *dumped* into that volume. To avoid this, the definition of dummy volumes can be useful. For example a shell around the target volume can be defined to help to ensure that the dose falls away sharply outside the target volume (Price et al. 2003b). Balancing all these tools is a new skill and indiscriminate application of them can result in an excessively complex plan that is difficult to deliver accurately. It is therefore advisable that, wherever possible, class solutions be developed including a table of acceptable dose volume limits.

In principle it would be possible to specify objectives directly in terms of tumour control probability and normal tissue complication probability (see Chapter 36). There has been much controversy about whether this is the right way to go, with those opposed to this saying that the biological parameters are not adequately known.

On the other hand the ranking of plans does not require such a high level of accuracy in terms of the absolute value of tumour control and several treatment planning system vendors are now offering the possibility of biological objectives; it is likely the use of these will grow (Grigorov et al. 2006). However, it is important to remember that if the complication probability for a particular organ is not specified in the plan design, the computer has no means of assessing the risk of that complication. This is especially important when doses are raised above the limits of normal experience.

When an IMRT plan has been satisfactorily computed and approved by the physician, the treatment control files can be generated. For MLC systems, these include leaf-sequence files for each gantry angle. Because IMRT involves complex beam shapes and control files, digital capability for plan transfer is essential to avoid possible mistakes during manual transfer. Depending on the individual clinic's information system, the files can be transferred by floppy disk or, preferably, directly transferred to the *record and verify* server through data-exchange software. If the department has a record and verify system that fully supports IMRT treatments, then many of these problems are reduced (and replaced by the need to verify the initial programming of the record and verification system).

44.3.5 PLAN VALIDATION AND QUALITY ASSURANCE

The implementation team should also design and implement an appropriate QA programme (see also Section 43.3.8). A number of useful papers are available to provide guidance on the QA programme (e.g. Losasso et al. 2001; Low 2002; Ezzell et al. 2003). The QA of IMRT consists of three parts: (1) commissioning and testing of the treatment planning and delivery systems, (2) routine QA of the delivery system and (3) patient specific dosimetry verification. The first part is to ensure the accuracy and performance of the inverse planning and IMRT delivery system. The second part is to ensure the normal operation of the dynamic or 'step & shoot' delivery system, and will involve additions to the daily, monthly and annual QA protocols (for example, the daily checks may be augmented to include a test of a dynamic treatment, including an output measurement that checks the stability of a

TABLE 44.2

Items to Be Tested before an IMRT Treatment

Item to be Verified	Method
Monitor units (absolute dose at a point)	Measurements in a phantom
MLC leaf sequences or fluence maps	Measurement with film or a digital imaging device
Dose distribution	Measurement with film or a dosimeter array (TLDs, ion chambers) in a phantom
Collision avoidance	A *dry run* before treatment

programmed MLC gap width). The third part is to ensure the accurate and safe treatment of the patient.

While IMRT produces conformal doses, the level of sophistication has been increased considerably. The goal of dosimetry validation is to verify that the planned dose distribution will be delivered to the patient (Wang et al. 1996; Low et al. 1998; Boyer et al. 1999). This ensures that the plan has been computed correctly and that the leaf sequence files and treatment parameters charted or stored in the record and verify system are correct and executable. To expedite IMRT delivery, an auto-sequencing delivery system is sometimes used. Such delivery systems (in different forms) are currently available from all major accelerator vendors. Items that need to be verified before the first treatment fraction are shown in Table 44.2.

Although the details of what is to be measured or calculated for dosimetric validation will depend on each clinic's needs and may change with experience, it is important to emphasise that new users will need to spend much more time validating IMRT plans than is common with conventional treatments. It is common practice to deliver the IMRT treatment plan to a phantom and to compare the doses measured in the phantom with those calculated by the planning system for that phantom (Paliwal et al. 2000; Ma et al. 2003; Budgell et al. 2005). Unless the phantom is similar in size and shape to the patient there may not be an area of uniform dose in the phantom that corresponds to the target in the patient. This will create difficulties for the dose measurement. Physical measurement will be necessary until reliable independent dose calculation methods are developed and validated (Ma et al., 2000; Pawlicki and Ma, 2001; Yang et al., 2003; see also Part F). However, it will never be possible to substitute dose check calculations (i.e. independent MU checking) for dose measurements entirely because the dose delivered is more dependent on the linac delivery system calibration than is the case with conventional treatments. For example, some calculation algorithms may not correctly model tongue-and-groove effects (Deng et al. 2001; Kamath et al. 2004; Que et al. 2004). Further discussion of these issues can be found in Section 39.3.9.3.

IMRT is a rapidly evolving modality, and the QA program for treatment planning will change with time so that it can handle various new issues that arise (Williams 2003). For example, two new approaches to reducing the time spent on quality control by automatic verification of the leaf positions include using the log of MLC leaf positions on Varian accelerators (Litzenberg et al. 2002; Stell et al. 2004), and using an electronic portal imaging device (Chang et al. 2004; van Esch et al. 2004). Electronic methods of measuring beam fluence are also being developed (Letourneau et al. 2004).

44.3.6 PATIENT IMMOBILISATION, POSITION VERIFICATION, AND TARGET LOCALISATION

The highly conformal nature of IMRT treatment requires satisfactory patient immobilisation to facilitate imaging and target localisation, and to ensure accurate dose delivery. New immobilisation devices may be necessary, such as supplementing thermoplastic masks with bite block fixation for head and neck treatments and using stereotactic body frames (see Chapter 45) and vacuum-fixation devices for abdominal treatments. Techniques to reduce the effect of organ motion due to breathing, such as respiratory gating (Hugo et al. 2003; Tsunashima et al. 2004; Jiang 2006; see also Section 29.3.3), may be desirable for thoracic and abdominal lesions. The inter-fraction reproducibility and reduction in intra-fraction organ motion that can be achieved with the immobilisation system should be investigated in advance to establish realistic margins for planning different sites. Devices to facilitate soft-tissue target localisation may be needed before each treatment to ensure the accuracy of isocentre placement (see below).

Verification of the patient position will be part of the initial plan validation. However, for subsequent IMRT treatments, patient setup and target localisation are the key to accurate dose delivery. The most critical requirement is that the treatment isocentre matches the planned isocentre. This can be done by comparing orthogonal films taken at simulation, digitally reconstructed radiographs (DRRs, see Section 32.4.5) from the planning system and portal images from the treatment unit. Wherever possible, portal images should be obtained for the fields used for treatment. It is useful to have a portal image showing the boundary of the intensity modulated aperture and compare this to the corresponding DRR from the planning system. Depending on the imaging system available, it may be possible to obtain a portal image of the modulated field superimposed on the patient's regional anatomy, but such images are often hard to interpret. In the case of dynamically delivered fields, one needs to consider how to acquire the MLC *envelope*, how to verify the position of a slit collimator, or how to operate an electronic portal imaging system.

Techniques based on bony structures provide adequate positioning accuracy for head and neck and spinal areas. Specially designed imaging systems based on ultrasound, CT, optical and radiographic detectors may be used for soft-tissue target localisation. The BAT ultrasound system (NOMOS Corp.) (Figure 44.1b) has been shown to be an effective and efficient target relocation device for prostate and other abdominal treatment setups (Langen et al. 2003; Trichter and Ennis 2003; Fuss et al. 2004), although Langen et al. found some operator variability. The positioning accuracy is between 2 mm and 5 mm (2σ) and between 3 min and 5 min is added to the treatment time. A CT scanner in the treatment room can provide a better than 2 mm (2σ) positioning accuracy for targets anywhere in the body (Figure 44.1c) using either bony structures or surgically inserted fiducial markers (Court and Dong 2003; Paskalev et al. 2004). At least, an extra 5 min will be needed to use this system before each treatment. A body localiser with vacuum fixation (Figure 44.1d) may reduce the movement of internal organs for high precision beam delivery for stereotactic body surgery or stereotactic IMRT (Wang et al. 2004). A combined optical and radiographic system can correlate the movements of the markers on the patient surface and the internal organs. Based on the real-time organ motion information, gated therapy or target tracking (Keall 2004; Keall et al. 2004; Murphy 2004) can be performed. A 5 mm (2σ) accuracy may be achieved using such real-time image guidance for breast, thoracic and abdominal treatments.

Because of the importance of patient positioning, the concept of using a diagnostic x-ray beam orthogonal to the therapy beam has been developed by both Elekta and Varian (see Section 40.1.5 and Figure 40.11). This can be used to obtain high quality verification images to facilitate, for example, matching of the position of seeds placed in the tumor volume, or to obtain cone-beam CT images allowing direct comparison of CT data with the planning CT scans (Jaffray et al. 2002). The use of such techniques can be expected to improve the accuracy and hence the outcome of radiotherapy (Ghilezan et al. 2004).

44.3.7 Treatment Delivery

As discussed in previous sections, there are several ways to deliver intensity-modulated beams for IMRT. Due to the increased complexity of intensity-modulated beam delivery the patient throughput on the treatment machine may decrease. IMRT treatments require more MUs, and often use more gantry angles (ports) than are used conventionally. For example, patients previously treated with lateral fields or a *four-field box* may now have many fields, including obliques or arcs. Because of limitations due to MLC leaf over-travel and collimator jaw over-travel, the maximum MLC field size for IMRT is usually smaller than the conventional maximum field size; two or more MLC fields may be required to deliver a conventional treatment field.

Studies have shown that the beam-on time ratio between IMRT and the conventional CFRT is generally 1–2 for prostate treatments and 1.5–5 for head and neck treatments, depending on the delivery system used (Bakai et al. 2003). The increase in beam-on-time is relevant to radiation protection issues both in terms of the increased leakage dose to the patient (see Section 61.5.2) and to the thickness of shielding required (see Sections 61.1.3 and 61.1.1.4). However, the beam-on time is only part of the overall treatment time and, where IMRT is able to simplify a complex conventional treatment, there may even be some reduction in the *overall* time taken to deliver the treatment. The overall time to deliver a fraction may be relevant to cell-killing (see Section 9.4) if this time exceeds that which corresponds to the high dose rate limit for which the LQ-expression is valid (Altman et al. 2006).

CHAPTER 45

STEREOTACTIC TECHNIQUES

Jim Warrington

CONTENTS

45.1 INTRODUCTION

The term *stereotactic* means the three-dimensional localisation of a particular point in space by a unique set of coordinates that relate to a fixed, external reference frame. Historically, such a frame has acted as a support for hollow probes that have carried either electrodes or biopsy needles to precise locations within either an animal or human brain, based on the analysis of plane and orthogonal x-ray films of the subject in the frame.

Since the 1950s, the stereotactic principle (Leksell 1951) has been adopted by neurosurgeons for the technique of *stereotactic radiosurgery* (SRS or RS), in which narrow beams of ^{60}Co γ-rays are focused onto a small target within the brain. Since the 1980s, adaptations of medical linear accelerators (linacs), used to produce similarly precise megavoltage photon beams, have made this technique accessible to many radiotherapy centres (Podgorsak et al. 1990; Tepper 1995). With the development of accurately relocatable head frames used in conjunction with CT based fiducial marker systems, precisely planned and accurately deliverable fractionated radiotherapy of the brain is readily achievable. The latter technique is referred to as *stereotactic radiotherapy* or SRT, and although usually applied solely to the brain, can in principle be applied anywhere in the body (SBRT 2006).

Stereotactic brachytherapy, typically using ^{125}I seeds implanted by stereotactically guided neurosurgical needles, is a complementary technique for very irregular lesions, but has the disadvantage, as with conventional neurosurgery, of being an invasive method, requiring hospitalisation of the patient. Radiosurgery procedures can usually be applied over a single day on an outpatient basis.

Stereotactic techniques using protons or heavy-ion beams (Kjellberg 1986; Steinberg et al. 1990) are described in Section 46.5.3.

45.2 CLINICAL INDICATIONS

Lesions suitable for treatment with single fraction radiosurgery range from benign arterovenous malformations (AVMs) to solitary, malignant brain metastases (Plowman 2000, 2002). Such lesions commonly receive doses in the range 15 Gy to 20 Gy. However, to reduce toxicity to sensitive structures, such as the brain stem and optic apparatus, many targets require fractionated treatments (Brada 1991; Shrieve et al. 1996). Such techniques are generally of a highly focused nature, thus they are often suitable for re-treating small recurrent tumours and as a boost to conventional brain and nasopharynx treatments. In these cases, *hypo-fractionated* regimens are sometimes used, typically involving five 5 Gy fractions. Most recent applications of SRT involve conventional fractionation of 25–35 fractions giving between 1.5 Gy and 2 Gy per fraction. Reduced PTV margins around the GTV (ICRU 1993b, 1999; see also Chapter 29 and Chapter 30) in SRT and RS reflect the improved accuracy of patient fixation, target localisation and treatment delivery, and the capability of making fine geometrical adjustments arising from electronic, megavoltage x-ray verification images.

45.3 RADIOSURGERY EQUIPMENT

45.3.1 GAMMA UNITS

Up until the mid-1980s, radiosurgery was almost exclusively delivered by the Gamma Knife treatment unit, pioneered by Lars Leksell (Figure 45.1) (Larsson et al. 1974; Leksell 1983; Walton et al. 1987; Lunsford et al. 1989).

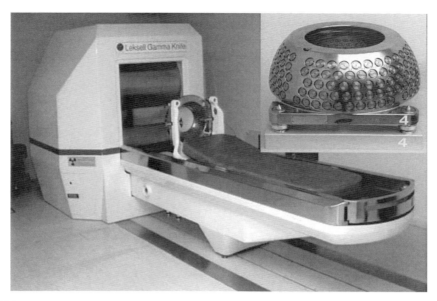

FIGURE 45.1
The ^{60}Co Gamma Knife radiosurgery unit. A tungsten secondary collimating shell is shown positioned over the patient's head prior to treatment, enabled by docking the couch inside the safe, which contains the primary collimator and ^{60}Co γ-ray sources.

This machine utilizes a *helmet* containing 201 separate ^{60}Co sources arranged in a 160° sectored array. These are focused onto a target with the aid of tungsten secondary collimating shells that define treatment spheres of 4 mm, 8 mm, 14 mm, or 18 mm in diameter. This unit represents the gold standard of geometrical precision, giving a dose delivery to within $\pm$ 0.4 mm (2 SD). It is robust and simple to operate, with mechanical movement restricted to a couch top that longitudinally docks into the treatment head (Figure 45.1). Translational adjustments to the patient's head frame align the planned isocentre with the focus of the cobalt-beam pencils. Its disadvantages lie in its high capital cost, its lack of versatility as a radiotherapy tool and its 201 radioactive sources which need replacing after 5–10 years of use.

45.4 LINEAR ACCELERATOR-BASED RADIOSURGERY

From the mid-1980s, linacs have been increasingly used to deliver radiosurgery (Betti and Derechinsky 1983, Colombo et al. 1985; Hartmann et al. 1985; Lutz et al. 1988; Podgorsak et al. 1988; Thomson et al. 1990) for lesions down to approximately 10 mm in diameter. Multiple noncoplanar arcs of circularly collimated x-ray beams directed to the linac isocentre can achieve a similarly focused treatment to that delivered by the sectored array of cobalt sources in the Gamma Knife.

The early linac-based systems used neuro-surgical floor stands (Figure 45.2), which provided accurately engineered, precise vernier movement of the patient's head support. This enabled precision movement of the patient so the centre of the patient's lesion could be accurately positioned at the treatment isocentre. During the last 10–15 years, linac manufacturers have introduced specialised, mechanically improved treatment gantries, patient support couches and beam-shaping devices capable of more precise treatment delivery. These specialised units (Friedman and Bova 1989; Yin et al. 2002) have almost comparable accuracy to the Gamma Knife, but offer greater flexibility for a broader range of treatment options. However, such linac-based facilities require carefully designed quality assurance programs to maintain the

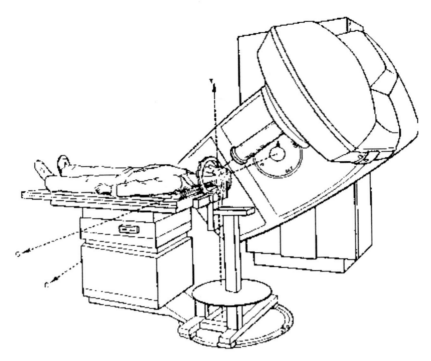

FIGURE 45.2
Early system for linac-based radiosurgery, after Lutz et al. 1988. (From Lutz, W., Winston, K. R., and Maleki, N., A system for stereotactic radiosurgery with a linear accelerator, *Int. J. Radiat. Oncol. Biol. Phys.*, 14, 373–381, 1988.

high precision needed. This, along with the linac's greater operational complexity, servicing needs and potential for breakdowns, all demand greater staffing resources.

The standard stereotactic beam delivery accessories for a linear accelerator are a set of quickly interchangeable, circular holed, tertiary collimators, which define circular cross-section beam pencils ranging typically from 10 mm to 50 mm in diameter in 2.5 mm steps. These can be augmented by the secondary collimators for more flexible conformation if the corresponding treatment planning software is available (Hacker et al. 1997). For fixed field, conformal SRT, the minimum requirement is a specialised blocking system (Figure 45.3), as this can be made more precise and less prone to collision problems compared to conventional blocking trays. Micro-multileaf collimators (Shiu et al. 1997; Cosgrove et al. 1999), although expensive, are very efficient and technically sophisticated for stereotactically guided treatments.

Other hardware requirements include such items as stereotactic frame mounting brackets, fiducial systems, setup boxes, a locking system for the couch top, stable precise room lasers, and a set of customised quality assurance tools for alignment and dosimetry checks (Warrington et al. 1994).

45.5 STEREOTACTIC FRAMES

45.5.1 NEUROSURGICAL FRAMES

Accurate fixing of the stereotactic frame to the patient's head has traditionally been achieved by means of three or four steel pins set into holes drilled into the patient's skull (e.g. the Leksell, Brainlab, or Brown-Roberts-Wells (BRW) surgical frames). Although frame fitting with this procedure requires at least a local anaesthetic, such frames remain the most reliable and stable platform for SRS. Their main disadvantage is that the whole planning

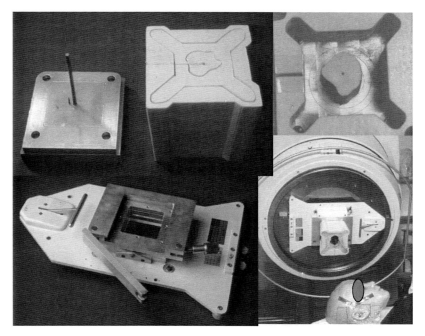

FIGURE 45.3
Stereotactic, conformal blocking system: clockwise from upper left; pouring jig and Styrofoam hot-wired pattern, jig before pouring alloy, block system mounted on linac head; coded block mounting device.

and treatment procedure usually has to be carried out in a single day. However, such frames have been removed and refitted for limited fractionation regimes (Schwade et al. 1990).

45.5.2 RELOCATABLE STEREOTACTIC FRAMES

For fractionated SRT treatments of up to 35 daily fractions, there is a need for simple but accurately relocatable fixation of the patient. This has led to developments of traditional thermoplastic masks (Lyman et al. 1989), or specialised frames, which locate onto the patient's upper dentition, as with the Radionics GTC (Gill–Thomas–Cosman) frame and variants (Gill et al. 1991; Kooy et al. 1994a; Adams et al. 1999a), shown in Figure 45.4a and the Medical Intelligence Head Arc frame (Sweeney et al. 1998) in Figure 45.4b. Others (Hariz et al. 1990;

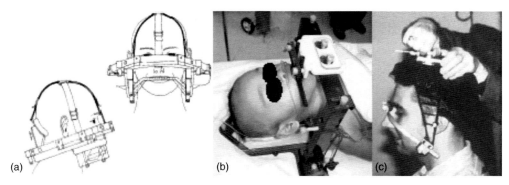

(a)   (b)   (c)

FIGURE 45.4
(a) Prototype Gill-Thomas (GTC) frame. (b) Head Arc vacuum bite-block system. (c) Fitting the Laitenen Stereodaptor.

Ashamalla et al. 2003) use the ear canals and bridge of the nose, as with the Laitinen Stereo-daptor (Figure 45.4c).

The repositional accuracy of a relocatable frame used for SRT should be less than 1 mm. Such precision is as much dependent on the skill and training of the staff routinely involved and the cooperation of the patient, as on the frame design itself. Relocation checks are best carried out by overlaying the bony anatomy from repeat sets of orthogonal x-ray, or low dose electronic portal images, and measuring the displacements of reference markers attached to the frame or field edges. Alternatively, nonradiographic checks, such as physical measurements to the surface of the head from a reference *depth helmet*, mounted onto the stereotactic frame, may be made. More recently, optical methods have been used, based on cameras in the treatment room, which track the patient's head position (Schulte et al. 2000; Liu et al. 2003).

45.6 STEREOTACTIC FIDUCIAL SYSTEMS

Central to the stereotactic method is the requirement for reference markers, or *fiducials*, to be imaged when attached to the stereotactic immobilisation system fixed to the patient (Siddon and Barth 1987). These markers are vital in providing accurate geometrical information on the coordinates of the planned isocentre(s). They are commonly in the form of various rod configurations attached to support rings, etchings or steel balls on the sides of plastic fiducial boxes or wires stretched between rigid spacers. The fundamental and ideal requirements of a stereotactic fiducial system are:

1. No significant scan artefacts generated or obscuring the anatomical images
2. Precise docking onto the patient's immobilisation system
3. An unambiguous (e.g. not window-dependent) and preferably simple marker arrangement, which enables manual checking of computed target coordinates, but has a sensitive, changing geometry from slice to slice
4. Capability of correcting for the effect of imaging slices being nonperpendicular to the cantilever scanner couch, with its tendency to sag from the weight of the patient as it passes through the scan plane

An example of a simple CT fiducial system is a bisected vee of apex angle 53.1°, etched onto the faces of a polymethylmethacrylate (PMMA, known as Perspex or Lucite) box (Figure 45.5a) that attaches to the stereotactic frame, which in turn is fixed to the patient's head. The stereotactic coordinates are obtained from simple linear measurements of the etching separations on the selected CT slice as shown. The BRW (later Cosman-Roberts-Wells, CRW) nine-rod system is also shown in Figure 45.5b with an associated CT slice and CT scan position.

If the scans are slightly oblique to the base frame due to a sagging, twisting CT couch, the unique coordinates of a selected target point can be computed from applying the known three-dimensional geometry of the system to the measurements in the vertical plane.

Although calculation software that generates the unique coordinates from the equations of fiducial markers is readily available, it is often preferable to adjust the frame tilt such that the patient CT scans are orthogonal to the couch. This facilitates ease of checking and retaining an intuitive feel for the precise isocentre setup geometry, which is ultimately transferred to the linac treatment couch.

For x-ray angiography, which is an accurate modality for visualising AVMs for radio-surgery, small 1 mm steel balls are incorporated into the four orthogonal faces of fiducial plates. This fiducial system can be particularly useful for radiographic checks on the relocation accuracy of stereotactic frames.

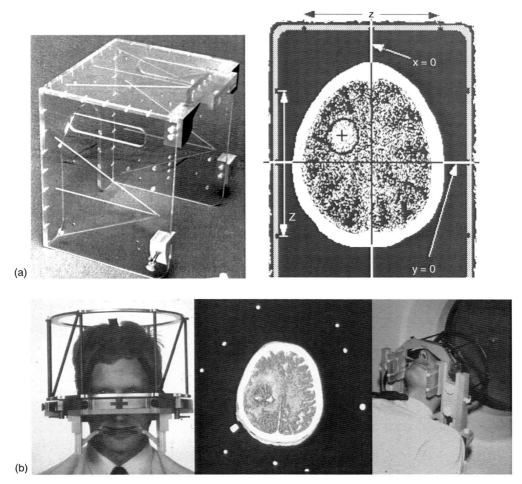

FIGURE 45.5
(a) Etched *bisected vee* fiducial box mountable on a stereotactic frame for CT scanning with coordinate measurements from a CT slice shown on the right. (b) BRW fiducial rod system and CT scan appearance.

For MRI or PET imaging, sealed tubes containing the appropriate liquid in similar configurations to x-ray CT localisers can be used to assist with the image registration process.

Knowledge of the precise magnification, resolution and distortion of the images used in planning stereotactic treatments is essential (Karger et al. 2003). Digital subtraction angiography (DSA) requires careful commissioning to establish the magnitude and geometry of any distortions that may arise.

45.7 STEREOTACTIC TREATMENT PLANNING

Commercially available stereotactic planning systems are often associated with combined neurosurgery, brachytherapy and radiosurgery packages. However, conventional radiotherapy planning systems increasingly offer an optional stereotactic module.

The ability of a planning computer to overlay or register images from various modalities such as x-ray CT, MRI, and PET (Plowman et al. 1997) is highly desirable. With MRI becoming an essential imaging tool for the brain, distortion-corrected image registration software is very important in high-precision stereotactic planning. X-ray CT scans represent the distortion-free anatomical baseline against which other modality images are co-registered and potentially *fused* (Kooy et al. 1994b; Alexander et al. 1995).

Real-time manipulation of the virtual patient, showing the juxtaposition of the target volume and sensitive structures, has become essential for SRT planning. Beam optimisation software is ever more important as the complexity of sensitive structure avoidance makes intuitive planning more difficult. Plan analysis tools for the rapid calculation of dose-volume data and assessment of optimised cost functions are therefore also required.

Stereotactic treatment planning not only requires the facility to plan multiple noncoplanar arcs focused to a single or several isocentres, but also to plan using multiple fixed noncoplanar conformal beams defined by either conformal blocking or multileaf collimators (MLCs). Table 45.1 shows some of the main planning options available for different target geometries.

Tumours with concavities wrapped around sensitive structures, such as the brain stem and optic apparatus, lend themselves to treatments using IMRT (see Chapter 43 and Chapter 44).

TABLE 45.1

Treatment Planning Options for Brain Lesions Using Linac-Based Stereotactic Radiotherapy

Lesion Shape	Treatment Planning Options	Planning Time (h)	Treatment Time (min)
Small sphere (up to 3 cm diameter)	4–6 multiple noncoplanar arcs, single isocentre. Single tertiary collimator	1	30
Large sphere (3–5 cm diameter)	2–4 multiple noncoplanar arcs, single isocentre. Single tertiary collimator	1	15–30
Ellipsoidal	6 multiple noncoplanar arcs, single isocentre, to include some shorter arc lengths with 1 or 2 collimator changes	3	30–45
Ellipsoidal	Double/triple isocentrer, 2–4 arcs per isocentre, 0–2 collimator changes. Up to 50% higher hot spot doses relative to prescription isodose surface	4	45–75
Ellipsoidal	Single isocentre, one tertiary collimator plus secondary jaw adjustments, 3–4 arcs	4	30
Ellipsoidal	4–6 fixed, noncoplanar fields using either blocks or multileaf collimation	4	15–30
Irregularly shaped with semi-enclosed sensitive structure	3–5 isocentres, 2–3 arcs per isocentre, evenly separated arc planes. Up to 50% increase in hot-spot doses relative to prescription isodose surface. Up to 3 different circular collimators	6	30–60
Irregularly shaped with semi-enclosed sensitive structure	4–6 fixed, noncoplanar fields using either conformal blocks or micro-leaf collimation. Limited scope for structure avoidance	6	15–30
Irregularly shaped with semi-enclosed sensitive structure	Conformal, dynamic arcing with micro-leaf collimator, 2–4 noncoplanar arcs (Grebe et al. 2001)	6	15–45
Irregularly shaped with semi-enclosed sensitive structure	3–9 fixed field, coplanar IMRT using step and shoot or dynamic mMLC[a]	8–16	15–45
Irregularly shaped with semi-enclosed sensitive structure	4–9 fixed field, noncoplanar IMRT using step and shoot or dynamic mMLC[a]	8–16	15–60

[a] Conformal compensators would provide a less sophisticated alternative.

These treatments, identified in the last two rows of Table 45.1, are designed by an inverse planning or optimisation process (Section 43.2 and Section 44.3.4), using the detailed specification of dose-volume constraints in and around the lesion to be treated. The inverse planning algorithms can either design conventional compensators or generate, via interpreters, complex MLC leaf sequences for each beam on the linear accelerator. These IMRT options are included in the table as the stereotactic process provides a good basis for such physically optimised methods (Baumert et al. 2003; Nakamura et al. 2003).

The planning physicist must be aware of the limitations of the given treatment machine geometry regarding potential collimator–couch–patient collisions. Customised software, which alerts the planner to such forbidden collision zones on a given linac, is invaluable (Nioutsikou et al. 2003).

Overlapping of beam entrance and exits should be avoided, and in general, arc planes (as defined by the couch angle) should have maximum separation according to the expression $180°/N$, where N is the number of arcs to be used. Avoidance of sensitive and sometimes previously treated structures means that optimisation software and biological TCP/NTCP modelling (see Chapter 36) are increasingly important developments.

The final dose distributions should be displayable in any required plane, preferably including adjacent anatomy. Plan comparisons require dose-volume histogram (DVH) analyses (see Chapter 35) to assist the clinical team in deciding the optimum treatment plan. Analysis of DVHs for various collimator arc regimes suggest that 3–5 arcs produce sufficient normal tissue sparing for RS and SRT (Graham et al. 1991). However, when planning treatments for benign conditions, it is relevant to consider increasing the number of arcs to reduce the exit dose from any given arc, or interrupting, for example, a sagittally orientated arc to avoid possibly sensitive exit doses (Shepherd et al. 1997). For larger, irregular target volumes (35 cm^3 to 70 cm^3), 4–6 fixed noncoplanar fields are likely to be a more appropriate technique. In this case, either conformal blocking or multileaf collimation of each portal, based on the beam's eye views (BEVs) of the target, is more likely to be used. Figure 45.6 shows a comparison of DVHs for three different beam configurations used in a stereotactically planned, elliptically shaped brain lesion (Laing et al. 1993). The sparing effect of conformal beam

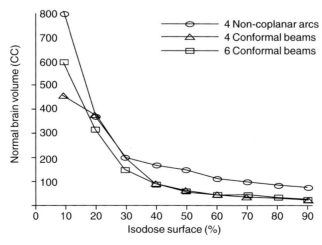

FIGURE 45.6
Cumulative dose volume histograms for normal brain with a 30 cm³ elliptical target volume. (After Laing, R. W. et al., *Radiother. Oncol.*, 28, 241–246, 1993.)

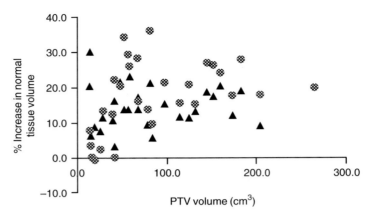

FIGURE 45.7

Increase in normal tissue volume irradiated to greater than 50% (triangles) and greater than 80% of dose prescription using conventional MLC (10 mm leaf width) compared to conformal blocks. (From Adams, E. J., Cosgrove, V. P., Shepherd, S. F. et al., *Radiother. Oncol.*, 51, 205–209, 1999. With permission.)

shaping for four and six fixed fields compared to a spherical treatment volume produced by single isocentre with multiple arcs and a circular collimator is clearly shown.

This graph shows only a simple comparison. Elliptical conformation can be achieved with multiple arcing techniques, having an adroit blend of different collimators and arc lengths. Alternatively, use of the secondary collimators to shield regions of circular fields can achieve satisfactory conformation for simpler, nonspherical target geometries. There are many comparisons of different stereotactic treatment regimes in the literature (e.g. Podgorsak et al. 1989; Serago et al. 1992a; Hall and Brenner 1993; Nedzi et al. 1993; Luxton et al. 1993; Philips et al. 1994; Cardinale et al. 1998; Verhey et al. 1998; Benedict et al. 2001). These are based on cumulative DVH comparisons.

Although the increasing availability of multileaf collimators enables efficient production of conformal beams, conventional 10 mm leaf systems are often too coarse for SRT. Figure 45.7 shows the results of DVH comparisons for fixed conformal fields shaped by MLC or lead alloy blocks for small to medium sized malignant brain lesions, planned stereotactically (Adams et al. 1999b). The added efficiency of the MLC device always has to be balanced against the gold standard of target conformation, which is obtained using customised lead alloy blocks.

Although micro-multileaf collimators bridge this gap in technology, they are expensive and generally limited to field sizes below 100 mm, and so have to be removed for conventional radiotherapy treatments on a non-dedicated accelerator. However, recent developments in technology, including the latest integrated MLCs in the linac head, which have leaf sizes of 5 mm or less at the isocentre (Varian 120 leaf, Elekta Beam Modulator), show a convergence of conformality (Monk et al. 2003).

45.8 BEAM DATA MEASUREMENT

The measurement of beam data for small, tertiary collimated, megavoltage photon beams typically involves up to five profiles (OARs), a central axis depth dose or TMR curve, output factors, often both in air and water, and build-up curves (see Section 23.2.1). A potential lack of lateral electronic equilibrium with these pencil beams requires small detectors having high spatial resolution (Rice et al. 1987; Serago et al. 1992b; McKerracher and Thwaites 1999; Yu et al. 2004). Profiles are therefore commonly measured with x-ray verification film, such as Kodak XV2 or EDR (Chetty and Charland 2002; Zhu et al. 2003), corrected for any dose/density nonlinearity (Figure 45.8).

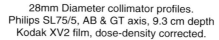

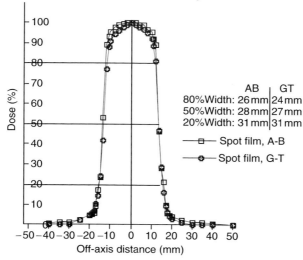

FIGURE 45.8
Profiles for a 28 mm diameter collimator at 93 mm depth from Kodak XV2 spot film, scanned across the AB (transverse) and GT (longitudinal) directions.

Small-volume (<0.2 cm^3) ionisation chambers are generally suitable for the depth-dose or TMR data down to 20 mm diameter fields. It is important to be aware of the diverging beam for depth-dose measurements of pencil beams, where the detector is exposed to different regions of the beam profile with depth. This could significantly distort the depth-dose curves for small fields. A computerised plotting tank system used in conjunction with a p-type electron diode or diamond detector (McKerracher and Thwaites 1999) is also an ideal means of acquiring these data down to about 15 mm diameter collimators. The setup for such measurements should ideally reflect the mean isocentre depth and scatter geometry of the patient. However, as with all beam data, measurements are often a compromise dictated by the needs of the treatment planning system and the size of the tank available. The smallest beams, below 20 mm diameter, for instance, should be measured with film, TLDs and a small diode, where the detector resolution is in the region of 2 mm or less. Overlap with the larger field sizes is recommended for such work so that discontinuities in data, such as output factors, can be investigated and reconciled. Note that tertiary collimators can also give rise to varying build-up depths to the peak dose.

45.9 TREATMENT SETUP

Prior to RS or a course of SRT treatment, it is important to be able to confirm the planned setup using the CT scanner, simulator or the treatment machine itself, depending on the type of lesion to be treated and the available facilities. Simulation of treatment fields, although often providing valuable confirmation of the planned isocentre position, can create unwanted confusion. This is due to small systematic and random errors from mechanical, optical and radiological differences between the simulator and treatment unit. Modern multislice CT scanners, increasingly used in radiotherapy for virtual simulation (Aird and Conway 2002), can now produce the high resolution data necessary for quality digitally-reconstructed radiographs (DRRs) for on-treatment reference images.

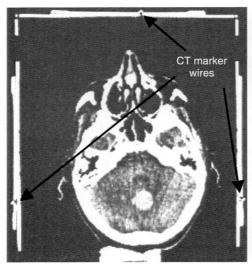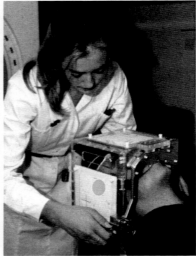

FIGURE 45.9
Check CT slice shown with wired setup plates on the fiducial box. The whole jig is shown being placed onto a patient in the stereotactic frame, on the right.

Translating the planned isocentre coordinates safely onto a simple, patient-specific setup system for the radiographers' daily use is a very important feature of the stereotactic process. The setup arrangements are generally of two types:

- Those involving manually setting three x, y, and z vernier scales to bring the patient's planned isocentre into coincidence with the linac isocentre. This type of system is commonly used in single-fraction radiosurgery and does not lend itself to repeated daily setups with its need for careful checking of finely adjusted vernier settings.
- Those using the setup of boxes with patient-specific plates or computer printouts with the isocentre crosses and beam projections clearly marked for room laser setup. These are more practical for daily fractionation regimes, but ideally require coding to avoid mistakes when a number of SRT patients are being treated in any given session.

If there is ready access to a CT scanner, a check CT slice with the laser setup crosses marked with radio-opaque markers or wires (Figure 45.9) is useful, as this can be directly compared to the corresponding planning slice and an additional frame relocation check. Figure 45.9 also shows a setup box with its attached, patient specific plates being placed on a stereotactic frame in its couch bracket.

Automatic or assisted setups on the linac couch are invaluable in limiting potential errors in setting up the patient. The uniquely positioned couch bracket, holding the stereotactic frame, enables couch position readouts to be included in the assisted setup parameters prior to making fine adjustments against the room lasers for the final setup.

45.10 PRETREATMENT QUALITY ASSURANCE

The nature of quality assurance checks (Tsai et al. 1991; Drzymala 1991; Warrington et al. 1994; AAPM 1995) depends on the precise technique used. Table 45.2 provides a summary of quality control procedures as an example of those undertaken in stereotactic radiotherapy, with the suggested tolerances in the right hand column. Figure 45.10 shows some of the equipment referred to.

TABLE 45.2

Frequency of Stereotactic Quality Assurance Tests

Test	Tolerance
Daily checks immediately before the stereotactic treatment session	
1. Movement of the light field centre cross for 360° collimator rotation	<0.5 mm
2. Comparison of the left lateral, right lateral and ceiling lasers with the corresponding light field centres at the cardinal gantry angles	<0.5 mm
3. Light field symmetry of the selected circular collimator about the field centre cross	<0.5 mm
4. (On-patient) fixation of the frame, using patient response, physical measurement (e.g. depth helmet or ruler measurements) and examination, followed by EPID checks after setup and comparison with DRRs	1 mm
Monthly	
1. Shadow films or EPID images of a radio-opaque object set at the room laser defined isocentre (Lutz et al. 1988) are taken at regular gantry intervals using a tertiary collimator attached to the treatment head (see Figure 38.5, Figure 45.10b, d). The films are measured on a scanning densitometer, and the displacements of the object centre from the irradiated field centres are determined	<1 mm
(Note: For radiosurgery of benign lesions, this check is often carried out prior to each single, large treatment fraction and may carry a 0.5-mm tolerance.)	
2. A single transverse 240° arc is executed over a water phantom, having a 0.2 cm³ Farmer chamber at the isocentre. The chamber dose is compared to a computer plan of the same *arcing treatment* for representative monitor units and collimation used on actual treatments (Perks et al. 1999; McKerracher and Thwaites 2002)	±2%
On an annual or six-monthly basis, this check can be broken down into separate arc segments and total doses to verify the consistency of arc calibration over the whole 360° range of gantry angles and monitor unit settings	
Annually or at commissioning	
1. A useful geometric and relative dosimetric check is to CT scan, plan and irradiate a *water-ball phantom* at the isocentre with a film cassette bisecting the two hemispheres (Figure 45.10b). The film is pierced before irradiation in the cassette to mark the laser setup and isocentre points. After irradiation, the film can be compared with the treatment plan, whether using fixed conformal fields or multiple arcs (Figure 45.10e)	
2. Similar to the above, with a humanoid head phantom mounted on the stereotactic frame, CT planning and treatment of a simulated lesion in the phantom is given. TLD pellets stacked along the three cardinal axes through the isocentre are measured and compared with the corresponding multiple-arc plan	<3% dose at isocentre 1 mm profile displacement
This is a very time-consuming procedure that often carries uncertainties in the positioning of the dosimeters in the sliced humanoid phantom head. Alternatively, measurements from verification film loaded into a similarly planned and treated phantom can be undertaken to give accurate geometrical information. Relative isodose information for the higher dose regions can be obtained from such film measurements provided the doses do not cause optical density saturation	
The more peripheral isodose information can be prone to large errors due to the enhanced response of film to scattered radiation. The new technique of gel dosimetry shows some promise for 3D verification of the isodose envelopes arising from conformal SRT (Cosgrove et al. 2000)	

Other checks on an annual basis should include relative output and profile checks, monitoring of the wear and tear on all stereotactic accessories including couch brackets, frames, setup boxes, fiducial jigs, collimators and mountings, couches and quality assurance tools.

Additional checks for the treatment planning computer and peripherals, portal imaging systems and scanners, as they relate to the stereotactic process, also need to be carefully incorporated into the established programmes. As with general radiotherapy QA protocols

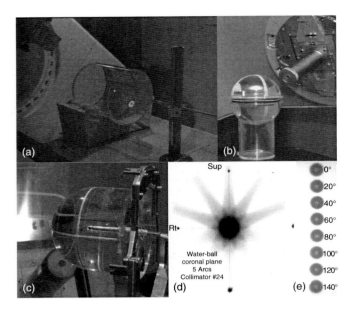

FIGURE 45.10
Quality assurance tools: (a) Gantry radiation isocentre apparatus; (b) water-ball phantom; (c) arc calibration phantom;
(d) developed film from water-ball phantom shown in (b); (e) developed strip film produced using the radiation
isocentre apparatus shown in (a).

(see Part H), stereotactic quality assurance should give a practical but focused and secure means
of quality control without overburdening the available physics and radiographer staff.

45.11 STEREOTACTIC CONFORMAL TREATMENTS AND QA

Low melting-point lead-alloy blocks, (Figure 45.3) based on beam's-eye-views (BEVs)
and manufactured with the aid of a computerised block cutter linked to the planning computer,
require careful quality checks prior to the first treatment fraction. As with the alternative micro-
multileaf collimator beam shaping, comparisons of the projected light fields onto hard copies of
the corresponding BEVs are a minimum required check. It is advantageous to have projections
of conformal fields marked on the setup box for a quick comparison of the light fields. This
reassuring facility is available in some stereotactic planning systems such that a composite
printout of all conformal field projections can be positioned on the setup surfaces or plates
of the treatment box.

Unambiguous labelling and configuring of these noncoplanar plans, particularly regarding
collimator twists, demands very careful checking. Automatic transfer to record and verify
systems create an even greater imperative for meticulous quality control prior to the first
fraction of treatment.

Electronic portal imaging devices (EPIDs) have become essential tools for completing the
quality assurance chain in the stereotactic process. Good quality DRRs from the CT planning
data set provide the reference images for the target-checking process. Note that there is limited
imager access for fields directed in cranio-caudal directions, and confusion often arises from
target-checking small, obliquely angled fields. It is often preferable to acquire orthogonal, AP
and lateral images on the first days of treatment to clearly establish the magnitude and direction
of any systematic errors in the isocentre position relative to the patient's anatomy. Once this has
been done, imaging can take place on a less frequent, weekly basis. The tertiary collimator will

often render EPID checks impractical, particularly for radiosurgery of small lesions. EPID imaging prior to the mounting of the tertiary collimator should be considered to obtain sufficient anatomical (bony) detail for overlay on reference DRRs. When mMLCs are being used, they can have their leaves retracted to enable such images to be taken.

45.12 ACCURACY AND ISODOSE PRESCRIPTION

The overall accuracy achievable with a linac-based radiosurgery system should be between 1 mm and 2 mm if sufficient emphasis is placed on the importance of quality assurance (Hartmann et al. 1994). However, localisation of the target volume often remains the greatest uncertainty, and it is important to view accuracy of SRT and RS in its full clinical perspective. Assessing the accuracy of a given system involves consideration of the cumulative effect of small (typically 0.5 mm to 1 mm) errors, which generally require a safety margin around the lesion of between 2 mm and 3 mm. Often with such treatments, a variable PTV margin is required with some compromise needed close to sensitive organs. This PTV is then covered by the *prescribed* isodose surface, commonly between 80 and 95% of the central dose. If multiple isocentres are used with arcing circular beams, dose uniformity is sacrificed to achieve more conformal target coverage. In this case, the prescribed isodose may be 50% of the maximum *hot spots* in the overlapping regions of adjacent spherical dose distributions, which correspond to multiple isocentre setups. Dose-volume statistics and radiobiological considerations may need to be applied to the PTV and organs-at-risk to evaluate the best plan. Ideally, single-isocentre treatments should be normalised and prescribed to 100% at the isocentre, ensuring that the required minimum encompasses the PTV.

45.13 STEREOTACTIC RADIOTHERAPY TO EXTRA-CRANIAL SITES

The stereotactic principle can be readily applied to other regions of the body, notably for head and neck treatments. The head and neck region can be accurately immobilised using either the frames described for upper head treatments, specialised head-and-neck frames or accurately made thermoplastic shells. Often, special emphasis is needed on the reproducible immobilisation of the shoulders and neck for the inclusion of nodal areas down to T1, for example. As with the treatment of the brain, better noncoplanar beam access is obtained if the head and neck system overhangs the end of the treatment couch and the headrest system is radiolucent.

An important development is the *stereotactic body frame* (also known as *extra-cranial stereotaxy*) (Lax et al. 1994; Blomgren et al. 1998). This system has the patient immobilised in an *air equivalent* vacuum bag containing expanded polystyrene granules, which in turn is contained within a rigid outer support box into which fiducial marker rods are embedded. The patient is CT scanned and treated in this box.

The concern with stereotactic radiotherapy applied to such sites as the abdomen and pelvis is the mobility of soft tissue. This causes a departure from the fundamental assumption in the stereotactic method of a fixed relationship between the support *frame* and the PTV. This potential for error requires, ideally, pretreatment soft-tissue monitoring with a CT scanner or ultrasound device, preferably in the treatment room. Breathing motion can also be a serious problem, requiring breathing control and gating of treatments (Hara et al. 2002; Jiang 2006). Fowler et al. (2004) have reviewed the successful use of large fractions (i.e. hypofractionation) when treating lung tumours in an extra-cranial stereotactic frame and a collection of very useful papers on this subject can be found in SBRT (2006). Provided that such facilities are available, the stereotactic body frame can provide a practical set of reference points against which inter- and

TABLE 45.3

Summary of Stereotactic Radiotherapy Procedures as Carried Out at the Royal Marsden Hospital

Procedure	Additional Details	Staff Involved	Time (h)
Seeing patient in clinic	Review of diagnosis, notes, clinical examination, discussion, deciding on treatment method, booking scans, planning and treatment slots and consenting patient	Clinicians and team	1–2
Fitting stereotactic fixation device to the patient	E.g. GTC frame: making dental mould, fitting frame including making headrest, checking reproducibility with physical, optical or radiographic measurements	Dentist or mould-room technician	2
CT scanning	Adjusting couch bracket tilt, checking fixation	Radiographer	0.45
MR scanning	Check similar patient head position in head coil (no frame)	Radiographer	0.45
Transfer of image data	Check slice set sequences, patient details, etc.	Physicist or radiographer	0.5
Localising and outlining MR/CT scans	Localising target volume and sensitive structures, fusing or registering images, growing PTV, outline adjustments	Clinician and physicist	1–2
Planning treatment	Viewing target and structures, deciding on technique and collimation, setting isocentre, applying appropriate beams, calculating isodoses, adjusting for optimal geometry/uniformity, obtaining printouts and DVH's	Physicist	2–6
Checking plan feasibility	Eliminate possible collisions by setting up on linac, on a scale model or running computer simulation	Physicist and radiographer	0.5
Obtaining plan approval and dose prescription	Presenting and discussing single or alternative plans to clinician with DVH analysis, if necessary, and write dose and fractionation prescription	Physicist and clinician	0.5–1
Labelling of plan/ setup box/MLC transfer/DRRs/ checking	Preparation of final plan with isocentre dose and coordinate check, DRR production and transfer of all files for auto setup, mMLC shapes and reference images	Physicist and radiographer	1–2
Conformal blocks	Transfer of BEV's to computerised block cutter, pouring of blocks and labelling/checking on unit. Output factor checks, if necessary	Physicist, mould room technician	4–6
MLC prescriptions	Transfer of plan data, checking prescription, monitor unit calculation and optical check of MLC shapes	Radiographers	1–2
CT slice check	Setup patient and take check slice on CT scanner. Check final position of isocentre against planning CT slice	Radiographers, physicist	0.5
Patient pretreatment	Show patient treatment machine and explain details of treatment	Radiographer	0.25
Quality assurance of treatment machine	Extra alignment checks on lasers and relation of mean radiation isocentre, arc calibrations, as appropriate	Physicist	0.5–1.5
Treatment of first fraction	Includes setting up patient, checking conformal fields optically, couch readouts and acquiring portal images	Radiographers and physicist	0.5–0.75
Treatment of each subsequent fraction	As above, according to protocols laid down for EPID verification imaging and isocentre adjustments for systematic error (e.g. after first 3 fractions)	Radiographers	0.25–0.5
Target checking	Checking relation of isocentre to patient's anatomy according to reference DRRs	Radiographer, physicist	0.5–1
Monitoring patient in weekly clinics	Review of patient progress (Also, discuss and review progress with rest of team in weekly audit with relevant clinician, etc.)	Clinicians (radiographer, physicist)	0.25–0.5

intra-fractional soft-tissue target movement can be measured and corrected for. However, in most centres limited to conventional portal imaging, the bony anatomy has to provide the practical baseline reference frame and an additional stereotactic reference system may not be helpful.

45.14 CONCLUSION

Stereotactic radiotherapy and radiosurgery techniques are increasingly being used where improved accuracy is required for treating well-defined targets inside the head. Their improved methods of patient fixation, multimodality imaging, fiducial marker-based treatment planning, treatment setup, verification and delivery represent a significant step forward in the evolution of high-precision radiotherapy. However, these techniques require the highest level of teamwork and continuity of training among different staff groups and are particularly demanding and time consuming for the radiotherapy physicist who has significant technical responsibilities at most stages of the stereotactic process. Good communication between physicists and radiotherapy engineers is particularly important after servicing or repair of the machines. Changes to various components on the linac or treatment couch, including major retuning of the beams, can profoundly affect the accuracy of stereotactic treatments.

Table 45.3 summarises the stereotactic planning and treatment process. The times involved can vary considerably depending on the individual patient case and the associated evolving technology. The impact of any technical problems that may arise are not included.

CHAPTER 46

PROTON BEAMS IN RADIOTHERAPY

Alejandro Mazal

CONTENTS

46.1 INTRODUCTION

There is significant worldwide interest in the medical use of proton beams with energies ranging between 65 and 250 MeV because of their favourable depth–dose properties (see Figure 46.1), and considerable technological progress has been made (Bonnet 1993; ICRU 1998; Sisterson 2004). This chapter presents the physical and technological basis of proton therapy (Wilson 1946; Fowler 1981; Bonnet 1993; Chu et al. 1993; Webb 1993; Linz 1995), as well as covering the terminology, procedures and tools now available.

46.2 PHYSICAL BASIS: THE INTERACTIONS OF PROTONS WITH MATTER

The interaction of protons with matter is the basis for the therapeutic potential of these beams, in particular because of their characteristic of high ionisation density at the end of the

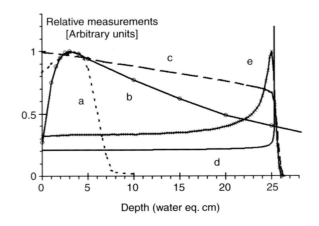

FIGURE 46.1
(a) Depth–dose curve for 15 MeV electrons; (b) depth–dose curve for 15 MV photons; (c) fluence vs. depth for 200 MeV protons; (d) stopping power vs. depth for a 200 MeV proton beam; (e) depth dose curve (Bragg peak) for a 200 MeV proton beam, being the result of the combination of (c) and (d). (Data measured at Orsay, France.)

range and their weak scattering. Protons are charged particles and are relatively 'heavy' compared to electrons (the mass of the proton, M, is 1836 times the mass, m, of the electron). Depending on their energy (represented by their velocity), they interact with the medium through various mechanisms: inelastic or elastic collisions with the nucleus, with the bound atomic electrons and, for certain processes, with the whole atom (see also Chapter 3). In clinical applications, protons are considered as having an *intermediate energy*, larger than $10^{-4}Mc^2$ and lower than Mc^2. A detailed discussion can be found in Bichsel (1968).

46.2.1   INELASTIC INTERACTIONS WITH THE NUCLEUS

Inelastic interactions with the nucleus result in either a significant deflection of the incident proton associated with a nuclear reaction, or bremsstrahlung[*]. In nuclear reactions like (p,n), (p,d) and others, the production of an unstable nucleus and some other processes lead to the disappearance of incident protons, with production of neutrons and recoil nuclei as well as activation of the medium with γ-ray production. These interactions create particular problems for radiation protection. The proportion of the nuclear interactions can be estimated by measuring the reduction in the *planar* fluence (see Section 6.3.5) of the protons at different depths in matter, before stopping at the end of the range (Figure 46.1), according to:

$$I = I_0 e^{-t/s} \tag{46.1}$$

where I_0 is the incident planar fluence, I the planar fluence after a thickness t of absorber, and s is the proton mean free path length given by

$$s = \frac{A^{1/3}}{0.032} \text{ in g cm}^{-2} \tag{46.2}$$

where A is the mass number of the absorber. In practice, for the high-energy proton beams used in clinical applications, a practical figure for the reduction in the proton planar fluence due to nuclear interactions in water or tissue is about 1% per cm.

46.2.2   INELASTIC INTERACTIONS WITH ATOMIC ELECTRONS

Inelastic interactions with bound atomic electrons represent the principal process by which protons lose energy along their trajectory at the energies and in the materials of clinical interest, creating atomic excitation or ionisation, as well as a small deflection of the incident proton (see also Chapter 3). Although it is possible for electrons to lose a large fraction of their energy and be deflected through large angles in an inelastic interaction[†], the energy transferred by protons at each interaction is always small, the maximum possible value being approximately $4m/M$, where m is the rest mass of the electron and M is the rest mass of the proton. The amount of energy transferred in each interaction has a probability distribution, which results in energy- and range straggling (see Section 3.4.2) after passing through a given thickness of absorber. This is the reason for the *finite* slope of the final part of the depth–fluence curve. The energy-straggling distribution is represented by a *Vavilov* distribution, which resembles a Gaussian at high energies (corresponding to very small energy losses) and with a *tail associated with larger energy losses* at low energies (Bichsel 1968). In practice, energy-and particularly range straggling

---

[*] Bremsstrahlung losses are negligible for the proton energies of interest in clinical applications. They only become important for energies larger than the rest mass energy of the particle, i.e. 938 MeV for protons, and even then this energy-loss mechanism very much less probable than bremsstrahlung for electrons (see Chapter 3).
[†] In practice, the probability for an electron to undergo for such large energy losses or deflections is very small but they are allowed in principle (see Chapter 3).

for clinical proton beams can be assumed to have nearly Gaussian distributions. The fluctuation in the pathlength of clinical proton beams in water is of the order of 1–1.3% (1σ) of the range (Bichsel 1968).

The average loss of energy by collision per unit distance along the path of a proton (dE/ds) is represented by the collision stopping power:

$$\left(\frac{dE}{ds}\right)_{col} = 4\pi z_{eff}^2 e^4 \frac{N_A Z}{A} \frac{1}{mv^2} \left\{ \ln\left(\frac{2mv^2}{I(1-\beta^2)}\right) - \beta^2 - \sum\left(\frac{C_i}{Z}\right) \right\} \qquad (46.3)$$

where the quantities not thus far defined are z_{eff} and e, the effective charge of the particle and the electronic charge, respectively. $N_A Z/A$ represents the number of electrons per gram; m and v are the mass of the electron and velocity of the particle, respectively; β is the ratio of the velocity of the particle to the velocity of light; I is the mean excitation energy of the atom of the absorbing material; and $\sum(C_i/Z)$ includes density and shell correction terms (see also Chapter 3).

When one considers only energy transfers below a given value, dE/ds is termed the *restricted stopping power* or *restricted linear energy transfer* (LET) (see Chapter 3 and Chapter 6). These two concepts are rather similar for charged heavy particles in the range of energies used in therapy. Tabulated values are available (ICRU 1993a). Stopping power is proportional to the inverse of the square of the velocity v together with a more complex function of the velocity and the mean excitation energy of the medium I, including also shell corrections (the final term) when v is of the same order as the velocity of the electrons in their orbits (see Chapter 3).

When the proton is almost at the end of its range its capacity to ionise increases rapidly since its velocity is low (Figure 43.1), giving rise to the *Bragg peak*. The amount of ionisation produced by a beam of protons is a combination of the proton fluence and the energy-transfer function i.e. the stopping power (see Section 6.6.2). The depth–dose curve for a broad beam of heavy charged particles is known as the *Bragg curve*. For clinical beams, the ratio of the dose at the entrance surface to the dose at the peak varies between 0.2 and 0.4. The dose beyond the peak is practically zero. The *median* range is approximately equal to the depth of the distal 80% of the Bragg curve. The most probable energy at the depth of the peak is approximately 10% of the initial kinetic energy of the particle.

The absence of bremsstrahlung in clinical proton beams means that there is no *tail* in the depth–dose curve beyond the end of the proton range (apart from that due to range straggling). In the case of heavier particles, secondary products of nuclear interactions give rise to a significant tail beyond the peak, with potentially large radiobiological effects, which can, in some cases, limit their clinical application (see Chapter 49).

46.2.3   Elastic Interactions with the Nucleus

Elastic interactions with the nucleus cause a deviation of the incident proton, with a negligible change in energy; this process is often called *Rutherford scattering*. The total cross section for this process decreases rapidly with the energy of the particle and the differential cross section decreases with increasing deflection angle (Bichsel 1968; see also Section 3.5). Thus only a very few of the protons undergo single, large deflections. Most collisions involve a *distant* interaction of the particle with a nucleus; the nuclear charge is partially screened by the atomic electrons, and the incident particle experiences only a small deflection. The multiplicity of small-angle deviations along the proton path is known as *multiple Coulomb scattering*. Molière (1947), Highland (1975) and others have evaluated this scattering. This subject has been reviewed by Gottschalk et al. (1993) who also considered scattering in thick absorbers[*].

---

[*] Multiple scattering is employed in the *condensed-history* Monte-Carlo technique—see Section 5.4.

The angular distribution of particles after traversing a thin foil can be represented to first order by a Gaussian, where the mean angle of multiple scattering is given by

$$\bar{\theta}_0 = 14.1 \frac{z}{p\text{v}} \left\{ \sqrt{\frac{L}{L_R}} \left(1 + \frac{1}{9} \log\left(\frac{L}{L_R}\right) \right) \right\} \tag{46.4}$$

where z, p and v are the charge number (charge/e, which for protons is unity), momentum and velocity of the incident proton, respectively; L is the thickness of the scatterer; L_R is the *radiation length* characteristic of the scattering material:

$$L_R \propto \frac{A}{NZ(Z + 1)} \log(183Z^{-1/3})$$

where N is Avogadro's number, Z is the atomic number and A the atomic weight of the target material.

The *scattering power* is the mean square scattering angle per unit thickness of an absorber traversed by the charged particles. It is analogous to the concept of stopping power (see Section 3.5).

The dependence of the stopping power and the scattering power on Z is the basis of the choice of materials for beam shaping: in practice high-Z materials are used to produce strong scattering minimising energy losses (e.g. lead scatterers to increase the beam width), and low-Z materials degrade the energy of the beam while minimising the scattering (e.g. graphite, PMMA, or water degraders to adapt the beam range to a given target). Combinations of these materials are used to achieve specific results: different beam attenuation with constant scattering (as in modulator wheels, see Section 46.3.1.1), different scattering with constant energy loss (as in scattering systems with complex shapes).

The dependence of multiple scattering on an inverse power of the proton (kinetic) energy explains the broadening of a proton (pencil) beam near the end of the range (Figure 46.2).

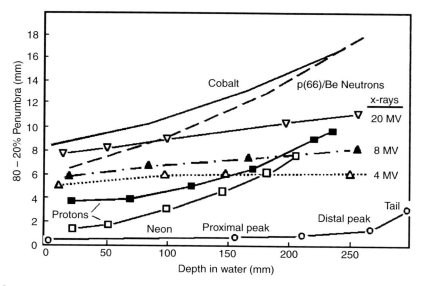

FIGURE 46.2

Lateral penumbra as a function of depth in water for beams of ^{60}Co γ-rays, neutrons, 4, 8 and 20 MV x-rays, protons and neon ions. Proton and photon penumbras are rather similar at large depth. Clinical light ion beams have intermediate penumbras between those presented for proton and neon beams. (From Schreuder, N. Personal communication; Mazal 1996a; Chu, W. T., Ludewigt, B. A., and Renner, T. R., *Rev. Sci. Instrum.*, 64, 2055–2122, 1993.)

For *small* proton beams (e.g. diameter 10 mm) with energy of the order of 100 MeV, the multiple scattering at the end of the range leads to the disappearance of the Bragg peak (Figure 46.9b) as a consequence of a lack of *lateral scattering equilibrium*: i.e. protons scattered *away* from the beam axis are not *replaced* by protons that would be scattered *towards* the beam axis as in the case of a *broad* beam (Vatnitsky et al. 1999a). Complex heterogeneities in the human body combined with this multiple scattering are the cause of beam degradation in clinical applications, even for broad beams (Urie et al. 1991; see Figure 46.9c).

46.3 TECHNOLOGICAL BASIS OF PROTON THERAPY

The principal components of a proton therapy facility are the accelerator with its associated systems, the beam transport system into one or more rooms, the shielding, the beam-shaping devices, the patient-positioning system and the control system.

46.3.1 TYPES OF PARTICLE ACCELERATOR

Accelerators producing proton beams for medical use (which generally have energies between 200 MeV and 250 MeV and beam intensities of several tens or hundreds of nanoamps) are commonplace in the fields of atomic and nuclear physics research. A review of the current accelerators and projected new systems for proton therapy, including economic and strategic considerations, has been made by Schwartz et al. (1995). Here we present the operating principles of these accelerators.

Static accelerators (van der Graaf, Tandems, etc.) are unable to produce sufficiently high energies. On the other hand, *linear* proton accelerators are able to provide an adequate beam, with a principle similar to the electron accelerators already widely used in medical applications (Chapter 11). These have been several proposals (Amaldi and Silari 1994) to build a proton linear accelerator for medical use (Figure 46.3). However, with current technology and considering the cost-performance ratio, *circular* accelerators represent a better solution.

46.3.2 THE CYCLOTRON

Figure 46.4 presents a schematic diagram of a cyclotron. The *ion source* injects the protons into the centre of the machine (e.g. plasma is produced by an electric arc from a tungsten filament in hydrogen gas). Two hollow semicircular electrodes, called 'Dees' because of their shape, are excited to a high-frequency, alternating high voltage. When a proton is in the electric field between the two electrodes, the sector of negative polarity exerts a force of attraction and, therefore, acceleration. A magnetic circuit and a set of coils

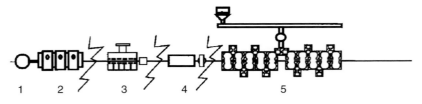

FIGURE 46.3
The principle of the proton linear accelerator: (1) source; (2) RFQ (radiofrequency quadrupole) for first bunching and acceleration; (3) associated cavities for acceleration at medium energies; (4) beam bending and focusing; (5) last series of tanks with accelerating cavities for high energy. (Modified from Benincasa, G. et al., *The TERA Project: The RITA Network and the Design of Compact Proton Accelerators*, Vol. 2, Part 1, Amaldi, U., Gandalfo, M., and Picardi, L., Eds., INFN Divisione Ricerca, Frascati, 1995.)

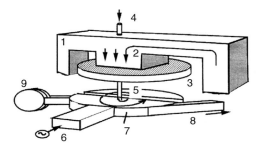

FIGURE 46.4
The principle of the cyclotron: (1) magnetic circuit; (2) magnetic field; (3) coils; (4) gas injection; (5) ion source; (6) high-frequency high voltage; (7) accelerating electrodes (Dees); (8) extraction; (9) vacuum systems.

are used to create a strong magnetic field *perpendicular* to the trajectory of the particle (e.g. 1–2 T). When the proton re-enters the cavity of the electrodes, no electric force acts on it. The magnetic field B bends the particle of charge q moving with a speed v into a circle of radius r. It emerges again into the space between the *Dees* exactly when the electric field has changed its direction, being accelerated again and going on a circular trajectory with a larger radius, and so on.

The operating principle of the cyclotron is thus expressed by equating the centrifugal force on a particle of mass M at a distance r to the force exerted on it by the magnetic field:

$$q\mathrm{v}B = \frac{M\mathrm{v}^2}{r}, \quad \text{or} \quad qBr = M\mathrm{v} \tag{46.5}$$

The charge, the magnetic field and the mass are constant, and therefore an increase in velocity corresponds to an increase of the radius of the trajectory, with an angular velocity ω that depends only on the magnetic field and the mass and charge of the particle:

$$\omega = \frac{\mathrm{v}}{r} = \frac{qB}{M} \tag{46.6}$$

The accelerating electric field between the *Dees* remains at this constant frequency, staying in phase with the passage of the particles. There are various other considerations in addition to the above guiding principles of operation. Among the principal ones are the problems of stability of phase, focusing and extraction. The magnetic field tends to decrease near the edges, curving the lines of force and imparting radial components to the field. This results in focusing forces that keep the particles oscillating around the central plane; the magnetic field must decrease with radius to keep these oscillations bounded.

The electric field between the *Dees* not only accelerates the protons, but also focuses and defocuses the beam with a net effect depending on the phase of the particle. To maintain synchronisation, the particle must arrive at the accelerating gap when the voltage is increasing, during a vertical defocusing stage. The beam diameter increases with radius at first, and after half the final radius the magnetic forces focus the beam again. Since the field must decrease with radius to keep the beam focused, there is also a change in the equilibrium between the centrifugal and the magnetic forces. If the field decreases more slowly than r^{-1}, the net effect is radial oscillations with a final focusing effect. A detailed analysis of these effects lies outside the scope of this chapter and can be studied in specialist references (e.g. Rosenblat 1968).

46.3.3 THE SYNCHROCYCLOTRON

When the particle energy is such that relativistic effects cause a change in the mass of the accelerated protons, the frequency of the electric field between the Dees of a cyclotron can no longer be constant (Equation 46.6). As the magnetic field in a cyclotron must decrease with radius to provide vertical focusing of the beam, this also affects the frequency of revolution of the particles. To maintain synchronicity between the electric field and the passage of the particles, a modulation of the frequency is necessary (reducing the frequency when the mass of the proton increases with energy at large radius, thus requiring a longer time to move through the Dee). This principle is applied in the synchrocyclotron. The frequency variation is obtained by using a variable capacitor in the high-frequency circuit. In practice this is achieved by a *rotary condenser*, which is a complex device. The disadvantage of the synchrocyclotron is the production of a pulsed beam instead of a continuous one, but the usual repetition frequencies of the *macropulses* (e.g. 450 Hz) are still well adapted to clinical use of the beam. The proton-therapy facilities at Harvard (Boston), Uppsala and Orsay are examples of the use of synchrocyclotrons in proton therapy.

46.3.4 THE ISOCHRONOUS CYCLOTRON WITH SEPARATE SECTORS

Another way to solve the relativistic effect on the particle mass could be to increase the magnetic field strength B towards the outside of the circular Dees (Equation 46.5 and Equation 46.6). This solution, however, has limitations in a cyclotron, because it is associated with beam defocusing. This, in turn, results in a loss of particles due to collisions with the walls of the vacuum chamber and the magnetic pole.

The required focus can be re-established by using alternating gradients of the magnetic field along the trajectory. The accelerator has several sectors in a spiral (alternation of *hills* and *valleys*—Figure 46.5). At the boundary of these sectors, the magnetic field plays a role equivalent to an optical lens, focusing and defocusing the beam. It is then possible to

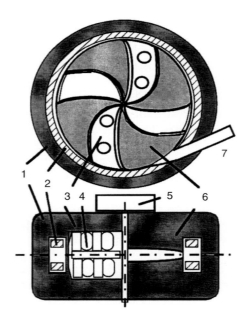

FIGURE 46.5
The isochronous spiral separate sectors cyclotron: (1) magnet yoke; (2) coils (may be superconducting); (3) valley; (4) electrodes, stems and resonators for acceleration in a valley; (5) radiofrequency generator and ion source; (6) magnet "Hill"; (7) extraction (e.g. Boston, Massachusetts).

increase the magnetic field with increasing radius by bringing the pole tips closer together (to compensate for the increase in mass of the particles) without losing focus, and thereby maintain a constant frequency for the accelerating potential (and therefore a continuous beam). Such machines are known as *isochronous cyclotrons*. The acceleration is provided by resonating radio-frequency cavities connected to the accelerating electrodes in the *valleys*. Examples of such an accelerator for proton therapy are the ones developed by *Ion Beam Applications* (IBA, Belgium) for centres in Boston, China, Korea and a superconducting version developed by *Accel* (Germany) for centres in Munich and Switzerland.

46.3.5 THE SYNCHROTRON

The diameter of a cyclotron increases with the energy and, in consequence, so does the weight and the cost of the magnetic circuit and other accessories. If we consider only the final trajectory of the particle, Equation 46.5 shows that the speed of the particle can be increased (taking also into account the change in mass) at constant radius r, with a variable magnetic field B in the annular region occupied by the acceleration chamber. This is the principle of operation of a synchrotron (Figure 46.6). The particles are accelerated up to several MeV before being injected into the ring. At each revolution, a high-frequency cavity produces an acceleration synchronised with the angular frequency of the particles. The deflecting dipoles modify the magnetic field to maintain the trajectory of the protons at a constant diameter, until they reach the desired energy. This type of accelerator thus makes it possible to produce a beam with variable energy, with the disadvantage that one obtains a pulsed beam of low pulse-repetition frequency. Typical acceleration cycles are between one and three seconds. Examples of synchrotrons developed for proton therapy are those manu- factured by *Optivus* for Loma Linda, in California (Slater et al. 1992) and by *Mitsubishi* and *Hitachi* for centres in Japan and the U.S.A. *Siemens* and other companies propose using a synchrotron for carbon-ion therapy (see Chapter 49) that can also produce protons.

46.3.6 NEW DEVELOPMENTS

New approaches to the acceleration and the transport of particles in the range of interest for therapy are being studied. High-intensity lasers (several hundred TW), which can deliver ultra-short pulses (approximately 50 femtoseconds at a wavelength of 0.82 µm, with an energy of 1 J and repetition rate of 10 Hz), will ionise a solid or gas target, creating a plasma state.

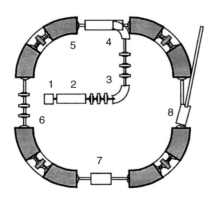

FIGURE 46.6
Principle of the synchrotron: (1) ion source; (2) RFQ; (3) bending dipoles and focusing quadrupoles; (4) injection; (5) bending dipoles with variable magnetic field; (6) linear section, focusing; (7) radiofrequency cavity for accelera- tion; (8) extraction. (e.g. Loma Linda, California).

Subsequent interaction of the pulse with the created plasma leads to generation of longitudinal electric fields that ultimately accelerate the charged particles. The generated electric fields can be greater than several hundreds of GV/m, arising from either electron expulsion from the target (in the case of solid targets) or excitation of the plasma waves (in the case of gas targets) induced by the light pressure of the high-intensity laser. These fields can be used to accelerate electrons and protons up to the energies required for clinical applications in a few millimeters (Malka et al. 2002). Fourkal et al. (2002) presented simulations showing that, under optimal interaction conditions, protons could be accelerated up to energies of 300 MeV by a petawatt laser field. Protons at energies between 10 and 50 MeV have been produced experimentally, but with very low repetition frequencies. Increasing the laser intensity, frequency and focalisation, as well as optimising the target design, have been proposed as solutions for developing compact devices for the next generation of particle accelerators. Other aspects like the beam energy spectra (Fourkal et al. 2003), intensity and stability need further improvements.

Recently, a new idea for accelerating protons and heavier ions (like carbon) has emerged, the so-called "non-scaling fixed field alternating gradient" accelerator (Keil et al. 2006). This combines the fixed magnetic field, high current capability and rapid repetition rate of the cyclotron with variable energy extraction of the synchrotron. A major research programme is underway to investigate whether this technology is feasible (http://gow.epsrc.ac.uk/ViewGrant.aspx?GrantRef=EP/E032869/1).

The use of plasma techniques or the transport of the beam by using a magnetic "guide" along a superconductor can be considered as promising ideas but their practical feasibility remains to be proven.

An intriguing idea is that of using a beam of *antiprotons*; as a result of antiproton-proton annihilation the height of the Bragg Peak is approximately double that of a proton beam of the same energy and entrance dose level (Hozscheiter et al. 2006; Hall 2006a).

46.3.7 General Technical Design Considerations

The extraction of the beam can be achieved by various methods. In a synchrocyclotron, electromagnetic channels cancel the main magnetic field and let the particles come out in a straight line. Electric deflection is often used in conventional cyclotrons. An electromagnet *kicker* can be used, particularly for synchrotrons[*].

To avoid disturbing the acceleration of the beam by absorption or scattering, a high vacuum (e.g. 10^{-6} mbar) is maintained in the accelerating chamber. This vacuum must also be maintained along the path of the beam[†].

Magnetic fields are limited in strength by the saturation of the iron in the magnetic circuit and by the current necessary in the coils. Superconducting coils and ironless magnetic circuits have been proposed to solve these problems.

Vacuum and cryogenic systems are critical parts which determine the capacity of a facility to return to clinical operation after a breakdown or maintenance. This is of paramount importance for medical applications.

46.3.8 Beam Transport

It is common practice for accelerator facilities devoted to physics research to have several beam lines. This is because of the cost and the size of accelerators, the multiple users involved in

---

[*] When negative ions are being accelerated a different approach can be used. It is possible to *strip off* the electrons from the ions with a thin target once they have reached the desired energy. The resulting positive ions will emerge from the accelerating chamber with a curved trajectory opposite to that of the orbit of the negative ions.
[†] For negative ions, the vacuum requirements are more severe in order to minimise interactions.

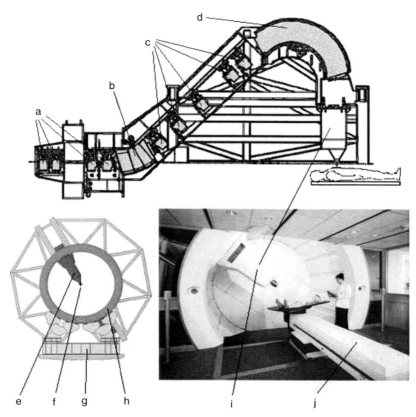

FIGURE 46.7
Isocentric rotational gantry for proton beam delivery. (a) and (c) quadrupole magnets; (b) 45° dipole magnet; (d) 135° dipole magnet (e) and (i) nozzle (see Figure 46.8); (f) isocenter; (g) gantry support; (h) rotating gantry structure; (j) patient positioner. (Modified from the original drawing, courtesy of IBA, Belgium.)

separate research programmes, the complex systems required for beam transport, measurement and focusing the beam, as well as the radiation-protection requirements.

For medical use, the diversity of the *clinical targets* leads to a requirement for different technical specifications for each application, and therefore to a need for specialised lines and rooms. Fast switching of the beam to different rooms can optimise the use of an installation, as the time for patient set-up is usually longer than the *beam-on* time. The beam must therefore be transported after extraction into several different treatment rooms along evacuated conduits. Electric and magnetic systems are used to modify the trajectory (e.g. dipoles) and to focus the beam in a similar way to optical lenses (e.g. quadrupoles). A particular case of optimisation of the beam-transport system, which acts as a constraint on the present development of proton therapy, is the design of compact and isocentric gantries (Figure 46.7), similar to those in use with electron linear accelerators (see Chapter 11), which allow the beam direction to be changed without moving the patient. The basic concepts related to beam delivery and gantries for proton therapy can be found in the following references: Coutrakon et al. (1991), Pedroni (1994), Renner et al. (1994), and Pedroni et al. (1995, 1999).

46.3.9 Radiation Protection and Neutron Shielding

The use of proton beams for radiotherapy requires specific approaches for radiation protection (Sullivan 1992). A review was carried out by Brassart et al. (1995) for the proton and neutron therapy centre in Nice. Shielding represents a significant component in the total cost of an installation. Whilst the primary beam of protons is easy to stop, the secondary particles resulting from the nuclear interactions, in particular the neutron yield, require a considerable thickness of material containing a high proportion of hydrogen to provide enough attenuation (Alsmiller et al. 1975; Tesch 1985; Hagan et al. 1988; Siebers et al. 1992; Mazal et al. 1997). Concrete is the most common material used for shielding as it represents a good compromise between attenuation, mechanical characteristics and cost.

In practice, the attenuation of the *equivalent dose* (see Section 59.1.2) is expressed by:

$$H(t) = H_0 \frac{\exp(-t/d)}{r^2} \qquad (46.7)$$

where $H(t)$ is the equivalent dose behind a thickness t of shielding, H_0 is the equivalent dose at 1 m with no shielding (or extrapolated to zero shielding thickness) and is known as the 'source term', d is the attenuation length for equivalent dose through the shield and r is the distance from the source to the surface, or point of interest beyond the shielding material.

For protons with energies in the range for clinical applications (up to 200 MeV) on targets such as water and aluminum, the attenuation length decreases from 1000 kg m^{-2} to 500 kg m^{-2} for angles from 0° to 90° relative to the beam direction. The source term H_0 is dependent on the angle to the beam direction; different values have been published (Siebers et al. 1992; Mazal et al. 1997) for the source term ranging from 1.3 to 10 at 0° and of the order of unity at 90° for the same kind of beam and targets. In practice proton-therapy facilities have concrete walls exceeding between 1.5 m and 2 m in thickness and up to 5 m in critical areas.

The *raw* protection of the patient (i.e. the dose outside the treatment field) must also be evaluated (Binns and Hough 1997), especially when the degradation of the beam energy is performed in the treatment room. Both the protons and the secondary particles *activate* the hardware close to the beam (Antoni 1996). Consideration to this must be given at each interface in the accelerator and beam lines to guarantee the protection of personnel (in particular from γ- and β-rays). The accelerating chamber, the transport system and colli- mators, whether made of brass or low-melting-point alloys (containing Bi, Pb, Sn, etc.), are examples of components with high levels of activation after the beam has been running. Some of these are often handled a short time after having been exposed to the beam, with the risk of irradiation or toxic ingestion of radioactivity. The activation in some materials decreases rapidly (of the order of ten minutes) immediately after the irradiation. However, at relatively long times after irradiation, brass contains isotopes of Co, Mn, Fe and Zn, while the standard shielding alloy, Cerrobend, contains a broad spectrum going up to Tl, Bi and, more critically, Po. The activation of air and of water is also a problem to be taken into account: air-ventilation systems and filters are necessary in all the rooms, as well as the use of closed cooling circuits for the accelerator. The possible contamination of the ground and the groundwater must be avoided or maintained within acceptable limits following inter- national and local regulations.

46.3.10 BEAM SHAPING: TOOLS AND DOSIMETRY

Figure 46.8 presents the principle of a beam line, at the end of the transport system. Various devices are necessary for beam shaping and dosimetry, and for adapting the dosimetric characteristics of the beam to the particular volume for each angle of incidence. They have been described in detail by Chu et al. (1993).

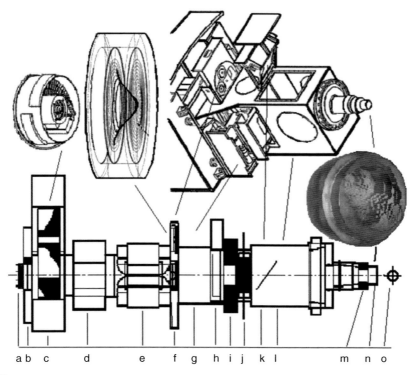

a b c   d   e   f g   h i j k l   m   n o

FIGURE 46.8
Typical passive and dynamic beam-shaping devices for proton therapy in a fixed beam line or in a "nozzle" of a gantry system: (a) vacuum window and first ion chamber (for monitoring of centering, beam intensity etc.) and/or SEM (secondary emission monitor); (b) first scattering foil; (c) modulator wheels; (d) first scanning magnet; (e) second scanning magnet; (f) second scatterers; (g) on axis, remote controlled x-ray tube for set-up verification (or film holder if tube opposite to beam direction); (h) and (i) collimator jaws, range verification devices; (j) second ion chamber (centering, flatness, monitor units); (k) light field for beam simulation; (l) snout holder; (m) patient aperture (or multi-leaf collimator); (n) patient compensator; (o) isocenter. Other components may be present like absorbers, intermediate antiscatter collimators, multiwire chambers. (Modified from original drawings from Harald Paganetti (personal communication) and IBA, Belgium.)

46.3.11 DEPTH–DOSE CURVES: RANGE AND MODULATION

The *raw* or *pristine* Bragg peak, i.e. that due to a monoenergetic beam (see Figure 46.1) must be adapted for clinical use. The maximum depth of the target volume determines the range necessary, and consequently the energy of the beam. This range can be changed either by using an absorber, or by a change of the energy produced in the case of a variable-energy accelerator.

The *raw* peak is too narrow for clinical applications. One solution is to superimpose, during the irradiation, a certain number of peaks placed at various depths with a *modulator* (Koehler et al. 1975) to ensure a homogeneous coverage of the target volume (Figure 46.9). With this approach, the dose at the surface will be increased relative to that of the resulting *spread out* Bragg peak (SOBP). This is one of the principal limitations of the use of a single beam of protons in therapy; the skin dose could be higher than that caused by a high-energy photon beam (see Chapter 22).

The modulation is produced by interposing in the beam a wheel with sectors of variable thickness. The angular opening of each sector is calculated to obtain a plateau over the desired range of depths. An almost homogeneous dose in this SOBP can be obtained independently of the original quality of the beam. However, the surface dose and especially the distal slope of the depth-dose curve are dependent on the energy dispersion of the original beam.

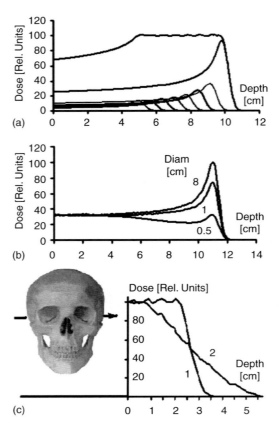

FIGURE 46.9
Modifications of proton depth–dose curves: (a) modulated or spread out Bragg peak (SOBP) as a result of a weighted sum of single energy peaks of different energies using a range modulator wheel, ridge filters or dynamic techniques; (b) single-energy depth doses for small beam sizes showing the effect of the lack of lateral proton equilibrium; (c) effect of inhomogeneities on the distal fall off of a SOBP: (1) after traversing a homogeneous phantom; (2) after the complex inhomogeneities in the base of the skull. (Figure (a) courtesy H. Kooy, Boston; Figure (b) courtesy C. Desblancs, Orsay; Figure (c) modified from Urie, M., Goitein, M., Holley, W. R., and Chen, G. T. Y., *Phys. Med. Biol.*, 31, 1–15, 1986.)

46.3.11.1   Lateral Dose Profiles and Penumbra

The proton beam transported from the extraction point of the accelerator to the treatment room generally has a narrow Gaussian profile. A larger beam size with the required homogeneity for clinical applications can be obtained by single or double *passive scatterers*, or by sweeping or scanning the beam (Koehler et al. 1977; Grusell et al. 1994b; Nauraye et al. 1995). The lateral penumbra at the entrance of the medium is strongly dependent on the characteristics and the localisation of the elements interposed in the beam, as well as on the position of the final collimator. At depth, this penumbra is largely determined by the multiple scattering in the medium. Figure 46.2 shows the variation of the penumbra width with depth under clinical conditions. Experimental results such as these are used in the development of computational models for treatment planning (Rosenwald et al. 1994).

46.3.11.2   The Shaping of the Dose Distribution around a Target Volume

The shaping of the dose distribution to conform to a target volume is achieved using various devices.

The dose to critical normal tissue is limited laterally by customised apertures. Close to the aperture, strong scattering creates *horns* in the profiles (Gottschalk et al. 1993). Monte Carlo simulations (van Luijk et al. 2001) show that protons that are incident on the entrance face of the collimator and leave it through the aperture wall make the largest contribution to scatter. There will be some very low-energy protons in this scatter component of the incident *beam*, and at very low energies the Relative Biological Effectiveness (RBE) (see Chapter 7) will be considerably greater than unity. However, by analyzing the energy distribution of protons at different positions in the patient (or a phantom) it has been shown that the net effect of such low-energy protons is to increase the normal-tissue damage probability by no more than 1% at worst. The width of the penumbra at the entrance surface is largely determined by scatter in air.

In the beam direction, isodoses are shaped by a 'compensator' (Wagner 1982; Urie et al. 1984; Sisterson et al. 1989; Petti 1997). The compensator causes each *ray* inside the beam to have a different range, adapted to the distal depth of the target volume in the direction of the ray. *Smearing* tools are used to modify ideally calculated compensators to take into account the risk of patient- and organ-intrafraction movements, as well as the consequences of multiple scattering (Urie et al. 1984).

46.3.11.3 Dynamic Techniques of Beam Shaping

Beam shaping can also be done dynamically (Kawachi et al. 1983; Scheib et al. 1994; Pedroni et al. 1995). Sweeping a small pencil beam in position, while varying its energy and intensity, allows control of the dose distribution without using the "passive" components described in Section 46.3.11.1 and Section 46.3.11.2. Using beam scanning, the dose can be *conformed* not only to the distal end of the target but also to the proximal end, thereby reducing high doses to normal tissue overlying the target volume. Beam scanning can be done continuously or point by point, allowing optimisation of the dose distribution using a limited number of beams without the use of patient- or beam-specific hardware and without attenuating the beam. This minimises the problem of radiation protection of the patient and of personnel. On the other hand, the beam control system and the dosimetry are more complex (Boon et al. 1998). In addition, as only a small part of the target volume is irradiated at any given instant with a high dose rate, this can give rise to additional problems of junctions and risks of possible movement. Various groups have presented the potential advantages of this approach, but its clinical use is still very limited with proton beams[*].

The use of dynamic techniques is also the basis of the development of optimisation techniques with intensity-modulated proton beams (IMPT) (Trofimov and Bortfeld 2003, see Figure 46.13c) following a similar approach to intensity-modulated photon beams (IMRT, see Chapter 43) but having an additional degree of freedom, because of the finite range of the charged particles. This results in a reduction of the integral dose to healthy tissues.

Fully automated 3D intensity-modulated proton therapy (IMPT) with a spot-scanning system in a compact gantry is employed at the Paul Scherrer Institute, Switzerland, for clinical applications such as base of the skull and thoracic chordomas and chondrosarcomas (Pedroni et al. 1999; Lomax et al 2001).

---

[*] Dynamic techniques are usually used in preference to passive systems when using heavier ions (Section 49.3.3) to minimise beam fragmentation.

46.4 TREATMENT IMPLEMENTATION

The clinical exploitation of the potential dosimetric advantages of proton beams is only possible if a full set of precision tools is available for the preparation and the delivery of each treatment. Four aspects will be considered:

1. Treatment planning software
2. Patient set-up aids
3. Dosimetry, including radiobiological considerations
4. Quality assurance for the whole procedure

46.4.1 TREATMENT PLANNING SYSTEMS

46.4.1.1 Calculation Models

The calculation of the dose distribution in the patient can be done by several methods: the three best known are ray tracing, the pencil beam and the Monte Carlo technique (see Part F). A short description of these techniques follows, based on the assumption that high resolution CT images of the patient anatomy and target are available. The planning process has been described by Bussiere and Adams (2003).

Ray Tracing. Tracing a straight line between the source and each calculation point makes possible the calculation of *radiological depths* into the body, taking into account heterogeneities (Lee et al. 1993; Belshi 1995; Schreuder et al. 1999; Russell et al. 2000). These depths are used to estimate the dose from *Bragg curves* stored in a beam library or represented by analytical expressions. However, this does not explicitly model the effect of multiple scattering. The lateral penumbra can be calculated with models similar to those used for photon and electron beams (Dutreix and Briot 1985; Rosenwald and Drouard 1991; Rosenwald et al. 1994). Parameters having a major influence on penumbra broadening are the depth (Figure 46.2), the distance from the final collimator, the presence of beam modifiers such as compensators (Urie et al. 1984), the air gap between the compensator and the patient surface (Sisterson et al. 1989), the characteristics and the position of beam-shaping devices and the original beam divergence. In spite of its limitations, this model is still used for fast calculations in clinical applications.

Pencil Beams. The multiple scattering of protons can be approximately taken into account by the modelling based on narrow beams with Gaussian profiles (see Chapter 27; Petti 1992, 1996, 1997; Lee et al. 1993; Russell et al. 1995, 2000; Hong et al. 1996; Carlsson et al. 1997; Deasy 1998; Schaffner et al. 1999; Szymanowski et al. 2001). Superposition (or convolution) of these elementary beams gives the distribution of dose for beams of greater cross-sectional area and irregular shape. The pencil-beam approach is a clear improvement over ray-tracing, and yields similar results to Monte-Carlo simulation (see Chapter 28) in the majority of clinical situations.

Monte Carlo Method. The Monte-Carlo technique of modeling the elementary interactions and then combining the "histories" of a large number of particles leads to more accurate calculations of the dose distribution in the presence of heterogeneities (Chapter 28; Petti 1996; Kohno et al. 2003; Paganetti et al. 2005). However, it requires powerful computational capabilities, a good knowledge of the proton interaction coefficients and some means of assigning tissue-composition data to the patient images acquired from the usual methods like CT scanning (see Chapter 5 and Chapter 28).

46.4.1.2 The Effect of Heterogeneities

When a proton beam penetrates the body, the heterogeneities (bone, air cavities, lung etc.) modify the dose distribution significantly compared to a homogeneous medium such as water (Goitein 1978; Urie et al. 1984, 1986; Zink 1987; Oozeer et al. 1993). This is due to the dependence on density, and to a lesser extent on atomic number, of the proton stopping power, and hence range, and the degree of lateral scattering. Two types of heterogeneities can be identified: *simple* and *complex*. A heterogeneity is *simple* when it can be modelled by taking into account its global density using a *radiological depth* equivalent to water, which involves a simple modification to the range of the beam, with no deformation of the distal edge, or the flat modulated. The upper part of the skull can, under certain conditions, be modelled using this principle. A heterogeneity is *complex* when its form, its size or its composition implies large variations in depth in contiguous rays, so the amount of multiple scattering is affected and, in consequence, so is the depth–dose curve. In clinical practice, beams are not planned in directions where a sharp heterogeneity must be traversed. The compensation of heterogeneities by the use of bolus or compensators must include the effect of uncertainties in the evaluation of the planned dose distribution. *Expanding* and *smearing* routines are applied when calculating compensators to ensure that the target volume can be covered by a specified dose given prespecified repositioning and motion uncertainties, taking into account the heterogeneities (Urie et al. 1984).

A stoichiometric (*stoicheion*: element, *metron*: measure) method has been described by Schneider et al. (1996) to improve the calibration of CT scans for treatment planning with protons taking into account the anatomical heterogeneities (see also Section 39.2.1). The relationship between Hounsfield CT units and proton stopping powers (or electron densities for photon beams) is based on measurements using tissue-equivalent materials with known composition.

46.4.1.3   Tools to Assist in the Visualisation and the Evaluation of the Treatment

The practical implementation of proton therapy requires some advanced treatment planning tools, most of which are common to any conformal technique of high-precision radiotherapy (see Part G; Zink 1987). Some examples are:

1. The determination of the various volumes of interest (i.e. the target volume and critical organs) requires tools to correlate images and contours from different imaging modalities (e.g. CT, MRI, and ultrasound), but the dose calculations are based on CT data.

2. The visualisation of the spatial orientation of structures (e.g. 'beam eye's view') and specific functions such as the automatic determination of the shapes of complex fields.

3. The use of 3D volume expansion algorithms and automatic margins (Belshi 1995).

4. The calculation of the dimensions of beam-shaping devices (e.g. compensators and filters), with the necessary documentation and interfaces for their fabrication and their quality control.

5. The calculation or measurement of dose-delivery parameters (Monitor Units). Many centres still calibrate each treatment field prior to treatment, or use their own database and/or empirical analytical fits to predict output factors. Kooy et al. (2003) have published a predictive model. It is based on (a) the inverse square of the distance between the entrance dose at the patient surface and the monitor chamber reading at the beam line; (b) theoretical derivation of the ratio between entrance dose and dose at the center of the SOBP (Bortfeld and Schlegel 1996); (c) experimental correction factors and analytical fitting to experimental data to correct for specific changes in the beam line (e.g. scatterers and modulators) for different irradiation conditions.

6. The calculation of dose distributions in easily reproducible conditions (e.g. with a flat surface) for quality control purposes.

7. Presentation of the calculated doses in various formats: isodoses superimposed on anatomical images, dose–volume histograms, doses on anatomical surfaces (e.g. observer's eye view).

8. Systematic analysis of uncertainties associated with collimator–compensator–patient misa-lignment, patient movements, beam range, etc. Among the suggested solutions, is the calculation of the distributions of nominal, maximum, and minimum doses taking into account these uncertainties (Goitein 1985; Urie et al. 1991).

9. Tools for comparison and for optimisation of treatment plans (Zink 1987; Brahme et al. 1989; Hoffmann et al. 2006) using dosimetric or *biological* figures of merit (see Chapter 36).

46.4.1.4 *Comparison with Photon Dose Distributions*

Comparative treatment planning has been extensively used to investigate potential improvements in the treatment of different clinical targets with protons, compared with photon beams, either with conventional conformal approaches or with intensity modulation. Lomax and coworkers amongst others (e.g. Lee et al. 1994; Isacsson et al. 1998; Johansson et al. 2002) have performed such studies for varying indications and lesion sites (Lomax et al. 1999) including paranasal sinus (Lomax et al. 2003a) and breast and regional nodes (Lomax et al. 2003b). In general, better conformation of dose distributions around the target volume can be achieved with protons or, when conformation of the high-dose regions is equivalent, better homogeneity in the dose distribution in the target and a significantly lower *integral dose* in healthy tissues is achieved with protons (Lee et al. 1994).

46.4.2 PATIENT SETUP

The precision and the reproducibility of the patient set-up required for proton therapy are of the order of millimeters, or even better for certain applications like ophthalmology (Verhey et al. 1982; Bonnett et al. 1993; Meyroneinc et al. 2001). To achieve this, attention must be paid to at least four aspects:

1. *The treatment planning software* must provide the specification of position needed to guarantee that the patient is set up according to the treatment plan. This can be achieved with digitally reconstructed radiographs (DRR see Section 32.4.5) but, in practice, anatomical reference marks or external fiducials having a fixed link with the target volume are also used. Markers may also be inserted into internal structures, such as tantalum clips sutured on the sclera and steel or gold marker seeds implanted in the skull.

2. *The patient immobilisation* must guarantee almost perfect reproducibility between the acqui-sition of the CT scan and each treatment session and minimise intrafraction movements. Present systems are based on masks, foam moulds, carbon fibre supports, orthopaedic devices and invasive or noninvasive stereotactic frames.

3. The *system of verification of patient positioning* must ensure that the reference fiducials are identified at each treatment session. Radiological systems installed in the treatment room must facilitate fast and reliable identification of the anatomical structures or reference marks. By establishing their position in space, relative to the isocentre of the beams or the field boundaries and comparing this position with the planned positions, the corrections necessary for patient repositioning can be calculated. Optical methods are also used based on video cameras and reflective reference marks or using real-time digital stereophotogrammetry tech-niques (Jones et al. 1994). The activation in phantoms of positron-emitting radionuclides induced by proton irradiation has been measured with on-line PET by Hishikawa et al. (2002) and Parodi et al. (2002), and correctly predicted by calculation. Thus in situ proton therapy monitoring by means of in-beam PET may be feasible, as already clinically implemented for the monitoring of carbon ion therapy at GSI Darmstadt (see Chapter 49).

4. *The patient positioning system* must enable corrections to the patient's position to be carried out with high precision. Present solutions are sometimes based on robotics concepts (Figure 46.10). These systems allow linear and angular corrections, in general with six degrees of freedom and accuracy better than 1 mm and 0.5°.

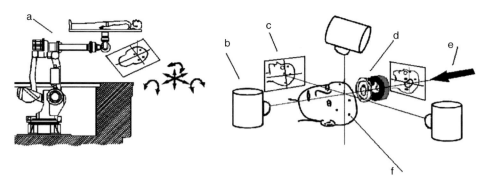

FIGURE 46.10
Patient set up: (a) robotic patient positioner; (b) x-ray tube; (c) films or electronic detector; (d) patient aperture and collimator; (e) proton beam; (f) implanted fiducial clips.

Such a level of precision and reliability can only be assured by a comprehensive quality assurance programme (Mazal et al. 1996b).

46.4.3 DOSIMETRY

Proton-beam dosimetry is carried out using several different detector systems: ionisation chambers, Faraday cups, calorimeters and semiconductors (Figure 46.11). It has many similarities with the dosimetry of electrons and photons, which is described in Part D.

The *ionisation chamber*, generally calibrated in a ^{60}Co gamma-ray beam, is well adapted for routine measurements and is also considered the reference for intercomparisons and calibration of a proton beam (see Chapter 15 and especially Section 18.7).

The *calorimeter* measures the energy deposited in an absorber by the associated temperature increase in the core of the detector. Sensitive thermistors are inserted in the core and connected to a measuring Wheatstone bridge and DC amplifiers to measure temperature rises of the order of several milli–Kelvin. The core is thermally insulated from the room temperature by means of a vacuum and different layers of 'equilibrium shells' or jackets which have electrical circuits to measure and maintain a stable temperature. Calorimeters in graphite (Palmans et al. 2004), tissue-equivalent material or distilled water have been built and used for proton and neutron beam calibrations, usually working in quasi-adiabatic mode. The total energy E imparted to the core is given by:

$$E = TC(1 - h)^{-1}$$

where T is the temperature increase, C is the total heat capacity of the core, and h the heat defect (the fraction of absorbed energy not converted into a temperature rise in the absorber). The heat defect is about 0.01–0.03 for tissue-equivalent calorimeters in proton beams used in clinical applications, and somewhat lower for water calorimeters, and in both cases increases slightly with increasing LET (Brede et al. 1997). Ohmic heating elements are also inserted in the core to allow electrical calibration of the calorimeter.

The main factor introducing uncertainties in the absolute measurement of absorbed dose is the heat defect. The experimental conditions for the use of a calorimeter are critical and require that the detector be in thermal equilibrium in the room. Consequently the initial time to achieve thermal equilibrium, together with the time for data acquisition and then to re-establish equilibrium after each irradiation, is long, limiting the use of calorimetry for routine calibrations of a clinical beam.

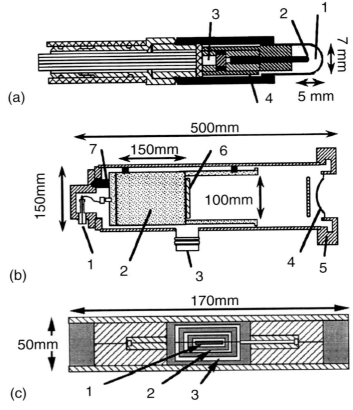

FIGURE 46.11
Detectors used for proton dosimetry: (a) Ionisation chamber: (1) gas cavity; (2) central electrode; (3) signal conductor; (4) vent to the atmosphere. (b) Faraday cup: (1) signal conductor; (2) absorber (e.g. copper, graphite); (3) vacuum; (4) entrance window; (5) steel cup; (6) permanent magnet; (7) insulators. (c) Tissue equivalent calorimeter: (1) core; (2) inner jacket; (3) outer jacket; (4) vacuum chamber.

The *Faraday cup* measures either the beam current directly or, by integration, a charge Q. Then the fluence, Φ, is given by $\Phi = Q/Ae$, where A is the beam area and e the electronic charge. It follows then (see Section 6.6) that the dose at the surface is

$$D_{med} = \Phi \left(\frac{S(E_z)}{\rho} \right)_{col,med}$$

where Φ refers to the fluence in air and $((S(E_z))/\rho)_{col,med}$ is the mass collision stopping power for the medium evaluated at (mean) energy E_z. The stopping power is strongly dependent on the beam energy, and any contamination (e.g. coming from the collimator) gives significant errors in dose calculation. Grusell et al. (1995) proposed the use of uncollimated beams to correlate ionisation chamber and Faraday cup measurement, with an estimated accuracy of 1.6% (1 SD). The Faraday cup method is only of practical use for *monoenergetic* proton beams at the surface of a phantom. However, it has the advantage of being completely independent of other methods.

Because of the uncertainties in these dosimetry systems it is essential to carry out dosimetric intercomparisons between detectors and between centres (Vatnitsky et al. 1996, 1999b; Delacroix et al. 1997) to standardise methods and results.

Several different protocols for absolute dose determination have been proposed: the American 'AAPM protocol' (AAPM 1986) and a 'European' code of practice (Vynckier et al. 1991, 1993) have been the basis for the development of international protocols such as ICRU 59 by the ICRU (ICRU 1998) and TRS 398 (IAEA 2000) by the International Atomic Energy Agency, to determine the absorbed dose in water according to well defined procedures (see Chapter 18). A thimble ionisation chamber is recommended as reference dosimeter, calibrated in a ^{60}Co beam in terms of air kerma or in terms of absorbed dose to water. In the latter case, the general equation is:

$$D_{\mathrm{w}} = MN_{\mathrm{D,w}}k$$

where M is the chamber reading corrected by the usual factors (temperature, pressure, etc.), $N_{\mathrm{D,w}}$ is the chamber calibration factor for absorbed dose to water and k is a factor to correct for differences between the quality of the calibrating beam and that of the user's beam.

Newhauser et al. (2002a, 2002b) applied the ICRU 59 protocol to calibrate proton beam lines for eye treatments and for radiosurgery. Differences of 1–2% were found compared to calibrations with a Faraday cup. Medin et al. (2000) compared the formalism and data in ICRU Report 59 and the IAEA TRS 398 Code of Practice including five different cylindrical ionisation chamber types commonly used in proton beam dosimetry. Differences of the order of 2% were found mainly because of differences in the recommended values of perturbation factors and $(W/e)_{\mathrm{air}}^{\mathrm{proton}}/(W/e)_{\mathrm{air}}^{\mathrm{cobalt-60}}$ (mean energies required to produce an ion pair in the chamber air cavity at the beam qualities of the proton beam and in the reference condition, respectively), depending on the chamber type, beam energy and calibration formalism.

Plane-parallel chambers are well suited to the measurement of depth–doses curves and also for calibrating beams in Gy/monitor unit (MU) (for the analogy with photon-beam therapy see Section 20.1.2 and Section 38.7.3). Palmans et al. (2002) presented experimental data and Monte Carlo calculations on perturbation correction factors for four plane-parallel ionization chamber types in 75 MeV modulated and nonmodulated proton beams. They confirmed, within the experimental uncertainties, that the overall perturbation correction factor for a plane-parallel chamber in a low-energy proton beam is unity, as had been assumed in IAEA TRS-398 and other dosimetry protocols.

Semiconductors are used to make *relative* measurements (depth–dose curves, profiles and isodose distributions) (Koehler 1967), because of their spatial resolution, high sensitivity and low cost. The dependence of the response of solid-state detectors on the particle energy (or LET) are not yet well known. De Angelis et al. (2002) studied the response of high doped Hi-pSi diodes to low-energy proton beams (26.7 and 12 MeV nominal energy). The diode response was linear with dose and the standard deviation of repeated readings was less than 2.5%. After performing corrections to the depth-dose measurements for dose rate and for the water-silicon mass collision stopping-power ratio, the difference, at the Bragg peak, with respect to the reference chamber was about 4%. They ascribed this to inadequate knowledge of the materials in front of the sensitive volume of the diode.

Radiological films are widely used in proton dosimetry, especially for measurements perpendicular to the beam axis, as in general the film response is nonlinear at high ionisation density such as at the end of the range of charged particles (Dutreix et al. 1978). Experimental data and a model for calculating the response of films to heavy charged particle irradiation as a function of particle type and energy have been published by Spielberger et al. (2001, 2002, 2003), based on the particle's track structure. These data enable three-dimensional dose verification to be carried out in ion beams with film. The feasibility of using *radiochromic films* for proton dosimetry (Vatnitsky 1997) for verification of an ophthalmic proton beam (Troja et al. 2000) and for verification of the proton Bragg peak in stereotactic radiosurgery with multiple beams has been studied (Vatnisky et al. 1999a). In this latter study, the authors validated the prescribed dose delivery within ±5%, one standard deviation, by comparing calculated doses with measured values.

Thermoluminescent detectors are also used for proton dose determination. TLD-100 detectors have been used to measure transverse and longitudinal distributions of 62 MeV unmodulated proton beams for a line devoted to eye treatments (Sabini et al. 2002). The analysis of the response of these detectors has elements in common with studies related to dosimetry in space (Olko et al. 2002).

Nichiporov et al. (1995) studied *alanine-based detectors* for their possible use as passive transfer dosimeters for clinical proton beams. Measurements from a number of experiments yielded agreement between the dosimetry standards of the Institute of Theoretical and Experimental Physics (Russia) and the National Institute of Standards and Technology (U.S.) to within 2.5%.

Significant efforts are being made to develop multidimensional detectors to enable the rapid acquisition of "dynamic" beams. A *scintillating screen* (Gd_2O_2S:Tb) mounted at the beam-exit side of a phantom and coupled to a low-noise, charge-coupled device (CCD) camera gives a fast and accurate two-dimensional image of the dose distribution at the screen position, with high spatial resolution (SD 1.3 mm) (Boon et al. 1998). In the Bragg peak, where the ionisation density is high, some quenching of the light output has been observed, though this is described well by existing models.

The dosimetric properties of *Ferrous sulphate gels* (see Section 17.3) irradiated with proton beams and evaluated in three dimensions using MRI have been studied by Bäck et al. (1999). The relative $1/T_1$ at the Bragg peak was 15–20% lower than the corresponding ionisation chamber data for the monoenergetic proton beam. The change in sensitivity with depth was explained in terms of a dependence on LET. BANG polymer gel dosimetry was used for the dosimetric verification of an ophthalmologic 68 MeV proton beam by Heufelder et al. (2003), with similar *quenching* effects of the Bragg maximum due to the high LET.

A 3D 'Magic cube' detector has been developed and used with proton beams (Brusasco et al. 1997; Amerio et al. 2002). It is based on a stack of 12 parallel-plate strip or pixel segmented ionisation chambers interleaved with interchangeable tissue-equivalent slabs. A *Very Large Scale Integration* (VLSI) electronic readout device is connected to a PC-based data acquisition system. Sixty four strips, 4 mm wide and 25 cm long, cover a sensitive area of 25×25 cm^2, while for the pixel chamber 1024 channels are available.

46.4.4   RADIOBIOLOGICAL CONSIDERATIONS

From a radiobiological point of view, the behaviour of protons is very close to that of photons and electrons (see Part B). The various studies carried out on this subject are based on in vitro or in vivo measurements, such as the technique of Withers based on the regeneration of the intestinal crypts in mice. With this technique, Gueulette et al. (1997, 2001) found in vivo RBE values, relative to ^{60}Co gamma radiation, which were slightly larger than the currently accepted value of 1.10 and independent of fractionation for one, three and 10 fractions. In this study, and in several other radiobiological and microdosimetric studies, variations of RBE with depth (Robertson 1994) have been found, with the RBE increasing at the end of the range of the particles. This is consistent with predictions from microdosimetric studies (Cosgrove et al. 1997; Tilly 2002; Tilly et al. 2005).

Paganetti et al. (2002) analysed published RBE values for in vitro and in vivo endpoints. The values for cell survival in vitro indicated a substantial spread between the diverse cell lines. The average value at the middle of the SOBP over all dose levels was approximately 1.2 in vitro and approximately 1.1 in vivo. Both in vitro and in vivo data indicated a statistically significant increase in RBE at lower doses per fraction, this increase being much smaller in vivo, and a measurable increase in RBE over the final few millimeters of the SOBP. The experimental in vivo data indicate that continued use of a generic RBE value of 1.1 (relative to ^{60}Co γ-rays) is reasonable (Raju 1980). Thus, in clinical practice, one commonly encounters the quantity *cobalt Gray equivalent*.

At present, there seems to be too much uncertainty in the RBE value for any one human tissue to propose tissue-specific or fraction-size-specific RBE values (Tilly 2002; Tilly et al. 2005). There are data which indicate enhanced biological effectiveness for low doses of protons of very low energy (< 10 MeV); these findings are also of interest for human interplanetary missions (Yang 1999).

46.5 CLINICAL APPLICATIONS

Details of the clinical basis of proton therapy are beyond the scope of this chapter: the interested reader should consult the extensive literature on this subject (e.g. Habrand et al. 1995, 1996; Loeffler et al. 1997; Krengli et al. 1998; Glimelius et al. 2005; Lundkvist et al. 2005). This includes clinical results and trial protocols for the treatment of skull-base and cervical-spine chordoma and low-grade chondrosarcoma, skull-base meningioma, pituitary tumours, paranasal sinus carcinoma, glioblastoma multiforme, artero-venous malformations, uveal melanoma, macular degeneration, retinoblastoma, thoracic spine-sacrum tumours, prostate carcinoma and other lesions. This section is limited to some technical and dosimetric considerations related to the common uses of proton therapy in treating ophthalmic and intracranial sites as well as radiosurgery or stereotactic radiotherapy.

46.5.1 OPHTHALMIC APPLICATIONS

The principal ophthalmic application is the treatment of uveal melanomas (Gragoudas et al. 1978; Austin-Seymour 1985; Pignol et al. 1994; Courdi et al. 1999; Munzenrider 2001; Desjardins et al. 2003; Damato et al. 2005a, 2005b). Other ophthalmic targets, such as angiomas and haemangiomas (Munzenrider 1994) have also been treated with protons with promising results. Dosimetric studies (Mazal et al. 1998; Adams et al. 1999c) and recent clinical trials have been initiated to treat age-related macular degeneration (Flaxel et al. 2000; Zur et al. 2001; Ciulla et al. 2002). A miscellaneous number of other ocular tumours are also treated with protons, such as retinoblastomas under general anesthesia and metastatic lesions (Munzenrider 1994).

For uveal melanomas, the differential diagnosis and the collection of data are carried out by a clinical examination of the fundus of the eye, eye biometry, A-scan and B-scan ultrasound images of the tumour and angiography. It is also important to assess the potential metastatic dissemination to other parts of the body. For special cases (e.g. invasion of optic nerve), CT or MRI scans are carried out. In some clinics CT or MRI scans are used for all cases.

Radio-opaque clips are sutured onto the sclera around the base of the tumour (rendered visible by transillumination) by the ophthalmic surgeon under general anaesthesia in the operating theatre. The clips (made of tantalum) have a diameter of 2.5 mm and a thickness of 0.5 mm, and will stay in place permanently. They are used for radiological verification of the position and the orientation of the eye at each treatment session (Figure 46.12). The immobilisation system is prepared (a thermoplastic masks with a byte block) and orthogonal x-ray films are taken at the first planning session. The patient is seated on a mechanical chair which enables the position of the eye to be adjusted with a reproducibility better than 1 mm and less than 1°. The information collected at the planning stage allows the reconstruction of the principal anatomical characteristics of the patient in a computer model of the eye. The model is adjusted according to the size of the eye. Based on the clinical data and the clip positions the tumour is drawn onto the eye model. Bondiau et al. (2003) have described a fully automatic technique for 3D reconstruction of a virtual eye from CT slices and merging of the CT slices with fundus retinography. These data can then be used for the simulation, planning and control of the proton therapy.

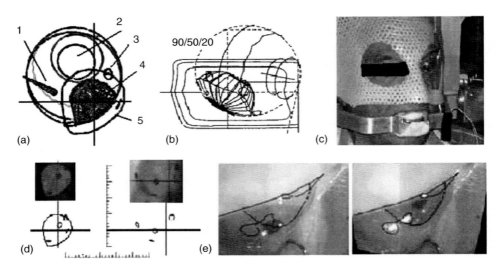

FIGURE 46.12

Image guided treatment of eye tumours with fixed proton beams. (a) Virtual model of the eye for planning, frontal beam's eye view: (1) optic nerve; (2) lens; (3) clips; (4) target; (5) aperture. (b) Isodoses on a lateral view from EYEPLAN. (c) Immobilisation devices: thermoplastic mask, byte block, eyelid retractors. (d) Orthogonal x-rays and clip coordinates for patient set-up. (e) Video verification of the gaze and eye position with audio coaching during treatment. (Figure (b) from Goitein, M. and Miller, T., *Med. Phys.*, 10, 275–283, 1983; Figure (e) is courtesy of the Centre de Protontherapy d'Orsay.)

By optimising the position of the eye and the direction of the gaze, an optimal plan of treatment based on a single anterior beam is produced. The EYEPLAN software, developed in Boston (Goitein and Miller 1983), established the definitive requirements of a treatment planning system. Some improvements have been added through a collaboration between the Paul Scherrer Institute in Villigen, Switzerland and Clatterbridge Centre for Oncology in the United Kingdom. This software is used in the majority of proton therapy centres in the world (Sheen 2001). The available functions include beam's eye view, isodoses in any plane of the eye (including complex surfaces like the fundus of the eye), dose–volume histograms and associated tables, files for the computerised fabrication of customised personal collimators and the documents necessary for patient set-up during the treatment, including the position of the clips as they will be seen on the x-ray images taken in the treatment room (or *x-ray eye view*).

The treatment plan is validated during a simulation session; the patient fixates a luminous reference point whose coordinates in space have been determined at the planning stage. Controlling the direction of gaze requires the active participation of the patient; except in particular situations, the eye is not fixed mechanically (e.g. with a sucker lens). Eyelid retractors are used to immobilise the eyelids during the session and to minimise the irradiation of the tear glands and eyelids. Additional measurements are carried out at the simulation session which includes determining the thickness of the eyelid and the curvature of the eye. Consideration is then given to possible modification of the irradiation parameters such as the range and the modulation of the beam, or the use of filters. The treatment is carried out in four or five sessions, delivering doses of around 15 Gy (cobalt-equivalent) per session (the value usually adopted for the RBE is 1.1).

At each session, orthogonal x-ray images are taken using film, Polaroid or fluoroscopic systems to verify the position of the eye relative to the proton beam. A conventional or an infrared video camera giving a magnified view of the eye allows the position of the eye to be monitored during the irradiation, so that the beam can be interrupted if the movements are out of tolerance. To achieve the precision required, specific procedures are observed: at each session a radiographer (radiation therapy technologist), a doctor and a physicist are usually present. Dosimetric characteristics of the beam such as the Bragg curve, beam profiles and Gy/MU are checked on each treatment day. These are time-consuming tasks and the procedures adopted depend on the experience and the beam characteristics at each centre.

46.5.2 INTRACRANIAL TUMOURS

Commonly occurring intracranial targets that are particularly suited to treatment with proton beams include chordomas and chondrosarcomas of the base of the skull (Austin-Seymour et al. 1985; Habrand et al. 1995). The approach to treatment is similar to that described in the Section 46.5.1 for ophthalmic treatments. Much of the development of proton therapy techniques was carried out with fixed beam lines at a time when isocentric gantrys were uncommon. Consequently the techniques developed for patient positioning and fixation were designed either to treat the patient in a chair or to facilitate repositioning the patient between beams. For seated patients, some centres have installed a specially constructed CT scanner which operates in the horizontal plane.

Fiducial reference markers made of gold or steel can be surgically implanted onto the skull surface and used in a similar way to the clips used for eye treatments. Implantation can be carried out under local anaesthetic. The markers can be visualised either radiographically or using a CT scanner or even cone-beam CT (see Section 40.1.5). Using two orthogonal x-ray images it is possible to calculate the translations and rotations necessary to correct the patient position. These corrections can be applied using a precision patient positioning system such as that shown in Figure 46.10 (Grusell et al. 1994c). Interfraction reproducibility of approximately ± 1 mm and $\pm 1°$ can be achieved. To achieve this level of accuracy a corresponding CT scan slice thickness must be used. If necessary the slice thickness can be smaller in the region of the fiducial markers. Ideally coregistered MR and CT images should be used.

Volumes of interest should be defined following the recommendations of ICRU 62 (1999) (see also Section 29.2), although these are not entirely suited to proton therapy. With proton beams, very precise demarcation of the beam is possible even in the direction of the beam and this should be borne in mind when considering the appropriate margins to use. However, uncertainties in the beam range may exist, especially in the presence of inhomogeneities. The definition of a proton beam requires the specification of the collimator, compensator, beam range and modulation. The treatment plan consists not only of the isodose distributions (see Figure 46.13) and dose-volume histograms as in photon therapy, but also data to allow the production of collimators and compensators.

A variety of different materials can be used for the collimators: e.g. low melting point alloy, lead or brass. Compensators are made of wax or Plexiglas usually with the aid of a computer numerical control (CNC) milling machine using 'cutting files' produced by the treatment planning system.

Before the start of treatment, quality control of all the parameters and devices must be carried out. These include profiles and depth dose curves of the beam to verify the range and modulation, monitor units and dose at the surface. The patient documentation, including the setup instructions must also be checked. Each day or before each treatment session, checks of the beam characteristics and beam calibration must be carried out. For proton beams the depth of treatment varies rapidly with beam energy and it is therefore particularly important to check this. Other considerations specific to proton beams include estimation of the neutron and gamma dose from the direct beam, which may be measured using radiation protection film dosimeters. Although much research work has been done on in vivo dosimetry with thermoluminescent dosimeters, their use is not yet routine.

Because of the complexity of the treatment and the difficulty of achieving the necessary precision, it is common practice for a radiation oncologist and a physicist to be present to oversee the treatment setup in addition to the radiographers.

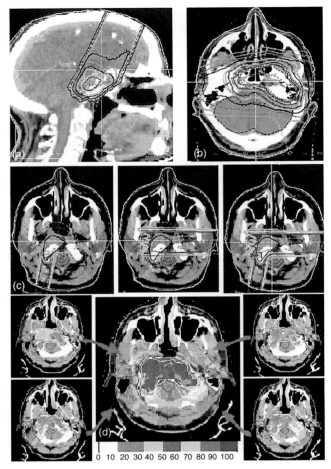

FIGURE 46.13
(See colour insert following page 590.) Treatment planning systems and dosimetry of intracranial targets with high energy proton beams: (a) noncoplanar beams, ray-tracing algorithm; (b) combined treatment with photons and protons, pencil beam algorithm; (c) "patching" techniques to boost targets surrounding a critical structure; (d) intensity modulated proton therapy with four coplanar scanned proton beam. (Figures (a–c) courtesy of Centre de Protontherapie d'Orsay with Isis system, Paris; Figure (d) is courtesy of Trofimov, A. and Bortfeld, T., Boston with KONRAD system, German Cancer Research Center (DKFZ), Heidelberg.)

46.5.3 STEREOTACTIC RADIOSURGERY

Proton therapy is ideally suited to single-fraction brain treatments using stereotactic methods (see Chapter 45). Targets include both tumours and benign conditions such as arteriovenous malformations (AVMs) and functional surgery (Kjellberg and Abe 1988; Fabrikant et al. 1992). The stereotactic radiotherapy programme established at the Lawrence Berkeley Laboratory (where protons were first used for radiotherapy treatment in 1954) was moved to the new proton facility at Loma Linda, California. In Uppsala, Sweden, Grusell et al. (1994a) have developed techniques for the treatment of AVMs and in Faure, South Africa, Jones et al. (1994) have established cross-fire techniques.

A variable number (between 3 and 10) of small noncoplanar beams with diameters between 5 mm and 40 mm are used in a similar manner to photon radiosurgery. The precise positioning of the patient is achieved with the aid of an invasive stereotactic frame (see Section 45.4.1). Early work was based on the use of the cross-fire technique taking advantage of the small lateral penumbra of the proton beam in the plateau region of the depth dose curve (see Figure 46.2). Subsequently more use has been made of the depth-dose characteristics associated with the Bragg peak to minimise the number of beams. Collimator design techniques are similar to those described for intracranial tumours in Section 46.5.2.

Compared to *photon-beam* stereotactic radiosurgery, protons offer better conformation to the target volume, excellent dose homogeneity within the target volume and a lower *integral dose*. However, for small diameter proton beams there is a loss of lateral scattering equilibrium. This manifests itself as a reduction in the proton fluence on the beam axis, as discussed in Section 46.2.3. The relative amplitude of the Bragg peak is reduced and it can even disappear when the beam diameter is only a few millimetres. These effects must be taken into account in the beam modelling and in the calculation of the dose and the design of beam-modifying devices.

46.5.4 Other Sites Suitable for Proton Therapy

In addition to the ophthalmic and intracranial tumours already described, there are many other sites where proton therapy is of clinical benefit. These include tumours in the vertebral and paravertebral regions, head and neck (paranasal and nasopharynx), central nervous system (meningiomas and gliomas), lung and pelvic tumours (prostate, sacral chordomas and sarcomas) (Munkel et al. 1994; Munzenrider 1994; Lundkvist et al. 2005).

The toxicity of paediatric radiotherapy can be significantly reduced through the use of proton beams while maintaining or improving tumour control. This is not only related to the improved degree of conformality of the high doses encompassing a given target, but also to the significant reduction of the *integral dose* to the healthy tissues, making protons a treatment of choice for paediatric tumours in a number of different locations (Hug et al. 2002; Noel et al. 2003; St Clair et al. 2004). The expected reduction in the probability of second cancer induction is of particular significance for young patients (Mu 2005; Hall 2006b).

46.6 ECONOMIC AND FUNCTIONAL CONSIDERATIONS

The hospital-based facility in Boston, which has three treatment rooms attached to a 230 MeV isochronous cyclotron and which became operational in 2001, cost a total of US\$46M. Two of the treatment rooms have a gantry, the third has two fixed beam lines dedicated to eye treatments and radiosurgery respectively. Each gantry costs around the same as the accelerator itself. Goitein and Jermann (2003) compared the cost per fraction of proton therapy to the cost per fraction for x-ray therapy. They estimated the current construction cost of a two-gantry proton facility complete with its associated equipment at €62.5M (US\$80M) compared to a two-linear-accelerator x-ray facility at €16.8M (US\$21.6M). Considering the construction, operating and business costs, the costs per fraction were estimated to be €1025 (US\$1320) for protons and €425 (US\$550) for x-rays. This represents a cost ratio of 2.4 ± 0.35 (85% confidence interval). If more proton facilities are built the costs of proton therapy are likely to fall due to economies of scale in the manufacture of proton accelerators and gantries. Goitein and Jermann estimated that were the costs of the capital investment to be written off, the cost ratio would be in the range of 1.3–1.6.

The investment necessary to convert an existing nuclear physics research accelerator to provide the facilities necessary for medical use is very dependent on the detailed local arrangements. In general this will require changes to the beam energy (e.g. degrading from high energies) and modifications to the beam-transport system and shielding and additional safety features. Facilities specific to the patient treatment include an isocentric gantry and patient support system together with the standard hospital clinical facilities. Physics support for dosimetry and a workshop to prepare the patient-specific hardware described in Section 46.5 will also be needed. In the case of the Orsay Centre in France, which did not include an isocentric gantry, the cost of the necessary conversions was similar to the cost of a modern linear accelerator with its associated equipment (about US$2M in 1991). However, such *converted* facilities require a large technical staff and a significant amount of maintenance.

The capacity of existing proton treatment centres range between 20 and 300 patients per year, as some share their activity with physics research programmes. In new, dedicated facilities it is expected that approximately 1000–2000 patients can be treated per year with one accelerator servicing several treatment rooms. The number of fractions per patient is strongly dependent of the clinical protocol in use: 1 fraction for stereotactic irradiations, 4–5 fractions for the ophthalmic treatments, 10 fractions for a boost after a photon treatment, and more than 25 fractions for a treatment given entirely with protons.

The estimation of the number of staff necessary is sometimes difficult, particularly when maintenance and development personnel are shared with a research centre. From the reported figures, the current average is 1 member of staff for every 10 patients treated per year. However, at dedicated treatment facilities treating a large number of patients the staff to patient ratio is much lower. Estimates of the cost of operation (excluding staff salaries) for these facilities vary between US$250 and US$400 per hour of operation. In facilities which share their activity with physics research, the running costs are probably higher than this.

46.7 CONCLUSION

There is worldwide interest in the development of proton therapy as a tool for conformal radiotherapy. Whereas the first treatment units were based around nuclear physics research facilities, the present trend is towards dedicated facilities integrated within a hospital, with modern equipment developed specifically for proton radiotherapy. Due to their inherent depth–dose qualities, proton beams can deliver the same dose to the target volume as photon beams, whilst lowering the integral dose delivered to the patient and, in some cases, reducing the probability of damaging critical organs. For some clinical indications, proton-beam radiotherapy is already established as the treatment of choice with evidence provided through clinical trials. For other indications, the potential benefit of protons still requires clinical validation, but it is expected that their advantages will be demonstrated. Proton-beam radiotherapy is seen as the radiotherapy technique of the future (Suit 1992; Suit et al. 2003; Lundkvist et al. 2005). Cost-effectiveness studies comparing protons with conformal techniques using photon beams (see Chapter 43 through Chapter 45) have been carried out (ANDEM 1995) but these must be kept up to date. Protons should also be compared with the use of heavy ions described in Chapter 49.

CHAPTER 47

TOTAL BODY IRRADIATION

*Philip Mayles**

CONTENTS

47.1 CLINICAL GOALS

Total body irradiation (TBI) is used primarily as part of the treatment for leukaemia, but *hemi-body* irradiation can also be used for pain control. For the treatment of leukaemia, the aim is to destroy the patient's leukaemic bone marrow and then to replace it with a bone marrow transplant. Systemic diseases such as leukaemia are usually treated with chemotherapy. However, TBI is able to make two contributions to therapy (Barrett 1995):

---

* Section 47.5.2 includes contributions by Ginette Marinello.

- Killing leukaemic cells
- Suppressing the immune system to combat rejection of subsequent bone marrow transplants.

Although the first of these is also the goal of the chemotherapy administered at the same time, radiotherapy has the advantage that there are no *sanctuary sites* (Doughty et al. 1987). Opinion is divided on the relative importance of the above two effects and it may also depend on the type of leukaemia being treated. The dose selected is to some extent determined by the philosophy being followed. If immunosuppression is considered the primary aim, then the required dose will be lower (Novitzky et al. 2004). Two reviews of this topic have been written by Vriesendorp (1990, 2003). It has been suggested that Tomotherapy (see Section 43.3.4) could also be used for TBI, which would allow sensitive organs to be spared (Hui et al. 2005), but there is a concern that this may spare leukaemic cells as well.

47.2 DOSE AND DOSE RATE

The dose-limiting organ is generally regarded as the lung, which can develop radiation-induced interstitial pneumonitis. Pneumonitis is also associated with chemotherapy and graft versus host disease, and it is difficult to establish an exact dose-response relationship. For single-fraction TBI there is evidence that dose rate is a significant factor. Barrett et al. (1983) argued that the lower rate of mortality due to interstitial pneumonitis, 4.5% at the Royal Marsden compared to 35% in the early days in Seattle (Thomas et al. 1977) might be due to the lower dose rate used (2.5 cGy/min compared to 5.5–8.0 cGy/min). However, this difference in mortality might have been associated with a slightly lower total dose or with patient selection. Kim et al. (1985) compared the results between patients treated on a Cobalt unit at 4.7–6.3 cGy/min and patients treated on an accelerator by a swept beam technique that was associated with instantaneous dose rates of 21–23.5 cGy/min. In the first group, 2/11 died of pneumonitis, compared to 8/11 in the second group. Ringdén et al. (1983) obtained a significantly higher death rate for patients treated at a dose rate of 7 cGy/min compared to 4 cGy/min. Beyzadeoglu et al. (2004) also found that less than 4 cGy/min produced significantly less interstitial pneumonitis, whereas the median total dose to the lung, which ranged from 8.9 to 10.9 Gy, was not a significant factor in their study. The need for an even lower dose rate is suggested by O'Donoghue (1986) based on radiobiological considerations. Treatments at these very low dose rates are very time-consuming and most centres that do not have a specialised facility have opted for fractionated treatments, although there is some evidence that single-fraction treatments may provide better leukaemic cell kill (Cossett et al. 1990). A single fraction dose of 10–10.5 Gy is generally considered to be equivalent to a fractionated total dose of around 12 Gy (a discussion of these issues can be found in Chapter 9). However, Beyzadeoglu et al. (2002) showed that there is a statistically significant increased incidence in cataract formation at dose rates above 4 cGy/min, even for fractionated TBI. In a study by Ozsahin et al. (1996), the effect of dose rate was apparently independent of fractionation. They reported that low dose rates (<4.8 cGy/min) were less effective in controlling the disease, but that high dose rates (>9 cGy/min) were more toxic to the lungs. They therefore recommended the use of a dose rate between these two limits. The effect of the dose rate in fractionated TBI has also been demonstrated by Carruthers and Wallington (2004), who found a significantly increased rate of pneumonitis when 15 cGy/min was used compared to 7.5 cGy/min. The effect of radiation dose rate can also be dependent on the chemotherapy regime. Safwat et al. (1996), using a mouse model, found that when TBI was combined with cyclophosphamide, low dose rates were more toxic to lung than high dose rates, whereas without chemotherapy, high dose rates were more toxic.

47.3 DOSE SPECIFICATION

The point at which the dose is specified is an important consideration. It has been rec-ommended (Dutreix and Broerse 1982; Leer et al. 1990) that the dose be specified at the centre of the abdomen, but some centres, such as the Royal Marsden, prefer to specify the dose to the lung since this is the dose-limiting organ. (In the survey carried out by Leer et al. (1990) only two out of 18 centres followed this latter practice.) In contrast, the American College of Radiologists practice guidelines (ACR 2001) do not specify the dose prescription point, although they suggest the abdomen at the level of the umbilicus. This choice will depend partly on whether the dose to the lungs is controlled by shielding. If no lung shielding is used, it makes sense to prescribe a maximum dose to the lung. In any case, it is recommended that the dose to head, neck, shoulder, thorax, abdomen, pelvis and ankle be recorded (Briot et al. 1990); the dose to upper thighs and knees may also be recorded. European guidelines on reporting TBI have been published (Sanchez-Doblado et al. 1995).

Dose homogeneity is regarded by some clinicians as important (Doughty et al. 1987), but by others (Barrett 1995) as less so because they consider that the aim is to give as much dose as possible to the areas most likely to contain disease. This can be taken further by the application of a boost dose to the brain or to the spleen (Lapidot et al. 1988).

47.4 AVAILABLE TECHNIQUES

Useful reviews of the many techniques in use are given by Quast (1987) and Thomas (1990). Apart from hospitals such as the Fred Hutchinson Cancer Research Center in Seattle and the Royal Marsden in the United Kingdom, which have specialist TBI facilities (see Section 47.4.1), most treatments are done using a horizontal beam with the patient located as far away from the source as possible. The size of the field available will depend on the distance from the source that can be achieved. On a conventional treatment machine, the head can be rotated to 45° to achieve the maximum field size. The patients may then be treated lying on their side with an anterior-posterior (AP) beam or lying supine with a lateral beam. In either case, it is necessary to rotate the patient so that an effectively parallel opposed beam will be achieved. Knees may be bent to reduce the length of the patient (see Figure 47.1). In some centres, the patient sits in a chair. From the point of view of reproducibility of the setup, the supine position is undoubtedly superior, but has the disadvantage that the lateral separation is always greater than the AP separation, which may be more of a problem when using lower-energy beams (e.g. cobalt-60).

Umek et al. (1996) describe a cobalt-60 technique in which the couch is moved during the treatment, thus reducing the need for extended source-skin distance. Couch speed is altered depending on the thickness of the patient to produce a relatively homogeneous dose distribution.

47.4.1 CUSTOM-DESIGNED FACILITIES

The Royal Marsden is one of three centres in the world that has a specially designed treat-ment facility for TBI (Edser 1988; Lewis and Rosenbloom 1988)[*]. Two Theratron-780 heads are mounted on a frame, which allows them to be up to eight metres apart in the vertical direction. This enables a routine treatment distance of 3.50 m to be used. In a cobalt-60

---

[*] The other two are the original centre in Seattle (Thomas et al. 1977) and the Joint Center in Boston, where there are two linear accelerators mounted to give a horizontal beam.

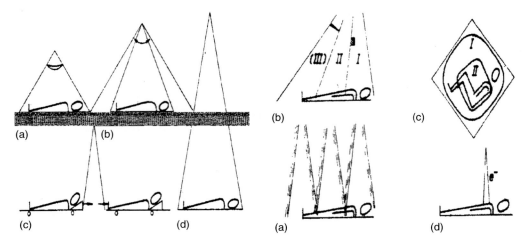

FIGURE 47.1
Alternative approaches to TBI. (From Quast, U., *Radiother. Oncol.*, 9, 91–106, 1987. With permission.)

Co beam, the dose rate drops towards the edge of the beam and so a compensating or flattening filter has been designed to reduce the dose at the centre of the beam relative to the periphery. The machine is equipped with a set of four attenuators, which are used to alter the machine dose rate as the source decays. One of the attenuators was designed to flatten the beam and is left in at all times; a second thick filter reduces the dose rate to about 0.2 Gy/min. Combinations of the other two filters allow the dose rate to be altered in four steps. One of the attenuators is equivalent to one year's decay and the other to two year's decay. This enables the dose rate to be kept to within 20% of its original value over a four-year cycle. The collimators are designed to provide a field length of 2 m at a treatment distance as small as 2 m.

The choice of a vertical AP beam was made on the grounds of patient reproducibility. However, when measurements were made, it was found that the dose rate at the patient's head was reduced because of loss of scatter, which could not be corrected by placing scattering material around the head. For this reason, patients had to be treated lying on their side as well. In practice, one of the problems of single-fraction TBI is that the patient may vomit, and it is easier to cope with this when the patient is not supine; also many patients find lying on their side more comfortable. Patients are treated either 50% supine and 50% on side or 33% supine and 67% on side for single-fraction TBI. For fractionated TBI, patients are less likely to vomit, but the dose distribution is improved by including the lateral treatment position. From the point of view of the reproducibility of dosimetry, however, it can be clearly demonstrated that the lateral position is less satisfactory.

47.4.2 USE OF A STANDARD LINEAR ACCELERATOR

Figure 47.1 illustrates a number of different ways of treating patients using an accelerator beam. When carrying out TBI on a linear accelerator, logistical considerations dictate that treatment is either fractionated or carried out at a weekend. Fractionation is the preferred method, as there is less concern about dose-rate effects (but see Section 47.2). If single-fraction TBI is to be used, it is important to lower the dose rate; this can usually be achieved by altering the pulse-repetition frequency (PRF) of the accelerator (see Section 11.3.2). The patient will normally be placed as far away from the source as possible to obtain a large field with a horizontal beam. Again, for reasons of setup reproducibility, patients are best treated supine rather than lying on their side. With a lateral beam and a supine patient, the dose to the head will normally be too high. This can be corrected either by including treatment with the patient lying on their side or by some form of compensating filter (see Section 47.5.4). For fractionated TBI, the latter option is clearly preferable because of the improved ease of setup, but if single fractions are being used the difficulty of maintaining the patient in exactly the same position for three hours or so may make the use of any form of compensation impractical. Doughty et al. (1987) have described the technique used at St. Bartholomew's Hospital, which aims to ensure that the dose is homogeneous within 5% throughout the body, although as mentioned above, it is questionable whether an increased dose to the legs is necessarily a bad thing.

Miralbell et al. (1994) have developed a technique in which the patient is supported in an upright position, thus allowing treatment with a horizontal AP beam. This has the advantage that the thickness of the patient is smaller in the AP direction and partial lung shielding is easier. Methods of treating with a vertical AP beam are shown in Figure 47.1.

Because for there are "horns" at the edge of the (conventionally flattened) beam, it may be necessary to use an *inverse* flattening filter (Doughty et al. 1987) to achieve a flat beam. The design of this additional flattening filter should be based on measurements at depth (e.g. 10 cm) rather than on in-air measurements. It can take the form of a series of concentric rings of Perspex, which can be mounted on the shadow tray of the accelerator. It is wise to check that the beam is flat before *each* treatment. This can conveniently be done using the diodes used for measurement (see Section 47.5.2).

47.5 DOSIMETRY CONSIDERATIONS

The dosimetry of TBI gives rise to a number of problems that are not normally found at shorter SSDs and with smaller field sizes. A thorough analysis has been presented by Briot et al. (1990). Measurements made in a cuboidal phantom may not be appropriate, as scatter from throughout the long thin patient will contribute to the dose. Attempts at dose calculation have proved difficult, although with high energy x-ray beams (Houdek and Piciotta 1987; Jani and Pennington 1991) the results have been more successful than with cobalt-60 beams. In addition to the problems of scatter, there is also the effect of the density of lung. Lung densities vary between patients, and since this is the region of greatest risk to the patient, it is important to achieve the highest possible accuracy. For these reasons it is common practice to carry out a *test dose* on a previous day so that measurements can be checked and monitor units calculated. For a cobalt-60 beam, the in-air scattered electrons associated with the very large field size (Kassaee et al. 2001) are sufficient to eliminate the build-up effect, but if an accelerator is being used it is necessary to use a Perspex™ sheet in front of the patient. It is not particularly critical how far from the patient this sheet is placed (Planskoy et al. 1996a). A 1 cm thick sheet is sufficient to bring the surface dose to 99% for an 8 MV beam.

47.5.1 CHOICE OF ENERGY

The energy chosen will often depend on what is available. There are advantages to being able to use a high-energy machine, as this reduces the difference between the dose at the mid-line of the patient and the dose at d_{max}. Table 47.1 compares the ratio of mid-line doses to peak doses for cobalt-60, 6-MV x-rays and 10-MV x-rays for a 30 cm thick patient. The table illustrates clearly the advantages of using higher energies.

47.5.2 DOSE MEASUREMENT

Because of the importance of accurate dosimetry in this potentially life-threatening treatment, it is usual to carry out in vivo dosimetry (van Dyk et al. 1986; ESTRO 1987;

TABLE 47.1

Effect of Energy on the Ratio of Maximum to Minimum Dose

	SSD = 2.5 m	SSD = 3.5 m
Cobalt-60	0.909	0.923
6-MV x-rays	0.940	0.953
10-MV x-rays	0.967	0.976

Indovina et al. 1989; Briot et al. 1990; Mege et al. 1994; Planskoy et al. 1996a, 1996b; Mangili et al. 1999). Doses delivered during TBI are difficult to evaluate accurately because of:

- The very large fields employed. These frequently require that the patient is positioned very close to the floor or to the walls of the treatment room, which may produce significant electron and photon backscattered components that reach the patient's skin (Lam et al. 1979; van Dyk et al. 1980; van Dam et al. 1988)
- The necessity to correct for tissue inhomogeneities (Marinello et al. 1982) and to monitor the dose to the organs at risk (lungs, liver, etc.)
- The difficulty of achieving a reproducible setup and the possibility of the patient movement during the treatment, especially for low dose-rate single fraction treatments.

The in vivo measurements can be used either as a check of the calculated doses or calculated doses may be used to check the measurements.

Because the treatment technique is designed for clinical reasons to have the minimum possible build-up region, both the skin and the in vivo detectors are almost in full-build up conditions. The build-up cap required can therefore be thin or may even be unnecessary with some dosimeters (consequently there is no risk of target underdosage due to the attenuation caused by in vivo detectors in TBI conditions).

In vivo dosimetry generally consists of entrance and exit dose measurements performed on different regions of the patient's skin (Figure 47.2). The points of measurement must be chosen such that they allow for both verification of the dose delivered at the point of dose specification (usually taken at mid-pelvis or mid-abdomen) and estimation of the homogeneity of the midline dose distribution in the cranio-caudal direction of the patient using the general methodology detailed in van Dam and Marinello (1994). Theoretically, the lung dose can also be deduced from entrance and exit dose measurements, but the problem of the inaccuracy of its estimation from a measurement at a single point has been pointed out by several authors (Marinello et al. 1982; Briot et al. 1990; Planskoy et al. 1996b).

Either diodes or thermoluminescent dosimeters (TLD) can be used. If the department has access to the necessary specialised and expensive readout system, alanine dosimeters are particularly suitable. The nondestructive readout of the electron spin resonance signal can be advantageously exploited to monitor the dose increment fraction-by-fraction in fractionated therapy, the final objective being to check the total dose, provided that the same dosimeter is applied each day at the same position on the patient (Indovina et al. 1989).

Diodes provide an immediate response, which enables the immediate correction of any patient movements. Care must be taken when making measurements behind shielding blocks, because low-energy scattered radiation contributes a substantial fraction of the dose, which can introduce inaccuracies in dose determination due to possible variation of dosimeter response with (lower) energy. For the low dose rates used in single-fraction TBI, it becomes very important to balance the diode measurement circuit accurately to minimise the *offset current*. It is also important to consider the effects of temperature (see Section 16.3.4.7) on the diodes, which will be in contact with the patient's skin for a long time and will therefore reach body temperature. This may affect both the offset current and the sensitivity of the diodes.

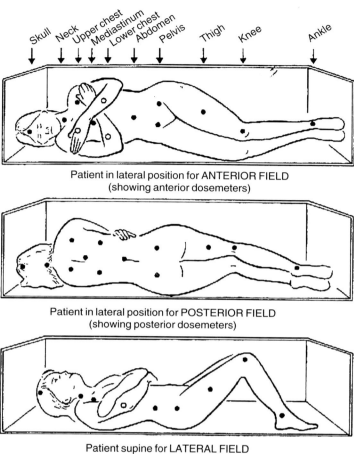

Patient in lateral position for ANTERIOR FIELD
(showing anterior dosemeters)

Patient in lateral position for POSTERIOR FIELD
(showing posterior dosemeters)

Patient supine for LATERAL FIELD
(showing lateral dosemeters)

FIGURE 47.2
Example of in vivo dosimetry practiced at the Royal Marsden Hospital for total body irradiation: filled circles show the position of dosimeters nearer irradiation sources and open circles the dosimeters shielded by the arms. (From Rosenbloom, M., Hickling, P. A., Chow, M., Chittenden, S., Machardy, J. et al., *J. Eur. Radiother.*, 3, 246–248, 1982.)

Diodes have directional asymmetry and it is important to consider this aspect when calibrating them for use in TBI, in which the contribution of electron contamination and scattered photons may be much greater than at standard SSDs and field sizes. The type of diode must be selected according to the energy in use, but diodes that are recommended for normal use at a lower energy may be appropriate because of the reduction in build-up (see above).

TLDs allow a large number of sites to be measured at the same time and give accurate results even behind shielding blocks because of their closeness to tissue equivalence over a wide energy range. With automatic TLD readers adapted for medical use, the dose readout can be obtained approximately 15–30 minutes after irradiation. Moreover, most TLD readers can be connected to a portable computer so that stored data can be processed immediately after readout and the results directly integrated into the patient's file.

Doses to the head, the lungs, the abdomen and the pelvis should be measured (see Section 47.3). Extremity doses should also be measured, at least occasionally. When using a horizontal accelerator beam, it is important to guard against the effect of low-energy back-scattered radiation from the wall of the treatment room. This can be reduced by placing a low-atomic-number absorber between the patient and the wall (van Dam et al. 1988).

Whatever method is being used, the calibration of the detectors must be done in the TBI geometry.

A useful approach to the assessment of the homogeneity of the dose distribution is to take a portal film (Signorotto et al. 1996). This can be scanned with a densitometer to obtain the dose distribution on the exit surface of the patient. If no compensators are used, this should ideally show a uniform field, but if the input dose distribution is not uniform, the exit dose will not bear such a straightforward relationship to the mid-line dose distribution. Some authors have proposed the use of a combination of in vivo measurements at specific points and portal dosimetry (films or electronic portal imaging) to overcome this problem (Essers and Mijnheer 1999; Mangili et al. 1999).

When making measurements in large fields with ionisation chambers it is important to bear in mind the possibility of stem and cable effects (van Dyk 1987; Planskoy et al. 1996a). The amount of charge arising from outside the chamber itself can be assessed by shielding the chamber (see Section 15.4.3). The expected decrease in dose due to the shielding can be calculated and compared to the actual measurement.

It is important to remember that the dose measured by surface dosimeters is the maximum dose; to obtain the mid-line dose, a correction must be applied. For treatments from a single portal, closer agreement to the mid-line dose may be obtained by using the geometric mean of the entry and exit doses (Mijnheer 1996). However, the best approach is to use a graph of the mid-line dose against patient separation. In comparing doses between treatment clinics, it is important to find out exactly how the dose is measured. For example, at the Royal Marsden the surface doses from the lateral and the anterior-posterior beams are added together, even though they do not represent the dose at the same point. This practice arose because of the intention to stay within lung tolerance. The difference between the dose calculated in this way and the mid-line dose is about 7%.

47.5.3 DOSE CALCULATION

A number of centres have sought to calculate the dose as opposed to measuring it (van Dyk 1985, 1987; van Dyk et al. 1986; Doughty et al. 1987; Houdek and Piciotta 1987; Thwaites et al. 1990; Jani and Pennington 1991) with varying degrees of success; a key paper is that by van Dyk (1987). A satisfactory approach is to make measurements at a constant Source Chamber Distance in a phantom designed to represent a full-size patient[*]. These measurements can be used to produce a graph of tissue-phantom ratios (see Section 23.2.1.6) relative to a measurement at 5 cm deep in a phantom. Then it simply remains to measure the absolute dose at a depth of 5 cm to be able to predict the dose at the mid-line in the abdomen. This is particularly convenient if the patient is always set up with the mid-line at a fixed distance from the source. To obtain doses in the region of the lung, it may be possible to take a few CT scans. For an 8 MV beam, this gives satisfactory results simply using the equivalent pathlength method (C. D. Lee unpublished results). Abraham et al. (2000) obtained satisfactory calculations of the lung dose using a standard treatment planning system with a full CT representation of the patient. The extended geometry of the TBI situation can create problems for standard planning systems (Well et al. 1991). To overcome these problems Planskoy et al. (1996a, 1996b) adapted the Bentley–Milan-based electron algorithm[†] of the *Target* system (International General Electric) to calculate dose distributions around the lung.

---

[*] This can be created using *solid water* close to the ionisation chamber and hardboard or large containers of water to make up the rest of the shape of the patient.
[†] The *electron* algorithm allowed more fan lines to be used than the *photon* algorithm; the words *photon* and *electron* here are essentially just labels. It is characteristic of the Bentley–Milan model that it simply reproduces the measured dose distribution in a structured way.

Doses at the extremities will be reduced due to lack of scatter, as will doses close to the lung. It is probably sensible to concentrate on calculating mid-line doses since these are less likely to be affected by the electron contamination experienced at the surface of the patient (van Dyk 1987). It is also important to remember that very young patients may also be treated and the scatter dose will be less in this case.

47.5.4 COMPENSATION

To achieve a uniform dose throughout the body, it is common practice to use various forms of compensation. Because skin sparing is not desirable, the use of bolus is not counter-indicated, and in the St. Bartholomew's technique (Doughty et al. 1987) bolus is extensively used with the aim of achieving a constant separation equivalent to that of the hips. The disadvantage is that the increase in the uniformity of surface dose measurements may be at the expense of the dose at mid-line. For the lung, *lead rubber* can be used, usually attached to the Perspex screen close to the patient, or brass compensators may be made to the thickness required. These may also be mounted on the head of the accelerator, which has the advantage that the borders are rather less sharp, in keeping with the fuzziness of the border of the lung. A simple and effective approach to lung shielding which is convenient for single-fraction TBI is to use the patient's arms (which is the technique used at the Royal Marsden).

Port films taken on the treatment machine provide a transmission map of the patient. These data can be used to design lung compensators (Hussein and Kennelly 1996) that accurately reflect the transmission through the patient. Another technique of compensation is described by Galvin et al. (1980).

There may be problems associated with beam *softening* towards the edge of an accelerator beam, which will mean that the attenuation of any absorber will be increased. van Dyk (1987) reports that (effective) linear attenuation coefficients may vary by about 10%, depending on the field size, and this should be borne in mind when designing flattening filters and compensators (Table M.3 can also be consulted).

47.6 SUMMARY

The main points are summarised as follows:

- TBI delivers a potentially lethal dose of radiation and thus the accuracy of the dosimetry is particularly important.

- Care must be taken to ensure that the dose is specified in the same way if dose comparisons are made between centres. The dose rates and fractionation used must also be considered.

- The concurrent chemotherapy has a synergistic effect with the radiation and the treatment regime must therefore be considered as a whole.

- The lung is the dose-limiting organ and thus particular care must be taken to ensure that the lung doses are correctly determined.

- Measurements must be made in *TBI conditions* rather than at standard SSDs.

- Measurements at depth are more reliable than surface measurements (although only the latter are possible on the patient). The effects of electron contamination close to the surface must be borne in mind.

- In vivo dosimetry is an essential check, even when calculation is used as the method of setting the monitor units.

- Calculation of doses is an important safeguard, even when monitor-unit settings are based on measurements.

- The possible effect of beam *softening* off-axis should be considered.

CHAPTER 48

TOTAL SKIN ELECTRON IRRADIATION

David Thwaites and Alan McKenzie [*]

CONTENTS

48.1 INTRODUCTION

The technique of total skin electron irradiation (TSEI), or total skin electron therapy (TSET), is used to treat a variety of cutaneous malignancies where most or all of the skin is involved. The aim is to treat the whole skin to as uniform a dose as possible, both circumferentially around the patients as well as along their length. The uniformity requirement comprises an acceptably uniform and high surface dose, uniform dose at d_{max} (that should be

---

[*] With contributions by Ginette Marinello.

at, or close to, the surface), and also a uniform penetration and dose delivery to the required therapeutic depth, selected to match the presenting disease and stage. At the same time, the treatment should spare other underlying organs and minimise the whole-body dose from the bremsstrahlung x-ray component of the beam.

The main use of TSEI is in the treatment of *mycosis fungoides*, a cutaneous T-cell lymphoma arising in a very small number of people per million each year, and related conditions (Szur 1975; Hoppe et al. 1977). However, a range of other rarely presenting malignant diseases requires a similar approach, including inflammatory breast disease and Kaposi's sarcoma. TSEI techniques have also been used for other dermatological conditions covering large areas that do not respond to alternative treatments (e.g. Kavanagh et al. 1997). There is no clear consensus on the prescription dose, being typically in the range of 20–40 Gy, delivered over a period from 2 to 6 weeks, or longer. Dose per fraction and fractionation schedules are often dictated by a compromise between clinical requirements and pragmatic technical and time restrictions. TSEI is not widely practised because of the rarity of the diseases for which it is appropriate, and also due to the technical complexity of commissioning the technique and of the treatment itself, coupled with long patient set-up and treatment times.

Some of the techniques and practical considerations necessary to implement TSEI are reviewed in AAPM report 23 (1987).

48.2 GENERAL CLINICAL AND PATIENT-RELATED PROBLEMS

The patient-related problems are mainly linked to the marked irregularities in surface shape and to the variations in the cross-sectional dimensions of different areas of the body, making it intrinsically difficult to achieve uniform distributions. Additionally there is always some self-shielding of one part of the body by others and some areas of the body may be omitted from treatment due to the patient position (e.g. the soles of the feet if the patient is standing). Other areas may always receive low doses because their position dictates some self-shielding and poor penetration because all beams are tangentially incident (e.g. the perineum or the top of the head). Such areas may require subsequent boosts with small-area fields, if deemed clinically necessary. A significant number of the patients referred for TSEI are elderly and may not be able to stand for long periods, or may have difficulty taking up complex positions, if these are required to minimise self-shielding. TSEI patients can present for treatment at various stages of their disease (e.g. for mycosis fungoides: with superficial involvement only, through a generalised plaque stage, to a tumoral stage). Consequently, variable depth penetration may be required for different patients. Some may present having previously received treatment to a number of areas with electrons or kilovoltage x-ray beams. There may be a requirement to treat only part of the body but a large area, or to shield specific areas (e.g. fingernails, toenails, and eyes). The same patient may return subsequently for repeated treatment or treatments.

48.3 PHYSICAL AND PRACTICAL REQUIREMENTS

48.3.1 DOSE DISTRIBUTION ALONG THE PATIENT

A uniform distribution, to around ±5%, is required. This covers the entire length of the patient, generally positioned with arms above the head (i.e. with dimensions of around 80 cm × 220 cm). This has been achieved in a variety of ways:

1. Using medium SSD (1.5–2 m) with the patient in a lying position and either adding large fields with wide penumbra (Thwaites 1988), translating the patient through the beam (Williams et al. 1979) or by arcing beams longitudinally (Wu et al. 1997);

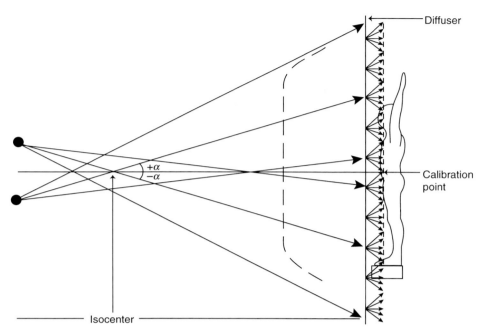

FIGURE 48.1
Schematic illustration of the dual-beam technique showing how two scattered distributions can combine to produce a uniform profile over the patient length.

2. Using longer SSDs (3–5 m) and the patient in a standing position. In this case the unmodified beam profile approaches a wide Gaussian, due to scattering in the air. A uniform distribution over the patient's height is achieved either with dual added fields angled both up and down from the horizontal by ± 10–$20°$, such that the combined distribution is uniform (Figure 48.1) e.g. the Stanford technique (Karzmark 1967; Page et al. 1970) or the Memorial technique (Edelstein et al. 1973; Holt and Perry 1982); or with a single field utilising a customised scattering filter to produce the uniform distribution (Podgorsak et al. 1983; Kim et al. 1984); or with vertical arcs (Sewchand et al. 1979).

The distribution may be affected by scatter from the floor toward the feet of a standing patient and this should be taken into consideration. A low platform of Styrofoam, or similar material, can be used to minimise this.

It is important to take account of the effect of the intervening air in these extended-distance treatments. Three metres of air will add a relatively small contribution to the mean-square angle of scatter, but this amounts to a significant lateral displacement over 3 m. Therefore, it is important not to restrict the aperture of the scattered radiation at the linear accelerator head, so that in-scattering of electrons from air outside the target area can help to maintain beam uniformity.

48.3.2   DOSE DISTRIBUTION AROUND THE PATIENT

A uniform distribution and penetration is required around the patient and this requires multiple beams. Some early reported techniques suggested the use of only two beams, but this gives very variable distributions across the patient, particularly at the edges, where the electrons are more and more obliquely incident, eventually being tangential with minimal penetration (see Figure 48.2 for two single opposed fields).

The distributions are improved by increasing the number of fields (or effective number) around the patient:

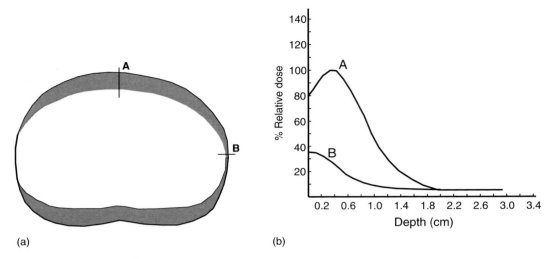

(a)   (b)

FIGURE 48.2
The effect of two single electron fields incident on the pelvis of a patient from the anterior and posterior directions.
(a) Section through the pelvis. The shaded area is the part of the patient that receives a dose greater than 10% of the
peak dose as measured with a film in the transverse plane of a phantom. (b) Depth dose curves along the two lines
A and B in (a). (Adapted from Kumar, P. P. and Patel, I. S., *Clin. Radiol.*, 33, 495–497, 1982.)

- Two fields with varying angulation of the patient (Williams et al. 1979)
- Four-fields also with changing angulation at different fractions (Thwaites 1988)
- Six-fields (Figure 48.3a, Stanford technique)
- Eight-fields (Figure 48.3b)
- Continuous (slow) rotation (McGill technique, Podgorsak et al. 1983).

For patients lying horizontally, this means positioning prone, supine, on left or right side
as necessary, with the aspect to be irradiated presented to a nominally zero gantry angle beam,
with angulation achieved either by small gantry-angle changes, or by modifying the angulation
of the couch surface using supporting wedges. For standing patients, a number of pre-set
positions can be selected on a rotatable platform. Using more fields improves the uniformity
of the dose distribution around the patient in terms of the magnitude of the surface dose, the
depth of penetration of the electrons and the depth (and magnitude) of the maximum dose.

48.3.3 BEAM ENERGY

The penetration needs to be matched to the required clinical depth. Generally, for superficial
involvement only, or where shallow plaques are present, this requires the therapeutic range to be
in the range 3–6 mm. The beam penetration of interest here is the final effective penetration
achieved by the total combination of all field positions and directions. This produces a significantly smaller depth of dose maximum and therapeutic range than for a single field, due to the
effects of oblique incidence, scattering, etc, although the practical range and bremsstrahlung
are similar, all other things being equal. The beam energies that are used for TSEI are generally
chosen to be in the range of 6–10 MeV as they emerge from the accelerator. These are
typically reduced to 3–6 MeV at the patient surface partly by the energy loss in the air (approximately 0.25 MeV/m) and partly by the addition of plastic (e.g. PMMA) energy-degrader sheets,
typically of 5–10 mm thickness, close to the patient* (i.e. approximately 20 cm). The depth

---

*These sheets also maintain a high surface dose and help to scatter the beam at this position, thus increasing the doses
in less directly accessible areas and improving the distribution.

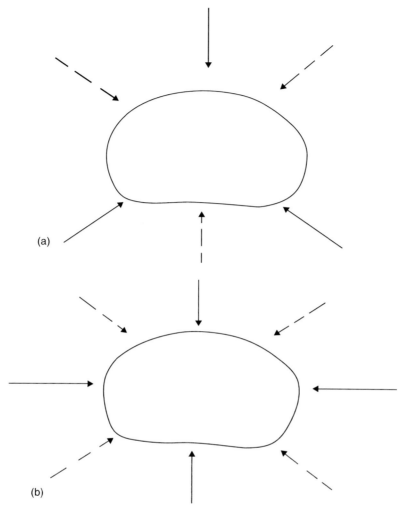

FIGURE 48.3
The beam directions around the patient in (a) six-field and (b) eight-field techniques. Typically, half the fields are delivered on alternate fractions (full lines) and the other half on the other fractions (dashed lines).

doses of the single field vary with angle of incidence across the beam (i.e. around the patient surface as illustrated in Figure 48.4; Bjarngard et al. 1977; Pla et al. 1988).

Using a dual-angled technique, the penetration will be reduced even on the central axis. The final effective depth-dose will depend on appropriate weighted addition of all possible penetrations at any point on the skin from all incident beams, dependent upon the technique being used (Kumar and Patel 1982) (see Figure 48.5). Changes in the effective penetration can be achieved by varying the thickness of the degrader.

The distribution around the patient and the penetration will vary somewhat with the size of patient cross-section and, therefore, with anatomical position on the patient. This will be more marked for shorter SSD techniques.

48.3.4 X-Ray Contamination

Typically, for beams of 6–10 MeV initial energy, the bremsstrahlung tail of the electron depth-dose curve is around 1% (0.5–2%) of the maximum dose. For TSEI treatments, the

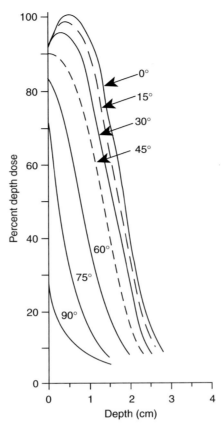

FIGURE 48.4
Percentage depth dose curves at different angles from normal incidence, for a nominal 4.5 MeV surface energy electron beam measured as part of TSEI commissioning. Each curve is normalised to the 100% value at 0°.

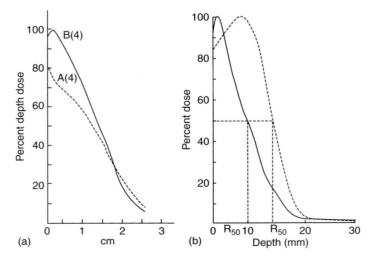

FIGURE 48.5
Illustration of the combined depth-dose curve at two extreme positions on a cylindrical phantom for (a) a four-field technique for the same beam as in Figure 48.4 and (b) for a dual-beam six-field technique.

x-ray component will irradiate the whole body and the contribution from each field direction will therefore be additive. The higher the number of fields employed, the lower the given dose on each individual field. For typical multiple-field irradiations, or for rotational techniques, the single-field percentage x-ray contribution is approximately doubled when measured at the patient midline (Holt and Perry 1982; Podgorsak et al. 1983) to approximately 2% (1–4%) of the total TSEI dose. One advantage of the dual-angled technique is to minimise the effect of the x-ray contribution, as the x-rays are predominantly forward-thrown (see Section 3.3) and the angling aims this mainly above and below the patient.

48.3.5　DOSE RATE AND TREATMENT TIME

Some shorter SSD techniques were dictated by the available dose rates available, as a long SSD reduces the dose rate faster than that due to the inverse square law alone because of scatter, and to such a degree that the treatment times became excessively long. High dose-rate modes are available on at least one electron energy on most current multi-modality accelerators. These typically provide dose rates up to a factor of nearly 10 times higher than normal at 100 cm SSD, with appropriate safety interlocks. This, in turn, allows dose rates to be achieved on the order of 1 Gy/min at 3–5 m SSD. However, for either short or long SSD techniques, the set-ups and treatments are complex and typically the treatment times are long in any case. As a result, compromises are often introduced in centres where the accelerators have a heavy workload and where too much time cannot be dedicated to one treatment. In these cases, not all fields are given at each fraction. For example, three of the six Stanford-technique directions (six separate fields in practice if dual-angling is employed) may be given on half the fractions, alternating with the other three on the other fractions. Main fields may be given on four fractions a week, with top-up fields to under-dosed areas on the fifth day etc. In this way, the uniform dose distribution is delivered over a cycle of treatment, with the treatment time being acceptable on any given day.

48.3.6　DOSIMETRY

The dosimetry of TSEI is reviewed in AAPM (1987). In addition to beam uniformity, depth doses, and x-ray contamination discussed above, the single beam must be calibrated (dose/MU) in TSEI conditions of distance, degrader etc., using a suitable ionisation chamber. Generally, at these beam energies, this should be a parallel-plate chamber (see Section 15.3.2). Cable effects must be checked carefully and minimised in large-field irradiations. The combined beam skin dose per MU must be ascertained as a measured ratio to the single-field value, using film or TLD on the surface of representative cylindrical phantoms. This value will depend on the technique, but can also be checked by calculation from a weighted addition of the contributions from the different beams (Pla et al. 1984).

For TLD measurements, the dosimeter set-up must be planned with care. The TLDs should be embedded level with the phantom surface rather than be placed on it, to measure appropriate values. For the reasons described in Section 40.2.3.1, thin thermoluminescent dosimeters are the best suited for this particular application (Marshall and Docherty 1971; Marinello et al. 1980; Cotento et al. 1984; Anacak et al. 2003). Small samples of monocoated silver-bromide or radiochromic films can also be used, but are less convenient. In any case the dosimeters, enclosed in thin protective sachets, are attached to different measurement points on the patient in order to check both the homogeneity of the dose distribution and the absolute dose (Figure 48.6). Care must be taken that there is no layer of air between the skin and the dosimeters because otherwise the results may not be representative of the surface dose.

The results of in vivo measurements taken during the first treatment fraction provide a check of the patient set up (always difficult for TSEI) and if necessary, corrective action can be taken for the subsequent treatment fractions. The series of in vivo measurements made during all the treatment fractions will enable the cumulative exposure to all areas of skin to be

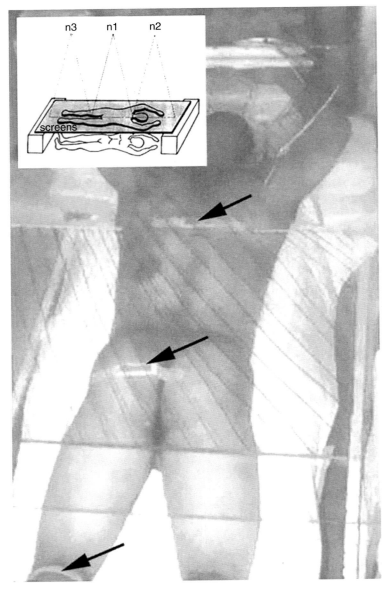

FIGURE 48.6
Example of in vivo dosimetry practised at Henri Mondor Hospital for total electron irradiation: TL dosimeters, shown in the figure by arrows, are placed on each beam entrance, along the 2 field junctions and at curvatures. (Marinello, G. and Le Bourgeois, J. P., Radiotherapie Oncologique, Editions Hermann, Paris, 1992.)

determined and, in particular, to point out the unavoidable zones of overdosage or under-dosage due to the patient's anatomy. For instance, the supplementary dose to be added to the interior of the thighs, which are always underdosed, can be deduced from in vivo measurements.

The Stanford Technique: Many techniques are variations of the 'Stanford' technique (Karzmark 1967; Page et al. 1970). Therefore, although the principles have been outlined above, it is worth reiterating a few practical points. In the original Stanford approach, a treatment plane 2 m high and 0.8 m wide was defined 3 m from the end of the linear accel-erator treatment head. Field coverage was achieved by placing a 1 mm thick aluminum scatterer at the end of the treatment head, together with a slab of tissue-equivalent material sufficiently thick to reduce the energy of the electrons to the required level at the patient treatment plane. The resulting beam spread is sufficient to cover the treatment plane laterally with acceptable uniformity, but, in the vertical plane, coverage is achieved by abutting dual beams, one above the other, with the central axis of the lower beam aimed below the patient's feet and that of the upper beam aimed above the patient's head, using gantry angles approximately 20° below and 20° above the horizontal. In practice the optimum angles would be tested in an individual centre by measuring the distribution in the overlap region using film for the specific conditions of the particular beam, distance and irradiation set-up.

As noted above, this dual-beam geometry also has the advantage of reducing the intensity of x-ray contamination in the treatment plane to less than 1% of the peak electron dose in the dual field, compared with around 2% on the central axis of each of the two component beams. This is an important consideration, because adequate dose uniformity around the patient is achieved with an arrangement of six dual beams at intervals of 60°. Because three beams generally contribute to the surface electron dose, but all six contribute to the x-ray dose, the percentage x-ray contamination for the complete treatment is effectively double that of one dual beam. For a 0.7% x-ray contamination in one dual beam, a typical prescribed skin dose of 36 Gy would result in an x-ray dose of $2 \times 0.7\% \times 36$ Gy $= 0.5$ Gy, which is clinically acceptable.

Total energy loss from the linear accelerator head to the patient, including the energy degradation of the scatterers, is typically 2–4 MeV. Because of the contribution of oblique beams, in a typical full six dual-beam treatment, where the incident electron energy at the treatment plane is 4 MeV, the depth of the 90% dose will be of the order of only 2–3 mm, and the dose maximum will be at a depth of 1 mm or less. Using the six dual beams, the mean dose in a cylindrical phantom at the depth of maximum dose will be typically 2.5–3 times that of only one of the six dual-beams measured at the centre of the patient treatment field at the depth of maximum dose.

Patients support themselves by holding a light frame or an overhead grip that arranges the arms in a position to minimise self-shielding on at least one arm. The arms are alternated in position on alternate fractions.

Throughout treatment consideration must be given to shielding parts of the body as the accumulate dose. Typically this will include nails, hands and feet, but additional areas may be selected to be excluded from treatment due to non-involvement. Internal eye shields positioned beneath the eyelids to protect the lenses pose problems of backscatter. Another choice may be to shield the eyelids and to boost the lids later with low-kV x-rays using an eye shield under the lids. Shielding must be done with 2–3 mm of lead placed close to or on the patient's skin, rather than close to the linear accelerator head. Dose should be monitored during treatment with a dosemeter such as an ionisation chamber placed close to the treatment plane (e.g. attached to the degrader sheet). Thermoluminescent dosemeters may be used to monitor dose on selected areas of the body if required. Careful assessment of the dose to under-dosed areas will provide the basis for calculating boost doses if these are deemed clinically necessary, but it can be noted that some centres only use boost doses in exceptional circumstances, yet report results that are apparently no worse than at other centres.

48.4　CONCLUSIONS

Total skin electron irradiation is a complex technique that can never fully meet its ideal aims. Dose uniformity to ±5% can be readily achieved in the basic distribution across an area large enough to cover the whole patient. However, the addition of these distributions around variable-size body cross-sections will produce further variations in dose of up to typically ±10% at best. Even in the best circumstances there will be under-dosed regions that will not be clearly defined, so that even when boost doses are added to these regions the variation will probably still be between $+10\%$ and -25%. Where under-dosed regions are not boosted, variations may be significantly greater.

All the techniques employed for TSEI are specialized and time-consuming, requiring extensive development and careful dosimetry, planning, execution, and quality control. Referral to centres specialising in TSEI is often the preferred option, thus balancing the effort required to implement and maintain such techniques against the number of patients for which it is provided.

CHAPTER 49

HIGH-LET MODALITIES

Roland Sabattier, Oliver Jäkel, and Alejandro Mazal

CONTENTS

49.1 THERAPY WITH NEUTRONS

49.1.1 INTRODUCTION

The interest in neutron beams in radiation therapy is based solely on their radio biological effectiveness (Raju 1980; Withers 1985; Griffin et al. 1988; see also Section 7.14). The depth-dose characteristics of the early neutron beams were poor, resembling those of kilovoltage x-ray and ^{60}Co γ-ray beams (Fowler 1975). In Table 49.1 it can be seen that the depth–dose

TABLE 49.1

Neutrons Generated in Different Facilities; the 50% Depth Is the Depth of the 50% Isodose on the Central Axis of the Beam

Site	Incident Particle	Particle Energy (MeV)	SSD (cm)	50% Depth (cm)	Depth of d_{max} (mm)
Louvain	Protons	65	162.5	17.6	16
Seattle	Protons	50	150	14.8	
Orleans	Protons	34	169	12.8	7
Detroit	Deuterons	48	183	13.5	9
Chiba	Deuterons	30	175	11.7	4
Essen	Deuterons	14	125	8.1	

Source: From International Commission on Radiation Units and Measurements (ICRU), Report 45, Clinical neutron dosimetry part I: Determination of absorbed dose in a patient treated by external beams of fast neutrons, ICRU, Bethesda, MD, 1989; Maughan, R. L. and Yudelev, M. *Med. Phys.*, 22, 1459–1465, 1995. With permission.

characteristics of the current neutron facilities resemble those of 4-MV to 8-MV x-ray beams. The very high density of ionisation (or high LET [linear energy transfer]) along the path of secondary particles in neutron-irradiated media is associated with a greater response of (radio-resistant) tumours and the reduced influence of conditions such as oxygen status, cell line, mitotic cycle and repair phenomena (see Section 7.14).

The first patients were treated with neutrons in the late 1930s. Currently, only about 10 facilities worldwide are involved in external-beam neutron therapy. Some of those programs were initiated two decades ago. Some old neutron facilities are ceasing operation, while there are few, if any, current proposals to build new facilities for fast neutron therapy. This small number of facilities is partly due to the relatively high cost of equipment compared to photon therapy but principally to the high frequency of late sequelae of the preliminary clinical trials done in the 1930s and some poor clinical results reported in the 1980s. Those results can now be clearly attributed to the difficulties experienced by all the pioneer teams in dealing with the radiobiological properties of neutron beams and the suboptimal technical quality available at the early facilities. These early facilities had beams of low energy and hence poor penetrating power and more recent higher energy installations did not have satisfactory facilities for producing irregular fields or the necessary advanced treatment planning systems (Scalliet 1991). Because of the reduction in the differential biological effects between tumour and normal tissues observed with high-LET particles, it is essential that neutron therapy be applied with conformal techniques.

49.1.2 FAST NEUTRON BEAMS

49.1.2.1 Fast Neutron Interactions in Biological Tissues

Neutrons, which are indirectly ionising, interact with nuclei of biological tissues through elastic and inelastic scattering, nuclear reactions and neutron capture. Ionisations are produced by charged secondary particles such as protons, α-particles and recoils of heavy nuclei (carbon, oxygen, and nitrogen ions) (Bewley 1989). Because of the relative masses of hydrogen, oxygen, carbon and nitrogen in soft tissues, the relative cross-section values and the mean energy transferred to nuclei by incident neutrons are such that hydrogen atoms are responsible for the major part of the energy transfer. One consequence of using neutron rather than photon beams is a smaller risk of bone necrosis but a higher risk of fibrosis in fatty (or *adipose*) tissues (due to their high hydrogen content).

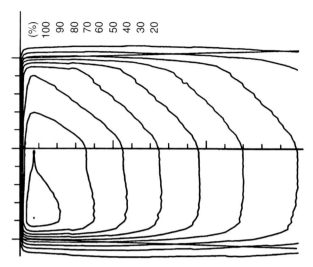

FIGURE 49.1
Partially wedged isodoses for a p(34)+Be neutron beam, field size: 10×10 cm. (From the Orléans facility, France.
With permission.)

Neutron dose distributions at a *macroscopic* level (on-axis depth doses and isodoses curves)
are quite similar to those of photons (Figure 49.1; see also Figure 46.1).

However, dose distributions at a *microscopic* level (i.e. subcellular level) are very different,
due to the high values of LET associated with the charged secondary particles generated by
neutrons (Figure 49.2). High neutron energies give rise to a broad spectrum of LET values.
This is because the mean energy of recoil protons increases, and the associated LET decreases,
with increasing neutron energy.

The secondary α particles produced by neutron interactions are especially important
because of their very high LET. The part of the LET spectrum above 250 keV μm^{-1} is due
to heavy recoils such as carbon or oxygen ions.

49.1.2.2 *Beam Production*

Many deuterium–tritium generators were used in the 1980s. Fusion reactions

$$d + t \rightarrow n + {}^4He$$

produce neutrons at 14 MeV to 15 MeV. These monoenergetic neutron beams had a percentage depth-dose value at 10 cm depth on the central axis of about 50% for a source-surface
distance (SSD) in the range 80 cm to 100 cm, with a typical dose rate of about 15 cGy min^{-1}
at 80 cm. The 80% to 20% penumbra at 10 cm depth was larger than 2.4 cm. These facilities,
which were designed to accelerate deuterium nuclei to energies of the order of 250 KeV, are no
longer used because of the poor physical characteristics of the neutron beams and the relatively
short lifetime of tritium (in the range 80 to a few hundred hours). Another important limiting
factor was the high activity of tritium (typically 20 TBq) which had to be contained*.

Current facilities are all based on cyclotrons accelerating either deuterons (stripping
reaction) or most frequently protons (inelastic reaction) on a beryllium target.

---

* The twenty-first century version of the deuterium–tritium neutron generator uses a sealed tube containing a solid
target made of titanium, scandium, or zirconium metal hydrides and an acceleration potential of 80–180 keV, but this
does not solve all the associated problems.

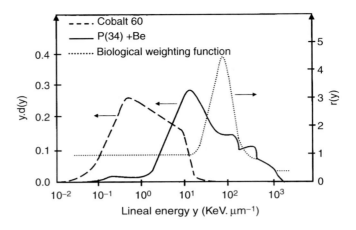

FIGURE 49.2

Comparison of microdosimetric spectra (dose distributions in lineal energy d(y) for a 2 μm simulated diameter—see footnote to Section 49.1.2.5) for ^{60}Co photons and p(34)+Be fast neutrons (left ordinate). A biological weighting function r(y) for an early effect in mice (right ordinate) is used to calculate the corresponding weighted distributions r(y)d(y) for radiobiological intercomparisons of various neutron beams. (From Wambersie, A. and Menzel, H. G., *Radiat. Prot. Dosim.*, 70, 517–527, 1997. With permission.)

Deuterons produce neutrons because the neutrons are *stripped* from the deuteron as it passes close to a nucleus whereas protons interact directly with the nucleus of the atom resulting in the ejection of a neutron. Beryllium is the most effective target for neutron production (Bewley et al. 1984) and is able to dissipate the heat generated (typically 1–2 kW) because of its high thermal conductivity and stability at high temperatures. Deuterons and protons interacting with beryllium give different neutron energy spectra and dose rates. The higher the energy of the incident particle, the higher the energy of the neutron beam with consequently better penetration.

The maximum energies used today in neutron therapy facilities are 66 MeV for protons and 50 MeV for deuterons (Jones 1999). For deuterons and protons of the same energy, the neutron dose rate is higher with deuterons but the neutron energy (and hence penetration) is greater with protons. An even better way to increase the penetration of the neutron beam is to use protons with semi-thick targets and hydrogenous filters (such as high-density polyethylene) to reduce the high fluence of low-energy neutrons. A disadvantage of the proton-induced reaction is the broader angle of emission of the neutrons, requiring more shielding around the target and the primary collimator. The large size of treatment heads results in SSDs generally greater than 150 cm. Details of some neutron facilities are given in Table 49.1. For these high-energy neutron beams the 80% to 20% penumbra width is less than 2 cm but is still larger than that obtained with high-energy x-rays.

As observed with high-energy photons, the on-axis depth dose increases from the surface to a maximum depth at which charged secondary-particle equilibrium is reached (see Section 6.5). The depth of maximum dose is less than that for megavoltage photon beams because of the shorter ranges of the secondary particles, as shown in Table 49.1. For D–T generators, the depth-dose maximum is at only 2 mm.

In practice there is a limit to the incident proton energy of about 70 MeV because of the decrease in the LET of the neutrons at higher energies, which reduces the radiobiological benefit. Seventy mega-electronvolt protons produce a neutron beam with depth-dose characteristics similar to those of an 8-MV x-ray beam.

49.1.2.3 The Treatment Head and Equipment

The treatment head of a neutron machine is larger than that of an x-ray machine due to the larger size of the magnets and the necessary shielding. They are, however, designed in a similar way: target, flattening filter, transmission ionisation chambers, and beam modifiers (see Section 11.4). Beam monitoring can also be performed by measuring the total electric charge of particles incident on the target.

Various collimation and shielding systems have been proposed, including compound shields: steel for high-energy neutrons, followed by polyethylene to absorb low-energy neutrons and an outer layer of lead to attenuate gamma rays emitted from the second layer. Borated wood collimators have also been used.

It is clearly established that the *shape* of the neutron field has to conform to that of the clinical target volume for each beam direction (i.e. the beams's eye view; Section 32.4.2) (Austin-Seymour et al. 1994; Russell et al. 1994). A minimum requirement is therefore to have shielding blocks adapted for each beam with transmission as low as for those used with high-energy x-rays (see Section 22.5). For the highest energies (typically above 50 MeV), the weight of the blocks needed to shape the fields is quite high, and the high levels of radioactivity induced in the shielding blocks makes it difficult to limit the doses received by the radiographers. Multi-leaf collimators (see Section 22.5.4) are therefore essential to optimise the use of high-energy neutron beams.

Many of the existing facilities are non-isocentric or not-fully-isocentric. An example of an isocentric facility with multi-leaf collimator is that in Seattle (U.S.A.), in clinical operation since 1985. In Detroit, a compact isocentric cyclotron with a multirod collimator has been used since 1990. For nonisocentric facilities, heavy and complex patient fixation devices have been designed to achieve the patient positions required for the different beam directions required by the treatment plan.

The ancillary equipment such as light-beam simulation and lasers is quite similar to those used for photon-beam therapy machines.

49.1.2.4 Dosimetry

The dosimetry protocols for therapeutic applications of fast neutron beams in U.S. and European centres were standardised in 1987 and the common protocol is described in ICRU Report 45 (1989) and by Mijnheer et al. (1987).

A clinical neutron beam always consists in practice of a mixed field of neutrons and photons. Their respective contributions to the total absorbed dose have to be determined separately because of the large difference in relative biological effectiveness (RBE) of each component. The photon component which is generated in the target, collimation devices or the patient, accounts for a few percent of the total dose[*]. For routine measurements, the neutron and photon dose components are separated by a double-detector method based on two instruments with different sensitivities to the two types of radiation: typically an A150 ionisation chamber and a Geiger–Müller (GM) counter. The GM counter is primarily sensitive to the photon component of the dose, but the neutron sensitivity of the counter, k_u, has to be taken into account. Data from microdosimetric measurements with tissue-equivalent proportional counters are used to derive a value of k_u.

The combined neutron and photon dose is measured with a tissue-equivalent plastic (Shonka A150) ionisation chamber filled with tissue-equivalent methane-based gas (to correspond to *Fano-theorem* conditions; see Section 6.7.7). Such homogeneous chambers are used because the conditions for a *Bragg–Gray* cavity (see Section 6.7.3) cannot be fulfilled in practice and because there is no *cavity theory* applicable to fast neutron dosimetry with inhomogeneous chambers. Chambers filled with gas are calibrated in terms of air kerma in a ^{60}Co γ-ray beam at a National Standards laboratory.

---

[*] The photon dose can be expressed as a percentage of the total dose at 2 cm and 10 cm depth. Typical percentages are, respectively, 5.2% and 7% (Orleans), 6.5% and 8% (Seattle), and 3.1–4.1% (Louvain) (ICRU 1989; Sabattier et al. 1990).

An A150 calorimeter is the most fundamental instrument for measuring absorbed dose in a neutron beam (see the Introduction to Part D) with the minimum uncertainty in the determination of the total absorbed dose. Caumes et al. (1985) described the calibration of an ionisation chamber in terms of absorbed dose in A150 determined using a calorimeter directly in a clinical neutron beam.

For clinical applications, dose distributions have to be measured in a water phantom but the dose should be specified in ICRU muscle as reference material. Kerma ratios can be calculated from neutron energy spectra. For neutron energies higher than 15 MeV a full calculation cannot be made and the dosimetry protocol (ICRU 1989) recommends a value of 0.95 for the K_{tissue}/K_{A150} ratio.

Kerma in fatty tissue with a high percentage of hydrogen is 13% to 15% higher than kerma in ICRU muscle (for the same neutron fluence). By contrast, kerma in bone is only 59% to 66% of the kerma in reference ICRU muscle.

49.1.2.5 Microdosimetry

Specification of radiation quality is important for clinical neutron beams because, although radiobiological parameters are dependent to some extent on the proportion of photons in the beam, the principal determinant of the RBE is the secondary charged-particle spectrum (ICRU 1983; Pihet 1989).

Microdosimetric measurements provide an effective way to describe the beam quality. Microdosimetric spectra (Figure 49.2) and data such as dose-mean lineal energy* (y_D) or dose-mean lineal energy corrected for saturation (y^*) can be obtained with A150 proportional counters containing tissue-equivalent gas at low pressure to simulate micrometer-sized volumes (Liu 2000). Relevant radiobiological data can be derived from variations of $yd(y)$ vs. y spectra or from the parameters y_D and y^* at different depths, or at off-axis points in a clinical neutron beam as well as at a reference point common to several facilities (Pihet et al. 1988; Menzel et al. 1990; Wambersie and Menzel 1997).

Proportional counters are frequently used to analyse a mixed photon–neutron field both inside and outside the beam and are therefore useful for radiation protection measurements around medical facilities or high-energy accelerators. A powerful technique for obtaining values of the quality factor Q for a beam of mixed energy is to use the values of $Q(L)$ as a function of LET, L, published in ICRP 60 (1990) to weight the $yd(y)$ vs. y spectrum, for any point of interest inside or outside the beam (Folkerts 1988; Vynckier et al. 1988).

49.1.2.6 Clinical Use of Fast Neutrons

For identical absorbed doses, dose distributions at a microscopic scale are vastly different according to whether one uses low-LET radiation such as photons and electrons or high-LET neutrons. Figure 49.3 shows the physical dose distribution, which is similar to that obtained with a photon beam, but the very high density of ionisation along the path of the secondary particles (high LET) results in high RBE values. RBE values for neutrons are much higher than those observed for high-energy photons or electrons (which are low-LET radiation; see Section 7.14). They are also higher than the RBE of protons and low energy x-rays. For fast neutrons, the RBE varies within wide limits (about 2 to 5) depending on the neutron energy spectrum, dose, and biological system (Wambersie and Menzel 1997) and therefore an

---

* *Lineal energy* is defined by ICRU as ε/l, where ε is the energy imparted to matter in a volume of interest by an energy deposition event and l is the mean chord length in that volume. This is not the same as the LET, (see Section 7.14) which is a *macroscopic* quantity. Lineal energy can be described in terms of the *frequency-mean* lineal energy which is the mean of the lineal energy distribution or the *dose-mean* lineal energy which is the mean of the distribution weighted for dose. The quality factor, Q, (see Section 59.1.2) expresses the relationship between the absorbed dose and the *equivalent dose* and is related to the lineal energy.

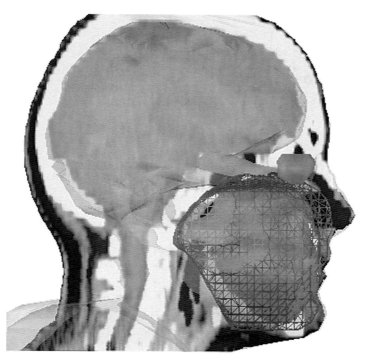

FIGURE 49.3
(See colour insert following page 590.) Example of 3D dose distribution for a large tumour of the maxillary sinus with irregular fields and partially wedged neutron beams at the Orléans neutron therapy facility. The wireframe represents the prescribed dose distribution around the clinical target volume.

accurate determination of the RBE relative to photons is necessary before starting clinical applications and when exchanging clinical results (Gueulette et al. 1984, 1996; Grégoire 1993). Some heavy-ion beams (e.g. carbon, see Section 49.3) have RBE values similar to those of neutron beams.

Differential biological effects (e.g. dependence on oxygenation) observed when using low-LET radiation (i.e. photons and electrons) are significantly reduced with high-LET radiation. Clinical interest in neutrons is based on this fact. Radiosensitivity is less dependent on cell type and other related factors with high-LET than with low-LET radiation. Environmental conditions (e.g. oxygen status), position in the cell cycle and repair mechanisms, all play a less important role than with photons. The large proportion of time spent in the radiation-tolerant resting phases of their cycle makes the cells in slowly growing tumours relatively resistant to low-LET radiation. The similar responses observed for all cells in all conditions under high-LET irradiation lead us to expect radio-resistant tumours to have a greater response to treatment with neutrons than with photons, for a similar response from the surrounding healthy tissue.

The most promising results reported for external neutron therapy are in the treatment of large, slow-growing or radioresistant tumours such as salivary gland tumours, locally advanced prostatic adenocarcinomas, low-grade soft-tissue sarcomas and some sarcomas of bone (Wambersie and Menzel 1996).

49.1.3 CONCLUSIONS

Tumour response to irradiation with fast neutrons is much less dependent on the differential biological effects currently observed with photons and electrons (see Part B). Basic radiobiological data justify neutron therapy for some radioresistant tumours but it must be emphasised that all treatments require a conformal approach. Conformal neutron therapy did not exist for the pioneers in the 1980s, but was also not available at some more recent facilities because of the difficulties of designing irregular fields and isocentric gantries. The use of fast neutrons necessitates investigations in the fields of radiobiology, microdosimetry and dosimetry. For more details, readers are encouraged to consult the historical review of fast neutron therapy including current trends and future needs by Wambersie and Menzel (1996).

49.2   BORON NEUTRON CAPTURE THERAPY

A potentially promising way to treat highly malignant tumours like glioblastomas and melanomas is to enhance the biological effects of neutrons on tumour cells using thermal boron neutron capture therapy (BNCT) in which ^{10}B atoms are incorporated within tumour cells and then irradiated with thermal neutrons (Zamenhof et al. 1975; Barth et al. 1996). Boron has a very large cross section for absorbing thermal neutrons. In consequence, the thermal neutron components of the beam spectrum produce a selective enhancement through the ^{10}B (n, α) ^{7}Li reaction. The secondary α and recoil ^{7}Li ion particles have an average total kinetic energy of 2.33 MeV, with ranges of 9 µm and 5 µm respectively, i.e. close to tumour-cell diameters, delivering a differential dose to the boron-loaded tumour cells. The thermal neutron capture in ^{10}B is two orders of magnitude greater than in tissue, but a minimal and differential boron concentration must be ensured at the tumour level (Bartelink et al. 1996). This can be achieved, for example, where a tumour breaks the blood–brain barrier allowing large boron compounds to selectively enter tumour cells in the brain.

Patients were treated by BNCT in the U.S.A. in the 1950s at two facilities (Brookhaven [Farr et al. 1954] and MIT [Asbury et al. 1972]) and the technique was improved in Japan in the late 1960s, giving rise to new projects like the European Centre at Petten, The Netherlands (Sauerwein and Zurlo 2002).

Two ways to implement Boron capture therapy are being investigated:

- BNCT with thermal neutrons from a reactor or from low-energy, high-current accelerators (Chadha et al. 1998).
- Boron neutron capture enhancement of fast neutron beams (BNCEFN) with the thermal component of the neutron beam (Pignol et al. 1999).

Barth et al. (2004) also report some success in combining BNCT with external beam therapy.

In parallel, the likely degree of success of BNCT is related to the design of boron compounds with a differential selectivity for tumour cells (Bartelink et al. 1996; Cumberlin 2002; Barth 2003).

49.3   LIGHT AND HEAVY IONS

49.3.1   INTRODUCTION

Nuclei heavier than protons are of interest for radiation therapy (Raju 1980; Kraft 2000) as they combine excellent dose-distribution properties similar to proton beams (Chapter 46) with radiobiological properties similar to neutron beams (Figure 49.4). If these two high-LET modalities are compared, the relationship of neutron therapy to heavy-ion therapy is akin to that of high-energy x-ray therapy to proton therapy (Chapter 46).

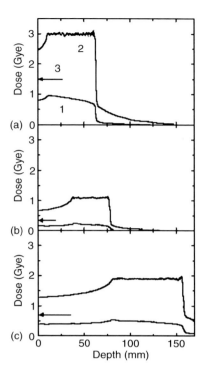

FIGURE 49.4
Modulated depth–dose distribution for a carbon-ion beam for various modulation depths and dose levels, corresponding to different RBE values in the center of the SOBP: (a) RBE=3.4; (b) RBE=5.5; (c) RBE=4. In the diagram, curve 1 shows the modulated *physical* depth-dose. Curve 2 shows that because of the increase of the RBE with depth, the resulting biologically effective dose is homogeneous throughout the target volume. It can be seen that the RBE is higher for lower dose levels and that the depth modulation is not constant. The arrow (3) indicates the effective skin dose. The latter differs from curve 1 because the biological effectiveness of ion beams depends on cell type.

The rationale for the use of ion beams in radiotherapy, including the practical clinical limitations, is as follows:

1. Multiple-scattering processes are less important for heavier ions compared to protons, due to the greater mass of the former. Consequently, range straggling as well as lateral scattering (see Figure 46.2) is reduced; the resulting lateral and distal dose gradients are therefore steeper than for proton beams. However, in contrast to protons, nuclear interactions of heavier ions with matter occur primarily as so-called projectile fragmentation. These light fragments are mainly produced close to the Bragg peak and the range of some fragments is greater than that of the primary ion. This leads to a tail of secondary particles beyond the Bragg peak (Figure 49.4), which becomes more pronounced for heavier ions. Due to this tail, ions heavier than neon are less advantageous for radiotherapy.

2. The LET is higher for heavier ions, and consequently the RBE is greater. For high-LET radiation the oxygen enhancement ratio (OER) is lower and the cell-survival curves become more and more linear so that fractionation effects become smaller. This can be expressed in terms of a higher α/β ratio (see Section 7.14). This is the rationale for hypo-fractionation using ion beams, i.e. treatment protocols with a smaller number of fractions and a shorter overall treatment time (see Chapter 9). However, the lack of fractionation sensitivity is limited to the high-LET region in the Bragg peak. Normal tissue in the entrance region, where the LET is lower, will still benefit from fractionation. For very high LET values (beyond 1000 keV cm^{-1}), the RBE decreases again, and this is another factor limiting the use of ions heavier than neon (Figure 49.5). As an example, for argon ions the RBE close to the surface is higher than that in the Spread-Out Bragg Peak (SOBP) region.

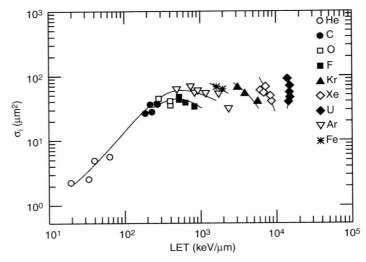

FIGURE 49.5
Variation of the inactivation cross section (and consequently RBE) for mammalian cells after irradiation with various heavy ions with different LET values. The increase with higher LET and for heavier ions is clearly visible. At very high LET values (reached with ions with $Z > 9$) a reduction in the inactivation cross section is observed. (Courtesy of G. Kraft.)

3. Due to the nuclear fragmentation, heavy ions produce some radioactive isotopes in the body which are positron emitters, creating the possibility of monitoring the stopping point of the projectile ions using online PET (Positron Emission Tomography) techniques (see Figure 49.6). The use of a radioactive beam (e.g. ^{11}C ions) for therapy is currently being investigated at Heavy Ion Medical Accelerator (HIMAC) to increase the PET signal (Li et al. 2004). However, the intensity of these beams is very low, which increases the treatment time.

4. The costs of beam production, beam transport (including an isocentric gantry) and shielding, are considerably higher than for protons.

5. Although some experience has been gained, patient numbers are still low and it has not yet been demonstrated in large clinical trials which tumours would benefit from heavy-ion therapy.

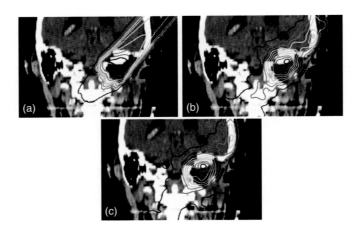

FIGURE 49.6
(See colour insert following page 590.) Example of the online PET imaging of the induced activity in a patient treated for a tumour of the skull base. The figure shows the absorbed dose distribution of one of the treatment fields (a), the expected activity as calculated from the treatment plan (b) and the measured PET activity (c). Due to *wash out* effects in soft tissue, the measured activity is in general somewhat lower than the calculated activity, while in the bony structures the measured and calculated activities agree very well. (Courtesy of W. Enghardt.)

The biological characteristics of helium beams ($Z=2$) are similar to those of proton beams. Beams of nuclei with atomic numbers up to neon ($Z=10$) are called *light* ions, including carbon ($Z=6$) and oxygen ($Z=8$). Beams with higher atomic number (silicon $Z=14$, argon $Z=18$, and so on) are called *heavy* ions. There is also interest in the use of *radioactive* beams, based on the optimal use of PET techniques for *online* dose-deposition monitoring.

49.3.2 Physical Parameters, Radiobiological Data and Treatment Planning

As the LET and RBE increase towards the Bragg peak, the *physical* dose has to be modulated in such a way that the resulting *biologically effective* dose is constant throughout the target volume. Modulated depth–dose curves for heavy-ion beams therefore show a significant decrease towards the distal end (Figure 49.4). While the entrance dose is lower and the distal gradient is higher than in proton beams, there is a tail of secondaries behind the Bragg peak that increases with the atomic number of the particle. Ions in the region around carbon are considered nowadays as optimal for clinical applications, since they combine very high RBE values with only a moderate fragmentation tail. An overview of the medical physics aspects of carbon-ion therapy can be found in Jäkel et al. (2003a).

Dosimetry intercomparisons have been performed between one of the Japanese facilities (HIMAC) and the German facility, Gesellschaft für Schwerionenforschung (GSI), using cylindrical ionisation chambers (Fukumura et al. 1998; Hartmann et al. 1999). The Japanese and the German dosimetry protocols are very similar to dosimetry protocols for proton beams (see Section 46.4.3). While the Japanese protocol is based on ionisation chambers calibrated in air kerma, the German protocol uses chambers calibrated in absorbed dose to water (see Section 18.2 and Section 18.3).

There are small differences between the values adopted by the above two institutions for (W_{air}/e) and for the water-to-air mass stopping-power ratio ($s_{w,air}$) for carbon ions. However, the product of these parameters differs only by about 0.5%, which is also the difference between the final dose determinations at the two institutions. The International Atomic Agency has published a dosimetry protocol for heavy-ion beams (IAEA 2000), which is based on chambers calibrated in absorbed dose to water and has been shown to be very similar to the German dosimetry protocol (Jäkel et al. 2003b). In spite of the agreement between the centres, the overall uncertainty in heavy-ion dosimetry due to a lack of data on W-values and stopping-power ratios is around 3%. There is agreement that cylindrical ionisation chambers (see Chapter 15) are the most suitable types for the dosimetry of carbon ions following international protocols such as TRS-398 (IAEA 2000).

It is mandatory that radiobiological modelling of the depth–dose modulation be included in the treatment planning process, in order to account for the variation of RBE and thus to achieve a homogeneous biological effect in the tumour. The RBE is dependent on the spectrum of secondary particles (which varies with depth), the absorbed dose, the fractionation scheme, the endpoint under consideration (e.g. cell killing) and the type of cells in the irradiated tissue. For a *broad beam* technique such as that used at HIMAC, the dependence of RBE on the depth-dose determines the shape of the depth modulators, which are defined for a particular fractionation regime. If a beam-scanning technique is used, with variable depth–dose modulation, the composition of the radiation field at every point has to be known to be able to model the biological effectiveness for the patient. In Figure 49.4, some examples of the varying depth modulation for a carbon ion beam are shown, as applied to the treatment of chondrosarcoma patients at GSI. The basis of the underlying biological modelling of these depth–dose curves is described in Krämer (2000). The clinical treatment planning aspects are discussed in Jäkel et al. (2001).

49.3.3 ACCELERATORS, BEAM-DELIVERY SYSTEMS, AND PET MONITORING

Due to the high energies needed for the treatment of deep-seated tumours using beams of heavy ions, synchrotrons are the usual choice for accelerators in a clinical facility.

At Berkeley, a heavy-ion linear accelerator (SUPERHILAC) served as an injector to a synchrotron (BEVATRON) in the Bevalac facility. This was primarily a research facility and the clinical program initiated in 1977 continued until the shutdown of the facility in 1992 using beams of helium, carbon, neon, argon, and silicon ions (Alonso 1994). Two treatment rooms with horizontal beams were available (Chu 1993). The majority of the patients were treated in a sitting position. For treatment planning, a CT scanner was modified to scan seated patients i.e. in horizontal planes.

The modulation of the depth-dose was performed using a fixed accelerator energy and *passive* range modulators, such as ridge filters and patient-specific compensators (or bolus on the skin surface). The beam was broadened using a double-scattering system which was upgraded to a *beam-wobbling* system in the 1980s. Patient-specific field collimators were used to confine the beam to the target volume. Just before the facility was shut down, a 2D magnetic beam scanning system together with a passive range modulator was installed, but this was used for the treatment of a single patient only.

Based on the experience in Berkeley, the first hospital-based ion facility, HIMAC, in Chiba, Japan became operational in 1994, offering ion beams from helium to argon, with a maximum range of 30 cm in tissue (energies up to 800 MeV/u*), a dose rate of 5 Gy/min (beam intensities of 2×10^9 particles/s), a field size of 22 cm, and vertical and horizontal beams (Kanai et al. 1999).

The accelerator complex consists of two different ion sources, followed by a radiofrequency (RFQ) and an *Alvarez* linac and a pair of superposed synchrotrons of about 40 m in diameter with a *flat top* intensity per pulse (20% duty factor) and a repetition rate of 0.5 Hz. The beams are fed to horizontal or vertical beam lines, in three different treatment rooms, a biological irradiation room and physical experimentation rooms. Patients can be treated either lying down or seated. At HIMAC the activation of tissue is measured *off-line* with a conventional PET scanner.

The beam delivery system is passive, as the Berkeley system was. It is based on a combination of wobbling magnets and scattering foils for beam spreading, *ridge* filters (instead of modulators) to spread out the Bragg peak, absorbers, a steel multileaf collimator and compensators.

At the research laboratory Gesellschaft für Schwerionenforschung (GSI) in Darmstadt, a therapy unit began clinical operation in 1997 (Kraft 2000). Only one treatment room is available, equipped with a horizontal beam line and a treatment couch. A treatment chair has also been installed.

The GSI synchrotron, Schwerionen-Synchrotron (SIS), can produce ion beams from protons through to uranium. However, only carbon ions are used for therapy. Energies up to several GeV/u are available, while the maximum energy used for therapy is 450 MeV/u.

The beam delivery system is dynamically controlled and allows 3D scanning of arbitrarily shaped volumes with a high spatial resolution (Haberer et al. 1993). Using a magnetic deflection system, the intensity-controlled raster scanner can deliver a monoenergetic pencil beam over an arbitrarily shaped area. To do so, a beam of 6 mm to 10 mm (full-width at half-maximum) is scanned over a regular grid of points with typically 2 mm to 3 mm spacing. After completion of a scan, the accelerator energy can be switched from pulse to pulse and

---

* u is the *unified atomic mass* or atomic mass unit, or Dalton (Da), which is the unit of mass used to express atomic masses and molecular masses. It is defined to be 1/12 of the mass of one atom of ^{12}C. Accordingly: $1u = 1/N_A$ gram $= 1/(1000N_A)$ kg (where N_A is Avogadro's number) $\approx 1.66053886 \times 10^{-27}$ kg

another scan can be performed with a different radiological depth. In total, 256 accelerator energies are available. The intensity control moves the beam to the next beam spot when a predefined number of particles has been reached. An on-line monitoring of the beam position and a feedback loop to the scanner is used to keep the beam stable at each scan spot.

In summary, the system allows a very high degree of dose conformation and, at the same time, makes any patient-specific beam-shaping hardware unnecessary. Furthermore a dedicated PET scanner is installed in the treatment room which enables the induced activity in the patient to be monitored *on-line* during the irradiation (Enghardt et al. 1999). An example of the capability of this PET camera is shown in Figure 49.6.

A second hospital-based ion therapy facility came into operation in Japan in 2001. For the first time, both proton and ion beams are available in the same centre. The Hyogo Ion Beam Medical Center (HIBMC) at Harima Science Garden City began operation with protons in 2001 and with carbon ions in 2002 (Kagawa et al. 2002; Itano et al. 2003). Six therapy rooms are available with seven treatment ports. Three rooms are dedicated to carbon ion beams: one with a vertical beam line, one with a horizontal one and one with a 45° oblique beam line. Two proton treatment rooms are equipped with commercially designed gantries. The beam delivery system is based on the HIMAC system.

49.3.4 CLINICAL PROGRAMS

The first experiences with the clinical application of ion beams were between 1977 and 1992 at the Lawrence Berkeley Laboratory (LBL). Treatments of skull-base tumours such as chordoma, low-grade chondrosarcoma and meningioma were performed, as well as tumours of the paranasal sinus, nasopharynx, bone, and soft-tissue sarcomas and prostate cancer (Castro 1994). The results of the clinical phase-I/II study at LBL have to be regarded in the light of the fact that there was at the time almost no clinical information available about the biological effectiveness of ion beams for the different types of tumour and normal tissues. The studies were defined as *dose-finding* studies. The majority of patients were treated with helium ions, which are very similar to protons in terms of biological effectiveness. Four hundred and thirty-three patients were treated with ions heavier than helium, the majority with neon ions, but carbon, argon and silicon ions were also used.

At the HIMAC in Chiba well over 2000 patients were treated with carbon ions during the first eleven years of operation. These were mainly tumours of the head and neck, lung, prostate, liver, bone and soft-tissue sarcoma, uterus, and esophagus (Tsujii et al. 2004).

Dose-finding studies for carbon-ion radiotherapy for prostate cancer, non-small-cell lung cancer (NSCLC) and soft tissue sarcomas (Tsujii et al. 2004) are ongoing at NIRS in Chiba, Japan. In two phase I/II trials for Stage I NSCLC, using different fractionation schemes (18 fractions in 6 weeks and 9 fractions in 3 weeks), a dose-escalation study was performed (Miyamoto et al. 2003). For unresectable bone and soft tissue sarcomas, a further phase-I/II trial was performed (Kamada et al. 2002), using 16 fractions over four weeks. These trials all show very promising results.

Between 1997 and the end of 2005, nearly 300 patients were treated with carbon-ion radiotherapy at GSI. The majority of patients were treated for chordoma and low-grade chondrosarcoma of the skull base using a dose of 60 GyE* in 20 fractions of 3.0 GyE. For these tumours carbon-ion radiotherapy is offered as a routine treatment to patients. Clinical phase I/II trials for the treatment of sacral and spinal chordoma as well as for adenoidcystic carcinoma are ongoing. For these treatments carbon ions are applied as a boost treatment (using six fractions of 3.0 GyE) after an intensity modulated radiation therapy (IMRT) treatment using photons. Although follow-up time is still limited, the control rates are

---

* GyE = cobalt gray equivalent.

very encouraging (Schulz-Ertner et al. 2004). Furthermore, no severe radiation-induced side effects (neither acute nor late) have been observed (Schulz-Ertner et al. 2002).

49.3.5　Economic Aspects, Future Developments and Conclusions

Although there are promising preliminary results from the clinical trials at HIMAC and GSI, the number of patients as well as the follow-up time is still limited. The clinical value of heavy-ion beams especially as compared to modern photon IMRT techniques (see Chapter 43 and Chapter 44) and proton radiotherapy (Chapter 46) has not yet been evaluated. To do so, hospital-based facilities must be established which enable randomised trials between heavy ions, photon IMRT and proton radiotherapy to be carried out for various indications, using similar levels of sophistication in treatment planning and beam delivery.

There are several facilities currently planned in Europe (Sisterson 2003a,b) that may serve this purpose. One is already under construction at the Heidelberg university hospital (Figure 49.7). The facility will feature a dedicated heavy-ion synchrotron, which will be a downsized version of the GSI synchrotron (Eickhoff et al. 2000; Bär et al. 2001). For the first time the facility will provide an isocentric gantry system (Spiller et al. 2000) as well as two fixed, horizontal beam lines. All treatment rooms will be equipped with the same beam-scanning method as is currently performed at GSI. The integration of beam scanning into the isocentric gantry is a technological highlight of the facility (Heeg et al. 2004). Furthermore, not only carbon ions but also protons, helium and oxygen ions will be available for radiotherapy. The facility is scheduled to be finished in 2006 and to start clinical operation in 2007,

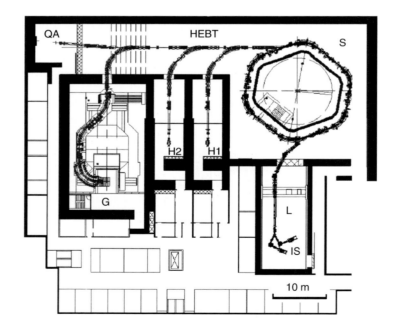

FIGURE 49.7
Layout of the clinical ion facility in Heidelberg, showing the two ion sources (IS) and Linac injector (L), synchrotron (S), high energy transfer beam line (HEBT), an irradiation room for dosimetry and quality assurance (QA), two treatment rooms with a fixed horizontal beam line (H1, H2) and the isocentric gantry (G) in the third treatment room. The shielding walls of concrete are also shown. In front of the entrance door to each treatment room, the patient changing rooms are shown. Offices are on the first floor of the building with all other technical equipment one floor lower. The size of the building is approximately 70 m × 70 m. It is connected to the radiotherapy department of the university hospital, where all diagnostic services are located. (Courtesy of Architects Nickl & Partners, Munich.)

the final capacity being 1000 patients treated per year. The projected capital investment in the facility is €72 M.

It is clear that the enormous capital investment and technical complications for a hospital-based ion facility make tumour therapy with ion beams more expensive as compared to conventional therapy (the treatment cost reimbursed currently for fully fractionated carbon-ion therapy at GSI is €20,000). However, if ion therapy for certain specific tumour types is proven to be superior to conventional therapy in terms of clinical outcome, it may come to be considered as a cost-effective therapy. This is due to the high costs associated with the treatment of recurrent tumours (Land et al. 2005).

Another clinical facility for protons and carbon ions is under construction in Pavia near Milan. It will feature a proton gantry but only fixed horizontal beam lines for carbon ions. The facility is planned to be operational in 2008. A great deal of effort has also been expended in the design and planning of some other European ion-beam centres, such as the MedAustron project in Wiener Neustadt (Austria), the Etoile project in Lyon (France) and the Nordic Light-Ion facility in Stockholm (Brahme 2004; Svensson et al. 2004).

When the first of these new hospital-based facilities comes into clinical operation, a number of questions have to be addressed: what tumour indications can be treated most effectively with ions? Which tumours are more effectively treated with IMRT or protons? Which type of ion is best suited for which tumour? Is ion therapy cost effective? It will certainly take many years of clinical, radiobiological and physical research to answer these questions and thus determine the role of ion beams in tumour therapy.

PART I: SPECIAL TECHNIQUES
REFERENCES

AAPM (American Association of Physicist in Medicine), Report 16, *Protocol for Heavy Charged-Particle Therapy Beam Dosimetry,* American Institute of Physics, New York, 1986.

AAPM (American Association of Physicist in Medicine), Report 23: *Total Skin Electron Therapy: Technique and Dosimetry*, American Institute of Physics, NewYork, 1987.

AAPM (American Association of Physicist in Medicine), Report 54: *Stereotactic Radiosurgery*, Report of Task Group 42, American Institute of Physics, New York, 1995.

Abraham, D., Colussi, V., Shina, D., Kinsell, T., and Sibata, C., TBI treatment planning using the ADAC pinnacle treatment planning system, *Med. Dosim.*, 25, 219–224, 2000.

ACR (American College of Radiologists), *ACR Practice Guideline for the Performance of Total Body Irradiation*, American College of Radiologists, Reston, VA, 2001.

Adams, E. J., Suter, B. L., Warrington, A. P., Black, P., Saran, F. H., and Brada, M., Design and implementation of a system for treating paediatric patients with stereotactically guided conformal radiotherapy, *Radiother. Oncol.*, 51(Suppl. 1), 8, 1999.

Adams, E. J., Cosgrove, V. P., Shepherd, S. F., Warrington, A. P., Bedford, J. L. et al., Comparison of a multi-leaf collimator with conformal blocks for the delivery of stereotactically guided conformal radiotherapy, *Radiother. Oncol.*, 51, 205–209, 1999.

Adams, J. A., Paiva, K. L., Munzenrider, J. E., Miller, J. W., and Gragoudas, E. S., Proton beam therapy for age-related macular degeneration: Development of a standard plan, *Med. Dosim.*, 24, 233–238, 1999.

Ahnesjö, A., Hårdemark, B., Isacsson, I., and Montelius, A., The IMRT information process-mastering the degrees of freedom in external beam therapy, *Phys. Med. Biol.*, 51, R381–R402, 2006.

Aird, E. G. and Conway, J., CT simulation for radiotherapy treatment planning, *Br. J. Radiol.*, 75, 937–949, 2002.

Alexander, E., Kooy, H. M., van Herk, M., Schwartz, M., Barnes, P. D. et al., Magnetic resonance image-directed stereotactic neurosurgery: Use of image fusion with computerized tomography to enhance spatial accuracy, *J. Neurosurg.*, 83, 271–276, 1995.

Alonso, J., Synchrotrons: The American experience, in *Hadrontherapy in Oncology, Proceedings of the 1st International Symposium on Hadrontherapy*, Elsevier, Amsterdam, pp. 266–281, 1994.

Alsmiller, R. G., Santoro, R. T., and Barish, J., Shielding calculations for a 200 MeV proton accelerator and comparisons with experimental data, *Particle Accelerators*, 7, 1–7, 1975.

Altman, M. B., Chmura, S. J., Deasy, J. O., and Roeske, J. C., Optimization of the temporal pattern of radiation: an IMRT based study, *Int. J. Radiat. Oncol. Biol. Phys.*, 66, 898–905, 2006.

Amaldi, U. and Silari, M., Eds., *The TERA Project and the Centre for Oncological Hadrontherapy*, Instituto Nazionale di Fisica Nucleare, Rome, 1994.

Amerio, S., Boriano, A., Bourhaleb, F., Cirio, R., Donetti, M., et al., The Magic cube and the pixel ionization chamber: Detectors for monitor and dosimetry of radiotherapy beams, http://www.to.infn.it/activities/experiments/tera/publications/pixel_chamber/siena_2002.pdf, 2002.

Anacak, Y., Arican, Z., Bar-Deroma, R., Tamir, A., and Kuten, A., Total skin electron irradiation: Evaluation of dose uniformity throughout the skin surface, *Med. Dosim.*, 28, 31–34, 2003.

ANDEM (Agence Nationale pour le Développement de l'Evaluation Médicale), Therapeutic use of cyclotrons in oncology: Clinical and economic evaluation, *Usage Thérapeutique des Cyclotrons en Cancérologie: Évaluation Clinique et Économique*, ANDEM, Paris, 1995.

Antoni, R., Radiation protection study of activation in a 200 MeV proton beam intended for clinical application (*Etude de la radioprotection liée à l'activation des pièces soumises à un faisceau de protons de 200 MeV destiné à des applications cliniques*) Mémoire Physique Fondamentale, Université Paris VII., Paris, 1996.

Asbury, A. K., Ojemann, R. G., Nielsen, S. L., and Sweet, W. H., Neuropathologic study of fourteen cases of malignant brain tumor treated by boron-10 slow neutron capture radiation, *J. Neuropathol. Exp. Neurol.*, 31, 278–303, 1972.

Ashamalla, H., Addeo, D., Ikoro, N. C., Ross, P., Cosma, M., and Nasr, N., Commissioning and clinical results utilizing the Gildenberg-Laitinen adapter device for x-ray in fractionated stereotactic radiotherapy, *Int. J. Radiat. Oncol. Biol. Phys.*, 56, 592–598, 2003.

Austin-Seymour, M., Munzenrider, J. E., Goitein, M., Gentry, R., Gragoudas, E. et al., Progress in low-LET heavy particle therapy: Intracranial and paracranial tumors and uveal melanomas, *Radiat. Res. Suppl.*, 8, S219–S226, 1985.

Austin-Seymour, M., Caplan, R., Russell, K., Laramore, G., Jacky, J. et al., Impact of a multileaf collimator on treatment morbidity in localized carcinoma of the prostate, *Int. J. Radiat. Oncol. Biol. Phys.*, 30, 1065–1071, 1994.

Bäck, S., Medin, J., Magnusson, P., Olsson, P., Grusell, E., and Olsson, L., Ferrous sulphate gel dosimetry and MRI for proton beam dose measurements, *Phys. Med. Biol.*, 44, 1983–1996, 1999.

Bakai, A., Alber, M., and Nusslin, F., Estimation of a radiation time prolongation factor for intensity-modulated radiotherapy, *Phys. Med. Biol.*, 48, N25–N29, 2003.

Bär, R., Eickhoff, H., and Haberer, T., Status and control requirements of the planned heavy ion tumor therapy accelerator facility HICAT, in *Proceedings of the 8th International Conference on Accelerator and Large Experimental Physics Control Systems* (ICALEPCS), 134–136, Stanford Linac Accelerator Center, Menlo Park, CA, http://www.slac.stanford.edu/econf/C011127/TUAP031.pdf, 2001.

Barrett, A., Total-body irradiation, in *The Oxford Textbook of Oncology*, Peckham, M., Pinedo, H. M., and Veronesi, U., Eds., Oxford University Press, Oxford, pp. 744–755, 1995.

Barrett, A., Depledge, M. H., and Powles, R. L., Interstitial pneumonitis following bone marrow transplantation after low dose rate total body irradiation, *Int. J. Radiat. Oncol. Biol. Phys.*, 9, 1029–1033, 1983.

Barrett, H. H., Barber, H. B., Ervin, P. A., Myers, K. J., Paxman, R. G. et al., New directions in coded aperture imaging, in *Information Processing in Medical Imaging: Proceedings of the 8th Conference*, Deconinck, F., Ed., Martinus Nijhoff, Dordrecht, pp. 106–129, 1984.

Bartelink, H., Mijnheer, B., and Dewit, L., Is there a future for neutron capture therapy?, *Int. J. Radiat. Oncol. Biol. Phys.*, 36, 267–268, 1996.

Barth, R. F., A critical assessment of boron neutron capture therapy: An overview, *J. Neurooncol.*, 62, 1–5, 2003.

Barth, R. F., Soloway, A. H., and Brugger, R. M., Boron neutron capture therapy of brain tumors: Past history, current status, and future potential, *Cancer Invest.*, 14, 534–550, 1996.

Barth, R. F., Grecula, J. C., Yang, W., Rotaru, J. H., Nawrocky, M. et al., Combination of boron neutron capture therapy and external beam radiotherapy for brain tumors, *Int. J. Radiat. Oncol. Biol. Phys.*, 58, 267–277, 2004.

Baumert, B. G., Norton, I. A., and Davis, J. B., Intensity-modulated stereotactic radiotherapy vs. stereotactic conformal radiotherapy for the treatment of meningioma located predominantly in the skull base, *Int. J. Radiat. Oncol. Biol. Phys.*, 57, 580–592, 2003.

Belshi, R., Use of proton beams for conformal radiotherapy of intracranial tumours, (*Utilisation des faisceaux de protons pour la radiothérapie conformationnelle des tumeurs intracrâniennes*), Thesis, Université Paul Sabattier, Toulouse, 1995.

Benedict, S. H., Cardinale, R. M., Wu, Q., Zwicker, R. D., Broaddus, W. C., and Mohan, R., Intensity-modulated stereotactic radiosurgery using dynamic micro-multileaf collimation, *Int. J. Radiat. Oncol. Biol. Phys.*, 50, 751–758, 2001.

Benincasa, G., Bourquin, P., Lombardi, A., Nonis, M., Orlandi, G., et al., High frequency proton linac, in Vol. 2, Part 1 of *The TERA Project: The RITA Network and the Design of Compact Proton Accelerators*, Amaldi, U., Gandalfo M., Picardi, L., Eds., INFN Divisione Ricerca, Frascati, pp. 215–256, 1995.

Bentley, R. and Milan, J., An interactive digital computer system for radiotherapy treatment planning, *Br. J. Radiol.*, 44, 826–833, 1971.

Bentzen, S. M., Theragnostic imaging for radiation oncology: dose-painting by numbers, *Lancet Oncol.*, 6, 112–117, 2005.

Betti, O. and Derechinsky, V., Multiple-beam stereotaxic irradiation, *Neurochirurgie*, 29, 295–298, 1983.

Betti, O., Munari, C., and Rosler, J. R., Stereotactic radiosurgery with the linear accelerator: Treatment of arteriovenous malformations, *Neurosurgery*, 24, 311–321, 1989.

Bewley, D. K.,, Institute of Physics Publishing, Bristol, 1989.

Bewley, D. K., Meulders, J. P., and Page, B. C., New neutron sources for radiotherapy, *Phys. Med. Biol.*, 29, 341–349, 1984.

Beyzadeoglu, M., Dirican, B., Oysul, K., Arpaci, F., and Pak, Y., Evaluation of fractionated total body irradiation and dose rate on cataractogenesis in bone marrow transplantation, *Haematologia (Budap)*, 32, 25–30, 2002.

Beyzadeoglu, M., Oysul, K., Dirican, B., Arpaci, F., Balkan, A. et al., Effect of dose-rate and lung dose in total body irradiation on interstitial pneumonitis after bone marrow transplantation, *Tohoku J. Exp. Med.*, 202, 255–263, 2004.

Bichsel, H., Charged-particle interactions, in *Radiation Dosimetry*, Attix, F. H. and Roesch, W. C., Eds., Vol. 1, Academic Press, New York, pp. 157–228, 1968.

Binns, P. J. and Hough, J. H., Secondary dose exposures during 200 MeV proton therapy. *Proceedings of the 8th Symposium on Neutron Dosimetry*, 13–17 Nov 1995, *Radiat. Prot. Dosim.*, 70, 441–444, 1997.

Bjarngard, B. E., Chen, G. T. Y., Piontek, R. W., and Svensson, G. K., Analysis of dose distributions in whole body superficial electron therapy, *Int. J. Radiat. Oncol. Biol. Phys.*, 2, 319–324, 1977.

Blomgren, H., Lax, I., Naslund, I., and Svanstrom, R., Stereotactic high dose fraction radiation therapy of extracranial tumors using an accelerator, Clinical experience of the first thirty-one patients, *Acta Oncol.*, 34, 861–870, 1995.

Blomgren, H., Lax, I., Göranson, H., Kraepelin, T., Nilsson, B. et al., Radiosurgery for tumours in the body: Clinical experience using a new method, *J. Radiosurg.*, 1, 63–75, 1998.

Bondiau, P. Y., Malandain, G., Chauvel, P., Peyrade, F., Courdi, A. et al., Automatic three-dimensional model for proton therapy of the eye: Preliminary results, *Med. Phys.*, 30, 1013–1020, 2003.

Bonnett, D. E., Current developments in proton therapy: A review, *Phys. Med. Biol.*, 38, 1393–1401, 1993.

Bonnett, D. E., Kacperek, A., Sheen, M. A., Goodall, R., and Saxton, T. E., The 62 MeV proton beam for the treatment of ocular melanoma at Clatterbridge, *Br. J. Radiol.*, 66, 907–914, 1993.

Boon, S. N., van Luijk, P., Schippers, J. M., Meertens, H., Denis, J. M. et al., Fast 2D phantom dosimetry for scanning proton beams, *Med. Phys.*, 25, 464–475, 1998.

Boon, S. N., van Luijk, P., Bohringer, T., Coray, A., Lomax, A. et al., Performance of a fluorescent screen and CCD camera as a two-dimensional dosimetry system for dynamic treatment techniques, *Med. Phys.*, 27, 2198–2208, 2000.

Bortfeld, T., IMRT: a review and preview, *Phys. Med. Biol.*, 51, R363–R379, 2006.

Bortfeld, T. and Schlegel, W., An analytical approximation of depth-dose distributions for therapeutic proton beam, *Phys. Med. Biol.*, 41, 1331–1339, 1996.

Bortfeld, T., Boyer, A. L., Schlegel, W., Kahler, D. L., and Waldron, T. J., Realization and verification of three-dimensional conformal radiotherapy with modulated fields, *Int. J. Radiat. Oncol. Biol. Phys.*, 30, 899–908, 1994a.

Bortfeld, T. R., Kahler, L., Waldron, T. J., and Boyer, A., X-ray field compensation with multileaf collimators, *Int. J. Radiat. Oncol. Biol. Phys.*, 28, 723–730, 1994b.

Bortfeld, T., Stein, J., and Preiser, K., Clinically relevant intensity modulation optimization using physical criteria, in *Proceedings of the 12th ICCR, Salt Lake City*, Medical Physics Publishing, Madison, 1–4, 1997.

Bortfeld, T., Jiang, S. B., and Rietzel, E., Effects of motion on the total dose distribution, *Semin. Radiat. Oncol.*, 14, 41–51, 2004.

Bortfeld, T., Schmidt-Ullrich, R., De Neve, W., and Wazer, D. E., Eds., *Image-guided IMRT*, Springer Verlag, Berlin, 2006.

Boyer, A. L., Xing, L., Ma, C.-M., Curran, B., Hill, R., Kania, A., and Bleier, A., Theoretical considerations of monitor unit calculations for intensity modulated beam treatment planning, *Med. Phys.*, 26, 187–195, 1999.

Brada, M., Radiosurgery for brain tumours, *Eur. J. Cancer.*, 27, 1545–1548, 1991.

Brahme, A., Optimization of stationary and moving beam radiation therapy techniques, *Radiother. Oncol.*, 12, 129–140, 1988.

Brahme, A., Recent advances in light ion radiation therapy, *Int. J. Radiat. Oncol. Phys. Biol.*, 58, 603–616, 2004.

Brahme, A., Källman, P., and Lind, B. K., Optimisation of proton and heavy ion therapy using an adaptive inversion algorithm, *Radiother. Oncol.*, 15, 189–197, 1989.

Brassart, N., Tuyn, J., Herault, J., and Chauvel, P., Specific aspects of radiation protection in relation to a cyclotron (*Aspects spécifiques de la radioprotection autour d'un cyclotron*), *Radioprotection*, 30, 411–422, 1995.

Brede, H. J., Hecker, O., and Hollnagel, R., Measurement of the heat defect in water and A-150 plastic for high energy protons, deuterons and alpha particles, *Proceedings of the 8th Symposium Neutron Dosimetry Paris*, November 1995, *Radiat. Prot. Dosim.*, 70, 505–508, 1997.

Briot, E., Dutreix, A., and Bridier, A., Dosimetry for total body irradiation, *Radiother. Oncol.*, 18(Suppl. 1), 16–29, 1990.

Brusasco, C., Cattai, A., Cirio, R., Dellacasa, G., Donetti, et al., Strip Ionization chambers as 3D detector for hadrontherapy, *Nucl. Instrum. Meth. Phys. Res. A.*, 389, 499–512, http://www.to.infn.it/activities/experiments/tera/publications/magic_cube/strip_ionization.pdf, 1997.

Budgell, G. J., Perrin, B. A., Mott, J. H., Fairfoul, J., and Mackay, R. I., Quantitative analysis of patient-specific dosimetric IMRT verification, *Phys. Med. Biol.*, 50, 103–119, 2005.

Busse, P. M., Harling, O. K., Palmer, M. R., Kiger, W. S., Kaplan, J. et al., A critical examination of the results from the Harvard-MIT NCT program phase I clinical trial of neutron capture therapy for intracranial disease, *J. Neurooncol.*, 62, 111–121, 2003.

Bussiere, M. R. and Adams, J. A., Treatment planning for conformal proton radiation therapy, *Technol. Cancer Res. Treat.*, 2, 389–399, 2003.

Cardinale, R. M., Benedict, S. H., Wu, Q., Zwicker, R. D., Gaballa, H. E., and Mohan, R., A comparison of three stereotactic radiotherapy techniques; Arcs vs. non coplanar fixed fields vs. intensity modulation, *Int. J. Radiat. Oncol. Biol. Phys.*, 42, 431–436, 1998.

Carlsson, A. K., Andreo, P., and Brahme, A., A Monte Carlo and analytical calculation of proton pencil beams for computerized treatment plan optimization, *Phys. Med. Biol.*, 42, 1033–1053, 1997.

Carruthers, S. A. and Wallington, M. M., Total body irradiation and pneumonitis risk: A review of outcomes, *Br. J. Cancer.*, 90, 2080–2084, 2004.

Castro, J. R., Heavy ion therapy: Bevalac epoch, in *Hadrontherapy in Oncology, Proceedings of the 1st International Symposium on Hadrontherapy*, Elsevier, Amsterdam, 208–216, 1994.

Caumes, J., Ostrowsky, A., Steinschaden, K., Mancaux, M., Cance, M., and Simoen. J.P.,Tissue equivalent calorimetry for direct calibration of ionisaiton chambers in neutron beams (*Calorimètre équivalent tissu pour l'étalonnage direct des chambres d'ionisation dans les faisceaux de neutrons*) *Proceedings of the 5th Symposium on Neutron Dosimetry*, Vol. 2, CEC, EUR 9762 EN Luxembourg, pp. 1215–1226, 1985.

Chadha, M., Capala, J., Coderre, J. A., Elowitz, E. H., Iwai, J. et al., Boron neutron-capture therapy (BNCT) for glioblastoma multiforme (GBM) using the epithermal neutron beam at the Brookhaven National Laboratory, *Int. J. Radiat. Oncol. Biol. Phys.*, 40, 829–834, 1998.

Chang, J., Obcemea, C.H., Sillanpaa, J., Mechalakos, J., and Burman C., Use of EPID for leaf position accuracy QA of dynamic multi-leaf collimator (DMLC) treatment, Med. Phys., 31, 2091–2096, 2004.

Chapman, J. D., Bradley, J. D., Eary, J. F., Haubner, R., Larson, S. M. et al., Molecular (functional) imaging for radiotherapy applications: An RTOG symposium, *Int. J. Radiat. Oncol. Biol. Phys.*, 55, 294–301, 2003.

Chen, Y., Xing, L., Luxton, G., Li, J. G., and Boyer, A. L., A multi-purpose quality assurance tool for MLC-based IMRT, *Proceedings of the XIIIth International Conference on the Use of Computers in Radiation Therapy* (ICCR), Schlegel, W., and Bortfeld, T., Eds., Springer, Heidelberg, 371–373, 2000.

Chen, L., Price, R. A., Nguyen, T. B., Wang, L., Li, J. S., Qin, L., Ding, M., Palacio, E., Ma, C. M., and Pollack, A., Dosimetric evaluation of MRI-based treatment planning for prostate cancer, *Phys. Med. Biol.*, 49, 5157–5170, 2004.

Chetty, I. J. and Charland, P. M., Investigation of Kodak extended dose range (EDR) film for megavoltage photon beam dosimetry, *Phys. Med. Biol.*, 47, 3629–3641, 2002.

Chu, W. T., Ludewigt, B. A., and Renner, T. R., Instrumentation for treatment of cancer using proton and light-ion beams, *Rev. Sci. Instrum.*, 64, 2055–2122, 1993.

Ciulla, T. A., Danis, R. P., Klein, S. B., Malinovsky, V. E., Soni, P. S. et al., Proton therapy for exudative age-related macular degeneration: A randomized, sham-controlled clinical trial, *Am. J. Ophthalmol.*, 134, 905–906, 2002.

Coleman, C. N., Radiation oncology–linking technology and biology in the treatment of cancer, *Acta Oncol.*, 41, 6–13, 2002.

Colombo, F., Benedetti, A., Pozza, F., Zanardo, Z., Avanzo, R. C. et al., Stereotactic radiosurgery utilizing a linear accelerator, *Appl. Neurophysiol.*, 48, 133–145, 1985.

Contento, G., Malisan, M. R., and Padovani, R., Response of thermoluminescence dosimeters to beta and skin dose assessment, *Phys. Med. Biol.*, 29, 661–678, 1984.

Convery, D. and Webb, S., Generation of discrete beam-intensity modulation by dynamic multileaf collimation under minimum leaf separation constraints, *Phys. Med. Biol.*, 43, 2521–2538, 1998.

Cosgrove, V. P., Delacroix, S., Green, S., Mazal, A., and Scott, M. C., Microdosimetric studies on the Orsay proton synchrocyclotron at 73 and 200 MeV, *Proceedings of the 8th Symposium on Neutron Dosimetry, Radiat. Prot. Dosim.*, 70, 493–496, 1997.

Cosgrove, V. P., Jahn, U., Pfaender, M., Bauer, S., Budach, V., and Wurm, R. E., Commissioning of a micro multi-leaf collimator and planning system for stereotactic radiosurgery, *Radiother. Oncol.*, 50, 325–336, 1999.

Cosgrove, V. P., Murphy, P. S., McJury, M., Adams, E. J., Warrington, A. P. et al., The reproducibility of polyacrylamide gel dosimetry applied to stereotactic conformal radiotherapy, *Phys. Med. Biol.*, 45, 1195–1210, 2000.

Cosset, J. M., Girinsky, T., Malaise, E., Chaillet, M. P., and Dutreix, J., Clinical basis for TBI fractionation, *Radiother. Oncol.*, 18(Suppl. 1), 60–67, 1990.

Courdi, A., Caujolle, J. P., Grange, J. D., Diallo-Rosier, L., Sahel, J. et al., Results of proton therapy of uveal melanomas treated in nice, *Int. J. Radiat. Oncol. Biol. Phys.*, 45, 5–11, 1999.

Court, L. E. and Dong, L., Automatic registration of the prostate for computed-tomography-guided radiotherapy, *Med. Phys.*, 30, 2751–2757, 2003.

Court, L. E., Dong, L., Lee, A. K., Cheung, R., Bonnen, M. D. et al., An automatic CT-guided adaptive radiation therapy technique by online modification of multileaf collimator leaf positions for prostate cancer, *Int. J. Radiat. Oncol. Biol. Phys.*, 62, 154–163, 2005.

Coutrakon, G., Bauman, M., Lesyna, D., Miller, D., Nusbaum, J. et al., A prototype beam delivery system for the proton medical accelerator at Loma Linda, *Med. Phys.*, 18, 1093–1099, 1991.

Cumberlin, R. L., Clinical research in neutron capture therapy, *Int. J. Radiat. Oncol. Biol. Phys.*, 54, 992–998, 2002.

Damato, B., Kacperek, A., Chopra, M., Sheen, M. A., Campbell, I. R., and Errington, R. D. Proton beam radiotherapy of iris melanoma, *Int. J. Radiat. Oncol. Biol. Phys.*, 63, 109–115, 2005a.

Damato, B., Kacperek, A., Chopra, M., Campbell, I. R., and Errington, R. D. Proton beam radiotherapy of choroidal melanoma: the Liverpool-Clatterbridge experience, *Int. J. Radiat. Oncol. Biol. Phys.*, 62, 1405–1411, 2005b.

Danciu, C. and Proimos, B. S., Gravity oriented absorbers in conformal radiotherapy for cervix cancer, *Radiother. Oncol.*, 51(Suppl. 1), S38, 1999.

Dawson, L. A. and Balter, J. M., Interventions to reduce organ motion effects in radiation delivery, *Semin. Radiat. Oncol.*, 14, 76–80, 2004.

De Angelis, C., Onori, S., Pacili, M., Cirrone, G. A., Cuttone, G. et al., Preliminary results on a dedicated silicon diode detector for proton dosimetry, *Radiat. Prot. Dosim.*, 101, 461–464, 2002.

De Santis, M., Caiazza, A., and Simili, A., Biological factors and therapeutic modulation in brain tumor radiotherapy, *Rays*, 27, 201–203, 2002.

Dearnaley, D. P., Khoo, V. S., Norman, A. R., Meyer, L., Nahum, A. et al., Comparison of radiation side-effects of conformal and conventional radiotherapy in prostate cancer: A randomised trial, *Lancet*, 353, 267–272, 1999.

Deasy, J. O., A proton dose calculation algorithm for conformal therapy simulations based on Moliere's theory of lateral deflections, *Med. Phys.*, 25, 476–483, 1998.

Delacroix, S., Bridier, A., Mazal, A., Daures, J., Ostrowsky, A. et al., Proton dosimetry comparison involving iono-metry and calorimetry, *Int. J. Radiat. Oncol. Biol. Phys.*, 37, 711–718, 1997.

Deng, J., Pawlicki, T., Chen, Y., Li, J., Jiang, S. B., and Ma, C. M., The MLC tongue-and-groove effect on IMRT dose distributions, *Phys. Med. Biol.*, 46, 1039–1060, 2001.

Desjardins, L., Levy, C., D'hermies, F., Frau, E., Schlienger, P. et al., Initial results of proton therapy in choroidal melanoma at the d'Orsey Center for Proton Therapy; the first 464 cases, *Cancer Radiother.*, 1, 222–226, 1997.

Desjardins, L., Lumbroso, L., Levy, C., Mazal, A., Delacroix, S. et al., Treatment of uveal melanoma with iodine 125 plaques or proton beam therapy: Indications and comparison of local recurrence rates, *J. Fr. Ophtalmol.*, 26, 269–276, 2003.

Djajaputra, D. and Li, S., Real-time 3D surface-image-guided beam setup in radiotherapy of breast cancer, *Med. Phys.*, 32, 65–75, 2005.

Dobbs, J., Logue, J., McNair, H., McNee, S., and Webb, S., *Development and Implementation of Conformal Radiotherapy in the United Kingdom*, IPEM/RCR/SCR, London, http://www.rcr.ac.uk/docs/oncology/pdf/conformal.pdf, 2002

Doughty, D., Lambert, G. D., Hirst, A., Marks, A. M., and Plowman, P. N., Improved total-body irradiation dosim-etry, *Br. J. Radiol.*, 60, 269–278, 1987.

Drzymala, R., Quality assurance for linac-based stereotactic radiosurgery in quality assurance in radiotherapy physics, in *Quality Assurance in Radiotherapy Physics: Proceedings of the American College of Medical Physics Symposium*, Starkshall, G. and Horton, J., Eds., Medical Physics Publishing, Madison WI, 1991.

Dutreix, A. and Briot. E., The development of a pencil-beam algorithm for clinical use at the Institut Gustave Roussy, in *The Computation of Dose Distributions in Electron Beam Radiotherapy*, Nahum, A. E., Ed., Medical Physics Publishing, Madison WI, pp. 242–270, 1985.

Dutreix, A. and Broerse, J. J., Summary of round table discussion on physical aspects of total body irradiation, *J. Eur. Radiother.*, 3, 262–264, 1982.

Dutreix, A., Bridier, A., and Wambersie, A., Is film dosimetry reliable for clinical applications of high energy neutron beams? in *Proceedings of the 3rd Symposium on Neutron Dosimetry in Biology and Medicine*, Burger G., Ebert, H. G., Eds., Euratom, Commission of the European Communities, Brussels, 1978.

Edelstein, G. R., Clark, T., and Holt, J. G., Dosimetry for total body electron beam therapy in the treatment of mycosis fungoides, *Radiology*, 108, 691–694, 1973.

Edser, E., The development of the double-headed cobalt machine for total body irradiation, *Br. J. Radiol.*, 61, 1192, 1988 (abstract).

Eickhoff, H., Böhne, D., Haberer, T., Schlitt, B., Debus, J. et al., The proposed dedicated ion beam facility for cancer therapy at the clinic, in *Proceedings of the EPAC 2000*, 2512–2514, http://accelconf.web.cern.ch/accelconf/e00/PAPERS/WEP3A13.pdf, 2000.

Ellis, F., Hall, E. J., and Oliver, R., A compensator for variation in tissue thickness for high energy beams, *Br. J. Radiol.*, 32, 421–422, 1959.

Enghardt, W., Debus, J., Haberer, T., Hasch, B. G., Hinz, R. et al. The application of PET to quality assurance of heavy-ion tumor therapy, *Strahlenther. Onkol.*, Suppl. II, 175, 33–36, 1999.

Essers, M. and Mijnheer, B. J., In vivo dosimetry during external photon beam therapy, *Int. J. Radiat. Oncol. Biol. Phys.*, 43, 245–259, 1999.

ESTRO (European Society for Therapeutic Radiology and Oncology), Fifth meeting European Society for Thera-peutic Radiology and Oncology, *Proceedings of the Symposium on Total Body Irradiation, Radiother. Oncol.*, 9, 85–135, 1987.

Evans, P. M., Hansen, V. N., and Swindell, W., The optimum intensities for multiple static multileaf collimator field compensator, *Med. Phys.*, 24, 1147–1156, 1997.

Ezzel, G. A., Galvin, J. M., Low, D., Palta, J. R., Rosen, I. et al., Guidance document on delivery, treatment planning and clinical implementation of IMRT: Report of the IMRT Subcommittee of the AAPM radiation therapy committee, *Med. Phys.*, 30, 2089–2115, 2003.

Fabrikant, J. I., Levy, R. P., Steinberg, G. K., Phillips, M. H., Frankel, K. A. et al., Charged-particle radiosurgery for intracranial vascular malformations, *Neurosurg. Clin. N., Am.*, 3, 99–139, 1992.

Farr, L. E., Sweet, W. H., Robertson, J. S., Foster, C. G., Locksley, H. B. et al., Neutron capture therapy with boron in the treatment of glioblastoma multiforme, *Am. J., Roentgenol. Radium. Ther. Nucl. Med.*, 71, 279–293, 1954.

Fenwick, J. D., Tome, W. A., Soisson, E. T., Mehta, M. M., and Mackie, T. R., Tomotherapy and other innovative IMRT delivery systems, *Semin. Radiat. Oncol.*, 16, 199–208, 2006.

Fiorino, C., Ardesi, A., del Vecchio, A., Cattaneo, G. M., Fossati, V. et al., Conformal technics by the movement of bars in pendular fields 2 The dosimetric aspects, *Radiol. Med.*, 84, 310–316, 1992.

Flaxel, C. J., Friedrichsen, E. J., Smith, J. O., Oeinck, S. C., Blacharski, P. A. et al., Proton beam irradiation of subfoveal choroidal neovascularisation in age-related macular degeneration, *Eye*, 14, 155–164, 2000.

Folkerts, K. H., Menzel, H. G., Schuhmacher, H., and Arend, E., TEPC radiation protection dosimetry in the environment of accelerators and at nuclear facilities, *Radiat. Prot. Dosim.*, 23, 261–264, 1988.

Followill, D., Geis, P., and Boyer, A., Estimates of whole-body dose equivalent produced by beam intensity modulated conformal therapy, *Int. J. Radiat. Oncol. Biol. Phys.*, 38, 667–672, 1997.

Ford, E. C., Mageras, G. S., Yorke, E., Rosenzweig, K. E., Wagman, R., and Ling, C. C., Evaluation of respiratory movement during gated radiotherapy using film and electronic portal imaging, *Int. J. Radiat. Oncol. Biol. Phys.*, 52, 522–531, 2002.

Ford, E. C., Mageras, G. S., Yorke, E., and Ling, C. C., Respiration-correlated spiral CT: A method of measuring respiratory-induced anatomic motion for radiation treatment planning, *Med. Phys.*, 30, 88–97, 2003.

Forrest, L. J., Mackie, T. R., Ruchala, K., Turek, M., Kapatoes, J. et al., The utility of megavoltage computed tomography images from a helical tomotherapy system for setup verification purposes, *Int. J. Radiat. Oncol. Biol. Phys.*, 60, 1639–1644, 2004.

Fourkal, E., Shahine, B., Ding, M., Li, J. S., Tajima, T., and Ma, C. M., Particle in cell simulation of laser-accelerated proton beams for radiation therapy, *Med. Phys.*, 29, 2788–2798, 2002.

Fourkal, E., Li, J. S., Ding, M., Tajima, T., and Ma, C. M., Particle selection for laser-accelerated proton therapy feasibility study, *Med. Phys.*, 30, 1660–1670, 2003.

Fowler, J. F., Dose fractionation schedules—biologic aspects and applications to high LET radiotherapy, *J. Can. Assoc. Radiol.*, 26, 40–43, 1975.

Fowler, J. F., *Nuclear Particles in Cancer Treatment*, Medical physics handbook 8, Adam Hilger, Bristol, 1981.

Fowler, J. F., Tomé, W. A., Fenwick, J. D., and Mehta, M. P., A challenge to traditional radiation oncology, *Int. J. Radiat. Oncol. Biol. Phys.*, 60, 1241–1256, 2004.

Friedman, W. A. and Bova, F. J., The University of Florida radiosurgery system, *Surg. Neurol.*, 32, 334–342, 1989.

Fukumura, A., Hiraoka, T., Omata, K., Tekeshita, M., Kawachi, K. et al., Carbon beam dosimetry intercomparison at HIMAC, *Phys. Med. Biol.*, 43, 3459–3463, 1998.

Fuss, M., Salter, B. J., Cavanaugh, S. X., Fuss, C., Sadeghi, A. et al., Daily ultrasound-based image-guided targeting for radiotherapy of upper abdominal malignancies, *Int. J. Radiat. Oncol. Biol. Phys.*, 59, 1245–1256, 2004.

Galvin, J. M., DAngin, G. J., and Walsh, G., The use of tissue compensation to improve dose uniformity for total body irradiation, *Int. J. Radiat. Oncol. Biol. Phys.*, 6, 676–771, 1980.

Garcia, R., Oozeer, R., Le Thanh, H., Chastel, D., Doyen, J. C. et al., Radiotherapy of lung cancer: The inspiration breath hold with spirometric monitoring, *Cancer Radiother.*, 6, 30–38, 2002.

Ghilezan, M., Yan, D., Liang, J., Jaffray, D., Wong, J., and Martinez, A., Online image-guided intensity-modulated radiotherapy for prostate cancer: How much improvement can we expect? A theoretical assessment of clinical benefits and potential dose escalation by improving precision and accuracy of radiation delivery, *Int. J. Radiat. Oncol. Biol. Phys.*, 60, 1602–1610, 2004.

Gill, S. S., Thomas, D. G., Warrington, A. P., and Brada, M., Relocatable frame for stereotactic external beam radiotherapy, *Int. J. Radiat. Oncol. Biol. Phys.*, 20, 599–603, 1991.

Giraud, P., Reboul, F., Clippe, S., Garcia, R., Carrie, C., Campana, F., Dubray, B., Rosenwald, J.-C., and Cosset, J. M., Respiration-gated radiotherapy: Current techniques and potential benefits, *Cancer Radiother.*, 7(Suppl. 1), 15s–25s, 2003.

Giraud, P., Simon, L., Saliou, M., Reboul, F., Garcia, R., Carrie, C., Lerolle, U., Rosenwald, J.-C., and Cosset, J. M., Respiratory gated radiotherapy: The 4D radiotherapy, *Bull. Cancer*, 92, 83–89, 2005.

Glimelius, B., Ask., A., Bjelkengren, G., Björk-Eriksson, T., Blomquist, E., Johansson, B., Karlsson, M., and Zackrisson B., Number of patients potentially eligible for proton therapy, *Acta. Oncol.*, 44, 836–849, 2005.

Goitein, M., A technique for calculating the influence of thin inhomogeneities on charged particle beams, *Med. Phys.*, 5, 258–264, 1978.

Goitein, M., Calculation of the uncertainty in the dose delivered during radiation therapy, *Med. Phys.*, 12, 608–612, 1985.

Goitein, M., Organ and tumor motion: An overview, *Semin. Radiat. Oncol.*, 14, 2–9, 2004.

Goitein, M. and Jermann, J., The relative costs of proton and x-ray radiation therapy, *Clin. Oncol. (R., Coll. Radiol.)*, 15, S37–S50, 2003.

Goitein, M. and Miller, T., Planning protontherapy of the eye, *Med. Phys.*, 10, 275–283, 1983.

Gottschalk, B., Koehler, A. M., Schneider, R. J., Sisterson, J. M., and Wagner, M. S., Multiple Coulomb scattering of 160 MeV protons, *Nucl. Instrum. Methods Phys. Res. B.*, 74, 467–490, 1993.

Gragoudas, E. S., Goitein, M., Koehler, A., Constable, I. J., Wagne, M. S. et al., Proton irradiation of choroidal melanomas: Preliminary results, *Arch. Ophthalmol.*, 96, 1583–1591, 1978.

Graham, J. D., Nahum, A. E., and Brada, M., Optimum technique for stereotactic radiotherapy by linear accelerator based on 3-dimensional dose distributions, *Radiother. Oncol.*, 22, 29–35, 1991.

Grebe, G., Pfaender, M., Roll, R., Luedemann, L., and Wurm, R. E., Dynamic arc radiosurgery and radiotherapy: Commissioning and verification of dose distributions, *Int. J. Radiat. Oncol. Biol. Phys.*, 49, 1451–1460, 2001. See also (Erratum in: *Int. J. Radiat. Oncol. Biol. Phys.* 51:865 (2001)).

Grégoire, V., Beauduin, M., Gueulette, J., De Coster, B. M., Octave-Prignot, M. et al., Radiobiological intercomparison of p(45)+Be and p(65)+Be neutron beams for lung tolerance in mice after single and fractionated irradiation, *Radiat. Res.*, 133, 27–32, 1993.

Grégoire, V., Bol, A., Geets, X., and Eng, J. L., Is PET-based treatment planning the new standard in modern radiotherapy? The head and neck paradigm, *Semin. Radiat. Oncol.*, 16, 232–238, 2006.

Griffin, T. W., Wambersie, A., Laramore, G., and Castro, M. D., High LET: Heavy particle trials, *Int. J. Radiat. Oncol. Biol. Phys.*, 14, S83–S92, 1988.

Grigorov, G. N., Chow, J. C. L., Grigorov, L., Jiang, R., and Barnett, R. B., IMRT: Improvement in treatment planning efficiency using NTCP calculation independent of the dose-volume-histogram, *Med. Phys.*, 33, 1250–1258, 2006.

Grusell, E., Montelius, A., Russell, K. R., Blomquist, E., Pellettieri, L. et al. Stereotactic proton beam irradiation of arteriovenous malformations and malignant tumors in the brain, In *Hadrontherapy in Oncology, Proceedings of the 1st International Symposium on Hadrontherapy*, Como, Italy, October 1993, Amaldi, U., Larsson, B., Eds., Elsevier, Amsterdam, 13–121, 1994a.

Grusell, E., Montelius, A., Brahme, A., Rikner, G., and Russell, K., A general solution to charged particle beam flattening using an optimized dual-scattering-foil technique, with application to proton therapy beams, *Phys. Med. Biol.*, 39, 2201–2216, 1994b.

Grusell, E., Montelius, A., Russell, K. R., Blomquist, E., Pellettieri, L. et al., Patient positioning for fractionated precision radiation treatment of targets in the head using fiducial markers, *Radiother. Oncol.*, 33, 68–72, 1994c.

Grusell, E., Isacsson, U., Montelius, A., and Medin, A., Faraday cup dosimetry in a proton therapy beam without collimation, *Phys. Med. Biol.*, 40, 1831–1840, 1995.

Gueulette, J., Octave-Prignot, M., and Wambersie, A., RBE of d(50) + Be neutrons, as a function of the dose per fraction, for early intestinal tolerance in mice, *J. Eur. Radiother.*, 5, 180–182, 1984.

Gueulette, J., Beauduin, M., Gregoire, V., Vynckier, S., De Coster, B. M. et al., RBE variation between fast neutron beams as a function of energy. Intercomparison involving 7 neutrontherapy facilities, *Bull. Cancer Radiother.*, 83(Suppl), 55s–63s, 1996.

Gueulette, J., Bohm, L., De Coster, B. M., Vynckier, S., Octave-Prignot, M. et al., RBE variation as a function of depth in the 200-MeV proton beam produced at the National Accelerator Centre in Faure (South Africa), *Radiother. Oncol.*, 42, 303–309, 1997.

Gueulette, J., Slabbert, J. P., Bohm, L., De Coster, B. M., Rosier, J. F. et al., Proton RBE for early intestinal tolerance in mice after fractionated irradiation, *Radiother. Oncol.*, 61, 177–184, 2001.

Haberer, T., Becher, W., Schardt, D., and Kraft, G., Magnetic scanning system for heavy ion therapy, *Nucl. Instrum. Methods Phys. Res. A.*, 330, 296–305, 1993.

Habrand, J. L., Schlienger, P. S., Schwartz, L., Pontvert, D., Lenir-Cohen-Solal, C. et al., Clinical applications of proton therapy, experiences and ongoing studies, *Radiat. Environ. Biophys.*, 34, 41–44, 1995.

Habrand, J. L., Desjardins, L., Mammar, H., and Mazal, A., Protontherapy, Clinical features, *Bull. Cancer Radiother.*, 83, 247–253, 1996.

Hacker, F. L., Kooy, H. M., Bellerive, M. R., Killoran, J. H., Leber, Z. H. et al., Beam shaping for conformal fractionated stereotactic radiotherapy: A modelling study, *Int. J. Radiat. Oncol. Biol. Phys.*, 38, 1113–1121, 1997.

Hagan, W. K., Colborn, B. L., Armstrong, T. W., and Allen, M., Radiation shielding calculations for a 70–250 MeV proton therapy facility, *Nucl. Sci. Eng.*, 98, 272–278, 1988.

Hall, E. J., Antiprotons for radiotherapy?, *Radioth. Oncol.*, 81, 231–232, 2006a.

Hall, E. J., Intensity-modulated radiation therapy, protons, and the risk of second cancers, *Int. J. Radiat. Oncol. Biol. Phys.*, 65, 1–7, 2006b.

Hall, E. J. and Brenner, D. J., The radiobiology of radiosurgery: Rationale for different treatment regimes for AVMs and malignancies, *Int. J. Radiat. Oncol. Biol.* Phys., 25, 381–385, 1993.

Hara, R., Itami, J., Kondo, T., Aruga, T., Abe, Y. et al., Stereotactic single high dose irradiation of lung tumors under respiratory gating, *Radiother. Oncol.*, 63, 159–163, 2002.

Hariz, M. I., Henriksson, R., Lofroth, P. O., Laitinen, L. V., and Saterborg, N. E., A non-invasive method for fractionated stereotactic irradiation of brain tumors with linear accelerator, *Radiother. Oncol.*, 17, 57–72, 1990.

Hartmann, G. H., Schlegel, W., Sturm, V., Kober, B., Pastyr, O., and Lorenz, W. J., Cerebral radiation surgery using moving field irradiation at a linear accelerator facility, *Int. J. Radiat. Oncol. Biol. Phys.*, 11, 1185–1192, 1985.

Hartmann, G. H., Bauer-Kirpes, B., Serago, C. F., and Lorenz, W. J., Precision and accuracy of stereotactic convergent beam irradiations from a linear accelerator, *Int. J. Radiat.. Oncol. Biol. Phys.*, 28, 481–492, 1994.

Hartmann, G., Heeg, P., Jäkel, O., Karger, C., and Krießbach, A., Determination of water absorbed dose in a carbon ion beam using thimble ionization chambers, *Phys. Med. Biol.*, 44, 1193–1206, 1999.

Heeg, P., Eickhoff, H., and Haberer, T., Die Konzeption der Heidelberger Ionentherapieanlage HICAT.Z, *Med. Phys.*, 14, 17–24, 2004.

Heufelder, J., Stiefel, S., Pfaender, M., Ludemann, L., Grebe, G., and Heese, J., Use of BANG polymer gel for dose measurements in a 68 MeV proton beam, *Med. Phys.*, 30, 1235–1240, 2003.

Highland, V. L., Some practical remarks on multiple scattering, *Nucl. Instrum. Meth.*, 129, 497–499, 1975.

Hishikawa, Y., Kagawa, K., Murakami, M., Sakai, H., Akagi, T., and Abe, M., Usefulness of positron-emission tomographic images after proton therapy, *Int. J. Radiat. Oncol. Biol. Phys.*, 53, 1388–1391, 2002.

Hoffmann, A. L., Siem, A. Y. D., den Hertog, D., Kaanders, J. H. A. M., and Huizenga, H., Derivative-free generation and interpolation of convex Pareto optimal IMRT plans, *Phys. Med. Biol.*, 51, 6349–6369, 2006.

Holt, J. G. and Perry, D. J., Some physical considerations in whole skin electron beam therapy, *Med. Phys.*, 9, 769–776, 1982.

Holzscheiter, M. H., Bassler, N., Agazaryan, N., Beyer. G., Blackmore, E., DeMarco, J. J., et al., The biological effectiveness of antiproton irradiation, *Radioth. Oncol.*, 81, 233–242, 2006.

Hong, L., Goitein, M., Bucciolini, M., Comiskey, R., Gottschalk, B., Rosenthal, S. et al., A pencil beam algorithm for proton dose calculations, *Phys. Med. Biol.*, 41, 1305–1330, 1996.

Hoppe, R. T., Fuks, Z., and Bagshaw, M. A., The rationale for curative radiotherapy in mycosis fungoides, *Int. J. Radiat. Oncol. Biol. Phys.*, 2, 843–851, 1977.

Hornick, D. C., Litzenberg, D. W., Lam, K. L., Balter, J. M., Hetrick, J., and Ten Haken, R. K., A tilt and roll device for automated correction of rotational setup errors, *Med. Phys.*, 25, 1739–1740, 1998.

Houdek, P. V. and Piciotta, V. J., Comparison of calculated and measured data for total body irradiation, *Phys. Med. Biol.*, 32, 1101–1108, 1987.

Hug, E. B., Sweeney, R. A., Nurre, P. M., Holloway, K. C., Slater, J. D., and Munzenrider, J. E., Proton radiotherapy in management of pediatric base of skull tumors, *Int. J. Radiat. Oncol. Biol. Phys.*, 52, 1017–1024, 2002.

Hugo, G. D., Agazaryan, N., and Solberg, T. D., An evaluation of gating window size, delivery method, and composite field dosimetry of respiratory-gated IMRT, *Med. Phys.*, 29, 2517–2525, 2002.

Hugo, G. D., Agazaryan, N., and Solberg, T. D., The effects of tumor motion on planning and delivery of respiratory-gated IMRT, *Med. Phys.*, 30, 1052–1066, 2003.

Hugo, G. D., Yan, D., and Liang, J., Population and patient-specific target margins for 4D adaptive radiotherapy to account for intra- and inter-fraction variation in lung tumour position, *Phys. Med. Biol.*, 52, 257–274, 2007.

Hui, S. K., Kapatoes, J., Fowler, J., Henderson, D., Olivera, G. et al., Feasibility study of helical tomotherapy for total body or total marrow irradiation, *Med. Phys.*, 32, 3214–3224, 2005.

Hussein, S. and Kennelly, G. M., Lung compensation in total body irradiation: A radiographic method, *Med. Phys.*, 23, 357–360, 1996.

IAEA (International Atomic Energy Agency), *Absorbed dose determination in External Beam Radiotherapy, An international code of practice for dosimetry based on standards of absorbed dosee to water*, IAEA Technical Report Series 398, IAEA, Vienna, 2000.

ICRP (International commission on Radiological Protetction), ICRP Publication 60, Recommendations of the International Commission on Radiological Protection, *Ann. ICRP*, 21, 1990.

ICRU (International Commission on Radiation Units and Measurements), Report 36, *Microdosimetry*. Bethesda, ICRU, MD, 1983.

ICRU (International Commission on Radiation Units and Measurements), Report 45, *Clinical neutron dosimetry part I: Determination of absorbed dose in a patient treated by external beams of fast neutrons*, ICRU, Bethesda, MD, 1989.

ICRU (International Commission on Radiation Units and Measurements), Report 49, *Stopping power and Ranges for proton and alpha particles*, ICRU, Bethesda, MD, 1993a.

ICRU (International Commission on Radiation Units and Measurements), Report 50, *Prescribing, Recording and Reporting Photon Beam Therapy*, Landberg, T., Chavaudra, J., Dobbs, H. J., Hanks, G., Johansson, K. A., Möller, T., Purdy, J., Eds., ICRU, Bethesda, MD, 1993b.

ICRU (International Commission on Radiation Units and Measurements), Report 59, *Clinical Proton Dosimetry—Part I: Beam Production, Beam Delivery and Measurement of Absorbed Dose*, ICRU, Bethesda, MD, 1998.

ICRU (International Commission on Radiation Units and Measurements), Report 62, *Prescribing, Recording and Reporting Photon Beam Therapy*, Supplement to ICRU Report 50, ICRU, Bethesda, MD, 1999.

Ilfield, D. N., Wright, K. A., and Salzman, F. A., Synchronous shielding and field shaping for megavolt irradiation of advanced cervical carcinoma, *Am. J., Roentgenol.*, 112, 792–796, 1971.

IMRTCWG (Intensity Modulated Radiation Therapy Collaborative Working Group), IMRT—current status and issues of interest, *Int. J. Radiat. Oncol. Biol. Phys.*, 51, 880–914, 2001.

Indovina, P. L., Benassi, M., Giacco, G. C., Primavera, A., and Rosati, A., In vivo ESR dosimetry in total body irradiation, *Strahlenther. Onkol.*, 165, 611–616, 1989.

Isacsson, U., Lennernas, B., Grusell, E., Jung, B., Montelius, A., and Glimelius, B., Comparative treatment planning between proton and x-ray therapy in esophageal cancer, *Int. J. Radiat. Oncol. Biol. Phys.*, 41, 441–450, 1998.

Itano, A., Akagi, T., Higashi, A., Fukushima, S., Fujita, A. et al., Operation of medical Accelerator PATRO at Hyogo Ion Beam Medical Center, in *Workshop on Accelerator Operation*, March 2003, KEK (Tsukuba). http://conference.kek.jp/wao2003/papers/10p2-2.pdf, 2003.

Jaffray, D. A., Drake, D. G., Moreau, M., Martinez, A. A., and Wong, J. W., A radiographic and tomographic imaging system integrated into a medical linear accelerator for localization of bone and soft-tissue targets, *Int. J. Radiat. Oncol. Biol. Phys.*, 45, 773–789, 1999.

Jaffray, D. A., Siewerdsen, J. H., Wong, J. W., and Martinez, A. A., Flat-panel cone-beam computed tomography for image-guided radiation therapy, *Int. J. Radiat. Oncol. Biol. Phys.*, 53, 1337–1349, 2002.

Jäkel, O., Krämer, M., Karger, C. P., and Debus, J., Treatment planning for heavy ion radio-therapy: Clinical implementation and application, *Phys. Med. Biol.*, 46, 1101–1116, 2001.

Jäkel, O., Schulz-Ertner, D., Karger, C. P., Nikoghosyan, A., and Debus, J., Heavy ion therapy: Status and perspectives, *Technol. Cancer Res. Treat.*, 2, 377–388, 2003a.

Jäkel, O., Hartmann, G. H., Herg, P., and Karger, C. P., Dosimetry of C12-ion beams at the German Heavy Ion Therapy Facility—Comparison between the currently used approach and the new TRS-398 In Standards and codes of practice in medical radiation dosimetry, Vol. 2, IAEA, Vienna, 303–310, http://www-pub.iaea.org/MTCD/publications/PDF/Pub1153/Start.pdf, 2003b.

Jani, S. K. and Pennington, E. C., Depth dose characteristics of 24-MV x-ray beams at extended SSD, *Med. Phys.*, 18, 292–294, 1991.

Jennings, A., The tracking cobalt project: From moving-beam therapy to three-dimensional programmed irradiation, in *Progress in Medical Radiation Physics*, Orton, C. G., Ed., Vol. 2, Plenum, New York, pp. 1–44, 1985.

Jiang, S. B., Radiotherapy of Mobile Tumours, *Semin. Radiat. Oncol.*, 16, 239–248, 2006.

Johansson, J., Isacsson, U., Lindman, H., Montelius, A., and Glimelius, B., Node-positive left-sided breast cancer patients after breast-conserving surgery: Potential outcomes of radiotherapy modalities and techniques, *Radiother. Oncol.*, 65, 89–98, 2002.

Johnson, L. S., Milliken, B. D., Hadley, S. W., Pelizzari, C. A., Haraf, D. J., and Chen, G. T., Initial clinical experience with a video-based patient positioning system, *Int. J. Radiat. Oncol. Biol. Phys.*, 45, 205–213, 1999.

Jones, D. T. L., Present status and future trends of heavy particle radiotherapy, in *Proceedings of the 15th International Conference on Cyclotrons and Their Applications*, Baron, E. and Lieuvin, M., Eds., Institute of Physics, Bristol, pp. 13–20, 1999.

Jones, D. T. L., Schreuder, A. N., Symons, J. E., and Yudelev. M., The NAC particle therapy facilities. in *Hadrontherapy in Oncology, Proceedings of the 1st International Symposium on Hadrontherapy*, Elsevier, Amsterdam, 307–328, 1994.

Kagawa, K., Murakami, M., Hishikawa, Y., Abe, M., Akagi, T. et al., Preclinical biological assessment of proton and carbon ion beams at Hyogo Ion Beam Medical Center, *Int. J. Radiat. Oncol. Biol. Phys.*, 54, 928–938, 2002.

Kamada, T., Tsujii, H., Tsuji, H., Yanagi, T. J., Mizoe, J. et al., Efficacy and safety of carbon ion radiotherapy in bone and soft tissue sarcomas, *J. Clin. Oncol.*, 20, 4466–4471, 2002.

Kamath, S., Sahni, S., Palta, J., Ranka, S., and Li, J., Optimal leaf sequencing with elimination of tongue-and-groove underdosage, *Phys. Med. Biol.*, 49, N7–N19, 2004.

Kanai, T., Endo, M., Minohara, S., Miyahara, N., Koyama-ito, H. et al., Biophysical characteristics of HIMAC clinical irradiation system for heavy-ion radiation therapy, *Int. J. Radiat. Oncol. Biol. Phys.*, 44, 201–210, 1999.

Karger, C. P., Hipp, P., Henze, M., Echner, G., Hoss, A. et al., Stereotactic imaging for radiotherapy: Accuracy of CT, MRI, PET and SPECT, *Phys. Med. Biol.*, 48, 211–221, 2003.

Karzmark, C. J., Large field superficial electron therapy with linear accelerators, *Br. J. Radiol.*, 40, 697–703, 1967.

Kassaee, A., Xiao, Y., Bloch, P., Goldwein, J., Rosenthal, D. I., and Bjarngard, B. E., Doses near the surface during total-body irradiation with 15 MV x-rays, *Int. J. Cancer*, 96(Suppl), 125–130, 2001.

Kavanagh, G. M., Matheson, L. M., Thwaites, D. I., and Hunter, J. A., Electron beam therapy in Arndt-Gottron's scleromyxoedema, *Br. J. Dermatol.*, 137, 152–153, 1997.

Kawachi, K., Kanai, T., Matsuzawa, H., and Inada, T., Three dimensional spot beam scanning methods for proton conformation therapy, *Acta Radiol. Oncol.*, 364, 81–88, 1983.

Keall, P., 4-Dimensional computed tomography imaging and treatment planning, *Semin. Radiat. Oncol.*, 14, 81–90, 2004.

Keall, P. J., Todor, A. D., Vedam, S. S., Bartee, C. L., Siebers, J. V. et al., On the use of EPID-based implanted marker tracking for 4D radiotherapy, *Med. Phys.*, 31, 3492–3499, 2004.

Keall, P. J., Mageras, G. S., Balter, J. M., Emery, R. S., Forster, K. M. et al., The management of respiratory motion in radiation oncology report of AAPM Task Group 76, *Med. Phys.*, 33, 3874–3900, 2006.

Keil, E, Sessler, A. M., and Trbojevic, D., Hadron cancer therapy complex employing non-scaling FFAG accelerator and fixed field gantry design, *Proceedings of EPAC 2006*, Edinburgh (http://accelcomf.web.cern.ch/accelconf/e06/PAPERS/TUPLS079.pdf) 2006.

Kieffer, J., The laminagraph and its variations: Applications and implications of the planigraphic principles, *Am. J. Roentgenol.*, 39, 497–513, 1938.

Kim, T. H., Pla, C., Pla, M., and Podgorsak, E. B., Clinical aspects of a rotational total skin electron irradiation technique, *Br. J. Radiol.*, 57, 501–506, 1984.

Kim, T. H., Rybka, W. B., Lehnert, S., Podgorsak, E. B., and Freeman, C. R., Interstitial pneumonitis following total body irradiation for bone marrow transplantation using two different dose rates, *Int. J. Radiat. Oncol. Biol. Phys.*, 11, 1285–1291, 1985.

King, C. R., Lehmann, J., Adler, J. R., and Hai, J., CyberKnife radiotherapy for localized prostate cancer: Rationale and technical feasibility, *Technol. Cancer Res. Treat.*, 2, 25–30, 2003.

Kini, V. R., Vedam, S. S., Keall, P. J., Patil, S., Chen, C., and Mohan, R., Patient training in respiratory-gated radiotherapy, *Med. Dosim.*, 28, 7–11, 2003.

Kjellberg, R. N., Stereotactic bragg peak proton beam radiosurgery for cerebral arteriovenous malformations, *Ann. Clin. Res.*, 18(Suppl. 47), 17–19, 1986.

Kjellberg, R. N., and Abe. M., Stereotactic Bragg peak proton beam therapy, in *Modern Stereotactic Neurosurgery*, Lunsford, L. D., Ed., Vienna, Springer, pp.463–470, 1988.

Koehler, A. M., Dosimetry of proton beams using small diodes, *Radiat. Res.*, Suppl. 7, 53–63, 1967.

Koehler, A. M., Schneider, R. J., and Sisterson, J. M., Range modulators for protons and heavy ions, *Nucl. Instrum. Methods.*, 131, 437–440, 1975.

Koehler, A. M., Schneider, R. J., and Sisterson, J. M., Flattening of proton dose distributions for large-field radiotherapy, *Med. Phys.*, 4, 297–301, 1977.

Kohno, R., Takada, Y., Sakae, T., Terunuma, T., Matsumoto, K. et al., Experimental evaluation of validity of simplified Monte Carlo method in proton dose calculations, *Phys. Med. Biol.*, 48, 1277–1288, 2003.

Kooy, H. M., Dunbar, S. F., Tarbell, N. J., Mannarino, E., Ferarro, N. et al., Adaptation and verification of the relocatable Gill-Thomas-Cosman frame in stereotactic radiotherapy, *Int. J. Radiat. Oncol. Biol. Phys.*, 30, 685–691, 1994.

Kooy, H. M., van Herk, M., Barnes, P. D., Alexander, E., Dunbar, S. F. et al., Image fusion for stereotactic radiotherapy and radiosurgery treatment planning, *Int. J. Radiat. Oncol. Biol. Phys.*, 28, 1229–1234, 1994.

Kooy, H. M., Schaefer, M., Rosenthal, S., and Bortfeld, T., Monitor unit calculations for range-modulated spread-out Bragg peak fields, *Phys. Med. Biol.*, 48, 2797–2808, 2003.

Kraft, G., Tumor therapy with heavy charged particles, *Prog. Part. Nucl. Phys.*, 45, S473–S544, 2000.

Krämer, M., Jäkel, O., Haberer, T., Kraft, G., Schardt, D., and Weber, U., Treatment planning for heavy ion radiotherapy: Physical beam model and dose optimization, *Phys. Med. Biol.*, 45, 3299–3317, 2000.

Krengli, M., Liebsch, N. J., Hug, E. B., and Orecchia, R., Review of current protocols for proton therapy in USA, *Tumori*, 84, 209–216, 1998.

Kubo, H. D., Len, P. M., Minohara, S., and Mostafavi, H., Breathing-synchronized radiotherapy program at the University of California Davis Cancer Center, *Med. Phys.*, 27, 346–353, 2000.

Kumar, P. P. and Patel, I. S., Comparison of dose distributions with different techniques of total skin electron beam therapy, *Clin. Radiol.*, 33, 495–497, 1982.

Laing, R. W., Bentley, R. E., Nahum, A. E., Warrington, A. P., and Brada, M., Stereotactic radiotherapy of irregular targets: A comparison between static conformal beams and non-coplanar arcs, *Radiother. Oncol.*, 28, 241–246, 1993.

Lam, W.-C., Lindskoug, B. A., Order, S., and Grant, D., The dosimetry of ^{60}Co total body irradiation, *Int. J. Radiat. Oncol. Biol. Phys.*, 5, 905–911, 1979.

Land, B., Jäkel, O., Combs, S., Schulz-Ertner, D., Cost-effectiveness of Carbon ion radiation therapy, in *Proceedings of the 9th Workshop on Charged Particles in Biology and Medicine*, 136–140, 2005.

Langen, K. M., Pouliot, J., Anezinos, C., Aubin, M., Gottschalk, A. R. et al., Evaluation of ultrasound-based prostate localization for image-guided radiotherapy, *Int. J. Radiat. Oncol. Biol. Phys.*, 57, 635–644, 2003.

Lapidot, T., Singer, T. S., Salomon, O., Terenzi, A., Schwartz, E., and Reisner, Y., Booster irradiation to the spleen following total body irradiation. A new immunosuppressive approach for allogeneic bone marrow transplantation, *J. Immunol.*, 141, 2619–2624, 1988.

Larsson, B., Liden, K., and Sarby, B., Irradiation of small structures through the intact skull, *Acta Radiol. Ther. Phys. Biol.*, 13, 512–534, 1974.

Lax, I. and Brahme, A., Rotation therapy using a novel high-gradient filter, *Radiology*, 145, 473–478, 1982.

Lax, I., Blomgren, H., Naslund, I., and Svanstrom, R., Stereotactic radiotherapy of malignancies in the abdomen Methodological aspects, *Acta Oncol.*, 33, 677–683, 1994.

Lee, M., Nahum, A. E., and Webb, S., An empirical method to build up a model of proton dose distribution for a radiotherapy treatment planning package, *Phys. Med. Biol.*, 38, 989–998, 1993.

Lee, M., Wynne, C., Webb, S., Nahum, A. E., and Dearnaley, D., A comparison of proton and megavoltage x-ray treatment planning for prostate cancer, *Radiother. Oncol.*, 33, 239–253, 1994.

Leer, J. W., Broerse, J. J., De Vroome, H., Chin, A., Noordijk, E. M., and Dutreix, A., Techniques applied for total body irradiation, *Radiother. Oncol.*, 18(Suppl. 1), 10–15, 1990.

Leksell, L., The stereotaxis method and radiosurgery of the brain, *Acta Chir. Scand.*, 102, 316–319, 1951.

Leksell, L., Stereotactic radiosurgery, *J. Neurol. Neurosurg. Psychiatry.*, 46, 797–803, 1983.

Letourneau, D., Gulam, M., Yan, D., Oldham, M., and Wong, J. W., Evaluation of a 2D diode array for IMRT quality assurance, *Radiother. Oncol.*, 70, 199–206, 2004.

Lewis, M. A. and Rosenbloom, M. E., A double-headed cobalt-60 unit for large-field irradiation, *Br. J. Radiol.*, 61, 1192, 1988 (abstract).

Li, Q., Komori, M., Kanai, T., Kitagawa, A., Urakabe, E. et al., The LET spectra at different penetration depths along secondary 9C and 11C beams, *Phys. Med. Biol.*, 49, 5119–5133, 2004.

Lind, B. and Brahme, A., Development of treatment techniques for radiotherapy optimization, *Int. J. Imaging Syst. Technol.*, 6, 33–42, 1995.

Ling, C. C., Humm, J., Larson, S., Amols, H., Fuks, Z. et al., Towards multidimensional radiotherapy (MD-CRT): Biological imaging and Biological conformality, *Int. J. Radiat. Oncol. Biol. Phys.*, 47, 551–560, 2000.

Linz, U., Ed, *Ion Beams in Tumor Therapy*, Chapman and Hall, Weinheim, 1995.

Litzenberg, D. W., Moran, J. M., and Fraass, B. A., Verification of dynamic and segmental IMRT delivery by dynamic log file analysis, *J. Appl.Clin. Med. Phys.*, 3, 63–72, 2002.

Liu, J. C., Evaluation of a microdosimetric-based neutron instrument, SLAC Publication 8258, Stanford Linear Accelerator Center, Stanford, CA, http://www.slac.stanford.edu/cgiwrap/getdoc/slacpub8258.pdf, 2000.

Liu, H., Yu, Y., Schell, M. C., O'Dell, W. G., Ruo, R., and Okunieff, P., Optimal marker placement in photogrammetry patient positioning system, *Med. Phys.*, 30, 103–110, 2003.

Loeffler, J. S., Smith, A. R., and Suit, H. D., The potential role of proton beams in radiation oncology, *Semin. Oncol.*, 24, 686–695, 1997.

Lomax, A. J., Bortfeld, T., Goitein, G., Debus, J., Dykstra, C. et al., A treatment planning inter-comparison of proton and intensity modulated photon radiotherapy, *Radiother. Oncol.*, 51, 257–271, 1999.

Lomax, A. J., Boehringer, T., Coray, A., Egger, E., Goitein, G. et al., Intensity modulated proton therapy: A clinical example, *Med. Phys.*, 28, 317–324, 2001.

Lomax, A. J., Cella, L., Weber, D., Kurtz, J. M., and Miralbell, R., Potential role of intensity-modulated photons and protons in the treatment of the breast and regional nodes, *Int. J. Radiat. Oncol. Biol. Phys.*, 55, 785–792, 2003a.

Lomax, A. J., Goitein, M., and Adams, J., Intensity modulation in radiotherapy: Photons versus protons in the paranasal sinus, *Radiother. Oncol.*, 66, 11–18, 2003b.

LoSasso, T., Chui, C. S., and Ling, C. C., Comprehensive quality assurance for the delivery of intensity modulated radiotherapy with a multileaf collimator used in the dynamic mode, *Med. Phys.*, 28, 2209–2219, 2001.

Low, D. A., Quality assurance of intensity-modulated radiotherapy, *Semin. Radiat. Oncol.*, 12, 219–228, 2002.

Low, D. A., Mutic, S., Dempsey, J. F., Gerber, R. L., Bosch, W. R. et al., Quantitative dosimetric verification of an IMRT planning and delivery system, *Radiother. Oncol.*, 49, 305–316, 1998.

Lundkvist, J., Ekman, M., Ericsson, S. R, Jonsson, B., and Glimelius, B., Proton therapy of cancer: potential clinical advantages and cost-effectiveness, *Acta Oncol.*, 44, 850–861, 2005.

Lunsford, L. D., Flickinger, J., Lindner, G., and Maitz, A., Stereotactic radiosurgery of the brain using the first United States 201 cobalt-60 source gamma knife, *Neurosurgery.*, 24, 151–159, 1989.

Lutz, W., Winston, K. R., and Maleki, N., A system for stereotactic radiosurgery with a linear accelerator, *Int. J. Radiat. Oncol. Biol. Phys.*, 14, 373–381, 1988.

Luxton, G., Petrovich, Z., Jozsef, G., Nedzi, L. A., and Apuzzo, M. L., Stereotactic radiosurgery: Principles and comparison of treatment methods, *Neurosurgery.*, 32, 241–259, 1993.

Lyman, J. T., Phillips, M. H., Frankel, K. A., and Fabrikant, J. I., Stereotactic frame for neuroradiology and charged particle Bragg peak radiosurgery of intracranial disorders, *Int. J. Radiat. Oncol. Biol. Phys.*, 16, 1615–1621, 1989.

Ma, C. M., Pawlicki, T., Jiang, S. B., Mok, E., Kapur, A. et al., Monte Carlo verification of IMRT dose distributions from a commercial treatment planning optimisation system, *Phys. Med. Biol.*, 45, 2483–2495, 2000.

Ma, C. M., Jiang, S. B., Pawlicki, T., Chen, Y., Li, J. S. et al., A quality assurance phantom for IMRT dose verification, *Phys. Med. Biol.*, 48, 561–572, 2003.

Mackie, T. R., Holmes, T., Swerdloff, S., Reckwerdt, P., Deasy, J. O. et al., Tomotherapy: A new concept for the delivery of dynamic conformal radiotherapy, *Med. Phys.*, 20, 1709–1719, 1993.

Mackie, T. R., Kapatoes, J., Ruchala, K., Lu, W., Wu, C. et al., Image guidance for precise conformal radiotherapy, *Int. J. Radiat. Oncol. Biol. Phys.*, 56, 89–105, 2003.

Mageras, G. S. and Yorke, E., Deep inspiration breath hold and respiratory gating strategies for reducing organ motion in radiation treatment, *Semin. Radiat. Oncol.*, 14, 65–75, 2004.

Mageras, G. S., Yorke, E., Rosenzweig, K., Braban, L., Keatley, E., Ford, E., Leibel, S. A., and Ling, C. C., Fluoro-scopic evaluation of diaphragmatic motion reduction with a respiratory gated radiotherapy system, *J. Appl. Clin. Med. Phys.*, 2 (4), 191–200, 2001.

Mah, K., Caldwell, C. B., Ung, Y. C., Danjoux, C. E., Balogh, J. M. et al., ^{18}FDT-PET/CT integration: Impact on target and critical organs in radiation therapy of non-small cell lung carcinoma, a prospective study, *Int. J. Radiat. Oncol. Biol. Phys.*, 52, 339–350, 2002.

Malinen, E., Søvik, Å., Hristov, D., Bruland, Ø. S., and Olsen, D. R., Adapting radiotherapy to hypoxic tumours, *Phys. Med. Biol.*, 51, 4903–4921, 2006.

Malka, V., Fritzler, S., Lefebvre, E., Aleonard, M. M., Burgy, F. et al., Electron acceleration by a wake field forced by an intense ultrashort laser pulse, *Science.*, 298, 1596–1600, 2002.

Mangili, P., Fiorini, C., Rosso, A., Cattaneo, G. M., Parisi, R. et al., In vivo dosimetry by diode semiconductors in combination with portal films during TBI: Reporting a 5-year clinical experience, *Radiother. Oncol.*, 52, 269–276, 1999.

Marinello, G. and Le Bourgeois, J. P., Skin irradiation (Irradiation cutanée totale), in *Radiotherapie Oncologique*, Le Bourgeois, J. P., Chavaudra, J., and Eschwege, F., Eds., Editions Hermann, Paris, pp. 531–537, 1992.

Marinello, G., Barret, C., and Le Bourgeois, J. P., Lithium borate discs for skin measurements: Application to total superficial electron beam therapy, *Nucl. Instrum. Methods.*, 175, 198–200, 1980.

Marinello, G., Barrie, A. M., and Le Bourgeois, J. P., Measurements and calculation of lung dose in total body irradiation performed with Cobalt-60, *J. Eur. Radioth.*, 3, 174–182, 1982.

Marinker, M. and Peckham, M., *Clinical Futures*, BMJ Books, London, 1998.

Marshall, M. and Docherty, J., Measurement of skin dose from low energy beta and gamma radiation using thermo-luminescent discs, *Phys. Med. Biol.*, 16, 503–510, 1971.

Maryanski, M. J., Ibbott, G. S., Eastman, P., Schulz, R. J., and Gore, J. C., Radiation dosimetry using MRI of polymer gels, *Med. Phys.*, 23, 699–705, 1996.

Maughan, R. L. and Yudelev, M., Physical characteristics of a clinical d(485) + Be neutron therapy beam produced by a superconducting cyclotron, *Med. Phys.*, 22, 1459–1465, 1995.

Mazal, A., Delacroix, S., Arianer, J., Clapier, F., Nauraye, C. et al., Protontherapy: Physical and technical basis, *Bull. Cancer Radiother.*, (*La protontherapie; Bases physiques et technologiques*), 83, 230–246, 1996a.

Mazal, A., Habrand, J. L., Lafortune, F., and Breteau, N., Quality assurance in protontherapy: A systematic approach in progress at Orsay, *Bull. Cancer Radiother.*, 83(Suppl), 179s–184s, 1996b.

Mazal, A., Gall, K., Bottollier-Depois, J. F., Michaud, S., Delacroix, D. et al., Shielding measurements for a proto-ntherapy beam of 200 MeV, *Proceedings of the 8th Symposium on Neutron Dosimetry*, 13–17 Nov 1995, *Radiat. Prot. Dosim.*, 70, 429–436, 1997.

Mazal, A., Schwartz, L., Lacroix, F., Mammar, H., Delacroix, S. et al., A preliminary comparative treatment planning study for radiotherapy of age-related maculopathy, *Radiother. Oncol.*, 47, 91–98, 1998.

McKerracher, C. and Thwaites, D. I., Assessment of new small-field detectors against standard field detectors for practical stereotactic beam data acquisition, *Phys. Med. Biol.*, 44, 2143–2160, 1999.

McKerracher, C. and Thwaites, D. I., Verification of the dose to the isocentre in stereotactic plans, *Radiother. Oncol.*, 64, 97–107, 2002.

Medin, J., Andreo, P., and Vynckier, S., Comparison of dosimetry recommendations for clinical proton beams, *Phys. Med. Biol.*, 45, 3195–3211, 2000. (Comment in Phys. Med. Biol., 45:L51–L54).

Meeks, S. L., Tome, W. A., Willoughby, T. R., Kupelian, P. A., Wagner, T. H., Buatti, J. M., and Bova, F. J., Optically guided patient positioning techniques, *Semin. Radiat. Oncol.*, 15, 192–201, 2005.

Mege, J. P., Marinello, G., Chossière, L., Piedbois, P., and Le Bourgeois, P., Correlation between measured and prescribed doses in vivo: Aplication to TBI before bone marrow transplant (*Correlation Entre Doses Prescrites and Doses Mesurées in vivo: Application Aux Irradiations Corporelles Totales Avant Greffe de Moelle*), *Bull. Cancer Radiother.*, 81, 468, 1994.

Menzel, H. G., Pihet, P., and Wambersie, A., Microdosimetric specification of radiation quality in neutron radiation therapy, *Int. J. Radiat. Biol.*, 57, 865–883, 1990.

Meyer, J. L. and Purdy, A., *3D Conformal Radiotherapy: A New era in the Irradiation of Cancer*, Karger, Basel, 1996.

Meyroneinc, S., Ferrand, R., Delacroix, S., Nauraye, C., Desblancs, C. et al., Which place for heavy charged particles in radiotherapy, CERN Accelerator School, *Particle Accelerators for Medicine and Industries*, http://cas.web.cern.ch/cas/Pruhonice/PDF/Meyroneinc.pdf, 2001.

Mijnheer, B. J., Present and future applications of in vivo dosimetry. *Proceedings of the 15th annual ESTRO Meting*, Vienna, *Radiother. Oncol.*, 40(Suppl. 1), S26, 1996.

Mijnheer, B. J., Wootton, P., Williams, J. R., Eenmaa, J., and Parnell, C. J., Uniformity in dosimetry protocols for therapeutic applications of fast neutron beams, *Med. Phys.*, 14, 1020–1026, 1987.

Miralbell, R., Rouzaud, M., Grob, E., Nouet, P., Bieri, S. et al., Can a total body irradiation technique be fast and reproducible?, *Int. J. Radiat. Oncol. Biol. Phys.*, 29, 1167–1173, 1994.

Miyamoto, T., Yamamoto, N., Nishimura, H., Koto, M., Tsujii, H. et al., Carbon ion radiotherapy for stage I non-small cell lung cancer, *Radiother. Oncol.*, 66, 127–140, 2003.

Mohan, R., Zhang, X., Wang, H., Kang, Y., Wang, X. et al., Use of deformed intensity distributions for on-line modification of image-guided IMRT to account for interfractional anatomic changes, *Int. J. Radiat. Oncol. Biol. Phys.*, 61, 1258–1266, 2005.

Molière, G. Z., Theory of scattering of high energy charged particles I., Single scattering in a shielded Coulomb field (*Theorie der Streuung schneller geladener Teilchen. I., Einzelstreuung am abgeschirmten Coulomb-Field*), *Z. Natur-forsch.*, 2a, 133–145, 1947.

Monk, J. E., Perks, J. R., Doughty, D., and Plowman, P. N., Comparison of a micro-multileaf collimator with a 5 mm leaf-width collimator for intracranial stereotactic radiotherapy, *Int. J. Radiat. Oncol. Biol. Phys.*, 57, 1443–1449, 2003.

Mould, R. F., *A Century of X-rays and Radioactivity in Medicine*, Institute of Physics Publishing, Bristol, Chapters 18 and 19, 1993.

Mould, R. F., Invited review: Röntgen and the discovery of x-rays, *Br. J. Radiol.*, 68, 1145–1176, 1995.

Mu, X., Clinical application of intensity and energy modulated radiotherapy with photon and electron beams, *PhD diss.*, Umeå University, 2005 (available from Dept. Radiation Sciences, Oncology, Umeå University, Umeå, Sweden.)

Munkel, G., Lomax, A., Scheib, S., Pedroni, E., Blattmann, H., Indications for protontherapy, in *Hadrontherapy in oncology, Proceedings of the 1st International Symposium on Hadrontherapy*, Como, Italy, October 1993, Elsevier, Amsterdam, 109–112, 1994.

Munzenrider, J. E., Proton therapy with the Harvard cyclotron, in Hadrontherapy in oncology, Amaldi, U. and Larsson, B., Eds., *Proceedings of the 1st International Symposium on Hadrontherapy, Como, Italy, October 1993*, Elsevier, Amsterdam, pp. 83–101, 1994.

Munzenrider, J. E., Uveal melanomas conservation treatment, *Hematol. Oncol. Clin. North Am.*, 15, 389–402, 2001.

Murphy, M. J., Tracking moving organs in real time, *Semin. Radiat. Oncol.*, 14, 91–100, 2004.

Mutic, S., Low, D. A., Klein, E. E., Dempsey, J. F., and Purdy, J. A., Room shielding for intensity modulated radiation therapy treatment facilities, *Int. J. Radiat. Oncol. Biol. Phys.*, 50, 239–246, 2001.

Nakamura, J. L., Pirzkall, A., Carol, M. P., Xia, P., Smith, V. et al., Comparison of intensity-modulated radiosurgery with gamma knife radiosurgery for challenging skull base lesions, *Int. J. Radiat. Oncol. Biol. Phys.*, 55, 99–109, 2003.

Nauraye, C., Mazal, A., Delacroix, S., Bridier, A., Chavaudra, J., and Rosenwald, J.-C., An experimental approach to the design of a scattering system for a proton therapy beam line dedicated to ophtalmological applications, *Int. J. Radiat. Oncol. Biol. Phys.*, 32, 1177–1183, 1995.

Nedzi, L. A., Kooy, H. M., Alexander, E., Svensson, G. K., and Loeffler, J. S., Dynamic field shaping for stereotactic radiosurgery: A modelling study, *Int. J. Radiat. Oncol. Biol. Phys.*, 25, 859–869, 1993.

Neicu, T., Shirato, H., Seppenwoolde, Y., and Jiang, S. B., Synchronized moving aperture radiation therapy (SMART): Average tumour trajectory for lung patients, *Phys. Med. Biol.*, 48, 587–598, 2003.

Nestle, U., Kremp, S., and Grosu, A. L., Practical integration of [18F]-FDG-PET and PET-CT in the planning of radiotherapy for non-small cell lung cancer (NSCLC): the technical basis, ICRU-target volumes, problems, perspectives, *Radiother. Oncol.*, 81, 209–225, 2006.

Newhauser, W. D., Burns, J., and Smith, A. R., Dosimetry for ocular proton beam therapy at the Harvard Cyclotron Laboratory based on the ICRU Report 59, *Med. Phys.*, 29, 1953–1961, 2002a.

Newhauser, W. D., Myers, K. D., Rosenthal, S. J., and Smith, A. R., Proton beam dosimetry for radiosurgery: Implementation of the ICRU Report 59 at the Harvard Cyclotron Laboratory, *Phys. Med. Biol.*, 47, 1369–1389, 2002b.

Nichiporov, D., Kostjuchenko, V., Puhl, J. M., Bensen, D. L., Desrosiers, M. F. et al., Investigation of applicability of alanine and radiochromic detectors to dosimetry of proton clinical beams, *Appl. Radiat. Isot.*, 46, 1355–1362, 1995.

Nioutsikou, E., Bedford, J. L., and Webb, S., Patient-specific planning for prevention of mechanical collisions during radiotherapy, *Phys. Med. Biol.*, 48, N313–N321, 2003.

Noel, G., Habrand, J. L., Helfre, S., Mammar, H., Kalifa, C. et al., Proton beam therapy in the management of central nervous system tumors in childhood: The preliminary experience of the Centre de Protontherapie d'Orsay, *Med. Pediatr. Oncol.*, 40, 309–315, 2003.

Novitzky, N., Thomas, V., Stubbings, H., Hale, G., and Waldmann, H., Radiotherapy-based conditioning is effective immunosuppression for patients undergoing transplantation with T-cell depleted stem cell grafts for severe aplasia, *Cytotherapy*, 6, 450–456, 2004.

O'Donoghue, J. A., Fractionated versus low dose-rate total body irradiation Radiobiological considerations in the selection of regimes, *Radiother. Oncol.*, 7, 241–247, 1986.

Olko, P., Bilski, P., Budzanowski, M., Waligorski, M. P., and Reitz, G., Modeling the response of thermoluminescence detectors exposed to low- and high-LET radiation fields, *J. Radiat. Res. (Tokyo)*, 43, S59–S62, 2002.

Oozeer, R., Mazal, A., Rosenwald, J.-C., Nauraye, C., Delacroix, S., Effect of heterogeneities in a proton beam intended for clinical applications: Preliminary study, (*Effets des hétérogénéités dans un faisceau de protons destiné aux applications cliniques: étude préliminaire*) *Proc XXXIIème Congrès de la Société Française des Physiciens d'Hôpital*, June 1993, 1993.

Ozsahin, M., Belkacemi, Y., Pene, F., Laporte, J., Rio, B. et al., Interstitial pneumonitis following autologous bone-marrow transplantation conditioned with cyclophosphamide and total-body irradiation, *Int. J. Radiat. Oncol. Biol. Phys.*, 34, 71–77, 1996.

Paganetti, H., Niemierko, A., Ancukiewicz, M., Gerweck, L. E., Goitein, M. et al., Relative biological effectiveness (RBE) values for proton beam therapy, *Int. J. Radiat. Oncol. Biol. Phys.*, 53, 407–421, 2002.

Paganetti, H., Jiang, H., and Trofimov, A., 4D Monte Carlo simulation of proton beam scanning: modelling of variations in time and space to study the interplay between scanning pattern and time-dependent patient geometry, *Phys. Med. Biol.*, 50, 983–990, 2005.

Page, V., Gardner, A., and Karzmark, C. J., Patient dosimetry in the treatment of large large superficial lesions, *Radiology*, 94, 635–641, 1970.

Paliwal, B., Tomé, W., Richardson, S., and Mackie, T. R., A spiral phantom for IMRT and tomotherapy treatment delivery verification, *Med. Phys.*, 27, 2503–2507, 2000.

Palmans, H., Verhaegen, F., Denis, J. M., and Vynckier, S., Dosimetry using plane-parallel ionization chambers in a 75 MeV clinical proton beam, *Phys. Med. Biol.*, 47, 2895–2905, 2002.

Palmans, H., Thomas, R., Simon, M., Duane, S., Kacperek, A. et al., A small-body portable graphite calorimeter for dosimetry in low-energy clinical proton beams, *Phys. Med. Biol.*, 49, 3737–3749, 2004.

Palta, J. R. and Mackie, T. R., *Intensity Modulated Radiation Therapy—the State of the Art*, Medical Physics Publishing, Madison, WI, 2003.

Parodi, K., Enghardt, W., and Haberer, T., In-beam PET measurements of beta+ radioactivity induced by proton beams, *Phys. Med. Biol.*, 47, 21–36, 2002.

Paskalev, K., Ma, C.-M., Jacob, R., Price, R. A., McNeeley, S., and Pollack, A., Daily target localization for prostate patients based on 3D image correlation, *Phys. Med. Biol.*, 49, 931–939, 2004.

Pawlicki, T. and Ma, C.-M., Monte Carlo dose modeling for MLC-based IMRT, *Med. Dosim.*, 26, 157–168, 2001.

Pedroni, E., Beam delivery, in *Hadrontherapy in oncology, Proceedings of the 1st International Symposium on Hadrontherapy*, Como, Italy, October 1993, Elsevier, Amsterdam, 434–452, 1994.

Pedroni, E., Bacher, R., Blattmann, H., Bohringer, T., Coray, A. et al., The 200-MeV proton therapy project at the Paul Scherrer Institute: Conceptual design and practical realization, *Med. Phys.*, 22, 37–53, 1995.

Pedroni, E., Bohringer, T., Coray, A., Egger, E., Grossmann, M. et al., Initial experience of using an active beam delivery technique at PSI, *Strahlenther. Oncol.*, 175(Suppl. 2), 18–20, 1999.

Perez, C. A., Bradley, J., Chao, C. K., Grigsby, P. W., Mutic, S., and Malyapa, R., Functional imaging in treatment planning in radiation therapy: A review, *Rays*, 27, 157–173, 2002.

Perks, J., Rosenberg, I., and Warrington, A. P., Dose quality assurance for stereotactic radiotherapy treatments, *Phys. Med. Biol.*, 44, P209–P215, 1999.

Petti, P. L., Differential pencil beam dose calculation for charged particles, *Med. Phys.*, 19, 137–149, 1992.

Petti, P. L., Evaluation of a pencil-beam dose calculation technique for charged particle radiotherapy, *Int. J. Radiat. Oncol. Biol. Phys.*, 35, 1049–1057, 1996.

Petti, P. L., New compensator design options for charged-particle radiotherapy, *Phys. Med. Biol.*, 42, 1289–1300, 1997.

Phillips, M. H., Stelzer, K. J., Griffin, T. W., Mayberg, M. R., and Winn, H. R., Stereotactic radiosurgery: A review and comparison of methods, *J. Clin. Oncol.*, 12, 1085–1099, 1994.

Pignol, J. P., Meyer, L., Methlin, A., Wagner, J. P., Abbe, J. C., and Sahel, J., Radiotherapy of ocular melanoma: Physical and radiobiological bases, current techniques and future prospects (*Radiothérapie des mélanomes oculaires: Bases physiques et radiobiologiques, techniques actuelles et perspectives d'avenir*), *Bull. Cancer Radiother.*, 81, 127–142, 1994.

Pignol, J. P., Paquis, P., Cuendet, P., Gibon, D., Diop, C. M., and Sabattier, R., Beam collimation and bolusing material optimizations for ^{10}boron neutron capture enhancement of fast neutron (BNCEFN): Definition of the optimum irradiation technique, *Int. J. Radiat. Oncol. Biol. Phys.*, 43, 1151–1159, 1999.

Pihet, P., Microdosimetric study of high energy neutron beams. Dosimetric and radiobiological applications. (Etude microdosimétrique de faisceaux de neutrons de haute énergie Applications dosimétriques et radiobiologiques), Thesis, Université catholique de Louvain, Louvain La Neuve, Belgium, 1989.

Pihet, P., Gueulette, J., Menzel, H. G., Grillmaier, R. E., and Wambersie, A., Use of microdosimetric data of clinical relevance in neutron therapy planning, *Radiat. Prot. Dosim.*, 23, 471–474, 1988.

Pla, C., Heese, R., Pla, M., and Podgorsak, E. B., Calculation of surface dose in rotational total skin electron irradiation, *Med. Phys.*, 11, 539–546, 1984.

Pla, M., Pla, E., and Podgorsak, E. B., The influence of beam parameters on percentage depth dose in electron arc therapy, *Med. Phys.*, 15, 49–55, 1988.

Planskoy, B., Bedford, A. M., Davis, F. M., Tapper, P. D., and Loverock, L. T., Physical aspects of total-body irradiation at the Middlesex Hospital (UCL group of hospitals) London 1988–1993: I Phantom measurements and planning methods, *Phys. Med. Biol.*, 41, 2307–2326, 1996a.

Planskoy, B., Bedford, A. M., Davis, F. M., Tapper, P. D., and Loverock, L. T., Physical aspects of total-body irradiation at the Middlesex Hospital (UCL group of hospitals) London 1988–1993: II In vivo planning and dosimetry, *Phys. Med. Biol.*, 41, 2327–2343, 1996b.

Plowman, P. N., Stereotactic intracranial radiotherapy/radiosurgery has come of age, *J. R. Coll. Physicians Lond.*, 34, 273–281, 2000.

Plowman, P. N., Radiation therapy for vascular malformations–a changing scene, *Clin. Oncol. (R. Coll. Radiol.)*, 14, 91–96, 2002.

Plowman, P. N., Saunders, C. A., and Maisey, M., On the usefulness of brain PET scanning to the paediatric neuro-oncologist, *Br. J..Neurosurg.*, 11, 525–532, 1997.

Podgorsak, E. B., Pla, C., Pla, M., Lefebvre, E. Y., and Heese, R., Physical aspects of a rotational total skin electron irradiation, *Med. Phys.*, 10, 159–168, 1983.

Podgorsak, E. B., Olivier, A., Pla, M., Hazel, J., de Lotbiniere, A., and Pike, B., Physical aspects of dynamic stereotactic radiosurgery, *Appl. Neurophysiol.*, 50, 263–268, 1987.

Podgorsak, E. B., Olivier, A., Pla, M., Lefebvre, P. Y., and Hazel, J., Dynamic stereotactic radiosurgery, *Int. J. Radiat. Oncol. Biol. Phys.*, 14, 115–126, 1988.

Podgorsak, E. B., Pike, G. B., Olivier, A., Pla, M., and Souhami, L., Radiosurgery with high energy photon beams: A comparison among techniques, *Int. J. Radiat. Oncol. Biol. Phys.*, 16, 857–865, 1989.

Podgorsak, E. B., Pike, G. B., Pla, M., Olivier, A., and Souhami, L., Radiosurgery with photon beams: Physical aspects and adequacy of linear accelerators, *Radiother. Oncol.*, 17, 349–358, 1990.

Price, R. A., Chibani, O., and Ma, C.-M., Shielding evaluation for IMRT implementation in an existing accelerator vault, *J. Appl. Clin. Med. Phys.*, 4 (3), 231–238, 2003a.

Price, R. A, Murphy, S., McNeeley, S. W., Ma, C.-M., Horwitz, E., Movsas, B., Raben, A., and Pollack, A., A method for increased dose conformity and segment reduction for SMLC delivered IMRT treatment of the prostate. *Int. J. Radiat. Oncol. Biol. Phys.*, 57, 843–852, 2003b.

Proimos, B., Synchronous field shaping in rotational megavolt therapy, *Radiology*, 74, 753–757, 1960.

Purdy, J. A., Current ICRU definitions of volumes: Limitations and future directions, *Semin. Radiat. Oncol.*, 14, 27–40, 2004.

Quast, U., Total body irradiation—review of treatment techniques in Europe, *Radiother. Oncol.*, 9, 91–106, 1987.

Que, W., Kung, J., and Dai, J., Tongue-and-groove effect in intensity modulated radiotherapy with static multileaf collimator fields, *Phys. Med. Biol.*, 49, 399–405, 2004.

Raaijmakers, A. J., Raaymakers, B. W., and Lagendijk, J. J., Integrating a MRI scanner with a 6 MV radiotherapy accelerator: Dose increase at tissue-air interfaces in a lateral magnetic field due to returning electrons, *Phys. Med. Biol.*, 50, 1363–1376, 2005.

Raaymakers, B. W., Lagendijk, J. J. W., Van der Heide, U. A., Overweg, J., Brown, K., et al., Integrating a MRI scanner with a radiotherapy accelerator: A new concept of precise on line radiotherapy guidance and treatment monitoring, *Proceedings of the 14th International Conference on the Use of the Computers in Radiation Therapy (Seoul, South Korea)*, Yi, B. Y. and Choi, E. K., Eds., Jeong, Seoul, pp.89–92, 2004.

Raju, M. R., *Heavy Particle Radiotherapy*, Academic Press, New York, 1980.

Renner, T. R., Chu, W. T., and Ludewigt, B. A., Advantages of beam-scanning and requirements of hadrontherapy facilities, in *Hadrontherapy in oncology, Proceedings of the 1st International Symposium on Hadrontherapy*, Como, Italy, October 1993, Elsevier, Amsterdam, pp.453–461, 1994.

Rice, R. K., Hansen, J. L., Svensson, G. K., and Siddon, R. L., Measurements of dose distributions in small beams of 6 MV x-rays, *Phys. Med. Biol.*, 32, 1087–1099, 1987.

Ringdén, O., Båryd, I., Johansson, B., Gahrton, G., Groth, C. G. et al., Increased mortality by septicemia, interstitial pneumonitis and pulmonary fibrosis among bone marrow transplant recipients receiving an increased mean dose rate of total irradiation, *Acta Radiol. Oncol.*, 22, 423–428, 1983.

Robertson, J. B., Eaddy, J. M., Archambeau, J. O., Coutrakon, G. B., Miller, D. W. et al., Variation of measured proton relative biological effectiveness as a function of initial proton energy, in *Hadrontherapy in Oncology, Proceedings of the 1st International Symposium on Hadrontherapy*, Como, Italy, October 1993, Elsevier, Amsterdam, pp.706–711, 1994.

Rosenblat, J. Particle acceleration, in *Methuen's Monographs on Physical Subjects*, Methuen, London, 1968.

Rosenbloom, M., Hickling, P. A., Chow, M., Chittenden, S., Machardy, J. et al., Total body irradiation at the Royal Marsden Hospital Sutton, *J. Eur. Radioth.*, 3, 246–248, 1982.

Rosenwald, J.-C., Drouard, J., Modelling of the penumbra region with extension of the concept of separation of primary and scatter to electron beams. (Modélisation de la zone de pénombre lorsqu'on étend le concept de séparation primaire-diffusé aux faisceaux d'électrons) Proc. XXX Congrès Soc. Française des Physiciens d'Hôpital, pp.21–31, 1991.

Rosenwald, J.-C., Oozeer, R., Belshi, R., Drouard, J., and Mazal, A., Penumbra representation for "primary-scatter" decomposition of electron and proton beams, In *Proceedings of the XIth International Conference on the Use of Computers*, in *Radiation Therapy*, Hounsell, A. R., Wilkinson, J. M., and Williams, P. C., Eds., Medical Physics Publishing, Madison, WI, pp.128–129, 1994.

Rowbottom, C. G., Webb, S., and Oldham, M., Is it possible to optimize a radiotherapy treatment plan?, *Int. J. Radiat. Oncol. Biol. Phys.*, 43, 698–699, 1999.

Russell, K. J., Caplan, R. J., Laramore, G. E., Burnison, C. M., Maor, M. H. et al., Photon versus fast neutron external beam radiotherapy in the treatment of locally advanced prostate cancer: Results of a randomized prospective trial, *Int. J. Radiat. Oncol. Biol. Phys.*, 28, 47–54, 1994.

Russell, K. R., Grusell, E., and Montelius, A., Dose calculations in proton beams: Range straggling corrections and energy scaling, *Phys. Med. Biol.*, 40, 1031–1043, 1995.

Russell, K. R., Isacsson, U., Saxner, M., Ahnesjö, A., Montelius, A. et al., Implementation of pencil kernel and depth penetration algorithms for treatment planning of proton beams, *Phys. Med. Biol.*, 45, 9–27, 2000.

SBRT, 2006, 3rd Acta Oncologica symposium on Stereotactic Body Radiotherapy, June 15–17, Copenhagen, *Acta Oncol.*, 45, 771–994, 2006.

Sabattier, R., Destembert, B., and Breteau, N., Technical status report of the Orléans neutrontherapy facility, *Strahlenther. Oncol.*, 166, 86–89, 1990.

Sabini, M. G., Raffaele, L., Bucciolini, M., Cirrone, G. A., Cuttone, G. et al., The use of thermoluminescent detectors for measurements of proton dose distribution, *Radiat. Prot. Dosimetry.*, 101, 453–456, 2002.

Safwat, A., Nielsen, O. S., El, S., and Overgaard, J., Effect of radiation dose rate and cyclophosphamide on pulmonary toxicity after total body irradiation in a mouse model, *Int. J. Radiat. Oncol. Biol. Phys.*, 34, 85–91, 1996.

Sánchez-Doblado, F., Quast, U., Arrans, R., Errazquin, L., Sánchez-Nieto, B., and Terrón J. A., Total body irradiation prior to bone marrow transplantation, *Report of the European group for Blood and Marrow Transplantation*, 1995.

Sauerwein, W. and Zurlo, Z., The EORTC Boron Neutron Capture Therapy (BNCT) Group: Achievements and future projects, *Eur. J., Cancer.*, 38(Suppl. 4), S31–S34, 2002.

Scalliet, P., The trouble with neutrons, *Eur. J., Cancer.*, 27, 225–230, 1991.

Schaffner, B., Pedroni, E., and Lomax, A., Dose calculation models for proton treatment planning using a dynamic beam delivery system: An attempt to include density heterogeneity effects in the analytical dose calculation, *Phys. Med. Biol.*, 44, 27–41, 1999.

Scheib, S., Pedroni, E., Lomax, A., Blattmann, H., Bohringer, T. et al., Spot scanning with protons at PSI: Experimental results and treatment planning, in *Hadrontherapy in oncology, Proceedings of the 1st International Symposium on Hadrontherapy*, Como, Italy, October 1993, Elsevier, Amsterdam, 471–480, 1994.

Scherrer, A., Kufer, K. H., Bortfeld, T., Monz, M., and Alonso, F., IMRT planning on adaptive volume structures–a decisive reduction in computational complexity, *Phys. Med. Biol.*, 50, 2033–2053, 2005.

Schlegel, W. and Mahr, A., *3D Conformal Radiation Therapy*, Springer, Heidelberg, 2001.

Schneider, U., Pedroni, E., and Lomax, A., The calibration of CT Hounsfield units for radiotherapy treatment planning, *Phys. Med. Biol.*, 41, 111–124, 1996. (Comments in Phys. Med. Biol. 41:1524–1527 and in Phys. Med. Biol. 48:861–874 [2003]).

Schreuder, A. N., Jones, D. T., Symons, J. E., De Kock, E. A., Hough, J. K. et al., The NAC proton treatment planning system, *Strahlenther. Oncol.*, 175(Suppl. 2), 10–12, 1999.

Schulte, R. W., Fargo, R. A., Meinass, H. J., Slater, J. D., and Slater, J. M., Analysis of head motion prior to and during proton beam therapy, *Int. J. Radiat. Oncol. Biol. Phys.*, 47, 1105–1110, 2000.

Schulz-Ertner, D., Haberer, T., Scholz, M., Thilmann, C., Wenz, F. et al., Acute radiation-induced toxicity of heavy ion radiotherapy delivered with intensity modulated pencil beam scanning in patients with base of skull tumors, *Radiother. Oncol.*, 64, 189–195, 2002.

Schulz-Ertner, D., Nikoghosyan, A., Thilmann, C., Haberer, T., Jäkel, O. et al., Results of carbonion radiotherapy in 152 patients, *Int. J. Radiat. Oncol. Biol. Phys.*, 58, 631–640, 2004.

Schwade, J. G., Houdek, P. V., Landy, H. J., Bujnoski, J. L., Lewin, A. A. et al., Small-field stereotactic external-beam radiation therapy of intracranial lesions: Fractionated treatment with a fixed-halo immobilization device, *Radiology*, 176, 563–565, 1990.

Schwartz, L. H., Laisne, A., Ivanov, A. A., Mandrillon, P., Chauvel, P. et al., What equipment should be used for the proton therapy of tomorrow?, *(Quel appareillage pour la protonthérapie de demain?) Bull. Cancer Radiother.*, 82, 365–369, 1995.

Schweikard, A., Glosser, G., Bodduluri, M., Murphy, M. J., and Adler, J. R., Robotic motion compensation for respiratory movement during radiosurgery, *Comput. Aided Surg.*, 5, 263–277, 2000.

Serago, C. F., Houdek, P. V., Bauer-Kirpes, B., Lewin, A. A., Abitbol, A. A. et al., Stereotactic radiosurgery: Dose-volume analysis of linear accelerator techniques, *Med. Phys.*, 19, 181–185, 1992.

Serago, C. F., Houdek, P. V., Hartmann, G. H., Saini, D. S., Serago, M. E., and Kaydee, A., Tissue maximum ratios (and other parameters) of small circular 4, 6, 10, 15 and 24 MV x-ray beams for radiosurgery, *Phys. Med. Biol.*, 37, 1943–1956, 1992.

Sewchand, W., Khan, F. M., and Williamson, J., Total body superficial electron beam therapy using a multiple field pendulum arc technique, *Radiology*, 130, 493–498, 1979.

Sheen, M., EYEPLAN User Manual v. 3.01 (Douglas Cyclotron Laboratory, Clatterbridge, UK) 2001.

Shepherd, S. F., Childs, P. J., Graham, J. D., Warrington, A. P., and Brada, M., Whole body doses from linear accelerator-based stereotactic radiotherapy, *Int. J. Radiat. Oncol. Biol. Phys.*, 38, 657–665, 1997.

Sherouse, G. W. Conformal radiotherapy without multileaf collimators, in *ESTRO Pre-meeting Workshop on "Challenges in Conformal Radiotherapy* ESTRO" (Brussels), 1997.

Shirato, H., Seppenwoolde, Y., Kitamura, K., Onimura, R., and Shimizu, S., Intrafractional tumor motion: Lung and liver, *Semin. Radiat. Oncol.*, 14, 10–18, 2004.

Shiu, A. S., Kooy, H. M., Ewton, J. R., Tung, S. S., Wong, J. et al., Comparison of miniature multileaf collimation (MMLC) with circular collimation for stereotactic treatment, *Int. J. Radiat. Oncol. Biol. Phys.*, 37, 679–688, 1997.

Shrieve, D. C., Kooy, H. M., Tarbell, N. J., and Loeffler, J. S., Fractionated stereotactic radiotherapy, in *Important Advances in Oncology*, DeVita, V. T., Hellman, S., and Rosenberg, S. A., Eds., Lippincott, Williams and Wilkins, Hagerstown, MD, pp.205–224, 1996.

Siddon, R. L. and Barth, N. H., Stereotaxic localization of intracranial targets, *Int. J. Radiat. Oncol. Biol. Phys.*, 13, 1241–1246, 1987.

Siebers, J. V., DeLuca, P. M., Pearson, D. W., and Coutrakon, G., Measurements of neutron dose equivalent and penetration in concrete for 230 MeV proton bombardment of Al, Fe, and Pb targets, *Radiat. Prot. Dosim.*, 44, 247–251, 1992.

Signorotto, P., Del Vecchio, A., Fiorino, C., Mangili, P., and Calandrino, R., Diodes and portal-films in TBI in-vivo dosimetry, *Proceedings of the 15th Annual ESTRO Meeting*, Vienna, *Radiother. Oncol.*, 40(Suppl. 1), S126, 1996.

Sisterson, J., Ed., Particles 18 (Harvard: Proton Therapy Cooperative Group), http://ptcog.mgh.harvard.edu/particles_archives.htm, 1996.

Sisterson, J., Ed., Status of ion beam therapy in 2002, in *CP680 Applications of accelerators to research and industry*, 17th International Conference, Duggan, J. L. and Morgan, I. L., Eds., American Institute of Physics, New York, 2003

Sisterson, J., Ed., Particles 31. (Harvard: Proton Therapy Cooperative Group) http://ptcog.mgh.harvard.edu/particles_archives.htm, 2003.

Sisterson, J., Ed., Particles 33 (Harvard: Proton Therapy Cooperative Group) http://ptcog.mgh.harvard.edu/particles_archives.htm, 2004.

Sisterson, J., Urie, M., Koehler, A., and Goitein, M., Distal penetration of proton beam: The effects of air gaps between compensating bolus and patient, *Phys. Med. Biol.*, 34, 1309–1315, 1989.

Slater, J. M., Archambeau, J. O., Miller, D., Notarus, M., Preston, W., and Slater, J. D., The proton treatment center at Loma Linda University Medical Center: Rational for description of its development, *Int. J. Radiat. Oncol. Biol. Phys.*, 22, 383–389, 1992.

Smith, A. R., *Radiation Therapy Physics*, Springer Verlag, Berlin, 1995.

Spielberger, B., Scholz, M., Kramer, M., and Kraft, G., Experimental investigations of the response of films to heavy-ion irradiation, *Phys. Med. Biol.*, 46, 2889–2897, 2001.

Spielberger, B., Scholz, M., Kramer, M., and Kraft, G., Calculation of the x-ray film response to heavy charged particle irradiation, *Phys. Med. Biol.*, 47, 4107–4120, 2002.

Spielberger, B., Kramer, M., and Kraft, G., Three-dimensional dose verification with x-ray films in conformal carbon ion therapy, *Phys. Med. Biol.*, 48, 497–505, 2003.

Spiller, P., Boehne, D., Dolinskii, A., Eickhoff, H., Franczak, B. et al., Gantry studies for the proposed heavy ion cancer therapy facility in Heidelberg, in *Proceedings of the 7th European Particle Accelerator Conference (EPAC)*, CERN, Vienna, 2551–2553, http://accelconf.web.cern.ch/accelconf/e00/PAPERS/WEP3A15.pdf, 2000.

Spirou, S. V. and Chui, C. S., Generation of arbitrary intensity profiles by dynamic jaws or multileaf collimators, *Med. Phys.*, 21, 1031–1041, 1994.

Starkschall, G., Forster, K. M., Kitamura, K., Cardenas, A., Tucker, S. L., and Stevens, C. W., Correlation of gross tumor volume excursion with potential benefits of respiratory gating, *Int. J. Radiat. Oncol. Biol. Phys.*, 60, 1291–1297, 2004.

Stathakis, S., Price, R. A., and Ma, C.-M., Shielding evaluation of existing accelerator vaults under IMRT conditions, *Med. Phys.*, 32, 448–454, 2005.

St Clair, W. H., Adams, J. A., Bues, M., Fullerton, B. C., La Shell, S. et al., Advantage of protons compared to conventional x-ray or IMRT in the treatment of a pediatric patient with medulloblastoma, *Int. J. Radiat. Oncol. Biol. Phys.*, 58, 727–734, 2004.

Stein, J., Bortfeld, T., Dorschel, B., and Schlegel, W., Dynamic x-ray compensation for conformal radiotherapy by means of multi-leaf collimation, *Radiother. Oncol.*, 32, 163–173, 1994.

Steinberg, G. K., Fabrikant, J. I., Marks, M. P., Levy, R. P., Frankel, K. A. et al., Stereotactic heavy-charged-particle Bragg-peak radiation for intracranial arteriovenous malformations, *N. Engl. J. Med.*, 323, 96–101, 1990.

Stell, A. M., Li, J. G., Zeidan, O. A., and Dempsey, J. F., An extensive log-file analysis of step-and-shoot intensity modulated radiation therapy segment delivery errors, *Med. Phys.*, 31, 1593–1602, 2004.

Sternick, E. S., *The Theory and Practice of Intensity Modulated Radiation Therapy*, Advanced Medical Publishing, Madison, 1997.

Suit, H. D., National Cancer Institute Proton Workshop: Potential clinical gains by use of superior radiation dose distribution, *Int. J. Radiat. Oncol. Biol. Phys.*, 22, 233–234, 1992.

Suit, H., Goldberg, S., Niemierko, A., Trofimov, A., Adams, J. et al., Proton beams to replace photon beams in radical dose treatments, *Acta Oncol.*, 42, 800–808, 2003.

Sullivan, A. H., *A Guide to Radiation and Radioactivity Levels Near High Energy Particle Accelerators*, Nuclear Technology Publishing, Ashford, 1992.

Svensson, H., Ringborg, U., Näslund, I., and Brahme, A., Development of light ion therapy at the Karolinska Hospital and Institute, *Radiother. Oncol.*, 73(Suppl. 2), S206–S210, 2004.

Svensson, R., Källman, P., and Brahme, A., An analytical solution for the dynamic control of multileaf collimators, *Phys. Med. Biol.*, 39, 37–61, 1994.

Svensson, R., Lind, B., and Brahme, A., A new compact treatment unit design combining narrow pencil beam scanning and segmental multileaf collimation, *Radiother. Oncol.*, 51(Suppl. 1), S21, 1999.

Sweeney, R., Bale, R., Vogele, M., Nevinny-Stickel, M., Bluhm, A. et al., Repositioning accuracy: Comparison of a noninvasive head holder with thermoplastic mask for fractionated radiotherapy and a case report, *Int. J. Radiat. Oncol. Biol. Phys.*, 41, 475–483, 1998.

Swerdloff, S., Mackie, T. R., Holmes, T., Method and Apparatus for Radiation Therapy, US Patent 5, 317, 616, 1994.

Swerdloff, S., Mackie, T. R., Holmes, T., Multi-leaf Radiation Attenuator for Radiation Therapy, US Patent 5, 351, 280 1994.

Szur, L., The treatment of mycosis fungoides and related conditions with particular emphasis on electron therapy, *Br. J. Cancer.*, 2(Suppl. 2), 368–378, 1975.

Szymanowski, H., Mazal, A., Nauraye, C., Biensan, S., Ferrand, R. et al., Experimental determination and verification of the parameters used in a proton pencil beam algorithm, *Med. Phys.*, 28, 975–987, 2001.

Takahashi, S. Conformation Radiotherapy: Rotation techniques as applied to radiography and radiotherapy of cancer, *Acta Radiologica* (Suppl. 242), 1965.

Ten Haken, R. K. and Lawrence, T. S., The Clinical Application of Intensity-Modulated Radiation Therapy, *Semin. Radiat. Oncol*, 16, 224–231, 2006.

Tepper, J. E., Ed., Stereotactic radiosurgery, *Semin. Radiat. Oncol.*, (Special Issue) 5 (3), 1995.

Tesch, K., A simple estimation of the lateral shielding for proton accelerators in the energy range 50 to 1000 MeV, *Radiat. Prot. Dosim.*, 11, 165–172, 1985.

Thomas, A., Isherwood, I., and Wells, P. N. T., *The Invisible Light—100 Years of Medical Radiology*, Blackwell Science, Oxford, 1995.

Thomas, E. D., Total body irradiation regimens for marrow grafting, *Int. J. Radiat. Oncol. Biol. Phys.*, 19, 1285–1288, 1990.

Thomas, E. D., Buckner, C. D., Banaji, M., Clift, R. A., Fefer, A. et al., One hundred patients with acute leukaemia treated by chemotherapy, total body irradiation and allogenic marrow transplantation, *Blood*, 49, 511–533, 1977.

Thomson, E. S., Gill, S. S., and Doughty, D., Stereotactic multiple arc radiotherapy, *Br. J. Radiol.*, 63, 745–751, 1990.

Thwaites, D. I., Total skin irradiation at short fsd, *Proceedings of the British Institute of Radiology—Radiology, 88, 46th Annual Congress, Br. J. Radiol.*, 61, 743, 1988.

Thwaites, D. I., Ritchie, G. L., and Parker, A. C., Total body irradiation for bone marrow transplantation in Edinburgh: Techniques, dosimetry and results, *Radiother. Oncol.*, 18(Suppl. 1), 143–145, 1990.

Tilly, N. *Radiobiological investigations of proton and light ion therapy. PhD Thesis*, Institute for Medical Radiation Physics, Stockholm University, 2002.

Tilly, N., Johansson, J., Isacsson, U., Media, J., Blomquist, E., Grusell, E., and Glimelius, B., The influence of RBE variations in a clinical proton treatment plan for a hypopharynx cancer, *Phys. Med. Biol.*, 50, 983–990, 2005.

Trichter, F. and Ennis, R. D., Prostate localization using transabdominal ultrasound imaging, *Int. J. Radiat. Oncol. Biol. Phys.*, 56, 1225–1233, 2003.

Trofimov, A. and Bortfeld, T., Beam delivery sequencing for intensity modulated proton therapy, *Phys. Med. Biol.*, 48, 1321–1331, 2003.

Troja, S. O., Egger, E., Francescon, P., Gueli, A. M., Kacperek, A. et al., 2D and 3D dose distribution determination in proton beam radiotherapy with GafChromic film detectors, *Technol. Health Care.*, 8, 155–164, 2000.

Tsai, J. S., Buck, B. A., Svensson, G. K., Alexander, E., Cheng, C. W. et al., Quality assurance in stereotactic radiosurgery using a standard linear accelerator, *Int. J. Radiat. Oncol. Biol. Phys.*, 21, 737–748, 1991.

Tsujii, H., Mizoe, J. E., Kamada, T., Baba, M., Kato, H. et al., Overview of clinical experiences on carbon ion radiotherapy at NIRS, *Radioth. Oncol.*, 73 (Suppl. 2), S41–S49, 2004.

Tsunashima, Y., Sakae, T., Shioyama, Y., Kagei, K., Terunuma, T. et al., Correlation between the respiratory waveform measured using a respiratory sensor and 3D tumor motion in gated radiotherapy, *Int. J. Radiat. Oncol. Biol. Phys.*, 60, 951–958, 2004.

Umek, B., Zwitter, M., and Habic, H., Total body irradiation with translation method, *Radiother. Oncol.*, 38, 253–255, 1996.

Urie, M., Goitein, M., and Wagner, M., Compensating for heterogeneities in proton radiation therapy, *Phys. Med. Biol.*, 29, 553–566, 1984.

Urie, M., Goitein, M., Holley, W. R., and Chen, G. T. Y., Degradation of the Bragg peak due to inhomogeneities, *Phys. Med. Biol.*, 31, 1–15, 1986.

Urie, M., Goitein, M., Doppke, K., Kutcher, J. G., LoSasso, T. et al., The role of uncertaintiy analysis in treatment planning, *Int. J. Radiat. Oncol. Biol. Phys.*, 21, 91–107, 1991.

van Dam, J., Marinello G., *Methods for In Vivo Dosimetry in External Radiotherapy*, ESTRO Booklet Number 1, Garant Editions, Apeldoorn, Leuven, 1994.

van Dam, J., Rijnders, A., Vanuytsel, L., and Zhang, H.-Z., Practical implications of back-scatter from outside the patient on the dose distribution during total body irradiation, *Radiother. Oncol.*, 13, 193–201, 1988.

van Dyk, J., Magna-field irradiation: Physical considerations, *Int. J. Radiat. Oncol. Biol. Phys.*, 11, 1285–1291, 1985.

van Dyk, J., Dosimetry for total body irradiation, *Radiother. Oncol.*, 9, 107–118, 1987.

van Dyk, J., *The modern technology of radiation oncology*, Medical Physics Publishing, Madison WI, 1999.

van Dyk, J., Leung, P. M., and Cunningham, J., Dosimetric considerations of very large Cobalt-60 fields, *Int. J. Radiat. Oncol. Biol. Phys.*, 6, 753–759, 1980.

van Dyk, J., Galvin, J. M., Glasgow, G. P., and Podgorsak E. B., *The physical aspects of total and half body photon irradiation*, Report of Task Group 29 of the Radiation Therapy Committee of the American Association of Physicists in Medicine, AAPM Report No 17, American Institute of Physics, New York, 1986.

van Esch, A., Depuydt, T., and Huyskens, D. P., The use of an aSi-based EPID for routine absolute dosimetric pre-treatment verification of dynamic IMRT fields, *Radiother. Oncol.*, 71, 223–234, 2004.

van Luijk, P., van t' Veld, A. A., Zelle, H. D., and Schippers, J. M., Collimator scatter and 2D dosimetry in small proton beams, *Phys. Med. Biol.*, 46, 653–670, 2001.

van Santvoort, J. P. C. and Heijmen, B. J. M., Dynamic multileaf collimation without tongue-and-groove underdosage effects, *Phys. Med. Biol.*, 41, 2091–2105, 1996.

Vatnitsky, S. M., Radiochromic film dosimetry for clinical proton beams, *Appl. Radiat. Isot.*, 48, 643–651, 1997.

Vatnitsky, S., Siebers, J., Miller, D., Moyers, M., Schaefer, M. et al., Proton dosimetry intercomparison, *Radiother. Oncol.*, 41, 169–177, 1996.

Vatnitsky, S., Schulte, R., Galindo, R., Meinass, H., and Miller, D., Radiochromic film dosimetry for verification of dose distributions delivered with proton-beam radiosurgery, *Phys. Med. Biol.*, 42, 1887–1898, 1997.

Vatnitsky, S. M., Miller, D. W., Moyers, M. F., Levy, R. P., Schulte, R. W. et al., Dosimetry techniques for narrow proton beam radiosurgery, *Phys. Med. Biol.*, 44, 2789–2801, 1999a.

Vatnitsky, S., Moyers, M., Miller, D., Abell, G., Slater, J. M. et al., Proton dosimetry intercomparison based on the ICRU report 59 protocol, *Radiother. Oncol.*, 51,273–279, (Erratum in: Radiother. Oncol. 52,281 [1999].), 1999b.

Verhey, L. J., Goitein, M., McNulty, P., Munzenrider, J., and Suit, H., Precise positioning of patients for radiation therapy, *Int. J. Radiat. Oncol. Biol. Phys.*, 8, 289–294, 1982.

Verhey, L. J., Smith, V., and Serago, C. F., Comparison of radiosurgery treatment modalities based on physical dose distributions, *Int. J. Radiat. Oncol. Biol. Phys.*, 40, 497–505, 1998.

Vriesendorp, H. M., Prediction of effects of therapeutic total body irradiation in man, *Radiother. Oncol.*, 18(Suppl. 1), 37–50, 1990.

Vriesendorp, H. M., Aims of conditioning, *Exp. Hematol.*, 31, 844–854, 2003.

Vynckier, S., Sabattier, R., Kunz, A., Menzel, H. G., and Wambersie, A., Determination of the dose equivalent and the quality factor in the environment of clinical neutron beams, *Radiat. Prot. Dosim.*, 23, 269–272, 1988.

Vynckier, S., Bonnett, D. E., and Jones, D., Code of practice for clinical proton dosimetry, *Radiother. Oncol.*, 20, 53–63, 1991.

Vynckier, S., Bonnett, D. E., and Jones, D., Supplement to the code of practice for clinical proton dosimetry, *Radiother. Oncol.*, 32, 174–179, 1993.

Wagner, M. S., Automated range compensation for proton therapy, *Med. Phys.*, 9, 749–752, 1982.

Walton, L., Bomford, C. K., and Ramsden, D., The Sheffield stereotactic radiosurgery unit: Physical characteristics and principles of operation, *Br. J. Radiol.*, 60, 897–906, 1987.

Wambersie, A. and Menzel, H. G., Present status, trends and needs in fast neutron therapy, *Bull. Cancer Radiother.*, 83(Suppl. 1), 68s–77s, 1996.

Wambersie, A. and Menzel, H. G., Specification of absorbed dose and radiation quality in heavy particle therapy (a review), *Radiat. Prot. Dosim.*, 70, 517–527, 1997.

Wambersie, A., Richard, R., and Breteau, N., Development of fast neutron therapy worldwide, *Acta Oncol.*, 33, 261–274, 1994.

Wang, L., Jacob, R., Chen, L., Feigenberg, S., Ma, C. -M. et al., Stereotactic IMRT for prostate cancer: Setup accuracy of a new stereotactic body localization system, *J. Appl. Clin. Med. Phys.*, 5 (2), 18–28, 2004.

Wang, X., Spirou, S., LoSasso, T., Stein, J., Chui, C. S., and Mohan, B., Dosimetric verification of intensity-modulated fields, *Med. Phys.*, 23, 317–327, 1996.

Warrington, A. P., Laing, R. W., and Brada, M., Quality assurance in fractionated stereotactic radiotherapy, *Radiother. Oncol.*, 30, 239–246, 1994.

Webb, S., *The Physics of Medical Imaging*, Institute of Physics Publishing, Bristol, 1988.

Webb, S., Optimisation of conformal radiotherapy dose distributions by simulated annealing, *Phys. Med. Biol.*, 34, 1349–1369, 1989.

Webb, S., *From the Watching of Shadows: The Origins of Radiological Tomography*, Institute of Physics Publishing, Bristol, 1990.

Webb, S., Optimization of conformal radiotherapy dose distributions by simulated annealing: 2 Inclusion of scatter in the 2D technique, *Phys. Med. Biol.*, 36, 1227–1237, 1991.

Webb, S., *The Physics of Three Dimensional Radiation Therapy: Conformal Radiotherapy, Radiosurgery and Treatment Planning*, Institute of Physics Publishing, Bristol, 1993.

Webb, S., Optimizing the planning of intensity-modulated radiotherapy, *Phys. Med. Biol.*, 39, 2229–2246, 1994.

Webb, S., Optimising radiation therapy inverse treatment planning using the simulated annealing technique, *Int. J. Imaging Syst. Technol.*, 6, 71–79, 1995.

Webb, S., *The Physics of Conformal Radiotherapy*, Institute of Physics Publishing, Bristol, 1997a.

Webb S., Inverse planning for IMRT: The role of simulated annealing, in *The Theory and Practice of Intensity Modulated Radiation Therapy*, Sternick, S. Ed., Advanced Medical Publishing, Madison, 1997b.

Webb, S., The physics of radiation treatment, *Physics World,* November, 39–43, 1998a.

Webb, S., Configuration options for intensity-modulated radiation therapy using multiple static fields shaped by a multileaf collimator, *Phys. Med. Biol.*, 43, 241–260, 1998b.

Webb, S., Configuration options for intensity-modulated radiation therapy using multiple static fields shaped by a multileaf collimator II: Constraints and limitations on 2D modulation, *Phys. Med. Biol.*, 43, 1481–1495, 1998c.

Webb, S., Conformal intensity-modulated radiotherapy (IMRT) delivered by robotic linac–testing IMRT to the limit? *Phys. Med. Biol.*, 44, 1639–1654, 1999.

Webb, S., *Intensity Modulated Radiation Therapy*, Institute of Physics Publishing, Bristol, 2000a.

Webb, S., Advances in three-dimensional conformal radiation therapy physics with intensity modulation, *Lancet Oncol.*, 1, 30–36, 2000b.

Webb, S., Conformal intensity-modulated radiotherapy (IMRT) delivered by robotic linac–conformality versus efficiency of dose delivery, *Phys. Med. Biol.*, 45, 1715–1730, 2000c.

Webb, S., A new concept of multileaf collimator (the shuttling MLC)—an interpreter for high-efficiency IMRT, *Phys. Med. Biol.*, 45, 3343–3358, 2000d.

Webb, S., Concepts for shuttling multileaf collimators for intensity-modulated radiation therapy, *Phys. Med. Biol.*, 46, 637–651, 2001.

Webb, S., Intensity-modulated radiation therapy using only jaws and a mask, *Phys. Med. Biol.*, 47, 257–275, 2002a.

Webb, S., Intensity-modulated radiation therapy using only jaws and a mask: II A simplified concept of relocatable single-bixel attenuators, *Phys. Med. Biol.*, 47, 1869–1879, 2002b.

Webb, S., The physical basis of IMRT and inverse planning, *Br. J. Radiol.*, 76, 678–689, 2003a.

Webb, S., Intensity modulated beam delivery and image guidance, in *Atti del Corso di Aggiornamento—Metodi per la Conformazione della dose in radioterapia*, Cirio, R., Ed., Istituto Scientifico Europeo, Arona, 2003b.

Webb, S., Historical perspective on IMRT, in *IMRT- The State of the Art*, Palta, J. R. and Mackie, T. R., Eds., Medical Physics Publishing, Madison WI, pp. 1–23, 2003c.

Webb, S., Direct aperture optimization for a variable aperture collimator for intensity modulated radiation therapy, *Phys. Med. Biol.*, 49, N47–N55, 2004a.

Webb, S., *Contemporary IMRT*, Institute of Physics Publishing, Bristol, 2004b.

Webb, S., Radiotherapy physics: The next 10 years of technical development, *Imaging Oncol.*, 1, 43–50, 2004c.

Webb, S. and Evans, P., Eds., Innovative Technologies in Radiation Therapy, *Semin. Radiat. Oncol.*, 16, 2006.

Webb, S., Bortfeld, T., Stein, J., and Convery, D., The effect of stair-step leaf transmission on the tongue-and-groove problem in dynamic radiotherapy with a multileaf collimator, *Phys. Med. Biol.*, 42, 595–602, 1997.

Webb, S., Hartmann, G., Eschner, G., and Schlegel, W., Intensity-modulated radiation therapy using a variable-aperture collimator, *Phys. Med. Biol.*, 48, 1223–1248, 2003.

Well, J. C., Christ, G., Nusslin, F., and Schmidt, F., Computer-assisted irradiation planning in whole-body irradiation, *Strahlenther. Onkol.*, 167, 295–300, 1991.

Williams, P. C., IMRT: Delivery technique and quality assurance, *Br. J. Radiol.*, 76, 766–776, 2003.

Williams, P. C., Hunter, R. D., and Jackson, M., Whole body electron therapy in mycosis fungoides: A successful translational technique achieved by modification of an established linear accelerator, *Br. J. Radiol.*, 52, 302–307, 1979.

Wilson, R., Radiological use of fast protons, *Radiology*, 47, 487–491, 1946.

Withers, H. R., Neutron radiobiology and clinical consequences, *Strahlentherapie*, 161, 739–745, 1985.

Wong, J. W., Sharpe, M. B., Jaffray, D. A., Kini, V. R., Robertson, J. M. et al., The use of active breathing control (ABC) to reduce margin for breathing motion, *Int. J. Radiat. Oncol. Biol. Phys.*, 44, 911–919, 1999.

Wu, J. M., Leung, S. W., Wang, C. J., and Chui, S., Lying-on position of total skin electron therapy, *Int. J. Radiat. Oncol. Biol. Phys.*, 39, 521–528, 1997.

Yan, D., Wong, J., Vicini, F., Michalski, J., and Pan, C., Adaptive modification of treatment planning to minimize the deleterious effects of treatment setup errors, *Int. J. Radiat. Oncol. Biol. Phys.*, 38, 197–206, 1997.

Yan, Y., Song, Y., and Boyer, L., An investigation of a video-based patient repositioning technique, *Int. J. Radiat. Oncol. Biol. Phys.*, 54, 606–614, 2002.

Yang, T. C., Proton radiobiology and uncertainties, *Radiat. Meas.*, 30, 383–392, 1999.

Yang, Y., Xing, L., Li, J. G., Palta, J., Chen, Y., Luxton, G., and Boyer, A., Independent dosimetric calculation with inclusion of head scatter and MLC transmission for IMRT, *Med. Phys.*, 30, 2937–2947, 2003.

Yin, F. F., Zhu, J., Yan, H., Gaun, H., Hammoud, R., Ryu, S., and Kim, J. H., Dosimetric characteristics of Novalis shaped beam surgery unit, *Med. Phys.*, 29, 1729–1738, 2002.

Yu, C. X., Symons, M., Du, M. N., Martinez, A. A., and Wong, J. W., A method for implementing dynamic photon beam intensity modulation using independent jaws and a multileaf collimator, *Phys. Med. Biol.*, 40, 769–787, 1995.

Yu, C. X., Li, X. A., Ma, L., Chen, D., and Naqvi, S., Clinical implementation of intensity-modulated arc therapy, *Int. J. Radiat. Oncol. Biol. Phys.*, 53, 453–463, 2002.

Yu, C., Jozsef, G., Apuzzo, M. L., and Petrovich, Z., Measurements of the relative output factors for CyberKnife collimators, *Neurosurgery*, 54, 157–161, 2004.

Zamenhof, R. G., Murray, B. W., Brownell, G. L., Wellum, G. R., and Tolpin, I., Boron neutron capture therapy for the treatment of cerebral gliomas. I., Theoretical evaluation of the efficacy of various neutron beams, *Med. Phys.*, 2, 47–60, 1975.

Zhang, T., Keller, H., O'Brien, M. J., Mackie, T. R., and Paliwal, B., Application of the spirometer in respiratory gated radiotherapy, *Med. Phys.*, 30, 3165–3171, 2003.

Zhu, X. R., Yoo, S., Jursinic, P. A., Grimm, D. F., Lopez, F. et al., Characteristics of sensitometric curves of radiographic films, *Med. Phys.*, 30, 912–919, 2003.

Zink, S., *Evaluation of Treatment Planning for Particle Beam Radiotherapy*, National Cancer Institute, Bethesda, 1987.

Zur, C., Caujolle, J. P., Chauvel, P., Darmon, J., Iborra-Brassart, N., and Gastaud, P., Proton therapy of occult neovessels in age-related macular degeneration, *J. Fr. Ophtalmol.*, 24, 949–954, 2001.

PART J

BRACHYTHERAPY

Editors: Philip Mayles and Jean-Claude Rosenwald

INTRODUCTION

This Part deals with all forms of sealed source brachytherapy. Chapter 50 provides a clinical introduction. This leads to a discussion of the sources available for brachytherapy (Chapter 51), and methods of afterloading are then considered (Chapter 52). Brachytherapy treatment planning can be carried out both manually and with a computer, and both aspects are considered in Chapter 53 and Chapter 54, respectively. Finally, in Chapter 55 there is a section on the radiobiology of brachytherapy that has a number of considerations that are peculiar to this form of treatment. Brachytherapy can be time consuming for the Radiation Oncologist and for this reason its use had declined. However, with the availability of afterloading systems that allow optimisation of the dose distribution by the use of stepping sources (see Chapter 52) interest is growing again (Mazeron 2005, Hoskin and Bownes 2006, Guedea et al. 2007). Bentzen et al. (2005) identified the need for more work to be done on the requirements for brachytherapy and this was the motivation for the patterns of care study reported by Guedea et al. (2007), which identified the need for more brachytherapy resources. Brachytherapy is the ultimate in conformal therapy especially with the possibilities for optimisation that are increasingly becoming available. However, no optimisation technique can overcome the problem of an implant that misses the intended target geometrically and it is important that physicists involved in brachytherapy have a firm grounding in the experience of the more traditional techniques so that the hazards of stepping outside the traditional framework are fully understood.

CHAPTER 50

CLINICAL INTRODUCTION TO BRACHYTHERAPY

Peter Blake

CONTENTS

50.1 HISTORY AND RATIONALE FOR BRACHYTHERAPY

The term brachytherapy refers to treating tumours from a short distance, in contrast to teletherapy, where tumours are treated at a long distance from the radiation source. Brachytherapy is usually achieved by placing radioactive sources on or in the tissue to be irradiated. In teletherapy, tissue absorption largely governs dose distribution, whereas the inverse square law is most responsible in brachytherapy. Whereas teletherapy is designed to produce a homogenous dose distribution in most cases, brachytherapy uses the

1093

inhomogeneous dose distributions found around sources to create a high dose in tumours, whilst producing a low dose in normal tissue.

Brachytherapy can use sources that are loaded within body cavities (intracavitary), sources that are implanted into tissues (interstitial), trains of sources within the lumen of organs, such as the bronchus or oesophagus (intraluminal), or sources supported in a mould over a tumour (superficial brachytherapy). The prime example of intracavitary brachytherapy is the use of radioactive sources within applicators in the uterus and vagina, whilst the best example of interstitial brachytherapy would be the use of radioactive needles in the treatment of carcinoma of the tongue. Intraluminal brachytherapy is a relatively new technique made possible by high dose rate micro-sources, which need only remain within the lumen for a few minutes, thus avoiding the problems of obstruction. Superficial brachytherapy has largely been replaced by the use of high energy electrons for tumours of the skin, but may still have a place in the treatment of tumours in the upper airways and oropharynx. It is also occasionally used in the treatment of vaginal and vulval cancers.

Historically, brachytherapy developed because of the inadequacy of external radiotherapy in treating deep-seated tumours. The gynaecological organs were particularly suitable for treatment by brachytherapy because of their accessibility through the vagina. Similarly, the treatment of head and neck cancer relied heavily on interstitial brachytherapy because of the ease of access through the mouth and nose. As a consequence, systems were developed for both intracavitary and interstitial therapy in several major radiotherapy centres, particularly Paris*, Stockholm and Manchester. In the first half of the 20th century, all of these centres used radium, which had the advantage of a long half-life, making recalibration and replacement of sources unnecessary. Radium produces a high energy emission that is not preferentially absorbed in bone by the photoelectric effect. Therefore, it can be used to deliver large doses to tumours adjacent to bone that would be at risk of osteonecrosis if treated with low-energy orthovoltage x-rays. This latter advantage became less important as high energy external-beam therapy improved in the 1950s and 1960s and, as the daughter product of radium was radon gas, radium was phased out of use and replaced with caesium, which has solid decay products. Various applicator designs were created to allow live-source implantation techniques to be replaced by either manual or remote after-loading, whereby the radioactive source was only introduced into the applicators after the applicators were correctly positioned, be they intracavitary or interstitial applicators. Today, newly developed systems allow radioactive sources to be remotely after-loaded by either cable-driven or pneumatic techniques, enabling the activity of the radioactive sources to be increased to allow low-, medium-, and high-dose-rate brachytherapy with complete staff protection.

50.1.1 Intracavitary Brachytherapy

50.1.1.1 Cervix

Intracavitary brachytherapy has largely been developed for the treatment of gynaecological cancers. Cervical cancer was one of the first tumours to be treated by brachytherapy and, in its early stages, can be treated extremely successfully. A tube is passed through the cervix into the uterus, where radioactive sources are loaded. In addition, further sources placed at the top of the vagina can increase the dose rate delivered to the cervix in relation to the dose rate to sensitive, normal tissues, such as the rectum and bladder. In this way, a tumouricidal dose can be delivered to a small volume without risking damage to the bowel and bladder. The techniques based on specifically designed applicators (see Figure 52.1 through Figure 52.3)developed in Paris, Stockholm, Manchester and subsequently, in Houston by Gilbert

---

* In French-speaking countries, brachytherapy is named "curiethérapie" after the name of Pierre and Marie Curie who discovered radium in 1898.

Fletcher, all rely on creating a three dimensional dose distribution using multiple sources in the uterus and vagina to create a high dose volume around the cervix, whilst sparing normal tissues. However, developments in micro-processor controlled, high-dose-rate after-loading now allow complex isodose patterns to be built up with a smaller number of applicators, or indeed, a single-line applicator. While such a system cannot produce as complex a dose distribution as multiple applicators, being constrained by the requirement to limit the high dose region close to the source, there may be circumstances in which the simplicity of treatment makes it an attractive option.

50.1.1.2 Endometrium

The treatment of choice for endometrial carcinoma confined to the body of the uterus is total hysterectomy, followed by adjuvant pelvic radiotherapy and vaginal brachytherapy, depending on histological prognostic factors. Vaginal brachytherapy can be achieved easily with either a line source or two *Manchester* style vaginal applicators. However, occasionally hysterectomy is not possible and the uterus remains in situ. Heyman developed a system of filling the uterus with small applicators, each containing a radium source (Heyman's capsules). These applicators were active and were removed by pulling on an attached string. Skill was needed to position these applicators in a configuration that would produce an even and tumouricidal dose to the tumour, as well as not entangle while being withdrawn from the uterus. This is a prime example of a skill virtually lost in brachytherapy, as there are very few clinicians who now have any expertise in this technique. Where the technique is still used, both low-and high-dose-rate after-loading systems have been developed.

However, with microprocessor-controlled high-dose-rate systems and CT-compatible applicators, isodose patterns can be created to match those of the classical Heyman system, delivering a homogeneous dose to the uterine wall.

50.1.1.3 Vagina

An obturator can treat the vagina by irradiating the mucosa evenly from a central line source or, in some circumstances, by producing higher doses in specific areas of the vaginal wall, using one or more peripherally loaded sources. Shields can be incorporated into vaginal obturators, although in general, they must be used with great caution if the tumour is not to be shielded unintentionally.

50.1.1.4 Other Sites

Occasionally, there is a role for intracavitary therapy in treating the nasopharynx, and techniques have been developed particularly for the postnasal space, whereby sources can be introduced into applicators via the nostrils. Similarly, in the treatment of antral tumours, an intracavitary applicator introduced into the maxillary antrum may be useful or, indeed, a mould may be used. However, as external-beam techniques have developed and as high energy electrons have become available, this is less commonly used. Intracavitary therapy may be of use in low rectal tumours with a rectal obturator, although again, this technique has largely been rendered obsolete by conformal external-beam techniques.

50.1.2 Interstitial Therapy

50.1.2.1 Head and Neck

In head and neck cancer, interstitial therapy played a very large part in improving cure rates in early tumours, due to the problems of delivering a tumouricidal dose to a volume without causing bone necrosis. However, the placing of these interstitial applicators, which usually take the form of active or manually after-loaded needles or hairpins, requires considerable expertise. This expertise has become quite scarce in many centres, as external-beam techniques have improved, decreasing the use of brachytherapy. Nevertheless, in centres with a tradition of brachytherapy, such as those in France, techniques have developed to replace manually after-loaded sources with remote after-loading sources using flexible plastic tubes. The tubes are introduced into the tumour volume and then brought out through the skin to be connected to the after-loading machine. This allows fractionated high-dose-rate regimens to be used, simplifying patient care during treatment.

As with gynaecological cancer treatments, brachytherapy alone can be an effective treatment for early stage disease of the head and neck, but is more commonly used as a boost following external-beam therapy to the primary tumour and draining lymph nodes. As such, it can be used for more advanced cancers that have undergone regression with the first phase of external-beam treatment. Equally, brachytherapy has a post-operative role when it is thought that the tumour bed may not be free of disease. An after-loading technique using plastic tubes laid into the tissues of the neck after node dissection is one of the most frequent uses of brachytherapy in head and neck cancer. Clinical guidelines for head and neck brachytherapy have been developed by the American Brachytherapy Society (Nag et al. 2001).

50.1.2.2 Breast

Following lumpectomy for breast cancer, interstitial implantation of the tumour bed may be used after external-beam therapy as the second phase of treatment. However, the wider availability of high energy electrons to provide a boost to this area and the difficulty in accurately delineating the tumour bed several weeks postoperatively, has meant that this technique has declined in the U.K. and elsewhere. However, it is still widely used in France, the U.S. and Italy. The use of an interstitial implant as the sole treatment in conservative treatment of early breast cancer is also being evaluated. This technique uses an inflatable balloon catheter to allow treatment of the tumour bed after mastectomy (Keisch et al. 2003).

50.1.2.3 Prostate Gland

Prostate gland work is an expanding area, in that early cancer of the prostate, with no evidence of spread beyond the capsule of the gland, may now be irradiated using permanently implanted ^{125}I seeds (Hilaris 1997; Vicini 1999; Maurer 2002). The technique requires that the seeds are distributed evenly throughout the entire gland using ultrasound-guided needles. The positioning of these seeds may be *pre-planned* using CT scans or controlled with real-time computer planning on the basis of a predefined loading pattern. There is considerable skill needed in obtaining a satisfactory dose distribution within the prostate gland without causing morbidity to the urethra. Nevertheless, this is one of the few areas of brachytherapy that is expanding rapidly at the present time.

50.1.3 INTRALUMINAL THERAPY: BRONCHUS AND OESOPHAGUS

Intraluminal therapy was largely not possible in the past because of the size of applicators and the need for them to remain in place for many hours or days when using low dose rates. Obstruction to the bronchus or oesophagus could not be tolerated for long enough to deliver an effective dose. However, as after-loading techniques have improved and sources are produced of higher activity and smaller size, the intraluminal applicators now need only be in place for a matter of minutes. This allows high doses of radiation to be delivered to the superficial tissues of the bronchus and oesophagus and early results from centres using this for both radical (with external-beam radiotherapy) and palliative treatment are encouraging. The integration of brachytherapy with endoscopic laser resection is problematic and there is some evidence that if both techniques are used, the risk of subsequent catastrophic bleeding, particularly in the bronchus, is much increased. The recommendations of the American Brachytherapy Society can be found in Gaspar et al. (1997).

50.1.4 SUPERFICIAL BRACHYTHERAPY

Superficial brachytherapy is little used these days, but has the potential for development with high dose rate, after-loading systems. The advantage of superficial brachytherapy in this situation is that high energy radiation can be delivered to superficial tumours with a very rapid fall-off within the tissue beneath, due to the inverse-square law. Although electrons may be more suitable for irradiating deeply invasive tumours to a depth of a few centimetres, brachytherapy is an attractive option for treating to depths of only a few millimetres. As such, it may well have a use in the treatment of vulval intra-epithelial neoplasia and other very superficial skin tumours. Occasionally, there are sites where it may be used internally, as in the vagina or oropharynx, when a custom-made mould is used. Treatment of ocular tumours using an applicator placed in contact with the sclera, with either a β-emitter (e.g. ruthenium-106) or with the low energy x-rays from iodine-125 seeds, is another area where superficial brachytherapy is still widely applied.

50.1.5 INTEGRATION OF BRACHYTHERAPY WITH EXTERNAL-BEAM THERAPY

In the early days, brachytherapy was prescribed on an entirely empirical basis, relating to the total number of milligrams of radium and the number of hours they remained in place. Prescriptions were therefore in milligram-hours. Accurate dose calculation in small volumes was not possible until relatively recently and, largely, dose distribution has been calculated retrospectively to empirically-derived treatment regimens and tolerance doses, rather than regimens having been derived from dosimetric principles. As a result, there was no easy way to integrate the dose from external-beam therapy and that from brachytherapy. In addition, the radio-biological processes that take place in high-dose-rate external-beam therapy are very different from those in low dose rate brachytherapy. Consequently, the biological effect on malignant and normal tissue is not directly related to the numerical sum of the external-beam and brachytherapy doses. However, mathematical models have been developed (see Section 55.8), which reflect the biological effectiveness of both external-beam and brachytherapy treatment regimens and the iso-effect curves of these treatments can then be summated. As of yet, this is not a standard method of either prescribing or recording brachytherapy treatments in association with external-beam therapy, but may well be the way forward.

ICRU Report 38 (ICRU 1985) recommends a system of reporting gynaecological brachytherapy in terms of the prescription, dose rate and doses to specific points. In addition, an estimate is made of the volume irradiated to a total dose of 60 Gy. However, when dose rates other than those achieved with radium systems of brachytherapy are used, an assumed equivalent dose has to be stated. This introduces a major problem in reporting therapy using ICRU 38, as there is no agreed equivalence of doses at different dose rates.

50.1.5.1 Practical Considerations

In the early days of combined external-beam and brachytherapy regimens for the treatment of gynaecological cancer, central shielding was used in the external phase with the intention that central tissues largely treated by brachytherapy and peripheral tissues at the pelvic side wall were largely treated with external-beam radiotherapy. However, there is a degree of uncertainty as to the combined dose around the edge of the shields, particularly if the brachytherapy applicators are not symmetrically placed in relation to the shield. As a consequence, there is a swing away from this philosophy towards using external-beam therapy first, with open fields, followed by brachytherapy. *Hot* and *cold* spots can be avoided this way. Some of the potential benefits of a very high dose that can be achieved around a brachytherapy source may be lost, however. Unfortunately, differences in technique of delivering external-beam radiotherapy and brachytherapy and in the sequencing and overall time of these treatments, make it virtually impossible to tease out an ideal treatment from historical data, in terms of tumour control and the avoidance of normal tissue damage. However, it is now clear that the overall treatment time of both the external-beam and brachytherapy contributions should not be excessively prolonged, due to the danger of accelerated tumour-cell re-population during treatment. Therefore, the advice for cancer of the cervix is that treatment is completed within a total of eight weeks.

50.1.6   Dose-Rate Effects in Brachytherapy

50.1.6.1   Medium- and High-Dose-Rate Brachytherapy

During low-dose-rate brachytherapy, there is repair of sub-lethal damage and, in general, this is more effective in normal tissues than in malignant tissues. There is less time for this repair to take place as brachytherapy dose-rates increase and, as a consequence, the treatment has to be fractionated to allow repair between fractions. In addition, it has been found that the total dose delivered to tissue also needs to be reduced. There are complex mathematical models dictating the size of these reductions. However, most centres use empirically derived regimens whereby, in changing from low-dose-rate therapy to higher dose rates, there is an overall reduction in total dose and a maximum tolerated fraction size.

Increasing the dose rate in gynaecological brachytherapy from $0.5\,\mathrm{cGy\,h^{-1}}$ to $1.75\,\mathrm{Gy\,h^{-1}}$, it has been found that a total dose reduction of 10% to 15% is needed in order to achieve an effect on tumour and normal tissues that is similar to classical radium systems. With the use of high-dose-rate brachytherapy with iridium or cobalt sources, where dose rates can exceed $1\,\mathrm{Gy\,min^{-1}}$ to point A, the total dose reduction appears to be of the order of 35% to 45% and the maximum tolerated fraction size at point A is 8 Gy (see Section 54.2.2 for the definition of point A). There have been few randomised trials in this area but those that have taken place, and those historical series that have been analysed, would seem to indicate the accuracy of this approach.

50.1.6.2   Pulsed Dose-Rate Brachytherapy

Because of the concerns relating to radio-biology of high-dose-rate brachytherapy, systems using pulsed brachytherapy have been developed. A high dose rate is delivered for a very short time in many fractions over the same total length of time that a low-dose-rate insertion would have taken. Typically, a treatment fraction may last a few minutes and be delivered hourly for between 5 and 6 days. Theoretically, this allows the normal tissues to undergo repair, yet has the advantage of segregating the patient for irradiation for only a small proportion of the total treatment time. However, the disadvantages are that the potential for high throughput of patients on treatment machines is lost with pulsed-dose-rate therapy as each patient requires a machine of their own for several days, and that the numerous source positionings in a course of pulsed-dose-rate therapy increase the risk of a malfunction. Nevertheless, pulsed-dose-rate brachytherapy is particularly attractive in interstitial treatment. In gynaecological brachytherapy, the geometrical stability of high-dose-rate applicators outweighs the radio-biological disadvantages, whereby one would expect to see a higher incidence of late effects for a lower incidence of tumour cure. However, in interstitial therapy the positioning of the sources in relation to sensitive normal tissues is such that distances are much shorter and this geometrical advantage is lost. Therefore, pulsed- dose- rate therapy may have a place in mimicking low-dose-rate brachytherapy in sites suitable for interstitial therapy, but is unlikely to be of benefit in gynaecological intracavitary treatment*.

50.1.7 CURRENT WEAKNESSES AND POTENTIAL DEVELOPMENTS

At present the main weakness in brachytherapy is the lack of agreement on the optimum technique, dose and dose-rate for most brachytherapy treatments at most sites. Differences have arisen because of the wide variety of radioactive isotopes, applicators, and after-loading systems used, and these differences are, if anything, growing wider. In addition, there is still little consistency in dose prescription or dose reporting, or in the integrated dosimetry with external-beam radiotherapy. These areas are being addressed, but documents such as ICRU 38 (ICRU 1985) became outmoded almost concurrently with their production, because of the development of high-dose-rate brachytherapy. Also, the technology is, in many cases, moving ahead more quickly than clinical practice can accommodate.

The use of *iso-effect dosimetry* is very attractive and should move forward over the next few years, but requires a consensus on the validity of mathematical modelling and the wider availability of sophisticated computing systems. A National Cancer Institute review of brachytherapy research (Cumberlin and Coleman 2002) identified a number of areas for development. These included improved understanding of the radiation biology, improved imaging including the use of MR compatible gynaecological applicators and standardization of reporting. The challenge of dealing with the shape and relative position changes that occur both during brachytherapy and between brachytherapy and external-beam treatment was also identified.

For expertise to be maintained in the multidisciplinary team involved in brachytherapy, both for the delivery of a sound service and for teaching and research, it is likely that brachytherapy services will have to become concentrated in a small number of cancer centres. These centres could then have access to a wide variety of brachytherapy techniques to allow treatment to be tailor-made for the patients' needs. A smaller number of centres are also more likely to have a consensus view on conjectural areas.

---

* Editors note: This is controversial. PDR brachytherapy for gynaecological treatments is more widely used in Continental Europe, especially in France, than in the U.K. Those radiation oncologists who prefer it, take the view that the radiobiological disadvantages of high-dose-rate brachytherapy are the determining factor.

CHAPTER 51

CALIBRATION AND QUALITY ASSURANCE OF BRACHYTHERAPY SOURCES

Colin H. Jones

CONTENTS

51.1 INTRODUCTION

Brachytherapy developed largely through the use of sealed radium and radon sources. In the 1950s, alternative artificially produced nuclides became available, and gradually radium and radon were replaced with ^{137}Cs, ^{192}Ir, ^{60}Co, ^{198}Au, and ^{125}I sources (Trott 1987; Godden 1988). Although radium and radon are no longer used, many of the techniques that are used currently are based on the clinical experience gained with those sources over more than sixty years. Table M.4 in Part M gives data for a number of brachytherapy sources in current use.

51.2 PARTICULAR SOURCES

51.2.1 RADIUM-226

Latterly, therapeutic sources of ^{226}Ra consisted of dried radium sulphate powder doubly encapsulated in hermetically sealed containers made of platinum–iridium alloy. An inert filler was used to facilitate the uniform distribution of the radioactive nuclide in the source container. The 1620-year half-life of radium corresponds to a decay of approximately 1% in 17 years, so the useful life of a radium source was, in practice, determined by its mechanical properties, rather than by its rate of radioactive decay. The spectrum of emission is complex, with the emission of α, β, and γ radiation. Many of the β-particles emitted by radium and its decay products present in a sealed source are of high energy. Heavy filtration (typically 0.5 mm or 1 mm of Pt) is therefore necessary for sources intended for brachytherapy, so that the practical minimum diameter of such a source is about 2 mm. A filtration thickness of 0.5 mm Pt reduces the dose contributed by β-particles to about 1% of that contributed by the γ-radiation. ^{226}Ra itself and several of its decay products, including ^{222}Ra, are α-emitters. Although α-radiation is absorbed by a small amount of filtration, rupture of a radium source implanted within a patient's body could result in radium being deposited in bone, and radon gas, which is soluble in tissue and of high toxicity, irradiating local cells to a high radiation dose. Another disadvantage of radium is that the radiation from a sealed source from ^{214}Pb (RaB) and ^{214}Bi (RaC) is of high energy, so that a relatively large thickness of lead, or other protective material, would be needed for radiation protection purposes. For these reasons, and also because the maximum concentration of radium is relatively low, high activity sources are not available for afterloading techniques, and radium has been replaced by less hazardous nuclides.

51.2.2 RADON-222

Radon shares many of the properties of radium, the emission in each case being from the same decay products in radioactive equilibrium. The practical maximum activity per unit volume is about 70 times greater than that for radium, and geometrically small radon sources of high activity are therefore possible. The half-life of radon is 3.825 days. Traditionally, radon sources were fabricated as seeds or as linear sources, usually in gold capillary tubing of 0.5 mm wall thickness with crimped ends.

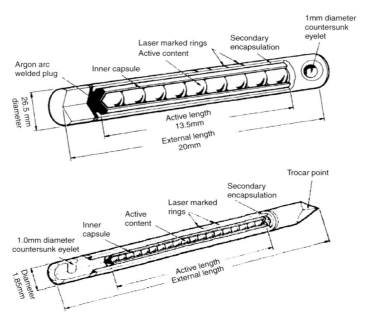

FIGURE 51.1
Typical construction of ^{137}Cs needle and tube.

51.2.3   CAESIUM-137

Up until the commercial availability and development of automatic afterloading equipment, ^{137}Cs sources of similar shape and equivalent activity were often used as a direct replacement for radium in conventional intracavitary and interstitial therapy. Figure 51.1 illustrates the construction of a typical ^{137}Cs tube and needle. ^{137}Cs is a product of nuclear fission and is produced as a by-product of nuclear fuel reprocessing. It emits β-rays and 0.662 MeV γ-rays and has a half-life of 30.17 years, long enough to allow clinical sources to be used for 10 years or so; a 2% annual reduction in source activity occurs. Caesium is a toxic alkali metal of low melting point; earlier sources were made from caesium chloride or caesium sulphate, which are soluble in water. Sources were originally doubly encapsulated in platinum-iridium and, latterly, in stainless steel. In the case of point sources or small capsule sources, the caesium is incorporated as a constituent in a glass or ceramic; in this form the caesium is relatively insoluble. ^{137}Cs glass has a maximum specific activity of about 450 GBq g^{-1} and an activity concentration of 1.2 GBq mm^{-3}, so that this form can be used for high-activity brachytherapy sources. ^{137}Cs incorporated into zirconium phosphate has a lower maximum concentration of about 180 MBq mm^{-3}; this is the form sometimes used in needles and tubes for interstitial therapy. Trace quantities of ^{134}Cs (usually less than 1%) emit β-particles, and sources require a filter equivalent to 0.1 mm of platinum.

Caesium seeds have also been used. Typical seeds are 5 mm long and have an external diameter of 0.5 mm. The active element is an extruded gold wire 3 mm long and 0.3 mm in diameter, within which the ^{137}Cs is uniformly distributed, and which is surrounded by a 0.076 mm Pt filter and a 0.1 mm stainless steel wall.

51.2.4   IRIDIUM-192

^{192}Ir has a half-life of 73.8 days and emits γ photons with energies ranging from 9 keV to 884.5 keV, but the weighted average energy of a ^{192}Ir brachytherapy source is 397 keV (Goetsch et al. 1991). With a high maximum-activity concentration of 330 GBq mm^{-3}, ^{192}Ir is suitable for high-activity afterloading sources; it is also available in the form of seeds and flexible wires. In wire form it is produced by reactor irradiation of a 75%/25% iridium/platinum alloy, which is usually provided as wire, clad with 0.1 mm of pure platinum. Pure iridium is very hard and brittle, and is difficult to fabricate. The iridium/platinum wire is available with 0.3 mm and 0.6 mm overall diameter. European manufacturers also produce wires with a 0.5 mm diameter. The wire should be cut to the required length by means of special cutters which are designed to minimise particulate contamination. For clinical use, wires are often sheathed in plastic tubing. Hairpin wires and single pins cut to length before or after activation are used for implantation directly into tissue with the aid of removable stainless steel guides.

In the U.S., ^{192}Ir seeds are used in place of wire sources. There are two types of ^{192}Ir seed sources available commercially, both having a physical length of about 3 mm—the primary difference being that of encapsulation. The first type has a 0.3 mm diameter core of 90% Pt/10% Ir with 0.1 mm Pt encapsulation giving an external diameter of 0.5 mm. The second type has a 0.1 mm diameter core of 70% Pt/30% Ir and is doubly encapsulated with two stainless steel sheaths, each 0.1 mm thick with an external diameter of 0.5 mm.

51.2.5 COBALT-60

This nuclide is used in some high-dose-rate remote-loading machines since activity concentrations of up to 130 GBq mm^{-3} are available; ^{60}Co emits 0.318 MeV β-rays, 1.17 MeV and 1.33 MeV γ-rays and has a half-life of 5.27 years. Because cobalt tends to be corrosive, it is usually nickel plated; encapsulation with 0.1 mm to 0.2 mm platinum-equivalent is necessary to filter the β-particles.

51.2.6 GOLD-198

This nuclide is produced by reactor irradiation of pure gold: activity concentrations of 7.4 GBq mm^{-3} can be readily produced. The active material is typically 0.5 mm in diameter and is sheathed in platinum of 0.1 mm to 0.2 mm wall thickness, which acts as an effective β filter. ^{198}Au emits β-rays and 412 keV γ-rays; it decays with a half-life of 2.7 days. Sources are available as grains, seeds and wires. Grains are usually 2.5 mm long and are supplied in magazines which fit directly into commercially available implantation guns. Gold wire has also been used in intracavitary treatments of the oesophagus.

51.2.7 IODINE-125

This nuclide has a photon energy of 0.03 MeV and has a 59.4-day half-life. The initial decay is by electron capture, which produces a 35.5 keV gamma, but this is followed by de-excitation from the K and L shells resulting in a principal energy of 27.4 keV (from the K shell). The spectrum is further complicated by characteristic x-rays from the encapsulation materials. The isotope is produced by the reactor irradiation of ^{124}Xe; activity concentrations as high as 3.7 GBq mm^{-3} can be obtained by encapsulating an iodide-activated ion exchange bead. Sealed sources of ^{125}I are being used as temporary or permanent implant sources in the management of prostatic, intra-ocular, and head-and-neck malignancies. Many designs of encapsulated sources are available, including the type 6711 (Amersham, now GE Healthcare) and Symmetra S06 (Bebig) seeds shown in Figure 51.2. The model 6711 seed consists of ^{125}I adsorbed onto the surface of a radio-opaque silver rod, which is encapsulated within titanium tubing. (The other widely used source from Amersham, the 6702, has now been discontinued.) The Bebig seed uses a gold marker and has a slightly higher effective energy (30 keV compared to 27.4 keV) as a result. In view of the low photon energy of ^{125}I emissions and

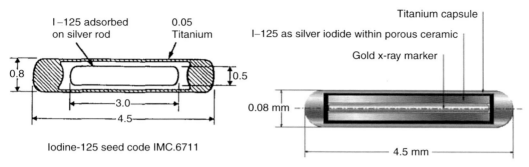

FIGURE 51.2
Construction of the 6711 (Amersham) and Symmetra S06 (Bebig) ^{125}I seeds.

the different source types that are available, the dosimetry of ^{125}I seeds has attracted much attention. Experimental studies (Burns and Raeside 1988; Nath et al. 1993) and Monte Carlo calculations (Burns and Raeside 1987) show that the transverse axial dose distributions of seed models are almost identical, but that the longitudinal axis dose distributions are significantly different. Details of all the ^{125}I seeds in common use in the U.S.A can be found at the web site http://adcl.mdanderson.org/rpc/BrachySeeds/SourceCharacteristics.htm together with references to treatment planning data for them. Further discussion of the dosimetry of ^{125}I seeds can be found in Section 53.4.2.

51.2.8 BETA SOURCES: STRONTIUM-90, YTTRIUM-90, AND RUTHENIUM-106

Although ^{90}Sr has a relatively low β energy of 546 keV maximum, it is in radioactive equilibrium with its daughter element, ^{90}Y, which has a maximum β energy of 2.27 MeV. This means that even though the half-life of ^{90}Y is only 64 h, the more energetic β-rays are present for the 29-year half-life of ^{90}Sr. It has been used in specially fabricated ophthalmic applicators designed to fit over the cornea of the eye (Jones and Dermentzoglou 1971). The applicators contain a ^{90}Sr refractory compound in rolled silver sheet embedded into an inactive silver shell. Depending on the design, the active discs allow radiation from either the front or back of the applicator, the thickness of silver used for screening the low-energy β-rays being 0.1 mm.

^{106}Ru is used in a form similar to ^{90}Sr applicators, but the 3.54 MeV β-particles are more penetrating and have a maximum range in tissue of 18 mm compared to those of ^{90}Sr which have a range of 12 mm. The applicators are designed for the treatment of intraocular tumours and treatment of ciliar body melanoma (Davelaar et al. 1992). A thin film of ^{106}Ru is contained within a sheet of pure silver 1 mm thick. The applicators are hemi-spherically shaped with a radius of 7.5 mm or 10 mm. ^{106}Ru has a half-life of 368 days.

^{90}Y can be formed by nuclear fission and has been used in the form of rods for pituitary ablation (Jones et al. 1963). The source consists of yttrium oxide, pressed and sintered into rod form 1 mm to 2 mm diameter, 4 mm to 5 mm long. Typically, two or three rods of 150 MBq activity are implanted permanently into the pituitary.

51.2.9 OTHER SOURCES

A number of other sources have been used or are being introduced which are described below. These sources are not widely used outside the U.S.

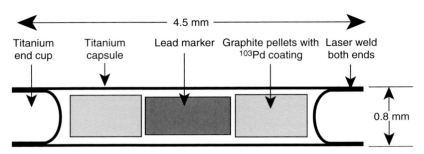

FIGURE 51.3
Theragenics model 200 ^{103}Pd seed.

51.2.9.1 Palladium-103

This nuclide has an average energy of 20.9 keV and a 17 day half-life, and is available in the form of seeds with strengths up to 110 MBq. The encapsulation consists of a titanium tube 0.8 mm in diameter and 4.5 mm long (0.06 mm wall), with laser welded end caps. The ^{103}Pd is contained in a palladium layer electroplated on the surfaces of two graphite pellets, one on either side of a central lead marker pellet (see Figure 51.3). Absorption of incident photons by the titanium wall and the lead marker, and the self-shielding by palladium, result in an anisotropic emission pattern with an axial fluence much lower than the fluence along the transverse axis (Chiu-Tsao and Anderson 1991). Nath et al. (1992) review some of the considerations in using ^{103}Pd compared to ^{125}I.

51.2.9.2 Samarium-145

^{145}Sm is produced by neutron irradiation of ^{144}Sm (96.5% enriched); decay is by electron capture with 140 keV x-rays per 100 disintegrations in the energy region 38 keV to 45 keV plus 13 γ-rays at 61 keV. The half-life is 340 days.

Fairchild et al. (1987) report the successful production of sources encapsulated in titanium tubes, 0.8 mm×4.5 mm, with strengths up to 111 MBq. Interest in the development of ^{145}Sm sources is due in part to the possibility of providing a source of low-energy photons with energies appropriate for maximising possible radiation enhancement with the thymidine analogue, iodinated deoxyuridine. By stimulating Auger cascades with photons above the K-absorption edge of iodine (33.2 keV), it is hoped to augment the effects of sensitisation. It is thought that such a technique might be especially effective in treating brain tumours, as the supporting normal brain tissue does not synthesise DNA.

51.2.9.3 Americium-241

Sealed sources of ^{241}Am emit primarily 60 keV γ-rays. For photons with energies less than 30 keV, the attenuation in water is much greater than the build-up of dose due to scatter, and the absorbed dose decreases with depth faster than expected from the inverse-square law. However, for ^{241}Am photons in water the multiple Compton scattering events compete effectively with the photoelectric effect. The Compton interactions produce a scattered photon with essentially the same energy as the incident photon (see Section 4.3.2) so that multiple photon scattering within the medium occurs, which produces a build up of energy absorption with depth, leading to the observed similarity of dose distributions in water to those produced by ^{137}Cs sources. However, the ^{241}Am γ-rays interact with high-atomic-number materials such as lead, principally through the photoelectric effect. The half-value thickness (HVT) of ^{241}Am is only 1/8 mm of lead, so shielding of sources can be achieved more easily than in the case of ^{137}Cs or ^{60}Co sources. The interest in ^{241}Am sources is based mainly on the fact that it could lead to more effective shielding of critical normal tissues at risk in intracavitary irradiation; it is suggested that in vivo shielding of bladder and rectum may also be possible by using high-atomic-number materials such as Hypaque solution and barium sulphate. However, ^{241}Am also decays by alpha emission with a half-life of 432.2 years, its exposure rate constant for photons is low, and self-absorption of photons in the source material and encapsulation is high. Nevertheless, experimental sources of ^{241}Am in the chemical form of AmO_2 mixed with aluminium powder encapsulated in titanium have been developed and investigated for clinical use (Nath et al. 1987).

51.2.9.4 Ytterbium-169

^{169}Yb has been introduced as a brachytherapy source (Mason et al. 1992); this nuclide is a lanthanide (rare earth) with a half-life of 32 days and a photon energy spectrum ranging from 49.8 keV to 307.7 keV with an average energy of 93 keV. The nuclide is produced by neutron activation of ytterbium oxide (Yb_2O_3) and very high concentrations—comparable to those of ^{192}Ir—may be produced. The only contaminant radionuclide present in any significant concentration is ^{175}Yb, which has a relatively short half-life of 4.2 days. These properties are appropriate for both high activity sources for high-dose-rate after-loading procedures, and permanently implantable interstitial sources. Furthermore, the photon energy is such that in multiple-source implants, due to its uniform dose build-up characteristics, the expected dose homogeneity in an ytterbium implant is likely to be superior to that obtainable with nuclides of either higher or lower energy. In comparison with ^{125}I, the ^{169}Yb photons are more penetrating, and the radiation distribution around the source has a less pronounced anisotropic effect. Since the half-life of ^{169}Yb is shorter than that of ^{125}I, the initial dose rates for a ^{169}Yb implant will be relatively higher. This may be clinically advantageous, since there is evidence that in some tumours, lower dose rates may impair disease control.

51.2.9.5 Californium-252

This is a man-made isotope produced by bombarding ^{239}Pu with neutrons in a high-flux reactor. The radionuclide decays by α emission with a half-life of 2.64 years and emits neutrons by spontaneous fission with a half-life of 85 years. It is the alpha emission that dictates the practical life of the source, but it is the neutron emission which makes the nuclide of such interest. The fission spectrum of neutrons has an average energy of about 2 MeV with a modal energy of 1 MeV. The associated γ-rays are within 0.5 MeV to 1 MeV. ^{252}Cf seeds have been used in preloaded plastic tubes with a 10 mm spacing between seeds. The seeds themselves have a 0.25 mm thickness of 10% iridium platinum alloy on their walls, which produces a strong angular dependence in the absorbed dose distribution for the β- and γ-rays. Monte Carlo calculation techniques and experimental measurements have been made to establish the neutron dose delivered to tissue from point and linear sources (Krishnaswamy 1972).

51.2.9.6 Caesium-131

^{131}Cs is a low energy x-ray emitter, with the most prominent peaks in the 29 keV to 34 keV region and has a half-life of 9.7 days. Brachytherapy seed sources (typically 4.5 mm long $\times$ 0.8 mm diameter) have been developed principally for prostate implantation. The shorter half-life is thought to offer advantages in biological effectiveness over ^{125}I and ^{103}Pd seed sources (Murphy et al. 2004).

51.3 SPECIFICATION OF BRACHYTHERAPY SOURCE STRENGTH

Historically, the strength of sealed radioactive sources was specified in terms of activity (in mCi or MBq) or equivalent mass of radium (mg RaEq). Radium tubes and needles were originally specified in terms of the actual mass of ^{226}Ra they contained, and sources utilising radium substitutes such as ^{137}Cs and ^{192}Ir have been specified in terms of equivalent mass of radium, to give the same dose rate at 10 mm from the source. The quantity *apparent* or *effective* activity has also been widely used. However, these methods of activity definition can be confusing when encapsulated sources are considered.

In 1973 Wambersie et al. (1973) proposed that source strength should be specified in terms of the quantity of radiation emitted. This would avoid errors due to uncertainty in the relationship between the source activity and the radiation emitted, and reduce uncertainties in corrections for filtration in sheathing material and absorption in the source material itself. To this end, various national and international bodies have recommended the use of air kerma to specify source strength (CFMRI 1983; BCRU 1984; ICRU 1985).

The recommended quantity is the *Reference Air Kerma Rate* (RAKR), defined by the ICRU (1985) as "the kerma rate to air, in air, at a reference distance of 1 m, corrected for air attenuation and scattering". For needles, tubes, wires and other rigid sources, the direction from the source centre to the reference point shall be at right angles to the long axis of the source. In the case of wire sources, it is recommended that the RAKR is specified for a 1 mm length of wire (BIR 1993). The SI unit of RAKR is Gy s^{-1} at 1 m, but for the purpose of source specification it is more convenient to use μGy h^{-1} at 1 m for LDR sources and μGy s^{-1} at 1 m and mGy h^{-1} at 1 m for HDR applications.

The American Association of Physicists in Medicine Task Group 32 (AAPM 1987) have recommended a slightly different quantity *air kerma strength*, S_k. This is defined as "the product of air kerma rate in free space and the square of the distance to the calibration point". It is specified that the calibration measurement must be performed with the distance between detector and source large enough that the source can be treated as a point source and the detector as a point detector. Generally, this condition is fulfilled if the distance from source to chamber is at least ten times the length of the source. The AAPM recommend that air kerma strength be expressed in units of μGy m^2 h^{-1} (or the numerically identical cGy cm^2 h^{-1}), which they denote by the symbol U. In practice, the air kerma strength and the RAKR are numerically identical although dimensionally different (Rivard et al. 2004)[*]. In this book, RAKR will be used.

The RAKR, $\dot{K}_R$, is related numerically to the activity, A in Bq, of the source as follows:

$$\dot{K}_R = \Gamma_\delta A / d_{\text{ref}}^2 \tag{51.1}$$

where Γ_δ is the air kerma rate constant (μGy h^{-1} Bq^{-1} m^2) for the specific radionuclide. d_{ref} (=1) in the denominator is required as described in the footnote[*]. Γ_δ and its (now obsolete) counterpart, the exposure rate constant, are difficult to measure precisely, due to the uncertainty in the determination of the activity, and their use should be avoided whenever possible. The RAKR can be related to the exposure rate $\dot{X}_R$ at a distance of 1 m:

$$\dot{K}_R = \dot{X}_R \frac{W_{\text{air}}}{e}(1-g)^{-1} \tag{51.2}$$

---

[*] The identity of RAKR and source strength depends theoretically on the inverse square law being obeyed between the measurement distance and 1 m. However, since the measurement of RAKR is made at a distance from the source large enough to consider it as a point source and includes a correction for air attenuation (IAEA 2002), this condition is fulfilled. There is then the dimensional difference. At present there is an ongoing discussion about the proper definition of reference air kerma rate (IAEA 2002). In the original CFMRI (1983) definition the units for RAKR were μGy m^2 h^{-1}, interpreting the expression "at 1 m" in the definition as meaning that the unit m^2 must be appended to the air kerma rate unit, but this was dropped in ICRU (1985). When using the RAKR to calculate the kerma rate at a distance different from 1 m, it is necessary to avoid the confusion resulting from the dimensional imbalance, which can be done either by insertion of the factor d_{ref} (=1 m) into some of the equations as required or by following the CFMRI interpretation. Although ICRU defines RAKR as being measured at 1 m, we have specifically included this as a reminder.

TABLE 51.1

Air Kerma Rate Constants

Nuclide	Air Kerma Rate Constant $(\mu Gy\ h^{-1}\ MBq^{-1}\ m^2)$
^{60}Co	0.30_6
^{137}Cs	0.077_2
^{192}Ir	0.108
^{125}I	0.033_7

Source: From ICRU (International Commission on Radiation Units and Measurements), Dose and volume specification for reporting interstitial therapy, ICRU Report 58, ICRU, Washington, DC, 1997. With permission.

where W_{air} is the average energy to create an ion pair in dry air and e is the charge of an electron (W_{air}/e has the value $33.97\ J\ C^{-1}$), and g is the fraction of the kinetic energy of the secondary charged particles which is converted to bremsstrahlung in air under calibration conditions. The term $(1-g)$ is taken as 0.997 for ^{60}Co, 0.999 for ^{137}Cs, and 1.000 for ^{192}Ir (NCS 1991).

For compliance with radiation protection requirements it is common to start from a source specification in terms of activity in MBq. For this purpose the RAKR in $\mu Gy\ h^{-1}$ should be divided by the air kerma rate constant as given in Table 51.1.

For some dose calculation methods conversion of the activity to traditional units is required. This is discussed in Section 53.6.

51.4 CALIBRATION OF BRACHYTHERAPY SOURCES

The strength of a brachytherapy source is specified in terms of the air kerma rate at a point in free space at a distance of 1 m from the source on the radial plane of symmetry (i.e. the plane which bisects the active length and the cylindrical axis of the source). In the case of a wire source, the output should be specified for a 10 mm length, but specified as the RAKR per mm.

Before being used clinically, sources should be measured by the user. In principle, there are two methods that are frequently used for calibrating sources. A calibrated well-type chamber can be used, and is usually the method of choice. An in-air air kerma measurement can also be made with an ionisation chamber at a known distance from the source. In the latter case, a correction for the chamber volume has to be made; when the source-to-chamber distance is small compared with the size of the chamber, this correction factor can be significant. The preferred measurement device for low-strength sources is a well chamber (i.e. an isotope calibrator) (IAEA 2002). The calibration of the well chamber should be traceable to a National Standards Laboratory and checked regularly with a source of known strength. For low-strength sources this is best achieved with the aid of one or more radioactive sources calibrated at a National Standards Laboratory. Air kerma rate calibration of small sources of ^{60}Co, ^{137}Cs, ^{226}Ra, and ^{192}Ir at the National Physical Laboratory, U.K., has been described by Rossiter et al. (1991). The overall estimated uncertainty (95% confidence level) was found to be $\pm 1.4\%$ for ^{60}Co, ^{137}Cs, and ^{226}Ra and $\pm 1.5\%$ for ^{192}Ir. In the U.S., the National Institute of Standards and Technology (NIST), formerly the National Bureau of Standards (NBS), has been instrumental in developing a wide range of traceable standards including those for ^{192}Ir, ^{125}I, and ^{103}Pd seeds.

The well chamber used for calibrating sources should respond linearly throughout its measuring range; its energy response must be known, and care must be taken to ensure that there is no drop in sensitivity when measuring high activities. Before use, the dependence of the chamber sensitivity on the position of the source within the chamber should be determined experimentally. Usually, the chamber sensitivity increases gradually with distance from the bottom of the well to a maximum where the response is relatively uniform, and then drops as the source approaches the chamber entrance*. A characteristic response of this form will require measurements to be made to ascertain how the chamber responds to sources of different lengths. A polymethyl methacrylate (PMMA, known as Perspex or Lucite) source holder should be used to ensure that sources are measured at the same position in the chamber; it is usual for the source to be positioned in the region of greatest sensitivity, and where change in response with source displacement is minimal. Ideally, the well chamber should be used in a location at least 300 mm from adjacent objects that might cause radiation scatter. The user should be aware that the response of the chamber is also dependent upon the source filtration and encapsulation (Williamson et al. 1982).

The calibration method will depend on the radionuclide and its activity. It might be necessary to have one or more long-half-life sources available for checking the reproducibility of the chamber. Chamber calibration checks with sources of known strength should be made every six months. If the well chamber is open to the atmosphere, temperature and pressure corrections must be made. Measurements made at the IAEA (2002) over a two-year period showed that the reproducibility of the well chamber response used with a calibrated ^{137}Cs source was generally within 0.5%.

Calibration measurements can also be made in air with an ion chamber. Although the RAKR is a quantity specified at 1 m, it is not always practical to make a measurement at this distance. The RAKR, $\dot{K}_R$, may be determined from measurements made in air using the equation:

$$\dot{K}_R = N_K \, (M_u/t) \, k_{air} \, k_{scatt} \, k_n \, (d/d_{ref})^2 \qquad (51.3)$$

where N_K is the air kerma calibration factor of the ionisation chamber at the actual photon energy; M_u is the measured charge collected during the time t and corrected for ambient temperature and pressure, recombination losses and transit effects during source transfer (in the case of remote afterloading systems); k_{air} is the correction for attenuation in air of the primary photons between the source and the chamber; k_{scatt} is the correction for the scattered radiation from the walls, floor, measurement set-up, air, etc.; k_n is the nonuniformity correction; d is the measurement distance, i.e. the distance between the centre of the source and the centre of the ionisation chamber; and d_{ref} is the reference distance of 1 m.

Methods of determining these factors and typical values are given in IAEA Technical Document 1274 (IAEA 2002). The document provides descriptions of methods for determining room scatter and data from investigations by Kondo and Randolph (1960) and Bielajew (1990) about chamber correction factors for ionisation chambers in point-source photon beams. The distance from source to chamber will depend upon source strength and the sensitivity of the ionisation chamber, but is typically between 5 cm and 20 cm for chambers up to 30 cm^3. For low dose-rate sources, large-volume ionisation chambers (30 cm^3 to 100 cm^3) are required to obtain an adequate signal. Leakage currents must be measured and allowed for if these are greater than 0.1% of the signal. The uncertainty of the non-uniformity correction factor for very large chambers is often large, and can limit their use.

In the case of high-dose-rate sources, ionisation chambers of about 1 cm^3 can be used (e.g. Farmer-type chambers of 0.6 cm^3). For ^{60}Co and ^{137}Cs calibrations, chamber factors are usually available, but this is not so for ^{192}Ir. Until such time as calibration laboratories are able to supply users with a factor for ^{192}Ir (as the UK National Physical Laboratory (NPL) is now able to do), the principle for determining the air kerma calibration for ^{192}Ir is to calibrate it at

---

* For some chambers, where the bottom part of the well is also collecting the ionisation, the maximum of response could be found there.

an appropriate x-ray quality (such as 250 kV) and at ^{137}Cs or ^{60}Co. With knowledge of the air kerma calibration factors at these energies, the air kerma calibration factor for ^{192}Ir is obtained by interpolation.

Air kerma rates (and/or source activity) should be measured and compared with the RAKR in the manufacturer's test report. Discrepancies greater than the accuracy limits specified by the manufacturer should be explored further. The most appropriate method of measuring individual sources will depend largely on the type of source. Some typical examples are described in the following subsections. For sources with shorter half-lives, such as ^{192}Ir, it is important to include a correction for source decay when calculating the required treatment time.

51.4.1   LONG HALF-LIFE SOURCES OF LOW ACTIVITY (^{226}Ra, ^{137}Cs, AND ^{60}Co)

Ideally, the user should have calibrated standard sources for each radionuclide to be measured, and for each type of source. This is rarely possible, and if the user has only one calibrated source, correction factors have to be used to allow for the energy response of the well chamber, variation in source geometry, and any differences in source encapsulation. The appropriate standard should be used to calibrate all other similar sources. This is best achieved by sequential placement of the standard source and the source to be calibrated in the same position in the well chamber, and can be achieved with the aid of a PMMA holder to locate the source centrally. It should be possible to determine the dose rate from individual sources (e.g. ^{137}Cs) within 5%. Corrections for radioactive decay should be made at suitably frequent intervals: the frequency recommended is every month for ^{60}Co sources and twice a year for ^{137}Cs sources.

51.4.2   SOURCES USED IN AUTOMATIC AFTERLOADING SYSTEMS (^{137}Cs, ^{192}Ir, AND ^{60}Co)

For automatic afterloading systems (see Section 52.3 and Section 52.4), transfer of the source into the measurement position will, during transit, irradiate the well chamber or ion chamber. If the measuring system does not employ an externally triggered electrometer to collect charge after the source has stopped moving, then a correction must be made to account for the dose delivered to the detector during the transit of the source to and from its measurement position. This is simply done by subtracting two readings taken for different intervals to eliminate the transit charge common to both.

There are typically four types of problem:

1. *Preloaded source trains (such as the ^{137}Cs Curietron type sources).* Ideally, the user should calibrate these sources before they are loaded into the flexible source holders. If this is not possible, the manufacturer should guarantee that the output of each source loaded into the source train is not different by more than 5% from that stated. When possible, the user should confirm the position of individual sources in a source train by means of autoradiography and radiography, and check the isodose pattern for each train using film, thermoluminescent dosimeters, or radiochromic dosimetric film. The location and relative activity of individual sources along a source train can also be recorded by means of a device employing a highly collimated detector.

2. *Multiple medium activity sources such as ^{137}Cs sources used in the LDR/MDR Selectron.* Upon receipt from the source supplier, the variation in activity between individual sources should be measured with a well chamber. The variation in source strength between sources in a batch should preferably be less than 3.5%. The user should

request a certificate of the homogeneity of the source activities from the manufacturer and confirm that the measured data agree with the manufacturer's data within acceptable limits. For dose distribution calculations, the weighted-mean air kerma rate should be determined.

3. *Single high-activity sources such as the ^{192}Ir source used in HDR remote afterloading equipment such as the HDR or PDR microSelectron, the GammaMed, and the Varian HDR afterloading machines.* High-activity sources can be measured in a calibrated well chamber or by measuring the source output at a known distance with a Farmer type ionisation chamber. If a calibrated well chamber designed to measure high-strength sources is available, the calibration process is straightforward. However, care should be taken to ensure that the HDR source does not cause a temperature increase inside the chamber (Goetsch et al. 1992). The response of the chamber should be checked at regular intervals using a source with a long half-life. The reproducibility of such measurements should be on the order of 0.5%.

When measurements are made with an ionisation chamber, corrections have to be made for room scatter, air attenuation, and the finite size of the chamber. It is also probable that the user will have to experimentally determine a calibration factor for the Farmer type chamber. This can be done in the following way:

Goetsch et al. (1991) showed that for ^{192}Ir, which emits γ photons with a mean weighted energy of 397 keV, a total wall thickness of 0.36 g cm^{-2} was required for build-up, and also to eliminate the effect of secondary electrons from the source and its capsule. If a total-chamber-wall thickness of 0.36 g cm^{-2} is not available, a ^{60}Co build-up cap can be used instead. Source output measurements are made with ion chamber (and build-up cap) at 250 kV. The chamber calibration factor $N_{K,Ir}$ for ^{192}Ir is then

$$N_{K,Ir} = (0.8\, A_{W,250\,kV}\, N_{K,250\,kV} + 0.2\, A_{W,Co}\, N_{K,Co})/A_{W,Ir} \qquad (51.4)$$

where A_W is the A-wall factor, which is a correction factor for all energies determined by applying the linear attenuation curves measured with different cap materials and thicknesses for each beam.

The response of the chamber alone is

$$N_{chamber} = N_K A_W \qquad (51.5)$$

The values correspond to the thickness of the chamber wall plus build-up caps of various thicknesses. If A_W is not available for the chamber in use, then A_W can be set to unity for each energy in the equation 51.4 and the calibration factor is determined from

$$N_{K,Ir} = 0.8 N_{K,250\,kV} + 0.2 N_{K,Co} \qquad (51.6)$$

With the use of this equation, the uncertainty in the air kerma factor for Ir increases by approximately 0.5%. It is now possible to obtain a direct calibration for a thimble chamber or a well chamber from the NPL.

The values for A_W for a number of different chambers are given in IAEA Technical Document 1274 (IAEA 2002). The source chamber distance is typically 10 cm to 20 cm. Table 51.2 lists typical factors for calibration in air.

Whatever methods are used for calibrating HDR sources, before being used clinically, new sources should be calibrated from two independent sets of measurements. The dose rate of an HDR source in clinical use should be checked at least every month, and preferably more often. A well chamber is a convenient method of making daily QA checks.

TABLE 51.2

Typical Factors for In-Air Calibration of ^{192}Ir HDR Sources

Typical A-Wall Factors	$A_{W,250kV}$	$A_{W,Ir}$	$A_{W,Co}$
NE 0.2 cm^3 Farmer 2515/3 Graphite/0.066 g cm^{-2} + PMMA cap 0.543 g cm^{-2}	0.994	0.982	0.986
NE 0.6 cm^3 Farmer 2571 Graphite/0.065 g cm^{-2} + PMMA cap 0.550 g cm^{-2}	0.998	0.989	0.989
NE 0.6 cm^3 Farmer 2581 A150/0.040 g cm^{-2} +PMMA cap 0.584 g cm^{-2}	0.986	0.988	0.987

Room scatter factor, $k_{scatt}=0.940$ at 1 m.

Nonuniformity correction factors for a Farmer-type ionisation chamber NE 2571, NE 2581:
At 100 mm distance	$k_n=1.009$
At 150 mm	$k_n=1.005$
At 200 mm	$k_n=1.004$
At 300 mm	$k_n=1.002$

Air attenuation factors:
At 100 mm distance	$k_{air}=1.001$
At 200 mm	$k_{air}=1.002$
At 300 mm	$k_{air}=1.004$

Source: From IAEA (International Atomic Energy Agency), Calibration of brachytherapy sources. Guidelines on standardised procedures for the calibration of brachytherapy sources at secondary standard dosimetry laboratories (SSDLs) and hospitals, Technical Document 1079, IAEA, Vienna, 1999*. With permission.

4. *Multiple high activity sources such as* ^{60}Co *sources used in the* HDR ^{60}Co *Selectron*. Techniques similar to those described in (3) can be used to determine the strength of individual sources and to determine a weighted-mean air kerma rate for dose distribution calculations.

51.4.3 SHORT HALF-LIFE SOURCES OF LOW ACTIVITY SUCH AS ^{198}Au, ^{192}Ir, AND ^{125}I SOURCES

A suitable long-half-life source should be selected as a reference source. Then a calibrated standard source of the appropriate short-half-life isotope should be obtained and compared with the reference source by sequential placement within the well chamber using the same chamber settings. In this way, it is possible to determine the relative sensitivity of the system to the two sources. The chamber can then be calibrated for the particular radionuclide and the long-lived reference source used to verify that the chamber is operating properly after the short-half-life source has decayed away.

If a short half-life nuclide is not available as a standard, it might be necessary to measure the source strength by measuring the air kerma rate of a typical source. Generally, the air kerma rate will be too low for high precision measurements and it will be necessary to measure the air kerma rate at a distance from multiple sources; the relative output from each source can be determined with the aid of the well chamber.

---

* The more recent publication (IAEA 2002) omits the values for the 0.2 cm^3 Farmer chamber, but other values are the same. Factors are also given for a number of other chambers based on the Monte Carlo calculations of Ferreira et al. (1999).

The reference source should be measured every time the chamber is used to calibrate the short-half-life sources. When appropriate, the measurement should be corrected for decay of the reference source.

When several small sources such as ^{125}I or ^{192}Ir seeds are used for an implant, the total activity of the batch should be known within 5% and the activity of individual sources should not differ by more than 15%. Well chamber checks of Rapid Strand ^{125}I seeds usually measure 5 seeds at a time (Butler et al. 1998). However the well chamber must be calibrated precisely for this geometry and for this type of source since, for the low energy radiation emitted from ^{125}I, the chamber response is very sensitive to any spectral change.

Wire sources should be checked to ensure that the linear activity is uniform with $\pm 5\%$. If the wire is bent during use it can cause small areas of low linear activity.

51.4.4 OPHTHALMIC APPLICATORS

The calibration of ophthalmic applicators is problematic because of the extremely rapid fall-off in dose rate. Soares et al. (2001) showed in an international calibration that there are differences in calibration for β-emitting plaques of the order of 14% and the uncertainty quoted by standards laboratories is higher than for other measurements. Kollaard et al. (2006) recommend that the calibration should be specified at a distance of 2 mm from the source surface. Methods available include alanine, radiochromic film, scintillation detectors and TLD. It is recommended that verification of the activity for ophthalmic applicators should be possible within 10% (Kollaard et al. 2006).

51.5 QUALITY ASSURANCE OF SEALED SOURCES

Quality assurance in brachytherapy includes procedures and tests to ensure source security (including systems to identify sources uniquely), maintain source integrity, and also provide patient dosimetry within clinically acceptable limits. The storage and use of radioactive sources must provide maximum patient and personnel safety. Whereas all of these matters are important, this section addresses only those aspects that are relevant to the calibration of individual sources. These include checks for leakage and contamination, and measurements to ascertain the distribution of radioactivity in the source. Quality assurance checks associated with the storage, manipulation, cleaning, and sterilisation of sources are not described, nor are the procedures related to the clinical dosimetry of multiple sources.

51.5.1 LEAKAGE AND CONTAMINATION CHECKS

These tests are to confirm that the source does not leak radioactivity, and that the surface of the source is not contaminated with free radioactivity. All sealed sources such as caesium tubes and needles must conform to BS 5288 (BSI 1976). Sources obtained from manufacturers are issued with leakage test certificates which describe the immersion and wipe tests that have been carried out. The upper limit of the safety level is 200 Bq. New encapsulated sources should be wiped with a swab or tissue moistened with water or ethanol, and measured with a Geiger-Müller or scintillation counter capable of detecting 200 Bq. The leakage from a sealed non-gaseous source or leak-proof container may be estimated by multiplying the total activity measured in a wipe test by a factor of ten. If the tests indicate that there is a free activity of 2 kBq, the source must be regarded as leaking, and must be sealed immediately in an airtight container and returned to the manufacturer. For encapsulated long-life sources, the maximum time between leakage tests should be 2 years, even though the sources might not be in clinical use; annual tests should be made on old sources that have been in use for several years.

Low-activity sealed sources used in automatic afterloading systems should be swab tested annually or, if this is not possible, source catheters should be checked regularly for contamination. It is difficult to swab test high-activity afterloading sources, and reliance must be placed upon catheter monitoring.

Iridium (^{192}Ir) and gold (^{198}Au) wires and seeds are not classified as sealed radiation sources because their cladding remains slightly radioactive. The principal hazards are those caused by scoring of the surface and particulate fragmentation when the wire is manipulated or cut with inappropriate equipment. Handling tools and cutting equipment should be monitored and decontaminated regularly. Only cutting equipment designed to cut cleanly should be used.

51.5.2   RADIOGRAPHY AND AUTORADIOGRAPHY

These tests provide information about the distribution of radioactive material within its container, and positional data about individual sources in radioactive source trains and applicators. Autoradiography is also useful for checking the uniformity of radioactive wires and ribbons of radioactive seeds. Although manufacturers often provide autoradiographs of sealed sources, users should carry out autoradiography on all new sources before clinical use. Autoradiographs should be evaluated to confirm that the source activity is distributed uniformly (within $\pm 5\%$) throughout the inner cell. Needle sources might have two or more cells; the source activity and distribution in each cell should be the same (within 10%) unless one or more of the cells is deliberately loaded differentially.

Radiography and autoradiography should be used to check the configuration of single and multiple sources in pre-loaded source trains. In the case of afterloading machines whose source configuration can be programmed, autoradiographic checks should be carried out at commissioning, and after machine service or catheter replacement (see Section 52.5.1 and Figure 52.8).

CHAPTER 52

AFTERLOADING EQUIPMENT FOR BRACHYTHERAPY

Margaret Bidmead and Colin H. Jones

CONTENTS

52.1 INTRODUCTION

Afterloading for brachytherapy was introduced in order to reduce the radiation dose to some radiotherapy staff groups, such as the radiotherapist performing the brachytherapy and the theatre staff attending the insertion. The first methods used manual afterloading, in which the inactive applicators or needles were positioned and checked before the active sources were introduced. With the advent of remote afterloading techniques, the radiation dose to other staff, such as nurses, source curators and technical staff preparing radioactive source trains, was reduced. The different techniques and their clinical applications will be discussed in the following sections.

52.2 MANUAL AFTERLOADING

52.2.1 INTRACAVITARY TECHNIQUES

Although manual afterloading has been used in various techniques since the very earliest days of brachytherapy, the method became popular principally after technical innovations by Henschke et al. (1963) and the introduction of miniature caesium sources (Horwitz et al. 1964) for intracavitary radiotherapy.

Afterloading can be used for all types of brachytherapy, including interstitial, intracavitary and mould-type treatments. The design constraints on interstitial afterloading devices are obviously greater than those for intracavitary techniques, in which applicators of relatively large volume can be accommodated in many of the body's cavities. Interstitial afterloading consists of inserting either needles or catheters (of small diameter) into the tumour volume; these are then loaded with radioactive wires or seeds. In the case of intracavitary therapy, especially in techniques for treating carcinoma of the cervix, applicator design has varied widely. Chassagne and Pierquin (1966), at the Institut Gustave Roussy, described a technique using a moulded plastic vaginal applicator as shown in Figure 52.1, which could then be afterloaded with ^{192}Ir wires or ^{137}Cs sources. This involves fabricating an acrylic mould for individual patients. Flexible source trains are inserted into the mould via flexible plastic tubes. The mould is such a good fit that patients may be allowed to get out of bed for exercise during treatment; local hygiene is assured by daily vaginal irrigation via a special additional tube. Figure 52.1 shows also a series of *Delouche* applicators. Unlike the moulds which are individualized, the Delouche applicators are manufactured as standard sizes (Delouche and Gest 1974). They are disposable. Both types of applicators may be afterloaded manually but they may also be connected to a remote-controlled system (see Section 52.3). In these methods, the dose fractionation is based essentially on the Paris low-dose-rate system.

Figure 52.2 illustrates another intracavitary method using a disposable manually afterloaded system together with a flexible spring container developed commercially by Amersham International (now GE Healthcare). The uterine and ovoid tubes are made of

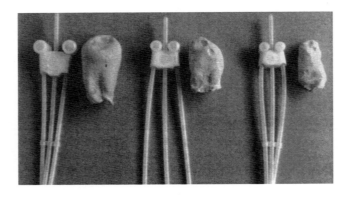

FIGURE 52.1
Different sizes of *Chassagne* vaginal moulds used at the Institut Gustave Roussy shown side by side together with the disposable *Delouche* applicators which are designed to fit different sizes of vaginal cavities using cylindrical plastic ovoids of 3 different diameters offering 6 possibilities for interspacing. The flexible plastic tubes which guide the sources up to the moulds are not shown. They are similar to the *Delouche* central tube (uterine catheter) or could be thinner if ^{192}Ir is being used.

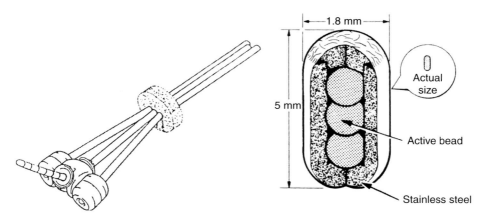

FIGURE 52.2
Amersham disposable manual afterloading system with detail of sources.

flexible polyvinyl tubing and the ovoids, spacers and washers are made of high-impact poly-styrene. The applicators, which are impregnated with barium, separate into sections that can be held together by a polystyrene disc through which the sections pass. Although the device is based on the Manchester system, in practice the principal axis of the ovoid sources lies at a different angle from that of the Manchester ovoids and allowance must be made for this difference when dose calculations are made.

52.2.2 Interstitial Techniques

Tumours of the oral cavity lend themselves to interstitial brachytherapy. If they are early stage, they can be treated with brachytherapy alone; otherwise, a combined external beam and interstitial approach can be used.

Small tongue tumours can be treated using the hairpin technique and the Paris system of dosimetry. Stainless steel slotted hairpin guides are inserted into the patient under general anaesthetic. The position of these guides is optimised whilst still inactive and x-ray films or fluoroscopic images are frequently acquired at this stage to assist in the positional optimisation. Separation between the hairpins is ideally between 12 mm and 15 mm. The active hairpins are prepared in the radioactive source preparation laboratory and transported to the theatre for insertion. The radiation oncologist inserts the sources and removes the hairpin guides at the same time. Each hairpin is sutured in place to ensure stability for the duration of the implant. Imaging in the form of orthogonal x-rays or CT scans is performed afterwards for the purposes of dose calculation.

If the target volume is too thick for a hairpin-type implant, then the *plastic loop* technique (using iridium wire) can be employed. Stainless steel needles are inserted through the skin in as parallel a fashion as possible, and run through the tumour before exiting through the skin. Nylon cord is then passed through the needle and plastic tube (of the same diameter as the needle) is pulled through into the channel made by the needle as the needle is removed. The skin surface is identified by radio-opaque washers which also act as anchorage for the plastic tubes. Dummy marker wires are then fed into the plastic tubes to act as rulers for the determination of active-wire length and position. Images are taken for dosimetric purposes and the active-wire lengths are cut and heat-sealed into small-diameter plastic tubes. The active wires are then loaded into the patient, anchored by lead washers and remain in situ for the calculated duration of the implant.

These techniques are particularly applicable to buccal mucosal tumours, lower alveoli, neck node recurrences and breast implantation (Pierquin and Marinello 1997).

52.3 REMOTE-CONTROLLED AFTERLOADING: LOW AND MEDIUM DOSE RATES

There are several commercially available remote-controlled afterloading systems. Sources can be driven by cable or by pneumatic transfer. The source can be in the form of a single capsule or a series of capsules in a flexible spring container attached to a cable or multiple spherical sources that are blown pneumatically into the patient. Table 52.1 compares the source characteristics and mechanisms for some of the equipment that is available.

There are several methods of achieving low dose-rate afterloading, each using a different applicator system and method of source transfer. Some of these methods are illustrated by reference to commercially available systems, all of which have the following features in common:

- The shielding of the machine head conforms to IEC (1989), which specifies an air kerma rate $1\ \mu Gy\ h^{-1}$ at 1 m and $10\ \mu Gy\ h^{-1}$ at 50 mm.
- The positioning of the source(s) within the applicator (or catheter) is reproducible within 1 mm.
- In the event of machine failure, source retraction takes place automatically.
- Treatment exposure is controlled and documented.

The *Curietron* system uses a number of preloaded source trains whose composition is specified at the time of purchase. They are made up of ^{137}Cs sealed sources, which are individually 1.65 mm diameter and 5.3 mm long, with an active length of 3.6 mm. The reference air kerma rate for these sources is between 30 and 100 $\mu Gy\ h^{-1}\ cm^{-1}$ at 1 m. A number of these sources, together with an appropriate number of spherical spacers, are loaded into a stainless steel spring which is 3 mm in diameter and 250 mm long. The end of this spring distal to the source train is sealed with a hook, which serves as identification of the source train and provides the means of connection to a drive cable. Up to 20 of these preloaded source trains may be stored in the source safe. For treatment, a selection of up to four source trains is transferred from the source safe (using the drive cable and a transfer sleeve) into the Curietron source projector. *Delouche* applicators or *Chassagne* moulds* are inserted into the patient and connected using connecting sleeves to the projector so that the sources can be remotely inserted into the applicators using the drive cables.

The original *Buchler* system is also driven by cable, but makes use of a single radioactive source, which is made to move linearly by means of a cam device. Variously shaped cams are available, so that the movement of the source along the applicator can be selected according to treatment requirements. During treatment, the position of the source is controlled by the cam (called a *programme disc*), which moves the source with constant frequency within the applicator over a path length of up to 200 mm. For gynaecological work, a three-channel system is available: one channel provides an oscillating source for the uterine tube and two channels with fixed source positions are used for loading of the ovoids. Machine design is such that different nuclides can be used. For example, the oscillating source could be ^{137}Cs, ^{192}Ir, or ^{60}Co, depending upon the source activity: sources used for irradiating the vagina are usually ^{137}Cs.

In the *Nucletron LDR/MDR Selectron* machine, the sources are in the form of 2.5 mm diameter spherical pellets. Source trains can be made up of a combination of 48 spacers and sources. The number of sources and their positions in each source train is independently programmable. Nucletron manufactured a three-channel and a six-channel LDR/MDR caesium-137 machine. The six-channel machine can be used to treat two patients

---

* *Fletcher-type* applicators can also be used with a slightly different type of source train and connector.

TABLE 52.1

Details of Sources Used in Remote Afterloading Equipment

Nuclide	Activity	Number of Sources	Type of Source	Source Diameter	Dose Rate	Transfer Mechanism
^{137}Cs	0.5–1.5 GBq	48	Pellet	2.5 mm	LDR/MDR	Pneumatic
^{137}Cs	0.75 GBq	variable	Capsules	1.7 mm	LDR	Cable
^{137}Cs	30–250 MBq/cm	15	Ribbon	< 2.3 mm	LDR/MDR	Pneumatic; Cable
^{192}Ir	400 GBq	1	Cylinder	1.1($\times$ 5) mm	HDR	Cable
^{192}Ir	20–40 GBq	1	Cylinder	1.1($\times$ 5) mm	PDR	Cable

simultaneously. As the radioactive sources are ^{137}Cs, they are usually replaced every 10 years. This is not due to the reduction in dose rate but because with multiple source transfers it is possible that the sources have become distorted and could be damaged or stick in the applicator tubes. The height of the source stack is measured regularly to check the distortion of the sources and pellets.

The sources are sorted magnetically and transferred pneumatically into the patient via stainless steel applicators that fit onto the end of pneumatic flexible tubing. The position of the source is continuously monitored and maintained at the programmed position by a pneumatic air check. A microprocessor controls all the hardware and processing functions of the Selectron. A self-diagnosis function gives appropriate warnings of any faults and sources are automatically withdrawn if a fault occurs.

52.3.1 APPLICATORS

The applicators for LDR Selectron intracavitary insertions for the treatment of carcinoma of the cervix were developed from the Manchester principles as shown in Figure 52.3.

Approximate source geometry was maintained, although the applicators are now made of stainless steel and can either be anchored together only at the distal end from the patient, or can be made to fix together in reproducible rigid geometry, as shown in Figure 52.4. The applicators are inserted into the patient under general anaesthetic and dummy sources are positioned inside the applicators for identification on x-ray. The dose distribution and treatment time is calculated from the applicator and source co-ordinates; it should be calculated on an individual basis for each patient, as the dose to rectal and bladder points varies from patient to patient. Once the treatment plan has been calculated, the Selectron is programmed with the active source positions and the plastic transfer tubes, into which the sources and dummy pellets are blown, are inserted into the steel applicators. Each applicator is uniquely interlocked to each channel so that it is impossible to place an incorrect catheter into the applicator. Treatment times are of the order of 10 h to 20 h and the patients are treated (often overnight) in a specially protected room on a ward. Nursing staff have a remote control with which to interrupt the treatment; the sources are automatically retracted and nursing procedures can be carried out with no radiation hazard. The machine is fail-safe, as the sources are maintained within the patient by air pressure so that if the power were to fail, pressure would drop and the sources would automatically return to the safe.

There are many other types of applicator produced for use with the LDR Selectron, but one of the more commonly used for the treatment of carcinoma of the vagina is the straight line source *Dobbie* type applicator, as shown in Figure 52.5.

Sources are loaded into the straight applicator and spacers of different diameters are used as appropriate to best fit the patient's anatomy. Dose is usually prescribed at 0.5 cm from the

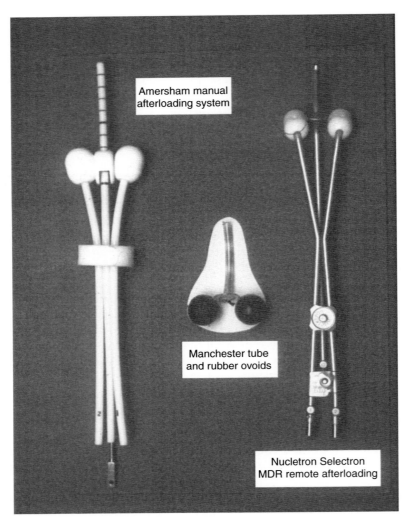

FIGURE 52.3
Amersham and Selectron applicators compared with the Manchester system.

surface of the applicator. Some centres use a uterine tube and Dobbie, differentially loaded, for treatment of carcinoma of the cervix, as illustrated in the lower part of Figure 52.5. This technique also lends itself to customisation in order to keep rectal doses within tolerance. However, care should be taken with the specification of dose calculation points (both for

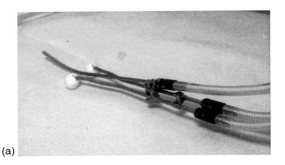

(a)

(b)

FIGURE 52.4
(a) Rigid Selectron applicators; (b) fixed-geometry applicators.

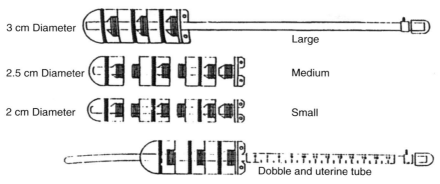

FIGURE 52.5
Dobbie type applicators.

prescription and critical organ dose) to ensure that the correct and safe loading patterns are used.

Carcinoma of the endometrium is difficult to treat, as the uterine wall is of variable thickness and is often considerably distorted. Heyman's capsules (Heyman 1929) were an attempt to produce an even dose distribution to the uterine wall by packing the uterus with

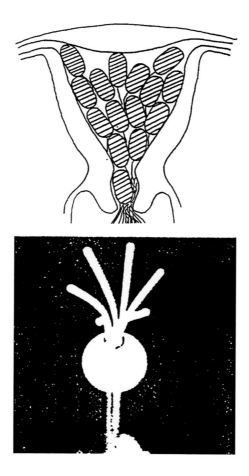

FIGURE 52.6
AP view of the applicator for clinical use with six afterloading tubes that expand due to the inherent stress of the material.

many small sources, but it is very difficult to achieve a satisfactory dose distribution. Devices developed for use with afterloading methods usually take the form of two curved tubes that curl into the uterine cavity, although multiple tubes like the ribs of an umbrella have also been used, as shown in Figure 52.6.

In most centres, the standard treatment is total hysterectomy, bilateral salpingo-oophorectomy and adjuvant radiotherapy, which is often given via an intracavitary insertion of two ovoids. The size of ovoid is selected to be the *best fit* in the patient. Doses are prescribed at 5 mm from the surface of the ovoids.

52.4 REMOTE-CONTROLLED AFTERLOADING: HIGH DOSE RATES

Historically, the TEM Cathetron was one of the first commercially available high-dose-rate remote afterloading machines. Using ^{60}Co sources sealed in tubes connected to catheters, treatment times were reduced from hours to minutes. The basic unit comprised a source safe and a set of stainless steel catheters, which could be connected independently to the source safe via flexible tubes. The safe, which was located in the treatment room, housed up to nine separate ^{60}Co sources, each in a source pencil; each pencil was capable of accommodating individual sources up to a maximum length of 100 mm. Typical source activities were 100 GBq to 200 GBq.

The catheters were connected to the flexible tubes to allow exposure of the source, which was controlled automatically from outside the treatment room by means of electro-mechanically driven cables. Much of the pioneering work in HDR brachytherapy was carried out with the Cathetron (O'Conell et al. 1967; Joslin et al. 1972). The machine was functionally reliable, but one of its limitations was that the source configurations, both in terms of source activity and source distribution, were limited to a choice of nine source pencils.

Liversage et al. (1967) devised source loadings to achieve the same relative contributions to point A (see Section 54.2.2) from the central tube and ovoids, as in the Manchester system. In producing isodose curves, allowance was made for the effects of oblique screening. Because the Cathetron source pencil comprises a number of short capsules, the γ-ray absorption through the hemispherical ends increases with obliquity rather less than would be the case if the sources were contained in a long steel cylindrical tube.

Santhamma and Das (1978) carried out experiments using TLD to determine the optimum source arrangement to achieve a *Manchester* distribution, although the vaginal colpostats are not in the true Manchester orientation with respect to the central uterine tube. The results verified theoretical calculations that took into account the filtration at the ends of the sources, the effective filtration of the ^{60}Co γ-rays in the sheathing of the source capsule, the thickness of the helical steel spring surrounding the source element and the wall of the stainless steel catheter.

In recent years, ^{60}Co has been replaced by ^{192}Ir for HDR work. The Nucletron HDR microSelectron and other machines such as the GammaMed 12i and the VariSource make use of a single 370 GBq ^{192}Ir source (1 mm×4 mm long), which is cable-driven into the patient. The source can be programmed to move in step-fashion through one or more catheters. The systems are designed to be used for intraluminal, interstitial, intracavitary and intraoperative brachytherapy. Table 52.2 shows the key features of the three systems mentioned above.

When using an HDR system for brachytherapy, the procedure is to insert the applicators or catheters and then to drive a simulator (dummy) source through them to test the overall source travel distance, and check for obstructions and incorrectly attached applicators, needles or catheters. Only after a successful test run can the ^{192}Ir source be exposed. The single, small, but highly active source can be programmed to travel consecutively through between 18 and 24 channels (depending on the make of machine) and to stop at up to 60 different dwell

TABLE 52.2

Key Features of Afterloading Systems

	GammaMed 12i	VariSource	HDR microSelectron
HDR	✓	✓	✓
PDR	✓	✗	✓
Number of channels	24	20	18
Adjustable height	✓	✓	✓
Stepwidth	1–10 mm	Variable	2.5, 5
Source diameter	0.9 mm	0.6 mm	1.1 mm
Check position	PC controlled with position-control panel	Camera and scale	Hand controllers for dummy source

positions in each channel. Dwell times at each position can be varied up to 999 s. A variety of rigid applicators, needles and flexible catheters from 1 mm outer diameter upward are available and provide means by which a large number of different techniques can be used. The position of the source can be controlled within 1 mm and can be programmed to travel in incremental steps (typically 2.5 mm, 5.0 mm, or 10.0 mm).

The relatively short half-life of ^{192}Ir leads to regular source replacement, usually every three months: the source exchange can be accomplished without any associated risk to the operator.

Treatments of gynaecological sites with HDR machines are similar to those with the LDR Selectron, but of course the treatment time is about 15 min, compared with 15 h. The dose reduction for the biological effect should be between 50% and 70% of the dose with the conventional Manchester system, and the treatment is given in several fractions. The rigid applicators are inserted in theatre and clamped in position; the clamps are attached to a long bar fixed to the operating table so that the applicators do not move during treatment. The applicators are much smaller in diameter than the LDR Selectron applicators, and can often be used without a general anaesthetic, so that the patient can have treatment as an outpatient.

Rectal retractors or packing are often used to give increased sparing to the rectum. The level of protection of the treatment room for an HDR microSelectron (which is often a protected operating theatre) will depend on the size and location of the room. Although ^{192}Ir gamma radiation is less penetrating than ^{137}Cs radiation and treatment times are much shorter, the activity of the HDR ^{192}Ir source is very much higher than those sources in the LDR Selectron machines. The patient is observed via TV monitors, whilst pulse and respiration are monitored. The source is driven out independently and automatically returns at the end of a pre-set time.

52.4.1 ENDOVASCULAR BRACHYTHERAPY

The endovascular application of brachytherapy concerns internal radiation of the wall of a blood vessel, usually an artery, with the aim of reducing a condition termed restenosis, a re-narrowing of a blood vessel (Nath et al. 1999, Pötter et al. 2001a).

There are currently two major sites of interest for endovascular therapy: the femoral and popliteal arteries are being investigated by the PARIS trial (Peripheral Artery Radiation Investigational Study) to demonstrate the safety and efficacy of delivering radiation to these sites; and the coronary arteries are being investigated in the PREVENT study (Proliferation Reduction with Vascular Energy Trial). Both the peripheral and coronary endovascular brachytherapy studies make use of a multiple balloon catheter system, which is introduced into the vessel over a guidewire and positioned, with the array of balloons spanning the target area.

The balloons are then inflated with fluid, which centres the catheter within the lumen of the artery, helping to optimise the dose distribution to the vessel wall. A closed-ended radiation sheath is inserted and a miniature radioactive source—either an ^{192}Ir gamma source (Nucletron system) or a pure beta emitter (GammaMed BetaMed)—is advanced, step-wise, through the catheter to the target site, where it remains for a pre-planned period of a few minutes before removal.

A typical system for coronary restenosis comprises three components: the afterloader, a flexible β-emitting ^{32}P source wire (diameter approximately 0.45 mm) and a centring catheter. The beta emitter and source strength are chosen to allow for treatment in a regularly-shielded cardiac catheterisation laboratory. The source wire can be used in multiple cases because it is protected from blood contact by the centring catheter, which contains a dedicated, dead-end lumen. The normal interventional procedure (angioplasty, stent placement, etc.) is performed and the guide wire is placed in position, followed by the pre-programmed radioactive source. The advantage of remote afterloading is that the procedure can be temporarily interrupted for cardiological procedures to take place; the afterloader automatically records the data relating to any interruptions and calculates the remaining treatment time needed to complete the procedure. This procedure is gradually being replaced by the use of drug-eluting stents.

52.4.2 PDR MICROSELECTRON

The pulsed-dose-rate (PDR) microSelectron is an afterloading machine based on similar mechanical principles to the HDR microSelectron. The machine offers variable, low-dose-rate pulsed brachytherapy: the source activity is in the range 20 GBq to 40 GBq. The pulse length is between 10 and 30 min per hour, according to source activity. This has advantages for the patient in terms of comfort and convenience and there may also be therapeutic advantages (Santhamma and Das 1978; Brenner and Hall 1991). There are a number of reports of the outcome of such treatments (Mangold et al. 2001; Peiffert et al. 2001; Sminia et al. 2001; Harms et al. 2002).

52.5 ASPECTS OF QUALITY ASSURANCE FOR AFTERLOADING SYSTEMS

There are four important elements that are common to all QA programs. These are:

1. A person experienced in brachytherapy applications of radiation physics should be made responsible for drawing up the quality assurance program and for ensuring compliance with it.
2. A detailed record of the program should be kept, specifying the procedures to be adhered to, the tests to be carried out and the frequency of the tests to be undertaken and recording the results of all tests in a traceable logbook.
3. Radioactive sources should only be used in compliance with local, national and/or international recommendations.
4. Incidents that have, or might have, affected the precision of treatment or the safe use of sources should be noted, so that the program can be modified in the light of experience. The overall program should be reviewed periodically.

Specific areas for quality assurance procedures are discussed in the following sections and are summarised in Table 52.3

52.5.1 SOURCE POSITIONING

When using afterloading techniques, source positioning and the reproducibility of source positions are very important. Autoradiographs of the sources positioned in the applicators should be compared with radiographs of dummy sources in the same applicators to identify

TABLE 52.3

Summary of Recommended Checks

Frequency	Test	Tolerance
Daily	Room safety, door interlocks, alarms	Functional
	Console functions, batteries, switches, printer	Functional
	Visual inspection of applicators	Undamaged
Monthly	Agreement of source and dummy positions	1 mm
	Reproducibility of source positioning	1 mm
Each source change	Calibration	3%
	Timer function	1%
	Position of source in applicators	1 mm
	Mechanical integrity of applicators	Functional
	Source transit time	
Annually	Dose calculation algorithm	3%, 1 mm
	Simulate emergency conditions	Functional
	Digitiser; verification with test phantom	<2 mm
	Review documentation, review procedures	Update when necessary
	Measurement verification vs. planning system	Within 5%

the relationship between the programmed source positions and the dummy markers that are used for treatment planning. Simple wax and Perspex phantoms can be used to check the reproducibility of applicator positioning, as shown in Figure 52.7 and Figure 52.8.

Lead-foil markers can be embedded into the surface of the phantom to provide marks and scales on the film. Once the applicators have been positioned, the film (e.g. Kodak XV2 film) is taped onto the phantom. Secondary electron emission from the lead-foil markers shows up on the autoradiographs, so that the source position reproducibility and source relationship to the applicator surface can be compared and recorded. All applicators to be used clinically should be tested in this way. Autoradiographs should be performed on various clinical source configurations. This should be done routinely after service or maintenance to check the function of the machine.

For high-dose-rate systems when dose rates are too high to allow reasonable exposure times, autoradiographs with film are not as precise, so radiation-sensitive paper such as Detex and Gafchromic material are used. The advantage of these is that they are not significantly light-sensitive, so they can be used in daylight. Exposure time needs to be on the order of 300 s, however, for a 370 GBq source.

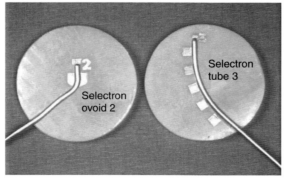

FIGURE 52.7
Wax phantoms for quality control of the source positions within the applicators.

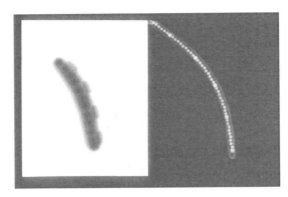

FIGURE 52.8
Autoradiograph of sources (right-hand side) and x-ray of dummy sources (left-hand side).

52.5.2 MACHINE FUNCTION

The machine function itself must be regularly checked and the checks documented. These checks will take place at different frequencies, some daily, weekly, or monthly and some three-monthly. These checks should include:

- Couplings and connectors
- Transit tubes and applicators
- Treatment timers—including a measurement of timer errors
- Fail-safe devices during power failure or air pressure failure (e.g. backup storage batteries)
- Limitations of source positions and treatment lengths
- Room and machine interlocks
- Programming and carrying out a simulated treatment
- Measurement of source transit time
- Measurement of radiation exposure levels adjacent to the intermediate and main source safes at the surface and at 1 m.

52.5.3 FACILITY CHECKS

Room protection should be sufficient to maintain acceptable dose levels in surrounding areas (see Section 61.2.1). These dose levels must be checked at commissioning and after structural alterations. Radiation warning signs should be prominently displayed. Independent radiation monitors should be available for regular use inside the treatment room. These should be activated when the source is exposed and tested each time the machine is to be used. There should be a hand-held dose monitor outside the room, which should be tested regularly for use in emergencies.

52.5.4 EMERGENCY PROCEDURES

Emergency procedures that will differ between various applications of the afterloading machine should be well documented and posted in a prominent position (see also Section 61.4.1). These should be tested at regular intervals. Equipment should be available for removal and storage of radioactive sources in the event of a source return failure. A procedure common to most situations is to maintain the integrity of the system (i.e. not to disconnect the applicators unless the exact position of the sources is known).

52.5.5 OTHER

Also essential to the safe operation and treatment of afterloading equipment are

- Training of personnel, including written instructions, demonstrations, etc.
- Patient instruction and explanation (including instructions to visitors)
- Comprehensive record keeping of checks performed.

CHAPTER 53

DOSE CALCULATION FOR BRACHYTHERAPY SOURCES

Philip Mayles

CONTENTS

53.1 INTRODUCTION

This chapter sets out the methods by which the dose delivered by brachytherapy sources can be calculated. It is important that the physicist has a clear appreciation of the theoretical basis of these calculations so that, whether applying a manual calculation system or using a computer algorithm, the correction factors that are implicit in the algorithm and those that must be applied explicitly are fully understood. The issue of dose calculation is considered first in a generic way, and then the calculation of dose for specific sources will be considered.

53.2 DOSE FROM A POINT SOURCE

The dose rate, $\dot{D}(r)$, at a distance r from a point source can be expressed in terms of the dose rate at r_{ref} from the source $\dot{D}_{ref}$ by the simple relationship:

$$\dot{D}(r) = \dot{D}_{ref} \frac{r_{ref}^2}{r^2} f(r) \tag{53.1}$$

The reference distance, r_{ref}, usually considered is 1 m.

$\dot{D}_{ref}$ is the absorbed dose rate to water* that would be achieved in vacuo, i.e. with no attenuation and scattering. As explained in Section 6.4, if $\dot{K}_R$ is the reference air kerma rate (RAKR) of the source (see Section 51.3), then

$$\dot{D}_{ref} = \dot{K}_R \left(\frac{\mu_{tr}}{\rho} \right)_{air}^{water} (1 - g) \tag{53.2}$$

where $(\mu_{tr}/\rho)_{air}^{water}$ is the ratio of the mass energy transfer coefficient for water and the mass energy transfer coefficient for air (BIR 1993) and $(1 - g)$ is the energy loss of secondary electrons to bremsstrahlung in water. The factor $(1 - g)$ may mostly be ignored (see Section 51.3) and will therefore be omitted from subsequent equations. Between the energies of 150 keV and 1.5 MeV, the ratio of the mass energy transfer coefficients lies in the range 1.107–1.112, and the BIR working party recommend the use of the value 1.11 for all common energies except ^{125}I, for which 1.02 is recommended.

The factor $f(r)$ is a correction factor for attenuation and scattering. Because of attenuation and scattering, the actual dose rate at 1 cm from the source may be either greater (for ^{192}Ir) or less (for ^{125}I) than that predicted by the straight conversion from air kerma to dose to water. At a specified distance, r, when this factor is greater than 1, it indicates that scattered radiation makes a significant contribution to the dose. The BIR working party recommended the use of the values of $f(r)$ given by Sakelliou et al. (1992) as given in Table 53.1. However, these data, like those of Dale (1982), were calculated with the incorrect decay scheme for ^{192}Ir published in the *Radiochemical Manual 1982* (Wilson 1966). Figure 53.1 shows the data for ^{192}Ir given in Table 53.1 in a graphical form. All the data have been renormalised to a value of 1.0 at 1 cm†.

Combining Equation 53.1 and Equation 53.2 and omitting the bremsstrahlung correction factor we obtain:

$$\dot{D}(r) = \dot{K}_R \left(\frac{\mu_{tr}}{\rho} \right)_{air}^{water} \left(\frac{100}{r} \right)^2 f(r) \tag{53.3}$$

where r is in centimetres and $\dot{K}_R$ and $\dot{D}(r)$ have the units $\mu Gy\ h^{-1}$. (Note that, conveniently, by replacing the term $(100/r)^2$ by $(1/r)^2$—i.e. by dividing both sides of the equation by 10 000—the units of $\dot{D}(r)$ become $cGy\ h^{-1}$). As before, $(\mu_{tr}/\rho)_{air}^{water}$ is 1.11 for ^{192}Ir and 1.02 for ^{125}I.

---

* As in external-beam therapy it is usual for reference dosimetry to be based on dose to water. If required the dose to particular tissue types can be calculated instead.
† For a point source, the normalised value of $f(r)$ is strictly equal to the radial dose function $g(r)$ used in the TG43 formalism (as in Equation 53.8).

TABLE 53.1

Polynomial Coefficients for the Variation of Dose Rate with Distance from the Source

Coefficient	a	b	c	d
Sakelliou				
Cobalt-60	1.000	−0.013350	−0.0003451	
Caesium-137	1.000	−0.005767	−0.0008628	
Gold-198	1.000	0.006678	−0.0015270	
Iridium-192	1.000	0.012500	−0.0018340	
Meisberger				
Cobalt-60	0.99423	−0.005318	−0.002610	0.0001327
Caesium-137	1.00910	−0.009015	−0.0003459	−0.00002817
Gold-198	1.03060	−0.008134	0.001111	−0.0001597
Iridium-192	1.01280	0.005019	−0.001178	−0.00002008
TG43				
Iridium-192	0.989054	0.00881319	0.00351778	−0.00146637[a]

Coefficients in the formulae for $f(r)$ in which $f(r) = a + br + cr^2 + dr^3$, as given by Sakelliou et al. (1992), compared to the long-established Meisberger et al. (1968) values of the coefficients. Also shown are the TG43 coefficients, $g(r)$, for ^{192}Ir, which differ from the others in being normalised to the value at 1 cm from the source.

[a] TG43 adds a fourth-order term with the coefficient 0.000092437.

Sources: From Sakelliou, L., Saklliariou, K., Sarigiannis, K., Angelopoulos, A., Perris, A., and Zarris, G., *Phys. Med. Biol.*, 37, 1859–1872, 1992; Meisberger, L. L., Keller, R. J., and Shalek, R. J., *Radiology*, 90, 953–957, 1968; Nath, R., Anderson, L. L., Luxton, G., Weaver, K. A., Williamson, J. F., and Meigooni, A. S., *Med. Phys.*, 22, 209–234.

53.3 EXTENSION TO A LINE SOURCE

Consider the activity spread out uniformly over a wire to form a line source. The geometry of this is represented by Figure 53.2, ignoring the attenuation of the wire for the present. The wire may be broken down into elementary lengths (dL) and the linear activity of the source in terms of the RAKR per cm must now be considered. Ignoring attenuation and scattering in water (i.e. assuming $f(r) = 1$), the elemental dose rate at P, d$\dot{D}_P$, is then given by

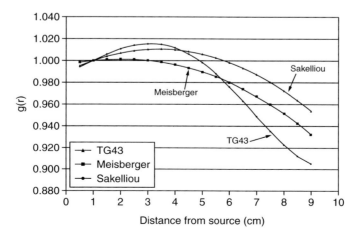

FIGURE 53.1

Graph showing the radial dose function, $g(r)$, for ^{192}Ir as calculated using the data in Table 53.1 (all data are normalised to 1.0 at 1 cm).

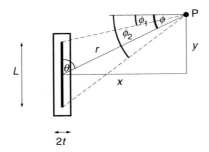

FIGURE 53.2
Calculation of dose rate due to a line source. L is the length of the active part of the source and t is the radius of the source (including its encapsulation).

$$dD_P = \left(\frac{\mu_{tr}}{\rho}\right)_{air}^{water} \dot{K}_R' \, dL \, \frac{1}{r^2} \tag{53.4}$$

where $\dot{K}_R'$ is the RAKR per unit length of the source in μGy h^{-1} cm^{-1} at 1 m, D_p is in cGy h^{-1} and r and L are in cm. The equation may be rewritten in polar coordinates using the fact that

$$r = x \sec \phi \text{ and } L = x \tan \phi, \quad dL = x \, (\sec^2 \phi) \, d\phi;$$

therefore:

$$\dot{D}_P = \left(\frac{\mu_{tr}}{\rho}\right)_{air}^{water} \dot{K}_R' \int_{-L/2}^{+L/2} \frac{1}{r^2} \, dL = \frac{\left(\frac{\mu_{tr}}{\rho}\right)_{air}^{water} \dot{K}_R'}{x} \int_{\phi_1}^{\phi_2} d\phi = \frac{\left(\frac{\mu_{tr}}{\rho}\right)_{air}^{water} \dot{K}_R'}{x} (\phi_2 - \phi_1) \tag{53.5}$$

where ϕ is in radians and x in cm. (Note that ϕ has been used here rather than the more usual θ to point out the different definitions used in this approach compared to that of TG43 described in Section 53.4.)

Now consider the effect of attenuation in the sheath, but for the time being, continue to ignore attenuation and scattering in the medium. If t is the radius of the sheath, then this attenuation is e$^{-\mu t \sec \phi}$, where μ is the linear attenuation coefficient of the material of the sheath (which is assumed to be the same as that of the solid wire). Equation 53.3 then becomes:

$$dD_P = \left(\frac{\mu_{tr}}{\rho}\right)_{air}^{water} \dot{K}_R' \, dL \, \frac{1}{r^2} \frac{e^{-\mu t \sec \phi}}{e^{-\mu t}} \tag{53.6}$$

Note that it is necessary to divide by e$^{-\mu t}$ because the definition of RAKR involves a measurement that includes the effect of attenuation in the direction perpendicular to the wire. Converting to polar coordinates gives the result:

$$\dot{D}_P = \frac{\left(\frac{\mu_{tr}}{\rho}\right)_{air}^{water} \dot{K}_R'}{xe^{-\mu t}} \int_{\phi_1}^{\phi_2} e^{-\mu t \sec \phi} d\phi \tag{53.7}$$

The integral term is known as a Sievert integral and can only be solved numerically.

Tabulations of the Sievert integral are available in a number of forms (see Section 54.1.3), such as the *Manchester system* tables (Paterson and Parker 1967), the French *escargot* curves (Schlienger et al. 1970) and the *Oxford cross-line curves* (Hall et al. 1966; Welsh et al. 1983). As is the case with Equation 53.6, the Manchester and Oxford tables ignore the effect of tissue attenuation and scattering.

With a computer, it is just as easy to divide the source into a number of small segments and treat each one as a point source (but allowing for the self-attenuation of the source) in which case Equation 53.2 can be used together with a correction for oblique filtration.

53.4 THE TG43 FORMALISM

In practice, because of the effects of scattering that are ignored, the Sievert integral calculation does not accurately represent the dose from encapsulated sources (Williamson 1996) along the axis of the source, even when $f(r)$ is included in the calculation. This is especially true for ^{125}I sources. To improve calculation accuracy, it is desirable to make use of Monte Carlo calculations using real source geometries, etc. To facilitate this, the AAPM (Nath et al. 1995) have proposed that the different components of the dose calculation be divided up into geometry, attenuation and scattering, and anisotropy. They have proposed the formula:

$$\dot{D}(r,\theta) = S_k \Lambda \frac{G(r,\theta)}{G(r_0,\theta_0)} g(r) F(r,\theta) \qquad (53.8)$$

where $\dot{D}(r,\theta)$ is the dose rate in water, expressed in cGy h^{-1} at a distance r in cm from a line source and θ is defined as in Figure 53.2; S_k is the air kerma source strength expressed in U (i.e. μGy h^{-1} m^2); Λ is the dose rate constant expressed in cGy h^{-1} U^{-1}; $G(r,\theta)/G(r_0,\theta_0)$ is the geometry factor; r_0, θ_0, the reference position, is defined by $r_0 = 1$ cm and $\theta_0 = 90°$, $g(r)$ is the radial dose function; and $F(r,\theta)$ is the anisotropy function.

The concept of air kerma strength is discussed in Section 51.3. It is numerically identical to the RAKR. The dose rate constant, Λ, is defined as the dose rate to water at the reference position from a unit air kerma strength source. It incorporates the effect of the source dimensions, of the source encapsulation and the effect of scatter. Using the formalism of Equation 53.2, it is equal to $(\mu_{tr}/\rho)_{air}^{water} \times f(1.0)$ for a point source[*]. The geometry factor $G(r,\theta)$ represents the inverse square law for point sources, whereas for a line source it is the ratio $(\phi_2 - \phi_1)/(L \times x) = (\phi_2 - \phi_1)/(L \times r \sin \theta)$, as shown in Equation 53.5. Note that $G(r,\theta)$ is calculated relative to the value at 1 cm. $g(r)$ is defined in the central axial plane (i.e. for $\theta = \theta_0 = 90°$). It includes the effect of water scatter and attenuation but excludes the geometrical factor. For more complex source-activity distributions $G(r,\theta)$ and $g(r)$ should be calculated, but the line-source approximation is usually adequate[†]. In any event, the calculation of $G(r,\theta)$ is a purely geometric one, excluding any attenuation in the source (or medium). The anisotropy factor, $F(r,\theta)$, describes the anisotropy due to attenuation within the source. Anisotropy factors must be derived for each individual type of source—factors for common sources are contained within the report. If anisotropy factors are not available, they may be calculated or measured.

53.4.1 APPLICATION OF TG43 TO ^{192}Ir Dosimetry

For ^{192}Ir, the TG43 report only provided data for a stainless-steel-encapsulated ^{192}Ir source, and no data were provided for ^{192}Ir wires or for the high dose-rate sources used in afterloading devices. However, since then a number of authors have applied the TG43 formalism to such sources (Ballester et al. 1997; Lliso et al. 2001, 2003)[‡]. Ballester et al. provide data for the application of TG43 to 0.3 mm diameter ^{192}Ir wire of lengths 10 mm and 50 mm.

---

[*] i.e. $1.11 \times f(1.0)$ for energies between 150 keV and 1.5 MeV and $1.02 \times f(1.0)$ for ^{125}I. For ^{125}I, the situation is complicated by the change in the standards laboratory calibration, as discussed in the next section.
[†] In the 2004 revision of TG43 (Rivard et al. 2004), $G(r,\theta)$ and $g(r)$ are replaced by $G_L(r,\theta)$ and $g_L(r)$ to emphasise that the line-source approximation is used for the geometry function.
[‡] A further revision of TG43 to deal with ^{192}Ir is planned.

However, because widely differing lengths of wire are used in practice, it may be best to restrict the use of TG43 use to the calculation of high dose-rate iridium sources. Details and references for a number of ^{192}Ir sources are given in the ESTRO booklet on brachytherapy quality control (Venselaar and Pérez-Calatayud 2004).

53.4.2 APPLICATION OF TG43 TO ^{125}I DOSIMETRY

The dosimetry of ^{125}I has always had a higher degree of uncertainty than that of isotopes emitting higher energy radiation. The use of TG43 data involved a decrease of ten percent in the dose calculated at a distance from a seed. This arose mostly from a change in the recommended values for the dose rate constant and the anisotropy factor (or equivalent quantities in older dosimetry protocols) for the Amersham sources which were almost exclusively used at that time (Kubo et al. 1998; Williamson et al. 1999). Bice et al. (1998) recommended that the reference dose for prostate implants be changed from 160 Gy to 144 Gy so as to maintain the same clinical outcome when using the TG53 protocol.

On January 1, 1999, a further complication arose because NIST changed its standard calibration for ^{125}I by 11.5% (Kubo et al. 1998; Williamson et al. 1999). This was because the earlier standard was affected by 4.5 keV titanium characteristic x-rays, which do not affect the dose in water beyond 1 mm from the source. The AAPM definition of source strength expressly excludes the effect of low-energy x-rays below 5 keV, depending on the application (Rivard et al. 2004). This had two repercussions. First, it was necessary to be certain which version of the standard was used by the supplier when specifying the source strength[*]. Second, if the new standard was used, it was necessary to compensate for this change by decreasing by the same amount the values of the dose rate constant to be used for dose calculation. To avoid misinterpretations, these new recommended Λ values were indicated by assigning to the Λ symbol the subscript "99std" (Williamson 1999) or "N99S" (Rivard et al. 2004). The value of Λ should be increased by 1.115 compared to values quoted based on the old standard (Williamson et al. 1999). Rivard et al. provide a set of consistent values. Further refinements of the values for dose rate constants to be used for different source designs have been achieved and the up-to-date recommended values for Λ can be obtained by following the links on the AAPM/RPC website http://adcl.mdanderson.org/rpc/BrachySeeds/SourceCharacteristies.htm or from Rivard et al. (2004).

The value of Λ depends on the design of the source, as this can affect the radiation that is actually emitted from the source capsule, resulting in differences in the dose at 10 mm, giving a range of Λ values between 0.94 cGy h^{-1} U^{-1} and 1.04 cGy h^{-1} U^{-1} for different source types (Rivard et al. 2004). For a specified source type such as the Amersham 6711 seed, the updated TG53 recommended value is 0.965 cGy h^{-1} U^{-1} which is consistent with the calculated and experimental values found more recently by Dolan et al. (2006) and well within the 1%–4% uncertainty usually quoted for Monte-Carlo calculated values or 6%–8% quoted for experimental values (Dolan et al. 2006). An analytical approach for a large range of source types and radionuclides is also possible and gives quite accurate results (Chen and Nath 2001).

53.5 DECAY CORRECTION

In the preceding calculations no mention has been made of radioactive decay (see Section 2.3.2). To correct for decay the formula:

$$\dot{K} = \dot{K}_0 \, e^{\left(\frac{-0.693t}{T}\right)} \tag{53.9}$$

---

[*] By now, all suppliers should be using the new standard.

can be used to calculate the RAKR at the time of treatment, where $\dot{K}$ is the RAKR t days after the RAKR $\dot{K}_0$ was measured (the *activity date*), and T is the half-life of the isotope (see Table M.4). For an HDR implant the activity time should be the time at which the irradiation takes place—this is normally done automatically within the software. For a manually afterloaded implant the dose rate, $\dot{D}$, at the start of the treatment can be calculated. The total dose is then given by:

$$D = \dot{D} \int_0^{t_1} e^{\frac{-0.693t}{T}} = \dot{D} \frac{T}{0.693}\left(1 - e^{\frac{-0.693\,t_1}{T}}\right)$$
(53.10)

with t_1 being the duration of the implant. For sources used in radiotherapy that have half lives around two months the correction for decay during a one week treatment will be about 4%. If T is large compared to t_1 the exponential term approximates to $1 + \frac{0.693\,t_1}{T}$ so Equation 53.10 simplifies to $D = \dot{D}t_1$ as one would expect. In practice it is convenient to make an initial calculation of the duration of the implant based on the RAKR at the time of the implant and to calculate the additional time $\varDelta t$ that will be required using the formula:

$$\varDelta t = \frac{-\ln\left(1 - 0.693\frac{t_1}{T}\right)}{0.693} T - t_1$$
(53.11)

Tables can be compiled of $\varDelta t$ as a function of t_1. The $\varDelta t$ correction is greater than 1% as soon as t_1 extends to more than about 2% of T. For a permanent implant the term in the brackets is 1.0 and the effective duration of the implant is the mean life (*1.44 T*) of the isotope, which is about 87 days for iodine-125 (see Section 2.3.2.2). However, as discussed by Wallner et al. (2002) the effects of such an extended treatment are affected by biological factors as well as the physical dose.

53.6 CONVERSION FACTORS FOR OLDER METHODS OF ACTIVITY SPECIFICATION

Brachytherapy was, for many years, exclusively delivered using ^{226}Ra sources. Because of this, data were widely available to calculate dose rates for radium sources. If source attenuation and scattering are ignored, these data can be used for other isotopes if source activity is specified in terms of milligrams radium equivalent (mg RaEq), and this unit was widely used and reasonably unambiguous. The phasing-out of radium and the need for compliance with radiation protection regulations couched in terms of units of activity led to the specification of activity in MBq or mCi. The almost-universal adoption of air kerma specification reduces the ambiguity in relation to modern dose-calculation methods, but where older computer software or data tables are used, it is necessary to convert RAKR into these older units. It may also be necessary to be familiar with these units in order to be able to understand the historical literature. For this reason, a brief discussion of these earlier units is included here. It is based on the paper by Williamson and Nath (1991), to which the reader is referred for further explanation.

53.6.1 THE ROENTGEN

Because many of the tables used for dose calculation were defined to provide the exposure rate in Roentgen, R, it is necessary to convert from values in R h^{-1} to air kerma rate and thence to absorbed dose rate. To convert from exposure rate in R h^{-1} to air kerma rate in cGy h^{-1}, one must multiply by the product of the average energy required to produce one ion pair in air (divided by the electron charge), W_{air}/e and $(1-g)$ where g is the fractional energy loss to bremsstrahlung (see Equation 6.9 in Section 6.3.2 and Equation 51.2). The former quantity has the value 33.97 J C^{-1} yielding a conversion factor of $0.876/(1-g)$ cGy R^{-1}. Subsequent conversion to dose requires the application of Equation 53.2. This leads to the value $1.11 \times (1-g) \times 0.876/(1-g) = 0.97$ cGy R^{-1} in water for ^{192}Ir.

53.6.2 RADIUM EQUIVALENCE

Because most early brachytherapy was delivered with ^{226}Ra and activity of radium was expressed in terms of the mass of radium in the source, the concept of mg RaEq has been widely used. An added complication is that radium was used in different encapsulations—from 0.5 mm Pt filtration up to 2 mm—but it is the former that was the basis for the Manchester system, and hence is what is referred to when defining equivalence. Equivalence is defined in terms of having the same RAKR. Since the exposure rate at 1 cm from 1 mg of ^{226}Ra filtered with 0.5 mm Pt is 8.25 R h^{-1}, the air kerma rate is $8.25 \times 0.876 = 7.227$ cGy h^{-1} at 1 cm which is an RAKR of 7.227 μGy h^{-1} at 1 m. So for any isotope, one may convert from mg RaEq to RAKR by multiplying by 7.227.

53.6.3 ACTIVITY UNITS

Conversion from activity quoted in MBq or mCi to RAKR is not straightforward, and the conversion factor depends on the isotope. It is, first of all, necessary to examine how the supplier defines activity. This may either be the *contained activity* or the *apparent activity*. The former is the activity that would be measured with any source cladding removed. The apparent activity is defined as the activity of an unfiltered point source of the same isotope which has the same RAKR as the filtered source. If the contained activity is specified, it is necessary to calculate the attenuation due to source filtration and then to multiply by the air kerma rate constant. If, as is more likely, the apparent activity is quoted, then the conversion factor is, in principle, simply the air kerma rate constant. However—for ^{192}Ir in particular—there is some confusion about the appropriate value of the air kerma rate constant, with values of 0.1136 μGy h^{-1} m^2 MBq^{-1} (4.80 R h^{-1} m^2 mCi^{-1}) and 0.1099 μGy h^{-1} m^2 MBq^{-1} (4.64 R h^{-1} m^2 mCi^{-1}) both appearing in the literature. The former was derived from the decay scheme published in the 1982 edition of the *Radiochemical Manual*, whereas the latter is that calculated with the current edition. Although one might suppose that the most appropriate value was the current best estimate, this is not necessarily the case. The manufacturer's calibration of ^{192}Ir was actually based on an air kerma measurement, and the activity was derived from this using the quoted air kerma rate constant which was, therefore, the factor to use to convert back from activity to RAKR. Furthermore, some commercial software asks for the activity in mCi or MBq and therefore carries out a calculation implicitly or explicitly based on the air kerma rate constant. In this case, when converting to mCi for RAKR, the factor used should be the air kerma rate constant used by the software supplier. The current best estimates of the air kerma rate constant are given in Table 51.1 but they should not be used for dose calculation without checking which coefficients are used by the source supplier or by the software designer to specify the source activity.

53.7 LIMITATIONS OF BRACHYTHERAPY DOSE CALCULATIONS

Unless a full Monte Carlo calculation is carried out, the accuracy of brachytherapy dose calculations will always involve some limitations. However, because dose values are dominated by the distance to the nearest source and dose gradients are very steep, algorithms based on the simple dose calculation methods described above are adequate for most practical purposes. However, it is important to bear in mind the factors that may limit accuracy.

53.7.1 UNKNOWN SOURCE ORIENTATION

For seed sources that are not contained in a guide tube, the sources are likely to be randomly orientated. A practical solution in this event is to treat them as point sources and to ignore the anisotropy factor. This will inevitably result in a calculated dose that is slightly different from that actually delivered. The TG43 report recommends that this problem be overcome by multiplying the dose rate in the transverse plane by the mean anisotropy factor averaged over all values of θ. This is called the *anisotropy* constant. Its value is given by TG43 as 0.98 for ^{192}Ir seeds, and 0.93 for the model 6711 ^{125}I seed (Nath et al. 1993).

53.7.2 SHIELDING BY OTHER SOURCES

None of the standard algorithms take into account shielding of sources by other sources or applicators. Markman et al. (2001) compared Monte Carlo calculations of the dose distribution with and without taking into account the interactions with the applicator and the other sources for the LDR Selectron and found differences around 10%. By pre-calculating the distribution around the entire applicator and its contents and then adding these together, they were able to reduce the error to around 4%.

53.7.3 APPLICATOR SHIELDING

Gynaecological applicators are usually made of stainless steel. This will attenuate the radiation by a small percentage. Standard source models do not take this into account, but an adequate correction (albeit not considering oblique filtration) may be made by reducing the source activity by a small amount. A factor of 2% has been found appropriate for Selectron ^{137}Cs sources.

53.7.4 INHOMOGENEITY CORRECTIONS

The tables of $g(r)$ given in TG43 are based on water-equivalent phantoms. Meigooni and Nath (1992) investigated the effect of a 20-mm-thick polystyrene cavity on the dose rate due to ^{103}Pd, ^{125}I, ^{241}Am, and ^{192}Ir. They found that the cavity made little difference to the dose calculated with an ^{192}Ir source, but for ^{103}Pd, ^{125}I and ^{241}Am, the dose beyond the cavity was greater by a factor of 130%, 55% and 10%, respectively. They proposed a simple calculation method that replaces $g(r)$ for a homogeneous medium with a corrected value. Consider a point source with an inhomogeneity which begins a distance r_1 from the source and ends at r_2, as illustrated in Figure 53.3. The value of $g(r)$ that is appropriate for use in Equation 53.8 is then:

$$g_{\text{medium}_1}(r) \qquad \qquad \text{for } r < r_1$$

$$g_{\text{medium}_2}(r)\frac{g_{\text{medium}_1}(r_1)}{g_{\text{medium}_2}(r_1)} \qquad \qquad \text{for } r_1 < r < r_2$$

$$g_{\text{medium}_1}(r)\frac{g_{\text{medium}_2}(r_2)g_{\text{medium}_1}(r_1)}{g_{\text{medium}_2}(r_1)g_{\text{medium}_1}(r_2)} \qquad \text{for } r > r_2$$

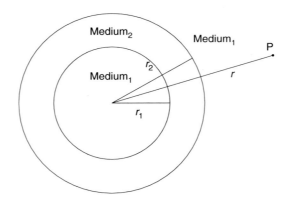

FIGURE 53.3
Calculation of dose at point P from a point source with an inhomogeneity. Medium$_2$ represents the inhomogeneity.

The appropriate value of $g_{\text{medium}}(r)$ can be obtained from the literature (e.g. Dale 1983; Meigooni and Nath 1992) or calculated *de novo*.

Other workers have proposed a method based on the separation of the primary and scatter contribution (Russell and Ahnesjö 1996; Williamson 1996; Anagnostopoulos et al. 2003). Anagnostopoulos et al. propose to replace $\Lambda g(r)$ in Equation 53.8 by the following formula:

$$\frac{\sum_i nE_i \left(\frac{\mu_{\text{en}}(E_i)}{\rho}\right)_{\text{medium}} e^{-\mu_{\text{medium}}(E_i)r}\left[1 + SPR_{\text{medium}}(E_i,r)\right]}{\sum_i n_iE_i \left(\frac{\mu_{\text{en}}(E_i)}{\rho}\right)_{\text{air}}}$$

where n_iE_i represents the energy spectrum of the isotope and SPR is the scatter-to-primary ratio in the medium, which is the ratio of the scatter dose divided by the primary dose at a distance r from the source. $\mu_{\text{medium}}(E_i)$ is the linear attenuation coefficient at the energy in question for the medium, and $\mu_{\text{en}}(E_i)/\rho$ is the corresponding mass energy-absorption coefficient. Although in principle these values need to be evaluated for each source and medium, the authors have shown that for ^{192}Ir, the scatter-to-primary ratio scales with density and is given by the equation:

$$SPR_{\text{medium}}(\rho r) = 0.121(\rho r) + 0.0068(\rho r)^2 \qquad (53.12)$$

where ρ is the density of the medium in g cm^{-3} and r is in cm. The values in Equation 53.12 are strictly the values for water but Anagnostopoulos et al. found that the equation worked well for other media.

CHAPTER 54

BRACHYTHERAPY TREATMENT PLANNING

Margaret Bidmead and Dorothy Ingham [*]

CONTENTS

---

[*] With a contribution from Jean-Claude Rosenwald (Section 54.4).

54.1 INTERSTITIAL IMPLANTS

54.1.1 INTRODUCTION

Radiation oncologists use interstitial implant techniques for the treatment of many different sites within the body. An interstitial implant administers a high dose of radiation to the tumour, whilst effectively sparing surrounding normal tissue and nearby critical organs. The dose homogeneity of such an implant will be different from that achievable with external-beam radiotherapy, as there is a very high dose region immediately around the implanted sources. This high dose region is within the target volume and is not clinically significant. In order to ensure adequate tumour coverage and sparing of critical organs, dosimetry systems are usually based on the minimum dose to a volume.

In the 1930s Paterson and Parker (1934, 1938) developed the Manchester System and Edith Quimby developed the Quimby system (Quimby and Castro 1953, Goodwin et al. 1970) for use with radium sources. These two systems had their own unique implantation rules and dose specification, but were completely self-consistent and have been used for many years to calculate dose to tissue. Godden (1988) provides an excellent description of the Manchester system and its distribution rules.

Once other isotopes such as caesium-137, cobalt-60 (Fletcher 1953, 1980) and iridium-192 (Pierquin et al. 1978 and Dutreix et al. 1982) became available, other dosimetry systems with their own implantation rules were developed. The advent of computerised dosimetry and the availability of multiple-plane isodose distributions provide radiation oncologists with more comprehensive information about the variation of dose within the treatment volume. Three-dimensional dose distributions are now possible, together with volumetric data on doses to critical organs.

Dosimetry systems have been developed for interstitial treatments to ensure reproducibility between implants and adequate coverage of tumour volumes, provided that the rules that have been developed are adhered to as closely as possible. For iridium implants, the system most often used is the Paris system developed by Pierquin et al. (1978).

Afterloading techniques, where inactive applicators can be optimally placed within tumours and loaded with radioactive sources once the dosimetry calculations have been performed, have allowed more flexibility and optimisation of dose distributions. The pioneers of these techniques for interstitial radiotherapy were Henschke in the U.S.A. and Pierquin in Paris.

54.1.2 PARIS PREDICTIVE DOSE-PLANNING SYSTEM

The Paris dosimetry system was developed for flexible wire implants, particularly iridium-192, and acts as a basis applicable for standard implants. There are often exceptions where special calculations are necessary, in addition to the standard dosimetry.

54.1.2.1 Basic Principles

The basic implantation rules of the Paris system are:

1. Active sources should be straight and parallel.

2. The sources should be placed equidistant from each other.

3. The linear activity of each source should be uniform and identical.

4. Each source should extend beyond the limits of the volume so that the dose calculated as the reference isodose is achieved over all the target volume.

5. The plane through the mid-points of the sources should be at right angles to the axis of each source.

6. In any one implant, sources should be equidistant, but from one implant to the other, the inter-source separation can be varied (between 5 mm and 20 mm)*.

7. For volume implants the distribution produced in cross-section through the central plane should be either an equilateral triangle or a square.

8. The maximum activity in any implant using iridium should not exceed 5.5 GBq (i.e. a maximum Reference Air Kerma Rate of 550 μGy h^{-1} at 1 m).

These are the implantation rules, but there are some additional guidelines that have been developed to assist the placement and choice of sources within the tumour volume, derived from clinical experience (Dutreix et al. 1982).

54.1.2.2 Single Source

The dose variation around a single source is dependent on its length. Very close to the wire the dose rate is determined by the local activity of the wire and is only marginally dependent on the length. Further away from the wire there is a relatively greater contribution from more distant parts of the wire. Dutreix et al. (1962) suggested that a high dose sleeve ("manchon de surdosage") where the dose rate was twice the treatment dose rate could be tolerated around a

---

* The maximum separation between sources is constrained by the requirement to limit the magnitude of the high dose *sleeves* around the wires (as explained in Section 54.1.2.2) and this limits the separation to about 20 mm. On the other hand, if the sources are too close together it may be difficult to keep the wires sufficiently parallel to each other along their length to avoid hot and cold spots. For single plane implants the separation of the wires determines the thickness of the treated volume (see Section 54.1.2.3).

wire provided that its diameter did not exceed 10 mm, although 8 mm would be safer. If we restrict this high dose sleeve to 8 mm the diameter of the volume treated by a single source varies from 12 mm for a 1 cm long source to 16 mm for a 10 cm long source. The *useful length* treated is 0.7 times the *radioactive length*.

54.1.2.3 Planar Implants

Varying the spacing between sources from one implant to another is possible in the Paris system, so that greater dose uniformity and optimal tumour coverage can be achieved. In general, the longer the wire, the greater the separation between wires with an upper limit of 2.0 cm. If the separation between the sources is s, and n is the number of wires of active length L; the dimensions of the treatment volume are length ℓ, thickness t, width w and lateral margin d: the relationships relating s and L to ℓ, w, t and d are given in Table 54.1, taken from Godden (1988). Figure 54.1 shows the target volume dimensions for single and double plane implants.

54.1.2.4 Nonplanar Implants

It is sometimes necessary to treat a cylindrical volume, such as vagina and rectum with straight and parallel radioactive sources. In these circumstances, the reference isodose curve for dose prescription is often chosen to limit dose to a critical organ in preference to a strict *Paris* type of dosimetry.

54.1.2.5 Dose Specification and Calculation

The *central plane* of an implant is defined as a plane perpendicular to the sources, which is at right angles to the long axis of the sources and is situated mid-way along their length, as shown in Figure 54.2. Dose calculations are based on the distribution of sources across this central plane.

The *basal dose rate* (BDR) provides a measure of the dose rate in the centre of the treated volume and acts as the basis for dosimetry. It is always calculated from the position of the sources in the central plane and is the minimum dose rate between a pair or group of sources. For a single plane implant, the dose minima occur at the midpoints between sources in the central plane. For an implant in two or more planes, triangles may be formed at the intersection of sources with the central plane and the basal dose (BD) points are taken where the bisectors of the triangles intersect. The BDR is then the mean of the dose rates at the midpoints of all component triangles as shown in Figure 54.3a and Figure 54.3b.

When squares are formed by the intersection of wires with the central plane, then the BDR is the mean of the dose rates at the midpoints of the squares as shown in Figure 54.3c and d. The value of the BDR at the different calculation points across the implant should be

TABLE 54.1

Relationships for Planar Implants

Implant Type	ℓ	w	t	d
Single plane	0.65L to 0.75L[a]	$(n-1)s+0.74s$	0.46s to 0.59s[b]	0.34s to 0.37s
Triangular double plane	0.7L	—	1.2s	0.15s
Square double plane	0.7L	$(n-1)s+0.74s$	1.5s	0.28s

The treatment volume has length ℓ, width w, thickness t and the lateral margin is d.

[a] Variation between 0.65 and 0.75 as the active length, L, increases from 10 mm to 100 mm.

[b] Variation between 0.46 and 0.59 as the active length, L, increases from 10 mm to 100 mm.

Source: From Godden, T. J., *Physical Aspects of Brachytherapy*, Medical Physics Handbook 19, Adam Hilger, Bristol and Philadelphia, 1988.

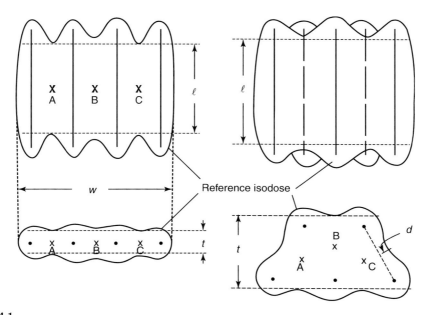

FIGURE 54.1
Positions of the basal-dose points (see Section 54.1.2.5), reference isodose and dimensions of the volume ($\ell \times w \times t$) defined for a single-plane and double-plane ^{192}Ir wire implant. (A cross indicates a basal-dose calculation point). (From Pierquin, B., Dutreix, A., Paine, C. H., Chassagne, D., Marinello, G., and Ash, D., *Acta Radiol. Oncol. Radiat. Phys. Biol.*, 17, 33–48, 1978.)

within ±10% of the mean value, for acceptable homogeneity. The BDR depends on the number, activity, and length of sources and their distance from the calculation point.

For hairpins, the value for active length is taken as the leg length as shown in Figure 54.4 in addition to one-half the loop length. This is valid as long as the total length of the hairpin is at

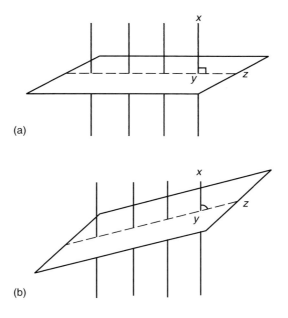

FIGURE 54.2
The central plane is correct in (a), where *XYZ* is a right angle and not in (b), where this angle is acute. In (b) an incorrect source separation would be employed for calculation of the basal dose rate.

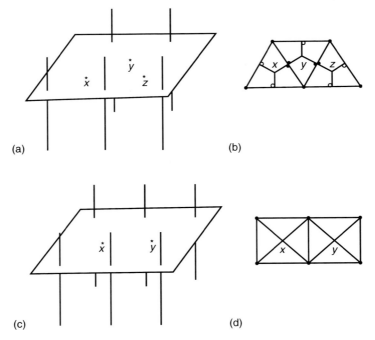

FIGURE 54.3
(a, b) Five-line-two-plane implant, where the sources are arranged in triangles. Such triangles should be equilateral. Whether or not they are strictly equilateral, BD is found at points x, y and z—where the perpendicular bisectors of the sides of the triangles intersect. (c, d) Six-line-two-plane implant where the sources are arranged in squares. BD is found in the centre of each square (at x and y).

least twice the separation between the legs. For example, if the leg length is 3 cm, the radioactive length for calculation purposes will be 3.6 cm (inter-leg spacing is 1.2 cm).

The *reference dose rate* (RDR) is the dose rate used as the basis for dose prescription and hence for calculating the total time of the implant. It is defined as 85% of the basal dose rate. The reference dose rate should encompass an envelope around the implant. Within the limits of source geometry, 85% leads to an acceptable compromise between too steep a dose gradient in passing from the margin of the treatment envelope towards the basal dose rate in its interior, and too great a ripple of the contour of the treatment envelope (see Figure 54.5).

The *treatment volume* is defined as the volume enclosed by the 85% reference isodose. The minimum dimensions of the isodose envelope produced should correspond as accurately as

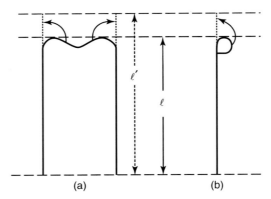

FIGURE 54.4
Radioactive length (ℓ') of a hairpin, which should be used for calculation purposes.

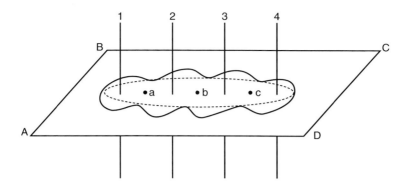

FIGURE 54.5
Four lines (1,2,3,4) transecting the central plane (A, B, C, D) on which dose calculation is carried out. The basal dose rate (BDR) is the mean of the dose rates at a, b and c. The reference dose rate (85% BDR) has an irregular contour (wavy solid line) and totally encloses the target volume (dotted line).

possible with those of the target volume, which it must enclose. The width, w, for a single plane is the maximum width of the reference isodose in the central plane. It is equal to the distance between the most lateral sources plus 37% of the separation between the sources added to each side (see column w in Table 54.1). The thickness, t, of the treatment volume varies with the length and the separation of the sources (see column t in Table 54.1).

54.1.3 DOSE CALCULATION FOR REAL IMPLANTS

Once the source positions within a tumour volume have been defined, whether for a pre-implant calculation or from an actual implant by the use of orthogonal or stereo radiographs, transverse CT, or manual measurement, the basal and reference dose rates can be calculated. It is necessary to know the length of the wire used in each source in addition to the distance between sources. Prior to an implant, a preliminary calculation is made on an *ideal* implant so that iridium wire of the correct source strength can be ordered. This preliminary calculation is often done by hand with the aid of dose data in the form of *crossline graphs* or *escargot curves*, and the source strength selected to give a reference dose rate of approximately 10 Gy per day (42 cGy/hr). The calculation to provide isodose distributions for the actual implant can then be done on a computer by reconstructing the actual source positions and checked using one of the following manual methods or an equivalent computerised lookup table.

54.1.3.1 Crossline Curves

Hall et al. (1966) derived crossline curves, such as those shown in Figure 54.6. The curve shown is for a ^{192}Ir wire of 5 cm length and shows dose rates at different distances from the axis of a line source. A different graph is necessary for each active length used in 1 cm intervals. Crossline 0 corresponds to the central plane of the Paris system. The curves labeled "2, 3, 4" also refer to dose rates at distance from the axis of the wire, but starting at a point on the wire which is 2 cm, 3 cm and 4 cm away from the central plane (towards the end of the wire), respectively. Data are based on wire of linear activity 1 mg Ra-equivalent per cm (RAKR 7.23 μGy h^{-1} at 1 m) and the dose rate is given in cGy/h.

The distance from each basal-dose point to each source is calculated, then the dose rates at these distances from each individual source are measured from the crossline graphs and summed for each point. The mean of all the basal dose points is taken and 85% of this mean is the reference dose rate. The prescription dose is assigned to this value.

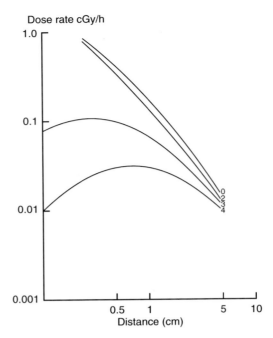

FIGURE 54.6
Crossline curves for a 5 cm source for a source of linear activity of 1 mg-Ra equivalent per cm. The dose rate is given in cGy h^{-1} at different distances from the source axis along crosslines at position 0, 2 cm, 3 cm and 4 cm from the source midplane.

The crosslines as provided give the exposure rate in Roentgen ignoring scatter and absorption in tissue and the encapsulation. The dose rate calculated should therefore be multiplied by correction factors as discussed by Welsh et al. (1983).

54.1.3.2 Escargot Curves

The method, adopted by Pierquin in Paris, makes use of *escargot*[*] curves described by Schlienger et al. (1970). To use these curves it is necessary to reconstruct the source intersections in the central plane. These were originally obtained from conventional radiological tomography acquired in planes perpendicular to the mean source axis. The escargot curves, as shown in Figure 54.7, are constructed with the appropriate magnification factor so that they can be placed directly onto the reconstruction, the centre of the curve being placed on each point where the source intersects the central plane. The graph is then turned so that the line appropriate for the length of the active wire passes through the point at which the dose is required. The dose rate in Gy h^{-1} is then read directly from the spiral line. These curves were initially based on linear activity of 1 mCi cm^{-1}, although a new method of specifying source activity was adopted in 1974 following the proposal of Dutreix et al. (1982) (see Section 51.3).

54.1.3.3 Calculation of Reference Dose Rate

Once the dose rate to each calculation point has been calculated, the basal dose rate is the mean of the values obtained at all the points of calculation. The actual reference dose rate at the time of application is 85% of the basal dose rate corrected for the actual linear activity of the wire used. The treatment time during which the therapy is given is then determined from

---

[*] The name "escargot" is the French word for snail.

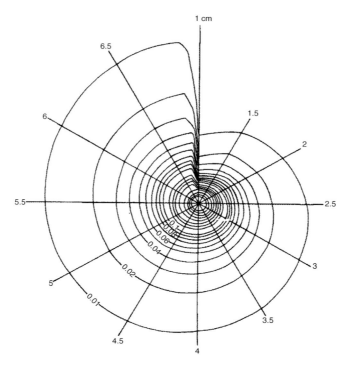

FIGURE 54.7
The escargot isodose lines in the central plane of an active wire. The figures on the curved lines give absorbed dose rate in Gy h^{-1} for a source linear activity of 1 mCi cm^{-1}. This single *escargot* may be used for wire lengths between 1 cm and 7 cm (shown at the tip of each radial line).

this reference dose rate. It may be necessary to apply a further correction to account for the decay of the source during the implant as discussed in Section 53.5.

There is a paper by Casebow (1985) giving tables for standard iridium-192 wire single- and double-plane implants, together with the dimension of the treatment volume within the 85% dose contour. Various workers, including Kwan et al. (1983), have investigated single- and double-plane implants using different source and plane separations (using iridium seeds). Having established the best variables for different thicknesses of implant, they were able to produce standard tables for square implants. These facilitate a rapid estimation of minimum and maximum target dose rates with rectangular implants being calculated on the basis of the *equivalent square* implant.

In flexible wire interstitial techniques, however, there are usually deviations from the ideal implant and often the implanted wires do not strictly comply with Paris system rules. This may produce increased areas of hot and cold spots if the dose is based solely on the basal dose rate, and therefore production of isodose distributions in multiple planes is essential for evaluation. To overcome the problems associated with nonideal implants, Mayles et al. (1985) developed a dose optimisation method, which achieves a clinically acceptable dose distribution by using sources of different linear activity, or by removing sources of the same linear activity at different times during the implant. For interstitial treatments of vagina and rectum, circular templates are frequently used, where the reference dose rate is often specified in a circular plane at a distance (such as 1 cm) from the sources. The dose is frequently limited by the dose to the critical organ in closest proximity.

The introduction of high-dose rate (HDR) or pulsed-dose rate (PDR) remote afterloading machines with single high activity sources (see Section 52.3 and Section 52.4) has increased the opportunity for dose optimisation as discussed later in this chapter.

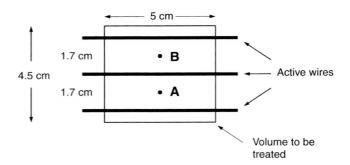

FIGURE 54.8
Example of a single-plane source arrangement of the Paris system used to treat a parallelepipedic volume which is 1.5 cm thick. The active length of the wires is 7 cm. The resulting dose distribution is shown in Figure 54.9.

54.1.3.4 *Example Iridium Implant Calculation*

An example of an iridium implant calculation based on the Paris system follows:

The volume to be treated is illustrated in Figure 54.8 and is 5 cm long and 4.5 cm wide, by 1 cm thick. The Paris system would use a single plane implant with the basal dose calculated at points A and B. From Table 54.1 if the separation of the wires is 1.7 cm and their length is 7 cm:

- The thickness of the treated volume would be: $0.55 \times 1.7 = 0.94$ cm
- The width of the treated volume would be: $2 \times 1.7 + 2 \times 0.37 \times 1.7 = 4.7$ cm
- The length of the treated volume would be $0.72 \times 7 = 5.0$ cm

For ^{192}Ir wire with a linear Reference Air Kerma Rate (RAKR) of 6.5 μGy h^{-1} cm^{-1} at 1 m (0.9 mg Ra Eq cm^{-1}), the total RAKR would be 137 μGy h^{-1} at 1 m. The basal dose rate is 21.6 cGy h^{-1} from the two closer wires and 5.3 cGy h^{-1} from the third wire, giving 48.5 cGy h^{-1}. The reference dose rate is therefore 41.2 cGy h^{-1}. The dose distribution in the central plane and a plane at 0.5 cm from the sources is shown in Figure 54.9[*].

Interstitial brachytherapy can be given on its own for the treatment of an early stage tumour, e.g. stage-I carcinoma of the tongue, where a typical treatment dose might be 65 Gy in 6.5 days to the 85% isodose. It may also be given as part of a combined radiotherapy approach where the dose delivered by the interstitial brachytherapy would be about 30–35 Gy, depending on the dose already delivered by the external-beam techniques.

54.1.4 ICRU RECOMMENDATIONS ON INTERSTITIAL BRACHYTHERAPY DOSIMETRY

The International Commission on Radiation Units and Measurements (ICRU) published their Report 58 giving recommendations on the reporting of Interstitial Brachytherapy (ICRU 1997). The general concepts are similar to those of the Paris system although expressed in terms of the target volumes defined by ICRU 50 (1993)—see Section 29.2. These recommendations allow implants following different systems to be described in the same terms.

---

[*] A comparison of this implant with a similar volume implanted with sources following Manchester system rules is given in a paper by Gillin et al. (1984).

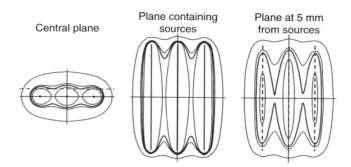

FIGURE 54.9
Dose distributions following the Paris system for the source arrangement shown in Figure 54.8. Dose distributions are shown in the central plane perpendicular to the wires (left), parallel to the wires (centre) and in a plane 5 mm from the source plane and parallel to it (right) - this plane is indicated by the dotted line in the left hand figure. The isodose lines shown are (from the outside in) 50%, 90%, 100% and 120% relative to the reference dose rate (RDR). The 100%, RDR, isodose is bold. The BDR is approximately 120% (actually 117.6%).

54.1.4.1 Description of Volumes

The *planning target volume* (PTV) is considered to be the same as the *clinical target volume* (CTV) in interstitial brachytherapy. It is important that the CTV should be defined prior to the implantation process rather than defining it based on what has actually been implanted. The radiation oncologist will determine an isodose surface, which represents the *minimum target dose* and should ideally encompass the CTV. This isodose surface defines the *treated volume*.

54.1.4.2 Reference Dose

The minimum target dose could potentially be set at any level to ensure that the CTV is covered adequately, but this would risk inappropriately high doses within the implant. In order to provide a more objective reference the *mean central dose* is defined. This is exactly equivalent to the *basal dose rate* of the Paris system. In order to specify the mean central dose it is necessary to identify the central plane of the implant exactly, as in Section 54.1.3. ICRU envisages that there might be a need in a complex implant to divide the implant into two parts and identify the central plane of each part. The mean central dose is defined as the arithmetic mean of the local minima between the sources on the central plane of the implant.

A second approach to ensuring that there is an objective reference is the requirement that the *total reference air kerma* (TRAK) be calculated for the implant. (This concept is discussed in detail in Section 54.2.6.5.)

54.1.4.3 High and Low Dose Volumes

An implant will have high dose regions around the individual sources. Since the size of these high dose regions will influence the toxicity of the implant, it is necessary to measure them. For this purpose, the high dose volume around each wire is defined as the volume encompassed by the isodose corresponding to 150% of the mean central dose around each wire. The maximum dimension of the largest high dose region in all planes should be reported. This will usually be in a direction parallel to the wires.

It is possible also that there may be part of the CTV that does not reach the minimum target dose. Any volume within the CTV, which receives less than 90% of the prescribed dose, is defined as a low dose volume. In a well-constructed implant prescribed to the minimum target dose (i.e. following the Paris system), there would be no low dose volume.

54.1.4.4 Dose Uniformity

The aim of brachytherapy is to deliver a relatively uniform dose to the CTV. However, it is to be accepted that dose rates close to the wires will be significantly higher. In order to express the uniformity of the implant two parameters are recommended:

- The spread in the individual minimum doses used to calculate the mean central dose in the central plane expressed as a percentage of the mean central dose.
- The dose homogeneity index: defined as the ratio of minimum target dose to the mean central dose.

An implant following the Paris system exactly would have a dose homogeneity index of 85%, but the report gives examples of two implants where this index is just over 70%.

54.1.4.5 Implant Duration

The overall time and the temporal pattern of irradiation should be recorded. The irradiation time is the time during which a radioactive source is present in the patient. The overall treatment time is the total time elapsed from the beginning of the first irradiation to the end of the last one. The *instantaneous dose rate* is the *dose per fraction* divided by the *irradiation time* for that fraction. The *average overall treatment dose rate* is the *total dose* divided by the *overall treatment time*.

54.2 GYNAECOLOGICAL INTRACAVITARY TREATMENTS

54.2.1 INTRODUCTION

Intracavitary brachytherapy involves the insertion of radioactive sources into natural body cavities. Many sites are treated in this way, e.g. rectum, antrum, oesophagus, nasopharynx, etc. The most widespread use of this technique has been in the treatment of gynaecological malignancies. However, this means that for extensive tumours, brachytherapy is only a part of the treatment and external-beam therapy is often given before (or occasionally after) the intracavitary irradiation.

The original brachytherapy techniques (the Stockholm system (Heyman 1929, 1935) and the Paris system (Regaud 1929)) evolved using sealed radium sources for the treatment of uterine cancer. Both methods used radium inserted into the uterine canal and the lateral fornices of the vagina, but there were important difference in terms of the activity, the distribution of radium used and the duration of treatment. The dosimetry systems were based on the concept of milligram hours, which is the total activity (in mg) multiplied by the time in hours. However, when intracavitary and external-beam treatment are combined, it becomes important to calculate the absorbed dose in tissue, taking into account the radiobiological effect of the different dose rates (see Section 55.4).

54.2.2 MANCHESTER RADIUM SYSTEM

The classical Manchester system was the first system to attempt reproducible dosimetry to a specified dose point. Tod and Meredith (1953) began to calculate "dose" in roentgen to various sites in the pelvis. It was decided that the limiting dose was not the dose to the rectum or bladder, but to an area where the uterine vessels cross the ureter, the *paracervical triangle* (Figure 54.10a). This area is also frequently the site of early infiltration of the disease. This led to the definition of Points A and B.

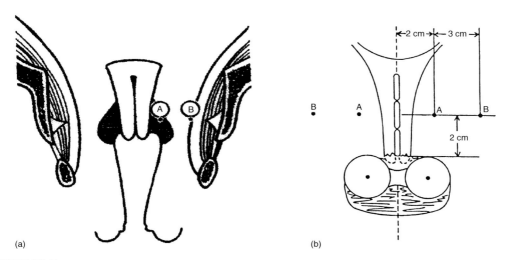

(a)   (b)

FIGURE 54.10
(a) Anatomical drawing of the pelvic area showing the approximate location of Points A and B; (b) Manchester system showing Points A and B positions and their relationship with the intracavitary sources.

- *Point A*, is defined as a point at 2 cm lateral to the centre of the uterine canal and 2 cm from the mucous membrane of the lateral fornix of the vagina in the plane of the uterus (Figure 54.10b). Point A is a geometric and not an anatomical point, its position can only be defined with respect to the applicators in an idealised, or reference geometry. It allows the treatment to be prescribed in terms of dose rather than milligram hours. The dose to Point A can be regarded as the minimum dose on the surface of a large volume of potentially tumour bearing tissue.

- *Point B* is defined to be 5 cm from the mid-line and 2 cm up from the mucous membrane of the lateral fornix. This gives a good indication of the lateral spread of the effective dose in addition to the dose near the pelvic wall near to the obturator nodes.

Loading rules were produced for the uterus and vaginal sources to enable the same dose to be given to Point A, regardless of the shape and size of the uterus and vagina, and consequently the applicators used.

To achieve a constant dose rate at Point A, the Manchester system defines the number of *units* of radium to be used in each ovoid and tube size. The radium loading in the uterine tubes is $(10+10+15)$ mg, $(10+15)$ mg and 20 mg for long, medium and short tubes. The ovoids are loaded with 22.5 mg, 20 mg and 17.5 mg of radium for large, medium and small ovoids respectively. When a short tube (~ 2 cm) is used, a dose reduction of at least 13% should be made to the dose prescribed to Point A. This is because the dose rate at Point A will not be representative of a standard insertion owing to the relative dose rates from the ovoids and the uterine tube at this point*.

The standard "dose" to Point A in the Manchester system was fixed at 53 R h^{-1} and the total "dose" for two equal fractions of treatment was 8000 R delivered in 144 h. The dose at Point B is between 25% and 30% of the dose at Point A. Typically the "dose" rate at Point A from the uterine tube was between 27 R h^{-1} and 34 R h^{-1} and from the ovoids about 19 R h^{-1}. These "dose" rates apply for an ideal insertion and often the disease does not allow perfect positioning. Point A is always perpendicular to the uterine canal and 2 cm

---

* In this case the most superior part of the uterine tube would be at the same distance from the mucosal surface of the lateral fornix as Point A and the dose distribution would not have the characteristic pear shape shown in Figure 54.12 - so the dose rate would be falling off faster than usual at Point A.

from it, whereas Point B remains constant relative to the midline of the patient. Wilkinson (1983) pointed out that even with displaced sources, the dose rate at Point A calculated for the ideal insertion is what is used to determine the treatment time, even if the perfect geometry is not achieved for that particular insertion.

54.2.3 EXTENSION OF THE MANCHESTER SYSTEM TO LOW-DOSE-RATE REMOTE AFTERLOADING

54.2.3.1 Geometrical Considerations

The principles of the Manchester system are still employed when using modern low-dose-rate (LDR), pulsed-dose-rate (PDR), or high-dose-rate (HDR) afterloading systems, though the geometry of the applicators used is more rigid. When one takes x-rays, it is easier to identify the flange on the uterine tube than the mucosal surface of the lateral fornix, as often the plastic *ovoids* cannot be seen on x-ray. These two identification points are not necessarily at the same position in the pelvis so there can be a variation in position of Point A, dependent on which of these points is chosen as the reference origin for Point A. This is particularly true if applicators that do not fit to the same rigid geometry each time are used. Care should be taken to report exactly how Point A has been identified, in addition to the dose prescribed. Figure 54.11 shows an AP x-ray of a typical uterine tube and ovoid application using a rigid LDR applicator system. The black dots on the image show the positions of the reconstruction points of the applicators and the position of the flange.

Point doses to Point A and Point B are usually calculated from orthogonal radiographs of the application. Tables are available of typical dose rates to Point A and Point B from standard (fixed geometry) applicators. However, it is now usual to carry out calculations of dose distributions on an individual patient basis. Most modern brachytherapy planning systems require applicator information, digitised from x-rays or from co-ordinates typed at a keyboard. Orthogonal x-rays are the simplest way to locate the applicators and sources, but sometimes a sufficiently penetrated lateral film is difficult to achieve. It is then necessary either to have another reconstruction method from a different set of films or to use a template of the applicator at the correct magnification and superimpose this on the lateral x-ray. (Different reconstruction techniques are discussed in Section 54.5.2).

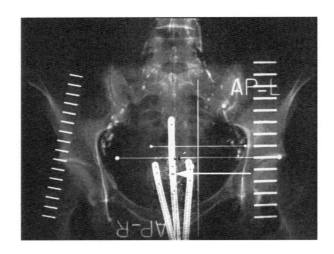

FIGURE 54.11

X-ray of typical uterine tube and ovoid application. The black dots which are just distinguishable on the central axis of the applicators show the positions of the points that will be digitised to allow reconstruction of the applicators and the position of the flange - the latter being the most inferior dot on the central tube level with the white arrow.

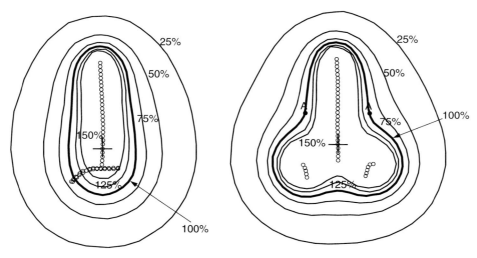

FIGURE 54.12
Lateral and AP isodose distributions for an LDR insertion. Isodoses are shown as percentages of the dose to point A.

Once the positions of the applicators have been defined, some computer systems will allow the applicator and patient points to be entered separately. Points A and B can be defined from the position of the applicators and only the patient points have to be digitised separately from the radiographs. Some computer systems will position the sources inside the applicators automatically, once the applicators have been defined and the active sources requested. This also speeds up the time spent on dosimetry, which is often important for HDR calculations where the patient is waiting for dosimetry to be calculated before treatment can commence. If it is necessary to modify the resulting dose distribution (for example, if the rectal dose is too high) then automatic positioning of the sources inside the applicators is a very useful feature which means that the sources do not have to be re-entered into the computer. Automatic source placement also assists in computer optimisation of dose distributions.

Many planning systems allow anatomy to be traced in from radiographs so that dose distributions can be superimposed in the correct position. Distributions can be selected in any plane and plotted at any magnification, to assist with the accurate positioning of shielding if required. Figure 54.12 shows typical AP and lateral isodose distributions, displayed in the plane of the uterine tube. Three dimensional brachytherapy planning is now a feature of some systems, where isodose surfaces and volumes irradiated are displayed. Input of data is often from CT data sets in combination with orthogonal x-rays.

54.2.3.2 Source Loading Pattern

When the Selectron LDR machine was first introduced, the aim was to reproduce (as closely as possible) the isodose patterns achieved with the traditional Manchester radium system (Wilkinson et al. 1983), by using standard loading patterns and treatment times. The afterloading source patterns that simulate the radium system were achieved by comparing dose distributions for sources arranged in the same reference geometry as is used for the Point A calculation. Similar dose distributions are expected with similarly distributed source activities, so the 10 mg, 15 mg and 20 mg radium tubes were replaced by 2, 3 and 4 caesium pellets. Where the ovoid activity required was 22.5 mg and 17.5 mg radium equivalents, it was not possible to keep exactly the same ratio; consequently, four pellets were used but the treatment times for the ovoid channels were adjusted accordingly. Now a more common practice is to alter the number of ovoid sources to correspond with the variation in ovoid diameter. Mini, small, medium and large ovoids contain 2, 3, 4 and 5 sources respectively, to give similar dose

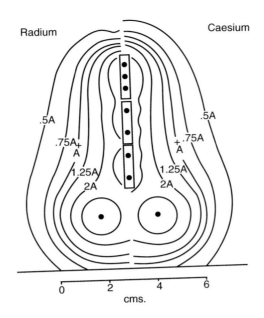

FIGURE 54.13
A comparison of the relative isodose distributions from radium G-type tubes and the corresponding Caesium-137 pellet-loading pattern for the reference geometry. The positions of the active pellets are shown, the intervening spaces being taken up by inactive pellets. Isodoses are shown as multiples of the Point A dose.

rate contributions from the ovoids of all sizes to Point A. This may be preferable to varying the dose rate to Point A during the treatment delivery.

An example of a relative dose distribution comparison for a 6 cm uterine tube and 2 medium ovoids is shown in Figure 54.13. There is no correction for the oblique filtration in the metal applicators or for the shielding effects when the pellets are in line. The errors here will only be significant at the tips of the applicators along the axis where the dose would be reduced by a factor of up to two. Any other small errors are considered insignificant when such steep dose gradients exist. The dose-rate to Point A when using 1.5 GBq (15 mg radium equivalent, RAKR 120 μGy h^{-1} at 1 m) caesium-137 pellets is about 3.5 times greater than the classical Manchester system so a dose-rate correction for the biological effect is applied as discussed in Section 55.4.3. Typical Selectron source positions for this insertion are given in Table 54.2 compared to the classical Manchester system loading.

54.2.3.3 Optimisation

There is also the possibility of optimising the dose distribution and tailoring it to individual patients using the LDR Selectron. Compared with external-beam treatment there are relatively few options for altering dose distributions in intracavitary therapy. However, some

TABLE 54.2

Equivalent Manchester and Selectron Loadings

	Manchester	Selectron Loading Positions
6-cm uterine tube		
Fundus	15 mg	3, 5, 7 (45 mg Ra equiv)
	10 mg	11, 15 (30 mg Ra equiv)
	10 mg	19, 23 (30 mg Ra equiv)
Medium Ovoids	2×20 mg	2×3, 4, 5, 6 (60 mg Ra equiv)

modification of the dose distribution to tailor it to the individual patient anatomy can be achieved by varying the number of active pellets and their position in each applicator. For example, if the dose to the rectum is considered to be too high, the ovoid sources can be moved from position 3 to position 4, 5, 6 or 7 (or removed completely), which results in a shift in the isodose curves. Because of the high dose gradients, the dose to the anterior rectal wall could be altered sufficiently by this change, whilst making very little difference to the dose higher in the pelvis. However, dose to the bladder may be increased unacceptably.

If the disease is concentrated around the cervix, the loading of the uterine applicator can be increased in this area. Care should always be taken with optimisation techniques because clinical data has been collected over many years using standard rules for source positioning and dose prescription. These rules no longer apply when dose distributions are tailor-made to fit treatment volumes. However, optimisation is being investigated because the potential gains in local tumour control with a reduction in normal tissue damage could be considerable.

Pulsed-dose-rate (PDR) intracavitary treatments give very similar dose distributions. They offer similar optimisation possibilities based on a single iridium-192 source which moves along the catheters as for the HDR treatments (see Section 54.2.7).

54.2.4 HDR Treatment of Gynaecological Carcinomas

Carcinoma of the cervix may be treated on a HDR machine using the narrow uterine tube and ovoids, which have built-in packing (an extra 5 mm of plastic incorporated instead of conventional packing) (Nag et al. 2000b). The applicators lock together in a fixed geometry and are fixed to the table during the treatment. With the narrow tubes, it is possible to use local anesthetic for the insertion of these applicators although this is technically difficult and not recommended. X-rays are taken before the first treatment, but because the applicators fit together in a standard, reproducible way, an atlas of pre-calculated dose distributions may be used to select the required dosimetry for the first insertion. A typical isodose distribution is shown in Figure 54.14.

The treatment times are calculated based on the actual activity of the source. Standard dwell positions and times are used (see Figure 54.14). Dose to a point 5 mm behind the back of the ovoids (which is the ICRU rectal point) is calculated to represent the minimum rectal dose[*]. The dwell positions and times are chosen to mimic the classical Manchester pear-shaped dose distribution as closely as possible.

Table 54.3 shows typical dwell positions to reproduce the dose distribution shown in Figure 54.14. Dwell position separation is 2.5 mm, i.e. one active dwell position every 5 mm except in the ovoid applicators.

After the first treatment, the actual dose distribution is calculated for that particular patient, including the dose to the rectum. If any alterations need to be made, these can then be calculated in preparation for the subsequent insertion(s). In order to avoid the second application of a general anaesthetic it is possible to insert a Schmidt sleeve at the first insertion which remains in place until the completion of all fractions.

For the radiotherapy treatment of carcinoma of the endometrium and carcinoma of the vagina, similar techniques may be used as for the LDR Selectron, but owing to the small diameter of the applicators through which the source travels it is possible to give this treatment on an outpatient basis, using only local anaesthetic. Dwell times are calculated from standard tables, which are produced to give a dose rate at the prescription distance from the source, using a specified activity. These tables are based on double-ovoid insertions of known

---

[*] The dwell weightings in the ovoids does sometimes have to be reduced, or a dwell position not used, to bring the rectal dose within tolerance.

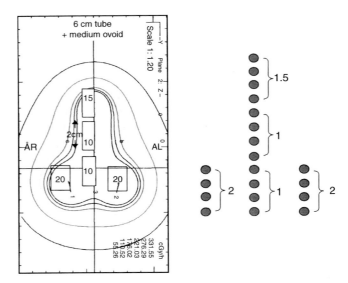

FIGURE 54.14
(**See colour insert following page 590.**) HDR source dwell position pattern with relative dwell weights to reproduce typical Manchester loading.

separation, or line-source simulation. The dose rate for each individual insertion is decayed for source activity. Radiographs are taken of every patient as a record of treatment and so that subsequent dosimetry and field matching can be achieved when necessary.

54.2.5 COMBINED INTRACAVITARY AND EXTERNAL-BEAM THERAPY

Intracavitary therapy is often used in combination with external-beam therapy, which is usually given before brachytherapy. When attempting combined dose distributions it is essential to convert the delivered doses into radiobiologically equivalent doses before combining the brachytherapy and external-beam components. When prescribing the intracavitary dose for this technique the dose given can be influenced by the tolerance of the bladder and rectum. It used to be the practice, for some stages of disease, to treat with intracavitary therapy first, followed by external-beam therapy. The external-beam treatment often incorporated shielding and compensation to spare the volume irradiated by the intracavitary treatment. If a wedge filter of some description was used consideration had to be given to the rapid fall-off in the dose from the intracavitary therapy, the maximum dose to Point A, and the desired dose to the rest of the pelvis. The wedge had to be reproducibly positioned with its thickest part over the position occupied by the uterine sources. The use of such wedges has been described by many authors (e.g. Tranter 1959, Jones et al. 1972). When using shaped wedges or shielding blocks

TABLE 54.3

Source Positions to Reproduce Manchester Loading Patterns

	Manchester	HDR Selectron Dwell Positions and Weights
6-cm uterine tube		
Fundus	15 mg	1, 3, 5, 7 Dwell time weight 1.5
	10 mg	9, 11, 13, 15 Dwell time weight 1.0
	10 mg	17, 19, 21, 23 Dwell time weight 1.0
Medium Ovoids	2×20 mg	3, 4, 5, 6 Dwell time weight 2.0

there were many difficulties associated with their reproducible positioning and the consequent matching of steep dose gradients.

1. When adding intracavitary to external-beam dose distributions, the pear shaped intracavitary volume is often at an inclined angle within the pelvis so the position of the shielding block must be matched to the projection of this distribution onto a horizontal plane through the patient.

2. For the ideal dose distribution, the uterus must be in the same position with and without the applicator inserted. This is frequently not the case and as the position of the block is defined from the radiographs of the intracavitary insertion, the volume shielded is not necessarily the volume irradiated by the sources, this results in areas of over- and under-dose.

3. If more than one intracavitary treatment is used, the summed dose distributions should be used and the doses to Points B are likely to be different. There are also inaccuracies in the day-to-day set up of external-beam treatments.

4. A more fundamental inaccuracy occurs from the use of radiographic techniques to calculate the true position of the sources. An accuracy of at best ± 3 mm can be achieved. In addition, the positioning of the block on the lead tray cannot be achieved to better than ± 1 mm which when magnified can result in a ± 2 mm uncertainty. This can result in significant areas of under- and over-dose due to the high dose-gradients in this region.

For the above reasons the use of such wedges and shielding devices is not so commonly practised nowadays.

54.2.6 ICRU RECOMMENDATIONS ON GYNAECOLOGICAL BRACHYTHERAPY

Intracavitary therapy is frequently used in conjunction with external-beam therapy for the treatment of gynaecological malignancies. For this reason, the ICRU have published Report 38 (1985)* on *Dose and volume specification for reporting intracavitary therapy in gynaecology* recommending that similar terminology and definitions be used, to that used in the ICRU Report 29 (1978) and anticipating the more recent ICRU Report 50 (1993) *Prescribing, recording and reporting photon beam therapy*. This is only possible to a limited extent, since the high dose gradients near intracavitary sources make the specification of absorbed dose to specific points within the target volume less meaningful than in external-beam therapy. The necessary modifications are made in the way in which various volumes are defined.

54.2.6.1 Target and Treatment Volumes

The *target* volume is that volume which contains tissues to be irradiated to a specified absorbed dose according to a specified time-dose pattern

The *treatment* volume is the volume enclosed by a reference isodose surface, which encompasses at least the target volume. In intracavitary therapy, the dose cannot be specified in terms of a percentage of the dose at any point within the target volume due to the steep dose gradient. It is therefore necessary firstly to agree upon a reference dose level, secondly to plan the treatment to include the *target volume* within that reference isodose surface and then to determine the dimensions of the *treatment volume*.

The *irradiated* volume is larger than the *treatment volume* and includes tissues irradiated to a significant absorbed dose.

---

* The ICRU report 38 does not include the recent possibilities offered by modern imaging techniques. It is currently under review (Pötter et al. 2001b) and will include some of the concepts developed in the recommendations issued by the GEC-ESTRO group (Haie-Meder et al. 2005; Pötter et al. 2006) or by the American Brachytherapy Society (Nag et al. 2004).

Using these definitions, the ICRU recommend that for an insertion of uterine and vaginal sources the specification of the intracavitary application should be in terms of the *treatment volume*, which the ICRU have defined as the volume enclosed by the 60 Gy reference isodose surface for classical low-dose-rate therapy[*]. If there is an external-beam component in addition to the brachytherapy, then the brachytherapy isodose must be selected to allow for the external-beam dose (e.g. if the external-beam contribution is 50 Gy, then the brachytherapy *treatment volume* is the volume enclosed by the isodose that is equivalent to 10 Gy at low dose rate).

This reference volume should be defined by the height, width and thickness of the pear-shaped volume enclosed by the 60 Gy absorbed dose level, measured in the oblique-frontal and (oblique-) sagittal planes containing the uterine sources as shown in Figure 54.15, taken from ICRU 38.

The dose delivered to a gynaecological malignancy is limited by the radiation dose that can be tolerated by the bladder and the rectum. The determination and specification of the bladder and rectal doses should either be measured or calculated. The recommended calculation points are shown in Figure 54.16. Chassagne and Horiot (1977) have described methods of determining points of reference for both organs.

54.2.6.2 Dose to the Bladder

A Foley catheter is inserted and partially filled with $7\ cm^3$ of radio-opaque fluid. The catheter is pulled down to bring the balloon against the urethra. On the AP radiograph, the reference point is taken as the centre of the balloon. On the lateral radiograph, the reference point is taken at the posterior surface of the balloon along an AP line drawn through the centre of the balloon.

54.2.6.3 Dose to the Rectum

The rectal point is located at 5 mm behind the posterior vaginal wall as shown in Figure 54.16. Knowledge of the dose to the rectum can mean that the sources or applicators need to be repositioned to keep the rectal dose to less than 65% of the dose to point A, and less than 65 Gy in total.

In practice, it is difficult to identify the posterior vaginal wall on the lateral radiograph. The rectal point calculation is therefore usually made at a point 5 mm posterior to the back of the ovoids or 5 mm posterior to the packing if there is any in place.

If rectal and bladder measurements are to be made, there are several pitfalls that should be noted:

1. Applicators are normally inserted in the lithotomy position and packing and measurement of dose points is much easier in this position. For low-, pulsed- or medium-dose-rate treatments, the treatment is always given with the patient lying flat so that the critical anatomy is in a different position.

2. When using afterloading systems with high activity sources, microsources have to be used to measure dose points. This means some radiation exposure to staff. Using microsources also means that it is necessary to measure dose rates of the order of 0.05 cGy s^{-1} (3 cGy h^{-1}) which is technically difficult. If using an ionisation chamber the measuring volume is of the order of 10.6 cm^3, which is large for clinical use. Other solid state and photovoltaic detectors are often temperature and energy dependent, so have to be used with care. Measurements have been made at various centres using TLD inserted into rectal catheters, which are left in place during the course of an insertion (see Section 40.2.3.5). However, this gives a retrospective average dose over the whole intracavitary treatment, and is not necessarily the maximum dose received by the anterior rectal wall.

[*] For medium- or high-dose-rate techniques the radiation oncologist must indicate the dose level he believes to be equivalent to 60 Gy, delivered at low dose rate.

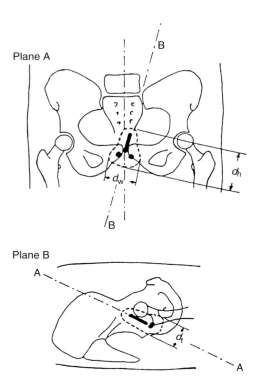

FIGURE 54.15
Geometry for measurement of the size of the pear-shaped 60 Gy isodose surface (broken line) in a typical treatment of cervix carcinoma using one uterine applicator and two vaginal applicators. Plane A is the oblique-frontal plane that contains the intrauterine device. The oblique-frontal plane is obtained by rotation of the frontal plane around a transverse axis. Plane B is the oblique-sagittal plane that contains the intrauterine device. The oblique-sagittal plane is obtained by the rotation of the sagittal plane around the AP axis. The height (d_h) and the width (d_w) of the reference volume are measured in plane A as the maximal sizes parallel and perpendicular to the uterine applicator, respectively. The thickness (d_t) of the reference volume is measured in plane B as the maximal size perpendicular to the uterine applicator. (From ICRU Report 38, 1985.)

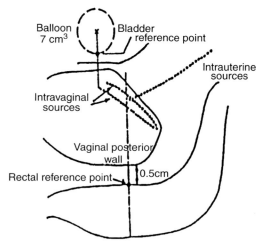

FIGURE 54.16
Determination of the reference points for bladder and rectum (see text). (From ICRU Report 38, 1985.)

3. When using a straight measuring probe it is very difficult not to distend the anatomy. It is also difficult to be sure that maximum doses are measured.

Some centres acquire CT scans of the patient with the applicators in place to define the maximum bladder and rectal points by superimposing isodose curves on the relevant transverse slices. This is a good method, but practically difficult to achieve and expensive on a routine basis. The use of CT and Magnetic Resonance Imaging (MRI) is increasing in brachytherapy dosimetry and allows volumetric dose calculations for bladder and rectum.

54.2.6.4 Dose to the Pelvic Wall

The points related to bony structures (Figure 54.17 and Figure 54.18) are the mid-external lymph nodes (R and L EXT), the low para-aortic area (R and L PARA) and the low common iliac lymph nodes (R and L COM). The pelvic wall reference points represent the absorbed dose at the distal part of the parametrium and at the obturator lymph nodes. The evaluation of these doses is useful when intracavitary therapy is combined with external-beam therapy.

Typical doses to the above pelvic points expressed as a percentage of the dose to Point A are shown in Table 54.4.

54.2.6.5 Total Reference Air Kerma

There are steep dose gradients in the region of the sources, so that specifying a treatment in terms of the dose at a point is not recommended. It is proposed that the total intracavitary treatment should be specified in terms of total air kerma. The reference air kerma rate (RAKR) of a source is the kerma rate to air, in air at a reference distance of one metre, corrected for air attenuation and scattering (see Section 51.3). The *total reference air kerma* (TRAK) is conceptually similar to specifying the number of milligram hours in traditional systems.

For example, a point source containing 1 mg radium with 0.5 mm Pt filtration, producing an exposure rate of 8.25 R h^{-1} at 1 cm would have an RAKR of 7.2 μGy h^{-1} at 1 m. (The conversion factor from exposure to air kerma used is 0.873 cGy R^{-1}).

A typical Manchester type insertion of 75 mg radium with 1 mm Pt filtration (difference in filtration reduces dose-rate by 0.93) would produce an air kerma rate of

$$75 \times 0.825 \times 0.873 \times 0.93 = 502 \ \mu\text{Gy h}^{-1} \text{at 1 m.}$$

The TRAK is the sum of the products of the RAKR and the duration of the application for each source. Assuming the sources are left in position for 72 h, then the TRAK is 3.6 cGy at 1 m for a typical *Manchester* insertion.

The TRAK is independent of the source geometry and proportional to the integral dose to the patient. It provides a useful quantity to compare treatments within one centre or between different centres and to explain possible differences in the delivered dose or in the treatment volume dimensions. It can also serve as a useful index for radiation protection of personnel. By application of the inverse square law it can be used to evaluate to a reasonable approximation, the total absorbed dose delivered during treatment at distances between 10 cm and 20 cm from the sources (Dutreix et al. 1982). However, it cannot be used to calculate absorbed doses in the tumour or target volume (i.e. close to the sources).

54.2.6.6 Time-Dose Pattern

The prescription method for a particular type of intracavitary system cannot necessarily be extended from one system to another. It is therefore important to record the specific time-dose pattern that has been used. With all intracavitary treatment, whether continuous low dose rate, or fractionated high dose rate the prescription should specify the total radiation dose, the length of

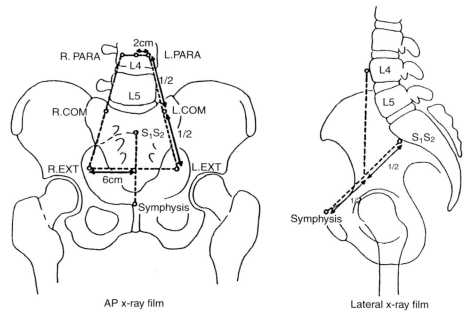

FIGURE 54.17
Determination of the lymphatic trapezoid. On the left is an anteroposterior view and on the right a lateral view.

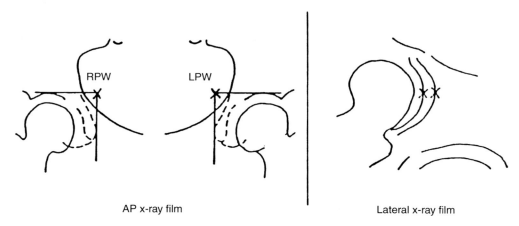

FIGURE 54.18
Determination of the right (RPW) and the left (LPW) pelvic wall reference points.

TABLE 54.4

Typical Doses Expressed as a Percentage of the Dose to Point A

A	R	BL	RPW LPW	R/L EXT	R/L COM	R/L PARA
100%	65%	60–80%	25%	25%	15%	5%

treatment and the number of times the treatment is to be repeated. Frequently the intracavitary treatment will be a part of a combined approach using external-beam therapy. The external-beam dose must also be recorded. For example, a typical radiotherapy prescription for the treatment of carcinoma of the cervix would be 50 Gy in 28 fractions over 5.5 weeks delivered by external therapy (8 MV or 10 MV x-rays) to the pelvis, using an anterior and two lateral wedged fields. This would be followed by 25 Gy delivered over 12 h to Point A from a uterine tube and two vaginal ovoids remotely afterloaded with caesium.

For a complete description of the treatment a more specific prescription detailing field sizes, wedge compensator and blocks, the length of the uterine tube and size of vaginal ovoids and the distribution of activity within the applicators is necessary.

There is often confusion between low and medium dose rates for caesium-137 Selectron dosimetry. The ICRU definitions of dose rate are as follows:

Low dose rate	0.4 Gy h^{-1} to 2 Gy h^{-1}
Medium dose rate	2 Gy h^{-1} to 12 Gy h^{-1}
High dose rate	Above 12 Gy h^{-1}

The classical Manchester system gives a dose rate to Point A of 53 cGy h^{-1}, which is clearly low-dose-rate therapy. Modern afterloading machines with higher activity caesium-137 sources give dose rates between 1.5 Gy h^{-1} and 2.0 Gy h^{-1}. This would be termed low dose rate by the ICRU definition, but clinical experience has shown that a dose reduction is needed to maintain the same isoeffect as standard therapy (Roberts et al. 2004). The suggestion is therefore that the definitions should be changed so that dose-rate classification would reflect the requirement for a reduction in dose compared to doses given by standard activity systems (Pötter et al. 2001b). There is a much clearer division between medium and high dose rate systems.

54.3 HIGH-DOSE-RATE TECHNIQUES FOR NON GYNAECOLOGICAL SITES

The small physical size of the iridium-192 source lends itself to use in other sites where a small diameter applicator can be used. The use for endovascular brachytherapy has already been described in Section 52.4.1. Other uses are described below.

54.3.1 CARCINOMA OF THE BRONCHUS AND OESOPHAGUS

Patients suitable for treatment with HDR brachytherapy are those with localised endo-bronchial lesions where both the proximal and distal extent of the tumour can be identified bronchoscopically. The treatment is largely palliative and results have shown that it is at least as effective as external-beam palliation, achieving the same relief of haemoptysis, cough and breathlessness but with less associated morbidity. The adverse effects on normal tissue are confined to those tissues within the immediate vicinity of the bronchus. It has the benefit to the patient of requiring fewer hospital visits and less treatment time (as the dose is given in a single fraction), whilst preserving the possibility of more than one fraction. Intraluminal therapy can also be used in combination with external-beam treatment. Many centres use this technique and a large clinical trial has been conducted at the Christie Hospital (Stout et al. 2000). The technique of treatment is usually to perform a bronchoscopy on the patient on the treatment couch (provided that x-ray facilities are available in the treatment room). The patient is screened while the catheter containing the guide wire with dummy sources is placed at the correct site for treatment. The bronchoscope is removed and the patient is re-screened to check the position of the catheter. The length of the tumour volume to be treated is determined, which includes a treatment margin of (usually) 10 mm at either end. An anterior radiograph and two oblique films may be taken for dosimetry purposes so that clear visualization of the dummy source and catheters are possible. The guide wire is removed and the dosimetry calculated from standard tables decayed for source activity. The procedure for treatment of carcinoma of the oesophagus is similar although the external diameter of the applicator is 10 mm. The dose is usually prescribed at 10 mm from the source.

54.3.2   CARCINOMA OF THE NASOPHARYNX

The small diameter of the flexible catheter and small source size lend themselves to insertion into small orifices. Two catheters are inserted through the nose. This technique has been used to deliver a boost treatment in nasopharyngeal carcinoma in several centres (Gao et al., 1990).

54.3.3   CARCINOMA OF THE BREAST

HDR is very useful for the treatment of breast carcinoma, using templates and rigid introducers. Lumpectomy is performed and before the skin is closed, a single metal introducer is placed along the tumour bed to obtain the correct template orientation. The skin is then closed, leaving the needle in situ. The remaining needles are inserted through the templates and the breast is slightly compressed. The needles are fixed in position. Measurements are taken which enable the active lengths to be calculated. 5 mm skin sparing is usually allowed deep to the skin surface (reproducible positioning is therefore very important). The dwell positions and times are calculated for the treatment, as the HDR has a single high activity source. Linear sources are simulated by stepping the source at 5 mm intervals along the needle. The dose distribution is calculated, based on the Paris system. The treatment is fractionated and treatment time is on the order of a few minutes. Once the machine has been programmed, a test run is performed to check dwell positions and times using film.

54.3.4   SPECIALISED MOULDS AND APPLICATORS

Moulds and applicators can be made for particular applications. Removable applicators, such as moulds for treating hard palate, are particularly useful as the patient can be treated on an outpatient basis. The limiting factor is usually the angle through which the source cable has to turn, although the dummy source is used for checking the travel before the active source is used for treatment. The mould is constructed to place the catheters with a few mm stand-off from the skin in order to increase the relative dose at depth within the patient. Dosimetry is based on either the Manchester or Paris system principles.

54.4   PROSTATE BRACHYTHERAPY

Transperineal source implantation, guided by trans-rectal ultrasound and remote afterloading techniques with HDR, PDR, or low dose rate ^{125}I or ^{103}Pd have recently become very popular as one of several methods for treating prostate cancer. These techniques lend themselves to more precise placement of radioactive sources within the prostate, allowing dose to urethra and rectum to be minimised. The advent of high-frequency rectal ultrasound probes has meant that the placing of the applicators or sources can be monitored both in the transverse and longitudinal planes. The planning and dosimetry process is similar whether using afterloading or seed implantation. However, permanent implants, in which iodine or palladium seeds are distributed within the prostate volume and never removed from the patient, are more widely used. Specific recommendations have been issued both in the United States and in Europe (Nag et al. 1999; Yu et al. 1999; Ash et al. 2000). What follows considers only such permanent implants although the process is similar for afterloading methods (Kovács et al. 2005).

Conduct of prostate implants may be carried out either as a two-stage process involving a planning visit and a treatment visit or as a single session. The two-stage process will be considered first. A volume study of the prostate is performed, several days before the actual implant, using transrectal ultrasound with a stepper to move the ultrasound probe in the longitudinal direction (normally with 5 mm increments). During this study, the prostate and critical organs (especially the urethra) are contoured at several levels. These data are used to determine the prostate volume and hence the optimal position of the needles with the aid of a dedicated treatment planning system. Fixed needle positions on a Cartesian grid (5 mm steps) are used, according to a template mechanically attached to the probe support (see Figure 54.19).

The positions of the active sources are then determined within each needle. Although this is often done without the use of *dosimetry rules* more controlled systems have been evolved. Examples are *peripheral* or *modified peripheral* loading or nomogram based implants, where the total activity is calculated according to the prostate volume and the loading is mostly peripheral with 75% of the activity distributed close to the external surface of the prostate (Stone et al. 1995, Stock et al. 1998). Additional guidelines are as follows:

- It is best to avoid having seeds at the same depth in adjacent needles.
- It is better to have a larger separation between seeds along a more central needle, than a peripheral one, where the prostate contour changes more rapidly.
- It is common practice to avoid placing seeds close to the prostate-rectal interface or close to the urethra.

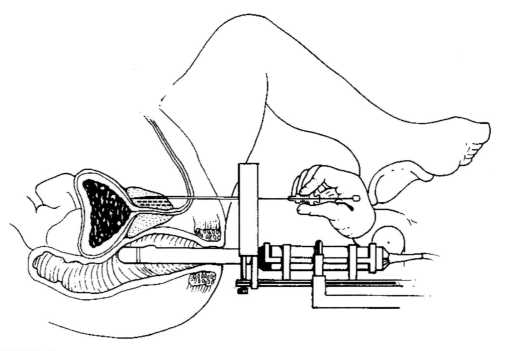

FIGURE 54.19
Typical patient set up for prostate implant. A trans-rectal ultrasound probe gives a real time image of the prostate and of the inserted needles, in either the transverse or sagittal plane. This allows control of the needle position and seed loading within the prostate capsule. The needles are guided by a template with a Cartesian grid of 5 mm spacing. (After Blasko, J. C., Mate, T., Sylvester, J. E., Grimm, P. D., and Cavanagh, W., *Semin. Radiat. Oncol.*, 12, 81–94, 2002.)

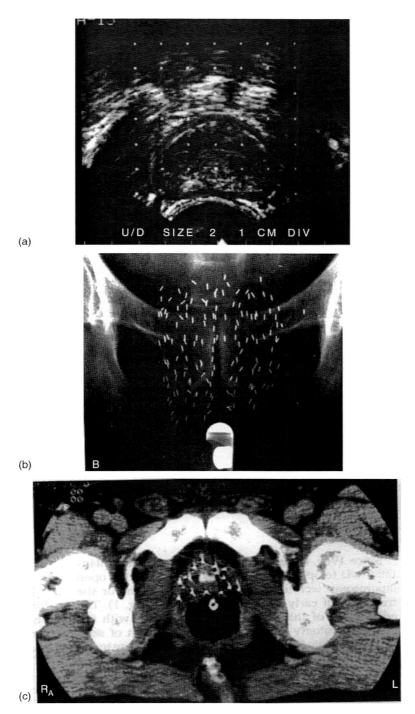

(a)

(b)

(c)

FIGURE 54.20
Transverse images of implanted seed using ultrasound, x-ray and CT.

One problem with the pre-implant volumetric study is that it is difficult to set the patient exactly in the same position for the actual implant. In addition, the shape of the prostate changes as the needles are inserted or removed, which causes the pre-planned theoretical seed distribution within the prostate to be difficult to be reproduced precisely. Finally, the patient has to come to the hospital and be anaesthetised twice. For all these reasons real-time dose planning is becoming more popular. Several approaches are possible but in all cases the treatment planning computer is brought into the operating room and is directly connected to the ultrasound machine so that 2D or 3D real time images of the prostate can be acquired. The basis for source placement is either a seed distribution computed automatically according to some predefined algorithm or predefined rules as indicated above. The dose distribution is calculated in real time as seeds are inserted. The seed insertion could be performed either line by line with preloaded needles which are then removed leaving the seeds retained within the prostate gland, or one after the other with a *Mick applicator* attached successively to each needle which is then retracted step by step.

Typical seed activities used are between 0.3 mCi and 0.4 mCi (0.4–0.5 μGy h^{-1} m^2) for ^{125}I and between 1 mCi and 1.4 mCi (1.3–1.8 μGy h^{-1} m^2) for ^{103}Pd. A typical distance between adjacent seeds is of the order of 1 cm. The details of dose calculation for seed sources are discussed in Section 53.2 and Section 53.4.2. For permanent implants, it is not possible to adjust the treatment time. The total dose is calculated assuming a constant activity and an overall equivalent time of the half-life multiplied by $\sqrt{2}$. The isodose distribution produced is then examined and compared with the contours of the prostate. The lowest value isodose that just encompasses the prostate contours on all levels is selected. This is called the *minimum peripheral dose*, which is generally selected to be the *prescribed dose*. When brachytherapy is used as monotherapy, the recommended prescribed dose is usually 145 Gy for ^{125}I and 125 Gy for ^{103}Pd. The limits of dose to the urethra and the rectum are not firmly established but it is recommended that the length of urethra receiving more that 200% of the prescribed dose should be limited.

Once the seeds have been implanted it is not possible to alter the outcome for that particular implant, but it is useful to analyse the dose distribution and assess the quality of the implant. Post-implant dosimetry is therefore recommended (Nag et al. 2000a). This is usually done from transverse CT slices performed several weeks after the implant (to allow for stabilisation of the post-implant oedema). Orthogonal x-ray films were originally used for reconstruction. Rosenthal and Nath (1983) developed a method of reconstructing an implant from uncorrelated images and this method was further refined by van der Laarse and Batterman (1984), using the fact that the seeds should appear in the same order along the y-axis on all three films. Films are now mostly used as a check of the overall seed distribution. Figure 54.20 shows typical images from ultrasound, x-ray and CT. Ash et al. (2000) recommend that the following dosimetric indexes should be reported:

- The value of the prescribed dose
- D_{90}, the dose that covers 90% of the prostate volume (which should be at least equal to the prescribed dose)
- V_{100}, the percentage of the prostate volume that has received the prescribed dose
- V_{150}, the volume that receives 50% more than the prescribed dose

54.5 DOSIMETRY CONSIDERATIONS FOR BRACHYTHERAPY IN GENERAL

54.5.1 SEALED SOURCE DOSIMETRY

The dose calculation around a sealed source has become progressively more accurate and more complicated over the years. Originally, a source was considered as a single point source using the specific gamma-ray constant to relate the exposure rate of the source to the specified source activity or strength. The encapsulation and filtration of the source was ignored and the *apparent* source strength used to approximate the effects of filtration and self-absorption. Many sources in brachytherapy are not actually point sources, but line sources and therefore the exposure rate at a distance from a line source can be calculated by integrating over the total source length. Filtration and self-absorption in the source then become much more significant and should be included in the dose model.

Various authors have attempted to calculate effective absorption coefficients and anistropy factors required to model linear sources. Sievert (1932) published extensive tables of the integral for values of exposure round a line source and these have been compared with Monte Carlo evaluations by Williamson (1996) for iridium and other sources. Monte Carlo calculations can be used not only to model the source, but also to predict the effects of different tissues on the absorbed dose.

Other calculation methods required the application of a tissue correction factor, such as the polynomial expressions used by Meisberger et al. (1968), Van Kleffens and Star (1979), and Dale (1986). A good review and presentation of the theory of sealed source dosimetry is provided by both Godden (1988) and Williamson et al. (1995) (see Chapter 53). Much work still has to be done on calculation of dose in different tissues and the effect of air cavities. The increased use of CT scanning for brachytherapy patients will provide improved data.

54.5.2 METHODS OF IMPLANT RECONSTRUCTION

Radiographic techniques are often used to define the spatial positions of the applicators and radioactive sources. The choice of reconstruction techniques is influenced by the availability of isocentric x-ray equipment, the complexity of the source arrangement and the site of the implant in relation to the patient anatomy.

Radiographs can be used to reconstruct three-dimensional models of the applicators and source positions, which in turn are used to produce multiplanar dose distributions.

1. Direct measurement can be used for single plane implants for superficial skin tumours, and when rigid plastic templates or jigs are used with rigid introducers in a standard pattern (e.g. breast implants.)

2. Orthogonal radiographs using views at right angles to each other, preferably parallel and perpendicular to the plane of the implant are the most commonly used method*. A pair of orthogonal films share one common co-ordinate axis, so the x and y dimensions are obtained from one film and the other film gives the y and z dimensions[†]. The y-axis is therefore common to both films. A base line common to both films, which is often a line perpendicular to the intersection of the radiographs and is usually parallel to the edge of the films, must be defined. The magnification of films must be taken into account and there are various devices such as rulers or rings to assist in this calculation. Dummy marker wires can be positioned in plastic tubes and applicators to determine directly the length of active wire to be loaded. They can also be used to define the magnification at the plane of the implant or insertion.

    For accurate results from this technique, the two films must be mutually perpendicular; the central ray should be perpendicular to the film and, as for the other methods, the patient must not move between exposures. It may also be necessary to correct for beam divergence.

3. An isocentric (variable angle) technique is useful for separating multiple sources (e.g. iodine seeds in a prostate, multiple catheters in a breast or head and neck implant) and for visualising sources where bone or other structures may obscure them, or where a good lateral film is not possible. A simulator should be used for this technique because the best plane can then be chosen whilst screening, so that all the sources or applicators are clearly visible and do not obscure one another. The highest accuracy with this method is when there is 90 degrees

---

* If image planes parallel to the plane of the implant are not easily achieved it is easy to rotate the reconstructed planes within the computer to achieve a more appropriate display.
[†] This is in the co-ordinate system recommended by the International Electrotechnical Commission where y is the caudo-cranial direction (see Section 11.5.4).

between the two beams. It is sometimes useful to take three films to assist in source identification, though only two are used for the actual reconstruction process.

4. Semi-orthogonal methods have been developed using a proprietary reconstruction box with markers. They are particularly suited for mobile x-ray systems, which do not have accurate scales or movements. The x-ray beams are approximately AP (or PA) and lateral and the radiographic images of the cross-wire markers on the AP and lateral box faces are used to reconstruct the spatial localisation of the x-ray foci. This method accepts beams whose central axes do not intersect and are not perpendicular to one another, as long as the projections of the cross-wires on the two corresponding box faces are visible on the radiographs. A typical set-up with a localisation box is shown in Figure 54.21.

5. Stereo shift films, which involve taking two AP films at a measured distance between them with a common reference point projected onto both films, may be used for source localisation when there are many sources.

6. To take a tomographic section through the central plane of the implant, either with x-ray tomography or more commonly with CT, is a useful method that gives both the relationship between sources on the central axis and associated anatomical structures, e.g. breast (lung), tongue (spinal cord), intracavitary techniques (bladder and rectum). It is now possible to input source and applicator positions directly from a series of transverse CT scans.

It is easier and safer to perform any of these radiographic techniques whilst afterloading applicators are inactive. When entering source positions into computers it is important to have the orientation of the radiographs correct for the particular computer's dosimetry system and to identify each end of the active sources correctly, in addition to the applicator itself. Dummy sources, consisting of radioopaque markers, inserted precisely where the sources are expected to be, may be used to enable the source positions to be digitised. There are also several computer

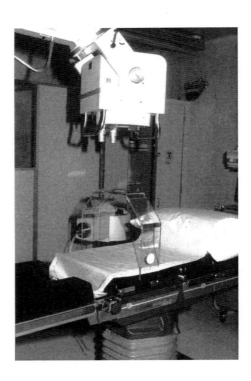

FIGURE 54.21
(**See colour insert following page 590.**) X-ray unit and localisation box used to reconstruct the 3D coordinates of sources. With such a box, it is not necessary to have an accurate x-ray set-up since the radioopaque markers of the box can be used to restore an orthogonal coordinate system.

programs which will automatically place active sources in specified positions within applicators once the co-ordinates of the applicator have been defined, and will automatically determine the central dosimetry plane after the sources have been reconstructed from the films.

A tracking technique may be useful to reconstruct dummy source positions on two radiographs, if a correspondence can be established between the points belonging to the same catheter as seen separately on both radiographs and there are sufficient points to allow good reconstruction (Rosenwald 1975).

Some inherent limitations in reconstruction techniques, which lead to inaccuracies, are the accuracy with which the magnification is determined; the limitation of the calculational accuracy obtainable with the equations used for deriving source position and, in the absence of a localisation box, the fact that the radiation beam may not be perpendicular to the films. Greater accuracy can usually be achieved by using radiotherapy simulators for applicator or source visualization. Quality control of the reconstruction methods may be carried out using geometrical phantoms (Roué et al. 2006).

54.5.3 Computerised Dosimetry

Dose computation algorithms have evolved over the years to calculate dose distributions around sources in a variety of planes and orientations. Stovall and Shalek (1972) and Shalek and Stovall (1975) described a number of computer techniques available at that time. These have since been extended to include the calculation of source coordinates from radiographs, and the sorting of multiple seeds and wires, particularly useful for prostate implants. Faults in the dose rate formula, including errors in the constants used can also introduce errors into the dose computation, although errors in the source-positional data supplied to the computer program due to inaccurate data entry from the localisation films are usually more significant. Kline et al. (1985) reported a comparison of reference dose point data with dose calculations carried out using a Theraplan computer for iridium wire. This paper highlights some of these pitfalls of brachytherapy dosimetry.

Welsh et al. (1983) have reported that, with their computer dosimetry system, repetition of the dose distribution calculation for the same wire arrangement may show differences of as much as 5% in the doses at individual points although the average discrepancy is much less than this (<3%). The differences have been shown to be due to small variations in the digitised wire position data and the position of the calculation plane. Wire positions should therefore be entered carefully and a representative dose rate chosen for the prescription of the dose to reduce the errors to less than 3%. Rigorous quality assurance checks should be performed on the input and output devices of the treatment planning system, in addition to QA checks on the sources and applicators in an attempt to minimise these errors.

To verify computer calculations a database of individually measured data or appropriate published data must be available. For external-beam treatments, data are usually measured for the individual linear accelerator to be used. However, brachytherapy is different in that experimental determination of dose rates at short, clinically relevant distances around brachytherapy sources is extremely difficult and there can be large uncertainties in the measurements. Several phantoms have been developed to check consistency of dose distributions, but absolute dosimetry is more of a problem. Some errors in computer dosimetry have been found to be because of incorrect input parameters, particularly the use of the correct exposure-rate constant or unit of source strength. It is important to know what internal factors the computer uses, whether tissue correction algorithms are used, if nonlinear sources can be modeled, and if self-absorption in sources, applicators and cables is calculated. There are also many different units associated with brachytherapy and although every book on brachytherapy provides a conversion scheme (see Section 53.5), the use of the correct units for a particular treatment-planning computer can be crucial. For a particular brachytherapy computer planning system users are constrained by the concepts in the software development as well as the validity of the calculation algorithms (IAEA 2004; ESTRO 2004). It is important to apply a source calibration methodology that is consistent with the assumptions made by the software designer. There are uncertainties in the value of the specific gamma rate constant but in some senses it is a *dummy variable* when source calibration is specified in terms of reference air kerma rate (see Section 51.3).

54.6 OPTIMISATION IN AFTERLOADING BRACHYTHERAPY

The objective of treatment planning for brachytherapy is to obtain an optimum dose distribution to deliver the desired target dose at the desired dose rate, whilst minimising the dose to surrounding normal tissue. In interstitial techniques, it is very rare to achieve an ideal implant, some method of optimisation that can be applied after the needles have been implied would therefore be beneficial.

History shows us that brachytherapy techniques have evolved around available radiation sources. Prominent examples are the Manchester system for radium-226 and the Paris system for iridium-192. We now have available the miniature stepping source and it is appropriate that a dosimetry system should evolve around this type of source.

Brachytherapy has always been a geometrically conformal therapy. The applicators placed within the treatment volume enable the delivery of a high dose of radiation to a small volume of tissue while sparing normal tissues. External-beam therapy in moving to conformal therapy uses imaging based treatment planning together with volume-dose analysis. Should brachytherapy be moving in this direction?

The established dosimetry systems have evolved empirically over a long period. Considerable clinical experience in their application has been accrued. Progress to new dosimetry systems should only proceed with care being taken in comparing the old with the new, using the computational tools currently available.

The miniature stepping sources can be used to mimic traditional source arrangements such as those of the Paris or Manchester systems or they can be used with variable dwell times in each position so producing *optimised* dose distributions. Implants can be optimised to produce as uniform a dose coverage as possible throughout the implanted volume or they can be optimised to encompass a volume outlined on a set of images. In the latter case the dose in the implanted volume may be very non-uniform.

54.6.1 IMAGE-BASED PLANNING

Conformal radiotherapy to the prostate can be delivered either by external-beam therapy or by means of a stepping source in implanted needles, as well as by a seed implant (see Section 54.3). CT can be used to define a target volume and any critical structures such as the rectum. Dose volume histograms (see Section 35.3) are easy to interpret for external-beam therapy (Figure 54.22). Deviation from the ideal, with all the target volume receiving the prescribed dose, is immediately apparent. In brachytherapy, the inevitable inverse-square falloff of dose around the sources results in an inherent inhomogeneity. This obscures regions of underdosage or overdosage, which may be present.

Clinical experience has demonstrated tolerance of significant high dose volumes within implants. The Paris system allows quite a range of needle separations resulting in significantly different high dose volumes (Pierquin et al. 1978). However, these treatments have involved conventional doses of radiation. As dose escalation is the aim of much conformal therapy the tolerance of significant high dose regions may be reduced and the greater uniformity produced

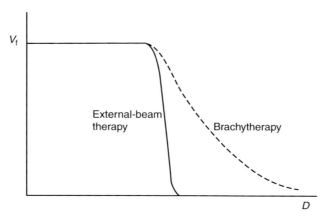

FIGURE 54.22
Representative cumulative dose volume histogram.

by optimised plans may be required. The presence of low dose regions within the target volume is undesirable as it could result in local recurrence. It is clear that dose volume analysis of a more sophisticated nature needs to be applied in considering the brachytherapy target volume.

The issues relating to dose to critical structures in brachytherapy are similar to those in external-beam therapy. In gynaecological brachytherapy, for example, the dose is limited by rectum, bladder and bowel tolerance. Indeed the widely used Manchester point A for dose prescriptions is intended to limit the dose to a normal tissue tolerance point (Paterson and Parker 1967). Most gynaecological applications use standard source loading patterns, planning being performed to determine critical organ doses usually by means of ICRU reference point doses (ICRU 1985) (Figure 54.16). However, more recent image based planning has demonstrated how unrepresentative these points can be of whole organ doses and dose volume analysis could play a useful role (Figure 54.23).

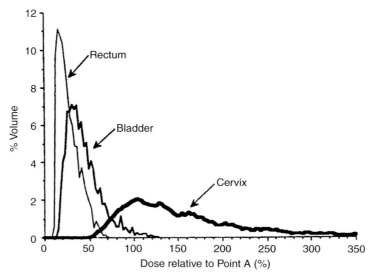

FIGURE 54.23
Representative dose volume histogram.

54.6.2   Approaches to Optimisation in Brachytherapy

Brachytherapy is an inherently conformal therapy due to the applicators being positioned within the treatment volume. This geometric conformality can be improved upon to produce a conformal dose distribution by optimisation of the source dwell positions and times.

The miniature stepping sources employed in modern afterloaders contain iridium-192 in a stainless steel capsule of about 1 mm diameter, the active source length being between 3 mm and 5 mm. The afterloader enables this single source to be stepped through the implanted catheters stopping at programmed positions for variable *dwell* times. Dwell positions have largely been predetermined by applicator placement, so dose optimisation involves varying the dwell times in each position. Optimisation can be performed to produce as uniform a dose as possible in dose points, which may be distributed throughout a volume or across a surface.

54.6.3   Surface Optimisation

Surface optimisation can be used when an outline of the target volume is available, dose points can then be placed across the surface of the volume. This technique is not directly comparable with any existing dosimetry system.

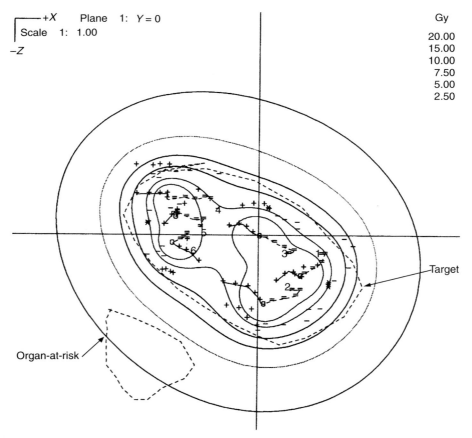

FIGURE 54.24
The inappropriate use of surface optimisation.

The type of situation where this technique has been employed is in the treatment of small solitary brain lesions. A limited number of catheters are implanted under image guidance. The surface of the treatment volume is defined by a number of points and the dwell times within the catheters optimised to give as uniform a dose as possible to these points and within the volume defined. This method must be applied with caution because depending on the distance from the catheters to the volume surface, some very high-dose regions can exist within the volume. Figure 54.24 demonstrates the inappropriate use of this type of optimisation in an implant to the rectum containing insufficient catheters to encompass the target volume. The dose distribution has been *optimised* to deliver the prescribed dose to the surface of the outlined target volume. This has resulted in massive regions of over dosage within the target volume.

54.6.4 VOLUME OPTIMISATION

If one considers a regular interstitial implant using the Paris system, the basal dose points are in the mid transverse plane only. Consider these points extended throughout the length of the treatment volume with optimisation of dwell times to deliver as uniform a dose as possible to these points. Mathematically this can be treated in the same way as the previous surface point example but the results will be amenable to comparison with the previously used Paris system.

An alternative volume optimisation technique involves using the other dwell positions within the catheters as dose points and optimising dwell times to produce as even a dose as possible within them (Edmundson 1990, 1991). This is a particularly useful technique in situations involving curved flexible catheters where the placement of suitable dose points can be difficult to perform.

Figure 54.25 compares a Paris-type distribution with an optimised distribution. The solid lines are for the optimised plan using an active source length of 50 mm, the dotted lines are the Paris distribution using an active source length of 60 mm.

One of the main advantages of the optimised distribution lies in not having any active source positions outside the target volume. Care needs to be taken in changing systems, as the distributions are also different in the transverse planes (Van der Laarse 1994). The increased weighting of outer catheters relative to inner catheters in optimised plans results in extended coverage by the reference isodose in the transverse directions. These optimised plans are essentially direct descendants of the Manchester interstitial system, which used more radium around the periphery than in the centre of the implant. They have the added advantage of being applicable in situations where the geometry of the implant deviates from the regular ideal.

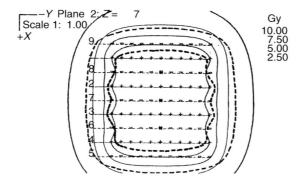

FIGURE 54.25
Comparison of Paris system based plan and an optimised plan. The Paris system distribution is represented by dotted lines and uses sources with 60 mm active length, whereas the active length required for the optimised plan is only 50 mm.

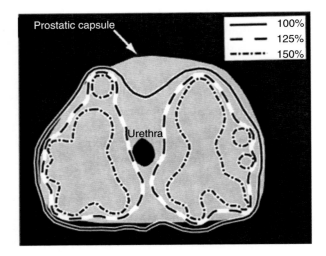

FIGURE 54.26
Typical isodose pattern in a plane through the middle of an HDR prostate implant. (From Mate, T. P., Gottesman, J. E., Hatton, J., Gribble, M., and Van Hollebeke, L., High dose-rate afterloading [192]Iridium prostate brachytherapy: Feasibility report, *Int. J. Radiat. Oncol. Biol. Phys.*, 41, 525–533, 1998.)

The ideal planning situation involves conforming the dose distribution to a target outlined on an image utilising the available optimisation techniques. HDR interstitial implants to the prostate are prime examples (Mate et al. 1998). The treatment catheters are inserted under ultrasound guidance and the planning is performed on these images or on post implant CT scans. The planning aims to deliver a higher dose to the periphery of the gland than to the urethra. Optimisation is used to obtain the best conformation. Figure 54.26 shows a typical mid gland dose distribution.

Optimisation cannot make a technically poor implant good but it can reduce the incidence of too hot or too cold regions within the treatment volume.

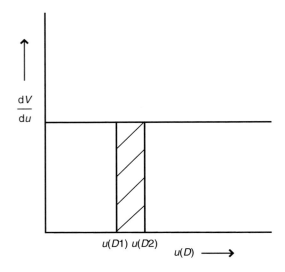

FIGURE 54.27
Natural dose volume histogram for a point source. The shaded area represents the volume receiving a dose larger than D_1 and smaller than D_2. $u(D)$ is defined in Equation 54.1.

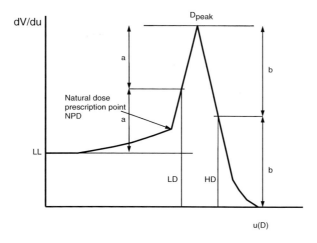

FIGURE 54.28
Idealised natural dose volume histogram for a multi-source implant. LL (lower limit) represents the value of *dV/du* at a distance from the sources large enough for them to be treated as a single point source. The natural dose prescription point is at the low dose base of the peak and represents the edge of the high dose volume. HD (high dose) and LD (low dose) represent the doses at half the maximum on the high dose and low dose sides of the peak respectively*.

54.6.5 DOSE VOLUME ANALYSIS

The rapid drop in dose rate around sources (due to the inverse-square law) renders integral dose volume histograms of little value in evaluating the dose distribution within the target volume, although they are important to show whether parts of the target will be underdosed. The presence of hot or cold regions that may be of clinical significance are masked by the inherent inhomogeneity. The use of differential histograms can reduce this problem. The process can be taken further in using *natural dose volume histograms* (Anderson 1986).

The dose distribution around a point source is a series of concentric spheres of equal dose rate varying as the inverse square of the radius. The volume receiving more than a specified dose D is then proportional to the square root of D raised to the power of 3. Anderson introduced a new variable, $u(D)$, to be used as a replacement for D on the DVH abscissa axis and defined as:

$$u(D) = D_b^{-3/2} - D^{-3/2} \qquad (54.1)$$

where D can be either dose or dose rate and D_b is the minimum dose that is of interest. The value chosen for D_b is not critical. If a dose volume histogram is created using u by plotting the volume element that receives $u(D)$ against $u(D)$, the differential histogram for a point source will be a horizontal line as shown in Figure 54.27.

Since the point source inverse square distance variation reflects the largest gradients encountered in an implant and is represented as a horizontal line in Figure 54.2, when there are multiple sources there will be areas of uniform dose and these will give rise to a peak in the natural dose volume histogram (as can be seen in Figure 54.31). Figure 54.28 is an idealised representation of this. Outside the implanted volume *dV/du* will decrease slowly until it eventually reaches a constant value equal to $\frac{4}{3}\pi D_{1cm}^{3/2}$, where $D_{1\ cm}$ is the dose, in cGy, at 1 cm if all the activity in the implant were concentrated at a point—calculated using Equation 54.1. Close to the wires the dose rate will be quite large but *dV/du* will tend towards 0.

---

* Some authors (e.g. Moerland et al. 2000) derive LD from half the peak in the same way as HD, but this gives an unnecessarily pessimistic view of the implant and is not what Anderson intended.

The histogram provides an immediate visual assessment of a treatment plan. A plan with regular applicator geometry and even dose coverage is characterised by a high narrow peak. Greater inhomogeneity is represented by a broader, shallower peak.

Anderson defined a low dose point, LD, and a high dose point, HD, determined as shown in Figure 54.28. The peak of the distribution corresponds approximately to the basal dose (see Section 54.1.2.5) of the Paris system (Langmack and Thomas 1995). The difference between LD and HD, and between HD and D_{peak}, relative to the prescribed dose provide an indication of the uniformity of the implant.

Three approaches to dose prescription using natural dose volume histograms are possible:

- The prescription dose, PD, can be derived from the histogram as the point at which the curve flattens out (NPD in Figure 54.28).
- The prescription dose, PD, can be defined as the minimum dose in the target volume and can then be compared to the value NPD.
- The prescription dose, PD, can be taken as $0.85\ D_{peak}$.

Anderson also described a uniformity index where:

$$U = \frac{V_{(PD\,to\,HD)}\big/\big(u(PD) - u(HD)\big)}{V_{(above\,PD)}\big/\big(u(PD) - u(\infty)\big)} \tag{54.2}$$

Thomadsen et al. (1994) modified this to provide a quality index which is independent of the prescription dose as follows:

$$U = \frac{V_{(LD\,to\,HD)}\big/\big(u(LD) - u(HD)\big)}{V_{(above\,LD)}\big/\big(u(LD) - u(\infty)\big)} \tag{54.3}$$

To illustrate the application of this concept in practice, a seventeen catheter prostate implant is considered. The implant is shown in Figure 54.29. Geometrical optimisation was used to achieve the dose prescribed at dose points placed around the surface of the target volume (indicated by the dotted line on the dose distribution.) The cumulative histogram

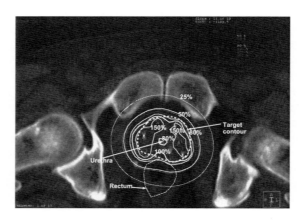

FIGURE 54.29
(See colour insert following page 590.) Seventeen-catheter prostate implant. The intended target volume is shown dotted and the thick line represents the prescription 100% isodose line. Note the sparing of the urethra in the centre of the prostate which is only given 80% of the dose.

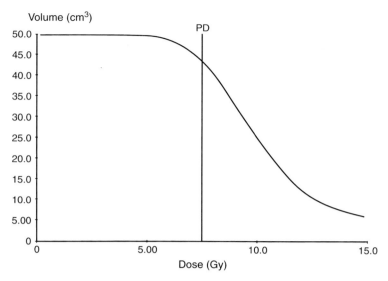

FIGURE 54.30
Cumulative histogram for the implant in Figure 54.29. PD is the prescription dose (100% on Figure 54.29).

for the target (Figure 54.30) shows whether the target was encompassed by the chosen dose rate. The marker PD in Figure 54.30 indicates the position of the prescription dose.

The natural dose volume histogram (Figure 54.31) shows the prescription dose compared to $0.85\,D_{peak}$ and the NPD at the inflection of the curve. It will be seen that the prescribed dose, PD, derived from the mean of the dose points is above the NPD but below $0.85\,D_{peak}$. This is because of the deliberate sparing of the urethra. One measure of the quality of the implant is the natural dose ratio (NDR) which is the ratio NPD/PD. If $NDR > 1$ it implies that the implant is overdosed, whereas if $NDR < 1$, as here, it implies that a lower prescription isodose could be chosen.

As an illustration of how natural dose volume histograms can demonstrate a poor quality implant, Figure 54.32 shows the histogram for the inadequate rectal implant which was shown

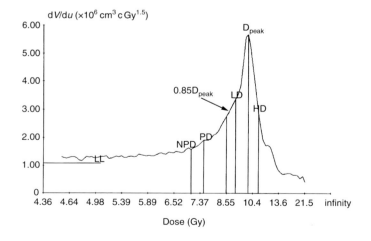

FIGURE 54.31
Natural dose volume histogram for the implant in Figure 54.29. D_{peak} is the dose corresponding to the peak of the histogram. PD is the prescription dose based on the mean of the dose points. NPD is the natural prescription dose. LL, LD and HD are as defined in Figure 54.28. Note that the prescription based on 85% of D_{peak} gives a higher dose.

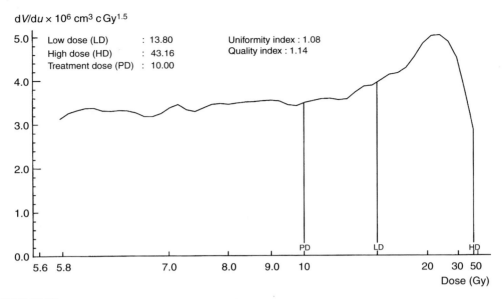

FIGURE 54.32

Natural dose volume histogram for the unsatisfactory implant of Figure 54.25. PD is the prescription dose, and LD and HD are as defined in Figure 54.28. The quality index is defined in Equation 54.3 and the uniformity index in Equation 54.2.

in Figure 54.25. The broad shallow peak demonstrates the dose inhomogeneity within the target volume. The peak dose occurs at a dose of greater than twice the prescription dose, which is indicated as PD on the histogram. The quality and uniformity indices are both low.

Geometrically conformal brachytherapy can be made dose conformal using optimisation techniques with or without image-based planning. Dose volume analysis enables detailed comparisons to be made between old and new techniques. It can also give detailed critical organ dose information rather than using doses to fixed reference points.

To use new techniques to the full extent, one needs to have a treatment volume defined. This is not always the case in brachytherapy but the increasing availability of CT and MRI should make it more common practice in the future (Nag et al. 2004; Haie-Meder et al. 2005; Pötter et al. 2006).

CHAPTER 55

RADIOBIOLOGY OF BRACHYTHERAPY

Roger G. Dale

CONTENTS

55.1 INTRODUCTION

The underlying radiobiology of brachytherapy is essentially the same as that for external-beam radiotherapy and this chapter should, therefore, be read in conjunction with Part B. For brachytherapy in particular, the all-pervasive nature of the inverse-square law means that there is a complex interplay between the physical and radiobiological aspects. This interplay influences the way in which brachytherapy treatments should be evaluated, both in absolute and comparative terms. In planning any brachytherapy procedure it is probably true to say that attention to physics should always take precedence over attention to radiobiology, because the radiobiological considerations are of little value when applied to treatments with poor source geometry or that are misplaced.

This section covers some aspects of this physics/radiobiology interaction. As the linear-quadratic (LQ) dose-effect model continues to be the standard method of evaluating radiobiological effects, it will be used here. There remain a number of improperly understood factors in radiobiology and quantitative radiobiological assessments, which although useful, should be used with a modicum of caution. They should be used to help inform the clinical decision-making process, never to override it.

55.2 α/β RATIOS

For any type of radiotherapy, one of the principal determinants of biological effect is the α/β ratio. Despite being sometimes referred to as *fractionation factors*, α/β ratios provide an indication of the sensitivity of a given tumour or organ to changes in dose-rate, as well as to changes in dose per fraction. α (in units of Gy^{-1}) and β (in units of Gy^{-2}) are the two radiosensitivity coefficients used in the LQ model and are respective measures of the initial slope and degree of downward curvature (*bendiness*) of the underlying cell-survival curve. The α/β value (in units of Gy) is generally higher for tumours (typical range 5–25 Gy) than for late-responding normal tissues (typical range 1–5 Gy). An important exception to this generalisation about the tumour values is prostate adenocarcinoma, for which the α/β ratio may be as low as around 1 Gy (Bentzen and Ritter 2005).

Tissues with lower values of α/β are said to have more sparing capacity than tissues with higher values. This means that the reduction in cell kill, brought about by a reduction in dose per fraction or dose rate, will be relatively greater in low α/β tissues. Therefore, compared to tumours, late-responding normal tissues would be preferentially spared in treatments involving small fraction sizes or low dose rates.

The principle involved is demonstrated in Figure 55.1 and Figure 55.2. It will be seen that the sparing effect resulting from lowering of dose rate exhibits itself as a *straightening out* of the cell-survival curves. It is thus analogous to the sparing effect associated with the use of multiple small-dose fractions. Excellent clinical results have often been claimed for traditional continuous-low-dose-rate (CLDR) treatments, delivered over periods of several days, and it is clear from the graphs that CLDR essentially emulates what might be expected of a hyperfractionated treatment. In both cases a relatively higher degree of radiobiological sparing is likely in the tissues with low α/β values. However, in most CLDR implants the sources are arranged within the tumour, where the dose and dose rate are maximised. Adjacent normal structures usually receive a lower dose, delivered at a lower dose rate. This is a physical characteristic of brachytherapy which may also explain the good clinical results attributed to CLDR.

55.3 REPAIR RATES (μ VALUES)

Sub-lethal DNA damage in cells may repair if allowed sufficient time. If further damage accrues before the earlier damage has repaired, then the sub-lethal damage becomes lethal.

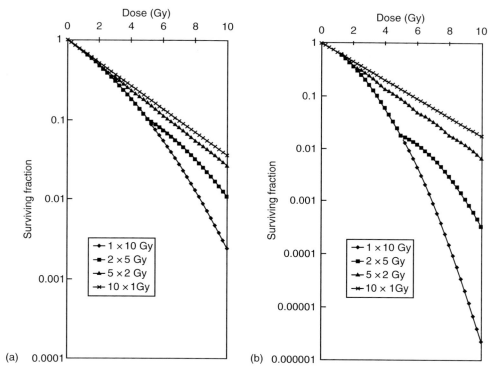

FIGURE 55.1
Effect of fractionation (a) on tumour (b) on late-responding normal tissue.

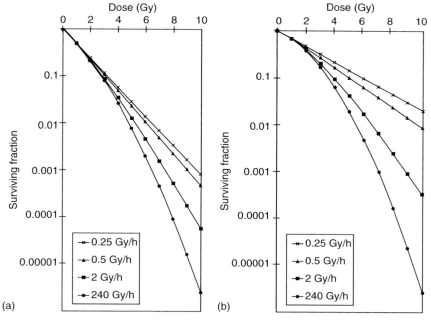

FIGURE 55.2
Effect of changing dose rate (a) on tumour (b) on normal tissue.

Although this is a simplistic explanation of the repair process, it can help in the understanding of several facets of clinical radiotherapy.

Average repair half-times for mammalian tissues are usually approximately 0.5–3 h, with increasing evidence that tumour repair half-lives are probably shorter than those for late-responding normal tissues. If the repair process is assumed to be mono-exponential, then the repair rate (μ) is related to the repair half-time ($T_{1/2}$) by:

$$\mu = \frac{\ln 2}{T_{1/2}}$$

In conventional external-beam radiotherapy, the sub-lethal damage remaining after each fraction is essentially fully repaired by the time of delivery of the next fraction. In CLDR treatments, the repair mechanism works in competition with the induction of further damage caused by the ongoing irradiation.

55.4 THE CONCEPT OF BIOLOGICAL DOSE

Biological dose is specified within the LQ model in terms of the biologically effective dose (BED). The unit of measure of BED, as for physical dose, is the Gray (Gy). In any particular irradiated tissue, BED is related to the logarithm of the surviving fraction of cells. Irrespective of the type of radiotherapy involved, BED is always calculated from the total physical dose (TD) in the following way:

Biologically Effective Dose (BED) = Total Physical Dose (TD) × Relative Effectiveness (RE)

i.e.   BED (in Gy) = TD (in Gy) × RE (dimensionless)

BED is used for inter-comparing the biological effects of various radiotherapy schedules. It is also useful in the inter-comparison of different types of radiotherapy. BED can be thought of as the notional physical dose that could be delivered if the radiation were administered in a very large number of tiny fractions, although this is not a perfect definition.

The RE factor takes account of the other physical details of the treatment (e.g. dose per fraction, dose rate, etc.), together with the radiobiological factors (α/β, μ, etc.) associated with the irradiated tissue or organ. Because the radiobiological factors are tissue-specific, it is usually the case that two adjacent tissues, each receiving the same total dose, dose per fraction, dose rate, etc., will be associated with different BEDs.

If the normal-tissue α/β value is less than that for the tumour then, provided the two tissues each receive the same dose and pattern of irradiation (i.e. dose rate and/or number of fractions), the normal-tissue BED will always exceed that for the tumour. This does not mean that the normal tissue is receiving a higher biological dose, it simply reflects what was discussed earlier, i.e. that there is greater potential to bring about preferential damage sparing in the normal tissue if the fractional dose (or dose rate) is reduced. The numerical difference between the BED and the physical dose is a measure of the potential capacity of a tissue to be spared radiation damage.

It is common to write both the BED symbol and its numerical value with a suffix indicating which value of α/β is being used in the calculation, i.e. BED_3, BED_{10}, $100~Gy_3$, $72~Gy_{10}$, etc.

55.4.1 EFFECT OF CELLULAR REPOPULATION ON BIOLOGICAL DOSE

Irrespective of the physical dose delivered, if the cells in an irradiated tissue repopulate during a course of treatment then the resultant biological dose to that tissue is less than it would be if no repopulation occurred. This is because some of the delivered dose will be *wasted* in combating the cellular repopulation that occurs during treatment, with the result that fewer cells are killed. The incorporation of the repopulation effect in BED equations is achieved through the use of a subtractive repopulation factor (RF) that takes account of the treatment time and the repopulation rate, i.e.:

$$BED = TD \times RE - RF$$

Repopulation is a significant phenomenon in many tumour types and is particularly important when treatment times are protracted. Repopulation is related to the overall treatment time (T days) via the relationship:

$$RF = kT$$

The parameter k may be thought of as the amount of biological dose which is *used up* each day in sterilising the new cells born that day. It is the hypothetical daily BED that must be delivered simply to stop the tumour growing further during treatment. For some squamous cell carcinomas k may be in excess of 0.9 Gy day^{-1}, but it is still not clear if k retains a fixed value throughout treatment or whether it is small at the beginning and increases after a lag time. For head and neck cancers the lag time may be of the order of 28 days, but there are few data available for other tumour sites (Roberts and Hendry 1993; Dale et al. 2002). If k does change during treatment, then an average value should be used in the RF, or an explicit allowance for the lag time prior to the onset of fast repopulation should be made.

55.4.2 FRACTIONATED HIGH DOSE-RATE (FHDR) BRACHYTHERAPY

For any fractionated treatment (teletherapy or brachytherapy) for which each fractional dose is given in a short period of time the relative effectiveness factor to use is:

$$RE = 1 + \frac{d}{\alpha/\beta}$$

where d is the dose per fraction. Hence, because $BED = (TD \times RE) - RF$, then:

$$BED = Nd\left[1 + \frac{d}{(\alpha/\beta)}\right] - kT \qquad (55.1)$$

where N is the number of fractions.

Equation 55.1 is valid only if the fractions are sufficiently well-spaced (usually a minimum of 6 h) to allow complete recovery of sub-lethal damage between successive fractions.

55.4.3 DOSE-RATE EFFECTS

For the delivery of any given physical dose, reducing the dose rate means that treatment time is proportionately extended. As treatment time becomes longer there is more time available for intra-cellular repair to occur. This phenomenon will reduce the biological effectiveness of the delivered dose and, for any given physical dose, BED is less than it would have been for a shorter exposure time. Hence the key characteristic of the dose-rate effect (a reduction in radiation effect as the dose rate is lowered) essentially results from more time being available for sub-lethal damage repair, rather than the reduced dose rate per se.

In addition to intra-cellular repair, the phenomena of cell-cycle progression and repopulation are also dose-rate related. The approximate time scales over which the principal effects operate are as follows:

1. Intra-cellular repair: from minutes to hours.
2. Cell cycle progression: tens or hundreds of hours.
3. Repopulation: days or weeks.

In terms of their quantitative impact, effects 1 and 3 are reasonably well understood and are probably the most important. Repopulation can occur in any treatment, but is likely to be more significant in fractionated high dose-rate (FHDR) treatments, as they will usually be of longer duration than CLDR treatments.

Cell-cycle progression can lead to an inverse dose-rate effect, whereby radiosensitivity increases, rather than decreases, as dose rate is reduced. Although the effect has been observed in vitro, its influence in clinical practice is not known with certainty. For CLDR brachytherapy, its impact is likely to be small on account of the inevitable dose-rate inhomogeneities that exist.

55.4.4 Continuous Low Dose-Rate (CLDR) Brachytherapy

For continuous brachytherapy treatments lasting up to a few days, the repopulation effect is thus usually negligible and may be omitted. Thus, $BED = TD \times RE$.

Because:

$$RE = 1 + \frac{2R}{\mu(\alpha/\beta)}$$

then:

$$BED = RT\left[1 + \frac{2R}{\mu(\alpha/\beta)}\right] \tag{55.2}$$

where R is the dose rate (Gy h^{-1}), T (h) is the treatment time and μ (h^{-1}) is the time constant for exponential repair. Equation 55.2 is a simplified form of a more complex expression, but may be used when T is more than about twelve hours (Dale 1985; Thames 1985).

Equation 55.2 assumes that the intra-cellular repair follows a mono-exponential pattern. There is evidence to suggest that several repair components may exist in some tissue types. If there are m exponential repair components, each with repair constant μ_i and fractional presence a_i such that:

$$\sum_i^m a_i = 1$$

then Equation 55.2 is better written as:

$$BED = RT\left[1 + \frac{2R}{(\alpha/\beta)} \sum_{i=1}^m \left(\frac{a_i}{\mu_i}\right)\right] \tag{55.3}$$

If the treatment is delivered via a permanent implant, the BED is given as:

$$BED = \frac{R_0}{\lambda}\left[1 + \frac{R_0}{(\mu + \lambda)(\alpha/\beta)}\right] \tag{55.4}$$

where R_0 is the initial dose rate (Gy h^{-1}) and λ is the radioactive decay constant (h^{-1}).

55.5 THE CONCEPT OF EQUIVALENCE

A particular type of treatment (Treatment A) will be associated with a particular pattern of biological effects in the tumour and critical normal tissues. If an alternative treatment (Treatment B) is given and produces exactly the same pattern of biological effects, then the two treatments can be said to be *equivalent*. Equivalence does not mean that both treatments produce the same amount of biological damage in all the irradiated structures; rather it means that Treatments A and B produce the same differential pattern of damage.

In reality, it is almost impossible to attain true equivalence between alternative treatments. It is possible to arrange for treatments A and B to produce a similar effect on the tumour (tumour-control rate unchanged) or a similar normal-tissue response (complication-rate unchanged), but rarely both. Generally two or more treatments can be designed to be equivalent only for one specific biological end-point. A more balanced assessment of the relative merits of alternative treatments can only be achieved by separately considering the radiobiological consequences for each critical structure, e.g. by using BED values to review the likely changes to the therapeutic index.

Considerations about equivalence typically arise when seeking to replace an existing treatment (CLDR, say) with a newer alternative. In some of the older literature the concept of equivalence between treatments was used fairly loosely and the potential shortcomings of the idea need to be appreciated.

55.6 INTERCOMPARISONS BETWEEN CLDR AND FHDR

The use of CLDR has been superseded in many Centres by FHDR brachytherapy. Modern high-dose-rate (HDR) afterloading systems offer the ability to treat patients in fully protected rooms, using fractions of short duration. However, practical FHDR treatment is only possible if the number of fractions is kept relatively small, which often necessitates a relatively large dose-per-fraction.

As conventional radiobiological wisdom indicates that large fractional doses are best avoided, does this mean that FHDR brachytherapy with small fraction numbers is an unsafe option?

The answer to this is that physical-dose gradients may sometimes work in the patient's favour, as the following examples demonstrate.

Example 1

A CLDR treatment involves delivering 60 Gy in 144 h. What total dose is required using six FHDR fractions to obtain a similar biological effect on the tumour? (Use generic α/β values of 10 Gy and 3 Gy, respectively for tumour and late-responding normal tissue, and $\mu = 0.5$ h^{-1} for both tissues. Neglect any repopulation effects.)

For the CLDR, dose-rate (R) is $60/144 = 0.417$ Gy h^{-1}. From Equation 55.2:

$$\text{BED}_{10} = 60\left[1 + \frac{2 \times 0.417}{0.5 \times 10}\right] = 70.0 \text{ Gy}_{10}$$

$$\text{BED}_3 = 60 \times \left[1 + \frac{2 \times 0.417}{0.5 \times 3}\right] = 93.4 \text{ Gy}_3$$

The six-fraction FHDR treatment that delivers the same tumour effect must achieve this same BED_{10} value, i.e.:

$$6d\left[1 + \frac{d}{10}\right] = 70.0 \, Gy_{10}$$

leading to $d=6.91$ Gy, i.e. a total FHDR physical dose of $6\times6.91=41.46$ Gy.
For such a treatment, the BED_3 value will be given by:

$$BED_3 = 41.46\left[1 + \frac{6.91}{3}\right] = 137.0 \, Gy_3$$

i.e. relative to the original CLDR treatment the critical late-responding normal tissues appear to be significantly overdosed (47% increase in BED_3) by this FHDR regime. This indeed suggests that attempting to treat in a small number of fractions, with correspondingly large dose per fraction, is a bad option.

Suppose, however, that it is possible to position the FHDR applicators such that a physical sparing of 0.8 is afforded to the critical normal tissues, i.e. the dose per fraction to such structures is reduced by 20%. The sparing factor of 0.8 operates on both dose per fraction and total dose, therefore the net BED_3 given is:

$$BED_3 = 41.46 \times 0.8 \times \left[1 + \frac{0.8 \times 6.91}{3}\right] = 94.3 \, Gy_3$$

This BED_3 value is only 1% higher than that associated with the original FHDR treatment. Thus, a modest degree of extra geometrical sparing (a physical effect) with FHDR has offset a potentially dangerous biological disadvantage.

Geometrical sparing of normal structures is usually more easily achieved in gynaecological than in other applications.

Example 2

A CLDR treatment involves delivering 60 Gy in 144 h. What TD is required using a permanent implant of ^{103}Pd (half-life of 17 days) to obtain a similar biological effect?

$$\text{Radioactive decay constant} \, (\lambda) = 0.693/17 = 0.04076 \, day^{-1} = 1.66 \times 10^{-3} \, h^{-1}$$

To find the required physical dose it is necessary to use Equation 55.4 to determine R_0, the initial dose rate that will give the required BED_{10} value which, from Example 1, is 70.0 Gy_{10}.

$$\frac{R_0}{1.699 \times 10^{-3}}\left[1 + \frac{R_0}{(0.5 + 0.001699) \times 10}\right] = 70.0$$

leading to $R_0=0.1162$ Gy h^{-1} and, hence, a total dose (R_0/λ) of 68.4 Gy (using the mean life as in Section 2.2.3.2.2).

For adjacent late-responding normal structures receiving this same dose (and dose rate), the associated BED_3 is:

$$BED_3 = 68.4 \times \left[1 + \frac{0.1162}{(0.5 + 0.001699) \times 3}\right] = 73.7 \, Gy_3$$

Thus, the ^{103}Pd implant can achieve a better therapeutic index (same BED_{10}, lower BED_3) than the original CLDR treatment without any physical sparing of the normal tissues. This is because, despite the requirement for a higher physical dose, the palladium delivers that dose over a longer period than the 6 days of the CLDR treatment, thus providing greater opportunity for normal-tissue sparing.

These calculations ignore the RF factor. When radionuclides of very long half-life are used, tumour repopulation can become a significant factor and may diminish the potential benefits (Dale 1989; Ling 1992; Armpilia et al. 2003).

55.7 PULSED BRACHYTHERAPY

Pulsed brachytherapy (usually wrongly called pulsed-dose-rate (PDR) brachytherapy) seeks to offer the same clinical advantages as CLDR but without the need for visitor and staff access to be restricted throughout the whole period of treatment. The intention with pulsed brachytherapy is to emulate the radiobiological effects associated with a CLDR treatment. The dose which would have been given by CLDR is delivered in the same overall time period, but as a series of short fractions (pulses) with radiation-free gaps between each pulse.

As the number of pulses and dose-per-pulse are inversely related, a treatment using a very large number of closely-spaced pulses becomes increasingly identical to a CLDR treatment as the number of pulses is increased. However, to realise one of the main practical benefits of pulsed brachytherapy treatment (freedom for the patient to receive visitors or move around between fractions), it is desirable to keep the pulse durations short and the inter-pulse times long. The essential radiobiological question is this: What is the maximum dose per pulse and maximum inter-pulse interval that can be allowed before the therapeutic ratio (relative to CLDR) becomes significantly compromised?

The question is not easily answered, although guidelines are emerging (Fowler and Mount 1992; Millar et al. 1996). The radiobiology (and the associated mathematics) of pulsed brachytherapy is complex and, more so than in the case of FHDR or CLDR, the reliability of the predictions is more dependent on the radiobiological assumptions that are made. This is particularly true in relation to the mechanism of sub-lethal damage repair; because the pulses are closely spaced, the repair between fractions is incomplete and this leads to a cascading amount of additional cell lethality as treatment progresses.

The overall indications are that, provided there is a modification to the total dose, intervals as long as 3 h may be adequate if the pulsed brachytherapy treatment is being used merely as one component of a radical treatment. If the entire treatment is to be given by pulsed brachytherapy then gaps of no longer than 1 h may be required (Visser et al. 1996).

55.8 DOSE GRADIENTS AND INTEGRATED BIOLOGICAL RESPONSE

In most brachytherapy applications the dose prescription point will be part of an isodose surface that encloses volumes of tissue where the doses will rise rapidly in the close vicinity of the sources. Clearly this means that the physical dose and its radiobiological counterpart, the BED, are only correctly defined at the prescription point. Just a few millimetres on either side of this point the dose-rate (and hence the delivered dose) may be very different.

Does this mean that the BED concept is meaningless when applied to brachytherapy? The answer is that although the dose gradients have a significant effect, the net increment in BED caused by the dose gradients can be quantified and is usually not more than 15–30% in most clinical cases. Therefore, calculated reference point BEDs can be corrected by the use of multiplying factors which take account of the specific treatment conditions. The issues involved are discussed in depth elsewhere (Dale et al. 1997), but Table 55.1 and Table 55.2 summarise the multiplying factors which apply in a range of circumstances. These multiplying factors are

TABLE 55.1

BED Multiplying Factors for FHDR Brachytherapy

Dose per fraction (Gy)	$N=1$	$N=2$	$N=4$	$N=6$	$N=8$	$N=10$
2	1.72	1.54	1.39	1.32	1.27	1.24
4	1.55	1.39	1.27	1.21	1.18	1.15
6	1.45	1.31	1.21	1.16	1.13	1.11
8	1.38	1.26	1.17	1.13	1.10	1.09
10	1.33	1.22	1.14	1.10	1.08	1.07

It is assumed that $\alpha/\beta=20$ Gy, $\alpha=0.35$ Gy^{-1}. N is the number of fractions. To obtain the BED, the BED calculated at the reference point is multiplied by the factor in the Table.

derived from a purely analytical approach based on ideal single source geometries, but direct assessment of integrated BED around clinical gynaecological applications indicates that the theoretical values are very representative of what happens in practice.

The integrated BED (i.e. the BED at the prescription point multiplied by the factor from the relevant table) is the biological dose associated with the uniform physical dose that would have to be applied to the same volume of tissue to yield the same surviving fraction of cells as does the brachytherapy. The concept of integrated biological response is particularly useful in cases where brachytherapy is combined with teletherapy, as it allows a more realistic summation of the respective biological doses delivered by the two components.

The multiplying factors are a function of dose per fraction and number of fractions (N). The BED at the dose reference point is calculated in the usual way, using the most appropriate radiobiological factors. The net BED within the volume enclosed by the reference iso-effect surface is found by multiplying the calculated BED by the value from the appropriate table. Unlike the BED, the multiplying factors are relatively insensitive to changes in the radiobiological parameters. A more detailed review of the relationship between the multiplying factors and the BEDs from which they are derived has been undertaken by Armpilia et al. (2006). The article considers CLDR, FHDR and pulsed brachytherapy treatments.

55.9 CONSEQUENCES OF DYNAMIC CHANGES DURING THE COURSE OF A BRACHYTHERAPY TREATMENT

During the course of a brachytherapy treatment the treated volume may shrink (due to cell depletion) or swell (due to oedema). If the treatment involves placing radioactive sources or

TABLE 55.2

BED Multiplying Factors for CLDR Brachytherapy

Dose-rate (Gy h^{-1})	$T=25$ h	$T=50$ h	$T=75$ h	$T=100$ h	$T=125$ h	$T=150$ h	$T=175$ h	$T=200$ h
0.2	1.48	1.35	1.28	1.25	1.21	1.19	1.17	1.16
0.4	1.35	1.24	1.19	1.16	1.14	1.12	1.11	1.10
0.6	1.28	1.19	1.15	1.12	1.10	1.09	1.08	1.08
0.8	1.24	1.16	1.12	1.10	1.09	1.07	1.07	1.06
1.0	1.21	1.14	1.10	1.08	1.07	1.06	1.06	1.05

It is assumed that $\alpha/\beta=20$ Gy, $\alpha=0.35$ Gy^{-1}, $\mu=1$ h^{-1}. T is the treatment time.

afterloading applicators directly within the tumour, then the dose may be significantly enhanced or reduced because of these respective effects (Dale and Jones 1994).

In the case of pseudo-exponential shrinkage, if the linear shrinkage rate is z (day^{-1}), then Equation 55.1 and Equation 55.2 are respectively modified to Equation 55.5 and Equation 55.6:

$$\text{BED} = N\mathrm{d}X\left[1 + \frac{\mathrm{d}X}{(\alpha/\beta)}\right] \qquad (55.5)$$

$$\text{BED} = RTY\left[1 + \frac{2RY}{\mu(\alpha/\beta)}\right] \qquad (55.6)$$

where $X = 1 + (N-1)zt$, $Y = 1 + zT$, and t is the average time interval between successive fractions.

Some imaging studies have demonstrated that, where brachytherapy is preceded by a course of chemotherapy, tumour shrinkage is indeed exponential with time (Tan et al. 1996).

In cases where brachytherapy causes short-term swelling rather than shrinkage, the same equations made be used but with the positive z replaced by an appropriate negative coefficient.

55.10 THE RBE EFFECT IN BRACHYTHERAPY

Although relative biological effectiveness (RBE) effects are usually ignored in brachytherapy assessments, they may be of major importance. Some brachytherapy radio-nuclides emit radiations that are more efficient in producing biological damage than are the conventional x-rays used in external-beam therapy (Scalliet and Wambersie 1987; Ling et al. 1995; Wuu and Zaider 1998). A smaller physical dose is thus required with the radionuclide to produce a particular biological effect, the increased efficiency being measured in terms of RBE. For example, ^{125}I, with an RBE of around 1.45, requires in principle only to deliver (1/1.45 = 0.69 Gy) to produce the same effect as 1 Gy of conventional x-rays.

However, allowance for RBE effects is made potentially problematic on account of the fact that they vary with dose (or dose rate), the RBE values being highest at very low dose (or dose rate). Fortunately, the incorporation of the RBE phenomenon into LQ equations is easily accomplished by changing the format of the expression for the relative effectiveness in all BED equations from $[1 + \ldots]$ to $[\text{RBE}_{\text{max}} + \ldots]$, where RBE_{max} is the maximum RBE occurring at zero dose. For example, Equation 55.4 is changed to:

$$\text{BED} = \frac{R_0}{\lambda}\left[\text{RBE}_{\text{max}} + \frac{R_0}{(\mu + \lambda)(\alpha/\beta)}\right]$$

This simple generic change to BED equations allows the variation of RBE with dose (or dose rate) to be properly allowed for and also results in BED values that may be used for direct inter-comparison with those associated with treatments using conventional x-rays. The incorporation of RBE into radiobiological equations is described by Dale and Jones (1999), and the practical implications relevant to permanent implant brachytherapy have been examined by Antipas et al. (2001) and Armpilia et al. (2003).

PART J: BRACHYTHERAPY
REFERENCES

AAPM (American Association of Physicists in Medicine), Report No. 21. *Specification of brachytherapy source strength*. Report of Task Group 32, American Institute of Physics, New York: 1987.

Anagnostopoulos, G., Baltas, D., Karaiskos, P., Pantelis, E., Papagiannis, P., and Sakelliou, L., An analytical dosimetry model as a step towards accounting for inhomogeneities and bounded geometries in ^{192}Ir brachytherapy treatment planning, *Phys. Med. Biol.*, 48, 1625–1647, 2003.

Anderson, L. L., A natural volume dose histogram in brachytherapy, *Med. Phys.*, 13(6), 898–903, 1986.

Antipas, V., Dale, R. G., and Coles, I. P., A theoretical investigation into the role of tumour radiosensitivity, clonogen repopulation, tumour shrinkage and radionuclide RBE in permanent brachytherapy implants of ^{125}I and ^{103}Pd, *Phys. Med. Biol.*, 46, 2557–2569, 2001.

Armpilia, C. I., Dale, R. G., Coles, I. P., Jones, B., and Antipas, V., The determination of radiobiologically optimized half-lives for radionuclides used in permanent brachytherapy implants, *Int. J. Radiat. Oncol. Biol. Phys.*, 55, 378–385, 2003.

Armpilia, C., Dale, R. G., Sandilos, P., and Vlachos, L., Radiobiological modelling of dose-gradient effects in low dose rate, high dose rate and pulsed brachytherapy, *Phys. Med. Biol.*, 51, 4399–4411, 2006.

Ash, D., Flynn, A., Battermann, J., de Reijke, T., Lavagnini, P. et al., ESTRO/EAU/EORTC recommendations on permanent seed implantation for localized prostate cancer, *Radiother. Oncol.*, 57(3), 315–321, 2000.

Ballester, F., Hernandez, C., Pérez-Calatayud, J., and Lliso, F., Monte Carlo calculation of dose rate distributions around ^{192}Ir wires, *Med. Phys.*, 24, 1221–1228, 1997.

BCRU (British Committee on Radiation Units and Measurements), Specification of brachytherapy sources. Memorandum from the British Committee on Radiation Units and Measurements, *Br. J. Radiol.*, 57, 941–942, 1984.

Bentzen, S. M. and Ritter, M. A., The alpha/beta ratio for prostate cancer: What is it, really?, *Radiother. Oncol.*, 76, 1–3, 2005.

Bentzen, S. M., Heeren, G., Cottier, B., Slotman, B., Glimelius, B., Lievens, Y., and van den Bogaert, W., Towards evidence-based guidelines for radiotherapy infrastructure and staffing needs in Europe: the ESTRO QUARTS project, *Radiother. Oncol.*, 75, 355–365, 2005.

Bice, W. S., Prestidge, B. R., Prete, J. J., and Dubois, D. F., Clinical impact of implementing the recommendations of AAPM Task Group 43 on permanent prostate brachytherapy using ^{125}I, *Int. J. Radiat. Oncol. Biol. Phys.*, 40, 1237–1241, 1998.

Bielajew, A. F., Correction factors for thick-walled ionisation chambers in point-source photon beams, *Phys. Med. Biol.*, 35, 501–516, 1990.

BIR (British Institute of Radiology), *Recommendations for brachytherapy dosimetry*. Report of a Joint Working Party of the BIR and the IPSM, BIR, London, 1993.

Blasko, J. C., Mate, T., Sylvester, J. E., Grimm, P. D., and Cavanagh, W., Brachytherapy for carcinoma of the prostate: Techniques, patient selection, and clinical outcomes, *Semin. Radiat. Oncol.*, 12, 81–94, 2002.

Brenner, D. J. and Hall, E. J., Conditions for the equivalence of continuous low dose rate brachytherapy, *Int. J. Radiol. Onc. Biol. Phys.*, 20, 181–190, 1991.

BSI (British Standard Institution), *BS 5288:1976 Specification. Sealed Radioactive Sources*, BSI, London, 1976.

Burns, G. S. and Raeside, D. E., Monte Carlo simulation of the dose distribution around ^{125}I seeds, *Med. Phys.*, 14, 420–424, 1987.

Burns, G. S. and Raeside, D. E., Two-dimensional dose distribution around a commercial ^{125}I seed, *Med. Phys.*, 15, 56–60, 1988.

Butler, W. M., Dorsey, A. T., Nelson, K. R., and Merrick, G. S., Quality assurance calibration of ^{125}I rapid strand in a sterile environment, *Int. J. Radiat. Oncol. Biol. Phys.*, 41, 217–222, 1998.

Casebow, M. P., Paris technique implant dosimetry tables for 0.3 and 0.6 mm diameter ^{192}Ir wire, *Br. J. Radiol.*, 58, 549–553, 1985.

CFMRI (Comité Français de Mesures des Rayonnements Ionisants), Recommandations pour la détermination des doses absorbées en curiethérapie, Rapport n°1, Bureau National de Métrologie, Ed., Chiron, Paris, 1983.

Chassagne, D. and Pierquin, B., Plesiocurietherapy of vaginal cancers by plastic mold with iridium-192 (non-radioactive preparation) [La plesio curie therapie du cancer du vagin par moulage plastique avec iridium-192—preparation non radioactive], *J. Radiol. Electrol. Med. Nucl. [J. de Radiologie et d'Electrologie]*, 47, 89–93, 1996.

Chassagne, D. and Horiot, J. C., Proposals for common definitions of reference points in gynecological brachytherapy, *J. Radiol. Electrol. Med. Nucl. [J De Radiologie et d'Electrologie]*, 58, 371–373, 1977.

Chen, Z. and Nath, R., Dose rate constant and energy spectrum of interstitial brachytherapy sources, *Med. Phys.*, 28, 86–96, 2001.

Chiu-Tsao, S. T. and Anderson, L. L., Thermoluminescent dosimetry for ^{103}Pd seeds (model 200) in solid water phantom, *Med. Phys.*, 18, 449–452, 1991.

Chiu-Tsao, S. T., Anderson, L. L., O'Brien, K., and Sana, R., Dose rate determination for ^{125}I seeds, *Med. Phys.*, 17, 815–825, 1990.

Cumberlin, R. L. and Coleman, C. N., New directions in brachytherapy, *Int. J. Radiat. Oncol. Biol. Phys.*, 53, 6–11, 2002.

Dale, R. G., A Monte Carlo derivation of parameters for use in the tissue dosimetry of medium and low energy nuclides, *Br. J. Radiol.*, 55, 748–757, 1982.

Dale, R. G., Some theoretical derivations relating to the tissue dosimetry of brachytherapy nuclides, with particular reference to iodine-125, *Med. Phys.*, 10, 176–183, 1983.

Dale, R. G., The application of the linear-quadratic dose-effect equation to fractionated and protracted radiotherapy, *Br. J. Radiol.*, 58, 515–528, 1985.

Dale, R. G., Revisions to radial dose function data for ^{125}I and ^{131}Cs, *Med. Phys.*, 13, 963–964, 1986.

Dale, R. G., Radiobiological assessment of permanent implants using tumour repopulation factors in the linear-quadratic model, *Br. J. Radiol.*, 62, 241–244, 1989.

Dale, R. G. and Jones, B., The effect of tumour shrinkage on biologically effective dose, and possible implications for fractionated high dose rate brachytherapy, *Radiother. Oncol.*, 33, 125–132, 1994.

Dale, R. G. and Jones, B., The assessment of RBE effects using the concept of biologically effective dose, *Int. J. Radiat. Oncol. Biol. Phys.*, 43, 639–645, 1999.

Dale, R. G., Coles, I. P., Deehan, C., and O'Donoghue, J. A., Calculation of integrated biological response in brachytherapy, *Int. J. Radiat. Oncol. Biol. Phys.*, 38, 633–642, 1997.

Dale, R. G., Hendry, J. H., Jones, B., Robertson, A. G., Deehan, C., and Sinclair, J. A., Practical methods for compensating for missed treatment days in radiotherapy, with particular reference to head and neck schedules, *Clin. Oncol.*, 14, 382–393, 2002.

Davelaar, J., Schaling, D. F., Hennen, L. A., and Broerse, J. J., Dosimetry of ruthenium-106 eye applicators, *Med. Phys.*, 19, 691–694, 1992.

Delouche, G. and Gest, J., Treatment of uterine cancer by Curietron, *Nouv. Presse Med.*, 3, 597–600, 1974.

Dolan, J., Li Z., and Williamson, J. F., Monte Carlo and experimental dosimetry of an ^{125}I brachytherapy seed, *Med. Phys.*, 33, 4675–4684, 2006.

Dutreix, A., Marinello, G., and Wambersie, A., *Dosimetrie en Curietherapie*, Masson, Paris, 1982.

Edmundson, G. K., Geometry based optimisation for stepping source implants, in *Brachytherapy HDR and LDR*, Martinez, A. A., Orton, C. G., and Mould, R. F., Eds., Nucletron, Vienendaal, pp. 184–192, 1990.

Edmundson, G. K., Geometry based optimization for stepping source implants, *Activity*, 5, 22–26, 1991.

ESTRO (European Society for Radiotherapy and Oncology), *A practical guide to quality control of brachytherapy equipment*, Physics for clinical radiotherapy ESTRO booklet No 8, Venselaar, J. and Pérez-Calatayud, J., Eds., ESTRO, Brussels, 2004.

Fairchild, R. G., Kalef-Ezra, J., Packer, S., Wielopolski, L., Laster, B. H. et al., Samarium-145: A new brachytherapy source, *Phys. Med. Biol.*, 32, 847–858, 1987.

Ferreira, I. H., de Almeida, C. E., Marre, D., Marechal, M. H., and Bridier, A., Monte Carlo calculations of the ionization chamber wall correction factors for ^{192}Ir and ^{60}Co gamma rays and 250 kV x-rays for use in calibration of ^{192}Ir HDR brachytherapy sources, *Phys. Med. Biol.*, 44, 1897–1904, 1999.

Fletcher, G. H., Cervical radium applicators with screening in the direction of bladder and rectum, *Radiology*, 60, 77–84, 1953.

Fletcher, G. H., *Textbook of Radiotherapy*, 3rd ed., Lea and Febiger, Philadelphia, 1980.

Fowler, J. and Mount, M., Pulsed brachytherapy: The conditions for no significant loss of therapeutic ratio compared with traditional low dose rate brachytherapy, *Int. J. Radiat. Oncol. Biol. Phys.*, 23, 661–669, 1992.

Gao, L., Guo-Zhen, X., Jie-hua, Y., Su-yan, L., Guang-li, X., et al., HDR brachytherapy for residual or recurrent nasopharyngeal carcinoma: preliminary experience, *Activity*, 4(4), 80–83, 1990.

Gaspar, L. E., Nag, S., Herskovic, A., Mantravadi, R., Speiser, B., American Brachytherapy Society (ABS) consensus guidelines for brachytherapy of esophageal cancer, *Int. J. Radiat. Oncol. Biol. Phys.*, 38, 127–132, 1997.

Gerbaulet, A., Pötter, R., Mazeron, J.-J., Meertens, H., and Van Limbergen, E. *The GEC ESTRO Handbook of Brachytherapy*. ISBN 90-804532-6, ESTRO (www.estro.be), Brussels, 2002.

Gillin, M. T., Kline, R. W., Wilson, J. F., and Cox, J. D., Single and double plane implants. A comparison of the Manchester system with the Paris System, *Int. J. Rad. Oncol. Biol. Phys.*, 10, 921–925, 1984.

Godden, T. J., *Physical Aspects of Brachytherapy*, Medical Physics Handbook 19, Adam Hilger, Bristol and Philadelphia, 1988.

Goetsch, S. J., Attix, F. H., Pearson, D. W., and Thomadsen, B. R., Calibration of ^{192}Ir high-dose-rate afterloading systems, *Med. Phys.*, 18, 462–467, 1991.

Goetsch, S. J., Attix, F. H., DeWerd, L. A., and Thomadsen, B. R., A new re-entrant ionization chamber for the calibration of iridium-192 high dose rate sources, *Int. J. Radiat. Oncol. Biol. Phys.*, 24, 167–170, 1992.

Goodwin, P. N., Quimby, E. H., and Morgan, R. H., Dosage with radioactive materials, in *Physical Foundation of Radiology*, 4th ed., Harper & Row, New York, pp. 241–284, 1970.

Guedea, F., Ellison, T., Venselaar, J., Borras, J. M., Hoskin, P. et al., Overview of brachytherapy resources in Europe: A survey of patterns of care study for brachytherapy in Europe, *Radiother. Oncol.*, 82, 50–54, 2007.

Haie-Meder, C., Pötter, R., Van Limbergen, E., Briot, E., De Brabandere, M. et al., Recommendations from Gynae-cological (GYN) GEC ESTRO working group (I): Concepts and terms in 3D image based 3D treatment planning in cervix cancer brachytherapy with emphasis on MRI assessment of GTV and CTV, *Radiother. Oncol.*, 74, 235–245, 2005.

Hall, E. J., Oliver, R., and Shepstone, B. J., Routine dosimetry with tantalum 182 and iridium 192 wires, *Acta Radiol. Ther. Phys. Biol.*, 4, 155–160, 1966.

Harms, W., Krempien, R., Hensley, F. W., Berns, C., Fritz, P., and Wannenmacher, M., 5-year results of pulsed dose rate brachytherapy applied as a boost after breast-conserving therapy in patients at high risk for local recurrence from breast cancer, *Strahlenther. Onkol.*, 178, 607–614, 2002.

Henschke, U. K., Hilaris, B. S., and Mabian, G. D., Afterloading in interstitial and intracavitary radiation therapy, *Am. J. Roent. Rad. Ther. Nucl. Med.*, 90, 386–395, 1963.

Heyman, J., The technique in the treatment of cancer uteri at Radiumhemmet, *Acta. Radiol.*, 10, 49, 1929.

Heyman, J., The so-called Stockholm method and the results of treatment of uterine cancer at the Radiumhemmet, *Acta Radiol.*, 16, 129–147, 1935.

Hilaris, B. S., Brachytherapy in cancer of the prostate: An historical perspective, *Semin. Surg. Oncol.*, 13, 399–405, 1997.

Horwitz, H., Kereiakes, J. G., Baker, G. K., Cluxton, S. E., and Barett, C. M., An afterloading system utilising caesium-137 for the treatment of carcinoma of the cervix, *Am. J. Roentg. Rad. Therapy and Nuc. Med.*, 91, 176–191, 1964.

Hoskin, P. J. and Bownes, P., Innovative Technologies in Radiation Therapy: Brachytherapy, *Semin. Radiat. Oncol.*, 16, 209–217, 2006.

IAEA (International Atomic Energy Agency), Technical Document No. 1079. *Calibration of brachytherapy sources. Guidelines on standardized procedures for the calibration of brachytherapy sources at Secondary Standard Dosimetry Laboratories (SSDLs) and hospitals*, IAEA, Vienna, 1999.

IAEA (International Atomic Energy Agency), Safety Report No. 17: *Lessons learned from accidental exposures in Radio-therapy*, IAEA, Vienna, 2000.

IAEA (International Atomic Energy Agency), Technical Document No. 1274. *Calibration of photon and beta ray sources used in brachytherapy. Guidelines on standardized procedures at Secondary Standards Dosimetry Laboratories (SSDLs) and hospitals* (Supersedes IAEA 1999), IAEA, Vienna, 2002.

IAEA (International Atomic Energy Agency), *Commissioning and quality assurance of computerized planning systems for radiation treatment of cancer*, IAEA Technical Report Series 430, IAEA, Vienna, 2004.

IEC (International Electrotechnical Commission), Medical electrical equipment, Section 2.17 *Specification for remote-controlled automatically driven gamma-ray afterloading equipment*. IEC Publication 601-2-17 (British standard BS EN 60601-2-17:1996) (New draft standard IEC 62C/343/CDV will replace it), IEC, Geneva, 1989.

ICRU (International Commission on Radiation Units and Measurements), *Dose specification for reporting external beam therapy*. ICRU Report 29, ICRU, Bethesda, MD, 1978.

ICRU (International Commission on Radiation Units and Measurements), *Dose and volume specification for reporting intracavitary therapy in gynaecology*. ICRU Report 38, ICRU, Bethesda, MD, 1985.

ICRU (International Commission on Radiation Units and Measurements), *Prescribing, recording and reporting photon beam therapy*, ICRU Report 50, ICRU, Bethesda, MD, 1993.

ICRU (International Commission on Radiation Units and Measurements), *Dose and volume specification for reporting interstitial therapy*. ICRU Report 58, ICRU, Washington, DC, 1997.

Jones, C. H. and Dermentzoglou, F., Practical aspects of ^{90}Sr ophthalmic applicator dosimetry, *Br. J. Radiol.*, 44, 203–210, 1971.

Jones, E., Mallard, J. R., and Elmanharawy, M. S., The experimental determination of the dose distribution around yttrium-90 sources suitable for pituitary implantation, *Phys. Med. Biol.*, 8, 59–73, 1963.

Jones, J. C., Milan, S., and Lillicrap, S. C., The planning of treatment of gynaecological cancer with combined intracavitary and external beam irradiation, *Br. J. Radiol.*, 45, 684–691, 1972.

Joslin, C. A. F., Smith, C. W., and Mallik, A., The treatment of cervix cancer using high activity cobalt 60 sources, *Br. J. Radiol.*, 45, 257–270, 1972.

Keisch, M., Vicini, F., Kuske, R. R., Hebert, M., White, J. et al., Initial clinical experience with the MammoSite breast brachytherapy applicator in women with early-stage breast cancer treated with breast-conserving therapy, *Int. J. Radiat. Oncol. Biol. Phys.*, 55, 289–293, 2003.

Kline, R. W., Gillin, M. T., Grimm, D. F., and Niroomand-Rad, A., Computer dosimetry of ^{192}Ir wire, *Med. Phys.*, 12, 634–638, 1985.

Kollaard, R. P., Dries, W. J. F., van Kleffens, H. J., Aalbers, T. H. L., van der Marel, H. et al., Recommendations on detectors and quality control procedures for brachytherapy beta sources, *Radiother. Oncol.*, 78, 223–229, 2006.

Kondo, S. and Randolph, M. L., Effect of finite size of ionization chambers on measurement of small photon sources, *Rad. Res.*, 13, 37–60, 1960.

Kovács, G., Pötter, R., Loch, T., Hammer, J., Kolkman-Deurloo, I. K. et al., GEC/ESTRO-EAU recommendations on temporary brachytherapy using stepping sources for localised prostate cancer, *Radiother. Oncol.*, 74, 137–148, 2005.

Krishnaswamy, V., Calculated depth dose tablets for californium-252 sources in tissue, *Phys. Med. Biol.*, 17, 56–63, 1972.

Kubo, H. D., Coursey, B. M., Hanson, W. F., Kline, R. W., Seltzer, S. M. et al., Report of the ad hoc committee of the AAPM radiation therapy committee on ^{125}I sealed source dosimetry, *Int. J. Radiat. Oncol. Biol. Phys.*, 40, 697–702, 1998.

Kwan, D. K., Kagan, A. R., Olch, A. J., Chan, P. Y. M., Hinte, B. L., and Wollin, M., Single and double plane iridium 192 interstitial implants, *Med. Phys.*, 10, 456–461, 1983.

Langmack, K. A. and Thomas, S. J., The application of dose-volume histograms to the Paris and Manchester systems of brachytherapy dosimetry, *Br. J. Radiol.*, 68, 42–48, 1995.

Ling, C. C., Permanent implants using Au-198, Pd-103 and I-125: Radiobiological considerations based on the linear quadratic model, *Int. J. Radiat. Oncol. Biol. Phys.*, 23, 81–87, 1992.

Ling, C. C., Li, W. X., and Anderson, L. L., The relative biological effectiveness of I-125 and Pd-103, *Int. J. Radiat. Oncol. Biol. Phys.*, 32, 373–378, 1995.

Liversage, W. E., Martin-Smith, P., and Ramsey, N. W., The treatment of uterine carcinoma using the Cathetron. Part II. Physical measurements, *Br. J. Radiol.*, 40, 887–894, 1967.

Lliso, F., Pérez-Calatayud, J., Carmona, V., Ballester, F., Lluch, J. L., Serrano, M. A., Limami, Y., and Casal, E., Fitted dosimetric parameters of high dose-rate ^{192}Ir sources according to the AAPM TG43 formalism, *Med. Phys.*, 28, 654–660, 2001.

Lliso, F., Pérez-Calatayud, J., Carmona, V., Ballester, F., Puchades, V., and Granero, D., Technical note: Fitted dosimetric parameters of high dose-rate ^{192}Ir sources according to the AAPM TG43 formalism, *Med. Phys.*, 30, 651–654, 2003.

Mangold, C. A., Rijnders, A., Georg, D., Van Limbergen, E., Potter, R., and Huyskens, D., Quality control in interstitial brachytherapy of the breast using pulsed dose rate: Treatment planning and dose delivery with an Ir-192 afterloading system, *Radiother. Oncol.*, 58, 43–51, 2001.

Markman, J., Williamson, J. F., Dempsey, J. F., and Low, D. A., On the validity of the superposition principle in dose calculations for intracavitary implants with shielded vaginal colpostats, *Med. Phys.*, 28, 147–155, 2001.

Mason, D. L., Battista, J. J., Barnett, R. B., and Porter, A. T., Ytterbium-169: Calculated physical properties of a new radiation source for brachytherapy, *Med. Phys.*, 19, 695–703, 1992.

Mate, T. P., Gottesman, J. E., Hatton, J., Gribble, M., and Van Hollebeke, L., High dose-rate afterloading ^{192}Iridium prostate brachytherapy: Feasibility report, *Int. J. Radiat. Oncol. Biol. Phys.*, 41, 525–533, 1998.

Maurer, U., Wiegel, T., Hinkelbein, W., and Eble, M. J., Interstitial brachytherapy with permanent seed implants in early prostate cancer, *Front Radiat. Ther. Oncol.*, 36, 166–170, 2002.

Mayles, W. P., Mayles, H. M., and Turner, P. C., Physical aspects of interstitial therapy using flexible iridium-192 wire, *Br. J. Radiol.*, 58, 529–535, 1985.

Mazeron, J. J., Brachytherapy: a new era, *Radiother. Oncol.*, 74, 223–225, 2005.

Meigooni, A. S. and Nath, R., Tissue inhomogeneity correction for brachytherapy sources in a heterogeneous phantom with cylindrical symmetry, *Med. Phys.*, 19, 401–407, 1992.

Meisberger, L. L., Keller, R. J., and Shalek, R. J., The effective attenuation in water of the gamma rays of gold 198, iridium 192, caesium 137, radium 226 and cobalt 60, *Radiology*, 90, 953–957, 1968.

Millar, W. T., Hendry, J. H., and Canney, P. A., The influence of the number of fractions and bi-exponential repair kinetics on biological equivalence in pulsed brachytherapy, *Br. J. Radiol.*, 69, 457–468, 1996.

Moerland, M. A., van der Laarse, R., Luthmann, R. W., Wijrdeman, H. K., and Battermann, J. J., The combined use of the natural and the cumulative dose-volume histograms in planning and evaluation of permanent prostatic seed implants, *Radiother. Oncol.*, 57, 279–284, 2000.

Nag, S., Beyer, D., Friedland, J., Grimm, P., and Nath, R., American Brachytherapy Society (ABS) recommendations for transperineal permanent brachytherapy of prostate cancer, *Int. J. Radiat. Oncol. Biol. Phys.*, 44, 789–799, 1999.

Nag, S., Bice, W., DeWyngaert, K., Prestidge, B., Stock, R., and Yu, Y., The American Brachytherapy Society recommendations for permanent prostate brachytherapy postimplant dosimetric analysis, *Int. J. Radiat. Oncol. Biol. Phys.*, 46, 221–230, 2000a.

Nag, S., Erickson, B., Thomadsen, B., Orton, C., Demanes, J. D., and Petereit, D., The American Brachytherapy Society recommendations for high-dose-rate brachytherapy for carcinoma of the cervix, *Int. J. Radiat. Oncol. Biol. Phys.*, 48, 201–211, 2000b.

Nag, S., Cano, E. R., Demanes, D. J., Puthawala, A. A., and Vikram, B., The American Brachytherapy Society recommendations for high-dose-rate brachytherapy for head-and-neck carcinoma, *Int. J. Radiat. Oncol. Biol. Phys.*, 50, 1190–1206, 2001.

Nag, S., Cardenes, H., Chang, S., Das, I. J., Erickson, B. et al., Proposed guidelines for image-based intracavitary brachytherapy for cervical carcinoma: Report from Image-Guided Brachytherapy Working Group, *Int. J. Radiat. Oncol. Biol. Phys.*, 60, 1160–1172, 2004.

Nath, R., Gray, L., and Park, C. H., Dose distributions around cylindrical 241Am sources for a clinical intracavitary applicator, *Med. Phys.*, 14, 809–817, 1987.

Nath, R., Meigooni, A. S., and Melillo, A., Some treatment planning considerations for ^{103}Pd and ^{125}I permanent interstitial implants, *Int. J. Radiat. Oncol. Biol. Phys.*, 22, 1131–1138, 1992.

Nath, R., Meigooni, A. S., Muench, P., and Melillo, A., Anisotropy functions for ^{103}Pd, ^{125}I, and ^{192}Ir interstitial brachytherapy sources, *Med. Phys.*, 20, 1465–1473, 1993.

Nath, R., Anderson, L. L., Luxton, G., Weaver, K. A., Williamson, J. F., and Meigooni, A. S., Dosimetry of interstitial brachytherapy sources: Recommendations of the AAPM Radiation Therapy Committee Task Group No. 43. American Association of Physicists in Medicine, *Med. Phys.*, 22, 209–234, 1995.

Nath, R., Amols, H., Coffey, C., Duggan, D., Jani, S. et al., Intravascular brachytherapy physics: Report of the AAPM Radiation Therapy Committee Task Group No. 60. American Association of Physicists in Medicine, *Med. Phys.*, 26, 119–152, 1999.

NCS (Netherlands Commission on Radiation Dosimetry), *Recommendations for dosimetry and quality control of radioactive sources used in brachytherapy*, NCS Report 4, The Netherlands Measurement Institute, Delft: 1991.

O'Connell, D., Joslin, C. A., Howard, N., Ramsey, N. W., and Liversage, W. E., The treatment of uterine carcinoma using the Cathetron. Part I. Technique, *Br. J. Radiol.*, 40, 882–887, 1967.

Parker, H. M., A dosage system for interstitial radium therapy. II. Physical aspects, *Br. J. Radiol.*, 11, 252–266, 1938.

Paterson, R. and Parker, H. M., A dosage system for gamma ray therapy, *Br. J. Radiol.*, 7, 592–632, 1934.

Paterson, R. and Parker, H. M., A dosage system for interstitial radium therapy, *Br. J. Radiol.*, 11, 313–339, 1938.

Paterson, R. and Parker, H. M., *Radium Dosage the Manchester System*, 2nd ed., Livingstone, Edinburgh, 1967.

Peiffert, D., Castelain, B., Thomas, L., Ardiet, J. M., Baillet, F., and Mazeron, J. J., Pulsed dose rate brachytherapy in head and neck cancers. Feasibility study of a French cooperative group, *Radiother. Oncol.*, 58, 71–75, 2001.

Pierquin, B. and Dutreix, A., Pour une nouvelle méthodologie en curiethérapie: le système de Paris, *Ann. Radiol.*, 9, 757–760, 1966.

Pierquin, B. and Marinello, G., *A practical manual of brachytherapy*, Medical Physics Publishing, Madison, WI, 1997.

Pierquin, B., Dutreix, A., Paine, C. H., Chassagne, D., Marinello, G., and Ash, D., The Paris system in interstitial radiation therapy, *Acta. Radiol. Oncol. Radiat. Phys. Biol.*, 17, 33–48, 1978.

Pötter, R., Van Limbergen, E., Dries, W., Popowski, Y., Coen, V. et al., Recommendations of the EVA GEC ESTRO Working Group: prescribing, recording, and reporting in endovascular brachytherapy. Quality assurance, equipment, personnel and education, *Radiother. Oncol.*, 59, 339–360, 2001a. (Erratum in *Radiother. Oncol.*, 60, 337–338, 2001.)

Pötter, R., Van Limbergen, E., Gerstner, N., and Wambersie, A., Survey of the use of the ICRU 38 in recording and reporting cervical cancer brachytherapy, *Radiother. Oncol.*, 58, 11–18, 2001b.

Pötter, R., Haie-Meder, C., Limbergen, E. V., Barillot, I., Brabandere, M. D. et al., Recommendations from gynaecological (GYN) GEC ESTRO working group (II): Concepts and terms in 3D image-based treatment planning in cervix cancer brachytherapy—3D dose volume parameters and aspects of 3D image-based anatomy, radiation physics, radiobiology, *Radiother. Oncol.*, 78, 67–77, 2006.

Quimby, E. H. and Castro, V., The calculation of dosage in interstitial radium therapy, *Am. J. Roentgenol. Radium. Ther. Nucl. Med.*, 70, 739–749, 1953.

Regaud, C., Radium therapy of cancer at the Radium Institute of Paris, *Am. J. Roentgenol.*, 21, 1, 1929.

Rivard, M. J., Coursey, B. M., DeWerd, L. A., Hanson, W. F., Huq, M. S. et al., Update of AAPM Task Group No. 43 Report: A revised AAPM protocol for brachytherapy dose calculations, *Med. Phys.*, 31, 633–674, 2004.

Roberts, S. A. and Hendry, J. H., The delay before onset of accelerated tumour cell repopulation during radiotherapy: A direct maximum-likelihood analysis of a collection of worldwide tumour-control datasets, *Radiother. Oncol.*, 29, 69–74, 1993.

Roberts, S. A., Hendry, J. H., Swindell, R., Wilkinson, J. M., and Hunter, R. D., Compensation for changes in dose-rate in radical low-dose-rate brachytherapy: A radiobiological analysis of a randomised clinical trial, *Radiother. Oncol.*, 70, 63–74, 2004.

Rosenthal, M. S. and Nath, R., An automatic seed identification technique for interstitial implants using three isocentric radiographs, *Med. Phys.*, 10, 475–479, 1983.

Rosenwald, J.-C., Automatic localization of curved wires in brachytherapy: The COUREP program, *Comput. Methods Programs Biomed.*, 4, 103–112, 1975.

Rossiter, M. J., Williams, T. T., and Bass, G. A., Air kerma rate calibration of small sources of ^{60}Co, ^{137}Cs, ^{226}Ra and ^{192}Ir, *Phys. Med. Biol.*, 36, 279–284, 1991.

Roué, A., Ferreira, I. H., Dam, J. V., Svensson, H., and Venselaar, J. L. M., The EQUAL-ESTRO audit on geometric reconstruction techniques in brachytherapy, *Radiother. Oncol.*, 78, 78–83, 2006.

Russell, K. R. and Ahnesjö, A., Dose calculation in brachytherapy for a ^{192}Ir source using a primary and scatter dose separation technique, *Phys. Med. Biol.*, 41, 1007–1024, 1996.

Sakelliou, L., Saklliariou, K., Sarigiannis, K., Angelopoulos, A., Perris, A., and Zarris, G., Dose rate distributions around ^{60}Co, ^{137}Cs, ^{198}Au, ^{241}Am, ^{125}I (models 6702 and 6711) brachytherapy sources and the nuclide ^{99}Tc$_m$, *Phys. Med. Biol.*, 37, 1859–1872, 2001.

Santhamma, A. V. and Das, K. R., Dosimetry of cathetron applicators in intracavitary therapy, *Br. J. Radiol.*, 51, 507–514, 1978.

Scalliet, P. and Wambersie, A., Which RBE for iodine-125 in clinical applications?, *Radiother. Oncol.*, 9, 221–230, 1987.

Schlienger, M., Rosenwald, J.-C., Miclutia, M., Quint, R., and Pierquin, B., Dosimetric control in brachycurie therapy by the means of escargot isodoses, *Acta Radiol. Ther. Phys. Biol.*, 9, 282–288, 1970.

Shalek, R. J. and Stovall, M., Use of computers for the calculation of dose from iridium 192 implants, in *Afterloading: 20 years of experience 1955–1975*, Hilaris, B. S., Ed., Memorial Sloan-Kettering, New York, pp. 87–89, 1975.

Sievert, R. M., A method for the measurement of x rays, radium radiation and cosmic radiation along with some investigations into its applicability to physics and medicine. [Eine Methode zur Messung von Röntgen, Radium- und Ultrastrahlung nebst einige Untersuchungen ueber die Anwendbarkeit derselben in der Physik und der Medizin.], *Acta Radiol.*, (Suppl. 14), 1932.

Sminia, P., Schneider, C. J., vanTienhoven, G., Koedooder, K., Blank, L. E., and González González, D., Office hours pulsed brachytherapy boost in breast cancer, *Radiother. Oncol.*, 59, 273–280, 2001.

Soares, C. G., Vynckier, S., Jarvinen, H., Cross, W. G., Sipila, P. et al., Dosimetry of beta-ray ophthalmic applicators: comparison of different measurement methods, *Med. Phys.*, 28, 1373–1384, 2001.

Stock, R. G., Stone, N. N., Tabert, A., Iannuzzi, C., and DeWyngaert, J. K., A dose-response study for I-125 prostate implants, *Int. J. Radiat. Oncol. Biol. Phys.*, 41, 101–108, 1998.

Stone, N. N., Stock, R. G., DeWyngaert, J. K., and Tabert, A., Prostate brachytherapy: Improvements in prostate volume measurements and dose distribution using interactive ultrasound guided implantation and three-dimensional dosimetry, *Radiat. Oncol. Invest.*, 3, 185–195, 1995.

Stout, R., Barber, P., Burt, P., Hopwood, P., Swindell, R., Hodgetts, J., and Lomax, L., Clinical and quality of life outcomes in the first United Kingdom randomized trial of endobronchial brachytherapy (intraluminal radiotherapy) vs. external beam radiotherapy in the palliative treatment of inoperable non-small cell lung cancer, *Radiother. Oncol.*, 56, 323–327, 2000.

Stovall, M. and Shalek, R. J., A review of computer techniques for dosimetry of interstitial and intracavitary radiotherapy, *Comput. Programs Biomed.*, 2, 125–136, 1972.

Tan, L. T., Jones, B., Green, J. A., Kingston, R. E., and Clark, P. I., Treatment of carcinomas of the uterine cervix which remain bulky after initial external beam radiotherapy: A pilot study using integrated cytotoxic chemotherapy prior to brachytherapy, *Br. J. Radiol.*, 69, 165–171, 1996.

Thames, H. D., An incomplete-repair model for survival after fractionated and continuous irradiations, *Int. J. Radiat. Biol. Relat Stud. Phys. Chem. Med.*, 47, 319–339, 1985.

Thomadsen, B. R., Houdek, P. V., van der Laarse, R., Edmunson, G., Kolkman-Deurloo, I. K., and Visser, A. G., Treatment planning and optimization, in *High Dose Rate Brachytherapy: A Textbook*, Futura, New York, pp. 79–145, 1994.

Tod, M. and Meredith, W. J., Treatment of cancer of the cervix uteri, a revised Manchester method, *Br. J. Radiol.*, 26, 252–257, 1953.

Tranter, F. W., A wedge filter for use in treatment of carcinoma of the cervix uteri with 4 MV x-rays, *Br. J. Radiol.*, 32, 350–352, 1959.

Trott, N.G., Ed., Radionuclides in brachytherapy: Radium and after. *Br. J. Radiol.* (Suppl. 21), 1987.

Van der Laarse, R., The stepping source dosimetry systems as an extension of the Paris system, in *Brachytherapy from Radium to Optimisation*, Nucletron, Veenendahl, pp. 319–330, 1994.

Van der Laarse, R. and Batterman, J. J., Treatment planning of radioactive seed implants with the Selectron treatment planning system, in *Brachytherapy*, Mould, R. F., Ed., Nucletron, Veenendahl, 295–301, 1984.

Van Kleffens, H. J. and Star, W. M., Application of stereo x-ray photogrammetry (SRM) in the determination of absorbed dose values during intracavitary radiation therapy, *Int. J. Radiat. Oncol. Biol. Phys.*, 5, 557–563, 1979.

Vicini, F. A., Kini, V. R., Edmundson, G., Gustafson, G. S., Stromberg, J., and Martinez, A., A comprehensive review of prostate cancer brachytherapy: Defining an optimal technique, *Int. J. Radiat. Oncol. Biol. Phys.*, 44, 483–491, 1999.

Visser, A. G., van den Aardweg, G. J., and Levendag, P. C., Pulsed dose rate and fractionated high dose rate brachytherapy: Choice of brachytherapy schedules to replace low dose rate treatments, *Int. J. Radiat. Oncol. Biol. Phys.*, 34, 497–505, 1996.

Wallner, K., Merrick, G., True, L., Cavanagh, W., Simpson, C., and Butler, W., I-125 versus Pd-103 for low-risk prostate cancer: morbidity outcomes from a prospective randomized multicenter trial, *Cancer J.*, 8, 67–73, 2002.

Wambersie, A. and Prignot, M., Replacement of radium by cesium-137 in gynecologic curietherapy, *J. Radiol. Electrol. Med. Nucl.*, 54, 261–270, 1973.

Welsh, A. D., Dixon-Brown, A., and Stedeford, J. B. H., Calculation of dose distribution for iridium 192 implants, *Acta Radiol. Oncol.*, 22, 331–336, 1983.

Wilkinson, J. M., Moore, C. J., Notley, H. M., and Hunter, R. D., The use of Selectron afterloading equipment to simulate and extend the Manchester system for intracavitary therapy of the cervix uteri, *Br. J. Radiol.*, 56, 409–414, 1983.

Williamson, J. F., The Sievert integral revisited: Evaluation and extension to ^{125}I, [169]Yb, and ^{192}Ir brachytherapy sources, *Int. J. Radiat. Oncol. Biol. Phys.*, 36, 1239–1250, 1996.

Williamson, J. F. and Nath, R., Clinical implementation of AAPM Task Group 32 recommendations on brachytherapy source strength specification, *Med. Phys.*, 18, 439–448, 1991.

Williamson, J. F., Khan, F. M., Sharma, S. C., and Fullerton, G. D., Methods for routine calibration of brachytherapy sources, *Radiology*, 142, 511–515, 1982.

Williamson, J. F., Thomadsen, B. R., and Nath, R., *Brachytherapy Physics Proceedings of the AAPM Summer School 1994*, Medical Physics Publishing, Madison, WI, 1995.

Williamson, J. F., Coursey, B. M., DeWerd, L. A., Hanson, W. F., Nath, R., and Ibbott, G., Guidance to users of Nycomed Amersham and North American Scientific, Inc., I-125 interstitial sources: Dosimetry and calibration changes: Recommendations of the American Association of Physicists in Medicine Radiation Therapy Committee Ad Hoc Subcommittee on Low-Energy Seed Dosimetry, *Med. Phys.*, 26, 570–573, 1999.

Wilson, B. J., Ed., The Radiochemical Manual. 2nd Edition, The Radiochemical Centre, Amersham, 1966.

Wuu, C. S. and Zaider, M. A., Calculation of the relative biological effectiveness of ^{125}I and [103]Pd brachytherapy sources using the concept of proximity function, *Med. Phys.*, 25, 2186–2189, 1995.

Yu, Y., Anderson, L. L., Li, Z., Mellenberg, D. E., Nath, R. et al., Permanent prostate seed implant brachytherapy: Report of the American Association of Physicists in Medicine Task Group No. 64, *Med. Phys.*, 26, 2054–2076, 1999.

ADDITIONAL READING

See also Williamson et al. (1995), Godden (1988), CFMRI (1983), Pierquin and Marinello (1997), and Paterson and Parker (1967) in references.

Dale, R. G. and Jones, B., The clinical radiobiology of brachytherapy, *Br. J. Radiol.*, 71, 465–483, 1998.

ICRU (International Commission on Radiation Units and Measurements), *Dosimetry of beta ray sources and low energy photon brachytherapy sources*, ICRU Report 65, ICRU, Bethesda, MD, 2001.

Martinez, A. A., Orton, C. G. and Mould, R. F., *Brachytherapy HDR and LDR 1990 Remote Afterloading: State of the Art*, Nucletron, Veenendahl, 1990.

Massey, J. B., Pointon, R. S., and Wilkinson, J. M., The Manchester system and the BCRU recommendation for brachytherapy source specification, *Brit. J. Radiol.*, 58, 911–913, 1985.

Mould, R. F., Ed., Brachytherapy 2. Proceedings of the 5th International Selectron Users Meeting, Nucletron, Veenendahl, 1989.

Williamson, J. F., Brachytherapy technology and physics practice since 1950: a half-century of progress, *Phys. Med. Biol.*, 51, R303–R325, 2006.

PART K

THERAPY WITH UNSEALED SOURCES

Editors: Philip Mayles and Jean-Claude Rosenwald

INTRODUCTION

Radionuclide therapy has been used to treat both benign and malignant diseases for many years. For example, ^{131}I is a standard form of treatment for thyrotoxicosis and thyroid cancer using the natural property of the thyroid gland to take up iodide, and radionuclide therapy with ^{32}P and ^{89}Sr has also been widely practiced to relieve bone pain. More recently, other short- and long-lived radionuclides have been introduced. This Part deals with the scientific basis of this unique form of radiotherapy. When compared to external-beam radiotherapy, therapy with unsealed sources has the potential to deliver larger internal radiation doses more selectively to target tissues. It also has the advantage of relative ease of use because it is often a noninvasive procedure with relatively few side-effects and, in many applications, can be performed on an outpatient basis.

Most early treatments involved the intravenous or oral administration of the therapeutic agent and were referred to as systemic radionuclide therapy. These systemic treatments relied on the preferential uptake and prolonged retention of the radiolabelled agent by the tumour, resulting in a high tumour-to-normal-tissue dose ratio.

With the improved availability of a broader range of therapy radionuclides (see Chapter 57) and the development of new tumour-specific pharmaceuticals (see Chapter 58), there has been a gradually increasing interest in the therapeutic applications of unsealed radioactive sources. The latest terminology—targeted radionuclide therapy—refers to the selective delivery of the therapy radionuclide to tumour cells using a biologically-targeted molecule to which the radionuclide is attached. This seeks to satisfy the fundamental aim of all forms of radiotherapy, that is, to maximise the therapeutic ratio, i.e. to deliver the highest possible radiation dose to the tumour while limiting the dose to normal tissues (see Section 43.1).

Both physical and biological mechanisms can be used to achieve tissue targeting. Improvements in *physical* targeting have been achieved by careful selection of the most appropriate radionuclide. The basis for selection is to match the physical range of the particles emitted to the size of the tumours to be treated together with careful consideration of the distribution of the radionuclide within the tumour at both the macroscopic and cellular level. Improved physical targeting has also been achieved by the introduction of alternative, more-direct routes of administration (e.g. intra-arterial, intra-cavitary, and intra-lesional). Improvements in *biological* targeting are being achieved by the development of new carrier molecules that can provide selective targeting of tumour sites, for example, on cell surface membrane receptors, in the cytoplasm, or in the nucleus. Biologically-targeted (or cell-directed) radiotherapy provides the opportunity to deliver radiation dose directly to tumour cells with almost complete sparing of normal tissues.

The traditional approach to the practice of radionuclide therapy was to prescribe a fixed amount of radioactivity. The amount was either based on previous experience or determined by gradually increasing the amount until unacceptable toxicity was observed. More recently, pre-therapy tracer studies have been introduced to measure the uptake of the radionuclide in different organs and tissues. This enables the treatment prescription to be given, not in terms of activity administered, but of absorbed dose to the tumour or in terms of the limiting dose to an organ-at-risk. Although this is a significant improvement when compared to the traditional approach, careful consideration must be given to the fact that dose estimates are often subject to large uncertainties. This is due to oversimplification of the models used in conventional internal dosimetry that will be described in Chapter 56. The latest approach at research centres with facilities for quantitative tomographic imaging and multi-modality image registration is to use much more realistic models so that patient-specific 3D dose-distribution maps can be produced (see Section 56.4). For details on the more practical (including radiation protection) aspects of radionuclide therapy, the reader is referred to Flower and Chittenden (2000).

CHAPTER 56

DOSIMETRY OF UNSEALED SOURCES

Maggie Flower and Jamal Zweit

CONTENTS

56.1 MEDICAL INTERNAL RADIATION DOSIMETRY (MIRD) IN THEORY

The conventional method of calculating absorbed dose delivered internally is known as the medical internal radiation dosimetry (MIRD) scheme (Loevinger et al. 1991). The body is considered as a set of source organs that have a significant uptake of the radiopharmaceutical, and a set of target organs that are irradiated by the source organs. It is important to note that the MIRD definition of target organs is very different from that used in external-beam radiotherapy (see Section 29.2). For unsealed source therapy, the target organs are any organs (or tissues) of interest for which the absorbed dose is to be estimated and hence can include normal tissues (e.g. those at risk) as well as tumours.

For a self-irradiating tumour with uniform uptake, the mean absorbed dose, D (in Gy), to the tumour is given by:

$$D = \frac{\tilde{A}}{m} \sum_i \Delta_i \phi_i \tag{56.1}$$

where $\tilde{A}$ is the cumulated activity in the tumour (in MBq h), m is the tumour mass (in g), Δ_i is the equilibrium absorbed-dose constant for each nuclear transition (in g Gy/(MBq h)), and ϕ_i is the absorbed fraction of energy. For more general cases, the absorbed dose $D_{t \leftarrow s}$ (in Gy) to a target organ from activity in a source organ is given by:

$$D_{t \leftarrow s} = \frac{\tilde{A}_s}{m_t} \sum_i \Delta_i \phi_i \tag{56.2}$$

where the subscripts t and s denote the target and source organs, respectively. The total dose to a single target organ is simply the sum of the absorbed doses from all the source organs.

The cumulated activity, $\tilde{A}_s$, is proportional to the total number of radioactive disintegrations which occur in the source organ, and depends on the activity administered; the uptake of, retention by, and excretion from the organ; and the physical decay of the radionuclide. $\tilde{A}_s$ is equal to the area under the activity-time curve for that organ.

It is helpful when considering both Δ_i and ϕ_i to separate the different emissions into nonpenetrating radiation (β-particles and electrons) and penetrating radiation (x- and γ-rays). Δ_i (in g Gy/(MBq h)) is given by:

$$\Delta_i = 0.576 \, n_i \bar{E}_i \tag{56.3}$$

where n_i is the mean number of the ith type of radiation emitted per disintegration, and $\bar{E}_i$ is the mean energy (in MeV) of each emission. Δ_i is found from nuclear decay tables for each type of radiation emitted. MIRD tables (MIRD 1975a) list the values for most radionuclides used in nuclear medicine.

For nonpenetrating radiation, ϕ_i is equal to zero when the target and source organs are geometrically separated, or is equal to 1.0 when they are the same organ. At a source-target interface, ϕ_i is set to 0.5 (e.g. when considering the absorbed dose to the wall of a body cavity, such as the bladder containing radioactive urine). For penetrating radiation, the fraction ϕ_i depends strongly on both the radiation energy and on the geometry of the source-target configuration. Values of ϕ_i derived from Monte Carlo calculations can be found in MIRD Pamphlet 5 (MIRD 1969) for various radionuclides and pairs of source and target organs.

Equation 56.2 can be simplified by the introduction of the mean dose per unit cumulated activity $S_{t \leftarrow s}$ (in Gy/(MBq h)) defined by:

$$S_{t \leftarrow s} = \frac{1}{m_t} \sum_i \Delta_i \phi_i \tag{56.4}$$

Values of the mean dose per unit cumulated activity (known as S values) have been tabulated for a variety of radionuclides and for different source–target configurations in both standard man (MIRD 1975b) and children (NRCP 1983). S values for a wider range of mathematical models of the human body including models for pregnant women are now available on the Radiation Dose Assessment Resource (RADAR) website (see References).

The traditional MIRD scheme, as described above, for calculating the dose from internally administered radionuclides makes the assumption that the activity is uniformly distributed in the source organ. It also assumes that the shape, size and position of the organs are as represented by the human phantom described in Pamphlet 5 (MIRD 1969). The traditional MIRD scheme does not permit the determination of a maximum or minimum dose to each target organ considered, but simply provides the mean dose to the target volume. The MIRD Committee has started to address the requirement for the dosimetry of nonuniform activity distributions by providing S values at the cellular (Goddu et al. 1997) and image-voxel level (Bolch et al. 1999). A more recent development is the availability of S values for dosimetry using α-particle emitters (Stinchcomb and Roeske 1999). Clairand et al. (2000) have developed mathematical models to allow consideration of the patient height, which they show affects the calculated dose.

56.2 MIRD IN PRACTICE

A program called MIRDOSE (Stabin 1996) is often used for calculating absorbed doses. This program is particularly useful when dealing with new radiopharmaceuticals or when special biokinetic models need to be considered. The MIRDOSE code has been renamed OLINDA (organ level internal dose assessment) after it was rewritten for greater computer platform independence. OLINDA is deployed on the RADAR website for rapid electronic access.

The total dose to a single target organ arising from radiation emitted by several source organs is simply the sum of the absorbed doses from all the source organs. Figure 56.1 is a flow chart that summarizes how MIRD is applied in practice. Sequential planar gamma camera imaging is often used to obtain activity-time curves needed for the determination of the cumulated activity. Alternatively sequential quantitative tomographic imaging (SPECT (Zanzonico et al. 1989) or PET (Ott et al. 1992)) can be used to assess the specific cumulated activity.

However, many problems can be encountered when trying to use the gamma camera for imaging therapy radionuclides, since camera design has been optimised for diagnostic imaging of tracer levels of radioactivity. These problems are associated with high count rates (especially during the first few days after a therapy administration), and poor resolution (as a result of collimator penetration of any high-energy photons and the broad energy spectrum of bremsstrahlung radiation). Special low-sensitivity collimators or lead sheets attached to the front face of the camera can be used to reduce the count rate. Other options are to perform a pre-therapy tracer study using a radioisotope more suited to gamma-camera imaging (e.g. ^{123}I instead of ^{131}I) or to use dual radionuclides, using a simultaneous injection of both imaging and therapy radionuclides (see Section 57.7).

In addition to the count-rate limitations of the gamma camera, the need to limit doses to staff and other patients also makes data acquisition at early time points difficult. A combination of early measurements during a tracer study and late measurements following therapy administration is a reasonable compromise. However, the kinetics of the radiopharmaceutical may not be identical in the tracer and therapy studies as a result of cell damage, so extrapolation of data can be subject to error.

It is difficult to achieve precise estimates of the magnitude and distribution of internal dose delivered from unsealed sources (Flux et al. 2002). Major sources of error are associated with the practical difficulties of obtaining enough time points for the accurate determination of $\tilde{A}_s$ and the accuracy of methods used to estimate m_t. A review of radionuclide therapy practices and facilities in Europe (Hoefnagel et al. 1999) showed that dosimetry was still not routinely performed in most centres and Fisher (2003) reported that this was also the case in the U.S. However, with the advent of improved imaging and counting techniques, estimates of internal dose are becoming more commonplace. These estimates are either *predictive* using a pre-treatment tracer study or *retrospective* using data acquired during therapy.

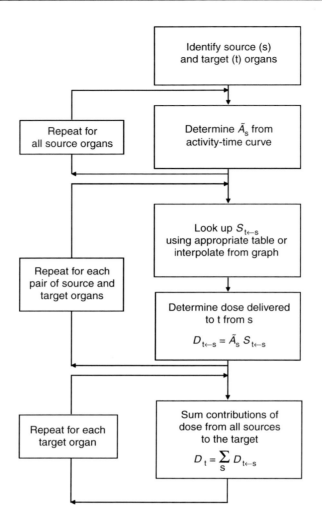

FIGURE 56.1
Flow chart illustrating how MIRD is applied in practice. (From Cormack, J., Towson, J. E. C., and Flower, M. A., Radiation protection and dosimetry in clinical practice, *Nuclear Medicine in Clinical Diagnosis and Treatment*, Murray, I. P. C. and Ell, P. J., Eds., 2nd Ed., Vol. 2, Churchill Livingstone, Edinburgh, 1998. With permission.)

56.3 DOSE-LIMITING ORGANS

The dose-limiting organs for new targeting agents are not necessarily known in advance and therefore detailed dosimetry and toxicity studies are required for each new carrier molecule considered for targeted radionuclide therapy. The organs at risk depend on both the route of administration and on the radiopharmaceutical. In systemic therapy the dose-limiting organ is often the bone marrow: if there is a risk of bone-marrow ablation, either a bone marrow harvest will be performed prior to therapy, for subsequent regrafting if necessary, or peripheral blood stem cell support is used. For intrathecal administrations, the dose-limiting organ is the spinal cord. If there is high uptake in the liver, this could lead to radiation hepatitis and chronic veno-occlusive liver disease. The bladder is also at risk from radioactivity in the urine, and patients should be hydrated and/or catheterised to reduce the dose to the bladder wall.

As the bone marrow is often the dose-limiting organ, bone-marrow dosimetry merits special attention. Various methods (Siegel et al. 1990) have been used to calculate the marrow dose but inaccuracies are caused by:

- Inadequate knowledge of the distribution of radionuclide within the marrow
- Treating the β-rays as nonpenetrating
- Ignoring the photon dose to the marrow from source organs other than the bone marrow
- Ignoring the effects of backscattering of electrons at the marrow-bone interface

In the estimation of absorbed dose to red marrow from nonspecific uptake of radiolabeled antibodies, it was recommended initially (Siegel et al. 1990) that the following assumptions be made:

- Marrow specific activity is one-third of blood specific activity.
- Clearance of activity from bone marrow is equal to that from blood.
- Marrow activity is uniformly distributed.

Subsequently, Sgouros (1993) presented a simple equation for calculating the red-marrow-to-blood activity concentration ratio given the haematocrit and red marrow extracellular fluid fraction of a patient. He also presented a series of equations that may be used to correct the activity concentration in bone-marrow biopsies for bone, yellow-marrow and blood contamination. Given the uncertainties associated with marrow dose estimates, the current method of choice for the assessment of bone-marrow toxicity from radioimmunotherapy (RIT) is the radiobiological endpoint method whereby the activity administered is slowly escalated until critical haematopoietic depression is reached.

56.4 PATIENT-SPECIFIC CALCULATIONS

In research centres, patient-specific dosimetry is being introduced, especially for RIT, to overcome some of the limitations of the MIRD system. The program MABDOS (Johnson 1988) starts with reference man but permits the inclusion of a spherical tumour at any site in the body. Also, with the increasing availability of multimodality imaging and image registration techniques, several groups (Giap et al. 1995; Tagesson et al. 1996; Flux et al. 1997; Kolbert et al. 1997; Thierens et al. 2001; Guy et al. 2003) have started to produce patient-specific 3D dose distributions resulting from radionuclide therapy. Dose volume histograms and tumour control probabilities (TCPs) can be derived from the 3D dose maps in the same way as for external-beam radiotherapy. The equivalent uniform dose (EUD), proposed initially for external-beam radiotherapy (see Chapter 36), can be applied to radionuclide therapy (O'Donoghue 1999). The EUD is based on the radiobiological linear quadratic (LQ) model and a 3D absorbed-dose distribution.

56.5 ALTERNATIVE DOSIMETRY METHODS

56.5.1 In Vivo Dosimetry

Two alternative methods of in vivo dosimetry are autoradiography (Puncher and Blower 1994) and the use of miniature thermoluminescent dosimeters (TLDs) (Wessels and Griffith 1986). These techniques have been developed and evaluated in tumour models and/or biopsy specimens and provide valuable insight into the optimal properties of therapy agents. An indirect method of in vivo dosimetry is biological dosimetry which involves the analysis of chromosomal aberrations in cultured blood cells (IAEA 1986). The results from in vivo studies can provide valuable input into theoretical calculations of dose distributions which can be performed at both the macroscopic and microscopic level. Evidence from microautoradiography and from miniature dosimeters suggests that the assumption of a uniform distribution of uptake is incorrect. There may be significant heterogeneity in the 10–100 µm range (Adelstein and Kassis 1987).

56.5.2 THEORETICAL APPROACHES

Dose distributions can be calculated using Monte Carlo techniques (Humm and Cobb 1990; Tagesson et al. 1996; Coulot et al. 2003; Descalle et al. 2003) or by integrating theoretically-derived (usually by Monte Carlo methods) dose point kernels over the spatial radionuclide distribution (Giap et al. 1995; Kolbert et al. 1997). An example of the latter is a theoretical model that has been used to predict TCP for spherical neuroblastoma tumours of different size (Nahum 1996) and used to demonstrate the advantages to be gained from using cocktails of β emitters to treat tumours with a broad range of sizes simultaneously (Zweit et al. 1997). This model was extended to include the radiobiological effects of cellular repair and proliferation (Atthey et al. 2000). Theoretical microdosimetry at the multicellular level (Coulot et al. 2003; Malaroda et al. 2003) is particularly valuable in assessing the dose distribution from inhomogeneous depositions of short-range emitters. For example, in RIT, cell killing will depend on whether or not the antibody binds to the cell-surface membrane or is internalized by the tumour cells (Humm and Cobb 1990; Hartman et al. 2000).

CHAPTER 57

RADIONUCLIDE SELECTION FOR UNSEALED SOURCE THERAPY

Maggie Flower, Jamal Zweit, and Mark Atthey

CONTENTS

57.1 INTRODUCTION

There are three basic considerations when choosing a radionuclide for targeted radiotherapy:

- Physical properties of the radionuclide (i.e. type, yield and energy of emissions, and physical half-life)
- Chemical properties in relation to: conjugation and binding stability; targeting molecule; and accumulation, retention and clearance profiles for the tissues of interest
- Production methods in relation to radioactivity levels and chemical and radiochemical purity suitable for clinical application and specific activity requirements when, for example, conjugated to a receptor ligand type molecule (Zweit 1996)

The types of radionuclide decay that are of greatest potential use in unsealed source therapy are those which involve the emission of β-particles, Auger and conversion electrons, and α-particles. These particles can be considered as nonpenetrating radiation in contrast with x- and γ-rays, which are penetrating radiation. Although the β-emitters ^{131}I and ^{32}P have been the most commonly used radionuclides in the past, their physical properties are not necessarily ideal. Other β-emitting radionuclides have been introduced into the clinic, especially over the last couple of decades. Emitters of Auger and conversion electrons and α-particles may be more widely used in the future (see Section 57.3 and Section 57.4).

57.2 BETA EMITTERS

There is a wide choice of β-emitting radionuclides for targeted radiotherapy and hence a flexibility in the choice of β range in tissue. Table 57.1 lists the physical properties of β emitters that have been used or are of potential use in radionuclide therapy. The radionuclides are listed in order of increasing β energy, and only the most abundant β and γ radiations are included. The values of Δ_p and Δ_{np} (the total equilibrium dose constants for penetrating and nonpenetrating radiations, respectively) listed in Table 57.1 are useful indicators of the γ dose relative to the β dose for each therapy radionuclide. These parameters were defined in Section 56.1.

Beta decay (see Section 2.2.2.4) gives rise to a continuous spectrum of β energies. Details of spectra can be found in Cross et al. (1983). For those radionuclides with more complex decay schemes, the resulting spectrum is a sum of the continuous spectra for each β emission. The average energy is approximately one-third of the maximum energy for each β emission. The maximum range, in water or soft tissue (in mm), is approximately equal to the maximum energy (in MeV) multiplied by five (Adelstein and Kassis 1987). More precise relationships between the range and the β energy have been given by Cole (1969) and by Prestwich et al. (1985). The maximum range is often calculated using the continuous slowing down approximation (CSDA) (see Section 3.4.3). This is then referred to as the CSDA range. However, the latter does not take account of the contorted paths of the β-particles as they interact with their surroundings, nor of the decay scheme, since the β-particle with maximum energy may not be very abundant. In addition to the maximum range, two other parameters (mean range and X_{90}) have been used, especially in the context of radionuclide therapy, to describe the distance over which most of the β energy is deposited. The mean range usually refers to the CSDA range for a β-particle with the mean energy. The X_{90} value, used by Simpkin and Mackie (1990) and Johnson and Yanch (1991), is defined as the radius of the sphere in which 90% of the emitted energy is absorbed. More recently, the X_{90} value has been referred to as the therapeutic range (Johnson et al. 1995). Both the mean range and the X_{90} value can be calculated from the full energy spectrum using Monte Carlo techniques (see Chapter 5). Some examples of these values are given in Table 57.2.

Pure β emitters (e.g. ^{32}P, ^{89}Sr and ^{90}Y) may be considered as ideal for delivering a high local dose while sparing tissues at a distance. However, radionuclides which emit γ-rays in addition to β-particles (e.g. ^{131}I, ^{153}Sm and ^{186}Re) have the advantage that external counting and imaging techniques can be applied more easily to assess the uptake and distribution of the therapy agent. The disadvantage is that the γ-rays increase the whole-body dose to the patient and, in some cases (e.g. for ^{131}I), their energy may not be well-matched to the gamma camera, so only poor resolution imaging is available. In this regard, the lower flux and energies of the γ-rays from, for example ^{153}Sm and ^{186}Re, are advantageous compared to ^{131}I. Imaging the bremsstrahlung radiation from pure β emitters is also a possibility for monitoring the bio-distribution of these therapy agents, but again the spatial resolution is poor (Smith et al. 1988; Siegel et al. 1995).

The range of the β-particles is very important in relation to the size of the tumour to be treated (Wheldon 1993; O'Donoghue et al. 1995). High-energy β-particles (see Table 57.1) provide radiation cross-fire, which can reduce the problem of heterogeneous uptake in large tumours. These long-range β-particles are also better suited to the treatment of large volumes. If the β emitter is uniformly distributed throughout the tumour, the dose rate will be uniform

TABLE 57.1

Physical Properties of β-Emitting Radionuclides for Targeted Radiotherapy

Radio-nuclide	$T_{1/2}$ (days)	n_{np}	E_{np} (MeV)	n_p	E_p (MeV)	Δ_{np} (g Gy $MBq^{-1} h^{-1}$)	Δ_p (g Gy $MBq^{-1} h^{-1}$)
Low-energy β emitters							
^{191}Os	15.4	1.00	0.038	0.26	0.129	0.078	0.046
^{35}S	87.4	1.00	0.049	—	—	0.028	—
^{33}P	25.4	1.00	0.077	—	—	0.044	—
^{45}Ca	163.0	1.00	0.077	—	—	0.044	—
^{199}Au	3.2	0.66	0.082	0.37	0.158	0.082	0.051
^{169}Er	9.3	0.55	0.101	—	—	0.060	—
^{67}Cu	2.6	0.57	0.121	0.49	0.185	0.089	0.066
^{47}Sc	3.4	0.68	0.143	0.68	0.159	0.093	0.062
^{177}Lu	6.7	0.79	0.149	0.11	0.208	0.085	0.020
^{161}Tb	6.9	0.67	0.154	0.22	0.025	0.113	0.020
^{105}Rh	1.4	0.75	0.179	0.19	0.319	0.088	0.045
Medium-energy β emitters							
^{131}I	8.0	0.89	0.192	0.81	0.364	0.109	0.219
^{153}Sm	2.0	0.43	0.229	0.28	0.103	0.156	0.035
^{77}As	1.6	0.97	0.232	0.02	0.239	0.131	0.005
^{143}Pr	13.6	1.00	0.314	—	—	0.181	—
^{198}Au	2.7	0.99	0.315	0.96	0.412	0.188	0.233
^{159}Gd	0.77	0.64	0.319	0.10	0.363	0.174	0.029
^{109}Pd	0.56	1.00	0.361	0.04	0.088	0.252	0.007
^{186}Re	3.8	0.73	0.362	0.09	0.137	0.198	0.012
^{111}Ag	7.5	0.93	0.363	0.07	0.342	0.204	0.015
^{149}Pm	2.2	0.97	0.370	0.03	0.286	0.210	0.006
High-energy β emitters							
^{165}Dy	0.10	0.83	0.455	0.04	0.095	0.257	0.015
^{89}Sr	50.5	1.00	0.583	—	—	0.336	—
^{32}P	14.3	1.00	0.695	—	—	0.400	—
^{166}Ho	1.1	0.51	0.695	0.06	0.081	0.400	0.017
^{188}Re	0.71	0.72	0.764	0.149	0.155	0.448	0.033
^{114m}In	49.5	0.99[a]	0.777[a]	0.15	0.190	0.526	0.056
^{142}Pr	0.80	0.96	0.833	0.04	1.576	0.465	0.034
^{90}Y	2.7	1.00	0.935	—	—	0.539	—
^{76}As	1.1	0.51	1.267	0.45	0.559	0.611	0.246

n and *E* are the mean number per decay and mean energy respectively of the most abundant radiations; Δ is the total equilibrium dose constant; subscripts np and p refer to nonpenetrating and penetrating radiations respectively.
[a] Emitted from daughter.
Source: From ICRP (International Commission on Radiological Protection), Radionuclide transformations. Energy and intensity of emissions, ICRP Publication 38, *Annals ICRP*, 11–13, 1983. With permission.

throughout this volume except for a reduction close to the outer edge where electronic equilibrium no longer exists. This is illustrated in Figure 57.1. For very small tumours (less than a few mm in diameter), a large fraction of the total energy from β particles emitted within the tumour will be deposited in surrounding tissues. Hence, for the treatment of micrometastases, short-range electrons and α emitters would be more appropriate, provided that the activity distribution is uniform on a microscopic scale.

One practical advantage of using pure β emitters is that there is little risk of irradiating other people, and patients can therefore often be treated in out-patient clinics (Zanzonico et al. 1999). Beta emitters with accompanying γ-rays, however, pose a significant radiation risk to others, and in some cases, an in-patient stay of several days is required before the activity level is sufficiently low for the patient to return home. Radiation protection considerations are discussed further in Section 61.4.2

TABLE 57.2

Examples of Maximum β Range and X_{90} Values

Radionuclide	Maximum Energy (MeV)	Maximum Range[a] (mm)	X_{90} (mm)
^{33}P	0.24	0.62	0.22
^{199}Au	0.45	1.57	0.36
^{67}Cu	0.57	2.15	0.63
^{47}Sc	0.60	2.27	0.73
^{177}Lu	0.49	1.75	0.63
^{131}I	0.80	4.27	0.90
^{153}Sm	0.81	3.39	1.15
^{143}Pr	0.93	4.06	1.66
^{186}Re	1.07	4.78	1.93
^{111}Ag	1.02	4.62	1.91
^{32}P	1.71	8.31	3.74
^{188}Re	2.11	10.51	4.89
^{90}Y	2.28	11.27	5.53

Data from β spectra provided by Manuel Bardiès and Monte Carlo code (using Electron Gamma Shower, version 4 (EGS4)) written by Mark Atthey.

[a]   CSDA range of particle with maximum energy using the formula from Prestwich et al. (1989).

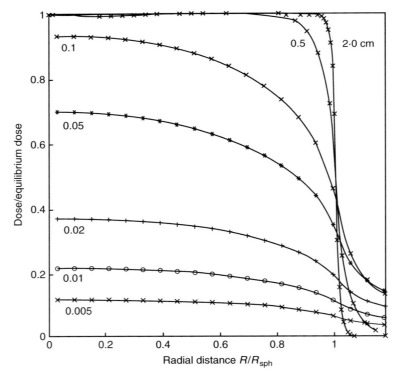

FIGURE 57.1

Distributions of beta dose to a sphere of radius R_{sph} uniformly filled with ^{131}I. The graph shows the beta dose (expressed as a fraction of the equilibrium dose) as a function of the scaled radial distance R/R_{sph} for various values (0.005–2.0 cm) of R_{sph}. (Data from Nahum, A. E., *Phys. Med. Biol.*, 41, 1957–1972, 1996.)

57.3 AUGER-ELECTRON EMITTERS

Radionuclides, which decay by electron capture or internal conversion, emit low-energy characteristic x-rays and Auger electrons (see Section 1.4.4). In some cases, low-energy conversion electrons are also emitted. Most Auger electrons have very short range (< 1 μm) and are therefore only of use in therapy if the source is attached to, or very close to, the cell nucleus. Table 57.3 lists the physical properties of Auger-electron emitters, some of which have been considered for radionuclide therapy. The total equilibrium dose constant for Auger electrons alone (Δ_{Auger}) has been listed separately from Δ_{np} for all nonpenetrating radiations. Once again, a comparison of the total equilibrium dose constants is a useful indication of radiation dose contributions from the different types of radiation. It indicates, for instance, that ^{117m}Sn emits both Auger electrons as well as conversion electrons and γ-rays, and hence has been proposed as a useful therapy radionuclide when conjugated to diethylenetriamine pentaacetic acid (DTPA) and used in palliative bone therapy (Krishnamurthy et al. 1997). The 159 keV γ-photons are well suited for imaging with the gamma camera.

The potential use of Auger-electron emitters for targeted radionuclide therapy has been discussed by Humm (1986) and O'Donoghue and Wheldon (1996). Biological efficacy is critically dependent on the sub-cellular and sub-nuclear localisation of Auger-electron emitters. Potential methods of targeting include the use of analogues of DNA precursors (e.g. iodo- and bromo-deoxyuridine) and of molecules that bind to DNA (e.g. oligonucleotides and other compounds

TABLE 57.3

Physical Properties of Auger-Electron Emitters

Radionuclide	$T_{1/2}$ (days)	Auger Electron Yield per Decay	Δ_{Auger} (g Gy MBq^{-1} h^{-1})	Δ_{np} (g Gy MBq^{-1} h^{-1})	Δ_p (g Gy MBq^{-1} h^{-1})
^{51}Cr	27.7	5.4	0.002	0.002	0.019
^{67}Ga	3.3	5.8	0.004	0.020	0.091
^{71}Ge	11.2	4.2	0.003	0.003	0.002
^{75}Se	120	4.4	0.003	0.008	0.227
^{77}Br	2.4	3.8	0.003	0.005	0.183
^{80m}Br	0.18	5.5	0.004	0.035	0.014
^{99m}Tc	0.25	1.2	0.001	0.009	0.073
^{103}Pd	2.8	2.8	0.003	0.003	0.008
^{103m}Rh	0.04	2.3	0.002	0.022	0.001
^{111}In	2.8	3.1	0.004	0.020	0.233
^{113m}In	0.07	0.9	0.001	0.077	0.148
^{115m}In	4.5	1.3	0.002	0.099	0.093
^{117m}Sn	13.6	2.8	0.004	0.093	0.091
^{119}Sb	1.6	4.4	0.005	0.015	0.013
^{123}I	0.55	2.9	0.004	0.016	0.098
^{125}I	60.1	5.4	0.007	0.011	0.024
^{131}Cs	9.7	2.7	0.004	0.004	0.013
^{161}Ho	0.38	4.5	0.002	0.013	0.001
^{165}Er	0.42	2.4	0.005	0.005	0.022
^{193m}Pt	4.3	4.0	0.006	0.079	0.007
^{195m}Pt	4.0	5.5	0.013	0.105	0.044
^{201}Tl	3.0	3.6	0.008	0.025	0.054
^{203}Pb	2.2	2.4	0.007	0.030	0.205

Δ is the equilibrium dose constant; subscripts np and p refer to nonpenetrating and penetrating radiations, respectively.
Source: Data from ICRP (International Commission on Radiological Protection), *Radionuclide transformations. Energy and intensity of emissions*, ICRP Publication 38, Annals ICRP, 11–13, 1983. With permission.

which internalise to the nucleus). Unfortunately, heterogeneity of radionuclide uptake is a serious limitation on the success of targeted therapy with Auger-electron emitters. Experimental studies (Neshasteh-Riz et al. 1998) have suggested that combinations of ^{131}IUdR (iodo-deoxyuridine) (with cross-fire β irradiation) and ^{123}IUdR or ^{125}IUdR may be more effective than Auger-electron emitters alone if proliferative heterogeneity of the tumour cell population exists.

57.4 ALPHA EMITTERS

The advantages of α emitters (e.g. ^{211}At, ^{213}Bi) for radionuclide therapy are their short range (typically 50–90 μm, i.e. several cell diameters), high linear energy transfer (LET) and hence increased biological effectiveness (see Section 7.14). ^{213}Bi has the advantage of being available from an ^{225}Ac/^{213}Bi generator system (McDevitt et al. 1999). A single α-particle can deposit approximately 0.25 Gy in a 10 μm diameter cell nucleus (Humm 1986). This offers the possibility of combining cell-specific targeting with radiation of similar range. Preclinical studies with a variety of ^{213}Bi- and ^{211}At-labelled radiopharmaceuticals have produced exciting results. Several approaches have been explored, including labelled colloids, monoclonal antibodies, metabolic precursors, receptor-avid ligands and other low molecular weight molecules (Vaidyanathan and Zalutsky 1996; McDevitt et al. 1998). However, patient studies are not recommended until highly specific tumour targeting can be guaranteed. One way to achieve this would be to find more direct routes of administration that confine the agent to the region of the tumour. Alternatively, α-emitting compounds could be used in vitro, e.g. to purge bone marrow and/or peripheral blood of tumour cells.

57.5 RADIOCHEMICAL CONJUGATION

Methods for incorporating the radionuclide into a carrier molecule (radiolabelling) vary from direct labelling, in which the radionuclide is directly attached to a chemical moiety on the molecule, to indirect methods using pre- and post-conjugate approaches. The latter can some-times involve complex reactions and purifications. In both scenarios the stability of attachment and the metabolic profile of the radiolabelled compound must be evaluated in preclinical studies. Radiochemistry aspects of radiolabelled therapeutic conjugates have been extensively reviewed (Wilbur 1992; Blower et al. 1996) and are beyond the scope of this chapter. The choice of labelling approach is governed by a number of factors, which include:

- The chemical properties of the radionuclide and conjugate molecule
- The site of attachment on the carrier molecule
- The overall pharmacokinetic profile of the targeting molecule, the conjugate, the radionuclide and other radiolabelled and unradiolabelled species which could be produced as a result of in vivo metabolism.

57.6 RADIOBIOLOGICAL CONSIDERATIONS

In contrast to fractionated high-dose-rate external-beam radiotherapy, radionuclide therapy delivers continuous irradiation at relatively low dose rates. In addition, the dose rate varies during therapy, increasing initially during the uptake phase, then decreasing at a rate which depends on both the physical half-life of the radionuclide and the biological clearance rate of the radiolabelled compound.

From radiobiological considerations (Wheldon and O'Donoghue 1990), continuous low-dose-rate irradiation may be viewed as being a particularly effective type of radiotherapy, as it ensures that the differential sparing between tumour and normal tissues is maximised. A given dose has a reduced biological effect if delivered at a low dose rate, but the magnitude of this reduction varies for different tissues. For early-responding tissues (bone marrow, epithelium) the effect is small; for late-responding tissues (vascular endothelium, nervous tissue, kidney) it can be large. Only at the later stages of radionuclide therapy, when the dose rate is low, will the advantageous effect of low dose rate (i.e. in sparing late-responding normal tissues) be maximised. Dale (1996) pointed out that the beneficial effect of low dose rate may be reduced by its lack of effect on cell proliferation. Further discussion of these issues can be found in Chapter 55.

The linear-quadratic (LQ) model for cell survival (see Part B), applied originally to external-beam therapy and brachytherapy, can also be applied to radionuclide therapy. When so applied, the model has to be modified to allow for the period of accumulation of the radionuclide in the tissues of interest and to take into consideration the proliferation and repair rate of normal and tumour tissues. Howell et al. (1998) have shown that the modified LQ model can be helpful when choosing a radionuclide for a particular type of therapy such as radioimmunotherapy. In particular, for a given therapeutic effect in a tumour, a longer-lived radionuclide (i.e. a lower overall dose rate) can result in a lower detrimental effect on the bone marrow than a short-lived radionuclide. The optimal physical half-life is about two to three times the biological clearance half-time of the therapeutic compound in the tumour.

57.7 MATCHED PAIRS FOR IMAGING/THERAPY

The dosimetry of radionuclide therapy relies on quantitative in vivo uptake measurements of the therapeutic agents. These measurements can be best achieved by high-resolution positron-emission tomography (PET) imaging if a positron- and β-emitting radioisotope of the same element is available. The use of a radioisotope of a different element, even when linked to the same carrier molecule, could influence the biodistribution. Pre-therapy imaging with an isotope of the same element is therefore essential. Some examples of PET radionuclides, which have analogues that are potentially suitable for therapy (see list in Table 57.1) are ^{64}Cu, ^{71}As, ^{72}As, ^{83}Sr, ^{86}Y, and ^{124}I. Of this group, ^{64}Cu looks the most promising. It is one of the very few radionuclides which combine β, positron, and Auger-electron emissions in its decay. The biological effects of ^{64}Cu emissions in tumour cells have been found to be at least as lethal as those from the β-emitter ^{67}Cu (Apelgot et al. 1989)[*]. This makes ^{64}Cu a good candidate radionuclide both for PET imaging and for targeted therapy (Obata et al. 2003). Also, the chemistry of copper is amenable to the development of compounds, which could potentially exploit certain targeting approaches such as redox mechanisms (Blower et al. 1996).

---

[*] In fact, there is no reason to believe that a given dose delivered by positrons will have a different biological effect to the same dose delivered by electrons of similar energy.

CHAPTER 58

RADIOPHARMACEUTICAL TARGETING FOR UNSEALED SOURCE THERAPY

Maggie Flower and Jamal Zweit

CONTENTS

58.1 INTRODUCTION

The number of available and potential radiopharmaceuticals for targeted radiotherapy has increased greatly since the early days of systemic radionuclide therapy. Table 58.1 provides some examples of radiopharmaceuticals together with current or potential applications.

TABLE 58.1

Examples of Radiopharmaceuticals for Targeted Radiotherapy

Mechanism	Radiopharmaceutical[a]	Application
Incorporation into DNA	^{125}I-IUdR	Choriocarcinoma, colorectal carcinoma
Metabolic	^{131}I-iodide	Thyrotoxicosis, goitre, differentiated thyroid carcinoma
Metabolic	^{131}I-mIBG	Neural crest tumours
Metabolic	^{32}P-phosphate	Polycythaemia vera, essential thrombo-cythaemia
Receptors	^{80m}Br-oestrogen, ^{125}I-tamoxifen; ^{111}In/^{90}Y/^{188}Re-octreotide	Breast; abdominal tumours
Immunologic	^{131}I/^{90}Y/^{177}Lu/^{153}Sm/^{186}Re/^{188}Re/^{67}Cu/^{211}At/^{212}Bi-antibodies	Radioimmunotherapy
Extracellular adsorption	^{32}P-orthophosphate, ^{89}Sr-chloride, ^{153}Sm-EDTMP, ^{186}Re- HEDP, ^{117m}Sn(4+)-DTPA	Relief of bone pain
Extracellular adsorption	^{166}Ho-DOTMP	Multiple myeloma
Radiolabelled cells	^{114m}In-lymphocytes	lymphoma
Intra-arterial	^{131}I-lipiodol, ^{90}Y-microspheres	Hepatic tumours
Intracavitary	^{32}P-phosphate, ^{90}Y-silicate colloid	Malignant effusions
Intra-articular	^{90}Y/^{198}Au-colloid, ^{153}Sm-PHYP, ^{165}Dy-FHMA, ^{169}Er-citrate colloid	Radiation synovectomy
Intravesical	^{67}Cu-595 antibody	Superficial bladder cancer
Intralesional	^{131}I-antitenascin; ^{111}In/^{90}Y/^{188}Re-octreotide	Glioma; abdominal tumours
Endovascular	^{188}Re-perrhenate	Balloon angioplasty

[a] IUdR, iodo-deoxyuridine; mIBG, metaiodobenzylguanidine; EDTMP, ethylene diamine tetramethylenephosphonate; HEDP, hydroxyethylidine diphosphonate; DTPA, diethylenetriamine pentaacetic acid; PHYP, particulate hydroxyapatite; FHMA, ferric hydroxide macroaggregates.

The following sections provide more information on the principles of the tumour-targeting mechanisms involved.

58.2 INTRACELLULAR MECHANISMS

58.2.1 INCORPORATION INTO DNA

Nucleosides (i.e. a subunit of a nucleic acid) radiolabelled with an Auger-electron emitter (e.g. ^{123}IUdR, ^{125}IUdR, ^{77}BrUdR and ^{80m}BrUdR) have been shown using experimental systems to produce extensive DNA damage in the cell nucleus. This damage is due to the fact that these nucleosides bind covalently to DNA so that the Auger-emitting radionuclide is in close proximity to the genome. The greater the proliferation of the cells, the higher the uptake of the radiolabelled nucleoside. To facilitate targeting to tumours and to reduce toxicity in normal cells with high proliferation rates (e.g. gut and bone marrow), the direct injection into tumours of nucleosides labelled with Auger-electron emitters is currently under investigation (Kassis et al. 1996).

58.2.2 METABOLIC PROCESSES

The oldest types of radionuclide therapy are the treatment of thyroid disorders with radioiodine (Clarke 1998) and the treatment of polycythaemia vera with radiophosphorus

(Berk et al. 1986). Both of these treatments make use of metabolic processes to deliver a therapeutic radiation dose to selective tissues. In radioiodine therapy, because iodine is a precursor of the thyroid hormone thyroxine, it is taken up into the follicular cells of the thyroid. In radiophosphorus therapy, incorporation of orthophosphate into the nucleic acids of rapidly proliferating cells is considered to be the targeting mechanism in the treatment of myeloproliferative diseases. The ^{32}P is selectively concentrated by the mitotically-active bone-marrow cells and later incorporated into the calcium phosphate of adjacent bone, from which the marrow is irradiated further.

Meta-iodobenzylguanidine (mIBG) has structural similarities to both guanethidine and the endogenous catacholamines (noradrenaline and adrenaline). ^{131}I-mIBG can therefore be incorporated into the cellular cytoplasm via the catecholamine pathways, and has been used to treat tumours which are derived from the neural crest (e.g. neuroblastoma and phaeochromo-cytoma) (Hoefnagel and Lewington 1998). There has been a lot of research into elucidating the uptake and retention mechanisms of different tumour cell types (Wafelman et al. 1994). Cellular uptake of mIBG occurs by two distinct mechanisms, active and passive. The passive mechanism is nonsaturable and occurs in all cells synthesising the noradrenaline transporter (NAT) molecule. Selective uptake of mIBG depends on cellular expression of the NAT gene (Mairs et al. 1994).

58.3 CELL-SURFACE RECEPTORS

It is possible to attach radionuclides to the cell membrane surface via receptor binding of hormones and antibodies. To achieve an effective radiation dose, the tumours must have sufficient and accessible binding sites. Also, the radiopharmaceutical must be of a high specific activity (amount of radioactivity per mass of the radiolabelled compound) to prevent saturation of the binding sites by unlabelled compound. Other variables to be considered are the receptor density, the receptor homogeneity, any loss of affinity for the receptor of the radiolabelled compound compared to the native (unlabelled) molecule, and whether or not cellular internalisation occurs.

58.3.1 HORMONES

Many tumours have hormone receptors—in the cytoplasm or at the cell membrane—to which radiolabelled hormones, analogues, or antihormone drugs may be targeted. Examples of hormone–receptor binding radiopharmaceuticals for treating breast cancer are ^{80m}Br-oestrogen and ^{125}I-tamoxifen (Hoefnagel 1991).

58.3.2 PEPTIDES

Radiolabelled octreotide, a synthetic 8-amino acid analogue of the naturally occurring regulatory peptide somatostatin, is a potential therapeutic agent. Early clinical treatments (Fjälling et al. 1996) used the Auger-electron emitter ^{111}In, but the β emitters ^{90}Y and ^{188}Re have also been considered (De Jjong et al. 1997; Zamora et al. 1997). One limitation of octreotide is that it only recognises two of the somatostatin subtypes (i.e. SSTR2 and SSTR5). A combination of multiple peptide receptors may be the best way of overcoming the dependence of binding and internalisation on the peptide subtype.

58.3.3 ANTIBODIES

The development of radiolabelled monoclonal antibodies (mAbs) that bind to tumour-associated antigens opened up the possibility of radioimmunotherapy (RIT) (Humm 1986;

Bruland 1995; Wilder et al. 1996). Experimental and clinical RIT studies have been carried out with a wide range of β emitters, and α emitters have also been considered (see Table 58.1). Despite the obvious potential, there are many difficulties with this type of therapy that have limited its success to date. These difficulties are mainly caused by nonspecific binding, slow clearance rates and moderate specificity. An enormous effort is being made to overcome these difficulties. The combination of biological targeting (using radiolabelled mAbs) and physical targeting (with more direct routes of administration, e.g. intrathecal (Papanastassiou et al. 1995), intravesical (Hughes et al. 1997), intralesional (Riva et al. 1994), and direct injection into surgically-created resection cavities (Akabani et al. 1999) may finally enable RIT to realise its inherent potential.

58.4 EXTRACELLULAR MECHANISMS

Radionuclides can be brought into close proximity to tumour cells by various extracellular mechanisms, e.g. by targeting the extracellular osteoid, by using radiolabelled cells, and by using more direct routes of administration.

58.4.1 BONE-SEEKING AGENTS

A variety of bone-seeking radiopharmaceuticals are now available for targeted radio-therapy of bone metastases (Lewington 2002). Although external-beam radiotherapy is of value in the management of well-localised pain, systemic targeted radiotherapy offers the advantage that multiple sites can be treated simultaneously, whilst minimising the normal-tissue absorbed dose. However, the exact mechanism of pain relief is not fully understood. Initial clinical experience was with ^{32}P orthophosphate and ^{89}Sr chloride (Porter et al. 1993). Comparisons have been made with newer agents ^{186}Re-HEDP (De Klerk et al. 1996), ^{153}Sm-EDTMP (Bayouth et al. 1994) and ^{117m}Sn (IV)-DTPA (Krishnamurthy et al. 1997). Early results showed that the speed of onset and duration of pain relief is governed by the half-life of the radionuclide. Early response correlated with shorter half-life, whereas prolonged duration correlated with longer half-life. This is due to the fact that if the half-life is shorter, then the initial dose rate is higher (assuming the same total dose). However, there are other issues to be considered with short half-lives, e.g. if a patient is too ill to undergo treatment, then it may not be possible to use the ordered radiopharmaceutical for another patient. Another application of a skeletal-targeted radiotherapeutic agent is ^{166}Ho-DOTMP, which was used in patients with multiple myeloma (Bayouth et al. 1995). With this agent, the long range of the high-energy β-particles allowed ablation of marrow from radioactivity localized in bone matrix.

58.4.2 RADIOLABELLED CELLS

^{114m}In-labelled lymphocytes have a potential therapeutic role in the management of lymphoma (Sharma et al. 1997). The principle here is that the malignant lymphocytes may retain their ability to migrate and recirculate into the spleen, liver, bone marrow and lymph nodes. After harvesting, labelling with ^{114m}In and reinjecting, the lymphocytes can carry and deliver a therapeutic radiation dose to the lymphoreticular system.

58.4.3 INTRA-ARTERIAL INFUSION

Intra-arterial infusion has been used for the treatment of hepatic malignancies. This is because hepatic tumours derive their blood supply almost entirely from the hepatic artery, whereas normal liver is supplied mainly by the portal venous system. Therefore, administration of the radiolabelled agent into the hepatic artery can be used to deliver high doses to small liver tumours. Various radiopharmaceuticals have been investigated, including ^{131}I Lipiodol (Leung et al. 1994), but these have now generally been replaced by ^{90}Y glass microspheres (Andrews et al. 1994; Ho et al. 1996). More recent studies (Dancey et al. 2000; Salem et al. 2002) have shown that significantly higher doses of radiation can be delivered to hepatocellular carcinoma (HCC) by intrahepatic arterial administration of ^{90}Y-microspheres than by external beam radiation. These groups have also demonstrated that this new alternative treatment appears to be beneficial in unresectable HCC with acceptable toxicity.

58.4.4 INTRACAVITARY ROUTES

A large number of intracavitary routes have been used for targeted radiotherapy (e.g. intraperitoneal, intrapleural, intrapericardial, intrathecal, intra-articular). The basic principle of this type of treatment also applies to intracystic and intravesical administration. In intracavitary treatments, the colloidal particles are assumed to plate out on the cavity/cyst/vessel wall after administration. Therefore, the therapy radionuclides are chosen to match the β range to the thickness of surface to be treated so that underlying tissues are spared. For example, in radiation synovectomy (Deutsch et al. 1993) ^{169}Er colloid (mean β-range in tissue $=0.3$ mm) is used mainly to treat joints in the fingers whereas ^{90}Y colloid (mean β-range $=3.6$ mm) is commonly used to treat knee joints.

58.4.5 INTRALESIONAL INSTILLATION

Direct instillation of a therapy agent is only appropriate for local control of isolated lesions, where the rationale is that the radiopharmaceutical will be retained within the tumour volume, thus improving the therapeutic ratio. Examples are intralesional administration of ^{131}I-labelled antitenascin MAb in patients with glioma (Riva et al. 1994), and infusional brachytherapy of nonresectable pancreatic cancer using ^{32}P chromic phosphate (Order et al. 1994).

58.4.6 ENDOVASCULAR CATHETERS

Local irradiation using a liquid-filled balloon is a new approach for the efficient and cost-effective reduction of restenosis after angioplasty (Kotzerke et al. 2000). The most attractive radionuclide for this purpose is the high-energy-β emitter ^{188}Re, which is available from a ^{188}W/^{188}Re generator system. It could be argued that this type of therapy should not be included here because the liquid source is *sealed* in the balloon rather than unsealed. However, this recent development in therapeutic nuclear medicine will result in the availability of generator-produced ^{188}Re for other unsealed-source therapy applications.

58.5 CURRENT STATUS OF THERAPY USING UNSEALED SOURCES

There has been a considerable development in the clinical practice of radionuclide therapy over the last couple of decades. This development has been driven by the production of new radionuclides, research into new targeting agents, and by the failures and limitations of other methods of treatment, especially for malignant disease. The success of therapy using unsealed sources depends on very careful matching of the physical, chemical and biological properties of the agent to the clinical problem to be addressed. In the treatment of cancer, this depends on

TABLE 58.2

Areas of Development in Radionuclide Therapy

Topic	Issues to Be Evaluated
Radionuclide	Availability, production method, activity levels, specific activity, matching of all physical properties to each specific clinical application
Carrier	Specificity of biological targeting, and dependence on targeting mechanism and/or route of administration
Therapeutic ratio	Possible pharmacological enhancement by vasoactive drugs, radiosensitising agents, pre-targeting (in RIT) to improve uptake and retention in tumour, and increased excretion from blood and normal tissues.
	Potentially more specific targeting pathways involved in tumour aggressiveness and progression
Dosimetry calculations	Accuracy of dose model, validity of assumptions made, propagation of errors, effect of heterogeneity of uptake in tumour
Normal tissue toxicity	Availability of bone-marrow harvest or peripheral stem cell support.
	Improvement in targeting and manipulation of normal tissue kinetics of carrier

factors such as tumour type, lesion size, tumour accessibility and on the presence or absence of metastatic disease. Table 58.2 provides a checklist of most of the issues that need to be considered in relation to this unique form of radiation therapy. Reviews of the status of unsealed-source therapy and molecular targeting with radionuclides can be found in Zanzonico (2000), Goldenberg (2002) and Britz-Cunningham and Adelstein (2003).

If targeted radiotherapy using unsealed sources is to become more successful, various combinations of therapeutic agents and treatment modalities will be involved, in addition to the more direct routes of administration already mentioned. Examples of such combinations are cocktails of:

- β emitters with different ranges to treat tumours of variable size (Wheldon and O'Donoghue 1990; Zweit et al. 1997)
- Radionuclides with different half-lives, e.g. to produce both rapid response and prolonged symptom relief in palliation of bone pain (Lewington 1993)
- Radionuclides with different types of emissions, i.e. β emitters combined with Auger-electron emitters or α emitters (O'Donoghue and Wheldon 1996; Neshasteh-Riz et al. 1998)
- Different mAbs to improve targeting in radioimmunotherapy (Wilder et al. 1996)

Therapeutic agents that use molecules with better targeting and pharmacokinetic profiles should also result in more successful treatments. For example, single-chain versions of mAbs, peptides and specifically targeted small metal or halogen complexes, can improve the therapeutic ratio by virtue of enhanced target uptake and retention and accelerated elimination from normal tissues (Blower et al. 1996; Wilder et al. 1996; Lewis et al. 1999a, 1999b).

The rationale behind the use of multiple treatment modalities in relation to tumour size is illustrated in the following example. In neuroblastoma therapy, total body irradiation or chemotherapy could be used to control micrometastases and small tumours less than 1 mm in size; ^{131}I-mIBG could treat tumours with dimensions between 1 mm and 1 cm; and external beam radiotherapy could be directed at easily detected, large (> 1 cm) tumour masses (Gaze et al. 1995).

PART K: THERAPY WITH UNSEALED SOURCES
REFERENCES

Adelstein, S. J. and Kassis, A. I., Radiobiologic implications of the microscopic distribution of energy from radionuclides, *Nucl. Med. Biol.*, 14, 165–169, 1987.

Akabani, G., Reist, C. J., Cokgor, I., Friedman, A. H., Friedman, H. S. et al., Dosimetry of ^{131}I-labeled 81C6 monoclonal antibody administered into surgically created resection cavities in patients with malignant brain tumors, *J. Nucl. Med.*, 40, 631–638, 1999.

Andrews, J. C., Walker, S. C., Ackermann, R. J., Cotton, L. A., Ensminger, W. D. et al., Hepatic radioembolization with yttrium-90 containing glass microspheres: Preliminary results and clinical follow-up, *J. Nucl. Med.*, 35, 1637–1646, 1994.

Apelgot, S., Coppey, J., Gaudemer, A., Grisvard, J., Guille, E. et al., Similar lethal effect in mammalian cells for 2 radioisotopes of copper with different decay schemes, Cu-64 and Cu-67, *Int. J. Radiat. Biol.*, 55, 365–384, 1989.

Atthey, M., Nahum, A. E., Flower, M. A., and McCready, V. R., Effects of cellular repair and proliferation on targeted radionuclide therapy, *Phys. Med. Biol.*, 45, N15–N20, 2000.

Babich, J. W., A study of the uptake and retention mechanisms of meta-iodobenzylguanidine in neuroblastoma cells, PhD diss., University of London, 1994.

Bayouth, J. E., Macey, D. J., Kasi, L. P., and Fossella, F. V., Dosimetry and toxicity of Samarium-153-EDTMP administered for bone pain due to skeletal metastases, *J. Nucl. Med.*, 35, 63–69, 1994.

Bayouth, J. E., Macey, D. J., Kasi, L. P., Garlich, J. R., McMillan, K. et al., Pharmacokinetics, dosimetry and toxicity of Holmium-166-DOTMP for bone marrow ablation multiple myeloma, *J. Nucl. Med.*, 36, 730–737, 1995.

Berk, P. D., Goldberg, J. D., Donovan, P. B., Fruchtman, S. M., Berlin, N. I., and Wasserman, L. R., Therapeutic recommendations in polycythemia vera based on Polycythemia Vera Study Group protocols, *Semin. Hematol.*, 23, 132–143, 1986.

Blower, P. J., Lewis, J. S., and Zweit, J., Copper radionuclides and radiopharmaceuticals in nuclear medicine, *Nucl. Med. Biol.*, 23, 957–980, 1996.

Bolch, W. E., Bouchet, L. G., Robertson, J. S., Wessels, B. W., and Siegel, J. A., MIRD pamphlet number 17: The dosimetry of nonuniform activity distributions—Radionuclide S values at the voxel level, *J. Nucl. Med.*, 40, 11S–36S, 1999.

Britz-Cunningham, S. H. and Adelstein, S. J., Molecular targeting with radionuclides: State of the science, *J. Nucl. Med.*, 44, 1945–1961, 2003.

Bruland, O. S., Cancer therapy with radiolabelled antibodies—an overview, *Acta Oncol.*, 34, 1085–1094, 1995.

Clairand, I., Bouchet, L. G., Ricard, M., Durigon, M., Di Paola, M., and Aubert, B., Improvement of internal dose calculations using mathematical models of different adult heights, *Phys. Med. Biol.*, 45, 2771–2785, 2000.

Clarke, S. E. M., Radioiodine therapy of the thyroid, in *Nuclear Medicine in Clinical Diagnosis and Treatment*, Murray, I. P. C. and Ell, P. J., Eds., 2nd ed., Churchill Livingstone, Edinburgh, pp. 1049–1062, 1998.

Cole, A., Absorption of 20 eV to 50,000 eV electron beams in air and plastic, *Radiat. Res.*, 38, 7–33, 1969.

Cormack, J., Towson, J. E. C., and Flower, M. A., Radiation protection and dosimetry in clinical practice, in *Nuclear Medicine in Clinical Diagnosis and Treatment*, Murray, I. P. C. and Ell, P. J., Eds., 2nd ed., Churchill Livingstone, Edinburgh, pp. 1651–1677, 1998.

Coulot, J., Ricard, M., and Aubert, B., Validation of the EGS usercode DOSE3D for internal beta dose calculation at the cellular and tissue levels, *Phys. Med. Biol.*, 48, 2591–2602, 2003.

Cross, W. G., Ing, H., and Freedman, N., A short atlas of beta-ray spectra, *Phys. Med. Biol.*, 28, 1251–1260, 1983.

Dale, R. G., Dose-rate effects in targeted radiotherapy, *Phys. Med. Biol.*, 41, 1871–1884, 1996.

Dancey, J. E., Shepherd, F. A., Paul, K., Sniderman, K. W., Houle, S. et al., Treatment of nonresectable hepatocellular carcinoma with intrahepatic ^{90}Y-microspheres, *J. Nucl. Med.*, 41, 1673–1681, 2000.

De Jjong, M., Bakker, W. H., Krenning, E. P., Breeman, W. A. P., van der Pluijm, M. E. et al., Yttrium-90 and indium-111 labelling, receptor binding and biodistribution of [DOTA[0],D-Phe[1],Tyr[3]]octreotide, a promising somatostatin analogue for radionuclide therapy, *Eur. J. Nucl. Med.*, 24, 368–371, 1997.

De Klerk, J. M. H., Van het Schip, A. D., Zonnenberg, B. A., Van Dijk, A., Quirijnen, J. M. S. P. et al., Phase 1 study of rhenium-186-HEDP in patients with bone metastases originating from breast cancer, *J. Nucl. Med.*, 37, 244–249, 1996.

Descalle, M. A., Hartmann Siantar, C. L., Dauffy, L., Nigg, D. W., Wemple, C. A. et al., Application of MINERVA Monte Carlo simulations to targeted radionuclide therapy, *Cancer Biother. Radiopharm.*, 18, 71–79, 2003.

Deutsch, E., Brodack, J. W., and Deutsch, K. F., Radiation synovectomy revisited, *Eur. J. Nucl. Med.*, 20, 1113–1127, 1993.

Fisher, D. R., Assessments for high dose radionuclide therapy treatment planning, *Radiat. Prot. Dosimetry*, 105, 581–586, 2003.

Fjälling, M., Andersson, P., Forssell-Aronsson, E., Grétarsdóttir, J., Johansson, V. et al., Systemic radionuclide therapy using indium-111-DTPA-D-Phe^1-octreotide in midgut carcinoid syndrome, *J. Nucl. Med.*, 37, 1519–1521, 1996.

Flower, M. A. and Chittenden, S. J., Unsealed source therapy, in *Radiotherapy Physics in Practice*, Williams, J. R. and Thwaites, D. I., Eds., 2nd ed., Oxford University Press, Oxford, pp. 297–313, 2000.

Flux, G. D., Guy, M. J., Beddows, R., Pryor, M., and Flower, M. A., Estimation and implications of random errors in whole-body dosimetry for targeted radionuclide therapy, *Phys. Med. Biol.*, 47, 3211–3223, 2002.

Flux, G. D., Webb, S., Ott, R. J., Chittenden, S. J., and Thomas, R., Three-dimensional dosimetry for intralesional radionuclide therapy using mathematical modelling and multimodality imaging, *J. Nucl. Med.*, 38, 1059–1066, 1997.

Gaze, M. N., Wheldon, T. E., O'Donoghue, J. A., Hilditch, T. E., McNee, S. G. et al., Multi-modality megatherapy with [^{131}I]metaiodobenzylguanidine, high dose melphalan and total body irradiation with bone marrow rescue: Feasibility study of a new strategy for advanced neuroblastoma, *Eur. J. Cancer*, 31, 252–256, 1995.

Giap, H. B., Macey, D. J., and Podoloff, D. A., Development of a SPECT-based three-dimensional treatment planning system for radioimmunotherapy, *J. Nucl. Med.*, 36, 1885–1894, 1995.

Goddu, S. M., Howell, R. W., Bouchet, L. G., Bolch, W. E., and Rao, D. V., *MIRD Cellular S values. Self-Absorbed Dose Per Cumulated Activity for Selected Radionuclides and Monoenergetic Electron and Alpha Particle Emitters Incorporated into Different Cell Compartments*, Society of Nuclear Medicine, Reston, VA, 1997.

Goldenberg, D. M., Targeted therapy of cancer with radiolabeled antibodies, *J. Nucl. Med.*, 43, 693–713, 2002.

Guy, M. J., Flux, G. D., Papavasileiou, P., Flower, M. A., and Ott, R. J., RMDP: A dedicated package for I-131 SPECT quantification, registration, and patient-specific dosimetry, *Cancer Biother. Radiopharm.*, 18, 61–70, 2003.

Hartman, T., Lundqvist, H., Westlin, J. E., and Carlsson, J., Radiation doses to the cell nucleus in single cells and cells in micrometastases in targeted therapy with ^{131}I labeled ligands or antibodies, *Int. J. Radiat. Oncol. Biol. Phys.*, 46, 1025–1036, 2000.

Ho, S., Lau, W. Y., Leung, T. W. T., Chan, M., Ngar, Y. K. et al., Partition model for estimating radiation doses from yttrium-90 microspheres in treating hepatic tumours, *Eur. J. Nucl. Med.*, 23, 947–952, 1996.

Hoefnagel, C. A., Radionuclide therapy revisited, *Eur. J. Nucl. Med.*, 18, 408–431, 1991.

Hoefnagel, C. A. and Lewington, V. J., MIBG therapy, in *Nuclear Medicine in Clinical Diagnosis and Treatment*, Murray, I. P. C. and Ell, P. J., Eds., 2nd ed., Churchill Livingstone, Edinburgh, pp. 1067–1081, 1998.

Hoefnagel, C. A., Clarke, S. E. M., Chatal, J. F., Lewington, V. J., Nilsson, S. et al., Radionuclide therapy practice and facilities in Europe, *Eur. J. Nucl. Med.*, 26, 277–282, 1999.

Howell, R. W., Goddu, S. M., and Rao, D. V., Proliferation and the advantage of longer-lived radionuclides in radioimmunotherapy, *Med. Phys.*, 25, 37–42, 1998.

Hughes, O. D. M., Bishop, M. C., Perkins, A. C., Frier, M., Price, M. R. et al., Preclinical evaluation of copper-67 labelled anti-muc1 mucin antibody c595 for therapeutic use in bladder cancer, *Eur. J. Nucl. Med.*, 24, 439–443, 1997.

Humm, J. L., Dosimetric aspects of radiolabeled antibodies for tumor therapy, *J. Nucl. Med.*, 27, 1490–1497, 1986.

Humm, J. L. and Cobb, L. M., Nonuniformity of tumor dose in radioimmunotherapy, *J. Nucl. Med.*, 31, 75–83, 1990.

IAEA (International Atomic Energy Agency), *Biological dosimetry. Chromosomal aberration analysis to dose assessment*, Technical Report Series No. 260, Vienna, IAEA, 1986.

ICRP (International Commission on Radiological Protection), *Radionuclide transformations. Energy and intensity of emissions*, ICRP Publication 38, Annals ICRP, 11–13, 1983.

Johnson, L. S. and Yanch, J. C., Absorbed dose profiles for radionuclides of frequent use in radiation synovectomy, *Arthritis and Rheumatism*, 34, 1521–1530, 1991.

Johnson, L. S., Yanch, J. C., Shortkroff, S., Barnes, C. L., Spitzer, A. I., and Sledge, C. B., Beta-particle dosimetry in radiation synovectomy, *Eur. J. Nucl. Med.*, 22, 977–988, 1995.

Johnson, T. K., MABDOS: A generalized program for internal radionuclide dosimetry, *Computer Methods Programs Biomed.*, 27, 159–167, 1988.

Kassis, A. I., Adelstein, S. J., and Mariani, G., Radiolabeled nucleoside analogs in cancer diagnosis and therapy, *Q. J. Nucl. Med.*, 40, 301–319, 1996.

Kolbert, K. S., Sgouros, G., Scott, A. M., Bronstein, J. E., Malane, R. A. et al., Implementation and evaluation of patient-specific three-dimensional internal dosimetry, *J. Nucl. Med.*, 38, 301–308, 1997.

Kotzerke, J., Hanke, H., and Höher, M., Endovascular brachytherapy for the prevention of restenosis after angioplasty, *Eur. J. Nucl. Med.*, 27, 223–236, 2000.

Krishnamurthy, G. T., Swailem, F. M., Srivastava, S. C., Atkins, H. L., and Simpson, L. J., Tin-117m(4+)DTPA: pharmacokinetics and imaging characteristics in patients with metastatic bone pain, *J. Nucl. Med.*, 38, 230–237, 1997.

Leung, W. T., Lau, W. Y., Ho, S., Chan, M., Leung, N. et al., Selective internal radiation-therapy with intraarterial iodine-131-lipiodol in inoperable hepatocellular-carcinoma, *J. Nucl. Med.*, 35, 1313–1318, 1994.

Lewington, V. J., Targeted radionuclide therapy for bone metastases, *Eur. J. Nucl. Med.*, 20, 66–74, 1993.

Lewington, V. J., A practical guide to targeted therapy for bone pain palliation, *Nucl. Med. Commun.*, 23, 833–836, 2002.

Lewis, J. S., Lewis, M. R., Srinivasan, A., Schmidt, M. A., Wang, J., and Anderson, C. J., Comparison of four Cu-64-labelled somatostatin analogues in vitro and in a tumour-bearing rat model: Evaluation of new derivatives for positron emission tomography imaging and targeted radiotherapy, *J. Med. Chem.*, 42, 1341–1347, 1999a.

Lewis, J. S., Srinivasan, A., Schmidt, M. A., and Anderson, C. J., In vitro and in vivo evaluation of Cu-64-TETA-Tyr3-octreotate, A new somatostatin analogue with improved target tissue uptake, *Nucl. Med. Biol.*, 26, 267–273, 1999b.

Loevinger, R., Budinger, T. F., and Watson, E. E., *MIRD Primer for Absorbed Dose Calculations*, Society of Nuclear Medicine, New York, 1991.

Mairs, R. J., Livingstone, A., Gaze, M. N., Wheldon, T. E., and Barrett, A., Prediction of accumulation of I-131-labeled metaiodobenzylguanidine in neuroblastoma cell-lines by means of reverse transcription and polymerase chain-reaction, *Br. J. Cancer.*, 70, 97–101, 1994.

Malaroda, A., Flux, G. D., Buffa, F. M., and Ott, R. J., Multicellular dosimetry in voxel geometry for targeted radionuclide therapy, *Cancer Biother. Radiopharm.*, 18, 451–461, 2003.

McDevitt, M. R., Sgouros, G., Finn, R. D., Humm, J. L., Jurcic, J. G. et al., Radioimmunotherapy with alpha-emitting nuclides, *Eur. J. Nucl. Med.*, 25, 1341–1351, 1998.

McDevitt, M. R., Finn, R. D., Sgouros, G., Ma, D., and Scheinberg, D. A., An ^{225}Ac/^{213}Bi generator system for therapeutic clinical applications: Construction and operation, *Appl. Radiat. Isot.*, 5, 895–904, 1999.

MIRD (Medical Internal Radiation Dose committee), *Absorbed fractions for photon dosimetry*, MIRD Pamphlet 5, Society of Nuclear Medicine, New York, 1969.

MIRD (Medical Internal Radiation Dose committee), *Radionuclide decay Schemes and nuclear parameters for use in radiation dose estimation*, MIRD Pamphlet 10, Society of Nuclear Medicine, New York, 1975a.

MIRD (Medical Internal Radiation Dose committee), *Absorbed dose per unit cumulated activity for selected radionuclides and organs*, MIRD Pamphlet 11, Society of Nuclear Medicine, New York, 1975b.

Nahum, A. E., Microdosimetry and radiocurability: Modelling targeted therapy with β-emitters, *Phys. Med. Biol.*, 41, 1957–1972, 1996.

NCRP (National Council on Radiation Protection and Measurements), *Protection in nuclear medicine and ultrasound diagnostic procedures in children*, NCRP Report No. 73, NRCP, Washington, DC, 1983.

Neshasteh-Riz, A., Mairs, R. J., Angerson, W. J., Stanton, P. D., Reeves, J. R. et al., Differential cytotoxicity of ^{123}IUdR, ^{125}IUdR and ^{131}IUdR to human glioma cells in monolayer or spheroid culture: Effect of proliferative hetergeneity and radiation cross-fire, *Br. J. Cancer.*, 77, 385–390, 1998.

Obata, A., Yoshimoto, M., Kasamatsu, S., Naiki, H., Takamatsu, S. et al., Intra-tumoral distribution of Cu-64-ATSM: A comparison study with FDG, *Nucl. Med. Biol.*, 30, 529–534, 2003.

O'Donoghue, J. A., Implications of nonuniform tumor doses for radioimmumotherapy, *J. Nucl. Med.*, 40, 1337–1341, 1999.

O'Donoghue, J. A. and Wheldon, T. E., Targeted radiotherapy using Auger electron emitters, *Phys. Med. Biol.*, 41, 1973–1992, 1996.

O'Donoghue, J. A., Bardiès, M., and Wheldon, T. E., Relationships between tumor size and curability for uniformly targeted therapy with beta-emitting radionuclides, *J. Nucl. Med.*, 36, 1902–1909, 1995.

Order, S. E., Siegel, J. A., Lustig, R. A., Principato, R., Zeiger, S. L. et al., Infusional brachytherapy in the treatment of non-resectable pancreatic cancer: A new radiation modality (preliminary report of the phase I study), *Antibody Immunoconjugates Radiopharm.*, 7, 11–27, 1994.

Ott, R. J., Tait, D., Flower, M. A., Babich, J. W., and Lambrecht, R. M., Treatment planning for ^{131}I-mIBG radio-therapy of neural crest tumours using ^{124}I-mIBG positron emission tomography, *Br. J. Radiol.*, 65, 787–791, 1992.

Papanastassiou, V., Pizer, B. L., Chandler, C. L., Zananiri, T. F., Kemshead, J. T., and Hopkins, K. I., Pharmacoki-netics and dose estimates following intrathecal administration of ^{131}I-monoclonal antibodies for the treatment of central nervous system malignancies, *Int. J. Radiat. Oncol. Biol. Phys.*, 31, 541–552, 1995.

Porter, A. T., McEwan, A. J. B., Powe, J. E., Reid, R., McGowan, D. G. et al., Results of a randomized phase-III trial to evaluate the efficacy of strontium-89 adjuvant to local field external beam irradiation in the management of endocrine resistant metastatic prostate cancer, *Int. J. Radiat. Oncol. Biol. Phys.*, 25, 805–813, 1993.

Prestwich, W. V., Chan, L. B., Kwok, C. S., and Wilson, B. C., Dose point kernels for beta-emitting radioisotopes, in *Proc. 4th Int. Radiopharmaceutical Dosimetry Symposium*, in Oak Ridge, US Government Printing Office, Washington, 545–561, 1985.

Prestwich, W. V., Nunes, J., and Kwok, C. S., Beta dose point kernels for radionuclides of potential use in radio-immunotherapy, *J. Nucl. Med.*, 30, 1036–1046, 1989.

Puncher, M. R. B. and Blower, P. J., Radionuclide targeting and dosimetry at the microscopic level: The role of microautoradiography, *Eur. J. Nucl. Med.*, 21, 1347–1365, 1994.

RADAR (Radiation Dose Assessment Resource), http://www.doseinfo-radar.com.

Riva, P., Arista, A., Tison, V., Sturiale, C., Franceschi, G. et al., Intralesional radioimmunotherapy of malignant gliomas—an effective treatment in recurrent tumors, *Cancer*, 73, 1076–1082, 1994.

Salem, R., Thurston, K. G., Carr, B. I., Goin, J. E., and Geschwind, J. F., Yttrium-90 microspheres: Radiation therapy for unresectable liver cancer, *J. Vasc. Interv. Radiol.*, 13, S223–S229, 2002.

Sgouros, G., Bone-marrow dosimetry for radioimmunotherapy—theoretical considerations, *J. Nucl. Med.*, 34, 689–694, 1993.

Sharma, H. L., Cowan, R. A., Murby, B., Owens, S., Nuttall, P. M. et al., Treatment of lymphoid cell malignancy with In-114m labelled autologous lymphocytes, *Anticancer Res.*, 3B, 1815–1821, 1997.

Siegel, J. A., Wessels, B. W., Watson, E. E., Stabin, M. G., Vriesendorp, H. M. et al., Bone marrow dosimetry and toxicity for radioimmunotherapy, *Antibody, Immunoconjugates, and Radiopharm.*, 3, 213–233, 1990.

Siegel, J. A., Zeiger, L. S., Order, S. E., and Wallner, P. E., Quantitative bremsstrahlung single photon emission computed tomographic imaging: use for volume, activity, and absorbed dose calculations, *Int. J. Radiat. Oncol. Biol. Phys.*, 31, 953–958, 1995.

Simpkin, D. J. and Mackie, T. R., EGS4 Monte Carlo determination of the beta dose kernel in water, *Med. Phys.*, 17, 179–186, 1990.

Smith, T., Crawley, J. C., Shawe, D. J., and Gumpel, J. M., SPECT using Bremsstrahlung to quantify ^{90}Y uptake in Baker's cysts: its application in radiation synovectomy of the knee, *Eur. J. Nucl. Med.*, 14, 498–503, 1988.

Stabin, M. G., MIRDOSE: Personal computer software for internal dose assessment in nuclear medicine, *J. Nucl. Med.*, 37, 538–546, 1996.

Stinchcomb, T. G. and Roeske, J. C., Values of "S", $<z_1>$, and $<z_1^2>$ for dosimetry using alpha-particle emitters, *Med. Phys.*, 26, 1960–1971, 1999.

Tagesson, M., Ljungberg, M., and Strand, S.-E., A Monte Carlo program converting activity distributions to absorbed dose distributions in a radionuclide treatment planning system, *Acta Oncol.*, 35, 367–372, 1996.

Thierens, H. M., Monsieurs, M. A., Brans, B., Van Driessche, T., Christiaens, I., and Dierckx, R. A., Dosimetry from organ to cellular dimensions, *Comput. Med. Imaging Graph.*, 25, 187–193, 2001.

Vaidyanathan, G. and Zalutsky, M. R., Targeted therapy using alpha emitters, *Phys. Med. Biol.*, 41, 1915–1931, 1996.

Wafelman, A. R., Hoefnagel, C. A., Maes, R. A. A., and Beijnen, J. H., Radioiodinated metaiodobenzylguanidine: A review of its biodistribution and pharmacokinetics, drug interactions, cytotoxicity and dosimetry, *Eur. J. Nucl. Med.*, 21, 545–559, 1994.

Wessels, B. W. and Griffith, M. H., Miniature thermoluminescent dosimeter absorbed dose measurements in tumor phantom models, *J. Nucl. Med.*, 27, 1308–1314, 1986.

Wheldon, T. E., Radionuclide therapy of cancer: Particle range and therapeutic effectiveness, *Nucl. Med. Commun.*, 14, 408–410, 1993.

Wheldon, T. E. and O'Donoghue, J. A., The radiobiology of targeted radiotherapy, *Int. J. Radiat. Biol.*, 58, 1–21, 1990.

Wilbur, D. S., Radiohalogenation of proteins: an overview of radionuclides, labeling methods, and reagents for conjugate labeling, *Bioconj. Chem.*, 3(6), 433–470, 1992.

Wilder, R. B., Denardo, G. L., and Denardo, S. J., Radioimmunotherapy—recent results and future directions, *J. Clin. Oncol.*, 14, 1383–1400, 1996.

Zamora, P. O., Bender, H., Gulhke, S., Marek, M. J., Knapp, F. F. et al., Pre-clinical experience with Re-188-rc-160, a radiolabeled somatostatin analog for use in peptide-targeted radiotherapy, *Anticancer Res.*, 17, 1803–1808, 1997.

Zanzonico, P. B., Internal radionuclide radiation dosimetry: A review of basic concepts and recent developments, *J. Nucl. Med.*, 41, 297–308, 2000.

Zanzonico, P. B., Bigler, R. E., Sgouros, G., and Strauss, A., Quantitative SPECT in radiation dosimetry, *Semin. Nucl. Med.*, 19, 47–61, 1989.

Zanzonico, P. B., Binkert, B. L., and Goldsmith, S. J., Bremsstrahlung radiation exposure from pure beta-ray emitters, *J. Nucl. Med.*, 40, 1024–1028, 1999.

Zweit, J., Radionuclides and carrier molecules for therapy, *Phys. Med. Biol.*, 41, 1905–1914, 1996.

Zweit, J., Nahum, A., Gianolini, S., Flower, M., Flux, G., and McCready, V., Radionuclide cocktails for therapy of tumours with a broad range of sizes, *J. Nucl. Med.*, 38(Suppl.), p. 226P, 1997.

PART L

RADIATION PROTECTION IN RADIOTHERAPY

Editors: Philip Mayles and Jean-Claude Rosenwald

INTRODUCTION

In considering radiation protection for radiotherapy, a balance must be achieved between mitigating risk and the cost of the measures required to do so. To assess this, an understanding of the risks associated with radiation treatment is required, and this is the subject of Chapter 59.

The use of radiation is governed by a regulatory framework, and Chapter 60 considers issues of regulation. This regulatory framework is different in each country, but the principles are common to all. This Part is mainly concerned with minimising the exposure of staff and members of the public that may also include the protection of the foetus on the rare occasions that a patient may require radiotherapy while pregnant. When dealing with issues of legislation, U.K. and European law is sometimes referred to in order to provide a coherent context, but wherever possible, the text has been written to be generally applicable. A summary of U.K. legislation can be found in Appendix L.1.

The final chapter (Chapter 61) deals with the application of these principles to practical radiotherapy. This includes the design of treatment rooms and certain practical issues relating to patient treatment.

An example of the calculations necessary in the design of a radiotherapy treatment room is included in Appendix L.2.

In reducing exposure to radiation there are three factors that interact; time, distance and shielding. The interplay of these factors is a common theme throughout Chapter 61. In an area where the dose rate is significant it is self evident that the dose received is directly proportional to the time spent there. The time should therefore be restricted to the minimum required to perform the necessary tasks. The fact that dose rates from a small source of radiation decrease as the inverse square of the distance from the source means that increasing the distance from the source can be very effective in reducing the dose rate. In brachytherapy, for example, the use of long forceps to handle sources is essential, but the distance factor is also an important part of shielding calculations for external-beam therapy. When time and distance are insufficient to reduce dose rates to a safe level, or in order to further improve staff protection, it is necessary to introduce shielding. This is obviously the case for the walls of a room in which teletherapy equipment or high-dose-rate brachytherapy equipment is located. Shielding is also needed for transportation of sources and for sealed or unsealed source handling. However, it should be borne in mind that individual protective shielding reduces mobility and may therefore conflict with the need to reduce the time. This becomes a consideration when deciding whether it is appropriate to wear protective clothing, which will encumber the operator.

With unsealed sources, personnel are exposed to the risk of both external exposure and contamination. Contamination may result in radioactive particles being incorporated into body tissues, either through superficial skin contamination or inhalation or ingestion. The protection against such risks is dealt with in Section 61.3.4.

Radiation protection should also address the issue of protection of the patient in terms of ensuring that the dose delivered to the target as well as to surrounding normal tissues is as intended. This is the role of the quality assurance measures described in Part H and is the reason that radiation safety regulators have an interest in the oversight of radiotherapy. It is well known that radiation is a carcinogen as well as being a cure for cancer and there is growing interest in second malignancies that may arise as a result of treatment (NRPB 2000; Bhatia et al. 2003). These issues are briefly considered at the end of the Chapter 61.

CHAPTER 59

THEORETICAL BACKGROUND TO RADIATION PROTECTION

Mike Rosenbloom

CONTENTS

59.1 BASIC CONCEPTS

59.1.1 DETERMINISTIC AND STOCHASTIC EFFECTS

In Part B, Gordon Steel describes the process whereby ionising radiation acts directly on the DNA of the nucleus causing cell death, as well as the mechanisms for subsequent repair. For large doses of radiation, the damage, which may occur early or late in the life of the organism, is more or less predictable and its severity may be determined by the magnitude of the dose and the timing of its delivery. For this reason, such damage is described as *deterministic*. In general, the dose–response function is sigmoid and exhibits a threshold. This threshold is large compared to the doses that will be experienced by staff and public, and with the rare exception of the foetus of a pregnant patient undergoing therapy, deterministic effects will not be discussed within this chapter*.

At low doses, cellular repair mechanisms will usually take place and the organism suffers no damage. Occasionally, the cellular repair process may be imperfect, with the result that cells retain their ability to replicate, but with coding damage to the DNA that causes the propagation of errors in succeeding generations of cells. The consequence of this might ultimately be cancer, or if the somatic cells of the gonads are involved, the production of genetic effects in successive generations.

The consequences of this damage may be catastrophic for the individual in whom it occurs. However, as it potentially requires only one nucleus to be damaged, the magnitude of the dose determines its *likelihood* rather than its *severity*. For this reason, damage of this nature is described as *stochastic* (i.e. determined by chance).

To estimate the effect of low doses of radiation, it is necessary to know the risk (i.e. the *probability*) of cancer induction or hereditary effects as a function of exposure to radiation.

59.1.2 EQUIVALENT DOSE

Before considering the effects of low doses of radiation on the human body, the appropriate units in which to express the dose will be discussed. The efficacy of ionising radiation will be determined not only by the biological sensitivity of an individual organ, but also by the density of ionisation produced in the vicinity of the cell nucleus. Particles such as α-particles and neutrons are densely ionising. This is represented in terms of the linear energy transfer (LET), which for these particles is high (see also Chapter 49). The *unrestricted linear energy transfer* is defined as dE/dl, where dE is the energy lost by a charged particle in traversing a distance dl. High-LET particles will be more damaging for a given physical dose because they are able to cause multiple damage within an individual cell. This leads to the definition of $Q(L)$ (ICRP 1990), which is the *quality factor* as a function of unrestricted linear energy transfer, L. ICRP defines this as 1 for $L < 10$; as $(0.32L - 2.2)$ for L between 10 and 100, and as $300/\sqrt{L}$ for higher values. For radiation protection purposes, a simpler quantity is required whose magnitude is likely to be related to the biological effects of the radiation. For this purpose, the *equivalent dose*[†] was defined in 1977.

---

* The statement that the damage at high doses is predictable and determined by the dose is simplistic. Damage to individual cells will be randomly determined. Organ damage results from damage to large numbers of individual cells and this smoothes out random fluctuations giving rise to the well known sigmoid curve. The gradient of the curve is an indication of the statistical nature of the process.

† The terms *equivalent dose* and *dose equivalent* are often used interchangeably. Strictly *dose equivalent* should be used for the product of absorbed dose and *quality factor*, where the quality factor is defined in terms of the LET. *Equivalent dose* should be used for the product of the absorbed dose and the *radiation weighting factor*, which is defined by ICRP in terms of the type and energy of the radiation, as in Equation 59.1. ICRP (2007) is proposing to replace the term *equivalent dose* with *radiation-weighted dose*.

The equivalent dose, H_T, to an organ, T, exposed to a mixture of radiations, R, is defined as

$$H_T = \sum_R W_R D_{T,R} \tag{59.1}$$

where W_R is a dimensionless weighting factor appropriate to radiation R, $D_{T,R}$ is the dose in Gy to the organ T, and H_T is the equivalent dose.

To differentiate the equivalent dose from the physical dose defined in Chapter 6, its unit has been given a special name, the sievert (Sv). When $D_{T,R}$ is expressed in gray (Gy), H_T is expressed in Sv. In radiation protection, it is often useful to refer to millisieverts (mSv) or microsieverts (μSv).

For photons and electrons, the value of W_R is 1 and consequently the physical dose (Gy) and the equivalent dose (Sv) are numerically identical. For protons, W_R is 5 and for α-particles 20. The value for neutrons varies continuously with energy between 5 (at very low and very high energies) and 20 (between 0.1 MeV and 2 MeV) (ICRP 1990). Ideally, a dosimeter calibrated to read in mSv should have an energy response that matches the weighting factor curve, but this is not easy to achieve in practice. The draft ICRP publication* (ICRP 2006d) recommends a numerical formula for neutrons of energies greater than 1.0 MeV:

$$W_R = 5.0 + 17.0 \exp\left[\frac{-(\ln(2E_n))^2}{6}\right] \tag{59.2}$$

where E_n is the energy of the neutrons. For energies below 1 MeV, the formula becomes:

$$W_R = 2.5 + 18.2 \exp\left[\frac{-(\ln E_n)^2}{6}\right] \tag{59.3}$$

59.1.3 EFFECTIVE DOSE

The discussion thus far has assumed that all body tissues are equally irradiated. This will often not be the case and it is then necessary to assess the likely effect taking doses to individual organs into effect. To facilitate this, the concept of *effective dose* has been defined. The effective dose, E, to the individual is defined as:

$$E = \sum_T W_T H_T \tag{59.4}$$

where, considering each organ, T, within the body, H_T is the equivalent dose defined in equation 59.1 and W_T is a weighting factor which takes into account the ability of that organ to contribute to the overall risk of harm. By definition:

$$\sum_T W_T = 1 \tag{59.5}$$

This double summation is used to calculate the effect of nonuniform irradiation on the body where individual organs have differing individual sensitivities to ionising radiation. Values of W_T are given in Table 59.1. The derivation of the values in this table will be explained in Section 59.2.8, but we first need to consider the issue of risk estimation.

---

* These formulae should not be used until the new ICRP recommendations are published.

TABLE 59.1

Weighting Factors for Different Body Organs Recommended by ICRP

Organ	Weight, W_T (ICRP 1990)	Proposed (ICRP 2006b) Weight, W_T
Gonads	0.20	0.08
Red bone marrow	0.12	0.12
Colon	0.12	0.12
Lung	0.12	0.12
Stomach	0.12	0.12
Bladder	0.05	0.04
Breast	0.05	0.12
Liver	0.05	0.04
Oesophagus	0.05	0.04
Thyroid	0.05	0.04
Skin	0.01	0.01
Brain	Included in remaining organs	0.01
Salivary glands	Not specifically included	0.01
Bone surfaces	0.01	0.01
Remaining organs	0.05[a]	0.12[b]

[a] The remaining organs are adrenals, brain, upper large intestine, small intestine, kidney, muscle, pancreas, spleen, thymus and uterus.

[b] The ICRP (2006c) list is adrenals, extrathoracic airways, gall bladder, heart wall, kidneys, lymphatic nodes, muscle, oral mucosa, pancreas, prostate, small intestine, spleen, thymus and uterus/cervix. The new weighting factors should not be used until the new recommendations are published by ICRP.

59.2 RISK ESTIMATION

The exponential attenuation of x- and γ-rays in passing through protective barriers means that it is impossible to completely eliminate exposure of staff and public arising from the use of ionising radiation in medicine and they will in any case be exposed to natural background radiation. Natural background radiation is responsible for a worldwide average dose of 2.4 mSv per year, of which 1.2 mSv is due to radon and its daughter products (Thorne 2003). The risk associated with the genetic effects of radiation is smaller than the risk of induction of cancer (UNSCEAR 2001), so it is the latter that is the principal consideration in determining dose limits. It is also important to ensure that the costs of reducing exposure to a given level are commensurate with the benefits (i.e. reduction in the excess incidence of cancer) so obtained. This cost benefit comparison cannot be made without knowledge of the risk of developing cancer as a function of exposure.

To calculate the risk associated with radiation it is necessary to have a model for the effect of the radiation. In a given unirradiated population, there will be a certain incidence of cancer that will depend on the population and will vary with factors such as age and gender. The system of risk estimation seeks to establish the excess risk of dying of cancer, together with an allowance for associated morbidity, resulting from a specific equivalent dose of radiation. Useful reviews of the issues involved are contained in the UNSCEAR reports (2000, 2001) and in the review article by Little (2003).

59.2.1 SOURCES OF DATA FOR RISK ESTIMATES

The qualitative evidence for the carcinogenic nature of ionising radiation comes from a variety of sources. These are listed in Table 59.2. Although most of these establish beyond doubt that ionising radiation is carcinogenic, very few of them provide good quantitative

TABLE 59.2

Sources of Data for Risk Estimations

Exposed human populations
   Atomic bombs
     Japanese survivors
     Marshall Islanders
   Medical therapy
     Pelvic radiotherapy (cervix)
     Spinal radiotherapy (ankylosing spondylitis)
     Neck and chest radiotherapy (thyroid)
     Treatment of Hodgkins Disease
     Scalp irradiation (tinea capitis)
     Breast radiotherapy
     Radium treatment
   Medical diagnosis
     Multiple fluoroscopies (breast)
     Prenatal irradiation
     Thorotrast injections
   Occupational
     Uranium miners
     Radium ingestion (dial painters)
     Registered radiation workers

data, and it is fair to say that almost our entire knowledge of risk estimates arises from the prolonged and ongoing studies of survivors of the atomic bombs dropped on the Japanese cities of Hiroshima and Nagasaki. These studies have relied on observations of the excess incidence of cancer and the correlation of these observations with estimates of exposure made on the basis of mathematical modelling and thermoluminescent dosimetry (using building materials exposed at the time of the explosions). These risk estimates are handicapped by the fact that the data are obtained from persons exposed to large doses of radiation delivered more or less instantaneously whereas the effects of small doses received over prolonged periods of exposure are of primary interest here. They are also handicapped by lack of precision in the dose estimates, the fact that many of those exposed are still alive and may still yet develop cancer, and the difference in the response to radiation between different population groups.

Studies of large groups of workers occupationally exposed to low levels of radiation are beginning to yield results. However, the data are subject to such large uncertainties that it is only possible to say that the risk coefficients are not inconsistent with those yielded by studies of atomic bomb survivors (Cardis et al. 1994; NRPB 2000). A more recent analysis with larger numbers in the U.K. was also consistent with these data (Muirhead et al. 1999). Data from many sources continue to be monitored by UNSCEAR (2000) and others.

Evidence for hereditary effects comes from animal studies with mice. There is no evidence of hereditary damage in the offspring of atom bomb survivors, but an estimate of the risk of genetic effects is made on the basis of these animal studies together with recent knowledge of the mutation rate in humans. These sources of data lead to an estimate of the doubling dose for mutations (i.e. the dose that will result in twice as many mutations as occur naturally) of 1 Sv (UNSCEAR 2001). This estimate of the doubling dose has been stable for a number of years, but the estimate of the genetic risk associated with the doubling dose of 1 Sv is subject to considerations such as whether a mutation is compatible with a live birth and whether its consequences are lethal. ICRP 60 (1990) gave the risk associated with doubling the number of mutations as 1 case in 100 live births (1%), but more recent risk estimates are as low as 1 in,

330 (BRER 2006). In any case, it is clear that the principal risks are of cancer induction rather than the genetic risk.

59.2.2 Dose and Dose-Rate Effectiveness Factor

Current risk estimates are based on the assumption that risks are given by a linear quadratic function of dose (or equivalent dose):

$$f(D) = (\alpha_1 D + \alpha_2 D^2) \exp[-(\beta_1 D + \beta_2 D^2)]$$

where $f(D)$ is the risk estimate at dose D, and α_1 and α_2 are coefficients of the linear and quadratic terms for the induction of stochastic effects. β_1 and β_2 are terms representing cell killing at higher doses (UNSCEAR 2000). There is some direct evidence from the observed incidence of leukaemia as a function of dose amongst the atomic bomb survivors that the response is indeed linear quadratic, but for solid cancers it is likely that a linear relation is the best fit to the data and the quadratic term is zero. At the low dose rates encountered in occupational and public exposure it is to be expected that repair processes can mitigate the effect of the radiation so that the effect is less than the effect at the high doses (>200 mSv) received by the atom bomb survivors. To express this reduction in effect at low doses, it is therefore appropriate to reduce the estimate of risk by a factor called the *dose and dose-rate effectiveness factor* (DDREF). The ICRP (1990) recommends that the risks estimated on the basis of high-dose irradiation should be reduced by approximately a factor of two.

59.2.3 Low Dose Threshold Theories

Because of the low incidence of radiation-induced cancer compared to the natural incidence of the disease, there is little direct evidence of the magnitude of the risk for doses below 200 mSv. The possibility of a threshold below which radiation is not harmful is a subject of much controversy. Studies of large cohorts, such as the study of British radiologists, have failed to show any increased risk associated with the low doses they received (Berrington et al. 2001). Some have argued that very small doses of radiation may even be beneficial (so-called *radiation hormesis*) (Cameron and Moulder 1998), bearing in mind the existence of natural background radiation. The Board on Radiation Effects Research (BRER 2006), ICRP (1990, 2006a) and the radiation protection community have, however, taken the view that in the absence of clear evidence to the contrary, a linear no-threshold model should be assumed. An excellent review of these issues is given by Johansson (2003). There is also a view that the effect of radiation at low doses is underestimated by the linear model (Hall 2004). The majority view of the Committee Examining Radiation Risks of Internal Emitters (CERRIE 2004) is that the linear no-threshold model is the most appropriate.

59.2.4 Relative and Absolute Risks

The risk of cancer induction can be expressed either as an *added* absolute risk or as the excess *multiplicative* risk relative to the normal risk of contracting cancer. The *excess absolute risk* (EAR) model assumes that a unit of exposure results in an absolute number of increased cancers. The *excess relative risk* (ERR) model defines risk as a multiplicative factor per unit exposure operating on the natural risk, in the absence of radiation, for a given cancer in a given population. The justification for this model derives from the possibility that cancer induction is a multi-step process of which radiation is not the only factor. In extrapolating from the risk measured in the group of Japanese atomic-bomb survivors (in which men of military age were under-represented) to other population groups in other parts of the world where the natural incidence of cancer may be different, the relative risk model has been preferred. Relative risks are also particularly appropriate when considering irradiation at a young age (BRER 2006). However, there is some evidence in particular cases, for example breast (Preston 2002) and leukaemia (Little 2003), that the excess absolute risk model provides a better correlation of cancer incidence and exposure than the excess relative risk model. In any case the aim is to estimate the absolute increase in risk in a given population.

59.2.5 EFFECT OF AGE AT EXPOSURE AND LIFETIME RISK

Cancer induction by radiation has a latent period which is generally taken to be 10 years for solid cancers and 2 years for leukaemia. It might be expected that the risk of dying of cancer following a single radiation episode would peak after a few years and would then revert to the background level. This has only been demonstrated for leukaemia (NCRP 1993); for other cancers, it appears that the effect of a radiation exposure continues indefinitely. The issue can only be resolved when the younger atomic bomb survivors have died. Because the onset of cancer may occur at any time after the irradiation, we need to consider the excess lifetime risk (ELR). However, because everyone has to die of something, the ELR for all causes is zero because, for example, an increase in cancer as cause of death will be associated with a reduction in heart disease as a cause of death. The UNSCEAR report (2000), therefore, recommends the *risk of exposure-induced death* (REID) as a measure of the impact of radiation exposure. It may also be useful to consider the *loss of life-expectancy* (LLE). It might be expected that because of the greater length of time during which the cancer has time to occur, the number of excess cancer deaths would be greater for those irradiated at a young age than for those irradiated late in life. However, although there is an increased risk for those exposed during childhood, for adults the REID is found to vary little with the age at exposure (UNSCEAR 2000), although this remains controversial. This can be explained on the basis that multiple events are necessary for cancer induction so that the latency for older people will be less than for younger ones (BRER 2006). This will be particularly relevant when considering second cancers associated with radiotherapy treatment (see Section 61.5.2).

59.2.6 THE CONCEPT OF DETRIMENT

To quantify the risk associated with radiation, it is necessary to have some measure of the damage done by the radiation. The negative impact of radiation varies in severity from death at one extreme, to the induction of an easily cured cancer at the other. The ICRP uses the concept of *detriment* as a measure of the total negative impact on health that would eventually be experienced by an exposed group and its descendants as a result of exposure to radiation. The definition of *detriment* in Report 26 (ICRP 1977) used the expected number of cases of a radiation-induced health effect, weighted by the severity of the effect. The weighting factor was 1 for death, with smaller factors for other less severe effects. Although the number of cancer deaths arising from a certain amount of radiation is relatively unambiguous, the impact of the nonfatal cancers is harder to quantify and consequently the detriment calculation can be complicated.

59.2.7 RISK COEFFICIENTS

To calculate the detriment associated with the exposure, it is necessary to know the probability that a given dose will result in harm. The probability of injury or harm (detriment) per unit equivalent dose is called the *risk coefficient* arising from a particular *risk factor*. Detriment values for whole-body exposure are given in Table 59.3. Also shown are the proposed values in

TABLE 59.3

Nominal Risk Coefficients (i.e. Probability) for Stochastic Effects

Exposed Population	Detriment (10^{-2} Sv^{-1})			
	Fatal Cancer	Nonfatal Cancer	Severe Hereditary Effects	Total
Adult workers	4.0 (4.1)^{a,b}	0.8	0.8 (0.1)^{a,b}	5.6 (4.2)^{a}
Whole population	5.0 (5.5)^{a,b}	1.0	1.3 (0.2)^{a,b}	7.3 (5.7)^{a}

[a] Figures in brackets are the proposed values from the report of ICRP Committee 1 (ICRP 2006c).
[b] In the proposed ICRP recommendations the detriment from fatal and nonfatal cancers is combined by estimating the detriment associated with a nonfatal cancer.
Source: From ICRP (International commission on Radiological Protection), ICRP Publication 60, Recommendations of the International Commission on Radiological Protection, *Ann. ICRP,* 21, 1990.

the draft ICRP report in which the detriment associated with fatal and nonfatal cancer is combined.

59.2.8 RISK COEFFICIENTS FOR PARTIAL BODY IRRADIATION

As described in Section 59.1.3, when only part of the body is irradiated we can use the effective dose to estimate the detriment. In practice, the values in Table 59.3 are derived by first considering the detriment for individual organs separately. Data from ICRP 26 (1977), ICRP 60 (1990) and the draft 2007 recommendations (ICRP 2007) are shown in Table 59.4.

The calculation starts with an estimate of the cancer incidence and the number of cancers that result in the death of the patient[*]. Patients who die are said to have suffered a relative detriment of 1.0. It is, however, recognised that cancer treatment is also traumatic and those who do not die also suffer some detriment. To represent this a weighting factor, or relative detriment, is required which will be used as a multiplier for the number of cancer survivors so that they can be added to the number who die to obtain an effective, or *lethality adjusted,* number of *fatalities.* This weighting factor was numerically equal to the lethality fraction in ICRP 60, but in the new draft it is proposed to modify this slightly for the more curable cancers to ensure that there is a minimum factor. Different cancers result in different numbers of years of life lost and a further weighting factor is applied representing the relative life lost. This produces a figure for the overall detriment for each cancer. The weighting factors in Table 59.1 are based on the relative detriment, although somewhat simplified.

The ICRP calculations based on effective dose are intended to be used to calculate the effect of irradiation on large numbers of people—either the adult population if the irradiation is occupational exposure, or the whole population. They are not intended to calculate the expected detriment to a particular individual. Effective dose is not appropriate for use when high doses are delivered to individual organs (ICRP 1990; CERRIE 2004).

Note the improvement in the cure rates of cancer between the ICRP 1990 and ICRP 2006c reports. The increase in the estimate of the number of breast cancers caused by radiation is also noteworthy. The current estimate of the impact of genetic effects is significantly reduced. This is because of the lack of evidence of such effects, perhaps associated with the unviability of mutated cells.

---

[*] In ICRP 60 and ICRP 26, the data are presented as shown in Table 59.4 in terms of the total number of fatal cancers. In the new draft the data are presented starting with cancer incidence and have been adapted in the table to be consistent with the earlier documents. The reader is referred to the reports themselves for a full discussion of the concepts used. A somewhat simplified discussion is given here to bring out the salient points and to assist the reader's understanding of the reports.

TABLE 59.4

Fatal Risk Coefficients (Probabilities) and Detriment for Individual Organs. Values Represent the Total Number of Deaths per Year among a Population of 10,000 of All Ages Receiving a Dose of 1 Sv

Organ	ICRP 26 (1977) Fatal Cancers	ICRP 60 (1990) Fatal Cancers[a]	Lethality	Detriment	Proposed ICRP Fatal Cancers[a]	Lethality	Detriment
Gonads		100[a]		133.3	16[a]	0.80	25.4
Ovary		10	0.67	14.6			9.9
Red bone marrow	20	50	0.99	104.0	28	0.67	61.5
Colon		85	0.55	102.7	31	0.48	47.9
Lung	20	85	0.95	80.3	101	0.89	90.2[b]
Stomach		110	0.90	100.0	65	0.83	67.7
Bladder		30	0.50	29.4	12	0.29	16.7
Breast	25	20	0.50	36.4	33	0.29	79.8
Liver		15	0.95	15.8	29	0.95	26.6
Oesophagus		30	0.95	24.2	14	0.93	13.1
Thyroid	5	8	0.10	15.2	2	0.07	12.7
Skin		2	0.00	4.0	2	0.002	4.0
Bone surfaces	5	5	0.70	6.5	3	0.45	5.1
Remaining organs	50	50	0.71	58.9	71	0.49	113.5
Total	125	500[a]		725.3	414[a]		574

We will use the instance of oesophagus cancer to show how the calculation works, using the figures from the draft document. If 1 Sv is given to 10,000 people, 17 will contract oesophagus cancer and of these 93% will die. The 1.9 survivors will have suffered severely from their treatment and will probably suffer lasting side effects. Thus instead of recording 15.8 lethal cancers the 1.1 survivors are said to have suffered a detriment of 0.937 so the effective number of deaths is 16.9. This is multiplied by the relative cancer free life lost (0.87) to give a detriment of 15.

a   Note that the values given for the gonads are for fatalities related to hereditary effects. These values have not been added to the totals.

b   As an example of how the calculation is carried out this figure is obtained as follows. If 1 Sv is given to 10,000 people, 114 will contract lung cancer and of these 89% will die. The 13 survivors will have suffered severely from their treatment and will probably suffer lasting side effects. Thus instead of recording 101 lethal cancers as the detriment, each of the survivors are said to have suffered a quality of life detriment of 0.1 + 0.89(1 − 0.1)=0.901; so the effective number of *lives* lost is 113. This is multiplied by the *relative life lost* (0.80) to give a detriment of 90.

Sources: From ICRP (International Commission on Radiological Protection), Publication 26, Recommendations of the International Commission on Radiological Protection, *Ann. ICRP, 1,* 1977; ICRP, ICRP Publication 60. Recommendations of the International Commission on Radiological Protection, *Ann. ICRP, 21,* 1990. The figures in the final three columns are from the draft of the revised ICRP recommendations and represent the trends of ICRP thinking. They should not be taken as definitive values. Committee 1 Task Group Report. Biological and Epidemiological Information on Health Risks Attributable to Ionising Radiation: A Summary of Judgements for the Purposes of Radiological Protection of Humans. February 2006. www.icrp.org/Health_risks.pdf on 24th August 2006.

59.2.9   EFFECT OF RADIATION ON THE DEVELOPING FOETUS

Studies of the offspring of mothers pregnant at the time of the Hiroshima and Nagasaki bombs have shown a number of deterministic effects on the brain of the developing foetus. These effects, which are summarised in NCRP Commentary Number 9 (NCRP 1994) and in ICRP Report 84 (2000), are strongly linked to the post-conception age of the foetus. They include mental retardation, loss of Intelligence Quotient (IQ) points and small head size. Their incidence is greatest in the period from 8 weeks to 15 weeks post conception, is smaller between 16 weeks and 25 weeks and very low, or absent, subsequently (Otake and Schull 1998). The effects occur at low doses, perhaps as low as 100 mSv, and their incidence is consistent with either linearity or a threshold dose of the same value. Prior to 2 weeks post conception, the most likely effect of radiation is reabsorption of the embryo. Growth retardation either as an adult or at birth can result from radiation at almost any subsequent time.

In utero exposure to x-rays for diagnostic purposes has long been linked to an excess risk of childhood cancer (Stewart et al. 1956). Wakeford and Little (2003) conclude that there is good evidence for an increased risk of cancer following foetal irradiation of 10 mSv. However, studies of Japanese children irradiated in utero by the atomic bombs indicate that the risks of irradiation in utero and in the first six years of life are similar (Delongchamp et al. 1997), but the studies also support the possibility of a higher risk in the first trimester of pregnancy. Risks of irradiation in childhood are up to 10 times greater than in adulthood (Little 2003).

59.3 PRINCIPLES OF RADIATION PROTECTION

59.3.1 PRACTICES AND INTERVENTIONS

In establishing the framework of radiation protection, the ICRP (1990, 1998) divides its areas of concern into *practices* and *interventions*. *Practices* cover those activities that lead to an increase in the overall exposure to radiation. However, much radiation exposure is either the result of naturally occurring activity (such as radon) or of accidental or inadvertent manmade contamination (such as the Chernobyl disaster). In these cases, any radiation protection activity has to be by way of an intervention in the lives of people who have no choice about their involvement. In these circumstances, ICRP recognizes that the human cost of removing the hazard may be greater than the benefit associated with the reduction in radiation dose.

In what follows, only those parts of radiation protection that involve the management of dose from practices will be considered.

59.3.2 BASIS FOR DOSE REGULATION

The assumption that there is no threshold dose below which damage cannot occur, leads us to the conclusion that no dose can be regarded as *safe* (however, see Section 59.2.3). On the other hand, it is not possible to reduce the dose to staff, patients and public to zero. The problem becomes one of finding an acceptable framework of compromise that allows the benefits patients derive from exposure to radiation to be balanced against the unavoidable potential harm incurred by staff and the public.

ICRP 60 (1990) provides us with the necessary framework to do this by means of the three principles of justification, optimisation and limitation:

- *Justification*: no practice shall be adopted unless its introduction produces a net benefit.
- *Optimisation*: all exposures shall be kept As Low As Reasonably Achievable* (ALARA), economic and social factors being taken into account.
- *Limitation*: the dose equivalent to individuals shall not exceed the limits recommended for the appropriate circumstances.

Optimisation describes the process whereby the net *benefit* is maximised after taking into account the *cost*, however measured, of using ionising radiation. This is discussed further in Section 59.3.4. The magnitude of dose limits is covered in Section 60.2.2. In the present Chapter the theoretical basis for these limits is considered.

---

* This is the wording used by both ICRP (1990, 2007) and by the EU legislation (ECD 1996a, 1997). U.K. legislation has changed this to As Low As Reasonably Practicable, ALARP (HS 1999a; HS 2000; IPEM 2002).

59.3.3 Limitation of Dose to Staff

Limits are set with a view toward ensuring that no deterministic effects can occur and that risks of stochastic effects are kept acceptably low.

The setting of limits for the avoidance of deterministic effects is a comparatively simple matter in that it requires ensuring that no organ or part of the body can receive a dose above an amount set at a level comfortably below the threshold for any relevant deterministic effect.

If there is no safe threshold dose, setting limits for adverse effects whose chance of occurrence is determined by exposure must involve a value judgement. For staff, it is assumed that the individual exposed gains some benefit, i.e. gainful employment, and the dose level is set at a level where some degree of risk is accepted. Additional limits have been set for occupationally exposed women of child-bearing age, or who are actually pregnant, to ensure that the foetus does not receive a dose greater than an ordinary member of the public.

59.3.4 Limitation of Dose to the General Public

For public exposure, there is no individual benefit; consequently, the dose limit must be set much lower than for staff. Where large numbers of people are being exposed the detriment is greater than when there are only a few. This low limit takes into account the genetic burden associated with the irradiation of the entire population.

The low value of the dose limits for the public would cause problems when certain groups of persons other than the patient are exposed as a consequence of medical treatment or diagnosis. These persons can be classified as *comforters and carers* and are discussed in Section 60.2.3.3.

59.3.5 Dose Constraints

The process of optimisation may involve the application of constraints to individual doses arising from any specific practice. They are prospective figures that are used in planning and represent maximum target values that will be tolerated from that single practice. Although they appear to function as limits in that they specify maximum values, they will usually be lower than the overall dose limit that will be applied to an individual and do not have the legal force of a limit. A constraint of (say) 0.3 mSv applied to the dose arising to a member of the public from a single practice should prevent a limit of 1 mSv being exceeded when all likely artificial sources of radiation have been taken into account (see Section 60.2.1).

Constraints will also be applied to individual doses where the object of optimisation is to minimise the collective dose, i.e. the sum of all doses to all individuals. Once again, a constraint of 0.3 mSv for any single practice should prevent all members of the public receiving the legal limit of 1 mSv. Further examples of the use of constraints in conjunction with limits will be seen in Section 60.2.3.3 and Section 61.4.2.2. The concept of constraints in radiation protection is similar to the considerations applied to the doses received by *organs-at-risk* in planning radiotherapy, although there are no legal limits in the case of medical exposures.

59.3.6 Risk Assessment

Risk assessment is an essential part of radiation protection practice. A radiological risk assessment should consider fully all risks arising from a single practice taking into account previous relevant knowledge and procedures. Particular attention should be given to the possibilities of failure of each safety control system, assessment of the consequences of such failure and the development of contingency plans to reduce the consequences of foreseeable failures to acceptable levels.

The U.K. Health and Safety Executive has published a simple guide to the five steps of the risk assessment process (HSE 1998) that involves:

- Identifying the hazards
- Deciding on who might be harmed and how
- Evaluating the risks and deciding whether the existing precautions are adequate and need to be improved
- Recording the findings
- Reviewing the assessment and revising it if necessary

The IAEA (1996) lists the following issues that must be included in a risk assessment:

- The nature and magnitude of potential exposures and the likelihood of their occurrence
- The limits and technical conditions for operation of the source
- The ways in which structures, systems, components and procedures related to protection or safety might fail, singly or in combination, or otherwise lead to potential exposures; and the consequences of such failures
- The ways in which changes in the environment could affect protection or safety
- The ways in which operating procedures related to protection or safety might be erroneous, and the consequences of such errors
- The protection and safety implications of any proposed modifications

The risk assessment should list the actions that are proposed to mitigate any identified risk. Where a risk assessment identifies the possibility of a radiation accident, contingency plans must be drawn up for such an eventuality.

CHAPTER 60

RADIATION PROTECTION REGULATION

Mike Rosenbloom and Philip Mayles

CONTENTS

60.1 THE REGULATORY FRAMEWORK

Many different organisations are concerned with radiation protection. The International Commission for Radiological Protection (ICRP), originally founded in 1928 by the International Society of Radiology (ISR), is an international body with charitable status whose remit is to function as an advisory body providing recommendations and guidance on radiation protection. It operates through a number of task groups convened for specific purposes and is composed of a main commission and four standing committees dealing with radiation effects, doses from radiation exposure, protection in medicine, and the application of ICRP recommendations[*]. A sister organisation, the International Commission on Radiological Units (ICRU), deals largely with issues related to the measurement of ionising radiation[*].

The International Atomic Energy Agency (IAEA) serves as the world's central inter-governmental forum for scientific and technical co-operation in the nuclear field, and as the international inspectorate for the application of nuclear safeguards and verification measures covering civilian nuclear programmes[*]. A specialised agency within the United Nations system, the IAEA came into being in 1957. It deals specifically with the safe generation of nuclear energy and defines many of the procedures and criteria required. The IAEA recommendations for radiation protection are contained in the Basic Safety Standards (BSS, IAEA 1996). The BSS represents an internationally agreed standard that has, in principle, been accepted by all member states and is based on the recommendations of the ICRP. The BSS sets out the requirements for a framework to regulate radiation safety in each country, including the need to investigate "any therapeutic treatment delivered to either the wrong patient or the wrong tissue, or using the wrong pharmaceutical, or with a dose or dose fractionation differing substantially from the values prescribed by the medical practitioner or which may lead to undue acute secondary effects". The IAEA will carry out accident investigations if requested by a member state (e.g. IAEA 2001).

In the U.S.A., radiation protection legislation is a responsibility of each individual state and significant variations are found between adjacent states. The National Council on Radiation Protection and Measurements (NCRP[*]) provides advice to legislators in a way similar to the Health Protection Agency (HPA)[*,†] in the U.K.

Within the European Union, requirements for legislation are imposed on member states by means of directives that specify an obligation to achieve certain objectives, but do not directly formulate the legislation by which these objectives will be attained. The publication of ICRP 26 (1977) led directly to the adoption by the European Community in 1980 of Council Directive 80/836/Euratom (ECD 1980) (subsequently amended by Council Directive 84/467/Euratom (ECD 1984b)). This Directive laid down basic safety standards for the protection of workers and public against the dangers of ionising radiation and became known as the Basic Safety Standards (BSS) Directive. A new Basic Safety Standard was adopted in 1996 and published as Council Directive 96/29/Euratom (ECD 1996a). In parallel with this, protection of the patient was covered by the 1984 Directive 84/466/Euratom (ECD 1984a) and later by the 1997 Medical Exposure Directive 97/43/Euratom (ECD 1997) known as the *patient Directive*. After an European Council of Ministers directive has been published, it is the responsibility of each member state to interpret the directive into its own legislative framework. In what follows the U.K. legislation will often be used as an illustration of the general principles. In addition, a note on the U.K. implementation of the directives is provided in Appendix L.1.

---

[*] Further information can be obtained from the respective web sites: ICRP at http://www.icrp.org; ICRU at http://www.icru.org; IAEA at http://www.iaea.or.at; NCRP at http://www.ncrp.com; and the HPA at http://www.hpa.org.uk/radiation.

[†] The Radiation Protection Division of the HPA has taken over the functions of the National Radiological Protection Board.

60.2 PROTECTION OF STAFF AND THE PUBLIC

60.2.1 DESIGNATED AREAS

Control of exposure to radiation is achieved partly by the *designation* of areas of work where there is a potential for such exposure.

The BSS (ECD 1996a; IAEA 1996) designates two levels of control: *controlled areas* that are areas "subject to special rules for the purposes of protection against ionising radiation and to which access is controlled" and *supervised areas*, where occupational exposure conditions need to be kept under review. The European Council Directive requires that member states should provide guidance on the circumstances that would require an area to be controlled. In Great Britain*, the guidance defines a controlled area as one in which it is likely that persons could be exposed to more than three tenths of any applicable dose limit (i.e. an effective dose greater than 6 mSv a year), but requires that specific arrangements should be made for all areas where there is a possibility of exposure exceeding the general public limit of 1 mSv (HS 2000). For an area to be unsupervised, the regulations (ECD 1996a; HS 1999a) set a limit of 1 mSv/year, but a limit of 0.3 mSv/year for any single source of radiation is recommended by the NRPB and the IPEM (2002). The Medical and Dental Guidance Notes (MDGN, IPEM 2002, see Appendix L.1) provide detailed recommendations on the designation of areas which are reproduced in Table 60.1. This table refers to three different ways of considering the dose limit. The *instantaneous dose rate* (IDR) is the dose rate averaged over 1 min. The *time averaged dose rate* (TADR) is the dose rate averaged over an 8 h day, taking into account use and workload but with an occupancy factor of 1 (for the use of the TADR see Section 61.1.3). Finally, TADR 2000 is the dose rate taking into account all factors, including occupancy averaged over 2000 h, which is considered to be the working year for an office worker. The 3 mSv/h for the TADR 2000 is directly derived from the 6 mSv per annum limit. 7.5 mSv/h[†] allows for an occupancy factor of 0.4. If the recommendation derived from the TADR and the TADR 2000 conflict, Appendix A of the MDGN provides a decision tree that determines the outcome depending on whether the area is open to the general public or not.

The definition of an area as controlled could be either in terms of prescribed dose rates for external radiation, prescribed levels of air concentration or surface contamination for potential internal radiation, or a combination of the two.

Entry within controlled areas, which must be adequately demarcated, is restricted to classified workers or those entering under a written system of work designed to restrict their exposure. Local rules (see Appendix L.3) must be in existence to ensure that all persons entering a designated area restrict their exposure and one or more radiation protection officers or supervisors must be appointed to ensure that work is performed in compliance with these local rules.

60.2.2 LOCAL RULES

The local rules must adequately describe the area and the procedures to be followed within that area. They should also contain the name of the radiation protection supervisor. All persons working under such local rules are required to have read them. Monitoring must be in place to demonstrate that doses are adequately restricted.

---

* Different rules apply in Northern Ireland.

[†] IPEM Report 75 recommended that new rooms should be designed to limit the IDR to 7.5 µSv/h. This figure was given expecting a tightening of the regulations, but is superseded by IPEM 2002.

TABLE 60.1

IPEM Guidance on Radiation Levels Defining Controlled and Supervised Areas

	Controlled (μSv/h)	Supervised (μSv/h)	Unsupervised (μSv/h)
Instantaneous dose rate (IDR)	>2000	>7.5	<7.5
Time averaged dose rate (TADR)	>7.5	>2.5	<0.5
Time averaged dose rate over 2,000 h (TADR 2000)	>3	>0.5	<0.15

Source: From IPEM (Institute of Physics and Engineering in Medicine), Medical and Dental Guidance Notes, A good practice guide to implement ionising radiation protection legislation in the clinical environment, IPEM, York, 2002.

60.2.3 DOSE LIMITS FOR STAFF AND PUBLIC

Dose limits are set in terms of effective dose in order to restrict stochastic effects and equivalent organ doses to limit deterministic effects. The widely accepted dose limits recommended by ICRP (1990) are set for specific groups as follows.

60.2.3.1 *Employees*

Annual limits recommended by the European Union (EU) and ICRP for employees over the age of 18 are an annual effective dose (see Section 59.1.3) of 20 mSv averaged over a five-year period*, but subject to an annual limit of 50 mSv. Annual limits for equivalent doses (see Section 59.1.2) are set as 150 mSv for the lens of the eye and 500 mSv for hands, forearms, feet, ankles and any area of skin averaged over 1 cm². These limits are reduced to three tenths of the above values for employees under 18 years of age. An employer is obliged to ensure, subsequent to notification of a pregnancy, that the equivalent dose to the foetus is unlikely to exceed 1 mSv in the remaining period of the pregnancy and that, whilst breastfeeding, an employee cannot incur significant radioactive contamination†.

The current limits on effective dose are lower than those previously defined because of the change in risk coefficients when the ICRP 60 recommendations replaced ICRP 26. Although there was no change in the estimate of the dose response for deterministic effects, the area over which skin doses can be averaged has been reduced from 100 cm² to 1 cm², which effectively lowers the skin dose limit.

60.2.3.2 *Other Persons*

Although there is no clearly defined threshold for negligible risk, analysis of general human behaviour indicates that individuals are prepared to neglect risk at a range of values according to the activity being undertaken. At currently accepted estimates of risk for the general public (see Table 59.3), a dose of 1 mSv leads to an average risk of 5×10^{-5} and has been recommended by ICRP as the annual public dose limit (excluding exposure from natural radiation). An annual dose constraint of 300 μSv is applied to public exposure from all new practices.

---

* In U.K. law, the possibility of averaging over 5 years is not automatically applicable. If more than 20 mSv are to be received in any one year the employer is required to demonstrate that it is not practicable to keep the annual dose below 20 mSv.
† U.K. legislation additionally requires that for women of reproductive capacity, the equivalent dose averaged throughout the abdomen is restricted to 13 mSv in any three-month period.

Because members of the public are not subject to the same systems of control and thus liable to accumulate doses from different sources of radiation and because the public also includes children, the dose limits for hands, forearms, feet, ankles and skin are reduced by a factor of ten as compared to limits for employees to 50 mSv and 15 mSv for the lens of the eye.

Annual effective dose limits are set to 1 mSv for persons other than an employee aged over 16. This limit might be impracticable where exposure arises as a result of prolonged proximity to persons treated with radioactive substances, and it may therefore be relaxed for individuals (including children) who are exposed to radiation from patients after leaving hospital. A limit of 5 mSv in five consecutive years is applied to such persons.

60.2.3.3 Classification and Monitoring of Staff

The requirements for monitoring of workers depends on the level of dose that they are expected to receive. The European Council Directive defines them as category A workers if there is a possibility that the annual effective dose received may be greater than 6 mSv or exceeds three-tenths of the limits for equivalent doses to specific organs and category B if the dose received is less*. Category A workers must be subject to systematic dose monitoring and medical surveillance to ensure that they are fit to undertake work with ionising radiation; whereas for category B workers, the monitoring only has to be sufficient to demonstrate that they are correctly classified as category B.

Monitoring nonclassified workers working in controlled areas is important to confirm that dose restriction is adequate, and it is therefore normal to insist that all persons working within a controlled area wear personal dosimeters. Health workers rarely need to be classified as category A workers, but it may be necessary in a few cases, particularly among those who work with radioactive materials. Some previously unclassified staff may need to be classified as a consequence of the limit for skin dose being defined as the dose averaged over 1 cm^2 rather than 100 cm^2, as was the case with the 1985 recommendations (ICRP 1985). The practice of monitoring hand or mid-finger doses may need to be replaced by monitoring fingertip doses.

60.2.3.4 Dose Limitation in Special Situations: Comforters and Carers

Individuals who are exposed as nonprofessionals in the care and comfort of persons undergoing exposure for medical purposes (known as *comforters and carers* in U.K. legislation) are not subject to the dose limit of 1 mSv for the general public, but all reasonable efforts must be taken to minimise their exposure. This is of particular relevance where young children are hospitalised for unsealed source radiotherapy and when it is desirable for their parents to be nearby. Such persons must have had the risks clearly explained to them and given their agreement to this exposure. It is a requirement of EU legislation that dose constraints be applied to such exposures and that the carer receives the dose knowingly and willingly.

Singleton et al. (2003) have produced guidance on the principles that should be followed and have assessed the magnitude of doses likely to be received in different therapy situations. They conclude that only rarely should it be necessary to invoke the possibility of exceeding a limit of 1 mSv, except in the case of therapy with isotope sources. Guidance on appropriate constraints is also available in the Medical and Dental Guidance Notes (IPEM 2002). A particularly common situation is when family members are exposed by patients who

---

* In Great Britain, category A workers are called *classified* workers.

return home following therapy with radioactive substances; this will be considered in detail in Section 61.4.2.2.

60.2.4 QUALIFIED EXPERT (OR RADIATION PROTECTION ADVISER)

Employers, when work with ionising radiation is about to take place, need to seek the advice of a *qualified expert** as to whether they need to appoint a qualified expert (ECD 1996a). In general, a qualified expert (RPA) will be required wherever there is a need to designate areas for work with ionising radiation (see Section 60.2.1) or there is any other significant use of ionising radiation. All hospitals giving radiotherapy will need a qualified expert, who will have a wide range of responsibilities including the evaluation of all procedures where the use of ionising radiation is involved and advising on appropriate shielding and other measures to limit dose.

60.2.5 RISK ASSESSMENT

An obligation to carry out risk assessment is implied by the Basic Safety Standards (ECD 1996a) and is an explicit requirement of U.K. law. A suitable record must be made of all risk assessments. Details of what should be considered are discussed in Section 59.3.6 and further advice can be found in the Medical and Dental Guidance Notes (IPEM 2002).

60.2.5.1 Critical Examination of Equipment

The European Council Directive requires that a critical examination of plans for new installations should be carried out by a qualified expert, whereas U.K. legislation requires that this critical examination is carried out once the equipment has been installed. Table 60.2 indicates those radiation protection issues that should be considered in respect of any new installation.

60.2.5.2 Contingency Planning

Employers are required to make a hazard assessment before any work with ionising radiation is undertaken. This assessment should be recorded and take account of the likelihood of any accident and the severity of its consequences. Where this assessment indicates that there is a risk of accident, procedures should be in place to minimise this risk and *contingency plans* should be drawn up to deal with possible accidents. These contingency plans should form part of the local rules.

Work with large teletherapy units carries the risk of a source being stuck in the exposed position. Contingency plans must include clear instructions on how the patient should be moved to safety and what data should be noted to assist in assessment of the dose received by the patient. Guidance should also be given on how the unit will subsequently be made safe.

Work with radioactive substances carries the risk of spillage; contingency plans should include instructions on summoning help, preventing the spread of contamination and making the area safe.

---

* Known in the U.K. as the *Radiation Protection Adviser* (RPA).

TABLE 60.2

Components Requiring Critical Examination Before Use

Interlocks	Warning Systems	Safety Design Features	Barriers
Door interlocks	Warning signals	Exposure termination	Maze design and entrance
Emergency *off* buttons	Entry warning signs	Tube or head leakage	doors where fitted
or switches	Beam on indications	Beam filtration and	Primary and secondary
Beam off/disable	Indication of selected	collimators	barriers (position and
buttons	parameters (e.g.	Protection of exposure	adequacy of protection
Microswitch interlocks	energy, HDR, tube)	switch against	for adjacent areas)
Alignment and filter	Unambiguous	accidental activation	Protective cubicle and
interlocks	labelling	Fluoroscopy	mobile screens
Last-man-out systems		(termination	Heating, ventilation,
		and maximum skin	air-conditioning and
		dose rate limitation)	cable penetrations
		Automatic Exposure	through barriers
		Control (AEC) (dose	Provision and adequacy
		rate termination,	of personal protection
		back-up timer,	(drapes, aprons, shields)
		chamber/mode	
		selection)	

Source: From IPEM (Institute of Physics and Engineering in Medicine), Medical and Dental Guidance Notes, A good practice guide to implement ionising radiation protection legislation in the clinical environment, IPEM, York, 2002.

60.3 PROTECTION OF THE PATIENT

60.3.1 JUSTIFICATION OF MEDICAL EXPOSURES

Whereas the Basic Safety Standards Directive (ECD 1996a) deals with protection of staff and the general public, a second Council Directive 97/43/Euratom (ECD 1997) deals with protection of the patient and is referred to as the Medical Exposure Directive.

There is no fixed limit for medical exposure of individuals. However, the Directive identifies the duties of the *prescriber*, who is the referring physician or other health professional, and the *practitioner*, who is the person (usually a radiation oncologist) who takes responsibility for the justification, optimisation, and clinical evaluation of the exposure*. The practitioner may delegate the *practical aspects* of the irradiation to other suitably qualified people (*operators* in the U.K.). The need to justify and optimise individual medical exposures has always been part of the standard practice of radiotherapy, but the Directive makes this a legal obligation and extends the concept to diagnostic exposures.

Justification requires that there is net benefit to the patient from the exposure. In the case of exposures for research purposes, where there may not be a net benefit to the individual patient, it is required that the justification for the exposures shall be considered by an ethical committee.

60.3.2 OPTIMISATION OF EXPOSURES

In addition to being justified, the directive requires that exposures must be optimised. Optimisation involves the selection of appropriate equipment and techniques based on adequate information. It also requires that there is appropriate quality assurance.

---

* Note that U.K. law uses slightly different terminology as discussed in Appendix L.1.

In determining the optimum approach, it is permissible to consider economic and social factors. The directive requires that exposures are individually planned and that doses to nontarget tissues are as low as reasonably achievable.

60.3.3 DIAGNOSTIC REFERENCE LEVELS

Diagnostic reference levels (DRLs) are dose levels in medical radiodiagnostic practices or, in the case of radiopharmaceuticals, levels of activity, for typical examinations for groups of standard-sized patients or standard phantoms for broadly defined types of equipment. These levels are expected not to be exceeded for standard procedures when good and normal practice regarding diagnostic and technical performance is applied, but they do not represent an absolute dose limit. ICRP describes diagnostic reference levels in Publication 73 (ICRP 1996) and has published further recommendations on the internet (http://www.icrp.org/docs/DRL_for_web.pdf)*. DRLs are an important control in diagnostic radiology and if they are regularly exceeded, an investigation should take place.

The IPEM recommends (2002) that DRLs are not appropriately applied to radiotherapy simulation or CT scanning because of the much higher doses that are going to be given therapeutically. However, the obligation to optimise exposures remains.

60.3.4 MEDICAL PHYSICS EXPERT

The European Council Directive creates the requirement for a medical physics expert (MPE) to be closely involved in radiotherapy treatment. The MPE is required to be appropriately trained and gives "advice on patient dosimetry, on the development and use of complex techniques and equipment, on optimization, on quality assurance, including quality control, and on other matters relating to radiation protection" (ECD 1997).

60.3.5 OTHER REQUIREMENTS

There are a number of other specific requirements of the European Council Directive:

- Written instructions for radioactive patients leaving a hospital must be provided to restrict the exposure of persons with whom they might come into contact.
- All female patients of child bearing age must be asked whether they might be pregnant or breast feeding where relevant.
- A clinical audit must be carried out.
- Equipment at any site should be limited to that which is necessary to prevent unnecessary proliferation of equipment.
- Adequate training for staff must be provided and recorded; training must include continuing professional development.
- Written procedures must exist for medical exposures together with written protocols for every type of radiological practice for each piece of equipment.

60.4 LICENSING ARRANGEMENTS

The IAEA Basic Safety Standards (1996) require that all radiation work is appropriately licensed. This is reflected in the EU requirement for prior authorisation of such work.

---

* A new ICRP report entitled "Radiological Protection in Medicine" was put out for consultation in January 2007.

The licensing arrangements differ from one country to another, but have the common goal of ensuring that the uses of radiation are appropriately controlled. Licensing is principally required for the storage, use and disposal of radioactive material, but it is also required for the administration of radioactive substances to patients.

60.4.1 STORAGE OF RADIOACTIVE MATERIAL

A license is required to permit the storage of radioactive material. Storage for use will have a different authorisation from storage as waste, and storage of waste materials is generally discouraged. Meticulous records of both sources-intended-for-use and classified-as-waste must be maintained. All sources must be regularly inspected, and sealed sources used for brachytherapy should be regularly leak tested; the frequency of testing will be dependent on use. The risk of possible use of radioactive material to create a dirty bomb increases the importance of such vigilance and this has led to a European Council Directive regarding the security of high-activity sealed sources (ECD 2003).

60.4.2 DISPOSAL OF RADIOACTIVE WASTE

As a consequence of the use of unsealed radioactive substances in the diagnosis or treatment of patients some radioactivity is likely to be discharged into the environment. The quantity discharged will depend partly on the physical and biological half-lives of the substance. Often this discharge can be reduced by storing waste to allow physical decay of short-lived nuclides. The discharge and accumulation will be regulated by an appropriate government agency. Any organisation discharging significant amounts of radioactivity into the environment or storing waste is required to seek authorisation to do so. Authorisation will be granted if the inspectors are satisfied that there is justification for the discharge and adequate accounting for radioactive waste. The authorisation will limit the rate of discharge and the time for which radioactive waste can be accumulated on site.

A number of routes of discharge can be identified; these include discharge into the atmosphere, discharge as low-level waste for burial at landfill sites, incineration on-site or at another site, and discharge into sewers. The request for authorisation will require a list of the amounts to be discharged by each route and a radiological impact assessment of each route identified.

The inpatient treatment of patients with radionuclides, particularly ^{131}I can result in gigabecquerel (GBq) quantities of nuclide being excreted in patients' urine. The comparatively long half-life of eight days means there will be little physical decay, and most iodine administered will reach the sewer unless means are adopted to prevent this.

Means to reduce (abate) discharge of ^{131}I into the environment rely on its radioactive decay during storage of the urine before it is discharged from the premises. This will require either direct collection of patients' urine for storage, or systems of tanks for storage of sewage from the treatment suite or ward (Leung and Nikolic 1998; Goddard 1999). Direct collection of urine can involve the use of bottles, chemical toilets, or toilets allowing rapid freezing of contents and subsequent transfer to large freezers for radioactive decay. Examples of all of these systems can be found throughout the world.

The operation of systems to allow the physical decay of ^{131}I before discharge to the environment can be either distasteful or expensive or both. A proper radiological assessment of direct discharge to the sewer will usually establish that the dose to the most exposed member of any critical group, be it recreational bathers, fishermen or sewerage workers, will be less than the IAEA *de minimis* dose (IAEA 1999) of 10 μSv/year with correspondingly lower population doses. There is therefore no justification for any other practice in most circumstances. In the event that water discharged from sewage treatment works is eventually abstracted downstream for drinking, the process will involve large dilutions and storage in reservoirs with the result that, once again, dose to the population does not justify abatement procedures.

A distinction should be drawn between the disposal of a radioactive source as waste and the transfer of the source to another organisation. For example, unused brachytherapy sources are usually returned to the manufacturer. In the case of permanent implants, such as iodine-125 prostate implants, the patient will be discharged from hospital with the implant in place. This is not classified as disposal as waste for regulatory purposes.

60.4.3 ADMINISTRATION OF RADIOACTIVE SUBSTANCES

The administration of radioactive substances to patients must also be controlled. Such substances may only be used under the direction of a suitably qualified doctor who holds the necessary certificate and works in an approved centre with minimum standards of staffing and equipment. However, the prescription of activities for therapy is left to the professional judgement of the clinician.

60.5 TRANSPORT OF RADIOACTIVE SUBSTANCES

The law distinguishes between the movement and transport of radioactive substances. Transport is the operative term whenever a public highway or place is involved. Because transportation of a particular consignment of goods may affect many different countries, the regulations governing transport are more closely harmonised and are based on the IAEA safety standards (IAEA 2000). The regulations do not apply to radioactive material within the body following diagnosis or therapy. The arrangements for travel from a hospital for patients who have been administered radioactive substances will be dealt with in Section 61.4.2.2.

The transport regulations are relevant to the delivery of radioactive materials required for treatment to a hospital. They are also relevant to the delivery and return of sealed sources used in teletherapy and brachytherapy.

All deliveries of radioactive substances must be made in special containers (Type A) designed and tested to meet stringent conditions to ensure that there will be no dispersal of radioactive material during transport in any reasonably foreseeable circumstances. These conditions can be relaxed to allow transport in *excepted* packages where only small quantities are involved, but most aspects of the transport of excepted packages are still subject to control under the transport regulations. Therapy quantities will always exceed the limits for transport as excepted packages and will always be made in Type A containers. Type A containers are divided into further categories (see Table 60.3) depending on the external dose rate. The labelling of these containers will give some indication of external dose rates and the restrictions placed on drivers for their handling. Apart from more relaxed conditions for parking of vehicles containing only white label packages, there is little difference in the requirements made for carriers of different categories of Type A packages. The *transport index* for a single container or group of containers is numerically 100 times the dose rate in mSv/h at 1 m (being originally based on the dose rate in mrem/h).

The regulations impose obligations on consignor and carrier and clarify training, general documentation, recording and packaging requirements. They also describe the precautions required for the custody of radioactive materials in transit and the prevention of exposure of persons not directly concerned with the transport of radioactive materials. The European Council Directive R6/35/EC (ECD 1996b) created a legal requirement for entities who consign or transport dangerous goods by road, rail or inland waterways (above the threshold quantities) to appoint a suitably qualified dangerous-goods safety adviser. When sealed sources are being transported between EU countries, a certificate must be obtained from the licensing

TABLE 60.3

Transport Categories for Carriage of Radioactive Materials

Category	Maximum Surface Dose Rate	Maximum Transport Index[a]	Exclusions
I (white)	$<5\ \mu\text{Sv h}^{-1}$	—	Fissile class II or III
II (yellow)	$<500\ \mu\text{Sv h}^{-1}$	1.0	Fissile class III
III (yellow)	$<2\ \text{mSv h}^{-1}$	10	
III (yellow)[b]	$<10\ \text{mSv h}^{-1}$	>10	

[a] The Transport Index is the dose rate in mSv/h at 1 m multiplied by 100

[b] Such packages must be transported under exclusive use, i.e. the consignor has sole responsibility for the loading of the conveyance or large container.

authority of the recipient country to confirm that the consignee is licensed to hold the source in question. This is to conform to Council Regulation 1493/93 (ECD 1993).

More stringent regulations apply to the transport of large teletherapy sources which must be transported in Type B packages. The responsibility for observation of these regulations will lie with the specialist companies delivering these sources.

CHAPTER 61

PRACTICAL RADIATION PROTECTION IN RADIOTHERAPY

Mike Rosenbloom[*]

CONTENTS

---

[*] With contributions from Philip Mayles (Section 61.5.2 through Section 61.5.4).

61.1   DESIGN OF EXTERNAL BEAM TREATMENT FACILITIES

The protection of staff directly concerned with the practice of radiotherapy as well as other staff and the public at large, whether as visitors to the hospital or as persons in the vicinity of the hospital, must be considered in the design of radiation treatment facilities. A discussion of the approaches available to reduce staff doses is given in the Introduction to Part L. Protection of staff and public is mainly achieved by the use of shielded rooms and warning devices that exclude persons other than patients from areas of high radiation dose.

Protection of personnel may not be the only requirement. Linear accelerators produce pulsed radiation and even where adequate protection is provided for the radiological safety of personnel, sensitive radiation equipment such as counters and gamma cameras may be affected by these pulses which will usually be intermittent and can thus lead to spurious readings or image artefacts.

In what follows, the general methodology for designing treatment rooms will be described. A more detailed treatment of the subject is to be found in IPEM Report 75 (IPEM 1997) or NCRP Report 151 (NCRP 2005) and data can be found in the *Handbook of Radiological Protection* (RSAC 1971) and the NCRP Report 49, Report 51 (NCRP 1976, 1977) and in BS4094 parts 1 and 2 (BSI 1996, 1971). An example calculation for a specific treatment room is given in Appendix L.2.

61.1.1 COMPONENTS OF RADIATION

Three principle components may require shielding. These are the primary beam, leakage radiation and scattered radiation. For high-energy accelerators, there may also be significant production of neutrons. Photon bombardment of parts of high-energy linear accelerators may lead to induced activity and this can be a hazard to maintenance staff.

61.1.1.1 Primary Radiation

The primary beam is the first consideration in designing shielding. Adequate shielding extending beyond the maximum diagonal width of the beam as defined by the collimation system of the equipment will be required to reduce the dose to acceptable levels outside the treatment room for persons within the geometrical projection of the primary beam.

Lower-energy equipment is often designed to be used without restriction of the beam direction. This requires that all external walls are of adequate thickness to attenuate the primary beam. Special attention should be paid to the door, which will usually require incorporated lead or other shielding materials of adequate thickness.

Almost all megavoltage equipment is now mounted isocentrically*. This restricts the area of wall that needs to be thick enough to attenuate the full primary beam. A primary barrier in the wall of the room is to be preferred, but may not always be possible for practical reasons. Installing such a barrier is usually possible for new installations. However, where new radiotherapy equipment producing more penetrating radiation is installed in treatment rooms originally designed for equipment of lower energy, it may not be possible to provide adequate attenuation of the primary beam for all gantry orientations. Beam stoppers—retractable or fixed—may be attached to the equipment such that they intercept the primary beam after it passes through the patient. However, they limit access to the patient for setting up and are best avoided wherever possible.

The expense of shielding can often be minimised by situating bunkers underground or against earth banks. Care will be needed to ensure that in future site development this shielding is not inadvertently removed.

Where there are particular weaknesses in room shielding, restrictions on use or limitation of access to external areas may be necessary.

61.1.1.2 Leakage Radiation

Leakage radiation is emitted by the equipment outside the useful beam of radiation. In orthovoltage or superficial units, this will arise entirely from the target and is a consequence of the side lobes of the polar emission of bremsstrahlung. It is attenuated by internal lead shielding. For orthovoltage equipment scattered radiation is not a problem because the walls must also be primary barriers (see Section 61.1.1.1).

For teleisotope equipment, leakage radiation arises from the source. In the *safe* position, leakage radiation is potentially significant for protection of staff within the room and adequate internal shielding must be provided. In some unit designs, particularly those in which the source moves into the treatment position, shielding of the source in the exposed position may be reduced to save space and weight. In such units, it is the leakage radiation while the source is exposed that must be used in calculating the thickness of the room walls (see also Section 12.3.1).

For linear accelerators, leakage radiation can arise from a variety of internal sources, including the waveguide, radio-frequency source, beam-transport system, target, filters, collimators, etc. It is a technical challenge to provide full protection against leakage radiation as the shielding materials required may be heavy and occupy considerable space. It is not uncommon to find leakage radiation from linear accelerators approaching the maximum allowed by regulatory authorities. This is usually set at 0.1% of the useful beam. Because

---

* Some specialised units, such as those designed specifically for total-body irradiation may not be mounted isocentrically, but the movement of these units is likely to be more restricted and this paragraph will apply in the same way.

TABLE 61.1

Leakage Limits for Teletherapy Treatment Units

Energy	5 cm From Surface	1 m From Target
X-ray generators		
<500 kVp	300 mSv/h	10 mSv/h
>500 kVp		0.1% of main beam[a]
Gamma units		
Off	0.2 mSv/h	0.02 mSv/h
On		10 mSv/h or 0.1% of main beam

[a]   Note that ICRP Report 33 (1982) recommended 0.2%; therefore, older units may have more leakage.

of the filtration effects of the internal shielding, it is often of an energy similar to the primary beam. In a conservative design it is assumed that this is the case and, if the same level of occupancy (see Section 61.1.1.5) is assumed, this will lead to a secondary barrier that is up to three tenth-value-thickness less thick than the primary barrier. (Section 61.1.1.7 discusses the validity of the assumption that leakage radiation is of the same energy as the primary beam.) Leakage radiation is usually assumed to be isotropic in design calculations. Table 61.1 shows the leakage limits generally imposed.

61.1.1.3   Scattered Radiation

Scatter may arise either from the walls of the room or from the patient. It will be of lower energy than the leakage radiation and it can be reasonably assumed in room design that any barrier providing protection against leakage radiation will also provide adequate protection against scattered radiation. Because scattering is by the Compton effect, for scatter through 90° the energy of the scattered radiation will be about 0.5 MeV and backscattered radiation will have an energy around 0.25 MeV (NCRP 1977) (see also Section 4.3.2.1).

Special consideration needs to be given to room access. A maze offering at least one or, if possible, two reflections from any surface irradiated by the primary beam will usually afford adequate protection against scatter. Sometimes space precludes the construction of an adequate maze and, in such situations, a protective door may be needed. Such doors are often heavy, will give the patient a feeling of being *shut in* and are to be avoided if possible[*]. The provision of an adequate maze may be problematic where an existing room is to be adapted to house a higher-energy unit. The calculation of the dose rate in the maze is covered in Section 61.1.2.3.

61.1.1.4   Neutrons and Induced Radioactivity

Neutrons are a particular problem with high energy (>10 MV) linear accelerators and a good discussion of the issues can be found in NCRP Report 79 (NCRP 1984). They can be produced by interactions between energetic electrons or photons and the nucleus (see Section 4.3.4). They can also be produced by fission induced by photons and electrons in elements heavier than bismuth. These processes result in a neutron flux while the beam is on but they also lead to induced radioactivity with the result that parts of the equipment struck by the generating electron beam (or by the bremsstrahlung) can emit radiation even after the beam is switched off. There is also a possibility of inducing radioactivity in the patient, patient support system, or treatment room fabric.

[*] It is unusual in the U.K. for rooms to have protective doors. In other countries such doors are the norm.

Most neutrons will arise within the treatment head, principally from the x-ray target. They will be approximately isotropic and penetrate the photon shielding in most directions. Photon shielding is usually made of lead or tungsten for which the only sources of energy loss will be inelastic scattering and $(n,2n)$ reactions. As a result, the effect of the head shielding is to degrade the *average* neutron energy as an exponential function of the shielding thickness.

Although the head shielding reduces the average energy of the neutrons, the number of particles will remain essentially unchanged. This is a result of the competing effects of a small amount ($\sim 15\%$) of attenuation and of the $(n,2n)$ reaction. Multiple scattering leads to a more or less isotropic emission of neutrons from the head, so that small gaps in the shielding will not lead to collimated beams of neutrons.

Within the treatment room, the treatment head can be regarded more or less as a point source of neutrons whose average energy has been degraded from around 4 MeV (for 15 MeV electrons) to around 1 MeV. The intensity of this component will fall off according to the inverse-square law. The concrete walls of the treatment room will contribute a fluence of scattered neutrons whose average energy will be about a quarter of this value. The fluence of the scattered neutrons will be more or less uniform within the room. These neutrons will contribute to patient dose.

Generally speaking, neutrons rarely contribute above 0.5% of the equivalent dose at the iso-centre, and the tenth-value-thickness for neutrons absorbed in concrete are lower than those for photons. As a consequence the shielding provided against photons will be adequate for neutrons.

However, a particular problem in shielding against neutrons arises from scatter down the maze. Three components will contribute to the dose along the maze:

1. The original high energy neutrons present in the treatment room
2. Thermal neutrons produced by repeated scatter
3. Neutron-capture γ-rays

The relative contributions of these components are difficult to separate, but regarding them as approximately equal is satisfactory for most purposes.

Mazes designed for lower-energy linear accelerators are unlikely to provide adequate protection against neutrons when a higher-energy linear accelerator is installed in the treatment room. A number of measures can be incorporated into mazes to improve their effectiveness against neutrons. Extra baffles will effectively increase the length of the maze, and lining the maze with polythene-loaded boron or lithium will reduce the energy of fast neutrons and provide capture for thermal neutrons. This will be more effective if placed at the inner end of the maze. Ultimately, a door may be needed. This door should incorporate neutron-absorbing material (i.e. containing highly hydrogenated components) such as boron with a layer of lead to absorb the neutron-capture γ-rays. Such doors will be heavy and require complicated and expensive drive mechanisms.

Induced radioactivity will be present within the accelerator structure, especially in the head (Powell et al. 1987). It may also be present in components of the beam-transport system. The intensity of this activity will increase with prolonged running of the linear accelerator and will decay with time after irradiation. Induced activity is a particular hazard to maintenance staff and care should always be taken to ensure that levels are acceptably low when maintenance or repair work on the components of the treatment head takes place after operation of linear accelerators at high energies. This may be more of a problem with the longer irradiation times associated with intensity modulated radiotherapy (IMRT) (Rawlinson et al. 2002; see also Section 61.5.2) and is also a particular problem for proton or neutron treatment facilities (Yudelev et al. 1995, 1997). Further details of radiation protection for proton facilities are given in Section 48.3.9.

Extensive measurements of the neutron components of various accelerators have been made by Followill et al. (2003).

61.1.2 Barrier Attenuation

In order to calculate the required thickness of material needed to attenuate the radiation to an acceptable level, information on barrier penetration is required. Attenuation depends on the energy of the radiation. The measurement of attenuation for kilovoltage x-ray beams was described in Section 21.2.2. There the aim was to avoid any scattered radiation reaching the detector, in order to measure the attenuation of the primary beam only. This is called the *narrow beam* attenuation. In radiation protection it is the *broad beam* attenuation that is relevant. A broad beam will generally be less attenuated than would be expected from the narrow beam situation because scattered radiation remains part of the beam. This distinction is sometimes emphasised by using the expression half-value thickness (HVT) for broad beam attenuation and half-value layer (HVL) for narrow beam attenuation and we will use this convention here. For radiation protection, where the attenuation required is several orders of magnitude it is usual to record the *tenth*-value-thickness, i.e. the thickness required to reduce the dose rate by a factor of 10. As the radiation is attenuated the mean energy of the radiation is reduced and the second and subsequent TVTs will be less than the first. This is in contrast to the effect described in Section 22.2.2 in which the mean energy is increased by filtration because the transmitted radiation is dominated by scatter. In attenuation tables (e.g. NRPB 2005) these may be designated as the first tenth-value-thickness (TVT_1) and the *equilibrium* tenth-value-thickness (TVT_e). Radiation that has been scattered before reaching the walls of the treatment room will have a lower energy that of the primary beam as discussed in Section 61.1.1.3 and different TVT values will be appropriate for this situation.

61.1.3 Treatment-Room Design Calculations

For a practical example of a shielding calculation the reader is referred to Appendix L.2. Here we will consider the guiding principles. The starting point in the design of a treatment room is consideration of its use. Constraints will differ depending on whether it is a new bunker or a replacement of an old machine with a new one. Ergonomic factors, such as the size of the room, the means of entry, especially for patients on trolleys, and ease of access to the patient should be considered as well as the relationship of the room with other facilities.

Design of the radiation protection should start with a consideration of the energy of the equipment and the overall workload, usually expressed as the product of the number of fields per unit time (day, week, or year) and the average dose per field. Conservative projections should always be used, and it is normal to neglect patient attenuation in all calculations. It is important to consider the locations of personnel to be protected.

For the primary beam, it may be acceptable, depending on national regulations, to take account of the fractional time that the beam will have a given orientation. This is called the *use factor*. Care should be taken that these factors are generous and take account of future evolution of treatment techniques. For example, factors of 0.5 for vertical beam orientations and 0.25 for other orientations would usually be appropriate, but the factors chosen should be justified individually taking into account any special uses of the room. Record and verify systems allow the assumptions made to be tested against real data. Because factors should be generous, their sum will be in excess of unity. Where there is no alternative, areas of low occupancy may have lower shielding: the thickness of the barriers can be reduced by the incorporation of a suitable *occupancy factor* into the barrier thickness calculation. In determining the occupancy factor, the proportion of time for which any person is present should be considered rather than the time spent by a particular individual. Thus, a corridor might have an occupancy factor of 0.25 (NCRP (2005) recommends 0.2), but a very busy corridor might have an occupancy factor of 1.

Economies in the cost of shielding can be obtained by situating high-energy installations in single-storey buildings with thinner roofs and carefully restricted personnel access. In such cases, particular account should be taken of *skyshine*, where penetrating radiation can be scattered by air back to the ground. It is also important to consider the occupation of nearby tall buildings, present or future, within the projection of the primary beam.

According to the U.K. regulations, the aim of shielding design is to reduce the time averaged dose rate over 2000 h (TADR2000)* to an acceptable level while ensuring that the instantaneous dose rate (IDR) is not excessive based on the criteria in Table 60.1. The NCRP recommendations are based on the aim to reduce the weekly dose to 0.02 mSv week^{-1} for uncontrolled areas and 0.1 mSv week^{-1} for controlled areas (NCRP 2005), calculating workloads in Gy week^{-1}. Modern accelerators can have dose rates up to 6 Gy/min or even 10 Gy/min. However, a high dose rate at the isocentre also reduces the beam on time. Assuming that fifty 2.4 Gy fractions can be delivered in an eight hour day and that the depth dose at the isocentre is about 50%, the TADR at the isocentre is about 0.5 Gy/min or 30 Gy/h. If a large number of IMRT treatments are being delivered, the beam will be on for longer and consideration therefore needs to be given to whether the shielding should be increased (Followill et al. 1997; Price et al. 2003; Stathakis et al. 2005; NRPB 2005). For the primary barrier the calculation given above is still approximately valid because the mean dose averaged over the area of an individual field is not increased; indeed, the number of patients that can be treated will probably be less. However, when calculating the protection required against leakage radiation, the *use factor* may need to be increased by up to three times. Consideration also needs to be given to the doses delivered during commissioning when the beam may be on continuously for some time. Temporary restrictions may be necessary.

61.1.3.1 Primary Barriers

The calculation of a typical barrier thickness against primary radiation can be based on the following equation:

$$\dot{D}_{\mathrm{p}} = \frac{TF_{\mathrm{u}}F_{\mathrm{o}}r_{\mathrm{iso}}^2}{r_{\mathrm{s}}^2}\dot{D}_{\mathrm{iso}}$$

where $\dot{D}_{\mathrm{p}}$ is the TADR2000 to be allowed in the area to be protected, as derived from Table 60.1; T is the fraction of the incident radiation transmitted through the barrier; F_{u} is the use factor (the fraction of treatments with the range of beam orientations potentially irradiating the point to be shielded); F_{o} is the occupancy factor; r_{iso} is the source to isocentre distance (usually 1 m); r_{s} is the distance from the source to the point of interest; and $\dot{D}_{\mathrm{iso}}$ is the TADR delivered to the isocentre (30 Gy/h).

Consider a point 6 m from the source (5 m from the isocentre) where the objective is to achieve a TADR2000 of 0.15 μSv/h, with an occupancy factor of 0.25 and a use factor of 0.3. The wall transmission ensuring the required attenuation will be 2.4×10^{-6}, which corresponds to about 2.2 m of concrete for a 6 MV beam (for which the tenth-value-thickness for standard 2.355 g/cm^{3†} concrete is 343 mm). For 15 MV, the thickness will be 25% greater.

61.1.3.2 Secondary Barriers

The calculation of the attenuation against leakage radiation is given by the equation:

$$\dot{D}_{\mathrm{L}} = \frac{TkF_{\mathrm{o}}r_{\mathrm{iso}}^2}{r_{\mathrm{s}}^2}\dot{D}_{\mathrm{iso}}$$

---

* For a detailed discussion of TADR and IDR (see Section 60.2.1).
$\dagger$ Shielding density is often quoted in t/m^3 in which the unit t is 1000 kg. This is referred to as a *metric ton* in some countries (e.g. the U.S.A.) and as a *tonne* in other countries (e.g. the U.K.) t/m^3 is numerically equal to g/cm^3.

where k (usually 10^{-3}) is the fractional leakage radiation. Note that there is no use factor and that r is usually taken to be the distance to the isocentre as representing the average position of the target. Attenuation against leakage radiation needs to be provided for all areas where staff, public, or patients might potentially be located. A critical factor is the energy of the leakage radiation. IPEM Report 75 (1997) provides evidence that the assumption that the energy of the leakage radiation is equal to the energy of the primary beam is somewhat conservative, which is consistent with the findings of Nelson and LaRiviere (1984). For a point 5 m from the isocentre the barrier thickness for a 6 MV accelerator calculated assuming the energy is equal to that of the primary beam would be 1.3 m, but a more realistic calculation might lead to 1.1 m. Where possible, it is wise to be conservative, as increasing the barrier thickness after the machine has been installed is likely to be very expensive.

61.1.3.3 Room Entrances: Protection against Scattered Radiation

As stated above, because the energy of the scattered radiation is substantially lower than that of the primary beam, any shielding designed for leakage radiation should also be sufficient to account for the scattered radiation. However, scattered radiation is important when considering the dose rate at the entrance of the maze.

Exact calculations of attenuation of radiation in a maze are best achieved with Monte Carlo calculations (Al Affan and Smith 1996; Al Affan et al. 1998; Al Affan 2000), but an approximate method proposed in NCRP Report 51 (1977) gives reasonable results, although Al Affan found that this method tends to underestimate the dose rate at the maze entrance. The formula is:

$$\dot{D}_m = \dot{D}_{iso} \frac{\alpha_1 A_1}{d_1^2} \frac{\alpha_2 A_2}{d_2^2} \frac{\alpha_3 A_3}{d_3^2} \frac{\alpha_4 A_4}{d_4^2} \tag{61.1}$$

where $\dot{D}_m$ is the dose rate at the entrance of the maze; α_1 is the reflection coefficient of the scattering material; A_1 is the corresponding area of the scattering material that is irradiated, which for the maximum field size is about $0.25\ m^2$. As a general rule, the remaining A_i (i.e. the area of the cross-section of the maze) are between $6\ m^2$ and $8\ m^2$. α_1 is about 0.0015 and $\alpha_2 A_2 \approx \alpha_3 A_3 \approx 0.1$, i.e. each bend in the maze introduces a 10% reduction in dose rate[*]. This is illustrated in Figure 61.1.

61.1.3.4 Room Entrances: Protection against Neutrons

At energies above 10 MV, it is necessary to consider the effect of neutrons. As discussed in Section 61.1.1.4, the dose rate due to neutron penetration through barriers is always less than that due to photon penetration, but neutrons undergo multiple reflections at the walls and therefore are less attenuated by mazes than photons. In addition, the tissue radiosensitivity to neutrons, as expressed by the equivalent dose in Sv, is 20 times higher than for photons (see Section 59.1.2). Methods for calculating the effect of mazes on neutron dose are discussed in IPEM Report 75 (1997). NCRP Report 51 (1977) provides graphs for the attenuation of the maze depending on its design (i.e. the number of bends) and its area and length. An alternative empirical method is described by Kersey (1979). He assumes that the neutrons are attenuated within the treatment room according to the inverse square law and that a 5 m length of maze provides an equivalent tenth-value-thickness of attenuation. However, for short mazes, 3 m may be more appropriate (IPEM 1997; Carinou et al. 1999). In assessing the neutron dose rate related to the photon dose rate in the primary beam, a

---

[*] More detailed information can be found in the *Handbook of Radiological Protection* (RSAC 1971) or IPEM Report 75 (1997). The mean energy of the primary beam may be taken as half the nominal energy for this purpose.

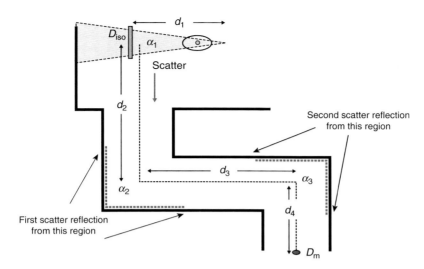

FIGURE 61.1
Diagram illustrating the reduction of dose rate resulting from scattered photons down a maze. (Figure courtesy of Dr. J. E. Shaw.)

figure of about 1.2 mSv/Gy at the isocentre is a reasonable assumption, although more exact figures can be found in the literature (McGinley 2002; Followill 2003; Waller 2003; Waller et al. 2003a). If the dose rate at the entrance of the maze is found to be unacceptable, lining the maze with borated polyethylene (e.g. Premadex®) or wood (Carinou et al. 1999; Waller 2003b) or the use of a neutron door containing polyethylene and borated plastic (BPE) sandwiched between lead (O'Brien 1985; NCRP 2005) can provide a solution.

61.1.3.5 Skyshine

The dose rate from skyshine, $\dot{D}_{sky}$, can be calculated by the following equation (IPEM 1997):

$$\dot{D}_{sky} = \frac{2.3 \times 10^{-2} A_{iso}^{1.3} T}{d_c^2 (d_r + 2)^2} \dot{D}_{iso}$$

where A_{iso}, is the beam area at the isocentre; $\dot{D}_{iso}$ is the dose rate at the isocentre; T is the roof transmission; d_c is the distance from the source to the roof; d_r is the off axis distance of the point of interest at roof level. All units of length are in metres.

If a roof is given adequate protection to reduce the primary beam penetration to a level such that short-term occupation will not lead to excessive dose, the residual penetrating primary radiation will not cause dangerous levels of skyshine at ground level. However, these levels may be high enough to interfere with sensitive radiation counting equipment especially that found in nuclear medicine departments.

61.1.4   CHOICE OF MATERIALS

The choice of the materials for room shielding will depend on factors including the energy of the radiation, available space, and cost. Representative data for some typical materials are given in Table 61.2.

At low energies, the photoelectric effect may favour the use of high-atomic-number materials, such as lead or barium as in barytes plaster. The space and weight savings might justify the use of such high-cost materials.

TABLE 61.2

Data for Typical Shielding Materials

	Standard Concrete	Barytes Concrete	Steel	Lead	Dry Earth	Ledite®
Density (g cm^{-3})	2.35	3.5	7.8	11.36	1.5[a]	5.6[a]
TVT 4 MV (in mm)						
Primary	290	205[a]	91	54	450[a]	150 [a]
Leakage	254	180[a]	79	47	400[a]	130[a]
TVT 6 MV (in mm)						
Primary	343	225[a]	98	55	540[a]	165[a]
Leakage	279	185[a]	80	45	440[a]	135[a]
TVT 10 MV (in mm)						
Primary	389	250[a]	105	56	610[a]	180[a]
Leakage	305	470[a]	85	46	480[a]	145[a]
TVT 15 MV (in mm)						
Primary	432	265[a]	108	57	680[a]	190[a]
Leakage	330	210[a]	87	50	520[a]	150[a]
TVT 20 MV (in mm)						
Primary	457	270[a]	111	55	715[a]	190[a]
Leakage	343	210[a]	88	49	540[a]	150[a]
Relative cost (approx)	1.0	2.6	2900	4000	0	4320

Note that costs (relative to standard concrete) are based on cost per unit mass of material only and are very approximate as they will depend on local conditions. This may be only a small proportion of the total cost of the building project and does not consider such factors as the reusability of Ledite®.

[a]   These values are interpolated from the data in IPEM Report 75 and should be used with caution.

Source: From IPEM (Institute of Physics and Engineering in Medicine), *The Design of Radiotherapy Treatment Room Facilities*, Stedeford, B., Morgan, H. M., and Mayles, W. P. M., Eds., IPEM Report No 75, IPEM, York, 1997.

At high-energies, mass attenuation coefficients become independent of atomic number. Full-density concrete (usual density 2.35 g/cm^3), either poured or in blocks, is the material of choice, although others including sand have been used. It is possible to obtain *high-density concrete* with a density of approximately 3.2 g/cm^3, achieved by loading with barium or other high-density material, although this is relatively more expensive and only worth considering where space is limited. Siting high-energy bunkers underground may enable the use of earth as a shielding material and significantly reduce the cost of shielding. The independence of attenuation on atomic number usually makes the cost of high-atomic-number materials unattractive except in circumstances where space is especially limited. If necessary, steel sheet can be used. At 6 MV, 10 mm of steel is equivalent to 35 mm of concrete and 6 mm of lead. Materials used for construction must have the necessary rigidity and steel is generally preferable to lead for this reason. Atomic International market a proprietary material called Ledite® (Barish 1993) that consists of scrap steel cast into interlocking concrete blocks. The attenuation of these blocks is about twice that of concrete. The blocks have the advantage that if the shielding is no longer needed or the department moves to another site, the walls can be dismantled and reused.

61.1.5   INTERLOCKS AND OTHER ISSUES

Interlocks are required to ensure that staff have left the treatment room before irradiation can begin. It is convenient to have two buttons, one inside the treatment room and one outside. These have to be pressed in sequence and within a set time before the radiation can be turned on. Emergency-off buttons one either side of the treatment room are required in case someone is left in the room when the machine is turned on (in spite of the door interlocks). Radiation warning lights should be displayed at the entrance to the maze and in the treatment room. These can follow a three stage process: the first stage appears when power is applied to the machine, the second when all the interlocks are complete and finally an indication that radiation is on.

Facilities are required to monitor the patient. The decrease in cost of closed circuit television has obviated the need for direct or indirect viewing arrangements such as mirrors and attenuating windows. Additionally there is a requirement for two way audio communication, although this may be less useful.

There is a need for electrical and other services to be taken through the radiation barrier inside ducts. Careful consideration must be given to the siting of these as it is possible that they will provide a route of escape for the radiation. Any break in the primary barrier should be avoided and ducts should cross the barrier at an angle. A particular requirement is for a tube to allow dosimetry cables to be passed into the room.

The lasers used to define the machine isocentre are often inset into the primary barrier in order to protect them from being knocked and this can present a weak spot in the barrier. If they are to be recessed, they should be mounted on a steel plate of equivalent thickness to the missing concrete (see Table 61.2).

61.1.6 UPGRADING EXISTING TREATMENT ROOMS

Installing new equipment into old rooms frequently turns out to be more complex than might be expected. Design guidelines for modern rooms have moved toward increasing the level of shielding specified. In addition, dose rates provided by modern accelerators, and hence throughput, have increased; there is also frequently a desire to install a machine with a higher energy. When installing new equipment, it is necessary to try to bring the shielding up to modern standards. Many older machines did not have full 360° rotation and consequently shielding in the roof area may be inadequate. Possible solutions to these problems include

- Building walls with interlocking concrete blocks to avoid pouring concrete
- Using steel or lead sheets or Ledite® in the walls
- Restricting use of the equipment
- Increasing the length of the maze or introducing baffles to increase the number of scattering angles
- Using a solid door, if there is not one already.

61.1.7 PROTECTION SURVEYS

It is important that after installation a full survey is conducted to verify the design calculations and to ensure that builders have followed instructions*. The first step in this process is to choose the correct survey instrument. In the case of linear accelerators, the pulsed nature of the radiation can give erroneous results with equipment such as Geiger-Müller tubes that count photons, in contrast to ionisation-based instruments that measure charge. Scintillation detectors can be used, provided that their response to

---

* For example, modern building regulations require that steel inside concrete should not protrude. This can be a problem with the shuttering used to create the frame into which concrete is poured and builders often use plastic sleeves around the tie rods so that the rods can be withdrawn once the concrete has been poured. Unfortunately this leaves a hole in the barrier, which must be filled with appropriate material.

pulsed radiation is verified. Film and TLD can be used for longer-term environmental surveys. Neutron measurements will be needed if the energy of the machine is above 10 MeV.

Before carrying out protection surveys, the expected instantaneous dose rates should be calculated so that they can be compared with measurements. Measurement should be made under conservative conditions (largest field size, diagonal collimator orientation, and maximum dose rate). Measurements in the primary beam projection should be made with no phantoms or other attenuating material in place, but, for scatter dose measurements, a water-equivalent scattering object should be placed in the beam. Measurements should cover all working conditions likely to be found, ensuring an adequate range of machine orientations. Full checks should also be made of all interlocks, displays, warning signs, and labels. Finally, the results should be reported to the *qualified expert* (Radiation Protection Adviser) if done by others.

61.1.7.1 Neutron Monitoring

Neutrons can be measured by activation, etched-track detectors, superheated-drop detectors, diodes, and ionisation detectors. Most detectors respond primarily to thermal neutrons (energy below 0.5 eV). Higher-energy neutrons are moderated to be detected as thermal neutrons. The ideal detector is not sensitive to photons because there will usually be a mixed photon and neutron beam; for this reason, the first three detector types are preferred. However, diodes and ionisation chambers are useful as direct-reading dosimeters. Neutron detectors should ideally have a sensitivity profile for different neutron energies that is matched to that of the equivalent dose (Section 59.1.2) rather than the physical dose.

Activation detectors make use of materials such as indium (half-life$=$54 min) or gold (half-life$=$2.7 days) that have high neutron absorptions. After exposure, the resulting activity can be counted to give a measure of the neutron fluence. Thermal neutrons can be distinguished from fast neutrons by the differential effects of the moderator that may or may not be incorporated into the detector.

Etched-track detectors make use of microscopic radiation damage caused by charged particles (protons) produced as secondary radiation when neutrons are captured in suitable materials (e.g. boron). Etching with caustic solution causes this damage to appear as pits that can be counted. Some etched-track detectors use polycarbonate foils, but the more recent development of CR-39 plastic or PADC (polyallyl diglycol carbonate) detects recoil protons from neutrons with an energy threshold of around 150 keV. Thermal neutrons are detected by interacting with nitrogen nuclei in the holder and this detector has good discrimination against photons (Spurney et al. 1996; Kralik and Turek 2004).

The superheated-drop detector uses superheated liquid drops suspended homogeneously in a viscoelastic gel. When neutrons interact with the gel, the drops nucleate to form vapour bubbles that can be detected by the naked eye. The dose is estimated by counting the number of bubbles, which is directly related to the dose in μSv, with approximately one bubble being produced per μSv (d'Errico and Alberts 1994).

Diodes are damaged by secondary heavy particles emitted by neutrons. Changes in electrical properties can therefore be related to neutron fluence (Yudelev et al. 2004). Care must be taken to avoid misreadings resulting from exposure to primary photons.

While ionisation chambers consisting of air-filled chambers respond primarily to photons, proportional counters filled with boron triflouride respond to thermal neutrons; surrounding them with suitable moderators enhances their sensitivity to higher-energy neutrons. Their sensitivity to pulsed photon fields precludes their use where such fields are present and confines their usefulness to measurements outside the treatment room or at the far end of the maze.

61.2 DESIGN OF FACILITIES FOR SEALED AND UNSEALED SOURCE THERAPY

Brachytherapy embraces the use of sources varying from iodine-125 seeds with activities on the order of tens of megabecquerels (MBq) and γ-ray energy of 25 keV to high-dose-rate afterloading devices with cobalt-60 sources of hundreds of gigabecquerels (GBq) and γ radiation of energy in excess of 1 MeV (see Part J). A wide range of treatment regimes are used from continuous irradiation lasting several days (or even many weeks in the case of permanent implants) to fractionated treatments with multiple sessions, each taking only a few minutes. Treatment may be given by sources that cannot be readily withdrawn from the patient or by using high-dose-rate remote afterloading equipment with sources that can be withdrawn instantaneously whenever this is required. Clearly, the shielding requirements will differ and the design brief will need to take these factors into account.

In the past, low-dose-rate treatments were given in open wards, but this is no longer acceptable. For medium- or high-dose-rate afterloading equipment, a specially designed facility is essential. In any event, appropriate controlled areas must be defined, except possibly for therapy with ^{125}I seeds.

61.2.1 REMOTE AFTERLOADING

The room shielding calculations will broadly follow the same principles as for external beam therapy (Section 61.1). However, radiation will be emitted more or less isotropically, with the result that shielding will need to cover all directions. On the other hand, the product of total activity and exposure time is likely to be such that, with the exception of high-dose-rate equipment, wall thicknesses are unlikely to be much in excess of 400 mm of concrete. These are comparable to the thicknesses required for freely mounted orthovoltage equipment. It should not be forgotten that shielding of floors and ceilings may also be required. If more than one room is required, the total amount of shielding can be minimised by locating them adjacent to each other. However, consideration must still be given to radiation transmitted through the dividing wall. This will affect staff attending one patient while the other is being treated. Some regulatory authorities also require that once a patient's treatment is complete, the patient should be treated as a member of the public for protection purposes.

High-dose-rate afterloading equipment will require concrete walls between 500 mm and 800 mm thick, depending upon the size of the room and the nuclide being used. Room doors will usually incorporate lead to take advantage of the greater mass attenuation resulting from photoelectric absorption. The thickness of lead, and thus the weight of the doors, will be minimised by ensuring that no radiation from sources within the patient can fall directly on them. This can be achieved by internal mazes.

Afterloading equipment can be located in modified or shared external-beam therapy rooms, but in this event it is essential that appropriate interlocks are installed to ensure that there is no confusion.

Staff safety will require that equipment is interlocked to room doors and that adequate warnings are displayed. In addition, it may be advisable to include dose-rate monitors. The dose rate in the vicinity of the patient or safe working times should be displayed. Closed-circuit television and two-way audio communication will reduce the need for staff entry.

Room design should include arrangements for visitors and contingency planning for medical emergencies that might require removal of the sources.

61.2.2 FACILITIES FOR UNSEALED-SOURCE THERAPY

Ideally, high-activity radiotherapy with unsealed sources, especially ^{131}I should be undertaken in specially designed wards with adequate shielding incorporated into the walls, floor, and ceiling.

The plans for one such three-bed, purpose-built unit are shown in Figure 61.2. Each room has its own *en-suite* bathroom and toilet, which minimises the potential for cross-contamination. The toilet may be connected to a storage facility to allow excreted radioactive substances to decay before being allowed into the main sewer (see Section 60.4.2), but this is no longer recommended by the ICRP (ICRP 2005a). All surfaces must be smooth and easy to clean. A whole-body counter (compensated Geiger-Müller tube) can be incorporated into the ceiling above each bed to enable the measurement of residual activity within the patient.

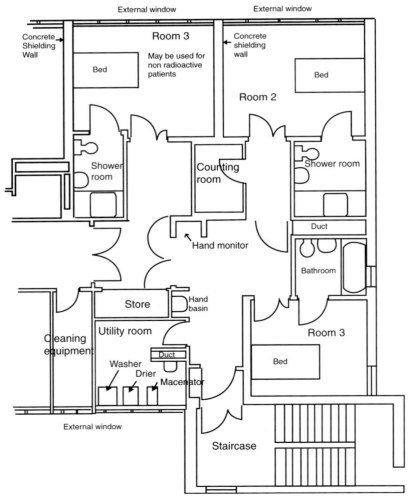

FIGURE 61.2
Design of ward for unsealed source therapy. The facility could also be used for manual-afterloading brachytherapy.

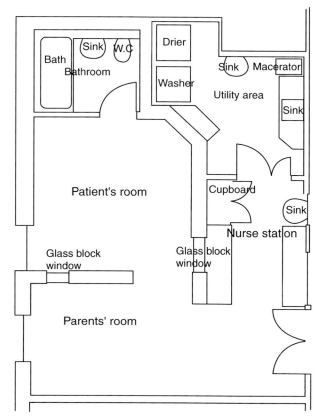

FIGURE 61.3
Design of specialist unit for paediatric unsealed source therapy.

In addition to the design of the patient rooms, radiation hygiene will be aided by the provision of a utility room with washing and drying machines (to allow separate washing of potentially-radioactive contaminated linen) and a macerator for disposal of lightweight contaminated waste (e.g. disposable crockery, nappies, and dressings).

The treatment of young children, especially with agents such as mIBG (see Section 58.1.2) poses special problems because they are likely to be emotionally upset. A facility allowing a parent to be with the child but adequately shielded will reduce the emotional trauma. One such facility is shown in Figure 61.3, where it can be seen that the parent has a good view of the child but is able to sleep separated only by a wall that is no longer than the minimum to ensure adequate protection against penetrating radiation.

In addition to these facilities appropriate radiopharmaceutical preparation facilities will be needed (see Section 61.3.4.1).

61.3 EQUIPMENT AND SOURCE HANDLING

61.3.1 TELETHERAPY ISOTOPE SOURCES

Teletherapy equipment with high-activity sources (usually cobalt-60) poses additional diffi- culties compared to accelerators in that the potential for accidental exposure persists when the power supply is removed. A range of additional measures is required to ensure that inadvertent actions cannot result in exposure to personnel when the equipment is not in use. These include the use of a radiation detector in the treatment room, leaving the equipment with the collimators closed, and perhaps a mechanical to secure the source mechanism in the safe position. Details of safety issues relating to cobalt-60 sources are discussed in Section 12.3. Regular checks should be carried out to ensure that the source is still properly encapsulated (Section 12.3.3).

Because they decay, sources will be changed periodically to prevent unacceptable prolongation of treatment time. This work will always be done by external contractors. Under the European Council Directive 90/641[*], it is the responsibility of the employer responsible for a controlled area to ensure, either directly or through contractual arrangements, that radiation protection of outside workers is adequately provided for[†]. Because the licensed source handlers will be classified radiation workers it is the responsibility of the hospital to ensure that their radiation record books are filled in when the operation is complete[‡]. Additional considerations are discussed in Section 12.4.

Local rules must include contingency plans for use in the event of the source not returning properly to the safe position. These are discussed in Section 12.3.4.

61.3.2 Afterloading Equipment

Although afterloading has made a considerable contribution to the radiological safety of staff, its use requires continued monitoring to ensure that this safety is achieved. In particular, all interlocks should be checked regularly and consistent retraction and insertion of sources should be verified. There will also be special procedures for source changing, although for high-dose-rate afterloading devices, this will normally be the responsibility of the manufacturer.

Provision of appropriate contingency plans for failure of the source transfer mechanism is important. For low-dose-rate afterloading systems, it is usual for the equipment to be left in the care of nursing staff and it is essential that the staff are properly trained to deal with emergencies. Consideration should be given to the provision of an emergency storage facility for the sources in the event of failure. However, the instructions for the use of this need to be carefully thought out, as damage may be inflicted on the applicators, making recovery of the sources more difficult than is necessary.

61.3.3 Manipulation of Low Activity Brachytherapy Sources

Modern afterloading equipment is the most effective means of minimising dose to the operator. However, implantation or manual afterloading of solid sources such as ^{198}Au grains, ^{125}I seeds, and ^{192}Ir wire, etc. is still a common practice. For such sources the principal considerations are

- Irradiation of staff while handling the sources (particularly the hands)
- Irradiation of medical and nursing staff while dealing with the patient
- The possibility of contamination from a leaking source
- Security of the sources
- Sterility of the sources

---

[*] And the Ionising Radiation Regulations 1999 in the U.K.
[†] An approach to this that is sometimes used is that the contractor becomes fully responsible for the controlled area for the period of the source change using the services of a *qualified expert* (RPA) appointed by them. In that event, they remain responsible for their own radiation records.
[‡] In some countries (e.g. France) records are maintained centrally which obviates the need for the hospital to do this.

Sources should be kept in shielded containers whenever possible and shielded source preparation facilities should be provided. Systems for handling sources should be established that make best use of the standard methods of dose reduction using time, distance, and shielding. The time needed to handle sources can be minimised by the provision of suitable handling equipment and the provision of an ergonomic environment. Distance can be increased by the use of long handled forceps, but self-gripping forceps should not be used as they may damage the encapsulation of the source. Shielding should be designed to allow rapid and easy work as otherwise there may be a significant time penalty. Automatic handling facilities should be considered wherever their use will reduce operator exposure.

Where sealed sources such as ^{137}Cs tubes are still used, they should be subject to regular inspection and leakage tests. These leakage tests should start when sources are first received and be repeated throughout their working life at intervals not exceeding two years. For older sources and those used in automatic afterloading equipment, annual testing is advisable. Sources should also be inspected for mechanical damage after cleaning or use. Identification marks should be checked. If any source fails any of these tests, it should be returned to the manufacturer and not re-used.

In principle, solid sources should not be cut or modified in any way. A common exception is in the use of ^{192}Ir wire, for which appropriate cutting facilities are available. Care should be taken that activity transferred to the cutter is not dispersed further and that short lengths of wire are not left in the handling area. The wire must be fully accounted for at all stages of its use up to its final disposal.

All equipment used in connection with sealed sources should be monitored for contamination. If contamination is found, its source should be identified and corrective action should be taken. The precautions described in Section 61.3.4.3 are also relevant to the use of sealed sources.

Apart from the precautions necessary for their safe handling, careful accounting is necessary to ensure that sources are not lost. Regular audit (monthly is normally appropriate) should be performed at suitable intervals with independent annual audit by a senior staff member*.

Sterility of sources is a particular problem for implanted seeds and these are usually delivered presterilised from the manufacturer. Resterilisation using steam sterilisation of ^{125}I sources is possible, but not ideal. Iridium wires are usually encapsulated in a catheter and do not therefore need to be sterilised. With hairpins, however, the wire is in contact with tissue. The pins are usually considered self-sterilising, but it is usual to soak them in chlorhexidine, followed by rinsing in sterile water before use.

61.3.4 UNSEALED SOURCES

Radiation protection for unsealed source therapy involves the same precautions as for small sealed sources, but the potential for contamination becomes a major consideration. Accidental spillage creates the potential for contamination of surfaces as well as of staff and patients and for ingestion or inhalation of radioactive substances. Such spills are common, particularly when syringes are used for injection.

61.3.4.1 Dispensing and Handling of Unsealed Radioactive Substances

The preparation of radiopharmaceuticals requires full pharmaceutical facilities as well as radiation protection. For this reason, many hospitals limit their activities to dispensing prepared solutions. It is important to be prepared for the possibility of spillage by ensuring that

---

* In the current international security climate, the safekeeping of sources attains an even greater importance.

appropriate containment is provided in the form of deep trays. Syringes used for drawing up radioactive liquids should be shielded. Therapeutic isotopes are likely to be β emitters for which a transparent polymethyl methacrylate (PMMA, known as Perspex™ or Lucite™) shield is appropriate because less bremsstrahlung is produced than with the lead or tungsten shields used for γ emitters. After dispensing is complete, surfaces should be checked for contamination with an appropriate detector (see below). The local rules given in Appendix L.3 indicate other precautions that should be taken.

During movement from the dispensing laboratory to the administration room (ideally nearby), the radiopharmaceutical must be well-shielded and packaged to avoid spills in the event of an accident. A sturdy trolley is useful for this purpose, but it may also be carried by hand in a shielded container. Two persons should be present during movement of therapy-level quantities of radiopharmaceuticals in case of accidents.

^{131}I poses particular problems. It is used in large quantities for the treatment of thyroid carcinoma. Typical activities are in the region 3 GBq to 10 GBq, but these activities will vary widely according to local practice and the stage of disease treated. Not only does it have a penetrating γ-ray (360 keV) requiring thick shielding, but in the elemental form it is also volatile. Because it is readily incorporated into normally functioning thyroid glands, it constitutes a particular hazard. Therapy quantities of ^{131}I should be handled in fume cupboards (or similar) behind substantial shielding. Obtaining iodine in the correct quantities for administration as a liquid or in capsule form will reduce the amount of handling required.

^{131}I is also used in labelled mIBG for the treatment of neural-crest tumours (see Section 58.1.2). Although often used at larger activities (up to 35 GBq) and thus more of a problem in regard to exposure to external radiation, it is only the free iodide, i.e. between 5 percent and 10 percent formed by radiolysis that will be taken up by the thyroid. Patients receiving mIBG therapy will have their thyroids blocked by administration of inactive iodide to prevent this uptake.

61.3.4.2 Monitors for Working with Unsealed Radioisotopes

Contamination and exposure-rate monitors (calibrated for each radionuclide, e.g. in terms of the annual limit of intake) must be readily available and used at all stages of the therapy procedure. For example, a thin-end-window Geiger counter is suitable for contamination monitoring of β emitters, and a side-window Geiger counter or scintillation detector is suitable for γ emitters. Suitably mounted and shielded monitors should be available for checking contamination of hands.

61.3.4.3 Monitoring of Staff Working with Unsealed Radioactive Substances

During preparation, dispensing, and administration of a therapy radiopharmaceutical, it is advisable to monitor finger dose; this may be carried out by using small lithium fluoride thermoluminescent dosimeters. To assess external dose, a personal monitoring badge must be worn. If shielding is used and a high dose rate is expected above the level of the shielding, it is advisable to wear two dosimeters on shielded and unshielded parts of body. These precautions are also appropriate for handling sealed sources.

An integrating personal dosimeter with direct readout and a dose-rate alarm is useful at high dose rates. In the case of suspected or possible inhalation of volatile radionuclides (e.g. ^{131}I), staff are advised to monitor both their chest (immediately after exposure) and thyroid (at 24 h) with a calibrated NaI scintillation detector. Regular whole-body counting will provide reassurance, but this should be tempered with the knowledge that the effective half-life of therapy agents is often measured in days.

61.3.4.4 Contingency Arrangements

Most incidents involving unsealed sources in hospitals do not warrant emergency action, but can be dealt with by trained staff working in the area using a *spill kit* (container of items useful for decontamination of personnel and equipment). However, contingency plans must be drawn up to cover the possibility of a serious incident. In such an event, the radiation protection supervisor for the area must be informed. Equipment for use in an emergency with unsealed sources (e.g. respirators, barriers for restricting access, portable monitors, and a list of useful telephone numbers) must be available. The effectiveness of contingency plans should be tested in training exercises using low-activity sources.

61.4 PATIENT TREATMENTS

61.4.1 Management of Brachytherapy Patients

61.4.1.1 Manual Afterloading

In manual afterloading procedures, the sources are permanently located in the patient until they are removed. This means that during hospitalization there is a potential hazard for staff or visitors entering the room and the additional risk of source loss. Appropriate warning notices must be displayed that indicate the activity implanted and any restrictions on visiting, etc. Consideration must always be given to the possibility that implanted sources may become displaced. Normal procedures may include x-raying patients to ensure that implanted sources are correctly positioned in the patient. All dressings, bedding, and other items removed from the vicinity of the patient should be monitored to ensure that sources have not been inadvertently removed. Shielded pots should be placed near the bedside of patients to receive displaced sources. Patients should be monitored after removal of sources to verify that the sources have been completely removed. Portable monitoring equipment and remote handling tools should be provided. Additional fixed radiation detectors, located at well chosen exits, are useful to make sure that sources are not inadvertently taken out of the hospital by patients, visitors or staff. Mobile shields can be used for staff safety and may also be appropriate for the safety of visitors. Typical thicknesses for such shields are between 20 mm and 30 mm of lead for ^{192}Ir, ^{198}Au or ^{137}Cs sources. Lead aprons as used in radiology are of no benefit for these sources. For ^{125}I sources, 0.1 mm of lead or equivalent thicknesses of other metals may be used (see attenuation data in Table M.4).

61.4.1.2 Automatic Afterloading

High-dose-rate afterloading will be carried out in a shielded room with appropriate interlocks in a similar way to external-beam therapy. The treatment lasts for a few minutes and there are no subsequent radiation protection issues for nursing staff. The ICRP has published recommendations on the safety of high-dose-rate brachytherapy (ICRP 2005b). For low-dose-rate or pulsed-dose-rate afterloading, the irradiation will take place while the patient is in bed and the nurses will be able to operate the equipment so that there is, in principle, no hazard to them when they are with the patient. However, it is possible when applicators are not correctly fitted for the applicators to become dislodged and a procedure to deal with the situation where the sources do not return to the safe position is required.

61.4.1.3 Patients with Permanent Implants Leaving Hospital

Patients who have had their prostates implanted with ^{125}I or ^{103}Pd will leave the hospital shortly after the implant. Task Group 57 of the ICRP has made recommendations for this situation (ICRP 2006b). It is unlikely that doses greater than 1 mSv/year will be received by relatives and others as a result of such implants. However, patients should be advised to restrict close contact with children and pregnant women for the first two months following the implant. There is the possibility of seeds being expelled in the urine or during ejaculation and patients should avoid sexual intercourse for a week following the implant and should use a condom for the first five ejaculations. Immediately following the implant, it is advisable to strain the urine to check for such seeds although the ICRP report points out that there will be little hazard if they are flushed down the toilet and straining is not advised after the first few days. Patients should be advised that if they find such seeds in the first week or two after the implant, they should avoid physical contact with the seed, placing it in a sealed container and returning them to the hospital. Patients and their carers should be made aware of the protection issues relating to cremation (see Section 61.4.3). The advice given is similar to that described in Section 61.4.2.3.

61.4.2 MANAGEMENT OF THE PATIENT UNDERGOING UNSEALED-SOURCE RADIOTHERAPY

61.4.2.1 In-Patient Care

During therapy with liquid radionuclides (such as ^{131}I or ^{90}Sr), as with solid sources, the patient is a source of external radiation. However, in addition, all body fluids of the patient (e.g. urine, perspiration, vomit, saliva, blood) will be radioactive, with consequent risk of contamination. Although practices may vary at different centres, the precautions described below should avoid contamination during the treatment of thyroid carcinoma with ^{131}I. During other types of therapy, if there is less risk of contamination, it may not be necessary to carry out all of these precautions.

Prior to administration of the radionuclide, plastic-backed absorbent paper should be taped to the floor around the patient's toilet and on other surfaces where contamination is most likely. After administration of the therapy agent, radiation measurements should be made around the treatment facility. Warning signs must then be clearly displayed to indicate the designation of the area (i.e. controlled or supervised). These signs should show:

- The nature and activity of the radio-nuclide used
- The date and time of administration
- Instructions to attending staff such as the use of protective clothing
- Recommended maximum time close to the patient (calculated taking account of the number of patients treated each year, the remaining activity within the patient, and the dose limits for staff)

Until discharge, in-patients should be identified with special signs (e.g. wearing a wristband), indicating that they are radioactive.

Mobile lead shields may be necessary for the protection of staff working close to the patient. They may also be placed at the door of the room to shield visitors and provide a demarcation. If the patient is catheterised, the urine bag should be placed in a lead pot (with lid) to reduce exposure to staff, and to minimise errors in measurements of whole-body counts.

The patient must be confined to their own suite while their activity exceeds levels that might give rise to excess exposure to others. As a guide, this level is reached when the product of the energy of the photons emitted per disintegration and the total activity in the body exceeds 150 MBq MeV, which is an activity of 400 MBq of ^{131}I (IPEM 2002). If patients are required to move to another part of the hospital (e.g. for a scan) they should be accompanied by staff suitably trained in radiation protection. Before leaving their room, the patient should shower and change into clean clothing. Overshoes should be worn and be removed on return.

Bed linen and clothing should be changed frequently and put into plastic bags. Each item should be monitored and, if contaminated, washed in the designated washing machine, dried, and re-monitored. Items still contaminated can be placed in labelled plastic bags and stored in a designated locked room until the remaining activity can be neglected. Contaminated waste should be macerated for disposal as liquid waste or bagged for future disposal as solid radioactive waste.

All staff entering the patient's room should wear plastic overshoes, aprons, and gloves. These are discarded in designated plastic bags on exit from the room and treated as radioactive waste. If staff who enter the patient's room handle potentially contaminated items, they should monitor themselves for contamination immediately afterwards. No item should be removed from the patient's room unless first checked for contamination by trained staff. If found to be contaminated, they should either be cleaned, disposed of, or put into store to allow for return after decay to acceptable levels.

Sampling of body fluids should be avoided if possible, especially in the first few days of therapy when they are likely to be most radioactive. Samples should be monitored and, if necessary, instructions on handling given to the laboratory testing the sample.

Regular assessments of the residual activity in the patient should be made to update the time allowed for staff to spend in the vicinity of the patient and to determine the time at which it is deemed safe to discharge the patient. These may either be based on calculations of the proportion of the initial activity remaining, or preferably on measurements with hand-held survey meters. If fitted, the whole-body counter in the ceiling can be used for this purpose (see Section 61.2.2).

Some patients may require particularly intensive care. In such cases, a detailed individual risk assessment should be carried out and appropriate precautions put into place. An example of the considerations required can be found in the paper by Greaves and Tindale (2001), which also considers the issues associated with the subsequent death of the patient (see also Section 61.4.3).

After discharge of an in-patient, the treatment room should be monitored for contamination and, if necessary, decontaminated by cleaning with proprietary detergents such as Decon™ (diluted 1:100). Moveable items that cannot be adequately decontaminated may be stored until the activity remaining is negligible. Nonremovable contamination can be covered with an adhesive plastic film (e.g. Fablon™).

61.4.2.2 *Patient Leaving Hospital*

Guidance on precautions to be taken when patients leave a hospital has been issued by the European Commission (EC 1998), the BIR (BIR 1999), and the IPEM (IPEM 2002). As discussed in Section 60.2.3.3, the statutory dose limits do not apply to those knowingly and willingly helping in the support and comfort of patients, but other basic principles such as the justification of practices and optimisation of radiation protection involving appropriate dose constraints* must be applied. In deciding on the appropriate activity levels, it is reasonable to consider the availability of appropriate facilities in the hospital, as well as the views of individual patients and carers (EC 1998). Many regulatory authorities prescribe activity levels at which patients may be discharged, but these vary widely. The IAEA (1996) give a limit of 1100 Mbq, but suggest that 400 MBq would be good practice. Within the European Union (EU), recommended activity levels for ^{131}I vary from 95 to 800 MBq, although the majority of states recommend between 400 and 600 MBq. While it is appropriate for general guidance

---

* See Section 59.3.5 for an explanation of the difference between *dose limits* and *dose constraints*.

TABLE 61.3

Threshold Activity Levels (MBq) for Discharging Patients after Unsealed Source Radiotherapy

Radio-nuclide	Return to Radiosensitive Work[a], Contact with Children, <10 MBq MeV	No Restrictions, Except Radio Sensitive Work[a] and Contact with Children, <50 MBq MeV	Travel by Public Transport, <150 MBq MeV	Travel by Private Transport, <300 MBq MeV
32P	300	1500	4500	9000
90Y	100	500	1500	3000
131I	30	150	400	800
198Au	30	150	400	800

[a] Examples of radiosensitive work include radionuclide assays, radiation monitoring, or work with radiosensitive materials.
Source: From NRPB (National Radiological Protection Board), *Guidance Notes for the Protection of Persons against Ionising Radiations Arising from Medical and Dental Use*, HMSO, London, 1988. (While this document is now obsolete the relative values remain useful.)

to be developed (depending on national regulations) the individual circumstances of each patient should also be considered.

In addition to any nursing requirements, patients, especially those treated with iodine for thyroid cancer, may be hospitalised because they are a potential source of ionising radiation for others, either as a consequence of external radiation or the risk of contamination. The patient will usually be anxious to return home and there will be pressures to release the bed for the treatment of others. Consequently, the patient will return home as soon as it is safe for this to happen.

The main factors that determine whether it is safe for the patient to be discharged from hospital are the radionuclide administered, the residual activity in the patient, the domestic circumstances, and the method of transport of the patient to home. Consideration must also be given to the state of health and nursing requirements of the patient. For example, a patient treated with a pure β emitter will not present an external radiation hazard, but might give rise to a contamination hazard if incontinent. It might be advisable to delay the discharge of such a patient, or to give special instructions (e.g. regarding handling of contaminated laundry).

Current legislation emphasises the need for practice to be based on risk estimates set against dose constraints, but Table 61.3, which is taken from the U.K. Guidance Notes of 1988 (NRPB 1988) provides a useful practical baseline. While the figure of 800 MBq remains the U.K. recommendation (IPEM 2002) for discharge of a patient receiving ^{131}I therapy from hospital, the EU document (EC 1998) is based around a more conservative figure of 400 MBq. The values for other isotopes are not available in the other recommendations, but are based on the product of activity and energy of the photons emitted per disintegration (Mountford 1997). No radiation protection precautions were recommended where this product did not exceed 10 MBq MeV. At levels greater than this, a patient might be discharged, but have to observe certain restrictions after leaving the hospital. In 1988, there was little available information about doses received by persons in the vicinity of radioactive patients and the table relates to compliance with the definition of a controlled area in the vicinity of radioactive patients. An activity energy product of 150 MBq MeV will require a controlled area within one metre of a patient.

Considerable progress has been made since the 1988 guidance notes were written in measuring doses to persons exposed to radiation from radioactive patients, particularly after treatment with ^{131}I (Monsieurs et al. 1998; Barrington et al. 1999; Reiners and Lassman 1999). The European Commission guidance (EC 1998) identified six groups of people:

TABLE 61.4

Proposed Dose Constraints [mSv] for Family and Close Friends per Treatment with ^{131}I

Groups of Persons	Dose Constraint (mSv)
Children under 10 years of age (including unborn children)	1
Adults up to about age 60 years	3
Adults older than about 60 years age	15

Consideration must be given to the possibility of exposure during the year from other sources and where this cannot be controlled (as for members of the public) a constraint of 0.3 mSv applies. The term "Family and Close Friends" used in this document is equivalent to "Comforters and Carers" used by ICRP.
Source: From EC (European Commission), Radiation Protection 97, Radiation protection following iodine-131 therapy, EU, Brussels, 1998, http://europa.eu.int/comm/energy/nuclear/radioprotection/publication/doc/097_en.pdf.

- Children up to 2 years of age (who are likely to spend more time in close contact with their parents and relatives)
- Children aged between 3 years and 10 years (because the risks of cancer induction when children are irradiated are 2–3 times greater than when adults are irradiated)
- Pregnant women (the foetus being equivalent to a young child)
- Partners
- Partners over 60 years of age (because their probability of contracting cancer or of passing on genetic damage is between 3 and 10 times lower)
- Other people—who must be treated as members of the public

This guidance (Table 61.4) proposes constraints for family and close friends based on risks for exposure at various ages. The guidance also provides detailed advice in a form suitable to be given to patients.

Those persons who come into contact with the patient, where they cannot be classed as comforters and carers, are still subject to limits (see also Section 60.2.3.4). For example, in the U.K. these limits are 1 mSv/year; the special circumstances of these exposures would allow averaging over 5 years, but there is no dispensation to allow exposure to the constraint of 15 mSv from a single administration. However, persons exposed as comforters and carers (including care at home) are not subject to limits and these constraints could, if necessary, be applied.

When discharging a patient to another institution (e.g. a nursing home or hospital) details of radionuclide administered, residual activity, and estimates of activity excreted in the urine must be supplied and care taken to ensure that appropriately trained staff are available (IPEM 2002).

61.4.2.3 Advice to Patients and Carers

Prior to the administration of radioactive substances, it is good practice to explain clearly to patients and their relatives what will happen to them when they are in the hospital, together with the restrictions that will be placed on visitors. This discussion should cover the arrangements for their discharge and give adequate information about radiation hygiene after leaving the hospital. Information should also be sought from the patient about their family circumstances and the nature of facilities at home. It is particularly important to explain restrictions on close contact with children.

On discharge from hospital, it is recommended that the patient be given a card with the following information (IPEM 2002):

TABLE 61.5

Activities of Therapy Nuclides above Which Burial or Cremation Are Not Recommended

Radionuclide	Burial (MBq)	Cremation (MBq)
^{131}I	400	400
^{125}I seeds	4000	—[a]
^{103}Pd seeds	15 000	—[a]
^{90}Y colloid	2000	70
^{198}Au seeds	4000	—[a]
^{198}Au colloid	400	100
^{32}P	2000	30
^{90}Sr	2000	200

[a]   The encapsulation of radioactive seeds is required to withstand the heat of cremation and it is therefore possible that the contents of the seeds will not be dispersed. A risk assessment should therefore be carried out to determine the point at which this will no longer present a significant risk. For ^{125}I seeds ICRP (2006b) recommends that cremation does not take place before one year has elapsed.

Source: From IPEM (Institute of Physics and Engineering in Medicine), *Medical and Dental Guidance Notes. A Good Practice Guide to Implement Ionising Radiation Protection Legislation in the Clinical Environment,* IPEM, York, 2002.

- Name and address of the patient
- Name, address and telephone number of the hospital
- Contact name in the hospital in case of difficulty
- The name of the prescribing physician
- The type and quantity of radionuclide administered
- The date of administration
- Restrictions to be followed, especially concerning contact with young children
- The period of time for which the restrictions apply

A new problem has arisen with the potential terrorist threat to use radioactive materials to contaminate public areas. As a result, many airports and other places have installed radiation detectors and these frequently detect patients who are radioactive. It is therefore wise to determine whether a patient is expecting to travel in the foreseeable future and to warn them of the need to justify their radioactive state.

61.4.3   DEATH OF PATIENTS CONTAINING RADIOACTIVE MATERIALS

If a patient should die soon after receiving therapy with radioactive material, the *qualified expert* (RPA) should be notified immediately. Precautions may be required (e.g. for post mortem and/or regarding disposal of the corpse by burial or cremation). Activity limits for burial or cremation for the most common therapy radionuclides are listed in Table 61.5. It may be necessary to consider removal of the organ containing radioactive material if a risk assessment shows that the activity of the seeds will present a risk to crematorium staff. This is likely to be the case if death occurs within one year of a prostate seed implant with ^{125}I. If a post mortem is necessary, appropriate precautions should be taken. Unless the minimum limit in Table 61.3 has been reached, it is recommended that embalming should not take place.

61.5 RISKS ASSOCIATED WITH RADIOTHERAPY TREATMENT

61.5.1 RADIOTHERAPY OF PATIENTS OF CHILD-BEARING AGE

If the patient being considered for radiotherapy is pregnant, the foetus would be at direct risk from radiation. There is also the possibility that radiation damage to the gonads of father or mother can result in sterility or genetic damage to the as-yet-unconceived child. If the mother is breast feeding, then therapy agents can be passed on in breast milk. The following paragraphs consider such issues.

61.5.1.1 External-Beam Therapy of Potentially Pregnant Women

Radiotherapy should be avoided when a woman is pregnant. Termination of the pregnancy, other methods of therapy, or postponement of radiotherapy are alternatives that may be considered. The risk to the foetus from radiotherapy (see Section 59.2.9) must be balanced against the corresponding risk from alternative therapies and the risk to the mother of hormonal changes in pregnancy. The appropriate course of action should be decided in discussion between the patient, the oncologist, and the physicist. There may therefore be occasions when it is decided to proceed with treatment. In such cases, the dose to the foetus must be estimated and precautions taken to minimise the dose.

Although the treatment of pregnant mothers is unusual, it is important that the physicist should be aware of the possibilities so that appropriate advice can be given when required. Task Group 36 of the AAPM has produced an extensive review of the subject (Stovall et al. 1995) and guidance is available in ICRP Report 84 (2000). The foetal dose will evidently depend on the proximity of the uterus to the target volume. The dose results from a combination of head leakage, external scatter, and scatter from the patient's own tissues. Extensive data are available in the literature for peripheral doses so that foetal dose can be calculated as a function of field size, beam energy, machine type, treatment volume, and distance from the field edge to within a factor of approximately two (O'Connor et al. 1987; Ngu et al. 1992; Sneed et al. 1995; Van der Giessen 1996, 1997). This is sufficiently accurate for the purpose of advising on the risks, bearing in mind the uncertainty inherent in risk estimates. A common site to be treated is the breast and it is argued that the dose received is unlikely to be sufficient to cause an unacceptably high risk to the foetus; certainly not to the extent of recommending abortion (Antypas et al. 1998; Greskovich and Macklis 2000; Fenig et al. 2001). The risk to the foetus of radiation induced cancer is around 15% for 1 Sv.

The dose delivered to the foetus depends on the type of treatment unit and on the use of accessories such as wedge filters. By an appropriate choice of treatment and perhaps some additional shielding, it should be possible to minimise the foetal dose. For example, intensity modulated radiotherapy should probably be avoided because the contribution of head leakage will be greater. Once the beam arrangements have been decided, phantom measurements should be made to confirm the predictions. During treatment, TLD measurements should be made to verify the calculations.

61.5.1.2 Unsealed-Source Therapy of Pregnant or Nursing Mothers

Radiopharmaceuticals used for therapy may be able to cross the placental barrier and lead to direct uptake by the foetus. There may also be irradiation of the foetus by penetrating radiation from uptake within the mother. The use of MIRD methodology (see Section 56.1 and Section 56.2) will enable an estimate of this dose. Clearly shielding the foetus is not an option in this case. Particular care needs to be taken with regard to the nursing mother because radioisotopes will find their way into breast milk. The risk of uptake from breast milk can be avoided by suspension of breast feeding. Guidance on the periods of suspension for individual isotopes can be found in Section 7 of the ARSAC guidance notes (ARSAC 1998), but for ^{131}I it is recommended that breast feeding is stopped altogether.

61.5.1.3 Irradiation of the Gonads

General sound advice is that gonads should be shielded from external radiation. Treatment of young adults with a potential for future healthy life and parenthood requires that dose to the gonads is minimised and that its magnitude be estimated. High doses to the testes or ovaries can result in temporary or permanent sterilisation. A woman's ova are formed early in life and radiation exposure may lead to genetic damage expressed in future generations. Testicular shields can be used to reduce the gonad dose, although the benefit is small. Modern technology for the storage of sperm and ova provides an alternative solution. It is also possible to move the ovaries surgically prior to treatment.

In the case of radionuclide therapy, future conception should be avoided until after all significant radioactivity has been eliminated from either the father's or the mother's body. The necessary period for this will vary with the nuclide and activity administered. Practical guidance is also available in the ARSAC guidance notes (ARSAC 1998).

61.5.2 SECOND CANCER RISK FROM RADIOTHERAPY

The requirement for optimisation and justification of radiotherapy treatments imposed by the European Council Directive on medical exposure (ECD 1997) increases the importance of an understanding of the risks associated with radiotherapy treatment and when carrying out research studies there is a requirement to estimate the effect of any radiation delivered (ICRP 1993). While issues such as hair loss and the formation of cataracts have long been well understood, stochastic risks such as the induction of second cancers have only recently been taken seriously. The growing literature on this subject has been reviewed by the National Radiological Protection Board (NRPB 2000) and by the Board on Radiation Effects Research (BRER 2006). It is concluded that doses to tissues in the target volume are likely to be sufficiently high as to sterilise any induced cancer, but that doses to surrounding tissues may cause second cancers. The evidence is that the risk level is the same as that derived from the atomic bomb data referred to in Section 59.2. However, Aird (2004) and Harrison (2004) draw attention to the problems of using the concept of effective dose (Section 59.1.3) and population-based statistics when dealing with individual radiotherapy patients. Direct proof of a causal relationship between radiotherapy and second cancers is difficult because of the conflicting factors that the patients may have an increased susceptibility to cancer whether for hereditary or lifestyle reasons.

A particular problem arises when young patients are being treated. Examples in which there is good evidence of second cancers are the contralateral breast (Gao et al. 2003) and patients treated with mantle fields for Hodgkin's disease who have a significantly increased risk of contracting breast cancer (Bhatia et al. 1996), particularly if irradiated below the age of 30 (Aisenberg et al. 1997). Fortunately, there is reasonable success in treating breast cancer (Horwich and Swerdlow 2004), but the high incidence has lead to a strong desire to find other methods of treatment. Khan and Haffty (2001) have evaluated the site of second tumours within the contralateral breast and found that the site is not correlated with the area receiving the highest dose, which may call into question the causal relationship in the case of breast cancer treatment.

There is also some evidence of second cancers following the treatment of older patients. For example Movsas et al. (1998) and Brenner et al. (2000) show a small increase in second cancers following prostate treatment. Yasui et al. (2003) point out the need to consider the natural incidence of cancer with age for the subjects of studies of second cancers. Waddington and McKenzie (2004) have analysed the expected impact of double exposure portal films and predict that these will be likely to cause cancer. However, the age of such patients may reduce the practical impact and any effect of cancer induction must be balanced against improvements in localisation increasing the potential for cancer cure. Nevertheless, it is preferable to obtain images using the actual treatment field where this is sufficient to obtain positional information and new techniques are becoming available to obtain portal images with reduced dose (Vetterli 2004).

A number of authors point out that IMRT involves a higher scatter dose and that IMRT treatment is likely to increase the number of induced second cancers (Verellen and Vanhavere 1999; Dorr and Hermann 2002; Followill et al. 2003; Hall and Wuu 2003; Gregoire and Maingon 2004). Followill et al. (1997) recommended that high-energy photon beams should not be used for IMRT because of the increased risk of second cancers predicted from the associated neutron dose. Price et al. (2003) found that a Varian 18 MV machine produced significantly more neutrons than the corresponding Siemens machine because the latter had a lower electron energy (see Section 22.2.2). Further work on this has been reported by Vanhavere et al. (2004). This risk of cancer induction must be set against the benefits of IMRT.

61.5.3 DETERMINISTIC RISK OF ORGAN DAMAGE

It has long been known that acute radiation damage can be caused to various organs, such as the optic nerve, the spinal cord, the lungs, liver, and kidneys. However, more recently, late effects of irradiation of the heart have been identified. In a meta-analysis of breast radiotherapy treatment Cuzick et al. (1994) found that, although radiotherapy improved local control, it did not improve survival because of the increase in deaths associated with cardiac problems. More direct evidence of cardiac damage has become available since then, and it is clear that dose to the heart should be minimised (Gagliardi et al. 2001; Adams et al. 2003).

61.5.4 PATIENTS WITH PACEMAKERS

Patients who have implanted cardiac pacemakers may also be at risk. Early reports of the assessment of damage to pacemakers suggested that there was not a problem (Walz et al. 1975), but modern pacemakers that use CMOS logic are more sensitive to damage than older types (Adamec 1982; Last 1998). Several publications give advice on this subject including the American Association of Physicists in Medicine (AAPM) guidance (Marbach et al. 1994) and more recent publications by Last (1998) and Solan et al. (2004). The general consensus has been that the dose to the pacemaker should be less than 2 Gy and that it should therefore never be in the direct beam. However, work by Mouton et al. (2002) has identified a pacemaker failure with a dose as low as 0.15 Gy. Solan et al. suggest that pulse and blood pressure monitoring should take place before and after each treatment and that pacemaker dependent patients should be monitored using the electrocardiograph at the first treatment, but this level of monitoring may be hard to implement. Manufacturers provide advice on the susceptibility of their pacemakers to radiation, but the studies by Mouton and Solan et al. call into question the reliability of such advice. More recently, implantable cardiac defibrillator devices have been introduced and these are believed to be more susceptible to the effects of radiation; it is suggested that the dose delivered to them should be limited to 1 Gy. There does not seem to be good evidence for the view that patients with pacemakers should only be treated on cobalt-60 units, although the AAPM recommend that such patients are not treated on betatrons. It would be prudent to establish the patient's degree of dependence on the pacemaker before commencing treatment and to base the level of monitoring on this. It would also be sensible to ensure that the pacemaker is checked after treatment and that the patient is warned of the possible dangers.

PART L: RADIATION PROTECTION IN
RADIOTHERAPY

APPENDIX L

RADIATION PROTECTION REGULATION IN THE UNITED KINGDOM

*Philip Mayles**

CONTENTS

---

* With contributions from Mike Rosenbloom.

L.1 U.K. REGULATORY FRAMEWORK

L.1.1 INTRODUCTION

As discussed in Chapter 60 all radiation protection legislation in the European Union (EU) is based on directives of the European Council of Ministers, which are then interpreted into the legal framework of individual states. Each state will interpret the legislation in a slightly different way. This Appendix, which describes the U.K. legislation*, is included both for the benefit of U.K. readers and to provide a comprehensive overview of a self-consistent legal framework. Attention is drawn to some small differences from the original Council Directives.

The legislation applies equally to therapy and diagnosis, but only the application to therapy will be considered here. The Appendix should be read in conjunction with Chapter 60. where a more international treatment is given. This summary of the legislation should not be regarded as a substitute for the original text. References are given to the specific regulations and their associated schedules.

Oversight of radiation protection issues is the responsibility of a number of government departments. The Secretary of State for Environment, Food and Rural Affairs has overall responsibility for radiation protection legislation, the Secretary of State for Work and Pensions for health and safety legislation and the Secretary of State for Health for protection of the patient. These functions are carried out through several agencies.

Because the process to introduce a new act of Parliament is complex and time consuming, all the radiation protection regulations were "laid before parliament" under the Health and Safety at Work Act 1974 and this, in spite of being a simpler process, took several years to achieve. The Health and Safety at Work Act 1974 led to the setting up of the Health and Safety Commission (HSC) and its operational arm the Health and Safety Executive (HSE). The HSC is a body of up to ten people appointed by the Secretary of State for Work and Pensions. It has a full time chair and part-time members including representatives from employers, unions, local authorities and public interest. Its function is to make arrangements to secure the health, safety and welfare of people at work and the public. It is the interface between ministers and the HSE. The HSE is formally a body of three people appointed by the HSC with the consent of the Secretary of State. It advises and assists the Commission in its functions and also has specific statutory responsibilities of its own, notably for the enforcement of health and safety law. However, the term *HSE* is also used to describe its staff (around 4000) that includes inspectors, policy advisers, technologists, and scientific and medical experts. The HSE ensures the safety of employees and public from the use of ionising radiation in hospitals and other places.

---

* What is described is strictly the legislation for England. Detailed legislative arrangements in England, Scotland and Wales are complex, but in radiation protection terms are similar. Some minor differences exist in Northern Ireland legislation.

The Environment Agency was formed in 1996 as an independent agency charged with protecting the environment from all manner of potentially damaging agents including ionising radiation. It incorporates the Nuclear Inspectorate that is responsible for the regulation of the disposal of radioactive waste (as well as for the supervision of the nuclear industry).

The Department of Health (DoH) regulates the medical exposure of persons to ionising radiation. A committee, the Administration of Radioactive Substances Advisory Committee (ARSAC), is concerned directly with all administration of radioactive substances to patients for medicinal purposes. The Health Protection Agency was set up in 2003 to take responsibility for advice on health protection issues. It now incorporates the National Radiological Protection Board (NRPB). From November 2006 the Commission for Healthcare Audit and Inspection (Healthcare Commission) has taken responsibility for inspections relating to protection of the patient.

L.1.2 IONISING RADIATION REGULATIONS 1999

The Ionising Radiation Regulations 1999 (IRR1999) (HS 1999a) are derived from the EU Basic Safety Standards (ECD 1996a) and also incorporate the Outside Workers Directive (ECD 1990) (replacing the Outside Workers Regulations 1993 (HS 1993a)). They set out the requirements for the protection of staff and the public and apply equally to hospitals as to other users of radiation. They are accompanied by an Approved Code of Practice (ACoP) (HSE 2000) which sets out ways of achieving compliance with the law. Those who choose to depart from the code must be prepared to show that their own approach is an equally valid way of meeting the legal requirements. Guidance on the application of IRR1999 in medicine is also provided by the Medical and Dental Guidance Notes (IPEM 2002).

L.1.2.1 Dose Limits

The dose limits for staff and the public are set out in Regulation 11 and Schedule 4 and are listed in Section 60.2.2.3. *Category A* workers who might receive more than 6 mSv per year as defined by the EU BSS (ECD 1996a) are called *classified* workers in the U.K. (Regulation 20). Classified workers must be subject to medical supervision (Regulation 24) and more elaborate record keeping (Regulation 21). Their dose records must be kept for at least 50 years by an approved dosimetry service.

L.1.2.2 Investigations

It is a requirement that an *investigation* is carried out (Regulation 8) the first time that an employee exceeds 15 mSv in a calendar year in order to verify that the dose is indeed as low as reasonable practicable. An investigation must always be carried out if an overexposure of a member of staff occurs (Regulation 25) or if an equipment malfunction results in the overexposure of a patient (Regulation 32(6)) or if a personal dosimeter is lost or damaged (Regulation 22).

L.1.2.3 Control of Areas

The Regulations define the requirements for *controlled* areas and *supervised* areas (Regulation 16). The EU legislation leaves the definition of these areas to be defined by the individual states. In the IRR 1999 a controlled area is defined as one in which it might be possible to receive an annual dose of 6 mSv and a supervised area as one where a dose of 1 mSv might be reached. Both controlled and supervised areas must be suitably signed (Regulation 18) and controlled areas be physically demarcated or if this is not practicable, delineated by some other suitable means.

L.1.2.4 Equipment

Before new radiation work is carried out the employer must notify the HSE (Regulation 6). Before equipment is installed a *risk assessment* must be carried out (Regulation 7) to identify any possibility of the equipment leading to a radiation accident. If such a risk is identified, contingency plans must be set in place to mitigate the risk (Regulation 12). The supplier is obliged to carry out a *critical examination* of the equipment either in conjunction with their own Radiological Protection Adviser (RPA, see Section L1.2.5) or with the hospital's RPA (Regulation 31). The hospital is required to ensure that quality control of the equipment (Regulation 32) and its associated interlocks (Regulation 10) is carried out. The RPA should be consulted about the suitability of the *quality assurance programme*.

When an equipment malfunction occurs which results in a patient receiving a dose that is *much greater than intended* the Regulations require that this be reported to the HSE (Regulation 32). The definition of "much greater than intended" is contained in the HSE document PM77 (HSE 2006) as 10% for a whole therapy course or 20% for a single fraction. The requirement to report such overdoses is contained also in the IAEA BSS (IAEA 1996), but is not a specific requirement of the European Council Directive. In other European countries the reporting requirement is less juridical.

L.1.2.5 Radiation Protection Adviser

The Regulations require the appointment of a *Radiation Protection Adviser* if ionising radiation is to be used (Regulation 13). The duty to comply with the law remains with the employing authority who is in principle at liberty to ignore the advice of the RPA (although this would only be done in rare circumstances). An RPA is required to have an appropriate certificate of competence the attainment of which requires a portfolio of evidence showing both academic understanding and practical application or the achievement of a national vocational qualification (NVQ). Schedule 5 of the Regulations lists the issues about which an RPA must be consulted:

- Controlled and supervised areas
- New installations including the radiation shielding and interlocks
- Calibration and correct use of monitoring equipment
- The testing of interlocks

The RPA may also advise on a wider range of issues including risk assessment, investigations of estimated doses, investigations of high exposures, and training. A fuller list is given in the ACoP.

L.1.2.6 Radiation Protection Supervisors

The day to day oversight of radiation work is the responsibility of *Radiation Protection Supervisors* (RPS) appointed by the employer (Regulation 17). They are responsible for the application of the *local rules* (see Section 60.2.2).

L.1.2.7 Training

The Regulations require that the employer ensures that staff are adequately trained (Regulation 14). Training must be appropriate to the work being done, but particular training must be given to those acting as RPS. Training is also required for those entering a controlled area under a *system of work*.

L.1.3 Ionising Radiation (Medical Exposures) Regulations 2000

The Ionising Radiation (Medical Exposures) Regulations (IR(ME)R 2000) (HS 2000*) derive from the European Council Medical Exposures Directive (ECD 1997). They set out the regulations governing patient safety. Guidance on the application of IR(ME)R 2000 is provided by the Medical and Dental Guidance Notes (IPEM 2002) which also covers the medical application of IRR1999. This was produce by a working group of the Institute of Physics and Engineering in Medicine (IPEM) in association with the NRPB (now HPA), the HSE, the Health Departments and the Environment Agencies and in cooperation with the Royal College of Radiologists and the College of Radiographers. As such, it has a similar (but perhaps slightly lower) standing in legal terms to the ACoP associated with IRR 1999.

L.1.3.1 Definition of Roles

The Regulations define (Regulation 2) four classes of professional contribution to radiation therapy (and diagnosis). The duties associated with these roles are defined in Regulations 5 and 9. These are:

1. The *referrer* who sends the patient for treatment and is required to provide sufficient clinical information to allow the justification of the treatment to be determined[†]
2. The *practitioner* who is responsible for the clinical oversight of the treatment and for justifying any radiation exposure[‡]
3. The *operator* carries out any "physical aspect" of the treatment process (including for example machine calibration or treatment planning)[¶]; the Regulations make clear that all operators are personally responsible for their own contribution to the patient treatment
4. The *medical physics expert* (MPE) who must be "closely involved" in all radiotherapy treatment

The Regulations do not define who may fulfil these roles, apart from requiring that they be health professionals and in the case of the referrer that they be state registered (HS 2006). However, they do require that each employer produces a list of named individuals together with records of their training (Regulation 11) (although not expressly for the referrer). Employers are required to define written procedures to ensure that patients are correctly identified (Schedule 1). Employees are required to follow the defined procedures (Regulation 5). Operators must identify the patient and ensure that the exposure has been *justified* (Regulation 6) before exposing the patient to radiation.

L.1.3.2 Optimisation

Regulation 7, which deals with optimisation of exposures, includes the requirement that therapeutic exposures are individually planned to ensure that non-target tissues are given doses *as low as reasonably practicable* and that the equipment selected for treatment should be appropriate. It is a requirement that the clinical outcome of every exposure is evaluated. It is common

---

* Some minor amendments to the regulations came into force in November 2006 (HS 2006).
† In the EU legislation the term used is the "prescriber", but this is not the interpretation applied in the U.K. regulations.
‡ The use of the word "practitioner" is unfortunate because of its wider use to describe level 2 in the four tier structure of health service professional groups. The term "justifier" may better describe the role.
¶ This role is not expressly recognised in the European Council Directive.

practice for therapeutic exposures to be defined to include the whole course of radiotherapy including the necessary planning and verification exposures.

L.1.3.3   Investigation of Errors

In parallel with IRR 1999, the Regulations require that a human error that leads to an exposure *much greater than intended* must be reported to the Healthcare Commission (Regulation 4), who are responsible for investigation (and potentially the prosecution of individuals in the event that the error resulted from a breach of the Regulations). The Regulations do not define "much greater than intended". A course of radiotherapy may not be compromised by a significant error (even 20%) in a single fraction and it is accepted that for this purpose the course should be treated as a whole.

Errors in positioning of the beam may be as significant as errors in the magnitude of the dose and where they have a potential impact on treatment outcome, should also be reported. Because the requirement for reporting of overdoses derives from a regulatory framework there is no statutory requirement for reporting underdoses under the Regulations, even though they may also compromise treatment outcome.

L.1.3.4   Training

The requirement for staff to be adequately trained is a central part of the Regulations as it is of the European Council Directive. Comprehensive records of staff training must be kept in association with the list of referrers, practitioners and operators (Regulation 11). A list of topics which training must cover is in Schedule 2.

L.1.3.5   Equipment

The employer is required to draw up a list of equipment held, to include information about the manufacturer, serial number and year of manufacture and installation (Regulation 10). There is a requirement that the equipment is limited to the amount necessary, implying that redundant equipment must be decommissioned.

L.1.4   RADIATION (EMERGENCY PREPAREDNESS AND PUBLIC INFORMATION) REGULATIONS 2001

The Radiation (Emergency Preparedness and Public Information) Regulations (REPPIR) (HS 2001) are primarily intended to ensure that the nuclear industry has appropriate contingency plans. It requires that where sources of activity greater than a certain level are held, the employer should have appropriate contingency plans in place. Nondispersible sources (i.e. sources which by virtue of their physical or chemical form cannot cause a radiation emergency affecting the public in any reasonably foreseeable event) such as ^{192}Ir and ^{60}Co therapy sources used in afterloading or teletherapy equipment, are exempt except when being transported and it is therefore unlikely that there would be a requirement under these Regulations in a hospital.

L.1.5   RADIOACTIVE SUBSTANCES ACT 1993 CONTROL OF SOURCES AND WASTE DISPOSAL

The holding and disposal of radioactive substances is governed by the Radioactive Substances Act 1993 (HS 1993b). The Environment Agency for England and Wales and the Scottish Environment Protection Agency grants a license to hold specified activities of radioactive sources and also licenses disposal of waste. Following the European Council Directive on the control of high activity sources (ECD 2003) the government has issued the High-Activity Sealed Radioactive Sources and Orphan Sources (HASS) Regulations 2005 (HS 2005) which require that for high activity sources (e.g. ^{60}Co $>$ 4 GBq and ^{192}Ir $>$ 10 GBq) appropriate arrangements are in place for source security and eventual disposal before an authorisation is given. Guidance has been issued on the implementation of these regulations (EA 2006). To allow for the replacement of therapy sources, the licensed activity should allow for the presence of both the new source and the decayed source at the same time. Use of dispersible radioactive substances in a hospital will generate radioactive waste in the form of both patient excretions and unused quantities of isotopes. An estimate must be made of the amount of all such radioactive waste and the routes by which it will reach the environment. Before applying for an authorisation for the disposal of this waste an *environmental risk assessment* must be made for each route of disposal. Advice on making such assessments has been issued by the Environment Agency (EA 2002a) and the NRPB (McDonnell 2004). A therapy source that is no longer in use becomes classified as waste and counts towards the amount of waste material that may be held. It is therefore necessary that disposal arrangements are set in place and it is better if the supplier will take back the decayed sources. For those holding an authorisation to dispose of waste the Environment Agency requires that they carry out an assessment to show that the "best practical means" (BPM) are applied to minimise the impact of waste disposals. The requirement derives from European Commission requirements (EC 2003). The Environment Agency has issued guidance on carrying out this assessment (EA 2002b).

L.1.6 ADMINISTRATION OF RADIOACTIVE SUBSTANCES TO PATIENTS

The Regulations relating to the administration of radioactive substances to patients are contained in three pieces of legislation: The Medicines (Administration of Radioactive Substances) Regulations 1978 (MARS Regulations) (DH 1978a), The Medicines (Radioactive Substances) Order 1978 (DH 1978b) and The Medicines (Administration of Radioactive Substances) Amendment Regulations 1995 (DH 1995). The first of these is the principal one and established the Administration of Radioactive Substances Committee which oversees the grant of licenses by ministers under the Regulations. Licenses are granted to individual consultants for up to five years. The actual administration of the radioactive substance may be carried out by another person under the supervision of a license holder, but if this is to happen the agreement of the license holder must be recorded in writing. The license covers specified uses of radioactive materials in a specified location. The application for the license requires the applicant to show how scientific and radiation protection support will be provided and the responsible scientists will have to sign the application. It is the responsibility of the clinician to ensure that a valid license is held for a specific procedure. ARSAC licenses may also be issued for two years for clinical research purposes. In such cases ethical approval must be given for each study. An ARSAC license is not required for teletherapy, but is required for any form of brachytherapy.

L.1.7 THE RADIOACTIVE MATERIAL (ROAD TRANSPORT) REGULATIONS 2002

The regulations governing the transport of radioactive substances are the Radioactive Material (Road Transport) Regulations 2002 (HS 2002a)*, the Carriage of Dangerous

---

* Rail transport has separate regulations (The Packaging, Labelling and Carriage of Radioactive Material by Rail Regulations 2002 (HS 2002b)) but these are unlikely to affect medical physicists.

Goods by Road (Driver Training) Regulations 1996 (HS 1996), the Transport of Dangerous Goods (Safety Advisers) Regulations 1999 (HS 1999b) and the Carriage of Dangerous Goods and Use of Transportable Pressure Equipment Regulations 2004 (HS 2004). As discussed in Section 60.5 transport regulations are more closely aligned across the EU than other parts of the legislation. The Regulations (HS 2002a) define the classification of packages for transport purposes (Regulation 23 and Schedule 11) as specified in Table 60.3 and also the design of different types of packages (Schedule 8 and Schedule 9) which must be used depending on the source activity. Contents limits for different types of package are contained in Schedule 4 which relates to the detailed list of activity levels given in Table I of Schedule 1. All packages other than Type A packages require an approval certificate (Schedule 10 and Schedule 12, Part IV). The documentation and vehicle labelling that is required when consigning radioactive packages (Regulation 48) is listed in detail in Schedule 6 which contains details of the wording that must be used, etc. These include UN numbers (Schedule 1, Table VIII). The Regulations require that a quality assurance programme (Regulation 18) is in place to ensure compliance with the Regulations. Small packages containing activities less than those listed in Table III of Schedule 1 (HS 2002a) can be defined as "excepted packages" under the Transport Regulations, but the conditions of such exemptions are rarely met. The Regulations are not relevant when the radioactive substance is contained in a patient.

HS 1996 requires the training of drivers of vehicles carrying radioactive packages. HS 1999 requires (Regulation 4.1) that where packages are consigned for transport the employer has appointed someone who has undertaken an approved course of training in the Transport Regulations (HS 1999b). This can be an onerous requirement and for this reason it is common practice for the transport company to take responsibility for the whole process including the consignment of the packages. However, an employer who only occasionally engages in the transport of dangerous goods within the United Kingdom, and if that transport does not create a significant risk to the health and safety of persons and to the environment, is exempt (Schedule 1, Section 4). This exemption is interpreted as not more than 10 Type A packages with a total transport index of up to 3 (IPEM 2002).

L.2 EXAMPLE WALL-THICKNESS CALCULATIONS

Figure L.1 shows a treatment room originally designed to accommodate a 4 MV accelerator in about 1970. Since then, the 4 MV machine has been replaced by a 5 MV machine and some thickening of the walls at A was carried out. It is now due to be used for a modern 6 MV accelerator. This room will be used to illustrate some of the considerations in designing a linear accelerator treatment room. (For further examples of treatment rooms, the reader is referred to IPEM Report 75 [IPEM 1997] and NCRP Report 151 [NCRP 2005].)

The dose at point A, which is located on the main hospital corridor, is considered first. The dose rate at the isocentre is 3.70 Gy/min. The distance to point A from the source is 5.97 m, so the unattenuated dose rate is:

$$\frac{3.70 \times 1^2}{5.97^2} = 0.104 \text{ Gy/min} \tag{L.1}$$

The tenth-value-thickness (TVT) for 4 MV is 290 mm and for 6 MV is 343 mm. Therefore, for 5 MV a value of 316.5 mm is appropriate (see Table 61.2). The wall is apparently 1.8 m thick, so the attenuated instantaneous dose rate is:

$$0.104 \times e^{\frac{-\ln(10) \times 1.8 \times 1000}{316.5}} \times 60 = 12.8 \text{ } \mu\text{Sv/h} \tag{L.2}$$

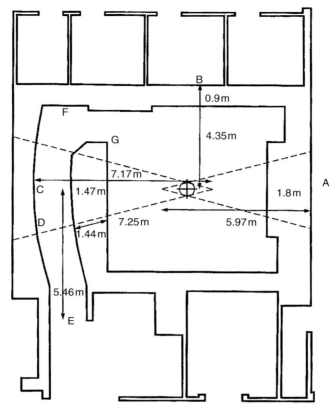

FIGURE L.1
Diagram of a 5 MV treatment room that is to be upgraded to a 6 MV room.

The dose rate actually measured at this point is 50 µSv/h*. This indicates that the density of the concrete in the wall was not quality controlled at the time of building; consequently, the effective thickness is only about 1.6 m. For the modern 6 MV accelerator, the dose rate is 6.0 Gy/min and the TVT for 6 MV is 343 mm, so the attenuated dose rate at A can be expected to increase to 200 µSv/h. To calculate the time-averaged dose rate (TADR), it is appropriate to consider:

- The duty cycle of the machine: about 1 in 10-calculated as the total dose delivered to the isocentre in a working day divided by the number of hours
- An orientation, or *use* factor: for standard use, a factor of 0.33 is appropriate
- An occupancy factor: taken as 1.0 for the TADR

This leads to a TADR of 1.7 µSv/h for the 5 MV machine and 6.6 µSv/h for the 6 MV machine. TADR2000 can include an occupancy factor for which a value of 0.3 for a busy corridor might be appropriate (noting that the occupancy does not need to refer to one individual). Thus, the calculated TADR2000 is 0.5 µSv/h and 2.0 µSv/h for the two machines, respectively.

---

* The use of Gy and Sv is somewhat arbitrary since for photons they are numerically equal. Here Gy is used to express therapy-level dose rates and Sv to express protection-level dose rates.

Note that this point needs to be a supervised area, but not a controlled area, on the basis of the TADR. It is not really practicable to have a supervised area in a corridor, but with the 5 MV machine and on the basis of the predicted dose rather the measured dose, the situation is almost acceptable. For the new machine, when current recommendations should be followed, at least an additional TVT is to be recommended. This could be achieved with 100 mm of steel or 180 mm of Ledite blocks.

At point B, leakage radiation is being considered. For this purpose, it is customary to measure distances from the isocentre rather than from the source and the nearest point is at 4.35 m. Leakage radiation is usually less than 0.1% of the dose rate in the beam, so the unattenuated dose rate at B is

$$\frac{3.70 \times 1^2}{4.35^2 \times 1000} = 0.00020 \text{ Gy/min.} \tag{L.3}$$

The wall thickness at that point is 0.9 m and the TVT is 266.5 mm for scattered radiation. Thus, the instantaneous dose rate is calculated as

$$0.00020 \times e^{\frac{-\ln(10) \times 0.9 \times 1000}{266.5}} \times 60 = 4.9 \text{ µSv/h.} \tag{L.4}$$

The dose rate actually measured at this point was 1.5 µSv/h. On the inside of the wall at this point, the dose rate was 20 mSv/h, as would be expected; it therefore appears that the leakage radiation is less penetrating than expected by a factor of three[*]. The reason for this is unclear.

Consider now the dose rate in the maze. At C, the dose rate can be calculated by analogy to point A as

$$\frac{3.70 \times 1^2}{7.17^2} \times e^{\frac{-\ln(10) \times 1.47 \times 1000}{316.5}} \times 60 = 98 \text{ µSv/h} \tag{L.5}$$

whereas at D, the dose rate is

$$\frac{3.70 \times 1^2}{7.25^2} \times e^{\frac{-\ln(10) \times 1.44 \times 1000}{316.5}} \times 60 = 119 \text{ µSv/h} \tag{L.6}$$

Going down the maze, the dose rate falls off. An approximate estimate of the dose rate at the maze entrance can be calculated using Equation 61.1. The scatter function (from IPEM 1997) for this condition is 0.022 and so the dose rate at the entrance of the maze, which is 5.46 m from the central axis of the primary beam is:

$$108.5 \times \frac{0.022 \times 3.5^2}{5.46^2} = 0.98 \text{ µSv/h} \tag{L.7}$$

In fact, the dose rate measured at the entrance of the maze is about 29 µSv/h with the largest field size. Measurements at point F showed a reflected dose rate of 32 mSv/h when a scattering phantom is placed at the isocentre; inspection of the figure shows that there is a direct line of sight from the isocentre to the wall, allowing radiation to access the maze entrance with only a single reflection. This was corrected by adding a nib to the maze wall at G. This reduced the dose rate at F to 1.65 mSv/h and the dose rate at the maze entrance to 3 µSv/h.

---

[*] A similar finding was recorded for the point C in Table III2.4 of IPEM report 75 (IPEM 1997). The calculated TVT is 232 mm rather than 266 mm, which endorses the use of a lower value to the TVT for leakage radiation.

L.3 LOCAL RULES FOR HANDLING RADIOACTIVE SOURCES

The following is an example of local rules that could be laminated and displayed in an unsealed source preparation lab:

1. Film badges must be worn at all times.

2. Laboratory coats must be worn and done up.

3. Eating, drinking, smoking, and application of cosmetics are forbidden.

4. Tissue handkerchiefs provided must be used, not your own, and hands monitored before use.

5. Open wounds and cuts must be covered with a waterproof dressing.

6. Rubber/PVC gloves must be worn. They must be removed before leaving the radiation areas to prevent possible spread of contamination.

7. Gloves and surfaces should be monitored frequently during, and immediately after, working. Contaminated gloves should be changed immediately.

8. Wash and monitor hands before leaving the working area.

9. Full use must be made of appropriate protective devices i.e. syringe shields, lead pots, handling tools, and shielding devices. Radioactive materials are to be handled at the greatest distance consistent with safety and minimising time of exposure.

10. Do not work alone in a controlled area unless another member of staff is aware of this and able to render assistance if required.

11. Where possible, secondary containment should be used for working with, and movement of, unsealed sources. In the event of a spill or suspected contamination, follow the spill procedures given in the full Local Rules.

12. Staff handling radioactive materials should not be interrupted.

13. Radioactive materials must be stored in designated places.

14. Waste disposal must be in accordance with Local Rules.

15. The local RPS is: . . . .

REFERENCES

Adamec, R., Haefliger, J. M., Killisch, J. P., Niederer, J., and Jaquet, P., Damaging effect of therapeutic radiation on programmable pacemakers, *Pacing Clin. Electrophysiol.*, 5, 146–150, 1982.

Adams, M. J., Lipshultz, S. E., Schwartz, C., Fajardo, L. F., Coen, V., and Constine, L. S., Radiation-associated cardiovascular disease: Manifestations and management, *Semin. Radiat. Oncol.*, 13, 346–356, 2003.

Aird, E. G., Second cancer risk, concomitant exposures and IRMER(2000), *Br. J. Radiol.*, 77, 983–985, 2004.

Aisenberg, A. C., Finkelstein, D. M., Doppke, K. P., Koerner, F. C., Boivin, J. F., and Willett, C. G., High risk of breast carcinoma after irradiation of young women with Hodgkin's disease, *Cancer*, 79, 1203–1210, 1997.

Al Affan, I. A., Estimation of the dose at the maze entrance for x-rays from radiotherapy linear accelerators, *Med. Phys.*, 27, 231–238, 2000.

Al-Affan, I. A. M. and Smith, C. W., Radiation quality of scattered photons at the maze entrance of radiotherapy rooms for photon beams of energy 0.5–30 MeV, *Radiat. Prot. Dosimetry*, 67, 299–302, 1996.

Al-Affan, I. A. M., Smith, C. W., Morgan, H. M., and Lillicrap, S. C., Dose rate and energy distribution of x-rays from a linear accelerator at the maze entrance of a radiotherapy room by measurement and monte carlo simulation, *Radiat. Prot. Dosimetry*, 78, 273–277, 1998.

Antypas, C., Sandilos, P., Kouvaris, J., Balafouta, E., Karinou, E. et al., Fetal dose evaluation during breast cancer radiotherapy, *Int. J. Radiat. Biol.*, 40, 995–999, 1998.

ARSAC (Administration of Radioactive Substances Advisory Committee), Notes for guidance on the clinical administration of radiopharmaceuticals and use of sealed radioactive sources. NRPB for Department of Health. (http://www.arsac.org.uk/notes_for_guidence/index.htm) 1998.

Barish, R. J., Evaluation of a new high-density shielding material, *Health Phys.*, 64, 412–416, 1993.

Barrington, S., O'Doherty, M. J., Kettle, A. G., Thomson, W. H., Mountford, P. J. et al., Radiation exposure of the families of outpatients treated with radioiodine (iodine-131) for hyperthyroidism, *Eur. J. Nucl. Med.*, 26, 686–692, 1999.

Berrington, A., Darby, S. C., Weiss, H. A., and Doll, R., 100 years of observation on British radiologists: Mortality from cancer and other causes 1897–1997, *Br. J. Radiol.*, 74, 507–519, 2001.

Bhatia, S., Robison, L. L., Oberlin, O., Greenberg, M., Bunin, G., Fossati-Bellani, F., and Meadows, A. T., Breast cancer and other second neoplasms after childhood Hodgkin's disease, *N. Engl. J. Med.*, 334, 745–751, 1996.

Bhatia, S., Yasui, Y., Robison, L. L., Birch, J. M., Bogue, M. K. et al., High risk of subsequent neoplasms continues with extended follow-up of childhood Hodgkin's disease: Report from the Late Effects Study Group, *J. Clin. Oncol.*, 21, 4386–4394, 2003.

BIR (British Institute of Radiology), Patients leaving hospital after administration of radioactive substances, *Br. J. Radiol.*, 72, 121–125, 1999.

Brenner, D. J., Curtis, R. E., Hall, E. J., and Ron, E., Second malignancies in prostate carcinoma patients after radiotherapy compared with surgery, *Cancer*, 88, 398–406, 2000.

BRER (Board on Radiation Effects Research), Health effects from exposure to low levels of ionizing radiations. Committee to Assess Health Risks from Exposure to Low Levels of Ionizing Radiations (BEIR VII) National Academy of Sciences, National Research Council, National Academy Press, Washington, DC, (http://books.nap.edu/catalog.php?record_id=11340) 2006.

BSI (British Standards Institution), *Recommendation for Data on Shielding from Ionizing Radiation. Shielding from Gamma Radiation*, BS 4094-1, BSI, London, 1966.

BSI (British Standards Institution), *Recommendation for Data on Shielding from Ionizing Radiation. Shielding from X-Radiation*, BS 4094-2, BSI, London, 1971.

Cameron, J. R. and Moulder, J. E., Proposition: Radiation hormesis should be elevated to a position of scientific respectability, *Med. Phys.*, 25, 1407–1410, 1998.

Cardis, E., Gilbert, G. S., Carpenter, L., Howe, G., Kato, I. et al., Direct estimates of cancer mortality due to low doses of ionising radiation: An international study. IARC study group on cancer risk among nuclear industry workers, *Lancet*, 344(8929), 1039–1043, 1994.

Carinou, E., Kamenopoulou, V., and Stamatelatos, I. E., Evaluation of neutron dose in the maze of medical electron accelerators, *Med. Phys.*, 26, 2520–2525, 1999.

CERRIE (Committee Examining Radiation Risks of Internal Emitters), Report of the Committee Examining Radiation Risks of Internal Emitters, CERRIE, London, http://www.cerrie.org, 2004.

Cuzick, J., Stewart, H., Rutqvist, L., Houghton, J., Edwards, R., Redmond, C., Peto, R., Baum, M., Fisher, B., and Host, H., Cause-specific mortality in long-term survivors of breast cancer who participated in trials of radiotherapy, *J. Clin. Oncol.*, 12, 447–453, 1994.

Delongchamp, R. R., Mabuchi, K., Yoshimoto, Y., and Preston, D. L., Cancer mortality among atomic bomb survivors exposed in utero or as young children, October 1950 - May 1992, *Radiat. Res.*, 147, 385–395, 1997.

d'Errico, F. and Alberts, W. G., Superheated-drop (bubble) neutron detectors and their compliance with ICRP-60, *Radiat. Prot. Dosimetry*, 54, 357–360, 1994.

DH (Department of Health), Statutory Instrument 1978 No. 1006 The Medicines (Administration of Radioactive Substances) Regulations 1978, 1978a.

DH (Department of Health), Statutory Instrument 1978 No. 1004 The Medicines (Administration of Radioactive Substances) Order 1978, 1978b.

DH (Department of Health), Statutory Instrument 1995 No. 2147 The Medicines (Administration of Radioactive Substances) Amendment Regulations 1995. http://www.legislation.hmso.gov.uk/si/si1995/Uksi_19952147_en_1.htm, 1995.

Doll, R., Effects of small doses of ionising radiation, *J. Radiol. Prot.*, 18, 163–174, 1998.

Dorr, W. and Herrmann, T., Second primary tumors after radiotherapy for malignancies. Treatment-related parameters, *Strahlenther. Onkol.*, 178, 357–362, 2002.

EA (Environment Agency), Authorisation of discharges of radioactive waste to the environment. Principles for the assessment of prospective public doses. Interim guidance. http://www.food.gov.uk/multimedia/pdfs/PublicDoseGuidance.pdf, 2002a.

EA (Environment Agency), Interim guidance on "best practicable means" for non-nuclear users of radioactive substances, (http://www.safeguard-international.com/bestpracticablemeans.pdf) 2002b.

EA (Environment Agency), The Environment Agency's interim guidance to users of sealed souces on the high-activity sealed radioactive sources and orphan sources regulations 2005. Versions 3. (http://publications.environment-agency.gov.uk/pdf/PMHO0206BKIA-e-e.pdf) 2006.

EC (European Commission), Radiation protection 97. Radiation protection following iodine-131 therapy, EU, Brussels, http://europa.eu.int/comm/energy/nuclear/radioprotection/publication/doc/097_en.pdf, 1998.

EC (European Commission), Evaluation of the application of the concepts of exemption and clearance for practices according to title III of Council Directive 96/29/Euratom of 13 May 1996 in EU Member States Volume 2: Appendices Issue N° 134 http://europa.eu.int/comm/energy/nuclear/radioprotection/publication/doc/134_appendice_en.pdf, 2003.

ECD (European Council Directive). Council directive 80/836/Euratom of 15 July 1980 amending the Directives laying down the basic safety standards for the health protection of the general public and workers against the dangers of ionizing radiation, EU, Brussels, 1980.

ECD (European Council Directive), Council directive 84/466/Euratom of 3 September 1984 laying down basic measures for the radiation protection of persons undergoing medical examination or treatment, EU, Brussels, 1984a.

ECD (European Council Directive), Council directive 84/467/Euratom of 3 September 1984 amending Directive 80/236/Euratom as regards the basic safety standards for the health protection of the general public and workers against the dangers of ionizing radiation, EU, Brussels, 1984b.

ECD (European Council Directive). Council directive 90/641/Euratom of 4 December 1990 on the operational protection of outside workers exposed to the risk of ionizing radiation during their activities in controlled areas, EU, Brussels, 1990.

ECD (European Council Directive), Council Regulation (Euratom) No 1493/93 of 8 June 1993 on shipment of radioactive substances between member states, EU, Brussels, http://europa.eu.int/comm/energy/nuclear/radioprotection/doc/legislation/931493_en.pdf, 1993.

ECD (European Council Directive), Council Directive 96/29/Euratom of 13 May 1996 laying down basic safety standards for the protection of the health of workers and the general public against the dangers arising from ionizing radiation, EU, Brussels, 1996a.

ECD (European), Council directive 97/43/Euratom of 30 June 1997 on health protection of individuals against the dangers of ionizing radiation in relation to medical exposure, EU, Brussels, 1997.

EC (European), Council Directive 2003/122/Euratom of 22 December 2003 on the control of high-activity sealed radioactive sources and orphan sources http://europa.eu.int/eur-lex/pri/en/oj/dat/2003/l_346/l_34620031231en00570064.pdf, 2003b.

Fenig, E., Mishaeli, M., Kalish, Y., and Lishner, M., Pregnancy and radiation, *Cancer Treat. Rev.*, 27, 1–7, 2001.

Followill, D., Geis, P., and Boyer, A., Estimates of whole-body dose equivalent produced by beam intensity modulated conformal therapy, *Int. J. Radiat. Oncol. Biol. Phys.*, 38, 667–672, 1997.

Followill, D. S., Stovall, M. S., Kry, S. F., and Ibbott, G. S., Neutron source strength measurements for Varian, Siemens, Elekta, and General Electric linear accelerators, *J. Appl. Clin. Med. Phys.*, 4, 189–194, 2003.

Gagliardi, G., Lax, I., and Rutqvist, L. E., Partial irradiation of the heart, *Semin. Radiat. Oncol.*, 11, 224–233, 2001.

Gao, X., Fisher, S. G., and Emami, B., Risk of second primary cancer in the contralateral breast in women treated for early-stage breast cancer: A population-based study, *Int. J. Radiat. Oncol. Biol. Phys.*, 56, 1038–1045, 2003.

Goddard, C., The use of delay tanks in the management of radioactive waste from thyroid therapy, *Nucl. Med. Commun.*, 20, 85–94, 1999.

Greaves, C. and Tindale, W., Radioiodine therapy: Care of the helpless patient and handling of the radioactive corpse, *J. Radiol. Prot.*, 21, 381–392, 2001.

Gregoire, V. and Maingon, P., Intensity-modulated radiation therapy in head and neck squamous cell carcinoma: An adaptation of 2-dimensional concepts or a reconsideration of current clinical practice, *Semin. Radiat. Oncol.*, 14, 110–120, 2004.

Greskovich, J. F. Jr., and Macklis, R. M., Radiation therapy in pregnancy: Risk calculation and risk minimization, *Semin. Oncol.*, 27, 633–645, 2000.

Hall, E. J., Henry S. Kaplan Distinguished Scientist Award 2003. The crooked shall be made straight; dose–response relationships for carcinogenesis, *Int. J. Radiat. Biol.*, 80, 327–337, 2004.

Hall, E. J. and Wuu, C. S., Radiation-induced second cancers: The impact of 3D-CRT and IMRT, *Int. J. Radiat. Oncol. Biol. Phys.*, 56, 83–88, 2003.

Harrison, R. M., Second cancers following radiotherapy: A suggested common dosimetry framework for therapeutic and concomitant exposures, *Br. J. Radiol.*, 77, 986–990, 2004.

Horwich, A. and Swerdlow, A. J., Second primary breast cancer after Hodgkin's disease, *Br. J. Cancer*, 90, 294–298, 2004.

HS (Health and Safety Legislation), *The Ionising Radiations Regulations 1985*, HMSO, London, 1985.

HS (Health and Safety Legislation), Statutory Instrument 1993 No. 2379? The Ionising Radiations (Outside Workers) Regulations 1993. http://www.legislation.hmso.gov.uk/si/si1993/Uksi_19932379_en_1.htm, 1993a.

HS (Health and Safety Legislation), Radioactive Substances Act 1993 (c.12). http://www.legislation.hmso.gov.uk/acts/acts1993/Ukpga_19930012_en_1.htm, 1993b.

HS (Health and Safety Legislation), Statutory Instrument 1996 No. 2094? The Carriage of Dangerous Goods by Road (Driver Training) Regulations 1996. http://www.hmso.gov.uk/si/si1996/Uksi_19962094_en_1.htm, 1996.

HS (Health and Safety Legislation), Statutory Instrument 1999 No. 3232. The Ionising Radiations Regulations 1999. http://www.legislation.hmso.gov.uk/si/si1999/19993232.htm, 1999.

HS (Health and Safety Legislation), Statutory Instrument 1999 No. 257. The Transport of Dangerous Goods (Safety Advisers) Regulations 1999. http://www.legislation.hmso.gov.uk/si/si1999/19990257.htm, 1999.

HS (Health and Safety Legislation), Statutory Instrument 2000 No. 1059. The Ionising Radiation (Medical Exposure) Regulations 2000. http://www.legislation.hmso.gov.uk/si/si2000/20001059.htm, 2000.

HS (Health and Safety Legislation), Statutory Instrument 2001 No. 2975. The Radiation (Emergency Preparedness and Public Information) Regulations 2001. http://www.legislation.hmso.gov.uk/si/si2001/20012975.htm, 2001.

HS (Health and Safety Legislation), Statutory Instrument 2002 No. 1093. The Radioactive Material (Road Transport) Regulations 2002. http://www.hmso.gov.uk/si/si2002/20021093.htm, 2002a.

HS (Health and Safety Legislation), Statutory Instrument 2002 No. 2099. The Packaging, Labelling and Carriage of Radioactive Material by Rail Regulations 2002. http://www.hmso.gov.uk/si/si2002/20022099.htm, 2002b.

HS (Health and Safety Legislation). Statutory Instrument 2004 No. 568. The Carriage of Dangerous Goods and Use of Transportable Pressure Equipment Regulations 2004. http://www.opsi.gov.uk/si/si2004/20040568.htm, 2004.

HS (Health and Safety Legislation). Statutory Instrument 2004 No. 2686. The High-activity Sealed Radioactive Sources and Orphan Sources Regulations 2005. http://www.opsi.gov.uk/si/si2005/20052686.htm, 2005.

HS (Health and Safety Legislation), Statutory Instrument 2006 No. 2523. The Ionising Radiation (Medical Exposure) (Amendent) Regulations 2006. http://www.opsi.gov.uk/si/si2006/20062523.htm, 2006.

HSE (Health and Safety Executive), Five steps to risk assessment. Health and safety guidance note 163, HMSO, London, http://www.hse.gov.uk/pubns/raindex.htm, 1998a.

HSE (Health and Safety Executive), Work with ionizing radiation. Ionising Radiations Regulations 1999. Approved Code of Practice and Guidance, HMSO, Norwich, 2000.

HSE (Health and Safety Executive), Equipment used in connection with medical exposure. Health and safety guidance note PM77 3rd ed., HMSO, London, 2006.

IAEA (International Atomic Energy Agency), *International Basic Safety Standards for protection against ionizing radiation and for the safety of radiation sources*, IAEA Safety Series Number 115, IAEA, Vienna, 1996.

IAEA (International Atomic Energy Agency), Application of radiological exclusion and exemption principles to sea disposal. The concept of 'de minimis' for radioactive substances under the London Convention 1972. IAEA TECDOC-1068, IAEA, Vienna. (http://www.pub.iaea.org/MTCD/publications/PDF/te_1068_prn.pdf), 1999.

IAEA (International Atomic Energy Agency), *The Regulations for the Safe Transport of Radioactive Material*, (1996 Edition (Revised)) TS-R-1 (ST-1, Revised). IAEA, Vienna, 2000.

IAEA (International Atomic Energy Agency), *Investigation of an accidental exposure of radiotherapy patients in Panama*, IAEA, Vienna, 2001.

ICRP (International Commission on Radiological Protection), ICRP Publication 26. Recommendations of the International Commission on Radiological Protection. *Ann. ICRP*, 1, 1977.

ICRP (International Commission on Radiological Protection), Report 33. *Protection against ionising radiation from external sources used in medicine*, Pergamon Press, Oxford, 1982.

ICRP (International Commission on Radiological Protection), Report 44. *Protection of the patient in radiotherapy*, Pergamon Press, Oxford, 1985.

ICRP (International Commission on Radiological Protection), ICRP Publication 60. Recommendations of the International Commission on Radiological Protection, *Ann. ICRP*, 21, 1990.

ICRP (International Commission on Radiological Protection), Report 62. *Radiological protection in biomedical research*, Pergamon Press, Oxford, 1993.

ICRP (International Commission on Radiological Protection), Report 73. *Radiological protection and safety in medicine*, Pergamon Press, Oxford, 1996.

ICRP (International Commission on Radiological Protection), Report 76. *Protection from potential exposures: Application to selected radiation sources*, Pergamon Press, Oxford, 1997.

ICRP (International Commission on Radiological Protection), Publication 77. Radiological protection policy for the disposal of radioactive waste, *Ann ICRP*, 27 (Suppl), 1998.

ICRP (International Commission on Radiological Protection), Publication 84. *Pregnancy and medical radiation*, Elsevier, Amsterdam, 2000.

ICRP (International Commission on Radiological Protection), Radiation and your patient: A guide for medical practitioners. ICRP Supporting Guidance 2, Pergamon Press, Oxford, http://www.icrp.org/educational_area.asp, 2001.

ICRP (International Commission on Radiological Protection), Publication 94. *Release of patients after therapy with unsealed radionuclides*, Elsevier, Amsterdam, 2005a.

ICRP (International Commission on Radiological Protection), Publication 97. *Prevention of high-dose-rate brachytherapy accidents*, Elsevier, Amsterdam, 2005b.

ICRP (International Commission on Radiological Protection), Publication 99. *Low dose extrapolation of radiation related cancer risk*, Elsevier, Amsterdam, 2006a.

ICRP (International Commission on Radiological Protection), Publication 98. *Radiation aspects of brachytherapy for prostate cancer*, Elsevier, Amsterdam, 2006b.

ICRP (International Commission on Radiological Protection) Committee 1 Task Group Report: C1 *Foundation Document Biological and Epidemiological Information on Health Risks Attributable to Ionising Radiation: A Summary of Judgements for the Purposes of Radiological Protection of Humans*, (Annex A of Main Recommendations), http://www.icrp.org/Health_risks.pdf, 2006c.

ICRP (International Commission on Radiological Protection) Committee 2 *Basis for dosimetric quantities used in radiological protection*, (Annex B of Main recommendations), http://www.icrp.org/Dosimetry.pdf, 2006d.

ICRP (International Commission on Radiological Protection) 2007 Draft *Recommendations of the international commission on radiological protection*. http://www.icrp.org/docs/ICRP_Draft_recommendations_12_January_2007.pdf, to be published by Elsevier, Amsterdam, 2007.

IEC (International Electrotechnical Commission). Report 61859, Guidelines for radiotherapy treatment rooms design, IEC, Geneva, 1997.

IPEM (Institute of Physics and Engineering in Medicine), *The Design of Radiotherapy Treatment Room Facilities*, Stedeford, B., Morgan, H. M., and Mayles, W. P. M. Eds., IPEM Report No 75, IPEM, New York, 1997.

IPEM (Institute of Physics and Engineering in Medicine), *Medical and Dental Guidance Notes. A Good Practice Guide to Implement Ionising Radiation Protection Legislation in the Clinical Environment*, IPEM, New York, 2002.

Johansson, L., Hormesis, an update of the present position, *Eur. J. Nucl. Med. Mol. Imaging*, 30, 921–933, 2003.

Kersey, R. W., Estimation of neutron and gamma radiation doses in the entrance mazes of SL75-20 linear accelerator treatment rooms, *Medicamundi*, 24, 151–155, 1979.

Khan, A. J. and Haffty, B. G., The location of contralateral breast cancers after radiation therapy, *Breast J.*, 7, 331–336, 2001.

Kralik, M. and Turek, K., Characterisation of neutron fields around high-energy x-ray radiotherapy machines, *Radiat. Prot. Dosimetry*, 110, 503–507, 2004.

Last, A., Radiotherapy in patients with cardiac pacemakers, *Br. J. Radiol.*, 71, 4–10, 1998.

Leung, P. M. K. and Nikolic, M., Disposal of therapeutic ^{131}I waste using a multiple holding tank system, *Health Phys.*, 75, 315–321, 1998.

Little, M. P., Risks associated with ionizing radiation, *Br. Med. Bull.*, 68, 259–275, 2003.

Marbach, J. R., Sontag, M. R., Van Dyk, J., and Wolbarst, A. B., Management of radiation oncology patients with implanted cardiac pacemakers: Report of AAPM Task Group No. 34. American Association of Physicists in Medicine, *Med. Phys.*, 21, 85–90, 1994.

McDonnell, C. E., NRPB Report. Radiological assessments for small users, NRPB, Chilton, 2004.

McGinley, P. H., *Shielding techniques for radiation oncology facilities*, 2nd ed., Medical Physics Publishing, Madison, WI, 2002.

Monsieurs, M., Thierens, H., Dierckx, R. A., Casier, K., De Beare, E. et al., Real-life radiation burden to relatives of patients treated with iodine-131: A study in eight centres in Flanders (Belgium), *Eur. J. Nucl. Med.*, 25, 1368–1376, 1998.

Mountford, P. J., Risk assessment of the nuclear medicine patient, *Br. J. Radiol.*, 70, 671–684, 1997.

Mouton, J., Haug, R., Bridier, A., Dodinot, B., and Eschwege, F., Influence of high-energy photon beam irradiation on pacemaker operation, *Phys. Med. Biol.*, 47, 2879–2893, 2002.

Movsas, B., Hanlon, A. L., Pinover, W., and Hanks, G. E., Is there an increased risk of second primaries following prostate irradiation? *Int. J. Radiat. Oncol. Biol. Phys.*, 41, 251–255, 1998.

Muirhead, C. R., Goodill, A. A., Haylock, R. G. E., Vokes, J., Little, M. P. et al., *Second Analysis of the National Registry for Radiation Workers: Occupational Exposure to Ionising Radiation and Mortality*, NRPB, Chilton, 1999.

NCRP (National Council on Radiation Protection and Measurements), Report 49. *Structural shielding design and evaluation for medical use of X rays and Gamma rays of energies up to 10 MeV.*, NCRP, Bethesda, MD, 1976.

NCRP (National Council on Radiation Protection and Measurements), Report 51. *Radiation protection design guidelines for 0.1–100 MeV particle accelerator facilities*, NCRP, Bethesda, MD, 1977.

NCRP (National Council on Radiation Protection and Measurements). Report 79. *Neutron contamination from medical electron accelerators*, NCRP, Bethesda, MD, 1984.

NCRP (National Council on Radiation Protection and Measurements). Report No 115. *Risk estimates for radiation protection*, NCRP, Bethesda, MD, 1993.

NCRP (National Council on Radiation Protection and Measurements), *Considerations regarding the unintended radiation exposure of the embryo, fetus or nursing child*. NCRP Commentary number 9, NCRP, Bethesda, MD, 1994.

NCRP (National Council on Radiation Protection and Measurements), Report No 151. *Structural Shielding Design and Evaluation for Megavoltage x- and Gamma-Ray Radiotherapy Facilities*, NCRP, Bethesda, MD, 2005.

Nelson, W. R. and LaRiviere, P. D., Primary and leakage radiation calculations at 6, 10 and 25 MeV, *Health Phys.*, 47, 811–818, 1984.

Ngu, S. L. C., Duval, P., and Collins, C., Foetal radiation dose in radiotherapy for breast cancer, *Australas. Radiol.*, 36, 321–322, 1992.

NRPB (National Radiological Protection Board), *Guidance Notes for the Protection of Persons against Ionising Radiations Arising from Medical and Dental Use*, HMSO, London, 1988.

NRPB (National Radiological Protection Board), Risks of second cancer in therapeutically irradiated populations. Comparison with cancer risks in the Japanese atomic bomb survivors and in other exposed groups. Report of an advisory group on ionizing radiation. *Documents of the NRPB*, 11 (1), NRPB, Chilton, 2000.

O'Brien, P., Michaels, H. B., Gillies, B., Aldrich, J. E., and Andrew, J. W., Radiation protection aspects of a new high-energy linear accelerator, *Med. Phys.*, 12, 101–107, 1985.

O'Connor, M., Blackwell, C. R., and Kisrow, K., Fetal dose for patients undergoing irradiation for Hodgkin's Disease, *Med. Dosimetry*, 12, 9–11, 1987.

Otake, M. and Schull, W. J., Radiation-related brain damage and growth retardation among the prenatally exposed atomic bomb survivors, *Int. J. Radiat. Biol.*, 74, 159–171, 1998.

Powell, N. L., Newing, A., Bullen, M. A., Sims, C., and Leaton, S. F., A radiation safety survey on a Clinac-20 linear accelerator, *Phys. Med. Biol.*, 32, 707–718, 1987.

Preston, D. L., Mattsson, A., Holmberg, E., Shore, R., Hildreth, N. G., and Boice, J. D., Radiation effects on breast cancer risk: A pooled analysis of eight cohorts, *Radiat. Res.*, 158, 220–235, 2002.

Price, R. A., Chibani, O., and Ma, C. M., Shielding evaluation for IMRT implementation in an existing accelerator vault, *J. Appl. Clin. Med. Phys.*, 4, 231–238, 2003.

Rawlinson, J. A., Islam, M. K., and Galbraith, D. M., Dose to radiation therapists from activation at high-energy accelerators used for conventional and intensity-modulated radiation therapy, *Med. Phys.*, 29, 598–608, 2002.

Reiners, C. and Lassman, M., Radioiodine (^{131}I) treatment of hyperthyroidism: Radiation protection and quality assurance, *Eur. J. Nucl. Med.*, 26, 683–685, 1999.

RSAC (Radioactive Substances Advisory Committee), *Handbook of Radiological Protection*, HMSO, London, 1971.

Singleton, M., Griffiths, C., Morrison, G., and Soanes, T., Dose constaints for comforters and carers. HSE Research Report 155, HSE, London, http://www.hse.gov.uk/research/rrhtm/rr155.htm, 2003.

Sneed, P. K., Albright, N. W., Wara, W. M., Prados, P. D., and Wilson, C. B., Fetal dose estimates for radiotherapy of brain tumours during pregnancy, *Int. J. Radiat. Biol.*, 32, 823–830, 1995.

Solan, A. N., Solan, M. J., Bednarz, G., and Goodkin, M. B., Treatment of patients with cardiac pacemakers and implantable cardioverter-defibrillators during radiotherapy, *Int. J. Radiat. Oncol. Biol. Phys.*, 59, 897–904, 2004.

Spurny, F., Johansson, L., Satherberg, A., Bednar, J., and Turek, K., The contribution of secondary heavy particles to the absorbed dose from high-energy photon beams, *Phys. Med. Biol.*, 41, 2643–2656, 1996.

Stathakis, S., Price, R. A., and Ma, C.-M., Shielding evaluation of existing accelerator vaults under IMRT conditions, *Med. Phys.*, 32, 448–454, 2005.

Stewart, A., Webb, J., Giles, D., and Hewitt, D., Malignant disease in childhood and diagnostic *irradiation in utero*, *Lancet*, 344, 345–346, 1956.

Stovall, M., Blackwell, C. R., Cundiff, J., Novack, D. H., Palta, J. R. et al., Fetal dose from radiotherapy with photon beams: Report of AAPM Radiation Therapy Committee Task Group number 36, *Med. Phys.*, 22, 63–83, 1995.

Thorne, M. C., Background radiation: Natural and man-made, *J. Radiol. Prot.*, 23, 29–42, 2003.

UNSCEAR (United Nations Scientific Committee on the Effects of Atomic Radiation), Sources and effects of ionizing radiation. Report to the General Assembly, United Nations, New York, http://www.unscear.org/unscear/en/publications/2000_2.html, 2000.

UNSCEAR (United Nations Scientific Committee on the Effects of Atomic Radiation), Hereditary effects of radiation. Report to the General Assembly, United Nations, New York, http://www.unscear.org/unscear/en/publications/2001.html, 2001.

Van der Giessen, P. H., A simple and generally applicable method to estimate the peripheral dose in radiation teletherapy with high energy X-rays or gamma radiation, *Int. J. Radiat. Biol.*, 35, 1059–1068, 1996.

Van der Giessen, P. H., Measurement of the peripheral dose for the tangential breast treatment technique with Co-60 gamma radiation and high energy X-rays, *Radiother. Oncol.*, 42, 257–264, 1997.

Vanhavere, F., Huyskens, D., and Struelens, L., Peripheral neutron and gamma doses in radiotherapy with an 18 MV linear accelerator, *Radiat. Prot. Dosimetry*, 110, 607–612, 2004.

Verellen, D. and Vanhavere, F., Risk assessment of radiation-induced malignancies based on whole-body equivalent dose estimates for IMRT treatment in the head and neck region, *Radiother. Oncol.*, 53, 199–203, 1999.

Vetterli, D., Riem, H., Aebersold, D. M., Greiner, R. H., Manser, P. et al., Introduction of a novel dose saving acquisition mode for the PortalVision aS500 EPID to facilitate on-line patient setup verification, *Med. Phys.*, 31, 828–831, 2004.

Waddington, S. P. and McKenzie, A. L., Assessment of effective dose from concomitant exposures required in verification of the target volume in radiotherapy, *Br. J. Radiol.*, 77, 557–561, 2004.

Wakeford, R. and Little, M. P., Risk coefficients for childhood cancer after intrauterine irradiation: A review, *Int. J. Radiat. Biol.*, 79, 293–309, 2003.

Waller, E. J., Neutron production associated with radiotherapy linear accelerators using intensity modulated radiation therapy mode, *Health Phys.*, 85, S75–S77, 2003.

Waller, E. J., Jamieson, R. J., Cole, D., Cousins, T., and Jammal, R. B., Experimental and computational determination of neutron dose equivalent around radiotherapy accelerators, *Radiat. Prot. Dosimetry*, 107, 225–232, 2003a.

Waller, E. J., Jamieson, R. J., Cole, D., Cousins, T., and Jammal, R. B., Effectiveness of customised neutron shielding in the maze of radiotherapy accelerators, *Radiat. Prot. Dosimetry*, 107, 233–238, 2003b.

Walz, B. J., Reder, R. E., Pastore, J. O., Littman, P., and Johnson, R., Cardiac pacemakers. Does radiation therapy affect performance, *J. Am. Med. Assoc.*, 234, 72–73, 1975.

Yasui, Y., Liu, Y., Neglia, J. P., Friedman, D. L., Bhatia, S. et al., A methodological issue in the analysis of second-primary cancer incidence in long-term survivors of childhood cancers, *Am. J. Epidemiol.*, 158, 1108–1113, 2003.

Yudelev, M., Maughan, R. L., and Dunlap, K., Shielding and radiation safety around a superconducting cyclotron neutron therapy facility, *Health Phys.*, 69, 130–136, 1995.

Yudelev, M., Maughan, R. L., Dunlap, K., Jordan, L. E., and Saxena, R., Dose equivalents to neutron therapy facility staff due to induced activation, *Health Phys.*, 72, 361–367, 1997.

Yudelev, M., Alyousef, K., Brandon, J., Perevertailo, V., Lerch, M. L., and Rosenfeld, A. B., Application of semi-conductors for dosimetry of fast-neutron therapy beam, *Radiat. Prot. Dosimetry*, 110, 573–578, 2004.

Useful links to radiation protection sources can be found at http://www.radiation.org.uk.

PART M

REFERENCE DATA

Editor: Alan Nahum

CONTENTS

1303

TABLES M.1 PHYSICAL CONSTANTS AND USEFUL DATA

Compiled by Jean-Claude Rosenwald

INTERNATIONAL SYSTEM OF UNITS (SI)*

TABLE M.1a

SI Basic Quantities and Units

Quantity	Name[a]	Symbol[b]
Length	metre	m
Mass	kilogram	kg
Time	second	s
Electric current	ampere	A
Thermodynamic temperature	kelvin	K
Amount of substance	mole	mol
Luminous intensity	candela	cd

[a] According to the SI system, the full names or the units are always treated like ordinary nouns (i.e. written in lower-case upright letters, in the plural if appropriate).
[b] The symbols of the units are printed in upright lower-case letters unless they are derived from a proper name, in which case the first letter is a capital letter (with the exception of the "litre" symbol which may written as "L").

TABLE M.1b

Useful SI-Derived Quantities and Units with Special Names and Symbols

Quantity	Name	Symbol	Expression in Other SI Units
Plane angle	radian	rad[a]	—
Solid angle	steradian	sr	—
Force	newton	N	$m\,kg\,s^{-2}$
Pressure	pascal	Pa	$N\,m^{-2}$
Energy	joule	J	$N\,m$
Power	watt	W	$J\,s^{-1}$
Electric charge	coulomb	C	$A\,s$
Electric potential difference	volt	V	$W\,A^{-1}$
Celsius temperature	degree Celsius[b]	°C	K
Activity	becquerel	Bq	s^{-1}
Absorbed dose, kerma	gray	Gy	$J\,kg^{-1}$
Dose equivalent	sievert	Sv	$J\,kg^{-1}$

[a] Radian (rad) should not be confused with the obsolete unit for abosorbed dose (rad).
[b] The degree Celsius is the special name for the kelvin used to express Celsius temperatures. The degree Celsius and the kelvin are equal in size, so that the numerical value of a temperature difference or temperature interval is the same when expressed in either degrees Celsius or in kelvins.

---

* A full description of the International System of Units may be found in the official brochure of the Bureau International des Poids et Mesures (see http://www.bipm.org/fr/si/si_brochure/).

TABLE M.1c

Conversions of Obsolete Units (All Coefficients Are Exact Values)

Quantity	Obsolete Unit	Coefficient	SI Unit
Activity	1 Ci (curie)	3.7×10^{10}	Bq
Exposure[a]	1 R (röntgen)	2.58×10^{-4}	$C \, kg^{-1}$ (in air)
Kerma or absorbed dose	1 rad	1×10^{-2}	Gy
Dose equivalent	1 rem	1×10^{-2}	Sv

To convert the *quantity* in *obsolete units* to the *quantity* in *SI units*, multiply by the relevant *coefficient*. Other obsolete units associated with source specification in brachytherapy are discussed in Section 53.6.
[a]   Air kerma is the modern replacement quantity for exposure (see Section 6.3.2).

TABLE M.1d

SI Prefixes

Decimal Multiples			Decimal Submultiples		
Factor	Name	Symbol	Factor	Name	Symbol
10^1	deka	da	10^{-1}	deci	d
10^2	hecto	h	10^{-2}	centi	c
10^3	kilo	k	10^{-3}	milli	m
10^6	mega	M	10^{-6}	micro	μ
10^9	giga	G	10^{-9}	nano	n
10^{12}	tera	T	10^{-12}	pico	p
10^{15}	peta	P	10^{-15}	femto	f
10^{18}	exa	E	10^{-18}	atto	a
10^{21}	zetta	Z	10^{-21}	zepto	z
10^{24}	yotta	Y	10^{-24}	yocto	y

USEFUL PHYSICAL CONSTANTS

TABLE M.1e

Constant	Symbol	Value	Unit	Comment
Speed of light in vacuo	c	299792458	$m \, s^{-1}$	Exact value
Avogadro's number	N_A	6.0221×10^{23}	mol^{-1}	
Atomic mass constant	m_u	1.6605×10^{-27}	kg	
Molar mass of carbon		12×10^{-3}	kg	Exact value
Planck's constant	h	6.6261×10^{-34}	J s	
Boltzmann's constant	k	1.3807×10^{-23}	$J \, K^{-1}$	
Elementary charge	e	1.6022×10^{-19}	C	Electron or proton charge
Standard atmosphere		101325	Pa	Exact value
Melting point of ice at 101 325 Pa		273.15	K	
		0	°C	
Standard acceleration of gravity	g_n	9.80665	$m \, s^{-2}$	Exact value
Standard air density		1.200	$kg \, m^{-3}$	At 20°C, 101 325 Pa, 50% relative air humidity
Density of dry air at 0°C	ρ_{air}	1.293	$kg \, m^{-3}$	At 273.15 K, 101 325 Pa
Mean energy expended in air per ion pair formed	W_{air}/e	33.97	$J \, C^{-1}$	

From Mohr and Taylor; CODATA Recommended Values of Fundamental Physical Constants: 2002, published in Review of Modern Physics 77, 1 (2005) and available at http://physics.nist.gov/constants. Most constant values are given here to 4 significant figures, which is sufficient for applications in radiotherapy physics. More precise statements of uncertainty are given by Mohr and Taylor. See also more accurate figures in Chapter 1.

ENERGY CONVERSION

TABLE M.1f

	Equivalence ⇒	Mass (m)	Wavelength (λ)	Frequency (ν)	Temperature (T)
	Relation ⇒	mc^2	hc/λ	$h\nu$	kT
Unit	J	kg	m	Hz	K
1 J	1	1.1127×10^{-17}	1.9858×10^{-25}	1.5092×10^{33}	7.24297×10^{22}
1 eV	1.6022×10^{-19}	1.7827×10^{-36}	1.2394×10^{-6}	2.4180×10^{14}	1.16045×10^{4}

This table shows how mass, wavelength, frequency, and temperature are related to energy. For example the energy equivalent to 1kg is $1/1.1127 \times 10^{-17}$ J.

PARTICLES

TABLE M.1g

Constant	Value	Unit	Comment
Unified atomic mass unit	1.6605×10^{-27}	kg	Symbol = u
Electron rest mass	9.1094×10^{-31}	kg	1/1823 u
Proton rest mass	1.6726×10^{-27}	kg	1.0073 u
Neutron rest mass	1.6749×10^{-27}	kg	1.0087 u
Alpha particle rest mass	6.6447×10^{-27}	kg	4.0015 u
Electron rest equivalent energy	0.51100	MeV	Derived from $E = mc^2$
Proton rest equivalent energy	938.272	MeV	Derived from $E = mc^2$

TABLES M.2 ELECTRON STOPPING POWERS, RANGES, AND RADIATION YIELDS

Compiled by Alan Nahum

Table M.2a through Table M.2z give stopping powers, ranges and radiation yields for electrons in a number of materials of medical interest. Elements have been placed first (a–f), then compounds (g–i), in order of increasing I-value (see below), then mixtures (j–q) and body tissues (r–z), in alphabetical order. The electron energies extend from 0.01 MeV (10 keV) up to 50 MeV, covering the range of interest in radiotherapy; at energies below 10 keV the validity of the Bethe-Bloch expression for the mass collision stopping power becomes questionable, especially in high atomic-number materials. The values have been obtained using the ESTAR computer code, accessed via http://physics.nist.gov/PhysRefData/Star/Text/ESTAR. html where full details of the calculational procedures are given (Berger et al. 2005).

Each table also contains the mass density ρ, the ratio $\langle Z/A \rangle$, the mean excitation energy I (see Section 3.2.2) and the atomic composition $(Z-f_w)$ of each substance, as given in the ESTAR database, where f_w is the fraction by weight of the element Z within the material. The mass stopping powers are given in MeV cm^2 g^{-1} which are the units most commonly used in radiotherapy, due to the beam energies also being expressed in MeV. To obtain linear stopping powers the values in the tables should be multiplied by the density ρ.

Collision stopping power is covered in detail in Section 3.2, radiative stopping power S_{rad}/ρ and radiation yield in Section 3.3, and the csda range R_{csda} or r_o in Section 3.4.3. The expression used to evaluate S_{col}/ρ in the ESTAR code is identical to Equation 3.7 given in Section 3.2.2 and hence the values in Table M.2a through Table M.2z can be verified by substituting $\langle Z/A \rangle$, I and the density-effect correction δ into Equation 3.7 for the desired electron kinetic energy. In this way exact values at intermediate energies can be computed if desired.

The numbers in Tables M.2a–z are identical to those given in ICRU (1984) for the same substance, with the exception of water; the reason for this difference is that ICRU (1984) uses the density-effect correction for water derived by Ashley (1982). For substances not included here, ICRU (1984) should be consulted, where tables of all the parameters required to evaluate S_{col}/ρ from Equation 3.7 can be found.

TABLE M.2a

Electrons in Carbon (Graphite)

$\langle Z/A \rangle$: 0.49954; ρ: 1.7000 g cm^{-3}; I: 78.0 eV; Composition ($Z-f_w$): 6–1.0000

Kinetic Energy (MeV)	Stopping Power (MeV cm^2 g^{-1})			CSDA Range (g cm^{-2})	Radiation Yield	Density-Effect Parameter δ
	Collision	Radiative	Total			
0.010	2.014×10^1	3.150×10^{-3}	2.014×10	2.820×10^{-4}	8.665×10^{-5}	1.920×10^{-3}
0.015	1.471×10^1	3.168×10^{-3}	1.471×10	5.767×10^{-4}	1.199×10^{-4}	3.073×10^{-3}
0.020	1.177×10^1	3.176×10^{-3}	1.177×10	9.595×10^{-4}	1.506×10^{-4}	4.347×10^{-3}
0.030	8.626	3.194×10^{-3}	8.629	1.966×10^{-3}	2.073×10^{-4}	7.237×10^{-3}
0.040	6.950	3.215×10^{-3}	6.953	3.267×10^{-3}	2.597×10^{-4}	1.055×10^{-2}
0.050	5.901	3.241×10^{-3}	5.904	4.835×10^{-3}	3.090×10^{-4}	1.425×10^{-2}
0.060	5.179	3.270×10^{-3}	5.183	6.648×10^{-3}	3.559×10^{-4}	1.832×10^{-2}
0.070	4.652	3.303×10^{-3}	4.655	8.688×10^{-3}	4.008×10^{-4}	2.271×10^{-2}
0.080	4.249	3.337×10^{-3}	4.252	1.094×10^{-2}	4.441×10^{-4}	2.740×10^{-2}
0.090	3.931	3.375×10^{-3}	3.935	1.339×10^{-2}	4.860×10^{-4}	3.237×10^{-2}
0.100	3.674	3.414×10^{-3}	3.677	1.602×10^{-2}	5.268×10^{-4}	3.760×10^{-2}
0.150	2.886	3.640×10^{-3}	2.890	3.156×10^{-2}	7.168×10^{-4}	6.694×10^{-2}
0.200	2.485	3.896×10^{-3}	2.489	5.032×10^{-2}	8.911×10^{-4}	1.003×10^{-1}
0.300	2.087	4.489×10^{-3}	2.092	9.462×10^{-2}	1.213×10^{-3}	1.740×10^{-1}
0.400	1.896	5.173×10^{-3}	1.901	1.450×10^{-1}	1.518×10^{-3}	2.524×10^{-1}
0.500	1.788	5.935×10^{-3}	1.794	1.993×10^{-1}	1.817×10^{-3}	3.321×10^{-1}
0.600	1.722	6.759×10^{-3}	1.729	2.561×10^{-1}	2.115×10^{-3}	4.114×10^{-1}
0.700	1.679	7.637×10^{-3}	1.687	3.147×10^{-1}	2.416×10^{-3}	4.891×10^{-1}
0.800	1.650	8.559×10^{-3}	1.659	3.746×10^{-1}	2.719×10^{-3}	5.648×10^{-1}
0.900	1.631	9.523×10^{-3}	1.640	4.352×10^{-1}	3.026×10^{-3}	6.382×10^{-1}
1.000	1.617	1.053×10^{-2}	1.627	4.964×10^{-1}	3.337×10^{-3}	7.091×10^{-1}
1.500	1.593	1.602×10^{-2}	1.609	8.062×10^{-1}	4.954×10^{-3}	1.028
2.000	1.597	2.213×10^{-2}	1.619	1.116	6.665×10^{-3}	1.295
3.000	1.621	3.561×10^{-2}	1.657	1.727	1.029×10^{-2}	1.720
4.000	1.647	5.026×10^{-2}	1.697	2.323	1.410×10^{-2}	2.051
5.000	1.669	6.576×10^{-2}	1.735	2.906	1.803×10^{-2}	2.323
6.000	1.689	8.193×10^{-2}	1.771	3.476	2.204×10^{-2}	2.555
7.000	1.706	9.865×10^{-2}	1.804	4.036	2.610×10^{-2}	2.758
8.000	1.720	1.158×10^{-1}	1.836	4.585	3.020×10^{-2}	2.939
9.000	1.733	1.334×10^{-1}	1.867	5.125	3.432×10^{-2}	3.104
10.000	1.745	1.513×10^{-1}	1.896	5.657	3.845×10^{-2}	3.256
15.000	1.787	2.444×10^{-1}	2.032	8.202	5.903×10^{-2}	3.879
20.000	1.816	3.417×10^{-1}	2.157	1.059×10	7.917×10^{-2}	4.361
30.000	1.852	5.435×10^{-1}	2.396	1.498×10	1.173×10^{-1}	5.088
40.000	1.877	7.508×10^{-1}	2.627	1.897×10	1.522×10^{-1}	5.628
50.000	1.895	9.617×10^{-1}	2.857	2.262×10	1.841×10^{-1}	6.057

TABLE M.2b

Electrons in Aluminium

$\langle Z/A \rangle$: 0.48181; ρ: 2.699 g cm^{-3}; I: 166.0 eV; Composition ($Z-f_w$): 13–1.0000

Kinetic Energy (MeV)	Stopping Power (MeV cm^2 g^{-1})			CSDA Range (g cm^{-2})	Radiation Yield	Density-Effect Parameter δ
	Collision	Radiative	Total			
0.010	1.649×10	6.559×10^{-3}	1.650×10	3.539×10^{-4}	2.132×10^{-4}	3.534×10^{-4}
0.015	1.220×10	6.798×10^{-3}	1.221×10	7.111×10^{-4}	3.017×10^{-4}	6.538×10^{-4}
0.020	9.844	6.926×10^{-3}	9.851	1.170×10^{-3}	3.840×10^{-4}	1.031×10^{-3}
0.030	7.287	7.059×10^{-3}	7.294	2.367×10^{-3}	5.353×10^{-4}	2.005×10^{-3}
0.040	5.909	7.133×10^{-3}	5.916	3.900×10^{-3}	6.736×10^{-4}	3.246×10^{-3}
0.050	5.039	7.191×10^{-3}	5.046	5.738×10^{-3}	8.022×10^{-4}	4.732×10^{-3}
0.060	4.439	7.243×10^{-3}	4.446	7.855×10^{-3}	9.232×10^{-4}	6.440×10^{-3}
0.070	3.998	7.295×10^{-3}	4.005	1.023×10^{-2}	1.038×10^{-3}	8.351×10^{-3}
0.080	3.661	7.350×10^{-3}	3.668	1.284×10^{-2}	1.147×10^{-3}	1.045×10^{-2}
0.090	3.394	7.411×10^{-3}	3.401	1.568×10^{-2}	1.252×10^{-3}	1.271×10^{-2}
0.100	3.177	7.476×10^{-3}	3.185	1.872×10^{-2}	1.353×10^{-3}	1.513×10^{-2}
0.150	2.513	7.865×10^{-3}	2.521	3.659×10^{-2}	1.816×10^{-3}	2.907×10^{-2}
0.200	2.174	8.344×10^{-3}	2.183	5.804×10^{-2}	2.231×10^{-3}	4.525×10^{-2}
0.300	1.839	9.487×10^{-3}	1.849	1.083×10^{-1}	2.982×10^{-3}	8.116×10^{-2}
0.400	1.680	1.082×10^{-2}	1.691	1.652×10^{-1}	3.678×10^{-3}	1.190×10^{-1}
0.500	1.592	1.230×10^{-2}	1.604	2.260×10^{-1}	4.349×10^{-3}	1.569×10^{-1}
0.600	1.540	1.390×10^{-2}	1.554	2.894×10^{-1}	5.009×10^{-3}	1.943×10^{-1}
0.700	1.507	1.560×10^{-2}	1.522	3.545×10^{-1}	5.664×10^{-3}	2.307×10^{-1}
0.800	1.486	1.739×10^{-2}	1.503	4.206×10^{-1}	6.319×10^{-3}	2.661×10^{-1}
0.900	1.473	1.925×10^{-2}	1.492	4.874×10^{-1}	6.976×10^{-3}	3.005×10^{-1}
1.000	1.465	2.119×10^{-2}	1.486	5.546×10^{-1}	7.636×10^{-3}	3.339×10^{-1}
1.500	1.460	3.177×10^{-2}	1.491	8.913×10^{-1}	1.101×10^{-2}	4.898×10^{-1}
2.000	1.475	4.350×10^{-2}	1.518	1.224	1.449×10^{-2}	6.349×10^{-1}
3.000	1.510	6.924×10^{-2}	1.580	1.869	2.173×10^{-2}	9.145×10^{-1}
4.000	1.540	9.702×10^{-2}	1.637	2.491	2.918×10^{-2}	1.183
5.000	1.564	1.263×10^{-1}	1.690	3.092	3.675×10^{-2}	1.433
6.000	1.583	1.567×10^{-1}	1.739	3.675	4.436×10^{-2}	1.661
7.000	1.599	1.879×10^{-1}	1.787	4.242	5.197×10^{-2}	1.868
8.000	1.613	2.200×10^{-1}	1.833	4.795	5.955×10^{-2}	2.055
9.000	1.625	2.526×10^{-1}	1.877	5.334	6.708×10^{-2}	2.226
10.000	1.636	2.858×10^{-1}	1.921	5.861	7.454×10^{-2}	2.384
15.000	1.676	4.574×10^{-1}	2.134	8.328	1.105×10^{-1}	3.016
20.000	1.704	6.357×10^{-1}	2.340	1.056×10	1.438×10^{-1}	3.484
30.000	1.743	1.003	2.746	1.450×10	2.027×10^{-1}	4.168
40.000	1.769	1.379	3.148	1.790×10	2.528×10^{-1}	4.669
50.000	1.789	1.761	3.550	2.089×10	2.959×10^{-1}	5.068

TABLE M.2c

Electrons in Silicon

$\langle Z/A \rangle$: 0.49848; ρ: 2.330 g cm^{-3}; I: 173.0 eV; Composition ($Z-f_w$): 14–1.0000

Kinetic Energy (MeV)	Stopping Power (MeV cm^2 g^{-1})			CSDA Range (g cm^{-2})	Radiation Yield	Density-Effect Parameter δ
	Collision	Radiative	Total			
0.010	1.689×10	7.255×10^{-3}	1.690×10	3.461×10^{-4}	2.289×10^{-4}	1.037×10^{-3}
0.015	1.251×10	7.555×10^{-3}	1.252×10	6.946×10^{-4}	3.252×10^{-4}	1.641×10^{-3}
0.020	1.010×10	7.720×10^{-3}	1.011×10	1.142×10^{-3}	4.151×10^{-4}	2.298×10^{-3}
0.030	7.480	7.892×10^{-3}	7.487	2.308×10^{-3}	5.807×10^{-4}	3.766×10^{-3}
0.040	6.067	7.988×10^{-3}	6.075	3.802×10^{-3}	7.322×10^{-4}	5.424×10^{-3}
0.050	5.175	8.061×10^{-3}	5.183	5.592×10^{-3}	8.731×10^{-4}	7.257×10^{-3}
0.060	4.559	8.123×10^{-3}	4.568	7.653×10^{-3}	1.006×10^{-3}	9.252×10^{-3}
0.070	4.107	8.185×10^{-3}	4.116	9.964×10^{-3}	1.131×10^{-3}	1.139×10^{-2}
0.080	3.761	8.248×10^{-3}	3.769	1.251×10^{-2}	1.251×10^{-3}	1.368×10^{-2}
0.090	3.487	8.317×10^{-3}	3.496	1.526×10^{-2}	1.366×10^{-3}	1.608×10^{-2}
0.100	3.265	8.389×10^{-3}	3.274	1.822×10^{-2}	1.476×10^{-3}	1.861×10^{-2}
0.150	2.583	8.821×10^{-3}	2.592	3.561×10^{-2}	1.981×10^{-3}	3.271×10^{-2}
0.200	2.236	9.349×10^{-3}	2.245	5.646×10^{-2}	2.433×10^{-3}	4.868×10^{-2}
0.300	1.892	1.062×10^{-2}	1.903	1.054×10^{-1}	3.249×10^{-3}	8.402×10^{-2}
0.400	1.729	1.209×10^{-2}	1.741	1.606×10^{-1}	4.003×10^{-3}	1.216×10^{-1}
0.500	1.638	1.374×10^{-2}	1.652	2.197×10^{-1}	4.728×10^{-3}	1.599×10^{-1}
0.600	1.585	1.551×10^{-2}	1.600	2.812×10^{-1}	5.441×10^{-3}	1.980×10^{-1}
0.700	1.551	1.740×10^{-2}	1.568	3.444×10^{-1}	6.148×10^{-3}	2.355×10^{-1}
0.800	1.529	1.938×10^{-2}	1.549	4.086×10^{-1}	6.855×10^{-3}	2.721×10^{-1}
0.900	1.516	2.145×10^{-2}	1.537	4.734×10^{-1}	7.564×10^{-3}	3.077×10^{-1}
1.000	1.507	2.360×10^{-2}	1.531	5.386×10^{-1}	8.275×10^{-3}	3.424×10^{-1}
1.500	1.502	3.533×10^{-2}	1.538	8.652×10^{-1}	1.190×10^{-2}	5.020×10^{-1}
2.000	1.518	4.833×10^{-2}	1.567	1.188	1.565×10^{-2}	6.439×10^{-1}
3.000	1.558	7.682×10^{-2}	1.634	1.812	2.340×10^{-2}	8.976×10^{-1}
4.000	1.591	1.076×10^{-1}	1.699	2.412	3.134×10^{-2}	1.131
5.000	1.618	1.399×10^{-1}	1.758	2.991	3.937×10^{-2}	1.351
6.000	1.639	1.735×10^{-1}	1.813	3.551	4.742×10^{-2}	1.557
7.000	1.657	2.081×10^{-1}	1.865	4.095	5.546×10^{-2}	1.748
8.000	1.672	2.435×10^{-1}	1.916	4.624	6.344×10^{-2}	1.925
9.000	1.685	2.795×10^{-1}	1.965	5.139	7.136×10^{-2}	2.088
10.000	1.697	3.161×10^{-1}	2.013	5.642	7.919×10^{-2}	2.239
15.000	1.740	5.057×10^{-1}	2.245	7.992	1.168×10^{-1}	2.858
20.000	1.769	7.023×10^{-1}	2.472	1.011×10	1.514×10^{-1}	3.323
30.000	1.809	1.108	2.917	1.383×10	2.123×10^{-1}	4.003
40.000	1.837	1.523	3.360	1.702×10	2.638×10^{-1}	4.501
50.000	1.858		3.802	1.982×10	3.077×10^{-1}	4.897

TABLE M.2d

Electrons in Copper

$\langle Z/A \rangle$: 0.45636; ρ: 8.960 g cm^{-3}; I: 322.0 eV; Composition $(Z-f_w)$: 29–1.0000

Kinetic Energy (MeV)	Stopping Power (MeV cm^2 g^{-1})			CSDA Range g cm^{-2}	Radiation Yield	Density-Effect Parameter δ
	Collision	Radiative	Total			
0.010	1.318×10	1.213×10^{-2}	1.319×10	4.601×10^{-4}	4.701×10^{-4}	1.244×10^{-3}
0.015	9.904	1.327×10^{-2}	9.917	9.028×10^{-4}	6.904×10^{-4}	1.938×10^{-3}
0.020	8.066	1.399×10^{-2}	8.080	1.465×10^{-3}	9.019×10^{-4}	2.683×10^{-3}
0.030	6.040	1.488×10^{-2}	6.055	2.914×10^{-3}	1.301×10^{-3}	4.334×10^{-3}
0.040	4.931	1.543×10^{-2}	4.947	4.754×10^{-3}	1.674×10^{-3}	6.220×10^{-3}
0.050	4.226	1.583×10^{-2}	4.242	6.946×10^{-3}	2.025×10^{-3}	8.365×10^{-3}
0.060	3.736	1.615×10^{-2}	3.753	9.459×10^{-3}	2.358×10^{-3}	1.080×10^{-2}
0.070	3.375	1.641×10^{-2}	3.392	1.227×10^{-2}	2.674×10^{-3}	1.354×10^{-2}
0.080	3.098	1.665×10^{-2}	3.114	1.535×10^{-2}	2.977×10^{-3}	1.664×10^{-2}
0.090	2.877	1.688×10^{-2}	2.894	1.868×10^{-2}	3.268×10^{-3}	2.013×10^{-2}
0.100	2.698	1.710×10^{-2}	2.715	2.225×10^{-2}	3.547×10^{-3}	2.404×10^{-2}
0.150	2.146	1.816×10^{-2}	2.164	4.314×10^{-2}	4.822×10^{-3}	5.053×10^{-2}
0.200	1.861	1.926×10^{-2}	1.881	6.807×10^{-2}	5.950×10^{-3}	8.595×10^{-2}
0.300	1.579	2.172×10^{-2}	1.601	1.263×10^{-1}	7.945×10^{-3}	1.604×10^{-1}
0.400	1.444	2.450×10^{-2}	1.469	1.918×10^{-1}	9.741×10^{-3}	2.302×10^{-1}
0.500	1.370	2.757×10^{-2}	1.398	2.617×10^{-1}	1.143×10^{-2}	2.958×10^{-1}
0.600	1.326	3.087×10^{-2}	1.357	3.345×10^{-1}	1.307×10^{-2}	3.581×10^{-1}
0.700	1.298	3.437×10^{-2}	1.333	4.089×10^{-1}	1.467×10^{-2}	4.173×10^{-1}
0.800	1.281	3.803×10^{-2}	1.319	4.843×10^{-1}	1.625×10^{-2}	4.739×10^{-1}
0.900	1.270	4.185×10^{-2}	1.312	5.604×10^{-1}	1.782×10^{-2}	5.280×10^{-1}
1.000	1.263	4.580×10^{-2}	1.309	6.367×10^{-1}	1.938×10^{-2}	5.799×10^{-1}
1.500	1.259	6.733×10^{-2}	1.327	1.017	2.720×10^{-2}	8.121×10^{-1}
2.000	1.273	9.103×10^{-2}	1.364	1.389	3.509×10^{-2}	1.011
3.000	1.305	1.425×10^{-1}	1.448	2.101	5.095×10^{-2}	1.343
4.000	1.334	1.976×10^{-1}	1.531	2.772	6.668×10^{-2}	1.617
5.000	1.358	2.552×10^{-1}	1.613	3.408	8.209×10^{-2}	1.850
6.000	1.378	3.146×10^{-1}	1.693	4.013	9.710×10^{-2}	2.052
7.000	1.396	3.756×10^{-1}	1.771	4.591	1.117×10^{-1}	2.229
8.000	1.411	4.378×10^{-1}	1.849	5.143	1.258×10^{-1}	2.388
9.000	1.424	5.009×10^{-1}	1.925	5.673	1.394×10^{-1}	2.532
10.000	1.436	5.650×10^{-1}	2.001	6.183	1.526×10^{-1}	2.664
15.000	1.482	8.949×10^{-1}	2.377	8.472	2.122×10^{-1}	3.194
20.000	1.513	1.236	2.749	1.043×10	2.628×10^{-1}	3.597
30.000	1.555	1.936	3.491	1.365×10	3.437×10^{-1}	4.209
40.000	1.582	2.650	4.233	1.624×10	4.059×10^{-1}	4.676
50.000	1.603	3.375	4.978	1.842×10	4.554×10^{-1}	5.054

TABLE M.2e

Electrons in Tungsten

$\langle Z/A \rangle$: 0.40250; ρ: 19.300 g cm^{-3}; I: 727.0 eV; Composition $(Z-f_w)$: 74–1.0000

Kinetic Energy (MeV)	Stopping Power (MeV cm^2 g^{-1})			CSDA Range (g cm^{-2})	Radiation Yield	Density-Effect Parameter δ
	Collision	Radiative	Total			
0.010	8.974	1.977×10^{-2}	8.993	7.490×10^{-4}	1.076×10^{-3}	9.911×10^{-4}
0.015	6.945	2.320×10^{-2}	6.968	1.387×10^{-3}	1.639×10^{-3}	1.544×10^{-3}
0.020	5.753	2.563×10^{-2}	5.779	2.179×10^{-3}	2.200×10^{-3}	2.133×10^{-3}
0.030	4.394	2.908×10^{-2}	4.423	4.181×10^{-3}	3.305×10^{-3}	3.417×10^{-3}
0.040	3.631	3.160×10^{-2}	3.662	6.681×10^{-3}	4.381×10^{-3}	4.834×10^{-3}
0.050	3.137	3.364×10^{-2}	3.171	9.627×10^{-3}	5.430×10^{-3}	6.378×10^{-3}
0.060	2.791	3.539×10^{-2}	2.826	1.298×10^{-2}	6.453×10^{-3}	8.041×10^{-3}
0.070	2.533	3.694×10^{-2}	2.570	1.669×10^{-2}	7.453×10^{-3}	9.817×10^{-3}
0.080	2.334	3.834×10^{-2}	2.373	2.075×10^{-2}	8.430×10^{-3}	1.170×10^{-2}
0.090	2.176	3.964×10^{-2}	2.216	2.511×10^{-2}	9.385×10^{-3}	1.369×10^{-2}
0.100	2.047	4.084×10^{-2}	2.088	2.977×10^{-2}	1.032×10^{-2}	1.577×10^{-2}
0.150	1.646	4.595×10^{-2}	1.692	5.668×10^{-2}	1.470×10^{-2}	2.755×10^{-2}
0.200	1.439	5.021×10^{-2}	1.489	8.835×10^{-2}	1.865×10^{-2}	4.131×10^{-2}
0.300	1.234	5.797×10^{-2}	1.292	1.611×10^{-1}	2.558×10^{-2}	7.370×10^{-2}
0.400	1.138	6.565×10^{-2}	1.203	2.416×10^{-1}	3.164×10^{-2}	1.111×10^{-1}
0.500	1.085	7.353×10^{-2}	1.159	3.265×10^{-1}	3.712×10^{-2}	1.515×10^{-1}
0.600	1.055	8.162×10^{-2}	1.136	4.137×10^{-1}	4.221×10^{-2}	1.932×10^{-1}
0.700	1.036	8.993×10^{-2}	1.126	5.022×10^{-1}	4.702×10^{-2}	2.352×10^{-1}
0.800	1.025	9.841×10^{-2}	1.124	5.911×10^{-1}	5.161×10^{-2}	2.768×10^{-1}
0.900	1.019	1.071×10^{-1}	1.126	6.800×10^{-1}	5.602×10^{-2}	3.176×10^{-1}
1.000	1.016	1.159×10^{-1}	1.132	7.686×10^{-1}	6.030×10^{-2}	3.575×10^{-1}
1.500	1.021	1.624×10^{-1}	1.183	1.201	8.022×10^{-2}	5.416×10^{-1}
2.000	1.037	2.117×10^{-1}	1.249	1.613	9.856×10^{-2}	7.015×10^{-1}
3.000	1.072	3.158×10^{-1}	1.388	2.372	1.321×10^{-1}	9.684×10^{-1}
4.000	1.101	4.248×10^{-1}	1.526	3.059	1.625×10^{-1}	1.188
5.000	1.126	5.372×10^{-1}	1.663	3.687	1.902×10^{-1}	1.378
6.000	1.146	6.523×10^{-1}	1.798	4.265	2.157×10^{-1}	1.544
7.000	1.163	7.697×10^{-1}	1.933	4.801	2.393×10^{-1}	1.694
8.000	1.178	8.890×10^{-1}	2.067	5.301	2.612×10^{-1}	1.830
9.000	1.191	1.010	2.201	5.770	2.816×10^{-1}	1.955
10.000	1.203	1.132	2.335	6.211	3.006×10^{-1}	2.070
15.000	1.247	1.759	3.006	8.094	3.800×10^{-1}	2.544
20.000	1.277	2.406	3.682	9.594	4.403×10^{-1}	2.910
30.000	1.316	3.735	5.051	1.190×10	5.270×10^{-1}	3.471
40.000	1.343	5.096	6.439	1.365×10	5.871×10^{-1}	3.901
50.000	1.362	6.477	7.840	1.506×10	6.316×10^{-1}	4.252

TABLE M.2f

Electrons in Lead

$\langle Z/A \rangle$: 0.39575; ρ: 11.35 g cm^{-3}; I: 823.0 eV; Composition (Z $- f_w$): 82–1.0000

Kinetic Energy (MeV)	Stopping Power (MeV cm^{2}g^{-1})			CSDA Range (g cm^{-2})	Radiation Yield	Density-Effect Parameter δ
	Collision	Radiative	Total			
0.010	8.428	2.045×10^{-2}	8.448	8.255×10^{-4}	1.191×10^{-3}	4.841×10^{-4}
0.015	6.561	2.421×10^{-2}	6.585	1.502×10^{-3}	1.810×10^{-3}	7.491×10^{-4}
0.020	5.453	2.693×10^{-2}	5.480	2.339×10^{-3}	2.432×10^{-3}	1.029×10^{-3}
0.030	4.182	3.086×10^{-2}	4.212	4.445×10^{-3}	3.664×10^{-3}	1.633×10^{-3}
0.040	3.463	3.376×10^{-2}	3.497	7.066×10^{-3}	4.872×10^{-3}	2.294×10^{-3}
0.050	2.997	3.613×10^{-2}	3.034	1.015×10^{-2}	6.055×10^{-3}	3.011×10^{-3}
0.060	2.670	3.817×10^{-2}	2.708	1.365×10^{-2}	7.214×10^{-3}	3.783×10^{-3}
0.070	2.426	3.998×10^{-2}	2.466	1.752×10^{-2}	8.349×10^{-3}	4.608×10^{-3}
0.080	2.237	4.162×10^{-2}	2.279	2.175×10^{-2}	9.461×10^{-3}	5.485×10^{-3}
0.090	2.087	4.313×10^{-2}	2.130	2.629×10^{-2}	1.055×10^{-2}	6.413×10^{-3}
0.100	1.964	4.454×10^{-2}	2.008	3.113×10^{-2}	1.162×10^{-2}	7.392×10^{-3}
0.150	1.583	5.054×10^{-2}	1.633	5.905×10^{-2}	1.664×10^{-2}	1.300×10^{-2}
0.200	1.387	5.555×10^{-2}	1.442	9.180×10^{-2}	2.118×10^{-2}	1.971×10^{-2}
0.300	1.193	6.460×10^{-2}	1.257	1.668×10^{-1}	2.917×10^{-2}	3.579×10^{-2}
0.400	1.102	7.340×10^{-2}	1.175	2.494×10^{-1}	3.614×10^{-2}	5.437×10^{-2}
0.500	1.053	8.228×10^{-2}	1.135	3.361×10^{-1}	4.241×10^{-2}	7.443×10^{-2}
0.600	1.026	9.132×10^{-2}	1.117	4.250×10^{-1}	4.820×10^{-2}	9.529×10^{-2}
0.700	1.009	1.005×10^{-1}	1.110	5.149×10^{-1}	5.363×10^{-2}	1.166×10^{-1}
0.800	1.000	1.098×10^{-1}	1.110	6.050×10^{-1}	5.877×10^{-2}	1.380×10^{-1}
0.900	9.957×10^{-1}	1.193×10^{-1}	1.115	6.949×10^{-1}	6.369×10^{-2}	1.595×10^{-1}
1.000	9.939×10^{-1}	1.290×10^{-1}	1.123	7.843×10^{-1}	6.842×10^{-2}	1.809×10^{-1}
1.500	1.004	1.792×10^{-1}	1.183	1.219	9.009×10^{-2}	2.854×10^{-1}
2.000	1.024	2.319×10^{-1}	1.256	1.629	1.096×10^{-1}	3.855×10^{-1}
3.000	1.063	3.427×10^{-1}	1.406	2.381	1.447×10^{-1}	5.743×10^{-1}
4.000	1.095	4.582×10^{-1}	1.553	3.057	1.761×10^{-1}	7.479×10^{-1}
5.000	1.120	5.773×10^{-1}	1.698	3.673	2.045×10^{-1}	9.061×10^{-1}
6.000	1.142	6.991×10^{-1}	1.841	4.239	2.304×10^{-1}	1.050
7.000	1.160	8.233×10^{-1}	1.983	4.762	2.543×10^{-1}	1.182
8.000	1.175	9.495×10^{-1}	2.125	5.249	2.765×10^{-1}	1.304
9.000	1.189	1.077	2.266	5.705	2.970×10^{-1}	1.417
10.000	1.201	1.206	2.407	6.133	3.162×10^{-1}	1.523
15.000	1.246	1.870	3.116	7.954	3.955×10^{-1}	1.964
20.000	1.277	2.554	3.830	9.399	4.555×10^{-1}	2.310
30.000	1.318	3.961	5.279	1.161×10	5.412×10^{-1}	2.841
40.000	1.345	5.402	6.747	1.329×10	6.002×10^{-1}	3.247
50.000	1.365	6.865	8.231	1.463×10	6.439×10^{-1}	3.579

TABLE M.2g

Electrons in Water (Liquid)

$\langle Z/A \rangle$: 0.55509; ρ: 1.000 g cm^{-3}; I: 75.0 eV, Composition $(Z - f_w)$: 1–0.111898; 8–0.888102

Kinetic Energy (MeV)	Stopping Power (MeV cm^{2}g^{-1})			CSDA Range (g cm^{-2})	Radiation Yield	Density-Effect Parameter δ
	Collision	Radiative	Total			
0.010	2.256×10	3.898×10^{-3}	2.256×10	2.515×10^{-4}	9.408×10^{-5}	0.000
0.015	1.647×10	3.944×10^{-3}	1.647×10	5.147×10^{-4}	1.316×10^{-4}	0.000
0.020	1.317×10	3.963×10^{-3}	1.318×10	8.566×10^{-4}	1.663×10^{-4}	0.000
0.030	9.653	3.984×10^{-3}	9.657	1.756×10^{-3}	2.301×10^{-4}	0.000
0.040	7.777	4.005×10^{-3}	7.781	2.919×10^{-3}	2.886×10^{-4}	0.000
0.050	6.603	4.031×10^{-3}	6.607	4.320×10^{-3}	3.435×10^{-4}	0.000
0.060	5.797	4.062×10^{-3}	5.801	5.940×10^{-3}	3.955×10^{-4}	0.000
0.070	5.207	4.098×10^{-3}	5.211	7.762×10^{-3}	4.453×10^{-4}	0.000
0.080	4.757	4.138×10^{-3}	4.761	9.773×10^{-3}	4.931×10^{-4}	0.000
0.090	4.402	4.181×10^{-3}	4.407	1.196×10^{-2}	5.393×10^{-4}	0.000
0.100	4.115	4.228×10^{-3}	4.119	1.431×10^{-2}	5.842×10^{-4}	0.000
0.150	3.238	4.494×10^{-3}	3.242	2.817×10^{-2}	7.926×10^{-4}	0.000
0.200	2.793	4.801×10^{-3}	2.798	4.488×10^{-2}	9.826×10^{-4}	0.000
0.300	2.355	5.514×10^{-3}	2.360	8.421×10^{-2}	1.331×10^{-3}	0.000
0.400	2.148	6.339×10^{-3}	2.154	1.288×10^{-1}	1.658×10^{-3}	0.000
0.500	2.034	7.257×10^{-3}	2.041	1.766×10^{-1}	1.976×10^{-3}	0.000
0.600	1.963	8.254×10^{-3}	1.972	2.265×10^{-1}	2.292×10^{-3}	2.938×10^{-2}
0.700	1.917	9.313×10^{-3}	1.926	2.778×10^{-1}	2.608×10^{-3}	7.435×10^{-2}
0.800	1.886	1.042×10^{-2}	1.896	3.302×10^{-1}	2.928×10^{-3}	1.267×10^{-1}
0.900	1.864	1.159×10^{-2}	1.876	3.832×10^{-1}	3.251×10^{-3}	1.835×10^{-1}
1.000	1.849	1.280×10^{-2}	1.862	4.367×10^{-1}	3.579×10^{-3}	2.428×10^{-1}
1.500	1.822	1.942×10^{-2}	1.841	7.075×10^{-1}	5.281×10^{-3}	5.437×10^{-1}
2.000	1.824	2.678×10^{-2}	1.850	9.785×10^{-1}	7.085×10^{-3}	8.218×10^{-1}
3.000	1.846	4.299×10^{-2}	1.889	1.514	1.092×10^{-2}	1.288
4.000	1.870	6.058×10^{-2}	1.931	2.037	1.495×10^{-2}	1.660
5.000	1.892	7.917×10^{-2}	1.971	2.550	1.911×10^{-2}	1.967
6.000	1.911	9.854×10^{-2}	2.010	3.052	2.336×10^{-2}	2.227
7.000	1.928	1.185×10^{-1}	2.047	3.545	2.766×10^{-2}	2.453
8.000	1.943	1.391×10^{-1}	2.082	4.030	3.200×10^{-2}	2.652
9.000	1.956	1.601×10^{-1}	2.116	4.506	3.636×10^{-2}	2.831
10.000	1.968	1.814×10^{-1}	2.149	4.975	4.072×10^{-2}	2.992
15.000	2.014	2.926×10^{-1}	2.306	7.219	6.243×10^{-2}	3.633
20.000	2.046	4.086×10^{-1}	2.454	9.320	8.355×10^{-2}	4.107
30.000	2.089	6.489×10^{-1}	2.738	1.317×10	1.233×10^{-1}	4.806
40.000	2.118	8.955×10^{-1}	3.013	1.665×10	1.594×10^{-1}	5.326
50.000	2.139	1.146	3.286	1.983×10	1.923×10^{-1}	5.741

TABLE M.2h

Electrons in Lithium Fluoride (LiF)

$\langle Z/A \rangle$: 0.46262; ρ : 2.635 g cm^{-3}; I: 94.0 eV; Composition (Z$-f_w$): 3–0.267585; 9–0.732415

Kinetic Energy (MeV)	Stopping Power (MeV cm^{2}g^{-1})			CSDA Range (g cm^{-2})	Radiation Yield	Density-Effect Parameter δ
	Collision	Radiative	Total			
0.010	1.796×10	3.678×10^{-3}	1.796×10	3.181×10^{-4}	1.117×10^{-4}	0.000
0.015	1.315×10	3.735×10^{-3}	1.316×10	6.480×10^{-4}	1.561×10^{-4}	0.000
0.020	1.055×10	3.762×10^{-3}	1.055×10	1.076×10^{-3}	1.973×10^{-4}	0.000
0.030	7.748	3.792×10^{-3}	7.751	2.198×10^{-3}	2.729×10^{-4}	0.000
0.040	6.252	3.815×10^{-3}	6.256	3.645×10^{-3}	3.423×10^{-4}	0.000
0.050	5.315	3.840×10^{-3}	5.319	5.386×10^{-3}	4.071×10^{-4}	0.000
0.060	4.670	3.867×10^{-3}	4.674	7.397×10^{-3}	4.684×10^{-4}	0.000
0.070	4.198	3.898×10^{-3}	4.202	9.659×10^{-3}	5.269×10^{-4}	0.000
0.080	3.838	3.932×10^{-3}	3.842	1.215×10^{-2}	5.831×10^{-4}	0.000
0.090	3.553	3.970×10^{-3}	3.557	1.486×10^{-2}	6.372×10^{-4}	0.000
0.100	3.323	4.011×10^{-3}	3.327	1.777×10^{-2}	6.896×10^{-4}	0.000
0.150	2.619	4.253×10^{-3}	2.623	3.492×10^{-2}	9.321×10^{-4}	0.000
0.200	2.261	4.540×10^{-3}	2.266	5.555×10^{-2}	1.153×10^{-3}	0.000
0.300	1.907	5.215×10^{-3}	1.912	1.041×10^{-1}	1.558×10^{-3}	2.136×10^{-2}
0.400	1.737	5.992×10^{-3}	1.743	1.591×10^{-1}	1.939×10^{-3}	6.095×10^{-2}
0.500	1.642	6.852×10^{-3}	1.649	2.183×10^{-1}	2.310×10^{-3}	1.098×10^{-1}
0.600	1.583	7.779×10^{-3}	1.591	2.801×10^{-1}	2.679×10^{-3}	1.648×10^{-1}
0.700	1.546	8.765×10^{-3}	1.555	3.437×10^{-1}	3.048×10^{-3}	2.236×10^{-1}
0.800	1.521	9.800×10^{-3}	1.530	4.086×10^{-1}	3.419×10^{-3}	2.846×10^{-1}
0.900	1.503	1.088×10^{-2}	1.514	4.743×10^{-1}	3.794×10^{-3}	3.467×10^{-1}
1.000	1.491	1.200×10^{-2}	1.504	5.406×10^{-1}	4.173×10^{-3}	4.093×10^{-1}
1.500	1.471	1.818×10^{-2}	1.489	8.756×10^{-1}	6.138×10^{-3}	7.142×10^{-1}
2.000	1.474	2.505×10^{-2}	1.499	1.210	8.214×10^{-3}	9.917×10^{-1}
3.000	1.493	4.021×10^{-2}	1.533	1.870	1.262×10^{-2}	1.461
4.000	1.513	5.666×10^{-2}	1.570	2.515	1.725×10^{-2}	1.839
5.000	1.531	7.402×10^{-2}	1.605	3.145	2.202×10^{-2}	2.154
6.000	1.547	9.211×10^{-2}	1.639	3.761	2.687×10^{-2}	2.422
7.000	1.560	1.108×10^{-1}	1.671	4.365	3.178×10^{-2}	2.655
8.000	1.572	1.299×10^{-1}	1.702	4.958	3.672×10^{-2}	2.861
9.000	1.583	1.494×10^{-1}	1.732	5.541	4.168×10^{-2}	3.046
10.000	1.592	1.693×10^{-1}	1.761	6.113	4.663×10^{-2}	3.214
15.000	1.629	2.723×10^{-1}	1.901	8.844	7.108×10^{-2}	3.881
20.000	1.654	3.797×10^{-1}	2.034	1.139×10	9.463×10^{-2}	4.374
30.000	1.688	6.014×10^{-1}	2.289	1.602×10	1.384×10^{-1}	5.099
40.000	1.711	8.289×10^{-1}	2.540	2.016×10	1.776×10^{-1}	5.633
50.000	1.728	1.060	2.788	2.392×10	2.129×10^{-1}	6.057

TABLE M.2i

Electrons in Lithium Tetraborate ($Li_2B_4O_7$)

$\langle Z/A \rangle$: 0.48485; ρ: 2.440 g cm^{-3}; I: 94.6 eV; Composition ($Z - f_w$): 3–0.082081; 5–0.255715; 8–0.662204

Kinetic Energy (MeV)	Stopping Power (MeV cm^{2}g^{-1})			CSDA Range (g cm^{-2})	Density-Effect Parameter δ	Radiation Yield
	Collision	Radiative	Total			
0.010	1.880×10	3.547×10^{-3}	1.880×10	3.040×10^{-4}	1.039×10^{-4}	0.000
0.015	1.377×10	3.582×10^{-3}	1.377×10	6.192×10^{-4}	1.442×10^{-4}	0.000
0.020	1.104×10	3.598×10^{-3}	1.104×10	1.028×10^{-3}	1.815×10^{-4}	0.000
0.030	8.111	3.617×10^{-3}	8.115	2.100×10^{-3}	2.500×10^{-4}	0.000
0.040	6.546	3.638×10^{-3}	6.550	3.482×10^{-3}	3.128×10^{-4}	0.000
0.050	5.565	3.662×10^{-3}	5.568	5.145×10^{-3}	3.716×10^{-4}	0.000
0.060	4.890	3.692×10^{-3}	4.893	7.066×10^{-3}	4.274×10^{-4}	0.000
0.070	4.396	3.725×10^{-3}	4.400	9.226×10^{-3}	4.808×10^{-4}	0.000
0.080	4.019	3.761×10^{-3}	4.022	1.161×10^{-2}	5.321×10^{-4}	0.000
0.090	3.721	3.801×10^{-3}	3.724	1.419×10^{-2}	5.816×10^{-4}	0.000
0.100	3.480	3.843×10^{-3}	3.483	1.697×10^{-2}	6.297×10^{-4}	0.000
0.150	2.742	4.086×10^{-3}	2.746	3.335×10^{-2}	8.528×10^{-4}	0.000
0.200	2.368	4.365×10^{-3}	2.373	5.306×10^{-2}	1.056×10^{-3}	0.000
0.300	1.999	5.013×10^{-3}	2.004	9.941×10^{-2}	1.428×10^{-3}	8.291×10^{-3}
0.400	1.819	5.763×10^{-3}	1.824	1.520×10^{-1}	1.779×10^{-3}	6.448×10^{-2}
0.500	1.717	6.598×10^{-3}	1.724	2.085×10^{-1}	2.121×10^{-3}	1.294×10^{-1}
0.600	1.655	7.504×10^{-3}	1.663	2.676×10^{-1}	2.463×10^{-3}	1.980×10^{-1}
0.700	1.615	8.467×10^{-3}	1.623	3.285×10^{-1}	2.806×10^{-3}	2.680×10^{-1}
0.800	1.588	9.479×10^{-3}	1.597	3.907×10^{-1}	3.152×10^{-3}	3.382×10^{-1}
0.900	1.569	1.054×10^{-2}	1.580	4.536×10^{-1}	3.501×10^{-3}	4.077×10^{-1}
1.000	1.557	1.164×10^{-2}	1.568	5.172×10^{-1}	3.856×10^{-3}	4.760×10^{-1}
1.500	1.534	1.766×10^{-2}	1.552	8.385×10^{-1}	5.695×10^{-3}	7.929×10^{-1}
2.000	1.538	2.435×10^{-2}	1.562	1.160	7.638×10^{-3}	1.067
3.000	1.560	3.909×10^{-2}	1.599	1.793	1.175×10^{-2}	1.515
4.000	1.583	5.508×10^{-2}	1.638	2.411	1.607×10^{-2}	1.870
5.000	1.603	7.197×10^{-2}	1.675	3.014	2.051×10^{-2}	2.163
6.000	1.621	8.958×10^{-2}	1.710	3.605	2.504×10^{-2}	2.415
7.000	1.636	1.078×10^{-1}	1.743	4.184	2.962×10^{-2}	2.635
8.000	1.649	1.264×10^{-1}	1.775	4.752	3.423×10^{-2}	2.831
9.000	1.661	1.455×10^{-1}	1.806	5.311	3.886×10^{-2}	3.008
10.000	1.671	1.649×10^{-1}	1.836	5.860	4.350×10^{-2}	3.170
15.000	1.710	2.660×10^{-1}	1.976	8.483	6.649×10^{-2}	3.823
20.000	1.737	3.716×10^{-1}	2.109	1.093×10	8.879×10^{-2}	4.314
30.000	1.772	5.900×10^{-1}	2.362	1.541×10	1.305×10^{-1}	5.042
40.000	1.796	8.142×10^{-1}	2.611	1.943×10	1.683×10^{-1}	5.581
50.000	1.814	1.042	2.856	2.309×10	2.025×10^{-1}	6.007

TABLE M.2j

Electrons in A-150, Tissue-Equivalent Plastic

$\langle Z/A \rangle$: 0.54903; ρ: 1.127^a g cm^{-3}; I: 65.1 eV; Composition $(Z - f_w)$: 1–0.101327; 6–0.775501; 7–0.035057; 8–0.052316; 9–0.017422; 20–0.018378

Kinetic Energy (MeV)	Stopping Power (MeV cm²g⁻¹)			CSDA Range (g cm⁻²)	Radiation Yield	Density-Effect Parameter δ
	Collision	Radiative	Total			
0.010	2.294×10	3.156×10^{-3}	2.295×10	2.463×10^{-4}	7.529×10^{-5}	0.000
0.015	1.671×10	3.188×10^{-3}	1.671×10	5.055×10^{-4}	1.050×10^{-4}	0.000
0.020	1.335×10	3.205×10^{-3}	1.336×10	8.426×10^{-4}	1.327×10^{-4}	0.000
0.030	9.769	3.232×10^{-3}	9.772	1.731×10^{-3}	1.838×10^{-4}	0.000
0.040	7.863	3.258×10^{-3}	7.866	2.880×10^{-3}	2.311×10^{-4}	0.000
0.050	6.671	3.287×10^{-3}	6.675	4.267×10^{-3}	2.756×10^{-4}	0.000
0.060	5.853	3.319×10^{-3}	5.857	5.871×10^{-3}	3.180×10^{-4}	0.000
0.070	5.256	3.352×10^{-3}	5.259	7.677×10^{-3}	3.586×10^{-4}	0.000
0.080	4.800	3.388×10^{-3}	4.803	9.669×10^{-3}	3.977×10^{-4}	0.000
0.090	4.441	3.427×10^{-3}	4.444	1.184×10^{-2}	4.356×10^{-4}	0.000
0.100	4.150	3.467×10^{-3}	4.153	1.417×10^{-2}	4.723×10^{-4}	0.000
0.150	3.262	3.697×10^{-3}	3.265	2.793×10^{-2}	6.436×10^{-4}	0.000
0.200	2.812	3.959×10^{-3}	2.816	4.452×10^{-2}	8.003×10^{-4}	0.000
0.300	2.369	4.563×10^{-3}	2.373	8.362×10^{-2}	1.089×10^{-3}	0.000
0.400	2.156	5.260×10^{-3}	2.161	1.280×10^{-1}	1.361×10^{-3}	2.970×10^{-2}
0.500	2.033	6.036×10^{-3}	2.039	1.757×10^{-1}	1.628×10^{-3}	9.751×10^{-2}
0.600	1.957	6.875×10^{-3}	1.964	2.258×10^{-1}	1.895×10^{-3}	1.708×10^{-1}
0.700	1.907	7.768×10^{-3}	1.915	2.774×10^{-1}	2.164×10^{-3}	2.464×10^{-1}
0.800	1.874	8.706×10^{-3}	1.883	3.300×10^{-1}	2.436×10^{-3}	3.223×10^{-1}
0.900	1.851	9.687×10^{-3}	1.860	3.835×10^{-1}	2.711×10^{-3}	3.975×10^{-1}
1.000	1.834	1.071×10^{-2}	1.845	4.375×10^{-1}	2.990×10^{-3}	4.712×10^{-1}
1.500	1.803	1.630×10^{-2}	1.819	7.111×10^{-1}	4.447×10^{-3}	8.100×10^{-1}
2.000	1.804	2.252×10^{-2}	1.827	9.856×10^{-1}	5.992×10^{-3}	1.099
3.000	1.827	3.625×10^{-2}	1.863	1.528	9.281×10^{-3}	1.561
4.000	1.852	5.116×10^{-2}	1.903	2.059	1.275×10^{-2}	1.922
5.000	1.874	6.695×10^{-2}	1.941	2.579	1.633×10^{-2}	2.218
6.000	1.894	8.341×10^{-2}	1.977	3.090	2.000×10^{-2}	2.470
7.000	1.911	1.004×10^{-1}	2.011	3.591	2.372×10^{-2}	2.691
8.000	1.926	1.179×10^{-1}	2.044	4.084	2.749×10^{-2}	2.887
9.000	1.939	1.358×10^{-1}	2.075	4.570	3.127×10^{-2}	3.064
10.000	1.951	1.540×10^{-1}	2.105	5.048	3.508×10^{-2}	3.227
15.000	1.995	2.488×10^{-1}	2.243	7.348	5.411×10^{-2}	3.885
20.000	2.024	3.478×10^{-1}	2.372	9.515	7.282×10^{-2}	4.385
30.000	2.063	5.532×10^{-1}	2.616	1.353×10	1.085×10^{-1}	5.126
40.000	2.089	7.643×10^{-1}	2.854	1.718×10	1.414×10^{-1}	5.671
50.000	2.109	9.792×10^{-1}	3.089	2.055×10	1.717×10^{-1}	6.101

^{a} The density of any sample of plastic used for the first time should be determined experimentally if high dosimetric accuracy is required (see Section 18.8).

TABLE M.2k

Electrons in Air, Dry (Near Sea Level)[a]

$\langle Z/A \rangle$: 0.49919; ρ: 1.2048×10^{-3} g cm^{-3}; I: 85.7 eV; Composition $(Z-f_w)$: 6–0.000124; 7–0.755267; 8–0.231781; 18–0.012827

Kinetic Energy (MeV)	Stopping Power (MeV cm^{2}g^{-1})			CSDA Range (g cm^{-2})	Radiation Yield	Density-Effect Parameter δ
	Collision	Radiative	Total			
0.010	1.975×10	3.897×10^{-3}	1.976×10	2.884×10^{-4}	1.082×10^{-4}	0.000
0.015	1.444×10	3.937×10^{-3}	1.445×10	5.886×10^{-4}	1.506×10^{-4}	0.000
0.020	1.157×10	3.954×10^{-3}	1.158×10	9.782×10^{-4}	1.898×10^{-4}	0.000
0.030	8.491	3.976×10^{-3}	8.495	2.002×10^{-3}	2.619×10^{-4}	0.000
0.040	6.848	3.998×10^{-3}	6.852	3.322×10^{-3}	3.280×10^{-4}	0.000
0.050	5.818	4.025×10^{-3}	5.822	4.913×10^{-3}	3.900×10^{-4}	0.000
0.060	5.110	4.057×10^{-3}	5.114	6.751×10^{-3}	4.488×10^{-4}	0.000
0.070	4.593	4.093×10^{-3}	4.597	8.817×10^{-3}	5.050×10^{-4}	0.000
0.080	4.197	4.133×10^{-3}	4.201	1.110×10^{-2}	5.590×10^{-4}	0.000
0.090	3.885	4.175×10^{-3}	3.889	1.357×10^{-2}	6.112×10^{-4}	0.000
0.100	3.633	4.222×10^{-3}	3.637	1.623×10^{-2}	6.618×10^{-4}	0.000
0.150	2.861	4.485×10^{-3}	2.865	3.193×10^{-2}	8.968×10^{-4}	0.000
0.200	2.469	4.789×10^{-3}	2.474	5.082×10^{-2}	1.111×10^{-3}	0.000
0.300	2.084	5.495×10^{-3}	2.089	9.528×10^{-2}	1.502×10^{-3}	0.000
0.400	1.902	6.311×10^{-3}	1.908	1.456×10^{-1}	1.869×10^{-3}	0.000
0.500	1.802	7.223×10^{-3}	1.809	1.995×10^{-1}	2.225×10^{-3}	0.000
0.600	1.743	8.210×10^{-3}	1.751	2.558×10^{-1}	2.577×10^{-3}	0.000
0.700	1.706	9.258×10^{-3}	1.715	3.136×10^{-1}	2.930×10^{-3}	0.000
0.800	1.683	1.036×10^{-2}	1.694	3.723×10^{-1}	3.283×10^{-3}	0.000
0.900	1.669	1.151×10^{-2}	1.681	4.316×10^{-1}	3.639×10^{-3}	0.000
1.000	1.661	1.271×10^{-2}	1.674	4.912×10^{-1}	3.997×10^{-3}	0.000
1.500	1.661	1.927×10^{-2}	1.680	7.901×10^{-1}	5.836×10^{-3}	0.000
2.000	1.684	2.656×10^{-2}	1.711	1.085	7.748×10^{-3}	0.000
3.000	1.740	4.260×10^{-2}	1.783	1.658	1.173×10^{-2}	0.000
4.000	1.790	5.999×10^{-2}	1.850	2.208	1.583×10^{-2}	0.000
5.000	1.833	7.838×10^{-2}	1.911	2.740	2.001×10^{-2}	0.000
6.000	1.870	9.754×10^{-2}	1.967	3.255	2.422×10^{-2}	0.000
7.000	1.902	1.173×10^{-1}	2.020	3.757	2.846×10^{-2}	0.000
8.000	1.931	1.376×10^{-1}	2.068	4.246	3.269×10^{-2}	0.000
9.000	1.956	1.584×10^{-1}	2.115	4.724	3.692×10^{-2}	0.000
10.000	1.979	1.795×10^{-1}	2.159	5.192	4.113×10^{-2}	0.000
15.000	2.069	2.895×10^{-1}	2.359	7.405	6.182×10^{-2}	0.000
20.000	2.134	4.042×10^{-1}	2.539	9.447	8.167×10^{-2}	0.000
30.000	2.226	6.417×10^{-1}	2.868	1.315×10	1.186×10^{-1}	7.563×10^{-3}
40.000	2.282	8.855×10^{-1}	3.167	1.646×10	1.520×10^{-1}	1.375×10^{-1}
50.000	2.319	1.133	3.452	1.948×10	1.825×10^{-1}	3.189×10^{-1}

[a] CRC Handbook of Chemistry and Physics (1979): 78.09% N_2, 20.95% O_2, 0.93% Ar and 0.03% CO_2

TABLE M.2l

Electrons in C-552 Air-Equivalent Plastic

$\langle Z/A \rangle$: 0.49969; ρ: 1.760[a] g cm^{-3}; I: 86.8 eV; Composition ($Z-f_w$): 1–0.024681; 6–0.501610; 8–0.004527; 9–0.465209; 14–0.003973

Kinetic Energy (MeV)	Stopping Power (MeV cm^{2}g^{-1})			CSDA Range (g cm^{-2})	Radiation Yield	Density-Effect Parameter δ
	Collision	Radiative	Total			
0.010	1.972×10	3.767×10^{-3}	1.972×10	2.890×10^{-4}	1.046×10^{-4}	0.000
0.015	1.442×10	3.812×10^{-3}	1.443×10	5.897×10^{-4}	1.458×10^{-4}	0.000
0.020	1.156×10	3.834×10^{-3}	1.156×10	9.797×10^{-4}	1.839×10^{-4}	0.000
0.030	8.482	3.860×10^{-3}	8.486	2.004×10^{-3}	2.541×10^{-4}	0.000
0.040	6.841	3.883×10^{-3}	6.844	3.326×10^{-3}	3.185×10^{-4}	0.000
0.050	5.813	3.909×10^{-3}	5.816	4.918×10^{-3}	3.789×10^{-4}	0.000
0.060	5.106	3.939×10^{-3}	5.109	6.758×10^{-3}	4.360×10^{-4}	0.000
0.070	4.589	3.973×10^{-3}	4.593	8.827×10^{-3}	4.907×10^{-4}	0.000
0.080	4.194	4.010×10^{-3}	4.198	1.111×10^{-2}	5.431×10^{-4}	0.000
0.090	3.882	4.050×10^{-3}	3.886	1.359×10^{-2}	5.938×10^{-4}	0.000
0.100	3.630	4.093×10^{-3}	3.634	1.625×10^{-2}	6.429×10^{-4}	0.000
0.150	2.859	4.347×10^{-3}	2.863	3.196×10^{-2}	8.705×10^{-4}	0.000
0.200	2.468	4.643×10^{-3}	2.473	5.086×10^{-2}	1.078×10^{-3}	0.000
0.300	2.083	5.336×10^{-3}	2.088	9.535×10^{-2}	1.458×10^{-3}	0.000
0.400	1.899	6.134×10^{-3}	1.906	1.457×10^{-1}	1.815×10^{-3}	1.228×10^{-2}
0.500	1.794	7.020×10^{-3}	1.801	1.998×10^{-1}	2.164×10^{-3}	6.801×10^{-2}
0.600	1.729	7.977×10^{-3}	1.737	2.564×10^{-1}	2.510×10^{-3}	1.308×10^{-1}
0.700	1.687	8.995×10^{-3}	1.696	3.147×10^{-1}	2.858×10^{-3}	1.972×10^{-1}
0.800	1.659	1.006×10^{-2}	1.669	3.742×10^{-1}	3.209×10^{-3}	2.650×10^{-1}
0.900	1.639	1.118×10^{-2}	1.650	4.345×10^{-1}	3.564×10^{-3}	3.331×10^{-1}
1.000	1.626	1.234×10^{-2}	1.638	4.953×10^{-1}	3.923×10^{-3}	4.005×10^{-1}
1.500	1.602	1.871×10^{-2}	1.620	8.030×10^{-1}	5.787×10^{-3}	7.168×10^{-1}
2.000	1.605	2.581×10^{-2}	1.630	1.111	7.758×10^{-3}	9.920×10^{-1}
3.000	1.627	4.144×10^{-2}	1.668	1.717	1.194×10^{-2}	1.440
4.000	1.650	5.840×10^{-2}	1.709	2.310	1.632×10^{-2}	1.794
5.000	1.671	7.633×10^{-2}	1.748	2.888	2.084×10^{-2}	2.086
6.000	1.690	9.501×10^{-2}	1.785	3.454	2.544×10^{-2}	2.335
7.000	1.705	1.143×10^{-1}	1.820	4.009	3.010×10^{-2}	2.552
8.000	1.719	1.341×10^{-1}	1.853	4.554	3.478×10^{-2}	2.745
9.000	1.731	1.543×10^{-1}	1.886	5.089	3.949×10^{-2}	2.920
10.000	1.742	1.748×10^{-1}	1.917	5.615	4.419×10^{-2}	3.080
15.000	1.783	2.817×10^{-1}	2.065	8.125	6.748×10^{-2}	3.724
20.000	1.811	3.931×10^{-1}	2.204	1.047×10	9.002×10^{-2}	4.210
30.000	1.848	6.235×10^{-1}	2.472	1.475×10	1.321×10^{-1}	4.932
40.000	1.873	8.601×10^{-1}	2.733	1.859×10	1.702×10^{-1}	5.466
50.000	1.892	1.101	2.992	2.209×10	2.045×10^{-1}	5.890

[a] The density of any sample of plastic used for the first time should be determined experimentally if high dosimetric accuracy is required (see Section 18.8).

TABLE M.2m

Electrons in Glass, Borosilicate (Pyrex)

$\langle Z/A \rangle$: 0.49707; ρ: 2.230 g cm^{-3}; I: 134.0 eV; Composition $(Z-f_w)$: 5–0.040066; 8–0.539559; 11–0.028191; 13–0.011644; 14–0.377220; 19–003321

Kinetic Energy (MeV)	Stopping Power (MeV cm^{2}g^{-1})			CSDA Range (g cm^{-2})	Radiation Yield	Density-Effect Parameter δ
	Collision	Radiative	Total			
0.010	1.787×10	5.400×10^{-3}	1.788×10	3.236×10^{-4}	1.632×10^{-4}	0.000
0.015	1.317×10	5.548×10^{-3}	1.318×10	6.540×10^{-4}	2.296×10^{-4}	0.000
0.020	1.060×10	5.626×10^{-3}	1.060×10	1.080×10^{-3}	2.914×10^{-4}	0.000
0.030	7.823	5.707×10^{-3}	7.828	2.193×10^{-3}	4.048×10^{-4}	0.000
0.040	6.331	5.759×10^{-3}	6.337	3.624×10^{-3}	5.087×10^{-4}	0.000
0.050	5.393	5.803×10^{-3}	5.399	5.341×10^{-3}	6.055×10^{-4}	0.000
0.060	4.746	5.847×10^{-3}	4.752	7.321×10^{-3}	6.968×10^{-4}	0.000
0.070	4.272	5.893×10^{-3}	4.278	9.543×10^{-3}	7.837×10^{-4}	0.000
0.080	3.909	5.943×10^{-3}	3.915	1.199×10^{-2}	8.667×10^{-4}	0.000
0.090	3.623	5.997×10^{-3}	3.629	1.465×10^{-2}	9.466×10^{-4}	0.000
0.100	3.390	6.055×10^{-3}	3.396	1.750×10^{-2}	1.024×10^{-3}	0.000
0.150	2.679	6.393×10^{-3}	2.685	3.427×10^{-2}	1.378×10^{-3}	0.000
0.200	2.318	6.796×10^{-3}	2.325	5.440×10^{-2}	1.698×10^{-3}	0.000
0.300	1.962	7.749×10^{-3}	1.970	1.016×10^{-1}	2.277×10^{-3}	0.000
0.400	1.793	8.857×10^{-3}	1.802	1.549×10^{-1}	2.814×10^{-3}	9.055×10^{-3}
0.500	1.698	1.009×10^{-2}	1.708	2.121×10^{-1}	3.333×10^{-3}	4.648×10^{-2}
0.600	1.640	1.143×10^{-2}	1.651	2.717×10^{-1}	3.846×10^{-3}	9.011×10^{-2}
0.700	1.603	1.285×10^{-2}	1.616	3.330×10^{-1}	4.359×10^{-3}	1.372×10^{-1}
0.800	1.579	1.434×10^{-2}	1.593	3.953×10^{-1}	4.873×10^{-3}	1.862×10^{-1}
0.900	1.563	1.590×10^{-2}	1.579	4.584×10^{-1}	5.391×10^{-3}	2.361×10^{-1}
1.000	1.552	1.751×10^{-2}	1.570	5.219×10^{-1}	5.913×10^{-3}	2.861×10^{-1}
1.500	1.538	2.636×10^{-2}	1.564	8.418×10^{-1}	8.599×10^{-3}	5.274×10^{-1}
2.000	1.547	3.617×10^{-2}	1.583	1.160	1.141×10^{-2}	7.451×10^{-1}
3.000	1.576	5.771×10^{-2}	1.634	1.782	1.729×10^{-2}	1.114
4.000	1.604	8.102×10^{-2}	1.685	2.384	2.339×10^{-2}	1.419
5.000	1.627	1.056×10^{-1}	1.733	2.970	2.961×10^{-2}	1.680
6.000	1.647	1.311×10^{-1}	1.778	3.539	3.590×10^{-2}	1.907
7.000	1.664	1.574×10^{-1}	1.821	4.095	4.222×10^{-2}	2.110
8.000	1.678	1.844×10^{-1}	1.863	4.638	4.853×10^{-2}	2.292
9.000	1.691	2.119×10^{-1}	1.903	5.169	5.482×10^{-2}	2.459
10.000	1.703	2.399×10^{-1}	1.943	5.689	6.109×10^{-2}	2.611
15.000	1.746	3.850×10^{-1}	2.131	8.145	9.159×10^{-2}	3.229
20.000	1.775	5.360×10^{-1}	2.311	1.040×10	1.204×10^{-1}	3.696
30.000	1.814	8.477×10^{-1}	2.662	1.442×10	1.725×10^{-1}	4.388
40.000	1.840	1.167	3.007	1.796×10	2.180×10^{-1}	4.901
50.000	1.860	1.491	3.351	2.111×10	2.579×10^{-1}	5.310

TABLE M.2n

Electrons in Glass, Lead

$\langle Z/A \rangle$: 0.42101; ρ: 6.220 g cm^{-3}; I: 526.4 eV; Composition (Z$-f_w$): 8–0.156453; 14–0.080866; 22–0.008092; 33–0.002651; 82–0.751938

Kinetic Energy (MeV)	Stopping Power (MeV cm^{2}g^{-1})			CSDA Range (g cm^{-2})	Radiation Yield	Density-Effect Parameter δ
	Collision	Radiative	Total			
0.010	1.049×10	1.674×10^{-2}	1.050×10	6.065×10^{-4}	7.624×10^{-4}	0.000
0.015	8.008	1.962×10^{-2}	8.028	1.157×10^{-3}	1.181×10^{-3}	0.000
0.020	6.584	2.168×10^{-2}	6.606	1.848×10^{-3}	1.602×10^{-3}	0.000
0.030	4.985	2.466×10^{-2}	5.010	3.608×10^{-3}	2.437×10^{-3}	0.000
0.040	4.098	2.686×10^{-2}	4.125	5.822×10^{-3}	3.258×10^{-3}	0.000
0.050	3.529	2.865×10^{-2}	3.558	8.443×10^{-3}	4.064×10^{-3}	0.000
0.060	3.131	3.020×10^{-2}	3.162	1.143×10^{-2}	4.854×10^{-3}	0.000
0.070	2.837	3.157×10^{-2}	2.869	1.476×10^{-2}	5.629×10^{-3}	0.000
0.080	2.610	3.282×10^{-2}	2.643	1.840×10^{-2}	6.390×10^{-3}	0.000
0.090	2.430	3.397×10^{-2}	2.464	2.232×10^{-2}	7.136×10^{-3}	0.000
0.100	2.283	3.504×10^{-2}	2.318	2.651×10^{-2}	7.868×10^{-3}	0.000
0.150	1.830	3.964×10^{-2}	1.869	5.081×10^{-2}	1.132×10^{-2}	0.000
0.200	1.597	4.351×10^{-2}	1.641	7.952×10^{-2}	1.447×10^{-2}	0.000
0.300	1.367	5.056×10^{-2}	1.418	1.457×10^{-1}	2.005×10^{-2}	7.930×10^{-3}
0.400	1.258	5.746×10^{-2}	1.316	2.192×10^{-1}	2.497×10^{-2}	3.553×10^{-2}
0.500	1.199	6.445×10^{-2}	1.263	2.970×10^{-1}	2.946×10^{-2}	6.527×10^{-2}
0.600	1.164	7.159×10^{-2}	1.236	3.771×10^{-1}	3.363×10^{-2}	9.612×10^{-2}
0.700	1.143	7.886×10^{-2}	1.222	4.585×10^{-1}	3.758×10^{-2}	1.274×10^{-1}
0.800	1.131	8.626×10^{-2}	1.217	5.406×10^{-1}	4.135×10^{-2}	1.589×10^{-1}
0.900	1.123	9.380×10^{-2}	1.217	6.227×10^{-1}	4.497×10^{-2}	1.901×10^{-1}
1.000	1.120	1.015×10^{-1}	1.221	7.048×10^{-1}	4.848×10^{-2}	2.210×10^{-1}
1.500	1.124	1.414×10^{-1}	1.266	1.108	6.485×10^{-2}	3.675×10^{-1}
2.000	1.142	1.836×10^{-1}	1.326	1.494	7.996×10^{-2}	4.998×10^{-1}
3.000	1.179	2.724×10^{-1}	1.452	2.214	1.078×10^{-1}	7.309×10^{-1}
4.000	1.211	3.652×10^{-1}	1.576	2.875	1.334×10^{-1}	9.302×10^{-1}
5.000	1.237	4.610×10^{-1}	1.698	3.486	1.571×10^{-1}	1.107
6.000	1.259	5.591×10^{-1}	1.818	4.055	1.792×10^{-1}	1.265
7.000	1.277	6.591×10^{-1}	1.936	4.588	1.999×10^{-1}	1.409
8.000	1.293	7.609×10^{-1}	2.054	5.089	2.194×10^{-1}	1.540
9.000	1.307	8.639×10^{-1}	2.171	5.563	2.377×10^{-1}	1.662
10.000	1.319	9.682×10^{-1}	2.287	6.012	2.550×10^{-1}	1.774
15.000	1.365	1.504	2.869	7.959	3.290×10^{-1}	2.244
20.000	1.396	2.056	3.453	9.546	3.872×10^{-1}	2.609
30.000	1.438	3.194	4.632	1.204×10	4.736×10^{-1}	3.166
40.000	1.467	4.358	5.825	1.396×10	5.354×10^{-1}	3.587
50.000	1.488	5.541	7.029	1.552×10	5.822×10^{-1}	3.928

TABLE M.2o

Electrons in Photographic Emulsion

$\langle Z/A \rangle$: 0.45453; ρ: 3.815 g cm^{-3}; I: 331.0 eV; Composition ($Z-f_w$): 1–0.014100; 6–0.072261; 7–0.019320; 8–0.066101; 16–0.001890; 35–0.349104; 47–0.474105; 53–0.003120

Kinetic Energy (MeV)	Stopping Power (MeV cm^{2}g^{-1})			CSDA Range (g cm^{-2})	Radiation Yield	Density-Effect Parameter δ
	Collision	Radiative	Total			
0.010	1.302×10	1.310×10^{-2}	1.304×10	4.665×10^{-4}	4.992×10^{-4}	0.000
0.015	9.798	1.463×10^{-2}	9.812	9.140×10^{-4}	7.494×10^{-4}	0.000
0.020	7.984	1.565×10^{-2}	8.000	1.482×10^{-3}	9.934×10^{-4}	0.000
0.030	5.983	1.697×10^{-2}	6.000	2.945×10^{-3}	1.462×10^{-3}	0.000
0.040	4.887	1.786×10^{-2}	4.905	4.801×10^{-3}	1.906×10^{-3}	0.000
0.050	4.190	1.854×10^{-2}	4.209	7.011×10^{-3}	2.330×10^{-3}	0.000
0.060	3.706	1.910×10^{-2}	3.725	9.543×10^{-3}	2.737×10^{-3}	0.000
0.070	3.349	1.959×10^{-2}	3.369	1.237×10^{-2}	3.128×10^{-3}	0.000
0.080	3.075	2.003×10^{-2}	3.095	1.547×10^{-2}	3.506×10^{-3}	0.000
0.090	2.857	2.044×10^{-2}	2.877	1.883×10^{-2}	3.870×10^{-3}	0.000
0.100	2.680	2.081×10^{-2}	2.701	2.242×10^{-2}	4.224×10^{-3}	0.000
0.150	2.136	2.251×10^{-2}	2.159	4.338×10^{-2}	5.851×10^{-3}	0.000
0.200	1.858	2.405×10^{-2}	1.883	6.833×10^{-2}	7.294×10^{-3}	0.000
0.300	1.585	2.721×10^{-2}	1.612	1.263×10^{-1}	9.822×10^{-3}	0.000
0.400	1.453	3.066×10^{-2}	1.484	1.913×10^{-1}	1.206×10^{-2}	3.297×10^{-2}
0.500	1.381	3.443×10^{-2}	1.415	2.604×10^{-1}	1.415×10^{-2}	7.178×10^{-2}
0.600	1.338	3.846×10^{-2}	1.376	3.322×10^{-1}	1.615×10^{-2}	1.112×10^{-1}
0.700	1.311	4.269×10^{-2}	1.354	4.055×10^{-1}	1.809×10^{-2}	1.505×10^{-1}
0.800	1.295	4.711×10^{-2}	1.342	4.797×10^{-1}	2.000×10^{-2}	1.895×10^{-1}
0.900	1.284	5.168×10^{-2}	1.336	5.544×10^{-1}	2.187×10^{-2}	2.278×10^{-1}
1.000	1.278	5.640×10^{-2}	1.335	6.293×10^{-1}	2.373×10^{-2}	2.652×10^{-1}
1.500	1.278	8.177×10^{-2}	1.360	1.001	3.289×10^{-2}	4.394×10^{-1}
2.000	1.294	1.094×10^{-1}	1.403	1.363	4.193×10^{-2}	5.933×10^{-1}
3.000	1.331	1.687×10^{-1}	1.499	2.053	5.974×10^{-2}	8.560×10^{-1}
4.000	1.363	2.317×10^{-1}	1.595	2.699	7.708×10^{-2}	1.077
5.000	1.390	2.973×10^{-1}	1.687	3.309	9.385×10^{-2}	1.268
6.000	1.412	3.649×10^{-1}	1.777	3.886	1.100×10^{-1}	1.437
7.000	1.431	4.340×10^{-1}	1.865	4.436	1.256×10^{-1}	1.590
8.000	1.448	5.045×10^{-1}	1.952	4.960	1.406×10^{-1}	1.729
9.000	1.462	5.760×10^{-1}	2.038	5.461	1.551×10^{-1}	1.856
10.000	1.475	6.485×10^{-1}	2.123	5.942	1.690×10^{-1}	1.975
15.000	1.523	1.022	2.545	8.089	2.312×10^{-1}	2.471
20.000	1.555	1.407	2.963	9.909	2.834×10^{-1}	2.858
30.000	1.598	2.202	3.800	1.288×10	3.659×10^{-1}	3.451
40.000	1.626	3.015	4.641	1.526×10	4.286×10^{-1}	3.899
50.000	1.648	3.840	5.488	1.724×10	4.781×10^{-1}	4.261

TABLE M.2p

Electrons in Polymethyl Methacrylate (PMMA or Perspex)

$\langle Z/A \rangle$: 0.53937; ρ: 1.190[a] g cm^{-3}; I: 74.0 eV; Composition ($Z-f_w$): 1–0.080541; 6–0.599846; 8–0.319613

Kinetic Energy (MeV)	Stopping Power (MeV cm^{2}g^{-1})			CSDA Range (g cm^{-2})	Radiation Yield	Density-Effect Parameter δ
	Collision	Radiative	Total			
0.010	2.198×10	3.332×10^{-3}	2.198×10	2.580×10^{-4}	8.330×10^{-5}	0.000
0.015	1.604×10	3.359×10^{-3}	1.604×10	5.282×10^{-4}	1.158×10^{-4}	0.000
0.020	1.283×10	3.372×10^{-3}	1.283×10	8.793×10^{-4}	1.460×10^{-4}	0.000
0.030	9.400	3.391×10^{-3}	9.404	1.803×10^{-3}	2.015×10^{-4}	0.000
0.040	7.573	3.413×10^{-3}	7.576	2.997×10^{-3}	2.526×10^{-4}	0.000
0.050	6.429	3.438×10^{-3}	6.433	4.436×10^{-3}	3.007×10^{-4}	0.000
0.060	5.644	3.468×10^{-3}	5.647	6.100×10^{-3}	3.464×10^{-4}	0.000
0.070	5.070	3.502×10^{-3}	5.073	7.972×10^{-3}	3.901×10^{-4}	0.000
0.080	4.631	3.538×10^{-3}	4.635	1.004×10^{-2}	4.322×10^{-4}	0.000
0.090	4.286	3.577×10^{-3}	4.289	1.228×10^{-2}	4.729×10^{-4}	0.000
0.100	4.006	3.619×10^{-3}	4.010	1.470×10^{-2}	5.125×10^{-4}	0.000
0.150	3.151	3.855×10^{-3}	3.155	2.894×10^{-2}	6.966×10^{-4}	0.000
0.200	2.719	4.126×10^{-3}	2.723	4.610×10^{-2}	8.650×10^{-4}	0.000
0.300	2.292	4.751×10^{-3}	2.297	8.653×10^{-2}	1.175×10^{-3}	0.000
0.400	2.090	5.474×10^{-3}	2.096	1.323×10^{-1}	1.466×10^{-3}	0.000
0.500	1.975	6.278×10^{-3}	1.981	1.815×10^{-1}	1.751×10^{-3}	4.112×10^{-2}
0.600	1.903	7.149×10^{-3}	1.910	2.330×10^{-1}	2.035×10^{-3}	1.005×10^{-1}
0.700	1.856	8.076×10^{-3}	1.864	2.860×10^{-1}	2.320×10^{-3}	1.650×10^{-1}
0.800	1.825	9.050×10^{-3}	1.834	3.401×10^{-1}	2.609×10^{-3}	2.321×10^{-1}
0.900	1.803	1.007×10^{-2}	1.813	3.950×10^{-1}	2.902×10^{-3}	3.001×10^{-1}
1.000	1.788	1.113×10^{-2}	1.799	4.504×10^{-1}	3.199×10^{-3}	3.679×10^{-1}
1.500	1.759	1.693×10^{-2}	1.776	7.308×10^{-1}	4.744×10^{-3}	6.887×10^{-1}
2.000	1.762	2.338×10^{-2}	1.785	1.012	6.383×10^{-3}	9.689×10^{-1}
3.000	1.784	3.761×10^{-2}	1.822	1.567	9.868×10^{-3}	1.425
4.000	1.809	5.307×10^{-2}	1.862	2.109	1.354×10^{-2}	1.783
5.000	1.832	6.943×10^{-2}	1.901	2.641	1.733×10^{-2}	2.077
6.000	1.851	8.648×10^{-2}	1.937	3.162	2.120×10^{-2}	2.327
7.000	1.868	1.041×10^{-1}	1.972	3.673	2.513×10^{-2}	2.545
8.000	1.883	1.222×10^{-1}	2.005	4.176	2.910×10^{-2}	2.739
9.000	1.896	1.407×10^{-1}	2.037	4.671	3.309×10^{-2}	2.914
10.000	1.908	1.596×10^{-1}	2.067	5.158	3.710×10^{-2}	3.073
15.000	1.952	2.577×10^{-1}	2.210	7.496	5.709×10^{-2}	3.716
20.000	1.982	3.603×10^{-1}	2.342	9.693	7.667×10^{-2}	4.202
30.000	2.022	5.728×10^{-1}	2.595	1.375×10	1.138×10^{-1}	4.927
40.000	2.049	7.912×10^{-1}	2.840	1.743×10	1.480×10^{-1}	5.463
50.000	2.069	1.013	3.082	2.081×10	1.792×10^{-1}	5.889

[a] The density of any sample of plastic used for the first time should be determined experimentally if high dosimetric accuracy is required (see Section 18.8).

TABLE M.2q

Electrons in Polystyrene

$\langle Z/A \rangle$: 0.53768; ρ: 1.060[a] g cm^{-3}; I: 68.7 eV; Composition $(Z - f_w)$: 1–0.077421; 6–0.922579

Kinetic Energy (MeV)	Stopping Power (MeV cm^{2}g^{-1})			CSDA Range (g cm^{-2})	Radiation Yield	Density-Effect Parameter δ
	Collision	Radiative	Total			
0.010	2.223×10	2.982×10^{-3}	2.224×10	2.546×10^{-4}	7.406×10^{-5}	0.000
0.015	1.621×10	2.999×10^{-3}	1.621×10	5.218×10^{-4}	1.027×10^{-4}	0.000
0.020	1.296×10	3.008×10^{-3}	1.296×10	8.694×10^{-4}	1.292×10^{-4}	0.000
0.030	9.485	3.027×10^{-3}	9.488	1.785×10^{-3}	1.782×10^{-4}	0.000
0.040	7.637	3.048×10^{-3}	7.640	2.968×10^{-3}	2.235×10^{-4}	0.000
0.050	6.481	3.074×10^{-3}	6.484	4.395×10^{-3}	2.662×10^{-4}	0.000
0.060	5.688	3.103×10^{-3}	5.691	6.047×10^{-3}	3.068×10^{-4}	0.000
0.070	5.108	3.135×10^{-3}	5.111	7.905×10^{-3}	3.458×10^{-4}	0.000
0.080	4.666	3.169×10^{-3}	4.669	9.955×10^{-3}	3.834×10^{-4}	0.000
0.090	4.317	3.206×10^{-3}	4.320	1.218×10^{-2}	4.197×10^{-4}	0.000
0.100	4.034	3.244×10^{-3}	4.038	1.458×10^{-2}	4.550×10^{-4}	0.000
0.150	3.172	3.463×10^{-3}	3.176	2.873×10^{-2}	6.199×10^{-4}	0.000
0.200	2.735	3.711×10^{-3}	2.739	4.579×10^{-2}	7.709×10^{-4}	0.000
0.300	2.305	4.284×10^{-3}	2.309	8.598×10^{-2}	1.050×10^{-3}	0.000
0.400	2.101	4.945×10^{-3}	2.106	1.315×10^{-1}	1.312×10^{-3}	2.731×10^{-3}
0.500	1.984	5.680×10^{-3}	1.989	1.805×10^{-1}	1.570×10^{-3}	5.420×10^{-2}
0.600	1.911	6.475×10^{-3}	1.918	2.318×10^{-1}	1.827×10^{-3}	1.152×10^{-1}
0.700	1.864	7.322×10^{-3}	1.871	2.846×10^{-1}	2.087×10^{-3}	1.810×10^{-1}
0.800	1.832	8.212×10^{-3}	1.840	3.385×10^{-1}	2.349×10^{-3}	2.492×10^{-1}
0.900	1.810	9.142×10^{-3}	1.819	3.932×10^{-1}	2.615×10^{-3}	3.179×10^{-1}
1.000	1.794	1.011×10^{-2}	1.804	4.484×10^{-1}	2.885×10^{-3}	3.862×10^{-1}
1.500	1.766	1.541×10^{-2}	1.781	7.281×10^{-1}	4.293×10^{-3}	7.064×10^{-1}
2.000	1.768	2.132×10^{-2}	1.789	1.008	5.788×10^{-3}	9.834×10^{-1}
3.000	1.791	3.435×10^{-2}	1.825	1.562	8.971×10^{-3}	1.431
4.000	1.816	4.852×10^{-2}	1.865	2.104	1.233×10^{-2}	1.782
5.000	1.839	6.353×10^{-2}	1.902	2.635	1.580×10^{-2}	2.070
6.000	1.859	7.919×10^{-2}	1.938	3.156	1.936×10^{-2}	2.316
7.000	1.876	9.539×10^{-2}	1.971	3.667	2.297×10^{-2}	2.531
8.000	1.891	1.120×10^{-1}	2.003	4.171	2.662×10^{-2}	2.722
9.000	1.904	1.290×10^{-1}	2.033	4.666	3.029×10^{-2}	2.896
10.000	1.916	1.464×10^{-1}	2.062	5.155	3.399×10^{-2}	3.054
15.000	1.960	2.367×10^{-1}	2.196	7.502	5.249×10^{-2}	3.702
20.000	1.989	3.311×10^{-1}	2.320	9.717	7.072×10^{-2}	4.196
30.000	2.027	5.270×10^{-1}	2.554	1.382×10	1.056×10^{-1}	4.933
40.000	2.053	7.284×10^{-1}	2.782	1.757×10	1.378×10^{-1}	5.478
50.000	2.073	9.334×10^{-1}	3.006	2.103×10	1.676×10^{-1}	5.908

[a] The density of any sample of plastic used for the first time should be determined experimentally if high dosimetric accuracy is required (see Section 18.8).

TABLE M.2r

Electrons in Adipose Tissue (ICRP)

$\langle Z/A \rangle$: 0.55847; ρ: 0.920 g cm^{-3}; I: 63.2 eV; Composition (Z $- f_w$): 1–0.119477; 6–0.637240; 7–0.007970; 8–0.232333; 11–0.000500; 12–0.000020; 15–0.000160; 16–0.000730; 17–0.001190; 19–0.000320; 20–0.000020; 26–0.000020; 30–0.000020

Kinetic Energy (MeV)	Stopping Power (MeV cm^{2}g^{-1})			CSDA Range (g cm^{-2})	Radiation Yield	Density-Effect Parameter δ
	Collision	Radiative	Total			
0.010	2.347×10	3.168×10^{-3}	2.347×10	2.406×10^{-4}	7.396×10^{-5}	0.000
0.015	1.709×10	3.194×10^{-3}	1.709×10	4.940×10^{-4}	1.031×10^{-4}	0.000
0.020	1.365×10	3.207×10^{-3}	1.366×10	8.237×10^{-4}	1.301×10^{-4}	0.000
0.030	9.984	3.227×10^{-3}	9.987	1.693×10^{-3}	1.800×10^{-4}	0.000
0.040	8.034	3.249×10^{-3}	8.037	2.818×10^{-3}	2.260×10^{-4}	0.000
0.050	6.816	3.275×10^{-3}	6.819	4.175×10^{-3}	2.693×10^{-4}	0.000
0.060	5.979	3.305×10^{-3}	5.983	5.745×10^{-3}	3.106×10^{-4}	0.000
0.070	5.369	3.338×10^{-3}	5.372	7.513×10^{-3}	3.501×10^{-4}	0.000
0.080	4.903	3.373×10^{-3}	4.906	9.464×10^{-3}	3.881×10^{-4}	0.000
0.090	4.535	3.411×10^{-3}	4.539	1.159×10^{-2}	4.250×10^{-4}	0.000
0.100	4.238	3.452×10^{-3}	4.241	1.387×10^{-2}	4.608×10^{-4}	0.000
0.150	3.330	3.681×10^{-3}	3.334	2.734×10^{-2}	6.277×10^{-4}	0.000
0.200	2.871	3.943×10^{-3}	2.875	4.359×10^{-2}	7.805×10^{-4}	0.000
0.300	2.418	4.547×10^{-3}	2.422	8.190×10^{-2}	1.062×10^{-3}	0.000
0.400	2.204	5.244×10^{-3}	2.209	1.253×10^{-1}	1.328×10^{-3}	0.000
0.500	2.081	6.020×10^{-3}	2.087	1.720×10^{-1}	1.588×10^{-3}	4.193×10^{-2}
0.600	2.005	6.860×10^{-3}	2.011	2.209×10^{-1}	1.848×10^{-3}	1.029×10^{-1}
0.700	1.954	7.753×10^{-3}	1.962	2.712×10^{-1}	2.109×10^{-3}	1.693×10^{-1}
0.800	1.921	8.692×10^{-3}	1.929	3.227×10^{-1}	2.374×10^{-3}	2.383×10^{-1}
0.900	1.897	9.674×10^{-3}	1.907	3.748×10^{-1}	2.642×10^{-3}	3.081×10^{-1}
1.000	1.880	1.070×10^{-2}	1.891	4.275×10^{-1}	2.915×10^{-3}	3.778×10^{-1}
1.500	1.849	1.629×10^{-2}	1.865	6.944×10^{-1}	4.334×10^{-3}	7.070×10^{-1}
2.000	1.850	2.252×10^{-2}	1.873	9.621×10^{-1}	5.842×10^{-3}	9.938×10^{-1}
3.000	1.872	3.626×10^{-2}	1.908	1.491	9.055×10^{-3}	1.460
4.000	1.897	5.120×10^{-2}	1.948	2.010	1.245×10^{-2}	1.825
5.000	1.920	6.701×10^{-2}	1.987	2.518	1.596×10^{-2}	2.125
6.000	1.939	8.350×10^{-2}	2.023	3.017	1.955×10^{-2}	2.379
7.000	1.956	1.005×10^{-1}	2.057	3.507	2.319×10^{-2}	2.601
8.000	1.972	1.181×10^{-1}	2.090	3.990	2.688×10^{-2}	2.798
9.000	1.985	1.360×10^{-1}	2.121	4.465	3.059×10^{-2}	2.976
10.000	1.997	1.542×10^{-1}	2.151	4.933	3.432×10^{-2}	3.138
15.000	2.042	2.492×10^{-1}	2.291	7.183	5.300×10^{-2}	3.790
20.000	2.073	3.485×10^{-1}	2.421	9.305	7.138×10^{-2}	4.282
30.000	2.113	5.544×10^{-1}	2.668	1.324×10	1.065×10^{-1}	5.012
40.000	2.141	7.661×10^{-1}	2.907	1.683×10	1.389×10^{-1}	5.552
50.000	2.161	9.815×10^{-1}	3.143	2.013×10	1.688×10^{-1}	5.979

TABLE M.2s

Electrons in Bone, Compact (ICRU)

$\langle Z/A \rangle$: 0.53010; ρ: 1.850 g cm^{-3}; I: 91.9 eV; Composition (Z$-f_w$): 1$-$0.063984; 6$-$0.278000; 7$-$0.027000; 8$-$0.410016; 12$-$0.002000; 15$-$0.070000; 16$-$0.002000; 20$-$0.147000

Kinetic Energy (MeV)	Stopping Power (MeV cm^{2}g^{-1})			CSDA Range (g cm^{-2})	Radiation Yield	Density-Effect Parameter δ
	Collision	Radiative	Total			
0.010	2.068×10	4.793×10^{-3}	2.068×10	2.761×10^{-4}	1.236×10^{-4}	0.000
0.015	1.514×10	4.942×10^{-3}	1.514×10	5.627×10^{-4}	1.756×10^{-4}	0.000
0.020	1.213×10	5.026×10^{-3}	1.214×10	9.343×10^{-4}	2.244×10^{-4}	0.000
0.030	8.912	5.121×10^{-3}	8.917	1.910×10^{-3}	3.148×10^{-4}	0.000
0.040	7.191	5.184×10^{-3}	7.196	3.167×10^{-3}	3.982×10^{-4}	0.000
0.050	6.112	5.237×10^{-3}	6.117	4.681×10^{-3}	4.764×10^{-4}	0.000
0.060	5.370	5.288×10^{-3}	5.375	6.431×10^{-3}	5.504×10^{-4}	0.000
0.070	4.827	5.338×10^{-3}	4.832	8.397×10^{-3}	6.210×10^{-4}	0.000
0.080	4.412	5.390×10^{-3}	4.418	1.056×10^{-2}	6.887×10^{-4}	0.000
0.090	4.085	5.445×10^{-3}	4.090	1.292×10^{-2}	7.540×10^{-4}	0.000
0.100	3.820	5.502×10^{-3}	3.826	1.545×10^{-2}	8.171×10^{-4}	0.000
0.150	3.010	5.826×10^{-3}	3.016	3.037×10^{-2}	1.108×10^{-3}	0.000
0.200	2.599	6.200×10^{-3}	2.605	4.831×10^{-2}	1.371×10^{-3}	0.000
0.300	2.194	7.075×10^{-3}	2.201	9.052×10^{-2}	1.848×10^{-3}	0.000
0.400	1.996	8.090×10^{-3}	2.004	1.384×10^{-1}	2.292×10^{-3}	6.259×10^{-2}
0.500	1.883	9.221×10^{-3}	1.893	1.898×10^{-1}	2.724×10^{-3}	1.344×10^{-1}
0.600	1.815	1.045×10^{-2}	1.825	2.437×10^{-1}	3.153×10^{-3}	2.080×10^{-1}
0.700	1.770	1.175×10^{-2}	1.782	2.992×10^{-1}	3.582×10^{-3}	2.819×10^{-1}
0.800	1.740	1.311×10^{-2}	1.753	3.558×10^{-1}	4.014×10^{-3}	3.550×10^{-1}
0.900	1.719	1.454×10^{-2}	1.734	4.132×10^{-1}	4.449×10^{-3}	4.267×10^{-1}
1.000	1.705	1.602×10^{-2}	1.721	4.711×10^{-1}	4.889×10^{-3}	4.966×10^{-1}
1.500	1.680	2.414×10^{-2}	1.705	7.638×10^{-1}	7.161×10^{-3}	8.162×10^{-1}
2.000	1.684	3.313×10^{-2}	1.718	1.056	9.547×10^{-3}	1.089
3.000	1.709	5.289×10^{-2}	1.762	1.631	1.457×10^{-2}	1.527
4.000	1.735	7.427×10^{-2}	1.809	2.191	1.981×10^{-2}	1.872
5.000	1.758	9.680×10^{-2}	1.855	2.737	2.517×10^{-2}	2.156
6.000	1.778	1.202×10^{-1}	1.898	3.270	3.060×10^{-2}	2.399
7.000	1.795	1.444×10^{-1}	1.939	3.791	3.607×10^{-2}	2.612
8.000	1.810	1.692×10^{-1}	1.979	4.301	4.156×10^{-2}	2.801
9.000	1.823	1.945×10^{-1}	2.018	4.802	4.705×10^{-2}	2.973
10.000	1.835	2.202×10^{-1}	2.055	5.293	5.252×10^{-2}	3.130
15.000	1.879	3.537×10^{-1}	2.233	7.625	7.939×10^{-2}	3.766
20.000	1.909	4.926×10^{-1}	2.402	9.783	1.051×10^{-1}	4.248
30.000	1.949	7.799×10^{-1}	2.729	1.369×10	1.523×10^{-1}	4.965
40.000	1.976	1.074	3.050	1.715×10	1.943×10^{-1}	5.496
50.000	1.996	1.374	3.369	2.027×10	2.316×10^{-1}	5.916

TABLE M.2t

Electrons in Bone, Cortical (ICRP)

$\langle Z/A \rangle$: 0.52130; ρ: 1.850 g cm^{-3}; I: 106.4 eV; Composition ($Z - f_w$): 1–0.047234; 6–0.144330; 7–0.041990; 8–0.446096; 12–0.002200; 15–0.104970; 16–0.003150; 20–0.209930; 30–0.000100

Kinetic Energy (MeV)	Stopping Power (MeV cm^{2}g^{-1})			CSDA Range (g cm^{-2})	Radiation Yield	Density-Effect Parameter δ
	Collision	Radiative	Total			
0.010	1.972×10	5.461×10^{-3}	1.972×10	2.909×10^{-4}	1.468×10^{-4}	0.000
0.015	1.447×10	5.664×10^{-3}	1.447×10	5.911×10^{-4}	2.095×10^{-4}	0.000
0.020	1.162×10	5.778×10^{-3}	1.162×10	9.795×10^{-4}	2.683×10^{-4}	0.000
0.030	8.546	5.907×10^{-3}	8.552	1.997×10^{-3}	3.775×10^{-4}	0.000
0.040	6.904	5.989×10^{-3}	6.910	3.308×10^{-3}	4.781×10^{-4}	0.000
0.050	5.873	6.054×10^{-3}	5.879	4.884×10^{-3}	5.723×10^{-4}	0.000
0.060	5.163	6.113×10^{-3}	5.169	6.703×10^{-3}	6.614×10^{-4}	0.000
0.070	4.643	6.171×10^{-3}	4.649	8.747×10^{-3}	7.463×10^{-4}	0.000
0.080	4.246	6.230×10^{-3}	4.252	1.100×10^{-2}	8.276×10^{-4}	0.000
0.090	3.932	6.292×10^{-3}	3.939	1.345×10^{-2}	9.058×10^{-4}	0.000
0.100	3.679	6.356×10^{-3}	3.685	1.607×10^{-2}	9.814×10^{-4}	0.000
0.150	2.901	6.719×10^{-3}	2.908	3.155×10^{-2}	1.329×10^{-3}	0.000
0.200	2.507	7.140×10^{-3}	2.514	5.015×10^{-2}	1.641×10^{-3}	0.000
0.300	2.119	8.129×10^{-3}	2.127	9.386×10^{-2}	2.206×10^{-3}	0.000
0.400	1.932	9.276×10^{-3}	1.941	1.433×10^{-1}	2.730×10^{-3}	3.416×10^{-2}
0.500	1.825	1.055×10^{-2}	1.836	1.964×10^{-1}	3.236×10^{-3}	9.154×10^{-2}
0.600	1.760	1.194×10^{-2}	1.772	2.519×10^{-1}	3.737×10^{-3}	1.532×10^{-1}
0.700	1.718	1.341×10^{-2}	1.732	3.090×10^{-1}	4.237×10^{-3}	2.167×10^{-1}
0.800	1.690	1.495×10^{-2}	1.705	3.672×10^{-1}	4.740×10^{-3}	2.807×10^{-1}
0.900	1.671	1.657×10^{-2}	1.688	4.262×10^{-1}	5.245×10^{-3}	3.443×10^{-1}
1.000	1.659	1.824×10^{-2}	1.677	4.857×10^{-1}	5.755×10^{-3}	4.071×10^{-1}
1.500	1.638	2.740×10^{-2}	1.665	7.857×10^{-1}	8.382×10^{-3}	6.995×10^{-1}
2.000	1.643	3.755×10^{-2}	1.681	1.085	1.113×10^{-2}	9.535×10^{-1}
3.000	1.670	5.981×10^{-2}	1.730	1.671	1.689×10^{-2}	1.368
4.000	1.697	8.386×10^{-2}	1.781	2.241	2.288×10^{-2}	1.697
5.000	1.720	1.092×10^{-1}	1.829	2.795	2.898×10^{-2}	1.970
6.000	1.740	1.355×10^{-1}	1.876	3.335	3.514×10^{-2}	2.203
7.000	1.758	1.627×10^{-1}	1.921	3.862	4.133×10^{-2}	2.408
8.000	1.773	1.904×10^{-1}	1.964	4.377	4.752×10^{-2}	2.591
9.000	1.787	2.188×10^{-1}	2.006	4.880	5.369×10^{-2}	2.757
10.000	1.799	2.476×10^{-1}	2.046	5.374	5.983	2.909
15.000	1.844	3.971×10^{-1}	2.241	7.707	8.974×10^{-2}	3.525
20.000	1.874	5.525×10^{-1}	2.427	9.850	1.180×10^{-1}	3.994
30.000	1.915	8.735×10^{-1}	2.788	1.369×10	1.694×10^{-1}	4.696
40.000	1.942	1.202	3.144	1.707×10	2.143×10^{-1}	5.217
50.000	1.962	1.537	3.498	2.008×10	2.538×10^{-1}	5.631

TABLE M.2u

Electrons in Lung Tissue (ICRP)

$\langle Z/A \rangle$: 0.54965; ρ: 1.050 g cm^{-3} (deflated)[a]; I: 75.3 eV; Composition $(Z - f_w)$:– 1–0.101278; 6–0.102310; 7–0.028650; 8–0.757072; 11–0.001840; 12–0.000730; 15–0.000800; 16–0.002250; 17–0.002660; 19–0.001940; 20–0.000090; 26–0.000370; 30–0.000010

Kinetic Energy (MeV)	Stopping Power (MeV cm^{2}g^{-1})			CSDA Range (g cm^{-2})	Radiation Yield	Density-Effect Parameter δ
	Collision	Radiative	Total			
0.010	2.232×10	3.842×10^{-3}	2.233×10	2.542×10^{-4}	9.375×10^{-4}	0.000
0.015	1.629×10	3.887×10^{-3}	1.630×10	5.202×10^{-4}	1.311×10^{-4}	0.000
0.020	1.304×10	3.908×10^{-3}	1.304×10	8.657×10^{-4}	1.657×10^{-4}	0.000
0.030	9.552	3.931×10^{-3}	9.556	1.775×10^{-3}	2.293×10^{-4}	0.000
0.040	7.696	3.954×10^{-3}	7.700	2.950×10^{-3}	2.877×10^{-4}	0.000
0.050	6.535	3.981×10^{-3}	6.539	4.365×10^{-3}	3.425×10^{-4}	0.000
0.060	5.737	4.012×10^{-3}	5.741	6.002×10^{-3}	3.945×10^{-4}	0.000
0.070	5.153	4.048×10^{-3}	5.157	7.844×10^{-3}	4.441×10^{-4}	0.000
0.080	4.708	4.087×10^{-3}	4.712	9.876×10^{-3}	4.919×10^{-4}	0.000
0.090	4.357	4.130×10^{-3}	4.361	1.208×10^{-2}	5.381×10^{-4}	0.000
0.100	4.073	4.176×10^{-3}	4.077	1.446×10^{-2}	5.829×10^{-4}	0.000
0.150	3.204	4.440×10^{-3}	3.209	2.847×10^{-2}	7.910×10^{-4}	0.000
0.200	2.764	4.743×10^{-3}	2.769	4.534×10^{-2}	9.808×10^{-4}	0.000
0.300	2.331	5.448×10^{-3}	2.336	8.509×10^{-2}	1.329×10^{-3}	0.000
0.400	2.126	6.263×10^{-3}	2.132	1.301×10^{-1}	1.655×10^{-3}	0.000
0.500	2.013	7.171×10^{-3}	2.020	1.784×10^{-1}	1.973×10^{-3}	3.564×10^{-3}
0.600	1.942	8.156×10^{-3}	1.950	2.288×10^{-1}	2.288×10^{-3}	4.469×10^{-2}
0.700	1.895	9.202×10^{-3}	1.904	2.808×10^{-1}	2.605×10^{-3}	9.560×10^{-2}
0.800	1.864	1.030×10^{-2}	1.874	3.337×10^{-1}	2.925×10^{-3}	1.523×10^{-1}
0.900	1.842	1.145×10^{-2}	1.854	3.874×10^{-1}	3.248×10^{-3}	2.124×10^{-1}
1.000	1.827	1.265×10^{-2}	1.840	4.416×10^{-1}	3.576×10^{-3}	2.741×10^{-1}
1.500	1.800	1.919×10^{-2}	1.819	7.156×10^{-1}	5.279×10^{-3}	5.807×10^{-1}
2.000	1.802	2.646×10^{-2}	1.828	9.899×10^{-1}	7.084×10^{-3}	8.593×10^{-1}
3.000	1.824	4.248×10^{-2}	1.867	1.532	1.092×10^{-2}	1.323
4.000	1.849	5.986×10^{-2}	1.908	2.061	1.495×10^{-2}	1.691
5.000	1.871	7.823×10^{-2}	1.949	2.580	1.911×10^{-2}	1.994
6.000	1.890	9.738×10^{-2}	1.987	3.088	2.335×10^{-2}	2.252
7.000	1.907	1.172×10^{-1}	2.024	3.587	2.765×10^{-2}	2.475
8.000	1.921	1.374×10^{-1}	2.059	4.076	3.198×10^{-2}	2.672
9.000	1.935	1.582×10^{-1}	2.093	4.558	3.634×10^{-2}	2.849
10.000	1.947	1.793×10^{-1}	2.126	5.032	4.070×10^{-2}	3.010
15.000	1.992	2.892×10^{-1}	2.281	7.301	6.239×10^{-2}	3.649
20.000	2.024	4.038×10^{-1}	2.427	9.425	8.349×10^{-2}	4.124
30.000	2.066	6.413×10^{-1}	2.707	1.332×10	1.232×10^{-1}	4.828
40.000	2.094	8.851×10^{-1}	2.979	1.684×10	1.593×10^{-1}	5.351
50.000	2.115	1.133	3.248	2.006×10	1.922×10^{-1}	5.768

[a] The density of lung encountered clinically is generally around 0.3 g cm^{-3}; however, for the purposes of evaluating $(S/\rho)_{col}$, which above ≈ 0.5 MeV is generally *not* independent of the density due to the *density effect* (see Section 3.2.3), the *local* density of lung tissue, i.e., 1.05 g cm^{-3}, is the more correct one to use; therefore the correct *linear stopping* power will be given by multiplying the values in this table by the clinical *bulk* lung density. The very small amount of air in inflated lung will make a negligible difference to the linear stopping power derived from these numbers, which strictly only apply to lung *tissue*.

TABLE M.2v

Electrons in Muscle, Skeletal (ICRP)

$\langle Z/A \rangle$: 0.549378; ρ: 1.040 g cm^{-3}; I: 75.3 eV; Composition ($Z - f_w$): 1–0.100637; 6–0.107830; 7–0.027680; 8–0.754773; 11–0.000750; 12–0.000190; 15–0.001800; 16–0.002410; 17–0.000790; 19–0.003020; 20–0.000030; 26–0.000040; 30–0.000050

Kinetic Energy (MeV)	Stopping Power (MeV cm^{2}g^{-1})			CSDA Range (g cm^{-2})	Radiation Yield	Density-Effect Parameter δ
	Collision	Radiative	Total			
0.010	2.231×10	3.835×10^{-3}	2.232×10	2.543×10^{-4}	9.365×10^{-4}	0.000
0.015	1.629×10	3.880×10^{-3}	1.629×10	5.204×10^{-4}	1.310×10^{-4}	0.000
0.020	1.303×10	3.901×10^{-3}	1.303×10	8.662×10^{-4}	1.655×10^{-4}	0.000
0.030	9.547	3.924×10^{-3}	9.551	1.776×10^{-3}	2.290×10^{-4}	0.000
0.040	7.692	3.946×10^{-3}	7.696	2.951×10^{-3}	2.874×10^{-4}	0.000
0.050	6.531	3.973×10^{-3}	6.535	4.367×10^{-3}	3.420×10^{-4}	0.000
0.060	5.734	4.004×10^{-3}	5.738	6.005×10^{-3}	3.939×10^{-4}	0.000
0.070	5.151	4.040×10^{-3}	5.155	7.848×10^{-3}	4.435×10^{-4}	0.000
0.080	4.706	4.079×10^{-3}	4.710	9.881×10^{-3}	4.912×10^{-4}	0.000
0.090	4.355	4.122×10^{-3}	4.359	1.209×10^{-2}	5.373×10^{-4}	0.000
0.100	4.071	4.168×10^{-3}	4.075	1.447×10^{-2}	5.820×10^{-4}	0.000
0.150	3.203	4.431×10^{-3}	3.207	2.848×10^{-2}	7.899×10^{-4}	0.000
0.200	2.763	4.734×10^{-3}	2.768	4.537×10^{-2}	9.794×10^{-4}	0.000
0.300	2.330	5.438×10^{-3}	2.335	8.513×10^{-2}	1.327×10^{-3}	0.000
0.400	2.125	6.252×10^{-3}	2.131	1.302×10^{-1}	1.653×10^{-3}	0.000
0.500	2.012	7.158×10^{-3}	2.019	1.785×10^{-1}	1.970×10^{-3}	2.283×10^{-3}
0.600	1.941	8.141×10^{-3}	1.949	2.290×10^{-1}	2.285×10^{-3}	4.259×10^{-2}
0.700	1.895	9.186×10^{-3}	1.904	2.809×10^{-1}	2.601×10^{-3}	9.285×10^{-2}
0.800	1.863	1.028×10^{-2}	1.874	3.339×10^{-1}	2.921×10^{-3}	1.491×10^{-1}
0.900	1.842	1.143×10^{-2}	1.853	3.876×10^{-1}	3.244×10^{-3}	2.087×10^{-1}
1.000	1.827	1.262×10^{-2}	1.839	4.418×10^{-1}	3.571×10^{-3}	2.701×10^{-1}
1.500	1.799	1.916×10^{-2}	1.818	7.158×10^{-1}	5.272×10^{-3}	5.754×10^{-1}
2.000	1.801	2.642×10^{-2}	1.828	9.903×10^{-1}	7.074×10^{-3}	8.534×10^{-1}
3.000	1.824	4.241×10^{-2}	1.866	1.532	1.090×10^{-2}	1.316
4.000	1.848	5.977×10^{-2}	1.908	2.062	1.493×10^{-2}	1.684
5.000	1.870	7.811×10^{-2}	1.948	2.580	1.908×10^{-2}	1.987
6.000	1.889	9.722×10^{-2}	1.987	3.089	2.332×10^{-2}	2.244
7.000	1.906	1.170×10^{-1}	2.023	3.587	2.761×10^{-2}	2.467
8.000	1.921	1.372×10^{-1}	2.058	4.077	3.194×10^{-2}	2.664
9.000	1.934	1.579×10^{-1}	2.092	4.559	3.629×10^{-2}	2.841
10.000	1.946	1.790×10^{-1}	2.125	5.033	4.065×10^{-2}	3.002
15.000	1.992	2.887×10^{-1}	2.281	7.303	6.231×10^{-2}	3.640
20.000	2.023	4.032×10^{-1}	2.427	9.428	8.339×10^{-2}	4.115
30.000	2.066	6.403×10^{-1}	2.706	1.333×10	1.230×10^{-1}	4.819
40.000	2.094	8.838×10^{-1}	2.978	1.685×10	1.592×10^{-1}	5.341
50.000	2.115	1.131	3.247	2.006×10	1.920×10^{-1}	5.758

TABLE M.2w

Electrons in Muscle, Striated (ICRU)

$\langle Z/A \rangle$: 0.55005; ρ: 1.040 g cm^{-3}; I: 74.7 eV; Composition $(Z - f_w)$: 1–0.101997; 6–0.123000; 7–0.035000; 8–0.729003; 11–0.000800; 12–0.002000; 15–0.002000; 16–0.005000; 16–0.005000; 19–0.003000

Kinetic Energy (MeV)	Stopping Power (MeV cm^{2}g^{-1})			CSDA Range (g cm^{-2})	Radiation Yield	Density-Effect Parameter δ
	Collision	Radiative	Total			
0.010	2.237×10	3.816×10^{-3}	2.238×10	2.536×10^{-4}	9.292×10^{-4}	0.000
0.015	1.633×10	3.862×10^{-3}	1.633×10	5.189×10^{-4}	1.300×10^{-4}	0.000
0.020	1.306×10	3.882×10^{-3}	1.307×10	8.638×10^{-4}	1.642×10^{-4}	0.000
0.030	9.572	3.905×10^{-3}	9.576	1.771×10^{-3}	2.273×10^{-4}	0.000
0.040	7.711	3.928×10^{-3}	7.715	2.943×10^{-3}	2.852×10^{-4}	0.000
0.050	6.547	3.955×10^{-3}	6.551	4.356×10^{-3}	3.396×10^{-4}	0.000
0.060	5.747	3.986×10^{-3}	5.751	5.990×10^{-3}	3.911×10^{-4}	0.000
0.070	5.163	4.022×10^{-3}	5.167	7.828×10^{-3}	4.404×10^{-4}	0.000
0.080	4.717	4.061×10^{-3}	4.721	9.856×10^{-3}	4.878×10^{-4}	0.000
0.090	4.365	4.104×10^{-3}	4.369	1.206×10^{-2}	5.336×10^{-4}	0.000
0.100	4.080	4.150×10^{-3}	4.084	1.443×10^{-2}	5.780×10^{-4}	0.000
0.150	3.210	4.412×10^{-3}	3.214	2.841×10^{-2}	7.845×10^{-4}	0.000
0.200	2.769	4.714×10^{-3}	2.774	4.526×10^{-2}	9.728×10^{-4}	0.000
0.300	2.335	5.415×10^{-3}	2.340	8.494×10^{-2}	1.318×10^{-3}	0.000
0.400	2.129	6.226×10^{-3}	2.136	1.299×10^{-1}	1.642×10^{-3}	0.000
0.500	2.016	7.129×10^{-3}	2.023	1.781×10^{-1}	1.957×10^{-3}	4.347×10^{-3}
0.600	1.945	8.108×10^{-3}	1.953	2.285×10^{-1}	2.271×10^{-3}	4.652×10^{-2}
0.700	1.898	9.148×10^{-3}	1.907	2.803×10^{-1}	2.585×10^{-3}	9.821×10^{-2}
0.800	1.867	1.024×10^{-2}	1.877	3.332×10^{-1}	2.903×10^{-3}	1.555×10^{-1}
0.900	1.845	1.139×10^{-2}	1.856	3.868×10^{-1}	3.224×10^{-3}	2.161×10^{-1}
1.000	1.830	1.257×10^{-2}	1.842	4.409×10^{-1}	3.549×10^{-3}	2.782×10^{-1}
1.500	1.802	1.908×10^{-2}	1.821	7.146×10^{-1}	5.242×10^{-3}	5.857×10^{-1}
2.000	1.804	2.632×10^{-2}	1.830	9.886×10^{-1}	7.035×10^{-3}	8.645×10^{-1}
3.000	1.826	4.225×10^{-2}	1.869	1.530	1.084×10^{-2}	1.328
4.000	1.851	5.954×10^{-2}	1.910	2.059	1.485×10^{-2}	1.696
5.000	1.873	7.782×10^{-2}	1.951	2.577	1.898×10^{-2}	1.999
6.000	1.892	9.686×10^{-2}	1.989	3.084	2.320×10^{-2}	2.256
7.000	1.909	1.165×10^{-1}	2.026	3.583	2.747×10^{-2}	2.479
8.000	1.924	1.367×10^{-1}	2.061	4.072	3.178×10^{-2}	2.676
9.000	1.937	1.574×10^{-1}	2.094	4.553	3.611×10^{-2}	2.853
10.000	1.949	1.784×10^{-1}	2.127	5.027	4.045×10^{-2}	3.014
15.000	1.995	2.877×10^{-1}	2.282	7.295	6.202×10^{-2}	3.653
20.000	2.026	4.018×10^{-1}	2.428	9.418	8.302×10^{-2}	4.129
30.000	2.068	6.381×10^{-1}	2.706	1.332×10	1.225×10^{-1}	4.833
40.000	2.097	8.807×10^{-1}	2.977	1.684×10	1.585×10^{-1}	5.357
50.000	2.118	1.127	3.245	2.005×10	1.913×10^{-1}	5.775

TABLE M.2x

Electrons in Skin (ICRP)

$\langle Z/A \rangle$: 0.54933; ρ: 1.100 g cm^{-3}; I: 72.7 eV; Composition $(Z-f_w)$: 1–0.100588; 6–0.228250; 7–0.046420; 8–0.619002; 11–0.000070; 12–0.000060; 15–0.000330; 16–0.001590; 17–0.002670; 19–0.000850; 20–0.000150; 26–0.000010; 30–0.000010

Kinetic Energy (MeV)	Stopping Power (MeV cm^{2}g^{-1})			CSDA Range (g cm^{-2})	Radiation Yield	Density-Effect Parameter δ
	Collision	Radiative	Total			
0.010	2.247×10	3.678×10^{-3}	2.247×10	2.523×10^{-4}	8.934×10^{-4}	0.000
0.015	1.639×10	3.718×10^{-3}	1.639×10	5.167×10^{-4}	1.248×10^{-4}	0.000
0.020	1.311×10	3.736×10^{-3}	1.311×10	8.603×10^{-4}	1.577×10^{-4}	0.000
0.030	9.602	3.758×10^{-3}	9.605	1.765×10^{-3}	2.181×10^{-4}	0.000
0.040	7.734	3.780×10^{-3}	7.738	2.934×10^{-3}	2.737×10^{-4}	0.000
0.050	6.565	3.806×10^{-3}	6.569	4.342×10^{-3}	3.259×10^{-4}	0.000
0.060	5.763	3.837×10^{-3}	5.767	5.972×10^{-3}	3.753×10^{-4}	0.000
0.070	5.176	3.872×10^{-3}	5.180	7.805×10^{-3}	4.227×10^{-4}	0.000
0.080	4.729	3.911×10^{-3}	4.733	9.828×10^{-3}	4.682×10^{-4}	0.000
0.090	4.376	3.953×10^{-3}	4.380	1.203×10^{-2}	5.123×10^{-4}	0.000
0.100	4.090	3.997×10^{-3}	4.094	1.439×10^{-2}	5.550×10^{-4}	0.000
0.150	3.217	4.252×10^{-3}	3.221	2.834×10^{-2}	7.538×10^{-4}	0.000
0.200	2.775	4.545×10^{-3}	2.779	4.516×10^{-2}	9.352×10^{-4}	0.000
0.300	2.339	5.225×10^{-3}	2.344	8.476×10^{-2}	1.268×10^{-3}	0.000
0.400	2.133	6.011×10^{-3}	2.139	1.296×10^{-1}	1.581×10^{-3}	0.000
0.500	2.018	6.886×10^{-3}	2.024	1.778×10^{-1}	1.885×10^{-3}	2.256×10^{-2}
0.600	1.945	7.835×10^{-3}	1.953	2.281×10^{-1}	2.189×10^{-3}	7.560×10^{-2}
0.700	1.897	8.843×10^{-3}	1.906	2.800×10^{-1}	2.494×10^{-3}	1.356×10^{-1}
0.800	1.865	9.902×10^{-3}	1.875	3.329×10^{-1}	2.802×10^{-3}	1.995×10^{-1}
0.900	1.843	1.101×10^{-2}	1.854	3.866×10^{-1}	3.114×10^{-3}	2.653×10^{-1}
1.000	1.827	1.216×10^{-2}	1.840	4.407×10^{-1}	3.430×10^{-3}	3.317×10^{-1}
1.500	1.798	1.847×10^{-2}	1.817	7.149×10^{-1}	5.075×10^{-3}	6.519×10^{-1}
2.000	1.800	2.549×10^{-2}	1.826	9.897×10^{-1}	6.818×10^{-3}	9.362×10^{-1}
3.000	1.822	4.094×10^{-2}	1.863	1.532	1.052×10^{-2}	1.403
4.000	1.847	5.771×10^{-2}	1.904	2.063	1.442×10^{-2}	1.772
5.000	1.869	7.545×10^{-2}	1.944	2.583	1.845×10^{-2}	2.075
6.000	1.888	9.393×10^{-2}	1.982	3.092	2.256×10^{-2}	2.331
7.000	1.905	1.130×10^{-1}	2.018	3.592	2.672×10^{-2}	2.554
8.000	1.920	1.326×10^{-1}	2.052	4.084	3.092×10^{-2}	2.751
9.000	1.933	1.527×10^{-1}	2.085	4.567	3.514×10^{-2}	2.929
10.000	1.945	1.731×10^{-1}	2.118	5.043	3.938×10^{-2}	3.090
15.000	1.990	2.792×10^{-1}	2.269	7.322	6.045×10^{-2}	3.733
20.000	2.021	3.901×10^{-1}	2.411	9.459	8.100×10^{-2}	4.213
30.000	2.062	6.197×10^{-1}	2.682	1.339×10	1.198×10^{-1}	4.926
40.000	2.090	8.556×10^{-1}	2.946	1.694×10	1.552×10^{-1}	5.455
50.000	2.111	1.095	3.206	2.020×10	1.875×10^{-1}	5.875

TABLE M.2y

Electrons in Soft Tissue (ICRP)

$\langle Z/A \rangle$: 0.55121; ρ: 1.000 g cm^{-3}; I: 72.3 eV; Composition $(Z - f_w)$: 1–0.104472; 6–0.232190; 7–0.024880; 8–0.630238;11–0.001130; 12–0.000130; 15–0.001330; 16–0.001990; 17–0.001340; 19–0.001990; 20–0.000230; 26–0.000050; 30–0.000030

Kinetic Energy (MeV)	Stopping Power (MeV cm^{2}g^{-1})			CSDA Range (g cm^{-2})	Radiation Yield	Density-Effect Parameter δ
	Collision	Radiative	Total			
0.010	2.257×10	3.680×10^{-3}	2.257×10	2.512×10^{-4}	8.894×10^{-4}	0.000
0.015	1.646×10	3.721×10^{-3}	1.647×10	5.144×10^{-4}	1.243×10^{-4}	0.000
0.020	1.317×10	3.740×10^{-3}	1.317×10	8.565×10^{-4}	1.571×10^{-4}	0.000
0.030	9.643	3.762×10^{-3}	9.647	1.757×10^{-3}	2.173×10^{-4}	0.000
0.040	7.767	3.785×10^{-3}	7.771	2.921×10^{-3}	2.728×10^{-4}	0.000
0.050	6.593	3.811×10^{-3}	6.597	4.324×10^{-3}	3.248×10^{-4}	0.000
0.060	5.787	3.842×10^{-3}	5.791	5.946×10^{-3}	3.742×10^{-4}	0.000
0.070	5.198	3.877×10^{-3}	5.202	7.772×10^{-3}	4.214×10^{-4}	0.000
0.080	4.749	3.916×10^{-3}	4.753	9.786×10^{-3}	4.668×10^{-4}	0.000
0.090	4.394	3.958×10^{-3}	4.398	1.198×10^{-2}	5.107×10^{-4}	0.000
0.100	4.107	4.002×10^{-3}	4.111	1.433×10^{-2}	5.533×10^{-4}	0.000
0.150	3.230	4.258×10^{-3}	3.235	2.822×10^{-2}	7.515×10^{-4}	0.000
0.200	2.786	4.551×10^{-3}	2.791	4.497×10^{-2}	9.324×10^{-4}	0.000
0.300	2.349	5.232×10^{-3}	2.354	8.441×10^{-2}	1.265×10^{-3}	0.000
0.400	2.142	6.018×10^{-3}	2.148	1.291×10^{-1}	1.576×10^{-3}	0.000
0.500	2.027	6.895×10^{-3}	2.034	1.770×10^{-1}	1.880×10^{-3}	7.184×10^{-3}
0.600	1.955	7.844×10^{-3}	1.963	2.271×10^{-1}	2.182×10^{-3}	5.251×10^{-2}
0.700	1.908	8.854×10^{-3}	1.917	2.787×10^{-1}	2.485×10^{-3}	1.066×10^{-1}
0.800	1.876	9.914×10^{-3}	1.886	3.314×10^{-1}	2.792×10^{-3}	1.657×10^{-1}
0.900	1.854	1.102×10^{-2}	1.865	3.847×10^{-1}	3.102×10^{-3}	2.276×10^{-1}
1.000	1.839	1.218×10^{-2}	1.851	4.385×10^{-1}	3.416×10^{-3}	2.908×10^{-1}
1.500	1.810	1.849×10^{-2}	1.829	7.110×10^{-1}	5.051×10^{-3}	6.012×10^{-1}
2.000	1.812	2.552×10^{-2}	1.838	9.839×10^{-1}	6.784×10^{-3}	8.807×10^{-1}
3.000	1.835	4.099×10^{-2}	1.876	1.523	1.047×10^{-2}	1.343
4.000	1.859	5.778×10^{-2}	1.917	2.050	1.434×10^{-2}	1.710
5.000	1.881	7.553×10^{-2}	1.957	2.566	1.835×10^{-2}	2.012
6.000	1.901	9.404×10^{-2}	1.995	3.073	2.243×10^{-2}	2.268
7.000	1.918	1.132×10^{-1}	2.031	3.569	2.658×10^{-2}	2.491
8.000	1.932	1.328×10^{-1}	2.065	4.058	3.076×10^{-2}	2.688
9.000	1.946	1.528×10^{-1}	2.099	4.538	3.496×10^{-2}	2.865
10.000	1.958	1.733×10^{-1}	2.131	5.011	3.917×10^{-2}	3.025
15.000	2.003	2.796×10^{-1}	2.283	7.276	6.014×10^{-2}	3.666
20.000	2.035	3.905×10^{-1}	2.425	9.401	8.060×10^{-2}	4.144
30.000	2.077	6.204×10^{-1}	2.697	1.331×10	1.192×10^{-1}	4.853
40.000	2.105	8.565×10^{-1}	2.961	1.684×10	1.545×10^{-1}	5.379
50.000	2.126	1.097	3.222	2.008×10	1.867×10^{-1}	5.799

TABLE M.2z

Electrons in Soft Tissue (ICRU 4-Component)

⟨Z/A⟩: 0.54975; ρ: 1.000 g cm^{-3}; I: 74.9 eV; Composition $(Z-f_w)$: 1–0.101172; 6–0.111000; 7–0.026000; 8–0.761828

Kinetic Energy (MeV)	Stopping Power (MeV cm²g⁻¹)			CSDA Range (g cm⁻²)	Radiation Yield	Density-Effect Parameter δ
	Collision	Radiative	Total			
0.010	2.235×10	3.795×10^{-3}	2.235×10	2.538×10^{-4}	9.262×10^{-4}	0.000
0.015	1.631×10	3.837×10^{-3}	1.632×10	5.195×10^{-4}	1.294×10^{-4}	0.000
0.020	1.305×10	3.855×10^{-3}	1.305×10	8.647×10^{-4}	1.635×10^{-4}	0.000
0.030	9.562	3.876×10^{-3}	9.566	1.773×10^{-3}	2.260×10^{-4}	0.000
0.040	7.704	3.897×10^{-3}	7.708	2.946×10^{-3}	2.835×10^{-4}	0.000
0.050	6.541	3.923×10^{-3}	6.545	4.361×10^{-3}	3.374×10^{-4}	0.000
0.060	5.742	3.953×10^{-3}	5.746	5.996×10^{-3}	3.885×10^{-4}	0.000
0.070	5.158	3.989×10^{-3}	5.162	7.836×10^{-3}	4.374×10^{-4}	0.000
0.080	4.712	4.028×10^{-3}	4.717	9.866×10^{-3}	4.844×10^{-4}	0.000
0.090	4.361	4.070×10^{-3}	4.365	1.207×10^{-2}	5.299×10^{-4}	0.000
0.100	4.076	4.116×10^{-3}	4.081	1.444×10^{-2}	5.740×10^{-4}	0.000
0.150	3.207	4.377×10^{-3}	3.211	2.844×10^{-2}	7.790×10^{-4}	0.000
0.200	2.767	4.677×10^{-3}	2.771	4.530×10^{-2}	9.660×10^{-4}	0.000
0.300	2.333	5.374×10^{-3}	2.338	8.502×10^{-2}	1.309×10^{-3}	0.000
0.400	2.128	6.179×10^{-3}	2.134	1.300×10^{-1}	1.631×10^{-3}	0.000
0.500	2.015	7.077×10^{-3}	2.022	1.783×10^{-1}	1.944×10^{-3}	0.000
0.600	1.944	8.050×10^{-3}	1.952	2.286×10^{-1}	2.255×10^{-3}	3.374×10^{-2}
0.700	1.898	9.084×10^{-3}	1.907	2.805×10^{-1}	2.568×10^{-3}	8.114×10^{-2}
0.800	1.867	1.017×10^{-2}	1.877	3.334×10^{-1}	2.883×10^{-3}	1.351×10^{-1}
0.900	1.846	1.131×10^{-2}	1.857	3.870×10^{-1}	3.202×10^{-3}	1.930×10^{-1}
1.000	1.831	1.249×10^{-2}	1.843	4.410×10^{-1}	3.525×10^{-3}	2.529×10^{-1}
1.500	1.803	1.896×10^{-2}	1.822	7.145×10^{-1}	5.205×10^{-3}	5.542×10^{-1}
2.000	1.806	2.615×10^{-2}	1.832	9.884×10^{-1}	6.985×10^{-3}	8.304×10^{-1}
3.000	1.828	4.199×10^{-2}	1.870	1.529	1.077×10^{-2}	1.292
4.000	1.853	5.919×10^{-2}	1.912	2.058	1.475×10^{-2}	1.660
5.000	1.875	7.736×10^{-2}	1.952	2.575	1.885×10^{-2}	1.963
6.000	1.894	9.629×10^{-2}	1.990	3.083	2.305×10^{-2}	2.220
7.000	1.911	1.159×10^{-1}	2.026	3.581	2.729×10^{-2}	2.443
8.000	1.925	1.359×10^{-1}	2.061	4.070	3.158×10^{-2}	2.640
9.000	1.939	1.565×10^{-1}	2.095	4.551	3.588×10^{-2}	2.817
10.000	1.950	1.774×10^{-1}	2.128	5.025	4.020×10^{-2}	2.978
15.000	1.996	2.861×10^{-1}	2.282	7.292	6.165×10^{-2}	3.616
20.000	2.028	3.996×10^{-1}	2.427	9.416	8.255×10^{-2}	4.090
30.000	2.070	6.347×10^{-1}	2.705	1.331×10	1.219×10^{-1}	4.792
40.000	2.099	8.761×10^{-1}	2.975	1.684×10	1.577×10^{-1}	5.314
50.000	2.120	1.122	3.241	2.006×10	1.904×10^{-1}	5.731

TABLE M.3 PHOTON INTERACTION COEFFICIENTS

Compiled by Alan Nahum[*]

Table M.3a through Table M.3z give interaction coefficients for photons in a number of materials of medical interest. The materials are identical to those included in Table M.2. Elements have been placed first (a–f), then compounds (g–i), then mixtures (j–q) and body tissues (r–z), in alphabetical order. The photon energies extend from 0.001 MeV (1 keV) up to 50 MeV, covering the range of interest in radiotherapy. The values for the partial coefficients (coherent, σ_{coh}/ρ; Compton, σ_C/ρ; photoelectric, τ/ρ; pair + triplet production, κ/ρ) and the total attenuation coefficient, μ/ρ, have been obtained using the XCOM computer code, accessed via http://physics.nist.gov/PhysRefData/Xcom/Text/XCOM.html where full details of the calculational procedures are given (Berger et al. 2005). The mass energy-transfer coefficient, μ_{tr}/ρ, mass energy-absorption coefficient, μ_{en}/ρ, and (1-g) were supplied by Steve Seltzer of NIST; these were evaluated using the procedures described in Seltzer (1993) and Hubbell and Seltzer (1995).

Each table also contains the mass density ρ and the atomic composition (by weight) of each substance. The units are $cm^2 g^{-1}$; to convert to $m^2 kg^{-1}$ the numbers should be multiplied by 0.1. To obtain linear coefficients the values in the tables must be multiplied by the density ρ. For substances not included here, ICRU (1989) and Hubbell and Seltzer (1995) can be consulted.

The various photon interaction processes are covered in detail in Chapter 4; the macroscopic quantities μ/ρ, μ_{tr}/ρ and μ_{en}/ρ are defined in Section 4.4.1 and Section 4.4.2, respectively.

---

[*] We acknowledge the generous assistance of Steve Seltzer, NIST, in providing values of the mass energy-transfer coefficient μ_{tr}/ρ, mass energy-absorption coefficient μ_{en}/ρ, and (1-g) for all the substances in the tables.

TABLE M.3a

Photons in Carbon (Graphite)

ρ:1.700 (g cm^{-3}), Composition ($Z-f_w$): 6−1.0000

K, L, M Edges	Energy (MeV)	Coherent σ_{coh}/ρ	Compton σ_C/ρ	Photoelectric τ/ρ	Pair + Triplet κ/ρ	Total Attenuation μ/ρ	Energy-Transfer μ_{tr}/ρ	Energy-Absorption μ_{en}/ρ	$(1-g)$
					Mass Coefficients (cm^2 g^{-1})				
	0.0010	1.08	1.26×10^{-2}	2.21×10^{3}	0.00	2.21×10^{3}	2.209×10^{3}	2.209×10^{3}	1.0000
	0.0015	9.59×10^{-1}	2.51×10^{-2}	6.99×10^{2}	0.00	7.00×10^{2}	6.990×10^{2}	6.990×10^{2}	0.9999
	0.0020	8.32×10^{-1}	3.86×10^{-2}	3.02×10^{2}	0.00	3.03×10^{2}	3.017×10^{2}	3.016×10^{2}	0.9999
	0.0030	6.13×10^{-1}	6.41×10^{-2}	8.96×10	0.00	9.03×10	8.964×10	8.963×10	0.9999
	0.0040	4.60×10^{-1}	8.45×10^{-2}	3.72×10	0.00	3.78×10	3.724×10	3.723×10	0.9999
	0.0050	3.59×10^{-1}	9.95×10^{-2}	1.87×10	0.00	1.91×10	1.866×10	1.866×10	0.9999
	0.0060	2.92×10^{-1}	1.10×10^{-1}	1.05×10	0.00	1.09×10	1.055×10	1.054×10	0.9999
	0.0080	2.10×10^{-1}	1.25×10^{-1}	4.24	0.00	4.58	4.243	4.243	0.9998
	0.0100	1.62×10^{-1}	1.35×10^{-1}	2.08	0.00	2.37	2.079	2.078	0.9998
	0.0150	9.79×10^{-2}	1.51×10^{-1}	5.59×10^{-1}	0.00	8.07×10^{-1}	5.628×10^{-1}	5.627×10^{-1}	0.9998
	0.0200	6.48×10^{-2}	1.60×10^{-1}	2.18×10^{-1}	0.00	4.42×10^{-1}	2.239×10^{-1}	2.238×10^{-1}	0.9998
	0.0300	3.36×10^{-2}	1.65×10^{-1}	5.71×10^{-2}	0.00	2.56×10^{-1}	6.616×10^{-2}	6.614×10^{-2}	0.9997
	0.0400	2.05×10^{-2}	1.65×10^{-1}	2.19×10^{-2}	0.00	2.08×10^{-1}	3.344×10^{-2}	3.343×10^{-2}	0.9997
	0.0500	1.37×10^{-2}	1.63×10^{-1}	1.04×10^{-2}	0.00	1.87×10^{-1}	2.398×10^{-2}	2.397×10^{-2}	0.9997
	0.0600	9.81×10^{-3}	1.60×10^{-1}	5.67×10^{-3}	0.00	1.75×10^{-1}	2.099×10^{-2}	2.098×10^{-2}	0.9997
	0.0800	5.71×10^{-3}	1.53×10^{-1}	2.17×10^{-3}	0.00	1.61×10^{-1}	2.038×10^{-2}	2.037×10^{-2}	0.9998
	0.1000	3.72×10^{-3}	1.47×10^{-1}	1.03×10^{-3}	0.00	1.51×10^{-1}	2.148×10^{-2}	2.147×10^{-2}	0.9997
	0.1500	1.68×10^{-3}	1.33×10^{-1}	2.71×10^{-4}	0.00	1.35×10^{-1}	2.450×10^{-2}	2.449×10^{-2}	0.9996
	0.2000	9.54×10^{-4}	1.22×10^{-1}	1.06×10^{-4}	0.00	1.23×10^{-1}	2.657×10^{-2}	2.655×10^{-2}	0.9995
	0.3000	4.26×10^{-4}	1.06×10^{-1}	2.98×10^{-5}	0.00	1.07×10^{-1}	2.872×10^{-2}	2.870×10^{-2}	0.9993
	0.4000	2.40×10^{-4}	9.52×10^{-2}	1.27×10^{-5}	0.00	9.55×10^{-2}	2.953×10^{-2}	2.950×10^{-2}	0.9991
	0.5000	1.54×10^{-4}	8.70×10^{-2}	6.84×10^{-6}	0.00	8.72×10^{-2}	2.973×10^{-2}	2.969×10^{-2}	0.9989
	0.6000	1.07×10^{-4}	8.05×10^{-2}	4.25×10^{-6}	0.00	8.06×10^{-2}	2.960×10^{-2}	2.956×10^{-2}	0.9987
	0.8000	6.02×10^{-5}	7.07×10^{-2}	2.14×10^{-6}	0.00	7.08×10^{-2}	2.890×10^{-2}	2.885×10^{-2}	0.9982
	1.0000	3.85×10^{-5}	6.36×10^{-2}	1.33×10^{-6}	0.00	6.36×10^{-2}	2.798×10^{-2}	2.792×10^{-2}	0.9978
	1.2500	2.47×10^{-5}	5.69×10^{-2}	8.35×10^{-7}	0.00	5.69×10^{-2}	2.676×10^{-2}	2.669×10^{-2}	0.9972
	1.5000	1.71×10^{-5}	5.17×10^{-2}	6.06×10^{-7}	0.00	5.18×10^{-2}	2.560×10^{-2}	2.551×10^{-2}	0.9966
	2.0000	9.63×10^{-6}	4.41×10^{-2}	3.83×10^{-7}	0.00	4.44×10^{-2}	2.357×10^{-2}	2.345×10^{-2}	0.9952
	3.0000	4.28×10^{-6}	3.47×10^{-2}	2.15×10^{-7}	0.00	3.56×10^{-2}	2.064×10^{-2}	2.048×10^{-2}	0.9922
	4.0000	2.41×10^{-6}	2.89×10^{-2}	1.48×10^{-7}	0.00	3.05×10^{-2}	1.871×10^{-2}	1.849×10^{-2}	0.9887
	5.0000	1.54×10^{-6}	2.50×10^{-2}	1.12×10^{-7}	0.00	2.71×10^{-2}	1.736×10^{-2}	1.710×10^{-2}	0.9850
	6.0000	1.07×10^{-6}	2.21×10^{-2}	9.03×10^{-8}	0.00	2.47×10^{-2}	1.638×10^{-2}	1.607×10^{-2}	0.9811
	8.0000	6.02×10^{-7}	1.81×10^{-2}	6.49×10^{-8}	0.00	2.15×10^{-2}	1.509×10^{-2}	1.468×10^{-2}	0.9731

10.0000	3.85×10^{-7}	1.54×10^{-2}	5.06×10^{-8}	0.00	1.96×10^{-2}	1.429×10^{-2}	1.380×10^{-2}	0.9651
15.0000	1.71×10^{-7}	1.14×10^{-2}	3.25×10^{-8}	0.00	1.70×10^{-2}	1.330×10^{-2}	1.258×10^{-2}	0.9458
20.0000	9.63×10^{-8}	9.14×10^{-3}	2.40×10^{-8}	0.00	1.58×10^{-2}	1.291×10^{-2}	1.198×10^{-2}	0.9278
30.0000	4.28×10^{-8}	6.65×10^{-3}	1.57×10^{-8}	0.00	1.47×10^{-2}	1.276×10^{-2}	1.142×10^{-2}	0.8949
40.0000	2.41×10^{-8}	5.29×10^{-3}	1.17×10^{-8}	0.00	1.44×10^{-2}	1.287×10^{-2}	1.113×10^{-2}	0.8654
50.0000	1.54×10^{-8}	4.41×10^{-3}	9.28×10^{-9}	0.00	1.43×10^{-2}	1.304×10^{-2}	1.093×10^{-2}	0.8385

TABLE M.3b

Photons in Aluminum

ρ: 2.699 (g cm^{-3}), Composition ($Z-f_w$): 13–1.0000

K, L, M Edges	Energy (MeV)	Coherent σ_{coh}/ρ	Compton σ_C/ρ	Photoelectric τ/ρ	Pair + Triplet κ/ρ	Total Attenuation μ/ρ	Energy-Transfer μ_{tr}/ρ	Energy-Absorption μ_{en}/ρ	$(1-g)$
	0.0010	2.26	1.43×10^{-2}	1.18×10^{3}	0.00	1.19×10^{3}	1.183×10^{3}	1.183×10^{3}	1
	0.0015	2.04	2.48×10^{-2}	4.00×10^{2}	0.00	4.02×10^{2}	4.002×10^{2}	4.001×10^{2}	1
	0.001560	2.01	2.59×10^{-2}	3.60×10^{2}	0.00	3.62×10^{2}	3.600×10^{2}	3.600×10^{2}	1
13 K	0.001560	2.01	2.59×10^{-2}	3.96×10^{3}	0.00	3.96×10^{3}	3.829×10^{3}	3.829×10^{3}	1
	0.0020	1.84	3.37×10^{-2}	2.26×10^{3}	0.00	2.26×10^{3}	2.204×10^{3}	2.204×10^{3}	1
	0.0030	1.52	4.73×10^{-2}	7.87×10^{2}	0.00	7.88×10^{2}	7.732×10^{2}	7.732×10^{2}	1
	0.0040	1.30	5.81×10^{-2}	3.59×10^{2}	0.00	3.60×10^{2}	3.546×10^{2}	3.545×10^{2}	0.9999
	0.0050	1.12	6.79×10^{-2}	1.92×10^{2}	0.00	1.93×10^{2}	1.903×10^{2}	1.902×10^{2}	0.9998
	0.0060	9.64×10^{-1}	7.70×10^{-2}	1.14×10^{2}	0.00	1.15×10^{2}	1.133×10^{2}	1.133×10^{2}	0.9997
	0.0080	7.23×10^{-1}	9.29×10^{-2}	4.95×10	0.00	5.03×10	4.920×10	4.918×10	0.9995
	0.0100	5.51×10^{-1}	1.06×10^{-1}	2.56×10	0.00	2.62×10	2.544×10	2.543×10	0.9994
	0.0150	3.14×10^{-1}	1.27×10^{-1}	7.51	0.00	7.96	7.493	7.487	0.9992
	0.0200	2.05×10^{-1}	1.37×10^{-1}	3.10	0.00	3.44	3.097	3.094	0.9990
	0.0300	1.10×10^{-1}	1.46×10^{-1}	8.72×10^{-1}	0.00	1.13	8.790×10^{-1}	8.779×10^{-1}	0.9988
	0.0400	6.86×10^{-2}	1.49×10^{-1}	3.50×10^{-1}	0.00	5.68×10^{-1}	3.606×10^{-1}	3.601×10^{-1}	0.9986
	0.0500	4.68×10^{-2}	1.50×10^{-1}	1.72×10^{-1}	0.00	3.68×10^{-1}	1.843×10^{-1}	1.840×10^{-1}	0.9985
	0.0600	3.39×10^{-2}	1.48×10^{-1}	9.56×10^{-2}	0.00	2.78×10^{-1}	1.101×10^{-1}	1.099×10^{-1}	0.9984
	0.0800	2.00×10^{-2}	1.44×10^{-1}	3.78×10^{-2}	0.00	2.02×10^{-1}	5.520×10^{-2}	5.511×10^{-2}	0.9984
	0.1000	1.32×10^{-2}	1.39×10^{-1}	1.84×10^{-2}	0.00	1.70×10^{-1}	3.801×10^{-2}	3.795×10^{-2}	0.9984
	0.1500	6.12×10^{-3}	1.27×10^{-1}	4.99×10^{-3}	0.00	1.38×10^{-1}	2.832×10^{-2}	2.827×10^{-2}	0.9984
	0.2000	3.50×10^{-3}	1.17×10^{-1}	2.00×10^{-3}	0.00	1.22×10^{-1}	2.750×10^{-2}	2.745×10^{-2}	0.9981
	0.3000	1.58×10^{-3}	1.02×10^{-1}	5.74×10^{-4}	0.00	1.04×10^{-1}	2.823×10^{-2}	2.816×10^{-2}	0.9976
	0.4000	8.93×10^{-4}	9.16×10^{-2}	2.48×10^{-4}	0.00	9.28×10^{-2}	2.870×10^{-2}	2.862×10^{-2}	0.9971
	0.5000	5.73×10^{-4}	8.37×10^{-2}	1.34×10^{-4}	0.00	8.45×10^{-2}	2.878×10^{-2}	2.868×10^{-2}	0.9966
	0.6000	3.99×10^{-4}	7.75×10^{-2}	8.40×10^{-5}	0.00	7.80×10^{-2}	2.863×10^{-2}	2.851×10^{-2}	0.9961
	0.8000	2.25×10^{-4}	6.81×10^{-2}	4.25×10^{-5}	0.00	6.84×10^{-2}	2.792×10^{-2}	2.778×10^{-2}	0.9951
	1.0000	1.44×10^{-4}	6.13×10^{-2}	2.64×10^{-5}	0.00	6.15×10^{-2}	2.702×10^{-2}	2.686×10^{-2}	0.9941
	1.2500	9.21×10^{-5}	5.48×10^{-2}	1.69×10^{-5}	3.13×10^{-5}	5.50×10^{-2}	2.583×10^{-2}	2.565×10^{-2}	0.9928
	1.5000	6.39×10^{-5}	4.98×10^{-2}	1.22×10^{-5}	1.71×10^{-4}	5.01×10^{-2}	2.472×10^{-2}	2.451×10^{-2}	0.9915
	2.0000	3.60×10^{-5}	4.25×10^{-2}	7.63×10^{-6}	6.75×10^{-4}	4.32×10^{-2}	2.291×10^{-2}	2.266×10^{-2}	0.9887
	3.0000	1.60×10^{-5}	3.35×10^{-2}	4.22×10^{-6}	1.93×10^{-3}	3.54×10^{-2}	2.059×10^{-2}	2.024×10^{-2}	0.9827

Mass Coefficients (cm^2 g^{-1})

4.0000	9.00×10^{-6}	2.79×10^{-2}	2.88×10^{-6}	3.15×10^{-3}	3.11×10^{-2}	1.928×10^{-2}	1.882×10^{-2}	0.9761
5.0000	5.76×10^{-6}	2.41×10^{-2}	2.18×10^{-6}	4.25×10^{-3}	2.84×10^{-2}	1.852×10^{-2}	1.795×10^{-2}	0.9692
6.0000	4.00×10^{-6}	2.13×10^{-2}	1.74×10^{-6}	5.25×10^{-3}	2.66×10^{-2}	1.807×10^{-2}	1.739×10^{-2}	0.9623
8.0000	2.25×10^{-6}	1.74×10^{-2}	1.24×10^{-6}	6.94×10^{-3}	2.44×10^{-2}	1.768×10^{-2}	1.678×10^{-2}	0.9488
10.0000	1.44×10^{-6}	1.48×10^{-2}	9.66×10^{-7}	8.34×10^{-3}	2.32×10^{-2}	1.763×10^{-2}	1.650×10^{-2}	0.9357
15.0000	6.40×10^{-7}	1.10×10^{-2}	6.19×10^{-7}	1.09×10^{-2}	2.19×10^{-2}	1.801×10^{-2}	1.631×10^{-2}	0.9052
20.0000	3.60×10^{-7}	8.82×10^{-3}	4.55×10^{-7}	1.29×10^{-2}	2.17×10^{-2}	1.861×10^{-2}	1.633×10^{-2}	0.8775
30.0000	1.60×10^{-7}	6.42×10^{-3}	2.97×10^{-7}	1.55×10^{-2}	2.20×10^{-2}	1.980×10^{-2}	1.641×10^{-2}	0.8284
40.0000	9.00×10^{-8}	5.10×10^{-3}	2.20×10^{-7}	1.74×10^{-2}	2.25×10^{-2}	2.084×10^{-2}	1.637×10^{-2}	0.7857
50.0000	5.76×10^{-8}	4.26×10^{-3}	1.75×10^{-7}	1.88×10^{-2}	2.31×10^{-2}	2.169×10^{-2}	1.622×10^{-2}	0.7478

TABLE M.3c

Photons in Silicon

ρ: 2.330 (g cm^{-3}), Composition ($Z-f_w$): 14−1.0000

K, L, M Edges	Energy (MeV)	Mass Coefficients (cm^2 g^{-1})							
		Coherent σ_{coh}/ρ	Compton σ_C/ρ	Photoelectric τ/ρ	Pair+Triplet κ/ρ	Total Attenuation μ/ρ	Energy-Transfer μ_{tr}/ρ	Energy-Absorption μ_{en}/ρ	$(1-g)$
	0.0010	2.53	1.32×10^{-2}	1.57×10^{3}	0.00	1.57×10^{3}	1.567×10^{3}	1.567×10^{3}	1
	0.0015	2.29	2.39×10^{-2}	5.33×10^{2}	0.00	5.36×10^{2}	5.332×10^{2}	5.331×10^{2}	1
	0.001839	2.12	3.08×10^{-2}	3.07×10^{2}	0.00	3.09×10^{2}	3.070×10^{2}	3.070×10^{2}	0.9999
14 K	0.001839	2.12	3.08×10^{-2}	3.19×10^{3}	0.00	3.19×10^{3}	3.059×10^{3}	3.059×10^{3}	1
	0.0020	2.05	3.39×10^{-2}	2.77×10^{3}	0.00	2.78×10^{3}	2.669×10^{3}	2.669×10^{3}	1
	0.0030	1.67	4.96×10^{-2}	9.77×10^{2}	0.00	9.78×10^{2}	9.517×10^{2}	9.516×10^{2}	1
	0.0040	1.40	6.13×10^{-2}	4.51×10^{2}	0.00	4.53×10^{2}	4.427×10^{2}	4.427×10^{2}	0.9999
	0.0050	1.21	7.11×10^{-2}	2.44×10^{2}	0.00	2.45×10^{2}	2.400×10^{2}	2.400×10^{2}	0.9998
	0.0060	1.05	7.98×10^{-2}	1.46×10^{2}	0.00	1.47×10^{2}	1.440×10^{2}	1.439×10^{2}	0.9997
	0.0080	8.04×10^{-1}	9.51×10^{-2}	6.38×10	0.00	6.47×10	6.316×10	6.313×10	0.9996
	0.0100	6.22×10^{-1}	1.08×10^{-1}	3.31×10	0.00	3.39×10	3.290×10	3.289×10	0.9994
	0.0150	3.59×10^{-1}	1.29×10^{-1}	9.85	0.00	1.03×10	9.802	9.794	0.9992
	0.0200	2.34×10^{-1}	1.40×10^{-1}	4.09	0.00	4.46	4.080	4.076	0.9990
	0.0300	1.25×10^{-1}	1.50×10^{-1}	1.16	0.00	1.44	1.166	1.165	0.9987
	0.0400	7.89×10^{-2}	1.53×10^{-1}	4.69×10^{-1}	0.00	7.01×10^{-1}	4.789×10^{-1}	4.782×10^{-1}	0.9985
	0.0500	5.40×10^{-2}	1.54×10^{-1}	2.31×10^{-1}	0.00	4.38×10^{-1}	2.434×10^{-1}	2.430×10^{-1}	0.9983
	0.0600	3.92×10^{-2}	1.53×10^{-1}	1.29×10^{-1}	0.00	3.21×10^{-1}	1.436×10^{-1}	1.434×10^{-1}	0.9982
	0.0800	2.32×10^{-2}	1.48×10^{-1}	5.12×10^{-2}	0.00	2.23×10^{-1}	6.909×10^{-2}	6.896×10^{-2}	0.9981
	0.1000	1.54×10^{-2}	1.43×10^{-1}	2.50×10^{-2}	0.00	1.84×10^{-1}	4.522×10^{-2}	4.513×10^{-2}	0.9981
	0.1500	7.13×10^{-3}	1.31×10^{-1}	6.81×10^{-3}	0.00	1.45×10^{-1}	3.092×10^{-2}	3.086×10^{-2}	0.9981
	0.2000	4.08×10^{-3}	1.21×10^{-1}	2.74×10^{-3}	0.00	1.28×10^{-1}	2.911×10^{-2}	2.905×10^{-2}	0.9979
	0.3000	1.84×10^{-3}	1.06×10^{-1}	7.88×10^{-4}	0.00	1.08×10^{-1}	2.940×10^{-2}	2.932×10^{-2}	0.9973
	0.4000	1.04×10^{-3}	9.48×10^{-2}	3.41×10^{-4}	0.00	9.61×10^{-2}	2.978×10^{-2}	2.968×10^{-2}	0.9967
	0.5000	6.70×10^{-4}	8.66×10^{-2}	1.85×10^{-4}	0.00	8.75×10^{-2}	2.983×10^{-2}	2.971×10^{-2}	0.9962
	0.6000	4.66×10^{-4}	8.02×10^{-2}	1.16×10^{-4}	0.00	8.08×10^{-2}	2.964×10^{-2}	2.951×10^{-2}	0.9956
	0.8000	2.62×10^{-4}	7.05×10^{-2}	5.85×10^{-5}	0.00	7.08×10^{-2}	2.890×10^{-2}	2.875×10^{-2}	0.9945
	1.0000	1.68×10^{-4}	6.34×10^{-2}	3.64×10^{-5}	0.00	6.36×10^{-2}	2.796×10^{-2}	2.778×10^{-2}	0.9935
	1.2500	1.08×10^{-4}	5.67×10^{-2}	2.33×10^{-5}	3.52×10^{-5}	5.69×10^{-2}	2.673×10^{-2}	2.652×10^{-2}	0.9921
	1.5000	7.47×10^{-5}	5.15×10^{-2}	1.68×10^{-5}	1.91×10^{-4}	5.18×10^{-2}	2.559×10^{-2}	2.535×10^{-2}	0.9907
	2.0000	4.20×10^{-5}	4.40×10^{-2}	1.05×10^{-5}	7.53×10^{-4}	4.48×10^{-2}	2.374×10^{-2}	2.345×10^{-2}	0.9877
	3.0000	1.87×10^{-5}	3.46×10^{-2}	5.80×10^{-6}	2.15×10^{-3}	3.68×10^{-2}	2.141×10^{-2}	2.101×10^{-2}	0.9812

4.0000	1.05×10^{-5}	2.89×10^{-2}	3.95×10^{-6}	3.51×10^{-3}	3.24×10^{-2}	2.015×10^{-2}	1.963×10^{-2}	0.9742
5.0000	6.73×10^{-6}	2.49×10^{-2}	2.98×10^{-6}	4.73×10^{-3}	2.97×10^{-2}	1.943×10^{-2}	1.878×10^{-2}	0.967
6.0000	4.67×10^{-6}	2.20×10^{-2}	2.39×10^{-6}	5.83×10^{-3}	2.79×10^{-2}	1.903×10^{-2}	1.827×10^{-2}	0.9598
8.0000	2.63×10^{-6}	1.80×10^{-2}	1.70×10^{-6}	7.71×10^{-3}	2.57×10^{-2}	1.875×10^{-2}	1.773×10^{-2}	0.9456
10.0000	1.68×10^{-6}	1.54×10^{-2}	1.32×10^{-6}	9.25×10^{-3}	2.46×10^{-2}	1.881×10^{-2}	1.753×10^{-2}	0.9319
15.0000	7.48×10^{-7}	1.14×10^{-2}	8.46×10^{-7}	1.22×10^{-2}	2.35×10^{-2}	1.940×10^{-2}	1.746×10^{-2}	0.9003
20.0000	4.21×10^{-7}	9.12×10^{-3}	6.22×10^{-7}	1.42×10^{-2}	2.34×10^{-2}	2.015×10^{-2}	1.757×10^{-2}	0.8716
30.0000	1.87×10^{-7}	6.64×10^{-3}	4.06×10^{-7}	1.72×10^{-2}	2.38×10^{-2}	2.158×10^{-2}	1.772×10^{-2}	0.8208
40.0000	1.05×10^{-7}	5.28×10^{-3}	3.01×10^{-7}	1.93×10^{-2}	2.45×10^{-2}	2.279×10^{-2}	1.770×10^{-2}	0.7768
50.0000	6.73×10^{-8}	4.41×10^{-3}	2.39×10^{-7}	2.08×10^{-2}	2.52×10^{-2}	2.378×10^{-2}	1.755×10^{-2}	0.7379

TABLE M.3d

Photons in Copper

ρ: 8.960 (g cm^{-3}), Composition ($Z-f_w$): 29–1.000

K, L, M Edges	Energy (MeV)	Mass Coefficients (cm^2 g^{-1})							(1−g)
		Coherent σ_{coh}/ρ	Compton σ_C/ρ	Photoelectric τ/ρ	Pair+Triplet κ/ρ	Total Attenuation μ/ρ	Energy-Transfer μ_{tr}/ρ	Energy-Absorption μ_{en}/ρ	
	0.0010	5.05	5.91×10^{-3}	1.06×10^4	0.00	1.06×10^4	1.049×10^4	1.049×10^4	1
	0.0010	5.03	6.36×10^{-3}	9.33×10^3	0.00	9.33×10^3	9.241×10^3	9.241×10^3	1
	0.001096	5.01	6.84×10^{-3}	8.24×10^3	0.00	8.25×10^3	8.186×10^3	8.186×10^3	1
29 L1	0.001096	5.01	6.84×10^{-3}	9.34×10^3	0.00	9.35×10^3	9.282×10^3	9.282×10^3	1
	0.0015	4.81	1.09×10^{-2}	4.41×10^3	0.00	4.42×10^3	4.393×10^3	4.393×10^3	1
	0.0020	4.53	1.59×10^{-2}	2.15×10^3	0.00	2.15×10^3	2.142×10^3	2.142×10^3	0.9999
	0.0030	3.95	2.59×10^{-2}	7.45×10^2	0.00	7.49×10^2	7.431×10^2	7.430×10^2	0.9999
	0.0040	3.40	3.53×10^{-2}	3.44×10^2	0.00	3.47×10^2	3.433×10^2	3.432×10^2	0.9998
	0.0050	2.91	4.39×10^{-2}	1.87×10^2	0.00	1.90×10^2	1.867×10^2	1.866×10^2	0.9997
	0.0060	2.50	5.18×10^{-2}	1.13×10^2	0.00	1.16×10^2	1.129×10^2	1.128×10^2	0.9996
	0.0080	1.87	6.57×10^{-2}	5.06×10	0.00	5.26×10	5.057×10	5.055×10	0.9995
	0.008979	1.65	7.16×10^{-2}	3.66×10	0.00	3.83×10	3.654×10	3.652×10	0.9994
29 K	0.008979	1.65	7.16×10^{-2}	2.77×10^2	0.00	2.78×10^2	1.825×10^2	1.824×10^2	0.9999
	0.0100	1.45	7.73×10^{-2}	2.14×10^2	0.00	2.16×10^2	1.484×10^2	1.484×10^2	0.9999
	0.0150	8.80×10^{-1}	9.76×10^{-2}	7.31×10	0.00	7.41×10	5.790×10	5.788×10	0.9996
	0.0200	6.06×10^{-1}	1.10×10^{-1}	3.31×10	0.00	3.38×10	2.791×10	2.788×10	0.9992
	0.0300	3.37×10^{-1}	1.23×10^{-1}	1.05×10	0.00	1.09×10	9.367	9.350	0.9982
	0.0400	2.12×10^{-1}	1.29×10^{-1}	4.52	0.00	4.86	4.174	4.163	0.9974
	0.0500	1.47×10^{-1}	1.31×10^{-1}	2.34	0.00	2.61	2.199	2.192	0.9967
	0.0600	1.08×10^{-1}	1.31×10^{-1}	1.35	0.00	1.59	1.295	1.290	0.9960
	0.0800	6.59×10^{-2}	1.29×10^{-1}	5.68×10^{-1}	0.00	7.63×10^{-1}	5.609×10^{-1}	5.581×10^{-1}	0.9951
	0.1000	4.45×10^{-2}	1.26×10^{-1}	2.88×10^{-1}	0.00	4.58×10^{-1}	2.966×10^{-1}	2.949×10^{-1}	0.9943
	0.1500	2.11×10^{-2}	1.17×10^{-1}	8.35×10^{-2}	0.00	2.22×10^{-1}	1.034×10^{-1}	1.027×10^{-1}	0.9932
	0.2000	1.23×10^{-2}	1.09×10^{-1}	3.49×10^{-2}	0.00	1.56×10^{-1}	5.824×10^{-2}	5.782×10^{-2}	0.9927
	0.3000	5.62×10^{-3}	9.58×10^{-2}	1.05×10^{-2}	0.00	1.12×10^{-1}	3.648×10^{-2}	3.619×10^{-2}	0.9918
	0.4000	3.21×10^{-3}	8.63×10^{-2}	4.66×10^{-3}	0.00	9.41×10^{-2}	3.151×10^{-2}	3.122×10^{-2}	0.9906
	0.5000	2.07×10^{-3}	7.90×10^{-2}	2.57×10^{-3}	0.00	8.36×10^{-2}	2.965×10^{-2}	2.933×10^{-2}	0.9893
	0.6000	1.44×10^{-3}	7.32×10^{-2}	1.62×10^{-3}	0.00	7.63×10^{-2}	2.861×10^{-2}	2.826×10^{-2}	0.9880
	0.8000	8.15×10^{-4}	6.44×10^{-2}	8.26×10^{-4}	0.00	6.61×10^{-2}	2.721×10^{-2}	2.681×10^{-2}	0.9854
	1.0000	5.23×10^{-4}	5.80×10^{-2}	5.14×10^{-4}	0.00	5.90×10^{-2}	2.607×10^{-2}	2.562×10^{-2}	0.9828
	1.2500	3.35×10^{-4}	5.19×10^{-2}	3.30×10^{-4}	8.02×10^{-5}	5.26×10^{-2}	2.478×10^{-2}	2.428×10^{-2}	0.9796

1.5000	2.33×10^{-4}	4.72×10^{-2}	2.38×10^{-4}	4.02×10^{-4}	4.80×10^{-2}	2.371×10^{-2}	2.316×10^{-2}	0.9765
2.0000	1.31×10^{-4}	4.03×10^{-2}	1.46×10^{-4}	1.51×10^{-3}	4.20×10^{-2}	2.227×10^{-2}	2.160×10^{-2}	0.9702
3.0000	5.83×10^{-5}	3.17×10^{-2}	7.92×10^{-5}	4.17×10^{-3}	3.60×10^{-2}	2.112×10^{-2}	2.023×10^{-2}	0.9575
4.0000	3.28×10^{-5}	2.64×10^{-2}	5.32×10^{-5}	6.67×10^{-3}	3.32×10^{-2}	2.106×10^{-2}	1.989×10^{-2}	0.9446
5.0000	2.10×10^{-5}	2.28×10^{-2}	3.98×10^{-5}	8.88×10^{-3}	3.18×10^{-2}	2.144×10^{-2}	1.998×10^{-2}	0.9320
6.0000	1.46×10^{-5}	2.02×10^{-2}	3.17×10^{-5}	1.08×10^{-2}	3.11×10^{-2}	2.203×10^{-2}	2.027×10^{-2}	0.9198
8.0000	8.20×10^{-6}	1.65×10^{-2}	2.24×10^{-5}	1.42×10^{-2}	3.07×10^{-2}	2.343×10^{-2}	2.100×10^{-2}	0.8967
10.0000	5.25×10^{-6}	1.41×10^{-2}	1.72×10^{-5}	1.69×10^{-2}	3.10×10^{-2}	2.484×10^{-2}	2.174×10^{-2}	0.8750
15.0000	2.33×10^{-6}	1.04×10^{-2}	1.09×10^{-5}	2.21×10^{-2}	3.25×10^{-2}	2.795×10^{-2}	2.309×10^{-2}	0.8262
20.0000	1.31×10^{-6}	8.35×10^{-3}	7.99×10^{-6}	2.57×10^{-2}	3.41×10^{-2}	3.048×10^{-2}	2.387×10^{-2}	0.7833
30.0000	5.83×10^{-7}	6.08×10^{-3}	5.19×10^{-6}	3.08×10^{-2}	3.69×10^{-2}	3.435×10^{-2}	2.443×10^{-2}	0.7112
40.0000	3.28×10^{-7}	4.83×10^{-3}	3.84×10^{-6}	3.44×10^{-2}	3.92×10^{-2}	3.716×10^{-2}	2.425×10^{-2}	0.6527
50.0000	2.10×10^{-7}	4.03×10^{-3}	3.05×10^{-6}	3.70×10^{-2}	4.10×10^{-2}	3.934×10^{-2}	2.377×10^{-2}	0.6042

TABLE M.3e

Photons in Tungsten

ρ: 19.30 (g cm^{-3}), Composition (Z − f$_w$): 74–1.000

K, L, M Edges	Energy (MeV)	Mass Coefficients (cm^2 g^{-1})							
		Coherent σ_{coh}/ρ	Compton σ_c/ρ	Photoelectric τ/ρ	Pair + Triplet κ/ρ	Total Attenuation μ/ρ	Energy-Transfer μ_{tr}/ρ	Energy-Absorption μ_{en}/ρ	(1-g)
	0.0010	1.14×10	4.34×10^{-3}	3.67×10^{3}	0.00	3.68×10^{3}	3.671×10^{3}	3.671×10^{3}	0.9999
	0.0015	1.10×10	7.51×10^{-3}	1.63×10^{3}	0.00	1.64×10^{3}	1.632×10^{3}	1.632×10^{3}	0.9998
	0.001809	1.06×10	9.38×10^{-3}	1.10×10^{3}	0.00	1.11×10^{3}	1.097×10^{3}	1.097×10^{3}	0.9998
74 M5	0.001809	1.06×10	9.38×10^{-3}	1.30×10^{3}	0.00	1.32×10^{3}	1.312×10^{3}	1.311×10^{3}	0.9998
	0.001872	1.06×10	9.75×10^{-3}	2.85×10^{3}	0.00	2.86×10^{3}	2.853×10^{3}	2.853×10^{3}	0.9999
74 M4	0.001872	1.06×10	9.75×10^{-3}	3.11×10^{3}	0.00	3.12×10^{3}	3.116×10^{3}	3.116×10^{3}	0.9999
	0.0020	1.04×10	1.05×10^{-2}	3.91×10^{3}	0.00	3.92×10^{3}	3.853×10^{3}	3.853×10^{3}	0.9999
	0.002281	1.01×10	1.22×10^{-2}	2.82×10^{3}	0.00	2.83×10^{3}	2.781×10^{3}	2.781×10^{3}	0.9999
74 M3	0.002281	1.01×10	1.22×10^{-2}	3.27×10^{3}	0.00	3.28×10^{3}	3.226×10^{3}	3.226×10^{3}	0.9999
	0.002575	9.82	1.39×10^{-2}	2.44×10^{3}	0.00	2.45×10^{3}	2.407×10^{3}	2.407×10^{3}	0.9999
74 M2	0.002575	9.82	1.39×10^{-2}	2.59×10^{3}	0.00	2.60×10^{3}	2.558×10^{3}	2.558×10^{3}	0.9999
	0.002820	9.55	1.53×10^{-2}	2.09×10^{3}	0.00	2.10×10^{3}	2.071×10^{3}	2.071×10^{3}	0.9999
74 M1	0.002820	9.55	1.53×10^{-2}	2.18×10^{3}	0.00	2.19×10^{3}	2.160×10^{3}	2.160×10^{3}	0.9999
	0.0030	9.36	1.63×10^{-2}	1.89×10^{3}	0.00	1.90×10^{3}	1.873×10^{3}	1.873×10^{3}	0.9999
	0.0040	8.37	2.19×10^{-2}	9.48×10^{2}	0.00	9.56×10^{2}	9.407×10^{2}	9.405×10^{2}	0.9998
	0.0050	7.48	2.71×10^{-2}	5.46×10^{2}	0.00	5.53×10^{2}	5.425×10^{2}	5.423×10^{2}	0.9997
	0.0060	6.70	3.20×10^{-2}	3.45×10^{2}	0.00	3.51×10^{2}	3.429×10^{2}	3.428×10^{2}	0.9996
	0.0080	5.42	4.07×10^{-2}	1.65×10^{2}	0.00	1.71×10^{2}	1.644×10^{2}	1.643×10^{2}	0.9993
	0.0100	4.45	4.79×10^{-2}	9.24×10	0.00	9.69×10	9.213×10	9.204×10	0.9990
	0.01021	4.36	4.86×10^{-2}	8.76×10	0.00	9.20×10	8.733×10	8.724×10	0.9990
74 L3	0.01021	4.36	4.86×10^{-2}	2.29×10^{2}	0.00	2.33×10^{2}	1.967×10^{2}	1.966×10^{2}	0.9996
	0.01154	3.86	5.28×10^{-2}	1.65×10^{2}	0.00	1.69×10^{2}	1.445×10^{2}	1.444×10^{2}	0.9995
74 L2	0.01154	3.86	5.28×10^{-2}	2.27×10^{2}	0.00	2.31×10^{2}	1.890×10^{2}	1.889×10^{2}	0.9996
	0.01210	3.68	5.44×10^{-2}	2.03×10^{2}	0.00	2.07×10^{2}	1.700×10^{2}	1.699×10^{2}	0.9996
74 L1	0.01210	3.68	5.44×10^{-2}	2.34×10^{2}	0.00	2.38×10^{2}	1.949×10^{2}	1.948×10^{2}	0.9996
	0.0150	2.89	6.20×10^{-2}	1.36×10^{2}	0.00	1.39×10^{2}	1.173×10^{2}	1.172×10^{2}	0.9994
	0.0200	2.04	7.25×10^{-2}	6.36×10	0.00	6.57×10	5.704×10	5.697×10	0.9989
	0.0300	1.20	8.64×10^{-2}	2.14×10	0.00	2.27×10	1.996×10	1.991×10	0.9975
	0.0400	7.94×10^{-1}	9.43×10^{-2}	9.78	0.00	1.07×10	9.278	9.242	0.9961
	0.0500	5.61×10^{-1}	9.86×10^{-2}	5.29	0.00	5.95	5.077	5.051	0.9948
	0.0600	4.21×10^{-1}	1.01×10^{-1}	3.19	0.00	3.71	3.090	3.070	0.9935

74 K

0.06953	3.32×10^{-1}	1.02×10^{-1}	2.12	0.00	2.55	2.066	2.050	0.9923
0.06953	3.32×10^{-1}	1.02×10^{-1}	1.08×10	0.00	1.12×10	3.228	3.212	0.9951
0.0800	2.64×10^{-1}	1.03×10^{-1}	7.44	0.00	7.81	2.894	2.880	0.9953
0.1000	1.81×10^{-1}	1.02×10^{-1}	4.15	0.00	4.44	2.113	2.100	0.9937
0.1500	8.88×10^{-2}	9.74×10^{-2}	1.40	0.00	1.58	9.489×10^{-1}	9.380×10^{-1}	0.9885
0.2000	5.29×10^{-2}	9.18×10^{-2}	6.40×10^{-1}	0.00	7.84×10^{-1}	4.997×10^{-1}	4.914×10^{-1}	0.9835
0.3000	2.51×10^{-2}	8.21×10^{-2}	2.17×10^{-1}	0.00	3.24×10^{-1}	2.027×10^{-1}	1.978×10^{-1}	0.9756
0.4000	1.46×10^{-2}	7.46×10^{-2}	1.03×10^{-1}	0.00	1.92×10^{-1}	1.136×10^{-1}	1.101×10^{-1}	0.9696
0.5000	9.54×10^{-3}	6.87×10^{-2}	5.96×10^{-2}	0.00	1.38×10^{-1}	7.719×10^{-2}	7.448×10^{-2}	0.9649
0.6000	6.71×10^{-3}	6.38×10^{-2}	3.88×10^{-2}	0.00	1.09×10^{-1}	5.910×10^{-2}	5.678×10^{-2}	0.9607
0.8000	3.84×10^{-3}	5.64×10^{-2}	2.04×10^{-2}	0.00	8.07×10^{-2}	4.227×10^{-2}	4.030×10^{-2}	0.9533
1.0000	2.48×10^{-3}	5.09×10^{-2}	1.28×10^{-2}	0.00	6.62×10^{-2}	3.463×10^{-2}	3.277×10^{-2}	0.9464
1.2500	1.60×10^{-3}	4.56×10^{-2}	8.28×10^{-3}	3.23×10^{-4}	5.58×10^{-2}	2.949×10^{-2}	2.767×10^{-2}	0.9380
1.5000	1.11×10^{-3}	4.15×10^{-2}	5.91×10^{-3}	1.51×10^{-3}	5.00×10^{-2}	2.673×10^{-2}	2.487×10^{-2}	0.9305
2.0000	6.29×10^{-4}	3.54×10^{-2}	3.57×10^{-3}	4.68×10^{-3}	4.43×10^{-2}	2.460×10^{-2}	2.257×10^{-2}	0.9175
3.0000	2.81×10^{-4}	2.79×10^{-2}	1.88×10^{-3}	1.07×10^{-2}	4.07×10^{-2}	2.501×10^{-2}	2.238×10^{-2}	0.8951
4.0000	1.58×10^{-4}	2.33×10^{-2}	1.23×10^{-3}	1.57×10^{-2}	4.04×10^{-2}	2.704×10^{-2}	2.365×10^{-2}	0.8745
5.0000	1.01×10^{-4}	2.01×10^{-2}	9.04×10^{-4}	1.99×10^{-2}	4.10×10^{-2}	2.937×10^{-2}	2.511×10^{-2}	0.8548
6.0000	7.03×10^{-5}	1.78×10^{-2}	7.10×10^{-4}	2.35×10^{-2}	4.21×10^{-2}	3.169×10^{-2}	2.649×10^{-2}	0.8360
8.0000	3.96×10^{-5}	1.46×10^{-2}	4.92×10^{-4}	2.96×10^{-2}	4.47×10^{-2}	3.605×10^{-2}	2.887×10^{-2}	0.8008
10.0000	2.53×10^{-5}	1.24×10^{-2}	3.75×10^{-4}	3.47×10^{-2}	4.75×10^{-2}	3.998×10^{-2}	3.072×10^{-2}	0.7684
15.0000	1.13×10^{-5}	9.17×10^{-3}	2.33×10^{-4}	4.44×10^{-2}	5.38×10^{-2}	4.814×10^{-2}	3.361×10^{-2}	0.6981
20.0000	6.33×10^{-6}	7.36×10^{-3}	1.69×10^{-4}	5.14×10^{-2}	5.89×10^{-2}	5.428×10^{-2}	3.475×10^{-2}	0.6402
30.0000	2.81×10^{-6}	5.36×10^{-3}	1.08×10^{-4}	6.10×10^{-2}	6.65×10^{-2}	6.309×10^{-2}	3.479×10^{-2}	0.5514
40.0000	1.58×10^{-6}	4.26×10^{-3}	7.98×10^{-5}	6.76×10^{-2}	7.20×10^{-2}	6.925×10^{-2}	3.373×10^{-2}	0.4870
50.0000	1.01×10^{-6}	3.56×10^{-3}	6.31×10^{-5}	7.25×10^{-2}	7.62×10^{-2}	7.389×10^{-2}	3.238×10^{-2}	0.4382

TABLE M.3f

Photons in Lead

ρ: 11.35 (g cm^{-3}), Composition ($Z - f_w$): 82–1.0000

K, L, M Edges	Energy (MeV)	Mass Coefficients (cm^2 g^{-1})							(1-g)
		Coherent σ_{coh}/ρ	Compton σ_C/ρ	Photoelectric τ/ρ	Pair + Triplet κ/ρ	Total Attenuation μ/ρ	Energy-Transfer μ_{tr}/ρ	Energy-Absorption μ_{en}/ρ	
	0.0010	1.25×10	3.59×10^{-3}	5.20×10^{3}	0.00	5.21×10^{3}	5.197×10^{3}	5.197×10^{3}	0.9999
	0.0015	1.20×10	6.60×10^{-3}	2.34×10^{3}	0.00	2.36×10^{3}	2.344×10^{3}	2.344×10^{3}	0.9998
	0.0020	1.14×10	9.62×10^{-3}	1.27×10^{3}	0.00	1.29×10^{3}	1.274×10^{3}	1.274×10^{3}	0.9998
	0.002484	1.09×10	1.24×10^{-2}	7.90×10^{2}	0.00	8.01×10^{2}	7.897×10^{2}	7.895×10^{2}	0.9997
82 M5	0.002484	1.09×10	1.24×10^{-2}	1.38×10^{3}	0.00	1.40×10^{3}	1.367×10^{3}	1.366×10^{3}	0.9999
	0.002586	1.08×10	1.30×10^{-2}	1.93×10^{3}	0.00	1.94×10^{3}	1.895×10^{3}	1.895×10^{3}	0.9999
82 M4	0.002586	1.08×10	1.30×10^{-2}	2.44×10^{3}	0.00	2.45×10^{3}	2.390×10^{3}	2.390×10^{3}	0.9999
	0.0030	1.03×10	1.52×10^{-2}	1.95×10^{3}	0.00	1.96×10^{3}	1.913×10^{3}	1.913×10^{3}	0.9999
	0.003066	1.02×10	1.56×10^{-2}	1.85×10^{3}	0.00	1.86×10^{3}	1.809×10^{3}	1.808×10^{3}	0.9999
82 M3	0.003066	1.02×10	1.56×10^{-2}	2.14×10^{3}	0.00	2.15×10^{3}	2.091×10^{3}	2.090×10^{3}	0.9999
	0.003554	9.65	1.81×10^{-2}	1.49×10^{3}	0.00	1.50×10^{3}	1.459×10^{3}	1.459×10^{3}	0.9999
82 M2	0.003554	9.65	1.81×10^{-2}	1.57×10^{3}	0.00	1.58×10^{3}	1.546×10^{3}	1.546×10^{3}	0.9999
	0.003851	9.34	1.96×10^{-2}	1.30×10^{3}	0.00	1.31×10^{3}	1.279×10^{3}	1.279×10^{3}	0.9998
82 M1	0.003851	9.34	1.96×10^{-2}	1.36×10^{3}	0.00	1.37×10^{3}	1.335×10^{3}	1.335×10^{3}	0.9999
	0.0040	9.18	2.04×10^{-2}	1.24×10^{3}	0.00	1.25×10^{3}	1.221×10^{3}	1.221×10^{3}	0.9998
	0.0050	8.21	2.52×10^{-2}	7.22×10^{2}	0.00	7.30×10^{2}	7.126×10^{2}	7.124×10^{2}	0.9997
	0.0060	7.36	2.97×10^{-2}	4.60×10^{2}	0.00	4.67×10^{2}	4.548×10^{2}	4.546×10^{2}	0.9996
	0.0080	6.00	3.81×10^{-2}	2.23×10^{2}	0.00	2.29×10^{2}	2.208×10^{2}	2.207×10^{2}	0.9994
	0.0100	4.98	4.54×10^{-2}	1.26×10^{2}	0.00	1.31×10^{2}	1.248×10^{2}	1.247×10^{2}	0.9991
	0.01304	3.85	5.44×10^{-2}	6.31×10	0.00	6.70×10	6.279×10	6.271×10	0.9986
82 L3	0.01304	3.85	5.44×10^{-2}	1.58×10^{2}	0.00	1.62×10^{2}	1.292×10^{2}	1.291×10^{2}	0.9993
	0.0150	3.31	5.92×10^{-2}	1.08×10^{2}	0.00	1.12×10^{2}	9.108×10	9.100×10	0.9992
	0.01520	3.26	5.96×10^{-2}	1.04×10^{2}	0.00	1.08×10^{2}	8.815×10	8.808×10	0.9992
82 L2	0.01520	3.26	5.96×10^{-2}	1.45×10^{2}	0.00	1.49×10^{2}	1.132×10^{2}	1.131×10^{2}	0.9993
	0.01586	3.10	6.11×10^{-2}	1.31×10^{2}	0.00	1.34×10^{2}	1.033×10^{2}	1.032×10^{2}	0.9993
82 L1	0.01586	3.10	6.11×10^{-2}	1.52×10^{2}	0.00	1.55×10^{2}	1.181×10^{2}	1.180×10^{2}	0.9994
	0.0200	2.34	6.90×10^{-2}	8.40×10	0.00	8.64×10	6.906×10	6.900×10	0.9990
	0.0300	1.38	8.23×10^{-2}	2.89×10	0.00	3.03×10	2.542×10	2.537×10	0.9978
	0.0400	9.20×10^{-1}	9.02×10^{-2}	1.33×10	0.00	1.44×10	1.216×10	1.211×10	0.9963
	0.0500	6.55×10^{-1}	9.48×10^{-2}	7.29	0.00	8.04	6.776	6.741	0.9948
	0.0600	4.90×10^{-1}	9.73×10^{-2}	4.43	0.00	5.02	4.178	4.150	0.9934

82 K

Energy								
0.0800	3.08×10^{-1}	9.92×10^{-2}	2.01	0.00	2.42	1.934	1.916	0.9907
0.08801	2.63×10^{-1}	9.93×10^{-2}	1.55	0.00	1.91	1.497	1.482	0.9896
0.08801	2.63×10^{-1}	9.93×10^{-2}	7.32	0.00	7.68	2.175	2.160	0.9929
0.1000	2.13×10^{-1}	9.89×10^{-2}	5.24	0.00	5.55	1.990	1.976	0.9932
0.1500	1.05×10^{-1}	9.48×10^{-2}	1.81	0.00	2.01	1.069	1.056	0.9885
0.2000	6.26×10^{-2}	8.97×10^{-2}	8.46×10^{-1}	0.00	9.99×10^{-1}	5.975×10^{-1}	5.874×10^{-1}	0.9831
0.3000	2.99×10^{-2}	8.04×10^{-2}	2.93×10^{-1}	0.00	4.03×10^{-1}	2.525×10^{-1}	2.458×10^{-1}	0.9736
0.4000	1.75×10^{-2}	7.31×10^{-2}	1.42×10^{-1}	0.00	2.32×10^{-1}	1.419×10^{-1}	1.371×10^{-1}	0.9661
0.5000	1.14×10^{-2}	6.73×10^{-2}	8.26×10^{-2}	0.00	1.61×10^{-1}	9.514×10^{-2}	9.135×10^{-2}	0.9601
0.6000	8.06×10^{-3}	6.26×10^{-2}	5.41×10^{-2}	0.00	1.25×10^{-1}	7.143×10^{-2}	6.822×10^{-2}	0.955
0.8000	4.62×10^{-3}	5.54×10^{-2}	2.87×10^{-2}	0.00	8.87×10^{-2}	4.911×10^{-2}	4.647×10^{-2}	0.9461
1.0000	2.99×10^{-3}	4.99×10^{-2}	1.81×10^{-2}	0.00	7.10×10^{-2}	3.896×10^{-2}	3.655×10^{-2}	0.9380
1.2500	1.93×10^{-3}	4.48×10^{-2}	1.17×10^{-2}	3.78×10^{-4}	5.88×10^{-2}	3.226×10^{-2}	2.996×10^{-2}	0.9287
1.5000	1.35×10^{-3}	4.07×10^{-2}	8.32×10^{-3}	1.81×10^{-3}	5.22×10^{-2}	2.873×10^{-2}	2.644×10^{-2}	0.9205
2.0000	7.63×10^{-4}	3.48×10^{-2}	5.03×10^{-3}	5.45×10^{-3}	4.61×10^{-2}	2.604×10^{-2}	2.362×10^{-2}	0.9071
3.0000	3.41×10^{-4}	2.74×10^{-2}	2.63×10^{-3}	1.19×10^{-2}	4.23×10^{-2}	2.629×10^{-2}	2.325×10^{-2}	0.8846
4.0000	1.92×10^{-4}	2.29×10^{-2}	1.72×10^{-3}	1.71×10^{-2}	4.20×10^{-2}	2.837×10^{-2}	2.451×10^{-2}	0.8638
5.0000	1.23×10^{-4}	1.98×10^{-2}	1.26×10^{-3}	2.16×10^{-2}	4.27×10^{-2}	3.082×10^{-2}	2.601×10^{-2}	0.8439
6.0000	8.54×10^{-5}	1.75×10^{-2}	9.89×10^{-4}	2.53×10^{-2}	4.39×10^{-2}	3.327×10^{-2}	2.745×10^{-2}	0.8250
8.0000	4.81×10^{-5}	1.43×10^{-2}	6.84×10^{-4}	3.17×10^{-2}	4.67×10^{-2}	3.788×10^{-2}	2.990×10^{-2}	0.7894
10.0000	3.08×10^{-5}	1.22×10^{-2}	5.20×10^{-4}	3.70×10^{-2}	4.97×10^{-2}	4.205×10^{-2}	3.182×10^{-2}	0.7567
15.0000	1.37×10^{-5}	9.02×10^{-3}	3.23×10^{-4}	4.72×10^{-2}	5.66×10^{-2}	5.073×10^{-2}	3.479×10^{-2}	0.6859
20.0000	7.70×10^{-6}	7.24×10^{-3}	2.33×10^{-4}	5.45×10^{-2}	6.21×10^{-2}	5.728×10^{-2}	3.596×10^{-2}	0.6278
30.0000	3.42×10^{-6}	5.27×10^{-3}	1.50×10^{-4}	6.48×10^{-2}	7.02×10^{-2}	6.668×10^{-2}	3.595×10^{-2}	0.5391
40.0000	1.92×10^{-6}	4.19×10^{-3}	1.10×10^{-4}	7.18×10^{-2}	7.61×10^{-2}	7.326×10^{-2}	3.482×10^{-2}	0.4753
50.0000	1.23×10^{-6}	3.50×10^{-3}	8.70×10^{-5}	7.70×10^{-2}	8.06×10^{-2}	7.818×10^{-2}	3.338×10^{-2}	0.4270

TABLE M.3g

Photons in Water, Liquid

ρ: 1.000 (g cm^{-3}), Composition ($Z - f_w$): 1−0.111898; 8−0.888102

K, L, M Edges	Energy (MeV)	Coherent σ_{coh}/ρ	Compton σ_C/ρ	Photoelectric τ/ρ	Pair+Triplet κ/ρ	Total Attenuation μ/ρ	Energy-Transfer μ_{tr}/ρ	Energy-Absorption μ_{en}/ρ	$(1-g)$
						Mass Coefficients (cm^2 g^{-1})			
	0.0010	1.37	1.32×10^{-2}	4.08×10^{3}	0.00	4.08×10^{3}	4.065×10^{3}	4.065×10^{3}	1
	0.0015	1.27	2.67×10^{-2}	1.37×10^{3}	0.00	1.38×10^{3}	1.372×10^{3}	1.372×10^{3}	1
	0.0020	1.15	4.18×10^{-2}	6.16×10^{2}	0.00	6.17×10^{2}	6.152×10^{2}	6.152×10^{2}	0.9999
	0.0030	9.09×10^{-1}	7.07×10^{-2}	1.92×10^{2}	0.00	1.93×10^{2}	1.917×10^{2}	1.917×10^{2}	0.9999
	0.0040	7.08×10^{-1}	9.43×10^{-2}	8.20×10	0.00	8.28×10	8.192×10	8.191×10	0.9999
	0.0050	5.58×10^{-1}	1.12×10^{-1}	4.19×10	0.00	4.26×10	4.189×10	4.188×10	0.9998
	0.0060	4.49×10^{-1}	1.26×10^{-1}	2.41×10	0.00	2.46×10	2.406×10	2.405×10	0.9998
	0.0080	3.10×10^{-1}	1.44×10^{-1}	9.92	0.00	1.04×10	9.918	9.915	0.9998
	0.0100	2.31×10^{-1}	1.55×10^{-1}	4.94	0.00	5.33	4.945	4.944	0.9998
	0.0150	1.33×10^{-1}	1.70×10^{-1}	1.37	0.00	1.67	1.374	1.374	0.9997
	0.0200	8.86×10^{-2}	1.77×10^{-1}	5.44×10^{-1}	0.00	8.10×10^{-1}	5.505×10^{-1}	5.503×10^{-1}	0.9997
	0.0300	4.69×10^{-2}	1.83×10^{-1}	1.46×10^{-1}	0.00	3.76×10^{-1}	1.557×10^{-1}	1.557×10^{-1}	0.9996
	0.0400	2.87×10^{-2}	1.83×10^{-1}	5.68×10^{-2}	0.00	2.68×10^{-1}	6.950×10^{-2}	6.947×10^{-2}	0.9996
	0.0500	1.94×10^{-2}	1.80×10^{-1}	2.72×10^{-2}	0.00	2.27×10^{-1}	4.225×10^{-2}	4.223×10^{-2}	0.9996
	0.0600	1.39×10^{-2}	1.77×10^{-1}	1.49×10^{-2}	0.00	2.06×10^{-1}	3.191×10^{-2}	3.190×10^{-2}	0.9996
	0.0800	8.16×10^{-3}	1.70×10^{-1}	5.77×10^{-3}	0.00	1.84×10^{-1}	2.598×10^{-2}	2.597×10^{-2}	0.9996
	0.1000	5.35×10^{-3}	1.63×10^{-1}	2.76×10^{-3}	0.00	1.71×10^{-1}	2.547×10^{-2}	2.546×10^{-2}	0.9996
	0.1500	2.44×10^{-3}	1.47×10^{-1}	7.31×10^{-4}	0.00	1.51×10^{-1}	2.765×10^{-2}	2.764×10^{-2}	0.9995
	0.2000	1.39×10^{-3}	1.35×10^{-1}	2.89×10^{-4}	0.00	1.37×10^{-1}	2.969×10^{-2}	2.967×10^{-2}	0.9994
	0.3000	6.22×10^{-4}	1.18×10^{-1}	8.16×10^{-5}	0.00	1.19×10^{-1}	3.195×10^{-2}	3.192×10^{-2}	0.9992
	0.4000	3.51×10^{-4}	1.06×10^{-1}	3.49×10^{-5}	0.00	1.06×10^{-1}	3.282×10^{-2}	3.279×10^{-2}	0.9989
	0.5000	2.25×10^{-4}	9.66×10^{-2}	1.88×10^{-5}	0.00	9.69×10^{-2}	3.303×10^{-2}	3.299×10^{-2}	0.9987
	0.6000	1.56×10^{-4}	8.94×10^{-2}	1.17×10^{-5}	0.00	8.96×10^{-2}	3.289×10^{-2}	3.284×10^{-2}	0.9984
	0.8000	8.79×10^{-5}	7.86×10^{-2}	5.92×10^{-6}	0.00	7.87×10^{-2}	3.212×10^{-2}	3.206×10^{-2}	0.9980
	1.0000	5.63×10^{-5}	7.07×10^{-2}	3.68×10^{-6}	0.00	7.07×10^{-2}	3.111×10^{-2}	3.103×10^{-2}	0.9975
	1.2500	3.60×10^{-5}	6.32×10^{-2}	2.33×10^{-6}	1.78×10^{-5}	6.32×10^{-2}	2.974×10^{-2}	2.965×10^{-2}	0.9969
	1.5000	2.50×10^{-5}	5.74×10^{-2}	1.69×10^{-6}	9.82×10^{-5}	5.75×10^{-2}	2.844×10^{-2}	2.833×10^{-2}	0.9962
	2.0000	1.41×10^{-5}	4.90×10^{-2}	1.06×10^{-6}	3.91×10^{-4}	4.94×10^{-2}	2.621×10^{-2}	2.608×10^{-2}	0.9948
	3.0000	6.26×10^{-6}	3.85×10^{-2}	5.94×10^{-7}	1.13×10^{-3}	3.97×10^{-2}	2.300×10^{-2}	2.281×10^{-2}	0.9916
	4.0000	3.52×10^{-6}	3.22×10^{-2}	4.08×10^{-7}	1.87×10^{-3}	3.40×10^{-2}	2.091×10^{-2}	2.066×10^{-2}	0.9880
	5.0000	2.25×10^{-6}	2.78×10^{-2}	3.09×10^{-7}	2.54×10^{-3}	3.03×10^{-2}	1.946×10^{-2}	1.915×10^{-2}	0.9840

6.0000	1.56×10^{-6}	2.45×10^{-2}	2.48×10^{-7}	3.16×10^{-3}	2.77×10^{-2}	1.843×10^{-2}	1.806×10^{-2}	0.9800
8.0000	8.80×10^{-7}	2.01×10^{-2}	1.78×10^{-7}	4.21×10^{-3}	2.43×10^{-2}	1.707×10^{-2}	1.658×10^{-2}	0.9716
10.0000	5.63×10^{-7}	1.71×10^{-2}	1.39×10^{-7}	5.09×10^{-3}	2.22×10^{-2}	1.626×10^{-2}	1.566×10^{-2}	0.9633
15.0000	2.50×10^{-7}	1.27×10^{-2}	8.91×10^{-8}	6.75×10^{-3}	1.94×10^{-2}	1.528×10^{-2}	1.441×10^{-2}	0.9432
20.0000	1.41×10^{-7}	1.02×10^{-2}	6.56×10^{-8}	7.98×10^{-3}	1.81×10^{-2}	1.495×10^{-2}	1.382×10^{-2}	0.9245
30.0000	6.26×10^{-8}	7.40×10^{-3}	4.29×10^{-8}	9.71×10^{-3}	1.71×10^{-2}	1.490×10^{-2}	1.327×10^{-2}	0.8904
40.0000	3.52×10^{-8}	5.88×10^{-3}	3.19×10^{-8}	1.09×10^{-2}	1.68×10^{-2}	1.510×10^{-2}	1.298×10^{-2}	0.8600
50.0000	2.25×10^{-8}	4.91×10^{-3}	2.53×10^{-8}	1.18×10^{-2}	1.67×10^{-2}	1.537×10^{-2}	1.279×10^{-2}	0.8323

TABLE M.3h

Photons in Lithium Fluoride

ρ: 2.635 (g cm^{-3}), Composition (Z − f$_w$): 3−0.267585; 9−0.732415

K, L, M Edges / Energy (MeV)	Mass Coefficients (cm^2 g^{-1})							(1-g)
	Coherent σ_{coh}/ρ	Compton σ_C/ρ	Photoelectric τ/ρ	Pair + Triplet κ/ρ	Total Attenuation μ/ρ	Energy-Transfer μ_{tr}/ρ	Energy-Absorption μ_{en}/ρ	
0.0010	1.30	1.29×10^{-2}	4.20×10^{3}	0.00	4.20×10^{3}	4.175×10^{3}	4.175×10^{3}	1
0.0015	1.21	2.22×10^{-2}	1.47×10^{3}	0.00	1.47×10^{3}	1.460×10^{3}	1.460×10^{3}	1
0.0020	1.11	3.10×10^{-2}	6.69×10^{2}	0.00	6.70×10^{2}	6.668×10^{2}	6.667×10^{2}	0.9999
0.0030	9.04×10^{-1}	4.85×10^{-2}	2.13×10^{2}	0.00	2.14×10^{2}	2.122×10^{2}	2.121×10^{2}	0.9999
0.0040	7.26×10^{-1}	6.49×10^{-2}	9.20×10	0.00	9.28×10	9.190×10	9.188×10	0.9998
0.0050	5.84×10^{-1}	7.93×10^{-2}	4.75×10	0.00	4.81×10	4.743×10	4.742×10	0.9998
0.0060	4.75×10^{-1}	9.14×10^{-2}	2.75×10	0.00	2.80×10	2.743×10	2.742×10	0.9998
0.0080	3.30×10^{-1}	1.09×10^{-1}	1.14×10	0.00	1.19×10	1.142×10	1.142×10	0.9997
0.0100	2.44×10^{-1}	1.21×10^{-1}	5.74	0.00	6.10	5.735	5.733	0.9997
0.0150	1.38×10^{-1}	1.37×10^{-1}	1.61	0.00	1.88	1.613	1.612	0.9996
0.0200	9.05×10^{-2}	1.45×10^{-1}	6.44×10^{-1}	0.00	8.79×10^{-1}	6.497×10^{-1}	6.494×10^{-1}	0.9996
0.0300	4.78×10^{-2}	1.50×10^{-1}	1.74×10^{-1}	0.00	3.73×10^{-1}	1.827×10^{-1}	1.826×10^{-1}	0.9995
0.0400	2.93×10^{-2}	1.51×10^{-1}	6.84×10^{-2}	0.00	2.49×10^{-1}	7.895×10^{-2}	7.891×10^{-2}	0.9995
0.0500	1.97×10^{-2}	1.49×10^{-1}	3.30×10^{-2}	0.00	2.02×10^{-1}	4.543×10^{-2}	4.541×10^{-2}	0.9995
0.0600	1.42×10^{-2}	1.47×10^{-1}	1.81×10^{-2}	0.00	1.79×10^{-1}	3.225×10^{-2}	3.223×10^{-2}	0.9995
0.0800	8.31×10^{-3}	1.41×10^{-1}	7.04×10^{-3}	0.00	1.56×10^{-1}	2.386×10^{-2}	2.385×10^{-2}	0.9995
0.1000	5.45×10^{-3}	1.35×10^{-1}	3.38×10^{-3}	0.00	1.44×10^{-1}	2.230×10^{-2}	2.229×10^{-2}	0.9995
0.1500	2.49×10^{-3}	1.23×10^{-1}	8.99×10^{-4}	0.00	1.26×10^{-1}	2.333×10^{-2}	2.332×10^{-2}	0.9994
0.2000	1.42×10^{-3}	1.13×10^{-1}	3.56×10^{-4}	0.00	1.14×10^{-1}	2.485×10^{-2}	2.484×10^{-2}	0.9993
0.3000	6.35×10^{-4}	9.83×10^{-2}	1.01×10^{-4}	0.00	9.90×10^{-2}	2.666×10^{-2}	2.663×10^{-2}	0.9990
0.4000	3.59×10^{-4}	8.81×10^{-2}	4.33×10^{-5}	0.00	8.85×10^{-2}	2.737×10^{-2}	2.734×10^{-2}	0.9987
0.5000	2.30×10^{-4}	8.05×10^{-2}	2.34×10^{-5}	0.00	8.08×10^{-2}	2.753×10^{-2}	2.749×10^{-2}	0.9984
0.6000	1.60×10^{-4}	7.45×10^{-2}	1.46×10^{-5}	0.00	7.47×10^{-2}	2.741×10^{-2}	2.736×10^{-2}	0.9981
0.8000	8.99×10^{-5}	6.55×10^{-2}	7.35×10^{-6}	0.00	6.56×10^{-2}	2.678×10^{-2}	2.671×10^{-2}	0.9976
1.0000	5.76×10^{-5}	5.89×10^{-2}	4.57×10^{-6}	0.00	5.89×10^{-2}	2.592×10^{-2}	2.585×10^{-2}	0.9970
1.2500	3.68×10^{-5}	5.27×10^{-2}	2.90×10^{-6}	1.67×10^{-5}	5.27×10^{-2}	2.470×10^{-2}	2.470×10^{-2}	0.9963
1.5000	2.56×10^{-5}	4.79×10^{-2}	2.10×10^{-6}	9.28×10^{-5}	4.80×10^{-2}	2.371×10^{-2}	2.361×10^{-2}	0.9956
2.0000	1.44×10^{-5}	4.08×10^{-2}	1.32×10^{-6}	3.70×10^{-4}	4.12×10^{-2}	2.186×10^{-2}	2.173×10^{-2}	0.9940
3.0000	6.40×10^{-6}	3.21×10^{-2}	7.37×10^{-7}	1.07×10^{-3}	3.32×10^{-2}	1.926×10^{-2}	1.907×10^{-2}	0.9903
4.0000	3.60×10^{-6}	2.68×10^{-2}	5.05×10^{-7}	1.77×10^{-3}	2.86×10^{-2}	1.758×10^{-2}	1.733×10^{-2}	0.9861
5.0000	2.30×10^{-6}	2.31×10^{-2}	3.83×10^{-7}	2.39×10^{-3}	2.55×10^{-2}	1.644×10^{-2}	1.614×10^{-2}	0.9817

6.0000	1.60×10^{-6}	2.05×10^{-2}	3.07×10^{-7}	2.97×10^{-3}	2.34×10^{-2}	1.564×10^{-2}	1.528×10^{-2}	0.9770
8.0000	9.00×10^{-7}	1.67×10^{-2}	2.20×10^{-7}	3.96×10^{-3}	2.07×10^{-2}	1.462×10^{-2}	1.414×10^{-2}	0.9677
10.0000	5.76×10^{-7}	1.43×10^{-2}	1.71×10^{-7}	4.78×10^{-3}	1.90×10^{-2}	1.403×10^{-2}	1.345×10^{-2}	0.9584
15.0000	2.56×10^{-7}	1.05×10^{-2}	1.10×10^{-7}	6.32×10^{-3}	1.69×10^{-2}	1.338×10^{-2}	1.253×10^{-2}	0.9362
20.0000	1.44×10^{-7}	8.47×10^{-3}	8.09×10^{-8}	7.46×10^{-3}	1.59×10^{-2}	1.323×10^{-2}	1.211×10^{-2}	0.9156
30.0000	6.40×10^{-8}	6.16×10^{-3}	5.29×10^{-8}	9.06×10^{-3}	1.52×10^{-2}	1.336×10^{-2}	1.174×10^{-2}	0.8785
40.0000	3.60×10^{-8}	4.90×10^{-3}	3.93×10^{-8}	1.02×10^{-2}	1.51×10^{-2}	1.365×10^{-2}	1.154×10^{-2}	0.8454
50.0000	2.30×10^{-8}	4.09×10^{-3}	3.13×10^{-8}	1.10×10^{-2}	1.51×10^{-2}	1.396×10^{-2}	1.138×10^{-2}	0.8155

TABLE M.3i

Photons in Lithium Tetraborate

ρ: 2.440 (g cm^{-3}), Composition ($Z-f_w$): 3–0.082081; 5–0.255715; 8–0.662204

Energy (MeV)	Coherent σ_{coh}/ρ	Compton σ_C/ρ	Photoelectric τ/ρ	Pair + Triplet κ/ρ	Total Attenuation μ/ρ	Energy-Transfer μ_{tr}/ρ	Energy-Absorption μ_{en}/ρ	$(1-g)$
0.0010	1.23	1.23×10^{-2}	3.37×10^{3}	0.00	3.37×10^{3}	3.363×10^{3}	3.363×10^{3}	1
0.0015	1.13	2.33×10^{-2}	1.13×10^{3}	0.00	1.13×10^{3}	1.124×10^{3}	1.124×10^{3}	0.9999
0.0020	1.01	3.51×10^{-2}	5.02×10^{2}	0.00	5.03×10^{2}	5.016×10^{2}	5.015×10^{2}	0.9999
0.0030	7.90×10^{-1}	5.74×10^{-2}	1.55×10^{2}	0.00	1.56×10^{2}	1.553×10^{2}	1.553×10^{2}	0.9999
0.0040	6.16×10^{-1}	7.60×10^{-2}	6.62×10	0.00	6.69×10	6.613×10	6.612×10	0.9998
0.0050	4.87×10^{-1}	9.07×10^{-2}	3.38×10	0.00	3.43×10	3.373×10	3.373×10	0.9998
0.0060	3.95×10^{-1}	1.02×10^{-1}	1.93×10	0.00	1.98×10	1.933×10	1.933×10	0.9998
0.0080	2.76×10^{-1}	1.18×10^{-1}	7.95	0.00	8.34	7.948	7.946	0.9998
0.0100	2.07×10^{-1}	1.29×10^{-1}	3.95	0.00	4.29	3.955	3.954	0.9997
0.0150	1.20×10^{-1}	1.45×10^{-1}	1.09	0.00	1.36	1.096	1.096	0.9997
0.0200	7.97×10^{-2}	1.53×10^{-1}	4.33×10^{-1}	0.00	6.65×10^{-1}	4.386×10^{-1}	4.384×10^{-1}	0.9997
0.0300	4.20×10^{-2}	1.59×10^{-1}	1.16×10^{-1}	0.00	3.17×10^{-1}	1.245×10^{-1}	1.244×10^{-1}	0.9996
0.0400	2.56×10^{-2}	1.59×10^{-1}	4.50×10^{-2}	0.00	2.30×10^{-1}	5.615×10^{-2}	5.612×10^{-2}	0.9996
0.0500	1.72×10^{-2}	1.57×10^{-1}	2.16×10^{-2}	0.00	1.96×10^{-1}	3.470×10^{-2}	3.469×10^{-2}	0.9996
0.0600	1.24×10^{-2}	1.54×10^{-1}	1.18×10^{-2}	0.00	1.79×10^{-1}	2.666×10^{-2}	2.665×10^{-2}	0.9996
0.0800	7.23×10^{-3}	1.48×10^{-1}	4.56×10^{-3}	0.00	1.60×10^{-1}	2.222×10^{-2}	2.221×10^{-2}	0.9996
0.1000	4.73×10^{-3}	1.42×10^{-1}	2.18×10^{-3}	0.00	1.49×10^{-1}	2.202×10^{-2}	2.201×10^{-2}	0.9996
0.1500	2.16×10^{-3}	1.29×10^{-1}	5.77×10^{-4}	0.00	1.31×10^{-1}	2.409×10^{-2}	2.408×10^{-2}	0.9995
0.2000	1.22×10^{-3}	1.18×10^{-1}	2.28×10^{-4}	0.00	1.20×10^{-1}	2.591×10^{-2}	2.589×10^{-2}	0.9994
0.3000	5.48×10^{-4}	1.03×10^{-1}	6.44×10^{-5}	0.00	1.04×10^{-1}	2.790×10^{-2}	2.787×10^{-2}	0.9991
0.4000	3.09×10^{-4}	9.24×10^{-2}	2.76×10^{-5}	0.00	9.27×10^{-2}	2.867×10^{-2}	2.863×10^{-2}	0.9989
0.5000	1.98×10^{-4}	8.44×10^{-2}	1.49×10^{-5}	0.00	8.46×10^{-2}	2.885×10^{-2}	2.881×10^{-2}	0.9986
0.6000	1.38×10^{-4}	7.81×10^{-2}	9.25×10^{-6}	0.00	7.82×10^{-2}	2.873×10^{-2}	2.868×10^{-2}	0.9984
0.8000	7.74×10^{-5}	6.86×10^{-2}	4.67×10^{-6}	0.00	6.87×10^{-2}	2.806×10^{-2}	2.800×10^{-2}	0.9979
1.0000	4.96×10^{-5}	6.17×10^{-2}	2.90×10^{-6}	0.00	6.18×10^{-2}	2.717×10^{-2}	2.710×10^{-2}	0.9973
1.2500	3.17×10^{-5}	5.52×10^{-2}	1.83×10^{-6}	1.62×10^{-5}	5.52×10^{-2}	2.598×10^{-2}	2.589×10^{-2}	0.9967
1.5000	2.20×10^{-5}	5.02×10^{-2}	1.33×10^{-6}	8.96×10^{-5}	5.03×10^{-2}	2.484×10^{-2}	2.474×10^{-2}	0.9960
2.0000	1.24×10^{-5}	4.28×10^{-2}	8.37×10^{-7}	3.57×10^{-4}	4.32×10^{-2}	2.290×10^{-2}	2.278×10^{-2}	0.9945
3.0000	5.51×10^{-6}	3.37×10^{-2}	4.68×10^{-7}	1.03×10^{-3}	3.47×10^{-2}	2.012×10^{-2}	1.994×10^{-2}	0.9910
4.0000	3.10×10^{-6}	2.81×10^{-2}	3.21×10^{-7}	1.70×10^{-3}	2.98×10^{-2}	1.832×10^{-2}	1.808×10^{-2}	0.9871
5.0000	1.98×10^{-6}	2.43×10^{-2}	2.44×10^{-7}	2.32×10^{-3}	2.66×10^{-2}	1.708×10^{-2}	1.678×10^{-2}	0.9829

Mass Coefficients (cm^2 g^{-1})

K, L, M Edges

6.0000	1.38×10^{-6}	2.14×10^{-2}	1.96×10^{-7}	2.88×10^{-3}	2.43×10^{-2}	1.619×10^{-2}	1.585×10^{-2}	0.9785
8.0000	7.75×10^{-7}	1.75×10^{-2}	1.40×10^{-7}	3.84×10^{-3}	2.14×10^{-2}	1.505×10^{-2}	1.459×10^{-2}	0.9697
10.0000	4.96×10^{-7}	1.49×10^{-2}	1.09×10^{-7}	4.64×10^{-3}	1.96×10^{-2}	1.437×10^{-2}	1.381×10^{-2}	0.9609
15.0000	2.20×10^{-7}	1.11×10^{-2}	7.03×10^{-8}	6.15×10^{-3}	1.72×10^{-2}	1.358×10^{-2}	1.276×10^{-2}	0.9397
20.0000	1.24×10^{-7}	8.87×10^{-3}	5.17×10^{-8}	7.26×10^{-3}	1.61×10^{-2}	1.334×10^{-2}	1.227×10^{-2}	0.9200
30.0000	5.51×10^{-8}	6.46×10^{-3}	3.38×10^{-8}	8.84×10^{-3}	1.53×10^{-2}	1.336×10^{-2}	1.182×10^{-2}	0.8844
40.0000	3.10×10^{-8}	5.13×10^{-3}	2.51×10^{-8}	9.93×10^{-3}	1.51×10^{-2}	1.358×10^{-2}	1.158×10^{-2}	0.8526
50.0000	1.98×10^{-8}	4.28×10^{-3}	2.00×10^{-8}	1.08×10^{-2}	1.51×10^{-2}	1.385×10^{-2}	1.141×10^{-2}	0.8237

TABLE M.3j

Photons in A-150 Tissue-Equivalent Plastic

ρ: 1.127 (g cm^{-3}), Composition ($Z-f_w$): 1−0.101327; 6−0.775501; 7−0.035057; 8−0.052316; 9−0.017422; 20−0.018378

K, L, M Edges	Energy (MeV)	Mass Coefficients (cm^2 g^{-1})							(1-g)
		Coherent σ_{coh}/ρ	Compton σ_C/ρ	Photoelectric τ/ρ	Pair + Triplet κ/ρ	Total Attenuation μ/ρ	Energy-Transfer μ_{tr}/ρ	Energy-Absorption μ_{en}/ρ	
	0.0010	1.09	1.61×10^{-2}	2.26×10^{3}	0.00	2.26×10^{3}	2.256×10^{3}	2.256×10^{3}	1
	0.0015	9.74×10^{-1}	3.18×10^{-2}	7.27×10^{2}	0.00	7.28×10^{2}	7.267×10^{2}	7.267×10^{2}	1
	0.0020	8.52×10^{-1}	4.86×10^{-2}	3.17×10^{2}	0.00	3.18×10^{2}	3.172×10^{2}	3.172×10^{2}	0.9999
	0.0030	6.38×10^{-1}	7.91×10^{-2}	9.58×10	0.00	9.65×10	9.577×10	9.576×10	0.9999
	0.0040	4.84×10^{-1}	1.02×10^{-1}	4.02×10	0.00	4.08×10	4.021×10	4.021×10	0.9999
	0.004038	4.80×10^{-1}	1.03×10^{-1}	3.91×10	0.00	3.97×10	3.907×10	3.907×10	0.9999
20 K	0.004038	4.80×10^{-1}	1.03×10^{-1}	5.57×10	0.00	5.63×10	5.326×10	5.326×10	0.9999
	0.0050	3.80×10^{-1}	1.19×10^{-1}	3.02×10	0.00	3.07×10	2.902×10	2.901×10	0.9999
	0.0060	3.09×10^{-1}	1.31×10^{-1}	1.77×10	0.00	1.81×10	1.709×10	1.708×10	0.9999
	0.0080	2.21×10^{-1}	1.46×10^{-1}	7.55	0.00	7.91	7.337	7.337	0.9999
	0.0100	1.70×10^{-1}	1.57×10^{-1}	3.86	0.00	4.19	3.771	3.771	0.9999
	0.0150	1.03×10^{-1}	1.71×10^{-1}	1.12	0.00	1.39	1.106	1.106	0.9998
	0.0200	6.79×10^{-2}	1.79×10^{-1}	4.60×10^{-1}	0.00	7.07×10^{-1}	4.606×10^{-1}	4.605×10^{-1}	0.9998
	0.0300	3.55×10^{-2}	1.84×10^{-1}	1.29×10^{-1}	0.00	3.48×10^{-1}	1.378×10^{-1}	1.378×10^{-1}	0.9997
	0.0400	2.17×10^{-2}	1.83×10^{-1}	5.19×10^{-2}	0.00	2.56×10^{-1}	6.413×10^{-2}	6.411×10^{-2}	0.9997
	0.0500	1.46×10^{-2}	1.80×10^{-1}	2.55×10^{-2}	0.00	2.20×10^{-1}	4.019×10^{-2}	4.018×10^{-2}	0.9997
	0.0600	1.05×10^{-2}	1.76×10^{-1}	1.42×10^{-2}	0.00	2.01×10^{-1}	3.095×10^{-2}	3.095×10^{-2}	0.9997
	0.0800	6.13×10^{-3}	1.69×10^{-1}	5.63×10^{-3}	0.00	1.80×10^{-1}	2.561×10^{-2}	2.561×10^{-2}	0.9997
	0.1000	4.01×10^{-3}	1.61×10^{-1}	2.75×10^{-3}	0.00	1.68×10^{-1}	2.521×10^{-2}	2.520×10^{-2}	0.9997
	0.1500	1.82×10^{-3}	1.46×10^{-1}	7.51×10^{-4}	0.00	1.48×10^{-1}	2.737×10^{-2}	2.737×10^{-2}	0.9996
	0.2000	1.04×10^{-3}	1.34×10^{-1}	3.03×10^{-4}	0.00	1.35×10^{-1}	2.938×10^{-2}	2.937×10^{-2}	0.9996
	0.3000	4.64×10^{-4}	1.17×10^{-1}	8.73×10^{-5}	0.00	1.17×10^{-1}	3.161×10^{-2}	3.159×10^{-2}	0.9994
	0.4000	2.62×10^{-4}	1.05×10^{-1}	3.79×10^{-5}	0.00	1.05×10^{-1}	3.247×10^{-2}	3.245×10^{-2}	0.9992
	0.5000	1.68×10^{-4}	9.56×10^{-2}	2.06×10^{-5}	0.00	9.58×10^{-2}	3.268×10^{-2}	3.265×10^{-2}	0.9990
	0.6000	1.17×10^{-4}	8.84×10^{-2}	1.29×10^{-5}	0.00	8.86×10^{-2}	3.253×10^{-2}	3.249×10^{-2}	0.9988
	0.8000	6.57×10^{-5}	7.77×10^{-2}	6.51×10^{-6}	0.00	7.78×10^{-2}	3.177×10^{-2}	3.172×10^{-2}	0.9984
	1.0000	4.20×10^{-5}	6.99×10^{-2}	4.05×10^{-6}	0.00	6.99×10^{-2}	3.076×10^{-2}	3.070×10^{-2}	0.998
	1.2500	2.69×10^{-5}	6.25×10^{-2}	2.58×10^{-6}	1.46×10^{-5}	6.25×10^{-2}	2.942×10^{-2}	2.934×10^{-2}	0.9974
	1.5000	1.87×10^{-5}	5.68×10^{-2}	1.87×10^{-6}	8.07×10^{-5}	5.69×10^{-2}	2.813×10^{-2}	2.805×10^{-2}	0.9969
	2.0000	1.05×10^{-5}	4.85×10^{-2}	1.16×10^{-6}	3.21×10^{-4}	4.88×10^{-2}	2.589×10^{-2}	2.578×10^{-2}	0.9957
	3.0000	4.67×10^{-6}	3.81×10^{-2}	6.42×10^{-7}	9.32×10^{-4}	3.91×10^{-2}	2.263×10^{-2}	2.247×10^{-2}	0.9929

4.0000	2.63×10^{-6}	3.18×10^{-2}	4.37×10^{-7}	1.54×10^{-3}	3.34×10^{-2}	2.046×10^{-2}	2.025×10^{-2}	0.9898
5.0000	1.68×10^{-6}	2.75×10^{-2}	3.30×10^{-7}	2.11×10^{-3}	2.96×10^{-2}	1.893×10^{-2}	1.867×10^{-2}	0.9864
6.0000	1.17×10^{-6}	2.43×10^{-2}	2.64×10^{-7}	2.63×10^{-3}	2.69×10^{-2}	1.781×10^{-2}	1.751×10^{-2}	0.9828
8.0000	6.57×10^{-7}	1.99×10^{-2}	1.88×10^{-7}	3.51×10^{-3}	2.34×10^{-2}	1.632×10^{-2}	1.592×10^{-2}	0.9755
10.0000	4.21×10^{-7}	1.69×10^{-2}	1.46×10^{-7}	4.26×10^{-3}	2.12×10^{-2}	1.538×10^{-2}	1.489×10^{-2}	0.9681
15.0000	1.87×10^{-7}	1.25×10^{-2}	9.34×10^{-8}	5.67×10^{-3}	1.82×10^{-2}	1.417×10^{-2}	1.346×10^{-2}	0.9500
20.0000	1.05×10^{-7}	1.00×10^{-2}	6.86×10^{-8}	6.71×10^{-3}	1.68×10^{-2}	1.366×10^{-2}	1.275×10^{-2}	0.9331
30.0000	4.67×10^{-8}	7.31×10^{-3}	4.48×10^{-8}	8.19×10^{-3}	1.55×10^{-2}	1.337×10^{-2}	1.206×10^{-2}	0.9021
40.0000	2.63×10^{-8}	5.81×10^{-3}	3.32×10^{-8}	9.22×10^{-3}	1.50×10^{-2}	1.340×10^{-2}	1.171×10^{-2}	0.8741
50.0000	1.68×10^{-8}	4.85×10^{-3}	2.64×10^{-8}	1.00×10^{-2}	1.49×10^{-2}	1.353×10^{-2}	1.148×10^{-2}	0.8486

TABLE M.3k

Photons in Air, Dry (Near Sea Level)

ρ: 1.760 (g cm^{-3}), Composition (Z−f_w): 6−0.000124; 7−0.755267; 8−0.231781; 18−0.012827

K, L, M Edges	Energy (MeV)	Coherent σ_{coh}/ρ	Compton σ_C/ρ	Photoelectric τ/ρ	Pair + Triplet κ/ρ	Total Attenuation μ/ρ	Energy-Transfer μ_{tr}/ρ	Energy-Absorption μ_{en}/ρ	$(1-\bar{g})$
					Mass Coefficients (cm^2 g^{-1})				
	0.0010	1.36	1.04×10^{-2}	3.60×10^{3}	0.00	3.61×10^{3}	3.599×10^{3}	3.599×10^{3}	1
	0.0015	1.25	2.12×10^{-2}	1.19×10^{3}	0.00	1.19×10^{3}	1.188×10^{3}	1.188×10^{3}	0.9999
	0.0020	1.12	3.34×10^{-2}	5.27×10^{2}	0.00	5.28×10^{2}	5.263×10^{2}	5.262×10^{2}	0.9999
	0.0030	8.63×10^{-1}	5.75×10^{-2}	1.62×10^{2}	0.00	1.62×10^{2}	1.615×10^{2}	1.614×10^{2}	0.9999
	0.003203	8.18×10^{-1}	6.20×10^{-2}	1.33×10^{2}	0.00	1.34×10^{2}	1.330×10^{2}	1.330×10^{2}	0.9999
18 K	0.003203	8.18×10^{-1}	6.20×10^{-2}	1.48×10^{2}	0.00	1.48×10^{2}	1.460×10^{2}	1.460×10^{2}	0.9999
	0.0040	6.65×10^{-1}	7.77×10^{-2}	7.72×10	0.00	7.79×10	7.637×10	7.636×10	0.9999
	0.0050	5.22×10^{-1}	9.33×10^{-2}	3.97×10	0.00	4.03×10	3.932×10	3.931×10	0.9998
	0.0060	4.21×10^{-1}	1.05×10^{-1}	2.29×10	0.00	2.34×10	2.271×10	2.270×10	0.9998
	0.0080	2.95×10^{-1}	1.21×10^{-1}	9.51	0.00	9.92	9.448	9.446	0.9998
	0.0100	2.22×10^{-1}	1.32×10^{-1}	4.77	0.00	5.12	4.743	4.742	0.9997
	0.0150	1.31×10^{-1}	1.47×10^{-1}	1.34	0.00	1.61	1.334	1.334	0.9997
	0.0200	8.75×10^{-2}	1.56×10^{-1}	5.35×10^{-1}	0.00	7.78×10^{-1}	5.391×10^{-1}	5.389×10^{-1}	0.9996
	0.0300	4.62×10^{-2}	1.62×10^{-1}	1.45×10^{-1}	0.00	3.54×10^{-1}	1.538×10^{-1}	1.537×10^{-1}	0.9996
	0.0400	2.83×10^{-2}	1.63×10^{-1}	5.71×10^{-2}	0.00	2.49×10^{-1}	6.836×10^{-2}	6.833×10^{-2}	0.9995
	0.0500	1.91×10^{-2}	1.61×10^{-1}	2.76×10^{-2}	0.00	2.08×10^{-1}	4.100×10^{-2}	4.098×10^{-2}	0.9995
	0.0600	1.37×10^{-2}	1.59×10^{-1}	1.52×10^{-2}	0.00	1.87×10^{-1}	3.042×10^{-2}	3.041×10^{-2}	0.9996
	0.0800	8.03×10^{-3}	1.52×10^{-1}	5.92×10^{-3}	0.00	1.66×10^{-1}	2.408×10^{-2}	2.407×10^{-2}	0.9996
	0.1000	5.26×10^{-3}	1.46×10^{-1}	2.85×10^{-3}	0.00	1.54×10^{-1}	2.326×10^{-2}	2.325×10^{-2}	0.9996
	0.1500	2.40×10^{-3}	1.32×10^{-1}	7.61×10^{-4}	0.00	1.36×10^{-1}	2.497×10^{-2}	2.496×10^{-2}	0.9995
	0.2000	1.36×10^{-3}	1.22×10^{-1}	3.03×10^{-4}	0.00	1.23×10^{-1}	2.674×10^{-2}	2.672×10^{-2}	0.9993
	0.3000	6.10×10^{-4}	1.06×10^{-1}	8.61×10^{-5}	0.00	1.07×10^{-1}	2.875×10^{-2}	2.949×10^{-2}	0.9991
	0.4000	3.44×10^{-4}	9.51×10^{-2}	3.70×10^{-5}	0.00	9.55×10^{-2}	2.953×10^{-2}	2.966×10^{-2}	0.9988
	0.5000	2.20×10^{-4}	8.69×10^{-2}	2.00×10^{-5}	0.00	8.71×10^{-2}	2.971×10^{-2}	2.953×10^{-2}	0.9985
	0.6000	1.53×10^{-4}	8.04×10^{-2}	1.25×10^{-5}	0.00	8.06×10^{-2}	2.958×10^{-2}	2.882×10^{-2}	0.9983
	0.8000	8.62×10^{-5}	7.06×10^{-2}	6.30×10^{-6}	0.00	7.07×10^{-2}	2.889×10^{-2}	2.789×10^{-2}	0.9977
	1.0000	5.52×10^{-5}	6.35×10^{-2}	3.92×10^{-6}	0.00	6.36×10^{-2}	2.797×10^{-2}	2.666×10^{-2}	0.9972
	1.2500	3.53×10^{-5}	5.68×10^{-2}	2.48×10^{-6}	1.78×10^{-5}	5.69×10^{-2}	2.675×10^{-2}	2.547×10^{-2}	0.9965
	1.5000	2.45×10^{-5}	5.16×10^{-2}	1.80×10^{-6}	9.85×10^{-5}	5.17×10^{-2}	2.557×10^{-2}	2.345×10^{-2}	0.9958
	2.0000	1.38×10^{-5}	4.41×10^{-2}	1.13×10^{-6}	3.92×10^{-4}	4.45×10^{-2}	2.359×10^{-2}	2.057×10^{-2}	0.9943
	3.0000	6.13×10^{-6}	3.47×10^{-2}	6.28×10^{-7}	1.13×10^{-3}	3.58×10^{-2}	2.076×10^{-2}		0.9909

4.0000	3.45×10^{-6}	2.89×10^{-2}	4.30×10^{-7}	1.87×10^{-3}	3.08×10^{-2}	1.894×10^{-2}	1.870×10^{-2}	0.9871
5.0000	2.21×10^{-6}	2.50×10^{-2}	3.26×10^{-7}	2.54×10^{-3}	2.75×10^{-2}	1.770×10^{-2}	1.740×10^{-2}	0.9831
6.0000	1.53×10^{-6}	2.21×10^{-2}	2.61×10^{-7}	3.15×10^{-3}	2.52×10^{-2}	1.683×10^{-2}	1.647×10^{-2}	0.9790
8.0000	8.63×10^{-7}	1.81×10^{-2}	1.87×10^{-7}	4.20×10^{-3}	2.23×10^{-2}	1.571×10^{-2}	1.525×10^{-2}	0.9708
10.0000	5.52×10^{-7}	1.54×10^{-2}	1.45×10^{-7}	5.07×10^{-3}	2.04×10^{-2}	1.506×10^{-2}	1.450×10^{-2}	0.9627
15.0000	2.45×10^{-7}	1.14×10^{-2}	9.33×10^{-8}	6.72×10^{-3}	1.81×10^{-2}	1.434×10^{-2}	1.353×10^{-2}	0.9436
20.0000	1.38×10^{-7}	9.13×10^{-3}	6.87×10^{-8}	7.92×10^{-3}	1.71×10^{-2}	1.415×10^{-2}	1.311×10^{-2}	0.9261
30.0000	6.14×10^{-8}	6.65×10^{-3}	4.49×10^{-8}	9.63×10^{-3}	1.63×10^{-2}	1.427×10^{-2}	1.277×10^{-2}	0.8948
40.0000	3.45×10^{-8}	5.29×10^{-3}	3.33×10^{-8}	1.08×10^{-2}	1.61×10^{-2}	1.456×10^{-2}	1.262×10^{-2}	0.8669
50.0000	2.21×10^{-8}	4.41×10^{-3}	2.65×10^{-8}	1.17×10^{-2}	1.61×10^{-2}	1.488×10^{-2}	1.252×10^{-2}	0.8415

TABLE M.3l

Photons in C-552 Air-Equivalent Plastic

ρ: 1.760 (g cm^{-3}), Composition (Z−f_w): 1−0.024681; 6−0.501610; 8−0.004527; 9−0.465209; 14−0.003973

K, L, M Edges	Energy (MeV)	Mass Coefficients (cm^2 g^{-1})							(1−g)
		Coherent σ_{coh}/ρ	Compton σ_C/ρ	Photoelectric τ/ρ	Pair + Triplet κ/ρ	Total Attenuation μ/ρ	Energy-Transfer μ_{tr}/ρ	Energy-Absorption μ_{en}/ρ	
	0.0010	1.32	1.07×10^{-2}	3.76×10^{3}	0.00	3.76×10^{3}	3.747×10^{3}	3.747×10^{3}	1
	0.0015	1.21	2.15×10^{-2}	1.28×10^{3}	0.00	1.28×10^{3}	1.276×10^{3}	1.276×10^{3}	1
	0.001839	1.13	2.96×10^{-2}	7.29×10^{2}	0.00	7.30×10^{2}	7.271×10^{2}	7.271×10^{2}	0.9999
14K	0.001839	1.13	2.96×10^{-2}	7.40×10^{2}	0.00	7.42×10^{2}	7.381×10^{2}	7.380×10^{2}	0.9999
	0.0020	1.09	3.36×10^{-2}	5.86×10^{2}	0.00	5.87×10^{2}	5.840×10^{2}	5.840×10^{2}	0.9999
	0.0030	8.57×10^{-1}	5.73×10^{-2}	1.84×10^{2}	0.00	1.85×10^{2}	1.833×10^{2}	1.832×10^{2}	0.9999
	0.0040	6.71×10^{-1}	7.73×10^{-2}	7.89×10	0.00	7.96×10	7.874×10	7.873×10	0.9999
	0.0050	5.33×10^{-1}	9.29×10^{-2}	4.05×10	0.00	4.11×10	4.042×10	4.041×10	0.9998
	0.0060	4.32×10^{-1}	1.05×10^{-1}	2.33×10	0.00	2.38×10	2.328×10	2.327×10	0.9998
	0.0080	3.04×10^{-1}	1.22×10^{-1}	9.64	0.00	1.01×10	9.634	9.632	0.9998
	0.0100	2.28×10^{-1}	1.33×10^{-1}	4.82	0.00	5.18	4.818	4.817	0.9997
	0.0150	1.32×10^{-1}	1.49×10^{-1}	1.34	0.00	1.62	1.346	1.346	0.9997
	0.0200	8.73×10^{-2}	1.57×10^{-1}	5.35×10^{-1}	0.00	7.79×10^{-1}	5.411×10^{-1}	5.409×10^{-1}	0.9996
	0.0300	4.60×10^{-2}	1.63×10^{-1}	1.44×10^{-1}	0.00	3.53×10^{-1}	1.532×10^{-1}	1.531×10^{-1}	0.9996
	0.0400	2.82×10^{-2}	1.64×10^{-1}	5.64×10^{-2}	0.00	2.48×10^{-1}	6.781×10^{-2}	6.778×10^{-2}	0.9995
	0.0500	1.90×10^{-2}	1.62×10^{-1}	2.71×10^{-2}	0.00	2.08×10^{-1}	4.062×10^{-2}	4.060×10^{-2}	0.9995
	0.0600	1.36×10^{-2}	1.59×10^{-1}	1.49×10^{-2}	0.00	1.87×10^{-1}	3.017×10^{-2}	3.016×10^{-2}	0.9996
	0.0800	7.98×10^{-3}	1.53×10^{-1}	5.77×10^{-3}	0.00	1.66×10^{-1}	2.396×10^{-2}	2.395×10^{-2}	0.9996
	0.1000	5.22×10^{-3}	1.46×10^{-1}	2.77×10^{-3}	0.00	1.54×10^{-1}	2.321×10^{-2}	2.320×10^{-2}	0.9996
	0.1500	2.38×10^{-3}	1.33×10^{-1}	7.35×10^{-4}	0.00	1.36×10^{-1}	2.496×10^{-2}	2.495×10^{-2}	0.9995
	0.2000	1.35×10^{-3}	1.22×10^{-1}	2.91×10^{-4}	0.00	1.23×10^{-1}	2.675×10^{-2}	2.674×10^{-2}	0.9993
	0.3000	6.06×10^{-4}	1.06×10^{-1}	8.23×10^{-5}	0.00	1.07×10^{-1}	2.877×10^{-2}	2.874×10^{-2}	0.9991
	0.4000	3.42×10^{-4}	9.52×10^{-2}	3.53×10^{-5}	0.00	9.56×10^{-2}	2.952×10^{-2}	2.952×10^{-2}	0.9988
	0.5000	2.19×10^{-4}	8.70×10^{-2}	1.90×10^{-5}	0.00	8.72×10^{-2}	2.974×10^{-2}	2.970×10^{-2}	0.9986
	0.6000	1.52×10^{-4}	8.05×10^{-2}	1.19×10^{-5}	0.00	8.06×10^{-2}	2.961×10^{-2}	2.956×10^{-2}	0.9983
	0.8000	8.57×10^{-5}	7.07×10^{-2}	5.99×10^{-6}	0.00	7.08×10^{-2}	2.892×10^{-2}	2.885×10^{-2}	0.9978
	1.0000	5.49×10^{-5}	6.36×10^{-2}	3.72×10^{-6}	0.00	6.37×10^{-2}	2.800×10^{-2}	2.792×10^{-2}	0.9973
	1.2500	3.51×10^{-5}	5.69×10^{-2}	2.36×10^{-6}	1.72×10^{-5}	5.69×10^{-2}	2.677×10^{-2}	2.668×10^{-2}	0.9966
	1.5000	2.44×10^{-5}	5.17×10^{-2}	1.71×10^{-6}	9.51×10^{-5}	5.18×10^{-2}	2.561×10^{-2}	2.551×10^{-2}	0.9959
	2.0000	1.37×10^{-5}	4.41×10^{-2}	1.08×10^{-6}	3.79×10^{-4}	4.45×10^{-2}	2.360×10^{-2}	2.347×10^{-2}	0.9943
	3.0000	6.10×10^{-6}	3.47×10^{-2}	6.00×10^{-7}	1.09×10^{-3}	3.58×10^{-2}	2.076×10^{-2}	2.057×10^{-2}	0.9908

4.0000	3.43×10^{-6}	2.89×10^{-2}	4.11×10^{-7}	1.81×10^{-3}	3.08×10^{-2}	1.892×10^{-2}	1.867×10^{-2}	0.9869
5.0000	2.20×10^{-6}	2.50×10^{-2}	3.12×10^{-7}	2.46×10^{-3}	2.75×10^{-2}	1.766×10^{-2}	1.735×10^{-2}	0.9826
6.0000	1.53×10^{-6}	2.21×10^{-2}	2.51×10^{-7}	3.05×10^{-3}	2.52×10^{-2}	1.676×10^{-2}	1.640×10^{-2}	0.9782
8.0000	8.58×10^{-7}	1.81×10^{-2}	1.79×10^{-7}	4.07×10^{-3}	2.21×10^{-2}	1.561×10^{-2}	1.513×10^{-2}	0.9693
10.0000	5.49×10^{-7}	1.54×10^{-2}	1.40×10^{-7}	4.92×10^{-3}	2.03×10^{-2}	1.494×10^{-2}	1.435×10^{-2}	0.9604
15.0000	2.44×10^{-7}	1.14×10^{-2}	8.97×10^{-8}	6.51×10^{-3}	1.79×10^{-2}	1.416×10^{-2}	1.330×10^{-2}	0.9391
20.0000	1.37×10^{-7}	9.14×10^{-3}	6.60×10^{-8}	7.69×10^{-3}	1.68×10^{-2}	1.394×10^{-2}	1.281×10^{-2}	0.9193
30.0000	6.10×10^{-8}	6.66×10^{-3}	4.32×10^{-8}	9.35×10^{-3}	1.60×10^{-2}	1.401×10^{-2}	1.238×10^{-2}	0.8834
40.0000	3.43×10^{-8}	5.29×10^{-3}	3.21×10^{-8}	1.05×10^{-2}	1.58×10^{-2}	1.427×10^{-2}	1.215×10^{-2}	0.8515
50.0000	2.20×10^{-8}	4.42×10^{-3}	2.55×10^{-8}	1.14×10^{-2}	1.58×10^{-2}	1.456×10^{-2}	1.197×10^{-2}	0.8225

TABLE M.3m

Photons in Glass, Borosilicate (Pyrex)

ρ: 2.230 (g cm^{-3}), Composition ($Z-f_w$): 5–0.040066; 8–0.539559; 11–0.028191; 13–0.011644; 14–0.377220; 19–003321

K, L, M Edges	Energy (MeV)	Coherent σ_{coh}/ρ	Compton σ_C/ρ	Photoelectric τ/ρ	Pair+Triplet κ/ρ	Total Attenuation μ/ρ	Energy-Transfer μ_{tr}/ρ	Energy-Absorption μ_{en}/ρ	$(1-g)$
				Mass Coefficients (cm^2 g^{-1})					
	0.0010	1.89	1.08×10^{-2}	3.16×10^{3}	0.00	3.16×10^{3}	3.155×10^{3}	3.155×10^{3}	1
	0.001072	1.86	1.21×10^{-2}	2.63×10^{3}	0.00	2.63×10^{3}	2.627×10^{3}	2.627×10^{3}	1
11 K	0.001072	1.86	1.21×10^{-2}	2.79×10^{3}	0.00	2.80×10^{3}	2.790×10^{3}	2.790×10^{3}	1
	0.0015	1.73	2.07×10^{-2}	1.15×10^{3}	0.00	1.15×10^{3}	1.148×10^{3}	1.148×10^{3}	0.9999
	0.001560	1.70	2.20×10^{-2}	1.04×10^{3}	0.00	1.04×10^{3}	1.033×10^{3}	1.033×10^{3}	0.9999
13 K	0.001560	1.70	2.20×10^{-2}	1.08×10^{3}	0.00	1.08×10^{3}	1.073×10^{3}	1.073×10^{3}	0.9999
	0.001839	1.61	2.78×10^{-2}	6.87×10^{2}	0.00	6.88×10^{2}	6.845×10^{2}	6.845×10^{2}	0.9999
14 K	0.001839	1.61	2.78×10^{-2}	1.77×10^{3}	0.00	1.78×10^{3}	1.723×10^{3}	1.723×10^{3}	1
	0.0020	1.55	3.12×10^{-2}	1.50×10^{3}	0.00	1.50×10^{3}	1.457×10^{3}	1.457×10^{3}	1
	0.0030	1.25	5.08×10^{-2}	5.11×10^{2}	0.00	5.12×10^{2}	5.012×10^{2}	5.011×10^{2}	0.9999
	0.003607	1.10	6.12×10^{-2}	3.07×10^{2}	0.00	3.09×10^{2}	3.036×10^{2}	3.036×10^{2}	0.9999
19 K	0.003607	1.10	6.12×10^{-2}	3.11×10^{2}	0.00	3.12×10^{2}	3.067×10^{2}	3.067×10^{2}	0.9999
	0.0040	1.02	6.73×10^{-2}	2.34×10^{2}	0.00	2.35×10^{2}	2.307×10^{2}	2.307×10^{2}	0.9999
	0.0050	8.46×10^{-1}	8.08×10^{-2}	1.25×10^{2}	0.00	1.26×10^{2}	1.235×10^{2}	1.235×10^{2}	0.9998
	0.0060	7.13×10^{-1}	9.19×10^{-2}	7.42×10	0.00	7.50×10	7.339×10	7.337×10	0.9998
	0.0080	5.24×10^{-1}	1.08×10^{-1}	3.21×10	0.00	3.27×10	3.180×10	3.178×10	0.9997
	0.0100	4.00×10^{-1}	1.20×10^{-1}	1.65×10	0.00	1.70×10	1.642×10	1.642×10	0.9996
	0.0150	2.31×10^{-1}	1.38×10^{-1}	4.85	0.00	5.22	4.830	4.828	0.9995
	0.0200	1.52×10^{-1}	1.48×10^{-1}	2.00	0.00	2.30	1.997	1.995	0.9993
	0.0300	8.09×10^{-2}	1.56×10^{-1}	5.62×10^{-1}	0.00	7.99×10^{-1}	5.689×10^{-1}	5.684×10^{-1}	0.9992
	0.0400	5.04×10^{-2}	1.58×10^{-1}	2.25×10^{-1}	0.00	4.34×10^{-1}	2.363×10^{-1}	2.361×10^{-1}	0.9991
	0.0500	3.43×10^{-2}	1.57×10^{-1}	1.11×10^{-1}	0.00	3.02×10^{-1}	1.237×10^{-1}	1.235×10^{-1}	0.9990
	0.0600	2.48×10^{-2}	1.55×10^{-1}	6.16×10^{-2}	0.00	2.42×10^{-1}	7.656×10^{-2}	7.648×10^{-2}	0.9990
	0.0800	1.46×10^{-2}	1.50×10^{-1}	2.44×10^{-2}	0.00	1.89×10^{-1}	4.235×10^{-2}	4.230×10^{-2}	0.9990
	0.1000	9.65×10^{-3}	1.44×10^{-1}	1.19×10^{-2}	0.00	1.66×10^{-1}	3.212×10^{-2}	3.209×10^{-2}	0.9990
	0.1500	4.45×10^{-3}	1.31×10^{-1}	3.22×10^{-3}	0.00	1.39×10^{-1}	2.730×10^{-2}	2.727×10^{-2}	0.9989
	0.2000	2.54×10^{-3}	1.21×10^{-1}	1.29×10^{-3}	0.00	1.25×10^{-1}	2.760×10^{-2}	2.757×10^{-2}	0.9987
	0.3000	1.14×10^{-3}	1.05×10^{-1}	3.71×10^{-4}	0.00	1.07×10^{-1}	2.890×10^{-2}	2.885×10^{-2}	0.9983
	0.4000	6.47×10^{-4}	9.46×10^{-2}	1.60×10^{-4}	0.00	9.54×10^{-2}	2.952×10^{-2}	2.946×10^{-2}	0.9979
	0.5000	4.15×10^{-4}	8.65×10^{-2}	8.68×10^{-5}	0.00	8.70×10^{-2}	2.965×10^{-2}	2.957×10^{-2}	0.9975
	0.6000	2.89×10^{-4}	8.00×10^{-2}	5.42×10^{-5}	0.00	8.04×10^{-2}	2.950×10^{-2}	2.941×10^{-2}	0.9971

0.8000	1.63×10^{-4}	7.03×10^{-2}	2.74×10^{-5}	0.00	7.05×10^{-2}	2.879×10^{-2}	2.868×10^{-2}	0.9963
1.0000	1.04×10^{-4}	6.33×10^{-2}	1.71×10^{-5}	0.00	6.34×10^{-2}	2.787×10^{-2}	2.774×10^{-2}	0.9956
1.2500	6.66×10^{-5}	5.66×10^{-2}	1.09×10^{-5}	2.55×10^{-5}	5.67×10^{-2}	2.664×10^{-2}	2.650×10^{-2}	0.9946
1.5000	4.63×10^{-5}	5.14×10^{-2}	7.88×10^{-6}	1.39×10^{-4}	5.16×10^{-2}	2.549×10^{-2}	2.533×10^{-2}	0.9935
2.0000	2.60×10^{-5}	4.39×10^{-2}	4.92×10^{-6}	5.51×10^{-4}	4.45×10^{-2}	2.357×10^{-2}	2.337×10^{-2}	0.9913
3.0000	1.16×10^{-5}	3.45×10^{-2}	2.72×10^{-6}	1.58×10^{-3}	3.61×10^{-2}	2.098×10^{-2}	2.069×10^{-2}	0.9864
4.0000	6.51×10^{-6}	2.88×10^{-2}	1.86×10^{-6}	2.59×10^{-3}	3.14×10^{-2}	1.941×10^{-2}	1.904×10^{-2}	0.9809
5.0000	4.17×10^{-6}	2.49×10^{-2}	1.40×10^{-6}	3.51×10^{-3}	2.84×10^{-2}	1.841×10^{-2}	1.795×10^{-2}	0.9752
6.0000	2.89×10^{-6}	2.20×10^{-2}	1.12×10^{-6}	4.33×10^{-3}	2.63×10^{-2}	1.775×10^{-2}	1.721×10^{-2}	0.9694
8.0000	1.63×10^{-6}	1.80×10^{-2}	8.02×10^{-7}	5.75×10^{-3}	2.37×10^{-2}	1.701×10^{-2}	1.629×10^{-2}	0.9577
10.0000	1.04×10^{-6}	1.53×10^{-2}	6.23×10^{-7}	6.92×10^{-3}	2.22×10^{-2}	1.668×10^{-2}	1.579×10^{-2}	0.9464
15.0000	4.63×10^{-7}	1.13×10^{-2}	3.99×10^{-7}	9.12×10^{-3}	2.05×10^{-2}	1.654×10^{-2}	1.522×10^{-2}	0.9197
20.0000	2.60×10^{-7}	9.10×10^{-3}	2.93×10^{-7}	1.07×10^{-2}	1.98×10^{-2}	1.678×10^{-2}	1.503×10^{-2}	0.8954
30.0000	1.16×10^{-7}	6.62×10^{-3}	1.91×10^{-7}	1.29×10^{-2}	1.96×10^{-2}	1.749×10^{-2}	1.490×10^{-2}	0.8518
40.0000	6.51×10^{-8}	5.26×10^{-3}	1.42×10^{-7}	1.46×10^{-2}	1.98×10^{-2}	1.818×10^{-2}	1.479×10^{-2}	0.8135
50.0000	4.17×10^{-8}	4.39×10^{-3}	1.13×10^{-7}	1.57×10^{-2}	2.01×10^{-2}	1.880×10^{-2}	1.465×10^{-2}	0.7792

TABLE M.3n

Photons in Glass, Lead

ρ: 6.220 (g cm^{-3}), Composition ($Z-f_w$): 8−0.156453; 14−0.080866; 22−0.008092; 33−0.002651; 82−0.751938

K, L, M Edges	Energy (MeV)	Coherent σ_{coh}/ρ	Compton σ_C/ρ	Photoelectric τ/ρ	Pair + Triplet κ/ρ	Total Attenuation μ/ρ	Energy-Transfer μ_{tr}/ρ	Energy-Absorption μ_{en}/ρ	$(1-g)$
	0.0010	9.89	5.20×10^{-3}	4.81×10^3	0.00	4.82×10^3	4.804×10^3	4.804×10^3	0.9999
	0.001323	9.63	8.14×10^{-3}	2.68×10^3	0.00	2.69×10^3	2.693×10^3	2.693×10^3	0.9999
33 L3	0.001323	9.63	8.14×10^{-3}	2.69×10^3	0.00	2.70×10^3	2.702×10^3	2.702×10^3	0.9999
	0.001359	9.60	8.48×10^{-3}	2.55×10^3	0.00	2.56×10^3	2.556×10^3	2.556×10^3	0.9999
33 L2	0.001359	9.60	8.48×10^{-3}	2.55×10^3	0.00	2.56×10^3	2.560×10^3	2.560×10^3	0.9999
	0.0015	9.47	9.85×10^{-3}	2.08×10^3	0.00	2.09×10^3	2.078×10^3	2.078×10^3	0.9999
33 L1	0.001526	9.45	1.01×10^{-2}	2.00×10^3	0.00	2.01×10^3	2.002×10^3	2.002×10^3	0.9999
	0.001526	9.45	1.01×10^{-2}	2.00×10^3	0.00	2.01×10^3	2.004×10^3	2.004×10^3	0.9999
14 K	0.001839	9.16	1.31×10^{-2}	1.33×10^3	0.00	1.34×10^3	1.330×10^3	1.330×10^3	0.9999
	0.001839	9.16	1.31×10^{-2}	1.56×10^3	0.00	1.57×10^3	1.552×10^3	1.552×10^3	0.9999
	0.0020	9.00	1.47×10^{-2}	1.31×10^3	0.00	1.32×10^3	1.298×10^3	1.298×10^3	0.9999
	0.002484	8.54	1.92×10^{-2}	7.96×10^2	0.00	8.04×10^2	7.899×10^2	7.897×10^2	0.9998
82 M5	0.002484	8.54	1.92×10^{-2}	1.24×10^3	0.00	1.25×10^3	1.224×10^3	1.223×10^3	0.9999
	0.002586	8.44	2.01×10^{-2}	1.63×10^3	0.00	1.64×10^3	1.600×10^3	1.600×10^3	0.9999
82 M4	0.002586	8.44	2.01×10^{-2}	2.01×10^3	0.00	2.02×10^3	1.973×10^3	1.973×10^3	0.9999
	0.0030	8.05	2.38×10^{-2}	1.59×10^3	0.00	1.60×10^3	1.555×10^3	1.555×10^3	0.9999
	0.003066	7.98	2.44×10^{-2}	1.50×10^3	0.00	1.51×10^3	1.469×10^3	1.469×10^3	0.9999
82 M3	0.003066	7.98	2.44×10^{-2}	1.72×10^3	0.00	1.73×10^3	1.681×10^3	1.681×10^3	0.9999
	0.003554	7.55	2.84×10^{-2}	1.19×10^3	0.00	1.20×10^3	1.170×10^3	1.170×10^3	0.9999
82 M2	0.003554	7.55	2.84×10^{-2}	1.26×10^3	0.00	1.27×10^3	1.236×10^3	1.235×10^3	0.9999
	0.003851	7.29	3.07×10^{-2}	1.04×10^3	0.00	1.05×10^3	1.021×10^3	1.020×10^3	0.9999
82 M1	0.003851	7.29	3.07×10^{-2}	1.08×10^3	0.00	1.09×10^3	1.062×10^3	1.062×10^3	0.9999
	0.0040	7.16	3.19×10^{-2}	9.87×10^2	0.00	9.94×10^2	9.711×10^2	9.709×10^2	0.9999
	0.004966	6.41	3.87×10^{-2}	5.81×10^2	0.00	5.88×10^2	5.734×10^2	5.733×10^2	0.9998
22 K	0.004966	6.41	3.87×10^{-2}	5.86×10^2	0.00	5.92×10^2	5.773×10^2	5.772×10^2	0.9998
	0.0050	6.39	3.89×10^{-2}	5.76×10^2	0.00	5.83×10^2	5.679×10^2	5.678×10^2	0.9998
	0.0060	5.72	4.52×10^{-2}	3.66×10^2	0.00	3.71×10^2	3.613×10^2	3.612×10^2	0.9997
	0.0080	4.65	5.56×10^{-2}	1.76×10^2	0.00	1.81×10^2	1.745×10^2	1.745×10^2	0.9995
	0.0100	3.85	6.40×10^{-2}	9.90×10	0.00	1.03×10^2	9.828×10	9.821×10	0.9993
	0.011870	3.27	7.03×10^{-2}	6.33×10	0.00	6.66×10	6.294×10	6.288×10	0.9991
33 K	0.011870	3.27	7.03×10^{-2}	6.37×10	0.00	6.70×10	6.314×10	6.308×10	0.9991

Shell	Energy								
82 L3	0.013040	2.97	7.38×10^{-2}	4.98×10	0.00	5.29×10	4.942×10	4.937×10	0.9990
	0.013040	2.97	7.38×10^{-2}	1.21×10^{2}	0.00	1.24×10^{2}	9.932×10	9.927×10	0.9995
	0.0150	2.55	7.88×10^{-2}	8.29×10	0.00	8.56×10	6.996×10	6.991×10	0.9994
82 L2	0.015200	2.51	7.93×10^{-2}	8.01×10	0.00	8.27×10	6.770×10	6.766×10	0.9994
	0.015200	2.51	7.93×10^{-2}	1.11×10^{2}	0.00	1.13×10^{2}	8.654×10	8.649×10	0.9995
82 L1	0.015860	2.39	8.08×10^{-2}	1.00×10^{2}	0.00	1.02×10^{2}	7.892×10	7.888×10	0.9995
	0.015860	2.39	8.08×10^{-2}	1.15×10^{2}	0.00	1.18×10^{2}	9.002×10	8.998×10	0.9996
	0.0200	1.80	8.85×10^{-2}	6.38×10	0.00	6.57×10	5.256×10	5.253×10	0.9993
	0.0300	1.06	1.01×10^{-1}	2.19×10	0.00	2.31×10	1.931×10	1.927×10	0.9984
	0.0400	7.05×10^{-1}	1.07×10^{-1}	1.01×10	0.00	1.09×10	9.224	9.200	0.9974
	0.0500	5.01×10^{-1}	1.10×10^{-1}	5.52	0.00	6.14	5.138	5.119	0.9964
	0.0600	3.75×10^{-1}	1.12×10^{-1}	3.36	0.00	3.84	3.167	3.153	0.9954
	0.0800	2.35×10^{-1}	1.12×10^{-1}	1.52	0.00	1.87	1.468	1.459	0.9936
82 K	0.088005	2.01×10^{-1}	1.11×10^{-1}	1.17	0.00	1.48	1.137	1.129	0.9929
	0.088005	2.01×10^{-1}	1.11×10^{-1}	5.51	0.00	5.82	1.647	1.639	0.9951
	0.1000	1.63×10^{-1}	1.10×10^{-1}	3.94	0.00	4.22	1.506	1.499	0.9953
	0.1500	8.01×10^{-2}	1.04×10^{-1}	1.37	0.00	1.55	8.107×10^{-1}	8.044×10^{-1}	0.9922
	0.2000	4.78×10^{-2}	9.74×10^{-2}	6.37×10^{-1}	0.00	7.82×10^{-1}	4.563×10^{-1}	4.511×10^{-1}	0.9884
	0.3000	2.28×10^{-2}	8.66×10^{-2}	2.20×10^{-1}	0.00	3.30×10^{-1}	1.971×10^{-1}	1.936×10^{-1}	0.9821
	0.4000	1.33×10^{-2}	7.85×10^{-2}	1.07×10^{-1}	0.00	1.98×10^{-1}	1.140×10^{-1}	1.114×10^{-1}	0.9771
	0.5000	8.71×10^{-3}	7.21×10^{-2}	6.21×10^{-2}	0.00	1.43×10^{-1}	7.892×10^{-2}	7.680×10^{-2}	0.9731
	0.6000	6.14×10^{-3}	6.70×10^{-2}	4.07×10^{-2}	0.00	1.14×10^{-1}	6.105×10^{-2}	5.919×10^{-2}	0.9697
	0.8000	3.52×10^{-3}	5.91×10^{-2}	2.16×10^{-2}	0.00	8.42×10^{-2}	4.408×10^{-2}	4.247×10^{-2}	0.9635
	1.0000	2.28×10^{-3}	5.33×10^{-2}	1.36×10^{-2}	0.00	6.91×10^{-2}	3.622×10^{-2}	3.469×10^{-2}	0.9577
	1.2500	1.47×10^{-3}	4.77×10^{-2}	8.79×10^{-3}	2.91×10^{-4}	5.83×10^{-2}	3.088×10^{-2}	2.936×10^{-2}	0.9508
	1.5000	1.03×10^{-3}	4.34×10^{-2}	6.26×10^{-3}	1.39×10^{-3}	5.21×10^{-2}	2.793×10^{-2}	2.638×10^{-2}	0.9445
	2.0000	5.80×10^{-4}	3.71×10^{-2}	3.79×10^{-3}	4.24×10^{-3}	4.57×10^{-2}	2.543×10^{-2}	2.374×10^{-2}	0.9334
	3.0000	2.59×10^{-4}	2.92×10^{-2}	1.98×10^{-3}	9.38×10^{-3}	4.08×10^{-2}	2.498×10^{-2}	2.283×10^{-2}	0.9138
	4.0000	1.46×10^{-4}	2.44×10^{-2}	1.30×10^{-3}	1.35×10^{-2}	3.94×10^{-2}	2.617×10^{-2}	2.344×10^{-2}	0.8958
	5.0000	9.36×10^{-5}	2.10×10^{-2}	9.50×10^{-4}	1.71×10^{-2}	3.92×10^{-2}	2.776×10^{-2}	2.440×10^{-2}	0.8788
	6.0000	6.50×10^{-5}	1.86×10^{-2}	7.44×10^{-4}	2.01×10^{-2}	3.96×10^{-2}	2.945×10^{-2}	2.540×10^{-2}	0.8626
	8.0000	3.66×10^{-5}	1.52×10^{-2}	5.15×10^{-4}	2.53×10^{-2}	4.11×10^{-2}	3.274×10^{-2}	2.725×10^{-2}	0.8322
	10.0000	2.34×10^{-5}	1.30×10^{-2}	3.91×10^{-4}	2.96×10^{-2}	4.30×10^{-2}	3.580×10^{-2}	2.878×10^{-2}	0.8040
	15.0000	1.04×10^{-5}	9.60×10^{-3}	2.43×10^{-4}	3.79×10^{-2}	4.77×10^{-2}	4.231×10^{-2}	3.138×10^{-2}	0.7417
	20.0000	5.86×10^{-6}	7.70×10^{-3}	1.76×10^{-4}	4.38×10^{-2}	5.16×10^{-2}	4.730×10^{-2}	3.258×10^{-2}	0.6888
	30.0000	2.60×10^{-6}	5.61×10^{-3}	1.13×10^{-4}	5.21×10^{-2}	5.77×10^{-2}	5.455×10^{-2}	3.296×10^{-2}	0.6041
	40.0000	1.46×10^{-6}	4.46×10^{-3}	8.28×10^{-5}	5.77×10^{-2}	6.22×10^{-2}	5.968×10^{-2}	3.223×10^{-2}	0.5400
	50.0000	9.37×10^{-7}	3.72×10^{-3}	6.55×10^{-5}	6.19×10^{-2}	6.57×10^{-2}	6.354×10^{-2}	3.114×10^{-2}	0.4900

TABLE M.3o

Photons in Photographic Emulsion

ρ: 3.815 (g cm^{-3}), Composition $(Z-f_w)$: 1–0.014100; 6–0.072261; 7–0.019320; 8–0.066101; 16–0.001890; 35–0.349104; 47–0.474105; 53–0.003120

K, L, M Edges	Energy (MeV)	Mass Coefficients (cm^2 g^{-1})							(1-g)
		Coherent σ_{coh}/ρ	Compton σ_C/ρ	Photoelectric τ/ρ	Pair + Triplet κ/ρ	Total Attenuation μ/ρ	Energy-Transfer μ_{tr}/ρ	Energy-Absorption μ_{en}/ρ	
	0.0010	5.99	6.60×10^{-3}	4.81×10^3	0.00	4.81×10^3	4.805×10^3	4.805×10^3	1
	0.001072	5.94	7.41×10^{-3}	4.08×10^3	0.00	4.09×10^3	4.094×10^3	4.093×10^3	1
53 M1	0.001072	5.94	7.41×10^{-3}	4.08×10^3	0.00	4.09×10^3	4.095×10^3	4.094×10^3	1
	0.0015	5.68	1.27×10^{-2}	1.85×10^3	0.00	1.86×10^3	1.854×10^3	1.854×10^3	0.9999
	0.001550	5.64	1.34×10^{-2}	1.71×10^3	0.00	1.72×10^3	1.714×10^3	1.714×10^3	0.9999
35 L3	0.001550	5.64	1.34×10^{-2}	2.89×10^3	0.00	2.90×10^3	2.867×10^3	2.867×10^3	1
	0.001596	5.61	1.40×10^{-2}	2.55×10^3	0.00	2.55×10^3	2.531×10^3	2.531×10^3	1
35 L2	0.001596	5.61	1.40×10^{-2}	3.07×10^3	0.00	3.08×10^3	3.048×10^3	3.048×10^3	1
35 L2	0.001782	5.48	1.64×10^{-2}	2.37×10^3	0.00	2.38×10^3	2.356×10^3	2.356×10^3	1
35 L1	0.001782	5.48	1.64×10^{-2}	2.56×10^3	0.00	2.56×10^3	2.537×10^3	2.536×10^3	1
	0.0020	5.32	1.92×10^{-2}	1.93×10^3	0.00	1.94×10^3	1.917×10^3	1.917×10^3	0.9999
	0.002472	4.97	2.53×10^{-2}	1.13×10^3	0.00	1.14×10^3	1.131×10^3	1.131×10^3	0.9999
16 K	0.002472	4.97	2.53×10^{-2}	1.14×10^3	0.00	1.14×10^3	1.134×10^3	1.134×10^3	0.9999
	0.0030	4.60	3.16×10^{-2}	6.97×10^2	0.00	7.02×10^2	6.934×10^2	6.933×10^2	0.9999
	0.003351	4.37	3.55×10^{-2}	5.23×10^2	0.00	5.27×10^2	5.210×10^2	5.210×10^2	0.9999
47 L3	0.003351	4.37	3.55×10^{-2}	9.42×10^2	0.00	9.47×10^2	9.196×10^2	9.196×10^2	0.9999
	0.003524	4.26	3.74×10^{-2}	8.30×10^2	0.00	8.34×10^2	8.106×10^2	8.105×10^2	0.9999
47 L2	0.003524	4.26	3.74×10^{-2}	1.03×10^3	0.00	1.03×10^3	9.988×10^2	9.987×10^2	0.9999
	0.003806	4.08	4.03×10^{-2}	8.49×10^2	0.00	8.53×10^2	8.252×10^2	8.252×10^2	0.9999
47 L1	0.003806	4.08	4.03×10^{-2}	9.37×10^2	0.00	9.41×10^2	9.093×10^2	9.092×10^2	0.9999
	0.0040	3.97	4.22×10^{-2}	8.30×10^2	0.00	8.34×10^2	8.062×10^2	8.062×10^2	0.9999
	0.004557	3.66	4.74×10^{-2}	5.92×10^2	0.00	5.96×10^2	5.783×10^2	5.783×10^2	0.9999
53 L3	0.004557	3.66	4.74×10^{-2}	5.93×10^2	0.00	5.97×10^2	5.798×10^2	5.797×10^2	0.9999
	0.004852	3.51	4.99×10^{-2}	5.04×10^2	0.00	5.08×10^2	4.925×10^2	4.925×10^2	0.9999
53 L2	0.004852	3.51	4.99×10^{-2}	5.05×10^2	0.00	5.09×10^2	4.932×10^2	4.931×10^2	0.9999
	0.0050	3.44	5.11×10^{-2}	4.67×10^2	0.00	4.71×10^2	4.563×10^2	4.562×10^2	0.9999
	0.005188	3.35	5.26×10^{-2}	4.24×10^2	0.00	4.27×10^2	4.147×10^2	4.146×10^2	0.9999
53 L1	0.005188	3.35	5.26×10^{-2}	4.24×10^2	0.00	4.28×10^2	4.150×10^2	4.149×10^2	0.9999
	0.0060	3.01	5.86×10^{-2}	2.90×10^2	0.00	2.93×10^2	2.838×10^2	2.838×10^2	0.9998
	0.0080	2.36	7.10×10^{-2}	1.34×10^2	0.00	1.36×10^2	1.320×10^2	1.320×10^2	0.9997
	0.0100	1.90	8.07×10^{-2}	7.29×10	0.00	7.49×10	7.208×10	7.205×10	0.9995

35 K	0.01347	1.38	9.35×10^{-2}	3.20×10	0.00	3.34×10	3.170×10	3.168×10	0.9993
	0.01347	1.38	9.35×10^{-2}	7.57×10	0.00	7.72×10	5.137×10	5.135×10	0.9996
	0.0150	1.21	9.79×10^{-2}	5.71×10	0.00	5.84×10	4.042×10	4.040×10	0.9995
	0.0200	8.37×10^{-1}	1.09×10^{-1}	2.63×10	0.00	2.73×10	2.043×10	2.042×10	0.9993
	0.02551	5.95×10^{-1}	1.17×10^{-1}	1.35×10	0.00	1.42×10	1.106×10	1.105×10	0.9989
47 K	0.02551	5.95×10^{-1}	1.17×10^{-1}	3.52×10	0.00	3.59×10	1.664×10	1.663×10	0.9993
	0.0300	4.71×10^{-1}	1.21×10^{-1}	2.30×10	0.00	2.36×10	1.257×10	1.256×10	0.9991
	0.03317	4.06×10^{-1}	1.23×10^{-1}	1.76×10	0.00	1.81×10	1.032×10	1.031×10	0.9990
53 K	0.03317	4.06×10^{-1}	1.23×10^{-1}	1.77×10	0.00	1.82×10	1.034×10	1.032×10	0.9990
	0.0400	3.05×10^{-1}	1.26×10^{-1}	1.06×10	0.00	1.10×10	6.920	6.910	0.9986
	0.0500	2.14×10^{-1}	1.29×10^{-1}	5.70	0.00	6.04	4.112	4.104	0.9979
	0.0600	1.58×10^{-1}	1.29×10^{-1}	3.41	0.00	3.69	2.618	2.611	0.9973
	0.0800	9.63×10^{-2}	1.28×10^{-1}	1.49	0.00	1.72	1.243	1.238	0.9960
	0.1000	6.51×10^{-2}	1.25×10^{-1}	7.81×10^{-1}	0.00	9.71×10^{-1}	6.874×10^{-1}	6.840×10^{-1}	0.9950
	0.1500	3.15×10^{-2}	1.16×10^{-1}	2.39×10^{-1}	0.00	3.86×10^{-1}	2.373×10^{-1}	2.356×10^{-1}	0.9930
	0.2000	1.85×10^{-2}	1.08×10^{-1}	1.03×10^{-1}	0.00	2.29×10^{-1}	1.194×10^{-1}	1.184×10^{-1}	0.9917
	0.3000	8.54×10^{-3}	9.50×10^{-2}	3.24×10^{-2}	0.00	1.36×10^{-1}	5.668×10^{-2}	5.612×10^{-2}	0.9901
	0.4000	4.90×10^{-3}	8.57×10^{-2}	1.47×10^{-2}	0.00	1.05×10^{-1}	4.090×10^{-2}	4.044×10^{-2}	0.9888
	0.5000	3.17×10^{-3}	7.85×10^{-2}	8.24×10^{-3}	0.00	8.99×10^{-2}	3.493×10^{-2}	3.449×10^{-2}	0.9874
	0.6000	2.22×10^{-3}	7.28×10^{-2}	5.25×10^{-3}	0.00	8.03×10^{-2}	3.198×10^{-2}	3.153×10^{-2}	0.9859
	0.8000	1.26×10^{-3}	6.41×10^{-2}	2.70×10^{-3}	0.00	6.81×10^{-2}	2.892×10^{-2}	2.842×10^{-2}	0.9829
	1.0000	8.08×10^{-4}	5.77×10^{-2}	1.69×10^{-3}	0.00	6.02×10^{-2}	2.709×10^{-2}	2.655×10^{-2}	0.9799
	1.2500	5.19×10^{-4}	5.16×10^{-2}	1.08×10^{-3}	1.15×10^{-4}	5.33×10^{-2}	2.542×10^{-2}	2.482×10^{-2}	0.9761
	1.5000	3.61×10^{-4}	4.69×10^{-2}	7.77×10^{-4}	5.46×10^{-4}	4.86×10^{-2}	2.419×10^{-2}	2.352×10^{-2}	0.9725
	2.0000	2.03×10^{-4}	4.01×10^{-2}	4.74×10^{-4}	1.94×10^{-3}	4.27×10^{-2}	2.271×10^{-2}	2.192×10^{-2}	0.9653
	3.0000	9.04×10^{-5}	3.15×10^{-2}	2.53×10^{-4}	5.11×10^{-3}	3.70×10^{-2}	2.184×10^{-2}	2.077×10^{-2}	0.9512
	4.0000	5.09×10^{-5}	2.63×10^{-2}	1.68×10^{-4}	8.01×10^{-3}	3.45×10^{-2}	2.211×10^{-2}	2.072×10^{-2}	0.9374
	5.0000	3.26×10^{-5}	2.27×10^{-2}	1.25×10^{-4}	1.06×10^{-2}	3.34×10^{-2}	2.279×10^{-2}	2.106×10^{-2}	0.9240
	6.0000	2.26×10^{-5}	2.01×10^{-2}	9.88×10^{-5}	1.27×10^{-2}	3.30×10^{-2}	2.363×10^{-2}	2.153×10^{-2}	0.9112
	8.0000	1.27×10^{-5}	1.64×10^{-2}	6.93×10^{-5}	1.65×10^{-2}	3.31×10^{-2}	2.547×10^{-2}	2.259×10^{-2}	0.8871
	10.0000	8.15×10^{-6}	1.40×10^{-2}	5.31×10^{-5}	1.96×10^{-2}	3.37×10^{-2}	2.727×10^{-2}	2.358×10^{-2}	0.8645
	15.0000	3.62×10^{-6}	1.04×10^{-2}	3.34×10^{-5}	2.55×10^{-2}	3.59×10^{-2}	3.117×10^{-2}	2.536×10^{-2}	0.8138
	20.0000	2.04×10^{-6}	8.32×10^{-3}	2.43×10^{-5}	2.97×10^{-2}	3.80×10^{-2}	3.424×10^{-2}	2.634×10^{-2}	0.7693
	30.0000	9.06×10^{-7}	6.06×10^{-3}	1.57×10^{-5}	3.55×10^{-2}	4.16×10^{-2}	3.886×10^{-2}	2.700×10^{-2}	0.6948
	40.0000	5.09×10^{-7}	4.81×10^{-3}	1.16×10^{-5}	3.95×10^{-2}	4.43×10^{-2}	4.216×10^{-2}	2.677×10^{-2}	0.6348
	50.0000	3.26×10^{-7}	4.02×10^{-3}	9.21×10^{-6}	4.25×10^{-2}	4.65×10^{-2}	4.469×10^{-2}	2.617×10^{-2}	0.5856

TABLE M.3p

Photons in Polymethyl Methacrylate (Perspex, Lucite)

ρ: 3.815 (g cm^{-3}), Composition $(Z-f_w)$: 1–0.080541; 6–0.599846; 8–0.319613

K, L, M Edges	Energy (MeV)	Mass Coefficients (cm^{2}g^{-1})							
		Coherent σ_{coh}/ρ	Compton σ_C/ρ	Photoelectric τ/ρ	Pair + Triplet κ/ρ	Total Attenuation μ/ρ	Energy-Transfer μ_{tr}/ρ	Energy-Absorption μ_{en}/ρ	$(1-g)$
	0.0010	1.16	1.44×10^{-2}	2.79×10^{3}	0.00	2.79×10^{3}	2.788×10^{3}	2.788×10^{3}	1
	0.0015	1.04	2.86×10^{-2}	9.14×10^{2}	0.00	9.15×10^{2}	9.131×10^{2}	9.131×10^{2}	1
	0.0020	9.23×10^{-1}	4.42×10^{-2}	4.03×10^{2}	0.00	4.04×10^{2}	4.024×10^{2}	4.024×10^{2}	0.9999
	0.0030	7.01×10^{-1}	7.31×10^{-2}	1.23×10^{2}	0.00	1.24×10^{2}	1.228×10^{2}	1.228×10^{2}	0.9999
	0.0040	5.35×10^{-1}	9.59×10^{-2}	5.18×10	0.00	5.25×10	5.182×10	5.181×10	0.9999
	0.0050	4.20×10^{-1}	1.13×10^{-1}	2.63×10	0.00	2.68×10	2.627×10	2.627×10	0.9999
	0.0060	3.39×10^{-1}	1.25×10^{-1}	1.50×10	0.00	1.55×10	1.498×10	1.498×10	0.9999
	0.0080	2.39×10^{-1}	1.41×10^{-1}	6.11	0.00	6.49	6.115	6.114	0.9998
	0.0100	1.81×10^{-1}	1.51×10^{-1}	3.02	0.00	3.36	3.027	3.026	0.9998
	0.0150	1.07×10^{-1}	1.66×10^{-1}	8.28×10^{-1}	0.00	1.10	8.326×10^{-1}	8.324×10^{-1}	0.9998
	0.0200	7.10×10^{-2}	1.74×10^{-1}	3.26×10^{-1}	0.00	5.71×10^{-1}	3.329×10^{-1}	3.328×10^{-1}	0.9997
	0.0300	3.72×10^{-2}	1.79×10^{-1}	8.67×10^{-2}	0.00	3.03×10^{-1}	9.649×10^{-2}	9.646×10^{-2}	0.9997
	0.0400	2.27×10^{-2}	1.79×10^{-1}	3.36×10^{-2}	0.00	2.35×10^{-1}	4.600×10^{-2}	4.599×10^{-2}	0.9997
	0.0500	1.52×10^{-2}	1.76×10^{-1}	1.61×10^{-2}	0.00	2.07×10^{-1}	3.068×10^{-2}	3.067×10^{-2}	0.9997
	0.0600	1.09×10^{-2}	1.73×10^{-1}	8.77×10^{-3}	0.00	1.92×10^{-1}	2.530×10^{-2}	2.530×10^{-2}	0.9997
	0.0800	6.38×10^{-3}	1.65×10^{-1}	3.38×10^{-3}	0.00	1.75×10^{-1}	2.303×10^{-2}	2.302×10^{-2}	0.9997
	0.1000	4.17×10^{-3}	1.58×10^{-1}	1.61×10^{-3}	0.00	1.64×10^{-1}	2.369×10^{-2}	2.368×10^{-2}	0.9997
	0.1500	1.89×10^{-3}	1.43×10^{-1}	4.25×10^{-4}	0.00	1.46×10^{-1}	2.658×10^{-2}	2.657×10^{-2}	0.9996
	0.2000	1.07×10^{-3}	1.32×10^{-1}	1.68×10^{-4}	0.00	1.33×10^{-1}	2.874×10^{-2}	2.872×10^{-2}	0.9995
	0.3000	4.81×10^{-4}	1.15×10^{-1}	4.72×10^{-5}	0.00	1.15×10^{-1}	3.102×10^{-2}	3.099×10^{-2}	0.9993
	0.4000	2.71×10^{-4}	1.03×10^{-1}	2.02×10^{-5}	0.00	1.03×10^{-1}	3.188×10^{-2}	3.185×10^{-2}	0.9991
	0.5000	1.74×10^{-4}	9.39×10^{-2}	1.09×10^{-5}	0.00	9.41×10^{-2}	3.210×10^{-2}	3.206×10^{-2}	0.9989
	0.6000	1.21×10^{-4}	8.69×10^{-2}	6.77×10^{-6}	0.00	8.70×10^{-2}	3.195×10^{-2}	3.191×10^{-2}	0.9987
	0.8000	6.79×10^{-5}	7.63×10^{-2}	3.42×10^{-6}	0.00	7.64×10^{-2}	3.121×10^{-2}	3.116×10^{-2}	0.9983
	1.0000	4.35×10^{-5}	6.87×10^{-2}	2.12×10^{-6}	0.00	6.87×10^{-2}	3.022×10^{-2}	3.015×10^{-2}	0.9978
	1.2500	2.78×10^{-5}	6.14×10^{-2}	1.34×10^{-6}	1.52×10^{-5}	6.14×10^{-2}	2.890×10^{-2}	2.882×10^{-2}	0.9973
	1.5000	1.93×10^{-5}	5.58×10^{-2}	9.72×10^{-7}	8.43×10^{-5}	5.59×10^{-2}	2.764×10^{-2}	2.755×10^{-2}	0.9967
	2.0000	1.09×10^{-5}	4.76×10^{-2}	6.12×10^{-7}	3.36×10^{-4}	4.80×10^{-2}	2.544×10^{-2}	2.533×10^{-2}	0.9954
	3.0000	4.83×10^{-6}	3.75×10^{-2}	3.42×10^{-7}	9.75×10^{-4}	3.84×10^{-2}	2.227×10^{-2}	2.210×10^{-2}	0.9925
	4.0000	2.72×10^{-6}	3.12×10^{-2}	2.35×10^{-7}	1.61×10^{-3}	3.29×10^{-2}	2.017×10^{-2}	1.995×10^{-2}	0.9891
	5.0000	1.74×10^{-6}	2.70×10^{-2}	1.79×10^{-7}	2.20×10^{-3}	2.92×10^{-2}	1.870×10^{-2}	1.843×10^{-2}	0.9855

6.0000	1.21×10^{-6}	2.39×10^{-2}	1.44×10^{-7}	2.73×10^{-3}	2.66×10^{-2}	1.763×10^{-2}	1.731×10^{-2}	0.9818
8.0000	6.80×10^{-7}	1.95×10^{-2}	1.03×10^{-7}	3.67×10^{-3}	2.32×10^{-2}	1.621×10^{-2}	1.579×10^{-2}	0.9741
10.0000	4.35×10^{-7}	1.66×10^{-2}	8.02×10^{-8}	4.43×10^{-3}	2.11×10^{-2}	1.534×10^{-2}	1.482×10^{-2}	0.9663
15.0000	1.93×10^{-7}	1.23×10^{-2}	5.16×10^{-8}	5.90×10^{-3}	1.82×10^{-2}	1.423×10^{-2}	1.348×10^{-2}	0.9475
20.0000	1.09×10^{-7}	9.87×10^{-3}	3.80×10^{-8}	6.98×10^{-3}	1.68×10^{-2}	1.379×10^{-2}	1.282×10^{-2}	0.9299
30.0000	4.83×10^{-8}	7.18×10^{-3}	2.49×10^{-8}	8.50×10^{-3}	1.57×10^{-2}	1.358×10^{-2}	1.219×10^{-2}	0.8978
40.0000	2.72×10^{-8}	5.71×10^{-3}	1.85×10^{-8}	9.57×10^{-3}	1.53×10^{-2}	1.367×10^{-2}	1.187×10^{-2}	0.8689
50.0000	1.74×10^{-8}	4.77×10^{-3}	1.47×10^{-8}	1.04×10^{-2}	1.51×10^{-2}	1.384×10^{-2}	1.166×10^{-2}	0.8426

TABLE M.3q

Photons in Polystyrene

ρ: 1.060 (g cm^{-3}), Composition (Z$-f_w$): 1$-$0.077421; 6$-$0.922579

K, L, M Edges / Energy (MeV)	Mass Coefficients (cm^2 g^{-1})							
	Coherent σ_{coh}/ρ	Compton σ_C/ρ	Photoelectric τ/ρ	Pair + Triplet κ/ρ	Total Attenuation μ/ρ	Energy-Transfer μ_{tr}/ρ	Energy-Absorption μ_{en}/ρ	$(1-g)$
0.0010	1.02	1.55×10^{-2}	2.04×10^{3}	0.00	2.04×10^{3}	2.038×10^{3}	2.038×10^{3}	1
0.0015	9.07×10^{-1}	3.08×10^{-2}	6.45×10^{2}	0.00	6.46×10^{2}	6.451×10^{2}	6.450×10^{2}	1
0.0020	7.87×10^{-1}	4.71×10^{-2}	2.78×10^{2}	0.00	2.79×10^{2}	2.783×10^{2}	2.783×10^{2}	0.9999
0.0030	5.78×10^{-1}	7.68×10^{-2}	8.27×10	0.00	8.34×10	8.272×10	8.271×10	0.9999
0.0040	4.33×10^{-1}	9.96×10^{-2}	3.44×10	0.00	3.49×10	3.436×10	3.435×10	0.9999
0.0050	3.38×10^{-1}	1.16×10^{-1}	1.72×10	0.00	1.77×10	1.722×10	1.721×10	0.9999
0.0060	2.74×10^{-1}	1.27×10^{-1}	9.73	0.00	1.01×10	9.731	9.730	0.9999
0.0080	1.96×10^{-1}	1.43×10^{-1}	3.91	0.00	4.25	3.916	3.915	0.9999
0.0100	1.51×10^{-1}	1.52×10^{-1}	1.92	0.00	2.22	1.919	1.918	0.9998
0.0150	9.12×10^{-2}	1.68×10^{-1}	5.15×10^{-1}	0.00	7.74×10^{-1}	5.201×10^{-1}	5.200×10^{-1}	0.9998
0.0200	6.03×10^{-2}	1.75×10^{-1}	2.01×10^{-1}	0.00	4.36×10^{-1}	2.076×10^{-1}	2.075×10^{-1}	0.9998
0.0300	3.13×10^{-2}	1.80×10^{-1}	5.26×10^{-2}	0.00	2.64×10^{-1}	6.248×10^{-2}	6.247×10^{-2}	0.9998
0.0400	1.90×10^{-2}	1.79×10^{-1}	2.02×10^{-2}	0.00	2.18×10^{-1}	3.265×10^{-2}	3.264×10^{-2}	0.9998
0.0500	1.27×10^{-2}	1.76×10^{-1}	9.62×10^{-3}	0.00	1.99×10^{-1}	2.422×10^{-2}	2.421×10^{-2}	0.9998
0.0600	9.11×10^{-3}	1.73×10^{-1}	5.23×10^{-3}	0.00	1.87×10^{-1}	2.173×10^{-2}	2.172×10^{-2}	0.9998
0.0800	5.30×10^{-3}	1.65×10^{-1}	2.00×10^{-3}	0.00	1.72×10^{-1}	2.160×10^{-2}	2.160×10^{-2}	0.9998
0.1000	3.45×10^{-3}	1.58×10^{-1}	9.52×10^{-4}	0.00	1.62×10^{-1}	2.296×10^{-2}	2.296×10^{-2}	0.9998
0.1500	1.56×10^{-3}	1.43×10^{-1}	2.50×10^{-4}	0.00	1.45×10^{-1}	2.633×10^{-2}	2.632×10^{-2}	0.9997
0.2000	8.86×10^{-4}	1.31×10^{-1}	9.81×10^{-5}	0.00	1.32×10^{-1}	2.858×10^{-2}	2.857×10^{-2}	0.9996
0.3000	3.96×10^{-4}	1.14×10^{-1}	2.75×10^{-5}	0.00	1.15×10^{-1}	3.090×10^{-2}	3.088×10^{-2}	0.9994
0.4000	2.23×10^{-4}	1.02×10^{-1}	1.17×10^{-5}	0.00	1.03×10^{-1}	3.178×10^{-2}	3.175×10^{-2}	0.9992
0.5000	1.43×10^{-4}	9.36×10^{-2}	6.31×10^{-6}	0.00	9.38×10^{-2}	3.199×10^{-2}	3.196×10^{-2}	0.9990
0.6000	9.93×10^{-5}	8.66×10^{-2}	3.92×10^{-6}	0.00	8.67×10^{-2}	3.185×10^{-2}	3.182×10^{-2}	0.9988
0.8000	5.58×10^{-5}	7.61×10^{-2}	1.98×10^{-6}	0.00	7.62×10^{-2}	3.111×10^{-2}	3.106×10^{-2}	0.9985
1.0000	3.57×10^{-5}	6.84×10^{-2}	1.23×10^{-6}	0.00	6.85×10^{-2}	3.012×10^{-2}	3.006×10^{-2}	0.9981
1.2500	2.29×10^{-5}	6.12×10^{-2}	7.70×10^{-7}	1.36×10^{-5}	6.12×10^{-2}	2.881×10^{-2}	2.874×10^{-2}	0.9976
1.5000	1.59×10^{-5}	5.56×10^{-2}	5.59×10^{-7}	7.58×10^{-5}	5.57×10^{-2}	2.755×10^{-2}	2.747×10^{-2}	0.9970
2.0000	8.94×10^{-6}	4.75×10^{-2}	3.53×10^{-7}	3.02×10^{-4}	4.78×10^{-2}	2.535×10^{-2}	2.524×10^{-2}	0.9959
3.0000	3.97×10^{-6}	3.73×10^{-2}	1.98×10^{-7}	8.78×10^{-4}	3.82×10^{-2}	2.214×10^{-2}	2.199×10^{-2}	0.9932
4.0000	2.24×10^{-6}	3.11×10^{-2}	1.36×10^{-7}	1.45×10^{-3}	3.26×10^{-2}	1.999×10^{-2}	1.979×10^{-2}	0.9901
5.0000	1.43×10^{-6}	2.69×10^{-2}	1.04×10^{-7}	2.00×10^{-3}	2.89×10^{-2}	1.848×10^{-2}	1.824×10^{-2}	0.9868

6.0000	9.94×10^{-7}	2.38×10^{-2}	8.34×10^{-8}	2.48×10^{-3}	2.63×10^{-2}	1.737×10^{-2}	1.708×10^{-2}	0.9834
8.0000	5.59×10^{-7}	1.94×10^{-2}	5.99×10^{-8}	3.33×10^{-3}	2.28×10^{-2}	1.588×10^{-2}	1.550×10^{-2}	0.9762
10.0000	3.58×10^{-7}	1.66×10^{-2}	4.67×10^{-8}	4.03×10^{-3}	2.06×10^{-2}	1.494×10^{-2}	1.448×10^{-2}	0.9690
15.0000	1.59×10^{-7}	1.23×10^{-2}	3.00×10^{-8}	5.38×10^{-3}	1.76×10^{-2}	1.371×10^{-2}	1.305×10^{-2}	0.9514
20.0000	8.94×10^{-8}	9.84×10^{-3}	2.21×10^{-8}	6.36×10^{-3}	1.62×10^{-2}	1.318×10^{-2}	1.232×10^{-2}	0.9348
30.0000	3.97×10^{-8}	7.16×10^{-3}	1.45×10^{-8}	7.77×10^{-3}	1.49×10^{-2}	1.286×10^{-2}	1.163×10^{-2}	0.9044
40.0000	2.24×10^{-8}	5.69×10^{-3}	1.08×10^{-8}	8.75×10^{-3}	1.44×10^{-2}	1.286×10^{-2}	1.128×10^{-2}	0.8770
50.0000	1.43×10^{-8}	4.75×10^{-3}	8.56×10^{-9}	9.50×10^{-3}	1.43×10^{-2}	1.296×10^{-2}	1.104×10^{-2}	0.8519

TABLE M.3r

Photons in Adipose Tissue (ICRP)

ρ: 0.920 (g cm^{-3}), Composition (Z $-$ f$_w$): 1$-$0.119477; 6$-$0.637240; 7$-$0.007970; 8$-$0.232333; 11$-$0.000500; 12$-$0.000020; 15$-$0.000160; 16$-$0.000730; 17$-$0.001190; 19$-$0.000320; 20$-$0.000020; 26$-$0.000020; 30$-$0.000020

K, L, M Edges	Energy (MeV)	Coherent σ_{coh}/ρ	Compton σ_C/ρ	Photoelectric τ/ρ	Pair + Triplet κ/ρ	Total Attenuation μ/ρ	Energy-Transfer μ_{tr}/ρ	Energy-Absorption μ_{en}/ρ	$(1-g)$
	0.0010	1.10	1.62×10^{-2}	2.51×10^{3}	0.00	2.51×10^{3}	2.505×10^{3}	2.505×10^{3}	1
	0.001020	1.09	1.67×10^{-2}	2.38×10^{3}	0.00	2.38×10^{3}	2.376×10^{3}	2.376×10^{3}	1
30 L3	0.001020	1.09	1.67×10^{-2}	2.38×10^{3}	0.00	2.38×10^{3}	2.376×10^{3}	2.376×10^{3}	1
	0.001043	1.09	1.74×10^{-2}	2.24×10^{3}	0.00	2.24×10^{3}	2.235×10^{3}	2.235×10^{3}	1
30 L2	0.001043	1.09	1.74×10^{-2}	2.24×10^{3}	0.00	2.24×10^{3}	2.235×10^{3}	2.235×10^{3}	1
	0.001072	1.08	1.82×10^{-2}	2.07×10^{3}	0.00	2.07×10^{3}	2.073×10^{3}	2.073×10^{3}	1
11 K	0.001072	1.08	1.82×10^{-2}	2.08×10^{3}	0.00	2.08×10^{3}	2.076×10^{3}	2.076×10^{3}	1
	0.001194	1.05	2.20×10^{-2}	1.54×10^{3}	0.00	1.55×10^{3}	1.546×10^{3}	1.545×10^{3}	1
30 L1	0.001194	1.05	2.20×10^{-2}	1.54×10^{3}	0.00	1.55×10^{3}	1.546×10^{3}	1.546×10^{3}	1
	0.001305	1.03	2.55×10^{-2}	1.21×10^{3}	0.00	1.21×10^{3}	1.207×10^{3}	1.207×10^{3}	1
12 K	0.001305	1.03	2.55×10^{-2}	1.21×10^{3}	0.00	1.21×10^{3}	1.207×10^{3}	1.207×10^{3}	1
	0.0015	9.87×10^{-1}	3.21×10^{-2}	8.18×10^{2}	0.00	8.19×10^{2}	8.173×10^{2}	8.173×10^{2}	1
	0.0020	8.69×10^{-1}	4.93×10^{-2}	3.59×10^{2}	0.00	3.60×10^{2}	3.590×10^{2}	3.590×10^{2}	0.9999
	0.002145	8.34×10^{-1}	5.42×10^{-2}	2.93×10^{2}	0.00	2.94×10^{2}	2.928×10^{2}	2.927×10^{2}	0.9999
15 K	0.002145	8.34×10^{-1}	5.42×10^{-2}	2.93×10^{2}	0.00	2.94×10^{2}	2.931×10^{2}	2.931×10^{2}	0.9999
	0.002472	7.61×10^{-1}	6.48×10^{-2}	1.94×10^{2}	0.00	1.95×10^{2}	1.936×10^{2}	1.936×10^{2}	0.9999
16 K	0.002472	7.61×10^{-1}	6.48×10^{-2}	1.95×10^{2}	0.00	1.96×10^{2}	1.949×10^{2}	1.949×10^{2}	0.9999
	0.002822	6.89×10^{-1}	7.55×10^{-2}	1.32×10^{2}	0.00	1.33×10^{2}	1.319×10^{2}	1.319×10^{2}	0.9999
17 K	0.002822	6.89×10^{-1}	7.55×10^{-2}	1.34×10^{2}	0.00	1.35×10^{2}	1.335×10^{2}	1.335×10^{2}	0.9999
	0.0030	6.55×10^{-1}	8.05×10^{-2}	1.12×10^{2}	0.00	1.12×10^{2}	1.115×10^{2}	1.115×10^{2}	0.9999
	0.003607	5.53×10^{-1}	9.60×10^{-2}	6.45×10	0.00	6.51×10	6.434×10	6.433×10	0.9999
19 K	0.003607	5.53×10^{-1}	9.60×10^{-2}	6.48×10	0.00	6.55×10	6.464×10	6.463×10	0.9999
	0.0040	4.98×10^{-1}	1.05×10^{-1}	4.75×10	0.00	4.81×10	4.736×10	4.602×10	0.9999
	0.004038	4.93×10^{-1}	1.05×10^{-1}	4.61×10	0.00	4.67×10	4.603×10	4.604×10	0.9999
20 K	0.004038	4.93×10^{-1}	1.05×10^{-1}	4.62×10	0.00	4.68×10	4.604×10	4.604×10	0.9999
	0.0050	3.90×10^{-1}	1.22×10^{-1}	2.41×10	0.00	2.46×10	2.406×10	2.406×10	0.9999
	0.0060	3.15×10^{-1}	1.34×10^{-1}	1.38×10	0.00	1.42×10	1.375×10	1.374×10	0.9999
	0.007112	2.57×10^{-1}	1.44×10^{-1}	8.13	0.00	8.53	8.116	8.115	0.9999
26 K	0.007112	2.57×10^{-1}	1.44×10^{-1}	8.14	0.00	8.54	8.121	8.120	0.9999
	0.0080	2.23×10^{-1}	1.50×10^{-1}	5.64	0.00	6.01	5.627	5.627	0.9999

Mass Coefficients (cm^2 g^{-1})

30 K

0.009659	1.77×10^{-1}	1.59×10^{-1}	3.12	0.00	3.45	3.115	3.115	0.9998
0.009659	1.77×10^{-1}	1.59×10^{-1}	3.12	0.00	3.46	3.118	3.117	0.9998
0.0100	1.69×10^{-1}	1.60×10^{-1}	2.80	0.00	3.13	2.794	2.794	0.9998
0.0150	1.00×10^{-1}	1.75×10^{-1}	7.69×10^{-1}	0.00	1.04	7.724×10^{-1}	7.723×10^{-1}	0.9998
0.0200	6.65×10^{-2}	1.82×10^{-1}	3.04×10^{-1}	0.00	5.53×10^{-1}	3.104×10^{-1}	3.104×10^{-1}	0.9998
0.0300	3.48×10^{-2}	1.87×10^{-1}	8.11×10^{-2}	0.00	3.03×10^{-1}	9.119×10^{-2}	9.117×10^{-2}	0.9997
0.0400	2.12×10^{-2}	1.86×10^{-1}	3.15×10^{-2}	0.00	2.39×10^{-1}	4.437×10^{-2}	4.436×10^{-2}	0.9997
0.0500	1.42×10^{-2}	1.83×10^{-1}	1.51×10^{-2}	0.00	2.12×10^{-1}	3.025×10^{-2}	3.024×10^{-2}	0.9997
0.0600	1.02×10^{-2}	1.79×10^{-1}	8.27×10^{-3}	0.00	1.98×10^{-1}	2.539×10^{-2}	2.538×10^{-2}	0.9998
0.0800	5.96×10^{-3}	1.71×10^{-1}	3.19×10^{-3}	0.00	1.81×10^{-1}	2.354×10^{-2}	2.354×10^{-2}	0.9998
0.1000	3.89×10^{-3}	1.64×10^{-1}	1.53×10^{-3}	0.00	1.69×10^{-1}	2.438×10^{-2}	2.438×10^{-2}	0.9997
0.1500	1.77×10^{-3}	1.48×10^{-1}	4.04×10^{-4}	0.00	1.51×10^{-1}	2.749×10^{-2}	2.748×10^{-2}	0.9997
0.2000	1.00×10^{-3}	1.36×10^{-1}	1.60×10^{-4}	0.00	1.37×10^{-1}	2.974×10^{-2}	2.973×10^{-2}	0.9996
0.3000	4.48×10^{-4}	1.19×10^{-1}	4.51×10^{-5}	0.00	1.19×10^{-1}	3.211×10^{-2}	3.209×10^{-2}	0.9994
0.4000	2.53×10^{-4}	1.06×10^{-1}	1.93×10^{-5}	0.00	1.07×10^{-1}	3.301×10^{-2}	3.298×10^{-2}	0.9992
0.5000	1.62×10^{-4}	9.72×10^{-2}	1.04×10^{-5}	0.00	9.74×10^{-2}	3.323×10^{-2}	3.320×10^{-2}	0.9990
0.6000	1.12×10^{-4}	9.00×10^{-2}	6.49×10^{-6}	0.00	9.01×10^{-2}	3.309×10^{-2}	3.305×10^{-2}	0.9988
0.8000	6.33×10^{-5}	7.90×10^{-2}	3.28×10^{-6}	0.00	7.91×10^{-2}	3.231×10^{-2}	3.226×10^{-2}	0.9984
1.0000	4.05×10^{-5}	7.11×10^{-2}	2.04×10^{-6}	0.00	7.11×10^{-2}	3.129×10^{-2}	3.123×10^{-2}	0.9980
1.2500	2.59×10^{-5}	6.36×10^{-2}	1.28×10^{-6}	1.45×10^{-5}	6.36×10^{-2}	2.992×10^{-2}	2.985×10^{-2}	0.9975
1.5000	1.80×10^{-5}	5.78×10^{-2}	9.32×10^{-7}	8.04×10^{-5}	5.79×10^{-2}	2.861×10^{-2}	2.853×10^{-2}	0.9970
2.0000	1.01×10^{-5}	4.93×10^{-2}	5.86×10^{-7}	3.20×10^{-4}	4.96×10^{-2}	2.633×10^{-2}	2.622×10^{-2}	0.9958
3.0000	4.51×10^{-6}	3.88×10^{-2}	3.28×10^{-7}	9.31×10^{-4}	3.97×10^{-2}	2.301×10^{-2}	2.285×10^{-2}	0.9931
4.0000	2.53×10^{-6}	3.24×10^{-2}	2.25×10^{-7}	1.55×10^{-3}	3.39×10^{-2}	2.079×10^{-2}	2.058×10^{-2}	0.9900
5.0000	1.62×10^{-6}	2.79×10^{-2}	1.71×10^{-7}	2.11×10^{-3}	3.00×10^{-2}	1.922×10^{-2}	1.897×10^{-2}	0.9867
6.0000	1.13×10^{-6}	2.47×10^{-2}	1.37×10^{-7}	2.62×10^{-3}	2.73×10^{-2}	1.808×10^{-2}	1.778×10^{-2}	0.9832
8.0000	6.34×10^{-7}	2.02×10^{-2}	9.83×10^{-8}	3.52×10^{-3}	2.37×10^{-2}	1.654×10^{-2}	1.615×10^{-2}	0.9760
10.0000	4.05×10^{-7}	1.72×10^{-2}	7.65×10^{-8}	4.26×10^{-3}	2.15×10^{-2}	1.558×10^{-2}	1.510×10^{-2}	0.9687
15.0000	1.80×10^{-7}	1.27×10^{-2}	4.91×10^{-8}	5.67×10^{-3}	1.84×10^{-2}	1.433×10^{-2}	1.362×10^{-2}	0.9510
20.0000	1.01×10^{-7}	1.02×10^{-2}	3.62×10^{-8}	6.71×10^{-3}	1.69×10^{-2}	1.379×10^{-2}	1.289×10^{-2}	0.9343
30.0000	4.51×10^{-8}	7.44×10^{-3}	2.37×10^{-8}	8.20×10^{-3}	1.56×10^{-2}	1.347×10^{-2}	1.217×10^{-2}	0.9037
40.0000	2.53×10^{-8}	5.91×10^{-3}	1.76×10^{-8}	9.23×10^{-3}	1.51×10^{-2}	1.349×10^{-2}	1.182×10^{-2}	0.8762
50.0000	1.62×10^{-8}	4.94×10^{-3}	1.40×10^{-8}	1.00×10^{-2}	1.50×10^{-2}	1.361×10^{-2}	1.158×10^{-2}	0.8510

TABLE M.3s

Photons in Bone, Compact (ICRU)

ρ: 1.850 (g cm^{-3}), Composition ($Z-f_w$): 1−0.063984; 6−0.278000; 7−0.027000; 8−0.410016; 12−0.002000; 15−0.070000; 16−0.002000; 20−0.147000

K, L, M Edges	Energy (MeV)	Coherent σ_{coh}/ρ	Compton σ_{C}/ρ	Photoelectric τ/ρ	Pair + Triplet κ/ρ	Total Attenuation μ/ρ	Energy-Transfer μ_{tr}/ρ	Energy-Absorption μ_{en}/ρ	$(1-g)$
				Mass Coefficients (cm^{2}g^{-1})					
	0.0010	1.70	1.35×10^{-2}	3.44×10^{3}	0.00	3.44×10^{3}	3.435×10^{3}	3.435×10^{3}	1
	0.001305	1.60	2.10×10^{-2}	1.69×10^{3}	0.00	1.69×10^{3}	1.688×10^{3}	1.688×10^{3}	1
12K	0.001305	1.60	2.10×10^{-2}	1.70×10^{3}	0.00	1.70×10^{3}	1.698×10^{3}	1.698×10^{3}	1
	0.0015	1.55	2.61×10^{-2}	1.16×10^{3}	0.00	1.17×10^{3}	1.163×10^{3}	1.163×10^{3}	1
	0.0020	1.39	3.96×10^{-2}	5.24×10^{2}	0.00	5.25×10^{2}	5.234×10^{2}	5.234×10^{2}	0.9999
	0.002145	1.34	4.35×10^{-2}	4.30×10^{2}	0.00	4.31×10^{2}	4.294×10^{2}	4.293×10^{2}	0.9999
15K	0.002145	1.34	4.35×10^{-2}	5.85×10^{2}	0.00	5.87×10^{2}	5.760×10^{2}	5.760×10^{2}	0.9999
	0.002472	1.25	5.20×10^{-2}	4.00×10^{2}	0.00	4.01×10^{2}	3.947×10^{2}	3.946×10^{2}	0.9999
16K	0.002472	1.25	5.20×10^{-2}	4.04×10^{2}	0.00	4.05×10^{2}	3.981×10^{2}	3.981×10^{2}	0.9999
	0.0030	1.10	6.49×10^{-2}	2.38×10^{2}	0.00	2.40×10^{2}	2.353×10^{2}	2.353×10^{2}	0.9999
	0.0040	8.81×10^{-1}	8.54×10^{-2}	1.06×10^{2}	0.00	1.07×10^{2}	1.048×10^{2}	1.048×10^{2}	0.9999
	0.004038	8.74×10^{-1}	8.61×10^{-2}	1.03×10^{2}	0.00	1.04×10^{2}	1.020×10^{2}	1.020×10^{2}	0.9999
20K	0.004038	8.74×10^{-1}	8.61×10^{-2}	2.36×10^{2}	0.00	2.37×10^{2}	2.155×10^{2}	2.155×10^{2}	0.9999
	0.0050	7.16×10^{-1}	1.01×10^{-1}	1.35×10^{2}	0.00	1.35×10^{2}	1.248×10^{2}	1.248×10^{2}	0.9999
	0.0060	5.94×10^{-1}	1.13×10^{-1}	8.18×10	0.00	8.25×10	7.667×10	7.667×10	0.9999
	0.0080	4.35×10^{-1}	1.29×10^{-1}	3.67×10	0.00	3.73×10	3.496×10	3.495×10	0.9999
	0.0100	3.37×10^{-1}	1.40×10^{-1}	1.95×10	0.00	1.99×10	1.869×10	1.869×10	0.9998
	0.0150	2.04×10^{-1}	1.56×10^{-1}	5.97	0.00	6.33	5.810	5.809	0.9997
	0.0200	1.36×10^{-1}	1.64×10^{-1}	2.53	0.00	2.83	2.486	2.485	0.9996
	0.0300	7.24×10^{-2}	1.71×10^{-1}	7.39×10^{-1}	0.00	9.82×10^{-1}	7.379×10^{-1}	7.374×10^{-1}	0.9994
	0.0400	4.50×10^{-2}	1.72×10^{-1}	3.04×10^{-1}	0.00	5.21×10^{-1}	3.127×10^{-1}	3.125×10^{-1}	0.9993
	0.0500	3.08×10^{-2}	1.70×10^{-1}	1.52×10^{-1}	0.00	3.52×10^{-1}	1.645×10^{-1}	1.644×10^{-1}	0.9993
	0.0600	2.24×10^{-2}	1.67×10^{-1}	8.56×10^{-2}	0.00	2.75×10^{-1}	1.011×10^{-1}	1.010×10^{-1}	0.9992
	0.0800	1.33×10^{-2}	1.61×10^{-1}	3.45×10^{-2}	0.00	2.09×10^{-1}	5.355×10^{-2}	5.351×10^{-2}	0.9992
	0.1000	8.79×10^{-3}	1.54×10^{-1}	1.70×10^{-2}	0.00	1.80×10^{-1}	3.859×10^{-2}	3.855×10^{-2}	0.9992
	0.1500	4.07×10^{-3}	1.40×10^{-1}	4.73×10^{-3}	0.00	1.49×10^{-1}	3.039×10^{-2}	3.037×10^{-2}	0.9991
	0.2000	2.33×10^{-3}	1.29×10^{-1}	1.92×10^{-3}	0.00	1.33×10^{-1}	2.998×10^{-2}	2.995×10^{-2}	0.9990
	0.3000	1.05×10^{-3}	1.13×10^{-1}	5.61×10^{-4}	0.00	1.14×10^{-1}	3.099×10^{-2}	3.095×10^{-2}	0.9987
	0.4000	5.95×10^{-4}	1.01×10^{-1}	2.45×10^{-4}	0.00	1.02×10^{-1}	3.155×10^{-2}	3.150×10^{-2}	0.9983
	0.5000	3.82×10^{-4}	9.22×10^{-2}	1.33×10^{-4}	0.00	9.28×10^{-2}	3.166×10^{-2}	3.160×10^{-2}	0.9980
	0.6000	2.66×10^{-4}	8.53×10^{-2}	8.35×10^{-5}	0.00	8.57×10^{-2}	3.148×10^{-2}	3.141×10^{-2}	0.9976

0.8000	1.50×10^{-4}	7.50×10^{-2}	4.24×10^{-5}	0.00	7.52×10^{-2}	3.071×10^{-2}	3.062×10^{-2}	0.9970
1.0000	9.60×10^{-5}	6.75×10^{-2}	2.64×10^{-5}	0.00	6.76×10^{-2}	2.972×10^{-2}	2.962×10^{-2}	0.9963
1.2500	6.15×10^{-5}	6.03×10^{-2}	1.69×10^{-5}	2.33×10^{-5}	6.04×10^{-2}	2.842×10^{-2}	2.829×10^{-2}	0.9955
1.5000	4.27×10^{-5}	5.48×10^{-2}	1.22×10^{-5}	1.27×10^{-4}	5.50×10^{-2}	2.718×10^{-2}	2.704×10^{-2}	0.9946
2.0000	2.40×10^{-5}	4.68×10^{-2}	7.57×10^{-6}	5.00×10^{-4}	4.73×10^{-2}	2.510×10^{-2}	2.492×10^{-2}	0.9928
3.0000	1.07×10^{-5}	3.68×10^{-2}	4.16×10^{-6}	1.43×10^{-3}	3.83×10^{-2}	2.220×10^{-2}	2.195×10^{-2}	0.9885
4.0000	6.01×10^{-6}	3.07×10^{-2}	2.82×10^{-6}	2.35×10^{-3}	3.31×10^{-2}	2.039×10^{-2}	2.006×10^{-2}	0.9838
5.0000	3.84×10^{-6}	2.65×10^{-2}	2.12×10^{-6}	3.18×10^{-3}	2.97×10^{-2}	1.919×10^{-2}	1.878×10^{-2}	0.9788
6.0000	2.67×10^{-6}	2.34×10^{-2}	1.70×10^{-6}	3.93×10^{-3}	2.74×10^{-2}	1.836×10^{-2}	1.788×10^{-2}	0.9737
8.0000	1.50×10^{-6}	1.92×10^{-2}	1.21×10^{-6}	5.22×10^{-3}	2.44×10^{-2}	1.735×10^{-2}	1.672×10^{-2}	0.9634
10.0000	9.61×10^{-7}	1.63×10^{-2}	9.34×10^{-7}	6.29×10^{-3}	2.26×10^{-2}	1.682×10^{-2}	1.603×10^{-2}	0.9533
15.0000	4.27×10^{-7}	1.21×10^{-2}	5.96×10^{-7}	8.31×10^{-3}	2.04×10^{-2}	1.633×10^{-2}	1.517×10^{-2}	0.9294
20.0000	2.40×10^{-7}	9.70×10^{-3}	4.37×10^{-7}	9.78×10^{-3}	1.95×10^{-2}	1.633×10^{-2}	1.482×10^{-2}	0.9073
30.0000	1.07×10^{-7}	7.06×10^{-3}	2.85×10^{-7}	1.19×10^{-2}	1.89×10^{-2}	1.673×10^{-2}	1.452×10^{-2}	0.8677
40.0000	6.01×10^{-8}	5.61×10^{-3}	2.11×10^{-7}	1.33×10^{-2}	1.89×10^{-2}	1.723×10^{-2}	1.434×10^{-2}	0.8325
50.0000	3.84×10^{-8}	4.68×10^{-3}	1.68×10^{-7}	1.44×10^{-2}	1.91×10^{-2}	1.770×10^{-2}	1.418×10^{-2}	0.8008

TABLE M.3t

Photons in Bone, Cortical (ICRP)

ρ: 1.850 (g cm^{-3}), Composition ($Z-f_w$): 1−0.047234; 6−0.144330; 7−0.041990; 8−0.446096; 12−0.002200; 15−0.104970; 16−0.003150; 20−0.209930; 30−0.000100

K, L, M Edges	Energy (MeV)	Coherent σ_{coh}/ρ	Compton σ_{C}/ρ	Photoelectric τ/ρ	Pair+Triplet κ/ρ	Total Attenuation μ/ρ	Energy-Transfer μ_{tr}/ρ	Energy-Absorption μ_{en}/ρ	$(1-g)$
					Mass Coefficients $\times10^{-2}$(cm^{2}g^{-1})				
	0.0010	1.94	1.28×10^{-2}	3.74×10^{3}	0.00	3.74×10^{3}	3.730×10^{3}	3.730×10^{3}	1
30 L3	0.001020	1.93	1.32×10^{-2}	3.55×10^{3}	0.00	3.55×10^{3}	3.545×10^{3}	3.545×10^{3}	1
	0.001020	1.93	1.32×10^{-2}	3.55×10^{3}	0.00	3.55×10^{3}	3.545×10^{3}	3.545×10^{3}	1
30 L2	0.001043	1.93	1.37×10^{-2}	3.34×10^{3}	0.00	3.35×10^{3}	3.343×10^{3}	3.343×10^{3}	1
	0.001043	1.93	1.37×10^{-2}	3.34×10^{3}	0.00	3.35×10^{3}	3.343×10^{3}	3.343×10^{3}	1
30 L1	0.001194	1.87	1.71×10^{-2}	2.34×10^{3}	0.00	2.34×10^{3}	2.340×10^{3}	2.340×10^{3}	1
	0.001194	1.87	1.71×10^{-2}	2.34×10^{3}	0.00	2.34×10^{3}	2.340×10^{3}	2.340×10^{3}	1
12 K	0.001305	1.84	1.97×10^{-2}	1.84×10^{3}	0.00	1.84×10^{3}	1.844×10^{3}	1.844×10^{3}	1
	0.001305	1.84	1.97×10^{-2}	1.85×10^{3}	0.00	1.86×10^{3}	1.855×10^{3}	1.855×10^{3}	1
	0.0015	1.77	2.43×10^{-2}	1.28×10^{3}	0.00	1.28×10^{3}	1.275×10^{3}	1.274×10^{3}	0.9999
	0.0020	1.60	3.66×10^{-2}	5.77×10^{2}	0.00	5.79×10^{2}	5.769×10^{2}	5.768×10^{2}	0.9999
15 K	0.002145	1.55	4.01×10^{-2}	4.74×10^{2}	0.00	4.76×10^{2}	4.739×10^{2}	4.739×10^{2}	0.9999
	0.002145	1.55	4.01×10^{-2}	7.08×10^{2}	0.00	7.09×10^{2}	6.939×10^{2}	6.938×10^{2}	0.9999
16 K	0.002472	1.44	4.79×10^{-2}	4.87×10^{2}	0.00	4.89×10^{2}	4.792×10^{2}	4.791×10^{2}	0.9999
	0.002472	1.44	4.79×10^{-2}	4.93×10^{2}	0.00	4.95×10^{2}	4.846×10^{2}	4.846×10^{2}	0.9999
	0.0030	1.29	5.98×10^{-2}	2.94×10^{2}	0.00	2.95×10^{2}	2.892×10^{2}	2.892×10^{2}	0.9999
	0.0040	1.04	7.89×10^{-2}	1.32×10^{2}	0.00	1.33×10^{2}	1.301×10^{2}	1.301×10^{2}	0.9999
20 K	0.004038	1.03	7.96×10^{-2}	1.28×10^{2}	0.00	1.29×10^{2}	1.267×10^{2}	1.267×10^{2}	0.9999
	0.004038	1.03	7.96×10^{-2}	3.18×10^{2}	0.00	3.19×10^{2}	2.888×10^{2}	2.888×10^{2}	0.9999
	0.0050	8.49×10^{-1}	9.39×10^{-2}	1.83×10^{2}	0.00	1.83×10^{2}	1.685×10^{2}	1.684×10^{2}	0.9999
	0.0060	7.10×10^{-1}	1.05×10^{-1}	1.11×10^{2}	0.00	1.12×10^{2}	1.040×10^{2}	1.040×10^{2}	0.9998
	0.0080	5.23×10^{-1}	1.22×10^{-1}	5.02×10	0.00	5.09×10	4.770×10	4.769×10	0.9998
30 K	0.009659	4.24×10^{-1}	1.31×10^{-1}	2.95×10	0.00	3.00×10	2.824×10	2.823×10	0.9998
	0.009659	4.24×10^{-1}	1.31×10^{-1}	2.95×10	0.00	3.01×10	2.825×10	2.824×10	0.9998
	0.0100	4.08×10^{-1}	1.33×10^{-1}	2.67×10	0.00	2.73×10	2.562×10	2.561×10	0.9998
	0.0150	2.47×10^{-1}	1.49×10^{-1}	8.23	0.00	8.63	8.008	8.005	0.9996
	0.0200	1.65×10^{-1}	1.58×10^{-1}	3.50	0.00	3.83	3.436	3.434	0.9995
	0.0300	8.82×10^{-2}	1.66×10^{-1}	1.03	0.00	1.28	1.021	1.020	0.9993
	0.0400	5.51×10^{-2}	1.67×10^{-1}	4.23×10^{-1}	0.00	6.45×10^{-1}	4.302×10^{-1}	4.299×10^{-1}	0.9991
	0.0500	3.78×10^{-2}	1.66×10^{-1}	2.11×10^{-1}	0.00	4.15×10^{-1}	2.234×10^{-1}	2.232×10^{-1}	0.999
	0.0600	2.75×10^{-2}	1.63×10^{-1}	1.19×10^{-1}	0.00	3.10×10^{-1}	1.343×10^{-1}	1.342×10^{-1}	0.999

0.0800	1.64×10^{-2}	1.57×10^{-1}	4.82×10^{-2}	0.00	2.22×10^{-1}	6.683×10^{-2}	6.676×10^{-2}	0.9989
0.1000	1.09×10^{-2}	1.51×10^{-1}	2.38×10^{-2}	0.00	1.86×10^{-1}	4.496×10^{-2}	4.491×10^{-2}	0.9989
0.1500	5.04×10^{-3}	1.38×10^{-1}	6.62×10^{-3}	0.00	1.49×10^{-1}	3.184×10^{-2}	3.181×10^{-2}	0.9989
0.2000	2.89×10^{-3}	1.27×10^{-1}	2.70×10^{-3}	0.00	1.32×10^{-1}	3.028×10^{-2}	3.024×10^{-2}	0.9987
0.3000	1.31×10^{-3}	1.11×10^{-1}	7.88×10^{-4}	0.00	1.13×10^{-1}	3.071×10^{-2}	3.065×10^{-2}	0.9983
0.4000	7.41×10^{-4}	9.92×10^{-2}	3.44×10^{-4}	0.00	1.00×10^{-1}	3.113×10^{-2}	3.106×10^{-2}	0.9979
0.5000	4.76×10^{-4}	9.07×10^{-2}	1.87×10^{-4}	0.00	9.13×10^{-2}	3.119×10^{-2}	3.111×10^{-2}	0.9975
0.6000	3.31×10^{-4}	8.39×10^{-2}	1.17×10^{-4}	0.00	8.44×10^{-2}	3.099×10^{-2}	3.090×10^{-2}	0.9971
0.8000	1.87×10^{-4}	7.37×10^{-2}	5.96×10^{-5}	0.00	7.40×10^{-2}	3.022×10^{-2}	3.011×10^{-2}	0.9963
1.0000	1.20×10^{-4}	6.63×10^{-2}	3.70×10^{-5}	0.00	6.65×10^{-2}	2.924×10^{-2}	2.911×10^{-2}	0.9956
1.2500	7.65×10^{-5}	5.93×10^{-2}	2.37×10^{-5}	2.70×10^{-5}	5.94×10^{-2}	2.795×10^{-2}	2.780×10^{-2}	0.9946
1.5000	5.32×10^{-5}	5.39×10^{-2}	1.71×10^{-5}	1.46×10^{-4}	5.41×10^{-2}	2.674×10^{-2}	2.657×10^{-2}	0.9936
2.0000	2.99×10^{-5}	4.60×10^{-2}	1.06×10^{-5}	5.75×10^{-4}	4.66×10^{-2}	2.472×10^{-2}	2.451×10^{-2}	0.9914
3.0000	1.33×10^{-5}	3.62×10^{-2}	5.84×10^{-6}	1.64×10^{-3}	3.79×10^{-2}	2.199×10^{-2}	2.170×10^{-2}	0.9866
4.0000	7.48×10^{-6}	3.02×10^{-2}	3.96×10^{-6}	2.68×10^{-3}	3.29×10^{-2}	2.033×10^{-2}	1.995×10^{-2}	0.9812
5.0000	4.79×10^{-6}	2.61×10^{-2}	2.98×10^{-6}	3.62×10^{-3}	2.97×10^{-2}	1.926×10^{-2}	1.879×10^{-2}	0.9756
6.0000	3.33×10^{-6}	2.31×10^{-2}	2.38×10^{-6}	4.48×10^{-3}	2.75×10^{-2}	1.856×10^{-2}	1.800×10^{-2}	0.9699
8.0000	1.87×10^{-6}	1.89×10^{-2}	1.69×10^{-6}	5.94×10^{-3}	2.48×10^{-2}	1.776×10^{-2}	1.702×10^{-2}	0.9585
10.0000	1.20×10^{-6}	1.61×10^{-2}	1.31×10^{-6}	7.14×10^{-3}	2.32×10^{-2}	1.739×10^{-2}	1.648×10^{-2}	0.9474
15.0000	5.32×10^{-7}	1.19×10^{-2}	8.36×10^{-7}	9.41×10^{-3}	2.13×10^{-2}	1.720×10^{-2}	1.585×10^{-2}	0.9212
20.0000	2.99×10^{-7}	9.54×10^{-3}	6.13×10^{-7}	1.10×10^{-2}	2.06×10^{-2}	1.742×10^{-2}	1.563×10^{-2}	0.8973
30.0000	1.33×10^{-7}	6.95×10^{-3}	3.99×10^{-7}	1.34×10^{-2}	2.03×10^{-2}	1.811×10^{-2}	1.548×10^{-2}	0.8545
40.0000	7.48×10^{-8}	5.52×10^{-3}	2.96×10^{-7}	1.50×10^{-2}	2.05×10^{-2}	1.881×10^{-2}	1.536×10^{-2}	0.8167
50.0000	4.79×10^{-8}	4.61×10^{-3}	2.35×10^{-7}	1.62×10^{-2}	2.08×10^{-2}	1.943×10^{-2}	1.521×10^{-2}	0.7828

TABLE M.3u

Photons in Lung Tissue (ICRP)

ρ:1.050[a] (g cm^{-3}), Composition $(Z-f_w)$:1–0.101278; 6–0.102310; 7–0.028650;8–0.757072; 11–0.001840; 12–0.000730; 15–0.000800; 16–0.002250; 17–0.002660; 19–0.001940; 20–0.000090; 26–0.000370; 30–0.000010

K, L, M Edges	Energy (MeV)	Mass Coefficients (cm^{2}g^{-1})							(1-g)
		Coherent σ_{coh}/ρ	Compton σ_C/ρ	Photoelectric τ/ρ	Pair + Triplet κ/ρ	Total Attenuation μ/ρ	Energy-Transfer μ_{tr}/ρ	Energy-Absorption μ_{en}/ρ	
	0.0010	1.35	1.33×10^{-2}	3.82×10^{3}	0.00	3.82×10^{3}	3.814×10^{3}	3.814×10^{3}	1
	0.001020	1.35	1.37×10^{-2}	3.63×10^{3}	0.00	3.63×10^{3}	3.623×10^{3}	3.623×10^{3}	1
30 L3	0.001020	1.35	1.37×10^{-2}	3.63×10^{3}	0.00	3.63×10^{3}	3.623×10^{3}	3.623×10^{3}	1
	0.001043	1.34	1.43×10^{-2}	3.42×10^{3}	0.00	3.42×10^{3}	3.415×10^{3}	3.415×10^{3}	1
30 L2	0.001043	1.34	1.43×10^{-2}	3.42×10^{3}	0.00	3.42×10^{3}	3.415×10^{3}	3.415×10^{3}	1
	0.001072	1.33	1.50×10^{-2}	3.18×10^{3}	0.00	3.18×10^{3}	3.174×10^{3}	3.174×10^{3}	1
11 K	0.001072	1.33	1.50×10^{-2}	3.19×10^{3}	0.00	3.19×10^{3}	3.184×10^{3}	3.184×10^{3}	1
	0.001194	1.31	1.81×10^{-2}	2.39×10^{3}	0.00	2.39×10^{3}	2.392×10^{3}	2.392×10^{3}	1
30 L1	0.001194	1.31	1.81×10^{-2}	2.39×10^{3}	0.00	2.39×10^{3}	2.392×10^{3}	2.392×10^{3}	1
	0.001305	1.29	2.12×10^{-2}	1.88×10^{3}	0.00	1.88×10^{3}	1.881×10^{3}	1.881×10^{3}	1
12 K	0.001305	1.29	2.12×10^{-2}	1.89×10^{3}	0.00	1.89×10^{3}	1.885×10^{3}	1.885×10^{3}	1
	0.0015	1.24	2.68×10^{-2}	1.29×10^{3}	0.00	1.29×10^{3}	1.290×10^{3}	1.290×10^{3}	1
	0.0020	1.12	4.18×10^{-2}	5.78×10^{2}	0.00	5.79×10^{2}	5.774×10^{2}	5.773×10^{2}	0.9999
	0.002145	1.09	4.62×10^{-2}	4.74×10^{2}	0.00	4.75×10^{2}	4.729×10^{2}	4.728×10^{2}	0.9999
15 K	0.002145	1.09	4.62×10^{-2}	4.75×10^{2}	0.00	4.76×10^{2}	4.745×10^{2}	4.745×10^{2}	0.9999
	0.002472	1.00	5.58×10^{-2}	3.17×10^{2}	0.00	3.18×10^{2}	3.162×10^{2}	3.162×10^{2}	0.9999
16 K	0.002472	1.00	5.58×10^{-2}	3.21×10^{2}	0.00	3.22×10^{2}	3.201×10^{2}	3.200×10^{2}	0.9999
	0.002822	9.22×10^{-1}	6.56×10^{-2}	2.19×10^{2}	0.00	2.20×10^{2}	2.183×10^{2}	2.183×10^{2}	0.9999
17 K	0.002822	9.22×10^{-1}	6.56×10^{-2}	2.23×10^{2}	0.00	2.24×10^{2}	2.219×10^{2}	2.219×10^{2}	0.9999
	0.0030	8.82×10^{-1}	7.04×10^{-2}	1.87×10^{2}	0.00	1.88×10^{2}	1.861×10^{2}	1.861×10^{2}	0.9999
	0.003607	7.57×10^{-1}	8.52×10^{-2}	1.09×10^{2}	0.00	1.10×10^{2}	1.085×10^{2}	1.084×10^{2}	0.9999
19 K	0.003607	7.57×10^{-1}	8.52×10^{-2}	1.11×10^{2}	0.00	1.12×10^{2}	1.103×10^{2}	1.103×10^{2}	0.9999
	0.0040	6.86×10^{-1}	9.36×10^{-2}	8.17×10	0.00	8.25×10	8.127×10	8.126×10	0.9999
	0.004038	6.80×10^{-1}	9.43×10^{-2}	7.94×10	0.00	8.02×10	7.902×10	7.901×10	0.9999
20 K	0.004038	6.80×10^{-1}	9.43×10^{-2}	7.95×10	0.00	8.03×10	7.909×10	7.908×10	0.9999
	0.0050	5.41×10^{-1}	1.11×10^{-1}	4.20×10	0.00	4.27×10	4.182×10	4.182×10	0.9998
	0.0060	4.36×10^{-1}	1.24×10^{-1}	2.42×10	0.00	2.48×10	2.413×10	2.412×10	0.9998
	0.007112	3.53×10^{-1}	1.35×10^{-1}	1.44×10	0.00	1.49×10	1.437×10	1.437×10	0.9998
26 K	0.007112	3.53×10^{-1}	1.35×10^{-1}	1.45×10	0.00	1.50×10	1.446×10	1.446×10	0.9998
	0.0080	3.03×10^{-1}	1.42×10^{-1}	1.01×10	0.00	1.06×10	1.009×10	1.009×10	0.9998

30 K

0.009659	2.37×10^{-1}	1.51×10^{-1}	5.67	0.00	6.06		5.640	5.639	0.9998
0.009659	2.37×10^{-1}	1.51×10^{-1}	5.67	0.00	6.06		5.641	5.640	0.9998
0.0100	2.27×10^{-1}	1.53×10^{-1}	5.09	0.00	5.47		5.065	5.063	0.9998
0.0150	1.32×10^{-1}	1.68×10^{-1}	1.43	0.00	1.72		1.424	1.424	0.9997
0.0200	8.74×10^{-2}	1.76×10^{-1}	5.70×10^{-1}	0.00	8.33×10^{-1}		5.752×10^{-1}	5.751×10^{-1}	0.9997
0.0300	4.63×10^{-2}	1.81×10^{-1}	1.55×10^{-1}	0.00	3.82×10^{-1}		1.641×10^{-1}	1.640×10^{-1}	0.9996
0.0400	2.83×10^{-2}	1.81×10^{-1}	6.07×10^{-2}	0.00	2.70×10^{-1}		7.314×10^{-2}	7.311×10^{-2}	0.9996
0.0500	1.91×10^{-2}	1.79×10^{-1}	2.93×10^{-2}	0.00	2.27×10^{-1}		4.408×10^{-2}	4.407×10^{-2}	0.9996
0.0600	1.37×10^{-2}	1.75×10^{-1}	1.61×10^{-2}	0.00	2.05×10^{-1}		3.290×10^{-2}	3.289×10^{-2}	0.9996
0.0800	8.06×10^{-3}	1.68×10^{-1}	6.27×10^{-3}	0.00	1.82×10^{-1}		2.627×10^{-2}	2.626×10^{-2}	0.9996
0.1000	5.28×10^{-3}	1.61×10^{-1}	3.01×10^{-3}	0.00	1.69×10^{-1}		2.550×10^{-2}	2.549×10^{-2}	0.9996
0.1500	2.41×10^{-3}	1.46×10^{-1}	8.04×10^{-4}	0.00	1.49×10^{-1}		2.746×10^{-2}	2.745×10^{-2}	0.9995
0.2000	1.37×10^{-3}	1.34×10^{-1}	3.19×10^{-4}	0.00	1.36×10^{-1}		2.943×10^{-2}	2.941×10^{-2}	0.9994
0.3000	6.14×10^{-4}	1.17×10^{-1}	9.07×10^{-5}	0.00	1.17×10^{-1}		3.165×10^{-2}	3.162×10^{-2}	0.9992
0.4000	3.46×10^{-4}	1.05×10^{-1}	3.90×10^{-5}	0.00	1.05×10^{-1}		3.251×10^{-2}	3.247×10^{-2}	0.9989
0.5000	2.22×10^{-4}	9.57×10^{-2}	2.10×10^{-5}	0.00	9.59×10^{-2}		3.271×10^{-2}	3.267×10^{-2}	0.9987
0.6000	1.54×10^{-4}	8.85×10^{-2}	1.31×10^{-5}	0.00	8.87×10^{-2}		3.257×10^{-2}	3.252×10^{-2}	0.9985
0.8000	8.68×10^{-5}	7.78×10^{-2}	6.63×10^{-6}	0.00	7.79×10^{-2}		3.181×10^{-2}	3.175×10^{-2}	0.998
1.0000	5.56×10^{-5}	7.00×10^{-2}	4.12×10^{-6}	0.00	7.00×10^{-2}		3.080×10^{-2}	3.072×10^{-2}	0.9975
1.2500	3.56×10^{-5}	6.26×10^{-2}	2.61×10^{-6}	1.76×10^{-5}	6.26×10^{-2}		2.945×10^{-2}	2.936×10^{-2}	0.9969
1.5000	2.47×10^{-5}	5.69×10^{-2}	1.89×10^{-6}	9.70×10^{-5}	5.70×10^{-2}		2.816×10^{-2}	2.806×10^{-2}	0.9962
2.0000	1.39×10^{-5}	4.85×10^{-2}	1.19×10^{-6}	3.86×10^{-4}	4.89×10^{-2}		2.595×10^{-2}	2.582×10^{-2}	0.9948
3.0000	6.18×10^{-6}	3.82×10^{-2}	6.62×10^{-7}	1.11×10^{-3}	3.93×10^{-2}		2.278×10^{-2}	2.258×10^{-2}	0.9916
4.0000	3.48×10^{-6}	3.18×10^{-2}	4.53×10^{-7}	1.84×10^{-3}	3.37×10^{-2}		2.070×10^{-2}	2.045×10^{-2}	0.988
5.0000	2.22×10^{-6}	2.75×10^{-2}	3.43×10^{-7}	2.51×10^{-3}	3.00×10^{-2}		1.927×10^{-2}	1.896×10^{-2}	0.984
6.0000	1.54×10^{-6}	2.43×10^{-2}	2.76×10^{-7}	3.12×10^{-3}	2.74×10^{-2}		1.824×10^{-2}	1.787×10^{-2}	0.98
8.0000	8.69×10^{-7}	1.99×10^{-2}	1.97×10^{-7}	4.16×10^{-3}	2.40×10^{-2}		1.689×10^{-2}	1.641×10^{-2}	0.9716
10.0000	5.56×10^{-7}	1.69×10^{-2}	1.54×10^{-7}	5.03×10^{-3}	2.20×10^{-2}		1.609×10^{-2}	1.550×10^{-2}	0.9633
15.0000	2.47×10^{-7}	1.25×10^{-2}	9.85×10^{-8}	6.68×10^{-3}	1.92×10^{-2}		1.512×10^{-2}	1.426×10^{-2}	0.9432
20.0000	1.39×10^{-7}	1.01×10^{-2}	7.25×10^{-8}	7.88×10^{-3}	1.79×10^{-2}		1.478×10^{-2}	1.367×10^{-2}	0.9245
30.0000	6.18×10^{-8}	7.32×10^{-3}	4.74×10^{-8}	9.60×10^{-3}	1.69×10^{-2}		1.474×10^{-2}	1.312×10^{-2}	0.8905
40.0000	3.48×10^{-8}	5.82×10^{-3}	3.52×10^{-8}	1.08×10^{-2}	1.66×10^{-2}		1.493×10^{-2}	1.284×10^{-2}	0.8601
50.0000	2.22×10^{-8}	4.86×10^{-3}	2.80×10^{-8}	1.17×10^{-2}	1.66×10^{-2}		1.519×10^{-2}	1.265×10^{-2}	0.8324

a The density of lung encountered clinically is generally around 0.3 g cm^{-3}; therefore the correct *linear* attenuation coefficient will be given, to a good approximation, by multiplying the values in this table by the clinical *bulk* lung density. The very small amount of air in inflated lung will make a negligible difference to the linear values derived from these numbers, which strictly only apply to lung *tissue*.

TABLE M.3v

Photons in Muscle, Skeletal (ICRP)

ρ: 1.040 (g cm^{-3}), Composition ($Z-f_w$): 1–0.100637; 6–0.107830; 7–0.027680; 8–0.754773; 11–0.000750; 12–0.000190; 15–0.001800; 16–0.002410; 17–0.000790; 19–0.003020; 20–0.000030; 26–0.000040; 30–0.000050

K, L, M Edges	Energy (MeV)	Coherent σ_{coh}/ρ	Compton σ_C/ρ	Photoelectric τ/ρ	Pair + Triplet κ/ρ	Total Attenuation μ/ρ	Energy-Transfer μ_{tr}/ρ	Energy-Absorption μ_{en}/ρ	(1-g)
					Mass Coefficients (cm^2 g^{-1})				
	0.0010	1.35	1.33×10^{-2}	3.82×10^{3}	0.00	3.82×10^{3}	3.809×10^{3}	3.809×10^{3}	1
	0.001020	1.34	1.37×10^{-2}	3.63×10^{3}	0.00	3.63×10^{3}	3.619×10^{3}	3.618×10^{3}	1
30 L3	0.001020	1.34	1.37×10^{-2}	3.63×10^{3}	0.00	3.63×10^{3}	3.619×10^{3}	3.619×10^{3}	1
	0.001043	1.34	1.43×10^{-2}	3.42×10^{3}	0.00	3.42×10^{3}	3.411×10^{3}	3.411×10^{3}	1
30 L2	0.001043	1.34	1.43×10^{-2}	3.42×10^{3}	0.00	3.42×10^{3}	3.411×10^{3}	3.411×10^{3}	1
	0.001072	1.33	1.50×10^{-2}	3.17×10^{3}	0.00	3.17×10^{3}	3.170×10^{3}	3.170×10^{3}	1
11 K	0.001072	1.33	1.50×10^{-2}	3.18×10^{3}	0.00	3.18×10^{3}	3.175×10^{3}	3.174×10^{3}	1
	0.001194	1.31	1.81×10^{-2}	2.38×10^{3}	0.00	2.39×10^{3}	2.384×10^{3}	2.384×10^{3}	1
30 L1	0.001194	1.31	1.81×10^{-2}	2.38×10^{3}	0.00	2.39×10^{3}	2.384×10^{3}	2.384×10^{3}	1
	0.001305	1.28	2.12×10^{-2}	1.88×10^{3}	0.00	1.88×10^{3}	1.875×10^{3}	1.875×10^{3}	1
12 K	0.001305	1.28	2.12×10^{-2}	1.88×10^{3}	0.00	1.88×10^{3}	1.876×10^{3}	1.876×10^{3}	1
	0.0015	1.24	2.68×10^{-2}	1.29×10^{3}	0.00	1.29×10^{3}	1.283×10^{3}	1.283×10^{3}	1
	0.0020	1.12	4.18×10^{-2}	5.75×10^{2}	0.00	5.76×10^{2}	5.740×10^{2}	5.740×10^{2}	0.9999
	0.002145	1.08	4.61×10^{-2}	4.71×10^{2}	0.00	4.72×10^{2}	4.701×10^{2}	4.701×10^{2}	0.9999
15 K	0.002145	1.08	4.61×10^{-2}	4.75×10^{2}	0.00	4.76×10^{2}	4.739×10^{2}	4.738×10^{2}	0.9999
	0.002472	1.00	5.58×10^{-2}	3.16×10^{2}	0.00	3.17×10^{2}	3.158×10^{2}	3.158×10^{2}	0.9999
16 K	0.002472	1.00	5.58×10^{-2}	3.21×10^{2}	0.00	3.22×10^{2}	3.200×10^{2}	3.199×10^{2}	0.9999
	0.002822	9.20×10^{-1}	6.56×10^{-2}	2.19×10^{2}	0.00	2.20×10^{2}	2.184×10^{2}	2.183×10^{2}	0.9999
17 K	0.002822	9.20×10^{-1}	6.56×10^{-2}	2.20×10^{2}	0.00	2.21×10^{2}	2.194×10^{2}	2.194×10^{2}	0.9999
	0.0030	8.80×10^{-1}	7.04×10^{-2}	1.84×10^{2}	0.00	1.85×10^{2}	1.839×10^{2}	1.839×10^{2}	0.9999
	0.003607	7.55×10^{-1}	8.51×10^{-2}	1.07×10^{2}	0.00	1.08×10^{2}	1.070×10^{2}	1.070×10^{2}	0.9999
19 K	0.003607	7.55×10^{-1}	8.51×10^{-2}	1.11×10^{2}	0.00	1.11×10^{2}	1.099×10^{2}	1.098×10^{2}	0.9999
	0.0040	6.84×10^{-1}	9.35×10^{-2}	8.14×10	0.00	8.22×10	8.099×10	8.098×10	0.9999
	0.004038	6.78×10^{-1}	9.43×10^{-2}	7.92×10	0.00	7.99×10	7.875×10	7.874×10	0.9999
20 K	0.004038	6.78×10^{-1}	9.43×10^{-2}	7.92×10	0.00	8.00×10	7.877×10	7.876×10	0.9999
	0.0050	5.39×10^{-1}	1.11×10^{-1}	4.19×10	0.00	4.25×10	4.166×10	4.165×10	0.9998
	0.0060	4.35×10^{-1}	1.24×10^{-1}	2.41×10	0.00	2.47×10	2.403×10	2.403×10	0.9998
	0.007112	3.52×10^{-1}	1.35×10^{-1}	1.44×10	0.00	1.49×10	1.432×10	1.431×10	0.9998
26 K	0.007112	3.52×10^{-1}	1.35×10^{-1}	1.44×10	0.00	1.49×10	1.433×10	1.432×10	0.9998
	0.0080	3.02×10^{-1}	1.42×10^{-1}	1.00×10	0.00	1.05×10	9.986	9.984	0.9998

30 K

Energy								
0.009659	2.36×10^{-1}	1.51×10^{-1}	5.59	0.00	5.98	5.578	5.577	0.9998
0.009659	2.36×10^{-1}	1.51×10^{-1}	5.60	0.00	5.99	5.584	5.583	0.9998
0.0100	2.26×10^{-1}	1.53×10^{-1}	5.03	0.00	5.41	5.013	5.012	0.9998
0.0150	1.31×10^{-1}	1.68×10^{-1}	1.41	0.00	1.71	1.408	1.407	0.9997
0.0200	8.72×10^{-2}	1.75×10^{-1}	5.63×10^{-1}	0.00	8.25×10^{-1}	5.679×10^{-1}	5.677×10^{-1}	0.9997
0.0300	4.61×10^{-2}	1.81×10^{-1}	1.52×10^{-1}	0.00	3.79×10^{-1}	1.619×10^{-1}	1.618×10^{-1}	0.9996
0.0400	2.83×10^{-2}	1.81×10^{-1}	5.97×10^{-2}	0.00	2.69×10^{-1}	7.221×10^{-2}	7.218×10^{-2}	0.9996
0.0500	1.90×10^{-2}	1.79×10^{-1}	2.88×10^{-2}	0.00	2.26×10^{-1}	4.360×10^{-2}	4.359×10^{-2}	0.9996
0.0600	1.37×10^{-2}	1.75×10^{-1}	1.58×10^{-2}	0.00	2.05×10^{-1}	3.263×10^{-2}	3.261×10^{-2}	0.9996
0.0800	8.03×10^{-3}	1.68×10^{-1}	6.15×10^{-3}	0.00	1.82×10^{-1}	2.615×10^{-2}	2.614×10^{-2}	0.9996
0.1000	5.26×10^{-3}	1.61×10^{-1}	2.96×10^{-3}	0.00	1.69×10^{-1}	2.543×10^{-2}	2.542×10^{-2}	0.9996
0.1500	2.40×10^{-3}	1.46×10^{-1}	7.88×10^{-4}	0.00	1.49×10^{-1}	2.743×10^{-2}	2.742×10^{-2}	0.9995
0.2000	1.36×10^{-3}	1.34×10^{-1}	3.13×10^{-4}	0.00	1.36×10^{-1}	2.941×10^{-2}	2.939×10^{-2}	0.9994
0.3000	6.11×10^{-4}	1.17×10^{-1}	8.88×10^{-5}	0.00	1.17×10^{-1}	3.163×10^{-2}	3.160×10^{-2}	0.9992
0.4000	3.45×10^{-4}	1.05×10^{-1}	3.81×10^{-5}	0.00	1.05×10^{-1}	3.249×10^{-2}	3.245×10^{-2}	0.9989
0.5000	2.21×10^{-4}	9.56×10^{-2}	2.06×10^{-5}	0.00	9.59×10^{-2}	3.270×10^{-2}	3.265×10^{-2}	0.9987
0.6000	1.54×10^{-4}	8.85×10^{-2}	1.28×10^{-5}	0.00	8.86×10^{-2}	3.255×10^{-2}	3.250×10^{-2}	0.9985
0.8000	8.65×10^{-5}	7.78×10^{-2}	6.49×10^{-6}	0.00	7.78×10^{-2}	3.179×10^{-2}	3.173×10^{-2}	0.998
1.0000	5.54×10^{-5}	6.99×10^{-2}	4.03×10^{-6}	0.00	7.00×10^{-2}	3.079×10^{-2}	3.071×10^{-2}	0.9975
1.2500	3.54×10^{-5}	6.25×10^{-2}	2.56×10^{-6}	1.75×10^{-5}	6.26×10^{-2}	2.944×10^{-2}	2.934×10^{-2}	0.9969
1.5000	2.46×10^{-5}	5.68×10^{-2}	1.85×10^{-6}	9.69×10^{-5}	5.70×10^{-2}	2.815×10^{-2}	2.804×10^{-2}	0.9962
2.0000	1.38×10^{-5}	4.85×10^{-2}	1.16×10^{-6}	3.85×10^{-4}	4.89×10^{-2}	2.594×10^{-2}	2.581×10^{-2}	0.9948
3.0000	6.15×10^{-6}	3.82×10^{-2}	6.48×10^{-7}	1.11×10^{-3}	3.93×10^{-2}	2.276×10^{-2}	2.257×10^{-2}	0.9916
4.0000	3.46×10^{-6}	3.18×10^{-2}	4.44×10^{-7}	1.84×10^{-3}	3.37×10^{-2}	2.069×10^{-2}	2.044×10^{-2}	0.988
5.0000	2.22×10^{-6}	2.75×10^{-2}	3.36×10^{-7}	2.51×10^{-3}	3.00×10^{-2}	1.925×10^{-2}	1.895×10^{-2}	0.9841
6.0000	1.54×10^{-6}	2.43×10^{-2}	2.70×10^{-7}	3.11×10^{-3}	2.74×10^{-2}	1.823×10^{-2}	1.786×10^{-2}	0.98
8.0000	8.65×10^{-7}	1.99×10^{-2}	1.93×10^{-7}	4.15×10^{-3}	2.40×10^{-2}	1.688×10^{-2}	1.640×10^{-2}	0.9717
10.0000	5.54×10^{-7}	1.69×10^{-2}	1.50×10^{-7}	5.02×10^{-3}	2.19×10^{-2}	1.608×10^{-2}	1.549×10^{-2}	0.9633
15.0000	2.46×10^{-7}	1.25×10^{-2}	9.65×10^{-8}	6.67×10^{-3}	1.92×10^{-2}	1.510×10^{-2}	1.425×10^{-2}	0.9433
20.0000	1.38×10^{-7}	1.01×10^{-2}	7.10×10^{-8}	7.87×10^{-3}	1.79×10^{-2}	1.477×10^{-2}	1.365×10^{-2}	0.9246
30.0000	6.15×10^{-8}	7.32×10^{-3}	4.64×10^{-8}	9.58×10^{-3}	1.69×10^{-2}	1.472×10^{-2}	1.311×10^{-2}	0.8906
40.0000	3.46×10^{-8}	5.82×10^{-3}	3.45×10^{-8}	1.08×10^{-2}	1.66×10^{-2}	1.491×10^{-2}	1.283×10^{-2}	0.8602
50.0000	2.22×10^{-8}	4.86×10^{-3}	2.74×10^{-8}	1.17×10^{-2}	1.65×10^{-2}	1.517×10^{-2}	1.263×10^{-2}	0.8326

TABLE M.3w

Photons in Muscle, Striated (ICRU)

ρ: 1.040 (g cm^{-3}), Composition ($Z-f_w$): 1–0.101997; 6–0.123000; 7–0.035000; 8–0.729003; 11–0.000800; 12–0.002000; 15–0.002000; 16–0.005000

K, L, M Edges	Energy (MeV)	Mass Coefficients (cm^2 g^{-1})							(1-g)
		Coherent σ_{coh}/ρ	Compton σ_C/ρ	Photoelectric τ/ρ	Pair + Triplet κ/ρ	Total Attenuation μ/ρ	Energy-Transfer μ_{tr}/ρ	Energy-Absorption μ_{en}/ρ	
	0.0010	1.34	1.34×10^{-2}	3.76×10^{3}	0.00	3.76×10^{3}	3.753×10^{3}	3.753×10^{3}	1
	0.001072	1.32	1.52×10^{-2}	3.12×10^{3}	0.00	3.13×10^{3}	3.122×10^{3}	3.122×10^{3}	1
11 K	0.001072	1.32	1.52×10^{-2}	3.13×10^{3}	0.00	3.13×10^{3}	3.127×10^{3}	3.127×10^{3}	1
	0.001305	1.28	2.14×10^{-2}	1.85×10^{3}	0.00	1.85×10^{3}	1.846×10^{3}	1.846×10^{3}	1
12 K	0.001305	1.28	2.14×10^{-2}	1.85×10^{3}	0.00	1.85×10^{3}	1.847×10^{3}	1.847×10^{3}	1
	0.0015	1.23	2.70×10^{-2}	1.27×10^{3}	0.00	1.27×10^{3}	1.263×10^{3}	1.263×10^{3}	1
	0.0020	1.11	4.21×10^{-2}	5.66×10^{2}	0.00	5.67×10^{2}	5.648×10^{2}	5.648×10^{2}	0.9999
	0.002145	1.08	4.65×10^{-2}	4.63×10^{2}	0.00	4.64×10^{2}	4.625×10^{2}	4.625×10^{2}	0.9999
15 K	0.002145	1.08	4.65×10^{-2}	4.68×10^{2}	0.00	4.69×10^{2}	4.667×10^{2}	4.667×10^{2}	0.9999
	0.002472	9.95×10^{-1}	5.62×10^{-2}	3.12×10^{2}	0.00	3.13×10^{2}	3.110×10^{2}	3.110×10^{2}	0.9999
16 K	0.002472	9.95×10^{-1}	5.62×10^{-2}	3.21×10^{2}	0.00	3.22×10^{2}	3.196×10^{2}	3.196×10^{2}	0.9999
	0.0030	8.73×10^{-1}	7.09×10^{-2}	1.84×10^{2}	0.00	1.85×10^{2}	1.831×10^{2}	1.831×10^{2}	0.9999
	0.003607	7.48×10^{-1}	8.57×10^{-2}	1.07×10^{2}	0.00	1.08×10^{2}	1.067×10^{2}	1.067×10^{2}	0.9999
19 K	0.003607	7.48×10^{-1}	8.57×10^{-2}	1.10×10^{2}	0.00	1.11×10^{2}	1.095×10^{2}	1.095×10^{2}	0.9999
	0.0040	6.78×10^{-1}	9.41×10^{-2}	8.12×10	0.00	8.20×10	8.074×10	8.073×10	0.9999
	0.0050	5.35×10^{-1}	1.12×10^{-1}	4.18×10	0.00	4.24×10	4.155×10	4.154×10	0.9998
	0.0060	4.31×10^{-1}	1.25×10^{-1}	2.41×10	0.00	2.46×10	2.399×10	2.398×10	0.9998
	0.0080	3.00×10^{-1}	1.42×10^{-1}	1.00×10	0.00	1.04×10	9.968	9.966	0.9998
	0.0100	2.24×10^{-1}	1.53×10^{-1}	5.01	0.00	5.39	5.001	5.000	0.9998
	0.0150	1.31×10^{-1}	1.68×10^{-1}	1.40	0.00	1.70	1.405	1.404	0.9997
	0.0200	8.67×10^{-2}	1.76×10^{-1}	5.61×10^{-1}	0.00	8.24×10^{-1}	5.668×10^{-1}	5.666×10^{-1}	0.9997
	0.0300	4.59×10^{-2}	1.81×10^{-1}	1.52×10^{-1}	0.00	3.79×10^{-1}	1.616×10^{-1}	1.615×10^{-1}	0.9996
	0.0400	2.81×10^{-2}	1.81×10^{-1}	5.96×10^{-2}	0.00	2.69×10^{-1}	7.212×10^{-2}	7.209×10^{-2}	0.9996
	0.0500	1.89×10^{-2}	1.79×10^{-1}	2.87×10^{-2}	0.00	2.26×10^{-1}	4.358×10^{-2}	4.356×10^{-2}	0.9996
	0.0600	1.36×10^{-2}	1.75×10^{-1}	1.58×10^{-2}	0.00	2.05×10^{-1}	3.262×10^{-2}	3.261×10^{-2}	0.9996
	0.0800	7.98×10^{-3}	1.68×10^{-1}	6.14×10^{-3}	0.00	1.82×10^{-1}	2.616×10^{-2}	2.615×10^{-2}	0.9996
	0.1000	5.23×10^{-3}	1.61×10^{-1}	2.95×10^{-3}	0.00	1.69×10^{-1}	2.545×10^{-2}	2.544×10^{-2}	0.9996
	0.1500	2.39×10^{-3}	1.46×10^{-1}	7.87×10^{-4}	0.00	1.49×10^{-1}	2.746×10^{-2}	2.745×10^{-2}	0.9995
	0.2000	1.36×10^{-3}	1.34×10^{-1}	3.12×10^{-4}	0.00	1.36×10^{-1}	2.944×10^{-2}	2.943×10^{-2}	0.9994
	0.3000	6.08×10^{-4}	1.17×10^{-1}	8.86×10^{-5}	0.00	1.18×10^{-1}	3.167×10^{-2}	3.164×10^{-2}	0.9992
	0.4000	3.43×10^{-4}	1.05×10^{-1}	3.81×10^{-5}	0.00	1.05×10^{-1}	3.253×10^{-2}	3.249×10^{-2}	0.9989

0.5000	2.20×10^{-4}	9.58×10^{-2}	2.05×10^{-5}	0.00	9.60×10^{-2}	3.274×10^{-2}	3.269×10^{-2}	0.9987
0.6000	1.53×10^{-4}	8.86×10^{-2}	1.28×10^{-5}	0.00	8.88×10^{-2}	3.259×10^{-2}	3.254×10^{-2}	0.9985
0.8000	8.60×10^{-5}	7.78×10^{-2}	6.47×10^{-6}	0.00	7.79×10^{-2}	3.183×10^{-2}	3.177×10^{-2}	0.998
1.0000	5.50×10^{-5}	7.00×10^{-2}	4.02×10^{-6}	0.00	7.01×10^{-2}	3.082×10^{-2}	3.075×10^{-2}	0.9975
1.2500	3.52×10^{-5}	6.26×10^{-2}	2.55×10^{-6}	1.74×10^{-5}	6.27×10^{-2}	2.947×10^{-2}	2.938×10^{-2}	0.9969
1.5000	2.45×10^{-5}	5.69×10^{-2}	1.85×10^{-6}	9.64×10^{-5}	5.70×10^{-2}	2.818×10^{-2}	2.808×10^{-2}	0.9963
2.0000	1.38×10^{-5}	4.86×10^{-2}	1.16×10^{-6}	3.84×10^{-4}	4.90×10^{-2}	2.597×10^{-2}	2.584×10^{-2}	0.9949
3.0000	6.12×10^{-6}	3.82×10^{-2}	6.47×10^{-7}	1.11×10^{-3}	3.93×10^{-2}	2.279×10^{-2}	2.260×10^{-2}	0.9917
4.0000	3.44×10^{-6}	3.19×10^{-2}	4.43×10^{-7}	1.83×10^{-3}	3.37×10^{-2}	2.071×10^{-2}	2.046×10^{-2}	0.9881
5.0000	2.20×10^{-6}	2.75×10^{-2}	3.35×10^{-7}	2.50×10^{-3}	3.00×10^{-2}	1.927×10^{-2}	1.896×10^{-2}	0.9841
6.0000	1.53×10^{-6}	2.43×10^{-2}	2.69×10^{-7}	3.10×10^{-3}	2.74×10^{-2}	1.823×10^{-2}	1.787×10^{-2}	0.9801
8.0000	8.60×10^{-7}	1.99×10^{-2}	1.93×10^{-7}	4.13×10^{-3}	2.40×10^{-2}	1.688×10^{-2}	1.641×10^{-2}	0.9718
10.0000	5.51×10^{-7}	1.69×10^{-2}	1.50×10^{-7}	5.00×10^{-3}	2.19×10^{-2}	1.607×10^{-2}	1.549×10^{-2}	0.9635
15.0000	2.45×10^{-7}	1.25×10^{-2}	9.63×10^{-8}	6.64×10^{-3}	1.92×10^{-2}	1.509×10^{-2}	1.424×10^{-2}	0.9435
20.0000	1.38×10^{-7}	1.01×10^{-2}	7.09×10^{-8}	7.83×10^{-3}	1.79×10^{-2}	1.475×10^{-2}	1.364×10^{-2}	0.9249
30.0000	6.12×10^{-8}	7.33×10^{-3}	4.63×10^{-8}	9.54×10^{-3}	1.69×10^{-2}	1.469×10^{-2}	1.309×10^{-2}	0.891
40.0000	3.44×10^{-8}	5.82×10^{-3}	3.44×10^{-8}	1.07×10^{-2}	1.65×10^{-2}	1.488×10^{-2}	1.280×10^{-2}	0.8607
50.0000	2.20×10^{-8}	4.86×10^{-3}	2.74×10^{-8}	1.17×10^{-2}	1.65×10^{-2}	1.513×10^{-2}	1.261×10^{-2}	0.8331

TABLE M.3x

Photons in Skin (ICRP)

ρ: 1.100 (g cm^{-3}), Composition ($Z - f_w$): 1–0.100588; 6–0.228250; 7–0.046420; 8–0.619002; 11–0.000070; 12–0.000060; 15–0.000330; 16–0.001590; 17–0.002670; 19–0.000850; 20–0.000150; 26–0.000010; 30–0.000010

K, L, M Edges	Energy (MeV)	Coherent σ_{coh}/ρ	Compton σ_C/ρ	Photoelectric τ/ρ	Pair+Triplet κ/ρ	Total Attenuation μ/ρ	Energy-Transfer μ_{tr}/ρ	Energy-Absorption μ_{en}/ρ	$(1-g)$
	0.0010	1.29	1.38×10^{-2}	3.52×10^{3}	0.00	3.52×10^{3}	3.507×10^{3}	3.507×10^{3}	1
	0.001020	1.28	1.43×10^{-2}	3.34×10^{3}	0.00	3.34×10^{3}	3.331×10^{3}	3.331×10^{3}	1
30 L3	0.001020	1.28	1.43×10^{-2}	3.34×10^{3}	0.00	3.34×10^{3}	3.331×10^{3}	3.331×10^{3}	1
	0.001043	1.28	1.48×10^{-2}	3.14×10^{3}	0.00	3.14×10^{3}	3.139×10^{3}	3.138×10^{3}	1
30 L2	0.001043	1.28	1.48×10^{-2}	3.14×10^{3}	0.00	3.14×10^{3}	3.139×10^{3}	3.139×10^{3}	1
	0.001072	1.27	1.56×10^{-2}	2.92×10^{3}	0.00	2.92×10^{3}	2.915×10^{3}	2.915×10^{3}	1
11 K	0.001072	1.27	1.56×10^{-2}	2.92×10^{3}	0.00	2.92×10^{3}	2.916×10^{3}	2.916×10^{3}	1
	0.001194	1.25	1.88×10^{-2}	2.19×10^{3}	0.00	2.19×10^{3}	2.186×10^{3}	2.186×10^{3}	1
30 L1	0.001194	1.25	1.88×10^{-2}	2.19×10^{3}	0.00	2.19×10^{3}	2.186×10^{3}	2.186×10^{3}	1
	0.001305	1.22	2.20×10^{-2}	1.72×10^{3}	0.00	1.72×10^{3}	1.717×10^{3}	1.717×10^{3}	1
12 K	0.001305	1.22	2.20×10^{-2}	1.72×10^{3}	0.00	1.72×10^{3}	1.717×10^{3}	1.717×10^{3}	1
	0.0015	1.18	2.77×10^{-2}	1.17×10^{3}	0.00	1.18×10^{3}	1.172×10^{3}	1.172×10^{3}	1
	0.0020	1.06	4.31×10^{-2}	5.23×10^{2}	0.00	5.24×10^{2}	5.226×10^{2}	5.226×10^{2}	0.9999
	0.002145	1.02	4.76×10^{-2}	4.28×10^{2}	0.00	4.29×10^{2}	4.276×10^{2}	4.276×10^{2}	0.9999
15 K	0.002145	1.02	4.76×10^{-2}	4.29×10^{2}	0.00	4.30×10^{2}	4.283×10^{2}	4.283×10^{2}	0.9999
	0.002472	9.45×10^{-1}	5.74×10^{-2}	2.85×10^{2}	0.00	2.86×10^{2}	2.848×10^{2}	2.848×10^{2}	0.9999
16 K	0.002472	9.45×10^{-1}	5.74×10^{-2}	2.88×10^{2}	0.00	2.89×10^{2}	2.876×10^{2}	2.876×10^{2}	0.9999
	0.002822	8.64×10^{-1}	6.73×10^{-2}	1.96×10^{2}	0.00	1.97×10^{2}	1.958×10^{2}	1.958×10^{2}	0.9999
17 K	0.002822	8.64×10^{-1}	6.73×10^{-2}	2.00×10^{2}	0.00	2.01×10^{2}	1.994×10^{2}	1.993×10^{2}	0.9999
	0.0030	8.26×10^{-1}	7.21×10^{-2}	1.68×10^{2}	0.00	1.69×10^{2}	1.671×10^{2}	1.671×10^{2}	0.9999
	0.003607	7.06×10^{-1}	8.70×10^{-2}	9.75×10	0.00	9.83×10	9.718×10	9.717×10	0.9999
19 K	0.003607	7.06×10^{-1}	8.70×10^{-2}	9.84×10	0.00	9.92×10	9.797×10	9.796×10	0.9999
	0.0040	6.39×10^{-1}	9.54×10^{-2}	7.24×10	0.00	7.31×10	7.209×10	7.208×10	0.9999
	0.004038	6.33×10^{-1}	9.61×10^{-2}	7.03×10	0.00	7.11×10	7.009×10	7.008×10	0.9999
20 K	0.004038	6.33×10^{-1}	9.61×10^{-2}	7.05×10	0.00	7.12×10	7.020×10	7.019×10	0.9999
	0.0050	5.03×10^{-1}	1.13×10^{-1}	3.71×10	0.00	3.78×10	3.701×10	3.701×10	0.9999
	0.0060	4.06×10^{-1}	1.26×10^{-1}	2.14×10	0.00	2.19×10	2.130×10	2.129×10	0.9999
	0.007112	3.28×10^{-1}	1.36×10^{-1}	1.27×10	0.00	1.32×10	1.266×10	1.265×10	0.9998
26 K	0.007112	3.28×10^{-1}	1.36×10^{-1}	1.27×10	0.00	1.32×10	1.266×10	1.266×10	0.9998
	0.0080	2.83×10^{-1}	1.43×10^{-1}	8.83	0.00	9.26	8.810	8.808	0.9998

30 K

0.009659	2.22×10^{-1}	1.52×10^{-1}	4.92	0.00	5.29	4.909	4.908	0.9998
0.009659	2.22×10^{-1}	1.52×10^{-1}	4.92	0.00	5.29	4.910	4.909	0.9998
0.0100	2.12×10^{-1}	1.54×10^{-1}	4.41	0.00	4.78	4.405	4.404	0.9998
0.0150	1.24×10^{-1}	1.69×10^{-1}	1.23	0.00	1.52	1.231	1.230	0.9997
0.0200	8.21×10^{-2}	1.76×10^{-1}	4.89×10^{-1}	0.00	7.48×10^{-1}	4.953×10^{-1}	4.951×10^{-1}	0.9997
0.0300	4.33×10^{-2}	1.82×10^{-1}	1.32×10^{-1}	0.00	3.57×10^{-1}	1.416×10^{-1}	1.415×10^{-1}	0.9996
0.0400	2.65×10^{-2}	1.81×10^{-1}	5.15×10^{-2}	0.00	2.59×10^{-1}	6.409×10^{-2}	6.407×10^{-2}	0.9996
0.0500	1.79×10^{-2}	1.79×10^{-1}	2.48×10^{-2}	0.00	2.21×10^{-1}	3.964×10^{-2}	3.962×10^{-2}	0.9996
0.0600	1.28×10^{-2}	1.75×10^{-1}	1.36×10^{-2}	0.00	2.02×10^{-1}	3.042×10^{-2}	3.041×10^{-2}	0.9996
0.0800	7.52×10^{-3}	1.68×10^{-1}	5.28×10^{-3}	0.00	1.81×10^{-1}	2.528×10^{-2}	2.527×10^{-2}	0.9997
0.1000	4.92×10^{-3}	1.61×10^{-1}	2.53×10^{-3}	0.00	1.68×10^{-1}	2.501×10^{-2}	2.500×10^{-2}	0.9996
0.1500	2.24×10^{-3}	1.46×10^{-1}	6.73×10^{-4}	0.00	1.49×10^{-1}	2.731×10^{-2}	2.730×10^{-2}	0.9996
0.2000	1.27×10^{-3}	1.34×10^{-1}	2.67×10^{-4}	0.00	1.36×10^{-1}	2.936×10^{-2}	2.934×10^{-2}	0.9994
0.3000	5.71×10^{-4}	1.17×10^{-1}	7.56×10^{-5}	0.00	1.17×10^{-1}	3.161×10^{-2}	3.159×10^{-2}	0.9992
0.4000	3.22×10^{-4}	1.05×10^{-1}	3.24×10^{-5}	0.00	1.05×10^{-1}	3.248×10^{-2}	3.245×10^{-2}	0.999
0.5000	2.06×10^{-4}	9.56×10^{-2}	1.75×10^{-5}	0.00	9.59×10^{-2}	3.269×10^{-2}	3.265×10^{-2}	0.9988
0.6000	1.43×10^{-4}	8.85×10^{-2}	1.09×10^{-5}	0.00	8.86×10^{-2}	3.255×10^{-2}	3.250×10^{-2}	0.9985
0.8000	8.07×10^{-5}	7.77×10^{-2}	5.50×10^{-6}	0.00	7.78×10^{-2}	3.179×10^{-2}	3.173×10^{-2}	0.9981
1.0000	5.17×10^{-5}	6.99×10^{-2}	3.42×10^{-6}	0.00	7.00×10^{-2}	3.078×10^{-2}	3.071×10^{-2}	0.9976
1.2500	3.31×10^{-5}	6.25×10^{-2}	2.17×10^{-6}	1.68×10^{-5}	6.26×10^{-2}	2.943×10^{-2}	2.935×10^{-2}	0.997
1.5000	2.30×10^{-5}	5.68×10^{-2}	1.57×10^{-6}	9.29×10^{-5}	5.69×10^{-2}	2.815×10^{-2}	2.804×10^{-2}	0.9964
2.0000	1.29×10^{-5}	4.85×10^{-2}	9.87×10^{-7}	3.70×10^{-4}	4.89×10^{-2}	2.593×10^{-2}	2.580×10^{-2}	0.9951
3.0000	5.74×10^{-6}	3.81×10^{-2}	5.50×10^{-7}	1.07×10^{-3}	3.92×10^{-2}	2.273×10^{-2}	2.255×10^{-2}	0.9919
4.0000	3.23×10^{-6}	3.18×10^{-2}	3.77×10^{-7}	1.77×10^{-3}	3.36×10^{-2}	2.063×10^{-2}	2.039×10^{-2}	0.9884
5.0000	2.07×10^{-6}	2.75×10^{-2}	2.86×10^{-7}	2.41×10^{-3}	2.99×10^{-2}	1.918×10^{-2}	1.888×10^{-2}	0.9846
6.0000	1.44×10^{-6}	2.43×10^{-2}	2.30×10^{-7}	3.00×10^{-3}	2.73×10^{-2}	1.813×10^{-2}	1.778×10^{-2}	0.9806
8.0000	8.08×10^{-7}	1.99×10^{-2}	1.65×10^{-7}	4.00×10^{-3}	2.39×10^{-2}	1.675×10^{-2}	1.629×10^{-2}	0.9725
10.0000	5.17×10^{-7}	1.69×10^{-2}	1.28×10^{-7}	4.84×10^{-3}	2.18×10^{-2}	1.591×10^{-2}	1.535×10^{-2}	0.9644
15.0000	2.30×10^{-7}	1.25×10^{-2}	8.22×10^{-8}	6.43×10^{-3}	1.90×10^{-2}	1.488×10^{-2}	1.406×10^{-2}	0.9448
20.0000	1.29×10^{-7}	1.01×10^{-2}	6.05×10^{-8}	7.59×10^{-3}	1.76×10^{-2}	1.451×10^{-2}	1.344×10^{-2}	0.9265
30.0000	5.74×10^{-8}	7.32×10^{-3}	3.96×10^{-8}	9.25×10^{-3}	1.66×10^{-2}	1.440×10^{-2}	1.286×10^{-2}	0.8931
40.0000	3.23×10^{-8}	5.81×10^{-3}	2.94×10^{-8}	1.04×10^{-2}	1.62×10^{-2}	1.455×10^{-2}	1.256×10^{-2}	0.8633
50.0000	2.07×10^{-8}	4.85×10^{-3}	2.34×10^{-8}	1.13×10^{-2}	1.61×10^{-2}	1.478×10^{-2}	1.236×10^{-2}	0.8361

TABLE M.3γ

Photons in Soft Tissue (ICRP)

ρ: 1.000 (g cm^{-3}), Composition ($Z-f_w$): 1–0.104472; 6–0.232190; 7–0.024880; 8–0.630238; 11–0.001130; 12–0.000130; 15–0.001330; 16–0.001340; 19–0.001990; 20–0.000230; 26–0.000050; 30–0.000030

K, L, M Edges	Energy (MeV)	Coherent σ_{coh}/ρ	Compton σ_{Cl}/ρ	Photoelectric τ/ρ	Pair+Triplet κ/ρ	Total Attenuation μ/ρ	Energy-Transfer μ_{tr}/ρ	Energy-Absorption μ_{en}/ρ	$(1-g)$
	0.0010	1.29	1.39×10^{-2}	3.51×10^{3}	0.00	3.51×10^{3}	3.501×10^{3}	3.501×10^{3}	1
	0.001020	1.28	1.44×10^{-2}	3.33×10^{3}	0.00	3.33×10^{3}	3.325×10^{3}	3.325×10^{3}	1
30 L3	0.001020	1.28	1.44×10^{-2}	3.33×10^{3}	0.00	3.33×10^{3}	3.325×10^{3}	3.325×10^{3}	1
	0.001043	1.28	1.50×10^{-2}	3.14×10^{3}	0.00	3.14×10^{3}	3.134×10^{3}	3.134×10^{3}	1
30 L2	0.001043	1.28	1.50×10^{-2}	3.14×10^{3}	0.00	3.14×10^{3}	3.134×10^{3}	3.134×10^{3}	1
	0.001072	1.27	1.57×10^{-2}	2.91×10^{3}	0.00	2.91×10^{3}	2.911×10^{3}	2.911×10^{3}	1
11 K	0.001072	1.27	1.57×10^{-2}	2.92×10^{3}	0.00	2.92×10^{3}	2.918×10^{3}	2.918×10^{3}	1
	0.001194	1.25	1.90×10^{-2}	2.19×10^{3}	0.00	2.19×10^{3}	2.188×10^{3}	2.188×10^{3}	1
30 L1	0.001194	1.25	1.90×10^{-2}	2.19×10^{3}	0.00	2.19×10^{3}	2.188×10^{3}	2.188×10^{3}	1
	0.001305	1.22	2.22×10^{-2}	1.72×10^{3}	0.00	1.72×10^{3}	1.719×10^{3}	1.719×10^{3}	1
12 K	0.001305	1.22	2.22×10^{-2}	1.72×10^{3}	0.00	1.72×10^{3}	1.719×10^{3}	1.719×10^{3}	1
	0.0015	1.18	2.80×10^{-2}	1.18×10^{3}	0.00	1.18×10^{3}	1.174×10^{3}	1.174×10^{3}	1
	0.0020	1.06	4.35×10^{-2}	5.25×10^{2}	0.00	5.26×10^{2}	5.238×10^{2}	5.238×10^{2}	0.9999
	0.002145	1.03	4.80×10^{-2}	4.29×10^{2}	0.00	4.30×10^{2}	4.287×10^{2}	4.287×10^{2}	0.9999
15 K	0.002145	1.03	4.80×10^{-2}	4.32×10^{2}	0.00	4.33×10^{2}	4.315×10^{2}	4.314×10^{2}	0.9999
	0.002472	9.46×10^{-1}	5.78×10^{-2}	2.88×10^{2}	0.00	2.89×10^{2}	2.872×10^{2}	2.871×10^{2}	0.9999
16 K	0.002472	9.46×10^{-1}	5.78×10^{-2}	2.91×10^{2}	0.00	2.92×10^{2}	2.906×10^{2}	2.906×10^{2}	0.9999
	0.002822	8.66×10^{-1}	6.78×10^{-2}	1.98×10^{2}	0.00	1.99×10^{2}	1.980×10^{2}	1.980×10^{2}	0.9999
17 K	0.002822	8.66×10^{-1}	6.78×10^{-2}	2.00×10^{2}	0.00	2.01×10^{2}	1.998×10^{2}	1.998×10^{2}	0.9999
	0.0030	8.27×10^{-1}	7.27×10^{-2}	1.68×10^{2}	0.00	1.69×10^{2}	1.674×10^{2}	1.674×10^{2}	0.9999
	0.003607	7.08×10^{-1}	8.76×10^{-2}	9.76×10	0.00	9.84×10	9.735×10	9.734×10	0.9999
19 K	0.003607	7.08×10^{-1}	8.76×10^{-2}	9.97×10	0.00	1.01×10^{2}	9.921×10	9.920×10	0.9999
	0.004000	6.41×10^{-1}	9.60×10^{-2}	7.34×10	0.00	7.41×10	7.306×10	7.305×10	0.9999
	0.004038	6.35×10^{-1}	9.68×10^{-2}	7.14×10	0.00	7.21×10	7.104×10	7.103×10	0.9999
20 K	0.004038	6.35×10^{-1}	9.68×10^{-2}	7.16×10	0.00	7.23×10	7.121×10	7.121×10	0.9999
	0.0050	5.04×10^{-1}	1.14×10^{-1}	3.78×10	0.00	3.84×10	3.760×10	3.760×10	0.9999
	0.0060	4.07×10^{-1}	1.27×10^{-1}	2.17×10	0.00	2.23×10	2.166×10	2.166×10	0.9998
	0.007112	3.30×10^{-1}	1.37×10^{-1}	1.29×10	0.00	1.34×10	1.289×10	1.289×10	0.9998
26 K	0.007112	3.30×10^{-1}	1.37×10^{-1}	1.30×10	0.00	1.34×10	1.290×10	1.290×10	0.9998
	0.0080	2.84×10^{-1}	1.44×10^{-1}	9.02	0.00	9.44	8.986	8.984	0.9998

Mass Coefficients (cm^2 g^{-1})

30 K

Energy								
0.009659	2.23×10^{-1}	1.53×10^{-1}	5.03	0.00	5.40	5.014	5.013	0.9998
0.009659	2.23×10^{-1}	1.53×10^{-1}	5.03	0.00	5.41	5.017	5.016	0.9998
0.0100	2.13×10^{-1}	1.55×10^{-1}	4.52	0.00	4.88	4.503	4.502	0.9998
0.0150	1.24×10^{-1}	1.69×10^{-1}	1.26	0.00	1.55	1.262	1.261	0.9997
0.0200	8.24×10^{-2}	1.77×10^{-1}	5.03×10^{-1}	0.00	7.63×10^{-1}	5.088×10^{-1}	5.087×10^{-1}	0.9997
0.0300	4.35×10^{-2}	1.82×10^{-1}	1.36×10^{-1}	0.00	3.62×10^{-1}	1.457×10^{-1}	1.456×10^{-1}	0.9996
0.0400	2.66×10^{-2}	1.82×10^{-1}	5.32×10^{-2}	0.00	2.62×10^{-1}	6.582×10^{-2}	6.579×10^{-2}	0.9996
0.0500	1.79×10^{-2}	1.79×10^{-1}	2.57×10^{-2}	0.00	2.23×10^{-1}	4.054×10^{-2}	4.052×10^{-2}	0.9996
0.0600	1.29×10^{-2}	1.76×10^{-1}	1.41×10^{-2}	0.00	2.03×10^{-1}	3.096×10^{-2}	3.095×10^{-2}	0.9996
0.0800	7.55×10^{-3}	1.69×10^{-1}	5.48×10^{-3}	0.00	1.82×10^{-1}	2.554×10^{-2}	2.553×10^{-2}	0.9997
0.1000	4.94×10^{-3}	1.62×10^{-1}	2.63×10^{-3}	0.00	1.69×10^{-1}	2.518×10^{-2}	2.517×10^{-2}	0.9996
0.1500	2.25×10^{-3}	1.46×10^{-1}	7.00×10^{-4}	0.00	1.49×10^{-1}	2.743×10^{-2}	2.742×10^{-2}	0.9996
0.2000	1.28×10^{-3}	1.34×10^{-1}	2.78×10^{-4}	0.00	1.36×10^{-1}	2.947×10^{-2}	2.945×10^{-2}	0.9994
0.3000	5.74×10^{-4}	1.17×10^{-1}	7.88×10^{-5}	0.00	1.18×10^{-1}	3.172×10^{-2}	3.170×10^{-2}	0.9992
0.4000	3.24×10^{-4}	1.05×10^{-1}	3.38×10^{-5}	0.00	1.05×10^{-1}	3.259×10^{-2}	3.256×10^{-2}	0.9990
0.5000	2.07×10^{-4}	9.60×10^{-2}	1.83×10^{-5}	0.00	9.62×10^{-2}	3.280×10^{-2}	3.276×10^{-2}	0.9988
0.6000	1.44×10^{-4}	8.88×10^{-2}	1.14×10^{-5}	0.00	8.89×10^{-2}	3.266×10^{-2}	3.261×10^{-2}	0.9985
0.8000	8.11×10^{-5}	7.80×10^{-2}	5.75×10^{-6}	0.00	7.81×10^{-2}	3.190×10^{-2}	3.184×10^{-2}	0.9981
1.0000	5.19×10^{-5}	7.02×10^{-2}	3.58×10^{-6}	0.00	7.02×10^{-2}	3.089×10^{-2}	3.081×10^{-2}	0.9976
1.2500	3.33×10^{-5}	6.27×10^{-2}	2.26×10^{-6}	1.68×10^{-5}	6.28×10^{-2}	2.953×10^{-2}	2.945×10^{-2}	0.9970
1.5000	2.31×10^{-5}	5.70×10^{-2}	1.64×10^{-6}	9.30×10^{-5}	5.71×10^{-2}	2.824×10^{-2}	2.814×10^{-2}	0.9964
2.0000	1.30×10^{-5}	4.87×10^{-2}	1.03×10^{-6}	3.70×10^{-4}	4.91×10^{-2}	2.602×10^{-2}	2.589×10^{-2}	0.9951
3.0000	5.77×10^{-6}	3.83×10^{-2}	5.75×10^{-7}	1.07×10^{-3}	3.94×10^{-2}	2.281×10^{-2}	2.263×10^{-2}	0.9920
4.0000	3.25×10^{-6}	3.19×10^{-2}	3.94×10^{-7}	1.77×10^{-3}	3.37×10^{-2}	2.070×10^{-2}	2.046×10^{-2}	0.9885
5.0000	2.08×10^{-6}	2.76×10^{-2}	2.98×10^{-7}	2.41×10^{-3}	3.00×10^{-2}	1.924×10^{-2}	1.894×10^{-2}	0.9847
6.0000	1.44×10^{-6}	2.44×10^{-2}	2.40×10^{-7}	3.00×10^{-3}	2.74×10^{-2}	1.818×10^{-2}	1.783×10^{-2}	0.9807
8.0000	8.12×10^{-7}	1.99×10^{-2}	1.72×10^{-7}	4.00×10^{-3}	2.39×10^{-2}	1.680×10^{-2}	1.634×10^{-2}	0.9727
10.0000	5.20×10^{-7}	1.70×10^{-2}	1.33×10^{-7}	4.85×10^{-3}	2.18×10^{-2}	1.596×10^{-2}	1.539×10^{-2}	0.9646
15.0000	2.31×10^{-7}	1.26×10^{-2}	8.57×10^{-8}	6.43×10^{-3}	1.90×10^{-2}	1.492×10^{-2}	1.410×10^{-2}	0.9451
20.0000	1.30×10^{-7}	1.01×10^{-2}	6.30×10^{-8}	7.60×10^{-3}	1.77×10^{-2}	1.454×10^{-2}	1.347×10^{-2}	0.9268
30.0000	5.77×10^{-8}	7.34×10^{-3}	4.12×10^{-8}	9.26×10^{-3}	1.66×10^{-2}	1.443×10^{-2}	1.289×10^{-2}	0.8936
40.0000	3.25×10^{-8}	5.83×10^{-3}	3.06×10^{-8}	1.04×10^{-2}	1.62×10^{-2}	1.458×10^{-2}	1.259×10^{-2}	0.8639
50.0000	2.08×10^{-8}	4.87×10^{-3}	2.43×10^{-8}	1.13×10^{-2}	1.62×10^{-2}	1.481×10^{-2}	1.239×10^{-2}	0.8368

TABLE M.3z

Photons in Soft Tissue (ICRU 4-Component)

ρ: 1.000 (g cm^{-3}), Composition $(Z-f_w)$: 1–0.101172; 6–0.111000; 7–0.026000; 8–0.761828

K, L, M Edges	Energy (MeV)	Coherent σ_{coh}/ρ	Compton σ_C/ρ	Photoelectric τ/ρ	Pair + Triplet κ/ρ	Total Attenuation μ/ρ	Energy-Transfer μ_{tr}/ρ	Energy-Absorption μ_{en}/ρ	(1−g)
					Mass Coefficients (cm^2 g^{-1})				
	0.0010	1.33	1.33×10^{-2}	3.83×10^{3}	0.00	3.83×10^{3}	3.818×10^{3}	3.818×10^{3}	1
	0.0015	1.23	2.68×10^{-2}	1.28×10^{3}	0.00	1.29×10^{3}	1.283×10^{3}	1.283×10^{3}	1
	0.0020	1.11	4.19×10^{-2}	5.74×10^{2}	0.00	5.76×10^{2}	5.736×10^{2}	5.736×10^{2}	0.9999
	0.0030	8.69×10^{-1}	7.05×10^{-2}	1.78×10^{2}	0.00	1.79×10^{2}	1.781×10^{2}	1.781×10^{2}	0.9999
	0.0040	6.75×10^{-1}	9.38×10^{-2}	7.60×10	0.00	7.68×10	7.599×10	7.598×10	0.9999
	0.0050	5.31×10^{-1}	1.11×10^{-1}	3.88×10	0.00	3.95×10	3.881×10	3.880×10	0.9998
	0.0060	4.28×10^{-1}	1.25×10^{-1}	2.23×10	0.00	2.28×10	2.226×10	2.226×10	0.9998
	0.0080	2.97×10^{-1}	1.42×10^{-1}	9.17	0.00	9.60	9.165	9.163	0.9998
	0.0100	2.21×10^{-1}	1.53×10^{-1}	4.56	0.00	4.94	4.565	4.564	0.9998
	0.0150	1.28×10^{-1}	1.68×10^{-1}	1.26	0.00	1.56	1.267	1.266	0.9997
	0.0200	8.53×10^{-2}	1.76×10^{-1}	5.01×10^{-1}	0.00	7.62×10^{-1}	5.072×10^{-1}	5.070×10^{-1}	0.9997
	0.0300	4.51×10^{-2}	1.81×10^{-1}	1.34×10^{-1}	0.00	3.60×10^{-1}	1.439×10^{-1}	1.438×10^{-1}	0.9996
	0.0400	2.76×10^{-2}	1.81×10^{-1}	5.22×10^{-2}	0.00	2.61×10^{-1}	6.476×10^{-2}	6.474×10^{-2}	0.9996
	0.0500	1.86×10^{-2}	1.79×10^{-1}	2.50×10^{-2}	0.00	2.22×10^{-1}	3.988×10^{-2}	3.987×10^{-2}	0.9996
	0.0600	1.34×10^{-2}	1.75×10^{-1}	1.37×10^{-2}	0.00	2.02×10^{-1}	3.053×10^{-2}	3.051×10^{-2}	0.9996
	0.0800	7.83×10^{-3}	1.68×10^{-1}	5.29×10^{-3}	0.00	1.81×10^{-1}	2.531×10^{-2}	2.530×10^{-2}	0.9996
	0.1000	5.13×10^{-3}	1.61×10^{-1}	2.53×10^{-3}	0.00	1.69×10^{-1}	2.502×10^{-2}	2.501×10^{-2}	0.9996
	0.1500	2.34×10^{-3}	1.46×10^{-1}	6.70×10^{-4}	0.00	1.49×10^{-1}	2.733×10^{-2}	2.732×10^{-2}	0.9995
	0.2000	1.33×10^{-3}	1.34×10^{-1}	2.65×10^{-4}	0.00	1.36×10^{-1}	2.938×10^{-2}	2.936×10^{-2}	0.9994
	0.3000	5.95×10^{-4}	1.17×10^{-1}	7.47×10^{-5}	0.00	1.17×10^{-1}	3.164×10^{-2}	3.161×10^{-2}	0.9992
	0.4000	3.36×10^{-4}	1.05×10^{-1}	3.20×10^{-5}	0.00	1.05×10^{-1}	3.250×10^{-2}	3.247×10^{-2}	0.9989
	0.5000	2.15×10^{-4}	9.57×10^{-2}	1.72×10^{-5}	0.00	9.59×10^{-2}	3.272×10^{-2}	3.267×10^{-2}	0.9987
	0.6000	1.49×10^{-4}	8.85×10^{-2}	1.07×10^{-5}	0.00	8.87×10^{-2}	3.257×10^{-2}	3.252×10^{-2}	0.9985
	0.8000	8.41×10^{-5}	7.78×10^{-2}	5.42×10^{-6}	0.00	7.79×10^{-2}	3.182×10^{-2}	3.175×10^{-2}	0.998
	1.0000	5.39×10^{-5}	7.00×10^{-2}	3.37×10^{-6}	0.00	7.00×10^{-2}	3.081×10^{-2}	3.073×10^{-2}	0.9976
	1.2500	3.45×10^{-5}	6.26×10^{-2}	2.13×10^{-6}	1.73×10^{-5}	6.26×10^{-2}	2.946×10^{-2}	2.937×10^{-2}	0.9969
	1.5000	2.39×10^{-5}	5.69×10^{-2}	1.55×10^{-6}	9.57×10^{-5}	5.70×10^{-2}	2.817×10^{-2}	2.806×10^{-2}	0.9963
	2.0000	1.35×10^{-5}	4.85×10^{-2}	9.72×10^{-7}	3.81×10^{-4}	4.89×10^{-2}	2.596×10^{-2}	2.582×10^{-2}	0.9949
	3.0000	5.99×10^{-6}	3.82×10^{-2}	5.43×10^{-7}	1.10×10^{-3}	3.93×10^{-2}	2.277×10^{-2}	2.258×10^{-2}	0.9917
	4.0000	3.37×10^{-6}	3.18×10^{-2}	3.73×10^{-7}	1.82×10^{-3}	3.37×10^{-2}	2.069×10^{-2}	2.044×10^{-2}	0.9881
	5.0000	2.16×10^{-6}	2.75×10^{-2}	2.83×10^{-7}	2.48×10^{-3}	3.00×10^{-2}	1.924×10^{-2}	1.894×10^{-2}	0.9843

6.0000	1.50×10^{-6}	2.43×10^{-2}	2.27×10^{-7}	3.08×10^{-3}	2.74×10^{-2}	1.821×10^{-2}	1.785×10^{-2}	0.9802
8.0000	8.42×10^{-7}	1.99×10^{-2}	1.63×10^{-7}	4.11×10^{-3}	2.40×10^{-2}	1.685×10^{-2}	1.638×10^{-2}	0.972
10.0000	5.39×10^{-7}	1.69×10^{-2}	1.27×10^{-7}	4.97×10^{-3}	2.19×10^{-2}	1.604×10^{-2}	1.546×10^{-2}	0.9637
15.0000	2.39×10^{-7}	1.25×10^{-2}	8.16×10^{-8}	6.60×10^{-3}	1.91×10^{-2}	1.505×10^{-2}	1.420×10^{-2}	0.9438
20.0000	1.35×10^{-7}	1.01×10^{-2}	6.00×10^{-8}	7.79×10^{-3}	1.78×10^{-2}	1.470×10^{-2}	1.360×10^{-2}	0.9253
30.0000	5.99×10^{-8}	7.32×10^{-3}	3.93×10^{-8}	9.49×10^{-3}	1.68×10^{-2}	1.463×10^{-2}	1.305×10^{-2}	0.8915
40.0000	3.37×10^{-8}	5.82×10^{-3}	2.92×10^{-8}	1.07×10^{-2}	1.65×10^{-2}	1.481×10^{-2}	1.276×10^{-2}	0.8613
50.0000	2.16×10^{-8}	4.86×10^{-3}	2.32×10^{-8}	1.16×10^{-2}	1.64×10^{-2}	1.507×10^{-2}	1.256×10^{-2}	0.8339

TABLE M.4 RADIOACTIVE NUCLIDES USED IN RADIOTHERAPY

Compiled by Philip Mayles

TABLE M.4

Radioactive Nuclides Used in Radiotherapy[a]

Nuclide[d]	Decay Process	Decays to[e]	Half Life	γ keV	β keV	Other keV	Probability	1st HVL water	1st HVT lead or β range in perspex[k]	Air Kerma Rate Constant[c] μGy h⁻¹ MBq⁻¹ at 1 m
¹⁹⁸Au	β-	¹⁹⁸Hg	**2.6943 d**	**411.8**			**0.956**	66 mm[f]	2.5 mm[g]	0.0565[h]
				675.9			0.008			
					960.7		0.990			
²⁴¹Am	α	²³⁷Np	432.7 y			5388 α	0.014	0.1 mm[i]	0.1 mm[i]	
						5443 α	0.128			
						5486 α	**0.852**			
						5512 α	0.002			
						5544 α	0.003			
				26.34			0.024			
				33.19			0.001			
				59.536			**0.359**			
				69.77			0.006			
²⁵²Cf	α	²⁴⁸Cm	2.645 y			5976.7 α	0.002			
						6075.7 α	0.152			
						6118.3 α	0.816			
	Spontaneous fission	Various	966.1 d							
⁶⁰Co	β-	⁶⁰Ni	**5.2718 y**	**1173.2**			**0.999**	110 mm[f]	13 mm[g]	0.309[h]
				1332.5			**1.000**			
					317.9		1.000			

Nuclide	Decay	Half-life	E (keV)	E_β (keV)	Intensity			
^{137}Cs	β-	30.17 y		511.5	0.946	81 mm^f	6.5 mm^f	0.078^h
				1173.2	0.054			
→^{137m}Ba	IT		**661.7**		**0.901**			
^{125}I	EC	59.43 d	35.5		0.067	24 mm^f	0.025 mm^g	0.033^h
			27.0^n		**1.125**			
	IT		**31.0**		**0.254**			
^{131}I	β-	8.04 d	80.2		0.026	63 mm^f	2.54 mm	0.055
			177.21		0.003			
			284.3		0.061			
			318.1		0.001			
			325.8		0.003			
			364.5		**0.812**			
			503.0		0.004			
			637.0		0.073			
			642.7		0.002			
			722.9		0.018			
				247.9	0.021			
				303.9	0.006			
				333.8	0.074			
				606.3	**0.894**			
				629.7	0.001			
				806.9	0.004			
^{131m}Xe	IT	11.87 d	163.9		0.020			
^{192}Ir	β- 95.2%	73.831 d	**205.8**		0.032	63 mm^f	2.5 mm^g	0.113^h
			296.0		**0.287**			
			308.5		**0.298**			
			316.5		**0.830**			
	EC 48%		**468.1**		**0.477**			
			484.6		0.031			
			588.6		0.045			
			604.4		0.081			
			612.5		0.053			
^{192}Os				79	0.001			
				256	0.056			
				536	0.416			
				672	0.481			

TABLE M.4 (Continued)

Radioactive Nuclides Used in Radiotherapy[a]

Nuclide[d]	Decay Process	Decays to[e]	Half Life	γ keV	β keV	Other keV	Probability	1st HVL water	1st HVT lead or β range in perspex[k]	Air Kerma Rate Constant[c] μGy h^{-1} MBq^{-1} at 1 m
^{32}P	β-	^{32}S	**14.27 d**		**1710.4**		**1.000**	0.9 mm[l]	6.9 mm[k]	
^{103}Pd → ^{103m}Rh	EC	^{103m}Rh	**16.98 d**	**39.756**				26 mm[f]	0.04 mm[m]	0.035[h]
	IT	^{103}Rh	56.115 m				0.001			
^{226}Ra[o]	α	^{222}Rn	1600 y			4784.5 α	0.945		12 mm[g]	0.195[h]
				186.0			0.328			
						4601.9 α	0.056			
→ ^{222}Rn	α	^{218}Po	**3.825 d**			5489.7 α	0.999		12 mm[g]	0.195[g]
				510			0.001			
→ ^{218}Po	α	^{214}Pb	3.06 m			6002.6 α	1.000			
→ ^{214}Pb	β-	^{214}Bi	26.9 m		678		0.482			
					735		0.430			
					1030		0.056			
				241.9			0.075			
				295.2			0.192			
				351.9			0.369			
→ ^{214}Bi	β-	^{214}Po	19.9 m		1066		0.059			
					1151		0.043			
					1423		0.085			
					1505		0.180			
					1540		0.182			
					1892		0.076			
					3270		0.161			
				609.3			0.469			
				768.4			0.050			
				1120.3			0.155			
				1238.1			0.061			
				1377.7			0.041			
				1764.5			0.162			
				2204.1			0.0525			

Attenuation[b]

Nuclide	Decay	Half-life	β (keV)	γ (keV)	α (keV)	Intensity	Range	Range	Range
→^{214}Po	α	0.0002 s			7687.1 α	1.000			
→^{210}Pb	β-	22.3 y	16.5			0.82			
			63.0			0.18			
→^{210}Bi	β-	5.01 d	1161.5			1.0			
→^{210}Po	α	138.4 d			5304.5 α	1.0			
^{106}Ru →^{106}Rh	β-	**1.008 y**	39.4			1.000	2.3 mm[l]	15.2 mm[k]	
^{106}Rh →^{106}Pd	β-	30.1 s		428.5		0.001			
				511.9		0.212			
				616.1		0.008			
				621.9		0.102			
				873.5		0.005			
				1050.4		0.016			
				1128.0		0.004			
				1194.5		0.001			
				2112.5		0.002			
			917			0.001			
			1539			0.005			
			1835			0.001			
			1979			0.018			
			2407			0.101			
			2413			0.007			
			3029			0.084			
			3541			**0.781**			
^{145}Sm	EC	**340 d**		**61.25**		**0.122**	34 mm[f]	0.13 mm[m]	
^{89}Sr	β-	**50.52 d**	1492			**1.000**	0.7 mm	5.9 mm	
^{90}Sr	β-	**28.64 y**	546.2			1.000	1 mm		
→^{90}Y	β-	2.6713 d	**2279.2**			**1.000**		9.5 mm[k]	
(^{99}Mo) →^{99m}Tc	β-	2.748 d		Various		Various			
→^{99m}Tc	IT	**6.01 h**		**140.5**		**0.890**	45 mm[f]	0.17 mm	
^{182}Ta	β-	**114.7 d**		**67.75**		0.408	66 mm[f]	1.7 mm[m]	0.16
				100.1		0.141			
				152.4		0.074			
				222.1		0.074			
				229.3		0.037			
				264.1		0.035			

TABLE M.4 (Continued)

Radioactive Nuclides Used in Radiotherapy[a]

Nuclide[d]	Decay Process	Decays to[e]	Half Life	γ keV	β keV	Other keV	Probability	1st HVL water	1st HVT lead or β range in perspex[k]	Air Kerma Rate Constant[c] μGy h⁻¹ MBq⁻¹ at 1 m
				1121.3			0.346			
				1189.1			0.161			
				1221.4			**0.268**			
				1231.0			0.113			
					260.4		0.293			
					303.4		0.001			
					326.1		0.019			
					370.8		0.006			
					439.8		0.201			
					482.5		0.021			
					524.4		0.394			
					556.2		0.009			
					592.2		0.038			
					1713.5		0.018			
⁹⁰Y	β-	⁹⁰Zr	**2.6713 d**		**2279.2**		**1.000**	1 mm	9.5 mm	
¹⁶⁹Yb	EC	¹⁶⁹Tm	**32.01 d**	8.401			0.004	43 mm[f]	0.3 mm[m]	0.027[c]
				63.119			**0.416**			
				93.613			0.026			
				109.78			0.174			
				118.19			0.019			
				130.52			0.115			
				177.21			0.223			
				197.95			**0.359**			
				261.07			0.017			
				307.73			0.099			

The most important emissions are shown in bold. In the Half-life column "d" and "y" are used for "days" and "years".

a   Unless indicated otherwise, data are from Longworth et al. (1998). While every effort to ensure the accuracy of the data has been made, the table is not intended as a definitive reference.

b Attenuation is quoted as an indication of penetrative ability and is the approximate value for a point source. The HVT in lead gives an indication of the amount of material needed for radiation protection purposes. When shielding β-emitters it is better to use a low atomic number material close to the source as lead will produce bremsstrahlung photons which will be more penetrating than the original β-particles.

c When calculating dose rates for radiotherapy purposes the calculation should be based on source strength or air kerma rate at 1 m (see Chapter 53). However, it is sometimes necessary to be able to convert from air kerma rate to MBq for regulatory purposes or to be able to calculate the expected dose rate for a particular activity for radiation protection purposes. In such circumstances these values may be used. Where data were not available the value has been calculated based on the decay scheme. The formula used is:

$$\Gamma = \frac{\sum_i f_i \times E_i \times (\mu_{en}/\rho)_{E_i}}{4\pi(1-g)} \times 3600$$

where Γ is the specific γ-ray constant (in Gy m^2 h^{-1} Bq^{-1}); f is the probability of emission, per disintegration, of a γ-ray of energy E (expressed in J); μ_{en}/ρ is the mass energy-absorption coefficient for air in m^2 kg^{-1}; 3600 is the number of seconds in an hour. Normally E is in MeV and it is therefore necessary to multiply by 1.602×10^{-13} to convert to joules and, if the activity is in MBq, to multiply by 10^6. Tables of mass energy-absorption coefficient are usually in cm^{2}g^{-1} (see Table M.3k) and these must be converted to m^2 kg^{-1} by dividing by 10^2. If these units are used the formula becomes:

$$\Gamma = \frac{4.59 \sum_i f_i \times E_i(\text{MeV}) \times (\mu_{en}/\rho)_{E_i}}{(1-g)} \quad \mu\text{Gy m}^2 \text{ h}^{-1} \text{MB}_q^{-1}$$

d This method may slightly underestimate the air kerma rate because it ignores bremsstrahlung contributions from the βs (Glasgow and Dillman 1979).

e Isotopes shown with an arrow indicate the daughter of the parent isotope and its associated decay products.

f This is the isotope which results from the initial decay process.

g The HVL for water is very dependent on the measurement conditions. For many sources (e.g. ^{192}Ir) scattered radiation often compensates for attenuation close to the source so that the dose rate in water is actually higher than in air. The data presented here have been calculated based on the linear attenuation coefficients for the individual energies. The values presented give an indication of the relative penetration, but should not be given greater weight. As discussed in Part J, in brachytherapy the dose rate is principally determined by the inverse square law.

g From Burger (2003).

h From BIR (1987).

i α-particles are stopped by less than a millimetre of tissue and so these parameters are not relevant. However, there is some γ-ray output which requires some shielding (HVT approximately 0.1 mm of lead).

j From Aird et al (2000).

k For β-particles the value quoted is calculated from the formula $R = 412\, E^{1.265 - 0.0954\ln E}$ where E is the maximum energy in MeV and R is expressed in mg cm^{-2} (Bull 1998) using a density of 1.15 g cm^{-3} for perspex (PMMA). R is the thickness of perspex required to completely absorb all β-particles.

l Based on nomogram Figure 6.3 in Bull (1998).

m This value is calculated from the attenuation coefficients in the same way as for the HVL in water (Note f) and similar reservations apply.

n Emissions at 27 keV and 31 keV are characteristic x rays from Te.

o The Radium decay series is very complex and is included for general information only. Less important emissions with less than 4% probability have been omitted.

PART M: REFERENCE DATA
REFERENCES

Aird, E. G., Williams, J. R., and Rembowska, A., in *Brachytherapy in Radiotherapy Physics in Practice*, Williams, J. R. and Thwaites, D. I., Eds., 2nd ed., OUP, Oxford, 2000.

Ashley, J. C., Density effect in liquid water, *Radiat. Res.*, 89, 32–37, 1982.

Berger, M. J., Coursey, J. S., Zucker, M. A., and Chang, J., *ESTAR, PSTAR, and ASTAR: Computer Programs for Calculating Stopping-Power and Range Tables for Electrons, Protons, and Helium Ions* (version 1.2.3). [Online] Available: http://physics.nist.gov/Star [2006, September 28]. National Institute of Standards and Technology, Gaithersburg, MD, 2005.

Berger, M. J., Hubbell, J. H., Seltzer, S. M., Chang, J., Coursey, J. S., Sukumar, R., and Zucker, D. S., *XCOM: Photon Cross Section Database* (version 1.3). Available on-line at: http://physics.nist.gov/xcom [2006, September 25]. National Institute of Standards and Technology, Gaithersburg, MD, 2005.

BIPM (Bureau International des Poids et Mesures) Le Système international d'unités—The International System of Units, 8ème édition (see http://www.bipm.org/fr/si/si_brochure/), 2006.

BIR (British Institute of Radiology), Radionuclides in brachytherapy: Radium and after, Trott, N. G., Ed., *Br. J. Radiol.*, Suppl. 21 (London: BIR), 1987.

Borg, J. and Rogers, D. W. O., Monte Carlo Calculations of Photon Spectra in air from 192Ir Sources *NRC Report PIRS-629r* (see http://www.irs.inms.nrc.ca/inms/irs/papers/PIRS629r/pirs629r.html), 1998.

Burger, J., Radioactive sources in brachytherapy, *Radiol. Oncol.*, 37, 127–131, 2003.

Bull, R., Radiological protection, in *The Radiochemical Manual*, Longworth, G., Carpenter, B., Bull, R., Toole, J., and Nichols, A., Eds.,, AEA Technology, Harwell, 1998. Crown copyright.

Glasgow, G. P. and Dillman, L. T., Specific gamma-ray constant and exposure rate constant of ^{192}Ir, *Med. Phys.*, 6, 49–52, 1979.

Glasgow, G. P. and Dillman, L. T., The specific gamma-ray constant and exposure rate constant of ^{182}Ta, *Med. Phys.*, 9, 250–253, 1982.

Hubbell, J. H. and Seltzer, S. M., *Tables of X-Ray Mass Attenuation Coefficients and Mass Energy-Absorption Coefficients 1 keV to 20 MeV for Elements Z = 1 to 92 and 48 Additional Substances of Dosimetric Interest*, National Institute of Standards and Technology Report NISTIR 5632, 1995.

ICRU (International Commission on Radiation Units and Measurements), Stopping powers for electrons and positrons, ICRU Report 37, ICRU, Bethesda, MD, 1984.

ICRU (International Commission on Radiation Units and Measurements), Tissue Substitutes in Radiation Dosimetry and Measurement, ICRU Report 44, ICRU, Bethesda, MD, 1989.

Mohr, P. J. and Taylor, B. N., CODATA recommended values of the fundamental physical constants: 2002, *Rev. Mod. Phys.*, 77(1), 1–107, 2005.

NIST Nuclear Data http://www.niehs.nih.gov/odhsb/radhyg/nucdata/125I.pdf, 2005.

Seltzer, S. M., Calculation of photon mass energy-transfer and mass energy-absorption coefficients, *Radiat. Res.*, 136, 147–170, 1993.

INDEX

R